Site Work & Landscape Costs with RSMeans data

Derrick Hale, PE, Senior Editor

GORDIAN®

2021
40th annual edition

Vice President, Data
Tim Duggan

Vice President, Product
Ted Kail

Product Manager
Jason Jordan

Principal Engineer
Bob Mewis

Engineers: Architectural Divisions
1, 3, 4, 5, 6, 7, 8, 9, 10, 11, 12, 13, 41
Matthew Doheny, Manager
Sam Babbitt
Sergeleen Edouard
Richard Goldin

Scott Keller
Thomas Lane

Engineers: Civil Divisions and Wages
2, 31, 32, 33, 34, 35, 44, 46
Derrick Hale, PE, Manager
Christopher Babbitt
Stephen Bell
Michael Lynch
Elisa Mello
David Yazbek

Engineers: Mechanical, Electrical, Plumbing & Conveying Divisions
14, 21, 22, 23, 25, 26, 27, 28, 48
Joseph Kelble, Manager
Brian Adams

Michelle Curran
Antonio D'Aulerio
Thomas Lyons
Jake MacDonald

Contributing Engineers
John Melin, PE, Manager
Paul Cowan
Barry Hutchinson
Gerard Lafond, PE
Matthew Sorrentino

Technical Project Lead
Kevin Souza

Production
Debra Panarelli, Manager
Jonathan Forgit
Sharon Larsen
Sheryl Rose
Janice Thalin

Data Quality
Joseph Ingargiola, Manager
David Byars
Audrey Considine
Ellen D'Amico

Cover Design
Blaire Collins

Innovation
Ray Diwakar, Vice President
Kedar Gaikwad
Todd Glowac
Srini Narla
Joseph Woughter

RSMeans data from Gordian
Construction Publishers & Consultants
1099 Hingham Street, Suite 201
Rockland, MA 02370
United States of America
1.800.448.8182
RSMeans.com

Copyright 2020 by The Gordian Group Inc.
All rights reserved.
Cover photo © iStock.com/Gang Zhou

Printed in the United States of America
ISSN 1520-5320
ISBN 978-1-950656-67-7

| 0281 | $319.00 per copy (in United States)
Price is subject to change without prior notice. |

Related Data and Services

Our engineers recommend the following products and services to complement *Site Work & Landscape Costs with RSMeans data:*

Annual Cost Data Books
2021 Building Construction Costs with RSMeans data
2021 Heavy Construction Costs with RSMeans data

Reference Books
Landscape Estimating Methods
Estimating Building Costs
RSMeans Estimating Handbook
Green Building: Project Planning & Estimating
How to Estimate with Means Data and CostWorks
Plan Reading & Material Takeoff
Project Scheduling and Management for Construction

Virtual, Instructor-led & On-site Training Offerings
Site Work & Heavy Construction Estimating
Training for our online estimating solution
Plan Reading & Material Takeoff
Unit Price Estimating

RSMeans data
For access to the latest cost data, an intuitive search, and an easy-to-use estimate builder, take advantage of the time savings available from our online application.

To learn more visit: **RSMeans.com/online**

Enterprise Solutions
Building owners, facility managers, building product manufacturers, and attorneys across the public and private sectors engage with RSMeans data Enterprise to solve unique challenges where trusted construction cost data is critical.

To learn more visit: **RSMeans.com/Enterprise**

Custom Built Data Sets
Building and Space Models: Quickly plan construction costs across multiple locations based on geography, project size, building system component, product options, and other variables for precise budgeting and cost control.

Predictive Analytics: Accurately plan future builds with custom graphical interactive dashboards, negotiate future costs of tenant build-outs, and identify and compare national account pricing.

Consulting
Building Product Manufacturing Analytics: Validate your claims and assist with new product launches.

Third-Party Legal Resources: Used in cases of construction cost or estimate disputes, construction product failure vs. installation failure, eminent domain, class action construction product liability, and more.

API
For resellers or internal application integration, RSMeans data is offered via API. Deliver Unit, Assembly, and Square Foot Model data within your interface. To learn more about how you can provide your customers with the latest in localized construction cost data visit: **RSMeans.com/API**

Table of Contents

Foreword

The Value of RSMeans data from Gordian

Since 1942, RSMeans data has been the industry-standard materials, labor, and equipment cost information database for contractors, facility owners and managers, architects, engineers, and anyone else that requires the latest localized construction cost information. More than 75 years later, the objective remains the same: to provide facility and construction professionals with the most current and comprehensive construction cost database possible.

With the constant influx of new construction methods and materials, in addition to ever-changing labor and material costs, last year's cost data is not reliable for today's designs, estimates, or budgets. Gordian's cost engineers apply real-world construction experience to identify and quantify new building products and methodologies, adjust productivity rates, and adjust costs to local market conditions across the nation. This adds up to more than 22,000 hours in cost research annually. This unparalleled construction cost expertise is why so many facility and construction professionals rely on RSMeans data year over year.

About Gordian

Gordian originated in the spirit of innovation and a strong commitment to helping clients reach and exceed their construction goals. In 1982, Gordian's chairman and founder, Harry H. Mellon, created Job Order Contracting while serving as chief engineer at the Supreme Headquarters Allied Powers Europe. Job Order Contracting is a unique indefinite delivery/indefinite quantity (IDIQ) process, which enables facility owners to complete a substantial number of repair, maintenance, and construction projects with a single, competitively awarded contract. Realizing facility and infrastructure owners across various industries could greatly benefit from the time and cost saving advantages of this innovative construction procurement solution, he established Gordian in 1990.

Continuing the commitment to provide the most relevant and accurate facility and construction data, software, and expertise in the industry, Gordian enhanced the fortitude of its data with the acquisition of RSMeans in 2014. And in an effort to expand its facility management capabilities, Gordian acquired Sightlines, the leading provider of facilities benchmarking data and analysis, in 2015.

Our Offerings

Gordian is the leader in facility and construction cost data, software, and expertise for all phases of the building life cycle. From planning to design, procurement, construction, and operations, Gordian's solutions help clients maximize efficiency, optimize cost savings, and increase building quality with its highly specialized data engineers, software, and unique proprietary data sets.

Our Commitment

At Gordian, we do more than talk about the quality of our data and the usefulness of its application. We stand behind all of our RSMeans data—from historical cost indexes to construction materials and techniques—to craft current costs and predict future trends. If you have any questions about our products or services, please call us toll-free at 800.448.8182 or visit our website at gordian.com.

How the Cost Data Is Built: An Overview

Unit Prices*

All cost data have been divided into 50 divisions according to the MasterFormat® system of classification and numbering.

Assemblies*

The cost data in this section have been organized in an "Assemblies" format. These assemblies are the functional elements of a building and are arranged according to the 7 elements of the UNIFORMAT II classification system. For a complete explanation of a typical "Assembly", see "RSMeans data: Assemblies—How They Work."

Residential Models*

Model buildings for four classes of construction—economy, average, custom, and luxury—are developed and shown with complete costs per square foot.

Commercial/Industrial/Institutional Models*

This section contains complete costs for 77 typical model buildings expressed as costs per square foot.

Green Commercial/Industrial/Institutional Models*

This section contains complete costs for 25 green model buildings expressed as costs per square foot.

References*

This section includes information on Equipment Rental Costs, Crew Listings, Historical Cost Indexes, City Cost Indexes, Location Factors, Reference Tables, and Change Orders, as well as a listing of abbreviations.

- **Equipment Rental Costs:** Included are the average costs to rent and operate hundreds of pieces of construction equipment.
- **Crew Listings:** This section lists all the crews referenced in the cost data. A crew is composed of more than one trade classification and/or the addition of power equipment to any trade classification. Power equipment is included in the cost of the crew. Costs are shown both with bare labor rates and with the installing contractor's overhead and profit added. For each, the total crew cost per eight-hour day and the composite cost per labor-hour are listed.

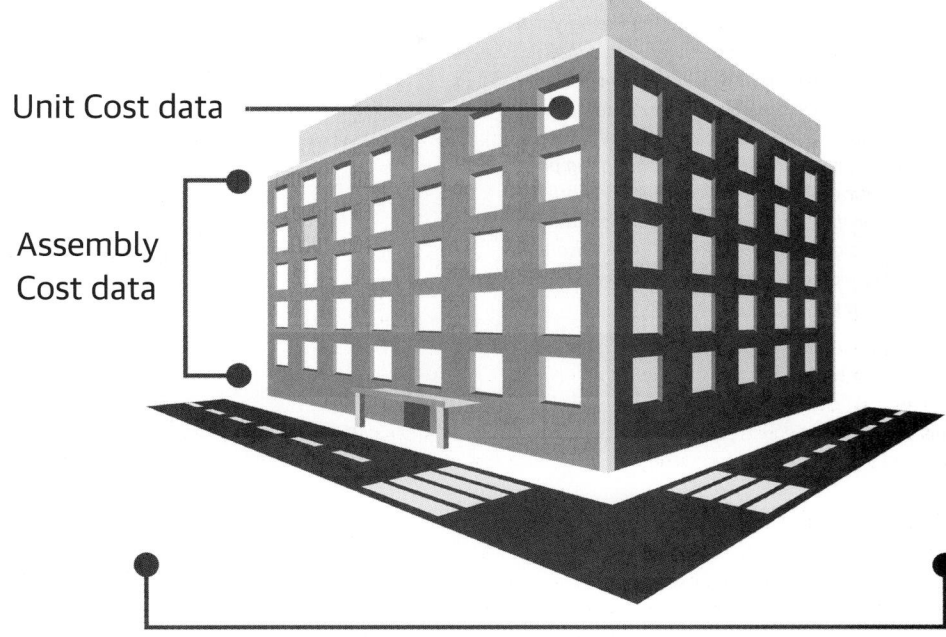

Unit Cost data

Assembly Cost data

Square Foot Models

- **Historical Cost Indexes:** These indexes provide you with data to adjust construction costs over time.
- **City Cost Indexes:** All costs in this data set are U.S. national averages. Costs vary by region. You can adjust for this by CSI Division to over 730 cities in 900+ 3-digit zip codes throughout the U.S. and Canada by using this data.
- **Location Factors:** You can adjust total project costs to over 730 cities in 900+ 3-digit zip codes throughout the U.S. and Canada by using the weighted number, which applies across all divisions.
- **Reference Tables:** At the beginning of selected major classifications in the Unit Prices are reference numbers indicators. These numbers refer you to related information in the Reference Section. In this section, you'll find reference tables, explanations, and estimating information that support how we develop the unit price data, technical data, and estimating procedures.
- **Change Orders:** This section includes information on the factors that influence the pricing of change orders.

- **Abbreviations:** A listing of abbreviations used throughout this information, along with the terms they represent, is included.

Index (printed versions only)

A comprehensive listing of all terms and subjects will help you quickly find what you need when you are not sure where it occurs in MasterFormat®.

Conclusion

This information is designed to be as comprehensive and easy to use as possible.

The Construction Specifications Institute (CSI) and Construction Specifications Canada (CSC) have produced the 2018 edition of MasterFormat®, a system of titles and numbers used extensively to organize construction information.

All unit prices in the RSMeans cost data are now arranged in the 50-division MasterFormat® 2018 system.

---

*   Not all information is available in all data sets

Note: The material prices in RSMeans cost data are "contractor's prices." They are the prices that contractors can expect to pay at the lumberyards, suppliers'/distributors' warehouses, etc. Small orders of specialty items would be higher than the costs shown, while very large orders, such as truckload lots, would be less. The variation would depend on the size, timing, and negotiating power of the contractor. The labor costs are primarily for new construction or major renovation rather than repairs or minor alterations. With reasonable exercise of judgment, the figures can be used for any building work.

Estimating with RSMeans data: Unit Prices

Following these steps will allow you to complete an accurate estimate using RSMeans data: Unit Prices.

1. Scope Out the Project

- Think through the project and identify the CSI divisions needed in your estimate.
- Identify the individual work tasks that will need to be covered in your estimate.
- The Unit Price data have been divided into 50 divisions according to CSI MasterFormat® 2018.
- In printed versions, the Unit Price Section Table of Contents on page 1 may also be helpful when scoping out your project.
- Experienced estimators find it helpful to begin with Division 2 and continue through completion. Division 1 can be estimated after the full project scope is known.

2. Quantify

- Determine the number of units required for each work task that you identified.
- Experienced estimators include an allowance for waste in their quantities. (Waste is not included in our Unit Price line items unless otherwise stated.)

3. Price the Quantities

- Use the search tools available to locate individual Unit Price line items for your estimate.
- Reference Numbers indicated within a Unit Price section refer to additional information that you may find useful.
- The crew indicates who is performing the work for that task. Crew codes are expanded in the Crew Listings in the Reference Section to include all trades and equipment that comprise the crew.
- The Daily Output is the amount of work the crew is expected to complete in one day.
- The Labor-Hours value is the amount of time it will take for the crew to install one unit of work.
- The abbreviated Unit designation indicates the unit of measure upon which the crew, productivity, and prices are based.
- Bare Costs are shown for materials, labor, and equipment needed to complete the Unit Price line item. Bare costs do not include waste, project overhead, payroll insurance, payroll taxes, main office overhead, or profit.
- The Total Incl O&P cost is the billing rate or invoice amount of the installing contractor or subcontractor who performs the work for the Unit Price line item.

4. Multiply

- Multiply the total number of units needed for your project by the Total Incl O&P cost for each Unit Price line item.
- Be careful that your take off unit of measure matches the unit of measure in the Unit column.
- The price you calculate is an estimate for a completed item of work.
- Keep scoping individual tasks, determining the number of units required for those tasks, matching each task with individual Unit Price line items, and multiplying quantities by Total Incl O&P costs.
- An estimate completed in this manner is priced as if a subcontractor, or set of subcontractors, is performing the work. The estimate does not yet include Project Overhead or Estimate Summary components such as general contractor markups on subcontracted work, general contractor office overhead and profit, contingency, and location factors.

5. Project Overhead

- Include project overhead items from Division 1–General Requirements.
- These items are needed to make the job run. They are typically, but not always, provided by the general contractor. Items include, but are not limited to, field personnel, insurance, performance bond, permits, testing, temporary utilities, field office and storage facilities, temporary scaffolding and platforms, equipment mobilization and demobilization, temporary roads and sidewalks, winter protection, temporary barricades and fencing, temporary security, temporary signs, field engineering and layout, final cleaning, and commissioning.
- Each item should be quantified and matched to individual Unit Price line items in Division 1, then priced and added to your estimate.

- An alternate method of estimating project overhead costs is to apply a percentage of the total project cost—usually 5% to 15% with an average of 10% (see General Conditions).
- Include other project related expenses in your estimate such as:
  - Rented equipment not itemized in the Crew Listings
  - Rubbish handling throughout the project (see section 02 41 19.19)

6. Estimate Summary

- Include sales tax as required by laws of your state or county.
- Include the general contractor's markup on self-performed work, usually 5% to 15% with an average of 10%.
- Include the general contractor's markup on subcontracted work, usually 5% to 15% with an average of 10%.
- Include the general contractor's main office overhead and profit:
  - RSMeans data provides general guidelines on the general contractor's main office overhead (see section 01 31 13.60 and Reference Number RO13113-50).
  - Markups will depend on the size of the general contractor's operations, projected annual revenue, the level of risk, and the level of competition in the local area and for this project in particular.
- Include a contingency, usually 3% to 5%, if appropriate.
- Adjust your estimate to the project's location by using the City Cost Indexes or the Location Factors in the Reference Section:
  - Look at the rules in "How to Use the City Cost Indexes" to see how to apply the Indexes for your location.
  - When the proper Index or Factor has been identified for the project's location, convert it to a multiplier by dividing it by 100, then multiply that multiplier by your estimated total cost. The original estimated total cost will now be adjusted up or down from the national average to a total that is appropriate for your location.

Editors' Note:
We urge you to spend time reading and understanding the supporting material. An accurate estimate requires experience, knowledge, and careful calculation. The more you know about how we at RSMeans developed the data, the more accurate your estimate will be. In addition, it is important to take into consideration the reference material such as Equipment Listings, Crew Listings, City Cost Indexes, Location Factors, and Reference Tables.

How to Use the Cost Data: The Details

What's Behind the Numbers? The Development of Cost Data

RSMeans data engineers continually monitor developments in the construction industry in order to ensure reliable, thorough, and up-to-date cost information. While overall construction costs may vary relative to general economic conditions, price fluctuations within the industry are dependent upon many factors. Individual price variations may, in fact, be opposite to overall economic trends. Therefore, costs are constantly tracked and complete updates are performed yearly. Also, new items are frequently added in response to changes in materials and methods.

Costs in U.S. Dollars

All costs represent U.S. national averages and are given in U.S. dollars. The City Cost Index (CCI) with RSMeans data can be used to adjust costs to a particular location. The CCI for Canada can be used to adjust U.S. national averages to local costs in Canadian dollars. No exchange rate conversion is necessary because it has already been factored in.

[G] The processes or products identified by the green symbol in our publications have been determined to be environmentally responsible and/or resource-efficient solely by RSMeans data engineering staff. The inclusion of the green symbol does not represent compliance with any specific industry association or standard.

Material Costs

RSMeans data engineers contact manufacturers, dealers, distributors, and contractors all across the U.S. and Canada to determine national average material costs. If you have access to current material costs for your specific location, you may wish to make adjustments to reflect differences from the national average. Included within material costs are fasteners for a normal installation. RSMeans data engineers use manufacturers' recommendations, written specifications, and/or standard construction practices for the sizing and spacing of fasteners. Adjustments to material costs may be required for your specific application or location. The manufacturer's warranty is assumed. Extended warranties are not included in the material costs. **Material costs do not include sales tax.**

Labor Costs

Labor costs are based upon a mathematical average of trade-specific wages in 30 major U.S. cities. The type of wage (union, open shop, or residential) is identified on the inside back cover of printed publications or selected by the estimator when using the electronic products. Markups for the wages can also be found on the inside back cover of printed publications and/or under the labor references found in the electronic products.

- If wage rates in your area vary from those used, or if rate increases are expected within a given year, labor costs should be adjusted accordingly.

Labor costs reflect productivity based on actual working conditions. In addition to actual installation, these figures include time spent during a normal weekday on tasks, such as material receiving and handling, mobilization at the site, site movement, breaks, and cleanup.

Productivity data is developed over an extended period so as not to be influenced by abnormal variations and reflects a typical average.

Equipment Costs

Equipment costs include not only rental but also operating costs for equipment under normal use. The operating costs include parts and labor for routine servicing, such as the repair and replacement of pumps, filters, and worn lines. Normal operating expendables, such as fuel, lubricants, tires, and electricity (where applicable), are also included. Extraordinary operating expendables with highly variable wear patterns, such as diamond bits and blades, are excluded. These costs are included under materials. Equipment rental rates are obtained from industry sources throughout North America—contractors, suppliers, dealers, manufacturers, and distributors.

Rental rates can also be treated as reimbursement costs for contractor-owned equipment. Owned equipment costs include depreciation, loan payments, interest, taxes, insurance, storage, and major repairs.

Equipment costs do not include operators' wages.

Equipment Cost/Day—The cost of equipment required for each crew is included in the Crew Listings in the Reference Section (small tools that are considered essential everyday tools are not listed out separately). The Crew Listings itemize specialized tools and heavy equipment along with labor trades. The daily cost of itemized equipment included in a crew is based on dividing the weekly bare rental rate by 5 (number of working days per week), then adding the hourly operating cost times 8 (the number of hours per day). This Equipment Cost/Day is shown in the last column of the Equipment Rental Costs in the Reference Section.

Mobilization, Demobilization—The cost to move construction equipment from an equipment yard or rental company to the job site and back again is not included in equipment costs. Mobilization (to the site) and demobilization (from the site) costs can be found in the Unit Price Section. If a piece of equipment is already at the job site, it is not appropriate to utilize mobilization or demobilization costs again in an estimate.

Overhead and Profit

Total Cost including O&P for the installing contractor is shown in the last column of the Unit Price and/or Assemblies. This figure is the sum of the bare material cost plus 10% for profit, the bare labor cost plus total overhead and profit, and the bare equipment cost plus 10% for profit. Details for the calculation of overhead and profit on labor are shown on the inside back cover of the printed product and in the Reference Section of the electronic product.

General Conditions

Cost data in this data set are presented in two ways: Bare Costs and Total Cost including O&P (Overhead and Profit). General Conditions, or General Requirements, of the contract should also be added to the Total Cost including O&P when applicable. Costs for General Conditions are listed in Division 1 of the Unit Price Section and in the Reference Section.

General Conditions for the installing contractor may range from 0% to 10% of the Total Cost including O&P. For the general or prime contractor, costs for General Conditions may range from 5% to 15% of the Total Cost including O&P, with a figure of 10% as the most typical allowance. If applicable, the Assemblies and Models sections use costs that include the installing contractor's overhead and profit (O&P).

Factors Affecting Costs

Costs can vary depending upon a number of variables. Here's a listing of some factors that affect costs and points to consider.

Quality—The prices for materials and the workmanship upon which productivity is based represent sound construction work. They are also in line with industry standard and manufacturer specifications and are frequently used by federal, state, and local governments.

Overtime—We have made no allowance for overtime. If you anticipate premium time or work beyond normal working hours, be sure to make an appropriate adjustment to your labor costs.

Productivity—The productivity, daily output, and labor-hour figures for each line item are based on an eight-hour work day in daylight hours in moderate temperatures and up to a 14' working height unless otherwise indicated. For work that extends beyond normal work hours or is performed under adverse conditions, productivity may decrease.

Size of Project—The size, scope of work, and type of construction project will have a significant impact on cost. Economies of scale can reduce costs for large projects. Unit costs can often run higher for small projects.

Location—Material prices are for metropolitan areas. However, in dense urban areas, traffic and site storage limitations may increase costs. Beyond a 20-mile radius of metropolitan areas, extra trucking or transportation charges may also increase the material costs slightly. On the other hand, lower wage rates may be in effect. Be sure to consider both of these factors when preparing an estimate, particularly if the job site is located in a central city or remote rural location. In addition, highly specialized subcontract items may require travel and per-diem expenses for mechanics.

Other Factors—

- season of year
- contractor management
- weather conditions
- local union restrictions
- building code requirements
- availability of:
  - adequate energy
  - skilled labor
  - building materials
- owner's special requirements/restrictions
- safety requirements
- environmental considerations
- access

Unpredictable Factors—General business conditions influence "in-place" costs of all items. Substitute materials and construction methods may have to be employed. These may affect the installed cost and/or life cycle costs. Such factors may be difficult to evaluate and cannot necessarily be predicted on the basis of the job's location in a particular section of the country. Thus, where these factors apply, you may find significant but unavoidable cost variations for which you will have to apply a measure of judgment to your estimate.

Rounding of Costs

In printed publications only, all unit prices in excess of $5.00 have been rounded to make them easier to use and still maintain adequate precision of the results.

How Subcontracted Items Affect Costs

A considerable portion of all large construction jobs is usually subcontracted. In fact, the percentage done by subcontractors is constantly increasing and may run over 90%. Since the workers employed by these companies do nothing else but install their particular products, they soon become experts in that line. As a result, installation by these firms is accomplished so efficiently that the total in-place cost, even with the general contractor's overhead and profit, is no more, and often less, than if the principal contractor had handled the installation. Companies that deal with construction specialties are anxious to have their products perform well and, consequently, the installation will be the best possible.

Contingencies

The allowance for contingencies generally provides for unforeseen construction difficulties. On alterations or repair jobs, 20% is not too much. If drawings are final and only field contingencies are being considered, 2% or 3% is probably sufficient and often nothing needs to be added. Contractually, changes in plans will be covered by extras. The contractor should consider inflationary price trends and possible material shortages during the course of the job. These escalation factors are dependent upon both economic conditions and the anticipated time between the estimate and actual construction. If drawings are not complete or approved, or a budget cost is wanted, it is wise to add 5% to 10%. Contingencies, then, are a matter of judgment.

Important Estimating Considerations

The productivity, or daily output, of each craftsman or crew assumes a well-managed job where tradesmen with the proper tools and equipment, along with the appropriate construction materials, are present. Included are daily set-up and cleanup time, break time, and plan layout time. Unless otherwise indicated, time for material movement on site (for items

that can be transported by hand) of up to 200' into the building and to the first or second floor is also included. If material has to be transported by other means, over greater distances, or to higher floors, an additional allowance should be considered by the estimator.

While horizontal movement is typically a sole function of distances, vertical transport introduces other variables that can significantly impact productivity. In an occupied building, the use of elevators (assuming access, size, and required protective measures are acceptable) must be understood at the time of the estimate. For new construction, hoist wait and cycle times can easily be 15 minutes and may result in scheduled access extending beyond the normal work day. Finally, all vertical transport will impose strict weight limits likely to preclude the use of any motorized material handling.

The productivity, or daily output, also assumes installation that meets manufacturer/designer/standard specifications. A time allowance for quality control checks, minor adjustments, and any task required to ensure proper function or operation is also included. For items that require connections to services, time is included for positioning, leveling, securing the unit, and making all the necessary connections (and start up where applicable) to ensure a complete installation. Estimating of the services themselves (electrical, plumbing, water, steam, hydraulics, dust collection, etc.) is separate.

In some cases, the estimator must consider the use of a crane and an appropriate crew for the installation of large or heavy items. For those situations where a crane is not included in the assigned crew and as part of the line item cost,

then equipment rental costs, mobilization and demobilization costs, and operator and support personnel costs must be considered.

Labor-Hours

The labor-hours expressed in this publication are derived by dividing the total daily labor-hours for the crew by the daily output. Based on average installation time and the assumptions listed above, the labor-hours include: direct labor, indirect labor, and nonproductive time. A typical day for a craftsman might include but is not limited to:

- Direct Work
  - ☐ Measuring and layout
  - ☐ Preparing materials
  - ☐ Actual installation
  - ☐ Quality assurance/quality control
- Indirect Work
  - ☐ Reading plans or specifications
  - ☐ Preparing space
  - ☐ Receiving materials
  - ☐ Material movement
  - ☐ Giving or receiving instruction
  - ☐ Miscellaneous
- Non-Work
  - ☐ Chatting
  - ☐ Personal issues
  - ☐ Breaks
  - ☐ Interruptions (i.e., sickness, weather, material or equipment shortages, etc.)

If any of the items for a typical day do not apply to the particular work or project situation, the estimator should make any necessary adjustments.

Final Checklist

Estimating can be a straightforward process provided you remember the basics. Here's a checklist of some of the steps you should remember to complete before finalizing your estimate.

Did you remember to:

- factor in the City Cost Index for your locale?
- take into consideration which items have been marked up and by how much?
- mark up the entire estimate sufficiently for your purposes?
- read the background information on techniques and technical matters that could impact your project time span and cost?
- include all components of your project in the final estimate?
- double check your figures for accuracy?
- call RSMeans data engineers if you have any questions about your estimate or the data you've used? Remember, Gordian stands behind all of our products, including our extensive RSMeans data solutions. If you have any questions about your estimate, about the costs you've used from our data, or even about the technical aspects of the job that may affect your estimate, feel free to call the Gordian RSMeans editors at 1.800.448.8182.

Unit Price Section

Table of Contents

1

Table of Contents (cont.)

RSMeans data: Unit Prices— How They Work

All RSMeans data: Unit Prices are organized in the same way.

03 30 Cast-In-Place Concrete

03 30 53 – Miscellaneous Cast-In-Place Concrete

03 30 53.40 Concrete In Place		Crew	Daily Output	Labor-Hours	Unit	Material	2021 Bare Costs Labor	Equipment	Total	Total Incl O&P
0010	**CONCRETE IN PLACE** R033053-60									
0020	Including forms (4 uses), Grade 60 rebar, concrete (Portland cement									
0050	Type I), placement and finishing unless otherwise indicated R033105-20									
0300	Beams (3500 psi), 5 kip/L.F., 10' span R033105-70	C-14A	15.62	12.804	C.Y.	420	700	29	1,149	1,550
0350	25' span	"	18.55	10.782		440	590	24.50	1,054.50	1,375
0500	Chimney foundations (5000 psi), over 5 C.Y.	C-14C	32.22	3.476		187	182	.83	369.83	480
0510	(3500 psi), under 5 C.Y.	"	23.71	4.724		223	247	1.13	471.13	615
0700	Columns, square (4000 psi), 12" x 12", up to 1% reinforcing by area	C-14A	11.96	16.722		475	915	37.50	1,427.50	1,950
3540	Equipment pad (3000 psi), 3' x 3' x 6" thick	C-14H	45	1.067	Ea.	55.50	57	.61	113.11	147
3550	4' x 4' x 6" thick		30	1.600		84	85.50	.91	170.41	220
3560	5' x 5' x 8" thick		18	2.667		150	143	1.52	294.52	380
3570	6' x 6' x 8" thick		14	3.429		204	184	1.95	389.95	500
3580	8' x 8' x 10" thick		8	6		415	320	3.42	738.42	945
3590	10' x 10' x 12" thick		5	9.600		720	515	5.45	1,240.45	1,550

It is important to understand the structure of RSMeans data: Unit Prices so that you can find information easily and use it correctly.

---

1 Line Numbers

Line Numbers consist of 12 characters, which identify a unique location in the database for each task. The first 6 or 8 digits conform to the Construction Specifications Institute MasterFormat® 2018. The remainder of the digits are a further breakdown in order to arrange items in understandable groups of similar tasks. Line numbers are consistent across all of our publications, so a line number in any of our products will always refer to the same item of work.

2 Descriptions

Descriptions are shown in a hierarchical structure to make them readable. In order to read a complete description, read up through the indents to the top of the section. Include everything that is above and to the left that is not contradicted by information below. For instance, the complete description for line 03 30 53.40 3550 is "Concrete in place, including forms (4 uses), Grade 60 rebar, concrete (Portland cement Type 1), placement and finishing unless otherwise indicated; Equipment pad (3,000 psi), 4' × 4' × 6" thick."

3 RSMeans data

When using **RSMeans data**, it is important to read through an entire section to ensure that you use the data that most closely matches your work. Note that sometimes there is additional information shown in the section that may improve your price. There are frequently lines that further describe, add to, or adjust data for specific situations.

4 Reference Information

Gordian's RSMeans engineers have created **reference** information to assist you in your estimate. **If** there is information that applies to a section, it will be indicated at the start of the section. The Reference Section is located in the back of the data set.

5 Crews

Crews include labor and/or equipment necessary to accomplish each task. In this case, Crew C-14H is used. Gordian's RSMeans staff selects a crew to represent the workers and equipment that are

typically used for that task. In this case, Crew C-14H consists of one carpenter foreman (outside), two carpenters, one rodman, one laborer, one cement finisher, and one gas engine vibrator. Details of all crews can be found in the Reference Section.

Crews - Standard

Crew No.	Bare Costs		Incl. Subs O & P		Cost Per Labor-Hour	
Crew C-14H	Hr.	Daily	Hr.	Daily	Bare Costs	Incl. O&P
1 Carpenter Foreman (outside)	$56.70	$453.60	$84.60	$676.80	$53.53	$79.68
2 Carpenters	54.70	875.20	81.65	1306.40		
1 Rodman (reinf.)	58.90	471.20	88.05	704.40		
1 Laborer	44.40	355.20	66.25	530.00		
1 Cement Finisher	51.80	414.40	75.90	607.20		
1 Gas Engine Vibrator		27.15		29.86	.57	.62
48 L.H., Daily Totals		$2596.75		$3854.67	$54.10	$80.31

6 Daily Output

The **Daily Output** is the amount of work that the crew can do in a normal 8-hour workday, including mobilization, layout, movement of materials, and cleanup. In this case, crew C-14H can install thirty 4' × 4' × 6" thick concrete pads in a day. Daily output is variable and based on many factors, including the size of the job, location, and environmental conditions. RSMeans data represents work done in daylight (or adequate lighting) and temperate conditions.

7 Labor-Hours

The figure in the **Labor-Hours** column is the amount of labor required to perform one unit of work—in this case the amount of labor required to construct one 4' × 4' equipment pad. This figure is calculated by dividing the number of hours of labor in the crew by the daily output (48 labor-hours divided by 30 pads = 1.6 hours of labor per pad). Multiply 1.6 times 60 to see the value in minutes: 60 × 1.6 = 96

minutes. Note: the labor-hour figure is not dependent on the crew size. A change in crew size will result in a corresponding change in daily output, but the labor-hours per unit of work will not change.

8 Unit of Measure

All RSMeans data: Unit Prices include the typical **Unit of Measure** used for estimating that item. For concrete-in-place the typical unit is cubic yards (C.Y.) or each (Ea.). For installing broadloom carpet it is square yard and for gypsum board it is square foot. The estimator needs to take special care that the unit in the data matches the unit in the take-off. Unit conversions may be found in the Reference Section.

9 Bare Costs

Bare Costs are the costs of materials, labor, and equipment that the installing contractor pays. They represent the cost, in U.S. dollars, for one unit of work. They do not include any markups for profit or labor burden.

10 Bare Total

The **Total column** represents the total bare cost for the installing contractor in U.S. dollars. In this case, the sum of $84 for material + $85.50 for labor + $.91 for equipment is $170.41.

11 Total Incl O&P

The **Total Incl O&P column** is the total cost, including overhead and profit, that the installing contractor will charge the customer. This represents the cost of materials plus 10% profit, the cost of labor plus labor burden and 10% profit, and the cost of equipment plus 10% profit. It does not include the general contractor's overhead and profit. Note: See the inside back cover of the printed product or the Reference Section of the electronic product for details on how the labor burden is calculated.

National Average

The RSMeans data in our print publications represent a "national average" cost. This data should be modified to the project location using the **City Cost Indexes** or **Location Factors** tables found in the Reference Section. Use the Location Factors to adjust estimate totals if the project covers multiple trades. Use the City Cost Indexes (CCI) for single trade projects or projects where a more detailed analysis is required. All figures in the two tables are derived from the same research. The last row of data in the CCI—the weighted average—is the same as the numbers reported for each location in the location factor table.

Project Name: Pre-Engineered Steel Building   **Architect: As Shown**

Location: Anywhere, USA   01/01/21   STD

Line Number	Description	Qty	Unit	Material	Labor	Equipment	SubContract	Estimate Total
03 30 53.40 3940	Strip footing, 12" x 24", reinforced	15	C.Y.	$2,625.00	$1,830.00	$8.40	$0.00	
03 30 53.40 3950	Strip footing, 12" x 36", reinforced	34	C.Y.	$5,644.00	$3,315.00	$15.30	$0.00	
03 11 13.65 3000	Concrete slab edge forms	500	L.F.	$225.00	$1,390.00	$0.00	$0.00	
03 22 11.10 0200	Welded wire fabric reinforcing	150	C.S.F.	$3,150.00	$4,575.00	$0.00	$0.00	
03 31 13.35 0300	Ready mix concrete, 4000 psi for slab on grade	278	C.Y.	$35,862.00	$0.00	$0.00	$0.00	
03 31 13.70 4300	Place, strike off & consolidate concrete slab	278	C.Y.	$0.00	$5,560.00	$136.22	$0.00	
03 35 13.30 0250	Machine float & trowel concrete slab	15,000	S.F.	$0.00	$10,350.00	$750.00	$0.00	
03 15 16.20 0140	Cut control joints in concrete slab	950	L.F.	$38.00	$437.00	$57.00	$0.00	
03 39 23.13 0300	Sprayed concrete curing membrane	150	C.S.F.	$1,830.00	$1,125.00	$0.00	$0.00	
Division 03	**Subtotal**			**$49,374.00**	**$28,582.00**	**$966.92**	**$0.00**	**$78,922.92**
08 36 13.10 2650	Manual 10' x 10' steel sectional overhead door	8	Ea.	$9,800.00	$3,880.00	$0.00	$0.00	
08 36 13.10 2860	Insulation and steel back panel for OH door	800	S.F.	$4,760.00	$0.00	$0.00	$0.00	
Division 08	**Subtotal**			**$14,560.00**	**$3,880.00**	**$0.00**	**$0.00**	**$18,440.00**
13 34 19.50 1100	Pre-Engineered Steel Building, 100' x 150' x 24'	15,000	SF Flr.	$0.00	$0.00	$0.00	$367,500.00	
13 34 19.50 6050	Framing for PESB door opening, 3' x 7'	4	Opng.	$0.00	$0.00	$0.00	$2,300.00	
13 34 19.50 6100	Framing for PESB door opening, 10' x 10'	8	Opng.	$0.00	$0.00	$0.00	$9,400.00	
13 34 19.50 6200	Framing for PESB window opening, 4' x 3'	6	Opng.	$0.00	$0.00	$0.00	$3,360.00	
13 34 19.50 5750	PESB door, 3' x 7', single leaf	4	Opng.	$2,740.00	$772.00	$0.00	$0.00	
13 34 19.50 7750	PESB sliding window, 4' x 3' with screen	6	Opng.	$2,940.00	$660.00	$68.70	$0.00	
13 34 19.50 6550	PESB gutter, eave type, 26 ga., painted	300	L.F.	$2,595.00	$906.00	$0.00	$0.00	
13 34 19.50 8650	PESB roof vent, 12" wide x 10' long	15	Ea.	$705.00	$3,615.00	$0.00	$0.00	
13 34 19.50 6900	PESB insulation, vinyl faced, 4" thick	27,400	S.F.	$14,522.00	$10,412.00	$0.00	$0.00	
Division 13	**Subtotal**			**$23,502.00**	**$16,365.00**	**$68.70**	**$382,560.00**	**$422,495.70**
			Subtotal	$87,436.00	$48,827.00	$1,035.62	$382,560.00	$519,858.62
Division 01	**General Requirements @ 7%**			6,120.52	3,417.89	72.49	26,779.20	
			Estimate Subtotal	$93,556.52	$52,244.89	$1,108.11	$409,339.20	$519,858.62
			Sales Tax @ 5%	4,677.83		55.41	10,233.48	
			Subtotal A	98,234.35	52,244.89	1,163.52	419,572.68	
			GC O & P	9,823.43	26,331.42	116.35	41,957.27	
			Subtotal B	108,057.78	78,576.31	1,279.87	461,529.95	$649,443.91
			Contingency @ 5%					32,472.20
			Subtotal C					$681,916.11
			Bond @ $12/1000 +10% O&P					9,001.29
			Subtotal D					$690,917.40
			Location Adjustment Factor		114.20			98,110.27
			Grand Total					**$789,027.67**

This estimate is based on an interactive spreadsheet. You are free to download it and adjust it to your methodology.
A copy of this spreadsheet is available at **RSMeans.com/2021books.**

Sample Estimate

This sample demonstrates the elements of an estimate, including a tally of the RSMeans data lines and a summary of the markups on a contractor's work to arrive at a total cost to the owner. The Location Factor with RSMeans data is added at the bottom of the estimate to adjust the cost of the work to a specific location.

---

1 Work Performed

The body of the estimate shows the RSMeans data selected, including the line number, a brief description of each item, its take-off unit and quantity, and the bare costs of materials, labor, and equipment. This estimate also includes a column titled "SubContract." This data is taken from the column "Total Incl O&P" and represents the total that a subcontractor would charge a general contractor for the work, including the sub's markup for overhead and profit.

2 Division 1, General Requirements

This is the first division numerically but the last division estimated. Division 1 includes project-wide needs provided by the general contractor. These requirements vary by project but may include temporary facilities and utilities, security, testing, project cleanup, etc. For small projects a percentage can be used—typically between 5% and 15% of project cost. For large projects the costs may be itemized and priced individually.

3 Sales Tax

If the work is subject to state or local sales taxes, the amount must be added to the estimate. Sales tax may be added to material costs, equipment costs, and subcontracted work. In this case, sales tax was added in all three categories. It was assumed that approximately half the subcontracted work would be material cost, so the tax was applied to 50% of the subcontract total.

4 GC O&P

This entry represents the general contractor's markup on material, labor, equipment, and subcontractor costs. Our standard markup on materials, equipment, and subcontracted work is 10%. In this estimate, the markup on the labor performed by the GC's workers uses "Skilled Workers Average" shown in Column F on the table "Installing Contractor's Overhead & Profit," which can be found on the inside back cover of the printed product or in the Reference Section of the electronic product.

5 Contingency

A factor for contingency may be added to any estimate to represent the cost of unknowns that may occur between the time that the estimate is performed and the time the project is constructed. The amount of the allowance will depend on the stage of design at which the estimate is done and the contractor's assessment of the risk involved. Refer to section 01 21 16.50 for contingency allowances.

6 Bonds

Bond costs should be added to the estimate. The figures here represent a typical performance bond, ensuring the owner that if the general contractor does not complete the obligations in the construction contract the bonding company will pay the cost for completion of the work.

7 Location Adjustment

Published prices are based on national average costs. If necessary, adjust the total cost of the project using a location factor from the "Location Factor" table or the "City Cost Index" table. Use location factors if the work is general, covering multiple trades. If the work is by a single trade (e.g., masonry) use the more specific data found in the "City Cost Indexes."

Estimating Tips
01 20 00 Price and Payment Procedures

- Allowances that should be added to estimates to cover contingencies and job conditions that are not included in the national average material and labor costs are shown in Section 01 21.
- When estimating historic preservation projects (depending on the condition of the existing structure and the owner's requirements), a 15–20% contingency or allowance is recommended, regardless of the stage of the drawings.

01 30 00 Administrative Requirements

- Before determining a final cost estimate, it is good practice to review all the items listed in Subdivisions 01 31 and 01 32 to make final adjustments for items that may need customizing to specific job conditions.
- Requirements for initial and periodic submittals can represent a significant cost to the General Requirements of a job. Thoroughly check the submittal specifications when estimating a project to determine any costs that should be included.

01 40 00 Quality Requirements

- All projects will require some degree of quality control. This cost is not included in the unit cost of construction listed in each division. Depending upon the terms of the contract, the various costs of inspection and testing can be the responsibility of either the owner or the contractor. Be sure to include the required costs in your estimate.

01 50 00 Temporary Facilities and Controls

- Barricades, access roads, safety nets, scaffolding, security, and many more requirements for the execution of a safe project are elements of direct cost. These costs can easily be overlooked when preparing an estimate. When looking through the major classifications of this subdivision, determine which items apply to each division in your estimate.
- Construction equipment rental costs can be found in the Reference Section in Section 01 54 33. Operators' wages are not included in equipment rental costs.
- Equipment mobilization and demobilization costs are not included in equipment rental costs and must be considered separately.

- The cost of small tools provided by the installing contractor for his workers is covered in the "Overhead" column on the "Installing Contractor's Overhead and Profit" table that lists labor trades, base rates, and markups. Therefore, it is included in the "Total Incl. O&P" cost of any unit price line item.

01 70 00 Execution and Closeout Requirements

- When preparing an estimate, thoroughly read the specifications to determine the requirements for Contract Closeout. Final cleaning, record documentation, operation and maintenance data, warranties and bonds, and spare parts and maintenance materials can all be elements of cost for the completion of a contract. Do not overlook these in your estimate.

Reference Numbers

Reference numbers are shown at the beginning of some major classifications. These numbers refer to related items in the Reference Section. The reference information may be an estimating procedure, an alternate pricing method, or technical information.

Note: Not all subdivisions listed here necessarily appear. ■

Same Data. Simplified.

Enjoy the convenience and efficiency of accessing your costs anywhere:

- **Skip the multiplier** by setting your location
- **Quickly search,** edit, favorite and share costs
- **Stay on top of price changes** with automatic updates

Discover more at rsmeans.com/online

---

01 11 Summary of Work

01 11 31 – Professional Consultants

01 11 31.10 Architectural Fees		Crew	Daily Output	Labor-Hours	Unit	Material	2021 Bare Costs Labor	Equipment	Total	Total Incl O&P
0010	**ARCHITECTURAL FEES**	R011110-10								
0020	For new construction									
0060	Minimum				Project				4.90%	4.90%
0090	Maximum				"				16%	16%

01 11 31.20 Construction Management Fees

0010	**CONSTRUCTION MANAGEMENT FEES**									
0020	$1,000,000 job, minimum				Project				4.50%	4.50%
0050	Maximum								7.50%	7.50%
0300	$50,000,000 job, minimum								2.50%	2.50%
0350	Maximum				↓				4%	4%

01 11 31.30 Engineering Fees

0010	**ENGINEERING FEES**	R011110-30								
0800	Landscaping & site development, minimum				Contrct				2.50%	2.50%
0900	Maximum				"				6%	6%

01 11 31.50 Models

0010	**MODELS**									
0500	2 story building, scaled 100' x 200', simple materials and details				Ea.	4,775			4,775	5,250
0510	Elaborate materials and details				"	31,400			31,400	34,500

01 11 31.75 Renderings

0010	**RENDERINGS** Color, matted, 20" x 30", eye level,									
0020	1 building, minimum				Ea.	2,375			2,375	2,600
0050	Average					3,425			3,425	3,775
0100	Maximum					5,600			5,600	6,150
1000	5 buildings, minimum					4,125			4,125	4,550
1100	Maximum					8,250			8,250	9,075
2000	Aerial perspective, color, 1 building, minimum					3,100			3,100	3,425
2100	Maximum					8,550			8,550	9,400
3000	5 buildings, minimum					6,650			6,650	7,300
3100	Maximum				↓	12,400			12,400	13,600

01 21 Allowances

01 21 16 – Contingency Allowances

01 21 16.50 Contingencies

0010	**CONTINGENCIES**, Add to estimate									
0020	Conceptual stage				Project				20%	20%
0050	Schematic stage								15%	15%
0100	Preliminary working drawing stage (Design Dev.)								10%	10%
0150	Final working drawing stage				↓				3%	3%

01 21 53 – Factors Allowance

01 21 53.50 Factors

0010	**FACTORS** Cost adjustments									
0100	Add to construction costs for particular job requirements									
1100	Equipment usage curtailment, add, minimum				Costs	1%	1%			
1150	Maximum				"	3%	10%			
1400	Material handling & storage limitation, add, minimum				Costs	1%	1%			
1450	Maximum					6%	7%			
2300	Temporary shoring and bracing, add, minimum					2%	5%			
2350	Maximum				↓	5%	12%			

01 21 Allowances

01 21 53 – Factors Allowance

01 21 53.60 Security Factors

		Crew	Daily Output	Labor-Hours	Unit	Material	2021 Bare Costs Labor	Equipment	Total	Total Incl O&P
0010	**SECURITY FACTORS** R012153-60									
0100	Additional costs due to security requirements									
0110	Daily search of personnel, supplies, equipment and vehicles									
0120	Physical search, inventory and doc of assets, at entry				Costs		30%			
0130	At entry and exit						50%			
0140	Physical search, at entry						6.25%			
0150	At entry and exit						12.50%			
0160	Electronic scan search, at entry						2%			
0170	At entry and exit						4%			
0180	Visual inspection only, at entry						.25%			
0190	At entry and exit						.50%			
0200	ID card or display sticker only, at entry						.12%			
0210	At entry and exit				▼		.25%			
0220	Day 1 as described below, then visual only for up to 5 day job duration									
0230	Physical search, inventory and doc of assets, at entry				Costs		5%			
0240	At entry and exit						10%			
0250	Physical search, at entry						1.25%			
0260	At entry and exit						2.50%			
0270	Electronic scan search, at entry						.42%			
0280	At entry and exit				▼		.83%			
0290	Day 1 as described below, then visual only for 6-10 day job duration									
0300	Physical search, inventory and doc of assets, at entry				Costs		2.50%			
0310	At entry and exit						5%			
0320	Physical search, at entry						.63%			
0330	At entry and exit						1.25%			
0340	Electronic scan search, at entry						.21%			
0350	At entry and exit				▼		.42%			
0360	Day 1 as described below, then visual only for 11-20 day job duration									
0370	Physical search, inventory and doc of assets, at entry				Costs		1.25%			
0380	At entry and exit						2.50%			
0390	Physical search, at entry						.31%			
0400	At entry and exit						.63%			
0410	Electronic scan search, at entry						.10%			
0420	At entry and exit				▼		.21%			
0430	Beyond 20 days, costs are negligible									
0440	Escort required to be with tradesperson during work effort				Costs		6.25%			

01 21 53.65 Infectious Disease Precautions

		Crew	Daily Output	Labor-Hours	Unit	Material	2021 Bare Costs Labor	Equipment	Total	Total Incl O&P
0010	**INFECTIOUS DISEASE PRECAUTIONS** R012153-65									
0100	Additional costs due to infectious disease precautions									
0120	Daily temperature checks				Costs		1%			
0130	Donning & doffing masks and gloves						1%			
0140	Washing hands						1%			
0150	Informational meetings						2%			
0160	Maintaining social distance						1%			
0170	Disinfecting tools or equipment				▼		3%			
0200	N95 rated masks				Ea.	1.40			1.40	1.54
0210	Surgical masks				"	.16			.16	.18
0220	Black nitrile disposable gloves				Pair	.18			.18	.20
0250	Hand washing station & 2 x weekly service				Week				158	174

For customer support on your Site Work & Landscape Costs with RSMeans Data, call 800.448.8182.

11

01 21 Allowances

01 21 57 – Overtime Allowance

01 21 57.50 Overtime		Crew	Daily Output	Labor-Hours	Unit	Material	2021 Bare Costs Labor	2021 Bare Costs Equipment	Total	Total Incl O&P
0010	**OVERTIME** for early completion of projects or where	R012909-90								
0020	labor shortages exist, add to usual labor, up to					Costs		100%			

01 21 63 – Taxes

01 21 63.10 Taxes

		Crew	Daily Output	Labor-Hours	Unit	Material	2021 Bare Costs Labor	2021 Bare Costs Equipment	Total	Total Incl O&P
0010	**TAXES**	R012909-80								
0020	Sales tax, State, average					%	5.08%			
0050	Maximum	R012909-85					7.50%			
0200	Social Security, on first $118,500 of wages							7.65%		
0300	Unemployment, combined Federal and State, minimum	R012909-86						.60%		
0350	Average							9.60%		
0400	Maximum							12%		

01 31 Project Management and Coordination

01 31 13 – Project Coordination

01 31 13.20 Field Personnel

		Crew	Daily Output	Labor-Hours	Unit	Material	2021 Bare Costs Labor	2021 Bare Costs Equipment	Total	Total Incl O&P
0010	**FIELD PERSONNEL**									
0020	Clerk, average				Week		495		495	750
0100	Field engineer, junior engineer						1,241		1,241	1,877
0120	Engineer						1,825		1,825	2,775
0140	Senior engineer						2,400		2,400	3,625
0160	General purpose laborer, average	1 Clab	.20	40			1,775		1,775	2,650
0180	Project manager, minimum						2,175		2,175	3,300
0200	Average						2,500		2,500	3,800
0220	Maximum						2,850		2,850	4,325
0240	Superintendent, minimum						2,125		2,125	3,225
0260	Average						2,325		2,325	3,525
0280	Maximum						2,650		2,650	4,025
0290	Timekeeper, average				Week		1,350		1,350	2,050

01 31 13.30 Insurance

		Crew	Daily Output	Labor-Hours	Unit	Material	2021 Bare Costs Labor	2021 Bare Costs Equipment	Total	Total Incl O&P
0010	**INSURANCE**	R013113-40								
0020	Builders risk, standard, minimum				Job				.24%	.24%
0050	Maximum	R013113-50							.80%	.80%
0200	All-risk type, minimum								.25%	.25%
0250	Maximum	R013113-60							.62%	.62%
0400	Contractor's equipment floater, minimum				Value				.50%	.50%
0450	Maximum				"				1.50%	1.50%
0600	Public liability, average				Job				2.02%	2.02%
0800	Workers' compensation & employer's liability, average									
0850	by trade, carpentry, general				Payroll		11.97%			
0900	Clerical						.38%			
0950	Concrete						10.84%			
1000	Electrical						4.91%			
1050	Excavation						7.81%			
1250	Masonry						13.40%			
1300	Painting & decorating						10.44%			
1350	Pile driving						12.06%			
1450	Plumbing						5.77%			
1600	Steel erection, structural						17.21%			
1650	Tile work, interior ceramic						8.10%			
1700	Waterproofing, brush or hand caulking						6.05%			

For customer support on your Site Work & Landscape Costs with RSMeans Data, call 800.448.8182.

01 31 Project Management and Coordination

01 31 13 – Project Coordination

01 31 13.30 Insurance

		Crew	Daily Output	Labor-Hours	Unit	Material	2021 Bare Costs Labor	Equipment	Total	Total Incl O&P
1800	Wrecking				Payroll		15.48%			
2000	Range of 35 trades in 50 states, excl. wrecking & clerical, min.						1.37%			
2100	Average						10.60%			
2200	Maximum						120.29%			

01 31 13.40 Main Office Expense

		Crew	Daily Output	Labor-Hours	Unit	Material	2021 Bare Costs Labor	Equipment	Total	Total Incl O&P
0010	**MAIN OFFICE EXPENSE** Average for General Contractors R013113-50									
0020	As a percentage of their annual volume									
0125	Annual volume under $1,000,000				% Vol.				17.50%	
0145	Up to $2,500,000								8%	
0150	Up to $4,000,000								6.80%	
0200	Up to $7,000,000								5.60%	
0250	Up to $10,000,000								5.10%	
0300	Over $10,000,000								3.90%	

01 31 13.50 General Contractor's Mark-Up

		Crew	Daily Output	Labor-Hours	Unit	Material	2021 Bare Costs Labor	Equipment	Total	Total Incl O&P
0010	**GENERAL CONTRACTOR'S MARK-UP** on Change Orders									
0200	Extra work, by subcontractors, add				%				10%	10%
0250	By General Contractor, add								15%	15%
0400	Omitted work, by subcontractors, deduct all but								5%	5%
0450	By General Contractor, deduct all but								7.50%	7.50%
0600	Overtime work, by subcontractors, add				%				15%	15%
0650	By General Contractor, add				"				10%	10%

01 31 13.80 Overhead and Profit

		Crew	Daily Output	Labor-Hours	Unit	Material	2021 Bare Costs Labor	Equipment	Total	Total Incl O&P
0010	**OVERHEAD & PROFIT** Allowance to add to items in this R013113-50									
0020	book that do not include Subs O&P, average				%				25%	
0100	Allowance to add to items in this book that R013113-55									
0110	do include Subs O&P, minimum				%				5%	5%
0150	Average								10%	10%
0200	Maximum								15%	15%
0300	Typical, by size of project, under $100,000								30%	
0350	$500,000 project								25%	
0400	$2,000,000 project								20%	
0450	Over $10,000,000 project								15%	

01 31 13.90 Performance Bond

		Crew	Daily Output	Labor-Hours	Unit	Material	2021 Bare Costs Labor	Equipment	Total	Total Incl O&P
0010	**PERFORMANCE BOND**									
0020	For buildings, minimum				Job				.60%	.60%
0100	Maximum								2.50%	2.50%
0190	Highways & bridges, new construction, minimum								1%	1%
0200	Maximum								1.50%	1.50%
0300	Highways & bridges, resurfacing, minimum								.40%	.40%
0350	Maximum								.94%	.94%

01 31 14 – Facilities Services Coordination

01 31 14.20 Lock Out/Tag Out

		Crew	Daily Output	Labor-Hours	Unit	Material	2021 Bare Costs Labor	Equipment	Total	Total Incl O&P
0010	**LOCK OUT/TAG OUT**									
0020	Miniature circuit breaker lock out device	1 Elec	220	.036	Ea.	24	2.32		26.32	30
0030	Miniature pin circuit breaker lock out device		220	.036		19.90	2.32		22.22	25.50
0040	Single circuit breaker lock out device		220	.036		21	2.32		23.32	26.50
0050	Multi-pole circuit breaker lock out device (15 to 225 Amp)		210	.038		19.90	2.43		22.33	25.50
0060	Large 3 pole circuit breaker lock out device (over 225 Amp)		210	.038		21	2.43		23.43	27
0080	Square D I-Line circuit breaker lock out device		210	.038		33	2.43		35.43	39.50
0090	Lock out disconnect switch, 30 to 100 Amp		330	.024		11	1.54		12.54	14.40
0100	100 to 400 Amp		330	.024		11	1.54		12.54	14.40

For customer support on your Site Work & Landscape Costs with RSMeans Data, call 800.448.8182.

13

01 31 Project Management and Coordination

01 31 14 – Facilities Services Coordination

01 31 14.20 Lock Out/Tag Out	Crew	Daily Output	Labor-Hours	Unit	Material	2021 Bare Costs Labor	Equipment	Total	Total Incl O&P
0110	Over 400 Amp	1 Elec	330	.024	Ea.	11	1.54		12.54	14.40
0120	Lock out hasp for multiple lockout tags		200	.040		5.60	2.55		8.15	9.95
0130	Electrical cord plug lock out device		220	.036		10.50	2.32		12.82	15
0140	Electrical plug prong lock out device (3-wire grounding plug)		220	.036		6.95	2.32		9.27	11.10
0150	Wall switch lock out		200	.040		25	2.55		27.55	32
0160	Fire alarm pull station lock out	1 Stpi	200	.040		18.75	2.73		21.48	24.50
0170	Sprinkler valve tamper and flow switch lock out device	1 Skwk	220	.036		16.60	2.08		18.68	21.50
0180	Lock out sign		330	.024		18.90	1.38		20.28	23
0190	Lock out tag		440	.018		5	1.04		6.04	7.05

01 32 Construction Progress Documentation

01 32 13 – Scheduling of Work

01 32 13.50 Scheduling of Work

0010	**SCHEDULING**								
0025	Scheduling, critical path, $50 million project, initial schedule				Ea.			25,750	28,325
0030	Monthly updates				"			3,348	3,682

01 32 33 – Photographic Documentation

01 32 33.50 Photographs

0010	**PHOTOGRAPHS**								
0020	8" x 10", 4 shots, 2 prints ea., std. mounting				Set	545		545	600
0100	Hinged linen mounts					550		550	605
0200	8" x 10", 4 shots, 2 prints each, in color					575		575	630
0300	For I.D. slugs, add to all above					5.10		5.10	5.60
0500	Aerial photos, initial fly-over, 5 shots, digital images					350		350	385
0550	10 shots, digital images, 1 print					595		595	655
0600	For each additional print from fly-over					310		310	340
0700	For full color prints, add					40%			
0750	Add for traffic control area				Ea.	360		360	395
0900	For over 30 miles from airport, add per				Mile	7.20		7.20	7.90
1500	Time lapse equipment, camera and projector, buy				Ea.	2,600		2,600	2,850
1550	Rent per month				"	1,325		1,325	1,450
1700	Cameraman and processing, black & white				Day	1,300		1,300	1,425
1720	Color				"	1,525		1,525	1,675

01 41 Regulatory Requirements

01 41 26 – Permit Requirements

01 41 26.50 Permits

0010	**PERMITS**								
0020	Rule of thumb, most cities, minimum				Job			.50%	.50%
0100	Maximum				"			2%	2%

01 45 23 – Testing and Inspecting Services

01 45 23.50 Testing	Crew	Daily Output	Labor-Hours	Unit	Material	2021 Bare Costs Labor	Equipment	Total	Total Incl O&P
0010 **TESTING** and Inspecting Services									
0080 Supervision of Earthwork per day	1 Skwk	1	8	Day		455		455	685
0082 Quality Control of Earthwork per day	1 Clab	1	8	"		355		355	530
0200 Asphalt testing, compressive strength Marshall stability, set of 3				Ea.				545	600
0250 Extraction, individual tests on sample								236	260
0300 Penetration								91	100
0350 Mix design, 5 specimens								182	200
0360 Additional specimen								36	40
0400 Specific gravity								53	58
0420 Swell test								64	70
0450 Water effect and cohesion, set of 6				Ea.				182	200
0470 Water effect and plastic flow								64	70
0600 Concrete testing, aggregates, abrasion, ASTM C 131								205	225
0650 Absorption, ASTM C 127								77	85
0800 Petrographic analysis, ASTM C 295								910	1,000
0900 Specific gravity, ASTM C 127								77	85
1000 Sieve analysis, washed, ASTM C 136								130	140
1050 Unwashed								130	140
1200 Sulfate soundness								182	200
1300 Weight per cubic foot								80	88
1500 Cement, physical tests, ASTM C 150								320	350
1600 Chemical tests, ASTM C 150								245	270
1800 Compressive test, cylinder, delivered to lab, ASTM C 39								36	40
1900 Picked up by lab, 30 minute round trip, add to above	1 Skwk	16	.500			28.50		28.50	43
1950 1 hour round trip, add to above		8	1			57		57	86
2000 2 hour round trip, add to above		4	2			114		114	172
2200 Compressive strength, cores (not incl. drilling), ASTM C 42								95	105
2250 Core drilling, 4" diameter (plus technician), up to 6" thick	B-89A	14	1.143		.80	58	8.70	67.50	97.50
2260 Technician for core drilling				Hr.				50	55
2300 Patching core holes, to 12" diameter	1 Cefi	22	.364	Ea.	35	18.85		53.85	66
2400 Drying shrinkage at 28 days								236	260
2500 Flexural test beams, ASTM C 78								136	150
2600 Mix design, one batch mix								259	285
2650 Added trial batches								120	132
2800 Modulus of elasticity, ASTM C 469								195	215
2900 Tensile test, cylinders, ASTM C 496								52	58
3000 Water-Cement ratio curve, 3 batches								141	155
3100 4 batches								186	205
3300 Masonry testing, absorption, per 5 brick, ASTM C 67								72	80
3350 Chemical resistance, per 2 brick								50	55
3400 Compressive strength, per 5 brick, ASTM C 67								95	105
3420 Efflorescence, per 5 brick, ASTM C 67								101	112
3440 Imperviousness, per 5 brick								87	96
3470 Modulus of rupture, per 5 brick								86	95
3500 Moisture, block only								68	75
3550 Mortar, compressive strength, set of 3								32	35
4100 Reinforcing steel, bend test				Ea.				68	75
4200 Tensile test, up to #8 bar								68	75
4220 #9 to #11 bar								114	125
4240 #14 bar and larger								182	200
4400 Soil testing, Atterberg limits, liquid and plastic limits								91	100
4510 Hydrometer analysis								155	170
4600 Sieve analysis, washed, ASTM D 422								136	150

01 45 Quality Control

01 45 23 – Testing and Inspecting Services

01 45 23.50 Testing		Crew	Daily Output	Labor-Hours	Unit	Material	2021 Bare Costs Labor	Equipment	Total	Total Incl O&P
4710	Consolidation test (ASTM D 2435), minimum				Ea.				455	500
4750	Moisture content, ASTM D 2216								18	20
4780	Permeability test, double ring infiltrometer								500	550
4800	Permeability, var. or constant head, undist., ASTM D 2434								264	290
4850	Recompacted								250	275
4900	Proctor compaction, 4" standard mold, ASTM D 698								230	253
4950	6" modified mold								110	120
5100	Shear tests, triaxial, minimum								410	450
5150	Maximum								545	600
5300	Direct shear, minimum, ASTM D 3080								320	350
5350	Maximum								410	450
5550	Technician for inspection, per day, earthwork								556	612
5570	Concrete								556	612
5650	Bolting								556	612
5750	Roofing								556	612
5790	Welding								556	612
5820	Non-destructive metal testing, dye penetrant				Day				556	612
5840	Magnetic particle								556	612
5860	Radiography								556	612
5880	Ultrasonic								556	612
6000	Welding certification, minimum				Ea.				91	100
6100	Maximum				"				364	400
7000	Underground storage tank									
7520	Volumetric tightness, °= 12,000 gal.				Ea.	470			470	515
7530	12,000 – 29,999 gal.					585			585	645
7540	= 30,000 gal.					665			665	730
7600	Vadose zone (soil gas) sampling, 10-40 samples, min.				Day				1,375	1,500
7610	Maximum				"				2,275	2,500
7700	Ground water monitoring incl. drilling 3 wells, min.				Total				4,550	5,000
7710	Maximum				"				6,375	7,000
8000	X-ray concrete slabs				Ea.				182	200

01 51 Temporary Utilities

01 51 13 – Temporary Electricity

01 51 13.50 Temporary Power Equip (Pro-Rated Per Job)		Crew	Daily Output	Labor-Hours	Unit	Material	2021 Bare Costs Labor	Equipment	Total	Total Incl O&P
0010	**TEMPORARY POWER EQUIP (PRO-RATED PER JOB)**									
0020	Service, overhead feed, 3 use									
0030	100 Amp	1 Elec	1.25	6.400	Ea.	405	410		815	1,050
0040	200 Amp		1	8		505	510		1,015	1,300
0050	400 Amp		.75	10.667		905	680		1,585	2,000
0060	600 Amp		.50	16		1,300	1,025		2,325	2,950
0100	Underground feed, 3 use									
0110	100 Amp	1 Elec	2	4	Ea.	415	255		670	835
0120	200 Amp		1.15	6.957		545	445		990	1,250
0130	400 Amp		1	8		925	510		1,435	1,775
0140	600 Amp		.75	10.667		1,125	680		1,805	2,250
0150	800 Amp		.50	16		1,625	1,025		2,650	3,325
0160	1000 Amp		.35	22.857		1,800	1,450		3,250	4,150
0170	1200 Amp		.25	32		2,000	2,050		4,050	5,225
0180	2000 Amp		.20	40		2,400	2,550		4,950	6,400
0200	Transformers, 3 use									

16

01 51 Temporary Utilities

01 51 13 – Temporary Electricity

01 51 13.50 Temporary Power Equip (Pro-Rated Per Job)		Crew	Daily Output	Labor-Hours	Unit	Material	2021 Bare Costs Labor	Equipment	Total	Total Incl O&P
0210	30 kVA	1 Elec	1	8	Ea.	1,650	510		2,160	2,575
0220	45 kVA		.75	10.667		1,975	680		2,655	3,175
0230	75 kVA		.50	16		3,325	1,025		4,350	5,175
0240	112.5 kVA		.40	20		4,000	1,275		5,275	6,325
0250	Feeder, PVC, CU wire in trench									
0260	60 Amp	1 Elec	96	.083	L.F.	2.42	5.30		7.72	10.55
0270	100 Amp		85	.094		4.53	6		10.53	13.90
0280	200 Amp		59	.136		10.30	8.65		18.95	24
0290	400 Amp		42	.190		27.50	12.15		39.65	48.50
0300	Feeder, PVC, aluminum wire in trench									
0310	60 Amp	1 Elec	96	.083	L.F.	1.84	5.30		7.14	9.90
0320	100 Amp		85	.094		2.01	6		8.01	11.10
0330	200 Amp		59	.136		5.50	8.65		14.15	18.90
0340	400 Amp		42	.190		8.70	12.15		20.85	27.50
0350	Feeder, EMT, CU wire									
0360	60 Amp	1 Elec	90	.089	L.F.	2.89	5.65		8.54	11.55
0370	100 Amp		80	.100		4.28	6.35		10.63	14.15
0380	200 Amp		60	.133		11.35	8.50		19.85	25
0390	400 Amp		35	.229		31.50	14.55		46.05	56.50
0400	Feeder, EMT, aluminum wire									
0410	60 Amp	1 Elec	90	.089	L.F.	3.48	5.65		9.13	12.25
0420	100 Amp		80	.100		4.51	6.35		10.86	14.40
0430	200 Amp		60	.133		7.95	8.50		16.45	21.50
0440	400 Amp		35	.229		16.40	14.55		30.95	39.50
0500	Equipment, 3 use									
0510	Spider box, 50 Amp	1 Elec	8	1	Ea.	800	63.50		863.50	970
0520	Lighting cord, 100'		8	1		118	63.50		181.50	225
0530	Light stanchion		8	1		138	63.50		201.50	247
0540	Temporary cords, 100', 3 use									
0550	Feeder cord, 50 Amp	1 Elec	16	.500	Ea.	202	32		234	271
0560	Feeder cord, 100 Amp		12	.667		1,225	42.50		1,267.50	1,425
0570	Tap cord, 50 Amp		12	.667		635	42.50		677.50	765
0580	Tap cord, 100 Amp		6	1.333		1,075	85		1,160	1,325
0590	Temporary cords, 50', 3 use									
0600	Feeder cord, 50 Amp	1 Elec	16	.500	Ea.	296	32		328	375
0610	Feeder cord, 100 Amp		12	.667		585	42.50		627.50	710
0620	Tap cord, 50 Amp		12	.667		273	42.50		315.50	365
0630	Tap cord, 100 Amp		6	1.333		590	85		675	775
0700	Connections									
0710	Compressor or pump									
0720	30 Amp	1 Elec	7	1.143	Ea.	17.30	73		90.30	127
0730	60 Amp		5.30	1.509		28	96		124	174
0740	100 Amp		4	2		73.50	127		200.50	270
0860	Office trailer									
0870	60 Amp	1 Elec	4.50	1.778	Ea.	57.50	113		170.50	231
0880	100 Amp		3	2.667		86.50	170		256.50	345
0890	200 Amp		2	4		325	255		580	735

01 51 13.80 Temporary Utilities

		Crew	Daily Output	Labor-Hours	Unit	Material	2021 Bare Costs Labor	Equipment	Total	Total Incl O&P
0010	**TEMPORARY UTILITIES**									
0350	Lighting, lamps, wiring, outlets, 40,000 S.F. building, 8 strings	1 Elec	34	.235	CSF Flr	3.55	15		18.55	26.50
0360	16 strings	"	17	.471		7.10	30		37.10	52.50
0400	Power for temp lighting only, 6.6 KWH, per month								.92	1.01

01 51 Temporary Utilities

01 51 13 – Temporary Electricity

01 51 13.80 Temporary Utilities	Crew	Daily Output	Labor-Hours	Unit	Material	2021 Bare Costs Labor	Equipment	Total	Total Incl O&P
0430	11.8 KWH, per month				CSF Flr				1.65	1.82
0450	23.6 KWH, per month								3.30	3.63
0600	Power for job duration incl. elevator, etc., minimum								53	58
0650	Maximum								110	121
0675	Temporary cooling				Ea.	1,050			1,050	1,150
0700	Temporary construction water bill per month, average				Month	78			78	86

01 52 Construction Facilities

01 52 13 – Field Offices and Sheds

01 52 13.20 Office and Storage Space

		Crew	Daily Output	Labor-Hours	Unit	Material	Labor	Equipment	Total	Total Incl O&P
0010	**OFFICE AND STORAGE SPACE**									
0020	Office trailer, furnished, no hookups, 20' x 8', buy	2 Skwk	1	16	Ea.	10,000	915		10,915	12,400
0250	Rent per month					192			192	211
0300	32' x 8', buy	2 Skwk	.70	22.857		15,000	1,300		16,300	18,500
0350	Rent per month					243			243	267
0400	50' x 10', buy	2 Skwk	.60	26.667	Ea.	30,500	1,525		32,025	35,800
0450	Rent per month					355			355	390
0500	50' x 12', buy	2 Skwk	.50	32		25,800	1,825		27,625	31,200
0550	Rent per month					450			450	500
0700	For air conditioning, rent per month, add					53			53	58.50
0800	For delivery, add per mile				Mile	11.95			11.95	13.15
0900	Bunk house trailer, 8' x 40' duplex dorm with kitchen, no hookups, buy	2 Carp	1	16	Ea.	89,000	875		89,875	99,500
0910	9 man with kitchen and bath, no hookups, buy		1	16		91,500	875		92,375	102,000
0920	18 man sleeper with bath, no hookups, buy		1	16		98,500	875		99,375	110,000
1000	Portable buildings, prefab, on skids, economy, 8' x 8'		265	.060	S.F.	29.50	3.30		32.80	37.50
1100	Deluxe, 8' x 12'		150	.107	"	33.50	5.85		39.35	45.50
1200	Storage boxes, 20' x 8', buy	2 Skwk	1.80	8.889	Ea.	3,100	510		3,610	4,200
1250	Rent per month					88			88	96.50
1300	40' x 8', buy	2 Skwk	1.40	11.429		4,600	655		5,255	6,050
1350	Rent per month					135			135	148

01 52 13.40 Field Office Expense

		Crew	Daily Output	Labor-Hours	Unit	Material	Labor	Equipment	Total	Total Incl O&P
0010	**FIELD OFFICE EXPENSE**									
0100	Office equipment rental average				Month	228			228	251
0120	Office supplies, average				"	90.50			90.50	99.50
0125	Office trailer rental, see Section 01 52 13.20									
0140	Telephone bill; avg. bill/month incl. long dist.				Month	87			87	96
0160	Lights & HVAC				"	163			163	179

01 54 Construction Aids

01 54 09 – Protection Equipment

01 54 09.60 Safety Nets

		Crew	Daily Output	Labor-Hours	Unit	Material	Labor	Equipment	Total	Total Incl O&P
0010	**SAFETY NETS**									
0020	No supports, stock sizes, nylon, 3-1/2" mesh				S.F.	3.16			3.16	3.48
0100	Polypropylene, 6" mesh					1.63			1.63	1.79
0200	Small mesh debris nets, 1/4" mesh, stock sizes					.57			.57	.63
0220	Combined 3-1/2" mesh and 1/4" mesh, stock sizes					4.92			4.92	5.40
0300	Rental, 4" mesh, stock sizes, 3 months					.88			.88	.97
0320	6 month rental					1.03			1.03	1.13
0340	12 months					1.42			1.42	1.56

01 54 Construction Aids

01 54 09 – Protection Equipment

01 54 09.60 Safety Nets

	Crew	Daily Output	Labor-Hours	Unit	Material	2021 Bare Costs Labor	2021 Bare Costs Equipment	Total	Total Incl O&P
6220 Safety supplies and first aid kits				Month	28			28	30.50

01 54 16 – Temporary Hoists

01 54 16.50 Weekly Forklift Crew

	Crew	Daily Output	Labor-Hours	Unit	Material	Labor	Equipment	Total	Total Incl O&P
0010 **WEEKLY FORKLIFT CREW**									
0100 All-terrain forklift, 45' lift, 35' reach, 9000 lb. capacity	A-3P	.20	40	Week		2,225	1,725	3,950	5,200

01 54 19 – Temporary Cranes

01 54 19.50 Daily Crane Crews

	Crew	Daily Output	Labor-Hours	Unit	Material	Labor	Equipment	Total	Total Incl O&P
0010 **DAILY CRANE CREWS** for small jobs, portal to portal R015433-15									
0100 12-ton truck-mounted hydraulic crane	A-3H	1	8	Day		490	735	1,225	1,525
0200 25-ton	A-3I	1	8	Day		490	810	1,300	1,625
0300 40-ton	A-3J	1	8			490	1,275	1,765	2,150
0400 55-ton	A-3K	1	16			910	1,525	2,435	3,025
0500 80-ton	A-3L	1	16			910	2,200	3,110	3,775
0900 If crane is needed on a Saturday, Sunday or Holiday									
0910 At time-and-a-half, add				Day		50%			
0920 At double time, add				"		100%			

01 54 19.60 Monthly Tower Crane Crew

	Crew	Daily Output	Labor-Hours	Unit	Material	Labor	Equipment	Total	Total Incl O&P
0010 **MONTHLY TOWER CRANE CREW**, excludes concrete footing									
0100 Static tower crane, 130' high, 106' jib, 6200 lb. capacity	A-3N	.05	176	Month		10,800	38,200	49,000	58,000

01 54 23 – Temporary Scaffolding and Platforms

01 54 23.70 Scaffolding

	Crew	Daily Output	Labor-Hours	Unit	Material	Labor	Equipment	Total	Total Incl O&P
0010 **SCAFFOLDING** R015423-10									
0015 Steel tube, regular, no plank, labor only to erect & dismantle									
0090 Building exterior, wall face, 1 to 5 stories, 6'-4" x 5' frames	3 Carp	8	3	C.S.F.		164		164	245
0460 Building interior, wall face area, up to 16' high	"	12	2			109		109	163
0906 Complete system for face of walls, no plank, material only rent/mo					81.50			81.50	90
0908 Interior spaces, no plank, material only rent/mo				C.C.F.	5.60			5.60	6.15
0910 Steel tubular, heavy duty shoring, buy									
0920 Frames 5' high 2' wide				Ea.	92.50			92.50	102
0925 5' high 4' wide					118			118	130
0930 6' high 2' wide					120			120	132
0935 6' high 4' wide					119			119	131
0940 Accessories									
0945 Cross braces				Ea.	19.20			19.20	21
0950 U-head, 8" x 8"					23			23	25
0955 J-head, 4" x 8"					16.70			16.70	18.35
0960 Base plate, 8" x 8"					18.05			18.05	19.90
0965 Leveling jack					36			36	39.50
1000 Steel tubular, regular, buy									
1100 Frames 3' high 5' wide				Ea.	54			54	59.50
1150 5' high 5' wide					71			71	78
1200 6'-4" high 5' wide					83.50			83.50	92
1350 7'-6" high 6' wide					156			156	171
1500 Accessories, cross braces					19.20			19.20	21
1550 Guardrail post					24			24	26.50
1600 Guardrail 7' section					8.15			8.15	9
1650 Screw jacks & plates					26			26	28.50
1700 Sidearm brackets					26			26	29
1750 8" casters					37.50			37.50	41
1800 Plank 2" x 10" x 16'-0"				Ea.	66.50			66.50	73.50
1900 Stairway section					298			298	330

For customer support on your Site Work & Landscape Costs with RSMeans Data, call 800.448.8182.

19

01 54 23 – Temporary Scaffolding and Platforms

01 54 23.70 Scaffolding

01 54 23.70 Scaffolding		Crew	Daily Output	Labor-Hours	Unit	Material	2021 Bare Costs Labor	Equipment	Total	Total Incl O&P
1910	Stairway starter bar				Ea.	32.50			32.50	35.50
1920	Stairway inside handrail					57.50			57.50	63.50
1930	Stairway outside handrail					91.50			91.50	101
1940	Walk-thru frame guardrail					42			42	46.50
2000	Steel tubular, regular, rent/mo.									
2100	Frames 3' high 5' wide				Ea.	9.90			9.90	10.90
2150	5' high 5' wide					9.90			9.90	10.90
2200	6'-4" high 5' wide					5.75			5.75	6.35
2250	7'-6" high 6' wide					9.90			9.90	10.90
2500	Accessories, cross braces					3.35			3.35	3.69
2550	Guardrail post					3.25			3.25	3.58
2600	Guardrail 7' section					11.10			11.10	12.20
2650	Screw jacks & plates					4.35			4.35	4.79
2700	Sidearm brackets					5.20			5.20	5.70
2750	8" casters					7.85			7.85	8.65
2800	Outrigger for rolling tower					9.10			9.10	10
2850	Plank 2" x 10" x 16'-0"					9.90			9.90	10.90
2900	Stairway section					32			32	35.50
2940	Walk-thru frame guardrail					8.25			8.25	9.10
3000	Steel tubular, heavy duty shoring, rent/mo.									
3250	5' high 2' & 4' wide				Ea.	8.45			8.45	9.25
3300	6' high 2' & 4' wide					8.45			8.45	9.25
3500	Accessories, cross braces					.91			.91	1
3600	U-head, 8" x 8"					2.49			2.49	2.74
3650	J-head, 4" x 8"					2.49			2.49	2.74
3700	Base plate, 8" x 8"					.91			.91	1
3750	Leveling jack					2.47			2.47	2.72
5700	Planks, 2" x 10" x 16'-0", labor only to erect & remove to 50' H	3 Carp	72	.333			18.25		18.25	27
5800	Over 50' high	4 Carp	80	.400			22		22	32.50
6000	Heavy duty shoring for elevated slab forms to 8'-2" high, floor area									
6100	Labor only to erect & dismantle	4 Carp	16	2	C.S.F.		109		109	163
6110	Materials only, rent/mo.				"	43.50			43.50	48
6500	To 14'-8" high									
6600	Labor only to erect & dismantle	4 Carp	10	3.200	C.S.F.		175		175	261
6610	Materials only, rent/mo				"	64			64	70

01 54 23.75 Scaffolding Specialties

01 54 23.75 Scaffolding Specialties		Crew	Daily Output	Labor-Hours	Unit	Material	2021 Bare Costs Labor	Equipment	Total	Total Incl O&P
0010	**SCAFFOLDING SPECIALTIES**									
1200	Sidewalk bridge, heavy duty steel posts & beams, including									
1210	parapet protection & waterproofing (material cost is rent/month)									
1220	8' to 10' wide, 2 posts	3 Carp	15	1.600	L.F.	65.50	87.50		153	203
1230	3 posts	"	10	2.400	"	96.50	131		227.50	300
1500	Sidewalk bridge using tubular steel scaffold frames including									
1510	planking (material cost is rent/month)	3 Carp	45	.533	L.F.	9.80	29		38.80	54.50
1600	For 2 uses per month, deduct from all above					50%				
1700	For 1 use every 2 months, add to all above					100%				
1900	Catwalks, 20" wide, no guardrails, 7' span, buy				Ea.	148			148	163
2000	10' span, buy					207			207	228
3800	Rolling ladders with handrails, 30" wide, buy, 2 step					289			289	320
4000	7 step					1,125			1,125	1,225
4050	10 step					1,375			1,375	1,500
4100	Rolling towers, buy, 5' wide, 7' long, 10' high					1,250			1,250	1,375
4200	For additional 5' high sections, to buy					180			180	198

01 54 Construction Aids

01 54 23 – Temporary Scaffolding and Platforms

01 54 23.75 Scaffolding Specialties		Crew	Daily Output	Labor-Hours	Unit	Material	2021 Bare Costs Labor	Equipment	Total	Total Incl O&P
4300	Complete incl. wheels, railings, outriggers,									
4350	21' high, to buy				Ea.	2,050			2,050	2,250
4400	Rent/month = 5% of purchase cost				"		103		103	113

01 54 36 – Equipment Mobilization

01 54 36.50 Mobilization

		Crew	Daily Output	Labor-Hours	Unit	Material	Labor	Equipment	Total	Total Incl O&P
0010	**MOBILIZATION** (Use line item again for demobilization) R015436-50									
0015	Up to 25 mi. haul dist. (50 mi. RT for mob/demob crew)									
1200	Small equipment, placed in rear of, or towed by pickup truck	A-3A	4	2	Ea.		111	44	155	214
1300	Equipment hauled on 3-ton capacity towed trailer	A-3Q	2.67	3			167	93	260	350
1400	20-ton capacity	B-34U	2	8			425	224	649	885
1500	40-ton capacity	B-34N	2	8			440	345	785	1,050
1600	50-ton capacity	B-34V	1	24			1,350	995	2,345	3,100
1700	Crane, truck-mounted, up to 75 ton (driver only)	1 Eqhv	4	2			123		123	183
1800	Over 75 ton (with chase vehicle)	A-3E	2.50	6.400			360	70.50	430.50	620
2400	Crane, large lattice boom, requiring assembly	B-34W	.50	144	↓		7,775	7,075	14,850	19,400
2500	For each additional 5 miles haul distance, add						10%	10%		
3000	For large pieces of equipment, allow for assembly/knockdown									
3001	For mob/demob of vibroflotation equip, see Section 31 45 13.10									
3100	For mob/demob of micro-tunneling equip, see Section 33 05 07.36									
3200	For mob/demob of pile driving equip, see Section 31 62 19.10									
3300	For mob/demob of caisson drilling equip, see Section 31 63 26.13									

01 55 Vehicular Access and Parking

01 55 23 – Temporary Roads

01 55 23.50 Roads and Sidewalks

		Crew	Daily Output	Labor-Hours	Unit	Material	Labor	Equipment	Total	Total Incl O&P
0010	**ROADS AND SIDEWALKS** Temporary									
0050	Roads, gravel fill, no surfacing, 4" gravel depth	B-14	715	.067	S.Y.	5.35	3.13	.30	8.78	10.85
0100	8" gravel depth	"	615	.078	"	10.65	3.64	.35	14.64	17.55
1000	Ramp, 3/4" plywood on 2" x 6" joists, 16" OC	2 Carp	300	.053	S.F.	2.44	2.92		5.36	7.05
1100	On 2" x 10" joists, 16" OC	"	275	.058	"	3.39	3.18		6.57	8.50

01 56 Temporary Barriers and Enclosures

01 56 13 – Temporary Air Barriers

01 56 13.60 Tarpaulins

		Crew	Daily Output	Labor-Hours	Unit	Material	Labor	Equipment	Total	Total Incl O&P
0010	**TARPAULINS**									
0020	Cotton duck, 10-13.13 oz./S.Y., 6' x 8'				S.F.	.74			.74	.81
0050	30' x 30'					.62			.62	.68
0100	Polyvinyl coated nylon, 14-18 oz., minimum					1.46			1.46	1.61
0150	Maximum					1.46			1.46	1.61
0200	Reinforced polyethylene 3 mils thick, white					.06			.06	.07
0300	4 mils thick, white, clear or black					.17			.17	.19
0400	5.5 mils thick, clear					.16			.16	.18
0500	White, fire retardant					.56			.56	.62
0600	12 mils, oil resistant, fire retardant					.51			.51	.56
0700	8.5 mils, black					.31			.31	.34
0710	Woven polyethylene, 6 mils thick					.06			.06	.07
0730	Polyester reinforced w/integral fastening system, 11 mils thick					.23			.23	.25
0740	Polyethylene, reflective, 23 mils thick				↓	1.34			1.34	1.47

For customer support on your Site Work & Landscape Costs with RSMeans Data, call 800.448.8182.

21

01 56 Temporary Barriers and Enclosures

01 56 13 – Temporary Air Barriers

01 56 13.90 Winter Protection	Crew	Daily Output	Labor-Hours	Unit	Material	2021 Bare Costs Labor	Equipment	Total	Total Incl O&P
0010 **WINTER PROTECTION**									
0100 Framing to close openings	2 Clab	500	.032	S.F.	.77	1.42		2.19	2.97
0200 Tarpaulins hung over scaffolding, 8 uses, not incl. scaffolding		1500	.011		.25	.47		.72	.99
0250 Tarpaulin polyester reinf. w/integral fastening system, 11 mils thick		1600	.010		.25	.44		.69	.94
0300 Prefab fiberglass panels, steel frame, 8 uses	▼	1200	.013	▼	3.06	.59		3.65	4.25

01 56 16 – Temporary Dust Barriers

01 56 16.10 Dust Barriers, Temporary

	Crew	Daily Output	Labor-Hours	Unit	Material	2021 Bare Costs Labor	Equipment	Total	Total Incl O&P
0010 **DUST BARRIERS, TEMPORARY**, erect and dismantle									
0020 Spring loaded telescoping pole & head, to 12', erect and dismantle	1 Clab	240	.033	Ea.		1.48		1.48	2.21
0025 Cost per day (based upon 250 days)				Day	.22			.22	.25
0030 To 21', erect and dismantle	1 Clab	240	.033	Ea.		1.48		1.48	2.21
0035 Cost per day (based upon 250 days)				Day	.73			.73	.81
0040 Accessories, caution tape reel, erect and dismantle	1 Clab	480	.017	Ea.		.74		.74	1.10
0045 Cost per day (based upon 250 days)				Day	.34			.34	.38
0060 Foam rail and connector, erect and dismantle	1 Clab	240	.033	Ea.		1.48		1.48	2.21
0065 Cost per day (based upon 250 days)				Day	.10			.10	.11
0070 Caution tape	1 Clab	384	.021	C.L.F.	2.55	.93		3.48	4.18
0080 Zipper, standard duty	1 Clab	60	.133	Ea.	10.75	5.90		16.65	20.50
0090 Heavy duty		48	.167	"	10.85	7.40		18.25	23
0100 Polyethylene sheet, 4 mil		37	.216	Sq.	2.70	9.60		12.30	17.25
0110 6 mil		37	.216	"	7.65	9.60		17.25	23
1000 Dust partition, 6 mil polyethylene, 1" x 3" frame	2 Carp	2000	.008	S.F.	.49	.44		.93	1.19
1080 2" x 4" frame		2000	.008	"	.60	.44		1.04	1.31
4000 Dust & infectious control partition, adj. to 10' high, obscured, 4' panel		90	.178	Ea.	580	9.70		589.70	655
4010 3' panel		90	.178		550	9.70		559.70	620
4020 2' panel		90	.178		430	9.70		439.70	490
4030 1' panel	1 Carp	90	.089		285	4.86		289.86	320
4040 6" panel	"	90	.089		265	4.86		269.86	299
4050 2' panel with HEPA filtered discharge port	2 Carp	90	.178		500	9.70		509.70	565
4060 3' panel with 32" door		90	.178		895	9.70		904.70	1,000
4070 4' panel with 36" door		90	.178		995	9.70		1,004.70	1,125
4080 4'-6" panel with 44" door		90	.178		1,200	9.70		1,209.70	1,350
4090 Hinged corner		80	.200		185	10.95		195.95	220
4100 Outside corner		80	.200		150	10.95		160.95	181
4110 T post		80	.200		150	10.95		160.95	181
4120 Accessories, ceiling grid clip		360	.044		7.45	2.43		9.88	11.85
4130 Panel locking clip		360	.044		5.15	2.43		7.58	9.30
4140 Panel joint closure strip		360	.044		8	2.43		10.43	12.45
4150 Screw jack		360	.044		6.65	2.43		9.08	10.95
4160 Digital pressure difference gauge					275			275	305
4180 Combination lockset	1 Carp	13	.615	▼	200	33.50		233.50	271
4185 Sealant tape, 2" wide	1 Clab	192	.042	C.L.F.	10.65	1.85		12.50	14.50
4190 System in place, including door and accessories									
4200 Based upon 25 uses	2 Carp	51	.314	L.F.	9.85	17.15		27	36.50
4210 Based upon 50 uses		51	.314		4.94	17.15		22.09	31
4230 Based upon 100 uses	▼	51	.314	▼	2.47	17.15		19.62	28

01 56 23 – Temporary Barricades

01 56 23.10 Barricades

	Crew	Daily Output	Labor-Hours	Unit	Material	2021 Bare Costs Labor	Equipment	Total	Total Incl O&P
0010 **BARRICADES**									
0020 5' high, 3 rail @ 2" x 8", fixed	2 Carp	20	.800	L.F.	9.45	44		53.45	76
0150 Movable	"	30	.533	"	8.15	29		37.15	52.50

For customer support on your Site Work & Landscape Costs with RSMeans Data, call 800.448.8182.

01 56 Temporary Barriers and Enclosures

01 56 23 – Temporary Barricades

01 56 23.10 Barricades		Crew	Daily Output	Labor-Hours	Unit	Material	2021 Bare Costs Labor	Equipment	Total	Total Incl O&P
0300	Stock units, 58' high, 8' wide, reflective, buy				Ea.	211			211	232
0350	With reflective tape, buy				"	410			410	455
0400	Break-a-way 3" PVC pipe barricade									
0410	with 3 ea. 1' x 4' reflectorized panels, buy				Ea.	146			146	161
0500	Barricades, plastic, 8" x 24" wide, foldable					57.50			57.50	63.50
0800	Traffic cones, PVC, 18" high					11.60			11.60	12.75
0850	28" high				Ea.	26			26	28.50
1000	Guardrail, wooden, 3' high, 1" x 6" on 2" x 4" posts	2 Carp	200	.080	L.F.	2.05	4.38		6.43	8.80
1100	2" x 6" on 4" x 4" posts	"	165	.097		3.46	5.30		8.76	11.70
1200	Portable metal with base pads, buy					14.35			14.35	15.80
1250	Typical installation, assume 10 reuses	2 Carp	600	.027		2.19	1.46		3.65	4.59
1300	Barricade tape, polyethylene, 7 mil, 3" wide x 300' long roll	1 Clab	128	.063	Ea.	7.65	2.78		10.43	12.55
3000	Detour signs, set up and remove									
3010	Reflective aluminum, MUTCD, 24" x 24", post mounted	1 Clab	20	.400	Ea.	3.46	17.75		21.21	30.50

01 56 26 – Temporary Fencing

01 56 26.50 Temporary Fencing

		Crew	Daily Output	Labor-Hours	Unit	Material	2021 Bare Costs Labor	Equipment	Total	Total Incl O&P
0010	**TEMPORARY FENCING**									
0020	Chain link, 11 ga., 4' high	2 Clab	400	.040	L.F.	3.05	1.78		4.83	6
0100	6' high		300	.053		5.80	2.37		8.17	9.90
0200	Rented chain link, 6' high, to 1000' (up to 12 mo.)		400	.040		3.44	1.78		5.22	6.45
0250	Over 1000' (up to 12 mo.)		300	.053		4.07	2.37		6.44	8
0350	Plywood, painted, 2" x 4" frame, 4' high	A-4	135	.178		9.45	9.25		18.70	24
0400	4" x 4" frame, 8' high	"	110	.218		17.05	11.35		28.40	35.50
0500	Wire mesh on 4" x 4" posts, 4' high	2 Carp	100	.160		12.25	8.75		21	26.50
0550	8' high	"	80	.200		18	10.95		28.95	36
0600	Plastic safety fence, light duty, 4' high, posts at 10'	B-1	500	.048		1.61	2.16		3.77	5
0610	Medium duty, 4' high, posts at 10'		500	.048		1.50	2.16		3.66	4.88
0620	Heavy duty, 4' high, posts at 10'		500	.048		1.70	2.16		3.86	5.10
0630	Reflective heavy duty, 4' high, posts at 10'		500	.048		6.35	2.16		8.51	10.25

01 56 29 – Temporary Protective Walkways

01 56 29.50 Protection

		Crew	Daily Output	Labor-Hours	Unit	Material	2021 Bare Costs Labor	Equipment	Total	Total Incl O&P
0010	**PROTECTION**									
0020	Stair tread, 2" x 12" planks, 1 use	1 Carp	75	.107	Tread	7.50	5.85		13.35	16.95
0100	Exterior plywood, 1/2" thick, 1 use		65	.123		3.01	6.75		9.76	13.35
0200	3/4" thick, 1 use		60	.133		4.15	7.30		11.45	15.45
2200	Sidewalks, 2" x 12" planks, 2 uses		350	.023	S.F.	1.25	1.25		2.50	3.24
2300	Exterior plywood, 2 uses, 1/2" thick		750	.011		.50	.58		1.08	1.42
2400	5/8" thick		650	.012		.58	.67		1.25	1.64
2500	3/4" thick		600	.013		.69	.73		1.42	1.85

For customer support on your Site Work & Landscape Costs with RSMeans Data, call 800.448.8182.

23

01 57 Temporary Controls

01 57 33 – Temporary Security

01 57 33.50 Watchman	Crew	Daily Output	Labor-Hours	Unit	Material	2021 Bare Costs Labor	2021 Bare Costs Equipment	Total	Total Incl O&P
0010 **WATCHMAN**									
0020 Service, monthly basis, uniformed person, minimum				Hr.				27.65	30.40
0100 Maximum								56	61.62
0200 Person and command dog, minimum								31	34
0300 Maximum								60	65

01 58 Project Identification

01 58 13 – Temporary Project Signage

01 58 13.50 Signs

	Crew	Daily Output	Labor-Hours	Unit	Material	2021 Bare Costs Labor	2021 Bare Costs Equipment	Total	Total Incl O&P
0010 **SIGNS**									
0020 High intensity reflectorized, no posts, buy				Ea.	22			22	24

01 66 Product Storage and Handling Requirements

01 66 19 – Material Handling

01 66 19.10 Material Handling

	Crew	Daily Output	Labor-Hours	Unit	Material	2021 Bare Costs Labor	2021 Bare Costs Equipment	Total	Total Incl O&P
0010 **MATERIAL HANDLING**									
0020 Above 2nd story, via stairs, per C.Y. of material per floor	2 Clab	145	.110	C.Y.		4.90		4.90	7.30
0030 Via elevator, per C.Y. of material		240	.067			2.96		2.96	4.42
0050 Distances greater than 200', per C.Y. of material per each addl 200'		300	.053			2.37		2.37	3.53

01 71 Examination and Preparation

01 71 23 – Field Engineering

01 71 23.13 Construction Layout

	Crew	Daily Output	Labor-Hours	Unit	Material	2021 Bare Costs Labor	2021 Bare Costs Equipment	Total	Total Incl O&P
0010 **CONSTRUCTION LAYOUT**									
1100 Crew for layout of building, trenching or pipe laying, 2 person crew	A-6	1	16	Day		880	34.50	914.50	1,375
1200 3 person crew	A-7	1	24			1,425	34.50	1,459.50	2,175
1400 Crew for roadway layout, 4 person crew	A-8	1	32			1,850	34.50	1,884.50	2,800

01 71 23.19 Surveyor Stakes

	Crew	Daily Output	Labor-Hours	Unit	Material	2021 Bare Costs Labor	2021 Bare Costs Equipment	Total	Total Incl O&P
0010 **SURVEYOR STAKES**									
0020 Hardwood, 1" x 1" x 48" long				C	70			70	77
0100 2" x 2" x 18" long					78			78	86
0150 2" x 2" x 24" long					150			150	165

01 74 Cleaning and Waste Management

01 74 13 – Progress Cleaning

01 74 13.20 Cleaning Up

	Crew	Daily Output	Labor-Hours	Unit	Material	2021 Bare Costs Labor	2021 Bare Costs Equipment	Total	Total Incl O&P
0010 **CLEANING UP**									
0020 After job completion, allow, minimum				Job				.30%	.30%
0040 Maximum				"				1%	1%
0200 Rubbish removal, see Section 02 41 19.19									

01 76 Protecting Installed Construction

01 76 13 – Temporary Protection of Installed Construction

01 76 13.20 Temporary Protection	Crew	Daily Output	Labor-Hours	Unit	Material	2021 Bare Costs Labor	Equipment	Total	Total Incl O&P
0010 **TEMPORARY PROTECTION**									
0020 Flooring, 1/8" tempered hardboard, taped seams	2 Carp	1500	.011	S.F.	.51	.58		1.09	1.43
0030 Peel away carpet protection	1 Clab	3200	.003	"	.11	.11		.22	.29

01 91 Commissioning

01 91 13 – General Commissioning Requirements

01 91 13.50 Building Commissioning

	Crew	Daily Output	Labor-Hours	Unit	Material	2021 Bare Costs Labor	Equipment	Total	Total Incl O&P
0010 **BUILDING COMMISSIONING**									
0100 Systems operation and verification during turnover				%				.25%	.25%
0150 Including all systems subcontractors								.50%	.50%
0200 Systems design assistance, operation, verification and training								.50%	.50%
0250 Including all systems subcontractors								1%	1%

For customer support on your Site Work & Landscape Costs with RSMeans Data, call 800.448.8182.

25

Division Notes

		CREW	DAILY OUTPUT	LABOR-HOURS	UNIT	BARE COSTS				TOTAL INCL O&P
						MAT.	LABOR	EQUIP.	TOTAL	

Estimating Tips

02 30 00 Subsurface Investigation

In preparing estimates on structures involving earthwork or foundations, all information concerning soil characteristics should be obtained. Look particularly for hazardous waste, evidence of prior dumping of debris, and previous stream beds.

02 40 00 Demolition and Structure Moving

The costs shown for selective demolition do not include rubbish handling or disposal. These items should be estimated separately using RSMeans data or other sources.

- Historic preservation often requires that the contractor remove materials from the existing structure, rehab them, and replace them. The estimator must be aware of any related measures and precautions that must be taken when doing selective demolition and cutting and patching. Requirements may include special handling and storage, as well as security.

- In addition to Subdivision 02 41 00, you can find selective demolition items in each division. Example: Roofing demolition is in Division 7.

- Absent of any other specific reference, an approximate demolish-in-place cost can be obtained by halving the new-install labor cost. To remove for reuse, allow the entire new-install labor figure.

02 40 00 Building Deconstruction

This section provides costs for the careful dismantling and recycling of most low-rise building materials.

02 50 00 Containment of Hazardous Waste

This section addresses on-site hazardous waste disposal costs.

02 80 00 Hazardous Material Disposal/Remediation

This subdivision includes information on hazardous waste handling, asbestos remediation, lead remediation, and mold remediation. See reference numbers R028213-20 and R028319-60 for further guidance in using these unit price lines.

02 90 00 Monitoring Chemical Sampling, Testing Analysis

This section provides costs for on-site sampling and testing hazardous waste.

Reference Numbers

Reference numbers are shown at the beginning of some major classifications. These numbers refer to related items in the Reference Section. The reference information may be an estimating procedure, an alternate pricing method, or technical information.

Note: Not all subdivisions listed here necessarily appear. ■

Same Data. Simplified.

Enjoy the convenience and efficiency of accessing your costs anywhere:

- **Skip the multiplier** by setting your location
- **Quickly search,** edit, favorite and share costs
- **Stay on top of price changes** with automatic updates

Discover more at rsmeans.com/online

02 21 Surveys

02 21 13 – Site Surveys

02 21 13.09 Topographical Surveys

02 21 13.09 Topographical Surveys	Crew	Daily Output	Labor-Hours	Unit	Material	2021 Bare Costs Labor	Equipment	Total	Total Incl O&P
0010 **TOPOGRAPHICAL SURVEYS**									
0020 Topographical surveying, conventional, minimum	A-7	3.30	7.273	Acre	35	430	10.40	475.40	695
0050 Average	"	1.95	12.308		52.50	730	17.60	800.10	1,175
0100 Maximum	A-8	.60	53.333	↓	70	3,075	57.50	3,202.50	4,750

02 21 13.13 Boundary and Survey Markers

02 21 13.13 Boundary and Survey Markers	Crew	Daily Output	Labor-Hours	Unit	Material	2021 Bare Costs Labor	Equipment	Total	Total Incl O&P
0010 **BOUNDARY AND SURVEY MARKERS**									
0300 Lot location and lines, large quantities, minimum	A-7	2	12	Acre	34.50	715	17.15	766.65	1,125
0320 Average	"	1.25	19.200		61	1,150	27.50	1,238.50	1,800
0400 Small quantities, maximum	A-8	1	32	↓	77	1,850	34.50	1,961.50	2,875
0600 Monuments, 3' long	A-7	10	2.400	Ea.	37.50	143	3.43	183.93	258
0800 Property lines, perimeter, cleared land	"	1000	.024	L.F.	.08	1.43	.03	1.54	2.26
0900 Wooded land	A-8	875	.037	"	.10	2.11	.04	2.25	3.30

02 21 13.16 Aerial Surveys

02 21 13.16 Aerial Surveys	Crew	Daily Output	Labor-Hours	Unit	Material	2021 Bare Costs Labor	Equipment	Total	Total Incl O&P
0010 **AERIAL SURVEYS**									
1500 Aerial surveying, including ground control, minimum fee, 10 acres				Total				4,700	4,700
1510 100 acres								9,400	9,400
1550 From existing photography, deduct				↓				1,625	1,625
1600 2' contours, 10 acres				Acre				470	470
1850 100 acres								94	94
2000 1000 acres								90	90
2050 10,000 acres				↓				85	85

02 31 Geophysical Investigations

02 31 23 – Electromagnetic Investigations

02 31 23.10 Ground Penetrating Radar

02 31 23.10 Ground Penetrating Radar	Crew	Daily Output	Labor-Hours	Unit	Material	2021 Bare Costs Labor	Equipment	Total	Total Incl O&P
0010 **GROUND PENETRATING RADAR**									
0100 Ground Penetrating Radar				Hr.		210		210	235

02 32 Geotechnical Investigations

02 32 13 – Subsurface Drilling and Sampling

02 32 13.10 Boring and Exploratory Drilling

02 32 13.10 Boring and Exploratory Drilling	Crew	Daily Output	Labor-Hours	Unit	Material	2021 Bare Costs Labor	Equipment	Total	Total Incl O&P
0010 **BORING AND EXPLORATORY DRILLING**									
0020 Borings, initial field stake out & determination of elevations	A-6	1	16	Day		880	34.50	914.50	1,375
0100 Drawings showing boring details				Total		335		335	425
0200 Report and recommendations from P.E.						775		775	970
0300 Mobilization and demobilization	B-55	4	6	↓		275	310	585	750
0350 For over 100 miles, per added mile		450	.053	Mile		2.45	2.76	5.21	6.70
0600 Auger holes in earth, no samples, 2-1/2" diameter		78.60	.305	L.F.		14	15.85	29.85	38.50
0650 4" diameter		67.50	.356			16.30	18.45	34.75	45
0800 Cased borings in earth, with samples, 2-1/2" diameter		55.50	.432		25	19.85	22.50	67.35	81.50
0850 4" diameter	↓	32.60	.736		28.50	34	38	100.50	124
1000 Drilling in rock, "BX" core, no sampling	B-56	34.90	.458	↓		23	45.50	68.50	84
1050 With casing & sampling	B-56	31.70	.505	L.F.	25	25	50	100	120
1200 "NX" core, no sampling		25.92	.617			31	61.50	92.50	114
1250 With casing and sampling	↓	25	.640	↓	26	32	63.50	121.50	146
1400 Borings, earth, drill rig and crew with truck mounted auger	B-55	1	24	Day		1,100	1,250	2,350	3,025
1450 Rock using crawler type drill	B-56	1	16	"		800	1,600	2,400	2,950
1500 For inner city borings, add, minimum								10%	10%
1510 Maximum								20%	20%

02 32 Geotechnical Investigations

02 32 19 – Exploratory Excavations

02 32 19.10 Test Pits

		Crew	Daily Output	Labor-Hours	Unit	Material	2021 Bare Costs Labor	2021 Bare Costs Equipment	Total	Total Incl O&P
0010	**TEST PITS**									
0020	Hand digging, light soil	1 Clab	4.50	1.778	C.Y.		79		79	118
0100	Heavy soil	"	2.50	3.200			142		142	212
0120	Loader-backhoe, light soil	B-11M	28	.571			29.50	8.40	37.90	53.50
0130	Heavy soil	"	20	.800			41.50	11.75	53.25	74.50
1000	Subsurface exploration, mobilization				Mile				6.75	8.40
1010	Difficult access for rig, add				Hr.				260	320
1020	Auger borings, drill rig, incl. samples				L.F.				26.50	33
1030	Hand auger								31.50	40
1050	Drill and sample every 5', split spoon								31.50	40
1060	Extra samples				Ea.				36	45.50

02 41 Demolition

02 41 13 – Selective Site Demolition

02 41 13.15 Hydrodemolition

		Crew	Daily Output	Labor-Hours	Unit	Material	2021 Bare Costs Labor	2021 Bare Costs Equipment	Total	Total Incl O&P
0010	**HYDRODEMOLITION**									
0015	Hydrodemolition, concrete pavement									
0120	Includes removal but excludes disposal of concrete									
0130	2" depth	B-5E	1000	.064	S.F.		3.15	2.75	5.90	7.70
0410	4" depth		800	.080			3.93	3.43	7.36	9.65
0420	6" depth		600	.107			5.25	4.58	9.83	12.85
0520	8" depth		300	.213			10.50	9.15	19.65	25.50

02 41 13.16 Hydroexcavation

		Crew	Daily Output	Labor-Hours	Unit	Material	2021 Bare Costs Labor	2021 Bare Costs Equipment	Total	Total Incl O&P
0010	**HYDROEXCAVATION**									
0015	Hydroexcavation									
0105	Assumes onsite disposal									
0110	Mobilization or Demobilization	B-6D	8	2.500	Ea.		127	160	287	365
0130	Normal Conditions		48	.417	B.C.Y.		21	26.50	47.50	61
0140	Adverse Conditions		30	.667	"		34	42.50	76.50	97.50
0160	Minimum labor/equipment charge		4	5	Ea.		253	320	573	730

02 41 13.17 Demolish, Remove Pavement and Curb

		Crew	Daily Output	Labor-Hours	Unit	Material	2021 Bare Costs Labor	2021 Bare Costs Equipment	Total	Total Incl O&P
0010	**DEMOLISH, REMOVE PAVEMENT AND CURB** R024119-10									
5010	Pavement removal, bituminous roads, up to 3" thick	B-38	690	.058	S.Y.		2.90	1.76	4.66	6.25
5050	4"-6" thick		420	.095			4.76	2.90	7.66	10.30
5100	Bituminous driveways		640	.063			3.12	1.90	5.02	6.75
5200	Concrete to 6" thick, hydraulic hammer, mesh reinforced	B-38	255	.157	S.Y.		7.85	4.77	12.62	16.95
5300	Rod reinforced		200	.200	"		10	6.10	16.10	21.50
5400	Concrete, 7"-24" thick, plain		33	1.212	C.Y.		60.50	37	97.50	131
5500	Reinforced		24	1.667	"		83	50.50	133.50	180
5600	With hand held air equipment, bituminous, to 6" thick	B-39	1900	.025	S.F.		1.18	.19	1.37	1.97
5700	Concrete to 6" thick, no reinforcing		1600	.030			1.40	.22	1.62	2.32
5800	Mesh reinforced		1400	.034			1.60	.25	1.85	2.66
5900	Rod reinforced		765	.063			2.92	.46	3.38	4.87
6000	Curbs, concrete, plain	B-6	360	.067	L.F.		3.21	.60	3.81	5.45
6100	Reinforced		275	.087			4.20	.79	4.99	7.10
6200	Granite		360	.067			3.21	.60	3.81	5.45
6300	Bituminous		528	.045			2.19	.41	2.60	3.71
6400	Wood		570	.042			2.03	.38	2.41	3.44
6500	Site demo, berms under 4" in height, bituminous		528	.045			2.19	.41	2.60	3.71
6600	4" or over in height		300	.080			3.85	.72	4.57	6.55

For customer support on your Site Work & Landscape Costs with RSMeans Data, call 800.448.8182.

29

02 41 13 – Selective Site Demolition

02 41 13.20 Selective Demo, Highway Guard Rails & Barriers	Crew	Daily Output	Labor-Hours	Unit	Material	2021 Bare Costs Labor	Equipment	Total	Total Incl O&P
0010 **SELECTIVE DEMOLITION, HIGHWAY GUARD RAILS & BARRIERS**									
0100 Guard rail, corrugated steel	B-6	600	.040	L.F.		1.92	.36	2.28	3.27
0200 End sections		40	.600	Ea.		29	5.40	34.40	49
0300 Wrap around		40	.600	"		29	5.40	34.40	49
0400 Timber 4" x 8"		600	.040	L.F.		1.92	.36	2.28	3.27
0500 Three 3/4" cables		600	.040	"		1.92	.36	2.28	3.27
0600 Wood posts		240	.100	Ea.		4.81	.90	5.71	8.15
0700 Guide rail, 6" x 6" box beam	B-80B	120	.267	L.F.		12.60	1.99	14.59	21
0800 Median barrier, 6" x 8" box beam		240	.133			6.30	.99	7.29	10.50
0850 Precast concrete 3'-6" high x 2' wide		300	.107			5.05	.80	5.85	8.40
0900 Impact barrier, MUTCD, barrel type	B-16	60	.533	Ea.		25	9.65	34.65	47.50
1000 Resilient guide fence and light shield 6' high	"	120	.267	L.F.		12.45	4.83	17.28	24
1100 Concrete posts, 6'-5" triangular	B-6	200	.120	Ea.		5.75	1.08	6.83	9.80
1200 Speed bumps 10-1/2" x 2-1/4" x 48"		300	.080	L.F.		3.85	.72	4.57	6.55
1300 Pavement marking channelizing		200	.120	Ea.		5.75	1.08	6.83	9.80
1400 Barrier and curb delineators		300	.080			3.85	.72	4.57	6.55
1500 Rumble strips 24" x 3-1/2" x 1/2"		150	.160			7.70	1.44	9.14	13.10

02 41 13.23 Utility Line Removal

	Crew	Daily Output	Labor-Hours	Unit	Material	Labor	Equipment	Total	Total Incl O&P
0010 **UTILITY LINE REMOVAL**									
0015 No hauling, abandon catch basin or manhole	B-6	7	3.429	Ea.		165	31	196	280
0020 Remove existing catch basin or manhole, masonry		4	6			289	54	343	490
0030 Catch basin or manhole frames and covers, stored		13	1.846			89	16.65	105.65	150
0040 Remove and reset		7	3.429			165	31	196	280
0900 Hydrants, fire, remove only	B-21A	5	8			440	86.50	526.50	750
0950 Remove and reset	B-21A	2	20	Ea.		1,100	216	1,316	1,875
2900 Pipe removal, sewer/water, no excavation, 12" diameter	B-6	175	.137	L.F.		6.60	1.24	7.84	11.20
2930 15"-18" diameter	B-12Z	150	.160			8	10.45	18.45	23.50
2960 21"-24" diameter		120	.200			10	13.05	23.05	29.50
3000 27"-36" diameter		90	.267			13.35	17.40	30.75	39
3200 Steel, welded connections, 4" diameter	B-6	160	.150			7.20	1.35	8.55	12.25
3300 10" diameter	"	80	.300			14.45	2.70	17.15	24.50

02 41 13.30 Minor Site Demolition

	Crew	Daily Output	Labor-Hours	Unit	Material	Labor	Equipment	Total	Total Incl O&P
0010 **MINOR SITE DEMOLITION** R024119-10									
0100 Roadside delineators, remove only	B-80	175	.183	Ea.		8.90	6	14.90	19.90
0110 Remove and reset	"	100	.320	"		15.60	10.50	26.10	35
0800 Guiderail, corrugated steel, remove only	B-80A	100	.240	L.F.		10.65	8.50	19.15	25.50
0850 Remove and reset	"	40	.600	"		26.50	21.50	48	63.50
0860 Guide posts, remove only	B-80B	120	.267	Ea.		12.60	1.99	14.59	21
0870 Remove and reset	B-55	50	.480	"		22	25	47	60.50
1000 Masonry walls, block, solid	B-5	1800	.031	C.F.		1.52	.83	2.35	3.19
1200 Brick, solid		900	.062			3.04	1.67	4.71	6.35
1400 Stone, with mortar		900	.062			3.04	1.67	4.71	6.35
1500 Dry set		1500	.037			1.82	1	2.82	3.82
1600 Median barrier, precast concrete, remove and store	B-3	430	.112	L.F.		5.50	5.35	10.85	14.15
1610 Remove and reset	"	390	.123	"		6.10	5.90	12	15.60
4000 Sidewalk removal, bituminous, 2" thick	B-6	350	.069	S.Y.		3.30	.62	3.92	5.60
4010 2-1/2" thick		325	.074			3.55	.67	4.22	6.05
4050 Brick, set in mortar		185	.130			6.25	1.17	7.42	10.60
4060 Dry set		270	.089			4.28	.80	5.08	7.30
4100 Concrete, plain, 4"		160	.150			7.20	1.35	8.55	12.25
4110 Plain, 5"		140	.171			8.25	1.54	9.79	14
4120 Plain, 6"		120	.200			9.60	1.80	11.40	16.35

02 41 Demolition

02 41 13 – Selective Site Demolition

02 41 13.30 Minor Site Demolition

		Crew	Daily Output	Labor-Hours	Unit	Material	2021 Bare Costs Labor	Equipment	Total	Total Incl O&P
4200	Mesh reinforced, concrete, 4"	B-6	150	.160	S.Y.		7.70	1.44	9.14	13.10
4210	5" thick		131	.183			8.80	1.65	10.45	14.95
4220	6" thick		112	.214			10.30	1.93	12.23	17.45
4300	Slab on grade removal, plain	B-5	45	1.244	C.Y.		61	33.50	94.50	127
4310	Mesh reinforced		33	1.697			83	45.50	128.50	174
4320	Rod reinforced		25	2.240			109	60	169	229
4400	For congested sites or small quantities, add up to								200%	200%
4450	For disposal on site, add	B-11A	232	.069			3.57	6.55	10.12	12.50
4500	To 5 miles, add	B-34D	76	.105			5.40	8.60	14	17.50

02 41 13.33 Railtrack Removal

		Crew	Daily Output	Labor-Hours	Unit	Material	2021 Bare Costs Labor	Equipment	Total	Total Incl O&P
0010	**RAILTRACK REMOVAL**									
3500	Railroad track removal, ties and track	B-13	330	.170	L.F.		8.20	1.78	9.98	14.15
3600	Ballast	B-14	500	.096	C.Y.		4.47	.43	4.90	7.15
3700	Remove and re-install, ties & track using new bolts & spikes	B-14	50	.960	L.F.		44.50	4.32	48.82	71.50
3800	Turnouts using new bolts and spikes	"	1	48	Ea.		2,225	216	2,441	3,575

02 41 13.34 Selective Demolition, Utility Materials

		Crew	Daily Output	Labor-Hours	Unit	Material	2021 Bare Costs Labor	Equipment	Total	Total Incl O&P
0010	**SELECTIVE DEMOLITION, UTILITY MATERIALS** R024119-10									
0015	Excludes excavation									
0020	See other utility items in Section 02 41 13.33									
0100	Fire hydrant extensions	B-20	14	1.714	Ea.		84.50		84.50	127
0200	Precast utility boxes up to 8' x 14' x 7'	B-13	2	28			1,350	293	1,643	2,350
0300	Handholes and meter pits	B-6	2	12			575	108	683	980
0400	Utility valves 4"-12"	B-20	4	6			296		296	445
0500	14"-24"	B-21	2	14			715	95.50	810.50	1,175

02 41 13.36 Selective Demolition, Utility Valves and Accessories

		Crew	Daily Output	Labor-Hours	Unit	Material	2021 Bare Costs Labor	Equipment	Total	Total Incl O&P
0010	**SELECTIVE DEMOLITION, UTILITY VALVES & ACCESSORIES**									
0015	Excludes excavation									
0100	Utility valves 4"-12" diam.	B-20	4	6	Ea.		296		296	445
0200	14"-24" diam.	B-21	2	14			715	95.50	810.50	1,175
0300	Crosses 4"-12" diam.	B-20	8	3			148		148	221
0400	14"-24" diam.	B-21	4	7			355	47.50	402.50	590
0500	Utility cut-in valves 4"-12" diam.	B-20	20	1.200			59		59	88.50
0600	Curb boxes	"	20	1.200			59		59	88.50

02 41 13.38 Selective Demo., Water & Sewer Piping & Fittings

		Crew	Daily Output	Labor-Hours	Unit	Material	2021 Bare Costs Labor	Equipment	Total	Total Incl O&P
0010	**SELECTIVE DEMOLITION, WATER & SEWER PIPING AND FITTINGS**									
0015	Excludes excavation									
0090	Concrete pipe 4"-10" diameter	B-6	250	.096	L.F.		4.62	.86	5.48	7.85
0100	42"-48" diameter	B-13B	96	.583			28	10.30	38.30	53.50
0200	60"-84" diameter	"	80	.700			34	12.40	46.40	64
0300	96" diameter	B-13C	80	.700			34	29	63	82.50
0400	108"-144" diameter	"	64	.875			42.50	36	78.50	103
0450	Concrete fittings 12" diameter	B-6	24	1	Ea.		48	9	57	81.50
0480	Concrete end pieces 12" diameter		200	.120	L.F.		5.75	1.08	6.83	9.80
0485	15" diameter		150	.160			7.70	1.44	9.14	13.10
0490	18" diameter		150	.160			7.70	1.44	9.14	13.10
0500	24"-36" diameter		100	.240			11.55	2.16	13.71	19.60
0600	Concrete fittings 24"-36" diameter		12	2	Ea.		96	18	114	163
0700	48"-84" diameter	B-13B	12	4.667			225	82.50	307.50	425
0800	96" diameter	"	8	7			340	124	464	640
0900	108"-144" diameter	B-13C	4	14			675	580	1,255	1,625
1000	Ductile iron pipe 4" diameter	B-21B	200	.200	L.F.		9.65	2.38	12.03	17
1100	6"-12" diameter		175	.229			11	2.72	13.72	19.45

02 41 Demolition

02 41 13 – Selective Site Demolition

02 41 13.38 Selective Demo., Water & Sewer Piping & Fittings	Crew	Daily Output	Labor-Hours	Unit	Material	2021 Bare Costs Labor	Equipment	Total	Total Incl O&P
1200	14"-24" diameter	B-21B	120	.333	L.F.		16.05	3.97	20.02	28.50
1300	Ductile iron fittings 4"-12" diameter		24	1.667	Ea.		80.50	19.85	100.35	142
1400	14"-16" diameter	↓	18	2.222	"		107	26.50	133.50	189
1500	18"-24" diameter	B-21B	12	3.333	Ea.		161	39.50	200.50	284
1600	Plastic pipe 3/4"-4" diameter	B-6	700	.034	L.F.		1.65	.31	1.96	2.80
1700	6"-8" diameter		500	.048			2.31	.43	2.74	3.92
1800	10"-18" diameter		300	.080			3.85	.72	4.57	6.55
1900	20"-36" diameter		200	.120			5.75	1.08	6.83	9.80
1910	42"-48" diameter		180	.133			6.40	1.20	7.60	10.85
1920	54"-60" diameter		160	.150	↓		7.20	1.35	8.55	12.25
2000	Plastic fittings 4"-8" diameter		75	.320	Ea.		15.40	2.88	18.28	26
2100	10"-14" diameter		50	.480			23	4.32	27.32	39.50
2200	16"-24" diameter		20	1.200			57.50	10.80	68.30	98
2210	30"-36" diameter		15	1.600			77	14.40	91.40	131
2220	42"-48" diameter	↓	12	2	↓		96	18	114	163
2300	Copper pipe 3/4"-2" diameter	Q-1	500	.032	L.F.		1.95		1.95	2.91
2400	2-1/2"-3" diameter		300	.053			3.25		3.25	4.85
2500	4"-6" diameter		200	.080	↓		4.87		4.87	7.25
2600	Copper fittings 3/4"-2" diameter		15	1.067	Ea.		65		65	97
2700	Cast iron pipe 4" diameter	↓	200	.080	L.F.		4.87		4.87	7.25
2800	5"-6" diameter	Q-2	200	.120			7.60		7.60	11.30
2900	8"-12" diameter	Q-3	200	.160	↓		10.30		10.30	15.40
3000	Cast iron fittings 4" diameter	Q-1	30	.533	Ea.		32.50		32.50	48.50
3100	5"-6" diameter	Q-2	30	.800			50.50		50.50	75.50
3200	8"-15" diameter	Q-3	30	1.067	↓		68.50		68.50	103
3300	Vent cast iron pipe 4"-8" diameter	Q-1	200	.080	L.F.		4.87		4.87	7.25
3400	10"-15" diameter	Q-3	200	.160	"		10.30		10.30	15.40
3500	Vent cast iron fittings 4"-8" diameter	Q-2	30	.800	Ea.		50.50		50.50	75.50
3600	10"-15" diameter	"	20	1.200	"		76		76	113

02 41 13.40 Selective Demolition, Metal Drainage Piping

		Crew	Daily Output	Labor-Hours	Unit	Material	2021 Bare Costs Labor	Equipment	Total	Total Incl O&P
0010	**SELECTIVE DEMOLITION, METAL DRAINAGE PIPING**									
0015	Excludes excavation									
0100	CMP pipe, aluminum, 6"-10" diam.	B-11M	800	.020	L.F.		1.03	.29	1.32	1.86
0110	12" diam.		600	.027			1.38	.39	1.77	2.49
0120	18" diam.		600	.027			1.38	.39	1.77	2.49
0140	Steel, 6"-10" diam.		800	.020			1.03	.29	1.32	1.86
0150	12" diam.		600	.027			1.38	.39	1.77	2.49
0160	18" diam.		400	.040			2.07	.59	2.66	3.73
0170	24" diam.	B-13	300	.187			9	1.96	10.96	15.60
0180	30"-36" diam.		250	.224			10.80	2.35	13.15	18.75
0190	48"-60" diam.	↓	200	.280			13.50	2.93	16.43	23
0200	72" diam.	B-13B	100	.560	↓		27	9.90	36.90	51.50
0210	CMP end sections, steel, 10"-18" diam.	B-11M	40	.400	Ea.		20.50	5.90	26.40	37.50
0220	24"-36" diam.	B-13	30	1.867			90	19.55	109.55	156
0230	48" diam.	"	20	2.800	↓		135	29.50	164.50	235
0240	60" diam.	B-13	10	5.600	Ea.		270	58.50	328.50	470
0250	72" diam.	B-13B	10	5.600			270	99	369	515
0260	CMP fittings, 8"-12" diam.	B-11M	60	.267			13.80	3.92	17.72	25
0270	18" diam.	"	40	.400			20.50	5.90	26.40	37.50
0280	24"-48" diam.	B-13	30	1.867			90	19.55	109.55	156
0290	60" diam.	"	20	2.800			135	29.50	164.50	235
0300	72" diam.	B-13B	10	5.600	↓		270	99	369	515

02 41 Demolition

02 41 13 – Selective Site Demolition

	02 41 13.40 Selective Demolition, Metal Drainage Piping	Crew	Daily Output	Labor-Hours	Unit	Material	2021 Bare Costs Labor	2021 Bare Costs Equipment	Total	Total Incl O&P
0310	Oval arch 17" x 13", 21" x 15", 15"-18" equivalent	B-11M	400	.040	L.F.		2.07	.59	2.66	3.73
0320	28" x 20", 24" equivalent	B-13	300	.187			9	1.96	10.96	15.60
0330	35" x 24", 42" x 29", 30"-36" equivalent		250	.224			10.80	2.35	13.15	18.75
0340	49" x 33", 57" x 38", 42"-48" equivalent	↓	200	.280	↓		13.50	2.93	16.43	23
0350	Oval arch 17" x 13" end piece, 15" equivalent	B-11M	40	.400	Ea.		20.50	5.90	26.40	37.50
0360	42" x 29" end piece, 36" equivalent	B-13	30	1.867	"		90	19.55	109.55	156

02 41 13.42 Selective Demolition, Manholes and Catch Basins

		Crew	Daily Output	Labor-Hours	Unit	Material	Labor	Equipment	Total	Total Incl O&P
0010	**SELECTIVE DEMOLITION, MANHOLES & CATCH BASINS**									
0015	Excludes excavation									
0100	Manholes, precast or brick over 8' deep	B-6	8	3	V.L.F.		144	27	171	245
0200	Cast in place 4'-8' deep	B-9	127	.315	SF Face		14.10	2.80	16.90	24
0300	Over 8' deep	"	100	.400	"		17.90	3.56	21.46	30.50
0400	Top, precast, 8" thick, 4'-6' diam.	B-6	8	3	Ea.		144	27	171	245
0500	Steps	1 Clab	60	.133	"		5.90		5.90	8.85

02 41 13.43 Selective Demolition, Box Culvert

		Crew	Daily Output	Labor-Hours	Unit	Material	Labor	Equipment	Total	Total Incl O&P
0010	**SELECTIVE DEMOLITION, BOX CULVERT**									
0015	Excludes excavation									
0100	Box culvert 8' x 6' x 3' to 8' x 8' x 8'	B-69	300	.160	L.F.		7.85	4.86	12.71	17.05
0200	8' x 10' x 3' to 8' x 12' x 8'	"	200	.240	"		11.75	7.30	19.05	25.50

02 41 13.44 Selective Demolition, Septic Tanks and Related Components

		Crew	Daily Output	Labor-Hours	Unit	Material	Labor	Equipment	Total	Total Incl O&P
0010	**SELECTIVE DEMOLITION, SEPTIC TANKS & RELATED COMPONENTS**									
0020	Excludes excavation and disposal									
0100	Septic tanks, precast, 1000-1250 gal.	B-21	8	3.500	Ea.		179	24	203	294
0200	1500 gal.		7	4			204	27.50	231.50	335
0300	2000-2500 gal.		5	5.600			286	38	324	470
0400	4000 gal.	↓	4	7			355	47.50	402.50	590
0500	Precast, 5000 gal., multiple sections	B-13	3	18.667			900	196	1,096	1,575
0600	15,000 gal.	B-13B	1.70	32.941			1,600	580	2,180	3,025
0800	40,000 gal.	"	.80	70			3,375	1,250	4,625	6,400
0900	Precast, 50,000 gal., 5 piece	B-13C	.60	93.333			4,500	3,850	8,350	11,000
1000	Cast-in-place, 75,000 gal.	B-13L	.50	32			1,975	4,200	6,175	7,550
1100	100,000 gal.	"	.30	53.333			3,275	7,000	10,275	12,600
1200	HDPE, 1000 gal.	B-21	9	3.111			159	21	180	262
1300	1500 gal.	↓	8	3.500			179	24	203	294
1400	Galley, 4' x 4' x 4'		16	1.750	↓		89.50	11.95	101.45	147
1500	Distribution boxes, concrete, 7 outlets	2 Clab	16	1	Ea.		44.50		44.50	66.50
1600	9 outlets	"	8	2			89		89	133
1700	Leaching chambers 13' x 3'-7" x 1'-4", standard	B-13	16	3.500			169	36.50	205.50	293
1800	8' x 4' x 1'-6", heavy duty		14	4			193	42	235	335
1900	13' x 3'-9" x 1'-6"		12	4.667			225	49	274	390
2100	20' x 4' x 1'-6"	↓	5	11.200			540	117	657	935
2200	Leaching pit 6'-6" x 6' deep	B-21	5	5.600			286	38	324	470
2300	6'-6" x 8' deep		4	7			355	47.50	402.50	590
2400	8' x 6' deep H20		4	7			355	47.50	402.50	590
2500	8' x 8' deep H20		3	9.333			475	63.50	538.50	785
2600	Velocity reducing pit, precast 6' x 3' deep	↓	4.70	5.957	↓		305	40.50	345.50	500

02 41 13.46 Selective Demolition, Steel Pipe With Insulation

		Crew	Daily Output	Labor-Hours	Unit	Material	Labor	Equipment	Total	Total Incl O&P
0010	**SELECTIVE DEMOLITION, STEEL PIPE WITH INSULATION**									
0020	Excludes excavation									
0100	Steel pipe, with insulation, 3/4"-4"	B-1A	400	.060	L.F.		2.70	.93	3.63	5.05
0200	5"-10"	B-1B	360	.089			4.37	2.36	6.73	9.10
0300	12"-16"	↓	240	.133	↓		6.55	3.54	10.09	13.70

02 41 Demolition

02 41 13 - Selective Site Demolition

02 41 13.46 Selective Demolition, Steel Pipe With Insulation

		Crew	Daily Output	Labor-Hours	Unit	Material	2021 Bare Costs Labor	Equipment	Total	Total Incl O&P
0400	18"-24"	B-1B	160	.200	L.F.		9.85	5.30	15.15	20.50
0450	26"-36"	↓	100	.320	↓		15.75	8.50	24.25	33
0500	Steel gland seal, with insulation, 3/4"-4"	B-1A	100	.240	Ea.		10.80	3.73	14.53	20.50
0600	5"-10"	B-1B	75	.427			21	11.30	32.30	44
0700	12"-16"		60	.533			26	14.15	40.15	54.50
0800	18"-24"		50	.640			31.50	17	48.50	65.50
0850	26"-36"	↓	40	.800			39.50	21	60.50	82
0900	Demo steel fittings with insulation, 3/4"-4"	B-1A	60	.400			18.05	6.20	24.25	34
1000	5"-10"	B-1B	40	.800			39.50	21	60.50	82
1100	12"-16"		30	1.067			52.50	28.50	81	109
1200	18"-24"		20	1.600			78.50	42.50	121	164
1300	26"-36"		15	2.133			105	56.50	161.50	219
1400	Steel pipe anchors, 5"-10"		40	.800			39.50	21	60.50	82
1500	12"-16"		30	1.067			52.50	28.50	81	109
1600	18"-24"		20	1.600			78.50	42.50	121	164
1700	26"-36"	↓	15	2.133	↓		105	56.50	161.50	219

02 41 13.48 Selective Demolition, Gasoline Containment Piping

		Crew	Daily Output	Labor-Hours	Unit	Material	2021 Bare Costs Labor	Equipment	Total	Total Incl O&P
0010	**SELECTIVE DEMOLITION, GASOLINE CONTAINMENT PIPING**									
0020	Excludes excavation									
0030	Excludes environmental site remediation									
0100	Gasoline plastic primary containment piping 2"-4"	Q-6	800	.030	L.F.		1.91		1.91	2.86
0200	Fittings 2"-4"		40	.600	Ea.		38.50		38.50	57
0300	Gasoline plastic secondary containment piping 3"-6"		800	.030	L.F.		1.91		1.91	2.86
0400	Fittings 3"-6"	↓	40	.600	Ea.		38.50		38.50	57

02 41 13.50 Selective Demolition, Natural Gas, PE Pipe

		Crew	Daily Output	Labor-Hours	Unit	Material	2021 Bare Costs Labor	Equipment	Total	Total Incl O&P
0010	**SELECTIVE DEMOLITION, NATURAL GAS, PE PIPE**									
0020	Excludes excavation									
0100	Natural gas coils, PE, 1-1/4"-3"	Q-6	800	.030	L.F.		1.91		1.91	2.86
0200	Joints, 40', PE, 3"-4"		800	.030			1.91		1.91	2.86
0300	6"-8"	↓	600	.040	↓		2.55		2.55	3.81

02 41 13.51 Selective Demolition, Natural Gas, Steel Pipe

		Crew	Daily Output	Labor-Hours	Unit	Material	2021 Bare Costs Labor	Equipment	Total	Total Incl O&P
0010	**SELECTIVE DEMOLITION, NATURAL GAS, STEEL PIPE**									
0020	Excludes excavation									
0100	Natural gas steel pipe 1"-4"	B-1A	800	.030	L.F.		1.35	.47	1.82	2.53
0200	5"-10"	B-1B	360	.089			4.37	2.36	6.73	9.10
0300	12"-16"		240	.133			6.55	3.54	10.09	13.70
0400	18"-24"	↓	160	.200	↓		9.85	5.30	15.15	20.50
0500	Natural gas steel fittings 1"-4"	B-1A	160	.150	Ea.		6.75	2.33	9.08	12.65
0600	5"-10"	B-1B	160	.200			9.85	5.30	15.15	20.50
0700	12"-16"		108	.296			14.55	7.85	22.40	30
0800	18"-24"	↓	70	.457	↓		22.50	12.15	34.65	47

02 41 13.52 Selective Demo, Natural Gas, Valves, Fittings, Regulators

		Crew	Daily Output	Labor-Hours	Unit	Material	2021 Bare Costs Labor	Equipment	Total	Total Incl O&P
0010	**SELECTIVE DEMO, NATURAL GAS, VALVES, FITTINGS, REGULATORS**									
0100	Gas stops 1-1/4"-2"	1 Plum	22	.364	Ea.		24.50		24.50	37
0200	Gas regulator 1-1/2"-2"	"	22	.364			24.50		24.50	37
0300	3"-4"	Q-1	22	.727			44.50		44.50	66
0400	Gas plug valve, 3/4"-2"	1 Plum	22	.364			24.50		24.50	37
0500	2-1/2"-3"	Q-1	10	1.600	↓		97.50		97.50	145

For customer support on your Site Work & Landscape Costs with RSMeans Data, call 800.448.8182.

02 41 Demolition

02 41 13 – Selective Site Demolition

02 41 13.54 Selective Demolition, Electric Ducts and Fittings	Crew	Daily Output	Labor-Hours	Unit	Material	2021 Bare Costs Labor	2021 Bare Costs Equipment	Total	Total Incl O&P
0010 **SELECTIVE DEMOLITION, ELECTRIC DUCTS & FITTINGS**									
0020 Excludes excavation									
0100 Plastic conduit, 1/2"-2"	1 Elec	600	.013	L.F.		.85		.85	1.26
0200 3"-6"	2 Elec	400	.040	"		2.55		2.55	3.79
0300 Fittings, 1/2"-2"	1 Elec	50	.160	Ea.		10.20		10.20	15.15
0400 3"-6"	"	40	.200	"		12.75		12.75	18.95

02 41 13.56 Selective Demolition, Electric Duct Banks

02 41 13.56 Selective Demolition, Electric Duct Banks	Crew	Daily Output	Labor-Hours	Unit	Material	2021 Bare Costs Labor	2021 Bare Costs Equipment	Total	Total Incl O&P
0010 **SELECTIVE DEMOLITION, ELECTRIC DUCT BANKS**									
0020 Excludes excavation									
0100 Hand holes sized to 4' x 4' x 4'	R-3	7	2.857	Ea.		181	27.50	208.50	300
0200 Manholes sized to 6' x 10' x 7'	B-13	6	9.333	"		450	98	548	780
0300 Conduit 1 @ 2" diameter, EB plastic, no concrete	2 Elec	1000	.016	L.F.		1.02		1.02	1.51
0400 2 @ 2" diameter		500	.032			2.04		2.04	3.03
0500 4 @ 2" diameter		250	.064			4.08		4.08	6.05
0600 1 @ 3" diameter		800	.020			1.27		1.27	1.89
0700 2 @ 3" diameter		400	.040			2.55		2.55	3.79
0800 4 @ 3" diameter		200	.080			5.10		5.10	7.55
0900 1 @ 4" diameter		800	.020			1.27		1.27	1.89
1000 2 @ 4" diameter		400	.040			2.55		2.55	3.79
1100 4 @ 4" diameter		200	.080			5.10		5.10	7.55
1200 6 @ 4" diameter		100	.160			10.20		10.20	15.15
1300 1 @ 5" diameter	2 Elec	500	.032	L.F.		2.04		2.04	3.03
1400 2 @ 5" diameter		250	.064			4.08		4.08	6.05
1500 4 @ 5" diameter		160	.100			6.35		6.35	9.45
1600 6 @ 5" diameter		100	.160			10.20		10.20	15.15
1700 1 @ 6" diameter		500	.032			2.04		2.04	3.03
1800 2 @ 6" diameter		250	.064			4.08		4.08	6.05
1900 4 @ 6" diameter		160	.100			6.35		6.35	9.45
2000 6 @ 6" diameter		100	.160			10.20		10.20	15.15
2100 Conduit 1 EB plastic, with concrete, 0.92 C.F./L.F.	B-9	150	.267			11.95	2.37	14.32	20.50
2200 2 EB plastic 1.52 C.F./L.F.		100	.400			17.90	3.56	21.46	30.50
2300 2 x 2 EB plastic 2.51 C.F./L.F.		80	.500			22.50	4.45	26.95	38.50
2400 2 x 3 EB plastic 3.56 C.F./L.F.		60	.667			30	5.95	35.95	51
2500 Conduit 2 @ 2" diameter, steel, no concrete	2 Elec	400	.040			2.55		2.55	3.79
2600 4 @ 2" diameter		200	.080			5.10		5.10	7.55
2700 2 @ 3" diameter		200	.080			5.10		5.10	7.55
2800 4 @ 3" diameter		100	.160			10.20		10.20	15.15
2900 2 @ 4" diameter		200	.080			5.10		5.10	7.55
3000 4 @ 4" diameter		100	.160			10.20		10.20	15.15
3100 6 @ 4" diameter		50	.320			20.50		20.50	30.50
3200 2 @ 5" diameter		120	.133			8.50		8.50	12.60
3300 4 @ 5" diameter		60	.267			17		17	25
3400 6 @ 5" diameter		40	.400			25.50		25.50	38
3500 2 @ 6" diameter		120	.133			8.50		8.50	12.60
3600 4 @ 6" diameter		60	.267			17		17	25
3700 6 @ 6" diameter		40	.400			25.50		25.50	38
3800 Conduit 2 steel, with concrete, 1.52 C.F./L.F.	B-9	80	.500			22.50	4.45	26.95	38.50
3900 2 x 2 EB steel 2.51 C.F./L.F.		60	.667			30	5.95	35.95	51
4000 2 x 3 steel 3.56 C.F./L.F.		50	.800			36	7.10	43.10	61.50
4100 Conduit fittings, PVC type EB, 2"-3"	1 Elec	30	.267	Ea.		17		17	25
4200 4"-6"	"	20	.400	"		25.50		25.50	38

For customer support on your Site Work & Landscape Costs with RSMeans Data, call 800.448.8182.

35

02 41 Demolition

02 41 13 – Selective Site Demolition

02 41 13.60 Selective Demolition Fencing		Crew	Daily Output	Labor-Hours	Unit	Material	2021 Bare Costs Labor	2021 Bare Costs Equipment	Total	Total Incl O&P
0010	**SELECTIVE DEMOLITION FENCING** R024119-10									
0700	Snow fence, 4' high	B-6	1000	.024	L.F.		1.15	.22	1.37	1.96
1600	Fencing, barbed wire, 3 strand	2 Clab	430	.037			1.65		1.65	2.47
1650	5 strand	"	280	.057			2.54		2.54	3.79
1700	Chain link, posts & fabric, 8'-10' high, remove only	B-6	445	.054			2.59	.49	3.08	4.40
1755	6' high, remove only	"	520	.046			2.22	.42	2.64	3.77
1760	Wood fence to 6', remove only, minimum	2 Clab	500	.032			1.42		1.42	2.12
1770	Maximum		250	.064			2.84		2.84	4.24
1775	Fencing, wood, average all types, 4'-6' high		432	.037	▼		1.64		1.64	2.45
1780	Fence post, wood, 4" x 4", 6'-8' high	▼	96	.167	Ea.		7.40		7.40	11.05
1785	Fence rail, 2" x 4", 8' long, remove only	2 Clab	7680	.002	L.F.		.09		.09	.14
1790	Remove and store	B-80	235	.136	"		6.65	4.48	11.13	14.85

02 41 13.62 Selective Demo., Chain Link Fences & Gates

		Crew	Daily Output	Labor-Hours	Unit	Material	2021 Bare Costs Labor	2021 Bare Costs Equipment	Total	Total Incl O&P
0010	**SELECTIVE DEMOLITION, CHAIN LINK FENCES & GATES**									
0100	Chain link, gates, 3'-4' width	B-6	30	.800	Ea.		38.50	7.20	45.70	65.50
0200	10'-12' width		16	1.500			72	13.50	85.50	123
0300	14' width		15	1.600			77	14.40	91.40	131
0400	20' width		10	2.400	▼		115	21.50	136.50	196
0500	18' width with overhead & cantilever		80	.300	L.F.		14.45	2.70	17.15	24.50
0510	Sliding		80	.300			14.45	2.70	17.15	24.50
0520	Cantilever to 40' wide		80	.300	▼		14.45	2.70	17.15	24.50
0530	Motor operators	2 Skwk	1	16	Ea.		915		915	1,375
0540	Transmitter systems	"	15	1.067	"		61		61	91.50
0600	Chain link, fence, 5' high	B-6	890	.027	L.F.		1.30	.24	1.54	2.20
0650	3'-4' high		1000	.024			1.15	.22	1.37	1.96
0675	12' high	▼	400	.060	▼		2.89	.54	3.43	4.89
0800	Chain link, fence, braces	B-1	2000	.012	Ea.		.54		.54	.81
0900	Privacy slats	"	2000	.012			.54		.54	.81
1000	Fence posts, steel, in concrete	B-6	80	.300	▼		14.45	2.70	17.15	24.50
1100	Fence fabric & accessories, fabric to 8' high		800	.030	L.F.		1.44	.27	1.71	2.45
1200	Barbed wire		5000	.005	"		.23	.04	.27	.39
1300	Extension arms & eye tops		300	.080	Ea.		3.85	.72	4.57	6.55
1400	Fence rails		2000	.012	L.F.		.58	.11	.69	.98
1500	Reinforcing wire	▼	5000	.005	"		.23	.04	.27	.39

02 41 13.64 Selective Demolition, Vinyl Fences and Gates

		Crew	Daily Output	Labor-Hours	Unit	Material	2021 Bare Costs Labor	2021 Bare Costs Equipment	Total	Total Incl O&P
0010	**SELECTIVE DEMOLITION, VINYL FENCES & GATES**									
0100	Vinyl fence up to 6' high	B-6	1000	.024	L.F.		1.15	.22	1.37	1.96
0200	Gates up to 6' high	"	40	.600	Ea.		29	5.40	34.40	49

02 41 13.66 Selective Demolition, Misc Metal Fences and Gates

		Crew	Daily Output	Labor-Hours	Unit	Material	2021 Bare Costs Labor	2021 Bare Costs Equipment	Total	Total Incl O&P
0010	**SELECTIVE DEMOLITION, MISC METAL FENCES & GATES**									
0100	Misc steel mesh fences, 4'-6' high	B-6	600	.040	L.F.		1.92	.36	2.28	3.27
0200	Kennels, 6'-12' long	2 Clab	8	2	Ea.		89		89	133
0300	Tops, 6'-12' long	"	30	.533	"		23.50		23.50	35.50
0400	Security fences, 12'-16' high	B-6	100	.240	L.F.		11.55	2.16	13.71	19.60
0500	Metal tubular picket fences, 4'-6' high		500	.048	"		2.31	.43	2.74	3.92
0600	Gates, 3'-4' wide	▼	20	1.200	Ea.		57.50	10.80	68.30	98

02 41 13.68 Selective Demolition, Wood Fences and Gates

		Crew	Daily Output	Labor-Hours	Unit	Material	2021 Bare Costs Labor	2021 Bare Costs Equipment	Total	Total Incl O&P
0010	**SELECTIVE DEMOLITION, WOOD FENCES & GATES**									
0100	Wood fence gates 3'-4' wide	2 Clab	20	.800	Ea.		35.50		35.50	53
0200	Wood fence, open rail, to 4' high		2560	.006	L.F.		.28		.28	.41
0300	to 8' high	▼	368	.043	"		1.93		1.93	2.88

02 41 Demolition

02 41 13 – Selective Site Demolition

02 41 13.68 Selective Demolition, Wood Fences and Gates	Crew	Daily Output	Labor-Hours	Unit	Material	2021 Bare Costs Labor	Equipment	Total	Total Incl O&P
0400 Post, in concrete	2 Clab	50	.320	Ea.		14.20		14.20	21

02 41 13.70 Selective Demolition, Rip-Rap and Rock Lining	Crew	Daily Output	Labor-Hours	Unit	Material	Labor	Equipment	Total	Total Incl O&P
0010 **SELECTIVE DEMOLITION, RIP-RAP & ROCK LINING** R024119-10									
0100 Slope protection, broken stone	B-13	62	.903	C.Y.		43.50	9.45	52.95	75.50
0200 3/8 to 1/4 C.Y. pieces		60	.933	S.Y.		45	9.80	54.80	78
0300 18" depth		60	.933	"		45	9.80	54.80	78
0400 Dumped stone		93	.602	Ton		29	6.30	35.30	50.50
0500 Gabions, 6"-12" deep		60	.933	S.Y.		45	9.80	54.80	78
0600 18"-36" deep		30	1.867	"		90	19.55	109.55	156

02 41 13.72 Selective Demo., Shore Protect/Mooring Struct.	Crew	Daily Output	Labor-Hours	Unit	Material	Labor	Equipment	Total	Total Incl O&P
0010 **SELECTIVE DEMOLITION, SHORE PROTECT/MOORING STRUCTURES**									
0100 Breakwaters, bulkheads, concrete, maximum R024119-10	B-9	12	3.333	L.F.		149	29.50	178.50	256
0200 Breakwaters, bulkheads, concrete, 12', minimum		10	4			179	35.50	214.50	305
0300 Maximum		9	4.444			199	39.50	238.50	340
0400 Steel, from shore	B-40B	54	.889			43.50	39.50	83	109
0500 from barge	B-76A	30	2.133			102	80.50	182.50	241
0600 Jetties, docks, floating	B-21B	600	.067	S.F.		3.21	.79	4	5.65
0700 Pier supported, 3"-4" decking		300	.133			6.45	1.59	8.04	11.35
0800 Floating, prefab, small boat, minimum		600	.067			3.21	.79	4	5.65
0900 Maximum		300	.133			6.45	1.59	8.04	11.35
1000 Floating, prefab, per slip, minimum		3.20	12.500	Ea.		605	149	754	1,075
1010 Maximum		2.80	14.286	"		690	170	860	1,200

02 41 13.74 Selective Demolition, Piles	Crew	Daily Output	Labor-Hours	Unit	Material	Labor	Equipment	Total	Total Incl O&P
0010 **SELECTIVE DEMOLITION, PILES**									
0100 Cast in place piles, corrugated, 8"-10"	B-19	600	.107	V.L.F.		6.10	3.41	9.51	13.05
0200 12"-14"		500	.128			7.30	4.09	11.39	15.65
0300 16"		400	.160			9.15	5.10	14.25	19.55
0400 fluted, 12"		600	.107			6.10	3.41	9.51	13.05
0500 14"-18"		500	.128			7.30	4.09	11.39	15.65
0600 end bearing, 12"		600	.107			6.10	3.41	9.51	13.05
0700 14"-18"		500	.128			7.30	4.09	11.39	15.65
0800 Precast prestressed piles, 12"-14" diam.		700	.091			5.20	2.92	8.12	11.15
0900 16"-24" diam.		600	.107			6.10	3.41	9.51	13.05
1000 36"-66" diam.		300	.213			12.20	6.80	19	26
1100 10"-14" thick		600	.107			6.10	3.41	9.51	13.05
1200 16"-24" thick	B-38	500	.080			4	2.43	6.43	8.65
1300 Pressure grouted pile, 5"	B-19	150	.427			24.50	13.65	38.15	52
1400 Steel piles, 8"-12" tip		600	.107			6.10	3.41	9.51	13.05
1500 H sections HP8 to HP12		600	.107			6.10	3.41	9.51	13.05
1600 HP14	B-19A	600	.107			6.10	4.87	10.97	14.65
1700 Steel pipe piles, 8"-12"	B-19	600	.107			6.10	3.41	9.51	13.05
1800 14"-18" plain		500	.128			7.30	4.09	11.39	15.65
1900 concrete filled, 14"-18"		400	.160			9.15	5.10	14.25	19.55
2000 Timber piles to 14" diam.		600	.107			6.10	3.41	9.51	13.05

02 41 13.76 Selective Demolition, Water Wells	Crew	Daily Output	Labor-Hours	Unit	Material	Labor	Equipment	Total	Total Incl O&P
0010 **SELECTIVE DEMOLITION, WATER WELLS**									
0100 Well, 40' deep with casing & gravel pack, 24"-36" diam.	B-23	.25	160	Ea.		7,175	6,475	13,650	17,800
0200 Riser pipe, 1-1/4", for observation well	"	300	.133	L.F.		5.95	5.40	11.35	14.85
0300 Pump, 1/2 to 5 HP up to 100' depth	Q-1	3	5.333	Ea.		325		325	485
0400 Up to 150' well 25 HP pump	Q-22	1.50	10.667			650	315	965	1,325
0500 Up to 500' well 30 HP pump	"	1	16			975	475	1,450	1,975
0600 Well screen 2"-8"	B-23	300	.133	V.L.F.		5.95	5.40	11.35	14.85

For customer support on your Site Work & Landscape Costs with RSMeans Data, call 800.448.8182.

37

02 41 Demolition

02 41 13 – Selective Site Demolition

02 41 13.76 Selective Demolition, Water Wells		Crew	Daily Output	Labor-Hours	Unit	Material	2021 Bare Costs Labor	2021 Bare Costs Equipment	Total	Total Incl O&P
0700	10"-16"	B-23	200	.200	V.L.F.		8.95	8.10	17.05	22.50
0800	18"-26"		150	.267			11.95	10.80	22.75	29.50
0900	Slotted PVC for wells 1-1/4"-8"		600	.067			2.99	2.70	5.69	7.45
1000	Well screen and casing 6"-16"		300	.133			5.95	5.40	11.35	14.85
1100	18"-26"		150	.267			11.95	10.80	22.75	29.50
1200	30"-36"		100	.400			17.90	16.20	34.10	44.50

02 41 13.78 Selective Demolition, Radio Towers		Crew	Daily Output	Labor-Hours	Unit	Material	2021 Bare Costs Labor	2021 Bare Costs Equipment	Total	Total Incl O&P
0010	**SELECTIVE DEMOLITION, RADIO TOWERS**									
0100	Radio tower, guyed, 50'	2 Skwk	16	1	Ea.		57		57	86
0200	190', 40 lb. section	K-2	.70	34.286			1,950	1,225	3,175	4,325
0300	200', 70 lb. section		.70	34.286			1,950	1,225	3,175	4,325
0400	300', 70 lb. section		.40	60			3,425	2,125	5,550	7,600
0500	270', 90 lb. section		.40	60			3,425	2,125	5,550	7,600
0600	400'		.30	80			4,575	2,825	7,400	10,100
0700	Self supported, 60'		.90	26.667			1,525	945	2,470	3,375
0800	120'		.80	30			1,725	1,075	2,800	3,800
0900	190'		.40	60			3,425	2,125	5,550	7,600

02 41 13.80 Selective Demo., Utility Poles and Cross Arms		Crew	Daily Output	Labor-Hours	Unit	Material	2021 Bare Costs Labor	2021 Bare Costs Equipment	Total	Total Incl O&P
0010	**SELECTIVE DEMOLITION, UTILITY POLES & CROSS ARMS**									
0100	Utility poles, wood, 20'-30' high	R-3	6	3.333	Ea.		212	32	244	350
0200	35'-45' high	"	5	4			254	38	292	415
0300	Cross arms, wood, 4'-6' long	1 Elec	5	1.600			102		102	151

02 41 13.82 Selective Removal, Pavement Lines and Markings		Crew	Daily Output	Labor-Hours	Unit	Material	2021 Bare Costs Labor	2021 Bare Costs Equipment	Total	Total Incl O&P
0010	**SELECTIVE REMOVAL, PAVEMENT LINES & MARKINGS**									
0015	Does not include traffic control costs									
0020	See other items in Section 32 17 23.13									
0100	Remove permanent painted traffic lines and markings	B-78A	500	.016	C.L.F.		.89	1.99	2.88	3.51
0200	Temporary traffic line tape	2 Clab	1500	.011	L.F.		.47		.47	.71
0300	Thermoplastic traffic lines and markings	B-79A	500	.024	C.L.F.		1.33	3.02	4.35	5.30
0400	Painted pavement markings	B-78B	500	.036	S.F.		1.64	.82	2.46	3.35

02 41 13.84 Selective Demolition, Walks, Steps and Pavers		Crew	Daily Output	Labor-Hours	Unit	Material	2021 Bare Costs Labor	2021 Bare Costs Equipment	Total	Total Incl O&P
0010	**SELECTIVE DEMOLITION, WALKS, STEPS AND PAVERS**									
0100	Splash blocks	1 Clab	300	.027	S.F.		1.18		1.18	1.77
0200	Tree grates	"	50	.160	Ea.		7.10		7.10	10.60
0300	Walks, limestone pavers	2 Clab	150	.107	S.F.		4.74		4.74	7.05
0400	Redwood sections		600	.027			1.18		1.18	1.77
0500	Redwood planks		480	.033			1.48		1.48	2.21
0600	Shale paver	2 Clab	300	.053	S.F.		2.37		2.37	3.53
0700	Tile thinset paver	"	675	.024	"		1.05		1.05	1.57
0800	Wood round	B-1	350	.069	Ea.		3.09		3.09	4.61
0900	Asphalt block	2 Clab	450	.036	S.F.		1.58		1.58	2.36
1000	Bluestone		450	.036			1.58		1.58	2.36
1100	Slate, 1" or thinner		675	.024			1.05		1.05	1.57
1200	Granite blocks		300	.053			2.37		2.37	3.53
1300	Precast patio blocks		450	.036			1.58		1.58	2.36
1400	Planter blocks		600	.027			1.18		1.18	1.77
1500	Brick paving, dry set		300	.053			2.37		2.37	3.53
1600	Mortar set		180	.089			3.95		3.95	5.90
1700	Dry set on edge		240	.067			2.96		2.96	4.42
1800	Steps, brick		200	.080	L.F.		3.55		3.55	5.30
1900	Railroad tie		150	.107			4.74		4.74	7.05
2000	Bluestone		180	.089			3.95		3.95	5.90

02 41 Demolition

02 41 13 − Selective Site Demolition

02 41 13.84 Selective Demolition, Walks, Steps and Pavers	Crew	Daily Output	Labor-Hours	Unit	Material	2021 Bare Costs Labor	Equipment	Total	Total Incl O&P
2100	Wood/steel edging for steps	2 Clab	1000	.016	L.F.		.71		.71	1.06
2200	Timber or railroad tie edging for steps	↓	400	.040	↓		1.78		1.78	2.65

02 41 13.86 Selective Demolition, Athletic Surfaces

		Crew	Daily Output	Labor-Hours	Unit	Material	Labor	Equipment	Total	Total Incl O&P
0010	**SELECTIVE DEMOLITION, ATHLETIC SURFACES**									
0100	Synthetic grass	2 Clab	2000	.008	S.F.		.36		.36	.53
0200	Surface coat latex rubber	"	2000	.008	"		.36		.36	.53
0300	Tennis court posts	B-11C	16	1	Ea.		51.50	13.50	65	92

02 41 13.88 Selective Demolition, Lawn Sprinkler Systems

		Crew	Daily Output	Labor-Hours	Unit	Material	Labor	Equipment	Total	Total Incl O&P
0010	**SELECTIVE DEMOLITION, LAWN SPRINKLER SYSTEMS**									
0100	Golf course sprinkler system, 9 hole	4 Clab	.10	320	Ea.		14,200		14,200	21,200
0200	Sprinkler system, 24' diam. @ 15' OC, per head	B-20	110	.218	Head		10.75		10.75	16.10
0300	60' diam. @ 24' OC, per head	"	52	.462	"		23		23	34
0400	Sprinkler heads, plastic	2 Clab	150	.107	Ea.		4.74		4.74	7.05
0500	Impact circle pattern, 28'-76' diam.		75	.213			9.45		9.45	14.15
0600	Pop-up, 42'-76' diam.		50	.320			14.20		14.20	21
0700	39'-99' diam.		50	.320			14.20		14.20	21
0800	Sprinkler valves		40	.400			17.75		17.75	26.50
0900	Valve boxes		40	.400			17.75		17.75	26.50
1000	Controls		2	8			355		355	530
1100	Backflow preventer		4	4			178		178	265
1200	Vacuum breaker	↓	4	4	↓		178		178	265

02 41 13.90 Selective Demolition, Retaining Walls

		Crew	Daily Output	Labor-Hours	Unit	Material	Labor	Equipment	Total	Total Incl O&P
0010	**SELECTIVE DEMOLITION, RETAINING WALLS**									
0020	See other retaining wall items in Section 32 32									
0100	Concrete retaining wall, 6' high, no reinforcing	B-13K	200	.080	L.F.		4.92	10.25	15.17	18.60
0200	8' high		150	.107			6.55	13.70	20.25	25
0300	10' high		150	.107			6.55	13.70	20.25	25
0400	With reinforcing, 6' high	↓	200	.080	↓		4.92	10.25	15.17	18.60
0500	8' high	B-13K	150	.107	L.F.		6.55	13.70	20.25	25
0600	10' high		120	.133			8.20	17.10	25.30	31
0700	20' high		60	.267	↓		16.40	34	50.40	62
0800	Concrete cribbing, 12' high, open/closed face	↓	150	.107	S.F.		6.55	13.70	20.25	25
0900	Interlocking segmental retaining wall	B-62	800	.030			1.44	.22	1.66	2.40
1000	Wall caps	"	600	.040			1.92	.30	2.22	3.20
1100	Metal bin retaining wall, 10' wide, 4'-12' high	B-13	1200	.047			2.25	.49	2.74	3.90
1200	10' wide, 16'-28' high		1000	.056	↓		2.70	.59	3.29	4.68
1300	Stone filled gabions, 6' x 3' x 1'		170	.329	Ea.		15.90	3.45	19.35	27.50
1400	6' x 3' x 1'-6"		75	.747			36	7.85	43.85	62.50
1500	6' x 3' x 3'		25	2.240			108	23.50	131.50	187
1600	9' x 3' x 1'		75	.747			36	7.85	43.85	62.50
1700	9' x 3' x 1'-6"		33	1.697			82	17.80	99.80	142
1800	9' x 3' x 3'		12	4.667			225	49	274	390
1900	12' x 3' x 1'		42	1.333			64.50	13.95	78.45	111
2000	12' x 3' x 1'-6"		20	2.800			135	29.50	164.50	235
2100	12' x 3' x 3'	↓	6	9.333	↓		450	98	548	780

02 41 13.92 Selective Demolition, Parking Appurtenances

		Crew	Daily Output	Labor-Hours	Unit	Material	Labor	Equipment	Total	Total Incl O&P
0010	**SELECTIVE DEMOLITION, PARKING APPURTENANCES**									
0100	Bumper rails, garage, 6" wide	B-6	300	.080	L.F.		3.85	.72	4.57	6.55
0200	12" channel rail		300	.080			3.85	.72	4.57	6.55
0300	Parking bumper, timber	↓	1000	.024	↓		1.15	.22	1.37	1.96
0400	Folding, with locks	B-1	100	.240	Ea.		10.80		10.80	16.15
0500	Flexible fixed garage stanchion	B-6	150	.160	↓		7.70	1.44	9.14	13.10

39

For customer support on your Site Work & Landscape Costs with RSMeans Data, call 800.448.8182.

02 41 13 – Selective Site Demolition

02 41 13.92 Selective Demolition, Parking Appurtenances	Crew	Daily Output	Labor-Hours	Unit	Material	2021 Bare Costs Labor	2021 Bare Costs Equipment	Total	Total Incl O&P
0600 Wheel stops, precast concrete	B-6	120	.200	Ea.		9.60	1.80	11.40	16.35
0700 Thermoplastic		120	.200			9.60	1.80	11.40	16.35
0800 Pipe bollards, 6"-12" diam.	↓	80	.300	↓		14.45	2.70	17.15	24.50

02 41 13.93 Selective Demolition, Site Furnishings

		Crew	Daily Output	Labor-Hours	Unit	Material	Labor	Equipment	Total	Total Incl O&P
0010	SELECTIVE DEMOLITION, SITE FURNISHINGS									
0100	Benches, all types	2 Clab	10	1.600	Ea.		71		71	106
0200	Trash receptacles, all types		80	.200			8.90		8.90	13.25
0300	Trash enclosures, steel or wood	↓	10	1.600	↓		71		71	106

02 41 13.94 Selective Demolition, Athletic Screening

		Crew	Daily Output	Labor-Hours	Unit	Material	Labor	Equipment	Total	Total Incl O&P
0010	SELECTIVE DEMOLITION, ATHLETIC SCREENING									
0020	See other items in Sections 02 41 13.60 & 02 41 13.62									
0100	Baseball backstops	B-6	2	12	Ea.		575	108	683	980
0200	Basketball goal, single		6	4			192	36	228	325
0300	Double		4	6	↓		289	54	343	490
0400	Tennis wire mesh, pair ends		5	4.800	Set		231	43.50	274.50	395
0500	Enclosed court		3	8	Ea.		385	72	457	655
0600	Wood or masonry handball court	↓	1	24	"		1,150	216	1,366	1,975

02 41 13.95 Selective Demo., Athletic/Playground Equipment

		Crew	Daily Output	Labor-Hours	Unit	Material	Labor	Equipment	Total	Total Incl O&P
0010	SELECTIVE DEMO., ATHLETIC/PLAYGROUND EQUIPMENT									
0020	See other items in Section 02 41 13.96									
0100	Bike rack	B-6	32	.750	Ea.		36	6.75	42.75	61.50
0200	Climber arch		16	1.500			72	13.50	85.50	123
0300	Fitness trail, 9 to 10 stations wood or metal		.50	48			2,300	430	2,730	3,925
0400	16 to 20 stations wood or metal		.25	96			4,625	865	5,490	7,825
0500	Monkey bars 14' long		8	3			144	27	171	245
0600	Parallel bars 10' long		8	3			144	27	171	245
0700	Post tether ball		32	.750			36	6.75	42.75	61.50
0800	Poles 10'-6" long		32	.750			36	6.75	42.75	61.50
0900	Pole ground sockets		80	.300			14.45	2.70	17.15	24.50
1000	See-saw, 2 units		12	2			96	18	114	163
1100	4 units		10	2.400			115	21.50	136.50	196
1200	6 units		8	3			144	27	171	245
1300	Shelter, fiberglass, 3 person		10	2.400			115	21.50	136.50	196
1400	Slides, 12' long		8	3			144	27	171	245
1500	20' long		6	4			192	36	228	325
1600	Swings, 4 seats		6	4			192	36	228	325
1700	8 seats		4	6			289	54	343	490
1800	Whirlers, 8'-10' diameter		6	4	↓		192	36	228	325
1900	Football or football/soccer goal posts		3	8	Pair		385	72	457	655
2000	Soccer goal posts	↓	4	6	"		289	54	343	490
2100	Playground surfacing, 4" depth	B-62	2000	.012	S.F.		.58	.09	.67	.96
2200	2" topping	"	3500	.007	"		.33	.05	.38	.55
2300	Platform/paddle tennis court, alum. deck & frame	B-1	.20	120	Court		5,400		5,400	8,075
2400	Aluminum deck and wood frame		.25	96	"		4,325		4,325	6,450
2500	Heater		2	12	Ea.		540		540	805
2600	Wood deck & frame		.25	96	Court		4,325		4,325	6,450
2700	Wood deck & steel frame		.25	96			4,325		4,325	6,450
2800	Steel deck & wood frame	↓	.25	96	↓		4,325		4,325	6,450

02 41 13.96 Selective Demo., Modular Playground Equipment

		Crew	Daily Output	Labor-Hours	Unit	Material	Labor	Equipment	Total	Total Incl O&P
0010	SELECTIVE DEMOLITION, MODULAR PLAYGROUND EQUIPMENT									
0100	Modular playground, 48" deck, square or triangle shape	B-1	2	12	Ea.		540		540	805
0200	Various posts	↓	18	1.333	↓		60		60	89.50

02 41 Demolition

02 41 13 – Selective Site Demolition

02 41 13.96 Selective Demo., Modular Playground Equipment

		Crew	Daily Output	Labor-Hours	Unit	Material	2021 Bare Costs Labor	2021 Bare Costs Equipment	Total	Total Incl O&P
0300	Roof 54" x 54"	B-1	18	1.333	Ea.		60		60	89.50
0400	Wheel chair transfer module		6	4			180		180	269
0500	Guardrail pipe		90	.267	L.F.		12		12	17.95
0600	Steps, 3 risers		16	1.500	Ea.		67.50		67.50	101
0700	Activity panel crawl through		8	3			135		135	202
0800	Alphabet/spelling panel		8	3			135		135	202
0900	Crawl tunnel, each 4' section		8	3			135		135	202
1000	Slide tunnel		16	1.500			67.50		67.50	101
1100	Spiral slide tunnel	B-1	12	2	Ea.		90		90	135
1200	Ladder, horizontal or vertical		10	2.400			108		108	161
1300	Corkscrew climber		6	4			180		180	269
1400	Fire pole		12	2			90		90	135
1500	Bridge, ring chamber		8	3			135		135	202
1600	Bridge, suspension		8	3			135		135	202

02 41 13.97 Selective Demolition, Traffic Light Systems

		Crew	Daily Output	Labor-Hours	Unit	Material	2021 Bare Costs Labor	2021 Bare Costs Equipment	Total	Total Incl O&P
0010	**SELECTIVE DEMOLITION, TRAFFIC LIGHT SYSTEMS**									
0100	Traffic signal, pedestrian cross walk	R-2	.50	112	Total		6,700	765	7,465	10,800
0200	8 signal intersection, 2 each direction	"	.25	224			13,400	1,525	14,925	21,600
0300	Each traffic phase controller	L-9	3	12			600		600	905
0400	Each semi-actuated detector		1.50	24			1,200		1,200	1,825
0500	Each fully-actuated detector		1.50	24			1,200		1,200	1,825
0600	Pedestrian push button system		2	18			905		905	1,350
0650	Each optically programmed head		3	12			600		600	905
0700	School signal flashing system		1	36	Signal		1,800		1,800	2,725
0800	Complete traffic light intersection system, without lane control	R-2	.25	224	Ea.		13,400	1,525	14,925	21,600
0900	With lane control		.20	280			16,700	1,900	18,600	27,000
1000	Traffic light, each left turn protection system		.66	84.848			5,075	580	5,655	8,175

02 41 13.98 Selective Demolition, Plantings

		Crew	Daily Output	Labor-Hours	Unit	Material	2021 Bare Costs Labor	2021 Bare Costs Equipment	Total	Total Incl O&P
0010	**SELECTIVE DEMOLITION, PLANTINGS**									
0100	Remove sod 1" deep, 3 M.S.F. or greater, level ground	B-63	40	1	M.S.F.		46.50	4.49	50.99	74.50
0200	Less than 3 M.S.F., level ground		27	1.481			69	6.65	75.65	110
0300	3 M.S.F. or greater, sloped ground		12	3.333			155	14.95	169.95	248
0400	Less than 3 M.S.F., sloped ground		8	5			233	22.50	255.50	375
0500	Edging, aluminum	B-1	800	.030	L.F.		1.35		1.35	2.02
0600	Brick horizontal		400	.060			2.70		2.70	4.04
0700	Brick vertical		200	.120			5.40		5.40	8.05
0800	Corrugated aluminum		1200	.020			.90		.90	1.35
0900	Granite 5" x 16"		600	.040			1.80		1.80	2.69
1000	Polyethylene grass barrier		1200	.020			.90		.90	1.35
1100	Precast 2" x 8" x 16"		800	.030			1.35		1.35	2.02
1200	Railroad ties		200	.120			5.40		5.40	8.05
1300	Wood		400	.060			2.70		2.70	4.04
1400	Steel strips including stakes		600	.040			1.80		1.80	2.69
1500	Planters, concrete, 48" diam.	2 Clab	30	.533	Ea.		23.50		23.50	35.50
1600	Concrete 7' diam.		20	.800			35.50		35.50	53
1700	Fiberglass 36" diam.		30	.533			23.50		23.50	35.50
1800	60" diam.		20	.800			35.50		35.50	53
1900	24" square		30	.533			23.50		23.50	35.50
2000	48" square		30	.533			23.50		23.50	35.50
2100	Wood 48" square	2 Clab	30	.533	Ea.		23.50		23.50	35.50
2200	48" diam.		20	.800			35.50		35.50	53
2300	72" diam.		20	.800			35.50		35.50	53

02 41 Demolition

02 41 13 - Selective Site Demolition

	02 41 13.98 Selective Demolition, Plantings	Crew	Daily Output	Labor-Hours	Unit	Material	2021 Bare Costs Labor	2021 Bare Costs Equipment	Total	Total Incl O&P
2400	Planter/bench, fiberglass 72"-96" square	2 Clab	10	1.600	Ea.		71		71	106
2500	Wood, 72" square		10	1.600			71		71	106
2600	Tree guying, stakes, 3"-4" caliper		40	.400			17.75		17.75	26.50
2700	anchors, 3" caliper		40	.400			17.75		17.75	26.50
2800	3"-6" caliper		30	.533			23.50		23.50	35.50
2900	6" caliper		20	.800			35.50		35.50	53
3000	8" caliper		15	1.067			47.50		47.50	70.50

02 41 16 - Structure Demolition

02 41 16.13 Building Demolition

		Crew	Daily Output	Labor-Hours	Unit	Material	2021 Bare Costs Labor	2021 Bare Costs Equipment	Total	Total Incl O&P
0010	**BUILDING DEMOLITION** Large urban projects, incl. 20 mi. haul R024119-10									
0011	No foundation or dump fees, C.F. is vol. of building standing									
0020	Steel	B-8	21500	.003	C.F.		.15	.13	.28	.38
0050	Concrete		15300	.004			.21	.19	.40	.53
0080	Masonry		20100	.003			.16	.14	.30	.40
0100	Mixture of types		20100	.003			.16	.14	.30	.40
0500	Small bldgs, or single bldgs, no salvage included, steel	B-3	14800	.003			.16	.16	.32	.41
0600	Concrete		11300	.004			.21	.20	.41	.53
0650	Masonry		14800	.003			.16	.16	.32	.41
0700	Wood		14800	.003			.16	.16	.32	.41
0750	For buildings with no interior walls, deduct								30%	30%
1000	Demolition single family house, one story, wood 1600 S.F.	B-3	1	48	Ea.		2,375	2,300	4,675	6,075
1020	3200 S.F.		.50	96			4,750	4,600	9,350	12,200
1200	Demolition two family house, two story, wood 2400 S.F.		.67	71.964			3,550	3,450	7,000	9,125
1220	4200 S.F.		.38	128			6,325	6,150	12,475	16,200
1300	Demolition three family house, three story, wood 3200 S.F.		.50	96			4,750	4,600	9,350	12,200
1320	5400 S.F.		.30	160			7,925	7,675	15,600	20,300
5000	For buildings with no interior walls, deduct								30%	30%

02 41 16.15 Explosive/Implosive Demolition

		Crew	Daily Output	Labor-Hours	Unit	Material	2021 Bare Costs Labor	2021 Bare Costs Equipment	Total	Total Incl O&P
0010	**EXPLOSIVE/IMPLOSIVE DEMOLITION** R024119-10									
0011	Large projects,									
0020	No disposal fee based on building volume, steel building	B-5B	16900	.003	C.F.		.16	.16	.32	.41
0100	Concrete building		16900	.003			.16	.16	.32	.41
0200	Masonry building		16900	.003			.16	.16	.32	.41
0400	Disposal of material, minimum	B-3	445	.108	C.Y.		5.35	5.20	10.55	13.65
0500	Maximum	"	365	.132	"		6.50	6.30	12.80	16.65

02 41 16.17 Building Demolition Footings and Foundations

		Crew	Daily Output	Labor-Hours	Unit	Material	2021 Bare Costs Labor	2021 Bare Costs Equipment	Total	Total Incl O&P
0010	**BUILDING DEMOLITION FOOTINGS AND FOUNDATIONS** R024119-10									
0200	Floors, concrete slab on grade,									
0240	4" thick, plain concrete	B-13L	5000	.003	S.F.		.20	.42	.62	.75
0280	Reinforced, wire mesh	B-13L	4000	.004	S.F.		.25	.53	.78	.95
0300	Rods		4500	.004			.22	.47	.69	.84
0400	6" thick, plain concrete		4000	.004			.25	.53	.78	.95
0420	Reinforced, wire mesh		3200	.005			.31	.66	.97	1.18
0440	Rods		3600	.004			.27	.58	.85	1.05
1000	Footings, concrete, 1' thick, 2' wide		300	.053	L.F.		3.28	7	10.28	12.60
1080	1'-6" thick, 2' wide		250	.064			3.93	8.40	12.33	15.10
1120	3' wide		200	.080			4.92	10.50	15.42	18.85
1140	2' thick, 3' wide		175	.091			5.60	12	17.60	21.50
1200	Average reinforcing, add								10%	10%
1220	Heavy reinforcing, add								20%	20%
2000	Walls, block, 4" thick	B-13L	8000	.002	S.F.		.12	.26	.38	.47
2040	6" thick		6000	.003			.16	.35	.51	.63

02 41 Demolition

02 41 16 – Structure Demolition

02 41 16.17 Building Demolition Footings and Foundations		Crew	Daily Output	Labor-Hours	Unit	Material	2021 Bare Costs Labor	Equipment	Total	Total Incl O&P
2080	8" thick	B-13L	4000	.004	S.F.		.25	.53	.78	.95
2100	12" thick	↓	3000	.005			.33	.70	1.03	1.26
2200	For horizontal reinforcing, add								10%	10%
2220	For vertical reinforcing, add								20%	20%
2400	Concrete, plain concrete, 6" thick	B-13L	4000	.004			.25	.53	.78	.95
2420	8" thick		3500	.005			.28	.60	.88	1.08
2440	10" thick		3000	.005			.33	.70	1.03	1.26
2500	12" thick	↓	2500	.006			.39	.84	1.23	1.52
2600	For average reinforcing, add								10%	10%
2620	For heavy reinforcing, add								20%	20%
4000	For congested sites or small quantities, add up to								200%	200%
4200	Add for disposal, on site	B-11A	232	.069	C.Y.		3.57	6.55	10.12	12.50
4250	To five miles	B-30	220	.109	"		5.90	8.45	14.35	18

02 41 16.33 Bridge Demolition

		Crew	Daily Output	Labor-Hours	Unit	Material	2021 Bare Costs Labor	Equipment	Total	Total Incl O&P
0010	**BRIDGE DEMOLITION**									
0100	Bridges, pedestrian, precast, 60'-150' long	B-21C	250	.224	S.F.		10.80	8.35	19.15	25.50
0200	Steel, 50'-160' long, 8'-10' wide	"	500	.112			5.40	4.17	9.57	12.65
0300	Laminated wood, 80'-130' long	C-12	300	.160	↓		8.70	1.59	10.29	14.75

02 41 19 – Selective Demolition

02 41 19.13 Selective Building Demolition

0010	**SELECTIVE BUILDING DEMOLITION**									
0020	Costs related to selective demolition of specific building components									
0025	are included under Common Work Results (XX 05)									
0030	in the component's appropriate division.									

02 41 19.16 Selective Demolition, Cutout

		Crew	Daily Output	Labor-Hours	Unit	Material	2021 Bare Costs Labor	Equipment	Total	Total Incl O&P
0010	**SELECTIVE DEMOLITION, CUTOUT** R024119-10									
0020	Concrete, elev. slab, light reinforcement, under 6 C.F.	B-9	65	.615	C.F.		27.50	5.45	32.95	47
0050	Light reinforcing, over 6 C.F.	"	75	.533	"		24	4.74	28.74	40.50
0200	Slab on grade to 6" thick, not reinforced, under 8 S.F.	B-9	85	.471	S.F.		21	4.18	25.18	36
0250	8-16 S.F.	"	175	.229	"		10.25	2.03	12.28	17.55
0255	For over 16 S.F. see Line 02 41 16.17 0400									
0600	Walls, not reinforced, under 6 C.F.	B-9	60	.667	C.F.		30	5.95	35.95	51
0650	6-12 C.F.	"	80	.500	"		22.50	4.45	26.95	38.50
0655	For over 12 C.F. see Line 02 41 16.17 2500									
1000	Concrete, elevated slab, bar reinforced, under 6 C.F.	B-9	45	.889	C.F.		40	7.90	47.90	68
1050	Bar reinforced, over 6 C.F.	"	50	.800	"		36	7.10	43.10	61.50
1200	Slab on grade to 6" thick, bar reinforced, under 8 S.F.		75	.533	S.F.		24	4.74	28.74	40.50
1250	8-16 S.F.	↓	150	.267	"		11.95	2.37	14.32	20.50
1255	For over 16 S.F. see Line 02 41 16.17 0440									
1400	Walls, bar reinforced, under 6 C.F.	B-9	50	.800	C.F.		36	7.10	43.10	61.50
1450	6-12 C.F.	"	70	.571	"		25.50	5.10	30.60	43.50
1455	For over 12 C.F. see Lines 02 41 16.17 2500 and 2600									
2000	Brick, to 4 S.F. opening, not including toothing									
2040	4" thick	B-9	30	1.333	Ea.		59.50	11.85	71.35	102
2060	8" thick		18	2.222			99.50	19.75	119.25	171
2080	12" thick		10	4			179	35.50	214.50	305
2400	Concrete block, to 4 S.F. opening, 2" thick		35	1.143			51	10.15	61.15	87.50
2420	4" thick		30	1.333			59.50	11.85	71.35	102
2440	8" thick		27	1.481			66.50	13.15	79.65	114
2460	12" thick		24	1.667			74.50	14.80	89.30	127
2600	Gypsum block, to 4 S.F. opening, 2" thick		80	.500			22.50	4.45	26.95	38.50
2620	4" thick	↓	70	.571			25.50	5.10	30.60	43.50

For customer support on your Site Work & Landscape Costs with RSMeans Data, call 800.448.8182.

43

02 41 Demolition

02 41 19 – Selective Demolition

02 41 19.16 Selective Demolition, Cutout	Crew	Daily Output	Labor-Hours	Unit	Material	2021 Bare Costs Labor	2021 Bare Costs Equipment	Total	Total Incl O&P
2640	8" thick	B-9	55	.727	Ea.		32.50	6.45	38.95	55.50
2800	Terra cotta, to 4 S.F. opening, 4" thick		70	.571			25.50	5.10	30.60	43.50
2840	8" thick		65	.615			27.50	5.45	32.95	47
2880	12" thick		50	.800			36	7.10	43.10	61.50
3000	Toothing masonry cutouts, brick, soft old mortar	1 Brhe	40	.200	V.L.F.		8.75		8.75	13.15
3100	Hard mortar		30	.267			11.65		11.65	17.55
3200	Block, soft old mortar		70	.114			4.99		4.99	7.55
3400	Hard mortar		50	.160			7		7	10.55
6000	Walls, interior, not including re-framing,									
6010	openings to 5 S.F.									
6100	Drywall to 5/8" thick	1 Clab	24	.333	Ea.		14.80		14.80	22
6200	Paneling to 3/4" thick		20	.400			17.75		17.75	26.50
6300	Plaster, on gypsum lath		20	.400			17.75		17.75	26.50
6340	On wire lath		14	.571			25.50		25.50	38
7000	Wood frame, not including re-framing, openings to 5 S.F.									
7200	Floors, sheathing and flooring to 2" thick	1 Clab	5	1.600	Ea.		71		71	106
7310	Roofs, sheathing to 1" thick, not including roofing		6	1.333			59		59	88.50
7410	Walls, sheathing to 1" thick, not including siding		7	1.143			50.50		50.50	75.50

02 41 19.18 Selective Demolition, Disposal Only

		Crew	Daily Output	Labor-Hours	Unit	Material	Labor	Equipment	Total	Total Incl O&P
0010	**SELECTIVE DEMOLITION, DISPOSAL ONLY** R024119-10									
0015	Urban bldg w/salvage value allowed									
0020	Including loading and 5 mile haul to dump									
0200	Steel frame	B-3	430	.112	C.Y.		5.50	5.35	10.85	14.15
0300	Concrete frame		365	.132			6.50	6.30	12.80	16.65
0400	Masonry construction		445	.108			5.35	5.20	10.55	13.65
0500	Wood frame		247	.194			9.60	9.35	18.95	24.50

02 41 19.19 Selective Demolition

		Crew	Daily Output	Labor-Hours	Unit	Material	Labor	Equipment	Total	Total Incl O&P
0010	**SELECTIVE DEMOLITION**, Rubbish Handling R024119-10									
0020	The following are to be added to the demolition prices									
0050	The following are components for a complete chute system									
0100	Top chute circular steel, 4' long, 18" diameter R024119-30	B-1C	15	1.600	Ea.	272	72	19.50	363.50	430
0102	23" diameter		15	1.600		295	72	19.50	386.50	455
0104	27" diameter		15	1.600		320	72	19.50	411.50	480
0106	30" diameter		15	1.600		340	72	19.50	431.50	505
0108	33" diameter		15	1.600		365	72	19.50	456.50	530
0110	36" diameter		15	1.600		385	72	19.50	476.50	555
0112	Regular chute, 18" diameter		15	1.600		204	72	19.50	295.50	355
0114	23" diameter		15	1.600		227	72	19.50	318.50	380
0116	27" diameter		15	1.600		249	72	19.50	340.50	405
0118	30" diameter		15	1.600		261	72	19.50	352.50	415
0120	33" diameter		15	1.600		295	72	19.50	386.50	455
0122	36" diameter		15	1.600		320	72	19.50	411.50	480
0124	Control door chute, 18" diameter		15	1.600		385	72	19.50	476.50	555
0126	23" diameter		15	1.600		410	72	19.50	501.50	580
0128	27" diameter		15	1.600		430	72	19.50	521.50	605
0130	30" diameter		15	1.600		455	72	19.50	546.50	630
0132	33" diameter		15	1.600		475	72	19.50	566.50	655
0134	36" diameter		15	1.600		500	72	19.50	591.50	680
0136	Chute liners, 14 ga., 18"-30" diameter		15	1.600		209	72	19.50	300.50	360
0138	33"-36" diameter		15	1.600		261	72	19.50	352.50	415
0140	17% thinner chute, 30" diameter		15	1.600		227	72	19.50	318.50	380
0142	33% thinner chute, 30" diameter		15	1.600		170	72	19.50	261.50	315

02 41 Demolition

02 41 19 – Selective Demolition

02 41 19.19 Selective Demolition		Crew	Daily Output	Labor-Hours	Unit	Material	2021 Bare Costs Labor	Equipment	Total	Total Incl O&P
0144	Top chute cover	1 Clab	24	.333	Ea.	143	14.80		157.80	179
0146	Door chute cover	"	24	.333		143	14.80		157.80	179
0148	Top chute trough	2 Clab	12	1.333		475	59		534	615
0150	Bolt down frame & counter weights, 250 lb.	B-1	4	6	Ea.	4,175	270		4,445	5,000
0152	500 lb.		4	6		6,175	270		6,445	7,200
0154	750 lb.		4	6		10,500	270		10,770	12,000
0156	1000 lb.		2.67	8.989		11,500	405		11,905	13,300
0158	1500 lb.		2.67	8.989		14,800	405		15,205	16,800
0160	Chute warning light system, 5 stories	B-1C	4	6		7,975	270	73	8,318	9,250
0162	10 stories	"	2	12		14,800	540	146	15,486	17,300
0164	Dust control device for dumpsters	1 Clab	8	1		166	44.50		210.50	250
0166	Install or replace breakaway cord		8	1		27	44.50		71.50	96
0168	Install or replace warning sign		16	.500		10	22		32	44
0600	Dumpster, weekly rental, 1 dump/week, 6 C.Y. capacity (2 tons)				Week	415			415	455
0700	10 C.Y. capacity (3 tons)					480			480	530
0725	20 C.Y. capacity (5 tons)					565			565	625
0800	30 C.Y. capacity (7 tons)					730			730	800
0840	40 C.Y. capacity (10 tons)					775			775	850
0900	Alternate pricing for dumpsters									
0910	Delivery, average for all sizes				Ea.	75			75	82.50
0920	Haul, average for all sizes					235			235	259
0930	Rent per day, average for all sizes					20			20	22
0940	Rent per month, average for all sizes					80			80	88
0950	Disposal fee per ton, average for all sizes				Ton	88			88	97
2000	Load, haul, dump and return, 0'-50' haul, hand carried	2 Clab	24	.667	C.Y.		29.50		29.50	44
2005	Wheeled		37	.432			19.20		19.20	28.50
2040	0'-100' haul, hand carried		16.50	.970			43		43	64
2045	Wheeled		25	.640			28.50		28.50	42.50
2050	Forklift	A-3R	25	.320			17.75	11.35	29.10	39
2080	Haul and return, add per each extra 100' haul, hand carried	2 Clab	35.50	.451			20		20	30
2085	Wheeled		54	.296			13.15		13.15	19.65
2120	For travel in elevators, up to 10 floors, add		140	.114			5.05		5.05	7.55
2130	0'-50' haul, incl. up to 5 riser stairs, hand carried		23	.696			31		31	46
2135	Wheeled		35	.457			20.50		20.50	30.50
2140	6-10 riser stairs, hand carried		22	.727			32.50		32.50	48
2145	Wheeled		34	.471			21		21	31
2150	11-20 riser stairs, hand carried		20	.800			35.50		35.50	53
2155	Wheeled		31	.516			23		23	34
2160	21-40 riser stairs, hand carried		16	1			44.50		44.50	66.50
2165	Wheeled		24	.667			29.50		29.50	44
2170	0'-100' haul, incl. 5 riser stairs, hand carried		15	1.067			47.50		47.50	70.50
2175	Wheeled		23	.696			31		31	46
2180	6-10 riser stairs, hand carried	2 Clab	14	1.143	C.Y.		50.50		50.50	75.50
2185	Wheeled		21	.762			34		34	50.50
2190	11-20 riser stairs, hand carried		12	1.333			59		59	88.50
2195	Wheeled		18	.889			39.50		39.50	59
2200	21-40 riser stairs, hand carried		8	2			89		89	133
2205	Wheeled		12	1.333			59		59	88.50
2210	Haul and return, add per each extra 100' haul, hand carried		35.50	.451			20		20	30
2215	Wheeled		54	.296			13.15		13.15	19.65
2220	For each additional flight of stairs, up to 5 risers, add		550	.029	Flight		1.29		1.29	1.93
2225	6-10 risers, add		275	.058			2.58		2.58	3.85
2230	11-20 risers, add		138	.116			5.15		5.15	7.70

R024119-20

For customer support on your Site Work & Landscape Costs with RSMeans Data, call 800.448.8182.

45

02 41 19 – Selective Demolition

02 41 19.19 Selective Demolition	Crew	Daily Output	Labor-Hours	Unit	Material	2021 Bare Costs Labor	Equipment	Total	Total Incl O&P
2235	21-40 risers, add	2 Clab	69	.232	Flight		10.30		10.30	15.35
3000	Loading & trucking, including 2 mile haul, chute loaded	B-16	45	.711	C.Y.		33	12.85	45.85	63.50
3040	Hand loading truck, 50' haul	"	48	.667			31	12.05	43.05	60
3080	Machine loading truck	B-17	120	.267			13.05	5.20	18.25	25
5000	Haul, per mile, up to 8 C.Y. truck	B-34B	1165	.007			.35	.50	.85	1.08
5100	Over 8 C.Y. truck	"	1550	.005			.26	.37	.63	.81

02 41 19.20 Selective Demolition, Dump Charges

		Crew	Daily Output	Labor-Hours	Unit	Material	Labor	Equipment	Total	Total Incl O&P
0010	**SELECTIVE DEMOLITION, DUMP CHARGES**	R024119-10								
0020	Dump charges, typical urban city, tipping fees only									
0100	Building construction materials				Ton	74			74	81
0200	Trees, brush, lumber					63			63	69.50
0300	Rubbish only					63			63	69.50
0500	Reclamation station, usual charge					74			74	81

02 41 19.21 Selective Demolition, Gutting

		Crew	Daily Output	Labor-Hours	Unit	Material	Labor	Equipment	Total	Total Incl O&P
0010	**SELECTIVE DEMOLITION, GUTTING**	R024119-10								
0020	Building interior, including disposal, dumpster fees not included									
0500	Residential building									
0560	Minimum	B-16	400	.080	SF Flr.		3.73	1.45	5.18	7.15
0580	Maximum	"	360	.089	"		4.14	1.61	5.75	7.95
0900	Commercial building									
1000	Minimum	B-16	350	.091	SF Flr.		4.26	1.65	5.91	8.15
1020	Maximum	"	250	.128	"		5.95	2.32	8.27	11.45

02 41 19.25 Selective Demolition, Saw Cutting

		Crew	Daily Output	Labor-Hours	Unit	Material	Labor	Equipment	Total	Total Incl O&P
0010	**SELECTIVE DEMOLITION, SAW CUTTING**	R024119-10								
0012	For concrete saw cutting, see Section 03 81									
0015	Asphalt, up to 3" deep	B-89	1050	.015	L.F.	.11	.79	1.02	1.92	2.43
0020	Each additional inch of depth	"	1800	.009		.04	.46	.59	1.09	1.38
1200	Masonry walls, hydraulic saw, brick, per inch of depth	B-89B	300	.053		.04	2.78	5.20	8.02	9.90
1220	Block walls, solid, per inch of depth	"	250	.064		.04	3.34	6.25	9.63	11.85
2000	Brick or masonry w/hand held saw, per inch of depth	A-1	125	.064		.05	2.84	.91	3.80	5.30
5000	Wood sheathing to 1" thick, on walls	1 Carp	200	.040	L.F.		2.19		2.19	3.27
5020	On roof	"	250	.032	"		1.75		1.75	2.61

02 41 19.27 Selective Demolition, Torch Cutting

		Crew	Daily Output	Labor-Hours	Unit	Material	Labor	Equipment	Total	Total Incl O&P
0010	**SELECTIVE DEMOLITION, TORCH CUTTING**	R024119-10								
0020	Steel, 1" thick plate	E-25	333	.024	L.F.	2.23	1.50	.04	3.77	4.81
0040	1" diameter bar	"	600	.013	Ea.	.37	.83	.02	1.22	1.72
1000	Oxygen lance cutting, reinforced concrete walls									
1040	12"-16" thick walls	1 Clab	10	.800	L.F.		35.50		35.50	53
1080	24" thick walls	"	6	1.333	"		59		59	88.50

02 42 Removal and Diversion of Construction Materials

02 42 10 – Building Deconstruction

02 42 10.10 Estimated Salvage Value or Savings

		Crew	Daily Output	Labor-Hours	Unit	Material	2021 Bare Costs Labor	2021 Bare Costs Equipment	Total	Total Incl O&P
0010	**ESTIMATED SALVAGE VALUE OR SAVINGS**									
0015	Excludes material handling, packaging, container costs and									
0020	transportation for salvage or disposal									
0050	All items in Section 02 42 10.10 are credit deducts and not costs									
0100	Copper wire salvage value	G			Lb.	1.70			1.70	1.87
0200	Brass salvage value	G				1.25			1.25	1.38
0250	Bronze					1.45			1.45	1.60
0300	Steel	G				.05			.05	.06
0304	304 Stainless Steel					.30			.30	.33
0316	316 Stainless Steel					.45			.45	.50
0400	Cast Iron	G				.06			.06	.07
0600	Aluminum siding	G				.33			.33	.36
0610	Aluminum wire					.47			.47	.52

02 42 10.20 Deconstruction of Building Components

		Crew	Daily Output	Labor-Hours	Unit	Material	2021 Bare Costs Labor	2021 Bare Costs Equipment	Total	Total Incl O&P
0010	**DECONSTRUCTION OF BUILDING COMPONENTS**									
0012	Buildings one or two stories only									
0015	Excludes material handling, packaging, container costs and									
0020	transportation for salvage or disposal									
0050	Deconstruction of plumbing fixtures									
0100	Wall hung or countertop lavatory	G	2 Clab	16	1	Ea.		44.50		44.50	66.50
0110	Single or double compartment kitchen sink	G		14	1.143			50.50		50.50	75.50
0120	Wall hung urinal	G		14	1.143			50.50		50.50	75.50
0130	Floor mounted	G		8	2			89		89	133
0140	Floor mounted water closet	G		16	1			44.50		44.50	66.50
0150	Wall hung	G		14	1.143			50.50		50.50	75.50
0160	Water fountain, free standing	G		16	1			44.50		44.50	66.50
0170	Wall hung or deck mounted	G		12	1.333			59		59	88.50
0180	Bathtub, steel or fiberglass	G		10	1.600			71		71	106
0190	Cast iron	G		8	2			89		89	133
0200	Shower, single	G	2 Clab	6	2.667	Ea.		118		118	177
0210	Group	G	"	7	2.286	"		101		101	151
0300	Deconstruction of electrical fixtures									
0310	Surface mounted incandescent fixtures	G	2 Clab	48	.333	Ea.		14.80		14.80	22
0320	Fluorescent, 2 lamp	G		32	.500			22		22	33
0330	4 lamp	G		24	.667			29.50		29.50	44
0340	Strip fluorescent, 1 lamp	G		40	.400			17.75		17.75	26.50
0350	2 lamp	G		32	.500			22		22	33
0400	Recessed drop-in fluorescent fixture, 2 lamp	G		27	.593			26.50		26.50	39.50
0410	4 lamp	G		18	.889			39.50		39.50	59
0500	Deconstruction of appliances									
0510	Cooking stoves	G	2 Clab	26	.615	Ea.		27.50		27.50	41
0520	Dishwashers	G	"	26	.615	"		27.50		27.50	41
0600	Deconstruction of millwork and trim									
0610	Cabinets, wood	G	2 Carp	40	.400	L.F.		22		22	32.50
0620	Countertops	G		100	.160	"		8.75		8.75	13.05
0630	Wall paneling, 1" thick	G		500	.032	S.F.		1.75		1.75	2.61
0640	Ceiling trim	G		500	.032	L.F.		1.75		1.75	2.61
0650	Wainscoting	G		500	.032	S.F.		1.75		1.75	2.61
0660	Base, 3/4"-1" thick	G		600	.027	L.F.		1.46		1.46	2.18
0700	Deconstruction of doors and windows									
0710	Doors, wrap, interior, wood, single, no closers	G	2 Carp	21	.762	Ea.	4	41.50		45.50	66.50
0720	Double	G		13	1.231		8	67.50		75.50	109

For customer support on your Site Work & Landscape Costs with RSMeans Data, call 800.448.8182.

47

02 42 Removal and Diversion of Construction Materials

02 42 10 – Building Deconstruction

02 42 10.20 Deconstruction of Building Components

			Crew	Daily Output	Labor-Hours	Unit	Material	2021 Bare Costs Labor	Equipment	Total	Total Incl O&P
0730	Solid core, single, exterior or interior	G	2 Carp	10	1.600	Ea.	4	87.50		91.50	135
0740	Double	G	↓	8	2	↓	8	109		117	172
0810	Windows, wrap, wood, single										
0812	with no casement or cladding	G	2 Carp	21	.762	Ea.	4	41.50		45.50	66.50
0820	with casement and/or cladding	G	"	18	.889	"	4	48.50		52.50	77
0900	Deconstruction of interior finishes										
0910	Drywall for recycling	G	2 Clab	1775	.009	S.F.		.40		.40	.60
0920	Plaster wall, first floor	G		1775	.009			.40		.40	.60
0930	Second floor	G	↓	1330	.012	↓		.53		.53	.80
1000	Deconstruction of roofing and accessories										
1010	Built-up roofs	G	2 Clab	570	.028	S.F.		1.25		1.25	1.86
1020	Gutters, fascia and rakes	G	"	1140	.014	L.F.		.62		.62	.93
2000	Deconstruction of wood components										
2010	Roof sheeting	G	2 Clab	570	.028	S.F.		1.25		1.25	1.86
2020	Main roof framing	G		760	.021	L.F.		.93		.93	1.39
2030	Porch roof framing	G	↓	445	.036			1.60		1.60	2.38
2040	Beams 4" x 8"	G	B-1	375	.064	↓		2.88		2.88	4.30
2050	4" x 10"	G	B-1	300	.080	L.F.		3.61		3.61	5.40
2055	4" x 12"	G		250	.096			4.33		4.33	6.45
2060	6" x 8"	G		250	.096			4.33		4.33	6.45
2065	6" x 10"	G		200	.120			5.40		5.40	8.05
2070	6" x 12"	G		170	.141			6.35		6.35	9.50
2075	8" x 12"	G		126	.190			8.60		8.60	12.80
2080	10" x 12"	G	↓	100	.240			10.80		10.80	16.15
2100	Ceiling joists	G	2 Clab	800	.020			.89		.89	1.33
2150	Wall framing, interior	G		1230	.013	↓		.58		.58	.86
2160	Sub-floor	G		2000	.008	S.F.		.36		.36	.53
2170	Floor joists	G		2000	.008	L.F.		.36		.36	.53
2200	Wood siding (no lead or asbestos)	G		1300	.012	S.F.		.55		.55	.82
2300	Wall framing, exterior	G		1600	.010	L.F.		.44		.44	.66
2400	Stair risers	G		53	.302	Ea.		13.40		13.40	20
2500	Posts	G	↓	800	.020	L.F.		.89		.89	1.33
3000	Deconstruction of exterior brick walls										
3010	Exterior brick walls, first floor	G	2 Clab	200	.080	S.F.		3.55		3.55	5.30
3020	Second floor	G		64	.250	"		11.10		11.10	16.55
3030	Brick chimney	G	↓	100	.160	C.F.		7.10		7.10	10.60
4000	Deconstruction of concrete										
4010	Slab on grade, 4" thick, plain concrete	G	B-9	500	.080	S.F.		3.58	.71	4.29	6.15
4020	Wire mesh reinforced	G		470	.085			3.81	.76	4.57	6.55
4030	Rod reinforced	G		400	.100			4.48	.89	5.37	7.70
4110	Foundation wall, 6" thick, plain concrete	G		160	.250			11.20	2.22	13.42	19.15
4120	8" thick	G		140	.286			12.80	2.54	15.34	22
4130	10" thick	G	↓	120	.333	↓		14.95	2.96	17.91	26
9000	Deconstruction process, support equipment as needed										
9010	Daily use, portal to portal, 12-ton truck-mounted hydraulic crane crew	G	A-3H	1	8	Day		490	735	1,225	1,525
9020	Daily use, skid steer and operator	G	A-3C	1	8			445	445	890	1,150
9030	Daily use, backhoe 48 HP, operator and labor	G	"	1	8	↓		445	445	890	1,150

02 42 10.30 Deconstruction Material Handling

			Crew	Daily Output	Labor-Hours	Unit	Material	2021 Bare Costs Labor	Equipment	Total	Total Incl O&P
0010	**DECONSTRUCTION MATERIAL HANDLING**										
0012	Buildings one or two stories only										
0100	Clean and stack brick on pallet	G	2 Clab	1200	.013	Ea.		.59		.59	.88
0200	Haul 50' and load rough lumber up to 2" x 8" size	G	↓	2000	.008	"		.36		.36	.53

02 42 Removal and Diversion of Construction Materials

02 42 10 – Building Deconstruction

02 42 10.30 Deconstruction Material Handling

02 42 10.30 Deconstruction Material Handling		Crew	Daily Output	Labor-Hours	Unit	Material	2021 Bare Costs Labor	Equipment	Total	Total Incl O&P
0210	Lumber larger than 2" x 8"	G	2 Clab	3200	.005	B.F.		.22		.22	.33
0300	Finish wood for recycling stack and wrap per pallet	G		8	2	Ea.	32	89		121	168
0350	Light fixtures			6	2.667		57.50	118		175.50	241
0375	Windows			6	2.667		54.50	118		172.50	237
0400	Miscellaneous materials			8	2		16	89		105	151
1000	See Section 02 41 19.19 for bulk material handling										

02 43 Structure Moving

02 43 13 – Structure Relocation

02 43 13.13 Building Relocation

			Crew	Daily Output	Labor-Hours	Unit	Material	2021 Bare Costs Labor	Equipment	Total	Total Incl O&P
0010	**BUILDING RELOCATION**										
0011	One day move, up to 24' wide										
0020	Reset on existing foundation					Total				11,500	11,500
0040	Wood or steel frame bldg., based on ground floor area	G	B-4	185	.259	S.F.		11.90	2.70	14.60	20.50
0060	Masonry bldg., based on ground floor area	G	"	137	.350			16.05	3.64	19.69	28
0200	For 24'-42' wide, add									15%	15%

02 56 Site Containment

02 56 13 – Waste Containment

02 56 13.10 Containment of Hazardous Waste

			Crew	Daily Output	Labor-Hours	Unit	Material	2021 Bare Costs Labor	Equipment	Total	Total Incl O&P
0010	**CONTAINMENT OF HAZARDOUS WASTE**										
0020	OSHA Hazard level C										
0030	OSHA Hazard level D decrease labor and equipment, deduct							45%	45%		
0035	OSHA Hazard level B increase labor and equipment, add							22%	22%		
0040	OSHA Hazard level A increase labor and equipment, add							71%	71%		
0100	Excavation of contaminated soil & waste										
0105	Includes one respirator filter and two disposable suits per work day										
0110	3/4 C.Y. excavator to 10' deep		B-12F	51	.314	B.C.Y.	1.56	16.60	13.75	31.91	42
0120	Labor crew to 6' deep		B-2	19	2.105		10.45	94.50		104.95	153
0130	6'-12' deep		"	12	3.333		16.60	149		165.60	241
0200	Move contaminated soil/waste up to 150' on-site with 2.5 C.Y. loader		B-10T	300	.040	L.C.Y.	.27	2.17	2.13	4.57	5.85
0210	300'		"	186	.065	"	.43	3.49	3.43	7.35	9.45
0300	Secure burial cell construction										
0310	Various liner and cover materials										
0400	Very low density polyethylene (VLDPE)										
0410	50 mil top cover		B-47H	4000	.008	S.F.	.71	.46	.21	1.38	1.70
0420	80 mil liner		"	4000	.008	"	1.10	.46	.21	1.77	2.13
0500	Chlorosulfonated polyethylene										
0510	36 mil hypalon top cover		B-47H	4000	.008	S.F.	2.71	.46	.21	3.38	3.90
0520	45 mil hypalon liner		"	4000	.008	"	3.29	.46	.21	3.96	4.54
0600	Polyvinyl chloride (PVC)										
0610	60 mil top cover		B-47H	4000	.008	S.F.	1.14	.46	.21	1.81	2.17
0620	80 mil liner		"	4000	.008	"	1.63	.46	.21	2.30	2.71
0700	Rough textured H.D. polyethylene (HDPE)										
0710	40 mil top cover		B-47H	4000	.008	S.F.	.60	.46	.21	1.27	1.58
0720	60 mil top cover			4000	.008		.81	.46	.21	1.48	1.81
0722	60 mil liner			4000	.008		.77	.46	.21	1.44	1.77
0730	80 mil liner			3800	.008		.99	.49	.22	1.70	2.07
1000	3/4" crushed stone, 6" deep ballast around liner		B-6	30	.800	L.C.Y.	25.50	38.50	7.20	71.20	94

For customer support on your Site Work & Landscape Costs with RSMeans Data, call 800.448.8182.

49

02 56 Site Containment

02 56 13 – Waste Containment

02 56 13.10 Containment of Hazardous Waste	Crew	Daily Output	Labor-Hours	Unit	Material	2021 Bare Costs Labor	Equipment	Total	Total Incl O&P
1100 Hazardous waste, ballast cover with common borrow material	B-63	56	.714	L.C.Y.	13.50	33.50	3.21	50.21	68
1110 Mixture of common borrow & topsoil		56	.714		20.50	33.50	3.21	57.21	75.50
1120 Bank sand		56	.714		18.75	33.50	3.21	55.46	73.50
1130 Medium priced clay		44	.909		26.50	42.50	4.08	73.08	97
1140 Mixture of common borrow & medium priced clay		56	.714		20	33.50	3.21	56.71	75

02 58 Snow Control

02 58 13 – Snow Fencing

02 58 13.10 Snow Fencing System

	Crew	Daily Output	Labor-Hours	Unit	Material	2021 Bare Costs Labor	Equipment	Total	Total Incl O&P
0010 **SNOW FENCING SYSTEM**									
7001 Snow fence on steel posts 10' OC, 4' high	B-1	500	.048	L.F.	.39	2.16		2.55	3.66

02 65 Underground Storage Tank Removal

02 65 10 – Underground Tank and Contaminated Soil Removal

02 65 10.30 Removal of Underground Storage Tanks

	Crew	Daily Output	Labor-Hours	Unit	Material	2021 Bare Costs Labor	Equipment	Total	Total Incl O&P
0010 **REMOVAL OF UNDERGROUND STORAGE TANKS** R026510-20									
0011 Petroleum storage tanks, non-leaking									
0100 Excavate & load onto trailer									
0110 3,000 gal. to 5,000 gal. tank [G]	B-14	4	12	Ea.		560	54	614	895
0120 6,000 gal. to 8,000 gal. tank [G]	B-3A	3	13.333			630	232	862	1,200
0130 9,000 gal. to 12,000 gal. tank [G]	"	2	20			945	350	1,295	1,775
0190 Known leaking tank, add				%				100%	100%
0200 Remove sludge, water and remaining product from bottom									
0201 of tank with vacuum truck									
0300 3,000 gal. to 5,000 gal. tank [G]	A-13	5	1.600	Ea.		89	149	238	296
0310 6,000 gal. to 8,000 gal. tank [G]		4	2			111	187	298	370
0320 9,000 gal. to 12,000 gal. tank [G]		3	2.667			148	249	397	495
0390 Dispose of sludge off-site, average				Gal.				6.25	6.80
0400 Insert inert solid CO_2 "dry ice" into tank									
0401 For cleaning/transporting tanks (1.5 lb./100 gal. cap) [G]	1 Clab	500	.016	Lb.	1	.71		1.71	2.16
0503 Disconnect and remove piping [G]	1 Plum	160	.050	L.F.		3.39		3.39	5.05
0603 Transfer liquids, 10% of volume [G]	"	1600	.005	Gal.		.34		.34	.51
0703 Cut accessway into underground storage tank [G]	1 Clab	5.33	1.501	Ea.		66.50		66.50	99.50
0813 Remove sludge, wash and wipe tank, 500 gal. [G]	1 Plum	8	1			67.50		67.50	101
0823 3,000 gal. [G]		6.67	1.199			81		81	121
0833 5,000 gal. [G]		6.15	1.301			88		88	131
0843 8,000 gal. [G]		5.33	1.501			102		102	152
0853 10,000 gal. [G]		4.57	1.751			119		119	177
0863 12,000 gal. [G]		4.21	1.900			129		129	192
1020 Haul tank to certified salvage dump, 100 miles round trip									
1023 3,000 gal. to 5,000 gal. tank				Ea.				760	830
1026 6,000 gal. to 8,000 gal. tank				Ea.				880	960
1029 9,000 gal. to 12,000 gal. tank				"				1,050	1,150
1100 Disposal of contaminated soil to landfill									
1110 Minimum				C.Y.				145	160
1111 Maximum				"				400	440
1120 Disposal of contaminated soil to									
1121 bituminous concrete batch plant									
1130 Minimum				C.Y.				80	88

02 65 Underground Storage Tank Removal

02 65 10 – Underground Tank and Contaminated Soil Removal

02 65 10.30 Removal of Underground Storage Tanks		Crew	Daily Output	Labor-Hours	Unit	Material	2021 Bare Costs Labor	Equipment	Total	Total Incl O&P
1131	Maximum				C.Y.				115	125
1203	Excavate, pull & load tank, backfill hole, 8,000 gal. +	G	B-12C	.50	32	Ea.		1,700	1,875	3,575	4,600
1213	Haul tank to certified dump, 100 miles rt, 8,000 gal. +	G	B-34K	1	8			410	865	1,275	1,575
1223	Excavate, pull & load tank, backfill hole, 500 gal.	G	B-11C	1	16			825	216	1,041	1,475
1233	Excavate, pull & load tank, backfill hole, 3,000-5,000 gal.	G	B-11M	.50	32			1,650	470	2,120	3,000
1243	Haul tank to certified dump, 100 miles rt, 500 gal.	G	B-34L	1	8			445	198	643	880
1253	Haul tank to certified dump, 100 miles rt, 3,000-5,000 gal.	G	B-34M	1	8			445	850	1,295	1,600
2010	Decontamination of soil on site incl poly tarp on top/bottom										
2011	Soil containment berm and chemical treatment										
2020	Minimum	G	B-11C	100	.160	C.Y.	6.25	8.25	2.16	16.66	21.50
2021	Maximum	G	"	100	.160		8.10	8.25	2.16	18.51	23.50
2050	Disposal of decontaminated soil, minimum									135	150
2055	Maximum									400	440

02 81 Transportation and Disposal of Hazardous Materials

02 81 20 – Hazardous Waste Handling

02 81 20.10 Hazardous Waste Cleanup/Pickup/Disposal

		Crew	Daily Output	Labor-Hours	Unit	Material	Labor	Equipment	Total	Total Incl O&P
0010	**HAZARDOUS WASTE CLEANUP/PICKUP/DISPOSAL**									
0100	For contractor rental equipment, i.e., dozer,									
0110	Front end loader, dump truck, etc., see 01 54 33 Reference Section									
1000	Solid pickup									
1100	55 gal. drums				Ea.				240	265
1120	Bulk material, minimum				Ton				190	210
1130	Maximum				"				595	655
1200	Transportation to disposal site									
1220	Truckload = 80 drums or 25 C.Y. or 18 tons									
1260	Minimum				Mile				3.95	4.45
1270	Maximum				"				7.25	7.98
3000	Liquid pickup, vacuum truck, stainless steel tank									
3100	Minimum charge, 4 hours									
3110	1 compartment, 2200 gallon				Hr.				140	155
3120	2 compartment, 5000 gallon				"				200	225
3400	Transportation in 6900 gallon bulk truck				Mile				7.95	8.75
3410	In teflon lined truck				"				10.20	11.25
5000	Heavy sludge or dry vacuumable material				Hr.				140	155
6000	Dumpsite disposal charge, minimum				Ton				140	155
6020	Maximum				"				415	455

02 91 Chemical Sampling, Testing and Analysis

02 91 10 – Monitoring, Sampling, Testing and Analysis

02 91 10.10 Monitoring, Chem. Sampling, Testing & Analysis

		Crew	Daily Output	Labor-Hours	Unit	Material	Labor	Equipment	Total	Total Incl O&P
0010	**MONITORING, CHEMICAL SAMPLING, TESTING AND ANALYSIS**									
0015	Field sampling of waste									
0100	Field samples, sample collection, sludge	1 Skwk	32	.250	Ea.		14.30		14.30	21.50
0110	Contaminated soils	"	32	.250	"		14.30		14.30	21.50
0200	Vials and bottles									
0210	32 oz. clear wide mouth jar (case of 12)				Ea.	39.50			39.50	43.50
0220	32 oz. Boston round bottle (case of 12)					40			40	44
0230	32 oz. HDPE bottle (case of 12)					30.50			30.50	33.50

02 91 Chemical Sampling, Testing and Analysis

02 91 10 – Monitoring, Sampling, Testing and Analysis

02 91 10.10 Monitoring, Chem. Sampling, Testing & Analysis	Crew	Daily Output	Labor-Hours	Unit	Material	2021 Bare Costs Labor	Equipment	Total	Total Incl O&P
0300	Laboratory analytical services									
0310	Laboratory testing 13 metals				Ea.	222			222	244
0312	13 metals + mercury					215			215	236
0314	8 metals					165			165	181
0316	Mercury only					61			61	67
0318	Single metal (only Cs, Li, Sr, Ta)					47.50			47.50	52.50
0320	Single metal (excludes Hg, Cs, Li, Sr, Ta)					47.50			47.50	52.50
0400	Hydrocarbons standard					190			190	209
0410	Hydrocarbons fingerprint					240			240	265
0500	Radioactivity gross alpha					160			160	176
0510	Gross alpha & beta					160			160	176
0520	Radium 226					107			107	118
0530	Radium 228					127			127	139
0540	Radon					155			155	170
0550	Uranium					140			140	153
0600	Volatile organics without GC/MS					161			161	178
0610	Volatile organics including GC/MS					310			310	340
0630	Synthetic organic compounds					1,075			1,075	1,200
0640	Herbicides					218			218	240
0650	Pesticides					156			156	171
0660	PCB's					128			128	141

Estimating Tips
General

- Carefully check all the plans and specifications. Concrete often appears on drawings other than structural drawings, including mechanical and electrical drawings for equipment pads. The cost of cutting and patching is often difficult to estimate. See Subdivision 03 81 for Concrete Cutting, Subdivision 02 41 19.16 for Cutout Demolition, Subdivision 03 05 05.10 for Concrete Demolition, and Subdivision 02 41 19.19 for Rubbish Handling (handling, loading, and hauling of debris).

- Always obtain concrete prices from suppliers near the job site. A volume discount can often be negotiated, depending upon competition in the area. Remember to add for waste, particularly for slabs and footings on grade.

03 10 00 Concrete Forming and Accessories

- A primary cost for concrete construction is forming. Most jobs today are constructed with prefabricated forms. The selection of the forms best suited for the job and the total square feet of forms required for efficient concrete forming and placing are key elements in estimating concrete construction. Enough forms must be available for erection to make efficient use of the concrete placing equipment and crew.

- Concrete accessories for forming and placing depend upon the systems used. Study the plans and specifications to ensure that all special accessory requirements have been included in the cost estimate, such as anchor bolts, inserts, and hangers.

- Included within costs for forms-in-place are all necessary bracing and shoring.

03 20 00 Concrete Reinforcing

- Ascertain that the reinforcing steel supplier has included all accessories, cutting, bending, and an allowance for lapping, splicing, and waste. A good rule of thumb is 10% for lapping, splicing, and waste. Also, 10% waste should be allowed for welded wire fabric.

- The unit price items in the subdivisions for Reinforcing In Place, Glass Fiber Reinforcing, and Welded Wire Fabric include the labor to install accessories such as beam and slab bolsters, high chairs, and bar ties and tie wire. The material cost for these accessories is not included; they may be obtained from the Accessories Subdivisions.

03 30 00 Cast-In-Place Concrete

- When estimating structural concrete, pay particular attention to requirements for concrete additives, curing methods, and surface treatments. Special consideration for climate, hot or cold, must be included in your estimate. Be sure to include requirements for concrete placing equipment and concrete finishing.

- For accurate concrete estimating, the estimator must consider each of the following major components individually: forms, reinforcing steel, ready-mix concrete, placement of the concrete, and finishing of the top surface. For faster estimating, Subdivision 03 30 53.40 for Concrete-In-Place can be used; here, various items of concrete work are presented that include the costs of all five major components (unless specifically stated otherwise).

03 40 00 Precast Concrete
03 50 00 Cast Decks and Underlayment

- The cost of hauling precast concrete structural members is often an important factor. For this reason, it is important to get a quote from the nearest supplier. It may become economically feasible to set up precasting beds on the site if the hauling costs are prohibitive.

Reference Numbers

Reference numbers are shown at the beginning of some major classifications. These numbers refer to related items in the Reference Section. The reference information may be an estimating procedure, an alternate pricing method, or technical information.

Note: Not all subdivisions listed here necessarily appear. ■

Same Data. Simplified.

Enjoy the convenience and efficiency of accessing your costs anywhere:

- **Skip the multiplier** by setting your location
- **Quickly search,** edit, favorite and share costs
- **Stay on top of price changes** with automatic updates

Discover more at rsmeans.com/online

---

03 01 Maintenance of Concrete

03 01 30 – Maintenance of Cast-In-Place Concrete

03 01 30.62 Concrete Patching

		Crew	Daily Output	Labor-Hours	Unit	Material	2021 Bare Costs Labor	Equipment	Total	Total Incl O&P
0010	**CONCRETE PATCHING**									
0100	Floors, 1/4" thick, small areas, regular grout	1 Cefi	170	.047	S.F.	1.53	2.44		3.97	5.25
0150	Epoxy grout	"	100	.080	"	11.15	4.14		15.29	18.30
2000	Walls, including chipping, cleaning and epoxy grout									
2100	1/4" deep	1 Cefi	65	.123	S.F.	6.55	6.40		12.95	16.55
2150	1/2" deep		50	.160		13.10	8.30		21.40	26.50
2200	3/4" deep	↓	40	.200	↓	19.65	10.35		30	36.50
2400	Walls, including chipping or sand blasting,									
2410	Priming, and two part polymer mix, 1/4" deep	1 Cefi	80	.100	S.F.	1.62	5.20		6.82	9.40
2420	1/2" deep		60	.133		3.24	6.90		10.14	13.65
2430	3/4" deep	↓	40	.200		4.87	10.35		15.22	20.50

03 05 Common Work Results for Concrete

03 05 05 – Selective Demolition for Concrete

03 05 05.10 Selective Demolition, Concrete

		Crew	Daily Output	Labor-Hours	Unit	Material	2021 Bare Costs Labor	Equipment	Total	Total Incl O&P
0010	**SELECTIVE DEMOLITION, CONCRETE**	R024119-10								
0012	Excludes saw cutting, torch cutting, loading or hauling									
0050	Break into small pieces, reinf. less than 1% of cross-sectional area	B-9	24	1.667	C.Y.		74.50	14.80	89.30	127
0060	Reinforcing 1% to 2% of cross-sectional area		16	2.500			112	22	134	192
0070	Reinforcing more than 2% of cross-sectional area	↓	8	5	↓		224	44.50	268.50	385
0150	Remove whole pieces, up to 2 tons per piece	E-18	36	1.111	Ea.		67	42.50	109.50	150
0160	2-5 tons per piece		30	1.333			80.50	51	131.50	180
0170	5-10 tons per piece		24	1.667			101	63.50	164.50	225
0180	10-15 tons per piece	↓	18	2.222			134	85	219	300
0250	Precast unit embedded in masonry, up to 1 C.F.	D-1	16	1			48.50		48.50	73.50
0260	1-2 C.F.		12	1.333			65		65	98
0270	2-5 C.F.		10	1.600			78		78	117
0280	5-10 C.F.	↓	8	2	↓		97.50		97.50	147
0990	For hydrodemolition see Section 02 41 13.15									

03 05 13 – Basic Concrete Materials

03 05 13.20 Concrete Admixtures and Surface Treatments

		Crew	Daily Output	Labor-Hours	Unit	Material	2021 Bare Costs Labor	Equipment	Total	Total Incl O&P
0010	**CONCRETE ADMIXTURES AND SURFACE TREATMENTS**									
0040	Abrasives, aluminum oxide, over 20 tons				Lb.	1.90			1.90	2.09
0050	1 to 20 tons					1.98			1.98	2.17
0070	Under 1 ton					2.05			2.05	2.26
0100	Silicon carbide, black, over 20 tons					2.97			2.97	3.27
0110	1 to 20 tons					3.09			3.09	3.40
0120	Under 1 ton				↓	3.21			3.21	3.53
0200	Air entraining agent, .7 to 1.5 oz. per bag, 55 gallon drum				Gal.	22			22	24.50
0220	5 gallon pail					34.50			34.50	38
0300	Bonding agent, acrylic latex, 250 S.F./gallon, 5 gallon pail					27			27	29.50
0320	Epoxy resin, 80 S.F./gallon, 4 gallon case				↓	62.50			62.50	69
0400	Calcium chloride, 50 lb. bags, T.L. lots				Ton	1,450			1,450	1,575
0420	Less than truckload lots				Bag	24			24	26.50
0500	Carbon black, liquid, 2 to 8 lb. per bag of cement				Lb.	16.05			16.05	17.65
0600	Concrete admixture, integral colors, dry pigment, 5 lb. bag				Ea.	31			31	34
0610	10 lb. bag					47			47	51.50
0620	25 lb. bag				↓	82			82	90.50
0920	Dustproofing compound, 250 S.F./gal., 5 gallon pail				Gal.	8			8	8.80
1010	Epoxy based, 125 S.F./gal., 5 gallon pail				"	65.50			65.50	72

03 05 Common Work Results for Concrete

03 05 13 – Basic Concrete Materials

03 05 13.20 Concrete Admixtures and Surface Treatments

		Crew	Daily Output	Labor-Hours	Unit	Material	2021 Bare Costs Labor	Equipment	Total	Total Incl O&P
1100	Hardeners, metallic, 55 lb. bags, natural (grey)				Lb.	1.04			1.04	1.15
1200	Colors					1.53			1.53	1.68
1300	Non-metallic, 55 lb. bags, natural grey					.60			.60	.66
1320	Colors					.85			.85	.93
1550	Release agent, for tilt slabs, 5 gallon pail				Gal.	18.55			18.55	20.50
1570	For forms, 5 gallon pail					13.90			13.90	15.30
1590	Concrete release agent for forms, 100% biodegradable, zero VOC, 5 gal. pail G					21.50			21.50	24
1595	55 gallon drum G					19.90			19.90	22
1600	Sealer, hardener and dustproofer, epoxy-based, 125 S.F./gal., 5 gallon unit					65.50			65.50	72
1620	3 gallon unit					72			72	79
1630	Sealer, solvent-based, 250 S.F./gal., 55 gallon drum					21			21	23
1640	5 gallon pail					41			41	45
1650	Sealer, water based, 350 S.F., 55 gallon drum					26.50			26.50	29
1660	5 gallon pail					36			36	39.50
1900	Set retarder, 100 S.F./gal., 1 gallon pail					6.35			6.35	6.95
2000	Waterproofing, integral 1 lb. per bag of cement				Lb.	1.04			1.04	1.14
2100	Powdered metallic, 40 lbs. per 100 S.F., standard colors					4.50			4.50	4.95
2120	Premium colors					6.30			6.30	6.95
3000	For colored ready-mix concrete, add to prices in section 03 31 13.35									
3100	Subtle shades, 5 lb. dry pigment per C.Y., add				C.Y.	31			31	34
3400	Medium shades, 10 lb. dry pigment per C.Y., add					47			47	51.50
3700	Deep shades, 25 lb. dry pigment per C.Y., add					82			82	90.50
6000	Concrete ready mix additives, recycled coal fly ash, mixed at plant G				Ton	58.50			58.50	64.50
6010	Recycled blast furnace slag, mixed at plant				"	88			88	96.50

03 05 13.25 Aggregate

0010	**AGGREGATE** R033105-20									
0100	Lightweight vermiculite or perlite, 4 C.F. bag, C.L. lots G				Bag	28			28	30.50
0150	L.C.L. lots G				"	31			31	34
0250	Sand & stone, loaded at pit, crushed bank gravel				Ton	21			21	23
0350	Sand, washed, for concrete				Ton	21.50			21.50	23.50
0400	For plaster or brick					21.50			21.50	23.50
0450	Stone, 3/4" to 1-1/2"					20			20	22
0470	Round, river stone					43.50			43.50	48
0500	3/8" roofing stone & 1/2" pea stone					32			32	35
0550	For trucking 10-mile round trip, add to the above	B-34B	117	.068			3.51	4.95	8.46	10.70
0600	For trucking 30-mile round trip, add to the above	"	72	.111			5.70	8.05	13.75	17.35
0850	Sand & stone, loaded at pit, crushed bank gravel				C.Y.	29.50			29.50	32.50
0950	Sand, washed, for concrete					30			30	33
1000	For plaster or brick					30			30	33
1050	Stone, 3/4" to 1-1/2"					38.50			38.50	42.50
1055	Round, river stone					45			45	49.50
1100	3/8" roofing stone & 1/2" pea stone					31			31	34
1150	For trucking 10-mile round trip, add to the above	B-34B	78	.103			5.25	7.45	12.70	16
1200	For trucking 30-mile round trip, add to the above	"	48	.167			8.55	12.05	20.60	26
1310	Onyx chips, 50 lb. bags				Cwt.	80			80	88
1330	Quartz chips, 50 lb. bags					34.50			34.50	37.50
1410	White marble, 3/8" to 1/2", 50 lb. bags					10.25			10.25	11.30
1430	3/4", bulk				Ton	179			179	197

03 05 13.30 Cement

0010	**CEMENT** R033105-20									
0240	Portland, Type I/II, T.L. lots, 94 lb. bags				Bag	14.70			14.70	16.15
0250	L.T.L./L.C.L. lots				"	19.95			19.95	22

For customer support on your Site Work & Landscape Costs with RSMeans Data, call 800.448.8182.

55

03 05 Common Work Results for Concrete

03 05 13 – Basic Concrete Materials

03 05 13.30 Cement		Crew	Daily Output	Labor-Hours	Unit	Material	2021 Bare Costs Labor	Equipment	Total	Total Incl O&P
0300	Trucked in bulk, per cwt.				Cwt.	9.10			9.10	10
0400	Type III, high early strength, T.L. lots, 94 lb. bags				Bag	11.80			11.80	13
0420	L.T.L. or L.C.L. lots					21.50			21.50	23.50
0500	White, type III, high early strength, T.L. or C.L. lots, bags					23.50			23.50	26
0520	L.T.L. or L.C.L. lots					43			43	47.50
0600	White, type I, T.L. or C.L. lots, bags					29.50			29.50	32.50
0620	L.T.L. or L.C.L. lots					31			31	34

03 05 13.80 Waterproofing and Dampproofing

		Crew	Daily Output	Labor-Hours	Unit	Material	Labor	Equipment	Total	Total Incl O&P
0010	**WATERPROOFING AND DAMPPROOFING**									
0050	Integral waterproofing, add to cost of regular concrete				C.Y.	6.25			6.25	6.85

03 05 13.85 Winter Protection

		Crew	Daily Output	Labor-Hours	Unit	Material	Labor	Equipment	Total	Total Incl O&P
0010	**WINTER PROTECTION**									
0012	For heated ready mix, add				C.Y.	5.85			5.85	6.40
0100	Temporary heat to protect concrete, 24 hours	2 Clab	50	.320	M.S.F.	181	14.20		195.20	220
0200	Temporary shelter for slab on grade, wood frame/polyethylene sheeting									
0201	Build or remove, light framing for short spans	2 Carp	10	1.600	M.S.F.	530	87.50		617.50	710
0210	Large framing for long spans	"	3	5.333	"	620	292		912	1,125
0500	Electrically heated pads, 110 volts, 15 watts/S.F., buy				S.F.	14.50			14.50	15.95
0600	20 watts/S.F., buy					19.25			19.25	21
0710	Electrically heated pads, 15 watts/S.F., 20 uses					.72			.72	.80

03 11 Concrete Forming

03 11 13 – Structural Cast-In-Place Concrete Forming

03 11 13.20 Forms In Place, Beams and Girders

			Crew	Daily Output	Labor-Hours	Unit	Material	Labor	Equipment	Total	Total Incl O&P
0010	**FORMS IN PLACE, BEAMS AND GIRDERS**	R031113-40									
3500	Bottoms only, to 30" wide, job-built plywood, 1 use	R031113-60	C-2	230	.209	SFCA	6.60	11.15		17.75	24
3550	2 use			265	.181		3.70	9.65		13.35	18.45
3600	3 use			280	.171		2.64	9.15		11.79	16.55
3650	4 use			290	.166		2.15	8.85		11	15.50
4000	Sides only, vertical, 36" high, job-built plywood, 1 use			335	.143		6.60	7.65		14.25	18.70
4050	2 use			405	.119		3.64	6.30		9.94	13.45
4100	3 use			430	.112		2.65	5.95		8.60	11.80
4150	4 use			445	.108		2.15	5.75		7.90	10.95
4500	Sloped sides, 36" high, 1 use			305	.157		6.60	8.40		15	19.75
4550	2 use			370	.130		3.69	6.90		10.59	14.35
4600	3 use			405	.119		2.64	6.30		8.94	12.35
4650	4 use			425	.113		2.14	6		8.14	11.35
5000	Upstanding beams, 36" high, 1 use			225	.213		7.65	11.35		19	25.50
5050	2 use			255	.188		4.28	10.05		14.33	19.70
5100	3 use			275	.175		3.09	9.30		12.39	17.30
5150	4 use			280	.171		2.51	9.15		11.66	16.40

03 11 13.25 Forms In Place, Columns

			Crew	Daily Output	Labor-Hours	Unit	Material	Labor	Equipment	Total	Total Incl O&P
0010	**FORMS IN PLACE, COLUMNS**	R031113-40									
0500	Round fiberglass, 4 use per mo., rent, 12" diameter		C-1	160	.200	L.F.	22.50	10.45		32.95	40.50
0550	16" diameter	R031113-60		150	.213		26	11.10		37.10	45
0600	18" diameter			140	.229		28.50	11.90		40.40	49.50
0650	24" diameter			135	.237		35.50	12.35		47.85	57.50
0700	28" diameter			130	.246		40	12.85		52.85	63
0800	30" diameter			125	.256		42	13.35		55.35	66
0850	36" diameter			120	.267		55	13.90		68.90	81.50

03 11 Concrete Forming

03 11 13 – Structural Cast-In-Place Concrete Forming

03 11 13.25 Forms In Place, Columns

		Crew	Daily Output	Labor-Hours	Unit	Material	2021 Bare Costs Labor	Equipment	Total	Total Incl O&P
1500	Round fiber tube, recycled paper, 1 use, 8" diameter **G**	C-1	155	.206	L.F.	2.94	10.75		13.69	19.30
1550	10" diameter **G**		155	.206		3.51	10.75		14.26	19.90
1600	12" diameter **G**		150	.213		4.14	11.10		15.24	21
1650	14" diameter **G**		145	.221		5.85	11.50		17.35	23.50
1700	16" diameter **G**		140	.229		6.30	11.90		18.20	25
1720	18" diameter **G**		140	.229		7.40	11.90		19.30	26
1750	20" diameter **G**		135	.237		7.65	12.35		20	27
1800	24" diameter **G**		130	.246		12.70	12.85		25.55	33
1850	30" diameter **G**		125	.256		17.60	13.35		30.95	39.50
1900	36" diameter **G**		115	.278		21.50	14.50		36	45
1950	42" diameter **G**	C-1	100	.320	L.F.	43	16.70		59.70	72.50
2000	48" diameter **G**	"	85	.376	"	89	19.65		108.65	128
2200	For seamless type, add					15%				
5000	Job-built plywood, 8" x 8" columns, 1 use	C-1	165	.194	SFCA	3.73	10.10		13.83	19.20
5050	2 use		195	.164		2.14	8.55		10.69	15.10
5100	3 use		210	.152		1.49	7.95		9.44	13.50
5150	4 use		215	.149		1.22	7.75		8.97	12.95
5500	12" x 12" columns, 1 use		180	.178		3.76	9.25		13.01	18
5550	2 use		210	.152		2.07	7.95		10.02	14.15
5600	3 use		220	.145		1.50	7.60		9.10	12.95
5650	4 use		225	.142		1.22	7.40		8.62	12.40
6000	16" x 16" columns, 1 use		185	.173		3.85	9		12.85	17.70
6050	2 use		215	.149		2.04	7.75		9.79	13.85
6100	3 use		230	.139		1.55	7.25		8.80	12.50
6150	4 use		235	.136		1.26	7.10		8.36	12
6500	24" x 24" columns, 1 use		190	.168		4.43	8.80		13.23	17.95
6550	2 use		216	.148		2.44	7.70		10.14	14.25
6600	3 use		230	.139		1.77	7.25		9.02	12.75
6650	4 use		238	.134		1.44	7		8.44	12.05
7000	36" x 36" columns, 1 use		200	.160		3.55	8.35		11.90	16.35
7050	2 use		230	.139		1.99	7.25		9.24	13
7100	3 use		245	.131		1.42	6.80		8.22	11.70
7150	4 use		250	.128		1.15	6.65		7.80	11.20
7400	Steel framed plywood, based on 50 uses of purchased									
7420	forms, and 4 uses of bracing lumber									
7500	8" x 8" column	C-1	340	.094	SFCA	2.29	4.91		7.20	9.80
7550	10" x 10"		350	.091		2.03	4.77		6.80	9.35
7600	12" x 12"		370	.086		1.72	4.51		6.23	8.65
7650	16" x 16"		400	.080		1.33	4.17		5.50	7.65
7700	20" x 20"		420	.076		1.18	3.97		5.15	7.25
7750	24" x 24"		440	.073		.84	3.79		4.63	6.55
7755	30" x 30"		440	.073		1.04	3.79		4.83	6.80
7760	36" x 36"		460	.070		.92	3.63		4.55	6.40

03 11 13.30 Forms In Place, Culvert

		Crew	Daily Output	Labor-Hours	Unit	Material	2021 Bare Costs Labor	Equipment	Total	Total Incl O&P
0010	**FORMS IN PLACE, CULVERT** R031113-40									
0015	5' to 8' square or rectangular, 1 use	C-1	170	.188	SFCA	5.65	9.80		15.45	21
0050	2 use R031113-60		180	.178		3.33	9.25		12.58	17.50
0100	3 use		190	.168		2.55	8.80		11.35	15.90
0150	4 use		200	.160		2.16	8.35		10.51	14.85

For customer support on your Site Work & Landscape Costs with RSMeans Data, call 800.448.8182.

57

03 11 13 – Structural Cast-In-Place Concrete Forming

03 11 13.35 Forms In Place, Elevated Slabs		Crew	Daily Output	Labor-Hours	Unit	Material	2021 Bare Costs Labor	Equipment	Total	Total Incl O&P
0010	**FORMS IN PLACE, ELEVATED SLABS** R031113-40									
1000	Flat plate, job-built plywood, to 15' high, 1 use R031113-60	C-2	470	.102	S.F.	5.75	5.45		11.20	14.50
1050	2 use	"	520	.092	"	3.16	4.92		8.08	10.85
1100	3 use	C-2	545	.088	S.F.	2.30	4.70		7	9.55
1150	4 use		560	.086		1.87	4.57		6.44	8.85
1500	15' to 20' high ceilings, 4 use		495	.097		1.90	5.15		7.05	9.80
1600	21' to 35' high ceilings, 4 use		450	.107		2.23	5.70		7.93	10.95
2000	Flat slab, drop panels, job-built plywood, to 15' high, 1 use		449	.107		6.75	5.70		12.45	15.95
2050	2 use		509	.094		3.72	5.05		8.77	11.60
2100	3 use		532	.090		2.70	4.81		7.51	10.15
2150	4 use		544	.088		2.20	4.70		6.90	9.40
2250	15' to 20' high ceilings, 4 use		480	.100		3.15	5.35		8.50	11.40
2350	20' to 35' high ceilings, 4 use		435	.110		3.47	5.90		9.37	12.60
3000	Floor slab hung from steel beams, 1 use		485	.099		3.61	5.30		8.91	11.85
3050	2 use		535	.090		2.66	4.78		7.44	10.05
3100	3 use		550	.087		2.34	4.65		6.99	9.50
3150	4 use		565	.085	↓	2.18	4.53		6.71	9.15
5000	Box out for slab openings, over 16" deep, 1 use		190	.253	SFCA	4.83	13.45		18.28	25.50
5050	2 use		240	.200	"	2.66	10.65		13.31	18.80
5500	Shallow slab box outs, to 10 S.F.		42	1.143	Ea.	19.20	61		80.20	112
5550	Over 10 S.F. (use perimeter)		600	.080	L.F.	2.56	4.27		6.83	9.15
6000	Bulkhead forms for slab, with keyway, 1 use, 2 piece		500	.096		2.35	5.10		7.45	10.25
6100	3 piece (see also edge forms)	↓	460	.104		2.68	5.55		8.23	11.25
6200	Slab bulkhead form, 4-1/2" high, exp metal, w/keyway & stakes [G]	C-1	1200	.027		.94	1.39		2.33	3.10
6210	5-1/2" high [G]		1100	.029		1.19	1.52		2.71	3.57
6215	7-1/2" high [G]		960	.033		1.40	1.74		3.14	4.13
6220	9-1/2" high [G]		840	.038	↓	1.56	1.99		3.55	4.68
6500	Curb forms, wood, 6" to 12" high, on elevated slabs, 1 use		180	.178	SFCA	2.63	9.25		11.88	16.75
6550	2 use		205	.156		1.45	8.15		9.60	13.75
6600	3 use		220	.145		1.05	7.60		8.65	12.45
6650	4 use		225	.142	↓	.86	7.40		8.26	12
7000	Edge forms to 6" high, on elevated slab, 4 use		500	.064	L.F.	.34	3.34		3.68	5.35
7070	7" to 12" high, 1 use		162	.198	SFCA	1.94	10.30		12.24	17.50
7080	2 use		198	.162		1.07	8.45		9.52	13.70
7090	3 use		222	.144		.78	7.50		8.28	12.05
7101	4 use		350	.091	↓	.63	4.77		5.40	7.80
7500	Depressed area forms to 12" high, 4 use		300	.107	L.F.	1.07	5.55		6.62	9.50
7550	12" to 24" high, 4 use		175	.183		1.46	9.55		11.01	15.85
8000	Perimeter deck and rail for elevated slabs, straight		90	.356		18.85	18.55		37.40	48
8050	Curved		65	.492	↓	26	25.50		51.50	67
8500	Void forms, round plastic, 8" high x 3" diameter [G]		450	.071	Ea.	1.62	3.71		5.33	7.35
8550	4" diameter [G]	↓	425	.075	"	2.41	3.93		6.34	8.50
8600	6" diameter [G]	C-1	400	.080	Ea.	4	4.17		8.17	10.60
8650	8" diameter [G]	"	375	.085	"	7.10	4.45		11.55	14.45

03 11 13.40 Forms In Place, Equipment Foundations

	03 11 13.40 Forms In Place, Equipment Foundations	Crew	Daily Output	Labor-Hours	Unit	Material	Labor	Equipment	Total	Total Incl O&P
0010	**FORMS IN PLACE, EQUIPMENT FOUNDATIONS** R031113-40									
0020	1 use	C-2	160	.300	SFCA	3.96	16		19.96	28.50
0050	2 use R031113-60		190	.253		2.18	13.45		15.63	22.50
0100	3 use		200	.240		1.59	12.80		14.39	21
0150	4 use	↓	205	.234	↓	1.30	12.50		13.80	20

58

For customer support on your Site Work & Landscape Costs with RSMeans Data, call 800.448.8182.

03 11 Concrete Forming

03 11 13 – Structural Cast-In-Place Concrete Forming

03 11 13.45 Forms In Place, Footings

		Crew	Daily Output	Labor-Hours	Unit	Material	2021 Bare Costs Labor	Equipment	Total	Total Incl O&P
0010	**FORMS IN PLACE, FOOTINGS** R031113-40									
0020	Continuous wall, plywood, 1 use	C-1	375	.085	SFCA	7.85	4.45		12.30	15.30
0050	2 use R031113-60		440	.073		4.32	3.79		8.11	10.40
0100	3 use		470	.068		3.14	3.55		6.69	8.75
0150	4 use		485	.066		2.56	3.44		6	7.95
0500	Dowel supports for footings or beams, 1 use		500	.064	L.F.	1.56	3.34		4.90	6.70
1000	Integral starter wall, to 4" high, 1 use		400	.080		1.58	4.17		5.75	7.95
1500	Keyway, 4 use, tapered wood, 2" x 4"	1 Carp	530	.015		.37	.83		1.20	1.64
1550	2" x 6"		500	.016		.53	.88		1.41	1.89
2000	Tapered plastic		530	.015		1.58	.83		2.41	2.97
2250	For keyway hung from supports, add		150	.053		1.56	2.92		4.48	6.05
3000	Pile cap, square or rectangular, job-built plywood, 1 use	C-1	290	.110	SFCA	4.48	5.75		10.23	13.55
3050	2 use		346	.092		2.46	4.82		7.28	9.90
3100	3 use		371	.086		1.79	4.50		6.29	8.65
3150	4 use		383	.084		1.46	4.36		5.82	8.10
4000	Triangular or hexagonal, 1 use		225	.142		5.25	7.40		12.65	16.85
4050	2 use		280	.114		2.89	5.95		8.84	12.10
4100	3 use		305	.105		2.10	5.45		7.55	10.45
4150	4 use		315	.102		1.71	5.30		7.01	9.80
5000	Spread footings, job-built lumber, 1 use		305	.105		3.28	5.45		8.73	11.75
5050	2 use		371	.086		1.82	4.50		6.32	8.70
5100	3 use		401	.080		1.31	4.16		5.47	7.65
5150	4 use		414	.077		1.06	4.03		5.09	7.15
6000	Supports for dowels, plinths or templates, 2' x 2' footing		25	1.280	Ea.	9.30	66.50		75.80	110
6050	4' x 4' footing		22	1.455		18.55	76		94.55	134
6100	8' x 8' footing		20	1.600		37	83.50		120.50	165
6150	12' x 12' footing		17	1.882		51	98		149	202
7000	Plinths, job-built plywood, 1 use		250	.128	SFCA	5.30	6.65		11.95	15.75
7100	4 use		270	.119	"	1.73	6.20		7.93	11.10

03 11 13.47 Forms In Place, Gas Station Forms

			Crew	Daily Output	Labor-Hours	Unit	Material	2021 Bare Costs Labor	Equipment	Total	Total Incl O&P
0010	**FORMS IN PLACE, GAS STATION FORMS**										
0050	Curb fascia, with template, 12 ga. steel, left in place, 9" high	G	1 Carp	50	.160	L.F.	13.80	8.75		22.55	28.50
1000	Sign or light bases, 18" diameter, 9" high	G	"	9	.889	Ea.	87	48.50		135.50	169
1050	30" diameter, 13" high	G	1 Carp	8	1	Ea.	138	54.50		192.50	234
2000	Island forms, 10' long, 9" high, 3'-6" wide	G	C-1	10	3.200		385	167		552	675
2050	4' wide	G		9	3.556		400	185		585	715
2500	20' long, 9" high, 4' wide	G		6	5.333		640	278		918	1,125
2550	5' wide	G		5	6.400		670	335		1,005	1,225

03 11 13.50 Forms In Place, Grade Beam

		Crew	Daily Output	Labor-Hours	Unit	Material	2021 Bare Costs Labor	Equipment	Total	Total Incl O&P
0010	**FORMS IN PLACE, GRADE BEAM** R031113-40									
0020	Job-built plywood, 1 use	C-2	530	.091	SFCA	3.97	4.83		8.80	11.55
0050	2 use R031113-60		580	.083		2.19	4.41		6.60	9
0100	3 use		600	.080		1.59	4.27		5.86	8.10
0150	4 use		605	.079		1.29	4.23		5.52	7.70

03 11 13.55 Forms In Place, Mat Foundation

		Crew	Daily Output	Labor-Hours	Unit	Material	2021 Bare Costs Labor	Equipment	Total	Total Incl O&P
0010	**FORMS IN PLACE, MAT FOUNDATION** R031113-40									
0020	Job-built plywood, 1 use	C-2	290	.166	SFCA	4.43	8.85		13.28	18.05
0050	2 use R031113-60		310	.155		1.80	8.25		10.05	14.30
0100	3 use		330	.145		1.20	7.75		8.95	12.90
0120	4 use		350	.137		1.06	7.30		8.36	12.05

03 11 Concrete Forming

03 11 13 – Structural Cast-In-Place Concrete Forming

03 11 13.65 Forms In Place, Slab On Grade

		Crew	Daily Output	Labor-Hours	Unit	Material	2021 Bare Costs Labor	Equipment	Total	Total Incl O&P
0010	**FORMS IN PLACE, SLAB ON GRADE** R031113-40									
1000	Bulkhead forms w/keyway, wood, 6" high, 1 use	C-1	510	.063	L.F.	1.65	3.27		4.92	6.70
1050	2 uses R031113-60		400	.080		.91	4.17		5.08	7.20
1100	4 uses		350	.091		.54	4.77		5.31	7.70
1400	Bulkhead form for slab, 4-1/2" high, exp metal, incl. keyway & stakes G		1200	.027		.94	1.39		2.33	3.10
1410	5-1/2" high G		1100	.029		1.19	1.52		2.71	3.57
1420	7-1/2" high G		960	.033		1.40	1.74		3.14	4.13
1430	9-1/2" high G		840	.038		1.56	1.99		3.55	4.68
2000	Curb forms, wood, 6" to 12" high, on grade, 1 use		215	.149	SFCA	2.99	7.75		10.74	14.90
2050	2 use		250	.128		1.65	6.65		8.30	11.75
2100	3 use		265	.121		1.20	6.30		7.50	10.70
2150	4 use		275	.116		.97	6.05		7.02	10.10
3000	Edge forms, wood, 4 use, on grade, to 6" high		600	.053	L.F.	.45	2.78		3.23	4.65
3050	7" to 12" high		435	.074	SFCA	1.14	3.83		4.97	6.95
3060	Over 12"		350	.091	"	1.43	4.77		6.20	8.65
3500	For depressed slabs, 4 use, to 12" high		300	.107	L.F.	1.18	5.55		6.73	9.60
3550	To 24" high		175	.183		1.53	9.55		11.08	15.95
4000	For slab blockouts, to 12" high, 1 use		200	.160		1.20	8.35		9.55	13.75
4050	To 24" high, 1 use		120	.267		1.51	13.90		15.41	22.50
4100	Plastic (extruded), to 6" high, multiple use, on grade		800	.040		6.50	2.09		8.59	10.25
5020	Wood, incl. wood stakes, 1" x 3"		900	.036		1.23	1.85		3.08	4.13
5050	2" x 4"		900	.036		1.31	1.85		3.16	4.21
6000	Trench forms in floor, wood, 1 use		160	.200	SFCA	3.15	10.45		13.60	19
6050	2 use		175	.183	"	1.73	9.55		11.28	16.15
6100	3 use	C-1	180	.178	SFCA	1.26	9.25		10.51	15.25
6150	4 use		185	.173	"	1.02	9		10.02	14.60
8760	Void form, corrugated fiberboard, 4" x 12", 4' long G		3000	.011	S.F.	3.85	.56		4.41	5.05
8770	6" x 12", 4' long		3000	.011		4.64	.56		5.20	5.95
8780	1/4" thick hardboard protective cover for void form	2 Carp	1500	.011		.94	.58		1.52	1.90

03 11 13.85 Forms In Place, Walls

		Crew	Daily Output	Labor-Hours	Unit	Material	2021 Bare Costs Labor	Equipment	Total	Total Incl O&P
0010	**FORMS IN PLACE, WALLS** R031113-10									
0100	Box out for wall openings, to 16" thick, to 10 S.F.	C-2	24	2	Ea.	42	107		149	205
0150	Over 10 S.F. (use perimeter) R031113-40	"	280	.171	L.F.	3.66	9.15		12.81	17.65
0250	Brick shelf, 4" W, add to wall forms, use wall area above shelf									
0260	1 use R031113-60	C-2	240	.200	SFCA	4.02	10.65		14.67	20.50
0300	2 use		275	.175		2.21	9.30		11.51	16.35
0350	4 use		300	.160		1.61	8.55		10.16	14.50
0500	Bulkhead, wood with keyway, 1 use, 2 piece		265	.181	L.F.	3.19	9.65		12.84	17.90
0600	Bulkhead forms with keyway, 1 piece expanded metal, 8" wall G	C-1	1000	.032		1.40	1.67		3.07	4.03
0610	10" wall G		800	.040		1.56	2.09		3.65	4.83
0620	12" wall G		525	.061		1.87	3.18		5.05	6.80
0700	Buttress, to 8' high, 1 use	C-2	350	.137	SFCA	5.60	7.30		12.90	17.05
0750	2 use		430	.112		3.07	5.95		9.02	12.30
0800	3 use		460	.104		2.24	5.55		7.79	10.75
0850	4 use		480	.100		1.84	5.35		7.19	10
1000	Corbel or haunch, to 12" wide, add to wall forms, 1 use		150	.320	L.F.	3.75	17.05		20.80	29.50
1050	2 use		170	.282		2.06	15.05		17.11	25
1100	3 use		175	.274		1.50	14.60		16.10	23.50
1150	4 use		180	.267		1.22	14.20		15.42	22.50
2000	Wall, job-built plywood, to 8' high, 1 use		370	.130	SFCA	4.22	6.90		11.12	14.95
2050	2 use		435	.110		2.72	5.90		8.62	11.80
2100	3 use		495	.097		1.98	5.15		7.13	9.90

For customer support on your Site Work & Landscape Costs with RSMeans Data, call 800.448.8182.

03 11 13 – Structural Cast-In-Place Concrete Forming

03 11 13.85 Forms In Place, Walls		Crew	Daily Output	Labor-Hours	Unit	Material	2021 Bare Costs Labor	Equipment	Total	Total Incl O&P
2150	4 use	C-2	505	.095	SFCA	1.61	5.05		6.66	9.30
2400	Over 8' to 16' high, 1 use		280	.171		4.66	9.15		13.81	18.75
2450	2 use		345	.139		2.03	7.40		9.43	13.30
2500	3 use		375	.128		1.45	6.80		8.25	11.80
2550	4 use		395	.122		1.19	6.50		7.69	10.95
2700	Over 16' high, 1 use		235	.204		4.14	10.90		15.04	21
2750	2 use		290	.166		2.28	8.85		11.13	15.65
2800	3 use		315	.152		1.66	8.10		9.76	13.95
2850	4 use		330	.145		1.35	7.75		9.10	13.10
4000	Radial, smooth curved, job-built plywood, 1 use		245	.196		3.77	10.45		14.22	19.75
4050	2 use		300	.160		2.07	8.55		10.62	15.05
4100	3 use		325	.148		1.51	7.85		9.36	13.40
4150	4 use	▼	335	.143	▼	1.23	7.65		8.88	12.75
4200	Below grade, job-built plywood, 1 use	C-2	225	.213	SFCA	4.23	11.35		15.58	21.50
4210	2 use		225	.213		2.33	11.35		13.68	19.55
4220	3 use		225	.213		1.96	11.35		13.31	19.15
4230	4 use		225	.213		1.38	11.35		12.73	18.50
4300	Curved, 2' chords, job-built plywood, to 8' high, 1 use		290	.166		3.29	8.85		12.14	16.75
4350	2 use		355	.135		1.81	7.20		9.01	12.75
4400	3 use		385	.125		1.32	6.65		7.97	11.35
4450	4 use		400	.120		1.07	6.40		7.47	10.75
4500	Over 8' to 16' high, 1 use		290	.166		1.38	8.85		10.23	14.65
4525	2 use		355	.135		.76	7.20		7.96	11.60
4550	3 use		385	.125		.55	6.65		7.20	10.50
4575	4 use		400	.120		.46	6.40		6.86	10.05
4600	Retaining wall, battered, job-built plyw'd, to 8' high, 1 use		300	.160		3.08	8.55		11.63	16.15
4650	2 use		355	.135		1.69	7.20		8.89	12.60
4700	3 use		375	.128		1.23	6.80		8.03	11.55
4750	4 use		390	.123		1	6.55		7.55	10.90
4900	Over 8' to 16' high, 1 use		240	.200		3.39	10.65		14.04	19.65
4950	2 use		295	.163		1.87	8.70		10.57	15
5000	3 use		305	.157		1.36	8.40		9.76	14
5050	4 use		320	.150		1.10	8		9.10	13.15
5100	Retaining wall form, plywood, smooth curve, 1 use		200	.240		4.94	12.80		17.74	24.50
5120	2 use		235	.204		2.72	10.90		13.62	19.25
5130	3 use		250	.192		1.98	10.25		12.23	17.45
5140	4 use	▼	260	.185		1.62	9.85		11.47	16.50
5500	For gang wall forming, 192 S.F. sections, deduct					10%	10%			
5550	384 S.F. sections, deduct					20%	20%			
7500	Lintel or sill forms, 1 use	1 Carp	30	.267		5.15	14.60		19.75	27.50
7520	2 use		34	.235		2.84	12.85		15.69	22.50
7540	3 use		36	.222		2.07	12.15		14.22	20.50
7560	4 use	▼	37	.216	▼	1.68	11.85		13.53	19.50
7800	Modular prefabricated plywood, based on 20 uses of purchased									
7820	forms, and 4 uses of bracing lumber									
7860	To 8' high	C-2	800	.060	SFCA	1.31	3.20		4.51	6.20
8060	Over 8' to 16' high		600	.080		1.40	4.27		5.67	7.90
8600	Pilasters, 1 use		270	.178		4.92	9.50		14.42	19.55
8620	2 use		330	.145		2.70	7.75		10.45	14.55
8640	3 use		370	.130		1.97	6.90		8.87	12.45
8660	4 use	▼	385	.125	▼	1.60	6.65		8.25	11.65
9010	Steel framed plywood, based on 50 uses of purchased									
9020	forms, and 4 uses of bracing lumber									

03 11 Concrete Forming

03 11 13 – Structural Cast-In-Place Concrete Forming

03 11 13.85 Forms In Place, Walls

		Crew	Daily Output	Labor-Hours	Unit	Material	2021 Bare Costs Labor	Equipment	Total	Total Incl O&P
9060	To 8' high	C-2	600	.080	SFCA	.74	4.27		5.01	7.15
9260	Over 8' to 16' high		450	.107		.74	5.70		6.44	9.30
9460	Over 16' to 20' high	↓	400	.120	↓	.74	6.40		7.14	10.35
9480	For battered walls, 1 side battered, add					10%	10%			
9485	For battered walls, 2 sides battered, add					15%	15%			

03 11 16 – Architectural Cast-in-Place Concrete Forming

03 11 16.13 Concrete Form Liners

		Crew	Daily Output	Labor-Hours	Unit	Material	2021 Bare Costs Labor	Equipment	Total	Total Incl O&P
0010	**CONCRETE FORM LINERS**									
5750	Liners for forms (add to wall forms), ABS plastic									
5800	Aged wood, 4" wide, 1 use	1 Carp	256	.031	SFCA	3.11	1.71		4.82	5.95
5820	2 use		256	.031		1.71	1.71		3.42	4.43
5830	3 use		256	.031		1.24	1.71		2.95	3.92
5840	4 use		256	.031		1.01	1.71		2.72	3.66
5900	Fractured rope rib, 1 use		192	.042		4.57	2.28		6.85	8.45
5925	2 use		192	.042		2.51	2.28		4.79	6.15
5950	3 use		192	.042		1.83	2.28		4.11	5.40
6000	4 use		192	.042		1.49	2.28		3.77	5.05
6100	Ribbed, 3/4" deep x 1-1/2" OC, 1 use		224	.036		5.05	1.95		7	8.45
6125	2 use		224	.036		2.77	1.95		4.72	5.95
6150	3 use		224	.036		2.01	1.95		3.96	5.15
6200	4 use		224	.036		1.63	1.95		3.58	4.72
6300	Rustic brick pattern, 1 use		224	.036		3.23	1.95		5.18	6.45
6325	2 use		224	.036		1.78	1.95		3.73	4.87
6350	3 use		224	.036		1.29	1.95		3.24	4.34
6400	4 use		224	.036		1.05	1.95		3	4.07
6500	3/8" striated, random, 1 use		224	.036		3.52	1.95		5.47	6.80
6525	2 use		224	.036		1.94	1.95		3.89	5.05
6550	3 use		224	.036		1.41	1.95		3.36	4.47
6600	4 use		224	.036		1.14	1.95		3.09	4.18
6850	Random vertical rustication, 1 use		384	.021		6.10	1.14		7.24	8.45
6900	2 use		384	.021		3.37	1.14		4.51	5.40
6925	3 use		384	.021		2.45	1.14		3.59	4.39
6950	4 use		384	.021		1.99	1.14		3.13	3.89
7050	Wood, beveled edge, 3/4" deep, 1 use		384	.021	L.F.	.14	1.14		1.28	1.85
7100	1" deep, 1 use		384	.021	"	.24	1.14		1.38	1.96
7200	4" wide aged cedar, 1 use		256	.031	SFCA	3.23	1.71		4.94	6.10
7300	4" variable depth rough cedar	↓	224	.036	"	4.52	1.95		6.47	7.90

03 11 19 – Insulating Concrete Forming

03 11 19.10 Insulating Forms, Left In Place

		Crew	Daily Output	Labor-Hours	Unit	Material	2021 Bare Costs Labor	Equipment	Total	Total Incl O&P
0010	**INSULATING FORMS, LEFT IN PLACE**									
0020	Forms include layout, exclude rebar, embedments, bucks for openings,									
0030	scaffolding, wall bracing, concrete, and concrete placing.									
0040	S.F. is for exterior face but includes forms for both faces of wall									
0100	Straight blocks or panels, molded, walls up to 4' high									
0110	4" core wall	4 Carp	1984	.016	S.F.	3.64	.88		4.52	5.30
0120	6" core wall		1808	.018		3.66	.97		4.63	5.50
0130	8" core wall		1536	.021		3.79	1.14		4.93	5.85
0140	10" core wall		1152	.028		4.32	1.52		5.84	7
0150	12" core wall	↓	992	.032	↓	4.91	1.76		6.67	8.05
0200	90 degree corner blocks or panels, molded, walls up to 4' high									
0210	4" core wall	4 Carp	1880	.017	S.F.	3.75	.93		4.68	5.50
0220	6" core wall	↓	1708	.019	↓	3.77	1.02		4.79	5.70

03 11 19 — Insulating Concrete Forming

03 11 19.10 Insulating Forms, Left In Place	Crew	Daily Output	Labor-Hours	Unit	Material	2021 Bare Costs Labor	Equipment	Total	Total Incl O&P
0230	8" core wall	4 Carp	1324	.024	S.F.	3.90	1.32		5.22	6.25
0240	10" core wall		987	.032		4.51	1.77		6.28	7.60
0250	12" core wall		884	.036		4.83	1.98		6.81	8.25
0300	45 degree corner blocks or panels, molded, walls up to 4' high									
0310	4" core wall	4 Carp	1880	.017	S.F.	3.95	.93		4.88	5.75
0320	6" core wall		1712	.019		4.08	1.02		5.10	6
0330	8" core wall		1324	.024		4.19	1.32		5.51	6.60
0400	T blocks or panels, molded, walls up to 4' high									
0420	6" core wall	4 Carp	1540	.021	S.F.	4.81	1.14		5.95	7
0430	8" core wall		1325	.024		4.86	1.32		6.18	7.30
0440	Non-standard corners or Ts requiring trimming & strapping		192	.167		4.91	9.10		14.01	19
0500	Radius blocks or panels, molded, walls up to 4' high, 6" core wall									
0520	5' to 10' diameter, molded blocks or panels	4 Carp	2400	.013	S.F.	5.95	.73		6.68	7.60
0530	10' to 15' diameter, requiring trimming and strapping, add		500	.064		4.47	3.50		7.97	10.15
0540	15'-1" to 30' diameter, requiring trimming and strapping, add		1200	.027		4.47	1.46		5.93	7.10
0550	30'-1" to 60' diameter, requiring trimming and strapping, add		1600	.020		4.47	1.09		5.56	6.55
0560	60'-1" to 100' diameter, requiring trimming and strapping, add		2800	.011		4.47	.63		5.10	5.85
0600	Additional labor for blocks/panels in higher walls (excludes scaffolding)									
0610	4'-1" to 9'-4" high, add						10%			
0620	9'-5" to 12'-0" high, add						20%			
0630	12'-1" to 20'-0" high, add						35%			
0640	Over 20'-0" high, add						55%			
0700	Taper block or panels, molded, single course									
0720	6" core wall	4 Carp	1600	.020	S.F.	3.93	1.09		5.02	5.95
0730	8" core wall	"	1392	.023	"	3.98	1.26		5.24	6.25
0800	ICF brick ledge (corbel) block or panels, molded, single course									
0820	6" core wall	4 Carp	1200	.027	S.F.	4.33	1.46		5.79	6.95
0830	8" core wall	"	1152	.028	"	4.43	1.52		5.95	7.15
0900	ICF curb (shelf) block or panels, molded, single course									
0930	8" core wall	4 Carp	688	.047	S.F.	3.74	2.54		6.28	7.90
0940	10" core wall	"	544	.059	"	4.27	3.22		7.49	9.50
0950	Wood form to hold back concrete to form shelf, 8" high	2 Carp	400	.040	L.F.	1.41	2.19		3.60	4.82
1000	ICF half height block or panels, molded, single course									
1010	4" core wall	4 Carp	1248	.026	S.F.	4.74	1.40		6.14	7.30
1020	6" core wall		1152	.028		4.76	1.52		6.28	7.50
1030	8" core wall		942	.034		4.81	1.86		6.67	8.05
1040	10" core wall		752	.043		5.30	2.33		7.63	9.30
1050	12" core wall		648	.049		5.25	2.70		7.95	9.80
1100	ICF half height block/panels, made by field sawing full height block/panels									
1110	4" core wall	4 Carp	800	.040	S.F.	1.82	2.19		4.01	5.25
1120	6" core wall		752	.043		1.83	2.33		4.16	5.50
1130	8" core wall		600	.053		1.90	2.92		4.82	6.45
1140	10" core wall		496	.065		2.16	3.53		5.69	7.65
1150	12" core wall		400	.080		2.46	4.38		6.84	9.25
1200	Additional insulation inserted into forms between ties									
1210	1 layer (2" thick)	4 Carp	14000	.002	S.F.	1.04	.13		1.17	1.33
1220	2 layers (4" thick)		7000	.005		2.08	.25		2.33	2.66
1230	3 layers (6" thick)		4622	.007		3.12	.38		3.50	4
1300	EPS window/door bucks, molded, permanent									
1310	4" core wall (9" wide)	2 Carp	200	.080	L.F.	3.49	4.38		7.87	10.40
1320	6" core wall (11" wide)		200	.080		3.58	4.38		7.96	10.50
1330	8" core wall (13" wide)		176	.091		3.75	4.97		8.72	11.55
1340	10" core wall (15" wide)		152	.105		4.88	5.75		10.63	13.95

For customer support on your Site Work & Landscape Costs with RSMeans Data, call 800.448.8182.

63

03 11 Concrete Forming

03 11 19 – Insulating Concrete Forming

03 11 19.10 Insulating Forms, Left In Place

		Crew	Daily Output	Labor-Hours	Unit	Material	2021 Bare Costs Labor	2021 Bare Costs Equipment	Total	Total Incl O&P
1350	12" core wall (17" wide)	2 Carp	152	.105	L.F.	5.45	5.75		11.20	14.60
1360	2" x 6" temporary buck bracing (includes installing and removing)	↓	400	.040	↓	1.06	2.19		3.25	4.43
1400	Wood window/door bucks (instead of EPS bucks), permanent									
1410	4" core wall (9" wide)	2 Carp	400	.040	L.F.	2.02	2.19		4.21	5.50
1420	6" core wall (11" wide)		400	.040		2.50	2.19		4.69	6
1430	8" core wall (13" wide)		350	.046		8.15	2.50		10.65	12.75
1440	10" core wall (15" wide)		300	.053		11.30	2.92		14.22	16.75
1450	12" core wall (17" wide)		300	.053		11.30	2.92		14.22	16.75
1460	2" x 6" temporary buck bracing (includes installing and removing)	↓	800	.020	↓	1.06	1.09		2.15	2.79
1500	ICF alignment brace (incl. stiff-back, diagonal kick-back, work platform									
1501	bracket & guard rail post), fastened to one face of wall forms @ 6' O.C.									
1510	1st tier up to 10' tall									
1520	Rental of ICF alignment brace set, per set				Week	10.10			10.10	11.10
1530	Labor (includes installing & removing)	2 Carp	30	.533	Ea.		29		29	43.50
1560	2nd tier from 10' to 20' tall (excludes mason's scaffolding up to 10' high)									
1570	Rental of ICF alignment brace set, per set				Week	10.10			10.10	11.10
1580	Labor (includes installing & removing)	4 Carp	30	1.067	Ea.		58.50		58.50	87
1600	2" x 10" wood plank for work platform, 16' long									
1610	Plank material cost pro-rated over 20 uses				Ea.	1.61			1.61	1.78
1620	Labor (includes installing & removing)	2 Carp	48	.333	"		18.25		18.25	27
1700	2" x 4" lumber for top & middle rails for work platform									
1710	Railing material cost pro-rated over 20 uses				Ea.	.03			.03	.04
1720	Labor (includes installing & removing)	2 Carp	2400	.007	L.F.		.36		.36	.54
1800	ICF accessories									
1810	Wire clip to secure forms in place	2 Carp	2100	.008	Ea.	.39	.42		.81	1.05
1820	Masonry anchor embedment (excludes ties by mason)		1600	.010		4.26	.55		4.81	5.50
1830	Ledger anchor embedment (excludes timber hanger & screws)	↓	128	.125	↓	7.65	6.85		14.50	18.60
1900	See section 01 54 23.70 for mason's scaffolding components									
1910	See section 03 15 19.05 for anchor bolt sleeves									
1920	See section 03 15 19.10 for anchor bolts									
1930	See section 03 15 19.20 for dovetail anchor components									
1940	See section 03 15 19.30 for embedded inserts									
1950	See section 03 21 05.10 for rebar accessories									
1960	See section 03 21 11.60 for reinforcing bars in place									
1970	See section 03 31 13.35 for ready-mix concrete material									
1980	See section 03 31 13.70 for placement and consolidation of concrete									
1990	See section 06 05 23.60 for timber connectors									

03 11 23 – Permanent Stair Forming

03 11 23.75 Forms In Place, Stairs

			Crew	Daily Output	Labor-Hours	Unit	Material	2021 Bare Costs Labor	2021 Bare Costs Equipment	Total	Total Incl O&P
0010	**FORMS IN PLACE, STAIRS**	R031113-40									
0015	(Slant length x width), 1 use		C-2	165	.291	S.F.	9.45	15.50		24.95	33.50
0050	2 use	R031113-60		170	.282		5.30	15.05		20.35	28.50
0100	3 use			180	.267		3.92	14.20		18.12	25.50
0150	4 use			190	.253	↓	3.24	13.45		16.69	23.50
1000	Alternate pricing method (1.0 L.F./S.F.), 1 use			100	.480	LF Rsr	9.45	25.50		34.95	48.50
1050	2 use			105	.457		5.30	24.50		29.80	42.50
1100	3 use			110	.436		3.92	23.50		27.42	39
1150	4 use			115	.417	↓	3.24	22.50		25.74	36.50
2000	Stairs, cast on sloping ground (length x width), 1 use			220	.218	S.F.	3.91	11.65		15.56	21.50
2025	2 use			232	.207		2.15	11.05		13.20	18.80
2050	3 use			244	.197		1.56	10.50		12.06	17.35
2100	4 use		↓	256	.188	↓	1.27	10		11.27	16.30

03 15 Concrete Accessories

03 15 05 – Concrete Forming Accessories

03 15 05.12 Chamfer Strips

		Crew	Daily Output	Labor-Hours	Unit	Material	2021 Bare Costs Labor	Equipment	Total	Total Incl O&P
0010	**CHAMFER STRIPS**									
2000	Polyvinyl chloride, 1/2" wide with leg	1 Carp	535	.015	L.F.	.76	.82		1.58	2.06
2200	3/4" wide with leg		525	.015		.84	.83		1.67	2.16
2400	1" radius with leg		515	.016		1.04	.85		1.89	2.41
2800	2" radius with leg		500	.016		1.78	.88		2.66	3.27
5000	Wood, 1/2" wide		535	.015		.12	.82		.94	1.35
5200	3/4" wide		525	.015		.14	.83		.97	1.39
5400	1" wide		515	.016		.24	.85		1.09	1.53

03 15 05.30 Hangers

					Unit	Material	Labor	Equipment	Total	Total Incl O&P
0010	**HANGERS**									
8500	Wire, black annealed, 15 gauge	G			Cwt.	150			150	165
8600	16 gauge				"	153			153	168

03 15 05.70 Shores

		Crew	Daily Output	Labor-Hours	Unit	Material	Labor	Equipment	Total	Total Incl O&P
0010	**SHORES**									
0020	Erect and strip, by hand, horizontal members									
0500	Aluminum joists and stringers	G	2 Carp	60	.267	Ea.		14.60		14.60	22
0600	Steel, adjustable beams	G		45	.356			19.45		19.45	29
0700	Wood joists			50	.320			17.50		17.50	26
0800	Wood stringers			30	.533			29		29	43.50
1000	Vertical members to 10' high	G		55	.291			15.90		15.90	24
1050	To 13' high	G		50	.320			17.50		17.50	26
1100	To 16' high	G		45	.356			19.45		19.45	29
1500	Reshoring	G		1400	.011	S.F.	.62	.63		1.25	1.61
1600	Flying truss system	G	C-17D	9600	.009	SFCA		.50	.08	.58	.84
1760	Horizontal, aluminum joists, 6-1/4" high x 5' to 21' span, buy	G				L.F.	16.05			16.05	17.65
1770	Beams, 7-1/4" high x 4' to 30' span	G				"	19.25			19.25	21
1810	Horizontal, steel beam, W8x10, 7' span, buy	G				Ea.	63			63	69
1830	10' span	G					73			73	80
1920	15' span	G					125			125	138
1940	20' span	G					176			176	194
1970	Steel stringer, W8x10, 4' to 16' span, buy	G				L.F.	7.30			7.30	8
3000	Rent for job duration, aluminum joist @ 2' OC, per mo.	G				SF Flr.	.40			.40	.44
3050	Steel W8x10	G					.18			.18	.20
3060	Steel adjustable	G					.18			.18	.20
3500	#1 post shore, steel, 5'-7" to 9'-6" high, 10,000# cap., buy	G				Ea.	158			158	174
3550	#2 post shore, 7'-3" to 12'-10" high, 7800# capacity	G					182			182	200
3600	#3 post shore, 8'-10" to 16'-1" high, 3800# capacity	G					199			199	219
5010	Frame shoring systems, steel, 12,000#/leg, buy										
5040	Frame, 2' wide x 6' high	G				Ea.	120			120	132
5250	X-brace	G				Ea.	19.20			19.20	21
5550	Base plate	G					18.05			18.05	19.90
5600	Screw jack	G					36			36	39.50
5650	U-head, 8" x 8"	G					23			23	25

03 15 05.75 Sleeves and Chases

		Crew	Daily Output	Labor-Hours	Unit	Material	Labor	Equipment	Total	Total Incl O&P
0010	**SLEEVES AND CHASES**									
0100	Plastic, 1 use, 12" long, 2" diameter	1 Carp	100	.080	Ea.	1.27	4.38		5.65	7.95
0150	4" diameter		90	.089		2.99	4.86		7.85	10.55
0200	6" diameter		75	.107		11.30	5.85		17.15	21
0250	12" diameter		60	.133		18.05	7.30		25.35	31
5000	Sheet metal, 2" diameter	G	100	.080		1.61	4.38		5.99	8.30
5100	4" diameter	G	90	.089		2.01	4.86		6.87	9.45
5150	6" diameter	G	75	.107		1.74	5.85		7.59	10.60

For customer support on your Site Work & Landscape Costs with RSMeans Data, call 800.448.8182.

65

03 15 05 – Concrete Forming Accessories

03 15 05.75 Sleeves and Chases

			Crew	Daily Output	Labor-Hours	Unit	Material	2021 Bare Costs Labor	2021 Bare Costs Equipment	Total	Total Incl O&P
5200	12" diameter	G	1 Carp	60	.133	Ea.	3.67	7.30		10.97	14.95
6000	Steel pipe, 2" diameter	G		100	.080		4.35	4.38		8.73	11.35
6100	4" diameter	G		90	.089		17.20	4.86		22.06	26
6150	6" diameter	G		75	.107		19.45	5.85		25.30	30
6200	12" diameter	G		60	.133		87	7.30		94.30	107

03 15 05.80 Snap Ties

			Crew	Daily Output	Labor-Hours	Unit	Material	2021 Bare Costs Labor	2021 Bare Costs Equipment	Total	Total Incl O&P
0010	**SNAP TIES**, 8-1/4" L&W (Lumber and wedge)										
0100	2250 lb., w/flat washer, 8" wall	G				C	75			75	82.50
0150	10" wall	G					132			132	145
0200	12" wall	G					189			189	208
0250	16" wall	G					167			167	184
0300	18" wall	G					174			174	191
0500	With plastic cone, 8" wall	G					64.50			64.50	71
0550	10" wall	G					67			67	73.50
0600	12" wall	G					72.50			72.50	79.50
0650	16" wall	G					79.50			79.50	87.50
0700	18" wall	G					81.50			81.50	90
1000	3350 lb., w/flat washer, 8" wall	G					163			163	179
1100	10" wall	G					178			178	196
1150	12" wall	G					189			189	208
1200	16" wall	G					209			209	230
1250	18" wall	G					218			218	240
1500	With plastic cone, 8" wall	G					129			129	142
1550	10" wall	G					141			141	155
1600	12" wall	G					152			152	167
1650	16" wall	G					165			165	182
1700	18" wall	G					167			167	184

03 15 05.85 Stair Tread Inserts

			Crew	Daily Output	Labor-Hours	Unit	Material	2021 Bare Costs Labor	2021 Bare Costs Equipment	Total	Total Incl O&P
0010	**STAIR TREAD INSERTS**										
0105	Cast nosing insert, abrasive surface, pre-drilled, includes screws										
0110	Aluminum, 3" wide x 3' long		1 Cefi	32	.250	Ea.	57.50	12.95		70.45	82.50
0120	4' long		1 Cefi	31	.258	Ea.	74.50	13.35		87.85	102
0130	5' long		"	30	.267	"	89.50	13.80		103.30	119
0135	Extruded nosing insert, black abrasive strips, continuous anchor										
0140	Aluminum, 3" wide x 3' long		1 Cefi	64	.125	Ea.	35.50	6.50		42	48.50
0150	4' long			60	.133		48	6.90		54.90	63
0160	5' long			56	.143		62.50	7.40		69.90	79.50
0165	Extruded nosing insert, black abrasive strips, pre-drilled, incl. screws										
0170	Aluminum, 3" wide x 3' long		1 Cefi	32	.250	Ea.	45	12.95		57.95	68.50
0180	4' long			31	.258		59	13.35		72.35	84.50
0190	5' long			30	.267		83.50	13.80		97.30	112

03 15 05.95 Wall and Foundation Form Accessories

			Crew	Daily Output	Labor-Hours	Unit	Material	2021 Bare Costs Labor	2021 Bare Costs Equipment	Total	Total Incl O&P
0010	**WALL AND FOUNDATION FORM ACCESSORIES**										
0020	Coil tie system										
0050	Coil ties 1/2", 6000 lb., to 8"	G				C	281			281	310
0150	24"	G					475			475	520
0300	3/4", 12,000 lb., to 8"	G					545			545	600
0400	24"	G					805			805	885
0900	Coil bolts, 1/2" diameter x 3" long	G					144			144	158
0940	12" long	G					410			410	455
1000	3/4" diameter x 3" long	G					315			315	345
1040	12" long	G					865			865	950

03 15 Concrete Accessories

03 15 05 – Concrete Forming Accessories

03 15 05.95 Wall and Foundation Form Accessories		Crew	Daily Output	Labor-Hours	Unit	Material	2021 Bare Costs Labor	Equipment	Total	Total Incl O&P
1300	Adjustable coil tie, 3/4" diameter, 20" long	G			C	1,200			1,200	1,325
1350	3/4" diameter, 36" long	G				1,425			1,425	1,550
1400	Tie cones, plastic, 1" setback length, 1/2" bolt diameter					87			87	95.50
1420	3/4" bolt diameter					171			171	188
1500	Welding coil tie, 1/2" diameter x 4" long	G				270			270	297
1550	3/4" diameter x 6" long	G				350			350	385
1700	Waler holders, 1/2" diameter	G				266			266	292
1750	3/4" diameter	G				296			296	325
1900	Flat washers, 4" x 5" x 1/4" for 3/4" diameter	G			▼	1,275			1,275	1,400
2000	Footings, turnbuckle form aligner	G			Ea.	13.50			13.50	14.85
2050	Spreaders for footer, adjustable	G			"	26.50			26.50	29.50
2100	Lagstud, threaded, 1/2" diameter	G			C.L.F.	149			149	164
2200	1" diameter	G			"	545			545	600
2300	Lagnuts, 1/2" diameter	G			C	39			39	43
2400	1" diameter	G				270			270	297
2600	Lagnuts with handle, 1/2" diameter	G				365			365	400
2700	1" diameter	G				1,025			1,025	1,125
2750	Plastic set back plugs for 3/4" diameter					99.50			99.50	110
2800	Rock anchors, 1/2" diameter	G			▼	1,325			1,325	1,475
2900	1" diameter	G			C	1,925			1,925	2,100
2950	Batter washer, 1/2" diameter	G			"	665			665	730
3000	Form oil, up to 1200 S.F./gallon coverage				Gal.	17.15			17.15	18.90
3050	Up to 800 S.F./gallon				"	21.50			21.50	23.50
3500	Form patches, 1-3/4" diameter				C	23.50			23.50	26
3550	2-3/4" diameter				"	40			40	44
4000	Nail stakes, 3/4" diameter, 18" long	G			Ea.	2.66			2.66	2.93
4050	24" long	G				2.64			2.64	2.90
4200	30" long	G				3.33			3.33	3.66
4250	36" long	G			▼	4.20			4.20	4.62
5000	Pencil rods, 1/4" diameter	G			C.L.F.	65.50			65.50	72
5200	Clamps for 1/4" pencil rods	G			Ea.	6.25			6.25	6.85
5300	Clamping jacks for 1/4" pencil rod clamp	G			"	101			101	111
6000	She-bolts, 7/8" x 20"	G			C	6,100			6,100	6,700
6150	1-1/4" x 20"	G				6,950			6,950	7,650
6300	Inside rods, threaded, 1/2" diameter x 6"	G				175			175	193
6350	1/2" diameter x 12"	G				310			310	340
6500	3/4" diameter x 6"	G				460			460	505
6550	3/4" diameter x 12"	G			▼	655			655	720
6700	For wing nuts, see taper ties									
7000	Taper tie system									
7100	Taper ties, 3/4" to 1/2" diameter x 30"	G			Ea.	59.50			59.50	65
7200	1-1/4" to 1" diameter x 30"	G			"	96			96	106
7300	Wing nuts, 1/2" diameter	G			C	695			695	765
7550	1-1/4" diameter	G				1,050			1,050	1,150
7700	Flat washers, 1/4" x 3" x 4", for 1/2" diam. bolt	G				310			310	340
7800	7/16" x 5" x 5", for 1" diam. bolt	G			▼	1,600			1,600	1,750
9000	Wood form accessories									
9100	Tie plate	G			C	640			640	700
9200	Corner washer	G				1,400			1,400	1,550
9300	Panel bolt	G				965			965	1,050
9400	Panel wedge	G				161			161	177
9500	Stud clamp	G			▼	690			690	760

For customer support on your Site Work & Landscape Costs with RSMeans Data, call 800.448.8182.

67

03 15 Concrete Accessories

03 15 13 – Waterstops

03 15 13.50 Waterstops	Crew	Daily Output	Labor-Hours	Unit	Material	2021 Bare Costs Labor	2021 Bare Costs Equipment	Total	Total Incl O&P
0010 **WATERSTOPS**, PVC and Rubber									
0020 PVC, ribbed 3/16" thick, 4" wide	1 Carp	155	.052	L.F.	1.66	2.82		4.48	6.05
0050 6" wide		145	.055		2.62	3.02		5.64	7.40
0500 With center bulb, 6" wide, 3/16" thick		135	.059		2.73	3.24		5.97	7.85
0550 3/8" thick	↓	130	.062	↓	3.16	3.37		6.53	8.50
0600 9" wide x 3/8" thick	1 Carp	125	.064	L.F.	4.17	3.50		7.67	9.85
0800 Dumbbell type, 6" wide, 3/16" thick		150	.053		4.87	2.92		7.79	9.70
0850 3/8" thick		145	.055		4.30	3.02		7.32	9.25
1000 9" wide, 3/8" thick, plain		130	.062		6.60	3.37		9.97	12.25
1050 Center bulb		130	.062		9.95	3.37		13.32	15.95
1250 Ribbed type, split, 3/16" thick, 6" wide		145	.055		2.73	3.02		5.75	7.50
1300 3/8" thick		130	.062		5.35	3.37		8.72	10.90
2000 Rubber, flat dumbbell, 3/8" thick, 6" wide		145	.055		5.60	3.02		8.62	10.65
2050 9" wide		135	.059		11.35	3.24		14.59	17.35
2500 Flat dumbbell split, 3/8" thick, 6" wide		145	.055		2.73	3.02		5.75	7.50
2550 9" wide		135	.059		5.35	3.24		8.59	10.75
3000 Center bulb, 1/4" thick, 6" wide		145	.055		6.95	3.02		9.97	12.15
3050 9" wide		135	.059		14.50	3.24		17.74	21
3500 Center bulb split, 3/8" thick, 6" wide		145	.055		5.40	3.02		8.42	10.45
3550 9" wide	↓	135	.059	↓	9	3.24		12.24	14.75
5000 Waterstop fittings, rubber, flat									
5010 Dumbbell or center bulb, 3/8" thick,									
5200 Field union, 6" wide	1 Carp	50	.160	Ea.	29	8.75		37.75	44.50
5250 9" wide		50	.160		32.50	8.75		41.25	49
5500 Flat cross, 6" wide		30	.267		43.50	14.60		58.10	69.50
5550 9" wide		30	.267		64.50	14.60		79.10	93
6000 Flat tee, 6" wide		30	.267		47.50	14.60		62.10	74
6050 9" wide		30	.267		62	14.60		76.60	90.50
6500 Flat ell, 6" wide		40	.200		46.50	10.95		57.45	68
6550 9" wide		40	.200		57.50	10.95		68.45	79.50
7000 Vertical tee, 6" wide		25	.320		18.25	17.50		35.75	46
7050 9" wide		25	.320		30.50	17.50		48	60
7500 Vertical ell, 6" wide		35	.229		22	12.50		34.50	43
7550 9" wide	↓	35	.229	↓	42.50	12.50		55	65

03 15 16 – Concrete Construction Joints

03 15 16.20 Control Joints, Saw Cut

	Crew	Daily Output	Labor-Hours	Unit	Material	Labor	Equipment	Total	Total Incl O&P
0010 **CONTROL JOINTS, SAW CUT**									
0100 Sawcut control joints in green concrete									
0120 1" depth	C-27	2000	.008	L.F.	.03	.41	.06	.50	.70
0140 1-1/2" depth		1800	.009		.04	.46	.06	.56	.79
0160 2" depth	↓	1600	.010	↓	.06	.52	.07	.65	.90
0180 Sawcut joint reservoir in cured concrete									
0182 3/8" wide x 3/4" deep, with single saw blade	C-27	1000	.016	L.F.	.04	.83	.11	.98	1.38
0184 1/2" wide x 1" deep, with double saw blades		900	.018		.08	.92	.13	1.13	1.58
0186 3/4" wide x 1-1/2" deep, with double saw blades	↓	800	.020		.17	1.04	.14	1.35	1.87
0190 Water blast joint to wash away laitance, 2 passes	C-29	2500	.003	↓		.14	.04	.18	.25
0200 Air blast joint to blow out debris and air dry, 2 passes	C-28	2000	.004	L.F.		.21	.02	.23	.32
0300 For backer rod, see Section 07 91 23.10									
0340 For joint sealant, see Section 03 15 16.30 or 07 92 13.20									
0900 For replacement of joint sealant, see Section 07 01 90.81									

68

For customer support on your Site Work & Landscape Costs with RSMeans Data, call 800.448.8182.

03 15 Concrete Accessories

03 15 16 – Concrete Construction Joints

03 15 16.30 Expansion Joints

		Crew	Daily Output	Labor-Hours	Unit	Material	2021 Bare Costs Labor	Equipment	Total	Total Incl O&P
0010	**EXPANSION JOINTS**									
0020	Keyed, cold, 24 ga., incl. stakes, 3-1/2" high ☐G	1 Carp	200	.040	L.F.	.89	2.19		3.08	4.25
0050	4-1/2" high ☐G		200	.040		.94	2.19		3.13	4.30
0100	5-1/2" high ☐G		195	.041		1.19	2.24		3.43	4.66
0150	7-1/2" high ☐G		190	.042		1.40	2.30		3.70	4.98
0160	9-1/2" high ☐G		185	.043		1.56	2.37		3.93	5.25
0300	Poured asphalt, plain, 1/2" x 1"	1 Clab	450	.018		.94	.79		1.73	2.21
0350	1" x 2"		400	.020		3.77	.89		4.66	5.50
0500	Neoprene, liquid, cold applied, 1/2" x 1"		450	.018		2.72	.79		3.51	4.17
0550	1" x 2"		400	.020		10.75	.89		11.64	13.15
0700	Polyurethane, poured, 2 part, 1/2" x 1"		400	.020		1.42	.89		2.31	2.89
0750	1" x 2"		350	.023		5.65	1.01		6.66	7.75
0900	Rubberized asphalt, hot or cold applied, 1/2" x 1"		450	.018		.34	.79		1.13	1.55
0950	1" x 2"		400	.020		1.29	.89		2.18	2.75
1100	Hot applied, fuel resistant, 1/2" x 1"		450	.018		.51	.79		1.30	1.74
1150	1" x 2"		400	.020		1.94	.89		2.83	3.46
2000	Premolded, bituminous fiber, 1/2" x 6"	1 Carp	375	.021		.41	1.17		1.58	2.19
2050	1" x 12"		300	.027		1.63	1.46		3.09	3.97
2140	Concrete expansion joint, recycled paper and fiber, 1/2" x 6" ☐G		390	.021		.43	1.12		1.55	2.14
2150	1/2" x 12" ☐G		360	.022		.86	1.22		2.08	2.76
2250	Cork with resin binder, 1/2" x 6"		375	.021		1.49	1.17		2.66	3.38
2300	1" x 12"		300	.027		3.65	1.46		5.11	6.20
2500	Neoprene sponge, closed cell, 1/2" x 6"		375	.021		2.41	1.17		3.58	4.39
2550	1" x 12"		300	.027		7.85	1.46		9.31	10.80
2750	Polyethylene foam, 1/2" x 6"		375	.021		.65	1.17		1.82	2.46
2800	1" x 12"		300	.027		3.44	1.46		4.90	5.95
3000	Polyethylene backer rod, 3/8" diameter		460	.017		.16	.95		1.11	1.60
3050	3/4" diameter		460	.017		.08	.95		1.03	1.50
3100	1" diameter		460	.017		.15	.95		1.10	1.59
3500	Polyurethane foam, with polybutylene, 1/2" x 1/2"		475	.017		1.25	.92		2.17	2.76
3550	1" x 1"		450	.018		3.18	.97		4.15	4.95
3750	Polyurethane foam, regular, closed cell, 1/2" x 6"		375	.021		.93	1.17		2.10	2.76
3800	1" x 12"		300	.027		3.31	1.46		4.77	5.80
4000	Polyvinyl chloride foam, closed cell, 1/2" x 6"		375	.021		2.52	1.17		3.69	4.51
4050	1" x 12"		300	.027		8.70	1.46		10.16	11.75
4100	Rod, 3/8" polyethylene/rubberized asphalt sealer, 1/4" x 3/4"	2 Clab	600	.027	L.F.	.33	1.18		1.51	2.13
4250	Rubber, gray sponge, 1/2" x 6"	1 Carp	375	.021		2.12	1.17		3.29	4.07
4300	1" x 12"		300	.027		7.60	1.46		9.06	10.60
4400	Redwood heartwood, 1" x 4"		400	.020		2.97	1.09		4.06	4.90
4450	1" x 6"		375	.021		2.70	1.17		3.87	4.71
5000	For installation in walls, add						75%			
5250	For installation in boxouts, add						25%			

03 15 19 – Cast-In Concrete Anchors

03 15 19.05 Anchor Bolt Accessories

		Crew	Daily Output	Labor-Hours	Unit	Material	2021 Bare Costs Labor	Equipment	Total	Total Incl O&P
0010	**ANCHOR BOLT ACCESSORIES**									
0015	For anchor bolts set in fresh concrete, see Section 03 15 19.10									
8150	Anchor bolt sleeve, plastic, 1" diameter bolts	1 Carp	60	.133	Ea.	10.45	7.30		17.75	22.50
8500	1-1/2" diameter		28	.286		13.10	15.65		28.75	38
8600	2" diameter		24	.333		15.05	18.25		33.30	43.50
8650	3" diameter		20	.400		27.50	22		49.50	63
8800	Templates, steel, 8" bolt spacing ☐G	2 Carp	16	1		10.75	54.50		65.25	93.50
8850	12" bolt spacing ☐G		15	1.067		11.20	58.50		69.70	99.50

03 15 19.05 Anchor Bolt Accessories

		Crew	Daily Output	Labor-Hours	Unit	Material	2021 Bare Costs Labor	2021 Bare Costs Equipment	Total	Total Incl O&P
8900	16" bolt spacing	G 2 Carp	14	1.143	Ea.	13.45	62.50		75.95	108
8950	24" bolt spacing	G	12	1.333		17.90	73		90.90	129
9100	Wood, 8" bolt spacing		16	1		1.40	54.50		55.90	83
9150	12" bolt spacing		15	1.067		1.85	58.50		60.35	89
9200	16" bolt spacing		14	1.143		2.32	62.50		64.82	96
9250	24" bolt spacing		16	1		3.24	54.50		57.74	85

03 15 19.10 Anchor Bolts

		Crew	Daily Output	Labor-Hours	Unit	Material	2021 Bare Costs Labor	2021 Bare Costs Equipment	Total	Total Incl O&P
0010	**ANCHOR BOLTS**									
0015	Made from recycled materials									
0025	Single bolts installed in fresh concrete, no templates									
0030	Hooked w/nut and washer, 1/2" diameter, 8" long	G 1 Carp	132	.061	Ea.	1.55	3.32		4.87	6.65
0040	12" long	G	131	.061		1.72	3.34		5.06	6.90
0050	5/8" diameter, 8" long	G	129	.062		3.40	3.39		6.79	8.80
0060	12" long	G	127	.063		4.18	3.45		7.63	9.75
0070	3/4" diameter, 8" long	G	127	.063		4.18	3.45		7.63	9.75
0080	12" long	G	125	.064		5.25	3.50		8.75	11
0090	2-bolt pattern, including job-built 2-hole template, per set									
0100	J-type, incl. hex nut & washer, 1/2" diameter x 6" long	G 1 Carp	21	.381	Set	7.20	21		28.20	39
0110	12" long	G	21	.381		7.90	21		28.90	39.50
0120	18" long	G	21	.381		8.90	21		29.90	41
0130	3/4" diameter x 8" long	G	20	.400		12.80	22		34.80	46.50
0140	12" long	G	20	.400		14.90	22		36.90	49
0150	18" long	G	20	.400		18.05	22		40.05	52.50
0160	1" diameter x 12" long	G	19	.421		28	23		51	65.50
0170	18" long	G 1 Carp	19	.421	Set	33	23		56	70.50
0180	24" long	G	19	.421		39	23		62	77.50
0190	36" long	G	18	.444		52	24.50		76.50	93.50
0200	1-1/2" diameter x 18" long	G	17	.471		49	25.50		74.50	92.50
0210	24" long	G	16	.500		57.50	27.50		85	105
0300	L-type, incl. hex nut & washer, 3/4" diameter x 12" long	G	20	.400		21.50	22		43.50	56
0310	18" long	G	20	.400		26.50	22		48.50	61.50
0320	24" long	G	20	.400		31	22		53	67
0330	30" long	G	20	.400		38.50	22		60.50	74.50
0340	36" long	G	20	.400		43	22		65	80
0350	1" diameter x 12" long	G	19	.421		34	23		57	72
0360	18" long	G	19	.421		41.50	23		64.50	80
0370	24" long	G	19	.421		50	23		73	89.50
0380	30" long	G	19	.421		58.50	23		81.50	98.50
0390	36" long	G	18	.444		66	24.50		90.50	109
0400	42" long	G	18	.444		79	24.50		103.50	124
0410	48" long	G	18	.444		88.50	24.50		113	134
0420	1-1/4" diameter x 18" long	G	18	.444		43	24.50		67.50	84
0430	24" long	G	18	.444		50	24.50		74.50	92
0440	30" long	G	17	.471		57.50	25.50		83	102
0450	36" long	G	17	.471		64.50	25.50		90	110
0460	42" long	G 2 Carp	32	.500		72.50	27.50		100	121
0470	48" long	G	32	.500		82	27.50		109.50	132
0480	54" long	G	31	.516		96	28		124	148
0490	60" long	G	31	.516		105	28		133	158
0500	1-1/2" diameter x 18" long	G	33	.485		48	26.50		74.50	92.50
0510	24" long	G	32	.500		55.50	27.50		83	102
0520	30" long	G	31	.516		62	28		90	110

03 15 19.10 Anchor Bolts		Crew	Daily Output	Labor-Hours	Unit	Material	2021 Bare Costs Labor	Equipment	Total	Total Incl O&P
0530	36" long	G	2 Carp	30	.533	Set	70.50	29		99.50	121
0540	42" long	G		30	.533		79.50	29		108.50	131
0550	48" long	G		29	.552		89	30		119	143
0560	54" long	G		28	.571		107	31.50		138.50	165
0570	60" long	G		28	.571		117	31.50		148.50	176
0580	1-3/4" diameter x 18" long	G		31	.516		80.50	28		108.50	131
0590	24" long	G		30	.533		93.50	29		122.50	147
0600	30" long	G		29	.552		108	30		138	164
0610	36" long	G		28	.571		123	31.50		154.50	182
0620	42" long	G		27	.593		137	32.50		169.50	200
0630	48" long	G		26	.615		150	33.50		183.50	217
0640	54" long	G		26	.615		186	33.50		219.50	255
0650	60" long	G		25	.640		200	35		235	273
0660	2" diameter x 24" long	G	▼	27	.593	▼	147	32.50		179.50	211
0670	30" long	G	2 Carp	27	.593	Set	166	32.50		198.50	231
0680	36" long	G		26	.615		181	33.50		214.50	251
0690	42" long	G		25	.640		202	35		237	275
0700	48" long	G		24	.667		231	36.50		267.50	310
0710	54" long	G		23	.696		274	38		312	355
0720	60" long	G		23	.696		295	38		333	380
0730	66" long	G		22	.727		315	40		355	405
0740	72" long	G	▼	21	.762	▼	345	41.50		386.50	440
1000	4-bolt pattern, including job-built 4-hole template, per set										
1100	J-type, incl. hex nut & washer, 1/2" diameter x 6" long	G	1 Carp	19	.421	Set	9.95	23		32.95	45.50
1110	12" long	G		19	.421		11.30	23		34.30	47
1120	18" long	G		18	.444		13.40	24.50		37.90	51
1130	3/4" diameter x 8" long	G		17	.471		21	25.50		46.50	62
1140	12" long	G		17	.471		25.50	25.50		51	66.50
1150	18" long	G		17	.471		31.50	25.50		57	73.50
1160	1" diameter x 12" long	G		16	.500		51.50	27.50		79	97.50
1170	18" long	G		15	.533		61	29		90	111
1180	24" long	G		15	.533		73.50	29		102.50	125
1190	36" long	G		15	.533		99	29		128	153
1200	1-1/2" diameter x 18" long	G		13	.615		94	33.50		127.50	154
1210	24" long	G		12	.667		111	36.50		147.50	177
1300	L-type, incl. hex nut & washer, 3/4" diameter x 12" long	G		17	.471		38.50	25.50		64	81
1310	18" long	G		17	.471		48.50	25.50		74	91.50
1320	24" long	G		17	.471		58	25.50		83.50	102
1330	30" long	G		16	.500		72.50	27.50		100	121
1340	36" long	G		16	.500		82	27.50		109.50	131
1350	1" diameter x 12" long	G		16	.500		63.50	27.50		91	111
1360	18" long	G		15	.533		78.50	29		107.50	130
1370	24" long	G		15	.533		95.50	29		124.50	149
1380	30" long	G		15	.533		112	29		141	167
1390	36" long	G		15	.533		128	29		157	184
1400	42" long	G		14	.571		154	31.50		185.50	216
1410	48" long	G		14	.571		172	31.50		203.50	236
1420	1-1/4" diameter x 18" long	G		14	.571		81.50	31.50		113	136
1430	24" long	G		14	.571		96	31.50		127.50	153
1440	30" long	G		13	.615		110	33.50		143.50	173
1450	36" long	G	▼	13	.615		125	33.50		158.50	188
1460	42" long	G	2 Carp	25	.640		141	35		176	208
1470	48" long	G		24	.667	▼	160	36.50		196.50	231

03 15 19 – Cast-In Concrete Anchors

03 15 19.10 Anchor Bolts

		Crew	Daily Output	Labor-Hours	Unit	Material	2021 Bare Costs Labor	Equipment	Total	Total Incl O&P
1480	54" long	2 Carp G	23	.696	Set	188	38		226	263
1490	60" long	G	23	.696		206	38		244	283
1500	1-1/2" diameter x 18" long	G	25	.640		91.50	35		126.50	154
1510	24" long	2 Carp G	24	.667	Set	106	36.50		142.50	172
1520	30" long	G	23	.696		119	38		157	188
1530	36" long	G	22	.727		136	40		176	210
1540	42" long	G	22	.727		155	40		195	230
1550	48" long	G	21	.762		173	41.50		214.50	253
1560	54" long	G	20	.800		210	44		254	297
1570	60" long	G	20	.800		230	44		274	320
1580	1-3/4" diameter x 18" long	G	22	.727		156	40		196	232
1590	24" long	G	21	.762		183	41.50		224.50	263
1600	30" long	G	21	.762		212	41.50		253.50	295
1610	36" long	G	20	.800		241	44		285	330
1620	42" long	G	19	.842		270	46		316	365
1630	48" long	G	18	.889		297	48.50		345.50	400
1640	54" long	G	18	.889		365	48.50		413.50	480
1650	60" long	G	17	.941		395	51.50		446.50	510
1660	2" diameter x 24" long	G	19	.842		290	46		336	390
1670	30" long	G	18	.889		325	48.50		373.50	435
1680	36" long	G	18	.889		360	48.50		408.50	470
1690	42" long	G	17	.941		400	51.50		451.50	515
1700	48" long	G	16	1		460	54.50		514.50	585
1710	54" long	G	15	1.067		545	58.50		603.50	685
1720	60" long	G	15	1.067		585	58.50		643.50	730
1730	66" long	G	14	1.143		625	62.50		687.50	785
1740	72" long	G	14	1.143		685	62.50		747.50	850
1990	For galvanized, add				Ea.	75%				

03 15 19.20 Dovetail Anchor System

		Crew	Daily Output	Labor-Hours	Unit	Material	2021 Bare Costs Labor	Equipment	Total	Total Incl O&P
0010	**DOVETAIL ANCHOR SYSTEM**									
0500	Dovetail anchor slot, galvanized, foam-filled, 26 ga.	1 Carp G	425	.019	L.F.	1.38	1.03		2.41	3.06
0600	24 ga.	G	400	.020		.30	1.09		1.39	1.96
0625	22 ga.	G	400	.020		.45	1.09		1.54	2.13
0900	Stainless steel, foam-filled, 26 ga.	G	375	.021		2.21	1.17		3.38	4.17
1200	Dovetail brick anchor, corrugated, galvanized, 3-1/2" long, 16 ga.	1 Bric G	10.50	.762	C	31.50	41		72.50	96
1300	12 ga.	G	10.50	.762		40	41		81	106
1500	Seismic, galvanized, 3-1/2" long, 16 ga.	G	10.50	.762		74.50	41		115.50	144
1600	12 ga.	G	10.50	.762		101	41		142	173
6000	Dovetail stone panel anchors, galvanized, 1/8" x 1" wide, 3-1/2" long	G	10.50	.762		90.50	41		131.50	162
6100	1/4" x 1" wide	G	10.50	.762		177	41		218	257

03 15 19.30 Inserts

		Crew	Daily Output	Labor-Hours	Unit	Material	2021 Bare Costs Labor	Equipment	Total	Total Incl O&P
0010	**INSERTS**									
6000	Thin slab, ferrule type									
6100	1/2" diameter bolt	1 Carp G	60	.133	Ea.	3.83	7.30		11.13	15.10
6150	5/8" diameter bolt	G	60	.133		4.73	7.30		12.03	16.10
6200	3/4" diameter bolt	G	60	.133		5.50	7.30		12.80	16.95
6250	1" diameter bolt	1 Carp G	60	.133	Ea.	8.85	7.30		16.15	20.50
9950	For galvanized inserts, add					30%				

03 15 Concrete Accessories

03 15 19 – Cast-In Concrete Anchors

03 15 19.45 Machinery Anchors		Crew	Daily Output	Labor-Hours	Unit	Material	2021 Bare Costs Labor	Equipment	Total	Total Incl O&P
0010	**MACHINERY ANCHORS**, heavy duty, incl. sleeve, floating base nut,									
0020	lower stud & coupling nut, fiber plug, connecting stud, washer & nut.									
0030	For flush mounted embedment in poured concrete heavy equip. pads.									
0200	Stud & bolt, 1/2" diameter	G	E-16	40	.400	Ea.	60.50	24.50	3.72	88.72	109
0300	5/8" diameter	G		35	.457		67	28	4.25	99.25	122
0500	3/4" diameter	G		30	.533		80.50	32.50	4.96	117.96	145
0600	7/8" diameter	G		25	.640		95	39	5.95	139.95	171
0800	1" diameter	G		20	.800		105	49	7.45	161.45	199
0900	1-1/4" diameter	G		15	1.067		135	65.50	9.90	210.40	260

03 21 Reinforcement Bars

03 21 05 – Reinforcing Steel Accessories

03 21 05.10 Rebar Accessories						Unit	Material	Labor	Equipment	Total	Total Incl O&P
0010	**REBAR ACCESSORIES**	R032110-70									
0030	Steel & plastic made from recycled materials										
0700	Bag ties, 16 ga., plain, 4" long	G				C	.71			.71	.78
0710	5" long	G					.82			.82	.90
0720	6" long	G					.93			.93	1.02
0730	7" long	G					1.02			1.02	1.12
1200	High chairs, individual (HC), 3" high, plain steel	G					50			50	55
1202	Galvanized	G					114			114	126
1204	Stainless tipped legs	G					470			470	515
1206	Plastic tipped legs	G					66			66	72.50
1210	5" high, plain	G					60			60	66
1212	Galvanized	G					147			147	162
1214	Stainless tipped legs	G					590			590	650
1216	Plastic tipped legs	G					76			76	83.50
1220	8" high, plain	G					83			83	91.50
1222	Galvanized	G					180			180	198
1224	Stainless tipped legs	G					635			635	695
1226	Plastic tipped legs	G					99			99	109
1230	12" high, plain	G					139			139	153
1232	Galvanized	G					360			360	395
1234	Stainless tipped legs	G					690			690	760
1236	Plastic tipped legs	G					154			154	169
1400	Individual high chairs, with plate (HCP), 5" high	G					139			139	153
1410	8" high	G					162			162	178
1500	Bar chair (BC), 1-1/2" high, plain steel	G					46			46	50.50
1520	Galvanized	G					52			52	57
1530	Stainless tipped legs	G				C	420			420	465
1540	Plastic tipped legs	G				"	49			49	54
1700	Continuous high chairs (CHC), legs 8" OC, 4" high, plain steel	G				C.L.F.	62			62	68
1705	Galvanized	G					62.50			62.50	69
1710	Stainless tipped legs	G					560			560	615
1715	Plastic tipped legs	G					78			78	86
1718	Epoxy dipped	G					93			93	102
1720	6" high, plain	G					85			85	93.50
1725	Galvanized	G					100			100	110
1730	Stainless tipped legs	G					580			580	640
1735	Plastic tipped legs	G					101			101	111

03 21 05 – Reinforcing Steel Accessories

03 21 05.10 Rebar Accessories		Crew	Daily Output	Labor-Hours	Unit	Material	2021 Bare Costs Labor	Equipment	Total	Total Incl O&P
1738	Epoxy dipped	G				C.L.F.	117			117	129
1740	8" high, plain	G					122			122	134
1745	Galvanized	G					154			154	169
1750	Stainless tipped legs	G					620			620	685
1755	Plastic tipped legs	G					138			138	152
1758	Epoxy dipped	G					153			153	168
1900	For continuous bottom wire runners, add	G					37			37	40.50
1940	For continuous bottom plate, add	G				▼	216			216	238
2200	Screed chair base, 1/2" coil thread diam., 2-1/2" high, plain steel	G				C	385			385	425
2210	Galvanized	G					430			430	475
2220	5-1/2" high, plain	G					465			465	510
2250	Galvanized	G					510			510	560
2300	3/4" coil thread diam., 2-1/2" high, plain steel	G					465			465	515
2310	Galvanized	G					535			535	590
2320	5-1/2" high, plain steel	G					585			585	645
2350	Galvanized	G					725			725	800
2400	Screed holder, 1/2" coil thread diam. for pipe screed, plain steel, 6" long	G					465			465	510
2420	12" long	G					710			710	785
2500	3/4" coil thread diam. for pipe screed, plain steel, 6" long	G					580			580	640
2520	12" long	G					925			925	1,025
2700	Screw anchor for bolts, plain steel, 3/4" diameter x 4" long	G					530			530	585
2720	1" diameter x 6" long	G					1,000			1,000	1,100
2740	1-1/2" diameter x 8" long	G					1,350			1,350	1,500
2800	Screw anchor eye bolts, 3/4" x 3" long	G					3,125			3,125	3,450
2820	1" x 3-1/2" long	G					4,050			4,050	4,450
2840	1-1/2" x 6" long	G					12,400			12,400	13,600
2900	Screw anchor bolts, 3/4" x 9" long	G					1,975			1,975	2,175
2920	1" x 12" long	G				▼	3,825			3,825	4,200
3800	Subgrade chairs, #4 bar head, 3-1/2" high	G				C	48			48	53
3850	12" high	G					54.50			54.50	59.50
3900	#6 bar head, 3-1/2" high	G					48			48	53
3950	12" high	G					54.50			54.50	59.50
4200	Subgrade stakes, no nail holes, 3/4" diameter, 12" long	G					370			370	410
4250	24" long	G					490			490	540
4300	7/8" diameter, 12" long	G					415			415	455
4350	24" long	G				▼	710			710	780
4500	Tie wire, 16 ga. annealed steel	G				Cwt.	153			153	168

03 21 11 – Plain Steel Reinforcement Bars

03 21 11.50 Reinforcing Steel, Mill Base Plus Extras

			Crew	Daily Output	Labor-Hours	Unit	Material	2021 Bare Costs Labor	Equipment	Total	Total Incl O&P
0010	**REINFORCING STEEL, MILL BASE PLUS EXTRAS**	R032110-10									
0150	Reinforcing, A615 grade 40, mill base	G				Ton	635			635	700
0200	Detailed, cut, bent, and delivered	G					905			905	995
0650	Reinforcing steel, A615 grade 60, mill base	G					645			645	705
0700	Detailed, cut, bent, and delivered	G				▼	910			910	1,000
1000	Reinforcing steel, extras, included in delivered price										
1005	Mill extra, added for delivery to shop					Ton	42.50			42.50	46.50
1010	Shop extra, added for handling & storage						48.50			48.50	53.50
1020	Shop extra, added for bending, limited percent of bars						39.50			39.50	43
1030	Average percent of bars						78.50			78.50	86.50
1050	Large percent of bars	R032110-70					157			157	173
1200	Shop extra, added for detailing, under 50 tons						58.50			58.50	64.50
1250	50 to 150 tons	R032110-80				▼	44			44	48.50

For customer support on your Site Work & Landscape Costs with RSMeans Data, call 800.448.8182.

03 21 Reinforcement Bars

03 21 11 – Plain Steel Reinforcement Bars

03 21 11.50 Reinforcing Steel, Mill Base Plus Extras

		Crew	Daily Output	Labor-Hours	Unit	Material	2021 Bare Costs Labor	Equipment	Total	Total Incl O&P
1300	150 to 500 tons				Ton	42			42	46
1350	Over 500 tons					39.50			39.50	43.50
1700	Shop extra, added for listing					6.10			6.10	6.70
2000	Mill extra, added for quantity, under 20 tons					26			26	28.50
2100	Shop extra, added for quantity, under 20 tons					39			39	43
2200	20 to 49 tons					29.50			29.50	32
2250	50 to 99 tons					19.55			19.55	21.50
2300	100 to 300 tons					11.70			11.70	12.90
2500	Shop extra, added for size, #3					163			163	180
2550	#4					81.50			81.50	90
2600	#5					41			41	45
2650	#6					36.50			36.50	40.50
2700	#7 to #11					49			49	54
2750	#14					61			61	67.50
2800	#18					69.50			69.50	76.50
2900	Shop extra, added for delivery to job				▼	16.15			16.15	17.75

03 21 11.60 Reinforcing In Place

			Crew	Daily Output	Labor-Hours	Unit	Material	2021 Bare Costs Labor	Equipment	Total	Total Incl O&P
0010	**REINFORCING IN PLACE**, 50-60 ton lots, A615 Grade 60	R032110-10									
0020	Includes labor, but not material cost, to install accessories										
0030	Made from recycled materials										
0100	Beams & Girders, #3 to #7	G	4 Rodm	1.60	20	Ton	1,250	1,175		2,425	3,125
0150	#8 to #18	G		2.70	11.852		1,250	700		1,950	2,425
0200	Columns, #3 to #7	G		1.50	21.333		1,250	1,250		2,500	3,250
0250	#8 to #18	G		2.30	13.913		1,250	820		2,070	2,600
0300	Spirals, hot rolled, 8" to 15" diameter	G		2.20	14.545		1,575	855		2,430	3,025
0320	15" to 24" diameter	G		2.20	14.545		1,525	855		2,380	2,950
0330	24" to 36" diameter	G		2.30	13.913		1,450	820		2,270	2,800
0340	36" to 48" diameter	G		2.40	13.333		1,375	785		2,160	2,675
0360	48" to 64" diameter	G		2.50	12.800		1,525	755		2,280	2,800
0380	64" to 84" diameter	R032110-70	G		2.60	12.308		1,575	725		2,300	2,825
0390	84" to 96" diameter		G		2.70	11.852		1,650	700		2,350	2,875
0400	Elevated slabs, #4 to #7	R032110-80	G		2.90	11.034		1,250	650		1,900	2,350
0500	Footings, #4 to #7	G		2.10	15.238		1,250	900		2,150	2,725
0550	#8 to #18	G		3.60	8.889		1,250	525		1,775	2,150
0600	Slab on grade, #3 to #7	G		2.30	13.913		1,250	820		2,070	2,600
0700	Walls, #3 to #7	G		3	10.667		1,250	630		1,880	2,325
0750	#8 to #18	G	▼	4	8	▼	1,250	470		1,720	2,075
0900	For other than 50-60 ton lots										
1000	Under 10 ton job, #3 to #7, add						25%	10%			
1010	#8 to #18, add						20%	10%			
1050	10-50 ton job, #3 to #7, add						10%				
1060	#8 to #18, add						5%				
1100	60-100 ton job, #3 to #7, deduct						5%				
1110	#8 to #18, deduct						10%				
1150	Over 100 ton job, #3 to #7, deduct						10%				
1160	#8 to #18, deduct						15%				
1200	Reinforcing in place, A615 Grade 75, add	G				Ton	129			129	142
1220	Grade 90, add						145			145	160
2000	Unloading & sorting, add to above		C-5	100	.560			33	5.85	38.85	55.50
2200	Crane cost for handling, 90 picks/day, up to 1.5 tons/bundle, add to above			135	.415			24.50	4.35	28.85	41.50
2210	1.0 ton/bundle			92	.609			35.50	6.40	41.90	60.50
2220	0.5 ton/bundle		▼	35	1.600	▼		94	16.75	110.75	158

For customer support on your Site Work & Landscape Costs with RSMeans Data, call 800.448.8182.

75

03 21 Reinforcement Bars

03 21 11 – Plain Steel Reinforcement Bars

03 21 11.60 Reinforcing In Place		Crew	Daily Output	Labor-Hours	Unit	Material	2021 Bare Costs Labor	Equipment	Total	Total Incl O&P
2400	Dowels, 2 feet long, deformed, #3	G 2 Rodm	520	.031	Ea.	.51	1.81		2.32	3.27
2410	#4	G	480	.033		.91	1.96		2.87	3.94
2420	#5	G	435	.037		1.42	2.17		3.59	4.81
2430	#6	G	360	.044		2.05	2.62		4.67	6.15
2450	Longer and heavier dowels, add	G 2 Rodm	725	.022	Lb.	.68	1.30		1.98	2.69
2500	Smooth dowels, 12" long, 1/4" or 3/8" diameter	G	140	.114	Ea.	.87	6.75		7.62	11
2520	5/8" diameter	G	125	.128		1.53	7.55		9.08	12.95
2530	3/4" diameter	G	110	.145		1.89	8.55		10.44	14.90
2600	Dowel sleeves for CIP concrete, 2-part system									
2610	Sleeve base, plastic, for 5/8" smooth dowel sleeve, fasten to edge form	1 Rodm	200	.040	Ea.	.53	2.36		2.89	4.10
2615	Sleeve, plastic, 12" long, for 5/8" smooth dowel, snap onto base		400	.020		1.54	1.18		2.72	3.45
2620	Sleeve base, for 3/4" smooth dowel sleeve		175	.046		.64	2.69		3.33	4.73
2625	Sleeve, 12" long, for 3/4" smooth dowel		350	.023		1.42	1.35		2.77	3.57
2630	Sleeve base, for 1" smooth dowel sleeve		150	.053		.68	3.14		3.82	5.45
2635	Sleeve, 12" long, for 1" smooth dowel		300	.027		1.37	1.57		2.94	3.86
2700	Dowel caps, visual warning only, plastic, #3 to #8	2 Rodm	800	.020		.39	1.18		1.57	2.19
2720	#8 to #18		750	.021		.89	1.26		2.15	2.85
2750	Impalement protective, plastic, #4 to #9		800	.020		1.22	1.18		2.40	3.10

03 21 13 – Galvanized Reinforcement Steel Bars

03 21 13.10 Galvanized Reinforcing

		Crew	Daily Output	Labor-Hours	Unit	Material	Labor	Equipment	Total	Total Incl O&P
0010	**GALVANIZED REINFORCING**									
0150	Add to plain steel rebar pricing for galvanized rebar				Ton	485			485	535

03 21 16 – Epoxy-Coated Reinforcement Steel Bars

03 21 16.10 Epoxy-Coated Reinforcing

		Crew	Daily Output	Labor-Hours	Unit	Material	Labor	Equipment	Total	Total Incl O&P
0010	**EPOXY-COATED REINFORCING**									
0100	Add to plain steel rebar pricing for epoxy-coated rebar				Ton	970			970	1,075

03 21 21 – Composite Reinforcement Bars

03 21 21.11 Glass Fiber-Reinforced Polymer Reinf. Bars

		Crew	Daily Output	Labor-Hours	Unit	Material	Labor	Equipment	Total	Total Incl O&P
0010	**GLASS FIBER-REINFORCED POLYMER REINFORCEMENT BARS**									
0020	Includes labor, but not material cost, to install accessories									
0050	#2 bar, .043 lb./L.F.	4 Rodm	9500	.003	L.F.	.41	.20		.61	.75
0100	#3 bar, .092 lb./L.F.		9300	.003		.66	.20		.86	1.03
0150	#4 bar, .160 lb./L.F.		9100	.004		.95	.21		1.16	1.36
0200	#5 bar, .258 lb./L.F.		8700	.004		1.17	.22		1.39	1.61
0250	#6 bar, .372 lb./L.F.		8300	.004		2.04	.23		2.27	2.58
0300	#7 bar, .497 lb./L.F.		7900	.004		2.52	.24		2.76	3.13
0350	#8 bar, .620 lb./L.F.		7400	.004		2.29	.25		2.54	2.90
0400	#9 bar, .800 lb./L.F.		6800	.005		4.15	.28		4.43	4.98
0450	#10 bar, 1.08 lb./L.F.		5800	.006		3.77	.32		4.09	4.64
0500	For bends, add per bend				Ea.	1.59			1.59	1.75

76

For customer support on your Site Work & Landscape Costs with RSMeans Data, call 800.448.8182.

03 22 Fabric and Grid Reinforcing

03 22 11 – Plain Welded Wire Fabric Reinforcing

03 22 11.10 Plain Welded Wire Fabric

03 22 11.10 Plain Welded Wire Fabric		Crew	Daily Output	Labor-Hours	Unit	Material	2021 Bare Costs Labor	2021 Bare Costs Equipment	Total	Total Incl O&P
0010	**PLAIN WELDED WIRE FABRIC** ASTM A185 R032205-30									
0020	Includes labor, but not material cost, to install accessories									
0030	Made from recycled materials									
0050	Sheets									
0100	6 x 6 - W1.4 x W1.4 (10 x 10) 21 lb./C.S.F. [G]	2 Rodm	35	.457	C.S.F.	14.95	27		41.95	57
0200	6 x 6 - W2.1 x W2.1 (8 x 8) 30 lb./C.S.F. [G]		31	.516		21	30.50		51.50	69
0300	6 x 6 - W2.9 x W2.9 (6 x 6) 42 lb./C.S.F. [G]		29	.552		33.50	32.50		66	85.50
0400	6 x 6 - W4 x W4 (4 x 4) 58 lb./C.S.F. [G]		27	.593		39	35		74	95
0500	4 x 4 - W1.4 x W1.4 (10 x 10) 31 lb./C.S.F. [G]		31	.516		24.50	30.50		55	72.50
0600	4 x 4 - W2.1 x W2.1 (8 x 8) 44 lb./C.S.F. [G]		29	.552		30.50	32.50		63	82
0650	4 x 4 - W2.9 x W2.9 (6 x 6) 61 lb./C.S.F. [G]		27	.593		49	35		84	106
0700	4 x 4 - W4 x W4 (4 x 4) 85 lb./C.S.F. [G]		25	.640		61	37.50		98.50	124
0750	Rolls									
0800	2 x 2 - #14 galv., 21 lb./C.S.F., beam & column wrap [G]	2 Rodm	6.50	2.462	C.S.F.	45.50	145		190.50	267
0900	2 x 2 - #12 galv. for gunite reinforcing [G]	"	6.50	2.462	"	63	145		208	287

03 22 13 – Galvanized Welded Wire Fabric Reinforcing

03 22 13.10 Galvanized Welded Wire Fabric

03 22 13.10 Galvanized Welded Wire Fabric		Crew	Daily Output	Labor-Hours	Unit	Material	2021 Bare Costs Labor	2021 Bare Costs Equipment	Total	Total Incl O&P
0010	**GALVANIZED WELDED WIRE FABRIC**									
0100	Add to plain welded wire pricing for galvanized welded wire				Lb.	.24			.24	.27

03 22 16 – Epoxy-Coated Welded Wire Fabric Reinforcing

03 22 16.10 Epoxy-Coated Welded Wire Fabric

03 22 16.10 Epoxy-Coated Welded Wire Fabric		Crew	Daily Output	Labor-Hours	Unit	Material	2021 Bare Costs Labor	2021 Bare Costs Equipment	Total	Total Incl O&P
0010	**EPOXY-COATED WELDED WIRE FABRIC**									
0100	Add to plain welded wire pricing for epoxy-coated welded wire				Lb.	.49			.49	.53

03 23 Stressed Tendon Reinforcing

03 23 05 – Prestressing Tendons

03 23 05.50 Prestressing Steel

03 23 05.50 Prestressing Steel		Crew	Daily Output	Labor-Hours	Unit	Material	2021 Bare Costs Labor	2021 Bare Costs Equipment	Total	Total Incl O&P
0010	**PRESTRESSING STEEL**									
3000	Slabs on grade, 0.5-inch diam. non-bonded strands, HDPE sheathed,									
3050	attached dead-end anchors, loose stressing-end anchors									
3100	25' x 30' slab, strands @ 36" OC, placing	2 Rodm	2940	.005	S.F.	.69	.32		1.01	1.23
3105	Stressing	C-4A	3750	.004			.25	.02	.27	.40
3110	42" OC, placing	2 Rodm	3200	.005		.61	.29		.90	1.11
3115	Stressing	C-4A	4040	.004			.23	.02	.25	.37
3120	48" OC, placing	2 Rodm	3510	.005		.53	.27		.80	.99
3125	Stressing	C-4A	4390	.004			.21	.02	.23	.34
3150	25' x 40' slab, strands @ 36" OC, placing	2 Rodm	3370	.005		.66	.28		.94	1.15
3155	Stressing	C-4A	4360	.004			.22	.02	.24	.34
3160	42" OC, placing	2 Rodm	3760	.004		.57	.25		.82	1
3165	Stressing	C-4A	4820	.003			.20	.02	.22	.31
3170	48" OC, placing	2 Rodm	4090	.004		.51	.23		.74	.90
3175	Stressing	C-4A	5190	.003			.18	.01	.19	.29
3200	30' x 30' slab, strands @ 36" OC, placing	2 Rodm	3260	.005		.66	.29		.95	1.16
3205	Stressing	C-4A	4190	.004			.22	.02	.24	.36
3210	42" OC, placing	2 Rodm	3530	.005	S.F.	.60	.27		.87	1.05
3215	Stressing	C-4A	4500	.004			.21	.02	.23	.33
3220	48" OC, placing	2 Rodm	3840	.004		.53	.25		.78	.95
3225	Stressing	C-4A	4850	.003			.19	.02	.21	.31
3230	30' x 40' slab, strands @ 36" OC, placing	2 Rodm	3780	.004		.64	.25		.89	1.07
3235	Stressing	C-4A	4920	.003			.19	.02	.21	.31

03 23 Stressed Tendon Reinforcing

03 23 05 – Prestressing Tendons

03 23 05.50 Prestressing Steel	Crew	Daily Output	Labor-Hours	Unit	Material	2021 Bare Costs Labor	Equipment	Total	Total Incl O&P
3240 42" OC, placing	2 Rodm	4190	.004	S.F.	.56	.22		.78	.95
3245 Stressing	C-4A	5410	.003			.17	.01	.18	.28
3250 48" OC, placing	2 Rodm	4520	.004		.50	.21		.71	.86
3255 Stressing	C-4A	5790	.003			.16	.01	.17	.25
3260 30' x 50' slab, strands @ 36" OC, placing	2 Rodm	4300	.004		.60	.22		.82	.99
3265 Stressing	C-4A	5650	.003			.17	.01	.18	.26
3270 42" OC, placing	2 Rodm	4720	.003		.54	.20		.74	.89
3275 Stressing	C-4A	6150	.003			.15	.01	.16	.24
3280 48" OC, placing	2 Rodm	5240	.003		.47	.18		.65	.79
3285 Stressing	C-4A	6760	.002	▼		.14	.01	.15	.22

03 24 Fibrous Reinforcing

03 24 05 – Reinforcing Fibers

03 24 05.30 Synthetic Fibers

				Unit	Material	Labor	Equipment	Total	Total Incl O&P
0010 **SYNTHETIC FIBERS**									
0100 Synthetic fibers, add to concrete				Lb.	6.15			6.15	6.75
0110 1-1/2 lb./C.Y.				C.Y.	9.50			9.50	10.45

03 24 05.70 Steel Fibers

				Unit	Material	Labor	Equipment	Total	Total Incl O&P
0010 **STEEL FIBERS**									
0140 ASTM A850, Type V, continuously deformed, 1-1/2" long x 0.045" diam.									
0150 Add to price of ready mix concrete	G			Lb.	1.20			1.20	1.32
0205 Alternate pricing, dosing at 5 lb./C.Y., add to price of RMC	G			C.Y.	6			6	6.60
0210 10 lb./C.Y.	G				12			12	13.20
0215 15 lb./C.Y.	G				18			18	19.80
0220 20 lb./C.Y.	G				24			24	26.50
0225 25 lb./C.Y.	G				30			30	33
0230 30 lb./C.Y.	G				36			36	39.50
0235 35 lb./C.Y.	G				42			42	46
0240 40 lb./C.Y.	G				48			48	53
0250 50 lb./C.Y.	G				60			60	66
0275 75 lb./C.Y.	G				90			90	99
0300 100 lb./C.Y.	G			▼	120			120	132

03 30 Cast-In-Place Concrete

03 30 53 – Miscellaneous Cast-In-Place Concrete

03 30 53.40 Concrete In Place

		Crew	Daily Output	Labor-Hours	Unit	Material	Labor	Equipment	Total	Total Incl O&P
0010 **CONCRETE IN PLACE**	R033053-60									
0020 Including forms (4 uses), Grade 60 rebar, concrete (Portland cement										
0050 Type I), placement and finishing unless otherwise indicated	R033105-20									
0300 Beams (3500 psi), 5 kip/L.F., 10' span	R033105-70	C-14A	15.62	12.804	C.Y.	420	700	29	1,149	1,550
0350 25' span		"	18.55	10.782		440	590	24.50	1,054.50	1,375
0500 Chimney foundations (5000 psi), over 5 C.Y.		C-14C	32.22	3.476		187	182	.83	369.83	480
0510 (3500 psi), under 5 C.Y.		"	23.71	4.724		223	247	1.13	471.13	615
0700 Columns, square (4000 psi), 12" x 12", up to 1% reinforcing by area		C-14A	11.96	16.722		475	915	37.50	1,427.50	1,950
0800 16" x 16", up to 1% reinforcing by area			16.22	12.330		375	675	27.50	1,077.50	1,450
0900 24" x 24", up to 1% reinforcing by area			23.66	8.453		310	460	19	789	1,050
1000 36" x 36", up to 1% reinforcing by area			33.69	5.936		268	325	13.35	606.35	795
1100 Columns, round (4000 psi), tied, 12" diameter, up to 1% reinforcing by area			20.97	9.537		420	520	21.50	961.50	1,275
1200 16" diameter, up to 1% reinforcing by area		▼	31.49	6.351		365	345	14.30	724.30	935

78

03 30 53 – Miscellaneous Cast-In-Place Concrete

03 30 53.40 Concrete In Place	Crew	Daily Output	Labor-Hours	Unit	Material	2021 Bare Costs Labor	Equipment	Total	Total Incl O&P
1300	20" diameter, up to 1% reinforcing by area	C-14A	41.04	4.873	C.Y.	335	266	10.95	611.95	780
1400	24" diameter, up to 1% reinforcing by area		51.85	3.857		340	211	8.70	559.70	700
1500	36" diameter, up to 1% reinforcing by area		75.04	2.665		330	146	6	482	590
3100	Elevated slabs, flat plate, including finish, not									
3110	including forms or reinforcing									
3150	Regular concrete (4000 psi), 4" slab	C-8	2613	.021	S.F.	1.72	1.05	.16	2.93	3.62
3200	6" slab		2585	.022		2.52	1.06	.16	3.74	4.52
3250	2-1/2" thick floor fill		2685	.021		1.12	1.02	.16	2.30	2.91
3300	Lightweight, 110 #/C.F., 2-1/2" thick floor fill		2585	.022		1.31	1.06	.16	2.53	3.19
3400	Cellular concrete, 1-5/8" fill, under 5000 S.F.		2000	.028		.89	1.37	.21	2.47	3.24
3450	Over 10,000 S.F.		2200	.025		.85	1.24	.19	2.28	3
3540	Equipment pad (3000 psi), 3' x 3' x 6" thick	C-14H	45	1.067	Ea.	55.50	57	.61	113.11	147
3550	4' x 4' x 6" thick		30	1.600		84	85.50	.91	170.41	220
3560	5' x 5' x 8" thick		18	2.667		150	143	1.52	294.52	380
3570	6' x 6' x 8" thick		14	3.429		204	184	1.95	389.95	500
3580	8' x 8' x 10" thick		8	6		415	320	3.42	738.42	945
3590	10' x 10' x 12" thick		5	9.600		720	515	5.45	1,240.45	1,550
3600	Flexural concrete on grade, direct chute, 500 psi, no forms, reinf, finish	C-8A	150	.320	C.Y.	124	15.10		139.10	160
3610	650 psi		150	.320		137	15.10		152.10	174
3620	750 psi		150	.320		206	15.10		221.10	250
3650	Pumped, 500 psi	C-8	70	.800		124	39	6.05	169.05	202
3660	650 psi		70	.800		137	39	6.05	182.05	216
3670	750 psi		70	.800		206	39	6.05	251.05	292
3800	Footings (3000 psi), spread under 1 C.Y.	C-14C	28	4		208	209	.96	417.96	540
3813	Install new concrete (3000 psi) light pole base, 24" diam. x 8'	C-1	2.66	12.030		345	625		970	1,325
3825	1 C.Y. to 5 C.Y.	C-14C	43	2.605		266	136	.63	402.63	495
3850	Over 5 C.Y.	"	75	1.493		245	78	.36	323.36	385
3900	Footings, strip (3000 psi), 18" x 9", unreinforced	C-14L	40	2.400		159	123	.67	282.67	360
3920	18" x 9", reinforced	C-14C	35	3.200		189	167	.77	356.77	460
3925	20" x 10", unreinforced	C-14L	45	2.133		153	109	.60	262.60	335
3930	20" x 10", reinforced	C-14C	40	2.800		178	146	.67	324.67	415
3935	24" x 12", unreinforced	C-14L	55	1.745		150	89.50	.49	239.99	299
3940	24" x 12", reinforced	C-14C	48	2.333		175	122	.56	297.56	375
3945	36" x 12", unreinforced	C-14L	70	1.371		144	70	.38	214.38	263
3950	36" x 12", reinforced	C-14C	60	1.867		166	97.50	.45	263.95	330
4000	Foundation mat (3000 psi), under 10 C.Y.		38.67	2.896		259	151	.70	410.70	510
4050	Over 20 C.Y.		56.40	1.986		225	104	.48	329.48	405
4200	Wall, free-standing (3000 psi), 8" thick, 8' high	C-14D	45.83	4.364		199	237	9.80	445.80	585
4250	14' high		27.26	7.337		237	400	16.50	653.50	875
4260	12" thick, 8' high		64.32	3.109		179	169	7	355	455
4270	14' high		40.01	4.999		191	272	11.25	474.25	630
4300	15" thick, 8' high		80.02	2.499		172	136	5.60	313.60	400
4350	12' high		51.26	3.902		172	212	8.80	392.80	515
4500	18' high		48.85	4.094		195	223	9.20	427.20	555
4520	Handicap access ramp (4000 psi), railing both sides, 3' wide	C-14H	14.58	3.292	L.F.	405	176	1.88	582.88	710
4525	5' wide		12.22	3.928		420	210	2.24	632.24	775
4530	With 6" curb and rails both sides, 3' wide		8.55	5.614		415	300	3.20	718.20	910
4535	5' wide		7.31	6.566		425	350	3.74	778.74	995
4650	Slab on grade (3500 psi), not including finish, 4" thick	C-14E	60.75	1.449	C.Y.	149	77.50	.45	226.95	279
4700	6" thick	"	92	.957	"	144	51	.30	195.30	234
4701	Thickened slab edge (3500 psi), for slab on grade poured									
4702	monolithically with slab; depth is in addition to slab thickness;									
4703	formed vertical outside edge, earthen bottom and inside slope									

79

For customer support on your Site Work & Landscape Costs with RSMeans Data, call 800.448.8182.

03 30 Cast-In-Place Concrete

03 30 53 – Miscellaneous Cast-In-Place Concrete

03 30 53.40 Concrete In Place	Crew	Daily Output	Labor-Hours	Unit	Material	2021 Bare Costs Labor	Equipment	Total	Total Incl O&P
4705	8" deep x 8" wide bottom, unreinforced	C-14L	2190	.044	L.F.	4.58	2.24	.01	6.83	8.40
4710	8" x 8", reinforced	C-14C	1670	.067		7.50	3.51	.02	11.03	13.50
4715	12" deep x 12" wide bottom, unreinforced	C-14L	1800	.053		9.05	2.73	.01	11.79	14.05
4720	12" x 12", reinforced	C-14C	1310	.086		14.40	4.47	.02	18.89	22.50
4725	16" deep x 16" wide bottom, unreinforced	C-14L	1440	.067		15	3.41	.02	18.43	21.50
4730	16" x 16", reinforced	C-14C	1120	.100		21.50	5.25	.02	26.77	31.50
4735	20" deep x 20" wide bottom, unreinforced	C-14L	1150	.083		22.50	4.27	.02	26.79	31
4740	20" x 20", reinforced	C-14C	920	.122		30.50	6.35	.03	36.88	43
4745	24" deep x 24" wide bottom, unreinforced	C-14L	930	.103		31.50	5.30	.03	36.83	42.50
4750	24" x 24", reinforced	C-14C	740	.151	▼	42	7.90	.04	49.94	58
4751	Slab on grade (3500 psi), incl. troweled finish, not incl. forms									
4760	or reinforcing, over 10,000 S.F., 4" thick	C-14F	3425	.021	S.F.	1.65	1.04	.01	2.70	3.36
4820	6" thick		3350	.021		2.41	1.07	.01	3.49	4.23
4840	8" thick		3184	.023		3.29	1.12	.01	4.42	5.30
4900	12" thick		2734	.026		4.94	1.31	.01	6.26	7.40
4950	15" thick		2505	.029	▼	6.20	1.42	.01	7.63	8.95
5000	Slab on grade (3000 psi), incl. broom finish, not incl. forms									
5001	or reinforcing, 4" thick	C-14G	2873	.019	S.F.	1.62	.95	.01	2.58	3.20
5010	6" thick		2590	.022		2.53	1.06	.01	3.60	4.35
5020	8" thick	▼	2320	.024	▼	3.30	1.18	.01	4.49	5.40
5200	Lift slab in place above the foundation, incl. forms, reinforcing,									
5210	concrete (4000 psi) and columns, over 20,000 S.F./floor	C-14B	2113	.098	S.F.	7.50	5.35	.21	13.06	16.50
5900	Pile caps (3000 psi), incl. forms and reinf., sq. or rect., under 10 C.Y.	C-14C	54.14	2.069	C.Y.	221	108	.50	329.50	405
5950	Over 10 C.Y.		75	1.493		202	78	.36	280.36	340
6000	Triangular or hexagonal, under 10 C.Y.		53	2.113		153	111	.51	264.51	335
6050	Over 10 C.Y.	▼	85	1.318		173	69	.32	242.32	293
6200	Retaining walls (3000 psi), gravity, 4' high see Section 32 32	C-14D	66.20	3.021		176	164	6.80	346.80	445
6250	10' high		125	1.600		166	87	3.60	256.60	315
6300	Cantilever, level backfill loading, 8' high		70	2.857		185	155	6.45	346.45	440
6350	16' high	▼	91	2.198	▼	179	119	4.95	302.95	380
6800	Stairs (3500 psi), not including safety treads, free standing, 3'-6" wide	C-14H	83	.578	LF Nose	7.60	31	.33	38.93	54.50
6850	Cast on ground		125	.384	"	6.05	20.50	.22	26.77	37.50
7000	Stair landings, free standing		200	.240	S.F.	6.10	12.85	.14	19.09	26
7050	Cast on ground	▼	475	.101	"	4.50	5.40	.06	9.96	13.05

03 31 Structural Concrete

03 31 13 – Heavyweight Structural Concrete

03 31 13.25 Concrete, Hand Mix

		Crew	Daily Output	Labor-Hours	Unit	Material	2021 Bare Costs Labor	Equipment	Total	Total Incl O&P
0010	**CONCRETE, HAND MIX** for small quantities or remote areas									
0050	Includes bulk local aggregate, bulk sand, bagged Portland									
0060	cement (Type I) and water, using gas powered cement mixer									
0125	2500 psi	C-30	135	.059	C.F.	4.35	2.63	1.09	8.07	9.90
0130	3000 psi		135	.059		4.71	2.63	1.09	8.43	10.35
0135	3500 psi		135	.059		4.92	2.63	1.09	8.64	10.55
0140	4000 psi		135	.059		5.15	2.63	1.09	8.87	10.85
0145	4500 psi		135	.059		5.45	2.63	1.09	9.17	11.15
0150	5000 psi	▼	135	.059	▼	5.85	2.63	1.09	9.57	11.55
0300	Using pre-bagged dry mix and wheelbarrow (80-lb. bag = 0.6 C.F.)									
0340	4000 psi	1 Clab	48	.167	C.F.	13.90	7.40		21.30	26.50

03 31 Structural Concrete

03 31 13 – Heavyweight Structural Concrete

03 31 13.30 Concrete, Volumetric Site-Mixed

		Crew	Daily Output	Labor-Hours	Unit	Material	2021 Bare Costs Labor	Equipment	Total	Total Incl O&P
0010	**CONCRETE, VOLUMETRIC SITE-MIXED**									
0015	Mixed on-site in volumetric truck									
0020	Includes local aggregate, sand, Portland cement (Type I) and water									
0025	Excludes all additives and treatments									
0100	3000 psi, 1 C.Y. mixed and discharged				C.Y.	205			205	226
0110	2 C.Y.					160			160	176
0120	3 C.Y.					145			145	160
0130	4 C.Y.					128			128	140
0140	5 C.Y.					125			125	138
0200	For truck holding/waiting time past first 2 on-site hours, add				Hr.	93			93	102
0210	For trip charge beyond first 20 miles, each way, add				Mile	3.68			3.68	4.05
0220	For each additional increase of 500 psi, add				Ea.	4.72			4.72	5.20

03 31 13.35 Heavyweight Concrete, Ready Mix

		Crew	Daily Output	Labor-Hours	Unit	Material	2021 Bare Costs Labor	Equipment	Total	Total Incl O&P
0010	**HEAVYWEIGHT CONCRETE, READY MIX**, delivered									
0012	Includes local aggregate, sand, Portland cement (Type I) and water									
0015	Excludes all additives and treatments R033105-20									
0020	2000 psi				C.Y.	109			109	120
0100	2500 psi					112			112	124
0150	3000 psi					124			124	137
0200	3500 psi					127			127	139
0300	4000 psi					129			129	142
0350	4500 psi					133			133	146
0400	5000 psi					137			137	151
0411	6000 psi					141			141	155
0412	8000 psi					149			149	163
0413	10,000 psi					156			156	172
0414	12,000 psi					164			164	180
1000	For high early strength (Portland cement Type III), add					10%				
1010	For structural lightweight with regular sand, add					25%				
1300	For winter concrete (hot water), add					5.85			5.85	6.40
1410	For mid-range water reducer, add					4.22			4.22	4.64
1420	For high-range water reducer/superplasticizer, add					6.65			6.65	7.30
1430	For retarder, add					4.22			4.22	4.64
1440	For non-Chloride accelerator, add					8.70			8.70	9.60
1450	For Chloride accelerator, per 1%, add					3.69			3.69	4.06
1460	For fiber reinforcing, synthetic (1 lb./C.Y.), add					8.30			8.30	9.15
1500	For Saturday delivery, add					7.70			7.70	8.45
1510	For truck holding/waiting time past 1st hour per load, add				Hr.	114			114	125
1520	For short load (less than 4 C.Y.), add per load				Ea.	97			97	107
2000	For all lightweight aggregate, add				C.Y.	45%				
4000	Flowable fill: ash, cement, aggregate, water									
4100	40-80 psi				C.Y.	77			77	84.50
4150	Structural: ash, cement, aggregate, water & sand									
4200	50 psi				C.Y.	77			77	85
4250	140 psi					78			78	85.50
4300	500 psi					80			80	88
4350	1000 psi					83.50			83.50	91.50

03 31 13.70 Placing Concrete

		Crew	Daily Output	Labor-Hours	Unit	Material	2021 Bare Costs Labor	Equipment	Total	Total Incl O&P
0010	**PLACING CONCRETE** R033105-70									
0020	Includes labor and equipment to place, level (strike off) and consolidate									
0050	Beams, elevated, small beams, pumped	C-20	60	1.067	C.Y.		50.50	7.95	58.45	84
0100	With crane and bucket	C-7	45	1.600			77	24	101	141

03 31 13 – Heavyweight Structural Concrete

03 31 13.70 Placing Concrete		Crew	Daily Output	Labor-Hours	Unit	Material	2021 Bare Costs Labor	Equipment	Total	Total Incl O&P
0200	Large beams, pumped	C-20	90	.711	C.Y.		33.50	5.30	38.80	56
0250	With crane and bucket	C-7	65	1.108			53	16.75	69.75	97.50
0400	Columns, square or round, 12" thick, pumped	C-20	60	1.067			50.50	7.95	58.45	84
0450	With crane and bucket	C-7	40	1.800			86.50	27.50	114	159
0600	18" thick, pumped	C-20	90	.711			33.50	5.30	38.80	56
0650	With crane and bucket	C-7	55	1.309			63	19.80	82.80	116
0800	24" thick, pumped	C-20	92	.696			33	5.20	38.20	54.50
0850	With crane and bucket	C-7	70	1.029			49.50	15.55	65.05	90.50
1000	36" thick, pumped	C-20	140	.457			21.50	3.41	24.91	36.50
1050	With crane and bucket	C-7	100	.720			34.50	10.90	45.40	63.50
1200	Duct bank, direct chute	C-6	155	.310			14.25	.35	14.60	21.50
1400	Elevated slabs, less than 6" thick, pumped	C-20	140	.457			21.50	3.41	24.91	36.50
1450	With crane and bucket	C-7	95	.758			36.50	11.45	47.95	66.50
1500	6" to 10" thick, pumped	C-20	160	.400			18.95	2.99	21.94	31.50
1550	With crane and bucket	C-7	110	.655			31.50	9.90	41.40	57.50
1600	Slabs over 10" thick, pumped	C-20	180	.356			16.85	2.66	19.51	28
1650	With crane and bucket	C-7	130	.554			26.50	8.40	34.90	49
1900	Footings, continuous, shallow, direct chute	C-6	120	.400			18.40	.45	18.85	28
1950	Pumped	C-20	150	.427			20	3.19	23.19	33.50
2000	With crane and bucket	C-7	90	.800			38.50	12.10	50.60	70.50
2100	Footings, continuous, deep, direct chute	C-6	140	.343			15.75	.39	16.14	24
2150	Pumped	C-20	160	.400			18.95	2.99	21.94	31.50
2200	With crane and bucket	C-7	110	.655			31.50	9.90	41.40	57.50
2400	Footings, spread, under 1 C.Y., direct chute	C-6	55	.873			40	.99	40.99	60.50
2450	Pumped	C-20	65	.985			46.50	7.35	53.85	77.50
2500	With crane and bucket	C-7	45	1.600			77	24	101	141
2600	Over 5 C.Y., direct chute	C-6	120	.400			18.40	.45	18.85	28
2650	Pumped	C-20	150	.427			20	3.19	23.19	33.50
2700	With crane and bucket	C-7	100	.720			34.50	10.90	45.40	63.50
2900	Foundation mats, over 20 C.Y., direct chute	C-6	350	.137			6.30	.15	6.45	9.55
2950	Pumped	C-20	400	.160	C.Y.		7.60	1.20	8.80	12.60
3000	With crane and bucket	C-7	300	.240			11.50	3.63	15.13	21
3200	Grade beams, direct chute	C-6	150	.320			14.70	.36	15.06	22.50
3250	Pumped	C-20	180	.356			16.85	2.66	19.51	28
3300	With crane and bucket	C-7	120	.600			29	9.10	38.10	53
3700	Pile caps, under 5 C.Y., direct chute	C-6	90	.533			24.50	.60	25.10	37
3750	Pumped	C-20	110	.582			27.50	4.35	31.85	46
3800	With crane and bucket	C-7	80	.900			43	13.65	56.65	79.50
3850	Pile cap, 5 C.Y. to 10 C.Y., direct chute	C-6	175	.274			12.60	.31	12.91	19.10
3900	Pumped	C-20	200	.320			15.15	2.39	17.54	25
3950	With crane and bucket	C-7	150	.480			23	7.25	30.25	42.50
4000	Over 10 C.Y., direct chute	C-6	215	.223			10.25	.25	10.50	15.55
4050	Pumped	C-20	240	.267			12.65	1.99	14.64	21
4100	With crane and bucket	C-7	185	.389			18.65	5.90	24.55	34.50
4300	Slab on grade, up to 6" thick, direct chute R033105-70	C-6	110	.436			20	.49	20.49	30.50
4350	Pumped	C-20	130	.492			23.50	3.68	27.18	38.50
4400	With crane and bucket	C-7	110	.655			31.50	9.90	41.40	57.50
4600	Over 6" thick, direct chute	C-6	165	.291			13.35	.33	13.68	20.50
4650	Pumped	C-20	185	.346			16.40	2.58	18.98	27.50
4700	With crane and bucket	C-7	145	.497			24	7.50	31.50	44
4900	Walls, 8" thick, direct chute	C-6	90	.533			24.50	.60	25.10	37
4950	Pumped	C-20	100	.640			30.50	4.78	35.28	50.50
5000	With crane and bucket	C-7	80	.900			43	13.65	56.65	79.50

03 31 Structural Concrete

03 31 13 – Heavyweight Structural Concrete

03 31 13.70 Placing Concrete

		Crew	Daily Output	Labor-Hours	Unit	Material	2021 Bare Costs Labor	Equipment	Total	Total Incl O&P
5050	12" thick, direct chute	C-6	100	.480	C.Y.		22	.54	22.54	33.50
5100	Pumped	C-20	110	.582			27.50	4.35	31.85	46
5200	With crane and bucket	C-7	90	.800			38.50	12.10	50.60	70.50
5300	15" thick, direct chute	C-6	105	.457			21	.52	21.52	32
5350	Pumped	C-20	120	.533			25.50	3.98	29.48	42
5400	With crane and bucket	C-7	95	.758			36.50	11.45	47.95	66.50
5600	Wheeled concrete dumping, add to placing costs above									
5610	Walking cart, 50' haul, add	C-18	32	.281	C.Y.		12.55	3.65	16.20	23
5620	150' haul, add		24	.375			16.75	4.87	21.62	30.50
5700	250' haul, add		18	.500			22.50	6.50	29	40.50
5800	Riding cart, 50' haul, add	C-19	80	.113			5	1.74	6.74	9.40
5810	150' haul, add		60	.150			6.70	2.32	9.02	12.55
5900	250' haul, add		45	.200			8.90	3.09	11.99	16.70
6000	Concrete in-fill for pan-type metal stairs and landings. Manual placement									
6010	includes up to 50' horizontal haul from point of concrete discharge.									
6100	Stair pan treads, 2" deep									
6110	Flights in 1st floor level up/down from discharge point	C-8A	3200	.015	S.F.		.71		.71	1.05
6120	2nd floor level		2500	.019			.91		.91	1.34
6130	3rd floor level		2000	.024			1.13		1.13	1.68
6140	4th floor level		1800	.027			1.26		1.26	1.87
6200	Intermediate stair landings, pan-type 4" deep									
6210	Flights in 1st floor level up/down from discharge point	C-8A	2000	.024	S.F.		1.13		1.13	1.68
6220	2nd floor level		1500	.032			1.51		1.51	2.24
6230	3rd floor level		1200	.040			1.89		1.89	2.80
6240	4th floor level		1000	.048			2.27		2.27	3.36

03 35 Concrete Finishing

03 35 13 – High-Tolerance Concrete Floor Finishing

03 35 13.30 Finishing Floors, High Tolerance

		Crew	Daily Output	Labor-Hours	Unit	Material	2021 Bare Costs Labor	Equipment	Total	Total Incl O&P
0010	**FINISHING FLOORS, HIGH TOLERANCE**									
0012	Finishing of fresh concrete flatwork requires that concrete									
0013	first be placed, struck off & consolidated									
0015	Basic finishing for various unspecified flatwork									
0100	Bull float only	C-10	4000	.006	S.F.		.30		.30	.44
0125	Bull float & manual float		2000	.012			.59		.59	.87
0150	Bull float, manual float & broom finish, w/edging & joints		1850	.013			.64		.64	.94
0200	Bull float, manual float & manual steel trowel		1265	.019			.94		.94	1.38
0210	For specified Random Access Floors in ACI Classes 1, 2, 3 and 4 to achieve									
0215	Composite Overall Floor Flatness and Levelness values up to FF35/FL25									
0250	Bull float, machine float & machine trowel (walk-behind)	C-10C	1715	.014	S.F.		.69	.05	.74	1.08
0300	Power screed, bull float, machine float & trowel (walk-behind)	C-10D	2400	.010			.49	.08	.57	.81
0350	Power screed, bull float, machine float & trowel (ride-on)	C-10E	4000	.006			.30	.06	.36	.51
0352	For specified Random Access Floors in ACI Classes 5, 6, 7 and 8 to achieve									
0354	Composite Overall Floor Flatness and Levelness values up to FF50/FL50									
0356	Add for two-dimensional restraightening after power float	C-10	6000	.004	S.F.		.20		.20	.29
0358	For specified Random or Defined Access Floors in ACI Class 9 to achieve									
0360	Composite Overall Floor Flatness and Levelness values up to FF100/FL100									
0362	Add for two-dimensional restraightening after bull float & power float	C-10	3000	.008	S.F.		.39		.39	.58
0364	For specified Superflat Defined Access Floors in ACI Class 9 to achieve									
0366	Minimum Floor Flatness and Levelness values of FF100/FL100									
0368	Add for 2-dim'l restraightening after bull float, power float, power trowel	C-10	2000	.012	S.F.		.59		.59	.87

For customer support on your Site Work & Landscape Costs with RSMeans Data, call 800.448.8182.

83

03 35 16 – Heavy-Duty Concrete Floor Finishing

03 35 16.30 Finishing Floors, Heavy-Duty		Crew	Daily Output	Labor-Hours	Unit	Material	2021 Bare Costs Labor	Equipment	Total	Total Incl O&P
0010	**FINISHING FLOORS, HEAVY-DUTY**									
1800	Floor abrasives, dry shake on fresh concrete, .25 psf, aluminum oxide	1 Cefi	850	.009	S.F.	.48	.49		.97	1.23
1850	Silicon carbide		850	.009		.74	.49		1.23	1.53
2000	Floor hardeners, dry shake, metallic, light service, .50 psf		850	.009		.56	.49		1.05	1.32
2050	Medium service, .75 psf		750	.011		.84	.55		1.39	1.73
2100	Heavy service, 1.0 psf		650	.012		1.12	.64		1.76	2.16
2150	Extra heavy, 1.5 psf		575	.014		1.68	.72		2.40	2.90
2300	Non-metallic, light service, .50 psf		850	.009		.14	.49		.63	.87
2350	Medium service, .75 psf		750	.011		.21	.55		.76	1.04
2400	Heavy service, 1.0 psf		650	.012		.28	.64		.92	1.24
2450	Extra heavy, 1.5 psf		575	.014		.43	.72		1.15	1.53
2800	Trap rock wearing surface, dry shake, for monolithic floors									
2810	2.0 psf	C-10B	1250	.032	S.F.	.02	1.52	.27	1.81	2.57
3800	Dustproofing, liquid, for cured concrete, solvent-based, 1 coat	1 Cefi	1900	.004		.19	.22		.41	.53
3850	2 coats		1300	.006		.69	.32		1.01	1.23
4000	Epoxy-based, 1 coat		1500	.005		.17	.28		.45	.58
4050	2 coats		1500	.005		.34	.28		.62	.77

03 35 19 – Colored Concrete Finishing

03 35 19.30 Finishing Floors, Colored

		Crew	Daily Output	Labor-Hours	Unit	Material	Labor	Equipment	Total	Total Incl O&P
0010	**FINISHING FLOORS, COLORED**									
3000	Floor coloring, dry shake on fresh concrete (0.6 psf)	1 Cefi	1300	.006	S.F.	.48	.32		.80	1
3050	(1.0 psf)	"	625	.013	"	.80	.66		1.46	1.85
3100	Colored dry shake powder only				Lb.	.80			.80	.88
3600	1/2" topping using 0.6 psf dry shake powdered color	C-10B	590	.068	S.F.	7.30	3.21	.57	11.08	13.40
3650	1.0 psf dry shake powdered color	"	590	.068	"	7.60	3.21	.57	11.38	13.75

03 35 23 – Exposed Aggregate Concrete Finishing

03 35 23.30 Finishing Floors, Exposed Aggregate

		Crew	Daily Output	Labor-Hours	Unit	Material	Labor	Equipment	Total	Total Incl O&P
0010	**FINISHING FLOORS, EXPOSED AGGREGATE**									
1600	Exposed local aggregate finish, seeded on fresh concrete, 3 lb./S.F.	1 Cefi	625	.013	S.F.	.20	.66		.86	1.19
1650	4 lb./S.F.	"	465	.017	"	.34	.89		1.23	1.68

03 35 29 – Tooled Concrete Finishing

03 35 29.30 Finishing Floors, Tooled

		Crew	Daily Output	Labor-Hours	Unit	Material	Labor	Equipment	Total	Total Incl O&P
0010	**FINISHING FLOORS, TOOLED**									
4400	Stair finish, fresh concrete, float finish	1 Cefi	275	.029	S.F.		1.51		1.51	2.21
4500	Steel trowel finish		200	.040			2.07		2.07	3.04
4600	Silicon carbide finish, dry shake on fresh concrete, .25 psf		150	.053		.48	2.76		3.24	4.57

03 35 29.60 Finishing Walls

		Crew	Daily Output	Labor-Hours	Unit	Material	Labor	Equipment	Total	Total Incl O&P
0010	**FINISHING WALLS**									
0020	Break ties and patch voids	1 Cefi	540	.015	S.F.	.04	.77		.81	1.17
0050	Burlap rub with grout		450	.018		.04	.92		.96	1.40
0100	Carborundum rub, dry		270	.030			1.53		1.53	2.25
0150	Wet rub		175	.046			2.37		2.37	3.47
0300	Bush hammer, green concrete	B-39	1000	.048	S.F.		2.24	.36	2.60	3.73
0350	Cured concrete	"	650	.074			3.44	.55	3.99	5.75
0500	Acid etch	1 Cefi	575	.014		.13	.72		.85	1.21
0600	Float finish, 1/16" thick	"	300	.027		.38	1.38		1.76	2.44
0700	Sandblast, light penetration	E-11	1100	.029		.52	1.41	.27	2.20	3.06
0750	Heavy penetration	"	375	.085		1.04	4.15	.81	6	8.45
0850	Grind form fins flush	1 Clab	700	.011	L.F.		.51		.51	.76

84

For customer support on your Site Work & Landscape Costs with RSMeans Data, call 800.448.8182.

03 35 Concrete Finishing

03 35 33 – Stamped Concrete Finishing

03 35 33.50 Slab Texture Stamping	Crew	Daily Output	Labor-Hours	Unit	Material	2021 Bare Costs Labor	Equipment	Total	Total Incl O&P
0010 **SLAB TEXTURE STAMPING**									
0050 Stamping requires that concrete first be placed, struck off, consolidated,									
0060 bull floated and free of bleed water. Decorative stamping tasks include:									
0100 Step 1 - first application of dry shake colored hardener	1 Cefi	6400	.001	S.F.	.43	.06		.49	.56
0110 Step 2 - bull float	↓	6400	.001			.06		.06	.09
0130 Step 3 - second application of dry shake colored hardener	↓	6400	.001		.21	.06		.27	.32
0140 Step 4 - bull float, manual float & steel trowel	3 Cefi	1280	.019			.97		.97	1.42
0150 Step 5 - application of dry shake colored release agent	1 Cefi	6400	.001		.10	.06		.16	.20
0160 Step 6 - place, tamp & remove mats	3 Cefi	2400	.010		.85	.52		1.37	1.69
0170 Step 7 - touch up edges, mat joints & simulated grout lines	1 Cefi	1280	.006			.32		.32	.47
0300 Alternate stamping estimating method includes all tasks above	4 Cefi	800	.040		1.58	2.07		3.65	4.78
0400 Step 8 - pressure wash @ 3000 psi after 24 hours	1 Cefi	1600	.005			.26		.26	.38
0500 Step 9 - roll 2 coats cure/seal compound when dry	"	800	.010	↓	.81	.52		1.33	1.65

03 37 Specialty Placed Concrete

03 37 13 – Shotcrete

03 37 13.30 Gunite (Dry-Mix)

03 37 13.30 Gunite (Dry-Mix)	Crew	Daily Output	Labor-Hours	Unit	Material	2021 Bare Costs Labor	Equipment	Total	Total Incl O&P
0010 **GUNITE (DRY-MIX)**									
0020 Typical in place, 1" layers, no mesh included	C-16	2000	.028	S.F.	.52	1.37	.20	2.09	2.82
0100 Mesh for gunite 2 x 2, #12	2 Rodm	800	.020		.63	1.18		1.81	2.45
0150 #4 reinforcing bars @ 6" each way	"	500	.032		2.07	1.88		3.95	5.10
0300 Typical in place, including mesh, 2" thick, flat surfaces	C-16	1000	.056		1.66	2.74	.39	4.79	6.30
0350 Curved surfaces		500	.112		1.66	5.50	.79	7.95	10.80
0500 4" thick, flat surfaces		750	.075		2.70	3.65	.52	6.87	8.95
0550 Curved surfaces	↓	350	.160		2.70	7.80	1.12	11.62	15.80
0900 Prepare old walls, no scaffolding, good condition	C-10	1000	.024			1.18		1.18	1.74
0950 Poor condition	"	275	.087			4.31		4.31	6.35
1100 For high finish requirement or close tolerance, add						50%			
1150 Very high				↓		110%			

03 37 13.60 Shotcrete (Wet-Mix)

03 37 13.60 Shotcrete (Wet-Mix)	Crew	Daily Output	Labor-Hours	Unit	Material	2021 Bare Costs Labor	Equipment	Total	Total Incl O&P
0010 **SHOTCRETE (WET-MIX)**									
0020 Wet mix, placed @ up to 12 C.Y./hour, 3000 psi	C-8C	80	.600	C.Y.	167	29	7.85	203.85	236
0100 Up to 35 C.Y./hour	C-8E	240	.200	C.Y.	151	9.55	2.72	163.27	183
1010 Fiber reinforced, 1" thick	C-8C	1740	.028	S.F.	1.23	1.34	.36	2.93	3.74
1020 2" thick		900	.053		2.45	2.58	.70	5.73	7.30
1030 3" thick		825	.058		3.68	2.82	.76	7.26	9.10
1040 4" thick	↓	750	.064	↓	4.91	3.10	.84	8.85	10.95

03 37 23 – Roller-Compacted Concrete

03 37 23.50 Concrete, Roller-Compacted

03 37 23.50 Concrete, Roller-Compacted	Crew	Daily Output	Labor-Hours	Unit	Material	2021 Bare Costs Labor	Equipment	Total	Total Incl O&P
0010 **CONCRETE, ROLLER-COMPACTED**									
0100 Mass placement, 1' lift, 1' layer	B-10C	1280	.009	C.Y.		.51	1.59	2.10	2.51
0200 2' lift, 6" layer	"	1600	.008			.41	1.28	1.69	2.01
0210 Vertical face, formed, 1' lift	B-11V	400	.060			2.66	.42	3.08	4.44
0220 6" lift	"	200	.120			5.35	.83	6.18	8.85
0300 Sloped face, nonformed, 1' lift	B-11L	384	.042			2.15	2.79	4.94	6.30
0360 6" lift	"	192	.083	↓		4.31	5.60	9.91	12.55
0400 Surface preparation, vacuum truck	B-6A	3280	.006	S.Y.		.31	.11	.42	.58
0450 Water clean	B-9A	3000	.008			.37	.16	.53	.73
0460 Water blast	B-9B	800	.030			1.40	.71	2.11	2.88
0500 Joint bedding placement, 1" thick	B-11C	975	.016	↓		.85	.22	1.07	1.50

03 37 Specialty Placed Concrete

03 37 23 – Roller-Compacted Concrete

03 37 23.50 Concrete, Roller-Compacted	Crew	Daily Output	Labor-Hours	Unit	Material	2021 Bare Costs Labor	Equipment	Total	Total Incl O&P
0510	Conveyance of materials, 18 C.Y. truck, 5 min. cycle	B-34F	2048	.004	C.Y.		.20	.46	.66	.81
0520	10 min. cycle		1024	.008			.40	.92	1.32	1.62
0540	15 min. cycle		680	.012			.60	1.39	1.99	2.43
0550	With crane and bucket	C-23A	1600	.025			1.25	1.66	2.91	3.68
0560	With 4 C.Y. loader, 4 min. cycle	B-10U	480	.025			1.35	2.02	3.37	4.24
0570	8 min. cycle		240	.050			2.71	4.03	6.74	8.45
0580	12 min. cycle		160	.075			4.06	6.05	10.11	12.70
0590	With belt conveyor	C-7D	600	.093			4.37	.34	4.71	6.90
0600	With 17 C.Y. scraper, 5 min. cycle	B-33J	1440	.006			.33	1.68	2.01	2.34
0610	10 min. cycle		720	.011			.66	3.37	4.03	4.68
0620	15 min. cycle		480	.017			.98	5.05	6.03	7
0630	20 min. cycle		360	.022			1.31	6.75	8.06	9.35
0640	Water cure, small job, < 500 C.Y.	B-94C	8	1	Hr.		44.50	11.50	56	79
0650	Large job, over 500 C.Y.	B-59A	8	3	"		140	58	198	273
0660	RCC paving, with asphalt paver including material	B-25C	1000	.048	C.Y.	77	2.38	2.39	81.77	90.50
0670	8" thick layers		4200	.011	S.Y.	20	.57	.57	21.14	23.50
0680	12" thick layers		2800	.017	"	30	.85	.85	31.70	35

03 39 Concrete Curing

03 39 13 – Water Concrete Curing

03 39 13.50 Water Curing

0010	WATER CURING									
0015	With burlap, 4 uses assumed, 7.5 oz.	2 Clab	55	.291	C.S.F.	13.75	12.90		26.65	34.50
0100	10 oz.	"	55	.291	"	25	12.90		37.90	46.50
0400	Curing blankets, 1" to 2" thick, buy				S.F.	.64			.64	.70

03 39 23 – Membrane Concrete Curing

03 39 23.13 Chemical Compound Membrane Concrete Curing

0010	CHEMICAL COMPOUND MEMBRANE CONCRETE CURING									
0300	Sprayed membrane curing compound	2 Clab	95	.168	C.S.F.	12.20	7.50		19.70	24.50
0700	Curing compound, solvent based, 400 S.F./gal., 55 gallon lots				Gal.	21			21	23
0720	5 gallon lots					41			41	45
0800	Curing compound, water based, 250 S.F./gal., 55 gallon lots					26.50			26.50	29
0820	5 gallon lots					36			36	39.50

03 39 23.23 Sheet Membrane Concrete Curing

0010	SHEET MEMBRANE CONCRETE CURING									
0200	Curing blanket, burlap/poly, 2-ply	2 Clab	70	.229	C.S.F.	20.50	10.15		30.65	37.50

03 41 Precast Structural Concrete

03 41 13 – Precast Concrete Hollow Core Planks

03 41 13.50 Precast Slab Planks

0010	PRECAST SLAB PLANKS									
0020	Prestressed roof/floor members, grouted, solid, 4" thick	C-11	2400	.030	S.F.	8.40	1.79	.97	11.16	13.05
0050	6" thick		2800	.026		8.55	1.54	.83	10.92	12.65
0100	Hollow, 8" thick		3200	.023		12.20	1.35	.73	14.28	16.30
0150	10" thick		3600	.020		9.90	1.20	.65	11.75	13.45
0200	12" thick		4000	.018		11.70	1.08	.58	13.36	15.15

03 41 Precast Structural Concrete

03 41 23 – Precast Concrete Stairs

03 41 23.50 Precast Stairs	Crew	Daily Output	Labor-Hours	Unit	Material	2021 Bare Costs Labor	Equipment	Total	Total Incl O&P
0010 **PRECAST STAIRS**									
0020 Precast concrete treads on steel stringers, 3' wide	C-12	75	.640	Riser	135	35	6.35	176.35	207
0200 Front entrance 4' wide with 48" platform, 2 risers		17	2.824	Flight	615	154	28	797	935
0250 5 risers		13	3.692		970	201	36.50	1,207.50	1,425
0300 Front entrance, 5' wide with 48" platform, 2 risers		16	3		690	163	29.50	882.50	1,025
1200 Basement entrance stairwell, 6 steps, incl. steel bulkhead door	B-51	22	2.182		1,900	99	9.05	2,008.05	2,250
1250 14 steps	"	11	4.364		3,425	198	18.05	3,641.05	4,075

03 41 33 – Precast Structural Pretensioned Concrete

03 41 33.10 Precast Beams

03 41 33.10 Precast Beams	Crew	Daily Output	Labor-Hours	Unit	Material	2021 Bare Costs Labor	Equipment	Total	Total Incl O&P
0010 **PRECAST BEAMS**									
0011 L-shaped, 20' span, 12" x 20"	C-11	32	2.250	Ea.	4,800	135	72.50	5,007.50	5,550
1200 Rectangular, 20' span, 12" x 20"		32	2.250		4,150	135	72.50	4,357.50	4,850
1250 18" x 36"		24	3		6,050	179	97	6,326	7,025
1300 24" x 44"		22	3.273		7,050	196	106	7,352	8,175
1400 30' span, 12" x 36"		24	3		6,875	179	97	7,151	7,950
1450 18" x 44"		20	3.600		9,550	215	116	9,881	11,000
1500 24" x 52"		16	4.500		11,600	269	145	12,014	13,400
1600 40' span, 12" x 52"		20	3.600		11,400	215	116	11,731	13,100
1650 18" x 52"		16	4.500		13,500	269	145	13,914	15,500
1700 24" x 52"		12	6		15,500	360	194	16,054	17,900
2000 "T" shaped, 20' span, 12" x 20"		32	2.250		5,775	135	72.50	5,982.50	6,625
2050 18" x 36"	C-11	24	3	Ea.	7,600	179	97	7,876	8,750
2100 24" x 44"		22	3.273		7,825	196	106	8,127	9,025
2200 30' span, 12" x 36"		24	3		10,400	179	97	10,676	11,800
2250 18" x 44"		20	3.600		13,100	215	116	13,431	14,900
2300 24" x 52"		16	4.500		12,800	269	145	13,214	14,600
2500 40' span, 12" x 52"		20	3.600		16,800	215	116	17,131	19,000
2550 18" x 52"		16	4.500		15,900	269	145	16,314	18,100
2600 24" x 52"		12	6		17,000	360	194	17,554	19,500

03 45 Precast Architectural Concrete

03 45 13 – Faced Architectural Precast Concrete

03 45 13.50 Precast Wall Panels

03 45 13.50 Precast Wall Panels	Crew	Daily Output	Labor-Hours	Unit	Material	2021 Bare Costs Labor	Equipment	Total	Total Incl O&P
0010 **PRECAST WALL PANELS**									
0050 Uninsulated, smooth gray									
0150 Low rise, 4' x 8' x 4" thick	C-11	320	.225	S.F.	33	13.45	7.25	53.70	64.50
0210 8' x 8', 4" thick		576	.125		32.50	7.45	4.04	43.99	52
0250 8' x 16' x 4" thick		1024	.070		32	4.20	2.27	38.47	44.50
0600 High rise, 4' x 8' x 4" thick		288	.250		33	14.95	8.05	56	68
0650 8' x 8' x 4" thick		512	.141		32.50	8.40	4.54	45.44	54
0700 8' x 16' x 4" thick		768	.094		32	5.60	3.03	40.63	47.50
0750 10' x 20', 6" thick		1400	.051		55	3.07	1.66	59.73	67
0800 Insulated panel, 2" polystyrene, add					1.23			1.23	1.35
0850 2" urethane, add					.78			.78	.86
1200 Finishes, white, add					3.43			3.43	3.77
1250 Exposed aggregate, add					.71			.71	.78
1300 Granite faced, domestic, add					28.50			28.50	31.50
1350 Brick faced, modular, red, add					8.95			8.95	9.85
2200 Fiberglass reinforced cement with urethane core									
2210 R20, 8' x 8', 5" plain finish	E-2	750	.075	S.F.	30	4.45	2.28	36.73	42.50

03 45 Precast Architectural Concrete

03 45 13 – Faced Architectural Precast Concrete

03 45 13.50 Precast Wall Panels	Crew	Daily Output	Labor-Hours	Unit	Material	2021 Bare Costs Labor	Equipment	Total	Total Incl O&P
2220 Exposed aggregate or brick finish	E-2	600	.093	S.F.	49.50	5.55	2.86	57.91	66

03 47 Site-Cast Concrete

03 47 13 – Tilt-Up Concrete

03 47 13.50 Tilt-Up Wall Panels

	Crew	Daily Output	Labor-Hours	Unit	Material	2021 Bare Costs Labor	Equipment	Total	Total Incl O&P
0010 **TILT-UP WALL PANELS**									
0015 Wall panel construction, walls only, 5-1/2" thick	C-14	1600	.090	S.F.	6.15	4.81	.91	11.87	14.95
0100 7-1/2" thick	"	1550	.093	"	7.85	4.96	.94	13.75	17.10

03 48 Precast Concrete Specialties

03 48 43 – Precast Concrete Trim

03 48 43.40 Precast Lintels

	Crew	Daily Output	Labor-Hours	Unit	Material	2021 Bare Costs Labor	Equipment	Total	Total Incl O&P
0010 **PRECAST LINTELS**, smooth gray, prestressed, stock units only									
0800 4" wide x 8" high x 4' long	D-10	28	1.143	Ea.	35.50	61.50	15.45	112.45	148
0850 8' long		24	1.333		81	71.50	18.05	170.55	216
1000 6" wide x 8" high x 4' long		26	1.231		60	66	16.65	142.65	183
1050 10' long	D-10	22	1.455	Ea.	156	78	19.65	253.65	310
1200 8" wide x 8" high x 4' long		24	1.333		58	71.50	18.05	147.55	191
1250 12' long		20	1.600		201	86	21.50	308.50	375
1275 For custom sizes, types, colors, or finishes of precast lintels, add					150%				

03 53 Concrete Topping

03 53 16 – Iron-Aggregate Concrete Topping

03 53 16.50 Floor Topping

	Crew	Daily Output	Labor-Hours	Unit	Material	2021 Bare Costs Labor	Equipment	Total	Total Incl O&P
0010 **FLOOR TOPPING**									
0400 Integral topping/finish, on fresh concrete, using 1:1:2 mix, 3/16" thick	C-10B	1000	.040	S.F.	.17	1.89	.34	2.40	3.35
0450 1/2" thick		950	.042		.44	1.99	.35	2.78	3.82
0500 3/4" thick		850	.047		.66	2.23	.39	3.28	4.46
0600 1" thick		750	.053		.88	2.53	.45	3.86	5.20
0800 Granolithic topping, on fresh or cured concrete, 1:1:1-1/2 mix, 1/2" thick		590	.068		.49	3.21	.57	4.27	5.90
0820 3/4" thick		580	.069		.74	3.27	.58	4.59	6.30
0850 1" thick		575	.070		.99	3.29	.58	4.86	6.60
0950 2" thick		500	.080		1.97	3.79	.67	6.43	8.50
1200 Heavy duty, 1:1:2, 3/4" thick, preshrunk, gray, 20 M.S.F.		320	.125		.93	5.90	1.05	7.88	10.95
1300 100 M.S.F.		380	.105		.66	4.99	.88	6.53	9.10

03 54 Cast Underlayment

03 54 16 – Hydraulic Cement Underlayment

03 54 16.50 Cement Underlayment	Crew	Daily Output	Labor-Hours	Unit	Material	2021 Bare Costs		Total	Total Incl O&P
						Labor	Equipment		
0010 **CEMENT UNDERLAYMENT**									
2510 Underlayment, P.C. based, self-leveling, 4100 psi, pumped, 1/4" thick	C-8	20000	.003	S.F.	2.09	.14	.02	2.25	2.52
2520 1/2" thick		19000	.003		4.19	.14	.02	4.35	4.84
2530 3/4" thick		18000	.003		6.30	.15	.02	6.47	7.15
2540 1" thick		17000	.003		8.35	.16	.02	8.53	9.45
2550 1-1/2" thick		15000	.004		12.55	.18	.03	12.76	14.10
2560 Hand placed, 1/2" thick	C-18	450	.020		4.19	.89	.26	5.34	6.25
2610 Topping, P.C. based, self-leveling, 6100 psi, pumped, 1/4" thick	C-8	20000	.003		3.40	.14	.02	3.56	3.96
2620 1/2" thick		19000	.003		6.80	.14	.02	6.96	7.75
2630 3/4" thick		18000	.003		10.20	.15	.02	10.37	11.50
2660 1" thick		17000	.003		13.60	.16	.02	13.78	15.20
2670 1-1/2" thick		15000	.004		20.50	.18	.03	20.71	23
2680 Hand placed, 1/2" thick	C-18	450	.020		6.80	.89	.26	7.95	9.10

03 62 Non-Shrink Grouting

03 62 13 – Non-Metallic Non-Shrink Grouting

03 62 13.50 Grout, Non-Metallic Non-Shrink

	Crew	Daily Output	Labor-Hours	Unit	Material	Labor	Equipment	Total	Total Incl O&P
0010 **GROUT, NON-METALLIC NON-SHRINK**									
0300 Non-shrink, non-metallic, 1" deep	1 Cefi	35	.229	S.F.	6.80	11.85		18.65	25
0350 2" deep	"	25	.320	"	13.60	16.60		30.20	39.50

03 62 16 – Metallic Non-Shrink Grouting

03 62 16.50 Grout, Metallic Non-Shrink

	Crew	Daily Output	Labor-Hours	Unit	Material	Labor	Equipment	Total	Total Incl O&P
0010 **GROUT, METALLIC NON-SHRINK**									
0020 Column & machine bases, non-shrink, metallic, 1" deep	1 Cefi	35	.229	S.F.	9.10	11.85		20.95	27.50
0050 2" deep	"	25	.320	"	18.25	16.60		34.85	44.50

03 63 Epoxy Grouting

03 63 05 – Grouting of Dowels and Fasteners

03 63 05.10 Epoxy Only

	Crew	Daily Output	Labor-Hours	Unit	Material	Labor	Equipment	Total	Total Incl O&P
0010 **EPOXY ONLY**									
1500 Chemical anchoring, epoxy cartridge, excludes layout, drilling, fastener									
1530 For fastener 3/4" diam. x 6" embedment	2 Skwk	72	.222	Ea.	4.89	12.70		17.59	24.50
1535 1" diam. x 8" embedment		66	.242		7.35	13.85		21.20	29
1540 1-1/4" diam. x 10" embedment		60	.267		14.70	15.25		29.95	39
1545 1-3/4" diam. x 12" embedment		54	.296		24.50	16.90		41.40	52.50
1550 14" embedment		48	.333		29.50	19.05		48.55	61
1555 2" diam. x 12" embedment		42	.381		39	22		61	75.50
1560 18" embedment		32	.500		49	28.50		77.50	97

03 81 Concrete Cutting

03 81 13 – Flat Concrete Sawing

03 81 13.50 Concrete Floor/Slab Cutting

		Daily Output	Labor-Hours	Unit	Material	2021 Bare Costs Labor	Equipment	Total	Total Incl O&P
0010	**CONCRETE FLOOR/SLAB CUTTING**								
0050	Includes blade cost, layout and set-up time								
0300	Saw cut concrete slabs, plain, up to 3" deep	B-89	1060	.015	L.F.	.10	.79	1.01	1.90	2.40
0320	Each additional inch of depth		3180	.005		.03	.26	.34	.63	.80
0400	Mesh reinforced, up to 3" deep		980	.016		.12	.85	1.09	2.06	2.60
0420	Each additional inch of depth		2940	.005		.04	.28	.36	.68	.86
0500	Rod reinforced, up to 3" deep		800	.020		.14	1.04	1.33	2.51	3.19
0520	Each additional inch of depth		2400	.007		.05	.35	.44	.84	1.06

03 81 13.75 Concrete Saw Blades

		Daily Output	Labor-Hours	Unit	Material	2021 Bare Costs Labor	Equipment	Total	Total Incl O&P
0010	**CONCRETE SAW BLADES**								
3000	Blades for saw cutting, included in cutting line items								
3020	Diamond, 12" diameter				Ea.	219			219	241
3040	18" diameter					345			345	380
3080	24" diameter					490			490	535
3120	30" diameter					865			865	950
3160	36" diameter					1,225			1,225	1,350
3200	42" diameter					1,850			1,850	2,050

03 81 16 – Track Mounted Concrete Wall Sawing

03 81 16.50 Concrete Wall Cutting

		Daily Output	Labor-Hours	Unit	Material	2021 Bare Costs Labor	Equipment	Total	Total Incl O&P
0010	**CONCRETE WALL CUTTING**								
0750	Includes blade cost, layout and set-up time								
0800	Concrete walls, hydraulic saw, plain, per inch of depth	B-89B	250	.064	L.F.	.03	3.34	6.25	9.62	11.85
0820	Rod reinforcing, per inch of depth	"	150	.107	"	.05	5.55	10.40	16	19.80

03 82 Concrete Boring

03 82 13 – Concrete Core Drilling

03 82 13.10 Core Drilling

		Daily Output	Labor-Hours	Unit	Material	2021 Bare Costs Labor	Equipment	Total	Total Incl O&P
0010	**CORE DRILLING**								
0015	Includes bit cost, layout and set-up time								
0020	Reinforced concrete slab, up to 6" thick								
0100	1" diameter core	B-89A	17	.941	Ea.	.20	48	7.15	55.35	79.50
0150	For each additional inch of slab thickness in same hole, add		1440	.011		.03	.56	.08	.67	.98
0200	2" diameter core		16.50	.970		.27	49	7.35	56.62	82.50
0250	For each additional inch of slab thickness in same hole, add		1080	.015		.04	.75	.11	.90	1.30
0300	3" diameter core		16	1		.41	51	7.60	59.01	85
0350	For each additional inch of slab thickness in same hole, add		720	.022		.07	1.13	.17	1.37	1.96
0500	4" diameter core		15	1.067		.50	54	8.10	62.60	90.50
0550	For each additional inch of slab thickness in same hole, add		480	.033		.08	1.69	.25	2.02	2.91
0700	6" diameter core		14	1.143		.80	58	8.70	67.50	97.50
0750	For each additional inch of slab thickness in same hole, add		360	.044		.13	2.26	.34	2.73	3.90
0900	8" diameter core		13	1.231		1.16	62.50	9.35	73.01	105
0950	For each additional inch of slab thickness in same hole, add		288	.056		.19	2.82	.42	3.43	4.90
1100	10" diameter core		12	1.333		1.43	67.50	10.15	79.08	114
1150	For each additional inch of slab thickness in same hole, add		240	.067		.24	3.38	.51	4.13	5.85
1300	12" diameter core		11	1.455		1.63	74	11.05	86.68	125
1350	For each additional inch of slab thickness in same hole, add		206	.078		.27	3.94	.59	4.80	6.85
1500	14" diameter core		10	1.600		2.02	81	12.15	95.17	138
1550	For each additional inch of slab thickness in same hole, add		180	.089		.34	4.51	.68	5.53	7.85
1700	18" diameter core		9	1.778		2.80	90	13.50	106.30	153
1750	For each additional inch of slab thickness in same hole, add		144	.111		.47	5.65	.84	6.96	9.90

For customer support on your Site Work & Landscape Costs with RSMeans Data, call 800.448.8182.

03 82 Concrete Boring

03 82 13 – Concrete Core Drilling

03 82 13.10 Core Drilling

		Crew	Daily Output	Labor-Hours	Unit	Material	2021 Bare Costs Labor	Equipment	Total	Total Incl O&P
1754	24" diameter core	B-89A	8	2	Ea.	4.07	102	15.20	121.27	173
1756	For each additional inch of slab thickness in same hole, add	▼	120	.133		.68	6.75	1.01	8.44	12
1760	For horizontal holes, add to above				▼		20%	20%		
1770	Prestressed hollow core plank, 8" thick									
1780	1" diameter core	B-89A	17.50	.914	Ea.	.26	46.50	6.95	53.71	77.50
1790	For each additional inch of plank thickness in same hole, add		3840	.004		.03	.21	.03	.27	.39
1794	2" diameter core		17.25	.928		.36	47	7.05	54.41	78.50
1796	For each additional inch of plank thickness in same hole, add		2880	.006		.04	.28	.04	.36	.52
1800	3" diameter core		17	.941		.55	48	7.15	55.70	80
1810	For each additional inch of plank thickness in same hole, add		1920	.008		.07	.42	.06	.55	.78
1820	4" diameter core		16.50	.970		.67	49	7.35	57.02	83
1830	For each additional inch of plank thickness in same hole, add		1280	.013		.08	.63	.10	.81	1.14
1840	6" diameter core		15.50	1.032		1.06	52.50	7.85	61.41	88.50
1850	For each additional inch of plank thickness in same hole, add	▼	960	.017	▼	.13	.85	.13	1.11	1.56
1860	8" diameter core	B-89A	15	1.067	Ea.	1.55	54	8.10	63.65	91.50
1870	For each additional inch of plank thickness in same hole, add		768	.021		.19	1.06	.16	1.41	1.97
1880	10" diameter core		14	1.143		1.90	58	8.70	68.60	98.50
1890	For each additional inch of plank thickness in same hole, add		640	.025		.24	1.27	.19	1.70	2.37
1900	12" diameter core		13.50	1.185		2.18	60	9	71.18	102
1910	For each additional inch of plank thickness in same hole, add	▼	548	.029	▼	.27	1.48	.22	1.97	2.76
3000	Bits for core drilling, included in drilling line items									
3010	Diamond, premium, 1" diameter				Ea.	78			78	86
3020	2" diameter					107			107	118
3030	3" diameter					165			165	182
3040	4" diameter					202			202	222
3060	6" diameter					320			320	350
3080	8" diameter					465			465	510
3110	10" diameter					570			570	630
3120	12" diameter					655			655	720
3140	14" diameter					805			805	890
3180	18" diameter					1,125			1,125	1,225
3240	24" diameter				▼	1,625			1,625	1,800

03 82 16 – Concrete Drilling

03 82 16.10 Concrete Impact Drilling

		Crew	Daily Output	Labor-Hours	Unit	Material	2021 Bare Costs Labor	Equipment	Total	Total Incl O&P
0010	**CONCRETE IMPACT DRILLING**									
0020	Includes bit cost, layout and set-up time, no anchors									
0050	Up to 4" deep in concrete/brick floors/walls									
0100	Holes, 1/4" diameter	1 Carp	75	.107	Ea.	.05	5.85		5.90	8.75
0150	For each additional inch of depth in same hole, add		430	.019		.01	1.02		1.03	1.53
0200	3/8" diameter		63	.127		.05	6.95		7	10.40
0250	For each additional inch of depth in same hole, add		340	.024		.01	1.29		1.30	1.93
0300	1/2" diameter		50	.160		.05	8.75		8.80	13.10
0350	For each additional inch of depth in same hole, add		250	.032		.01	1.75		1.76	2.62
0400	5/8" diameter		48	.167		.08	9.10		9.18	13.70
0450	For each additional inch of depth in same hole, add		240	.033		.02	1.82		1.84	2.74
0500	3/4" diameter		45	.178		.11	9.70		9.81	14.60
0550	For each additional inch of depth in same hole, add		220	.036		.03	1.99		2.02	3
0600	7/8" diameter		43	.186		.13	10.20		10.33	15.35
0650	For each additional inch of depth in same hole, add		210	.038		.03	2.08		2.11	3.15
0700	1" diameter		40	.200		.21	10.95		11.16	16.60
0750	For each additional inch of depth in same hole, add		190	.042		.05	2.30		2.35	3.50
0800	1-1/4" diameter	▼	38	.211		.35	11.50		11.85	17.60

For customer support on your Site Work & Landscape Costs with RSMeans Data, call 800.448.8182.

91

03 82 Concrete Boring

03 82 16 – Concrete Drilling

03 82 16.10 Concrete Impact Drilling	Crew	Daily Output	Labor-Hours	Unit	Material	2021 Bare Costs Labor	2021 Bare Costs Equipment	Total	Total Incl O&P
0850 For each additional inch of depth in same hole, add	1 Carp	180	.044	Ea.	.09	2.43		2.52	3.73
0900 1-1/2" diameter	↓	35	.229	↓	.37	12.50		12.87	19.05
0950 For each additional inch of depth in same hole, add	1 Carp	165	.048	Ea.	.09	2.65		2.74	4.06
1000 For ceiling installations, add						40%			

Estimating Tips
04 05 00 Common Work Results for Masonry

- The terms mortar and grout are often used interchangeably—and incorrectly. Mortar is used to bed masonry units, seal the entry of air and moisture, provide architectural appearance, and allow for size variations in the units. Grout is used primarily in reinforced masonry construction and to bond the masonry to the reinforcing steel. Common mortar types are M (2500 psi), S (1800 psi), N (750 psi), and O (350 psi), and they conform to ASTM C270. Grout is either fine or coarse and conforms to ASTM C476, and in-place strengths generally exceed 2500 psi. Mortar and grout are different components of masonry construction and are placed by entirely different methods. An estimator should be aware of their unique uses and costs.

- Mortar is included in all assembled masonry line items. The mortar cost, part of the assembled masonry material cost, includes all ingredients, all labor, and all equipment required. Please see reference number RO040513-10.

- Waste, specifically the loss/droppings of mortar and the breakage of brick and block, is included in all unit cost lines that include mortar and masonry units in this division. A factor of 25% is added for mortar and 3% for brick and concrete masonry units.

- Scaffolding or staging is not included in any of the Division 4 costs. Refer to Subdivision 01 54 23 for scaffolding and staging costs.

04 20 00 Unit Masonry

- The most common types of unit masonry are brick and concrete masonry. The major classifications of brick are building brick (ASTM C62), facing brick (ASTM C216), glazed brick, fire brick, and pavers. Many varieties of texture and appearance can exist within these classifications, and the estimator would be wise to check local custom and availability within the project area. For repair and remodeling jobs, matching the existing brick may be the most important criteria.

- Brick and concrete block are priced by the piece and then converted into a price per square foot of wall. Openings less than two square feet are generally ignored by the estimator because any savings in units used are offset by the cutting and trimming required.

- It is often difficult and expensive to find and purchase small lots of historic brick. Costs can vary widely. Many design issues affect costs, selection of mortar mix, and repairs or replacement of masonry materials. Cleaning techniques must be reflected in the estimate.

- All masonry walls, whether interior or exterior, require bracing. The cost of bracing walls during construction should be included by the estimator, and this bracing must remain in place until permanent bracing is complete. Permanent bracing of masonry walls is accomplished by masonry itself, in the form of pilasters or abutting wall corners, or by anchoring the walls to the structural frame. Accessories in the form of anchors, anchor slots, and ties are used, but their supply and installation can be by different trades. For instance, anchor slots on spandrel beams and columns are supplied and welded in place by the steel fabricator, but the ties from the slots into the masonry are installed by the bricklayer. Regardless of the installation method, the estimator must be certain that these accessories are accounted for in pricing.

Reference Numbers
Reference numbers are shown at the beginning of some major classifications. These numbers refer to related items in the Reference Section. The reference information may be an estimating procedure, an alternate pricing method, or technical information.

Note: Not all subdivisions listed here necessarily appear. ■

Same Data. Simplified.
Enjoy the convenience and efficiency of accessing your costs anywhere:
- **Skip the multiplier** by setting your location
- **Quickly search,** edit, favorite and share costs
- **Stay on top of price changes** with automatic updates

Discover more at rsmeans.com/online

---

04 01 20 – Maintenance of Unit Masonry

04 01 20.10 Patching Masonry

		Crew	Daily Output	Labor-Hours	Unit	Material	2021 Bare Costs Labor	Equipment	Total	Total Incl O&P
0010	**PATCHING MASONRY**									
0500	CMU patching, includes chipping, cleaning and epoxy									
0520	1/4" deep	1 Cefi	65	.123	S.F.	4.23	6.40		10.63	14
0540	3/8" deep		50	.160		6.35	8.30		14.65	19.15
0580	1/2" deep		40	.200		8.45	10.35		18.80	24.50

04 01 20.20 Pointing Masonry

		Crew	Daily Output	Labor-Hours	Unit	Material	2021 Bare Costs Labor	Equipment	Total	Total Incl O&P
0010	**POINTING MASONRY**									
0300	Cut and repoint brick, hard mortar, running bond	1 Bric	80	.100	S.F.	.57	5.35		5.92	8.75
0320	Common bond		77	.104		.57	5.60		6.17	9.05
0360	Flemish bond		70	.114		.60	6.15		6.75	9.90
0400	English bond		65	.123		.60	6.60		7.20	10.60
0600	Soft old mortar, running bond		100	.080		.57	4.30		4.87	7.10
0620	Common bond		96	.083		.57	4.48		5.05	7.40
0640	Flemish bond		90	.089		.60	4.77		5.37	7.85
0680	English bond		82	.098		.60	5.25		5.85	8.55
0700	Stonework, hard mortar		140	.057	L.F.	.76	3.07		3.83	5.45
0720	Soft old mortar		160	.050	"	.76	2.69		3.45	4.89
1000	Repoint, mask and grout method, running bond		95	.084	S.F.	.76	4.52		5.28	7.65
1020	Common bond		90	.089		.76	4.77		5.53	8.05
1040	Flemish bond		86	.093		.80	5		5.80	8.45
1060	English bond		77	.104		.80	5.60		6.40	9.30
2000	Scrub coat, sand grout on walls, thin mix, brushed		120	.067		3.89	3.58		7.47	9.65
2020	Troweled		98	.082		5.40	4.38		9.78	12.50

04 05 Common Work Results for Masonry

04 05 05 – Selective Demolition for Masonry

04 05 05.10 Selective Demolition

		Crew	Daily Output	Labor-Hours	Unit	Material	2021 Bare Costs Labor	Equipment	Total	Total Incl O&P
0010	**SELECTIVE DEMOLITION** R024119-10									
1000	Chimney, 16" x 16", soft old mortar	1 Clab	55	.145	C.F.		6.45		6.45	9.65
1020	Hard mortar		40	.200			8.90		8.90	13.25
1030	16" x 20", soft old mortar		55	.145			6.45		6.45	9.65
1040	Hard mortar		40	.200			8.90		8.90	13.25
1050	16" x 24", soft old mortar		55	.145			6.45		6.45	9.65
1060	Hard mortar		40	.200			8.90		8.90	13.25
1080	20" x 20", soft old mortar		55	.145			6.45		6.45	9.65
1100	Hard mortar		40	.200			8.90		8.90	13.25
1110	20" x 24", soft old mortar		55	.145			6.45		6.45	9.65
1120	Hard mortar		40	.200			8.90		8.90	13.25
1140	20" x 32", soft old mortar		55	.145			6.45		6.45	9.65
1160	Hard mortar	1 Clab	40	.200	C.F.		8.90		8.90	13.25
1200	48" x 48", soft old mortar		55	.145			6.45		6.45	9.65
1220	Hard mortar		40	.200			8.90		8.90	13.25
1250	Metal, high temp steel jacket, 24" diameter	E-2	130	.431	V.L.F.		25.50	13.20	38.70	54
1260	60" diameter	"	60	.933			55.50	28.50	84	117
1280	Flue lining, up to 12" x 12"	1 Clab	200	.040			1.78		1.78	2.65
1282	Up to 24" x 24"	"	150	.053			2.37		2.37	3.53
4000	Fireplace, brick, 30" x 24" opening									
4020	Soft old mortar	1 Clab	2	4	Ea.		178		178	265
4040	Hard mortar		1.25	6.400			284		284	425
4100	Stone, soft old mortar		1.50	5.333			237		237	355
4120	Hard mortar		1	8			355		355	530

04 05 13 – Masonry Mortaring

04 05 13.10 Cement

		Crew	Daily Output	Labor-Hours	Unit	Material	2021 Bare Costs Labor	Equipment	Total	Total Incl O&P
0010	**CEMENT**									
0100	Masonry, 70 lb. bag, T.L. lots				Bag	11.75			11.75	12.90
0150	L.T.L. lots					12.45			12.45	13.70
0200	White, 70 lb. bag, T.L. lots					20.50			20.50	22.50
0250	L.T.L. lots				↓	17.10			17.10	18.80

04 05 13.20 Lime

		Crew	Daily Output	Labor-Hours	Unit	Material	2021 Bare Costs Labor	Equipment	Total	Total Incl O&P
0010	**LIME**									
0020	Masons, hydrated, 50 lb. bag, T.L. lots				Bag	11.05			11.05	12.15
0050	L.T.L. lots					12.15			12.15	13.40
0200	Finish, double hydrated, 50 lb. bag, T.L. lots					12.05			12.05	13.25
0250	L.T.L. lots				↓	13.25			13.25	14.55

04 05 13.23 Surface Bonding Masonry Mortaring

		Crew	Daily Output	Labor-Hours	Unit	Material	2021 Bare Costs Labor	Equipment	Total	Total Incl O&P
0010	**SURFACE BONDING MASONRY MORTARING**									
0020	Gray or white colors, not incl. block work	1 Bric	540	.015	S.F.	.20	.80		1	1.42

04 05 13.30 Mortar

			Crew	Daily Output	Labor-Hours	Unit	Material	2021 Bare Costs Labor	Equipment	Total	Total Incl O&P
0010	**MORTAR**	R040513-10									
0020	With masonry cement										
0100	Type M, 1:1:6 mix		1 Brhe	143	.056	C.F.	6.85	2.44		9.29	11.25
0200	Type N, 1:3 mix			143	.056		5.45	2.44		7.89	9.70
0300	Type O, 1:3 mix			143	.056		4.82	2.44		7.26	9
0400	Type PM, 1:1:6 mix, 2500 psi			143	.056		6.80	2.44		9.24	11.15
0500	Type S, 1/2:1:4 mix	↓		143	.056	↓	6.60	2.44		9.04	10.95
2000	With Portland cement and lime										
2100	Type M, 1:1/4:3 mix		1 Brhe	143	.056	C.F.	9.45	2.44		11.89	14.10
2200	Type N, 1:1:6 mix, 750 psi			143	.056		7.30	2.44		9.74	11.75
2300	Type O, 1:2:9 mix (Pointing Mortar)			143	.056		8.85	2.44		11.29	13.40
2400	Type PL, 1:1/2:4 mix, 2500 psi			143	.056		6.10	2.44		8.54	10.40
2600	Type S, 1:1/2:4 mix, 1800 psi	↓		143	.056		9.50	2.44		11.94	14.15
2650	Pre-mixed, type S or N					↓	6.40			6.40	7.05
2700	Mortar for glass block		1 Brhe	143	.056	C.F.	13	2.44		15.44	18
2900	Mortar for fire brick, dry mix, 10 lb. pail					Ea.	25			25	27.50

04 05 13.91 Masonry Restoration Mortaring

		Crew	Daily Output	Labor-Hours	Unit	Material	2021 Bare Costs Labor	Equipment	Total	Total Incl O&P
0010	**MASONRY RESTORATION MORTARING**									
0020	Masonry restoration mix				Lb.	.73			.73	.80
0050	White				"	1.20			1.20	1.32

04 05 13.93 Mortar Pigments

			Crew	Daily Output	Labor-Hours	Unit	Material	2021 Bare Costs Labor	Equipment	Total	Total Incl O&P
0010	**MORTAR PIGMENTS**, 50 lb. bags (2 bags per M bricks)	R040513-10									
0020	Color admixture, range 2 to 10 lb. per bag of cement, light colors					Lb.	6.65			6.65	7.30
0050	Medium colors						7.65			7.65	8.40
0100	Dark colors					↓	15.85			15.85	17.40

04 05 13.95 Sand

		Crew	Daily Output	Labor-Hours	Unit	Material	2021 Bare Costs Labor	Equipment	Total	Total Incl O&P
0010	**SAND**, screened and washed at pit									
0020	For mortar, per ton				Ton	21.50			21.50	23.50
0050	With 10 mile haul					43			43	47
0100	With 30 mile haul				↓	71.50			71.50	78.50
0200	Screened and washed, at the pit				C.Y.	30			30	33
0250	With 10 mile haul					59.50			59.50	65.50
0300	With 30 mile haul				↓	99			99	109

04 05 Common Work Results for Masonry

04 05 13 – Masonry Mortaring

04 05 13.98 Mortar Admixtures	Crew	Daily Output	Labor-Hours	Unit	Material	2021 Bare Costs Labor	Equipment	Total	Total Incl O&P
0010 MORTAR ADMIXTURES									
0020 Waterproofing admixture, per quart (1 qt. to 2 bags of masonry cement)				Qt.	3.66			3.66	4.03

04 05 16 – Masonry Grouting

04 05 16.30 Grouting

	Crew	Daily Output	Labor-Hours	Unit	Material	2021 Bare Costs Labor	Equipment	Total	Total Incl O&P
0010 GROUTING									
0011 Bond beams & lintels, 8" deep, 6" thick, 0.15 C.F./L.F.	D-4	1480	.022	L.F.	.83	1.06	.13	2.02	2.65
0020 8" thick, 0.2 C.F./L.F.		1400	.023		1.33	1.12	.14	2.59	3.30
0050 10" thick, 0.25 C.F./L.F.		1200	.027		1.39	1.31	.16	2.86	3.66
0060 12" thick, 0.3 C.F./L.F.		1040	.031		1.66	1.51	.18	3.35	4.30
0200 Concrete block cores, solid, 4" thk., by hand, 0.067 C.F./S.F. of wall	D-8	1100	.036	S.F.	.37	1.81		2.18	3.13
0210 6" thick, pumped, 0.175 C.F./S.F.	D-4	720	.044		.97	2.18	.26	3.41	4.64
0250 8" thick, pumped, 0.258 C.F./S.F.		680	.047		1.43	2.31	.28	4.02	5.35
0300 10" thick, pumped, 0.340 C.F./S.F.		660	.048		1.88	2.38	.29	4.55	5.95
0350 12" thick, pumped, 0.422 C.F./S.F.		640	.050		2.34	2.46	.30	5.10	6.60
0500 Cavity walls, 2" space, pumped, 0.167 C.F./S.F. of wall		1700	.019		.93	.93	.11	1.97	2.53
0550 3" space, 0.250 C.F./S.F.		1200	.027		1.39	1.31	.16	2.86	3.66
0600 4" space, 0.333 C.F./S.F.		1150	.028		1.85	1.37	.17	3.39	4.26
0700 6" space, 0.500 C.F./S.F.		800	.040		2.77	1.97	.24	4.98	6.25
0800 Door frames, 3' x 7' opening, 2.5 C.F. per opening		60	.533	Opng.	13.85	26	3.17	43.02	58
0850 6' x 7' opening, 3.5 C.F. per opening		45	.711	"	19.40	35	4.23	58.63	78.50
2000 Grout, C476, for bond beams, lintels and CMU cores		350	.091	C.F.	5.55	4.49	.54	10.58	13.45

04 05 19 – Masonry Anchorage and Reinforcing

04 05 19.05 Anchor Bolts

	Crew	Daily Output	Labor-Hours	Unit	Material	2021 Bare Costs Labor	Equipment	Total	Total Incl O&P
0010 ANCHOR BOLTS									
0015 Installed in fresh grout in CMU bond beams or filled cores, no templates									
0020 Hooked, with nut and washer, 1/2" diameter, 8" long	1 Bric	132	.061	Ea.	1.55	3.25		4.80	6.60
0030 12" long		131	.061		1.72	3.28		5	6.85
0040 5/8" diameter, 8" long		129	.062		3.40	3.33		6.73	8.75
0050 12" long		127	.063		4.18	3.38		7.56	9.70
0060 3/4" diameter, 8" long		127	.063		4.18	3.38		7.56	9.70
0070 12" long		125	.064		5.25	3.44		8.69	10.95

04 05 19.16 Masonry Anchors

	Crew	Daily Output	Labor-Hours	Unit	Material	2021 Bare Costs Labor	Equipment	Total	Total Incl O&P
0010 MASONRY ANCHORS									
0020 For brick veneer, galv., corrugated, 7/8" x 7", 22 ga.	1 Bric	10.50	.762	C	15.95	41		56.95	79
0100 24 ga.		10.50	.762		10.40	41		51.40	73
0150 16 ga.		10.50	.762		33	41		74	97.50
0200 Buck anchors, galv., corrugated, 16 ga., 2" bend, 8" x 2"		10.50	.762		58.50	41		99.50	126
0250 8" x 3"		10.50	.762		63.50	41		104.50	131
0300 Adjustable, rectangular, 4-1/8" wide									
0350 Anchor and tie, 3/16" wire, mill galv.									
0400 2-3/4" eye, 3-1/4" tie	1 Bric	1.05	7.619	M	480	410		890	1,150
0500 4-3/4" tie		1.05	7.619		520	410		930	1,200
0520 5-1/2" tie		1.05	7.619		565	410		975	1,225
0550 4-3/4" eye, 3-1/4" tie		1.05	7.619		525	410		935	1,200
0570 4-3/4" tie		1.05	7.619		580	410		990	1,250
0580 5-1/2" tie		1.05	7.619		680	410		1,090	1,375
0660 Cavity wall, Z-type, galvanized, 6" long, 1/8" diam.		10.50	.762	C	24.50	41		65.50	88.50
0670 3/16" diameter		10.50	.762		30.50	41		71.50	95
0680 1/4" diameter		10.50	.762		40	41		81	106
0850 8" long, 3/16" diameter		10.50	.762		28	41		69	92
0855 1/4" diameter		10.50	.762		50.50	41		91.50	117

04 05 Common Work Results for Masonry

04 05 19 – Masonry Anchorage and Reinforcing

04 05 19.16 Masonry Anchors		Crew	Daily Output	Labor-Hours	Unit	Material	2021 Bare Costs Labor	Equipment	Total	Total Incl O&P
1000	Rectangular type, galvanized, 1/4" diameter, 2" x 6"	1 Bric	10.50	.762	C	80	41		121	150
1050	4" x 6"		10.50	.762		91	41		132	162
1100	3/16" diameter, 2" x 6"		10.50	.762		54	41		95	121
1150	4" x 6"		10.50	.762		55.50	41		96.50	123
1200	Mesh wall tie, 1/2" mesh, hot dip galvanized									
1400	16 ga., 12" long, 3" wide	1 Bric	9	.889	C	91.50	47.50		139	173
1420	6" wide		9	.889		134	47.50		181.50	220
1440	12" wide		8.50	.941		212	50.50		262.50	310
1500	Rigid partition anchors, plain, 8" long, 1" x 1/8"		10.50	.762		191	41		232	273
1550	1" x 1/4"		10.50	.762		325	41		366	420
1580	1-1/2" x 1/8"		10.50	.762		227	41		268	310
1600	1-1/2" x 1/4"		10.50	.762		360	41		401	455
1650	2" x 1/8"	1 Bric	10.50	.762	C	335	41		376	425
1700	2" x 1/4"	"	10.50	.762	"	395	41		436	495
2000	Column flange ties, wire, galvanized									
2300	3/16" diameter, up to 3" wide	1 Bric	10.50	.762	C	80.50	41		121.50	150
2350	To 5" wide		10.50	.762		87.50	41		128.50	158
2400	To 7" wide		10.50	.762		94.50	41		135.50	166
2600	To 9" wide		10.50	.762		100	41		141	172
2650	1/4" diameter, up to 3" wide		10.50	.762		112	41		153	185
2700	To 5" wide		10.50	.762		133	41		174	208
2800	To 7" wide		10.50	.762		149	41		190	226
2850	To 9" wide		10.50	.762		162	41		203	240
2900	For hot dip galvanized, add					35%				
4000	Channel slots, 1-3/8" x 1/2" x 8"									
4100	12 ga., plain	1 Bric	10.50	.762	C	254	41		295	340
4150	16 ga., galvanized	"	10.50	.762	"	143	41		184	219
4200	Channel slot anchors									
4300	16 ga., galvanized, 1-1/4" x 3-1/2"				C	64			64	70.50
4350	1-1/4" x 5-1/2"					74.50			74.50	82
4400	1-1/4" x 7-1/2"					68.50			68.50	75.50
4500	1/8" plain, 1-1/4" x 3-1/2"					143			143	158
4550	1-1/4" x 5-1/2"					153			153	168
4600	1-1/4" x 7-1/2"					167			167	184
4700	For corrugation, add					77			77	85
4750	For hot dip galvanized, add					35%				
5000	Dowels									
5100	Plain, 1/4" diameter, 3" long				C	49			49	54
5150	4" long					54.50			54.50	59.50
5200	6" long					66.50			66.50	73
5300	3/8" diameter, 3" long					65			65	71.50
5350	4" long					80.50			80.50	88.50
5400	6" long					93			93	102
5500	1/2" diameter, 3" long					94.50			94.50	104
5550	4" long					112			112	123
5600	6" long					146			146	161
5700	5/8" diameter, 3" long					130			130	143
5750	4" long					159			159	175
5800	6" long					219			219	241
6000	3/4" diameter, 3" long					161			161	177
6100	4" long					205			205	225
6150	6" long					293			293	320
6300	For hot dip galvanized, add					35%				

For customer support on your Site Work & Landscape Costs with RSMeans Data, call 800.448.8182.

97

04 05 19.26 Masonry Reinforcing Bars

	04 05 19.26 Masonry Reinforcing Bars	Crew	Daily Output	Labor-Hours	Unit	Material	2021 Bare Costs Labor	Equipment	Total	Total Incl O&P
0010	**MASONRY REINFORCING BARS** R040519-50									
0015	Steel bars A615, placed horiz., #3 & #4 bars	1 Bric	450	.018	Lb.	.62	.95		1.57	2.12
0020	#5 & #6 bars		800	.010		.62	.54		1.16	1.49
0050	Placed vertical, #3 & #4 bars		350	.023		.62	1.23		1.85	2.53
0060	#5 & #6 bars		650	.012		.62	.66		1.28	1.68
0200	Joint reinforcing, regular truss, to 6" wide, mill std galvanized		30	.267	C.L.F.	23	14.30		37.30	47
0250	12" wide		20	.400		29.50	21.50		51	65
0400	Cavity truss with drip section, to 6" wide		30	.267		23.50	14.30		37.80	47
0450	12" wide		20	.400		26.50	21.50		48	62
0500	Joint reinforcing, ladder type, mill std galvanized									
0600	9 ga. sides, 9 ga. ties, 4" wall	1 Bric	30	.267	C.L.F.	22	14.30		36.30	45.50
0650	6" wall		30	.267		22	14.30		36.30	45.50
0700	8" wall		25	.320		26.50	17.20		43.70	55
0750	10" wall		20	.400		26	21.50		47.50	61
0800	12" wall		20	.400		27.50	21.50		49	62.50
1000	Truss type									
1100	9 ga. sides, 9 ga. ties, 4" wall	1 Bric	30	.267	C.L.F.	24.50	14.30		38.80	48.50
1150	6" wall		30	.267		26	14.30		40.30	50
1200	8" wall		25	.320		27.50	17.20		44.70	56.50
1250	10" wall		20	.400		23.50	21.50		45	58.50
1300	12" wall		20	.400		25	21.50		46.50	60
1500	3/16" sides, 9 ga. ties, 4" wall		30	.267		28.50	14.30		42.80	53
1550	6" wall		30	.267		36	14.30		50.30	61.50
1600	8" wall		25	.320		32.50	17.20		49.70	62
1650	10" wall		20	.400		39	21.50		60.50	75.50
1700	12" wall		20	.400		38	21.50		59.50	74.50
2000	3/16" sides, 3/16" ties, 4" wall		30	.267		40.50	14.30		54.80	66
2050	6" wall		30	.267		41.50	14.30		55.80	67.50
2100	8" wall		25	.320		40.50	17.20		57.70	70.50
2150	10" wall		20	.400		45	21.50		66.50	82
2200	12" wall		20	.400		46.50	21.50		68	84
2500	Cavity truss type, galvanized									
2600	9 ga. sides, 9 ga. ties, 4" wall	1 Bric	25	.320	C.L.F.	35.50	17.20		52.70	65
2650	6" wall		25	.320		40.50	17.20		57.70	70.50
2700	8" wall		20	.400		48	21.50		69.50	85
2750	10" wall		15	.533		43	28.50		71.50	90.50
2800	12" wall		15	.533		52	28.50		80.50	100
3000	3/16" sides, 9 ga. ties, 4" wall		25	.320		45	17.20		62.20	75.50
3050	6" wall		25	.320		33	17.20		50.20	62
3100	8" wall		20	.400		60.50	21.50		82	99
3150	10" wall	1 Bric	15	.533	C.L.F.	43	28.50		71.50	90
3200	12" wall	"	15	.533	"	43	28.50		71.50	90
3500	For hot dip galvanizing, add				Ton	485			485	535

04 05 23.13 Masonry Control and Expansion Joints

	04 05 23.13 Masonry Control and Expansion Joints	Crew	Daily Output	Labor-Hours	Unit	Material	2021 Bare Costs Labor	Equipment	Total	Total Incl O&P
0010	**MASONRY CONTROL AND EXPANSION JOINTS**									
0020	Rubber, for double wythe 8" minimum wall (Brick/CMU)	1 Bric	400	.020	L.F.	2.55	1.07		3.62	4.43
0025	"T" shaped		320	.025		1.37	1.34		2.71	3.53
0030	Cross-shaped for CMU units		280	.029		1.68	1.53		3.21	4.16
0050	PVC, for double wythe 8" minimum wall (Brick/CMU)		400	.020		1.64	1.07		2.71	3.42
0120	"T" shaped		320	.025		1.08	1.34		2.42	3.21
0160	Cross-shaped for CMU units		280	.029		1.28	1.53		2.81	3.72

For customer support on your Site Work & Landscape Costs with RSMeans Data, call 800.448.8182.

04 05 Common Work Results for Masonry

04 05 23 – Masonry Accessories

04 05 23.95 Wall Plugs		Crew	Daily Output	Labor-Hours	Unit	Material	2021 Bare Costs Labor	Equipment	Total	Total Incl O&P
0010	**WALL PLUGS** (for nailing to brickwork)									
0020	25 ga., galvanized, plain	1 Bric	10.50	.762	C	52	41		93	119
0050	Wood filled	"	10.50	.762	"	121	41		162	196

04 21 Clay Unit Masonry

04 21 13 – Brick Masonry

04 21 13.13 Brick Veneer Masonry

		Crew	Daily Output	Labor-Hours	Unit	Material	2021 Bare Costs Labor	Equipment	Total	Total Incl O&P
0010	**BRICK VENEER MASONRY**, T.L. lots, excl. scaff., grout & reinforcing R042110-20									
0015	Material costs incl. 3% brick and 25% mortar waste									
0020	Standard, select common, 4" x 2-2/3" x 8" (6.75/S.F.)	D-8	1.50	26.667	M	705	1,325		2,030	2,775
0050	Red, 4" x 2-2/3" x 8", running bond		1.50	26.667		750	1,325		2,075	2,825
0100	Full header every 6th course (7.88/S.F.) R042110-50		1.45	27.586		750	1,375		2,125	2,900
0150	English, full header every 2nd course (10.13/S.F.)		1.40	28.571		750	1,425		2,175	2,975
0200	Flemish, alternate header every course (9.00/S.F.)		1.40	28.571		750	1,425		2,175	2,975
0250	Flemish, alt. header every 6th course (7.13/S.F.)		1.45	27.586		750	1,375		2,125	2,900
0300	Full headers throughout (13.50/S.F.)		1.40	28.571		745	1,425		2,170	2,975
0350	Rowlock course (13.50/S.F.)		1.35	29.630		745	1,475		2,220	3,050
0400	Rowlock stretcher (4.50/S.F.)		1.40	28.571		760	1,425		2,185	2,975
0450	Soldier course (6.75/S.F.)		1.40	28.571		750	1,425		2,175	2,975
0500	Sailor course (4.50/S.F.)		1.30	30.769		760	1,525		2,285	3,125
0601	Buff or gray face, running bond (6.75/S.F.)		1.50	26.667		750	1,325		2,075	2,825
0700	Glazed face, 4" x 2-2/3" x 8", running bond		1.40	28.571		2,475	1,425		3,900	4,875
0750	Full header every 6th course (7.88/S.F.)		1.35	29.630		2,375	1,475		3,850	4,825
1000	Jumbo, 6" x 4" x 12" (3.00/S.F.)		1.30	30.769		1,975	1,525		3,500	4,475
1051	Norman, 4" x 2-2/3" x 12" (4.50/S.F.)		1.45	27.586		1,225	1,375		2,600	3,425
1100	Norwegian, 4" x 3-1/5" x 12" (3.75/S.F.)		1.40	28.571		2,075	1,425		3,500	4,450
1150	Economy, 4" x 4" x 8" (4.50/S.F.)		1.40	28.571		1,050	1,425		2,475	3,325
1201	Engineer, 4" x 3-1/5" x 8" (5.63/S.F.)		1.45	27.586		710	1,375		2,085	2,850
1251	Roman, 4" x 2" x 12" (6.00/S.F.)		1.50	26.667		1,375	1,325		2,700	3,525
1300	S.C.R., 6" x 2-2/3" x 12" (4.50/S.F.)		1.40	28.571		1,375	1,425		2,800	3,675
1350	Utility, 4" x 4" x 12" (3.00/S.F.)		1.08	37.037		1,900	1,850		3,750	4,850
1360	For less than truck load lots, add					15%				
1400	For battered walls, add						30%			
1450	For corbels, add						75%			
1500	For curved walls, add						30%			
1550	For pits and trenches, deduct						20%			
1999	Alternate method of figuring by square foot									
2000	Standard, sel. common, 4" x 2-2/3" x 8" (6.75/S.F.)	D-8	230	.174	S.F.	4.75	8.65		13.40	18.20
2020	Red, 4" x 2-2/3" x 8", running bond		220	.182		5.05	9.05		14.10	19.20
2050	Full header every 6th course (7.88/S.F.)		185	.216		5.90	10.75		16.65	22.50
2100	English, full header every 2nd course (10.13/S.F.)		140	.286		7.55	14.20		21.75	30
2150	Flemish, alternate header every course (9.00/S.F.)		150	.267		6.75	13.25		20	27.50
2200	Flemish, alt. header every 6th course (7.13/S.F.)		205	.195		5.35	9.70		15.05	20.50
2250	Full headers throughout (13.50/S.F.)		105	.381		10.05	18.95		29	39.50
2300	Rowlock course (13.50/S.F.)		100	.400		10.05	19.90		29.95	41
2350	Rowlock stretcher (4.50/S.F.)		310	.129		3.41	6.40		9.81	13.40
2400	Soldier course (6.75/S.F.)		200	.200		5.05	9.95		15	20.50
2450	Sailor course (4.50/S.F.)		290	.138		3.41	6.85		10.26	14.10
2600	Buff or gray face, running bond (6.75/S.F.)		220	.182		5.35	9.05		14.40	19.50
2700	Glazed face brick, running bond		210	.190		16	9.45		25.45	32

04 21 13.13 Brick Veneer Masonry

		Crew	Daily Output	Labor-Hours	Unit	Material	2021 Bare Costs Labor	Equipment	Total	Total Incl O&P
2750	Full header every 6th course (7.88/S.F.)	D-8	170	.235	S.F.	18.70	11.70		30.40	38
3000	Jumbo, 6" x 4" x 12" running bond (3.00/S.F.)		435	.092		5.40	4.57		9.97	12.85
3050	Norman, 4" x 2-2/3" x 12" running bond (4.5/S.F.)		320	.125		6.70	6.20		12.90	16.75
3100	Norwegian, 4" x 3-1/5" x 12" (3.75/S.F.)		375	.107		7.70	5.30		13	16.45
3150	Economy, 4" x 4" x 8" (4.50/S.F.)		310	.129		4.72	6.40		11.12	14.85
3200	Engineer, 4" x 3-1/5" x 8" (5.63/S.F.)		260	.154		3.99	7.65		11.64	15.90
3250	Roman, 4" x 2" x 12" (6.00/S.F.)		250	.160		8.10	7.95		16.05	21
3300	S.C.R., 6" x 2-2/3" x 12" (4.50/S.F.)		310	.129		6.15	6.40		12.55	16.40
3350	Utility, 4" x 4" x 12" (3.00/S.F.)		360	.111		5.55	5.50		11.05	14.40
3360	For less than truck load lots, add					.10%				
3400	For cavity wall construction, add						15%			
3450	For stacked bond, add						10%			
3500	For interior veneer construction, add						15%			
3550	For curved walls, add						30%			

04 21 13.14 Thin Brick Veneer

		Crew	Daily Output	Labor-Hours	Unit	Material	2021 Bare Costs Labor	Equipment	Total	Total Incl O&P
0010	**THIN BRICK VENEER**									
0015	Material costs incl. 3% brick and 25% mortar waste									
0020	On & incl. metal panel support sys, modular, 2-2/3" x 5/8" x 8", red	D-7	92	.174	S.F.	13.85	8.10		21.95	27
0100	Closure, 4" x 5/8" x 8"		110	.145		9.45	6.80		16.25	20.50
0110	Norman, 2-2/3" x 5/8" x 12"		110	.145		9.85	6.80		16.65	20.50
0120	Utility, 4" x 5/8" x 12"		125	.128		9.20	5.95		15.15	18.85
0130	Emperor, 4" x 3/4" x 16"		175	.091		10.40	4.27		14.67	17.70
0140	Super emperor, 8" x 3/4" x 16"		195	.082		11.40	3.83		15.23	18.15
0150	For L shaped corners with 4" return, add				L.F.	9.25			9.25	10.20
0200	On masonry/plaster back-up, modular, 2-2/3" x 5/8" x 8", red	D-7	137	.117	S.F.	8.80	5.45		14.25	17.65
0210	Closure, 4" x 5/8" x 8"		165	.097		4.40	4.52		8.92	11.45
0220	Norman, 2-2/3" x 5/8" x 12"		165	.097		4.77	4.52		9.29	11.85
0230	Utility, 4" x 5/8" x 12"		185	.086		4.12	4.03		8.15	10.45
0240	Emperor, 4" x 3/4" x 16"		260	.062		5.30	2.87		8.17	10.05
0250	Super emperor, 8" x 3/4" x 16"		285	.056		6.35	2.62		8.97	10.80
0260	For L shaped corners with 4" return, add				L.F.	10			10	11
0270	For embedment into pre-cast concrete panels, add				S.F.	14.40			14.40	15.85

04 21 13.18 Columns

		Crew	Daily Output	Labor-Hours	Unit	Material	2021 Bare Costs Labor	Equipment	Total	Total Incl O&P
0010	**COLUMNS**, solid, excludes scaffolding, grout and reinforcing									
0050	Brick, 8" x 8", 9 brick/V.L.F.	D-1	56	.286	V.L.F.	6.50	13.90		20.40	28
0100	12" x 8", 13.5 brick/V.L.F.		37	.432		9.75	21		30.75	42.50
0200	12" x 12", 20 brick/V.L.F.		25	.640		14.50	31		45.50	63
0300	16" x 12", 27 brick/V.L.F.		19	.842		19.55	41		60.55	83.50
0400	16" x 16", 36 brick/V.L.F.		14	1.143		26	55.50		81.50	113
0500	20" x 16", 45 brick/V.L.F.		11	1.455		32.50	71		103.50	143
0600	20" x 20", 56 brick/V.L.F.		9	1.778		40.50	86.50		127	175
0700	24" x 20", 68 brick/V.L.F.		7	2.286		49	111		160	222
0800	24" x 24", 81 brick/V.L.F.		6	2.667		58.50	130		188.50	261
1000	36" x 36", 182 brick/V.L.F.		3	5.333		132	260		392	535

04 21 13.35 Common Building Brick

		Crew	Daily Output	Labor-Hours	Unit	Material	2021 Bare Costs Labor	Equipment	Total	Total Incl O&P
0010	**COMMON BUILDING BRICK**, C62, T.L. lots, material only									
0020	Standard				M	610			610	670
0050	Select				"	580			580	640

04 21 13.45 Face Brick

		Crew	Daily Output	Labor-Hours	Unit	Material	2021 Bare Costs Labor	Equipment	Total	Total Incl O&P
0010	**FACE BRICK** Material Only, C216, T.L. lots									
0300	Standard modular, 4" x 2-2/3" x 8"				M	625			625	690
0450	Economy, 4" x 4" x 8"					910			910	1,000

04 21 Clay Unit Masonry

04 21 13 – Brick Masonry

04 21 13.45 Face Brick

		Crew	Daily Output	Labor-Hours	Unit	Material	2021 Bare Costs Labor	Equipment	Total	Total Incl O&P
0510	Economy, 4" x 4" x 12"				M	1,450			1,450	1,575
0550	Jumbo, 6" x 4" x 12"					1,625			1,625	1,800
0610	Jumbo, 8" x 4" x 12"					1,625			1,625	1,800
0650	Norwegian, 4" x 3-1/5" x 12"					1,875			1,875	2,075
0710	Norwegian, 6" x 3-1/5" x 12"					1,575			1,575	1,750
0850	Standard glazed, plain colors, 4" x 2-2/3" x 8"					2,200			2,200	2,425
1000	Deep trim shades, 4" x 2-2/3" x 8"				M	2,375			2,375	2,625
1080	Jumbo utility, 4" x 4" x 12"					1,675			1,675	1,850
1120	4" x 8" x 8"					1,975			1,975	2,150
1140	4" x 8" x 16"					6,875			6,875	7,550
1260	Engineer, 4" x 3-1/5" x 8"					580			580	640
1350	King, 4" x 2-3/4" x 10"					575			575	635
1770	Standard modular, double glazed, 4" x 2-2/3" x 8"					3,000			3,000	3,300
1850	Jumbo, colored glazed ceramic, 6" x 4" x 12"					2,800			2,800	3,075
2050	Jumbo utility, glazed, 4" x 4" x 12"					5,750			5,750	6,325
2100	4" x 8" x 8"					6,775			6,775	7,450
2150	4" x 16" x 8"					7,225			7,225	7,950
3050	Used brick					495			495	540
3150	Add for brick to match existing work, minimum					5%				
3200	Maximum					50%				

04 22 Concrete Unit Masonry

04 22 10 – Concrete Masonry Units

04 22 10.10 Concrete Block

		Crew	Daily Output	Labor-Hours	Unit	Material	2021 Bare Costs Labor	Equipment	Total	Total Incl O&P
0010	**CONCRETE BLOCK** Material Only	R042210-20								
0020	2" x 8" x 16" solid, normal-weight, 2,000 psi				Ea.	1.21			1.21	1.33
0050	3,500 psi					1.40			1.40	1.54
0100	5,000 psi					1.50			1.50	1.65
0150	Lightweight, std.					1.49			1.49	1.64
0300	3" x 8" x 16" solid, normal-weight, 2,000 psi					.96			.96	1.06
0350	3,500 psi					1.40			1.40	1.54
0400	5,000 psi					1.64			1.64	1.80
0450	Lightweight, std.					1.47			1.47	1.62
0600	4" x 8" x 16" hollow, normal-weight, 2,000 psi					1.51			1.51	1.66
0650	3,500 psi					1.43			1.43	1.57
0700	5,000 psi					1.82			1.82	2
0750	Lightweight, std.					1.43			1.43	1.57
1300	Solid, normal-weight, 2,000 psi					1.89			1.89	2.08
1350	3,500 psi					1.67			1.67	1.84
1400	5,000 psi					1.65			1.65	1.82
1450	Lightweight, std.					1.54			1.54	1.69
1600	6" x 8" x 16" hollow, normal-weight, 2,000 psi					1.93			1.93	2.12
1650	3,500 psi					2.08			2.08	2.29
1700	5,000 psi					2.24			2.24	2.46
1750	Lightweight, std.					2.27			2.27	2.50
2300	Solid, normal-weight, 2,000 psi					2.67			2.67	2.94
2350	3,500 psi					1.65			1.65	1.82
2400	5,000 psi					2.25			2.25	2.48
2450	Lightweight, std.				Ea.	2.60			2.60	2.86
2600	8" x 8" x 16" hollow, normal-weight, 2,000 psi					1.73			1.73	1.90
2650	3,500 psi					2.49			2.49	2.74

04 22 10 – Concrete Masonry Units

04 22 10.10 Concrete Block	Crew	Daily Output	Labor-Hours	Unit	Material	2021 Bare Costs Labor	Equipment	Total	Total Incl O&P
2700	5,000 psi				Ea.	2.64			2.64	2.90
2750	Lightweight, std.					2.91			2.91	3.20
3200	Solid, normal-weight, 2,000 psi					2.77			2.77	3.05
3250	3,500 psi					2.98			2.98	3.28
3300	5,000 psi					3.25			3.25	3.58
3350	Lightweight, std.					2.36			2.36	2.60
3400	10" x 8" x 16" hollow, normal-weight, 2,000 psi					1.73			1.73	1.90
3410	3,500 psi					2.49			2.49	2.74
3420	5,000 psi					2.64			2.64	2.90
3430	Lightweight, std.					2.91			2.91	3.20
3480	Solid, normal-weight, 2,000 psi					2.77			2.77	3.05
3490	3,500 psi					2.98			2.98	3.28
3500	5,000 psi					3.25			3.25	3.58
3510	Lightweight, std.					2.36			2.36	2.60
3600	12" x 8" x 16" hollow, normal-weight, 2,000 psi					3.25			3.25	3.58
3650	3,500 psi					3.06			3.06	3.37
3700	5,000 psi					3.66			3.66	4.03
3750	Lightweight, std.					2.70			2.70	2.97
4300	Solid, normal-weight, 2,000 psi					4.13			4.13	4.54
4350	3,500 psi					3.71			3.71	4.08
4400	5,000 psi					3.57			3.57	3.93
4500	Lightweight, std.					3.44			3.44	3.78

04 22 10.14 Concrete Block, Back-Up

		Crew	Daily Output	Labor-Hours	Unit	Material	Labor	Equipment	Total	Total Incl O&P
0010	**CONCRETE BLOCK, BACK-UP**, C90, 2000 psi									
0020	Normal weight, 8" x 16" units, tooled joint 1 side									
0050	Not-reinforced, 2000 psi, 2" thick	D-8	475	.084	S.F.	1.72	4.19		5.91	8.20
0200	4" thick		460	.087		2.20	4.32		6.52	8.90
0300	6" thick		440	.091		2.81	4.52		7.33	9.90
0350	8" thick		400	.100		2.74	4.97		7.71	10.50
0400	10" thick		330	.121		3.29	6		9.29	12.70
0450	12" thick	D-9	310	.155		4.74	7.55		12.29	16.55

04 22 10.16 Concrete Block, Bond Beam

		Crew	Daily Output	Labor-Hours	Unit	Material	Labor	Equipment	Total	Total Incl O&P
0010	**CONCRETE BLOCK, BOND BEAM**, C90, 2000 psi									
0020	Not including grout or reinforcing									
0125	Regular block, 6" thick	D-8	584	.068	L.F.	2.76	3.40		6.16	8.20
0130	8" high, 8" thick	"	565	.071		2.93	3.52		6.45	8.55
0150	12" thick	D-9	510	.094		4.35	4.58		8.93	11.70
0525	Lightweight, 6" thick	D-8	592	.068		3.11	3.36		6.47	8.45
0530	8" high, 8" thick	"	575	.070		3.66	3.46		7.12	9.20
0550	12" thick	D-9	520	.092	L.F.	4.91	4.50		9.41	12.15
2000	Including grout and 2 #5 bars									
2100	Regular block, 8" high, 8" thick	D-8	300	.133	L.F.	5.55	6.65		12.20	16.10
2150	12" thick	D-9	250	.192		7.70	9.35		17.05	22.50
2500	Lightweight, 8" high, 8" thick	D-8	305	.131		6.30	6.50		12.80	16.70
2550	12" thick	D-9	255	.188		8.25	9.15		17.40	23

04 22 10.19 Concrete Block, Insulation Inserts

		Crew	Daily Output	Labor-Hours	Unit	Material	Labor	Equipment	Total	Total Incl O&P
0010	**CONCRETE BLOCK, INSULATION INSERTS**									
0100	Styrofoam, plant installed, add to block prices									
0200	8" x 16" units, 6" thick				S.F.	1.28			1.28	1.41
0250	8" thick					1.65			1.65	1.82
0300	10" thick					1.34			1.34	1.47
0350	12" thick					1.71			1.71	1.88

04 22 Concrete Unit Masonry

04 22 10 – Concrete Masonry Units

04 22 10.19 Concrete Block, Insulation Inserts	Crew	Daily Output	Labor-Hours	Unit	Material	2021 Bare Costs Labor	Equipment	Total	Total Incl O&P
0500	8" x 8" units, 8" thick				S.F.	1.63			1.63	1.79
0550	12" thick				↓	1.62			1.62	1.78

04 22 10.23 Concrete Block, Decorative

		Crew	Daily Output	Labor-Hours	Unit	Material	Labor	Equipment	Total	Total Incl O&P
0010	**CONCRETE BLOCK, DECORATIVE**, C90, 2000 psi									
0020	Embossed, simulated brick face									
0100	8" x 16" units, 4" thick	D-8	400	.100	S.F.	3.17	4.97		8.14	11
0200	8" thick		340	.118		3.29	5.85		9.14	12.40
0250	12" thick	↓	300	.133	↓	4.82	6.65		11.47	15.30
0400	Embossed both sides									
0500	8" thick	D-8	300	.133	S.F.	5.55	6.65		12.20	16.10
0550	12" thick	"	275	.145	"	6.75	7.25		14	18.30
1000	Fluted high strength									
1100	8" x 16" x 4" thick, flutes 1 side,	D-8	345	.116	S.F.	4.17	5.75		9.92	13.30
1150	Flutes 2 sides		335	.119		4.72	5.95		10.67	14.15
1200	8" thick	↓	300	.133		5.30	6.65		11.95	15.80
1250	For special colors, add				↓	.66			.66	.72
1400	Deep grooved, smooth face									
1450	8" x 16" x 4" thick	D-8	345	.116	S.F.	2.64	5.75		8.39	11.60
1500	8" thick	"	300	.133	"	4.54	6.65		11.19	15
2000	Formblock, incl. inserts & reinforcing									
2100	8" x 16" x 8" thick	D-8	345	.116	S.F.	4.49	5.75		10.24	13.65
2150	12" thick	"	310	.129	"	5.85	6.40		12.25	16.05
2500	Ground face									
2600	8" x 16" x 4" thick	D-8	345	.116	S.F.	4.30	5.75		10.05	13.45
2650	6" thick		325	.123		4.92	6.10		11.02	14.60
2700	8" thick	↓	300	.133		5.70	6.65		12.35	16.25
2750	12" thick	D-9	265	.181	↓	7.30	8.80		16.10	21.50
2900	For special colors, add, minimum					15%				
2950	For special colors, add, maximum					45%				
4000	Slump block									
4100	4" face height x 16" x 4" thick	D-1	165	.097	S.F.	5.40	4.72		10.12	13
4150	6" thick		160	.100		6.65	4.87		11.52	14.65
4200	8" thick		155	.103		9.50	5.05		14.55	18
4250	10" thick		140	.114		12.15	5.55		17.70	22
4300	12" thick		130	.123		13.45	6		19.45	24
4400	6" face height x 16" x 6" thick		155	.103		6.20	5.05		11.25	14.35
4450	8" thick		150	.107		7.65	5.20		12.85	16.25
4500	10" thick		130	.123		14.55	6		20.55	25
4550	12" thick	↓	120	.133	↓	14.55	6.50		21.05	26
5000	Split rib profile units, 1" deep ribs, 8 ribs									
5100	8" x 16" x 4" thick	D-8	345	.116	S.F.	4.90	5.75		10.65	14.10
5150	6" thick		325	.123		5.65	6.10		11.75	15.40
5200	8" thick	↓	300	.133		6.85	6.65		13.50	17.55
5250	12" thick	D-9	275	.175		7.85	8.50		16.35	21.50
5400	For special deeper colors, 4" thick, add					1.42			1.42	1.57
5450	12" thick, add					1.38			1.38	1.52
5600	For white, 4" thick, add					1.42			1.42	1.57
5650	6" thick, add					1.38			1.38	1.52
5700	8" thick, add					1.48			1.48	1.63
5750	12" thick, add				↓	1.42			1.42	1.57
6000	Split face									
6100	8" x 16" x 4" thick	D-8	350	.114	S.F.	3.86	5.70		9.56	12.80

04 22 10.23 Concrete Block, Decorative

		Crew	Daily Output	Labor-Hours	Unit	Material	2021 Bare Costs Labor	Equipment	Total	Total Incl O&P
6150	6" thick	D-8	325	.123	S.F.	4.38	6.10		10.48	14
6200	8" thick	↓	300	.133		5.60	6.65		12.25	16.15
6250	12" thick	D-9	270	.178		6.25	8.65		14.90	19.95
6300	For scored, add					.39			.39	.43
6400	For special deeper colors, 4" thick, add					.66			.66	.72
6450	6" thick, add					.84			.84	.92
6500	8" thick, add					.71			.71	.78
6550	12" thick, add					.84			.84	.92
6650	For white, 4" thick, add					1.42			1.42	1.57
6700	6" thick, add					1.29			1.29	1.42
6750	8" thick, add					1.41			1.41	1.55
6800	12" thick, add				↓	1.38			1.38	1.52
7000	Scored ground face, 2 to 5 scores									
7100	8" x 16" x 4" thick	D-8	340	.118	S.F.	8.85	5.85		14.70	18.55
7150	6" thick		310	.129		10.15	6.40		16.55	21
7200	8" thick	↓	290	.138		11.50	6.85		18.35	23
7250	12" thick	D-9	265	.181	↓	17.10	8.80		25.90	32
8000	Hexagonal face profile units, 8" x 16" units									
8100	4" thick, hollow	D-8	340	.118	S.F.	4.18	5.85		10.03	13.40
8200	Solid		340	.118		5.30	5.85		11.15	14.60
8300	6" thick, hollow		310	.129		4.19	6.40		10.59	14.25
8350	8" thick, hollow	↓	290	.138		4.17	6.85		11.02	14.95
8500	For stacked bond, add				↓		26%			
8550	For high rise construction, add per story	D-8	67.80	.590	M.S.F.		29.50		29.50	44
8600	For scored block, add					10%				
8650	For honed or ground face, per face, add				Ea.	1.19			1.19	1.31
8700	For honed or ground end, per end, add				"	1.54			1.54	1.69
8750	For bullnose block, add					10%				
8800	For special color, add					13%				

04 22 10.24 Concrete Block, Exterior

		Crew	Daily Output	Labor-Hours	Unit	Material	2021 Bare Costs Labor	Equipment	Total	Total Incl O&P
0010	**CONCRETE BLOCK, EXTERIOR**, C90, 2000 psi R042210-20									
0020	Reinforced alt courses, tooled joints 2 sides									
0100	Normal weight, 8" x 16" x 6" thick	D-8	395	.101	S.F.	2.56	5.05		7.61	10.40
0200	8" thick		360	.111		4.25	5.50		9.75	12.95
0250	10" thick	↓	290	.138		4.58	6.85		11.43	15.40
0300	12" thick	D-9	250	.192		5.35	9.35		14.70	20
0500	Lightweight, 8" x 16" x 6" thick	D-8	450	.089		3.64	4.42		8.06	10.65
0600	8" thick		430	.093		3.55	4.62		8.17	10.85
0650	10" thick	↓	395	.101		4.60	5.05		9.65	12.65
0700	12" thick	D-9	350	.137	↓	5.05	6.70		11.75	15.60

04 22 10.26 Concrete Block Foundation Wall

		Crew	Daily Output	Labor-Hours	Unit	Material	2021 Bare Costs Labor	Equipment	Total	Total Incl O&P
0010	**CONCRETE BLOCK FOUNDATION WALL**, C90/C145									
0050	Normal-weight, cut joints, horiz joint reinf, no vert reinf.									
0200	Hollow, 8" x 16" x 6" thick	D-8	455	.088	S.F.	3.38	4.37		7.75	10.30
0250	8" thick		425	.094		3.33	4.68		8.01	10.70
0300	10" thick	↓	350	.114		3.87	5.70		9.57	12.80
0350	12" thick	D-9	300	.160		5.35	7.80		13.15	17.65
0500	Solid, 8" x 16" block, 6" thick	D-8	440	.091		4.22	4.52		8.74	11.45
0550	8" thick	"	415	.096		4.50	4.79		9.29	12.15
0600	12" thick	D-9	350	.137	↓	6.35	6.70		13.05	17
1000	Reinforced, #4 vert @ 48"									
1100	Hollow, 8" x 16" block, 4" thick	D-8	455	.088	S.F.	3.42	4.37		7.79	10.35

04 22 Concrete Unit Masonry

04 22 10 – Concrete Masonry Units

04 22 10.26 Concrete Block Foundation Wall

		Crew	Daily Output	Labor-Hours	Unit	Material	2021 Bare Costs Labor	2021 Bare Costs Equipment	Total	Total Incl O&P
1125	6" thick	D-8	445	.090	S.F.	4.47	4.47		8.94	11.65
1150	8" thick		415	.096		4.88	4.79		9.67	12.55
1200	10" thick	↓	340	.118		5.90	5.85		11.75	15.25
1250	12" thick	D-9	290	.166		7.80	8.05		15.85	21
1500	Solid, 8" x 16" block, 6" thick	D-8	430	.093		4.22	4.62		8.84	11.60
1600	8" thick	"	405	.099	↓	4.50	4.91		9.41	12.35
1650	12" thick	D-9	340	.141	S.F.	6.35	6.90		13.25	17.30

04 22 10.32 Concrete Block, Lintels

		Crew	Daily Output	Labor-Hours	Unit	Material	2021 Bare Costs Labor	2021 Bare Costs Equipment	Total	Total Incl O&P
0010	**CONCRETE BLOCK, LINTELS**, C90, normal weight									
0100	Including grout and horizontal reinforcing									
0200	8" x 8" x 8", 1 #4 bar	D-4	300	.107	L.F.	4.90	5.25	.63	10.78	14
0250	2 #4 bars		295	.108		5.20	5.35	.65	11.20	14.40
0400	8" x 16" x 8", 1 #4 bar		275	.116		4.31	5.70	.69	10.70	14.10
0450	2 #4 bars		270	.119		4.59	5.85	.71	11.15	14.60
1000	12" x 8" x 8", 1 #4 bar		275	.116		6.45	5.70	.69	12.84	16.45
1100	2 #4 bars		270	.119		6.75	5.85	.71	13.31	16.95
1150	2 #5 bars		270	.119		7.05	5.85	.71	13.61	17.30
1200	2 #6 bars		265	.121		7.40	5.95	.72	14.07	17.85
1500	12" x 16" x 8", 1 #4 bar		250	.128		7.60	6.30	.76	14.66	18.65
1600	2 #3 bars		245	.131		7.60	6.40	.78	14.78	18.90
1650	2 #4 bars		245	.131		7.85	6.40	.78	15.03	19.15
1700	2 #5 bars	↓	240	.133	↓	8.20	6.55	.79	15.54	19.70

04 22 10.34 Concrete Block, Partitions

		Crew	Daily Output	Labor-Hours	Unit	Material	2021 Bare Costs Labor	2021 Bare Costs Equipment	Total	Total Incl O&P
0010	**CONCRETE BLOCK, PARTITIONS**, excludes scaffolding R042210-20									
1000	Lightweight block, tooled joints, 2 sides, hollow									
1100	Not reinforced, 8" x 16" x 4" thick	D-8	440	.091	S.F.	2	4.52		6.52	9
1150	6" thick		410	.098		3.10	4.85		7.95	10.70
1200	8" thick		385	.104		3.97	5.15		9.12	12.15
1250	10" thick	↓	370	.108		4.22	5.35		9.57	12.75
1300	12" thick	D-9	350	.137		4.02	6.70		10.72	14.45
1500	Reinforced alternate courses, 4" thick	D-8	435	.092		2.17	4.57		6.74	9.30
1600	6" thick		405	.099		3.26	4.91		8.17	11
1650	8" thick		380	.105		4.17	5.25		9.42	12.50
1700	10" thick	↓	365	.110		4.41	5.45		9.86	13.05
1750	12" thick	D-9	345	.139	↓	4.23	6.80		11.03	14.85
4000	Regular block, tooled joints, 2 sides, hollow									
4100	Not reinforced, 8" x 16" x 4" thick	D-8	430	.093	S.F.	2.09	4.62		6.71	9.25
4150	6" thick		400	.100		2.71	4.97		7.68	10.50
4200	8" thick		375	.107		2.64	5.30		7.94	10.90
4250	10" thick	↓	360	.111		3.19	5.50		8.69	11.80
4300	12" thick	D-9	340	.141		4.64	6.90		11.54	15.45
4500	Reinforced alternate courses, 8" x 16" x 4" thick	D-8	425	.094		2.28	4.68		6.96	9.55
4550	6" thick		395	.101		2.91	5.05		7.96	10.80
4600	8" thick		370	.108		2.85	5.35		8.20	11.25
4650	10" thick	↓	355	.113		4.68	5.60		10.28	13.60
4700	12" thick	D-9	335	.143	↓	4.83	7		11.83	15.80

For customer support on your Site Work & Landscape Costs with RSMeans Data, call 800.448.8182.

105

04 27 10 – Multiple-Wythe Masonry

04 27 10.30 Brick Walls

		Crew	Daily Output	Labor-Hours	Unit	Material	2021 Bare Costs Labor	Equipment	Total	Total Incl O&P
0010	**BRICK WALLS**, including mortar, excludes scaffolding R042110-20									
0020	Estimating by number of brick									
0140	Face brick, 4" thick wall, 6.75 brick/S.F.	D-8	1.45	27.586	M	740	1,375		2,115	2,900
0150	Common brick, 4" thick wall, 6.75 brick/S.F. R042110-50		1.60	25		720	1,250		1,970	2,675
0204	8" thick, 13.50 brick/S.F.		1.80	22.222		740	1,100		1,840	2,500
0250	12" thick, 20.25 brick/S.F.		1.90	21.053		745	1,050		1,795	2,400
0304	16" thick, 27.00 brick/S.F.		2	20		755	995		1,750	2,325
0500	Reinforced, face brick, 4" thick wall, 6.75 brick/S.F.		1.40	28.571		760	1,425		2,185	3,000
0520	Common brick, 4" thick wall, 6.75 brick/S.F.		1.55	25.806		745	1,275		2,020	2,750
0550	8" thick, 13.50 brick/S.F.		1.75	22.857		765	1,125		1,890	2,550
0600	12" thick, 20.25 brick/S.F.		1.85	21.622		770	1,075		1,845	2,475
0650	16" thick, 27.00 brick/S.F.		1.95	20.513		775	1,025		1,800	2,375
0790	Alternate method of figuring by square foot									
0800	Face brick, 4" thick wall, 6.75 brick/S.F.	D-8	215	.186	S.F.	4.99	9.25		14.24	19.45
0850	Common brick, 4" thick wall, 6.75 brick/S.F.		240	.167		4.87	8.30		13.17	17.85
0900	8" thick, 13.50 brick/S.F.		135	.296		10	14.75		24.75	33
1000	12" thick, 20.25 brick/S.F.		95	.421		15.05	21		36.05	48
1050	16" thick, 27.00 brick/S.F.		75	.533		20.50	26.50		47	62.50
1200	Reinforced, face brick, 4" thick wall, 6.75 brick/S.F.		210	.190		5.15	9.45		14.60	19.90
1220	Common brick, 4" thick wall, 6.75 brick/S.F.		235	.170		5	8.45		13.45	18.25
1250	8" thick, 13.50 brick/S.F.		130	.308		10.30	15.30		25.60	34.50
1260	8" thick, 2.25 brick/S.F.		130	.308		1.89	15.30		17.19	25
1300	12" thick, 20.25 brick/S.F.		90	.444		15.55	22		37.55	50.50
1350	16" thick, 27.00 brick/S.F.		70	.571		21	28.50		49.50	66

04 27 10.40 Steps

		Crew	Daily Output	Labor-Hours	Unit	Material	2021 Bare Costs Labor	Equipment	Total	Total Incl O&P
0010	**STEPS**									
0012	Entry steps, select common brick	D-1	.30	53.333	M	580	2,600		3,180	4,575

04 41 10 – Dry Placed Stone

04 41 10.10 Rough Stone Wall

			Crew	Daily Output	Labor-Hours	Unit	Material	2021 Bare Costs Labor	Equipment	Total	Total Incl O&P
0011	**ROUGH STONE WALL**, Dry										
0012	Dry laid (no mortar), under 18" thick	G	D-1	60	.267	C.F.	14.75	13		27.75	36
0100	Random fieldstone, under 18" thick	G	D-12	60	.533		14.75	26		40.75	56
0150	Over 18" thick	G	"	63	.508		17.70	25		42.70	57
0500	Field stone veneer	G	D-8	120	.333	S.F.	14	16.55		30.55	40.50

04 43 10 – Masonry with Natural and Processed Stone

04 43 10.05 Ashlar Veneer

		Crew	Daily Output	Labor-Hours	Unit	Material	2021 Bare Costs Labor	Equipment	Total	Total Incl O&P
0011	**ASHLAR VENEER** +/- 4" thk, random or random rectangular									
0150	Sawn face, split joints, low priced stone	D-8	140	.286	S.F.	13	14.20		27.20	36
0200	Medium priced stone	"	130	.308	"	15.55	15.30		30.85	40
0300	High priced stone	D-8	120	.333	S.F.	18.75	16.55		35.30	45.50
0600	Seam face, split joints, medium price stone		125	.320		18.45	15.90		34.35	44.50
0700	High price stone		120	.333		17.95	16.55		34.50	45
1000	Split or rock face, split joints, medium price stone		125	.320		10.45	15.90		26.35	35.50
1100	High price stone		120	.333		18.30	16.55		34.85	45

04 43 Stone Masonry

04 43 10 — Masonry with Natural and Processed Stone

04 43 10.10 Bluestone	Crew	Daily Output	Labor-Hours	Unit	Material	2021 Bare Costs Labor	Equipment	Total	Total Incl O&P
0010 **BLUESTONE**, cut to size									
0100 Paving, natural cleft, to 4', 1" thick	D-8	150	.267	S.F.	8.15	13.25		21.40	29
0150 1-1/2" thick		145	.276		8.10	13.70		21.80	29.50
0200 Smooth finish, 1" thick		150	.267		7.85	13.25		21.10	28.50
0250 1-1/2" thick		145	.276		9.15	13.70		22.85	30.50
0300 Thermal finish, 1" thick		150	.267		7.60	13.25		20.85	28.50
0350 1-1/2" thick		145	.276		8.10	13.70		21.80	29.50
0500 Sills, natural cleft, 10" wide to 6' long, 1-1/2" thick	D-11	70	.343	L.F.	13.50	17.50		31	41.50
0550 2" thick		63	.381		14.75	19.45		34.20	46
0600 Smooth finish, 1-1/2" thick		70	.343		14.45	17.50		31.95	42.50
0650 2" thick		63	.381		17.10	19.45		36.55	48.50
0800 Thermal finish, 1-1/2" thick		70	.343		12.75	17.50		30.25	40.50
0850 2" thick		63	.381		10.95	19.45		30.40	41.50
1000 Stair treads, natural cleft, 12" wide, 6' long, 1-1/2" thick	D-10	115	.278		15.60	14.95	3.76	34.31	44
1050 2" thick		105	.305		17	16.35	4.12	37.47	47.50
1100 Smooth finish, 1-1/2" thick		115	.278		14.60	14.95	3.76	33.31	42.50
1150 2" thick		105	.305		15.45	16.35	4.12	35.92	46
1300 Thermal finish, 1-1/2" thick		115	.278		15	14.95	3.76	33.71	43
1350 2" thick		105	.305		14.30	16.35	4.12	34.77	45
2000 Coping, finished top & 2 sides, 12" to 6'									
2100 Natural cleft, 1-1/2" thick	D-10	115	.278	L.F.	12.95	14.95	3.76	31.66	41
2150 2" thick		105	.305		15.05	16.35	4.12	35.52	45.50
2200 Smooth finish, 1-1/2" thick		115	.278		12.90	14.95	3.76	31.61	41
2250 2" thick		105	.305		15.05	16.35	4.12	35.52	45.50
2300 Thermal finish, 1-1/2" thick		115	.278		12.95	14.95	3.76	31.66	41
2350 2" thick		105	.305		15.05	16.35	4.12	35.52	45.50

04 43 10.45 Granite	Crew	Daily Output	Labor-Hours	Unit	Material	2021 Bare Costs Labor	Equipment	Total	Total Incl O&P
0010 **GRANITE**, cut to size									
0050 Veneer, polished face, 3/4" to 1-1/2" thick									
0150 Low price, gray, light gray, etc.	D-10	130	.246	S.F.	25	13.20	3.33	41.53	51
0180 Medium price, pink, brown, etc.		130	.246		28.50	13.20	3.33	45.03	55
0220 High price, red, black, etc.		130	.246		43	13.20	3.33	59.53	70.50
0300 1-1/2" to 2-1/2" thick, veneer									
0350 Low price, gray, light gray, etc.	D-10	130	.246	S.F.	29	13.20	3.33	45.53	55.50
0500 Medium price, pink, brown, etc.	"	130	.246	"	34	13.20	3.33	50.53	61
0550 High price, red, black, etc.	D-10	130	.246	S.F.	53	13.20	3.33	69.53	81.50
0700 2-1/2" to 4" thick, veneer									
0750 Low price, gray, light gray, etc.	D-10	110	.291	S.F.	39	15.60	3.93	58.53	71
0850 Medium price, pink, brown, etc.		110	.291		44.50	15.60	3.93	64.03	77
0950 High price, red, black, etc.		110	.291		58.50	15.60	3.93	78.03	92
1000 For bush hammered finish, deduct					5%				
1050 Coarse rubbed finish, deduct					10%				
1100 Honed finish, deduct					5%				
1150 Thermal finish, deduct					18%				
1800 Carving or bas-relief, from templates or plaster molds									
1850 Low price, gray, light gray, etc.	D-10	80	.400	C.F.	164	21.50	5.40	190.90	219
1875 Medium price, pink, brown, etc.		80	.400		345	21.50	5.40	371.90	420
1900 High price, red, black, etc.		80	.400		530	21.50	5.40	556.90	620
2000 Intricate or hand finished pieces									
2010 Mouldings, radius cuts, bullnose edges, etc.									
2050 Add for low price gray, light gray, etc.					30%				
2075 Add for medium price, pink, brown, etc.					165%				

For customer support on your Site Work & Landscape Costs with RSMeans Data, call 800.448.8182.

107

04 43 10 – Masonry with Natural and Processed Stone

04 43 10.45 Granite		Crew	Daily Output	Labor-Hours	Unit	Material	2021 Bare Costs Labor	Equipment	Total	Total Incl O&P
2100	Add for high price red, black, etc.					300%				
2450	For radius under 5', add				L.F.	100%				
2500	Steps, copings, etc., finished on more than one surface									
2550	Low price, gray, light gray, etc.	D-10	50	.640	C.F.	92	34.50	8.65	135.15	162
2575	Medium price, pink, brown, etc.		50	.640		120	34.50	8.65	163.15	193
2600	High price, red, black, etc.		50	.640		147	34.50	8.65	190.15	223
2700	Pavers, Belgian block, 8"-13" long, 4"-6" wide, 4"-6" deep	D-11	120	.200	S.F.	25	10.20		35.20	42.50
2800	Pavers, 4" x 4" x 4" blocks, split face and joints									
2850	Low price, gray, light gray, etc.	D-11	80	.300	S.F.	12.10	15.30		27.40	36.50
2875	Medium price, pink, brown, etc.		80	.300		19.45	15.30		34.75	44.50
2900	High price, red, black, etc.		80	.300		27	15.30		42.30	52.50
3000	Pavers, 4" x 4" x 4", thermal face, sawn joints									
3050	Low price, gray, light gray, etc.	D-11	65	.369	S.F.	24.50	18.85		43.35	55.50
3075	Medium price, pink, brown, etc.		65	.369		28.50	18.85		47.35	60
3100	High price, red, black, etc.		65	.369		32.50	18.85		51.35	64
4000	Soffits, 2" thick, low price, gray, light gray	D-13	35	1.371		38	71.50	12.35	121.85	163
4050	Medium price, pink, brown, etc.		35	1.371		61	71.50	12.35	144.85	188
4100	High price, red, black, etc.		35	1.371		84	71.50	12.35	167.85	213
4200	Low price, gray, light gray, etc.		35	1.371		63.50	71.50	12.35	147.35	191
4250	Medium price, pink, brown, etc.		35	1.371		86.50	71.50	12.35	170.35	216
4300	High price, red, black, etc.		35	1.371		109	71.50	12.35	192.85	241
5000	Reclaimed or antique									
5010	Treads, up to 12" wide	D-10	100	.320	L.F.	24	17.15	4.33	45.48	57.50
5020	Up to 18" wide		100	.320		45.50	17.15	4.33	66.98	81
5030	Capstone, size varies		50	.640		26.50	34.50	8.65	69.65	90
5040	Posts		30	1.067	V.L.F.	31.50	57	14.40	102.90	136

04 43 10.50 Lightweight Natural Stone

			Crew	Daily Output	Labor-Hours	Unit	Material	2021 Bare Costs Labor	Equipment	Total	Total Incl O&P
0011	**LIGHTWEIGHT NATURAL STONE** Lava type										
0100	Veneer, rubble face, sawed back, irregular shapes	G	D-10	130	.246	S.F.	9.15	13.20	3.33	25.68	33.50
0200	Sawed face and back, irregular shapes	G		130	.246	"	9.15	13.20	3.33	25.68	33.50
1000	Reclaimed or antique, barn or foundation stone			1	32	Ton	180	1,725	435	2,340	3,250

04 43 10.55 Limestone

		Crew	Daily Output	Labor-Hours	Unit	Material	2021 Bare Costs Labor	Equipment	Total	Total Incl O&P
0010	**LIMESTONE**, cut to size									
0020	Veneer facing panels									
0500	Texture finish, light stick, 4-1/2" thick, 5' x 12'	D-4	300	.107	S.F.	26	5.25	.63	31.88	37.50
1400	Sugarcube, textured finish, 4-1/2" thick, 5' x 12'	D-10	275	.116		38.50	6.25	1.57	46.32	53
1450	5" thick, 5' x 14' panels		275	.116		39	6.25	1.57	46.82	54
2000	Coping, sugarcube finish, top & 2 sides		30	1.067	C.F.	66.50	57	14.40	137.90	175
2100	Sills, lintels, jambs, trim, stops, sugarcube finish, simple		20	1.600		66.50	86	21.50	174	226
2150	Detailed		20	1.600		66.50	86	21.50	174	226
2300	Steps, extra hard, 14" wide, 6" rise		50	.640	L.F.	29	34.50	8.65	72.15	92.50
3000	Quoins, plain finish, 6" x 12" x 12"	D-12	25	1.280	Ea.	54.50	63		117.50	155
3050	6" x 16" x 24"	"	25	1.280	"	73	63		136	175

04 43 10.60 Marble

		Crew	Daily Output	Labor-Hours	Unit	Material	2021 Bare Costs Labor	Equipment	Total	Total Incl O&P
0011	**MARBLE**, ashlar, split face, +/- 4" thick, random									
0040	Lengths 1' to 4' & heights 2" to 7-1/2", average	D-8	175	.229	S.F.	16.85	11.35		28.20	35.50
1000	Facing, polished finish, cut to size, 3/4" to 7/8" thick									
1050	Carrara or equal	D-10	130	.246	S.F.	35	13.20	3.33	51.53	62
1100	Arabescato or equal		130	.246		37	13.20	3.33	53.53	64
1300	1-1/4" thick, Botticino Classico or equal		125	.256		22.50	13.75	3.46	39.71	49.50
1350	Statuarietto or equal		125	.256		37.50	13.75	3.46	54.71	66
1500	2" thick, Crema Marfil or equal		120	.267		48	14.30	3.61	65.91	78.50

04 43 10 – Masonry with Natural and Processed Stone

04 43 10.60 Marble	Crew	Daily Output	Labor-Hours	Unit	Material	2021 Bare Costs Labor	Equipment	Total	Total Incl O&P
1550 Cafe Pinta or equal	D-10	120	.267	S.F.	66.50	14.30	3.61	84.41	98.50
1700 Rubbed finish, cut to size, 4" thick									
1740 Average	D-10	100	.320	S.F.	41.50	17.15	4.33	62.98	76.50
1780 Maximum	"	100	.320	"	77.50	17.15	4.33	98.98	116
2500 Flooring, polished tiles, 12" x 12" x 3/8" thick									
2510 Thin set, Giallo Solare or equal	D-11	90	.267	S.F.	14.75	13.60		28.35	36.50
2600 Sky Blue or equal		90	.267		11.35	13.60		24.95	33
2700 Mortar bed, Giallo Solare or equal		65	.369		14.85	18.85		33.70	45
2740 Sky Blue or equal		65	.369		11.35	18.85		30.20	41
3210 Stairs, risers, 7/8" thick x 6" high	D-10	115	.278	L.F.	14.95	14.95	3.76	33.66	43
3360 Treads, 12" wide x 1-1/4" thick	D-10	115	.278	L.F.	46	14.95	3.76	64.71	77
3500 Thresholds, 3' long, 7/8" thick, 4" to 5" wide, plain	D-12	24	1.333	Ea.	34.50	65.50		100	137
3550 Beveled	"	24	1.333	"	76	65.50		141.50	183

04 43 10.75 Sandstone or Brownstone

	Crew	Daily Output	Labor-Hours	Unit	Material	2021 Bare Costs Labor	Equipment	Total	Total Incl O&P
0011 **SANDSTONE OR BROWNSTONE**									
0100 Sawed face veneer, 2-1/2" thick, to 2' x 4' panels	D-10	130	.246	S.F.	21.50	13.20	3.33	38.03	47.50
0150 4" thick, to 3'-6" x 8' panels		100	.320		21.50	17.15	4.33	42.98	55
0300 Split face, random sizes		100	.320		15	17.15	4.33	36.48	47.50

04 43 10.80 Slate

	Crew	Daily Output	Labor-Hours	Unit	Material	2021 Bare Costs Labor	Equipment	Total	Total Incl O&P
0010 **SLATE**									
0040 Pennsylvania - blue gray to black									
0050 Vermont - unfading green, mottled green & purple, gray & purple									
0100 Virginia - blue black									
0200 Exterior paving, natural cleft, 1" thick									
0250 6" x 6" Pennsylvania	D-12	100	.320	S.F.	7.05	15.75		22.80	31.50
0300 Vermont		100	.320		11.40	15.75		27.15	36
0350 Virginia		100	.320		14.70	15.75		30.45	39.50
0500 24" x 24", Pennsylvania		120	.267		13.55	13.10		26.65	34.50
0550 Vermont		120	.267		28.50	13.10		41.60	51
0600 Virginia		120	.267		21.50	13.10		34.60	43.50
0700 18" x 30" Pennsylvania		120	.267		15.35	13.10		28.45	36.50
0750 Vermont		120	.267		28.50	13.10		41.60	51
0800 Virginia		120	.267		19	13.10		32.10	41
1000 Interior flooring, natural cleft, 1/2" thick									
1100 6" x 6" Pennsylvania	D-12	100	.320	S.F.	4.16	15.75		19.91	28
1150 Vermont		100	.320		10.05	15.75		25.80	34.50
1200 Virginia		100	.320		11.60	15.75		27.35	36.50
1300 24" x 24" Pennsylvania		120	.267		8.10	13.10		21.20	28.50
1350 Vermont		120	.267		23	13.10		36.10	45
1400 Virginia		120	.267		15.35	13.10		28.45	36.50
1500 18" x 24" Pennsylvania		120	.267		8.10	13.10		21.20	28.50
1550 Vermont		120	.267		16.85	13.10		29.95	38.50
1600 Virginia		120	.267		15.55	13.10		28.65	37
2000 Facing panels, 1-1/4" thick, to 4' x 4' panels									
2100 Natural cleft finish, Pennsylvania	D-10	180	.178	S.F.	35	9.55	2.40	46.95	55.50
2110 Vermont		180	.178		28.50	9.55	2.40	40.45	48
2120 Virginia		180	.178		34.50	9.55	2.40	46.45	55
2150 Sand rubbed finish, surface, add					10.60			10.60	11.65
2200 Honed finish, add					7.65			7.65	8.45
2500 Ribbon, natural cleft finish, 1" thick, to 9 S.F.	D-10	80	.400		13.65	21.50	5.40	40.55	53
2550 Sand rubbed finish		80	.400		18.50	21.50	5.40	45.40	58.50
2600 Honed finish		80	.400		17.20	21.50	5.40	44.10	57

For customer support on your Site Work & Landscape Costs with RSMeans Data, call 800.448.8182.

109

04 43 Stone Masonry

04 43 10 – Masonry with Natural and Processed Stone

04 43 10.80 Slate		Crew	Daily Output	Labor-Hours	Unit	Material	2021 Bare Costs Labor	Equipment	Total	Total Incl O&P
2700	1-1/2" thick	D-10	78	.410	S.F.	17.75	22	5.55	45.30	58.50
2750	Sand rubbed finish		78	.410		23.50	22	5.55	51.05	65
2800	Honed finish		78	.410		22	22	5.55	49.55	63.50
2850	2" thick		76	.421		21.50	22.50	5.70	49.70	64
2900	Sand rubbed finish		76	.421		29.50	22.50	5.70	57.70	73
2950	Honed finish		76	.421		27	22.50	5.70	55.20	70.50
3100	Stair landings, 1" thick, black, clear	D-1	65	.246		21	12		33	41
3200	Ribbon	"	65	.246		23	12		35	43
3500	Stair treads, sand finish, 1" thick x 12" wide									
3550	Under 3 L.F.	D-10	85	.376	L.F.	23	20	5.10	48.10	61
3600	3 L.F. to 6 L.F.	"	120	.267	"	25	14.30	3.61	42.91	52.50
3700	Ribbon, sand finish, 1" thick x 12" wide									
3750	To 6 L.F.	D-10	120	.267	L.F.	21	14.30	3.61	38.91	48.50
4000	Stools or sills, sand finish, 1" thick, 6" wide	D-12	160	.200		12	9.85		21.85	28
4100	Honed finish		160	.200		11.45	9.85		21.30	27.50
4200	10" wide		90	.356		18.50	17.50		36	47
4250	Honed finish		90	.356		17.20	17.50		34.70	45.50
4400	2" thick, 6" wide		140	.229		19.30	11.25		30.55	38
4450	Honed finish		140	.229		18.40	11.25		29.65	37.50
4600	10" wide		90	.356		30	17.50		47.50	59.50
4650	Honed finish		90	.356		28.50	17.50		46	58
4800	For lengths over 3', add					25%				

04 54 Refractory Brick Masonry

04 54 10 – Refractory Brick Work

04 54 10.10 Fire Brick

		Crew	Daily Output	Labor-Hours	Unit	Material	2021 Bare Costs Labor	Equipment	Total	Total Incl O&P
0010	**FIRE BRICK**									
0012	Low duty, 2000°F, 9" x 2-1/2" x 4-1/2"	D-1	.60	26.667	M	1,750	1,300		3,050	3,875
0050	High duty, 3000°F	"	.60	26.667	"	3,150	1,300		4,450	5,400

04 54 10.20 Fire Clay

		Crew	Daily Output	Labor-Hours	Unit	Material	2021 Bare Costs Labor	Equipment	Total	Total Incl O&P
0010	**FIRE CLAY**									
0020	Gray, high duty, 100 lb. bag				Bag	28.50			28.50	31
0050	100 lb. drum, premixed (400 brick per drum)				Drum	44			44	48.50

04 57 Masonry Fireplaces

04 57 10 – Brick or Stone Fireplaces

04 57 10.10 Fireplace

		Crew	Daily Output	Labor-Hours	Unit	Material	2021 Bare Costs Labor	Equipment	Total	Total Incl O&P
0010	**FIREPLACE**									
0100	Brick fireplace, not incl. foundations or chimneys									
0110	30" x 29" opening, incl. chamber, plain brickwork	D-1	.40	40	Ea.	630	1,950		2,580	3,625
0200	Fireplace box only (110 brick)	"	2	8	"	162	390		552	765
0300	For elaborate brickwork and details, add					35%	35%			
0400	For hearth, brick & stone, add	D-1	2	8	Ea.	214	390		604	820

110

For customer support on your Site Work & Landscape Costs with RSMeans Data, call 800.448.8182.

04 72 Cast Stone Masonry

04 72 10 – Cast Stone Masonry Features

04 72 10.10 Coping

	Crew	Daily Output	Labor-Hours	Unit	Material	2021 Bare Costs Labor	Equipment	Total	Total Incl O&P
0010 **COPING**, stock units									
0050 Precast concrete, 10" wide, 4" tapers to 3-1/2", 8" wall	D-1	75	.213	L.F.	25	10.40		35.40	43
0100 12" wide, 3-1/2" tapers to 3", 10" wall		70	.229		27	11.15		38.15	46.50
0110 14" wide, 4" tapers to 3-1/2", 12" wall		65	.246		36.50	12		48.50	58
0150 16" wide, 4" tapers to 3-1/2", 14" wall		60	.267		39	13		52	62.50
0250 Precast concrete corners		40	.400	Ea.	51	19.50		70.50	85.50
0300 Limestone for 12" wall, 4" thick		90	.178	L.F.	16.90	8.65		25.55	31.50
0350 6" thick		80	.200		23.50	9.75		33.25	40.50
0500 Marble, to 4" thick, no wash, 9" wide		90	.178		10.40	8.65		19.05	24.50
0550 12" wide		80	.200		18.50	9.75		28.25	35
0700 Terra cotta, 9" wide		90	.178		8.10	8.65		16.75	22
0750 12" wide		80	.200		8.20	9.75		17.95	23.50
0800 Aluminum, for 12" wall		80	.200		9.45	9.75		19.20	25

04 72 20 – Cultured Stone Veneer

04 72 20.10 Cultured Stone Veneer Components

	Crew	Daily Output	Labor-Hours	Unit	Material	2021 Bare Costs Labor	Equipment	Total	Total Incl O&P
0010 **CULTURED STONE VENEER COMPONENTS**									
0110 On wood frame and sheathing substrate, random sized cobbles, corner stones	D-8	70	.571	V.L.F.	8.45	28.50		36.95	52.50
0120 Field stones		140	.286	S.F.	5.90	14.20		20.10	28
0130 Random sized flats, corner stones		70	.571	V.L.F.	8.65	28.50		37.15	52.50
0140 Field stones		140	.286	S.F.	9.20	14.20		23.40	31.50
0150 Horizontal lined ledgestones, corner stones		75	.533	V.L.F.	8.35	26.50		34.85	49
0160 Field stones		150	.267	S.F.	6.30	13.25		19.55	27
0170 Random shaped flats, corner stones		65	.615	V.L.F.	8.40	30.50		38.90	55
0180 Field stones		150	.267	S.F.	6.15	13.25		19.40	27
0190 Random shaped/textured face, corner stones		65	.615	V.L.F.	8.45	30.50		38.95	55.50
0200 Field stones		130	.308	S.F.	6.10	15.30		21.40	30
0210 Random shaped river rock, corner stones		65	.615	V.L.F.	8.45	30.50		38.95	55.50
0220 Field stones		130	.308	S.F.	6.10	15.30		21.40	30
0240 On concrete or CMU substrate, random sized cobbles, corner stones		70	.571	V.L.F.	7.60	28.50		36.10	51.50
0250 Field stones		140	.286	S.F.	5.45	14.20		19.65	27.50
0260 Random sized flats, corner stones		70	.571	V.L.F.	7.80	28.50		36.30	51.50
0270 Field stones		140	.286	S.F.	8.75	14.20		22.95	31
0280 Horizontal lined ledgestones, corner stones		75	.533	V.L.F.	7.50	26.50		34	48.50
0290 Field stones		150	.267	S.F.	5.85	13.25		19.10	26.50
0300 Random shaped flats, corner stones		70	.571	V.L.F.	7.50	28.50		36	51.50
0310 Field stones		140	.286	S.F.	5.75	14.20		19.95	28
0320 Random shaped/textured face, corner stones		65	.615	V.L.F.	7.60	30.50		38.10	54.50
0330 Field stones		130	.308	S.F.	5.70	15.30		21	29.50
0340 Random shaped river rock, corner stones		65	.615	V.L.F.	7.60	30.50		38.10	54.50
0350 Field stones	D-8	130	.308	S.F.	5.70	15.30		21	29.50
0360 Cultured stone veneer, #15 felt weather resistant barrier	1 Clab	3700	.002	Sq.	5.70	.10		5.80	6.40
0370 Expanded metal lath, diamond, 2.5 lb./S.Y., galvanized	1 Lath	85	.094	S.Y.	3.87	5.10		8.97	11.70
0390 Water table or window sill, 18" long	1 Bric	80	.100	Ea.	9.25	5.35		14.60	18.25

For customer support on your Site Work & Landscape Costs with RSMeans Data, call 800.448.8182.

111

Division Notes

	CREW	DAILY OUTPUT	LABOR-HOURS	UNIT	BARE COSTS				TOTAL INCL O&P
					MAT.	LABOR	EQUIP.	TOTAL	

Estimating Tips

05 05 00 Common Work Results for Metals

- Nuts, bolts, washers, connection angles, and plates can add a significant amount to both the tonnage of a structural steel job and the estimated cost. As a rule of thumb, add 10% to the total weight to account for these accessories.

- Type 2 steel construction, commonly referred to as "simple construction," consists generally of field-bolted connections with lateral bracing supplied by other elements of the building, such as masonry walls or x-bracing. The estimator should be aware, however, that shop connections may be accomplished by welding or bolting. The method may be particular to the fabrication shop and may have an impact on the estimated cost.

05 10 00 Structural Steel

- Steel items can be obtained from two sources: a fabrication shop or a metals service center. Fabrication shops can fabricate items under more controlled conditions than crews in the field can. They are also more efficient and can produce items more economically. Metal service centers serve as a source of long mill shapes to both fabrication shops and contractors.

- Most line items in this structural steel subdivision, and most items in 05 50 00 Metal Fabrications, are indicated as being shop fabricated. The bare material cost for these shop fabricated items is the "Invoice Cost" from the shop and includes the mill base price of steel plus mill extras, transportation to the shop, shop drawings and detailing where warranted, shop fabrication and handling, sandblasting and a shop coat of primer paint, all necessary structural bolts, and delivery to the job site. The bare labor cost and bare equipment cost for these shop fabricated items are for field installation or erection.

- Line items in Subdivision 05 12 23.40 Lightweight Framing, and other items scattered in Division 5, are indicated as being field fabricated. The bare material cost for these field fabricated items is the "Invoice Cost" from the metals service center and includes the mill base price of steel plus mill extras, transportation to the metals service center, material handling, and delivery of long lengths of mill shapes to the job site. Material costs for structural bolts and welding rods should be added to the estimate. The bare labor cost and bare equipment cost for these items are for both field fabrication and field installation or erection, and include time for cutting, welding, and drilling in the fabricated metal items. Drilling into concrete and fasteners to fasten field fabricated items to other work is not included and should be added to the estimate.

05 20 00 Steel Joist Framing

- In any given project the total weight of open web steel joists is determined by the loads to be supported and the design. However, economies can be realized in minimizing the amount of labor used to place the joists. This is done by maximizing the joist spacing and therefore minimizing the number of joists required to be installed on the job. Certain spacings and locations may be required by the design, but in other cases maximizing the spacing and keeping it as uniform as possible will keep the costs down.

05 30 00 Steel Decking

- The takeoff and estimating of a metal deck involve more than the area of the floor or roof and the type of deck specified or shown on the drawings. Many different sizes and types of openings may exist. Small openings for individual pipes or conduits may be drilled after the floor/roof is installed, but larger openings may require special deck lengths as well as reinforcing or structural support. The estimator should determine who will be supplying this reinforcing. Additionally, some deck terminations are part of the deck package, such as screed angles and pour stops, and others will be part of the steel contract, such as angles attached to structural members and cast-in-place angles and plates. The estimator must ensure that all pieces are accounted for in the complete estimate.

05 50 00 Metal Fabrications

- The most economical steel stairs are those that use common materials, standard details, and most importantly, a uniform and relatively simple method of field assembly. Commonly available A36/A992 channels and plates are very good choices for the main stringers of the stairs, as are angles and tees for the carrier members. Risers and treads are usually made by specialty shops, and it is most economical to use a typical detail in as many places as possible. The stairs should be pre-assembled and shipped directly to the site. The field connections should be simple and straightforward enough to be accomplished efficiently, and with minimum equipment and labor.

Reference Numbers

Reference numbers are shown at the beginning of some major classifications. These numbers refer to related items in the Reference Section. The reference information may be an estimating procedure, an alternate pricing method, or technical information.

Note: Not all subdivisions listed here necessarily appear. ■

Same Data. Simplified.

Enjoy the convenience and efficiency of accessing your costs anywhere:

- **Skip the multiplier** by setting your location
- **Quickly search,** edit, favorite and share costs
- **Stay on top of price changes** with automatic updates

Discover more at rsmeans.com/online

05 01 Maintenance of Metals

05 01 10 – Maintenance of Structural Metal Framing

05 01 10.51 Cleaning of Structural Metal Framing

		Crew	Daily Output	Labor-Hours	Unit	Material	2021 Bare Costs Labor	Equipment	Total	Total Incl O&P
0010	**CLEANING OF STRUCTURAL METAL FRAMING**									
6125	Steel surface treatments, PDCA guidelines									
6170	Wire brush, hand (SSPC-SP2)	1 Psst	400	.020	S.F.	.02	.94		.96	1.54
6180	Power tool (SSPC-SP3)	"	700	.011		.07	.54		.61	.95
6215	Pressure washing, up to 5,000 psi, 5,000-15,000 S.F./day	1 Pord	10000	.001			.04		.04	.06
6220	Steam cleaning, 600 psi @ 300 °F, 1,250-2,500 S.F./day		2000	.004			.19		.19	.28
6225	Water blasting, up to 25,000 psi, 1,750-3,500 S.F./day		2500	.003			.15		.15	.22
6230	Brush-off blast (SSPC-SP7)	E-11	1750	.018		.17	.89	.17	1.23	1.75
6235	Com'l blast (SSPC-SP6), loose scale, fine pwder rust, 2.0#/S.F. sand		1200	.027		.35	1.30	.25	1.90	2.66
6240	Tight mill scale, little/no rust, 3.0#/S.F. sand		1000	.032		.52	1.55	.30	2.37	3.30
6245	Exist coat blistered/pitted, 4.0#/S.F. sand		875	.037		.69	1.78	.35	2.82	3.89
6250	Exist coat badly pitted/nodules, 6.7#/S.F. sand		825	.039		1.16	1.88	.37	3.41	4.58
6255	Near white blast (SSPC-SP10), loose scale, fine rust, 5.6#/S.F. sand		450	.071		.97	3.45	.67	5.09	7.15
6260	Tight mill scale, little/no rust, 6.9#/S.F. sand		325	.098		1.19	4.78	.93	6.90	9.75
6265	Exist coat blistered/pitted, 9.0#/S.F. sand		225	.142		1.55	6.90	1.34	9.79	13.90
6270	Exist coat badly pitted/nodules, 11.3#/S.F. sand		150	.213		1.95	10.35	2.02	14.32	20.50

05 05 Common Work Results for Metals

05 05 05 – Selective Demolition for Metals

05 05 05.10 Selective Demolition, Metals

		Crew	Daily Output	Labor-Hours	Unit	Material	2021 Bare Costs Labor	Equipment	Total	Total Incl O&P
0010	**SELECTIVE DEMOLITION, METALS** R024119-10									
0015	Excludes shores, bracing, cutting, loading, hauling, dumping									
0020	Remove nuts only up to 3/4" diameter	1 Sswk	480	.017	Ea.		1.01		1.01	1.56
0030	7/8" to 1-1/4" diameter		240	.033			2.01		2.01	3.11
0040	1-3/8" to 2" diameter		160	.050			3.02		3.02	4.67
0060	Unbolt and remove structural bolts up to 3/4" diameter		240	.033			2.01		2.01	3.11
0070	7/8" to 2" diameter		160	.050			3.02		3.02	4.67
0140	Light weight framing members, remove whole or cut up, up to 20 lb.		240	.033			2.01		2.01	3.11
0150	21-40 lb.	2 Sswk	210	.076			4.59		4.59	7.10
0160	41-80 lb.	3 Sswk	180	.133			8.05		8.05	12.45
0170	81-120 lb.	4 Sswk	150	.213			12.85		12.85	19.90
0230	Structural members, remove whole or cut up, up to 500 lb.	E-19	48	.500			29.50	32	61.50	80.50
0240	1/4-2 tons	E-18	36	1.111	Ea.		67	42.50	109.50	150
0250	2-5 tons	E-24	30	1.067			64	19.55	83.55	120
0260	5-10 tons	E-20	24	2.667			159	88.50	247.50	340
0270	10-15 tons	E-2	18	3.111			186	95	281	390
0340	Fabricated item, remove whole or cut up, up to 20 lb.	1 Sswk	96	.083			5.05		5.05	7.80
0350	21-40 lb.	2 Sswk	84	.190			11.50		11.50	17.75
0360	41-80 lb.	3 Sswk	72	.333			20		20	31
0370	81-120 lb.	4 Sswk	60	.533			32		32	50
0380	121-500 lb.	E-19	48	.500			29.50	32	61.50	80.50
0390	501-1000 lb.	"	36	.667			39.50	42.50	82	107
0500	Steel roof decking, uncovered, bare	B-2	5000	.008	S.F.		.36		.36	.53

05 05 13 – Shop-Applied Coatings for Metal

05 05 13.50 Paints and Protective Coatings

		Crew	Daily Output	Labor-Hours	Unit	Material	2021 Bare Costs Labor	Equipment	Total	Total Incl O&P
0010	**PAINTS AND PROTECTIVE COATINGS**									
5900	Galvanizing structural steel in shop, under 1 ton				Ton	700			700	770
5950	1 ton to 20 tons					670			670	735
6000	Over 20 tons					645			645	710

05 05 19 – Post-Installed Concrete Anchors

05 05 19.10 Chemical Anchors		Crew	Daily Output	Labor-Hours	Unit	Material	2021 Bare Costs Labor	Equipment	Total	Total Incl O&P
0010	**CHEMICAL ANCHORS**									
0020	Includes layout & drilling									
1430	Chemical anchor, w/rod & epoxy cartridge, 3/4" diameter x 9-1/2" long	B-89A	27	.593	Ea.	11.90	30	4.50	46.40	63
1435	1" diameter x 11-3/4" long		24	.667		24	34	5.05	63.05	82.50
1440	1-1/4" diameter x 14" long		21	.762		38.50	38.50	5.80	82.80	106
1445	1-3/4" diameter x 15" long		20	.800		70.50	40.50	6.10	117.10	145
1450	18" long		17	.941		84.50	48	7.15	139.65	172
1455	2" diameter x 18" long		16	1		148	51	7.60	206.60	247
1460	24" long		15	1.067		194	54	8.10	256.10	305

05 05 19.20 Expansion Anchors										
0010	**EXPANSION ANCHORS**									
0100	Anchors for concrete, brick or stone, no layout and drilling									
0200	Expansion shields, zinc, 1/4" diameter, 1-5/16" long, single	G 1 Carp	90	.089	Ea.	.31	4.86		5.17	7.60
0300	1-3/8" long, double	G	85	.094		.42	5.15		5.57	8.15
0600	1/2" diameter, 2-1/16" long, single	G	80	.100		1.23	5.45		6.68	9.50
1000	3/4" diameter, 2-3/4" long, single	G	70	.114		3.75	6.25		10	13.50
1100	3-15/16" long, double	G	65	.123		5.55	6.75		12.30	16.15
3000	Toggle bolts, bright steel, 1/8" diameter, 2" long	G	85	.094		.33	5.15		5.48	8.05
3400	1/4" diameter, 3" long	G	75	.107		.44	5.85		6.29	9.20
3600	3/8" diameter, 3" long	G	70	.114		.73	6.25		6.98	10.15
5000	Screw anchors for concrete, masonry,									
5100	stone & tile, no layout or drilling included									
6600	Lead, #6 & #8, 3/4" long	G 1 Carp	260	.031	Ea.	.22	1.68		1.90	2.75
6700	#10 - #14, 1-1/2" long	G	200	.040		.52	2.19		2.71	3.84
6800	#16 & #18, 1-1/2" long	G	160	.050		.70	2.74		3.44	4.85
6900	Plastic, #6 & #8, 3/4" long		260	.031		.05	1.68		1.73	2.57
8000	Wedge anchors, not including layout or drilling									
8050	Carbon steel, 1/4" diameter, 1-3/4" long	G 1 Carp	150	.053	Ea.	.76	2.92		3.68	5.20
8100	3-1/4" long	G	140	.057		1	3.13		4.13	5.75
8150	3/8" diameter, 2-1/4" long	G	145	.055		.43	3.02		3.45	4.98
8200	5" long	G	140	.057		.76	3.13		3.89	5.50
8250	1/2" diameter, 2-3/4" long	G	140	.057		1.14	3.13		4.27	5.95
8300	7" long	G	125	.064		1.96	3.50		5.46	7.40
8350	5/8" diameter, 3-1/2" long	G	130	.062		2.13	3.37		5.50	7.35
8400	8-1/2" long	G	115	.070		4.54	3.81		8.35	10.70
8450	3/4" diameter, 4-1/4" long	G	115	.070		3.45	3.81		7.26	9.50
8500	10" long	G	95	.084		7.85	4.61		12.46	15.55
8550	1" diameter, 6" long	G	100	.080		5.65	4.38		10.03	12.75
8575	9" long	G	85	.094		7.35	5.15		12.50	15.75
8600	12" long	G	75	.107		7.95	5.85		13.80	17.40
8650	1-1/4" diameter, 9" long	G	70	.114		33.50	6.25		39.75	46.50
8700	12" long	G	60	.133		43	7.30		50.30	58.50
8750	For type 303 stainless steel, add					350%				
8800	For type 316 stainless steel, add					450%				
8950	Self-drilling concrete screw, hex washer head, 3/16" diam. x 1-3/4" long	G 1 Carp	300	.027	Ea.	.21	1.46		1.67	2.41
8960	2-1/4" long	G	250	.032		.29	1.75		2.04	2.93
8970	Phillips flat head, 3/16" diam. x 1-3/4" long	G	300	.027		.20	1.46		1.66	2.40
8980	2-1/4" long	G	250	.032		.31	1.75		2.06	2.95

For customer support on your Site Work & Landscape Costs with RSMeans Data, call 800.448.8182.

115

05 05 Common Work Results for Metals

05 05 21 – Fastening Methods for Metal

05 05 21.10 Cutting Steel

		Crew	Daily Output	Labor-Hours	Unit	Material	2021 Bare Costs Labor	Equipment	Total	Total Incl O&P
0010	**CUTTING STEEL**									
0020	Hand burning, incl. preparation, torch cutting & grinding, no staging									
0050	Steel to 1/4" thick	E-25	400	.020	L.F.	.64	1.25	.03	1.92	2.68
0100	1/2" thick		320	.025		1.04	1.56	.04	2.64	3.60
0150	3/4" thick		260	.031		1.61	1.92	.05	3.58	4.80
0200	1" thick		200	.040		2.23	2.49	.06	4.78	6.40

05 05 21.15 Drilling Steel

		Crew	Daily Output	Labor-Hours	Unit	Material	2021 Bare Costs Labor	Equipment	Total	Total Incl O&P
0010	**DRILLING STEEL**									
1910	Drilling & layout for steel, up to 1/4" deep, no anchor									
1920	Holes, 1/4" diameter	1 Sswk	112	.071	Ea.	.06	4.31		4.37	6.70
1925	For each additional 1/4" depth, add		336	.024		.06	1.44		1.50	2.29
1930	3/8" diameter		104	.077		.08	4.64		4.72	7.30
1935	For each additional 1/4" depth, add		312	.026		.08	1.55		1.63	2.48
1940	1/2" diameter	1 Sswk	96	.083	Ea.	.08	5.05		5.13	7.90
1945	For each additional 1/4" depth, add		288	.028		.08	1.68		1.76	2.68
1950	5/8" diameter		88	.091		.12	5.50		5.62	8.65
1955	For each additional 1/4" depth, add		264	.030		.12	1.83		1.95	2.96
1960	3/4" diameter		80	.100		.16	6.05		6.21	9.55
1965	For each additional 1/4" depth, add		240	.033		.16	2.01		2.17	3.29
1970	7/8" diameter		72	.111		.19	6.70		6.89	10.55
1975	For each additional 1/4" depth, add		216	.037		.19	2.23		2.42	3.67
1980	1" diameter		64	.125		.30	7.55		7.85	12
1985	For each additional 1/4" depth, add		192	.042		.30	2.51		2.81	4.22
1990	For drilling up, add						40%			

05 05 21.90 Welding Steel

		Crew	Daily Output	Labor-Hours	Unit	Material	2021 Bare Costs Labor	Equipment	Total	Total Incl O&P
0010	**WELDING STEEL**, Structural R050521-20									
0020	Field welding, 1/8" E6011, cost per welder, no operating engineer	E-14	8	1	Hr.	5.45	62.50	18.60	86.55	123
0200	With 1/2 operating engineer	E-13	8	1.500		5.45	90	18.60	114.05	165
0300	With 1 operating engineer	E-12	8	2		5.45	118	18.60	142.05	206
0500	With no operating engineer, 2# weld rod per ton	E-14	8	1	Ton	5.45	62.50	18.60	86.55	123
0600	8# E6011 per ton	"	2	4		22	249	74.50	345.50	490
0800	With one operating engineer per welder, 2# E6011 per ton	E-12	8	2		5.45	118	18.60	142.05	206
0900	8# E6011 per ton	"	2	8		22	470	74.50	566.50	820
1200	Continuous fillet, down welding									
1300	Single pass, 1/8" thick, 0.1#/L.F.	E-14	150	.053	L.F.	.27	3.32	.99	4.58	6.55
1400	3/16" thick, 0.2#/L.F.		75	.107		.54	6.65	1.98	9.17	13.10
1500	1/4" thick, 0.3#/L.F.		50	.160		.82	9.95	2.97	13.74	19.55
1610	5/16" thick, 0.4#/L.F.		38	.211		1.09	13.10	3.91	18.10	26
1800	3 passes, 3/8" thick, 0.5#/L.F.		30	.267		1.36	16.60	4.96	22.92	32.50
2010	4 passes, 1/2" thick, 0.7#/L.F.		22	.364		1.90	22.50	6.75	31.15	44.50
2600	For vertical joint welding, add						20%			
2700	Overhead joint welding, add						300%			
2900	For semi-automatic welding, obstructed joints, deduct						5%			
3000	Exposed joints, deduct						15%			
4000	Cleaning and welding plates, bars, or rods									
4010	to existing beams, columns, or trusses	E-14	12	.667	L.F.	1.36	41.50	12.40	55.26	79.50

05 05 23 – Metal Fastenings

05 05 23.10 Bolts and Hex Nuts

		Crew	Daily Output	Labor-Hours	Unit	Material	2021 Bare Costs Labor	Equipment	Total	Total Incl O&P
0010	**BOLTS & HEX NUTS**, Steel, A307									
0100	1/4" diameter, 1/2" long G	1 Sswk	140	.057	Ea.	.14	3.45		3.59	5.50
0200	1" long G		140	.057		.17	3.45		3.62	5.55

116

05 05 23 – Metal Fastenings

05 05 23.10 Bolts and Hex Nuts		Crew	Daily Output	Labor-Hours	Unit	Material	2021 Bare Costs Labor	Equipment	Total	Total Incl O&P
0400	3" long	G 1 Sswk	130	.062	Ea.	.41	3.71		4.12	6.20
0600	3/8" diameter, 1" long	G	130	.062		.17	3.71		3.88	5.95
0800	3" long	G	120	.067		.26	4.02		4.28	6.50
1100	1/2" diameter, 1-1/2" long	G 1 Sswk	120	.067	Ea.	.39	4.02		4.41	6.65
1400	6" long	G	110	.073		1.04	4.39		5.43	7.95
1600	5/8" diameter, 1-1/2" long	G	120	.067		.82	4.02		4.84	7.10
2000	8" long	G	105	.076		2.42	4.59		7.01	9.75
2200	3/4" diameter, 2" long	G	120	.067		1.18	4.02		5.20	7.50
2600	10" long	G	85	.094		4.44	5.70		10.14	13.70
2800	1" diameter, 3" long	G	105	.076		3.11	4.59		7.70	10.50
3000	12" long	G	75	.107		7.75	6.45		14.20	18.50
3100	For galvanized, add					75%				
3200	For stainless, add					350%				

05 05 23.30 Lag Screws

		Crew	Daily Output	Labor-Hours	Unit	Material	Labor	Equipment	Total	Total Incl O&P
0010	**LAG SCREWS**									
0020	Steel, 1/4" diameter, 2" long	G 1 Carp	200	.040	Ea.	.13	2.19		2.32	3.41
0100	3/8" diameter, 3" long	G	150	.053		.44	2.92		3.36	4.83
0200	1/2" diameter, 3" long	G	130	.062		1.08	3.37		4.45	6.20
0300	5/8" diameter, 3" long	G	120	.067		1.72	3.65		5.37	7.35

05 05 23.35 Machine Screws

		Crew	Daily Output	Labor-Hours	Unit	Material	Labor	Equipment	Total	Total Incl O&P
0010	**MACHINE SCREWS**									
0020	Steel, round head, #8 x 1" long	G 1 Carp	4.80	1.667	C	3.98	91		94.98	140
0110	#8 x 2" long	G	2.40	3.333		14.70	182		196.70	288
0200	#10 x 1" long	G	4	2		6.25	109		115.25	170
0300	#10 x 2" long	G	2	4		10.90	219		229.90	335

05 05 23.50 Powder Actuated Tools and Fasteners

		Crew	Daily Output	Labor-Hours	Unit	Material	Labor	Equipment	Total	Total Incl O&P
0010	**POWDER ACTUATED TOOLS & FASTENERS**									
0020	Stud driver, .22 caliber, single shot				Ea.	146			146	160
0100	.27 caliber, semi automatic, strip				"	625			625	685
0300	Powder load, single shot, .22 cal, power level 2, brown				C	7.85			7.85	8.60
0400	Strip, .27 cal, power level 4, red					12			12	13.20
0600	Drive pin, .300 x 3/4" long	G 1 Carp	4.80	1.667		16.95	91		107.95	155
0700	.300 x 3" long with washer	G "	4	2		22	109		131	187

05 05 23.80 Vibration and Bearing Pads

		Crew	Daily Output	Labor-Hours	Unit	Material	Labor	Equipment	Total	Total Incl O&P
0010	**VIBRATION & BEARING PADS**									
0300	Laminated synthetic rubber impregnated cotton duck, 1/2" thick	2 Sswk	24	.667	S.F.	83	40		123	153
0400	1" thick		20	.800		155	48		203	246
0600	Neoprene bearing pads, 1/2" thick		24	.667		30	40		70	95
0700	1" thick		20	.800		60	48		108	141
0900	Fabric reinforced neoprene, 5000 psi, 1/2" thick		24	.667		12.25	40		52.25	75.50
1000	1" thick		20	.800		24.50	48		72.50	102
1200	Felt surfaced vinyl pads, cork and sisal, 5/8" thick		24	.667		6.30	40		46.30	69
1300	1" thick		20	.800		11.35	48		59.35	87
1500	Teflon bonded to 10 ga. carbon steel, 1/32" layer		24	.667		58.50	40		98.50	127
1600	3/32" layer		24	.667		88	40		128	159
1800	Bonded to 10 ga. stainless steel, 1/32" layer		24	.667		104	40		144	177
1900	3/32" layer	2 Sswk	24	.667	S.F.	126	40		166	201
2100	Circular machine leveling pad & stud				Kip	7.20			7.20	7.90

For customer support on your Site Work & Landscape Costs with RSMeans Data, call 800.448.8182.

117

05 05 23 – Metal Fastenings

05 05 23.90 Welding Rod	Crew	Daily Output	Labor-Hours	Unit	Material	2021 Bare Costs Labor	Equipment	Total	Total Incl O&P
0010 **WELDING ROD**									
0020 Steel, type 6011, 1/8" diam., less than 500#				Lb.	2.72			2.72	2.99
0100 500# to 2,000#					2.45			2.45	2.70
0200 2,000# to 5,000#					2.30			2.30	2.53
0300 5/32" diam., less than 500#					3.94			3.94	4.33
0310 500# to 2,000#					3.55			3.55	3.91
0320 2,000# to 5,000#					3.34			3.34	3.67
0400 3/16" diam., less than 500#					2.87			2.87	3.16
0500 500# to 2,000#					2.59			2.59	2.85
0600 2,000# to 5,000#					2.43			2.43	2.68
0620 Steel, type 6010, 1/8" diam., less than 500#					5.70			5.70	6.30
0630 500# to 2,000#					5.15			5.15	5.65
0640 2,000# to 5,000#					4.83			4.83	5.30
0650 Steel, type 7018 Low Hydrogen, 1/8" diam., less than 500#					2.99			2.99	3.28
0660 500# to 2,000#					2.69			2.69	2.96
0670 2,000# to 5,000#					2.53			2.53	2.78
0700 Steel, type 7024 Jet Weld, 1/8" diam., less than 500#					3.72			3.72	4.09
0710 500# to 2,000#					3.35			3.35	3.69
0720 2,000# to 5,000#					3.15			3.15	3.46
1550 Aluminum, type 4043 TIG, 1/8" diam., less than 10#					7			7	7.70
1560 10# to 60#					6.30			6.30	6.90
1570 Over 60#					5.90			5.90	6.50
1600 Aluminum, type 5356 TIG, 1/8" diam., less than 10#					6.55			6.55	7.20
1610 10# to 60#					5.90			5.90	6.50
1620 Over 60#					5.55			5.55	6.10
1900 Cast iron, type 8 Nickel, 1/8" diam., less than 500#					46.50			46.50	51.50
1910 500# to 1,000#					42			42	46
1920 Over 1,000#					39.50			39.50	43.50
2000 Stainless steel, type 316/316L, 1/8" diam., less than 500#					7.20			7.20	7.90
2100 500# to 1,000#					6.50			6.50	7.15
2220 Over 1,000#					6.10			6.10	6.70

05 12 23 – Structural Steel for Buildings

05 12 23.15 Columns, Lightweight

	Crew	Daily Output	Labor-Hours	Unit	Material	2021 Bare Costs Labor	Equipment	Total	Total Incl O&P
0010 **COLUMNS, LIGHTWEIGHT**									
1000 Lightweight units (lally), 3-1/2" diameter	E-2	780	.072	L.F.	5.40	4.28	2.20	11.88	14.90
1050 4" diameter	"	900	.062	"	9.25	3.71	1.90	14.86	18
5800 Adjustable jack post, 8' maximum height, 2-3/4" diameter G				Ea.	45.50			45.50	50
5850 4" diameter G				Ea.	72.50			72.50	80

05 12 23.17 Columns, Structural

	Crew	Daily Output	Labor-Hours	Unit	Material	2021 Bare Costs Labor	Equipment	Total	Total Incl O&P
0010 **COLUMNS, STRUCTURAL** R051223-10									
0015 Made from recycled materials									
0020 Shop fab'd for 100-ton, 1-2 story project, bolted connections									
0800 Steel, concrete filled, extra strong pipe, 3-1/2" diameter	E-2	660	.085	L.F.	43.50	5.05	2.60	51.15	58.50
0830 4" diameter		780	.072		48.50	4.28	2.20	54.98	62.50
0890 5" diameter		1020	.055		58	3.27	1.68	62.95	70.50
0930 6" diameter		1200	.047		76.50	2.78	1.43	80.71	90
0940 8" diameter		1100	.051		76.50	3.04	1.56	81.10	90.50
1100 For galvanizing, add				Lb.	.33			.33	.37

05 12 Structural Steel Framing

05 12 23 – Structural Steel for Buildings

05 12 23.17 Columns, Structural

		Crew	Daily Output	Labor-Hours	Unit	Material	2021 Bare Costs Labor	Equipment	Total	Total Incl O&P
1500	Steel pipe, extra strong, no concrete, 3" to 5" diameter ⑥	E-2	16000	.004	Lb.	1.32	.21	.11	1.64	1.89
1600	6" to 12" diameter ⑥		14000	.004		1.32	.24	.12	1.68	1.95
3300	Structural tubing, square, A500GrB, 4" to 6" square, light section ⑥		11270	.005		1.32	.30	.15	1.77	2.07
3600	Heavy section ⑥		32000	.002		1.32	.10	.05	1.47	1.67
5100	Structural tubing, rect., 5" to 6" wide, light section ⑥		8000	.007		1.32	.42	.21	1.95	2.33
5200	Heavy section ⑥		12000	.005		1.32	.28	.14	1.74	2.04
5300	7" to 10" wide, light section ⑥		15000	.004		1.32	.22	.11	1.65	1.92
5400	Heavy section ⑥		18000	.003	▼	1.32	.19	.10	1.61	1.83
6800	W Shape, A992 steel, 2 tier, W8 x 24 ⑥		1080	.052	L.F.	35	3.09	1.59	39.68	45
6900	W8 x 48 ⑥		1032	.054		69.50	3.24	1.66	74.40	83.50
7050	W10 x 68 ⑥		984	.057		98.50	3.39	1.74	103.63	116
7100	W10 x 112 ⑥		960	.058		163	3.48	1.78	168.26	186
7200	W12 x 87 ⑥		984	.057		126	3.39	1.74	131.13	146
7250	W12 x 120 ⑥		960	.058		174	3.48	1.78	179.26	199
7350	W14 x 74 ⑥		984	.057		107	3.39	1.74	112.13	125
7450	W14 x 176 ⑥	▼	912	.061	▼	255	3.66	1.88	260.54	289
8090	For projects 75 to 99 tons, add				%	10%				
8092	50 to 74 tons, add					20%				
8094	25 to 49 tons, add					30%	10%			
8096	10 to 24 tons, add					50%	25%			
8098	2 to 9 tons, add					75%	50%			
8099	Less than 2 tons, add				▼	100%	100%			

05 12 23.18 Corner Guards

		Crew	Daily Output	Labor-Hours	Unit	Material	2021 Bare Costs Labor	Equipment	Total	Total Incl O&P
0010	**CORNER GUARDS**									
0020	Steel angle w/anchors, 1" x 1" x 1/4", 1.5#/L.F.	2 Carp	160	.100	L.F.	7.45	5.45		12.90	16.35
0100	2" x 2" x 1/4" angles, 3.2#/L.F.		150	.107		11.50	5.85		17.35	21.50
0200	3" x 3" x 5/16" angles, 6.1#/L.F.		140	.114		19	6.25		25.25	30.50
0300	4" x 4" x 5/16" angles, 8.2#/L.F.	▼	120	.133		15.75	7.30		23.05	28
0350	For angles drilled and anchored to masonry, add					15%	120%			
0370	Drilled and anchored to concrete, add				▼	20%	170%			
0400	For galvanized angles, add				L.F.	35%				
0450	For stainless steel angles, add				"	100%				

05 12 23.20 Curb Edging

		Crew	Daily Output	Labor-Hours	Unit	Material	2021 Bare Costs Labor	Equipment	Total	Total Incl O&P
0010	**CURB EDGING**									
0020	Steel angle w/anchors, shop fabricated, on forms, 1" x 1", 0.8#/L.F. ⑥	E-4	350	.091	L.F.	1.69	5.55	.43	7.67	10.95
0100	2" x 2" angles, 3.92#/L.F. ⑥		330	.097		6.65	5.90	.45	13	16.90
0200	3" x 3" angles, 6.1#/L.F. ⑥		300	.107		10.90	6.50	.50	17.90	22.50
0300	4" x 4" angles, 8.2#/L.F. ⑥		275	.116		14.20	7.05	.54	21.79	27
1000	6" x 4" angles, 12.3#/L.F. ⑥		250	.128		20.50	7.80	.60	28.90	35.50
1050	Steel channels with anchors, on forms, 3" channel, 5#/L.F. ⑥		290	.110		8.35	6.70	.51	15.56	20
1100	4" channel, 5.4#/L.F. ⑥		270	.119		8.95	7.20	.55	16.70	21.50
1200	6" channel, 8.2#/L.F. ⑥		255	.125		14.20	7.65	.58	22.43	28
1300	8" channel, 11.5#/L.F. ⑥		225	.142		19.40	8.65	.66	28.71	35.50
1400	10" channel, 15.3#/L.F. ⑥		180	.178		25.50	10.80	.83	37.13	45.50
1500	12" channel, 20.7#/L.F. ⑥	▼	140	.229		34	13.90	1.06	48.96	60
2000	For curved edging, add				▼	35%	10%			

05 12 23.40 Lightweight Framing

		Crew	Daily Output	Labor-Hours	Unit	Material	2021 Bare Costs Labor	Equipment	Total	Total Incl O&P
0010	**LIGHTWEIGHT FRAMING** R051223-35									
0015	Made from recycled materials									
0400	Angle framing, field fabricated, 4" and larger ⑥	E-3	440	.055	Lb.	.77	3.33	.34	4.44	6.35
0450	Less than 4" angles R051223-45 ⑥		265	.091		.79	5.50	.56	6.85	10.05
0600	Channel framing, field fabricated, 8" and larger ⑥	▼	500	.048	▼	.79	2.93	.30	4.02	5.75

05 12 Structural Steel Framing

05 12 23 – Structural Steel for Buildings

	05 12 23.40 Lightweight Framing		Crew	Daily Output	Labor-Hours	Unit	Material	2021 Bare Costs Labor	Equipment	Total	Total Incl O&P
0650	Less than 8" channels	G	E-3	335	.072	Lb.	.79	4.37	.44	5.60	8.10
1000	Continuous slotted channel framing system, shop fab, simple framing	G	2 Sswk	2400	.007		4.09	.40		4.49	5.10
1200	Complex framing	G	"	1600	.010		4.62	.60		5.22	6.05
1300	Cross bracing, rods, shop fabricated, 3/4" diameter	G	E-3	700	.034		1.58	2.09	.21	3.88	5.20
1310	7/8" diameter	G		850	.028		1.58	1.72	.18	3.48	4.59
1320	1" diameter	G		1000	.024		1.58	1.46	.15	3.19	4.16
1330	Angle, 5" x 5" x 3/8"	G		2800	.009		1.58	.52	.05	2.15	2.61
1350	Hanging lintels, shop fabricated	G	↓	850	.028		1.58	1.72	.18	3.48	4.59
1380	Roof frames, shop fabricated, 3'-0" square, 5' span	G	E-2	4200	.013		1.58	.80	.41	2.79	3.41
1400	Tie rod, not upset, 1-1/2" to 4" diameter, with turnbuckle	G	2 Sswk	800	.020		1.71	1.21		2.92	3.76
1420	No turnbuckle	G		700	.023		1.65	1.38		3.03	3.94
1500	Upset, 1-3/4" to 4" diameter, with turnbuckle	G		800	.020		1.71	1.21		2.92	3.76
1520	No turnbuckle	G	↓	700	.023	↓	1.65	1.38		3.03	3.94

05 12 23.45 Lintels

			Crew	Daily Output	Labor-Hours	Unit	Material	2021 Bare Costs Labor	Equipment	Total	Total Incl O&P
0010	**LINTELS**										
0015	Made from recycled materials										
0020	Plain steel angles, shop fabricated, under 500 lb.	G	1 Bric	550	.015	Lb.	1.02	.78		1.80	2.30
0100	500 to 1,000 lb.	G		640	.013		.99	.67		1.66	2.10
0200	1,000 to 2,000 lb.	G	↓	640	.013	↓	.96	.67		1.63	2.07
0300	2,000 to 4,000 lb.	G	1 Bric	640	.013	Lb.	.94	.67		1.61	2.04
0500	For built-up angles and plates, add to above	G					1.32			1.32	1.45
0700	For engineering, add to above						.13			.13	.15
0900	For galvanizing, add to above, under 500 lb.						.39			.39	.42
0950	500 to 2,000 lb.						.35			.35	.39
1000	Over 2,000 lb.					↓	.33			.33	.37
2000	Steel angles, 3-1/2" x 3", 1/4" thick, 2'-6" long	G	1 Bric	47	.170	Ea.	14.25	9.15		23.40	29.50
2100	4'-6" long	G		26	.308		25.50	16.50		42	53
2600	4" x 3-1/2", 1/4" thick, 5'-0" long	G		21	.381		32.50	20.50		53	67
2700	9'-0" long	G	↓	12	.667	↓	59	36		95	119

05 12 23.60 Pipe Support Framing

			Crew	Daily Output	Labor-Hours	Unit	Material	2021 Bare Costs Labor	Equipment	Total	Total Incl O&P
0010	**PIPE SUPPORT FRAMING**										
0020	Under 10#/L.F., shop fabricated	G	E-4	3900	.008	Lb.	1.77	.50	.04	2.31	2.75
0200	10.1 to 15#/L.F.	G		4300	.007		1.74	.45	.03	2.22	2.66
0400	15.1 to 20#/L.F.	G		4800	.007		1.71	.41	.03	2.15	2.55
0600	Over 20#/L.F.	G	↓	5400	.006	↓	1.69	.36	.03	2.08	2.45

05 12 23.65 Plates

			Crew	Daily Output	Labor-Hours	Unit	Material	2021 Bare Costs Labor	Equipment	Total	Total Incl O&P
0010	**PLATES**	R051223-80									
0015	Made from recycled materials										
0020	For connections & stiffener plates, shop fabricated										
0050	1/8" thick (5.1 lb./S.F.)	G				S.F.	6.75			6.75	7.40
0100	1/4" thick (10.2 lb./S.F.)	G					13.45			13.45	14.80
0300	3/8" thick (15.3 lb./S.F.)	G					20			20	22
0400	1/2" thick (20.4 lb./S.F.)	G					27			27	29.50
0450	3/4" thick (30.6 lb./S.F.)	G					40.50			40.50	44.50
0500	1" thick (40.8 lb./S.F.)	G				↓	54			54	59
2000	Steel plate, warehouse prices, no shop fabrication										
2100	1/4" thick (10.2 lb./S.F.)	G				S.F.	8.10			8.10	8.90
2210	1/4" steel plate, welded in place		E-18	528	.076		8.10	4.58	2.89	15.57	19.15
2220	1/2" steel plate, welded in place			480	.083		16.15	5.05	3.18	24.38	29
2230	3/4" steel plate, welded in place			384	.104		24	6.30	3.97	34.27	40.50
2240	1" steel plate, welded in place			320	.125		32.50	7.55	4.77	44.82	52.50
2250	1-1/2" steel plate, welded in place		↓	256	.156	↓	48.50	9.45	5.95	63.90	74.50

05 12 Structural Steel Framing

05 12 23 – Structural Steel for Buildings

05 12 23.65 **Plates**		Crew	Daily Output	Labor-Hours	Unit	Material	2021 Bare Costs Labor	Equipment	Total	Total Incl O&P
2260	2" steel plate, welded in place	E-18	192	.208	S.F.	64.50	12.60	7.95	85.05	99

05 12 23.75 Structural Steel Members

			Crew	Daily Output	Labor-Hours	Unit	Material	Labor	Equipment	Total	Total Incl O&P
0010	**STRUCTURAL STEEL MEMBERS**	R051223-10									
0015	Made from recycled materials										
0020	Shop fab'd for 100-ton, 1-2 story project, bolted connections										
0300	W 8 x 10	G	E-2	600	.093	L.F.	14.50	5.55	2.86	22.91	27.50
0500	x 31	G		550	.102		45	6.05	3.11	54.16	62
0600	W 10 x 12	G		600	.093		17.40	5.55	2.86	25.81	31
0700	x 22	G	↓	600	.093	↓	32	5.55	2.86	40.41	46.50
0900	x 49	G	E-2	550	.102	L.F.	71	6.05	3.11	80.16	90.50
1100	W 12 x 16	G		880	.064		23	3.80	1.95	28.75	33.50
1700	x 72	G		640	.088		104	5.20	2.68	111.88	126
1900	W 14 x 26	G		990	.057		37.50	3.37	1.73	42.60	48.50
2500	x 120	G	↓	720	.078		174	4.64	2.38	181.02	202
3300	W 18 x 35	G	E-5	960	.083		51	5	1.94	57.94	66
4100	W 21 x 44	G		1064	.075		64	4.51	1.75	70.26	79
4900	W 24 x 55	G		1110	.072		80	4.33	1.68	86.01	96.50
5900	x 94	G		1190	.067		136	4.04	1.56	141.60	158
6100	W 30 x 99	G		1200	.067		144	4	1.55	149.55	166
6700	W 33 x 118	G		1176	.068		171	4.08	1.58	176.66	196
7300	W 36 x 135	G	↓	1170	.068		196	4.11	1.59	201.70	223
8490	For projects 75 to 99 tons, add						10%				
8492	50 to 74 tons, add						20%				
8494	25 to 49 tons, add						30%	10%			
8496	10 to 24 tons, add						50%	25%			
8498	2 to 9 tons, add						75%	50%			
8499	Less than 2 tons, add					↓	100%	100%			

05 12 23.77 Structural Steel Projects

			Crew	Daily Output	Labor-Hours	Unit	Material	Labor	Equipment	Total	Total Incl O&P
0010	**STRUCTURAL STEEL PROJECTS**	R050516-30									
0015	Made from recycled materials										
0020	Shop fab'd for 100-ton, 1-2 story project, bolted connections										
0200	Apartments, nursing homes, etc., 1 to 2 stories R050523-10	G	E-5	10.30	7.767	Ton	2,650	465	181	3,296	3,825
1300	Industrial bldgs., 1 story, beams & girders, steel bearing	G		12.90	6.202		2,650	370	144	3,164	3,625
1400	Masonry bearing R051223-10	G	↓	10	8		2,650	480	186	3,316	3,850
2800	Power stations, fossil fuels, simple connections	G	E-6	11	11.636		2,650	700	197	3,547	4,200
2900	Moment/composite connections R051223-20	G	"	5.70	22.456		3,950	1,350	380	5,680	6,850
5390	For projects 75 to 99 tons, add						10%				
5392	50 to 74 tons, add R051223-25						20%				
5394	25 to 49 tons, add						30%	10%			
5396	10 to 24 tons, add R051223-30						50%	25%			
5398	2 to 9 tons, add						75%	50%			
5399	Less than 2 tons, add					↓	100%	100%			

05 12 23.78 Structural Steel Secondary Members

			Crew	Daily Output	Labor-Hours	Unit	Material	Labor	Equipment	Total	Total Incl O&P
0010	**STRUCTURAL STEEL SECONDARY MEMBERS**										
0015	Made from recycled materials										
0020	Shop fabricated for 20-ton girt/purlin framing package, materials only										
0100	Girts/purlins, C/Z-shapes, includes clips and bolts										
0110	6" x 2-1/2" x 2-1/2", 16 ga., 3.0 lb./L.F.					L.F.	3.56			3.56	3.92
0115	14 ga., 3.5 lb./L.F.						4.16			4.16	4.57
0120	8" x 2-3/4" x 2-3/4", 16 ga., 3.4 lb./L.F.					↓	4.04			4.04	4.44
0125	14 ga., 4.1 lb./L.F.					L.F.	4.87			4.87	5.35
0130	12 ga., 5.6 lb./L.F.					↓	6.65			6.65	7.30

05 12 Structural Steel Framing

05 12 23 – Structural Steel for Buildings

05 12 23.78 Structural Steel Secondary Members	Crew	Daily Output	Labor-Hours	Unit	Material	2021 Bare Costs Labor	2021 Bare Costs Equipment	Total	Total Incl O&P
0135	10" x 3-1/2" x 3-1/2", 14 ga., 4.7 lb./L.F.				L.F.	5.60			5.60	6.15
0140	12 ga., 6.7 lb./L.F.					7.95			7.95	8.75
0145	12" x 3-1/2" x 3-1/2", 14 ga., 5.3 lb./L.F.					6.30			6.30	6.90
0150	12 ga., 7.4 lb./L.F.				↓	8.80			8.80	9.65
0200	Eave struts, C-shape, includes clips and bolts									
0210	6" x 4" x 3", 16 ga., 3.1 lb./L.F.				L.F.	3.68			3.68	4.05
0215	14 ga., 3.9 lb./L.F.					4.63			4.63	5.10
0220	8" x 4" x 3", 16 ga., 3.5 lb./L.F.					4.16			4.16	4.57
0225	14 ga., 4.4 lb./L.F.					5.20			5.20	5.75
0230	12 ga., 6.2 lb./L.F.					7.35			7.35	8.10
0235	10" x 5" x 3", 14 ga., 5.2 lb./L.F.					6.15			6.15	6.80
0240	12 ga., 7.3 lb./L.F.					8.65			8.65	9.55
0245	12" x 5" x 4", 14 ga., 6.0 lb./L.F.					7.10			7.10	7.85
0250	12 ga., 8.4 lb./L.F.				↓	9.95			9.95	10.95
0300	Rake/base angle, excludes concrete drilling and expansion anchors									
0310	2" x 2", 14 ga., 1.0 lb./L.F.	2 Sswk	640	.025	L.F.	1.19	1.51		2.70	3.64
0315	3" x 2", 14 ga., 1.3 lb./L.F.		535	.030		1.54	1.80		3.34	4.49
0320	3" x 3", 14 ga., 1.6 lb./L.F.		500	.032		1.90	1.93		3.83	5.10
0325	4" x 3", 14 ga., 1.8 lb./L.F.	↓	480	.033	↓	2.14	2.01		4.15	5.45
0600	Installation of secondary members, erection only									
0610	Girts, purlins, eave struts, 16 ga., 6" deep	E-18	100	.400	Ea.		24	15.25	39.25	54
0615	8" deep		80	.500			30	19.10	49.10	67.50
0620	14 ga., 6" deep		80	.500			30	19.10	49.10	67.50
0625	8" deep		65	.615			37	23.50	60.50	83
0630	10" deep		55	.727			44	28	72	98
0635	12" deep		50	.800			48.50	30.50	79	108
0640	12 ga., 8" deep		50	.800			48.50	30.50	79	108
0645	10" deep		45	.889			53.50	34	87.50	120
0650	12" deep	↓	40	1	↓		60.50	38	98.50	135
0900	For less than 20-ton job lots									
0905	For 15 to 19 tons, add				%	10%				
0910	For 10 to 14 tons, add					25%				
0915	For 5 to 9 tons, add					50%	50%	50%		
0920	For 1 to 4 tons, add					75%	75%	75%		
0925	For less than 1 ton, add				↓	100%	100%	100%		

05 15 Wire Rope Assemblies

05 15 16 – Steel Wire Rope Assemblies

05 15 16.05 Accessories for Steel Wire Rope

			Crew	Daily Output	Labor-Hours	Unit	Material	2021 Bare Costs Labor	2021 Bare Costs Equipment	Total	Total Incl O&P
0010	**ACCESSORIES FOR STEEL WIRE ROPE**										
0015	Made from recycled materials										
1500	Thimbles, heavy duty, 1/4"	G	E-17	160	.100	Ea.	.45	6.15		6.60	10
1510	1/2"	G		160	.100		1.98	6.15		8.13	11.70
1520	3/4"	G		105	.152		4.50	9.35		13.85	19.40
1530	1"	G		52	.308		9	18.85		27.85	39
1540	1-1/4"	G		38	.421		13.85	26		39.85	55.50
1550	1-1/2"	G		13	1.231		39	75.50		114.50	160
1560	1-3/4"	G		8	2		80.50	123		203.50	279
1570	2"	G		6	2.667		117	163		280	380
1580	2-1/4"	G		4	4		158	245		403	555
1600	Clips, 1/4" diameter	G	↓	160	.100		1.97	6.15		8.12	11.65

05 15 Wire Rope Assemblies

05 15 16 – Steel Wire Rope Assemblies

05 15 16.05 Accessories for Steel Wire Rope		Crew	Daily Output	Labor-Hours	Unit	Material	2021 Bare Costs Labor	Equipment	Total	Total Incl O&P
1610	3/8" diameter	G	E-17	160	.100	Ea.	2.16	6.15		8.31	11.90
1620	1/2" diameter	G		160	.100		3.47	6.15		9.62	13.30
1630	3/4" diameter	G		102	.157		5.65	9.60		15.25	21
1640	1" diameter	G		64	.250		9.40	15.35		24.75	34
1650	1-1/4" diameter	G		35	.457		15.40	28		43.40	60.50
1670	1-1/2" diameter	G		26	.615		21	37.50		58.50	81.50
1680	1-3/4" diameter	G		16	1		48.50	61.50		110	148
1690	2" diameter	G		12	1.333		54	81.50		135.50	185
1700	2-1/4" diameter	G		10	1.600		79	98		177	239
1800	Sockets, open swage, 1/4" diameter	G		160	.100		32	6.15		38.15	44.50
1810	1/2" diameter	G		77	.208		46	12.75		58.75	70
1820	3/4" diameter	G		19	.842		71	51.50		122.50	159
1830	1" diameter	G		9	1.778		127	109		236	310
1840	1-1/4" diameter	G		5	3.200		177	196		373	500
1850	1-1/2" diameter	G		3	5.333		390	325		715	930
1860	1-3/4" diameter	G		3	5.333		690	325		1,015	1,250
1870	2" diameter	G		1.50	10.667		1,050	655		1,705	2,150
1900	Closed swage, 1/4" diameter	G		160	.100		19.10	6.15		25.25	30.50
1910	1/2" diameter	G		104	.154		33	9.45		42.45	51
1920	3/4" diameter	G		32	.500		49.50	30.50		80	102
1930	1" diameter	G		15	1.067		86.50	65.50		152	197
1940	1-1/4" diameter	G		7	2.286		130	140		270	360
1950	1-1/2" diameter	G		4	4		236	245		481	640
1960	1-3/4" diameter	G		3	5.333		350	325		675	890
1970	2" diameter	G		2	8		675	490		1,165	1,500
2000	Open spelter, galv., 1/4" diameter	G		160	.100		75	6.15		81.15	91.50
2010	1/2" diameter	G		70	.229		78	14		92	108
2020	3/4" diameter	G		26	.615		117	37.50		154.50	188
2030	1" diameter	G		10	1.600		325	98		423	510
2040	1-1/4" diameter	G		5	3.200		465	196		661	815
2050	1-1/2" diameter	G		4	4		985	245		1,230	1,450
2060	1-3/4" diameter	G		2	8		1,725	490		2,215	2,650
2070	2" diameter	G		1.20	13.333		1,975	815		2,790	3,450
2080	2-1/2" diameter	G		1	16		3,650	980		4,630	5,525
2100	Closed spelter, galv., 1/4" diameter	G		160	.100		32.50	6.15		38.65	45.50
2110	1/2" diameter	G		88	.182		35	11.15		46.15	55.50
2120	3/4" diameter	G		30	.533		52.50	32.50		85	109
2130	1" diameter	G		13	1.231		112	75.50		187.50	240
2140	1-1/4" diameter	G		7	2.286		179	140		319	415
2150	1-1/2" diameter	G		6	2.667		385	163		548	680
2160	1-3/4" diameter	G		2.80	5.714		515	350		865	1,100
2170	2" diameter	G		2	8		635	490		1,125	1,450
2200	Jaw & jaw turnbuckles, 1/4" x 4"	G		160	.100		8.40	6.15		14.55	18.75
2250	1/2" x 6"	G		96	.167		10.65	10.20		20.85	27.50
2260	1/2" x 9"	G		77	.208		14.20	12.75		26.95	35.50
2270	1/2" x 12"	G		66	.242		15.95	14.85		30.80	40.50
2300	3/4" x 6"	G		38	.421		21	26		47	63
2310	3/4" x 9"	G		30	.533		23	32.50		55.50	76
2320	3/4" x 12"	G		28	.571		29.50	35		64.50	86.50
2330	3/4" x 18"	G		23	.696		35.50	42.50		78	105
2350	1" x 6"	G		17	.941		40.50	57.50		98	134
2360	1" x 12"	G		13	1.231		44.50	75.50		120	166
2370	1" x 18"	G		10	1.600		66.50	98		164.50	225

05 15 Wire Rope Assemblies

05 15 16 – Steel Wire Rope Assemblies

05 15 16.05 Accessories for Steel Wire Rope		Crew	Daily Output	Labor-Hours	Unit	Material	2021 Bare Costs			Total	Total Incl O&P
							Labor	Equipment			
2380	1" x 24"	G	E-17	9	1.778	Ea.	73	109		182	250
2400	1-1/4" x 12"	G		7	2.286		74.50	140		214.50	299
2410	1-1/4" x 18"	G		6.50	2.462		92	151		243	335
2420	1-1/4" x 24"	G		5.60	2.857		124	175		299	410
2450	1-1/2" x 12"	G		5.20	3.077		305	189		494	630
2460	1-1/2" x 18"	G		4	4		325	245		570	740
2470	1-1/2" x 24"	G		3.20	5		440	305		745	960
2500	1-3/4" x 18"	G		3.20	5		665	305		970	1,200
2510	1-3/4" x 24"	G		2.80	5.714		755	350		1,105	1,375
2550	2" x 24"	G		1.60	10		1,025	615		1,640	2,075

05 15 16.50 Steel Wire Rope		Crew	Daily Output	Labor-Hours	Unit	Material	Labor	Equipment		Total	Total Incl O&P
0010	**STEEL WIRE ROPE**										
0015	Made from recycled materials										
0020	6 x 19, bright, fiber core, 5000' rolls, 1/2" diameter	G				L.F.	.77			.77	.85
0050	Steel core	G					1.01			1.01	1.11
0100	Fiber core, 1" diameter	G					2.59			2.59	2.85
0150	Steel core	G					2.96			2.96	3.25
0300	6 x 19, galvanized, fiber core, 1/2" diameter	G					1.13			1.13	1.25
0350	Steel core	G					1.30			1.30	1.43
0400	Fiber core, 1" diameter	G					3.32			3.32	3.65
0450	Steel core	G				L.F.	3.48			3.48	3.83
0500	6 x 7, bright, IPS, fiber core, <500 L.F. w/acc., 1/4" diameter	G	E-17	6400	.003		.46	.15		.61	.75
0510	1/2" diameter	G		2100	.008		1.12	.47		1.59	1.95
0520	3/4" diameter	G		960	.017		2.03	1.02		3.05	3.81
0550	6 x 19, bright, IPS, IWRC, <500 L.F. w/acc., 1/4" diameter	G		5760	.003		.85	.17		1.02	1.20
0560	1/2" diameter	G		1730	.009		1.38	.57		1.95	2.39
0570	3/4" diameter	G		770	.021		2.39	1.27		3.66	4.60
0580	1" diameter	G		420	.038		4.05	2.34		6.39	8.05
0590	1-1/4" diameter	G		290	.055		6.70	3.38		10.08	12.65
0600	1-1/2" diameter	G		192	.083		8.25	5.10		13.35	17
0610	1-3/4" diameter	G	E-18	240	.167		13.15	10.05	6.35	29.55	37
0620	2" diameter	G		160	.250		16.90	15.10	9.55	41.55	52
0630	2-1/4" diameter	G		160	.250		22.50	15.10	9.55	47.15	58.50
0650	6 x 37, bright, IPS, IWRC, <500 L.F. w/acc., 1/4" diameter	G	E-17	6400	.003		1.03	.15		1.18	1.38
0660	1/2" diameter	G		1730	.009		1.75	.57		2.32	2.81
0670	3/4" diameter	G		770	.021		2.83	1.27		4.10	5.10
0680	1" diameter	G		430	.037		4.49	2.28		6.77	8.45
0690	1-1/4" diameter	G		290	.055		6.80	3.38		10.18	12.70
0700	1-1/2" diameter	G		190	.084		9.70	5.15		14.85	18.65
0710	1-3/4" diameter	G	E-18	260	.154		15.40	9.30	5.85	30.55	37.50
0720	2" diameter	G		200	.200		20	12.10	7.65	39.75	49
0730	2-1/4" diameter	G		160	.250		26.50	15.10	9.55	51.15	62.50
0800	6 x 19 & 6 x 37, swaged, 1/2" diameter	G	E-17	1220	.013		4.34	.80		5.14	6
0810	9/16" diameter	G		1120	.014		5.05	.88		5.93	6.90
0820	5/8" diameter	G		930	.017		6	1.05		7.05	8.20
0830	3/4" diameter	G		640	.025		7.60	1.53		9.13	10.75
0840	7/8" diameter	G		480	.033		9.60	2.04		11.64	13.70
0850	1" diameter	G		350	.046		11.70	2.80		14.50	17.25
0860	1-1/8" diameter	G		288	.056		14.40	3.41		17.81	21
0870	1-1/4" diameter	G		230	.070		17.45	4.26		21.71	26
0880	1-3/8" diameter	G		192	.083		20	5.10		25.10	30
0890	1-1/2" diameter	G	E-18	300	.133		24.50	8.05	5.10	37.65	45

124

05 15 Wire Rope Assemblies

05 15 16 – Steel Wire Rope Assemblies

05 15 16.60 Galvanized Steel Wire Rope and Accessories		Crew	Daily Output	Labor-Hours	Unit	Material	2021 Bare Costs Labor	Equipment	Total	Total Incl O&P
0010	**GALVANIZED STEEL WIRE ROPE & ACCESSORIES**									
0015	Made from recycled materials									
3000	Aircraft cable, galvanized, 7 x 7 x 1/8"	G	E-17	5000	.003	L.F.	.16	.20		.36	.48
3100	Clamps, 1/8"	G	"	125	.128	Ea.	1.41	7.85		9.26	13.70

05 15 16.70 Temporary Cable Safety Railing

		Crew	Daily Output	Labor-Hours	Unit	Material	Labor	Equipment	Total	Total Incl O&P
0010	**TEMPORARY CABLE SAFETY RAILING**, Each 100' strand incl.									
0020	2 eyebolts, 1 turnbuckle, 100' cable, 2 thimbles, 6 clips									
0025	Made from recycled materials									
0100	One strand using 1/4" cable & accessories	G	2 Sswk	4	4	C.L.F.	79.50	241		320.50	460
0200	1/2" cable & accessories	G	2 Sswk	2	8	C.L.F.	157	480		637	920

05 21 Steel Joist Framing

05 21 13 – Deep Longspan Steel Joist Framing

05 21 13.50 Deep Longspan Joists

		Crew	Daily Output	Labor-Hours	Unit	Material	Labor	Equipment	Total	Total Incl O&P
0010	**DEEP LONGSPAN JOISTS**									
3015	Made from recycled materials									
3500	For less than 40-ton job lots									
3502	For 30 to 39 tons, add					%	10%			
3504	20 to 29 tons, add						20%			
3506	10 to 19 tons, add						30%			
3507	5 to 9 tons, add						50%	25%		
3508	1 to 4 tons, add						75%	50%		
3509	Less than 1 ton, add						100%	100%		
4010	SLH series, 40-ton job lots, bolted cross bridging, shop primer									
4040	Spans to 200' (shipped in 3 pieces)	G	E-7	13	6.154	Ton	2,550	370	155	3,075	3,550
6100	For less than 40-ton job lots									
6102	For 30 to 39 tons, add					%	10%			
6104	20 to 29 tons, add						20%			
6106	10 to 19 tons, add						30%			
6107	5 to 9 tons, add						50%	25%		
6108	1 to 4 tons, add						75%	50%		
6109	Less than 1 ton, add						100%	100%		

05 21 16 – Longspan Steel Joist Framing

05 21 16.50 Longspan Joists

		Crew	Daily Output	Labor-Hours	Unit	Material	Labor	Equipment	Total	Total Incl O&P
0010	**LONGSPAN JOISTS**									
2000	LH series, 40-ton job lots, bolted cross bridging, shop primer									
2015	Made from recycled materials									
2600	For less than 40-ton job lots									
2602	For 30 to 39 tons, add					%	10%			
2604	20 to 29 tons, add						20%			
2606	10 to 19 tons, add						30%			
2607	5 to 9 tons, add						50%	25%		
2608	1 to 4 tons, add						75%	50%		
2609	Less than 1 ton, add						100%	100%		

05 21 19 – Open Web Steel Joist Framing

05 21 19.10 Open Web Joists		Crew	Daily Output	Labor-Hours	Unit	Material	2021 Bare Costs Labor	Equipment	Total	Total Incl O&P
0010	**OPEN WEB JOISTS**									
0015	Made from recycled materials									
0440	K series, 30' to 50' spans	G E-7	17	4.706	Ton	1,950	283	118	2,351	2,725
0800	For less than 40-ton job lots									
0802	For 30 to 39 tons, add				%	10%				
0804	20 to 29 tons, add					20%				
0806	10 to 19 tons, add					30%				
0807	5 to 9 tons, add				%	50%	25%			
0808	1 to 4 tons, add					75%	50%			
0809	Less than 1 ton, add					100%	100%			
1010	CS series, 40-ton job lots, horizontal bridging, shop primer									
1040	Spans to 30'	G E-7	12	6.667	Ton	1,975	400	168	2,543	2,975
1500	For less than 40-ton job lots									
1502	For 30 to 39 tons, add				%	10%				
1504	20 to 29 tons, add					20%				
1506	10 to 19 tons, add					30%				
1507	5 to 9 tons, add					50%	25%			
1508	1 to 4 tons, add					75%	50%			
1509	Less than 1 ton, add					100%	100%			

05 31 13 – Steel Floor Decking

05 31 13.50 Floor Decking

05 31 13.50 Floor Decking		Crew	Daily Output	Labor-Hours	Unit	Material	2021 Bare Costs Labor	Equipment	Total	Total Incl O&P
0010	**FLOOR DECKING** R053100-10									
0015	Made from recycled materials									
5100	Non-cellular composite decking, galvanized, 1-1/2" deep, 16 ga.	G E-4	3500	.009	S.F.	4.90	.56	.04	5.50	6.30
5120	18 ga.	G	3650	.009		3.02	.53	.04	3.59	4.19
5140	20 ga.	G	3800	.008		3.57	.51	.04	4.12	4.75
5200	2" deep, 22 ga.	G	3860	.008		2.41	.50	.04	2.95	3.47
5300	20 ga.	G	3600	.009		3.43	.54	.04	4.01	4.66
5400	18 ga.	G	3380	.009		3.15	.58	.04	3.77	4.41
5500	16 ga.	G	3200	.010		4.84	.61	.05	5.50	6.30
5700	3" deep, 22 ga.	G	3200	.010		2.63	.61	.05	3.29	3.88
5800	20 ga.	G	3000	.011		3.77	.65	.05	4.47	5.20
5900	18 ga.	G	2850	.011		3.15	.68	.05	3.88	4.59
6000	16 ga.	G	2700	.012		5.50	.72	.06	6.28	7.25

05 31 23 – Steel Roof Decking

05 31 23.50 Roof Decking

05 31 23.50 Roof Decking		Crew	Daily Output	Labor-Hours	Unit	Material	2021 Bare Costs Labor	Equipment	Total	Total Incl O&P
0010	**ROOF DECKING**									
0015	Made from recycled materials									
2100	Open type, 1-1/2" deep, Type B, wide rib, galv., 22 ga., under 50 sq.	G E-4	4500	.007	S.F.	2.72	.43	.03	3.18	3.70
2400	Over 500 squares	G	5100	.006		1.96	.38	.03	2.37	2.77
2600	20 ga., under 50 squares	G	3865	.008		3.03	.50	.04	3.57	4.15
2700	Over 500 squares	G	4300	.007		2.18	.45	.03	2.66	3.14
2900	18 ga., under 50 squares	G	3800	.008		3.89	.51	.04	4.44	5.10
3000	Over 500 squares	G	4300	.007		2.80	.45	.03	3.28	3.82
3050	16 ga., under 50 squares	G	3700	.009		5.25	.53	.04	5.82	6.65
3100	Over 500 squares	G	4200	.008		3.79	.46	.04	4.29	4.93
3200	3" deep, Type N, 22 ga., under 50 squares	G	3600	.009		3.75	.54	.04	4.33	5
3260	over 500 squares	G	4000	.008		2.70	.49	.04	3.23	3.76

05 31 Steel Decking

05 31 23 – Steel Roof Decking

05 31 23.50 Roof Decking		Crew	Daily Output	Labor-Hours	Unit	Material	2021 Bare Costs Labor	Equipment	Total	Total Incl O&P
3300	20 ga., under 50 squares	E-4	3400	.009	S.F.	4.09	.57	.04	4.70	5.45
3360	over 500 squares		3800	.008		2.95	.51	.04	3.50	4.07
3400	18 ga., under 50 squares		3200	.010		5.30	.61	.05	5.96	6.85
3460	over 500 squares		3600	.009		3.83	.54	.04	4.41	5.10
3500	16 ga., under 50 squares		3000	.011		7.05	.65	.05	7.75	8.80
3560	over 500 squares		3400	.009		5.05	.57	.04	5.66	6.50
3700	4-1/2" deep, Type J, 20 ga., over 50 squares		2700	.012		3.99	.72	.06	4.77	5.55
4100	6" deep, Type H, 18 ga., over 50 squares		2000	.016		6.40	.97	.07	7.44	8.65
4500	7-1/2" deep, Type H, 18 ga., over 50 squares		1690	.019		7.60	1.15	.09	8.84	10.25

05 31 33 – Steel Form Decking

05 31 33.50 Form Decking

		Crew	Daily Output	Labor-Hours	Unit	Material	Labor	Equipment	Total	Total Incl O&P
0010	**FORM DECKING**									
0015	Made from recycled materials									
6100	Slab form, steel, 28 ga., 9/16" deep, Type UFS, uncoated	E-4	4000	.008	S.F.	1.91	.49	.04	2.44	2.89
6200	Galvanized	"	4000	.008	"	1.69	.49	.04	2.22	2.65

05 35 Raceway Decking Assemblies

05 35 13 – Steel Cellular Decking

05 35 13.50 Cellular Decking

		Crew	Daily Output	Labor-Hours	Unit	Material	Labor	Equipment	Total	Total Incl O&P
0010	**CELLULAR DECKING**									
0015	Made from recycled materials									
0200	Cellular units, galv, 1-1/2" deep, Type BC, 20-20 ga., over 15 squares	E-4	1460	.022	S.F.	11.70	1.33	.10	13.13	15.05
0250	18-20 ga.		1420	.023		9.20	1.37	.10	10.67	12.35
0300	18-18 ga.		1390	.023		9.40	1.40	.11	10.91	12.65
0320	16-18 ga.		1360	.024		16.25	1.43	.11	17.79	20
0340	16-16 ga.		1330	.024		18.10	1.46	.11	19.67	22.50
0400	3" deep, Type NC, galvanized, 20-20 ga.		1375	.023		12.90	1.41	.11	14.42	16.45

05 51 Metal Stairs

05 51 13 – Metal Pan Stairs

05 51 13.50 Pan Stairs

		Crew	Daily Output	Labor-Hours	Unit	Material	Labor	Equipment	Total	Total Incl O&P
0010	**PAN STAIRS**, shop fabricated, steel stringers									
0015	Made from recycled materials									
0200	Metal pan tread for concrete in-fill, picket rail, 3'-6" wide	E-4	35	.914	Riser	600	55.50	4.25	659.75	750
0300	4'-0" wide		30	1.067		590	65	4.96	659.96	755
0350	Wall rail, both sides, 3'-6" wide		53	.604		515	36.50	2.81	554.31	625
1500	Landing, steel pan, conventional		160	.200	S.F.	79.50	12.15	.93	92.58	107
1600	Pre-erected		255	.125	"	117	7.65	.58	125.23	140
1700	Pre-erected, steel pan tread, 3'-6" wide, 2 line pipe rail	E-2	87	.644	Riser	495	38.50	19.70	553.20	625

05 51 16 – Metal Floor Plate Stairs

05 51 16.50 Floor Plate Stairs

		Crew	Daily Output	Labor-Hours	Unit	Material	Labor	Equipment	Total	Total Incl O&P
0010	**FLOOR PLATE STAIRS**, shop fabricated, steel stringers									
0015	Made from recycled materials									
0400	Cast iron tread and pipe rail, 3'-6" wide	E-4	35	.914	Riser	565	55.50	4.25	624.75	715
0500	Checkered plate tread, industrial, 3'-6" wide	E-4	28	1.143	Riser	325	69.50	5.30	399.80	475
0550	Circular, for tanks, 3'-0" wide	"	33	.970		370	59	4.51	433.51	500
0600	For isolated stairs, add						100%			
0800	Custom steel stairs, 3'-6" wide, economy	E-4	35	.914		495	55.50	4.25	554.75	635

For customer support on your Site Work & Landscape Costs with RSMeans Data, call 800.448.8182.

127

05 51 Metal Stairs

05 51 16 – Metal Floor Plate Stairs

05 51 16.50 Floor Plate Stairs		Crew	Daily Output	Labor-Hours	Unit	Material	2021 Bare Costs Labor	Equipment	Total	Total Incl O&P
0810	Medium priced	G E-4	30	1.067	Riser	655	65	4.96	724.96	825
0900	Deluxe	G	20	1.600		820	97.50	7.45	924.95	1,075
1100	For 4' wide stairs, add					5%	5%			
1300	For 5' wide stairs, add					10%	10%			

05 51 19 – Metal Grating Stairs

05 51 19.50 Grating Stairs

		Crew	Daily Output	Labor-Hours	Unit	Material	Labor	Equipment	Total	Total Incl O&P
0010	**GRATING STAIRS**, shop fabricated, steel stringers, safety nosing on treads									
0015	Made from recycled materials									
0020	Grating tread and pipe railing, 3'-6" wide	G E-4	35	.914	Riser	390	55.50	4.25	449.75	520
0100	4'-0" wide	G "	30	1.067	"	385	65	4.96	454.96	530

05 51 33 – Metal Ladders

05 51 33.13 Vertical Metal Ladders

		Crew	Daily Output	Labor-Hours	Unit	Material	Labor	Equipment	Total	Total Incl O&P
0010	**VERTICAL METAL LADDERS**, shop fabricated									
0015	Made from recycled materials									
0020	Steel, 20" wide, bolted to concrete, with cage	G E-4	50	.640	V.L.F.	107	39	2.98	148.98	180
0100	Without cage	G	85	.376		42	23	1.75	66.75	83.50
0300	Aluminum, bolted to concrete, with cage	G	50	.640		106	39	2.98	147.98	180
0400	Without cage	G	85	.376		51	23	1.75	75.75	93.50

05 51 33.16 Inclined Metal Ladders

		Crew	Daily Output	Labor-Hours	Unit	Material	Labor	Equipment	Total	Total Incl O&P
0010	**INCLINED METAL LADDERS**, shop fabricated									
0015	Made from recycled materials									
3900	Industrial ships ladder, steel, 24" W, grating treads, 2 line pipe rail	G E-4	30	1.067	Riser	181	65	4.96	250.96	305
4000	Aluminum	G "	30	1.067	"	310	65	4.96	379.96	445

05 52 Metal Railings

05 52 13 – Pipe and Tube Railings

05 52 13.50 Railings, Pipe

		Crew	Daily Output	Labor-Hours	Unit	Material	Labor	Equipment	Total	Total Incl O&P
0010	**RAILINGS, PIPE**, shop fab'd, 3'-6" high, posts @ 5' OC									
0015	Made from recycled materials									
0020	Aluminum, 2 rail, satin finish, 1-1/4" diameter	G E-4	160	.200	L.F.	50.50	12.15	.93	63.58	76
0030	Clear anodized	G	160	.200		60	12.15	.93	73.08	86
0040	Dark anodized	G	160	.200		67.50	12.15	.93	80.58	94.50
0080	1-1/2" diameter, satin finish	G	160	.200		57.50	12.15	.93	70.58	83
0090	Clear anodized	G	160	.200		64.50	12.15	.93	77.58	91
0100	Dark anodized	G	160	.200		71	12.15	.93	84.08	98
0140	Aluminum, 3 rail, 1-1/4" diam., satin finish	G	137	.234		65	14.20	1.09	80.29	94.50
0150	Clear anodized	G	137	.234		90	14.20	1.09	105.29	122
0160	Dark anodized	G	137	.234		89	14.20	1.09	104.29	121
0200	1-1/2" diameter, satin finish	G	137	.234		85.50	14.20	1.09	100.79	117
0210	Clear anodized	G E-4	137	.234	L.F.	87.50	14.20	1.09	102.79	120
0220	Dark anodized	G	137	.234		107	14.20	1.09	122.29	141
0500	Steel, 2 rail, on stairs, primed, 1-1/4" diameter	G	160	.200		36	12.15	.93	49.08	60
0520	1-1/2" diameter	G	160	.200		31	12.15	.93	44.08	54
0540	Galvanized, 1-1/4" diameter	G	160	.200		46.50	12.15	.93	59.58	71
0560	1-1/2" diameter	G	160	.200		55.50	12.15	.93	68.58	81
0580	Steel, 3 rail, primed, 1-1/4" diameter	G	137	.234		51.50	14.20	1.09	66.79	79.50
0600	1-1/2" diameter	G	137	.234		56.50	14.20	1.09	71.79	85.50
0620	Galvanized, 1-1/4" diameter	G	137	.234		72	14.20	1.09	87.29	103
0640	1-1/2" diameter	G	137	.234		87	14.20	1.09	102.29	119
0700	Stainless steel, 2 rail, 1-1/4" diam., #4 finish	G	137	.234		146	14.20	1.09	161.29	184

05 52 Metal Railings

05 52 13 – Pipe and Tube Railings

05 52 13.50 Railings, Pipe

		Crew	Daily Output	Labor-Hours	Unit	Material	2021 Bare Costs Labor	Equipment	Total	Total Incl O&P
0720	High polish	E-4	137	.234	L.F.	236	14.20	1.09	251.29	283
0740	Mirror polish		137	.234		295	14.20	1.09	310.29	350
0760	Stainless steel, 3 rail, 1-1/2" diam., #4 finish		120	.267		219	16.20	1.24	236.44	267
0770	High polish		120	.267		365	16.20	1.24	382.44	425
0780	Mirror finish		120	.267		445	16.20	1.24	462.44	515
0900	Wall rail, alum. pipe, 1-1/4" diam., satin finish		213	.150		24	9.15	.70	33.85	41.50
0905	Clear anodized		213	.150		33.50	9.15	.70	43.35	52
0910	Dark anodized		213	.150		39	9.15	.70	48.85	58
0915	1-1/2" diameter, satin finish		213	.150		29.50	9.15	.70	39.35	47.50
0920	Clear anodized		213	.150		37	9.15	.70	46.85	56
0925	Dark anodized		213	.150		47	9.15	.70	56.85	66.50
0930	Steel pipe, 1-1/4" diameter, primed		213	.150		21	9.15	.70	30.85	38
0935	Galvanized		213	.150		30	9.15	.70	39.85	48
0940	1-1/2" diameter		176	.182		17.60	11.05	.85	29.50	37.50
0945	Galvanized		213	.150		30	9.15	.70	39.85	48
0955	Stainless steel pipe, 1-1/2" diam., #4 finish		107	.299		117	18.20	1.39	136.59	159
0960	High polish		107	.299		238	18.20	1.39	257.59	291
0965	Mirror polish		107	.299		281	18.20	1.39	300.59	340
2000	2-line pipe rail (1-1/2" T&B) with 1/2" pickets @ 4-1/2" OC,									
2005	attached handrail on brackets									
2010	42" high aluminum, satin finish, straight & level	E-4	120	.267	L.F.	390	16.20	1.24	407.44	455
2050	42" high steel, primed, straight & level	"	120	.267		148	16.20	1.24	165.44	188
4000	For curved and level rails, add						10%	10%		
4100	For sloped rails for stairs, add						30%	30%		

05 52 16 – Industrial Railings

05 52 16.50 Railings, Industrial

		Crew	Daily Output	Labor-Hours	Unit	Material	2021 Bare Costs Labor	Equipment	Total	Total Incl O&P
0010	**RAILINGS, INDUSTRIAL**, shop fab'd, 3'-6" high, posts @ 5' OC									
0020	2 rail, 3'-6" high, 1-1/2" pipe	E-4	255	.125	L.F.	40.50	7.65	.58	48.73	57
0100	2" angle rail	"	255	.125		36.50	7.65	.58	44.73	52.50
0200	For 4" high kick plate, 10 ga., add						5.70		5.70	6.30
0300	1/4" thick, add						7.45		7.45	8.20
0500	For curved level rails, add						10%	10%		
0550	For sloped rails for stairs, add						30%	30%		

05 53 Metal Gratings

05 53 13 – Bar Gratings

05 53 13.10 Floor Grating, Aluminum

		Crew	Daily Output	Labor-Hours	Unit	Material	2021 Bare Costs Labor	Equipment	Total	Total Incl O&P
0010	**FLOOR GRATING, ALUMINUM**, field fabricated from panels									
0015	Made from recycled materials									
0110	Bearing bars @ 1-3/16" OC, cross bars @ 4" OC,									
0111	Up to 300 S.F., 1" x 1/8" bar	E-4	900	.036	S.F.	24.50	2.16	.17	26.83	30.50
0112	Over 300 S.F.		850	.038		22.50	2.29	.18	24.97	28
0113	1-1/4" x 1/8" bar, up to 300 S.F.		800	.040		21.50	2.43	.19	24.12	27.50
0114	Over 300 S.F.		1000	.032		19.45	1.95	.15	21.55	24.50
0122	1-1/4" x 3/16" bar, up to 300 S.F.		750	.043		30.50	2.59	.20	33.29	37.50
0124	Over 300 S.F.		1000	.032		28	1.95	.15	30.10	33.50
0132	1-1/2" x 3/16" bar, up to 300 S.F.		700	.046		44.50	2.78	.21	47.49	53.50
0134	Over 300 S.F.		1000	.032		40.50	1.95	.15	42.60	47.50
0136	1-3/4" x 3/16" bar, up to 300 S.F.		500	.064		48	3.89	.30	52.19	59.50
0138	Over 300 S.F.		1000	.032		43.50	1.95	.15	45.60	51

For customer support on your Site Work & Landscape Costs with RSMeans Data, call 800.448.8182.

05 53 13 – Bar Gratings

	05 53 13.10 Floor Grating, Aluminum		Crew	Daily Output	Labor-Hours	Unit	Material	2021 Bare Costs Labor	Equipment	Total	Total Incl O&P
0146	2-1/4" x 3/16" bar, up to 300 S.F.	G	E-4	600	.053	S.F.	66	3.24	.25	69.49	78
0148	Over 300 S.F.	G		1000	.032		60	1.95	.15	62.10	69
0162	Cross bars @ 2" OC, 1" x 1/8", up to 300 S.F.	G		600	.053		29	3.24	.25	32.49	37.50
0164	Over 300 S.F.	G		1000	.032		26.50	1.95	.15	28.60	32
0172	1-1/4" x 3/16" bar, up to 300 S.F.	G		600	.053		46.50	3.24	.25	49.99	56.50
0174	Over 300 S.F.	G		1000	.032		42.50	1.95	.15	44.60	49.50
0182	1-1/2" x 3/16" bar, up to 300 S.F.	G		600	.053		56	3.24	.25	59.49	67
0184	Over 300 S.F.	G		1000	.032		50.50	1.95	.15	52.60	59
0186	1-3/4" x 3/16" bar, up to 300 S.F.	G		600	.053		72	3.24	.25	75.49	84.50
0188	Over 300 S.F.	G		1000	.032		65	1.95	.15	67.10	75
0200	For straight cuts, add					L.F.	4.56			4.56	5
0212	Bearing bars @ 15/16" OC, 1" x 1/8", up to 300 S.F.	G	E-4	520	.062	S.F.	32	3.74	.29	36.03	41.50
0214	Over 300 S.F.	G		920	.035		29	2.11	.16	31.27	35.50
0222	1-1/4" x 3/16", up to 300 S.F.	G		520	.062		50.50	3.74	.29	54.53	61.50
0224	Over 300 S.F.	G		920	.035		46	2.11	.16	48.27	54
0232	1-1/2" x 3/16", up to 300 S.F.	G		520	.062		63	3.74	.29	67.03	75.50
0234	Over 300 S.F.	G		920	.035		57.50	2.11	.16	59.77	66.50
0300	For curved cuts, add					L.F.	5.60			5.60	6.15
0400	For straight banding, add	G					5.80			5.80	6.40
0500	For curved banding, add	G					7.05			7.05	7.75
0600	For aluminum checkered plate nosings, add	G					7.55			7.55	8.35
0700	For straight toe plate, add	G					11.45			11.45	12.55
0800	For curved toe plate, add	G					13.35			13.35	14.65
1000	For cast aluminum abrasive nosings, add	G					11.10			11.10	12.20
1400	Extruded I bars are 10% less than 3/16" bars										
1600	Heavy duty, all extruded plank, 3/4" deep, 1.8 #/S.F.	G	E-4	1100	.029	S.F.	28.50	1.77	.14	30.41	34.50
1700	1-1/4" deep, 2.9 #/S.F.	G		1000	.032		37.50	1.95	.15	39.60	44
1800	1-3/4" deep, 4.2 #/S.F.	G		925	.035		43.50	2.10	.16	45.76	51.50
1900	2-1/4" deep, 5.0 #/S.F.	G		875	.037		70	2.22	.17	72.39	80.50
2100	For safety serrated surface, add						15%				

05 53 13.70 Floor Grating, Steel

			Crew	Daily Output	Labor-Hours	Unit	Material	2021 Bare Costs Labor	Equipment	Total	Total Incl O&P
0010	**FLOOR GRATING, STEEL**, field fabricated from panels										
0015	Made from recycled materials										
0300	Platforms, to 12' high, rectangular	G	E-4	3150	.010	Lb.	3.29	.62	.05	3.96	4.63
0400	Circular	G	"	2300	.014	"	4.11	.85	.06	5.02	5.90
0410	Painted bearing bars @ 1-3/16"										
0412	Cross bars @ 4" OC, 3/4" x 1/8" bar, up to 300 S.F.	G	E-2	500	.112	S.F.	8.65	6.70	3.43	18.78	23.50
0414	Over 300 S.F.	G		750	.075		7.85	4.45	2.28	14.58	17.95
0422	1-1/4" x 3/16", up to 300 S.F.	G		400	.140		13.85	8.35	4.28	26.48	33
0424	Over 300 S.F.	G		600	.093		12.60	5.55	2.86	21.01	25.50
0432	1-1/2" x 3/16", up to 300 S.F.	G		400	.140		16	8.35	4.28	28.63	35
0434	Over 300 S.F.	G		600	.093		14.55	5.55	2.86	22.96	27.50
0436	1-3/4" x 3/16", up to 300 S.F.	G		400	.140		22	8.35	4.28	34.63	42
0438	Over 300 S.F.	G		600	.093		20	5.55	2.86	28.41	33.50
0452	2-1/4" x 3/16", up to 300 S.F.	G		300	.187		26.50	11.15	5.70	43.35	52.50
0454	Over 300 S.F.	G		450	.124		24	7.40	3.81	35.21	42
0462	Cross bars @ 2" OC, 3/4" x 1/8", up to 300 S.F.	G		500	.112		16.30	6.70	3.43	26.43	32
0464	Over 300 S.F.	G		750	.075		13.60	4.45	2.28	20.33	24.50
0472	1-1/4" x 3/16", up to 300 S.F.	G		400	.140		19.80	8.35	4.28	32.43	39.50
0474	Over 300 S.F.	G		600	.093		16.50	5.55	2.86	24.91	30
0482	1-1/2" x 3/16", up to 300 S.F.	G		400	.140		22.50	8.35	4.28	35.13	42
0484	Over 300 S.F.	G		600	.093		18.55	5.55	2.86	26.96	32

05 53 Metal Gratings

05 53 13 – Bar Gratings

05 53 13.70 Floor Grating, Steel

		Crew	Daily Output	Labor-Hours	Unit	Material	2021 Bare Costs Labor	Equipment	Total	Total Incl O&P
0486	1-3/4" x 3/16", up to 300 S.F.	G E-2	400	.140	S.F.	33.50	8.35	4.28	46.13	54.50
0488	Over 300 S.F.	G	600	.093		28	5.55	2.86	36.41	42.50
0502	2-1/4" x 3/16", up to 300 S.F.	G	300	.187		33	11.15	5.70	49.85	59.50
0504	Over 300 S.F.	G	450	.124		27.50	7.40	3.81	38.71	45.50
0601	Painted bearing bars @ 15/16" OC, cross bars @ 4" OC,									
0612	Up to 300 S.F., 3/4" x 3/16" bars	G E-4	850	.038	S.F.	12.20	2.29	.18	14.67	17.15
0622	1-1/4" x 3/16" bars	G	600	.053		17.05	3.24	.25	20.54	24
0632	1-1/2" x 3/16" bars	G	550	.058		19.40	3.54	.27	23.21	27.50
0636	1-3/4" x 3/16" bars	G	450	.071		26.50	4.32	.33	31.15	36
0652	2-1/4" x 3/16" bars	G E-2	300	.187		26.50	11.15	5.70	43.35	52.50
0662	Cross bars @ 2" OC, up to 300 S.F., 3/4" x 3/16"	G	500	.112		17	6.70	3.43	27.13	32.50
0672	1-1/4" x 3/16" bars	G	400	.140		22	8.35	4.28	34.63	41.50
0682	1-1/2" x 3/16" bars	G	400	.140		24.50	8.35	4.28	37.13	44.50
0686	1-3/4" x 3/16" bars	G	300	.187		31	11.15	5.70	47.85	57.50
0690	For galvanized grating, add					25%				
0800	For straight cuts, add				L.F.	6.50			6.50	7.15
0900	For curved cuts, add					8.25			8.25	9.10
1000	For straight banding, add	G				7.05			7.05	7.75
1100	For curved banding, add	G				9.20			9.20	10.15
1200	For checkered plate nosings, add	G				8.10			8.10	8.90
1300	For straight toe or kick plate, add	G				14.50			14.50	15.95
1400	For curved toe or kick plate, add	G				16.45			16.45	18.05
1500	For abrasive nosings, add	G				10.75			10.75	11.85
1600	For safety serrated surface, bearing bars @ 1-3/16" OC, add					15%				
1700	Bearing bars @ 15/16" OC, add					25%				
2000	Stainless steel gratings, close spaced, 1" x 1/8" bars, up to 300 S.F.	G E-4	450	.071	S.F.	53.50	4.32	.33	58.15	66
2100	Standard spacing, 3/4" x 1/8" bars	G	500	.064		104	3.89	.30	108.19	120
2200	1-1/4" x 3/16" bars	G	400	.080		130	4.86	.37	135.23	151

05 53 16 – Plank Gratings

05 53 16.50 Grating Planks

		Crew	Daily Output	Labor-Hours	Unit	Material	2021 Bare Costs Labor	Equipment	Total	Total Incl O&P
0010	**GRATING PLANKS**, field fabricated from planks									
0020	Aluminum, 9-1/2" wide, 14 ga., 2" rib	G E-4	950	.034	L.F.	36	2.05	.16	38.21	43
0200	Galvanized steel, 9-1/2" wide, 14 ga., 2-1/2" rib	G	950	.034		19.55	2.05	.16	21.76	25
0300	4" rib	G	950	.034		21	2.05	.16	23.21	26.50
0500	12 ga., 2-1/2" rib	G	950	.034		22	2.05	.16	24.21	27.50
0600	3" rib	G	950	.034		26.50	2.05	.16	28.71	32.50
0800	Stainless steel, type 304, 16 ga., 2" rib	G	950	.034		40.50	2.05	.16	42.71	48
0900	Type 316	G	950	.034		56	2.05	.16	58.21	65.50

05 53 19 – Expanded Metal Gratings

05 53 19.10 Expanded Grating, Aluminum

		Crew	Daily Output	Labor-Hours	Unit	Material	2021 Bare Costs Labor	Equipment	Total	Total Incl O&P
0010	**EXPANDED GRATING, ALUMINUM**									
1200	Expanded aluminum, .65 #/S.F.	G E-4	1050	.030	S.F.	22	1.85	.14	23.99	27

05 53 19.20 Expanded Grating, Steel

		Crew	Daily Output	Labor-Hours	Unit	Material	2021 Bare Costs Labor	Equipment	Total	Total Incl O&P
0010	**EXPANDED GRATING, STEEL**									
2400	Expanded steel grating, at ground, 3.0 #/S.F.	G E-4	900	.036	S.F.	8.70	2.16	.17	11.03	13.10
2500	3.14 #/S.F.	G	900	.036		6.50	2.16	.17	8.83	10.70
2600	4.0 #/S.F.	G	850	.038		7.95	2.29	.18	10.42	12.50
2650	4.27 #/S.F.	G	850	.038		8.70	2.29	.18	11.17	13.30
2700	5.0 #/S.F.	G	800	.040		14.95	2.43	.19	17.57	20.50
2800	6.25 #/S.F.	G	750	.043		19.20	2.59	.20	21.99	25
2900	7.0 #/S.F.	G	700	.046		21	2.78	.21	23.99	28

05 53 Metal Gratings

05 53 19 – Expanded Metal Gratings

05 53 19.20 Expanded Grating, Steel		Crew	Daily Output	Labor-Hours	Unit	Material	2021 Bare Costs Labor	Equipment	Total	Total Incl O&P
3100	For flattened expanded steel grating, add				S.F.	8%				
3300	For elevated installation above 15', add				↓		15%			

05 53 19.30 Grating Frame

05 53 19.30 Grating Frame		Crew	Daily Output	Labor-Hours	Unit	Material	2021 Bare Costs Labor	Equipment	Total	Total Incl O&P
0010	**GRATING FRAME**, field fabricated									
0020	Aluminum, for gratings 1" to 1-1/2" deep	[G] 1 Sswk	70	.114	L.F.	3.28	6.90		10.18	14.25
0100	For each corner, add	[G]			Ea.	4.89			4.89	5.40

05 54 Metal Floor Plates

05 54 13 – Floor Plates

05 54 13.20 Checkered Plates

05 54 13.20 Checkered Plates		Crew	Daily Output	Labor-Hours	Unit	Material	2021 Bare Costs Labor	Equipment	Total	Total Incl O&P
0010	**CHECKERED PLATES**, steel, field fabricated									
0015	Made from recycled materials									
0020	1/4" & 3/8", 2000 to 5000 S.F., bolted	[G] E-4	2900	.011	Lb.	1.05	.67	.05	1.77	2.26
0100	Welded	[G]	4400	.007	"	1.02	.44	.03	1.49	1.84
0300	Pit or trench cover and frame, 1/4" plate, 2' to 3' wide	[G] ↓	100	.320	S.F.	12.65	19.45	1.49	33.59	45.50
0400	For galvanizing, add	[G]			Lb.	.30			.30	.33
0500	Platforms, 1/4" plate, no handrails included, rectangular	[G] E-4	4200	.008	↓	3.49	.46	.04	3.99	4.60
0600	Circular	[G] "	2500	.013	↓	4.37	.78	.06	5.21	6.10

05 54 13.70 Trench Covers

05 54 13.70 Trench Covers		Crew	Daily Output	Labor-Hours	Unit	Material	2021 Bare Costs Labor	Equipment	Total	Total Incl O&P
0010	**TRENCH COVERS**, field fabricated									
0020	Cast iron grating with bar stops and angle frame, to 18" wide	[G] 1 Sswk	20	.400	L.F.	198	24		222	256
0100	Frame only (both sides of trench), 1" grating	[G]	45	.178		1.22	10.70		11.92	17.95
0150	2" grating	[G] ↓	35	.229	↓	2.58	13.80		16.38	24.50
0200	Aluminum, stock units, including frames and									
0210	3/8" plain cover plate, 4" opening	[G] E-4	205	.156	L.F.	15.15	9.50	.73	25.38	32
0300	6" opening	[G]	185	.173		19.05	10.50	.80	30.35	38
0400	10" opening	[G]	170	.188		33.50	11.45	.88	45.83	55.50
0500	16" opening	[G] ↓	155	.206		38.50	12.55	.96	52.01	62.50
0700	Add per inch for additional widths to 24"	[G]				1.68			1.68	1.85
0900	For custom fabrication, add					50%				
1100	For 1/4" plain cover plate, deduct					12%				
1500	For cover recessed for tile, 1/4" thick, deduct					12%				
1600	3/8" thick, add					5%				
1800	For checkered plate cover, 1/4" thick, deduct					12%				
1900	3/8" thick, add					2%				
2100	For slotted or round holes in cover, 1/4" thick, add					3%				
2200	3/8" thick, add					4%				
2300	For abrasive cover, add				↓	12%				

For customer support on your Site Work & Landscape Costs with RSMeans Data, call 800.448.8182.

05 55 13 – Metal Stair Treads

05 55 13.50 Stair Treads		Crew	Daily Output	Labor-Hours	Unit	Material	2021 Bare Costs Labor	Equipment	Total	Total Incl O&P
0010	**STAIR TREADS**, stringers and bolts not included									
3000	Diamond plate treads, steel, 1/8" thick									
3005	Open riser, black enamel									
3010	9" deep x 36" long	G	2 Sswk	48	.333	Ea.	114	20		134	157
3020	42" long	G		48	.333		121	20		141	164
3030	48" long	G		48	.333		126	20		146	169
3040	11" deep x 36" long	G		44	.364		121	22		143	167
3050	42" long	G		44	.364		128	22		150	174
3060	48" long	G		44	.364		136	22		158	183
3110	Galvanized, 9" deep x 36" long	G		48	.333		181	20		201	230
3120	42" long	G		48	.333		192	20		212	242
3130	48" long	G		48	.333		204	20		224	255
3140	11" deep x 36" long	G		44	.364		187	22		209	240
3150	42" long	G		44	.364		204	22		226	258
3160	48" long	G		44	.364		212	22		234	267
3200	Closed riser, black enamel									
3210	12" deep x 36" long	G	2 Sswk	40	.400	Ea.	138	24		162	190
3220	42" long	G		40	.400		150	24		174	203
3230	48" long	G		40	.400		158	24		182	212
3240	Galvanized, 12" deep x 36" long	G		40	.400		218	24		242	278
3250	42" long	G		40	.400		238	24		262	300
3260	48" long	G		40	.400		249	24		273	310
4000	Bar grating treads									
4005	Steel, 1-1/4" x 3/16" bars, anti-skid nosing, black enamel									
4010	8-5/8" deep x 30" long	G	2 Sswk	48	.333	Ea.	56	20		76	92.50
4020	36" long	G		48	.333		65.50	20		85.50	104
4030	48" long	G		48	.333		101	20		121	142
4040	10-15/16" deep x 36" long	G		44	.364		73	22		95	114
4050	48" long	G		44	.364		103	22		125	147
4060	Galvanized, 8-5/8" deep x 30" long	G		48	.333		64.50	20		84.50	102
4070	36" long	G		48	.333		76.50	20		96.50	115
4080	48" long	G		48	.333		112	20		132	154
4090	10-15/16" deep x 36" long	G		44	.364		89.50	22		111.50	133
4100	48" long	G		44	.364		116	22		138	161
4200	Aluminum, 1-1/4" x 3/16" bars, serrated, with nosing									
4210	7-5/8" deep x 18" long	G	2 Sswk	52	.308	Ea.	54	18.55		72.55	88
4220	24" long	G		52	.308		62	18.55		80.55	97
4230	30" long	G		52	.308		72	18.55		90.55	108
4240	36" long	G		52	.308		186	18.55		204.55	234
4250	8-13/16" deep x 18" long	G		48	.333		71	20		91	109
4260	24" long	G		48	.333		117	20		137	160
4270	30" long	G		48	.333		117	20		137	160
4280	36" long	G		48	.333		205	20		225	256
4290	10" deep x 18" long	G		44	.364		137	22		159	185
4300	30" long	G		44	.364		182	22		204	234
4310	36" long	G		44	.364		221	22		243	277
5000	Channel grating treads									
5005	Steel, 14 ga., 2-1/2" thick, galvanized									
5010	9" deep x 36" long	G	2 Sswk	48	.333	Ea.	103	20		123	144
5020	48" long	G	"	48	.333	"	137	20		157	181

For customer support on your Site Work & Landscape Costs with RSMeans Data, call 800.448.8182.

133

05 55 Metal Stair Treads and Nosings

05 55 19 – Metal Stair Tread Covers

05 55 19.50 Stair Tread Covers for Renovation		Crew	Daily Output	Labor-Hours	Unit	Material	2021 Bare Costs			Total	Total Incl O&P
							Labor	Equipment			
0010	**STAIR TREAD COVERS FOR RENOVATION**										
0205	Extruded tread cover with nosing, pre-drilled, includes screws										
0210	Aluminum with black abrasive strips, 9" wide x 3' long	1 Carp	24	.333	Ea.	105	18.25			123.25	142
0220	4' long		22	.364		137	19.90			156.90	181
0230	5' long		20	.400		186	22			208	238
0240	11" wide x 3' long		24	.333		147	18.25			165.25	189
0250	4' long		22	.364		201	19.90			220.90	251
0260	5' long		20	.400		240	22			262	297
0305	Black abrasive strips with yellow front strips										
0310	Aluminum, 9" wide x 3' long	1 Carp	24	.333	Ea.	127	18.25			145.25	167
0320	4' long		22	.364		170	19.90			189.90	217
0330	5' long		20	.400		218	22			240	272
0340	11" wide x 3' long		24	.333		162	18.25			180.25	205
0350	4' long		22	.364		206	19.90			225.90	257
0360	5' long		20	.400		269	22			291	330
0405	Black abrasive strips with photoluminescent front strips										
0410	Aluminum, 9" wide x 3' long	1 Carp	24	.333	Ea.	161	18.25			179.25	204
0420	4' long		22	.364		182	19.90			201.90	231
0430	5' long		20	.400		256	22			278	315
0440	11" wide x 3' long		24	.333		182	18.25			200.25	228
0450	4' long		22	.364		243	19.90			262.90	297
0460	5' long		20	.400		305	22			327	370

05 56 Metal Castings

05 56 13 – Metal Construction Castings

05 56 13.50 Construction Castings

05 56 13.50 Construction Castings			Crew	Daily Output	Labor-Hours	Unit	Material	Labor	Equipment	Total	Total Incl O&P
0010	**CONSTRUCTION CASTINGS**										
0020	Manhole covers and frames, see Section 33 44 13.13										
0100	Column bases, cast iron, 16" x 16", approx. 65 lb.	G	E-4	46	.696	Ea.	139	42.50	3.23	184.73	222
0200	32" x 32", approx. 256 lb.	G		23	1.391	"	505	84.50	6.45	595.95	695
0600	Miscellaneous C.I. castings, light sections, less than 150 lb.	G		3200	.010	Lb.	9.05	.61	.05	9.71	11
1100	Heavy sections, more than 150 lb.	G		4200	.008		5.25	.46	.04	5.75	6.55
1300	Special low volume items	G		3200	.010		11.25	.61	.05	11.91	13.40
1500	For ductile iron, add						100%				

05 58 Formed Metal Fabrications

05 58 21 – Formed Chain

05 58 21.05 Alloy Steel Chain

05 58 21.05 Alloy Steel Chain			Crew	Daily Output	Labor-Hours	Unit	Material	Labor	Equipment	Total	Total Incl O&P
0010	**ALLOY STEEL CHAIN**, Grade 80, for lifting										
0015	Self-colored, cut lengths, 1/4"	G	E-17	4	4	C.L.F.	294	245		539	705
0020	3/8"	G		2	8		1,350	490		1,840	2,225
0030	1/2"	G		1.20	13.333		2,425	815		3,240	3,925
0040	5/8"	G		.72	22.222		3,175	1,350		4,525	5,575
0050	3/4"	G	E-18	.48	83.333		1,750	5,025	3,175	9,950	13,200
0060	7/8"	G		.40	100		2,950	6,050	3,825	12,825	16,700
0070	1"	G		.35	114		5,450	6,900	4,350	16,700	21,400
0080	1-1/4"	G		.24	167		13,500	10,100	6,350	29,950	37,300
0110	Hook, Grade 80, Clevis slip, 1/4"	G				Ea.	30			30	33
0120	3/8"	G					48			48	53

05 58 Formed Metal Fabrications

05 58 21 – Formed Chain

05 58 21.05 Alloy Steel Chain		Crew	Daily Output	Labor-Hours	Unit	Material	2021 Bare Costs Labor	Equipment	Total	Total Incl O&P
0130	1/2"	G			Ea.	74.50			74.50	82
0140	5/8"	G				101			101	112
0150	3/4"	G				129			129	142
0160	Hook, Grade 80, eye/sling w/hammerlock coupling, 15 ton	G				430			430	475
0170	22 ton	G				1,975			1,975	2,175
0180	37 ton	G				3,575			3,575	3,925

05 58 25 – Formed Lamp Posts

05 58 25.40 Lamp Posts

		Crew	Daily Output	Labor-Hours	Unit	Material	2021 Bare Costs Labor	Equipment	Total	Total Incl O&P
0010	**LAMP POSTS**									
0020	Aluminum, 7' high, stock units, post only	G	1 Carp	16	.500	Ea.	70.50	27.50		98	119
0100	Mild steel, plain	G	"	16	.500	"	100	27.50		127.50	151

05 71 Decorative Metal Stairs

05 71 13 – Fabricated Metal Spiral Stairs

05 71 13.50 Spiral Stairs

		Crew	Daily Output	Labor-Hours	Unit	Material	2021 Bare Costs Labor	Equipment	Total	Total Incl O&P
0010	**SPIRAL STAIRS**									
1805	Shop fabricated, custom ordered									
1810	Aluminum, 5'-0" diameter, plain units	G	E-4	45	.711	Riser	640	43	3.31	686.31	775
1900	Cast iron, 4'-0" diameter, plain units	G		45	.711		770	43	3.31	816.31	915
1920	Fancy units	G		25	1.280		1,650	78	5.95	1,733.95	1,950
2000	Steel, industrial checkered plate, 4' diameter	G		45	.711		610	43	3.31	656.31	740
2200	6' diameter	G		40	.800		655	48.50	3.72	707.22	800
3100	Spiral stair kits, 12 stacking risers to fit exact floor height									
3110	Steel, flat metal treads, primed, 3'-6" diameter	G	2 Carp	1.60	10	Flight	1,300	545		1,845	2,250
3120	4'-0" diameter	G		1.45	11.034		1,400	605		2,005	2,450
3130	4'-6" diameter	G		1.35	11.852		1,675	650		2,325	2,825
3140	5'-0" diameter	G		1.25	12.800		1,875	700		2,575	3,125
3210	Galvanized, 3'-6" diameter	G		1.60	10		2,125	545		2,670	3,150
3220	4'-0" diameter	G		1.45	11.034		2,225	605		2,830	3,350
3230	4'-6" diameter	G		1.35	11.852		2,500	650		3,150	3,725
3240	5'-0" diameter	G		1.25	12.800		2,800	700		3,500	4,150
3310	Checkered plate tread, primed, 3'-6" diameter	G		1.45	11.034		1,550	605		2,155	2,625
3320	4'-0" diameter	G		1.35	11.852		1,675	650		2,325	2,800
3330	4'-6" diameter	G		1.25	12.800		1,950	700		2,650	3,200
3340	5'-0" diameter	G		1.15	13.913		2,150	760		2,910	3,500
3410	Galvanized, 3'-6" diameter	G		1.45	11.034		2,375	605		2,980	3,525
3420	4'-0" diameter	G		1.35	11.852		2,475	650		3,125	3,700
3430	4'-6" diameter	G		1.25	12.800		2,675	700		3,375	4,000
3440	5'-0" diameter	G		1.15	13.913		3,125	760		3,885	4,575
3510	Red oak covers on flat metal treads, 3'-6" diameter			1.35	11.852		1,800	650		2,450	2,950
3520	4'-0" diameter			1.25	12.800		1,950	700		2,650	3,200
3530	4'-6" diameter			1.15	13.913		2,225	760		2,985	3,550
3540	5'-0" diameter			1.05	15.238		2,625	835		3,460	4,150

05 73 Decorative Metal Railings

05 73 16 – Wire Rope Decorative Metal Railings

05 73 16.10 Cable Railings		Crew	Daily Output	Labor-Hours	Unit	Material	2021 Bare Costs Labor	Equipment	Total	Total Incl O&P
0010	**CABLE RAILINGS**, with 316 stainless steel 1 x 19 cable, 3/16" diameter									
0015	Made from recycled materials									
0100	1-3/4" diameter stainless steel posts x 42" high, cables 4" OC	G 2 Sswk	25	.640	L.F.	39	38.50		77.50	103

05 73 23 – Ornamental Railings

05 73 23.50 Railings, Ornamental		Crew	Daily Output	Labor-Hours	Unit	Material	2021 Bare Costs Labor	Equipment	Total	Total Incl O&P
0010	**RAILINGS, ORNAMENTAL**, 3'-6" high, posts @ 6' OC									
0020	Bronze or stainless, hand forged, plain	G 2 Sswk	24	.667	L.F.	265	40		305	355
0100	Fancy	G	18	.889		530	53.50		583.50	670
0200	Aluminum, panelized, plain	G	24	.667		13.60	40		53.60	77
0300	Fancy	G	18	.889		26	53.50		79.50	112
0400	Wrought iron, hand forged, plain	G	24	.667		75	40		115	145
0500	Fancy	G	18	.889		233	53.50		286.50	340
0550	Steel, panelized, plain	G	24	.667		21.50	40		61.50	86
0560	Fancy	G	18	.889		34	53.50		87.50	120
0600	Composite metal/wood/glass, plain		18	.889		154	53.50		207.50	252
0700	Fancy		12	1.333		310	80.50		390.50	465

Estimating Tips
06 05 00 Common Work Results for Wood, Plastics, and Composites

- Common to any wood-framed structure are the accessory connector items such as screws, nails, adhesives, hangers, connector plates, straps, angles, and hold-downs. For typical wood-framed buildings, such as residential projects, the aggregate total for these items can be significant, especially in areas where seismic loading is a concern. For floor and wall framing, the material cost is based on 10 to 25 lbs. of accessory connectors per MBF. Hold-downs, hangers, and other connectors should be taken off by the piece.

  Included with material costs are fasteners for a normal installation. Gordian's RSMeans engineers use manufacturers' recommendations, written specifications, and/or standard construction practice for the sizing and spacing of fasteners. Prices for various fasteners are shown for informational purposes only. Adjustments should be made if unusual fastening conditions exist.

06 10 00 Carpentry

- Lumber is a traded commodity and therefore sensitive to supply and demand in the marketplace. Even with "budgetary" estimating of wood-framed projects, it is advisable to call local suppliers for the latest market pricing.

- The common quantity unit for wood-framed projects is "thousand board feet" (MBF). A board foot is a volume of wood—1" x 1' x 1' or 144 cubic inches. Board-foot quantities are generally calculated using nominal material dimensions—dressed sizes are ignored. Board foot per lineal foot of any stick of lumber can be calculated by dividing the nominal cross-sectional area by 12. As an example, 2,000 lineal feet of 2 x 12 equates to 4 MBF by dividing the nominal area, 2 x 12, by 12, which equals 2, and multiplying that by 2,000 to give 4,000 board feet. This simple rule applies to all nominal dimensioned lumber.

- Waste is an issue of concern at the quantity takeoff for any area of construction. Framing lumber is sold in even foot lengths, i.e., 8', 10', 12', 14', 16', and depending on spans, wall heights, and the grade of lumber, waste is inevitable. A rule of thumb for lumber waste is 5–10% depending on material quality and the complexity of the framing.

- Wood in various forms and shapes is used in many projects, even where the main structural framing is steel, concrete, or masonry. Plywood as a back-up partition material and 2x boards used as blocking and cant strips around roof edges are two common examples. The estimator should ensure that the costs of all wood materials are included in the final estimate.

06 20 00 Finish Carpentry

- It is necessary to consider the grade of workmanship when estimating labor costs for erecting millwork and an interior finish. In practice, there are three grades: premium, custom, and economy. The RSMeans daily output for base and case moldings is in the range of 200 to 250 L.F. per carpenter per day. This is appropriate for most average custom-grade projects. For premium projects, an adjustment to productivity of 25–50% should be made, depending on the complexity of the job.

Reference Numbers

Reference numbers are shown at the beginning of some major classifications. These numbers refer to related items in the Reference Section. The reference information may be an estimating procedure, an alternate pricing method, or technical information.

Note: Not all subdivisions listed here necessarily appear. ■

Same Data. Simplified.

Enjoy the convenience and efficiency of accessing your costs anywhere:

- **Skip the multiplier** by setting your location
- **Quickly search,** edit, favorite and share costs
- **Stay on top of price changes** with automatic updates

Discover more at rsmeans.com/online

06 05 05 – Selective Demolition for Wood, Plastics, and Composites

06 05 05.10 Selective Demolition Wood Framing	Crew	Daily Output	Labor-Hours	Unit	Material	2021 Bare Costs Labor	Equipment	Total	Total Incl O&P
0010 **SELECTIVE DEMOLITION WOOD FRAMING** R024119-10									
0100 Timber connector, nailed, small	1 Clab	96	.083	Ea.		3.70		3.70	5.50
0110 Medium		60	.133			5.90		5.90	8.85
0120 Large		48	.167			7.40		7.40	11.05
0130 Bolted, small		48	.167			7.40		7.40	11.05
0140 Medium		32	.250			11.10		11.10	16.55
0150 Large		24	.333			14.80		14.80	22
3162 Alternate pricing method	B-1	1.10	21.818	M.B.F.		985		985	1,475

06 05 23 – Wood, Plastic, and Composite Fastenings

06 05 23.60 Timber Connectors

06 05 23.60 Timber Connectors	Crew	Daily Output	Labor-Hours	Unit	Material	2021 Bare Costs Labor	Equipment	Total	Total Incl O&P
0010 **TIMBER CONNECTORS**									
0020 Add up cost of each part for total cost of connection									
0100 Connector plates, steel, with bolts, straight	2 Carp	75	.213	Ea.	31	11.65		42.65	51.50
0110 Tee, 7 ga.		50	.320		45.50	17.50		63	76
0120 T-Strap, 14 ga., 12" x 8" x 2"		50	.320		45.50	17.50		63	76
0150 Anchor plates, 7 ga., 9" x 7"		75	.213		31	11.65		42.65	51.50
0200 Bolts, machine, sq. hd. with nut & washer, 1/2" diameter, 4" long	1 Carp	140	.057		.74	3.13		3.87	5.50
0300 7-1/2" long		130	.062		1.37	3.37		4.74	6.50
0500 3/4" diameter, 7-1/2" long		130	.062		3.37	3.37		6.74	8.70
0610 Machine bolts, w/nut, washer, 3/4" diameter, 15" L, HD's & beam hangers		95	.084		6.35	4.61		10.96	13.85
0800 Drilling bolt holes in timber, 1/2" diameter		450	.018	Inch		.97		.97	1.45
0900 1" diameter		350	.023	"		1.25		1.25	1.87

06 11 Wood Framing

06 11 10 – Framing with Dimensional, Engineered or Composite Lumber

06 11 10.14 Posts and Columns

06 11 10.14 Posts and Columns	Crew	Daily Output	Labor-Hours	Unit	Material	2021 Bare Costs Labor	Equipment	Total	Total Incl O&P
0010 **POSTS AND COLUMNS**									
0100 4" x 4"	2 Carp	390	.041	L.F.	2.31	2.24		4.55	5.90
0150 4" x 6"		275	.058		2.89	3.18		6.07	7.95
0200 4" x 8"		220	.073		6.50	3.98		10.48	13.10
0250 6" x 6"		215	.074		6.85	4.07		10.92	13.65
0300 6" x 8"		175	.091		9.65	5		14.65	18.10
0350 6" x 10"		150	.107		9.90	5.85		15.75	19.60

06 11 10.20 Framing, Light

06 11 10.20 Framing, Light	Crew	Daily Output	Labor-Hours	Unit	Material	2021 Bare Costs Labor	Equipment	Total	Total Incl O&P
0010 **FRAMING, LIGHT**									
0020 Average cost for all light framing	2 Carp	1.05	15.238	M.B.F.	1,150	835		1,985	2,525
2870 Redwood joists, 2" x 6"		1.50	10.667		9,375	585		9,960	11,200
2880 2" x 8"		1.75	9.143		5,475	500		5,975	6,775
2890 2" x 10"		1.75	9.143		5,950	500		6,450	7,300
4000 Platform framing, 2" x 4"		.70	22.857		1,050	1,250		2,300	3,025
4100 2" x 6"		.76	21.053		1,075	1,150		2,225	2,900
4200 Pressure treated 6" x 6"		.80	20		2,025	1,100		3,125	3,850
4220 8" x 8"		.60	26.667		3,225	1,450		4,675	5,725
4500 Cedar post, columns & girts, 4" x 4"		.52	30.769		3,950	1,675		5,625	6,850
4510 4" x 6"		.55	29.091		4,225	1,600		5,825	7,000
4520 6" x 6"		.65	24.615		5,625	1,350		6,975	8,175
4530 8" x 8"		.70	22.857		5,450	1,250		6,700	7,875
4540 Redwood post, columns & girts, 4" x 4"		.52	30.769		8,450	1,675		10,125	11,800
4550 4" x 6"		.55	29.091		8,950	1,600		10,550	12,200
4560 6" x 6"		.65	24.615		11,800	1,350		13,150	14,900

06 11 Wood Framing

06 11 10 – Framing with Dimensional, Engineered or Composite Lumber

06 11 10.20 Framing, Light	Crew	Daily Output	Labor-Hours	Unit	Material	2021 Bare Costs Labor	Equipment	Total	Total Incl O&P
4570	8" x 8"	2 Carp	.70	22.857	M.B.F.	12,600	1,250		13,850	15,800

06 11 10.34 Sleepers

0010	**SLEEPERS**									
0100	On concrete, treated, 1" x 2"	2 Carp	2350	.007	L.F.	.45	.37		.82	1.05
0150	1" x 3"		2000	.008		.73	.44		1.17	1.46
0200	2" x 4"		1500	.011		.97	.58		1.55	1.93
0250	2" x 6"		1300	.012		1.40	.67		2.07	2.54

06 11 10.38 Treated Lumber Framing Material

0010	**TREATED LUMBER FRAMING MATERIAL**									
0100	2" x 4"				M.B.F.	1,350			1,350	1,475
0110	2" x 6"					1,250			1,250	1,375
0120	2" x 8"					1,425			1,425	1,575
0130	2" x 10"					1,200			1,200	1,325
0140	2" x 12"					1,825			1,825	2,025
0200	4" x 4"					1,675			1,675	1,850
0210	4" x 6"					1,450			1,450	1,600
0220	4" x 8"					2,300			2,300	2,550

06 11 10.42 Furring

0010	**FURRING**									
0012	Wood strips, 1" x 2", on walls, on wood	1 Carp	550	.015	L.F.	.42	.80		1.22	1.65
0015	On wood, pneumatic nailed		710	.011		.42	.62		1.04	1.38
0300	On masonry		495	.016		.45	.88		1.33	1.82
0400	On concrete		260	.031		.45	1.68		2.13	3.01
0600	1" x 3", on walls, on wood		550	.015		.62	.80		1.42	1.87
0605	On wood, pneumatic nailed		710	.011		.62	.62		1.24	1.60
0700	On masonry		495	.016		.67	.88		1.55	2.06
0800	On concrete		260	.031		.67	1.68		2.35	3.25

06 13 Heavy Timber Construction

06 13 33 – Heavy Timber Pier Construction

06 13 33.50 Jetties, Docks, Fixed

		Crew	Daily Output	Labor-Hours	Unit	Material	2021 Bare Costs Labor	Equipment	Total	Total Incl O&P
0010	**JETTIES, DOCKS, FIXED**									
0020	Pile supported, treated wood,									
0030	5' x 20' platform, freshwater	F-3	115	.348	S.F.	22	19.50	4.14	45.64	57.50
0060	Saltwater		115	.348		50.50	19.50	4.14	74.14	89
0100	6' x 20' platform, freshwater		110	.364		19.40	20.50	4.33	44.23	57
0160	Saltwater		110	.364		45	20.50	4.33	69.83	85
0200	8' x 20' platform, freshwater		105	.381		16.55	21.50	4.53	42.58	55
0260	Saltwater		105	.381		39.50	21.50	4.53	65.53	80.50
0420	5' x 30' platform, freshwater		100	.400		21	22.50	4.76	48.26	62
0460	Saltwater		100	.400		63	22.50	4.76	90.26	108
0500	For Greenhart lumber, add					40%				
0550	Diagonal planking, add								25%	25%

06 13 33.52 Jetties, Piers

0010	**JETTIES, PIERS**, Municipal with 3" x 12" framing and 3" decking,									
0020	wood piles and cross bracing, alternate bents battered	B-76	60	1.200	S.F.	91	68.50	62.50	222	273
0200	Treated piles, not including mobilization									
0210	50' long, CCA treated, shore driven	B-19	540	.119	V.L.F.	25.50	6.75	3.79	36.04	43
0220	Barge driven	B-76	320	.225		25.50	12.80	11.70	50	61
0230	ACZA treated, shore driven	B-19	540	.119		16.70	6.75	3.79	27.24	33

For customer support on your Site Work & Landscape Costs with RSMeans Data, call 800.448.8182.

139

06 13 Heavy Timber Construction

06 13 33 – Heavy Timber Pier Construction

06 13 33.52 Jetties, Piers		Crew	Daily Output	Labor-Hours	Unit	Material	2021 Bare Costs Labor	Equipment	Total	Total Incl O&P
0240	Barge driven	B-76	320	.225	V.L.F.	16.70	12.80	11.70	41.20	51
0250	30' long, CCA treated, shore driven	B-19	540	.119		18.50	6.75	3.79	29.04	35
0260	Barge driven	B-76	320	.225		18.50	12.80	11.70	43	53
0270	ACZA treated, shore driven	B-19	540	.119		8.05	6.75	3.79	18.59	23.50
0280	Barge driven	B-76	320	.225	↓	8.05	12.80	11.70	32.55	41.50
0300	Mobilization, barge, by tug boat	B-83	25	.640	Mile		33	29	62	81.50
0350	Standby time for shore pile driving crew	B-19	8	8	Hr.		455	256	711	975
0360	Standby time for barge driving rig	B-76	8	9	"		515	470	985	1,300

06 18 Glued-Laminated Construction

06 18 13 – Glued-Laminated Beams

06 18 13.20 Laminated Framing

		Crew	Daily Output	Labor-Hours	Unit	Material	2021 Bare Costs Labor	Equipment	Total	Total Incl O&P
0010	**LAMINATED FRAMING**									
0200	Straight roof beams, 20' clear span, beams 8' OC	F-3	2560	.016	SF Flr.	3.25	.88	.19	4.32	5.10
0500	40' clear span, beams 8' OC	"	3200	.013		6.15	.70	.15	7	8
0800	60' clear span, beams 8' OC	F-4	2880	.017		10.55	.92	.34	11.81	13.40
1700	Radial arches, 60' clear span, frames 8' OC		1920	.025		13.85	1.39	.52	15.76	17.90
2000	100' clear span, frames 8' OC		1600	.030		14.30	1.66	.62	16.58	18.85
2300	120' clear span, frames 8' OC	↓	1440	.033		19	1.85	.69	21.54	24.50
2600	Bowstring trusses, 20' OC, 40' clear span	F-3	2400	.017		8.60	.93	.20	9.73	11.05
2700	60' clear span	F-4	3600	.013		7.75	.74	.28	8.77	9.90
2800	100' clear span	"	4000	.012	↓	10.95	.67	.25	11.87	13.30
4300	Alternate pricing method: (use nominal footage of									
4310	components). Straight beams, camber less than 6"	F-3	3.50	11.429	M.B.F.	4,050	640	136	4,826	5,575
4600	Curved members, radius over 32'		2.50	16		4,450	895	190	5,535	6,450
4700	Radius 10' to 32'	↓	3	13.333	↓	4,425	745	159	5,329	6,150
6000	Laminated veneer members, southern pine or western species									
6050	1-3/4" wide x 5-1/2" deep	2 Carp	480	.033	L.F.	6.10	1.82		7.92	9.40
6100	9-1/2" deep		480	.033		5.20	1.82		7.02	8.40
6150	14" deep		450	.036		8.15	1.94		10.09	11.90
6200	18" deep	↓	450	.036	↓	11.30	1.94		13.24	15.30
6300	Parallel strand members, southern pine or western species									
6350	1-3/4" wide x 9-1/4" deep	2 Carp	480	.033	L.F.	5.40	1.82		7.22	8.65
6400	11-1/4" deep		450	.036		6.45	1.94		8.39	10
6450	14" deep		400	.040		9.50	2.19		11.69	13.70
6500	3-1/2" wide x 9-1/4" deep		480	.033		26	1.82		27.82	31
6550	11-1/4" deep		450	.036		28.50	1.94		30.44	34.50
6600	14" deep		400	.040		26	2.19		28.19	32
6650	7" wide x 9-1/4" deep		450	.036		48	1.94		49.94	56
6700	11-1/4" deep		420	.038		58.50	2.08		60.58	67.50
6750	14" deep	↓	400	.040	↓	70	2.19		72.19	80.50

06 44 39 – Wood Posts and Columns

06 44 39.20 Columns		Crew	Daily Output	Labor-Hours	Unit	Material	2021 Bare Costs Labor	Equipment	Total	Total Incl O&P
0010	**COLUMNS**									
4000	Rough sawn cedar posts, 4" x 4"	2 Carp	250	.064	V.L.F.	4.12	3.50		7.62	9.80
4100	4" x 6"		235	.068		6.85	3.72		10.57	13.10
4200	6" x 6"		220	.073		13.60	3.98		17.58	21
4300	8" x 8"		200	.080		21.50	4.38		25.88	30

For customer support on your Site Work & Landscape Costs with RSMeans Data, call 800.448.8182.

141

Division Notes

		CREW	DAILY OUTPUT	LABOR-HOURS	UNIT	BARE COSTS				TOTAL INCL O&P
						MAT.	LABOR	EQUIP.	TOTAL	

Estimating Tips

07 10 00 Dampproofing and Waterproofing

- Be sure of the job specifications before pricing this subdivision. The difference in cost between waterproofing and dampproofing can be great. Waterproofing will hold back standing water. Dampproofing prevents the transmission of water vapor. Also included in this section are vapor retarding membranes.

07 20 00 Thermal Protection

- Insulation and fireproofing products are measured by area, thickness, volume, or R-value. Specifications may give only what the specific R-value should be in a certain situation. The estimator may need to choose the type of insulation to meet that R-value.

07 30 00 Steep Slope Roofing
07 40 00 Roofing and Siding Panels

- Many roofing and siding products are bought and sold by the square. One square is equal to an area that measures 100 square feet.

  This simple change in unit of measure could create a large error if the estimator is not observant. Accessories necessary for a complete installation must be figured into any calculations for both material and labor.

07 50 00 Membrane Roofing
07 60 00 Flashing and Sheet Metal
07 70 00 Roofing and Wall Specialties and Accessories

- The items in these subdivisions compose a roofing system. No one component completes the installation, and all must be estimated. Built-up or single-ply membrane roofing systems are made up of many products and installation trades. Wood blocking at roof perimeters or penetrations, parapet coverings, reglets, roof drains, gutters, downspouts, sheet metal flashing, skylights, smoke vents, and roof hatches all need to be considered along with the roofing material. Several different installation trades will need to work together on the roofing system. Inherent difficulties in the scheduling and coordination of various trades must be accounted for when estimating labor costs.

07 90 00 Joint Protection

- To complete the weather-tight shell, the sealants and caulkings must be estimated. Where different materials meet—at expansion joints, at flashing penetrations, and at hundreds of other locations throughout a construction project—caulking and sealants provide another line of defense against water penetration. Often, an entire system is based on the proper location and placement of caulking or sealants. The detailed drawings that are included as part of a set of architectural plans show typical locations for these materials. When caulking or sealants are shown at typical locations, this means the estimator must include them for all the locations where this detail is applicable. Be careful to keep different types of sealants separate, and remember to consider backer rods and primers if necessary.

Reference Numbers

Reference numbers are shown at the beginning of some major classifications. These numbers refer to related items in the Reference Section. The reference information may be an estimating procedure, an alternate pricing method, or technical information.

Note: Not all subdivisions listed here necessarily appear. ■

Same Data. Simplified.

Enjoy the convenience and efficiency of accessing your costs anywhere:

- **Skip the multiplier** by setting your location
- **Quickly search,** edit, favorite and share costs
- **Stay on top of price changes** with automatic updates

Discover more at rsmeans.com/online

07 01 90 – Maintenance of Joint Protection

07 01 90.81 Joint Sealant Replacement	Crew	Daily Output	Labor-Hours	Unit	Material	2021 Bare Costs Labor	Equipment	Total	Total Incl O&P
0010	**JOINT SEALANT REPLACEMENT**									
0050	Control joints in concrete floors/slabs									
0100	Option 1 for joints with hard dry sealant									
0110	Step 1: Sawcut to remove 95% of old sealant									
0112	1/4" wide x 1/2" deep, with single saw blade	C-27	4800	.003	L.F.	.01	.17	.02	.20	.29
0114	3/8" wide x 3/4" deep, with single saw blade		4000	.004		.02	.21	.03	.26	.36
0116	1/2" wide x 1" deep, with double saw blades		3600	.004		.05	.23	.03	.31	.42
0118	3/4" wide x 1-1/2" deep, with double saw blades		3200	.005		.10	.26	.04	.40	.53
0120	Step 2: Water blast joint faces and edges	C-29	2500	.003			.14	.04	.18	.25
0130	Step 3: Air blast joint faces and edges	C-28	2000	.004			.21	.02	.23	.32
0140	Step 4: Sand blast joint faces and edges	E-11	2000	.016			.78	.15	.93	1.37
0150	Step 5: Air blast joint faces and edges	C-28	2000	.004			.21	.02	.23	.32
0200	Option 2 for joints with soft pliable sealant									
0210	Step 1: Plow joint with rectangular blade	B-62	2600	.009	L.F.		.44	.07	.51	.74
0220	Step 2: Sawcut to re-face joint faces									
0222	1/4" wide x 1/2" deep, with single saw blade	C-27	2400	.007	L.F.	.02	.35	.05	.42	.58
0224	3/8" wide x 3/4" deep, with single saw blade		2000	.008		.03	.41	.06	.50	.71
0226	1/2" wide x 1" deep, with double saw blades		1800	.009		.06	.46	.06	.58	.81
0228	3/4" wide x 1-1/2" deep, with double saw blades		1600	.010		.13	.52	.07	.72	.98
0230	Step 3: Water blast joint faces and edges	C-29	2500	.003			.14	.04	.18	.25
0240	Step 4: Air blast joint faces and edges	C-28	2000	.004			.21	.02	.23	.32
0250	Step 5: Sand blast joint faces and edges	E-11	2000	.016			.78	.15	.93	1.37
0260	Step 6: Air blast joint faces and edges	C-28	2000	.004			.21	.02	.23	.32
0290	For saw cutting new control joints, see Section 03 15 16.20									
8910	For backer rod, see Section 07 91 23.10									
8920	For joint sealant, see Section 03 15 16.30 or 07 92 13.20									

07 11 Dampproofing

07 11 13 – Bituminous Dampproofing

07 11 13.10 Bituminous Asphalt Coating

		Crew	Daily Output	Labor-Hours	Unit	Material	2021 Bare Costs Labor	Equipment	Total	Total Incl O&P
0010	**BITUMINOUS ASPHALT COATING**									
0030	Brushed on, below grade, 1 coat	1 Rofc	665	.012	S.F.	.22	.58		.80	1.18
0100	2 coat		500	.016		.44	.77		1.21	1.73
0300	Sprayed on, below grade, 1 coat		830	.010		.22	.46		.68	.99
0400	2 coat		500	.016		.43	.77		1.20	1.72
0500	Asphalt coating, with fibers				Gal.	9.55			9.55	10.50
0600	Troweled on, asphalt with fibers, 1/16" thick	1 Rofc	500	.016	S.F.	.42	.77		1.19	1.71
0700	1/8" thick		400	.020		.73	.96		1.69	2.37
1000	1/2" thick		350	.023		2.39	1.10		3.49	4.42

07 11 16 – Cementitious Dampproofing

07 11 16.20 Cementitious Parging

		Crew	Daily Output	Labor-Hours	Unit	Material	2021 Bare Costs Labor	Equipment	Total	Total Incl O&P
0010	**CEMENTITIOUS PARGING**									
0020	Portland cement, 2 coats, 1/2" thick	D-1	250	.064	S.F.	.49	3.12		3.61	5.25
0100	Waterproofed Portland cement, 1/2" thick, 2 coats	"	250	.064	"	2.26	3.12		5.38	7.20

07 12 Built-up Bituminous Waterproofing

07 12 13 – Built-Up Asphalt Waterproofing

07 12 13.20 Membrane Waterproofing	Crew	Daily Output	Labor-Hours	Unit	Material	2021 Bare Costs Labor	Equipment	Total	Total Incl O&P
0010 **MEMBRANE WATERPROOFING**									
0012 On slabs, 1 ply, felt, mopped	G-1	3000	.019	S.F.	.48	.84	.19	1.51	2.10
0015 On walls, 1 ply, felt, mopped		3000	.019		.48	.84	.19	1.51	2.10
0100 On slabs, 1 ply, glass fiber fabric, mopped		2100	.027		.53	1.20	.27	2	2.82
0105 On walls, 1 ply, glass fiber fabric, mopped		2100	.027		.53	1.20	.27	2	2.82
0300 On slabs, 2 ply, felt, mopped		2500	.022		.96	1.01	.23	2.20	2.93
0305 On walls, 2 ply, felt, mopped		2500	.022		.96	1.01	.23	2.20	2.93
0400 On slabs, 2 ply, glass fiber fabric, mopped		1650	.034		1.13	1.53	.35	3.01	4.09
0405 On walls, 2 ply, glass fiber fabric, mopped		1650	.034		1.13	1.53	.35	3.01	4.09
0600 On slabs, 3 ply, felt, mopped		2100	.027		1.44	1.20	.27	2.91	3.82
0605 On walls, 3 ply, felt, mopped		2100	.027		1.44	1.20	.27	2.91	3.82
0700 On slabs, 3 ply, glass fiber fabric, mopped		1550	.036		1.59	1.63	.37	3.59	4.79
0705 On walls, 3 ply, glass fiber fabric, mopped	▼	1550	.036		1.59	1.63	.37	3.59	4.79
0710 Asphaltic hardboard protection board, 1/8" thick	2 Rofc	500	.032		.77	1.54		2.31	3.35
0715 1/4" thick		450	.036		1.40	1.71		3.11	4.32
1000 EPS membrane protection board, 1/4" thick		3500	.005		.34	.22		.56	.73
1050 3/8" thick		3500	.005		.38	.22		.60	.77
1060 1/2" thick		3500	.005	▼	.42	.22		.64	.82
1070 Fiberglass fabric, black, 20/10 mesh	▼	116	.138	Sq.	15.45	6.65		22.10	28

07 13 Sheet Waterproofing

07 13 53 – Elastomeric Sheet Waterproofing

07 13 53.10 Elastomeric Sheet Waterproofing and Access.	Crew	Daily Output	Labor-Hours	Unit	Material	2021 Bare Costs Labor	Equipment	Total	Total Incl O&P
0010 **ELASTOMERIC SHEET WATERPROOFING AND ACCESS.**									
0090 EPDM, plain, 45 mils thick	2 Rofc	580	.028	S.F.	.74	1.33		2.07	2.98
0100 60 mils thick		570	.028		.95	1.35		2.30	3.23
0300 Nylon reinforced sheets, 45 mils thick		580	.028		.87	1.33		2.20	3.11
0400 60 mils thick	▼	570	.028	▼	1.01	1.35		2.36	3.30
0600 Vulcanizing splicing tape for above, 2" wide				C.L.F.	61.50			61.50	67.50
0700 4" wide				"	320			320	350
0900 Adhesive, bonding, 60 S.F./gal.				Gal.	30.50			30.50	33.50
1000 Splicing, 75 S.F./gal.				"	37.50			37.50	41.50
1200 Neoprene sheets, plain, 45 mils thick	2 Rofc	580	.028	S.F.	2.25	1.33		3.58	4.64
1300 60 mils thick		570	.028		1.59	1.35		2.94	3.94
1500 Nylon reinforced, 45 mils thick		580	.028		2.36	1.33		3.69	4.76
1600 60 mils thick		570	.028		2.64	1.35		3.99	5.10
1800 120 mils thick	▼	500	.032	▼	5.70	1.54		7.24	8.80
1900 Adhesive, splicing, 150 S.F./gal. per coat				Gal.	37.50			37.50	41.50
2100 Fiberglass reinforced, fluid applied, 1/8" thick	2 Rofc	500	.032	S.F.	1.63	1.54		3.17	4.29
2200 Polyethylene and rubberized asphalt sheets, 60 mils thick		550	.029		1.26	1.40		2.66	3.66
2400 Polyvinyl chloride sheets, plain, 10 mils thick		580	.028		.15	1.33		1.48	2.33
2500 20 mils thick		570	.028		.26	1.35		1.61	2.48
2700 30 mils thick	▼	560	.029	▼	.21	1.38		1.59	2.46
3000 Adhesives, trowel grade, 40-100 S.F./gal.				Gal.	25.50			25.50	28
3100 Brush grade, 100-250 S.F./gal.				"	23			23	25.50
3300 Bitumen modified polyurethane, fluid applied, 55 mils thick	2 Rofc	665	.024	S.F.	1.18	1.16		2.34	3.18

07 16 Cementitious and Reactive Waterproofing

07 16 16 – Crystalline Waterproofing

07 16 16.20 Cementitious Waterproofing	Crew	Daily Output	Labor-Hours	Unit	Material	2021 Bare Costs Labor	Equipment	Total	Total Incl O&P
0010 **CEMENTITIOUS WATERPROOFING**									
0020 1/8" application, sprayed on	G-2A	1000	.024	S.F.	.58	1.03	.64	2.25	2.97
0050 4 coat cementitious metallic slurry	1 Cefi	1.20	6.667	C.S.F.	72	345		417	585

07 17 Bentonite Waterproofing

07 17 13 – Bentonite Panel Waterproofing

07 17 13.10 Bentonite

	Crew	Daily Output	Labor-Hours	Unit	Material	2021 Bare Costs Labor	Equipment	Total	Total Incl O&P
0010 **BENTONITE**									
0020 Panels, 4' x 4', 3/16" thick	1 Rofc	625	.013	S.F.	2.22	.62		2.84	3.44
0100 Rolls, 3/8" thick, with geotextile fabric both sides	"	550	.015	"	1.51	.70		2.21	2.80
0300 Granular bentonite, 50 lb. bags (.625 C.F.)				Bag	16.60			16.60	18.25
0400 3/8" thick, troweled on	1 Rofc	475	.017	S.F.	.83	.81		1.64	2.23
0500 Drain board, expanded polystyrene, 1-1/2" thick	1 Rohe	1600	.005		.48	.18		.66	.82
0510 2" thick		1600	.005		.64	.18		.82	.99
0520 3" thick		1600	.005		.96	.18		1.14	1.35
0530 4" thick		1600	.005		1.28	.18		1.46	1.70
0600 With filter fabric, 1-1/2" thick		1600	.005		.55	.18		.73	.89
0625 2" thick		1600	.005		.71	.18		.89	1.07
0650 3" thick		1600	.005		1.03	.18		1.21	1.42
0675 4" thick		1600	.005		1.35	.18		1.53	1.77

07 19 Water Repellents

07 19 19 – Silicone Water Repellents

07 19 19.10 Silicone Based Water Repellents

	Crew	Daily Output	Labor-Hours	Unit	Material	2021 Bare Costs Labor	Equipment	Total	Total Incl O&P
0010 **SILICONE BASED WATER REPELLENTS**									
0020 Water base liquid, roller applied	2 Rofc	7000	.002	S.F.	.43	.11		.54	.65
0200 Silicone or stearate, sprayed on CMU, 1 coat	1 Rofc	4000	.002		.40	.10		.50	.61
0300 2 coats	"	3000	.003		.81	.13		.94	1.10

07 21 Thermal Insulation

07 21 13 – Board Insulation

07 21 13.10 Rigid Insulation

		Crew	Daily Output	Labor-Hours	Unit	Material	2021 Bare Costs Labor	Equipment	Total	Total Incl O&P
0010 **RIGID INSULATION**, for walls										
0040 Fiberglass, 1.5#/C.F., unfaced, 1" thick, R4.1	G	1 Carp	1000	.008	S.F.	.36	.44		.80	1.05
0060 1-1/2" thick, R6.2	G		1000	.008		.44	.44		.88	1.13
0080 2" thick, R8.3	G		1000	.008		.49	.44		.93	1.19
0120 3" thick, R12.4	G		800	.010		.59	.55		1.14	1.47
0370 3#/C.F., unfaced, 1" thick, R4.3	G		1000	.008		.57	.44		1.01	1.28
0390 1-1/2" thick, R6.5	G		1000	.008		.73	.44		1.17	1.45
0400 2" thick, R8.7	G		890	.009		1.25	.49		1.74	2.11
0420 2-1/2" thick, R10.9	G		800	.010		1.04	.55		1.59	1.96
0440 3" thick, R13	G		800	.010		1.63	.55		2.18	2.61
0520 Foil faced, 1" thick, R4.3	G		1000	.008		.80	.44		1.24	1.53
0540 1-1/2" thick, R6.5	G		1000	.008		1.18	.44		1.62	1.95
0560 2" thick, R8.7	G		890	.009		1.49	.49		1.98	2.37
0580 2-1/2" thick, R10.9	G		800	.010		1.75	.55		2.30	2.75
0600 3" thick, R13	G		800	.010		1.85	.55		2.40	2.86

07 21 Thermal Insulation

07 21 13 – Board Insulation

07 21 13.10 Rigid Insulation

		Crew	Daily Output	Labor-Hours	Unit	Material	2021 Bare Costs Labor	Equipment	Total	Total Incl O&P
1600	Isocyanurate, 4' x 8' sheet, foil faced, both sides									
1610	1/2" thick	G	1 Carp	800	.010	S.F.	.29	.55		.84	1.14
1620	5/8" thick	G		800	.010		.62	.55		1.17	1.50
1630	3/4" thick	G		800	.010		.44	.55		.99	1.30
1640	1" thick	G		800	.010		.62	.55		1.17	1.50
1650	1-1/2" thick	G		730	.011		.76	.60		1.36	1.73
1660	2" thick	G		730	.011		.78	.60		1.38	1.75
1670	3" thick	G		730	.011		3.13	.60		3.73	4.33
1680	4" thick	G		730	.011		2.60	.60		3.20	3.75
1700	Perlite, 1" thick, R2.77	G		800	.010		.46	.55		1.01	1.33
1750	2" thick, R5.55	G		730	.011		.83	.60		1.43	1.80
1900	Extruded polystyrene, 25 psi compressive strength, 1" thick, R5	G		800	.010		.61	.55		1.16	1.49
1940	2" thick, R10	G		730	.011		1.23	.60		1.83	2.24
1960	3" thick, R15	G		730	.011		1.75	.60		2.35	2.82
2100	Expanded polystyrene, 1" thick, R3.85	G		800	.010		.32	.55		.87	1.17
2120	2" thick, R7.69	G		730	.011		.64	.60		1.24	1.59
2140	3" thick, R11.49	G		730	.011		.96	.60		1.56	1.95

07 21 13.13 Foam Board Insulation

		Crew	Daily Output	Labor-Hours	Unit	Material	2021 Bare Costs Labor	Equipment	Total	Total Incl O&P
0010	**FOAM BOARD INSULATION**									
0600	Polystyrene, expanded, 1" thick, R4	G	1 Carp	680	.012	S.F.	.32	.64		.96	1.31
0700	2" thick, R8	G	"	675	.012	"	.64	.65		1.29	1.67

07 21 23 – Loose-Fill Insulation

07 21 23.20 Masonry Loose-Fill Insulation

		Crew	Daily Output	Labor-Hours	Unit	Material	2021 Bare Costs Labor	Equipment	Total	Total Incl O&P
0010	**MASONRY LOOSE-FILL INSULATION**, vermiculite or perlite									
0100	In cores of concrete block, 4" thick wall, .115 C.F./S.F.	G	D-1	4800	.003	S.F.	.76	.16		.92	1.07
0200	6" thick wall, .175 C.F./S.F.	G		3000	.005		1.15	.26		1.41	1.65
0300	8" thick wall, .258 C.F./S.F.	G		2400	.007		1.70	.32		2.02	2.35
0400	10" thick wall, .340 C.F./S.F.	G		1850	.009		2.23	.42		2.65	3.09
0500	12" thick wall, .422 C.F./S.F.	G		1200	.013		2.77	.65		3.42	4.03
0600	Poured cavity wall, vermiculite or perlite, water repellent	G		250	.064	C.F.	6.55	3.12		9.67	11.95
0700	Foamed in place, urethane in 2-5/8" cavity	G	G-2A	1035	.023	S.F.	1.45	.99	.62	3.06	3.84
0800	For each 1" added thickness, add	G	"	2372	.010	"	.55	.43	.27	1.25	1.59

07 25 Weather Barriers

07 25 10 – Weather Barriers or Wraps

07 25 10.10 Weather Barriers

		Crew	Daily Output	Labor-Hours	Unit	Material	2021 Bare Costs Labor	Equipment	Total	Total Incl O&P
0010	**WEATHER BARRIERS**									
0400	Asphalt felt paper, #15		1 Carp	37	.216	Sq.	5.70	11.85		17.55	24
0401	Per square foot		"	3700	.002	S.F.	.06	.12		.18	.24
0450	Housewrap, exterior, spun bonded polypropylene									
0470	Small roll		1 Carp	3800	.002	S.F.	.15	.12		.27	.33
0480	Large roll		"	4000	.002	"	.18	.11		.29	.35
2100	Asphalt felt roof deck vapor barrier, class 1 metal decks		1 Rofc	37	.216	Sq.	24.50	10.40		34.90	44
2200	For all other decks		"	37	.216		18.65	10.40		29.05	37.50
2800	Asphalt felt, 50% recycled content, 15 lb., 4 sq./roll		1 Carp	36	.222		5.95	12.15		18.10	24.50
2810	30 lb., 2 sq./roll		"	36	.222		10.20	12.15		22.35	29.50
3000	Building wrap, spun bonded polyethylene		2 Carp	8000	.002	S.F.	.14	.11		.25	.31

For customer support on your Site Work & Landscape Costs with RSMeans Data, call 800.448.8182.

147

07 26 Vapor Retarders

07 26 13 – Above-Grade Vapor Retarders

07 26 13.10 Vapor Retarders		Crew	Daily Output	Labor-Hours	Unit	Material	2021 Bare Costs Labor	Equipment	Total	Total Incl O&P
0010	**VAPOR RETARDERS**									
0020	Aluminum and kraft laminated, foil 1 side	G	1 Carp	37	.216	Sq.	17.65	11.85		29.50	37
0100	Foil 2 sides	G		37	.216		18.40	11.85		30.25	37.50
0600	Polyethylene vapor barrier, standard, 2 mil	G		37	.216		3.03	11.85		14.88	21
0700	4 mil	G		37	.216		2.70	11.85		14.55	20.50
0900	6 mil	G		37	.216		7.65	11.85		19.50	26
1200	10 mil	G		37	.216		8.15	11.85		20	26.50
1300	Clear reinforced, fire retardant, 8 mil	G		37	.216		26.50	11.85		38.35	46.50
1350	Cross laminated type, 3 mil	G		37	.216		12.85	11.85		24.70	32
1400	4 mil	G		37	.216		14.25	11.85		26.10	33.50
1800	Reinf. waterproof, 2 mil polyethylene backing, 1 side			37	.216		12	11.85		23.85	31
1900	2 sides			37	.216		15.15	11.85		27	34.50
2400	Waterproofed kraft with sisal or fiberglass fibers			37	.216		18.90	11.85		30.75	38.50

07 31 Shingles and Shakes

07 31 13 – Asphalt Shingles

07 31 13.10 Asphalt Roof Shingles

07 31 13.10 Asphalt Roof Shingles		Crew	Daily Output	Labor-Hours	Unit	Material	2021 Bare Costs Labor	Equipment	Total	Total Incl O&P
0010	**ASPHALT ROOF SHINGLES**									
0100	Standard strip shingles									
0150	Inorganic, class A, 25 year	1 Rofc	5.50	1.455	Sq.	75	70		145	197
0155	Pneumatic nailed		7	1.143		75	55		130	172
0200	30 year		5	1.600		115	77		192	251
0205	Pneumatic nailed		6.25	1.280		115	61.50		176.50	226
0250	Standard laminated multi-layered shingles									
0300	Class A, 240-260 lb./square	1 Rofc	4.50	1.778	Sq.	113	85.50		198.50	264
0305	Pneumatic nailed		5.63	1.422		113	68.50		181.50	236
0350	Class A, 250-270 lb./square		4	2		113	96.50		209.50	281
0355	Pneumatic nailed		5	1.600		113	77		190	250
0400	Premium, laminated multi-layered shingles									
0450	Class A, 260-300 lb./square	1 Rofc	3.50	2.286	Sq.	143	110		253	335
0455	Pneumatic nailed		4.37	1.831		143	88		231	300
0500	Class A, 300-385 lb./square		3	2.667		370	129		499	620
0505	Pneumatic nailed		3.75	2.133		370	103		473	575
0800	#15 felt underlayment		64	.125		5.70	6.05		11.75	16
0825	#30 felt underlayment		58	.138		9.60	6.65		16.25	21.50
0850	Self adhering polyethylene and rubberized asphalt underlayment		22	.364		68.50	17.55		86.05	104
0900	Ridge shingles		330	.024	L.F.	2.44	1.17		3.61	4.57
0905	Pneumatic nailed		412.50	.019	"	2.44	.93		3.37	4.20
1000	For steep roofs (7 to 12 pitch or greater), add						50%			

07 31 26 – Slate Shingles

07 31 26.10 Slate Roof Shingles

07 31 26.10 Slate Roof Shingles		Crew	Daily Output	Labor-Hours	Unit	Material	2021 Bare Costs Labor	Equipment	Total	Total Incl O&P
0010	**SLATE ROOF SHINGLES**									
0100	Buckingham Virginia black, 3/16" - 1/4" thick	G	1 Rots	1.75	4.571	Sq.	555	220		775	965

07 32 Roof Tiles

07 32 13 – Clay Roof Tiles

07 32 13.10 Clay Tiles

		Crew	Daily Output	Labor-Hours	Unit	Material	2021 Bare Costs Labor	Equipment	Total	Total Incl O&P
0010	**CLAY TILES**, including accessories									
0300	Flat shingle, interlocking, 15", 166 pcs./sq., fireflashed blend	3 Rots	6	4	Sq.	555	193		748	925
0500	Terra cotta red		6	4		655	193		848	1,025
0600	Roman pan and top, 18", 102 pcs./sq., fireflashed blend		5.50	4.364		515	210		725	905
0640	Terra cotta red	1 Rots	2.40	3.333		595	161		756	915
1100	Barrel mission tile, 18", 166 pcs./sq., fireflashed blend	3 Rots	5.50	4.364		430	210		640	810
1140	Terra cotta red		5.50	4.364		435	210		645	820
1700	Scalloped edge flat shingle, 14", 145 pcs./sq., fireflashed blend		6	4		1,125	193		1,318	1,550
1800	Terra cotta red		6	4		1,100	193		1,293	1,550
3010	#15 felt underlayment	1 Rofc	64	.125		5.70	6.05		11.75	16
3020	#30 felt underlayment		58	.138		9.60	6.65		16.25	21.50
3040	Polyethylene and rubberized asph. underlayment		22	.364		68.50	17.55		86.05	104

07 32 16 – Concrete Roof Tiles

07 32 16.10 Concrete Tiles

		Crew	Daily Output	Labor-Hours	Unit	Material	2021 Bare Costs Labor	Equipment	Total	Total Incl O&P
0010	**CONCRETE TILES**									
0020	Corrugated, 13" x 16-1/2", 90 per sq., 950 lb./sq.									
0050	Earthtone colors, nailed to wood deck	1 Rots	1.35	5.926	Sq.	118	286		404	595
0150	Blues		1.35	5.926		106	286		392	580
0200	Greens		1.35	5.926		117	286		403	595
0250	Premium colors		1.35	5.926		117	286		403	595
0500	Shakes, 13" x 16-1/2", 90 per sq., 950 lb./sq.									
0600	All colors, nailed to wood deck	1 Rots	1.50	5.333	Sq.	172	257		429	605
1500	Accessory pieces, ridge & hip, 10" x 16-1/2", 8 lb. each	"	120	.067	Ea.	3.74	3.21		6.95	9.30
1700	Rake, 6-1/2" x 16-3/4", 9 lb. each					3.74			3.74	4.11
1800	Mansard hip, 10" x 16-1/2", 9.2 lb. each					3.74			3.74	4.11
1900	Hip starter, 10" x 16-1/2", 10.5 lb. each					10.35			10.35	11.40
2000	3 or 4 way apex, 10" each side, 11.5 lb. each					11.85			11.85	13.05

07 32 19 – Metal Roof Tiles

07 32 19.10 Metal Roof Tiles

		Crew	Daily Output	Labor-Hours	Unit	Material	2021 Bare Costs Labor	Equipment	Total	Total Incl O&P
0010	**METAL ROOF TILES**									
0020	Accessories included, .032" thick aluminum, mission tile	1 Carp	2.50	3.200	Sq.	795	175		970	1,125
0200	Spanish tiles	"	3	2.667	"	565	146		711	840

07 41 Roof Panels

07 41 13 – Metal Roof Panels

07 41 13.10 Aluminum Roof Panels

		Crew	Daily Output	Labor-Hours	Unit	Material	2021 Bare Costs Labor	Equipment	Total	Total Incl O&P
0010	**ALUMINUM ROOF PANELS**									
0020	Corrugated or ribbed, .0155" thick, natural	G-3	1200	.027	S.F.	.77	1.46		2.23	3.05
0300	Painted		1200	.027		1.13	1.46		2.59	3.44
0400	Corrugated, .018" thick, on steel frame, natural finish		1200	.027		1.33	1.46		2.79	3.66
0600	Painted		1200	.027		1.66	1.46		3.12	4.03
0700	Corrugated, on steel frame, natural, .024" thick		1200	.027		1.93	1.46		3.39	4.32
0800	Painted		1200	.027		2.20	1.46		3.66	4.62
0900	.032" thick, natural		1200	.027		3.14	1.46		4.60	5.65
1200	Painted		1200	.027		3.41	1.46		4.87	5.95
1300	V-Beam, on steel frame construction, .032" thick, natural		1200	.027		2.75	1.46		4.21	5.25
1500	Painted		1200	.027		4.52	1.46		5.98	7.15
1600	.040" thick, natural		1200	.027		4	1.46		5.46	6.60
1800	Painted		1200	.027		5.35	1.46		6.81	8.10
1900	.050" thick, natural		1200	.027		4.11	1.46		5.57	6.70

07 41 Roof Panels

07 41 13 – Metal Roof Panels

07 41 13.10 Aluminum Roof Panels		Crew	Daily Output	Labor-Hours	Unit	Material	2021 Bare Costs Labor	Equipment	Total	Total Incl O&P
2100	Painted	G-3	1200	.027	S.F.	4.94	1.46		6.40	7.65
2200	For roofing on wood frame, deduct		4600	.007	↓	.08	.38		.46	.66
2400	Ridge cap, .032" thick, natural		800	.040	L.F.	3.38	2.20		5.58	7

07 41 13.20 Steel Roofing Panels

		Crew	Daily Output	Labor-Hours	Unit	Material	Labor	Equipment	Total	Total Incl O&P
0010	**STEEL ROOFING PANELS**									
0012	Corrugated or ribbed, on steel framing, 30 ga. galv	G-3	1100	.029	S.F.	1.89	1.60		3.49	4.48
0100	28 ga.		1050	.030		1.54	1.67		3.21	4.20
0300	26 ga.		1000	.032		1.95	1.76		3.71	4.79
0400	24 ga.		950	.034		3.07	1.85		4.92	6.15
0600	Colored, 28 ga.		1050	.030		1.92	1.67		3.59	4.62
0700	26 ga.		1000	.032	↓	2.17	1.76		3.93	5.05

07 41 33 – Plastic Roof Panels

07 41 33.10 Fiberglass Panels

		Crew	Daily Output	Labor-Hours	Unit	Material	Labor	Equipment	Total	Total Incl O&P
0010	**FIBERGLASS PANELS**									
0012	Corrugated panels, roofing, 8 oz./S.F.	G-3	1000	.032	S.F.	2.57	1.76		4.33	5.45
0100	12 oz./S.F.		1000	.032		4.65	1.76		6.41	7.75
0300	Corrugated siding, 6 oz./S.F.		880	.036		2.53	2		4.53	5.80
0400	8 oz./S.F.		880	.036		2.57	2		4.57	5.85
0500	Fire retardant		880	.036		4	2		6	7.40
0600	12 oz. siding, textured		880	.036		3.78	2		5.78	7.15
0700	Fire retardant		880	.036		4.85	2		6.85	8.35
0900	Flat panels, 6 oz./S.F., clear or colors		880	.036		3.06	2		5.06	6.35
1100	Fire retardant, class A		880	.036		3.63	2		5.63	7
1300	8 oz./S.F., clear or colors		880	.036		2.41	2		4.41	5.65
1700	Sandwich panels, fiberglass, 1-9/16" thick, panels to 20 S.F.		180	.178		36.50	9.75		46.25	54.50
1900	As above, but 2-3/4" thick, panels to 100 S.F.		265	.121	↓	25.50	6.65		32.15	38

07 42 Wall Panels

07 42 13 – Metal Wall Panels

07 42 13.20 Aluminum Siding Panels

		Crew	Daily Output	Labor-Hours	Unit	Material	Labor	Equipment	Total	Total Incl O&P
0010	**ALUMINUM SIDING PANELS**									
0012	Corrugated, on steel framing, .019" thick, natural finish	G-3	775	.041	S.F.	1.92	2.27		4.19	5.50
0100	Painted		775	.041		1.85	2.27		4.12	5.45
0400	Farm type, .021" thick on steel frame, natural		775	.041		1.75	2.27		4.02	5.35
0600	Painted		775	.041		1.85	2.27		4.12	5.45
0700	Industrial type, corrugated, on steel, .024" thick, mill		775	.041		2.41	2.27		4.68	6.05
0900	Painted		775	.041		2.56	2.27		4.83	6.25
1000	.032" thick, mill		775	.041		2.06	2.27		4.33	5.70
1200	Painted		775	.041		3.12	2.27		5.39	6.85
2750	.050" thick, natural		775	.041		3.50	2.27		5.77	7.25
2760	Painted		775	.041		4.04	2.27		6.31	7.85
3300	For siding on wood frame, deduct from above		2800	.011	↓	.09	.63		.72	1.04
3400	Screw fasteners, aluminum, self tapping, neoprene washer, 1"				M	208			208	229
3600	Stitch screws, self tapping, with neoprene washer, 5/8"				"	156			156	172
3630	Flashing, sidewall, .032" thick	G-3	800	.040	L.F.	3.20	2.20		5.40	6.80
3650	End wall, .040" thick		800	.040		3.74	2.20		5.94	7.40
3670	Closure strips, corrugated, .032" thick		800	.040		.97	2.20		3.17	4.37
3680	Ribbed, 4" or 8", .032" thick		800	.040		1.02	2.20		3.22	4.42
3690	V-beam, .040" thick		800	.040	↓	1.49	2.20		3.69	4.94
3800	Horizontal, colored clapboard, 8" wide, plain	2 Carp	515	.031	S.F.	3.13	1.70		4.83	6

07 42 Wall Panels

07 42 13 – Metal Wall Panels

07 42 13.20 Aluminum Siding Panels	Crew	Daily Output	Labor-Hours	Unit	Material	2021 Bare Costs Labor	Equipment	Total	Total Incl O&P
3810	Insulated	2 Carp	515	.031	S.F.	3.02	1.70		4.72	5.85

07 42 13.30 Steel Siding

		Crew	Daily Output	Labor-Hours	Unit	Material	Labor	Equipment	Total	Total Incl O&P
0010	**STEEL SIDING**									
0020	Beveled, vinyl coated, 8" wide	1 Carp	265	.030	S.F.	1.90	1.65		3.55	4.55
0050	10" wide	"	275	.029		1.86	1.59		3.45	4.43
0080	Galv, corrugated or ribbed, on steel frame, 30 ga.	G-3	800	.040		1.18	2.20		3.38	4.60
0100	28 ga.		795	.040		1.28	2.21		3.49	4.73
0300	26 ga.		790	.041		1.66	2.23		3.89	5.15
0400	24 ga.		785	.041		1.67	2.24		3.91	5.20
0600	22 ga.		770	.042		1.81	2.28		4.09	5.40
0700	Colored, corrugated/ribbed, on steel frame, 10 yr. finish, 28 ga.		800	.040		2.32	2.20		4.52	5.85
0900	26 ga.		795	.040		2.04	2.21		4.25	5.55
1000	24 ga.		790	.041		2.28	2.23		4.51	5.85
1020	20 ga.		785	.041		2.90	2.24		5.14	6.55

07 46 Siding

07 46 23 – Wood Siding

07 46 23.10 Wood Board Siding

		Crew	Daily Output	Labor-Hours	Unit	Material	Labor	Equipment	Total	Total Incl O&P
0010	**WOOD BOARD SIDING**									
3200	Wood, cedar bevel, A grade, 1/2" x 6"	1 Carp	295	.027	S.F.	4.42	1.48		5.90	7.05
3300	1/2" x 8"		330	.024		7.50	1.33		8.83	10.25
3500	3/4" x 10", clear grade		375	.021		7.35	1.17		8.52	9.80
3600	"B" grade		375	.021		4.38	1.17		5.55	6.55
3800	Cedar, rough sawn, 1" x 4", A grade, natural		220	.036		8.80	1.99		10.79	12.60
3900	Stained		220	.036		8.75	1.99		10.74	12.55
4100	1" x 12", board & batten, #3 & Btr., natural		420	.019		4.83	1.04		5.87	6.85
4200	Stained		420	.019		5.25	1.04		6.29	7.35
4400	1" x 8" channel siding, #3 & Btr., natural		330	.024		5.25	1.33		6.58	7.80
4500	Stained		330	.024		6	1.33		7.33	8.60
4700	Redwood, clear, beveled, vertical grain, 1/2" x 4"		220	.036		4.99	1.99		6.98	8.45
4750	1/2" x 6"		295	.027		5.40	1.48		6.88	8.15
4800	1/2" x 8"		330	.024		5.55	1.33		6.88	8.10
5000	3/4" x 10"		375	.021		5.35	1.17		6.52	7.65
5400	White pine, rough sawn, 1" x 8", natural		330	.024		2.41	1.33		3.74	4.63

07 46 29 – Plywood Siding

07 46 29.10 Plywood Siding Options

		Crew	Daily Output	Labor-Hours	Unit	Material	Labor	Equipment	Total	Total Incl O&P
0010	**PLYWOOD SIDING OPTIONS**									
0900	Plywood, medium density overlaid, 3/8" thick	2 Carp	750	.021	S.F.	1.20	1.17		2.37	3.06
1000	1/2" thick		700	.023		1.59	1.25		2.84	3.62
1100	3/4" thick		650	.025		2.62	1.35		3.97	4.89
1600	Texture 1-11, cedar, 5/8" thick, natural		675	.024		2.58	1.30		3.88	4.78
1900	Texture 1-11, fir, 5/8" thick, natural		675	.024		1.38	1.30		2.68	3.46
2050	Texture 1-11, S.Y.P., 5/8" thick, natural		675	.024		1.44	1.30		2.74	3.52
2200	Rough sawn cedar, 3/8" thick, natural		675	.024		1.28	1.30		2.58	3.35
2500	Rough sawn fir, 3/8" thick, natural		675	.024		1.02	1.30		2.32	3.06

07 46 33 – Plastic Siding

07 46 33.10 Vinyl Siding

		Crew	Daily Output	Labor-Hours	Unit	Material	Labor	Equipment	Total	Total Incl O&P
0010	**VINYL SIDING**									
3995	Clapboard profile, woodgrain texture, .048 thick, double 4	2 Carp	495	.032	S.F.	1.10	1.77		2.87	3.85
4000	Double 5		550	.029		1.10	1.59		2.69	3.59

07 46 33 – Plastic Siding

07 46 33.10 Vinyl Siding		Crew	Daily Output	Labor-Hours	Unit	Material	2021 Bare Costs Labor	Equipment	Total	Total Incl O&P
4005	Single 8	2 Carp	495	.032	S.F.	1.87	1.77		3.64	4.69
4010	Single 10		550	.029		2.24	1.59		3.83	4.84
4015	.044 thick, double 4		495	.032		1.09	1.77		2.86	3.84
4020	Double 5		550	.029		1.11	1.59		2.70	3.60
4025	.042 thick, double 4		495	.032		1.19	1.77		2.96	3.95
4030	Double 5		550	.029		1.19	1.59		2.78	3.69
4035	Cross sawn texture, .040 thick, double 4		495	.032		.76	1.77		2.53	3.47
4040	Double 5		550	.029		.73	1.59		2.32	3.18
4045	Smooth texture, .042 thick, double 4		495	.032		.97	1.77		2.74	3.70
4050	Double 5		550	.029		.81	1.59		2.40	3.27
4055	Single 8		495	.032		.73	1.77		2.50	3.44
4060	Cedar texture, .044 thick, double 4		495	.032		1.15	1.77		2.92	3.90
4065	Double 6		600	.027		1.49	1.46		2.95	3.82
4070	Dutch lap profile, woodgrain texture, .048 thick, double 5		550	.029		1.22	1.59		2.81	3.72
4075	.044 thick, double 4.5		525	.030		1.10	1.67		2.77	3.70
4080	.042 thick, double 4.5		525	.030		.91	1.67		2.58	3.49
4085	.040 thick, double 4.5		525	.030		.75	1.67		2.42	3.31
4100	Shake profile, 10" wide		400	.040		3.76	2.19		5.95	7.40
4105	Vertical pattern, .046 thick, double 5		550	.029		2.01	1.59		3.60	4.59
4110	.044 thick, triple 3		550	.029		1.72	1.59		3.31	4.27
4115	.040 thick, triple 4		550	.029		1.74	1.59		3.33	4.29
4120	.040 thick, triple 2.66		550	.029		1.92	1.59		3.51	4.49
4125	Insulation, fan folded extruded polystyrene, 1/4"		2000	.008		.30	.44		.74	.98
4130	3/8"		2000	.008		.33	.44		.77	1.01
4135	Accessories, J channel, 5/8" pocket		700	.023	L.F.	.64	1.25		1.89	2.57
4140	3/4" pocket		695	.023		.72	1.26		1.98	2.67
4145	1-1/4" pocket		680	.024		1.22	1.29		2.51	3.26
4150	Flexible, 3/4" pocket		600	.027		3.35	1.46		4.81	5.85
4155	Under sill finish trim		500	.032		.69	1.75		2.44	3.37
4160	Vinyl starter strip		700	.023		.75	1.25		2	2.69
4165	Aluminum starter strip		700	.023		.30	1.25		1.55	2.20
4170	Window casing, 2-1/2" wide, 3/4" pocket		510	.031		1.95	1.72		3.67	4.70
4175	Outside corner, woodgrain finish, 4" face, 3/4" pocket		700	.023		2.62	1.25		3.87	4.75
4180	5/8" pocket		700	.023		2.02	1.25		3.27	4.09
4185	Smooth finish, 4" face, 3/4" pocket		700	.023		2.66	1.25		3.91	4.79
4190	7/8" pocket		690	.023		1.92	1.27		3.19	4
4195	1-1/4" pocket		700	.023		1.52	1.25		2.77	3.54
4200	Soffit and fascia, 1' overhang, solid		120	.133		5.55	7.30		12.85	17
4205	Vented		120	.133		5.55	7.30		12.85	17
4207	18" overhang, solid		110	.145		6.55	7.95		14.50	19.10
4208	Vented		110	.145		6.55	7.95		14.50	19.10
4210	2' overhang, solid		100	.160		7.55	8.75		16.30	21.50
4215	Vented		100	.160		7.55	8.75		16.30	21.50
4217	3' overhang, solid		100	.160		9.55	8.75		18.30	23.50
4218	Vented		100	.160		9.55	8.75		18.30	23.50
4220	Colors for siding and soffits, add				S.F.	.15			.15	.17
4225	Colors for accessories and trim, add				L.F.	.29			.29	.32

07 46 33.20 Polypropylene Siding

		Crew	Daily Output	Labor-Hours	Unit	Material	Labor	Equipment	Total	Total Incl O&P
0010	**POLYPROPYLENE SIDING**									
4090	Shingle profile, random grooves, double 7	2 Carp	400	.040	S.F.	3.54	2.19		5.73	7.15
4092	Cornerpost for above	1 Carp	365	.022	L.F.	15.65	1.20		16.85	19
4095	Triple 5	2 Carp	400	.040	S.F.	3.72	2.19		5.91	7.35

07 46 Siding

07 46 33 – Plastic Siding

07 46 33.20 Polypropylene Siding	Crew	Daily Output	Labor-Hours	Unit	Material	2021 Bare Costs Labor	Equipment	Total	Total Incl O&P
4097	Cornerpost for above	1 Carp	365	.022	L.F.	12.50	1.20		13.70	15.55
5000	Staggered butt, double 7"	2 Carp	400	.040	S.F.	4.15	2.19		6.34	7.85
5002	Cornerpost for above	1 Carp	365	.022	L.F.	14.50	1.20		15.70	17.75
5010	Half round, double 6-1/4"	2 Carp	360	.044	S.F.	4.31	2.43		6.74	8.35
5020	Shake profile, staggered butt, double 9"	"	510	.031	"	3.84	1.72		5.56	6.80
5022	Cornerpost for above	1 Carp	365	.022	L.F.	11.65	1.20		12.85	14.65
5030	Straight butt, double 7"	2 Carp	400	.040	S.F.	4.31	2.19		6.50	8
5032	Cornerpost for above	1 Carp	365	.022	L.F.	14.65	1.20		15.85	17.95
6000	Accessories, J channel, 5/8" pocket	2 Carp	700	.023		.64	1.25		1.89	2.57
6010	3/4" pocket		695	.023		.72	1.26		1.98	2.67
6020	1-1/4" pocket		680	.024		1.22	1.29		2.51	3.26
6030	Aluminum starter strip		700	.023		.30	1.25		1.55	2.20

07 51 Built-Up Bituminous Roofing

07 51 13 – Built-Up Asphalt Roofing

07 51 13.10 Built-Up Roofing Components

		Crew	Daily Output	Labor-Hours	Unit	Material	2021 Bare Costs Labor	Equipment	Total	Total Incl O&P
0010	**BUILT-UP ROOFING COMPONENTS**									
0012	Asphalt saturated felt, #30, 2 sq./roll	1 Rofc	58	.138	Sq.	9.60	6.65		16.25	21.50
0200	#15, 4 sq./roll, plain or perforated, not mopped		58	.138		5.70	6.65		12.35	17.05
0300	Roll roofing, smooth, #65		15	.533		10	25.50		35.50	52.50
0500	#90		12	.667		35	32		67	90.50
0520	Mineralized		12	.667		36	32		68	91.50
0540	D.C. (double coverage), 19" selvage edge		10	.800		48	38.50		86.50	116
0580	Adhesive (lap cement)				Gal.	7.85			7.85	8.65
0800	Steep, flat or dead level asphalt, 10 ton lots, packaged				Ton	1,075			1,075	1,175

07 51 13.20 Built-Up Roofing Systems

		Crew	Daily Output	Labor-Hours	Unit	Material	2021 Bare Costs Labor	Equipment	Total	Total Incl O&P
0010	**BUILT-UP ROOFING SYSTEMS**									
0120	Asphalt flood coat with gravel/slag surfacing, not including									
0140	Insulation, flashing or wood nailers									
0200	Asphalt base sheet, 3 plies #15 asphalt felt, mopped	G-1	22	2.545	Sq.	127	115	26	268	355
0350	On nailable decks		21	2.667		118	120	27.50	265.50	355
0500	4 plies #15 asphalt felt, mopped		20	2.800		157	126	28.50	311.50	410
0550	On nailable decks		19	2.947		138	133	30	301	400
0700	Coated glass base sheet, 2 plies glass (type IV), mopped		22	2.545		124	115	26	265	350
0850	3 plies glass, mopped		20	2.800		150	126	28.50	304.50	400
0950	On nailable decks		19	2.947		140	133	30	303	400

07 51 13.30 Cants

		Crew	Daily Output	Labor-Hours	Unit	Material	2021 Bare Costs Labor	Equipment	Total	Total Incl O&P
0010	**CANTS**									
0012	Lumber, treated, 4" x 4" cut diagonally	1 Rofc	325	.025	L.F.	1.89	1.19		3.08	4
0300	Mineral or fiber, trapezoidal, 1" x 4" x 48"		325	.025		.36	1.19		1.55	2.32
0400	1-1/2" x 5-5/8" x 48"		325	.025		.47	1.19		1.66	2.44

07 65 Flexible Flashing

07 65 10 – Sheet Metal Flashing

07 65 10.10 Sheet Metal Flashing and Counter Flashing	Crew	Daily Output	Labor-Hours	Unit	Material	2021 Bare Costs Labor	Equipment	Total	Total Incl O&P
0010 **SHEET METAL FLASHING AND COUNTER FLASHING**									
0011 Including up to 4 bends									
0020 Aluminum, mill finish, .013" thick	1 Rofc	145	.055	S.F.	1.30	2.66		3.96	5.75
0030 .016" thick		145	.055		1.16	2.66		3.82	5.60
0060 .019" thick		145	.055		1.56	2.66		4.22	6.05
0100 .032" thick		145	.055		1.53	2.66		4.19	6
0200 .040" thick		145	.055		2.38	2.66		5.04	6.95
0300 .050" thick		145	.055		3.25	2.66		5.91	7.90
0400 Painted finish, add					.31			.31	.34

07 65 12 – Fabric and Mastic Flashings

07 65 12.10 Fabric and Mastic Flashing and Counter Flashing	Crew	Daily Output	Labor-Hours	Unit	Material	2021 Bare Costs Labor	Equipment	Total	Total Incl O&P
0010 **FABRIC AND MASTIC FLASHING AND COUNTER FLASHING**									
1300 Asphalt flashing cement, 5 gallon				Gal.	9.60			9.60	10.55
4900 Fabric, asphalt-saturated cotton, specification grade	1 Rofc	35	.229	S.Y.	3.32	11		14.32	21.50
5000 Utility grade		35	.229		1.45	11		12.45	19.45
5300 Close-mesh fabric, saturated, 17 oz./S.Y.		35	.229		2.21	11		13.21	20.50
5500 Fiberglass, resin-coated		35	.229		1.14	11		12.14	19.10

07 71 Roof Specialties

07 71 19 – Manufactured Gravel Stops and Fasciae

07 71 19.10 Gravel Stop

	Crew	Daily Output	Labor-Hours	Unit	Material	2021 Bare Costs Labor	Equipment	Total	Total Incl O&P
0010 **GRAVEL STOP**									
0020 Aluminum, .050" thick, 4" face height, mill finish	1 Shee	145	.055	L.F.	6.95	3.61		10.56	13.10
0080 Duranodic finish	"	145	.055	"	8.40	3.61		12.01	14.70

07 71 23 – Manufactured Gutters and Downspouts

07 71 23.10 Downspouts

	Crew	Daily Output	Labor-Hours	Unit	Material	2021 Bare Costs Labor	Equipment	Total	Total Incl O&P
0010 **DOWNSPOUTS**									
0020 Aluminum, embossed, .020" thick, 2" x 3"	1 Shee	190	.042	L.F.	.98	2.76		3.74	5.25
0100 Enameled		190	.042		1.40	2.76		4.16	5.70
0300 .024" thick, 2" x 3"		180	.044		2.23	2.91		5.14	6.85
0400 3" x 4"		140	.057		2.30	3.74		6.04	8.20
0600 Round, corrugated aluminum, 3" diameter, .020" thick		190	.042		1.93	2.76		4.69	6.30
0700 4" diameter, .025" thick		140	.057		6.85	3.74		10.59	13.20
4800 Steel, galvanized, round, corrugated, 2" or 3" diameter, 28 ga.		190	.042		2.82	2.76		5.58	7.25
4900 4" diameter, 28 ga.		145	.055		2.09	3.61		5.70	7.75
5100 5" diameter, 26 ga.		130	.062		3.87	4.03		7.90	10.30
5400 6" diameter, 28 ga.		105	.076		3.74	4.99		8.73	11.60
5700 Rectangular, corrugated, 28 ga., 2" x 3"		190	.042		2.43	2.76		5.19	6.85
5800 3" x 4"		145	.055		1.98	3.61		5.59	7.65
6000 Rectangular, plain, 28 ga., galvanized, 2" x 3"		190	.042		3.98	2.76		6.74	8.55
6100 3" x 4"		145	.055		5.10	3.61		8.71	11.05

07 71 23.30 Gutters

	Crew	Daily Output	Labor-Hours	Unit	Material	2021 Bare Costs Labor	Equipment	Total	Total Incl O&P
0010 **GUTTERS**									
0012 Aluminum, stock units, 5" K type, .027" thick, plain	1 Shee	125	.064	L.F.	2.95	4.19		7.14	9.55
0100 Enameled		125	.064		2.97	4.19		7.16	9.55
0300 5" K type, .032" thick, plain		125	.064		3.45	4.19		7.64	10.10
0400 Enameled		125	.064		3.57	4.19		7.76	10.25
0700 Copper, half round, 16 oz., stock units, 4" wide		125	.064		9.60	4.19		13.79	16.85
0900 5" wide		125	.064		7.80	4.19		11.99	14.90
1000 6" wide		118	.068		10.40	4.44		14.84	18.15

07 71 23.30 Gutters		Crew	Daily Output	Labor-Hours	Unit	Material	2021 Bare Costs Labor	Equipment	Total	Total Incl O&P
1200	K type, 16 oz., stock, 5" wide	1 Shee	125	.064	L.F.	8.15	4.19		12.34	15.25
1300	6" wide		125	.064		8.70	4.19		12.89	15.90
1500	Lead coated copper, 16 oz., half round, stock, 4" wide		125	.064		17.70	4.19		21.89	26
1600	6" wide		118	.068		17.80	4.44		22.24	26.50
1800	K type, stock, 5" wide		125	.064		18.65	4.19		22.84	27
1900	6" wide		125	.064		21	4.19		25.19	30
2100	Copper clad stainless steel, K type, 5" wide		125	.064		7.45	4.19		11.64	14.45
2200	6" wide		125	.064		10.25	4.19		14.44	17.60
2400	Steel, galv, half round or box, 28 ga., 5" wide, plain		125	.064		2.61	4.19		6.80	9.15
2500	Enameled		125	.064		2.23	4.19		6.42	8.75
2700	26 ga., stock, 5" wide		125	.064		2.60	4.19		6.79	9.15
2800	6" wide		125	.064		2.65	4.19		6.84	9.20
3000	Vinyl, O.G., 4" wide	1 Carp	115	.070		1.43	3.81		5.24	7.25
3100	5" wide		115	.070		1.74	3.81		5.55	7.60
3200	4" half round, stock units		115	.070		1.40	3.81		5.21	7.25
3250	Joint connectors				Ea.	3.24			3.24	3.56
3300	Wood, clear treated cedar, fir or hemlock, 3" x 4"	1 Carp	100	.080	L.F.	10.85	4.38		15.23	18.50
3400	4" x 5"	"	100	.080	"	24	4.38		28.38	33
5000	Accessories, end cap, K type, aluminum 5"	1 Shee	625	.013	Ea.	.74	.84		1.58	2.07
5010	6"		625	.013		1.50	.84		2.34	2.91
5020	Copper, 5"		625	.013		3.44	.84		4.28	5.05
5030	6"		625	.013		3.62	.84		4.46	5.25
5040	Lead coated copper, 5"		625	.013		13.20	.84		14.04	15.75
5050	6"		625	.013		14.05	.84		14.89	16.75
5060	Copper clad stainless steel, 5"		625	.013		3.58	.84		4.42	5.20
5070	6"		625	.013		3.58	.84		4.42	5.20
5080	Galvanized steel, 5"		625	.013		1.58	.84		2.42	3
5090	6"		625	.013		2.74	.84		3.58	4.27
5100	Vinyl, 4"	1 Carp	625	.013		6.55	.70		7.25	8.25
5110	5"	"	625	.013		2.99	.70		3.69	4.34
5120	Half round, copper, 4"	1 Shee	625	.013		4.66	.84		5.50	6.40
5130	5"		625	.013		5.30	.84		6.14	7.05
5140	6"		625	.013		8.55	.84		9.39	10.70
5150	Lead coated copper, 5"		625	.013		14.85	.84		15.69	17.60
5160	6"		625	.013		22	.84		22.84	26
5170	Copper clad stainless steel, 5"		625	.013		5.05	.84		5.89	6.80
5180	6"		625	.013		4.62	.84		5.46	6.35
5190	Galvanized steel, 5"		625	.013		2.82	.84		3.66	4.36
5200	6"		625	.013		3.50	.84		4.34	5.10
5210	Outlet, aluminum, 2" x 3"		420	.019		.73	1.25		1.98	2.68
5220	3" x 4"		420	.019		1.19	1.25		2.44	3.19
5230	2-3/8" round		420	.019		.67	1.25		1.92	2.62
5240	Copper, 2" x 3"		420	.019		7.80	1.25		9.05	10.50
5250	3" x 4"		420	.019		8.80	1.25		10.05	11.55
5260	2-3/8" round		420	.019		4.91	1.25		6.16	7.30
5270	Lead coated copper, 2" x 3"		420	.019		26	1.25		27.25	31
5280	3" x 4"		420	.019		30	1.25		31.25	35
5290	2-3/8" round		420	.019		26.50	1.25		27.75	31
5300	Copper clad stainless steel, 2" x 3"		420	.019		7.30	1.25		8.55	9.95
5310	3" x 4"		420	.019		8.35	1.25		9.60	11.05
5320	2-3/8" round		420	.019		4.91	1.25		6.16	7.30
5330	Galvanized steel, 2" x 3"		420	.019		3.81	1.25		5.06	6.05
5340	3" x 4"		420	.019		6.15	1.25		7.40	8.65

07 71 Roof Specialties

07 71 23 – Manufactured Gutters and Downspouts

	07 71 23.30 Gutters	Crew	Daily Output	Labor-Hours	Unit	Material	2021 Bare Costs Labor	2021 Bare Costs Equipment	Total	Total Incl O&P
5350	2-3/8" round	1 Shee	420	.019	Ea.	4.99	1.25		6.24	7.40
5360	K type mitres, aluminum		65	.123		4.94	8.05		12.99	17.60
5370	Copper		65	.123		17.60	8.05		25.65	31.50
5380	Lead coated copper		65	.123		56	8.05		64.05	73.50
5390	Copper clad stainless steel		65	.123		28	8.05		36.05	42.50
5400	Galvanized steel		65	.123		28	8.05		36.05	43
5420	Half round mitres, copper		65	.123		71	8.05		79.05	90.50
5430	Lead coated copper		65	.123		91.50	8.05		99.55	113
5440	Copper clad stainless steel		65	.123		57.50	8.05		65.55	75
5450	Galvanized steel		65	.123		34	8.05		42.05	49.50
5460	Vinyl mitres and outlets		65	.123		10.95	8.05		19	24
5470	Sealant		940	.009	L.F.	.01	.56		.57	.85
5480	Soldering		96	.083	"	.34	5.45		5.79	8.60

07 71 29 – Manufactured Roof Expansion Joints

07 71 29.10 Expansion Joints

		Crew	Daily Output	Labor-Hours	Unit	Material	2021 Bare Costs Labor	2021 Bare Costs Equipment	Total	Total Incl O&P
0010	**EXPANSION JOINTS**									
0300	Butyl or neoprene center with foam insulation, metal flanges									
0400	Aluminum, .032" thick for openings to 2-1/2"	1 Rofc	165	.048	L.F.	14.90	2.34		17.24	20
0600	For joint openings to 3-1/2"		165	.048		12.25	2.34		14.59	17.30
0610	For joint openings to 5"		165	.048		14.60	2.34		16.94	19.85
0620	For joint openings to 8"		165	.048		17.85	2.34		20.19	23.50
0700	Copper, 16 oz. for openings to 2-1/2"		165	.048		25	2.34		27.34	31.50
0900	For joint openings to 3-1/2"		165	.048		19.45	2.34		21.79	25.50
0910	For joint openings to 5"		165	.048		24.50	2.34		26.84	31
0920	For joint openings to 8"		165	.048		26.50	2.34		28.84	33
1000	Galvanized steel, 26 ga. for openings to 2-1/2"		165	.048		12.35	2.34		14.69	17.40
1200	For joint openings to 3-1/2"		165	.048		10.40	2.34		12.74	15.25
1210	For joint openings to 5"		165	.048		15.05	2.34		17.39	20.50
1220	For joint openings to 8"		165	.048		16.20	2.34		18.54	21.50
1810	For joint openings to 5"		165	.048		13.85	2.34		16.19	19
1820	For joint openings to 8"		165	.048		22	2.34		24.34	28.50
1900	Neoprene, double-seal type with thick center, 4-1/2" wide		125	.064		14.45	3.08		17.53	21
1950	Polyethylene bellows, with galv steel flat flanges		100	.080		7.75	3.86		11.61	14.75
1960	With galvanized angle flanges		100	.080		7.80	3.86		11.66	14.85
2000	Roof joint with extruded aluminum cover, 2"	1 Shee	115	.070		33.50	4.55		38.05	44
2700	Wall joint, closed cell foam on PVC cover, 9" wide	1 Rofc	125	.064		6.25	3.08		9.33	11.85
2800	12" wide	"	115	.070		7.05	3.35		10.40	13.20

07 71 43 – Drip Edge

07 71 43.10 Drip Edge, Rake Edge, Ice Belts

		Crew	Daily Output	Labor-Hours	Unit	Material	2021 Bare Costs Labor	2021 Bare Costs Equipment	Total	Total Incl O&P
0010	**DRIP EDGE, RAKE EDGE, ICE BELTS**									
0020	Aluminum, .016" thick, 5" wide, mill finish	1 Carp	400	.020	L.F.	.80	1.09		1.89	2.51
0100	White finish		400	.020		.64	1.09		1.73	2.33
0200	8" wide, mill finish		400	.020		1.40	1.09		2.49	3.17
0300	Ice belt, 28" wide, mill finish		100	.080		7.95	4.38		12.33	15.30
0400	Galvanized, 5" wide		400	.020		.69	1.09		1.78	2.39
0500	8" wide, mill finish		400	.020		.89	1.09		1.98	2.61

07 72 Roof Accessories

07 72 53 – Snow Guards

07 72 53.10 Snow Guard Options	Crew	Daily Output	Labor-Hours	Unit	Material	2021 Bare Costs Labor	2021 Bare Costs Equipment	Total	Total Incl O&P
0010 **SNOW GUARD OPTIONS**									
0100 Slate & asphalt shingle roofs, fastened with nails	1 Rofc	160	.050	Ea.	12.45	2.41		14.86	17.60
0200 Standing seam metal roofs, fastened with set screws		48	.167		17.80	8.05		25.85	32.50
0300 Surface mount for metal roofs, fastened with solder		48	.167		7.75	8.05		15.80	21.50
0400 Double rail pipe type, including pipe		130	.062	L.F.	36	2.97		38.97	44.50

07 76 Roof Pavers

07 76 16 – Roof Decking Pavers

07 76 16.10 Roof Pavers and Supports	Crew	Daily Output	Labor-Hours	Unit	Material	2021 Bare Costs Labor	2021 Bare Costs Equipment	Total	Total Incl O&P
0010 **ROOF PAVERS AND SUPPORTS**									
1000 Roof decking pavers, concrete blocks, 2" thick, natural	1 Clab	115	.070	S.F.	3.57	3.09		6.66	8.55
1100 Colors		115	.070	"	3.56	3.09		6.65	8.55
1200 Support pedestal, bottom cap		960	.008	Ea.	2.65	.37		3.02	3.47
1300 Top cap		960	.008		4.86	.37		5.23	5.90
1400 Leveling shims, 1/16"		1920	.004		1.21	.19		1.40	1.61
1500 1/8"		1920	.004		1.99	.19		2.18	2.47
1600 Buffer pad		960	.008		4.50	.37		4.87	5.50
1700 PVC legs (4" SDR 35)		2880	.003	Inch	.18	.12		.30	.38
2000 Alternate pricing method, system in place		101	.079	S.F.	8	3.52		11.52	14

07 81 Applied Fireproofing

07 81 16 – Cementitious Fireproofing

07 81 16.10 Sprayed Cementitious Fireproofing	Crew	Daily Output	Labor-Hours	Unit	Material	2021 Bare Costs Labor	2021 Bare Costs Equipment	Total	Total Incl O&P
0010 **SPRAYED CEMENTITIOUS FIREPROOFING**									
0050 Not including canvas protection, normal density									
0100 Per 1" thick, on flat plate steel	G-2	3000	.008	S.F.	.59	.37	.06	1.02	1.27
0200 Flat decking		2400	.010		.59	.46	.08	1.13	1.43
0400 Beams		1500	.016		.59	.74	.13	1.46	1.89
0500 Corrugated or fluted decks		1250	.019		.88	.89	.15	1.92	2.46
0700 Columns, 1-1/8" thick		1100	.022		.66	1.01	.17	1.84	2.42

07 91 Preformed Joint Seals

07 91 23 – Backer Rods

07 91 23.10 Backer Rods	Crew	Daily Output	Labor-Hours	Unit	Material	2021 Bare Costs Labor	2021 Bare Costs Equipment	Total	Total Incl O&P
0010 **BACKER RODS**									
0030 Backer rod, polyethylene, 1/4" diameter	1 Bric	4.60	1.739	C.L.F.	2.77	93.50		96.27	144
0050 1/2" diameter		4.60	1.739		19.90	93.50		113.40	163
0070 3/4" diameter		4.60	1.739		7.70	93.50		101.20	149
0090 1" diameter		4.60	1.739		15.45	93.50		108.95	158

07 91 26 – Joint Fillers

07 91 26.10 Joint Fillers	Crew	Daily Output	Labor-Hours	Unit	Material	2021 Bare Costs Labor	2021 Bare Costs Equipment	Total	Total Incl O&P
0010 **JOINT FILLERS**									
4360 Butyl rubber filler, 1/4" x 1/4"	1 Bric	290	.028	L.F.	.23	1.48		1.71	2.48
4365 1/2" x 1/2"		250	.032		.92	1.72		2.64	3.60
4370 1/2" x 3/4"		210	.038		1.38	2.05		3.43	4.60
4375 3/4" x 3/4"		230	.035		2.07	1.87		3.94	5.10
4380 1" x 1"		180	.044		2.76	2.39		5.15	6.65

For customer support on your Site Work & Landscape Costs with RSMeans Data, call 800.448.8182.

157

07 91 26 – Joint Fillers

07 91 26.10 Joint Fillers

		Crew	Daily Output	Labor-Hours	Unit	Material	2021 Bare Costs Labor	2021 Bare Costs Equipment	Total	Total Incl O&P
4390	For coloring, add					12%				
4980	Polyethylene joint backing, 1/4" x 2"	1 Bric	2.08	3.846	C.L.F.	14.85	207		221.85	325
4990	1/4" x 6"		1.28	6.250	"	33.50	335		368.50	540
5600	Silicone, room temp vulcanizing foam seal, 1/4" x 1/2"		1312	.006	L.F.	.45	.33		.78	.99
5610	1/2" x 1/2"		656	.012		.91	.65		1.56	1.99
5620	1/2" x 3/4"		442	.018		1.36	.97		2.33	2.95
5630	3/4" x 3/4"		328	.024		2.04	1.31		3.35	4.21
5640	1/8" x 1"		1312	.006		.45	.33		.78	.99
5650	1/8" x 3"		442	.018		1.36	.97		2.33	2.95
5670	1/4" x 3"		295	.027		2.72	1.46		4.18	5.20
5680	1/4" x 6"		148	.054		5.45	2.90		8.35	10.35
5690	1/2" x 6"		82	.098		10.85	5.25		16.10	19.85
5700	1/2" x 9"		52.50	.152		16.30	8.20		24.50	30.50
5710	1/2" x 12"		33	.242		21.50	13		34.50	43.50

07 92 13 – Elastomeric Joint Sealants

07 92 13.20 Caulking and Sealant Options

		Crew	Daily Output	Labor-Hours	Unit	Material	2021 Bare Costs Labor	2021 Bare Costs Equipment	Total	Total Incl O&P
0010	**CAULKING AND SEALANT OPTIONS**									
0050	Latex acrylic based, bulk				Gal.	30.50			30.50	33.50
0055	Bulk in place 1/4" x 1/4" bead	1 Bric	300	.027	L.F.	.10	1.43		1.53	2.27
0060	1/4" x 3/8"		294	.027		.16	1.46		1.62	2.38
0065	1/4" x 1/2"		288	.028		.21	1.49		1.70	2.48
0075	3/8" x 3/8"		284	.028		.24	1.51		1.75	2.54
0080	3/8" x 1/2"		280	.029		.32	1.53		1.85	2.66
0085	3/8" x 5/8"		276	.029		.40	1.56		1.96	2.78
0095	3/8" x 3/4"		272	.029		.48	1.58		2.06	2.90
0100	1/2" x 1/2"		275	.029		.42	1.56		1.98	2.82
0105	1/2" x 5/8"		269	.030		.53	1.60		2.13	2.99
0110	1/2" x 3/4"		263	.030		.64	1.63		2.27	3.16
0115	1/2" x 7/8"		256	.031		.74	1.68		2.42	3.35
0120	1/2" x 1"		250	.032		.85	1.72		2.57	3.52
0125	3/4" x 3/4"		244	.033		.96	1.76		2.72	3.70
0130	3/4" x 1"		225	.036		1.28	1.91		3.19	4.28
0135	1" x 1"		200	.040		1.70	2.15		3.85	5.10
0190	Cartridges				Gal.	41.50			41.50	45.50
0200	11 fl. oz. cartridge				Ea.	3.57			3.57	3.93
0400	1/4" x 1/4"	1 Bric	300	.027	L.F.	.15	1.43		1.58	2.32
0410	1/4" x 3/8"		294	.027		.22	1.46		1.68	2.44
0500	1/4" x 1/2"		288	.028		.29	1.49		1.78	2.57
0510	3/8" x 3/8"		284	.028		.33	1.51		1.84	2.64
0520	3/8" x 1/2"		280	.029		.44	1.53		1.97	2.79
0530	3/8" x 5/8"		276	.029		.55	1.56		2.11	2.94
0540	3/8" x 3/4"		272	.029		.65	1.58		2.23	3.10
0600	1/2" x 1/2"		275	.029		.58	1.56		2.14	2.99
0610	1/2" x 5/8"		269	.030		.73	1.60		2.33	3.21
0620	1/2" x 3/4"		263	.030		.88	1.63		2.51	3.42
0630	1/2" x 7/8"		256	.031		1.02	1.68		2.70	3.65
0640	1/2" x 1"		250	.032		1.17	1.72		2.89	3.87
0800	3/4" x 3/4"		244	.033		1.31	1.76		3.07	4.09
0900	3/4" x 1"		225	.036		1.75	1.91		3.66	4.81

07 92 Joint Sealants

07 92 13 – Elastomeric Joint Sealants

07 92 13.20 Caulking and Sealant Options		Crew	Daily Output	Labor-Hours	Unit	Material	2021 Bare Costs Labor	Equipment	Total	Total Incl O&P
1000	1" x 1"	1 Bric	200	.040	L.F.	2.19	2.15		4.34	5.65
3200	Polyurethane, 1 or 2 component				Gal.	54.50			54.50	60
3300	Cartridges				"	65			65	71.50
3500	Bulk, in place, 1/4" x 1/4"	1 Bric	300	.027	L.F.	.18	1.43		1.61	2.36
3510	1/4" x 3/8"		294	.027		.27	1.46		1.73	2.49
3520	3/8" x 3/8"		284	.028		.40	1.51		1.91	2.72
3530	3/8" x 1/2"		280	.029		.53	1.53		2.06	2.89
3540	3/8" x 5/8"		276	.029		.67	1.56		2.23	3.07
3550	3/8" x 3/4"		272	.029		.80	1.58		2.38	3.26
3560	1/2" x 1/2"		275	.029		.71	1.56		2.27	3.13
3570	1/2" x 5/8"		269	.030		.88	1.60		2.48	3.38
3580	1/2" x 3/4"		263	.030		1.07	1.63		2.70	3.64
3590	1/2" x 7/8"		256	.031		1.24	1.68		2.92	3.90
3655	1/2" x 1/4"		288	.028		.35	1.49		1.84	2.64
3800	3/4" x 3/8"		272	.029		.80	1.58		2.38	3.26
3850	3/4" x 3/4"		244	.033		1.61	1.76		3.37	4.42
3875	3/4" x 1"		225	.036		2.10	1.91		4.01	5.20
3900	1" x 1/2"		250	.032		1.42	1.72		3.14	4.15
4000	For coloring, add					15%				

07 95 Expansion Control

07 95 13 – Expansion Joint Cover Assemblies

07 95 13.50 Expansion Joint Assemblies

		Crew	Daily Output	Labor-Hours	Unit	Material	2021 Bare Costs Labor	Equipment	Total	Total Incl O&P
0010	**EXPANSION JOINT ASSEMBLIES**									
0200	Floor cover assemblies, 1" space, aluminum	1 Sswk	38	.211	L.F.	22.50	12.70		35.20	44.50
0300	Bronze		38	.211		53	12.70		65.70	78
2000	Roof closures, aluminum, flat roof, low profile, 1" space		57	.140		18.65	8.45		27.10	33.50
2300	Roof to wall, low profile, 1" space		57	.140		25	8.45		33.45	40

For customer support on your Site Work & Landscape Costs with RSMeans Data, call 800.448.8182.

159

Division Notes

	CREW	DAILY OUTPUT	LABOR-HOURS	UNIT	BARE COSTS				TOTAL INCL O&P
					MAT.	LABOR	EQUIP.	TOTAL	

Estimating Tips
08 10 00 Doors and Frames

All exterior doors should be addressed for their energy conservation (insulation and seals).

- Most metal doors and frames look alike, but there may be significant differences among them. When estimating these items, be sure to choose the line item that most closely compares to the specification or door schedule requirements regarding:
  - □ type of metal
  - □ metal gauge
  - □ door core material
  - □ fire rating
  - □ finish
- Wood and plastic doors vary considerably in price. The primary determinant is the veneer material. Lauan, birch, and oak are the most common veneers. Other variables include the following:
  - □ hollow or solid core
  - □ fire rating
  - □ flush or raised panel
  - □ finish
- Door pricing includes bore for cylindrical locksets and mortise for hinges.

08 30 00 Specialty Doors and Frames

- There are many varieties of special doors, and they are usually priced per each. Add frames, hardware, or operators required for a complete installation.

08 40 00 Entrances, Storefronts, and Curtain Walls

- Glazed curtain walls consist of the metal tube framing and the glazing material. The cost data in this subdivision is presented for the metal tube framing alone or the composite wall. If your estimate requires a detailed takeoff of the framing, be sure to add the glazing cost and any tints.

08 50 00 Windows

- Steel windows are unglazed and aluminum can be glazed or unglazed. Some metal windows are priced without glass. Refer to 08 80 00 Glazing for glass pricing. The grade C indicates commercial grade windows, usually ASTM C-35.
- All wood windows and vinyl are priced preglazed. The glazing is insulating glass. Add the cost of screens and grills if required and not already included.

08 70 00 Hardware

- Hardware costs add considerably to the cost of a door. The most efficient method to determine the hardware requirements for a project is to review the door and hardware schedule together. One type of door may have different hardware, depending on the door usage.
- Door hinges are priced by the pair, with most doors requiring 1-1/2 pairs per door. The hinge prices do not include installation labor because it is included in door installation.

Hinges are classified according to the frequency of use, base material, and finish.

08 80 00 Glazing

- Different openings require different types of glass. The most common types are:
  - □ float
  - □ tempered
  - □ insulating
  - □ impact-resistant
  - □ ballistic-resistant
- Most exterior windows are glazed with insulating glass. Entrance doors and window walls, where the glass is less than 18" from the floor, are generally glazed with tempered glass. Interior windows and some residential windows are glazed with float glass.
- Coastal communities require the use of impact-resistant glass, dependent on wind speed.
- The insulation or 'u' value is a strong consideration, along with solar heat gain, to determine total energy efficiency.

Reference Numbers

Reference numbers are shown at the beginning of some major classifications. These numbers refer to related items in the Reference Section. The reference information may be an estimating procedure, an alternate pricing method, or technical information.

Note: Not all subdivisions listed here necessarily appear. ■

08 05 Common Work Results for Openings

08 05 05 – Selective Demolition for Openings

08 05 05.10 Selective Demolition Doors

		Crew	Daily Output	Labor-Hours	Unit	Material	2021 Bare Costs Labor	Equipment	Total	Total Incl O&P
0010	**SELECTIVE DEMOLITION DOORS** R024119-10									
0200	Doors, exterior, 1-3/4" thick, single, 3' x 7' high	1 Clab	16	.500	Ea.		22		22	33
0220	Double, 6' x 7' high		12	.667			29.50		29.50	44
0500	Interior, 1-3/8" thick, single, 3' x 7' high		20	.400			17.75		17.75	26.50
0520	Double, 6' x 7' high		16	.500			22		22	33
0700	Bi-folding, 3' x 6'-8" high		20	.400			17.75		17.75	26.50
0720	6' x 6'-8" high		18	.444			19.75		19.75	29.50
0900	Bi-passing, 3' x 6'-8" high		16	.500			22		22	33
0940	6' x 6'-8" high		14	.571			25.50		25.50	38
1100	Door demo, floor door	2 Sswk	5	3.200			193		193	299
1500	Remove and reset, hollow core	1 Carp	8	1			54.50		54.50	81.50
1520	Solid		6	1.333			73		73	109
2000	Frames, including trim, metal		8	1			54.50		54.50	81.50
2200	Wood	2 Carp	32	.500			27.50		27.50	41
3000	Special doors, counter doors		6	2.667			146		146	218
3100	Double acting		10	1.600			87.50		87.50	131
3200	Floor door (trap type), or access type		8	2			109		109	163
3300	Glass, sliding, including frames		12	1.333			73		73	109
3400	Overhead, commercial, 12' x 12' high		4	4			219		219	325
3440	up to 20' x 16' high		3	5.333			292		292	435
3445	up to 35' x 30' high		1	16			875		875	1,300
3500	Residential, 9' x 7' high		8	2			109		109	163
3540	16' x 7' high		7	2.286			125		125	187
3620	Large		2.50	6.400			350		350	525
3700	Roll-up grille		5	3.200			175		175	261
3800	Revolving door, no wire or connections		2	8			440		440	655
3900	Storefront swing door		3	5.333			292		292	435
7550	Hangar door demo	2 Sswk	330	.048	S.F.		2.92		2.92	4.52

08 05 05.20 Selective Demolition of Windows

		Crew	Daily Output	Labor-Hours	Unit	Material	2021 Bare Costs Labor	Equipment	Total	Total Incl O&P
0010	**SELECTIVE DEMOLITION OF WINDOWS** R024119-10									
0200	Aluminum, including trim, to 12 S.F.	1 Clab	16	.500	Ea.		22		22	33
0240	To 25 S.F.		11	.727			32.50		32.50	48
0280	To 50 S.F.		5	1.600			71		71	106
0600	Glass, up to 10 S.F./window		200	.040	S.F.		1.78		1.78	2.65
0620	Over 10 S.F./window		150	.053	"		2.37		2.37	3.53
1000	Steel, including trim, to 12 S.F.		13	.615	Ea.		27.50		27.50	41
1020	To 25 S.F.		9	.889			39.50		39.50	59
1040	To 50 S.F.		4	2			89		89	133
2000	Wood, including trim, to 12 S.F.		22	.364			16.15		16.15	24
2020	To 25 S.F.		18	.444			19.75		19.75	29.50
2060	To 50 S.F.		13	.615			27.50		27.50	41
2065	To 180 S.F.		8	1			44.50		44.50	66.50
5083	Remove unit skylight	1 Carp	1.54	5.202			285		285	425

08 12 Metal Frames

08 12 13 – Hollow Metal Frames

08 12 13.13 Standard Hollow Metal Frames		Crew	Daily Output	Labor-Hours	Unit	Material	2021 Bare Costs Labor	Equipment	Total	Total Incl O&P
0010	**STANDARD HOLLOW METAL FRAMES**									
0020	16 ga., up to 5-3/4" jamb depth									
0025	3'-0" x 6'-8" single	G	2 Carp	16	1	Ea.	228	54.50		282.50	335
0028	3'-6" wide, single	G		16	1		269	54.50		323.50	380
0030	4'-0" wide, single	G		16	1		305	54.50		359.50	415
0040	6'-0" wide, double	G		14	1.143		299	62.50		361.50	425
0045	8'-0" wide, double	G		14	1.143		245	62.50		307.50	365
3600	up to 5-3/4" jamb depth, 4'-0" x 7'-0" single			15	1.067		216	58.50		274.50	325
3620	6'-0" wide, double			12	1.333		209	73		282	340
3640	8'-0" wide, double			12	1.333		315	73		388	460
4400	8-3/4" deep, 4'-0" x 7'-0", single			15	1.067		460	58.50		518.50	590
4440	8'-0" wide, double			12	1.333		440	73		513	590
4900	For welded frames, add						69.50			69.50	76.50
6200	8-3/4" deep, 4'-0" x 7'-0" single		2 Carp	15	1.067		271	58.50		329.50	385
6300	For "A" label use same price as "B" label										

08 13 Metal Doors

08 13 13 – Hollow Metal Doors

08 13 13.13 Standard Hollow Metal Doors

08 13 13.13 Standard Hollow Metal Doors			Crew	Daily Output	Labor-Hours	Unit	Material	2021 Bare Costs Labor	Equipment	Total	Total Incl O&P
0010	**STANDARD HOLLOW METAL DOORS**	R081313-20									
0015	Flush, full panel, hollow core										
0017	When noted doors are prepared but do not include glass or louvers										
0020	1-3/8" thick, 20 ga., 2'-0" x 6'-8"	G	2 Carp	20	.800	Ea.	390	44		434	495
0060	3'-0" x 6'-8"	G		17	.941		400	51.50		451.50	515
0100	3'-0" x 7'-0"	G		17	.941		405	51.50		456.50	520
0120	For vision lite, add						108			108	119
0140	For narrow lite, add						114			114	126
0320	Half glass, 20 ga., 2'-0" x 6'-8"	G	2 Carp	20	.800		775	44		819	915
0360	3'-0" x 6'-8"	G		17	.941		610	51.50		661.50	745
0400	3'-0" x 7'-0"	G		17	.941		475	51.50		526.50	600
0410	1-3/8" thick, 18 ga., 2'-0" x 6'-8"	G		20	.800		525	44		569	640
0420	3'-0" x 6'-8"	G		17	.941		530	51.50		581.50	655
0425	3'-0" x 7'-0"	G		17	.941		470	51.50		521.50	590
0450	For vision lite, add						108			108	119
0452	For narrow lite, add						114			114	126
0460	Half glass, 18 ga., 2'-0" x 6'-8"	G	2 Carp	20	.800		710	44		754	845
0470	3'-0" x 6'-8"	G		17	.941		720	51.50		771.50	865
0475	3'-0" x 7'-0"	G		17	.941		595	51.50		646.50	730
0500	Hollow core, 1-3/4" thick, full panel, 20 ga., 2'-8" x 6'-8"	G		18	.889		450	48.50		498.50	565
0520	3'-0" x 6'-8"	G		17	.941		435	51.50		486.50	555
0640	3'-0" x 7'-0"	G		17	.941		470	51.50		521.50	590
1000	18 ga., 2'-8" x 6'-8"	G		17	.941		485	51.50		536.50	605
1020	3'-0" x 6'-8"	G		16	1		495	54.50		549.50	625
1120	3'-0" x 7'-0"	G		17	.941		495	51.50		546.50	620
1212	For vision lite, add						108			108	119
1214	For narrow lite, add						114			114	126
1230	Half glass, 20 ga., 2'-8" x 6'-8"	G	2 Carp	20	.800		575	44		619	700
1240	3'-0" x 6'-8"	G		18	.889		545	48.50		593.50	675
1280	Embossed panel, 1-3/4" thick, poly core, 20 ga., 3'-0" x 7'-0"	G		18	.889		730	48.50		778.50	875
1290	Half glass, 1-3/4" thick, poly core, 20 ga., 3'-0" x 7'-0"	G		18	.889		665	48.50		713.50	805
1320	18 ga., 2'-8" x 6'-8"	G		18	.889		615	48.50		663.50	755

For customer support on your Site Work & Landscape Costs with RSMeans Data, call 800.448.8182.

163

08 13 Metal Doors

08 13 13 – Hollow Metal Doors

08 13 13.13 Standard Hollow Metal Doors		Crew	Daily Output	Labor-Hours	Unit	Material	2021 Bare Costs Labor	Equipment	Total	Total Incl O&P
1340	3'-0" x 6'-8"	**G** 2 Carp	17	.941	Ea.	610	51.50		661.50	745
1720	Insulated, 1-3/4" thick, full panel, 18 ga., 3'-0" x 6'-8"	**G**	15	1.067		535	58.50		593.50	675
1760	3'-0" x 7'-0"	**G**	15	1.067		560	58.50		618.50	700
1820	Half glass, 18 ga., 3'-0" x 6'-8"	**G**	16	1		625	54.50		679.50	765
1860	3'-0" x 7'-0"	**G**	16	1		690	54.50		744.50	840
2000	For vision lite, add					108			108	119
2010	For narrow lite, add					114			114	126
8100	For bottom louver, add					243			243	267

08 13 13.15 Metal Fire Doors

		Crew	Daily Output	Labor-Hours	Unit	Material	Labor	Equipment	Total	Total Incl O&P
0010	**METAL FIRE DOORS**									
0015	Steel, flush, "B" label, 90 minutes									
0020	Full panel, 20 ga., 2'-0" x 6'-8"	2 Carp	20	.800	Ea.	420	44		464	530
0060	3'-0" x 6'-8"		17	.941		425	51.50		476.50	540
0080	3'-0" x 7'-0"		17	.941		440	51.50		491.50	555
0140	18 ga., 3'-0" x 6'-8"		16	1		585	54.50		639.50	720
0220	For "A" label, 3 hour, 18 ga., use same price as "B" label									
0240	For vision lite, add				Ea.	165			165	182
0520	Flush, "B" label, 90 minutes, egress core, 20 ga., 2'-0" x 6'-8"	2 Carp	18	.889		685	48.50		733.50	825
0580	3'-0" x 7'-0"		16	1		715	54.50		769.50	870
0640	Flush, "A" label, 3 hour, egress core, 18 ga., 3'-0" x 6'-8"		15	1.067		745	58.50		803.50	905
0680	3'-0" x 7'-0"		15	1.067		415	58.50		473.50	540

08 14 Wood Doors

08 14 16 – Flush Wood Doors

08 14 16.09 Smooth Wood Doors

		Crew	Daily Output	Labor-Hours	Unit	Material	Labor	Equipment	Total	Total Incl O&P
0010	**SMOOTH WOOD DOORS**									
0015	Flush, interior, hollow core									
0180	3'-0" x 6'-8"	2 Carp	17	.941	Ea.	97.50	51.50		149	184
0202	1-3/4", 2'-0" x 6'-8"	"	17	.941		69	51.50		120.50	153
0430	For 7'-0" high, add					30			30	33
0480	For prefinishing, clear, add					63			63	69.50
1720	H.P. plastic laminate, 1-3/8", 2'-0" x 6'-8"	2 Carp	16	1		298	54.50		352.50	410
1780	3'-0" x 6'-8"		15	1.067		310	58.50		368.50	425
2020	Particle core, lauan face, 1-3/8", 2'-6" x 6'-8"		15	1.067		105	58.50		163.50	202
2080	3'-0" x 7'-0"		13	1.231		133	67.50		200.50	246
2110	1-3/4", 3'-0" x 7'-0"		13	1.231		168	67.50		235.50	284

08 14 16.20 Wood Fire Doors

		Crew	Daily Output	Labor-Hours	Unit	Material	Labor	Equipment	Total	Total Incl O&P
0010	**WOOD FIRE DOORS**									
0020	Particle core, 7 face plys, "B" label,									
0040	1 hour, birch face, 1-3/4" x 2'-6" x 6'-8"	2 Carp	14	1.143	Ea.	470	62.50		532.50	615
0080	3'-0" x 6'-8"		13	1.231		535	67.50		602.50	690
0090	3'-0" x 7'-0"		12	1.333		500	73		573	660
0440	M.D. overlay on hardboard, 2'-6" x 6'-8"		15	1.067		375	58.50		433.50	495
0480	3'-0" x 6'-8"		14	1.143		415	62.50		477.50	550
0740	90 minutes, birch face, 1-3/4" x 2'-6" x 6'-8"		14	1.143		345	62.50		407.50	475
0780	3'-0" x 6'-8"		13	1.231		440	67.50		507.50	585
0790	3'-0" x 7'-0"		12	1.333		460	73		533	615
2200	Custom architectural "B" label, flush, 1-3/4" thick, birch,									
2260	3'-0" x 7'-0"	2 Carp	14	1.143	Ea.	355	62.50		417.50	485

08 14 Wood Doors

08 14 33 – Stile and Rail Wood Doors

08 14 33.20 Wood Doors Residential

		Crew	Daily Output	Labor-Hours	Unit	Material	2021 Bare Costs Labor	Equipment	Total	Total Incl O&P
0010	**WOOD DOORS RESIDENTIAL**									
1000	Entrance door, colonial, 1-3/4" x 6'-8" x 2'-8" wide	2 Carp	16	1	Ea.	635	54.50		689.50	775
1300	Flush, birch, solid core, 1-3/4" x 6'-8" x 2'-8" wide	"	16	1	"	261	54.50		315.50	370
7310	Passage doors, flush, no frame included									
7700	Birch, hollow core, 1-3/8" x 6'-8" x 1'-6" wide	2 Carp	18	.889	Ea.	81.50	48.50		130	162
7740	2'-6" wide	"	18	.889	"	88	48.50		136.50	169

08 31 Access Doors and Panels

08 31 13 – Access Doors and Frames

08 31 13.20 Bulkhead/Cellar Doors

		Crew	Daily Output	Labor-Hours	Unit	Material	2021 Bare Costs Labor	Equipment	Total	Total Incl O&P
0010	**BULKHEAD/CELLAR DOORS**									
0020	Steel, not incl. sides, 44" x 62"	1 Carp	5.50	1.455	Ea.	720	79.50		799.50	910
0100	52" x 73"		5.10	1.569		900	86		986	1,125
0500	With sides and foundation plates, 57" x 45" x 24"		4.70	1.702		895	93		988	1,125
0600	42" x 49" x 51"	↓	4.30	1.860	↓	610	102		712	820

08 31 13.30 Commercial Floor Doors

		Crew	Daily Output	Labor-Hours	Unit	Material	2021 Bare Costs Labor	Equipment	Total	Total Incl O&P
0010	**COMMERCIAL FLOOR DOORS**									
0020	Aluminum tile, steel frame, one leaf, 2' x 2' opng.	2 Sswk	3.50	4.571	Opng.	740	276		1,016	1,225
0050	3'-6" x 3'-6" opening	"	3.50	4.571	"	1,575	276		1,851	2,150

08 32 Sliding Glass Doors

08 32 19 – Sliding Wood-Framed Glass Doors

08 32 19.10 Sliding Wood Doors

		Crew	Daily Output	Labor-Hours	Unit	Material	2021 Bare Costs Labor	Equipment	Total	Total Incl O&P
0010	**SLIDING WOOD DOORS**									
0020	Wood, tempered insul. glass, 6' wide, premium	2 Carp	4	4	Ea.	1,550	219		1,769	2,025
0100	Economy		4	4		1,225	219		1,444	1,675
0150	8' wide, wood, premium		3	5.333		1,975	292		2,267	2,600
0200	Economy		3	5.333		1,450	292		1,742	2,025
0235	10' wide, wood, premium		2.50	6.400		2,900	350		3,250	3,725
0240	Economy		2.50	6.400		2,475	350		2,825	3,250
0250	12' wide, wood, premium		2.50	6.400		3,200	350		3,550	4,050
0300	Economy	↓	2.50	6.400	↓	2,650	350		3,000	3,450

08 32 19.15 Sliding Glass Vinyl-Clad Wood Doors

			Crew	Daily Output	Labor-Hours	Unit	Material	2021 Bare Costs Labor	Equipment	Total	Total Incl O&P
0010	**SLIDING GLASS VINYL-CLAD WOOD DOORS**										
0020	Glass, sliding vinyl-clad, insul. glass, 6'-0" x 6'-8"	G	2 Carp	4	4	Opng.	1,625	219		1,844	2,100
0025	6'-0" x 6'-10" high	G		4	4		1,775	219		1,994	2,275
0030	6'-0" x 8'-0" high	G		4	4		2,175	219		2,394	2,725
0100	8'-0" x 6'-10" high	G		4	4		2,325	219		2,544	2,875
0500	4 leaf, 9'-0" x 6'-10" high	G		3	5.333		3,350	292		3,642	4,125
0600	12'-0" x 6'-10" high	G	↓	3	5.333	↓	4,375	292		4,667	5,225

08 36 Panel Doors

08 36 13 – Sectional Doors

08 36 13.10 Overhead Commercial Doors

		Crew	Daily Output	Labor-Hours	Unit	Material	2021 Bare Costs Labor	Equipment	Total	Total Incl O&P
0010	**OVERHEAD COMMERCIAL DOORS**									
1000	Stock, sectional, heavy duty, wood, 1-3/4" thick, 8' x 8' high	2 Carp	2	8	Ea.	1,225	440		1,665	1,975
1200	12' x 12' high		1.50	10.667		2,800	585		3,385	3,950
1300	Chain hoist, 14' x 14' high		1.30	12.308		3,800	675		4,475	5,175
1500	20' x 8' high		1.30	12.270		2,925	670		3,595	4,200
2300	Fiberglass and aluminum, heavy duty, sectional, 12' x 12' high		1.50	10.667		3,400	585		3,985	4,625
2600	Steel, 24 ga. sectional, manual, 8' x 8' high		2	8		1,175	440		1,615	1,950
2800	Chain hoist, 20' x 14' high	↓	.70	22.857	↓	3,700	1,250		4,950	5,950
2850	For 1-1/4" rigid insulation and 26 ga. galv.									
2860	back panel, add				S.F.	5.95			5.95	6.55
2900	For electric trolley operator, 1/3 HP, to 12' x 12', add	1 Carp	2	4	Ea.	1,150	219		1,369	1,575
2950	Over 12' x 12', 1/2 HP, add	"	1	8	"	1,250	440		1,690	2,025

08 41 Entrances and Storefronts

08 41 13 – Aluminum-Framed Entrances and Storefronts

08 41 13.20 Tube Framing

		Crew	Daily Output	Labor-Hours	Unit	Material	2021 Bare Costs Labor	Equipment	Total	Total Incl O&P
0010	**TUBE FRAMING**, For window walls and storefronts, aluminum stock									
1000	Flush tube frame, mill finish, 1/4" glass, 1-3/4" x 4", open header	2 Glaz	80	.200	L.F.	15.40	10.55		25.95	32.50
1050	Open sill		82	.195		13.75	10.25		24	30.50
1100	Closed back header		83	.193		24	10.15		34.15	41.50
1200	Vertical mullion, one piece		75	.213		21.50	11.25		32.75	40.50
1300	90° or 180° vertical corner post		75	.213		36	11.25		47.25	56.50
5000	Flush tube frame, mill fin., thermal brk., 2-1/4" x 4-1/2", open header		74	.216		16.40	11.40		27.80	35
5050	Open sill		75	.213		14.20	11.25		25.45	32.50
5100	Vertical mullion, one piece		69	.232		18.85	12.20		31.05	39
5200	90° or 180° vertical corner post	↓	69	.232	↓	19.10	12.20		31.30	39

08 41 26 – All-Glass Entrances and Storefronts

08 41 26.10 Window Walls Aluminum, Stock

		Crew	Daily Output	Labor-Hours	Unit	Material	2021 Bare Costs Labor	Equipment	Total	Total Incl O&P
0010	**WINDOW WALLS ALUMINUM, STOCK**, including glazing									
0020	Minimum	H-2	160	.150	S.F.	58.50	7.50		66	75
0050	Average		140	.171		80.50	8.55		89.05	101
0100	Maximum	↓	110	.218	↓	190	10.90		200.90	225

08 42 Entrances

08 42 26 – All-Glass Entrances

08 42 26.10 Swinging Glass Doors

		Crew	Daily Output	Labor-Hours	Unit	Material	2021 Bare Costs Labor	Equipment	Total	Total Incl O&P
0010	**SWINGING GLASS DOORS**									
0020	Including hardware, 1/2" thick, tempered, 3' x 7' opening	2 Glaz	2	8	Opng.	2,425	420		2,845	3,275
0100	6' x 7' opening	"	1.40	11.429	"	4,625	600		5,225	6,000

08 43 Storefronts

08 43 13 – Aluminum-Framed Storefronts

08 43 13.20 Storefront Systems	Crew	Daily Output	Labor-Hours	Unit	Material	2021 Bare Costs Labor	Equipment	Total	Total Incl O&P
0010 **STOREFRONT SYSTEMS**, aluminum frame clear 3/8" plate glass									
0020 incl. 3' x 7' door with hardware (400 sq. ft. max. wall)									
0500 Wall height to 12' high, commercial grade	2 Glaz	150	.107	S.F.	24	5.60		29.60	35
0700 Monumental grade		115	.139		48.50	7.35		55.85	64
1000 6' x 7' door with hardware, commercial grade		135	.119		73.50	6.25		79.75	90.50
1200 Monumental grade		100	.160		118	8.40		126.40	143

08 44 Curtain Wall and Glazed Assemblies

08 44 13 – Glazed Aluminum Curtain Walls

08 44 13.10 Glazed Curtain Walls

	Crew	Daily Output	Labor-Hours	Unit	Material	2021 Bare Costs Labor	Equipment	Total	Total Incl O&P
0010 **GLAZED CURTAIN WALLS**, aluminum, stock, including glazing									
0020 Minimum	H-1	205	.156	S.F.	38.50	8.80		47.30	56
0150 Average, double glazed	"	180	.178	"	92.50	10.05		102.55	117

08 51 Metal Windows

08 51 13 – Aluminum Windows

08 51 13.10 Aluminum Sash

	Crew	Daily Output	Labor-Hours	Unit	Material	2021 Bare Costs Labor	Equipment	Total	Total Incl O&P
0010 **ALUMINUM SASH**									
0020 Stock, grade C, glaze & trim not incl., casement	2 Sswk	200	.080	S.F.	43.50	4.82		48.32	55
0100 Fixed casement	"	200	.080	"	18.60	4.82		23.42	28

08 51 13.20 Aluminum Windows

	Crew	Daily Output	Labor-Hours	Unit	Material	2021 Bare Costs Labor	Equipment	Total	Total Incl O&P
0010 **ALUMINUM WINDOWS**, incl. frame and glazing, commercial grade									
1000 Stock units, casement, 3'-1" x 3'-2" opening	2 Sswk	10	1.600	Ea.	405	96.50		501.50	595
3000 Single-hung, 2' x 3' opening, enameled, standard glazed		10	1.600		225	96.50		321.50	395
3100 Insulating glass		10	1.600		273	96.50		369.50	450
3890 Awning type, 3' x 3' opening, standard glass		14	1.143		440	69		509	590
4000 Sliding aluminum, 3' x 2' opening, standard glazed		10	1.600		246	96.50		342.50	420
5500 Sliding, with thermal barrier and screen, 6' x 4', 2 track		8	2		725	121		846	980
5700 4 track		8	2		915	121		1,036	1,175

08 51 23 – Steel Windows

08 51 23.10 Steel Sash

	Crew	Daily Output	Labor-Hours	Unit	Material	2021 Bare Costs Labor	Equipment	Total	Total Incl O&P
0010 **STEEL SASH** Custom units, glazing and trim not included									
0100 Casement, 100% vented	2 Sswk	200	.080	S.F.	70	4.82		74.82	84.50
0200 50% vented		200	.080		57	4.82		61.82	70.50
0300 Fixed		200	.080		29.50	4.82		34.32	40
2000 Industrial security sash, 50% vented		200	.080		62	4.82		66.82	75.50
2100 Fixed		200	.080		49.50	4.82		54.32	62

08 51 66 – Metal Window Screens

08 51 66.10 Screens

	Crew	Daily Output	Labor-Hours	Unit	Material	2021 Bare Costs Labor	Equipment	Total	Total Incl O&P
0010 **SCREENS**									
0020 For metal sash, aluminum or bronze mesh, flat screen	2 Sswk	1200	.013	S.F.	4.74	.80		5.54	6.45
0800 Security screen, aluminum frame with stainless steel cloth	"	1200	.013	"	25	.80		25.80	28.50

08 52 Wood Windows

08 52 10 – Plain Wood Windows

08 52 10.40 Casement Window		Crew	Daily Output	Labor-Hours	Unit	Material	2021 Bare Costs Labor	2021 Bare Costs Equipment	Total	Total Incl O&P
0010	CASEMENT WINDOW, including frame, screen and grilles									
0100	2'-0" x 3'-0" H, double insulated glass	G	1 Carp	10	.800	Ea.	278	44		322	370
0200	2'-0" x 4'-6" high, double insulated glass	G		9	.889		390	48.50		438.50	505
0522	Vinyl-clad, premium, double insulated glass, 2'-0" x 3'-0"	G		10	.800		266	44		310	360
0525	2'-0" x 5'-0"	G		8	1		355	54.50		409.50	470

08 52 10.50 Double-Hung		Crew	Daily Output	Labor-Hours	Unit	Material	Labor	Equipment	Total	Total Incl O&P
0010	DOUBLE-HUNG, Including frame, screens and grilles									
0100	2'-0" x 3'-0" high, low E insul. glass	G	1 Carp	10	.800	Ea.	192	44		236	277
0300	4'-0" x 4'-6" high, low E insulated glass	G	"	8	1	"	320	54.50		374.50	435

08 52 10.70 Sliding Windows		Crew	Daily Output	Labor-Hours	Unit	Material	Labor	Equipment	Total	Total Incl O&P
0010	SLIDING WINDOWS									
0100	3'-0" x 3'-0" high, double insulated	G	1 Carp	10	.800	Ea.	292	44		336	385
0200	4'-0" x 3'-6" high, double insulated	G	"	9	.889	"	425	48.50		473.50	540

08 52 13 – Metal-Clad Wood Windows

08 52 13.10 Awning Windows, Metal-Clad		Crew	Daily Output	Labor-Hours	Unit	Material	Labor	Equipment	Total	Total Incl O&P
0010	AWNING WINDOWS, METAL-CLAD									
2000	Metal-clad, awning deluxe, double insulated glass, 34" x 22"		1 Carp	9	.889	Ea.	251	48.50		299.50	350
2100	40" x 22"		"	9	.889	"	300	48.50		348.50	405

08 52 13.20 Casement Windows, Metal-Clad		Crew	Daily Output	Labor-Hours	Unit	Material	Labor	Equipment	Total	Total Incl O&P
0010	CASEMENT WINDOWS, METAL-CLAD									
0100	Metal-clad, deluxe, dbl. insul. glass, 2'-0" x 3'-0" high	G	1 Carp	10	.800	Ea.	305	44		349	400
0130	2'-0" x 5'-0" high	G	"	8	1	"	365	54.50		419.50	485

08 52 13.30 Double-Hung Windows, Metal-Clad		Crew	Daily Output	Labor-Hours	Unit	Material	Labor	Equipment	Total	Total Incl O&P
0010	DOUBLE-HUNG WINDOWS, METAL-CLAD									
0100	Metal-clad, deluxe, dbl. insul. glass, 2'-6" x 3'-0" high	G	1 Carp	10	.800	Ea.	296	44		340	390
0160	3'-0" x 4'-6" high	G	"	9	.889	"	360	48.50		408.50	470

08 52 13.35 Picture and Sliding Windows Metal-Clad		Crew	Daily Output	Labor-Hours	Unit	Material	Labor	Equipment	Total	Total Incl O&P
0010	PICTURE AND SLIDING WINDOWS METAL-CLAD									
2400	Metal-clad, dlx sliding, dbl. insul. glass, 3'-0" x 3'-0" high	G	1 Carp	10	.800	Ea.	330	44		374	425
2420	4'-0" x 3'-6" high	G	"	9	.889	"	400	48.50		448.50	515

08 52 16 – Plastic-Clad Wood Windows

08 52 16.70 Vinyl-Clad, Premium, DBL. Insul. Glass		Crew	Daily Output	Labor-Hours	Unit	Material	Labor	Equipment	Total	Total Incl O&P
0010	VINYL-CLAD, PREMIUM, DBL. INSUL. GLASS									
1000	Sliding, 3'-0" x 3'-0"	G	1 Carp	10	.800	Ea.	655	44		699	785
1100	5'-0" x 4'-0"	G	"	9	.889	"	945	48.50		993.50	1,125

08 71 Door Hardware

08 71 20.40 Lockset		Crew	Daily Output	Labor-Hours	Unit	Material	Labor	Equipment	Total	Total Incl O&P
0010	LOCKSET, Standard duty									
0020	Non-keyed, passage, w/sect. trim		1 Carp	12	.667	Ea.	84	36.50		120.50	147
0100	Privacy			12	.667		75.50	36.50		112	138
0400	Keyed, single cylinder function			10	.800		196	44		240	281
0500	Lever handled, keyed, single cylinder function			10	.800		162	44		206	244

08 71 20.41 Dead Locks		Crew	Daily Output	Labor-Hours	Unit	Material	Labor	Equipment	Total	Total Incl O&P
0010	DEAD LOCKS									
0011	Mortise heavy duty outside key (security item)		1 Carp	9	.889	Ea.	193	48.50		241.50	285
0020	Double cylinder			9	.889		192	48.50		240.50	284
1000	Tubular, standard duty, outside key			10	.800		56	44		100	128

For customer support on your Site Work & Landscape Costs with RSMeans Data, call 800.448.8182.

08 71 Door Hardware

08 52 16 – Plastic-Clad Wood Windows

08 71 20.41 Dead Locks	Crew	Daily Output	Labor-Hours	Unit	Material	2021 Bare Costs Labor	Equipment	Total	Total Incl O&P
1010 Double cylinder	1 Carp	10	.800	Ea.	71.50	44		115.50	144

08 71 20.50 Door Stops

	Crew	Daily Output	Labor-Hours	Unit	Material	Labor	Equipment	Total	Total Incl O&P
0010 **DOOR STOPS**									
0020 Holder & bumper, floor or wall	1 Carp	32	.250	Ea.	36	13.70		49.70	60
1300 Wall bumper, 4" diameter, with rubber pad, aluminum		32	.250		13.80	13.70		27.50	35.50
1600 Door bumper, floor type, aluminum		32	.250		4.48	13.70		18.18	25.50
1900 Plunger type, door mounted	↓	32	.250	↓	33	13.70		46.70	57

08 71 20.65 Thresholds

	Crew	Daily Output	Labor-Hours	Unit	Material	Labor	Equipment	Total	Total Incl O&P
0010 **THRESHOLDS**									
0011 Threshold 3' long saddles aluminum	1 Carp	48	.167	L.F.	19.55	9.10		28.65	35
0100 Aluminum, 8" wide, 1/2" thick		12	.667	Ea.	64	36.50		100.50	125
0700 Rubber, 1/2" thick, 5-1/2" wide		20	.400		41	22		63	77.50
0800 2-3/4" wide	↓	20	.400	↓	49.50	22		71.50	87

08 71 20.90 Hinges

	Crew	Daily Output	Labor-Hours	Unit	Material	Labor	Equipment	Total	Total Incl O&P
0010 **HINGES**									
0012 Full mortise, avg. freq., steel base, USP, 4-1/2" x 4-1/2"				Pr.	48			48	53
0080 US10A					61.50			61.50	67.50
0400 Brass base, 4-1/2" x 4-1/2", US10					90.50			90.50	99.50
0480 US10B					73.50			73.50	80.50
0950 Full mortise, high frequency, steel base, 3-1/2" x 3-1/2", US26D					31.50			31.50	34.50
1000 4-1/2" x 4-1/2", USP					65			65	71.50
1400 Brass base, 3-1/2" x 3-1/2", US4					47.50			47.50	52.50
1430 4-1/2" x 4-1/2", US10				↓	60.50			60.50	66.50
3000 Half surface, half mortise, or full surface, average frequency									
3010 Steel base, 4-1/2" x 4-1/2", USP				Pr.	64.50			64.50	71
3400 Brass base, 4-1/2" x 4-1/2", US10				"	205			205	225

08 71 20.91 Special Hinges

	Crew	Daily Output	Labor-Hours	Unit	Material	Labor	Equipment	Total	Total Incl O&P
0010 **SPECIAL HINGES**									
1500 Non template, full mortise, average frequency									
1510 Steel base, 4" x 4", USP				Pr.	32			32	35.50
4600 Spring hinge, single acting, 6" flange, steel				Ea.	50			50	55
4700 Brass				"	96.50			96.50	106

08 71 21 – Astragals

08 71 21.10 Exterior Mouldings, Astragals

	Crew	Daily Output	Labor-Hours	Unit	Material	Labor	Equipment	Total	Total Incl O&P
0010 **EXTERIOR MOULDINGS, ASTRAGALS**									
2000 "L" extrusion, neoprene bulbs	1 Carp	75	.107	L.F.	4.37	5.85		10.22	13.50
2400 Spring hinged security seal, with cam	"	75	.107	"	6.65	5.85		12.50	16
3000 One piece stile protection									
3020 Neoprene fabric loop, nail on aluminum strips	1 Carp	60	.133	L.F.	1.23	7.30		8.53	12.25
3600 Spring bronze strip, nail on type		105	.076		2.35	4.17		6.52	8.80
4000 Extruded aluminum retainer, flush mount, pile insert	↓	105	.076	↓	2.72	4.17		6.89	9.20
5000 Two piece overlapping astragal, extruded aluminum retainer									
5010 Pile insert	1 Carp	60	.133	L.F.	3.16	7.30		10.46	14.40
5500 Interlocking aluminum, 5/8" x 1" neoprene bulb insert	"	45	.178	"	4.29	9.70		13.99	19.20

08 71 25 – Door Weatherstripping

08 71 25.10 Mechanical Seals, Weatherstripping

	Crew	Daily Output	Labor-Hours	Unit	Material	Labor	Equipment	Total	Total Incl O&P
0010 **MECHANICAL SEALS, WEATHERSTRIPPING**									
1000 Doors, wood frame, interlocking, for 3' x 7' door, zinc	1 Carp	3	2.667	Opng.	53.50	146		199.50	277
1300 6' x 7' opening, zinc	"	2	4	"	56.50	219		275.50	385
1700 Wood frame, spring type, bronze									

For customer support on your Site Work & Landscape Costs with RSMeans Data, call 800.448.8182.

169

08 71 Door Hardware

08 71 25 – Door Weatherstripping

08 71 25.10 Mechanical Seals, Weatherstripping	Crew	Daily Output	Labor-Hours	Unit	Material	2021 Bare Costs Labor	Equipment	Total	Total Incl O&P
1800	3' x 7' door	1 Carp	7.60	1.053	Opng.	22.50	57.50		80	111

08 75 Window Hardware

08 75 30 – Window Weatherstripping

08 75 30.10 Mechanical Weather Seals

		Crew	Daily Output	Labor-Hours	Unit	Material	Labor	Equipment	Total	Total Incl O&P
0010	**MECHANICAL WEATHER SEALS**, Window, double-hung, 3' x 5'									
0020	Zinc	1 Carp	7.20	1.111	Opng.	19.55	61		80.55	112
0500	As above but heavy duty, zinc	"	4.60	1.739	"	20	95		115	164

08 81 Glass Glazing

08 81 10 – Float Glass

08 81 10.10 Various Types and Thickness of Float Glass

		Crew	Daily Output	Labor-Hours	Unit	Material	Labor	Equipment	Total	Total Incl O&P
0010	**VARIOUS TYPES AND THICKNESS OF FLOAT GLASS**									
0020	3/16" plain	2 Glaz	130	.123	S.F.	7.10	6.50		13.60	17.45
0200	Tempered, clear	"	130	.123	"	21	6.50		27.50	33

08 81 30 – Insulating Glass

08 81 30.10 Reduce Heat Transfer Glass

			Crew	Daily Output	Labor-Hours	Unit	Material	Labor	Equipment	Total	Total Incl O&P
0010	**REDUCE HEAT TRANSFER GLASS**										
0015	2 lites 1/8" float, 1/2" thk under 15 S.F.										
0020	Clear	G	2 Glaz	95	.168	S.F.	27.50	8.85		36.35	43.50
0200	2 lites 3/16" float, for 5/8" thk unit, 15 to 30 S.F., clear	G		90	.178		17	9.35		26.35	32.50
0400	1" thk, dbl. glazed, 1/4" float, 30 to 70 S.F., clear	G	↓	75	.213	↓	17.35	11.25		28.60	36

08 83 Mirrors

08 83 13 – Mirrored Glass Glazing

08 83 13.10 Mirrors

		Crew	Daily Output	Labor-Hours	Unit	Material	Labor	Equipment	Total	Total Incl O&P
0010	**MIRRORS**, No frames, wall type, 1/4" plate glass, polished edge									
0100	Up to 5 S.F.	2 Glaz	125	.128	S.F.	9.80	6.75		16.55	21
1000	Float glass, up to 10 S.F., 1/8" thick		160	.100		6.85	5.25		12.10	15.40
1100	3/16" thick	↓	150	.107	↓	8.85	5.60		14.45	18.10

08 88 Special Function Glazing

08 88 56 – Ballistics-Resistant Glazing

08 88 56.10 Laminated Glass

		Crew	Daily Output	Labor-Hours	Unit	Material	Labor	Equipment	Total	Total Incl O&P
0010	**LAMINATED GLASS**									
0020	Clear float .03" vinyl 1/4" thick	2 Glaz	90	.178	S.F.	13.45	9.35		22.80	28.50
0200	.06" vinyl, 1/2" thick	"	65	.246	"	28	12.95		40.95	50.50

Estimating Tips

General

- Room Finish Schedule: A complete set of plans should contain a room finish schedule. If one is not available, it would be well worth the time and effort to obtain one.

09 20 00 Plaster and Gypsum Board

- Lath is estimated by the square yard plus a 5% allowance for waste. Furring, channels, and accessories are measured by the linear foot. An extra foot should be allowed for each accessory miter or stop.

- Plaster is also estimated by the square yard. Deductions for openings vary by preference, from zero deduction to 50% of all openings over 2 feet in width. The estimator should allow one extra square foot for each linear foot of horizontal interior or exterior angle located below the ceiling level. Also, double the areas of small radius work.

- Drywall accessories, studs, track, and acoustical caulking are all measured by the linear foot. Drywall taping is figured by the square foot. Gypsum wallboard is estimated by the square foot. No material deductions should be made for door or window openings under 32 S.F.

09 60 00 Flooring

- Tile and terrazzo areas are taken off on a square foot basis. Trim and base materials are measured by the linear foot. Accent tiles are listed per each. Two basic methods of installation are used. Mud set is approximately 30% more expensive than thin set.

The cost of grout is included with tile unit price lines unless otherwise noted. In terrazzo work, be sure to include the linear footage of embedded decorative strips, grounds, machine rubbing, and power cleanup.

- Wood flooring is available in strip, parquet, or block configuration. The latter two types are set in adhesives with quantities estimated by the square foot. The laying pattern will influence labor costs and material waste. In addition to the material and labor for laying wood floors, the estimator must make allowances for sanding and finishing these areas, unless the flooring is prefinished.

- Sheet flooring is measured by the square yard. Roll widths vary, so consideration should be given to use the most economical width, as waste must be figured into the total quantity. Consider also the installation methods available—direct glue down or stretched. Direct glue-down installation is assumed with sheet carpet unit price lines unless otherwise noted.

09 70 00 Wall Finishes

- Wall coverings are estimated by the square foot. The area to be covered is measured—length by height of the wall above the baseboards—to calculate the square footage of each wall. This figure is divided by the number of square feet in the single roll which is being used. Deduct, in full, the areas of openings such as doors and windows. Where a pattern match is required allow 25–30% waste.

09 80 00 Acoustic Treatment

- Acoustical systems fall into several categories. The takeoff of these materials should be by the square foot of area with a 5% allowance for waste. Do not forget about scaffolding, if applicable, when estimating these systems.

09 90 00 Painting and Coating

- New line items created for cut-ins with reference diagram.

- A major portion of the work in painting involves surface preparation. Be sure to include cleaning, sanding, filling, and masking costs in the estimate.

- Protection of adjacent surfaces is not included in painting costs. When considering the method of paint application, an important factor is the amount of protection and masking required. These must be estimated separately and may be the determining factor in choosing the method of application.

Reference Numbers

Reference numbers are shown at the beginning of some major classifications. These numbers refer to related items in the Reference Section. The reference information may be an estimating procedure, an alternate pricing method, or technical information.

Note: Not all subdivisions listed here necessarily appear. ■

Same Data. Simplified.

Enjoy the convenience and efficiency of accessing your costs anywhere:

- **Skip the multiplier** by setting your location
- **Quickly search,** edit, favorite and share costs
- **Stay on top of price changes** with automatic updates

Discover more at rsmeans.com/online

09 05 05 – Selective Demolition for Finishes

09 05 05.10 Selective Demolition, Ceilings	Crew	Daily Output	Labor-Hours	Unit	Material	2021 Bare Costs Labor	Equipment	Total	Total Incl O&P
0010 **SELECTIVE DEMOLITION, CEILINGS** R024119-10									
0200 Ceiling, gypsum wall board, furred and nailed or screwed	2 Clab	800	.020	S.F.		.89		.89	1.33
0220 On metal frame		760	.021			.93		.93	1.39
0240 On suspension system, including system		720	.022			.99		.99	1.47
1000 Plaster, lime and horse hair, on wood lath, incl. lath		700	.023			1.01		1.01	1.51
1020 On metal lath		570	.028			1.25		1.25	1.86
1100 Gypsum, on gypsum lath		720	.022			.99		.99	1.47
1120 On metal lath		500	.032			1.42		1.42	2.12
1200 Suspended ceiling, mineral fiber, 2' x 2' or 2' x 4'		1500	.011			.47		.47	.71
1250 On suspension system, incl. system		1200	.013			.59		.59	.88
1500 Tile, wood fiber, 12" x 12", glued		900	.018			.79		.79	1.18
1540 Stapled		1500	.011			.47		.47	.71
1580 On suspension system, incl. system		760	.021			.93		.93	1.39
2000 Wood, tongue and groove, 1" x 4"		1000	.016			.71		.71	1.06
2040 1" x 8"		1100	.015			.65		.65	.96
2400 Plywood or wood fiberboard, 4' x 8' sheets	↓	1200	.013			.59		.59	.88
2500 Remove & refinish textured ceiling	1 Plas	222	.036	↓	.03	1.80		1.83	2.72

09 05 05.20 Selective Demolition, Flooring

	Crew	Daily Output	Labor-Hours	Unit	Material	2021 Bare Costs Labor	Equipment	Total	Total Incl O&P
0010 **SELECTIVE DEMOLITION, FLOORING** R024119-10									
0200 Brick with mortar	2 Clab	475	.034	S.F.		1.50		1.50	2.23
0400 Carpet, bonded, including surface scraping		2000	.008			.36		.36	.53
0440 Scrim applied		8000	.002			.09		.09	.13
0480 Tackless		9000	.002	↓		.08		.08	.12
0500 Carpet pad		3500	.005	S.Y.		.20		.20	.30
0550 Carpet tile, releasable adhesive		5000	.003	S.F.		.14		.14	.21
0560 Permanent adhesive		1850	.009			.38		.38	.57
0600 Composition, acrylic or epoxy	↓	400	.040			1.78		1.78	2.65
0700 Concrete, scarify skin	A-1A	225	.036			2.03	.93	2.96	4.07
0800 Resilient, sheet goods	2 Clab	1400	.011			.51		.51	.76
0820 For gym floors	"	900	.018	↓		.79		.79	1.18
0850 Vinyl or rubber cove base	1 Clab	1000	.008	L.F.		.36		.36	.53
0860 Vinyl or rubber cove base, molded corner	"	1000	.008	Ea.		.36		.36	.53
0870 For glued and caulked installation, add to labor						50%			
0900 Vinyl composition tile, 12" x 12"	2 Clab	1000	.016	S.F.		.71		.71	1.06
2000 Tile, ceramic, thin set		675	.024			1.05		1.05	1.57
2020 Mud set		625	.026			1.14		1.14	1.70
2200 Marble, slate, thin set		675	.024			1.05		1.05	1.57
2220 Mud set		625	.026			1.14		1.14	1.70
2600 Terrazzo, thin set		450	.036			1.58		1.58	2.36
2620 Mud set		425	.038			1.67		1.67	2.49
2640 Terrazzo, cast in place	↓	300	.053			2.37		2.37	3.53
8000 Remove flooring, bead blast, simple floor plan	A-1A	1000	.008		1.68	.46	.21	2.35	2.77
8100 Complex floor plan		400	.020		1.68	1.14	.52	3.34	4.14
8150 Mastic only	↓	1500	.005	↓	1.68	.30	.14	2.12	2.46

09 05 05.30 Selective Demolition, Walls and Partitions

	Crew	Daily Output	Labor-Hours	Unit	Material	2021 Bare Costs Labor	Equipment	Total	Total Incl O&P
0010 **SELECTIVE DEMOLITION, WALLS AND PARTITIONS** R024119-10									
0100 Brick, 4" to 12" thick	B-9	220	.182	C.F.		8.15	1.62	9.77	13.95
0200 Concrete block, 4" thick		1150	.035	S.F.		1.56	.31	1.87	2.67
0280 8" thick	↓	1050	.038			1.71	.34	2.05	2.92
1000 Gypsum wallboard, nailed or screwed	1 Clab	1000	.008			.36		.36	.53
1010 2 layers		400	.020			.89		.89	1.33
1020 Glued and nailed	↓	900	.009	↓		.39		.39	.59

09 05 Common Work Results for Finishes

09 05 05 – Selective Demolition for Finishes

09 05 05.30 Selective Demolition, Walls and Partitions	Crew	Daily Output	Labor-Hours	Unit	Material	2021 Bare Costs Labor	Equipment	Total	Total Incl O&P
2200	Metal or wood studs, finish 2 sides, fiberboard	B-1	520	.046	S.F.		2.08		2.08	3.10
2250	Lath and plaster		260	.092			4.16		4.16	6.20
2300	Gypsum wallboard		520	.046			2.08		2.08	3.10
2350	Plywood		450	.053			2.40		2.40	3.59
3760	Tile, ceramic, on walls, thin set	1 Clab	300	.027			1.18		1.18	1.77
3800	Toilet partitions, slate or marble		5	1.600	Ea.		71		71	106
3820	Metal or plastic		8	1	"		44.50		44.50	66.50
5000	Wallcovering, vinyl	1 Pape	700	.011	S.F.		.53		.53	.78
5010	With release agent		1500	.005			.25		.25	.37
5025	Wallpaper, 2 layers or less, by hand		250	.032			1.48		1.48	2.20
5035	3 layers or more		165	.048			2.24		2.24	3.33

09 21 Plaster and Gypsum Board Assemblies

09 21 13 – Plaster Assemblies

09 21 13.10 Plaster Partition Wall

		Crew	Daily Output	Labor-Hours	Unit	Material	2021 Bare Costs Labor	Equipment	Total	Total Incl O&P
0010	**PLASTER PARTITION WALL**									
0400	Stud walls, 3.4 lb. metal lath, 3 coat gypsum plaster, 2 sides									
0600	2" x 4" wood studs, 16" OC	J-2	315	.152	S.F.	3.33	7.45	.36	11.14	15.10
0700	2-1/2" metal studs, 25 ga., 12" OC		325	.148		2.92	7.20	.35	10.47	14.30
0800	3-5/8" metal studs, 25 ga., 16" OC		320	.150		2.91	7.30	.35	10.56	14.45
0900	Gypsum lath, 2 coat vermiculite plaster, 2 sides									
1000	2" x 4" wood studs, 16" OC	J-2	355	.135	S.F.	2.85	6.60	.32	9.77	13.30
1200	2-1/2" metal studs, 25 ga., 12" OC		365	.132		2.54	6.40	.31	9.25	12.65
1300	3-5/8" metal studs, 25 ga., 16" OC		360	.133		2.53	6.50	.31	9.34	12.80

09 21 16 – Gypsum Board Assemblies

09 21 16.33 Partition Wall

		Crew	Daily Output	Labor-Hours	Unit	Material	2021 Bare Costs Labor	Equipment	Total	Total Incl O&P
0010	**PARTITION WALL** Stud wall, 8' to 12' high									
3200	5/8", interior, gypsum board, standard, tape & finish 2 sides									
3400	Installed on and including 2" x 4" wood studs, 16" OC	2 Carp	300	.053	S.F.	1.59	2.92		4.51	6.10
3800	Metal studs, NLB, 25 ga., 16" OC, 3-5/8" wide		340	.047		1.29	2.57		3.86	5.25
4800	Water resistant, on 2" x 4" wood studs, 16" OC		300	.053		1.73	2.92		4.65	6.25
5200	Metal studs, NLB, 25 ga. 16" OC, 3-5/8" wide		340	.047		1.43	2.57		4	5.40
6000	Fire resistant, 2 layers, 2 hr., on 2" x 4" wood studs, 16" OC		205	.078		2.39	4.27		6.66	9
6400	Metal studs, NLB, 25 ga., 16" OC, 3-5/8" wide		245	.065		2.09	3.57		5.66	7.65
9600	Partitions, for work over 8' high, add		1530	.010			.57		.57	.85

09 23 Gypsum Plastering

09 23 20 – Gypsum Plaster

09 23 20.10 Gypsum Plaster On Walls and Ceilings

		Crew	Daily Output	Labor-Hours	Unit	Material	2021 Bare Costs Labor	Equipment	Total	Total Incl O&P
0010	**GYPSUM PLASTER ON WALLS AND CEILINGS**									
0020	80# bag, less than 1 ton				Bag	16.20			16.20	17.85
0100	Over 1 ton				"	14.30			14.30	15.75
0300	2 coats, no lath included, on walls	J-1	105	.381	S.Y.	4.01	18.20	1.08	23.29	32.50
0600	On and incl. 3/8" gypsum lath on steel, on walls	J-2	97	.495		7.35	24	1.17	32.52	45.50
0900	3 coats, no lath included, on walls	J-1	87	.460		5.70	22	1.30	29	40
1200	On and including painted metal lath, on wood studs	J-2	86	.558		10.85	27	1.32	39.17	54

For customer support on your Site Work & Landscape Costs with RSMeans Data, call 800.448.8182.

173

09 24 Cement Plastering

09 24 23 – Cement Stucco

09 24 23.40 Stucco

		Crew	Daily Output	Labor-Hours	Unit	Material	2021 Bare Costs Labor	Equipment	Total	Total Incl O&P
0010	**STUCCO**									
0015	3 coats 7/8" thick, float finish, with mesh, on wood frame	J-2	63	.762	S.Y.	7.70	37	1.80	46.50	65.50
0100	On masonry construction, no mesh incl.	J-1	67	.597		3.10	28.50	1.69	33.29	48
0300	For trowel finish, add	1 Plas	170	.047			2.35		2.35	3.49
1000	Stucco, with bonding agent, 3 coats, on walls, no mesh incl.	J-1	200	.200		4.31	9.55	.57	14.43	19.55

09 26 Veneer Plastering

09 26 13 – Gypsum Veneer Plastering

09 26 13.80 Thin Coat Plaster

		Crew	Daily Output	Labor-Hours	Unit	Material	2021 Bare Costs Labor	Equipment	Total	Total Incl O&P
0010	**THIN COAT PLASTER**									
0012	1 coat veneer, not incl. lath	J-1	3600	.011	S.F.	.10	.53	.03	.66	.93
1000	In 50 lb. bags				Bag	13.65			13.65	15.05

09 29 Gypsum Board

09 29 10 – Gypsum Board Panels

09 29 10.30 Gypsum Board

		Crew	Daily Output	Labor-Hours	Unit	Material	2021 Bare Costs Labor	Equipment	Total	Total Incl O&P
0010	**GYPSUM BOARD** on walls & ceilings R092910-10									
0150	3/8" thick, on walls, standard, no finish included	2 Carp	2000	.008	S.F.	.35	.44		.79	1.04
0300	1/2" thick, on walls, standard, no finish included		2000	.008		.33	.44		.77	1.01
0350	Taped and finished (level 4 finish)		965	.017		.38	.91		1.29	1.77
0390	With compound skim coat (level 5 finish)		775	.021		.43	1.13		1.56	2.17
2000	5/8" thick, on walls, standard, no finish included		2000	.008		.36	.44		.80	1.05
2090	With compound skim coat (level 5 finish)		775	.021		.46	1.13		1.59	2.20
2200	Water resistant, no finish included		2000	.008		.43	.44		.87	1.12
2290	With compound skim coat (level 5 finish)		775	.021		.53	1.13		1.66	2.28
2510	Mold resistant, no finish included		2000	.008		.45	.44		.89	1.15
2530	With compound skim coat (level 5 finish)		775	.021		.55	1.13		1.68	2.30
4000	Fireproofing, beams or columns, 2 layers, 1/2" thick, incl finish		330	.048		.83	2.65		3.48	4.88
4010	Mold resistant		330	.048		.97	2.65		3.62	5.05
4050	5/8" thick		300	.053		.85	2.92		3.77	5.30
4060	Mold resistant		300	.053		.99	2.92		3.91	5.45
6010	1/2" thick on walls, multi-layer, lightweight, no finish included		1500	.011		1.66	.58		2.24	2.70
6015	Taped and finished (level 4 finish)		725	.022		1.71	1.21		2.92	3.68
6025	5/8" thick on walls, for wood studs, no finish included		1500	.011		2.81	.58		3.39	3.96
6035	With compound skim coat (level 5 finish)		580	.028		2.91	1.51		4.42	5.45
6040	For metal stud, no finish included		1500	.011		1.86	.58		2.44	2.92
6050	With compound skim coat (level 5 finish)		580	.028		1.96	1.51		3.47	4.41
6055	Abuse resist, no finish included		1500	.011		1.62	.58		2.20	2.65
6065	With compound skim coat (level 5 finish)		580	.028		1.72	1.51		3.23	4.15
6070	Shear rated, no finish included		1500	.011		1.63	.58		2.21	2.66
6080	With compound skim coat (level 5 finish)		580	.028		1.73	1.51		3.24	4.16
6085	For SCIF applications, no finish included		1500	.011		1.75	.58		2.33	2.80
6095	With compound skim coat (level 5 finish)		580	.028		1.85	1.51		3.36	4.29
6100	1-3/8" thick on walls, THX certified, no finish included		1500	.011		9.50	.58		10.08	11.30
6110	With compound skim coat (level 5 finish)		580	.028		9.60	1.51		11.11	12.85
6115	5/8" thick on walls, score & snap installation, no finish included		2000	.008		1.86	.44		2.30	2.70
6125	With compound skim coat (level 5 finish)		775	.021		1.96	1.13		3.09	3.85
7020	5/8" thick on ceilings, for wood joists, no finish included		1200	.013		2.81	.73		3.54	4.18
7030	With compound skim coat (level 5 finish)		410	.039		2.91	2.13		5.04	6.40

09 29 Gypsum Board

09 29 10 – Gypsum Board Panels

09 29 10.30 Gypsum Board	Crew	Daily Output	Labor-Hours	Unit	Material	2021 Bare Costs Labor	Equipment	Total	Total Incl O&P
7035	For metal joists, no finish included	2 Carp	1200	.013	S.F.	1.86	.73		2.59	3.14
7045	With compound skim coat (level 5 finish)		410	.039		1.96	2.13		4.09	5.35
7050	Abuse resist, no finish included		1200	.013		1.62	.73		2.35	2.87
7060	With compound skim coat (level 5 finish)		410	.039		1.72	2.13		3.85	5.10
7065	Shear rated, no finish included		1200	.013		1.63	.73		2.36	2.88
7075	With compound skim coat (level 5 finish)		410	.039		1.73	2.13		3.86	5.10
7080	For SCIF applications, no finish included		1200	.013		1.75	.73		2.48	3.02
7090	With compound skim coat (level 5 finish)		410	.039		1.85	2.13		3.98	5.25
8010	5/8" thick on ceilings, score & snap installation, no finish included		1600	.010		1.86	.55		2.41	2.87
8020	With compound skim coat (level 5 finish)		545	.029		1.96	1.61		3.57	4.56

09 31 Thin-Set Tiling

09 31 13 – Thin-Set Ceramic Tiling

09 31 13.10 Thin-Set Ceramic Tile

		Crew	Daily Output	Labor-Hours	Unit	Material	2021 Bare Costs Labor	Equipment	Total	Total Incl O&P
0010	THIN-SET CERAMIC TILE									
3300	Ceramic tile, porcelain type, 1 color, color group 2, 1" x 1"	D-7	183	.087	S.F.	5.75	4.08		9.83	12.30
3310	2" x 2" or 2" x 1"		190	.084		4.04	3.93		7.97	10.20
5400	Walls, interior, 4-1/4" x 4-1/4" tile		190	.084		3.02	3.93		6.95	9.05
5700	8-1/2" x 4-1/4" tile		190	.084		6.40	3.93		10.33	12.75

09 32 Mortar-Bed Tiling

09 32 13 – Mortar-Bed Ceramic Tiling

09 32 13.10 Ceramic Tile

		Crew	Daily Output	Labor-Hours	Unit	Material	2021 Bare Costs Labor	Equipment	Total	Total Incl O&P
0010	CERAMIC TILE									
6300	Exterior walls, frostproof, 4-1/4" x 4-1/4"	D-7	102	.157	S.F.	7.45	7.30		14.75	18.90
6400	1-3/8" x 1-3/8"	"	93	.172	"	6.95	8.05		15	19.40

09 32 16 – Mortar-Bed Quarry Tiling

09 32 16.10 Quarry Tile

		Crew	Daily Output	Labor-Hours	Unit	Material	2021 Bare Costs Labor	Equipment	Total	Total Incl O&P
0010	QUARRY TILE									
0100	Base, cove or sanitary, to 5" high, 1/2" thick	D-7	110	.145	L.F.	6.65	6.80		13.45	17.20
0700	Floors, 1,000 S.F. lots, red, 4" x 4" x 1/2" thick		120	.133	S.F.	9.40	6.20		15.60	19.40
0900	6" x 6" x 1/2" thick		140	.114	"	10.15	5.35		15.50	18.95

09 34 Waterproofing-Membrane Tiling

09 34 13 – Waterproofing-Membrane Ceramic Tiling

09 34 13.10 Ceramic Tile Waterproofing Membrane

		Crew	Daily Output	Labor-Hours	Unit	Material	2021 Bare Costs Labor	Equipment	Total	Total Incl O&P
0010	CERAMIC TILE WATERPROOFING MEMBRANE									
0020	On floors, including thinset									
0030	Fleece laminated polyethylene grid, 1/8" thick	D-7	250	.064	S.F.	2.13	2.99		5.12	6.70
0040	5/16" thick	"	250	.064	"	2.63	2.99		5.62	7.25
0050	On walls, including thinset									
0060	Fleece laminated polyethylene sheet, 8 mil thick	D-7	480	.033	S.F.	2.43	1.56		3.99	4.94
0070	Accessories, including thinset									
0080	Joint and corner sheet, 4 mils thick, 5" wide	1 Tilf	240	.033	L.F.	1.34	1.72		3.06	4
0090	7-1/4" wide		180	.044		1.72	2.30		4.02	5.25
0100	10" wide		120	.067		2.09	3.45		5.54	7.35
0110	Pre-formed corners, inside		32	.250	Ea.	9.15	12.95		22.10	29

09 34 Waterproofing-Membrane Tiling

09 34 13 – Waterproofing-Membrane Ceramic Tiling

09 34 13.10 Ceramic Tile Waterproofing Membrane		Crew	Daily Output	Labor-Hours	Unit	Material	2021 Bare Costs Labor	Equipment	Total	Total Incl O&P
0120	Outside	1 Tilf	32	.250	Ea.	7.65	12.95		20.60	27.50
0130	2" flanged floor drain with 6" stainless steel grate		16	.500	↓	390	26		416	470
0140	EPS, sloped shower floor		480	.017	S.F.	11.60	.86		12.46	14
0150	Curb	↓	32	.250	L.F.	17.55	12.95		30.50	38

09 51 Acoustical Ceilings

09 51 23 – Acoustical Tile Ceilings

09 51 23.30 Suspended Ceilings, Complete

		Crew	Daily Output	Labor-Hours	Unit	Material	2021 Bare Costs Labor	Equipment	Total	Total Incl O&P
0010	**SUSPENDED CEILINGS, COMPLETE**, incl. standard									
0100	suspension system but not incl. 1-1/2" carrier channels									
0600	Fiberglass ceiling board, 2' x 4' x 5/8", plain faced	1 Carp	500	.016	S.F.	3.57	.88		4.45	5.25
1800	Tile, Z bar suspension, 5/8" mineral fiber tile	"	150	.053	"	2.62	2.92		5.54	7.25

09 63 Masonry Flooring

09 63 13 – Brick Flooring

09 63 13.10 Miscellaneous Brick Flooring

		Crew	Daily Output	Labor-Hours	Unit	Material	2021 Bare Costs Labor	Equipment	Total	Total Incl O&P
0010	**MISCELLANEOUS BRICK FLOORING**									
0020	Acid-proof shales, red, 8" x 3-3/4" x 1-1/4" thick	D-7	.43	37.209	M	615	1,725		2,340	3,200
0050	2-1/4" thick	D-1	.40	40		1,000	1,950		2,950	4,025
0200	Acid-proof clay brick, 8" x 3-3/4" x 2-1/4" thick G	"	.40	40	↓	1,025	1,950		2,975	4,050
0260	Cast ceramic, pressed, 4" x 8" x 1/2", unglazed	D-7	100	.160	S.F.	7.40	7.45		14.85	19.05
0270	Glazed		100	.160		9.85	7.45		17.30	22
0280	Hand molded flooring, 4" x 8" x 3/4", unglazed		95	.168		9.75	7.85		17.60	22.50
0290	Glazed		95	.168		10.75	7.85		18.60	23.50
0300	8" hexagonal, 3/4" thick, unglazed		85	.188		9.35	8.80		18.15	23
0310	Glazed	↓	85	.188		16.90	8.80		25.70	31.50
0400	Heavy duty industrial, cement mortar bed, 2" thick, not incl. brick	D-1	80	.200		.90	9.75		10.65	15.70
0450	Acid-proof joints, 1/4" wide	"	65	.246		1.70	12		13.70	19.90
0500	Pavers, 8" x 4", 1" to 1-1/4" thick, red	D-7	95	.168		4.30	7.85		12.15	16.25
0510	Ironspot	"	95	.168		5.30	7.85		13.15	17.35
0540	1-3/8" to 1-3/4" thick, red	D-1	95	.168		4.15	8.20		12.35	16.90
0560	Ironspot		95	.168		5.25	8.20		13.45	18.15
0580	2-1/4" thick, red		90	.178		4.23	8.65		12.88	17.70
0590	Ironspot		90	.178		6.55	8.65		15.20	20.50
0700	Paver, adobe brick, 6" x 12", 1/2" joint G	↓	42	.381		1.42	18.55		19.97	29.50
0710	Mexican red, 12" x 12" G	1 Tilf	48	.167		2.08	8.60		10.68	14.90
0720	Saltillo, 12" x 12" G	"	48	.167	↓	1.73	8.60		10.33	14.50
0800	For sidewalks and patios with pavers, see Section 32 14 16.10									
0870	For epoxy joints, add	D-1	600	.027	S.F.	3.22	1.30		4.52	5.50
0880	For Furan underlayment, add	"	600	.027		2.66	1.30		3.96	4.89
0890	For waxed surface, steam cleaned, add	A-1H	1000	.008	↓	.21	.36	.08	.65	.84

176

For customer support on your Site Work & Landscape Costs with RSMeans Data, call 800.448.8182.

09 65 Resilient Flooring

09 65 13 - Resilient Base and Accessories

09 65 13.13 Resilient Base

		Crew	Daily Output	Labor-Hours	Unit	Material	2021 Bare Costs Labor	Equipment	Total	Total Incl O&P
0010	**RESILIENT BASE**									
0800	1/8" rubber base, 2-1/2" H, straight or cove, standard colors	1 Tilf	315	.025	L.F.	1.13	1.31		2.44	3.16
1100	4" high	"	315	.025	"	1.58	1.31		2.89	3.66

09 65 13.23 Resilient Stair Treads and Risers

		Crew	Daily Output	Labor-Hours	Unit	Material	2021 Bare Costs Labor	Equipment	Total	Total Incl O&P
0010	**RESILIENT STAIR TREADS AND RISERS**									
0300	Rubber, molded tread, 12" wide, 5/16" thick, black	1 Tilf	115	.070	L.F.	14.70	3.60		18.30	21.50
1800	Risers, 7" high, 1/8" thick, flat		250	.032		8.75	1.65		10.40	12.05
2100	Vinyl, molded tread, 12" wide, colors, 1/8" thick		115	.070		5.95	3.60		9.55	11.75
2400	Riser, 7" high, 1/8" thick, coved	↓	175	.046	↓	2.92	2.36		5.28	6.65

09 65 19 - Resilient Tile Flooring

09 65 19.19 Vinyl Composition Tile Flooring

		Crew	Daily Output	Labor-Hours	Unit	Material	2021 Bare Costs Labor	Equipment	Total	Total Incl O&P
0010	**VINYL COMPOSITION TILE FLOORING**									
7000	Vinyl composition tile, 12" x 12", 1/16" thick	1 Tilf	500	.016	S.F.	.88	.83		1.71	2.18
7350	1/8" thick, marbleized	"	500	.016	"	2.15	.83		2.98	3.58

09 65 19.23 Vinyl Tile Flooring

		Crew	Daily Output	Labor-Hours	Unit	Material	2021 Bare Costs Labor	Equipment	Total	Total Incl O&P
0010	**VINYL TILE FLOORING**									
7500	Vinyl tile, 12" x 12", 3/32" thick, standard colors/patterns	1 Tilf	500	.016	S.F.	3.51	.83		4.34	5.05
7550	1/8" thick, standard colors/patterns	"	500	.016	"	1.28	.83		2.11	2.62

09 66 Terrazzo Flooring

09 66 16 - Terrazzo Floor Tile

09 66 16.10 Tile or Terrazzo Base

		Crew	Daily Output	Labor-Hours	Unit	Material	2021 Bare Costs Labor	Equipment	Total	Total Incl O&P
0010	**TILE OR TERRAZZO BASE**									
0020	Scratch coat only	1 Mstz	150	.053	S.F.	.48	2.76		3.24	4.56
0500	Scratch and brown coat only	"	75	.107	"	.96	5.50		6.46	9.10

09 91 Painting

09 91 13 - Exterior Painting

09 91 13.60 Siding Exterior

		Crew	Daily Output	Labor-Hours	Unit	Material	2021 Bare Costs Labor	Equipment	Total	Total Incl O&P
0010	**SIDING EXTERIOR**, Alkyd (oil base)									
0450	Steel siding, oil base, paint 1 coat, brushwork	2 Pord	2015	.008	S.F.	.10	.37		.47	.66
0800	Paint 2 coats, brushwork		1300	.012		.20	.57		.77	1.07
1200	Stucco, rough, oil base, paint 2 coats, brushwork		1300	.012		.20	.57		.77	1.07
1400	Roller		1625	.010		.21	.46		.67	.92
1800	Texture 1-11 or clapboard, oil base, primer coat, brushwork		1300	.012		.14	.57		.71	1
2400	Paint 2 coats, brushwork	↓	810	.020		.30	.92		1.22	1.69
8000	For latex paint, deduct				↓	10%				

09 91 23 - Interior Painting

09 91 23.52 Miscellaneous, Interior

		Crew	Daily Output	Labor-Hours	Unit	Material	2021 Bare Costs Labor	Equipment	Total	Total Incl O&P
0010	**MISCELLANEOUS, INTERIOR**									
2400	Floors, conc./wood, oil base, primer/sealer coat, brushwork	2 Pord	1950	.008	S.F.	.09	.38		.47	.66
2450	Roller	"	5200	.003		.09	.14		.23	.31
3800	Grilles, per side, oil base, primer coat, brushwork	1 Pord	520	.015		.15	.71		.86	1.23
3900	Spray	"	1140	.007	↓	.28	.33		.61	.79
5000	Pipe, 1"-4" diameter, primer or sealer coat, oil base, brushwork	2 Pord	1250	.013	L.F.	.09	.59		.68	.98
5350	Paint 2 coats, brushwork		775	.021		.27	.96		1.23	1.71
6300	13"-16" diameter, primer or sealer coat, brushwork		310	.052		.36	2.40		2.76	3.95
6500	Paint 2 coats, brushwork	↓	195	.082	↓	1.07	3.81		4.88	6.85

09 91 Painting

09 91 23 – Interior Painting

09 91 23.52 Miscellaneous, Interior

09 91 23.52 Miscellaneous, Interior	Crew	Daily Output	Labor-Hours	Unit	Material	2021 Bare Costs Labor	2021 Bare Costs Equipment	Total	Total Incl O&P
8900	Trusses and wood frames, primer coat, oil base, brushwork	1 Pord	800	.010	S.F.	.08	.46		.54	.77
9220	Paint 2 coats, brushwork		500	.016		.24	.74		.98	1.37
9260	Stain, brushwork, wipe off		600	.013		.10	.62		.72	1.03
9280	Varnish, 3 coats, brushwork	↓	275	.029	↓	.30	1.35		1.65	2.33

09 91 23.72 Walls and Ceilings, Interior

09 91 23.72 Walls and Ceilings, Interior	Crew	Daily Output	Labor-Hours	Unit	Material	2021 Bare Costs Labor	2021 Bare Costs Equipment	Total	Total Incl O&P
0010	**WALLS AND CEILINGS, INTERIOR**									
0100	Concrete, drywall or plaster, latex, primer or sealer coat									
0150	Smooth finish, cut-in by brush	1 Pord	1150	.007	L.F.	.02	.32		.34	.50
0200	Brushwork		1150	.007	S.F.	.07	.32		.39	.55
0590	Paint 2 coats, smooth finish, cut-in by brush		680	.012	L.F.	.04	.55		.59	.85
0800	Brushwork		680	.012	S.F.	.16	.55		.71	.99
0880	Spray	↓	1625	.005	"	.15	.23		.38	.51

09 91 23.74 Walls and Ceilings, Interior, Zero VOC Latex

09 91 23.74 Walls and Ceilings, Interior, Zero VOC Latex		Crew	Daily Output	Labor-Hours	Unit	Material	2021 Bare Costs Labor	2021 Bare Costs Equipment	Total	Total Incl O&P
0010	**WALLS AND CEILINGS, INTERIOR, ZERO VOC LATEX**										
0100	Concrete, dry wall or plaster, latex, primer or sealer coat										
0190	Smooth finish, cut-in by brush		1 Pord	1150	.007	L.F.	.02	.32		.34	.50
0200	Brushwork	G		1150	.007	S.F.	.08	.32		.40	.56
0790	Paint 2 coats, smooth finish, cut-in by brush			680	.012	L.F.	.04	.55		.59	.85
0800	Brushwork	G		680	.012	S.F.	.04	.55		.59	.85
0880	Spray	G	↓	1625	.005	"	.14	.23		.37	.49

09 91 23.75 Dry Fall Painting

09 91 23.75 Dry Fall Painting	Crew	Daily Output	Labor-Hours	Unit	Material	2021 Bare Costs Labor	2021 Bare Costs Equipment	Total	Total Incl O&P
0010	**DRY FALL PAINTING**									
0100	Sprayed on walls, gypsum board or plaster									
0220	One coat	1 Pord	2600	.003	S.F.	.09	.14		.23	.30
0250	Two coats		1560	.005		.17	.24		.41	.54
0280	Concrete or textured plaster, one coat		1560	.005		.09	.24		.33	.44
0310	Two coats		1300	.006		.17	.29		.46	.61
0340	Concrete block, one coat		1560	.005		.09	.24		.33	.44
0370	Two coats		1300	.006		.17	.29		.46	.61
0400	Wood, one coat		877	.009		.09	.42		.51	.72
0430	Two coats	↓	650	.012	↓	.17	.57		.74	1.04
0440	On ceilings, gypsum board or plaster									
0470	One coat	1 Pord	1560	.005	S.F.	.09	.24		.33	.44
0500	Two coats		1300	.006		.17	.29		.46	.61
0530	Concrete or textured plaster, one coat		1560	.005		.09	.24		.33	.44
0560	Two coats		1300	.006		.17	.29		.46	.61
0570	Structural steel, bar joists or metal deck, one coat		1560	.005		.09	.24		.33	.44
0580	Two coats	↓	1040	.008	↓	.17	.36		.53	.72

09 96 High-Performance Coatings

09 96 23 – Graffiti-Resistant Coatings

09 96 23.10 Graffiti-Resistant Treatments

09 96 23.10 Graffiti-Resistant Treatments	Crew	Daily Output	Labor-Hours	Unit	Material	2021 Bare Costs Labor	2021 Bare Costs Equipment	Total	Total Incl O&P
0010	**GRAFFITI-RESISTANT TREATMENTS**, sprayed on walls									
0100	Non-sacrificial, permanent non-stick coating, clear, on metals	1 Pord	2000	.004	S.F.	1.98	.19		2.17	2.46
0200	Concrete		2000	.004		2.25	.19		2.44	2.76
0300	Concrete block		2000	.004		2.91	.19		3.10	3.48
0400	Brick		2000	.004		3.30	.19		3.49	3.91
0500	Stone		2000	.004		3.30	.19		3.49	3.91
0600	Unpainted wood		2000	.004		3.81	.19		4	4.47
2000	Semi-permanent cross linking polymer primer, on metals	↓	2000	.004	↓	.61	.19		.80	.95

09 96 High-Performance Coatings

09 96 23 - Graffiti-Resistant Coatings

09 96 23.10 Graffiti-Resistant Treatments	Crew	Daily Output	Labor-Hours	Unit	Material	2021 Bare Costs Labor	Equipment	Total	Total Incl O&P
2100	Concrete	1 Pord	2000	.004	S.F.	.73	.19		.92	1.08
2200	Concrete block		2000	.004		.91	.19		1.10	1.28
2300	Brick		2000	.004		.73	.19		.92	1.08
2400	Stone		2000	.004		.73	.19		.92	1.08
2500	Unpainted wood		2000	.004		1.01	.19		1.20	1.39
3000	Top coat, on metals		2000	.004		.56	.19		.75	.89
3100	Concrete		2000	.004		.64	.19		.83	.98
3200	Concrete block		2000	.004		.89	.19		1.08	1.26
3300	Brick		2000	.004		.74	.19		.93	1.10
3400	Stone		2000	.004		.74	.19		.93	1.10
3500	Unpainted wood		2000	.004		.89	.19		1.08	1.26
5000	Sacrificial, water based, on metal		2000	.004		.32	.19		.51	.64
5100	Concrete		2000	.004		.32	.19		.51	.64
5200	Concrete block		2000	.004		.32	.19		.51	.64
5300	Brick		2000	.004		.32	.19		.51	.64
5400	Stone		2000	.004		.32	.19		.51	.64
5500	Unpainted wood		2000	.004		.32	.19		.51	.64
8000	Cleaner for use after treatment									
8100	Towels or wipes, per package of 30				Ea.	.72			.72	.80
8200	Aerosol spray, 24 oz. can				"	19			19	21

09 96 46 - Intumescent Painting

09 96 46.10 Coatings, Intumescent

		Crew	Daily Output	Labor-Hours	Unit	Material	2021 Bare Costs Labor	Equipment	Total	Total Incl O&P
0010	**COATINGS, INTUMESCENT**, spray applied									
0100	On exterior structural steel, 0.25" d.f.t.	1 Pord	475	.017	S.F.	.50	.78		1.28	1.72
0150	0.51" d.f.t.		350	.023		.50	1.06		1.56	2.13
0200	0.98" d.f.t.		280	.029		.50	1.33		1.83	2.53
0300	On interior structural steel, 0.108" d.f.t.		300	.027		.49	1.24		1.73	2.37
0350	0.310" d.f.t.		150	.053		.49	2.48		2.97	4.20
0400	0.670" d.f.t.		100	.080		.49	3.72		4.21	6.05

09 97 Special Coatings

09 97 10 - Coatings and Paints

09 97 10.10 Coatings and Paints

		Crew	Daily Output	Labor-Hours	Unit	Material	2021 Bare Costs Labor	Equipment	Total	Total Incl O&P
0010	**COATINGS & PAINTS** in 5 gallon lots									
0050	For 100 gallons or more, deduct					10%				
0100	Paint, exterior alkyd (oil base)									
0200	Flat				Gal.	44.50			44.50	49
0300	Gloss					56			56	61.50
0400	Primer					34			34	37.50
0500	Latex (water base)									
0600	Acrylic stain				Gal.	54			54	59.50
0700	Gloss enamel					30			30	33
0800	Flat					30			30	33
0900	Primer					24			24	26.50
1000	Semi-gloss					60			60	66
2400	Masonry, exterior									
2500	Alkali-resistant primer				Gal.	23.50			23.50	26
2600	Block filler, epoxy					30			30	33
2700	Latex					18.25			18.25	20
2800	Latex, flat					29			29	31.50

179

09 97 10.10 Coatings and Paints	Crew	Daily Output	Labor-Hours	Unit	Material	2021 Bare Costs Labor	Equipment	Total	Total Incl O&P
2900 Semi-gloss				Gal.	34.50			34.50	38
4000 Metal									
4100 Galvanizing paint				Gal.	81			81	89
4200 High heat					66.50			66.50	73
4300 Heat-resistant					38.50			38.50	42.50
4400 Machinery enamel, alkyd					54.50			54.50	59.50
4500 Metal pretreatment (polyvinyl butyral)					137			137	151
4600 Rust inhibitor, ferrous metal					40			40	44
4700 Zinc chromate					162			162	178
4800 Zinc rich primer					160			160	175
5500 Coatings									
5600 Heavy duty									
5700 Acrylic urethane				Gal.	51			51	56
5800 Chlorinated rubber					70			70	77
5900 Coal tar epoxy					76.50			76.50	84
6000 Polyamide epoxy, finish					72			72	79.50
6100 Primer					47.50			47.50	52.50
6200 Silicone alkyd					63			63	69
6300 2 component solvent based acrylic epoxy					108			108	119
6400 Polyester epoxy					100			100	110
6500 Vinyl					37.50			37.50	41
6600 Special/Miscellaneous									
6700 Aluminum				Gal.	42			42	46
6900 Dry fall out, flat					25.50			25.50	28
7000 Fire retardant, intumescent					51			51	56
7100 Linseed oil					26.50			26.50	29
7200 Shellac					51			51	56
7300 Swimming pool, epoxy or urethane base					53			53	58
7400 Rubber base					60			60	66
7500 Texture paint					27.50			27.50	30.50
7600 Turpentine					44			44	48
7700 Water repellent, 5% silicone					35			35	38.50

09 97 13 – Steel Coatings

09 97 13.23 Exterior Steel Coatings

0010 **EXTERIOR STEEL COATINGS**									
6100 Cold galvanizing, brush in field	1 Psst	1100	.007	S.F.	.20	.34		.54	.77
6510 Paints & protective coatings, sprayed in field									
6520 Alkyds, primer	2 Psst	3600	.004	S.F.	.09	.21		.30	.43
6540 Gloss topcoats		3200	.005		.11	.24		.35	.50
6560 Silicone alkyd		3200	.005		.18	.24		.42	.58
6610 Epoxy, primer		3000	.005		.24	.25		.49	.66
6630 Intermediate or topcoat		2800	.006		.29	.27		.56	.75
6650 Enamel coat		2800	.006		.36	.27		.63	.83
6810 Latex primer		3600	.004		.06	.21		.27	.41
6830 Topcoats		3200	.005		.14	.24		.38	.54
6910 Universal primers, one part, phenolic, modified alkyd		2000	.008		.46	.38		.84	1.11
6940 Two part, epoxy spray		2000	.008		.43	.38		.81	1.09
7000 Zinc rich primers, self cure, spray, inorganic		1800	.009		.80	.42		1.22	1.55
7010 Epoxy, spray, organic		1800	.009		.29	.42		.71	.99
7020 Above one story, spray painting simple structures, add						25%			
7030 Intricate structures, add						50%			

Estimating Tips
General

- The items in this division are usually priced per square foot or each.

- Many items in Division 10 require some type of support system or special anchors that are not usually furnished with the item. The required anchors must be added to the estimate in the appropriate division.

- Some items in Division 10, such as lockers, may require assembly before installation. Verify the amount of assembly required. Assembly can often exceed installation time.

10 20 00 Interior Specialties

- Support angles and blocking are not included in the installation of toilet compartments, shower/dressing compartments, or cubicles. Appropriate line items from Division 5 or 6 may need to be added to support the installations.

- Toilet partitions are priced by the stall. A stall consists of a side wall, pilaster, and door with hardware. Toilet tissue holders and grab bars are extra.

- The required acoustical rating of a folding partition can have a significant impact on costs. Verify the sound transmission coefficient rating of the panel priced against the specification requirements.

- Grab bar installation does not include supplemental blocking or backing to support the required load. When grab bars are installed at an existing facility, provisions must be made to attach the grab bars to a solid structure.

Reference Numbers

Reference numbers are shown at the beginning of some major classifications. These numbers refer to related items in the Reference Section. The reference information may be an estimating procedure, an alternate pricing method, or technical information.

Note: Not all subdivisions listed here necessarily appear. ■

Same Data. Simplified.

Enjoy the convenience and efficiency of accessing your costs anywhere:

- **Skip the multiplier** by setting your location
- **Quickly search,** edit, favorite and share costs
- **Stay on top of price changes** with automatic updates

Discover more at rsmeans.com/online

10 05 Common Work Results for Specialties

10 05 05 – Selective Demolition for Specialties

10 05 05.10 Selective Demolition, Specialties	Crew	Daily Output	Labor-Hours	Unit	Material	2021 Bare Costs Labor	Equipment	Total	Total Incl O&P
0010 **SELECTIVE DEMOLITION, SPECIALTIES**									
3500 Flagpole, groundset, to 70' high, excluding base/foundation	K-1	1	16	Ea.		830	850	1,680	2,150
3555 To 30' high	"	2.50	6.400	"		330	340	670	870
4000 Removal of traffic signs, including supports									
4020 To 10 S.F.	B-80B	16	2	Ea.		94.50	14.90	109.40	157
4030 11 S.F. to 20 S.F.	"	5	6.400			300	47.50	347.50	505
4040 21 S.F. to 40 S.F.	B-14	1.80	26.667			1,250	120	1,370	1,975
4050 41 S.F. to 100 S.F.	B-13	1.30	43.077			2,075	450	2,525	3,600
4070 Remove traffic posts to 12'-0" high	B-6	100	.240			11.55	2.16	13.71	19.60
4310 Signs, street, reflective aluminum, including post and bracket	1 Clab	60	.133			5.90		5.90	8.85

10 13 Directories

10 13 10 – Building Directories

10 13 10.10 Directory Boards

	Crew	Daily Output	Labor-Hours	Unit	Material	2021 Bare Costs Labor	Equipment	Total	Total Incl O&P
0010 **DIRECTORY BOARDS**									
0900 Outdoor, weatherproof, black plastic, 36" x 24"	2 Carp	2	8	Ea.	685	440		1,125	1,400
1000 36" x 36"	"	1.50	10.667	"	263	585		848	1,150

10 14 Signage

10 14 19 – Dimensional Letter Signage

10 14 19.10 Exterior Signs

	Crew	Daily Output	Labor-Hours	Unit	Material	2021 Bare Costs Labor	Equipment	Total	Total Incl O&P
0010 **EXTERIOR SIGNS**									
0020 Letters, 2" high, 3/8" deep, cast bronze	1 Carp	24	.333	Ea.	31.50	18.25		49.75	61.50
0140 1/2" deep, cast aluminum		18	.444		31.50	24.50		56	71
0160 Cast bronze		32	.250		34	13.70		47.70	58
0300 6" high, 5/8" deep, cast aluminum		24	.333		28	18.25		46.25	57.50
0400 Cast bronze		24	.333		61	18.25		79.25	94.50
0600 8" high, 3/4" deep, cast aluminum		14	.571		33.50	31.50		65	83.50
0700 Cast bronze		20	.400		99	22		121	142
0900 10" high, 1" deep, cast aluminum		18	.444		62	24.50		86.50	105
1000 Cast bronze		18	.444		105	24.50		129.50	153
1200 12" high, 1-1/4" deep, cast aluminum		12	.667		58	36.50		94.50	118
1500 Cast bronze		18	.444		166	24.50		190.50	219
1600 14" high, 2-5/16" deep, cast aluminum		12	.667		108	36.50		144.50	174
1800 Fabricated stainless steel, 6" high, 2" deep		20	.400		44	22		66	81
1900 12" high, 3" deep		18	.444		84	24.50		108.50	129
2100 18" high, 3" deep		12	.667		93	36.50		129.50	158
2200 24" high, 4" deep		10	.800		185	44		229	270
2700 Acrylic, on high density foam, 12" high, 2" deep		20	.400		20.50	22		42.50	55
2800 18" high, 2" deep		18	.444		145	24.50		169.50	196
3900 Plaques, custom, 20" x 30", for up to 450 letters, cast aluminum	2 Carp	4	4		1,825	219		2,044	2,350
4000 Cast bronze		4	4		1,975	219		2,194	2,475
4200 30" x 36", up to 900 letters, cast aluminum		3	5.333		2,975	292		3,267	3,700
4300 Cast bronze		3	5.333		4,175	292		4,467	5,000
4500 36" x 48", for up to 1300 letters, cast bronze		2	8		4,425	440		4,865	5,525
4800 Signs, reflective alum. directional signs, dbl. face, 2-way, w/bracket		30	.533		99.50	29		128.50	154
4900 4-way		30	.533		200	29		229	264
5100 Exit signs, 24 ga. alum., 14" x 12" surface mounted	1 Carp	30	.267		52.50	14.60		67.10	79.50
5200 10" x 7"		20	.400		28.50	22		50.50	64

For customer support on your Site Work & Landscape Costs with RSMeans Data, call 800.448.8182.

10 14 Signage

10 14 19 - Dimensional Letter Signage

10 14 19.10 Exterior Signs

		Crew	Daily Output	Labor-Hours	Unit	Material	2021 Bare Costs Labor	Equipment	Total	Total Incl O&P
5400	Bracket mounted, double face, 12" x 10"	1 Carp	30	.267	Ea.	52	14.60		66.60	79.50
5500	Sticky back, stock decals, 14" x 10"	1 Clab	50	.160		22.50	7.10		29.60	35
6400	Replacement sign faces, 6" or 8"	"	50	.160		64	7.10		71.10	80.50
8000	Internally illuminated, custom									
8100	On pedestal, 84" x 30" x 12"	L-7	.50	56	Ea.	2,025	2,975		5,000	6,650

10 14 26 - Post and Panel/Pylon Signage

10 14 26.10 Post and Panel Signage

		Crew	Daily Output	Labor-Hours	Unit	Material	2021 Bare Costs Labor	Equipment	Total	Total Incl O&P
0010	**POST AND PANEL SIGNAGE**, Alum., incl. 2 posts,									
0011	2-sided panel (blank), concrete footings per post									
0025	24" W x 18" H	B-1	25	.960	Ea.	123	43.50		166.50	201
0050	36" W x 24" H		25	.960		975	43.50		1,018.50	1,150
0100	57" W x 36" H		25	.960		2,125	43.50		2,168.50	2,400
0150	81" W x 46" H		25	.960		475	43.50		518.50	585

10 14 43 - Photoluminescent Signage

10 14 43.10 Photoluminescent Signage

		Crew	Daily Output	Labor-Hours	Unit	Material	2021 Bare Costs Labor	Equipment	Total	Total Incl O&P
0010	**PHOTOLUMINESCENT SIGNAGE**									
0011	Photoluminescent exit sign, plastic, 7"x10"	1 Carp	32	.250	Ea.	23.50	13.70		37.20	46.50
0022	Polyester		32	.250		15.05	13.70		28.75	37
0033	Aluminum		32	.250		52.50	13.70		66.20	78.50
0044	Plastic, 10"x14"		32	.250		53	13.70		66.70	79
0055	Polyester		32	.250		55	13.70		68.70	81
0066	Aluminum		32	.250		121	13.70		134.70	154
0111	Directional exit sign, plastic, 10"x14"		32	.250		55	13.70		68.70	81
0122	Polyester		32	.250		55	13.70		68.70	81
0133	Aluminum		32	.250		67.50	13.70		81.20	94.50
0153	Kick plate exit sign, 10"x34"		28	.286		305	15.65		320.65	365
0173	Photoluminescent tape, 1"x60'		300	.027	L.F.	.53	1.46		1.99	2.77
0183	2"x60'		290	.028	"	1.05	1.51		2.56	3.40
0211	Photoluminescent directional arrows		42	.190	Ea.	12.75	10.40		23.15	29.50

10 14 53 - Traffic Signage

10 14 53.20 Traffic Signs

		Crew	Daily Output	Labor-Hours	Unit	Material	2021 Bare Costs Labor	Equipment	Total	Total Incl O&P
0010	**TRAFFIC SIGNS**									
0012	Stock, 24" x 24", no posts, .080" alum. reflectorized	B-80	70	.457	Ea.	97	22.50	15.05	134.55	157
0100	High intensity		70	.457		28.50	22.50	15.05	66.05	81.50
0300	30" x 30", reflectorized		70	.457		68.50	22.50	15.05	106.05	125
0400	High intensity		70	.457		182	22.50	15.05	219.55	250
0600	Guide and directional signs, 12" x 18", reflectorized		70	.457		42.50	22.50	15.05	80.05	96.50
0700	High intensity		70	.457		66.50	22.50	15.05	104.05	123
0900	18" x 24", stock signs, reflectorized		70	.457		32.50	22.50	15.05	70.05	86
1000	High intensity		70	.457		48.50	22.50	15.05	86.05	104
1200	24" x 24", stock signs, reflectorized		70	.457		38.50	22.50	15.05	76.05	92.50
1300	High intensity		70	.457		54.50	22.50	15.05	92.05	110
1500	Add to above for steel posts, galvanized, 10'-0" upright, bolted		200	.160		44.50	7.80	5.25	57.55	66.50
1600	12'-0" upright, bolted		140	.229		39.50	11.15	7.50	58.15	68.50
1800	Highway road signs, aluminum, over 20 S.F., reflectorized		350	.091	S.F.	13	4.46	3.01	20.47	24.50
2000	High intensity		350	.091		13	4.46	3.01	20.47	24.50
2200	Highway, suspended over road, 80 S.F. min., reflectorized		165	.194		18.75	9.45	6.40	34.60	41.50
2300	High intensity		165	.194		18.60	9.45	6.40	34.45	41.50
2350	Roadway delineators and reference markers		500	.064	Ea.	16.70	3.12	2.10	21.92	25.50
2360	Delineator post only, 6'		500	.064		17.60	3.12	2.10	22.82	26.50
2400	Highway sign bridge structure, 45' to 80'								33,900	33,900

For customer support on your Site Work & Landscape Costs with RSMeans Data, call 800.448.8182.

183

10 14 Signage

10 14 53 – Traffic Signage

10 14 53.20 Traffic Signs	Crew	Daily Output	Labor- Hours	Unit	Material	2021 Bare Costs Labor	Equipment	Total	Total Incl O&P
2410 Cantilever structure, add				Ea.				7,000	7,000
5200 Remove and relocate signs, including supports									
5210 To 10 S.F.	B-80B	5	6.400	Ea.	385	300	47.50	732.50	930
5220 11 S.F. to 20 S.F.	"	1.70	18.824		860	890	140	1,890	2,425
5230 21 S.F. to 40 S.F.	B-14	.56	85.714		905	4,000	385	5,290	7,375
5240 41 S.F. to 100 S.F.	B-13	.32	175	↓	1,475	8,450	1,825	11,750	16,300
8000 For temporary barricades and lights, see Section 01 56 23.10									
9000 Replace directional sign	B-80	6	5.333	Ea.	182	260	175	617	785

10 22 Partitions

10 22 16 – Folding Gates

10 22 16.10 Security Gates

		Crew	Daily Output	Labor- Hours	Unit	Material	2021 Bare Costs Labor	Equipment	Total	Total Incl O&P
0010	**SECURITY GATES**									
0300	Scissors type folding gate, ptd. steel, single, 6-1/2' high, 5-1/2' wide	2 Sswk	4	4	Opng.	170	241		411	560
0400	7-1/2' wide		4	4		264	241		505	665
0600	Double gate, 8' high, 8' wide		2.50	6.400		375	385		760	1,000
0750	14' wide	↓	2	8	↓	385	480		865	1,175

10 44 Fire Protection Specialties

10 44 16 – Fire Extinguishers

10 44 16.13 Portable Fire Extinguishers

		Crew	Daily Output	Labor- Hours	Unit	Material	2021 Bare Costs Labor	Equipment	Total	Total Incl O&P
0010	**PORTABLE FIRE EXTINGUISHERS**									
0140	CO_2, with hose and "H" horn, 10 lb.				Ea.	335			335	370
0160	15 lb.				"	395			395	435
1000	Dry chemical, pressurized									
1040	Standard type, portable, painted, 2-1/2 lb.				Ea.	43.50			43.50	48
1060	5 lb.					64			64	70.50
1080	10 lb.					98.50			98.50	109
1100	20 lb.					230			230	253
1120	30 lb.					430			430	475
1300	Standard type, wheeled, 150 lb.					1,750			1,750	1,925
2000	ABC all purpose type, portable, 2-1/2 lb.					40.50			40.50	44.50
2060	5 lb.					48.50			48.50	53.50
2080	9-1/2 lb.					73			73	80.50
2100	20 lb.					170			170	187
5000	Pressurized water, 2-1/2 gallon, stainless steel					98.50			98.50	108
5060	With anti-freeze					117			117	128
9400	Installation of extinguishers, 12 or more, on nailable surface	1 Carp	30	.267	↓		14.60		14.60	22
9420	On masonry or concrete	"	15	.533	↓		29		29	43.50

10 44 16.16 Wheeled Fire Extinguisher Units

		Crew	Daily Output	Labor- Hours	Unit	Material	2021 Bare Costs Labor	Equipment	Total	Total Incl O&P
0010	**WHEELED FIRE EXTINGUISHER UNITS**									
0350	CO_2, portable, with swivel horn									
0360	Wheeled type, cart mounted, 50 lb.				Ea.	1,425			1,425	1,575
0400	100 lb.				"	3,650			3,650	4,000
2200	ABC all purpose type									
2300	Wheeled, 45 lb.				Ea.	770			770	845
2360	150 lb.				"	1,950			1,950	2,150

10 57 Wardrobe and Closet Specialties

10 57 23 - Closet and Utility Shelving

10 57 23.19 Wood Closet and Utility Shelving	Crew	Daily Output	Labor-Hours	Unit	Material	2021 Bare Costs Labor	Equipment	Total	Total Incl O&P
0010 **WOOD CLOSET AND UTILITY SHELVING**									
0020 Pine, clear grade, no edge band, 1" x 8"	1 Carp	115	.070	L.F.	4.38	3.81		8.19	10.50
0100 1" x 10"		110	.073		5.45	3.98		9.43	11.95
0200 1" x 12"		105	.076		6.55	4.17		10.72	13.40

10 73 Protective Covers

10 73 13 - Awnings

10 73 13.10 Awnings, Fabric

		Crew	Daily Output	Labor-Hours	Unit	Material	2021 Bare Costs Labor	Equipment	Total	Total Incl O&P
0010	**AWNINGS, FABRIC**									
0020	Including acrylic canvas and frame, standard design									
0100	Door and window, slope, 3' high, 4' wide	1 Carp	4.50	1.778	Ea.	755	97		852	975
0110	6' wide		3.50	2.286		970	125		1,095	1,250
0120	8' wide		3	2.667		1,175	146		1,321	1,525
0200	Quarter round convex, 4' wide		3	2.667		1,175	146		1,321	1,525
0210	6' wide		2.25	3.556		1,525	194		1,719	1,975
0220	8' wide		1.80	4.444		1,850	243		2,093	2,425
0300	Dome, 4' wide		7.50	1.067		450	58.50		508.50	585
0310	6' wide		3.50	2.286		1,025	125		1,150	1,300
0320	8' wide		2	4		1,800	219		2,019	2,300
0350	Elongated dome, 4' wide		1.33	6.015		1,700	330		2,030	2,375
0360	6' wide		1.11	7.207		2,025	395		2,420	2,825
0370	8' wide		1	8		2,375	440		2,815	3,250
1000	Entry or walkway, peak, 12' long, 4' wide	2 Carp	.90	17.778		5,325	970		6,295	7,300
1010	6' wide		.60	26.667		8,200	1,450		9,650	11,200
1020	8' wide		.40	40		11,300	2,200		13,500	15,700
1100	Radius with dome end, 4' wide		1.10	14.545		4,025	795		4,820	5,625
1110	6' wide		.70	22.857		6,475	1,250		7,725	9,000
1120	8' wide		.50	32		9,225	1,750		10,975	12,700
2000	Retractable lateral arm awning, manual									
2010	To 12' wide, 8'-6" projection	2 Carp	1.70	9.412	Ea.	1,200	515		1,715	2,100
2020	To 14' wide, 8'-6" projection		1.10	14.545		1,400	795		2,195	2,750
2030	To 19' wide, 8'-6" projection		.85	18.824		1,900	1,025		2,925	3,625
2040	To 24' wide, 8'-6" projection		.67	23.881		2,400	1,300		3,700	4,600
2050	Motor for above, add	1 Carp	2.67	3		1,025	164		1,189	1,375
3000	Patio/deck canopy with frame									
3010	12' wide, 12' projection	2 Carp	2	8	Ea.	1,700	440		2,140	2,525
3020	16' wide, 14' projection	"	1.20	13.333		2,650	730		3,380	4,000
9000	For fire retardant canvas, add					7%				
9010	For lettering or graphics, add					35%				
9020	For painted or coated acrylic canvas, deduct					8%				
9030	For translucent or opaque vinyl canvas, add					10%				
9040	For 6 or more units, deduct					20%	15%			

10 73 16 - Canopies

10 73 16.20 Metal Canopies

		Crew	Daily Output	Labor-Hours	Unit	Material	2021 Bare Costs Labor	Equipment	Total	Total Incl O&P
0010	**METAL CANOPIES**									
0020	Wall hung, .032", aluminum, prefinished, 8' x 10'	K-2	1.30	18.462	Ea.	3,100	1,050	655	4,805	5,750
0300	8' x 20'		1.10	21.818		4,225	1,250	775	6,250	7,400
0500	10' x 10'		1.30	18.462		3,075	1,050	655	4,780	5,750
0700	10' x 20'		1.10	21.818		4,775	1,250	775	6,800	8,000
1000	12' x 20'		1	24		6,000	1,375	850	8,225	9,625

10 73 Protective Covers

10 73 16 – Canopies

10 73 16.20 Metal Canopies

		Crew	Daily Output	Labor-Hours	Unit	Material	2021 Bare Costs Labor	2021 Bare Costs Equipment	Total	Total Incl O&P
1360	12' x 30'	K-2	.80	30	Ea.	9,025	1,725	1,075	11,825	13,800
1700	12' x 40'	↓	.60	40		12,700	2,275	1,425	16,400	19,100
1900	For free standing units, add				↓	20%	10%			
7000	Carport, baked vinyl finish, .032", 20' x 10', no foundations, flat panel	K-2	4	6	Car	4,000	345	213	4,558	5,175
7250	Insulated flat panel		2	12	"	4,925	685	425	6,035	6,950
7500	Walkway cover, to 12' wide, stl., vinyl finish, .032", no fndtns., flat		250	.096	S.F.	20.50	5.50	3.40	29.40	34.50
7750	Arched	↓	200	.120	"	51	6.85	4.25	62.10	71

10 74 Manufactured Exterior Specialties

10 74 46 – Window Wells

10 74 46.10 Area Window Wells

		Crew	Daily Output	Labor-Hours	Unit	Material	2021 Bare Costs Labor	2021 Bare Costs Equipment	Total	Total Incl O&P
0010	**AREA WINDOW WELLS**, Galvanized steel									
0020	20 ga., 3'-2" wide, 1' deep	1 Sswk	29	.276	Ea.	17.95	16.65		34.60	45
0100	2' deep		23	.348		29.50	21		50.50	65
0300	16 ga., 3'-2" wide, 1' deep		29	.276		27.50	16.65		44.15	55.50
0400	3' deep		23	.348		48	21		69	85.50
0600	Welded grating for above, 15 lb., painted		45	.178		103	10.70		113.70	130
0700	Galvanized	↓	45	.178	↓	126	10.70		136.70	155

10 75 Flagpoles

10 75 16 – Ground-Set Flagpoles

10 75 16.10 Flagpoles

		Crew	Daily Output	Labor-Hours	Unit	Material	2021 Bare Costs Labor	2021 Bare Costs Equipment	Total	Total Incl O&P
0010	**FLAGPOLES**, ground set									
0050	Not including base or foundation									
0100	Aluminum, tapered, ground set 20' high	K-1	2	8	Ea.	1,250	415	425	2,090	2,475
0200	25' high		1.70	9.412		1,300	485	500	2,285	2,700
0300	30' high		1.50	10.667		2,050	550	565	3,165	3,700
0400	35' high		1.40	11.429		1,950	590	605	3,145	3,700
0500	40' high		1.20	13.333		2,225	690	710	3,625	4,250
0600	50' high		1	16		4,025	830	850	5,705	6,575
0700	60' high		.90	17.778		4,900	920	945	6,765	7,825
0800	70' high		.80	20		8,250	1,025	1,075	10,350	11,800
1100	Counterbalanced, internal halyard, 20' high		1.80	8.889		3,275	460	470	4,205	4,800
1200	30' high		1.50	10.667		2,900	550	565	4,015	4,650
1300	40' high		1.30	12.308		6,925	635	655	8,215	9,300
1400	50' high		1	16		9,650	830	850	11,330	12,800
2820	Aluminum, electronically operated, 30' high		1.40	11.429		4,675	590	605	5,870	6,700
2840	35' high		1.30	12.308		5,325	635	655	6,615	7,550
2860	39' high		1.10	14.545		6,150	755	775	7,680	8,725
2880	45' high		1	16		6,250	830	850	7,930	9,025
2900	50' high		.90	17.778		8,250	920	945	10,115	11,500
3000	Fiberglass, tapered, ground set, 23' high		2	8		825	415	425	1,665	2,000
3100	29'-7" high		1.50	10.667		1,275	550	565	2,390	2,850
3200	36'-1" high		1.40	11.429		1,700	590	605	2,895	3,425
3300	39'-5" high		1.20	13.333		2,025	690	710	3,425	4,025
3400	49'-2" high		1	16		4,925	830	850	6,605	7,575
3500	59' high	↓	.90	17.778	↓	4,350	920	945	6,215	7,200
4300	Steel, direct imbedded installation									
4400	Internal halyard, 20' high	K-1	2.50	6.400	Ea.	1,350	330	340	2,020	2,350

10 75 Flagpoles

10 75 16 – Ground-Set Flagpoles

10 75 16.10 Flagpoles		Crew	Daily Output	Labor-Hours	Unit	Material	2021 Bare Costs Labor	Equipment	Total	Total Incl O&P
4500	25' high	K-1	2.50	6.400	Ea.	2,175	330	340	2,845	3,275
4600	30' high		2.30	6.957		2,525	360	370	3,255	3,725
4700	40' high		2.10	7.619		3,500	395	405	4,300	4,875
4800	50' high		1.90	8.421		4,975	435	445	5,855	6,600
5000	60' high		1.80	8.889		7,200	460	470	8,130	9,125
5100	70' high		1.60	10		9,125	520	530	10,175	11,400
5200	80' high		1.40	11.429		12,000	590	605	13,195	14,800
5300	90' high		1.20	13.333		18,800	690	710	20,200	22,400
5500	100' high	▼	1	16	▼	17,200	830	850	18,880	21,100
6400	Wood poles, tapered, clear vertical grain fir with tilting									
6410	base, not incl. foundation, 4" butt, 25' high	K-1	1.90	8.421	Ea.	1,875	435	445	2,755	3,200
6800	6" butt, 30' high	"	1.30	12.308	"	3,675	635	655	4,965	5,725
7300	Foundations for flagpoles, including									
7400	excavation and concrete, to 35' high poles	C-1	10	3.200	Ea.	725	167		892	1,050
7600	40' to 50' high		3.50	9.143		1,350	475		1,825	2,175
7700	Over 60' high	▼	2	16	▼	1,675	835		2,510	3,100

10 75 23 – Wall-Mounted Flagpoles

10 75 23.10 Flagpoles		Crew	Daily Output	Labor-Hours	Unit	Material	2021 Bare Costs Labor	Equipment	Total	Total Incl O&P
0010	**FLAGPOLES**, structure mounted									
0100	Fiberglass, vertical wall set, 19'-8" long	K-1	1.50	10.667	Ea.	1,125	550	565	2,240	2,700
0200	23' long		1.40	11.429		1,675	590	605	2,870	3,375
0300	26'-3" long		1.30	12.308		1,625	635	655	2,915	3,450
0800	19'-8" long outrigger		1.30	12.308		1,325	635	655	2,615	3,150
1300	Aluminum, vertical wall set, tapered, with base, 20' high		1.20	13.333		1,300	690	710	2,700	3,225
1400	29'-6" high		1	16		3,075	830	850	4,755	5,525
2400	Outrigger poles with base, 12' long		1.30	12.308		1,400	635	655	2,690	3,200
2500	14' long	▼	1	16	▼	1,400	830	850	3,080	3,675

10 88 Scales

10 88 05 – Commercial Scales

10 88 05.10 Scales		Crew	Daily Output	Labor-Hours	Unit	Material	2021 Bare Costs Labor	Equipment	Total	Total Incl O&P
0010	**SCALES**									
0700	Truck scales, incl. steel weigh bridge,									
0800	not including foundation, pits									
1550	Digital, electronic, 100 ton capacity, steel deck 12' x 10' platform	3 Carp	.20	120	Ea.	14,900	6,575		21,475	26,200
1600	40' x 10' platform		.14	171		29,600	9,375		38,975	46,600
1640	60' x 10' platform		.13	185		41,400	10,100		51,500	60,500
1680	70' x 10' platform	▼	.12	200		41,600	10,900		52,500	62,000
2000	For standard automatic printing device, add					1,625			1,625	1,800
2100	For remote reading electronic system, add					3,150			3,150	3,475
2300	Concrete foundation pits for above, 8' x 6', 5 C.Y. required	C-1	.50	64		1,125	3,325		4,450	6,225
2400	14' x 6' platform, 10 C.Y. required		.35	91.429		1,650	4,775		6,425	8,950
2600	50' x 10' platform, 30 C.Y. required		.25	128		2,225	6,675		8,900	12,400
2700	70' x 10' platform, 40 C.Y. required	▼	.15	213		4,875	11,100		15,975	22,000
2750	Crane scales, dial, 1 ton capacity					1,250			1,250	1,375
2780	5 ton capacity					1,700			1,700	1,875
2800	Digital, 1 ton capacity					340			340	375
2850	10 ton capacity					775			775	850
3800	Portable, beam type, capacity 1000 lb., platform 18" x 24"					915			915	1,000
3900	Dial type, capacity 2000 lb., platform 24" x 24"					1,650			1,650	1,825

10 88 Scales

10 88 05 – Commercial Scales

10 88 05.10 Scales		Crew	Daily Output	Labor-Hours	Unit	Material	2021 Bare Costs Labor	Equipment	Total	Total Incl O&P
4000	Digital type, capacity 1000 lb., platform 24" x 30"				Ea.	2,650			2,650	2,925
4100	Portable contractor truck scales, 50 ton cap., 40' x 10' platform					30,900			30,900	34,000
4200	60' x 10' platform					37,000			37,000	40,700
4400	Heavy-Duty Steel Deck Truck Scales 20' x 10'	3 Carp	.20	120		18,300	6,575		24,875	29,900
4500	Heavy-Duty Steel Deck Truck Scales 80' x 10'		.10	240		49,100	13,100		62,200	73,500
4600	Heavy-Duty Steel Deck Truck Scales 160' x 10'		.04	600		106,000	32,800		138,800	165,500
4700	Heavy-Duty Steel Deck Truck Scales 20' x 12'		.18	137		21,400	7,500		28,900	34,700
4800	Heavy-Duty Steel Deck Truck Scales 80' x 12'		.09	267		61,000	14,600		75,600	89,500
4900	Heavy-Duty Steel Deck Truck Scales 160' x 12'		.04	600		123,500	32,800		156,300	184,500
5000	Heavy-Duty Steel Deck Truck Scales 20' x 14'		.16	150		28,400	8,200		36,600	43,400
5100	Heavy-Duty Steel Deck Truck Scales 80' x 14'		.06	400		67,000	21,900		88,900	106,500
5200	Heavy-Duty Steel Deck Truck Scales 160' x 14'		.03	800		131,000	43,800		174,800	210,000

Estimating Tips
General

- The items in this division are usually priced per square foot or each. Many of these items are purchased by the owner for installation by the contractor. Check the specifications for responsibilities and include time for receiving, storage, installation, and mechanical and electrical hookups in the appropriate divisions.

- Many items in Division 11 require some type of support system that is not usually furnished with the item. Examples of these systems include blocking for the attachment of casework and support angles for ceiling-hung projection screens. The required blocking or supports must be added to the estimate in the appropriate division.

- Some items in Division 11 may require assembly or electrical hookups. Verify the amount of assembly required or the need for a hard electrical connection and add the appropriate costs.

Reference Numbers

Reference numbers are shown at the beginning of some major classifications. These numbers refer to related items in the Reference Section. The reference information may be an estimating procedure, an alternate pricing method, or technical information.

Same Data. Simplified.

Enjoy the convenience and efficiency of accessing your costs anywhere:

- **Skip the multiplier** by setting your location
- **Quickly search,** edit, favorite and share costs
- **Stay on top of price changes** with automatic updates

Discover more at rsmeans.com/online

11 11 Vehicle Service Equipment

11 11 13 – Compressed-Air Vehicle Service Equipment

11 11 13.10 Compressed Air Equipment		Crew	Daily Output	Labor-Hours	Unit	Material	2021 Bare Costs Labor	Equipment	Total	Total Incl O&P
0010	**COMPRESSED AIR EQUIPMENT**									
0030	Compressors, electric, 1-1/2 HP, standard controls	L-4	1.50	16	Ea.	1,125	835		1,960	2,475
0550	Dual controls		1.50	16		1,050	835		1,885	2,400
0600	5 HP, 115/230 volt, standard controls		1	24		1,000	1,250		2,250	2,975
0650	Dual controls		1	24		1,325	1,250		2,575	3,325

11 11 19 – Vehicle Lubrication Equipment

11 11 19.10 Lubrication Equipment

		Crew	Daily Output	Labor-Hours	Unit	Material	Labor	Equipment	Total	Total Incl O&P
0010	**LUBRICATION EQUIPMENT**									
3000	Lube equipment, 3 reel type, with pumps, not including piping	L-4	.50	48	Set	10,400	2,500		12,900	15,300
3100	Hose reel, including hose, oil/lube, 1000 psi	2 Sswk	2	8	Ea.	295	480		775	1,075
3200	Grease, 5000 psi		2	8		455	480		935	1,250
3300	Air, 50', 160 psi		2	8		90	480		570	845
3350	25', 160 psi		2	8		54	480		534	805

11 11 33 – Vehicle Spray Painting Equipment

11 11 33.10 Spray Painting Equipment

		Crew	Daily Output	Labor-Hours	Unit	Material	Labor	Equipment	Total	Total Incl O&P
0010	**SPRAY PAINTING EQUIPMENT**									
4000	Spray painting booth, 26' long, complete	L-4	.40	60	Ea.	18,100	3,125		21,225	24,600

11 12 Parking Control Equipment

11 12 13 – Parking Key and Card Control Units

11 12 13.10 Parking Control Units

		Crew	Daily Output	Labor-Hours	Unit	Material	Labor	Equipment	Total	Total Incl O&P
0010	**PARKING CONTROL UNITS**									
5100	Card reader	1 Elec	2	4	Ea.	895	255		1,150	1,375
5120	Proximity with customer display	2 Elec	1	16		6,100	1,025		7,125	8,225
6000	Parking control software, basic functionality	1 Elec	.50	16		23,000	1,025		24,025	26,800
6020	Multi-function	"	.20	40		90,000	2,550		92,550	103,000

11 12 16 – Parking Ticket Dispensers

11 12 16.10 Ticket Dispensers

		Crew	Daily Output	Labor-Hours	Unit	Material	Labor	Equipment	Total	Total Incl O&P
0010	**TICKET DISPENSERS**									
5900	Ticket spitter with time/date stamp, standard	2 Elec	2	8	Ea.	6,525	510		7,035	7,925
5920	Mag stripe encoding	"	2	8	"	18,900	510		19,410	21,600

11 12 26 – Parking Fee Collection Equipment

11 12 26.13 Parking Fee Coin Collection Equipment

		Crew	Daily Output	Labor-Hours	Unit	Material	Labor	Equipment	Total	Total Incl O&P
0010	**PARKING FEE COIN COLLECTION EQUIPMENT**									
5200	Cashier booth, average	B-22	1	30	Ea.	10,400	1,550	287	12,237	14,100
5300	Collector station, pay on foot	2 Elec	.20	80		111,500	5,100		116,600	130,500
5320	Credit card only	"	.50	32		20,600	2,050		22,650	25,700

11 12 26.23 Fee Equipment

		Crew	Daily Output	Labor-Hours	Unit	Material	Labor	Equipment	Total	Total Incl O&P
0010	**FEE EQUIPMENT**									
5600	Fee computer	1 Elec	1.50	5.333	Ea.	13,000	340		13,340	14,800

11 12 33 – Parking Gates

11 12 33.13 Lift Arm Parking Gates

		Crew	Daily Output	Labor-Hours	Unit	Material	Labor	Equipment	Total	Total Incl O&P
0010	**LIFT ARM PARKING GATES**									
5000	Barrier gate with programmable controller	2 Elec	3	5.333	Ea.	3,325	340		3,665	4,175
5020	Industrial		3	5.333		5,175	340		5,515	6,175
5050	Non-programmable, with reader and 12' arm		3	5.333		1,725	340		2,065	2,400
5500	Exit verifier		1	16		19,800	1,025		20,825	23,300
5700	Full sign, 4" letters	1 Elec	2	4		1,225	255		1,480	1,725

11 12 Parking Control Equipment

11 12 33 – Parking Gates

11 12 33.13 Lift Arm Parking Gates	Crew	Daily Output	Labor-Hours	Unit	Material	2021 Bare Costs Labor	Equipment	Total	Total Incl O&P
5800	Inductive loop	2 Elec	4	4	Ea.	208	255		463	610
5950	Vehicle detector, microprocessor based	1 Elec	3	2.667	↓	465	170		635	765

11 13 Loading Dock Equipment

11 13 13 – Loading Dock Bumpers

11 13 13.10 Dock Bumpers

		Crew	Daily Output	Labor-Hours	Unit	Material	2021 Bare Costs Labor	Equipment	Total	Total Incl O&P
0010	**DOCK BUMPERS** Bolts not included									
0020	2" x 6" to 4" x 8", average	1 Carp	.30	26.667	M.B.F.	1,425	1,450		2,875	3,750
0050	Bumpers, lam. rubber blocks 4-1/2" thick, 10" high, 14" long		26	.308	Ea.	70	16.85		86.85	103
0300	36" long		17	.471		365	25.50		390.50	440
0500	12" high, 14" long		25	.320		69.50	17.50		87	103
0600	36" long		15	.533		111	29		140	166
0800	Laminated rubber blocks 6" thick, 10" high, 14" long		22	.364		79.50	19.90		99.40	117
0900	36" long		13	.615		208	33.50		241.50	280
0920	Extruded rubber bumpers, T section, 22" x 22" x 3" thick		41	.195		75.50	10.65		86.15	99
0940	Molded rubber bumpers, 24" x 12" x 3" thick	↓	20	.400	↓	71.50	22		93.50	111

11 14 Pedestrian Control Equipment

11 14 13 – Pedestrian Gates

11 14 13.19 Turnstiles

		Crew	Daily Output	Labor-Hours	Unit	Material	2021 Bare Costs Labor	Equipment	Total	Total Incl O&P
0010	**TURNSTILES**									
0020	One way, 4 arm, 46" diameter, economy, manual	2 Carp	5	3.200	Ea.	1,875	175		2,050	2,325
0100	Electric		1.20	13.333		2,225	730		2,955	3,550
0420	Three arm, 24" opening, light duty, manual		2	8		2,850	440		3,290	3,775
0450	Heavy duty		1.50	10.667		4,900	585		5,485	6,275
0460	Manual, with registering & controls, light duty		2	8		4,850	440		5,290	5,975
0470	Heavy duty	↓	1.50	10.667	↓	5,000	585		5,585	6,375

11 14 19 – Portable Posts and Railings

11 14 19.13 Portable Posts and Railings

		Crew	Daily Output	Labor-Hours	Unit	Material	2021 Bare Costs Labor	Equipment	Total	Total Incl O&P
0010	**PORTABLE POSTS AND RAILINGS**									
0020	Portable for pedestrian traffic control, standard				Ea.	38.50			38.50	42.50
0300	Deluxe posts				"	248			248	272
0600	Ropes for above posts, plastic covered, 1-1/2" diameter				L.F.	20.50			20.50	22.50
0700	Chain core				"	14.80			14.80	16.30
1500	Portable security or safety barrier, black with 7' yellow strap				Ea.	213			213	235
1510	12' yellow strap					251			251	276
1550	Sign holder, standard design				↓	92.50			92.50	102

11 66 Athletic Equipment

11 66 43 – Interior Scoreboards

11 66 43.10 Scoreboards		Crew	Daily Output	Labor-Hours	Unit	Material	2021 Bare Costs Labor	Equipment	Total	Total Incl O&P
0010	**SCOREBOARDS**									
7000	Baseball, minimum	R-3	1.30	15.385	Ea.	4,200	975	147	5,322	6,225
7200	Maximum		.05	400		18,000	25,400	3,825	47,225	61,500
7300	Football, minimum		.86	23.256		5,175	1,475	222	6,872	8,150
7400	Maximum	↓	.20	100	↓	25,600	6,350	955	32,905	38,600

11 68 Play Field Equipment and Structures

11 68 13 – Playground Equipment

11 68 13.10 Free-Standing Playground Equipment

			Crew	Daily Output	Labor-Hours	Unit	Material	2021 Bare Costs Labor	Equipment	Total	Total Incl O&P
0010	**FREE-STANDING PLAYGROUND EQUIPMENT** See also individual items										
0200	Bike rack, 10' long, permanent	G	B-1	12	2	Ea.	550	90		640	740
0240	Climber, arch, 6' high, 12' long, 5' wide			4	6		710	270		980	1,175
0260	Fitness trail, with signs, 9 to 10 stations, treated pine, minimum			.25	96		6,075	4,325		10,400	13,200
0270	Maximum			.17	141		22,000	6,375		28,375	33,700
0280	Metal, minimum			.25	96		11,300	4,325		15,625	18,900
0285	Maximum			.17	141		22,300	6,375		28,675	34,000
0300	Redwood, minimum			.25	96		12,300	4,325		16,625	20,000
0310	Maximum			.17	141		12,600	6,375		18,975	23,400
0320	16 to 20 stations, treated pine, minimum			.17	141		15,300	6,375		21,675	26,400
0330	Maximum			.13	185		16,100	8,325		24,425	30,200
0340	Metal, minimum			.17	141		13,500	6,375		19,875	24,300
0350	Maximum			.13	185		20,200	8,325		28,525	34,700
0360	Redwood, minimum			.17	141		20,100	6,375		26,475	31,600
0370	Maximum			.13	185		31,900	8,325		40,225	47,500
0392	Upper body warm-up station			2.60	9.231		3,700	415		4,115	4,700
0394	Bench stepper station			2.60	9.231		1,725	415		2,140	2,525
0396	Standing push up station			2.60	9.231		1,050	415		1,465	1,775
0398	Upper body stretch station			2.60	9.231		2,200	415		2,615	3,050
0400	Horizontal monkey ladder, 14' long, 6' high			4	6		1,000	270		1,270	1,500
0590	Parallel bars, 10' long			4	6		645	270		915	1,125
0600	Posts, tether ball set, 2-3/8" OD			12	2	↓	460	90		550	640
0800	Poles, multiple purpose, 10'-6" long			12	2	Pr.	275	90		365	440
1000	Ground socket for movable posts, 2-3/8" post			10	2.400		113	108		221	285
1100	3-1/2" post			10	2.400	↓	228	108		336	410
1300	See-saw, spring, steel, 2 units			6	4	Ea.	850	180		1,030	1,200
1400	4 units			4	6		1,600	270		1,870	2,175
1500	6 units			3	8		2,350	360		2,710	3,150
1700	Shelter, fiberglass golf tee, 3 person			4.60	5.217		4,225	235		4,460	5,000
1900	Slides, stainless steel bed, 12' long, 6' high			3	8		4,400	360		4,760	5,400
2000	20' long, 10' high			2	12		7,850	540		8,390	9,425
2200	Swings, plain seats, 8' high, 4 seats			2	12		1,725	540		2,265	2,700
2300	8 seats			1.30	18.462		2,725	830		3,555	4,250
2500	12' high, 4 seats			2	12		2,150	540		2,690	3,150
2600	8 seats			1.30	18.462		4,800	830		5,630	6,525
2800	Whirlers, 8' diameter			3	8		2,625	360		2,985	3,425
2900	10' diameter	↓		3	8	↓	6,425	360		6,785	7,625

11 68 13.20 Modular Playground

			Crew	Daily Output	Labor-Hours	Unit	Material	2021 Bare Costs Labor	Equipment	Total	Total Incl O&P
0010	**MODULAR PLAYGROUND** Basic components										
0100	Deck, square, steel, 48" x 48"		B-1	1	24	Ea.	710	1,075		1,785	2,400
0110	Recycled polyurethane			1	24		725	1,075		1,800	2,425
0120	Triangular, steel, 48" side			1	24	↓	830	1,075		1,905	2,550

11 68 Play Field Equipment and Structures

11 68 13 – Playground Equipment

11 68 13.20 Modular Playground

		Crew	Daily Output	Labor-Hours	Unit	Material	2021 Bare Costs Labor	Equipment	Total	Total Incl O&P
0130	Post, steel, 5" square	B-1	18	1.333	L.F.	55	60		115	150
0140	Aluminum, 2-3/8" square		20	1.200		60.50	54		114.50	147
0150	5" square		18	1.333		51	60		111	146
0160	Roof, square poly, 54" side		18	1.333	Ea.	2,100	60		2,160	2,400
0170	Wheelchair transfer module, for 3' high deck		3	8	"	3,675	360		4,035	4,575
0180	Guardrail, pipe, 36" high		60	.400	L.F.	298	18.05		316.05	355
0190	Steps, deck-to-deck, three 8" steps		8	3	Ea.	1,425	135		1,560	1,775
0200	Activity panel, crawl through panel		2	12		615	540		1,155	1,475
0210	Alphabet/spelling panel		2	12		660	540		1,200	1,525
0360	With guardrails		3	8		1,875	360		2,235	2,600
0370	Crawl tunnel, straight, 56" long		4	6		1,600	270		1,870	2,175
0380	90°, 4' long		4	6		1,925	270		2,195	2,500
1200	Slide, tunnel, for 56" high deck		8	3		2,300	135		2,435	2,725
1210	Straight, poly		8	3		605	135		740	865
1220	Stainless steel, 54" high deck		6	4		995	180		1,175	1,375
1230	Curved, poly, 40" high deck		6	4		970	180		1,150	1,350
1240	Spiral slide, 56"-72" high		5	4.800		5,500	216		5,716	6,375
1300	Ladder, vertical, for 24"-72" high deck		5	4.800		690	216		906	1,075
1310	Horizontal, 8' long		5	4.800		1,150	216		1,366	1,575
1320	Corkscrew climber, 6' high		3	8		1,525	360		1,885	2,225
1330	Fire pole for 72" high deck		6	4		425	180		605	740
1340	Bridge, ring climber, 8' long		4	6		2,575	270		2,845	3,250
1350	Suspension		4	6	L.F.	425	270		695	870

11 68 16 – Play Structures

11 68 16.10 Handball/Squash Court

		Crew	Daily Output	Labor-Hours	Unit	Material	2021 Bare Costs Labor	Equipment	Total	Total Incl O&P
0010	**HANDBALL/SQUASH COURT**, outdoor									
0900	Handball or squash court, outdoor, wood	2 Carp	.50	32	Ea.	5,000	1,750		6,750	8,100
1000	Masonry handball/squash court	D-1	.30	53.333	"	26,600	2,600		29,200	33,100

11 68 16.30 Platform/Paddle Tennis Court

		Crew	Daily Output	Labor-Hours	Unit	Material	2021 Bare Costs Labor	Equipment	Total	Total Incl O&P
0010	**PLATFORM/PADDLE TENNIS COURT** Complete with lighting, etc.									
0100	Aluminum slat deck with aluminum frame	B-1	.08	300	Court	70,000	13,500		83,500	97,000
0500	Aluminum slat deck with wood frame	C-1	.12	267		86,000	13,900		99,900	115,000
0800	Aluminum deck heater, add	B-1	1.18	20.339		3,025	915		3,940	4,700
0900	Douglas fir planking with wood frame 2" x 6" x 30'	C-1	.12	267		80,500	13,900		94,400	109,500
1000	Plywood deck with steel frame		.12	267		80,500	13,900		94,400	109,500
1100	Steel slat deck with wood frame		.12	267		47,800	13,900		61,700	73,000

11 68 33 – Athletic Field Equipment

11 68 33.13 Football Field Equipment

		Crew	Daily Output	Labor-Hours	Unit	Material	2021 Bare Costs Labor	Equipment	Total	Total Incl O&P
0010	**FOOTBALL FIELD EQUIPMENT**									
0020	Goal posts, steel, football, double post	B-1	1.50	16	Pr.	4,825	720		5,545	6,375
0100	Deluxe, single post		1.50	16		3,925	720		4,645	5,400
0300	Football, convertible to soccer		1.50	16		3,175	720		3,895	4,550
0500	Soccer, regulation		2	12		2,025	540		2,565	3,025

For customer support on your Site Work & Landscape Costs with RSMeans Data, call 800.448.8182.

193

11 82 Facility Solid Waste Handling Equipment

11 82 19 – Packaged Incinerators

11 82 19.10 Packaged Gas Fired Incinerators	Crew	Daily Output	Labor-Hours	Unit	Material	2021 Bare Costs Labor	Equipment	Total	Total Incl O&P
0010 **PACKAGED GAS FIRED INCINERATORS**									
4750 Large municipal incinerators, incl. stack, minimum	Q-3	.25	128	Ton/day	21,700	8,250		29,950	36,200
4850 Maximum	"	.10	320	"	58,000	20,600		78,600	94,500

11 82 26 – Facility Waste Compactors

11 82 26.10 Compactors

	Crew	Daily Output	Labor-Hours	Unit	Material	2021 Bare Costs Labor	Equipment	Total	Total Incl O&P
0010 **COMPACTORS**									
0020 Compactors, 115 volt, 250 lbs./hr., chute fed	L-4	1	24	Ea.	12,500	1,250		13,750	15,600
1400 For handling hazardous waste materials, 55 gallon drum packer, std.					21,800			21,800	24,000
1410 55 gallon drum packer w/HEPA filter					26,200			26,200	28,800
1420 55 gallon drum packer w/charcoal & HEPA filter					34,900			34,900	38,400
1430 All of the above made explosion proof, add					2,025			2,025	2,225
6000 Transfer station compactor, with power unit									
6050 and pedestal, not including pit, 50 tons/hour				Ea.	221,000			221,000	243,000

Estimating Tips
General

- The items in this division are usually priced per square foot or each. Most of these items are purchased by the owner and installed by the contractor. Do not assume the items in Division 12 will be purchased and installed by the contractor. Check the specifications for responsibilities and include receiving, storage, installation, and mechanical and electrical hookups in the appropriate divisions.

- Some items in this division require some type of support system that is not usually furnished with the item. Examples of these systems include blocking for the attachment of casework and heavy drapery rods. The required blocking must be added to the estimate in the appropriate division.

Reference Numbers

Reference numbers are shown at the beginning of some major classifications. These numbers refer to related items in the Reference Section. The reference information may be an estimating procedure, an alternate pricing method, or technical information.

Same Data. Simplified.

Enjoy the convenience and efficiency of accessing your costs anywhere:

- **Skip the multiplier** by setting your location
- **Quickly search,** edit, favorite and share costs
- **Stay on top of price changes** with automatic updates

Discover more at rsmeans.com/online

---

12 48 Rugs and Mats

12 48 13 - Entrance Floor Mats and Frames

12 48 13.13 Entrance Floor Mats	Crew	Daily Output	Labor-Hours	Unit	Material	2021 Bare Costs Labor	Equipment	Total	Total Incl O&P
0010 **ENTRANCE FLOOR MATS**									
0020　Recessed, black rubber, 3/8" thick, solid	1 Clab	155	.052	S.F.	2.42	2.29		4.71	6.10

12 92 Interior Planters and Artificial Plants

12 92 13 - Interior Artificial Plants

12 92 13.10 Plants

		Unit	Material			Total	Total Incl O&P
0010	**PLANTS** Permanent only, weighted						
0020	Preserved or polyester leaf, natural wood						
0030	Or molded trunk. For pots see Section 12 92 33.10						
0100	Plants, acuba, 5' high	Ea.	370			370	405
0200	Apidistra, 4' high		320			320	355
0300	Beech, variegated, 5' high		315			315	345
0400	Birds nest fern, 6' high		565			565	620
0500	Croton, 4' high		138			138	152
0600	Diffenbachia, 3' high		127			127	140
0700	Ficus Benjamina, 3' high		151			151	166
0720	6' high		380			380	420
0760	Nitida, 3' high		340			340	375
0780	6' high		595			595	655
0800	Helicona		315			315	345
1200	Palm, green date fan, 5' high		325			325	360
1220	7' high		415			415	455
1280	Green chamdora date, 5' high		297			297	325
1300	7' high		455			455	500
1400	Green giant, 7' high		705			705	775
1600	Rubber plant, 5' high		239			239	263
1800	Schefflera, 3' high		164			164	181
1820	4' high		197			197	217
1840	5' high		310			310	345
1860	6' high		425			425	470
1880	7' high		640			640	705
1900	Spathiphyllum, 3' high		227			227	250
4000	Trees, polyester or preserved, with						
4020	Natural trunks						
4100	Acuba, 10' high	Ea.	1,775			1,775	1,975
4120	12' high		2,500			2,500	2,750
4140	14' high		3,475			3,475	3,825
4400	Bamboo, 10' high		780			780	855
4420	12' high		900			900	990
4800	Beech, 10' high		2,400			2,400	2,625
4820	12' high		2,750			2,750	3,025
4840	14' high		3,700			3,700	4,075
5000	Birch, 10' high		2,400			2,400	2,625
5020	12' high		2,750			2,750	3,025
5040	14' high		2,975			2,975	3,275
5060	16' high		3,700			3,700	4,075
5080	18' high		4,175			4,175	4,600
5500	Ficus Benjamina, 10' high		1,750			1,750	1,925
5520	12' high		2,250			2,250	2,475
5540	14' high		2,525			2,525	2,775
5560	16' high		3,525			3,525	3,875

For customer support on your Site Work & Landscape Costs with RSMeans Data, call 800.448.8182.

12 92 Interior Planters and Artificial Plants

12 92 13 – Interior Artificial Plants

12 92 13.10 Plants	Crew	Daily Output	Labor-Hours	Unit	Material	2021 Bare Costs Labor	Equipment	Total	Total Incl O&P
5580	18' high				Ea.	5,275			5,275	5,800
6000	Magnolia, 10' high					2,575			2,575	2,825
6020	12' high					3,650			3,650	4,025
6040	14' high					5,000			5,000	5,500
6500	Maple, 10' high					276			276	305
6520	12' high					415			415	455
6540	14' high					510			510	560
7000	Palms, chamadora, 10' high					505			505	555
7020	12' high					640			640	705
7040	Date, 10' high					690			690	760
7060	12' high					795			795	875

12 92 33 – Interior Planters

12 92 33.10 Planters

		Crew	Daily Output	Labor-Hours	Unit	Material	Labor	Equipment	Total	Total Incl O&P
0010	**PLANTERS**									
1000	Fiberglass, hanging, 12" diameter, 7" high				Ea.	119			119	131
1100	15" diameter, 7" high					172			172	190
1200	36" diameter, 8" high					238			238	262
1500	Rectangular, 48" long, 16" high, 15" wide					485			485	535
1550	16" high, 24" wide					530			530	585
1600	24" high, 24" wide					780			780	855
1650	60" long, 30" high, 28" wide					990			990	1,100
1700	72" long, 16" high, 15" wide					810			810	890
1750	21" high, 24" wide					1,075			1,075	1,175
1800	30" high, 24" wide					1,175			1,175	1,300
2000	Round, 12" diameter, 13" high					101			101	111
2050	25" high					254			254	279
2150	14" diameter, 15" high					183			183	201
2200	16" diameter, 16" high					244			244	268
2250	18" diameter, 19" high					240			240	264
2300	23" high					375			375	415
2350	20" diameter, 16" high					242			242	266
2400	18" high					266			266	292
2450	21" high					345			345	375
2500	22" diameter, 10" high					242			242	266
2550	24" diameter, 16" high					285			285	315
2600	19" high					340			340	375
2650	25" high					395			395	435
2700	36" high					615			615	675
2750	48" high					855			855	940
2800	30" diameter, 16" high					340			340	370
2850	18" high					340			340	375
2900	21" high					380			380	420
3000	24" high					390			390	430
3350	27" high					430			430	475
3400	36" diameter, 16" high					390			390	430
3450	18" high					405			405	450
3500	21" high					440			440	485
3550	24" high					470			470	515
3600	27" high					510			510	560
3650	30" high					550			550	610
3700	48" diameter, 16" high					705			705	775
3750	21" high					760			760	840

12 92 33.10 Planters		Crew	Daily Output	Labor-Hours	Unit	Material	2021 Bare Costs Labor	Equipment	Total	Total Incl O&P
3800	24" high				Ea.	855			855	940
3850	27" high					895			895	985
3900	30" high					975			975	1,075
3950	36" high					1,025			1,025	1,125
4000	60" diameter, 16" high					995			995	1,100
4100	21" high					1,125			1,125	1,250
4150	27" high					1,250			1,250	1,400
4200	30" high					1,400			1,400	1,525
4250	33" high					1,600			1,600	1,750
4300	36" high					1,925			1,925	2,125
4400	39" high					2,075			2,075	2,300
5000	Square, 10" side, 20" high					209			209	230
5100	14" side, 15" high					258			258	284
5200	18" side, 19" high					400			400	440
5300	20" side, 16" high					315			315	350
5320	18" high					485			485	535
5340	21" high					390			390	425
5400	24" side, 16" high					360			360	400
5420	21" high					560			560	620
5440	25" high					535			535	590
5460	30" side, 16" high					500			500	550
5480	24" high					810			810	890
5490	27" high					600			600	660
5500	Round, 36" diameter, 16" high					600			600	660
5510	18" high					835			835	915
5520	21" high					920			920	1,000
5530	24" high					1,025			1,025	1,125
5540	27" high					1,150			1,150	1,275
5550	30" high					1,600			1,600	1,750
5800	48" diameter, 16" high					845			845	925
5820	21" high					945			945	1,050
5840	24" high					1,175			1,175	1,300
5860	27" high					1,275			1,275	1,400
5880	30" high					330			330	365
5900	60" diameter, 16" high					1,025			1,025	1,150
5920	21" high					1,200			1,200	1,325
5940	27" high					1,325			1,325	1,450
5960	30" high					1,475			1,475	1,625
5980	36" high					1,600			1,600	1,775
6000	Metal bowl, 32" diameter, 8" high, minimum					580			580	635
6050	Maximum					820			820	905
6100	Rectangle, 30" long x 12" wide, 6" high, minimum					445			445	490
6200	Maximum					600			600	660
6300	36" long x 12" wide, 6" high, minimum					875			875	965
6400	Maximum					540			540	595
6500	Square, 15" side, minimum					665			665	735
6600	Maximum					960			960	1,050
6700	20" side, minimum					1,375			1,375	1,525
6800	Maximum					605			605	665
6900	Round, 6" diameter x 6" high, minimum					64			64	70.50
7000	Maximum					76			76	83.50
7100	8" diameter x 8" high, minimum					85			85	93.50
7200	Maximum					91.50			91.50	101

12 92 33 – Interior Planters

12 92 33.10 Planters	Crew	Daily Output	Labor-Hours	Unit	Material	2021 Bare Costs Labor	Equipment	Total	Total Incl O&P
7300	10" diameter x 11" high, minimum				Ea.	109			109	120
7400	Maximum					161			161	177
7420	12" diameter x 13" high, minimum					101			101	111
7440	Maximum					211			211	232
7500	14" diameter x 15" high, minimum					136			136	150
7550	Maximum					239			239	263
7580	16" diameter x 17" high, minimum					147			147	162
7600	Maximum					262			262	288
7620	18" diameter x 19" high, minimum					178			178	196
7640	Maximum					320			320	350
7680	22" diameter x 20" high, minimum					224			224	247
7700	Maximum					410			410	450
7750	24" diameter x 21" high, minimum					292			292	320
7800	Maximum					545			545	600
7850	31" diameter x 18" high, minimum					635			635	700
7900	Maximum					1,625			1,625	1,775
7950	38" diameter x 24" high, minimum					1,075			1,075	1,175
8000	Maximum					2,700			2,700	2,975
8050	48" diameter x 24" high, minimum					1,425			1,425	1,550
8150	Maximum					2,800			2,800	3,075
8750	Wood, fiberglass liner, square									
8780	14" square, 15" high, minimum				Ea.	430			430	470
8800	Maximum					530			530	580
8820	24" square, 16" high, minimum					530			530	580
8840	Maximum					700			700	770
8860	36" square, 21" high, minimum					680			680	745
8880	Maximum					1,025			1,025	1,125
9000	Rectangle, 36" long x 12" wide, 10" high, minimum					480			480	530
9050	Maximum					625			625	685
9100	48" long x 12" wide, 10" high, minimum					520			520	570
9120	Maximum					655			655	720
9200	48" long x 12" wide, 24" high, minimum					625			625	685
9300	Maximum					915			915	1,000
9400	Plastic cylinder, molded, 10" diameter, 10" high					47			47	51.50
9500	11" diameter, 11" high					89			89	98
9600	13" diameter, 12" high					145			145	159
9700	16" diameter, 14" high					168			168	185

12 93 Interior Public Space Furnishings

12 93 13 – Bicycle Racks

12 93 13.10 Bicycle Racks	Crew	Daily Output	Labor-Hours	Unit	Material	2021 Bare Costs Labor	Equipment	Total	Total Incl O&P
0010	**BICYCLE RACKS**									
0020	Single side, grid, 1-5/8" OD stl. pipe, w/1/2" bars, galv, 5 bike cap	2 Clab	10	1.600	Ea.	199	71		270	325
0025	Powder coat finish		10	1.600		109	71		180	226
0030	Single side, grid, 1-5/8" OD stl. pipe, w/1/2" bars, galv, 9 bike cap		8	2		299	89		388	465
0035	Powder coat finish		8	2		141	89		230	288
0040	Single side, grid, 1-5/8" OD stl. pipe, w/1/2" bars, galv, 18 bike cap		4	4		350	178		528	650
0045	Powder coat finish		4	4		380	178		558	680
0050	S curve, 1-7/8" OD stl. pipe, 11 ga., galv, 5 bike cap		10	1.600		147	71		218	268
0055	Powder coat finish		10	1.600		147	71		218	268
0060	S curve, 1-7/8" OD stl. pipe, 11 ga., galv, 7 bike cap		9	1.778		262	79		341	405

12 93 13 – Bicycle Racks

12 93 13.10 Bicycle Racks		Crew	Daily Output	Labor-Hours	Unit	Material	2021 Bare Costs Labor	Equipment	Total	Total Incl O&P
0065	Powder coat finish	2 Clab	9	1.778	Ea.	206	79		285	345
0070	S curve, 1-7/8" OD stl. pipe, 11 ga., galv, 9 bike cap		8	2		335	89		424	505
0075	Powder coat finish		8	2		269	89		358	430
0080	S curve, 1-7/8" OD stl. pipe, 11 ga., galv, 11 bike cap		6	2.667		425	118		543	640
0085	Powder coat finish		6	2.667		425	118		543	640

12 93 23 – Trash and Litter Receptacles

12 93 23.10 Trash Receptacles

		Crew	Daily Output	Labor-Hours	Unit	Material	2021 Bare Costs Labor	Equipment	Total	Total Incl O&P
0010	**TRASH RECEPTACLES**									
0020	Fiberglass, 2' square, 18" high	2 Clab	30	.533	Ea.	655	23.50		678.50	755
0100	2' square, 2'-6" high		30	.533		940	23.50		963.50	1,050
0300	Circular, 2' diameter, 18" high		30	.533		575	23.50		598.50	665
0400	2' diameter, 2'-6" high		30	.533		615	23.50		638.50	710
1000	Alum. frame, hardboard panels, steel drum base,									
1020	30 gal. capacity, silk screen on plastic finish	2 Clab	25	.640	Ea.	435	28.50		463.50	525
1040	Aggregate finish		25	.640		555	28.50		583.50	655
1100	50 gal. capacity, silk screen on plastic finish		20	.800		655	35.50		690.50	775
1140	Aggregate finish		20	.800		700	35.50		735.50	825
1200	Formed plastic liner, 14 gal., silk screen on plastic finish		40	.400		365	17.75		382.75	430
1240	Aggregate finish		40	.400		395	17.75		412.75	460
1300	30 gal. capacity, silk screen on plastic finish		35	.457		470	20.50		490.50	550
1340	Aggregate finish		35	.457		515	20.50		535.50	595
1400	Redwood slats, plastic liner, leg base, 14 gal. capacity,									
1420	Varnish w/routed message	2 Clab	40	.400	Ea.	385	17.75		402.75	445
2000	Concrete, precast, 2' to 2-1/2' wide, 3' high, sandblasted	"	15	1.067	"	625	47.50		672.50	760
3000	Galv. steel frame and panels, leg base, poly bag retainer,									
3020	40 gal. capacity, silk screen on enamel finish	2 Clab	25	.640	Ea.	630	28.50		658.50	740
3040	Aggregate finish		25	.640		695	28.50		723.50	810
3200	Formed plastic liner, 50 gal., silk screen on enamel finish		20	.800		1,050	35.50		1,085.50	1,225
3240	Aggregate finish		20	.800		1,200	35.50		1,235.50	1,375
4000	Perforated steel, pole mounted, 12" diam., 10 gal., painted		25	.640		166	28.50		194.50	225
4040	Redwood slats		25	.640		380	28.50		408.50	460
4100	22 gal. capacity, painted		25	.640		171	28.50		199.50	231
4140	Redwood slats		25	.640		595	28.50		623.50	695
4500	Galvanized steel street basket, 52 gal. capacity, unpainted		40	.400		445	17.75		462.75	515

12 93 23.20 Trash Closure

		Crew	Daily Output	Labor-Hours	Unit	Material	2021 Bare Costs Labor	Equipment	Total	Total Incl O&P
0010	**TRASH CLOSURE**									
0020	Steel with pullover cover, 2'-3" wide, 4'-7" high, 6'-2" long	2 Clab	5	3.200	Ea.	2,375	142		2,517	2,800
0100	10'-1" long		4	4		3,075	178		3,253	3,650
0300	Wood, 10' wide, 6' high, 10' long		1.20	13.333		2,125	590		2,715	3,225

Estimating Tips

General

- The items and systems in this division are usually estimated, purchased, supplied, and installed as a unit by one or more subcontractors. The estimator must ensure that all parties are operating from the same set of specifications and assumptions, and that all necessary items are estimated and will be provided. Many times the complex items and systems are covered, but the more common ones, such as excavation or a crane, are overlooked for the very reason that everyone assumes nobody could miss them. The estimator should be the central focus and be able to ensure that all systems are complete.

- It is important to consider factors such as site conditions, weather, shape and size of building, as well as labor availability as they may impact the overall cost of erecting special structures and systems included in this division.

- Another area where problems can develop in this division is at the interface between systems.

The estimator must ensure, for instance, that anchor bolts, nuts, and washers are estimated and included for the air-supported structures and pre-engineered buildings to be bolted to their foundations. Utility supply is a common area where essential items or pieces of equipment can be missed or overlooked because each subcontractor may feel it is another's responsibility. The estimator should also be aware of certain items which may be supplied as part of a package but installed by others, and ensure that the installing contractor's estimate includes the cost of installation. Conversely, the estimator must also ensure that items are not costed by two different subcontractors, resulting in an inflated overall estimate.

13 30 00 Special Structures

- The foundations and floor slab, as well as rough mechanical and electrical, should be estimated, as this work is required for the assembly and erection of the structure. Generally, as noted in the data set, the pre-engineered building comes

as a shell. Pricing is based on the size and structural design parameters stated in the reference section. Additional features, such as windows and doors with their related structural framing, must also be included by the estimator. Here again, the estimator must have a clear understanding of the scope of each portion of the work and all the necessary interfaces.

Reference Numbers

Reference numbers are shown at the beginning of some major classifications. These numbers refer to related items in the Reference Section. The reference information may be an estimating procedure, an alternate pricing method, or technical information.

Note: Not all subdivisions listed here necessarily appear. ■

Same Data. Simplified.

Enjoy the convenience and efficiency of accessing your costs anywhere:

- **Skip the multiplier** by setting your location
- **Quickly search,** edit, favorite and share costs
- **Stay on top of price changes** with automatic updates

Discover more at rsmeans.com/online

13 05 05 – Selective Demolition for Special Construction

13 05 05.25 Selective Demolition, Geodesic Domes	Crew	Daily Output	Labor-Hours	Unit	Material	2021 Bare Costs Labor	Equipment	Total	Total Incl O&P
0010 **SELECTIVE DEMOLITION, GEODESIC DOMES**									
0050 Shell only, interlocking plywood panels, 30' diameter	F-5	3.20	10	Ea.		550		550	825
0060 34' diameter		2.30	13.913			770		770	1,150
0070 39' diameter		2	16			885		885	1,325
0080 45' diameter	F-3	2.20	18.182			1,025	216	1,241	1,775
0090 55' diameter		2	20			1,125	238	1,363	1,925
0100 60' diameter		2	20			1,125	238	1,363	1,925
0110 65' diameter		1.60	25			1,400	298	1,698	2,425

13 05 05.45 Selective Demolition, Lightning Protection

13 05 05.45 Selective Demolition, Lightning Protection	Crew	Daily Output	Labor-Hours	Unit	Material	Labor	Equipment	Total	Total Incl O&P
0010 **SELECTIVE DEMOLITION, LIGHTNING PROTECTION**									
0020 Air terminal & base, copper, 3/8" diam. x 10", to 75' H	1 Clab	16	.500	Ea.		22		22	33
0030 1/2" diam. x 12", over 75' H		16	.500			22		22	33
0050 Aluminum, 1/2" diam. x 12", to 75' H		16	.500			22		22	33
0060 5/8" diam. x 12", over 75' H		16	.500			22		22	33
0070 Cable, copper, 220 lb. per thousand feet, to 75' H		640	.013	L.F.		.56		.56	.83
0080 375 lb. per thousand feet, over 75' H		460	.017			.77		.77	1.15
0090 Aluminum, 101 lb. per thousand feet, to 75' H		560	.014			.63		.63	.95
0100 199 lb. per thousand feet, over 75' H		480	.017			.74		.74	1.10
0110 Arrester, 175 V AC, to ground		16	.500	Ea.		22		22	33
0120 650 V AC, to ground		13	.615	"		27.50		27.50	41

13 05 05.50 Selective Demolition, Pre-Engineered Steel Buildings

13 05 05.50 Selective Demolition, Pre-Engineered Steel Buildings	Crew	Daily Output	Labor-Hours	Unit	Material	Labor	Equipment	Total	Total Incl O&P
0010 **SELECTIVE DEMOLITION, PRE-ENGINEERED STEEL BUILDINGS**									
0500 Pre-engd. steel bldgs., rigid frame, clear span & multi post, excl. salvage									
0550 3,500 to 7,500 S.F.	L-10	1000	.024	SF Flr.		1.47	.48	1.95	2.77
0600 7,501 to 12,500 S.F.		1500	.016			.98	.32	1.30	1.85
0650 12,501 S.F. or greater		1650	.015			.89	.29	1.18	1.68
0700 Pre-engd. steel building components									
0710 Entrance canopy, including frame 4' x 4'	E-24	8	4	Ea.		240	73.50	313.50	450
0720 4' x 8'	"	7	4.571			274	84	358	510
0730 HM doors, self framing, single leaf	2 Skwk	8	2			114		114	172
0740 Double leaf		5	3.200			183		183	275
0760 Gutter, eave type		600	.027	L.F.		1.52		1.52	2.29
0770 Sash, single slide, double slide or fixed		24	.667	Ea.		38		38	57.50
0780 Skylight, fiberglass, to 30 S.F.		16	1			57		57	86
0785 Roof vents, circular, 12" to 24" diameter		12	1.333			76		76	115
0790 Continuous, 10' long		8	2			114		114	172
0900 Shelters, aluminum frame									
0910 Acrylic glazing, 3' x 9' x 8' high	2 Skwk	2	8	Ea.		455		455	685
0920 9' x 12' x 8' high	"	1.50	10.667	"		610		610	915

13 05 05.60 Selective Demolition, Silos

13 05 05.60 Selective Demolition, Silos	Crew	Daily Output	Labor-Hours	Unit	Material	Labor	Equipment	Total	Total Incl O&P
0010 **SELECTIVE DEMOLITION, SILOS**									
0020 Conc stave, indstrl, conical/sloping bott, excl fndtn, 12' diam., 35' H	E-24	.18	178	Ea.		10,700	3,250	13,950	19,900
0030 16' diam., 45' H		.12	267			16,000	4,900	20,900	29,900
0040 25' diam., 75' H		.08	400			24,000	7,325	31,325	44,900
0050 Steel, factory fabricated, 30,000 gal. cap, painted or epoxy lined	L-5	2	28			1,700	293	1,993	2,950

13 05 05.75 Selective Demolition, Storage Tanks

13 05 05.75 Selective Demolition, Storage Tanks		Crew	Daily Output	Labor-Hours	Unit	Material	Labor	Equipment	Total	Total Incl O&P
0010 **SELECTIVE DEMOLITION, STORAGE TANKS**										
0500 Steel tank, single wall, above ground, not incl. fdn., pumps or piping										
0510 Single wall, 275 gallon	R024119-10	Q-1	3	5.333	Ea.		325		325	485
0520 550 thru 2,000 gallon		B-34P	2	12			705	535	1,240	1,625
0530 5,000 thru 10,000 gallon		B-34Q	2	12			715	565	1,280	1,700
0540 15,000 thru 30,000 gallon		B-34S	2	16			1,000	1,550	2,550	3,200

202

13 05 Common Work Results for Special Construction

13 05 05 – Selective Demolition for Special Construction

13 05 05.75 Selective Demolition, Storage Tanks	Crew	Daily Output	Labor-Hours	Unit	Material	2021 Bare Costs Labor	Equipment	Total	Total Incl O&P
0600	Steel tank, double wall, above ground not incl. fdn., pumps & piping								
0620	500 thru 2,000 gallon	B-34P	2	12	Ea.		705	535	1,240	1,625

13 05 05.90 Selective Demolition, Tension Structures	Crew	Daily Output	Labor-Hours	Unit	Material	2021 Bare Costs Labor	Equipment	Total	Total Incl O&P
0010	**SELECTIVE DEMOLITION, TENSION STRUCTURES**									
0020	Steel/alum. frame, fabric shell, 60' clear span, 6,000 S.F.	B-41	2000	.022	SF Flr.		1.01	.15	1.16	1.68
0030	12,000 S.F.		2200	.020			.92	.14	1.06	1.52
0040	80' clear span, 20,800 S.F.		2440	.018			.83	.13	.96	1.38
0050	100' clear span, 10,000 S.F.	L-5	4350	.013			.78	.13	.91	1.35
0060	26,000 S.F.		4600	.012			.74	.13	.87	1.28
0070	36,000 S.F.		5000	.011			.68	.12	.80	1.18
0080	120' clear span, 24,000 S.F.		6000	.009			.57	.10	.67	.98
0090	150' clear span, 30,000 S.F.		7000	.008			.49	.08	.57	.84
0100	200' clear span, 40,000 S.F.	E-6	16000	.008			.48	.14	.62	.89
0110	For roll-up door, 12' x 14'	L-2	2	8	Ea.	385			385	580

13 11 Swimming Pools

13 11 13 – Below-Grade Swimming Pools

13 11 13.50 Swimming Pools	Crew	Daily Output	Labor-Hours	Unit	Material	2021 Bare Costs Labor	Equipment	Total	Total Incl O&P
0010	**SWIMMING POOLS** Residential in-ground, vinyl lined									
0020	Concrete sides, w/equip, sand bottom	B-52	300	.187	SF Surf	29.50	9.50	1.91	40.91	48.50
0100	Metal or polystyrene sides	B-14	410	.117		24.50	5.45	.53	30.48	35.50
0200	Add for vermiculite bottom					1.87			1.87	2.06
0500	Gunite bottom and sides, white plaster finish									
0600	12' x 30' pool	B-52	145	.386	SF Surf	54.50	19.60	3.95	78.05	93.50
0720	16' x 32' pool		155	.361		49	18.35	3.70	71.05	85.50
0750	20' x 40' pool		250	.224		44	11.35	2.29	57.64	68
0810	Concrete bottom and sides, tile finish									
0820	12' x 30' pool	B-52	80	.700	SF Surf	55	35.50	7.15	97.65	121
0830	16' x 32' pool		95	.589		45.50	30	6.05	81.55	101
0840	20' x 40' pool		130	.431		36.50	22	4.41	62.91	77.50
1100	Motel, gunite with plaster finish, incl. medium									
1150	capacity filtration & chlorination	B-52	115	.487	SF Surf	67.50	24.50	4.98	96.98	117
1200	Municipal, gunite with plaster finish, incl. high									
1250	capacity filtration & chlorination	B-52	100	.560	SF Surf	87	28.50	5.75	121.25	144
1350	Add for formed gutters				L.F.	128			128	141
1360	Add for stainless steel gutters				"	380			380	415
1700	Filtration and deck equipment only, as % of total				Total				20%	20%
1800	Automatic vacuum, hand tools, etc., 20' x 40' pool				SF Pool				.56	.62
1900	5,000 S.F. pool				"				.11	.12
3000	Painting pools, preparation + 3 coats, 20' x 40' pool, epoxy	2 Pord	.33	48.485	Total	1,300	2,250		3,550	4,775
3100	Rubber base paint, 18 gallons	"	.33	48.485		1,250	2,250		3,500	4,725
3500	42' x 82' pool, 75 gallons, epoxy paint	3 Pord	.14	171		5,500	7,975		13,475	17,900
3600	Rubber base paint	"	.14	171		5,150	7,975		13,125	17,500

13 11 23 – On-Grade Swimming Pools

13 11 23.50 Swimming Pools	Crew	Daily Output	Labor-Hours	Unit	Material	2021 Bare Costs Labor	Equipment	Total	Total Incl O&P
0010	**SWIMMING POOLS** Residential above ground, steel construction									
0100	Round, 15' diam.	B-80A	3	8	Ea.	880	355	283	1,518	1,800
0120	18' diam.		2.50	9.600		935	425	340	1,700	2,025
0140	21' diam.		2	12		1,050	535	425	2,010	2,425
0160	24' diam.		1.80	13.333		1,200	590	470	2,260	2,700

13 11 Swimming Pools

13 11 23 – On-Grade Swimming Pools

13 11 23.50 Swimming Pools

		Crew	Daily Output	Labor-Hours	Unit	Material	2021 Bare Costs Labor	Equipment	Total	Total Incl O&P
0180	27' diam.	B-80A	1.50	16	Ea.	1,500	710	565	2,775	3,325
0200	30' diam.		1	24		1,725	1,075	850	3,650	4,425
0220	Oval, 12' x 24'		2.30	10.435		1,525	465	370	2,360	2,775
0240	15' x 30'		1.80	13.333		1,700	590	470	2,760	3,275
0260	18' x 33'		1	24		1,925	1,075	850	3,850	4,650

13 11 46 – Swimming Pool Accessories

13 11 46.50 Swimming Pool Equipment

		Crew	Daily Output	Labor-Hours	Unit	Material	2021 Bare Costs Labor	Equipment	Total	Total Incl O&P
0010	**SWIMMING POOL EQUIPMENT**									
0020	Diving stand, stainless steel, 3 meter	2 Carp	.40	40	Ea.	19,800	2,200		22,000	25,100
0300	1 meter		2.70	5.926		11,400	325		11,725	13,000
0600	Diving boards, 16' long, aluminum		2.70	5.926		4,500	325		4,825	5,425
0700	Fiberglass		2.70	5.926		3,525	325		3,850	4,350
0800	14' long, aluminum		2.70	5.926		4,125	325		4,450	5,025
0850	Fiberglass		2.70	5.926		3,300	325		3,625	4,125
1100	Bulkhead, movable, PVC, 8'-2" wide	2 Clab	8	2		2,775	89		2,864	3,175
1120	7'-9" wide		8	2		2,900	89		2,989	3,325
1140	7'-3" wide		8	2		2,300	89		2,389	2,650
1160	6'-9" wide		8	2		2,300	89		2,389	2,650
1200	Ladders, heavy duty, stainless steel, 2 tread	2 Carp	7	2.286		1,075	125		1,200	1,350
1500	4 tread		6	2.667		900	146		1,046	1,225
1800	Lifeguard chair, stainless steel, fixed		2.70	5.926		3,225	325		3,550	4,000
1900	Portable					3,050			3,050	3,375
2100	Lights, underwater, 12 volt, with transformer, 300 watt	1 Elec	1	8		465	510		975	1,275
2200	110 volt, 500 watt, standard		1	8		291	510		801	1,075
2400	Low water cutoff type		1	8		315	510		825	1,100
3000	Pool covers, reinforced vinyl	3 Clab	1800	.013	S.F.	1.28	.59		1.87	2.29
3050	Automatic, electric								9.15	10.10
3100	Vinyl, for winter, 400 S.F. max pool surface	3 Clab	3200	.008		.23	.33		.56	.75
3200	With water tubes, 400 S.F. max pool surface	"	3000	.008		.36	.36		.72	.93
3250	Sealed air bubble polyethylene solar blanket, 16 mils					.37			.37	.41
3300	Slides, tubular, fiberglass, aluminum handrails & ladder, 5'-0", straight	2 Carp	1.60	10	Ea.	4,025	545		4,570	5,250
3320	8'-0", curved		3	5.333		8,000	292		8,292	9,225
3400	10'-0", curved		1	16		22,800	875		23,675	26,300
3420	12'-0", straight with platform		1.20	13.333		18,400	730		19,130	21,300

13 12 Fountains

13 12 13 – Exterior Fountains

13 12 13.10 Outdoor Fountains

		Crew	Daily Output	Labor-Hours	Unit	Material	2021 Bare Costs Labor	Equipment	Total	Total Incl O&P
0010	**OUTDOOR FOUNTAINS**									
0100	Outdoor fountain, 48" high with bowl and figures	2 Clab	2	8	Ea.	385	355		740	955
0200	Commercial, concrete or cast stone, 40-60" H, simple		2	8		975	355		1,330	1,600
0220	Average		2	8		1,900	355		2,255	2,600
0240	Ornate		2	8		4,300	355		4,655	5,275
0260	Metal, 72" high		2	8		1,525	355		1,880	2,200
0280	90" high		2	8		2,200	355		2,555	2,950
0300	120" high		2	8		5,275	355		5,630	6,325
0320	Resin or fiberglass, 40-60" H, wall type		2	8		690	355		1,045	1,300
0340	Waterfall type		2	8		1,175	355		1,530	1,800

13 18 Ice Rinks

13 18 13 – Ice Rink Floor Systems

13 18 13.50 Ice Skating	Crew	Daily Output	Labor-Hours	Unit	Material	2021 Bare Costs Labor	Equipment	Total	Total Incl O&P
0010 **ICE SKATING** Equipment incl. refrigeration, plumbing & cooling									
0020 coils & concrete slab, 85' x 200' rink									
0300 55° system, 5 mos., 100 ton				Total	594,500			594,500	654,000
0700 90° system, 12 mos., 135 ton				"	676,000			676,000	744,000
1200 Subsoil heating system (recycled from compressor), 85' x 200'	Q-7	.27	119	Ea.	43,700	7,700		51,400	59,500
1300 Subsoil insulation, 2 lb. polystyrene with vapor barrier, 85' x 200'	2 Carp	.14	114	"	31,600	6,250		37,850	44,000

13 18 16 – Ice Rink Dasher Boards

13 18 16.50 Ice Rink Dasher Boards	Crew	Daily Output	Labor-Hours	Unit	Material	2021 Bare Costs Labor	Equipment	Total	Total Incl O&P
0010 **ICE RINK DASHER BOARDS**									
1000 Dasher boards, 1/2" H.D. polyethylene faced steel frame, 3' acrylic									
1020 screen at sides, 5' acrylic ends, 85' x 200'	F-5	.06	533	Ea.	136,000	29,400		165,400	194,000
1100 Fiberglass & aluminum construction, same sides and ends	"	.06	533	"	152,000	29,400		181,400	211,000

13 31 Fabric Structures

13 31 13 – Air-Supported Fabric Structures

13 31 13.09 Air Supported Tank Covers	Crew	Daily Output	Labor-Hours	Unit	Material	2021 Bare Costs Labor	Equipment	Total	Total Incl O&P
0010 **AIR SUPPORTED TANK COVERS**, vinyl polyester									
0100 Scrim, double layer, with hardware, blower, standby & controls									
0200 Round, 75' diameter	B-2	4500	.009	S.F.	12.80	.40		13.20	14.70
0300 100' diameter		5000	.008		11.65	.36		12.01	13.35
0400 150' diameter		5000	.008		9.20	.36		9.56	10.70
0500 Rectangular, 20' x 20'		4500	.009		25	.40		25.40	27.50
0600 30' x 40'		4500	.009		25	.40		25.40	27.50
0700 50' x 60'		4500	.009		25	.40		25.40	27.50
0800 For single wall construction, deduct, minimum					.87			.87	.96
0900 Maximum					2.49			2.49	2.74
1000 For maximum resistance to atmosphere or cold, add					1.25			1.25	1.38
1100 For average shipping charges, add				Total	2,150			2,150	2,350

13 31 13.13 Single-Walled Air-Supported Structures	Crew	Daily Output	Labor-Hours	Unit	Material	2021 Bare Costs Labor	Equipment	Total	Total Incl O&P
0010 **SINGLE-WALLED AIR-SUPPORTED STRUCTURES**									
0020 Site preparation, incl. anchor placement and utilities	B-11B	1000	.016	SF Flr.	1.27	.80	.43	2.50	3.06
0030 For concrete, see Section 03 30 53.40									
0050 Warehouse, polyester/vinyl fabric, 28 oz., over 10 yr. life, welded									
0060 Seams, tension cables, primary & auxiliary inflation system,									
0070 airlock, personnel doors and liner									
0100 5,000 S.F.	4 Clab	5000	.006	SF Flr.	28	.28		28.28	31
0250 12,000 S.F.	"	6000	.005		19.10	.24		19.34	21.50
0400 24,000 S.F.	8 Clab	12000	.005		14.10	.24		14.34	15.85
0500 50,000 S.F.	"	12500	.005		12.45	.23		12.68	14.05
0700 12 oz. reinforced vinyl fabric, 5 yr. life, sewn seams,									
0710 accordion door, including liner									
0750 3,000 S.F.	4 Clab	3000	.011	SF Flr.	13.85	.47		14.32	15.90
0800 12,000 S.F.	"	6000	.005		11.80	.24		12.04	13.30
0850 24,000 S.F.	8 Clab	12000	.005		10	.24		10.24	11.35
0950 Deduct for single layer					1.13			1.13	1.24
1000 Add for welded seams					1.56			1.56	1.72
1050 Add for double layer, welded seams included					3.13			3.13	3.44
1250 Tedlar/vinyl fabric, 28 oz., with liner, over 10 yr. life,									
1260 incl. overhead and personnel doors									
1300 3,000 S.F.	4 Clab	3000	.011	SF Flr.	26	.47		26.47	29

For customer support on your Site Work & Landscape Costs with RSMeans Data, call 800.448.8182.

205

13 31 Fabric Structures

13 31 13 – Air-Supported Fabric Structures

13 31 13.13 Single-Walled Air-Supported Structures

		Crew	Daily Output	Labor-Hours	Unit	Material	2021 Bare Costs Labor	Equipment	Total	Total Incl O&P
1450	12,000 S.F.	4 Clab	6000	.005	SF Flr.	17.65	.24		17.89	19.75
1550	24,000 S.F.	8 Clab	12000	.005		14.10	.24		14.34	15.90
1700	Deduct for single layer				↓	2.08			2.08	2.29
2250	Greenhouse/shelter, woven polyethylene with liner, 2 yr. life,									
2260	sewn seams, including doors									
2300	3,000 S.F.	4 Clab	3000	.011	SF Flr.	16.75	.47		17.22	19.15
2350	12,000 S.F.	"	6000	.005		14.65	.24		14.89	16.45
2450	24,000 S.F.	8 Clab	12000	.005	↓	12.55	.24		12.79	14.15
2550	Deduct for single layer					1.03			1.03	1.13
2600	Tennis/gymnasium, polyester/vinyl fabric, 28 oz., over 10 yr. life,									
2610	including thermal liner, heat and lights									
2650	7,200 S.F.	4 Clab	6000	.005	SF Flr.	24	.24		24.24	27
2750	13,000 S.F.	"	6500	.005		19.10	.22		19.32	21.50
2850	Over 24,000 S.F.	8 Clab	12000	.005		17.45	.24		17.69	19.55
2860	For low temperature conditions, add					1.24			1.24	1.36
2870	For average shipping charges, add				Total	5,825			5,825	6,425
2900	Thermal liner, translucent reinforced vinyl				SF Flr.	1.24			1.24	1.36
2950	Metalized mylar fabric and mesh, double liner				"	2.48			2.48	2.73
3050	Stadium/convention center, teflon coated fiberglass, heavy weight,									
3060	over 20 yr. life, incl. thermal liner and heating system									
3100	Minimum	9 Clab	26000	.003	SF Flr.	61	.12		61.12	67
3110	Maximum	"	19000	.004	"	75	.17		75.17	83.50
3400	Doors, air lock, 15' long, 10' x 10'	2 Carp	.80	20	Ea.	21,300	1,100		22,400	25,100
3600	15' x 15'	"	.50	32		33,500	1,750		35,250	39,500
3700	For each added 5' length, add					5,975			5,975	6,575
3900	Revolving personnel door, 6' diameter, 6'-6" high	2 Carp	.80	20	↓	16,800	1,100		17,900	20,000

13 31 23 – Tensioned Fabric Structures

13 31 23.50 Tension Structures

		Crew	Daily Output	Labor-Hours	Unit	Material	2021 Bare Costs Labor	Equipment	Total	Total Incl O&P
0010	**TENSION STRUCTURES** Rigid steel/alum. frame, vinyl coated poly									
0100	Fabric shell, 60' clear span, not incl. foundations or floors									
0200	6,000 S.F.	B-41	1000	.044	SF Flr.	17.90	2.02	.31	20.23	23
0300	12,000 S.F.		1100	.040		21.50	1.84	.28	23.62	26.50
0400	80' to 99' clear span, 20,800 S.F.	↓	1220	.036		19.60	1.66	.25	21.51	24.50
0410	100' to 119' clear span, 10,000 S.F.	L-5	2175	.026		17.75	1.56	.27	19.58	22.50
0430	26,000 S.F.		2300	.024		16.40	1.48	.26	18.14	20.50
0450	36,000 S.F.		2500	.022		16.75	1.36	.23	18.34	21
0460	120' to 149' clear span, 24,000 S.F.		3000	.019		17.15	1.13	.20	18.48	21
0470	150' to 199' clear span, 30,000 S.F.	↓	6000	.009		18.80	.57	.10	19.47	21.50
0480	200' clear span, 40,000 S.F.	E-6	8000	.016	↓	23	.96	.27	24.23	27
0500	For roll-up door, 12' x 14', add	L-2	1	16	Ea.	6,775	775		7,550	8,625
0600	6,000 S.F.	B-41	1200	.037	SF Flr.	13	1.68	.26	14.94	17.10
0650	12,000 S.F.		1200	.037		15.25	1.68	.26	17.19	19.60
1000	Tension structure, PVC fabric, add		2500	.018		4	.81	.12	4.93	5.75
1200	Tension structure, HDPE fabric, add		2500	.018		2.50	.81	.12	3.43	4.10
1300	Tension structure, PTFE fabric, add	↓	2000	.022	↓	7.50	1.01	.15	8.66	9.95

13 33 Geodesic Structures

13 33 13 – Geodesic Domes

13 33 13.35 Geodesic Domes

		Crew	Daily Output	Labor-Hours	Unit	Material	2021 Bare Costs Labor	2021 Bare Costs Equipment	Total	Total Incl O&P
0010	**GEODESIC DOMES** Shell only, interlocking plywood panels R133423-30									
0400	30' diameter	F-5	1.60	20	Ea.	33,500	1,100		34,600	38,600
0500	33' diameter		1.14	28.070		24,300	1,550		25,850	29,000
0600	40' diameter		1	32		27,900	1,775		29,675	33,300
0700	45' diameter	F-3	1.13	35.556		28,900	2,000	425	31,325	35,200
0750	56' diameter		1	40		58,000	2,250	475	60,725	68,000
0800	60' diameter		1	40		63,000	2,250	475	65,725	73,500
0850	67' diameter		.80	50		93,500	2,800	595	96,895	108,000
1100	Aluminum panel, with 6" insulation									
1200	100' diameter				SF Flr.	25			25	27.50
1300	500' diameter				"	26			26	28.50
1600	Aluminum framed, plexiglass closure panels									
1700	40' diameter				SF Flr.	61.50			61.50	67.50
1800	200' diameter				"	60.50			60.50	66.50
2100	Aluminum framed, aluminum closure panels									
2200	40' diameter				SF Flr.	20			20	22
2300	100' diameter					31			31	34
2400	200' diameter					20.50			20.50	22.50
2700	Aluminum framed, fiberglass sandwich panel closure									
2800	6' diameter	2 Carp	150	.107	SF Flr.	28.50	5.85		34.35	39.50
2900	28' diameter	"	350	.046	"	25.50	2.50		28	32

13 34 Fabricated Engineered Structures

13 34 13 – Glazed Structures

13 34 13.13 Greenhouses

		Crew	Daily Output	Labor-Hours	Unit	Material	2021 Bare Costs Labor	2021 Bare Costs Equipment	Total	Total Incl O&P
0010	**GREENHOUSES**, Shell only, stock units, not incl. 2' stub walls,									
0020	foundation, floors, heat or compartments									
0300	Residential type, free standing, 8'-6" long x 7'-6" wide	2 Carp	59	.271	SF Flr.	24	14.85		38.85	48
0400	10'-6" wide		85	.188		43.50	10.30		53.80	63.50
0600	13'-6" wide		108	.148		46	8.10		54.10	62.50
0700	17'-0" wide		160	.100		42.50	5.45		47.95	54.50
0900	Lean-to type, 3'-10" wide		34	.471		42.50	25.50		68	85.50
1000	6'-10" wide		58	.276		26	15.10		41.10	51
1050	8'-0" wide		60	.267		75	14.60		89.60	105
1100	Wall mounted to existing window, 3' x 3'	1 Carp	4	2	Ea.	2,625	109		2,734	3,075
1120	4' x 5'	"	3	2.667	"	2,425	146		2,571	2,875
1200	Deluxe quality, free standing, 7'-6" wide	2 Carp	55	.291	SF Flr.	85.50	15.90		101.40	118
1220	10'-6" wide		81	.198		68.50	10.80		79.30	91
1240	13'-6" wide		104	.154		59.50	8.40		67.90	77.50
1260	17'-0" wide		150	.107		46	5.85		51.85	59
1400	Lean-to type, 3'-10" wide		31	.516		85	28		113	136
1420	6'-10" wide		55	.291		82.50	15.90		98.40	115
1440	8'-0" wide		97	.165		63.50	9		72.50	83.50
1500	Commercial, custom, truss frame, incl. equip., plumbing, elec.,									
1550	benches and controls, under 2,000 S.F.				SF Flr.	14.20			14.20	15.65
1700	Over 5,000 S.F.				"	12.70			12.70	14
2000	Institutional, custom, rigid frame, including compartments and									
2050	multi-controls, under 500 S.F.				SF Flr.	27			27	30
2150	Over 2,000 S.F.				"	12.35			12.35	13.60
3700	For 1/4" tempered glass, add				SF Surf	2.16			2.16	2.38
3900	Cooling, 1,200 CFM exhaust fan, add				Ea.	325			325	355

For customer support on your Site Work & Landscape Costs with RSMeans Data, call 800.448.8182.

207

13 34 13 – Glazed Structures

13 34 13.13 Greenhouses

13 34 13.13 Greenhouses	Crew	Daily Output	Labor-Hours	Unit	Material	2021 Bare Costs Labor	2021 Bare Costs Equipment	Total	Total Incl O&P
4000	7,850 CFM				Ea.	850			850	935
4200	For heaters, 10 MBH, add					245			245	270
4300	60 MBH, add					730			730	805
4500	For benches, 2' x 8', add					188			188	207
4600	4' x 10', add					495			495	545
4800	For ventilation & humidity control w/4 integrated outlets, add				Total	288			288	315
4900	For environmental controls and automation, 8 outputs, 9 stages, add				"	1,325			1,325	1,450
5100	For humidification equipment, add				Ea.	325			325	360
5200	For vinyl shading, add				S.F.	.39			.39	.43
6000	Geodesic hemisphere, 1/8" plexiglass glazing									
6050	8' diameter	2 Carp	2	8	Ea.	5,275	440		5,715	6,450
6150	24' diameter		.35	45.714		16,100	2,500		18,600	21,500
6250	48' diameter		.20	80		33,600	4,375		37,975	43,400

13 34 13.19 Swimming Pool Enclosures

13 34 13.19 Swimming Pool Enclosures	Crew	Daily Output	Labor-Hours	Unit	Material	2021 Bare Costs Labor	2021 Bare Costs Equipment	Total	Total Incl O&P
0010	SWIMMING POOL ENCLOSURES Translucent, free standing R131113-20									
0020	not including foundations, heat or light									
0200	Economy	2 Carp	200	.080	SF Hor.	42	4.38		46.38	52.50
0600	Deluxe	"	70	.229		115	12.50		127.50	145
0700	For motorized roof, 40% opening, solid roof, add					23			23	25
0800	Skylight type roof, add					13.95			13.95	15.35

13 34 16 – Grandstands and Bleachers

13 34 16.13 Grandstands

13 34 16.13 Grandstands	Crew	Daily Output	Labor-Hours	Unit	Material	2021 Bare Costs Labor	2021 Bare Costs Equipment	Total	Total Incl O&P
0010	GRANDSTANDS Permanent, municipal, including foundation									
0300	Steel, economy				Seat	25			25	27.50
0400	Steel, deluxe					27.50			27.50	30
0900	Composite, steel, wood and plastic, stock design, economy					34.50			34.50	38
1000	Deluxe					82.50			82.50	91

13 34 16.53 Bleachers

13 34 16.53 Bleachers	Crew	Daily Output	Labor-Hours	Unit	Material	2021 Bare Costs Labor	2021 Bare Costs Equipment	Total	Total Incl O&P
0010	BLEACHERS									
0020	Bleachers, outdoor, portable, 5 tiers, 42 seats	2 Sswk	120	.133	Seat	67	8.05		75.05	86
0100	5 tiers, 54 seats		80	.200		69	12.05		81.05	94.50
0200	10 tiers, 104 seats		120	.133		89.50	8.05		97.55	111
0300	10 tiers, 144 seats		80	.200		64.50	12.05		76.55	89
0500	Permanent bleachers, aluminum seat, steel frame, 24" row									
0600	8 tiers, 80 seats	2 Sswk	60	.267	Seat	74.50	16.10		90.60	107
0700	8 tiers, 160 seats		48	.333		74	20		94	113
0925	15 tiers, 154 to 165 seats		60	.267		73.50	16.10		89.60	106
0975	15 tiers, 214 to 225 seats		60	.267		66	16.10		82.10	97.50
1050	15 tiers, 274 to 285 seats		60	.267		68.50	16.10		84.60	100
1200	Seat backs only, 30" row, fiberglass		160	.100		25	6.05		31.05	37
1300	Steel and wood		160	.100		27	6.05		33.05	39
1400	NOTE: average seating is 1.5' in width									

13 34 19 – Metal Building Systems

13 34 19.50 Pre-Engineered Steel Buildings

13 34 19.50 Pre-Engineered Steel Buildings	Crew	Daily Output	Labor-Hours	Unit	Material	2021 Bare Costs Labor	2021 Bare Costs Equipment	Total	Total Incl O&P
0010	PRE-ENGINEERED STEEL BUILDINGS R133419-10									
0100	Clear span rigid frame, 26 ga. colored roofing and siding									
0150	20' to 29' wide, 10' eave height	E-2	425	.132	SF Flr.	8.25	7.85	4.03	20.13	25.50
0160	14' eave height		350	.160		10.05	9.55	4.89	24.49	31
0170	16' eave height		320	.175		9.45	10.45	5.35	25.25	32.50
0180	20' eave height		275	.204		10.35	12.15	6.25	28.75	37
0190	24' eave height		240	.233		12.70	13.90	7.15	33.75	43.50

13 34 19.50 Pre-Engineered Steel Buildings	Crew	Daily Output	Labor-Hours	Unit	Material	2021 Bare Costs Labor	Equipment	Total	Total Incl O&P
0200	30' to 49' wide, 10' eave height	E-2	535	.105	SF Flr.	6.65	6.25	3.20	16.10	20.50
0300	14' eave height		450	.124		7.15	7.40	3.81	18.36	23.50
0400	16' eave height		415	.135		7.65	8.05	4.13	19.83	25.50
0500	20' eave height		360	.156		8.30	9.30	4.76	22.36	28.50
0600	24' eave height		320	.175		9.15	10.45	5.35	24.95	32
0700	50' to 100' wide, 10' eave height		770	.073		5.60	4.34	2.22	12.16	15.25
0900	16' eave height		600	.093		6.50	5.55	2.86	14.91	18.85
1000	20' eave height		490	.114		7.05	6.80	3.50	17.35	22
1100	24' eave height		435	.129		7.70	7.70	3.94	19.34	24.50
1200	Clear span tapered beam frame, 26 ga. colored roofing/siding									
1300	30' to 39' wide, 10' eave height	E-2	535	.105	SF Flr.	7.50	6.25	3.20	16.95	21.50
1400	14' eave height		450	.124		8.30	7.40	3.81	19.51	24.50
1500	16' eave height		415	.135		8.70	8.05	4.13	20.88	26.50
1600	20' eave height		360	.156		9.60	9.30	4.76	23.66	30
1700	40' wide, 10' eave height		600	.093		6.65	5.55	2.86	15.06	19.05
1800	14' eave height		510	.110		7.40	6.55	3.36	17.31	22
1900	16' eave height		475	.118		7.75	7.05	3.61	18.41	23
2000	20' eave height		415	.135		8.50	8.05	4.13	20.68	26
2100	50' to 79' wide, 10' eave height		770	.073		6.25	4.34	2.22	12.81	15.95
2200	14' eave height		675	.083		6.75	4.95	2.54	14.24	17.85
2300	16' eave height		635	.088		7	5.25	2.70	14.95	18.70
2400	20' eave height		490	.114		8.55	6.80	3.50	18.85	23.50
2410	80' to 100' wide, 10' eave height		935	.060		5.55	3.57	1.83	10.95	13.55
2420	14' eave height		750	.075		6.10	4.45	2.28	12.83	16
2430	16' eave height		685	.082		6.35	4.88	2.50	13.73	17.20
2440	20' eave height		560	.100		6.80	5.95	3.06	15.81	20
2460	101' to 120' wide, 10' eave height		950	.059		5.10	3.52	1.80	10.42	13
2470	14' eave height		770	.073		5.65	4.34	2.22	12.21	15.30
2480	16' eave height		675	.083		6	4.95	2.54	13.49	17
2490	20' eave height		560	.100		6.40	5.95	3.06	15.41	19.55
2500	Single post 2-span frame, 26 ga. colored roofing and siding									
2600	80' wide, 14' eave height	E-2	740	.076	SF Flr.	5.65	4.51	2.31	12.47	15.65
2700	16' eave height		695	.081		6	4.81	2.46	13.27	16.65
2800	20' eave height		625	.090		6.50	5.35	2.74	14.59	18.35
2900	24' eave height		570	.098		7.10	5.85	3.01	15.96	20
3000	100' wide, 14' eave height		835	.067		5.45	4	2.05	11.50	14.40
3100	16' eave height		795	.070		5.05	4.20	2.15	11.40	14.35
3200	20' eave height		730	.077		6.20	4.58	2.35	13.13	16.40
3300	24' eave height		670	.084		6.85	4.98	2.56	14.39	17.95
3400	120' wide, 14' eave height		870	.064		8.80	3.84	1.97	14.61	17.70
3500	16' eave height		830	.067		5.65	4.02	2.06	11.73	14.65
3600	20' eave height		765	.073		6.70	4.37	2.24	13.31	16.55
3700	24' eave height		705	.079		7.35	4.74	2.43	14.52	18
3800	Double post 3-span frame, 26 ga. colored roofing and siding									
3900	150' wide, 14' eave height	E-2	925	.061	SF Flr.	4.46	3.61	1.85	9.92	12.50
4000	16' eave height		890	.063		4.65	3.75	1.92	10.32	12.95
4100	20' eave height		820	.068		5.10	4.07	2.09	11.26	14.15
4200	24' eave height		765	.073		5.55	4.37	2.24	12.16	15.30
4300	Triple post 4-span frame, 26 ga. colored roofing and siding									
4400	160' wide, 14' eave height	E-2	970	.058	SF Flr.	4.35	3.44	1.77	9.56	12
4500	16' eave height		930	.060		4.56	3.59	1.84	9.99	12.55
4600	20' eave height		870	.064		4.48	3.84	1.97	10.29	13
4700	24' eave height		815	.069		5.05	4.10	2.10	11.25	14.20

13 34 19 – Metal Building Systems

13 34 19.50 Pre-Engineered Steel Buildings	Crew	Daily Output	Labor-Hours	Unit	Material	2021 Bare Costs Labor	Equipment	Total	Total Incl O&P
4800	200' wide, 14' eave height	E-2	1030	.054	SF Flr.	4	3.24	1.66	8.90	11.20
4900	16' eave height		995	.056		4.14	3.36	1.72	9.22	11.60
5000	20' eave height		935	.060		4.57	3.57	1.83	9.97	12.50
5100	24' eave height	↓	885	.063	↓	5.10	3.77	1.94	10.81	13.55
7635	Insulation installation, over the purlin, second layer, up to 4" thick, add						90%			
7640	Insulation installation, between the purlins, up to 4" thick, add						100%			

13 34 23 – Fabricated Structures

13 34 23.15 Domes

		Crew	Daily Output	Labor-Hours	Unit	Material	Labor	Equipment	Total	Total Incl O&P
0010	**DOMES**									
1500	Domes, bulk storage, shell only, dual radius hemisphere, arch, steel									
1600	framing, corrugated steel covering, 150' diameter	E-2	550	.102	SF Flr.	33.50	6.05	3.11	42.66	49.50
1700	400' diameter	"	720	.078		27.50	4.64	2.38	34.52	39.50
1800	Wood framing, wood decking, to 400' diameter	F-4	400	.120	↓	28	6.65	2.48	37.13	43.50
1900	Radial framed wood (2" x 6"), 1/2" thick									
2000	plywood, asphalt shingles, 50' diameter	F-3	2000	.020	SF Flr.	67.50	1.12	.24	68.86	76
2100	60' diameter		1900	.021		63	1.18	.25	64.43	71
2200	72' diameter		1800	.022		52	1.25	.26	53.51	59
2300	116' diameter		1730	.023		33	1.30	.28	34.58	38
2400	150' diameter	↓	1500	.027	↓	35	1.49	.32	36.81	41

13 34 23.16 Fabricated Control Booths

		Crew	Daily Output	Labor-Hours	Unit	Material	Labor	Equipment	Total	Total Incl O&P
0010	**FABRICATED CONTROL BOOTHS**									
0100	Guard House, prefab conc. w/bullet resistant doors & windows, roof & wiring									
0110	8' x 8', Level III	L-10	1	24	Ea.	59,500	1,475	475	61,450	68,500
0120	8' x 8', Level IV	"	1	24	"	77,000	1,475	475	78,950	87,500

13 34 23.25 Garage Costs

		Crew	Daily Output	Labor-Hours	Unit	Material	Labor	Equipment	Total	Total Incl O&P
0010	**GARAGE COSTS**									
0300	Residential, wood, 12' x 20', one car prefab shell, stock, economy	2 Carp	1	16	Total	6,425	875		7,300	8,375
0350	Custom		.67	23.881		7,225	1,300		8,525	9,900
0400	Two car, 24' x 20', economy		.67	23.881		11,900	1,300		13,200	15,000
0450	Custom	↓	.50	32	↓	13,800	1,750		15,550	17,800

13 34 23.30 Garden House

		Crew	Daily Output	Labor-Hours	Unit	Material	Labor	Equipment	Total	Total Incl O&P
0010	**GARDEN HOUSE** Prefab wood, no floors or foundations									
0100	6' x 6'	2 Carp	200	.080	SF Flr.	50.50	4.38		54.88	62
0300	8' x 12'	"	48	.333	"	32	18.25		50.25	62

13 34 23.45 Kiosks

		Crew	Daily Output	Labor-Hours	Unit	Material	Labor	Equipment	Total	Total Incl O&P
0010	**KIOSKS**									
0020	Round, advertising type, 5' diameter, 7' high, aluminum wall, illuminated				Ea.	22,600			22,600	24,800
0100	Aluminum wall, non-illuminated					21,800			21,800	24,000
0500	Rectangular, 5' x 9', 7'-6" high, aluminum wall, illuminated					24,900			24,900	27,300
0600	Aluminum wall, non-illuminated				↓	25,500			25,500	28,100

13 34 23.60 Portable Booths

		Crew	Daily Output	Labor-Hours	Unit	Material	Labor	Equipment	Total	Total Incl O&P
0010	**PORTABLE BOOTHS** Prefab. aluminum with doors, windows, ext. roof									
0100	lights wiring & insulation, 15 S.F. building, OD, painted				S.F.	283			283	310
0300	30 S.F. building					265			265	292
0400	50 S.F. building					194			194	213
0600	80 S.F. building					165			165	181
0700	100 S.F. building				↓	151			151	166
0900	Acoustical booth, 27 Db @ 1,000 Hz, 15 S.F. floor				Ea.	4,375			4,375	4,800
1000	7' x 7'-6", including light & ventilation					9,000			9,000	9,900
1200	Ticket booth, galv. steel, not incl. foundations., 4' x 4'					4,025			4,025	4,425
1300	4' x 6'				↓	6,700			6,700	7,375

13 34 Fabricated Engineered Structures

13 34 23 – Fabricated Structures

13 34 23.70 Shelters

		Crew	Daily Output	Labor-Hours	Unit	Material	2021 Bare Costs Labor	2021 Bare Costs Equipment	Total	Total Incl O&P
0010	**SHELTERS**									
0020	Aluminum frame, acrylic glazing, 3' x 9' x 8' high	2 Sswk	1.14	14.035	Ea.	3,700	845		4,545	5,375
0100	9' x 12' x 8' high	"	.73	21.918	"	7,300	1,325		8,625	10,100

13 34 43 – Aircraft Hangars

13 34 43.50 Hangars

		Crew	Daily Output	Labor-Hours	Unit	Material	2021 Bare Costs Labor	2021 Bare Costs Equipment	Total	Total Incl O&P
0010	**HANGARS** Prefabricated steel T hangars, galv. steel roof &									
0100	walls, incl. electric bi-folding doors									
0110	not including floors or foundations, 4 unit	E-2	1275	.044	SF Flr.	14.25	2.62	1.34	18.21	21
0130	8 unit	"	1063	.053	"	12.40	3.14	1.61	17.15	20
1200	Alternate pricing method:									
1300	Galv. roof and walls, electric bi-folding doors, 4 plane	E-2	1.06	52.830	Plane	17,800	3,150	1,625	22,575	26,100
1500	8 plane		.91	61.538		13,200	3,675	1,875	18,750	22,200
1600	With bottom rolling doors, 4 plane		1.25	44.800		19,200	2,675	1,375	23,250	26,700
1800	8 plane		.97	57.732		13,900	3,450	1,775	19,125	22,400

13 34 53 – Agricultural Structures

13 34 53.50 Silos

		Crew	Daily Output	Labor-Hours	Unit	Material	2021 Bare Costs Labor	2021 Bare Costs Equipment	Total	Total Incl O&P
0010	**SILOS**									
0500	Steel, factory fab., 30,000 gallon cap., painted, economy	L-5	1	56	Ea.	24,800	3,400	585	28,785	33,200
0700	Deluxe	"	.50	112	"	39,400	6,800	1,175	47,375	55,000

13 47 Facility Protection

13 47 13 – Cathodic Protection

13 47 13.16 Cathodic Prot. for Underground Storage Tanks

		Crew	Daily Output	Labor-Hours	Unit	Material	2021 Bare Costs Labor	2021 Bare Costs Equipment	Total	Total Incl O&P
0010	**CATHODIC PROTECTION FOR UNDERGROUND STORAGE TANKS**									
1000	Anodes, magnesium type, 9 #	R-15	18.50	2.595	Ea.	39.50	162	15.25	216.75	300
1010	17 #		13	3.692		81.50	230	21.50	333	460
1020	32 #		10	4.800		120	300	28	448	610
1030	48 #		7.20	6.667		154	415	39	608	830
1100	Graphite type w/epoxy cap, 3" x 60" (32 #)	R-22	8.40	4.438		142	259		401	540
1110	4" x 80" (68 #)		6	6.213		245	360		605	810
1120	6" x 72" (80 #)		5.20	7.169		1,500	420		1,920	2,275
1130	6" x 36" (45 #)		9.60	3.883		750	226		976	1,150
2000	Rectifiers, silicon type, air cooled, 28 V/10 A	R-19	3.50	5.714		2,350	365		2,715	3,150
2010	20 V/20 A		3.50	5.714		2,375	365		2,740	3,150
2100	Oil immersed, 28 V/10 A		3	6.667		2,400	425		2,825	3,275
2110	20 V/20 A		3	6.667		3,175	425		3,600	4,125
3000	Anode backfill, coke breeze	R-22	3850	.010	Lb.	.27	.56		.83	1.14
4000	Cable, HMWPE, No. 8		2.40	15.533	M.L.F.	435	905		1,340	1,825
4010	No. 6		2.40	15.533		625	905		1,530	2,050
4020	No. 4		2.40	15.533		950	905		1,855	2,400
4030	No. 2		2.40	15.533		1,700	905		2,605	3,225
4040	No. 1		2.20	16.945		2,000	990		2,990	3,675
4050	No. 1/0		2.20	16.945		2,925	990		3,915	4,675
4060	No. 2/0		2.20	16.945		4,800	990		5,790	6,750
4070	No. 4/0		2	18.640		8,500	1,075		9,575	11,000
5000	Test station, 7 terminal box, flush curb type w/lockable cover	R-19	12	1.667	Ea.	79.50	106		185.50	246
5010	Reference cell, 2" diam. PVC conduit, cplg., plug, set flush	"	4.80	4.167	"	158	266		424	570

For customer support on your Site Work & Landscape Costs with RSMeans Data, call 800.448.8182.

211

13 53 Meteorological Instrumentation

13 53 09 - Weather Instrumentation

13 53 09.50 Weather Station	Crew	Daily Output	Labor-Hours	Unit	Material	2021 Bare Costs Labor	Equipment	Total	Total Incl O&P
0010 **WEATHER STATION**									
0020 Remote recording, solar powered, with rain gauge & display, 400' range				Ea.	890			890	980
0100 1 mile range				"	1,700			1,700	1,875

For customer support on your Site Work & Landscape Costs with RSMeans Data, call 800.448.8182.

Estimating Tips
22 10 00 Plumbing Piping and Pumps

This subdivision is primarily basic pipe and related materials. The pipe may be used by any of the mechanical disciplines, i.e., plumbing, fire protection, heating, and air conditioning.

Note: CPVC plastic piping approved for fire protection is located in 21 11 13.

- The labor adjustment factors listed in Subdivision 22 01 02.20 apply throughout Divisions 21, 22, and 23. CAUTION: the correct percentage may vary for the same items. For example, the percentage add for the basic pipe installation should be based on the maximum height that the installer must install for that particular section. If the pipe is to be located 14' above the floor but it is suspended on threaded rod from beams, the bottom flange of which is 18' high (4' rods), then the height is actually 18' and the add is 20%. The pipe cover, however, does not have to go above the 14' and so the add should be 10%.

- Most pipe is priced first as straight pipe with a joint (coupling, weld, etc.) every 10' and a hanger usually every 10'. There are exceptions with hanger spacing such as for cast iron pipe (5')

and plastic pipe (3 per 10'). Following each type of pipe there are several lines listing sizes and the amount to be subtracted to delete couplings and hangers. This is for pipe that is to be buried or supported together on trapeze hangers. The reason that the couplings are deleted is that these runs are usually long, and frequently longer lengths of pipe are used. By deleting the couplings, the estimator is expected to look up and add back the correct reduced number of couplings.

- When preparing an estimate, it may be necessary to approximate the fittings. Fittings usually run between 25% and 50% of the cost of the pipe. The lower percentage is for simpler runs, and the higher number is for complex areas, such as mechanical rooms.

- For historic restoration projects, the systems must be as invisible as possible, and pathways must be sought for pipes, conduit, and ductwork. While installations in accessible spaces (such as basements and attics) are relatively straightforward to estimate, labor costs may be more difficult to determine when delivery systems must be concealed.

22 40 00 Plumbing Fixtures

- Plumbing fixture costs usually require two lines: the fixture itself and its "rough-in, supply, and waste."

- In the Assemblies Section (Plumbing D2010) for the desired fixture, the System Components Group at the center of the page shows the fixture on the first line. The rest of the list (fittings, pipe, tubing, etc.) will total up to what we refer to in the Unit Price section as "Rough-in, supply, waste, and vent." Note that for most fixtures we allow a nominal 5' of tubing to reach from the fixture to a main or riser.

- Remember that gas- and oil-fired units need venting.

Reference Numbers

Reference numbers are shown at the beginning of some major classifications. These numbers refer to related items in the Reference Section. The reference information may be an estimating procedure, an alternate pricing method, or technical information.

Note: Not all subdivisions listed here necessarily appear. ■

Same Data. Simplified.

Enjoy the convenience and efficiency of accessing your costs anywhere:

- **Skip the multiplier** by setting your location
- **Quickly search,** edit, favorite and share costs
- **Stay on top of price changes** with automatic updates

Discover more at rsmeans.com/online

Note: "Powered in part by CINX™, based on licensed proprietary information of Harrison Publishing House, Inc."

Note: Trade Service, in part, has been used as a reference source for some of the material prices used in Division 22.

22 01 Operation and Maintenance of Plumbing

22 01 02 – Labor Adjustments

22 01 02.20 Labor Adjustment Factors	Crew	Daily Output	Labor-Hours	Unit	Material	2021 Bare Costs Labor	Equipment	Total	Total Incl O&P
0010 **LABOR ADJUSTMENT FACTORS** (For Div. 21, 22 and 23) R220102-20									
0100 Labor factors: The below are reasonable suggestions, but									
0110 each project must be evaluated for its own peculiarities, and									
0120 the adjustments be increased or decreased depending on the									
0130 severity of the special conditions.									
1000 Add to labor for elevated installation (Above floor level)									
1080 10' to 14.5' high						10%			
1100 15' to 19.5' high						20%			
1120 20' to 24.5' high						25%			
1140 25' to 29.5' high						35%			
1160 30' to 34.5' high						40%			
1180 35' to 39.5' high						50%			
1200 40' and higher						55%			
2000 Add to labor for crawl space									
2100 3' high						40%			
2140 4' high						30%			
3000 Add to labor for multi-story building									
3010 For new construction (No elevator available)									
3100 Add for floors 3 thru 10						5%			
3110 Add for floors 11 thru 15						10%			
3120 Add for floors 16 thru 20						15%			
3130 Add for floors 21 thru 30						20%			
3140 Add for floors 31 and up						30%			
3170 For existing structure (Elevator available)									
3180 Add for work on floor 3 and above						2%			
4000 Add to labor for working in existing occupied buildings									
4100 Hospital						35%			
4140 Office building						25%			
4180 School						20%			
4220 Factory or warehouse						15%			
4260 Multi dwelling						15%			
5000 Add to labor, miscellaneous									
5100 Cramped shaft						35%			
5140 Congested area						15%			
5180 Excessive heat or cold						30%			
9000 Labor factors: The above are reasonable suggestions, but									
9010 each project should be evaluated for its own peculiarities.									
9100 Other factors to be considered are:									
9140 Movement of material and equipment through finished areas									
9180 Equipment room									
9220 Attic space									
9260 No service road									
9300 Poor unloading/storage area									
9340 Congested site area/heavy traffic									

22 05 Common Work Results for Plumbing

22 05 05 – Selective Demolition for Plumbing

22 05 05.10 Plumbing Demolition

		Crew	Daily Output	Labor-Hours	Unit	Material	2021 Bare Costs Labor	2021 Bare Costs Equipment	Total	Total Incl O&P
0010	**PLUMBING DEMOLITION**									
1020	Fixtures, including 10' piping									
1100	Bathtubs, cast iron	1 Plum	4	2	Ea.		135		135	202
1120	Fiberglass		6	1.333			90.50		90.50	135
1140	Steel		5	1.600			108		108	162
1200	Lavatory, wall hung		10	.800			54		54	81
1220	Counter top		8	1			67.50		67.50	101
1300	Sink, single compartment		8	1			67.50		67.50	101
1320	Double compartment		7	1.143			77.50		77.50	115
1400	Water closet, floor mounted		8	1			67.50		67.50	101
1420	Wall mounted		7	1.143			77.50		77.50	115
1500	Urinal, floor mounted		4	2			135		135	202
1520	Wall mounted		7	1.143			77.50		77.50	115
1600	Water fountains, free standing		8	1			67.50		67.50	101
1620	Wall or deck mounted		6	1.333	▼		90.50		90.50	135
2000	Piping, metal, up thru 1-1/2" diameter		200	.040	L.F.		2.71		2.71	4.04
2050	2" thru 3-1/2" diameter	▼	150	.053			3.61		3.61	5.40
2100	4" thru 6" diameter	2 Plum	100	.160			10.85		10.85	16.15
2150	8" thru 14" diameter	"	60	.267			18.05		18.05	27
2153	16" thru 20" diameter	Q-18	70	.343			22	1.54	23.54	34
2155	24" thru 26" diameter		55	.436			28	1.95	29.95	43.50
2156	30" thru 36" diameter	▼	40	.600			38.50	2.69	41.19	60
2160	Plastic pipe with fittings, up thru 1-1/2" diameter	1 Plum	250	.032			2.17		2.17	3.23
2162	2" thru 3" diameter	"	200	.040			2.71		2.71	4.04
2164	4" thru 6" diameter	Q-1	200	.080			4.87		4.87	7.25
2166	8" thru 14" diameter		150	.107			6.50		6.50	9.70
2168	16" diameter	▼	100	.160	▼		9.75		9.75	14.55
2212	Deduct for salvage, aluminum scrap				Ton	455			455	500
2214	Brass scrap					2,175			2,175	2,400
2216	Copper scrap					3,725			3,725	4,100
2218	Lead scrap					910			910	1,000
2220	Steel scrap				▼				204	224
2250	Water heater, 40 gal.	1 Plum	6	1.333	Ea.		90.50		90.50	135

22 05 23 – General-Duty Valves for Plumbing Piping

22 05 23.20 Valves, Bronze

		Crew	Daily Output	Labor-Hours	Unit	Material	2021 Bare Costs Labor	2021 Bare Costs Equipment	Total	Total Incl O&P
0010	**VALVES, BRONZE** R220523-80									
1020	Angle, 150 lb., rising stem, threaded									
1030	1/8"	1 Plum	24	.333	Ea.	172	22.50		194.50	224
1040	1/4"		24	.333		172	22.50		194.50	224
1050	3/8"		24	.333		171	22.50		193.50	222
1060	1/2"		22	.364		171	24.50		195.50	225
1070	3/4"		20	.400		235	27		262	300
1080	1"		19	.421		325	28.50		353.50	405
1100	1-1/2"	▼	13	.615	▼	550	41.50		591.50	665
1102	Soldered same price as threaded									
1110	2"	1 Plum	11	.727	Ea.	865	49		914	1,025
1750	Check, swing, class 150, regrinding disc, threaded									
1800	1/8"	1 Plum	24	.333	Ea.	79.50	22.50		102	121
1830	1/4"		24	.333		81.50	22.50		104	124
1840	3/8"		24	.333		87.50	22.50		110	130
1850	1/2"		24	.333		89.50	22.50		112	132
1860	3/4"	▼	20	.400	▼	124	27		151	177

22 05 23 – General-Duty Valves for Plumbing Piping

22 05 23.20 Valves, Bronze		Crew	Daily Output	Labor-Hours	Unit	Material	2021 Bare Costs Labor	2021 Bare Costs Equipment	Total	Total Incl O&P
1870	1″	1 Plum	19	.421	Ea.	176	28.50		204.50	237
1880	1-1/4″		15	.533		257	36		293	335
1890	1-1/2″		13	.615		294	41.50		335.50	385
1900	2″		11	.727		425	49		474	540
1910	2-1/2″	Q-1	15	1.067		985	65		1,050	1,175
2000	For 200 lb., add					5%	10%			
2040	For 300 lb., add					15%	15%			
2850	Gate, N.R.S., soldered, 125 psi									
2900	3/8″	1 Plum	24	.333	Ea.	76.50	22.50		99	118
2920	1/2″		24	.333		76.50	22.50		99	118
2940	3/4″		20	.400		84.50	27		111.50	134
2950	1″		19	.421		122	28.50		150.50	177
2960	1-1/4″		15	.533		186	36		222	259
2970	1-1/2″		13	.615		207	41.50		248.50	290
2980	2″		11	.727		293	49		342	395
2990	2-1/2″	Q-1	15	1.067		640	65		705	800
3000	3″	″	13	1.231		780	75		855	965
3850	Rising stem, soldered, 300 psi									
3950	1″	1 Plum	19	.421	Ea.	242	28.50		270.50	310
3980	2″	″	11	.727		650	49		699	790
4000	3″	Q-1	13	1.231		2,150	75		2,225	2,450
4250	Threaded, class 150									
4310	1/4″	1 Plum	24	.333	Ea.	97.50	22.50		120	141
4320	3/8″		24	.333		97.50	22.50		120	141
4330	1/2″		24	.333		88	22.50		110.50	131
4340	3/4″		20	.400		99.50	27		126.50	150
4350	1″		19	.421		133	28.50		161.50	190
4360	1-1/4″		15	.533		181	36		217	253
4370	1-1/2″		13	.615		228	41.50		269.50	315
4380	2″		11	.727		305	49		354	410
4390	2-1/2″	Q-1	15	1.067		715	65		780	880
4400	3″	″	13	1.231		995	75		1,070	1,200
4500	For 300 psi, threaded, add					100%	15%			
4540	For chain operated type, add					15%				
4850	Globe, class 150, rising stem, threaded									
4920	1/4″	1 Plum	24	.333	Ea.	131	22.50		153.50	179
4940	3/8″		24	.333		129	22.50		151.50	176
4950	1/2″		24	.333		129	22.50		151.50	176
4960	3/4″		20	.400		172	27		199	231
4970	1″		19	.421		271	28.50		299.50	340
4980	1-1/4″		15	.533		430	36		466	530
4990	1-1/2″		13	.615		525	41.50		566.50	635
5000	2″		11	.727		785	49		834	940
5010	2-1/2″	Q-1	15	1.067		1,575	65		1,640	1,825
5020	3″	″	13	1.231		2,250	75		2,325	2,575
5120	For 300 lb. threaded, add					50%	15%			
5600	Relief, pressure & temperature, self-closing, ASME, threaded									
5640	3/4″	1 Plum	28	.286	Ea.	267	19.35		286.35	320
5650	1″		24	.333		425	22.50		447.50	505
5660	1-1/4″		20	.400		855	27		882	985
5670	1-1/2″		18	.444		1,650	30		1,680	1,850
5680	2″		16	.500		1,800	34		1,834	2,025
5950	Pressure, poppet type, threaded									

22 05 Common Work Results for Plumbing

22 05 23 – General-Duty Valves for Plumbing Piping

22 05 23.20 Valves, Bronze		Crew	Daily Output	Labor-Hours	Unit	Material	2021 Bare Costs Labor	Equipment	Total	Total Incl O&P
6000	1/2"	1 Plum	30	.267	Ea.	95.50	18.05		113.55	132
6040	3/4"	"	28	.286	"	111	19.35		130.35	151
6400	Pressure, water, ASME, threaded									
6440	3/4"	1 Plum	28	.286	Ea.	58	19.35		77.35	93
6450	1"		24	.333		360	22.50		382.50	430
6460	1-1/4"		20	.400		565	27		592	660
6470	1-1/2"		18	.444		780	30		810	905
6480	2"		16	.500		1,125	34		1,159	1,300
6490	2-1/2"		15	.533		4,425	36		4,461	4,925
6900	Reducing, water pressure									
6920	300 psi to 25-75 psi, threaded or sweat									
6940	1/2"	1 Plum	24	.333	Ea.	560	22.50		582.50	650
6950	3/4"		20	.400		585	27		612	685
6960	1"		19	.421		905	28.50		933.50	1,050
6970	1-1/4"		15	.533		1,550	36		1,586	1,750
6980	1-1/2"		13	.615		2,325	41.50		2,366.50	2,625
8350	Tempering, water, sweat connections									
8400	1/2"	1 Plum	24	.333	Ea.	119	22.50		141.50	165
8440	3/4"	"	20	.400	"	175	27		202	234
8650	Threaded connections									
8700	1/2"	1 Plum	24	.333	Ea.	167	22.50		189.50	218
8740	3/4"		20	.400		1,050	27		1,077	1,200
8750	1"		19	.421		1,200	28.50		1,228.50	1,350
8760	1-1/4"		15	.533		1,825	36		1,861	2,075
8770	1-1/2"		13	.615		2,000	41.50		2,041.50	2,250
8780	2"		11	.727		3,000	49		3,049	3,375

22 05 23.60 Valves, Plastic		Crew	Daily Output	Labor-Hours	Unit	Material	2021 Bare Costs Labor	Equipment	Total	Total Incl O&P
0010	**VALVES, PLASTIC**									
1150	Ball, PVC, socket or threaded, true union									
1230	1/2"	1 Plum	26	.308	Ea.	38	21		59	73
1240	3/4"		25	.320		45	21.50		66.50	82
1250	1"		23	.348		54	23.50		77.50	94
1260	1-1/4"		21	.381		85	26		111	132
1270	1-1/2"		20	.400		85	27		112	134
1280	2"		17	.471		134	32		166	196
1290	2-1/2"	Q-1	26	.615		189	37.50		226.50	263
1300	3"		24	.667		272	40.50		312.50	360
1310	4"		20	.800		460	48.50		508.50	580
1360	For PVC, flanged, add						100%	15%		
3150	Ball check, PVC, socket or threaded									
3200	1/4"	1 Plum	26	.308	Ea.	47.50	21		68.50	83.50
3220	3/8"		26	.308		47.50	21		68.50	83.50
3240	1/2"		26	.308		46.50	21		67.50	82
3250	3/4"		25	.320		52	21.50		73.50	89.50
3260	1"		23	.348		65	23.50		88.50	107
3270	1-1/4"		21	.381		109	26		135	159
3280	1-1/2"		20	.400		109	27		136	161
3290	2"		17	.471		149	32		181	211
3310	3"	Q-1	24	.667		415	40.50		455.50	515
3320	4"	"	20	.800		585	48.50		633.50	715
3360	For PVC, flanged, add						50%	15%		

217

For customer support on your Site Work & Landscape Costs with RSMeans Data, call 800.448.8182.

22 05 76 – Facility Drainage Piping Cleanouts

22 05 76.10 Cleanouts

	22 05 76.10 Cleanouts	Crew	Daily Output	Labor-Hours	Unit	Material	2021 Bare Costs Labor	Equipment	Total	Total Incl O&P
0010	**CLEANOUTS**									
0060	Floor type									
0080	Round or square, scoriated nickel bronze top									
0100	2" pipe size	1 Plum	10	.800	Ea.	440	54		494	560
0120	3" pipe size		8	1		460	67.50		527.50	605
0140	4" pipe size	↓	6	1.333	↓	615	90.50		705.50	815

22 05 76.20 Cleanout Tees

	22 05 76.20 Cleanout Tees	Crew	Daily Output	Labor-Hours	Unit	Material	2021 Bare Costs Labor	Equipment	Total	Total Incl O&P
0010	**CLEANOUT TEES**									
0100	Cast iron, B&S, with countersunk plug									
0200	2" pipe size	1 Plum	4	2	Ea.	133	135		268	350
0220	3" pipe size		3.60	2.222		205	150		355	450
0240	4" pipe size	↓	3.30	2.424		310	164		474	585
0260	5" pipe size	Q-1	5.50	2.909		610	177		787	935
0280	6" pipe size	"	5	3.200		895	195		1,090	1,275
0300	8" pipe size	Q-3	5	6.400	↓	1,225	410		1,635	1,975
0500	For round smooth access cover, same price									
0600	For round scoriated access cover, same price									
0700	For square smooth access cover, add				Ea.	60%				
4000	Plastic, tees and adapters. Add plugs									
4010	ABS, DWV									
4020	Cleanout tee, 1-1/2" pipe size	1 Plum	15	.533	Ea.	17.30	36		53.30	73
4030	2" pipe size	Q-1	27	.593		24	36		60	80.50
4040	3" pipe size		21	.762		35.50	46.50		82	109
4050	4" pipe size	↓	16	1		77.50	61		138.50	176
4100	Cleanout plug, 1-1/2" pipe size	1 Plum	32	.250		3.59	16.95		20.54	29.50
4110	2" pipe size	Q-1	56	.286		3.93	17.40		21.33	30.50
4120	3" pipe size		36	.444		6.40	27		33.40	47.50
4130	4" pipe size	↓	30	.533		11.05	32.50		43.55	60.50
4180	Cleanout adapter fitting, 1-1/2" pipe size	1 Plum	32	.250		5.30	16.95		22.25	31.50
4190	2" pipe size	Q-1	56	.286		7.35	17.40		24.75	34
4200	3" pipe size		36	.444		18.55	27		45.55	61
4210	4" pipe size		30	.533	↓	35	32.50		67.50	87
5000	PVC, DWV									
5010	Cleanout tee, 1-1/2" pipe size	1 Plum	15	.533	Ea.	14.20	36		50.20	69.50
5020	2" pipe size	Q-1	27	.593		16.55	36		52.55	72
5030	3" pipe size		21	.762		32	46.50		78.50	105
5040	4" pipe size	↓	16	1		57	61		118	154
5090	Cleanout plug, 1-1/2" pipe size	1 Plum	32	.250		3.32	16.95		20.27	29
5100	2" pipe size	Q-1	56	.286		3.73	17.40		21.13	30
5110	3" pipe size		36	.444		6.65	27		33.65	48
5120	4" pipe size		30	.533		9.85	32.50		42.35	59.50
5130	6" pipe size	↓	24	.667		31.50	40.50		72	95
5170	Cleanout adapter fitting, 1-1/2" pipe size	1 Plum	32	.250		4.47	16.95		21.42	30.50
5180	2" pipe size	Q-1	56	.286		6	17.40		23.40	32.50
5190	3" pipe size		36	.444		16.10	27		43.10	58
5200	4" pipe size		30	.533		26.50	32.50		59	77.50
5210	6" pipe size	↓	24	.667	↓	90.50	40.50		131	160

22 11 Facility Water Distribution

22 11 13 – Facility Water Distribution Piping

22 11 13.23 Pipe/Tube, Copper

		Crew	Daily Output	Labor-Hours	Unit	Material	2021 Bare Costs Labor	Equipment	Total	Total Incl O&P
0010	**PIPE/TUBE, COPPER**, Solder joints									
1000	Type K tubing, couplings & clevis hanger assemblies 10' OC									
1100	1/4" diameter	1 Plum	84	.095	L.F.	5	6.45		11.45	15.10
1120	3/8" diameter		82	.098		5.05	6.60		11.65	15.40
1140	1/2" diameter		78	.103		5.70	6.95		12.65	16.60
1160	5/8" diameter		77	.104		4.02	7.05		11.07	14.90
1180	3/4" diameter		74	.108		9.45	7.30		16.75	21.50
1200	1" diameter		66	.121		12.20	8.20		20.40	25.50
1220	1-1/4" diameter		56	.143		15.30	9.65		24.95	31.50
1240	1-1/2" diameter		50	.160		19.15	10.85		30	37
1260	2" diameter		40	.200		29.50	13.55		43.05	52.50
1280	2-1/2" diameter	Q-1	60	.267		46	16.25		62.25	75.50
1300	3" diameter	"	54	.296		63.50	18.05		81.55	96.50
1950	To delete cplgs. & hngrs., 1/4"-1" pipe, subtract					27%	60%			
1960	1-1/4"-3" pipe, subtract					14%	52%			

22 11 13.25 Pipe/Tube Fittings, Copper

		Crew	Daily Output	Labor-Hours	Unit	Material	2021 Bare Costs Labor	Equipment	Total	Total Incl O&P
0010	**PIPE/TUBE FITTINGS, COPPER**, Wrought unless otherwise noted									
0040	Solder joints, copper x copper									
0070	90° elbow, 1/4"	1 Plum	22	.364	Ea.	5.05	24.50		29.55	42.50
0090	3/8"		22	.364		4.89	24.50		29.39	42.50
0100	1/2"		20	.400		1.77	27		28.77	42.50
0110	5/8"		19	.421		3.39	28.50		31.89	46
0120	3/4"		19	.421		3.89	28.50		32.39	47
0130	1"		16	.500		9.55	34		43.55	61
0140	1-1/4"		15	.533		15.80	36		51.80	71.50
0150	1-1/2"		13	.615		22	41.50		63.50	86.50
0160	2"		11	.727		40	49		89	118
0180	3"	Q-1	11	1.455		104	88.50		192.50	247
0200	4"	"	9	1.778		228	108		336	415
0220	6"	Q-2	9	2.667		1,200	168		1,368	1,575
0250	45° elbow, 1/4"	1 Plum	22	.364		9	24.50		33.50	47
0270	3/8"		22	.364		7.70	24.50		32.20	45.50
0280	1/2"		20	.400		3.18	27		30.18	44
0290	5/8"		19	.421		13.50	28.50		42	57.50
0300	3/4"		19	.421		5.40	28.50		33.90	48.50
0310	1"		16	.500		13.55	34		47.55	65.50
0320	1-1/4"		15	.533		18.20	36		54.20	74
0340	2"		11	.727		37.50	49		86.50	115
0450	Tee, 1/4"		14	.571		10.10	38.50		48.60	68.50
0470	3/8"		14	.571		8.10	38.50		46.60	66.50
0480	1/2"		13	.615		2.96	41.50		44.46	65.50
0490	5/8"		12	.667		18.50	45		63.50	88
0500	3/4"		12	.667		7.15	45		52.15	75.50
0510	1"		10	.800		11.50	54		65.50	93.50
0520	1-1/4"		9	.889		29.50	60		89.50	123
0530	1-1/2"		8	1		44.50	67.50		112	150
0540	2"		7	1.143		71	77.50		148.50	193
0560	3"	Q-1	7	2.286		188	139		327	415
0580	4"	"	5	3.200		440	195		635	775
0600	6"	Q-2	6	4		1,850	253		2,103	2,400
0612	Tee, reducing on the outlet, 1/4"	1 Plum	15	.533		27.50	36		63.50	84.50
0613	3/8"		15	.533		23.50	36		59.50	80

22 11 13 – Facility Water Distribution Piping

22 11 13.25 Pipe/Tube Fittings, Copper

		Crew	Daily Output	Labor-Hours	Unit	Material	2021 Bare Costs Labor	Equipment	Total	Total Incl O&P
0614	1/2"	1 Plum	14	.571	Ea.	21.50	38.50		60	81.50
0615	5/8"		13	.615		43.50	41.50		85	110
0616	3/4"		12	.667		9.95	45		54.95	78.50
0617	1"		11	.727		47.50	49		96.50	126
0618	1-1/4"		10	.800		46.50	54		100.50	133
0619	1-1/2"		9	.889		49.50	60		109.50	145
0620	2"		8	1		81	67.50		148.50	190
0621	2-1/2"	Q-1	9	1.778		223	108		331	405
0622	3"		8	2		253	122		375	460
0623	4"		6	2.667		455	162		617	740
0624	5"		5	3.200		1,975	195		2,170	2,475
0625	6"	Q-2	7	3.429		3,475	217		3,692	4,125
0626	8"	"	6	4		13,400	253		13,653	15,100
0630	Tee, reducing on the run, 1/4"	1 Plum	15	.533		31	36		67	88.50
0631	3/8"		15	.533		35.50	36		71.50	93
0632	1/2"		14	.571		30	38.50		68.50	90.50
0633	5/8"		13	.615		12.25	41.50		53.75	75.50
0634	3/4"		12	.667		24	45		69	94
0635	1"		11	.727		38	49		87	116
0636	1-1/4"		10	.800		61.50	54		115.50	149
0637	1-1/2"		9	.889		106	60		166	206
0638	2"		8	1		135	67.50		202.50	250
0639	2-1/2"	Q-1	9	1.778		288	108		396	475
0640	3"		8	2		400	122		522	620
0641	4"		6	2.667		860	162		1,022	1,175
0642	5"		5	3.200		2,400	195		2,595	2,950
0643	6"	Q-2	7	3.429		3,650	217		3,867	4,350
0644	8"	"	6	4		12,600	253		12,853	14,200
0650	Coupling, 1/4"	1 Plum	24	.333		1.78	22.50		24.28	35.50
0670	3/8"		24	.333		2.31	22.50		24.81	36
0680	1/2"		22	.364		1.92	24.50		26.42	39
0690	5/8"		21	.381		5.05	26		31.05	44
0700	3/4"		21	.381		3.85	26		29.85	42.50
0710	1"		18	.444		7.65	30		37.65	53.50
0715	1-1/4"		17	.471		13.40	32		45.40	62.50
0716	1-1/2"		15	.533		17.70	36		53.70	73.50
0718	2"		13	.615		29.50	41.50		71	94.50

22 11 13.44 Pipe, Steel

		Crew	Daily Output	Labor-Hours	Unit	Material	2021 Bare Costs Labor	Equipment	Total	Total Incl O&P
0010	**PIPE, STEEL** R221113-50									
0020	All pipe sizes are to Spec. A-53 unless noted otherwise									
0050	Schedule 40, threaded, with couplings, and clevis hanger									
0060	assemblies sized for covering, 10' OC									
0540	Black, 1/4" diameter	1 Plum	66	.121	L.F.	7.85	8.20		16.05	21
0550	3/8" diameter		65	.123		10.25	8.35		18.60	24
0560	1/2" diameter		63	.127		2.69	8.60		11.29	15.80
0570	3/4" diameter		61	.131		3.27	8.90		12.17	16.85
0580	1" diameter		53	.151		3.73	10.20		13.93	19.35
0590	1-1/4" diameter	Q-1	89	.180		4.82	10.95		15.77	21.50
0600	1-1/2" diameter		80	.200		5.65	12.20		17.85	24.50
0610	2" diameter		64	.250		7.60	15.25		22.85	31
0620	2-1/2" diameter		50	.320		14.30	19.50		33.80	44.50
0630	3" diameter		43	.372		18.65	22.50		41.15	54.50

22 11 13 - Facility Water Distribution Piping

22 11 13.44 Pipe, Steel

		Crew	Daily Output	Labor-Hours	Unit	Material	2021 Bare Costs Labor	Equipment	Total	Total Incl O&P
0640	3-1/2" diameter	Q-1	40	.400	L.F.	28	24.50		52.50	67
0650	4" diameter	↓	36	.444	↓	28	27		55	71
1220	To delete coupling & hanger, subtract									
1230	1/4" diam. to 3/4" diam.					31%	56%			
1240	1" diam. to 1-1/2" diam.					23%	51%			
1250	2" diam. to 4" diam.					23%	41%			
1280	All pipe sizes are to Spec. A-53 unless noted otherwise									
1281	Schedule 40, threaded, with couplings and clevis hanger									
1282	assemblies sized for covering, 10' OC									
1290	Galvanized, 1/4" diameter	1 Plum	66	.121	L.F.	10.85	8.20		19.05	24
1300	3/8" diameter		65	.123		14.30	8.35		22.65	28
1310	1/2" diameter		63	.127		3.07	8.60		11.67	16.20
1320	3/4" diameter		61	.131		3.76	8.90		12.66	17.40
1330	1" diameter	↓	53	.151		4.45	10.20		14.65	20
1340	1-1/4" diameter	Q-1	89	.180		5.80	10.95		16.75	22.50
1350	1-1/2" diameter		80	.200		6.80	12.20		19	25.50
1360	2" diameter		64	.250		9.20	15.25		24.45	32.50
1370	2-1/2" diameter		50	.320		16.10	19.50		35.60	46.50
1380	3" diameter		43	.372		21	22.50		43.50	57
1390	3-1/2" diameter		40	.400		32.50	24.50		57	72.50
1400	4" diameter	↓	36	.444	↓	32.50	27		59.50	76.50
1750	To delete coupling & hanger, subtract									
1760	1/4" diam. to 3/4" diam.					31%	56%			
1770	1" diam. to 1-1/2" diam.					23%	51%			
1780	2" diam. to 4" diam.					23%	41%			

22 11 13.45 Pipe Fittings, Steel, Threaded

		Crew	Daily Output	Labor-Hours	Unit	Material	2021 Bare Costs Labor	Equipment	Total	Total Incl O&P
0010	**PIPE FITTINGS, STEEL, THREADED**									
0020	Cast iron									
1300	Extra heavy weight, black									
1310	Couplings, steel straight									
1320	1/4"	1 Plum	19	.421	Ea.	5.70	28.50		34.20	49
1330	3/8"		19	.421		6.20	28.50		34.70	49.50
1340	1/2"		19	.421		8.40	28.50		36.90	51.50
1350	3/4"		18	.444		8.95	30		38.95	55
1360	1"	↓	15	.533		11.40	36		47.40	66.50
1370	1-1/4"	Q-1	26	.615		18.30	37.50		55.80	76
1380	1-1/2"		24	.667		18.30	40.50		58.80	80.50
1390	2"		21	.762		28	46.50		74.50	100
1400	2-1/2"		18	.889		41.50	54		95.50	127
1410	3"		14	1.143		49	69.50		118.50	158
1420	3-1/2"		12	1.333		66	81		147	194
1430	4"	↓	10	1.600	↓	77.50	97.50		175	231
1510	90° elbow, straight									
1520	1/2"	1 Plum	15	.533	Ea.	43.50	36		79.50	102
1530	3/4"		14	.571		44	38.50		82.50	106
1540	1"	↓	13	.615		53.50	41.50		95	121
1550	1-1/4"	Q-1	22	.727		79.50	44.50		124	154
1560	1-1/2"		20	.800		98.50	48.50		147	181
1580	2"		18	.889		122	54		176	215
1590	2-1/2"		14	1.143		282	69.50		351.50	415
1600	3"		10	1.600		395	97.50		492.50	580
1610	4"	↓	6	2.667	↓	835	162		997	1,150

22 11 13.45 Pipe Fittings, Steel, Threaded		Crew	Daily Output	Labor-Hours	Unit	Material	2021 Bare Costs Labor	Equipment	Total	Total Incl O&P
1650	45° elbow, straight									
1660	1/2"	1 Plum	15	.533	Ea.	59	36		95	119
1670	3/4"		14	.571		59.50	38.50		98	123
1680	1"		13	.615		71.50	41.50		113	141
1690	1-1/4"	Q-1	22	.727		118	44.50		162.50	196
1700	1-1/2"		20	.800		130	48.50		178.50	216
1710	2"		18	.889		185	54		239	284
1720	2-1/2"		14	1.143		305	69.50		374.50	440
1800	Tee, straight									
1810	1/2"	1 Plum	9	.889	Ea.	67.50	60		127.50	165
1820	3/4"		9	.889		67.50	60		127.50	165
1830	1"		8	1		81.50	67.50		149	191
1840	1-1/4"	Q-1	14	1.143		122	69.50		191.50	238
1850	1-1/2"		13	1.231		157	75		232	284
1860	2"		11	1.455		195	88.50		283.50	345
1870	2-1/2"		9	1.778		415	108		523	615
1880	3"		6	2.667		580	162		742	880
1890	4"		4	4		1,125	244		1,369	1,625
4000	Standard weight, black									
4010	Couplings, steel straight, merchants									
4030	1/4"	1 Plum	19	.421	Ea.	1.53	28.50		30.03	44
4040	3/8"		19	.421		1.83	28.50		30.33	44.50
4050	1/2"		19	.421		1.95	28.50		30.45	44.50
4060	3/4"		18	.444		2.49	30		32.49	47.50
4070	1"		15	.533		3.49	36		39.49	58
4080	1-1/4"	Q-1	26	.615		4.45	37.50		41.95	61
4090	1-1/2"		24	.667		5.65	40.50		46.15	66.50
4100	2"		21	.762		8.05	46.50		54.55	78.50
4110	2-1/2"		18	.889		26.50	54		80.50	111
4120	3"		14	1.143		37.50	69.50		107	145
4130	3-1/2"		12	1.333		66.50	81		147.50	194
4140	4"		10	1.600		66.50	97.50		164	218
4200	Standard weight, galvanized									
4210	Couplings, steel straight, merchants									
4230	1/4"	1 Plum	19	.421	Ea.	1.76	28.50		30.26	44.50
4240	3/8"		19	.421		2.25	28.50		30.75	45
4250	1/2"		19	.421		2.39	28.50		30.89	45
4260	3/4"		18	.444		3	30		33	48.50
4270	1"		15	.533		4.19	36		40.19	58.50
4280	1-1/4"	Q-1	26	.615		5.35	37.50		42.85	62
4290	1-1/2"		24	.667		6.65	40.50		47.15	68
4300	2"		21	.762		9.95	46.50		56.45	80.50
4310	2-1/2"		18	.889		25	54		79	108
4320	3"		14	1.143		35.50	69.50		105	143
4330	3-1/2"		12	1.333		76.50	81		157.50	205
4340	4"		10	1.600		76.50	97.50		174	229
5000	Malleable iron, 150 lb.									
5020	Black									
5040	90° elbow, straight									
5060	1/4"	1 Plum	16	.500	Ea.	6.25	34		40.25	57.50
5070	3/8"		16	.500		6.25	34		40.25	57.50
5080	1/2"		15	.533		4.32	36		40.32	59
5090	3/4"		14	.571		5.20	38.50		43.70	63.50

222

For customer support on your Site Work & Landscape Costs with RSMeans Data, call 800.448.8182.

22 11 13 – Facility Water Distribution Piping

22 11 13.45 Pipe Fittings, Steel, Threaded		Crew	Daily Output	Labor-Hours	Unit	Material	2021 Bare Costs Labor	Equipment	Total	Total Incl O&P
5100	1"	1 Plum	13	.615	Ea.	9.05	41.50		50.55	72
5110	1-1/4"	Q-1	22	.727		14.95	44.50		59.45	82.50
5120	1-1/2"		20	.800		19.65	48.50		68.15	94
5130	2"		18	.889		34	54		88	119
5140	2-1/2"		14	1.143		75.50	69.50		145	187
5150	3"		10	1.600		110	97.50		207.50	267
5160	3-1/2"		8	2		320	122		442	530
5170	4"		6	2.667		211	162		373	475
5250	45° elbow, straight									
5270	1/4"	1 Plum	16	.500	Ea.	9.40	34		43.40	61
5280	3/8"		16	.500		9.40	34		43.40	61
5290	1/2"		15	.533		7.15	36		43.15	62
5300	3/4"		14	.571		8.85	38.50		47.35	67
5310	1"		13	.615		11.10	41.50		52.60	74.50
5320	1-1/4"	Q-1	22	.727		19.65	44.50		64.15	87.50
5330	1-1/2"		20	.800		24.50	48.50		73	99.50
5340	2"		18	.889		37	54		91	122
5350	2-1/2"		14	1.143		106	69.50		175.50	221
5360	3"		10	1.600		138	97.50		235.50	297
5370	3-1/2"		8	2		244	122		366	450
5380	4"		6	2.667		271	162		433	540
5450	Tee, straight									
5470	1/4"	1 Plum	10	.800	Ea.	9.05	54		63.05	91
5480	3/8"		10	.800		9.05	54		63.05	91
5490	1/2"		9	.889		5.80	60		65.80	96.50
5500	3/4"		9	.889		8.30	60		68.30	99
5510	1"		8	1		14.20	67.50		81.70	117
5520	1-1/4"	Q-1	14	1.143		23	69.50		92.50	130
5530	1-1/2"		13	1.231		28.50	75		103.50	144
5540	2"		11	1.455		49	88.50		137.50	186
5550	2-1/2"		9	1.778		105	108		213	278
5560	3"		6	2.667		155	162		317	410
5570	3-1/2"		5	3.200		375	195		570	705
5580	4"		4	4		375	244		619	775
5650	Coupling									
5670	1/4"	1 Plum	19	.421	Ea.	7.75	28.50		36.25	51
5680	3/8"		19	.421		7.75	28.50		36.25	51
5690	1/2"		19	.421		6	28.50		34.50	49
5700	3/4"		18	.444		7	30		37	52.50
5710	1"		15	.533		10.50	36		46.50	65.50
5720	1-1/4"	Q-1	26	.615		12.05	37.50		49.55	69.50
5730	1-1/2"		24	.667		16.30	40.50		56.80	78.50
5740	2"		21	.762		24	46.50		70.50	96
5750	2-1/2"		18	.889		66.50	54		120.50	154
5760	3"		14	1.143		90	69.50		159.50	203
5770	3-1/2"		12	1.333		171	81		252	310
5780	4"		10	1.600		181	97.50		278.50	345
6000	For galvanized elbows, tees, and couplings, add					20%				

22 11 13.47 Pipe Fittings, Steel

		Crew	Daily Output	Labor-Hours	Unit	Material	2021 Bare Costs Labor	Equipment	Total	Total Incl O&P
0010	**PIPE FITTINGS, STEEL**, flanged, welded & special									
0620	Gasket and bolt set, 150 lb., 1/2" pipe size	1 Plum	20	.400	Ea.	3.57	27		30.57	44.50
0622	3/4" pipe size		19	.421		3.73	28.50		32.23	46.50

For customer support on your Site Work & Landscape Costs with RSMeans Data, call 800.448.8182.

223

22 11 13 – Facility Water Distribution Piping

22 11 13.47 Pipe Fittings, Steel

22 11 13.47 Pipe Fittings, Steel		Crew	Daily Output	Labor-Hours	Unit	Material	2021 Bare Costs Labor	Equipment	Total	Total Incl O&P
0624	1" pipe size	1 Plum	18	.444	Ea.	4.59	30		34.59	50
0626	1-1/4" pipe size		17	.471		4.53	32		36.53	52.50
0628	1-1/2" pipe size		15	.533		4.29	36		40.29	58.50
0630	2" pipe size		13	.615		9.05	41.50		50.55	72
0640	2-1/2" pipe size		12	.667		8.40	45		53.40	77
0650	3" pipe size		11	.727		9.70	49		58.70	84
0660	3-1/2" pipe size		9	.889		15.75	60		75.75	107
0670	4" pipe size		8	1		16.80	67.50		84.30	120
0680	5" pipe size		7	1.143		28.50	77.50		106	147
0690	6" pipe size		6	1.333		30	90.50		120.50	168
0700	8" pipe size		5	1.600		31	108		139	196
0710	10" pipe size		4.50	1.778		58.50	120		178.50	244
0720	12" pipe size		4.20	1.905		62.50	129		191.50	261
0730	14" pipe size		4	2		52.50	135		187.50	260
0740	16" pipe size		3	2.667		69.50	181		250.50	345
0750	18" pipe size		2.70	2.963		135	201		336	445
0760	20" pipe size		2.30	3.478		219	235		454	590
0780	24" pipe size	↓	1.90	4.211	↓	273	285		558	725

22 11 13.48 Pipe, Fittings and Valves, Steel, Grooved-Joint

22 11 13.48 Pipe, Fittings and Valves, Steel, Grooved-Joint		Crew	Daily Output	Labor-Hours	Unit	Material	2021 Bare Costs Labor	Equipment	Total	Total Incl O&P
0010	**PIPE, FITTINGS AND VALVES, STEEL, GROOVED-JOINT**									
0012	Fittings are ductile iron. Steel fittings noted.									
0020	Pipe includes coupling & clevis type hanger assemblies, 10' OC									
1000	Schedule 40, black									
1040	3/4" diameter	1 Plum	71	.113	L.F.	5.85	7.65		13.50	17.85
1050	1" diameter		63	.127		5.40	8.60		14	18.80
1060	1-1/4" diameter		58	.138		6.95	9.35		16.30	21.50
1070	1-1/2" diameter		51	.157		7.85	10.60		18.45	24.50
1080	2" diameter	↓	40	.200		9.30	13.55		22.85	30
1090	2-1/2" diameter	Q-1	57	.281		14.90	17.10		32	42
1100	3" diameter		50	.320		18.35	19.50		37.85	49
1110	4" diameter		45	.356		26	21.50		47.50	61.50
1120	5" diameter	↓	37	.432		31.50	26.50		58	74
1130	6" diameter	Q-2	42	.571		35	36		71	92.50
1140	8" diameter		37	.649		87.50	41		128.50	157
1150	10" diameter		31	.774		120	49		169	205
1160	12" diameter	↓	27	.889	↓	135	56		191	232
1740	To delete coupling & hanger, subtract									
1750	3/4" diam. to 2" diam.					65%	27%			
1760	2-1/2" diam. to 5" diam.					41%	18%			
1770	6" diam. to 12" diam.					31%	13%			
1780	14" diam. to 24" diam.					35%	10%			
1800	Galvanized									
1840	3/4" diameter	1 Plum	71	.113	L.F.	6.30	7.65		13.95	18.35
1850	1" diameter		63	.127		6.30	8.60		14.90	19.75
1860	1-1/4" diameter		58	.138		8.15	9.35		17.50	23
1870	1-1/2" diameter		51	.157		9.25	10.60		19.85	26
1880	2" diameter	↓	40	.200		11.40	13.55		24.95	32.50
1890	2-1/2" diameter	Q-1	57	.281		17.05	17.10		34.15	44.50
1900	3" diameter		50	.320		21	19.50		40.50	52.50
1910	4" diameter		45	.356		30	21.50		51.50	65.50
1920	5" diameter	↓	37	.432		28	26.50		54.50	70
1930	6" diameter	Q-2	42	.571	↓	36.50	36		72.50	94

22 11 Facility Water Distribution

22 11 13 – Facility Water Distribution Piping

22 11 13.48 Pipe, Fittings and Valves, Steel, Grooved-Joint	Crew	Daily Output	Labor-Hours	Unit	Material	2021 Bare Costs Labor	Equipment	Total	Total Incl O&P
1940	8" diameter	Q-2	37	.649	L.F.	55	41		96	122
1950	10" diameter		31	.774		119	49		168	204
1960	12" diameter	↓	27	.889	↓	144	56		200	242
2540	To delete coupling & hanger, subtract									
2550	3/4" diam. to 2" diam.					36%	27%			
2560	2-1/2" diam. to 5" diam.					19%	18%			
2570	6" diam. to 12" diam.					14%	13%			
4690	Tee, painted									
4700	3/4" diameter	1 Plum	38	.211	Ea.	102	14.25		116.25	134
4740	1" diameter		33	.242		78.50	16.40		94.90	111
4750	1-1/4" diameter		27	.296		78.50	20		98.50	117
4760	1-1/2" diameter		22	.364		78.50	24.50		103	124
4770	2" diameter	↓	17	.471		78.50	32		110.50	134
4780	2-1/2" diameter	Q-1	27	.593		78.50	36		114.50	141
4790	3" diameter		22	.727		107	44.50		151.50	184
4800	4" diameter		17	.941		163	57.50		220.50	265
4810	5" diameter	↓	13	1.231		380	75		455	525
4820	6" diameter	Q-2	17	1.412		435	89		524	615
4830	8" diameter		14	1.714		955	108		1,063	1,200
4840	10" diameter		12	2		1,250	126		1,376	1,575
4850	12" diameter	↓	10	2.400		1,625	152		1,777	2,000
4900	For galvanized tees, add				↓	24%				
4939	Couplings									
4940	Flexible, standard, painted									
4950	3/4" diameter	1 Plum	100	.080	Ea.	29	5.40		34.40	40
4960	1" diameter		100	.080		29	5.40		34.40	40
4970	1-1/4" diameter		80	.100		37.50	6.75		44.25	51
4980	1-1/2" diameter		67	.119		40.50	8.10		48.60	56.50
4990	2" diameter	↓	50	.160		43.50	10.85		54.35	64
5000	2-1/2" diameter	Q-1	80	.200		50.50	12.20		62.70	73.50
5010	3" diameter		67	.239		56	14.55		70.55	83
5020	3-1/2" diameter		57	.281		80	17.10		97.10	114
5030	4" diameter		50	.320		80.50	19.50		100	118
5040	5" diameter	↓	40	.400		121	24.50		145.50	170
5050	6" diameter	Q-2	50	.480		143	30.50		173.50	203
5070	8" diameter		42	.571		231	36		267	310
5090	10" diameter		35	.686		375	43.50		418.50	480
5110	12" diameter	↓	32	.750		430	47.50		477.50	540
5200	For galvanized couplings, add				↓	33%				
5750	Flange, w/groove gasket, black steel									
5754	See Line 22 11 13.47 0620 for gasket & bolt set									
5760	ANSI class 125 and 150, painted									
5780	2" pipe size	1 Plum	23	.348	Ea.	167	23.50		190.50	218
5790	2-1/2" pipe size	Q-1	37	.432		207	26.50		233.50	268
5800	3" pipe size		31	.516		223	31.50		254.50	293
5820	4" pipe size		23	.696		298	42.50		340.50	390
5830	5" pipe size	↓	19	.842		345	51.50		396.50	455
5840	6" pipe size	Q-2	23	1.043		375	66		441	515
5850	8" pipe size		17	1.412		425	89		514	605
5860	10" pipe size		14	1.714		670	108		778	900
5870	12" pipe size	↓	12	2	↓	880	126		1,006	1,150
8000	Butterfly valve, 2 position handle, with standard trim									
8010	1-1/2" pipe size	1 Plum	30	.267	Ea.	360	18.05		378.05	420

For customer support on your Site Work & Landscape Costs with RSMeans Data, call 800.448.8182.

225

22 11 13 – Facility Water Distribution Piping

22 11 13.48 Pipe, Fittings and Valves, Steel, Grooved-Joint		Crew	Daily Output	Labor-Hours	Unit	Material	2021 Bare Costs Labor	Equipment	Total	Total Incl O&P
8020	2" pipe size	1 Plum	23	.348	Ea.	360	23.50		383.50	430
8030	3" pipe size	Q-1	30	.533		515	32.50		547.50	620
8050	4" pipe size	"	22	.727		565	44.50		609.50	690
8070	6" pipe size	Q-2	23	1.043		1,150	66		1,216	1,350
8080	8" pipe size		18	1.333		1,550	84		1,634	1,825
8090	10" pipe size		15	1.600		2,525	101		2,626	2,925
8200	With stainless steel trim									
8240	1-1/2" pipe size	1 Plum	30	.267	Ea.	455	18.05		473.05	525
8250	2" pipe size	"	23	.348		455	23.50		478.50	535
8270	3" pipe size	Q-1	30	.533		610	32.50		642.50	725
8280	4" pipe size	"	22	.727		665	44.50		709.50	795
8300	6" pipe size	Q-2	23	1.043		1,250	66		1,316	1,475
8310	8" pipe size		18	1.333		3,075	84		3,159	3,500
8320	10" pipe size		15	1.600		5,125	101		5,226	5,775
9000	Cut one groove, labor									
9010	3/4" pipe size	Q-1	152	.105	Ea.		6.40		6.40	9.55
9020	1" pipe size		140	.114			6.95		6.95	10.40
9030	1-1/4" pipe size		124	.129			7.85		7.85	11.75
9040	1-1/2" pipe size		114	.140			8.55		8.55	12.75
9050	2" pipe size		104	.154			9.35		9.35	14
9060	2-1/2" pipe size		96	.167			10.15		10.15	15.15
9070	3" pipe size		88	.182			11.10		11.10	16.55
9080	3-1/2" pipe size		83	.193			11.75		11.75	17.55
9090	4" pipe size		78	.205			12.50		12.50	18.65
9100	5" pipe size		72	.222			13.55		13.55	20
9110	6" pipe size		70	.229			13.95		13.95	21
9120	8" pipe size		54	.296			18.05		18.05	27
9130	10" pipe size		38	.421			25.50		25.50	38.50
9140	12" pipe size		30	.533			32.50		32.50	48.50
9210	Roll one groove									
9220	3/4" pipe size	Q-1	266	.060	Ea.		3.66		3.66	5.45
9230	1" pipe size		228	.070			4.28		4.28	6.40
9240	1-1/4" pipe size		200	.080			4.87		4.87	7.25
9250	1-1/2" pipe size		178	.090			5.50		5.50	8.15
9260	2" pipe size		116	.138			8.40		8.40	12.55
9270	2-1/2" pipe size		110	.145			8.85		8.85	13.25
9280	3" pipe size		100	.160			9.75		9.75	14.55
9290	3-1/2" pipe size		94	.170			10.35		10.35	15.50
9300	4" pipe size		86	.186			11.35		11.35	16.90
9310	5" pipe size		84	.190			11.60		11.60	17.30
9320	6" pipe size		80	.200			12.20		12.20	18.20
9330	8" pipe size		66	.242			14.75		14.75	22
9340	10" pipe size		58	.276			16.80		16.80	25
9350	12" pipe size		46	.348			21		21	31.50

22 11 13.74 Pipe, Plastic

		Crew	Daily Output	Labor-Hours	Unit	Material	2021 Bare Costs Labor	Equipment	Total	Total Incl O&P
0010	**PIPE, PLASTIC**									
0020	Fiberglass reinforced, couplings 10' OC, clevis hanger assy's, 3 per 10'									
0080	General service									
0120	2" diameter	Q-1	59	.271	L.F.	11.65	16.50		28.15	37.50
0140	3" diameter		52	.308		15.40	18.75		34.15	45
0150	4" diameter		48	.333		21	20.50		41.50	53.50
0160	6" diameter		39	.410		39	25		64	80

22 11 Facility Water Distribution

22 11 13 - Facility Water Distribution Piping

22 11 13.74 Pipe, Plastic

		Crew	Daily Output	Labor-Hours	Unit	Material	2021 Bare Costs Labor	Equipment	Total	Total Incl O&P
0170	8" diameter	Q-2	49	.490	L.F.	66	31		97	119
0180	10" diameter		41	.585		98	37		135	163
0190	12" diameter	↓	36	.667	↓	130	42		172	206
1800	PVC, couplings 10' OC, clevis hanger assemblies, 3 per 10'									
1820	Schedule 40									
1860	1/2" diameter	1 Plum	54	.148	L.F.	4.17	10.05		14.22	19.55
1870	3/4" diameter		51	.157		4.77	10.60		15.37	21
1880	1" diameter		46	.174		3.35	11.75		15.10	21
1890	1-1/4" diameter		42	.190		3.36	12.90		16.26	23
1900	1-1/2" diameter	↓	36	.222		4.90	15.05		19.95	28
1910	2" diameter	Q-1	59	.271		4.85	16.50		21.35	30
1920	2-1/2" diameter		56	.286		7.85	17.40		25.25	34.50
1930	3" diameter		53	.302		10.05	18.40		28.45	38.50
1940	4" diameter		48	.333		13	20.50		33.50	45
1950	5" diameter		43	.372		25	22.50		47.50	61.50
1960	6" diameter	↓	39	.410	↓	25	25		50	65
2340	To delete coupling & hangers, subtract									
2360	1/2" diam. to 1-1/4" diam.					65%	74%			
2370	1-1/2" diam. to 6" diam.					44%	57%			
4100	DWV type, schedule 40, couplings 10' OC, clevis hanger assy's, 3 per 10'									
4210	ABS, schedule 40, foam core type									
4212	Plain end black									
4214	1-1/2" diameter	1 Plum	39	.205	L.F.	2.79	13.90		16.69	23.50
4216	2" diameter	Q-1	62	.258		3.27	15.70		18.97	27
4218	3" diameter		56	.286		5.65	17.40		23.05	32
4220	4" diameter		51	.314		8.05	19.10		27.15	37.50
4222	6" diameter	↓	42	.381	↓	19.75	23		42.75	56
4240	To delete coupling & hangers, subtract									
4244	1-1/2" diam. to 6" diam.					43%	48%			
4400	PVC									
4410	1-1/4" diameter	1 Plum	42	.190	L.F.	3.42	12.90		16.32	23
4420	1-1/2" diameter	"	36	.222		2.35	15.05		17.40	25
4460	2" diameter	Q-1	59	.271		3.13	16.50		19.63	28
4470	3" diameter		53	.302		5.35	18.40		23.75	33.50
4480	4" diameter		48	.333		7.40	20.50		27.90	38.50
4490	6" diameter	↓	39	.410	↓	15.75	25		40.75	55
4510	To delete coupling & hangers, subtract									
4520	1-1/4" diam. to 1-1/2" diam.					48%	60%			
4530	2" diam. to 8" diam.					42%	54%			
4532	to delete hangers, 2" diam. to 8" diam.	Q-1	50	.320	L.F.	2.65	19.50		22.15	32
5300	CPVC, socket joint, couplings 10' OC, clevis hanger assemblies, 3 per 10'									
5302	Schedule 40									
5304	1/2" diameter	1 Plum	54	.148	L.F.	5.30	10.05		15.35	21
5305	3/4" diameter		51	.157		6.45	10.60		17.05	23
5306	1" diameter		46	.174		5.30	11.75		17.05	23.50
5307	1-1/4" diameter		42	.190		6.50	12.90		19.40	26.50
5308	1-1/2" diameter	↓	36	.222		6.35	15.05		21.40	29.50
5309	2" diameter	Q-1	59	.271		9.35	16.50		25.85	35
5310	2-1/2" diameter		56	.286		15.20	17.40		32.60	42.50
5311	3" diameter	↓	53	.302	↓	19.80	18.40		38.20	49.50
5318	To delete coupling & hangers, subtract									
5319	1/2" diam. to 3/4" diam.					37%	77%			
5320	1" diam. to 1-1/4" diam.					27%	70%			

For customer support on your Site Work & Landscape Costs with RSMeans Data, call 800.448.8182.

227

22 11 Facility Water Distribution

22 11 13 – Facility Water Distribution Piping

22 11 13.74 Pipe, Plastic	Crew	Daily Output	Labor-Hours	Unit	Material	2021 Bare Costs Labor	Equipment	Total	Total Incl O&P
5321 1-1/2" diam. to 3" diam.					21%	57%			
5360 CPVC, threaded, couplings 10' OC, clevis hanger assemblies, 3 per 10'									
5380 Schedule 40									
5460 1/2" diameter	1 Plum	54	.148	L.F.	6.15	10.05		16.20	21.50
5470 3/4" diameter		51	.157		7.90	10.60		18.50	24.50
5480 1" diameter		46	.174		6.85	11.75		18.60	25
5490 1-1/4" diameter		42	.190		7.65	12.90		20.55	27.50
5500 1-1/2" diameter		36	.222		7.35	15.05		22.40	30.50
5510 2" diameter	Q-1	59	.271		10.55	16.50		27.05	36
5520 2-1/2" diameter		56	.286		16.50	17.40		33.90	44
5530 3" diameter		53	.302		21.50	18.40		39.90	51.50
5730 To delete coupling & hangers, subtract									
5740 1/2" diam. to 3/4" diam.					37%	77%			
5750 1" diam. to 1-1/4" diam.					27%	70%			
5760 1-1/2" diam. to 3" diam.					21%	57%			
7280 PEX, flexible, no couplings or hangers									
7282 Note: For labor costs add 25% to the couplings and fittings labor total.									
7300 Non-barrier type, hot/cold tubing rolls									
7310 1/4" diameter x 100'				L.F.	.56			.56	.62
7350 3/8" diameter x 100'					.57			.57	.63
7360 1/2" diameter x 100'					.84			.84	.92
7370 1/2" diameter x 500'					.74			.74	.81
7380 1/2" diameter x 1000'					.64			.64	.70
7400 3/4" diameter x 100'					1.05			1.05	1.16
7410 3/4" diameter x 500'					1.21			1.21	1.33
7420 3/4" diameter x 1000'					1.21			1.21	1.33
7460 1" diameter x 100'					2.09			2.09	2.30
7470 1" diameter x 300'					2.09			2.09	2.30
7480 1" diameter x 500'					2.11			2.11	2.32
7500 1-1/4" diameter x 100'					3.56			3.56	3.92
7510 1-1/4" diameter x 300'					3.56			3.56	3.92
7540 1-1/2" diameter x 100'					4.88			4.88	5.35
7550 1-1/2" diameter x 300'					4.95			4.95	5.45
7596 Most sizes available in red or blue									

22 11 13.76 Pipe Fittings, Plastic

	Crew	Daily Output	Labor-Hours	Unit	Material	2021 Bare Costs Labor	Equipment	Total	Total Incl O&P
0010 **PIPE FITTINGS, PLASTIC**									
0030 Epoxy resin, fiberglass reinforced, general service									
0100 3"	Q-1	20.80	.769	Ea.	114	47		161	196
0110 4"		16.50	.970		123	59		182	223
0120 6"		10.10	1.584		228	96.50		324.50	395
0130 8"	Q-2	9.30	2.581		420	163		583	705
0140 10"		8.50	2.824		525	178		703	845
0150 12"		7.60	3.158		755	200		955	1,125
0380 Couplings									
0410 2"	Q-1	33.10	.483	Ea.	28	29.50		57.50	75
0420 3"		20.80	.769		32.50	47		79.50	106
0430 4"		16.50	.970		45	59		104	138
0440 6"		10.10	1.584		104	96.50		200.50	258
0450 8"	Q-2	9.30	2.581		176	163		339	435
0460 10"		8.50	2.824		242	178		420	535
0470 12"		7.60	3.158		345	200		545	680
0473 High corrosion resistant couplings, add					30%				

22 11 13.76 Pipe Fittings, Plastic		Crew	Daily Output	Labor-Hours	Unit	Material	2021 Bare Costs Labor	Equipment	Total	Total Incl O&P
2100	PVC schedule 80, socket joint									
2110	90° elbow, 1/2"	1 Plum	30.30	.264	Ea.	2.90	17.85		20.75	29.50
2130	3/4"		26	.308		3.71	21		24.71	35
2140	1"		22.70	.352		6	24		30	42
2150	1-1/4"		20.20	.396		7.80	27		34.80	48.50
2160	1-1/2"		18.20	.440		8.55	30		38.55	54
2170	2"	Q-1	33.10	.483		10.35	29.50		39.85	55.50
2180	3"		20.80	.769		26.50	47		73.50	99
2190	4"		16.50	.970		41.50	59		100.50	134
2200	6"		10.10	1.584		118	96.50		214.50	274
2210	8"	Q-2	9.30	2.581		325	163		488	600
2250	45° elbow, 1/2"	1 Plum	30.30	.264		5.50	17.85		23.35	32.50
2270	3/4"		26	.308		8.35	21		29.35	40
2280	1"		22.70	.352		12.50	24		36.50	49.50
2290	1-1/4"		20.20	.396		15.55	27		42.55	57
2300	1-1/2"		18.20	.440		18.85	30		48.85	65
2310	2"	Q-1	33.10	.483		24.50	29.50		54	71
2320	3"		20.80	.769		61	47		108	137
2330	4"		16.50	.970		112	59		171	212
2340	6"		10.10	1.584		142	96.50		238.50	300
2350	8"	Q-2	9.30	2.581		305	163		468	585
2400	Tee, 1/2"	1 Plum	20.20	.396		8.20	27		35.20	49
2420	3/4"		17.30	.462		8.60	31.50		40.10	56
2430	1"		15.20	.526		10.75	35.50		46.25	65
2440	1-1/4"		13.50	.593		29	40		69	91.50
2450	1-1/2"		12.10	.661		29.50	45		74.50	99.50
2460	2"	Q-1	20	.800		37	48.50		85.50	113
2470	3"		13.90	1.151		49	70		119	159
2480	4"		11	1.455		58	88.50		146.50	196
2490	6"		6.70	2.388		198	146		344	435
2500	8"	Q-2	6.20	3.871		460	245		705	870
2510	Flange, socket, 150 lb., 1/2"	1 Plum	55.60	.144		15.50	9.75		25.25	31.50
2514	3/4"		47.60	.168		16.60	11.40		28	35.50
2518	1"		41.70	.192		18.45	13		31.45	40
2522	1-1/2"		33.30	.240		19.45	16.25		35.70	46
2526	2"	Q-1	60.60	.264		26	16.10		42.10	52.50
2530	4"		30.30	.528		56	32		88	110
2534	6"		18.50	.865		87.50	52.50		140	175
2538	8"	Q-2	17.10	1.404		157	88.50		245.50	305
2550	Coupling, 1/2"	1 Plum	30.30	.264		5.45	17.85		23.30	32.50
2570	3/4"		26	.308		7.55	21		28.55	39.50
2580	1"		22.70	.352		7.90	24		31.90	44
2590	1-1/4"		20.20	.396		10.85	27		37.85	52
2600	1-1/2"		18.20	.440		16.45	30		46.45	62.50
2610	2"	Q-1	33.10	.483		17.45	29.50		46.95	63
2620	3"		20.80	.769		35.50	47		82.50	109
2630	4"		16.50	.970		73.50	59		132.50	169
2640	6"		10.10	1.584		95.50	96.50		192	249
2650	8"	Q-2	9.30	2.581		130	163		293	385
2660	10"		8.50	2.824		410	178		588	720
2670	12"		7.60	3.158		475	200		675	820
2700	PVC (white), schedule 40, socket joints									
2810	2"	Q-1	36.40	.440	Ea.	3.29	27		30.29	43.50

22 11 13.76 Pipe Fittings, Plastic		Crew	Daily Output	Labor-Hours	Unit	Material	2021 Bare Costs Labor	2021 Bare Costs Equipment	Total	Total Incl O&P
2840	4"	Q-1	18.20	.879	Ea.	21.50	53.50		75	104
4500	DWV, ABS, non pressure, socket joints									
4540	1/4 bend, 1-1/4"	1 Plum	20.20	.396	Ea.	8.15	27		35.15	49
4560	1-1/2"	"	18.20	.440		6.35	30		36.35	51.50
4570	2"	Q-1	33.10	.483		6.05	29.50		35.55	50.50
4580	3"	"	20.80	.769		24.50	47		71.50	97
4650	1/8 bend, same as 1/4 bend									
4800	Tee, sanitary									
4820	1-1/4"	1 Plum	13.50	.593	Ea.	10.85	40		50.85	72
4830	1-1/2"	"	12.10	.661		9.40	45		54.40	77.50
4840	2"	Q-1	20	.800		14.45	48.50		62.95	88.50
4850	3"		13.90	1.151		39.50	70		109.50	149
4860	4"		11	1.455		70	88.50		158.50	209
4862	Tee, sanitary, reducing, 2" x 1-1/2"		22	.727		12.60	44.50		57.10	80
4864	3" x 2"		15.30	1.046		31.50	63.50		95	130
4868	4" x 3"		12.10	1.322		70.50	80.50		151	198
4870	Combination Y and 1/8 bend									
4872	1-1/2"	1 Plum	12.10	.661	Ea.	24.50	45		69.50	94
4874	2"	Q-1	20	.800		25	48.50		73.50	100
4876	3"		13.90	1.151		59	70		129	170
4878	4"		11	1.455		123	88.50		211.50	268
4880	3" x 1-1/2"		15.50	1.032		66.50	63		129.50	168
4882	4" x 3"		12.10	1.322		88.50	80.50		169	218
4900	Wye, 1-1/4"	1 Plum	13.50	.593		12.50	40		52.50	74
4902	1-1/2"	"	12.10	.661		15.55	45		60.55	84
4904	2"	Q-1	20	.800		20.50	48.50		69	95
4906	3"		13.90	1.151		43.50	70		113.50	153
4908	4"		11	1.455		89	88.50		177.50	230
4910	6"		6.70	2.388		268	146		414	510
4918	3" x 1-1/2"		15.50	1.032		38.50	63		101.50	137
4920	4" x 3"		12.10	1.322		77	80.50		157.50	205
4922	6" x 4"		6.90	2.319		242	141		383	475
4930	Double wye, 1-1/2"	1 Plum	9.10	.879		48	59.50		107.50	142
4932	2"	Q-1	16.60	.964		57	58.50		115.50	151
4934	3"		10.40	1.538		128	93.50		221.50	281
4936	4"		8.25	1.939		251	118		369	450
4940	2" x 1-1/2"		16.80	.952		54.50	58		112.50	147
4942	3" x 2"		10.60	1.509		89.50	92		181.50	236
4944	4" x 3"		8.45	1.893		208	115		323	400
4946	6" x 4"		7.25	2.207		285	134		419	515
4950	Reducer bushing, 2" x 1-1/2"		36.40	.440		4.99	27		31.99	45.50
4952	3" x 1-1/2"		27.30	.586		22	35.50		57.50	77.50
4954	4" x 2"		18.20	.879		40.50	53.50		94	125
4956	6" x 4"		11.10	1.441		114	88		202	256
4960	Couplings, 1-1/2"	1 Plum	18.20	.440		3.26	30		33.26	48
4962	2"	Q-1	33.10	.483		4.09	29.50		33.59	48.50
4963	3"		20.80	.769		12.30	47		59.30	83.50
4964	4"		16.50	.970		20.50	59		79.50	111
4966	6"		10.10	1.584		86	96.50		182.50	239
4970	2" x 1-1/2"		33.30	.480		8.75	29.50		38.25	53
4972	3" x 1-1/2"		21	.762		26	46.50		72.50	98
4974	4" x 3"		16.70	.958		38.50	58.50		97	130
4978	Closet flange, 4"	1 Plum	32	.250		23.50	16.95		40.45	51

22 11 13.76 Pipe Fittings, Plastic	Crew	Daily Output	Labor-Hours	Unit	Material	2021 Bare Costs Labor	Equipment	Total	Total Incl O&P
4980	4" x 3"	1 Plum	34	.235	Ea.	25.50	15.95		41.45	52
5000	DWV, PVC, schedule 40, socket joints									
5040	1/4 bend, 1-1/4"	1 Plum	20.20	.396	Ea.	16.55	27		43.55	58
5060	1-1/2"	"	18.20	.440		4.75	30		34.75	50
5070	2"	Q-1	33.10	.483		7.45	29.50		36.95	52
5080	3"		20.80	.769		22	47		69	94
5090	4"		16.50	.970		43.50	59		102.50	136
5100	6"		10.10	1.584		152	96.50		248.50	310
5105	8"	Q-2	9.30	2.581		197	163		360	460
5110	1/4 bend, long sweep, 1-1/2"	1 Plum	18.20	.440		11	30		41	56.50
5112	2"	Q-1	33.10	.483		12.25	29.50		41.75	57.50
5114	3"		20.80	.769		28.50	47		75.50	101
5116	4"		16.50	.970		53.50	59		112.50	147
5150	1/8 bend, 1-1/4"	1 Plum	20.20	.396		11.30	27		38.30	52.50
5215	8"	Q-2	9.30	2.581		87	163		250	340
5250	Tee, sanitary 1-1/4"	1 Plum	13.50	.593		17.75	40		57.75	79.50
5254	1-1/2"	"	12.10	.661		8.30	45		53.30	76
5255	2"	Q-1	20	.800		12.20	48.50		60.70	86
5256	3"	"	13.90	1.151		32	70		102	141
5374	Coupling, 1-1/4"	1 Plum	20.20	.396		10.70	27		37.70	52
5376	1-1/2"	"	18.20	.440		2.23	30		32.23	47
5378	2"	Q-1	33.10	.483		3.06	29.50		32.56	47.50
5380	3"		20.80	.769		10.65	47		57.65	82
5390	4"		16.50	.970		18.10	59		77.10	108
5450	Solvent cement for PVC, industrial grade, per quart				Qt.	32			32	35
5500	CPVC, Schedule 80, threaded joints									
5540	90° elbow, 1/4"	1 Plum	32	.250	Ea.	12.80	16.95		29.75	39.50
5560	1/2"		30.30	.264		7.40	17.85		25.25	34.50
5570	3/4"		26	.308		11.10	21		32.10	43
5580	1"		22.70	.352		15.60	24		39.60	52.50
5590	1-1/4"		20.20	.396		30	27		57	73
5600	1-1/2"		18.20	.440		32.50	30		62.50	80
5610	2"	Q-1	33.10	.483		43.50	29.50		73	91.50
5620	2-1/2"		24.20	.661		134	40.50		174.50	208
5630	3"		20.80	.769		144	47		191	228
5660	45° elbow same as 90° elbow									
5700	Tee, 1/4"	1 Plum	22	.364	Ea.	25	24.50		49.50	64.50
5704	3/4"		17.30	.462		36	31.50		67.50	86
5706	1"		15.20	.526		38.50	35.50		74	95.50
5708	1-1/4"		13.50	.593		39	40		79	103
5710	1-1/2"		12.10	.661		40.50	45		85.50	112
5712	2"	Q-1	20	.800		45	48.50		93.50	122
5714	2-1/2"		16.20	.988		222	60		282	335
5716	3"		13.90	1.151		258	70		328	390
5730	Coupling, 1/4"	1 Plum	32	.250		16.30	16.95		33.25	43.50
5732	1/2"		30.30	.264		13.40	17.85		31.25	41.50
5734	3/4"		26	.308		21.50	21		42.50	55
5736	1"		22.70	.352		24.50	24		48.50	62.50
5738	1-1/4"		20.20	.396		26	27		53	68.50
5740	1-1/2"		18.20	.440		28	30		58	75.50
5742	2"	Q-1	33.10	.483		33	29.50		62.50	80.50
5744	2-1/2"		24.20	.661		59	40.50		99.50	125
5746	3"		20.80	.769		69	47		116	146

For customer support on your Site Work & Landscape Costs with RSMeans Data, call 800.448.8182.

231

22 11 13 - Facility Water Distribution Piping

22 11 13.76 Pipe Fittings, Plastic		Crew	Daily Output	Labor-Hours	Unit	Material	2021 Bare Costs Labor	Equipment	Total	Total Incl O&P
5900	CPVC, Schedule 80, socket joints									
5904	90° elbow, 1/4"	1 Plum	32	.250	Ea.	12.15	16.95		29.10	39
5906	1/2"		30.30	.264		4.77	17.85		22.62	32
5908	3/4"		26	.308		6.10	21		27.10	37.50
5910	1"		22.70	.352		9.65	24		33.65	46
5912	1-1/4"		20.20	.396		21	27		48	63
5914	1-1/2"	↓	18.20	.440		23.50	30		53.50	70
5916	2"	Q-1	33.10	.483		28	29.50		57.50	75
5918	2-1/2"		24.20	.661		65	40.50		105.50	132
5920	3"	↓	20.80	.769		73.50	47		120.50	151
5930	45° elbow, 1/4"	1 Plum	32	.250		18.10	16.95		35.05	45.50
5932	1/2"		30.30	.264		5.85	17.85		23.70	33
5934	3/4"		26	.308		8.40	21		29.40	40.50
5936	1"		22.70	.352		13.40	24		37.40	50.50
5938	1-1/4"		20.20	.396		26.50	27		53.50	69
5940	1-1/2"	↓	18.20	.440		27	30		57	74
5942	2"	Q-1	33.10	.483		30.50	29.50		60	77.50
5944	2-1/2"		24.20	.661		62	40.50		102.50	128
5946	3"	↓	20.80	.769		79.50	47		126.50	158
5960	Tee, 1/4"	1 Plum	22	.364		11.20	24.50		35.70	49.50
5964	3/4"		17.30	.462		11.40	31.50		42.90	59
5966	1"		15.20	.526		13.95	35.50		49.45	68.50
5968	1-1/4"		13.50	.593		29.50	40		69.50	92.50
5970	1-1/2"	↓	12.10	.661		33.50	45		78.50	104
5972	2"	Q-1	20	.800		37.50	48.50		86	114
5974	2-1/2"		16.20	.988		95.50	60		155.50	195
5976	3"	↓	13.90	1.151		95.50	70		165.50	210
5990	Coupling, 1/4"	1 Plum	32	.250		12.95	16.95		29.90	40
5992	1/2"		30.30	.264		5.05	17.85		22.90	32
5994	3/4"		26	.308		7.05	21		28.05	39
5996	1"		22.70	.352		9.50	24		33.50	46
5998	1-1/4"		20.20	.396		14.20	27		41.20	55.50
6000	1-1/2"	↓	18.20	.440		17.85	30		47.85	64
6002	2"	Q-1	33.10	.483		21	29.50		50.50	67
6004	2-1/2"		24.20	.661		46	40.50		86.50	111
6006	3"	↓	20.80	.769	↓	50	47		97	126
6360	Solvent cement for CPVC, commercial grade, per quart				Qt.	55			55	60.50
7340	PVC flange, slip-on, Sch 80 std., 1/2"	1 Plum	22	.364	Ea.	16	24.50		40.50	54.50
7350	3/4"		21	.381		17.05	26		43.05	57.50
7360	1"		18	.444		19	30		49	66
7370	1-1/4"		17	.471		19.60	32		51.60	69
7380	1-1/2"	↓	16	.500		20	34		54	72.50
7390	2"	Q-1	26	.615		26.50	37.50		64	85.50
7400	2-1/2"		24	.667		41	40.50		81.50	106
7410	3"		18	.889		45.50	54		99.50	131
7420	4"		15	1.067		57.50	65		122.50	160
7430	6"	↓	10	1.600		90.50	97.50		188	245
7440	8"	Q-2	11	2.182		162	138		300	385
7550	Union, schedule 40, socket joints, 1/2"	1 Plum	19	.421		5.25	28.50		33.75	48.50
7560	3/4"		18	.444		5.95	30		35.95	51.50
7570	1"		15	.533		6.10	36		42.10	61
7580	1-1/4"		14	.571		18.80	38.50		57.30	78
7590	1-1/2"	↓	13	.615	↓	20.50	41.50		62	84.50

22 11 13 – Facility Water Distribution Piping

22 11 13.76 Pipe Fittings, Plastic	Crew	Daily Output	Labor-Hours	Unit	Material	2021 Bare Costs Labor	Equipment	Total	Total Incl O&P
7600	2"	Q-1	20	.800	Ea.	27.50	48.50		76	103

22 11 13.78 Pipe, High Density Polyethylene Plastic (HDPE)

		Crew	Daily Output	Labor-Hours	Unit	Material	2021 Bare Costs Labor	Equipment	Total	Total Incl O&P
0010	**PIPE, HIGH DENSITY POLYETHYLENE PLASTIC (HDPE)**									
0020	Not incl. hangers, trenching, backfill, hoisting or digging equipment.									
0030	Standard length is 40', add a weld for each joint									
0035	For HDPE weld machine see 015433401685 in equipment rental									
0040	Single wall									
0050	Straight									
0054	1" diameter DR 11				L.F.	.91			.91	1
0058	1-1/2" diameter DR 11					1.17			1.17	1.29
0062	2" diameter DR 11					1.93			1.93	2.12
0066	3" diameter DR 11					2.34			2.34	2.57
0070	3" diameter DR 17					1.87			1.87	2.06
0074	4" diameter DR 11					3.92			3.92	4.31
0078	4" diameter DR 17					3.92			3.92	4.31
0082	6" diameter DR 11					9.75			9.75	10.70
0086	6" diameter DR 17					6.40			6.40	7.05
0090	8" diameter DR 11					16.15			16.15	17.80
0094	8" diameter DR 26					7.60			7.60	8.35
0098	10" diameter DR 11					25.50			25.50	28
0102	10" diameter DR 26					11.70			11.70	12.85
0106	12" diameter DR 11					37			37	40.50
0110	12" diameter DR 26					17.55			17.55	19.30
0114	16" diameter DR 11					56.50			56.50	62
0118	16" diameter DR 26					25.50			25.50	28
0122	18" diameter DR 11					72			72	79.50
0126	18" diameter DR 26					33			33	36.50
0130	20" diameter DR 11					87.50			87.50	96.50
0134	20" diameter DR 26					39			39	43
0138	22" diameter DR 11					107			107	118
0142	22" diameter DR 26					48.50			48.50	53.50
0146	24" diameter DR 11					127			127	139
0150	24" diameter DR 26					56.50			56.50	62
0154	28" diameter DR 17					117			117	129
0158	28" diameter DR 26					78			78	86
0162	30" diameter DR 21					109			109	120
0166	30" diameter DR 26					89.50			89.50	98.50
0170	36" diameter DR 26					89.50			89.50	98.50
0174	42" diameter DR 26					173			173	191
0178	48" diameter DR 26					228			228	251
0182	54" diameter DR 26					286			286	315
0300	90° elbow									
0304	1" diameter DR 11				Ea.	5.65			5.65	6.20
0308	1-1/2" diameter DR 11					7.05			7.05	7.75
0312	2" diameter DR 11					7.05			7.05	7.75
0316	3" diameter DR 11					14.10			14.10	15.55
0320	3" diameter DR 17					14.10			14.10	15.55
0324	4" diameter DR 11					19.75			19.75	22
0328	4" diameter DR 17					19.75			19.75	22
0332	6" diameter DR 11					45			45	49.50
0336	6" diameter DR 17					45			45	49.50
0340	8" diameter DR 11					112			112	123

For customer support on your Site Work & Landscape Costs with RSMeans Data, call 800.448.8182.

233

22 11 Facility Water Distribution

22 11 13 - Facility Water Distribution Piping

22 11 13.78 Pipe, High Density Polyethylene Plastic (HDPE)	Crew	Daily Output	Labor-Hours	Unit	Material	2021 Bare Costs Labor	Equipment	Total	Total Incl O&P
0344	8" diameter DR 26				Ea.	99			99	109
0348	10" diameter DR 11					415			415	460
0352	10" diameter DR 26					385			385	425
0356	12" diameter DR 11					440			440	485
0360	12" diameter DR 26					400			400	440
0364	16" diameter DR 11					520			520	575
0368	16" diameter DR 26					515			515	565
0372	18" diameter DR 11					655			655	720
0376	18" diameter DR 26					615			615	680
0380	20" diameter DR 11					770			770	845
0384	20" diameter DR 26					745			745	820
0388	22" diameter DR 11					795			795	875
0392	22" diameter DR 26					770			770	845
0396	24" diameter DR 11					900			900	990
0400	24" diameter DR 26					875			875	960
0404	28" diameter DR 17					1,100			1,100	1,200
0408	28" diameter DR 26					1,025			1,025	1,125
0412	30" diameter DR 17					1,550			1,550	1,700
0416	30" diameter DR 26					1,400			1,400	1,550
0420	36" diameter DR 26					1,800			1,800	1,975
0424	42" diameter DR 26					2,300			2,300	2,550
0428	48" diameter DR 26					2,700			2,700	2,975
0432	54" diameter DR 26				▼	6,425			6,425	7,050
0500	45° elbow									
0512	2" diameter DR 11				Ea.	5.65			5.65	6.25
0516	3" diameter DR 11					14.10			14.10	15.55
0520	3" diameter DR 17					14.10			14.10	15.55
0524	4" diameter DR 11					19.75			19.75	22
0528	4" diameter DR 17					19.75			19.75	22
0532	6" diameter DR 11					45			45	49.50
0536	6" diameter DR 17					45			45	49.50
0540	8" diameter DR 11					112			112	123
0544	8" diameter DR 26					65			65	71.50
0548	10" diameter DR 11					415			415	460
0552	10" diameter DR 26					385			385	425
0556	12" diameter DR 11					440			440	485
0560	12" diameter DR 26					400			400	440
0564	16" diameter DR 11					231			231	254
0568	16" diameter DR 26					218			218	240
0572	18" diameter DR 11					245			245	270
0576	18" diameter DR 26					225			225	247
0580	20" diameter DR 11					370			370	410
0584	20" diameter DR 26					355			355	390
0588	22" diameter DR 11					515			515	565
0592	22" diameter DR 26					490			490	535
0596	24" diameter DR 11					630			630	690
0600	24" diameter DR 26					605			605	665
0604	28" diameter DR 17					715			715	785
0608	28" diameter DR 26					695			695	765
0612	30" diameter DR 17					885			885	975
0616	30" diameter DR 26					860			860	945
0620	36" diameter DR 26					1,100			1,100	1,200
0624	42" diameter DR 26				▼	1,400			1,400	1,550

For customer support on your Site Work & Landscape Costs with RSMeans Data, call 800.448.8182.

22 11 13.78 Pipe, High Density Polyethylene Plastic (HDPE)	Crew	Daily Output	Labor-Hours	Unit	Material	2021 Bare Costs Labor	Equipment	Total	Total Incl O&P
0628	48" diameter DR 26				Ea.	1,525			1,525	1,675
0632	54" diameter DR 26				▼	2,050			2,050	2,250
0700	Tee									
0704	1" diameter DR 11				Ea.	7.35			7.35	8.10
0708	1-1/2" diameter DR 11					10.35			10.35	11.35
0712	2" diameter DR 11					8.85			8.85	9.75
0716	3" diameter DR 11					16.25			16.25	17.85
0720	3" diameter DR 17					16.25			16.25	17.85
0724	4" diameter DR 11					23.50			23.50	26
0728	4" diameter DR 17					23.50			23.50	26
0732	6" diameter DR 11					59			59	65
0736	6" diameter DR 17					59			59	65
0740	8" diameter DR 11					146			146	161
0744	8" diameter DR 17					146			146	161
0748	10" diameter DR 11					435			435	480
0752	10" diameter DR 17					435			435	480
0756	12" diameter DR 11					580			580	640
0760	12" diameter DR 17					580			580	640
0764	16" diameter DR 11					305			305	335
0768	16" diameter DR 17					253			253	278
0772	18" diameter DR 11					435			435	475
0776	18" diameter DR 17					355			355	390
0780	20" diameter DR 11					530			530	580
0784	20" diameter DR 17					435			435	475
0788	22" diameter DR 11					680			680	745
0792	22" diameter DR 17					535			535	590
0796	24" diameter DR 11					840			840	925
0800	24" diameter DR 17					710			710	780
0804	28" diameter DR 17					1,375			1,375	1,500
0812	30" diameter DR 17					1,575			1,575	1,725
0820	36" diameter DR 17					2,600			2,600	2,850
0824	42" diameter DR 26					2,900			2,900	3,175
0828	48" diameter DR 26				▼	3,100			3,100	3,425
1000	Flange adptr, w/back-up ring and 1/2 cost of plated bolt set									
1004	1" diameter DR 11				Ea.	28			28	31
1008	1-1/2" diameter DR 11					28			28	31
1012	2" diameter DR 11					17.70			17.70	19.50
1016	3" diameter DR 11					20.50			20.50	23
1020	3" diameter DR 17					20.50			20.50	23
1024	4" diameter DR 11					28			28	31
1028	4" diameter DR 17					28			28	31
1032	6" diameter DR 11					40			40	44
1036	6" diameter DR 17					40			40	44
1040	8" diameter DR 11					57.50			57.50	63.50
1044	8" diameter DR 26					57.50			57.50	63.50
1048	10" diameter DR 11					91.50			91.50	101
1052	10" diameter DR 26					91.50			91.50	101
1056	12" diameter DR 11					134			134	148
1060	12" diameter DR 26					134			134	148
1064	16" diameter DR 11					288			288	315
1068	16" diameter DR 26					288			288	315
1072	18" diameter DR 11					375			375	410
1076	18" diameter DR 26					375			375	410

235

For customer support on your Site Work & Landscape Costs with RSMeans Data, call 800.448.8182.

22 11 13 – Facility Water Distribution Piping

22 11 13.78 Pipe, High Density Polyethylene Plastic (HDPE)	Crew	Daily Output	Labor-Hours	Unit	Material	2021 Bare Costs Labor	Equipment	Total	Total Incl O&P
1080	20" diameter DR 11				Ea.	520			520	570
1084	20" diameter DR 26					520			520	570
1088	22" diameter DR 11					560			560	620
1092	22" diameter DR 26					560			560	615
1096	24" diameter DR 17					610			610	670
1100	24" diameter DR 32.5					610			610	670
1104	28" diameter DR 15.5					825			825	910
1108	28" diameter DR 32.5					825			825	910
1112	30" diameter DR 11					960			960	1,050
1116	30" diameter DR 21					960			960	1,050
1120	36" diameter DR 26					1,050			1,050	1,150
1124	42" diameter DR 26					1,200			1,200	1,325
1128	48" diameter DR 26					1,450			1,450	1,600
1132	54" diameter DR 26					1,775			1,775	1,950
1200	Reducer									
1208	2" x 1-1/2" diameter DR 11				Ea.	8.50			8.50	9.35
1212	3" x 2" diameter DR 11					8.50			8.50	9.35
1216	4" x 2" diameter DR 11					9.90			9.90	10.85
1220	4" x 3" diameter DR 11					12.70			12.70	14
1224	6" x 4" diameter DR 11					29.50			29.50	32.50
1228	8" x 6" diameter DR 11					45			45	49.50
1232	10" x 8" diameter DR 11					77.50			77.50	85.50
1236	12" x 8" diameter DR 11					127			127	140
1240	12" x 10" diameter DR 11					102			102	112
1244	14" x 12" diameter DR 11					113			113	124
1248	16" x 14" diameter DR 11					144			144	158
1252	18" x 16" diameter DR 11					175			175	193
1256	20" x 18" diameter DR 11					345			345	380
1260	22" x 20" diameter DR 11					425			425	470
1264	24" x 22" diameter DR 11					480			480	530
1268	26" x 24" diameter DR 11					565			565	620
1272	28" x 24" diameter DR 11					720			720	790
1276	32" x 28" diameter DR 17					930			930	1,025
1280	36" x 32" diameter DR 17					1,275			1,275	1,400
4000	Welding labor per joint, not including welding machine									
4010	Pipe joint size (cost based on thickest wall for each diam.)									
4030	1" pipe size	4 Skwk	273	.117	Ea.		6.70		6.70	10.05
4040	1-1/2" pipe size		175	.183			10.45		10.45	15.70
4050	2" pipe size		128	.250			14.30		14.30	21.50
4060	3" pipe size		100	.320			18.25		18.25	27.50
4070	4" pipe size		77	.416			23.50		23.50	35.50
4080	6" pipe size	5 Skwk	63	.635			36.50		36.50	54.50
4090	8" pipe size		48	.833			47.50		47.50	71.50
4100	10" pipe size		40	1			57		57	86
4110	12" pipe size	6 Skwk	41	1.171			67		67	101
4120	16" pipe size		34	1.412			80.50		80.50	121
4130	18" pipe size		32	1.500			85.50		85.50	129
4140	20" pipe size	8 Skwk	37	1.730			99		99	149
4150	22" pipe size		35	1.829			104		104	157
4160	24" pipe size		34	1.882			107		107	162
4170	28" pipe size		33	1.939			111		111	167
4180	30" pipe size		32	2			114		114	172
4190	36" pipe size		31	2.065			118		118	177

22 11 13 - Facility Water Distribution Piping

22 11 13.78 Pipe, High Density Polyethylene Plastic (HDPE)	Crew	Daily Output	Labor-Hours	Unit	Material	2021 Bare Costs Labor	Equipment	Total	Total Incl O&P
4200	42" pipe size	8 Skwk	30	2.133	Ea.		122		122	183
4210	48" pipe size	9 Skwk	33	2.182			125		125	187
4220	54" pipe size	"	31	2.323			133		133	200
5000	Dual wall contained pipe									
5040	Straight									
5054	1" DR 11 x 3" DR 11				L.F.	5.35			5.35	5.85
5058	1" DR 11 x 4" DR 11					5.60			5.60	6.20
5062	1-1/2" DR 11 x 4" DR 17					6.20			6.20	6.80
5066	2" DR 11 x 4" DR 17					6.55			6.55	7.20
5070	2" DR 11 x 6" DR 17					9.90			9.90	10.90
5074	3" DR 11 x 6" DR 17					11.10			11.10	12.20
5078	3" DR 11 x 6" DR 26					9.45			9.45	10.40
5086	3" DR 17 x 8" DR 17					12.05			12.05	13.30
5090	4" DR 11 x 8" DR 17					16.75			16.75	18.40
5094	4" DR 17 x 8" DR 26					13.10			13.10	14.40
5098	6" DR 11 x 10" DR 17					26			26	28.50
5102	6" DR 17 x 10" DR 26					17.65			17.65	19.45
5106	6" DR 26 x 10" DR 26					16.55			16.55	18.20
5110	8" DR 17 x 12" DR 26					25.50			25.50	28
5114	8" DR 26 x 12" DR 32.5					23.50			23.50	25.50
5118	10" DR 17 x 14" DR 26					38			38	42
5122	10" DR 17 x 16" DR 26					27.50			27.50	30
5126	10" DR 26 x 16" DR 26					36.50			36.50	40
5130	12" DR 26 x 16" DR 26					38			38	42
5134	12" DR 17 x 18" DR 26					53			53	58.50
5138	12" DR 26 x 18" DR 26					50.50			50.50	55.50
5142	14" DR 26 x 20" DR 32.5					51			51	56.50
5146	16" DR 26 x 22" DR 32.5					56.50			56.50	62.50
5150	18" DR 26 x 24" DR 32.5					70.50			70.50	77.50
5154	20" DR 32.5 x 28" DR 32.5					79.50			79.50	87.50
5158	22" DR 32.5 x 30" DR 32.5					79.50			79.50	87.50
5162	24" DR 32.5 x 32" DR 32.5					83			83	91
5166	36" DR 32.5 x 42" DR 32.5					97			97	107
5300	Force transfer coupling									
5354	1" DR 11 x 3" DR 11				Ea.	185			185	203
5358	1" DR 11 x 4" DR 17					217			217	239
5362	1-1/2" DR 11 x 4" DR 17					208			208	228
5366	2" DR 11 x 4" DR 17					208			208	228
5370	2" DR 11 x 6" DR 17					355			355	390
5374	3" DR 11 x 6" DR 17					380			380	420
5378	3" DR 11 x 6" DR 26					380			380	420
5382	3" DR 11 x 8" DR 11					380			380	420
5386	3" DR 11 x 8" DR 17					415			415	460
5390	4" DR 11 x 8" DR 17					425			425	470
5394	4" DR 17 x 8" DR 26					430			430	475
5398	6" DR 11 x 10" DR 17					510			510	560
5402	6" DR 17 x 10" DR 26					485			485	535
5406	6" DR 26 x 10" DR 26					665			665	730
5410	8" DR 17 x 12" DR 26					705			705	775
5414	8" DR 26 x 12" DR 32.5					815			815	895
5418	10" DR 17 x 14" DR 26					960			960	1,050
5422	10" DR 17 x 16" DR 26					975			975	1,075
5426	10" DR 26 x 16" DR 26					1,075			1,075	1,200

237

For customer support on your Site Work & Landscape Costs with RSMeans Data, call 800.448.8182.

22 11 13.78 Pipe, High Density Polyethylene Plastic (HDPE)	Crew	Daily Output	Labor-Hours	Unit	Material	2021 Bare Costs Labor	Equipment	Total	Total Incl O&P
5430	12" DR 26 x 16" DR 26				Ea.	1,325			1,325	1,450
5434	12" DR 17 x 18" DR 26					1,325			1,325	1,450
5438	12" DR 26 x 18" DR 26					1,450			1,450	1,600
5442	14" DR 26 x 20" DR 32.5					1,600			1,600	1,775
5446	16" DR 26 x 22" DR 32.5					1,775			1,775	1,950
5450	18" DR 26 x 24" DR 32.5					1,975			1,975	2,175
5454	20" DR 32.5 x 28" DR 32.5					2,150			2,150	2,375
5458	22" DR 32.5 x 30" DR 32.5					2,375			2,375	2,625
5462	24" DR 32.5 x 32" DR 32.5					2,625			2,625	2,900
5466	36" DR 32.5 x 42" DR 32.5					2,825			2,825	3,100
5600	90° elbow									
5654	1" DR 11 x 3" DR 11				Ea.	202			202	222
5658	1" DR 11 x 4" DR 17					238			238	262
5662	1-1/2" DR 11 x 4" DR 17					245			245	270
5666	2" DR 11 x 4" DR 17					240			240	264
5670	2" DR 11 x 6" DR 17					345			345	380
5674	3" DR 11 x 6" DR 17					375			375	415
5678	3" DR 17 x 6" DR 26					380			380	420
5682	3" DR 17 x 8" DR 11					510			510	560
5686	3" DR 17 x 8" DR 17					545			545	600
5690	4" DR 11 x 8" DR 17					460			460	510
5694	4" DR 17 x 8" DR 26					445			445	490
5698	6" DR 11 x 10" DR 17					805			805	890
5702	6" DR 17 x 10" DR 26					820			820	900
5706	6" DR 26 x 10" DR 26					800			800	880
5710	8" DR 17 x 12" DR 26					1,075			1,075	1,175
5714	8" DR 26 x 12" DR 32.5					885			885	975
5718	10" DR 17 x 14" DR 26					1,050			1,050	1,150
5722	10" DR 17 x 16" DR 26					1,025			1,025	1,125
5726	10" DR 26 x 16" DR 26					1,025			1,025	1,125
5730	12" DR 26 x 16" DR 26					1,050			1,050	1,150
5734	12" DR 17 x 18" DR 26					1,150			1,150	1,250
5738	12" DR 26 x 18" DR 26					1,150			1,150	1,275
5742	14" DR 26 x 20" DR 32.5					1,250			1,250	1,375
5746	16" DR 26 x 22" DR 32.5					1,300			1,300	1,425
5750	18" DR 26 x 24" DR 32.5					1,350			1,350	1,475
5754	20" DR 32.5 x 28" DR 32.5					1,400			1,400	1,525
5758	22" DR 32.5 x 30" DR 32.5					1,425			1,425	1,575
5762	24" DR 32.5 x 32" DR 32.5					1,475			1,475	1,625
5766	36" DR 32.5 x 42" DR 32.5					1,525			1,525	1,700
5800	45° elbow									
5804	1" DR 11 x 3" DR 11				Ea.	234			234	258
5808	1" DR 11 x 4" DR 17					192			192	211
5812	1-1/2" DR 11 x 4" DR 17					195			195	215
5816	2" DR 11 x 4" DR 17					186			186	205
5820	2" DR 11 x 6" DR 17					267			267	294
5824	3" DR 11 x 6" DR 17					287			287	315
5828	3" DR 17 x 6" DR 26					289			289	320
5832	3" DR 17 x 8" DR 11					370			370	405
5836	3" DR 17 x 8" DR 17					415			415	455
5840	4" DR 11 x 8" DR 17					340			340	375
5844	4" DR 17 x 8" DR 26					345			345	380
5848	6" DR 11 x 10" DR 17					535			535	590

22 11 13.78 Pipe, High Density Polyethylene Plastic (HDPE)	Crew	Daily Output	Labor-Hours	Unit	Material	2021 Bare Costs Labor	Equipment	Total	Total Incl O&P
5852	6" DR 17 x 10" DR 26				Ea.	545			545	600
5856	6" DR 26 x 10" DR 26					610			610	675
5860	8" DR 17 x 12" DR 26					725			725	800
5864	8" DR 26 x 12" DR 32.5					680			680	750
5868	10" DR 17 x 14" DR 26					770			770	845
5872	10" DR 17 x 16" DR 26					740			740	815
5876	10" DR 26 x 16" DR 26					785			785	865
5880	12" DR 26 x 16" DR 26					710			710	780
5884	12" DR 17 x 18" DR 26					1,000			1,000	1,100
5888	12" DR 26 x 18" DR 26					895			895	985
5892	14" DR 26 x 20" DR 32.5					970			970	1,075
5896	16" DR 26 x 22" DR 32.5					1,000			1,000	1,100
5900	18" DR 26 x 24" DR 32.5					1,075			1,075	1,175
5904	20" DR 32.5 x 28" DR 32.5					1,050			1,050	1,175
5908	22" DR 32.5 x 30" DR 32.5					1,100			1,100	1,200
5912	24" DR 32.5 x 32" DR 32.5					1,150			1,150	1,250
5916	36" DR 32.5 x 42" DR 32.5					1,175			1,175	1,300
6000	Access port with 4" riser									
6050	1" DR 11 x 4" DR 17				Ea.	279			279	305
6054	1-1/2" DR 11 x 4" DR 17					281			281	310
6058	2" DR 11 x 6" DR 17					340			340	375
6062	3" DR 11 x 6" DR 17					350			350	380
6066	3" DR 17 x 6" DR 26					335			335	370
6070	3" DR 17 x 8" DR 11					350			350	380
6074	3" DR 17 x 8" DR 17					360			360	395
6078	4" DR 11 x 8" DR 17					370			370	410
6082	4" DR 17 x 8" DR 26					355			355	390
6086	6" DR 11 x 10" DR 17					435			435	475
6090	6" DR 17 x 10" DR 26					395			395	435
6094	6" DR 26 x 10" DR 26					415			415	455
6098	8" DR 17 x 12" DR 26					445			445	490
6102	8" DR 26 x 12" DR 32.5					440			440	480
6200	End termination with vent plug									
6204	1" DR 11 x 3" DR 11				Ea.	94.50			94.50	104
6208	1" DR 11 x 4" DR 17					138			138	152
6212	1-1/2" DR 11 x 4" DR 17					138			138	152
6216	2" DR 11 x 4" DR 17					138			138	152
6220	2" DR 11 x 6" DR 17					280			280	310
6224	3" DR 11 x 6" DR 17					280			280	310
6228	3" DR 17 x 6" DR 26					280			280	310
6232	3" DR 17 x 8" DR 11					271			271	298
6236	3" DR 17 x 8" DR 17					305			305	335
6240	4" DR 11 x 8" DR 17					360			360	400
6244	4" DR 17 x 8" DR 26					340			340	375
6248	6" DR 11 x 10" DR 17					395			395	435
6252	6" DR 17 x 10" DR 26					365			365	400
6256	6" DR 26 x 10" DR 26					530			530	585
6260	8" DR 17 x 12" DR 26					555			555	610
6264	8" DR 26 x 12" DR 32.5					665			665	735
6268	10" DR 17 x 14" DR 26					885			885	970
6272	10" DR 17 x 16" DR 26					835			835	915
6276	10" DR 26 x 16" DR 26					940			940	1,025
6280	12" DR 26 x 16" DR 26					1,050			1,050	1,150

For customer support on your Site Work & Landscape Costs with RSMeans Data, call 800.448.8182.

239

22 11 13.78 Pipe, High Density Polyethylene Plastic (HDPE)	Crew	Daily Output	Labor-Hours	Unit	Material	2021 Bare Costs Labor	Equipment	Total	Total Incl O&P
6284	12" DR 17 x 18" DR 26				Ea.	1,225			1,225	1,325
6288	12" DR 26 x 18" DR 26					1,325			1,325	1,450
6292	14" DR 26 x 20" DR 32.5					1,475			1,475	1,625
6296	16" DR 26 x 22" DR 32.5					1,650			1,650	1,825
6300	18" DR 26 x 24" DR 32.5					1,875			1,875	2,050
6304	20" DR 32.5 x 28" DR 32.5					2,075			2,075	2,275
6308	22" DR 32.5 x 30" DR 32.5					2,325			2,325	2,550
6312	24" DR 32.5 x 32" DR 32.5					2,600			2,600	2,875
6316	36" DR 32.5 x 42" DR 32.5				▼	2,875			2,875	3,150
6600	Tee									
6604	1" DR 11 x 3" DR 11				Ea.	235			235	259
6608	1" DR 11 x 4" DR 17					255			255	281
6612	1-1/2" DR 11 x 4" DR 17					264			264	290
6616	2" DR 11 x 4" DR 17					240			240	263
6620	2" DR 11 x 6" DR 17					385			385	420
6624	3" DR 11 x 6" DR 17					450			450	500
6628	3" DR 17 x 6" DR 26					430			430	470
6632	3" DR 17 x 8" DR 11					525			525	580
6636	3" DR 17 x 8" DR 17					570			570	625
6640	4" DR 11 x 8" DR 17					685			685	755
6644	4" DR 17 x 8" DR 26					690			690	755
6648	6" DR 11 x 10" DR 17					895			895	985
6652	6" DR 17 x 10" DR 26					895			895	985
6656	6" DR 26 x 10" DR 26					1,100			1,100	1,200
6660	8" DR 17 x 12" DR 26					1,100			1,100	1,200
6664	8" DR 26 x 12" DR 32.5					1,425			1,425	1,550
6668	10" DR 17 x 14" DR 26					1,650			1,650	1,825
6672	10" DR 17 x 16" DR 26					1,825			1,825	2,025
6676	10" DR 26 x 16" DR 26					1,825			1,825	2,025
6680	12" DR 26 x 16" DR 26					2,100			2,100	2,300
6684	12" DR 17 x 18" DR 26					2,100			2,100	2,300
6688	12" DR 26 x 18" DR 26					2,375			2,375	2,600
6692	14" DR 26 x 20" DR 32.5					2,675			2,675	2,950
6696	16" DR 26 x 22" DR 32.5					3,025			3,025	3,325
6700	18" DR 26 x 24" DR 32.5					3,450			3,450	3,775
6704	20" DR 32.5 x 28" DR 32.5					3,875			3,875	4,275
6708	22" DR 32.5 x 30" DR 32.5					4,400			4,400	4,850
6712	24" DR 32.5 x 32" DR 32.5					4,975			4,975	5,475
6716	36" DR 32.5 x 42" DR 32.5				▼	5,725			5,725	6,275
6800	Wye									
6816	2" DR 11 x 4" DR 17				Ea.	585			585	640
6820	2" DR 11 x 6" DR 17					720			720	795
6824	3" DR 11 x 6" DR 17					765			765	840
6828	3" DR 17 x 6" DR 26					765			765	840
6832	3" DR 17 x 8" DR 11					815			815	895
6836	3" DR 17 x 8" DR 17					885			885	970
6840	4" DR 11 x 8" DR 17					970			970	1,075
6844	4" DR 17 x 8" DR 26					610			610	670
6848	6" DR 11 x 10" DR 17					1,500			1,500	1,650
6852	6" DR 17 x 10" DR 26					935			935	1,025
6856	6" DR 26 x 10" DR 26					1,300			1,300	1,425
6860	8" DR 17 x 12" DR 26					1,825			1,825	2,000
6864	8" DR 26 x 12" DR 32.5				▼	1,500			1,500	1,650

For customer support on your Site Work & Landscape Costs with RSMeans Data, call 800.448.8182.

22 11 Facility Water Distribution

22 11 13 – Facility Water Distribution Piping

22 11 13.78 Pipe, High Density Polyethylene Plastic (HDPE)	Crew	Daily Output	Labor-Hours	Unit	Material	2021 Bare Costs Labor	Equipment	Total	Total Incl O&P
6868	10" DR 17 x 14" DR 26				Ea.	1,600			1,600	1,750
6872	10" DR 17 x 16" DR 26					1,725			1,725	1,900
6876	10" DR 26 x 16" DR 26					1,875			1,875	2,050
6880	12" DR 26 x 16" DR 26					2,025			2,025	2,225
6884	12" DR 17 x 18" DR 26					2,175			2,175	2,375
6888	12" DR 26 x 18" DR 26					2,350			2,350	2,575
6892	14" DR 26 x 20" DR 32.5					2,525			2,525	2,775
6896	16" DR 26 x 22" DR 32.5					2,650			2,650	2,925
6900	18" DR 26 x 24" DR 32.5					3,025			3,025	3,350
6904	20" DR 32.5 x 28" DR 32.5					3,175			3,175	3,500
6908	22" DR 32.5 x 30" DR 32.5					3,425			3,425	3,775
6912	24" DR 32.5 x 32" DR 32.5					3,700			3,700	4,075
9000	Welding labor per joint, not including welding machine									
9010	Pipe joint size, outer pipe (cost based on the thickest walls)									
9020	Straight pipe									
9050	3" pipe size	4 Skwk	96	.333	Ea.		19.05		19.05	28.50
9060	4" pipe size	"	77	.416			23.50		23.50	35.50
9070	6" pipe size	5 Skwk	60	.667			38		38	57.50
9080	8" pipe size	"	40	1			57		57	86
9090	10" pipe size	6 Skwk	41	1.171			67		67	101
9100	12" pipe size		39	1.231			70.50		70.50	106
9110	14" pipe size		38	1.263			72		72	109
9120	16" pipe size		35	1.371			78.50		78.50	118
9130	18" pipe size	8 Skwk	45	1.422			81		81	122
9140	20" pipe size		42	1.524			87		87	131
9150	22" pipe size		40	1.600			91.50		91.50	137
9160	24" pipe size		38	1.684			96		96	145
9170	28" pipe size		37	1.730			99		99	149
9180	30" pipe size		36	1.778			102		102	153
9190	32" pipe size		35	1.829			104		104	157
9200	42" pipe size		32	2			114		114	172

22 11 19 – Domestic Water Piping Specialties

22 11 19.38 Water Supply Meters

		Crew	Daily Output	Labor-Hours	Unit	Material	2021 Bare Costs Labor	Equipment	Total	Total Incl O&P
0010	**WATER SUPPLY METERS**									
1000	Detector, serves dual systems such as fire and domestic or									
1020	process water, wide range cap., UL and FM approved									
1100	3" mainline x 2" by-pass, 400 GPM	Q-1	3.60	4.444	Ea.	8,425	271		8,696	9,675
1140	4" mainline x 2" by-pass, 700 GPM	"	2.50	6.400		8,425	390		8,815	9,850
1180	6" mainline x 3" by-pass, 1,600 GPM	Q-2	2.60	9.231		12,900	585		13,485	15,100
1220	8" mainline x 4" by-pass, 2,800 GPM		2.10	11.429		19,100	720		19,820	22,100
1260	10" mainline x 6" by-pass, 4,400 GPM		2	12		27,300	760		28,060	31,200
1300	10" x 12" mainlines x 6" by-pass, 5,400 GPM		1.70	14.118		37,000	890		37,890	42,000
2000	Domestic/commercial, bronze									
2020	Threaded									
2060	5/8" diameter, to 20 GPM	1 Plum	16	.500	Ea.	56.50	34		90.50	113
2080	3/4" diameter, to 30 GPM		14	.571		103	38.50		141.50	172
2100	1" diameter, to 50 GPM		12	.667		157	45		202	241
2300	Threaded/flanged									
2340	1-1/2" diameter, to 100 GPM	1 Plum	8	1	Ea.	385	67.50		452.50	520
2360	2" diameter, to 160 GPM	"	6	1.333	"	520	90.50		610.50	705
2600	Flanged, compound									
2640	3" diameter, 320 GPM	Q-1	3	5.333	Ea.	2,650	325		2,975	3,400

22 11 19.38 Water Supply Meters

		Crew	Daily Output	Labor-Hours	Unit	Material	2021 Bare Costs Labor	2021 Bare Costs Equipment	Total	Total Incl O&P
2660	4" diameter, to 500 GPM	Q-1	1.50	10.667	Ea.	4,250	650		4,900	5,650
2680	6" diameter, to 1,000 GPM		1	16		6,875	975		7,850	9,025
2700	8" diameter, to 1,800 GPM		.80	20		10,700	1,225		11,925	13,600
7000	Turbine									
7260	Flanged									
7300	2" diameter, to 160 GPM	1 Plum	7	1.143	Ea.	645	77.50		722.50	825
7320	3" diameter, to 450 GPM	Q-1	3.60	4.444		1,100	271		1,371	1,625
7340	4" diameter, to 650 GPM	"	2.50	6.400		1,825	390		2,215	2,575
7360	6" diameter, to 1,800 GPM	Q-2	2.60	9.231		3,025	585		3,610	4,200
7380	8" diameter, to 2,500 GPM		2.10	11.429		5,175	720		5,895	6,775
7400	10" diameter, to 5,500 GPM		1.70	14.118		6,950	890		7,840	8,975

22 11 19.42 Backflow Preventers

		Crew	Daily Output	Labor-Hours	Unit	Material	2021 Bare Costs Labor	2021 Bare Costs Equipment	Total	Total Incl O&P
0010	**BACKFLOW PREVENTERS**, Includes valves									
0020	and four test cocks, corrosion resistant, automatic operation									
1000	Double check principle									
1010	Threaded, with ball valves									
1020	3/4" pipe size	1 Plum	16	.500	Ea.	240	34		274	315
1030	1" pipe size		14	.571		335	38.50		373.50	430
1040	1-1/2" pipe size		10	.800		640	54		694	785
1050	2" pipe size		7	1.143		690	77.50		767.50	875
1080	Threaded, with gate valves									
1100	3/4" pipe size	1 Plum	16	.500	Ea.	1,275	34		1,309	1,450
1120	1" pipe size		14	.571		1,300	38.50		1,338.50	1,475
1140	1-1/2" pipe size		10	.800		1,150	54		1,204	1,350
1160	2" pipe size		7	1.143		2,050	77.50		2,127.50	2,375
1200	Flanged, valves are gate									
1210	3" pipe size	Q-1	4.50	3.556	Ea.	3,250	217		3,467	3,900
1220	4" pipe size	"	3	5.333		2,900	325		3,225	3,675
1230	6" pipe size	Q-2	3	8		4,525	505		5,030	5,725
1240	8" pipe size		2	12		7,725	760		8,485	9,625
1250	10" pipe size		1	24		11,600	1,525		13,125	15,100
1300	Flanged, valves are OS&Y									
1380	3" pipe size	Q-1	4.50	3.556	Ea.	2,975	217		3,192	3,600
1400	4" pipe size	"	3	5.333		3,700	325		4,025	4,525
1420	6" pipe size	Q-2	3	8		5,275	505		5,780	6,550
4000	Reduced pressure principle									
4100	Threaded, bronze, valves are ball									
4120	3/4" pipe size	1 Plum	16	.500	Ea.	450	34		484	545
4140	1" pipe size		14	.571		480	38.50		518.50	590
4150	1-1/4" pipe size		12	.667		1,175	45		1,220	1,375
4160	1-1/2" pipe size		10	.800		970	54		1,024	1,150
4180	2" pipe size		7	1.143		1,550	77.50		1,627.50	1,825
5000	Flanged, bronze, valves are OS&Y									
5060	2-1/2" pipe size	Q-1	5	3.200	Ea.	5,225	195		5,420	6,050
5080	3" pipe size		4.50	3.556		5,725	217		5,942	6,625
5100	4" pipe size		3	5.333		7,350	325		7,675	8,575
5120	6" pipe size	Q-2	3	8		10,400	505		10,905	12,200
5200	Flanged, iron, valves are gate									
5210	2-1/2" pipe size	Q-1	5	3.200	Ea.	3,200	195		3,395	3,800
5220	3" pipe size		4.50	3.556		2,475	217		2,692	3,050
5230	4" pipe size		3	5.333		3,675	325		4,000	4,500
5240	6" pipe size	Q-2	3	8		5,025	505		5,530	6,275

22 11 Facility Water Distribution

22 11 19 – Domestic Water Piping Specialties

22 11 19.42 Backflow Preventers		Crew	Daily Output	Labor-Hours	Unit	Material	2021 Bare Costs Labor	Equipment	Total	Total Incl O&P
5250	8" pipe size	Q-2	2	12	Ea.	9,275	760		10,035	11,300
5260	10" pipe size	▼	1	24	▼	14,100	1,525		15,625	17,800
5600	Flanged, iron, valves are OS&Y									
5660	2-1/2" pipe size	Q-1	5	3.200	Ea.	3,825	195		4,020	4,525
5680	3" pipe size		4.50	3.556		3,250	217		3,467	3,900
5700	4" pipe size	▼	3	5.333		4,725	325		5,050	5,675
5720	6" pipe size	Q-2	3	8		5,650	505		6,155	6,950
5740	8" pipe size		2	12		9,950	760		10,710	12,000
5760	10" pipe size	▼	1	24	▼	19,200	1,525		20,725	23,400

22 11 19.54 Water Hammer Arresters/Shock Absorbers

		Crew	Daily Output	Labor-Hours	Unit	Material	2021 Bare Costs Labor	Equipment	Total	Total Incl O&P
0010	**WATER HAMMER ARRESTERS/SHOCK ABSORBERS**									
0490	Copper									
0500	3/4" male IPS for 1 to 11 fixtures	1 Plum	12	.667	Ea.	31	45		76	102
0600	1" male IPS for 12 to 32 fixtures		8	1		52	67.50		119.50	158
0700	1-1/4" male IPS for 33 to 60 fixtures		8	1		52.50	67.50		120	159
0800	1-1/2" male IPS for 61 to 113 fixtures		8	1		77	67.50		144.50	186
0900	2" male IPS for 114 to 154 fixtures		8	1		118	67.50		185.50	231
1000	2-1/2" male IPS for 155 to 330 fixtures	▼	4	2	▼	330	135		465	565

22 11 19.64 Hydrants

		Crew	Daily Output	Labor-Hours	Unit	Material	2021 Bare Costs Labor	Equipment	Total	Total Incl O&P
0010	**HYDRANTS**									
0050	Wall type, moderate climate, bronze, encased									
0200	3/4" IPS connection	1 Plum	16	.500	Ea.	930	34		964	1,075
0300	1" IPS connection		14	.571		2,475	38.50		2,513.50	2,750
0500	Anti-siphon type, 3/4" connection	▼	16	.500	▼	865	34		899	1,000
1000	Non-freeze, bronze, exposed									
1100	3/4" IPS connection, 4" to 9" thick wall	1 Plum	14	.571	Ea.	530	38.50		568.50	640
1120	10" to 14" thick wall		12	.667		465	45		510	580
1140	15" to 19" thick wall		12	.667		590	45		635	720
1160	20" to 24" thick wall	▼	10	.800	▼	805	54		859	965
1200	For 1" IPS connection, add					15%	10%			
1240	For 3/4" adapter type vacuum breaker, add				Ea.	93.50			93.50	103
1280	For anti-siphon type, add				"	198			198	218
2000	Non-freeze bronze, encased, anti-siphon type									
2100	3/4" IPS connection, 5" to 9" thick wall	1 Plum	14	.571	Ea.	1,325	38.50		1,363.50	1,500
2120	10" to 14" thick wall		12	.667		1,725	45		1,770	1,975
2140	15" to 19" thick wall		12	.667		1,850	45		1,895	2,100
2160	20" to 24" thick wall	▼	10	.800	▼	1,900	54		1,954	2,175
2200	For 1" IPS connection, add					10%	10%			
3000	Ground box type, bronze frame, 3/4" IPS connection									
3080	Non-freeze, all bronze, polished face, set flush									
3100	2' depth of bury	1 Plum	8	1	Ea.	1,200	67.50		1,267.50	1,425
3120	3' depth of bury		8	1		1,300	67.50		1,367.50	1,525
3140	4' depth of bury		8	1		1,400	67.50		1,467.50	1,625
3160	5' depth of bury		7	1.143		1,500	77.50		1,577.50	1,775
3180	6' depth of bury		7	1.143		1,575	77.50		1,652.50	1,850
3200	7' depth of bury		6	1.333		1,800	90.50		1,890.50	2,100
3220	8' depth of bury		5	1.600		1,900	108		2,008	2,225
3240	9' depth of bury		4	2		2,000	135		2,135	2,400
3260	10' depth of bury	▼	4	2		2,075	135		2,210	2,500
3400	For 1" IPS connection, add					15%	10%			
3450	For 1-1/4" IPS connection, add					325%	14%			
3500	For 1-1/2" IPS connection, add	▼			▼	370%	18%			

22 11 19 – Domestic Water Piping Specialties

22 11 19.64 Hydrants	Crew	Daily Output	Labor-Hours	Unit	Material	2021 Bare Costs Labor	Equipment	Total	Total Incl O&P
3550	For 2" IPS connection, add				Ea.	445%	24%			
3600	For tapped drain port in box, add				↓	120			120	132
4000	Non-freeze, CI body, bronze frame & scoriated cover									
4010	with hose storage									
4100	2' depth of bury	1 Plum	7	1.143	Ea.	2,350	77.50		2,427.50	2,700
4120	3' depth of bury		7	1.143		2,450	77.50		2,527.50	2,825
4140	4' depth of bury		7	1.143		2,525	77.50		2,602.50	2,900
4160	5' depth of bury		6.50	1.231		2,575	83.50		2,658.50	2,950
4180	6' depth of bury		6	1.333		2,625	90.50		2,715.50	3,000
4200	7' depth of bury		5.50	1.455		2,700	98.50		2,798.50	3,125
4220	8' depth of bury		5	1.600		2,775	108		2,883	3,200
4240	9' depth of bury		4.50	1.778		2,600	120		2,720	3,025
4260	10' depth of bury	↓	4	2		2,975	135		3,110	3,475
4280	For 1" IPS connection, add					570			570	625
4300	For tapped drain port in box, add				↓	126			126	139
5000	Moderate climate, all bronze, polished face									
5020	and scoriated cover, set flush									
5100	3/4" IPS connection	1 Plum	16	.500	Ea.	880	34		914	1,025
5120	1" IPS connection	"	14	.571		2,000	38.50		2,038.50	2,250
5200	For tapped drain port in box, add				↓	126			126	139
6000	Ground post type, all non-freeze, all bronze, aluminum casing									
6010	guard, exposed head, 3/4" IPS connection									
6100	2' depth of bury	1 Plum	8	1	Ea.	1,400	67.50		1,467.50	1,650
6120	3' depth of bury		8	1		1,475	67.50		1,542.50	1,725
6140	4' depth of bury		8	1		1,650	67.50		1,717.50	1,925
6160	5' depth of bury		7	1.143		1,550	77.50		1,627.50	1,825
6180	6' depth of bury		7	1.143		1,900	77.50		1,977.50	2,200
6200	7' depth of bury		6	1.333		1,775	90.50		1,865.50	2,075
6220	8' depth of bury		5	1.600		1,875	108		1,983	2,225
6240	9' depth of bury		4	2		2,000	135		2,135	2,400
6260	10' depth of bury	↓	4	2	↓	2,100	135		2,235	2,525
6300	For 1" IPS connection, add					40%	10%			
6350	For 1-1/4" IPS connection, add					140%	14%			
6400	For 1-1/2" IPS connection, add					225%	18%			
6450	For 2" IPS connection, add					315%	24%			

22 11 23 – Domestic Water Pumps

22 11 23.10 General Utility Pumps

		Crew	Daily Output	Labor-Hours	Unit	Material	2021 Bare Costs Labor	Equipment	Total	Total Incl O&P
0010	**GENERAL UTILITY PUMPS**									
2000	Single stage									
3000	Double suction,									
3190	75 HP, to 2,500 GPM	Q-3	.28	114	Ea.	16,900	7,375		24,275	29,600
3220	100 HP, to 3,000 GPM		.26	123		21,000	7,925		28,925	34,900
3240	150 HP, to 4,000 GPM	↓	.24	133	↓	24,700	8,600		33,300	40,000

244

For customer support on your Site Work & Landscape Costs with RSMeans Data, call 800.448.8182.

22 12 Facility Potable-Water Storage Tanks

22 12 21 – Facility Underground Potable-Water Storage Tanks

22 12 21.13 Fiberglass, Undrgrnd Pot.-Water Storage Tanks	Crew	Daily Output	Labor-Hours	Unit	Material	2021 Bare Costs Labor	Equipment	Total	Total Incl O&P
0010 **FIBERGLASS, UNDERGROUND POTABLE-WATER STORAGE TANKS**									
0020 Excludes excavation, backfill & piping									
0030 Single wall									
2000 600 gallon capacity	B-21B	3.75	10.667	Ea.	4,325	515	127	4,967	5,650
2010 1,000 gallon capacity		3.50	11.429		5,450	550	136	6,136	6,975
2020 2,000 gallon capacity		3.25	12.308		7,675	595	146	8,416	9,475
2030 4,000 gallon capacity		3	13.333		10,100	645	159	10,904	12,200
2040 6,000 gallon capacity		2.65	15.094		11,500	730	180	12,410	13,900
2050 8,000 gallon capacity		2.30	17.391		14,100	840	207	15,147	17,000
2060 10,000 gallon capacity		2	20		15,400	965	238	16,603	18,700
2070 12,000 gallon capacity		1.50	26.667		22,100	1,275	315	23,690	26,700
2080 15,000 gallon capacity		1	40		24,200	1,925	475	26,600	30,100
2090 20,000 gallon capacity		.75	53.333		31,600	2,575	635	34,810	39,300
2100 25,000 gallon capacity		.50	80		48,000	3,850	950	52,800	59,500
2110 30,000 gallon capacity		.35	114		58,000	5,500	1,350	64,850	73,500
2120 40,000 gallon capacity		.30	133		79,000	6,425	1,575	87,000	98,000

22 13 Facility Sanitary Sewerage

22 13 16 – Sanitary Waste and Vent Piping

22 13 16.20 Pipe, Cast Iron

	Crew	Daily Output	Labor-Hours	Unit	Material	2021 Bare Costs Labor	Equipment	Total	Total Incl O&P
0010 **PIPE, CAST IRON**, Soil, on clevis hanger assemblies, 5' OC									
0020 Single hub, service wt., lead & oakum joints 10' OC									
2120 2" diameter	Q-1	63	.254	L.F.	13.90	15.45		29.35	38.50
2140 3" diameter		60	.267		18.65	16.25		34.90	45
2160 4" diameter		55	.291		24.50	17.75		42.25	53
2180 5" diameter	Q-2	76	.316		29	19.95		48.95	62
2200 6" diameter	"	73	.329		41.50	21		62.50	76.50
2220 8" diameter	Q-3	59	.542		66	35		101	125
2240 10" diameter		54	.593		108	38		146	176
2260 12" diameter		48	.667		156	43		199	235
2320 For service weight, double hub, add					10%				
2340 For extra heavy, single hub, add					48%	4%			
2360 For extra heavy, double hub, add					71%	4%			
2400 Lead for caulking (1#/diam. in.)	Q-1	160	.100	Lb.	1.02	6.10		7.12	10.20
2420 Oakum for caulking (1/8#/diam. in.)	"	40	.400	"	4.15	24.50		28.65	41
2960 To delete hangers, subtract									
2970 2" diam. to 4" diam.					16%	19%			
2980 5" diam. to 8" diam.					14%	14%			
2990 10" diam. to 15" diam.					13%	19%			
3000 Single hub, service wt., push-on gasket joints 10' OC									
3010 2" diameter	Q-1	66	.242	L.F.	15.15	14.75		29.90	38.50
3020 3" diameter		63	.254		20.50	15.45		35.95	45.50
3030 4" diameter		57	.281		26.50	17.10		43.60	54.50
3040 5" diameter	Q-2	79	.304		32.50	19.20		51.70	64
3050 6" diameter	"	75	.320		45	20		65	79.50
3060 8" diameter	Q-3	62	.516		73.50	33.50		107	130
3070 10" diameter		56	.571		121	37		158	188
3080 12" diameter		49	.653		172	42		214	252
3082 15" diameter		40	.800		250	51.50		301.50	350
3100 For service weight, double hub, add					65%				
3110 For extra heavy, single hub, add					48%	4%			

22 13 Facility Sanitary Sewerage

22 13 16 – Sanitary Waste and Vent Piping

22 13 16.20 Pipe, Cast Iron		Crew	Daily Output	Labor-Hours	Unit	Material	2021 Bare Costs Labor	Equipment	Total	Total Incl O&P
3120	For extra heavy, double hub, add					29%	4%			
3130	To delete hangers, subtract									
3140	2" diam. to 4" diam.					12%	21%			
3150	5" diam. to 8" diam.					10%	16%			
3160	10" diam. to 15" diam.					9%	21%			
4000	No hub, couplings 10' OC									
4100	1-1/2" diameter	Q-1	71	.225	L.F.	9.95	13.75		23.70	31.50
4120	2" diameter		67	.239		10.35	14.55		24.90	33
4140	3" diameter		64	.250		13.55	15.25		28.80	37.50
4160	4" diameter		58	.276		17.85	16.80		34.65	44.50
4180	5" diameter	Q-2	83	.289		21.50	18.25		39.75	51
4200	6" diameter	"	79	.304		30.50	19.20		49.70	62
4220	8" diameter	Q-3	69	.464		55	30		85	105
4240	10" diameter		61	.525		90	34		124	150
4244	12" diameter		58	.552		136	35.50		171.50	202
4248	15" diameter		52	.615		204	39.50		243.50	284
4280	To delete hangers, subtract									
4290	1-1/2" diam. to 6" diam.					22%	47%			
4300	8" diam. to 10" diam.					21%	44%			
4310	12" diam. to 15" diam.					19%	40%			

22 13 16.30 Pipe Fittings, Cast Iron

		Crew	Daily Output	Labor-Hours	Unit	Material	Labor	Equipment	Total	Total Incl O&P
0010	**PIPE FITTINGS, CAST IRON**, Soil									
0040	Hub and spigot, service weight, lead & oakum joints									
0080	1/4 bend, 2"	Q-1	16	1	Ea.	25.50	61		86.50	119
0120	3"		14	1.143		34	69.50		103.50	142
0140	4"		13	1.231		53.50	75		128.50	171
0160	5"	Q-2	18	1.333		74.50	84		158.50	208
0180	6"	"	17	1.412		93	89		182	235
0200	8"	Q-3	11	2.909		280	187		467	590
0220	10"		10	3.200		410	206		616	760
0224	12"		9	3.556		555	229		784	950
0266	Closet bend, 3" diameter with flange 10" x 16"	Q-1	14	1.143		141	69.50		210.50	259
0268	16" x 16"		12	1.333		157	81		238	294
0270	Closet bend, 4" diameter, 2-1/2" x 4" ring, 6" x 16"		13	1.231		123	75		198	247
0280	8" x 16"		13	1.231		107	75		182	230
0290	10" x 12"		12	1.333		100	81		181	231
0300	10" x 18"		11	1.455		147	88.50		235.50	294
0310	12" x 16"		11	1.455		124	88.50		212.50	268
0330	16" x 16"		10	1.600		161	97.50		258.50	325
0340	1/8 bend, 2"		16	1		18.15	61		79.15	111
0350	3"		14	1.143		28.50	69.50		98	136
0360	4"		13	1.231		41.50	75		116.50	158
0380	5"	Q-2	18	1.333		58.50	84		142.50	191
0400	6"	"	17	1.412		70.50	89		159.50	211
0420	8"	Q-3	11	2.909		211	187		398	510
0440	10"		10	3.200		305	206		511	645
0460	12"		9	3.556		575	229		804	970
0500	Sanitary tee, 2"	Q-1	10	1.600		35.50	97.50		133	184
0540	3"		9	1.778		57.50	108		165.50	226
0620	4"		8	2		70.50	122		192.50	260
0700	5"	Q-2	12	2		141	126		267	345
0800	6"	"	11	2.182		159	138		297	380

246

For customer support on your Site Work & Landscape Costs with RSMeans Data, call 800.448.8182.

22 13 Facility Sanitary Sewerage

22 13 16 – Sanitary Waste and Vent Piping

22 13 16.30 Pipe Fittings, Cast Iron		Crew	Daily Output	Labor-Hours	Unit	Material	2021 Bare Costs Labor	Equipment	Total	Total Incl O&P
0880	8"	Q-3	7	4.571	Ea.	420	295		715	905
1000	Tee, 2"	Q-1	10	1.600		51.50	97.50		149	202
1060	3"		9	1.778		76.50	108		184.50	247
1120	4"		8	2		98.50	122		220.50	290
1200	5"	Q-2	12	2		209	126		335	420
1300	6"	"	11	2.182		206	138		344	430
1380	8"	Q-3	7	4.571		320	295		615	790
1400	Combination Y and 1/8 bend									
1420	2"	Q-1	10	1.600	Ea.	44.50	97.50		142	194
1460	3"		9	1.778		68	108		176	237
1520	4"		8	2		93.50	122		215.50	285
1540	5"	Q-2	12	2		177	126		303	385
1560	6"		11	2.182		224	138		362	455
1580	8"		7	3.429		555	217		772	935
1582	12"	Q-3	6	5.333		1,150	345		1,495	1,800
1600	Double Y, 2"	Q-1	8	2		79	122		201	269
1610	3"		7	2.286		99	139		238	315
1620	4"		6.50	2.462		129	150		279	365
1630	5"	Q-2	9	2.667		236	168		404	510
1640	6"	"	8	3		340	190		530	655
1650	8"	Q-3	5.50	5.818		815	375		1,190	1,450
1660	10"		5	6.400		1,900	410		2,310	2,725
1670	12"		4.50	7.111		2,175	460		2,635	3,075
1740	Reducer, 3" x 2"	Q-1	15	1.067		25	65		90	125
1750	4" x 2"		14.50	1.103		28.50	67		95.50	132
1760	4" x 3"		14	1.143		32.50	69.50		102	140
1770	5" x 2"		14	1.143		69	69.50		138.50	180
1780	5" x 3"		13.50	1.185		73	72		145	188
1790	5" x 4"		13	1.231		42	75		117	158
1800	6" x 2"		13.50	1.185		68	72		140	183
1810	6" x 3"		13	1.231		66.50	75		141.50	185
1830	6" x 4"		12.50	1.280		65.50	78		143.50	188
1840	6" x 5"		11	1.455		70.50	88.50		159	210
1880	8" x 3"	Q-2	13.50	1.778		127	112		239	310
1900	8" x 4"		13	1.846		109	117		226	294
1920	8" x 5"		12	2		115	126		241	315
1940	8" x 6"		12	2		112	126		238	315
1960	Increaser, 2" x 3"	Q-1	15	1.067		63	65		128	166
1980	2" x 4"		14	1.143		60	69.50		129.50	170
2000	2" x 5"		13	1.231		66	75		141	185
2020	3" x 4"		13	1.231		65.50	75		140.50	185
2040	3" x 5"		13	1.231		66	75		141	185
2060	3" x 6"		12	1.333		88	81		169	218
2070	4" x 5"		13	1.231		78	75		153	198
2080	4" x 6"		12	1.333		89	81		170	219
2090	4" x 8"	Q-2	13	1.846		184	117		301	375
2100	5" x 6"	Q-1	11	1.455		132	88.50		220.50	277
2110	5" x 8"	Q-2	12	2		208	126		334	420
2120	6" x 8"		12	2		214	126		340	425
2130	6" x 10"		8	3		380	190		570	705
2140	8" x 10"		6.50	3.692		390	233		623	780
2150	10" x 12"		5.50	4.364		695	276		971	1,175
2500	Y, 2"	Q-1	10	1.600		32.50	97.50		130	181

For customer support on your Site Work & Landscape Costs with RSMeans Data, call 800.448.8182.

247

22 13 16.30 Pipe Fittings, Cast Iron		Crew	Daily Output	Labor-Hours	Unit	Material	2021 Bare Costs Labor	Equipment	Total	Total Incl O&P
2510	3"	Q-1	9	1.778	Ea.	60	108		168	228
2520	4"	↓	8	2		80.50	122		202.50	271
2530	5"	Q-2	12	2		143	126		269	345
2540	6"	"	11	2.182		186	138		324	410
2550	8"	Q-3	7	4.571		455	295		750	940
2560	10"		6	5.333		735	345		1,080	1,325
2570	12"		5	6.400		1,750	410		2,160	2,550
2580	15"	↓	4	8		3,700	515		4,215	4,850
3000	For extra heavy, add				↓	44%	4%			
3600	Hub and spigot, service weight gasket joint									
3605	Note: gaskets and joint labor have									
3606	been included with all listed fittings.									
3610	1/4 bend, 2"	Q-1	20	.800	Ea.	38.50	48.50		87	115
3620	3"	↓	17	.941		50.50	57.50		108	141
3630	4"	↓	15	1.067		74	65		139	179
3640	5"	Q-2	21	1.143		107	72		179	225
3650	6"	"	19	1.263		126	80		206	258
3660	8"	Q-3	12	2.667		355	172		527	645
3670	10"		11	2.909		535	187		722	870
3680	12"	↓	10	3.200		715	206		921	1,100
3700	Closet bend, 3" diameter with ring 10" x 16"	Q-1	17	.941		157	57.50		214.50	259
3710	16" x 16"		15	1.067		173	65		238	288
3730	Closet bend, 4" diameter, 1" x 4" ring, 6" x 16"		15	1.067		143	65		208	254
3740	8" x 16"		15	1.067		128	65		193	237
3750	10" x 12"		14	1.143		120	69.50		189.50	236
3760	10" x 18"		13	1.231		168	75		243	296
3770	12" x 16"		13	1.231		144	75		219	271
3780	16" x 16"		12	1.333		182	81		263	320
3800	1/8 bend, 2"		20	.800		31	48.50		79.50	107
3810	3"		17	.941		45	57.50		102.50	135
3820	4"	↓	15	1.067		62	65		127	166
3830	5"	Q-2	21	1.143		91	72		163	208
3840	6"	"	19	1.263		104	80		184	233
3850	8"	Q-3	12	2.667		284	172		456	565
3860	10"		11	2.909		430	187		617	755
3870	12"	↓	10	3.200		735	206		941	1,125
3900	Sanitary tee, 2"	Q-1	12	1.333		61	81		142	188
3910	3"		10	1.600		90	97.50		187.50	244
3920	4"	↓	9	1.778		112	108		220	285
3930	5"	Q-2	13	1.846		205	117		322	400
3940	6"	"	11	2.182		226	138		364	455
3950	8"	Q-3	8.50	3.765		570	243		813	985
3980	Tee, 2"	Q-1	12	1.333		77	81		158	206
3990	3"		10	1.600		109	97.50		206.50	265
4000	4"	↓	9	1.778		139	108		247	315
4010	5"	Q-2	13	1.846		273	117		390	475
4020	6"	"	11	2.182		272	138		410	505
4030	8"	Q-3	8	4	↓	465	258		723	895
4060	Combination Y and 1/8 bend									
4070	2"	Q-1	12	1.333	Ea.	70	81		151	198
4080	3"		10	1.600		100	97.50		197.50	255
4090	4"	↓	9	1.778		135	108		243	310
4100	5"	Q-2	13	1.846	↓	242	117		359	440

22 13 16.30 Pipe Fittings, Cast Iron		Crew	Daily Output	Labor-Hours	Unit	Material	2021 Bare Costs Labor	Equipment	Total	Total Incl O&P
4110	6"	Q-2	11	2.182	Ea.	291	138		429	525
4120	8"	Q-3	8	4		700	258		958	1,150
4121	12"	"	7	4.571		1,475	295		1,770	2,075
4160	Double Y, 2"	Q-1	10	1.600		117	97.50		214.50	274
4170	3"		8	2		148	122		270	345
4180	4"	↓	7	2.286		191	139		330	420
4190	5"	Q-2	10	2.400		330	152		482	590
4200	6"	"	9	2.667		435	168		603	730
4210	8"	Q-3	6	5.333		1,025	345		1,370	1,650
4220	10"		5	6.400		2,275	410		2,685	3,125
4230	12"	↓	4.50	7.111		2,650	460		3,110	3,600
4260	Reducer, 3" x 2"	Q-1	17	.941		54	57.50		111.50	145
4270	4" x 2"		16.50	.970		62	59		121	156
4280	4" x 3"		16	1		69.50	61		130.50	167
4290	5" x 2"		16	1		114	61		175	216
4300	5" x 3"		15.50	1.032		121	63		184	227
4310	5" x 4"		15	1.067		94.50	65		159.50	201
4320	6" x 2"		15.50	1.032		114	63		177	219
4330	6" x 3"		15	1.067		116	65		181	225
4336	6" x 4"		14	1.143		119	69.50		188.50	235
4340	6" x 5"	↓	13	1.231		136	75		211	262
4360	8" x 3"	Q-2	15	1.600		217	101		318	390
4370	8" x 4"		15	1.600		203	101		304	375
4380	8" x 5"		14	1.714		220	108		328	405
4390	8" x 6"	↓	14	1.714		218	108		326	400
4430	Increaser, 2" x 3"	Q-1	17	.941		79	57.50		136.50	173
4440	2" x 4"		16	1		80.50	61		141.50	180
4450	2" x 5"		15	1.067		98.50	65		163.50	205
4460	3" x 4"		15	1.067		86	65		151	192
4470	3" x 5"		15	1.067		98.50	65		163.50	205
4480	3" x 6"		14	1.143		121	69.50		190.50	237
4490	4" x 5"		15	1.067		110	65		175	218
4500	4" x 6"	↓	14	1.143		122	69.50		191.50	239
4510	4" x 8"	Q-2	15	1.600		257	101		358	435
4520	5" x 6"	Q-1	13	1.231		165	75		240	294
4530	5" x 8"	Q-2	14	1.714		281	108		389	470
4540	6" x 8"		14	1.714		286	108		394	475
4550	6" x 10"		10	2.400		510	152		662	785
4560	8" x 10"		8.50	2.824		520	178		698	835
4570	10" x 12"	↓	7.50	3.200		860	202		1,062	1,250
4600	Y, 2"	Q-1	12	1.333		58	81		139	185
4610	3"		10	1.600		93	97.50		190.50	247
4620	4"	↓	9	1.778		122	108		230	296
4630	5"	Q-2	13	1.846		207	117		324	400
4640	6"	"	11	2.182		252	138		390	485
4650	8"	Q-3	8	4		600	258		858	1,050
4660	10"		7	4.571		985	295		1,280	1,525
4670	12"		6	5.333		2,075	345		2,420	2,800
4672	15"	↓	5	6.400	↓	4,075	410		4,485	5,125
4900	For extra heavy, add					44%	4%			
4940	Gasket and making push-on joint									
4950	2"	Q-1	40	.400	Ea.	12.70	24.50		37.20	50.50
4960	3"	↓	35	.457	↓	16.30	28		44.30	59.50

22 13 16.30 Pipe Fittings, Cast Iron		Crew	Daily Output	Labor-Hours	Unit	Material	2021 Bare Costs Labor	Equipment	Total	Total Incl O&P
4970	4"	Q-1	32	.500	Ea.	20.50	30.50		51	68
4980	5"	Q-2	43	.558		32	35.50		67.50	88
4990	6"	"	40	.600		33	38		71	93
5000	8"	Q-3	32	1		73	64.50		137.50	176
5010	10"		29	1.103		127	71		198	245
5020	12"		25	1.280		162	82.50		244.50	300
5022	15"	▼	21	1.524	▼	193	98		291	360
5030	Note: gaskets and joint labor have									
5040	been included with all listed fittings.									
5990	No hub									
6000	Cplg. & labor required at joints not incl. in fitting									
6010	price. Add 1 coupling per joint for installed price									
6020	1/4 bend, 1-1/2"				Ea.	13.15			13.15	14.50
6060	2"					14.30			14.30	15.70
6080	3"					19.95			19.95	22
6120	4"					29.50			29.50	32.50
6140	5"					71.50			71.50	78.50
6160	6"					71.50			71.50	78.50
6180	8"					201			201	221
6184	1/4 bend, long sweep, 1-1/2"					33.50			33.50	37
6186	2"					31.50			31.50	34.50
6188	3"					38			38	41.50
6189	4"					60			60	66
6190	5"					117			117	128
6191	6"					133			133	146
6192	8"					345			345	380
6193	10"					670			670	735
6200	1/8 bend, 1-1/2"					11			11	12.10
6210	2"					12.30			12.30	13.55
6212	3"					16.50			16.50	18.15
6214	4"					21.50			21.50	24
6216	5"					45			45	49.50
6218	6"					48			48	53
6220	8"					138			138	152
6222	10"					263			263	289
6380	Sanitary tee, tapped, 1-1/2"					26			26	28.50
6382	2" x 1-1/2"					23			23	25.50
6384	2"					25			25	27.50
6386	3" x 2"					37			37	40.50
6388	3"					63.50			63.50	70
6390	4" x 1-1/2"					32.50			32.50	36
6392	4" x 2"					37			37	41
6393	4"					37			37	41
6394	6" x 1-1/2"					85.50			85.50	94
6396	6" x 2"					90			90	99.50
6459	Sanitary tee, 1-1/2"					18.40			18.40	20.50
6460	2"					19.75			19.75	21.50
6470	3"					24.50			24.50	27
6472	4"					46			46	50.50
6474	5"					108			108	119
6476	6"					110			110	121
6478	8"					445			445	490
6730	Y, 1-1/2"					18.70			18.70	20.50

22 13 16.30 Pipe Fittings, Cast Iron	Crew	Daily Output	Labor-Hours	Unit	Material	2021 Bare Costs Labor	Equipment	Total	Total Incl O&P
6740	2"				Ea.	18.25			18.25	20
6750	3"					26.50			26.50	29.50
6760	4"					42.50			42.50	47
6762	5"					101			101	111
6764	6"					113			113	125
6768	8"					267			267	293
6769	10"					590			590	650
6770	12"					1,175			1,175	1,275
6771	15"					2,725			2,725	2,975
6791	Y, reducing, 3" x 2"					19.75			19.75	21.50
6792	4" x 2"					28.50			28.50	31
6793	5" x 2"					62.50			62.50	69
6794	6" x 2"					69.50			69.50	76.50
6795	6" x 4"					90.50			90.50	99.50
6796	8" x 4"					155			155	171
6797	8" x 6"					191			191	210
6798	10" x 6"					430			430	475
6799	10" x 8"					515			515	565
6800	Double Y, 2"					29			29	32
6920	3"					53.50			53.50	59
7000	4"					109			109	120
7100	6"					192			192	211
7120	8"					545			545	600
7200	Combination Y and 1/8 bend									
7220	1-1/2"				Ea.	19.90			19.90	22
7260	2"					21			21	23
7320	3"					33			33	36
7400	4"					63			63	69.50
7480	5"					129			129	142
7500	6"					174			174	191
7520	8"					405			405	445
7800	Reducer, 3" x 2"					10.20			10.20	11.25
7820	4" x 2"					15.55			15.55	17.10
7840	4" x 3"					15.55			15.55	17.10
7842	6" x 3"					42			42	46
7844	6" x 4"					42			42	46
7846	6" x 5"					43			43	47
7848	8" x 2"					66.50			66.50	73
7850	8" x 3"					61.50			61.50	67.50
7852	8" x 4"					64.50			64.50	71
7854	8" x 5"					73			73	80.50
7856	8" x 6"					72			72	79
7858	10" x 4"					127			127	140
7860	10" x 6"					134			134	148
7862	10" x 8"					158			158	173
7864	12" x 4"					263			263	289
7866	12" x 6"					282			282	310
7868	12" x 8"					290			290	320
7870	12" x 10"					295			295	325
7872	15" x 4"					550			550	605
7874	15" x 6"					520			520	570
7876	15" x 8"					600			600	655
7878	15" x 10"					620			620	680

22 13 16.30 Pipe Fittings, Cast Iron

		Crew	Daily Output	Labor-Hours	Unit	Material	2021 Bare Costs Labor	Equipment	Total	Total Incl O&P
7880	15" x 12"				Ea.	625			625	690
8000	Coupling, standard (by CISPI Mfrs.)									
8020	1-1/2"	Q-1	48	.333	Ea.	13.90	20.50		34.40	46
8040	2"		44	.364		15.50	22		37.50	50
8080	3"		38	.421		18.20	25.50		43.70	58.50
8120	4"		33	.485		24	29.50		53.50	70.50
8160	5"	Q-2	44	.545		60.50	34.50		95	118
8180	6"	"	40	.600		61.50	38		99.50	124
8200	8"	Q-3	33	.970		100	62.50		162.50	204
8220	10"	"	26	1.231		124	79.50		203.50	254
8300	Coupling, cast iron clamp & neoprene gasket (by MG)									
8310	1-1/2"	Q-1	48	.333	Ea.	9.70	20.50		30.20	41
8320	2"		44	.364		13.05	22		35.05	47.50
8330	3"		38	.421		12.80	25.50		38.30	52.50
8340	4"		33	.485		19.45	29.50		48.95	65.50
8350	5"	Q-2	44	.545		36	34.50		70.50	91
8360	6"	"	40	.600		34.50	38		72.50	94.50
8380	8"	Q-3	33	.970		112	62.50		174.50	217
8400	10"	"	26	1.231		175	79.50		254.50	310
8600	Coupling, stainless steel, heavy duty									
8620	1-1/2"	Q-1	48	.333	Ea.	5.35	20.50		25.85	36.50
8630	2"		44	.364		5.80	22		27.80	39.50
8640	2" x 1-1/2"		44	.364		13.05	22		35.05	47.50
8650	3"		38	.421		6.05	25.50		31.55	45
8660	4"		33	.485		6.80	29.50		36.30	51.50
8670	4" x 3"		33	.485		18.75	29.50		48.25	64.50
8680	5"	Q-2	44	.545		14.90	34.50		49.40	68
8690	6"	"	40	.600		16.40	38		54.40	74.50
8700	8"	Q-3	33	.970		28	62.50		90.50	124
8710	10"		26	1.231		35	79.50		114.50	157
8712	12"		22	1.455		72	93.50		165.50	219
8715	15"		18	1.778		112	115		227	294

22 13 19 – Sanitary Waste Piping Specialties

22 13 19.13 Sanitary Drains

		Crew	Daily Output	Labor-Hours	Unit	Material	2021 Bare Costs Labor	Equipment	Total	Total Incl O&P
0010	**SANITARY DRAINS**									
0400	Deck, auto park, CI, 13" top									
0440	3", 4", 5", and 6" pipe size	Q-1	8	2	Ea.	2,100	122		2,222	2,500
0480	For galvanized body, add				"	1,300			1,300	1,425
0800	Promenade, heelproof grate, CI, 14" top									
0840	2", 3", and 4" pipe size	Q-1	10	1.600	Ea.	845	97.50		942.50	1,075
0860	5" and 6" pipe size		9	1.778		1,050	108		1,158	1,300
0880	8" pipe size		8	2		1,200	122		1,322	1,475
0940	For galvanized body, add					675			675	740
0960	With polished bronze top, 2"-3"-4" diam.					1,550			1,550	1,700
1200	Promenade, heelproof grate, CI, lateral, 14" top									
1240	2", 3" and 4" pipe size	Q-1	10	1.600	Ea.	1,100	97.50		1,197.50	1,350
1260	5" and 6" pipe size		9	1.778		1,250	108		1,358	1,525
1280	8" pipe size		8	2		1,425	122		1,547	1,750
1340	For galvanized body, add					550			550	605
1360	For polished bronze top, add					1,075			1,075	1,175
1500	Promenade, slotted grate, CI, 11" top									
1540	2", 3", 4", 5", and 6" pipe size	Q-1	12	1.333	Ea.	555	81		636	730

22 13 Facility Sanitary Sewerage

22 13 19 – Sanitary Waste Piping Specialties

22 13 19.13 Sanitary Drains

		Crew	Daily Output	Labor-Hours	Unit	Material	2021 Bare Costs Labor	Equipment	Total	Total Incl O&P
1600	For galvanized body, add				Ea.	385			385	425
1640	With polished bronze top				↓	1,125			1,125	1,250
2000	Floor, medium duty, CI, deep flange, 7" diam. top									
2040	2" and 3" pipe size	Q-1	12	1.333	Ea.	355	81		436	510
2080	For galvanized body, add					162			162	178
2120	With polished bronze top				↓	440			440	485
2400	Heavy duty, with sediment bucket, CI, 12" diam. loose grate									
2420	2", 3", 4", 5", and 6" pipe size	Q-1	9	1.778	Ea.	1,050	108		1,158	1,300
2460	With polished bronze top				"	1,575			1,575	1,725
2500	Heavy duty, cleanout & trap w/bucket, CI, 15" top									
2540	2", 3", and 4" pipe size	Q-1	6	2.667	Ea.	6,950	162		7,112	7,875
2560	For galvanized body, add					2,800			2,800	3,100
2580	With polished bronze top				↓	11,200			11,200	12,300

22 13 23 – Sanitary Waste Interceptors

22 13 23.10 Interceptors

		Crew	Daily Output	Labor-Hours	Unit	Material	2021 Bare Costs Labor	Equipment	Total	Total Incl O&P
0010	**INTERCEPTORS**									
0150	Grease, fabricated steel, 4 GPM, 8 lb. fat capacity	1 Plum	4	2	Ea.	1,750	135		1,885	2,125
0200	7 GPM, 14 lb. fat capacity		4	2		2,300	135		2,435	2,725
1000	10 GPM, 20 lb. fat capacity		4	2		2,875	135		3,010	3,350
1040	15 GPM, 30 lb. fat capacity		4	2		4,025	135		4,160	4,625
1060	20 GPM, 40 lb. fat capacity	↓	3	2.667		5,200	181		5,381	6,000
1120	50 GPM, 100 lb. fat capacity	Q-1	2	8		9,050	485		9,535	10,700
1160	100 GPM, 200 lb. fat capacity	"	2	8	↓	21,700	485		22,185	24,600
1580	For seepage pan, add					7%				
3000	Hair, cast iron, 1-1/4" and 1-1/2" pipe connection	1 Plum	8	1	Ea.	625	67.50		692.50	785
3100	For chrome-plated cast iron, add					470			470	520
4000	Oil, fabricated steel, 10 GPM, 2" pipe size	1 Plum	4	2		3,700	135		3,835	4,275
4100	15 GPM, 2" or 3" pipe size		4	2		5,075	135		5,210	5,800
4120	20 GPM, 2" or 3" pipe size	↓	3	2.667		6,700	181		6,881	7,650
4220	100 GPM, 3" pipe size	Q-1	2	8		20,500	485		20,985	23,200
6000	Solids, precious metals recovery, CI, 1-1/4" to 2" pipe	1 Plum	4	2		730	135		865	1,000
6100	Dental lab., large, CI, 1-1/2" to 2" pipe	"	3	2.667	↓	2,550	181		2,731	3,100

22 13 29 – Sanitary Sewerage Pumps

22 13 29.13 Wet-Pit-Mounted, Vertical Sewerage Pumps

		Crew	Daily Output	Labor-Hours	Unit	Material	2021 Bare Costs Labor	Equipment	Total	Total Incl O&P
0010	**WET-PIT-MOUNTED, VERTICAL SEWERAGE PUMPS**									
0020	Controls incl. alarm/disconnect panel w/wire. Excavation not included									
0260	Simplex, 9 GPM at 60 PSIG, 91 gal. tank				Ea.	3,750			3,750	4,100
0300	Unit with manway, 26" ID, 18" high					4,150			4,150	4,550
0340	26" ID, 36" high					4,075			4,075	4,475
0380	43" ID, 4' high					4,325			4,325	4,750
0600	Simplex, 9 GPM at 60 PSIG, 150 gal. tank, indoor					4,350			4,350	4,775
0700	Unit with manway, 26" ID, 36" high					4,975			4,975	5,475
0740	26" ID, 4' high					5,125			5,125	5,625
2000	Duplex, 18 GPM at 60 PSIG, 150 gal. tank, indoor					8,050			8,050	8,850
2060	Unit with manway, 43" ID, 4' high					9,575			9,575	10,500
2400	For core only				↓	2,025			2,025	2,225
3000	Indoor residential type installation									
3020	Simplex, 9 GPM at 60 PSIG, 91 gal. HDPE tank				Ea.	3,750			3,750	4,125

For customer support on your Site Work & Landscape Costs with RSMeans Data, call 800.448.8182.

253

22 13 29.14 Sewage Ejector Pumps		Crew	Daily Output	Labor-Hours	Unit	Material	2021 Bare Costs			Total	Total Incl O&P
							Labor	Equipment			
0010	**SEWAGE EJECTOR PUMPS**, With operating and level controls										
0100	Simplex system incl. tank, cover, pump 15' head										
0500	37 gal. PE tank, 12 GPM, 1/2 HP, 2" discharge	Q-1	3.20	5	Ea.	515	305			820	1,025
0510	3" discharge		3.10	5.161		550	315			865	1,075
0530	87 GPM, .7 HP, 2" discharge		3.20	5		795	305			1,100	1,325
0540	3" discharge		3.10	5.161		860	315			1,175	1,425
0600	45 gal. coated stl. tank, 12 GPM, 1/2 HP, 2" discharge		3	5.333		925	325			1,250	1,500
0610	3" discharge		2.90	5.517		960	335			1,295	1,550
0630	87 GPM, .7 HP, 2" discharge		3	5.333		1,175	325			1,500	1,775
0640	3" discharge		2.90	5.517		1,250	335			1,585	1,875
0660	134 GPM, 1 HP, 2" discharge		2.80	5.714		1,275	350			1,625	1,925
0680	3" discharge		2.70	5.926		1,350	360			1,710	2,025
0700	70 gal. PE tank, 12 GPM, 1/2 HP, 2" discharge		2.60	6.154		995	375			1,370	1,650
0710	3" discharge		2.40	6.667		1,075	405			1,480	1,775
0730	87 GPM, .7 HP, 2" discharge		2.50	6.400		1,325	390			1,715	2,025
0740	3" discharge		2.30	6.957		1,375	425			1,800	2,125
0760	134 GPM, 1 HP, 2" discharge		2.20	7.273		1,400	445			1,845	2,200
0770	3" discharge		2	8		1,500	485			1,985	2,375
0800	75 gal. coated stl. tank, 12 GPM, 1/2 HP, 2" discharge		2.40	6.667		1,100	405			1,505	1,825
0810	3" discharge		2.20	7.273		1,150	445			1,595	1,925
0830	87 GPM, .7 HP, 2" discharge		2.30	6.957		1,400	425			1,825	2,150
0840	3" discharge		2.10	7.619		1,475	465			1,940	2,300
0860	134 GPM, 1 HP, 2" discharge		2	8		1,500	485			1,985	2,375
0880	3" discharge		1.80	8.889		1,575	540			2,115	2,525
1040	Duplex system incl. tank, covers, pumps										
1060	110 gal. fiberglass tank, 24 GPM, 1/2 HP, 2" discharge	Q-1	1.60	10	Ea.	2,000	610			2,610	3,100
1080	3" discharge		1.40	11.429		2,100	695			2,795	3,350
1100	174 GPM, .7 HP, 2" discharge		1.50	10.667		2,575	650			3,225	3,825
1120	3" discharge		1.30	12.308		2,675	750			3,425	4,075
1140	268 GPM, 1 HP, 2" discharge		1.20	13.333		2,800	810			3,610	4,300
1160	3" discharge		1	16		2,900	975			3,875	4,650
1260	135 gal. coated stl. tank, 24 GPM, 1/2 HP, 2" discharge	Q-2	1.70	14.118		2,050	890			2,940	3,600
2000	3" discharge		1.60	15		2,175	950			3,125	3,825
2640	174 GPM, .7 HP, 2" discharge		1.60	15		2,700	950			3,650	4,400
2660	3" discharge		1.50	16		2,850	1,000			3,850	4,625
2700	268 GPM, 1 HP, 2" discharge		1.30	18.462		2,925	1,175			4,100	4,975
3040	3" discharge		1.10	21.818		3,125	1,375			4,500	5,475
3060	275 gal. coated stl. tank, 24 GPM, 1/2 HP, 2" discharge		1.50	16		2,575	1,000			3,575	4,350
3080	3" discharge		1.40	17.143		2,600	1,075			3,675	4,500
3100	174 GPM, .7 HP, 2" discharge		1.40	17.143		3,325	1,075			4,400	5,275
3120	3" discharge		1.30	18.462		3,575	1,175			4,750	5,700
3140	268 GPM, 1 HP, 2" discharge		1.10	21.818		3,650	1,375			5,025	6,050
3160	3" discharge		.90	26.667		3,850	1,675			5,525	6,750
3260	Pump system accessories, add										
3300	Alarm horn and lights, 115 V mercury switch	Q-1	8	2	Ea.	102	122			224	294
3340	Switch, mag. contactor, alarm bell, light, 3 level control		5	3.200		520	195			715	860
3380	Alternator, mercury switch activated		4	4		935	244			1,179	1,400

22 13 63.10 Graywater Recovery Systems

		Crew	Daily Output	Labor-Hours	Unit	Material	Labor	Equipment		Total	Total Incl O&P
0010	**GRAYWATER RECOVERY SYSTEMS**, for ext. irrigation										
2010	Residential system, sink to toilet	Q-1	2	8	Ea.	320	485			805	1,075
2020	Toilet top sink	1 Plum	8	1		170	67.50			237.50	288
3010	Prepackaged residential system, 6 people	Q-2	5	4.800		2,325	305			2,630	3,000

254

For customer support on your Site Work & Landscape Costs with RSMeans Data, call 800.448.8182.

22 13 Facility Sanitary Sewerage

22 13 29 – Sanitary Sewerage Pumps

22 13 63.10 Graywater Recovery Systems		Crew	Daily Output	Labor-Hours	Unit	Material	2021 Bare Costs Labor	Equipment	Total	Total Incl O&P
3020	9 people	Q-2	4	6	Ea.	2,625	380		3,005	3,475
3030	12 people	↓	3.50	6.857		2,825	435		3,260	3,775
3040	for irrigation supply	1 Plum	3	2.667		755	181		936	1,100
4010	Prepackaged commercial system, 1,530 gal.	Q-3	4.25	7.529		30,400	485		30,885	34,100
4020	3,060 gal.		3.50	9.143		56,000	590		56,590	62,500
4030	4,590 gal.	↓	2.50	12.800		70,000	825		70,825	78,000
5010	Lift station 75 GPM	Q-1	3	5.333	↓	1,550	325		1,875	2,175
6010	System components									

22 14 Facility Storm Drainage

22 14 23 – Storm Drainage Piping Specialties

22 14 23.33 Backwater Valves

		Crew	Daily Output	Labor-Hours	Unit	Material	2021 Bare Costs Labor	Equipment	Total	Total Incl O&P
0010	**BACKWATER VALVES**, CI Body									
6980	Bronze gate and automatic flapper valves									
7000	3" and 4" pipe size	Q-1	13	1.231	Ea.	2,675	75		2,750	3,050
7100	5" and 6" pipe size	"	13	1.231	"	4,100	75		4,175	4,625
7240	Bronze flapper valve, bolted cover									
7260	2" pipe size	Q-1	16	1	Ea.	785	61		846	950
7280	3" pipe size		14.50	1.103		1,225	67		1,292	1,450
7300	4" pipe size	↓	13	1.231		1,150	75		1,225	1,375
7320	5" pipe size	Q-2	18	1.333		1,800	84		1,884	2,125
7340	6" pipe size	"	17	1.412		2,550	89		2,639	2,950
7360	8" pipe size	Q-3	10	3.200		3,650	206		3,856	4,325
7380	10" pipe size	"	9	3.556	↓	5,975	229		6,204	6,925
7500	For threaded cover, same cost									
7540	Revolving disk type, same cost as flapper type									

22 14 26 – Facility Storm Drains

22 14 26.19 Facility Trench Drains

		Crew	Daily Output	Labor-Hours	Unit	Material	2021 Bare Costs Labor	Equipment	Total	Total Incl O&P
0010	**FACILITY TRENCH DRAINS**									
5980	Trench, floor, heavy duty, modular, CI, 12" x 12" top									
6000	2", 3", 4", 5" & 6" pipe size	Q-1	8	2	Ea.	1,175	122		1,297	1,475
6100	For unit with polished bronze top		8	2		1,725	122		1,847	2,075
6200	For 12" extension section, CI top		8	2		1,175	122		1,297	1,475
6240	For 12" extension section, polished bronze top	↓	8	2	↓	1,900	122		2,022	2,250
6600	Trench, floor, for cement concrete encasement									
6610	Not including trenching or concrete									
6640	Polyester polymer concrete									
6650	4" internal width, with grate									
6660	Light duty steel grate	Q-1	120	.133	L.F.	69.50	8.10		77.60	88.50
6670	Medium duty steel grate		115	.139		125	8.50		133.50	150
6680	Heavy duty iron grate	↓	110	.145	↓	123	8.85		131.85	149
6700	12" internal width, with grate									
6770	Heavy duty galvanized grate	Q-1	80	.200	L.F.	168	12.20		180.20	203
6800	Fiberglass									
6810	8" internal width, with grate									
6820	Medium duty galvanized grate	Q-1	115	.139	L.F.	113	8.50		121.50	137
6830	Heavy duty iron grate	"	110	.145	"	199	8.85		207.85	232

For customer support on your Site Work & Landscape Costs with RSMeans Data, call 800.448.8182.

255

22 14 29 – Sump Pumps

22 14 29.13 Wet-Pit-Mounted, Vertical Sump Pumps	Crew	Daily Output	Labor-Hours	Unit	Material	2021 Bare Costs Labor	Equipment	Total	Total Incl O&P
0010 **WET-PIT-MOUNTED, VERTICAL SUMP PUMPS**									
0400 Molded PVC base, 21 GPM at 15' head, 1/3 HP	1 Plum	5	1.600	Ea.	138	108		246	315
0800 Iron base, 21 GPM at 15' head, 1/3 HP		5	1.600		155	108		263	330
1200 Solid brass, 21 GPM at 15' head, 1/3 HP		5	1.600		250	108		358	435
2000 Sump pump, single stage									
2010 25 GPM, 1 HP, 1-1/2" discharge	Q-1	1.80	8.889	Ea.	4,000	540		4,540	5,200
2020 75 GPM, 1-1/2 HP, 2" discharge		1.50	10.667		4,225	650		4,875	5,625
2030 100 GPM, 2 HP, 2-1/2" discharge		1.30	12.308		4,300	750		5,050	5,850
2040 150 GPM, 3 HP, 3" discharge		1.10	14.545		4,300	885		5,185	6,050
2050 200 GPM, 3 HP, 3" discharge		1	16		4,550	975		5,525	6,450
2060 300 GPM, 10 HP, 4" discharge	Q-2	1.20	20		4,925	1,275		6,200	7,275
2070 500 GPM, 15 HP, 5" discharge		1.10	21.818		5,600	1,375		6,975	8,200
2080 800 GPM, 20 HP, 6" discharge		1	24		6,600	1,525		8,125	9,550
2090 1,000 GPM, 30 HP, 6" discharge		.85	28.235		7,250	1,775		9,025	10,700
2100 1,600 GPM, 50 HP, 8" discharge		.72	33.333		11,300	2,100		13,400	15,600
2110 2,000 GPM, 60 HP, 8" discharge	Q-3	.85	37.647		11,600	2,425		14,025	16,300
2202 For general purpose float switch, copper coated float, add	Q-1	5	3.200		109	195		304	410

22 14 29.16 Submersible Sump Pumps

	Crew	Daily Output	Labor-Hours	Unit	Material	2021 Bare Costs Labor	Equipment	Total	Total Incl O&P
0010 **SUBMERSIBLE SUMP PUMPS**									
7000 Sump pump, automatic									
7100 Plastic, 1-1/4" discharge, 1/4 HP	1 Plum	6.40	1.250	Ea.	173	84.50		257.50	315
7140 1/3 HP		6	1.333		229	90.50		319.50	385
7160 1/2 HP		5.40	1.481		250	100		350	425
7180 1-1/2" discharge, 1/2 HP		5.20	1.538		350	104		454	540
7500 Cast iron, 1-1/4" discharge, 1/4 HP		6	1.333		243	90.50		333.50	400
7540 1/3 HP		6	1.333		285	90.50		375.50	450
7560 1/2 HP		5	1.600		345	108		453	540

22 14 53 – Rainwater Storage Tanks

22 14 53.13 Fiberglass, Rainwater Storage Tank

	Crew	Daily Output	Labor-Hours	Unit	Material	2021 Bare Costs Labor	Equipment	Total	Total Incl O&P
0010 **FIBERGLASS, RAINWATER STORAGE TANK**									
2000 600 gallon	B-21B	3.75	10.667	Ea.	4,325	515	127	4,967	5,650
2010 1,000 gallon		3.50	11.429		5,450	550	136	6,136	6,975
2020 2,000 gallon		3.25	12.308		7,675	595	146	8,416	9,475
2030 4,000 gallon		3	13.333		10,100	645	159	10,904	12,200
2040 6,000 gallon		2.65	15.094		11,500	730	180	12,410	13,900
2050 8,000 gallon		2.30	17.391		14,100	840	207	15,147	17,000
2060 10,000 gallon		2	20		15,400	965	238	16,603	18,700
2070 12,000 gallon		1.50	26.667		22,100	1,275	315	23,690	26,700
2080 15,000 gallon		1	40		24,200	1,925	475	26,600	30,100
2090 20,000 gallon		.75	53.333		31,600	2,575	635	34,810	39,300
2100 25,000 gallon		.50	80		48,000	3,850	950	52,800	59,500
2110 30,000 gallon		.35	114		58,000	5,500	1,350	64,850	73,500
2120 40,000 gallon		.30	133		79,000	6,425	1,575	87,000	98,000

22 41 Residential Plumbing Fixtures

22 41 39 – Residential Faucets, Supplies and Trim

22 41 39.10 Faucets and Fittings	Crew	Daily Output	Labor-Hours	Unit	Material	2021 Bare Costs Labor	Equipment	Total	Total Incl O&P
0010 **FAUCETS AND FITTINGS**									
5000 Sillcock, compact, brass, IPS or copper to hose	1 Plum	24	.333	Ea.	12.90	22.50		35.40	47.50

22 47 Drinking Fountains and Water Coolers

22 47 13 – Drinking Fountains

22 47 13.10 Drinking Water Fountains

		Crew	Daily Output	Labor-Hours	Unit	Material	2021 Bare Costs Labor	Equipment	Total	Total Incl O&P
0010	**DRINKING WATER FOUNTAINS**, For connection to cold water supply									
1000	Wall mounted, non-recessed									
1400	Bronze, with no back	1 Plum	4	2	Ea.	1,125	135		1,260	1,450
1800	Cast aluminum, enameled, for correctional institutions		4	2		1,300	135		1,435	1,650
2000	Fiberglass, 12" back, single bubbler unit		4	2		1,800	135		1,935	2,175
2040	Dual bubbler		3.20	2.500		2,775	169		2,944	3,300
2400	Precast stone, no back		4	2		1,125	135		1,260	1,450
2700	Stainless steel, single bubbler, no back		4	2		870	135		1,005	1,150
2740	With back		4	2		1,125	135		1,260	1,450
2780	Dual handle, ADA compliant		4	2		785	135		920	1,075
2820	Dual level, ADA compliant		3.20	2.500		1,450	169		1,619	1,850
3300	Vitreous china									
3340	7" back	1 Plum	4	2	Ea.	670	135		805	935
3940	For vandal-resistant bottom plate, add					72.50			72.50	80
3960	For freeze-proof valve system, add	1 Plum	2	4		830	271		1,101	1,325
3980	For rough-in, supply and waste, add	"	2.21	3.620		370	245		615	770
4000	Wall mounted, semi-recessed									
4200	Poly-marble, single bubbler	1 Plum	4	2	Ea.	950	135		1,085	1,250
4600	Stainless steel, satin finish, single bubbler		4	2		1,700	135		1,835	2,075
4900	Vitreous china, single bubbler		4	2		1,075	135		1,210	1,375
5980	For rough-in, supply and waste, add		1.83	4.372		370	296		666	845
6000	Wall mounted, fully recessed									
6400	Poly-marble, single bubbler	1 Plum	4	2	Ea.	1,750	135		1,885	2,125
6800	Stainless steel, single bubbler		4	2		1,350	135		1,485	1,675
7560	For freeze-proof valve system, add		2	4		1,000	271		1,271	1,500
7580	For rough-in, supply and waste, add		1.83	4.372		370	296		666	845
7600	Floor mounted, pedestal type									
7780	ADA compliant unit	1 Plum	2	4	Ea.	1,500	271		1,771	2,050
8400	Stainless steel, architectural style		2	4		2,400	271		2,671	3,025
8600	Enameled iron, heavy duty service, 2 bubblers		2	4		3,475	271		3,746	4,225
8660	4 bubblers		2	4		4,175	271		4,446	4,975
8880	For freeze-proof valve system, add		2	4		680	271		951	1,150
8900	For rough-in, supply and waste, add		1.83	4.372		370	296		666	845
9100	Deck mounted									
9500	Stainless steel, circular receptor	1 Plum	4	2	Ea.	455	135		590	700
9760	White enameled steel, 14" x 9" receptor		4	2		395	135		530	635
9860	White enameled cast iron, 24" x 16" receptor		3	2.667		515	181		696	835
9980	For rough-in, supply and waste, add		1.83	4.372		370	296		666	845

22 51 Swimming Pool Plumbing Systems

22 51 19 – Swimming Pool Water Treatment Equipment

22 51 19.50 Swimming Pool Filtration Equipment	Crew	Daily Output	Labor-Hours	Unit	Material	2021 Bare Costs Labor	Equipment	Total	Total Incl O&P
0010 **SWIMMING POOL FILTRATION EQUIPMENT**									
0900 Filter system, sand or diatomite type, incl. pump, 6,000 gal./hr.	2 Plum	1.80	8.889	Total	2,325	600		2,925	3,450
1020 Add for chlorination system, 800 S.F. pool		3	5.333	Ea.	257	360		617	825
1040 5,000 S.F. pool		3	5.333	"	1,950	360		2,310	2,700

22 52 Fountain Plumbing Systems

22 52 16 – Fountain Pumps

22 52 16.10 Fountain Water Pumps

	Crew	Daily Output	Labor-Hours	Unit	Material	2021 Bare Costs Labor	Equipment	Total	Total Incl O&P
0010 **FOUNTAIN WATER PUMPS**									
0100 Pump w/controls									
0200 Single phase, 100' cord, 1/2 HP pump	2 Skwk	4.40	3.636	Ea.	1,325	208		1,533	1,750
0300 3/4 HP pump		4.30	3.721		1,325	212		1,537	1,800
0400 1 HP pump		4.20	3.810		1,775	218		1,993	2,275
0500 1-1/2 HP pump		4.10	3.902		3,250	223		3,473	3,900
0600 2 HP pump		4	4		4,325	228		4,553	5,125
0700 Three phase, 200' cord, 5 HP pump		3.90	4.103		6,175	234		6,409	7,150
0800 7-1/2 HP pump		3.80	4.211		12,900	240		13,140	14,600
0900 10 HP pump		3.70	4.324		14,500	247		14,747	16,300
1000 15 HP pump		3.60	4.444		20,200	254		20,454	22,600
2000 DESIGN NOTE: Use two horsepower per surface acre.									

22 52 33 – Fountain Ancillary

22 52 33.10 Fountain Miscellaneous

	Crew	Daily Output	Labor-Hours	Unit	Material	2021 Bare Costs Labor	Equipment	Total	Total Incl O&P
0010 **FOUNTAIN MISCELLANEOUS**									
1300 Lights w/mounting kits, 200 watt	2 Skwk	18	.889	Ea.	1,200	51		1,251	1,375
1400 300 watt		18	.889		1,375	51		1,426	1,575
1500 500 watt		18	.889		1,600	51		1,651	1,825
1600 Color blender		12	1.333		605	76		681	780

22 66 Chemical-Waste Systems for Lab. and Healthcare Facilities

22 66 53 – Laboratory Chemical-Waste and Vent Piping

22 66 53.30 Glass Pipe

	Crew	Daily Output	Labor-Hours	Unit	Material	2021 Bare Costs Labor	Equipment	Total	Total Incl O&P
0010 **GLASS PIPE**, Borosilicate, couplings & clevis hanger assemblies, 10' OC									
0020 Drainage									
1100 1-1/2" diameter	Q-1	52	.308	L.F.	13.60	18.75		32.35	43
1120 2" diameter		44	.364		17.90	22		39.90	52.50
1140 3" diameter		39	.410		22.50	25		47.50	62.50
1160 4" diameter		30	.533		43	32.50		75.50	95.50
1180 6" diameter		26	.615		77	37.50		114.50	141

22 66 53.40 Pipe Fittings, Glass

	Crew	Daily Output	Labor-Hours	Unit	Material	2021 Bare Costs Labor	Equipment	Total	Total Incl O&P
0010 **PIPE FITTINGS, GLASS**									
0020 Drainage, beaded ends									
0040 Coupling & labor required at joints not incl. in fitting									
0050 price. Add 1 per joint for installed price									
0070 90° bend or sweep, 1-1/2"				Ea.	35.50			35.50	39
0110 4"				"	118			118	130

22 66 Chemical-Waste Systems for Lab. and Healthcare Facilities

22 66 53 – Laboratory Chemical-Waste and Vent Piping

22 66 53.60 Corrosion Resistant Pipe

		Crew	Daily Output	Labor-Hours	Unit	Material	2021 Bare Costs Labor	Equipment	Total	Total Incl O&P
0010	**CORROSION RESISTANT PIPE**, No couplings or hangers									
0020	Iron alloy, drain, mechanical joint									
1000	1-1/2" diameter	Q-1	70	.229	L.F.	75.50	13.95		89.45	104
1100	2" diameter		66	.242		77.50	14.75		92.25	108
1120	3" diameter		60	.267		85	16.25		101.25	118
1140	4" diameter		52	.308		108	18.75		126.75	147
1980	Iron alloy, drain, B&S joint									
2000	2" diameter	Q-1	54	.296	L.F.	90.50	18.05		108.55	127
2100	3" diameter		52	.308		88	18.75		106.75	125
2120	4" diameter		48	.333		107	20.50		127.50	149
2140	6" diameter	Q-2	59	.407		162	25.50		187.50	217
2160	8" diameter	"	54	.444		330	28		358	405
2980	Plastic, epoxy, fiberglass filament wound, B&S joint									
3000	2" diameter	Q-1	62	.258	L.F.	12	15.70		27.70	36.50
3100	3" diameter		51	.314		14.05	19.10		33.15	44
3120	4" diameter		45	.356		20	21.50		41.50	54.50
3140	6" diameter		32	.500		28	30.50		58.50	76.50
3160	8" diameter	Q-2	38	.632		44	40		84	108
3980	Polyester, fiberglass filament wound, B&S joint									
4000	2" diameter	Q-1	62	.258	L.F.	13.05	15.70		28.75	38
4100	3" diameter		51	.314		17.15	19.10		36.25	47.50
4120	4" diameter		45	.356		25	21.50		46.50	60
4140	6" diameter		32	.500		36.50	30.50		67	86
4160	8" diameter	Q-2	38	.632		87	40		127	155
4980	Polypropylene, acid resistant, fire retardant, Schedule 40									
5000	1-1/2" diameter	Q-1	68	.235	L.F.	7.40	14.35		21.75	29.50
5100	2" diameter		62	.258		15.40	15.70		31.10	40.50
5120	3" diameter		51	.314		20.50	19.10		39.60	51
5140	4" diameter		45	.356		26	21.50		47.50	61
5160	6" diameter		32	.500		52	30.50		82.50	103
5980	Proxylene, fire retardant, Schedule 40									
6000	1-1/2" diameter	Q-1	68	.235	L.F.	11.40	14.35		25.75	34
6100	2" diameter		62	.258		15.60	15.70		31.30	40.50
6120	3" diameter		51	.314		28.50	19.10		47.60	59.50
6140	4" diameter		45	.356		40	21.50		61.50	76.50
6820	For Schedule 80, add					35%	2%			

22 66 53.70 Pipe Fittings, Corrosion Resistant

		Crew	Daily Output	Labor-Hours	Unit	Material	2021 Bare Costs Labor	Equipment	Total	Total Incl O&P
0010	**PIPE FITTINGS, CORROSION RESISTANT**									
0030	Iron alloy									
0050	Mechanical joint									
0060	1/4 bend, 1-1/2"	Q-1	12	1.333	Ea.	128	81		209	262
0080	2"		10	1.600		209	97.50		306.50	375
0090	3"		9	1.778		251	108		359	440
0100	4"		8	2		288	122		410	495
0160	Tee and Y, sanitary, straight									
0170	1-1/2"	Q-1	8	2	Ea.	139	122		261	335
0180	2"		7	2.286		186	139		325	415
0190	3"		6	2.667		288	162		450	555
0200	4"		5	3.200		530	195		725	875
3000	Epoxy, filament wound									
3030	Quick-lock joint									
3040	90° elbow, 2"	Q-1	28	.571	Ea.	99	35		134	161

For customer support on your Site Work & Landscape Costs with RSMeans Data, call 800.448.8182.

259

22 66 Chemical-Waste Systems for Lab. and Healthcare Facilities

22 66 53 – Laboratory Chemical-Waste and Vent Piping

22 66 53.70 Pipe Fittings, Corrosion Resistant	Crew	Daily Output	Labor-Hours	Unit	Material	2021 Bare Costs Labor	Equipment	Total	Total Incl O&P
3060	3"	Q-1	16	1	Ea.	113	61		174	215
3070	4"	↓	13	1.231	↓	154	75		229	282
4000	Polypropylene, acid resistant									
4020	Non-pressure, electrofusion joints									
4060	2"	Q-1	28	.571	Ea.	56.50	35		91.50	114
4090	4"	"	14	1.143	"	99	69.50		168.50	213

Estimating Tips

The labor adjustment factors listed in Subdivision 22 01 02.20 also apply to Division 23.

23 10 00 Facility Fuel Systems

- The prices in this subdivision for above- and below-ground storage tanks do not include foundations or hold-down slabs, unless noted. The estimator should refer to Divisions 3 and 31 for foundation system pricing. In addition to the foundations, required tank accessories, such as tank gauges, leak detection devices, and additional manholes and piping, must be added to the tank prices.

23 50 00 Central Heating Equipment

- When estimating the cost of an HVAC system, check to see who is responsible for providing and installing the temperature control system. It is possible to overlook controls, assuming that they would be included in the electrical estimate.
- When looking up a boiler, be careful on specified capacity. Some

manufacturers rate their products on output while others use input.

- Include HVAC insulation for pipe, boiler, and duct (wrap and liner).
- Be careful when looking up mechanical items to get the correct pressure rating and connection type (thread, weld, flange).

23 70 00 Central HVAC Equipment

- Combination heating and cooling units are sized by the air conditioning requirements. (See Reference No. R236000-20 for the preliminary sizing guide.)
- A ton of air conditioning is nominally 400 CFM.
- Rectangular duct is taken off by the linear foot for each size, but its cost is usually estimated by the pound. Remember that SMACNA standards now base duct on internal pressure.
- Prefabricated duct is estimated and purchased like pipe: straight sections and fittings.
- Note that cranes or other lifting equipment are not included on any

lines in Division 23. For example, if a crane is required to lift a heavy piece of pipe into place high above a gym floor, or to put a rooftop unit on the roof of a four-story building, etc., it must be added. Due to the potential for extreme variation—from nothing additional required to a major crane or helicopter—we feel that including a nominal amount for "lifting contingency" would be useless and detract from the accuracy of the estimate. When using equipment rental cost data from RSMeans, do not forget to include the cost of the operator(s).

Reference Numbers

Reference numbers are shown at the beginning of some major classifications. These numbers refer to related items in the Reference Section. The reference information may be an estimating procedure, an alternate pricing method, or technical information.

Note: Not all subdivisions listed here necessarily appear. ■

Same Data. Simplified.

Enjoy the convenience and efficiency of accessing your costs anywhere:

- **Skip the multiplier** by setting your location
- **Quickly search,** edit, favorite and share costs
- **Stay on top of price changes** with automatic updates

Discover more at rsmeans.com/online

23 05 Common Work Results for HVAC

23 05 05 – Selective Demolition for HVAC

23 05 05.10 HVAC Demolition

		Crew	Daily Output	Labor-Hours	Unit	Material	2021 Bare Costs Labor	Equipment	Total	Total Incl O&P
0010	**HVAC DEMOLITION**									
0100	Air conditioner, split unit, 3 ton	Q-5	2	8	Ea.		490		490	735
0150	Package unit, 3 ton	Q-6	3	8	"		510		510	760
0298	Boilers									
0300	Electric, up thru 148 kW	Q-19	2	12	Ea.		745		745	1,125
0310	150 thru 518 kW	"	1	24			1,500		1,500	2,225
0320	550 thru 2,000 kW	Q-21	.40	80			5,100		5,100	7,600
0330	2,070 kW and up	"	.30	107			6,800		6,800	10,100
0340	Gas and/or oil, up thru 150 MBH	Q-7	2.20	14.545			945		945	1,400
0350	160 thru 2,000 MBH		.80	40			2,600		2,600	3,875
0360	2,100 thru 4,500 MBH		.50	64			4,175		4,175	6,225
0370	4,600 thru 7,000 MBH		.30	107			6,950		6,950	10,400
0380	7,100 thru 12,000 MBH		.16	200			13,000		13,000	19,400
0390	12,200 thru 25,000 MBH		.12	267			17,300		17,300	25,900
1000	Ductwork, 4" high, 8" wide	1 Clab	200	.040	L.F.		1.78		1.78	2.65
1100	6" high, 8" wide		165	.048			2.15		2.15	3.21
1200	10" high, 12" wide		125	.064			2.84		2.84	4.24
1300	12"-14" high, 16"-18" wide		85	.094			4.18		4.18	6.25
1400	18" high, 24" wide		67	.119			5.30		5.30	7.90
1500	30" high, 36" wide		56	.143			6.35		6.35	9.45
1540	72" wide		50	.160			7.10		7.10	10.60
3000	Mechanical equipment, light items. Unit is weight, not cooling.	Q-5	.90	17.778	Ton		1,100		1,100	1,625
3600	Heavy items	"	1.10	14.545	"		895		895	1,325
5090	Remove refrigerant from system	1 Stpi	40	.200	Lb.		13.65		13.65	20.50

23 05 23 – General-Duty Valves for HVAC Piping

23 05 23.30 Valves, Iron Body

		Crew	Daily Output	Labor-Hours	Unit	Material	2021 Bare Costs Labor	Equipment	Total	Total Incl O&P
0010	**VALVES, IRON BODY** R220523-80									
1020	Butterfly, wafer type, gear actuator, 200 lb.									
1030	2"	1 Plum	14	.571	Ea.	118	38.50		156.50	188
1040	2-1/2"	Q-1	9	1.778		120	108		228	294
1050	3"		8	2		124	122		246	320
1060	4"		5	3.200		138	195		333	445
1070	5"	Q-2	5	4.800		155	305		460	625
1080	6"	"	5	4.800		176	305		481	650
1650	Gate, 125 lb., N.R.S.									
2150	Flanged									
2200	2"	1 Plum	5	1.600	Ea.	915	108		1,023	1,150
2240	2-1/2"	Q-1	5	3.200		935	195		1,130	1,325
2260	3"		4.50	3.556		1,050	217		1,267	1,475
2280	4"		3	5.333		1,500	325		1,825	2,125
2300	6"	Q-2	3	8		2,300	505		2,805	3,275
3550	OS&Y, 125 lb., flanged									
3600	2"	1 Plum	5	1.600	Ea.	625	108		733	845
3660	3"	Q-1	4.50	3.556		690	217		907	1,075
3680	4"	"	3	5.333		1,000	325		1,325	1,575
3700	6"	Q-2	3	8		1,650	505		2,155	2,575
3900	For 175 lb., flanged, add					200%	10%			
5450	Swing check, 125 lb., threaded									
5500	2"	1 Plum	11	.727	Ea.	615	49		664	750
5540	2-1/2"	Q-1	15	1.067		790	65		855	965
5550	3"		13	1.231		855	75		930	1,050
5560	4"		10	1.600		1,350	97.50		1,447.50	1,650

For customer support on your Site Work & Landscape Costs with RSMeans Data, call 800.448.8182.

23 05 Common Work Results for HVAC

23 05 23 - General-Duty Valves for HVAC Piping

23 05 23.30 Valves, Iron Body

		Crew	Daily Output	Labor-Hours	Unit	Material	2021 Bare Costs Labor	2021 Bare Costs Equipment	Total	Total Incl O&P
5950	Flanged									
6000	2"	1 Plum	5	1.600	Ea.	535	108		643	745
6040	2-1/2"	Q-1	5	3.200		490	195		685	825
6050	3"		4.50	3.556		520	217		737	900
6060	4"		3	5.333		820	325		1,145	1,375
6070	6"	Q-2	3	8		1,400	505		1,905	2,275

23 05 23.80 Valves, Steel

		Crew	Daily Output	Labor-Hours	Unit	Material	2021 Bare Costs Labor	2021 Bare Costs Equipment	Total	Total Incl O&P
0010	**VALVES, STEEL** R220523-80									
0800	Cast									
1350	Check valve, swing type, 150 lb., flanged									
1370	1"	1 Plum	10	.800	Ea.	880	54		934	1,050
1400	2"	"	8	1		850	67.50		917.50	1,025
1440	2-1/2"	Q-1	5	3.200		980	195		1,175	1,375
1450	3"		4.50	3.556		1,025	217		1,242	1,450
1460	4"		3	5.333		1,625	325		1,950	2,275
1540	For 300 lb., flanged, add					50%	15%			
1548	For 600 lb., flanged, add					110%	20%			
1950	Gate valve, 150 lb., flanged									
2000	2"	1 Plum	8	1	Ea.	865	67.50		932.50	1,050
2040	2-1/2"	Q-1	5	3.200		1,175	195		1,370	1,600
2050	3"		4.50	3.556		1,225	217		1,442	1,675
2060	4"		3	5.333		1,700	325		2,025	2,350
2070	6"	Q-2	3	8		2,675	505		3,180	3,675
3650	Globe valve, 150 lb., flanged									
3700	2"	1 Plum	8	1	Ea.	1,050	67.50		1,117.50	1,275
3740	2-1/2"	Q-1	5	3.200		1,350	195		1,545	1,775
3750	3"		4.50	3.556		1,425	217		1,642	1,875
3760	4"		3	5.333		1,950	325		2,275	2,625
3770	6"	Q-2	3	8		3,075	505		3,580	4,150
5150	Forged									
5650	Check valve, class 800, horizontal									
5698	Threaded									
5700	1/4"	1 Plum	24	.333	Ea.	86.50	22.50		109	129
5720	3/8"		24	.333		86.50	22.50		109	129
5730	1/2"		24	.333		86.50	22.50		109	129
5740	3/4"		20	.400		92.50	27		119.50	143
5750	1"		19	.421		109	28.50		137.50	163
5760	1-1/4"		15	.533		214	36		250	289

23 05 93 - Testing, Adjusting, and Balancing for HVAC

23 05 93.10 Balancing, Air

		Crew	Daily Output	Labor-Hours	Unit	Material	2021 Bare Costs Labor	2021 Bare Costs Equipment	Total	Total Incl O&P
0010	**BALANCING, AIR** (Subcontractor's quote incl. material and labor)									
0900	Heating and ventilating equipment									
1000	Centrifugal fans, utility sets				Ea.				420	420
1100	Heating and ventilating unit								630	630
1200	In-line fan								630	630
1300	Propeller and wall fan								119	119
1400	Roof exhaust fan								280	280
2000	Air conditioning equipment, central station								910	910
2100	Built-up low pressure unit								840	840
2200	Built-up high pressure unit								980	980
2300	Built-up high pressure dual duct								1,550	1,550
2400	Built-up variable volume								1,825	1,825

For customer support on your Site Work & Landscape Costs with RSMeans Data, call 800.448.8182.

263

23 05 Common Work Results for HVAC

23 05 93 – Testing, Adjusting, and Balancing for HVAC

23 05 93.10 Balancing, Air		Crew	Daily Output	Labor-Hours	Unit	Material	2021 Bare Costs Labor	Equipment	Total	Total Incl O&P
2500	Multi-zone A.C. and heating unit				Ea.				630	630
2600	For each zone over one, add								140	140
2700	Package A.C. unit								350	350
2800	Rooftop heating and cooling unit								490	490
3000	Supply, return, exhaust, registers & diffusers, avg. height ceiling								84	84
3100	High ceiling								126	126
3200	Floor height								70	70
3300	Off mixing box								56	56
3500	Induction unit								91	91
3600	Lab fume hood								420	420
3700	Linear supply								210	210
3800	Linear supply high								245	245
4000	Linear return								70	70
4100	Light troffers								84	84
4200	Moduline - master								84	84
4300	Moduline - slaves								42	42
4400	Regenerators								560	560
4500	Taps into ceiling plenums								105	105
4600	Variable volume boxes								84	84

23 05 93.20 Balancing, Water

		Crew	Daily Output	Labor-Hours	Unit	Material	2021 Bare Costs Labor	Equipment	Total	Total Incl O&P
0010	**BALANCING, WATER** (Subcontractor's quote incl. material and labor)									
0050	Air cooled condenser				Ea.				256	256
0080	Boiler								515	515
0100	Cabinet unit heater								88	88
0200	Chiller								620	620
0300	Convector								73	73
0400	Converter								365	365
0500	Cooling tower								475	475
0600	Fan coil unit, unit ventilator								132	132
0700	Fin tube and radiant panels								146	146
0800	Main and duct re-heat coils								135	135
0810	Heat exchanger								135	135
0900	Main balancing cocks								110	110
1000	Pumps								320	320
1100	Unit heater								102	102

23 05 93.50 Piping, Testing

		Crew	Daily Output	Labor-Hours	Unit	Material	2021 Bare Costs Labor	Equipment	Total	Total Incl O&P
0010	**PIPING, TESTING**									
0100	Nondestructive testing									
0110	Nondestructive hydraulic pressure test, isolate & 1 hr. hold									
0120	1" - 4" pipe									
0140	0-250 L.F.	1 Stpi	1.33	6.015	Ea.		410		410	615
0160	250-500 L.F.	"	.80	10			685		685	1,025
0180	500-1000 L.F.	Q-5	1.14	14.035			865		865	1,300
0200	1000-2000 L.F.		.80	20			1,225		1,225	1,825
0320	0-250 L.F.		1	16			985		985	1,475
0340	250-500 L.F.		.73	21.918			1,350		1,350	2,025
0360	500-1000 L.F.		.53	30.189			1,850		1,850	2,775
0380	1000-2000 L.F.		.38	42.105			2,600		2,600	3,875
1000	Pneumatic pressure test, includes soaping joints									
1120	1" - 4" pipe									
1140	0-250 L.F.	Q-5	2.67	5.993	Ea.	13.05	370		383.05	565
1160	250-500 L.F.		1.33	12.030		26	740		766	1,125

23 05 Common Work Results for HVAC

23 05 93 – Testing, Adjusting, and Balancing for HVAC

23 05 93.50 Piping, Testing		Crew	Daily Output	Labor-Hours	Unit	Material	2021 Bare Costs Labor	Equipment	Total	Total Incl O&P
1180	500-1000 L.F.	Q-5	.80	20	Ea.	39	1,225		1,264	1,875
1200	1000-2000 L.F.	↓	.50	32	↓	52	1,975		2,027	3,000
1300	6" - 10" pipe									
1320	0-250 L.F.	Q-5	1.33	12.030	Ea.	13.05	740		753.05	1,125
1340	250-500 L.F.		.67	23.881		26	1,475		1,501	2,225
1360	500-1000 L.F.		.40	40		52	2,450		2,502	3,725
1380	1000-2000 L.F.	↓	.25	64	↓	65.50	3,950		4,015.50	5,950
2000	X-Ray of welds									
2110	2" diam.	1 Stpi	8	1	Ea.	16.50	68.50		85	120
2120	3" diam.		8	1		16.50	68.50		85	120
2130	4" diam.		8	1		25	68.50		93.50	129
2140	6" diam.		8	1		25	68.50		93.50	129
2150	8" diam.		6.60	1.212		25	83		108	151
2160	10" diam.	↓	6	1.333	↓	33	91		124	173
3000	Liquid penetration of welds									
3110	2" diam.	1 Stpi	14	.571	Ea.	9.95	39		48.95	69.50
3120	3" diam.		13.60	.588		9.95	40		49.95	71
3130	4" diam.		13.40	.597		9.95	41		50.95	72
3140	6" diam.		13.20	.606		9.95	41.50		51.45	73
3150	8" diam.		13	.615		14.95	42		56.95	79.50
3160	10" diam.	↓	12.80	.625	↓	14.95	42.50		57.45	80.50

23 09 23.10 Control Components/DDC Systems

		Crew	Daily Output	Labor-Hours	Unit	Material	Labor	Equipment	Total	Total Incl O&P
0010	**CONTROL COMPONENTS/DDC SYSTEMS** (Sub's quote incl. M & L)									
0100	Analog inputs									
0110	Sensors (avg. 50' run in 1/2" EMT)									
0120	Duct temperature				Ea.				465	465
0130	Space temperature								665	665
0140	Duct humidity, +/- 3%								595	595
0150	Space humidity, +/- 2%								1,150	1,150
0160	Duct static pressure								680	680
0170	CFM/transducer								920	920
0172	Water temperature								940	940
0174	Water flow								3,450	3,450
0176	Water pressure differential								975	975
0177	Steam flow								2,400	2,400
0178	Steam pressure								1,025	1,025
0180	KW/transducer								1,350	1,350
0182	KWH totalization (not incl. elec. meter pulse xmtr.)								625	625
0190	Space static pressure				↓				1,075	1,075
1000	Analog outputs (avg. 50' run in 1/2" EMT)									
1010	P/I transducer				Ea.				635	635
1020	Analog output, matl. in MUX								305	305
1030	Pneumatic (not incl. control device)								645	645
1040	Electric (not incl. control device)				↓				380	380
2000	Status (alarms)									
2100	Digital inputs (avg. 50' run in 1/2" EMT)									
2110	Freeze				Ea.				435	435
2120	Fire								395	395
2130	Differential pressure (air)	2 Elec	3	5.333		283	340		623	815
2140	Differential pressure (water)								975	975
2150	Current sensor								435	435
2160	Duct high temperature thermostat				↓				570	570

For customer support on your Site Work & Landscape Costs with RSMeans Data, call 800.448.8182.

265

23 09 23.10 Control Components/DDC Systems	Crew	Daily Output	Labor-Hours	Unit	Material	2021 Bare Costs Labor	Equipment	Total	Total Incl O&P
2170	Duct smoke detector				Ea.					705	705
2200	Digital output (avg. 50' run in 1/2" EMT)									
2210	Start/stop				Ea.				340	340
2220	On/off (maintained contact)				"				585	585
3000	Controller MUX panel, incl. function boards									
3100	48 point				Ea.				5,275	5,275
3110	128 point				"				7,225	7,225
3200	DDC controller (avg. 50' run in conduit)									
3210	Mechanical room									
3214	16 point controller (incl. 120 volt/1 phase power supply)				Ea.				3,275	3,275
3229	32 point controller (incl. 120 volt/1 phase power supply)				"				5,425	5,425
3230	Includes software programming and checkout									
3260	Space									
3266	VAV terminal box (incl. space temp. sensor)				Ea.				840	840
3280	Host computer (avg. 50' run in conduit)									
3281	Package complete with PC, keyboard,									
3282	printer, monitor, basic software				Ea.				3,150	3,150
4000	Front end costs									
4100	Computer (P.C.) with software program				Ea.				6,350	6,350
4200	Color graphics software								3,925	3,925
4300	Color graphics slides								490	490
4350	Additional printer								980	980
4400	Communications trunk cable				L.F.				3.80	3.80
4500	Engineering labor (not incl. dftg.)				Point				94	94
4600	Calibration labor								120	120
4700	Start-up, checkout labor								120	120
4800	Programming labor, as req'd									
5000	Communications bus (data transmission cable)									
5010	#18 twisted shielded pair in 1/2" EMT conduit				C.L.F.				380	380
8000	Applications software									
8050	Basic maintenance manager software (not incl. data base entry)				Ea.				1,950	1,950
8100	Time program				Point				6.85	6.85
8120	Duty cycle								13.65	13.65
8140	Optimum start/stop								41.50	41.50
8160	Demand limiting								20.50	20.50
8180	Enthalpy program								41.50	41.50
8200	Boiler optimization				Ea.				1,225	1,225
8220	Chiller optimization				"				1,625	1,625
8240	Custom applications									
8260	Cost varies with complexity									

23 13 13 – Facility Underground Fuel-Oil, Storage Tanks

23 13 13.09 Single-Wall Steel Fuel-Oil Tanks	Crew	Daily Output	Labor-Hours	Unit	Material	2021 Bare Costs Labor	Equipment	Total	Total Incl O&P
0010 **SINGLE-WALL STEEL FUEL-OIL TANKS**									
5000 Tanks, steel ugnd., sti-p3, not incl. hold-down bars									
5500 Excavation, pad, pumps and piping not included									
5510 Single wall, 500 gallon capacity, 7 ga. shell	Q-5	2.70	5.926	Ea.	2,000	365		2,365	2,750
5520 1,000 gallon capacity, 7 ga. shell	"	2.50	6.400		2,900	395		3,295	3,775
5530 2,000 gallon capacity, 1/4" thick shell	Q-7	4.60	6.957		4,175	455		4,630	5,275
5535 2,500 gallon capacity, 7 ga. shell	Q-5	3	5.333		5,625	330		5,955	6,675
5540 5,000 gallon capacity, 1/4" thick shell	Q-7	3.20	10		13,200	650		13,850	15,500
5560 10,000 gallon capacity, 1/4" thick shell		2	16		10,800	1,050		11,850	13,400
5580 15,000 gallon capacity, 5/16" thick shell		1.70	18.824		11,100	1,225		12,325	14,000
5600 20,000 gallon capacity, 5/16" thick shell		1.50	21.333		27,000	1,400		28,400	31,800
5610 25,000 gallon capacity, 3/8" thick shell		1.30	24.615		30,900	1,600		32,500	36,400
5620 30,000 gallon capacity, 3/8" thick shell		1.10	29.091		34,800	1,900		36,700	41,000
5630 40,000 gallon capacity, 3/8" thick shell		.90	35.556		45,200	2,325		47,525	53,000
5640 50,000 gallon capacity, 3/8" thick shell	▼	.80	40	▼	51,500	2,600		54,100	60,500

23 13 13.23 Glass-Fiber-Reinfcd-Plastic, Fuel-Oil, Storage

	Crew	Daily Output	Labor-Hours	Unit	Material	2021 Bare Costs Labor	Equipment	Total	Total Incl O&P
0010 **GLASS-FIBER-REINFCD-PLASTIC, UNDERGRND. FUEL-OIL, STORAGE**									
0210 Fiberglass, underground, single wall, UL listed, not including									
0220 manway or hold-down strap									
0225 550 gallon capacity	Q-5	2.67	5.993	Ea.	4,225	370		4,595	5,200
0230 1,000 gallon capacity	"	2.46	6.504		5,450	400		5,850	6,575
0240 2,000 gallon capacity	Q-7	4.57	7.002		8,725	455		9,180	10,300
0245 3,000 gallon capacity		3.90	8.205		8,850	535		9,385	10,500
0250 4,000 gallon capacity		3.55	9.014		11,800	585		12,385	13,900
0255 5,000 gallon capacity		3.20	10		10,800	650		11,450	12,800
0260 6,000 gallon capacity		2.67	11.985		10,100	780		10,880	12,300
0270 8,000 gallon capacity		2.29	13.974		15,800	910		16,710	18,700
0280 10,000 gallon capacity		2	16		15,900	1,050		16,950	19,100
0282 12,000 gallon capacity		1.88	17.021		16,800	1,100		17,900	20,200
0284 15,000 gallon capacity		1.68	19.048		24,300	1,250		25,550	28,600
0290 20,000 gallon capacity		1.45	22.069		33,800	1,425		35,225	39,400
0300 25,000 gallon capacity		1.28	25		58,000	1,625		59,625	66,500
0320 30,000 gallon capacity		1.14	28.070		87,500	1,825		89,325	99,000
0340 40,000 gallon capacity		.89	35.955		125,000	2,350		127,350	140,500
0360 48,000 gallon capacity	▼	.81	39.506		185,000	2,575		187,575	207,500
0500 For manway, fittings and hold-downs, add					20%	15%			
1000 For helical heating coil, add	Q-5	2.50	6.400	▼	5,625	395		6,020	6,800
2210 Fiberglass, underground, single wall, UL listed, including									
2220 hold-down straps, no manways									
2225 550 gallon capacity	Q-5	2	8	Ea.	4,675	490		5,165	5,875
2230 1,000 gallon capacity	"	1.88	8.511		5,900	525		6,425	7,250
2240 2,000 gallon capacity	Q-7	3.55	9.014		9,175	585		9,760	11,000
2250 4,000 gallon capacity		2.90	11.034		12,200	720		12,920	14,600
2260 6,000 gallon capacity		2	16		11,000	1,050		12,050	13,700
2270 8,000 gallon capacity		1.78	17.978		16,700	1,175		17,875	20,100
2280 10,000 gallon capacity		1.60	20		16,800	1,300		18,100	20,500
2282 12,000 gallon capacity		1.52	21.053		17,700	1,375		19,075	21,600
2284 15,000 gallon capacity		1.39	23.022		25,200	1,500		26,700	29,900
2290 20,000 gallon capacity		1.14	28.070		35,200	1,825		37,025	41,400
2300 25,000 gallon capacity		.96	33.333		60,000	2,175		62,175	69,000
2320 30,000 gallon capacity	▼	.80	40	▼	90,500	2,600		93,100	103,500

23 21 Hydronic Piping and Pumps

23 21 23 – Hydronic Pumps

23 21 23.13 In-Line Centrifugal Hydronic Pumps

		Crew	Daily Output	Labor-Hours	Unit	Material	2021 Bare Costs Labor	2021 Bare Costs Equipment	Total	Total Incl O&P
0010	**IN-LINE CENTRIFUGAL HYDRONIC PUMPS**									
0600	Bronze, sweat connections, 1/40 HP, in line									
0640	3/4" size	Q-1	16	1	Ea.	273	61		334	390
1000	Flange connection, 3/4" to 1-1/2" size									
1040	1/12 HP	Q-1	6	2.667	Ea.	680	162		842	985
1060	1/8 HP		6	2.667		1,250	162		1,412	1,600
1100	1/3 HP		6	2.667		1,375	162		1,537	1,775
1140	2" size, 1/6 HP		5	3.200		1,775	195		1,970	2,250
1180	2-1/2" size, 1/4 HP	↓	5	3.200	↓	2,275	195		2,470	2,800
2000	Cast iron, flange connection									
2040	3/4" to 1-1/2" size, in line, 1/12 HP	Q-1	6	2.667	Ea.	480	162		642	765
2100	1/3 HP		6	2.667		865	162		1,027	1,200
2140	2" size, 1/6 HP		5	3.200		950	195		1,145	1,350
2180	2-1/2" size, 1/4 HP		5	3.200		1,150	195		1,345	1,550
2220	3" size, 1/4 HP	↓	4	4	↓	1,150	244		1,394	1,625
2600	For nonferrous impeller, add					3%				

23 55 Fuel-Fired Heaters

23 55 13 – Fuel-Fired Duct Heaters

23 55 13.16 Gas-Fired Duct Heaters

		Crew	Daily Output	Labor-Hours	Unit	Material	2021 Bare Costs Labor	2021 Bare Costs Equipment	Total	Total Incl O&P
0010	**GAS-FIRED DUCT HEATERS**, Includes burner, controls, stainless steel									
0020	heat exchanger. Gas fired, electric ignition									
1000	Outdoor installation, with power venter									
1020	75 MBH output	Q-5	4	4	Ea.	3,450	246		3,696	4,175
1060	120 MBH output		4	4		4,275	246		4,521	5,075
1100	187 MBH output		3	5.333		5,625	330		5,955	6,700
1140	300 MBH output		1.80	8.889		6,475	545		7,020	7,925
1180	450 MBH output	↓	1.40	11.429	↓	7,550	705		8,255	9,350

Estimating Tips
26 05 00 Common Work Results for Electrical

- Conduit should be taken off in three main categories—power distribution, branch power, and branch lighting—so the estimator can concentrate on systems and components, therefore making it easier to ensure all items have been accounted for.
- For cost modifications for elevated conduit installation, add the percentages to labor according to the height of installation and only to the quantities exceeding the different height levels, not to the total conduit quantities. Refer to subdivision 26 01 02.20 for labor adjustment factors.
- Remember that aluminum wiring of equal ampacity is larger in diameter than copper and may require larger conduit.
- If more than three wires at a time are being pulled, deduct percentages from the labor hours of that grouping of wires.
- When taking off grounding systems, identify separately the type and size of wire, and list each unique type of ground connection.

- The estimator should take the weights of materials into consideration when completing a takeoff. Topics to consider include: How will the materials be supported? What methods of support are available? How high will the support structure have to reach? Will the final support structure be able to withstand the total burden? Is the support material included or separate from the fixture, equipment, and material specified?
- Do not overlook the costs for equipment used in the installation. If scaffolding or highlifts are available in the field, contractors may use them in lieu of the proposed ladders and rolling staging.

26 20 00 Low-Voltage Electrical Transmission

- Supports and concrete pads may be shown on drawings for the larger equipment, or the support system may be only a piece of plywood for the back of a panelboard. In either case, they must be included in the costs.

26 40 00 Electrical and Cathodic Protection

- When taking off cathodic protection systems, identify the type and size of cable, and list each unique type of anode connection.

26 50 00 Lighting

- Fixtures should be taken off room by room using the fixture schedule, specifications, and the ceiling plan. For large concentrations of lighting fixtures in the same area, deduct the percentages from labor hours.

Reference Numbers

Reference numbers are shown at the beginning of some major classifications. These numbers refer to related items in the Reference Section. The reference information may be an estimating procedure, an alternate pricing method, or technical information.

Note: Not all subdivisions listed here necessarily appear. ■

Same Data. Simplified.

Enjoy the convenience and efficiency of accessing your costs anywhere:

- **Skip the multiplier** by setting your location
- **Quickly search,** edit, favorite and share costs
- **Stay on top of price changes** with automatic updates

Discover more at rsmeans.com/online

Note: "Powered in part by CINX™, based on licensed proprietary information of Harrison Publishing House, Inc."

Note: The following companies, in part, have been used as a reference source for some of the material prices used in Division 26:

Electriflex

Trade Service

26 05 05 – Selective Demolition for Electrical

26 05 05.10 Electrical Demolition		Crew	Daily Output	Labor-Hours	Unit	Material	2021 Bare Costs Labor	Equipment	Total	Total Incl O&P
0010	**ELECTRICAL DEMOLITION**									
0020	Electrical demolition, conduit to 10' high, incl. fittings & hangers									
0100	Rigid galvanized steel, 1/2" to 1" diameter	1 Elec	242	.033	L.F.		2.11		2.11	3.13
0120	1-1/4" to 2"	"	200	.040			2.55		2.55	3.79
0140	2-1/2" to 3-1/2"	2 Elec	302	.053			3.37		3.37	5
0160	4" to 6"	"	160	.100			6.35		6.35	9.45
0200	Electric metallic tubing (EMT), 1/2" to 1"	1 Elec	394	.020			1.29		1.29	1.92
0220	1-1/4" to 1-1/2"		326	.025			1.56		1.56	2.32
0240	2" to 3"	↓	236	.034			2.16		2.16	3.21
0260	3-1/2" to 4"	2 Elec	310	.052	↓		3.29		3.29	4.89
1300	Transformer, dry type, 1 phase, incl. removal of									
1320	supports, wire & conduit terminations									
1340	1 kVA	1 Elec	7.70	1.039	Ea.		66		66	98.50
1360	5 kVA		4.70	1.702			108		108	161
1380	10 kVA	↓	3.60	2.222			142		142	210
1400	37.5 kVA	2 Elec	3	5.333			340		340	505
1420	75 kVA	"	2.50	6.400	↓		410		410	605
1440	3 phase to 600 V, primary									
1460	3 kVA	1 Elec	3.87	2.067	Ea.		132		132	196
1520	75 kVA	2 Elec	2.69	5.948			380		380	565
1550	300 kVA	R-3	1.80	11.111			705	106	811	1,175
1570	750 kVA	"	1.10	18.182	↓		1,150	174	1,324	1,925

26 05 13 – Medium-Voltage Cables

26 05 13.16 Medium-Voltage, Single Cable

26 05 13.16 Medium-Voltage, Single Cable		Crew	Daily Output	Labor-Hours	Unit	Material	2021 Bare Costs Labor	Equipment	Total	Total Incl O&P
0010	**MEDIUM-VOLTAGE, SINGLE CABLE** Splicing & terminations not included									
0040	Copper, XLP shielding, 5 kV, #6 R260533-22	2 Elec	4.40	3.636	C.L.F.	168	232		400	530
0050	#4		4.40	3.636		218	232		450	585
0100	#2		4	4		232	255		487	635
0200	#1		4	4		330	255		585	745
0400	1/0		3.80	4.211		370	268		638	810
0600	2/0		3.60	4.444		400	283		683	860
0800	4/0	↓	3.20	5		610	320		930	1,150
1000	250 kcmil	3 Elec	4.50	5.333		725	340		1,065	1,300
1200	350 kcmil		3.90	6.154		945	390		1,335	1,625
1400	500 kcmil		3.60	6.667		1,000	425		1,425	1,725
1600	15 kV, ungrounded neutral, #1	2 Elec	4	4		360	255		615	775
1800	1/0		3.80	4.211		430	268		698	875
2000	2/0		3.60	4.444		490	283		773	960
2200	4/0	↓	3.20	5		650	320		970	1,200
2400	250 kcmil	3 Elec	4.50	5.333		720	340		1,060	1,300
2600	350 kcmil		3.90	6.154		910	390		1,300	1,575
2800	500 kcmil		3.60	6.667		1,025	425		1,450	1,750
3000	25 kV, grounded neutral, 1/0	2 Elec	3.60	4.444		595	283		878	1,075
3200	2/0		3.40	4.706		655	300		955	1,175
3400	4/0	↓	3	5.333		825	340		1,165	1,400
3600	250 kcmil	3 Elec	4.20	5.714		1,025	365		1,390	1,675
3800	350 kcmil		3.60	6.667		1,200	425		1,625	1,950
3900	500 kcmil	↓	3.30	7.273		1,400	465		1,865	2,250
4000	35 kV, grounded neutral, 1/0	2 Elec	3.40	4.706		610	300		910	1,125
4200	2/0		3.20	5		745	320		1,065	1,300
4400	4/0	↓	2.80	5.714		935	365		1,300	1,575
4600	250 kcmil	3 Elec	3.90	6.154		1,100	390		1,490	1,775

26 05 13.16 Medium-Voltage, Single Cable		Crew	Daily Output	Labor-Hours	Unit	Material	2021 Bare Costs Labor	Equipment	Total	Total Incl O&P
4800	350 kcmil	3 Elec	3.30	7.273	C.L.F.	1,325	465		1,790	2,150
5000	500 kcmil		3	8		1,550	510		2,060	2,450
5050	Aluminum, XLP shielding, 5 kV, #2	2 Elec	5	3.200		263	204		467	595
5070	#1		4.40	3.636		272	232		504	645
5090	1/0		4	4		315	255		570	725
5100	2/0		3.80	4.211		350	268		618	790
5150	4/0		3.60	4.444		420	283		703	880
5200	250 kcmil	3 Elec	4.80	5		505	320		825	1,025
5220	350 kcmil		4.50	5.333		595	340		935	1,150
5240	500 kcmil		3.90	6.154		765	390		1,155	1,425
5260	750 kcmil		3.60	6.667		1,000	425		1,425	1,725
5300	15 kV aluminum, XLP, #1	2 Elec	4.40	3.636		335	232		567	715
5320	1/0		4	4		350	255		605	765
5340	2/0		3.80	4.211		375	268		643	815
5360	4/0		3.60	4.444		460	283		743	925
5380	250 kcmil	3 Elec	4.80	5		550	320		870	1,075
5400	350 kcmil		4.50	5.333		615	340		955	1,175
5420	500 kcmil		3.90	6.154		855	390		1,245	1,525
5440	750 kcmil		3.60	6.667		1,125	425		1,550	1,875
6010	Copper, XLP shielding, 5 kV, buried in trench, #6	R-15	2.90	16.552	M.L.F.	1,675	1,025	97	2,797	3,475
6020	#4		2.90	16.552		2,175	1,025	97	3,297	4,025
6030	#2		2.90	16.552		2,325	1,025	97	3,447	4,175
6040	1/0		2.60	18.462		3,700	1,150	108	4,958	5,925
6050	2/0		2.60	18.462		3,975	1,150	108	5,233	6,225
6060	3/0		2.40	20		5,250	1,250	117	6,617	7,750
6070	4/0		2.40	20		6,100	1,250	117	7,467	8,700
6080	250 kcmil		2.20	21.818		7,275	1,350	128	8,753	10,200
6090	350 kcmil		2.15	22.326		9,425	1,400	131	10,956	12,600
6100	500 kcmil		2.05	23.415		10,100	1,450	137	11,687	13,400
6200	Installed on poles, #6	R-16	5.50	11.636		1,675	730	51	2,456	2,975
6210	#4		5.10	12.549		2,175	785	55	3,015	3,625
6220	#2		4.05	15.802		2,325	990	69.50	3,384.50	4,100
6230	1/0		3.35	19.104		3,700	1,200	84	4,984	5,950
6240	2/0		2.85	22.456		3,975	1,400	99	5,474	6,575
6250	3/0		2.50	25.600		5,250	1,600	113	6,963	8,275
6260	4/0		2.15	29.767		6,100	1,875	131	8,106	9,650
6270	250 kcmil		1.70	37.647		7,275	2,350	166	9,791	11,700
6280	350 kcmil		1.35	47.407		9,425	2,975	209	12,609	15,100
6290	500 kcmil		1.05	60.952		10,100	3,825	268	14,193	17,100
6300	15 kV, pulled in duct, #1	R-15	1.35	35.556		3,600	2,225	209	6,034	7,500
6310	1/0		1.35	35.556		4,300	2,225	209	6,734	8,250
6320	2/0		1.35	35.556		4,900	2,225	209	7,334	8,925
6330	3/0		1.30	36.923		5,300	2,300	217	7,817	9,500
6340	4/0		1.30	36.923		6,500	2,300	217	9,017	10,800
6350	250 kcmil		1.25	38.400		7,200	2,400	225	9,825	11,700
6360	350 kcmil		1.20	40		9,100	2,500	235	11,835	14,000
6370	500 kcmil		1.20	40		10,200	2,500	235	12,935	15,300
6400	Direct burial, #1		2.40	20		3,600	1,250	117	4,967	5,950
6410	1/0		2.40	20		4,300	1,250	117	5,667	6,700
6420	2/0		2.40	20		4,900	1,250	117	6,267	7,375
6430	3/0		2.20	21.818		5,300	1,350	128	6,778	8,000
6440	4/0		2.20	21.818		6,500	1,350	128	7,978	9,325
6450	250 kcmil		2	24		7,200	1,500	141	8,841	10,300

271

For customer support on your Site Work & Landscape Costs with RSMeans Data, call 800.448.8182.

26 05 13 – Medium-Voltage Cables

26 05 13.16 Medium-Voltage, Single Cable		Crew	Daily Output	Labor-Hours	Unit	Material	2021 Bare Costs Labor	Equipment	Total	Total Incl O&P
6460	350 kcmil	R-15	1.90	25.263	M.L.F.	9,100	1,575	148	10,823	12,500
6470	500 kcmil	↓	1.85	25.946		10,200	1,625	152	11,977	13,900
6500	Installed on poles, #1	R-16	3.85	16.623		3,600	1,050	73	4,723	5,600
6510	1/0		3.35	19.104		4,300	1,200	84	5,584	6,600
6520	2/0		2.85	22.456		4,900	1,400	99	6,399	7,600
6530	3/0		2.50	25.600		5,300	1,600	113	7,013	8,325
6540	4/0		2.15	29.767		6,500	1,875	131	8,506	10,100
6550	250 kcmil		1.85	34.595		7,200	2,175	152	9,527	11,300
6560	350 kcmil		1.60	40		9,100	2,500	176	11,776	13,900
6570	500 kcmil	↓	1.35	47.407	↓	10,200	2,975	209	13,384	16,000
6600	URD, 15 kV, 3/C, direct burial, #2	R-22	1500	.025	L.F.	12.65	1.45		14.10	16.05
7000	Copper, shielded, 35 kV, 1/0	R-16	1.10	58.182	M.L.F.	6,075	3,650	256	9,981	12,400
7020	3/0		1.05	60.952		7,100	3,825	268	11,193	13,800
7100	69 kV, 500 kcmil	↓	.90	71.111	↓	26,700	4,450	315	31,465	36,300

26 05 19 – Low-Voltage Electrical Power Conductors and Cables

26 05 19.55 Non-Metallic Sheathed Cable

26 05 19.55 Non-Metallic Sheathed Cable		Crew	Daily Output	Labor-Hours	Unit	Material	2021 Bare Costs Labor	Equipment	Total	Total Incl O&P
0010	**NON-METALLIC SHEATHED CABLE** 600 volt R260533-22									
0100	Copper with ground wire (Romex)									
0150	#14, 2 conductor	1 Elec	2.70	2.963	C.L.F.	17.40	189		206.40	299
0200	3 conductor		2.40	3.333		24.50	212		236.50	340
0220	4 conductor		2.20	3.636		41	232		273	390
0250	#12, 2 conductor		2.50	3.200		23.50	204		227.50	330
0300	3 conductor		2.20	3.636		37	232		269	385
0320	4 conductor		2	4		60.50	255		315.50	445
0350	#10, 2 conductor		2.20	3.636		45.50	232		277.50	395
0400	3 conductor		1.80	4.444		57.50	283		340.50	485
0420	4 conductor		1.60	5		96	320		416	580
0430	#8, 2 conductor		1.60	5		77.50	320		397.50	560
0450	3 conductor		1.50	5.333		110	340		450	625
0500	#6, 3 conductor	↓	1.40	5.714		385	365		750	960
0520	#4, 3 conductor	2 Elec	2.40	6.667		405	425		830	1,075
0540	#2, 3 conductor	"	2.20	7.273	↓	630	465		1,095	1,375
0550	SE type SER aluminum cable, 3 RHW and									
0600	1 bare neutral, 3 #8 & 1 #8	1 Elec	1.60	5	C.L.F.	71	320		391	555
0650	3 #6 & 1 #6	"	1.40	5.714		74.50	365		439.50	620
0700	3 #4 & 1 #6	2 Elec	2.40	6.667		82	425		507	720
0750	3 #2 & 1 #4		2.20	7.273		128	465		593	830
0800	3 #1/0 & 1 #2		2	8		173	510		683	945
0850	3 #2/0 & 1 #1		1.80	8.889		202	565		767	1,050
0900	3 #4/0 & 1 #2/0		1.60	10		325	635		960	1,300
1000	URD - triplex underground distribution cable, alum. 2 #4 + #4 neutral		2.80	5.714		83.50	365		448.50	630
1010	2 #2 + #4 neutral		2.65	6.038		102	385		487	680
1020	2 #2 + #2 neutral		2.55	6.275		103	400		503	710
1030	2 1/0 + #2 neutral		2.40	6.667		124	425		549	765
1040	2 1/0 + 1/0 neutral		2.30	6.957		140	445		585	815
1050	2 2/0 + #1 neutral		2.20	7.273		169	465		634	875
1060	2 2/0 + 2/0 neutral		2.10	7.619		171	485		656	910
1070	2 3/0 + 1/0 neutral		1.95	8.205		187	525		712	980
1080	2 3/0 + 3/0 neutral		1.95	8.205		203	525		728	1,000
1090	2 4/0 + 2/0 neutral		1.85	8.649		192	550		742	1,025
1100	2 4/0 + 4/0 neutral	↓	1.85	8.649		241	550		791	1,075
1450	UF underground feeder cable, copper with ground, #14, 2 conductor	1 Elec	4	2		19.70	127		146.70	211

272

26 05 Common Work Results for Electrical

26 05 19 – Low-Voltage Electrical Power Conductors and Cables

26 05 19.55 Non-Metallic Sheathed Cable		Crew	Daily Output	Labor-Hours	Unit	Material	2021 Bare Costs Labor	Equipment	Total	Total Incl O&P
1500	#12, 2 conductor	1 Elec	3.50	2.286	C.L.F.	34	146		180	254
1550	#10, 2 conductor		3	2.667		53	170		223	310
1600	#14, 3 conductor		3.50	2.286		36.50	146		182.50	256
1650	#12, 3 conductor		3	2.667		47	170		217	305
1700	#10, 3 conductor		2.50	3.200		78.50	204		282.50	390
1710	#8, 3 conductor		2	4		144	255		399	540
1720	#6, 3 conductor		1.80	4.444		212	283		495	655
2400	SEU service entrance cable, copper 2 conductors, #8 + #8 neutral		1.50	5.333		97	340		437	610
2600	#6 + #8 neutral		1.30	6.154		135	390		525	730
2800	#6 + #6 neutral		1.30	6.154		155	390		545	750
3000	#4 + #6 neutral	2 Elec	2.20	7.273		215	465		680	925
3200	#4 + #4 neutral		2.20	7.273		249	465		714	965
3400	#3 + #5 neutral		2.10	7.619		283	485		768	1,025
3600	#3 + #3 neutral		2.10	7.619		335	485		820	1,100
3800	#2 + #4 neutral		2	8		355	510		865	1,150
4000	#1 + #1 neutral		1.90	8.421		395	535		930	1,225
4200	1/0 + 1/0 neutral		1.80	8.889		870	565		1,435	1,800
4400	2/0 + 2/0 neutral		1.70	9.412		840	600		1,440	1,825
4600	3/0 + 3/0 neutral		1.60	10		1,100	635		1,735	2,175
4620	4/0 + 4/0 neutral		1.45	11.034		785	705		1,490	1,925
4800	Aluminum 2 conductors, #8 + #8 neutral	1 Elec	1.60	5		54	320		374	535
5000	#6 + #6 neutral	"	1.40	5.714		58.50	365		423.50	605
5100	#4 + #6 neutral	2 Elec	2.50	6.400		74	410		484	685
5200	#4 + #4 neutral		2.40	6.667		78	425		503	715
5300	#2 + #4 neutral		2.30	6.957		107	445		552	780
5400	#2 + #2 neutral		2.20	7.273		107	465		572	810
5450	1/0 + #2 neutral		2.10	7.619		179	485		664	915
5500	1/0 + 1/0 neutral		2	8		167	510		677	940
5550	2/0 + #1 neutral		1.90	8.421		200	535		735	1,025
5600	2/0 + 2/0 neutral		1.80	8.889		187	565		752	1,050
5800	3/0 + 1/0 neutral		1.70	9.412		288	600		888	1,200
6000	3/0 + 3/0 neutral		1.70	9.412		237	600		837	1,150
6200	4/0 + 2/0 neutral		1.60	10		221	635		856	1,200
6400	4/0 + 4/0 neutral		1.60	10		237	635		872	1,200
6500	Service entrance cap for copper SEU									
6600	100 amp	1 Elec	12	.667	Ea.	7	42.50		49.50	70.50
6700	150 amp	"	10	.800	"	12.95	51		63.95	90

26 05 26 – Grounding and Bonding for Electrical Systems

26 05 26.80 Grounding

		Crew	Daily Output	Labor-Hours	Unit	Material	Labor	Equipment	Total	Total Incl O&P
0010	**GROUNDING**									
0030	Rod, copper clad, 8' long, 1/2" diameter	1 Elec	5.50	1.455	Ea.	17	92.50		109.50	157
0040	5/8" diameter		5.50	1.455		15	92.50		107.50	155
0050	3/4" diameter		5.30	1.509		28.50	96		124.50	175
0080	10' long, 1/2" diameter		4.80	1.667		21	106		127	181
0090	5/8" diameter		4.60	1.739		20	111		131	187
0100	3/4" diameter		4.40	1.818		52	116		168	229
0130	15' long, 3/4" diameter		4	2		63	127		190	259
0150	Coupling, bronze, 1/2" diameter					9.30			9.30	10.25
0160	5/8" diameter					6.85			6.85	7.55
0170	3/4" diameter					19.20			19.20	21
0190	Drive studs, 1/2" diameter					16.60			16.60	18.30
0210	5/8" diameter					26			26	28.50

For customer support on your Site Work & Landscape Costs with RSMeans Data, call 800.448.8182.

273

26 05 26.80 Grounding	Crew	Daily Output	Labor-Hours	Unit	Material	2021 Bare Costs Labor	2021 Bare Costs Equipment	Total	Total Incl O&P
0220	3/4" diameter				Ea.	29			29	32
0230	Clamp, bronze, 1/2" diameter	1 Elec	32	.250		5.30	15.95		21.25	29.50
0240	5/8" diameter		32	.250		2.46	15.95		18.41	26
0250	3/4" diameter		32	.250		3.52	15.95		19.47	27.50
0260	Wire ground bare armored, #8-1 conductor		2	4	C.L.F.	77.50	255		332.50	465
0270	#6-1 conductor		1.80	4.444		150	283		433	585
0280	#4-1 conductor		1.60	5		148	320		468	640
0320	Bare copper wire, #14 solid		14	.571		6.90	36.50		43.40	61.50
0330	#12		13	.615		12.10	39		51.10	72
0340	#10		12	.667		16.55	42.50		59.05	81
0350	#8		11	.727		25	46.50		71.50	96.50
0360	#6		10	.800		39.50	51		90.50	119
0370	#4		8	1		80	63.50		143.50	183
0380	#2		5	1.600		126	102		228	290
0390	Bare copper wire, stranded, #8		11	.727		27.50	46.50		74	99.50
0400	#6		10	.800		43.50	51		94.50	124
0450	#4	2 Elec	16	1		64	63.50		127.50	165
0600	#2		10	1.600		112	102		214	274
0650	#1		9	1.778		122	113		235	305
0700	1/0		8	2		145	127		272	350
0750	2/0		7.20	2.222		192	142		334	420
0800	3/0		6.60	2.424		275	154		429	535
1000	4/0		5.70	2.807		305	179		484	600
1200	250 kcmil	3 Elec	7.20	3.333		375	212		587	730
1210	300 kcmil		6.60	3.636		760	232		992	1,175
1220	350 kcmil		6	4		540	255		795	975
1230	400 kcmil		5.70	4.211		715	268		983	1,200
1240	500 kcmil		5.10	4.706		710	300		1,010	1,225
1260	750 kcmil		3.60	6.667		1,250	425		1,675	2,000
1270	1,000 kcmil		3	8		1,200	510		1,710	2,050
1360	Bare aluminum, stranded, #6	1 Elec	9	.889		13.70	56.50		70.20	99
1370	#4	2 Elec	16	1		33.50	63.50		97	131
1380	#2		13	1.231		55.50	78.50		134	177
1390	#1		10.60	1.509		40	96		136	187
1400	1/0		9	1.778		80.50	113		193.50	257
1410	2/0		8	2		95.50	127		222.50	294
1420	3/0		7.20	2.222		136	142		278	360
1430	4/0		6.60	2.424		134	154		288	375
1440	250 kcmil	3 Elec	9.30	2.581		64	164		228	315
1450	300 kcmil		8.70	2.759		106	176		282	380
1460	400 kcmil		7.50	3.200		92	204		296	405
1470	500 kcmil		6.90	3.478		145	222		367	490
1480	600 kcmil		6	4		182	255		437	580
1490	700 kcmil		5.70	4.211		201	268		469	620
1500	750 kcmil		5.10	4.706		223	300		523	690
1510	1,000 kcmil		4.80	5		227	320		547	725
1800	Water pipe ground clamps, heavy duty									
2000	Bronze, 1/2" to 1" diameter	1 Elec	8	1	Ea.	24	63.50		87.50	121
2100	1-1/4" to 2" diameter		8	1		44.50	63.50		108	144
2200	2-1/2" to 3" diameter		6	1.333		66	85		151	199
2740	4/0 wire to building steel		7	1.143		2.74	73		75.74	111
2750	4/0 wire to motor frame		7	1.143		2.74	73		75.74	111
2760	4/0 wire to 4/0 wire		7	1.143		2.74	73		75.74	111

26 05 26 – Grounding and Bonding for Electrical Systems

26 05 26.80 Grounding		Crew	Daily Output	Labor-Hours	Unit	Material	2021 Bare Costs Labor	Equipment	Total	Total Incl O&P
2770	4/0 wire to #4 wire	1 Elec	7	1.143	Ea.	2.74	73		75.74	111
2780	4/0 wire to #8 wire		7	1.143		2.74	73		75.74	111
2790	Mold, reusable, for above					142			142	156
2800	Brazed connections, #6 wire	1 Elec	12	.667		16	42.50		58.50	80.50
3000	#2 wire		10	.800		21.50	51		72.50	99
3100	3/0 wire		8	1		32	63.50		95.50	130
3200	4/0 wire		7	1.143		36.50	73		109.50	149
3400	250 kcmil wire		5	1.600		42.50	102		144.50	198
3600	500 kcmil wire		4	2		52.50	127		179.50	247
3700	Insulated ground wire, copper #14		13	.615	C.L.F.	7.15	39		46.15	66.50
3710	#12		11	.727		17.95	46.50		64.45	89
3720	#10		10	.800		25	51		76	103
3730	#8		8	1		26	63.50		89.50	123
3740	#6		6.50	1.231		42	78.50		120.50	162
3750	#4	2 Elec	10.60	1.509		65	96		161	215
3770	#2		9	1.778		99.50	113		212.50	277
3780	#1		8	2		122	127		249	325
3790	1/0		6.60	2.424		151	154		305	395
3800	2/0		5.80	2.759		186	176		362	465
3810	3/0		5	3.200		232	204		436	560
3820	4/0		4.40	3.636		290	232		522	665
3830	250 kcmil	3 Elec	6	4		335	255		590	750
3840	300 kcmil		5.70	4.211		405	268		673	845
3850	350 kcmil		5.40	4.444		470	283		753	940
3860	400 kcmil		5.10	4.706		515	300		815	1,000
3870	500 kcmil		4.80	5		675	320		995	1,225
3880	600 kcmil		3.90	6.154		1,625	390		2,015	2,375
3890	750 kcmil		3.30	7.273		1,050	465		1,515	1,875
3900	1,000 kcmil		2.70	8.889		1,900	565		2,465	2,925
3960	Insulated ground wire, aluminum, #6	1 Elec	8	1		16.25	63.50		79.75	112
3970	#4	2 Elec	13	1.231		20.50	78.50		99	139
3980	#2		10.60	1.509		28	96		124	174
3990	#1		9	1.778		39	113		152	211
4000	1/0		8	2		43.50	127		170.50	237
4010	2/0		7.20	2.222		53	142		195	269
4020	3/0		6.60	2.424		65.50	154		219.50	300
4030	4/0		6.20	2.581		70	164		234	320
4040	250 kcmil	3 Elec	8.70	2.759		84	176		260	355
4050	300 kcmil		8.10	2.963		122	189		311	415
4060	350 kcmil		7.50	3.200		121	204		325	440
4070	400 kcmil		6.90	3.478		146	222		368	490
4080	500 kcmil		6	4		157	255		412	555
4090	600 kcmil		5.70	4.211		209	268		477	630
4100	700 kcmil		5.10	4.706		238	300		538	705
4110	750 kcmil		4.80	5		240	320		560	740
5000	Copper electrolytic ground rod system									
5010	Includes augering hole, mixing bentonite clay,									
5020	Installing rod, and terminating ground wire									
5100	Straight vertical type, 2" diam.									
5120	8.5' long, clamp connection	1 Elec	2.67	2.996	Ea.	755	191		946	1,125
5130	With exothermic weld connection		1.95	4.103		725	261		986	1,200
5140	10' long		2.35	3.404		740	217		957	1,125
5150	With exothermic weld connection		1.78	4.494		925	286		1,211	1,450

26 05 26 – Grounding and Bonding for Electrical Systems

26 05 26.80 Grounding		Crew	Daily Output	Labor-Hours	Unit	Material	2021 Bare Costs Labor	2021 Bare Costs Equipment	Total	Total Incl O&P
5160	12' long	1 Elec	2.16	3.704	Ea.	1,000	236		1,236	1,450
5170	With exothermic weld connection		1.67	4.790		1,100	305		1,405	1,650
5180	20' long		1.74	4.598		1,450	293		1,743	2,025
5190	With exothermic weld connection	↓	1.40	5.714		1,375	365		1,740	2,050
5195	40' long with exothermic weld connection	2 Elec	2	8	↓	3,025	510		3,535	4,100
5200	L-shaped, 2" diam.									
5220	4' vert. x 10' horiz., clamp connection	1 Elec	5.33	1.501	Ea.	1,300	95.50		1,395.50	1,575
5230	With exothermic weld connection	"	3.08	2.597	"	1,300	165		1,465	1,700
5300	Protective box at grade level, with breather slots									
5320	Round 12" long, fiberlyte	1 Elec	32	.250	Ea.	150	15.95		165.95	189
5330	Concrete	"	16	.500		113	32		145	172
5400	Bentonite clay, 50# bag, 1 per 10' of rod					52			52	57
5500	Equipotential earthing bar	1 Elec	2	4	↓	136	255		391	530

26 05 33 – Raceway and Boxes for Electrical Systems

26 05 33.13 Conduit

		Crew	Daily Output	Labor-Hours	Unit	Material	2021 Bare Costs Labor	2021 Bare Costs Equipment	Total	Total Incl O&P
0010	**CONDUIT** To 10' high, includes 2 terminations, 2 elbows,									
0020	11 beam clamps, and 11 couplings per 100 L.F.									
9100	PVC, schedule 40, 1/2" diameter	1 Elec	190	.042	L.F.	.90	2.68		3.58	4.98
9110	3/4" diameter		145	.055		.95	3.51		4.46	6.25
9120	1" diameter		125	.064		1.37	4.08		5.45	7.55
9130	1-1/4" diameter		110	.073		1.77	4.63		6.40	8.85
9140	1-1/2" diameter		100	.080		1.88	5.10		6.98	9.60
9150	2" diameter		90	.089		2.40	5.65		8.05	11.05
9160	2-1/2" diameter	↓	65	.123		4.61	7.85		12.46	16.70
9170	3" diameter	2 Elec	110	.145		3.94	9.25		13.19	18.10
9180	3-1/2" diameter		100	.160		5.70	10.20		15.90	21.50
9190	4" diameter		90	.178		5.35	11.30		16.65	23
9200	5" diameter		70	.229		9.25	14.55		23.80	31.50
9210	6" diameter	↓	60	.267	↓	9.35	17		26.35	35.50
9220	Elbows, 1/2" diameter	1 Elec	50	.160	Ea.	.49	10.20		10.69	15.70
9225	3/4" diameter		42	.190		.55	12.15		12.70	18.65
9230	1" diameter		35	.229		.86	14.55		15.41	22.50
9235	1-1/4" diameter		28	.286		1.30	18.20		19.50	28.50
9240	1-1/2" diameter		20	.400		2.27	25.50		27.77	40.50
9245	2" diameter	R-1A	36.40	.440		3.08	25		28.08	41
9250	2-1/2" diameter		26.70	.599		5.15	34.50		39.65	56.50
9255	3" diameter		22.90	.699		8.40	40		48.40	69
9260	3-1/2" diameter		22.20	.721		9.70	41.50		51.20	72
9265	4" diameter		18.20	.879		17.05	50.50		67.55	94
9270	5" diameter		12.10	1.322		18.35	76		94.35	133
9275	6" diameter	↓	11.10	1.441		36	82.50		118.50	163
9312	Couplings, 1/2" diameter	1 Elec	50	.160		.11	10.20		10.31	15.25
9314	3/4" diameter		42	.190		.20	12.15		12.35	18.25
9316	1" diameter		35	.229		.31	14.55		14.86	22
9318	1-1/4" diameter		28	.286		.41	18.20		18.61	27.50
9320	1-1/2" diameter	↓	20	.400		.63	25.50		26.13	38.50
9322	2" diameter	R-1A	36.40	.440		.80	25		25.80	38.50
9324	2-1/2" diameter		26.70	.599		2.10	34.50		36.60	53.50
9326	3" diameter		22.90	.699		2.04	40		42.04	61.50
9328	3-1/2" diameter		22.20	.721		2.35	41.50		43.85	64
9330	4" diameter		18.20	.879		3.06	50.50		53.56	78.50
9332	5" diameter	↓	12.10	1.322	↓	8.80	76		84.80	123

26 05 33.13 Conduit		Crew	Daily Output	Labor-Hours	Unit	Material	2021 Bare Costs Labor	2021 Bare Costs Equipment	Total	Total Incl O&P
9334	6" diameter	R-1A	11.10	1.441	Ea.	12.35	82.50		94.85	137
9335	See note on line 26 05 33.13 9995									
9340	Field bends, 45° & 90°, 1/2" diameter	1 Elec	45	.178	Ea.		11.30		11.30	16.85
9350	3/4" diameter		40	.200			12.75		12.75	18.95
9360	1" diameter		35	.229			14.55		14.55	21.50
9370	1-1/4" diameter		32	.250			15.95		15.95	23.50
9380	1-1/2" diameter		27	.296			18.85		18.85	28
9390	2" diameter		20	.400			25.50		25.50	38
9400	2-1/2" diameter		16	.500			32		32	47.50
9410	3" diameter		13	.615			39		39	58.50
9420	3-1/2" diameter		12	.667			42.50		42.50	63
9430	4" diameter		10	.800			51		51	75.50
9440	5" diameter		9	.889			56.50		56.50	84
9450	6" diameter		8	1			63.50		63.50	94.50
9460	PVC adapters, 1/2" diameter		50	.160		.21	10.20		10.41	15.40
9470	3/4" diameter		42	.190		.35	12.15		12.50	18.45
9480	1" diameter		38	.211		.38	13.40		13.78	20.50
9490	1-1/4" diameter		35	.229		.61	14.55		15.16	22
9500	1-1/2" diameter		32	.250		.81	15.95		16.76	24.50
9510	2" diameter		27	.296		1.02	18.85		19.87	29
9520	2-1/2" diameter		23	.348		1.67	22		23.67	35
9530	3" diameter		18	.444		2.13	28.50		30.63	44.50
9540	3-1/2" diameter		13	.615		2.80	39		41.80	61.50
9550	4" diameter		11	.727		4.33	46.50		50.83	74
9560	5" diameter		8	1		8.65	63.50		72.15	104
9570	6" diameter		6	1.333		10.30	85		95.30	137
9580	PVC-LB, LR or LL fittings & covers									
9590	1/2" diameter	1 Elec	20	.400	Ea.	2.96	25.50		28.46	41.50
9600	3/4" diameter		16	.500		4.25	32		36.25	52
9610	1" diameter		12	.667		4.65	42.50		47.15	68
9620	1-1/4" diameter		9	.889		7.40	56.50		63.90	92
9630	1-1/2" diameter		7	1.143		8.65	73		81.65	118
9640	2" diameter		6	1.333		16.25	85		101.25	144
9650	2-1/2" diameter		6	1.333		47	85		132	178
9660	3" diameter		5	1.600		56.50	102		158.50	214
9670	3-1/2" diameter		4	2		45.50	127		172.50	240
9680	4" diameter		3	2.667		63.50	170		233.50	320
9690	PVC-tee fitting & cover									
9700	1/2"	1 Elec	14	.571	Ea.	4.99	36.50		41.49	59.50
9710	3/4"		13	.615		5.65	39		44.65	64.50
9720	1"		10	.800		6.20	51		57.20	82.50
9730	1-1/4"		9	.889		10.65	56.50		67.15	96
9740	1-1/2"		8	1		11.35	63.50		74.85	107
9750	2"		7	1.143		19.35	73		92.35	130
9760	PVC-reducers, 3/4" x 1/2" diameter					1.30			1.30	1.43
9770	1" x 1/2" diameter					2.45			2.45	2.70
9780	1" x 3/4" diameter					2.48			2.48	2.73
9790	1-1/4" x 3/4" diameter					2.49			2.49	2.74
9800	1-1/4" x 1" diameter					3.80			3.80	4.18
9810	1-1/2" x 1-1/4" diameter					4.63			4.63	5.10
9820	2" x 1-1/4" diameter					5.15			5.15	5.65
9830	2-1/2" x 2" diameter					15			15	16.50
9840	3" x 2" diameter					14.90			14.90	16.35

For customer support on your Site Work & Landscape Costs with RSMeans Data, call 800.448.8182.

277

26 05 33 – Raceway and Boxes for Electrical Systems

26 05 33.13 Conduit

		Crew	Daily Output	Labor-Hours	Unit	Material	2021 Bare Costs Labor	2021 Bare Costs Equipment	Total	Total Incl O&P
9850	4" x 3" diameter				Ea.	17.95			17.95	19.75
9860	Cement, quart					17.50			17.50	19.25
9870	Gallon					111			111	123
9880	Heat bender, to 6" diameter					1,625			1,625	1,800
9900	Add to labor for higher elevated installation									
9905	10' to 14.5' high, add						10%			
9910	15' to 20' high, add						20%			
9920	20' to 25' high, add						25%			
9930	25' to 30' high, add						35%			
9940	30' to 35' high, add						40%			
9950	35' to 40' high, add						50%			
9960	Over 40' high, add						55%			
9995	Do not include labor when adding couplings to a fitting installation									

26 05 33.18 Pull Boxes

		Crew	Daily Output	Labor-Hours	Unit	Material	2021 Bare Costs Labor	2021 Bare Costs Equipment	Total	Total Incl O&P
0010	**PULL BOXES**									
2100	Pull box, NEMA 3R, type SC, raintight & weatherproof									
2150	6" L x 6" W x 6" D	1 Elec	10	.800	Ea.	19.50	51		70.50	97
2200	8" L x 6" W x 6" D		8	1		40.50	63.50		104	139
2250	10" L x 6" W x 6" D		7	1.143		32.50	73		105.50	144
2300	12" L x 12" W x 6" D		5	1.600		46.50	102		148.50	202
2350	16" L x 16" W x 6" D		4.50	1.778		92.50	113		205.50	270
2400	20" L x 20" W x 6" D		4	2		89	127		216	287
2450	24" L x 18" W x 8" D		3	2.667		148	170		318	415
2500	24" L x 24" W x 10" D		2.50	3.200		190	204		394	515
2550	30" L x 24" W x 12" D		2	4		860	255		1,115	1,325
2600	36" L x 36" W x 12" D		1.50	5.333		455	340		795	1,000
2800	Cast iron, pull boxes for surface mounting									
3000	NEMA 4, watertight & dust tight									
3050	6" L x 6" W x 6" D	1 Elec	4	2	Ea.	283	127		410	500
3100	8" L x 6" W x 6" D		3.20	2.500		470	159		629	755
3150	10" L x 6" W x 6" D		2.50	3.200		490	204		694	845
3200	12" L x 12" W x 6" D		2.30	3.478		715	222		937	1,125
3250	16" L x 16" W x 6" D		1.30	6.154		805	390		1,195	1,475
3300	20" L x 20" W x 6" D		.80	10		980	635		1,615	2,025
3350	24" L x 18" W x 8" D		.70	11.429		2,575	730		3,305	3,900
3400	24" L x 24" W x 10" D		.50	16		4,825	1,025		5,850	6,825
3450	30" L x 24" W x 12" D		.40	20		6,350	1,275		7,625	8,875
3500	36" L x 36" W x 12" D		.20	40		2,375	2,550		4,925	6,375

26 05 39 – Underfloor Raceways for Electrical Systems

26 05 39.30 Conduit In Concrete Slab

		Crew	Daily Output	Labor-Hours	Unit	Material	2021 Bare Costs Labor	2021 Bare Costs Equipment	Total	Total Incl O&P
0010	**CONDUIT IN CONCRETE SLAB** Including terminations, R337119-30									
0020	fittings and supports									
3230	PVC, schedule 40, 1/2" diameter	1 Elec	270	.030	L.F.	.36	1.89		2.25	3.19
3250	3/4" diameter		230	.035		.37	2.22		2.59	3.70
3270	1" diameter		200	.040		.42	2.55		2.97	4.25
3300	1-1/4" diameter		170	.047		.73	3		3.73	5.25
3330	1-1/2" diameter		140	.057		.76	3.64		4.40	6.25
3350	2" diameter		120	.067		.90	4.25		5.15	7.30
3370	2-1/2" diameter		90	.089		1.67	5.65		7.32	10.25
3400	3" diameter	2 Elec	160	.100		2.14	6.35		8.49	11.80
3430	3-1/2" diameter		120	.133		2.65	8.50		11.15	15.50
3440	4" diameter		100	.160		3.07	10.20		13.27	18.55

26 05 Common Work Results for Electrical

26 05 39 – Underfloor Raceways for Electrical Systems

26 05 39.30 Conduit In Concrete Slab

		Crew	Daily Output	Labor-Hours	Unit	Material	Labor	Equipment	Total	Total Incl O&P
3450	5" diameter	2 Elec	80	.200	L.F.	7.10	12.75		19.85	27
3460	6" diameter	↓	60	.267	↓	6.10	17		23.10	31.50
3530	Sweeps, 1" diameter, 30" radius	1 Elec	32	.250	Ea.	11.50	15.95		27.45	36
3550	1-1/4" diameter		24	.333		10.30	21		31.30	43
3570	1-1/2" diameter		21	.381		11.20	24.50		35.70	48.50
3600	2" diameter		18	.444		13.45	28.50		41.95	57
3630	2-1/2" diameter		14	.571		88.50	36.50		125	151
3650	3" diameter		10	.800		50	51		101	131
3670	3-1/2" diameter		8	1		49.50	63.50		113	149
3700	4" diameter		7	1.143		38.50	73		111.50	150
3710	5" diameter	↓	6	1.333		58	85		143	190
3730	Couplings, 1/2" diameter					.17			.17	.19
3750	3/4" diameter					.16			.16	.18
3770	1" diameter					.24			.24	.26
3800	1-1/4" diameter					.47			.47	.52
3830	1-1/2" diameter					.46			.46	.51
3850	2" diameter					.60			.60	.66
3870	2-1/2" diameter					1.27			1.27	1.40
3900	3" diameter					2.13			2.13	2.34
3930	3-1/2" diameter					2.08			2.08	2.29
3950	4" diameter					2.58			2.58	2.84
3960	5" diameter					6.15			6.15	6.75
3970	6" diameter					8.65			8.65	9.50
4030	End bells, 1" diameter, PVC	1 Elec	60	.133		3.19	8.50		11.69	16.10
4050	1-1/4" diameter		53	.151		4.98	9.60		14.58	19.80
4100	1-1/2" diameter		48	.167		3.38	10.60		13.98	19.50
4150	2" diameter		34	.235		4.70	15		19.70	27.50
4170	2-1/2" diameter		27	.296		3.69	18.85		22.54	32
4200	3" diameter		20	.400		5.45	25.50		30.95	44
4250	3-1/2" diameter		16	.500		7.35	32		39.35	55.50
4300	4" diameter		14	.571		6.40	36.50		42.90	61
4310	5" diameter		12	.667		10.05	42.50		52.55	74
4320	6" diameter		9	.889	↓	12.10	56.50		68.60	97.50
4350	Rigid galvanized steel, 1/2" diameter		200	.040	L.F.	2.32	2.55		4.87	6.35
4400	3/4" diameter		170	.047		2.48	3		5.48	7.15
4450	1" diameter		130	.062		3.73	3.92		7.65	9.90
4500	1-1/4" diameter		110	.073		5.25	4.63		9.88	12.70
4600	1-1/2" diameter		100	.080		5.90	5.10		11	14.05
4800	2" diameter	↓	90	.089	↓	7.20	5.65		12.85	16.35

26 05 39.40 Conduit In Trench

		Crew	Daily Output	Labor-Hours	Unit	Material	Labor	Equipment	Total	Total Incl O&P
0010	**CONDUIT IN TRENCH** Includes terminations and fittings R337119-30									
0020	Does not include excavation or backfill, see Section 31 23 16									
0200	Rigid galvanized steel, 2" diameter	1 Elec	150	.053	L.F.	6.90	3.40		10.30	12.65
0400	2-1/2" diameter	"	100	.080		13.60	5.10		18.70	22.50
0600	3" diameter	2 Elec	160	.100		15.45	6.35		21.80	26.50
0800	3-1/2" diameter		140	.114		20.50	7.30		27.80	33.50
1000	4" diameter		100	.160		22	10.20		32.20	39
1200	5" diameter		80	.200		48	12.75		60.75	72
1400	6" diameter	↓	60	.267	↓	59.50	17		76.50	90.50

279

For customer support on your Site Work & Landscape Costs with RSMeans Data, call 800.448.8182.

26 12 Medium-Voltage Transformers

26 12 19 – Pad-Mounted, Liquid-Filled, Medium-Voltage Transformers

26 12 19.10 Transformer, Oil-Filled		Crew	Daily Output	Labor-Hours	Unit	Material	2021 Bare Costs Labor	Equipment	Total	Total Incl O&P
0010	**TRANSFORMER, OIL-FILLED** primary delta or Y,									
0050	Pad mounted 5 kV or 15 kV, with taps, 277/480 V secondary, 3 phase									
0100	150 kVA	R-3	.65	30.769	Ea.	9,925	1,950	294	12,169	14,100
0200	300 kVA	"	.45	44.444	"	17,100	2,825	425	20,350	23,500

26 24 Switchboards and Panelboards

26 24 16 – Panelboards

26 24 16.30 Panelboards Commercial Applications

		Crew	Daily Output	Labor-Hours	Unit	Material	2021 Bare Costs Labor	Equipment	Total	Total Incl O&P
0010	**PANELBOARDS COMMERCIAL APPLICATIONS**									
0050	NQOD, w/20 amp 1 pole bolt-on circuit breakers									
0100	3 wire, 120/240 volts, 100 amp main lugs									
0150	10 circuits	1 Elec	1	8	Ea.	950	510		1,460	1,800
0200	14 circuits		.88	9.091		1,075	580		1,655	2,025
0250	18 circuits		.75	10.667		1,150	680		1,830	2,275
0300	20 circuits		.65	12.308		1,250	785		2,035	2,550
0350	225 amp main lugs, 24 circuits	2 Elec	1.20	13.333		1,425	850		2,275	2,825
0400	30 circuits		.90	17.778		1,650	1,125		2,775	3,475
0450	36 circuits		.80	20		1,900	1,275		3,175	4,000
0500	38 circuits		.72	22.222		2,050	1,425		3,475	4,350
0550	42 circuits		.66	24.242		2,125	1,550		3,675	4,650
0600	4 wire, 120/208 volts, 100 amp main lugs, 12 circuits	1 Elec	1	8		1,025	510		1,535	1,875
0650	16 circuits		.75	10.667		1,150	680		1,830	2,250
0700	20 circuits		.65	12.308		1,325	785		2,110	2,625
0750	24 circuits		.60	13.333		415	850		1,265	1,700
0800	30 circuits		.53	15.094		1,650	960		2,610	3,225
0850	225 amp main lugs, 32 circuits	2 Elec	.90	17.778		1,875	1,125		3,000	3,750
0900	34 circuits		.84	19.048		1,925	1,225		3,150	3,900
0950	36 circuits		.80	20		1,950	1,275		3,225	4,050
1000	42 circuits		.68	23.529		2,200	1,500		3,700	4,650
1010	400 amp main lugs, 42 circs		.68	23.529		2,200	1,500		3,700	4,650

26 27 Low-Voltage Distribution Equipment

26 27 13 – Electricity Metering

26 27 13.10 Meter Centers and Sockets

		Crew	Daily Output	Labor-Hours	Unit	Material	2021 Bare Costs Labor	Equipment	Total	Total Incl O&P
0010	**METER CENTERS AND SOCKETS**									
0100	Sockets, single position, 4 terminal, 100 amp	1 Elec	3.20	2.500	Ea.	73	159		232	320
0200	150 amp		2.30	3.478		55	222		277	390
0300	200 amp		1.90	4.211		106	268		374	515
0500	Double position, 4 terminal, 100 amp		2.80	2.857		234	182		416	525
0600	150 amp		2.10	3.810		275	243		518	660
0700	200 amp		1.70	4.706		515	300		815	1,025
1100	Meter centers and sockets, three phase, single pos, 7 terminal, 100 amp		2.80	2.857		157	182		339	440
1200	200 amp		2.10	3.810		264	243		507	650
1400	400 amp		1.70	4.706		840	300		1,140	1,375
2000	Meter center, main fusible switch, 1P 3W 120/240 V									
2030	400 amp	2 Elec	1.60	10	Ea.	1,000	635		1,635	2,050
2040	600 amp		1.10	14.545		1,600	925		2,525	3,125
2050	800 amp		.90	17.778		5,775	1,125		6,900	8,025
2060	Rainproof 1P 3W 120/240 V, 400 A		1.60	10		2,000	635		2,635	3,150

26 27 Low-Voltage Distribution Equipment

26 27 13 – Electricity Metering

26 27 13.10 Meter Centers and Sockets		Crew	Daily Output	Labor-Hours	Unit	Material	2021 Bare Costs Labor	Equipment	Total	Total Incl O&P
2070	600 amp	2 Elec	1.10	14.545	Ea.	3,475	925		4,400	5,200
2080	800 amp		.90	17.778		5,450	1,125		6,575	7,650
2100	3P 4W 120/208 V, 400 amp		1.60	10		1,075	635		1,710	2,125
2110	600 amp		1.10	14.545		1,800	925		2,725	3,350
2120	800 amp		.90	17.778		2,450	1,125		3,575	4,375
2130	Rainproof 3P 4W 120/208 V, 400 amp		1.60	10		2,275	635		2,910	3,475
2140	600 amp		1.10	14.545		3,675	925		4,600	5,425
2150	800 amp		.90	17.778		7,925	1,125		9,050	10,400
2170	Main circuit breaker, 1P 3W 120/240 V									
2180	400 amp	2 Elec	1.60	10	Ea.	1,475	635		2,110	2,550
2190	600 amp		1.10	14.545		1,900	925		2,825	3,450
2200	800 amp		.90	17.778		3,450	1,125		4,575	5,450
2210	1,000 amp		.80	20		4,850	1,275		6,125	7,250
2220	1,200 amp		.76	21.053		4,875	1,350		6,225	7,350
2230	1,600 amp		.68	23.529		8,500	1,500		10,000	11,600
2240	Rainproof 1P 3W 120/240 V, 400 amp		1.60	10		3,075	635		3,710	4,325
2250	600 amp		1.10	14.545		4,800	925		5,725	6,675
2260	800 amp		.90	17.778		5,600	1,125		6,725	7,850
2270	1,000 amp		.80	20		7,725	1,275		9,000	10,400
2280	1,200 amp		.76	21.053		10,400	1,350		11,750	13,500
2300	3P 4W 120/208 V, 400 amp		1.60	10		3,750	635		4,385	5,075
2310	600 amp		1.10	14.545		5,725	925		6,650	7,675
2320	800 amp		.90	17.778		6,800	1,125		7,925	9,175
2330	1,000 amp		.80	20		8,950	1,275		10,225	11,800
2340	1,200 amp		.76	21.053		11,400	1,350		12,750	14,600
2350	1,600 amp		.68	23.529		10,300	1,500		11,800	13,500
2360	Rainproof 3P 4W 120/208 V, 400 amp		1.60	10		4,075	635		4,710	5,425
2370	600 amp		1.10	14.545		5,725	925		6,650	7,675
2380	800 amp		.90	17.778		6,800	1,125		7,925	9,175
2390	1,000 amp		.76	21.053		8,950	1,350		10,300	11,900

26 28 Low-Voltage Circuit Protective Devices

26 28 16 – Enclosed Switches and Circuit Breakers

26 28 16.10 Circuit Breakers

		Crew	Daily Output	Labor-Hours	Unit	Material	Labor	Equipment	Total	Total Incl O&P
0010	**CIRCUIT BREAKERS** (in enclosure)									
2300	Enclosed (NEMA 7), explosion proof, 600 volt 3 pole, 50 amp	1 Elec	2.30	3.478	Ea.	1,975	222		2,197	2,500
2350	100 amp		1.50	5.333		2,050	340		2,390	2,750
2400	150 amp		1	8		4,925	510		5,435	6,175
2450	250 amp	2 Elec	1.60	10		6,150	635		6,785	7,725
2500	400 amp	"	1.20	13.333		6,200	850		7,050	8,075

26 28 16.20 Safety Switches

		Crew	Daily Output	Labor-Hours	Unit	Material	Labor	Equipment	Total	Total Incl O&P
0010	**SAFETY SWITCHES**									
0100	General duty 240 volt, 3 pole NEMA 1, fusible, 30 amp	1 Elec	3.20	2.500	Ea.	65.50	159		224.50	310
0200	60 amp		2.30	3.478		111	222		333	450
0300	100 amp		1.90	4.211		193	268		461	610
0400	200 amp		1.30	6.154		415	390		805	1,025
0500	400 amp	2 Elec	1.80	8.889		1,050	565		1,615	2,000
0600	600 amp	"	1.20	13.333		1,950	850		2,800	3,400
2900	Heavy duty, 240 volt, 3 pole NEMA 1 fusible									
2910	30 amp	1 Elec	3.20	2.500	Ea.	111	159		270	360
3000	60 amp		2.30	3.478		179	222		401	525

For customer support on your Site Work & Landscape Costs with RSMeans Data, call 800.448.8182.

281

26 28 Low-Voltage Circuit Protective Devices

26 28 16 – Enclosed Switches and Circuit Breakers

26 28 16.20 Safety Switches		Crew	Daily Output	Labor-Hours	Unit	Material	2021 Bare Costs Labor	2021 Bare Costs Equipment	Total	Total Incl O&P
3300	100 amp	1 Elec	1.90	4.211	Ea.	284	268		552	710
3500	200 amp	↓	1.30	6.154		495	390		885	1,125
3700	400 amp	2 Elec	1.80	8.889		1,200	565		1,765	2,175
3900	600 amp	"	1.20	13.333	↓	2,375	850		3,225	3,850

26 32 Packaged Generator Assemblies

26 32 13 – Engine Generators

26 32 13.13 Diesel-Engine-Driven Generator Sets

		Crew	Daily Output	Labor-Hours	Unit	Material	2021 Bare Costs Labor	2021 Bare Costs Equipment	Total	Total Incl O&P
0010	**DIESEL-ENGINE-DRIVEN GENERATOR SETS**									
2000	Diesel engine, including battery, charger,									
2010	muffler & day tank, 30 kW	R-3	.55	36.364	Ea.	11,800	2,300	345	14,445	16,800
2100	50 kW		.42	47.619		25,200	3,025	455	28,680	32,800
2200	75 kW		.35	57.143		19,500	3,625	545	23,670	27,500
2300	100 kW		.31	64.516		33,600	4,100	615	38,315	43,800
2400	125 kW		.29	68.966		33,300	4,375	660	38,335	43,800
2500	150 kW		.26	76.923		24,300	4,875	735	29,910	34,800
2501	Generator set, dsl eng in alum encl, incl btry, chgr, muf & day tank,150 kW		.26	76.923		41,500	4,875	735	47,110	53,500
2600	175 kW		.25	80		44,800	5,075	765	50,640	57,500
2700	200 kW		.24	83.333		45,400	5,300	795	51,495	58,500
2800	250 kW		.23	86.957		60,500	5,525	830	66,855	76,000
2900	300 kW		.22	90.909		55,500	5,775	870	62,145	70,500
3000	350 kW		.20	100		70,000	6,350	955	77,305	87,500
3100	400 kW		.19	105		60,500	6,675	1,000	68,175	78,000
3200	500 kW		.18	111		97,500	7,050	1,050	105,600	119,000
3220	600 kW	↓	.17	118		124,500	7,475	1,125	133,100	149,500
3240	750 kW	R-13	.38	111	↓	185,500	6,800	645	192,945	215,000

26 32 13.16 Gas-Engine-Driven Generator Sets

		Crew	Daily Output	Labor-Hours	Unit	Material	2021 Bare Costs Labor	2021 Bare Costs Equipment	Total	Total Incl O&P
0010	**GAS-ENGINE-DRIVEN GENERATOR SETS**									
0020	Gas or gasoline operated, includes battery,									
0050	charger & muffler									
0200	7.5 kW	R-3	.83	24.096	Ea.	7,525	1,525	230	9,280	10,800
0300	11.5 kW		.71	28.169		10,700	1,775	269	12,744	14,700
0400	20 kW		.63	31.746		10,600	2,025	305	12,930	15,000
0500	35 kW		.55	36.364		15,000	2,300	345	17,645	20,300
0520	60 kW	↓	.50	40		14,900	2,550	380	17,830	20,600
0600	80 kW	R-13	.40	105		23,600	6,475	615	30,690	36,200
0700	100 kW		.33	127		25,100	7,825	745	33,670	40,000
0800	125 kW		.28	150		27,800	9,225	880	37,905	45,200
0900	185 kW	↓	.25	168	↓	73,000	10,300	985	84,285	97,000

26 36 Transfer Switches

26 36 13 – Manual Transfer Switches

26 36 13.10 Non-Automatic Transfer Switches	Crew	Daily Output	Labor-Hours	Unit	Material	2021 Bare Costs Labor	Equipment	Total	Total Incl O&P
0010 **NON-AUTOMATIC TRANSFER SWITCHES** enclosed									
0100 Manual operated, 480 volt 3 pole, 30 amp	1 Elec	2.30	3.478	Ea.	1,525	222		1,747	2,000
0200 100 amp	"	1.30	6.154		3,475	390		3,865	4,400
0250 200 amp	2 Elec	2	8		4,600	510		5,110	5,800
0300 400 amp	"	1.60	10		4,525	635		5,160	5,950
1000 250 volt 3 pole, 30 amp	1 Elec	2.30	3.478		1,000	222		1,222	1,425
1150 100 amp	"	1.30	6.154		2,850	390		3,240	3,700
1200 200 amp	2 Elec	2	8		4,300	510		4,810	5,475
1500 Electrically operated, 480 volt 3 pole, 60 amp	1 Elec	1.90	4.211		1,600	268		1,868	2,175
1600 100 amp	"	1.30	6.154		1,600	390		1,990	2,350
1650 200 amp	2 Elec	2	8		2,650	510		3,160	3,675
1700 400 amp	"	1.60	10		3,700	635		4,335	5,025
2000 250 volt 3 pole, 30 amp	1 Elec	2.30	3.478		545	222		767	930
2050 60 amp	"	1.90	4.211		680	268		948	1,150
2150 200 amp	2 Elec	2	8		1,275	510		1,785	2,150
2200 400 amp	"	1.60	10		3,375	635		4,010	4,650
2500 NEMA 3R, 480 volt 3 pole, 60 amp	1 Elec	1.80	4.444		2,575	283		2,858	3,250
2550 100 amp	"	1.20	6.667		3,150	425		3,575	4,075
2650 400 amp	2 Elec	1.40	11.429		4,050	730		4,780	5,550

26 51 Interior Lighting

26 51 13 – Interior Lighting Fixtures, Lamps, and Ballasts

26 51 13.70 Residential Fixtures

	Crew	Daily Output	Labor-Hours	Unit	Material	2021 Bare Costs Labor	Equipment	Total	Total Incl O&P
0010 **RESIDENTIAL FIXTURES**									
2000 Incandescent, exterior lantern, wall mounted, 60 watt	1 Elec	16	.500	Ea.	64.50	32		96.50	119
2100 Post light, 150 W, with 7' post		4	2		254	127		381	470
2500 Lamp holder, weatherproof with 150 W PAR		16	.500		35.50	32		67.50	86.50
2550 With reflector and guard		12	.667		63	42.50		105.50	132

26 51 13.90 Ballast, Replacement HID

	Crew	Daily Output	Labor-Hours	Unit	Material	2021 Bare Costs Labor	Equipment	Total	Total Incl O&P
0010 **BALLAST, REPLACEMENT HID**									
7510 Multi-tap 120/208/240/277 V									
7550 High pressure sodium, 70 watt	1 Elec	10	.800	Ea.	56	51		107	138
7560 100 watt		9.40	.851		63	54		117	150
7570 150 watt		9	.889		67	56.50		123.50	158
7580 250 watt		8.50	.941		98.50	60		158.50	197
7590 400 watt		7	1.143		110	73		183	229
7600 1,000 watt		6	1.333		197	85		282	340
7610 Metal halide, 175 watt		8	1		58.50	63.50		122	159
7620 250 watt		8	1		73	63.50		136.50	175
7630 400 watt		7	1.143		76	73		149	192
7640 1,000 watt		6	1.333		141	85		226	281
7650 1,500 watt		5	1.600		228	102		330	400

For customer support on your Site Work & Landscape Costs with RSMeans Data, call 800.448.8182.

283

26 56 13.10 Lighting Poles		Crew	Daily Output	Labor-Hours	Unit	Material	2021 Bare Costs Labor	Equipment	Total	Total Incl O&P
0010	**LIGHTING POLES**									
2800	Light poles, anchor base									
2820	not including concrete bases									
2840	Aluminum pole, 8' high	1 Elec	4	2	Ea.	365	127		492	595
2850	10' high		4	2		400	127		527	630
2860	12' high		3.80	2.105		430	134		564	670
2870	14' high		3.40	2.353		445	150		595	715
2880	16' high	▼	3	2.667		635	170		805	950
3000	20' high	R-3	2.90	6.897		605	440	66	1,111	1,400
3200	30' high		2.60	7.692		1,550	490	73.50	2,113.50	2,525
3400	35' high		2.30	8.696		1,575	550	83	2,208	2,650
3600	40' high	▼	2	10		2,000	635	95.50	2,730.50	3,250
3800	Bracket arms, 1 arm	1 Elec	8	1		137	63.50		200.50	246
4000	2 arms		8	1		273	63.50		336.50	395
4200	3 arms		5.30	1.509		410	96		506	595
4400	4 arms		4.80	1.667		550	106		656	765
4500	Steel pole, galvanized, 8' high		3.80	2.105		720	134		854	995
4510	10' high		3.70	2.162		325	138		463	565
4520	12' high		3.40	2.353		810	150		960	1,125
4530	14' high		3.10	2.581		865	164		1,029	1,200
4540	16' high		2.90	2.759		915	176		1,091	1,250
4550	18' high	▼	2.70	2.963		965	189		1,154	1,350
4600	20' high	R-3	2.60	7.692		1,150	490	73.50	1,713.50	2,050
4800	30' high		2.30	8.696		1,275	550	83	1,908	2,300
5000	35' high		2.20	9.091		1,375	575	87	2,037	2,450
5200	40' high	▼	1.70	11.765		1,500	745	112	2,357	2,875
5400	Bracket arms, 1 arm	1 Elec	8	1		219	63.50		282.50	335
5600	2 arms		8	1		295	63.50		358.50	420
5800	3 arms		5.30	1.509		244	96		340	410
6000	4 arms	▼	5.30	1.509		335	96		431	515
6100	Fiberglass pole, 1 or 2 fixtures, 20' high	R-3	4	5		970	315	48	1,333	1,600
6200	30' high		3.60	5.556		1,200	355	53	1,608	1,900
6300	35' high		3.20	6.250		3,325	395	59.50	3,779.50	4,325
6400	40' high	▼	2.80	7.143		2,250	455	68	2,773	3,225
6420	Wood pole, 4-1/2" x 5-1/8", 8' high	1 Elec	6	1.333		390	85		475	550
6430	10' high		6	1.333		460	85		545	630
6440	12' high		5.70	1.404		570	89.50		659.50	765
6450	15' high		5	1.600		665	102		767	880
6460	20' high	▼	4	2		810	127		937	1,075
6461	Light poles, anchor base, w/o conc base, pwdr ct stl, 16' H	2 Elec	3.10	5.161		915	330		1,245	1,500
6462	20' high	R-3	2.90	6.897		1,150	440	66	1,656	1,975
6463	30' high		2.30	8.696		1,275	550	83	1,908	2,300
6464	35' high	▼	2.40	8.333		1,375	530	79.50	1,984.50	2,375
6470	Light pole conc base, max 6' buried, 2' exposed, 18" diam., average cost	C-6	6	8	▼	200	370	9.05	579.05	775
7300	Transformer bases, not including concrete bases									
7320	Maximum pole size, steel, 40' high	1 Elec	2	4	Ea.	1,450	255		1,705	1,950
7340	Cast aluminum, 30' high		3	2.667		765	170		935	1,100
7350	40' high	▼	2.50	3.200	▼	1,150	204		1,354	1,575
8000	Line cover protective devices									
8010	line cover									
8015	Refer to 26 01 02.20 for labor adjustment factors as they apply									
8100	MVLC 1" W x 1-1/2" H, 14/5'	1 Elec	60	.133	Ea.	4.84	8.50		13.34	17.90
8110	8'	▼	58	.138	▼	5.40	8.80		14.20	19

26 56 Exterior Lighting

26 56 13 – Lighting Poles and Standards

26 56 13.10 Lighting Poles

		Crew	Daily Output	Labor-Hours	Unit	Material	2021 Bare Costs Labor	Equipment	Total	Total Incl O&P
8120	MVLC 1" W x 1-1/2" H, 18/5'	1 Elec	60	.133	Ea.	6.15	8.50		14.65	19.35
8130	8'		58	.138		6.90	8.80		15.70	20.50
8140	MVLC 1" W x 1-1/2" H, 38/5'		60	.133		10.70	8.50		19.20	24.50
8150	8'		58	.138		13.25	8.80		22.05	27.50

26 56 19 – LED Exterior Lighting

26 56 19.55 Roadway LED Luminaire

			Crew	Daily Output	Labor-Hours	Unit	Material	Labor	Equipment	Total	Total Incl O&P
0010	**ROADWAY LED LUMINAIRE**										
0100	LED fixture, 72 LEDs, 120 V AC or 12 V DC, equal to 60 watt	G	1 Elec	2.70	2.963	Ea.	575	189		764	910
0110	108 LEDs, 120 V AC or 12 V DC, equal to 90 watt	G		2.70	2.963		765	189		954	1,125
0120	144 LEDs, 120 V AC or 12 V DC, equal to 120 watt	G		2.70	2.963		940	189		1,129	1,300
0130	252 LEDs, 120 V AC or 12 V DC, equal to 210 watt	G	2 Elec	4.40	3.636		1,300	232		1,532	1,775
0140	Replaces high pressure sodium fixture, 75 watt	G	1 Elec	2.70	2.963		410	189		599	730
0150	125 watt	G		2.70	2.963		470	189		659	795
0160	150 watt	G		2.70	2.963		545	189		734	880
0170	175 watt	G		2.70	2.963		805	189		994	1,175
0180	200 watt	G		2.70	2.963		770	189		959	1,125
0190	250 watt	G	2 Elec	4.40	3.636		880	232		1,112	1,325
0200	320 watt	G	"	4.40	3.636		960	232		1,192	1,400

26 56 19.60 Parking LED Lighting

			Crew	Daily Output	Labor-Hours	Unit	Material	Labor	Equipment	Total	Total Incl O&P
0010	**PARKING LED LIGHTING**										
0100	Round pole mounting, 88 lamp watts	G	1 Elec	2	4	Ea.	1,275	255		1,530	1,775

26 56 21 – HID Exterior Lighting

26 56 21.20 Roadway Luminaire

		Crew	Daily Output	Labor-Hours	Unit	Material	Labor	Equipment	Total	Total Incl O&P
0010	**ROADWAY LUMINAIRE**									
2650	Roadway area luminaire, low pressure sodium, 135 watt	1 Elec	2	4	Ea.	730	255		985	1,175
2700	180 watt	"	2	4		880	255		1,135	1,350
2750	Metal halide, 400 watt	2 Elec	4.40	3.636		640	232		872	1,050
2760	1,000 watt		4	4		775	255		1,030	1,225
2780	High pressure sodium, 400 watt		4.40	3.636		770	232		1,002	1,200
2790	1,000 watt		4	4		875	255		1,130	1,350

26 56 23 – Area Lighting

26 56 23.10 Exterior Fixtures

		Crew	Daily Output	Labor-Hours	Unit	Material	Labor	Equipment	Total	Total Incl O&P
0010	**EXTERIOR FIXTURES** With lamps									
0200	Wall mounted, incandescent, 100 watt	1 Elec	8	1	Ea.	168	63.50		231.50	280
0400	Quartz, 500 watt		5.30	1.509		61.50	96		157.50	211
0420	1,500 watt		4.20	1.905		100	121		221	290
1100	Wall pack, low pressure sodium, 35 watt		4	2		207	127		334	415
1150	55 watt		4	2		243	127		370	455
1160	High pressure sodium, 70 watt		4	2		208	127		335	420
1170	150 watt		4	2		219	127		346	430
1175	High pressure sodium, 250 watt		4	2		219	127		346	430
1180	Metal halide, 175 watt		4	2		223	127		350	435
1190	250 watt		4	2		253	127		380	470
1195	400 watt		4	2		410	127		537	645
1250	Induction lamp, 40 watt		4	2		520	127		647	765
1260	80 watt		4	2		600	127		727	850
1278	LED, poly lens, 26 watt		4	2		1,675	127		1,802	2,050
1280	110 watt		4	2		820	127		947	1,100
1500	LED, glass lens, 13 watt		4	2		1,675	127		1,802	2,050

For customer support on your Site Work & Landscape Costs with RSMeans Data, call 800.448.8182.

285

26 56 23 – Area Lighting

26 56 23.55 Exterior LED Fixtures		Crew	Daily Output	Labor-Hours	Unit	Material	2021 Bare Costs Labor	Equipment	Total	Total Incl O&P
0010	**EXTERIOR LED FIXTURES**									
0100	Wall mounted, indoor/outdoor, 12 watt	G	1 Elec	10	.800	Ea.	203	51		254	300
0110	32 watt	G		10	.800		460	51		511	580
0120	66 watt	G		10	.800		345	51		396	455
0200	outdoor, 110 watt	G		10	.800		1,200	51		1,251	1,375
0210	220 watt	G		10	.800		305	51		356	410
0300	modular, type IV, 120 V, 50 lamp watts	G		9	.889		1,425	56.50		1,481.50	1,650
0310	101 lamp watts	G		9	.889		1,325	56.50		1,381.50	1,525
0320	126 lamp watts	G		9	.889		1,650	56.50		1,706.50	1,875
0330	202 lamp watts	G		9	.889		1,900	56.50		1,956.50	2,150
0340	240 V, 50 lamp watts	G		8	1		1,225	63.50		1,288.50	1,450
0350	101 lamp watts	G		8	1		1,375	63.50		1,438.50	1,600
0360	126 lamp watts	G		8	1		1,300	63.50		1,363.50	1,525
0370	202 lamp watts	G		8	1		1,925	63.50		1,988.50	2,200
0400	wall pack, glass, 13 lamp watts	G		4	2		450	127		577	685
0410	poly w/photocell, 26 lamp watts	G		4	2		335	127		462	560
0420	50 lamp watts	G		4	2		785	127		912	1,050
0430	replacement, 40 watts	G		4	2		122	127		249	325
0440	60 watts	G		4	2		154	127		281	360

26 56 26 – Landscape Lighting

26 56 26.20 Landscape Fixtures

26 56 26.20 Landscape Fixtures		Crew	Daily Output	Labor-Hours	Unit	Material	2021 Bare Costs Labor	Equipment	Total	Total Incl O&P
0010	**LANDSCAPE FIXTURES**									
0012	Incl. conduit, wire, trench									
0030	Bollards									
0040	Incandescent, 24"	1 Elec	2.50	3.200	Ea.	365	204		569	710
0050	36"		2	4		475	255		730	900
0060	42"		2	4		505	255		760	935
0070	H.I.D., 24"		2.50	3.200		505	204		709	860
0080	36"		2	4		600	255		855	1,050
0090	42"		2	4		820	255		1,075	1,275
0100	Concrete, 18" diam.		1.20	6.667		1,650	425		2,075	2,450
0110	24" diam.		.75	10.667		2,075	680		2,755	3,275
0120	Dry niche									
0130	300 W 120 Volt	1 Elec	4	2	Ea.	1,075	127		1,202	1,375
0140	1000 W 120 Volt		2	4		1,450	255		1,705	1,950
0150	300 W 12 Volt		4	2		1,075	127		1,202	1,375
0160	Low voltage									
0170	Recessed uplight	1 Elec	2.20	3.636	Ea.	445	232		677	835
0180	Walkway		4	2		380	127		507	610
0190	Malibu - 5 light set		3	2.667		276	170		446	555
0200	Mushroom 24" pier		4	2		283	127		410	500
0210	Recessed, adjustable									
0220	Incandescent, 150 W	1 Elec	2.50	3.200	Ea.	695	204		899	1,075
0230	300 W	"	2.50	3.200	"	835	204		1,039	1,225
0250	Recessed uplight									
0260	Incandescent, 50 W	1 Elec	2.50	3.200	Ea.	565	204		769	925
0270	150 W		2.50	3.200		585	204		789	950
0280	300 W		2.50	3.200		655	204		859	1,025
0310	Quartz 500 W		2.50	3.200		580	204		784	945
0400	Recessed wall light									
0410	Incandescent 100 W	1 Elec	4	2	Ea.	214	127		341	425
0420	Fluorescent		4	2		195	127		322	405

26 56 Exterior Lighting

26 56 26 – Landscape Lighting

26 56 26.20 Landscape Fixtures

		Crew	Daily Output	Labor-Hours	Unit	Material	2021 Bare Costs Labor	Equipment	Total	Total Incl O&P
0430	H.I.D. 100 W	1 Elec	3	2.667	Ea.	400	170		570	690
0500	Step lights									
0510	Incandescent	1 Elec	5	1.600	Ea.	164	102		266	330
0520	Fluorescent	"	5	1.600	"	171	102		273	340
0600	Tree lights, surface adjustable									
0610	Incandescent 50 W	1 Elec	3	2.667	Ea.	345	170		515	625
0620	Incandescent 100 W	↓	3	2.667		169	170		339	440
0630	Incandescent 150 W	↓	2	4	↓	460	255		715	885
0700	Underwater lights									
0710	150 W 120 Volt	1 Elec	6	1.333	Ea.	1,025	85		1,110	1,250
0720	300 W 120 Volt		6	1.333		1,325	85		1,410	1,575
0730	1000 W 120 Volt		4	2		1,675	127		1,802	2,025
0740	50 W 12 Volt		6	1.333		1,250	85		1,335	1,500
0750	300 W 12 Volt	↓	6	1.333	↓	1,125	85		1,210	1,375
0800	Walkway, adjustable									
0810	Fluorescent, 2'	1 Elec	3	2.667	Ea.	465	170		635	760
0820	Fluorescent, 4'		3	2.667		495	170		665	795
0830	Fluorescent, 8'		2	4		965	255		1,220	1,425
0840	Incandescent, 50 W		4	2		440	127		567	675
0850	150 W	↓	4	2	↓	470	127		597	705
0900	Wet niche									
0910	300 W 120 Volt	1 Elec	2.50	3.200	Ea.	960	204		1,164	1,350
0920	1000 W 120 Volt		1.50	5.333		1,150	340		1,490	1,775
0930	300 W 12 Volt	↓	2.50	3.200	↓	1,100	204		1,304	1,500
0960	Landscape lights, solar powered, 6" diam. x 12" H, one piece w/stake [G]	1 Clab	80	.100		31	4.44		35.44	40.50
7380	Landscape recessed uplight, incl. housing, ballast, transformer									
7390	& reflector, not incl. conduit, wire, trench									
7420	Incandescent, 250 watt	1 Elec	5	1.600	Ea.	685	102		787	905
7440	Quartz, 250 watt		5	1.600		650	102		752	865
7460	500 watt	↓	4	2	↓	670	127		797	925

26 56 26.50 Landscape LED Fixtures

		Crew	Daily Output	Labor-Hours	Unit	Material	2021 Bare Costs Labor	Equipment	Total	Total Incl O&P
0010	**LANDSCAPE LED FIXTURES**									
0100	12 volt alum bullet hooded-BLK	1 Elec	5	1.600	Ea.	35.50	102		137.50	190
0200	12 volt alum bullet hooded-BRZ		5	1.600		98.50	102		200.50	260
0300	12 volt alum bullet hooded-GRN		5	1.600		98.50	102		200.50	260
1000	12 volt alum large bullet hooded-BLK		5	1.600		72.50	102		174.50	231
1100	12 volt alum large bullet hooded-BRZ		5	1.600		72.50	102		174.50	231
1200	12 volt alum large bullet hooded-GRN		5	1.600		72.50	102		174.50	231
2000	12 volt large bullet landscape light fixture		5	1.600		72.50	102		174.50	231
2100	12 volt alum light large bullet		5	1.600		72.50	102		174.50	231
2200	12 volt alum bullet light	↓	5	1.600		72.50	102		174.50	231

26 56 33 – Walkway Lighting

26 56 33.10 Walkway Luminaire

		Crew	Daily Output	Labor-Hours	Unit	Material	2021 Bare Costs Labor	Equipment	Total	Total Incl O&P
0010	**WALKWAY LUMINAIRE**									
6500	Bollard light, lamp & ballast, 42" high with polycarbonate lens									
6800	Metal halide, 175 watt	1 Elec	3	2.667	Ea.	980	170		1,150	1,325
6900	High pressure sodium, 70 watt		3	2.667		1,050	170		1,220	1,400
7000	100 watt		3	2.667		1,050	170		1,220	1,400
7100	150 watt		3	2.667		1,025	170		1,195	1,375
7200	Incandescent, 150 watt		3	2.667		605	170		775	915
7810	Walkway luminaire, square 16", metal halide 250 watt		2.70	2.963		780	189		969	1,125
7820	High pressure sodium, 70 watt	↓	3	2.667		890	170		1,060	1,225

26 56 Exterior Lighting

26 56 33 – Walkway Lighting

26 56 33.10 Walkway Luminaire		Crew	Daily Output	Labor-Hours	Unit	Material	2021 Bare Costs Labor	Equipment	Total	Total Incl O&P
7830	100 watt	1 Elec	3	2.667	Ea.	910	170		1,080	1,250
7840	150 watt		3	2.667		910	170		1,080	1,250
7850	200 watt		3	2.667		915	170		1,085	1,250
7910	Round 19", metal halide, 250 watt		2.70	2.963		1,250	189		1,439	1,650
7920	High pressure sodium, 70 watt		3	2.667		1,375	170		1,545	1,750
7930	100 watt		3	2.667		1,375	170		1,545	1,750
7940	150 watt		3	2.667		1,375	170		1,545	1,775
7950	250 watt		2.70	2.963		1,325	189		1,514	1,725
8000	Sphere 14" opal, incandescent, 200 watt		4	2		395	127		522	625
8020	Sphere 18" opal, incandescent, 300 watt		3.50	2.286		480	146		626	740
8040	Sphere 16" clear, high pressure sodium, 70 watt		3	2.667		830	170		1,000	1,175
8050	100 watt		3	2.667		885	170		1,055	1,225
8100	Cube 16" opal, incandescent, 300 watt		3.50	2.286		525	146		671	790
8120	High pressure sodium, 70 watt		3	2.667		765	170		935	1,100
8130	100 watt		3	2.667		785	170		955	1,125
8230	Lantern, high pressure sodium, 70 watt		3	2.667		690	170		860	1,000
8240	100 watt		3	2.667		740	170		910	1,075
8250	150 watt		3	2.667		695	170		865	1,025
8260	250 watt		2.70	2.963		970	189		1,159	1,350
8270	Incandescent, 300 watt		3.50	2.286		510	146		656	780
8330	Reflector 22" w/globe, high pressure sodium, 70 watt		3	2.667		505	170		675	805
8340	100 watt		3	2.667		510	170		680	810
8350	150 watt		3	2.667		515	170		685	820
8360	250 watt	▼	2.70	2.963	▼	660	189		849	1,000

26 56 33.55 Walkway LED Luminaire

0010	**WALKWAY LED LUMINAIRE**									
0100	Pole mounted, 86 watts, 4,350 lumens	G	1 Elec	3	2.667	Ea.	1,350	170		1,520	1,725
0110	4,630 lumens	G		3	2.667		2,225	170		2,395	2,700
0120	80 watts, 4,000 lumens	G	▼	3	2.667	▼	2,050	170		2,220	2,500

26 56 36 – Flood Lighting

26 56 36.20 Floodlights

0010	**FLOODLIGHTS** with ballast and lamp,									
1290	floor mtd, mount with swivel bracket									
1300	Induction lamp, 40 watt	1 Elec	3	2.667	Ea.	380	170		550	670
1310	80 watt		3	2.667		770	170		940	1,100
1320	150 watt	▼	3	2.667	▼	1,375	170		1,545	1,775
1400	Pole mounted, pole not included									
1950	Metal halide, 175 watt	1 Elec	2.70	2.963	Ea.	188	189		377	485
2000	400 watt	2 Elec	4.40	3.636		186	232		418	550
2200	1,000 watt		4	4		1,000	255		1,255	1,475
2210	1,500 watt	▼	3.70	4.324		450	275		725	905
2250	Low pressure sodium, 55 watt	1 Elec	2.70	2.963		420	189		609	745
2270	90 watt		2	4		600	255		855	1,050
2290	180 watt		2	4		625	255		880	1,075
2340	High pressure sodium, 70 watt		2.70	2.963		255	189		444	560
2360	100 watt		2.70	2.963		262	189		451	570
2380	150 watt	▼	2.70	2.963		295	189		484	605
2400	400 watt	2 Elec	4.40	3.636		300	232		532	675
2600	1,000 watt	"	4	4		605	255		860	1,050
9005	Solar powered floodlight, w/motion det, incl. batt pack for cloudy days	G	1 Elec	8	1		74	63.50		137.50	176
9020	Replacement battery pack	G	"	8	1	▼	20	63.50		83.50	117

For customer support on your Site Work & Landscape Costs with RSMeans Data, call 800.448.8182.

26 56 Exterior Lighting

26 56 36 – Flood Lighting

26 56 36.55 LED Floodlights

		Crew	Daily Output	Labor-Hours	Unit	Material	2021 Bare Costs Labor	Equipment	Total	Total Incl O&P
0010	**LED FLOODLIGHTS** with ballast and lamp,									
0020	Pole mounted, pole not included									
0100	11 watt	G	1 Elec	4	2	Ea.	320	127		447	540
0110	46 watt	G		4	2		267	127		394	480
0120	90 watt	G		4	2		2,225	127		2,352	2,650
0130	288 watt	G		4	2		705	127		832	965

26 61 Lighting Systems and Accessories

26 61 23 – Lamps Applications

26 61 23.10 Lamps

		Crew	Daily Output	Labor-Hours	Unit	Material	2021 Bare Costs Labor	Equipment	Total	Total Incl O&P
0010	**LAMPS**									
0520	Very high output, 4' long, 110 watt	1 Elec	.90	8.889	C	2,375	565		2,940	3,475
0525	8' long, 195 watt energy saver	G	.70	11.429		1,875	730		2,605	3,125
0550	8' long, 215 watt		.70	11.429		1,550	730		2,280	2,775
0600	Mercury vapor, mogul base, deluxe white, 100 watt		.30	26.667		14,600	1,700		16,300	18,600
0650	175 watt		.30	26.667		2,575	1,700		4,275	5,375
0700	250 watt		.30	26.667		11,100	1,700		12,800	14,700
0800	400 watt		.30	26.667		10,700	1,700		12,400	14,300
0900	1,000 watt		.20	40		10,800	2,550		13,350	15,700
1000	Metal halide, mogul base, 175 watt		.30	26.667		1,125	1,700		2,825	3,750
1100	250 watt		.30	26.667		2,825	1,700		4,525	5,625
1200	400 watt		.30	26.667		2,550	1,700		4,250	5,325
1300	1,000 watt		.20	40		3,725	2,550		6,275	7,875
1320	1,000 watt, 125,000 initial lumens		.20	40		6,875	2,550		9,425	11,400
1330	1,500 watt		.20	40		8,450	2,550		11,000	13,100
1350	High pressure sodium, 70 watt		.30	26.667		1,800	1,700		3,500	4,500
1360	100 watt		.30	26.667		1,875	1,700		3,575	4,575
1370	150 watt		.30	26.667		1,650	1,700		3,350	4,325
1380	250 watt		.30	26.667		1,625	1,700		3,325	4,300
1400	400 watt		.30	26.667		1,625	1,700		3,325	4,300
1450	1,000 watt		.20	40		4,575	2,550		7,125	8,800
1500	Low pressure sodium, 35 watt		.30	26.667		9,075	1,700		10,775	12,500
1550	55 watt		.30	26.667		4,975	1,700		6,675	8,000
1600	90 watt		.30	26.667		14,100	1,700		15,800	18,000
1650	135 watt		.20	40		9,175	2,550		11,725	13,900
1700	180 watt		.20	40		17,000	2,550		19,550	22,500
1750	Quartz line, clear, 500 watt		1.10	7.273		750	465		1,215	1,525
1760	1,500 watt		.20	40		2,275	2,550		4,825	6,275
1762	Spot, MR 16, 50 watt		1.30	6.154		2,125	390		2,515	2,925
1800	Incandescent, interior, A21, 100 watt		1.60	5		390	320		710	905
2500	Exterior, PAR 38, 75 watt		1.30	6.154		1,400	390		1,790	2,125
2600	PAR 38, 150 watt		1.30	6.154		2,150	390		2,540	2,925
2700	PAR 46, 200 watt		1.10	7.273		1,750	465		2,215	2,625
2800	PAR 56, 300 watt		1.10	7.273		1,575	465		2,040	2,450

Division Notes

		CREW	DAILY OUTPUT	LABOR-HOURS	UNIT	BARE COSTS				TOTAL INCL O&P
						MAT.	LABOR	EQUIP.	TOTAL	

Estimating Tips
31 05 00 Common Work Results for Earthwork

- Estimating the actual cost of performing earthwork requires careful consideration of the variables involved. This includes items such as type of soil, whether water will be encountered, dewatering, whether banks need bracing, disposal of excavated earth, and length of haul to fill or spoil sites, etc. If the project has large quantities of cut or fill, consider raising or lowering the site to reduce costs, while paying close attention to the effect on site drainage and utilities.

- If the project has large quantities of fill, creating a borrow pit on the site can significantly lower the costs.

- It is very important to consider what time of year the project is scheduled for completion. Bad weather can create large cost overruns from dewatering, site repair, and lost productivity from cold weather.

Reference Numbers

Reference numbers are shown at the beginning of some major classifications. These numbers refer to related items in the Reference Section. The reference information may be an estimating procedure, an alternate pricing method, or technical information.

Note: Not all subdivisions listed here necessarily appear. ■

Same Data. Simplified.

Enjoy the convenience and efficiency of accessing your costs anywhere:

- **Skip the multiplier** by setting your location
- **Quickly search,** edit, favorite and share costs
- **Stay on top of price changes** with automatic updates

Discover more at rsmeans.com/online

---

31 05 Common Work Results for Earthwork

31 05 13 – Soils for Earthwork

31 05 13.10 Borrow

31 05 13.10 Borrow		Crew	Daily Output	Labor-Hours	Unit	Material	2021 Bare Costs Labor	2021 Bare Costs Equipment	Total	Total Incl O&P
0010	**BORROW**									
0020	Spread, 200 HP dozer, no compaction, 2 mile RT haul									
0200	Common borrow	B-15	600	.047	C.Y.	13.50	2.45	4.46	20.41	23.50
0700	Screened loam		600	.047		29.50	2.45	4.46	36.41	41
0800	Topsoil, weed free	↓	600	.047		27	2.45	4.46	33.91	38.50
0900	For 5 mile haul, add	B-34B	200	.040	↓		2.05	2.90	4.95	6.25

31 05 16 – Aggregates for Earthwork

31 05 16.10 Borrow

		Crew	Daily Output	Labor-Hours	Unit	Material	2021 Bare Costs Labor	2021 Bare Costs Equipment	Total	Total Incl O&P
0010	**BORROW**									
0020	Spread, with 200 HP dozer, no compaction, 2 mile RT haul									
0100	Bank run gravel	B-15	600	.047	L.C.Y.	29.50	2.45	4.46	36.41	41
0300	Crushed stone (1.40 tons per C.Y.), 1-1/2"		600	.047		25.50	2.45	4.46	32.41	37
0320	3/4"		600	.047		25.50	2.45	4.46	32.41	37
0340	1/2"		600	.047		19.60	2.45	4.46	26.51	30
0360	3/8"		600	.047		33	2.45	4.46	39.91	45
0400	Sand, washed, concrete		600	.047		33	2.45	4.46	39.91	44.50
0500	Dead or bank sand		600	.047		18.75	2.45	4.46	25.66	29
0600	Select structural fill		600	.047		30	2.45	4.46	36.91	41.50
0610	Import 10 mile RT select structural fill	↓	408	.069		30	3.60	6.55	40.15	45.50
0900	For 5 mile haul, add	B-34B	200	.040	↓		2.05	2.90	4.95	6.25
1000	For flowable fill, see Section 03 31 13.35									

31 05 19 – Geosynthetics for Earthwork

31 05 19.53 Reservoir Liners

		Crew	Daily Output	Labor-Hours	Unit	Material	2021 Bare Costs Labor	2021 Bare Costs Equipment	Total	Total Incl O&P
0010	**RESERVOIR LINERS**									
0011	Membrane lining									
1100	30 mil, LLDPE	B-63B	1850	.017	S.F.	.39	.82	.24	1.45	1.92
1200	60 mil, HDPE		1600	.020		.68	.95	.28	1.91	2.47
1300	120 mil thick	↓	1440	.022	↓	1.12	1.05	.31	2.48	3.14

31 05 23 – Cement and Concrete for Earthwork

31 05 23.30 Plant Mixed Bituminous Concrete

		Crew	Daily Output	Labor-Hours	Unit	Material	2021 Bare Costs Labor	2021 Bare Costs Equipment	Total	Total Incl O&P
0010	**PLANT MIXED BITUMINOUS CONCRETE**									
0020	Asphaltic concrete plant mix (145 lb./C.F.)				Ton	68.50			68.50	75
0040	Asphaltic concrete less than 300 tons add trucking costs									
0200	All weather patching mix, hot				Ton	66.50			66.50	73
0250	Cold patch					99.50			99.50	109
0300	Berm mix					73.50			73.50	81
0400	Base mix					68.50			68.50	75
0500	Binder mix					68.50			68.50	75
0600	Sand or sheet mix				↓	65			65	71.50

31 05 23.40 Recycled Plant Mixed Bituminous Concrete

			Crew	Daily Output	Labor-Hours	Unit	Material	2021 Bare Costs Labor	2021 Bare Costs Equipment	Total	Total Incl O&P
0010	**RECYCLED PLANT MIXED BITUMINOUS CONCRETE**										
0200	Reclaimed pavement in stockpile	G				Ton	21.50			21.50	23.50
0400	Recycled pavement, at plant, ratio old:new, 70:30	G					35			35	38.50
0600	Ratio old:new, 30:70	G				↓	53			53	58.50

31 06 Schedules for Earthwork

31 06 60 – Schedules for Special Foundations and Load Bearing Elements

31 06 60.14 Piling Special Costs

		Crew	Daily Output	Labor-Hours	Unit	Material	2021 Bare Costs Labor	2021 Bare Costs Equipment	Total	Total Incl O&P
0010	**PILING SPECIAL COSTS**									
0011	Piling special costs, pile caps, see Section 03 30 53.40									
0500	Cutoffs, concrete piles, plain	1 Pile	5.50	1.455	Ea.		81.50		81.50	126
0600	With steel thin shell, add		38	.211			11.75		11.75	18.20
0700	Steel pile or "H" piles		19	.421			23.50		23.50	36.50
0800	Wood piles		38	.211			11.75		11.75	18.20
0900	Pre-augering up to 30' deep, average soil, 24" diameter	B-43	180	.267	L.F.		13.05	4.27	17.32	24
0920	36" diameter		115	.417			20.50	6.70	27.20	38
0960	48" diameter		70	.686			33.50	11	44.50	62
0980	60" diameter		50	.960			47	15.35	62.35	87
1000	Testing, any type piles, test load is twice the design load									
1050	50 ton design load, 100 ton test				Ea.				14,000	15,500
1100	100 ton design load, 200 ton test								20,000	22,000
1150	150 ton design load, 300 ton test								26,000	28,500
1200	200 ton design load, 400 ton test								28,000	31,000
1250	400 ton design load, 800 ton test								32,000	35,000
1500	Wet conditions, soft damp ground									
1600	Requiring mats for crane, add								40%	40%
1700	Barge mounted driving rig, add								30%	30%

31 06 60.15 Mobilization

		Crew	Daily Output	Labor-Hours	Unit	Material	2021 Bare Costs Labor	2021 Bare Costs Equipment	Total	Total Incl O&P
0010	**MOBILIZATION**									
0020	Set up & remove, air compressor, 600 CFM	A-5	3.30	5.455	Ea.		245	15.05	260.05	380
0100	1,200 CFM	"	2.20	8.182			365	22.50	387.50	575
0200	Crane, with pile leads and pile hammer, 75 ton	B-19	.60	107			6,100	3,400	9,500	13,000
0300	150 ton	"	.36	178			10,200	5,675	15,875	21,800
0500	Drill rig, for caissons, to 36", minimum	B-43	2	24			1,175	385	1,560	2,175
0520	Maximum		.50	96			4,700	1,525	6,225	8,700
0600	Up to 84"		1	48			2,350	770	3,120	4,350
0800	Auxiliary boiler, for steam small	A-5	1.66	10.843			485	30	515	760
0900	Large	"	.83	21.687			975	60	1,035	1,525
1100	Rule of thumb: complete pile driving set up, small	B-19	.45	142			8,125	4,550	12,675	17,400
1200	Large	"	.27	237			13,500	7,575	21,075	28,900
1300	Mobilization by water for barge driving rig									
1310	Minimum				Ea.				7,000	7,700
1320	Maximum				"				45,000	49,500
1500	Mobilization, barge, by tug boat	B-83	25	.640	Mile		33	29	62	81.50
1600	Standby time for shore pile driving crew	B-19	8	8	Hr.		455	256	711	975
1700	Standby time for barge driving rig	B-76	8	9	"		515	470	985	1,300

31 11 Clearing and Grubbing

31 11 10 – Clearing and Grubbing Land

31 11 10.10 Clear and Grub Site

		Crew	Daily Output	Labor-Hours	Unit	Material	2021 Bare Costs Labor	2021 Bare Costs Equipment	Total	Total Incl O&P
0010	**CLEAR AND GRUB SITE**									
0020	Cut & chip light trees to 6" diam.	B-7	1	48	Acre		2,275	1,600	3,875	5,125
0150	Grub stumps and remove	B-30	2	12			645	925	1,570	2,000
0160	Clear & grub brush including stumps	"	.58	41.379			2,225	3,200	5,425	6,850
0200	Cut & chip medium trees to 12" diam.	B-7	.70	68.571			3,225	2,275	5,500	7,325
0250	Grub stumps and remove	B-30	1	24			1,300	1,850	3,150	3,975
0260	Clear & grub dense brush including stumps	"	.47	51.064			2,750	3,950	6,700	8,450
0300	Cut & chip heavy trees to 24" diam.	B-7	.30	160			7,550	5,325	12,875	17,200
0350	Grub stumps and remove	B-30	.50	48			2,575	3,700	6,275	7,925

31 11 Clearing and Grubbing

31 11 10 – Clearing and Grubbing Land

31 11 10.10 Clear and Grub Site		Crew	Daily Output	Labor-Hours	Unit	Material	2021 Bare Costs Labor	Equipment	Total	Total Incl O&P
0400	If burning is allowed, deduct cut & chip				Acre				40%	40%
3000	Chipping stumps, to 18" deep, 12" diam.	B-86	20	.400	Ea.		23.50	9.45	32.95	45.50
3040	18" diameter		16	.500			29.50	11.85	41.35	57
3080	24" diameter		14	.571			33.50	13.50	47	65
3100	30" diameter		12	.667			39.50	15.75	55.25	76
3120	36" diameter		10	.800			47	18.90	65.90	91.50
3160	48" diameter		8	1			59	23.50	82.50	114
5000	Tree thinning, feller buncher, conifer									
5080	Up to 8" diameter	B-93	240	.033	Ea.		1.97	2.62	4.59	5.80
5120	12" diameter		160	.050			2.95	3.93	6.88	8.70
5240	Hardwood, up to 4" diameter		240	.033			1.97	2.62	4.59	5.80
5280	8" diameter		180	.044			2.62	3.49	6.11	7.75
5320	12" diameter		120	.067			3.93	5.25	9.18	11.60
7000	Tree removal, congested area, aerial lift truck									
7040	8" diameter	B-85	7	5.714	Ea.		278	111	389	540
7080	12" diameter		6	6.667			325	130	455	630
7120	18" diameter		5	8			390	156	546	750
7160	24" diameter		4	10			485	195	680	940
7240	36" diameter		3	13.333			650	260	910	1,250
7280	48" diameter		2	20			975	390	1,365	1,875

31 13 Selective Tree and Shrub Removal and Trimming

31 13 13 – Selective Tree and Shrub Removal

31 13 13.10 Selective Clearing

31 13 13.10 Selective Clearing		Crew	Daily Output	Labor-Hours	Unit	Material	2021 Bare Costs Labor	Equipment	Total	Total Incl O&P
0010	**SELECTIVE CLEARING**									
0020	Clearing brush with brush saw	A-1C	.25	32	Acre		1,425	209	1,634	2,350
0100	By hand	1 Clab	.12	66.667			2,950		2,950	4,425
0300	With dozer, ball and chain, light clearing	B-11A	2	8			415	760	1,175	1,450
0400	Medium clearing		1.50	10.667			550	1,025	1,575	1,950
0500	With dozer and brush rake, light		10	1.600			82.50	152	234.50	290
0550	Medium brush to 4" diameter		8	2			103	190	293	365
0600	Heavy brush to 4" diameter		6.40	2.500			129	238	367	455
1000	Brush mowing, tractor w/rotary mower, no removal									
1020	Light density	B-84	2	4	Acre		236	186	422	555
1040	Medium density		1.50	5.333			315	247	562	740
1080	Heavy density		1	8			470	370	840	1,125

31 13 13.20 Selective Tree Removal

31 13 13.20 Selective Tree Removal		Crew	Daily Output	Labor-Hours	Unit	Material	2021 Bare Costs Labor	Equipment	Total	Total Incl O&P
0010	**SELECTIVE TREE REMOVAL**									
0011	With tractor, large tract, firm									
0020	level terrain, no boulders, less than 12" diam. trees									
0300	300 HP dozer, up to 400 trees/acre, up to 25% hardwoods	B-10M	.75	16	Acre		865	2,375	3,240	3,925
0340	25% to 50% hardwoods		.60	20			1,075	2,975	4,050	4,900
0370	75% to 100% hardwoods		.45	26.667			1,450	3,975	5,425	6,525
0400	500 trees/acre, up to 25% hardwoods		.60	20			1,075	2,975	4,050	4,900
0440	25% to 50% hardwoods		.48	25			1,350	3,725	5,075	6,125
0470	75% to 100% hardwoods		.36	33.333			1,800	4,950	6,750	8,150
0500	More than 600 trees/acre, up to 25% hardwoods		.52	23.077			1,250	3,425	4,675	5,625
0540	25% to 50% hardwoods		.42	28.571			1,550	4,250	5,800	6,975
0570	75% to 100% hardwoods		.31	38.710			2,100	5,750	7,850	9,450
0900	Large tract clearing per tree									
1500	300 HP dozer, to 12" diameter, softwood	B-10M	320	.038	Ea.		2.03	5.60	7.63	9.20

For customer support on your Site Work & Landscape Costs with RSMeans Data, call 800.448.8182.

31 13 Selective Tree and Shrub Removal and Trimming

31 13 13 – Selective Tree and Shrub Removal

31 13 13.20 Selective Tree Removal		Crew	Daily Output	Labor-Hours	Unit	Material	2021 Bare Costs Labor	Equipment	Total	Total Incl O&P
1550	Hardwood	B-10M	100	.120	Ea.		6.50	17.85	24.35	29.50
1600	12" to 24" diameter, softwood		200	.060			3.25	8.95	12.20	14.65
1650	Hardwood		80	.150			8.10	22.50	30.60	36.50
1700	24" to 36" diameter, softwood		100	.120			6.50	17.85	24.35	29.50
1750	Hardwood		50	.240			13	35.50	48.50	59
1800	36" to 48" diameter, softwood		70	.171			9.30	25.50	34.80	42
1850	Hardwood	↓	35	.343	↓		18.55	51	69.55	83.50
2000	Stump removal on site by hydraulic backhoe, 1-1/2 C.Y.									
2040	4" to 6" diameter	B-17	60	.533	Ea.		26	10.40	36.40	50.50
2050	8" to 12" diameter	B-30	33	.727			39	56	95	121
2100	14" to 24" diameter		25	.960			51.50	74	125.50	159
2150	26" to 36" diameter	↓	16	1.500	↓		81	116	197	249
3000	Remove selective trees, on site using chain saws and chipper,									
3050	not incl. stumps, up to 6" diameter	B-7	18	2.667	Ea.		126	88.50	214.50	286
3100	8" to 12" diameter		12	4			189	133	322	425
3150	14" to 24" diameter		10	4.800			226	160	386	515
3200	26" to 36" diameter	↓	8	6			283	199	482	640
3300	Machine load, 2 mile haul to dump, 12" diam. tree	A-3B	8	2	↓		110	152	262	330

31 14 Earth Stripping and Stockpiling

31 14 13 – Soil Stripping and Stockpiling

31 14 13.23 Topsoil Stripping and Stockpiling		Crew	Daily Output	Labor-Hours	Unit	Material	2021 Bare Costs Labor	Equipment	Total	Total Incl O&P
0010	**TOPSOIL STRIPPING AND STOCKPILING**									
0020	200 HP dozer, ideal conditions	B-10B	2300	.005	C.Y.		.28	.66	.94	1.15
0100	Adverse conditions	"	1150	.010			.56	1.32	1.88	2.29
0200	300 HP dozer, ideal conditions	B-10M	3000	.004			.22	.60	.82	.97
0300	Adverse conditions	"	1650	.007			.39	1.08	1.47	1.78
0400	400 HP dozer, ideal conditions	B-10X	3900	.003			.17	.72	.89	1.04
0500	Adverse conditions	"	2000	.006			.32	1.40	1.72	2.02
0600	Clay, dry and soft, 200 HP dozer, ideal conditions	B-10B	1600	.008			.41	.95	1.36	1.65
0700	Adverse conditions	"	800	.015			.81	1.90	2.71	3.30
1000	Medium hard, 300 HP dozer, ideal conditions	B-10M	2000	.006			.32	.89	1.21	1.46
1100	Adverse conditions	"	1100	.011			.59	1.62	2.21	2.67
1200	Very hard, 400 HP dozer, ideal conditions	B-10X	2600	.005			.25	1.08	1.33	1.56
1300	Adverse conditions	"	1340	.009	↓		.48	2.09	2.57	3.02
1400	Loam or topsoil, remove and stockpile on site									
1420	6" deep, 200' haul	B-10B	865	.014	C.Y.		.75	1.76	2.51	3.05
1430	300' haul		520	.023			1.25	2.92	4.17	5.10
1440	500' haul		225	.053	↓		2.89	6.75	9.64	11.75
1450	Alternate method: 6" deep, 200' haul		5090	.002	S.Y.		.13	.30	.43	.52
1460	500' haul	↓	1325	.009	"		.49	1.15	1.64	1.99
1500	Loam or topsoil, remove/stockpile on site									
1510	By hand, 6" deep, 50' haul, less than 100 S.Y.	B-1	100	.240	S.Y.		10.80		10.80	16.15
1520	By skid steer, 6" deep, 100' haul, 101-500 S.Y.	B-62	500	.048			2.31	.36	2.67	3.84
1530	100' haul, 501-900 S.Y.	"	900	.027			1.28	.20	1.48	2.13
1540	200' haul, 901-1,100 S.Y.	B-63	1000	.040			1.86	.18	2.04	2.98
1550	By dozer, 200' haul, 1,101-4,000 S.Y.	B-10B	4000	.003	↓		.16	.38	.54	.66

For customer support on your Site Work & Landscape Costs with RSMeans Data, call 800.448.8182.

295

31 22 Grading

31 22 13 – Rough Grading

31 22 13.20 Rough Grading Sites	Crew	Daily Output	Labor-Hours	Unit	Material	2021 Bare Costs Labor	Equipment	Total	Total Incl O&P
0010 **ROUGH GRADING SITES**									
0100 Rough grade sites 400 S.F. or less, hand labor	B-1	2	12	Ea.		540		540	805
0120 410-1,000 S.F.	"	1	24			1,075		1,075	1,625
0130 1,100-3,000 S.F., skid steer & labor	B-62	1.50	16			770	120	890	1,275
0140 3,100-5,000 S.F.	"	1	24			1,150	180	1,330	1,925
0150 5,100-8,000 S.F.	B-63	1	40			1,875	180	2,055	2,975
0160 8,100-10,000 S.F.	"	.75	53.333			2,475	239	2,714	3,975
0170 8,100-10,000 S.F., dozer	B-10L	1	12			650	405	1,055	1,425
0200 Rough grade open sites 10,000-20,000 S.F., grader	B-11L	1.80	8.889			460	595	1,055	1,350
0210 20,100-25,000 S.F.		1.40	11.429			590	765	1,355	1,725
0220 25,100-30,000 S.F.		1.20	13.333			690	895	1,585	2,000
0230 30,100-35,000 S.F.		1	16			825	1,075	1,900	2,400
0240 35,100-40,000 S.F.		.90	17.778			920	1,200	2,120	2,675
0250 40,100-45,000 S.F.		.80	20			1,025	1,350	2,375	3,025
0260 45,100-50,000 S.F.		.72	22.222			1,150	1,500	2,650	3,375
0270 50,100-75,000 S.F.		.50	32			1,650	2,150	3,800	4,825
0280 75,100-100,000 S.F.		.36	44.444			2,300	2,975	5,275	6,700

31 22 16 – Fine Grading

31 22 16.10 Finish Grading

		Crew	Daily Output	Labor-Hours	Unit	Material	Labor	Equipment	Total	Total Incl O&P
0010	**FINISH GRADING**									
0011	Finish grading granular subbase for highway paving	B-32C	8000	.006	S.Y.		.31	.35	.66	.86
0012	Finish grading area to be paved with grader, small area	B-11L	400	.040			2.07	2.68	4.75	6.05
0100	Large area		2000	.008			.41	.54	.95	1.21
0200	Grade subgrade for base course, roadways		3500	.005			.24	.31	.55	.69
1020	For large parking lots	B-32C	5000	.010			.50	.57	1.07	1.36
1050	For small irregular areas	"	2000	.024			1.25	1.42	2.67	3.42
1100	Fine grade for slab on grade, machine	B-11L	1040	.015			.80	1.03	1.83	2.32
1150	Hand grading	B-18	700	.034			1.55	.24	1.79	2.57
1200	Fine grade granular base for sidewalks and bikeways	B-62	1200	.020			.96	.15	1.11	1.59
2550	Hand grade select gravel	2 Clab	60	.267	C.S.F.		11.85		11.85	17.65
3000	Hand grade select gravel, including compaction, 4" deep	B-18	555	.043	S.Y.		1.95	.30	2.25	3.24
3100	6" deep		400	.060			2.70	.41	3.11	4.50
3120	8" deep		300	.080			3.61	.55	4.16	6
3300	Finishing grading slopes, gentle	B-11L	8900	.002			.09	.12	.21	.27
3310	Steep slopes		7100	.002			.12	.15	.27	.34
3312	Steep slopes, large quantities		64	.250	M.S.F.		12.95	16.75	29.70	37.50
3500	Finish grading lagoon bottoms		4	4			207	268	475	605
3600	Fine grade, top of lagoon banks for compaction		30	.533			27.50	36	63.50	80.50

31 23 Excavation and Fill

31 23 16 – Excavation

31 23 16.13 Excavating, Trench

		Crew	Daily Output	Labor-Hours	Unit	Material	Labor	Equipment	Total	Total Incl O&P
0010	**EXCAVATING, TRENCH** R312316-40									
0011	Or continuous footing									
0020	Common earth with no sheeting or dewatering included									
0050	1' to 4' deep, 3/8 C.Y. excavator	B-11C	150	.107	B.C.Y.		5.50	1.44	6.94	9.80
0060	1/2 C.Y. excavator	B-11M	200	.080			4.14	1.18	5.32	7.45
0062	3/4 C.Y. excavator	B-12F	270	.059			3.14	2.60	5.74	7.55
0090	4' to 6' deep, 1/2 C.Y. excavator	B-11M	200	.080			4.14	1.18	5.32	7.45
0100	5/8 C.Y. excavator	B-12Q	250	.064			3.39	2.42	5.81	7.70

31 23 16.13 Excavating, Trench	Crew	Daily Output	Labor-Hours	Unit	Material	2021 Bare Costs Labor	Equipment	Total	Total Incl O&P
0110	3/4 C.Y. excavator	B-12F	300	.053	B.C.Y.		2.82	2.34	5.16	6.80
0120	1 C.Y. hydraulic excavator	B-12A	400	.040			2.12	2.08	4.20	5.45
0130	1-1/2 C.Y. excavator	B-12B	540	.030			1.57	1.29	2.86	3.76
0300	1/2 C.Y. excavator, truck mounted	B-12J	200	.080			4.23	4.25	8.48	11
0500	6' to 10' deep, 3/4 C.Y. excavator	B-12F	225	.071			3.76	3.12	6.88	9.05
0510	1 C.Y. excavator	B-12A	400	.040			2.12	2.08	4.20	5.45
0600	1 C.Y. excavator, truck mounted	B-12K	400	.040			2.12	2.46	4.58	5.85
0610	1-1/2 C.Y. excavator	B-12B	600	.027			1.41	1.16	2.57	3.38
0620	2-1/2 C.Y. excavator	B-12S	1000	.016			.85	1.57	2.42	2.98
0900	10' to 14' deep, 3/4 C.Y. excavator	B-12F	200	.080			4.23	3.51	7.74	10.15
0910	1 C.Y. excavator	B-12A	360	.044			2.35	2.31	4.66	6.05
1000	1-1/2 C.Y. excavator	B-12B	540	.030			1.57	1.29	2.86	3.76
1020	2-1/2 C.Y. excavator	B-12S	900	.018			.94	1.74	2.68	3.32
1030	3 C.Y. excavator	B-12D	1400	.011			.60	1.56	2.16	2.62
1040	3-1/2 C.Y. excavator	"	1800	.009			.47	1.21	1.68	2.03
1300	14' to 20' deep, 1 C.Y. excavator	B-12A	320	.050			2.65	2.60	5.25	6.80
1310	1-1/2 C.Y. excavator	B-12B	480	.033			1.76	1.45	3.21	4.22
1320	2-1/2 C.Y. excavator	B-12S	765	.021			1.11	2.05	3.16	3.90
1330	3 C.Y. excavator	B-12D	1000	.016			.85	2.18	3.03	3.66
1335	3-1/2 C.Y. excavator	"	1150	.014			.74	1.90	2.64	3.19
1340	20' to 24' deep, 1 C.Y. excavator	B-12A	288	.056			2.94	2.89	5.83	7.55
1342	1-1/2 C.Y. excavator	B-12B	432	.037			1.96	1.61	3.57	4.69
1344	2-1/2 C.Y. excavator	B-12S	685	.023			1.24	2.29	3.53	4.36
1346	3 C.Y. excavator	B-12D	900	.018			.94	2.43	3.37	4.07
1348	3-1/2 C.Y. excavator	"	1050	.015			.81	2.08	2.89	3.49
1352	4' to 6' deep, 1/2 C.Y. excavator w/trench box	B-13H	188	.085			4.50	5.15	9.65	12.35
1354	5/8 C.Y. excavator	"	235	.068			3.60	4.13	7.73	9.90
1356	3/4 C.Y. excavator	B-13G	282	.057			3	2.91	5.91	7.70
1358	1 C.Y. excavator	B-13D	376	.043			2.25	2.53	4.78	6.15
1360	1-1/2 C.Y. excavator	B-13E	508	.032			1.67	1.60	3.27	4.25
1362	6' to 10' deep, 3/4 C.Y. excavator w/trench box	B-13G	212	.075			3.99	3.87	7.86	10.20
1370	1 C.Y. excavator	B-13D	376	.043			2.25	2.53	4.78	6.15
1371	1-1/2 C.Y. excavator	B-13E	564	.028			1.50	1.44	2.94	3.83
1372	2-1/2 C.Y. excavator	B-13J	940	.017			.90	1.79	2.69	3.31
1374	10' to 14' deep, 3/4 C.Y. excavator w/trench box	B-13G	188	.085			4.50	4.37	8.87	11.50
1375	1 C.Y. excavator	B-13D	338	.047			2.51	2.82	5.33	6.85
1376	1-1/2 C.Y. excavator	B-13E	508	.032			1.67	1.60	3.27	4.25
1377	2-1/2 C.Y. excavator	B-13J	845	.019			1	2	3	3.68
1378	3 C.Y. excavator	B-13F	1316	.012			.64	1.75	2.39	2.89
1380	3-1/2 C.Y. excavator	"	1692	.009			.50	1.36	1.86	2.25
1381	14' to 20' deep, 1 C.Y. excavator w/trench box	B-13D	301	.053			2.81	3.16	5.97	7.65
1382	1-1/2 C.Y. excavator	B-13E	451	.035			1.88	1.81	3.69	4.79
1383	2-1/2 C.Y. excavator	B-13J	720	.022			1.18	2.34	3.52	4.33
1384	3 C.Y. excavator	B-13F	940	.017			.90	2.45	3.35	4.04
1385	3-1/2 C.Y. excavator	"	1081	.015			.78	2.13	2.91	3.51
1386	20' to 24' deep, 1 C.Y. excavator w/trench box	B-13D	271	.059			3.13	3.51	6.64	8.50
1387	1-1/2 C.Y. excavator	B-13E	406	.039			2.09	2.01	4.10	5.30
1388	2-1/2 C.Y. excavator	B-13J	645	.025			1.31	2.61	3.92	4.84
1389	3 C.Y. excavator	B-13F	846	.019			1	2.72	3.72	4.48
1390	3-1/2 C.Y. excavator	"	987	.016			.86	2.33	3.19	3.85
1391	Shoring by S.F./day trench wall protected loose mat., 4' W	B-6	3200	.008	SF Wall	.82	.36	.07	1.25	1.52
1392	Rent shoring per week per S.F. wall protected, loose mat., 4' W					2.28			2.28	2.51
1395	Hydraulic shoring, S.F. trench wall protected stable mat., 4' W	2 Clab	2700	.006		.24	.26		.50	.66

31 23 16.13 Excavating, Trench

	Crew	Daily Output	Labor-Hours	Unit	Material	2021 Bare Costs Labor	2021 Bare Costs Equipment	Total	Total Incl O&P
1397 semi-stable material, 4' W	2 Clab	2400	.007	SF Wall	.37	.30		.67	.85
1398 Rent hydraulic shoring per day/S.F. wall, stable mat., 4' W					.37			.37	.41
1399 semi-stable material			↓		.50			.50	.55
1400 By hand with pick and shovel 2' to 6' deep, light soil	1 Clab	8	1	B.C.Y.		44.50		44.50	66.50
1500 Heavy soil	"	4	2	"		89		89	133
1700 For tamping backfilled trenches, air tamp, add	A-1G	100	.080	E.C.Y.		3.55	.55	4.10	5.90
1900 Vibrating plate, add	B-18	180	.133	"		6	.92	6.92	9.95
2100 Trim sides and bottom for concrete pours, common earth	↓	1500	.016	S.F.		.72	.11	.83	1.20
2300 Hardpan	↓	600	.040	"		1.80	.28	2.08	2.99
2400 Pier and spread footing excavation, add to above				B.C.Y.				30%	30%
3000 Backfill trench, F.E. loader, wheel mtd., 1 C.Y. bucket									
3020 Minimal haul	B-10R	400	.030	L.C.Y.		1.62	.76	2.38	3.26
3040 100' haul	↓	200	.060			3.25	1.53	4.78	6.50
3060 200' haul	↓	100	.120	↓		6.50	3.06	9.56	13.05
3080 2-1/4 C.Y. bucket, minimum haul	B-10T	600	.020			1.08	1.06	2.14	2.78
3090 100' haul		300	.040	↓		2.17	2.13	4.30	5.55
3100 200' haul	↓	150	.080	↓		4.33	4.26	8.59	11.15
4000 For backfill with dozer, see Section 31 23 23.14									
4010 For compaction of backfill, see Section 31 23 23.23									
5020 Loam & sandy clay with no sheeting or dewatering included									
5050 1' to 4' deep, 3/8 C.Y. tractor loader/backhoe	B-11C	162	.099	B.C.Y.		5.10	1.33	6.43	9.05
5060 1/2 C.Y. excavator	B-11M	216	.074			3.83	1.09	4.92	6.90
5070 3/4 C.Y. excavator	B-12F	292	.055			2.90	2.40	5.30	6.95
5080 4' to 6' deep, 1/2 C.Y. excavator	B-11M	216	.074			3.83	1.09	4.92	6.90
5090 5/8 C.Y. excavator	B-12Q	276	.058			3.07	2.19	5.26	7
5100 3/4 C.Y. excavator	B-12F	324	.049			2.61	2.17	4.78	6.30
5110 1 C.Y. excavator	B-12A	432	.037			1.96	1.93	3.89	5.05
5120 1-1/2 C.Y. excavator	B-12B	583	.027			1.45	1.19	2.64	3.48
5130 1/2 C.Y. excavator, truck mounted	B-12J	216	.074			3.92	3.94	7.86	10.20
5140 6' to 10' deep, 3/4 C.Y. excavator	B-12F	243	.066			3.49	2.89	6.38	8.40
5150 1 C.Y. excavator	B-12A	432	.037			1.96	1.93	3.89	5.05
5160 1 C.Y. excavator, truck mounted	B-12K	432	.037			1.96	2.28	4.24	5.45
5170 1-1/2 C.Y. excavator	B-12B	648	.025			1.31	1.07	2.38	3.13
5180 2-1/2 C.Y. excavator	B-12S	1080	.015			.78	1.45	2.23	2.77
5190 10' to 14' deep, 3/4 C.Y. excavator	B-12F	216	.074			3.92	3.25	7.17	9.40
5200 1 C.Y. excavator	B-12A	389	.041			2.18	2.14	4.32	5.60
5210 1-1/2 C.Y. excavator	B-12B	583	.027			1.45	1.19	2.64	3.48
5220 2-1/2 C.Y. hydraulic backhoe	B-12S	970	.016			.87	1.62	2.49	3.08
5230 3 C.Y. excavator	B-12D	1512	.011			.56	1.44	2	2.42
5240 3-1/2 C.Y. excavator	"	1944	.008			.44	1.12	1.56	1.89
5250 14' to 20' deep, 1 C.Y. excavator	B-12A	346	.046			2.45	2.41	4.86	6.30
5260 1-1/2 C.Y. excavator	B-12B	518	.031			1.63	1.34	2.97	3.92
5270 2-1/2 C.Y. excavator	B-12S	826	.019			1.03	1.90	2.93	3.62
5280 3 C.Y. excavator	B-12D	1080	.015			.78	2.02	2.80	3.39
5290 3-1/2 C.Y. excavator	"	1242	.013			.68	1.76	2.44	2.95
5300 20' to 24' deep, 1 C.Y. excavator	B-12A	311	.051			2.72	2.68	5.40	7
5310 1-1/2 C.Y. excavator	B-12B	467	.034			1.81	1.49	3.30	4.34
5320 2-1/2 C.Y. excavator	B-12S	740	.022			1.14	2.12	3.26	4.04
5330 3 C.Y. excavator	B-12D	972	.016			.87	2.25	3.12	3.77
5340 3-1/2 C.Y. excavator	"	1134	.014			.75	1.93	2.68	3.23
5352 4' to 6' deep, 1/2 C.Y. excavator w/trench box	B-13H	205	.078			4.13	4.73	8.86	11.35
5354 5/8 C.Y. excavator	"	257	.062			3.30	3.77	7.07	9.05
5356 3/4 C.Y. excavator	B-13G	308	.052			2.75	2.67	5.42	7.05

298

31 23 Excavation and Fill

31 23 16 – Excavation

31 23 16.13 Excavating, Trench	Crew	Daily Output	Labor-Hours	Unit	Material	2021 Bare Costs		Total	Total Incl O&P
						Labor	Equipment		
5358	1 C.Y. excavator	B-13D	410	.039	B.C.Y.		2.07	2.32	4.39	5.65
5360	1-1/2 C.Y. excavator	B-13E	554	.029			1.53	1.47	3	3.90
5362	6' to 10' deep, 3/4 C.Y. excavator w/trench box	B-13G	231	.069			3.67	3.55	7.22	9.35
5364	1 C.Y. excavator	B-13D	410	.039			2.07	2.32	4.39	5.65
5366	1-1/2 C.Y. excavator	B-13E	616	.026			1.37	1.32	2.69	3.51
5368	2-1/2 C.Y. excavator	B-13J	1015	.016			.83	1.66	2.49	3.07
5370	10' to 14' deep, 3/4 C.Y. excavator w/trench box	B-13G	205	.078			4.13	4	8.13	10.55
5372	1 C.Y. excavator	B-13D	370	.043			2.29	2.57	4.86	6.25
5374	1-1/2 C.Y. excavator	B-13E	554	.029			1.53	1.47	3	3.90
5376	2-1/2 C.Y. excavator	B-13J	914	.018			.93	1.84	2.77	3.41
5378	3 C.Y. excavator	B-13F	1436	.011			.59	1.60	2.19	2.64
5380	3-1/2 C.Y. excavator	"	1847	.009			.46	1.25	1.71	2.05
5382	14' to 20' deep, 1 C.Y. excavator w/trench box	B-13D	329	.049			2.57	2.89	5.46	7
5384	1-1/2 C.Y. excavator	B-13E	492	.033			1.72	1.66	3.38	4.39
5386	2-1/2 C.Y. excavator	B-13J	780	.021			1.09	2.16	3.25	4
5388	3 C.Y. excavator	B-13F	1026	.016			.83	2.24	3.07	3.70
5390	3-1/2 C.Y. excavator	"	1180	.014			.72	1.95	2.67	3.22
5392	20' to 24' deep, 1 C.Y. excavator w/trench box	B-13D	295	.054			2.87	3.23	6.10	7.85
5394	1-1/2 C.Y. excavator	B-13E	444	.036			1.91	1.84	3.75	4.86
5396	2-1/2 C.Y. excavator	B-13J	695	.023			1.22	2.43	3.65	4.49
5398	3 C.Y. excavator	B-13F	923	.017			.92	2.50	3.42	4.11
5400	3-1/2 C.Y. excavator	"	1077	.015			.79	2.14	2.93	3.52
6020	Sand & gravel with no sheeting or dewatering included									
6050	1' to 4' deep, 3/8 C.Y. excavator	B-11C	165	.097	B.C.Y.		5	1.31	6.31	8.90
6060	1/2 C.Y. excavator	B-11M	220	.073			3.76	1.07	4.83	6.80
6070	3/4 C.Y. excavator	B-12F	297	.054			2.85	2.36	5.21	6.85
6080	4' to 6' deep, 1/2 C.Y. excavator	B-11M	220	.073			3.76	1.07	4.83	6.80
6090	5/8 C.Y. excavator	B-12Q	275	.058			3.08	2.20	5.28	7
6100	3/4 C.Y. excavator	B-12F	330	.048			2.57	2.13	4.70	6.15
6110	1 C.Y. excavator	B-12A	440	.036			1.92	1.89	3.81	4.95
6120	1-1/2 C.Y. excavator	B-12B	594	.027			1.43	1.17	2.60	3.42
6130	1/2 C.Y. excavator, truck mounted	B-12J	220	.073			3.85	3.87	7.72	10
6140	6' to 10' deep, 3/4 C.Y. excavator	B-12F	248	.065			3.41	2.83	6.24	8.20
6150	1 C.Y. excavator	B-12A	440	.036			1.92	1.89	3.81	4.95
6160	1 C.Y. excavator, truck mounted	B-12K	440	.036			1.92	2.24	4.16	5.35
6170	1-1/2 C.Y. excavator	B-12B	660	.024			1.28	1.05	2.33	3.07
6180	2-1/2 C.Y. excavator	B-12S	1100	.015			.77	1.42	2.19	2.72
6190	10' to 14' deep, 3/4 C.Y. excavator	B-12F	220	.073			3.85	3.19	7.04	9.25
6200	1 C.Y. excavator	B-12A	396	.040			2.14	2.10	4.24	5.50
6210	1-1/2 C.Y. excavator	B-12B	594	.027			1.43	1.17	2.60	3.42
6220	2-1/2 C.Y. excavator	B-12S	990	.016			.86	1.58	2.44	3.02
6230	3 C.Y. excavator	B-12D	1540	.010			.55	1.42	1.97	2.38
6240	3-1/2 C.Y. excavator	"	1980	.008			.43	1.10	1.53	1.85
6250	14' to 20' deep, 1 C.Y. excavator	B-12A	352	.045			2.41	2.37	4.78	6.20
6260	1-1/2 C.Y. excavator	B-12B	528	.030			1.60	1.32	2.92	3.84
6270	2-1/2 C.Y. excavator	B-12S	840	.019			1.01	1.87	2.88	3.55
6280	3 C.Y. excavator	B-12D	1100	.015			.77	1.99	2.76	3.33
6290	3-1/2 C.Y. excavator	"	1265	.013			.67	1.73	2.40	2.90
6300	20' to 24' deep, 1 C.Y. excavator	B-12A	317	.050			2.67	2.63	5.30	6.85
6310	1-1/2 C.Y. excavator	B-12B	475	.034			1.78	1.46	3.24	4.27
6320	2-1/2 C.Y. excavator	B-12S	755	.021			1.12	2.08	3.20	3.95
6330	3 C.Y. excavator	B-12D	990	.016			.86	2.21	3.07	3.71
6340	3-1/2 C.Y. excavator	"	1155	.014			.73	1.89	2.62	3.17

31 23 16.13 Excavating, Trench	Crew	Daily Output	Labor-Hours	Unit	Material	2021 Bare Costs Labor	Equipment	Total	Total Incl O&P
6352	4' to 6' deep, 1/2 C.Y. excavator w/trench box	B-13H	209	.077	B.C.Y.		4.05	4.64	8.69	11.15
6354	5/8 C.Y. excavator	"	261	.061			3.24	3.72	6.96	8.95
6356	3/4 C.Y. excavator	B-13G	314	.051			2.70	2.61	5.31	6.90
6358	1 C.Y. excavator	B-13D	418	.038			2.03	2.28	4.31	5.50
6360	1-1/2 C.Y. excavator	B-13E	564	.028			1.50	1.44	2.94	3.83
6362	6' to 10' deep, 3/4 C.Y. excavator w/trench box	B-13G	236	.068			3.59	3.48	7.07	9.20
6364	1 C.Y. excavator	B-13D	418	.038			2.03	2.28	4.31	5.50
6366	1-1/2 C.Y. excavator	B-13E	627	.026			1.35	1.30	2.65	3.44
6368	2-1/2 C.Y. excavator	B-13J	1035	.015			.82	1.63	2.45	3.01
6370	10' to 14' deep, 3/4 C.Y. excavator w/trench box	B-13G	209	.077			4.05	3.93	7.98	10.35
6372	1 C.Y. excavator	B-13D	376	.043			2.25	2.53	4.78	6.15
6374	1-1/2 C.Y. excavator	B-13E	564	.028			1.50	1.44	2.94	3.83
6376	2-1/2 C.Y. excavator	B-13J	930	.017			.91	1.81	2.72	3.35
6378	3 C.Y. excavator	B-13F	1463	.011			.58	1.57	2.15	2.59
6380	3-1/2 C.Y. excavator	"	1881	.009			.45	1.22	1.67	2.02
6382	14' to 20' deep, 1 C.Y. excavator w/trench box	B-13D	334	.048			2.54	2.85	5.39	6.90
6384	1-1/2 C.Y. excavator	B-13E	502	.032			1.69	1.62	3.31	4.30
6386	2-1/2 C.Y. excavator	B-13J	790	.020			1.07	2.13	3.20	3.95
6388	3 C.Y. excavator	B-13F	1045	.015			.81	2.20	3.01	3.63
6390	3-1/2 C.Y. excavator	"	1202	.013			.70	1.92	2.62	3.16
6392	20' to 24' deep, 1 C.Y. excavator w/trench box	B-13D	301	.053			2.81	3.16	5.97	7.65
6394	1-1/2 C.Y. excavator	B-13E	452	.035			1.87	1.80	3.67	4.77
6396	2-1/2 C.Y. excavator	B-13J	710	.023			1.19	2.37	3.56	4.39
6398	3 C.Y. excavator	B-13F	941	.017			.90	2.45	3.35	4.03
6400	3-1/2 C.Y. excavator	"	1097	.015			.77	2.10	2.87	3.46
7020	Dense hard clay with no sheeting or dewatering included									
7050	1' to 4' deep, 3/8 C.Y. excavator	B-11C	132	.121	B.C.Y.		6.25	1.64	7.89	11.15
7060	1/2 C.Y. excavator	B-11M	176	.091			4.70	1.34	6.04	8.45
7070	3/4 C.Y. excavator	B-12F	238	.067			3.56	2.95	6.51	8.55
7080	4' to 6' deep, 1/2 C.Y. excavator	B-11M	176	.091			4.70	1.34	6.04	8.45
7090	5/8 C.Y. excavator	B-12Q	220	.073			3.85	2.75	6.60	8.75
7100	3/4 C.Y. excavator	B-12F	264	.061			3.21	2.66	5.87	7.70
7110	1 C.Y. excavator	B-12A	352	.045			2.41	2.37	4.78	6.20
7120	1-1/2 C.Y. excavator	B-12B	475	.034			1.78	1.46	3.24	4.27
7130	1/2 C.Y. excavator, truck mounted	B-12J	176	.091			4.81	4.83	9.64	12.45
7140	6' to 10' deep, 3/4 C.Y. excavator	B-12F	198	.081			4.28	3.54	7.82	10.30
7150	1 C.Y. excavator	B-12A	352	.045			2.41	2.37	4.78	6.20
7160	1 C.Y. excavator, truck mounted	B-12K	352	.045			2.41	2.80	5.21	6.65
7170	1-1/2 C.Y. excavator	B-12B	528	.030			1.60	1.32	2.92	3.84
7180	2-1/2 C.Y. excavator	B-12S	880	.018			.96	1.78	2.74	3.39
7190	10' to 14' deep, 3/4 C.Y. excavator	B-12F	176	.091			4.81	3.99	8.80	11.55
7200	1 C.Y. excavator	B-12A	317	.050			2.67	2.63	5.30	6.85
7210	1-1/2 C.Y. excavator	B-12B	475	.034			1.78	1.46	3.24	4.27
7220	2-1/2 C.Y. excavator	B-12S	790	.020			1.07	1.98	3.05	3.78
7230	3 C.Y. excavator	B-12D	1232	.013			.69	1.77	2.46	2.97
7240	3-1/2 C.Y. excavator	"	1584	.010			.53	1.38	1.91	2.32
7250	14' to 20' deep, 1 C.Y. excavator	B-12A	282	.057			3	2.95	5.95	7.75
7260	1-1/2 C.Y. excavator	B-12B	422	.038			2.01	1.65	3.66	4.80
7270	2-1/2 C.Y. excavator	B-12S	675	.024			1.25	2.32	3.57	4.42
7280	3 C.Y. excavator	B-12D	880	.018			.96	2.48	3.44	4.16
7290	3-1/2 C.Y. excavator	"	1012	.016			.84	2.16	3	3.62
7300	20' to 24' deep, 1 C.Y. excavator	B-12A	254	.063			3.33	3.28	6.61	8.60
7310	1-1/2 C.Y. excavator	B-12B	380	.042			2.23	1.83	4.06	5.35

31 23 Excavation and Fill

31 23 16 – Excavation

31 23 16.13 Excavating, Trench	Crew	Daily Output	Labor-Hours	Unit	Material	Labor	Equipment	Total	Total Incl O&P
							2021 Bare Costs		
7320	2-1/2 C.Y. excavator	B-12S	605	.026	B.C.Y.		1.40	2.59	3.99	4.94
7330	3 C.Y. excavator	B-12D	792	.020			1.07	2.76	3.83	4.62
7340	3-1/2 C.Y. excavator	"	924	.017	↓		.92	2.36	3.28	3.97

31 23 16.14 Excavating, Utility Trench

		Crew	Daily Output	Labor-Hours	Unit	Material	Labor	Equipment	Total	Total Incl O&P
0010	**EXCAVATING, UTILITY TRENCH**									
0011	Common earth									
0050	Trenching with chain trencher, 12 HP, operator walking									
0100	4" wide trench, 12" deep	B-53	800	.010	L.F.		.56	.20	.76	1.05
0150	18" deep		750	.011			.59	.21	.80	1.11
0200	24" deep		700	.011			.63	.23	.86	1.20
0300	6" wide trench, 12" deep		650	.012			.68	.24	.92	1.29
0350	18" deep		600	.013			.74	.26	1	1.39
0400	24" deep		550	.015			.81	.29	1.10	1.52
0450	36" deep		450	.018			.99	.35	1.34	1.86
0600	8" wide trench, 12" deep		475	.017			.93	.33	1.26	1.76
0650	18" deep		400	.020			1.11	.40	1.51	2.09
0700	24" deep		350	.023			1.27	.45	1.72	2.39
0750	36" deep	↓	300	.027	↓		1.48	.53	2.01	2.79
0830	Fly wheel trencher, 18" wide trench, 6' deep, light soil	B-54A	1992	.005	B.C.Y.		.27	.57	.84	1.03
0840	Medium soil		1594	.006			.34	.72	1.06	1.29
0850	Heavy soil	↓	1295	.007			.41	.88	1.29	1.59
0860	24" wide trench, 9' deep, light soil	B-54B	4981	.002			.11	.25	.36	.44
0870	Medium soil		4000	.003			.14	.31	.45	.55
0880	Heavy soil	↓	3237	.003	↓		.17	.38	.55	.68
1000	Backfill by hand including compaction, add									
1050	4" wide trench, 12" deep	A-1G	800	.010	L.F.		.44	.07	.51	.74
1100	18" deep		530	.015			.67	.10	.77	1.11
1150	24" deep		400	.020			.89	.14	1.03	1.48
1300	6" wide trench, 12" deep		540	.015			.66	.10	.76	1.09
1350	18" deep		405	.020			.88	.13	1.01	1.46
1400	24" deep		270	.030			1.32	.20	1.52	2.18
1450	36" deep		180	.044			1.97	.30	2.27	3.27
1600	8" wide trench, 12" deep		400	.020			.89	.14	1.03	1.48
1650	18" deep		265	.030			1.34	.21	1.55	2.23
1700	24" deep		200	.040			1.78	.27	2.05	2.95
1750	36" deep	↓	135	.059	↓		2.63	.40	3.03	4.38
2000	Chain trencher, 40 HP operator riding									
2050	6" wide trench and backfill, 12" deep	B-54	1200	.007	L.F.		.37	.38	.75	.96
2100	18" deep		1000	.008			.44	.45	.89	1.16
2150	24" deep		975	.008			.46	.46	.92	1.19
2200	36" deep		900	.009			.49	.50	.99	1.29
2250	48" deep		750	.011			.59	.60	1.19	1.54
2300	60" deep		650	.012			.68	.69	1.37	1.78
2400	8" wide trench and backfill, 12" deep		1000	.008			.44	.45	.89	1.16
2450	18" deep		950	.008			.47	.47	.94	1.22
2500	24" deep		900	.009			.49	.50	.99	1.29
2550	36" deep		800	.010			.56	.56	1.12	1.45
2600	48" deep		650	.012			.68	.69	1.37	1.78
2700	12" wide trench and backfill, 12" deep		975	.008			.46	.46	.92	1.19
2750	18" deep		860	.009			.52	.52	1.04	1.35
2800	24" deep		800	.010			.56	.56	1.12	1.45
2850	36" deep	↓	725	.011	↓		.61	.62	1.23	1.59

31 23 16.14 Excavating, Utility Trench

	Crew	Daily Output	Labor-Hours	Unit	Material	2021 Bare Costs Labor	Equipment	Total	Total Incl O&P
3000	16" wide trench and backfill, 12" deep	B-54	835	.010	L.F.		.53	.54	1.07	1.38
3050	18" deep		750	.011			.59	.60	1.19	1.54
3100	24" deep		700	.011			.63	.64	1.27	1.66
3200	Compaction with vibratory plate, add								35%	35%
5100	Hand excavate and trim for pipe bells after trench excavation									
5200	8" pipe	1 Clab	155	.052	L.F.		2.29		2.29	3.42
5300	18" pipe	"	130	.062	"		2.73		2.73	4.08
5400	For clay or till, add up to								150%	150%
6000	Excavation utility trench, rock material									
6050	Note: assumption teeth change every 100' of trench									
6100	Chain trencher, 6" wide, 30" depth, rock saw	B-54D	400	.040	L.F.	20.50	2.07	1.09	23.66	27
6200	Chain trencher, 18" wide, 8' depth, rock trencher	B-54E	280	.057	"	33	2.95	3.63	39.58	44.50

31 23 16.15 Excavating, Utility Trench, Plow

	Crew	Daily Output	Labor-Hours	Unit	Material	2021 Bare Costs Labor	Equipment	Total	Total Incl O&P
0010	**EXCAVATING, UTILITY TRENCH, PLOW**									
0100	Single cable, plowed into fine material	B-54C	3800	.004	L.F.		.22	.30	.52	.65
0200	Two cable		3200	.005			.26	.36	.62	.78
0300	Single cable, plowed into coarse material		2000	.008			.41	.57	.98	1.25

31 23 16.16 Structural Excavation for Minor Structures

	Crew	Daily Output	Labor-Hours	Unit	Material	2021 Bare Costs Labor	Equipment	Total	Total Incl O&P
0010	**STRUCTURAL EXCAVATION FOR MINOR STRUCTURES** R312316-40									
0015	Hand, pits to 6' deep, sandy soil	1 Clab	8	1	B.C.Y.		44.50		44.50	66.50
0020	Normal soil	B-2	16	2.500			112		112	167
0030	Medium clay		12	3.333			149		149	223
0040	Heavy clay		8	5			224		224	335
0050	Loose rock		6	6.667			299		299	445
0100	Heavy soil or clay	1 Clab	4	2			89		89	133
0200	Pits to 2' deep, normal soil	B-2	24	1.667			74.50		74.50	111
0210	Sand and gravel		24	1.667			74.50		74.50	111
0220	Medium clay		18	2.222			99.50		99.50	149
0230	Heavy clay		12	3.333			149		149	223
0300	Pits 6' to 12' deep, sandy soil	1 Clab	5	1.600			71		71	106
0500	Heavy soil or clay		3	2.667			118		118	177
0700	Pits 12' to 18' deep, sandy soil		4	2			89		89	133
0900	Heavy soil or clay		2	4			178		178	265
1000	Hand trimming, bottom of excavation	B-2	2400	.017	S.F.		.75		.75	1.11
1010	Slopes and sides		2400	.017	"		.75		.75	1.11
1030	Around obstructions		8	5	B.C.Y.		224		224	335
1100	Hand loading trucks from stock pile, sandy soil	1 Clab	12	.667			29.50		29.50	44
1300	Heavy soil or clay	"	8	1			44.50		44.50	66.50
1500	For wet or muck hand excavation, add to above								50%	50%
1550	Excavation rock by hand/air tool	B-9	3.40	11.765	B.C.Y.		525	105	630	900
6000	Machine excavation, for spread and mat footings, elevator pits,									
6001	and small building foundations									
6030	Common earth, hydraulic backhoe, 1/2 C.Y. bucket	B-12E	55	.291	B.C.Y.		15.40	8.30	23.70	32
6035	3/4 C.Y. bucket	B-12F	90	.178			9.40	7.80	17.20	22.50
6040	1 C.Y. bucket	B-12A	108	.148			7.85	7.70	15.55	20
6050	1-1/2 C.Y. bucket	B-12B	144	.111			5.90	4.83	10.73	14.05
6060	2 C.Y. bucket	B-12C	200	.080			4.23	4.71	8.94	11.50
6070	Sand and gravel, 3/4 C.Y. bucket	B-12F	100	.160			8.45	7	15.45	20.50
6080	1 C.Y. bucket	B-12A	120	.133			7.05	6.95	14	18.15
6090	1-1/2 C.Y. bucket	B-12B	160	.100			5.30	4.35	9.65	12.70
6100	2 C.Y. bucket	B-12C	220	.073			3.85	4.29	8.14	10.45
6110	Clay, till, or blasted rock, 3/4 C.Y. bucket	B-12F	80	.200			10.60	8.75	19.35	25.50

31 23 Excavation and Fill

31 23 16 – Excavation

31 23 16.16 Structural Excavation for Minor Structures

		Crew	Daily Output	Labor-Hours	Unit	Material	2021 Bare Costs Labor	Equipment	Total	Total Incl O&P
6120	1 C.Y. bucket	B-12A	95	.168	B.C.Y.		8.90	8.75	17.65	23
6130	1-1/2 C.Y. bucket	B-12B	130	.123			6.50	5.35	11.85	15.60
6140	2 C.Y. bucket	B-12C	175	.091			4.84	5.40	10.24	13.15
6230	Sandy clay & loam, hydraulic backhoe, 1/2 C.Y. bucket	B-12E	60	.267			14.10	7.60	21.70	29.50
6235	3/4 C.Y. bucket	B-12F	98	.163			8.65	7.15	15.80	21
6240	1 C.Y. bucket	B-12A	116	.138			7.30	7.20	14.50	18.80
6250	1-1/2 C.Y. bucket	B-12B	156	.103			5.45	4.46	9.91	13
6260	2 C.Y. bucket	B-12C	216	.074			3.92	4.36	8.28	10.65
6280	3-1/2 C.Y. bucket	B-12D	300	.053			2.82	7.30	10.12	12.20
9010	For mobilization or demobilization, see Section 01 54 36.50									
9020	For dewatering, see Section 31 23 19.20									
9022	For larger structures, see Bulk Excavation, Section 31 23 16.42									
9024	For loading onto trucks, add								15%	15%
9026	For hauling, see Section 31 23 23.20									
9030	For sheeting or soldier bms/lagging, see Section 31 52 16.10									
9040	For trench excavation of strip ftgs, see Section 31 23 16.13									

31 23 16.26 Rock Removal

0010	**ROCK REMOVAL**									
0015	Drilling only rock, 2" hole for rock bolts	B-47	316	.076	L.F.		3.70	5.15	8.85	11.15
0800	2-1/2" hole for pre-splitting		250	.096			4.68	6.50	11.18	14.15
4600	Quarry operations, 2-1/2" to 3-1/2" diameter		240	.100			4.88	6.80	11.68	14.70

31 23 16.30 Drilling and Blasting Rock

0010	**DRILLING AND BLASTING ROCK**									
0020	Rock, open face, under 1,500 C.Y.	B-47	225	.107	B.C.Y.	4.69	5.20	7.25	17.14	21
0100	Over 1,500 C.Y.		300	.080		4.69	3.90	5.45	14.04	16.95
0200	Areas where blasting mats are required, under 1,500 C.Y.		175	.137		4.69	6.70	9.30	20.69	25.50
0250	Over 1,500 C.Y.		250	.096		4.69	4.68	6.50	15.87	19.30
0300	Bulk drilling and blasting, can vary greatly, average								9.65	12.20
0500	Pits, average								25.50	31.50
1300	Deep hole method, up to 1,500 C.Y.	B-47	50	.480		4.69	23.50	32.50	60.69	76
1400	Over 1,500 C.Y.		66	.364		4.69	17.75	24.50	46.94	58.50
1900	Restricted areas, up to 1,500 C.Y.		13	1.846		4.69	90	125	219.69	277
2000	Over 1,500 C.Y.		20	1.200		4.69	58.50	81.50	144.69	182
2200	Trenches, up to 1,500 C.Y.		22	1.091		13.60	53	74	140.60	176
2300	Over 1,500 C.Y.		26	.923		13.60	45	62.50	121.10	151
2500	Pier holes, up to 1,500 C.Y.		22	1.091		4.69	53	74	131.69	166
2600	Over 1,500 C.Y.		31	.774		4.69	38	52.50	95.19	120
2800	Boulders under 1/2 C.Y., loaded on truck, no hauling	B-100	80	.150			8.10	11.55	19.65	25
2900	Boulders, drilled, blasted	B-47	100	.240		4.69	11.70	16.30	32.69	40.50
3100	Jackhammer operators with foreman compressor, air tools	B-9	1	40	Day		1,800	355	2,155	3,075
3300	Track drill, compressor, operator and foreman	B-47	1	24	"		1,175	1,625	2,800	3,550
3500	Blasting caps				Ea.	6.70			6.70	7.35
3700	Explosives					.55			.55	.61
3800	Blasting mats, for purchase, no mobilization, 10' x 15' x 12"					1,275			1,275	1,400
3900	Blasting mats, rent, for first day					205			205	225
4000	Per added day					68.50			68.50	75.50
4200	Preblast survey for 6 room house, individual lot, minimum	A-6	2.40	6.667			365	14.35	379.35	565
4300	Maximum	"	1.35	11.852			650	25.50	675.50	1,000
4500	City block within zone of influence, minimum	A-8	25200	.001	S.F.		.07		.07	.11
4600	Maximum	"	15100	.002	"		.12		.12	.18
5000	Excavate and load boulders, less than 0.5 C.Y.	B-10T	80	.150	B.C.Y.		8.10	8	16.10	21
5020	0.5 C.Y. to 1 C.Y.	B-10U	100	.120			6.50	9.70	16.20	20.50

For customer support on your Site Work & Landscape Costs with RSMeans Data, call 800.448.8182.

303

31 23 16.30 Drilling and Blasting Rock

		Crew	Daily Output	Labor-Hours	Unit	Material	2021 Bare Costs Labor	2021 Bare Costs Equipment	Total	Total Incl O&P
5200	Excavate and load blasted rock, 3 C.Y. power shovel	B-12T	1530	.010	B.C.Y.		.55	1.36	1.91	2.33
5400	Haul boulders, 25 ton off-highway dump, 1 mile round trip	B-34E	330	.024			1.24	4.32	5.56	6.60
5420	2 mile round trip		275	.029			1.49	5.20	6.69	7.95
5440	3 mile round trip		225	.036			1.82	6.35	8.17	9.75
5460	4 mile round trip		200	.040			2.05	7.15	9.20	10.90
5600	Bury boulders on site, less than 0.5 C.Y., 300 HP dozer									
5620	150' haul	B-10M	310	.039	B.C.Y.		2.10	5.75	7.85	9.45
5640	300' haul		210	.057			3.09	8.50	11.59	13.95
5800	0.5 to 1 C.Y., 300 HP dozer, 150' haul		300	.040			2.17	5.95	8.12	9.80
5820	300' haul		200	.060			3.25	8.95	12.20	14.65

31 23 16.32 Ripping

		Crew	Daily Output	Labor-Hours	Unit	Material	2021 Bare Costs Labor	2021 Bare Costs Equipment	Total	Total Incl O&P
0010	**RIPPING** R312316-40									
0020	Ripping, trap rock, soft, 300 HP dozer, ideal conditions	B-11S	700	.017	B.C.Y.		.93	2.68	3.61	4.33
1500	Adverse conditions		660	.018			.98	2.84	3.82	4.60
1600	Medium hard, 300 HP dozer, ideal conditions		600	.020			1.08	3.13	4.21	5.05
1700	Adverse conditions		540	.022			1.20	3.48	4.68	5.60
2000	Very hard, 410 HP dozer, ideal conditions	B-11T	350	.034			1.86	8.40	10.26	12
2100	Adverse conditions	"	310	.039			2.10	9.50	11.60	13.55
2200	Shale, soft, 300 HP dozer, ideal conditions	B-11S	1500	.008			.43	1.25	1.68	2.03
2300	Adverse conditions	"	1350	.009			.48	1.39	1.87	2.25
2310	Grader rear ripper, 180 HP ideal conditions	B-11J	740	.022			1.12	1.57	2.69	3.40
2320	Adverse conditions	"	630	.025			1.31	1.85	3.16	3.99
2400	Medium hard, 300 HP dozer, ideal conditions	B-11S	1200	.010			.54	1.56	2.10	2.53
2500	Adverse conditions	"	1080	.011			.60	1.74	2.34	2.81
2510	Grader rear ripper, 180 HP ideal conditions	B-11J	625	.026			1.32	1.86	3.18	4.02
2520	Adverse conditions	"	530	.030			1.56	2.20	3.76	4.75
2600	Very hard, 410 HP dozer, ideal conditions	B-11T	800	.015			.81	3.68	4.49	5.25
2700	Adverse conditions	"	720	.017			.90	4.09	4.99	5.85
2800	Till, boulder clay/hardpan, soft, 300 HP dozer, ideal conditions	B-11S	7000	.002			.09	.27	.36	.43
2810	Adverse conditions	"	6300	.002			.10	.30	.40	.48
2815	Grader rear ripper, 180 HP ideal conditions	B-11J	1500	.011			.55	.78	1.33	1.67
2816	Adverse conditions	"	1275	.013			.65	.91	1.56	1.97
2820	Medium hard, 300 HP dozer, ideal conditions	B-11S	6000	.002			.11	.31	.42	.50
2830	Adverse conditions	"	5400	.002			.12	.35	.47	.56
2835	Grader rear ripper, 180 HP ideal conditions	B-11J	740	.022			1.12	1.57	2.69	3.40
2836	Adverse conditions	"	630	.025			1.31	1.85	3.16	3.99
2840	Very hard, 410 HP dozer, ideal conditions	B-11T	5000	.002			.13	.59	.72	.84
2850	Adverse conditions	"	4500	.003			.14	.65	.79	.94
3000	Dozing ripped material, 200 HP, 100' haul	B-10B	700	.017			.93	2.17	3.10	3.77
3050	300' haul	"	250	.048			2.60	6.10	8.70	10.55
3200	300 HP, 100' haul	B-10M	1150	.010			.56	1.55	2.11	2.55
3250	300' haul	"	400	.030			1.62	4.46	6.08	7.35
3400	410 HP, 100' haul	B-10X	1680	.007			.39	1.67	2.06	2.42
3450	300' haul	"	600	.020			1.08	4.68	5.76	6.75

31 23 16.35 Hydraulic Rock Breaking and Loading

0010	**HYDRAULIC ROCK BREAKING AND LOADING**									
0020	Solid rock mass excavation									
0022	Excavator/breaker and excavator/bucket (into trucks)									
0024	Double costs for trenching									
0026	Mobilization costs to be repeated for demobilization									
0100	110 HP excavator w/4,000 ft lb. breaker									
0120	Mobilization 2/20 ton									

31 23 Excavation and Fill

31 23 16 – Excavation

31 23 16.35 Hydraulic Rock Breaking and Loading

		Crew	Daily Output	Labor-Hours	Unit	Material	Labor	Equipment	Total	Total Incl O&P
							2021 Bare Costs			
0130	6,000 psi rock	B-13K	640	.025	B.C.Y.		1.54	3.21	4.75	5.80
0140	12,000 psi rock		240	.067			4.10	8.55	12.65	15.50
0150	18,000 psi rock		80	.200			12.30	25.50	37.80	46.50
0200	170 HP excavator w/5,000 ft lb. breaker									
0220	Mobilization 30 ton & 20 ton									
0240	12,000 psi rock	B-13L	640	.025	B.C.Y.		1.54	3.29	4.83	5.90
0250	18,000 psi rock		240	.067			4.10	8.75	12.85	15.75
0260	24,000 psi rock		80	.200			12.30	26.50	38.80	47.50
0300	270 HP excavator w/8,000 ft lb. breaker									
0320	Mobilization 40 ton & 30 ton									
0350	18,000 psi rock	B-13M	480	.033	B.C.Y.		2.05	6.65	8.70	10.35
0360	24,000 psi rock		240	.067			4.10	13.25	17.35	20.50
0370	30,000 psi rock		80	.200			12.30	40	52.30	62
0400	350 HP excavator w/12,000 ft lb. breaker									
0420	Mobilization 50 ton & 30 ton (may require disassembly & third hauler)									
0450	18,000 psi rock	B-13N	640	.025	B.C.Y.		1.54	5.90	7.44	8.75
0460	24,000 psi rock		400	.040			2.46	9.40	11.86	14
0470	30,000 psi rock		240	.067			4.10	15.70	19.80	23.50

31 23 16.42 Excavating, Bulk Bank Measure

		Crew	Daily Output	Labor-Hours	Unit	Material	Labor	Equipment	Total	Total Incl O&P
0010	**EXCAVATING, BULK BANK MEASURE** R312316-40									
0011	Common earth piled									
0020	For loading onto trucks, add								15%	15%
0050	For mobilization and demobilization, see Section 01 54 36.50 R312316-45									
0100	For hauling, see Section 31 23 23.20									
0200	Excavator, hydraulic, crawler mtd., 1 C.Y. cap. = 100 C.Y./hr.	B-12A	800	.020	B.C.Y.		1.06	1.04	2.10	2.72
0250	1-1/2 C.Y. cap. = 125 C.Y./hr.	B-12B	1000	.016			.85	.70	1.55	2.03
0260	2 C.Y. cap. = 165 C.Y./hr.	B-12C	1320	.012			.64	.71	1.35	1.75
0300	3 C.Y. cap. = 260 C.Y./hr.	B-12D	2080	.008			.41	1.05	1.46	1.77
0305	3-1/2 C.Y. cap. = 300 C.Y./hr.	"	2400	.007			.35	.91	1.26	1.53
0310	Wheel mounted, 1/2 C.Y. cap. = 40 C.Y./hr.	B-12E	320	.050			2.65	1.43	4.08	5.50
0360	3/4 C.Y. cap. = 60 C.Y./hr.	B-12F	480	.033			1.76	1.46	3.22	4.24
0500	Clamshell, 1/2 C.Y. cap. = 20 C.Y./hr.	B-12G	160	.100			5.30	5.50	10.80	13.95
0550	1 C.Y. cap. = 35 C.Y./hr.	B-12H	280	.057			3.02	4.36	7.38	9.30
0950	Dragline, 1/2 C.Y. cap. = 30 C.Y./hr.	B-12I	240	.067			3.53	4.48	8.01	10.20
1000	3/4 C.Y. cap. = 35 C.Y./hr.	"	280	.057			3.02	3.84	6.86	8.75
1050	1-1/2 C.Y. cap. = 65 C.Y./hr.	B-12P	520	.031			1.63	2.49	4.12	5.15
1100	3 C.Y. cap. = 112 C.Y./hr.	B-12V	900	.018			.94	2.27	3.21	3.89
1200	Front end loader, track mtd., 1-1/2 C.Y. cap. = 70 C.Y./hr.	B-10N	560	.021			1.16	1.02	2.18	2.85
1250	2-1/4 C.Y. cap. = 95 C.Y./hr.	B-10O	760	.016			.85	1.22	2.07	2.61
1300	3 C.Y. cap. = 130 C.Y./hr.	B-10P	1040	.012			.62	1.10	1.72	2.14
1350	5 C.Y. cap. = 160 C.Y./hr.	B-10Q	1280	.009			.51	1.14	1.65	2.01
1500	Wheel mounted, 3/4 C.Y. cap. = 45 C.Y./hr.	B-10R	360	.033			1.80	.85	2.65	3.62
1550	1-1/2 C.Y. cap. = 80 C.Y./hr.	B-10S	640	.019			1.01	.69	1.70	2.27
1600	2-1/4 C.Y. cap. = 100 C.Y./hr.	B-10T	800	.015			.81	.80	1.61	2.09
1601	3 C.Y. cap. = 140 C.Y./hr.	"	1120	.011			.58	.57	1.15	1.49
1650	5 C.Y. cap. = 185 C.Y./hr.	B-10U	1480	.008			.44	.65	1.09	1.37
1800	Hydraulic excavator, truck mtd. 1/2 C.Y. = 30 C.Y./hr.	B-12J	240	.067			3.53	3.54	7.07	9.15
1850	48" bucket, 1 C.Y. = 45 C.Y./hr.	B-12K	360	.044			2.35	2.73	5.08	6.50
3700	Shovel, 1/2 C.Y. cap. = 55 C.Y./hr.	B-12L	440	.036			1.92	1.99	3.91	5.05
3750	3/4 C.Y. cap. = 85 C.Y./hr.	B-12M	680	.024			1.25	1.59	2.84	3.61
3800	1 C.Y. cap. = 120 C.Y./hr.	B-12N	960	.017			.88	1.28	2.16	2.73
3850	1-1/2 C.Y. cap. = 160 C.Y./hr.	B-12O	1280	.013			.66	1.03	1.69	2.12

31 23 16.42 Excavating, Bulk Bank Measure	Crew	Daily Output	Labor-Hours	Unit	Material	2021 Bare Costs Labor	Equipment	Total	Total Incl O&P
3900 3 C.Y. cap. = 250 C.Y./hr.	B-12T	2000	.008	B.C.Y.		.42	1.04	1.46	1.78
4000 For soft soil or sand, deduct								15%	15%
4100 For heavy soil or stiff clay, add								60%	60%
4200 For wet excavation with clamshell or dragline, add								100%	100%
4250 All other equipment, add								50%	50%
4400 Clamshell in sheeting or cofferdam, minimum	B-12H	160	.100			5.30	7.65	12.95	16.30
4450 Maximum	"	60	.267			14.10	20.50	34.60	43.50
5000 Excavating, bulk bank measure, sandy clay & loam piled									
5020 For loading onto trucks, add								15%	15%
5100 Excavator, hydraulic, crawler mtd., 1 C.Y. cap. = 120 C.Y./hr.	B-12A	960	.017	B.C.Y.		.88	.87	1.75	2.27
5150 1-1/2 C.Y. cap. = 150 C.Y./hr.	B-12B	1200	.013			.71	.58	1.29	1.69
5300 2 C.Y. cap. = 195 C.Y./hr.	B-12C	1560	.010			.54	.60	1.14	1.47
5400 3 C.Y. cap. = 300 C.Y./hr.	B-12D	2400	.007			.35	.91	1.26	1.53
5500 3-1/2 C.Y. cap. = 350 C.Y./hr.	"	2800	.006			.30	.78	1.08	1.31
5610 Wheel mounted, 1/2 C.Y. cap. = 44 C.Y./hr.	B-12E	352	.045			2.41	1.30	3.71	5
5660 3/4 C.Y. cap. = 66 C.Y./hr.	B-12F	528	.030			1.60	1.33	2.93	3.85
8000 For hauling excavated material, see Section 31 23 23.20									

31 23 16.43 Excavating, Large Volume Projects

	Crew	Daily Output	Labor-Hours	Unit	Material	Labor	Equipment	Total	Total Incl O&P
0010 **EXCAVATING, LARGE VOLUME PROJECTS**, various materials									
0050 Minimum project size 200,000 B.C.Y.									
0080 For hauling, see Section 31 23 23.20									
0100 Unrestricted means continuous loading into hopper or railroad cars									
0200 Loader, 8 C.Y. bucket, 110% fill factor, unrestricted operation	B-14J	5280	.002	L.C.Y.		.12	.43	.55	.66
0250 10 C.Y. bucket	B-14K	7040	.002			.09	.38	.47	.56
0300 8 C.Y. bucket, 105% fill factor	B-14J	5040	.002			.13	.45	.58	.69
0350 10 C.Y. bucket	B-14K	6720	.002			.10	.40	.50	.58
0400 8 C.Y. bucket, 100% fill factor	B-14J	4800	.003			.14	.48	.62	.72
0450 10 C.Y. bucket	B-14K	6400	.002			.10	.42	.52	.62
0500 8 C.Y. bucket, 95% fill factor	B-14J	4560	.003			.14	.50	.64	.76
0550 10 C.Y. bucket	B-14K	6080	.002			.11	.45	.56	.65
0600 8 C.Y. bucket, 90% fill factor	B-14J	4320	.003			.15	.53	.68	.80
0650 10 C.Y. bucket	B-14K	5760	.002			.11	.47	.58	.69
0700 8 C.Y. bucket, 85% fill factor	B-14J	4080	.003			.16	.56	.72	.86
0750 10 C.Y. bucket	B-14K	5440	.002			.12	.50	.62	.73
0800 8 C.Y. bucket, 80% fill factor	B-14J	3840	.003			.17	.60	.77	.90
0850 10 C.Y. bucket	B-14K	5120	.002			.13	.53	.66	.77
0900 8 C.Y. bucket, 75% fill factor	B-14J	3600	.003			.18	.63	.81	.97
0950 10 C.Y. bucket	B-14K	4800	.003			.14	.56	.70	.82
1000 8 C.Y. bucket, 70% fill factor	B-14J	3360	.004			.19	.68	.87	1.04
1050 10 C.Y. bucket	B-14K	4480	.003			.14	.60	.74	.88
1100 8 C.Y. bucket, 65% fill factor	B-14J	3120	.004			.21	.73	.94	1.12
1150 10 C.Y. bucket	B-14K	4160	.003			.16	.65	.81	.95
1200 8 C.Y. bucket, 60% fill factor	B-14J	2880	.004			.23	.79	1.02	1.21
1250 10 C.Y. bucket	B-14K	3840	.003			.17	.70	.87	1.03
2100 Restricted loading trucks									
2200 Loader, 8 C.Y. bucket, 110% fill factor, loading trucks	B-14J	4400	.003	L.C.Y.		.15	.52	.67	.79
2250 10 C.Y. bucket	B-14K	5940	.002			.11	.46	.57	.66
2300 8 C.Y. bucket, 105% fill factor	B-14J	4200	.003			.15	.54	.69	.83
2350 10 C.Y. bucket	B-14K	5670	.002			.11	.48	.59	.69
2400 8 C.Y. bucket, 100% fill factor	B-14J	4000	.003			.16	.57	.73	.87
2450 10 C.Y. bucket	B-14K	5400	.002			.12	.50	.62	.73
2500 8 C.Y. bucket, 95% fill factor	B-14J	3800	.003			.17	.60	.77	.91

31 23 Excavation and Fill

31 23 16 – Excavation

31 23 16.43 Excavating, Large Volume Projects

		Crew	Daily Output	Labor-Hours	Unit	Material	2021 Bare Costs Labor	Equipment	Total	Total Incl O&P
2550	10 C.Y. bucket	B-14K	5130	.002	L.C.Y.		.13	.53	.66	.77
2600	8 C.Y. bucket, 90% fill factor	B-14J	3600	.003			.18	.63	.81	.97
2650	10 C.Y. bucket	B-14K	4860	.002			.13	.56	.69	.81
2700	8 C.Y. bucket, 85% fill factor	B-14J	3400	.004			.19	.67	.86	1.02
2750	10 C.Y. bucket	B-14K	4590	.003			.14	.59	.73	.86
2800	8 C.Y. bucket, 80% fill factor	B-14J	3200	.004			.20	.71	.91	1.09
2850	10 C.Y. bucket	B-14K	4320	.003			.15	.63	.78	.91
2900	8 C.Y. bucket, 75% fill factor	B-14J	3000	.004			.22	.76	.98	1.16
2950	10 C.Y. bucket	B-14K	4050	.003			.16	.67	.83	.97
4000	8 C.Y. bucket, 70% fill factor	B-14J	2800	.004			.23	.82	1.05	1.25
4050	10 C.Y. bucket	B-14K	3780	.003			.17	.72	.89	1.05
4100	8 C.Y. bucket, 65% fill factor	B-14J	2600	.005			.25	.88	1.13	1.34
4150	10 C.Y. bucket	B-14K	3510	.003			.19	.77	.96	1.13
4200	8 C.Y. bucket, 60% fill factor	B-14J	2400	.005			.27	.95	1.22	1.45
4250	10 C.Y. bucket	B-14K	3240	.004	▼		.20	.84	1.04	1.22
4290	Continuous excavation with no truck loading									
4300	Excavator, 4.5 C.Y. bucket, 100% fill factor, no truck loading	B-14A	4000	.003	B.C.Y.		.17	.86	1.03	1.20
4320	6 C.Y. bucket	B-14B	5900	.002			.11	.59	.70	.82
4330	7 C.Y. bucket	B-14C	6900	.002			.10	.50	.60	.69
4340	Shovel, 7 C.Y. bucket	B-14F	8900	.001			.08	.47	.55	.62
4360	12 C.Y. bucket	B-14G	14800	.001			.05	.41	.46	.52
4400	Excavator, 4.5 C.Y. bucket, 95% fill factor	B-14A	3800	.003			.18	.91	1.09	1.26
4420	6 C.Y. bucket	B-14B	5605	.002			.12	.63	.75	.87
4430	7 C.Y. bucket	B-14C	6555	.002			.10	.53	.63	.73
4440	Shovel, 7 C.Y. bucket	B-14F	8455	.001			.08	.49	.57	.66
4460	12 C.Y. bucket	B-14G	14060	.001			.05	.43	.48	.54
4500	Excavator, 4.5 C.Y. bucket, 90% fill factor	B-14A	3600	.003			.19	.96	1.15	1.33
4520	6 C.Y. bucket	B-14B	5310	.002			.13	.66	.79	.92
4530	7 C.Y. bucket	B-14C	6210	.002			.11	.56	.67	.78
4540	Shovel, 7 C.Y. bucket	B-14F	8010	.002			.08	.52	.60	.69
4560	12 C.Y. bucket	B-14G	13320	.001			.05	.45	.50	.57
4600	Excavator, 4.5 C.Y. bucket, 85% fill factor	B-14A	3400	.004			.20	1.01	1.21	1.41
4620	6 C.Y. bucket	B-14B	5015	.002			.13	.70	.83	.97
4630	7 C.Y. bucket	B-14C	5865	.002			.11	.59	.70	.82
4640	Shovel, 7 C.Y. bucket	B-14F	7565	.002			.09	.55	.64	.73
4660	12 C.Y. bucket	B-14G	12580	.001			.05	.48	.53	.61
4700	Excavator, 4.5 C.Y. bucket, 80% fill factor	B-14A	3200	.004			.21	1.08	1.29	1.50
4720	6 C.Y. bucket	B-14B	4720	.003			.14	.74	.88	1.03
4730	7 C.Y. bucket	B-14C	5520	.002			.12	.63	.75	.87
4740	Shovel, 7 C.Y. bucket	B-14F	7120	.002			.09	.58	.67	.78
4760	12 C.Y. bucket	B-14G	11840	.001	▼		.06	.51	.57	.64
5290	Excavation with truck loading									
5300	Excavator, 4.5 C.Y. bucket, 100% fill factor, with truck loading	B-14A	3400	.004	B.C.Y.		.20	1.01	1.21	1.41
5320	6 C.Y. bucket	B-14B	5000	.002			.13	.70	.83	.97
5330	7 C.Y. bucket	B-14C	5800	.002			.12	.60	.72	.83
5340	Shovel, 7 C.Y. bucket	B-14F	7500	.002			.09	.55	.64	.74
5360	12 C.Y. bucket	B-14G	12500	.001			.05	.48	.53	.61
5400	Excavator, 4.5 C.Y. bucket, 95% fill factor	B-14A	3230	.004			.21	1.07	1.28	1.48
5420	6 C.Y. bucket	B-14B	4750	.003			.14	.74	.88	1.02
5430	7 C.Y. bucket	B-14C	5510	.002			.12	.63	.75	.87
5440	Shovel, 7 C.Y. bucket	B-14F	7125	.002			.09	.58	.67	.78
5460	12 C.Y. bucket	B-14G	11875	.001			.06	.51	.57	.64
5500	Excavator, 4.5 C.Y. bucket, 90% fill factor	B-14A	3060	.004	▼		.22	1.13	1.35	1.57

31 23 16.43 Excavating, Large Volume Projects

		Crew	Daily Output	Labor-Hours	Unit	Material	2021 Bare Costs Labor	2021 Bare Costs Equipment	Total	Total Incl O&P
5520	6 C.Y. bucket	B-14B	4500	.003	B.C.Y.		.15	.78	.93	1.08
5530	7 C.Y. bucket	B-14C	5220	.002			.13	.67	.80	.92
5540	Shovel, 7 C.Y. bucket	B-14F	6750	.002			.10	.61	.71	.83
5560	12 C.Y. bucket	B-14G	11250	.001			.06	.54	.60	.68
5600	Excavator, 4.5 C.Y. bucket, 85% fill factor	B-14A	2890	.004			.23	1.19	1.42	1.66
5620	6 C.Y. bucket	B-14B	4250	.003			.16	.82	.98	1.14
5630	7 C.Y. bucket	B-14C	4930	.002			.14	.70	.84	.98
5640	Shovel, 7 C.Y. bucket	B-14F	6375	.002			.10	.65	.75	.88
5660	12 C.Y. bucket	B-14G	10625	.001			.06	.57	.63	.71
5700	Excavator, 4.5 C.Y. bucket, 80% fill factor	B-14A	2720	.004			.25	1.27	1.52	1.77
5720	6 C.Y. bucket	B-14B	4000	.003			.17	.88	1.05	1.21
5740	Shovel, 7 C.Y. bucket	B-14F	6000	.002			.11	.69	.80	.93
5760	12 C.Y. bucket	B-14G	10000	.001	↓		.07	.60	.67	.76
6000	Self propelled scraper, 1/4 push dozer, average productivity									
6100	31 C.Y., 1,500' haul	B-33K	2046	.007	B.C.Y.		.38	2.15	2.53	2.93
6200	44 C.Y.	B-33H	2640	.005	"		.29	2.04	2.33	2.68

31 23 16.46 Excavating, Bulk, Dozer

		Crew	Daily Output	Labor-Hours	Unit	Material	2021 Bare Costs Labor	2021 Bare Costs Equipment	Total	Total Incl O&P
0010	**EXCAVATING, BULK, DOZER**									
0011	Open site									
2000	80 HP, 50' haul, sand & gravel	B-10L	460	.026	B.C.Y.		1.41	.88	2.29	3.07
2010	Sandy clay & loam		440	.027			1.48	.92	2.40	3.21
2020	Common earth		400	.030			1.62	1.01	2.63	3.54
2040	Clay		250	.048			2.60	1.62	4.22	5.65
2200	150' haul, sand & gravel		230	.052			2.82	1.76	4.58	6.15
2210	Sandy clay & loam		220	.055			2.95	1.84	4.79	6.45
2220	Common earth		200	.060			3.25	2.03	5.28	7.05
2240	Clay		125	.096			5.20	3.25	8.45	11.30
2400	300' haul, sand & gravel		120	.100			5.40	3.38	8.78	11.75
2410	Sandy clay & loam		115	.104			5.65	3.53	9.18	12.30
2420	Common earth		100	.120			6.50	4.06	10.56	14.15
2440	Clay	↓	65	.185			10	6.25	16.25	22
3000	105 HP, 50' haul, sand & gravel	B-10W	700	.017			.93	.92	1.85	2.39
3010	Sandy clay & loam		680	.018			.96	.94	1.90	2.46
3020	Common earth		610	.020			1.06	1.05	2.11	2.75
3040	Clay		385	.031			1.69	1.66	3.35	4.34
3200	150' haul, sand & gravel		310	.039			2.10	2.07	4.17	5.40
3210	Sandy clay & loam		300	.040			2.17	2.14	4.31	5.60
3220	Common earth		270	.044			2.41	2.37	4.78	6.20
3240	Clay		170	.071			3.82	3.77	7.59	9.85
3300	300' haul, sand & gravel		140	.086			4.64	4.58	9.22	11.95
3310	Sandy clay & loam		135	.089			4.81	4.75	9.56	12.35
3320	Common earth		120	.100			5.40	5.35	10.75	13.90
3340	Clay	↓	100	.120			6.50	6.40	12.90	16.75
4000	200 HP, 50' haul, sand & gravel	B-10B	1400	.009			.46	1.09	1.55	1.88
4010	Sandy clay & loam		1360	.009			.48	1.12	1.60	1.94
4020	Common earth		1230	.010			.53	1.24	1.77	2.15
4040	Clay		770	.016			.84	1.97	2.81	3.43
4200	150' haul, sand & gravel		595	.020			1.09	2.55	3.64	4.44
4210	Sandy clay & loam		580	.021			1.12	2.62	3.74	4.55
4220	Common earth		516	.023			1.26	2.95	4.21	5.10
4240	Clay		325	.037			2	4.68	6.68	8.15
4400	300' haul, sand & gravel		310	.039			2.10	4.90	7	8.50

31 23 16.46 Excavating, Bulk, Dozer

		Crew	Daily Output	Labor-Hours	Unit	Material	2021 Bare Costs Labor	Equipment	Total	Total Incl O&P
4410	Sandy clay & loam	B-10B	300	.040	B.C.Y.		2.17	5.05	7.22	8.80
4420	Common earth		270	.044			2.41	5.65	8.06	9.80
4440	Clay		170	.071			3.82	8.95	12.77	15.55
5000	300 HP, 50' haul, sand & gravel	B-10M	1900	.006			.34	.94	1.28	1.54
5010	Sandy clay & loam		1850	.006			.35	.96	1.31	1.58
5020	Common earth		1650	.007			.39	1.08	1.47	1.78
5040	Clay		1025	.012			.63	1.74	2.37	2.86
5200	150' haul, sand & gravel		920	.013			.71	1.94	2.65	3.18
5210	Sandy clay & loam		895	.013			.73	1.99	2.72	3.27
5220	Common earth		800	.015			.81	2.23	3.04	3.66
5240	Clay		500	.024			1.30	3.57	4.87	5.85
5400	300' haul, sand & gravel		470	.026			1.38	3.80	5.18	6.25
5410	Sandy clay & loam		455	.026			1.43	3.92	5.35	6.45
5420	Common earth		410	.029			1.58	4.35	5.93	7.15
5440	Clay		250	.048			2.60	7.15	9.75	11.70
5500	460 HP, 50' haul, sand & gravel	B-10X	1930	.006			.34	1.45	1.79	2.10
5506	Sandy clay & loam		1880	.006			.35	1.49	1.84	2.15
5510	Common earth		1680	.007			.39	1.67	2.06	2.42
5520	Clay		1050	.011			.62	2.67	3.29	3.86
5530	150' haul, sand & gravel		1290	.009			.50	2.18	2.68	3.14
5535	Sandy clay & loam		1250	.010			.52	2.25	2.77	3.24
5540	Common earth		1120	.011			.58	2.51	3.09	3.62
5550	Clay		700	.017			.93	4.01	4.94	5.80
5560	300' haul, sand & gravel		660	.018			.98	4.25	5.23	6.15
5565	Sandy clay & loam		640	.019			1.01	4.39	5.40	6.35
5570	Common earth		575	.021			1.13	4.88	6.01	7.05
5580	Clay		350	.034			1.86	8	9.86	11.55
6000	700 HP, 50' haul, sand & gravel	B-10V	3500	.003			.19	1.48	1.67	1.91
6006	Sandy clay & loam		3400	.004			.19	1.52	1.71	1.95
6010	Common earth		3035	.004			.21	1.71	1.92	2.20
6020	Clay		1925	.006			.34	2.69	3.03	3.46
6030	150' haul, sand & gravel		2025	.006			.32	2.56	2.88	3.29
6035	Sandy clay & loam		1960	.006			.33	2.64	2.97	3.39
6040	Common earth		1750	.007			.37	2.96	3.33	3.80
6050	Clay		1100	.011			.59	4.70	5.29	6.10
6060	300' haul, sand & gravel		1030	.012			.63	5	5.63	6.50
6065	Sandy clay & loam		1005	.012			.65	5.15	5.80	6.60
6070	Common earth		900	.013			.72	5.75	6.47	7.45
6080	Clay		550	.022			1.18	9.40	10.58	12.10
6090	For dozer with ripper, see Section 31 23 16.32									

31 23 16.48 Excavation, Bulk, Dragline

		Crew	Daily Output	Labor-Hours	Unit	Material	2021 Bare Costs Labor	Equipment	Total	Total Incl O&P
0010	**EXCAVATION, BULK, DRAGLINE**									
0011	Excavate and load on truck, bank measure									
0012	Bucket drag line, 3/4 C.Y., sand/gravel	B-12I	440	.036	B.C.Y.		1.92	2.44	4.36	5.55
0100	Light clay		310	.052			2.73	3.47	6.20	7.90
0110	Heavy clay		280	.057			3.02	3.84	6.86	8.75
0120	Unclassified soil		250	.064			3.39	4.30	7.69	9.80
0200	1-1/2 C.Y. bucket, sand/gravel	B-12P	575	.028			1.47	2.26	3.73	4.68
0210	Light clay		440	.036			1.92	2.95	4.87	6.10
0220	Heavy clay		352	.045			2.41	3.68	6.09	7.65
0230	Unclassified soil		300	.053			2.82	4.32	7.14	8.95
0300	3 C.Y., sand/gravel	B-12V	720	.022			1.18	2.83	4.01	4.87

31 23 16 – Excavation

31 23 16.48 Excavation, Bulk, Dragline		Crew	Daily Output	Labor-Hours	Unit	Material	2021 Bare Costs Labor	Equipment	Total	Total Incl O&P
0310	Light clay	B-12V	700	.023	B.C.Y.		1.21	2.91	4.12	5
0320	Heavy clay		600	.027			1.41	3.40	4.81	5.85
0330	Unclassified soil	↓	550	.029	↓		1.54	3.71	5.25	6.40

31 23 16.50 Excavation, Bulk, Scrapers		Crew	Daily Output	Labor-Hours	Unit	Material	2021 Bare Costs Labor	Equipment	Total	Total Incl O&P
0010	**EXCAVATION, BULK, SCRAPERS** R312316-40									
0100	Elev. scraper 11 C.Y., sand & gravel 1,500' haul, 1/4 dozer	B-33F	690	.020	B.C.Y.		1.11	2.18	3.29	4.06
0150	3,000' haul R312316-45		610	.023			1.26	2.47	3.73	4.59
0200	5,000' haul		505	.028			1.52	2.98	4.50	5.55
0300	Common earth, 1,500' haul		600	.023			1.28	2.51	3.79	4.67
0350	3,000' haul		530	.026			1.45	2.84	4.29	5.30
0400	5,000' haul		440	.032			1.74	3.42	5.16	6.35
0410	Sandy clay & loam, 1,500' haul		648	.022			1.18	2.32	3.50	4.33
0420	3,000' haul		572	.024			1.34	2.63	3.97	4.89
0430	5,000' haul		475	.029			1.62	3.17	4.79	5.90
0500	Clay, 1,500' haul		375	.037			2.05	4.01	6.06	7.45
0550	3,000' haul		330	.042			2.33	4.56	6.89	8.45
0600	5,000' haul	↓	275	.051	↓		2.79	5.45	8.24	10.15
1000	Self propelled scraper, 14 C.Y., 1/4 push dozer									
1050	Sand and gravel, 1,500' haul	B-33D	920	.015	B.C.Y.		.83	3.12	3.95	4.67
1100	3,000' haul		805	.017			.95	3.57	4.52	5.35
1200	5,000' haul		645	.022			1.19	4.45	5.64	6.65
1300	Common earth, 1,500' haul		800	.018			.96	3.59	4.55	5.40
1350	3,000' haul		700	.020			1.10	4.10	5.20	6.15
1400	5,000' haul		560	.025			1.37	5.15	6.52	7.70
1420	Sandy clay & loam, 1,500' haul		864	.016			.89	3.32	4.21	4.97
1430	3,000' haul		786	.018			.98	3.65	4.63	5.50
1440	5,000' haul		605	.023			1.27	4.74	6.01	7.10
1500	Clay, 1,500' haul		500	.028			1.54	5.75	7.29	8.60
1550	3,000' haul		440	.032			1.74	6.50	8.24	9.80
1600	5,000' haul	↓	350	.040			2.19	8.20	10.39	12.25
2000	21 C.Y., 1/4 push dozer, sand & gravel, 1,500' haul	B-33E	1180	.012			.65	2.63	3.28	3.86
2100	3,000' haul		910	.015			.84	3.41	4.25	5
2200	5,000' haul		750	.019			1.02	4.14	5.16	6.10
2300	Common earth, 1,500' haul		1030	.014			.75	3.01	3.76	4.42
2350	3,000' haul		790	.018			.97	3.93	4.90	5.75
2400	5,000' haul		650	.022			1.18	4.77	5.95	7
2420	Sandy clay & loam, 1,500' haul		1112	.013			.69	2.79	3.48	4.10
2430	3,000' haul		854	.016			.90	3.63	4.53	5.35
2440	5,000' haul		702	.020			1.09	4.42	5.51	6.50
2500	Clay, 1,500' haul		645	.022			1.19	4.81	6	7.05
2550	3,000' haul		495	.028			1.55	6.25	7.80	9.20
2600	5,000' haul	↓	405	.035			1.90	7.65	9.55	11.25
2700	Towed, 10 C.Y., 1/4 push dozer, sand & gravel, 1,500' haul	B-33B	560	.025			1.37	4.27	5.64	6.75
2720	3,000' haul		450	.031			1.71	5.30	7.01	8.40
2730	5,000' haul		365	.038			2.10	6.55	8.65	10.35
2750	Common earth, 1,500' haul		420	.033			1.83	5.70	7.53	8.95
2770	3,000' haul		400	.035			1.92	6	7.92	9.45
2780	5,000' haul		310	.045			2.48	7.70	10.18	12.20
2785	Sandy clay & loam, 1,500' haul		454	.031			1.69	5.25	6.94	8.30
2790	3,000' haul		432	.032			1.78	5.55	7.33	8.75
2795	5,000' haul		340	.041			2.26	7.05	9.31	11.10
2800	Clay, 1,500' haul	↓	315	.044	↓		2.44	7.60	10.04	12

For customer support on your Site Work & Landscape Costs with RSMeans Data, call 800.448.8182.

31 23 Excavation and Fill

31 23 16 – Excavation

31 23 16.50 Excavation, Bulk, Scrapers

		Crew	Daily Output	Labor-Hours	Unit	Material	2021 Bare Costs Labor	Equipment	Total	Total Incl O&P
2820	3,000' haul	B-33B	300	.047	B.C.Y.		2.56	8	10.56	12.55
2840	5,000' haul	▼	225	.062			3.41	10.65	14.06	16.80
2900	15 C.Y., 1/4 push dozer, sand & gravel, 1,500' haul	B-33C	800	.018			.96	3.01	3.97	4.74
2920	3,000' haul		640	.022			1.20	3.77	4.97	5.95
2940	5,000' haul		520	.027			1.48	4.63	6.11	7.30
2960	Common earth, 1,500' haul		600	.023			1.28	4.02	5.30	6.35
2980	3,000' haul		560	.025			1.37	4.30	5.67	6.75
3000	5,000' haul		440	.032			1.74	5.50	7.24	8.65
3005	Sandy clay & loam, 1,500' haul		648	.022			1.18	3.72	4.90	5.85
3010	3,000' haul		605	.023			1.27	3.98	5.25	6.25
3015	5,000' haul		475	.029			1.62	5.05	6.67	8
3020	Clay, 1,500' haul		450	.031			1.71	5.35	7.06	8.45
3040	3,000' haul		420	.033			1.83	5.75	7.58	9
3060	5,000' haul	▼	320	.044	▼		2.40	7.55	9.95	11.85

31 23 19 – Dewatering

31 23 19.10 Cut Drainage Ditch

		Crew	Daily Output	Labor-Hours	Unit	Material	2021 Bare Costs Labor	Equipment	Total	Total Incl O&P
0010	**CUT DRAINAGE DITCH**									
0020	Cut drainage ditch common earth, 30' wide x 1' deep	B-11L	6000	.003	L.F.		.14	.18	.32	.41
0200	Clay and till		4200	.004			.20	.26	.46	.57
0250	Clean wet drainage ditch, 30' wide	▼	10000	.002	▼		.08	.11	.19	.24

31 23 19.20 Dewatering Systems

		Crew	Daily Output	Labor-Hours	Unit	Material	2021 Bare Costs Labor	Equipment	Total	Total Incl O&P
0010	**DEWATERING SYSTEMS**									
0020	Excavate drainage trench, 2' wide, 2' deep	B-11C	90	.178	C.Y.		9.20	2.40	11.60	16.35
0100	2' wide, 3' deep, with backhoe loader	"	135	.119			6.15	1.60	7.75	10.90
0200	Excavate sump pits by hand, light soil	1 Clab	7.10	1.127			50		50	74.50
0300	Heavy soil	"	3.50	2.286	▼		101		101	151
0500	Pumping 8 hrs., attended 2 hrs./day, incl. 20 L.F.									
0550	of suction hose & 100 L.F. discharge hose									
0600	2" diaphragm pump used for 8 hrs.	B-10H	4	3	Day		162	25	187	270
0620	Add per additional pump							72	72	79
0650	4" diaphragm pump used for 8 hrs.	B-10I	4	3			162	37.50	199.50	283
0670	Add per additional pump							116	116	127
0800	8 hrs. attended, 2" diaphragm pump	B-10H	1	12			650	99	749	1,075
0820	Add per additional pump							72	72	79
0900	3" centrifugal pump	B-10J	1	12			650	92	742	1,075
0920	Add per additional pump							79	79	86.50
1000	4" diaphragm pump	B-10I	1	12			650	149	799	1,125
1020	Add per additional pump							116	116	127
1100	6" centrifugal pump	B-10K	1	12			650	297	947	1,300
1120	Add per additional pump				▼			340	340	375
1300	CMP, incl. excavation 3' deep, 12" diameter	B-6	115	.209	L.F.	12	10.05	1.88	23.93	30
1400	18" diameter		100	.240	"	18.75	11.55	2.16	32.46	40
1600	Sump hole construction, incl. excavation and gravel, pit		1250	.019	C.F.	1.12	.92	.17	2.21	2.81
1700	With 12" gravel collar, 12" pipe, corrugated, 16 ga.		70	.343	L.F.	23.50	16.50	3.09	43.09	54
1800	15" pipe, corrugated, 16 ga.		55	.436		31	21	3.93	55.93	70
1900	18" pipe, corrugated, 16 ga.		50	.480		35.50	23	4.32	62.82	79
2000	24" pipe, corrugated, 14 ga.		40	.600	▼	43	29	5.40	77.40	96
2200	Wood lining, up to 4' x 4', add	▼	300	.080	SFCA	15.95	3.85	.72	20.52	24
9950	See Section 31 23 19.40 for wellpoints									
9960	See Section 31 23 19.30 for deep well systems									
9970	See Section 22 11 23 for pumps									

31 23 19 – Dewatering

31 23 19.30 Wells

		Crew	Daily Output	Labor-Hours	Unit	Material	2021 Bare Costs Labor	Equipment	Total	Total Incl O&P
0010	**WELLS**									
0011	For dewatering 10' to 20' deep, 2' diameter									
0020	with steel casing, minimum	B-6	165	.145	V.L.F.	37.50	7	1.31	45.81	53.50
0050	Average		98	.245		46	11.80	2.21	60.01	70.50
0100	Maximum	↓	49	.490	↓	52	23.50	4.41	79.91	97.50
0300	For dewatering pumps see 01 54 33 in Reference Section									
0500	For domestic water wells, see Section 33 21 13.10									

31 23 19.40 Wellpoints

		Crew	Daily Output	Labor-Hours	Unit	Material	2021 Bare Costs Labor	Equipment	Total	Total Incl O&P
0010	**WELLPOINTS** R312319-90									
0011	For equipment rental, see 01 54 33 in Reference Section									
0100	Installation and removal of single stage system									
0110	Labor only, 0.75 labor-hours per L.F.	1 Clab	10.70	.748	LF Hdr		33		33	49.50
0200	2.0 labor-hours per L.F.	"	4	2	"		89		89	133
0400	Pump operation, 4 @ 6 hr. shifts									
0410	Per 24 hr. day	4 Eqlt	1.27	25.197	Day		1,400		1,400	2,075
0500	Per 168 hr. week, 160 hr. straight, 8 hr. double time		.18	178	Week		9,875		9,875	14,700
0550	Per 4.3 week month	↓	.04	800	Month		44,400		44,400	66,000
0600	Complete installation, operation, equipment rental, fuel &									
0610	removal of system with 2" wellpoints 5' OC									
0700	100' long header, 6" diameter, first month	4 Eqlt	3.23	9.907	LF Hdr	160	550		710	995
0800	Thereafter, per month		4.13	7.748		128	430		558	780
1000	200' long header, 8" diameter, first month		6	5.333		153	296		449	610
1100	Thereafter, per month		8.39	3.814		72	212		284	395
1300	500' long header, 8" diameter, first month		10.63	3.010		56	167		223	310
1400	Thereafter, per month		20.91	1.530		40	85		125	171
1600	1,000' long header, 10" diameter, first month		11.62	2.754		48	153		201	281
1700	Thereafter, per month	↓	41.81	.765	↓	24	42.50		66.50	90
1900	Note: above figures include pumping 168 hrs. per week,									
1910	the pump operator, and one stand-by pump.									

31 23 23 – Fill

31 23 23.13 Backfill

		Crew	Daily Output	Labor-Hours	Unit	Material	2021 Bare Costs Labor	Equipment	Total	Total Incl O&P
0010	**BACKFILL** R312323-30									
0015	By hand, no compaction, light soil	1 Clab	14	.571	L.C.Y.		25.50		25.50	38
0100	Heavy soil		11	.727	"		32.50		32.50	48
0300	Compaction in 6" layers, hand tamp, add to above	↓	20.60	.388	E.C.Y.		17.25		17.25	25.50
0400	Roller compaction operator walking, add	B-10A	100	.120			6.50	1.67	8.17	11.55
0500	Air tamp, add	B-9D	190	.211			9.45	1.73	11.18	15.95
0600	Vibrating plate, add	A-1D	60	.133			5.90	.53	6.43	9.45
0800	Compaction in 12" layers, hand tamp, add to above	1 Clab	34	.235			10.45		10.45	15.60
0900	Roller compaction operator walking, add	B-10A	150	.080			4.33	1.11	5.44	7.65
1000	Air tamp, add	B-9	285	.140			6.30	1.25	7.55	10.75
1100	Vibrating plate, add	A-1E	90	.089	↓		3.95	1.84	5.79	7.90

31 23 23.14 Backfill, Structural

		Crew	Daily Output	Labor-Hours	Unit	Material	2021 Bare Costs Labor	Equipment	Total	Total Incl O&P
0010	**BACKFILL, STRUCTURAL**									
0011	Dozer or F.E. loader									
0020	From existing stockpile, no compaction									
1000	55 HP wheeled loader, 50' haul, common earth	B-11C	200	.080	L.C.Y.		4.14	1.08	5.22	7.35
2000	80 HP, 50' haul, sand & gravel	B-10L	1100	.011			.59	.37	.96	1.29
2010	Sandy clay & loam		1070	.011			.61	.38	.99	1.32
2020	Common earth		975	.012			.67	.42	1.09	1.45
2040	Clay	↓	850	.014	↓		.76	.48	1.24	1.67

31 23 23 – Fill

31 23 23.14 Backfill, Structural		Crew	Daily Output	Labor-Hours	Unit	Material	2021 Bare Costs Labor	Equipment	Total	Total Incl O&P
2200	150' haul, sand & gravel	B-10L	550	.022	L.C.Y.		1.18	.74	1.92	2.57
2210	Sandy clay & loam		535	.022			1.21	.76	1.97	2.64
2220	Common earth		490	.024			1.33	.83	2.16	2.89
2240	Clay		425	.028			1.53	.95	2.48	3.33
2400	300' haul, sand & gravel		370	.032			1.76	1.10	2.86	3.83
2410	Sandy clay & loam		360	.033			1.80	1.13	2.93	3.93
2420	Common earth		330	.036			1.97	1.23	3.20	4.28
2440	Clay		290	.041			2.24	1.40	3.64	4.88
3000	105 HP, 50' haul, sand & gravel	B-10W	1350	.009			.48	.47	.95	1.24
3010	Sandy clay & loam		1325	.009			.49	.48	.97	1.26
3020	Common earth		1225	.010			.53	.52	1.05	1.37
3040	Clay		1100	.011			.59	.58	1.17	1.52
3200	150' haul, sand & gravel		670	.018			.97	.96	1.93	2.50
3210	Sandy clay & loam		655	.018			.99	.98	1.97	2.56
3220	Common earth		610	.020			1.06	1.05	2.11	2.75
3240	Clay		550	.022			1.18	1.17	2.35	3.04
3300	300' haul, sand & gravel		465	.026			1.40	1.38	2.78	3.60
3310	Sandy clay & loam		455	.026			1.43	1.41	2.84	3.68
3320	Common earth		415	.029			1.57	1.54	3.11	4.03
3340	Clay		370	.032			1.76	1.73	3.49	4.53
4000	200 HP, 50' haul, sand & gravel	B-10B	2500	.005			.26	.61	.87	1.06
4010	Sandy clay & loam		2435	.005			.27	.62	.89	1.09
4020	Common earth		2200	.005			.30	.69	.99	1.20
4040	Clay		1950	.006			.33	.78	1.11	1.36
4200	150' haul, sand & gravel		1225	.010			.53	1.24	1.77	2.15
4210	Sandy clay & loam		1200	.010			.54	1.27	1.81	2.20
4220	Common earth		1100	.011			.59	1.38	1.97	2.40
4240	Clay		975	.012			.67	1.56	2.23	2.70
4400	300' haul, sand & gravel		805	.015			.81	1.89	2.70	3.28
4410	Sandy clay & loam		790	.015			.82	1.92	2.74	3.35
4420	Common earth		735	.016			.88	2.07	2.95	3.59
4440	Clay		660	.018			.98	2.30	3.28	4
5000	300 HP, 50' haul, sand & gravel	B-10M	3170	.004			.20	.56	.76	.93
5010	Sandy clay & loam		3110	.004			.21	.57	.78	.94
5020	Common earth		2900	.004			.22	.62	.84	1.01
5040	Clay		2700	.004			.24	.66	.90	1.09
5200	150' haul, sand & gravel		2200	.005			.30	.81	1.11	1.33
5210	Sandy clay & loam		2150	.006			.30	.83	1.13	1.36
5220	Common earth		1950	.006			.33	.92	1.25	1.51
5240	Clay		1700	.007			.38	1.05	1.43	1.73
5400	300' haul, sand & gravel		1500	.008			.43	1.19	1.62	1.96
5410	Sandy clay & loam		1470	.008			.44	1.21	1.65	2
5420	Common earth		1350	.009			.48	1.32	1.80	2.17
5440	Clay		1225	.010			.53	1.46	1.99	2.39
6000	For compaction, see Section 31 23 23.23									
6010	For trench backfill, see Sections 31 23 16.13 and 31 23 16.14									

31 23 23.15 Borrow, Loading and/or Spreading

		Crew	Daily Output	Labor-Hours	Unit	Material	Labor	Equipment	Total	Total Incl O&P
0010	**BORROW, LOADING AND/OR SPREADING**									
4000	Common earth, shovel, 1 C.Y. bucket	B-12N	840	.019	B.C.Y.	18.25	1.01	1.46	20.72	23
4010	1-1/2 C.Y. bucket	B-12O	1135	.014		18.25	.75	1.16	20.16	22.50
4020	3 C.Y. bucket	B-12T	1800	.009		18.25	.47	1.16	19.88	22
4030	Front end loader, wheel mounted									

31 23 23.15 Borrow, Loading and/or Spreading	Crew	Daily Output	Labor- Hours	Unit	Material	2021 Bare Costs Labor	Equipment	Total	Total Incl O&P
4050 3/4 C.Y. bucket	B-10R	550	.022	B.C.Y.	18.25	1.18	.56	19.99	22.50
4060 1-1/2 C.Y. bucket	B-10S	970	.012		18.25	.67	.46	19.38	21.50
4070 3 C.Y. bucket	B-10T	1575	.008		18.25	.41	.41	19.07	21
4080 5 C.Y. bucket	B-10U	2600	.005		18.25	.25	.37	18.87	21
5000 Select granular fill, shovel, 1 C.Y. bucket	B-12N	925	.017		30	.92	1.33	32.25	36
5010 1-1/2 C.Y. bucket	B-12O	1250	.013		30	.68	1.06	31.74	35
5020 3 C.Y. bucket	B-12T	1980	.008		30	.43	1.05	31.48	35
5030 Front end loader, wheel mounted									
5050 3/4 C.Y. bucket	B-10R	800	.015	B.C.Y.	30	.81	.38	31.19	34.50
5060 1-1/2 C.Y. bucket	B-10S	1065	.011		30	.61	.41	31.02	34.50
5070 3 C.Y. bucket	B-10T	1735	.007		30	.37	.37	30.74	34
5080 5 C.Y. bucket	B-10U	2850	.004		30	.23	.34	30.57	33.50
6000 Clay, till, or blasted rock, shovel, 1 C.Y. bucket	B-12N	715	.022		13.50	1.18	1.72	16.40	18.50
6010 1-1/2 C.Y. bucket	B-12O	965	.017		13.50	.88	1.37	15.75	17.65
6020 3 C.Y. bucket	B-12T	1530	.010		13.50	.55	1.36	15.41	17.20
6030 Front end loader, wheel mounted									
6035 3/4 C.Y. bucket	B-10R	465	.026	B.C.Y.	13.50	1.40	.66	15.56	17.65
6040 1-1/2 C.Y. bucket	B-10S	825	.015		13.50	.79	.53	14.82	16.60
6045 3 C.Y. bucket	B-10T	1340	.009		13.50	.48	.48	14.46	16.10
6050 5 C.Y. bucket	B-10U	2200	.005		13.50	.30	.44	14.24	15.75
6060 Front end loader, track mounted									
6065 1-1/2 C.Y. bucket	B-10N	715	.017	B.C.Y.	13.50	.91	.80	15.21	17.10
6070 3 C.Y. bucket	B-10P	1190	.010		13.50	.55	.96	15.01	16.70
6075 5 C.Y. bucket	B-10Q	1835	.007		13.50	.35	.79	14.64	16.25
7000 Topsoil or loam from stockpile, shovel, 1 C.Y. bucket	B-12N	840	.019		27	1.01	1.46	29.47	33
7010 1-1/2 C.Y. bucket	B-12O	1135	.014		27	.75	1.16	28.91	32.50
7020 3 C.Y. bucket	B-12T	1800	.009		27	.47	1.16	28.63	32
7030 Front end loader, wheel mounted									
7050 3/4 C.Y. bucket	B-10R	550	.022	B.C.Y.	27	1.18	.56	28.74	32.50
7060 1-1/2 C.Y. bucket	B-10S	970	.012		27	.67	.46	28.13	31.50
7070 3 C.Y. bucket	B-10T	1575	.008		27	.41	.41	27.82	31
7080 5 C.Y. bucket	B-10U	2600	.005		27	.25	.37	27.62	31
9000 Hauling only, excavated or borrow material, see Section 31 23 23.20									

31 23 23.16 Fill By Borrow and Utility Bedding

	Crew	Daily Output	Labor- Hours	Unit	Material	Labor	Equipment	Total	Total Incl O&P
0010 **FILL BY BORROW AND UTILITY BEDDING**									
0049 Utility bedding, for pipe & conduit, not incl. compaction									
0050 Crushed or screened bank run gravel	B-6	150	.160	L.C.Y.	33.50	7.70	1.44	42.64	49.50
0100 Crushed stone 3/4" to 1/2"		150	.160		25.50	7.70	1.44	34.64	41.50
0200 Sand, dead or bank		150	.160		18.75	7.70	1.44	27.89	33.50
0500 Compacting bedding in trench	A-1D	90	.089	E.C.Y.		3.95	.35	4.30	6.30
0600 If material source exceeds 2 miles, add for extra mileage.									
0610 See Section 31 23 23.20 for hauling mileage add.									

31 23 23.17 General Fill

	Crew	Daily Output	Labor- Hours	Unit	Material	Labor	Equipment	Total	Total Incl O&P
0010 **GENERAL FILL**									
0011 Spread dumped material, no compaction									
0020 By dozer	B-10B	1000	.012	L.C.Y.		.65	1.52	2.17	2.64
0100 By hand	1 Clab	12	.667	"		29.50		29.50	44
0150 Spread fill, from stockpile with 2-1/2 C.Y. F.E. loader									
0170 130 HP, 300' haul	B-10P	600	.020	L.C.Y.		1.08	1.91	2.99	3.71
0190 With dozer 300 HP, 300' haul	B-10M	600	.020	"		1.08	2.98	4.06	4.88
0400 For compaction of embankment, see Section 31 23 23.23									
0500 Gravel fill, compacted, under floor slabs, 4" deep	B-37	10000	.005	S.F.	.60	.22	.03	.85	1.01

314

For customer support on your Site Work & Landscape Costs with RSMeans Data, call 800.448.8182.

31 23 Excavation and Fill

31 23 23 – Fill

31 23 23.17 General Fill

		Crew	Daily Output	Labor-Hours	Unit	Material	2021 Bare Costs Labor	2021 Bare Costs Equipment	Total	Total Incl O&P
0600	6" deep	B-37	8600	.006	S.F.	.89	.26	.03	1.18	1.40
0700	9" deep		7200	.007		1.49	.31	.04	1.84	2.14
0800	12" deep		6000	.008	↓	2.08	.37	.04	2.49	2.90
1000	Alternate pricing method, 4" deep		120	.400	E.C.Y.	44.50	18.65	2.16	65.31	79.50
1100	6" deep		160	.300		44.50	13.95	1.62	60.07	72
1200	9" deep		200	.240		44.50	11.20	1.29	56.99	67
1300	12" deep	↓	220	.218	↓	44.50	10.15	1.18	55.83	65.50
1500	For fill under exterior paving, see Section 32 11 23.23									
1600	For flowable fill, see Section 03 31 13.35									

31 23 23.20 Hauling

		Crew	Daily Output	Labor-Hours	Unit	Material	2021 Bare Costs Labor	2021 Bare Costs Equipment	Total	Total Incl O&P
0010	**HAULING**									
0011	Excavated or borrow, loose cubic yards									
0012	no loading equipment, including hauling, waiting, loading/dumping									
0013	time per cycle (wait, load, travel, unload or dump & return)									
0014	8 C.Y. truck, 15 MPH avg., cycle 0.5 miles, 10 min. wait/ld./uld.	B-34A	320	.025	L.C.Y.		1.28	1.27	2.55	3.32
0016	cycle 1 mile		272	.029			1.51	1.50	3.01	3.91
0018	cycle 2 miles		208	.038			1.97	1.96	3.93	5.10
0020	cycle 4 miles		144	.056			2.85	2.83	5.68	7.35
0022	cycle 6 miles		112	.071			3.66	3.64	7.30	9.50
0024	cycle 8 miles		88	.091			4.66	4.63	9.29	12.05
0026	20 MPH avg., cycle 0.5 mile		336	.024			1.22	1.21	2.43	3.16
0028	cycle 1 mile		296	.027			1.39	1.38	2.77	3.58
0030	cycle 2 miles		240	.033			1.71	1.70	3.41	4.43
0032	cycle 4 miles		176	.045			2.33	2.32	4.65	6.05
0034	cycle 6 miles		136	.059			3.02	3	6.02	7.80
0036	cycle 8 miles		112	.071			3.66	3.64	7.30	9.50
0044	25 MPH avg., cycle 4 miles		192	.042			2.14	2.12	4.26	5.55
0046	cycle 6 miles		160	.050			2.57	2.55	5.12	6.65
0048	cycle 8 miles		128	.063			3.21	3.18	6.39	8.30
0050	30 MPH avg., cycle 4 miles		216	.037			1.90	1.89	3.79	4.92
0052	cycle 6 miles		176	.045			2.33	2.32	4.65	6.05
0054	cycle 8 miles		144	.056			2.85	2.83	5.68	7.35
0114	15 MPH avg., cycle 0.5 mile, 15 min. wait/ld./uld.		224	.036			1.83	1.82	3.65	4.74
0116	cycle 1 mile		200	.040			2.05	2.04	4.09	5.30
0118	cycle 2 miles		168	.048			2.44	2.43	4.87	6.30
0120	cycle 4 miles		120	.067			3.42	3.40	6.82	8.85
0122	cycle 6 miles		96	.083			4.28	4.25	8.53	11.05
0124	cycle 8 miles		80	.100			5.15	5.10	10.25	13.25
0126	20 MPH avg., cycle 0.5 mile		232	.034			1.77	1.76	3.53	4.57
0128	cycle 1 mile		208	.038			1.97	1.96	3.93	5.10
0130	cycle 2 miles		184	.043			2.23	2.22	4.45	5.75
0132	cycle 4 miles		144	.056			2.85	2.83	5.68	7.35
0134	cycle 6 miles		112	.071			3.66	3.64	7.30	9.50
0136	cycle 8 miles		96	.083			4.28	4.25	8.53	11.05
0144	25 MPH avg., cycle 4 miles		152	.053			2.70	2.68	5.38	7
0146	cycle 6 miles		128	.063			3.21	3.18	6.39	8.30
0148	cycle 8 miles		112	.071			3.66	3.64	7.30	9.50
0150	30 MPH avg., cycle 4 miles		168	.048			2.44	2.43	4.87	6.30
0152	cycle 6 miles		144	.056			2.85	2.83	5.68	7.35
0154	cycle 8 miles		120	.067			3.42	3.40	6.82	8.85
0214	15 MPH avg., cycle 0.5 mile, 20 min. wait/ld./uld.		176	.045			2.33	2.32	4.65	6.05
0216	cycle 1 mile	↓	160	.050	↓		2.57	2.55	5.12	6.65

315

31 23 23 – Fill

	31 23 23.20 Hauling	Crew	Daily Output	Labor-Hours	Unit	Material	2021 Bare Costs Labor	Equipment	Total	Total Incl O&P
0218	cycle 2 miles	B-34A	136	.059	L.C.Y.		3.02	3	6.02	7.80
0220	cycle 4 miles		104	.077			3.95	3.92	7.87	10.20
0222	cycle 6 miles		88	.091			4.66	4.63	9.29	12.05
0224	cycle 8 miles		72	.111			5.70	5.65	11.35	14.75
0226	20 MPH avg., cycle 0.5 mile		176	.045			2.33	2.32	4.65	6.05
0228	cycle 1 mile		168	.048			2.44	2.43	4.87	6.30
0230	cycle 2 miles		144	.056			2.85	2.83	5.68	7.35
0232	cycle 4 miles		120	.067			3.42	3.40	6.82	8.85
0234	cycle 6 miles		96	.083			4.28	4.25	8.53	11.05
0236	cycle 8 miles		88	.091			4.66	4.63	9.29	12.05
0244	25 MPH avg., cycle 4 miles		128	.063			3.21	3.18	6.39	8.30
0246	cycle 6 miles		112	.071			3.66	3.64	7.30	9.50
0248	cycle 8 miles		96	.083			4.28	4.25	8.53	11.05
0250	30 MPH avg., cycle 4 miles		136	.059			3.02	3	6.02	7.80
0252	cycle 6 miles		120	.067			3.42	3.40	6.82	8.85
0254	cycle 8 miles		104	.077			3.95	3.92	7.87	10.20
0314	15 MPH avg., cycle 0.5 mile, 25 min. wait/ld./uld.		144	.056			2.85	2.83	5.68	7.35
0316	cycle 1 mile		128	.063			3.21	3.18	6.39	8.30
0318	cycle 2 miles		112	.071			3.66	3.64	7.30	9.50
0320	cycle 4 miles		96	.083			4.28	4.25	8.53	11.05
0322	cycle 6 miles		80	.100			5.15	5.10	10.25	13.25
0324	cycle 8 miles		64	.125			6.40	6.35	12.75	16.60
0326	20 MPH avg., cycle 0.5 mile		144	.056			2.85	2.83	5.68	7.35
0328	cycle 1 mile		136	.059			3.02	3	6.02	7.80
0330	cycle 2 miles		120	.067			3.42	3.40	6.82	8.85
0332	cycle 4 miles		104	.077			3.95	3.92	7.87	10.20
0334	cycle 6 miles		88	.091			4.66	4.63	9.29	12.05
0336	cycle 8 miles		80	.100			5.15	5.10	10.25	13.25
0344	25 MPH avg., cycle 4 miles		112	.071			3.66	3.64	7.30	9.50
0346	cycle 6 miles		96	.083			4.28	4.25	8.53	11.05
0348	cycle 8 miles		88	.091			4.66	4.63	9.29	12.05
0350	30 MPH avg., cycle 4 miles		112	.071			3.66	3.64	7.30	9.50
0352	cycle 6 miles		104	.077			3.95	3.92	7.87	10.20
0354	cycle 8 miles		96	.083			4.28	4.25	8.53	11.05
0414	15 MPH avg., cycle 0.5 mile, 30 min. wait/ld./uld.		120	.067			3.42	3.40	6.82	8.85
0416	cycle 1 mile		112	.071			3.66	3.64	7.30	9.50
0418	cycle 2 miles		96	.083			4.28	4.25	8.53	11.05
0420	cycle 4 miles		80	.100			5.15	5.10	10.25	13.25
0422	cycle 6 miles		72	.111			5.70	5.65	11.35	14.75
0424	cycle 8 miles		64	.125			6.40	6.35	12.75	16.60
0426	20 MPH avg., cycle 0.5 mile		120	.067			3.42	3.40	6.82	8.85
0428	cycle 1 mile		112	.071			3.66	3.64	7.30	9.50
0430	cycle 2 miles		104	.077			3.95	3.92	7.87	10.20
0432	cycle 4 miles		88	.091			4.66	4.63	9.29	12.05
0434	cycle 6 miles		80	.100			5.15	5.10	10.25	13.25
0436	cycle 8 miles		72	.111			5.70	5.65	11.35	14.75
0444	25 MPH avg., cycle 4 miles		96	.083			4.28	4.25	8.53	11.05
0446	cycle 6 miles		88	.091			4.66	4.63	9.29	12.05
0448	cycle 8 miles		80	.100			5.15	5.10	10.25	13.25
0450	30 MPH avg., cycle 4 miles		96	.083			4.28	4.25	8.53	11.05
0452	cycle 6 miles		88	.091			4.66	4.63	9.29	12.05
0454	cycle 8 miles		80	.100			5.15	5.10	10.25	13.25
0514	15 MPH avg., cycle 0.5 mile, 35 min. wait/ld./uld.		104	.077			3.95	3.92	7.87	10.20

31 23 23.20 Hauling		Crew	Daily Output	Labor-Hours	Unit	Material	2021 Bare Costs		Total	Total Incl O&P
							Labor	Equipment		
0516	cycle 1 mile	B-34A	96	.083	L.C.Y.		4.28	4.25	8.53	11.05
0518	cycle 2 miles		88	.091			4.66	4.63	9.29	12.05
0520	cycle 4 miles		72	.111			5.70	5.65	11.35	14.75
0522	cycle 6 miles		64	.125			6.40	6.35	12.75	16.60
0524	cycle 8 miles		56	.143			7.35	7.30	14.65	18.95
0526	20 MPH avg., cycle 0.5 mile		104	.077			3.95	3.92	7.87	10.20
0528	cycle 1 mile		96	.083			4.28	4.25	8.53	11.05
0530	cycle 2 miles		96	.083			4.28	4.25	8.53	11.05
0532	cycle 4 miles		80	.100			5.15	5.10	10.25	13.25
0534	cycle 6 miles		72	.111			5.70	5.65	11.35	14.75
0536	cycle 8 miles		64	.125			6.40	6.35	12.75	16.60
0544	25 MPH avg., cycle 4 miles		88	.091			4.66	4.63	9.29	12.05
0546	cycle 6 miles		80	.100			5.15	5.10	10.25	13.25
0548	cycle 8 miles		72	.111			5.70	5.65	11.35	14.75
0550	30 MPH avg., cycle 4 miles		88	.091			4.66	4.63	9.29	12.05
0552	cycle 6 miles		80	.100			5.15	5.10	10.25	13.25
0554	cycle 8 miles		72	.111			5.70	5.65	11.35	14.75
1014	12 C.Y. truck, cycle 0.5 mile, 15 MPH avg., 15 min. wait/ld./uld.	B-34B	336	.024			1.22	1.72	2.94	3.73
1016	cycle 1 mile		300	.027			1.37	1.93	3.30	4.17
1018	cycle 2 miles		252	.032			1.63	2.30	3.93	4.96
1020	cycle 4 miles		180	.044			2.28	3.22	5.50	6.95
1022	cycle 6 miles		144	.056			2.85	4.02	6.87	8.70
1024	cycle 8 miles		120	.067			3.42	4.83	8.25	10.40
1025	cycle 10 miles		96	.083			4.28	6.05	10.33	13.05
1026	20 MPH avg., cycle 0.5 mile		348	.023			1.18	1.66	2.84	3.59
1028	cycle 1 mile		312	.026			1.32	1.86	3.18	4.01
1030	cycle 2 miles		276	.029			1.49	2.10	3.59	4.53
1032	cycle 4 miles		216	.037			1.90	2.68	4.58	5.80
1034	cycle 6 miles		168	.048			2.44	3.45	5.89	7.45
1036	cycle 8 miles		144	.056			2.85	4.02	6.87	8.70
1038	cycle 10 miles		120	.067			3.42	4.83	8.25	10.40
1040	25 MPH avg., cycle 4 miles		228	.035			1.80	2.54	4.34	5.50
1042	cycle 6 miles		192	.042			2.14	3.02	5.16	6.50
1044	cycle 8 miles		168	.048			2.44	3.45	5.89	7.45
1046	cycle 10 miles		144	.056			2.85	4.02	6.87	8.70
1050	30 MPH avg., cycle 4 miles		252	.032			1.63	2.30	3.93	4.96
1052	cycle 6 miles		216	.037			1.90	2.68	4.58	5.80
1054	cycle 8 miles		180	.044			2.28	3.22	5.50	6.95
1056	cycle 10 miles		156	.051			2.63	3.71	6.34	8
1060	35 MPH avg., cycle 4 miles		264	.030			1.55	2.19	3.74	4.73
1062	cycle 6 miles		228	.035			1.80	2.54	4.34	5.50
1064	cycle 8 miles		204	.039			2.01	2.84	4.85	6.15
1066	cycle 10 miles		180	.044			2.28	3.22	5.50	6.95
1068	cycle 20 miles		120	.067			3.42	4.83	8.25	10.40
1069	cycle 30 miles		84	.095			4.89	6.90	11.79	14.90
1070	cycle 40 miles		72	.111			5.70	8.05	13.75	17.35
1072	40 MPH avg., cycle 6 miles		240	.033			1.71	2.41	4.12	5.20
1074	cycle 8 miles		216	.037			1.90	2.68	4.58	5.80
1076	cycle 10 miles		192	.042			2.14	3.02	5.16	6.50
1078	cycle 20 miles		120	.067			3.42	4.83	8.25	10.40
1080	cycle 30 miles		96	.083			4.28	6.05	10.33	13.05
1082	cycle 40 miles		72	.111			5.70	8.05	13.75	17.35
1084	cycle 50 miles		60	.133			6.85	9.65	16.50	21

31 23 23.20 Hauling		Crew	Daily Output	Labor-Hours	Unit	Material	2021 Bare Costs Labor	Equipment	Total	Total Incl O&P
1094	45 MPH avg., cycle 8 miles	B-34B	216	.037	L.C.Y.		1.90	2.68	4.58	5.80
1096	cycle 10 miles		204	.039			2.01	2.84	4.85	6.15
1098	cycle 20 miles		132	.061			3.11	4.39	7.50	9.50
1100	cycle 30 miles		108	.074			3.80	5.35	9.15	11.60
1102	cycle 40 miles		84	.095			4.89	6.90	11.79	14.90
1104	cycle 50 miles		72	.111			5.70	8.05	13.75	17.35
1106	50 MPH avg., cycle 10 miles		216	.037			1.90	2.68	4.58	5.80
1108	cycle 20 miles		144	.056			2.85	4.02	6.87	8.70
1110	cycle 30 miles		108	.074			3.80	5.35	9.15	11.60
1112	cycle 40 miles		84	.095			4.89	6.90	11.79	14.90
1114	cycle 50 miles		72	.111			5.70	8.05	13.75	17.35
1214	15 MPH avg., cycle 0.5 mile, 20 min. wait/ld./uld.		264	.030			1.55	2.19	3.74	4.73
1216	cycle 1 mile		240	.033			1.71	2.41	4.12	5.20
1218	cycle 2 miles		204	.039			2.01	2.84	4.85	6.15
1220	cycle 4 miles		156	.051			2.63	3.71	6.34	8
1222	cycle 6 miles		132	.061			3.11	4.39	7.50	9.50
1224	cycle 8 miles		108	.074			3.80	5.35	9.15	11.60
1225	cycle 10 miles		96	.083			4.28	6.05	10.33	13.05
1226	20 MPH avg., cycle 0.5 mile		264	.030			1.55	2.19	3.74	4.73
1228	cycle 1 mile		252	.032			1.63	2.30	3.93	4.96
1230	cycle 2 miles		216	.037			1.90	2.68	4.58	5.80
1232	cycle 4 miles		180	.044			2.28	3.22	5.50	6.95
1234	cycle 6 miles		144	.056			2.85	4.02	6.87	8.70
1236	cycle 8 miles		132	.061			3.11	4.39	7.50	9.50
1238	cycle 10 miles		108	.074			3.80	5.35	9.15	11.60
1240	25 MPH avg., cycle 4 miles		192	.042			2.14	3.02	5.16	6.50
1242	cycle 6 miles		168	.048			2.44	3.45	5.89	7.45
1244	cycle 8 miles		144	.056			2.85	4.02	6.87	8.70
1246	cycle 10 miles		132	.061			3.11	4.39	7.50	9.50
1250	30 MPH avg., cycle 4 miles		204	.039			2.01	2.84	4.85	6.15
1252	cycle 6 miles		180	.044			2.28	3.22	5.50	6.95
1254	cycle 8 miles		156	.051			2.63	3.71	6.34	8
1256	cycle 10 miles		144	.056			2.85	4.02	6.87	8.70
1260	35 MPH avg., cycle 4 miles		216	.037			1.90	2.68	4.58	5.80
1262	cycle 6 miles		192	.042			2.14	3.02	5.16	6.50
1264	cycle 8 miles		168	.048			2.44	3.45	5.89	7.45
1266	cycle 10 miles		156	.051			2.63	3.71	6.34	8
1268	cycle 20 miles		108	.074			3.80	5.35	9.15	11.60
1269	cycle 30 miles		72	.111			5.70	8.05	13.75	17.35
1270	cycle 40 miles		60	.133			6.85	9.65	16.50	21
1272	40 MPH avg., cycle 6 miles		192	.042			2.14	3.02	5.16	6.50
1274	cycle 8 miles		180	.044			2.28	3.22	5.50	6.95
1276	cycle 10 miles		156	.051			2.63	3.71	6.34	8
1278	cycle 20 miles		108	.074			3.80	5.35	9.15	11.60
1280	cycle 30 miles		84	.095			4.89	6.90	11.79	14.90
1282	cycle 40 miles		72	.111			5.70	8.05	13.75	17.35
1284	cycle 50 miles		60	.133			6.85	9.65	16.50	21
1294	45 MPH avg., cycle 8 miles		180	.044			2.28	3.22	5.50	6.95
1296	cycle 10 miles		168	.048			2.44	3.45	5.89	7.45
1298	cycle 20 miles		120	.067			3.42	4.83	8.25	10.40
1300	cycle 30 miles		96	.083			4.28	6.05	10.33	13.05
1302	cycle 40 miles		72	.111			5.70	8.05	13.75	17.35
1304	cycle 50 miles		60	.133			6.85	9.65	16.50	21

31 23 23.20 Hauling		Crew	Daily Output	Labor-Hours	Unit	Material	2021 Bare Costs Labor	Equipment	Total	Total Incl O&P
1306	50 MPH avg., cycle 10 miles	B-34B	180	.044	L.C.Y.		2.28	3.22	5.50	6.95
1308	cycle 20 miles		132	.061			3.11	4.39	7.50	9.50
1310	cycle 30 miles		96	.083			4.28	6.05	10.33	13.05
1312	cycle 40 miles		84	.095			4.89	6.90	11.79	14.90
1314	cycle 50 miles		72	.111			5.70	8.05	13.75	17.35
1414	15 MPH avg., cycle 0.5 mile, 25 min. wait/ld./uld.		204	.039			2.01	2.84	4.85	6.15
1416	cycle 1 mile		192	.042			2.14	3.02	5.16	6.50
1418	cycle 2 miles		168	.048			2.44	3.45	5.89	7.45
1420	cycle 4 miles		132	.061			3.11	4.39	7.50	9.50
1422	cycle 6 miles		120	.067			3.42	4.83	8.25	10.40
1424	cycle 8 miles		96	.083			4.28	6.05	10.33	13.05
1425	cycle 10 miles		84	.095			4.89	6.90	11.79	14.90
1426	20 MPH avg., cycle 0.5 mile		216	.037			1.90	2.68	4.58	5.80
1428	cycle 1 mile		204	.039			2.01	2.84	4.85	6.15
1430	cycle 2 miles		180	.044			2.28	3.22	5.50	6.95
1432	cycle 4 miles		156	.051			2.63	3.71	6.34	8
1434	cycle 6 miles		132	.061			3.11	4.39	7.50	9.50
1436	cycle 8 miles		120	.067			3.42	4.83	8.25	10.40
1438	cycle 10 miles		96	.083			4.28	6.05	10.33	13.05
1440	25 MPH avg., cycle 4 miles		168	.048			2.44	3.45	5.89	7.45
1442	cycle 6 miles		144	.056			2.85	4.02	6.87	8.70
1444	cycle 8 miles		132	.061			3.11	4.39	7.50	9.50
1446	cycle 10 miles		108	.074			3.80	5.35	9.15	11.60
1450	30 MPH avg., cycle 4 miles		168	.048			2.44	3.45	5.89	7.45
1452	cycle 6 miles		156	.051			2.63	3.71	6.34	8
1454	cycle 8 miles		132	.061			3.11	4.39	7.50	9.50
1456	cycle 10 miles		120	.067			3.42	4.83	8.25	10.40
1460	35 MPH avg., cycle 4 miles		180	.044			2.28	3.22	5.50	6.95
1462	cycle 6 miles		156	.051			2.63	3.71	6.34	8
1464	cycle 8 miles		144	.056			2.85	4.02	6.87	8.70
1466	cycle 10 miles		132	.061			3.11	4.39	7.50	9.50
1468	cycle 20 miles		96	.083			4.28	6.05	10.33	13.05
1469	cycle 30 miles		72	.111			5.70	8.05	13.75	17.35
1470	cycle 40 miles		60	.133			6.85	9.65	16.50	21
1472	40 MPH avg., cycle 6 miles		168	.048			2.44	3.45	5.89	7.45
1474	cycle 8 miles		156	.051			2.63	3.71	6.34	8
1476	cycle 10 miles		144	.056			2.85	4.02	6.87	8.70
1478	cycle 20 miles		96	.083			4.28	6.05	10.33	13.05
1480	cycle 30 miles		84	.095			4.89	6.90	11.79	14.90
1482	cycle 40 miles		60	.133			6.85	9.65	16.50	21
1484	cycle 50 miles		60	.133			6.85	9.65	16.50	21
1494	45 MPH avg., cycle 8 miles		156	.051			2.63	3.71	6.34	8
1496	cycle 10 miles		144	.056			2.85	4.02	6.87	8.70
1498	cycle 20 miles		108	.074			3.80	5.35	9.15	11.60
1500	cycle 30 miles		84	.095			4.89	6.90	11.79	14.90
1502	cycle 40 miles		72	.111			5.70	8.05	13.75	17.35
1504	cycle 50 miles		60	.133			6.85	9.65	16.50	21
1506	50 MPH avg., cycle 10 miles		156	.051			2.63	3.71	6.34	8
1508	cycle 20 miles		120	.067			3.42	4.83	8.25	10.40
1510	cycle 30 miles		96	.083			4.28	6.05	10.33	13.05
1512	cycle 40 miles		72	.111			5.70	8.05	13.75	17.35
1514	cycle 50 miles		60	.133			6.85	9.65	16.50	21
1614	15 MPH avg., cycle 0.5 mile, 30 min. wait/ld./uld.		180	.044			2.28	3.22	5.50	6.95

319

For customer support on your Site Work & Landscape Costs with RSMeans Data, call 800.448.8182.

31 23 23.20 Hauling		Crew	Daily Output	Labor-Hours	Unit	Material	2021 Bare Costs Labor	2021 Bare Costs Equipment	Total	Total Incl O&P
1616	cycle 1 mile	B-34B	168	.048	L.C.Y.		2.44	3.45	5.89	7.45
1618	cycle 2 miles		144	.056			2.85	4.02	6.87	8.70
1620	cycle 4 miles		120	.067			3.42	4.83	8.25	10.40
1622	cycle 6 miles		108	.074			3.80	5.35	9.15	11.60
1624	cycle 8 miles		84	.095			4.89	6.90	11.79	14.90
1625	cycle 10 miles		84	.095			4.89	6.90	11.79	14.90
1626	20 MPH avg., cycle 0.5 mile		180	.044			2.28	3.22	5.50	6.95
1628	cycle 1 mile		168	.048			2.44	3.45	5.89	7.45
1630	cycle 2 miles		156	.051			2.63	3.71	6.34	8
1632	cycle 4 miles		132	.061			3.11	4.39	7.50	9.50
1634	cycle 6 miles		120	.067			3.42	4.83	8.25	10.40
1636	cycle 8 miles		108	.074			3.80	5.35	9.15	11.60
1638	cycle 10 miles		96	.083			4.28	6.05	10.33	13.05
1640	25 MPH avg., cycle 4 miles		144	.056			2.85	4.02	6.87	8.70
1642	cycle 6 miles		132	.061			3.11	4.39	7.50	9.50
1644	cycle 8 miles		108	.074			3.80	5.35	9.15	11.60
1646	cycle 10 miles		108	.074			3.80	5.35	9.15	11.60
1650	30 MPH avg., cycle 4 miles		144	.056			2.85	4.02	6.87	8.70
1652	cycle 6 miles		132	.061			3.11	4.39	7.50	9.50
1654	cycle 8 miles		120	.067			3.42	4.83	8.25	10.40
1656	cycle 10 miles		108	.074			3.80	5.35	9.15	11.60
1660	35 MPH avg., cycle 4 miles		156	.051			2.63	3.71	6.34	8
1662	cycle 6 miles		144	.056			2.85	4.02	6.87	8.70
1664	cycle 8 miles		132	.061			3.11	4.39	7.50	9.50
1666	cycle 10 miles		120	.067			3.42	4.83	8.25	10.40
1668	cycle 20 miles		84	.095			4.89	6.90	11.79	14.90
1669	cycle 30 miles		72	.111			5.70	8.05	13.75	17.35
1670	cycle 40 miles		60	.133			6.85	9.65	16.50	21
1672	40 MPH avg., cycle 6 miles		144	.056			2.85	4.02	6.87	8.70
1674	cycle 8 miles		132	.061			3.11	4.39	7.50	9.50
1676	cycle 10 miles		120	.067			3.42	4.83	8.25	10.40
1678	cycle 20 miles		96	.083			4.28	6.05	10.33	13.05
1680	cycle 30 miles		72	.111			5.70	8.05	13.75	17.35
1682	cycle 40 miles		60	.133			6.85	9.65	16.50	21
1684	cycle 50 miles		48	.167			8.55	12.05	20.60	26
1694	45 MPH avg., cycle 8 miles		144	.056			2.85	4.02	6.87	8.70
1696	cycle 10 miles		132	.061			3.11	4.39	7.50	9.50
1698	cycle 20 miles		96	.083			4.28	6.05	10.33	13.05
1700	cycle 30 miles		84	.095			4.89	6.90	11.79	14.90
1702	cycle 40 miles		60	.133			6.85	9.65	16.50	21
1704	cycle 50 miles		60	.133			6.85	9.65	16.50	21
1706	50 MPH avg., cycle 10 miles		132	.061			3.11	4.39	7.50	9.50
1708	cycle 20 miles		108	.074			3.80	5.35	9.15	11.60
1710	cycle 30 miles		84	.095			4.89	6.90	11.79	14.90
1712	cycle 40 miles		72	.111			5.70	8.05	13.75	17.35
1714	cycle 50 miles		60	.133			6.85	9.65	16.50	21
2000	Hauling, 8 C.Y. truck, small project cost per hour	B-34A	8	1	Hr.		51.50	51	102.50	133
2100	12 C.Y. truck	B-34B	8	1			51.50	72.50	124	156
2150	16.5 C.Y. truck	B-34C	8	1			51.50	79.50	131	164
2175	18 C.Y. 8 wheel truck	B-34I	8	1			51.50	94	145.50	181
2200	20 C.Y. truck	B-34D	8	1			51.50	81.50	133	167
2300	Grading at dump, or embankment if required, by dozer	B-10B	1000	.012	L.C.Y.		.65	1.52	2.17	2.64
2310	Spotter at fill or cut, if required	1 Clab	8	1	Hr.		44.50		44.50	66.50

31 23 23 - Fill

31 23 23.20 Hauling		Crew	Daily Output	Labor-Hours	Unit	Material	2021 Bare Costs Labor	2021 Bare Costs Equipment	Total	Total Incl O&P
2500	Dust control, light	B-59	1	8	Day		410	465	875	1,125
2510	Heavy	"	.50	16			820	930	1,750	2,250
2600	Haul road maintenance	B-86A	1	8			470	1,075	1,545	1,875
3014	16.5 C.Y. truck, 15 min. wait/ld./uld., 15 MPH, cycle 0.5 mile	B-34C	462	.017	L.C.Y.		.89	1.38	2.27	2.85
3016	cycle 1 mile		413	.019			.99	1.54	2.53	3.19
3018	cycle 2 miles		347	.023			1.18	1.84	3.02	3.79
3020	cycle 4 miles		248	.032			1.65	2.57	4.22	5.30
3022	cycle 6 miles		198	.040			2.07	3.22	5.29	6.65
3024	cycle 8 miles		165	.048			2.49	3.86	6.35	7.95
3025	cycle 10 miles		132	.061			3.11	4.83	7.94	9.95
3026	20 MPH avg., cycle 0.5 mile		479	.017			.86	1.33	2.19	2.74
3028	cycle 1 mile		429	.019			.96	1.49	2.45	3.06
3030	cycle 2 miles		380	.021			1.08	1.68	2.76	3.46
3032	cycle 4 miles		281	.028			1.46	2.27	3.73	4.68
3034	cycle 6 miles		231	.035			1.78	2.76	4.54	5.70
3036	cycle 8 miles		198	.040			2.07	3.22	5.29	6.65
3038	cycle 10 miles		165	.048			2.49	3.86	6.35	7.95
3040	25 MPH avg., cycle 4 miles		314	.025			1.31	2.03	3.34	4.18
3042	cycle 6 miles		264	.030			1.55	2.41	3.96	4.98
3044	cycle 8 miles		231	.035			1.78	2.76	4.54	5.70
3046	cycle 10 miles		198	.040			2.07	3.22	5.29	6.65
3050	30 MPH avg., cycle 4 miles		347	.023			1.18	1.84	3.02	3.79
3052	cycle 6 miles		281	.028			1.46	2.27	3.73	4.68
3054	cycle 8 miles		248	.032			1.65	2.57	4.22	5.30
3056	cycle 10 miles		215	.037			1.91	2.97	4.88	6.10
3060	35 MPH avg., cycle 4 miles		363	.022			1.13	1.76	2.89	3.62
3062	cycle 6 miles		314	.025			1.31	2.03	3.34	4.18
3064	cycle 8 miles		264	.030			1.55	2.41	3.96	4.98
3066	cycle 10 miles		248	.032			1.65	2.57	4.22	5.30
3068	cycle 20 miles		149	.054			2.75	4.28	7.03	8.85
3070	cycle 30 miles		116	.069			3.54	5.50	9.04	11.35
3072	cycle 40 miles		83	.096			4.94	7.70	12.64	15.85
3074	40 MPH avg., cycle 6 miles		330	.024			1.24	1.93	3.17	3.99
3076	cycle 8 miles		281	.028			1.46	2.27	3.73	4.68
3078	cycle 10 miles		264	.030			1.55	2.41	3.96	4.98
3080	cycle 20 miles		165	.048			2.49	3.86	6.35	7.95
3082	cycle 30 miles		132	.061			3.11	4.83	7.94	9.95
3084	cycle 40 miles		99	.081			4.15	6.45	10.60	13.30
3086	cycle 50 miles		83	.096			4.94	7.70	12.64	15.85
3094	45 MPH avg., cycle 8 miles		297	.027			1.38	2.15	3.53	4.43
3096	cycle 10 miles		281	.028			1.46	2.27	3.73	4.68
3098	cycle 20 miles		182	.044			2.25	3.50	5.75	7.20
3100	cycle 30 miles		132	.061			3.11	4.83	7.94	9.95
3102	cycle 40 miles		116	.069			3.54	5.50	9.04	11.35
3104	cycle 50 miles		99	.081			4.15	6.45	10.60	13.30
3106	50 MPH avg., cycle 10 miles		281	.028			1.46	2.27	3.73	4.68
3108	cycle 20 miles		198	.040			2.07	3.22	5.29	6.65
3110	cycle 30 miles		149	.054			2.75	4.28	7.03	8.85
3112	cycle 40 miles		116	.069			3.54	5.50	9.04	11.35
3114	cycle 50 miles		99	.081			4.15	6.45	10.60	13.30
3214	20 min. wait/ld./uld., 15 MPH, cycle 0.5 mile		363	.022			1.13	1.76	2.89	3.62
3216	cycle 1 mile		330	.024			1.24	1.93	3.17	3.99
3218	cycle 2 miles		281	.028			1.46	2.27	3.73	4.68

31 23 23.20 Hauling		Crew	Daily Output	Labor-Hours	Unit	Material	2021 Bare Costs Labor	2021 Bare Costs Equipment	Total	Total Incl O&P
3220	cycle 4 miles	B-34C	215	.037	L.C.Y.		1.91	2.97	4.88	6.10
3222	cycle 6 miles		182	.044			2.25	3.50	5.75	7.20
3224	cycle 8 miles		149	.054			2.75	4.28	7.03	8.85
3225	cycle 10 miles		132	.061			3.11	4.83	7.94	9.95
3226	20 MPH avg., cycle 0.5 mile		363	.022			1.13	1.76	2.89	3.62
3228	cycle 1 mile		347	.023			1.18	1.84	3.02	3.79
3230	cycle 2 miles		297	.027			1.38	2.15	3.53	4.43
3232	cycle 4 miles		248	.032			1.65	2.57	4.22	5.30
3234	cycle 6 miles		198	.040			2.07	3.22	5.29	6.65
3236	cycle 8 miles		182	.044			2.25	3.50	5.75	7.20
3238	cycle 10 miles		149	.054			2.75	4.28	7.03	8.85
3240	25 MPH avg., cycle 4 miles		264	.030			1.55	2.41	3.96	4.98
3242	cycle 6 miles		231	.035			1.78	2.76	4.54	5.70
3244	cycle 8 miles		198	.040			2.07	3.22	5.29	6.65
3246	cycle 10 miles		182	.044			2.25	3.50	5.75	7.20
3250	30 MPH avg., cycle 4 miles		281	.028			1.46	2.27	3.73	4.68
3252	cycle 6 miles		248	.032			1.65	2.57	4.22	5.30
3254	cycle 8 miles		215	.037			1.91	2.97	4.88	6.10
3256	cycle 10 miles		198	.040			2.07	3.22	5.29	6.65
3260	35 MPH avg., cycle 4 miles		297	.027			1.38	2.15	3.53	4.43
3262	cycle 6 miles		264	.030			1.55	2.41	3.96	4.98
3264	cycle 8 miles		231	.035			1.78	2.76	4.54	5.70
3266	cycle 10 miles		215	.037			1.91	2.97	4.88	6.10
3268	cycle 20 miles		149	.054			2.75	4.28	7.03	8.85
3270	cycle 30 miles		99	.081			4.15	6.45	10.60	13.30
3272	cycle 40 miles		83	.096			4.94	7.70	12.64	15.85
3274	40 MPH avg., cycle 6 miles		264	.030			1.55	2.41	3.96	4.98
3276	cycle 8 miles		248	.032			1.65	2.57	4.22	5.30
3278	cycle 10 miles		215	.037			1.91	2.97	4.88	6.10
3280	cycle 20 miles		149	.054			2.75	4.28	7.03	8.85
3282	cycle 30 miles		116	.069			3.54	5.50	9.04	11.35
3284	cycle 40 miles		99	.081			4.15	6.45	10.60	13.30
3286	cycle 50 miles		83	.096			4.94	7.70	12.64	15.85
3294	45 MPH avg., cycle 8 miles		248	.032			1.65	2.57	4.22	5.30
3296	cycle 10 miles		231	.035			1.78	2.76	4.54	5.70
3298	cycle 20 miles		165	.048			2.49	3.86	6.35	7.95
3300	cycle 30 miles		132	.061			3.11	4.83	7.94	9.95
3302	cycle 40 miles		99	.081			4.15	6.45	10.60	13.30
3304	cycle 50 miles		83	.096			4.94	7.70	12.64	15.85
3306	50 MPH avg., cycle 10 miles		248	.032			1.65	2.57	4.22	5.30
3308	cycle 20 miles		182	.044			2.25	3.50	5.75	7.20
3310	cycle 30 miles		132	.061			3.11	4.83	7.94	9.95
3312	cycle 40 miles		116	.069			3.54	5.50	9.04	11.35
3314	cycle 50 miles		99	.081			4.15	6.45	10.60	13.30
3414	25 min. wait/ld./uld., 15 MPH, cycle 0.5 mile		281	.028			1.46	2.27	3.73	4.68
3416	cycle 1 mile		264	.030			1.55	2.41	3.96	4.98
3418	cycle 2 miles		231	.035			1.78	2.76	4.54	5.70
3420	cycle 4 miles		182	.044			2.25	3.50	5.75	7.20
3422	cycle 6 miles		165	.048			2.49	3.86	6.35	7.95
3424	cycle 8 miles		132	.061			3.11	4.83	7.94	9.95
3425	cycle 10 miles		116	.069			3.54	5.50	9.04	11.35
3426	20 MPH avg., cycle 0.5 mile		297	.027			1.38	2.15	3.53	4.43
3428	cycle 1 mile		281	.028			1.46	2.27	3.73	4.68

31 23 23.20 Hauling		Crew	Daily Output	Labor-Hours	Unit	Material	2021 Bare Costs Labor	2021 Bare Costs Equipment	Total	Total Incl O&P
3430	cycle 2 miles	B-34C	248	.032	L.C.Y.		1.65	2.57	4.22	5.30
3432	cycle 4 miles		215	.037			1.91	2.97	4.88	6.10
3434	cycle 6 miles		182	.044			2.25	3.50	5.75	7.20
3436	cycle 8 miles		165	.048			2.49	3.86	6.35	7.95
3438	cycle 10 miles		132	.061			3.11	4.83	7.94	9.95
3440	25 MPH avg., cycle 4 miles		231	.035			1.78	2.76	4.54	5.70
3442	cycle 6 miles		198	.040			2.07	3.22	5.29	6.65
3444	cycle 8 miles		182	.044			2.25	3.50	5.75	7.20
3446	cycle 10 miles		165	.048			2.49	3.86	6.35	7.95
3450	30 MPH avg., cycle 4 miles		231	.035			1.78	2.76	4.54	5.70
3452	cycle 6 miles		215	.037			1.91	2.97	4.88	6.10
3454	cycle 8 miles		182	.044			2.25	3.50	5.75	7.20
3456	cycle 10 miles		165	.048			2.49	3.86	6.35	7.95
3460	35 MPH avg., cycle 4 miles		248	.032			1.65	2.57	4.22	5.30
3462	cycle 6 miles		215	.037			1.91	2.97	4.88	6.10
3464	cycle 8 miles		198	.040			2.07	3.22	5.29	6.65
3466	cycle 10 miles		182	.044			2.25	3.50	5.75	7.20
3468	cycle 20 miles		132	.061			3.11	4.83	7.94	9.95
3470	cycle 30 miles		99	.081			4.15	6.45	10.60	13.30
3472	cycle 40 miles		83	.096			4.94	7.70	12.64	15.85
3474	40 MPH, cycle 6 miles		231	.035			1.78	2.76	4.54	5.70
3476	cycle 8 miles		215	.037			1.91	2.97	4.88	6.10
3478	cycle 10 miles		198	.040			2.07	3.22	5.29	6.65
3480	cycle 20 miles		132	.061			3.11	4.83	7.94	9.95
3482	cycle 30 miles		116	.069			3.54	5.50	9.04	11.35
3484	cycle 40 miles		83	.096			4.94	7.70	12.64	15.85
3486	cycle 50 miles		83	.096			4.94	7.70	12.64	15.85
3494	45 MPH avg., cycle 8 miles		215	.037			1.91	2.97	4.88	6.10
3496	cycle 10 miles		198	.040			2.07	3.22	5.29	6.65
3498	cycle 20 miles		149	.054			2.75	4.28	7.03	8.85
3500	cycle 30 miles		116	.069			3.54	5.50	9.04	11.35
3502	cycle 40 miles		99	.081			4.15	6.45	10.60	13.30
3504	cycle 50 miles		83	.096			4.94	7.70	12.64	15.85
3506	50 MPH avg., cycle 10 miles		215	.037			1.91	2.97	4.88	6.10
3508	cycle 20 miles		165	.048			2.49	3.86	6.35	7.95
3510	cycle 30 miles		132	.061			3.11	4.83	7.94	9.95
3512	cycle 40 miles		99	.081			4.15	6.45	10.60	13.30
3514	cycle 50 miles		83	.096			4.94	7.70	12.64	15.85
3614	30 min. wait/ld./uld., 15 MPH, cycle 0.5 mile		248	.032			1.65	2.57	4.22	5.30
3616	cycle 1 mile		231	.035			1.78	2.76	4.54	5.70
3618	cycle 2 miles		198	.040			2.07	3.22	5.29	6.65
3620	cycle 4 miles		165	.048			2.49	3.86	6.35	7.95
3622	cycle 6 miles		149	.054			2.75	4.28	7.03	8.85
3624	cycle 8 miles		116	.069			3.54	5.50	9.04	11.35
3625	cycle 10 miles		116	.069			3.54	5.50	9.04	11.35
3626	20 MPH avg., cycle 0.5 mile		248	.032			1.65	2.57	4.22	5.30
3628	cycle 1 mile		231	.035			1.78	2.76	4.54	5.70
3630	cycle 2 miles		215	.037			1.91	2.97	4.88	6.10
3632	cycle 4 miles		182	.044			2.25	3.50	5.75	7.20
3634	cycle 6 miles		165	.048			2.49	3.86	6.35	7.95
3636	cycle 8 miles		149	.054			2.75	4.28	7.03	8.85
3638	cycle 10 miles		132	.061			3.11	4.83	7.94	9.95
3640	25 MPH avg., cycle 4 miles		198	.040			2.07	3.22	5.29	6.65

31 23 23.20 Hauling		Crew	Daily Output	Labor-Hours	Unit	Material	2021 Bare Costs Labor	2021 Bare Costs Equipment	Total	Total Incl O&P
3642	cycle 6 miles	B-34C	182	.044	L.C.Y.		2.25	3.50	5.75	7.20
3644	cycle 8 miles		149	.054			2.75	4.28	7.03	8.85
3646	cycle 10 miles		149	.054			2.75	4.28	7.03	8.85
3650	30 MPH avg., cycle 4 miles		198	.040			2.07	3.22	5.29	6.65
3652	cycle 6 miles		182	.044			2.25	3.50	5.75	7.20
3654	cycle 8 miles		165	.048			2.49	3.86	6.35	7.95
3656	cycle 10 miles		149	.054			2.75	4.28	7.03	8.85
3660	35 MPH avg., cycle 4 miles		215	.037			1.91	2.97	4.88	6.10
3662	cycle 6 miles		198	.040			2.07	3.22	5.29	6.65
3664	cycle 8 miles		182	.044			2.25	3.50	5.75	7.20
3666	cycle 10 miles		165	.048			2.49	3.86	6.35	7.95
3668	cycle 20 miles		116	.069			3.54	5.50	9.04	11.35
3670	cycle 30 miles		99	.081			4.15	6.45	10.60	13.30
3672	cycle 40 miles		83	.096			4.94	7.70	12.64	15.85
3674	40 MPH, cycle 6 miles		198	.040			2.07	3.22	5.29	6.65
3676	cycle 8 miles		182	.044			2.25	3.50	5.75	7.20
3678	cycle 10 miles		165	.048			2.49	3.86	6.35	7.95
3680	cycle 20 miles		132	.061			3.11	4.83	7.94	9.95
3682	cycle 30 miles		99	.081			4.15	6.45	10.60	13.30
3684	cycle 40 miles		83	.096			4.94	7.70	12.64	15.85
3686	cycle 50 miles		66	.121			6.20	9.65	15.85	19.95
3694	45 MPH avg., cycle 8 miles		198	.040			2.07	3.22	5.29	6.65
3696	cycle 10 miles		182	.044			2.25	3.50	5.75	7.20
3698	cycle 20 miles		132	.061			3.11	4.83	7.94	9.95
3700	cycle 30 miles		116	.069			3.54	5.50	9.04	11.35
3702	cycle 40 miles		83	.096			4.94	7.70	12.64	15.85
3704	cycle 50 miles		83	.096			4.94	7.70	12.64	15.85
3706	50 MPH avg., cycle 10 miles		182	.044			2.25	3.50	5.75	7.20
3708	cycle 20 miles		149	.054			2.75	4.28	7.03	8.85
3710	cycle 30 miles		116	.069			3.54	5.50	9.04	11.35
3712	cycle 40 miles		99	.081			4.15	6.45	10.60	13.30
3714	cycle 50 miles		83	.096			4.94	7.70	12.64	15.85
4014	20 C.Y. truck, 15 min. wait/ld./uld., 15 MPH, cycle 0.5 mile	B-34D	560	.014			.73	1.17	1.90	2.38
4016	cycle 1 mile		500	.016			.82	1.31	2.13	2.67
4018	cycle 2 miles		420	.019			.98	1.55	2.53	3.17
4020	cycle 4 miles		300	.027			1.37	2.18	3.55	4.44
4022	cycle 6 miles		240	.033			1.71	2.72	4.43	5.55
4024	cycle 8 miles		200	.040			2.05	3.26	5.31	6.65
4025	cycle 10 miles		160	.050			2.57	4.08	6.65	8.35
4026	20 MPH avg., cycle 0.5 mile		580	.014			.71	1.13	1.84	2.30
4028	cycle 1 mile		520	.015			.79	1.26	2.05	2.56
4030	cycle 2 miles		460	.017			.89	1.42	2.31	2.89
4032	cycle 4 miles		340	.024			1.21	1.92	3.13	3.91
4034	cycle 6 miles		280	.029			1.47	2.33	3.80	4.75
4036	cycle 8 miles		240	.033			1.71	2.72	4.43	5.55
4038	cycle 10 miles		200	.040			2.05	3.26	5.31	6.65
4040	25 MPH avg., cycle 4 miles		380	.021			1.08	1.72	2.80	3.50
4042	cycle 6 miles		320	.025			1.28	2.04	3.32	4.16
4044	cycle 8 miles		280	.029			1.47	2.33	3.80	4.75
4046	cycle 10 miles		240	.033			1.71	2.72	4.43	5.55
4050	30 MPH avg., cycle 4 miles		420	.019			.98	1.55	2.53	3.17
4052	cycle 6 miles		340	.024			1.21	1.92	3.13	3.91
4054	cycle 8 miles		300	.027			1.37	2.18	3.55	4.44

31 23 23 – Fill

31 23 23.20 Hauling		Crew	Daily Output	Labor-Hours	Unit	Material	2021 Bare Costs Labor	2021 Bare Costs Equipment	Total	Total Incl O&P
4056	cycle 10 miles	B-34D	260	.031	L.C.Y.		1.58	2.51	4.09	5.10
4060	35 MPH avg., cycle 4 miles		440	.018			.93	1.48	2.41	3.02
4062	cycle 6 miles		380	.021			1.08	1.72	2.80	3.50
4064	cycle 8 miles		320	.025			1.28	2.04	3.32	4.16
4066	cycle 10 miles		300	.027			1.37	2.18	3.55	4.44
4068	cycle 20 miles		180	.044			2.28	3.63	5.91	7.40
4070	cycle 30 miles		140	.057			2.93	4.66	7.59	9.55
4072	cycle 40 miles		100	.080			4.10	6.55	10.65	13.35
4074	40 MPH avg., cycle 6 miles		400	.020			1.03	1.63	2.66	3.33
4076	cycle 8 miles		340	.024			1.21	1.92	3.13	3.91
4078	cycle 10 miles		320	.025			1.28	2.04	3.32	4.16
4080	cycle 20 miles		200	.040			2.05	3.26	5.31	6.65
4082	cycle 30 miles		160	.050			2.57	4.08	6.65	8.35
4084	cycle 40 miles		120	.067			3.42	5.45	8.87	11.10
4086	cycle 50 miles		100	.080			4.10	6.55	10.65	13.35
4094	45 MPH avg., cycle 8 miles		360	.022			1.14	1.81	2.95	3.69
4096	cycle 10 miles		340	.024			1.21	1.92	3.13	3.91
4098	cycle 20 miles		220	.036			1.87	2.97	4.84	6.05
4100	cycle 30 miles		160	.050			2.57	4.08	6.65	8.35
4102	cycle 40 miles		140	.057			2.93	4.66	7.59	9.55
4104	cycle 50 miles		120	.067			3.42	5.45	8.87	11.10
4106	50 MPH avg., cycle 10 miles		340	.024			1.21	1.92	3.13	3.91
4108	cycle 20 miles		240	.033			1.71	2.72	4.43	5.55
4110	cycle 30 miles		180	.044			2.28	3.63	5.91	7.40
4112	cycle 40 miles		140	.057			2.93	4.66	7.59	9.55
4114	cycle 50 miles		120	.067			3.42	5.45	8.87	11.10
4214	20 min. wait/ld./uld., 15 MPH, cycle 0.5 mile		440	.018			.93	1.48	2.41	3.02
4216	cycle 1 mile		400	.020			1.03	1.63	2.66	3.33
4218	cycle 2 miles		340	.024			1.21	1.92	3.13	3.91
4220	cycle 4 miles		260	.031			1.58	2.51	4.09	5.10
4222	cycle 6 miles		220	.036			1.87	2.97	4.84	6.05
4224	cycle 8 miles		180	.044			2.28	3.63	5.91	7.40
4225	cycle 10 miles		160	.050			2.57	4.08	6.65	8.35
4226	20 MPH avg., cycle 0.5 mile		440	.018			.93	1.48	2.41	3.02
4228	cycle 1 mile		420	.019			.98	1.55	2.53	3.17
4230	cycle 2 miles		360	.022			1.14	1.81	2.95	3.69
4232	cycle 4 miles		300	.027			1.37	2.18	3.55	4.44
4234	cycle 6 miles		240	.033			1.71	2.72	4.43	5.55
4236	cycle 8 miles		220	.036			1.87	2.97	4.84	6.05
4238	cycle 10 miles		180	.044			2.28	3.63	5.91	7.40
4240	25 MPH avg., cycle 4 miles		320	.025			1.28	2.04	3.32	4.16
4242	cycle 6 miles		280	.029			1.47	2.33	3.80	4.75
4244	cycle 8 miles		240	.033			1.71	2.72	4.43	5.55
4246	cycle 10 miles		220	.036			1.87	2.97	4.84	6.05
4250	30 MPH avg., cycle 4 miles		340	.024			1.21	1.92	3.13	3.91
4252	cycle 6 miles		300	.027			1.37	2.18	3.55	4.44
4254	cycle 8 miles		260	.031			1.58	2.51	4.09	5.10
4256	cycle 10 miles		240	.033			1.71	2.72	4.43	5.55
4260	35 MPH avg., cycle 4 miles		360	.022			1.14	1.81	2.95	3.69
4262	cycle 6 miles		320	.025			1.28	2.04	3.32	4.16
4264	cycle 8 miles		280	.029			1.47	2.33	3.80	4.75
4266	cycle 10 miles		260	.031			1.58	2.51	4.09	5.10
4268	cycle 20 miles		180	.044			2.28	3.63	5.91	7.40

31 23 23.20 Hauling		Crew	Daily Output	Labor-Hours	Unit	Material	2021 Bare Costs Labor	2021 Bare Costs Equipment	Total	Total Incl O&P
4270	cycle 30 miles	B-34D	120	.067	L.C.Y.		3.42	5.45	8.87	11.10
4272	cycle 40 miles		100	.080			4.10	6.55	10.65	13.35
4274	40 MPH avg., cycle 6 miles		320	.025			1.28	2.04	3.32	4.16
4276	cycle 8 miles		300	.027			1.37	2.18	3.55	4.44
4278	cycle 10 miles		260	.031			1.58	2.51	4.09	5.10
4280	cycle 20 miles		180	.044			2.28	3.63	5.91	7.40
4282	cycle 30 miles		140	.057			2.93	4.66	7.59	9.55
4284	cycle 40 miles		120	.067			3.42	5.45	8.87	11.10
4286	cycle 50 miles		100	.080			4.10	6.55	10.65	13.35
4294	45 MPH avg., cycle 8 miles		300	.027			1.37	2.18	3.55	4.44
4296	cycle 10 miles		280	.029			1.47	2.33	3.80	4.75
4298	cycle 20 miles		200	.040			2.05	3.26	5.31	6.65
4300	cycle 30 miles		160	.050			2.57	4.08	6.65	8.35
4302	cycle 40 miles		120	.067			3.42	5.45	8.87	11.10
4304	cycle 50 miles		100	.080			4.10	6.55	10.65	13.35
4306	50 MPH avg., cycle 10 miles		300	.027			1.37	2.18	3.55	4.44
4308	cycle 20 miles		220	.036			1.87	2.97	4.84	6.05
4310	cycle 30 miles		180	.044			2.28	3.63	5.91	7.40
4312	cycle 40 miles		140	.057			2.93	4.66	7.59	9.55
4314	cycle 50 miles		120	.067			3.42	5.45	8.87	11.10
4414	25 min. wait/ld./uld., 15 MPH, cycle 0.5 mile		340	.024			1.21	1.92	3.13	3.91
4416	cycle 1 mile		320	.025			1.28	2.04	3.32	4.16
4418	cycle 2 miles		280	.029			1.47	2.33	3.80	4.75
4420	cycle 4 miles		220	.036			1.87	2.97	4.84	6.05
4422	cycle 6 miles		200	.040			2.05	3.26	5.31	6.65
4424	cycle 8 miles		160	.050			2.57	4.08	6.65	8.35
4425	cycle 10 miles		140	.057			2.93	4.66	7.59	9.55
4426	20 MPH avg., cycle 0.5 mile		360	.022			1.14	1.81	2.95	3.69
4428	cycle 1 mile		340	.024			1.21	1.92	3.13	3.91
4430	cycle 2 miles		300	.027			1.37	2.18	3.55	4.44
4432	cycle 4 miles		260	.031			1.58	2.51	4.09	5.10
4434	cycle 6 miles		220	.036			1.87	2.97	4.84	6.05
4436	cycle 8 miles		200	.040			2.05	3.26	5.31	6.65
4438	cycle 10 miles		160	.050			2.57	4.08	6.65	8.35
4440	25 MPH avg., cycle 4 miles		280	.029			1.47	2.33	3.80	4.75
4442	cycle 6 miles		240	.033			1.71	2.72	4.43	5.55
4444	cycle 8 miles		220	.036			1.87	2.97	4.84	6.05
4446	cycle 10 miles		200	.040			2.05	3.26	5.31	6.65
4450	30 MPH avg., cycle 4 miles		280	.029			1.47	2.33	3.80	4.75
4452	cycle 6 miles		260	.031			1.58	2.51	4.09	5.10
4454	cycle 8 miles		220	.036			1.87	2.97	4.84	6.05
4456	cycle 10 miles		200	.040			2.05	3.26	5.31	6.65
4460	35 MPH avg., cycle 4 miles		300	.027			1.37	2.18	3.55	4.44
4462	cycle 6 miles		260	.031			1.58	2.51	4.09	5.10
4464	cycle 8 miles		240	.033			1.71	2.72	4.43	5.55
4466	cycle 10 miles		220	.036			1.87	2.97	4.84	6.05
4468	cycle 20 miles		160	.050			2.57	4.08	6.65	8.35
4470	cycle 30 miles		120	.067			3.42	5.45	8.87	11.10
4472	cycle 40 miles		100	.080			4.10	6.55	10.65	13.35
4474	40 MPH, cycle 6 miles		280	.029			1.47	2.33	3.80	4.75
4476	cycle 8 miles		260	.031			1.58	2.51	4.09	5.10
4478	cycle 10 miles		240	.033			1.71	2.72	4.43	5.55
4480	cycle 20 miles		160	.050			2.57	4.08	6.65	8.35

31 23 23.20 Hauling	Crew	Daily Output	Labor-Hours	Unit	Material	2021 Bare Costs Labor	2021 Bare Costs Equipment	Total	Total Incl O&P
4482	cycle 30 miles	B-34D	140	.057	L.C.Y.		2.93	4.66	7.59	9.55
4484	cycle 40 miles		100	.080			4.10	6.55	10.65	13.35
4486	cycle 50 miles		100	.080			4.10	6.55	10.65	13.35
4494	45 MPH avg., cycle 8 miles		260	.031			1.58	2.51	4.09	5.10
4496	cycle 10 miles		240	.033			1.71	2.72	4.43	5.55
4498	cycle 20 miles		180	.044			2.28	3.63	5.91	7.40
4500	cycle 30 miles		140	.057			2.93	4.66	7.59	9.55
4502	cycle 40 miles		120	.067			3.42	5.45	8.87	11.10
4504	cycle 50 miles		100	.080			4.10	6.55	10.65	13.35
4506	50 MPH avg., cycle 10 miles		260	.031			1.58	2.51	4.09	5.10
4508	cycle 20 miles		200	.040			2.05	3.26	5.31	6.65
4510	cycle 30 miles		160	.050			2.57	4.08	6.65	8.35
4512	cycle 40 miles		120	.067			3.42	5.45	8.87	11.10
4514	cycle 50 miles		100	.080			4.10	6.55	10.65	13.35
4614	30 min. wait/ld./uld., 15 MPH, cycle 0.5 mile		300	.027			1.37	2.18	3.55	4.44
4616	cycle 1 mile		280	.029			1.47	2.33	3.80	4.75
4618	cycle 2 miles		240	.033			1.71	2.72	4.43	5.55
4620	cycle 4 miles		200	.040			2.05	3.26	5.31	6.65
4622	cycle 6 miles		180	.044			2.28	3.63	5.91	7.40
4624	cycle 8 miles		140	.057			2.93	4.66	7.59	9.55
4625	cycle 10 miles		140	.057			2.93	4.66	7.59	9.55
4626	20 MPH avg., cycle 0.5 mile		300	.027			1.37	2.18	3.55	4.44
4628	cycle 1 mile		280	.029			1.47	2.33	3.80	4.75
4630	cycle 2 miles		260	.031			1.58	2.51	4.09	5.10
4632	cycle 4 miles		220	.036			1.87	2.97	4.84	6.05
4634	cycle 6 miles		200	.040			2.05	3.26	5.31	6.65
4636	cycle 8 miles		180	.044			2.28	3.63	5.91	7.40
4638	cycle 10 miles		160	.050			2.57	4.08	6.65	8.35
4640	25 MPH avg., cycle 4 miles		240	.033			1.71	2.72	4.43	5.55
4642	cycle 6 miles		220	.036			1.87	2.97	4.84	6.05
4644	cycle 8 miles		180	.044			2.28	3.63	5.91	7.40
4646	cycle 10 miles		180	.044			2.28	3.63	5.91	7.40
4650	30 MPH avg., cycle 4 miles		240	.033			1.71	2.72	4.43	5.55
4652	cycle 6 miles		220	.036			1.87	2.97	4.84	6.05
4654	cycle 8 miles		200	.040			2.05	3.26	5.31	6.65
4656	cycle 10 miles		180	.044			2.28	3.63	5.91	7.40
4660	35 MPH avg., cycle 4 miles		260	.031			1.58	2.51	4.09	5.10
4662	cycle 6 miles		240	.033			1.71	2.72	4.43	5.55
4664	cycle 8 miles		220	.036			1.87	2.97	4.84	6.05
4666	cycle 10 miles		200	.040			2.05	3.26	5.31	6.65
4668	cycle 20 miles		140	.057			2.93	4.66	7.59	9.55
4670	cycle 30 miles		120	.067			3.42	5.45	8.87	11.10
4672	cycle 40 miles		100	.080			4.10	6.55	10.65	13.35
4674	40 MPH, cycle 6 miles		240	.033			1.71	2.72	4.43	5.55
4676	cycle 8 miles		220	.036			1.87	2.97	4.84	6.05
4678	cycle 10 miles		200	.040			2.05	3.26	5.31	6.65
4680	cycle 20 miles		160	.050			2.57	4.08	6.65	8.35
4682	cycle 30 miles		120	.067			3.42	5.45	8.87	11.10
4684	cycle 40 miles		100	.080			4.10	6.55	10.65	13.35
4686	cycle 50 miles		80	.100			5.15	8.15	13.30	16.65
4694	45 MPH avg., cycle 8 miles		220	.036			1.87	2.97	4.84	6.05
4696	cycle 10 miles		220	.036			1.87	2.97	4.84	6.05
4698	cycle 20 miles		160	.050			2.57	4.08	6.65	8.35

31 23 23.20 Hauling		Crew	Daily Output	Labor-Hours	Unit	Material	2021 Bare Costs Labor	2021 Bare Costs Equipment	Total	Total Incl O&P
4700	cycle 30 miles	B-34D	140	.057	L.C.Y.		2.93	4.66	7.59	9.55
4702	cycle 40 miles		100	.080			4.10	6.55	10.65	13.35
4704	cycle 50 miles		100	.080			4.10	6.55	10.65	13.35
4706	50 MPH avg., cycle 10 miles		220	.036			1.87	2.97	4.84	6.05
4708	cycle 20 miles		180	.044			2.28	3.63	5.91	7.40
4710	cycle 30 miles		140	.057			2.93	4.66	7.59	9.55
4712	cycle 40 miles		120	.067			3.42	5.45	8.87	11.10
4714	cycle 50 miles		100	.080			4.10	6.55	10.65	13.35
5000	22 C.Y. off-road, 15 min. wait/ld./uld., 5 MPH, cycle 2,000'	B-34F	528	.015			.78	1.79	2.57	3.13
5010	cycle 3,000'		484	.017			.85	1.95	2.80	3.42
5020	cycle 4,000'		440	.018			.93	2.15	3.08	3.75
5030	cycle 0.5 mile		506	.016			.81	1.87	2.68	3.27
5040	cycle 1 mile		374	.021			1.10	2.53	3.63	4.42
5050	cycle 2 miles		264	.030			1.55	3.58	5.13	6.25
5060	10 MPH, cycle 2,000'		594	.013			.69	1.59	2.28	2.78
5070	cycle 3,000'		572	.014			.72	1.65	2.37	2.89
5080	cycle 4,000'		528	.015			.78	1.79	2.57	3.13
5090	cycle 0.5 mile		572	.014			.72	1.65	2.37	2.89
5100	cycle 1 mile		506	.016			.81	1.87	2.68	3.27
5110	cycle 2 miles		374	.021			1.10	2.53	3.63	4.42
5120	cycle 4 miles		264	.030			1.55	3.58	5.13	6.25
5130	15 MPH, cycle 2,000'		638	.013			.64	1.48	2.12	2.59
5140	cycle 3,000'		594	.013			.69	1.59	2.28	2.78
5150	cycle 4,000'		572	.014			.72	1.65	2.37	2.89
5160	cycle 0.5 mile		616	.013			.67	1.54	2.21	2.69
5170	cycle 1 mile		550	.015			.75	1.72	2.47	3.01
5180	cycle 2 miles		462	.017			.89	2.05	2.94	3.58
5190	cycle 4 miles		330	.024			1.24	2.87	4.11	5
5200	20 MPH, cycle 2 miles		506	.016			.81	1.87	2.68	3.27
5210	cycle 4 miles		374	.021			1.10	2.53	3.63	4.42
5220	25 MPH, cycle 2 miles		528	.015			.78	1.79	2.57	3.13
5230	cycle 4 miles		418	.019			.98	2.26	3.24	3.96
5300	20 min. wait/ld./uld., 5 MPH, cycle 2,000'		418	.019			.98	2.26	3.24	3.96
5310	cycle 3,000'		396	.020			1.04	2.39	3.43	4.18
5320	cycle 4,000'		352	.023			1.17	2.69	3.86	4.70
5330	cycle 0.5 mile		396	.020			1.04	2.39	3.43	4.18
5340	cycle 1 mile		330	.024			1.24	2.87	4.11	5
5350	cycle 2 miles		242	.033			1.70	3.91	5.61	6.85
5360	10 MPH, cycle 2,000'		462	.017			.89	2.05	2.94	3.58
5370	cycle 3,000'		440	.018			.93	2.15	3.08	3.75
5380	cycle 4,000'		418	.019			.98	2.26	3.24	3.96
5390	cycle 0.5 mile		462	.017			.89	2.05	2.94	3.58
5400	cycle 1 mile		396	.020			1.04	2.39	3.43	4.18
5410	cycle 2 miles		330	.024			1.24	2.87	4.11	5
5420	cycle 4 miles		242	.033			1.70	3.91	5.61	6.85
5430	15 MPH, cycle 2,000'		484	.017			.85	1.95	2.80	3.42
5440	cycle 3,000'		462	.017			.89	2.05	2.94	3.58
5450	cycle 4,000'		462	.017			.89	2.05	2.94	3.58
5460	cycle 0.5 mile		484	.017			.85	1.95	2.80	3.42
5470	cycle 1 mile		440	.018			.93	2.15	3.08	3.75
5480	cycle 2 miles		374	.021			1.10	2.53	3.63	4.42
5490	cycle 4 miles		286	.028			1.43	3.31	4.74	5.80
5500	20 MPH, cycle 2 miles		396	.020			1.04	2.39	3.43	4.18

328

For customer support on your Site Work & Landscape Costs with RSMeans Data, call 800.448.8182.

31 23 23 – Fill

31 23 23.20 Hauling	Crew	Daily Output	Labor-Hours	Unit	Material	2021 Bare Costs Labor	Equipment	Total	Total Incl O&P
5510	cycle 4 miles	B-34F	330	.024	L.C.Y.		1.24	2.87	4.11	5
5520	25 MPH, cycle 2 miles		418	.019			.98	2.26	3.24	3.96
5530	cycle 4 miles		352	.023			1.17	2.69	3.86	4.70
5600	25 min. wait/ld./uld., 5 MPH, cycle 2,000'		352	.023			1.17	2.69	3.86	4.70
5610	cycle 3,000'		330	.024			1.24	2.87	4.11	5
5620	cycle 4,000'		308	.026			1.33	3.07	4.40	5.35
5630	cycle 0.5 mile		330	.024			1.24	2.87	4.11	5
5640	cycle 1 mile		286	.028			1.43	3.31	4.74	5.80
5650	cycle 2 miles		220	.036			1.87	4.30	6.17	7.50
5660	10 MPH, cycle 2,000'		374	.021			1.10	2.53	3.63	4.42
5670	cycle 3,000'		374	.021			1.10	2.53	3.63	4.42
5680	cycle 4,000'		352	.023			1.17	2.69	3.86	4.70
5690	cycle 0.5 mile		374	.021			1.10	2.53	3.63	4.42
5700	cycle 1 mile		330	.024			1.24	2.87	4.11	5
5710	cycle 2 miles		286	.028			1.43	3.31	4.74	5.80
5720	cycle 4 miles		220	.036			1.87	4.30	6.17	7.50
5730	15 MPH, cycle 2,000'		396	.020			1.04	2.39	3.43	4.18
5740	cycle 3,000'		374	.021			1.10	2.53	3.63	4.42
5750	cycle 4,000'		374	.021			1.10	2.53	3.63	4.42
5760	cycle 0.5 mile		374	.021			1.10	2.53	3.63	4.42
5770	cycle 1 mile		352	.023			1.17	2.69	3.86	4.70
5780	cycle 2 miles		308	.026			1.33	3.07	4.40	5.35
5790	cycle 4 miles		242	.033			1.70	3.91	5.61	6.85
5800	20 MPH, cycle 2 miles		330	.024			1.24	2.87	4.11	5
5810	cycle 4 miles		286	.028			1.43	3.31	4.74	5.80
5820	25 MPH, cycle 2 miles		352	.023			1.17	2.69	3.86	4.70
5830	cycle 4 miles		308	.026			1.33	3.07	4.40	5.35
6000	34 C.Y. off-road, 15 min. wait/ld./uld., 5 MPH, cycle 2,000'	B-34G	816	.010			.50	2.43	2.93	3.42
6010	cycle 3,000'		748	.011			.55	2.65	3.20	3.73
6020	cycle 4,000'		680	.012			.60	2.91	3.51	4.10
6030	cycle 0.5 mile		782	.010			.52	2.53	3.05	3.56
6040	cycle 1 mile		578	.014			.71	3.42	4.13	4.83
6050	cycle 2 miles		408	.020			1.01	4.85	5.86	6.85
6060	10 MPH, cycle 2,000'		918	.009			.45	2.16	2.61	3.04
6070	cycle 3,000'		884	.009			.46	2.24	2.70	3.15
6080	cycle 4,000'		816	.010			.50	2.43	2.93	3.42
6090	cycle 0.5 mile		884	.009			.46	2.24	2.70	3.15
6100	cycle 1 mile		782	.010			.52	2.53	3.05	3.56
6110	cycle 2 miles		578	.014			.71	3.42	4.13	4.83
6120	cycle 4 miles		408	.020			1.01	4.85	5.86	6.85
6130	15 MPH, cycle 2,000'		986	.008			.42	2.01	2.43	2.83
6140	cycle 3,000'		918	.009			.45	2.16	2.61	3.04
6150	cycle 4,000'		884	.009			.46	2.24	2.70	3.15
6160	cycle 0.5 mile		952	.008			.43	2.08	2.51	2.93
6170	cycle 1 mile		850	.009			.48	2.33	2.81	3.28
6180	cycle 2 miles		714	.011			.57	2.77	3.34	3.91
6190	cycle 4 miles		510	.016			.80	3.88	4.68	5.45
6200	20 MPH, cycle 2 miles		782	.010			.52	2.53	3.05	3.56
6210	cycle 4 miles		578	.014			.71	3.42	4.13	4.83
6220	25 MPH, cycle 2 miles		816	.010			.50	2.43	2.93	3.42
6230	cycle 4 miles		646	.012			.64	3.06	3.70	4.32
6300	20 min. wait/ld./uld., 5 MPH, cycle 2,000'		646	.012			.64	3.06	3.70	4.32
6310	cycle 3,000'		612	.013			.67	3.23	3.90	4.56

329

31 23 23.20 Hauling		Crew	Daily Output	Labor-Hours	Unit	Material	2021 Bare Costs		Total	Total Incl O&P
							Labor	Equipment		
6320	cycle 4,000'	B-34G	544	.015	L.C.Y.		.75	3.64	4.39	5.15
6330	cycle 0.5 mile		612	.013			.67	3.23	3.90	4.56
6340	cycle 1 mile		510	.016			.80	3.88	4.68	5.45
6350	cycle 2 miles		374	.021			1.10	5.30	6.40	7.45
6360	10 MPH, cycle 2,000'		714	.011			.57	2.77	3.34	3.91
6370	cycle 3,000'		680	.012			.60	2.91	3.51	4.10
6380	cycle 4,000'		646	.012			.64	3.06	3.70	4.32
6390	cycle 0.5 mile		714	.011			.57	2.77	3.34	3.91
6400	cycle 1 mile		612	.013			.67	3.23	3.90	4.56
6410	cycle 2 miles		510	.016			.80	3.88	4.68	5.45
6420	cycle 4 miles		374	.021			1.10	5.30	6.40	7.45
6430	15 MPH, cycle 2,000'		748	.011			.55	2.65	3.20	3.73
6440	cycle 3,000'		714	.011			.57	2.77	3.34	3.91
6450	cycle 4,000'		714	.011			.57	2.77	3.34	3.91
6460	cycle 0.5 mile		748	.011			.55	2.65	3.20	3.73
6470	cycle 1 mile		680	.012			.60	2.91	3.51	4.10
6480	cycle 2 miles		578	.014			.71	3.42	4.13	4.83
6490	cycle 4 miles		442	.018			.93	4.48	5.41	6.30
6500	20 MPH, cycle 2 miles		612	.013			.67	3.23	3.90	4.56
6510	cycle 4 miles		510	.016			.80	3.88	4.68	5.45
6520	25 MPH, cycle 2 miles		646	.012			.64	3.06	3.70	4.32
6530	cycle 4 miles		544	.015			.75	3.64	4.39	5.15
6600	25 min. wait/ld./uld., 5 MPH, cycle 2,000'		544	.015			.75	3.64	4.39	5.15
6610	cycle 3,000'		510	.016			.80	3.88	4.68	5.45
6620	cycle 4,000'		476	.017			.86	4.16	5.02	5.85
6630	cycle 0.5 mile		510	.016			.80	3.88	4.68	5.45
6640	cycle 1 mile		442	.018			.93	4.48	5.41	6.30
6650	cycle 2 miles		340	.024			1.21	5.80	7.01	8.20
6660	10 MPH, cycle 2,000'		578	.014			.71	3.42	4.13	4.83
6670	cycle 3,000'		578	.014			.71	3.42	4.13	4.83
6680	cycle 4,000'		544	.015			.75	3.64	4.39	5.15
6690	cycle 0.5 mile		578	.014			.71	3.42	4.13	4.83
6700	cycle 1 mile		510	.016			.80	3.88	4.68	5.45
6710	cycle 2 miles		442	.018			.93	4.48	5.41	6.30
6720	cycle 4 miles		340	.024			1.21	5.80	7.01	8.20
6730	15 MPH, cycle 2,000'		612	.013			.67	3.23	3.90	4.56
6740	cycle 3,000'		578	.014			.71	3.42	4.13	4.83
6750	cycle 4,000'		578	.014			.71	3.42	4.13	4.83
6760	cycle 0.5 mile		612	.013			.67	3.23	3.90	4.56
6770	cycle 1 mile		544	.015			.75	3.64	4.39	5.15
6780	cycle 2 miles		476	.017			.86	4.16	5.02	5.85
6790	cycle 4 miles		374	.021			1.10	5.30	6.40	7.45
6800	20 MPH, cycle 2 miles		510	.016			.80	3.88	4.68	5.45
6810	cycle 4 miles		442	.018			.93	4.48	5.41	6.30
6820	25 MPH, cycle 2 miles		544	.015			.75	3.64	4.39	5.15
6830	cycle 4 miles		476	.017			.86	4.16	5.02	5.85
7000	42 C.Y. off-road, 20 min. wait/ld./uld., 5 MPH, cycle 2,000'	B-34H	798	.010			.51	2.43	2.94	3.44
7010	cycle 3,000'		756	.011			.54	2.56	3.10	3.63
7020	cycle 4,000'		672	.012			.61	2.88	3.49	4.08
7030	cycle 0.5 mile		756	.011			.54	2.56	3.10	3.63
7040	cycle 1 mile		630	.013			.65	3.08	3.73	4.35
7050	cycle 2 miles		462	.017			.89	4.19	5.08	5.95
7060	10 MPH, cycle 2,000'		882	.009			.47	2.20	2.67	3.12

31 23 Excavation and Fill

31 23 23 – Fill

31 23 23.20 Hauling		Crew	Daily Output	Labor-Hours	Unit	Material	2021 Bare Costs Labor	Equipment	Total	Total Incl O&P
7070	cycle 3,000'	B-34H	840	.010	L.C.Y.		.49	2.31	2.80	3.27
7080	cycle 4,000'		798	.010			.51	2.43	2.94	3.44
7090	cycle 0.5 mile		882	.009			.47	2.20	2.67	3.12
7100	cycle 1 mile		798	.010			.51	2.43	2.94	3.44
7110	cycle 2 miles		630	.013			.65	3.08	3.73	4.35
7120	cycle 4 miles		462	.017			.89	4.19	5.08	5.95
7130	15 MPH, cycle 2,000'		924	.009			.44	2.10	2.54	2.97
7140	cycle 3,000'		882	.009			.47	2.20	2.67	3.12
7150	cycle 4,000'		882	.009			.47	2.20	2.67	3.12
7160	cycle 0.5 mile		882	.009			.47	2.20	2.67	3.12
7170	cycle 1 mile		840	.010			.49	2.31	2.80	3.27
7180	cycle 2 miles		714	.011			.57	2.71	3.28	3.85
7190	cycle 4 miles		546	.015			.75	3.55	4.30	5
7200	20 MPH, cycle 2 miles		756	.011			.54	2.56	3.10	3.63
7210	cycle 4 miles		630	.013			.65	3.08	3.73	4.35
7220	25 MPH, cycle 2 miles		798	.010			.51	2.43	2.94	3.44
7230	cycle 4 miles		672	.012			.61	2.88	3.49	4.08
7300	25 min. wait/ld./uld., 5 MPH, cycle 2,000'		672	.012			.61	2.88	3.49	4.08
7310	cycle 3,000'		630	.013			.65	3.08	3.73	4.35
7320	cycle 4,000'		588	.014			.70	3.30	4	4.67
7330	cycle 0.5 mile		630	.013			.65	3.08	3.73	4.35
7340	cycle 1 mile		546	.015			.75	3.55	4.30	5
7350	cycle 2 miles		378	.021			1.09	5.15	6.24	7.25
7360	10 MPH, cycle 2,000'		714	.011			.57	2.71	3.28	3.85
7370	cycle 3,000'		714	.011			.57	2.71	3.28	3.85
7380	cycle 4,000'		672	.012			.61	2.88	3.49	4.08
7390	cycle 0.5 mile		714	.011			.57	2.71	3.28	3.85
7400	cycle 1 mile		630	.013			.65	3.08	3.73	4.35
7410	cycle 2 miles		546	.015			.75	3.55	4.30	5
7420	cycle 4 miles		378	.021			1.09	5.15	6.24	7.25
7430	15 MPH, cycle 2,000'		756	.011			.54	2.56	3.10	3.63
7440	cycle 3,000'		714	.011			.57	2.71	3.28	3.85
7450	cycle 4,000'		714	.011			.57	2.71	3.28	3.85
7460	cycle 0.5 mile		714	.011			.57	2.71	3.28	3.85
7470	cycle 1 mile		672	.012			.61	2.88	3.49	4.08
7480	cycle 2 miles		588	.014			.70	3.30	4	4.67
7490	cycle 4 miles		462	.017			.89	4.19	5.08	5.95
7500	20 MPH, cycle 2 miles		630	.013			.65	3.08	3.73	4.35
7510	cycle 4 miles		546	.015			.75	3.55	4.30	5
7520	25 MPH, cycle 2 miles		672	.012			.61	2.88	3.49	4.08
7530	cycle 4 miles		588	.014			.70	3.30	4	4.67
8000	60 C.Y. off-road, 20 min. wait/ld./uld., 5 MPH, cycle 2,000'	B-34J	1140	.007			.36	2.43	2.79	3.21
8010	cycle 3,000'		1080	.007			.38	2.56	2.94	3.39
8020	cycle 4,000'		960	.008			.43	2.88	3.31	3.81
8030	cycle 0.5 mile		1080	.007			.38	2.56	2.94	3.39
8040	cycle 1 mile		900	.009			.46	3.08	3.54	4.06
8050	cycle 2 miles		660	.012			.62	4.20	4.82	5.55
8060	10 MPH, cycle 2,000'		1260	.006			.33	2.20	2.53	2.91
8070	cycle 3,000'		1200	.007			.34	2.31	2.65	3.05
8080	cycle 4,000'		1140	.007			.36	2.43	2.79	3.21
8090	cycle 0.5 mile		1260	.006			.33	2.20	2.53	2.91
8100	cycle 1 mile		1080	.007			.38	2.56	2.94	3.39
8110	cycle 2 miles		900	.009			.46	3.08	3.54	4.06

31 23 23.20 Hauling		Crew	Daily Output	Labor-Hours	Unit	Material	2021 Bare Costs Labor	2021 Bare Costs Equipment	Total	Total Incl O&P
8120	cycle 4 miles	B-34J	660	.012	L.C.Y.		.62	4.20	4.82	5.55
8130	15 MPH, cycle 2,000'		1320	.006			.31	2.10	2.41	2.77
8140	cycle 3,000'		1260	.006			.33	2.20	2.53	2.91
8150	cycle 4,000'		1260	.006			.33	2.20	2.53	2.91
8160	cycle 0.5 mile		1320	.006			.31	2.10	2.41	2.77
8170	cycle 1 mile		1200	.007			.34	2.31	2.65	3.05
8180	cycle 2 miles		1020	.008			.40	2.71	3.11	3.59
8190	cycle 4 miles		780	.010			.53	3.55	4.08	4.70
8200	20 MPH, cycle 2 miles		1080	.007			.38	2.56	2.94	3.39
8210	cycle 4 miles		900	.009			.46	3.08	3.54	4.06
8220	25 MPH, cycle 2 miles		1140	.007			.36	2.43	2.79	3.21
8230	cycle 4 miles		960	.008			.43	2.88	3.31	3.81
8300	25 min. wait/ld./uld., 5 MPH, cycle 2,000'		960	.008			.43	2.88	3.31	3.81
8310	cycle 3,000'		900	.009			.46	3.08	3.54	4.06
8320	cycle 4,000'		840	.010			.49	3.30	3.79	4.36
8330	cycle 0.5 mile		900	.009			.46	3.08	3.54	4.06
8340	cycle 1 mile		780	.010			.53	3.55	4.08	4.70
8350	cycle 2 miles		600	.013			.68	4.62	5.30	6.10
8360	10 MPH, cycle 2,000'		1020	.008			.40	2.71	3.11	3.59
8370	cycle 3,000'		1020	.008			.40	2.71	3.11	3.59
8380	cycle 4,000'		960	.008			.43	2.88	3.31	3.81
8390	cycle 0.5 mile		1020	.008			.40	2.71	3.11	3.59
8400	cycle 1 mile		900	.009			.46	3.08	3.54	4.06
8410	cycle 2 miles		780	.010			.53	3.55	4.08	4.70
8420	cycle 4 miles		600	.013			.68	4.62	5.30	6.10
8430	15 MPH, cycle 2,000'		1080	.007			.38	2.56	2.94	3.39
8440	cycle 3,000'		1020	.008			.40	2.71	3.11	3.59
8450	cycle 4,000'		1020	.008			.40	2.71	3.11	3.59
8460	cycle 0.5 mile		1080	.007			.38	2.56	2.94	3.39
8470	cycle 1 mile		960	.008			.43	2.88	3.31	3.81
8480	cycle 2 miles		840	.010			.49	3.30	3.79	4.36
8490	cycle 4 miles		660	.012			.62	4.20	4.82	5.55
8500	20 MPH, cycle 2 miles		900	.009			.46	3.08	3.54	4.06
8510	cycle 4 miles		780	.010			.53	3.55	4.08	4.70
8520	25 MPH, cycle 2 miles		960	.008			.43	2.88	3.31	3.81
8530	cycle 4 miles		840	.010			.49	3.30	3.79	4.36
9014	18 C.Y. truck, 8 wheels,15 min. wait/ld./uld.,15 MPH, cycle 0.5 mi.	B-34I	504	.016			.81	1.50	2.31	2.86
9016	cycle 1 mile		450	.018			.91	1.67	2.58	3.20
9018	cycle 2 miles		378	.021			1.09	1.99	3.08	3.81
9020	cycle 4 miles		270	.030			1.52	2.79	4.31	5.35
9022	cycle 6 miles		216	.037			1.90	3.49	5.39	6.70
9024	cycle 8 miles		180	.044			2.28	4.19	6.47	8
9025	cycle 10 miles		144	.056			2.85	5.25	8.10	10
9026	20 MPH avg., cycle 0.5 mile		522	.015			.79	1.44	2.23	2.77
9028	cycle 1 mile		468	.017			.88	1.61	2.49	3.08
9030	cycle 2 miles		414	.019			.99	1.82	2.81	3.48
9032	cycle 4 miles		324	.025			1.27	2.33	3.60	4.45
9034	cycle 6 miles		252	.032			1.63	2.99	4.62	5.70
9036	cycle 8 miles		216	.037			1.90	3.49	5.39	6.70
9038	cycle 10 miles		180	.044			2.28	4.19	6.47	8
9040	25 MPH avg., cycle 4 miles		342	.023			1.20	2.20	3.40	4.21
9042	cycle 6 miles		288	.028			1.43	2.62	4.05	5
9044	cycle 8 miles		252	.032			1.63	2.99	4.62	5.70

31 23 23.20 Hauling		Crew	Daily Output	Labor-Hours	Unit	Material	2021 Bare Costs		Total	Total Incl O&P
							Labor	Equipment		
9046	cycle 10 miles	B-34I	216	.037	L.C.Y.		1.90	3.49	5.39	6.70
9050	30 MPH avg., cycle 4 miles		378	.021			1.09	1.99	3.08	3.81
9052	cycle 6 miles		324	.025			1.27	2.33	3.60	4.45
9054	cycle 8 miles		270	.030			1.52	2.79	4.31	5.35
9056	cycle 10 miles		234	.034			1.75	3.22	4.97	6.15
9060	35 MPH avg., cycle 4 miles		396	.020			1.04	1.90	2.94	3.64
9062	cycle 6 miles		342	.023			1.20	2.20	3.40	4.21
9064	cycle 8 miles		288	.028			1.43	2.62	4.05	5
9066	cycle 10 miles		270	.030			1.52	2.79	4.31	5.35
9068	cycle 20 miles		162	.049			2.53	4.65	7.18	8.90
9070	cycle 30 miles		126	.063			3.26	6	9.26	11.45
9072	cycle 40 miles		90	.089			4.56	8.35	12.91	16
9074	40 MPH avg., cycle 6 miles		360	.022			1.14	2.09	3.23	4
9076	cycle 8 miles		324	.025			1.27	2.33	3.60	4.45
9078	cycle 10 miles		288	.028			1.43	2.62	4.05	5
9080	cycle 20 miles		180	.044			2.28	4.19	6.47	8
9082	cycle 30 miles		144	.056			2.85	5.25	8.10	10
9084	cycle 40 miles		108	.074			3.80	7	10.80	13.40
9086	cycle 50 miles		90	.089			4.56	8.35	12.91	16
9094	45 MPH avg., cycle 8 miles		324	.025			1.27	2.33	3.60	4.45
9096	cycle 10 miles		306	.026			1.34	2.46	3.80	4.72
9098	cycle 20 miles		198	.040			2.07	3.81	5.88	7.30
9100	cycle 30 miles		144	.056			2.85	5.25	8.10	10
9102	cycle 40 miles		126	.063			3.26	6	9.26	11.45
9104	cycle 50 miles		108	.074			3.80	7	10.80	13.40
9106	50 MPH avg., cycle 10 miles		324	.025			1.27	2.33	3.60	4.45
9108	cycle 20 miles		216	.037			1.90	3.49	5.39	6.70
9110	cycle 30 miles		162	.049			2.53	4.65	7.18	8.90
9112	cycle 40 miles		126	.063			3.26	6	9.26	11.45
9114	cycle 50 miles		108	.074			3.80	7	10.80	13.40
9214	20 min. wait/ld./uld.,15 MPH, cycle 0.5 mi.		396	.020			1.04	1.90	2.94	3.64
9216	cycle 1 mile		360	.022			1.14	2.09	3.23	4
9218	cycle 2 miles		306	.026			1.34	2.46	3.80	4.72
9220	cycle 4 miles		234	.034			1.75	3.22	4.97	6.15
9222	cycle 6 miles		198	.040			2.07	3.81	5.88	7.30
9224	cycle 8 miles		162	.049			2.53	4.65	7.18	8.90
9225	cycle 10 miles		144	.056			2.85	5.25	8.10	10
9226	20 MPH avg., cycle 0.5 mile		396	.020			1.04	1.90	2.94	3.64
9228	cycle 1 mile		378	.021			1.09	1.99	3.08	3.81
9230	cycle 2 miles		324	.025			1.27	2.33	3.60	4.45
9232	cycle 4 miles		270	.030			1.52	2.79	4.31	5.35
9234	cycle 6 miles		216	.037			1.90	3.49	5.39	6.70
9236	cycle 8 miles		198	.040			2.07	3.81	5.88	7.30
9238	cycle 10 miles		162	.049			2.53	4.65	7.18	8.90
9240	25 MPH avg., cycle 4 miles		288	.028			1.43	2.62	4.05	5
9242	cycle 6 miles		252	.032			1.63	2.99	4.62	5.70
9244	cycle 8 miles		216	.037			1.90	3.49	5.39	6.70
9246	cycle 10 miles		198	.040			2.07	3.81	5.88	7.30
9250	30 MPH avg., cycle 4 miles		306	.026			1.34	2.46	3.80	4.72
9252	cycle 6 miles		270	.030			1.52	2.79	4.31	5.35
9254	cycle 8 miles		234	.034			1.75	3.22	4.97	6.15
9256	cycle 10 miles		216	.037			1.90	3.49	5.39	6.70
9260	35 MPH avg., cycle 4 miles		324	.025			1.27	2.33	3.60	4.45

For customer support on your Site Work & Landscape Costs with RSMeans Data, call 800.448.8182.

333

31 23 23 – Fill

31 23 23.20 Hauling		Crew	Daily Output	Labor-Hours	Unit	Material	2021 Bare Costs		Total	Total Incl O&P
							Labor	Equipment		
9262	cycle 6 miles	B-34I	288	.028	L.C.Y.		1.43	2.62	4.05	5
9264	cycle 8 miles		252	.032			1.63	2.99	4.62	5.70
9266	cycle 10 miles		234	.034			1.75	3.22	4.97	6.15
9268	cycle 20 miles		162	.049			2.53	4.65	7.18	8.90
9270	cycle 30 miles		108	.074			3.80	7	10.80	13.40
9272	cycle 40 miles		90	.089			4.56	8.35	12.91	16
9274	40 MPH avg., cycle 6 miles		288	.028			1.43	2.62	4.05	5
9276	cycle 8 miles		270	.030			1.52	2.79	4.31	5.35
9278	cycle 10 miles		234	.034			1.75	3.22	4.97	6.15
9280	cycle 20 miles		162	.049			2.53	4.65	7.18	8.90
9282	cycle 30 miles		126	.063			3.26	6	9.26	11.45
9284	cycle 40 miles		108	.074			3.80	7	10.80	13.40
9286	cycle 50 miles		90	.089			4.56	8.35	12.91	16
9294	45 MPH avg., cycle 8 miles		270	.030			1.52	2.79	4.31	5.35
9296	cycle 10 miles		252	.032			1.63	2.99	4.62	5.70
9298	cycle 20 miles		180	.044			2.28	4.19	6.47	8
9300	cycle 30 miles		144	.056			2.85	5.25	8.10	10
9302	cycle 40 miles		108	.074			3.80	7	10.80	13.40
9304	cycle 50 miles		90	.089			4.56	8.35	12.91	16
9306	50 MPH avg., cycle 10 miles		270	.030			1.52	2.79	4.31	5.35
9308	cycle 20 miles		198	.040			2.07	3.81	5.88	7.30
9310	cycle 30 miles		144	.056			2.85	5.25	8.10	10
9312	cycle 40 miles		126	.063			3.26	6	9.26	11.45
9314	cycle 50 miles		108	.074			3.80	7	10.80	13.40
9414	25 min. wait/ld./uld.,15 MPH, cycle 0.5 mi.		306	.026			1.34	2.46	3.80	4.72
9416	cycle 1 mile		288	.028			1.43	2.62	4.05	5
9418	cycle 2 miles		252	.032			1.63	2.99	4.62	5.70
9420	cycle 4 miles		198	.040			2.07	3.81	5.88	7.30
9422	cycle 6 miles		180	.044			2.28	4.19	6.47	8
9424	cycle 8 miles		144	.056			2.85	5.25	8.10	10
9425	cycle 10 miles		126	.063			3.26	6	9.26	11.45
9426	20 MPH avg., cycle 0.5 mile		324	.025			1.27	2.33	3.60	4.45
9428	cycle 1 mile		306	.026			1.34	2.46	3.80	4.72
9430	cycle 2 miles		270	.030			1.52	2.79	4.31	5.35
9432	cycle 4 miles		234	.034			1.75	3.22	4.97	6.15
9434	cycle 6 miles		198	.040			2.07	3.81	5.88	7.30
9436	cycle 8 miles		180	.044			2.28	4.19	6.47	8
9438	cycle 10 miles		144	.056			2.85	5.25	8.10	10
9440	25 MPH avg., cycle 4 miles		252	.032			1.63	2.99	4.62	5.70
9442	cycle 6 miles		216	.037			1.90	3.49	5.39	6.70
9444	cycle 8 miles		198	.040			2.07	3.81	5.88	7.30
9446	cycle 10 miles		180	.044			2.28	4.19	6.47	8
9450	30 MPH avg., cycle 4 miles		252	.032			1.63	2.99	4.62	5.70
9452	cycle 6 miles		234	.034			1.75	3.22	4.97	6.15
9454	cycle 8 miles		198	.040			2.07	3.81	5.88	7.30
9456	cycle 10 miles		180	.044			2.28	4.19	6.47	8
9460	35 MPH avg., cycle 4 miles		270	.030			1.52	2.79	4.31	5.35
9462	cycle 6 miles		234	.034			1.75	3.22	4.97	6.15
9464	cycle 8 miles		216	.037			1.90	3.49	5.39	6.70
9466	cycle 10 miles		198	.040			2.07	3.81	5.88	7.30
9468	cycle 20 miles		144	.056			2.85	5.25	8.10	10
9470	cycle 30 miles		108	.074			3.80	7	10.80	13.40
9472	cycle 40 miles		90	.089			4.56	8.35	12.91	16

31 23 23 – Fill

31 23 23.20 Hauling		Crew	Daily Output	Labor-Hours	Unit	Material	2021 Bare Costs Labor	2021 Bare Costs Equipment	Total	Total Incl O&P
9474	40 MPH avg., cycle 6 miles	B-34I	252	.032	L.C.Y.		1.63	2.99	4.62	5.70
9476	cycle 8 miles		234	.034			1.75	3.22	4.97	6.15
9478	cycle 10 miles		216	.037			1.90	3.49	5.39	6.70
9480	cycle 20 miles		144	.056			2.85	5.25	8.10	10
9482	cycle 30 miles		126	.063			3.26	6	9.26	11.45
9484	cycle 40 miles		90	.089			4.56	8.35	12.91	16
9486	cycle 50 miles		90	.089			4.56	8.35	12.91	16
9494	45 MPH avg., cycle 8 miles		234	.034			1.75	3.22	4.97	6.15
9496	cycle 10 miles		216	.037			1.90	3.49	5.39	6.70
9498	cycle 20 miles		162	.049			2.53	4.65	7.18	8.90
9500	cycle 30 miles		126	.063			3.26	6	9.26	11.45
9502	cycle 40 miles		108	.074			3.80	7	10.80	13.40
9504	cycle 50 miles		90	.089			4.56	8.35	12.91	16
9506	50 MPH avg., cycle 10 miles		234	.034			1.75	3.22	4.97	6.15
9508	cycle 20 miles		180	.044			2.28	4.19	6.47	8
9510	cycle 30 miles		144	.056			2.85	5.25	8.10	10
9512	cycle 40 miles		108	.074			3.80	7	10.80	13.40
9514	cycle 50 miles		90	.089			4.56	8.35	12.91	16
9614	30 min. wait/ld./uld.,15 MPH, cycle 0.5 mi.		270	.030			1.52	2.79	4.31	5.35
9616	cycle 1 mile		252	.032			1.63	2.99	4.62	5.70
9618	cycle 2 miles		216	.037			1.90	3.49	5.39	6.70
9620	cycle 4 miles		180	.044			2.28	4.19	6.47	8
9622	cycle 6 miles		162	.049			2.53	4.65	7.18	8.90
9624	cycle 8 miles		126	.063			3.26	6	9.26	11.45
9625	cycle 10 miles		126	.063			3.26	6	9.26	11.45
9626	20 MPH avg., cycle 0.5 mile		270	.030			1.52	2.79	4.31	5.35
9628	cycle 1 mile		252	.032			1.63	2.99	4.62	5.70
9630	cycle 2 miles		234	.034			1.75	3.22	4.97	6.15
9632	cycle 4 miles		198	.040			2.07	3.81	5.88	7.30
9634	cycle 6 miles		180	.044			2.28	4.19	6.47	8
9636	cycle 8 miles		162	.049			2.53	4.65	7.18	8.90
9638	cycle 10 miles		144	.056			2.85	5.25	8.10	10
9640	25 MPH avg., cycle 4 miles		216	.037			1.90	3.49	5.39	6.70
9642	cycle 6 miles		198	.040			2.07	3.81	5.88	7.30
9644	cycle 8 miles		180	.044			2.28	4.19	6.47	8
9646	cycle 10 miles		162	.049			2.53	4.65	7.18	8.90
9650	30 MPH avg., cycle 4 miles		216	.037			1.90	3.49	5.39	6.70
9652	cycle 6 miles		198	.040			2.07	3.81	5.88	7.30
9654	cycle 8 miles		180	.044			2.28	4.19	6.47	8
9656	cycle 10 miles		162	.049			2.53	4.65	7.18	8.90
9660	35 MPH avg., cycle 4 miles		234	.034			1.75	3.22	4.97	6.15
9662	cycle 6 miles		216	.037			1.90	3.49	5.39	6.70
9664	cycle 8 miles		198	.040			2.07	3.81	5.88	7.30
9666	cycle 10 miles		180	.044			2.28	4.19	6.47	8
9668	cycle 20 miles		126	.063			3.26	6	9.26	11.45
9670	cycle 30 miles		108	.074			3.80	7	10.80	13.40
9672	cycle 40 miles		90	.089			4.56	8.35	12.91	16
9674	40 MPH avg., cycle 6 miles		216	.037			1.90	3.49	5.39	6.70
9676	cycle 8 miles		198	.040			2.07	3.81	5.88	7.30
9678	cycle 10 miles		180	.044			2.28	4.19	6.47	8
9680	cycle 20 miles		144	.056			2.85	5.25	8.10	10
9682	cycle 30 miles		108	.074			3.80	7	10.80	13.40
9684	cycle 40 miles		90	.089			4.56	8.35	12.91	16

31 23 23.20 Hauling		Crew	Daily Output	Labor-Hours	Unit	Material	2021 Bare Costs		Total	Total Incl O&P
							Labor	Equipment		
9686	cycle 50 miles	B-34I	72	.111	L.C.Y.		5.70	10.45	16.15	20
9694	45 MPH avg., cycle 8 miles		216	.037			1.90	3.49	5.39	6.70
9696	cycle 10 miles		198	.040			2.07	3.81	5.88	7.30
9698	cycle 20 miles		144	.056			2.85	5.25	8.10	10
9700	cycle 30 miles		126	.063			3.26	6	9.26	11.45
9702	cycle 40 miles		108	.074			3.80	7	10.80	13.40
9704	cycle 50 miles		90	.089			4.56	8.35	12.91	16
9706	50 MPH avg., cycle 10 miles		198	.040			2.07	3.81	5.88	7.30
9708	cycle 20 miles		162	.049			2.53	4.65	7.18	8.90
9710	cycle 30 miles		126	.063			3.26	6	9.26	11.45
9712	cycle 40 miles		108	.074			3.80	7	10.80	13.40
9714	cycle 50 miles		90	.089			4.56	8.35	12.91	16

31 23 23.23 Compaction

		Crew	Daily Output	Labor-Hours	Unit	Material	Labor	Equipment	Total	Total Incl O&P
0010	**COMPACTION**　R312323-30									
5000	Riding, vibrating roller, 6" lifts, 2 passes	B-10Y	3000	.004	E.C.Y.		.22	.19	.41	.53
5020	3 passes		2300	.005			.28	.25	.53	.70
5040	4 passes		1900	.006			.34	.31	.65	.85
5050	8" lifts, 2 passes		4100	.003			.16	.14	.30	.40
5060	12" lifts, 2 passes		5200	.002			.12	.11	.23	.31
5080	3 passes		3500	.003			.19	.17	.36	.46
5100	4 passes		2600	.005			.25	.22	.47	.62
5600	Sheepsfoot or wobbly wheel roller, 6" lifts, 2 passes	B-10G	2400	.005			.27	.57	.84	1.02
5620	3 passes		1735	.007			.37	.79	1.16	1.42
5640	4 passes		1300	.009			.50	1.05	1.55	1.89
5680	12" lifts, 2 passes		5200	.002			.12	.26	.38	.48
5700	3 passes		3500	.003			.19	.39	.58	.71
5720	4 passes		2600	.005			.25	.52	.77	.95
6000	Towed sheepsfoot or wobbly wheel roller, 6" lifts, 2 passes	B-10D	10000	.001			.06	.19	.25	.31
6020	3 passes		2000	.006			.32	.97	1.29	1.55
6030	4 passes		1500	.008			.43	1.30	1.73	2.08
6050	12" lifts, 2 passes		6000	.002			.11	.32	.43	.52
6060	3 passes		4000	.003			.16	.49	.65	.78
6070	4 passes		3000	.004			.22	.65	.87	1.03
6200	Vibrating roller, 6" lifts, 2 passes	B-10C	2600	.005			.25	.78	1.03	1.23
6210	3 passes		1735	.007			.37	1.18	1.55	1.85
6220	4 passes		1300	.009			.50	1.57	2.07	2.47
6250	12" lifts, 2 passes		5200	.002			.12	.39	.51	.62
6260	3 passes		3465	.003			.19	.59	.78	.93
6270	4 passes		2600	.005			.25	.78	1.03	1.23
7000	Walk behind, vibrating plate 18" wide, 6" lifts, 2 passes	A-1D	200	.040			1.78	.16	1.94	2.83
7020	3 passes		185	.043			1.92	.17	2.09	3.05
7040	4 passes		140	.057			2.54	.23	2.77	4.04
7200	12" lifts, 2 passes, 21" wide	A-1E	560	.014			.63	.30	.93	1.28
7220	3 passes		375	.021			.95	.44	1.39	1.90
7240	4 passes		280	.029			1.27	.59	1.86	2.54
7500	Vibrating roller 24" wide, 6" lifts, 2 passes	B-10A	420	.029			1.55	.40	1.95	2.75
7520	3 passes		280	.043			2.32	.60	2.92	4.12
7540	4 passes		210	.057			3.09	.79	3.88	5.50
7600	12" lifts, 2 passes		840	.014			.77	.20	.97	1.37
7620	3 passes		560	.021			1.16	.30	1.46	2.06
7640	4 passes		420	.029			1.55	.40	1.95	2.75
8000	Rammer tamper, 6" to 11", 4" lifts, 2 passes	A-1F	130	.062			2.73	.36	3.09	4.48

31 23 Excavation and Fill

31 23 23 – Fill

31 23 23.23 Compaction	Crew	Daily Output	Labor-Hours	Unit	Material	2021 Bare Costs Labor	Equipment	Total	Total Incl O&P
8050	3 passes	A-1F	97	.082	E.C.Y.		3.66	.48	4.14	6
8100	4 passes		65	.123			5.45	.72	6.17	8.95
8200	8" lifts, 2 passes		260	.031			1.37	.18	1.55	2.24
8250	3 passes		195	.041			1.82	.24	2.06	2.99
8300	4 passes		130	.062			2.73	.36	3.09	4.48
8400	13" to 18", 4" lifts, 2 passes	A-1G	390	.021			.91	.14	1.05	1.51
8450	3 passes		290	.028			1.22	.19	1.41	2.04
8500	4 passes		195	.041			1.82	.28	2.10	3.03
8600	8" lifts, 2 passes		780	.010			.46	.07	.53	.76
8650	3 passes		585	.014			.61	.09	.70	1.01
8700	4 passes		390	.021			.91	.14	1.05	1.51
9000	Water, 3,000 gal. truck, 3 mile haul	B-45	1888	.008		1.23	.47	.44	2.14	2.54
9010	6 mile haul		1444	.011		1.23	.61	.58	2.42	2.89
9020	12 mile haul		1000	.016		1.23	.88	.83	2.94	3.59
9030	6,000 gal. wagon, 3 mile haul	B-59	2000	.004		1.23	.21	.23	1.67	1.92
9040	6 mile haul	"	1600	.005		1.23	.26	.29	1.78	2.05

31 23 23.24 Compaction, Structural

		Crew	Daily Output	Labor-Hours	Unit	Material	Labor	Equipment	Total	Total Incl O&P
0010	**COMPACTION, STRUCTURAL**									
0020	Steel wheel tandem roller, 5 tons	B-10E	8	1.500	Hr.		81	32.50	113.50	157
0050	Air tamp, 6" to 8" lifts, common fill	B-9	250	.160	E.C.Y.		7.15	1.42	8.57	12.25
0060	Select fill	"	300	.133	"		5.95	1.19	7.14	10.20
0100	10 tons	B-10F	8	1.500	Hr.		81	31	112	155
0300	Sheepsfoot or wobbly wheel roller, 8" lifts, common fill	B-10G	1300	.009	E.C.Y.		.50	1.05	1.55	1.89
0400	Select fill	"	1500	.008	"		.43	.91	1.34	1.65
0420	Static roller, riding 33" diam.	B-10A	2800	.004	S.Y.		.23	.06	.29	.42
0600	Vibratory plate, 8" lifts, common fill	A-1D	200	.040	E.C.Y.		1.78	.16	1.94	2.83
0700	Select fill	"	216	.037	"		1.64	.15	1.79	2.61

31 23 23.25 Compaction, Airports

		Crew	Daily Output	Labor-Hours	Unit	Material	Labor	Equipment	Total	Total Incl O&P
0010	**COMPACTION, AIRPORTS**									
0100	Airport subgrade, compaction									
0110	non cohesive soils, 85% proctor, 12" depth	B-10G	15600	.001	S.Y.		.04	.09	.13	.16
0200	24" depth		7800	.002			.08	.17	.25	.31
0300	36" depth		5200	.002			.12	.26	.38	.48
0400	48" depth		3900	.003			.17	.35	.52	.63
0500	60" depth		3120	.004			.21	.44	.65	.79
0600	90% proctor, 12" depth		7200	.002			.09	.19	.28	.34
0700	24" depth		3600	.003			.18	.38	.56	.69
0800	36" depth		2400	.005			.27	.57	.84	1.02
0900	48" depth		1800	.007			.36	.76	1.12	1.37
1000	60" depth		1440	.008			.45	.95	1.40	1.71
1100	95% proctor, 12" depth	B-10D	6000	.002			.11	.32	.43	.52
1200	24" depth		3000	.004			.22	.65	.87	1.03
1300	36" depth		2000	.006			.32	.97	1.29	1.55
1400	42" depth		1715	.007			.38	1.14	1.52	1.81
1500	100% proctor, 12" depth		4500	.003			.14	.43	.57	.70
1550	18" depth		3000	.004			.22	.65	.87	1.03
1600	24" depth		2250	.005			.29	.87	1.16	1.38
1700	cohesive soils, 80% proctor, 12" depth	B-10G	7200	.002			.09	.19	.28	.34
1800	24" depth	B-10D	15000	.001			.04	.13	.17	.20
1900	36" depth		10000	.001			.06	.19	.25	.31
2000	85% proctor, 12" depth		6000	.002			.11	.32	.43	.52
2100	18" depth		4000	.003			.16	.49	.65	.78

For customer support on your Site Work & Landscape Costs with RSMeans Data, call 800.448.8182.

337

31 23 Excavation and Fill

31 23 23 – Fill

31 23 23.25 Compaction, Airports

		Crew	Daily Output	Labor-Hours	Unit	Material	2021 Bare Costs Labor	Equipment	Total	Total Incl O&P
2200	24" depth	B-10D	3000	.004	S.Y.	.22	.65		.87	1.03
2300	27" depth		2670	.004		.24	.73		.97	1.16
2400	90% proctor, 6" depth		10400	.001		.06	.19		.25	.30
2500	9" depth		6935	.002		.09	.28		.37	.45
2600	12" depth		5200	.002		.12	.37		.49	.60
2700	15" depth		4160	.003		.16	.47		.63	.74
2800	18" depth		3470	.003		.19	.56		.75	.90
2900	95% proctor, 6" depth		1500	.008		.43	1.30		1.73	2.08
3000	7" depth		1285	.009		.51	1.52		2.03	2.42
3100	8" depth		1125	.011		.58	1.73		2.31	2.76
3200	9" depth		1000	.012		.65	1.95		2.60	3.11

31 25 Erosion and Sedimentation Controls

31 25 14 – Stabilization Measures for Erosion and Sedimentation Control

31 25 14.16 Rolled Erosion Control Mats and Blankets

			Crew	Daily Output	Labor-Hours	Unit	Material	2021 Bare Costs Labor	Equipment	Total	Total Incl O&P
0010	**ROLLED EROSION CONTROL MATS AND BLANKETS**										
0020	Jute mesh, 100 S.Y. per roll, 4' wide, stapled	G	B-80A	2400	.010	S.Y.	1	.44	.35	1.79	2.15
0060	Polyethylene 3 dimensional geomatrix, 50 mil thick	G		700	.034		3.47	1.52	1.21	6.20	7.45
0062	120 mil thick	G		515	.047		7.45	2.07	1.65	11.17	13.10
0068	Slope stakes (placed @ 3' - 5' intervals)	G				Ea.	.16			.16	.18
0070	Paper biodegradable mesh	G	B-1	2500	.010	S.Y.	.14	.43		.57	.80
0080	Paper mulch	G	B-64	20000	.001		.14	.04	.02	.20	.23
0100	Plastic netting, stapled, 2" x 1" mesh, 20 mil	G	B-1	2500	.010		.31	.43		.74	.99
0120	Revegetation mat, webbed	G	2 Clab	1000	.016		2.72	.71		3.43	4.05
0160	Underdrain fabric, 18" x 100' roll	G	"	32	.500	Roll	26	22		48	61.50
0200	Polypropylene mesh, stapled, 6.5 oz./S.Y.	G	B-1	2500	.010	S.Y.	1.83	.43		2.26	2.66
0300	Tobacco netting, or jute mesh #2, stapled	G	"	2500	.010		.26	.43		.69	.94
0400	Soil sealant, liquid sprayed from truck	G	B-81	5000	.005		.35	.25	.11	.71	.88
0600	Straw in polymeric netting, biodegradable log		A-2	1000	.024	L.F.	2.25	1.10	.20	3.55	4.34
0705	Sediment Log, Filter Sock, 9"			1000	.024		2.25	1.10	.20	3.55	4.34
0710	Sediment Log, Filter Sock, 12"			1000	.024		3.50	1.10	.20	4.80	5.70
1000	Silt fence, install and remove	G	B-62	650	.037		.44	1.78	.28	2.50	3.43
1400	Barriers, w/degradable component, 3' H, incl. wd stakes	G	2 Clab	1600	.010		.80	.44		1.24	1.54

31 31 Soil Treatment

31 31 16 – Termite Control

31 31 16.13 Chemical Termite Control

			Crew	Daily Output	Labor-Hours	Unit	Material	2021 Bare Costs Labor	Equipment	Total	Total Incl O&P
0010	**CHEMICAL TERMITE CONTROL**										
0020	Slab and walls, residential		1 Skwk	1200	.007	SF Flr.	.32	.38		.70	.92
0030	SS mesh, no chemicals, avg 1,400 S.F. home, min	G		1000	.008		.32	.46		.78	1.04
0040	Max	G		500	.016		.32	.91		1.23	1.72
0100	Commercial, minimum			2496	.003		.33	.18		.51	.64
0200	Maximum			1645	.005		.50	.28		.78	.96
0400	Insecticides for termite control, minimum			14	.571	Gal.	2.14	32.50		34.64	51.50
0500	Maximum			11	.727	"	4.29	41.50		45.79	67
3000	Soil poisoning (sterilization)		1 Clab	8000	.001	S.F.	.35	.04		.39	.46
3100	Herbicide application from truck		B-59	14520	.001	S.Y.	3.15	.03	.03	3.21	3.55

31 32 Soil Stabilization

31 32 13 – Soil Mixing Stabilization

31 32 13.13 Asphalt Soil Stabilization

		Crew	Daily Output	Labor-Hours	Unit	Material	2021 Bare Costs Labor	Equipment	Total	Total Incl O&P
0010	**ASPHALT SOIL STABILIZATION**									
0011	Including scarifying and compaction									
0020	Asphalt, 1-1/2" deep, 1/2 gal./S.Y.	B-75	4000	.014	S.Y.	1.13	.76	1.37	3.26	3.88
0040	3/4 gal./S.Y.		4000	.014		1.69	.76	1.37	3.82	4.50
0100	3" deep, 1 gal./S.Y.		3500	.016		2.25	.86	1.57	4.68	5.50
0140	1-1/2 gal./S.Y.		3500	.016		3.38	.86	1.57	5.81	6.70
0200	6" deep, 2 gal./S.Y.		3000	.019		4.50	1.01	1.83	7.34	8.45
0240	3 gal./S.Y.		3000	.019		6.75	1.01	1.83	9.59	10.95
0300	8" deep, 2-2/3 gal./S.Y.		2800	.020		6	1.08	1.96	9.04	10.35
0340	4 gal./S.Y.		2800	.020		9	1.08	1.96	12.04	13.65
0500	12" deep, 4 gal./S.Y.		5000	.011		9	.60	1.10	10.70	12
0540	6 gal./S.Y.		2600	.022		13.50	1.16	2.11	16.77	18.90

31 32 13.16 Cement Soil Stabilization

		Crew	Daily Output	Labor-Hours	Unit	Material	2021 Bare Costs Labor	Equipment	Total	Total Incl O&P
0010	**CEMENT SOIL STABILIZATION**									
0011	Including scarifying and compaction									
1020	Cement, 4% mix, by volume, 6" deep	B-74	1100	.058	S.Y.	1.97	3.12	5.50	10.59	12.90
1030	8" deep		1050	.061		2.57	3.27	5.75	11.59	14.05
1060	12" deep		960	.067		3.85	3.58	6.30	13.73	16.55
1100	6% mix, 6" deep		1100	.058		2.83	3.12	5.50	11.45	13.80
1120	8" deep		1050	.061		3.68	3.27	5.75	12.70	15.30
1160	12" deep		960	.067		5.55	3.58	6.30	15.43	18.40
1200	9% mix, 6" deep		1100	.058		4.28	3.12	5.50	12.90	15.40
1220	8" deep		1050	.061		5.55	3.27	5.75	14.57	17.35
1260	12" deep		960	.067		8.40	3.58	6.30	18.28	21.50
1300	12% mix, 6" deep		1100	.058		5.55	3.12	5.50	14.17	16.80
1320	8" deep		1050	.061		7.45	3.27	5.75	16.47	19.45
1360	12" deep		960	.067		11.15	3.58	6.30	21.03	24.50

31 32 13.19 Lime Soil Stabilization

		Crew	Daily Output	Labor-Hours	Unit	Material	2021 Bare Costs Labor	Equipment	Total	Total Incl O&P
0010	**LIME SOIL STABILIZATION**									
0011	Including scarifying and compaction									
2020	Hydrated lime, for base, 2% mix by weight, 6" deep	B-74	1800	.036	S.Y.	.76	1.91	3.36	6.03	7.35
2030	8" deep		1700	.038		1.01	2.02	3.56	6.59	8.05
2060	12" deep		1550	.041		1.51	2.22	3.90	7.63	9.25
2100	4% mix, 6" deep		1800	.036		1.51	1.91	3.36	6.78	8.20
2120	8" deep		1700	.038		2.03	2.02	3.56	7.61	9.15
2160	12" deep		1550	.041		3.02	2.22	3.90	9.14	10.95
2200	6% mix, 6" deep		1800	.036		2.27	1.91	3.36	7.54	9.05
2220	8" deep		1700	.038		3.04	2.02	3.56	8.62	10.25
2260	12" deep		1550	.041		4.54	2.22	3.90	10.66	12.60

31 32 13.30 Calcium Chloride

		Crew	Daily Output	Labor-Hours	Unit	Material	2021 Bare Costs Labor	Equipment	Total	Total Incl O&P
0010	**CALCIUM CHLORIDE**									
0020	Calcium chloride, delivered, 100 lb. bags, truckload lots				Ton	705			705	775
0030	Solution, 4 lb. flake per gallon, tank truck delivery				Gal.	1.69			1.69	1.86

31 32 19 – Geosynthetic Soil Stabilization and Layer Separation

31 32 19.16 Geotextile Soil Stabilization

		Crew	Daily Output	Labor-Hours	Unit	Material	2021 Bare Costs Labor	Equipment	Total	Total Incl O&P
0010	**GEOTEXTILE SOIL STABILIZATION**									
1500	Geotextile fabric, woven, 200 lb. tensile strength	2 Clab	2500	.006	S.Y.	.93	.28		1.21	1.44
1510	Heavy duty, 600 lb. tensile strength		2400	.007		1.75	.30		2.05	2.37
1550	Non-woven, 120 lb. tensile strength		2500	.006		.77	.28		1.05	1.27

For customer support on your Site Work & Landscape Costs with RSMeans Data, call 800.448.8182.

339

31 32 36.16 Grouted Soil Nailing

	Crew	Daily Output	Labor-Hours	Unit	Material	2021 Bare Costs Labor	2021 Bare Costs Equipment	Total	Total Incl O&P
0010 **GROUTED SOIL NAILING**									
0020 Soil nailing does not include guniting of surfaces									
0030 See Section 03 37 13.30 for guniting of surfaces									
0035 Layout and vertical and horizontal control per day	A-6	1	16	Day		880	34.50	914.50	1,375
0038 Layout and vertical and horizontal control average holes per day	"	60	.267	Ea.		14.65	.57	15.22	22.50
0050 For grade 150 soil nail add $1.37/L.F. to base material cost									
0060 Material delivery add $1.83 to $2.00 per truck mile for shipping									
0090 Average soil nailing, grade 75, 15 min. setup per hole & 80'/hr. drilling									
0100 Soil nailing, drill hole, install #8 nail, grout 20' depth average	B-47G	16	2	Ea.	147	95.50	122	364.50	440
0110 25' depth average		14.20	2.254		184	107	137	428	515
0120 30' depth average		12.80	2.500		365	119	153	637	745
0130 35' depth average		11.60	2.759		420	132	168	720	845
0140 40' depth average		10.70	2.991		480	143	182	805	945
0150 45' depth average		9.90	3.232		540	154	197	891	1,050
0160 50' depth average		9.10	3.516		625	168	215	1,008	1,175
0170 55' depth average		8.50	3.765		685	180	230	1,095	1,275
0180 60' depth average		8	4		745	191	244	1,180	1,375
0190 65' depth average		7.50	4.267		805	203	260	1,268	1,475
0200 70' depth average		7.10	4.507		860	215	275	1,350	1,575
0210 75' depth average		6.70	4.776		920	228	291	1,439	1,675
0290 Average soil nailing, grade 75, 15 min. setup per hole & 90'/hr. drilling									
0300 Soil nailing, drill hole, install #8 nail, grout 20' depth average	B-47G	16.60	1.928	Ea.	147	92	118	357	430
0310 25' depth average		15	2.133		184	102	130	416	495
0320 30' depth average		13.70	2.336		365	111	142	618	725
0330 35' depth average		12.30	2.602		420	124	159	703	825
0340 40' depth average		11.40	2.807		480	134	171	785	920
0350 45' depth average		10.70	2.991		540	143	182	865	1,000
0360 50' depth average		9.80	3.265		625	156	199	980	1,125
0370 55' depth average		9.20	3.478		685	166	212	1,063	1,225
0380 60' depth average		8.70	3.678		745	175	224	1,144	1,325
0390 65' depth average		8.10	3.951		805	188	241	1,234	1,425
0400 70' depth average		7.70	4.156		860	198	254	1,312	1,525
0410 75' depth average		7.40	4.324		920	206	264	1,390	1,625
0490 Average soil nailing, grade 75, 15 min. setup per hole & 100'/hr. drilling									
0500 Soil nailing, drill hole, install #8 nail, grout 20' depth average	B-47G	17.80	1.798	Ea.	147	85.50	110	342.50	410
0510 25' depth average		16	2		184	95.50	122	401.50	480
0520 30' depth average		14.60	2.192		365	105	134	604	705
0530 35' depth average		13.30	2.406		420	115	147	682	795
0540 40' depth average		12.30	2.602		480	124	159	763	890
0550 45' depth average		11.40	2.807		540	134	171	845	985
0560 50' depth average		10.70	2.991		625	143	182	950	1,100
0570 55' depth average		10	3.200		685	153	195	1,033	1,200
0580 60' depth average		9.40	3.404		745	162	208	1,115	1,275
0590 65' depth average		8.90	3.596		805	171	219	1,195	1,375
0600 70' depth average		8.40	3.810		860	182	232	1,274	1,475
0610 75' depth average		8	4		920	191	244	1,355	1,575
0690 Average soil nailing, grade 75, 15 min. setup per hole & 110'/hr. drilling									
0700 Soil nailing, drill hole, install #8 nail, grout 20' depth average	B-47G	18.50	1.730	Ea.	147	82.50	106	335.50	400
0710 25' depth average		16.60	1.928		184	92	118	394	470
0720 30' depth average		15.50	2.065		365	98.50	126	589.50	685
0730 35' depth average		14.10	2.270		420	108	138	666	780
0740 40' depth average		13	2.462		480	117	150	747	870

For customer support on your Site Work & Landscape Costs with RSMeans Data, call 800.448.8182.

31 32 36.16 Grouted Soil Nailing		Crew	Daily Output	Labor-Hours	Unit	Material	2021 Bare Costs Labor	Equipment	Total	Total Incl O&P
0750	45' depth average	B-47G	12	2.667	Ea.	540	127	163	830	965
0760	50' depth average		11.40	2.807		625	134	171	930	1,075
0770	55' depth average		10.70	2.991		685	143	182	1,010	1,175
0780	60' depth average		10	3.200		745	153	195	1,093	1,250
0790	65' depth average		9.40	3.404		805	162	208	1,175	1,350
0800	70' depth average		9.10	3.516		860	168	215	1,243	1,425
0810	75' depth average		8.60	3.721		920	177	227	1,324	1,550
0890	Average soil nailing, grade 75, 15 min. setup per hole & 120'/hr. drilling									
0900	Soil nailing, drill hole, install #8 nail, grout 20' depth average	B-47G	19.20	1.667	Ea.	147	79.50	102	328.50	395
0910	25' depth average		17.10	1.871		184	89	114	387	460
0920	30' depth average		16	2		365	95.50	122	582.50	675
0930	35' depth average		14.60	2.192		420	105	134	659	770
0940	40' depth average		13.70	2.336		480	111	142	733	855
0950	45' depth average		12.60	2.540		540	121	155	816	945
0960	50' depth average		12	2.667		625	127	163	915	1,050
0970	55' depth average		11.20	2.857		685	136	174	995	1,150
0980	60' depth average		10.70	2.991		745	143	182	1,070	1,225
0990	65' depth average		10	3.200		805	153	195	1,153	1,325
1000	70' depth average		9.60	3.333		860	159	203	1,222	1,400
1010	75' depth average		9.10	3.516		920	168	215	1,303	1,500
1190	Difficult soil nailing, grade 75, 20 min. setup per hole & 80'/hr. drilling									
1200	Soil nailing, drill hole, install #8 nail, grout 20' depth difficult	B-47G	13.70	2.336	Ea.	147	111	142	400	485
1210	25' depth difficult		12.30	2.602		184	124	159	467	560
1220	30' depth difficult		11.20	2.857		365	136	174	675	795
1230	35' depth difficult		10.40	3.077		420	147	188	755	890
1240	40' depth difficult		9.60	3.333		480	159	203	842	990
1250	45' depth difficult		8.90	3.596		540	171	219	930	1,100
1260	50' depth difficult		8.30	3.855		625	184	235	1,044	1,225
1270	55' depth difficult		7.90	4.051		685	193	247	1,125	1,300
1280	60' depth difficult		7.40	4.324		745	206	264	1,215	1,425
1290	65' depth difficult		7	4.571		805	218	279	1,302	1,525
1300	70' depth difficult		6.60	4.848		860	231	296	1,387	1,625
1310	75' depth difficult		6.30	5.079		920	242	310	1,472	1,725
1390	Difficult soil nailing, grade 75, 20 min. setup per hole & 90'/hr. drilling									
1400	Soil nailing, drill hole, install #8 nail, grout 20' depth difficult	B-47G	14.10	2.270	Ea.	147	108	138	393	475
1410	25' depth difficult		13	2.462		184	117	150	451	540
1420	30' depth difficult		12	2.667		365	127	163	655	770
1430	35' depth difficult		10.90	2.936		420	140	179	739	870
1440	40' depth difficult		10.20	3.137		480	150	191	821	965
1450	45' depth difficult		9.60	3.333		540	159	203	902	1,050
1460	50' depth difficult		8.90	3.596		625	171	219	1,015	1,175
1470	55' depth difficult		8.40	3.810		685	182	232	1,099	1,275
1480	60' depth difficult		8	4		745	191	244	1,180	1,375
1490	65' depth difficult		7.50	4.267		805	203	260	1,268	1,475
1500	70' depth difficult		7.20	4.444		860	212	271	1,343	1,575
1510	75' depth difficult		6.90	4.638		920	221	283	1,424	1,675
1590	Difficult soil nailing, grade 75, 20 min. setup per hole & 100'/hr drilling									
1600	Soil nailing, drill hole, install #8 nail, grout 20' depth difficult	B-47G	15	2.133	Ea.	147	102	130	379	455
1610	25' depth difficult		13.70	2.336		184	111	142	437	525
1620	30' depth difficult		12.60	2.540		365	121	155	641	750
1630	35' depth difficult		11.70	2.735		420	130	167	717	845
1640	40' depth difficult		10.90	2.936		480	140	179	799	935
1650	45' depth difficult		10.20	3.137		540	150	191	881	1,025

31 32 Soil Stabilization

31 32 36 – Soil Nailing

31 32 36.16 Grouted Soil Nailing		Crew	Daily Output	Labor-Hours	Unit	Material	2021 Bare Costs Labor	Equipment	Total	Total Incl O&P
1660	50' depth difficult	B-47G	9.60	3.333	Ea.	625	159	203	987	1,150
1670	55' depth difficult		9.10	3.516		685	168	215	1,068	1,225
1680	60' depth difficult		8.60	3.721		745	177	227	1,149	1,325
1690	65' depth difficult		8.10	3.951		805	188	241	1,234	1,425
1700	70' depth difficult		7.70	4.156		860	198	254	1,312	1,525
1710	75' depth difficult		7.40	4.324		920	206	264	1,390	1,625
1790	Difficult soil nailing, grade 75, 20 min. setup per hole & 110'/hr drilling									
1800	Soil nailing, drill hole, install #8 nail, grout 20' depth difficult	B-47G	15.50	2.065	Ea.	147	98.50	126	371.50	450
1810	25' depth difficult		14.10	2.270		184	108	138	430	515
1820	30' depth difficult		13.30	2.406		365	115	147	627	730
1830	35' depth difficult		12.30	2.602		420	124	159	703	825
1840	40' depth difficult		11.40	2.807		480	134	171	785	920
1850	45' depth difficult		10.70	2.991		540	143	182	865	1,000
1860	50' depth difficult		10.20	3.137		625	150	191	966	1,125
1870	55' depth difficult		9.60	3.333		685	159	203	1,047	1,200
1880	60' depth difficult		9.10	3.516		745	168	215	1,128	1,300
1890	65' depth difficult		8.60	3.721		805	177	227	1,209	1,400
1900	70' depth difficult		8.30	3.855		860	184	235	1,279	1,475
1910	75' depth difficult		7.90	4.051		920	193	247	1,360	1,575
1990	Difficult soil nailing, grade 75, 20 min. setup per hole & 120'/hr drilling									
2000	Soil nailing, drill hole, install #8 nail, grout 20' depth difficult	B-47G	16	2	Ea.	147	95.50	122	364.50	440
2010	25' depth difficult		14.60	2.192		184	105	134	423	505
2020	30' depth difficult		13.70	2.336		365	111	142	618	725
2030	35' depth difficult		12.60	2.540		420	121	155	696	815
2040	40' depth difficult		12	2.667		480	127	163	770	900
2050	45' depth difficult		11.20	2.857		540	136	174	850	990
2060	50' depth difficult		10.70	2.991		625	143	182	950	1,100
2070	55' depth difficult		10	3.200		685	153	195	1,033	1,200
2080	60' depth difficult		9.60	3.333		745	159	203	1,107	1,275
2090	65' depth difficult		9.10	3.516		805	168	215	1,188	1,375
2100	70' depth difficult		8.70	3.678		860	175	224	1,259	1,450
2110	75' depth difficult		8.30	3.855		920	184	235	1,339	1,550
2990	Very difficult soil nailing, grade 75, 25 min. setup & 80'/hr. drilling									
3000	Soil nailing, drill, install #8 nail, grout 20' depth very difficult	B-47G	12	2.667	Ea.	147	127	163	437	530
3010	25' depth very difficult		10.90	2.936		184	140	179	503	610
3020	30' depth very difficult		10	3.200		365	153	195	713	845
3030	35' depth very difficult		9.40	3.404		420	162	208	790	935
3040	40' depth very difficult		8.70	3.678		480	175	224	879	1,050
3050	45' depth very difficult		8.10	3.951		540	188	241	969	1,150
3060	50' depth very difficult		7.60	4.211		625	201	257	1,083	1,275
3070	55' depth very difficult		7.30	4.384		685	209	267	1,161	1,350
3080	60' depth very difficult		6.90	4.638		745	221	283	1,249	1,450
3090	65' depth very difficult		6.50	4.923		805	235	300	1,340	1,575
3100	70' depth very difficult		6.20	5.161		860	246	315	1,421	1,650
3110	75' depth very difficult		5.90	5.424		920	259	330	1,509	1,775
3190	Very difficult soil nailing, grade 75, 25 min. setup & 90'/hr. drilling									
3200	Soil nailing, drill, install #8 nail, grout 20' depth very difficult	B-47G	12.30	2.602	Ea.	147	124	159	430	520
3210	25' depth very difficult		11.40	2.807		184	134	171	489	590
3220	30' depth very difficult		10.70	2.991		365	143	182	690	815
3230	35' depth very difficult		9.80	3.265		420	156	199	775	915
3240	40' depth very difficult		9.20	3.478		480	166	212	858	1,000
3250	45' depth very difficult		8.70	3.678		540	175	224	939	1,100
3260	50' depth very difficult		8.10	3.951		625	188	241	1,054	1,225

For customer support on your Site Work & Landscape Costs with RSMeans Data, call 800.448.8182.

31 32 36.16 Grouted Soil Nailing	Crew	Daily Output	Labor-Hours	Unit	Material	2021 Bare Costs Labor	Equipment	Total	Total Incl O&P
3270	55' depth very difficult	B-47G	7.70	4.156	Ea.	685	198	254	1,137	1,325
3280	60' depth very difficult		7.40	4.324		745	206	264	1,215	1,425
3290	65' depth very difficult		7	4.571		805	218	279	1,302	1,525
3300	70' depth very difficult		6.70	4.776		860	228	291	1,379	1,600
3310	75' depth very difficult		6.40	5		920	238	305	1,463	1,725
3390	Very difficult soil nailing, grade 75, 25 min. setup & 100'/hr. drilling									
3400	Soil nailing, drill, install #8 nail, grout 20' depth very difficult	B-47G	13	2.462	Ea.	147	117	150	414	500
3410	25' depth very difficult		12	2.667		184	127	163	474	570
3420	30' depth very difficult		11.20	2.857		365	136	174	675	795
3430	35' depth very difficult		10.40	3.077		420	147	188	755	890
3440	40' depth very difficult		9.80	3.265		480	156	199	835	980
3450	45' depth very difficult		9.20	3.478		540	166	212	918	1,075
3460	50' depth very difficult		8.70	3.678		625	175	224	1,024	1,200
3470	55' depth very difficult		8.30	3.855		685	184	235	1,104	1,275
3480	60' depth very difficult		7.90	4.051		745	193	247	1,185	1,375
3490	65' depth very difficult		7.50	4.267		805	203	260	1,268	1,475
3500	70' depth very difficult		7.20	4.444		860	212	271	1,343	1,575
3510	75' depth very difficult		6.90	4.638		920	221	283	1,424	1,675
3590	Very difficult soil nailing, grade 75, 25 min. setup & 110'/hr. drilling									
3600	Soil nailing, drill, install #8 nail, grout 20' depth very difficult	B-47G	13.30	2.406	Ea.	147	115	147	409	495
3610	25' depth very difficult		13.70	2.336		184	111	142	437	525
3620	30' depth very difficult		11.70	2.735		365	130	167	662	780
3630	35' depth very difficult		10.90	2.936		420	140	179	739	870
3640	40' depth very difficult		10.20	3.137		480	150	191	821	965
3650	45' depth very difficult		9.60	3.333		540	159	203	902	1,050
3660	50' depth very difficult		9.20	3.478		625	166	212	1,003	1,175
3670	55' depth very difficult		8.70	3.678		685	175	224	1,084	1,250
3680	60' depth very difficult		8.30	3.855		745	184	235	1,164	1,350
3690	65' depth very difficult		7.90	4.051		805	193	247	1,245	1,450
3700	70' depth very difficult		7.60	4.211		860	201	257	1,318	1,525
3710	75' depth very difficult		7.30	4.384		920	209	267	1,396	1,625
3790	Very difficult soil nailing, grade 75, 25 min. setup & 120'/hr. drilling									
3800	Soil nailing, drill, install #8 nail, grout 20' depth very difficult	B-47G	13.70	2.336	Ea.	147	111	142	400	485
3810	25' depth very difficult		12.60	2.540		184	121	155	460	555
3820	30' depth very difficult		12	2.667		365	127	163	655	770
3830	35' depth very difficult		11.20	2.857		420	136	174	730	860
3840	40' depth very difficult		10.70	2.991		480	143	182	805	945
3850	45' depth very difficult		10	3.200		540	153	195	888	1,050
3860	50' depth very difficult		9.60	3.333		625	159	203	987	1,150
3870	55' depth very difficult		9.10	3.516		685	168	215	1,068	1,225
3880	60' depth very difficult		8.70	3.678		745	175	224	1,144	1,325
3890	65' depth very difficult		8.30	3.855		805	184	235	1,224	1,425
3900	70' depth very difficult		8	4		860	191	244	1,295	1,500
3910	75' depth very difficult		7.60	4.211		920	201	257	1,378	1,600
3990	Severe soil nailing, grade 75, 30 min. setup per hole & 80'/hr. drilling									
4000	Soil nailing, drill hole, install #8 nail, grout 20' depth severe	B-47G	10.70	2.991	Ea.	147	143	182	472	575
4010	25' depth severe		9.80	3.265		184	156	199	539	655
4020	30' depth severe		9.10	3.516		365	168	215	748	885
4030	35' depth severe		8.60	3.721		420	177	227	824	980
4040	40' depth severe		8	4		480	191	244	915	1,075
4050	45' depth severe		7.50	4.267		540	203	260	1,003	1,175
4060	50' depth severe		7.10	4.507		625	215	275	1,115	1,300
4070	55' depth severe		6.80	4.706		685	224	287	1,196	1,400

31 32 36.16 Grouted Soil Nailing		Crew	Daily Output	Labor-Hours	Unit	Material	2021 Bare Costs Labor	Equipment	Total	Total Incl O&P
4080	60' depth severe	B-47G	6.40	5	Ea.	745	238	305	1,288	1,500
4090	65' depth severe		6.10	5.246		805	250	320	1,375	1,600
4100	70' depth severe		5.80	5.517		860	263	335	1,458	1,700
4110	75' depth severe		5.60	5.714		920	272	350	1,542	1,825
4190	Severe soil nailing, grade 75, 30 min. setup per hole & 90'/hr. drilling									
4200	Soil nailing, drill hole, install #8 nail, grout 20' depth severe	B-47G	10.90	2.936	Ea.	147	140	179	466	570
4210	25' depth severe		10.20	3.137		184	150	191	525	635
4220	30' depth severe		9.60	3.333		365	159	203	727	860
4230	35' depth severe		8.90	3.596		420	171	219	810	960
4240	40' depth severe		8.40	3.810		480	182	232	894	1,050
4250	45' depth severe		8	4		540	191	244	975	1,150
4260	50' depth severe		7.50	4.267		625	203	260	1,088	1,275
4270	55' depth severe		7.20	4.444		685	212	271	1,168	1,375
4280	60' depth severe		6.90	4.638		745	221	283	1,249	1,450
4290	65' depth severe		6.50	4.923		805	235	300	1,340	1,575
4300	70' depth severe		6.20	5.161		860	246	315	1,421	1,650
4310	75' depth severe		6	5.333		920	254	325	1,499	1,775
4390	Severe soil nailing, grade 75, 30 min. setup per hole & 100'/hr. drilling									
4400	Soil nailing, drill hole, install #8 nail, grout 20' depth severe	B-47G	11.40	2.807	Ea.	147	134	171	452	550
4410	25' depth severe		10.70	2.991		184	143	182	509	615
4420	30' depth severe		10	3.200		365	153	195	713	845
4430	35' depth severe		9.40	3.404		420	162	208	790	935
4440	40' depth severe		8.90	3.596		480	171	219	870	1,025
4450	45' depth severe		8.40	3.810		540	182	232	954	1,125
4460	50' depth severe		8	4		625	191	244	1,060	1,225
4470	55' depth severe		7.60	4.211		685	201	257	1,143	1,325
4480	60' depth severe		7.30	4.384		745	209	267	1,221	1,425
4490	65' depth severe		7	4.571		805	218	279	1,302	1,525
4500	70' depth severe		6.70	4.776		860	228	291	1,379	1,600
4510	75' depth severe		6.40	5		920	238	305	1,463	1,725
4590	Severe soil nailing, grade 75, 30 min. setup per hole & 110'/hr. drilling									
4600	Soil nailing, drill hole, install #8 nail, grout 20' depth severe	B-47G	11.70	2.735	Ea.	147	130	167	444	540
4610	25' depth severe		10.90	2.936		184	140	179	503	610
4620	30' depth severe		10.40	3.077		365	147	188	700	825
4630	35' depth severe		9.80	3.265		420	156	199	775	915
4640	40' depth severe		9.20	3.478		480	166	212	858	1,000
4650	45' depth severe		8.70	3.678		540	175	224	939	1,100
4660	50' depth severe		8.40	3.810		625	182	232	1,039	1,200
4670	55' depth severe		8	4		685	191	244	1,120	1,300
4680	60' depth severe		7.60	4.211		745	201	257	1,203	1,400
4690	65' depth severe		7.30	4.384		805	209	267	1,281	1,500
4700	70' depth severe		7.10	4.507		860	215	275	1,350	1,575
4710	75' depth severe		6.80	4.706		920	224	287	1,431	1,675
4790	Severe soil nailing, grade 75, 30 min. setup per hole & 120'/hr. drilling									
4800	Soil nailing, drill hole, install #8 nail, grout 20' depth severe	B-47G	12	2.667	Ea.	147	127	163	437	530
4810	25' depth severe		11.20	2.857		184	136	174	494	595
4820	30' depth severe		10.70	2.991		365	143	182	690	815
4830	35' depth severe		10	3.200		420	153	195	768	910
4840	40' depth severe		9.60	3.333		480	159	203	842	990
4850	45' depth severe		9.10	3.516		540	168	215	923	1,075
4860	50' depth severe		8.70	3.678		625	175	224	1,024	1,200
4870	55' depth severe		8.30	3.855		685	184	235	1,104	1,275
4880	60' depth severe		8	4		745	191	244	1,180	1,375

For customer support on your Site Work & Landscape Costs with RSMeans Data, call 800.448.8182.

31 32 Soil Stabilization

31 32 36 – Soil Nailing

31 32 36.16 Grouted Soil Nailing		Crew	Daily Output	Labor-Hours	Unit	Material	2021 Bare Costs Labor	Equipment	Total	Total Incl O&P
4890	65' depth severe	B-47G	7.60	4.211	Ea.	805	201	257	1,263	1,475
4900	70' depth severe		7.40	4.324		860	206	264	1,330	1,550
4910	75' depth severe		7.10	4.507		920	215	275	1,410	1,650

31 36 Gabions

31 36 13 – Gabion Boxes

31 36 13.10 Gabion Box Systems

		Crew	Daily Output	Labor-Hours	Unit	Material	Labor	Equipment	Total	Total Incl O&P
0010	**GABION BOX SYSTEMS**									
0400	Gabions, galvanized steel mesh mats or boxes, stone filled, 6" deep	B-13	200	.280	S.Y.	17.45	13.50	2.93	33.88	42.50
0500	9" deep		163	.344		26	16.60	3.60	46.20	57
0600	12" deep		153	.366		35.50	17.65	3.84	56.99	69.50
0700	18" deep		102	.549		45	26.50	5.75	77.25	95.50
0800	36" deep		60	.933		73	45	9.80	127.80	158

31 37 Riprap

31 37 13 – Machined Riprap

31 37 13.10 Riprap and Rock Lining

		Crew	Daily Output	Labor-Hours	Unit	Material	Labor	Equipment	Total	Total Incl O&P
0010	**RIPRAP AND ROCK LINING**									
0011	Random, broken stone									
0100	Machine placed for slope protection	B-12G	62	.258	L.C.Y.	32	13.65	14.15	59.80	71
0110	3/8 to 1/4 C.Y. pieces, grouted	B-13	80	.700	S.Y.	67.50	34	7.35	108.85	133
0200	18" minimum thickness, not grouted	"	53	1.057	"	19.85	51	11.05	81.90	110
0300	Dumped, 50 lb. average	B-11A	800	.020	Ton	28.50	1.03	1.90	31.43	35
0350	100 lb. average		700	.023		28.50	1.18	2.17	31.85	35.50
0370	300 lb. average		600	.027		28.50	1.38	2.53	32.41	36.50

31 41 Shoring

31 41 13 – Timber Shoring

31 41 13.10 Building Shoring

		Crew	Daily Output	Labor-Hours	Unit	Material	Labor	Equipment	Total	Total Incl O&P
0010	**BUILDING SHORING**									
0020	Shoring, existing building, with timber, no salvage allowance	B-51	2.20	21.818	M.B.F.	1,250	990	90.50	2,330.50	2,950
1000	On cribbing with 35 ton screw jacks, per box and jack	"	3.60	13.333	Jack	65	605	55	725	1,025

31 41 16 – Sheet Piling

31 41 16.10 Sheet Piling Systems

			Crew	Daily Output	Labor-Hours	Unit	Material	Labor	Equipment	Total	Total Incl O&P
0010	**SHEET PILING SYSTEMS**										
0020	Sheet piling, 50,000 psi steel, not incl. wales, 22 psf, left in place		B-40	10.81	5.920	Ton	1,900	340	325	2,565	2,950
0100	Drive, extract & salvage	R314116-40		6	10.667		530	610	590	1,730	2,150
0300	20' deep excavation, 27 psf, left in place			12.95	4.942		1,900	282	273	2,455	2,800
0400	Drive, extract & salvage	R314116-45		6.55	9.771		530	560	540	1,630	2,025
0600	25' deep excavation, 38 psf, left in place			19	3.368		1,900	192	186	2,278	2,575
0700	Drive, extract & salvage			10.50	6.095		530	350	335	1,215	1,475
0900	40' deep excavation, 38 psf, left in place			21.20	3.019		1,900	172	166	2,238	2,525
1000	Drive, extract & salvage			12.25	5.224		530	298	288	1,116	1,350
1200	15' deep excavation, 22 psf, left in place			983	.065	S.F.	22	3.72	3.59	29.31	33.50
1300	Drive, extract & salvage			545	.117		5.95	6.70	6.50	19.15	24
1500	20' deep excavation, 27 psf, left in place			960	.067		27.50	3.81	3.68	34.99	40.50
1600	Drive, extract & salvage			485	.132		7.70	7.55	7.30	22.55	28

345

31 41 Shoring

31 41 16 – Sheet Piling

31 41 16.10 Sheet Piling Systems	Crew	Daily Output	Labor-Hours	Unit	Material	2021 Bare Costs Labor	Equipment	Total	Total Incl O&P
1800	25' deep excavation, 38 psf, left in place	B-40	1000	.064	S.F.	41	3.66	3.53	48.19	54.50
1900	Drive, extract & salvage	↓	553	.116	↓	10.60	6.60	6.40	23.60	29
2100	Rent steel sheet piling and wales, first month				Ton	330			330	365
2200	Per added month					33			33	36.50
2300	Rental piling left in place, add to rental					1,225			1,225	1,350
2500	Wales, connections & struts, 2/3 salvage					530			530	585
2700	High strength piling, 60,000 psi, add					190			190	208
2800	65,000 psi, add					284			284	315
3000	Tie rod, not upset, 1-1/2" to 4" diameter with turnbuckle					2,375			2,375	2,600
3100	No turnbuckle					1,675			1,675	1,850
3300	Upset, 1-3/4" to 4" diameter with turnbuckle					2,500			2,500	2,750
3400	No turnbuckle				↓	2,125			2,125	2,325
3600	Lightweight, 18" to 28" wide, 7 ga., 9.22 psf, and									
3610	9 ga., 8.6 psf, minimum				Lb.	.72			.72	.79
3700	Average					.92			.92	1.01
3750	Maximum				↓	1.01			1.01	1.11
3900	Wood, solid sheeting, incl. wales, braces and spacers,									
3910	drive, extract & salvage, 8' deep excavation	B-31	330	.121	S.F.	3.34	5.70	.78	9.82	13.05
4000	10' deep, 50 S.F./hr. in & 150 S.F./hr. out		300	.133		3.44	6.25	.85	10.54	14
4100	12' deep, 45 S.F./hr. in & 135 S.F./hr. out		270	.148		3.54	6.95	.95	11.44	15.30
4200	14' deep, 42 S.F./hr. in & 126 S.F./hr. out		250	.160		3.64	7.50	1.02	12.16	16.35
4300	16' deep, 40 S.F./hr. in & 120 S.F./hr. out		240	.167		3.76	7.80	1.07	12.63	16.95
4400	18' deep, 38 S.F./hr. in & 114 S.F./hr. out		230	.174		3.88	8.15	1.11	13.14	17.65
4500	20' deep, 35 S.F./hr. in & 105 S.F./hr. out		210	.190		4.01	8.95	1.22	14.18	19.05
4520	Left in place, 8' deep, 55 S.F./hr.		440	.091		6	4.26	.58	10.84	13.60
4540	10' deep, 50 S.F./hr.		400	.100		6.35	4.69	.64	11.68	14.65
4560	12' deep, 45 S.F./hr.		360	.111		6.70	5.20	.71	12.61	15.90
4565	14' deep, 42 S.F./hr.		335	.119		7.10	5.60	.76	13.46	17
4570	16' deep, 40 S.F./hr.		320	.125		7.50	5.85	.80	14.15	17.90
4580	18' deep, 38 S.F./hr.		305	.131		8	6.15	.84	14.99	18.85
4590	20' deep, 35 S.F./hr.		280	.143	↓	8.60	6.70	.91	16.21	20.50
4700	Alternate pricing, left in place, 8' deep		1.76	22.727	M.B.F.	1,350	1,075	145	2,570	3,225
4800	Drive, extract and salvage, 8' deep	↓	1.32	30.303	"	1,200	1,425	194	2,819	3,675
5000	For treated lumber add cost of treatment to lumber									

31 43 Concrete Raising

31 43 13 – Pressure Grouting

31 43 13.13 Concrete Pressure Grouting	Crew	Daily Output	Labor-Hours	Unit	Material	2021 Bare Costs Labor	Equipment	Total	Total Incl O&P
0010	**CONCRETE PRESSURE GROUTING**									
0020	Grouting, pressure, cement & sand, 1:1 mix, minimum	B-61	124	.323	Bag	21.50	15.15	2.62	39.27	49
0100	Maximum		51	.784	"	21.50	37	6.35	64.85	85.50
0200	Cement and sand, 1:1 mix, minimum		250	.160	C.F.	43	7.50	1.30	51.80	60
0300	Maximum		100	.400		64.50	18.80	3.25	86.55	103
0310	Grouting, cement and sand, 1:2 mix, geothermal wells		250	.160		17.20	7.50	1.30	26	31.50
0320	Bentonite mix 1:4		250	.160		8.90	7.50	1.30	17.70	22.50
0400	Epoxy cement grout, minimum		137	.292		805	13.75	2.37	821.12	910
0500	Maximum	↓	57	.702	↓	805	33	5.70	843.70	940
0700	Alternate pricing method: (Add for materials)									
0710	5 person crew and equipment	B-61	1	40	Day		1,875	325	2,200	3,150

31 43 Concrete Raising

31 43 19 – Mechanical Jacking

31 43 19.10 Slabjacking	Crew	Daily Output	Labor-Hours	Unit	Material	2021 Bare Costs Labor	Equipment	Total	Total Incl O&P
0010 **SLABJACKING**									
0100 4" thick slab	D-4	1500	.021	S.F.	.39	1.05	.13	1.57	2.15
0150 6" thick slab		1200	.027		.55	1.31	.16	2.02	2.74
0200 8" thick slab		1000	.032		.65	1.57	.19	2.41	3.29
0250 10" thick slab		900	.036		.70	1.75	.21	2.66	3.63
0300 12" thick slab		850	.038		.76	1.85	.22	2.83	3.87

31 45 Vibroflotation and Densification

31 45 13 – Vibroflotation

31 45 13.10 Vibroflotation Densification

	Crew	Daily Output	Labor-Hours	Unit	Material	2021 Bare Costs Labor	Equipment	Total	Total Incl O&P
0010 **VIBROFLOTATION DENSIFICATION**									
0900 Vibroflotation compacted sand cylinder, minimum	B-60	750	.075	V.L.F.		3.84	3.04	6.88	9.10
0950 Maximum		325	.172			8.85	7.05	15.90	21
1100 Vibro replacement compacted stone cylinder, minimum		500	.112			5.75	4.57	10.32	13.60
1150 Maximum		250	.224			11.50	9.15	20.65	27.50
1300 Mobilization and demobilization, minimum		.47	119	Total		6,125	4,850	10,975	14,500
1400 Maximum		.14	400	"		20,600	16,300	36,900	48,600

31 48 Underpinning

31 48 13 – Underpinning Piers

31 48 13.10 Underpinning Foundations

	Crew	Daily Output	Labor-Hours	Unit	Material	2021 Bare Costs Labor	Equipment	Total	Total Incl O&P
0010 **UNDERPINNING FOUNDATIONS**									
0011 Including excavation,									
0020 forming, reinforcing, concrete and equipment									
0100 5' to 16' below grade, 100 to 500 C.Y.	B-52	2.30	24.348	C.Y.	345	1,225	249	1,819	2,500
0200 Over 500 C.Y.		2.50	22.400		310	1,125	229	1,664	2,300
0400 16' to 25' below grade, 100 to 500 C.Y.		2	28		380	1,425	286	2,091	2,850
0500 Over 500 C.Y.		2.10	26.667		355	1,350	273	1,978	2,725
0700 26' to 40' below grade, 100 to 500 C.Y.		1.60	35		410	1,775	360	2,545	3,500
0800 Over 500 C.Y.		1.80	31.111		380	1,575	320	2,275	3,125
0900 For under 50 C.Y., add					10%	40%			
1000 For 50 C.Y. to 100 C.Y., add					5%	20%			

31 52 Cofferdams

31 52 16 – Timber Cofferdams

31 52 16.10 Cofferdams

	Crew	Daily Output	Labor-Hours	Unit	Material	2021 Bare Costs Labor	Equipment	Total	Total Incl O&P
0010 **COFFERDAMS**									
0011 Incl. mobilization and temporary sheeting									
0020 Shore driven	B-40	960	.067	S.F.	27	3.81	3.68	34.49	40
0060 Barge driven	"	550	.116	"	27	6.65	6.40	40.05	47
0080 Soldier beams & lagging H-piles with 3" wood sheeting									
0090 horizontal between piles, including removal of wales & braces									
0100 No hydrostatic head, 15' deep, 1 line of braces, minimum	B-50	545	.206	S.F.	12.50	11.15	4.79	28.44	36
0200 Maximum		495	.226		13.85	12.30	5.30	31.45	40
0400 15' to 22' deep with 2 lines of braces, 10" H, minimum		360	.311		14.70	16.90	7.25	38.85	49.50
0500 Maximum		330	.339		16.65	18.40	7.90	42.95	55
0700 23' to 35' deep with 3 lines of braces, 12" H, minimum		325	.345		19.20	18.70	8.05	45.95	58.50

For customer support on your Site Work & Landscape Costs with RSMeans Data, call 800.448.8182.

347

31 52 Cofferdams

31 52 16 – Timber Cofferdams

31 52 16.10 Cofferdams	Crew	Daily Output	Labor-Hours	Unit	Material	2021 Bare Costs Labor	Equipment	Total	Total Incl O&P
0800	Maximum	B-50	295	.380	S.F.	21	20.50	8.85	50.35	64.50
1000	36' to 45' deep with 4 lines of braces, 14" H, minimum		290	.386		21.50	21	9	51.50	65.50
1100	Maximum		265	.423		22.50	23	9.85	55.35	71
1300	No hydrostatic head, left in place, 15' deep, 1 line of braces, min.		635	.176		16.65	9.55	4.11	30.31	37.50
1400	Maximum		575	.195		17.85	10.55	4.54	32.94	40.50
1600	15' to 22' deep with 2 lines of braces, minimum		455	.246		25	13.35	5.75	44.10	54.50
1700	Maximum		415	.270		27.50	14.65	6.30	48.45	60
1900	23' to 35' deep with 3 lines of braces, minimum		420	.267		29.50	14.45	6.20	50.15	61.50
2000	Maximum		380	.295		33	16	6.85	55.85	68
2200	36' to 45' deep with 4 lines of braces, minimum		385	.291		35.50	15.80	6.80	58.10	70.50
2300	Maximum	↓	350	.320		41.50	17.35	7.45	66.30	80.50
2350	Lagging only, 3" thick wood between piles 8' OC, minimum	B-46	400	.120		2.77	6.05	.10	8.92	12.35
2370	Maximum		250	.192		4.16	9.70	.17	14.03	19.50
2400	Open sheeting no bracing, for trenches to 10' deep, min.		1736	.028		1.25	1.40	.02	2.67	3.52
2450	Maximum	↓	1510	.032		1.39	1.60	.03	3.02	4
2500	Tie-back method, add to open sheeting, add, minimum								20%	20%
2550	Maximum				↓				60%	60%
2700	Tie-backs only, based on tie-backs total length, minimum	B-46	86.80	.553	L.F.	19.55	28	.48	48.03	64.50
2750	Maximum		38.50	1.247	"	34.50	63	1.08	98.58	135
3500	Tie-backs only, typical average, 25' long		2	24	Ea.	860	1,200	21	2,081	2,825
3600	35' long	↓	1.58	30.380	"	1,150	1,525	26.50	2,701.50	3,600
4500	Trench box, 7' deep, 16' x 8', see 01 54 33 in Reference Section				Day				191	210
4600	20' x 10', see 01 54 33 in Reference Section				"				240	265
5200	Wood sheeting, in trench, jacks at 4' OC, 8' deep	B-1	800	.030	S.F.	.96	1.35		2.31	3.08
5250	12' deep		700	.034		1.13	1.55		2.68	3.56
5300	15' deep	↓	600	.040	↓	1.56	1.80		3.36	4.41
6000	See also Section 31 41 16.10									

31 56 Slurry Walls

31 56 23 – Lean Concrete Slurry Walls

31 56 23.20 Slurry Trench

		Crew	Daily Output	Labor-Hours	Unit	Material	2021 Bare Costs Labor	Equipment	Total	Total Incl O&P
0010	**SLURRY TRENCH**									
0011	Excavated slurry trench in wet soils									
0020	backfilled with 3,000 psi concrete, no reinforcing steel									
0050	Minimum	C-7	333	.216	C.F.	9.20	10.35	3.27	22.82	29
0100	Maximum		200	.360	"	15.45	17.25	5.45	38.15	48.50
0200	Alternate pricing method, minimum		150	.480	S.F.	18.45	23	7.25	48.70	63
0300	Maximum	↓	120	.600		27.50	29	9.10	65.60	83.50
0500	Reinforced slurry trench, minimum	B-48	177	.316		13.85	15.75	5.90	35.50	45
0600	Maximum	"	69	.812	↓	46	40.50	15.20	101.70	128
0800	Haul for disposal, 2 mile haul, excavated material, add	B-34B	99	.081	C.Y.		4.15	5.85	10	12.65
0900	Haul bentonite castings for disposal, add	"	40	.200	"		10.25	14.50	24.75	31.50

31 62 13 – Concrete Piles

31 62 13.23 Prestressed Concrete Piles	Crew	Daily Output	Labor-Hours	Unit	Material	2021 Bare Costs Labor	Equipment	Total	Total Incl O&P
0010 **PRESTRESSED CONCRETE PILES**, 200 piles									
0020 Unless specified otherwise, not incl. pile caps or mobilization									
2200 Precast, prestressed, 50' long, cylinder, 12" diam., 2-3/8" wall	B-19	720	.089	V.L.F.	36.50	5.10	2.84	44.44	51
2300 14" diameter, 2-1/2" wall		680	.094		38	5.40	3.01	46.41	53.50
2500 16" diameter, 3" wall		640	.100		51.50	5.70	3.20	60.40	68.50
2600 18" diameter, 3-1/2" wall	B-19A	600	.107		65	6.10	4.87	75.97	86
2800 20" diameter, 4" wall		560	.114		56.50	6.55	5.20	68.25	77.50
2900 24" diameter, 5" wall		520	.123		80.50	7.05	5.60	93.15	105
2920 36" diameter, 5-1/2" wall	B-19	400	.160		99	9.15	5.10	113.25	129
2940 54" diameter, 6" wall		340	.188		250	10.75	6	266.75	298
2950 60" diameter, 6" wall		280	.229		257	13.05	7.30	277.35	310
2960 66" diameter, 6-1/2" wall		220	.291		320	16.60	9.30	345.90	385
3100 Precast, prestressed, 40' long, 10" thick, square		700	.091		14.50	5.20	2.92	22.62	27
3200 12" thick, square		680	.094		26	5.40	3.01	34.41	40
3275 Shipping for 60' long concrete piles, add to material cost					2.36			2.36	2.60
3400 14" thick, square	B-19	600	.107		29.50	6.10	3.41	39.01	45.50
3500 Octagonal		640	.100		28.50	5.70	3.20	37.40	43.50
3700 16" thick, square		560	.114		31	6.55	3.65	41.20	48
3800 Octagonal		600	.107		34.50	6.10	3.41	44.01	51
4000 18" thick, square	B-19A	520	.123		53	7.05	5.60	65.65	75.50
4100 Octagonal	B-19	560	.114		46	6.55	3.65	56.20	64.50
4300 20" thick, square	B-19A	480	.133		49	7.60	6.10	62.70	72.50
4400 Octagonal	B-19	520	.123		51.50	7.05	3.93	62.48	71.50
4600 24" thick, square	B-19A	440	.145		70	8.30	6.65	84.95	97
4700 Octagonal	B-19	480	.133		60.50	7.60	4.26	72.36	83.50
4730 Precast, prestressed, 60' long, 10" thick, square		700	.091		15.25	5.20	2.92	23.37	28
4740 12" thick, square (60' long)		680	.094		27	5.40	3.01	35.41	41
4750 Mobilization for 10,000 L.F. pile job, add		3300	.019			1.11	.62	1.73	2.37
4800 25,000 L.F. pile job, add		8500	.008			.43	.24	.67	.92
4850 Mobilization by water for barge driving rig, add								100%	100%

31 62 16 – Steel Piles

31 62 16.13 Steel Piles	Crew	Daily Output	Labor-Hours	Unit	Material	2021 Bare Costs Labor	Equipment	Total	Total Incl O&P
0010 **STEEL PILES**									
0100 Step tapered, round, concrete filled									
0110 8" tip, 12" butt, 60 ton capacity, 30' depth	B-19	760	.084	V.L.F.	17.75	4.81	2.69	25.25	30
0120 60' depth with extension		740	.086		37.50	4.94	2.76	45.20	52
0130 80' depth with extensions		700	.091		58.50	5.20	2.92	66.62	75.50
0250 "H" Sections, 50' long, HP8 x 36		640	.100		17.75	5.70	3.20	26.65	31.50
0400 HP10 x 42		610	.105		21	6	3.35	30.35	36
0500 HP10 x 57		610	.105		27	6	3.35	36.35	43
0700 HP12 x 53		590	.108		28	6.20	3.47	37.67	44.50
0800 HP12 x 74	B-19A	590	.108		34.50	6.20	4.95	45.65	53
1000 HP14 x 73		540	.119		36.50	6.75	5.40	48.65	56.50
1100 HP14 x 89		540	.119		46.50	6.75	5.40	58.65	67.50
1300 HP14 x 102		510	.125		53	7.15	5.75	65.90	76
1400 HP14 x 117		510	.125		61.50	7.15	5.75	74.40	85
1600 Splice on standard points, not in leads, 8" or 10"	1 Sswl	5	1.600	Ea.	117	96.50		213.50	278
1700 12" or 14"		4	2		165	121		286	370
1900 Heavy duty points, not in leads, 10" wide		4	2		187	121		308	395
2100 14" wide		3.50	2.286		231	138		369	465

For customer support on your Site Work & Landscape Costs with RSMeans Data, call 800.448.8182.

349

31 62 19.10 Wood Piles	Crew	Daily Output	Labor-Hours	Unit	Material	2021 Bare Costs Labor	2021 Bare Costs Equipment	Total	Total Incl O&P
0010 **WOOD PILES**									
0011 Friction or end bearing, not including									
0050 mobilization or demobilization									
0100 ACZA treated piles, 1.0 lb./C.F., up to 30' long, 12" butts, 8" points	B-19	625	.102	V.L.F.	15.05	5.85	3.27	24.17	29
0200 30' to 39' long, 12" butts, 8" points		700	.091		15.90	5.20	2.92	24.02	28.50
0300 40' to 49' long, 12" butts, 7" points		720	.089		16.75	5.10	2.84	24.69	29.50
0400 50' to 59' long, 13" butts, 7" points		800	.080		23.50	4.57	2.56	30.63	36
0500 60' to 69' long, 13" butts, 7" points		840	.076		27	4.35	2.43	33.78	39
0600 70' to 80' long, 13" butts, 6" points	▼	840	.076	▼	27	4.35	2.43	33.78	39.50
0800 ACZA Treated piles, 1.5 lb./C.F.									
0810 friction or end bearing, ASTM class B									
1000 Up to 30' long, 12" butts, 8" points	B-19	625	.102	V.L.F.	17.85	5.85	3.27	26.97	32
1100 30' to 39' long, 12" butts, 8" points		700	.091		18.50	5.20	2.92	26.62	31.50
1200 40' to 49' long, 12" butts, 7" points		720	.089		25.50	5.10	2.84	33.44	39.50
1300 50' to 59' long, 13" butts, 7" points	▼	800	.080		27.50	4.57	2.56	34.63	40
1400 60' to 69' long, 13" butts, 6" points	B-19A	840	.076		25	4.35	3.48	32.83	38
1500 70' to 80' long, 13" butts, 6" points	"	840	.076	▼	26	4.35	3.48	33.83	39
1600 ACZA treated piles, 2.5 lb./C.F.									
1610 8" butts, 10' long	B-19	400	.160	V.L.F.	8.05	9.15	5.10	22.30	28.50
1620 11' to 16' long		500	.128		8.05	7.30	4.09	19.44	24.50
1630 17' to 20' long		575	.111		8.05	6.35	3.56	17.96	22.50
1640 10" butts, 10' to 16' long		500	.128		23.50	7.30	4.09	34.89	41.50
1650 17' to 20' long		575	.111		23.50	6.35	3.56	33.41	39.50
1660 21' to 40' long		700	.091		23.50	5.20	2.92	31.62	37
1670 12" butts, 10' to 20' long		575	.111		29	6.35	3.56	38.91	45.50
1680 21' to 35' long		650	.098		29.50	5.60	3.15	38.25	44
1690 36' to 40' long		700	.091		30	5.20	2.92	38.12	44
1695 14" butts, to 40' long	▼	700	.091	▼	16.70	5.20	2.92	24.82	29.50
1700 Boot for pile tip, minimum	1 Pile	27	.296	Ea.	47	16.55		63.55	77.50
1800 Maximum		21	.381		142	21.50		163.50	189
2000 Point for pile tip, minimum		20	.400		47	22.50		69.50	86.50
2100 Maximum	▼	15	.533		170	30		200	233
2300 Splice for piles over 50' long, minimum	B-46	35	1.371		59	69	1.19	129.19	171
2400 Maximum		20	2.400	▼	77.50	121	2.09	200.59	272
2600 Concrete encasement with wire mesh and tube	▼	331	.145	V.L.F.	77.50	7.30	.13	84.93	97
2700 Mobilization for 10,000 L.F. pile job, add	B-19	3300	.019			1.11	.62	1.73	2.37
2800 25,000 L.F. pile job, add	"	8500	.008	▼		.43	.24	.67	.92
2900 Mobilization by water for barge driving rig, add								100%	100%

31 62 19.30 Marine Wood Piles

31 62 19.30 Marine Wood Piles	Crew	Daily Output	Labor-Hours	Unit	Material	2021 Bare Costs Labor	2021 Bare Costs Equipment	Total	Total Incl O&P
0010 **MARINE WOOD PILES**									
0040 Friction or end bearing, not including									
0050 mobilization or demobilization									
0055 Marine piles include production adjusted									
0060 due to thickness of silt layers									
0090 Shore driven piles with silt layer									
1000 10' silt, shore driven, ACZA treated 1.0lbs/C.F., 50', 13" butts, 7" points	B-19	952	.067	V.L.F.	19.60	3.84	2.15	25.59	29.50
1050 55', 7" points		936	.068		22.50	3.90	2.18	28.58	33.50
1075 60', 7" points		969	.066		23.50	3.77	2.11	29.38	34
1100 65', 7" points		958	.067		26.50	3.82	2.13	32.45	37.50
1125 70', 6" points		948	.068		27	3.86	2.16	33.02	38
1150 75', 6" points		940	.068		27	3.89	2.18	33.07	38.50
1175 80', 6" points	▼	933	.069	▼	29.50	3.92	2.19	35.61	41

31 62 Driven Piles

31 62 19 – Timber Piles

31 62 19.30 Marine Wood Piles		Crew	Daily Output	Labor-Hours	Unit	Material	2021 Bare Costs Labor	Equipment	Total	Total Incl O&P
1200	10' silt, treated 1.5lbs/C.F., 50', 7" points	B-19	952	.067	V.L.F.	24.50	3.84	2.15	30.49	35
1225	55', 7" points		936	.068		27.50	3.90	2.18	33.58	38.50
1250	60', 7" points		969	.066		28.50	3.77	2.11	34.38	39
1275	65', 7" points		958	.067		32	3.82	2.13	37.95	43
1300	70', 6" points		948	.068		32	3.86	2.16	38.02	44
1325	75', 6" points		940	.068		32.50	3.89	2.18	38.57	44
1350	80', 6" points		933	.069		35	3.92	2.19	41.11	47.50
1400	15' silt, shore driven, ACZA treated 1.0lbs/C.F., 50', 13" butts, 7" points		1053	.061		20.50	3.47	1.94	25.91	30
1450	55', 7" points		1023	.063		22.50	3.57	2	28.07	32.50
1475	60', 7" points		1050	.061		23.50	3.48	1.95	28.93	33.50
1500	65', 7" points		1030	.062		26.50	3.55	1.99	32.04	37
1525	70', 6" points		1013	.063		27	3.61	2.02	32.63	37
1550	75', 6" points		1000	.064		27	3.66	2.04	32.70	38
1575	80', 6" points		988	.065		29.50	3.70	2.07	35.27	40.50
1600	15' silt, treated 1.5lbs/C.F., 50', 7" points		1053	.061		24.50	3.47	1.94	29.91	34.50
1625	55', 7" points		1023	.063		27.50	3.57	2	33.07	37.50
1650	60', 7" points		1050	.061		28.50	3.48	1.95	33.93	38.50
1675	65', 7" points		1030	.062		32	3.55	1.99	37.54	42.50
1700	70', 6" points		1014	.063		32	3.60	2.02	37.62	43
1725	75', 6" points		1000	.064		32.50	3.66	2.04	38.20	43.50
1750	80', 6" points		988	.065		35	3.70	2.07	40.77	47
1800	20' silt, shore driven, ACZA treated 1.0lbs/C.F., 50', 13" butts, 7" points		1177	.054		20.50	3.11	1.74	25.35	29
1850	55', 7" points		1128	.057		22.50	3.24	1.81	27.55	32
1875	60', 7" points		1145	.056		23.50	3.19	1.79	28.48	33
1900	65', 7" points		1114	.057		26.50	3.28	1.84	31.62	36.50
1925	70', 6" points		1089	.059		27	3.36	1.88	32.24	36.50
1950	75', 6" points		1068	.060		27	3.42	1.91	32.33	37.50
1975	80', 6" points		1050	.061		29.50	3.48	1.95	34.93	40
2000	20' silt, treated 1.5lbs/C.F., 50', 7" points		1177	.054		24.50	3.11	1.74	29.35	33.50
2025	55', 7" points		1128	.057		27.50	3.24	1.81	32.55	37
2050	60', 7" points		1145	.056		28.50	3.19	1.79	33.48	38
2075	65', 7" points		1114	.057		32	3.28	1.84	37.12	42
2100	70', 6" points		1089	.059		32	3.36	1.88	37.24	42.50
2125	75', 6" points		1068	.060		32.50	3.42	1.91	37.83	43
2150	80', 6" points		1050	.061		35	3.48	1.95	40.43	46.50
2200	25' silt, shore driven, ACZA treated 1.0lbs/C.F., 60', 13" butts, 7" points		1260	.051		23.50	2.90	1.62	28.02	32
2225	65', 7" points		1213	.053		26.50	3.01	1.69	31.20	36
2250	70', 6" points		1176	.054		27	3.11	1.74	31.85	36
2275	75', 6" points		1146	.056		27	3.19	1.78	31.97	37
2300	80', 6" points		1120	.057		29.50	3.26	1.83	34.59	39.50
2350	25' silt, treated 1.5lbs/C.F., 60', 6" points		1260	.051		28.50	2.90	1.62	33.02	37
2375	65', 6" points		1213	.053		32	3.01	1.69	36.70	41.50
2400	70', 6" points		1176	.054		32	3.11	1.74	36.85	42
2425	75', 6" points		1146	.056		32.50	3.19	1.78	37.47	42.50
2450	80', 6" points		1120	.057		35	3.26	1.83	40.09	46
2500	30' silt, shore driven, ACZA treated 1.0lbs/C.F., 70', 13" butts, 6" points		1278	.050		27	2.86	1.60	31.46	35.50
2525	75', 6" points		1235	.052		27	2.96	1.66	31.62	36.50
2550	80', 6" points		1200	.053		29.50	3.05	1.70	34.25	39
2600	30' silt, treated 1.5lbs/C.F., 70', 6" points		1278	.050		32	2.86	1.60	36.46	41.50
2625	75', 6" points		1235	.052		32.50	2.96	1.66	37.12	42
2650	80', 6" points		1200	.053		35	3.05	1.70	39.75	45.50
2700	35' silt, shore driven, ACZA treated 1.0lbs/C.F., 70', 13" butts, 6" points		1400	.046		27	2.61	1.46	31.07	35
2725	75', 6" points		1340	.048		27	2.73	1.53	31.26	36

31 62 19.30 Marine Wood Piles		Crew	Daily Output	Labor-Hours	Unit	Material	2021 Bare Costs Labor	Equipment	Total	Total Incl O&P
2750	80', 6" points	B-19	1292	.050	V.L.F.	29.50	2.83	1.58	33.91	38.50
2800	35' silt, treated 1.5lbs/C.F., 70', 6" points		1400	.046		32	2.61	1.46	36.07	41
2825	75', 6" points		1340	.048		32.50	2.73	1.53	36.76	41.50
2850	80', 6" points		1292	.050		35	2.83	1.58	39.41	45
2875	40' silt, shore driven, ACZA treated 1.0lbs/C.F., 75', 13" butts, 6" points		1465	.044		27	2.49	1.40	30.89	35.50
2900	80', 6" points		1400	.046		29.50	2.61	1.46	33.57	38
2925	40' silt, treated 1.5lbs/C.F., 75', 6" points		1465	.044		32.50	2.49	1.40	36.39	41
2950	80', 6" points		1400	.046		35	2.61	1.46	39.07	44.50
2990	Barge driven piles with silt layer									
3000	10' silt, barge driven, ACZA treated 1.5lbs/C.F. 50', 13" butts, 7" points	B-19B	762	.084	V.L.F.	24.50	4.80	3.82	33.12	38.50
3050	55', 7" points		749	.085		27.50	4.88	3.89	36.27	41.50
3075	60', 7" points		775	.083		28.50	4.72	3.76	36.98	42.50
3100	65', 7" points		766	.084		32	4.77	3.80	40.57	46.50
3125	70', 6" points		759	.084		32	4.82	3.84	40.66	47
3150	75', 6" points		752	.085		32.50	4.86	3.87	41.23	47
3175	80', 6" points		747	.086		35	4.89	3.90	43.79	50.50
3200	10' silt, treated 2.5lbs/C.F., 50', 7" points		762	.084		31	4.80	3.82	39.62	45.50
3225	55', 7" points		749	.085		31.50	4.88	3.89	40.27	46
3250	60', 7" points		775	.083		32	4.72	3.76	40.48	46.50
3275	65', 7" points		766	.084		39	4.77	3.80	47.57	54.50
3300	70', 6" points		759	.084		39.50	4.82	3.84	48.16	55
3325	75', 6" points		752	.085		39.50	4.86	3.87	48.23	55
3350	80', 6" points		747	.086		46	4.89	3.90	54.79	62
3400	15' silt, barge driven, ACZA treated 1.5lbs/C.F. 50', 13" butts, 7" points		842	.076		24.50	4.34	3.46	32.30	37.50
3450	55', 7" points		819	.078		27.50	4.46	3.56	35.52	40.50
3475	60', 7" points		840	.076		28.50	4.35	3.47	36.32	41.50
3500	65', 7" points		824	.078		32	4.44	3.54	39.98	45.50
3525	70', 6" points		811	.079		32	4.51	3.59	40.10	46.50
3550	75', 6" points		800	.080		32.50	4.57	3.64	40.71	46.50
3575	80', 6" points		791	.081		35	4.62	3.68	43.30	50
3600	15' silt, treated 2.5lbs/C.F., 50', 7" points		842	.076		31	4.34	3.46	38.80	44.50
3625	55', 7" points		819	.078		31.50	4.46	3.56	39.52	45
3650	60', 7" points		840	.076		32	4.35	3.47	39.82	45.50
3675	65', 7" points		824	.078		39	4.44	3.54	46.98	53.50
3700	70', 6" points		811	.079		39.50	4.51	3.59	47.60	54.50
3725	75', 6" points		800	.080		39.50	4.57	3.64	47.71	54.50
3750	80', 6" points		791	.081		46	4.62	3.68	54.30	61.50
3800	20' silt, barge driven, ACZA treated 1.5lbs/C.F. 50', 13" butts, 7" points		941	.068		24.50	3.88	3.10	31.48	36.50
3850	55', 7" points		903	.071		27.50	4.05	3.23	34.78	39.50
3875	60', 7" points		916	.070		28.50	3.99	3.18	35.67	40.50
3900	65', 7" points		891	.072		32	4.10	3.27	39.37	45
3925	70', 6" points		871	.073		32	4.20	3.35	39.55	45.50
3950	75', 6" points		854	.075		32.50	4.28	3.41	40.19	46
3975	80', 6" points		840	.076		35	4.35	3.47	42.82	49.50
4000	20' silt, treated 2.5lbs/C.F., 50', 7" points		941	.068		31	3.88	3.10	37.98	43.50
4025	55', 7" points		903	.071		31.50	4.05	3.23	38.78	44
4050	60', 7" points		916	.070		32	3.99	3.18	39.17	44.50
4075	65', 7" points		891	.072		39	4.10	3.27	46.37	53
4100	70', 6" points		871	.073		39.50	4.20	3.35	47.05	53.50
4125	75', 6" points		854	.075		39.50	4.28	3.41	47.19	54
4150	80', 6" points		840	.076		46	4.35	3.47	53.82	61
4200	25' silt, barge driven, ACZA treated 1.5lbs/C.F. 60', 3" butts, 7" points		1008	.063		28.50	3.63	2.89	35.02	39.50
4225	65', 7" point		971	.066		32	3.76	3	38.76	44

For customer support on your Site Work & Landscape Costs with RSMeans Data, call 800.448.8182.

31 62 Driven Piles

31 62 19 – Timber Piles

31 62 19.30 Marine Wood Piles		Crew	Daily Output	Labor-Hours	Unit	Material	2021 Bare Costs		Total	Total Incl O&P
							Labor	Equipment		
4250	70', 6" points	B-19B	941	.068	V.L.F.	32	3.88	3.10	38.98	45
4275	75', 6" points		916	.070		32.50	3.99	3.18	39.67	45
4300	80', 6" points		896	.071		35	4.08	3.25	42.33	49
4350	25' silt, treated 2.5lbs/C.F., 60', 7" points		1008	.063		32	3.63	2.89	38.52	43.50
4375	65', 7" points		971	.066		39	3.76	3	45.76	52
4400	70', 6" points		941	.068		39.50	3.88	3.10	46.48	53
4425	75', 6" points		916	.070		39.50	3.99	3.18	46.67	53
4450	80', 6" points		896	.071		46	4.08	3.25	53.33	60.50
4500	30' silt, barge driven, ACZA treated 1.5lbs/C.F. 70', 13" butts, 6" points		1023	.063		32	3.57	2.85	38.42	44
4525	75', 6" points		988	.065		32.50	3.70	2.95	39.15	44.50
4550	80', 6" points		960	.067		35	3.81	3.04	41.85	48
4600	30' silt, treated 2.5lbs/C.F., 70', 6" points		1023	.063		39.50	3.57	2.85	45.92	52
4625	75', 6" points		988	.065		39.50	3.70	2.95	46.15	52.50
4650	80', 6" points		960	.067		46	3.81	3.04	52.85	59.50
4700	35' silt, barge driven, ACZA treated 1.5lbs/C.F. 70', 13" butts, 6" points		1120	.057		32	3.26	2.60	37.86	43.50
4725	75', 6" points		1072	.060		32.50	3.41	2.72	38.63	43.50
4750	80', 6" points		1034	.062		35	3.53	2.82	41.35	47.50
4800	35' silt, treated 2.5lbs/C.F., 70', 6" points		1120	.057		39.50	3.26	2.60	45.36	51.50
4825	75', 6" points		1072	.060		39.50	3.41	2.72	45.63	51.50
4850	80', 6" points		1034	.062		46	3.53	2.82	52.35	59
4875	40' silt, barge driven, ACZA treated 1.5lbs/C.F. 75', 13" butts, 6" points		1172	.055		32.50	3.12	2.49	38.11	43
4900	80', 6" points		1120	.057		35	3.26	2.60	40.86	47
4925	40' silt, treated 2.5lbs/C.F., 75', 6" points		1172	.055		39.50	3.12	2.49	45.11	51
4950	80', 6" points	▼	1120	.057		46	3.26	2.60	51.86	58.50
5000	Piles, wood, treated, added undriven length, 13" butt					39.50			39.50	43.50
5020	treated pile, 13" butts				▼	28.50			28.50	31

31 62 23 – Composite Piles

31 62 23.10 Recycled Plastic and Fiberglass Piles

			Crew	Daily Output	Labor-Hours	Unit	Material	Labor	Equipment	Total	Total Incl O&P
0010	**RECYCLED PLASTIC AND FIBERGLASS PILES**, 200 PILES										
5000	Marine pilings, recycled plastic w/fiberglass reinf, up to 90' L, 8" diam.	G	B-19	500	.128	V.L.F.	31.50	7.30	4.09	42.89	50
5010	10" diam.	G		500	.128		35	7.30	4.09	46.39	54
5020	13" diam.	G		400	.160		55	9.15	5.10	69.25	80
5030	16" diam.	G		400	.160		91.50	9.15	5.10	105.75	120
5040	20" diam.	G		350	.183		90.50	10.45	5.85	106.80	122
5050	24" diam.	G	▼	350	.183	▼	107	10.45	5.85	123.30	140

31 62 23.12 Marine Recycled Plastic and Fiberglass Piles

			Crew	Daily Output	Labor-Hours	Unit	Material	Labor	Equipment	Total	Total Incl O&P
0010	**MARINE RECYCLED PLASTIC AND FIBERGLASS PILES**, 200 PILES										
0040	Friction or end bearing, not including										
0050	mobilization or demobilization										
0055	Marine piles include production adjusted										
0060	due to thickness of silt layers										
0090	Shore driven piles with silt layer										
0100	Marine pilings, recycled plastic w/fiberglass, 50' long, 8" diam.,10' silt	G	B-19	595	.108	V.L.F.	31.50	6.15	3.44	41.09	47.50
0110	50' long, 10" diam.	G		595	.108		35	6.15	3.44	44.59	51.50
0120	50' long, 13" diam.	G		476	.134		55	7.70	4.30	67	77
0130	50' long, 16" diam.	G		476	.134		91.50	7.70	4.30	103.50	116
0140	50' long, 20" diam.	G		417	.153		90.50	8.75	4.90	104.15	118
0150	50' long, 24" diam.	G		417	.153		107	8.75	4.90	120.65	137
0210	55' long, 8" diam.	G		585	.109		31.50	6.25	3.50	41.25	48
0220	55' long, 10" diam.	G		585	.109		35	6.25	3.50	44.75	52
0230	55' long, 13" diam.	G		468	.137		55	7.80	4.37	67.17	77
0240	55' long, 16" diam.	G	▼	468	.137	▼	91.50	7.80	4.37	103.67	117

For customer support on your Site Work & Landscape Costs with RSMeans Data, call 800.448.8182.

353

31 62 23.12 Marine Recycled Plastic and Fiberglass Piles		Crew	Daily Output	Labor-Hours	Unit	Material	2021 Bare Costs Labor	Equipment	Total	Total Incl O&P
0250	55' long, 20" diam.	G	B-19	410	.156	V.L.F.	90.50	8.90	4.99	104.39	119
0260	55' long, 24" diam.	G		410	.156		107	8.90	4.99	120.89	137
0310	60' long, 8" diam.	G		576	.111		31.50	6.35	3.55	41.40	48
0320	60' long, 10" diam.	G		577	.111		35	6.35	3.54	44.89	52
0330	60' long, 13" diam.	G		462	.139		55	7.90	4.43	67.33	77.50
0340	60' long, 16" diam.	G		462	.139		91.50	7.90	4.43	103.83	117
0350	60' long, 20" diam.	G		404	.158		90.50	9.05	5.05	104.60	119
0360	60' long, 24" diam.	G		404	.158		107	9.05	5.05	121.10	137
0410	65' long, 8" diam.	G		570	.112		31.50	6.40	3.59	41.49	48.50
0420	65' long, 10" diam.	G		570	.112		35	6.40	3.59	44.99	52.50
0430	65' long, 13" diam.	G		456	.140		55	8	4.48	67.48	77.50
0440	65' long, 16" diam.	G		456	.140		91.50	8	4.48	103.98	117
0450	65' long, 20" diam.	G		399	.160		90.50	9.15	5.10	104.75	119
0460	65' long, 24" diam.	G		399	.160		107	9.15	5.10	121.25	138
0510	70' long, 8" diam.	G		565	.113		31.50	6.45	3.62	41.57	48.50
0520	70' long, 10" diam.	G		565	.113		35	6.45	3.62	45.07	52.50
0530	70' long, 13" diam.	G		452	.142		55	8.10	4.52	67.62	78
0540	70' long, 16" diam.	G		452	.142		91.50	8.10	4.52	104.12	117
0550	70' long, 20" diam.	G		395	.162		90.50	9.25	5.20	104.95	119
0560	70' long, 24" diam.	G		399	.160		107	9.15	5.10	121.25	138
0610	75' long, 8" diam.	G		560	.114		31.50	6.55	3.65	41.70	48.50
0620	75' long, 10" diam.	G		560	.114		35	6.55	3.65	45.20	52.50
0630	75' long, 13" diam.	G		448	.143		55	8.15	4.56	67.71	78
0640	75' long, 16" diam.	G		448	.143		91.50	8.15	4.56	104.21	117
0650	75' long, 20" diam.	G		392	.163		90.50	9.30	5.20	105	119
0660	75' long, 24" diam.	G		392	.163		107	9.30	5.20	121.50	138
0710	80' long, 8" diam.	G		556	.115		31.50	6.55	3.68	41.73	48.50
0720	80' long, 10" diam.	G		556	.115		35	6.55	3.68	45.23	52.50
0730	80' long, 13" diam.	G		444	.144		55	8.25	4.61	67.86	78
0740	80' long, 16" diam.	G		444	.144		91.50	8.25	4.61	104.36	118
0750	80' long, 20" diam.	G		389	.165		90.50	9.40	5.25	105.15	120
0760	80' long, 24" diam.	G		389	.165		107	9.40	5.25	121.65	138
1000	Marine pilings, recycled plastic w/fiberglass, 50' long, 8" diam.,15' silt	G		658	.097		31.50	5.55	3.11	40.16	46.50
1010	50' long, 10" diam.	G		658	.097		35	5.55	3.11	43.66	50.50
1020	50' long, 13" diam.	G		526	.122		55	6.95	3.89	65.84	75.50
1030	50' long, 16" diam.	G		526	.122		91.50	6.95	3.89	102.34	115
1040	50' long, 20" diam.	G		461	.139		90.50	7.95	4.44	102.89	116
1050	50' long, 24" diam.	G		461	.139		107	7.95	4.44	119.39	135
1210	55' long, 8" diam.	G		640	.100		31.50	5.70	3.20	40.40	46.50
1220	55' long, 10" diam.	G		640	.100		35	5.70	3.20	43.90	50.50
1230	55' long, 13" diam.	G		512	.125		55	7.15	3.99	66.14	76
1240	55' long, 16" diam.	G		512	.125		91.50	7.15	3.99	102.64	115
1250	55' long, 20" diam.	G		448	.143		90.50	8.15	4.56	103.21	117
1260	55' long, 24" diam.	G		448	.143		107	8.15	4.56	119.71	135
1310	60' long, 8" diam.	G		625	.102		31.50	5.85	3.27	40.62	47
1320	60' long, 10" diam.	G		625	.102		35	5.85	3.27	44.12	51
1330	60' long, 13" diam.	G		500	.128		55	7.30	4.09	66.39	76
1340	60' long, 16" diam.	G		500	.128		91.50	7.30	4.09	102.89	116
1350	60' long, 20" diam.	G		438	.146		90.50	8.35	4.67	103.52	117
1360	60' long, 24" diam.	G		438	.146		107	8.35	4.67	120.02	136
1410	65' long, 8" diam.	G		613	.104		31.50	5.95	3.34	40.79	47.50
1420	65' long, 10" diam.	G		613	.104		35	5.95	3.34	44.29	51.50
1430	65' long, 13" diam.	G		491	.130		55	7.45	4.16	66.61	76.50

31 62 23 – Composite Piles

31 62 23.12 Marine Recycled Plastic and Fiberglass Piles		Crew	Daily Output	Labor-Hours	Unit	Material	2021 Bare Costs Labor	Equipment	Total	Total Incl O&P
1440	65' long, 16" diam.	G	B-19	491	.130	V.L.F.	91.50	7.45	4.16	103.11	116
1450	65' long, 20" diam.	G		429	.149		90.50	8.50	4.77	103.77	118
1460	65' long, 24" diam.	G		429	.149		107	8.50	4.77	120.27	136
1510	70' long, 8" diam.	G		603	.106		31.50	6.05	3.39	40.94	47.50
1520	70' long, 10" diam.	G		603	.106		35	6.05	3.39	44.44	51.50
1530	70' long, 13" diam.	G		483	.133		55	7.55	4.23	66.78	76.50
1540	70' long, 16" diam.	G		483	.133		91.50	7.55	4.23	103.28	116
1550	70' long, 20" diam.	G		422	.152		90.50	8.65	4.85	104	118
1560	70' long, 24" diam.	G		422	.152		107	8.65	4.85	120.50	137
1610	75' long, 8" diam.	G		595	.108		31.50	6.15	3.44	41.09	47.50
1620	75' long, 10" diam.	G		595	.108		35	6.15	3.44	44.59	51.50
1630	75' long, 13" diam.	G		476	.134		55	7.70	4.30	67	77
1640	75' long, 16" diam.	G		476	.134		91.50	7.70	4.30	103.50	116
1650	75' long, 20" diam.	G		417	.153		90.50	8.75	4.90	104.15	118
1660	75' long, 24" diam.	G		417	.153		107	8.75	4.90	120.65	137
1710	80' long, 8" diam.	G		588	.109		31.50	6.20	3.48	41.18	48
1720	80' long, 10" diam.	G		588	.109		35	6.20	3.48	44.68	52
1730	80' long, 13" diam.	G		471	.136		55	7.75	4.34	67.09	77
1740	80' long, 16" diam.	G		471	.136		91.50	7.75	4.34	103.59	117
1750	80' long, 20" diam.	G		412	.155		90.50	8.85	4.96	104.31	118
1760	80' long, 24" diam.	G		412	.155		107	8.85	4.96	120.81	137
2000	Marine pilings, recycled plastic w/fiberglass, 50' long, 8" diam., 20' silt	G		735	.087		31.50	4.97	2.78	39.25	45
2010	50' long, 10" diam.	G		735	.087		35	4.97	2.78	42.75	49
2020	50' long, 13" diam.	G		588	.109		55	6.20	3.48	64.68	74
2030	50' long, 16" diam.	G		588	.109		91.50	6.20	3.48	101.18	113
2040	50' long, 20" diam.	G		515	.124		90.50	7.10	3.97	101.57	115
2050	50' long, 24" diam.	G		515	.124		107	7.10	3.97	118.07	133
2210	55' long, 8" diam.	G		705	.091		31.50	5.20	2.90	39.60	45.50
2220	55' long, 10" diam.	G		705	.091		35	5.20	2.90	43.10	49.50
2230	55' long, 13" diam.	G		564	.113		55	6.50	3.63	65.13	74.50
2240	55' long, 16" diam.	G		564	.113		91.50	6.50	3.63	101.63	114
2250	55' long, 20" diam.	G		494	.130		90.50	7.40	4.14	102.04	115
2260	55' long, 24" diam.	G		494	.130		107	7.40	4.14	118.54	134
2310	60' long, 8" diam.	G		682	.094		31.50	5.35	3	39.85	46
2320	60' long, 10" diam.	G		682	.094		35	5.35	3	43.35	50
2330	60' long, 13" diam.	G		546	.117		55	6.70	3.75	65.45	75
2340	60' long, 16" diam.	G		546	.117		91.50	6.70	3.75	101.95	114
2350	60' long, 20" diam.	G		477	.134		90.50	7.65	4.29	102.44	116
2360	60' long, 24" diam.	G		477	.134		107	7.65	4.29	118.94	134
2410	65' long, 8" diam.	G		663	.097		31.50	5.50	3.08	40.08	46.50
2420	65' long, 10" diam.	G		663	.097		35	5.50	3.08	43.58	50.50
2430	65' long, 13" diam.	G		531	.121		55	6.90	3.85	65.75	75
2440	65' long, 16" diam.	G		531	.121		91.50	6.90	3.85	102.25	115
2450	65' long, 20" diam.	G		464	.138		90.50	7.90	4.41	102.81	116
2460	65' long, 24" diam.	G		464	.138		107	7.90	4.41	119.31	135
2510	70' long, 8" diam.	G		648	.099		31.50	5.65	3.16	40.31	46.50
2520	70' long, 10" diam.	G		648	.099		35	5.65	3.16	43.81	50.50
2530	70' long, 13" diam.	G		518	.124		55	7.05	3.95	66	75.50
2540	70' long, 16" diam.	G		518	.124		91.50	7.05	3.95	102.50	115
2550	70' long, 20" diam.	G		454	.141		90.50	8.05	4.50	103.05	117
2560	70' long, 24" diam.	G		454	.141		107	8.05	4.50	119.55	135
2610	75' long, 8" diam.	G		636	.101		31.50	5.75	3.22	40.47	47
2620	75' long, 10" diam.	G		636	.101		35	5.75	3.22	43.97	51

31 62 23.12 Marine Recycled Plastic and Fiberglass Piles		Crew	Daily Output	Labor-Hours	Unit	Material	2021 Bare Costs Labor	Equipment	Total	Total Incl O&P
2630	75' long, 13" diam.	G	B-19	508	.126	V.L.F.	55	7.20	4.03	66.23	76
2640	75' long, 16" diam.	G		508	.126		91.50	7.20	4.03	102.73	115
2650	75' long, 20" diam.	G		445	.144		90.50	8.20	4.60	103.30	117
2660	75' long, 24" diam.	G		445	.144		107	8.20	4.60	119.80	136
2710	80' long, 8" diam.	G		625	.102		31.50	5.85	3.27	40.62	47
2720	80' long, 10" diam.	G		625	.102		35	5.85	3.27	44.12	51
2730	80' long, 13" diam.	G		500	.128		55	7.30	4.09	66.39	76
2740	80' long, 16" diam.	G		500	.128		91.50	7.30	4.09	102.89	116
2750	80' long, 20" diam.	G		437	.146		90.50	8.35	4.68	103.53	117
2760	80' long, 24" diam.	G		437	.146		107	8.35	4.68	120.03	136
4990	Barge driven piles with silt layer										
5100	Marine pilings, recycled plastic w/fiberglass, 50' long, 8" diam., 10' silt	G	B-19B	476	.134	V.L.F.	31.50	7.70	6.10	45.30	53
5110	50' long, 10" diam.	G		476	.134		35	7.70	6.10	48.80	57
5120	50' long, 13" diam.	G		381	.168		55	9.60	7.65	72.25	83.50
5130	50' long, 16" diam.	G		381	.168		91.50	9.60	7.65	108.75	123
5140	50' long, 20" diam.	G		333	.192		90.50	11	8.75	110.25	126
5150	50' long, 24" diam.	G		333	.192		107	11	8.75	126.75	144
5210	55' long, 8" diam.	G		468	.137		31.50	7.80	6.25	45.55	53.50
5220	55' long, 10" diam.	G		468	.137		35	7.80	6.25	49.05	57.50
5230	55' long, 13" diam.	G		374	.171		55	9.75	7.80	72.55	84
5240	55' long, 16" diam.	G		374	.171		91.50	9.75	7.80	109.05	123
5250	55' long, 20" diam.	G		328	.195		90.50	11.15	8.90	110.55	126
5260	55' long, 24" diam.	G		328	.195		107	11.15	8.90	127.05	145
5310	60' long, 8" diam.	G		461	.139		31.50	7.95	6.30	45.75	53.50
5320	60' long, 10" diam.	G		461	.139		35	7.95	6.30	49.25	57.50
5330	60' long, 13" diam.	G		369	.173		55	9.90	7.90	72.80	84.50
5340	60' long, 16" diam.	G		369	.173		91.50	9.90	7.90	109.30	124
5350	60' long, 20" diam.	G		323	.198		90.50	11.30	9	110.80	127
5360	60' long, 24" diam.	G		323	.198		107	11.30	9	127.30	145
5410	65' long, 8" diam.	G		456	.140		31.50	8	6.40	45.90	54
5420	65' long, 10" diam.	G		456	.140		35	8	6.40	49.40	58
5430	65' long, 13" diam.	G		365	.175		55	10	8	73	84.50
5440	65' long, 16" diam.	G		365	.175		91.50	10	8	109.50	124
5450	65' long, 20" diam.	G		319	.201		90.50	11.45	9.15	111.10	127
5460	65' long, 24" diam.	G		319	.201		107	11.45	9.15	127.60	146
5510	70' long, 8" diam.	G		452	.142		31.50	8.10	6.45	46.05	54
5520	70' long, 10" diam.	G		452	.142		35	8.10	6.45	49.55	58
5530	70' long, 13" diam.	G		361	.177		55	10.10	8.05	73.15	85
5540	70' long, 16" diam.	G		361	.177		91.50	10.10	8.05	109.65	124
5550	70' long, 20" diam.	G		316	.203		90.50	11.55	9.20	111.25	127
5560	70' long, 24" diam.	G		316	.203		107	11.55	9.20	127.75	146
5610	75' long, 8" diam.	G		448	.143		31.50	8.15	6.50	46.15	54
5620	75' long, 10" diam.	G		448	.143		35	8.15	6.50	49.65	58
5630	75' long, 13" diam.	G		358	.179		55	10.20	8.15	73.35	85
5640	75' long, 16" diam.	G		358	.179		91.50	10.20	8.15	109.85	125
5650	75' long, 20" diam.	G		313	.204		90.50	11.70	9.30	111.50	128
5660	75' long, 24" diam.	G		313	.204		107	11.70	9.30	128	146
5710	80' long, 8" diam.	G		444	.144		31.50	8.25	6.55	46.30	54.50
5720	80' long, 10" diam.	G		444	.144		35	8.25	6.55	49.80	58.50
5730	80' long, 13" diam.	G		356	.180		55	10.25	8.20	73.45	85
5740	80' long, 16" diam.	G		356	.180		91.50	10.25	8.20	109.95	125
5750	80' long, 20" diam.	G		311	.206		90.50	11.75	9.35	111.60	128
5760	80' long, 24" diam.	G		311	.206		107	11.75	9.35	128.10	146

31 62 Driven Piles

31 62 23 – Composite Piles

31 62 23.12 Marine Recycled Plastic and Fiberglass Piles		Crew	Daily Output	Labor-Hours	Unit	Material	2021 Bare Costs Labor	Equipment	Total	Total Incl O&P
6100	Marine pilings, recycled plastic w/fiberglass, 50' long, 8" diam.,15' silt	G	B-19B	526	.122	V.L.F.	31.50	6.95	5.55	44	51
6110	50' long, 10" diam.	G		526	.122		35	6.95	5.55	47.50	55
6120	50' long, 13" diam.	G		421	.152		55	8.70	6.90	70.60	81.50
6130	50' long, 16" diam.	G		421	.152		91.50	8.70	6.90	107.10	121
6140	50' long, 20" diam.	G		368	.174		90.50	9.95	7.90	108.35	123
6150	50' long, 24" diam.	G		368	.174		107	9.95	7.90	124.85	142
6210	55' long, 8" diam.	G		512	.125		31.50	7.15	5.70	44.35	51.50
6220	55' long, 10" diam.	G		512	.125		35	7.15	5.70	47.85	55.50
6230	55' long, 13" diam.	G		409	.156		55	8.95	7.10	71.05	82
6240	55' long, 16" diam.	G		409	.156		91.50	8.95	7.10	107.55	121
6250	55' long, 20" diam.	G		358	.179		90.50	10.20	8.15	108.85	124
6260	55' long, 24" diam.	G		358	.179		107	10.20	8.15	125.35	143
6310	60' long, 8" diam.	G		500	.128		31.50	7.30	5.85	44.65	52
6320	60' long, 10" diam.	G		500	.128		35	7.30	5.85	48.15	56
6330	60' long, 13" diam.	G		400	.160		55	9.15	7.30	71.45	82.50
6340	60' long, 16" diam.	G		400	.160		91.50	9.15	7.30	107.95	122
6350	60' long, 20" diam.	G		350	.183		90.50	10.45	8.35	109.30	125
6360	60' long, 24" diam.	G		350	.183		107	10.45	8.35	125.80	143
6410	65' long, 8" diam.	G		491	.130		31.50	7.45	5.95	44.90	52.50
6420	65' long, 10" diam.	G		491	.130		35	7.45	5.95	48.40	56.50
6430	65' long, 13" diam.	G		392	.163		55	9.30	7.45	71.75	83
6440	65' long, 16" diam.	G		392	.163		91.50	9.30	7.45	108.25	122
6450	65' long, 20" diam.	G		343	.187		90.50	10.65	8.50	109.65	125
6460	65' long, 24" diam.	G		343	.187		107	10.65	8.50	126.15	144
6510	70' long, 8" diam.	G		483	.133		31.50	7.55	6.05	45.10	52.50
6520	70' long, 10" diam.	G		483	.133		35	7.55	6.05	48.60	56.50
6530	70' long, 13" diam.	G		386	.166		55	9.45	7.55	72	83.50
6540	70' long, 16" diam.	G		386	.166		91.50	9.45	7.55	108.50	123
6550	70' long, 20" diam.	G		338	.189		90.50	10.80	8.60	109.90	126
6560	70' long, 24" diam.	G		338	.189		107	10.80	8.60	126.40	144
6610	75' long, 8" diam.	G		476	.134		31.50	7.70	6.10	45.30	53
6620	75' long, 10" diam.	G		476	.134		35	7.70	6.10	48.80	57
6630	75' long, 13" diam.	G		381	.168		55	9.60	7.65	72.25	83.50
6640	75' long, 16" diam.	G		381	.168		91.50	9.60	7.65	108.75	123
6650	75' long, 20" diam.	G		333	.192		90.50	11	8.75	110.25	126
6660	75' long, 24" diam.	G		333	.192		107	11	8.75	126.75	144
6710	80' long, 8" diam.	G		471	.136		31.50	7.75	6.20	45.45	53
6720	80' long, 10" diam.	G		471	.136		35	7.75	6.20	48.95	57
6730	80' long, 13" diam.	G		376	.170		55	9.70	7.75	72.45	84
6740	80' long, 16" diam.	G		376	.170		91.50	9.70	7.75	108.95	123
6750	80' long, 20" diam.	G		329	.195		90.50	11.10	8.85	110.45	126
6760	80' long, 24" diam.	G		329	.195		107	11.10	8.85	126.95	145
7100	Marine pilings, recycled plastic w/fiberglass, 50' long, 8" diam., 20' silt	G		588	.109		31.50	6.20	4.96	42.66	49.50
7110	50' long, 10" diam.	G		588	.109		35	6.20	4.96	46.16	53.50
7120	50' long, 13" diam.	G		471	.136		55	7.75	6.20	68.95	79
7130	50' long, 16" diam.	G		471	.136		91.50	7.75	6.20	105.45	119
7140	50' long, 20" diam.	G		412	.155		90.50	8.85	7.05	106.40	121
7150	50' long, 24" diam.	G		412	.155		107	8.85	7.05	122.90	139
7210	55' long, 8" diam.	G		564	.113		31.50	6.50	5.15	43.15	50
7220	55' long, 10" diam.	G		564	.113		35	6.50	5.15	46.65	54
7230	55' long, 13" diam.	G		451	.142		55	8.10	6.45	69.55	80
7240	55' long, 16" diam.	G		451	.142		91.50	8.10	6.45	106.05	119
7250	55' long, 20" diam.	G		395	.162		90.50	9.25	7.40	107.15	122

31 62 Driven Piles

31 62 23 – Composite Piles

31 62 23.12 Marine Recycled Plastic and Fiberglass Piles

Line	Description		Crew	Daily Output	Labor-Hours	Unit	Material	2021 Bare Costs Labor	2021 Bare Costs Equipment	Total	Total Incl O&P
7260	55' long, 24" diam.	G	B-19B	395	.162	V.L.F.	107	9.25	7.40	123.65	140
7310	60' long, 8" diam.	G		545	.117		31.50	6.70	5.35	43.55	50.50
7320	60' long, 10" diam.	G		545	.117		35	6.70	5.35	47.05	54.50
7330	60' long, 13" diam.	G		436	.147		55	8.40	6.70	70.10	80.50
7340	60' long, 16" diam.	G		436	.147		91.50	8.40	6.70	106.60	120
7350	60' long, 20" diam.	G		382	.168		90.50	9.55	7.65	107.70	123
7360	60' long, 24" diam.	G		382	.168		107	9.55	7.65	124.20	141
7410	65' long, 8" diam.	G		531	.121		31.50	6.90	5.50	43.90	51
7420	65' long, 10" diam.	G		531	.121		35	6.90	5.50	47.40	55
7430	65' long, 13" diam.	G		424	.151		55	8.60	6.85	70.45	81
7440	65' long, 16" diam.	G		424	.151		91.50	8.60	6.85	106.95	121
7450	65' long, 20" diam.	G		371	.173		90.50	9.85	7.85	108.20	123
7460	65' long, 24" diam.	G		371	.173		107	9.85	7.85	124.70	142
7510	70' long, 8" diam.	G		518	.124		31.50	7.05	5.65	44.20	51.50
7520	70' long, 10" diam.	G		518	.124		35	7.05	5.65	47.70	55.50
7530	70' long, 13" diam.	G		415	.154		55	8.80	7	70.80	81.50
7540	70' long, 16" diam.	G		415	.154		91.50	8.80	7	107.30	121
7550	70' long, 20" diam.	G		363	.176		90.50	10.05	8.05	108.60	124
7560	70' long, 24" diam.	G		363	.176		107	10.05	8.05	125.10	142
7610	75' long, 8" diam.	G		508	.126		31.50	7.20	5.75	44.45	52
7620	75' long, 10" diam.	G		508	.126		35	7.20	5.75	47.95	56
7630	75' long, 13" diam.	G		407	.157		55	9	7.15	71.15	82
7640	75' long, 16" diam.	G		407	.157		91.50	9	7.15	107.65	122
7650	75' long, 20" diam.	G		356	.180		90.50	10.25	8.20	108.95	124
7660	75' long, 24" diam.	G		356	.180		107	10.25	8.20	125.45	143
7710	80' long, 8" diam.	G		500	.128		31.50	7.30	5.85	44.65	52
7720	80' long, 10" diam.	G		500	.128		35	7.30	5.85	48.15	56
7730	80' long, 13" diam.	G		400	.160		55	9.15	7.30	71.45	82.50
7740	80' long, 16" diam.	G		400	.160		91.50	9.15	7.30	107.95	122
7750	80' long, 20" diam.	G		350	.183		90.50	10.45	8.35	109.30	125
7760	80' long, 24" diam.	G		350	.183		107	10.45	8.35	125.80	143

31 62 23.13 Concrete-Filled Steel Piles

Line	Description	Crew	Daily Output	Labor-Hours	Unit	Material	2021 Bare Costs Labor	2021 Bare Costs Equipment	Total	Total Incl O&P
0010	**CONCRETE-FILLED STEEL PILES** no mobilization or demobilization									
2600	Pipe piles, 50' L, 8" diam., 29 lb./L.F., no concrete	B-19	500	.128	V.L.F.	21	7.30	4.09	32.39	38.50
2700	Concrete filled		460	.139		26	7.95	4.45	38.40	45.50
2900	10" diameter, 34 lb./L.F., no concrete		500	.128		25.50	7.30	4.09	36.89	43.50
3000	Concrete filled		450	.142		31	8.10	4.54	43.64	51.50
3200	12" diameter, 44 lb./L.F., no concrete		475	.135		33	7.70	4.30	45	53
3300	Concrete filled		415	.154		37.50	8.80	4.93	51.23	60
3500	14" diameter, 46 lb./L.F., no concrete		430	.149		35.50	8.50	4.76	48.76	57
3600	Concrete filled		355	.180		40.50	10.30	5.75	56.55	66.50
3800	16" diameter, 52 lb./L.F., no concrete		385	.166		43	9.50	5.30	57.80	68
3900	Concrete filled		335	.191		55	10.90	6.10	72	84
4100	18" diameter, 59 lb./L.F., no concrete		355	.180		49.50	10.30	5.75	65.55	76.50
4200	Concrete filled		310	.206		63	11.80	6.60	81.40	94
4400	Splices for pipe piles, stl., not in leads, 8" diameter	1 Sswl	5	1.600	Ea.	90	96.50		186.50	248
4410	10" diameter		4.75	1.684		88.50	102		190.50	255
4430	12" diameter		4.50	1.778		111	107		218	288
4500	14" diameter		4.25	1.882		157	114		271	350
4600	16" diameter		4	2		177	121		298	380
4650	18" diameter		3.75	2.133		293	129		422	520
4710	Steel pipe pile backing rings, w/spacer, 8" diameter		12	.667		9.30	40		49.30	72.50

For customer support on your Site Work & Landscape Costs with RSMeans Data, call 800.448.8182.

31 62 Driven Piles

31 62 23 – Composite Piles

31 62 23.13 Concrete-Filled Steel Piles		Crew	Daily Output	Labor-Hours	Unit	Material	2021 Bare Costs Labor	Equipment	Total	Total Incl O&P
4720	10" diameter	1 Sswl	12	.667	Ea.	12.25	40		52.25	75.50
4730	12" diameter		10	.800		14.75	48		62.75	90.50
4740	14" diameter		9	.889		18.85	53.50		72.35	104
4750	16" diameter		8	1		22.50	60.50		83	118
4760	18" diameter		6	1.333		24.50	80.50		105	151
4800	Points, standard, 8" diameter		4.61	1.735		102	105		207	275
4840	10" diameter		4.45	1.798		134	108		242	315
4880	12" diameter		4.25	1.882		167	114		281	360
4900	14" diameter		4.05	1.975		224	119		343	430
5000	16" diameter		3.37	2.374		315	143		458	565
5050	18" diameter		3.50	2.286		400	138		538	655
5200	Points, heavy duty, 10" diameter		2.90	2.759		256	166		422	540
5240	12" diameter		2.95	2.712		315	164		479	600
5260	14" diameter		2.95	2.712		310	164		474	600
5280	16" diameter		2.95	2.712		375	164		539	665
5290	18" diameter		2.80	2.857		445	172		617	755
5500	For reinforcing steel, add	↓	1150	.007	Lb.	.80	.42		1.22	1.53
5700	For thick wall sections, add				"	.70			.70	.77
6020	Steel pipe pile end plates, 8" diameter	1 Sswl	14	.571	Ea.	31.50	34.50		66	88
6050	10" diameter		14	.571		41.50	34.50		76	99
6100	12" diameter		12	.667		53.50	40		93.50	121
6150	14" diameter		10	.800		62	48		110	143
6200	16" diameter		9	.889		78	53.50		131.50	169
6250	18" diameter		8	1		103	60.50		163.50	207
6300	Steel pipe pile shoes, 8" diameter		12	.667		66	40		106	135
6350	10" diameter		12	.667		84	40		124	155
6400	12" diameter		10	.800		99.50	48		147.50	185
6450	14" diameter		9	.889		147	53.50		200.50	244
6500	16" diameter		8	1		153	60.50		213.50	263
6550	18" diameter	↓	6	1.333	↓	223	80.50		303.50	370

31 63 Bored Piles

31 63 26 – Drilled Caissons

31 63 26.13 Fixed End Caisson Piles

		Crew	Daily Output	Labor-Hours	Unit	Material	2021 Bare Costs Labor	Equipment	Total	Total Incl O&P
0010	**FIXED END CAISSON PILES** R316326-60									
0015	Including excavation, concrete, 50 lb. reinforcing									
0020	per C.Y., not incl. mobilization, boulder removal, disposal									
0100	Open style, machine drilled, to 50' deep, in stable ground, no									
0110	casings or ground water, 18" diam., 0.065 C.Y./L.F.	B-43	200	.240	V.L.F.	10.10	11.75	3.84	25.69	33
0200	24" diameter, 0.116 C.Y./L.F.		190	.253		18.05	12.35	4.04	34.44	43
0300	30" diameter, 0.182 C.Y./L.F.		150	.320		28	15.65	5.10	48.75	60
0400	36" diameter, 0.262 C.Y./L.F.		125	.384		40.50	18.80	6.15	65.45	80
0500	48" diameter, 0.465 C.Y./L.F.		100	.480		72.50	23.50	7.70	103.70	123
0600	60" diameter, 0.727 C.Y./L.F.		90	.533		113	26	8.55	147.55	172
0700	72" diameter, 1.05 C.Y./L.F.		80	.600		163	29.50	9.60	202.10	235
0800	84" diameter, 1.43 C.Y./L.F.	↓	75	.640	↓	222	31.50	10.25	263.75	305
1000	For bell excavation and concrete, add									
1020	4' bell diameter, 24" shaft, 0.444 C.Y.	B-43	20	2.400	Ea.	55.50	117	38.50	211	279
1040	6' bell diameter, 30" shaft, 1.57 C.Y.		5.70	8.421		195	410	135	740	980
1060	8' bell diameter, 36" shaft, 3.72 C.Y.		2.40	20		465	980	320	1,765	2,300
1080	9' bell diameter, 48" shaft, 4.48 C.Y.		2	24		560	1,175	385	2,120	2,800

For customer support on your Site Work & Landscape Costs with RSMeans Data, call 800.448.8182.

359

31 63 26.13 Fixed End Caisson Piles		Crew	Daily Output	Labor-Hours	Unit	Material	2021 Bare Costs Labor	Equipment	Total	Total Incl O&P
1100	10' bell diameter, 60" shaft, 5.24 C.Y.	B-43	1.70	28.235	Ea.	650	1,375	450	2,475	3,250
1120	12' bell diameter, 72" shaft, 8.74 C.Y.		1	48		1,100	2,350	770	4,220	5,550
1140	14' bell diameter, 84" shaft, 13.6 C.Y.	↓	.70	68.571	↓	1,700	3,350	1,100	6,150	8,050
1200	Open style, machine drilled, to 50' deep, in wet ground, pulled									
1300	casing and pumping, 18" diameter, 0.065 C.Y./L.F.	B-48	160	.350	V.L.F.	10.10	17.45	6.55	34.10	44.50
1400	24" diameter, 0.116 C.Y./L.F.		125	.448		18.05	22.50	8.40	48.95	62.50
1500	30" diameter, 0.182 C.Y./L.F.		85	.659		28	33	12.30	73.30	93.50
1600	36" diameter, 0.262 C.Y./L.F.	↓	60	.933		40.50	46.50	17.45	104.45	134
1700	48" diameter, 0.465 C.Y./L.F.	B-49	55	1.600		72.50	83.50	29.50	185.50	238
1800	60" diameter, 0.727 C.Y./L.F.		35	2.514		113	131	46.50	290.50	375
1900	72" diameter, 1.05 C.Y./L.F.		30	2.933		163	153	54.50	370.50	470
2000	84" diameter, 1.43 C.Y./L.F.	↓	25	3.520	↓	222	184	65.50	471.50	595
2100	For bell excavation and concrete, add									
2120	4' bell diameter, 24" shaft, 0.444 C.Y.	B-48	19.80	2.828	Ea.	55.50	141	53	249.50	330
2140	6' bell diameter, 30" shaft, 1.57 C.Y.		5.70	9.825		195	490	184	869	1,150
2160	8' bell diameter, 36" shaft, 3.72 C.Y.	↓	2.40	23.333		465	1,175	435	2,075	2,725
2180	9' bell diameter, 48" shaft, 4.48 C.Y.	B-49	3.30	26.667		560	1,400	495	2,455	3,250
2200	10' bell diameter, 60" shaft, 5.24 C.Y.		2.80	31.429		650	1,650	585	2,885	3,825
2220	12' bell diameter, 72" shaft, 8.74 C.Y.		1.60	55		1,100	2,875	1,025	5,000	6,650
2240	14' bell diameter, 84" shaft, 13.6 C.Y.	↓	1	88	↓	1,700	4,600	1,625	7,925	10,600
2300	Open style, machine drilled, to 50' deep, in soft rocks and									
2400	medium hard shales, 18" diameter, 0.065 C.Y./L.F.	B-49	50	1.760	V.L.F.	10.10	92	32.50	134.60	185
2500	24" diameter, 0.116 C.Y./L.F.		30	2.933		18.05	153	54.50	225.55	310
2600	30" diameter, 0.182 C.Y./L.F.		20	4.400		28	230	81.50	339.50	465
2700	36" diameter, 0.262 C.Y./L.F.		15	5.867		40.50	305	109	454.50	625
2800	48" diameter, 0.465 C.Y./L.F.		10	8.800		72.50	460	163	695.50	950
2900	60" diameter, 0.727 C.Y./L.F.		7	12.571		113	655	233	1,001	1,375
3000	72" diameter, 1.05 C.Y./L.F.		6	14.667		163	765	272	1,200	1,625
3100	84" diameter, 1.43 C.Y./L.F.	↓	5	17.600	↓	222	920	325	1,467	1,975
3200	For bell excavation and concrete, add									
3220	4' bell diameter, 24" shaft, 0.444 C.Y.	B-49	10.90	8.073	Ea.	55.50	420	150	625.50	860
3240	6' bell diameter, 30" shaft, 1.57 C.Y.		3.10	28.387		195	1,475	525	2,195	3,025
3260	8' bell diameter, 36" shaft, 3.72 C.Y.		1.30	67.692		465	3,525	1,250	5,240	7,175
3280	9' bell diameter, 48" shaft, 4.48 C.Y.		1.10	80		560	4,175	1,475	6,210	8,525
3300	10' bell diameter, 60" shaft, 5.24 C.Y.		.90	97.778		650	5,100	1,825	7,575	10,400
3320	12' bell diameter, 72" shaft, 8.74 C.Y.		.60	147	↓	1,100	7,675	2,725	11,500	15,700
3340	14' bell diameter, 84" shaft, 13.6 C.Y.		.40	220	↓	1,700	11,500	4,075	17,275	23,700
3600	For rock excavation, sockets, add, minimum		120	.733	C.F.		38.50	13.60	52.10	72.50
3650	Average		95	.926			48.50	17.20	65.70	91.50
3700	Maximum	↓	48	1.833	↓		96	34	130	182
3900	For 50' to 100' deep, add				V.L.F.				7%	7%
4000	For 100' to 150' deep, add								25%	25%
4100	For 150' to 200' deep, add				↓				30%	30%
4200	For casings left in place, add				Lb.	1.39			1.39	1.53
4300	For other than 50 lb. reinf. per C.Y., add or deduct				"	1.31			1.31	1.44
4400	For steel I-beam cores, add	B-49	8.30	10.602	Ton	2,100	555	197	2,852	3,375
4500	Load and haul excess excavation, 2 miles	B-34B	178	.045	L.C.Y.		2.31	3.25	5.56	7.05
4600	For mobilization, 50 mile radius, rig to 36"	B-43	2	24	Ea.		1,175	385	1,560	2,175
4650	Rig to 84"	B-48	1.75	32			1,600	600	2,200	3,025
4700	For low headroom, add								50%	50%
4750	For difficult access, add								25%	25%
5000	Bottom inspection	1 Skwk	1.20	6.667	↓		380		380	575

31 63 Bored Piles

31 63 26 – Drilled Caissons

31 63 26.16 Concrete Caissons for Marine Construction	Crew	Daily Output	Labor-Hours	Unit	Material	2021 Bare Costs Labor	Equipment	Total	Total Incl O&P
0010 **CONCRETE CAISSONS FOR MARINE CONSTRUCTION**									
0100 Caissons, incl. mobilization and demobilization, up to 50 miles									
0200 Uncased shafts, 30 to 80 tons cap., 17" diam., 10' depth	B-44	88	.727	V.L.F.	25	41	23.50	89.50	116
0300 25' depth		165	.388		17.80	22	12.55	52.35	66.50
0400 80 to 150 ton capacity, 22" diameter, 10' depth		80	.800		31	45	26	102	131
0500 20' depth		130	.492		25	27.50	15.90	68.40	87
0700 Cased shafts, 10 to 30 ton capacity, 10-5/8" diam., 20' depth		175	.366		17.80	20.50	11.80	50.10	64
0800 30' depth		240	.267		16.60	14.95	8.60	40.15	50.50
0850 30 to 60 ton capacity, 12" diameter, 20' depth		160	.400		25	22.50	12.90	60.40	75.50
0900 40' depth		230	.278		19.15	15.60	9	43.75	55
1000 80 to 100 ton capacity, 16" diameter, 20' depth		160	.400		35.50	22.50	12.90	70.90	87
1100 40' depth		230	.278		33	15.60	9	57.60	70.50
1200 110 to 140 ton capacity, 17-5/8" diameter, 20' depth		160	.400		38.50	22.50	12.90	73.90	90
1300 40' depth		230	.278		35.50	15.60	9	60.10	73
1400 140 to 175 ton capacity, 19" diameter, 20' depth		130	.492		41.50	27.50	15.90	84.90	105
1500 40' depth	↓	210	.305	↓	38.50	17.10	9.85	65.45	79
1700 Over 30' long, L.F. cost tends to be lower									
1900 Maximum depth is about 90'									

31 63 29 – Drilled Concrete Piers and Shafts

31 63 29.13 Uncased Drilled Concrete Piers

31 63 29.13 Uncased Drilled Concrete Piers	Crew	Daily Output	Labor-Hours	Unit	Material	2021 Bare Costs Labor	Equipment	Total	Total Incl O&P
0010 **UNCASED DRILLED CONCRETE PIERS**									
0020 Unless specified otherwise, not incl. pile caps or mobilization									
0050 Cast in place augered piles, no casing or reinforcing									
0060 8" diameter	B-43	540	.089	V.L.F.	4.26	4.35	1.42	10.03	12.75
0065 10" diameter		480	.100		6.75	4.89	1.60	13.24	16.50
0070 12" diameter		420	.114		9.55	5.60	1.83	16.98	21
0075 14" diameter		360	.133		12.85	6.50	2.13	21.48	26.50
0080 16" diameter		300	.160		17.35	7.85	2.56	27.76	33.50
0085 18" diameter	↓	240	.200	↓	21.50	9.80	3.20	34.50	41.50
0100 Cast in place, thin wall shell pile, straight sided,									
0110 not incl. reinforcing, 8" diam., 16 ga., 5.8 lb./L.F.	B-19	700	.091	V.L.F.	9.60	5.20	2.92	17.72	21.50
0200 10" diameter, 16 ga. corrugated, 7.3 lb./L.F.		650	.098		12.60	5.60	3.15	21.35	26
0300 12" diameter, 16 ga. corrugated, 8.7 lb./L.F.		600	.107		16.30	6.10	3.41	25.81	31
0400 14" diameter, 16 ga. corrugated, 10.0 lb./L.F.		550	.116		19.20	6.65	3.72	29.57	35
0500 16" diameter, 16 ga. corrugated, 11.6 lb./L.F.	↓	500	.128	↓	23.50	7.30	4.09	34.89	41.50
0800 Cast in place friction pile, 50' long, fluted,									
0810 tapered steel, 4,000 psi concrete, no reinforcing									
0900 12" diameter, 7 ga.	B-19	600	.107	V.L.F.	29.50	6.10	3.41	39.01	45.50
1000 14" diameter, 7 ga.		560	.114		32	6.55	3.65	42.20	49.50
1100 16" diameter, 7 ga.		520	.123		38	7.05	3.93	48.98	56.50
1200 18" diameter, 7 ga.	↓	480	.133	↓	44	7.60	4.26	55.86	65
1300 End bearing, fluted, constant diameter,									
1320 4,000 psi concrete, no reinforcing									
1340 12" diameter, 7 ga.	B-19	600	.107	V.L.F.	31	6.10	3.41	40.51	47
1360 14" diameter, 7 ga.		560	.114		39	6.55	3.65	49.20	56.50
1380 16" diameter, 7 ga.		520	.123		45	7.05	3.93	55.98	64.50
1400 18" diameter, 7 ga.	↓	480	.133	↓	49.50	7.60	4.26	61.36	71

31 63 29.20 Cast In Place Piles, Adds

31 63 29.20 Cast In Place Piles, Adds	Crew	Daily Output	Labor-Hours	Unit	Material	2021 Bare Costs Labor	Equipment	Total	Total Incl O&P
0010 **CAST IN PLACE PILES, ADDS**									
1500 For reinforcing steel, add				Lb.	1.24			1.24	1.37
1700 For ball or pedestal end, add	B-19	11	5.818	C.Y.	151	330	186	667	875

31 63 Bored Piles

31 63 29 – Drilled Concrete Piers and Shafts

31 63 29.20 Cast In Place Piles, Adds	Crew	Daily Output	Labor-Hours	Unit	Material	2021 Bare Costs Labor	2021 Bare Costs Equipment	Total	Total Incl O&P
1900	For lengths above 60', concrete, add	B-19	11	5.818	C.Y.	158	330	186	674	885
2000	For steel thin shell, pipe only				Lb.	1.54			1.54	1.69

31 63 33 – Drilled Micropiles

31 63 33.10 Drilled Micropiles Metal Pipe

	31 63 33.10 Drilled Micropiles Metal Pipe	Crew	Daily Output	Labor-Hours	Unit	Material	2021 Bare Costs Labor	2021 Bare Costs Equipment	Total	Total Incl O&P
0010	**DRILLED MICROPILES METAL PIPE**									
0011	No mobilization or demobilization									
5000	Pressure grouted pin pile, 5" diam., cased, up to 50 ton,									
5040	End bearing, less than 20'	B-48	160	.350	V.L.F.	42	17.45	6.55	66	79.50
5080	More than 40'		240	.233		38.50	11.65	4.36	54.51	64
5120	Friction, loose sand and gravel		240	.233		42	11.65	4.36	58.01	68.50
5160	Dense sand and gravel		240	.233		42	11.65	4.36	58.01	68.50
5200	Uncased, up to 10 ton capacity, 20'		200	.280		23	13.95	5.25	42.20	52

Estimating Tips

32 01 00 Operations and Maintenance of Exterior Improvements

- Recycling of asphalt pavement is becoming very popular and is an alternative to removal and replacement. It can be a good value engineering proposal if removed pavement can be recycled, either at the project site or at another site that is reasonably close to the project site. Sections on repair of flexible and rigid pavement are included.

32 10 00 Bases, Ballasts, and Paving

- When estimating paving, keep in mind the project schedule. Also note that prices for asphalt and concrete are generally higher in the cold seasons. Lines for pavement markings, including tactile warning systems and fence lines, are included.

32 90 00 Planting

- The timing of planting and guarantee specifications often dictate the costs for establishing tree and shrub growth and a stand of grass or ground cover. Establish the work performance schedule to coincide with the local planting season. Maintenance and growth guarantees can add 20–100% to the total landscaping cost and can be contractually cumbersome. The cost to replace trees and shrubs can be as high as 5% of the total cost, depending on the planting zone, soil conditions, and time of year.

Reference Numbers

Reference numbers are shown at the beginning of some major classifications. These numbers refer to related items in the Reference Section. The reference information may be an estimating procedure, an alternate pricing method, or technical information.

Note: Not all subdivisions listed here necessarily appear. ■

Same Data. Simplified.

Enjoy the convenience and efficiency of accessing your costs anywhere:

- **Skip the multiplier** by setting your location
- **Quickly search,** edit, favorite and share costs
- **Stay on top of price changes** with automatic updates

Discover more at rsmeans.com/online

32 01 Operation and Maintenance of Exterior Improvements

32 01 11 – Paving Cleaning

32 01 11.51 Rubber and Paint Removal From Paving

32 01 11.51 Rubber and Paint Removal From Paving	Crew	Daily Output	Labor-Hours	Unit	Material	2021 Bare Costs Labor	Equipment	Total	Total Incl O&P
0010 **RUBBER AND PAINT REMOVAL FROM PAVING**									
0015 Does not include traffic control costs									
0020 See other items in Section 32 17 23.13									
0100 Remove permanent painted traffic lines and markings	B-78A	500	.016	C.L.F.		.89	1.99	2.88	3.51
0200 Temporary traffic line tape	2 Clab	1500	.011	L.F.		.47		.47	.71
0300 Thermoplastic traffic lines and markings	B-79A	500	.024	C.L.F.		1.33	3.02	4.35	5.30
0400 Painted pavement markings	B-78B	500	.036	S.F.		1.64	.82	2.46	3.35
1000 Prepare and clean, 25,000 S.Y.	B-91A	7.36	2.174	Mile		106	114	220	284

32 01 13 – Flexible Paving Surface Treatment

32 01 13.61 Slurry Seal (Latex Modified)

	Crew	Daily Output	Labor-Hours	Unit	Material	Labor	Equipment	Total	Total Incl O&P
0010 **SLURRY SEAL (LATEX MODIFIED)**									
0011 Chip seal, slurry seal, and microsurfacing, see section 32 12 36									

32 01 13.62 Asphalt Surface Treatment

	Crew	Daily Output	Labor-Hours	Unit	Material	Labor	Equipment	Total	Total Incl O&P
0010 **ASPHALT SURFACE TREATMENT**									
3000 Pavement overlay, polypropylene									
3040 6 oz./S.Y., ideal conditions	B-63	10000	.004	S.Y.	.83	.19	.02	1.04	1.21
3190 Prime coat, bituminous, 0.28 gallon/S.Y.	B-45	2400	.007	C.S.F.	8.80	.37	.35	9.52	10.65
3200 Tack coat, emulsion, 0.05 gal./S.Y., 1,000 S.Y.		2500	.006	S.Y.	.35	.35	.33	1.03	1.29
3240 10,000 S.Y.		10000	.002		.28	.09	.08	.45	.53
3275 10,000 S.Y.		10000	.002		.53	.09	.08	.70	.80
3280 0.15 gal./S.Y., 1,000 S.Y.		2500	.006		.96	.35	.33	1.64	1.96
3320 10,000 S.Y.		10000	.002		.77	.09	.08	.94	1.07
3780 Rubberized asphalt (latex) seal		5000	.003		1.44	.18	.17	1.79	2.02
5400 Thermoplastic coal-tar, Type I, small or irregular area	B-90	2400	.027		4.23	1.31	.93	6.47	7.65
5450 Roadway or large area		8000	.008		4.23	.39	.28	4.90	5.55
5500 Type II, small or irregular area		2400	.027		5.40	1.31	.93	7.64	8.95
6000 Gravel surfacing on asphalt, screened and rolled	B-11L	160	.100	C.Y.	23.50	5.15	6.70	35.35	41
7000 For subbase treatment, see Section 31 32 13									

32 01 13.64 Sand Seal

	Crew	Daily Output	Labor-Hours	Unit	Material	Labor	Equipment	Total	Total Incl O&P
0010 **SAND SEAL**									
2080 Sand sealing, sharp sand, asphalt emulsion, small area	B-91	10000	.006	S.Y.	1.57	.34	.23	2.14	2.48
2120 Roadway or large area	"	18000	.004	"	1.35	.19	.13	1.67	1.91
3000 Sealing random cracks, min. 1/2" wide, to 1-1/2", 1,000 L.F.	B-77	2800	.014	L.F.	1.56	.65	.38	2.59	3.11
3040 10,000 L.F.		4000	.010	"	1.08	.46	.27	1.81	2.17
3080 Alternate method, 1,000 L.F.		200	.200	Gal.	40	9.15	5.35	54.50	63
3120 10,000 L.F.		325	.123	"	32	5.60	3.30	40.90	47
3200 Multi-cracks (flooding), 1 coat, small area	B-92	460	.070	S.Y.	2.45	3.12	2.64	8.21	10.25
3320 Large area		1425	.022	"	15.25	1.01	.85	17.11	19.25
3360 Alternate method, small area		115	.278	Gal.	14.95	12.50	10.55	38	46.50
3400 Large area		715	.045	"	14.05	2.01	1.70	17.76	20.50
3500 Sealing, roads, resealing joints in concrete	B-77	525	.076	L.F.	.47	3.48	2.05	6	7.95

32 01 13.66 Fog Seal

	Crew	Daily Output	Labor-Hours	Unit	Material	Labor	Equipment	Total	Total Incl O&P
0010 **FOG SEAL**									
0012 Sealcoating, 2 coat coal tar pitch emulsion over 10,000 S.Y.	B-45	5000	.003	S.Y.	.83	.18	.17	1.18	1.35
0030 1,000 to 10,000 S.Y.	"	3000	.005		.83	.29	.28	1.40	1.66
0100 Under 1,000 S.Y.	B-1	1050	.023		.83	1.03		1.86	2.45
0300 Petroleum resistant, over 10,000 S.Y.	B-45	5000	.003		1.52	.18	.17	1.87	2.11
0320 1,000 to 10,000 S.Y.	"	3000	.005		1.52	.29	.28	2.09	2.42
0400 Under 1,000 S.Y.	B-1	1050	.023		1.52	1.03		2.55	3.21
0600 Non-skid pavement renewal, over 10,000 S.Y.	B-45	5000	.003		1.31	.18	.17	1.66	1.88
0620 1,000 to 10,000 S.Y.	"	3000	.005		1.31	.29	.28	1.88	2.19

32 01 Operation and Maintenance of Exterior Improvements

32 01 13 – Flexible Paving Surface Treatment

32 01 13.66 Fog Seal

		Crew	Daily Output	Labor-Hours	Unit	Material	2021 Bare Costs Labor	Equipment	Total	Total Incl O&P
0700	Under 1,000 S.Y.	B-1	1050	.023	S.Y.	1.31	1.03		2.34	2.98
0800	Prepare and clean surface for above	A-2	8545	.003	↓		.13	.02	.15	.22
1000	Hand seal asphalt curbing	B-1	4420	.005	L.F.	.64	.24		.88	1.07
1900	Asphalt surface treatment, single course, small area									
1901	0.30 gal./S.Y. asphalt material, 20 lb./S.Y. aggregate	B-91	5000	.013	S.Y.	1.48	.68	.46	2.62	3.14
1910	Roadway or large area		10000	.006	"	1.36	.34	.23	1.93	2.25
1920	Bituminous surface treatment, 12' wide roadway, single course		7	9.143	Mile	7,500	485	325	8,310	9,325
1950	Asphalt surface treatment, dbl. course for small area		3000	.021	S.Y.	2.53	1.13	.76	4.42	5.30
1960	Roadway or large area		6000	.011		2.28	.56	.38	3.22	3.76
1980	Asphalt surface treatment, single course, for shoulders	↓	7500	.009	↓	1.23	.45	.31	1.99	2.36

32 01 16 – Flexible Paving Rehabilitation

32 01 16.71 Cold Milling Asphalt Paving

		Crew	Daily Output	Labor-Hours	Unit	Material	Labor	Equipment	Total	Total Incl O&P
0010	**COLD MILLING ASPHALT PAVING**									
5200	Cold planing & cleaning, 1" to 3" asphalt pavmt., over 25,000 S.Y.	B-71	6000	.009	S.Y.		.48	.77	1.25	1.56
5280	5,000 S.Y. to 10,000 S.Y.		4000	.014	"		.71	1.16	1.87	2.34
5285	5,000 S.Y. to 10,000 S.Y.	↓	36000	.002	S.F.		.08	.13	.21	.26
5300	Asphalt pavement removal from conc. base, no haul									
5320	Rip, load & sweep 1" to 3"	B-70	8000	.007	S.Y.		.36	.29	.65	.85
5330	3" to 6" deep	"	5000	.011			.57	.47	1.04	1.36
5340	Profile grooving, asphalt pavement load & sweep, 1" deep	B-71	12500	.004			.23	.37	.60	.75
5350	3" deep		9000	.006			.32	.52	.84	1.04
5360	6" deep		5000	.011	↓		.57	.93	1.50	1.87
5400	Mixing material in windrow, 180 HP grader	B-11L	9400	.002	C.Y.		.09	.11	.20	.26
5450	For cold laid asphalt pavement, see Section 32 12 16.19									
7400	(Over 3") 5,000-25,000 S.F.	B-71	7500	.007	S.F.		.38	.62	1	1.25
7410	25,001-100,000 S.F.		27000	.002			.11	.17	.28	.35
7420	More than 100,000 S.F.	↓	50000	.001	↓		.06	.09	.15	.19

32 01 16.73 In Place Cold Reused Asphalt Paving

		Crew	Daily Output	Labor-Hours	Unit	Material	Labor	Equipment	Total	Total Incl O&P
0010	**IN PLACE COLD REUSED ASPHALT PAVING**									
5000	Reclamation, pulverizing and blending with existing base									
5040	Aggregate base, 4" thick pavement, over 15,000 S.Y.	B-73	2400	.027	S.Y.		1.43	1.90	3.33	4.23
5080	5,000 S.Y. to 15,000 S.Y.		2200	.029			1.56	2.07	3.63	4.60
5120	8" thick pavement, over 15,000 S.Y.		2200	.029			1.56	2.07	3.63	4.60
5160	5,000 S.Y. to 15,000 S.Y.	↓	2000	.032			1.72	2.27	3.99	5.05

32 01 16.74 In Place Hot Reused Asphalt Paving

			Crew	Daily Output	Labor-Hours	Unit	Material	Labor	Equipment	Total	Total Incl O&P
0010	**IN PLACE HOT REUSED ASPHALT PAVING**										
5500	Recycle asphalt pavement at site										
5520	Remove, rejuvenate and spread 4" deep	G	B-72	2500	.026	S.Y.	4.61	1.33	3.33	9.27	10.70
5521	6" deep	G	"	2000	.032	"	6.75	1.66	4.16	12.57	14.50

32 01 17 – Flexible Paving Repair

32 01 17.10 Repair of Asphalt Pavement Holes

		Crew	Daily Output	Labor-Hours	Unit	Material	Labor	Equipment	Total	Total Incl O&P
0010	**REPAIR OF ASPHALT PAVEMENT HOLES** (cold patch)									
0100	Flexible pavement repair holes, roadway, light traffic, 1 C.F. size each	B-37A	24	1	Ea.	9.50	46	14.80	70.30	95.50
0150	Group of two, 1 C.F. size each		16	1.500	Set	19	69	22	110	149
0200	Group of three, 1 C.F. size each	↓	12	2	"	28.50	91.50	29.50	149.50	201
0300	Medium traffic, 1 C.F. size each	B-37B	24	1.333	Ea.	9.50	60.50	14.80	84.80	117
0350	Group of two, 1 C.F. size each		16	2	Set	19	91	22	132	182
0400	Group of three, 1 C.F. size each	↓	12	2.667	"	28.50	121	29.50	179	245
0500	Highway/heavy traffic, 1 C.F. size each	B-37C	24	1.333	Ea.	9.50	62	23	94.50	129
0550	Group of two, 1 C.F. size each		16	2	Set	19	93	34.50	146.50	198
0600	Group of three, 1 C.F. size each		12	2.667	"	28.50	124	46	198.50	269

For customer support on your Site Work & Landscape Costs with RSMeans Data, call 800.448.8182.

365

32 01 17 – Flexible Paving Repair

32 01 17.10 Repair of Asphalt Pavement Holes

		Crew	Daily Output	Labor-Hours	Unit	Material	2021 Bare Costs Labor	2021 Bare Costs Equipment	Total	Total Incl O&P
0700	Add police officer and car for traffic control				Hr.	56.50			56.50	62
1000	Flexible pavement repair holes, parking lot, bag material, 1 C.F. size each	B-37D	18	.889	Ea.	60	41.50	6.25	107.75	135
1010	Economy bag material, 1 C.F. size each		18	.889	"	37.50	41.50	6.25	85.25	110
1100	Group of two, 1 C.F. size each		12	1.333	Set	120	62	9.35	191.35	235
1110	Group of two, economy bag, 1 C.F. size each		12	1.333		75.50	62	9.35	146.85	186
1200	Group of three, 1 C.F. size each		10	1.600		180	74.50	11.20	265.70	320
1210	Economy bag, group of three, 1 C.F. size each		10	1.600		113	74.50	11.20	198.70	247
1300	Flexible pavement repair holes, parking lot, 1 C.F. single hole	A-3A	4	2	Ea.	60	111	44	215	280
1310	Economy material, 1 C.F. single hole		4	2	"	37.50	111	44	192.50	255
1400	Flexible pavement repair holes, parking lot, 1 C.F. four holes		2	4	Set	240	222	88.50	550.50	690
1410	Economy material, 1 C.F. four holes		2	4	"	151	222	88.50	461.50	595
1500	Flexible pavement repair holes, large parking lot, bulk matl, 1 C.F. size	B-37A	32	.750	Ea.	9.50	34.50	11.10	55.10	74

32 01 17.20 Repair of Asphalt Pavement Patches

		Crew	Daily Output	Labor-Hours	Unit	Material	2021 Bare Costs Labor	2021 Bare Costs Equipment	Total	Total Incl O&P
0010	**REPAIR OF ASPHALT PAVEMENT PATCHES**									
0100	Flexible pavement patches, roadway, light traffic, sawcut 10-25 S.F.	B-89	12	1.333	Ea.		69.50	89	158.50	202
0150	Sawcut 26-60 S.F.		10	1.600			83.50	107	190.50	242
0200	Sawcut 61-100 S.F.		8	2			104	133	237	305
0210	Sawcut groups of small size patches		24	.667			35	44.50	79.50	101
0220	Large size patches		16	1			52	66.50	118.50	152
0300	Flexible pavement patches, roadway, light traffic, digout 10-25 S.F.	B-6	16	1.500			72	13.50	85.50	123
0350	Digout 26-60 S.F.		12	2			96	18	114	163
0400	Digout 61-100 S.F.		8	3			144	27	171	245
0450	Add 8 C.Y. truck, small project debris haulaway	B-34A	8	1	Hr.		51.50	51	102.50	133
0460	Add 12 C.Y. truck, small project debris haulaway	B-34B	8	1			51.50	72.50	124	156
0480	Add flagger for non-intersection medium traffic	1 Clab	8	1			44.50		44.50	66.50
0490	Add flasher truck for intersection medium traffic or heavy traffic	A-2B	8	1			49	25	74	101
0500	Flexible pavement patches, roadway, repave, cold, 15 S.F., 4" D	B-37	8	6	Ea.	45	279	32.50	356.50	500
0510	6" depth		8	6		68	279	32.50	379.50	525
0520	Repave, cold, 20 S.F., 4" depth		8	6		60	279	32.50	371.50	515
0530	6" depth		8	6		90.50	279	32.50	402	550
0540	Repave, cold, 25 S.F., 4" depth		8	6		74.50	279	32.50	386	535
0550	6" depth		8	6		113	279	32.50	424.50	575
0600	Repave, cold, 30 S.F., 4" depth		8	6		89.50	279	32.50	401	550
0610	6" depth		8	6		136	279	32.50	447.50	600
0640	Repave, cold, 40 S.F., 4" depth		8	6		119	279	32.50	430.50	580
0650	6" depth		8	6		181	279	32.50	492.50	650
0680	Repave, cold, 50 S.F., 4" depth		8	6		149	279	32.50	460.50	615
0690	6" depth		8	6		226	279	32.50	537.50	700
0720	Repave, cold, 60 S.F., 4" depth		8	6		179	279	32.50	490.50	650
0730	6" depth		8	6		272	279	32.50	583.50	750
0800	Repave, cold, 70 S.F., 4" depth		7	6.857		209	320	37	566	745
0810	6" depth		7	6.857		315	320	37	672	865
0820	Repave, cold, 75 S.F., 4" depth		7	6.857		223	320	37	580	760
0830	6" depth		7	6.857		340	320	37	697	890
0900	Repave, cold, 80 S.F., 4" depth		6	8		239	375	43	657	865
0910	6" depth		6	8		360	375	43	778	1,000
0940	Repave, cold, 90 S.F., 4" depth		6	8		269	375	43	687	900
0950	6" depth		6	8		410	375	43	828	1,050
0980	Repave, cold, 100 S.F., 4" depth		6	8		299	375	43	717	935
0990	6" depth		6	8		450	375	43	868	1,100
1000	Add flasher truck for paving operations in medium or heavy traffic	A-2B	8	1	Hr.		49	25	74	101
1100	Prime coat for repair 15-40 S.F., 25% overspray	B-37A	48	.500	Ea.	8.75	23	7.40	39.15	52.50

32 01 17 – Flexible Paving Repair

32 01 17.20 Repair of Asphalt Pavement Patches

		Crew	Daily Output	Labor-Hours	Unit	Material	2021 Bare Costs Labor	2021 Bare Costs Equipment	Total	Total Incl O&P
1150	41-60 S.F.	B-37A	40	.600	Ea.	17.50	27.50	8.90	53.90	70
1175	61-80 S.F.		32	.750		23.50	34.50	11.10	69.10	89.50
1200	81-100 S.F.		32	.750		29	34.50	11.10	74.60	95.50
1210	Groups of patches w/25% overspray		3600	.007	S.F.	.29	.31	.10	.70	.89
1300	Flexible pavement repair patches, street repave, hot, 60 S.F., 4" D	B-37	8	6	Ea.	97.50	279	32.50	409	560
1310	6" depth		8	6		148	279	32.50	459.50	615
1320	Repave, hot, 70 S.F., 4" depth		7	6.857		114	320	37	471	640
1330	6" depth		7	6.857		172	320	37	529	705
1340	Repave, hot, 75 S.F., 4" depth		7	6.857		121	320	37	478	650
1350	6" depth		7	6.857		184	320	37	541	720
1360	Repave, hot, 80 S.F., 4" depth		6	8		130	375	43	548	745
1370	6" depth		6	8		197	375	43	615	820
1380	Repave, hot, 90 S.F., 4" depth		6	8		146	375	43	564	765
1390	6" depth		6	8		222	375	43	640	845
1400	Repave, hot, 100 S.F., 4" depth		6	8		162	375	43	580	780
1410	6" depth		6	8		246	375	43	664	875
1420	Pave hot groups of patches, 4" depth		900	.053	S.F.	1.62	2.48	.29	4.39	5.80
1430	6" depth		900	.053	"	2.46	2.48	.29	5.23	6.75
1500	Add 8 C.Y. truck for hot asphalt paving operations	B-34A	1	8	Day		410	410	820	1,075
1550	Add flasher truck for hot paving operations in med/heavy traffic	A-2B	8	1	Hr.		49	25	74	101
2000	Add police officer and car for traffic control				"	56.50			56.50	62

32 01 17.61 Sealing Cracks In Asphalt Paving

		Crew	Daily Output	Labor-Hours	Unit	Material	2021 Bare Costs Labor	2021 Bare Costs Equipment	Total	Total Incl O&P
0010	**SEALING CRACKS IN ASPHALT PAVING**									
0100	Sealing cracks in asphalt paving, 1/8" wide x 1/2" depth, slow set	B-37A	2000	.012	L.F.	.18	.55	.18	.91	1.21
0110	1/4" wide x 1/2" depth		2000	.012		.20	.55	.18	.93	1.24
0130	3/8" wide x 1/2" depth		2000	.012		.23	.55	.18	.96	1.27
0140	1/2" wide x 1/2" depth		1800	.013		.26	.61	.20	1.07	1.42
0150	3/4" wide x 1/2" depth		1600	.015		.32	.69	.22	1.23	1.62
0160	1" wide x 1/2" depth		1600	.015		.38	.69	.22	1.29	1.69
0165	1/8" wide x 1" depth, rapid set	B-37F	2000	.016		.19	.73	.27	1.19	1.59
0170	1/4" wide x 1" depth		2000	.016		.37	.73	.27	1.37	1.79
0175	3/8" wide x 1" depth		1800	.018		.56	.81	.30	1.67	2.14
0180	1/2" wide x 1" depth		1800	.018		.74	.81	.30	1.85	2.35
0185	5/8" wide x 1" depth		1800	.018		.93	.81	.30	2.04	2.55
0190	3/4" wide x 1" depth		1600	.020		1.11	.91	.33	2.35	2.95
0195	1" wide x 1" depth		1600	.020		1.48	.91	.33	2.72	3.36
0200	1/8" wide x 2" depth, rapid set		2000	.016		.37	.73	.27	1.37	1.79
0210	1/4" wide x 2" depth		2000	.016		.74	.73	.27	1.74	2.20
0220	3/8" wide x 2" depth		1800	.018		1.11	.81	.30	2.22	2.75
0230	1/2" wide x 2" depth		1800	.018		1.48	.81	.30	2.59	3.16
0240	5/8" wide x 2" depth		1800	.018		1.85	.81	.30	2.96	3.57
0250	3/4" wide x 2" depth		1600	.020		2.23	.91	.33	3.47	4.18
0260	1" wide x 2" depth		1600	.020		2.97	.91	.33	4.21	5
0300	Add flagger for non-intersection medium traffic	1 Clab	8	1	Hr.		44.50		44.50	66.50
0400	Add flasher truck for intersection medium traffic or heavy traffic	A-2B	8	1	"		49	25	74	101

32 01 19 – Rigid Paving Surface Treatment

32 01 19.61 Sealing of Joints In Rigid Paving

		Crew	Daily Output	Labor-Hours	Unit	Material	2021 Bare Costs Labor	2021 Bare Costs Equipment	Total	Total Incl O&P
0010	**SEALING OF JOINTS IN RIGID PAVING**									
1000	Sealing, concrete pavement, preformed elastomeric	B-77	500	.080	L.F.	3.58	3.65	2.15	9.38	11.75
2000	Waterproofing, membrane, tar and fabric, small area	B-63	233	.172	S.Y.	15.10	8	.77	23.87	29.50
2500	Large area		1435	.028		13.30	1.30	.13	14.73	16.75
3000	Preformed rubberized asphalt, small area		100	.400		19.50	18.65	1.80	39.95	51.50

32 01 Operation and Maintenance of Exterior Improvements

32 01 19 – Rigid Paving Surface Treatment

32 01 19.61 Sealing of Joints In Rigid Paving

	Crew	Daily Output	Labor-Hours	Unit	Material	2021 Bare Costs Labor	Equipment	Total	Total Incl O&P
3500	Large area	B-63	367	.109	S.Y.	17.75	5.10	.49	23.34	27.50

32 01 29 – Rigid Paving Repair

32 01 29.61 Partial Depth Patching of Rigid Pavement

		Crew	Daily Output	Labor-Hours	Unit	Material	2021 Bare Costs Labor	Equipment	Total	Total Incl O&P
0010	**PARTIAL DEPTH PATCHING OF RIGID PAVEMENT**									
0100	Rigid pavement repair, roadway, light traffic, 25% pitting 15 S.F.	B-37F	16	2	Ea.	24	91	33	148	199
0110	Pitting 20 S.F.		16	2		32.50	91	33	156.50	208
0120	Pitting 25 S.F.		16	2		40.50	91	33	164.50	217
0130	Pitting 30 S.F.		16	2		48.50	91	33	172.50	226
0140	Pitting 40 S.F.		16	2		64.50	91	33	188.50	244
0150	Pitting 50 S.F.		12	2.667		80.50	121	44.50	246	320
0160	Pitting 60 S.F.		12	2.667		97	121	44.50	262.50	335
0170	Pitting 70 S.F.		12	2.667		113	121	44.50	278.50	355
0180	Pitting 75 S.F.		12	2.667		121	121	44.50	286.50	365
0190	Pitting 80 S.F.		8	4		129	182	66.50	377.50	485
0200	Pitting 90 S.F.		8	4		145	182	66.50	393.50	505
0210	Pitting 100 S.F.		8	4		161	182	66.50	409.50	520
0300	Roadway, light traffic, 50% pitting 15 S.F.		12	2.667		48.50	121	44.50	214	283
0310	Pitting 20 S.F.		12	2.667		64.50	121	44.50	230	300
0320	Pitting 25 S.F.		12	2.667		80.50	121	44.50	246	320
0330	Pitting 30 S.F.		12	2.667		97	121	44.50	262.50	335
0340	Pitting 40 S.F.		12	2.667		129	121	44.50	294.50	370
0350	Pitting 50 S.F.		8	4		161	182	66.50	409.50	520
0360	Pitting 60 S.F.		8	4		194	182	66.50	442.50	560
0370	Pitting 70 S.F.		8	4		226	182	66.50	474.50	595
0380	Pitting 75 S.F.		8	4		242	182	66.50	490.50	610
0390	Pitting 80 S.F.		6	5.333		258	243	88.50	589.50	740
0400	Pitting 90 S.F.		6	5.333		290	243	88.50	621.50	780
0410	Pitting 100 S.F.		6	5.333		325	243	88.50	656.50	815
1000	Rigid pavement repair, light traffic, surface patch, 2" deep, 15 S.F.		8	4		97	182	66.50	345.50	450
1010	Surface patch, 2" deep, 20 S.F.		8	4		129	182	66.50	377.50	485
1020	Surface patch, 2" deep, 25 S.F.		8	4		161	182	66.50	409.50	520
1030	Surface patch, 2" deep, 30 S.F.		8	4		194	182	66.50	442.50	560
1040	Surface patch, 2" deep, 40 S.F.		8	4		258	182	66.50	506.50	630
1050	Surface patch, 2" deep, 50 S.F.		6	5.333		325	243	88.50	656.50	815
1060	Surface patch, 2" deep, 60 S.F.		6	5.333		385	243	88.50	716.50	885
1070	Surface patch, 2" deep, 70 S.F.		6	5.333		450	243	88.50	781.50	955
1080	Surface patch, 2" deep, 75 S.F.		6	5.333		485	243	88.50	816.50	990
1090	Surface patch, 2" deep, 80 S.F.		5	6.400		515	291	106	912	1,125
1100	Surface patch, 2" deep, 90 S.F.		5	6.400		580	291	106	977	1,200
1200	Surface patch, 2" deep, 100 S.F.		5	6.400		645	291	106	1,042	1,250
2000	Add flagger for non-intersection medium traffic	1 Clab	8	1	Hr.		44.50		44.50	66.50
2100	Add flasher truck for intersection medium traffic or heavy traffic	A-2B	8	1			49	25	74	101
2200	Add police officer and car for traffic control					56.50			56.50	62

32 01 29.70 Full Depth Patching of Rigid Pavement

		Crew	Daily Output	Labor-Hours	Unit	Material	2021 Bare Costs Labor	Equipment	Total	Total Incl O&P
0010	**FULL DEPTH PATCHING OF RIGID PAVEMENT**									
0015	Pavement preparation includes sawcut, remove pavement and replace									
0020	6" of subbase and one layer of reinforcement									
0030	Trucking and haulaway of debris excluded									
0100	Rigid pavement replace, light traffic, prep, 15 S.F., 6" depth	B-37E	16	3.500	Ea.	24	173	58.50	255.50	350
0110	Replacement preparation 20 S.F., 6" depth		16	3.500		32	173	58.50	263.50	355
0115	25 S.F., 6" depth		16	3.500		40	173	58.50	271.50	365
0120	30 S.F., 6" depth		14	4		48	197	66.50	311.50	420

32 01 29 – Rigid Paving Repair

32 01 29.70 Full Depth Patching of Rigid Pavement		Crew	Daily Output	Labor-Hours	Unit	Material	2021 Bare Costs		Total	Total Incl O&P
							Labor	Equipment		
0125	35 S.F., 6" depth	B-37E	14	4	Ea.	56	197	66.50	319.50	430
0130	40 S.F., 6" depth		14	4		63.50	197	66.50	327	440
0135	45 S.F., 6" depth		14	4		71.50	197	66.50	335	445
0140	50 S.F., 6" depth		12	4.667		79.50	230	78	387.50	520
0145	55 S.F., 6" depth		12	4.667		87.50	230	78	395.50	525
0150	60 S.F., 6" depth		10	5.600		95.50	276	93.50	465	620
0155	65 S.F., 6" depth		10	5.600		104	276	93.50	473.50	625
0160	70 S.F., 6" depth		10	5.600		112	276	93.50	481.50	635
0165	75 S.F., 6" depth		10	5.600		120	276	93.50	489.50	645
0170	80 S.F., 6" depth		8	7		127	345	117	589	785
0175	85 S.F., 6" depth		8	7		135	345	117	597	790
0180	90 S.F., 6" depth		8	7		143	345	117	605	800
0185	95 S.F., 6" depth		8	7		151	345	117	613	810
0190	100 S.F., 6" depth	▼	8	7	▼	159	345	117	621	820
0200	Pavement preparation for 8" rigid paving same as 6"									
0290	Pavement preparation for 9" rigid paving same as 10"									
0300	Rigid pavement replace, light traffic, prep, 15 S.F., 10" depth	B-37E	16	3.500	Ea.	40.50	173	58.50	272	365
0302	Pavement preparation 10" includes sawcut, remove pavement and									
0304	replace 6" of subbase and two layers of reinforcement									
0306	Trucking and haulaway of debris excluded									
0310	Replacement preparation 20 S.F., 10" depth	B-37E	16	3.500	Ea.	54	173	58.50	285.50	380
0315	25 S.F., 10" depth		16	3.500		68	173	58.50	299.50	395
0320	30 S.F., 10" depth		14	4		81.50	197	66.50	345	455
0325	35 S.F., 10" depth		14	4		61.50	197	66.50	325	435
0330	40 S.F., 10" depth		14	4		108	197	66.50	371.50	485
0335	45 S.F., 10" depth		12	4.667		122	230	78	430	565
0340	50 S.F., 10" depth		12	4.667		136	230	78	444	580
0345	55 S.F., 10" depth		12	4.667		149	230	78	457	595
0350	60 S.F., 10" depth		10	5.600		163	276	93.50	532.50	690
0355	65 S.F., 10" depth		10	5.600		176	276	93.50	545.50	705
0360	70 S.F., 10" depth		10	5.600		190	276	93.50	559.50	720
0365	75 S.F., 10" depth		8	7		203	345	117	665	865
0370	80 S.F., 10" depth		8	7		217	345	117	679	880
0375	85 S.F., 10" depth		8	7		230	345	117	692	895
0380	90 S.F., 10" depth		8	7		244	345	117	706	910
0385	95 S.F., 10" depth		8	7		258	345	117	720	925
0390	100 S.F., 10" depth	▼	8	7	▼	271	345	117	733	940
0500	Add 8 C.Y. truck, small project debris haulaway	B-34A	8	1	Hr.		51.50	51	102.50	133
0550	Add 12 C.Y. truck, small project debris haulaway	B-34B	8	1			51.50	72.50	124	156
0600	Add flagger for non-intersection medium traffic	1 Clab	8	1			44.50		44.50	66.50
0650	Add flasher truck for intersection medium traffic or heavy traffic	A-2B	8	1			49	25	74	101
0700	Add police officer and car for traffic control				▼	56.50			56.50	62
0990	Concrete will be replaced using quick set mix with aggregate									
1000	Rigid pavement replace, light traffic, repour, 15 S.F., 6" depth	B-37F	18	1.778	Ea.	173	81	29.50	283.50	345
1010	20 S.F., 6" depth		18	1.778		230	81	29.50	340.50	405
1015	25 S.F., 6" depth		18	1.778		288	81	29.50	398.50	470
1020	30 S.F., 6" depth		16	2		345	91	33	469	555
1025	35 S.F., 6" depth		16	2		405	91	33	529	620
1030	40 S.F., 6" depth		16	2		460	91	33	584	680
1035	45 S.F., 6" depth		16	2		520	91	33	644	745
1040	50 S.F., 6" depth		16	2		575	91	33	699	810
1045	55 S.F., 6" depth		12	2.667		635	121	44.50	800.50	925
1050	60 S.F., 6" depth	▼	12	2.667	▼	690	121	44.50	855.50	990

369

For customer support on your Site Work & Landscape Costs with RSMeans Data, call 800.448.8182.

32 01 29 – Rigid Paving Repair

32 01 29.70 Full Depth Patching of Rigid Pavement		Crew	Daily Output	Labor-Hours	Unit	Material	2021 Bare Costs		Total	Total Incl O&P
							Labor	Equipment		
1055	65 S.F., 6" depth	B-37F	12	2.667	Ea.	750	121	44.50	915.50	1,050
1060	70 S.F., 6" depth		12	2.667		805	121	44.50	970.50	1,125
1065	75 S.F., 6" depth		10	3.200		865	146	53	1,064	1,225
1070	80 S.F., 6" depth		10	3.200		920	146	53	1,119	1,275
1075	85 S.F., 6" depth		8	4		980	182	66.50	1,228.50	1,425
1080	90 S.F., 6" depth		8	4		1,025	182	66.50	1,273.50	1,500
1085	95 S.F., 6" depth		8	4		1,100	182	66.50	1,348.50	1,550
1090	100 S.F., 6" depth	↓	8	4	↓	1,150	182	66.50	1,398.50	1,625
1099	Concrete will be replaced using 4,500 psi concrete ready mix									
1100	Rigid pavement replace, light traffic, repour, 15 S.F., 6" depth	A-2	12	2	Ea.	360	91.50	16.55	468.05	550
1110	20-50 S.F., 6" depth		12	2		360	91.50	16.55	468.05	550
1120	55-65 S.F., 6" depth		10	2.400		400	110	19.85	529.85	625
1130	70-80 S.F., 6" depth		10	2.400		430	110	19.85	559.85	660
1140	85-100 S.F., 6" depth		8	3		505	138	25	668	790
1200	Repour 15-40 S.F., 8" depth		12	2		360	91.50	16.55	468.05	550
1210	45-50 S.F., 8" depth		12	2		400	91.50	16.55	508.05	595
1220	55-60 S.F., 8" depth		10	2.400		430	110	19.85	559.85	660
1230	65-80 S.F., 8" depth		10	2.400		505	110	19.85	634.85	740
1240	85-90 S.F., 8" depth		8	3		540	138	25	703	830
1250	95-100 S.F., 8" depth		8	3		575	138	25	738	870
1300	Repour 15-30 S.F., 10" depth		12	2		360	91.50	16.55	468.05	550
1310	35-40 S.F., 10" depth		12	2		400	91.50	16.55	508.05	595
1320	45 S.F., 10" depth		12	2		430	91.50	16.55	538.05	630
1330	50-65 S.F., 10" depth		10	2.400		505	110	19.85	634.85	740
1340	70 S.F., 10" depth		10	2.400		540	110	19.85	669.85	780
1350	75-80 S.F., 10" depth		10	2.400		575	110	19.85	704.85	820
1360	85-95 S.F., 10" depth		8	3		645	138	25	808	945
1370	100 S.F., 10" depth	↓	8	3	↓	685	138	25	848	990

32 01 30 – Operation and Maintenance of Site Improvements

32 01 30.10 Site Maintenance

		Crew	Daily Output	Labor-Hours	Unit	Material	Labor	Equipment	Total	Total Incl O&P
0010	**SITE MAINTENANCE**									
0800	Flower bed maintenance									
0810	Cultivate bed, no mulch	1 Clab	14	.571	M.S.F.		25.50		25.50	38
0830	Fall clean-up of flower bed, including pick-up mulch for re-use		1	8			355		355	530
0840	Fertilize flower bed, dry granular 3 lb./M.S.F.		85	.094		1.26	4.18		5.44	7.65
1130	Police, hand pickup		30	.267			11.85		11.85	17.65
1140	Vacuum (outside)		48	.167			7.40		7.40	11.05
1200	Spring prepare		2	4			178		178	265
1300	Weed mulched bed		20	.400			17.75		17.75	26.50
1310	Unmulched bed	↓	8	1	↓		44.50		44.50	66.50
3000	Lawn maintenance									
3010	Aerate lawn, 18" cultivating width, walk behind	A-1K	95	.084	M.S.F.		3.74	.97	4.71	6.65
3040	48" cultivating width	B-66	750	.011			.59	.36	.95	1.27
3060	72" cultivating width	"	1100	.007	↓		.40	.24	.64	.87
3100	Edge lawn, by hand at walks	1 Clab	16	.500	C.L.F.		22		22	33
3150	At planting beds		7	1.143			50.50		50.50	75.50
3200	Using gas powered edger at walks		88	.091			4.04		4.04	6
3250	At planting beds		24	.333	↓		14.80		14.80	22
3260	Vacuum, 30" gas, outdoors with hose		96	.083	M.L.F.		3.70		3.70	5.50
3400	Weed lawn, by hand	↓	3	2.667	M.S.F.		118		118	177
3800	Lawn bed preparation see Section 32 91 13.23									
4510	Power rake	1 Clab	45	.178	M.S.F.		7.90		7.90	11.80

32 01 30 – Operation and Maintenance of Site Improvements

32 01 30.10 Site Maintenance	Crew	Daily Output	Labor-Hours	Unit	Material	2021 Bare Costs Labor	2021 Bare Costs Equipment	Total	Total Incl O&P
4700	Seeding lawn, see Section 32 92 19.14								
4750	Sodding, see Section 32 92 23.10								
5900	Road & walk maintenance								
5910	Asphaltic concrete paving, cold patch, 2" thick	B-37	350	.137	S.Y.	15.05	6.40	.74	22.19	27
5913	3" thick	"	260	.185	"	26.50	8.60	1	36.10	43
5915	De-icing roads and walks								
5920	Calcium chloride in truckload lots see Section 31 32 13.30								
6000	Ice melting comp., 90% calc. chlor., effec. to -30°F								
6010	50-80 lb. poly bags, med. applic. 19 lb./M.S.F., by hand	1 Clab	60	.133	M.S.F.	203	5.90		208.90	233
6050	With hand operated rotary spreader		110	.073		203	3.23		206.23	229
6100	Rock salt, med. applic. on road & walkway, by hand		60	.133		112	5.90		117.90	132
6110	With hand operated rotary spreader		110	.073		112	3.23		115.23	128
6130	Hosing, sidewalks & other paved areas		30	.267			11.85		11.85	17.65
6260	Sidewalk, brick pavers, steam cleaning	A-1H	950	.008	S.F.	.11	.37	.08	.56	.77
6400	Sweep walk by hand	1 Clab	15	.533	M.S.F.		23.50		23.50	35.50
6410	Power vacuum		100	.080			3.55		3.55	5.30
6420	Drives & parking areas with power vacuum		120	.067			2.96		2.96	4.42
6600	Shrub maintenance								
6640	Shrub bed fertilize dry granular 3 lb./M.S.F.	1 Clab	85	.094	M.S.F.	1.26	4.18		5.44	7.65
6800	Weed, by handhoe		8	1			44.50		44.50	66.50
6810	Spray out		32	.250			11.10		11.10	16.55
6820	Spray after mulch		48	.167			7.40		7.40	11.05
7100	Tree maintenance								
7200	Fertilize, tablets, slow release, 30 gram/tree	1 Clab	100	.080	Ea.	2.29	3.55		5.84	7.80
7420	Pest control, spray		24	.333		24.50	14.80		39.30	49
7430	Systemic		48	.167		25	7.40		32.40	38.50

32 01 30.20 Snow Removal	Crew	Daily Output	Labor-Hours	Unit	Material	2021 Bare Costs Labor	2021 Bare Costs Equipment	Total	Total Incl O&P
0010	**SNOW REMOVAL**								
0020	Plowing, 12 ton truck, 2"-4" deep	B-34A	250	.032	M.S.F.		1.64	1.63	3.27	4.24
0040	4"-10" deep		200	.040			2.05	2.04	4.09	5.30
0060	10"-15" deep		150	.053			2.74	2.72	5.46	7.10
0080	Pickup truck, 2"-4" deep	A-3A	175	.046			2.54	1.01	3.55	4.89
0100	4"-10" deep		130	.062			3.42	1.36	4.78	6.60
0120	10"-15" deep		75	.107			5.90	2.36	8.26	11.40
0140	Load and haul snow, 1 mile round trip	A-3B	230	.070	C.Y.		3.84	5.30	9.14	11.55
0160	2 mile round trip		175	.091			5.05	6.95	12	15.15
0180	3 mile round trip		150	.107			5.90	8.10	14	17.75
0200	4 mile round trip		130	.123			6.80	9.35	16.15	20.50
0220	5 mile round trip		120	.133			7.35	10.15	17.50	22
0240	Clearing with wheeled skid steer loader, 1 C.Y.	A-3C	240	.033			1.85	1.86	3.71	4.81
0260	Spread sand and salt mix	B-34A	375	.021	M.S.F.	7.45	1.09	1.09	9.63	11.05
0280	Sidewalks and drives, by hand	1 Clab	1200	.007	C.F.		.30		.30	.44
0300	Power, 24" blower	A-1M	7000	.001	"		.05	.01	.06	.09
0320	2"-4" deep, single driveway (10' x 50')		16	.500	Ea.		22	4.27	26.27	37.50
0340	Double driveway (20' x 50')		16	.500			22	4.27	26.27	37.50
0360	4"-10" deep, single driveway		16	.500			22	4.27	26.27	37.50
0380	Double driveway		16	.500			22	4.27	26.27	37.50
0400	10"-15" deep, single driveway		12	.667			29.50	5.70	35.20	50.50
0420	Double driveway		12	.667			29.50	5.70	35.20	50.50
0440	For heavy wet snow, add								20%	20%

For customer support on your Site Work & Landscape Costs with RSMeans Data, call 800.448.8182.

371

32 01 90.13 Fertilizing

		Crew	Daily Output	Labor-Hours	Unit	Material	2021 Bare Costs Labor	Equipment	Total	Total Incl O&P
0010	**FERTILIZING**									
0100	Dry granular, 4 lb./M.S.F., hand spread	1 Clab	24	.333	M.S.F.	2.35	14.80		17.15	24.50
0110	Push rotary	"	140	.057		2.35	2.54		4.89	6.40
0120	Tractor towed spreader, 8'	B-66	500	.016		2.35	.89	.54	3.78	4.50
0130	12' spread		800	.010		2.35	.56	.33	3.24	3.79
0140	Truck whirlwind spreader	↓	1200	.007		2.35	.37	.22	2.94	3.39
0180	Water soluble, hydro spread, 1.5 lb./M.S.F.	B-64	600	.027		2.41	1.24	.67	4.32	5.25
0190	Add for weed control				↓	.31			.31	.34

32 01 90.19 Mowing

		Crew	Daily Output	Labor-Hours	Unit	Material	2021 Bare Costs Labor	Equipment	Total	Total Incl O&P
0010	**MOWING**									
1650	Mowing brush, tractor with rotary mower									
1660	Light density	B-84	22	.364	M.S.F.		21.50	16.85	38.35	50.50
1670	Medium density		13	.615			36.50	28.50	65	85.50
1680	Heavy density		9	.889			52.50	41	93.50	124
2000	Mowing, brush/grass, tractor, rotary mower, highway/airport median	↓	13	.615	↓		36.50	28.50	65	85.50
2010	Traffic safety flashing truck for highway/airport median mowing	A-2B	1	8	Day		390	198	588	805
4050	Lawn mowing, power mower, 18" - 22"	1 Clab	65	.123	M.S.F.		5.45		5.45	8.15
4100	22" - 30"		110	.073			3.23		3.23	4.82
4150	30" - 32"	↓	140	.057			2.54		2.54	3.79
4160	Riding mower, 36" - 44"	B-66	300	.027			1.48	.89	2.37	3.19
4170	48" - 58"	"	480	.017	↓		.93	.56	1.49	1.99
4175	Mowing with tractor & attachments									
4180	3 gang reel, 7'	B-66	930	.009	M.S.F.		.48	.29	.77	1.03
4190	5 gang reel, 12'		1200	.007			.37	.22	.59	.80
4200	Cutter or sickle-bar, 5', rough terrain		210	.038			2.11	1.27	3.38	4.55
4210	Cutter or sickle-bar, 5', smooth terrain		340	.024	↓		1.31	.79	2.10	2.82
4220	Drainage channel, 5' sickle bar		5	1.600	Mile		89	53.50	142.50	191
4250	Lawn mower, rotary type, sharpen (all sizes)	1 Clab	10	.800	Ea.		35.50		35.50	53
4260	Repair or replace part		7	1.143	"		50.50		50.50	75.50
5000	Edge trimming with weed whacker	↓	5760	.001	L.F.		.06		.06	.09

32 01 90.23 Pruning

		Crew	Daily Output	Labor-Hours	Unit	Material	2021 Bare Costs Labor	Equipment	Total	Total Incl O&P
0010	**PRUNING**									
0020	1-1/2" caliper	1 Clab	84	.095	Ea.		4.23		4.23	6.30
0030	2" caliper		70	.114			5.05		5.05	7.55
0040	2-1/2" caliper		50	.160			7.10		7.10	10.60
0050	3" caliper	↓	30	.267			11.85		11.85	17.65
0060	4" caliper, by hand	2 Clab	21	.762			34		34	50.50
0070	Aerial lift equipment	B-85	38	1.053			51.50	20.50	72	99
0100	6" caliper, by hand	2 Clab	12	1.333			59		59	88.50
0110	Aerial lift equipment	B-85	20	2			97.50	39	136.50	188
0200	9" caliper, by hand	2 Clab	7.50	2.133			94.50		94.50	141
0210	Aerial lift equipment	B-85	12.50	3.200			156	62.50	218.50	300
0300	12" caliper, by hand	2 Clab	6.50	2.462			109		109	163
0310	Aerial lift equipment	B-85	10.80	3.704			180	72.50	252.50	350
0400	18" caliper by hand	2 Clab	5.60	2.857			127		127	189
0410	Aerial lift equipment	B-85	9.30	4.301			209	84	293	410
0500	24" caliper, by hand	2 Clab	4.60	3.478			154		154	230
0510	Aerial lift equipment	B-85	7.70	5.195			253	101	354	490
0600	30" caliper, by hand	2 Clab	3.70	4.324			192		192	286
0610	Aerial lift equipment	B-85	6.20	6.452			315	126	441	610
0700	36" caliper, by hand	2 Clab	2.70	5.926			263		263	395
0710	Aerial lift equipment	B-85	4.50	8.889	↓		435	173	608	835

32 01 Operation and Maintenance of Exterior Improvements

32 01 90 – Operation and Maintenance of Planting

32 01 90.23 Pruning

		Crew	Daily Output	Labor-Hours	Unit	Material	2021 Bare Costs Labor	Equipment	Total	Total Incl O&P
0800	48" caliper, by hand	2 Clab	1.70	9.412	Ea.		420		420	625
0810	Aerial lift equipment	B-85	2.80	14.286	↓		695	279	974	1,350

32 01 90.24 Shrub Pruning

		Crew	Daily Output	Labor-Hours	Unit	Material	2021 Bare Costs Labor	Equipment	Total	Total Incl O&P
0010	**SHRUB PRUNING**									
6700	Prune, shrub bed	1 Clab	7	1.143	M.S.F.		50.50		50.50	75.50
6710	Shrub under 3' height		190	.042	Ea.		1.87		1.87	2.79
6720	4' height		90	.089			3.95		3.95	5.90
6730	Over 6'		50	.160			7.10		7.10	10.60
7350	Prune trees from ground		20	.400			17.75		17.75	26.50
7360	High work	↓	8	1	↓		44.50		44.50	66.50

32 01 90.26 Watering

		Crew	Daily Output	Labor-Hours	Unit	Material	2021 Bare Costs Labor	Equipment	Total	Total Incl O&P
0010	**WATERING**									
4900	Water lawn or planting bed with hose, 1" of water	1 Clab	16	.500	M.S.F.		22		22	33
4910	50' soaker hoses, in place		82	.098			4.33		4.33	6.45
4920	60' soaker hoses, in place		89	.090	↓		3.99		3.99	5.95
7500	Water trees or shrubs, under 1" caliper		32	.250	Ea.		11.10		11.10	16.55
7550	1" - 3" caliper		17	.471			21		21	31
7600	3" - 4" caliper		12	.667			29.50		29.50	44
7650	Over 4" caliper	↓	10	.800	↓		35.50		35.50	53
9000	For sprinkler irrigation systems, see Section 32 84 23.10									

32 01 90.29 Topsoil Preservation

		Crew	Daily Output	Labor-Hours	Unit	Material	2021 Bare Costs Labor	Equipment	Total	Total Incl O&P
0010	**TOPSOIL PRESERVATION**									
0100	Weed planting bed	1 Clab	800	.010	S.Y.		.44		.44	.66

32 06 Schedules for Exterior Improvements

32 06 10 – Schedules for Bases, Ballasts, and Paving

32 06 10.10 Sidewalks, Driveways and Patios

		Crew	Daily Output	Labor-Hours	Unit	Material	2021 Bare Costs Labor	Equipment	Total	Total Incl O&P
0010	**SIDEWALKS, DRIVEWAYS AND PATIOS** No base									
0020	Asphaltic concrete, 2" thick	B-37	720	.067	S.Y.	7.25	3.11	.36	10.72	13
0100	2-1/2" thick	"	660	.073	"	9.15	3.39	.39	12.93	15.60
0110	Bedding for brick or stone, mortar, 1" thick	D-1	300	.053	S.F.	.91	2.60		3.51	4.92
0120	2" thick	"	200	.080		2.29	3.90		6.19	8.35
0130	Sand, 2" thick	B-18	8000	.003		.33	.14	.02	.49	.58
0140	4" thick	"	4000	.006	↓	.66	.27	.04	.97	1.18
0300	Concrete, 3,000 psi, CIP, 6 x 6 - W1.4 x W1.4 mesh,									
0310	broomed finish, no base, 4" thick	B-24	600	.040	S.F.	2.22	2.01		4.23	5.45
0350	5" thick		545	.044		2.77	2.22		4.99	6.35
0400	6" thick	↓	510	.047		3.23	2.37		5.60	7.05
0440	For other finishes, see Section 03 35 29.30									
0450	For bank run gravel base, 4" thick, add	B-18	2500	.010	S.F.	.68	.43	.07	1.18	1.47
0520	8" thick, add	"	1600	.015		1.38	.68	.10	2.16	2.64
0550	Exposed aggregate finish, add to above, minimum	B-24	1875	.013		.13	.64		.77	1.09
0600	Maximum		455	.053		.44	2.65		3.09	4.41
0700	Patterned surface, add to above, minimum		1200	.020			1.01		1.01	1.49
0710	Maximum	↓	500	.048	↓		2.41		2.41	3.58
0850	Splash block, precast concrete	1 Clab	150	.053	Ea.	11.35	2.37		13.72	16.05
0950	Concrete tree grate, 5' square	B-6	25	.960		425	46	8.65	479.65	550
0955	Tree well & cover, concrete, 3' square		25	.960		246	46	8.65	300.65	350
0960	Cast iron tree grate with frame, 2 piece, round, 5' diameter		25	.960		1,125	46	8.65	1,179.65	1,325
0980	Square, 5' side	↓	25	.960	↓	1,125	46	8.65	1,179.65	1,325

For customer support on your Site Work & Landscape Costs with RSMeans Data, call 800.448.8182.

373

32 06 10.10 Sidewalks, Driveways and Patios

		Crew	Daily Output	Labor-Hours	Unit	Material	2021 Bare Costs Labor	Equipment	Total	Total Incl O&P
1000	Crushed stone, 1" thick, white marble	2 Clab	1700	.009	S.F.	.50	.42		.92	1.17
1050	Bluestone		1700	.009		.19	.42		.61	.83
1070	Granite chips	↓	1700	.009		.20	.42		.62	.84
1200	For 2" asphaltic conc. base and tack coat, add to above	B-37	7200	.007		1.11	.31	.04	1.46	1.72
1660	Limestone pavers, 3" thick	D-1	72	.222		9.95	10.80		20.75	27
1670	4" thick		70	.229		13.15	11.15		24.30	31
1680	5" thick	↓	68	.235		16.45	11.45		27.90	35.50
1700	Redwood, prefabricated, 4' x 4' sections	2 Carp	316	.051		7.35	2.77		10.12	12.25
1750	Redwood planks, 1" thick, on sleepers	"	240	.067		7.35	3.65		11	13.55
1830	1-1/2" thick	B-28	167	.144		6.75	7.35		14.10	18.40
1840	2" thick		167	.144		8.70	7.35		16.05	20.50
1850	3" thick		150	.160		12.80	8.20		21	26.50
1860	4" thick		150	.160		16.80	8.20		25	31
1870	5" thick	↓	150	.160	↓	21.50	8.20		29.70	36
2100	River or beach stone, stock	B-1	18	1.333	Ton	31.50	60		91.50	124
2150	Quarried	"	18	1.333	"	55	60		115	150
2160	Load, dump, and spread stone with skid steer, 100' haul	B-62	24	1	C.Y.		48	7.50	55.50	80
2165	200' haul		18	1.333			64	9.95	73.95	106
2168	300' haul	↓	12	2	↓		96	14.95	110.95	159
2170	Shale paver, 2-1/4" thick	D-1	200	.080	S.F.	3.78	3.90		7.68	10
2200	Coarse washed sand bed, 1" thick	B-62	1350	.018	S.Y.	.95	.86	.13	1.94	2.48
2250	Stone dust, 4" thick	"	900	.027	"	5.95	1.28	.20	7.43	8.70
2300	Tile thinset pavers, 3/8" thick	D-1	300	.053	S.F.	3.80	2.60		6.40	8.10
2350	3/4" thick	"	280	.057	"	6.05	2.78		8.83	10.85
2400	Wood rounds, cypress	B-1	175	.137	Ea.	10.20	6.20		16.40	20.50

32 06 10.20 Steps

		Crew	Daily Output	Labor-Hours	Unit	Material	2021 Bare Costs Labor	Equipment	Total	Total Incl O&P
0010	**STEPS**									
0011	Incl. excav., borrow & concrete base as required									
0100	Brick steps	B-24	35	.686	LF Riser	18.55	34.50		53.05	71.50
0200	Railroad ties	2 Clab	25	.640		3.86	28.50		32.36	47
0300	Bluestone treads, 12" x 2" or 12" x 1-1/2"	B-24	30	.800	↓	45	40		85	109
0500	Concrete, cast in place, see Section 03 30 53.40									
0600	Precast concrete, see Section 03 41 23.50									
4000	Edging, redwood, 2" x 4"	2 Carp	330	.048	L.F.	3.11	2.65		5.76	7.40
4025	Steel edge strips, incl. stakes, 1/4" x 5"	B-1	390	.062		4.68	2.77		7.45	9.30
4050	Edging, landscape timber or railroad ties, 6" x 8"	2 Carp	170	.094	↓	2.74	5.15		7.89	10.70

32 11 Base Courses
32 11 23 - Aggregate Base Courses

32 11 23.23 Base Course Drainage Layers

		Crew	Daily Output	Labor-Hours	Unit	Material	2021 Bare Costs Labor	Equipment	Total	Total Incl O&P
0010	**BASE COURSE DRAINAGE LAYERS**									
0011	For Soil Stabilization, see Section 31 32									
0012	For roadways and large areas									
0050	Crushed 3/4" stone base, compacted, 3" deep	B-36C	5200	.008	S.Y.	2.50	.42	.83	3.75	4.29
0100	6" deep		5000	.008		5	.44	.86	6.30	7.10
0200	9" deep		4600	.009		7.50	.48	.93	8.91	10
0300	12" deep	↓	4200	.010		10	.52	1.02	11.54	12.95
0301	Crushed 1-1/2" stone base, compacted to 4" deep	B-36B	6000	.011		5.90	.56	.83	7.29	8.25
0302	6" deep		5400	.012		8.85	.63	.92	10.40	11.70
0303	8" deep		4500	.014		11.85	.75	1.10	13.70	15.35
0304	12" deep	↓	3800	.017	↓	17.75	.89	1.31	19.95	22.50

32 11 Base Courses

32 11 23 – Aggregate Base Courses

32 11 23.23 Base Course Drainage Layers

		Crew	Daily Output	Labor-Hours	Unit	Material	2021 Bare Costs Labor	Equipment	Total	Total Incl O&P
0350	Bank run gravel, spread and compacted									
0370	6" deep	B-32	6000	.005	S.Y.	5.70	.30	.47	6.47	7.25
0390	9" deep	↓	4900	.007		8.55	.36	.58	9.49	10.65
0400	12" deep	↓	4200	.008	↓	11.45	.42	.68	12.55	13.95
0600	Cold laid asphalt pavement, see Section 32 12 16.19									
1500	Alternate method to figure base course									
1510	Crushed stone, 3/4", compacted, 3" deep	B-36C	435	.092	E.C.Y.	25.50	5.05	9.85	40.40	47
1511	6" deep	B-36B	835	.077		25.50	4.05	5.95	35.50	41
1512	9" deep		1150	.056		25.50	2.94	4.31	32.75	37.50
1513	12" deep		1400	.046		25.50	2.41	3.54	31.45	36
1520	Crushed stone, 1-1/2", compacted, 4" deep		665	.096		25.50	5.10	7.45	38.05	44.50
1521	6" deep		900	.071		25.50	3.76	5.50	34.76	40
1522	8" deep		1000	.064		25.50	3.38	4.96	33.84	39
1523	12" deep	↓	1265	.051		25.50	2.67	3.92	32.09	37
1530	Gravel, bank run, compacted, 6" deep	B-36C	835	.048		29.50	2.63	5.15	37.28	42
1531	9" deep		1150	.035		29.50	1.91	3.73	35.14	39.50
1532	12" deep	↓	1400	.029	↓	29.50	1.57	3.07	34.14	38
2010	Crushed stone, 3/4" maximum size, 3" deep	B-36	540	.074	Ton	16.70	3.75	3.38	23.83	27.50
2011	6" deep		1625	.025		16.70	1.25	1.12	19.07	21.50
2012	9" deep		1785	.022		16.70	1.13	1.02	18.85	21
2013	12" deep		1950	.021		16.70	1.04	.94	18.68	21
2020	Crushed stone, 1-1/2" maximum size, 4" deep		720	.056		16.70	2.81	2.54	22.05	25.50
2021	6" deep		815	.049		16.70	2.49	2.24	21.43	24.50
2022	8" deep		835	.048		16.70	2.43	2.19	21.32	24.50
2023	12" deep	↓	975	.041		16.70	2.08	1.87	20.65	23.50
2030	Bank run gravel, 6" deep	B-32A	875	.027		19.75	1.48	1.99	23.22	26
2031	9" deep		970	.025		19.75	1.34	1.80	22.89	25.50
2032	12" deep	↓	1060	.023	↓	19.75	1.23	1.65	22.63	25
6000	Stabilization fabric, polypropylene, 6 oz./S.Y.	B-6	10000	.002	S.Y.	1.17	.12	.02	1.31	1.48
6900	For small and irregular areas, add						50%	50%		
7000	Prepare and roll sub-base, small areas to 2,500 S.Y.	B-32A	1500	.016	S.Y.		.87	1.16	2.03	2.57
8000	Large areas over 2,500 S.Y.	"	3500	.007			.37	.50	.87	1.10
8050	For roadways	B-32	4000	.008	↓		.44	.71	1.15	1.44

32 11 26 – Asphaltic Base Courses

32 11 26.13 Plant Mix Asphaltic Base Courses

		Crew	Daily Output	Labor-Hours	Unit	Material	2021 Bare Costs Labor	Equipment	Total	Total Incl O&P
0010	**PLANT MIX ASPHALTIC BASE COURSES**									
0011	For roadways and large paved areas									
0500	Bituminous concrete, 4" thick	B-25	4545	.019	S.Y.	15.35	.94	.60	16.89	18.95
0550	6" thick		3700	.024		22.50	1.15	.74	24.39	27.50
0560	8" thick		3000	.029		30	1.42	.91	32.33	36
0570	10" thick	↓	2545	.035	↓	37	1.68	1.08	39.76	44.50
1600	Macadam base, crushed stone or slag, dry-bound	B-36D	1400	.023	E.C.Y.	46.50	1.28	2.52	50.30	55.50
1610	Water-bound	B-36C	1400	.029	"	46.50	1.57	3.07	51.14	56.50
2000	Alternate method to figure base course									
2005	Bituminous concrete, 4" thick	B-25	1000	.088	Ton	68.50	4.27	2.74	75.51	84.50
2006	6" thick		1220	.072		68.50	3.50	2.25	74.25	82.50
2007	8" thick		1320	.067		68.50	3.24	2.08	73.82	82
2008	10" thick	↓	1400	.063	↓	68.50	3.05	1.96	73.51	81.50
8900	For small and irregular areas, add						50%	50%		

For customer support on your Site Work & Landscape Costs with RSMeans Data, call 800.448.8182.

375

32 11 Base Courses

32 11 26 – Asphaltic Base Courses

32 11 26.19 Bituminous-Stabilized Base Courses

	32 11 26.19 Bituminous-Stabilized Base Courses	Crew	Daily Output	Labor-Hours	Unit	Material	2021 Bare Costs Labor	Equipment	Total	Total Incl O&P
0010	**BITUMINOUS-STABILIZED BASE COURSES**									
0020	For roadways and large paved areas									
0700	Liquid application to gravel base, asphalt emulsion	B-45	6000	.003	Gal.	5.10	.15	.14	5.39	5.95
0800	Prime and seal, cut back asphalt		6000	.003	"	6.05	.15	.14	6.34	7
1000	Macadam penetration crushed stone, 2 gal./S.Y., 4" thick		6000	.003	S.Y.	10.20	.15	.14	10.49	11.60
1100	6" thick, 3 gal./S.Y.		4000	.004		15.35	.22	.21	15.78	17.40
1200	8" thick, 4 gal./S.Y.		3000	.005		20.50	.29	.28	21.07	23.50
8900	For small and irregular areas, add						50%	50%		

32 12 Flexible Paving

32 12 16 – Asphalt Paving

32 12 16.13 Plant-Mix Asphalt Paving

	32 12 16.13 Plant-Mix Asphalt Paving	Crew	Daily Output	Labor-Hours	Unit	Material	2021 Bare Costs Labor	Equipment	Total	Total Incl O&P
0010	**PLANT-MIX ASPHALT PAVING**									
0020	For highways and large paved areas, excludes hauling									
0025	See Section 31 23 23.20 for hauling costs									
0080	Binder course, 1-1/2" thick	B-25	7725	.011	S.Y.	5.60	.55	.35	6.50	7.35
0120	2" thick		6345	.014		7.45	.67	.43	8.55	9.65
0130	2-1/2" thick		5620	.016		9.30	.76	.49	10.55	11.85
0160	3" thick		4905	.018		11.15	.87	.56	12.58	14.15
0170	3-1/2" thick		4520	.019		13	.95	.61	14.56	16.40
0200	4" thick		4140	.021		14.85	1.03	.66	16.54	18.60
0300	Wearing course, 1" thick	B-25B	10575	.009		3.68	.45	.28	4.41	5.05
0340	1-1/2" thick		7725	.012		6.20	.61	.39	7.20	8.15
0380	2" thick		6345	.015		8.30	.75	.47	9.52	10.80
0420	2-1/2" thick		5480	.018		10.25	.87	.54	11.66	13.20
0460	3" thick		4900	.020		12.25	.97	.61	13.83	15.55
0470	3-1/2" thick		4520	.021		14.35	1.05	.66	16.06	18.10
0480	4" thick		4140	.023		16.40	1.15	.72	18.27	20.50
0600	Porous pavement, 1-1/2" open graded friction course	B-25	7725	.011		15.30	.55	.35	16.20	18.05
0800	Alternate method of figuring paving costs									
0810	Binder course, 1-1/2" thick	B-25	630	.140	Ton	68.50	6.80	4.35	79.65	90
0811	2" thick		690	.128		68.50	6.20	3.97	78.67	88.50
0812	3" thick		800	.110		68.50	5.35	3.42	77.27	86.50
0813	4" thick		900	.098		68.50	4.75	3.04	76.29	85.50
0850	Wearing course, 1" thick	B-25B	575	.167		72	8.25	5.20	85.45	97.50
0851	1-1/2" thick		630	.152		72	7.55	4.74	84.29	96
0852	2" thick		690	.139		72	6.90	4.33	83.23	94.50
0853	2-1/2" thick		765	.125		72	6.20	3.90	82.10	93
0854	3" thick		800	.120		72	5.95	3.73	81.68	92.50
1000	Pavement replacement over trench, 2" thick	B-21	90	.311	S.Y.	7.65	15.90	2.12	25.67	35
1050	4" thick		70	.400		15.15	20.50	2.73	38.38	50
1080	6" thick		55	.509		24	26	3.47	53.47	69.50
1100	Turnouts and driveway entrances to highways									
1110	Binder course, 1-1/2" thick	B-25	315	.279	Ton	68.50	13.55	8.70	90.75	105
1120	2" thick		345	.255		68.50	12.40	7.95	88.85	102
1130	3" thick		400	.220		68.50	10.70	6.85	86.05	98.50
1140	4" thick		450	.196		68.50	9.50	6.10	84.10	96
1150	Binder course, 1-1/2" thick		3860	.023	S.Y.	5.60	1.11	.71	7.42	8.60
1160	2" thick		3175	.028		7.45	1.35	.86	9.66	11.10
1170	2-1/2" thick		2810	.031		9.30	1.52	.97	11.79	13.55

32 12 16.13 Plant-Mix Asphalt Paving

		Crew	Daily Output	Labor-Hours	Unit	Material	2021 Bare Costs Labor	Equipment	Total	Total Incl O&P
1180	3" thick	B-25	2455	.036	S.Y.	11.15	1.74	12	14.01	16.10
1190	3-1/2" thick		2260	.039		13	1.89	21	16.10	18.45
1195	4" thick		2070	.043		14.85	2.06	32	18.23	21
1200	Wearing course, 1" thick	B-25B	290	.331	Ton	72	16.35	10 30	98.65	115
1210	1-1/2" thick		315	.305		72	15.05	50	96.55	112
1220	2" thick		345	.278		72	13.75	65	94.40	110
1230	2-1/2" thick		385	.249		72	12.35	75	92.10	106
1240	3" thick		400	.240		72	11.85	45	91.30	105
1250	Wearing course, 1" thick		5290	.018	S.Y.	3.68	.90	56	5.14	6
1260	1-1/2" thick		3860	.025		6.20	1.23	77	8.20	9.50
1270	2" thick		3175	.030		8.30	1.49	94	10.73	12.40
1280	2-1/2" thick		2740	.035		10.25	1.73	09	13.07	15.10
1285	3" thick		2450	.039		12.25	1.94	22	15.41	17.70
1290	3-1/2" thick		2260	.042		14.35	2.10	32	17.77	20.50
1295	4" thick		2070	.046		16.40	2.29	44	20.13	23
1400	Bridge approach less than 300 tons									
1410	Binder course, 1-1/2" thick	B-25	330	.267	Ton	68.50	12.95	8 30	89.75	103
1420	2" thick		360	.244		68.50	11.85	7 60	87.95	101
1430	3" thick		420	.210		68.50	10.15	50	85.15	97.50
1440	4" thick		470	.187		68.50	9.10	85	83.45	95
1450	Binder course, 1-1/2" thick		4055	.022	S.Y.	5.60	1.05	68	7.33	8.45
1460	2" thick		3330	.026		7.45	1.28	82	9.55	10.95
1470	2-1/2" thick		2950	.030		9.30	1.45	93	11.68	13.40
1480	3" thick		2575	.034		11.15	1.66	06	13.87	15.90
1490	3-1/2" thick		2370	.037		13	1.80	16	15.96	18.25
1495	4" thick		2175	.040		14.85	1.96	26	18.07	20.50
1500	Wearing course, 1" thick	B-25B	300	.320	Ton	72	15.80	95	97.75	114
1510	1-1/2" thick		330	.291		72	14.40	05	95.45	111
1520	2" thick		360	.267		72	13.20	30	93.50	108
1530	2-1/2" thick		400	.240		72	11.85	45	91.30	105
1540	3" thick		420	.229		72	11.30	10	90.40	104
1550	Wearing course, 1" thick		5555	.017	S.Y.	3.68	.85	54	5.07	5.90
1560	1-1/2" thick		4055	.024		6.20	1.17	74	8.11	9.35
1570	2" thick		3330	.029		8.30	1.43	90	10.63	12.25
1580	2-1/2" thick		2875	.033		10.25	1.65	04	12.94	14.90
1585	3" thick		2570	.037		12.25	1.85	16	15.26	17.50
1590	3-1/2" thick		2375	.040		14.35	2	26	17.61	20
1595	4" thick		2175	.044		16.40	2.18	37	19.95	23
1600	Bridge approach 300-800 tons									
1610	Binder course, 1-1/2" thick	B-25	390	.226	Ton	68.50	10.95		86.45	99
1620	2" thick		430	.205		68.50	9.95	35	84.80	97
1630	3" thick		500	.176		68.50	8.55	50	82.55	94
1640	4" thick		560	.157		68.50	7.65	89	81.04	92
1650	Binder course, 1-1/2" thick		4815	.018	S.Y.	5.60	.89	57	7.06	8.10
1660	2" thick		3955	.022		7.45	1.08	69	9.22	10.50
1670	2-1/2" thick		3500	.025		9.30	1.22	78	11.30	12.90
1680	3" thick		3060	.029		11.15	1.40	90	13.45	15.30
1690	3-1/2" thick		2820	.031		13	1.52	97	15.49	17.65
1695	4" thick		2580	.034		14.85	1.66	06	17.57	20
1700	Wearing course, 1" thick	B-25B	360	.267	Ton	72	13.20	30	93.50	108
1710	1-1/2" thick		390	.246		72	12.15	65	91.80	106
1720	2" thick		430	.223		72	11.05	95	90	104
1730	2-1/2" thick		475	.202		72	10	30	88.30	101

32 12 16.13 Plant-Mix Asphalt Paving		Crew	Daily Output	Labor-Hours	Unit	Material	2021 Bare Costs Labor	Equipment	Total	Total Incl O&P
1740	3" thick	B-25B	500	.192	Ton	72	9.50	5.95	87.45	100
1750	Wearing course, 1" thick		6590	.015	S.Y.	3.68	.72	.45	4.85	5.60
1760	1-1/2" thick		4815	.020		6.20	.99	.62	7.81	8.95
1770	2" thick		3955	.024		8.30	1.20	.76	10.26	11.75
1780	2-1/2" thick		3415	.028		10.25	1.39	.87	12.51	14.35
1785	3" thick		3055	.031		12.25	1.55	.98	14.78	16.85
1790	3-1/2" thick		2820	.034		14.35	1.68	1.06	17.09	19.45
1795	4" thick		2580	.037		16.40	1.84	1.16	19.40	22
1800	Bridge approach over 800 tons									
1810	Binder course, 1-1/2" thick	B-25	525	.168	Ton	68.50	8.15	5.20	81.85	93
1820	2" thick		575	.153		68.50	7.45	4.76	80.71	91.50
1830	3" thick		670	.131		68.50	6.40	4.09	78.99	89
1840	4" thick		750	.117		68.50	5.70	3.65	77.85	87.50
1850	Binder course, 1-1/2" thick		6450	.014	S.Y.	5.60	.66	.42	6.68	7.60
1860	2" thick		5300	.017		7.45	.81	.52	8.78	9.90
1870	2-1/2" thick		4690	.019		9.30	.91	.58	10.79	12.20
1880	3" thick		4095	.021		11.15	1.04	.67	12.86	14.55
1890	3-1/2" thick		3775	.023		13	1.13	.73	14.86	16.80
1895	4" thick		3455	.025		14.85	1.24	.79	16.88	19.05
1900	Wearing course, 1" thick	B-25B	480	.200	Ton	72	9.90	6.20	88.10	101
1910	1-1/2" thick		525	.183		72	9.05	5.70	86.75	99.50
1920	2" thick		575	.167		72	8.25	5.20	85.45	97.50
1930	2-1/2" thick		640	.150		72	7.40	4.67	84.07	95.50
1940	3" thick		670	.143		72	7.10	4.46	83.56	95
1950	Wearing course, 1" thick		8830	.011	S.Y.	3.68	.54	.34	4.56	5.20
1960	1-1/2" thick		6450	.015		6.20	.74	.46	7.40	8.40
1970	2" thick		5300	.018		8.30	.90	.56	9.76	11.10
1980	2-1/2" thick		4575	.021		10.25	1.04	.65	11.94	13.55
1985	3" thick		4090	.023		12.25	1.16	.73	14.14	16
1990	3-1/2" thick		3775	.025		14.35	1.26	.79	16.40	18.55
1995	4" thick		3455	.028		16.40	1.37	.86	18.63	21
2000	Intersections									
2010	Binder course, 1-1/2" thick	B-25	440	.200	Ton	68.50	9.70	6.25	84.45	96.50
2020	2" thick		485	.181		68.50	8.80	5.65	82.95	94.50
2030	3" thick		560	.157		68.50	7.65	4.89	81.04	92
2040	4" thick		630	.140		68.50	6.80	4.35	79.65	90
2050	Binder course, 1-1/2" thick		5410	.016	S.Y.	5.60	.79	.51	6.90	7.90
2060	2" thick		4440	.020		7.45	.96	.62	9.03	10.25
2070	2-1/2" thick		3935	.022		9.30	1.09	.70	11.09	12.60
2080	3" thick		3435	.026		11.15	1.24	.80	13.19	15
2090	3-1/2" thick		3165	.028		13	1.35	.87	15.22	17.25
2095	4" thick		2900	.030		14.85	1.47	.94	17.26	19.60
2100	Wearing course, 1" thick	B-25B	400	.240	Ton	72	11.85	7.45	91.30	105
2110	1-1/2" thick		440	.218		72	10.80	6.80	89.60	103
2120	2" thick		485	.198		72	9.80	6.15	87.95	101
2130	2-1/2" thick		535	.179		72	8.85	5.60	86.45	99
2140	3" thick		560	.171		72	8.45	5.35	85.80	98
2150	Wearing course, 1" thick		7400	.013	S.Y.	3.68	.64	.40	4.72	5.45
2160	1-1/2" thick		5410	.018		6.20	.88	.55	7.63	8.70
2170	2" thick		4440	.022		8.30	1.07	.67	10.04	11.50
2180	2-1/2" thick		3835	.025		10.25	1.24	.78	12.27	14
2185	3" thick		3430	.028		12.25	1.38	.87	14.50	16.45
2190	3-1/2" thick		3165	.030		14.35	1.50	.94	16.79	19.10

32 12 Flexible Paving

32 12 16 – Asphalt Paving

32 12 16.13 Plant-Mix Asphalt Paving

		Crew	Daily Output	Labor-Hours	Unit	Material	2021 Bare Costs Labor	Equipment	Total	Total Incl O&P
2195	4" thick	B-25B	2900	.033	S.Y.	16.40	1.64	.03	19.07	21.50
3000	Prime coat, emulsion, 0.30 gal./S.Y., 1000 S.Y.	B-45	2500	.006		2.10	.35	.33	2.78	3.21
3100	Tack coat, emulsion, 0.10 gal./S.Y., 1000 S.Y.	"	2500	.006		.70	.35	.33	1.38	1.67

32 12 16.14 Asphaltic Concrete Paving

		Crew	Daily Output	Labor-Hours	Unit	Material	2021 Bare Costs Labor	Equipment	Total	Total Incl O&P
0011	**ASPHALTIC CONCRETE PAVING**, parking lots & driveways									
0015	No asphalt hauling included									
0018	Use 6.05 C.Y. per inch per M.S.F. for hauling									
0020	6" stone base, 2" binder course, 1" topping	B-25C	9000	.005	S.F.	2.06	.26	.27	2.59	2.94
0025	2" binder course, 2" topping		9000	.005		2.49	.26	.27	3.02	3.42
0030	3" binder course, 2" topping		9000	.005		2.91	.26	.27	3.44	3.88
0035	4" binder course, 2" topping		9000	.005		3.32	.26	.27	3.85	4.33
0040	1-1/2" binder course, 1" topping		9000	.005		1.85	.26	.27	2.38	2.72
0042	3" binder course, 1" topping		9000	.005		2.47	.26	.27	3	3.40
0045	3" binder course, 3" topping		9000	.005		3.35	.26	.27	3.88	4.37
0050	4" binder course, 3" topping		9000	.005		3.76	.26	.27	4.29	4.82
0055	4" binder course, 4" topping		9000	.005		4.19	.26	.27	4.72	5.30
0300	Binder course, 1-1/2" thick		35000	.001		.62	.07	.07	.76	.86
0400	2" thick		25000	.002		.80	.10	.10	1	1.13
0500	3" thick		15000	.003		1.24	.16	.16	1.56	1.79
0600	4" thick		10800	.004		1.63	.22	.22	2.07	2.36
0800	Sand finish course, 3/4" thick		41000	.001		.31	.06	.06	.43	.49
0900	1" thick		34000	.001		.38	.07	.07	.52	.60
1000	Fill pot holes, hot mix, 2" thick	B-16	4200	.008		.82	.36	.14	1.32	1.58
1100	4" thick		3500	.009		1.20	.43	.17	1.80	2.14
1120	6" thick		3100	.010		1.61	.48	.19	2.28	2.70
1140	Cold patch, 2" thick	B-51	3000	.016		1.24	.73	.07	2.04	2.53
1160	4" thick		2700	.018		2.37	.81	.07	3.25	3.90
1180	6" thick		1900	.025		3.69	1.15	.10	4.94	5.85
2000	From 60% recycled content, base course 3" thick [G]	B-25	800	.110	Ton	40	5.35	3.42	48.77	55.50
3000	Prime coat, emulsion, 0.30 gal./S.Y., 1000 S.Y.	B-45	2500	.006	S.Y.	2.10	.35	.33	2.78	3.21
3100	Tack coat, emulsion, 0.10 gal./S.Y., 1000 S.Y.	"	2500	.006	"	.70	.35	.33	1.38	1.67

32 12 16.19 Cold-Mix Asphalt Paving

		Crew	Daily Output	Labor-Hours	Unit	Material	2021 Bare Costs Labor	Equipment	Total	Total Incl O&P
0010	**COLD-MIX ASPHALT PAVING** 0.5 gal. asphalt/S.Y. per in depth									
0020	Well graded granular aggregate									
0100	Blade mixed in windrows, spread & compacted 4" course	B-90A	1600	.035	S.Y.	23	1.86	1.71	26.57	30
0200	Traveling plant mixed in windrows, compacted 4" course	B-90B	3000	.016		23	.83	.76	24.59	27.50
0300	Rotary plant mixed in place, compacted 4" course	"	3500	.014		23	.71	.65	24.36	27.50
0400	Central stationary plant, mixed, compacted 4" course	B-36	7200	.006		46	.28	.25	46.53	51

32 12 36 – Seal Coats

32 12 36.13 Chip Seal

		Crew	Daily Output	Labor-Hours	Unit	Material	2021 Bare Costs Labor	Equipment	Total	Total Incl O&P
0010	**CHIP SEAL**									
0011	Excludes crack repair and flush coat									
1000	Fine - PMCRS-2h (20lbs/sy, 1/4" (No.10), .30gal/sy app. rate)									
1010	Small, irregular areas	B-91	5000	.013	S.Y.	1.10	.68	.46	2.24	2.72
1020	Parking Lot	B-91D	15000	.007		1.10	.36	.23	1.69	1.99
1030	Roadway	"	30000	.003		1.10	.18	.11	1.39	1.59
1090	For Each .5% Latex Additive, Add					.23			.23	.25
1100	Medium Fine - PMCRS-2h (25lbs/sy, 5/16" (No.8), .35gal/sy app. rate)									
1110	Small, irregular areas	B-91	4000	.016	S.Y.	1.29	.84	.57	2.70	3.30
1120	Parking Lot	B-91D	12000	.009		1.29	.44	.28	2.01	2.38
1130	Roadway	"	24000	.004		1.29	.22	.14	1.65	1.90
1190	For Each .5% Latex Additive, Add					.28			.28	.31

32 12 36.13 Chip Seal

		Crew	Daily Output	Labor-Hours	Unit	Material	2021 Bare Costs Labor	Equipment	Total	Total Incl O&P
1200	Medium - PMCRS-2h (30lbs/sy, 3/8" (No.6), .40gal/sy app. rate)									
1210	Small, irregular areas	B-91	3330	.019	S.Y.	1.41	1.01	.69	3.11	3.82
1220	Parking Lot	B-91D	10000	.010		1.41	.53	.34	2.28	2.71
1230	Roadway	"	20000	.005		1.41	.27	.17	1.85	2.14
1290	For Each .5% Latex Additive, Add				↓	.34			.34	.37
1300	Course - PMCRS-2h (30lbs/sy, 1/2" (No.4), .40gal/sy app. rate)									
1310	Small, irregular areas	B-91	2500	.026	S.Y.	1.39	1.35	.92	3.66	4.55
1320	Parking Lot	B-91D	7500	.014		1.39	.71	.45	2.55	3.08
1330	Roadway	"	15000	.007		1.39	.36	.23	1.98	2.30
1390	For Each .5% Latex Additive, Add				↓	.34			.34	.37
1400	Double - PMCRS-2h (Course Base with Fine Top)									
1410	Small, irregular areas	B-91	2000	.032	S.Y.	2.36	1.69	1.14	5.19	6.40
1420	Parking Lot	B-91D	6000	.017		2.36	.89	.57	3.82	4.54
1430	Roadway	"	12000	.009		2.36	.44	.28	3.08	3.57
1490	For Each .5% Latex Additive, Add				↓	.56			.56	.62

32 12 36.14 Flush Coat

		Crew	Daily Output	Labor-Hours	Unit	Material	2021 Bare Costs Labor	Equipment	Total	Total Incl O&P
0010	**FLUSH COAT**									
0011	Fog Seal with Sand Cover (18gal/sy, 6lbs/sy)									
1010	Small, irregular areas	B-91	2000	.032	S.Y.	.52	1.69	1.14	3.35	4.36
1020	Parking lot		6000	.011		.52	.56	.38	1.46	1.84
1030	Roadway	↓	12000	.005	↓	.52	.28	.19	.99	1.21

32 12 36.33 Slurry Seal

		Crew	Daily Output	Labor-Hours	Unit	Material	2021 Bare Costs Labor	Equipment	Total	Total Incl O&P
0010	**SLURRY SEAL**									
0011	Includes sweeping and cleaning of area									
1000	Type I-PMCQS-1h-EAS (12lbs/sy, 1/8", 20% asphalt emulsion)									
1010	Small, irregular areas	B-90	8000	.008	S.Y.	.98	.39	.28	1.65	1.97
1020	Parking Lot		25000	.003		.98	.13	.09	1.20	1.36
1030	Roadway	↓	50000	.001	↓	.98	.06	.04	1.08	1.21
1090	For Each .5% Latex Additive, Add					.15			.15	.16
1100	Type II-PMCQS-1h-EAS (15lbs/sy, 1/4", 18% asphalt emulsion)									
1110	Small, irregular areas	B-90	6000	.011	S.Y.	1.10	.52	.37	1.99	2.40
1120	Parking lot		20000	.003		1.10	.16	.11	1.37	1.56
1130	Roadway		40000	.002		1.10	.08	.06	1.24	1.39
1190	For Each .5% Latex Additive, Add				↓	.18			.18	.20
1200	Type III-PMCQS-1h-EAS (25lbs/sy, 3/8", 15% asphalt emulsion)									
1210	Small, irregular areas	B-90	4000	.016	S.Y.	1.51	.79	.56	2.86	3.44
1220	Parking lot		12000	.005		1.51	.26	.19	1.96	2.25
1230	Roadway	↓	24000	.003	↓	1.51	.13	.09	1.73	1.96
1290	For Each .5% Latex Additive, Add					.29			.29	.32

32 12 36.36 Microsurfacing

		Crew	Daily Output	Labor-Hours	Unit	Material	2021 Bare Costs Labor	Equipment	Total	Total Incl O&P
0010	**MICROSURFACING**									
1100	Type II-MSE (20lbs/sy, 1/4", 18% microsurfacing emulsion)									
1110	Small, irregular areas	B-90	5000	.013	S.Y.	1.65	.63	.44	2.72	3.24
1120	Parking lot		15000	.004		1.65	.21	.15	2.01	2.28
1130	Roadway	↓	30000	.002	↓	1.65	.10	.07	1.82	2.05
1200	Type IIIa-MSE (32lbs/sy, 3/8", 15% microsurfacing emulsion)									
1210	Small, irregular areas	B-90	3000	.021	S.Y.	2.29	1.05	.74	4.08	4.89
1220	Parking lot		9000	.007		2.29	.35	.25	2.89	3.31
1230	Roadway	↓	18000	.004	↓	2.29	.17	.12	2.58	2.92

32 13 Rigid Paving

32 13 13 – Concrete Paving

32 13 13.25 Concrete Pavement, Highways	Crew	Daily Output	Labor-Hours	Unit	Material	2021 Bare Costs Labor	Equipment	Total	Total Incl O&P
0010 **CONCRETE PAVEMENT, HIGHWAYS**									
0015 Including joints, finishing and curing									
0020 Fixed form, 12' pass, unreinforced, 6" thick	B-26	3000	.029	S.Y.	26.50	1.44	1.19	29.13	33
0030 7" thick		2850	.031		32.50	1.52	1.25	35.27	39.50
0100 8" thick		2750	.032		36.50	1.58	1.30	39.38	44
0110 8" thick, small area		1375	.064		36.50	3.15	2.60	42.25	47.50
0200 9" thick		2500	.035		41.50	1.73	1.43	44.66	49.50
0300 10" thick		2100	.042		45.50	2.06	1.70	49.26	55
0310 10" thick, small area		1050	.084		45.50	4.13	3.41	53.04	60
0400 12" thick		1800	.049		52	2.41	1.99	56.40	63.50
0410 Conc. pavement, w/jt., fnsh.& curing, fix form, 24' pass, unreinforced, 6"T		6000	.015		25	.72	.60	26.32	29
0420 7" thick		5700	.015		30.50	.76	.63	31.89	35.50
0430 8" thick		5500	.016		34.50	.79	.65	35.94	40
0440 9" thick		5000	.018		39.50	.87	.72	41.09	45
0450 10" thick		4200	.021		43.50	1.03	.85	45.38	50
0460 12" thick		3600	.024		50	1.20	.99	52.19	58
0470 15" thick		3000	.029		66.50	1.44	1.19	69.13	76.50
0500 Fixed form 12' pass, 15" thick		1500	.059		66.50	2.89	2.38	71.77	80
0510 For small irregular areas, add				%	10%	100%	100%		
0520 Welded wire fabric, sheets for rigid paving 2.33 lb./S.Y.	2 Rodm	389	.041	S.Y.	1.34	2.42		3.76	5.10
0530 Reinforcing steel for rigid paving 12 lb./S.Y.		666	.024		7.85	1.42		9.27	10.70
0540 Reinforcing steel for rigid paving 18 lb./S.Y.		444	.036		11.75	2.12		13.87	16.05
0610 For under 10' pass, add				%	10%	100%	100%		
0620 Slip form, 12' pass, unreinforced, 6" thick	B-26A	5600	.016	S.Y.	25.50	.77	.66	26.93	30
0622 7" thick		5600	.016		31	.77	.66	32.43	36
0624 8" thick		5300	.017		35	.82	.70	36.52	40.50
0626 9" thick		4820	.018		40	.90	.77	41.67	46
0628 10" thick		4050	.022		44	1.07	.91	45.98	50.50
0630 12" thick		3470	.025		50.50	1.25	1.06	52.81	58.50
0632 15" thick		2890	.030		64	1.50	1.28	66.78	73.50
0640 Slip form, 24' pass, unreinforced, 6" thick		11200	.008		25	.39	.33	25.72	28.50
0642 7" thick		11200	.008		29.50	.39	.33	30.22	33.50
0644 8" thick		10600	.008		33.50	.41	.35	34.26	38
0646 9" thick		9640	.009		38.50	.45	.38	39.33	43.50
0648 10" thick		8100	.011		42.50	.53	.46	43.49	48
0650 12" thick		6940	.013		49	.62	.53	50.15	55.50
0652 15" thick		5780	.015		61.50	.75	.64	62.89	70
0700 Finishing, broom finish small areas	2 Cefi	120	.133			6.90		6.90	10.10
0730 Transverse expansion joints, incl. premolded bit. jt. filler	C-1	150	.213	L.F.	2.08	11.10		13.18	18.90
0740 Transverse construction joint using bulkhead	"	73	.438	"	3.02	23		26.02	37.50
0750 Longitudinal joint tie bars, grouted	B-23	70	.571	Ea.	4.68	25.50	23	53.18	68.50
1000 Curing, with sprayed membrane by hand	2 Clab	1500	.011	S.Y.	1.10	.47		1.57	1.92

32 14 Unit Paving

32 14 13 – Precast Concrete Unit Paving

32 14 13.13 Interlocking Precast Concrete Unit Paving

		Daily	Labor-			2021 Bare Costs			Total
32 14 13.13 Interlocking Precast Concrete Unit Paving	Crew	Output	Hours	Unit	Material	Labor	Equipment	Total	Incl O&P
0010 **INTERLOCKING PRECAST CONCRETE UNIT PAVING**									
0020 "V" blocks for retaining soil	D-1	205	.078	S.F.	11.20	3.80		15	18.10

32 14 13.16 Precast Concrete Unit Paving Slabs

		Daily	Labor-			2021 Bare Costs			Total
0010 **PRECAST CONCRETE UNIT PAVING SLABS**									
0710 Precast concrete patio blocks, 2-3/8" thick, colors, 8" x 16"	D-1	265	.060	S.F.	11.15	2.94		14.09	16.70
0715 12" x 12"		300	.053		14.50	2.60		17.10	19.85
0720 16" x 16"		335	.048		15.95	2.33		18.28	21
0730 24" x 24"		510	.031		19.85	1.53		21.38	24.50
0740 Green, 8" x 16"		265	.060		14.35	2.94		17.29	20
0750 Exposed local aggregate, natural	2 Bric	250	.064		8.80	3.44		12.24	14.90
0800 Colors		250	.064		9.90	3.44		13.34	16.10
0850 Exposed granite or limestone aggregate		250	.064		9.10	3.44		12.54	15.20
0900 Exposed white tumblestone aggregate		250	.064		11.15	3.44		14.59	17.45

32 14 13.18 Precast Concrete Plantable Pavers

		Daily	Labor-			2021 Bare Costs			Total
0010 **PRECAST CONCRETE PLANTABLE PAVERS** (50% grass)									
0015 Subgrade preparation and grass planting not included									
0100 Precast concrete plantable pavers with topsoil, 24" x 16"	B-63	800	.050	S.F.	4.52	2.33	.22	7.07	8.70
0200 Less than 600 S.F. or irregular area	"	500	.080	"	4.52	3.73	.36	8.61	10.90
0300 3/4" crushed stone base for plantable pavers, 6" depth	B-62	1000	.024	S.Y.	3.90	1.15	.18	5.23	6.20
0400 8" depth		900	.027		5.20	1.28	.20	6.68	7.85
0500 10" depth		800	.030		6.50	1.44	.22	8.16	9.55
0600 12" depth		700	.034		7.80	1.65	.26	9.71	11.30
0700 Hydro seeding plantable pavers	B-81A	20	.800	M.S.F.	9.75	37.50	48.50	95.75	120
0800 Apply fertilizer and seed to plantable pavers	1 Clab	8	1	"	48	44.50		92.50	120

32 14 16 – Brick Unit Paving

32 14 16.10 Brick Paving

		Daily	Labor-			2021 Bare Costs			Total
0010 **BRICK PAVING**									
0012 4" x 8" x 1-1/2", without joints (4.5 bricks/S.F.)	D-1	110	.145	S.F.	2.74	7.10		9.84	13.65
0100 Grouted, 3/8" joint (3.9 bricks/S.F.)		90	.178		2.26	8.65		10.91	15.55
0200 4" x 8" x 2-1/4", without joints (4.5 bricks/S.F.)		110	.145		2.61	7.10		9.71	13.50
0300 Grouted, 3/8" joint (3.9 bricks/S.F.)		90	.178		2.26	8.65		10.91	15.55
0400 4" x 8" x 2-1/4", dry, on edge (8/S.F.)		140	.114		4.64	5.55		10.19	13.50
0450 Grouted, 3/8" joint (6.5/S.F.)		85	.188		3.77	9.15		12.92	17.95
0455 Pervious brick paving, 4" x 8" x 3-1/4", without joints (4.5 bricks/S.F.)		110	.145		2.83	7.10		9.93	13.75
0500 Bedding, asphalt, 3/4" thick	B-25	5130	.017		.68	.83	.53	2.04	2.58
0540 Course washed sand bed, 1" thick	B-18	5000	.005		.37	.22	.03	.62	.77
0580 Mortar, 1" thick	D-1	300	.053		.76	2.60		3.36	4.75
0620 2" thick		200	.080		1.52	3.90		5.42	7.55
2000 Brick pavers, laid on edge, 7.2/S.F.		70	.229		4.74	11.15		15.89	22
2500 For 4" thick concrete bed and joints, add		595	.027		1.34	1.31		2.65	3.44
2800 For steam cleaning, add	A-1H	950	.008		.10	.37	.08	.55	.76

32 14 23 – Asphalt Unit Paving

32 14 23.10 Asphalt Blocks

		Daily	Labor-			2021 Bare Costs			Total
0010 **ASPHALT BLOCKS**									
0020 Rectangular, 6" x 12" x 1-1/4", w/bed & neopr. adhesive	D-1	135	.119	S.F.	10.10	5.75		15.85	19.80
0100 3" thick		130	.123		14.15	6		20.15	24.50
0300 Hexagonal tile, 8" wide, 1-1/4" thick		135	.119		10.10	5.75		15.85	19.80
0400 2" thick		130	.123		14.15	6		20.15	24.50
0500 Square, 8" x 8", 1-1/4" thick		135	.119		10.10	5.75		15.85	19.80
0600 2" thick		130	.123		14.15	6		20.15	24.50
0900 For exposed aggregate (ground finish), add					.61			.61	.67

32 14 Unit Paving

32 14 23 – Asphalt Unit Paving

32 14 23.10 Asphalt Blocks	Crew	Daily Output	Labor-Hours	Unit	Material	2021 Bare Costs Labor	Equipment	Total	Total Incl O&P
0910	For colors, add				S.F.	.61			.61	.67

32 14 40 – Stone Paving

32 14 40.10 Stone Pavers

	32 14 40.10 Stone Pavers	Crew	Daily Output	Labor-Hours	Unit	Material	2021 Bare Costs Labor	Equipment	Total	Total Incl O&P
0010	**STONE PAVERS**									
1100	Flagging, bluestone, irregular, 1" thick,	D-1	81	.198	S.F.	9.65	9.60		19.25	25
1110	1-1/2" thick		90	.178		11.40	8.65		20.05	25.50
1120	Pavers, 1/2" thick		110	.145		16.10	7.10		23.20	28.50
1130	3/4" thick		95	.168		20.50	8.20		28.70	35
1140	1" thick		81	.198		22	9.60		31.60	38.50
1150	Snapped random rectangular, 1" thick		92	.174		14.65	8.45		23.10	29
1200	1-1/2" thick		85	.188		17.60	9.15		26.75	33
1250	2" thick		83	.193		20.50	9.40		29.90	36.50
1300	Slate, natural cleft, irregular, 3/4" thick		92	.174		9.70	8.45		18.15	23.50
1310	1" thick		85	.188		11.30	9.15		20.45	26.50
1350	Random rectangular, gauged, 1/2" thick		105	.152		21	7.40		28.40	34
1400	Random rectangular, butt joint, gauged, 1/4" thick	↓	150	.107		22.50	5.20		27.70	33
1450	For sand rubbed finish, add				↓	10			10	11
1500	For interior setting, add					25%			25%	25%
1550	Granite blocks, 3-1/2" x 3-1/2" x 3-1/2"	D-1	92	.174	S.F.	21.50	8.45		29.95	37
1560	4" x 4" x 4"		95	.168		23	8.20		31.20	37.50
1600	4" to 12" long, 3" to 5" wide, 3" to 5" thick		98	.163		18.05	7.95		26	32
1650	6" to 15" long, 3" to 6" wide, 3" to 5" thick	↓	105	.152		9.65	7.40		17.05	22
2000	Granite paver, sawn with thermal finish, white, 4" x 4" x 2"	D-2	80	.550		22.50	27.50		50	66
2005	4" x 4" x 3"		80	.550		37	27.50		64.50	82
2010	4" x 4" x 4"		80	.550		38	27.50		65.50	83.50
2015	4" x 8" x 2"		170	.259		27	13		40	49
2020	4" x 8" x 3"		170	.259		34	13		47	57
2025	4" x 8" x 4"		170	.259		37.50	13		50.50	61
2030	8" x 8" x 2"		300	.147		27.50	7.35		34.85	41.50
2035	8" x 8" x 3"		300	.147		36	7.35		43.35	51
2040	8" x 8" x 4"		300	.147		31.50	7.35		38.85	45.50
2045	12" x 12" x 2"		725	.061		25	3.04		28.04	32
2050	12" x 12" x 3"		725	.061		32	3.04		35.04	40
2055	12" x 12" x 4"	↓	725	.061	↓	35	3.04		38.04	43

32 16 Curbs, Gutters, Sidewalks, and Driveways

32 16 13 – Curbs and Gutters

32 16 13.13 Cast-in-Place Concrete Curbs and Gutters

	32 16 13.13 Cast-in-Place Concrete Curbs and Gutters	Crew	Daily Output	Labor-Hours	Unit	Material	2021 Bare Costs Labor	Equipment	Total	Total Incl O&P
0010	**CAST-IN-PLACE CONCRETE CURBS AND GUTTERS**									
0290	Forms only, no concrete									
0300	Concrete, wood forms, 6" x 18", straight	C-2	500	.096	L.F.	4	5.10		9.10	12.05
0400	6" x 18", radius	"	200	.240	"	4.21	12.80		17.01	23.50
0402	Forms and concrete complete									
0404	Concrete, wood forms, 6" x 18", straight & concrete	C-2A	500	.096	L.F.	7.50	5.05		12.55	15.80
0406	6" x 18", radius		200	.240		7.70	12.70		20.40	27.50
0410	Steel forms, 6" x 18", straight		700	.069		7.30	3.62		10.92	13.40
0411	6" x 18", radius	↓	400	.120		7.15	6.35		13.50	17.30
0415	Machine formed, 6" x 18", straight	B-69A	2000	.024		5.15	1.16	.62	6.93	8.05
0416	6" x 18", radius	"	900	.053	↓	5.20	2.58	1.37	9.15	11.10

For customer support on your Site Work & Landscape Costs with RSMeans Data, call 800.448.8182.

383

32 16 Curbs, Gutters, Sidewalks, and Driveways

32 16 13 – Curbs and Gutters

32 16 13.13 Cast-in-Place Concrete Curbs and Gutters

		Crew	Daily Output	Labor-Hours	Unit	Material	2021 Bare Costs Labor	Equipment	Total	Total Incl O&P
0421	Curb and gutter, straight									
0422	with 6" high curb and 6" thick gutter, wood forms									
0430	24" wide, 0.055 C.Y./L.F.	C-2A	375	.128	L.F.	21	6.75		27.75	33
0435	30" wide, 0.066 C.Y./L.F.		340	.141		23	7.45		30.45	36
0440	Steel forms, 24" wide, straight		700	.069		9.70	3.62		13.32	16.05
0441	Radius		500	.096		8.55	5.05		13.60	17
0442	30" wide, straight		700	.069		11.35	3.62		14.97	17.90
0443	Radius		500	.096		9.70	5.05		14.75	18.20
0445	Machine formed, 24" wide, straight	B-69A	2000	.024		6.95	1.16	.62	8.73	10.05
0446	Radius		900	.053		6.95	2.58	1.37	10.90	13
0447	30" wide, straight		2000	.024		8.10	1.16	.62	9.88	11.30
0448	Radius		900	.053		8.10	2.58	1.37	12.05	14.25
0451	Median mall, machine formed, 2' x 9" high, straight		2200	.022		6.95	1.06	.56	8.57	9.85
0452	Radius	B-69B	900	.053		6.95	2.58	.89	10.42	12.45
0453	4' x 9" high, straight		2000	.024		13.95	1.16	.40	15.51	17.50
0454	Radius		800	.060		13.95	2.90	1	17.85	21

32 16 13.23 Precast Concrete Curbs and Gutters

		Crew	Daily Output	Labor-Hours	Unit	Material	2021 Bare Costs Labor	Equipment	Total	Total Incl O&P
0010	**PRECAST CONCRETE CURBS AND GUTTERS**									
0550	Precast, 6" x 18", straight	B-29	700	.080	L.F.	9.05	3.86	1.22	14.13	17.05
0600	6" x 18", radius	"	325	.172	"	9.65	8.30	2.62	20.57	26

32 16 13.33 Asphalt Curbs

		Crew	Daily Output	Labor-Hours	Unit	Material	2021 Bare Costs Labor	Equipment	Total	Total Incl O&P
0010	**ASPHALT CURBS**									
0012	Curbs, asphaltic, machine formed, 8" wide, 6" high, 40 L.F./ton	B-27	1000	.032	L.F.	1.84	1.44	.25	3.53	4.45
0100	8" wide, 8" high, 30 L.F./ton		900	.036		2.46	1.60	.28	4.34	5.40
0150	Asphaltic berm, 12" W, 3" to 6" H, 35 L.F./ton, before pavement		700	.046		.04	2.05	.36	2.45	3.51
0200	12" W, 1-1/2" to 4" H, 60 L.F./ton, laid with pavement	B-2	1050	.038		.02	1.71		1.73	2.58

32 16 13.43 Stone Curbs

		Crew	Daily Output	Labor-Hours	Unit	Material	2021 Bare Costs Labor	Equipment	Total	Total Incl O&P
0010	**STONE CURBS**									
1000	Granite, split face, straight, 5" x 16"	D-13	275	.175	L.F.	16.45	9.10	1.57	27.12	33.50
1100	6" x 18"	"	250	.192		21.50	10	1.73	33.23	41
1300	Radius curbing, 6" x 18", over 10' radius	B-29	260	.215		26.50	10.40	3.27	40.17	48
1400	Corners, 2' radius	"	80	.700	Ea.	89	34	10.65	133.65	160
1600	Edging, 4-1/2" x 12", straight	D-13	300	.160	L.F.	8.25	8.35	1.44	18.04	23
1800	Curb inlets (guttermouth) straight	B-29	41	1.366	Ea.	198	66	21	285	340
2000	Indian granite (Belgian block)									
2100	Jumbo, 10-1/2" x 7-1/2" x 4", grey	D-1	150	.107	L.F.	8.30	5.20		13.50	16.95
2150	Pink		150	.107		9.80	5.20		15	18.60
2200	Regular, 9" x 4-1/2" x 4-1/2", grey		160	.100		4.77	4.87		9.64	12.60
2250	Pink		160	.100		6.50	4.87		11.37	14.50
2300	Cubes, 4" x 4" x 4", grey		175	.091		2.62	4.45		7.07	9.60
2350	Pink		175	.091		4.04	4.45		8.49	11.15
2400	6" x 6" x 6", pink		155	.103		13.50	5.05		18.55	22.50
2500	Alternate pricing method for Indian granite									
2550	Jumbo, 10-1/2" x 7-1/2" x 4" (30 lb.), grey				Ton	470			470	520
2600	Pink					570			570	625
2650	Regular, 9" x 4-1/2" x 4-1/2" (20 lb.), grey					335			335	370
2700	Pink					455			455	500
2750	Cubes, 4" x 4" x 4" (5 lb.), grey					320			320	350
2800	Pink					520			520	570
2850	6" x 6" x 6" (25 lb.), pink					525			525	580
2900	For pallets, add					22			22	24

32 17 Paving Specialties

32 17 13 – Parking Bumpers

32 17 13.13 Metal Parking Bumpers

		Crew	Daily Output	Labor-Hours	Unit	Material	2021 Bare Costs Labor	Equipment	Total	Total Incl O&P
0010	**METAL PARKING BUMPERS**									
0015	Bumper rails for garages, 12 ga. rail, 6" wide, with steel									
0020	posts 12'-6" OC, minimum	E-4	190	.168	L.F.	19.75	10.25	.78	30.78	38
0030	Average		165	.194		24.50	11.80	.90	37.20	46
0100	Maximum		140	.229		29.50	13.90	1.06	44.46	55
0300	12" channel rail, minimum		160	.200		24.50	12.15	.93	37.58	47
0400	Maximum	↓	120	.267	↓	37	16.20	1.24	54.44	67
1300	Pipe bollards, conc. filled/paint, 8' L x 4' D hole, 6" diam.	B-6	20	1.200	Ea.	680	57.50	10.80	748.30	845
1400	8" diam.		15	1.600		745	77	14.40	836.40	950
1500	12" diam.	↓	12	2		1,025	96	18	1,139	1,300
2030	Folding with individual padlocks	B-2	50	.800	↓	182	36		218	254

32 17 13.16 Plastic Parking Bumpers

		Crew	Daily Output	Labor-Hours	Unit	Material	2021 Bare Costs Labor	Equipment	Total	Total Incl O&P
0010	**PLASTIC PARKING BUMPERS**									
1200	Thermoplastic, 6" x 10" x 6'-0"	B-2	120	.333	Ea.	46	14.95		60.95	73

32 17 13.19 Precast Concrete Parking Bumpers

		Crew	Daily Output	Labor-Hours	Unit	Material	2021 Bare Costs Labor	Equipment	Total	Total Incl O&P
0010	**PRECAST CONCRETE PARKING BUMPERS**									
1000	Wheel stops, precast concrete incl. dowels, 6" x 10" x 6'-0"	B-2	120	.333	Ea.	58	14.95		72.95	86
1100	8" x 13" x 6'-0"	"	120	.333	"	86.50	14.95		101.45	118

32 17 13.26 Wood Parking Bumpers

		Crew	Daily Output	Labor-Hours	Unit	Material	2021 Bare Costs Labor	Equipment	Total	Total Incl O&P
0010	**WOOD PARKING BUMPERS**									
0020	Parking barriers, timber w/saddles, treated type									
0100	4" x 4" for cars	B-2	520	.077	L.F.	3.91	3.45		7.36	9.45
0200	6" x 6" for trucks		520	.077	"	8.20	3.45		11.65	14.15
0600	Flexible fixed stanchion, 2' high, 3" diameter	↓	100	.400	Ea.	44.50	17.90		62.40	75.50

32 17 23 – Pavement Markings

32 17 23.13 Painted Pavement Markings

		Crew	Daily Output	Labor-Hours	Unit	Material	2021 Bare Costs Labor	Equipment	Total	Total Incl O&P
0010	**PAINTED PAVEMENT MARKINGS**									
0020	Acrylic waterborne, white or yellow, 4" wide, less than 3,000 L.F.	B-78	20000	.002	L.F.	.12	.11	.05	.28	.35
0030	3,000-16,000 L.F.		20000	.002		.14	.11	.05	.30	.37
0040	over 16,000 L.F.		20000	.002		.12	.11	.05	.28	.35
0200	6" wide, less than 3,000 L.F.		11000	.004		.18	.20	.10	.48	.60
0220	3,000-16,000 L.F.		11000	.004		.21	.20	.10	.51	.64
0230	over 16,000 L.F.		11000	.004		.18	.20	.10	.48	.60
0500	8" wide, less than 3,000 L.F.		10000	.005		.23	.22	.11	.56	.71
0520	3,000-16,000 L.F.		10000	.005		.28	.22	.11	.61	.76
0530	over 16,000 L.F.		10000	.005		.23	.22	.11	.56	.71
0600	12" wide, less than 3,000 L.F.		4000	.012		.35	.55	.27	1.17	1.50
0604	3,000-16,000 L.F.		4000	.012		.42	.55	.27	1.24	1.57
0608	over 16,000 L.F.		4000	.012	↓	.35	.55	.27	1.17	1.50
0620	Arrows or gore lines		2300	.021	S.F.	.21	.95	.47	1.63	2.17
0640	Temporary paint, white or yellow, less than 3,000 L.F.		15000	.003	L.F.	.13	.15	.07	.35	.44
0642	3,000-16,000 L.F.		15000	.003		.10	.15	.07	.32	.41
0644	Over 16,000 L.F.	↓	15000	.003		.09	.15	.07	.31	.40
0660	Removal	1 Clab	300	.027			1.18		1.18	1.77
0680	Temporary tape	2 Clab	1500	.011		.52	.47		.99	1.28
0710	Thermoplastic, white or yellow, 4" wide, less than 6,000 L.F.	B-79	15000	.003		.45	.12	.12	.69	.81
0715	6,000 L.F. or more		15000	.003		.36	.12	.12	.60	.71
0730	6" wide, less than 6,000 L.F.		14000	.003		.68	.13	.13	.94	1.07
0735	6,000 L.F. or more		14000	.003		.54	.13	.13	.80	.92
0740	8" wide, less than 6,000 L.F.		12000	.003		.90	.15	.15	1.20	1.38
0745	6,000 L.F. or more	↓	12000	.003	↓	.72	.15	.15	1.02	1.18

32 17 23.13 Painted Pavement Markings

		Crew	Daily Output	Labor-Hours	Unit	Material	2021 Bare Costs Labor	2021 Bare Costs Equipment	Total	Total Incl O&P
0750	12" wide, less than 6,000 L.F.	B-79	6000	.007	L.F.	1.33	.30	.30	1.93	2.24
0755	6,000 L.F. or more		6000	.007	↓	1.06	.30	.30	1.66	1.95
0760	Arrows		660	.061	S.F.	.74	2.77	2.70	6.21	7.90
0770	Gore lines		2500	.016		.74	.73	.71	2.18	2.69
0780	Letters	↓	660	.061	↓	.75	2.77	2.70	6.22	7.95
0782	Thermoplastic material, small users				Ton	2,075			2,075	2,275
0784	Glass beads, highway use, add				Lb.	.67			.67	.74
0786	Thermoplastic material, highway departments				Ton	1,900			1,900	2,100
1000	Airport painted markings									
1050	Traffic safety flashing truck for airport painting	A-2B	1	8	Day		390	198	588	805
1100	Painting, white or yellow, taxiway markings	B-78	4000	.012	S.F.	.42	.55	.27	1.24	1.57
1110	with 12 lb. beads per 100 S.F.		4000	.012		.50	.55	.27	1.32	1.66
1200	Runway markings		3500	.014		.42	.62	.31	1.35	1.73
1210	with 12 lb. beads per 100 S.F.		3500	.014		.50	.62	.31	1.43	1.82
1300	Pavement location or direction signs		2500	.019		.42	.87	.44	1.73	2.24
1310	with 12 lb. beads per 100 S.F.		2500	.019	↓	.50	.87	.44	1.81	2.33
1350	Mobilization airport pavement painting	↓	4	12	Ea.		545	273	818	1,125
1400	Paint markings or pavement signs removal daytime	B-78B	400	.045	S.F.		2.05	1.02	3.07	4.18
1500	Removal nighttime		335	.054	"		2.45	1.22	3.67	5
1600	Mobilization pavement paint removal	↓	4	4.500	Ea.		205	102	307	415

32 17 23.14 Pavement Parking Markings

		Crew	Daily Output	Labor-Hours	Unit	Material	2021 Bare Costs Labor	2021 Bare Costs Equipment	Total	Total Incl O&P
0010	**PAVEMENT PARKING MARKINGS**									
0790	Layout of pavement marking	A-2	25000	.001	L.F.		.04	.01	.05	.08
0800	Lines on pvmt., parking stall, paint, white, 4" wide	B-78B	400	.045	Stall	5.20	2.05	1.02	8.27	9.90
0825	Parking stall, small quantities	2 Pord	80	.200		10.40	9.30		19.70	25.50
0830	Lines on pvmt., parking stall, thermoplastic, white, 4" wide	B-79	300	.133	↓	21.50	6.10	5.95	33.55	39
1000	Street letters and numbers	B-78B	1600	.011	S.F.	.88	.51	.25	1.64	2.02
1100	Pavement marking letter, 6"	2 Pord	400	.040	Ea.	11.75	1.86		13.61	15.70
1110	12" letter		272	.059		16.10	2.73		18.83	22
1120	24" letter		160	.100		32.50	4.65		37.15	43
1130	36" letter		84	.190		52.50	8.85		61.35	71
1140	42" letter		84	.190		65	8.85		73.85	84.50
1150	72" letter		40	.400		42.50	18.60		61.10	74.50
1200	Handicap symbol	↓	40	.400		33.50	18.60		52.10	64
1210	Handicap parking sign 12" x 18" and post	A-2	12	2	↓	133	91.50	16.55	241.05	300
1300	Pavement marking, thermoplastic tape including layout, 4" width	B-79B	320	.025	L.F.	2.77	1.11	.54	4.42	5.30
1310	12" width		192	.042	"	10.40	1.85	.90	13.15	15.20
1320	Letters including layout, 4"		240	.033	Ea.	18.50	1.48	.72	20.70	23.50
1330	6"		160	.050		20.50	2.22	1.09	23.81	27
1340	12"		120	.067		23.50	2.96	1.45	27.91	32
1350	48"		64	.125		88.50	5.55	2.71	96.76	109
1360	96"		32	.250		112	11.10	5.45	128.55	146
1380	4 letter words, 8' tall	↓	8	1	↓	420	44.50	21.50	486	555
1385	Sample words: BUMP, FIRE, LANE									

32 17 23.33 Plastic Pavement Markings

		Crew	Daily Output	Labor-Hours	Unit	Material	2021 Bare Costs Labor	2021 Bare Costs Equipment	Total	Total Incl O&P
0010	**PLASTIC PAVEMENT MARKINGS**									
0020	Thermoplastic markings with glass beads									
0100	Thermoplastic striping, 6 lb. beads per 100 S.F., white or yellow, 4" wide	B-79	15000	.003	L.F.	.46	.12	.12	.70	.82
0110	8 lb. beads per 100 S.F.		15000	.003		.47	.12	.12	.71	.82
0120	10 lb. beads per 100 S.F.		15000	.003		.47	.12	.12	.71	.83
0130	12 lb. beads per 100 S.F.		15000	.003		.48	.12	.12	.72	.83
0200	6 lb. beads per 100 S.F., white or yellow, 6" wide	↓	14000	.003	↓	.70	.13	.13	.96	1.09

32 17 Paving Specialties

32 17 23 – Pavement Markings

32 17 23.33 Plastic Pavement Markings	Crew	Daily Output	Labor-Hours	Unit	Material	2021 Bare Costs Labor	Equipment	Total	Total Incl O&P
0210	8 lb. beads per 100 S.F.	B-79	14000	.003	L.F.	.70	.13	.13	.96	1.10
0220	10 lb. beads per 100 S.F.		14000	.003		.71	.13	.13	.97	1.11
0230	12 lb. beads per 100 S.F.		14000	.003		.72	.13	.13	.98	1.12
0300	6 lb. beads per 100 S.F., white or yellow, 8" wide		12000	.003		.93	.15	.15	1.23	1.41
0310	8 lb. beads per 100 S.F.		12000	.003		.94	.15	.15	1.24	1.42
0320	10 lb. beads per 100 S.F.		12000	.003		.94	.15	.15	1.24	1.43
0330	12 lb. beads per 100 S.F.		12000	.003		.95	.15	.15	1.25	1.44
0400	6 lb. beads per 100 S.F., white or yellow, 12" wide		6000	.007		1.39	.30	.30	1.99	2.31
0410	8 lb. beads per 100 S.F.		6000	.007		1.40	.30	.30	2	2.32
0420	10 lb. beads per 100 S.F.		6000	.007		1.42	.30	.30	2.02	2.34
0430	12 lb. beads per 100 S.F.		6000	.007		1.43	.30	.30	2.03	2.35
0500	Gore lines with 6 lb. beads per 100 S.F.		2500	.016	S.F.	.78	.73	.71	2.22	2.74
0510	8 lb. beads per 100 S.F.		2500	.016		.79	.73	.71	2.23	2.75
0520	10 lb. beads per 100 S.F.		2500	.016		.81	.73	.71	2.25	2.77
0530	12 lb. beads per 100 S.F.		2500	.016		.82	.73	.71	2.26	2.78
0600	Arrows with 6 lb. beads per 100 S.F.		660	.061		.78	2.77	2.70	6.25	7.95
0610	8 lb. beads per 100 S.F.		660	.061		.79	2.77	2.70	6.26	7.95
0620	10 lb. beads per 100 S.F.		660	.061		.81	2.77	2.70	6.28	8
0630	12 lb. beads per 100 S.F.		660	.061		.82	2.77	2.70	6.29	8
0700	Letters with 6 lb. beads per 100 S.F.		660	.061		.79	2.77	2.70	6.26	7.95
0710	8 lb. beads per 100 S.F.		660	.061		.80	2.77	2.70	6.27	8
0720	10 lb. beads per 100 S.F.		660	.061		.82	2.77	2.70	6.29	8
0730	12 lb. beads per 100 S.F.		660	.061		.83	2.77	2.70	6.30	8
1000	Airport thermoplastic markings									
1050	Traffic safety flashing truck for airport markings	A-2B	1	8	Day		390	198	588	805
1100	Thermoplastic taxiway markings with 24 lb. beads per 100 S.F.	B-79	2500	.016	S.F.	.88	.73	.71	2.32	2.84
1200	Runway markings		2100	.019		.88	.87	.85	2.60	3.19
1250	Location or direction signs		1500	.027		.88	1.22	1.19	3.29	4.09
1350	Mobilization airport pavement marking		4	10	Ea.		455	445	900	1,175
1400	Thermostatic markings or pavement signs removal daytime	B-78B	400	.045	S.F.		2.05	1.02	3.07	4.18
1500	Removal nighttime		335	.054	"		2.45	1.22	3.67	5
1600	Mobilization thermostatic marking removal		4	4.500	Ea.		205	102	307	415
1700	Thermoplastic road marking material, small users				Ton	2,075			2,075	2,275
1710	Highway departments					1,900			1,900	2,100
1720	Thermoplastic airport marking material, small users					2,175			2,175	2,375
1730	Large users					1,875			1,875	2,050
1800	Glass beads material for highway use, type II				Lb.	.67			.67	.74
1900	Glass beads material for airport use, type III				"	.67			.67	.74

32 17 26 – Tactile Warning Surfacing

32 17 26.10 Tactile Warning Surfacing

		Crew	Daily Output	Labor-Hours	Unit	Material	Labor	Equipment	Total	Total Incl O&P
0010	**TACTILE WARNING SURFACING**									
0100	Detectable warning pad, ADA	2 Clab	400	.040	S.F.	20.50	1.78		22.28	25

For customer support on your Site Work & Landscape Costs with RSMeans Data, call 800.448.8182.

387

32 18 Athletic and Recreational Surfacing

32 18 13 – Synthetic Grass Surfacing

32 18 13.10 Artificial Grass Surfacing

		Crew	Daily Output	Labor-Hours	Unit	Material	2021 Bare Costs Labor	Equipment	Total	Total Incl O&P
0010	**ARTIFICIAL GRASS SURFACING**									
0015	Not including asphalt base or drainage,									
0020	but including cushion pad, over 50,000 S.F.									
0200	1/2" pile and 5/16" cushion pad, standard	C-17	3200	.025	S.F.	6.75	1.44		8.19	9.60
0300	Deluxe		2560	.031		6.50	1.80		8.30	9.85
0500	1/2" pile and 5/8" cushion pad, standard		2844	.028		6.25	1.62		7.87	9.35
0600	Deluxe		2327	.034		6.35	1.98		8.33	9.90
0700	Acrylic emulsion texture coat, sand filled	B-1	5200	.005	S.Y.	2.92	.21		3.13	3.52
0710	Rubber filled	"	5200	.005	"	3.07	.21		3.28	3.69
0720	Acrylic emulsion color coat									
0730	Brown, red, or tan	B-1	6000	.004	S.Y.	3.22	.18		3.40	3.81
0740	Green		6000	.004		3.22	.18		3.40	3.81
0750	Blue		6000	.004		3.11	.18		3.29	3.69
0760	Rubber acrylic cushion base coat		2600	.009		5.30	.42		5.72	6.40
0800	For asphaltic concrete base, 2-1/2" thick,									
0900	with 6" crushed stone sub-base, add	B-25	12000	.007	S.F.	1.91	.36	.23	2.50	2.89

32 18 16 – Synthetic Resilient Surfacing

32 18 16.13 Playground Protective Surfacing

		Crew	Daily Output	Labor-Hours	Unit	Material	2021 Bare Costs Labor	Equipment	Total	Total Incl O&P
0010	**PLAYGROUND PROTECTIVE SURFACING**									
0100	Resilient rubber surface, 4" thick, black	2 Skwk	300	.053	S.F.	13.25	3.05		16.30	19.20
0150	2" thick topping, colors	"	2800	.006		6.45	.33		6.78	7.60
0200	Wood chip mulch, 6" deep	1 Clab	300	.027		.99	1.18		2.17	2.86

32 18 16.16 Epoxy Acrylic Coatings

		Crew	Daily Output	Labor-Hours	Unit	Material	2021 Bare Costs Labor	Equipment	Total	Total Incl O&P
0010	**EPOXY ACRYLIC COATINGS**									
0100	Surface Prep, Sand blasting to SSPC-SP6, 2.0#/S.F. sand	E-11A	1200	.027	S.F.	.35	1.30	.50	2.15	2.93
0150	Layout and line striping by others									
0200	Painting epoxy, Foundation/Prime coat, large area (100 S.F. per Gal)	E-11B	2400	.010	S.F.	.37	.46	.16	.99	1.30
0210	Epoxy intermediate coat		2800	.009		.37	.40	.14	.91	1.17
0220	Epoxy top coat		2800	.009		.37	.40	.14	.91	1.17
0300	Painting epoxy, Foundation/Prime coat, small areas (100 S.F. per Gal)		1400	.017		.37	.79	.27	1.43	1.94
0310	Epoxy intermediate coat		1500	.016		.37	.74	.25	1.36	1.84
0320	Epoxy top coat		1500	.016		.37	.74	.25	1.36	1.84
0330	Adhesive Promoter		1400	.017		.27	.79	.27	1.33	1.84
0340	Sealer Concentrate		2400	.010		.15	.46	.16	.77	1.06
0400	Install debris tarp for construction work under 600 S.F.	2 Clab	2100	.008		.66	.34		1	1.23
0410	600 S.F. or more	"	2400	.007		.56	.30		.86	1.06

32 18 23 – Athletic Surfacing

32 18 23.33 Running Track Surfacing

		Crew	Daily Output	Labor-Hours	Unit	Material	2021 Bare Costs Labor	Equipment	Total	Total Incl O&P
0010	**RUNNING TRACK SURFACING**									
0020	Running track, asphalt concrete pavement, 2-1/2"	B-37	300	.160	S.Y.	12.90	7.45	.86	21.21	26
0102	Surface, latex rubber system, 1/2" thick, black	B-20	115	.209		58	10.30		68.30	79.50
0152	Colors		115	.209		71	10.30		81.30	94
0302	Urethane rubber system, 1/2" thick, black		110	.218		45	10.75		55.75	65.50
0402	Color coating		110	.218		55.50	10.75		66.25	77

32 18 23.53 Tennis Court Surfacing

		Crew	Daily Output	Labor-Hours	Unit	Material	2021 Bare Costs Labor	Equipment	Total	Total Incl O&P
0010	**TENNIS COURT SURFACING**									
0020	Tennis court, asphalt, incl. base, 2-1/2" thick, one court	B-37	450	.107	S.Y.	34.50	4.97	.57	40.04	46
0200	Two courts		675	.071		20.50	3.31	.38	24.19	28
0300	Clay courts		360	.133		45.50	6.20	.72	52.42	60
0400	Pulverized natural greenstone with 4" base, fast dry		250	.192		33	8.95	1.03	42.98	50.50
0800	Rubber-acrylic base resilient pavement		600	.080		67	3.73	.43	71.16	80

For customer support on your Site Work & Landscape Costs with RSMeans Data, call 800.448.8182.

32 18 Athletic and Recreational Surfacing

32 18 23 – Athletic Surfacing

32 18 23.53 Tennis Court Surfacing

		Crew	Daily Output	Labor-Hours	Unit	Material	2021 Bare Costs Labor	Equipment	Total	Total Incl O&P
1000	Colored sealer, acrylic emulsion, 3 coats	2 Clab	800	.020	S.Y.	7.35	.89		8.24	9.45
1100	3 coats, 2 colors	"	900	.018		10.25	.79		11.04	12.50
1200	For preparing old courts, add	1 Clab	825	.010			.43		.43	.64
1400	Posts for nets, 3-1/2" diameter with eye bolts	B-1	3.40	7.059	Pr.	263	320		583	765
1500	With pulley & reel		3.40	7.059	"	845	320		1,165	1,400
1700	Net, 42' long, nylon thread with binder		50	.480	Ea.	283	21.50		304.50	345
1800	All metal		6.50	3.692	"	560	166		726	865
2000	Paint markings on asphalt, 2 coats	1 Pord	1.78	4.494	Court	232	209		441	565
2200	Complete court with fence, etc., asphaltic conc., minimum	B-37	.20	240		34,100	11,200	1,300	46,600	55,500
2300	Maximum		.16	300		69,500	14,000	1,625	85,125	99,000
2800	Clay courts, minimum		.20	240		37,800	11,200	1,300	50,300	59,500
2900	Maximum		.16	300		63,000	14,000	1,625	78,625	91,500

32 31 Fences and Gates

32 31 11 – Gate Operators

32 31 11.10 Gate Operators

		Crew	Daily Output	Labor-Hours	Unit	Material	2021 Bare Costs Labor	Equipment	Total	Total Incl O&P
0010	**GATE OPERATORS**									
7810	Motor operators for gates (no elec wiring), 3' wide swing	2 Skwk	.50	32	Ea.	1,050	1,825		2,875	3,900
7815	Up to 20' wide swing		.50	32		1,350	1,825		3,175	4,250
7820	Up to 45' sliding		.50	32		3,050	1,825		4,875	6,100
7825	Overhead gate, 6' to 18' wide, sliding/cantilever		45	.356	L.F.	330	20.50		350.50	395
7830	Gate operators, digital receiver		7	2.286	Ea.	67.50	131		198.50	270
7835	Two button transmitter		24	.667		23.50	38		61.50	83
7840	3 button station		14	1.143		39	65.50		104.50	141
7845	Master slave system		4	4		176	228		404	540

32 31 13 – Chain Link Fences and Gates

32 31 13.10 Chain Link Gates and Fences

		Crew	Daily Output	Labor-Hours	Unit	Material	2021 Bare Costs Labor	Equipment	Total	Total Incl O&P
0010	**CHAIN LINK GATES AND FENCES**									
4750	Gate, transom for 10' fence, galv. steel, single, 3' x 7'	B-80A	52	.462	Ea.	410	20.50	16.35	446.85	500
4752	4' x 7'		10	2.400		440	107	85	632	740
4754	3' x 10'		8	3		400	133	106	639	755
4756	4' x 10'		10	2.400		455	107	85	647	755
4758	Double transom, 10' x 7'	B-80B	10	3.200		655	151	24	830	970
4760	12' x 7'		6	5.333		705	252	40	997	1,200
4762	14' x 7'		5	6.400		765	300	47.50	1,112.50	1,350
4764	10' x 10'		4	8		950	375	59.50	1,384.50	1,675
4766	12' x 10'		7	4.571		1,050	216	34	1,300	1,500
4768	14' x 10'		7	4.571		1,125	216	34	1,375	1,575
4780	Vinyl-clad, single transom, 3' x 7'		10	3.200		415	151	24	590	710
4782	4' x 7'		10	3.200		445	151	24	620	740
4784	3' x 10'		8	4		585	189	30	804	960
4786	4' x 10'		8	4		795	189	30	1,014	1,200
4788	Double transom, 10' x 7'		10	3.200		920	151	24	1,095	1,250
4790	12' x 7'		6	5.333		940	252	40	1,232	1,450
4792	14' x 7'		5	6.400		975	300	47.50	1,322.50	1,575
4794	10' x 10'		4	8		945	375	59.50	1,379.50	1,675
4798	12' x 12'		7	4.571		1,325	216	34	1,575	1,825
4799	14' x 14'		7	4.571		1,800	216	34	2,050	2,350

For customer support on your Site Work & Landscape Costs with RSMeans Data, call 800.448.8182.

389

32 31 13.20 Fence, Chain Link Industrial	Crew	Daily Output	Labor-Hours	Unit	Material	2021 Bare Costs Labor	Equipment	Total	Total Incl O&P
0010 **FENCE, CHAIN LINK INDUSTRIAL**									
0011 Schedule 40, including concrete									
0020 3 strands barb wire, 2" post @ 10' OC, set in concrete, 6' H									
0200 9 ga. wire, galv. steel, in concrete	B-80C	240	.100	L.F.	19.40	4.59	1.05	25.04	29.50
0248 Fence, add for vinyl coated fabric				S.F.	.71			.71	.78
0300 Aluminized steel	B-80C	240	.100	L.F.	26.50	4.59	1.05	32.14	37
0301 Fence, wrought iron		240	.100		43	4.59	1.05	48.64	55
0303 Fence, commercial 4' high		240	.100		32.50	4.59	1.05	38.14	43.50
0304 Fence, commercial 6' high		240	.100		58	4.59	1.05	63.64	72
0500 6 ga. wire, galv. steel		240	.100		25	4.59	1.05	30.64	35.50
0600 Aluminized steel		240	.100		34	4.59	1.05	39.64	45.50
0800 6 ga. wire, 6' high but omit barbed wire, galv. steel		250	.096		22.50	4.40	1.01	27.91	32
0900 Aluminized steel, in concrete		250	.096		30.50	4.40	1.01	35.91	41.50
0920 8' H, 6 ga. wire, 2-1/2" line post, galv. steel, in concrete		180	.133		35.50	6.10	1.41	43.01	49.50
0940 Aluminized steel, in concrete		180	.133	↓	43.50	6.10	1.41	51.01	58.50
1400 Gate for 6' high fence, 1-5/8" frame, 3' wide, galv. steel		10	2.400	Ea.	208	110	25.50	343.50	420
1500 Aluminized steel, in concrete	↓	10	2.400	"	224	110	25.50	359.50	440
2000 5'-0" high fence, 9 ga., no barbed wire, 2" line post, in concrete									
2010 10' OC, 1-5/8" top rail, in concrete									
2100 Galvanized steel, in concrete	B-80C	300	.080	L.F.	20.50	3.67	.84	25.01	29
2200 Aluminized steel, in concrete		300	.080	"	20	3.67	.84	24.51	28.50
2400 Gate, 4' wide, 5' high, 2" frame, galv. steel, in concrete		10	2.400	Ea.	229	110	25.50	364.50	445
2500 Aluminized steel, in concrete		10	2.400	"	183	110	25.50	318.50	395
3100 Overhead slide gate, chain link, 6' high, to 18' wide, in concrete	↓	38	.632	L.F.	97	29	6.65	132.65	158
3105 8' high, in concrete	B-80	30	1.067		102	52	35	189	229
3108 10' high, in concrete		24	1.333		156	65	44	265	315
3110 Cantilever type, in concrete		48	.667		189	32.50	22	243.50	281
3120 8' high, in concrete		24	1.333		182	65	44	291	345
3130 10' high, in concrete	↓	18	1.778	↓	211	86.50	58.50	356	425
5000 Double swing gates, incl. posts & hardware, in concrete									
5010 5' high, 12' opening, in concrete	B-80C	3.40	7.059	Opng.	620	325	74.50	1,019.50	1,250
5020 20' opening, in concrete		2.80	8.571		720	395	90.50	1,205.50	1,475
5060 6' high, 12' opening, in concrete		3.20	7.500		455	345	79	879	1,100
5070 20' opening, in concrete	↓	2.60	9.231		620	425	97.50	1,142.50	1,425
5080 8' high, 12' opening, in concrete	B-80	2.13	15.002		475	730	495	1,700	2,175
5090 20' opening, in concrete		1.45	22.069		750	1,075	725	2,550	3,225
5100 10' high, 12' opening, in concrete		1.31	24.427		865	1,200	805	2,870	3,625
5110 20' opening, in concrete		1.03	31.068		1,100	1,525	1,025	3,650	4,600
5120 12' high, 12' opening, in concrete		1.05	30.476		1,450	1,475	1,000	3,925	4,900
5130 20' opening, in concrete	↓	.85	37.647	↓	1,350	1,825	1,250	4,425	5,575
5190 For aluminized steel, add					20%				
7055 Braces, galv. steel	B-80A	960	.025	L.F.	2.90	1.11	.89	4.90	5.80
7056 Aluminized steel	"	960	.025	"	3.48	1.11	.89	5.48	6.45
7071 Privacy slats, vertical, vinyl	1 Clab	500	.016	S.F.	1.77	.71		2.48	3.01
7072 Redwood		450	.018		1.51	.79		2.30	2.84
7073 Diagonal, aluminum	↓	300	.027	↓	4.88	1.18		6.06	7.10

32 31 13.25 Fence, Chain Link Residential

	Crew	Daily Output	Labor-Hours	Unit	Material	2021 Bare Costs Labor	Equipment	Total	Total Incl O&P
0010 **FENCE, CHAIN LINK RESIDENTIAL**									
0011 Schedule 20, 11 ga. wire, 1-5/8" post									
0020 10' OC, 1-3/8" top rail, 2" corner post, galv. stl. 3' high	B-80C	500	.048	L.F.	4.95	2.20	.51	7.66	9.30
0050 4' high		400	.060		5.60	2.75	.63	8.98	10.95
0100 6' high	↓	200	.120	↓	7.25	5.50	1.26	14.01	17.60

32 31 13 – Chain Link Fences and Gates

32 31 13.25 Fence, Chain Link Residential

		Crew	Daily Output	Labor-Hours	Unit	Material	2021 Bare Costs Labor	Equipment	Total	Total Incl O&P
0150	Add for gate 3' wide, 1-3/8" frame, 3' high	B-80C	12	2	Ea.	88	91.50	21	200.50	257
0170	4' high		10	2.400		95	110	25.50	230.50	296
0190	6' high		10	2.400		104	110	25.50	239.50	305
0200	Add for gate 4' wide, 1-3/8" frame, 3' high		9	2.667		99.50	122	28	249.50	325
0220	4' high		9	2.667		107	122	28	257	330
0240	6' high		8	3		115	138	31.50	284.50	370
0350	Aluminized steel, 11 ga. wire, 3' high		500	.048	L.F.	8.60	2.20	.51	11.31	13.35
0380	4' high		400	.060		8.65	2.75	.63	12.03	14.30
0400	6' high		200	.120		11.20	5.50	1.26	17.96	22
0450	Add for gate 3' wide, 1-3/8" frame, 3' high		12	2	Ea.	109	91.50	21	221.50	280
0470	4' high		10	2.400		100	110	25.50	235.50	300
0490	6' high		10	2.400		133	110	25.50	268.50	340
0500	Add for gate 4' wide, 1-3/8" frame, 3' high		10	2.400		113	110	25.50	248.50	315
0520	4' high		9	2.667		122	122	28	272	350
0540	6' high		8	3		129	138	31.50	298.50	380
0620	Vinyl covered, 9 ga. wire, 3' high		500	.048	L.F.	7.45	2.20	.51	10.16	12.05
0640	4' high		400	.060		8.20	2.75	.63	11.58	13.85
0660	6' high		200	.120		9.05	5.50	1.26	15.81	19.60
0720	Add for gate 3' wide, 1-3/8" frame, 3' high		12	2	Ea.	106	91.50	21	218.50	276
0740	4' high		10	2.400		113	110	25.50	248.50	315
0760	6' high		10	2.400		131	110	25.50	266.50	335
0780	Add for gate 4' wide, 1-3/8" frame, 3' high		10	2.400		98.50	110	25.50	234	300
0800	4' high		9	2.667		111	122	28	261	335
0820	6' high		8	3		141	138	31.50	310.50	395
7076	Fence, for small jobs 100 L.F. fence or less w/or w/o gate, add				L.F.	20%				

32 31 13.26 Tennis Court Fences and Gates

		Crew	Daily Output	Labor-Hours	Unit	Material	2021 Bare Costs Labor	Equipment	Total	Total Incl O&P
0010	**TENNIS COURT FENCES AND GATES**									
0860	Tennis courts, 11 ga. wire, 2-1/2" post set									
0870	in concrete, 10' OC, 1-5/8" top rail									
0900	10' high	B-80	190	.168	L.F.	22	8.20	5.55	35.75	42.50
0920	12' high		170	.188	"	23	9.20	6.20	38.40	45.50
1000	Add for gate 4' wide, 1-5/8" frame 7' high		10	3.200	Ea.	230	156	105	491	600
1040	Aluminized steel, 11 ga. wire 10' high		190	.168	L.F.	19.70	8.20	5.55	33.45	40
1100	12' high		170	.188	"	25.50	9.20	6.20	40.90	48.50
1140	Add for gate 4' wide, 1-5/8" frame, 7' high		10	3.200	Ea.	239	156	105	500	610
1250	Vinyl covered, 9 ga. wire, 10' high		190	.168	L.F.	21	8.20	5.55	34.75	41.50
1300	12' high		170	.188	"	24.50	9.20	6.20	39.90	47.50
1310	Fence, CL, tennis court, transom gate, single, galv., 4' x 7'	B-80A	8.72	2.752	Ea.	390	122	97.50	609.50	715
1400	Add for gate 4' wide, 1-5/8" frame, 7' high	B-80	10	3.200	"	325	156	105	586	705

32 31 13.30 Fence, Chain Link, Gates and Posts

		Crew	Daily Output	Labor-Hours	Unit	Material	2021 Bare Costs Labor	Equipment	Total	Total Incl O&P
0010	**FENCE, CHAIN LINK, GATES & POSTS**									
0011	(1/3 post length in ground)									
0013	For Concrete, See Section 03 31									
6580	Line posts, galvanized, 2-1/2" OD, set in conc., 4'	B-80	80	.400	Ea.	49.50	19.50	13.15	82.15	98
6585	5'		76	.421		55	20.50	13.85	89.35	106
6590	6'		74	.432		57	21	14.20	92.20	110
6595	7'		72	.444		76.50	21.50	14.60	112.60	133
6600	8'		69	.464		84	22.50	15.25	121.75	143
6635	Vinyl coated, 2-1/2" OD, set in conc., 4'		79	.405		61.50	19.75	13.30	94.55	112
6640	5'		77	.416		65	20.50	13.65	99.15	117
6645	6'		74	.432		82.50	21	14.20	117.70	138
6650	7'		72	.444		77.50	21.50	14.60	113.60	134

32 31 Fences and Gates

32 31 13 – Chain Link Fences and Gates

32 31 13.30 Fence, Chain Link, Gates and Posts		Crew	Daily Output	Labor-Hours	Unit	Material	2021 Bare Costs Labor	Equipment	Total	Total Incl O&P
6655	8'	B-80	69	.464	Ea.	98.50	22.50	15.25	136.25	159
6660	End gate post, steel, 3" OD, set in conc., 4'		68	.471		52.50	23	15.50	91	109
6665	5'		65	.492		58.50	24	16.20	98.70	118
6670	6'		63	.508		60.50	25	16.70	102.20	122
6675	7'		61	.525		56.50	25.50	17.25	99.25	119
6680	8'		59	.542		69	26.50	17.85	113.35	135
6685	Vinyl, 4'		68	.471		39	23	15.50	77.50	94.50
6690	5'		65	.492		39	24	16.20	79.20	97
6695	6'		63	.508		60	25	16.70	101.70	121
6700	7'		61	.525		46.50	25.50	17.25	89.25	108
6705	8'		59	.542		75	26.50	17.85	119.35	142
6710	Corner post, galv. steel, 4" OD, set in conc., 4'		65	.492		91.50	24	16.20	131.70	154
6715	6'		63	.508		121	25	16.70	162.70	188
6720	7'		61	.525		120	25.50	17.25	162.75	189
6725	8'		65	.492		185	24	16.20	225.20	258
6730	Vinyl, 5'		65	.492		44	24	16.20	84.20	102
6735	6'		63	.508		85.50	25	16.70	127.20	149
6740	7'		61	.525		69	25.50	17.25	111.75	133
6745	8'		59	.542		89.50	26.50	17.85	133.85	158
7031	For corner, end & pull post bracing, add					20%	15%			
7770	Gates, sliding w/overhead support, 4' high	B-80B	35	.914	L.F.	136	43	6.80	185.80	222
7775	5' high		32	1		151	47	7.45	205.45	245
7780	6' high		28	1.143		150	54	8.55	212.55	255
7785	7' high		25	1.280		183	60.50	9.55	253.05	300
7790	8' high		23	1.391		191	65.50	10.40	266.90	320
7795	Cantilever, manual, exp. roller (pr), 40' wide x 8' high	B-22	1	30	Ea.	6,000	1,550	287	7,837	9,225
7800	30' wide x 8' high		1	30		4,750	1,550	287	6,587	7,875
7805	24' wide x 8' high		1	30		3,100	1,550	287	4,937	6,075
7900	Auger fence post hole, 3' deep, medium soil, by hand	1 Clab	30	.267			11.85		11.85	17.65
7925	By machine	B-80	175	.183			8.90	6	14.90	19.90
7950	Rock, with jackhammer	B-9	32	1.250			56	11.10	67.10	96
7975	With rock drill	B-47C	65	.246			12.30	27.50	39.80	49

32 31 13.33 Chain Link Backstops

	CHAIN LINK BACKSTOPS									
0010	**CHAIN LINK BACKSTOPS**									
0015	Backstops, baseball, prefabricated, 30' wide, 12' high & 1 overhang	B-1	1	24	Ea.	2,700	1,075		3,775	4,600
0100	40' wide, 12' high & 2 overhangs	"	.75	32		7,125	1,450		8,575	9,975
0300	Basketball, steel, single goal	B-13	3.04	18.421		1,750	890	193	2,833	3,450
0400	Double goal	"	1.92	29.167		2,400	1,400	305	4,105	5,050
0600	Tennis, wire mesh with pair of ends	B-1	2.48	9.677	Set	2,650	435		3,085	3,550
0700	Enclosed court	"	1.30	18.462	Ea.	9,075	830		9,905	11,200

32 31 13.40 Fence, Fabric and Accessories

	FENCE, FABRIC & ACCESSORIES									
0010	**FENCE, FABRIC & ACCESSORIES**									
1000	Fabric, 9 ga., galv., 1.2 oz. coat, 2" chain link, 4'	B-80A	304	.079	L.F.	3.64	3.51	2.80	9.95	12.35
1150	5'		285	.084		4.38	3.74	2.98	11.10	13.70
1200	6'		266	.090		9.20	4.01	3.20	16.41	19.65
1250	7'		247	.097		10.20	4.31	3.44	17.95	21.50
1300	8'		228	.105		13.05	4.67	3.73	21.45	25.50
1400	9 ga., fused, 4'		304	.079		4.23	3.51	2.80	10.54	13
1450	5'		285	.084		4.79	3.74	2.98	11.51	14.15
1500	6'		266	.090		4.78	4.01	3.20	11.99	14.75
1550	7'		247	.097		6.05	4.31	3.44	13.80	16.90
1600	8'		228	.105		11	4.67	3.73	19.40	23

32 31 Fences and Gates

32 31 13 – Chain Link Fences and Gates

32 31 13.40 Fence, Fabric and Accessories	Crew	Daily Output	Labor-Hours	Unit	Material	Labor	2021 Bare Costs Equipment	Total	Total Incl O&P
1650 Barbed wire, galv., cost per strand	B-80A	2280	.011	L.F.	.15	.47	.37	.99	1.28
1700 Vinyl coated		2280	.011	↓	.14	.47	.37	.98	1.26
1750 Extension arms, 3 strands		143	.168	Ea.	4.19	7.45	5.95	17.59	22.50
1800 6 strands, 2-3/8"		119	.202		12.45	8.95	7.15	28.55	35
1850 Eye tops, 2-3/8"		143	.168	↓	1.49	7.45	5.95	14.89	19.30
1900 Top rail, incl. tie wires, 1-5/8", galv.		912	.026	L.F.	5.50	1.17	.93	7.60	8.80
1950 Vinyl coated		912	.026		5.35	1.17	.93	7.45	8.65
2100 Rail, middle/bottom, w/tie wire, 1-5/8", galv.		912	.026		4.95	1.17	.93	7.05	8.20
2150 Vinyl coated		912	.026		5.25	1.17	.93	7.35	8.50
2200 Reinforcing wire, coiled spring, 7 ga. galv.		2279	.011		.11	.47	.37	.95	1.23
2250 9 ga., vinyl coated		2282	.011	↓	.60	.47	.37	1.44	1.77
2300 Steel T-post, galvanized with clips, 5', common earth, flat		200	.120	Ea.	10.30	5.35	4.25	19.90	24
2310 Clay		176	.136		10.30	6.05	4.83	21.18	25.50
2320 Soil & rock		144	.167		10.30	7.40	5.90	23.60	29
2330 5-1/2', common earth, flat		200	.120		11.10	5.35	4.25	20.70	25
2340 Clay		176	.136		11.10	6.05	4.83	21.98	26.50
2350 Soil & rock		144	.167		11.10	7.40	5.90	24.40	30
2360 6', common earth, flat		200	.120		11.10	5.35	4.25	20.70	25
2370 Clay		176	.136		11.10	6.05	4.83	21.98	26.50
2375 Soil & rock		144	.167		11.10	7.40	5.90	24.40	30
2385 6-1/2', common earth, flat		200	.120		12.90	5.35	4.25	22.50	27
2390 Clay		176	.136		12.90	6.05	4.83	23.78	28.50
2395 Soil & rock		144	.167		12.90	7.40	5.90	26.20	32
2400 7', common earth, flat		184	.130		12.85	5.80	4.62	23.27	28
2410 Clay		160	.150		12.85	6.65	5.30	24.80	30
2420 Soil & rock		128	.188		12.85	8.35	6.65	27.85	34
2430 8', common earth, flat		184	.130		14.55	5.80	4.62	24.97	30
2440 Clay		160	.150		14.55	6.65	5.30	26.50	32
2450 Soil & rock		128	.188		14.55	8.35	6.65	29.55	35.50
2460 9', common earth, flat		184	.130		17.45	5.80	4.62	27.87	33
2470 Clay		160	.150		17.45	6.65	5.30	29.40	35
2480 Soil & rock		128	.188		17.45	8.35	6.65	32.45	39
2485 10', common earth, flat		168	.143		17.90	6.35	5.05	29.30	34.50
2490 Clay		144	.167		17.90	7.40	5.90	31.20	37.50
2495 Soil & rock		112	.214		17.90	9.50	7.60	35	42.50
2500 11', common earth, flat		168	.143		21	6.35	5.05	32.40	38
2510 Clay		144	.167		21	7.40	5.90	34.30	40.50
2530 Soil & rock		112	.214		21	9.50	7.60	38.10	45.50
2540 12', common earth, flat		168	.143		22	6.35	5.05	33.40	39
2550 Clay		144	.167		22	7.40	5.90	35.30	41.50
2560 Soil & rock		112	.214		22	9.50	7.60	39.10	46.50
2600 Steel T-post, galvanized with clips, 5', common earth, hills		180	.133		10.30	5.90	4.72	20.92	25.50
2610 Clay		160	.150		10.30	6.65	5.30	22.25	27
2620 Soil & rock		130	.185		10.30	8.20	6.55	25.05	31
2630 5-1/2', common earth, hills		180	.133		11.10	5.90	4.72	21.72	26.50
2640 Clay		160	.150		11.10	6.65	5.30	23.05	28
2650 Soil & rock		130	.185		11.10	8.20	6.55	25.85	31.50
2660 6', common earth, hills		180	.133		11.10	5.90	4.72	21.72	26.50
2670 Clay		160	.150		11.10	6.65	5.30	23.05	28
2675 Soil & rock		130	.185		11.10	8.20	6.55	25.85	31.50
2685 6-1/2', common earth, hills		180	.133		12.90	5.90	4.72	23.52	28.50
2690 Clay		160	.150		12.90	6.65	5.30	24.85	30
2695 Soil & rock		130	.185	↓	12.90	8.20	6.55	27.65	33.50

For customer support on your Site Work & Landscape Costs with RSMeans Data, call 800.448.8182.

393

32 31 Fences and Gates

32 31 13 – Chain Link Fences and Gates

32 31 13.40 Fence, Fabric and Accessories		Crew	Daily Output	Labor-Hours	Unit	Material	2021 Bare Costs Labor	Equipment	Total	Total Incl O&P
2700	7', common earth, hills	B-80A	166	.145	Ea.	12.85	6.40	5.10	24.35	29.50
2710	Clay		144	.167		12.85	7.40	5.90	26.15	31.50
2720	Soil & rock		116	.207		12.85	9.20	7.35	29.40	36
2730	8', common earth, hills		166	.145		14.55	6.40	5.10	26.05	31.50
2740	Clay		144	.167		14.55	7.40	5.90	27.85	33.50
2750	Soil & rock		116	.207		14.55	9.20	7.35	31.10	38
2760	9', common earth, hills		166	.145		17.45	6.40	5.10	28.95	34.50
2770	Clay		144	.167		17.45	7.40	5.90	30.75	37
2780	Soil & rock		116	.207		17.45	9.20	7.35	34	41
2785	10', common earth, hills		152	.158		17.90	7	5.60	30.50	36.50
2790	Clay		130	.185		17.90	8.20	6.55	32.65	39
2795	Soil & rock		101	.238		17.90	10.55	8.40	36.85	44.50
2800	11', common earth, hills		152	.158		21	7	5.60	33.60	39.50
2810	Clay		130	.185		21	8.20	6.55	35.75	42.50
2830	Soil & rock		101	.238		21	10.55	8.40	39.95	48
2840	12', common earth, hills		152	.158		22	7	5.60	34.60	40.50
2850	Clay		130	.185		22	8.20	6.55	36.75	43.50
2860	Soil & rock		101	.238		22	10.55	8.40	40.95	49

32 31 13.53 High-Security Chain Link Fences, Gates and Sys.

		Crew	Daily Output	Labor-Hours	Unit	Material	Labor	Equipment	Total	Total Incl O&P
0010	**HIGH-SECURITY CHAIN LINK FENCES, GATES AND SYSTEMS**									
0100	Fence, chain link, security, 7' H, standard FE-7, incl. excavation & posts	B-80C	480	.050	L.F.	51.50	2.29	.53	54.32	60.50
0200	Fence, barbed wire, security, 7' H, with 3 wire barbed wire arm	"	400	.060	"	9	2.75	.63	12.38	14.70
0300	Complete systems, including material and installation									
0310	Taunt wire fence detection system				M.L.F.				25,100	27,600
0410	Microwave fence detection system								41,300	45,400
0510	Passive magnetic fence detection system								19,500	21,400
0610	Infrared fence detection system								12,900	14,400
0710	Strain relief fence detection system								25,100	27,600
0810	Electro-shock fence detection system								35,900	39,500
0910	Photo-electric fence detection system								16,300	18,000

32 31 13.64 Chain Link Terminal Post

		Crew	Daily Output	Labor-Hours	Unit	Material	Labor	Equipment	Total	Total Incl O&P
0010	**CHAIN LINK TERMINAL POST**									
0110	16 ga., steel, 2-1/2" x 6' x 0.065 wall, incl. post cap, excavation	B-80C	80	.300	Ea.	13.40	13.75	3.16	30.31	38.50
0120	2-1/2" x 7'-6" x 0.065 wall		80	.300		17.65	13.75	3.16	34.56	43.50
0130	2-1/2" x 8'-6" x 0.095 wall		80	.300		26.50	13.75	3.16	43.41	53.50
0210	16 ga., steel, 2-1/2" x 6' x 0.065 wall, incl. floor flange	B-80A	80	.300		32.50	13.30	10.65	56.45	67.50
0220	2-1/2" x 8' x 0.065 wall		80	.300		33	13.30	10.65	56.95	67.50
0230	4" x 10' x 0.160 wall		80	.300		120	13.30	10.65	143.95	164
0240	4" x 12' x 0.160 wall		80	.300		133	13.30	10.65	156.95	178
0310	16 ga., steel, 4" x 11' x 0.226 wall, incl. post cap, excavation	B-80C	80	.300		178	13.75	3.16	194.91	220
0320	4" x 13'-6" x 0.226 wall		80	.300		207	13.75	3.16	223.91	252
0330	4" x 21' x 0.226 wall		80	.300		282	13.75	3.16	298.91	335

32 31 13.65 Chain Link Line Post

		Crew	Daily Output	Labor-Hours	Unit	Material	Labor	Equipment	Total	Total Incl O&P
0010	**CHAIN LINK LINE POST**									
0110	16 ga., steel, 1-5/8" x 6' x 0.065 wall, incl. post cap and excavation	B-80C	80	.300	Ea.	10.05	13.75	3.16	26.96	35
0120	1-5/8" x 7'-6" x 0.065 wall		80	.300		13	13.75	3.16	29.91	38.50
0130	2" x 8'-6" x 0.095 wall		80	.300		20	13.75	3.16	36.91	46
0210	16 ga., steel, 2" x 6' x 0.065 wall, incl. post top and floor flange	B-80A	80	.300		33.50	13.30	10.65	57.45	68.50
0220	2" x 8' x 0.065 wall		80	.300		37	13.30	10.65	60.95	72
0230	3" x 10' x 0.160 wall		80	.300		83	13.30	10.65	106.95	123
0240	3" x 12' x 0.160 wall		80	.300		94.50	13.30	10.65	118.45	136
0410	16 ga., steel, 3" x 11' x 0.203 wall, incl. post cap and excavation	B-80C	80	.300		123	13.75	3.16	139.91	159

32 31 Fences and Gates

32 31 13 – Chain Link Fences and Gates

32 31 13.65 Chain Link Line Post		Crew	Daily Output	Labor-Hours	Unit	Material	2021 Bare Costs Labor	Equipment	Total	Total Incl O&P
0420	3" x 13'-6" x 0.203 wall	B-80C	80	.300	Ea.	147	13.75	3.16	163.91	186
0430	3" x 21' x 0.203 wall	▼	80	.300	▼	186	13.75	3.16	202.91	229

32 31 13.66 Chain Link Top Rail										
0010	**CHAIN LINK TOP RAIL**									
0110	Fence, rail tubing, 16 ga., 1-3/8" x 21' x 0.065 wall incl. hardware	B-80A	266	.090	L.F.	1.22	4.01	3.20	8.43	10.85
0120	1-5/8" x 21' x 0.065 wall, swedge		266	.090		1.49	4.01	3.20	8.70	11.15
0130	1-5/8" x 21' x 0.111 wall		266	.090		2.25	4.01	3.20	9.46	12
0140	2" x 18' x 0.145 wall	▼	266	.090	▼	5	4.01	3.20	12.21	15

32 31 13.68 Chain Link Fabric										
0010	**CHAIN LINK FABRIC**									
0110	Fence, fabric, steel, galv., 11-1/2 ga., 2-1/4" mesh, 4' H, incl. hardware	B-80A	266	.090	L.F.	2.67	4.01	3.20	9.88	12.45
0120	5' H		266	.090		3	4.01	3.20	10.21	12.80
0130	6' H		266	.090		3.38	4.01	3.20	10.59	13.25
0210	Fence, fabric, steel, galv., 11 ga., 2" mesh, 6' H, incl. hardware		266	.090		4	4.01	3.20	11.21	13.90
0220	8' H		266	.090		5.25	4.01	3.20	12.46	15.25
0230	10' H		266	.090		9.85	4.01	3.20	17.06	20.50
0240	12' H		266	.090		11.60	4.01	3.20	18.81	22.50
0310	Fence, fabric, steel, galv., 9 ga., 2" mesh, 8' H, incl. hardware		266	.090		5.30	4.01	3.20	12.51	15.30
0320	10' H		266	.090		6.15	4.01	3.20	13.36	16.30
0330	12' H		266	.090		7.20	4.01	3.20	14.41	17.40
0340	14' H		266	.090		7.50	4.01	3.20	14.71	17.75
0410	Fence, fabric, steel, galv., 9 ga., 1" mesh, 8' H, incl. hardware		266	.090		11.50	4.01	3.20	18.71	22
0420	10' H		266	.090		18.45	4.01	3.20	25.66	30
0430	12' H		266	.090		22	4.01	3.20	29.21	33.50
0440	14' H	▼	266	.090	▼	25.50	4.01	3.20	32.71	37.50

32 31 13.70 Chain Link Barbed Wire										
0010	**CHAIN LINK BARBED WIRE**									
0110	Fence, barbed wire, 12-1/2 ga., 1320' roll incl. barb arm and brace band	B-80A	266	.090	L.F.	.86	4.01	3.20	8.07	10.45

32 31 13.80 Residential Chain Link Gate										
0010	**RESIDENTIAL CHAIN LINK GATE**									
0110	Residential 4' gate, single incl. hardware and concrete	B-80C	10	2.400	Ea.	191	110	25.50	326.50	400
0120	5'		10	2.400		201	110	25.50	336.50	415
0130	6'		10	2.400		213	110	25.50	348.50	425
0510	Residential 4' gate, double incl. hardware and concrete		10	2.400		290	110	25.50	425.50	510
0520	5'		10	2.400		305	110	25.50	440.50	525
0530	6'	▼	10	2.400	▼	350	110	25.50	485.50	575

32 31 13.82 Internal Chain Link Gate										
0010	**INTERNAL CHAIN LINK GATE**									
0110	Internal 6' gate, single incl. post flange, hardware and concrete	B-80C	10	2.400	Ea.	340	110	25.50	475.50	565
0120	8'		10	2.400		380	110	25.50	515.50	605
0130	10'		10	2.400		505	110	25.50	640.50	745
0510	Internal 6' gate, double incl. post flange, hardware and concrete		10	2.400		575	110	25.50	710.50	820
0520	8'		10	2.400		645	110	25.50	780.50	900
0530	10'	▼	10	2.400	▼	805	110	25.50	940.50	1,075

32 31 13.84 Industrial Chain Link Gate										
0010	**INDUSTRIAL CHAIN LINK GATE**									
0110	Industrial 8' gate, single incl. hardware and concrete	B-80C	10	2.400	Ea.	525	110	25.50	660.50	765
0120	10'		10	2.400		590	110	25.50	725.50	840
0510	Industrial 8' gate, double incl. hardware and concrete		10	2.400		790	110	25.50	925.50	1,050
0520	10'	▼	10	2.400	▼	895	110	25.50	1,030.50	1,175

For customer support on your Site Work & Landscape Costs with RSMeans Data, call 800.448.8182.

395

32 31 Fences and Gates

32 31 13 – Chain Link Fences and Gates

32 31 13.88 Chain Link Transom

		Crew	Daily Output	Labor-Hours	Unit	Material	2021 Bare Costs Labor	Equipment	Total	Total Incl O&P
0010	**CHAIN LINK TRANSOM**									
0110	Add for, single transom, 3' wide, incl. components & hardware	B-80C	10	2.400	Ea.	119	110	25.50	254.50	325
0120	Add for, double transom, 6' wide, incl. components & hardware	"	10	2.400	"	127	110	25.50	262.50	330

32 31 19 – Decorative Metal Fences and Gates

32 31 19.10 Decorative Fence

		Crew	Daily Output	Labor-Hours	Unit	Material	2021 Bare Costs Labor	Equipment	Total	Total Incl O&P
0010	**DECORATIVE FENCE**									
5300	Tubular picket, steel, 6' sections, 1-9/16" posts, 4' high	B-80C	300	.080	L.F.	38.50	3.67	.84	43.01	49
5400	2" posts, 5' high		240	.100		44	4.59	1.05	49.64	56
5600	2" posts, 6' high		200	.120		53	5.50	1.26	59.76	68
5700	Staggered picket, 1-9/16" posts, 4' high		300	.080		45	3.67	.84	49.51	56
5800	2" posts, 5' high		240	.100		45	4.59	1.05	50.64	57.50
5900	2" posts, 6' high		200	.120		59	5.50	1.26	65.76	74.50
6200	Gates, 4' high, 3' wide	B-1	10	2.400	Ea.	335	108		443	530
6300	5' high, 3' wide		10	2.400		246	108		354	430
6400	6' high, 3' wide		10	2.400		370	108		478	565
6500	4' wide		10	2.400		330	108		438	520

32 31 23 – Plastic Fences and Gates

32 31 23.10 Fence, Vinyl

		Crew	Daily Output	Labor-Hours	Unit	Material	2021 Bare Costs Labor	Equipment	Total	Total Incl O&P
0010	**FENCE, VINYL**									
0011	White, steel reinforced, stainless steel fasteners									
0020	Picket, 4" x 4" posts @ 6'-0" OC, 3' high	B-1	140	.171	L.F.	24	7.75		31.75	38
0030	4' high		130	.185		26.50	8.30		34.80	41.50
0040	5' high		120	.200		37	9		46	54.50
0100	Board (semi-privacy), 5" x 5" posts @ 7'-6" OC, 5' high		130	.185		25.50	8.30		33.80	40.50
0120	6' high		125	.192		28.50	8.65		37.15	44.50
0200	Basket weave, 5" x 5" posts @ 7'-6" OC, 5' high		160	.150		25.50	6.75		32.25	38
0220	6' high		150	.160		28.50	7.20		35.70	42.50
0300	Privacy, 5" x 5" posts @ 7'-6" OC, 5' high		130	.185		26	8.30		34.30	41
0320	6' high		150	.160		24.50	7.20		31.70	38
0350	Gate, 5' high		9	2.667	Ea.	355	120		475	570
0360	6' high		9	2.667		400	120		520	620
0400	For posts set in concrete, add		25	.960		9.45	43.50		52.95	75
0500	Post and rail fence, 2 rail		150	.160	L.F.	6.30	7.20		13.50	17.70
0510	3 rail		150	.160		8.15	7.20		15.35	19.75
0515	4 rail		150	.160		11.25	7.20		18.45	23

32 31 23.20 Fence, Recycled Plastic

			Crew	Daily Output	Labor-Hours	Unit	Material	2021 Bare Costs Labor	Equipment	Total	Total Incl O&P
0010	**FENCE, RECYCLED PLASTIC**										
9015	Fence rail, made from recycled plastic, various colors, 2 rail	G	B-1	150	.160	L.F.	11.35	7.20		18.55	23.50
9018	3 rail	G		150	.160		10.25	7.20		17.45	22
9020	4 rail	G		150	.160		15.85	7.20		23.05	28
9030	Fence pole, made from recycled plastic, various colors, 7'	G		96	.250	Ea.	60.50	11.25		71.75	83.50
9040	Stockade fence, made from recycled plastic, various colors, 4' high	G	B-80C	160	.150	L.F.	32.50	6.90	1.58	40.98	48
9050	6' high	G		160	.150	"	37	6.90	1.58	45.48	52.50
9060	6' pole	G		96	.250	Ea.	43	11.45	2.64	57.09	67
9070	9' pole	G		96	.250	"	42.50	11.45	2.64	56.59	66.50
9080	Picket fence, made from recycled plastic, various colors, 3' high	G		160	.150	L.F.	23.50	6.90	1.58	31.98	38
9090	4' high	G		160	.150	"	35	6.90	1.58	43.48	50.50
9100	3' high gate	G		8	3	Ea.	24.50	138	31.50	194	268
9110	4' high gate	G		8	3		31.50	138	31.50	201	276
9120	5' high pole	G		96	.250		45.50	11.45	2.64	59.59	70.50
9130	6' high pole	G		96	.250		42	11.45	2.64	56.09	66.50

32 31 Fences and Gates

32 31 23 – Plastic Fences and Gates

32 31 23.20 Fence, Recycled Plastic		Crew	Daily Output	Labor- Hours	Unit	Material	2021 Bare Costs Labor	Equipment	Total	Total Incl O&P
9140	Pole cap only	G			Ea.	4.89			4.89	5.40
9150	Keeper pins only	G			↓	.41			.41	.45

32 31 26 – Wire Fences and Gates

32 31 26.10 Fences, Misc. Metal

		Crew	Daily Output	Labor- Hours	Unit	Material	Labor	Equipment	Total	Total Incl O&P
0010	**FENCES, MISC. METAL**									
0012	Chicken wire, posts @ 4', 1" mesh, 4' high	B-80C	410	.059	L.F.	4.32	2.69	.62	7.63	9.45
0100	2" mesh, 6' high		350	.069		3.75	3.15	.72	7.62	9.60
0200	Galv. steel, 12 ga., 2" x 4" mesh, posts 5' OC, 3' high		300	.080		2.79	3.67	.84	7.30	9.50
0300	5' high		300	.080		3.30	3.67	.84	7.81	10.05
0400	14 ga., 1" x 2" mesh, 3' high		300	.080		3.10	3.67	.84	7.61	9.85
0500	5' high	↓	300	.080	↓	4.34	3.67	.84	8.85	11.20
1000	Kennel fencing, 1-1/2" mesh, 6' long, 3'-6" wide, 6'-2" high	2 Clab	4	4	Ea.	505	178		683	820
1050	12' long		4	4		715	178		893	1,050
1200	Top covers, 1-1/2" mesh, 6' long		15	1.067		146	47.50		193.50	232
1250	12' long	↓	12	1.333	↓	219	59		278	330
4492	Security fence, prison grade, barbed wire, set in concrete, 10' high	B-80	22	1.455	L.F.	64.50	71	48	183.50	230
4494	Security fence, prison grade, razor wire, set in concrete, 10' high		18	1.778		65.50	86.50	58.50	210.50	266
4500	Security fence, prison grade, set in concrete, 12' high		25	1.280		56	62.50	42	160.50	201
4600	16' high		20	1.600		87	78	52.50	217.50	270
4990	Security fence, prison grade, set in concrete, 10' high	↓	25	1.280	↓	56	62.50	42	160.50	201

32 31 26.20 Wire Fencing, General

		Crew	Daily Output	Labor- Hours	Unit	Material	Labor	Equipment	Total	Total Incl O&P
0010	**WIRE FENCING, GENERAL**									
0015	Barbed wire, galvanized, domestic steel, hi-tensile 15-1/2 ga.				M.L.F.	155			155	170
0020	Standard, 12-3/4 ga.					169			169	186
0210	Barbless wire, 2-strand galvanized, 12-1/2 ga.				↓	169			169	186
0500	Helical razor ribbon, stainless steel, 18" diam. x 18" spacing				C.L.F.	160			160	176
0600	Hardware cloth galv., 1/4" mesh, 23 ga., 2' wide				C.S.F.	46			46	50.50
0700	3' wide					34.50			34.50	38
0900	1/2" mesh, 19 ga., 2' wide					39.50			39.50	43.50
1000	4' wide					41.50			41.50	45.50
1200	Chain link fabric, steel, 2" mesh, 6 ga., galvanized					58.50			58.50	64
1300	9 ga., galvanized					56.50			56.50	62
1350	Vinyl coated					29			29	31.50
1360	Aluminized					221			221	243
1400	2-1/4" mesh, 11-1/2 ga., galvanized					45			45	49.50
1600	1-3/4" mesh (tennis courts), 11-1/2 ga. (core), vinyl coated					63			63	69.50
1700	9 ga., galvanized				↓	89			89	98
2100	Welded wire fabric, galvanized, 1" x 2", 14 ga.	2 Carp	1600	.010	S.F.	.67	.55		1.22	1.56
2200	2" x 4", 12-1/2 ga.				C.S.F.	44			44	48.50

32 31 29 – Wood Fences and Gates

32 31 29.10 Fence, Wood

		Crew	Daily Output	Labor- Hours	Unit	Material	Labor	Equipment	Total	Total Incl O&P
0010	**FENCE, WOOD**									
0011	Basket weave, 3/8" x 4" boards, 2" x 4"									
0020	stringers on spreaders, 4" x 4" posts									
0050	No. 1 cedar, 6' high	B-80C	160	.150	L.F.	21.50	6.90	1.58	29.98	35.50
0070	Treated pine, 6' high	"	150	.160	"	30	7.35	1.69	39.04	46
0200	Board fence, 1" x 4" boards, 2" x 4" rails, 4" x 4" post									
0220	Preservative treated, 2 rail, 3' high	B-80C	145	.166	L.F.	13.95	7.60	1.74	23.29	28.50
0240	4' high		135	.178		12.25	8.15	1.87	22.27	27.50
0260	3 rail, 5' high		130	.185		11.55	8.45	1.95	21.95	27.50
0300	6' high	↓	125	.192	↓	16.40	8.80	2.02	27.22	33.50

For customer support on your Site Work & Landscape Costs with RSMeans Data, call 800.448.8182.

397

32 31 29.10 Fence, Wood

	32 31 29.10 Fence, Wood	Crew	Daily Output	Labor-Hours	Unit	Material	2021 Bare Costs Labor	Equipment	Total	Total Incl O&P
0320	No. 2 grade western cedar, 2 rail, 3' high	B-80C	145	.166	L.F.	14.15	7.60	1.74	23.49	29
0340	4' high		135	.178		12.30	8.15	1.87	22.32	28
0360	3 rail, 5' high		130	.185		14.55	8.45	1.95	24.95	31
0400	6' high		125	.192		15.35	8.80	2.02	26.17	32.50
0420	No. 1 grade cedar, 2 rail, 3' high		145	.166		13.90	7.60	1.74	23.24	28.50
0440	4' high		135	.178		14.95	8.15	1.87	24.97	30.50
0460	3 rail, 5' high		130	.185		18.25	8.45	1.95	28.65	35
0500	6' high		125	.192		21.50	8.80	2.02	32.32	39.50
0540	Shadow box, 1" x 6" board, 2" x 4" rail, 4" x 4" post									
0560	Pine, pressure treated, 3 rail, 6' high	B-80C	150	.160	L.F.	32.50	7.35	1.69	41.54	49
0600	Gate, 3'-6" wide		8	3	Ea.	128	138	31.50	297.50	380
0620	No. 1 cedar, 3 rail, 4' high		130	.185	L.F.	19.15	8.45	1.95	29.55	36
0640	6' high		125	.192		26.50	8.80	2.02	37.32	44.50
0860	Open rail fence, split rails, 2 rail, 3' high, no. 1 cedar		160	.150		9.35	6.90	1.58	17.83	22.50
0870	No. 2 cedar		160	.150		9.75	6.90	1.58	18.23	22.50
0880	3 rail, 4' high, no. 1 cedar		150	.160		13.45	7.35	1.69	22.49	27.50
0890	No. 2 cedar		150	.160		9.10	7.35	1.69	18.14	23
0920	Rustic rails, 2 rail, 3' high, no. 1 cedar		160	.150		12.30	6.90	1.58	20.78	25.50
0930	No. 2 cedar		160	.150		13.95	6.90	1.58	22.43	27.50
0940	3 rail, 4' high		150	.160		12.95	7.35	1.69	21.99	27
0950	No. 2 cedar		150	.160		7.85	7.35	1.69	16.89	21.50
0960	Picket fence, gothic, pressure treated pine									
1000	2 rail, 3' high	B-80C	140	.171	L.F.	10.45	7.85	1.81	20.11	25
1020	3 rail, 4' high		130	.185	"	12.55	8.45	1.95	22.95	28.50
1040	Gate, 3'-6" wide		9	2.667	Ea.	75	122	28	225	297
1060	No. 2 cedar, 2 rail, 3' high		140	.171	L.F.	12	7.85	1.81	21.66	27
1100	3 rail, 4' high		130	.185	"	10.40	8.45	1.95	20.80	26
1120	Gate, 3'-6" wide		9	2.667	Ea.	82.50	122	28	232.50	305
1140	No. 1 cedar, 2 rail, 3' high		140	.171	L.F.	14.45	7.85	1.81	24.11	29.50
1160	3 rail, 4' high		130	.185	"	20.50	8.45	1.95	30.90	37.50
1170	Gate, 3'-6" wide		9	2.667	Ea.	272	122	28	422	515
1200	Rustic picket, molded pine, 2 rail, 3' high		140	.171	L.F.	11.10	7.85	1.81	20.76	26
1220	No. 1 cedar, 2 rail, 3' high		140	.171		11.95	7.85	1.81	21.61	27
1240	Stockade fence, no. 1 cedar, 3-1/4" rails, 6' high		160	.150		14.05	6.90	1.58	22.53	27.50
1260	8' high		155	.155		20.50	7.10	1.63	29.23	35
1270	Gate, 3'-6" wide		9	2.667	Ea.	277	122	28	427	520
1300	No. 2 cedar, treated wood rails, 6' high		160	.150	L.F.	14.75	6.90	1.58	23.23	28
1320	Gate, 3'-6" wide		8	3	Ea.	95.50	138	31.50	265	345
1360	Treated pine, treated rails, 6' high		160	.150	L.F.	14.85	6.90	1.58	23.33	28.50
1400	8' high		150	.160	"	20.50	7.35	1.69	29.54	35.50
1420	Gate, 3'-6" wide		9	2.667	Ea.	103	122	28	253	325

32 31 29.20 Fence, Wood Rail

	32 31 29.20 Fence, Wood Rail	Crew	Daily Output	Labor-Hours	Unit	Material	2021 Bare Costs Labor	Equipment	Total	Total Incl O&P
0010	**FENCE, WOOD RAIL**									
0012	Picket, No. 2 cedar, Gothic, 2 rail, 3' high	B-1	160	.150	L.F.	8.60	6.75		15.35	19.55
0050	Gate, 3'-6" wide	B-80C	9	2.667	Ea.	95	122	28	245	320
0400	3 rail, 4' high		150	.160	L.F.	9.65	7.35	1.69	18.69	23.50
0500	Gate, 3'-6" wide		9	2.667	Ea.	85.50	122	28	235.50	310
5000	Fence rail, redwood, 2" x 4", merch. grade, 8'	B-1	2400	.010	L.F.	2.96	.45		3.41	3.93
5050	Select grade, 8'		2400	.010	"	6.80	.45		7.25	8.15
6000	Fence post, select redwood, earth packed & treated, 4" x 4" x 6'		96	.250	Ea.	13.75	11.25		25	32
6010	4" x 4" x 8'		96	.250		26.50	11.25		37.75	46
6020	Set in concrete, 4" x 4" x 6'		50	.480		27	21.50		48.50	62

32 31 Fences and Gates

32 31 29 – Wood Fences and Gates

32 31 29.20 Fence, Wood Rail	Crew	Daily Output	Labor-Hours	Unit	Material	2021 Bare Costs Labor	Equipment	Total	Total Incl O&P
6030	4" x 4" x 8'	B-1	50	.480	Ea.	25	21.50		46.50	60
6040	Wood post, 4' high, set in concrete, incl. concrete		50	.480		15	21.50		36.50	49
6050	Earth packed		96	.250		16.80	11.25		28.05	35.50
6060	6' high, set in concrete, incl. concrete		50	.480		18.85	21.50		40.35	53.50
6070	Earth packed		96	.250		12.20	11.25		23.45	30

32 32 Retaining Walls

32 32 13 – Cast-in-Place Concrete Retaining Walls

32 32 13.10 Retaining Walls, Cast Concrete

		Crew	Daily Output	Labor-Hours	Unit	Material	2021 Bare Costs Labor	Equipment	Total	Total Incl O&P
0010	**RETAINING WALLS, CAST CONCRETE**									
1800	Concrete gravity wall with vertical face including excavation & backfill									
1850	No reinforcing									
1900	6' high, level embankment	C-17C	36	2.306	L.F.	84	133	15.20	232.20	310
2000	33° slope embankment		32	2.594		106	150	17.10	273.10	360
2200	8' high, no surcharge		27	3.074		113	177	20.50	310.50	415
2300	33° slope embankment		24	3.458		137	199	23	359	475
2500	10' high, level embankment		19	4.368		162	252	29	443	590
2600	33° slope embankment		18	4.611		224	266	30.50	520.50	680
2800	Reinforced concrete cantilever, incl. excavation, backfill & reinf.									
2900	6' high, 33° slope embankment	C-17C	35	2.371	L.F.	83	137	15.65	235.65	315
3000	8' high, 33° slope embankment		29	2.862		96	165	18.85	279.85	375
3100	10' high, 33° slope embankment		20	4.150		124	239	27.50	390.50	525
3200	20' high, 500 lb./L.F. surcharge		7.50	11.067		375	640	73	1,088	1,450
3500	Concrete cribbing, incl. excavation and backfill									
3700	12' high, open face	B-13	210	.267	S.F.	41.50	12.85	2.79	57.14	68
3900	Closed face	"	210	.267	"	39	12.85	2.79	54.64	65.50
4100	Concrete filled slurry trench, see Section 31 56 23.20									

32 32 23 – Segmental Retaining Walls

32 32 23.13 Segmental Conc. Unit Masonry Retaining Walls

		Crew	Daily Output	Labor-Hours	Unit	Material	2021 Bare Costs Labor	Equipment	Total	Total Incl O&P
0010	**SEGMENTAL CONC. UNIT MASONRY RETAINING WALLS**									
7100	Segmental retaining wall system, incl. pins and void fill									
7120	base and backfill not included									
7140	Large unit, 8" high x 18" wide x 20" deep, 3 plane split	B-62	300	.080	S.F.	15.15	3.85	.60	19.60	23
7150	Straight split		300	.080		15.10	3.85	.60	19.55	23
7160	Medium, lt. wt., 8" high x 18" wide x 12" deep, 3 plane split		400	.060		7.80	2.89	.45	11.14	13.40
7170	Straight split		400	.060		10.85	2.89	.45	14.19	16.70
7180	Small unit, 4" x 18" x 10" deep, 3 plane split		400	.060		16.90	2.89	.45	20.24	23.50
7190	Straight split		400	.060		13.05	2.89	.45	16.39	19.15
7200	Cap unit, 3 plane split		300	.080		15.30	3.85	.60	19.75	23.50
7210	Cap unit, straight split		300	.080		15.30	3.85	.60	19.75	23.50
7250	Geo-grid soil reinforcement 4' x 50'	2 Clab	22500	.001		.80	.03		.83	.93
7255	Geo-grid soil reinforcement 6' x 150'	"	22500	.001		.62	.03		.65	.73

32 32 26 – Metal Crib Retaining Walls

32 32 26.10 Metal Bin Retaining Walls

		Crew	Daily Output	Labor-Hours	Unit	Material	2021 Bare Costs Labor	Equipment	Total	Total Incl O&P
0010	**METAL BIN RETAINING WALLS**									
0011	Aluminized steel bin, excavation									
0020	and backfill not included, 10' wide									
0100	4' high, 5.5' deep	B-13	650	.086	S.F.	28.50	4.16	.90	33.56	38.50
0200	8' high, 5.5' deep		615	.091		33	4.40	.95	38.35	43.50
0300	10' high, 7.7' deep		580	.097		36.50	4.66	1.01	42.17	48.50

For customer support on your Site Work & Landscape Costs with RSMeans Data, call 800.448.8182.

399

32 32 Retaining Walls

32 32 26 – Metal Crib Retaining Walls

32 32 26.10 Metal Bin Retaining Walls	Crew	Daily Output	Labor-Hours	Unit	Material	2021 Bare Costs Labor	Equipment	Total	Total Incl O&P
0400	12' high, 7.7' deep	B-13	530	.106	S.F.	39.50	5.10	1.11	45.71	52.50
0500	16' high, 7.7' deep		515	.109		42	5.25	1.14	48.39	55
0600	16' high, 9.9' deep		500	.112		46.50	5.40	1.17	53.07	61
0700	20' high, 9.9' deep		470	.119		52.50	5.75	1.25	59.50	67.50
0800	20' high, 12.1' deep		460	.122		45.50	5.90	1.28	52.68	60
0900	24' high, 12.1' deep		455	.123		48.50	5.95	1.29	55.74	64
1000	24' high, 14.3' deep		450	.124		68.50	6	1.30	75.80	86
1100	28' high, 14.3' deep	↓	440	.127	↓	71.50	6.15	1.33	78.98	89
1300	For plain galvanized bin type walls, deduct					10%				

32 32 29 – Timber Retaining Walls

32 32 29.10 Landscape Timber Retaining Walls	Crew	Daily Output	Labor-Hours	Unit	Material	Labor	Equipment	Total	Total Incl O&P
0010	**LANDSCAPE TIMBER RETAINING WALLS**									
0100	Treated timbers, 6" x 6"	1 Clab	265	.030	L.F.	2.84	1.34		4.18	5.10
0110	6" x 8'	"	200	.040	"	7.40	1.78		9.18	10.80
0120	Drilling holes in timbers for fastening, 1/2"	1 Carp	450	.018	Inch		.97		.97	1.45
0130	5/8"	"	450	.018	"		.97		.97	1.45
0140	Reinforcing rods for fastening, 1/2"	1 Clab	312	.026	L.F.	.46	1.14		1.60	2.20
0150	5/8"	"	312	.026	"	.71	1.14		1.85	2.48
0160	Reinforcing fabric	2 Clab	2500	.006	S.Y.	2.25	.28		2.53	2.90
0170	Gravel backfill		28	.571	C.Y.	26.50	25.50		52	67.50
0180	Perforated pipe, 4" diameter with silt sock	↓	1200	.013	L.F.	1.26	.59		1.85	2.27
0190	Galvanized 60d common nails	1 Clab	625	.013	Ea.	.15	.57		.72	1.01
0200	20d common nails	"	3800	.002	"	.03	.09		.12	.18

32 32 36 – Gabion Retaining Walls

32 32 36.10 Stone Gabion Retaining Walls	Crew	Daily Output	Labor-Hours	Unit	Material	Labor	Equipment	Total	Total Incl O&P
0010	**STONE GABION RETAINING WALLS**									
4300	Stone filled gabions, not incl. excavation,									
4310	Stone, delivered, 3' wide									
4340	Galvanized, 6' long, 1' high	B-13	113	.496	Ea.	90.50	24	5.20	119.70	141
4400	1'-6" high		50	1.120		106	54	11.75	171.75	210
4490	3'-0" high		13	4.308		167	208	45	420	545
4590	9' long, 1' high		50	1.120		178	54	11.75	243.75	288
4650	1'-6" high		22	2.545		143	123	26.50	292.50	370
4690	3'-0" high		6	9.333		330	450	98	878	1,150
4890	12' long, 1' high		28	2		166	96.50	21	283.50	350
4950	1'-6" high		13	4.308		193	208	45	446	570
4990	3'-0" high		3	18.667		385	900	196	1,481	2,000
5200	PVC coated, 6' long, 1' high		113	.496		97	24	5.20	126.20	148
5250	1'-6" high		50	1.120		112	54	11.75	177.75	217
5300	3' high		13	4.308		151	208	45	404	525
5500	9' long, 1' high		50	1.120		117	54	11.75	182.75	221
5550	1'-6" high		22	2.545		136	123	26.50	285.50	360
5600	3' high		6	9.333		259	450	98	807	1,075
5800	12' long, 1' high		28	2		186	96.50	21	303.50	370
5850	1'-6" high		13	4.308		226	208	45	479	610
5900	3' high		3	18.667	↓	288	900	196	1,384	1,875
6000	Galvanized, 6' long, 1' high		75	.747	C.Y.	77	36	7.85	120.85	148
6010	1'-6" high		50	1.120		106	54	11.75	171.75	210
6020	3'-0" high		25	2.240		83.50	108	23.50	215	279
6030	9' long, 1' high		50	1.120		178	54	11.75	243.75	288
6040	1'-6" high		33.30	1.682		95	81	17.60	193.60	245
6050	3'-0" high	↓	16.70	3.353	↓	109	162	35	306	400

32 32 Retaining Walls

32 32 36 – Gabion Retaining Walls

32 32 36.10 Stone Gabion Retaining Walls

32 32 36.10 Stone Gabion Retaining Walls	Crew	Daily Output	Labor-Hours	Unit	Material	2021 Bare Costs Labor	Equipment	Total	Total Incl O&P
6060	12' long, 1' high	B-13	37.50	1.493	C.Y.	124	72	15.65	211.65	262
6070	1'-6" high		25	2.240		96.50	108	23.50	228	293
6080	3'-0" high		12.50	4.480		96.50	216	47	359.50	485
6100	PVC coated, 6' long, 1' high		75	.747		145	36	7.85	188.85	223
6110	1'-6" high		50	1.120		112	54	11.75	177.75	217
6120	3' high		25	2.240		75.50	108	23.50	207	270
6130	9' long, 1' high		50	1.120		117	54	11.75	182.75	221
6140	1'-6" high		33.30	1.682		90.50	81	17.60	189.10	240
6150	3' high		16.67	3.359		86.50	162	35	283.50	375
6160	12' long, 1' high		37.50	1.493		139	72	15.65	226.65	278
6170	1'-6" high		25	2.240		100	108	23.50	231.50	297
6180	3' high		12.50	4.480		72	216	47	335	455

32 32 53 – Stone Retaining Walls

32 32 53.10 Retaining Walls, Stone

32 32 53.10 Retaining Walls, Stone	Crew	Daily Output	Labor-Hours	Unit	Material	2021 Bare Costs Labor	Equipment	Total	Total Incl O&P
0010	**RETAINING WALLS, STONE**									
0015	Including excavation, concrete footing and									
0020	stone 3' below grade. Price is exposed face area.									
0200	Decorative random stone, to 6' high, 1'-6" thick, dry set	D-1	35	.457	S.F.	71	22.50		93.50	112
0300	Mortar set		40	.400		73	19.50		92.50	110
0500	Cut stone, to 6' high, 1'-6" thick, dry set		35	.457		73.50	22.50		96	115
0600	Mortar set		40	.400		74	19.50		93.50	111
0800	Random stone, 6' to 10' high, 2' thick, dry set		45	.356		75.50	17.30		92.80	109
0900	Mortar set		50	.320		90	15.60		105.60	122
1100	Cut stone, 6' to 10' high, 2' thick, dry set		45	.356		75.50	17.30		92.80	109
1200	Mortar set		50	.320		90	15.60		105.60	123
5100	Setting stone, dry		100	.160	C.F.		7.80		7.80	11.75
5600	With mortar		120	.133	"		6.50		6.50	9.80

32 33 Site Furnishings

32 33 33 – Site Manufactured Planters

32 33 33.10 Planters

32 33 33.10 Planters	Crew	Daily Output	Labor-Hours	Unit	Material	2021 Bare Costs Labor	Equipment	Total	Total Incl O&P
0010	**PLANTERS**									
0012	Concrete, sandblasted, precast, 48" diameter, 24" high	2 Clab	15	1.067	Ea.	665	47.50		712.50	800
0100	Fluted, precast, 7' diameter, 36" high		10	1.600		1,700	71		1,771	1,950
0300	Fiberglass, circular, 36" diameter, 24" high		15	1.067		755	47.50		802.50	900
0400	60" diameter, 24" high		10	1.600		1,200	71		1,271	1,425
0600	Square, 24" side, 36" high		15	1.067		480	47.50		527.50	600
0700	48" side, 36" high		15	1.067		790	47.50		837.50	940
0900	Planter/bench, 72" square, 36" high		5	3.200		3,175	142		3,317	3,700
1000	96" square, 27" high		5	3.200		3,625	142		3,767	4,175
1200	Wood, square, 48" side, 24" high		15	1.067		1,525	47.50		1,572.50	1,750
1300	Circular, 48" diameter, 30" high		10	1.600		1,125	71		1,196	1,325
1500	72" diameter, 30" high		10	1.600		1,975	71		2,046	2,250
1600	Planter/bench, 72"		5	3.200		3,900	142		4,042	4,475

32 33 43 – Site Seating and Tables

32 33 43.13 Site Seating

32 33 43.13 Site Seating	Crew	Daily Output	Labor-Hours	Unit	Material	2021 Bare Costs Labor	Equipment	Total	Total Incl O&P
0010	**SITE SEATING**									
0012	Seating, benches, park, precast conc., w/backs, wood rails, 4' long	2 Clab	5	3.200	Ea.	670	142		812	950
0100	8' long		4	4		1,175	178		1,353	1,550
0300	Fiberglass, without back, one piece, 4' long		10	1.600		1,700	71		1,771	1,975

32 33 43.13 Site Seating		Crew	Daily Output	Labor-Hours	Unit	Material	2021 Bare Costs Labor	Equipment	Total	Total Incl O&P
0400	8' long	2 Clab	7	2.286	Ea.	365	101		466	555
0500	Steel barstock pedestals w/backs, 2" x 3" wood rails, 4' long		10	1.600		1,325	71		1,396	1,550
0510	8' long		7	2.286		1,850	101		1,951	2,175
0520	3" x 8" wood plank, 4' long		10	1.600		1,500	71		1,571	1,750
0530	8' long		7	2.286		1,425	101		1,526	1,725
0540	Backless, 4" x 4" wood plank, 4' square		10	1.600		1,275	71		1,346	1,500
0550	8' long		7	2.286		1,250	101		1,351	1,525
0600	Aluminum pedestals, with backs, aluminum slats, 8' long		8	2		505	89		594	695
0610	15' long		5	3.200		1,025	142		1,167	1,325
0620	Portable, aluminum slats, 8' long		8	2		450	89		539	630
0630	15' long		5	3.200		655	142		797	930
0800	Cast iron pedestals, back & arms, wood slats, 4' long		8	2		465	89		554	645
0820	8' long		5	3.200		1,050	142		1,192	1,350
0840	Backless, wood slats, 4' long		8	2		760	89		849	970
0860	8' long		5	3.200		1,300	142		1,442	1,650
1700	Steel frame, fir seat, 10' long		10	1.600		405	71		476	550

32 34 10.10 Bridges, Highway		Crew	Daily Output	Labor-Hours	Unit	Material	2021 Bare Costs Labor	Equipment	Total	Total Incl O&P
0010	**BRIDGES, HIGHWAY**									
0020	Structural steel, rolled beams	E-5	8.50	9.412	Ton	2,650	565	219	3,434	4,000
0500	Built up, plate girders	E-6	10.50	12.190	"	3,325	730	206	4,261	5,000
1000	Concrete in place, no reinforcing, abutment footings	C-17B	30	2.733	C.Y.	273	157	12.95	442.95	550
1100	Walls, stems and wing walls		20	4.100		370	236	19.40	625.40	780
1200	Decks		17	4.824		555	278	23	856	1,050
1300	Sidewalks and parapets		11	7.455		995	430	35.50	1,460.50	1,775
2000	Reinforcing, in place	4 Rodm	3	10.667	Ton	2,925	630		3,555	4,175
2050	Galvanized coated		3	10.667		3,875	630		4,505	5,200
2100	Epoxy coated		3	10.667		3,675	630		4,305	4,975
3000	Expansion dams, steel, double upset 4" x 8" angles welded to									
3010	double 8" x 8" angles, 1-3/4" compression seal	C-22	30	1.400	L.F.	610	83	2.45	695.45	795
3040	Double 8" x 8" angles only, 1-3/4" compression seal		35	1.200		460	71	2.10	533.10	615
3050	Galvanized		35	1.200		565	71	2.10	638.10	730
3060	Double 8" x 6" angles only, 1-3/4" compression seal		35	1.200		350	71	2.10	423.10	495
3100	Double 10" channels, 1-3/4" compression seal		35	1.200		320	71	2.10	393.10	460
3420	For 3" compression seal, add					59			59	65
3440	For double slotted extrusions with seal strip, add					290			290	320
3490	For galvanizing, add				Lb.	.45			.45	.50
4000	Approach railings, steel, galv. pipe, 2 line	C-22	140	.300	L.F.	238	17.75	.53	256.28	289
4200	Bridge railings, steel, galv. pipe, 3 line w/screen		85	.494		294	29.50	.86	324.36	370
4220	4 line w/screen		75	.560		360	33	.98	393.98	445
4300	Aluminum, pipe, 3 line w/screen		95	.442		276	26	.77	302.77	345
8000	For structural excavation, see Section 31 23 16.16									
8010	For dewatering, see Section 31 23 19.20									

32 34 Fabricated Bridges

32 34 13 – Fabricated Pedestrian Bridges

32 34 13.10 Bridges, Pedestrian

		Crew	Daily Output	Labor-Hours	Unit	Material	2021 Bare Costs Labor	Equipment	Total	Total Incl O&P
0010	**BRIDGES, PEDESTRIAN**									
0011	Spans over streams, roadways, etc.									
0020	including erection, not including foundations									
0050	Precast concrete, complete in place, 8' wide, 60' span	E-2	215	.260	S.F.	146	15.55	7.95	169.50	193
0100	100' span		185	.303		159	18.05	9.25	186.30	213
0150	120' span		160	.350		173	21	10.70	204.70	235
0200	150' span		145	.386		180	23	11.80	214.80	247
0300	Steel, trussed or arch spans, compl. in place, 8' wide, 40' span		320	.175		129	10.45	5.35	144.80	164
0400	50' span		395	.142		116	8.45	4.34	128.79	145
0500	60' span		465	.120		116	7.20	3.68	126.88	142
0600	80' span		570	.098		138	5.85	3.01	146.86	164
0700	100' span		465	.120		193	7.20	3.68	203.88	228
0800	120' span		365	.153		245	9.15	4.69	258.84	288
0900	150' span		310	.181		260	10.75	5.55	276.30	310
1000	160' span		255	.220		260	13.10	6.70	279.80	315
1100	10' wide, 80' span		640	.088		129	5.20	2.68	136.88	153
1200	120' span		415	.135		167	8.05	4.13	179.18	201
1300	150' span		445	.126		188	7.50	3.85	199.35	222
1400	200' span		205	.273		200	16.30	8.35	224.65	254
1600	Wood, laminated type, complete in place, 80' span	C-12	203	.236		104	12.85	2.34	119.19	137
1700	130' span	"	153	.314		109	17.10	3.11	129.21	149

32 35 Screening Devices

32 35 16 – Sound Barriers

32 35 16.10 Traffic Barriers, Highway Sound Barriers

		Crew	Daily Output	Labor-Hours	Unit	Material	2021 Bare Costs Labor	Equipment	Total	Total Incl O&P
0010	**TRAFFIC BARRIERS, HIGHWAY SOUND BARRIERS**									
0020	Highway sound barriers, not including footing									
0100	Precast concrete, concrete columns @ 30' OC, 8" T, 8' H	C-12	400	.120	L.F.	175	6.55	1.19	182.74	204
0110	12' H		265	.181		263	9.85	1.80	274.65	305
0120	16' H		200	.240		350	13.05	2.38	365.43	405
0130	20' H		160	.300		440	16.35	2.97	459.32	510
0400	Lt. wt. composite panel, cementitious face, st. posts @ 12' OC, 8' H	B-80B	190	.168		182	7.95	1.26	191.21	213
0410	12' H		125	.256		272	12.10	1.91	286.01	320
0420	16' H		95	.337		365	15.90	2.51	383.41	425
0430	20' H		75	.427		455	20	3.18	478.18	535

32 84 Planting Irrigation

32 84 13 – Drip Irrigation

32 84 13.10 Subsurface Drip Irrigation

			Crew	Daily Output	Labor-Hours	Unit	Material	2021 Bare Costs Labor	Equipment	Total	Total Incl O&P
0010	**SUBSURFACE DRIP IRRIGATION**										
0011	Looped grid, pressure compensating										
0100	Preinserted PE emitter, line, hand bury, irregular area, small	G	3 Skwk	1200	.020	L.F.	.26	1.14		1.40	2.01
0150	Medium	G		1800	.013		.26	.76		1.02	1.44
0200	Large	G		2520	.010		.26	.54		.80	1.11
0250	Rectangular area, small	G		2040	.012		.26	.67		.93	1.30
0300	Medium	G		2640	.009		.26	.52		.78	1.07
0350	Large	G		3600	.007		.26	.38		.64	.86
0400	Install in trench, irregular area, small	G		4050	.006		.26	.34		.60	.80
0450	Medium	G		7488	.003		.26	.18		.44	.57

For customer support on your Site Work & Landscape Costs with RSMeans Data, call 800.448.8182.

403

32 84 Planting Irrigation

32 84 13 – Drip Irrigation

32 84 13.10 Subsurface Drip Irrigation		Crew	Daily Output	Labor-Hours	Unit	Material	2021 Bare Costs Labor	Equipment	Total	Total Incl O&P
0500	Large	G	3 Skwk 16560	.001	L.F.	.26	.08		.34	.41
0550	Rectangular area, small	G	8100	.003		.26	.17		.43	.54
0600	Medium		21960	.001		.26	.06		.32	.38
0650	Large	G	33264	.001		.26	.04		.30	.35
0700	Trenching and backfill	G	B-53 500	.016			.89	.32	1.21	1.67
0800	Vinyl tubing, 1/4", material only	G				.08			.08	.09
0850	Supply tubing, 1/2", material only, 100' coil	G				.12			.12	.13
0900	500' coil	G				.11			.11	.12
0950	Compression fittings	G	1 Skwk 90	.089	Ea.	1.57	5.10		6.67	9.40
1000	Barbed fittings, 1/4"	G	360	.022		.13	1.27		1.40	2.05
1100	Flush risers	G	60	.133		4.21	7.60		11.81	16.10
1150	Flush ends, figure eight	G	180	.044		.61	2.54		3.15	4.49
1300	Auto flush, spring loaded	G	90	.089		2.28	5.10		7.38	10.15
1350	Volumetric	G	90	.089		8.05	5.10		13.15	16.50
1400	Air relief valve, inline with compensation tee, 1/2"	G	45	.178		16.70	10.15		26.85	33.50
1450	1"	G	30	.267		18.55	15.25		33.80	43.50
1500	Round box for flush ends, 6"	G	30	.267		7.95	15.25		23.20	32
1600	Screen filter, 3/4" screen	G	12	.667		9.10	38		47.10	67.50
1650	1" disk	G	8	1		28	57		85	117
1700	1-1/2" disk	G	4	2		112	114		226	295
1750	2" disk	G	3	2.667		151	152		303	395
1800	Typical installation 18" OC, small, minimum				S.F.				1.53	2.35
1850	Maximum								1.78	2.75
1900	Large, minimum								1.48	2.28
2000	Maximum								1.71	2.64
2100	For non-pressure compensating systems, deduct								10%	10%
2150	For supply piping, see Section 33 11 13.45									
2680	Install PVC or hose bib to PE drip adapter fittings		1 Skwk 40	.200	Ea.	1.70	11.40		13.10	19.05
2690	Install three outlet manual manifold		4	2		50	114		164	227
2700	Install four zone control module		4	2		50	114		164	227
2710	Install six zone control module		4	2		106	114		220	288

32 84 23 – Underground Sprinklers

32 84 23.10 Sprinkler Irrigation System		Crew	Daily Output	Labor-Hours	Unit	Material	2021 Bare Costs Labor	Equipment	Total	Total Incl O&P
0010	**SPRINKLER IRRIGATION SYSTEM**									
0011	For lawns									
0800	Residential system, custom, 1" supply		B-20 2000	.012	S.F.	.28	.59		.87	1.20
0900	1-1/2" supply		" 1800	.013	"	.42	.66		1.08	1.44
1020	Pop up spray head w/risers, hi-pop, full circle pattern, 4"		2 Skwk 76	.211	Ea.	4.24	12		16.24	23
1030	1/2 circle pattern, 4"		76	.211		7.10	12		19.10	26
1040	6", full circle pattern		76	.211		20.50	12		32.50	40.50
1050	1/2 circle pattern, 6"		76	.211		10.90	12		22.90	30
1060	12", full circle pattern		76	.211		15.65	12		27.65	35.50
1070	1/2 circle pattern, 12"		76	.211		17.50	12		29.50	37.50
1080	Pop up bubbler head w/risers, hi-pop bubbler head, 4"		76	.211		5.05	12		17.05	23.50
1110	Impact full/part circle sprinklers, 28'-54' @ 25-60 psi		37	.432		23	24.50		47.50	62.50
1120	Spaced 37'-49' @ 25-50 psi		37	.432		19.40	24.50		43.90	58.50
1130	Spaced 43'-61' @ 30-60 psi		37	.432		65	24.50		89.50	109
1140	Spaced 54'-78' @ 40-80 psi		37	.432		110	24.50		134.50	158
1145	Impact rotor pop-up full/part commercial circle sprinklers									
1150	Spaced 42'-65' @ 35-80 psi		2 Skwk 25	.640	Ea.	20.50	36.50		57	77.50
1160	Spaced 48'-76' @ 45-85 psi		" 25	.640	"	19.20	36.50		55.70	76
1165	Impact rotor pop-up part. circle comm., 53'-75', 55-100 psi, w/accessories									

For customer support on your Site Work & Landscape Costs with RSMeans Data, call 800.448.8182.

32 84 Planting Irrigation

32 84 23 – Underground Sprinklers

32 84 23.10 Sprinkler Irrigation System	Crew	Daily Output	Labor-Hours	Unit	Material	2021 Bare Costs Labor	Equipment	Total	Total Incl O&P
1180 Sprinkler, premium, pop-up rotator, 50'-100'	2 Skwk	25	.640	Ea.	96.50	36.50		133	161
1250 Plastic case, 2 nozzle, metal cover		25	.640		117	36.50		153.50	184
1260 Rubber cover		25	.640		102	36.50		138.50	167
1270 Iron case, 2 nozzle, metal cover		22	.727		147	41.50		188.50	225
1280 Rubber cover		22	.727		146	41.50		187.50	224
1282 Impact rotor pop-up full circle commercial, 39'-99', 30-100 psi									
1284 Plastic case, metal cover	2 Skwk	25	.640	Ea.	93.50	36.50		130	158
1286 Rubber cover		25	.640		136	36.50		172.50	205
1288 Iron case, metal cover		22	.727		180	41.50		221.50	262
1290 Rubber cover		22	.727		183	41.50		224.50	265
1292 Plastic case, 2 nozzle, metal cover		22	.727		135	41.50		176.50	211
1294 Rubber cover		22	.727		138	41.50		179.50	214
1296 Iron case, 2 nozzle, metal cover		20	.800		158	45.50		203.50	243
1298 Rubber cover		20	.800		187	45.50		232.50	275
1305 Electric remote control valve, plastic, 3/4"		18	.889		20.50	51		71.50	99
1310 1"		18	.889		23.50	51		74.50	103
1320 1-1/2"		18	.889		98.50	51		149.50	185
1330 2"		18	.889		117	51		168	205
1335 Quick coupling valves, brass, locking cover									
1340 Inlet coupling valve, 3/4"	2 Skwk	18.75	.853	Ea.	23.50	48.50		72	99.50
1350 1"		18.75	.853		28	48.50		76.50	105
1360 Controller valve boxes, 6" round boxes		18.75	.853		5.30	48.50		53.80	79.50
1370 10" round boxes		14.25	1.123		11	64		75	109
1380 12" square box		9.75	1.641		42	93.50		135.50	188
1388 Electromech. control, 14 day 3-60 min., auto start to 23/day									
1390 4 station	2 Skwk	1.04	15.385	Ea.	68.50	880		948.50	1,400
1400 7 station		.64	25		178	1,425		1,603	2,350
1410 12 station		.40	40		202	2,275		2,477	3,650
1420 Dual programs, 18 station		.24	66.667		267	3,800		4,067	6,025
1430 23 station		.16	100		310	5,700		6,010	8,950
1435 Backflow preventer, bronze, 0-175 psi, w/valves, test cocks									
1440 3/4"	2 Skwk	6	2.667	Ea.	90	152		242	330
1450 1"		6	2.667		103	152		255	340
1460 1-1/2"		6	2.667		282	152		434	540
1470 2"		6	2.667		395	152		547	665
1475 Pressure vacuum breaker, brass, 15-150 psi									
1480 3/4"	2 Skwk	6	2.667	Ea.	49.50	152		201.50	284
1490 1"		6	2.667		57.50	152		209.50	292
1500 1-1/2"		6	2.667		85	152		237	325
1510 2"		6	2.667		159	152		311	405

32 91 13.16 Mulching

		Crew	Daily Output	Labor-Hours	Unit	Material	2021 Bare Costs Labor	Equipment	Total	Total Incl O&P
0010	**MULCHING**									
0100	Aged barks, 3" deep, hand spread	1 Clab	100	.080	S.Y.	4.01	3.55		7.56	9.70
0150	Skid steer loader	B-63	13.50	2.963	M.S.F.	445	138	13.30	596.30	710
0160	Skid steer loader	"	1500	.027	S.Y.	4.01	1.24	.12	5.37	6.40
0200	Hay, 1" deep, hand spread	1 Clab	475	.017	"	.40	.75		1.15	1.56
0250	Power mulcher, small	B-64	180	.089	M.S.F.	44.50	4.14	2.22	50.86	57.50
0350	Large	B-65	530	.030	"	44.50	1.41	1.03	46.94	52
0370	Fiber mulch recycled newsprint hand spread	1 Clab	500	.016	S.Y.	.25	.71		.96	1.34
0380	Power mulcher small	B-64	200	.080	M.S.F.	28	3.73	2	33.73	38.50
0390	Power mulcher large	B-65	600	.027	"	28	1.24	.91	30.15	33.50
0400	Humus peat, 1" deep, hand spread	1 Clab	700	.011	S.Y.	3.51	.51		4.02	4.62
0450	Push spreader	"	2500	.003	"	3.51	.14		3.65	4.07
0550	Tractor spreader	B-66	700	.011	M.S.F.	390	.63	.38	391.01	430
0600	Oat straw, 1" deep, hand spread	1 Clab	475	.017	S.Y.	.62	.75		1.37	1.80
0650	Power mulcher, small	B-64	180	.089	M.S.F.	69	4.14	2.22	75.36	84.50
0700	Large	B-65	530	.030	"	69	1.41	1.03	71.44	79
0750	Add for asphaltic emulsion	B-45	1770	.009	Gal.	5.85	.50	.47	6.82	7.65
0760	Pine straw, 1" deep, hand spread	1 Clab	1950	.004	S.F.	.15	.18		.33	.44
0800	Peat moss, 1" deep, hand spread		900	.009	S.Y.	5.15	.39		5.54	6.25
0850	Push spreader		2500	.003	"	5.15	.14		5.29	5.85
0950	Tractor spreader	B-66	700	.011	M.S.F.	570	.63	.38	571.01	630
1000	Polyethylene film, 6 mil	2 Clab	2000	.008	S.Y.	.61	.36		.97	1.20
1010	4 mil		2300	.007		.53	.31		.84	1.04
1020	1-1/2 mil		2500	.006		.31	.28		.59	.76
1050	Filter fabric weed barrier		2000	.008		.89	.36		1.25	1.51
1100	Redwood nuggets, 3" deep, hand spread	1 Clab	150	.053		3.61	2.37		5.98	7.50
1150	Skid steer loader	B-63	13.50	2.963	M.S.F.	400	138	13.30	551.30	660
1200	Stone mulch, hand spread, ceramic chips, economy	1 Clab	125	.064	S.Y.	7.70	2.84		10.54	12.70
1250	Deluxe	"	95	.084	"	12.35	3.74		16.09	19.20
1300	Granite chips	B-1	10	2.400	C.Y.	77.50	108		185.50	247
1400	Marble chips		10	2.400		232	108		340	415
1600	Pea gravel		28	.857		110	38.50		148.50	178
1700	Quartz		10	2.400		197	108		305	375
1760	Landscape mulch, 100% recycld tires, var colors, hand spread, 3" deep		10	2.400		300	108		408	490
1800	Tar paper, 15 lb. felt	1 Clab	800	.010	S.Y.	.51	.44		.95	1.22
1900	Wood chips, 2" deep, hand spread	"	220	.036	"	1.33	1.61		2.94	3.87
1950	Skid steer loader	B-63	20.30	1.970	M.S.F.	148	92	8.85	248.85	310

32 91 13.23 Structural Soil Mixing

		Crew	Daily Output	Labor-Hours	Unit	Material	2021 Bare Costs Labor	Equipment	Total	Total Incl O&P
0010	**STRUCTURAL SOIL MIXING**									
0100	Rake topsoil, site material, harley rock rake, ideal	B-6	33	.727	M.S.F.		35	6.55	41.55	59
0200	Adverse	"	7	3.429			165	31	196	280
0250	By hand (raking)	1 Clab	10	.800			35.50		35.50	53
0300	Screened loam, york rake and finish, ideal	B-62	24	1			48	7.50	55.50	80
0400	Adverse	"	20	1.200			57.50	9	66.50	96
0450	By hand (raking)	1 Clab	15	.533			23.50		23.50	35.50
1000	Remove topsoil & stock pile on site, 75 HP dozer, 6" deep, 50' haul	B-10L	30	.400			21.50	13.55	35.05	47.50
1050	300' haul		6.10	1.967			106	66.50	172.50	232
1100	12" deep, 50' haul		15.50	.774			42	26	68	91.50
1150	300' haul		3.10	3.871			210	131	341	455
1200	200 HP dozer, 6" deep, 50' haul	B-10B	125	.096			5.20	12.15	17.35	21
1250	300' haul		30.70	.391			21	49.50	70.50	86
1300	12" deep, 50' haul		62	.194			10.50	24.50	35	42.50

32 91 Planting Preparation

32 91 13 – Soil Preparation

32 91 13.23 Structural Soil Mixing		Crew	Daily Output	Labor-Hours	Unit	Material	2021 Bare Costs Labor	Equipment	Total	Total Incl O&P
1350	300' haul	B-10B	15.40	.779	M.S.F.		42	98.50	140.50	172
1400	Alternate method, 75 HP dozer, 50' haul	B-10L	860	.014	C.Y.		.76	.47	1.23	1.65
1450	300' haul	"	114	.105			5.70	3.56	9.26	12.40
1500	200 HP dozer, 50' haul	B-10B	2660	.005			.24	.57	.81	.99
1600	300' haul	"	570	.021			1.14	2.67	3.81	4.63
1800	Rolling topsoil, hand push roller	1 Clab	3200	.003	S.F.		.11		.11	.17
1850	Tractor drawn roller	B-66	10666	.001	"		.04	.03	.07	.09
1900	Remove rocks & debris from grade, by hand	B-62	80	.300	M.S.F.		14.45	2.24	16.69	24
1920	With rock picker	B-10S	140	.086			4.64	3.15	7.79	10.35
2000	Root raking and loading, residential, no boulders	B-6	53.30	.450			21.50	4.06	25.56	37
2100	With boulders		32	.750			36	6.75	42.75	61.50
2200	Municipal, no boulders		200	.120			5.75	1.08	6.83	9.80
2300	With boulders		120	.200			9.60	1.80	11.40	16.35
2400	Large commercial, no boulders	B-10B	400	.030			1.62	3.80	5.42	6.60
2500	With boulders	"	240	.050			2.71	6.35	9.06	11
2600	Rough grade & scarify subsoil to receive topsoil, common earth									
2610	200 HP dozer with scarifier	B-11A	80	.200	M.S.F.		10.35	19	29.35	36.50
2620	180 HP grader with scarifier	B-11L	110	.145			7.50	9.75	17.25	22
2700	Clay and till, 200 HP dozer with scarifier	B-11A	50	.320			16.55	30.50	47.05	58
2710	180 HP grader with scarifier	B-11L	40	.400			20.50	27	47.50	60.50
3000	Scarify subsoil, residential, skid steer loader w/scarifiers, 50 HP	B-66	32	.250			13.90	8.35	22.25	29.50
3050	Municipal, skid steer loader w/scarifiers, 50 HP	"	120	.067			3.70	2.23	5.93	7.95
3100	Large commercial, 75 HP, dozer w/scarifier	B-10L	240	.050			2.71	1.69	4.40	5.90
3200	Grader with scarifier, 135 HP	B-11L	280	.057			2.95	3.83	6.78	8.60
3500	Screen topsoil from stockpile, vibrating screen, wet material (organic)	B-10P	200	.060	C.Y.		3.25	5.75	9	11.15
3550	Dry material	"	300	.040			2.17	3.82	5.99	7.45
3600	Mixing with conditioners, manure and peat	B-10R	550	.022			1.18	.56	1.74	2.37
3650	Mobilization add for 2 days or less operation	B-34K	3	2.667	Job		137	289	426	525
3800	Spread conditioned topsoil, 6" deep, by hand	B-1	360	.067	S.Y.	6	3		9	11.10
3850	300 HP dozer	B-10M	27	.444	M.S.F.	650	24	66	740	825
3900	4" deep, by hand	B-1	470	.051	S.Y.	5.40	2.30		7.70	9.40
3920	300 HP dozer	B-10M	34	.353	M.S.F.	485	19.10	52.50	556.60	620
3940	180 HP grader	B-11L	37	.432	"	485	22.50	29	536.50	600
4000	Spread soil conditioners, alum. sulfate, 1#/S.Y., hand push spreader	1 Clab	17500	.001	S.Y.	19.20	.02		19.22	21
4050	Tractor spreader	B-66	700	.011	M.S.F.	2,125	.63	.38	2,126.01	2,350
4100	Fertilizer, 0.2#/S.Y., push spreader	1 Clab	17500	.001	S.Y.	.08	.02		.10	.12
4150	Tractor spreader	B-66	700	.011	M.S.F.	9.35	.63	.38	10.36	11.60
4200	Ground limestone, 1#/S.Y., push spreader	1 Clab	17500	.001	S.Y.	.16	.02		.18	.21
4250	Tractor spreader	B-66	700	.011	M.S.F.	17.80	.63	.38	18.81	21
4400	Manure, 18#/S.Y., push spreader	1 Clab	2500	.003	S.Y.	7.40	.14		7.54	8.35
4450	Tractor spreader	B-66	280	.029	M.S.F.	825	1.59	.96	827.55	910
4500	Perlite, 1" deep, push spreader	1 Clab	17500	.001	S.Y.	11.05	.02		11.07	12.20
4550	Tractor spreader	B-66	700	.011	M.S.F.	1,225	.63	.38	1,226.01	1,350
4600	Vermiculite, push spreader	1 Clab	17500	.001	S.Y.	8.45	.02		8.47	9.35
4650	Tractor spreader	B-66	700	.011	M.S.F.	940	.63	.38	941.01	1,025
5000	Spread topsoil, skid steer loader and hand dress	B-62	270	.089	C.Y.	27	4.28	.66	31.94	37
5100	Articulated loader and hand dress	B-100	320	.038		27	2.03	2.89	31.92	36
5200	Articulated loader and 75 HP dozer	B-10M	500	.024		27	1.30	3.57	31.87	36
5300	Road grader and hand dress	B-11L	1000	.016		27	.83	1.07	28.90	32.50
6000	Tilling topsoil, 20 HP tractor, disk harrow, 2" deep	B-66	450	.018	M.S.F.		.99	.59	1.58	2.12
6050	4" deep		360	.022			1.23	.74	1.97	2.66
6100	6" deep		270	.030			1.64	.99	2.63	3.54
6150	26" rototiller, 2" deep	A-1J	1250	.006	S.Y.		.28	.04	.32	.47

For customer support on your Site Work & Landscape Costs with RSMeans Data, call 800.448.8182.

407

32 91 Planting Preparation

32 91 13 – Soil Preparation

32 91 13.23 Structural Soil Mixing

	32 91 13.23 Structural Soil Mixing	Crew	Daily Output	Labor-Hours	Unit	Material	2021 Bare Costs Labor	Equipment	Total	Total Incl O&P
6200	4" deep	A-1J	1000	.008	S.Y.		.36	.05	.41	.59
6250	6" deep	↓	750	.011	↓		.47	.07	.54	.79

32 91 13.26 Planting Beds

		Crew	Daily Output	Labor-Hours	Unit	Material	2021 Bare Costs Labor	Equipment	Total	Total Incl O&P
0010	**PLANTING BEDS**									
0100	Backfill planting pit, by hand, on site topsoil	2 Clab	18	.889	C.Y.		39.50		39.50	59
0200	Prepared planting mix, by hand	"	24	.667			29.50		29.50	44
0300	Skid steer loader, on site topsoil	B-62	340	.071			3.40	.53	3.93	5.65
0400	Prepared planting mix	"	410	.059			2.82	.44	3.26	4.68
1000	Excavate planting pit, by hand, sandy soil	2 Clab	16	1			44.50		44.50	66.50
1100	Heavy soil or clay	"	8	2			89		89	133
1200	1/2 C.Y. backhoe, sandy soil	B-11C	150	.107			5.50	1.44	6.94	9.80
1300	Heavy soil or clay	"	115	.139			7.20	1.88	9.08	12.75
2000	Mix planting soil, incl. loam, manure, peat, by hand	2 Clab	60	.267		47.50	11.85		59.35	69.50
2100	Skid steer loader	B-62	150	.160	↓	47.50	7.70	1.20	56.40	65
3000	Pile sod, skid steer loader	"	2800	.009	S.Y.		.41	.06	.47	.68
3100	By hand	2 Clab	400	.040			1.78		1.78	2.65
4000	Remove sod, F.E. loader	B-10S	2000	.006			.32	.22	.54	.72
4100	Sod cutter	B-12K	3200	.005			.26	.31	.57	.73
4200	By hand	2 Clab	240	.067	↓		2.96		2.96	4.42
6000	For planting bed edging, see Section 32 94 13.20									

32 91 19 – Landscape Grading

32 91 19.13 Topsoil Placement and Grading

		Crew	Daily Output	Labor-Hours	Unit	Material	2021 Bare Costs Labor	Equipment	Total	Total Incl O&P
0010	**TOPSOIL PLACEMENT AND GRADING**									
0300	Fine grade, base course for paving, see Section 32 11 23.23									
0400	Spread from pile to rough finish grade, F.E. loader, 1-1/2 C.Y.	B-10S	200	.060	C.Y.		3.25	2.21	5.46	7.25
0500	Up to 200' radius, by hand	1 Clab	14	.571			25.50		25.50	38
0600	Top dress by hand, 1 C.Y. for 600 S.F.	"	11.50	.696	↓	29.50	31		60.50	78.50
0700	Furnish and place, truck dumped, screened, 4" deep	B-10S	1300	.009	S.Y.	3.69	.50	.34	4.53	5.15
0800	6" deep	"	820	.015	"	4.73	.79	.54	6.06	6.95
0900	Fine grading and seeding, incl. lime, fertilizer & seed,									
1000	With equipment	B-14	1000	.048	S.Y.	.49	2.24	.22	2.95	4.12

32 92 Turf and Grasses

32 92 19 – Seeding

32 92 19.13 Mechanical Seeding

		Crew	Daily Output	Labor-Hours	Unit	Material	2021 Bare Costs Labor	Equipment	Total	Total Incl O&P
0010	**MECHANICAL SEEDING**									
0020	Mechanical seeding, 215 lb./acre	B-66	1.50	5.333	Acre	585	296	178	1,059	1,275
0100	44 lb./M.S.Y.	"	2500	.003	S.Y.	.24	.18	.11	.53	.64
0101	44 lb./M.S.Y.	1 Clab	13950	.001	S.F.	.03	.03		.06	.07
0300	Fine grading and seeding incl. lime, fertilizer & seed,									
0310	with equipment	B-14	1000	.048	S.Y.	.47	2.24	.22	2.93	4.10
0400	Fertilizer hand push spreader, 35 lb./M.S.F.	1 Clab	200	.040	M.S.F.	9	1.78		10.78	12.55
0600	Limestone hand push spreader, 50 lb./M.S.F.		180	.044		5.45	1.97		7.42	8.95
0800	Grass seed hand push spreader, 4.5 lb./M.S.F.	↓	180	.044	↓	26.50	1.97		28.47	32.50
1000	Hydro or air seeding for large areas, incl. seed and fertilizer	B-81	8900	.003	S.Y.	.60	.14	.06	.80	.94
1100	With wood fiber mulch added	"	8900	.003	"	2.20	.14	.06	2.40	2.70
1300	Seed only, over 100 lb., field seed, minimum				Lb.	2.52			2.52	2.77
1400	Maximum					1.67			1.67	1.84
1500	Lawn seed, minimum					1.34			1.34	1.47
1600	Maximum				↓	3.01			3.01	3.31

32 92 Turf and Grasses

32 92 19 – Seeding

32 92 19.13 Mechanical Seeding		Crew	Daily Output	Labor-Hours	Unit	Material	2021 Bare Costs Labor	Equipment	Total	Total Incl O&P
1800	Aerial operations, seeding only, field seed	B-58	50	.480	Acre	820	23	46.50	889.50	985
1900	Lawn seed		50	.480		435	23	46.50	504.50	565
2100	Seed and liquid fertilizer, field seed		50	.480		905	23	46.50	974.50	1,075
2200	Lawn seed		50	.480		520	23	46.50	589.50	655

32 92 19.14 Seeding, Athletic Fields

		Crew	Daily Output	Labor-Hours	Unit	Material	Labor	Equipment	Total	Total Incl O&P
0010	**SEEDING, ATHLETIC FIELDS** R329219-50									
0020	Seeding, athletic fields, athletic field mix, 8#/M.S.F. push spreader	1 Clab	8	1	M.S.F.	8.85	44.50		53.35	76.50
0100	Tractor spreader	B-66	52	.154		8.85	8.55	5.15	22.55	28
0200	Hydro or air seeding, with mulch and fertilizer	B-81	80	.300		9.75	15.45	7.10	32.30	41.50
0400	Birdsfoot trefoil, 0.45#/M.S.F., push spreader	1 Clab	8	1		4.41	44.50		48.91	71.50
0500	Tractor spreader	B-66	52	.154		4.41	8.55	5.15	18.11	23
0600	Hydro or air seeding, with mulch and fertilizer	B-81	80	.300		8.50	15.45	7.10	31.05	40
0800	Bluegrass, 4#/M.S.F., common, push spreader	1 Clab	8	1		9.50	44.50		54	77
0900	Tractor spreader	B-66	52	.154		9.50	8.55	5.15	23.20	29
1000	Hydro or air seeding, with mulch and fertilizer	B-81	80	.300		15.70	15.45	7.10	38.25	48
1100	Baron, push spreader	1 Clab	8	1		16.90	44.50		61.40	85
1200	Tractor spreader	B-66	52	.154		16.90	8.55	5.15	30.60	37
1300	Hydro or air seeding, with mulch and fertilizer	B-81	80	.300		23	15.45	7.10	45.55	56.50
1500	Clover, 0.67#/M.S.F., white, push spreader	1 Clab	8	1		3.06	44.50		47.56	70
1600	Tractor spreader	B-66	52	.154		3.06	8.55	5.15	16.76	21.50
1700	Hydro or air seeding, with mulch and fertilizer	B-81	80	.300		16.85	15.45	7.10	39.40	49.50
1800	Ladino, push spreader	1 Clab	8	1		3	44.50		47.50	70
1900	Tractor spreader	B-66	52	.154		3	8.55	5.15	16.70	21.50
2000	Hydro or air seeding, with mulch and fertilizer	B-81	80	.300		13.20	15.45	7.10	35.75	45.50
2200	Fescue 5.5#/M.S.F., tall, push spreader	1 Clab	8	1		10.55	44.50		55.05	78
2300	Tractor spreader	B-66	52	.154		10.55	8.55	5.15	24.25	30
2400	Hydro or air seeding, with mulch and fertilizer	B-81	80	.300		35	15.45	7.10	57.55	69.50
2500	Chewing, push spreader	1 Clab	8	1		8.20	44.50		52.70	75.50
2600	Tractor spreader	B-66	52	.154		8.20	8.55	5.15	21.90	27.50
2700	Hydro or air seeding, with mulch and fertilizer	B-81	80	.300		27	15.45	7.10	49.55	60.50
2800	Creeping, push spreader	1 Clab	8	1		7.60	44.50		52.10	75
2810	Tractor spreader	B-66	26	.308		7.60	17.10	10.30	35	45
2820	Hydro or air seeding, with mulch and fertilizer	B-81	80	.300		25	15.45	7.10	47.55	58.50
2900	Crown vetch, 4#/M.S.F., push spreader	1 Clab	8	1		132	44.50		176.50	212
3000	Tractor spreader	B-66	52	.154		132	8.55	5.15	145.70	163
3100	Hydro or air seeding, with mulch and fertilizer	B-81	80	.300		181	15.45	7.10	203.55	231
3300	Rye, 10#/M.S.F., annual, push spreader	1 Clab	8	1		18.90	44.50		63.40	87.50
3400	Tractor spreader	B-66	52	.154		18.90	8.55	5.15	32.60	39.50
3500	Hydro or air seeding, with mulch and fertilizer	B-81	80	.300		41.50	15.45	7.10	64.05	76.50
3600	Fine textured, push spreader	1 Clab	8	1		15.15	44.50		59.65	83
3700	Tractor spreader	B-66	52	.154		15.15	8.55	5.15	28.85	35
3800	Hydro or air seeding, with mulch and fertilizer	B-81	80	.300		33.50	15.45	7.10	56.05	67.50
4000	Shade mix, 6#/M.S.F., push spreader	1 Clab	8	1		10.15	44.50		54.65	77.50
4100	Tractor spreader	B-66	52	.154		10.15	8.55	5.15	23.85	29.50
4200	Hydro or air seeding, with mulch and fertilizer	B-81	80	.300		22.50	15.45	7.10	45.05	55.50
4400	Slope mix, 6#/M.S.F., push spreader	1 Clab	8	1		11.25	44.50		55.75	79
4500	Tractor spreader	B-66	52	.154		11.25	8.55	5.15	24.95	31
4600	Hydro or air seeding, with mulch and fertilizer	B-81	80	.300		28	15.45	7.10	50.55	62
4800	Turf mix, 4#/M.S.F., push spreader	1 Clab	8	1		12.80	44.50		57.30	80.50
4900	Tractor spreader	B-66	52	.154		12.80	8.55	5.15	26.50	32.50
5000	Hydro or air seeding, with mulch and fertilizer	B-81	80	.300		32	15.45	7.10	54.55	66
5200	Utility mix, 7#/M.S.F., push spreader	1 Clab	8	1		9.40	44.50		53.90	77

32 92 Turf and Grasses

32 92 19 – Seeding

32 92 19.14 Seeding, Athletic Fields	Crew	Daily Output	Labor-Hours	Unit	Material	2021 Bare Costs Labor	2021 Bare Costs Equipment	Total	Total Incl O&P
5300 Tractor spreader	B-66	52	.154	M.S.F.	9.40	8.55	5.15	23.10	28.50
5400 Hydro or air seeding, with mulch and fertilizer	B-81	80	.300		35.50	15.45	7.10	58.05	70
5600 Wildflower, 0.10#/M.S.F., push spreader	1 Clab	8	1		1.63	44.50		46.13	68.50
5700 Tractor spreader	B-66	52	.154		1.63	8.55	5.15	15.33	20
5800 Hydro or air seeding, with mulch and fertilizer	B-81	80	.300	↓	8.95	15.45	7.10	31.50	40.50
7000 Apply fertilizer, 800 lb./acre	B-66	4	2	Ton	780	111	67	958	1,100
7025 Fertilizer, mechanical spread	1 Clab	1.75	4.571	Acre	5.45	203		208.45	310
7050 Apply limestone, 800 lb./acre	B-66	4.25	1.882	Ton	195	104	63	362	440
7060 Limestone, mechanical spread	1 Clab	1.74	4.598	Acre	87	204		291	400
7100 Apply mulch, see Section 32 91 13.16									

32 92 23 – Sodding

32 92 23.10 Sodding Systems	Crew	Daily Output	Labor-Hours	Unit	Material	2021 Bare Costs Labor	2021 Bare Costs Equipment	Total	Total Incl O&P
0010 **SODDING SYSTEMS**									
0020 Sodding, 1" deep, bluegrass sod, on level ground, over 8 M.S.F.	B-63	22	1.818	M.S.F.	212	85	8.15	305.15	370
0200 4 M.S.F.		17	2.353		400	110	10.55	520.55	615
0300 1,000 S.F.		13.50	2.963		335	138	13.30	486.30	590
0500 Sloped ground, over 8 M.S.F.		6	6.667		212	310	30	552	730
0600 4 M.S.F.		5	8		400	375	36	811	1,025
0700 1,000 S.F.		4	10		335	465	45	845	1,125
1000 Bent grass sod, on level ground, over 6 M.S.F.		20	2		330	93	9	432	515
1100 3 M.S.F.		18	2.222		360	104	10	474	560
1200 Sodding 1,000 S.F. or less		14	2.857		380	133	12.85	525.85	635
1500 Sloped ground, over 6 M.S.F.		15	2.667		330	124	11.95	465.95	565
1600 3 M.S.F.		13.50	2.963		360	138	13.30	511.30	615
1700 1,000 S.F.	↓	12	3.333	↓	380	155	14.95	549.95	670

32 92 26 – Sprigging

32 92 26.13 Stolonizing	Crew	Daily Output	Labor-Hours	Unit	Material	2021 Bare Costs Labor	2021 Bare Costs Equipment	Total	Total Incl O&P
0010 **STOLONIZING**									
0100 6" OC, by hand	1 Clab	4	2	M.S.F.	52.50	89		141.50	191
0110 Walk behind sprig planter	"	80	.100		52.50	4.44		56.94	64.50
0120 Towed sprig planter	B-66	350	.023		52.50	1.27	.76	54.53	60.50
0130 9" OC, by hand	1 Clab	5.20	1.538		24.50	68.50		93	129
0140 Walk behind sprig planter	"	92	.087		24.50	3.86		28.36	33
0150 Towed sprig planter	B-66	420	.019		24.50	1.06	.64	26.20	29.50
0160 12" OC, by hand	1 Clab	6	1.333		13.05	59		72.05	103
0170 Walk behind sprig planter	"	110	.073		13.05	3.23		16.28	19.15
0180 Towed sprig planter	B-66	500	.016		13.05	.89	.54	14.48	16.25
0200 Broadcast, by hand, 2 Bu. per M.S.F.	1 Clab	15	.533		6.15	23.50		29.65	42.50
0210 4 Bu. per M.S.F.		10	.800		10.55	35.50		46.05	64.50
0220 6 Bu. per M.S.F.	↓	6.50	1.231		15.15	54.50		69.65	98
0300 Hydro planter, 6 Bu. per M.S.F.	B-64	100	.160		15.15	7.45	4	26.60	32.50
0320 Manure spreader planting 6 Bu. per M.S.F.	B-66	200	.040	↓	15.15	2.22	1.34	18.71	21.50

32 93 13.30 Ground Covers	Crew	Daily Output	Labor-Hours	Unit	Material	2021 Bare Costs Labor	Equipment	Total	Total Incl O&P
0010 GROUND COVERS									
0011 Vines and climbing plants									
1000 Achillea tomentosa (Woolly Yarrow), Z4, cont									
1010 1 qt.				Ea.	6			6	6.60
1020 2 gal.				"	18.50			18.50	20.50
1100 Aegopodium podagaria (Snow-on-the-Mountain Goutweed), Z3, cont									
1110 1 qt.				Ea.	11.30			11.30	12.45
1120 1 gal.					14			14	15.40
1130 Flat/24 plants					22.50			22.50	25
1200 Ajuga reptans (Carpet Bugle), Z4, cont									
1210 4" pot				Ea.	4.86			4.86	5.35
1220 1 qt.					14			14	15.40
1230 1 gal.					12.85			12.85	14.15
1240 Flat/24 plants					51.50			51.50	57
1300 Akebia quinata (Five-Leaf Akebia), Z4, cont									
1310 1 gal.				Ea.	16.25			16.25	17.90
1320 5 gal.				"	58			58	64
1400 Alchemilla vulgaris (Common Lady's Mantle), Z3, cont									
1410 1 qt.				Ea.	35			35	38.50
1420 1 gal.				"	27			27	30
1500 Ampelopsis brevipedunculata elegans (Ampelopsis), Z4, cont									
1510 1 gal.				Ea.	19.60			19.60	21.50
1520 3 gal.				"	36.50			36.50	40
1600 Arctostaphylos uva-ursi (Bearberry), Z2, cont									
1610 3" pot				Ea.	53			53	58.50
1620 1 gal.					16.40			16.40	18.05
1630 5 gal.					31.50			31.50	34.50
1700 Aristolochia durior (Dutchman's Pipe), Z5, cont									
1710 2 gal.				Ea.	28.50			28.50	31.50
1720 5 gal.				"	34			34	37
1800 Armeria maritima (Thrift, Sea Pink), Z6, cont									
1810 1 qt.				Ea.	12			12	13.20
1820 1 gal.				"	14.80			14.80	16.30
1900 Artemisia schmitdiana (Silver Mound/Satiny Wormwood), Z3, cont									
1910 1 qt.				Ea.	7			7	7.70
1920 1 gal.				"	10			10	11
2000 Artemisia stellerana (Dusty Miller), Z2, cont									
2010 1 qt.				Ea.	8.40			8.40	9.20
2020 1 gal.				"	6			6	6.60
2100 Asarum europaeum (European Wild Ginger), Z4, cont									
2110 4" pot				Ea.	14.10			14.10	15.50
2120 1 gal.				"	33.50			33.50	37
2200 Baccharis pilularis (Dwarf Coyote Bush), Z7, cont									
2210 1 qt.				Ea.	6.90			6.90	7.60
2220 1 gal.				"	7.65			7.65	8.45
2300 Bergenia cordifolia (Heartleaf Bergenia), Z3, cont									
2310 1 qt.				Ea.	15.55			15.55	17.10
2320 1 gal.				"	26.50			26.50	29
2400 Bougainvillea glabra, Z10, cont									
2410 1 gal.				Ea.	18.20			18.20	20
2420 5 gal.				"	26			26	28.50
2500 Calluna vulgaris cultivars (Scotch Heather), Z4, cont									
2510 3" pot				Ea.	8			8	8.80

411

32 93 13.30 Ground Covers	Crew	Daily Output	Labor-Hours	Unit	Material	2021 Bare Costs Labor	Equipment	Total	Total Incl O&P
2520	1 qt.				Ea.	16.15			16.15	17.75
2530	1 gal.					38.50			38.50	42.50
2540	2 gal.					74			74	81.50
2600	Campsis radicans (Trumpet Creeper), Z5, cont									
2610	1 gal.				Ea.	9.60			9.60	10.55
2620	2 gal.					21.50			21.50	23.50
2630	5 gal.					31.50			31.50	34.50
2700	Celastrus scandens (American Bittersweet), Z2, cont									
2710	1 gal.				Ea.	14.90			14.90	16.35
2720	2 gal.				"	51			51	56.50
2800	Cerastium tomentosum (Snow-in-Summer), Z2, cont									
2810	1 qt.				Ea.	19.20			19.20	21
2820	1 gal.				"	12.95			12.95	14.25
2900	Ceratostigma (Dwarf Plumbago), Z6, cont									
2910	1 qt.				Ea.	13.90			13.90	15.30
2920	1 gal.				"	7.50			7.50	8.25
3000	Cissus antartica (Kangaroo Vine), Z8, cont									
3010	1 gal.				Ea.	7.90			7.90	8.65
3020	5 gal.				"	16.75			16.75	18.45
3100	Cistus crispus (Wrinkleleaf Rock Rose), Z7, cont									
3110	1 gal.				Ea.	24.50			24.50	26.50
3120	5 gal.				"	19.50			19.50	21.50
3200	Clematis (Hybrids & Clones), Z6, cont									
3210	1 gal.				Ea.	14.60			14.60	16.05
3220	2 gal.					18.85			18.85	20.50
3230	5 gal.					29			29	32
3300	Cocculus laurifolius (Laurel-leaf Snailseed), Z8, cont									
3310	1 gal.				Ea.	8.15			8.15	8.95
3320	5 gal.				"	22			22	24.50
3400	Conuallaria majalis (Lily-of-the-Valley), Z3, cont									
3410	1 qt.				Ea.	8.40			8.40	9.25
3420	1 gal.				"	23.50			23.50	25.50
3500	Coprosma prostrata (Prostrate Coprosma), Z7, cont									
3510	1 gal.				Ea.	12.05			12.05	13.25
3520	5 gal.				"	16.95			16.95	18.65
3600	Coronilla varia (Crown Vetch), Z3, cont									
3610	2-1/4" pot				Ea.	4.70			4.70	5.15
3620	Flat/50 plants				"	82			82	90
3700	Cotoneaster dammeri (Bearberry Cotoneaster), Z5, cont									
3710	1 gal.				Ea.	53.50			53.50	58.50
3720	2 gal.					13.95			13.95	15.35
3730	5 gal.					12.35			12.35	13.60
3800	Cytisus decumbens (Prostrate Broom), Z5, cont									
3810	1 gal.				Ea.	23			23	25.50
3820	2 gal.				"	16.95			16.95	18.65
3900	Dianthus plumarius (Grass Pinks), Z4, cont									
3910	1 qt.				Ea.	10.75			10.75	11.80
3920	1 gal.				"	36.50			36.50	40
4000	Dicentra formosa (Western Bleeding Heart), Z3, cont									
4010	2 qt.				Ea.	12.05			12.05	13.25
4020	1 gal.				"	14.35			14.35	15.80
4100	Distictis buccinatoria (Blood Red Trumpet Vine), Z9, cont									
4110	1 gal.				Ea.	6.50			6.50	7.15

412

32 93 13 – Ground Covers

32 93 13.30 Ground Covers	Crew	Daily Output	Labor-Hours	Unit	Material	2021 Bare Costs Labor	Equipment	Total	Total Incl O&P
4120	5 gal.				Ea.	46.50			46.50	51.50
4200	Epimedium grandiflorum (Barrenwort, Bishopshat), Z3, cont									
4210	1 qt.				Ea.	19			19	21
4220	1 gal.				"	13.70			13.70	15.10
4300	Euonymus fortunei colorata (Purple-Leaf Wintercreeper), Z5, cont									
4310	1 qt.				Ea.	15			15	16.50
4320	1 gal.					22			22	24
4330	Flat/24 plants					38.50			38.50	42
4400	Euonymus fortunei kewensis (Euonymus), Z5, cont									
4410	1 gal.				Ea.	19.65			19.65	21.50
4420	2 gal.				"	23			23	25
4500	Fragaria vesca (Alpine Strawberry), Z6, cont									
4510	4" pot				Ea.	13.95			13.95	15.35
4520	1 qt.					15.55			15.55	17.10
4530	1 gal.					29			29	32
4600	Galium odoratum (Sweet Woodruff), Z4, cont									
4610	4" pot				Ea.	9.95			9.95	10.95
4620	1 gal.				"	50			50	55
4700	Gardenia jasminoides "Radicans" (Creeping Gardenia), Z8, cont									
4710	1 gal.				Ea.	22.50			22.50	24.50
4720	5 gal.				"	25			25	27.50
4800	Gelsemium sempervirens (Carolina Jessamine), Z7, cont									
4810	1 gal.				Ea.	28.50			28.50	31
4820	5 gal.				"	18			18	19.80
4900	Genista pilosa (Silky-leaf Woadwaxen), Z5, cont									
4910	1 gal.				Ea.	13.40			13.40	14.75
4920	2 gal.				"	28			28	30.50
5000	Hardenbergia violacea (Happy Wanderer), Z9, cont									
5010	1 gal.				Ea.	6.50			6.50	7.15
5020	5 gal.				"	20.50			20.50	22.50
5100	Hedera helix varieties (English Ivy), Z5, cont									
5110	3" pot				Ea.	5.05			5.05	5.60
5120	5 gal.					14			14	15.40
5130	Flat/50 plants					54.50			54.50	60
5200	Helianthemum nummularium (Sun-Rose), Z5, cont									
5210	1 qt.				Ea.	7.80			7.80	8.55
5220	1 gal.				"	10			10	11
5300	Hemerocallis hybrids (Assorted Daylillies), Z4, cont									
5310	2 qt.				Ea.	8.70			8.70	9.55
5320	1 gal.					8.10			8.10	8.90
5330	5 gal.					13.75			13.75	15.15
5400	Hosta assorted (Plantain Lily), Z4, cont									
5410	2 qt.				Ea.	7			7	7.70
5420	1 gal.					22.50			22.50	25
5430	2 gal.					52.50			52.50	58
5500	Hydrangea petiolaris (Climbing Hydrangea), Z4, cont									
5510	1 gal.				Ea.	6.50			6.50	7.15
5520	2 gal.					25			25	27.50
5530	5 gal.					18.50			18.50	20.50
5600	Hypericum calycinum (St. John's Wort), Z6, cont									
5610	1 qt.				Ea.	12.50			12.50	13.75
5620	1 gal.					27			27	29.50
5630	5 gal.					50			50	55

32 93 Plants

32 93 13 – Ground Covers

32 93 13.30 Ground Covers	Crew	Daily Output	Labor-Hours	Unit	Material	2021 Bare Costs Labor	Equipment	Total	Total Incl O&P
5700	Iberis sempervirens (Candytuft), Z4, cont									
5710	1 qt.				Ea.	9.30			9.30	10.20
5720	1 gal.				"	13.65			13.65	15
5800	Jasminum nudiflorum (Winter Jasmine), Z5, cont									
5810	1 gal.				Ea.	25			25	27.50
5820	2 gal.					26			26	28.50
5830	5 gal.				↓	30			30	33
5900	Juniperus horizontalis varieties (Prostrate Juniper), Z2, cont									
5910	1 gal.				Ea.	30			30	33
5920	2 gal.					45			45	49.50
5930	5 gal.				↓	95			95	105
6000	Lamium maculatum (Dead Nettle), Z6, cont									
6010	1 qt.				Ea.	7.10			7.10	7.80
6020	1 gal.					12.50			12.50	13.75
6030	Flat/24 plants				↓	58			58	64
6100	Lantana camara (Common Lantana), Z8, cont									
6110	3" pot				Ea.	15.65			15.65	17.25
6120	1 gal.				"	6			6	6.60
6200	Lantana montevidensis (Trailing Lantana), Z10, cont									
6210	1 gal.				Ea.	9.35			9.35	10.30
6220	5 gal.				"	22			22	24
6300	Lavandula angustifolia (English Lavender), Z5, cont									
6310	4" pot				Ea.	12			12	13.20
6320	1 qt.					11.65			11.65	12.80
6330	1 gal.				↓	4.50			4.50	4.95
6400	Lirope spicata (Creeping Lilyturf), Z6, cont									
6410	2 qt.				Ea.	15.95			15.95	17.55
6420	1 gal.					1.50			1.50	1.65
6430	5 gal.				↓	16.25			16.25	17.90
6500	Lonicera heckrotti (Gold Flame Honeysuckle), Z5, cont									
6510	1 gal.				Ea.	42.50			42.50	47
6520	2 gal.				"	57			57	62.50
6600	Lonicera japonica "Halliana" (Halls Honeysuckle), Z4, cont									
6610	3" pot				Ea.	5.75			5.75	6.35
6620	1 gal.					13.05			13.05	14.35
6630	2 gal.				↓	15			15	16.50
6700	Lysimachia nummularia aurea (Creeping Jenny), Z3, cont									
6710	2 qt.				Ea.	6.95			6.95	7.65
6720	1 gal.				"	9.50			9.50	10.45
6800	Macfadyena unguis-cali (Cat Claw Vine), Z8, cont									
6810	1 gal.				Ea.	4.50			4.50	4.95
6820	5 gal.				"	18			18	19.80
6900	Myoporum parvifolium (Prostrate Myoporum), Z9, cont									
6910	1 gal.				Ea.	6			6	6.60
6920	5 gal.				"	18			18	19.80
7000	Nepeta (Persian Catmint), Z5, cont									
7010	1 qt.				Ea.	3.85			3.85	4.24
7020	1 gal.				"	30			30	33
7100	Ophiupogon japonicus (Mondo-Grass), Z7, cont									
7110	4" pot				Ea.	4.45			4.45	4.90
7120	1 gal.					10			10	11
7130	5 gal.				↓	14.50			14.50	15.95
7200	Pachysandra terminalis (Japanese Spurge), Z5, cont									

32 93 Plants

32 93 13 – Ground Covers

32 93 13.30 Ground Covers	Crew	Daily Output	Labor-Hours	Unit	Material	2021 Bare Costs Labor	Equipment	Total	Total Incl O&P
7210	2-1/2" pot				Ea.	8.95			8.95	9.85
7220	1 gal.					33			33	36
7230	Tray/50 plants					60			60	66
7240	Tray/100 plants				↓	119			119	131
7300	Parthenocissus quinquefolia (Virginia Creeper), Z3, cont									
7310	3" pot				Ea.	12.40			12.40	13.65
7320	1 gal.				"	19.15			19.15	21
7400	Parthenocissus tricuspidata veitchi (Boston Ivy), Z4, cont									
7410	3" pot				Ea.	13.80			13.80	15.20
7420	1 gal.					42			42	46
7430	5 gal.				↓	50.50			50.50	55.50
7500	Phlox subulata (Moss Phlox, Creeping Phlox), Z4, cont									
7510	1 qt.				Ea.	17.20			17.20	18.90
7520	1 gal.				"	26			26	28.50
7600	Polygonum auberti (Silver Fleece Vine), Z4, cont									
7610	1 qt.				Ea.	15.30			15.30	16.85
7620	1 gal.				"	27			27	30
7700	Polygonum cuspidatum compactum (Fleece Flower), Z4, cont									
7710	1 qt.				Ea.	19.60			19.60	21.50
7720	1 gal.				"	29			29	32
7800	Potentilla verna (Spring Cinquefoil), Z6, cont									
7810	1 qt.				Ea.	5.10			5.10	5.60
7820	1 gal.				"	10.50			10.50	11.55
7900	Rosmarinus officinalis "Prostratus" (Creeping Rosemary), Z8, cont									
7910	3" pot				Ea.	10.65			10.65	11.70
7920	1 gal.				"	37			37	41
8000	Sagina subulata (Irish Moss, Scotch Moss), Z6, cont									
8010	3" pot				Ea.	10.95			10.95	12
8020	4" pot					11.95			11.95	13.15
8030	1 qt.				↓	13			13	14.30
8100	Sedum (Stonecrop), Z4, cont									
8110	1 qt.				Ea.	8.95			8.95	9.85
8120	1 gal.				"	14.75			14.75	16.25
8200	Solanum jasminoides (Jasmine Nightshade), Z9, cont									
8210	1 gal.				Ea.	15			15	16.50
8220	5 gal.				"	32			32	35
8300	Stachys byzantina (Lamb's ears), Z4, cont									
8310	1 qt.				Ea.	14.25			14.25	15.70
8320	1 gal.				"	17.90			17.90	19.70
8400	Thymus pseudolanuginosus (Wooly Thyme), Z5, cont									
8410	3" pot				Ea.	8.60			8.60	9.45
8420	1 qt.					9.05			9.05	9.95
8430	1 gal.				↓	19.25			19.25	21
8500	Vaccinium crassifolium procumbent (Creeping Blueberry), Z7, cont									
8510	1 gal.				Ea.	17			17	18.70
8520	3 gal.				"	46			46	50.50
8600	Verbena peruviana "St. Paul", Z8, cont									
8610	2 qt.				Ea.	7.50			7.50	8.25
8620	1 gal.				"	16.05			16.05	17.65
8700	Veronica repens (Creeping Speedwell), Z5, cont									
8710	1 qt.				Ea.	10.90			10.90	12
8720	1 gal.				"	12.30			12.30	13.55
8800	Vinca major variegata (Variegated Greater Periwinkle), Z7, cont									

32 93 Plants

32 93 13 – Ground Covers

32 93 13.30 Ground Covers	Crew	Daily Output	Labor-Hours	Unit	Material	2021 Bare Costs Labor	Equipment	Total	Total Incl O&P
8810	2-1/4" pot				Ea.	8.10			8.10	8.90
8820	4" pot					5			5	5.50
8830	Bare Root				↓	1.56			1.56	1.72
8900	Vinca minor (Periwinkle), Z4, cont									
8910	2-1/4" pot				Ea.	2.40			2.40	2.64
8920	4" pot					10.45			10.45	11.50
8930	Bare Root				↓	16.95			16.95	18.65
9000	Waldsteinia sibirica (Barren Strawberry), Z4, cont									
9010	1 qt.				Ea.	12.50			12.50	13.75
9020	1 gal.				"	12.95			12.95	14.25
9100	Wisteria sinensis (Chinese Wisteria), Z5, cont									
9110	1 gal.				Ea.	42.50			42.50	47
9120	3 gal.				"	50			50	55

32 93 13.40 Ornamental Grasses

		Crew	Daily Output	Labor-Hours	Unit	Material	Labor	Equipment	Total	Total Incl O&P
0010	**ORNAMENTAL GRASSES**									
2000	Acorus gramineus (Japanese Sweet Flag), Z7, cont									
2010	1 gal.				Ea.	7			7	7.70
2100	Arundo donax (Giant Reed), Z7, cont									
2110	1 qt.				Ea.	11.95			11.95	13.15
2120	1 gal.				"	54.50			54.50	60
2200	Calamagrostis acutiflora stricta (Feather Reed Grass), Z5, cont									
2210	1 qt.				Ea.	8.45			8.45	9.30
2220	1 gal.				↓	29			29	32
2230	2 gal.				↓	55			55	60.50
2300	Carex elata "Bowles Golden" (Variegated Sedge), Z5, cont									
2310	1 qt.				Ea.	11.50			11.50	12.65
2320	1 gal.				"	13.60			13.60	14.95
2400	Carex morrowii "aureo variegata" (Japanese Sedge), Z5, cont									
2410	1 qt.				Ea.	18.95			18.95	21
2420	1 gal.					13.55			13.55	14.90
2430	2 gal.				↓	25			25	27.50
2500	Cortaderia selloana (Pampas Grass), Z8, cont									
2510	1 gal.				Ea.	9			9	9.90
2520	2 gal.					45			45	49.50
2530	3 gal.					25			25	27.50
2540	5 gal.				↓	95			95	105
2600	Deschampsia caespitosa (Tufted Hair Grass), Z4, cont									
2610	1 qt.				Ea.	30			30	33
2620	1 gal.				"	7.90			7.90	8.70
2700	Elymus glaucus (Blue Wild Rye), Z4, cont									
2710	1 qt.				Ea.	7.40			7.40	8.15
2720	1 gal.				"	8			8	8.80
2800	Erianthus ravennae (Ravenna Grass), Z6, cont									
2810	1 gal.				Ea.	24.50			24.50	27
2820	2 gal.					29.50			29.50	32.50
2830	7 gal.				↓	47			47	51.50
2900	Festuca cineria "Sea Urchin" (Sea Urchin Blue Fescue), Z6, cont									
2910	1 qt.				Ea.	15.95			15.95	17.55
2920	1 gal.					10			10	11
2930	2 gal.				↓	15			15	16.50
3000	Glyceria maxima variegata (Manna Grass), Z5, cont									
3010	1 qt.				Ea.	14.90			14.90	16.40

416

For customer support on your Site Work & Landscape Costs with RSMeans Data, call 800.448.8182.

32 93 13 – Ground Covers

32 93 13.40 Ornamental Grasses	Crew	Daily Output	Labor-Hours	Unit	Material	2021 Bare Costs Labor	Equipment	Total	Total Incl O&P
3020	1 gal.				Ea.	22.50			22.50	24.50
3100	Hakonechloa macra aureola (Variegated Hakonechloa), Z7, cont									
3110	1 qt.				Ea.	18.90			18.90	21
3120	1 gal.				"	21.50			21.50	23.50
3200	Helictotrichon sempervirens (Blue Oat Grass), Z4, cont									
3210	1 qt.				Ea.	7.20			7.20	7.95
3220	1 gal.					7.70			7.70	8.50
3230	2 gal.				↓	18.75			18.75	20.50
3300	Imperata cylindrica rubra (Japanese Blood Grass), Z7, cont									
3310	1 qt.				Ea.	6.90			6.90	7.60
3320	1 gal.					9.20			9.20	10.15
3330	2 gal.				↓	9			9	9.90
3400	Juncus effusus spiralis (Corkscrew Rush), Z4, cont									
3410	1 qt.				Ea.	7.75			7.75	8.50
3420	2 gal.				"	14.30			14.30	15.70
3500	Koeleria glauca (Blue Hair Grass), Z6, cont									
3510	1 qt.				Ea.	8.35			8.35	9.20
3520	1 gal.					8.90			8.90	9.80
3530	3 gal.				↓	23			23	25.50
3600	Miscanthus sinensis gracillimus (Maiden Grass), Z6, cont									
3610	1 gal.				Ea.	8.45			8.45	9.30
3620	2 gal.					14			14	15.40
3630	5 gal.				↓	24			24	26.50
3700	Miscanthus sinensis zebrinus (Zebra Grass), Z6, cont									
3710	1 gal.				Ea.	9			9	9.90
3720	2 gal.					12.05			12.05	13.25
3730	5 gal.				↓	22			22	24
3800	Molinia caerulea arundinacea (Tall Purple Moor Grass), Z5, cont									
3810	1 gal.				Ea.	19.35			19.35	21.50
3820	3 gal.				"	12.95			12.95	14.25
3900	Molinia caerulea variegata (Variegated Moor Grass), Z5, cont									
3910	1 qt.				Ea.	10.10			10.10	11.10
3920	1 gal.				"	19.35			19.35	21.50
4000	Pennisetum alopecuroides (Fountain Grass), Z6, cont									
4010	1 qt.				Ea.	6.90			6.90	7.60
4020	1 gal.					12.30			12.30	13.55
4030	2 gal.				↓	33.50			33.50	37
4100	Phalaris arundinacea "Picta" (Ribbon Grass), Z4, cont									
4110	1 qt.				Ea.	6.90			6.90	7.60
4120	1 gal.					11.60			11.60	12.75
4130	2 gal.				↓	13.30			13.30	14.60
4200	Sesleria autumnalis (Autumn Moor Grass), Z5, cont									
4210	1 qt.				Ea.	7.20			7.20	7.95
4220	1 gal.				"	14.35			14.35	15.75
4300	Spartina pectinata "Aureomarginata" (Cord Grass), Z5, cont									
4310	1 gal.				Ea.	14.35			14.35	15.80
4320	2 gal.				"	20.50			20.50	23
4400	Stipa gigantea (Giant Feather Grass), Z7, cont									
4410	1 gal.				Ea.	10.10			10.10	11.15
4420	2 gal.				"	13.50			13.50	14.85

32 93 23.10 Perennials	Crew	Daily Output	Labor-Hours	Unit	Material	2021 Bare Costs Labor	Equipment	Total	Total Incl O&P
0010 **PERENNIALS**									
0100 Achillea filipendulina (Fernleaf Yarrow), Z2, cont									
0110 1 gal.				Ea.	24.50			24.50	27
0120 2 gal.					12.35			12.35	13.60
0130 3 gal.				↓	16.50			16.50	18.15
0200 Agapanthus africanus (Lily-of-the-Nile), Z9, cont									
0210 1 gal.				Ea.	14			14	15.40
0220 5 gal.				"	21			21	23
0300 Amsonia ciliata (Blue Star), Z7, cont									
0310 1 qt.				Ea.	30			30	33
0320 1 gal.				"	28			28	31
0400 Amsonia tabernaemontana (Willow Amsonia), Z3, cont									
0410 1 qt.				Ea.	15			15	16.50
0420 1 gal.				"	20			20	22
0500 Anchusa azurea (Italian Bugloss), Z3, cont									
0510 4" pot				Ea.	8.95			8.95	9.85
0520 1 qt.				↓	25.50			25.50	28
0530 1 gal.					34.50			34.50	38
0600 Anemone x hybrida (Japanese Anemone), Z5, cont									
0610 4" pot				Ea.	10.05			10.05	11.05
0620 1 qt.					10			10	11
0630 2 gal.				↓	16.40			16.40	18.05
0700 Anthemis tinctoria (Golden Marguerite), Z3, cont									
0710 4" pot				Ea.	4.55			4.55	5
0720 1 qt.					7.95			7.95	8.75
0730 1 gal.				↓	15			15	16.50
0800 Aonitum napellus (Aconite Monkshood), Z2, cont									
0810 4" pot				Ea.	5.45			5.45	6
0820 1 qt.					8.70			8.70	9.60
0830 1 gal.				↓	17.40			17.40	19.15
0900 Aquilegia species and hybrids (Columbine), Z2, cont									
0910 2" pot				Ea.	3.83			3.83	4.21
0920 1 qt.					10.75			10.75	11.85
0930 1 gal.				↓	11.60			11.60	12.75
1000 Aruncus dioicus (Goatsbeard), Z4, cont									
1010 1 qt.				Ea.	51			51	56.50
1020 1 gal.					16.90			16.90	18.60
1030 2 gal.				↓	16.10			16.10	17.75
1100 Asclepias tuberosa (Butterfly Flower), Z4, cont									
1110 1 gal.				Ea.	18.55			18.55	20.50
1120 2 gal.				"	29.50			29.50	32.50
1200 Aster species and hybrids (Aster, Michaelemas Daisy), Z4, cont									
1210 1 gal.				Ea.	14.95			14.95	16.45
1220 2 gal.					22			22	24
1230 3 gal.				↓	31.50			31.50	35
1300 Astilbe species and hybrids (False Spirea), Z4, cont									
1310 4" pot				Ea.	7.50			7.50	8.25
1320 1 gal.					10.75			10.75	11.80
1330 2 gal.				↓	12.40			12.40	13.60
1400 Aubrieta deltoidea (Purple Rock Cress), Z4, cont									
1410 4" pot				Ea.	14.95			14.95	16.45
1420 1 qt.				↓	9			9	9.90

32 93 23.10 Perennials	Crew	Daily Output	Labor-Hours	Unit	Material	2021 Bare Costs Labor	Equipment	Total	Total Incl O&P
1430	1 gal.				Ea.	18.40			18.40	20
1500	Aurinia saxatilis (Basket of Gold), Z3, cont									
1510	4" pot				Ea.	10.90			10.90	11.95
1520	1 qt.					4.48			4.48	4.93
1530	1 gal.				↓	18.65			18.65	20.50
1600	Baptisia australis (False Indigo), Z3, cont									
1610	1 qt.				Ea.	7.30			7.30	8.05
1620	1 gal.				"	11.30			11.30	12.45
1700	Begonia grandis (Evans Begonia), Z6, cont									
1710	1 qt.				Ea.	14.65			14.65	16.15
1720	1 gal.				"	26			26	28.50
1800	Belamcanda chinensis (Blackberry Lily), Z5, cont									
1810	1 qt.				Ea.	11.95			11.95	13.15
1820	1 gal.				"	23			23	25
1900	Boltonia asteroides "Snowbank" (White Boltonia), Z3, cont									
1910	1 qt.				Ea.	12.30			12.30	13.55
1920	1 gal.				"	28			28	30.50
2000	Brunnera macrophylla (Siberian Bugloss), Z3, cont									
2010	1 qt.				Ea.	8.30			8.30	9.10
2020	1 gal.				"	9.20			9.20	10.10
2100	Caltha palustris (Marsh Marigold), Z3, cont									
2110	1 qt.				Ea.	14			14	15.40
2120	1 gal.				"	26.50			26.50	29
2200	Campanula carpatica "Blue Chips" (Bellflower), Z3, cont									
2210	1 gal.				Ea.	10.30			10.30	11.30
2220	2 gal.				"	16.25			16.25	17.85
2300	Ceratostigma plumbaginoides (Blue Plumbago), Z6, cont									
2310	1 qt.				Ea.	17.35			17.35	19.10
2320	1 gal.				"	25			25	27.50
2400	Chrysanthemum hybrids (Hardy Chrysanthemum), Z5, cont									
2410	1 qt.				Ea.	12.95			12.95	14.25
2420	6" to 8" pot					20			20	22
2430	1 gal.				↓	24			24	26.50
2500	Chrysogonum virginianum (Golden Star), Z6, cont									
2510	1 gal.				Ea.	5.25			5.25	5.80
2520	2 gal.				"	10.95			10.95	12.05
2600	Cimicifuga racemosa (Bugbane), Z5, cont									
2610	4" pot				Ea.	11.40			11.40	12.55
2620	1 qt.					15.60			15.60	17.15
2630	1 gal.				↓	17.95			17.95	19.75
2700	Coreopsis verticillata (Threadleaf Coreopsis), Z5, cont									
2710	2" pot				Ea.	12.10			12.10	13.30
2720	1 gal.				"	15.90			15.90	17.50
2800	Delphinium elatum and hybrids (Delphinium), Z2, cont									
2810	1 qt.				Ea.	9.15			9.15	10.05
2820	1 gal.					35			35	38.50
2830	2 gal.				↓	11.55			11.55	12.70
2900	Dianthus species and hybrids (Cottage Pink/Scotch Pink), Z3, cont									
2910	1 gal.				Ea.	10.75			10.75	11.85
2920	2 gal.				"	12.70			12.70	14
3000	Dicentra spectabilis (Bleeding Heart), Z3, cont									
3010	2 qt.				Ea.	18.15			18.15	20
3020	1 gal.				↓	14.85			14.85	16.35

For customer support on your Site Work & Landscape Costs with RSMeans Data, call 800.448.8182.

419

32 93 23.10 Perennials	Crew	Daily Output	Labor-Hours	Unit	Material	2021 Bare Costs Labor	Equipment	Total	Total Incl O&P
3030	2 gal.				Ea.	24.50			24.50	27
3100	Dictamnus (Gas Plant), Z3, cont									
3110	1 gal.				Ea.	35			35	38.50
3120	2 gal.					40.50			40.50	44.50
3130	3 gal.					52.50			52.50	57.50
3200	Digitalis purpurea (Foxglove), Z4, cont									
3210	1 gal.				Ea.	21			21	23
3220	2 gal.					50.50			50.50	56
3230	3 gal.					56			56	61.50
3300	Doronicum cordatun (Leopard's Bane), Z4, cont									
3310	2 qt.				Ea.	9.15			9.15	10.10
3320	1 gal.				"	15.55			15.55	17.10
3400	Echinacea purpurea (Purple Coneflower), Z3, cont									
3410	2 qt.				Ea.	6.45			6.45	7.10
3420	1 gal.					20.50			20.50	23
3430	2 gal.					21.50			21.50	23.50
3500	Echinops exaltatus (Globe Thistle), Z3, cont									
3510	1 qt.				Ea.	9.55			9.55	10.50
3520	1 gal.				"	14			14	15.40
3600	Erigeron speciosus (Oregon Fleabane), Z3, cont									
3610	1 qt.				Ea.	8.90			8.90	9.75
3620	1 gal.				"	13.80			13.80	15.20
3700	Euphorbia epithymoides (Cushion Spurge), Z4, cont									
3710	1 qt.				Ea.	8.65			8.65	9.55
3720	1 gal.				"	15			15	16.50
3800	Filipendula hexapetala (Meadowsweet), Z5, cont									
3810	1 qt.				Ea.	8.30			8.30	9.10
3820	1 gal.				"	10.35			10.35	11.40
3900	Gaillardia aristata (Blanket Flower), Z2, cont									
3910	1 qt.				Ea.	9.95			9.95	10.95
3920	1 gal.					9			9	9.90
3930	2 gal.					14.25			14.25	15.70
4000	Geranium sanguineum (Bloodred Geranium), Z3, cont									
4010	2" pot				Ea.	4.74			4.74	5.20
4020	1 gal.				"	10.90			10.90	12
4100	Geum x borisii (Boris Avens), Z3, cont									
4110	1 qt.				Ea.	8.20			8.20	9
4120	1 gal.				"	10.75			10.75	11.80
4200	Gypsophila paniculata (Baby's Breath), Z3, cont									
4210	2 qt.				Ea.	9.65			9.65	10.60
4220	1 gal.					10.65			10.65	11.70
4230	3 gal.					38.50			38.50	42.50
4300	Helenium autumnale (Sneezeweed), Z3, cont									
4310	2 qt.				Ea.	9.45			9.45	10.40
4320	1 gal.					12.35			12.35	13.55
4330	2 gal.					19.80			19.80	22
4400	Helianthus salicifolius (Willow Leaf Sunflower), Z5, cont									
4410	1 qt.				Ea.	8.10			8.10	8.90
4420	1 gal.					22			22	24
4430	2 gal.					23			23	25.50
4500	Heliopsis helianthoides (False Sunflower), Z3, cont									
4510	2 qt.				Ea.	9.45			9.45	10.40
4520	1 gal.					10.30			10.30	11.35

32 93 Plants

32 93 23 – Plants and Bulbs

32 93 23.10 Perennials	Crew	Daily Output	Labor-Hours	Unit	Material	2021 Bare Costs Labor	Equipment	Total	Total Incl O&P
4530	2 gal.				Ea.	16.70			16.70	18.35
4600	Helleborus niger (Christmas Rose), Z3, cont									
4610	2 qt.				Ea.	14.15			14.15	15.55
4620	1 gal.				"	22.50			22.50	25
4700	Hemerocallis hybrids (Diploid) (Day Lily), Z3, cont									
4710	1 qt.				Ea.	9			9	9.90
4720	1 gal.					10.15			10.15	11.20
4730	3 gal.				▼	18.70			18.70	20.50
4800	Hemerocallis hybrids (Tetraploid), Z3, cont									
4810	1 qt.				Ea.	8.80			8.80	9.65
4820	1 gal.					14.50			14.50	15.95
4830	3 gal.				▼	22			22	24.50
4900	Hibiscus moscheutos (Rose Mallow), Z5, cont									
4910	1 qt.				Ea.	8.15			8.15	8.95
4920	2 gal.					44			44	48.50
4930	3 gal.				▼	59			59	65
5000	Hosta species and hybrids (Plantain Lily), Z3, cont									
5010	1 gal.				Ea.	10.60			10.60	11.65
5020	2 gal.					11.60			11.60	12.75
5030	3 gal.				▼	22			22	24
5100	Iris hybrids (Pacific Coast Iris), Z6, cont									
5110	1 gal.				Ea.	10.65			10.65	11.70
5120	3 gal.				"	15.65			15.65	17.25
5200	Iris sibiricu and hybrids (Siberian Iris), Z3, cont									
5210	1 gal.				Ea.	10			10	11
5220	4 gal.					24			24	26
5230	3" pot				▼	6.20			6.20	6.80
5300	Kniphofia uvaria and hybrids (Torch Lily), Z7, cont									
5310	1 qt.				Ea.	7.05			7.05	7.75
5320	1 gal.				"	10.50			10.50	11.55
5400	Liatris spicata (Blazing Star, Gayfeather), Z3, cont									
5410	1 gal.				Ea.	10.25			10.25	11.25
5420	2 gal.				"	14.45			14.45	15.90
5500	Ligularia dentata (Golden Groundsel), Z5, cont									
5510	2 qt.				Ea.	9.95			9.95	10.95
5520	2 gal.				"	14.10			14.10	15.50
5600	Linum perenne (Perennial Flax), Z4, cont									
5610	1 qt.				Ea.	8.30			8.30	9.10
5620	1 gal.				"	10.65			10.65	11.70
5700	Lobelia cardinalis (Cardinal Flower), Z3, cont									
5710	1 qt.				Ea.	8.30			8.30	9.10
5720	1 gal.				"	10.65			10.65	11.70
5800	Lupinus "Rusell Hybrids" (Russell Lupines), Z3, cont									
5810	1 gal.				Ea.	12.15			12.15	13.35
5820	2 gal.					13			13	14.30
5830	3 gal.				▼	15.40			15.40	16.90
5900	Lychnis chalcedonica (Maltese Cross), Z3, cont									
5910	1 qt.				Ea.	8.05			8.05	8.85
5920	1 gal.				"	9			9	9.90
6000	Lysimachia clethroides (Gooseneck Loosestrife), Z3, cont									
6010	1 qt.				Ea.	8.20			8.20	9
6020	1 gal.				"	11.45			11.45	12.60
6100	Lythrum salicaria (Purple Loosestrife), Z3, cont									

32 93 23.10 Perennials	Crew	Daily Output	Labor-Hours	Unit	Material	2021 Bare Costs Labor	Equipment	Total	Total Incl O&P
6110	1 qt.				Ea.	8.15			8.15	9
6120	1 gal.				"	12.05			12.05	13.25
6200	Macleaya cordata (Plume Poppy), Z3, cont								
6210	1 qt.				Ea.	8.15			8.15	9
6220	1 gal.				"	12.15			12.15	13.40
6300	Mertensia virginica (Virginia Bluebells), Z3, cont								
6310	1 qt.				Ea.	8.05			8.05	8.90
6320	1 gal.				"	13.85			13.85	15.25
6400	Paeonia hybrids (Herbaceous Peony), Z5, cont								
6410	1 gal.				Ea.	10.65			10.65	11.70
6420	2 gal.					15.45			15.45	17
6430	3 gal.				↓	118			118	130
6500	Papaver orientalis (Oriental Poppy), Z3, cont								
6510	1 qt.				Ea.	24			24	26.50
6520	1 gal.				"	9.75			9.75	10.75
6600	Penstemon azureus (Azure Penstemon), Z8, cont								
6610	2" pot				Ea.	5.60			5.60	6.15
6620	1 gal.				"	10.35			10.35	11.35
6700	Phlox paniculata (Garden Phlox), Z4, cont								
6710	1 qt.				Ea.	8.60			8.60	9.45
6720	1 gal.				"	13			13	14.30
6800	Physostegia virginiana (False Dragonhead), Z3, cont								
6810	4" pot				Ea.	5.50			5.50	6.05
6820	1 gal.				"	9			9	9.90
6900	Platycodon grandiflorus (Balloon Flower), Z3, cont								
6910	2" pot				Ea.	5.85			5.85	6.45
6920	1 gal.				"	11.15			11.15	12.25
7000	Pulmonaria saccharata (Bethlehem Sage), Z4, cont								
7010	4" pot				Ea.	5.90			5.90	6.50
7020	1 qt.					8.30			8.30	9.10
7030	1 gal.				↓	10.65			10.65	11.70
7100	Rudbeckia fulgida "Goldstrum" (Black-eyed Susan), Z4, cont								
7110	1 qt.				Ea.	7.90			7.90	8.70
7120	1 gal.					10.85			10.85	11.95
7130	2 gal.				↓	13.90			13.90	15.30
7200	Salvia x superba (Perennial Salvia), Z5, cont								
7210	4" pot				Ea.	4.66			4.66	5.15
7220	1 qt.					10.10			10.10	11.15
7230	1 gal.				↓	9.60			9.60	10.55
7300	Scabiosa caucasica (Pincushion Flower), Z3, cont								
7310	1 qt.				Ea.	10.65			10.65	11.75
7320	1 gal.				"	12.20			12.20	13.40
7400	Sedum spectabile (Showy Sedum), Z3, cont								
7410	4" pot				Ea.	9.95			9.95	10.95
7420	2" pot					7.40			7.40	8.15
7430	1 gal.				↓	10.90			10.90	12
7500	Solidago hybrids (Goldenrod), Z3, cont								
7510	1 qt.				Ea.	8.30			8.30	9.10
7520	1 gal.				"	10.65			10.65	11.70
7700	Trollius europaeus (Common Globeflower), Z4, cont								
7710	1 qt.				Ea.	8.20			8.20	9.05
7720	1 gal.				"	11.45			11.45	12.55
7800	Veronica hybrids (Speedwell), Z4, cont								

32 93 Plants

32 93 23 – Plants and Bulbs

32 93 23.10 Perennials		Crew	Daily Output	Labor-Hours	Unit	Material	2021 Bare Costs Labor	Equipment	Total	Total Incl O&P
7810	1 gal.				Ea.	13.15			13.15	14.45
7820	2 gal.				"	18.95			18.95	21

32 93 33 – Shrubs

32 93 33.40 Shrubs

		Crew	Daily Output	Labor-Hours	Unit	Material	Labor	Equipment	Total	Total Incl O&P
0010	**SHRUBS**, Temperate Z2 to 6									
1000	Abelia grandiflora (Abelia), Z6, cont									
1002	15" to 18"				Ea.	42			42	46
1003	18" to 21"				"	43			43	47
1050	Acanthopanax sieboldianus (Five-Leaved Azalea), Z5, cont/B&B									
1051	15" to 18"				Ea.	32			32	35.50
1052	3' to 4'					40.50			40.50	44.50
1053	4' to 5'					41			41	45
1100	Aronia arbutifolia brilliantissima (Brilliant Chokeberry), Z5, B&B									
1101	2' to 3'				Ea.	31.50			31.50	35
1102	3' to 4'					41			41	45
1103	4' to 5'					44			44	48.50
1104	5' to 6'					71.50			71.50	78.50
1150	Aronia melancarpa "Elata" (Elata Black Chokeberry), Z3, B&B									
1151	2' to 3'				Ea.	15.70			15.70	17.25
1152	3' to 4'				"	19.40			19.40	21.50
1200	Azalea arborescens (Sweet Azalea), Z5, cont									
1201	15" to 18"				Ea.	22.50			22.50	24.50
1202	18" to 24"				"	39			39	43
1250	Azalea Evergreen Hybrids (Azalea Evergreen), Z6, cont/B&B									
1251	1 gal.				Ea.	6.15			6.15	6.80
1252	2 gal.					10.75			10.75	11.80
1253	3 gal.					15.40			15.40	16.90
1254	5 gal.					21.50			21.50	23.50
1255	15" to 18"					14.25			14.25	15.70
1256	18" to 24"					27.50			27.50	30
1257	2' to 2-1/2'					42			42	46
1258	2-1/2' to 3'					52.50			52.50	57.50
1300	Azalea exbury, Ilam Hybrids (Azalea exbury), Z5, cont/B&B									
1301	1 gal.				Ea.	6.90			6.90	7.60
1302	2 gal.					14.50			14.50	15.95
1303	15" to 18"					15.85			15.85	17.40
1304	18" to 24"					27.50			27.50	30
1305	2' to 2-1/2'					52.50			52.50	57.50
1306	2-1/2' to 3'					17.50			17.50	19.25
1307	4' to 5'					78.50			78.50	86.50
1350	Azalea gandavensis (Ghent Azalea), Z4, B&B									
1351	15" to 18"				Ea.	34.50			34.50	38
1352	18" to 24"					39.50			39.50	43.50
1353	2' to 3'					43			43	47.50
1354	3' to 4'					46			46	50.50
1355	4' to 5'					68.50			68.50	75
1356	5' to 6'					94.50			94.50	104
1400	Azalea kaempferi (Torch Azalea), Z6, cont/B&B									
1401	12" to 15"				Ea.	10.95			10.95	12.05
1402	15" to 18"					16.65			16.65	18.30
1403	18" to 24"					20.50			20.50	22.50
1404	2' to 3'					37.50			37.50	41

For customer support on your Site Work & Landscape Costs with RSMeans Data, call 800.448.8182.

423

32 93 Plants

32 93 33 – Shrubs

32 93 33.40 Shrubs		Crew	Daily Output	Labor-Hours	Unit	Material	2021 Bare Costs Labor	Equipment	Total	Total Incl O&P
1450	Berberis julianae "Nana" (Dwarf Wintergreen Barberry), Z6, B&B									
1451	18" to 24"				Ea.	24.50			24.50	27
1452	2' to 2-1/2'					22			22	24
1453	2-1/2' to 3'				▼	24.50			24.50	27
1500	Berberis mentorensis (Mentor Barberry), Z6, cont/B&B									
1501	1 gal.				Ea.	8.65			8.65	9.50
1502	5 gal.					16.75			16.75	18.45
1503	2' to 3'					22.50			22.50	24.50
1504	3' to 4'				▼	29.50			29.50	32.50
1550	Berberis thumbergii (Crimson Pygmy Barberry), Z5, cont									
1551	10" to 12"				Ea.	9.15			9.15	10.10
1552	12" to 15"					12.55			12.55	13.85
1553	15" to 18"				▼	21			21	23
1600	Berberis thunbergi (Japanese Green Barberry), Z5, cont/B&B									
1601	9" to 12"				Ea.	3.77			3.77	4.15
1602	12" to 15"					12.25			12.25	13.50
1603	15" to 18"					13.55			13.55	14.90
1604	18" to 24"					21.50			21.50	24
1605	2' to 3'				▼	37			37	40.50
1650	Berberis verruculosa (Barberry), Z5, cont/B&B									
1651	15" to 18"				Ea.	19.10			19.10	21
1652	18" to 24"				"	30			30	33
1700	Buddleia davidii (Orange-Eye Butterfly Bush), Z5, cont									
1701	9" to 12"				Ea.	12			12	13.20
1702	15" to 18"				"	17.85			17.85	19.60
1750	Buxus microphylla japonica (Japanese Boxwood), Z5, cont/B&B									
1751	1 gal.				Ea.	4.10			4.10	4.51
1752	8" to 10"					10.05			10.05	11.05
1753	12" to 15"					14.55			14.55	16
1754	15" to 18"					32.50			32.50	35.50
1755	18" to 24"					52			52	57
1756	24" to 30"					74			74	81.50
1757	30" to 36"				▼	84			84	92.50
1800	Buxus sempervirens (Common Boxwood), Z5, cont/B&B									
1801	10" to 12"				Ea.	18			18	19.80
1802	15" to 18"					21.50			21.50	23.50
1803	18" to 24"					18			18	19.80
1804	24" to 30"				▼	31.50			31.50	35
1850	Callicarpa japonica (Japanese Beautyberry), Z6, B&B									
1851	2' to 3'				Ea.	21			21	23
1852	3' to 4'					25			25	27.50
1853	4' to 5'				▼	37			37	40.50
1900	Caragana arborescens (Siberian Peashrub), Z2, B&B									
1901	18" to 24"				Ea.	13.95			13.95	15.35
1902	2' to 3'					24			24	26.50
1903	5' to 6'					66			66	72.50
1904	6' to 8'				▼	80.50			80.50	88.50
1950	Caragana arborescens (Walkers Weeping Peashrub), Z3, B&B									
1951	1-1/2" to 1-3/4" Cal.				Ea.	125			125	138
1952	1-3/4" to 2" Cal.				"	159			159	175
2000	Caryopteris x clandonensis (Bluebeard), Z5, cont									
2001	12" to 15"				Ea.	20			20	22
2002	1 gal.				▼	21.50			21.50	24

424

32 93 33 – Shrubs

32 93 33.40 Shrubs		Crew	Daily Output	Labor-Hours	Unit	Material	2021 Bare Costs Labor	Equipment	Total	Total Incl O&P
2003	18" to 24"				Ea.	31.50			31.50	35
2050	Chaenomeles speciosa (Flowering Quince), Z5, cont/B&B									
2051	10" to 12"				Ea.	13.45			13.45	14.80
2052	2' to 3'					17.55			17.55	19.30
2053	3' to 4'				▼	23			23	25.50
2100	Clethra alnifolia rosea (Pink Summersweet), Z4, B&B									
2101	2' to 3'				Ea.	21.50			21.50	23.50
2102	3' to 4'					32			32	35
2103	4' to 5'				▼	36			36	39.50
2150	Cornus alba siberica (Siberian Dogwood), Z3, B&B									
2151	3' to 4'				Ea.	25.50			25.50	28
2152	4' to 5'					33			33	36.50
2153	5' to 6'					62.50			62.50	69
2154	6' to 7'				▼	56.50			56.50	62
2200	Cornus alternifolia (Alternate Leaved Dogwood), Z3, B&B									
2201	4' to 5'				Ea.	74			74	81.50
2202	5' to 6'					81.50			81.50	89.50
2203	6' to 8'				▼	97			97	107
2250	Cornus mas (Cornelian Cherry), Z5, B&B									
2251	2' to 3'				Ea.	34			34	37.50
2252	3' to 4'					82			82	90
2253	4' to 5'					85			85	93.50
2254	5' to 6'				▼	106			106	117
2300	Cornus racemosa (Gray Dogwood), Z4, B&B									
2301	18" to 24"				Ea.	27.50			27.50	30.50
2302	2' to 3'					37.50			37.50	41.50
2303	3' to 4'				▼	58			58	63.50
2350	Cornus stolonifera flaviramea (Goldentwig Dogwood), Z3, cont/B&B									
2351	2' to 3'				Ea.	27			27	29.50
2352	3' to 4'					40			40	44
2353	4' to 5'					46.50			46.50	51
2354	5' to 6'					56			56	61.50
2355	6' to 7'				▼	65			65	71.50
2400	Corylus americana (Filbert, American Hazelnut), Z4, B&B									
2401	2-1/2' to 3'				Ea.	15.90			15.90	17.45
2402	3' to 4'					22			22	24
2403	4' to 5'				▼	46			46	50.50
2450	Cotinus coggygria (Smoke Tree), Z5, B&B									
2451	5 gal.				Ea.	43			43	47.50
2452	6' to 8'					66			66	72.50
2453	8' to 10'				▼	101			101	111
2500	Cotoneaster acutifolius (Peking Cotoneaster), Z4, B&B									
2501	2' to 3'				Ea.	13.75			13.75	15.15
2502	3' to 4'					19.35			19.35	21.50
2503	4' to 5'				▼	24.50			24.50	27
2550	Cotoneaster adpressa (Creeping Cotoneaster), Z5, cont									
2551	15" to 18" spread				Ea.	12.85			12.85	14.10
2600	Cotoneaster adpressa praecox (Early Cotoneaster), Z5, cont									
2601	12" to 15"				Ea.	20.50			20.50	22.50
2602	15" to 18"					32.50			32.50	35.50
2603	18" to 24"				▼	28			28	30.50
2650	Cotoneaster divaricatus (Spreading Cotoneaster), Z5, cont/B&B									
2651	1 gal.				Ea.	23.50			23.50	26

32 93 33.40 Shrubs		Crew	Daily Output	Labor-Hours	Unit	Material	2021 Bare Costs Labor	Equipment	Total	Total Incl O&P
2652	2 gal.				Ea.	25.50			25.50	28
2653	2' to 3'					22.50			22.50	24.50
2654	3' to 4'					38			38	41.50
2655	4' to 5'				↓	33.50			33.50	37
2700	Cotoneaster horizontalis (Rock Cotoneaster), Z5, cont									
2701	12" to 15"				Ea.	13.65			13.65	15
2702	15" to 18"					32.50			32.50	36
2703	18" to 21"				↓	41			41	45
2750	Cytisus praecox (Warminster Broom), Z6, cont									
2751	1 gal.				Ea.	6.15			6.15	6.80
2752	2 gal.					13			13	14.30
2753	5 gal.				↓	24.50			24.50	26.50
2800	Daphne burkwoodi somerset (Burkwood Daphne), Z6, cont									
2801	15" to 18"				Ea.	25			25	27.50
2802	18" to 24"				"	42			42	46.50
2850	Deutzia gracilis (Slender Deutzia), Z5, cont									
2851	1 gal.				Ea.	10			10	11
2852	2 gal.				"	45			45	49.50
2900	Diervilla sessilifolia (Southern Bush Honeysuckle), Z5, B&B									
2901	18" to 24"				Ea.	29			29	31.50
2902	2' to 3'					26			26	29
2903	3' to 4'				↓	39			39	43
2950	Elaeagnus angustifolia (Russian Olive), Z3, B&B									
2951	5 gal.				Ea.	27			27	29.50
3000	Elaeagnus umbellata (Autumn Olive), Z3, B&B									
3001	3' to 4'				Ea.	41.50			41.50	45.50
3002	4' to 5'					101			101	111
3003	5' to 6'				↓	191			191	210
3050	Enkianthus campanulatus (Bellflowertree), Z5, cont/B&B									
3051	2' to 3'				Ea.	50			50	55
3052	3' to 4'					56.50			56.50	62
3053	4' to 5'					60			60	65.50
3054	5' to 6'				↓	32.50			32.50	35.50
3100	Euonymus alatus (Winged Burning Bush), Z4, B&B									
3101	18" to 24"				Ea.	18.90			18.90	21
3102	2' to 2-1/2'					25			25	27.50
3103	2-1/2' to 3'					48			48	52.50
3104	3' to 3-1/2'					63			63	69
3105	3-1/2' to 4'					65.50			65.50	72
3106	4' to 5'					67			67	73.50
3107	5' to 6'					74			74	81
3108	6' to 8'				↓	84			84	92
3150	Euonymus alatus compacta (Dwarf Winged Burning Bush), Z4, B&B									
3151	15" to 18"				Ea.	29			29	32
3152	18" to 24"					27.50			27.50	30.50
3153	2' to 2-1/2'					33			33	36.50
3154	2-1/2' to 3'					30			30	33
3155	3' to 3-1/2'					51.50			51.50	56.50
3156	3-1/2' to 4'					59			59	65
3157	4' to 5'					61.50			61.50	67.50
3158	5' to 6'				↓	49.50			49.50	54.50
3200	Euonymus europaeus aldenhamensis (Spindletree), Z4, B&B									
3201	3' to 4'				Ea.	19.80			19.80	22

32 93 33.40 Shrubs		Crew	Daily Output	Labor-Hours	Unit	Material	2021 Bare Costs Labor	Equipment	Total	Total Incl O&P
3202	4' to 5'				Ea.	20			20	22
3203	5' to 6'					32			32	35.50
3204	6' to 8'				↓	65			65	71.50
3250	Euonymus fortunei (Wintercreeper), Z5, cont/B&B									
3251	1 gal.				Ea.	11.90			11.90	13.10
3252	2 gal.					15.90			15.90	17.45
3253	5 gal.					26.50			26.50	29.50
3254	24" to 30"				↓	24			24	26.50
3300	Euonymus fortunei "Emerald 'n Gold" (Wintercreeper), Z5, cont									
3301	10" to 12"				Ea.	4.49			4.49	4.94
3302	12" to 15"					8.30			8.30	9.15
3303	15" to 18"					23			23	25.50
3304	5 gal.				↓	22.50			22.50	25
3350	Euonymus fortunei emerald gaiety (Gaiety Wintercreeper), Z5, cont									
3351	10" to 12"				Ea.	6.95			6.95	7.65
3352	12" to 15"					15.60			15.60	17.15
3353	15" to 18"					25			25	27.50
3354	18" to 21"				↓	29			29	32
3400	Forsythia intermedia (Border Goldenbells), Z5, cont/B&B									
3401	5 gal.				Ea.	22			22	24
3402	2' to 3'					14.40			14.40	15.85
3403	3' to 4'					17.65			17.65	19.45
3404	4' to 5'					41			41	45.50
3405	5' to 6'					48			48	53
3406	6' to 8'				↓	54			54	59.50
3450	Forsythia ovata robusta (Korean Forsythia), Z5, B&B									
3451	2' to 3'				Ea.	34			34	37.50
3452	3' to 4'					50			50	55
3453	4' to 5'				↓	69			69	76
3500	Hamamelis vernalis (Vernal Witch-Hazel), Z4, B&B									
3501	2' to 3'				Ea.	41.50			41.50	45.50
3502	3' to 4'					51			51	56.50
3503	4' to 5'				↓	69.50			69.50	76.50
3550	Hibiscus syriacus (Rose of Sharon), Z6, cont/B&B									
3551	18" to 24"				Ea.	18.50			18.50	20.50
3552	2' to 3'					20.50			20.50	22.50
3553	3' to 4'					30.50			30.50	33.50
3554	4' to 5'					98			98	108
3555	5' to 6'				↓	56			56	61.50
3600	Hydrangea arborescens (Smooth Hydrangea), Z4, cont									
3601	15" to 18"				Ea.	29			29	32
3602	18" to 24"				"	16.60			16.60	18.30
3650	Hydrangea macrophylla nikko blue (Blue Hydrangea), Z6, cont									
3651	18" to 24"				Ea.	17.80			17.80	19.60
3652	24" to 30"				"	40			40	44
3700	Hydrangea paniculata grandiflora (Peegee Hydrangea), Z5, B&B									
3701	2' to 3'				Ea.	38			38	41.50
3702	3' to 4'				"	148			148	163
3750	Hydrangea quercifolia (Oakleaf Hydrangea), Z6, cont									
3751	18" to 24"				Ea.	28.50			28.50	31.50
3800	Hypericum hidcote (St. John's Wort), Z6, cont									
3801	12" to 15"				Ea.	28.50			28.50	31.50
3850	Ilex cornuta "Carissa" (Carissa Holly), Z7, cont									

32 93 33 – Shrubs

32 93 33.40 Shrubs		Crew	Daily Output	Labor-Hours	Unit	Material	2021 Bare Costs Labor	Equipment	Total	Total Incl O&P
3851	1 gal.				Ea.	24.50			24.50	27
3852	2 gal.					27			27	29.50
3853	5 gal.				↓	50			50	55
3900	Ilex cornuta rotunda (Dwarf Chinese Holly), Z6, cont									
3901	1 gal.				Ea.	35			35	38.50
3902	2 gal.					41			41	45
3903	5 gal.				↓	54.50			54.50	60
3950	Ilex crenata "Helleri" (Hellers Japanese Holly), Z6, cont									
3951	24" to 30"				Ea.	24			24	26
3952	30" to 36"					41.50			41.50	46
3953	3' to 4'				↓	49.50			49.50	54.50
4000	Ilex crenata "Hetzi" (Hetzi Japanese Holly), Z6, cont/B&B									
4001	2 gal.				Ea.	19.70			19.70	21.50
4002	3 gal.					33.50			33.50	37
4003	2' to 3'					86			86	95
4004	3' to 4'				↓	102			102	112
4050	Ilex glabra compacta (Compact Inkberry), Z4, B&B									
4051	15" to 18"				Ea.	24			24	26.50
4052	18" to 24"					32.50			32.50	36
4053	2' to 2-1/2'				↓	41.50			41.50	45.50
4100	Ilex meserve blue holly hybrids (Blue Holly), Z5, cont/B&B									
4101	15" to 18"				Ea.	24			24	26.50
4102	18" to 24"					23			23	25
4103	24" to 30"					42.50			42.50	46.50
4104	2-1/2' to 3'					52			52	57
4105	3-1/2' to 4'					59.50			59.50	65.50
4106	4' to 5'					107			107	117
4107	5' to 6'				↓	137			137	151
4150	Ilex opaca (American Holly), Z6, B&B									
4151	2' to 3'				Ea.	38			38	42
4152	3' to 4'					72			72	79
4153	4' to 5'					105			105	116
4154	5' to 6'					174			174	191
4155	6' to 7'					227			227	249
4156	7' to 8'					325			325	355
4157	8' to 10'					415			415	455
4158	10' to 12'					830			830	910
4159	12' to 14'				↓	900			900	990
4200	Ilex verticillata female (Winterberry), Z4, B&B									
4201	2' to 3'				Ea.	45.50			45.50	50
4202	3' to 4'				"	38			38	41.50
4250	Ilex verticillata male (Winterberry), Z4, B&B									
4251	2' to 3'				Ea.	29.50			29.50	32
4252	3' to 4'					38			38	41.50
4253	4' to 5'				↓	62			62	68
4300	Ilex x attenuata "Fosteri" (Foster Holly), Z6, B&B									
4301	4' to 5'				Ea.	135			135	149
4302	5' to 6'					205			205	225
4303	6' to 7'					273			273	300
4304	7' to 8'				↓	565			565	620
4350	Itea virginica (Virginia Sweetspire), Z6, B&B									
4351	21" to 30"				Ea.	46.50			46.50	51.50
4400	Kalmia latifolia (Mountain Laurel), Z5, cont/B&B									

428

32 93 33 – Shrubs

32 93 33.40 Shrubs		Crew	Daily Output	Labor-Hours	Unit	Material	2021 Bare Costs Labor	Equipment	Total	Total Incl O&P
4401	15" to 18"				Ea.	56.50			56.50	62.50
4402	18" to 24"					75.50			75.50	83
4403	2' to 3'					73.50			73.50	81
4404	4' to 5'					124			124	136
4405	5' to 6'					70			70	77
4406	6' to 8'					230			230	253
4450	Kerria japonica (Kerria), Z6, cont									
4451	12" to 15"				Ea.	16.20			16.20	17.85
4452	5 gal.				"	31			31	34
4500	Kolkwitzia amabilis (Beautybush), Z5, cont/B&B									
4501	2 gal.				Ea.	18.05			18.05	19.85
4502	15" to 18"					27.50			27.50	30
4503	2' to 3'					30			30	33
4504	3' to 4'					39.50			39.50	43.50
4505	4' to 5'					50			50	55
4506	5' to 6'					75.50			75.50	83
4507	6' to 8'					84.50			84.50	93
4550	Leucothoe axillaris (Coast Leucothoe), Z6, cont									
4551	15" to 18"				Ea.	20.50			20.50	22.50
4552	18" to 24"					24			24	26
4553	3 gal.					38.50			38.50	42.50
4600	Leucothoe fontanesiana (Drooping Leucothoe), Z6, cont/B&B									
4601	2 gal.				Ea.	35.50			35.50	39
4602	12" to 15"				"	24			24	26.50
4650	Ligustrum obtusifolium regelianum (Regal Privet), Z4, cont/B&B									
4651	18" to 24"				Ea.	30			30	33
4652	2' to 3'					15			15	16.50
4653	Bare root, 18" to 24"					9.35			9.35	10.25
4654	2' to 2-1/2'					10.35			10.35	11.35
4700	Ligustrum amurense (Amur Privet), Z4, Bare Root									
4701	2' to 3'				Ea.	15.30			15.30	16.85
4702	3' to 4'				"	29.50			29.50	32.50
4750	Lindera benzoin (Spicebush), Z5, cont/B&B									
4751	18" to 24"				Ea.	20.50			20.50	23
4752	24" to 30"				"	38.50			38.50	42.50
4800	Lonicera fragrantissima (Winter Honeysuckle), Z6, cont/B&B									
4801	2' to 3'				Ea.	32.50			32.50	36
4802	3' to 4'				"	42.50			42.50	46.50
4850	Lonicera xylosteum (Emerald Mound Honeysuckle), Z4, cont									
4851	15" to 18"				Ea.	21			21	23
4852	3 gal.					28.50			28.50	31.50
4853	5 gal.					41.50			41.50	45.50
4900	Mahonia aquifolium (Oregon Grape Holly), Z6, cont/B&B									
4901	1 gal.				Ea.	5.60			5.60	6.15
4902	2 gal.					17.80			17.80	19.55
4903	5 gal.					21.50			21.50	24
4950	Myrica pensylvanica (Northern Bayberry), Z2, B&B									
4951	15" to 18"				Ea.	20			20	22
4952	2' to 2-1/2'					31			31	34
4953	2-1/2' to 3'					45			45	49.50
5000	Osmanthus heterophyllus "Gulftide" (Gulftide Sweet Holly), Z6, B&B									
5001	18" to 24"				Ea.	22.50			22.50	25
5002	2' to 2-1/2'					43.50			43.50	48

429

For customer support on your Site Work & Landscape Costs with RSMeans Data, call 800.448.8182.

32 93 33.40 Shrubs		Crew	Daily Output	Labor-Hours	Unit	Material	2021 Bare Costs Labor	Equipment	Total	Total Incl O&P
5003	2-1/2' to 3'				Ea.	48			48	53
5004	3' to 3-1/2'				↓	53.50			53.50	58.50
5050	Paxistima canbyi (Canby Paxistima), Z5, B&B									
5051	6" to 9"				Ea.	8.65			8.65	9.50
5052	9" to 12"					11.50			11.50	12.65
5053	12" to 15"				↓	14.60			14.60	16.05
5100	Philadelphus coronarius (Sweet Mockorange), Z5, cont/B&B									
5101	2' to 3'				Ea.	26.50			26.50	29
5102	3' to 4'					54.50			54.50	59.50
5103	4' to 5'					15			15	16.50
5104	5' to 6'				↓	15			15	16.50
5150	Philadelphus virginalis (Hybrid Mockorange), Z6, cont/B&B									
5151	18" to 24"				Ea.	19.25			19.25	21
5152	2' to 3'					28			28	31
5153	3' to 4'					47			47	51.50
5154	4' to 5'				↓	99.50			99.50	109
5200	Physocarpus opulifolius luteus (Golden Ninebark), Z3, cont/B&B									
5201	5 gal.				Ea.	35			35	38.50
5202	3' to 4'					25			25	27.50
5203	4' to 5'				↓	23.50			23.50	25.50
5250	Pieris floribunda (Mountain Andromeda), Z5, cont									
5251	12" to 15" spread				Ea.	18.70			18.70	20.50
5252	15" to 18" spread				"	35			35	38.50
5300	Pieris japonica (Japanese Andromeda), Z6, cont/B&B									
5301	9" to 12"				Ea.	16.30			16.30	17.95
5302	12" to 15"					30			30	33
5303	18" to 24"					35			35	38.50
5304	24" to 30"					64			64	70.50
5305	2-1/2' to 3'				↓	85			85	93.50
5350	Potentilla fruticosa (Shrubby Cinquefoil), Z2, cont/B&B									
5351	1 gal.				Ea.	35			35	38.50
5352	10" to 12"					14.20			14.20	15.60
5353	12" to 15"					10			10	11
5354	15" to 18"					14.10			14.10	15.55
5355	18" to 24"					20.50			20.50	23
5356	2' to 3'				↓	46.50			46.50	51.50
5400	Prunus cistena (Purple-Leaf Sand Cherry), Z3, cont/B&B									
5401	5 gal.				Ea.	12.95			12.95	14.20
5402	18" to 24"					21			21	23
5403	2' to 3'					4.95			4.95	5.45
5404	3' to 4'					24			24	26.50
5405	4' to 5'					30			30	33
5406	5' to 6'				↓	46.50			46.50	51.50
5450	Prunus glandulosa albo plena (Flowering Almond), Z6, cont/B&B									
5451	15" to 18"				Ea.	33.50			33.50	36.50
5452	18" to 24"					130			130	143
5453	2' to 3'					150			150	165
5454	3' to 4'				↓	160			160	176
5500	Prunus glandulosa sinensis (Pink Flowering Almond), Z6, cont/B&B									
5501	5 gal.				Ea.	98.50			98.50	108
5502	18" to 24"				"	33			33	36
5503	2' to 3'				Ea.	15.65			15.65	17.25
5550	Prunus laurocerasus schiphaewsis (Cherry Laurel), Z5, B&B									

32 93 33.40 Shrubs	Crew	Daily Output	Labor-Hours	Unit	Material	2021 Bare Costs Labor	Equipment	Total	Total Incl O&P
5551	5 gal.				Ea.	19.80			19.80	22
5552	18" to 24"					47			47	51.50
5553	24" to 30"					49.50			49.50	54.50
5600	Prunus maritima (Beach Plum), Z4, cont/B&B									
5601	1 gal.				Ea.	49			49	54
5602	18" to 24"					52			52	57
5603	2' to 3'					71			71	78
5650	Pyracantha coccinea (LaLande Firethorn), Z6, cont/B&B									
5651	1 gal.				Ea.	33			33	36.50
5652	5 gal.					34			34	37.50
5653	5 gal. Espalier					13.90			13.90	15.30
5654	10 gal. Espalier					17.55			17.55	19.30
5700	Rhamus frangula columnaris (Tallhedge), Z2, B&B									
5701	2' to 3'				Ea.	23.50			23.50	25.50
5702	3' to 4'					50			50	55
5703	4' to 5'					37.50			37.50	41
5704	5' to 6'					51			51	56
5750	Rhododendron carolinianum (Carolina Rhododendron), Z5, cont/B&B									
5751	15" to 18"				Ea.	21			21	23
5752	18" to 24"					38.50			38.50	42
5753	24" to 30"					50.50			50.50	55.50
5800	Rhododendron catawbiense hybrids (Rhododendron), Z5, cont/B&B									
5801	1 gal.				Ea.	17.55			17.55	19.30
5802	2 gal.					25			25	27.50
5803	5 gal.					31.50			31.50	34.50
5804	12" to 15"					36.50			36.50	40
5805	15" to 18"					32.50			32.50	35.50
5806	18" to 24"					47			47	52
5807	24" to 30"					63.50			63.50	70
5808	30" to 36"					69.50			69.50	76.50
5809	36" to 42"					69.50			69.50	76.50
5850	Rhododendron PJM (PJM Rhododendron), Z4, cont/B&B									
5851	15" to 18"				Ea.	50			50	55
5852	18" to 24"					75			75	82.50
5853	2' to 2-1/2'					90			90	99
5854	2-1/2' to 3'					78			78	86
5855	3' to 3-1/2'					94			94	103
5900	Rhus aromatica "Gro-Low" (Fragrant Sumac), Z3, cont									
5901	15" to 18"				Ea.	60			60	66
5902	3 gal.					12.50			12.50	13.75
5903	5 gal.					10			10	11
5950	Rhus glabra cismontana (Dwarf Smooth Sumac), Z2, cont									
5951	1 gal.				Ea.	46.50			46.50	51
5952	5 gal.				"	23.50			23.50	26
6000	Rhus typhina lacianata (Cut-Leaf Staghorn Sumac), Z4, cont/B&B									
6001	5 gal.				Ea.	26			26	28.50
6002	7 gal.					30.50			30.50	33.50
6003	6' to 7'					62.50			62.50	69
6050	Ribes alpinum (Alpine Currant), Z3, cont/B&B									
6051	18" to 24"				Ea.	12.75			12.75	14.05
6052	2' to 3'				"	22.50			22.50	25
6100	Rosa hugonis (Father Hugo Rose), Z6, cont									
6101	1 gal.				Ea.	10.80			10.80	11.90

32 93 33.40 Shrubs		Crew	Daily Output	Labor-Hours	Unit	Material	2021 Bare Costs Labor	Equipment	Total	Total Incl O&P
6102	2 gal.				Ea.	15.50			15.50	17.05
6150	Rosa meidiland hybrids, Z6, cont									
6151	9" to 12"				Ea.	19.05			19.05	21
6152	12" to 15"				"	24			24	26.50
6200	Rosa multiflora (Japanese Rose), Z6, cont									
6201	1 gal.				Ea.	4.50			4.50	4.95
6202	2 gal.				"	11.55			11.55	12.70
6250	Rosa rugosa and hybrids, Z4, cont									
6251	18" to 24"				Ea.	23			23	25.50
6252	2' to 3'				"	22.50			22.50	25
6300	Rosa: hybrid teas, grandifloras and climbers, Zna, cont									
6301	2 gal. Non Pat.				Ea.	20.50			20.50	22.50
6302	2 gal. Patented				"	17			17	18.70
6350	Salix caprea (French Pussy Willow), Z5, B&B									
6351	4' to 5'				Ea.	33			33	36
6352	5' to 6'				"	43.50			43.50	48
6400	Salix discolor (Pussy Willow), Z2, cont									
6401	2' to 3'				Ea.	25			25	27.50
6450	Sambucus canadensis (American Elder), Z4, B&B									
6451	2' to 3'				Ea.	26.50			26.50	29
6500	Skimmia japonica (Japanese Skimmia), Z6, cont									
6501	1 gal.				Ea.	9.35			9.35	10.30
6502	2 gal.					15.45			15.45	17
6503	5 gal.					26			26	28.50
6550	Spiraea arguta compacta (Garland Spirea), Z5, B&B									
6551	3' to 4'				Ea.	25			25	27.50
6552	4' to 5'				"	40			40	44
6600	Spirea bumalda (Bumalda Spirea), Z5, cont/B&B									
6601	1 gal.				Ea.	8.75			8.75	9.65
6602	5 gal.					33			33	36.50
6603	15" to 18"					17.80			17.80	19.55
6604	18" to 24"					18.85			18.85	20.50
6650	Spirea japonica (Japanese Spirea), Z5, cont									
6651	15" to 18"				Ea.	14.25			14.25	15.70
6652	18" to 24"					28.50			28.50	31.50
6653	3 gal.					18.50			18.50	20.50
6654	5 gal.					21			21	23.50
6700	Spirea nipponica snowmound (Snowmound Spirea), Z5, cont/B&B									
6701	15" to 18"				Ea.	14.05			14.05	15.50
6702	18" to 24"					21			21	23
6703	2' to 3'					31.50			31.50	35
6704	3' to 4'					45.50			45.50	50
6705	4' to 5'					66			66	72.50
6750	Spirea vanhouttei (Van Houtte Spirea), Z4, cont/B&B									
6751	15" to 18"				Ea.	14.35			14.35	15.80
6752	2' to 3'					30.50			30.50	34
6753	3' to 4'					40.50			40.50	44.50
6754	4' to 5'					39.50			39.50	43.50
6755	5' to 6'					55.50			55.50	61
6800	Symphoricarpos albus (Common Snowberry), Z4, cont/B&B									
6801	18" to 24"				Ea.	10.70			10.70	11.80
6802	2' to 3'					17.30			17.30	19.05
6803	3' to 4'					20.50			20.50	22.50

32 93 33.40 Shrubs		Crew	Daily Output	Labor-Hours	Unit	Material	2021 Bare Costs Labor	Equipment	Total	Total Incl O&P
6850	Syringa chinensis (Roven Lilac), Z6, B&B									
6851	3' to 4'				Ea.	26.50			26.50	29
6852	4' to 5'					50.50			50.50	55.50
6853	5' to 6'				▼	86			86	94.50
6900	Syringa hugo koster (Hugo Koster Lilac), Z4, B&B									
6901	2' to 3'				Ea.	29.50			29.50	32.50
6902	3' to 4'					31.50			31.50	35
6903	4' to 5'					97			97	107
6904	5' to 6'				▼	49.50			49.50	54.50
6950	Syringa meyeri (Meyer Lilac), Z6, cont/B&B									
6951	12" to 15"				Ea.	57			57	62.50
6952	15" to 18"					51			51	56
6953	18" to 24"					69			69	76
6954	2' to 3'					75.50			75.50	83
6955	3' to 4'				▼	117			117	129
7000	Syringa persica (Persian Lilac), Z5, cont									
7001	5 gal.				Ea.	35			35	38.50
7050	Syringa prestonae hybrids (Preston Lilac), Z3, B&B									
7051	2' to 3'				Ea.	95			95	105
7052	3' to 4'					44.50			44.50	48.50
7053	4' to 5'					120			120	132
7054	5' to 6'					162			162	179
7055	6' to 7'				▼	168			168	185
7100	Syringa vulgaris and Mixed Hybrids (Common Lilac), Z4, cont/B&B									
7101	5 gal.				Ea.	95			95	105
7102	3' to 4'					95			95	105
7103	4' to 5'					200			200	220
7104	5' to 6'					52			52	57
7105	6' to 8'				▼	63			63	69.50
7200	Vaccinium corymbosum (Highbush Blueberry), Z4, cont/B&B									
7201	15" to 18"				Ea.	57			57	62.50
7202	18" to 24"					12			12	13.20
7203	2' to 3'					33.50			33.50	36.50
7204	3' to 4'				▼	41			41	45
7250	Viburnum bodnantense (Bodnant Viburnum), Z5, cont/B&B									
7251	18" to 24"				Ea.	11.35			11.35	12.45
7252	3' to 4'					20.50			20.50	22.50
7253	4' to 5'					39.50			39.50	43.50
7254	5' to 6'				▼	68.50			68.50	75.50
7300	Viburnum carlcephalum (Fragrant Snowball), Z6, cont/B&B									
7301	2' to 3'				Ea.	13.50			13.50	14.85
7302	3' to 4'					22			22	24
7303	4' to 5'				▼	66.50			66.50	73.50
7350	Viburnum carlesi (Korean Spice Viburnum), Z4, B&B									
7351	18" to 24"				Ea.	66			66	72.50
7352	2' to 2-1/2'					66			66	72.50
7353	2-1/2' to 3'					85			85	93.50
7354	3' to 4'				▼	37			37	40.50
7400	Viburnum dilatatum (Linden Viburnum), Z6, B&B									
7401	3' to 4'				Ea.	37			37	40.50
7402	4' to 5'					75			75	82.50
7403	5' to 6'				▼	38			38	42
7450	Viburnum juddii (Judd Viburnum), Z5, B&B									

32 93 Plants

32 93 33 – Shrubs

32 93 33.40 Shrubs	Crew	Daily Output	Labor-Hours	Unit	Material	2021 Bare Costs Labor	Equipment	Total	Total Incl O&P
7451	18" to 24"				Ea.	85			85	93.50
7452	2' to 2-1/2'					22			22	24.50
7453	2-1/2' to 3'				↓	37			37	40.50
7454	3' to 4'				Ea.	55.50			55.50	61.50
7500	Viburnum opulus sterile (Common Snowball), Z4, cont/B&B									
7501	2 gal.				Ea.	16.95			16.95	18.65
7502	5 gal.					45			45	49.50
7503	2' to 3'					37			37	40.50
7504	3' to 4'				↓	37			37	40.50
7550	Viburnum plicatum (Japanese Snowball), Z6, cont/B&B									
7551	18" to 21"				Ea.	95			95	105
7552	5' to 6'				"	130			130	143
7600	Viburnum plicatum var. tomentosum (Viburnum), Z5, cont/B&B									
7601	18" to 21"				Ea.	52.50			52.50	58
7602	2' to 3'					35			35	38.50
7603	3' to 4'					37			37	40.50
7604	4' to 5'				↓	25			25	27.50
7650	Viburnum prunifolium (Blackhaw), Z4, B&B									
7651	3' to 4'				Ea.	47			47	51.50
7652	4' to 5'					86			86	94.50
7653	5' to 6'					120			120	132
7654	6' to 7'				↓	61			61	67
7700	Viburnum rhytidophyllum (Leatherleaf Viburnum), Z6, B&B									
7701	2' to 3'				Ea.	44			44	48.50
7702	3' to 4'					25			25	27.50
7703	4' to 5'				↓	95			95	105
7750	Viburnum sieboldi (Siebold Viburnum), Z5, B&B									
7751	3' to 4'				Ea.	94.50			94.50	104
7752	4' to 5'					54			54	59.50
7753	5' to 6'					89.50			89.50	98.50
7754	6' to 7'				↓	86.50			86.50	95
7800	Viburnum trilobum (American Cranberry Bush), Z2, cont/B&B									
7801	2' to 3'				Ea.	19.90			19.90	22
7802	3' to 4'					22			22	24
7803	5' to 6'					75			75	82.50
7804	6' to 8'					50			50	55
7805	8' to 10'				↓	95			95	105
7850	Weigela florida and Hybrids (Old Fashioned Weigela), Z6, cont/B&B									
7851	2' to 3'				Ea.	15			15	16.50
7852	3' to 4'					23.50			23.50	26
7853	4' to 5'					55			55	60.50
7854	5' to 6'				↓	75			75	82.50
7900	Xanthorhiza simplicissima (Yellowroot), Z4, cont									
7901	4" pot				Ea.	8			8	8.80
7902	1 gal.				"	17.85			17.85	19.65
7950	Yucca filamentosa (Adams Needle), Z4, cont									
7951	9" to 12"				Ea.	3.99			3.99	4.39
7952	15" to 18"					23			23	25.50
7953	18" to 24"				↓	10.45			10.45	11.50

434

For customer support on your Site Work & Landscape Costs with RSMeans Data, call 800.448.8182.

32 93 33.50 Shrubs	Crew	Daily Output	Labor-Hours	Unit	Material	2021 Bare Costs Labor	Equipment	Total	Total Incl O&P
0010 **SHRUBS**, Warm temperate									
0011 & subtropical, Z7 to 10									
1000 Aspidistra elatior (Cast Iron Plant), Z7, cont									
1010 1 gal.				Ea.	6.45			6.45	7.10
1020 2 gal.				"	18.40			18.40	20.50
1100 Aucuba japonica "Variegata" (Gold Dust Plant), Z7, cont									
1110 1 gal.				Ea.	9.75			9.75	10.70
1120 3 gal.					62.50			62.50	68.50
1130 5 gal.				▼	36			36	39.50
1200 Azara microphylla (Boxleaf Azara), Z8, cont									
1210 5 gal.				Ea.	29.50			29.50	32
1220 15 gal.				"	65.50			65.50	72
1300 Boronia megastigma (Sweet Boronia), Z9, cont									
1310 1 gal.				Ea.	8.65			8.65	9.50
1320 5 gal.				"	25			25	27.50
1400 Brunfelsia pauciflora calycina (Brazil Raisin-tree), Z10, cont									
1410 1 gal.				Ea.	8.05			8.05	8.85
1420 5 gal.				"	24			24	26.50
1500 Calliandra haematocephala (Pink Powerpuff), Z10, cont									
1510 1 gal.				Ea.	7.45			7.45	8.20
1520 5 gal.					16.10			16.10	17.70
1530 15 gal.				▼	64.50			64.50	71
1600 Callistemon citrinus (Lemon Bottlebrush), Z9, cont									
1610 1 gal.				Ea.	5.35			5.35	5.90
1620 5 gal.					75			75	82.50
1630 10 gal.					48.50			48.50	53.50
1640 20 gal.				▼	82.50			82.50	91
1700 Camellia japonica (Japanese Camellia), Z7, cont									
1710 1 gal.				Ea.	9.30			9.30	10.25
1720 3 gal.					12.95			12.95	14.25
1730 5 gal.				▼	29			29	32
1800 Camellia sasanqua (Sasanqua Camellia), Z7, cont									
1810 1 gal.				Ea.	11.05			11.05	12.20
1820 3 gal.					12.60			12.60	13.85
1830 5 gal.				▼	27.50			27.50	30
1900 Carissa grandiflora (Natal Plum), Z9, cont									
1910 1 gal.				Ea.	6.20			6.20	6.85
1920 5 gal.					10.10			10.10	11.10
1930 15 gal.				▼	61			61	67
2000 Carpenteria californica (Evergreen Mockorange), Z8, cont									
2010 1 gal.				Ea.	9.35			9.35	10.30
2020 5 gal.				"	30			30	33
2100 Cassia artemisioides (Feathery Cassia), Z9, cont									
2110 1 gal.				Ea.	6.25			6.25	6.90
2120 5 gal.					14			14	15.40
2130 15 gal.				▼	69			69	76
2200 Ceanothus concha (Concha Wild Lilac), Z8, cont									
2210 1 gal.				Ea.	5.80			5.80	6.35
2220 5 gal.				"	46			46	50.50
2300 Choisa ternata (Mexican Orange), Z7, cont									
2310 1 gal.				Ea.	10.35			10.35	11.35
2320 5 gal.				"	25.50			25.50	28

435

32 93 33.50 Shrubs		Crew	Daily Output	Labor-Hours	Unit	Material	2021 Bare Costs Labor	Equipment	Total	Total Incl O&P
2400	Cistus Hybridus (White Rockrose), Z7, cont									
2410	1 gal.				Ea.	6.40			6.40	7.05
2420	5 gal.				"	13.75			13.75	15.10
2500	Coprosma repens (Mirror Plant), Z9, cont									
2510	1 gal.				Ea.	5.70			5.70	6.30
2520	5 gal.				"	15			15	16.50
2600	Correa pulchella "Carminebells" (Australian Fuchsia), Z9, cont									
2610	1 gal.				Ea.	7			7	7.70
2620	5 gal.				"	18			18	19.80
2700	Cyrtomium falcatum (Japanese Holly Fern), Z10, cont									
2710	1 gal.				Ea.	9.35			9.35	10.30
2720	5 gal.				"	21.50			21.50	23.50
2800	Daphne odora "Marginata" (Variegated Winter Daphne), Z7, cont									
2810	2 gal.				Ea.	32.50			32.50	36
2820	5 gal.					65.50			65.50	72
2830	15 gal.					103			103	114
2900	Dizygotheca elegantissima (False Aralia), Z10, cont									
2910	1 gal.				Ea.	24			24	26
2920	5 gal.					15.70			15.70	17.25
2930	15 gal.					64.50			64.50	71
3000	Dodonaea viscosa (Akeake), Z10, cont									
3010	1 gal.				Ea.	5.50			5.50	6.05
3020	5 gal.					13.10			13.10	14.40
3030	15 gal.					42.50			42.50	46.50
3100	Eleagnus ebbingei "Gilt Edge" (Variegated Silverberry), Z7, cont									
3110	1 gal.				Ea.	4.50			4.50	4.95
3120	5 gal.					13.20			13.20	14.50
3130	15 gal.					49			49	54
3200	Eleagnus pungens (Silverberry), Z7, cont									
3210	1 gal.				Ea.	4.53			4.53	4.98
3220	3 gal.					10.55			10.55	11.60
3230	5 gal.					13.15			13.15	14.50
3300	Escallonia rubra (Red Escallonia), Z8, cont									
3310	1 gal.				Ea.	5.55			5.55	6.15
3320	5 gal.					75			75	82.50
3330	15 gal.					75			75	82.50
3400	Euonymus japonica "Aureo-Variegata" (Gold Euonymus), Z7, cont									
3410	1 gal.				Ea.	6.30			6.30	6.90
3420	2 gal.					18.70			18.70	20.50
3430	5 gal.					31			31	34
3500	Fatsia japonica (Japanese Fatsia), Z8, cont									
3510	1 gal.				Ea.	25.50			25.50	28
3520	3 gal.					9.70			9.70	10.65
3530	5 gal.					11.30			11.30	12.45
3540	15 gal.					50			50	55
3600	Fremontodendron californicum (Pacific "Fremontia"), Z9, cont									
3610	2 gal.				Ea.	24.50			24.50	27
3620	5 gal.					44			44	48
3630	15 gal.					123			123	135
3700	Gardenia jasminoides (Cape Jasmine), Z8, cont									
3710	1 gal.				Ea.	8.40			8.40	9.25
3720	3 gal.					17.30			17.30	19.05
3730	5 gal.					19.70			19.70	21.50

32 93 33.50 Shrubs		Crew	Daily Output	Labor-Hours	Unit	Material	2021 Bare Costs Labor	Equipment	Total	Total Incl O&P
3800	Garrya elliptica (Silk Tassel), Z8, cont									
3810	1 gal.				Ea.	10.40			10.40	11.40
3820	5 gal.				"	23			23	25.50
3900	Grewia caffra (Lavender Starflower), Z9, cont									
3910	5 gal.				Ea.	14.75			14.75	16.25
3920	15 gal.				"	28.50			28.50	31.50
4000	Hebe buxifolia (Boxleaf Hebe), Z7, cont									
4010	1 gal.				Ea.	5.05			5.05	5.55
4020	5 gal.				"	13.85			13.85	15.25
4100	Heteromeles arbutifolia (Toyon), Z9, cont									
4110	1 gal.				Ea.	12.10			12.10	13.30
4120	5 gal.					27			27	29.50
4130	15 gal.				↓	81			81	89.50
4200	Hibiscus rosa-sinensis (Chinese Hibiscus), Z9, cont									
4210	1 gal.				Ea.	5.75			5.75	6.35
4220	5 gal.					17			17	18.70
4230	15 gal.				↓	46			46	50.50
4300	Ilex vomitoria (Yaupon Holly), Z7, cont									
4310	1 gal.				Ea.	9.95			9.95	10.90
4320	3 gal.					36.50			36.50	40.50
4330	10 gal.				↓	95.50			95.50	105
4400	Illicium parvifolium (Florida Anise Tree), Z8, cont									
4410	1 gal.				Ea.	9.45			9.45	10.35
4420	5 gal.				"	25			25	27.50
4500	Lagerstroemia indica (Crapemyrtle), Z7, cont									
4510	1 gal.				Ea.	4.78			4.78	5.25
4520	5 gal.					19.60			19.60	21.50
4530	3' to 4'					61.50			61.50	67.50
4540	4' to 5'					97			97	107
4550	5' to 6'				↓	102			102	112
4600	Ligustrum japonicum texanum (Wax Leaf Privet), Z7, cont									
4610	1 gal.				Ea.	7.05			7.05	7.80
4620	2 gal.					13.65			13.65	15
4630	5 gal.				↓	23			23	25
4700	Ligustrum lucidum (Glossy Privet), Z7, cont									
4710	1 gal.				Ea.	7.10			7.10	7.80
4720	5 gal.					13.75			13.75	15.10
4730	15 gal.				↓	78			78	86
4800	Ligustrum sinensis "Variegatum" (Variegated Privet), Z7, cont									
4810	1 gal.				Ea.	18.90			18.90	21
4820	2 gal.					13.45			13.45	14.80
4830	3 gal.				↓	16.70			16.70	18.35
4900	Mahonia lomariifolia (Burmese Grape), Z8, cont									
4910	1 gal.				Ea.	7.80			7.80	8.55
4920	5 gal.					19.65			19.65	21.50
4930	15 gal.				↓	60			60	66
5000	Myoporum laetum (Coast Sandalwood), Z9, cont									
5010	1 gal.				Ea.	7.05			7.05	7.75
5020	5 gal.					21.50			21.50	23.50
5030	15 gal.				↓	62			62	68.50
5100	Myrica californica (Pacific Wax Myrtle), Z7, cont									
5110	1 gal.				Ea.	11.95			11.95	13.10
5120	5 gal.				"	38.50			38.50	42.50

For customer support on your Site Work & Landscape Costs with RSMeans Data, call 800.448.8182.

437

32 93 Plants

32 93 33 – Shrubs

32 93 33.50 Shrubs	Crew	Daily Output	Labor-Hours	Unit	Material	2021 Bare Costs Labor	Equipment	Total	Total Incl O&P
5200 Myrtus communis "Compacta" (Dwarf Roman Myrtle), Z8, cont									
5210 1 gal.				Ea.	9.65			9.65	10.65
5220 5 gal.				"	34			34	37.50
5300 Nandina domestica (Heavenly Bamboo), Z7, cont									
5310 1 gal.				Ea.	4.50			4.50	4.95
5320 2 gal.					16.50			16.50	18.15
5330 5 gal.				▼	26.50			26.50	29
5400 Nerium oleander (Oleander), Z8, cont									
5410 1 gal.				Ea.	4.75			4.75	5.25
5420 3 gal.					9.50			9.50	10.45
5430 5 gal.				▼	14.50			14.50	15.95
5500 Osmanthus fragrans (Fragrant Tea Olive), Z8, cont									
5510 1 gal.				Ea.	6.30			6.30	6.95
5520 3 gal.				"	14.65			14.65	16.10
5600 Philodendron selloum (Selloum), Z10, cont									
5610 1 gal.				Ea.	4.50			4.50	4.95
5620 5 gal.					14			14	15.40
5630 15 gal.				▼	50			50	55
5700 Photinia fraseri (Christmas Berry), Z7, cont									
5710 1 gal.				Ea.	10.70			10.70	11.75
5720 2 gal.					17.80			17.80	19.60
5730 5 gal.					35			35	38.50
5740 15 gal.				▼	75			75	82.50
5800 Phyllostachys aurea (Golden Bamboo), Z7, cont									
5810 1 gal.				Ea.	9.45			9.45	10.40
5820 5 gal.					25			25	27.50
5830 15 gal.				▼	78.50			78.50	86.50
5900 Pittosporum tobira (Japanese Pittosporum), Z8, cont									
5910 1 gal.				Ea.	27.50			27.50	30
5920 3 gal.					9.30			9.30	10.25
5930 5 gal.				▼	13.75			13.75	15.15
6000 Plumbago auriculata (Cape Plumbago), Z9, cont									
6010 1 gal.				Ea.	4.50			4.50	4.95
6020 5 gal.					12.25			12.25	13.50
6030 15 gal.				▼	55.50			55.50	61
6100 Prunus carolina (Carolina Cherry Laurel), Z7, cont									
6110 1 gal.				Ea.	6			6	6.60
6120 3 gal.					51.50			51.50	57
6130 15 gal.				▼	56.50			56.50	62.50
6200 Punica granatum "nana" (Dwarf Pomegranate), Z7, cont									
6210 1 gal.				Ea.	5			5	5.50
6220 5 gal.					22.50			22.50	25
6230 7 gal.				▼	35			35	38.50
6300 Pyracantha koidzumii "Santa Cruz" (Pyracantha), Z8, cont									
6310 1 gal.				Ea.	5.85			5.85	6.40
6320 5 gal.					12.90			12.90	14.20
6330 5 gal. Espalier				▼	70			70	77
6400 Raphiolepis indica (India Hawthorne), Z8, cont									
6410 1 gal.				Ea.	14.90			14.90	16.35
6420 2 gal.					43			43	47.50
6430 5 gal.				▼	55.50			55.50	61.50
6500 Rhamnus alaternus (Italian Buckthorn), Z7, cont									
6510 1 gal.				Ea.	6			6	6.60

438

32 93 Plants

32 93 33 – Shrubs

32 93 33.50 Shrubs	Crew	Daily Output	Labor-Hours	Unit	Material	2021 Bare Costs Labor	Equipment	Total	Total Incl O&P
6520	5 gal.				Ea.	16			16	17.60
6530	15 gal.				↓	55			55	60.50
6600	Sarcococca hookerana humilis (Himalayan Sarcococca), Z8, cont									
6610	1 gal.				Ea.	53			53	58.50
6620	2 gal.				"	10.15			10.15	11.20
6700	Sarcococca ruscifolia (Fragrant Sarcococca), Z7, cont									
6710	1 gal.				Ea.	6.35			6.35	7
6720	5 gal.				"	14			14	15.40
6800	Solanum rantonnetii (Paraguay Nightshade), Z9, cont									
6810	1 gal.				Ea.	6.75			6.75	7.40
6820	5 gal.				"	17.70			17.70	19.50
6900	Ternstroemia gymnanthera (Japanese Cleyera), Z7, cont									
6910	1 gal.				Ea.	7.85			7.85	8.65
6920	3 gal.					26			26	28.50
6930	5 gal.				↓	33			33	36
7000	Viburnum davidii (David Viburnum), Z8, cont									
7010	1 gal.				Ea.	6.80			6.80	7.50
7020	3 gal.				"	12.95			12.95	14.25
7100	Viburnum japonicum (Japanese Viburnum), Z8, cont									
7110	1 gal.				Ea.	15.45			15.45	17
7120	3 gal.				"	14.15			14.15	15.55
7200	Viburnum tinus (Spring Bouquet Laurustinis), Z7, cont									
7210	1 gal.				Ea.	6.05			6.05	6.65
7220	5 gal.				"	10.80			10.80	11.85
7300	Vitex agnus castus (Chastetree), Z7, cont									
7310	5 gal.				Ea.	36			36	39.50
7400	Xylosma congestum (Shiny Xylosma), Z8, cont									
7410	1 gal.				Ea.	4.50			4.50	4.95
7420	5 gal.				↓	14			14	15.40
7430	15 gal.				↓	50			50	55
7500	Yucca aloifolia (Spanish Bayonet), Z7, cont									
7510	1 gal.				Ea.	7.20			7.20	7.95
7520	3 gal.				"	17.70			17.70	19.45

32 93 33.60 Cacti

		Crew	Daily Output	Labor-Hours	Unit	Material	2021 Bare Costs Labor	Equipment	Total	Total Incl O&P
0010	**CACTI**									
1000	Column cactus, Cereus species, Z8, cont									
1020	5 gal.				Ea.	30			30	33
1030	regular size, 15 gal.					79			79	87
1032	large size, 15 gal.				↓	79			79	87
1100	Golden barrel cactus, Echinocactus grusonii, Z8, cont									
1130	14"				Ea.	68			68	74.50
1200	Orchid cactus, Epiphyllum varieties, Z8, cont									
1210	1 gal.				Ea.	10.45			10.45	11.50
1212	2 gal.				"	21			21	23
1300	Cluster cactus, Mammillaria species, Z8, cont									
1310	1 gal.				Ea.	9.90			9.90	10.90
1312	2 gal.					18.30			18.30	20
1320	5 gal.				↓	30			30	33
1400	Burbank spineless Prickly Pear, Opuntia fiscus-indica, Z8, cont									
1420	5 gal.				Ea.	25			25	27.50
1430	15 gal.				"	52.50			52.50	58
1500	Bunny ears, Opuntia microdasys, Z8, cont									

32 93 33.60 Cacti		Crew	Daily Output	Labor-Hours	Unit	Material	2021 Bare Costs Labor	2021 Bare Costs Equipment	Total	Total Incl O&P
1510	1 gal.				Ea.	4.72			4.72	5.20
1520	5 gal.				"	19.60			19.60	21.50
1600	Prickly Pear, Opuntia species, Z8, cont									
1610	1 gal.				Ea.	6.50			6.50	7.15
1620	5 gal.					18			18	19.80
1630	15 gal.				↓	52.50			52.50	57.50
1700	Giant Saguaro Cactus, Carnegia gigantea, Z8, cont									
1713	3 gal.				Ea.	45			45	49.50
1720	5 gal.				"	24.50			24.50	27

32 93 43 – Trees

32 93 43.10 Planting		Crew	Daily Output	Labor-Hours	Unit	Material	2021 Bare Costs Labor	2021 Bare Costs Equipment	Total	Total Incl O&P
0010	**PLANTING**									
0011	Trees, shrubs and ground cover									
0100	Light soil									
0110	Bare root seedlings, 3" to 5" height	1 Clab	960	.008	Ea.		.37		.37	.55
0120	6" to 10"		520	.015			.68		.68	1.02
0130	11" to 16"		370	.022			.96		.96	1.43
0140	17" to 24"		210	.038			1.69		1.69	2.52
0200	Potted, 2-1/4" diameter		840	.010			.42		.42	.63
0210	3" diameter		700	.011			.51		.51	.76
0220	4" diameter		620	.013			.57		.57	.85
0300	Container, 1 gallon	2 Clab	84	.190	↓		8.45		8.45	12.60
0310	2 gallon	2 Clab	52	.308	Ea.		13.65		13.65	20.50
0320	3 gallon		40	.400			17.75		17.75	26.50
0330	5 gallon		29	.552			24.50		24.50	36.50
0400	Bagged and burlapped, 12" diameter ball, by hand	↓	19	.842			37.50		37.50	56
0410	Backhoe/loader, 48 HP	B-6	40	.600			29	5.40	34.40	49
0415	15" diameter, by hand	2 Clab	16	1			44.50		44.50	66.50
0416	Backhoe/loader, 48 HP	B-6	30	.800			38.50	7.20	45.70	65.50
0420	18" diameter by hand	2 Clab	12	1.333			59		59	88.50
0430	Backhoe/loader, 48 HP	B-6	27	.889			43	8	51	73
0440	24" diameter by hand	2 Clab	9	1.778			79		79	118
0450	Backhoe/loader, 48 HP	B-6	21	1.143			55	10.30	65.30	93.50
0470	36" diameter, backhoe/loader, 48 HP	"	17	1.412	↓		68	12.70	80.70	115
0550	Medium soil									
0560	Bare root seedlings, 3" to 5"	1 Clab	672	.012	Ea.		.53		.53	.79
0561	6" to 10"		364	.022			.98		.98	1.46
0562	11" to 16"		260	.031			1.37		1.37	2.04
0563	17" to 24"		145	.055			2.45		2.45	3.66
0570	Potted, 2-1/4" diameter		590	.014			.60		.60	.90
0572	3" diameter		490	.016			.72		.72	1.08
0574	4" diameter	↓	435	.018			.82		.82	1.22
0590	Container, 1 gallon	2 Clab	59	.271			12.05		12.05	17.95
0592	2 gallon		36	.444			19.75		19.75	29.50
0594	3 gallon		28	.571			25.50		25.50	38
0595	5 gallon		20	.800			35.50		35.50	53
0600	Bagged and burlapped, 12" diameter ball, by hand	↓	13	1.231			54.50		54.50	81.50
0605	Backhoe/loader, 48 HP	B-6	28	.857			41	7.70	48.70	70
0607	15" diameter, by hand	2 Clab	11.20	1.429			63.50		63.50	94.50
0608	Backhoe/loader, 48 HP	B-6	21	1.143			55	10.30	65.30	93.50
0610	18" diameter, by hand	2 Clab	8.50	1.882			83.50		83.50	125
0615	Backhoe/loader, 48 HP	B-6	19	1.263	↓		61	11.40	72.40	103

32 93 Plants

32 93 43 – Trees

32 93 43.10 Planting

		Crew	Daily Output	Labor-Hours	Unit	Material	2021 Bare Costs Labor	2021 Bare Costs Equipment	Total	Total Incl O&P
0620	24" diameter, by hand	2 Clab	6.30	2.540	Ea.		113		113	168
0625	Backhoe/loader, 48 HP	B-6	14.70	1.633			78.50	14.70	93.20	133
0630	36" diameter, backhoe/loader, 48 HP	"	12	2	↓		96	18	114	163
0700	Heavy or stoney soil									
0710	Bare root seedlings, 3" to 5"	1 Clab	470	.017	Ea.		.76		.76	1.13
0711	6" to 10"		255	.031			1.39		1.39	2.08
0712	11" to 16"		182	.044			1.95		1.95	2.91
0713	17" to 24"		101	.079			3.52		3.52	5.25
0720	Potted, 2-1/4" diameter		360	.022			.99		.99	1.47
0722	3" diameter		343	.023			1.04		1.04	1.55
0724	4" diameter	↓	305	.026	↓		1.16		1.16	1.74
0730	Container, 1 gallon	2 Clab	41.30	.387	Ea.		17.20		17.20	25.50
0732	2 gallon		25.20	.635			28		28	42
0734	3 gallon		19.60	.816			36		36	54
0735	5 gallon		14	1.143			50.50		50.50	75.50
0750	Bagged and burlapped, 12" diameter ball, by hand	↓	9.10	1.758			78		78	116
0751	Backhoe/loader	B-6	19.60	1.224			59	11.05	70.05	100
0752	15" diameter, by hand	2 Clab	7.80	2.051			91		91	136
0753	Backhoe/loader, 48 HP	B-6	14.70	1.633			78.50	14.70	93.20	133
0754	18" diameter, by hand	2 Clab	5.60	2.857			127		127	189
0755	Backhoe/loader, 48 HP	B-6	13.30	1.805			87	16.25	103.25	147
0756	24" diameter, by hand	2 Clab	4.40	3.636			161		161	241
0757	Backhoe/loader, 48 HP	B-6	10.30	2.330			112	21	133	190
0758	36" diameter backhoe/loader, 48 HP	"	8.40	2.857			137	25.50	162.50	234
2000	Stake out tree and shrub locations	2 Clab	220	.073	↓		3.23		3.23	4.82

32 93 43.30 Trees, Deciduous

		Crew	Daily Output	Labor-Hours	Unit	Material	2021 Bare Costs Labor	2021 Bare Costs Equipment	Total	Total Incl O&P
0010	**TREES, DECIDUOUS**									
0011	Z2 to 6									
0100	Acer campestre (Hedge Maple), Z4, B&B									
0110	4' to 5'				Ea.	56			56	61.50
0120	5' to 6'					85			85	93.50
0130	1-1/2" to 2" Cal.					147			147	162
0140	2" to 2-1/2" Cal.					180			180	198
0150	2-1/2" to 3" Cal.				↓	310			310	340
0200	Acer ginnala (Amur Maple), Z2, cont/BB									
0210	8' to 10'				Ea.	143			143	157
0220	10' to 12'					315			315	345
0230	12' to 14'				↓	277			277	305
0300	Acer griseum (Paperbark Maple), Z5, B&B									
0310	1-1/2" to 1-3/4" Cal.				Ea.	183			183	201
0320	1-3/4" to 2" Cal.				"	276			276	305
0400	Acer palmatum (Japanese Maple), Z6, cont/BB									
0410	2-1/2' to 3'				Ea.	100			100	110
0420	4' to 5'					89			89	98
0430	5' to 6'					111			111	122
0440	6' to 7'					117			117	129
0450	7' to 8'					194			194	213
0460	8' to 10'					276			276	305
0470	10' to 12'				↓	145			145	160
0500	Acer palmatum atropurpureum (Bloodgood Japan Maple), Z5, B&B									
0510	3' to 3-1/2'				Ea.	79			79	87
0520	3-1/2' to 4'				↓	111			111	122

32 93 43.30 Trees, Deciduous	Crew	Daily Output	Labor-Hours	Unit	Material	2021 Bare Costs Labor	Equipment	Total	Total Incl O&P
0530	4' to 4-1/2'				Ea.	153			153	168
0540	4-1/2' to 5'					156			156	172
0550	5' to 6'				▼	180			180	198
0600	Acer platanoides (Norway Maple), Z4, B&B									
0610	8' to 10'				Ea.	170			170	187
0620	1-1/2" to 2" Cal.					119			119	131
0630	2" to 2-1/2" Cal.					145			145	159
0640	2-1/2" to 3" Cal.					178			178	195
0650	3" to 3-1/2" Cal.					214			214	235
0660	Bare root, 8' to 10'					300			300	330
0670	10' to 12'					320			320	350
0680	12' to 14'				▼	300			300	330
0700	Acer platanoides columnare (Column maple), Z4, B&B									
0710	2" to 2-1/2" Cal.				Ea.	215			215	237
0720	2-1/2" to 3" Cal.					168			168	185
0730	3" to 3-1/2" Cal.					258			258	284
0740	3-1/2" to 4" Cal.					287			287	315
0750	4" to 4-1/2" Cal.					325			325	355
0760	4-1/2" to 5" Cal.					395			395	435
0770	5" to 5-1/2" Cal.				▼	520			520	575
0800	Acer rubrum (Red Maple), Z4, B&B									
0810	1-1/2" to 2" Cal.				Ea.	129			129	142
0820	2" to 2-1/2" Cal.					176			176	194
0830	2-1/2" to 3" Cal.					259			259	285
0840	Bare Root, 8' to 10'					224			224	247
0850	10' to 12'				▼	265			265	292
0900	Acer saccharum (Sugar Maple), Z3, B&B									
0910	1-1/2" to 2" Cal.				Ea.	133			133	146
0920	2" to 2-1/2" Cal.					185			185	204
0930	2-1/2" to 3" Cal.					204			204	224
0940	3" to 3-1/2" Cal.					235			235	258
0950	3-1/2" to 4" Cal.					257			257	283
0960	Bare Root, 8' to 10'					127			127	139
0970	10' to 12'					216			216	237
0980	12' to 14'				▼	330			330	365
1000	Aesculus carnea biroti (Ruby Horsechestnut), Z3, B&B									
1010	2' to 3'				Ea.	82			82	90
1020	3' to 4'					120			120	132
1030	4' to 5'					140			140	154
1040	5' to 6'				▼	164			164	180
1100	Amelanchier canadensis (shadblow), Z4, B&B									
1110	3' to 4'				Ea.	47			47	51.50
1120	4' to 5'				Ea.	76.50			76.50	84.50
1130	5' to 6'					111			111	123
1140	6' to 8'					124			124	136
1150	8' to 10'				▼	174			174	191
1200	Betula maximowicziana (monarch birch), Z5, B&B									
1210	6' to 8'				Ea.	42.50			42.50	46.50
1220	1-1/2" to 2" Cal.					168			168	185
1230	2" to 2-1/2" Cal.				▼	209			209	229
1300	Betula nigra (river birch), Z5, B&B									
1310	1-1/2" to 2" Cal.				Ea.	80.50			80.50	88.50
1320	2" to 2-1/2" Cal.				▼	119			119	131

32 93 Plants

32 93 43 – Trees

32 93 43.30 Trees, Deciduous		Crew	Daily Output	Labor-Hours	Unit	Material	2021 Bare Costs Labor	Equipment	Total	Total Incl O&P
1330	2-1/2" to 3" Cal.				Ea.	166			166	182
1340	3" to 3-1/2" Cal.				↓	259			259	285
1400	Betula papyrifera (canoe or paper birch), Z2, B&B									
1410	4' to 5'				Ea.	77			77	84.50
1420	1-1/2" to 2" Cal.					127			127	140
1430	2" to 2-1/2" Cal.					132			132	145
1440	2-1/2" to 3" Cal.					159			159	175
1450	3" to 3-1/2" Cal.					167			167	184
1460	3-1/2" to 4" Cal.					187			187	206
1470	4" to 4-1/2" Cal.					185			185	203
1480	4-1/2" to 5" Cal.				↓	231			231	254
1500	Castanea mollissima (Chinese Chestnut), Z5, B&B									
1510	2' to 3'				Ea.	20.50			20.50	22.50
1520	6' to 8'					26			26	28.50
1530	8' to 10'					111			111	122
1540	10' to 12'				↓	140			140	154
1600	Catalpa speciosa (northern catalpa), Z5, B&B									
1610	6' to 8'				Ea.	74			74	81.50
1620	2" Cal.				"	260			260	286
1700	Cercidiphyllum japonica (katsura tree), Z5, B&B									
1710	6' to 8'				Ea.	117			117	129
1720	1-1/2" to 2" Cal.					110			110	121
1730	2" to 2-1/2" Cal.					102			102	112
1740	2-1/2" to 3" Cal.					133			133	146
1750	3-1/2" to 4" Cal.					153			153	168
1760	4" to 4-1/2" Cal.				↓	156			156	172
1800	Cercis canadensis (eastern redbud), Z5, B&B									
1810	5 gal.				Ea.	72.50			72.50	79.50
1820	7 gal.					67			67	73.50
1830	4' to 5'				↓	72.50			72.50	80
1840	5' to 6'				Ea.	72			72	79.50
1850	6' to 8'				"	73.50			73.50	81
1900	Chionanthus virginicus (fringetree), Z4, B&B									
1910	5 gal.				Ea.	39.50			39.50	43.50
1920	5' to 6'					89			89	98
1930	6' to 8'				↓	109			109	120
2000	Cladrastis lutea (yellowood), Z4, B&B									
2010	1-1/2" to 2" Cal.				Ea.	129			129	141
2020	2" to 2-1/2" Cal.					138			138	152
2030	2-1/2" to 3" Cal.					213			213	234
2040	3" to 3-1/2" Cal.					267			267	294
2050	3-1/2" to 4" Cal.				↓	310			310	340
2100	Cornus florida (white flowering dogwood), Z5, B&B									
2110	4' to 5'				Ea.	45			45	49.50
2120	5' to 6'					98.50			98.50	108
2130	6' to 7'				↓	86			86	94.50
2200	Cornus florida rubra (pink flowering dogwood), Z6, B&B									
2210	4' to 5'				Ea.	58			58	64
2220	5' to 6'					90			90	99
2230	6' to 7'					148			148	162
2240	7' to 8'				↓	120			120	132
2300	Cornus kousa (Kousa Dogwood), Z5, B&B									
2310	3' to 4'				Ea.	52.50			52.50	57.50

443

32 93 43.30 Trees, Deciduous	Crew	Daily Output	Labor-Hours	Unit	Material	2021 Bare Costs Labor	Equipment	Total	Total Incl O&P
2320	4' to 5'				Ea.	50.50			50.50	55.50
2330	5' to 6'					118			118	129
2340	6' to 8'					191			191	210
2350	8' to 10'				↓	248			248	273
2400	Crataegus crus-galli (Cockspur Hawthorn), Z5, B&B									
2410	55 gal.				Ea.	127			127	140
2420	5' to 6'				"	92			92	101
2500	Crataegus oxyacantha superba (Crimson Hawthorn), Z5, B&B									
2510	5' to 6'				Ea.	81.50			81.50	90
2520	6' to 8'					84.50			84.50	93
2530	1-1/2" to 2" Cal.					90.50			90.50	99.50
2540	2" to 2-1/2" Cal.					87.50			87.50	96.50
2550	2-1/2" to 3" Cal.					107			107	117
2560	3" to 3-1/2" Cal.					216			216	238
2570	3-1/2" to 4" Cal.					235			235	259
2580	4" to 5" Cal.				↓	278			278	305
2600	Crataegus phaenopyrum (Washington Hawthorn), Z5, B&B									
2610	5' to 6'				Ea.	169			169	186
2620	6' to 7'				Ea.	218			218	239
2630	1-3/4" to 2" Cal.					124			124	137
2640	2" to 2-1/2" Cal.					149			149	164
2650	2-1/2" to 3" Cal.					214			214	235
2660	3" to 3-1/2" Cal.				↓	278			278	305
2700	Cydonia oblonga orange (Orange Quince), Z5, B&B									
2710	5' to 6'				Ea.	53			53	58
2720	6' to 8'				"	94			94	103
2800	Diospyros virginiana (Persimmon), Z4, B&B									
2810	3' to 4'				Ea.	31			31	34.50
2820	4' to 5'					24.50			24.50	27
2830	1-1/2" to 2" Cal.					129			129	142
2840	2" to 2-1/2" Cal.					245			245	270
2850	2-1/2" to 3" Cal.				↓	305			305	335
2900	Eleagnus angustifolia (Russian Olive, tree form), Z3, B&B									
2910	1-1/2" to 2" Cal.				Ea.	178			178	196
2920	2" to 2-1/2" Cal.					420			420	460
2930	2-1/2" to 3" Cal.				↓	620			620	680
3000	Eucommia ulmoides (Hardy Rubber Tree), Z4, B&B									
3010	6' to 8'				Ea.	104			104	114
3020	8' to 10'					121			121	133
3030	1-1/2" to 2" Cal.					117			117	129
3040	2" to 2-1/2" Cal.					144			144	158
3050	2-1/2" to 3" Cal.				↓	179			179	197
3100	Fagus grandiflora (American Beech), Z3, B&B									
3110	1-1/2" to 1-3/4" Cal.				Ea.	223			223	246
3120	1-3/4" to 2" Cal.					241			241	265
3130	2" to 2-1/2" Cal.				↓	375			375	415
3200	Fagus sylvatica (European Beech), Z5, B&B									
3210	6' to 8'				Ea.	217			217	239
3220	1-1/2" to 2" Cal.					380			380	420
3230	2" to 2-1/2" Cal.				↓	257			257	282
3300	Fagus sylvatica pendula (Weeping European Beech), Z4, B&B									
3310	1-3/4" to 2" cal.				Ea.	84			84	92.50
3320	2" to 2-1/2" Cal.				↓	120			120	132

32 93 43.30 Trees, Deciduous	Crew	Daily Output	Labor-Hours	Unit	Material	2021 Bare Costs Labor	Equipment	Total	Total Incl O&P
3330	2-1/2" to 3" Cal.				Ea.	130			130	143
3340	3" to 3-1/2" Cal.					420			420	460
3400	Franklinia alatamaha (Franklin Tree), Z5, B&B									
3410	3 gal.				Ea.	49			49	54
3420	7 gal.					70			70	77
3430	15 gal.					102			102	112
3440	5' to 6'					250			250	274
3450	6' to 8'					152			152	167
3500	Fraxinus americana (Autumn Purple Ash), Z3, B&B									
3510	5 gal.				Ea.	47.50			47.50	52.50
3520	1-1/2" to 2" Cal.					153			153	168
3530	2" to 2-1/2" Cal.					205			205	225
3600	Fraxinus pensylvanica (Seedless Green Ash), Z3, B&B									
3610	5 gal.				Ea.	32			32	35
3615	20 gal.					91.50			91.50	100
3620	55 gal.					122			122	134
3625	6' to 8'					75.50			75.50	83
3630	1-1/2" to 2" Cal.					114			114	125
3635	2" to 2-1/2" Cal.					149			149	164
3640	2-1/2" to 3" Cal.					158			158	174
3645	3" to 3-1/2" Cal.					204			204	224
3650	3-1/2" to 4" Cal.					256			256	281
3655	4" to 4-1/2" Cal.					281			281	310
3660	4-1/2" to 5" Cal.					485			485	530
3665	Bare root, 8' to 10'					91.50			91.50	101
3670	10' to 12'					151			151	166
3675	12' to 14'					153			153	168
3700	Ginkgo biloba (Maidenhair Tree), Z5, B&B									
3710	5 gal.				Ea.	69.50			69.50	76.50
3720	1-1/2" to 2" Cal.					156			156	171
3730	2" to 2-1/2" Cal.					235			235	258
3740	2-1/2" to 3" Cal.					276			276	305
3750	3" to 3-1/2" Cal.					360			360	400
3800	Gleditsia triacanthos inermis (Thornless Honeylocust), Z5, B&B									
3810	20 gal.				Ea.	68			68	75
3815	55 gal.					84			84	92.50
3820	8' to 10'					151			151	166
3825	1-1/2" to 2" Cal.					168			168	184
3830	2" to 2-1/2" Cal.					269			269	296
3835	2-1/2" to 3" Cal.					310			310	340
3840	3" to 3-1/2" Cal.					310			310	340
3845	3-1/2" to 4" Cal.					350			350	385
3850	4" to 4-1/2" Cal.					475			475	525
3855	4-1/2" to 5" Cal.					475			475	525
3860	Bare root, 10' to 12'					210			210	231
3865	12' to 14'					240			240	264
3870	14' to 16'				Ea.	310			310	340
3900	Gymnocladus dioicus (Kentucky Coffee Tree), Z5, B&B									
3910	1-3/4" Cal.				Ea.	129			129	142
3920	2" Cal.					222			222	244
3930	2-1/2" Cal.					290			290	320
4000	Halesia carolina (Silverbell), Z6, B&B									
4010	4' to 5'				Ea.	75			75	82.50

32 93 43.30 Trees, Deciduous		Crew	Daily Output	Labor-Hours	Unit	Material	2021 Bare Costs Labor	Equipment	Total	Total Incl O&P
4020	5' to 6'				Ea.	70.50			70.50	77.50
4030	6' to 8'					81.50			81.50	90
4040	8' to 10'					107			107	117
4050	10' to 12'				▼	170			170	187
4100	Hippophae rhamnoides (Common Sea Buckthorn), Z3, B&B									
4110	3 gal.				Ea.	20.50			20.50	22.50
4120	5' to 6'					92.50			92.50	102
4130	6' to 8'				▼	101			101	111
4200	Juglans cinera (Butternut), Z4, B&B									
4210	4' to 5'				Ea.	33.50			33.50	37
4220	5' to 6'					42			42	46
4230	6' to 8'				▼	63			63	69.50
4300	Juglans nigra (Black Walnut), Z5, B&B									
4310	4' to 5'				Ea.	23			23	25
4320	5' to 6'					51.50			51.50	56.50
4330	1-1/2" to 2" Cal.					179			179	197
4340	2" to 2-1/2" Cal.					183			183	202
4350	2-1/2" to 3" Cal.				▼	220			220	242
4400	Koelreuteria paniculata (Goldenrain Tree), Z6, B&B									
4410	5 gal.				Ea.	22			22	24
4420	15 gal.					58			58	64
4430	20 gal.					182			182	200
4440	55 gal.					217			217	239
4450	1-1/2" to 2" Cal.					209			209	230
4460	2" to 2-1/2" Cal.					335			335	370
4470	2-1/2" to 3" Cal.				▼	495			495	545
4500	Laburnum x watereri (Goldenchain tree), Z5, B&B									
4510	6' to 8'				Ea.	65.50			65.50	72
4520	8' to 9'					55			55	60.50
4530	1-3/4" to 2" Cal.					120			120	132
4540	2" to 2-1/2" Cal.					122			122	135
4550	2-1/2" to 3" Cal.				▼	179			179	197
4600	Larix decidua pendula (Weeping European Larch), Z2, B&B									
4610	3' to 4'				Ea.	105			105	116
4620	4' to 5'				Ea.	87.50			87.50	96.50
4630	5' to 6'					147			147	162
4640	6' to 7'				▼	224			224	247
4700	Larix kaempferi (Japanese Larch), Z4, B&B									
4710	5' to 6'				Ea.	173			173	190
4720	6' to 8'					154			154	170
4730	8' to 10'				▼	173			173	190
4800	Liquidambar styraciflua (Sweetgum), Z6, B&B									
4810	6' to 8'				Ea.	50			50	54.50
4820	8' to 10'					76			76	83.50
4830	1-1/2" to 2" Cal.					112			112	123
4840	2" to 2-1/2" Cal.					132			132	145
4850	2-1/2" to 3" Cal.				▼	142			142	156
4900	Liriodendron tulipifera (Tuliptree), Z5, B&B									
4910	7' to 8'				Ea.	62.50			62.50	68.50
4920	1-1/2" to 2" Cal.					155			155	170
4930	2" to 2-1/2" Cal.					176			176	193
4940	2-1/2" to 3" Cal.				▼	206			206	227
5000	Maclura pomifera (Osage Orange), Z5, B&B									

32 93 43.30 **Trees, Deciduous**	Crew	Daily Output	Labor-Hours	Unit	Material	2021 Bare Costs Labor	Equipment	Total	Total Incl O&P
5010	3' to 4'				Ea.	34.50			34.50	37.50
5020	4' to 5'					60			60	66
5030	6' to 8'					77			77	84.50
5100	Magnolia loebneri merrill (Merrill Magnolia), Z5, B&B									
5110	4' to 5'				Ea.	240			240	264
5120	5' to 6'					90.50			90.50	99.50
5130	6' to 7'					164			164	181
5140	7' to 8'					186			186	204
5200	Magnolia soulangena (Saucer Magnolia), Z6, B&B									
5210	3' to 4'				Ea.	75			75	82.50
5220	4' to 5'					85			85	93.50
5230	5' to 6'					90			90	99
5240	6' to 7'					144			144	158
5300	Magnolia stellata (Star Magnolia), Z6, B&B									
5310	2' to 3'				Ea.	47			47	51.50
5320	3' to 4'					70			70	77
5330	4' to 5'					100			100	110
5340	5' to 6'					95			95	104
5400	Magnolia virginiana (Sweetbay Magnolia), Z5, B&B									
5410	2' to 3'				Ea.	40			40	44
5420	3' to 4'					60			60	66
5430	4' to 5'					45			45	49.50
5440	5' to 6'				Ea.	130			130	143
5500	Magnolia x galaxy (Galaxy Magnolia), Z6, B&B									
5510	5 gal.				Ea.	35			35	38.50
5520	7 gal.					81.50			81.50	89.50
5530	2-1/2" to 3" Cal.					187			187	206
5600	Malus (Crabapple), Z5, B&B									
5610	5 gal.				Ea.	14.70			14.70	16.15
5615	15 gal.					29.50			29.50	32.50
5620	20 gal.					42			42	46.50
5625	55 gal.					93			93	102
5630	6' to 8'					81			81	89
5635	1-1/2" to 2" Cal.					110			110	121
5640	2" to 2-1/2" Cal.					136			136	150
5645	2-1/2" to 3" Cal.					235			235	259
5650	3" to 3-1/2" Cal.					234			234	257
5655	3-1/2" to 4" Cal.					266			266	293
5660	4" to 5" Cal.					365			365	400
5665	Bare root, 5' to 6'					66.50			66.50	73.50
5700	Metasequoia glyptostroboides (Dawn Redwood), Z5, B&B									
5710	5' to 6'				Ea.	100			100	110
5720	6' to 8'					120			120	132
5730	8' to 10'					259			259	285
5800	Morus alba (White Mulberry), Z4, B&B									
5810	5 gal.				Ea.	35			35	38.50
5900	Morus alba tatarica (Russian Mulberry), Z4, B&B									
5910	6' to 8'				Ea.	69			69	76
5920	10' to 12'					142			142	156
5930	2-1/2" to 3" Cal.					395			395	435
6000	Nyssa sylvatica (Tupelo), Z4, B&B									
6010	5' to 6'				Ea.	70.50			70.50	78
6020	1-1/2" to 2" Cal.					233			233	256

32 93 43.30 Trees, Deciduous	Crew	Daily Output	Labor-Hours	Unit	2021 Bare Costs			Total	Total Incl O&P
					Material	Labor	Equipment	Total	
6030	2" to 2-1/2" Cal.				Ea.	229			229	252
6040	2-1/2" to 3" Cal.				↓	475			475	520
6100	Ostrya virginiana (Hop Hornbeam), Z5, B&B									
6110	2" to 2-1/2" Cal.				Ea.	165			165	181
6120	2-1/2" to 3" Cal.				"	285			285	315
6200	Oxydendron arboreum (Sourwood), Z4, B&B									
6210	4' to 5'				Ea.	104			104	114
6220	5' to 6'					81.50			81.50	90
6230	6' to 7'					145			145	160
6240	7' to 8'				↓	140			140	154
6300	Parrotia persica (Persian Parrotia), Z6, B&B									
6310	5 gal.				Ea.	31			31	34
6320	3' to 4'				"	102			102	112
6400	Phellodendron amurense (Amur Corktree), Z3, B&B									
6410	6' to 8'				Ea.	81.50			81.50	89.50
6420	8' to 10'					119			119	131
6430	2" to 2-1/2" Cal.					170			170	187
6440	2-1/2" to 3" Cal.					212			212	233
6450	3" to 3-1/2" Cal.				↓	262			262	288
6500	Platanus acerifolia (London Plane Tree), Z5, B&B									
6510	15 gal.				Ea.	59			59	65
6520	20 gal.					78			78	85.50
6530	8' to 10'					121			121	133
6540	1-1/2" to 2" Cal.					179			179	196
6550	2" to 2-1/2" Cal.					157			157	172
6560	2-1/2" to 3" Cal.					252			252	277
6570	3" to 3-1/2" Cal.				↓	222			222	244
6600	Platanus occidentalis (Buttonwood, Sycamore), Z4, B&B									
6610	30" box				Ea.	246			246	270
6620	36" box					335			335	365
6630	42" box					870			870	960
6640	1-1/2" to 2" Cal.					125			125	137
6650	2" to 2-1/2" Cal.					185			185	203
6660	2-1/2" to 3" Cal.					213			213	235
6670	3" to 3-1/2" Cal.				↓	213			213	235
6700	Populus alba pyramidalis (Bolleana Poplar), Z3, cont/BB									
6710	5 gal.				Ea.	41			41	45.50
6720	15 gal.					47.50			47.50	52
6730	20 gal.				↓	61			61	67
6800	Populus canadensis (Hybrid Black Poplar), Z5, B&B									
6810	6' to 8'				Ea.	201			201	221
6820	1-1/2" to 2" Cal.					217			217	239
6830	2" to 2-1/2" Cal.					300			300	330
6900	Populus nigra italica (Lombardy Poplar), Z4, cont									
6910	5 gal.				Ea.	24			24	26.50
6920	20 gal.					62.50			62.50	68.50
6930	6' to 8' potted				↓	73			73	80
7000	Prunus cerasifera pissardi (Flowering Plum), Z5, cont/BB									
7010	5 gal.				Ea.	33.50			33.50	37
7020	15 gal.					65			65	71.50
7030	20 gal.				↓	167			167	184
7040	5' to 6'				Ea.	87.50			87.50	96.50
7050	6' to 8'				↓	160			160	176

32 93 43.30 **Trees, Deciduous**		Crew	Daily Output	Labor-Hours	Unit	Material	2021 Bare Costs Labor	2021 Bare Costs Equipment	Total	Total Incl O&P
7060	1-1/2" to 2" Cal.				Ea.	169			169	186
7070	2" to 2-1/2" Cal.				↓	350			350	385
7100	Prunus sargenti columnaris (Columnar Cherry), Z4, cont/BB									
7110	6' to 8'				Ea.	74			74	81.50
7120	1-1/2" to 2" Cal.					214			214	235
7130	2" to 2-1/2" Cal.					241			241	265
7140	2-1/2" to 3" Cal.				↓	465			465	515
7200	Prunus serrulata kwanzan (Flowering Cherry), Z6, cont/BB									
7210	20 gal.				Ea.	173			173	190
7220	1-1/2" to 2" Cal.					120			120	132
7230	2" to 2-1/2" Cal.					184			184	202
7240	3" to 3-1/2" Cal.					325			325	355
7250	3-1/2" to 4" Cal.					470			470	520
7260	4" to 5" Cal.				↓	400			400	440
7300	Prunus subhirtella pendula (Weeping Cherry), Z5, B&B									
7310	1-1/4" to 1-1/2" Cal.				Ea.	109			109	120
7320	1-1/2" to 1-3/4" Cal.					109			109	120
7330	1-3/4" to 2" Cal.					151			151	167
7340	2" to 2-1/2" Cal.					216			216	237
7350	2-1/2" to 3" Cal.				↓	248			248	273
7400	Prunus yedoensis (Yoshino Cherry), Z5, B&B									
7410	5 gal.				Ea.	146			146	160
7420	1-1/2" to 2" Cal.					220			220	242
7430	2" to 2-1/2" Cal.					213			213	235
7440	2-1/2" to 3" Cal.				↓	330			330	365
7500	Pyrus calleryana aristocrat (Aristocrat Flwrng Pear), Z5, cont/BB									
7510	15 gal.				Ea.	36.50			36.50	40
7520	20 gal.					64.50			64.50	71
7530	6' to 8'					54.50			54.50	60
7540	1-1/2" to 2" Cal.					125			125	137
7550	2" to 2-1/2" Cal.					153			153	168
7560	2-1/2" to 3" Cal.					293			293	320
7570	3" to 3-1/2" Cal.					201			201	221
7580	3-1/2" to 4" Cal.				↓	265			265	291
7600	Pyrus calleryana bradford (Bradford Flowering Pear), Z5, cont/BB									
7610	5 gal.				Ea.	22.50			22.50	24.50
7620	8' to 9'					80			80	88
7630	1-1/2" to 2" Cal.					161			161	177
7640	2" to 2-1/2" Cal.					227			227	250
7650	2-1/2" to 3" Cal.					310			310	345
7660	3" to 3-1/2" Cal.					455			455	500
7670	3-1/2" to 4" Cal.				↓	680			680	745
7700	Quercus acutissima (Sawtooth Oak), Z5, B&B									
7710	6' to 8'				Ea.	62			62	68
7720	8' to 10'					82.50			82.50	91
7730	10' to 12'					120			120	132
7740	1-1/2" to 2" Cal.					81.50			81.50	90
7750	2" to 2-1/2" Cal.				↓	118			118	130
7800	Quercus coccinea (Scarlet Oak), Z4, B&B									
7810	1-1/2" to 2" Cal.				Ea.	211			211	232
7820	2" to 2-1/2" Cal.				"	203			203	223
7900	Quercus palustris (Pin Oak), Z5, B&B									
7910	5 gal.				Ea.	35.50			35.50	39

32 93 43.30 Trees, Deciduous	Crew	Daily Output	Labor-Hours	Unit	Material	2021 Bare Costs Labor	Equipment	Total	Total Incl O&P
7920	1-1/2" to 2" Cal.				Ea.	162			162	178
7930	2" to 2-1/2" Cal.					167			167	184
7940	2-1/2" to 3" Cal.					214			214	236
7950	3" to 3-1/2" Cal.					259			259	285
7960	3-1/2" to 4" Cal.					292			292	320
7970	4" to 4-1/2" Cal.					415			415	455
7980	4-1/2" to 5" Cal.					360			360	400
7985	6" to 7" Cal.					690			690	760
7990	8" to 9" Cal.					975			975	1,075
8000	Quercus prinus (Chesnut Oak), Z4, B&B									
8010	6' to 8'				Ea.	83.50			83.50	91.50
8020	8' to 10'				"	209			209	229
8100	Quercus robur fastigiata (Columnar English Oak), Z4, B&B									
8110	6' to 8'				Ea.	63.50			63.50	70
8120	8' to 10'					101			101	111
8130	1-1/2" to 2" Cal.					147			147	161
8140	2" to 2-1/2" Cal.					203			203	224
8150	2-1/2" to 3" Cal.					252			252	277
8200	Quercus rubra (Red Oak), Z4, cont/BB									
8210	5 gal.				Ea.	11.35			11.35	12.50
8220	1-1/2" to 2" Cal.					102			102	112
8230	2" to 2-1/2" Cal.					127			127	139
8240	2-1/2" to 3" Cal.					210			210	231
8300	Robinia pseudoacacia (Black Locust), Z5, cont									
8310	24" box				Ea.	212			212	233
8320	30" box					300			300	330
8330	36" box					675			675	745
8400	Salix babylonica (Weeping Willow), Z5, cont/BB									
8410	24" box				Ea.	128			128	141
8420	36" box					430			430	475
8430	1-1/2" to 2" Cal.					127			127	139
8440	2" to 2-1/2" Cal.					216			216	237
8450	2-1/2" to 3" Cal.					223			223	245
8500	Sophora japonica "Regent" (Regent Scholartree), Z4, B&B									
8510	1-3/4" to 2" Cal.				Ea.	325			325	360
8520	2" to 2-1/2" Cal.					375			375	415
8530	2-1/2" to 3" Cal.					455			455	500
8600	Sorbus alnifolia (Korean Mountain Ash), Z5, B&B									
8610	8' to 10'				Ea.	87			87	95.50
8620	1-1/2" to 2" Cal.					123			123	135
8630	2" to 2-1/2" Cal.					141			141	155
8640	2-1/2" to 3" Cal.					172			172	189
8700	Sorbus aucuparia (European Mountain Ash), Z2, cont/BB									
8710	15 gal.				Ea.	65.50			65.50	72
8715	20 gal.					198			198	217
8720	1-1/2" to 2" Cal.					145			145	159
8725	2" to 2-1/2" Cal.					255			255	280
8730	2-1/2" to 3" Cal.					335			335	365
8735	3" to 3-1/2" Cal.					545			545	600
8740	3-1/2" to 4" Cal.					555			555	610
8745	4" to 5" Cal.					720			720	795
8750	5" to 6" Cal.					905			905	995
8755	Bare root, 8' to 10'					109			109	120

32 93 43.30 Trees, Deciduous	Crew	Daily Output	Labor-Hours	Unit	Material	2021 Bare Costs Labor	Equipment	Total	Total Incl O&P
8760	10' to 12'				Ea.	105			105	116
8800	Stewartia pseudocamellia (Japanese Stewartia), Z5, B&B									
8810	2' to 3'				Ea.	109			109	120
8820	3' to 4'					132			132	145
8830	6' to 8'				↓	530			530	580
8900	Styrax japonica (Japanese Snowbell), Z6, B&B									
8910	4' to 5'				Ea.	141			141	156
8920	5' to 6'					170			170	186
8930	6' to 8'				↓	288			288	315
9000	Syringa japonica (Japanese Tree, Lilac), Z4, B&B									
9010	5' to 6'				Ea.	95.50			95.50	105
9020	6' to 8'					93.50			93.50	103
9030	8' to 10'					227			227	250
9040	10' to 12'				↓	365			365	400
9100	Taxodium distichum (Bald Cypress), Z4, B&B									
9110	5' to 6'				Ea.	127			127	139
9120	6' to 7'					142			142	156
9130	7' to 8'					112			112	123
9140	8' to 10'				↓	186			186	204
9200	Tilia americana "Redmond" (Redmond Linden), Z3, B&B									
9210	8' to 10'				Ea.	161			161	177
9220	10' to 12'					213			213	234
9230	1-1/2" to 2" Cal.					220			220	242
9240	2" to 2-1/2" Cal.					229			229	252
9250	2-1/2" to 3" Cal.				↓	217			217	239
9300	Tilia cordata greenspire (Littleleaf Linden), Z4, B&B									
9310	1-1/2" to 2" Cal.				Ea.	243			243	267
9315	2" to 2-1/2" Cal.					282			282	310
9320	2-1/2" to 3" Cal.					360			360	395
9325	3" to 3-1/2" Cal.					415			415	455
9330	3-1/2" to 4" Cal.					470			470	515
9335	4" to 4-1/2" Cal.					360			360	395
9340	4-1/2" to 5" Cal.					530			530	585
9345	5" to 5-1/2" Cal.					610			610	670
9350	Bare root, 10' to 12'					214			214	235
9355	12' to 14'				↓	320			320	350
9400	Tilia tomentosa (Silver Linden), Z4, B&B									
9410	6' to 8'				Ea.	155			155	170
9420	8' to 10'					181			181	199
9430	10' to 12'					218			218	240
9440	2" to 2-1/2" Cal.					264			264	290
9450	2-1/2" to 3" Cal.				↓	330			330	365
9500	Ulmus americana (American Elm), Z3, B&B									
9510	6' to 8'				Ea.	69			69	76
9600	Zeklova serrata (Japanese Keaki Tree), Z4, B&B									
9610	1-1/2" to 2" Cal.				Ea.	184			184	203
9620	2" to 2-1/2" Cal.					185			185	203
9630	2-1/2" to 3" Cal.				↓	206			206	226

32 93 43.40 Trees, Conifers	Crew	Daily Output	Labor-Hours	Unit	Material	2021 Bare Costs Labor	Equipment	Total	Total Incl O&P
0010 **TREES, CONIFERS**									
0011	Z2 to 7								
1000	Abies balsamea nana (Dwarf Balsam Fir), Z3, cont								
1001	1 gal.				Ea.	50			50	55
1002	15" to 18"					41.50			41.50	45.50
1003	18" to 24"					92.50			92.50	102
1050	Abies concolor (White Fir), Z4, B&B								
1051	2' to 3'				Ea.	51			51	56
1052	3' to 4'				Ea.	69			69	76
1053	4' to 5'					85			85	93.50
1054	5' to 6'					156			156	171
1055	6' to 8'					221			221	243
1100	Abies concolor violacea, Z4, cont/BB								
1101	5' to 6'				Ea.	172			172	189
1102	6' to 7'				"	227			227	250
1150	Abies fraseri (Fraser Balsam Fir), Z5, B&B								
1151	5' to 6'				Ea.	156			156	171
1152	6' to 7'				"	221			221	243
1200	Abies iasiocarpa arizonica (Cork Fir), Z5, cont/BB								
1201	2' to 3'				Ea.	46.50			46.50	51
1202	3' to 4'					75.50			75.50	83
1203	4' to 5'					117			117	128
1204	5' to 6'					143			143	157
1205	6' to 7'					175			175	193
1206	8' to 10'					243			243	267
1207	10' to 12'					350			350	385
1250	Abies iasiocarpa compacta (Dwarf Alpine Fir), Z5, cont/BB								
1251	18" to 24"				Ea.	120			120	132
1252	2' to 3'					117			117	129
1253	3' to 4'					142			142	156
1254	4' to 5'					240			240	264
1255	5' to 6'					285			285	315
1300	Abies koreana (Korean Fir), Z5, B&B								
1301	10' to 12'				Ea.	480			480	530
1302	12' to 14'				"	580			580	640
1350	Abies pinsapo glauca (Blue Spanish Fir), Z6, cont								
1351	2' to 3' spread				Ea.	96.50			96.50	106
1352	3' to 4'					151			151	166
1353	4' to 5'					168			168	185
1400	Abies procera (nobilis) glauca (Noble Fir), Z5, B&B								
1401	15" to 18" spread				Ea.	53			53	58.50
1402	18" to 24" spread					68.50			68.50	75
1403	2' to 3'					95.50			95.50	105
1404	3' to 4'					124			124	137
1450	Abies veitchi (Veitch Fir), Z3, cont/BB								
1451	4' to 5'				Ea.	132			132	145
1452	5' to 6'				"	177			177	195
1500	Cedrus atlantica glauca (Blue Atlas Cedar), Z6, cont/BB								
1501	2' to 3'				Ea.	116			116	128
1502	3' to 4'				Ea.	165			165	182
1503	4' to 5'					228			228	251
1504	5' to 6'					135			135	149

32 93 43.40 Trees, Conifers	Crew	Daily Output	Labor-Hours	Unit	Material	2021 Bare Costs Labor	Equipment	Total	Total Incl O&P
1505	7' to 8'				Ea.	187			187	206
1550	Cedrus deodara (Indian Cedar), Z7, cont/BB									
1551	5 gal.				Ea.	29.50			29.50	32.50
1552	6' to 8'					256			256	282
1553	8' to 10'					238			238	262
1554	10' to 12'					420			420	465
1555	12' to 14'					590			590	650
1556	14' to 16'					760			760	835
1600	Cedrus deodara kashmir (Hardy Deodar Cedar), Z6, cont/BB									
1601	3' to 4'				Ea.	171			171	189
1602	4' to 5'					240			240	264
1603	5' to 6'					286			286	315
1650	Cedrus deodara shalimar (Hardy Deodar Cedar), Z6, cont/BB									
1651	2' to 3'				Ea.	134			134	148
1652	3' to 4'				"	152			152	168
1700	Cedrus libani sargenti (Weeping Cedar of Lebanon), Z6, cont									
1701	2' to 3'				Ea.	116			116	127
1702	3' to 4'				"	171			171	188
1750	Chamaecyparis lawsoniana (Dwarf Lawson's Cypress), Z6, cont									
1751	18" to 24"				Ea.	91			91	100
1752	2' to 2-1/2'				"	125			125	138
1800	Chamaecyparis nootkatensis glauca (Alaska Cedar), Z6, cont/BB									
1801	2' to 3'				Ea.	104			104	114
1802	3' to 4'					131			131	144
1803	4' to 5'					204			204	224
1804	5' to 6'					246			246	270
1850	Chamaecyparis nootkatensis lutea (Gold Cedar), Z6, cont/BB									
1851	3' to 4'				Ea.	145			145	159
1852	4' to 5'					177			177	195
1853	5' to 6'					157			157	173
1900	Chamaecyparis nootkatensis pendula (Alaska Cedar), Z6, cont/BB									
1901	2' to 3'				Ea.	102			102	112
1902	4' to 5'					156			156	172
1903	5' to 6'					158			158	174
1904	6' to 7'					197			197	216
1950	Chamaecyparis obtusa coralliformis (Torulosa), Z5, cont/BB									
1951	1 gal.				Ea.	5.20			5.20	5.70
1952	2 gal.				"	13.30			13.30	14.65
1953	5 gal.				Ea.	15.20			15.20	16.75
1954	7 gal.					31.50			31.50	35
1955	4' to 5'					78			78	86
1956	7' to 8'					107			107	118
2000	Chamaecyparis obtusa crippsi (Gold Hinoki Cypress), Z6, cont/BB									
2001	1 gal.				Ea.	40			40	44
2002	5 gal.					42			42	46
2003	2' to 3'					60			60	66
2004	3' to 4'					66.50			66.50	73
2005	4' to 5'					100			100	110
2050	Chamaecyparis obtusa filiciodes (Fernspray Cypress), Z5, cont/BB									
2051	5 gal.				Ea.	23			23	25.50
2052	2' to 3'					100			100	110
2053	3' to 4'					132			132	145
2100	Chamaecyparis obtusa gracilis (Hinoki Cypress), Z5, cont/BB									

32 93 43.40 Trees, Conifers		Crew	Daily Output	Labor-Hours	Unit	Material	2021 Bare Costs Labor	Equipment	Total	Total Incl O&P
2101	2' to 3'				Ea.	87.50			87.50	96
2102	3' to 4'					130			130	144
2103	10' to 12'				↓	285			285	315
2150	Chamaecyparis obtusa gracilis compacta (Cypress), Z5, cont/BB									
2151	15" to 18"				Ea.	53			53	58
2152	18" to 24"					79			79	87
2153	2' to 3'					101			101	111
2154	3' to 4'				↓	199			199	219
2200	Chamaecyparis obtusa gracilis nana (Hinoki Cypress), Z5, cont/BB									
2201	10" to 12"				Ea.	25.50			25.50	28
2202	12" to 15"					35			35	38.50
2203	15" to 18"					70.50			70.50	77.50
2204	18" to 24"					35			35	38.50
2250	Chamaecyparis obtusa lycopodioides (Cypress), Z5, cont/BB									
2251	18" to 24"				Ea.	23			23	25
2252	2' to 3'					76.50			76.50	84
2253	3' to 4'					124			124	137
2254	4' to 5'					145			145	159
2255	5' to 6'					191			191	210
2256	6' to 7'					187			187	206
2257	7' to 8'				↓	216			216	238
2300	Chamaecyparis obtusa magnifica (Hinoki Cypress), Z5, cont/BB									
2301	2' to 3'				Ea.	70			70	77
2302	3' to 4'					61.50			61.50	67.50
2303	4' to 5'					163			163	180
2304	5' to 6'				↓	224			224	246
2305	6' to 7'				Ea.	243			243	267
2350	Chamaecyparis pisifera (Sawara Cypress), Z5, cont/BB									
2351	4' to 5'				Ea.	48.50			48.50	53
2352	5' to 6'					67.50			67.50	74
2353	6' to 7'					132			132	145
2354	7' to 8'				↓	143			143	158
2400	Chamaecyparis pisifera aurea (Sawara Cypress), Z4, cont/BB									
2401	4' to 5'				Ea.	62			62	68
2402	5' to 6'					75.50			75.50	83
2403	6' to 7'					146			146	160
2404	7' to 8'				↓	164			164	180
2450	Chamaecyparis pisifera boulevard (Moss Cypress), Z5, cont/BB									
2451	18" to 24"				Ea.	27.50			27.50	30
2452	2' to 3'					61.50			61.50	67.50
2453	3' to 4'					97.50			97.50	107
2454	4' to 5'					123			123	135
2455	5' to 6'					176			176	193
2456	6' to 7'				↓	193			193	213
2500	Chamaecyparis pisifera compacta (Sawara Cypress), Z4, cont									
2501	15" to 18" spread				Ea.	35			35	38.50
2502	18" to 24" spread					52.50			52.50	57.50
2503	2' to 2-1/2' spread					73			73	80.50
2504	2-1/2' to 3' spread				↓	108			108	119
2550	Chamaecyparis pisifera filifera (Thread Cypress), Z4, cont/BB									
2551	18" to 24"				Ea.	47.50			47.50	52.50
2552	2' to 3'					75.50			75.50	83
2553	3' to 4'				↓	107			107	117

32 93 43 – Trees

32 93 43.40 Trees, Conifers	Crew	Daily Output	Labor-Hours	Unit	Material	2021 Bare Costs Labor	Equipment	Total	Total Incl O&P
2554	4' to 5'				Ea.	180			180	198
2555	5' to 6'					157			157	172
2556	6' to 7'				▼	221			221	243
2600	Chamaecyparis pisifera filifera aurea (Thread Cypress), Z4, cont									
2601	8" to 10"				Ea.	18			18	19.80
2602	10" to 12"					24.50			24.50	27
2603	15" to 18"				▼	30			30	33
2650	Chamaecyparis pisifera filifera nana (Thread Cypress), Z5, cont									
2651	15" to 18"				Ea.	44.50			44.50	49
2652	18" to 24"					63.50			63.50	70
2653	2' to 2-1/2'					77			77	84.50
2654	2-1/2' to 3'					92			92	101
2655	3' to 4'					122			122	134
2671	Gold spangle, 12" to 15"				▼	36			36	39.50
2672	18" to 24"				Ea.	39			39	43
2673	Bare root, 2' to 3'				"	75.50			75.50	83
2700	Chamaecyparis pisifera plumosa aurea (Plumed Cypress), Z6, B&B									
2701	18" to 24"				Ea.	67			67	73.50
2702	3' to 4'					98.50			98.50	108
2703	4' to 5'				▼	103			103	113
2750	Chamaecyparis pisifera plumosa compress., Z6, cont/BB									
2751	12" to 15"				Ea.	31.50			31.50	34.50
2752	15" to 18"					40			40	44
2753	18" to 24"					55.50			55.50	61.50
2754	2' to 2-1/2'					67.50			67.50	74
2755	2-1/2' to 3'				▼	72			72	79
2800	Chamaecyparis pisifera squarrosa (Cypress), Z5, cont/BB									
2801	1 gal.				Ea.	9.45			9.45	10.40
2802	2 gal.					17.35			17.35	19.05
2803	4' to 5'					22			22	24.50
2804	5' to 6'				▼	75			75	82.50
2850	Chamaecyparis pisifera squarrosa pygmaea (Cypress), Z6, cont									
2851	12" to 15"				Ea.	31			31	34.50
2852	15" to 18"					40			40	44
2853	18" to 24"				▼	52			52	57
2900	Chamaecyparis thyoides andelyensis (White Cedar), Z5, cont/BB									
2901	2' to 3'				Ea.	11.05			11.05	12.20
2902	5 gal.					71			71	78
2903	4' to 5'					110			110	121
2904	5' to 6'					165			165	181
2905	6' to 7'				▼	176			176	193
2950	Cryptomeria japonica cristata (Japanese Cedar), Z6, B&B									
2951	5' to 6'				Ea.	113			113	124
2952	6' to 8'				"	184			184	203
3000	Cryptomeria japonica lobbi (Lobb Cryptomeria), Z6, B&B									
3001	6' to 8'				Ea.	182			182	200
3002	8' to 10'					241			241	265
3003	10' to 12'				▼	340			340	375
3050	Cupressocyparis leylandi blue (Leylandi Cypress), Z6, B&B									
3051	5' to 6'				Ea.	166			166	182
3052	6' to 7'				"	163			163	179
3100	Cupressocyparis leylandii (Leylandi Cypress), Z6, cont/BB									
3101	1 gal.				Ea.	9.60			9.60	10.55

32 93 43.40 Trees, Conifers		Crew	Daily Output	Labor-Hours	Unit	Material	2021 Bare Costs Labor	Equipment	Total	Total Incl O&P
3102	5 gal.				Ea.	31.50			31.50	35
3103	4' to 5'				▼	90			90	99
3104	5' to 6'				Ea.	166			166	183
3105	6' to 7'				↓	164			164	181
3106	Box, 24"				▼	18.40			18.40	20.50
3150	Cupressus arizonica (Arizona Cypress), Z6, cont									
3151	5 gal.				Ea.	49.50			49.50	54.50
3152	7 gal.				"	57			57	62.50
3200	Cupressus macrocarpa (Monterey Cypress), Z8, cont									
3201	1 gal.				Ea.	6.65			6.65	7.35
3202	5 gal.				↓	17.75			17.75	19.55
3203	15 gal.				▼	61			61	67
3250	Cupressus sempervirens (Italian Cypress), Z7, cont									
3251	5 gal.				Ea.	27.50			27.50	30.50
3252	15 gal.				"	56.50			56.50	62
3300	Juniperus chinensis (Green Clmnr Chinese Juniper), Z4, cont/BB									
3301	24" to 30"				Ea.	44			44	48
3302	30" to 36"					47.50			47.50	52.50
3303	3' to 4'					79			79	87
3304	4' to 5'					85			85	93.50
3305	5' to 6'				▼	88.50			88.50	97
3350	Juniperus chinensis "Armstrongii" (Armstrong Juniper), Z4, cont									
3351	1 gal.				Ea.	6			6	6.60
3352	2 gal.				↓	95			95	105
3353	5 gal.				▼	28.50			28.50	31
3400	Juniperus chinensis "Blue Point" (Blue Point Juniper), Z4, cont/BB									
3401	1 gal.				Ea.	9.65			9.65	10.60
3402	5 gal.					23			23	25.50
3403	4' to 5'					64.50			64.50	71
3404	5' to 6'					136			136	149
3405	6' to 8'				▼	153			153	168
3450	Juniperus chinensis "Keteleeri" (Keteleer Juniper), Z5, cont/BB									
3451	3-1/2' to 4'				Ea.	74			74	81.50
3452	4' to 4-1/2'				"	155			155	170
3500	Juniperus chinensis "Sargenti Glauca" (Sargent Juniper), Z4, cont									
3501	12" to 15"				Ea.	15.90			15.90	17.50
3502	15" to 18"					20.50			20.50	22.50
3503	2' to 2-1/2'					47			47	51.50
3504	2-1/2' to 3'				▼	46			46	51
3550	Juniperus chinensis "Sea Green" (Sea Green Juniper), Z4, cont									
3555	15" to 18"				Ea.	18.65			18.65	20.50
3556	18" to 24"					28.50			28.50	31.50
3557	2' to 2-1/2'				▼	50.50			50.50	55.50
3558	2-1/2' to 3'				Ea.	50			50	55
3600	Juniperus chinensis blaauwi (Blaauwi Juniper), Z4, B&B									
3601	18" to 24"				Ea.	29			29	32
3602	2' to 3'					46			46	50.50
3603	3' to 4'					63			63	69
3604	4' to 5'					133			133	147
3605	5' to 6'				▼	180			180	198
3650	Juniperus chinensis columnaris (Chinese Juniper), Z4, B&B									
3651	4' to 5'				Ea.	135			135	149
3652	5' to 6'				▼	152			152	167

For customer support on your Site Work & Landscape Costs with RSMeans Data, call 800.448.8182.

32 93 Plants

32 93 43 – Trees

32 93 43.40 Trees, Conifers	Crew	Daily Output	Labor-Hours	Unit	Material	2021 Bare Costs Labor	Equipment	Total	Total Incl O&P
3653	6' to 7'				Ea.	172			172	189
3700	Juniperus chinensis densaerecta (Spartan Juniper), Z5, cont									
3701	1 gal.				Ea.	4.66			4.66	5.15
3702	5 gal.					200			200	220
3703	15 gal.					48			48	53
3704	3' to 4'					56			56	61.50
3705	4' to 5'					101			101	112
3706	5' to 6'					169			169	186
3707	6' to 8'					199			199	219
3750	Juniperus chinensis Hetzii (Hetz Juniper), Z4, cont/BB									
3751	2-1/2' to 3'				Ea.	46			46	50.50
3752	3' to 3-1/2'					78.50			78.50	86
3753	2' to 2-1/2'					69			69	76
3754	4' to 4-1/2'					108			108	119
3755	4-1/2' to 5'					181			181	199
3800	Juniperus chinensis iowa (Chinese Juniper), Z4, cont/BB									
3801	24" to 30"				Ea.	49			49	54
3802	3' to 4'					66			66	73
3803	4' to 5'					115			115	126
3804	5' to 6'					156			156	171
3850	Juniperus chinensis mountbatten (Chinese Juniper), Z4, cont/BB									
3851	2' to 3'				Ea.	53			53	58
3852	3' to 4'					70			70	77
3853	4' to 5'					112			112	123
3854	5' to 6'					149			149	164
3855	6' to 7'					161			161	177
3856	7' to 8'					189			189	208
3857	8' to 10'					242			242	266
3900	Juniperus chinensis obelisk (Chinese Juniper), Z5, cont/BB									
3901	2' to 3'				Ea.	50			50	55
3902	3' to 4'				"	59			59	65
3903	4' to 5'				Ea.	120			120	132
3904	5' to 6'					185			185	204
3905	6' to 7'					167			167	184
3950	Juniperus chinensis pfitzeriana (Pfitzer Juniper), Z4, cont/BB									
3951	1 gal.				Ea.	6.15			6.15	6.75
3952	2 gal.					27.50			27.50	30
3953	5 gal.					28.50			28.50	31.50
3954	9" to 12"					11.20			11.20	12.35
3955	15" to 18"					17.50			17.50	19.25
3956	18" to 24"					23.50			23.50	26
3957	2' to 2-1/2'					34.50			34.50	38
3958	2-1/2' to 3'					48.50			48.50	53
3959	3' to 3-1/2'					49			49	54
3970	Juniperus chinensis pfitzeriana aurea (Pfitzer Juniper), Z4, cont									
3971	1 gal.				Ea.	9.50			9.50	10.45
3972	2 gal.					26.50			26.50	29
3973	5 gal.					28.50			28.50	31.50
4000	Juniperus chinensis pfitzeriana compact (Pfitzer Juniper), Z4, cont									
4001	5 gal.				Ea.	26.50			26.50	29.50
4002	15" to 18"					21			21	23.50
4003	18" to 24"					26			26	29
4004	2' to 2-1/2'					39.50			39.50	43.50

32 93 Plants

32 93 43 – Trees

32 93 43.40 Trees, Conifers	Crew	Daily Output	Labor-Hours	Unit	Material	2021 Bare Costs Labor	Equipment	Total	Total Incl O&P
4005 2-1/2' to 3'				Ea.	43.50			43.50	48
4050 Juniperus chinensis robusta green (Green Juniper), Z5, cont/BB									
4051 15" to 18"				Ea.	17.60			17.60	19.40
4052 2' to 3'					38			38	42
4053 3' to 4'					59			59	65
4054 4' to 5'					125			125	138
4055 5' to 6'					99.50			99.50	110
4056 6' to 7'					162			162	179
4057 7' to 8'					189			189	208
4058 8' to 10'					189			189	208
4100 Juniperus chinensis torulosa (Hollywood Juniper), Z6, cont/BB									
4101 1 gal.				Ea.	40			40	44
4102 5 gal.					99.50			99.50	109
4103 15 gal.					67.50			67.50	74
4104 4' to 5'					92			92	101
4105 5' to 6'					173			173	190
4150 Juniperus horizontalis "Bar Harbor" (Bar Harbor Juniper), Z4, cont									
4151 9" to 12"				Ea.	16.65			16.65	18.30
4152 15" to 18"				"	25			25	27.50
4153 18" to 21"				Ea.	31			31	34
4154 5 gal.				"	31.50			31.50	34.50
4200 Juniperus horizontalis "Wiltoni" (Blue Rug Juniper), Z3, cont									
4201 12" to 15"				Ea.	20			20	22
4202 15" to 18"					25.50			25.50	28
4203 18" to 24"					26			26	29
4204 24" to 30"					34.50			34.50	38
4250 Juniperus horizontalis Plumosa Compacta (Juniper), Z3, cont									
4251 18" to 24"				Ea.	27			27	29.50
4252 24" to 30"					33.50			33.50	37
4253 30" to 36"					47			47	51.50
4254 3' to 3-1/2'					90			90	99
4300 Juniperus procumbens "Nana" (Procumbens Juniper), Z4, cont									
4301 10" to 12"				Ea.	12.50			12.50	13.75
4302 12" to 15"					21			21	23.50
4303 15" to 18"					26			26	29
4304 18" to 24"					30			30	33
4350 Juniperus sabina "Broadmoor" (Broadmoor Juniper), Z3, cont									
4351 10" to 12"				Ea.	21.50			21.50	23.50
4352 12" to 15"					21			21	23
4353 15" to 18"					28			28	31
4354 18" to 24"					35.50			35.50	39
4400 Juniperus sabina Tamariscifolia New Blue (Tam Juniper), Z4, cont									
4401 9" to 12"				Ea.	8.65			8.65	9.55
4402 15" to 18"					16.05			16.05	17.65
4403 18" to 21"					18.75			18.75	20.50
4450 Juniperus scopulorum "Gray Gleam" (Gray Gleam Juniper), Z3, cont									
4451 3-1/2' to 4'				Ea.	77			77	84.50
4452 4' to 4-1/2'				"	101			101	111
4500 Juniperus scopulorum blue heaven (Blue Heaven Juniper), Z4, B&B									
4501 4' to 5'				Ea.	107			107	118
4502 5' to 6'					139			139	153
4503 6' to 7'					164			164	180
4504 7' to 8'					232			232	255

For customer support on your Site Work & Landscape Costs with RSMeans Data, call 800.448.8182.

32 93 Plants

32 93 43 – Trees

32 93 43.40 Trees, Conifers	Crew	Daily Output	Labor-Hours	Unit	Material	2021 Bare Costs Labor	Equipment	Total	Total Incl O&P
4550	Juniperus scopulorum cologreen (Cologreen Juniper), Z5, cont									
4551	5 gal.				Ea.	38			38	42
4600	Juniperus scopulorum hillburn's silver globe (Juniper), Z5, BB									
4601	3' to 4'				Ea.	81			81	89
4602	4' to 5'				"	108			108	119
4650	Juniperus scopulorum pathfinder (Pathfinder Juniper), Z5, B&B									
4651	4' to 5'				Ea.	98.50			98.50	108
4652	5' to 6'					137			137	150
4653	7' to 8'				↓	166			166	183
4700	Juniperus scopulorum (Rocky Mountain Juniper), Z5, cont/BB									
4701	20 gal.				Ea.	127			127	140
4702	3' to 4'					33.50			33.50	37
4703	4' to 5'					85			85	93.50
4704	5' to 6'				↓	103			103	113
4750	Juniperus scopulorum welchii (Welchi Juniper), Z3, cont									
4751	5 gal.				Ea.	91			91	100
4752	15 gal.				"	118			118	130
4800	Juniperus scopulorum wichita blue (Wichita Blue Juniper), Z5, cont									
4801	3' to 4'				Ea.	83			83	91.50
4802	4' to 5'				"	151			151	167
4850	Juniperus squamata "Parsonii" (Parson's Juniper), Z4, cont									
4851	1 gal.				Ea.	8.50			8.50	9.35
4852	2 gal.					17.20			17.20	18.90
4853	3 gal.					17.20			17.20	18.95
4854	5 gal.				↓	21.50			21.50	23.50
4900	Juniperus squamata Blue Star (Blue Juniper), Z4, cont/BB									
4901	12" to 15"				Ea.	25			25	27.50
4902	15" to 18"				"	30			30	33
4950	Juniperus squamata meyeri (Meyer's Juniper), Z5, cont/BB									
4951	3 gal.				Ea.	26.50			26.50	29
4952	5 gal.					34.50			34.50	37.50
4953	2' to 3'				↓	40.50			40.50	44.50
5000	Juniperus virginiana (Eastern Red Cedar), Z2, B&B									
5001	2' to 3'				Ea.	46			46	50.50
5002	3' to 4'					50.50			50.50	55.50
5003	4' to 5'					88			88	96.50
5004	5' to 6'				↓	150			150	165
5050	Juniperus virginiana burkii (Burki Juniper), Z2, cont									
5051	4' to 4-1/2'				Ea.	112			112	123
5052	4-1/2' to 5'				"	125			125	137
5100	Juniperus virginiana canaerti (Canaert Red Cedar), Z4, B&B									
5101	2' to 3'				Ea.	50			50	55
5102	3' to 4'					60			60	66
5103	4' to 5'					65.50			65.50	72.50
5104	5' to 6'				↓	125			125	138
5150	Juniperus virginiana glauca (Silver Cedar), Z4, cont/BB									
5151	2' to 3'				Ea.	45			45	49.50
5152	3' to 4'				Ea.	46			46	50.50
5153	4' to 5'				"	90.50			90.50	99.50
5200	Juniperus virginiana globusa (Globe Red Cedar), Z4, cont/BB									
5201	2' to 3'				Ea.	48.50			48.50	53.50
5250	Juniperus virginiana gray owl (Gray Owl Juniper), Z4, cont									
5251	18" to 24"				Ea.	25.50			25.50	28

32 93 43.40 Trees, Conifers	Crew	Daily Output	Labor-Hours	Unit	Material	2021 Bare Costs Labor	Equipment	Total	Total Incl O&P
5252	2' to 2-1/2'				Ea.	60			60	66
5300	Juniperus virginiana skyrocket (Skyrocket Juniper), Z5, cont/BB									
5301	5 gal.				Ea.	65.50			65.50	72
5302	15 gal.					79			79	87
5303	2' to 3'					78.50			78.50	86.50
5304	3' to 4'					74			74	81.50
5305	4' to 5'					86.50			86.50	95
5306	5' to 6'					135			135	148
5307	6' to 8'					133			133	146
5350	Picea abies (Norway Spruce), Z2, cont/BB									
5351	2' to 3'				Ea.	32			32	35
5352	3' to 4'					62			62	68
5353	4' to 5'					104			104	115
5354	5' to 6'					187			187	206
5355	6' to 7'					225			225	248
5356	7' to 8'					189			189	208
5400	Picea abies clanbrasiliana (Barry Spruce), Z4, cont/BB									
5401	2' to 3'				Ea.	47.50			47.50	52.50
5402	3' to 4'				"	63			63	69.50
5450	Picea abies gregoryana (Gregory's Dwarf Spruce), Z4, cont/BB									
5451	18" to 24"				Ea.	77			77	84.50
5452	2' to 2-1/2'				"	132			132	145
5500	Picea abies nidiformis (Nest Spruce), Z4, cont/BB									
5501	12" to 15"				Ea.	21			21	23
5502	15" to 18"					30			30	33
5503	18" to 24"					34.50			34.50	38
5504	2' to 3'					55			55	60.50
5505	3' to 4'					75			75	82.50
5506	4' to 5'					107			107	117
5550	Picea abies pendula (Weeping Norway Spruce), Z4, cont/BB									
5551	2' to 3'				Ea.	47			47	52
5552	3' to 4'					76			76	83.50
5553	4' to 5'					110			110	121
5554	6' to 7'					238			238	262
5555	7' to 8'					208			208	229
5600	Picea abies remonti (Remont Spruce), Z4, cont/BB									
5601	2' to 3'				Ea.	87			87	95.50
5650	Picea glauca (White or Canadian Spruce), Z3, cont/BB									
5651	3' to 4'				Ea.	74			74	81.50
5652	4' to 5'					95.50			95.50	105
5653	5' to 6'					155			155	171
5654	6' to 7'					182			182	200
5655	7' to 8'					192			192	211
5700	Picea glauca conica (Dwarf Alberta Spruce), Z3, cont/BB									
5701	9" to 12"				Ea.	18.95			18.95	21
5702	15" to 18"					37			37	41
5703	18" to 24"					62.50			62.50	69
5704	2' to 2-1/2'					79			79	86.50
5705	2-1/2' to 3'					25			25	27.50
5750	Picea omorika (Serbian Spruce), Z4, cont/BB									
5751	2' to 3'				Ea.	58			58	63.50
5752	3' to 4'					100			100	110
5753	4' to 5'					146			146	161

32 93 43.40 Trees, Conifers	Crew	Daily Output	Labor-Hours	Unit	Material	2021 Bare Costs Labor	Equipment	Total	Total Incl O&P
5754	5' to 6'				Ea.	178			178	196
5755	6' to 7'					208			208	229
5800	Picea omorika nana (Dwarf Serbian Spruce), Z4, cont									
5801	9" to 12"				Ea.	25.50			25.50	28
5802	12" to 15"					42.50			42.50	46.50
5803	18" to 24"					51.50			51.50	56.50
5850	Picea omorika pendula (Weeping Serbian Spruce), Z4, cont/BB									
5851	3' to 4'				Ea.	107			107	117
5900	Picea pungens (Colorado Spruce), Z3, cont/BB									
5901	18" to 24"				Ea.	43.50			43.50	48
5902	2' to 3'					59			59	65
5903	3' to 4'					88.50			88.50	97.50
5904	4' to 5'					149			149	164
5905	5' to 6'					145			145	160
5906	6' to 7'					75			75	82.50
5907	7' to 8'					208			208	229
5950	Picea pungens glauca (Blue Colorado Spruce), Z3, cont/BB									
5951	18" to 24"				Ea.	16			16	17.60
5952	2' to 3'					75			75	82.50
5953	3' to 4'					44			44	48
5954	4' to 5'					125			125	138
5955	5' to 6'					175			175	193
5956	6' to 7'					207			207	228
6000	Picea pungens hoopsi (Hoops Blue Spruce), Z3, cont/BB									
6001	2' to 3'				Ea.	79.50			79.50	87.50
6002	3' to 4'					99.50			99.50	109
6003	4' to 5'					400			400	440
6004	5' to 6'					420			420	460
6005	14' to 16'					3,200			3,200	3,525
6006	16' to 18'					400			400	440
6007	18' to 20'					300			300	330
6050	Picea pungens hunnewelliana (Dwarf Blue Spruce), Z4, cont/BB									
6051	12" to 15"				Ea.	32.50			32.50	35.50
6052	15" to 18"					64.50			64.50	71
6053	18" to 24"					66			66	72.50
6054	2' to 2-1/2'					117			117	129
6100	Picea pungens kosteriana (Koster's Blue Spruce), Z3, cont/BB									
6101	2' to 3'				Ea.	71.50			71.50	78.50
6102	3' to 4'					25			25	27.50
6103	4' to 5'					60			60	66
6104	5' to 6'					89.50			89.50	98.50
6105	6' to 7'					132			132	145
6106	7' to 8'					89.50			89.50	98.50
6150	Picea pungens montgomery (Dwarf Blue Spruce), Z4, cont/BB									
6151	12" to 15"				Ea.	40			40	44
6152	15" to 18"					55			55	60.50
6153	18" to 24"					100			100	110
6154	4' to 5'					123			123	135
6155	5' to 6'					130			130	142
6200	Pinus albicaulis (White Bark Pine), Z4, cont/BB									
6201	2' to 3'				Ea.	70			70	77
6202	3' to 4'					90			90	99
6203	5' to 6'					49.50			49.50	54.50

32 93 43.40 Trees, Conifers		Crew	Daily Output	Labor-Hours	Unit	Material	2021 Bare Costs Labor	Equipment	Total	Total Incl O&P
6204	6' to 8'				Ea.	49.50			49.50	54.50
6205	8' to 10'					69.50			69.50	76.50
6206	10' to 12'					69.50			69.50	76.50
6207	12' to 14'					110			110	120
6208	14' to 16'				↓	570			570	625
6250	Pinus aristata (Bristlecone Pine), Z5, cont/BB									
6251	15" to 18"				Ea.	40			40	44
6252	18" to 24"					57.50			57.50	63
6253	2' to 3'					64.50			64.50	71
6254	3' to 4'					178			178	196
6255	4' to 5'				↓	350			350	385
6300	Pinus banksiana (Jack Pine), Z3, cont/BB									
6301	5' to 6'				Ea.	38			38	41.50
6302	6' to 8'					47.50			47.50	52
6303	8' to 10'				↓	57			57	62.50
6350	Pinus bungeana (Lacebark Pine), Z5, cont/BB									
6351	2' to 3'				Ea.	40			40	44
6352	3' to 4'					78.50			78.50	86.50
6353	4' to 5'					30			30	33
6354	5' to 6'					176			176	193
6355	6' to 7'					305			305	335
6356	7' to 8'				↓	39			39	43
6400	Pinus canariensis (Canary Island Pine), Z8, cont									
6401	1 gal.				Ea.	7.50			7.50	8.25
6402	5 gal.					28.50			28.50	31
6403	15 gal.				↓	65			65	71.50
6450	Pinus cembra (Swiss Stone Pine), Z3, cont/BB									
6451	1 gal.				Ea.	40			40	44
6452	2 gal.					55			55	60.50
6453	18" to 24"					78			78	86
6454	2' to 3'				↓	93.50			93.50	103
6500	Pinus densiflora umbraculifera (Tanyosho Pine), Z5, cont/BB									
6501	18" to 24"				Ea.	40			40	44
6502	2' to 3'					169			169	186
6503	3' to 4'					207			207	228
6504	4' to 5'					259			259	285
6505	5' to 6'					330			330	360
6506	7' to 8'				↓	315			315	345
6550	Pinus flexilis (Limber Pine), Z4, B&B									
6551	5 gal.				Ea.	39.50			39.50	43.50
6552	4' to 5'					141			141	155
6553	5' to 6'				↓	191			191	210
6600	Pinus koraiensis (Korean Pine), Z4, cont/BB									
6601	3' to 4'				Ea.	265			265	292
6602	4' to 5'					335			335	370
6603	5' to 6'				↓	405			405	450
6650	Pinus lambertiana (Sugar Pine), Z6, cont/BB									
6651	3' to 4'				Ea.	51			51	56
6652	4' to 5'					64.50			64.50	71
6653	7' to 8'					160			160	176
6654	8' to 10'				↓	231			231	254
6700	Pinus monticola (Western White Pine), Z6, B&B									
6701	5' to 6'				Ea.	44			44	48

32 93 43 – Trees

32 93 43.40 Trees, Conifers	Crew	Daily Output	Labor-Hours	Unit	Material	2021 Bare Costs Labor	Equipment	Total	Total Incl O&P
6702	6' to 7'				Ea.	54.50			54.50	60
6703	7' to 8'					65.50			65.50	72
6704	8' to 10'				▼	109			109	120
6750	Pinus mugo var. mugo (Mugho Pine), Z3, cont/BB									
6751	12" to 15"				Ea.	24.50			24.50	27
6752	15" to 18"					24.50			24.50	27
6753	18" to 24"					25			25	27
6754	2' to 2-1/2'					57			57	62.50
6755	2-1/2' to 3'					79.50			79.50	87.50
6756	3' to 3-1/2' spread					89			89	98
6757	3-1/2' to 4'					61			61	67
6758	1 gal.					61			61	67
6759	2 gal.				▼	102			102	112
6800	Pinus nigra (Austrian Pine), Z5, cont/BB									
6801	5 gal.				Ea.	56			56	61.50
6802	3' to 4'					75			75	82.50
6803	4' to 5'					56			56	61.50
6804	5' to 6'					99.50			99.50	109
6805	6' to 7'					185			185	203
6806	7' to 8'					69.50			69.50	76.50
6807	8' to 9'					223			223	245
6808	9' to 10'					167			167	183
6809	10' to 12'				▼	13			13	14.30
6850	Pinus parviflora (Japanese White Pine), Z5, cont									
6851	12' to 14'				Ea.	56			56	61.50
6900	Pinus parviflora glauca (Silver Japanese White Pine), Z5, cont/BB									
6901	5 gal.				Ea.	65.50			65.50	72
6902	3' to 4'					50			50	55
6903	4' to 5'					167			167	184
6904	5' to 6'					310			310	340
6905	6' to 7'				▼	350			350	385
6950	Pinus ponderosa (Western Yellow Pine), Z6, B&B									
6951	16' to 20'				Ea.	985			985	1,075
7000	Pinus radiata (Monterey Pine), Z8, cont									
7001	1 gal.				Ea.	8.90			8.90	9.80
7002	5 gal.					26			26	28.50
7003	15 gal.					100			100	110
7004	24" box				▼	325			325	355
7050	Pinus resinosa (Red Pine), Z3, cont/BB									
7051	2' to 3'				Ea.	42			42	46.50
7052	3' to 4'					63			63	69.50
7053	4' to 5'				▼	78.50			78.50	86.50
7054	5' to 6'				Ea.	153			153	169
7055	6' to 8'					177			177	195
7056	8' to 10'					248			248	272
7057	10' to 12'					400			400	440
7058	12' to 14'				▼	450			450	495
7100	Pinus rigida (Pitch Pine), Z5, cont/BB									
7101	2' to 3'				Ea.	41.50			41.50	46
7150	Pinus strobus (White Pine), Z3, cont/BB									
7151	2' to 3'				Ea.	21.50			21.50	23.50
7152	3' to 4'					52.50			52.50	57.50
7153	4' to 5'				▼	74			74	81.50

32 93 43.40 Trees, Conifers		Crew	Daily Output	Labor-Hours	Unit	Material	2021 Bare Costs Labor	Equipment	Total	Total Incl O&P
7154	5' to 6'				Ea.	109			109	120
7155	6' to 8'					124			124	137
7156	8' to 10'					234			234	257
7157	10' to 12'					340			340	375
7158	12' to 14'				↓	420			420	465
7200	Pinus strobus fastigiata (Upright White Pine), Z3, B&B									
7201	5' to 6'				Ea.	160			160	176
7250	Pinus strobus nana (Dwarf White Pine), Z3, cont/BB									
7251	5 gal.				Ea.	59.50			59.50	65.50
7252	15" to 18" spread					54			54	59.50
7253	18" to 24" spread					65.50			65.50	72
7254	2' to 2-1/2' spread					59.50			59.50	65.50
7255	2-1/2' to 3' spread					71.50			71.50	78.50
7256	3' to 4' spread					118			118	130
7257	4' to 5' spread					310			310	340
7258	5' to 6' spread				↓	405			405	445
7300	Pinus strobus pendula (Weeping White Pine), Z3, cont/BB									
7301	3' to 4'				Ea.	84.50			84.50	93
7302	4' to 5'					94			94	104
7303	6' to 8'					212			212	233
7304	8' to 10'					390			390	425
7305	10' to 12'				↓	470			470	520
7350	Pinus sylvestris (Scotch Pine), Z3, cont/BB									
7351	Seedlings				Ea.	7.20			7.20	7.95
7352	3' to 4'					75.50			75.50	83
7353	4' to 5'					80.50			80.50	88.50
7354	5' to 6'					152			152	167
7355	6' to 7'					205			205	226
7356	7' to 8'				↓	185			185	204
7400	Pinus sylvestris beauvronensis (Mini Scotch Pine), Z4, cont/BB									
7401	12" to 15"				Ea.	59			59	65
7402	15" to 18"					66			66	72.50
7403	2' to 2-1/2'				↓	100			100	110
7450	Pinus sylvestris fastigiata (Pyramidal Scotch Pine), Z3, cont/BB									
7451	5 gal.				Ea.	71.50			71.50	78.50
7452	7 gal.				"	111			111	122
7500	Pinus thumbergi (Japanese Black Pine), Z5, cont/BB									
7501	1 gal.				Ea.	7.60			7.60	8.35
7502	5 gal.					22.50			22.50	24.50
7503	7 gal.					49			49	54
7504	2' to 3'					48			48	52.50
7505	3' to 4'					66			66	72.50
7506	4' to 5'					89			89	98
7507	5' to 6'					160			160	176
7508	6' to 7'					209			209	230
7509	7' to 8'				↓	204			204	224
7550	Podocarpus macrophyllus "Maki" (Podocarpus), Z7, cont/BB									
7551	1 gal.				Ea.	9.75			9.75	10.75
7552	5 gal.					30			30	33
7553	7 gal.					48.50			48.50	53
7554	15 gal.					69			69	76
7555	4' to 5'				↓	75			75	82.50
7600	Pseudotsuga menziesii (Douglas Fir), Z5, cont/BB									

464

32 93 43.40 Trees, Conifers		Crew	Daily Output	Labor-Hours	Unit	Material	2021 Bare Costs Labor	Equipment	Total	Total Incl O&P
7601	5 gal.				Ea.	29.50			29.50	32.50
7602	3' to 4'					57			57	62.50
7603	4' to 5'					84			84	92.50
7604	5' to 6'					128			128	141
7605	6' to 7'					187			187	206
7650	Sciadopitys verticillata (Umbrella Pine), Z6, B&B									
7651	2 gal.				Ea.	38.50			38.50	42.50
7652	2' to 2-1/2'					63			63	69.50
7653	6' to 7'					325			325	355
7654	7' to 8'					445			445	490
7700	Sequoia sempervirens (Coast Redwood), Z7, cont									
7701	5 gal.				Ea.	35			35	38.50
7702	15 gal.					48.50			48.50	53
7703	24" box					172			172	189
7750	Taxus baccata adpressa fowle (Midget Boxleaf Yew), Z5, cont/BB									
7751	15" to 18"				Ea.	29.50			29.50	32
7752	18" to 24"					27			27	30
7753	2' to 2-1/2'					49.50			49.50	54.50
7754	2-1/2' to 3'					63			63	69.50
7755	3' to 4'				Ea.	70.50			70.50	77.50
7756	4' to 5'				"	137			137	150
7800	Taxus baccata repandens (Spreading English Yew), Z5, B&B									
7801	12" to 15"				Ea.	31.50			31.50	34.50
7802	15" to 18"					57			57	62.50
7803	2' to 2-1/2'					86.50			86.50	95
7804	2-1/2' to 3'					106			106	116
7805	3' to 3-1/2'					127			127	140
7850	Taxus cuspidata (Spreading Japanese Yew), Z4, cont/BB									
7851	15" to 18"				Ea.	59			59	65
7852	18" to 24"					63.50			63.50	70
7853	2' to 2-1/2'					81			81	89
7854	2-1/2' to 3'					102			102	112
7855	3' to 3-1/2'					124			124	136
7900	Taxus cuspidata capitata (Upright Japanese Yew), Z4, cont/BB									
7901	15" to 18"				Ea.	46.50			46.50	51
7902	18" to 24"					59			59	64.50
7903	2' to 2-1/2'					79			79	87
7904	2-1/2' to 3'					93			93	103
7905	3' to 4'					121			121	133
7906	4' to 5'					181			181	199
7907	5' to 6'					315			315	350
7908	6' to 7'					390			390	430
7909	7' to 8'					450			450	495
7950	Taxus cuspidata densiformis (Dense Spreading Yew), Z5, cont/BB									
7951	15" to 18"				Ea.	51			51	56
7952	18" to 24"					55.50			55.50	61
7953	2' to 2-1/2'					73			73	80
7954	2-1/2' to 3'					78.50			78.50	86
7955	3' to 3-1/2'					86			86	94.50
8000	Taxus cuspidata hicksi (Hick's Yew), Z4, cont/BB									
8001	15" to 18"				Ea.	50			50	55
8002	18" to 24"					76.50			76.50	84
8003	2' to 2-1/2'					80.50			80.50	89

For customer support on your Site Work & Landscape Costs with RSMeans Data, call 800.448.8182.

465

32 93 43.40 Trees, Conifers		Crew	Daily Output	Labor-Hours	Unit	Material	2021 Bare Costs Labor	Equipment	Total	Total Incl O&P
8004	2-1/2' to 3'				Ea.	92			92	101
8050	Taxus cuspidata nana (Dwarf Yew), Z4, cont/BB									
8051	15" to 18"				Ea.	34.50			34.50	38
8052	18" to 24"					72			72	79
8053	3' to 4'					92.50			92.50	102
8054	4' to 5'					116			116	127
8055	5' to 6'					118			118	129
8100	Thuja occidentalis "Smaragd" (Emerald Arborvitae), Z3, cont/BB									
8101	3' to 3-1/2'				Ea.	53			53	58
8102	3-1/2' to 4'					50			50	55
8103	4' to 4-1/2'					73			73	80
8104	4-1/2' to 5'					75.50			75.50	83
8105	5' to 6'					87.50			87.50	96.50
8150	Thuja occidentalis "Woodwardii" (Globe Arborvitae), Z2, B&B									
8151	15" to 18"				Ea.	21			21	23
8152	18" to 24"					27			27	30
8153	2' to 2-1/2'					31			31	34
8154	2-1/2' to 3'					40			40	44
8200	Thuja occidentalis douglasi pyramidalis (Arborvitae), Z4, cont/BB									
8201	2' to 3'				Ea.	50			50	55
8202	3' to 4'					62			62	68
8203	4' to 5'					87.50			87.50	96
8204	5' to 6'					97			97	107
8205	6' to 7'					155			155	170
8206	7' to 8'					163			163	179
8250	Thuja occidentalis nigra (Dark American Arborvitae), Z4, cont/BB									
8251	2' to 3'				Ea.	24			24	26.50
8252	3' to 4'					44.50			44.50	49
8253	4' to 5'					54.50			54.50	60
8254	5' to 6'					80			80	88
8255	6' to 8'					145			145	160
8256	8' to 10'					160			160	176
8300	Thuja occidentalis techney (Mission Arborvitae), Z4, cont/BB									
8301	2' to 3'				Ea.	44			44	48.50
8302	3' to 4'					51.50			51.50	57
8303	4' to 5'					62.50			62.50	68.50
8304	5' to 6'					108			108	119
8305	6' to 7'					125			125	137
8306	7' to 8'					164			164	181
8350	Thuja orientalis "aurea nana"' (Gold Arborvitae), Z4, cont									
8351	1 gal.				Ea.	9.90			9.90	10.90
8352	5 gal.				"	19.55			19.55	21.50
8400	Thuja orientalis "blue cone" (Blue Cone Arborvitae), Z6, cont									
8401	1 gal.				Ea.	7.45			7.45	8.20
8402	5 gal.				"	25.50			25.50	28
8450	Tsuga canadensis (Canadian Hemlock), Z4, cont/BB									
8451	2' to 2-1/2'				Ea.	58			58	63.50
8452	2-1/2' to 3'				"	73			73	80.50
8453	3' to 3-1/2'				Ea.	88			88	97
8454	3-1/2' to 4'					96.50			96.50	106
8455	4' to 5'					130			130	143
8456	5' to 5-1/2'					172			172	189
8457	5-1/2' to 6'					177			177	195

32 93 43.40 Trees, Conifers

		Crew	Daily Output	Labor-Hours	Unit	Material	2021 Bare Costs Labor	Equipment	Total	Total Incl O&P
8458	6' to 7'				Ea.	198			198	218
8459	7' to 8'				↓	210			210	231
8500	Tsuga canadensis sargentii (Sargents Weeping Hemlock), Z8, cont									
8501	1 gal.				Ea.	7.65			7.65	8.40
8502	3 gal.					22			22	24.50
8503	30" to 36"					126			126	138
8504	3' to 4'					131			131	144
8505	4' to 5'					170			170	187
8506	5' to 6'				↓	173			173	191
8550	Tsuga caroliniana (Caroline Hemlock), Z5, B&B									
8551	2-1/2' to 3'				Ea.	92			92	101
8552	3' to 3-1/2'					111			111	122
8553	3-1/2' to 4'					122			122	134
8554	4' to 5'				↓	170			170	187
8600	Tsuga diversifolia (Northern Japanese Hemlock), Z6, cont/BB									
8601	3' to 4'				Ea.	127			127	140
8602	4' to 5'				"	156			156	171

32 93 43.50 Trees and Palms

		Crew	Daily Output	Labor-Hours	Unit	Material	2021 Bare Costs Labor	Equipment	Total	Total Incl O&P
0010	**TREES & PALMS**									
0011	Warm temperate and subtropical									
0100	Acacia baileyana (Bailey Acacia), Z9, cont									
0110	1 gal.				Ea.	21			21	23
0120	5 gal.					18			18	19.80
0130	15 gal.				↓	92.50			92.50	102
0200	Acacia latifolia (Broadleaf Acacia (multistem)), Z10, cont									
0210	1 gal.				Ea.	8.35			8.35	9.15
0220	5 gal.					26			26	29
0230	15 gal.				↓	74			74	81
0300	Acacia melanoxylon (Blackwood Acacia), Z9, cont									
0310	1 gal.				Ea.	8.35			8.35	9.15
0320	5 gal.					31.50			31.50	34.50
0330	15 gal.				↓	70.50			70.50	77.50
0400	Agonis flexuosa (Willow Myrtle), Z9, cont									
0410	5 gal.				Ea.	18			18	19.80
0420	15 gal.				"	59			59	65
0500	Albizzia julibrissin (Silk Tree), Z7, cont									
0510	5 gal.				Ea.	225			225	248
0520	15 gal.				Ea.	48			48	52.50
0530	20 gal.					86.50			86.50	95
0540	24" box				↓	35			35	38.50
0600	Arbutus menziesii (Pacific Madrone), Z7, cont									
0610	1 gal.				Ea.	8			8	8.80
0620	5 gal.					65			65	71.50
0630	15 gal.				↓	61			61	67
0700	Arbutus unedo (Strawberry Tree), Z7, cont									
0710	1 gal.				Ea.	19.95			19.95	22
0720	5 gal.					24.50			24.50	27
0730	15 gal.				↓	79			79	87
0800	Archontophoenix cunninghamiana (King Palm), Z10, cont									
0810	5 gal.				Ea.	18.50			18.50	20.50
0820	15 gal.					65			65	71.50
0830	24" box				↓	225			225	248

For customer support on your Site Work & Landscape Costs with RSMeans Data, call 800.448.8182.

467

32 93 43.50 Trees and Palms	Crew	Daily Output	Labor-Hours	Unit	Material	2021 Bare Costs Labor	Equipment	Total	Total Incl O&P
0900 Arecastrum romanzoffianum (Queen Palm), Z10, cont									
0910 5 gal.				Ea.	17.50			17.50	19.25
0920 15 gal.					40			40	44
0930 24" box				↓	150			150	165
1000 Bauhinia variegata (Orchid Tree), Z9, cont									
1010 1 gal.				Ea.	6			6	6.60
1020 5 gal.					19.50			19.50	21.50
1030 24" box				↓	175			175	193
1100 Brachychiton populneus (Kurrajong), Z9, cont									
1110 5 gal.				Ea.	18			18	19.80
1120 15 gal.					44			44	48
1130 24" box				↓	170			170	187
1200 Callistemon citroides (Lemon Bottlebrush (Tree Form)), Z9, cont									
1210 15 gal.				Ea.	55			55	60.50
1220 24" box					175			175	193
1230 36" box				↓	570			570	630
1300 Casuarina stricta (Mountain She-Oak), Z9, cont									
1310 5 gal.				Ea.	23.50			23.50	26
1320 15 gal.				"	70			70	77
1400 Ceratoma siliqua (Carob), Z10, cont									
1410 5 gal.				Ea.	19.15			19.15	21
1420 15 gal.					47.50			47.50	52
1430 24" box				↓	213			213	234
1500 Chamaerops humilis (Mediterranean Fan Palm), Z9, cont									
1510 5 gal.				Ea.	19.60			19.60	21.50
1520 15 gal.				"	78.50			78.50	86.50
1530 24" box				Ea.	233			233	257
1600 Chrysobalanus icaco (Coco-plum), Z10, cont									
1610 1 gal.				Ea.	3.41			3.41	3.75
1620 3 gal.					4.48			4.48	4.93
1630 10 gal.					24.50			24.50	27
1640 20 gal.				↓	61			61	67
1700 Cinnamonum camphora (Camphor Tree), Z9, cont									
1710 5 gal.				Ea.	25.50			25.50	28
1720 15 gal.					65			65	71.50
1730 24" box				↓	225			225	248
1800 Cornus nattali (California Laurel), Z7, B&B									
1810 5' to 6'				Ea.	26.50			26.50	29.50
1820 6' to 8'					86			86	94.50
1830 8' to 10'				↓	134			134	148
2000 Eriobotrya japonica (Loquat), Z8, cont									
2010 3 gal.				Ea.	26			26	28.50
2020 10 gal.					39.50			39.50	43.50
2030 15 gal.					86			86	94.50
2040 15 gal. Espalier				↓	130			130	143
2100 Erythrina caffra (Kaffirboom Coral Tree), Z10, cont									
2110 5 gal.				Ea.	18.50			18.50	20.50
2120 15 gal.					59			59	65
2130 24" box				↓	175			175	193
2200 Eucalyptus camaldulensis (Red Gum), Z9, cont									
2210 5 gal.				Ea.	25			25	27.50
2220 15 gal.					48.50			48.50	53.50
2230 24" box				↓	159			159	175

32 93 43.50 Trees and Palms	Crew	Daily Output	Labor-Hours	Unit	Material	2021 Bare Costs Labor	Equipment	Total	Total Incl O&P
2300 Eucalyptus ficifolia (Flaming Gum), Z10, cont									
2310 5 gal.				Ea.	20.50			20.50	22.50
2320 15 gal.					44			44	48.50
2330 24" box				↓	171			171	188
2400 Feijoa sellowiana (Guava), Z8, cont									
2410 1 gal.				Ea.	4.50			4.50	4.95
2420 5 gal.					14.50			14.50	15.95
2430 15 gal.				↓	45			45	49.50
2500 Ficus benjamina (Benjamin Fig), Z10, cont									
2510 1 gal.				Ea.	28.50			28.50	31
2520 5 gal.					18.50			18.50	20.50
2530 24" box				↓	217			217	239
2600 Ficus retusa nitida (India Laurel Fig), Z10, cont									
2610 5 gal.				Ea.	17.95			17.95	19.75
2620 15 gal.				"	56			56	61.50
2630 24" box				Ea.	185			185	204
2700 Fraxinus uhdei (Shamel Ash), Z9, cont									
2710 5 gal.				Ea.	18.25			18.25	20
2720 15 gal.					55			55	60.50
2730 24" box					157			157	173
2740 36" box				↓	480			480	530
2800 Fraxinus velutina (Velvet Ash), Z7, cont									
2810 5 gal.				Ea.	19.55			19.55	21.50
2820 15 gal.					54			54	59.50
2830 24" box				↓	145			145	159
2900 Gordonia lasianthus (Loblolly Bay Gordonia), Z8, cont									
2910 5 gal.				Ea.	35			35	38.50
2920 7 gal.				"	58.50			58.50	64
3000 Grevillea robusta (Silk Oak), Z10, cont									
3010 5 gal.				Ea.	17.35			17.35	19.10
3020 15 gal.				"	48.50			48.50	53.50
3100 Hakea suaveolens (Sea Urchin Tree), Z10, cont									
3110 1 gal.				Ea.	4.50			4.50	4.95
3120 5 gal.					14			14	15.40
3130 15 gal.				↓	45.50			45.50	50
3200 Harpephyllum caffrum (Kaffir Plum), Z10, cont									
3210 5 gal.				Ea.	19.10			19.10	21
3220 15 gal.					49.50			49.50	54.50
3230 24" box				↓	192			192	211
3300 Hymenosporum flavum (Sweetshade), Z10, cont									
3310 5 gal.				Ea.	18.65			18.65	20.50
3320 15 gal.					46.50			46.50	51
3330 24" box				↓	112			112	123
3400 Jacaranda mimosifolia (Jacaranda), Z10, cont									
3410 5 gal.				Ea.	18.35			18.35	20
3420 15 gal.					56.50			56.50	62.50
3430 24" box				↓	186			186	204
3500 Jatropha curcas (Barbados Nut), Z10, cont/BB									
3510 3 gal.				Ea.	8			8	8.80
3520 10 gal.					34			34	37.50
3530 6' to 7'				↓	50			50	55
3600 Laurus nobilis (Sweet Bay), Z8, cont									
3610 1 gal.				Ea.	6			6	6.60

32 93 43.50 Trees and Palms	Crew	Daily Output	Labor-Hours	Unit	Material	2021 Bare Costs Labor	Equipment	Total	Total Incl O&P
3620	5 gal.				Ea.	14			14	15.40
3630	15 gal.					61			61	67
3640	24" box				↓	206			206	226
3700	Leptospermum laevigatum (Australian Tea Tree), Z9, cont									
3710	1 gal.				Ea.	4.50			4.50	4.95
3720	5 gal.					23.50			23.50	25.50
3730	15 gal.				↓	56.50			56.50	62
3800	Leptospermum scoparium Ruby Glow (Tea Tree), Z9, cont									
3810	1 gal.				Ea.	4.50			4.50	4.95
3820	5 gal.				"	14			14	15.40
3900	Magnolia grandiflora (Southern Magnolia), Z7, B&B									
3910	4' to 5'				Ea.	36.50			36.50	40
3920	5' to 6'					100			100	110
3930	6' to 7'					95.50			95.50	105
3940	7' to 8'				↓	135			135	149
4000	Melaleuca linariifolia (Flaxleaf Paperbark), Z9, cont									
4010	5 gal.				Ea.	18.50			18.50	20.50
4020	15 gal.					68.50			68.50	75.50
4030	24" box				↓	210			210	231
4100	Maytenus boaria (Mayten Tree), Z9, cont									
4110	5 gal.				Ea.	18.50			18.50	20.50
4120	15 gal.				"	78.50			78.50	86.50
4200	Metrosideros excelsus (New Zealand Christmas Tree), Z10, cont									
4210	5 gal.				Ea.	22.50			22.50	25
4220	15 gal.					50			50	55
4230	24" box				↓	176			176	193
4300	Olea europea (Olive), Z9, cont									
4310	1 gal.				Ea.	10			10	11
4320	5 gal.					32			32	35.50
4330	15 gal.				↓	73			73	80
4400	Parkinsonia aculeata (Jerusalem Thorn), Z9, cont									
4410	5 gal.				Ea.	22			22	24
4420	15 gal.					53.50			53.50	59
4430	24" box				↓	154			154	170
4500	Pistachia chinensis (Chinese Pistache), Z9, cont									
4510	5 gal.				Ea.	24			24	26.50
4520	15 gal.					60			60	66
4530	24" box				↓	166			166	183
4600	Pyrus Kawakamii (Evergreen Pear), Z8, cont									
4610	5 gal.				Ea.	29			29	32
4620	15 gal.					58			58	64
4630	15 gal. Espalier					55			55	60.50
4640	24" box				↓	185			185	204
4700	Quercus agrifolia (Coast Live Oak), Z9, cont									
4710	5 gal.				Ea.	26.50			26.50	29.50
4720	15 gal.					55			55	60.50
4730	20 gal.				↓	200			200	220
4800	Sapium sebiferum (Chinese Tallow Tree), Z8, cont									
4810	5 gal.				Ea.	20			20	22
4820	15 gal.					48			48	52.50
4830	24" box				↓	177			177	194
4900	Schinus molle (Pepper Tree), Z9, cont									
4910	5 gal.				Ea.	26.50			26.50	29.50

32 93 43.50 Trees and Palms		Crew	Daily Output	Labor-Hours	Unit	Material	2021 Bare Costs Labor	Equipment	Total	Total Incl O&P
4920	15 gal.				Ea.	66.50			66.50	73
4930	24" box					206			206	226
5000	Strelitzia nicolai (Giant Bird of Paradise), Z10, cont									
5010	5 gal.				Ea.	16.25			16.25	17.90
5020	15 gal.					65.50			65.50	72
5030	24" box					225			225	248
5100	Syzygium paniculatum (Brush Cherry), Z9, cont									
5110	1 gal.				Ea.	6.50			6.50	7.15
5120	5 gal.					20			20	22
5130	15 gal.					51.50			51.50	56.50
5200	Tabebuia chrysotricha (Golden Trumpet Tree), Z10, cont									
5210	5 gal.				Ea.	19.05			19.05	21
5220	15 gal.					51			51	56
5230	24" box					191			191	210
5300	Thevetia peruviana (Yellow Oleander), Z10, cont									
5310	1 gal.				Ea.	9.20			9.20	10.15
5320	5 gal.					16.10			16.10	17.75
5330	15 gal.					54			54	59.50
5400	Tipuana tipi (Common Tiputree), Z10, cont									
5410	1 gal.				Ea.	6.20			6.20	6.80
5420	5 gal.					18.50			18.50	20.50
5430	15 gal.					48			48	52.50
5440	24" box					165			165	182
5500	Tristania conferta (Brisbane Box), Z10, cont									
5510	5 gal.				Ea.	16.05			16.05	17.65
5520	15 gal.				"	48			48	53
5600	Umbellularia californica (California Laurel), Z7, cont									
5610	1 gal.				Ea.	7			7	7.70
5620	5 gal.					22.50			22.50	25
5630	15 gal.					64.50			64.50	71

32 93 43.60 Trees, Fruits and Nuts

		Crew	Daily Output	Labor-Hours	Unit	Material	Labor	Equipment	Total	Total Incl O&P
0010	**TREES, FRUITS AND NUTS**									
1000	Apple, different varieties, Z4, cont									
1010	5 gal.				Ea.	28.50			28.50	31.50
1020	Standard, 2-3 yr.					35.50			35.50	39
1030	Semidwarf, 2-3 yr.					30.50			30.50	33.50
1500	Apricot, Z5, cont/BB									
1510	5 gal.				Ea.	22.50			22.50	25
1520	Standard, 2-3 yr.					30.50			30.50	33.50
1530	Semidwarf, 2-3 yr.					29			29	32
2000	Carya ovata, Z4, bare root									
2010	Bare root, 2 yr.				Ea.	135			135	148
2500	Castanea mollissima (Chinese Chestnut), Z4, bare root									
2510	Bare root, 2 yr.				Ea.	49.50			49.50	54.50
3000	Cherry, Z4, B&B									
3010	Standard, 2-3 yr.				Ea.	31.50			31.50	34.50
3020	Semidwarf, 2-3 yr.				"	33.50			33.50	36.50
3500	Corylus americana (Filbert, American Hazelnut), Z5, B&B									
3510	2' to 3'				Ea.	13			13	14.30
3520	3' to 4'				"	20.50			20.50	22.50
4500	Grape different varieties, Z5, potted									
4510	Potted				Ea.	14.20			14.20	15.60

32 93 Plants

32 93 43 – Trees

32 93 43.60 Trees, Fruits and Nuts

	32 93 43.60 Trees, Fruits and Nuts	Crew	Daily Output	Labor-Hours	Unit	Material	2021 Bare Costs Labor	Equipment	Total	Total Incl O&P
5000	Juglans cinera (Butternut), Z4, bare root									
5010	Bare root, 2 yr.				Ea.	37			37	41
5500	Juglans nigra (Black Walnut), Z4, bare root									
5510	Bare root, 2'				Ea.	30			30	33
6000	Peach, Z6, cont/BB									
6010	5 gal.				Ea.	65			65	71.50
6020	Standard, 2-3 yr.					28			28	31
6030	Semidwarf, 2-3 yr.				▼	28.50			28.50	31.50
6500	Pear, Z5, B&B									
6510	Standard, 2-3 yr.				Ea.	23.50			23.50	25.50
6520	Semidwarf, 2-3 yr.				"	26.50			26.50	29
7000	Plum, Z5, cont or B&B									
7010	5 gal.				Ea.	36.50			36.50	40
7020	Standard, 2-3 yr.					23			23	25
7030	Semidwarf, 2-3 yr.				▼	30			30	33
7500	Raspberry (Everbearing), Z4									
7510	package of 5				Ea.	23			23	25.50
8000	Strawberry (Hybrid), Z5									
8010	25 per box				Ea.	13.35			13.35	14.65
8020	50 per box				"	30.50			30.50	33.50

32 94 Planting Accessories

32 94 13 – Landscape Edging

32 94 13.20 Edging

	32 94 13.20 Edging	Crew	Daily Output	Labor-Hours	Unit	Material	2021 Bare Costs Labor	Equipment	Total	Total Incl O&P
0010	**EDGING**									
0050	Aluminum alloy, including stakes, 1/8" x 4", mill finish	B-1	390	.062	L.F.	2.27	2.77		5.04	6.65
0051	Black paint		390	.062		2.63	2.77		5.40	7.05
0052	Black anodized		390	.062		3.04	2.77		5.81	7.50
0060	3/16" x 4", mill finish		380	.063		3.27	2.85		6.12	7.85
0061	Black paint		380	.063		3.77	2.85		6.62	8.40
0062	Black anodized		380	.063		4.40	2.85		7.25	9.10
0070	1/8" x 5-1/2" mill finish		370	.065		3.29	2.92		6.21	8
0071	Black paint		370	.065		3.90	2.92		6.82	8.65
0072	Black anodized		370	.065		4.47	2.92		7.39	9.30
0080	3/16" x 5-1/2" mill finish		360	.067		4.40	3		7.40	9.30
0081	Black paint		360	.067		4.97	3		7.97	9.95
0082	Black anodized	▼	360	.067		5.75	3		8.75	10.80
0100	Brick, set horizontally, 1-1/2 bricks per L.F.	D-1	370	.043		1.75	2.11		3.86	5.10
0150	Set vertically, 3 bricks per L.F.	"	135	.119		4.04	5.75		9.79	13.15
0200	Corrugated aluminum, roll, 4" wide	1 Carp	650	.012		2.08	.67		2.75	3.29
0250	6" wide	"	550	.015	▼	2.60	.80		3.40	4.05
0300	Concrete, cast in place, see Section 03 30 53.40									
0350	Granite, 5" x 16", straight	B-29	300	.187	L.F.	16.45	9	2.84	28.29	34.50
0400	Polyethylene grass barrier, 5" x 1/8"	D-1	400	.040		1.26	1.95		3.21	4.33
0410	5" x 1/4"		400	.040		1.45	1.95		3.40	4.53
0420	5" x 5/32"		400	.040		1.68	1.95		3.63	4.78
0430	6" x 3/32"		400	.040		1.35	1.95		3.30	4.42
0500	Precast scallops, green, 2" x 8" x 16"		400	.040		2.69	1.95		4.64	5.90
0550	2" x 8" x 16" other than green	▼	400	.040		2.15	1.95		4.10	5.30
0600	Railroad ties, 6" x 8"	2 Carp	170	.094		2.74	5.15		7.89	10.70
0650	7" x 9"		136	.118	▼	3.04	6.45		9.49	12.95

472

32 94 Planting Accessories

32 94 13 – Landscape Edging

32 94 13.20 Edging		Crew	Daily Output	Labor-Hours	Unit	Material	2021 Bare Costs Labor	Equipment	Total	Total Incl O&P
0750	Redwood 2" x 4"	2 Carp	330	.048	L.F.	5.45	2.65		8.10	9.95
0780	Landscape timbers, 100% recycled plastic, var colors, 4" x 4" x 8' G		250	.064		7.25	3.50		10.75	13.25
0790	6" x 6" x 8' G		250	.064		12.20	3.50		15.70	18.65
0800	Steel edge strips, incl. stakes, 1/4" x 5"	B-1	390	.062		4.68	2.77		7.45	9.30
0850	3/16" x 4"	"	390	.062		3.70	2.77		6.47	8.20
0900	Hardwood, pressure treated, 4" x 6"	2 Carp	250	.064		2.01	3.50		5.51	7.45
0940	6" x 6"		200	.080		2.87	4.38		7.25	9.70
0980	6" x 8"		170	.094		3.28	5.15		8.43	11.30
1000	Pine, pressure treated, 1" x 4"		500	.032		.69	1.75		2.44	3.37
1040	2" x 4"		330	.048		1.16	2.65		3.81	5.25
1080	4" x 6"		250	.064		3.86	3.50		7.36	9.50
1100	6" x 6"		200	.080		5.80	4.38		10.18	12.95
1140	6" x 8"		170	.094		7.60	5.15		12.75	16.05
1200	Edging, lawn, made from recycled tires, black, 1" x 6"	B-1	390	.062	L.F.	1.23	2.77		4	5.50

32 94 50 – Tree Guying

32 94 50.10 Tree Guying Systems		Crew	Daily Output	Labor-Hours	Unit	Material	2021 Bare Costs Labor	Equipment	Total	Total Incl O&P
0010	**TREE GUYING SYSTEMS**									
0015	Tree guying including stakes, guy wire and wrap									
0100	Less than 3" caliper, 2 stakes	2 Clab	35	.457	Ea.	14.50	20.50		35	46.50
0200	3" to 4" caliper, 3 stakes	"	21	.762		21	34		55	73.50
0300	6" to 8" caliper, 3 stakes	B-1	8	3		40	135		175	246
1000	Including arrowhead anchor, cable, turnbuckles and wrap									
1100	Less than 3" caliper, 3" anchors	2 Clab	20	.800	Ea.	25.50	35.50		61	81
1200	3" to 6" caliper, 4" anchors		15	1.067		35	47.50		82.50	109
1300	6" caliper, 6" anchors		12	1.333		29	59		88	121
1400	8" caliper, 8" anchors		9	1.778		125	79		204	255
2000	Tree guard, preformed plastic, 36" high		168	.095		1.98	4.23		6.21	8.50
2010	Snow fence		140	.114		32	5.05		37.05	42.50

32 96 Transplanting

32 96 23 – Plant and Bulb Transplanting

32 96 23.23 Planting		Crew	Daily Output	Labor-Hours	Unit	Material	2021 Bare Costs Labor	Equipment	Total	Total Incl O&P
0010	**PLANTING**									
0012	Moving shrubs on site, 12" ball	B-62	28	.857	Ea.	41	6.40		47.40	68.50
0100	24" ball	"	22	1.091	"	52.50	8.15		60.65	87.50

32 96 23.43 Moving Trees		Crew	Daily Output	Labor-Hours	Unit	Material	2021 Bare Costs Labor	Equipment	Total	Total Incl O&P
0010	**MOVING TREES**, On site									
0300	Moving trees on site, 36" ball	B-6	3.75	6.400	Ea.		310	57.50	367.50	525
0400	60" ball	"	1	24	"		1,150	216	1,366	1,975

32 96 43 – Tree Transplanting

32 96 43.20 Tree Removal		Crew	Daily Output	Labor-Hours	Unit	Material	2021 Bare Costs Labor	Equipment	Total	Total Incl O&P
0010	**TREE REMOVAL**									
0100	Dig & lace, shrubs, broadleaf evergreen, 18"-24" high	B-1	55	.436	Ea.		19.65		19.65	29.50
0200	2'-3'	"	35	.686			31		31	46
0300	3'-4'	B-6	30	.800			38.50	7.20	45.70	65.50
0400	4'-5'	"	20	1.200			57.50	10.80	68.30	98
1000	Deciduous, 12"-15"	B-1	110	.218			9.85		9.85	14.65
1100	18"-24"		65	.369			16.65		16.65	25
1200	2'-3'		55	.436			19.65		19.65	29.50
1300	3'-4'	B-6	50	.480			23	4.32	27.32	39.50

32 96 43.20 Tree Removal		Crew	Daily Output	Labor-Hours	Unit	Material	2021 Bare Costs Labor	Equipment	Total	Total Incl O&P
2000	Evergreen, 18"-24"	B-1	55	.436	Ea.		19.65		19.65	29.50
2100	2' to 2.5'		50	.480			21.50		21.50	32.50
2200	2.5' to 3'		35	.686			31		31	46
2300	3' to 3.5'		20	1.200			54		54	80.50
3000	Trees, deciduous, small, 2'-3'		55	.436			19.65		19.65	29.50
3100	3'-4'	B-6	50	.480			23	4.32	27.32	39.50
3200	4'-5'		35	.686			33	6.20	39.20	56
3300	5'-6'		30	.800			38.50	7.20	45.70	65.50
4000	Shade, 5'-6'	B-6	50	.480	Ea.		23	4.32	27.32	39.50
4100	6'-8'		35	.686			33	6.20	39.20	56
4200	8'-10'		25	.960			46	8.65	54.65	78.50
4300	2" caliper		12	2			96	18	114	163
5000	Evergreen, 4'-5'		35	.686			33	6.20	39.20	56
5100	5'-6'		25	.960			46	8.65	54.65	78.50
5200	6'-7'		19	1.263			61	11.40	72.40	103
5300	7'-8'		15	1.600			77	14.40	91.40	131
5400	8'-10'		11	2.182			105	19.65	124.65	179

Estimating Tips

33 10 00 Water Utilities
33 30 00 Sanitary Sewerage Utilities
33 40 00 Storm Drainage Utilities

- Never assume that the water, sewer, and drainage lines will go in at the early stages of the project. Consider the site access needs before dividing the site in half with open trenches, loose pipe, and machinery obstructions. Always inspect the site to establish that the site drawings are complete. Check off all existing utilities on your drawings as you locate them. Be especially careful with underground utilities because appurtenances are sometimes buried during regrading or repaving operations. If you find any discrepancies, mark up the site plan for further research. Differing site conditions can be very costly if discovered later in the project.

- See also Section 33 01 00 for restoration of pipe where removal/replacement may be undesirable. Use of new types of piping materials can reduce the overall project cost. Owners/design engineers should consider the installing contractor as a valuable source of current information on utility products and local conditions that could lead to significant cost savings.

Reference Numbers

Reference numbers are shown at the beginning of some major classifications. These numbers refer to related items in the Reference Section. The reference information may be an estimating procedure, an alternate pricing method, or technical information.

Note: Not all subdivisions listed here necessarily appear. ■

Same Data. Simplified.

Enjoy the convenience and efficiency of accessing your costs anywhere:

- **Skip the multiplier** by setting your location
- **Quickly search,** edit, favorite and share costs
- **Stay on top of price changes** with automatic updates

Discover more at rsmeans.com/online

---

33 01 10.10 Corrosion Resistance		Crew	Daily Output	Labor-Hours	Unit	Material	2021 Bare Costs Labor	Equipment	Total	Total Incl O&P
0010	**CORROSION RESISTANCE**									
0012	Wrap & coat, add to pipe, 4" diameter				L.F.	2.50			2.50	2.75
0020	5" diameter					2.78			2.78	3.06
0040	6" diameter					3.71			3.71	4.08
0060	8" diameter					5.05			5.05	5.55
0080	10" diameter					6.20			6.20	6.80
0100	12" diameter					6.65			6.65	7.30
0120	14" diameter					8.80			8.80	9.70
0140	16" diameter					10.40			10.40	11.45
0160	18" diameter					10.95			10.95	12
0180	20" diameter					12.35			12.35	13.60
0200	24" diameter					14.85			14.85	16.30
0220	Small diameter pipe, 1" diameter, add					.80			.80	.88
0240	2" diameter					1.10			1.10	1.21
0260	2-1/2" diameter					1.52			1.52	1.67
0280	3" diameter				▼	1.94			1.94	2.13
0300	Fittings, field covered, add				S.F.	9.85			9.85	10.85
0500	Coating, bituminous, per diameter inch, 1 coat, add				L.F.	.19			.19	.21
0540	3 coat					.62			.62	.68
0560	Coal tar epoxy, per diameter inch, 1 coat, add					.25			.25	.28
0600	3 coat					.69			.69	.76
1000	Polyethylene H.D. extruded, 0.025" thk., 1/2" diameter, add					.10			.10	.11
1020	3/4" diameter					.17			.17	.19
1040	1" diameter					.20			.20	.22
1060	1-1/4" diameter					.26			.26	.29
1080	1-1/2" diameter					.29			.29	.32
1100	0.030" thk., 2" diameter					.40			.40	.44
1120	2-1/2" diameter					.52			.52	.57
1140	0.035" thk., 3" diameter					.55			.55	.61
1160	3-1/2" diameter				▼	.64			.64	.70
1180	4" diameter				L.F.	.75			.75	.83
1200	0.040" thk., 5" diameter					.92			.92	1.01
1220	6" diameter					1.10			1.10	1.21
1240	8" diameter					1.40			1.40	1.54
1260	10" diameter					1.80			1.80	1.98
1280	12" diameter					2.16			2.16	2.38
1300	0.060" thk., 14" diameter					2.85			2.85	3.14
1320	16" diameter					3.27			3.27	3.60
1340	18" diameter					3.25			3.25	3.58
1360	20" diameter				▼	3.62			3.62	3.98
1380	Fittings, field wrapped, add				S.F.	2.36			2.36	2.60

33 01 10.20 Pipe Repair

33 01 10.20 Pipe Repair		Crew	Daily Output	Labor-Hours	Unit	Material	2021 Bare Costs Labor	Equipment	Total	Total Incl O&P
0010	**PIPE REPAIR**									
0020	Not including excavation or backfill									
0100	Clamp, stainless steel, lightweight, for steel pipe									
0110	3" long, 1/2" diameter pipe	1 Plum	34	.235	Ea.	16.15	15.95		32.10	42
0120	3/4" diameter pipe		32	.250		16.70	16.95		33.65	44
0130	1" diameter pipe		30	.267		17.65	18.05		35.70	46.50
0140	1-1/4" diameter pipe		28	.286		18.65	19.35		38	49.50
0150	1-1/2" diameter pipe		26	.308		19.50	21		40.50	52.50
0160	2" diameter pipe		24	.333		21.50	22.50		44	57
0170	2-1/2" diameter pipe		23	.348		23	23.50		46.50	60

For customer support on your Site Work & Landscape Costs with RSMeans Data, call 800.448.8182.

33 01 10 - Operation and Maintenance of Water Utilities

33 01 10.20 Pipe Repair		Crew	Daily Output	Labor-Hours	Unit	Material	2021 Bare Costs Labor	Equipment	Total	Total Incl O&P
0180	3" diameter pipe	1 Plum	22	.364	Ea.	25	24.50		49.50	64.50
0190	3-1/2" diameter pipe	↓	21	.381		26.50	26		52.50	67.50
0200	4" diameter pipe	B-20	44	.545		26.50	27		53.50	69.50
0210	5" diameter pipe		42	.571		36.50	28		64.50	82
0220	6" diameter pipe		38	.632		36.50	31		67.50	86.50
0230	8" diameter pipe		30	.800		43	39.50		82.50	107
0240	10" diameter pipe		28	.857		232	42.50		274.50	320
0250	12" diameter pipe		24	1		244	49.50		293.50	340
0260	14" diameter pipe		22	1.091		268	54		322	375
0270	16" diameter pipe		20	1.200		150	59		209	254
0280	18" diameter pipe		18	1.333		305	65.50		370.50	435
0290	20" diameter pipe		16	1.500		320	74		394	465
0300	24" diameter pipe	↓	14	1.714		355	84.50		439.50	520
0360	For 6" long, add					100%	40%			
0370	For 9" long, add					200%	100%			
0380	For 12" long, add					300%	150%			
0390	For 18" long, add				↓	500%	200%			
0400	Pipe freezing for live repairs of systems 3/8" to 6"									
0410	Note: Pipe freezing can also be used to install a valve into a live system									
0420	Pipe freezing each side 3/8"	2 Skwk	8	2	Ea.	590	114		704	820
0425	Pipe freezing each side 3/8", second location same kit		8	2		34	114		148	210
0430	Pipe freezing each side 3/4"		8	2		525	114		639	750
0435	Pipe freezing each side 3/4", second location same kit		8	2		34	114		148	210
0440	Pipe freezing each side 1-1/2"		6	2.667		525	152		677	810
0445	Pipe freezing each side 1-1/2" , second location same kit		6	2.667		34	152		186	267
0450	Pipe freezing each side 2"		6	2.667		830	152		982	1,150
0455	Pipe freezing each side 2", second location same kit		6	2.667		34	152		186	267
0460	Pipe freezing each side 2-1/2" to 3"		6	2.667		955	152		1,107	1,275
0465	Pipe freezing each side 2-1/2" to 3", second location same kit		6	2.667		68	152		220	305
0470	Pipe freezing each side 4"		4	4		1,575	228		1,803	2,075
0475	Pipe freezing each side 4", second location same kit		4	4		71	228		299	425
0480	Pipe freezing each side 5" to 6"		4	4		4,850	228		5,078	5,700
0485	Pipe freezing each side 5" to 6", second location same kit	↓	4	4		214	228		442	580
0490	Pipe freezing extra 20 lb. CO$_2$ cylinders (3/8" to 2" - 1 ea, 3" - 2 ea)					240			240	264
0500	Pipe freezing extra 50 lb. CO$_2$ cylinders (4" - 2 ea, 5"-6" - 6 ea)				↓	575			575	630
1000	Clamp, stainless steel, with threaded service tap									
1040	Full seal for iron, steel, PVC pipe									
1100	6" long, 2" diameter pipe	1 Plum	17	.471	Ea.	91	32		123	148
1110	2-1/2" diameter pipe		16	.500		94.50	34		128.50	155
1120	3" diameter pipe		15.60	.513		185	34.50		219.50	256
1130	3-1/2" diameter pipe	↓	15	.533		115	36		151	181
1140	4" diameter pipe	B-20	32	.750		123	37		160	191
1150	6" diameter pipe		28	.857		219	42.50		261.50	305
1160	8" diameter pipe		21	1.143		345	56.50		401.50	465
1170	10" diameter pipe		20	1.200		400	59		459	530
1180	12" diameter pipe	↓	17	1.412		425	69.50		494.50	575
1200	8" long, 2" diameter pipe	1 Plum	11.72	.683		124	46		170	206
1210	2-1/2" diameter pipe		11	.727		129	49		178	216
1220	3" diameter pipe		10.75	.744		174	50.50		224.50	266
1230	3-1/2" diameter pipe	↓	10.34	.774		136	52.50		188.50	227
1240	4" diameter pipe	B-20	22	1.091		605	54		659	745
1250	6" diameter pipe		19.31	1.243		213	61.50		274.50	325
1260	8" diameter pipe	↓	14.48	1.657		241	81.50		322.50	385

For customer support on your Site Work & Landscape Costs with RSMeans Data, call 800.448.8182.

477

33 01 10.20 Pipe Repair		Crew	Daily Output	Labor-Hours	Unit	Material	2021 Bare Costs Labor	2021 Bare Costs Equipment	Total	Total Incl O&P
1270	10" diameter pipe	B-20	13.80	1.739	Ea.	310	85.50		395.50	470
1280	12" diameter pipe	↓	11.72	2.048		345	101		446	530
1300	12" long, 2" diameter pipe	1 Plum	9.44	.847		207	57.50		264.50	315
1310	2-1/2" diameter pipe	"	8.89	.900	↓	211	61		272	325
1320	3" diameter pipe	1 Plum	8.67	.923	Ea.	273	62.50		335.50	395
1330	3-1/2" diameter pipe	"	8.33	.960		231	65		296	350
1340	4" diameter pipe	B-20	17.78	1.350		267	66.50		333.50	395
1350	6" diameter pipe		15.56	1.542		310	76		386	455
1360	8" diameter pipe		11.67	2.057		355	101		456	545
1370	10" diameter pipe		11.11	2.160		455	107		562	665
1380	12" diameter pipe	↓	9.44	2.542		520	125		645	765
1400	20" long, 2" diameter pipe	1 Plum	8.10	.988		251	67		318	375
1410	2-1/2" diameter pipe		7.62	1.050		277	71		348	410
1420	3" diameter pipe		7.43	1.077		345	73		418	490
1430	3-1/2" diameter pipe	↓	7.14	1.120		335	76		411	485
1440	4" diameter pipe	B-20	15.24	1.575		330	77.50		407.50	475
1450	6" diameter pipe		13.33	1.800		505	89		594	690
1460	8" diameter pipe		10	2.400		555	118		673	785
1470	10" diameter pipe		9.52	2.521		675	124		799	930
1480	12" diameter pipe	↓	8.10	2.963	↓	795	146		941	1,100
1600	Clamp, stainless steel, single section									
1640	Full seal for iron, steel, PVC pipe									
1700	6" long, 2" diameter pipe	1 Plum	17	.471	Ea.	94	32		126	152
1710	2-1/2" diameter pipe		16	.500		94.50	34		128.50	155
1720	3" diameter pipe		15.60	.513		110	34.50		144.50	173
1730	3-1/2" diameter pipe	↓	15	.533		115	36		151	181
1740	4" diameter pipe	B-20	32	.750		122	37		159	190
1750	6" diameter pipe		27	.889		152	44		196	234
1760	8" diameter pipe		21	1.143		179	56.50		235.50	282
1770	10" diameter pipe		20	1.200		269	59		328	385
1780	12" diameter pipe	↓	17	1.412		274	69.50		343.50	405
1800	8" long, 2" diameter pipe	1 Plum	11.72	.683		124	46		170	206
1805	2-1/2" diameter pipe		11.03	.725		129	49		178	216
1810	3" diameter pipe		10.76	.743		136	50.50		186.50	224
1815	3-1/2" diameter pipe	↓	10.34	.774		136	52.50		188.50	227
1820	4" diameter pipe	B-20	22.07	1.087		153	53.50		206.50	249
1825	6" diameter pipe		19.31	1.243		183	61.50		244.50	293
1830	8" diameter pipe		14.48	1.657		214	81.50		295.50	360
1835	10" diameter pipe		13.79	1.740		286	86		372	445
1840	12" diameter pipe	↓	11.72	2.048		395	101		496	580
1850	12" long, 2" diameter pipe	1 Plum	9.44	.847		207	57.50		264.50	315
1855	2-1/2" diameter pipe		8.89	.900		192	61		253	300
1860	3" diameter pipe		8.67	.923		219	62.50		281.50	335
1865	3-1/2" diameter pipe	↓	8.33	.960		231	65		296	350
1870	4" diameter pipe	B-20	17.78	1.350		243	66.50		309.50	365
1875	6" diameter pipe		15.56	1.542		292	76		368	435
1880	8" diameter pipe		11.67	2.057		345	101		446	530
1885	10" diameter pipe		11.11	2.160		450	107		557	655
1890	12" diameter pipe	↓	9.44	2.542		520	125		645	760
1900	20" long, 2" diameter pipe	1 Plum	8.10	.988		251	67		318	375
1905	2-1/2" diameter pipe		7.62	1.050		277	71		348	410
1910	3" diameter pipe		7.43	1.077		320	73		393	460
1915	3-1/2" diameter pipe	↓	7.14	1.120		335	76		411	485

33 01 Operation and Maintenance of Utilities

33 01 10 – Operation and Maintenance of Water Utilities

33 01 10.20 Pipe Repair

		Crew	Daily Output	Labor-Hours	Unit	Material	2021 Bare Costs Labor	Equipment	Total	Total Incl O&P
1920	4" diameter pipe	B-20	15.24	1.575	Ea.	400	77.50		477.50	555
1925	6" diameter pipe		13.33	1.800		485	89		574	670
1930	8" diameter pipe		10	2.400		555	118		673	790
1935	10" diameter pipe		9.52	2.521		690	124		814	940
1940	12" diameter pipe		8.10	2.963		750	146		896	1,050
2000	Clamp, stainless steel, two section									
2040	Full seal, for iron, steel, PVC pipe									
2100	8" long, 4" diameter pipe	B-20	24	1	Ea.	241	49.50		290.50	340
2110	6" diameter pipe		20	1.200		277	59		336	395
2120	8" diameter pipe		13	1.846		315	91		406	485
2130	10" diameter pipe		12	2		320	98.50		418.50	505
2140	12" diameter pipe		10	2.400		410	118		528	625
2200	10" long, 4" diameter pipe		16	1.500		315	74		389	460
2210	6" diameter pipe		13	1.846		355	91		446	525
2220	8" diameter pipe		9	2.667		395	131		526	630
2230	10" diameter pipe		8	3		520	148		668	790
2240	12" diameter pipe		7	3.429		595	169		764	910
2242	Clamp, stainless steel, three section									
2244	Full seal, for iron, steel, PVC pipe									
2250	10" long, 14" diameter pipe	B-20	6.40	3.750	Ea.	860	185		1,045	1,225
2260	16" diameter pipe		6	4		870	197		1,067	1,250
2270	18" diameter pipe		5	4.800		955	237		1,192	1,400
2280	20" diameter pipe		4.60	5.217		955	257		1,212	1,425
2290	24" diameter pipe		4	6		1,525	296		1,821	2,125
2320	For 12" long, add to 10"					15%	25%			
2330	For 20" long, add to 10"					70%	55%			
8000	For internal cleaning and inspection, see Section 33 01 30.11									
8100	For pipe testing, see Section 23 05 93.50									

33 01 30 – Operation and Maintenance of Sewer Utilities

33 01 30.11 Television Inspection of Sewers

		Crew	Daily Output	Labor-Hours	Unit	Material	2021 Bare Costs Labor	Equipment	Total	Total Incl O&P
0010	**TELEVISION INSPECTION OF SEWERS**									
0100	Pipe internal cleaning & inspection, cleaning, pressure pipe systems									
0120	Pig method, lengths 1000' to 10,000'									
0140	4" diameter thru 24" diameter, minimum				L.F.				3.60	4.14
0160	Maximum				"				18	21
6000	Sewage/sanitary systems									
6100	Power rodder with header & cutters									
6110	Mobilization charge, minimum				Total				695	800
6120	Mobilization charge, maximum				"				9,125	10,600
6140	Cleaning 4"-12" diameter				L.F.				6	6.60
6190	14"-24" diameter								8	8.80
6240	30" diameter								9.60	10.50
6250	36" diameter								12	13.20
6260	48" diameter								16	17.60
6270	60" diameter								24	26.50
6280	72" diameter								48	53
9000	Inspection, television camera with video									
9060	up to 500 linear feet				Total				715	820

33 01 30.23 Pipe Bursting

		Crew	Daily Output	Labor-Hours	Unit	Material	2021 Bare Costs Labor	Equipment	Total	Total Incl O&P
0010	**PIPE BURSTING**									
0011	300' runs, replace with HDPE pipe									
0020	Not including excavation, backfill, shoring, or dewatering									
0100	6" to 15" diameter, minimum				L.F.				200	220
0200	Maximum								550	605
0300	18" to 36" diameter, minimum								650	715
0400	Maximum				↓				1,050	1,150
0500	Mobilize and demobilize, minimum				Job				3,000	3,300
0600	Maximum				"				32,100	35,400

33 01 30.72 Cured-In-Place Pipe Lining

		Crew	Daily Output	Labor-Hours	Unit	Material	2021 Bare Costs Labor	Equipment	Total	Total Incl O&P
0010	**CURED-IN-PLACE PIPE LINING**									
0011	Not incl. bypass or cleaning									
0020	Less than 10,000 L.F., urban, 6" to 10"	C-17E	130	.615	L.F.	9.45	35.50	.28	45.23	63.50
0050	10" to 12"		125	.640		11.65	37	.29	48.94	68.50
0070	12" to 16"		115	.696		11.95	40	.32	52.27	73.50
0100	16" to 20"		95	.842		14	48.50	.39	62.89	89
0200	24" to 36"		90	.889		15.40	51	.41	66.81	94.50
0300	48" to 72"		80	1		24	57.50	.46	81.96	114
0500	Rural, 6" to 10"		180	.444		9.45	25.50	.20	35.15	49
0550	10" to 12"		175	.457		11.65	26.50	.21	38.36	52.50
0570	12" to 16"		160	.500		12.45	29	.23	41.68	57.50
0600	16" to 20"		135	.593		14.45	34	.27	48.72	67.50
0700	24" to 36"		125	.640		15.75	37	.29	53.04	73
0800	48" to 72"		100	.800		24	46	.37	70.37	96
1000	Greater than 10,000 L.F., urban, 6" to 10"	↓	160	.500	↓	9.45	29	.23	38.68	54
1050	10" to 12"	C-17E	155	.516	L.F.	11.65	29.50	.24	41.39	57.50
1070	12" to 16"		140	.571		11.95	33	.26	45.21	63
1100	16" to 20"		120	.667		14.45	38.50	.31	53.26	73.50
1200	24" to 36"		115	.696		15.75	40	.32	56.07	77.50
1300	48" to 72"		95	.842		24	48.50	.39	72.89	100
1500	Rural, 6" to 10"		215	.372		9.45	21.50	.17	31.12	42.50
1550	10" to 12"		210	.381		11.65	22	.18	33.83	46
1570	12" to 16"		185	.432		11.95	25	.20	37.15	51
1600	16" to 20"		150	.533		14.45	30.50	.25	45.20	62
1700	24" to 36"		140	.571		15.75	33	.26	49.01	67
1800	48" to 72"	↓	120	.667	↓	25	38.50	.31	63.81	85.50
2000	Cured in place pipe, non-pressure, flexible felt resin									
2100	6" diameter				L.F.				25.50	28
2200	8" diameter								26.50	29
2300	10" diameter								29	32
2400	12" diameter								34.50	38
2500	15" diameter								59	65
2600	18" diameter								82	90
2700	21" diameter								105	115
2800	24" diameter								177	195
2900	30" diameter								195	215
3000	36" diameter								205	225
3100	48" diameter				↓				218	240

33 01 Operation and Maintenance of Utilities

33 01 30 – Operation and Maintenance of Sewer Utilities

33 01 30.74 Sliplining, Excludes Cleaning

		Crew	Daily Output	Labor-Hours	Unit	Material	2021 Bare Costs Labor	2021 Bare Costs Equipment	Total	Total Incl O&P
0010	**SLIPLINING, excludes cleaning** and video inspection									
0020	Pipe relined with one pipe size smaller than original (4" for 6")									
0100	6" diameter, original size	B-6B	600	.080	L.F.	5.60	3.61	1.67	10.88	13.45
0150	8" diameter, original size		600	.080		10	3.61	1.67	15.28	18.25
0200	10" diameter, original size		600	.080		13.25	3.61	1.67	18.53	22
0250	12" diameter, original size		400	.120		17.15	5.40	2.50	25.05	29.50
0300	14" diameter, original size	↓	400	.120		18.05	5.40	2.50	25.95	30.50
0350	16" diameter, original size	B-6C	300	.160		21.50	7.20	7.90	36.60	43
0400	18" diameter, original size	"	300	.160	↓	27	7.20	7.90	42.10	49
1000	Pipe HDPE lining, make service line taps	B-6	4	6	Ea.	101	289	54	444	600

33 01 30.75 Cured-In-place Pipe Lining

		Crew	Daily Output	Labor-Hours	Unit	Material	2021 Bare Costs Labor	2021 Bare Costs Equipment	Total	Total Incl O&P
0010	**CURED-IN-PLACE PIPE LINING**									
0050	Bypass, Sewer	C-17E	500	.160	L.F.		9.20	.07	9.27	13.95
0090	Mobilize - (Any Size) 1 Mob per Job.		1	80	Job		4,600	37	4,637	6,975
0091	Relocate - (°12") (8hrs, 3 per day)		3	26.667	Ea.		1,525	12.25	1,537.25	2,325
0092	Relocate - (12" to 24") (8hrs, 2 per day)		2	40			2,300	18.40	2,318.40	3,475
0093	Relocate - (24" to 42") (8hrs, 1.5 per day)	↓	1.50	53.333	↓		3,075	24.50	3,099.50	4,650
0094	Relocate - (42") (8hrs, 1.5 per day)	C-17E	1	80	Ea.		4,600	37	4,637	6,975
0095	Additional Travel, $2.00 per Mile		3362.50	.024	Mile		1.37	.01	1.38	2.07
0096	Demobilize - (Any Size) 1 Demob per Job.		1	80	Job		4,600	37	4,637	6,975
0097	Refrigeration Truck (12", Cost per Each Location)		3	26.667	Ea.		1,525	12.25	1,537.25	2,325
0098	Refrigeration Truck (12"-24", Each Pipe / Location)		3	26.667			1,525	12.25	1,537.25	2,325
0099	Refrigeration Truck (24"-42", Each Pipe / Location)		1.50	53.333			3,075	24.50	3,099.50	4,650
0100	Refrigeration Truck (42", Each Pipe / Location)		1	80	↓		4,600	37	4,637	6,975
1210	12", 6mm, Epoxy, 250k, 2ft		6	13.333	L.F.	48	765	6.15	819.15	1,200
1211	12", 6mm, Epoxy, 250k, 4ft		12	6.667		48	385	3.07	436.07	630
1212	12", 6mm, Epoxy, 250k, 10ft		30	2.667		48	153	1.23	202.23	285
1213	12", 6mm, Epoxy, 250k, 20ft		60	1.333		48	76.50	.61	125.11	169
1214	12", 6mm, Epoxy, 250k, 50ft		150	.533		48	30.50	.25	78.75	99.50
1215	12", 6mm, Epoxy, 250k, 100ft		300	.267		48	15.35	.12	63.47	76
1216	12", 6mm, Epoxy, 250k, 300ft		900	.089		48	5.10	.04	53.14	60.50
1217	12", 6mm, Epoxy, 250k, 328ft	↓	1000	.080		48	4.60	.04	52.64	60
1218	12", 6mm, Epoxy, 250k, Transitions, add					20%				
1219	12", 6mm, Epoxy, 250k, 45 and 90 deg, add					50%				
1230	12", 5mm, Epoxy, 400k, 2ft	C-17E	6	13.333		40	765	6.15	811.15	1,200
1231	12", 5mm, Epoxy, 400k, 4ft		12	6.667		40	385	3.07	428.07	620
1232	12", 5mm, Epoxy, 400k, 10ft		30	2.667		40	153	1.23	194.23	276
1233	12", 5mm, Epoxy, 400k, 20ft		60	1.333		40	76.50	.61	117.11	160
1234	12", 5mm, Epoxy, 400k, 50ft		150	.533		40	30.50	.25	70.75	90.50
1235	12", 5mm, Epoxy, 400k, 100ft		300	.267		40	15.35	.12	55.47	67
1236	12", 5mm, Epoxy, 400k, 300ft		900	.089		40	5.10	.04	45.14	51.50
1237	12", 5mm, Epoxy, 400k, 328ft	↓	1000	.080		40	4.60	.04	44.64	51
1238	12", 5mm, Epoxy, 400k, Transitions, add					20%				
1239	12", 5mm, Epoxy, 400k, 45 and 90 deg, add					50%				
2410	24", 12mm, Epoxy, 250k, 2ft	C-17E	4	20		192	1,150	9.20	1,351.20	1,950
2411	24", 12mm, Epoxy, 250k, 4ft		8	10		192	575	4.60	771.60	1,075
2412	24", 12mm, Epoxy, 250k, 10ft		20	4		192	230	1.84	423.84	560
2413	24", 12mm, Epoxy, 250k, 20ft		40	2		192	115	.92	307.92	385
2414	24", 12mm, Epoxy, 250k, 50ft		100	.800		192	46	.37	238.37	281
2415	24", 12mm, Epoxy, 250k, 100ft		200	.400		192	23	.18	215.18	247
2416	24", 12mm, Epoxy, 250k, 300ft		600	.133		192	7.65	.06	199.71	224
2417	24", 12mm, Epoxy, 250k, 328ft	↓	656	.122		192	7	.06	199.06	223

For customer support on your Site Work & Landscape Costs with RSMeans Data, call 800.448.8182.

481

33 01 Operation and Maintenance of Utilities

33 01 30 – Operation and Maintenance of Sewer Utilities

33 01 30.75 Cured-In-place Pipe Lining	Crew	Daily Output	Labor-Hours	Unit	Material	2021 Bare Costs Labor	Equipment	Total	Total Incl O&P
2418	24", 12mm, Epoxy, 250k, Transitions, add				L.F.	20%				
2419	24", 12mm, Epoxy, 250k, 45 and 90 deg, add					50%				
2430	24", 10.5mm, Epoxy, 400k, 2ft	C-17E	4	20		168	1,150	9.20	1,327.20	1,925
2431	24", 10.5mm, Epoxy, 400k, 4ft		8	10		168	575	4.60	747.60	1,050
2432	24", 10.5mm, Epoxy, 400k, 10ft		20	4		168	230	1.84	399.84	530
2433	24", 10.5mm, Epoxy, 400k, 20ft		40	2		168	115	.92	283.92	360
2434	24", 10.5mm, Epoxy, 400k, 50ft		100	.800		168	46	.37	214.37	254
2435	24", 10.5mm, Epoxy, 400k, 100ft		200	.400		168	23	.18	191.18	220
2436	24", 10.5mm, Epoxy, 400k, 300ft		600	.133		168	7.65	.06	175.71	197
2437	24", 10.5mm, Epoxy, 400k, 328ft		656	.122		168	7	.06	175.06	196
2438	24", 10.5mm, Epoxy, 400k, Transitions, add					20%				
2439	24", 10.5mm, Epoxy, 400k, 45 and 90 deg, add					50%				

33 05 Common Work Results for Utilities

33 05 07 – Trenchless Installation of Utility Piping

33 05 07.13 Utility Directional Drilling

		Crew	Daily Output	Labor-Hours	Unit	Material	2021 Bare Costs Labor	Equipment	Total	Total Incl O&P
0010	**UTILITY DIRECTIONAL DRILLING**									
0011	Excluding access and splice pits (if required) and conduit (required)									
0012	Drilled hole diameters shown should be 50% larger than conduit									
0013	Assume access to H2O & removal of spoil/drilling mud as non-hazard material									
0014	Actual production rates adj. to account for risk factors for each soil type									
0100	Sand, silt, clay, common earth									
0110	Mobilization or demobilization	B-82A	1	32	Ea.		1,600	2,500	4,100	5,125
0120	6" diameter		180	.178	L.F.		8.90	13.90	22.80	28.50
0130	12" diameter		70	.457			23	36	59	73.50
0140	18" diameter		50	.640			32	50	82	103
0150	24" diameter		35	.914			45.50	71.50	117	147
0210	Mobilization or demobilization	B-82B	1	32	Ea.		1,600	2,600	4,200	5,250
0220	6" diameter		125	.256	L.F.		12.80	21	33.80	42
0230	12" diameter		50	.640			32	52	84	105
0240	18" diameter		30	1.067			53.50	87	140.50	175
0250	24" diameter		25	1.280			64	104	168	211
0260	30" diameter		20	1.600			80	130	210	262
0270	36" diameter		15	2.133			107	174	281	350
0300	Hard rock (solid bed, 24,000+ psi)									
0310	Mobilization or demobilization	B-82C	1	32	Ea.		1,600	2,775	4,375	5,425
0320	6" diameter		30	1.067	L.F.		53.50	93	146.50	182
0330	12" diameter		12	2.667			133	232	365	455
0340	18" diameter		8	4			200	350	550	685
0344	20" diameter		7.50	4.267			213	370	583	730
0350	24" diameter		7	4.571			228	400	628	775
0360	30" diameter	B-82C	6	5.333	L.F.		266	465	731	905
0370	36" diameter	"	4.50	7.111	"		355	620	975	1,200

33 05 07.23 Utility Boring and Jacking

		Crew	Daily Output	Labor-Hours	Unit	Material	2021 Bare Costs Labor	Equipment	Total	Total Incl O&P
0010	**UTILITY BORING AND JACKING**									
0011	Casing only, 100' minimum,									
0020	not incl. jacking pits or dewatering									
0100	Roadwork, 1/2" thick wall, 24" diameter casing	B-42	20	3.200	L.F.	142	159	53.50	354.50	455
0200	36" diameter		16	4		219	199	66.50	484.50	615
0300	48" diameter		15	4.267		305	212	71	588	735
0500	Railroad work, 24" diameter		15	4.267		142	212	71	425	555

33 05 Common Work Results for Utilities

33 05 07 – Trenchless Installation of Utility Piping

33 05 07.23 Utility Boring and Jacking

		Crew	Daily Output	Labor-Hours	Unit	Material	2021 Bare Costs Labor	Equipment	Total	Total Incl O&P
0600	36" diameter	B-42	14	4.571	L.F.	219	228	76	523	665
0700	48" diameter	↓	12	5.333		305	265	89	659	835
0900	For ledge, add								20%	20%
1000	Small diameter boring, 3", sandy soil	B-82	900	.018		23	.89	.20	24.09	27
1040	Rocky soil	"	500	.032	↓	23	1.60	.37	24.97	28.50
1100	Prepare jacking pits, incl. mobilization & demobilization, minimum				Ea.				3,225	3,700
1101	Maximum				"				22,000	25,500

33 05 07.36 Microtunneling

		Crew	Daily Output	Labor-Hours	Unit	Material	2021 Bare Costs Labor	Equipment	Total	Total Incl O&P
0010	**MICROTUNNELING**									
0011	Not including excavation, backfill, shoring,									
0020	or dewatering, average 50'/day, slurry method									
0100	24" to 48" outside diameter, minimum				L.F.				965	965
0110	Adverse conditions, add				"				500	500
1000	Rent microtunneling machine, average monthly lease				Month				97,500	107,000
1010	Operating technician				Day				630	705
1100	Mobilization and demobilization, minimum				Job				41,200	45,900
1110	Maximum				"				445,500	490,500

33 05 61 – Concrete Manholes

33 05 61.10 Storm Drainage Manholes, Frames and Covers

		Crew	Daily Output	Labor-Hours	Unit	Material	2021 Bare Costs Labor	Equipment	Total	Total Incl O&P
0010	**STORM DRAINAGE MANHOLES, FRAMES & COVERS**									
0020	Excludes footing, excavation, backfill (See line items for frame & cover)									
0050	Brick, 4' inside diameter, 4' deep	D-1	1	16	Ea.	700	780		1,480	1,950
0100	6' deep		.70	22.857		1,000	1,125		2,125	2,775
0150	8' deep		.50	32		1,300	1,550		2,850	3,775
0175	12' deep		.33	47.904	↓	1,900	2,325		4,225	5,600
0200	Add for 1' depth increase		4	4	V.L.F.	176	195		371	485
0400	Concrete blocks (radial), 4' ID, 4' deep		1.50	10.667	Ea.	430	520		950	1,250
0500	6' deep		1	16		570	780		1,350	1,800
0600	8' deep		.70	22.857	↓	710	1,125		1,835	2,450
0700	For depths over 8', add	↓	5.50	2.909	V.L.F.	73	142		215	293
0800	Concrete, cast in place, 4' x 4', 8" thick, 4' deep	C-14H	2	24	Ea.	680	1,275	13.70	1,968.70	2,650
0900	6' deep	C-14H	1.50	32	Ea.	985	1,725	18.25	2,728.25	3,650
1000	8' deep		1	48	"	1,400	2,575	27.50	4,002.50	5,400
1100	For depths over 8', add	↓	8	6	V.L.F.	166	320	3.42	489.42	665
1110	Precast, 4' ID, 4' deep	B-22	4.10	7.317	Ea.	905	380	70	1,355	1,625
1120	6' deep		3	10		1,075	515	95.50	1,685.50	2,050
1130	8' deep		2	15	↓	1,900	775	143	2,818	3,400
1140	For depths over 8', add	↓	16	1.875	V.L.F.	107	97	17.90	221.90	283
1150	5' ID, 4' deep	B-6	3	8	Ea.	1,875	385	72	2,332	2,725
1160	6' deep		2	12		3,250	575	108	3,933	4,550
1170	8' deep		1.50	16	↓	2,750	770	144	3,664	4,325
1180	For depths over 8', add		12	2	V.L.F.	470	96	18	584	680
1190	6' ID, 4' deep		2	12	Ea.	2,100	575	108	2,783	3,275
1200	6' deep		1.50	16		4,600	770	144	5,514	6,350
1210	8' deep		1	24	↓	5,375	1,150	216	6,741	7,900
1220	For depths over 8', add	↓	8	3	V.L.F.	425	144	27	596	715
1250	Slab tops, precast, 8" thick									
1300	4' diameter manhole	B-6	8	3	Ea.	288	144	27	459	560
1400	5' diameter manhole		7.50	3.200		685	154	29	868	1,000
1500	6' diameter manhole	↓	7	3.429		715	165	31	911	1,075
3800	Steps, heavyweight cast iron, 7" x 9"	1 Bric	40	.200		17.25	10.75		28	35
3900	8" x 9"	↓	40	.200		20.50	10.75		31.25	39

33 05 Common Work Results for Utilities

33 05 61 – Concrete Manholes

33 05 61.10 Storm Drainage Manholes, Frames and Covers	Crew	Daily Output	Labor-Hours	Unit	Material	2021 Bare Costs Labor	Equipment	Total	Total Incl O&P
3928	12" x 10-1/2"	1 Bric	40	.200	Ea.	31	10.75		41.75	50
4000	Standard sizes, galvanized steel		40	.200		24	10.75		34.75	42.50
4100	Aluminum		40	.200		31	10.75		41.75	50
4150	Polyethylene	↓	40	.200		32.50	10.75		43.25	52
4210	Rubber boot 6" diam. or smaller	1 Clab	32	.250		105	11.10		116.10	133
4215	8" diam.		24	.333		118	14.80		132.80	152
4220	10" diam.		19	.421		148	18.70		166.70	191
4225	12" diam.		16	.500		176	22		198	227
4230	16" diam.		15	.533		236	23.50		259.50	296
4235	18" diam.		15	.533		261	23.50		284.50	325
4240	24" diam.		14	.571		290	25.50		315.50	360
4245	30" diam.	↓	12	.667	↓	380	29.50		409.50	465

33 05 63 – Concrete Vaults and Chambers

33 05 63.13 Precast Concrete Utility Structures

		Crew	Daily Output	Labor-Hours	Unit	Material	2021 Bare Costs Labor	Equipment	Total	Total Incl O&P
0010	**PRECAST CONCRETE UTILITY STRUCTURES**, 6" thick									
0040	4' x 6' x 6' high, ID	B-13	2	28	Ea.	1,475	1,350	293	3,118	3,975
0050	5' x 10' x 6' high, ID		2	28		1,825	1,350	293	3,468	4,350
0100	6' x 10' x 6' high, ID		2	28		1,900	1,350	293	3,543	4,450
0150	5' x 12' x 6' high, ID		2	28		2,000	1,350	293	3,643	4,550
0200	6' x 12' x 6' high, ID		1.80	31.111		2,250	1,500	325	4,075	5,075
0250	6' x 13' x 6' high, ID	↓	1.50	37.333	↓	2,950	1,800	390	5,140	6,375
0300	8' x 14' x 7' high, ID	B-13	1	56	Ea.	3,200	2,700	585	6,485	8,175
0350	Hand hole, precast concrete, 1-1/2" thick									
0400	1'-0" x 2'-0" x 1'-9", ID, light duty	B-1	4	6	Ea.	525	270		795	985
0450	4'-6" x 3'-2" x 2'-0", OD, heavy duty	B-6	3	8		1,600	385	72	2,057	2,400
0460	Meter pit, 4' x 4', 4' deep		2	12		1,550	575	108	2,233	2,675
0470	6' deep		1.60	15		2,200	720	135	3,055	3,650
0480	8' deep		1.40	17.143		2,900	825	154	3,879	4,600
0490	10' deep		1.20	20		3,675	960	180	4,815	5,675
0500	15' deep		1	24		5,375	1,150	216	6,741	7,900
0510	6' x 6', 4' deep		1.40	17.143		2,625	825	154	3,604	4,275
0520	6' deep		1.20	20		3,925	960	180	5,065	5,950
0530	8' deep		1	24		5,225	1,150	216	6,591	7,725
0540	10' deep		.80	30		6,550	1,450	270	8,270	9,650
0550	15' deep	↓	.60	40	↓	9,950	1,925	360	12,235	14,200

33 05 71 – Cleanouts

		Crew	Daily Output	Labor-Hours	Unit	Material	2021 Bare Costs Labor	Equipment	Total	Total Incl O&P
0300	Cleanout Access Housing, X-Hvy Dty, Ductile Iron	B-20	10	2.400	Ea.	1,400	118		1,518	1,700
0306	Vandal Proof, Add					96.50			96.50	106
0313	Galvanized, Add	↓			↓	775			775	850

33 05 97 – Identification and Signage for Utilities

33 05 97.05 Utility Connection

		Crew	Daily Output	Labor-Hours	Unit	Material	2021 Bare Costs Labor	Equipment	Total	Total Incl O&P
0010	**UTILITY CONNECTION**									
0020	Water, sanitary, stormwater, gas, single connection	B-14	1	48	Ea.	4,925	2,225	216	7,366	9,000
0030	Telecommunication	"	3	16	"	420	745	72	1,237	1,650

33 05 97.10 Utility Accessories

		Crew	Daily Output	Labor-Hours	Unit	Material	2021 Bare Costs Labor	Equipment	Total	Total Incl O&P
0010	**UTILITY ACCESSORIES**									
0400	Underground tape, detectable, reinforced, alum. foil core, 2"	1 Clab	150	.053	C.L.F.	3	2.37		5.37	6.85

33 05 Common Work Results for Utilities

33 05 97 – Identification and Signage for Utilities

33 05 97.15 Pipe Casing	Crew	Daily Output	Labor-Hours	Unit	Material	2021 Bare Costs Labor	Equipment	Total	Total Incl O&P
0010 **PIPE CASING**									
1000 Spacers									
1010 Poly, std config, std runners, for hot/cold carrier pipe, 8" diameter band	2 Clab	60	.267	Ea.	37.50	11.85		49.35	59
1020 12" diameter band		60	.267		53.50	11.85		65.35	76.50
1030 16" diameter band		60	.267		69.50	11.85		81.35	94
1040 20" diameter band		40	.400		94	17.75		111.75	130
1050 24" diameter band		40	.400		118	17.75		135.75	157
1060 30" diameter band		30	.533		143	23.50		166.50	193
1070 36" diameter band		30	.533		168	23.50		191.50	220
1110 Carbon steel, 14 ga., poly coat, std config, 4 runners, 8" diameter band		60	.267		79	11.85		90.85	105
1120 12" diameter band		60	.267		89	11.85		100.85	116
1130 6 runners, 16" diameter band		60	.267		105	11.85		116.85	134
1140 20" diameter band		40	.400		144	17.75		161.75	186
1150 24" diameter band		40	.400		191	17.75		208.75	237
1160 30" diameter band		30	.533		270	23.50		293.50	335
1170 36" diameter band	2 Clab	30	.533	Ea.	375	23.50		398.50	450
1210 Stainless steel, 14 ga., grade 304, std config, 4 runners, 8" diameter band		60	.267		99.50	11.85		111.35	127
1220 12" diameter band		60	.267		117	11.85		128.85	147
1230 6 runners, 16" diameter band		60	.267		141	11.85		152.85	173
1240 20" diameter band		40	.400		191	17.75		208.75	237
1250 24" diameter band		40	.400		252	17.75		269.75	305
1260 30" diameter band		30	.533		315	23.50		338.50	380
1270 36" diameter band		30	.533		475	23.50		498.50	555
2000 End seals									
2010 Pull-on, synthetic rubber, stainless steel bands, 8" carrier pipe	2 Clab	60	.267	Ea.	58.50	11.85		70.35	81.50
2020 12" carrier pipe		60	.267		106	11.85		117.85	135
2030 16" carrier pipe		60	.267		147	11.85		158.85	180
2040 20" carrier pipe		40	.400		198	17.75		215.75	245
2050 24" carrier pipe		40	.400		242	17.75		259.75	293
2060 30" carrier pipe		30	.533		325	23.50		348.50	390
2070 36" carrier pipe		30	.533		370	23.50		393.50	440

33 11 Groundwater Sources

33 11 13 – Potable Water Supply Wells

33 11 13.10 Wells and Accessories

	Crew	Daily Output	Labor-Hours	Unit	Material	2021 Bare Costs Labor	Equipment	Total	Total Incl O&P
0010 **WELLS & ACCESSORIES**									
0011 Domestic									
0100 Drilled, 4" to 6" diameter	B-23	120	.333	L.F.		14.95	13.50	28.45	37.50
0200 8" diameter	"	95.20	.420	"		18.80	17	35.80	46.50
0400 Gravel pack well, 40' deep, incl. gravel & casing, complete									
0500 24" diameter casing x 18" diameter screen	B-23	.13	308	Total	38,700	13,800	12,400	64,900	77,000
0600 36" diameter casing x 18" diameter screen		.12	333	"	39,800	14,900	13,500	68,200	81,000
0800 Observation wells, 1-1/4" riser pipe		163	.245	V.L.F.	18.15	11	9.95	39.10	47.50
0900 For flush Buffalo roadway box, add	1 Skwk	16.60	.482	Ea.	52	27.50		79.50	99
1200 Test well, 2-1/2" diameter, up to 50' deep (15 to 50 GPM)	B-23	1.51	26.490	"	895	1,175	1,075	3,145	3,925
1300 Over 50' deep, add	"	121.80	.328	L.F.	24	14.70	13.30	52	63
1400 Remove & reset pump, minimum	B-21	4	7	Ea.		355	47.50	402.50	590
1420 Maximum	"	2	14	"		715	95.50	810.50	1,175
1500 Pumps, installed in wells to 100' deep, 4" submersible									
1510 1/2 HP	Q-1	3.22	4.969	Ea.	580	305		885	1,100

33 11 13.10 Wells and Accessories	Crew	Daily Output	Labor-Hours	Unit	Material	2021 Bare Costs Labor	2021 Bare Costs Equipment	Total	Total Incl O&P
1520	3/4 HP	Q-1	2.66	6.015	Ea.	980	365		1,345	1,625
1600	1 HP	▼	2.29	6.987		805	425		1,230	1,525
1700	1-1/2 HP	Q-22	1.60	10		2,250	610	297	3,157	3,700
1800	2 HP		1.33	12.030		1,800	735	360	2,895	3,500
1900	3 HP		1.14	14.035		2,050	855	415	3,320	4,000
2000	5 HP	▼	1.14	14.035	▼	2,575	855	415	3,845	4,550
2050	Remove and install motor only, 4 HP	Q-22	1.14	14.035	Ea.	1,525	855	415	2,795	3,400
3000	Pump, 6" submersible, 25' to 150' deep, 25 HP, 103 to 400 GPM		.89	17.978		9,325	1,100	535	10,960	12,500
3100	25' to 500' deep, 30 HP, 104 to 400 GPM	▼	.73	21.918		11,000	1,325	650	12,975	14,800
5000	Wells to 180' deep, 4" submersible, 1 HP	B-21	1.10	25.455		2,425	1,300	174	3,899	4,800
5500	2 HP		1.10	25.455		3,825	1,300	174	5,299	6,350
6000	3 HP		1	28		4,450	1,425	191	6,066	7,225
7000	5 HP	▼	.90	31.111	▼	6,425	1,600	212	8,237	9,650
8000	Steel well casing	B-23A	3020	.008	Lb.	1.29	.40	.29	1.98	2.33
8110	Well screen assembly, stainless steel, 2" diameter		273	.088	L.F.	95.50	4.39	3.23	103.12	115
8120	3" diameter		253	.095		141	4.74	3.48	149.22	166
8130	4" diameter		200	.120		181	6	4.40	191.40	213
8140	5" diameter		168	.143		170	7.15	5.25	182.40	203
8150	6" diameter		126	.190		190	9.50	7	206.50	231
8160	8" diameter		98.50	.244		253	12.15	8.95	274.10	305
8170	10" diameter		73	.329		315	16.40	12.05	343.45	390
8180	12" diameter		62.50	.384		370	19.15	14.10	403.25	455
8190	14" diameter		54.30	.442		415	22	16.20	453.20	510
8200	16" diameter		48.30	.497		460	25	18.25	503.25	560
8210	18" diameter		39.20	.612		580	30.50	22.50	633	710
8220	20" diameter		31.20	.769		655	38.50	28	721.50	815
8230	24" diameter		23.80	1.008		810	50.50	37	897.50	1,000
8240	26" diameter		21	1.143		1,025	57	42	1,124	1,250
8244	Well casing or drop pipe, PVC, 1/2" diameter		550	.044		1.30	2.18	1.60	5.08	6.45
8245	3/4" diameter		550	.044		1.31	2.18	1.60	5.09	6.45
8246	1" diameter		550	.044		1.35	2.18	1.60	5.13	6.50
8247	1-1/4" diameter		520	.046		1.70	2.30	1.69	5.69	7.15
8248	1-1/2" diameter		490	.049		1.80	2.45	1.80	6.05	7.60
8249	1-3/4" diameter		380	.063		1.84	3.15	2.32	7.31	9.25
8250	2" diameter		280	.086		2.11	4.28	3.14	9.53	12.20
8252	3" diameter		260	.092		4.18	4.61	3.39	12.18	15.20
8254	4" diameter		205	.117		4.43	5.85	4.30	14.58	18.30
8255	5" diameter		170	.141		4.45	7.05	5.20	16.70	21
8256	6" diameter		130	.185		8.50	9.20	6.75	24.45	30.50
8258	8" diameter		100	.240		13	12	8.80	33.80	42
8260	10" diameter		73	.329		21	16.40	12.05	49.45	61
8262	12" diameter		62	.387		23.50	19.35	14.20	57.05	70.50
8300	Slotted PVC, 1-1/4" diameter		521	.046		2.88	2.30	1.69	6.87	8.45
8310	1-1/2" diameter		488	.049		2.98	2.46	1.80	7.24	8.90
8320	2" diameter		273	.088		3.86	4.39	3.23	11.48	14.35
8330	3" diameter		253	.095		6.80	4.74	3.48	15.02	18.35
8340	4" diameter	▼	200	.120	▼	7.25	6	4.40	17.65	21.50
8350	5" diameter	B-23A	168	.143	L.F.	15.65	7.15	5.25	28.05	33.50
8360	6" diameter		126	.190		16.80	9.50	7	33.30	40.50
8370	8" diameter	▼	98.50	.244		24.50	12.15	8.95	45.60	55
8400	Artificial gravel pack, 2" screen, 6" casing	B-23B	174	.138		4.52	6.90	6.40	17.82	22.50
8405	8" casing		111	.216		6.20	10.80	10.05	27.05	34
8410	10" casing	▼	74.50	.322	▼	8.50	16.10	15	39.60	50

33 11 Groundwater Sources

33 11 13 – Potable Water Supply Wells

33 11 13.10 Wells and Accessories

		Crew	Daily Output	Labor-Hours	Unit	Material	2021 Bare Costs Labor	Equipment	Total	Total Incl O&P
8415	12" casing	B-23B	60	.400	L.F.	13.60	19.95	18.60	52.15	65.50
8420	14" casing		50.20	.478		13.90	24	22	59.90	75.50
8425	16" casing		40.70	.590		18.20	29.50	27.50	75.20	94
8430	18" casing		36	.667		22.50	33.50	31	87	108
8435	20" casing		29.50	.814		24.50	40.50	38	103	129
8440	24" casing		25.70	.934		28.50	46.50	43.50	118.50	149
8445	26" casing		24.60	.976		31.50	48.50	45.50	125.50	157
8450	30" casing		20	1.200		36	60	56	152	191
8455	36" casing		16.40	1.463		39.50	73	68	180.50	228
8500	Develop well		8	3	Hr.	305	150	139	594	710
8550	Pump test well		8	3		96.50	150	139	385.50	480
8560	Standby well	B-23A	8	3		99	150	110	359	455
8570	Standby, drill rig		8	3			150	110	260	345
8580	Surface seal well, concrete filled		1	24	Ea.	1,075	1,200	880	3,155	3,925
8590	Well test pump, install & remove	B-23	1	40			1,800	1,625	3,425	4,450
8600	Well sterilization, chlorine	2 Clab	1	16		125	710		835	1,200
8610	Well water pressure switch	1 Clab	12	.667		97.50	29.50		127	151
8630	Well water pressure switch with manual reset	"	12	.667		130	29.50		159.50	186
9950	See Section 31 23 19.40 for wellpoints									
9960	See Section 31 23 19.30 for drainage wells									

33 11 13.20 Water Supply Wells, Pumps

		Crew	Daily Output	Labor-Hours	Unit	Material	2021 Bare Costs Labor	Equipment	Total	Total Incl O&P
0010	**WATER SUPPLY WELLS, PUMPS**									
0011	With pressure control									
1000	Deep well, jet, 42 gal. galvanized tank									
1040	3/4 HP	1 Plum	.80	10	Ea.	1,125	675		1,800	2,250
3000	Shallow well, jet, 30 gal. galvanized tank									
3040	1/2 HP	1 Plum	2	4	Ea.	905	271		1,176	1,400

33 14 Water Utility Transmission and Distribution

33 14 13 – Public Water Utility Distribution Piping

33 14 13.10 Water Supply, Concrete Pipe

		Crew	Daily Output	Labor-Hours	Unit	Material	2021 Bare Costs Labor	Equipment	Total	Total Incl O&P
0010	**WATER SUPPLY, CONCRETE PIPE**									
0020	Not including excavation or backfill									
3000	Prestressed Concrete Pipe (PCCP), 150 psi, 12" diameter	B-13	192	.292	L.F.	94	14.10	3.06	111.16	128
3010	24" diameter	"	128	.438		94	21	4.59	119.59	141
3040	36" diameter	B-13B	96	.583		146	28	10.30	184.30	214
3050	48" diameter	B-13B	64	.875	L.F.	209	42.50	15.45	266.95	310
3070	72" diameter		60	.933		410	45	16.50	471.50	535
3080	84" diameter		40	1.400		530	67.50	25	622.50	710
3090	96" diameter	B-13C	40	1.400		775	67.50	58	900.50	1,025
3100	108" diameter		32	1.750		1,100	84.50	72	1,256.50	1,400
3102	120" diameter		16	3.500		1,625	169	144	1,938	2,175
3104	144" diameter		16	3.500		1,925	169	144	2,238	2,525
3110	Prestressed Concrete Pipe (PCCP), 150 psi, elbow, 90°, 12" diameter	B-13	24	2.333	Ea.	835	113	24.50	972.50	1,125
3140	24" diameter	"	6	9.333		1,700	450	98	2,248	2,650
3150	36" diameter	B-13B	4	14		3,175	675	248	4,098	4,775
3160	48" diameter		3	18.667		6,300	900	330	7,530	8,650
3180	72" diameter		1.60	35		17,900	1,700	620	20,220	22,900
3190	84" diameter		1.30	43.077		34,400	2,075	760	37,235	41,800
3200	96" diameter		1	56		41,400	2,700	990	45,090	50,500
3210	108" diameter	B-13C	.66	84.848		43,700	4,100	3,500	51,300	58,000

33 14 Water Utility Transmission and Distribution

33 14 13 – Public Water Utility Distribution Piping

33 14 13.10 Water Supply, Concrete Pipe		Crew	Daily Output	Labor-Hours	Unit	Material	2021 Bare Costs Labor	Equipment	Total	Total Incl O&P
3220	120" diameter	B-13C	.40	140	Ea.	56,500	6,750	5,775	69,025	78,500
3225	144" diameter	↓	.30	184		69,500	8,900	7,600	86,000	98,000
3230	Prestressed Concrete Pipe (PCCP), 150 psi, elbow, 45°, 12" diameter	B-13	24	2.333		520	113	24.50	657.50	765
3250	24" diameter	"	6	9.333		1,250	450	98	1,798	2,150
3260	36" diameter	B-13B	4	14		2,350	675	248	3,273	3,875
3270	48" diameter		3	18.667		4,175	900	330	5,405	6,325
3290	72" diameter		1.60	35		11,100	1,700	620	13,420	15,400
3300	84" diameter		1.30	42.945		17,000	2,075	760	19,835	22,600
3310	96" diameter	↓	1	56		21,100	2,700	990	24,790	28,300
3320	108" diameter	B-13C	.66	84.337		26,500	4,075	3,475	34,050	39,100
3330	120" diameter	↓	.40	140		34,800	6,750	5,775	47,325	54,500
3340	144" diameter	↓	.30	184	↓	43,000	8,900	7,600	59,500	69,000

33 14 13.15 Water Supply, Ductile Iron Pipe

		Crew	Daily Output	Labor-Hours	Unit	Material	2021 Bare Costs Labor	Equipment	Total	Total Incl O&P
0010	**WATER SUPPLY, DUCTILE IRON PIPE**									
0011	Cement lined									
0020	Not including excavation or backfill									
2000	Pipe, class 50 water piping, 18' lengths									
2020	Mechanical joint, 4" diameter	B-21	200	.140	L.F.	47	7.15	.95	55.10	63.50
2040	6" diameter		160	.175		55	8.95	1.19	65.14	75
2060	8" diameter		133.33	.210		50	10.70	1.43	62.13	72.50
2080	10" diameter		114.29	.245		65.50	12.50	1.67	79.67	92.50
2100	12" diameter		105.26	.266		88.50	13.60	1.81	103.91	120
2120	14" diameter		100	.280		104	14.30	1.91	120.21	138
2140	16" diameter		72.73	.385		106	19.65	2.63	128.28	148
2160	18" diameter		68.97	.406		141	20.50	2.77	164.27	189
2170	20" diameter		57.14	.490		142	25	3.34	170.34	198
2180	24" diameter	↓	47.06	.595	↓	157	30.50	4.06	191.56	223
2260	30" diameter	B-21	66.67	.420	L.F.	116	21.50	2.86	140.36	163
2270	36" diameter		58.82	.476		148	24.50	3.25	175.75	202
2300	Mechanical restrained joint, no fittings, 4" diameter		173.91	.161		46.50	8.20	1.10	55.80	64.50
3000	Push-on joint, 4" diameter		400	.070		24	3.57	.48	28.05	32.50
3020	6" diameter		333.33	.084		23.50	4.29	.57	28.36	33
3040	8" diameter		200	.140		33	7.15	.95	41.10	48.50
3060	10" diameter		181.82	.154		51	7.85	1.05	59.90	69.50
3080	12" diameter		160	.175		54	8.95	1.19	64.14	74
3100	14" diameter		133.33	.210		54	10.70	1.43	66.13	77
3120	16" diameter		114.29	.245		57.50	12.50	1.67	71.67	83.50
3140	18" diameter		100	.280		64	14.30	1.91	80.21	93.50
3160	20" diameter		88.89	.315		66.50	16.10	2.15	84.75	99.50
3180	24" diameter	↓	76.92	.364	↓	92.50	18.60	2.48	113.58	133
6170	Cap, 4" diameter	B-20	32	.750	Ea.	70	37		107	133
6180	6" diameter		25.60	.938		108	46		154	187
6190	8" diameter	↓	21.33	1.125		149	55.50		204.50	247
6198	10" diameter	B-21	16.84	1.663		264	85	11.35	360.35	430
6200	12" diameter		16.84	1.663		264	85	11.35	360.35	430
6205	15" diameter		11	2.545		830	130	17.35	977.35	1,125
6206	16" diameter		11	2.545		830	130	17.35	977.35	1,125
6210	18" diameter		11	2.545		835	130	17.35	982.35	1,125
6220	24" diameter		9.41	2.976		1,675	152	20.50	1,847.50	2,075
6230	30" diameter		8	3.500		3,300	179	24	3,503	3,925
6240	36" diameter	↓	6.90	4.058	↓	3,525	207	27.50	3,759.50	4,250
8000	Piping, fittings, mechanical joint, AWWA C110									

33 14 Water Utility Transmission and Distribution

33 14 13 – Public Water Utility Distribution Piping

33 14 13.15 Water Supply, Ductile Iron Pipe

		Crew	Daily Output	Labor-Hours	Unit	Material	Labor	Equipment	Total	Total Incl O&P
8006	90° bend, 4" diameter	B-20	16	1.500	Ea.	173	74		247	300
8020	6" diameter	"	12.80	1.875		256	92.50		348.50	420
8040	8" diameter	B-21	10.67	2.624		500	134	17.90	651.90	770
8060	10" diameter		11.43	2.450		695	125	16.70	836.70	965
8080	12" diameter		10.53	2.659		985	136	18.15	1,139.15	1,300
8100	14" diameter		10	2.800		1,325	143	19.10	1,487.10	1,700
8120	16" diameter		7.27	3.851		1,700	197	26.50	1,923.50	2,175
8140	18" diameter		6.90	4.058		2,375	207	27.50	2,609.50	2,975
8160	20" diameter		5.71	4.904		2,975	250	33.50	3,258.50	3,675
8180	24" diameter		4.70	5.957		4,675	305	40.50	5,020.50	5,650
8200	Wye or tee, 4" diameter	B-20	10.67	2.249		475	111		586	685
8220	6" diameter	"	8.53	2.814		645	139		784	920
8240	8" diameter	B-21	7.11	3.938		1,000	201	27	1,228	1,425
8260	10" diameter		7.62	3.675		975	188	25	1,188	1,375
8280	12" diameter		7.02	3.989		1,900	204	27	2,131	2,425
8300	14" diameter		6.67	4.198		2,150	214	28.50	2,392.50	2,725
8320	16" diameter		4.85	5.773		2,475	295	39.50	2,809.50	3,200
8340	18" diameter		4.60	6.087		4,600	310	41.50	4,951.50	5,550
8360	20" diameter		3.81	7.349		7,425	375	50	7,850	8,800
8380	24" diameter		3.14	8.917		10,100	455	61	10,616	11,800
8398	45° bend, 4" diameter	B-20	16	1.500		201	74		275	330
8400	6" diameter	"	12.80	1.875		279	92.50		371.50	445
8405	8" diameter	B-21	10.67	2.624		410	134	17.90	561.90	670
8410	12" diameter		10.53	2.659		870	136	18.15	1,024.15	1,175
8420	16" diameter		7.27	3.851		1,675	197	26.50	1,898.50	2,150
8430	20" diameter		5.71	4.904		2,400	250	33.50	2,683.50	3,050
8440	24" diameter		4.70	5.957		3,375	305	40.50	3,720.50	4,200
8450	Decreaser, 6" x 4" diameter	B-20	14.22	1.688		231	83		314	380
8460	8" x 6" diameter	B-21	11.64	2.406		345	123	16.40	484.40	580
8470	10" x 6" diameter		13.33	2.101		435	107	14.35	556.35	655
8480	12" x 6" diameter		12.70	2.205		615	113	15.05	743.05	860
8490	16" x 6" diameter		10	2.800		995	143	19.10	1,157.10	1,325
8500	20" x 6" diameter		8.42	3.325		1,800	170	22.50	1,992.50	2,250
8552	For water utility valves see Section 33 14 19									
8700	Joint restraint, ductile iron mechanical joints									
8710	4" diameter	B-20	32	.750	Ea.	26.50	37		63.50	85
8720	6" diameter	"	25.60	.938		34.50	46		80.50	107
8730	8" diameter	B-21	21.33	1.313		52	67	8.95	127.95	167
8740	10" diameter		18.28	1.532		89.50	78	10.45	177.95	227
8750	12" diameter		16.84	1.663		99	85	11.35	195.35	248
8760	14" diameter		16	1.750		141	89.50	11.95	242.45	300
8770	16" diameter		11.64	2.406		145	123	16.40	284.40	360
8780	18" diameter		11.03	2.539		202	130	17.30	349.30	435
8785	20" diameter		9.14	3.063		248	156	21	425	530
8790	24" diameter		7.53	3.718		340	190	25.50	555.50	685
9600	Steel sleeve with tap, 4" diameter	B-20	3	8		550	395		945	1,200
9620	6" diameter		2	12		590	590		1,180	1,525
9630	8" diameter		2	12		750	590		1,340	1,700

For customer support on your Site Work & Landscape Costs with RSMeans Data, call 800.448.8182.

489

33 14 13.20 Water Supply, Polyethylene Pipe, C901

		Crew	Daily Output	Labor-Hours	Unit	Material	2021 Bare Costs Labor	Equipment	Total	Total Incl O&P
0010	**WATER SUPPLY, POLYETHYLENE PIPE, C901**									
0020	Not including excavation or backfill									
1000	Piping, 160 psi, 3/4" diameter	Q-1A	525	.019	L.F.	.96	1.30		2.26	3
1120	1" diameter		485	.021		.66	1.40		2.06	2.83
1140	1-1/2" diameter		450	.022		2.06	1.51		3.57	4.53
1160	2" diameter		365	.027		2.72	1.87		4.59	5.75
2000	Fittings, insert type, nylon, 160 & 250 psi, cold water									
2220	Clamp ring, stainless steel, 3/4" diameter	Q-1A	345	.029	Ea.	5.55	1.97		7.52	9.05
2240	1" diameter		321	.031		5.55	2.12		7.67	9.25
2260	1-1/2" diameter		285	.035		5.55	2.39		7.94	9.65
2280	2" diameter		255	.039		5.55	2.67		8.22	10.10
2300	Coupling, 3/4" diameter		66	.152		1.40	10.30		11.70	16.95
2320	1" diameter		57	.175		1.88	11.95		13.83	19.90
2340	1-1/2" diameter		51	.196		4.56	13.35		17.91	25
2360	2" diameter		48	.208		5.80	14.20		20	27.50
2400	Elbow, 90°, 3/4" diameter		66	.152		2.29	10.30		12.59	17.90
2420	1" diameter		57	.175		2.78	11.95		14.73	21
2440	1-1/2" diameter		51	.196		7.15	13.35		20.50	28
2460	2" diameter		48	.208		9.55	14.20		23.75	31.50
2500	Tee, 3/4" diameter		42	.238		2.78	16.20		18.98	27
2520	1" diameter		39	.256		4.44	17.45		21.89	31
2540	1-1/2" diameter		33	.303		10.70	20.50		31.20	43
2560	2" diameter		30	.333		14.50	22.50		37	50

33 14 13.25 Water Supply, Polyvinyl Chloride Pipe

		Crew	Daily Output	Labor-Hours	Unit	Material	2021 Bare Costs Labor	Equipment	Total	Total Incl O&P
0010	**WATER SUPPLY, POLYVINYL CHLORIDE PIPE**									
0020	Not including excavation or backfill, unless specified									
2100	PVC pipe, Class 150, 1-1/2" diameter	Q-1A	750	.013	L.F.	.64	.91		1.55	2.06
2120	2" diameter		686	.015		.81	.99		1.80	2.37
2140	2-1/2" diameter		500	.020		1.18	1.36		2.54	3.33
2160	3" diameter	B-20	430	.056		1.77	2.75		4.52	6.05
2180	4" diameter		375	.064		3.79	3.16		6.95	8.90
2200	6" diameter		316	.076		5.50	3.74		9.24	11.65
2210	8" diameter		260	.092		7.05	4.55		11.60	14.55
3010	AWWA C905, PR 100, DR 25									
3030	14" diameter	B-21	213	.131	L.F.	13.75	6.70	.90	21.35	26
3040	16" diameter		200	.140		18.80	7.15	.95	26.90	32.50
3050	18" diameter		160	.175		23	8.95	1.19	33.14	39.50
3060	20" diameter		133	.211		28.50	10.75	1.44	40.69	48.50
3070	24" diameter		107	.262		42	13.35	1.78	57.13	68.50
3080	30" diameter		80	.350		70.50	17.85	2.39	90.74	107
3090	36" diameter		80	.350		109	17.85	2.39	129.24	149
3100	42" diameter		60	.467		148	24	3.18	175.18	201
3200	48" diameter		60	.467		188	24	3.18	215.18	246
3960	Pressure pipe, class 200, ASTM 2241, SDR 21, 3/4" diameter	Q-1A	1000	.010		.21	.68		.89	1.25
3980	1" diameter		900	.011		.25	.76		1.01	1.41
4000	1-1/2" diameter		750	.013		1.16	.91		2.07	2.64
4010	2" diameter		686	.015		.86	.99		1.85	2.43
4020	2-1/2" diameter		500	.020		2.16	1.36		3.52	4.41
4030	3" diameter	B-20	430	.056		2.30	2.75		5.05	6.65
4040	4" diameter		375	.064		3.78	3.16		6.94	8.90
4050	6" diameter		316	.076		8	3.74		11.74	14.40
4060	8" diameter	B-20	260	.092	L.F.	8.45	4.55		13	16.10

33 14 13.25 Water Supply, Polyvinyl Chloride Pipe	Crew	Daily Output	Labor-Hours	Unit	Material	2021 Bare Costs Labor	Equipment	Total	Total Incl O&P
4090	Including trenching to 3' deep, 3/4" diameter	Q-1C	300	.080	L.F.	.21	4.82	6.30	11.33	14.40
4100	1" diameter		280	.086		.25	5.15	6.75	12.15	15.45
4110	1-1/2" diameter		260	.092		1.16	5.55	7.30	14.01	17.60
4120	2" diameter		220	.109		.86	6.60	8.60	16.06	20
4130	2-1/2" diameter		200	.120		2.16	7.25	9.45	18.86	23.50
4140	3" diameter		175	.137		2.30	8.25	10.80	21.35	27
4150	4" diameter		150	.160		3.78	9.65	12.65	26.08	32.50
4160	6" diameter		125	.192		8	11.55	15.15	34.70	42.50
4165	Fittings									
4170	Elbow, 90°, 3/4"	Q-1A	114	.088	Ea.	1.22	5.95		7.17	10.25
4180	1"		100	.100		3.19	6.80		9.99	13.65
4190	1-1/2"		80	.125		19.80	8.50		28.30	34.50
4200	2"		72	.139		14.20	9.45		23.65	30
4210	3"	B-20	46	.522		20.50	25.50		46	61
4220	4"		36	.667		35	33		68	87.50
4230	6"		24	1		64.50	49.50		114	145
4240	8"		14	1.714		126	84.50		210.50	265
4250	Elbow, 45°, 3/4"	Q-1A	114	.088		1.19	5.95		7.14	10.20
4260	1"		100	.100		9.65	6.80		16.45	21
4270	1-1/2"		80	.125		16.25	8.50		24.75	30.50
4280	2"		72	.139		19.60	9.45		29.05	35.50
4290	2-1/2"		54	.185		22.50	12.60		35.10	43.50
4300	3"	B-20	46	.522		28.50	25.50		54	69.50
4310	4"		36	.667		48.50	33		81.50	103
4320	6"		24	1		89	49.50		138.50	172
4330	8"		14	1.714		126	84.50		210.50	265
4340	Tee, 3/4"	Q-1A	76	.132		3.25	8.95		12.20	16.95
4350	1"		66	.152		12.50	10.30		22.80	29
4360	1-1/2"		54	.185		17.65	12.60		30.25	38
4370	2"		48	.208		21	14.20		35.20	44
4380	2-1/2"		36	.278		23	18.90		41.90	53.50
4390	3"	B-20	30	.800		24.50	39.50		64	86
4400	4"		24	1		38.50	49.50		88	117
4410	6"		14.80	1.622		84.50	80		164.50	213
4420	8"		9	2.667		149	131		280	360
4430	Coupling, 3/4"	Q-1A	114	.088		1.69	5.95		7.64	10.75
4440	1"		100	.100		7.35	6.80		14.15	18.25
4450	1-1/2"		80	.125		9	8.50		17.50	22.50
4460	2"		72	.139		9.15	9.45		18.60	24
4470	2-1/2"		54	.185		17.35	12.60		29.95	38
4480	3"	B-20	46	.522		12.50	25.50		38	52.50
4490	4"		36	.667		23	33		56	74
4500	6"		24	1		32.50	49.50		82	110
4510	8"	B-20	14	1.714	Ea.	68.50	84.50		153	203
4520	Pressure pipe Class 150, SDR 18, AWWA C900, 4" diameter		380	.063	L.F.	3.79	3.11		6.90	8.85
4530	6" diameter		316	.076		5.50	3.74		9.24	11.65
4540	8" diameter	B-21	264	.106		7.05	5.40	.72	13.17	16.65
4550	10" diameter		220	.127		9.55	6.50	.87	16.92	21
4560	12" diameter		186	.151		13	7.70	1.03	21.73	27
8000	Fittings with rubber gasket									
8003	Class 150, DR 18									
8006	90° bend , 4" diameter	B-20	100	.240	Ea.	42	11.85		53.85	63.50
8020	6" diameter	"	90	.267		74.50	13.15		87.65	102

491

For customer support on your Site Work & Landscape Costs with RSMeans Data, call 800.448.8182.

33 14 13.25 Water Supply, Polyvinyl Chloride Pipe	Crew	Daily Output	Labor-Hours	Unit	Material	2021 Bare Costs Labor	Equipment	Total	Total Incl O&P
8040	8" diameter	B-21	80	.350	Ea.	144	17.85	2.39	164.24	188
8060	10" diameter		50	.560		280	28.50	3.82	312.32	355
8080	12" diameter		30	.933		370	47.50	6.35	423.85	485
8100	Tee, 4" diameter		90	.311		57.50	15.90	2.12	75.52	90
8120	6" diameter		80	.350		129	17.85	2.39	149.24	171
8140	8" diameter		70	.400		184	20.50	2.73	207.23	236
8160	10" diameter		40	.700		257	35.50	4.77	297.27	340
8180	12" diameter		20	1.400		360	71.50	9.55	441.05	515
8200	45° bend, 4" diameter	B-20	100	.240		42	11.85		53.85	63.50
8220	6" diameter	"	90	.267		72.50	13.15		85.65	99.50
8240	8" diameter	B-21	50	.560		138	28.50	3.82	170.32	199
8260	10" diameter		50	.560		157	28.50	3.82	189.32	220
8280	12" diameter		30	.933		239	47.50	6.35	292.85	340
8300	Reducing tee 6" x 4"		100	.280		126	14.30	1.91	142.21	162
8320	8" x 6"		90	.311		223	15.90	2.12	241.02	271
8330	10" x 6"		90	.311		195	15.90	2.12	213.02	241
8340	10" x 8"		90	.311		209	15.90	2.12	227.02	256
8350	12" x 6"		90	.311		263	15.90	2.12	281.02	315
8360	12" x 8"		90	.311		291	15.90	2.12	309.02	345
8400	Tapped service tee (threaded type) 6" x 6" x 3/4"		100	.280		95	14.30	1.91	111.21	129
8430	6" x 6" x 1"		90	.311		95	15.90	2.12	113.02	131
8440	6" x 6" x 1-1/2"		90	.311		95	15.90	2.12	113.02	131
8450	6" x 6" x 2"		90	.311		95	15.90	2.12	113.02	131
8460	8" x 8" x 3/4"		90	.311		140	15.90	2.12	158.02	180
8470	8" x 8" x 1"		90	.311		140	15.90	2.12	158.02	180
8480	8" x 8" x 1-1/2"		90	.311		140	15.90	2.12	158.02	180
8490	8" x 8" x 2"		90	.311		140	15.90	2.12	158.02	180
8500	Repair coupling 4"	B-20	100	.240		26	11.85		37.85	46
8520	6" diameter		90	.267		40	13.15		53.15	63.50
8540	8" diameter		50	.480		96	23.50		119.50	142
8560	10" diameter		50	.480		202	23.50		225.50	258
8580	12" diameter		50	.480		256	23.50		279.50	320
8600	Plug end 4"	B-20	100	.240	Ea.	22.50	11.85		34.35	42.50
8620	6" diameter		90	.267		40.50	13.15		53.65	64
8640	8" diameter		50	.480		69	23.50		92.50	112
8660	10" diameter		50	.480		110	23.50		133.50	157
8680	12" diameter		50	.480		148	23.50		171.50	199
8700	PVC pipe, joint restraint									
8710	4" diameter	B-20	32	.750	Ea.	46.50	37		83.50	107
8720	6" diameter		25.60	.938		57.50	46		103.50	133
8730	8" diameter		21.33	1.125		83.50	55.50		139	175
8740	10" diameter		18.28	1.313		133	64.50		197.50	243
8750	12" diameter		16.84	1.425		140	70.50		210.50	259
8760	14" diameter		16	1.500		190	74		264	320
8770	16" diameter		11.64	2.062		256	102		358	435
8780	18" diameter		11.03	2.176		315	107		422	505
8785	20" diameter		9.14	2.626		445	129		574	685
8790	24" diameter		7.53	3.187		515	157		672	800

33 14 13 – Public Water Utility Distribution Piping

33 14 13.35 Water Supply, HDPE	Crew	Daily Output	Labor-Hours	Unit	Material	2021 Bare Costs Labor	Equipment	Total	Total Incl O&P
0010 **WATER SUPPLY, HDPE**									
0011 Butt fusion joints, SDR 21 40' lengths not including excavation or backfill									
0100 4" diameter	B-22A	400	.100	L.F.	3.20	5.10	2.01	10.31	13.35
0200 6" diameter		380	.105		5.60	5.35	2.11	13.06	16.50
0300 8" diameter		320	.125		10	6.35	2.51	18.86	23.50
0400 10" diameter		300	.133		13.25	6.75	2.68	22.68	27.50
0500 12" diameter		260	.154		17.15	7.80	3.09	28.04	34
0600 14" diameter	B-22B	220	.182		18.05	9.25	6.65	33.95	41
0700 16" diameter		180	.222		21.50	11.30	8.15	40.95	49.50
0800 18" diameter		140	.286		27	14.50	10.50	52	62.50
0850 20" diameter		130	.308		34	15.60	11.30	60.90	73.50
0900 24" diameter		100	.400		70	20.50	14.70	105.20	124
1000 Fittings									
1100 Elbows, 90 degrees									
1200 4" diameter	B-22A	32	1.250	Ea.	16.55	63.50	25	105.05	141
1300 6" diameter		28	1.429		47	72.50	28.50	148	192
1400 8" diameter		24	1.667		94.50	84.50	33.50	212.50	267
1500 10" diameter		18	2.222		260	113	44.50	417.50	505
1600 12" diameter		12	3.333		283	169	67	519	635
1700 14" diameter	B-22B	9	4.444		625	226	163	1,014	1,200
1800 16" diameter		6	6.667		925	340	245	1,510	1,800
1900 18" diameter		4	10		1,250	510	365	2,125	2,550
2000 24" diameter		3	13.333		1,375	675	490	2,540	3,075
2100 Tees									
2200 4" diameter	B-22A	30	1.333	Ea.	22.50	67.50	27	117	156
2300 6" diameter	B-22A	26	1.538	Ea.	68.50	78	31	177.50	227
2400 8" diameter		22	1.818		131	92.50	36.50	260	320
2500 10" diameter		15	2.667		173	135	53.50	361.50	450
2600 12" diameter		10	4		505	203	80.50	788.50	950
2700 14" diameter	B-22B	8	5		595	254	184	1,033	1,225
2800 16" diameter		6	6.667		700	340	245	1,285	1,550
2900 18" diameter		4	10		765	510	365	1,640	2,000
3000 24" diameter		2	20		935	1,025	735	2,695	3,350
4100 Caps									
4110 4" diameter	B-22A	34	1.176	Ea.	15.60	59.50	23.50	98.60	132
4120 6" diameter		30	1.333		35	67.50	27	129.50	169
4130 8" diameter		26	1.538		58.50	78	31	167.50	216
4150 10" diameter		20	2		146	102	40	288	355
4160 12" diameter		14	2.857		176	145	57.50	378.50	475

33 14 13.40 Water Supply, Black Steel Pipe

33 14 13.40 Water Supply, Black Steel Pipe	Crew	Daily Output	Labor-Hours	Unit	Material	Labor	Equipment	Total	Total Incl O&P
0010 **WATER SUPPLY, BLACK STEEL PIPE**									
0011 Not including excavation or backfill									
1000 Pipe, black steel, plain end, welded, 1/4" wall thk, 8" diam.	B-35A	208	.269	L.F.	23	14.40	10.15	47.55	57.50
1010 10" diameter		204	.275		28.50	14.65	10.35	53.50	65
1020 12" diameter		195	.287		32.50	15.35	10.85	58.70	71
1030 18" diameter		175	.320		79.50	17.10	12.10	108.70	126
1040 5/16" wall thickness, 12" diameter		195	.287		65.50	15.35	10.85	91.70	107
1050 18" diameter		175	.320		103	17.10	12.10	132.20	152
1060 36" diameter		28.96	1.934		139	103	73	315	390
1070 3/8" wall thickness, 18" diameter		43.20	1.296		213	69.50	49	331.50	390
1080 24" diameter		36	1.556		162	83	59	304	365
1090 30" diameter		30.40	1.842		184	98.50	69.50	352	425

33 14 13 – Public Water Utility Distribution Piping

33 14 13.40 Water Supply, Black Steel Pipe

		Crew	Daily Output	Labor-Hours	Unit	Material	2021 Bare Costs Labor	Equipment	Total	Total Incl O&P
1100	1/2" wall thickness, 36" diameter	B-35A	26.08	2.147	L.F.	305	115	81	501	595
1110	48" diameter		21.68	2.583		560	138	97.50	795.50	930
1135	7/16" wall thickness, 48" diameter		20.80	2.692		330	144	102	576	690
1140	5/8" wall thickness, 48" diameter		21.68	2.583		510	138	97.50	745.50	875

33 14 13.45 Water Supply, Copper Pipe

		Crew	Daily Output	Labor-Hours	Unit	Material	2021 Bare Costs Labor	Equipment	Total	Total Incl O&P
0010	**WATER SUPPLY, COPPER PIPE**									
0020	Not including excavation or backfill									
2000	Tubing, type K, 20' joints, 3/4" diameter	Q-1	400	.040	L.F.	7.95	2.44		10.39	12.40
2200	1" diameter		320	.050		10.90	3.05		13.95	16.55
3000	1-1/2" diameter		265	.060		16.85	3.68		20.53	24
3020	2" diameter		230	.070		26	4.24		30.24	35
3040	2-1/2" diameter		146	.110		39.50	6.70		46.20	53.50
3060	3" diameter		134	.119		54.50	7.30		61.80	71
4012	4" diameter		95	.168		90	10.25		100.25	114
4014	5" diameter		80	.200		48.50	12.20		60.70	71.50
4016	6" diameter	Q-2	80	.300	L.F.	57.50	18.95		76.45	91.50
4018	8" diameter	"	80	.300	"	134	18.95		152.95	177
5000	Tubing, type L									
5108	2" diameter	Q-1	230	.070	L.F.	7.95	4.24		12.19	15.10
6010	3" diameter		134	.119		37	7.30		44.30	52
6012	4" diameter		95	.168		22	10.25		32.25	39.50
6016	6" diameter	Q-2	80	.300		49.50	18.95		68.45	83
7165	Fittings, brass, corporation stops, no lead, 3/4" diameter	1 Plum	19	.421	Ea.	76.50	28.50		105	127
7166	1" diameter		16	.500		100	34		134	161
7167	1-1/2" diameter		13	.615		213	41.50		254.50	296
7168	2" diameter		11	.727		335	49		384	445
7170	Curb stops, no lead, 3/4" diameter		19	.421		100	28.50		128.50	153
7171	1" diameter		16	.500		145	34		179	210
7172	1-1/2" diameter		13	.615		275	41.50		316.50	360
7173	2" diameter		11	.727		310	49		359	415
7180	Curb box, cast iron, 1/2" to 1" curb stops		12	.667		40.50	45		85.50	112
7200	1-1/4" to 2" curb stops		8	1		92	67.50		159.50	202
7220	Saddles, 3/4" & 1" diameter, add					66			66	73
7240	1-1/2" to 2" diameter, add					116			116	127
7250	For copper fittings, see Section 22 11 13.25									

33 14 13.90 Water Supply, Thrust Blocks

		Crew	Daily Output	Labor-Hours	Unit	Material	2021 Bare Costs Labor	Equipment	Total	Total Incl O&P
0010	**WATER SUPPLY, THRUST BLOCKS**									
0015	Piping, not including excavation or backfill									
0110	Thrust block for 90 degree elbow, 4" diameter	C-30	41	.195	Ea.	23.50	8.65	3.59	35.74	42.50
0115	6" diameter		23	.348		40	15.45	6.40	61.85	74
0120	8" diameter		14	.571		61.50	25.50	10.50	97.50	117
0125	10" diameter		9	.889		88	39.50	16.35	143.85	174
0130	12" diameter		7	1.143		119	50.50	21	190.50	230
0135	14" diameter		5	1.600		160	71	29.50	260.50	315
0140	16" diameter		4	2		202	89	37	328	395
0145	18" diameter		3	2.667		248	118	49	415	505
0150	20" diameter		2.50	3.200		300	142	59	501	605
0155	24" diameter		2	4		430	178	73.50	681.50	820
0210	Thrust block for tee or deadend, 4" diameter		65	.123		16.10	5.45	2.26	23.81	28.50
0215	6" diameter		35	.229		28.50	10.15	4.20	42.85	51
0220	8" diameter		21	.381		44	16.90	7	67.90	81
0225	10" diameter		14	.571		63.50	25.50	10.50	99.50	119

33 14 Water Utility Transmission and Distribution

33 14 13 – Public Water Utility Distribution Piping

33 14 13.90 Water Supply, Thrust Blocks

		Crew	Daily Output	Labor-Hours	Unit	Material	2021 Bare Costs Labor	2021 Bare Costs Equipment	Total	Total Incl O&P
0230	12" diameter	C-30	10	.800	Ea.	86.50	35.50	14.70	136.70	164
0235	14" diameter		7	1.143		116	50.50	21	187.50	226
0240	16" diameter		5	1.600		146	71	29.50	246.50	299
0245	18" diameter		4	2		179	89	37	305	370
0250	20" diameter		3.50	2.286		216	101	42	359	435
0255	24" diameter	C-30	2.50	3.200	Ea.	305	142	59	506	610

33 14 17 – Site Water Utility Service Laterals

33 14 17.15 Tapping, Crosses and Sleeves

		Crew	Daily Output	Labor-Hours	Unit	Material	2021 Bare Costs Labor	2021 Bare Costs Equipment	Total	Total Incl O&P
0010	**TAPPING, CROSSES AND SLEEVES**									
4000	Drill and tap pressurized main (labor only)									
4100	6" main, 1" to 2" service	Q-1	3	5.333	Ea.		325		325	485
4150	8" main, 1" to 2" service	"	2.75	5.818	"		355		355	530
4500	Tap and insert gate valve									
4600	8" main, 4" branch	B-21	3.20	8.750	Ea.		445	59.50	504.50	735
4650	6" branch		2.70	10.370			530	70.50	600.50	870
4700	10" main, 4" branch		2.70	10.370			530	70.50	600.50	870
4750	6" branch		2.35	11.915			610	81.50	691.50	1,000
4800	12" main, 6" branch		2.35	11.915			610	81.50	691.50	1,000
4850	8" branch		2.35	11.915			610	81.50	691.50	1,000
7000	Tapping crosses, sleeves, valves; with rubber gaskets									
7020	Crosses, 4" x 4"	B-21	37	.757	Ea.	1,400	38.50	5.15	1,443.65	1,625
7030	6" x 4"		25	1.120		1,675	57	7.65	1,739.65	1,950
7040	6" x 6"		25	1.120		1,675	57	7.65	1,739.65	1,950
7060	8" x 6"		21	1.333		2,050	68	9.10	2,127.10	2,350
7080	8" x 8"		21	1.333		2,225	68	9.10	2,302.10	2,550
7100	10" x 6"		21	1.333		4,050	68	9.10	4,127.10	4,575
7120	10" x 10"		21	1.333		4,400	68	9.10	4,477.10	4,925
7140	12" x 6"		18	1.556		4,050	79.50	10.60	4,140.10	4,600
7160	12" x 12"		18	1.556		5,125	79.50	10.60	5,215.10	5,775
7180	14" x 6"		16	1.750		10,100	89.50	11.95	10,201.45	11,200
7200	14" x 14"		16	1.750		10,700	89.50	11.95	10,801.45	11,900
7220	16" x 6"		14	2		10,900	102	13.65	11,015.65	12,100
7240	16" x 10"		14	2		11,000	102	13.65	11,115.65	12,300
7260	16" x 16"		14	2		11,600	102	13.65	11,715.65	12,900
7280	18" x 6"		10	2.800		16,200	143	19.10	16,362.10	18,100
7300	18" x 12"		10	2.800		16,400	143	19.10	16,562.10	18,200
7320	18" x 18"		10	2.800		16,800	143	19.10	16,962.10	18,700
7340	20" x 6"		8	3.500		13,100	179	24	13,303	14,800
7360	20" x 12"		8	3.500		14,100	179	24	14,303	15,800
7380	20" x 20"		8	3.500		20,600	179	24	20,803	22,900
7400	24" x 6"		6	4.667		17,000	238	32	17,270	19,100
7420	24" x 12"		6	4.667		17,200	238	32	17,470	19,300
7440	24" x 18"		6	4.667		26,100	238	32	26,370	29,100
7460	24" x 24"		6	4.667		26,700	238	32	26,970	29,800
7600	Cut-in sleeves with rubber gaskets, 4"		18	1.556		640	79.50	10.60	730.10	835
7620	6"		12	2.333		745	119	15.90	879.90	1,025
7640	8"		10	2.800		950	143	19.10	1,112.10	1,275
7660	10"		10	2.800		1,325	143	19.10	1,487.10	1,675
7680	12"		9	3.111		1,800	159	21	1,980	2,225
7800	Cut-in valves with rubber gaskets, 4"		18	1.556		460	79.50	10.60	550.10	640
7820	6"		12	2.333		600	119	15.90	734.90	855
7840	8"		10	2.800		910	143	19.10	1,072.10	1,225

For customer support on your Site Work & Landscape Costs with RSMeans Data, call 800.448.8182.

495

33 14 17 – Site Water Utility Service Laterals

33 14 17.15 Tapping, Crosses and Sleeves

		Crew	Daily Output	Labor-Hours	Unit	Material	2021 Bare Costs Labor	Equipment	Total	Total Incl O&P
7860	10"	B-21	10	2.800	Ea.	1,375	143	19.10	1,537.10	1,725
7880	12"		9	3.111		1,725	159	21	1,905	2,125
7900	Tapping valve 4", MJ, ductile iron		18	1.556		670	79.50	10.60	760.10	865
7920	6", MJ, ductile iron		12	2.333		885	119	15.90	1,019.90	1,175
7940	Tapping valve 8", MJ, ductile iron		10	2.800		850	143	19.10	1,012.10	1,175
7960	Tapping valve 10", MJ, ductile iron		10	2.800		1,175	143	19.10	1,337.10	1,525
7980	Tapping valve 12", MJ, ductile iron		8	3.500		2,125	179	24	2,328	2,625
8000	Sleeves with rubber gaskets, 4" x 4"		37	.757		1,125	38.50	5.15	1,168.65	1,300
8010	6" x 4"		25	1.120		1,050	57	7.65	1,114.65	1,250
8020	6" x 6"		25	1.120		1,325	57	7.65	1,389.65	1,575
8030	8" x 4"		21	1.333		950	68	9.10	1,027.10	1,150
8040	8" x 6"		21	1.333		1,650	68	9.10	1,727.10	1,900
8060	8" x 8"		21	1.333		1,450	68	9.10	1,527.10	1,675
8070	10" x 4"		21	1.333		885	68	9.10	962.10	1,075
8080	10" x 6"		21	1.333		3,200	68	9.10	3,277.10	3,600
8090	10" x 8"		21	1.333		1,400	68	9.10	1,477.10	1,625
8100	10" x 10"		21	1.333		2,500	68	9.10	2,577.10	2,850
8110	12" x 4"		18	1.556		3,175	79.50	10.60	3,265.10	3,600
8120	12" x 6"		18	1.556		3,175	79.50	10.60	3,265.10	3,600
8130	12" x 8"		18	1.556		3,375	79.50	10.60	3,465.10	3,825
8135	12" x 10"		18	1.556		2,850	79.50	10.60	2,940.10	3,250
8140	12" x 12"		18	1.556		3,175	79.50	10.60	3,265.10	3,625
8160	14" x 6"		16	1.750		1,600	89.50	11.95	1,701.45	1,925
8180	14" x 14"		16	1.750		2,775	89.50	11.95	2,876.45	3,200
8200	16" x 6"		14	2		1,750	102	13.65	1,865.65	2,100
8220	16" x 10"		14	2		3,825	102	13.65	3,940.65	4,375
8240	16" x 16"		14	2		9,325	102	13.65	9,440.65	10,500
8260	18" x 6"		10	2.800		1,250	143	19.10	1,412.10	1,600
8280	18" x 12"		10	2.800		3,800	143	19.10	3,962.10	4,425
8300	18" x 18"		10	2.800		12,300	143	19.10	12,462.10	13,800
8320	20" x 6"		8	3.500		2,525	179	24	2,728	3,075
8340	20" x 12"		8	3.500		3,575	179	24	3,778	4,225
8360	20" x 20"		8	3.500		15,100	179	24	15,303	16,900
8380	24" x 6"		6	4.667		2,100	238	32	2,370	2,700
8400	24" x 12"		6	4.667		9,600	238	32	9,870	11,000
8420	24" x 18"		6	4.667		11,200	238	32	11,470	12,700
8440	24" x 24"	▼	6	4.667		12,600	238	32	12,870	14,300
8800	Hydrant valve box, 6' long	B-20	20	1.200		232	59		291	345
8820	8' long		18	1.333		237	65.50		302.50	360
8830	Valve box w/lid 4' deep		14	1.714		98.50	84.50		183	235
8840	Valve box and large base w/lid	▼	14	1.714	▼	340	84.50		424.50	495

33 14 19 – Valves and Hydrants for Water Utility Service

33 14 19.10 Valves

		Crew	Daily Output	Labor-Hours	Unit	Material	2021 Bare Costs Labor	Equipment	Total	Total Incl O&P
0010	**VALVES**, water distribution									
0011	See Sections 22 05 23.20 and 22 05 23.60									
3000	Butterfly valves with boxes, cast iron, mechanical joint									
3100	4" diameter	B-6	6	4	Ea.	665	192	36	893	1,050
3140	6" diameter		6	4		875	192	36	1,103	1,300
3180	8" diameter		6	4		1,100	192	36	1,328	1,550
3300	10" diameter		6	4		1,500	192	36	1,728	1,975
3340	12" diameter		6	4		2,025	192	36	2,253	2,575
3400	14" diameter		4	6		4,325	289	54	4,668	5,275

33 14 19 – Valves and Hydrants for Water Utility Service

33 14 19.10 Valves		Crew	Daily Output	Labor-Hours	Unit	Material	2021 Bare Costs Labor	Equipment	Total	Total Incl O&P
3440	16" diameter	B-6	4	6	Ea.	6,150	289	54	6,493	7,250
3460	18" diameter		4	6		6,550	289	54	6,893	7,700
3480	20" diameter		4	6		10,100	289	54	10,443	11,600
3500	24" diameter		4	6		14,400	289	54	14,743	16,400
3510	30" diameter		4	6		12,000	289	54	12,343	13,700
3520	36" diameter		4	6		14,900	289	54	15,243	16,900
3530	42" diameter		4	6		18,900	289	54	19,243	21,300
3540	48" diameter	▼	4	6	▼	24,300	289	54	24,643	27,300
3600	With lever operator									
3610	4" diameter	B-6	6	4	Ea.	565	192	36	793	950
3614	6" diameter		6	4		780	192	36	1,008	1,175
3616	8" diameter		6	4		1,000	192	36	1,228	1,425
3618	10" diameter		6	4		1,400	192	36	1,628	1,875
3620	12" diameter		6	4		1,950	192	36	2,178	2,450
3622	14" diameter		4	6		4,000	289	54	4,343	4,900
3624	16" diameter		4	6		5,800	289	54	6,143	6,875
3626	18" diameter		4	6		6,225	289	54	6,568	7,325
3628	20" diameter		4	6		9,725	289	54	10,068	11,200
3630	24" diameter	▼	4	6	▼	14,100	289	54	14,443	16,000
3700	Check valves, flanged									
3710	4" diameter	B-6	6	4	Ea.	820	192	36	1,048	1,225
3712	5" diameter		6	4		1,050	192	36	1,278	1,475
3714	6" diameter		6	4		1,400	192	36	1,628	1,850
3716	8" diameter		6	4		2,625	192	36	2,853	3,225
3718	10" diameter		6	4		4,475	192	36	4,703	5,250
3720	12" diameter		6	4		7,450	192	36	7,678	8,525
3726	18" diameter	▼	4	6	▼	36,600	289	54	36,943	40,800
3730	24" diameter	B-6	4	6	Ea.	73,500	289	54	73,843	81,500
3800	Gate valves, C.I., 125 psi, mechanical joint, w/boxes									
3810	4" diameter	B-6	6	4	Ea.	1,100	192	36	1,328	1,550
3814	6" diameter		6	4		1,750	192	36	1,978	2,250
3816	8" diameter		6	4		3,050	192	36	3,278	3,675
3818	10" diameter		6	4		5,475	192	36	5,703	6,350
3820	12" diameter		6	4		7,450	192	36	7,678	8,500
3822	14" diameter		4	6		15,300	289	54	15,643	17,400
3824	16" diameter		4	6		21,600	289	54	21,943	24,300
3826	18" diameter		4	6		26,200	289	54	26,543	29,300
3828	20" diameter		4	6		36,400	289	54	36,743	40,500
3830	24" diameter		4	6		54,000	289	54	54,343	59,500
3831	30" diameter		4	6		58,500	289	54	58,843	65,000
3832	36" diameter	▼	4	6		91,500	289	54	91,843	101,000
3880	Sleeve, for tapping mains, 8" x 4", add					950			950	1,050
3884	10" x 6", add					3,200			3,200	3,500
3888	12" x 6", add					3,175			3,175	3,475
3892	12" x 8", add					3,375			3,375	3,700

33 14 19.20 Valves

		Crew	Daily Output	Labor-Hours	Unit	Material	2021 Bare Costs Labor	Equipment	Total	Total Incl O&P
0010	**VALVES**									
0011	Special trim or use									
0150	Altitude valve, single acting, modulating, 2-1/2" diameter	B-6	6	4	Ea.	3,600	192	36	3,828	4,275
0155	3" diameter		6	4		3,625	192	36	3,853	4,325
0160	4" diameter		6	4		4,150	192	36	4,378	4,900
0165	6" diameter	▼	6	4		5,250	192	36	5,478	6,100

33 14 19.20 Valves		Crew	Daily Output	Labor-Hours	Unit	Material	2021 Bare Costs Labor	Equipment	Total	Total Incl O&P
0170	8" diameter	B-6	6	4	Ea.	7,800	192	36	8,028	8,900
0175	10" diameter		6	4		8,150	192	36	8,378	9,300
0180	12" diameter		6	4		12,400	192	36	12,628	14,000
0250	Altitude valve, single acting, non-modulating, 2-1/2" diameter		6	4		3,575	192	36	3,803	4,275
0255	3" diameter		6	4		3,725	192	36	3,953	4,425
0260	4" diameter		6	4		4,275	192	36	4,503	5,050
0265	6" diameter		6	4		5,350	192	36	5,578	6,200
0270	8" diameter		6	4		7,725	192	36	7,953	8,825
0275	10" diameter		6	4		8,550	192	36	8,778	9,725
0280	12" diameter		6	4		13,200	192	36	13,428	14,800
0350	Altitude valve, double acting, non-modulating, 2-1/2" diameter		6	4		4,375	192	36	4,603	5,125
0355	3" diameter		6	4		4,550	192	36	4,778	5,325
0360	4" diameter		6	4		5,350	192	36	5,578	6,225
0365	6" diameter		6	4		6,025	192	36	6,253	6,950
0370	8" diameter		6	4		7,675	192	36	7,903	8,775
0375	10" diameter		6	4		10,400	192	36	10,628	11,800
0380	12" diameter		6	4		15,800	192	36	16,028	17,700
1000	Air release valve for water, 1/2" inlet	1 Plum	16	.500	Ea.	98	34		132	159
1005	3/4" inlet		16	.500		98	34		132	159
1010	1" inlet		14	.571		98	38.50		136.50	166
1020	2" inlet		9	.889		98	60		158	198
1100	Air release & vacuum valve for water, 1/2" inlet		16	.500		330	34		364	415
1105	3/4" inlet		16	.500		345	34		379	430
1110	1" inlet		14	.571		520	38.50		558.50	635
1120	2" inlet		9	.889		885	60		945	1,075
1130	3" inlet	Q-1	8	2		1,775	122		1,897	2,125
1140	4" inlet		5	3.200		2,375	195		2,570	2,900
1150	6" inlet		5	3.200		4,625	195		4,820	5,375
1160	8" inlet		5	3.200		11,400	195		11,595	12,900
1170	10" inlet		5	3.200		12,800	195		12,995	14,400
9000	Valves, gate valve, N.R.S. PIV with post, 4" diameter	B-6	6	4		2,625	192	36	2,853	3,200
9020	6" diameter		6	4		3,275	192	36	3,503	3,925
9040	8" diameter		6	4		4,550	192	36	4,778	5,350
9060	10" diameter		6	4		7,000	192	36	7,228	8,025
9080	12" diameter		6	4		8,950	192	36	9,178	10,200
9100	14" diameter		6	4		16,600	192	36	16,828	18,600
9120	OS&Y, 4" diameter		6	4		925	192	36	1,153	1,350
9140	6" diameter		6	4		1,300	192	36	1,528	1,750
9160	8" diameter		6	4		1,900	192	36	2,128	2,400
9180	10" diameter		6	4		2,925	192	36	3,153	3,550
9200	12" diameter		6	4		4,800	192	36	5,028	5,600
9220	14" diameter		4	6		4,975	289	54	5,318	5,975
9400	Check valves, rubber disc, 2-1/2" diameter		6	4		430	192	36	658	795
9420	3" diameter		6	4		590	192	36	818	975
9440	4" diameter		6	4		820	192	36	1,048	1,225
9480	6" diameter		6	4		1,400	192	36	1,628	1,850
9500	8" diameter		6	4		2,625	192	36	2,853	3,225
9520	10" diameter		6	4		4,475	192	36	4,703	5,250
9540	12" diameter		6	4		7,450	192	36	7,678	8,525
9542	14" diameter		4	6		7,900	289	54	8,243	9,175
9700	Detector check valves, reducing, 4" diameter		6	4		1,500	192	36	1,728	1,975
9720	6" diameter		6	4		2,425	192	36	2,653	3,000
9740	8" diameter		6	4		3,525	192	36	3,753	4,200

33 14 Water Utility Transmission and Distribution

33 14 19 – Valves and Hydrants for Water Utility Service

33 14 19.20 Valves

		Crew	Daily Output	Labor-Hours	Unit	Material	2021 Bare Costs Labor	Equipment	Total	Total Incl O&P
9760	10" diameter	B-6	6	4	Ea.	6,400	192	36	6,628	7,350
9800	Galvanized, 4" diameter		6	4		2,050	192	36	2,278	2,600
9820	6" diameter		6	4		2,825	192	36	3,053	3,450
9840	8" diameter		6	4		3,800	192	36	4,028	4,500
9860	10" diameter		6	4		6,925	192	36	7,153	7,925

33 14 19.30 Fire Hydrants

		Crew	Daily Output	Labor-Hours	Unit	Material	2021 Bare Costs Labor	Equipment	Total	Total Incl O&P
0010	**FIRE HYDRANTS**									
0020	Mechanical joints unless otherwise noted									
1000	Fire hydrants, two way; excavation and backfill not incl.									
1100	4-1/2" valve size, depth 2'-0"	B-21	10	2.800	Ea.	3,075	143	19.10	3,237.10	3,625
1120	2'-6"		10	2.800		3,075	143	19.10	3,237.10	3,625
1140	3'-0"		10	2.800		3,200	143	19.10	3,362.10	3,750
1160	3'-6"		9	3.111		3,200	159	21	3,380	3,775
1170	4'-0"		9	3.111		3,325	159	21	3,505	3,925
1200	4'-6"		9	3.111		3,325	159	21	3,505	3,925
1220	5'-0"		8	3.500		3,450	179	24	3,653	4,100
1240	5'-6"		8	3.500		3,450	179	24	3,653	4,100
1260	6'-0"		7	4		3,575	204	27.50	3,806.50	4,275
1280	6'-6"		7	4		3,575	204	27.50	3,806.50	4,275
1300	7'-0"		6	4.667		3,675	238	32	3,945	4,450
1340	8'-0"		6	4.667		3,725	238	32	3,995	4,500
1420	10'-0"		5	5.600		3,975	286	38	4,299	4,850
2000	5-1/4" valve size, depth 2'-0"		10	2.800		3,325	143	19.10	3,487.10	3,900
2080	4'-0"		9	3.111		3,650	159	21	3,830	4,275
2160	6'-0"		7	4		3,975	204	27.50	4,206.50	4,700
2240	8'-0"		6	4.667		4,350	238	32	4,620	5,200
2320	10'-0"		5	5.600		4,675	286	38	4,999	5,625
2350	For threeway valves, add					7%				
2400	Lower barrel extensions with stems, 1'-0"	B-20	14	1.714		340	84.50		424.50	500
2440	2'-0"		13	1.846		415	91		506	590
2480	3'-0"		12	2		1,375	98.50		1,473.50	1,650
2520	4'-0"		10	2.400		785	118		903	1,025
3200	Yard hydrant, flush, non freeze, 4' depth of bury, 3/4" conn.	2 Plum	4	4		157	271		428	580
4000	Above ground, 1" connection	1 Plum	8	1		148	67.50		215.50	264
5000	Indicator post									
5020	Adjustable, valve size 4" to 14", 4' bury	B-21	10	2.800	Ea.	1,400	143	19.10	1,562.10	1,775
5060	8' bury		7	4		1,625	204	27.50	1,856.50	2,100
5080	10' bury		6	4.667		2,125	238	32	2,395	2,750
5100	12' bury		5	5.600		1,750	286	38	2,074	2,400
5120	14' bury		4	7		1,675	355	47.50	2,077.50	2,450
5500	Non-adjustable, valve size 4" to 14", 3' bury		10	2.800		870	143	19.10	1,032.10	1,200
5520	3'-6" bury		10	2.800		870	143	19.10	1,032.10	1,200
5540	4' bury		9	3.111		870	159	21	1,050	1,225

33 16 Water Utility Storage Tanks

33 16 11 – Elevated Composite Water Storage Tanks

33 16 11.50 Elevated Water Storage Tanks

		Crew	Daily Output	Labor-Hours	Unit	Material	2021 Bare Costs Labor	Equipment	Total	Total Incl O&P
0010	**ELEVATED WATER STORAGE TANKS**									
0011	Not incl. pipe, pumps or foundation									
3000	Elevated water tanks, 100' to bottom capacity line, incl. painting									
3010	50,000 gallons				Ea.				185,000	204,000
3300	100,000 gallons								280,000	307,500
3400	250,000 gallons								751,500	826,500
3600	500,000 gallons								1,336,000	1,470,000
3700	750,000 gallons								1,622,000	1,783,500
3900	1,000,000 gallons								2,322,000	2,556,000

33 16 23 – Ground-Level Steel Water Storage Tanks

33 16 23.13 Steel Water Storage Tanks

		Crew	Daily Output	Labor-Hours	Unit	Material	2021 Bare Costs Labor	Equipment	Total	Total Incl O&P
0010	**STEEL WATER STORAGE TANKS**									
0910	Steel, ground level, ht./diam. less than 1, not incl. fdn., 100,000 gallons				Ea.				202,000	244,500
1000	250,000 gallons								295,500	324,000
1200	500,000 gallons								417,000	458,500
1250	750,000 gallons								538,000	591,500
1300	1,000,000 gallons								558,000	725,500
1500	2,000,000 gallons								1,043,000	1,148,000
1600	4,000,000 gallons								2,121,000	2,333,000
1800	6,000,000 gallons								3,095,000	3,405,000
1850	8,000,000 gallons								4,068,000	4,475,000
1910	10,000,000 gallons								5,050,000	5,554,500
2100	Steel standpipes, ht./diam. more than 1,100' to overflow, no fdn.									
2200	500,000 gallons				Ea.				546,500	600,500
2400	750,000 gallons								722,500	794,500
2500	1,000,000 gallons								1,060,500	1,167,000
2700	1,500,000 gallons								1,749,000	1,923,000
2800	2,000,000 gallons								2,327,000	2,559,000

33 16 36 – Ground-Level Reinforced Concrete Water Storage Tanks

33 16 36.16 Prestressed Conc. Water Storage Tanks

		Crew	Daily Output	Labor-Hours	Unit	Material	2021 Bare Costs Labor	Equipment	Total	Total Incl O&P
0010	**PRESTRESSED CONC. WATER STORAGE TANKS**									
0020	Not including fdn., pipe or pumps, 250,000 gallons				Ea.				299,000	329,500
0100	500,000 gallons								487,000	536,000
0300	1,000,000 gallons								707,000	807,500
0400	2,000,000 gallons								1,072,000	1,179,000
0600	4,000,000 gallons								1,706,000	1,877,000
0700	6,000,000 gallons								2,266,000	2,493,000
0750	8,000,000 gallons								2,924,000	3,216,000
0800	10,000,000 gallons								3,533,000	3,886,000

33 16 56 – Ground-Level Plastic Water Storage Cisterns

33 16 56.23 Plastic-Coated Fabric Pillow Water Tanks

		Crew	Daily Output	Labor-Hours	Unit	Material	2021 Bare Costs Labor	Equipment	Total	Total Incl O&P
0010	**PLASTIC-COATED FABRIC PILLOW WATER TANKS**									
7000	Water tanks, vinyl coated fabric pillow tanks, freestanding, 5,000 gallons	4 Clab	4	8	Ea.	3,850	355		4,205	4,750
7100	Supporting embankment not included, 25,000 gallons	6 Clab	2	24	Ea.	13,400	1,075		14,475	16,300
7200	50,000 gallons	8 Clab	1.50	42.667		20,500	1,900		22,400	25,400
7300	100,000 gallons	9 Clab	.90	80		36,700	3,550		40,250	45,600
7400	150,000 gallons		.50	144		43,300	6,400		49,700	57,000
7500	200,000 gallons		.40	180		70,000	8,000		78,000	89,000
7600	250,000 gallons		.30	240		87,000	10,700		97,700	111,500

33 31 Sanitary Sewerage Piping

33 31 11 – Public Sanitary Sewerage Gravity Piping

33 31 11.10 Sewage Collection, Valves	Crew	Daily Output	Labor-Hours	Unit	Material	2021 Bare Costs Labor	Equipment	Total	Total Incl O&P
0010	**SEWAGE COLLECTION, VALVES**									
1300	PVC swing check backwater valve									
1310	3" size	2 Skwk	15	1.067	Ea.	43	61		104	139
1315	4" size		15	1.067		1,275	61		1,336	1,500
1320	6" size	↓	13	1.231	↓	1,750	70.50		1,820.50	2,025
2000	Sewer relief valve									
2010	Plug style popper for outside cleanout									
2025	2" size	2 Skwk	56	.286	Ea.	3.54	16.30		19.84	28.50
2030	3" size		36	.444		3.86	25.50		29.36	42.50
2040	4" size	↓	30	.533	↓	5	30.50		35.50	51.50
3000	Air release valves for sewer pipes									
3020	2" size inlet connection	1 Plum	9	.889	Ea.	450	60		510	585
3030	3" size inlet connection	Q-1	8	2		950	122		1,072	1,225
3040	4" size inlet connection	"	5	3.200	↓	875	195		1,070	1,250
3100	Air release valves for sewer pipes with backwash valve and fittings									
3120	2" size inlet connection	1 Plum	9	.889	Ea.	550	60		610	695
3130	3" size inlet connection	Q-1	8	2		1,375	122		1,497	1,675
3140	4" size inlet connection	"	5	3.200	↓	1,525	195		1,720	1,975
3200	Air release & vacuum valves for sewer pipes									
3220	2" size inlet connection	1 Plum	9	.889	Ea.	820	60		880	990
3230	3" size inlet connection	Q-1	8	2	"	1,350	122		1,472	1,650

33 31 11.13 Sewage Collection, Vent Cast Iron Pipe

33 31 11.13 Sewage Collection, Vent Cast Iron Pipe	Crew	Daily Output	Labor-Hours	Unit	Material	2021 Bare Costs Labor	Equipment	Total	Total Incl O&P
0010	**SEWAGE COLLECTION, VENT CAST IRON PIPE**									
0020	Not including excavation or backfill									
2022	Sewage vent cast iron, B&S, 4" diameter	Q-1	66	.242	L.F.	24.50	14.75		39.25	49
2024	5" diameter	Q-2	88	.273		30.50	17.25		47.75	59
2026	6" diameter	"	84	.286		42	18.05		60.05	73
2028	8" diameter	Q-3	70	.457		67.50	29.50		97	118
2030	10" diameter		66	.485		113	31		144	171
2032	12" diameter		57	.561		162	36		198	232
2034	15" diameter	↓	49	.653	↓	232	42		274	320
8001	Fittings, bends and elbows									
8110	4" diameter	Q-1	13	1.231	Ea.	74	75		149	194
8112	5" diameter	Q-2	18	1.333		107	84		191	243
8114	6" diameter	"	17	1.412		126	89		215	272
8116	8" diameter	Q-3	11	2.909		355	187		542	670
8118	10" diameter		10	3.200		535	206		741	900
8120	12" diameter		9	3.556		715	229		944	1,125
8122	15" diameter	↓	7	4.571	↓	2,125	295		2,420	2,800
8500	Wyes and tees									
8510	4" diameter	Q-1	8	2	Ea.	122	122		244	315
8512	5" diameter	Q-2	12	2		207	126		333	415
8514	6" diameter	"	11	2.182		252	138		390	485
8516	8" diameter	Q-3	7	4.571		600	295		895	1,100
8518	10" diameter		6	5.333		985	345		1,330	1,600
8520	12" diameter		4	8		2,075	515		2,590	3,050
8522	15" diameter	↓	3	10.667	↓	4,075	685		4,760	5,525

33 31 11 – Public Sanitary Sewerage Gravity Piping

33 31 11.20 Sewage Collection, Plastic Pipe

		Crew	Daily Output	Labor-Hours	Unit	Material	2021 Bare Costs Labor	2021 Bare Costs Equipment	Total	Total Incl O&P
0010	**SEWAGE COLLECTION, PLASTIC PIPE**									
0020	Not including excavation & backfill									
3000	Piping, HDPE Corrugated Type S with watertight gaskets, 4" diameter	B-20	425	.056	L.F.	1.09	2.78		3.87	5.35
3020	6" diameter		400	.060		2.98	2.96		5.94	7.70
3040	8" diameter		380	.063		3.93	3.11		7.04	9
3060	10" diameter		370	.065		5.50	3.20		8.70	10.85
3080	12" diameter		340	.071		7.05	3.48		10.53	12.95
3100	15" diameter		300	.080		11.20	3.94		15.14	18.25
3120	18" diameter	B-21	275	.102		14.80	5.20	.69	20.69	25
3130	21" diameter		250	.112		22	5.70	.76	28.46	34
3140	24" diameter		250	.112		22	5.70	.76	28.46	34
3160	30" diameter		200	.140		28	7.15	.95	36.10	43
3180	36" diameter		180	.156		43.50	7.95	1.06	52.51	61
3200	42" diameter		175	.160		43.50	8.15	1.09	52.74	61.50
3220	48" diameter		170	.165		59.50	8.40	1.12	69.02	79.50
3240	54" diameter		160	.175		82.50	8.95	1.19	92.64	106
3260	60" diameter		150	.187		170	9.55	1.27	180.82	203
3300	Watertight elbows 12" diameter	B-20	11	2.182	Ea.	76.50	108		184.50	246
3320	15" diameter	"	9	2.667		117	131		248	325
3340	18" diameter	B-21	9	3.111		192	159	21	372	475
3350	21" diameter		9	3.111		305	159	21	485	595
3360	24" diameter		9	3.111		405	159	21	585	705
3380	30" diameter		8	3.500		645	179	24	848	1,000
3400	36" diameter		8	3.500		830	179	24	1,033	1,200
3420	42" diameter		6	4.667		1,050	238	32	1,320	1,550
3440	48" diameter		6	4.667		2,050	238	32	2,320	2,650
3460	Watertight tee 12" diameter	B-20	7	3.429		115	169		284	380
3480	15" diameter	"	6	4		298	197		495	625
3500	18" diameter	B-21	6	4.667		435	238	32	705	870
3520	24" diameter		5	5.600		830	286	38	1,154	1,375
3540	30" diameter		5	5.600		1,400	286	38	1,724	2,025
3560	36" diameter		4	7		1,300	355	47.50	1,702.50	2,025
3580	42" diameter		4	7		1,475	355	47.50	1,877.50	2,225
3600	48" diameter		4	7		1,575	355	47.50	1,977.50	2,325

33 31 11.25 Sewage Collection, Polyvinyl Chloride Pipe

		Crew	Daily Output	Labor-Hours	Unit	Material	2021 Bare Costs Labor	2021 Bare Costs Equipment	Total	Total Incl O&P
0010	**SEWAGE COLLECTION, POLYVINYL CHLORIDE PIPE**									
0020	Not including excavation or backfill									
2000	20' lengths, SDR 35, B&S, 4" diameter	B-20	375	.064	L.F.	2.21	3.16		5.37	7.15
2040	6" diameter		350	.069		4.26	3.38		7.64	9.75
2080	13' lengths, SDR 35, B&S, 8" diameter		335	.072		8.80	3.53		12.33	15
2120	10" diameter	B-21	330	.085		12.05	4.33	.58	16.96	20.50
2160	12" diameter		320	.088		17.05	4.47	.60	22.12	26
2170	14" diameter		280	.100		19.45	5.10	.68	25.23	30
2200	15" diameter		240	.117		22	5.95	.80	28.75	34
2250	16" diameter		220	.127		29	6.50	.87	36.37	42
2300	18" diameter		200	.140		34.50	7.15	.95	42.60	50
2350	20" diameter		195	.144		40.50	7.35	.98	48.83	56.50
2400	21" diameter		190	.147		48.50	7.50	1.01	57.01	66
2500	24" diameter		180	.156		63	7.95	1.06	72.01	82.50
2550	27" diameter		170	.165		80	8.40	1.12	89.52	102

502

For customer support on your Site Work & Landscape Costs with RSMeans Data, call 800.448.8182.

33 31 11.25 Sewage Collection, Polyvinyl Chloride Pipe	Crew	Daily Output	Labor-Hours	Unit	Material	2021 Bare Costs Labor	Equipment	Total	Total Incl O&P
2600	30" diameter	B-21	150	.187	L.F.	104	9.55	1.27	114.82	130
2650	36" diameter		130	.215		151	11	1.47	163.47	184
3040	Fittings, bends or elbows, 4" diameter	2 Skwk	24	.667	Ea.	8.60	38		46.60	67
3080	6" diameter		24	.667		15.80	38		53.80	75
3120	Tees, 4" diameter		16	1		19.20	57		76.20	107
3160	6" diameter		16	1		23.50	57		80.50	112
3200	Wyes, 4" diameter		16	1		18	57		75	106
3240	6" diameter		16	1		63.50	57		120.50	156
3661	Cap, 4"		48	.333		7.05	19.05		26.10	36.50
3670	6"		48	.333		25	19.05		44.05	56
3671	8"		40	.400		22.50	23		45.50	59
3672	10"		24	.667		69.50	38		107.50	134
3673	12"		20	.800		105	45.50		150.50	184
3674	15"	2 Skwk	18	.889	Ea.	169	51		220	263
3675	18"		18	.889		246	51		297	350
3676	21"		18	.889		685	51		736	830
3677	24"		18	.889		615	51		666	755
3745	Elbow, 90 degree, 8" diameter		20	.800		43.50	45.50		89	117
3750	10" diameter		12	1.333		199	76		275	335
3755	12" diameter		10	1.600		190	91.50		281.50	345
3760	15" diameter		9	1.778		400	102		502	595
3765	18" diameter		9	1.778		680	102		782	900
3770	21" diameter		9	1.778		1,175	102		1,277	1,425
3775	24" diameter		9	1.778		1,550	102		1,652	1,875
3795	Elbow, 45 degree, 8" diameter		20	.800		39	45.50		84.50	111
3805	12" diameter		10	1.600		146	91.50		237.50	298
3810	15" diameter		9	1.778		325	102		427	515
3815	18" diameter		9	1.778		525	102		627	730
3820	21" diameter		9	1.778		885	102		987	1,125
3825	24" diameter		9	1.778		1,200	102		1,302	1,475
3875	Elbow, 22-1/2 degree, 8" diameter		20	.800		39	45.50		84.50	112
3880	10" diameter		12	1.333		107	76		183	233
3885	12" diameter		12	1.333		140	76		216	269
3890	15" diameter		9	1.778		370	102		472	560
3895	18" diameter		9	1.778		540	102		642	750
3900	21" diameter		9	1.778		765	102		867	1,000
3905	24" diameter		9	1.778		1,100	102		1,202	1,375
3910	Tee, 8" diameter		13	1.231		52.50	70.50		123	164
3915	10" diameter		8	2		192	114		306	385
3920	12" diameter		7	2.286		293	131		424	520
3925	15" diameter		6	2.667		455	152		607	730
3930	18" diameter		6	2.667		755	152		907	1,050
3935	21" diameter		6	2.667		1,825	152		1,977	2,225
3940	24" diameter		6	2.667		3,250	152		3,402	3,775
4000	Piping, DWV PVC, no exc./bkfill., 10' L, Sch 40, 4" diameter	B-20	375	.064	L.F.	4.78	3.16		7.94	9.95
4010	6" diameter		350	.069		11.20	3.38		14.58	17.35
4020	8" diameter		335	.072		17.85	3.53		21.38	25
4030	Fittings, 1/4 bend DWV PVC, 4" diameter		19	1.263	Ea.	43.50	62.50		106	141
4040	6" diameter		12	2		197	98.50		295.50	365
4050	8" diameter		11	2.182		197	108		305	375
4060	1/8 bend DWV PVC, 4" diameter		19	1.263		36	62.50		98.50	133
4070	6" diameter		12	2		133	98.50		231.50	295
4080	8" diameter		11	2.182		87	108		195	257

33 31 Sanitary Sewerage Piping

33 31 11 – Public Sanitary Sewerage Gravity Piping

33 31 11.25 Sewage Collection, Polyvinyl Chloride Pipe

		Crew	Daily Output	Labor-Hours	Unit	Material	2021 Bare Costs Labor	Equipment	Total	Total Incl O&P
4090	Tee DWV PVC, 4" diameter	B-20	12	2	Ea.	58.50	98.50		157	213
4100	6" diameter		10	2.400		235	118		353	435
4110	8" diameter		9	2.667		520	131		651	770

33 31 11.30 Piping, Drainage and Sewage, Vitrified Clay C700

		Crew	Daily Output	Labor-Hours	Unit	Material	2021 Bare Costs Labor	Equipment	Total	Total Incl O&P
0010	**PIPING, DRAINAGE & SEWAGE, VITRIFIED CLAY C700**									
0020	Not including excavation or backfill,									
0025	Piping									
4030	Extra strength, compression joints, C425									
5000	4" diameter x 4' long	B-20	265	.091	L.F.	5.10	4.46		9.56	12.30
5020	6" diameter x 5' long	"	200	.120		7.30	5.90		13.20	16.90
5040	8" diameter x 5' long	B-21	200	.140		11.45	7.15	.95	19.55	24.50
5060	10" diameter x 5' long		190	.147		17.10	7.50	1.01	25.61	31
5080	12" diameter x 6' long		150	.187		16.50	9.55	1.27	27.32	34
5100	15" diameter x 7' long		110	.255		31	13	1.74	45.74	55.50
5120	18" diameter x 7' long		88	.318		51	16.25	2.17	69.42	83
5140	24" diameter x 7' long		45	.622		79.50	32	4.24	115.74	140
5160	30" diameter x 7' long	B-22	31	.968		82.50	50	9.25	141.75	176
5180	36" diameter x 7' long	"	20	1.500		121	77.50	14.35	212.85	266

33 31 11.37 Centrif. Cst. Fbgs-Reinf. Polymer Mort. Util. Pipe

		Crew	Daily Output	Labor-Hours	Unit	Material	2021 Bare Costs Labor	Equipment	Total	Total Incl O&P
0010	**CENTRIF. CST. FBGS-REINF. POLYMER MORT. UTIL. PIPE**									
0020	Not including excavation or backfill									
3348	48" diameter	B-13B	40	1.400	L.F.	181	67.50	25	273.50	325
3354	54" diameter		40	1.400		295	67.50	25	387.50	455
3363	63" diameter		20	2.800		365	135	49.50	549.50	655
3418	18" diameter		60	.933		128	45	16.50	189.50	226
3424	24" diameter		60	.933		141	45	16.50	202.50	240
3436	36" diameter		40	1.400		159	67.50	25	251.50	305
3444	44" diameter		40	1.400		175	67.50	25	267.50	320
3448	48" diameter		40	1.400		195	67.50	25	287.50	345

33 32 Sanitary Sewerage Equipment

33 32 13 – Packaged Wastewater Pumping Stations

33 32 13.13 Packaged Sewage Lift Stations

		Crew	Daily Output	Labor-Hours	Unit	Material	2021 Bare Costs Labor	Equipment	Total	Total Incl O&P
0010	**PACKAGED SEWAGE LIFT STATIONS**									
2500	Sewage lift station, 200,000 GPD	E-8	.20	520	Ea.	237,500	31,100	11,500	280,100	321,500
2510	500,000 GPD		.15	684		283,000	40,900	15,200	339,100	390,500
2520	800,000 GPD		.13	813		317,500	48,500	18,000	384,000	444,000

33 34 Onsite Wastewater Disposal

33 34 13 – Septic Tanks

33 34 13.13 Concrete Septic Tanks

		Crew	Daily Output	Labor-Hours	Unit	Material	2021 Bare Costs Labor	Equipment	Total	Total Incl O&P
0010	**CONCRETE SEPTIC TANKS**									
0011	Not including excavation or piping									
0015	Septic tanks, precast, 1,000 gallon	B-21	8	3.500	Ea.	1,000	179	24	1,203	1,400
0020	1,250 gallon		8	3.500		1,500	179	24	1,703	1,950
0060	1,500 gallon		7	4		1,675	204	27.50	1,906.50	2,175
0100	2,000 gallon		5	5.600		2,550	286	38	2,874	3,300
0140	2,500 gallon	B-21	5	5.600	Ea.	2,375	286	38	2,699	3,100
0180	4,000 gallon	"	4	7		6,825	355	47.50	7,227.50	8,125

33 34 Onsite Wastewater Disposal

33 34 13 – Septic Tanks

33 34 13.13 Concrete Septic Tanks

		Crew	Daily Output	Labor-Hours	Unit	Material	2021 Bare Costs Labor	Equipment	Total	Total Incl O&P
0200	5,000 gallon	B-13	3.50	16	Ea.	7,350	770	168	8,288	9,425
0220	5,000 gallon, 4 piece	"	3	18.667		9,000	900	196	10,096	11,500
0300	15,000 gallon, 4 piece	B-13B	1.70	32.941		25,700	1,600	580	27,880	31,300
0400	25,000 gallon, 4 piece		1.10	50.909		48,400	2,450	900	51,750	57,500
0500	40,000 gallon, 4 piece		.80	70		53,500	3,375	1,250	58,125	65,500
0520	50,000 gallon, 5 piece	B-13C	.60	93.333		61,500	4,500	3,850	69,850	79,000
0640	75,000 gallon, cast in place	C-14C	.25	448		75,000	23,400	108	98,508	117,500
0660	100,000 gallon	"	.15	747		93,000	39,000	179	132,179	160,000
0800	Septic tanks, precast, 2 compartment, 1,000 gallon	B-21	8	3.500		705	179	24	908	1,075
0805	1,250 gallon		8	3.500		1,075	179	24	1,278	1,475
0810	1,500 gallon		7	4		1,600	204	27.50	1,831.50	2,075
0815	2,000 gallon		5	5.600		2,575	286	38	2,899	3,325
0820	2,500 gallon		5	5.600		3,250	286	38	3,574	4,050
0890	Septic tanks, butyl rope sealant				L.F.	.41			.41	.45
0900	Concrete riser 24" x 12" with standard lid	1 Clab	6	1.333	Ea.	88.50	59		147.50	186
0905	24" x 12" with heavy duty lid		6	1.333		112	59		171	212
0910	24" x 8" with standard lid		6	1.333		83	59		142	180
0915	24" x 8" with heavy duty lid		6	1.333		107	59		166	207
0917	24" x 12" extension		12	.667		59.50	29.50		89	110
0920	24" x 8" extension		12	.667		54.50	29.50		84	104
0950	HDPE riser 20" x 12" with standard lid		8	1		126	44.50		170.50	206
0951	HDPE 20" x 12" with heavy duty lid		8	1		138	44.50		182.50	219
0955	HDPE 24" x 12" with standard lid		8	1		133	44.50		177.50	213
0956	HDPE 24" x 12" with heavy duty lid		8	1		158	44.50		202.50	241
0960	HDPE 20" x 6" extension		48	.167		32	7.40		39.40	46
0962	HDPE 20" x 12" extension		48	.167		56.50	7.40		63.90	73
0965	HDPE 24" x 6" extension		48	.167		42	7.40		49.40	57
0967	HDPE 24" x 12" extension		48	.167		68	7.40		75.40	86
1150	Leaching field chambers, 13' x 3'-7" x 1'-4", standard	B-13	16	3.500		475	169	36.50	680.50	820
1200	Heavy duty, 8' x 4' x 1'-6"		14	4		325	193	42	560	690
1300	13' x 3'-9" x 1'-6"		12	4.667		1,400	225	49	1,674	1,925
1350	20' x 4' x 1'-6"		5	11.200		1,250	540	117	1,907	2,300
1400	Leaching pit, precast concrete, 3' diameter, 3' deep	B-21	8	3.500		710	179	24	913	1,075
1500	6' diameter, 3' section		4.70	5.957		1,100	305	40.50	1,445.50	1,725
1600	Leaching pit, 6'-6" diameter, 6' deep		5	5.600		880	286	38	1,204	1,425
1620	8' deep		4	7		1,150	355	47.50	1,552.50	1,875
1700	8' diameter, H-20 load, 6' deep		4	7		1,400	355	47.50	1,802.50	2,125
1720	8' deep		3	9.333		2,575	475	63.50	3,113.50	3,600
2000	Velocity reducing pit, precast conc., 6' diameter, 3' deep		4.70	5.957		1,875	305	40.50	2,220.50	2,550

33 34 13.33 Polyethylene Septic Tanks

		Crew	Daily Output	Labor-Hours	Unit	Material	2021 Bare Costs Labor	Equipment	Total	Total Incl O&P
0010	**POLYETHYLENE SEPTIC TANKS**									
0015	High density polyethylene, 1,000 gallon	B-21	8	3.500	Ea.	1,600	179	24	1,803	2,050
0020	1,250 gallon		8	3.500		1,325	179	24	1,528	1,750
0025	1,500 gallon		7	4		1,400	204	27.50	1,631.50	1,850
1015	Septic tanks, HDPE, 2 compartment, 1,000 gallon		8	3.500		1,475	179	24	1,678	1,925
1020	1,500 gallon		7	4		1,600	204	27.50	1,831.50	2,075

33 34 16 – Septic Tank Effluent Filters

33 34 16.13 Septic Tank Gravity Effluent Filters

		Crew	Daily Output	Labor-Hours	Unit	Material	2021 Bare Costs Labor	Equipment	Total	Total Incl O&P
0010	**SEPTIC TANK GRAVITY EFFLUENT FILTERS**									
3000	Effluent filter, 4" diameter	1 Skwk	8	1	Ea.	36	57		93	126
3020	6" diameter		7	1.143		43.50	65.50		109	146
3030	8" diameter		7	1.143		254	65.50		319.50	375

For customer support on your Site Work & Landscape Costs with RSMeans Data, call 800.448.8182.

505

33 34 Onsite Wastewater Disposal

33 34 16 – Septic Tank Effluent Filters

33 34 16.13 Septic Tank Gravity Effluent Filters		Crew	Daily Output	Labor-Hours	Unit	Material	2021 Bare Costs Labor	Equipment	Total	Total Incl O&P
3040	8" diameter, very fine	1 Skwk	7	1.143	Ea.	505	65.50		570.50	655
3050	10" diameter, very fine		6	1.333		240	76		316	380
3060	10" diameter		6	1.333		274	76		350	415
3080	12" diameter		6	1.333		660	76		736	840
3090	15" diameter		5	1.600		1,100	91.50		1,191.50	1,325

33 34 51 – Drainage Field Systems

33 34 51.10 Drainage Field Excavation and Fill

		Crew	Daily Output	Labor-Hours	Unit	Material	2021 Bare Costs Labor	Equipment	Total	Total Incl O&P
0010	**DRAINAGE FIELD EXCAVATION AND FILL**									
2200	Septic tank & drainage field excavation with 3/4 C.Y. backhoe	B-12F	145	.110	C.Y.		5.85	4.84	10.69	14
2400	4' trench for disposal field, 3/4 C.Y. backhoe	"	335	.048	L.F.		2.53	2.09	4.62	6.05
2600	Gravel fill, run of bank	B-6	150	.160	C.Y.	26.50	7.70	1.44	35.64	42.50
2800	Crushed stone, 3/4"	"	150	.160	"	41	7.70	1.44	50.14	58

33 34 51.13 Utility Septic Tank Tile Drainage Field

		Crew	Daily Output	Labor-Hours	Unit	Material	2021 Bare Costs Labor	Equipment	Total	Total Incl O&P
0010	**UTILITY SEPTIC TANK TILE DRAINAGE FIELD**									
0015	Distribution box, concrete, 5 outlets	2 Clab	20	.800	Ea.	75	35.50		110.50	136
0020	7 outlets		16	1		93.50	44.50		138	170
0025	9 outlets		8	2		530	89		619	720
0115	Distribution boxes, HDPE, 5 outlets		20	.800		75	35.50		110.50	136
0117	6 outlets		15	1.067		75	47.50		122.50	153
0118	7 outlets		15	1.067		74.50	47.50		122	153
0120	8 outlets		10	1.600		79	71		150	193
0240	Distribution boxes, outlet flow leveler	1 Clab	50	.160		2.99	7.10		10.09	13.90
0260	Bull run valve 4" diam.	2 Skwk	16	1		66.50	57		123.50	160
0261	Bull run valve key					15.70			15.70	17.30
0300	Precast concrete, galley, 4' x 4' x 4'	B-21	16	1.750		246	89.50	11.95	347.45	420
0350	HDPE infiltration chamber 12" H x 15" W	2 Clab	300	.053	L.F.	7.05	2.37		9.42	11.30
0351	12" H x 15" W end cap	1 Clab	32	.250	Ea.	19.05	11.10		30.15	37.50
0355	chamber 12" H x 22" W	2 Clab	300	.053	L.F.	6.70	2.37		9.07	10.90
0356	12" H x 22" W end cap	1 Clab	32	.250	Ea.	23.50	11.10		34.60	42
0360	chamber 13" H x 34" W	2 Clab	300	.053	L.F.	14.45	2.37		16.82	19.40
0361	13" H x 34" W end cap	1 Clab	32	.250	Ea.	51	11.10		62.10	72.50
0365	chamber 16" H x 34" W	2 Clab	300	.053	L.F.	19.40	2.37		21.77	25
0366	16" H x 34" W end cap	1 Clab	32	.250	Ea.	16.90	11.10		28	35
0370	chamber 8" H x 16" W	2 Clab	300	.053	L.F.	10	2.37		12.37	14.55
0371	8" H x 16" W end cap	1 Clab	32	.250	Ea.	11.20	11.10		22.30	29

33 41 Subdrainage

33 41 16 – Subdrainage Piping

33 41 16.10 Piping, Subdrainage, Vitrified Clay

		Crew	Daily Output	Labor-Hours	Unit	Material	2021 Bare Costs Labor	Equipment	Total	Total Incl O&P
0010	**PIPING, SUBDRAINAGE, VITRIFIED CLAY**									
0020	Not including excavation and backfill									
1000	Foundation drain, 6" diameter	B-14	315	.152	L.F.	6.35	7.10	.69	14.14	18.35
1010	8" diameter		290	.166		9.50	7.70	.74	17.94	23
1020	12" diameter		275	.175		20.50	8.15	.79	29.44	35.50
3000	Perforated, 5' lengths, C700, 4" diameter		400	.120		5.30	5.60	.54	11.44	14.80
3020	6" diameter		315	.152		6.85	7.10	.69	14.64	18.90
3040	8" diameter		290	.166		7.55	7.70	.74	15.99	20.50
3060	12" diameter		275	.175		20.50	8.15	.79	29.44	36

33 41 Subdrainage

33 41 16 – Subdrainage Piping

33 41 16.25 Piping, Subdrainage, Corrugated Metal	Crew	Daily Output	Labor-Hours	Unit	Material	2021 Bare Costs Labor	Equipment	Total	Total Incl O&P
0010 **PIPING, SUBDRAINAGE, CORRUGATED METAL**									
0021 Not including excavation and backfill									
2010 Aluminum, perforated									
2020 6" diameter, 18 ga.	B-20	380	.063	L.F.	6.35	3.11		9.46	11.65
2200 8" diameter, 16 ga.	"	370	.065		10.45	3.20		13.65	16.30
2220 10" diameter, 16 ga.	B-21	360	.078		13.05	3.97	.53	17.55	21
2240 12" diameter, 16 ga.		285	.098		14.65	5	.67	20.32	24.50
2260 18" diameter, 16 ga.	↓	205	.137	↓	22	6.95	.93	29.88	35.50
3000 Uncoated galvanized, perforated									
3020 6" diameter, 18 ga.	B-20	380	.063	L.F.	6.75	3.11		9.86	12.10
3200 8" diameter, 16 ga.	"	370	.065		8.10	3.20		11.30	13.75
3220 10" diameter, 16 ga.	B-21	360	.078		8.60	3.97	.53	13.10	16
3240 12" diameter, 16 ga.		285	.098		9.60	5	.67	15.27	18.80
3260 18" diameter, 16 ga.	↓	205	.137	↓	14.65	6.95	.93	22.53	27.50
4000 Steel, perforated, asphalt coated									
4020 6" diameter, 18 ga.	B-20	380	.063	L.F.	6	3.11		9.11	11.25
4030 8" diameter, 18 ga.	"	370	.065		9.30	3.20		12.50	15.05
4040 10" diameter, 16 ga.	B-21	360	.078		9.40	3.97	.53	13.90	16.90
4050 12" diameter, 16 ga.		285	.098		9.40	5	.67	15.07	18.60
4060 18" diameter, 16 ga.	↓	205	.137	↓	16.85	6.95	.93	24.73	30

33 41 16.30 Piping, Subdrainage, Plastic	Crew	Daily Output	Labor-Hours	Unit	Material	2021 Bare Costs Labor	Equipment	Total	Total Incl O&P
0010 **PIPING, SUBDRAINAGE, PLASTIC**									
0020 Not including excavation and backfill									
2100 Perforated PVC, 4" diameter	B-14	314	.153	L.F.	2.21	7.10	.69	10	13.80
2110 6" diameter		300	.160		4.26	7.45	.72	12.43	16.60
2120 8" diameter		290	.166		8	7.70	.74	16.44	21
2130 10" diameter		280	.171		11.15	8	.77	19.92	25
2140 12" diameter	↓	270	.178	↓	15.75	8.30	.80	24.85	30.50

33 41 16.35 Piping, Subdrain., Corr. Plas. Tubing, Perf. or Plain	Crew	Daily Output	Labor-Hours	Unit	Material	2021 Bare Costs Labor	Equipment	Total	Total Incl O&P
0010 **PIPING, SUBDRAINAGE, CORR. PLASTIC TUBING, PERF. OR PLAIN**									
0020 In rolls, not including excavation and backfill									
0030 3" diameter	2 Clab	1200	.013	L.F.	.63	.59		1.22	1.57
0040 4" diameter		1200	.013		.81	.59		1.40	1.77
0041 With silt sock		1200	.013		1.26	.59		1.85	2.27
0050 5" diameter		900	.018		1.34	.79		2.13	2.65
0060 6" diameter		900	.018		1.82	.79		2.61	3.18
0080 8" diameter	↓	700	.023	↓	5.30	1.01		6.31	7.30
0200 Fittings									
0230 Elbows, 3" diameter	1 Clab	32	.250	Ea.	4.54	11.10		15.64	21.50
0240 4" diameter		32	.250		5.40	11.10		16.50	22.50
0250 5" diameter		32	.250		6.40	11.10		17.50	23.50
0260 6" diameter		32	.250		9.30	11.10		20.40	27
0280 8" diameter		32	.250		32.50	11.10		43.60	52.50
0330 Tees, 3" diameter		27	.296		4.21	13.15		17.36	24.50
0340 4" diameter		27	.296		5.55	13.15		18.70	26
0350 5" diameter		27	.296		7.65	13.15		20.80	28
0360 6" diameter		27	.296		11.75	13.15		24.90	32.50
0370 6" x 6" x 4"		27	.296		56.50	13.15		69.65	81.50
0380 8" diameter		27	.296		36.50	13.15		49.65	60
0390 8" x 8" x 6"		27	.296		86	13.15		99.15	114
0430 End cap, 3" diameter		32	.250		2.05	11.10		13.15	18.80
0440 4" diameter	↓	32	.250	↓	2.45	11.10		13.55	19.25

33 41 Subdrainage

33 41 16 – Subdrainage Piping

33 41 16.35 Piping, Subdrain., Corr. Plas. Tubing, Perf. or Plain	Crew	Daily Output	Labor-Hours	Unit	Material	2021 Bare Costs Labor	Equipment	Total	Total Incl O&P
0460	6" diameter	1 Clab	32	.250	Ea.	5.10	11.10		16.20	22
0480	8" diameter		32	.250		16.75	11.10		27.85	35
0530	Coupler, 3" diameter		32	.250		3.07	11.10		14.17	19.95
0540	4" diameter		32	.250		2.07	11.10		13.17	18.85
0550	5" diameter		32	.250		3.85	11.10		14.95	21
0560	6" diameter		32	.250		3.85	11.10		14.95	21
0580	8" diameter	↓	32	.250		7	11.10		18.10	24.50
0590	Heavy duty highway type, add					10%				
0660	Reducer, 6" to 4"	1 Clab	32	.250		5.95	11.10		17.05	23
0680	8" to 6"		32	.250		51.50	11.10		62.60	73.50
0730	"Y" fitting, 3" diameter	↓	27	.296	↓	4.47	13.15		17.62	24.50
0740	4" diameter	1 Clab	27	.296	Ea.	9.10	13.15		22.25	29.50
0750	5" diameter		27	.296		9.80	13.15		22.95	30.50
0760	6" diameter		27	.296		13.70	13.15		26.85	35
0780	8" diameter	↓	27	.296	↓	65.50	13.15		78.65	91.50
0860	Silt sock only for above tubing, 6" diameter				L.F.	3.59			3.59	3.95
0880	8" diameter				"	5.90			5.90	6.50

33 41 23 – Drainage Layers

33 41 23.19 Geosynthetic Drainage Layers

		Crew	Daily Output	Labor-Hours	Unit	Material	2021 Bare Costs Labor	Equipment	Total	Total Incl O&P
0010	**GEOSYNTHETIC DRAINAGE LAYERS**									
0100	Fabric, laid in trench, polypropylene, ideal conditions	2 Clab	2400	.007	S.Y.	1.56	.30		1.86	2.16
0110	Adverse conditions		1600	.010	"	1.56	.44		2	2.38
0170	Fabric ply bonded to 3 dimen. nylon mat, 0.4" thick, ideal conditions		2000	.008	S.F.	.27	.36		.63	.83
0180	Adverse conditions		1200	.013	"	.34	.59		.93	1.25
0185	Soil drainage mat on vertical wall, 0.44" thick		265	.060	S.Y.	1.75	2.68		4.43	5.95
0188	0.25" thick		300	.053	"	.85	2.37		3.22	4.46
0190	0.8" thick, ideal conditions		2400	.007	S.F.	.27	.30		.57	.74
0200	Adverse conditions		1600	.010	"	.41	.44		.85	1.11
0300	Drainage material, 3/4" gravel fill in trench	B-6	260	.092	C.Y.	32	4.44	.83	37.27	42.50
0400	Pea stone	"	260	.092	"	32	4.44	.83	37.27	42.50

33 42 Stormwater Conveyance

33 42 11 – Stormwater Gravity Piping

33 42 11.40 Piping, Storm Drainage, Corrugated Metal

		Crew	Daily Output	Labor-Hours	Unit	Material	2021 Bare Costs Labor	Equipment	Total	Total Incl O&P
0010	**PIPING, STORM DRAINAGE, CORRUGATED METAL**									
0020	Not including excavation or backfill									
2000	Corrugated metal pipe, galvanized									
2020	Bituminous coated with paved invert, 20' lengths									
2040	8" diameter, 16 ga.	B-14	330	.145	L.F.	7.25	6.80	.65	14.70	18.80
2060	10" diameter, 16 ga.		260	.185		7.55	8.60	.83	16.98	22
2080	12" diameter, 16 ga.		210	.229		12	10.65	1.03	23.68	30
2100	15" diameter, 16 ga.		200	.240		12.70	11.20	1.08	24.98	32
2120	18" diameter, 16 ga.		190	.253		18.75	11.75	1.14	31.64	39.50
2140	24" diameter, 14 ga.	↓	160	.300		20.50	13.95	1.35	35.80	45.50
2160	30" diameter, 14 ga.	B-13	120	.467		26.50	22.50	4.89	53.89	68
2180	36" diameter, 12 ga.		120	.467		29.50	22.50	4.89	56.89	71.50
2200	48" diameter, 12 ga.	↓	100	.560		40	27	5.85	72.85	91
2220	60" diameter, 10 ga.	B-13B	75	.747		70	36	13.20	119.20	146
2240	72" diameter, 8 ga.	"	45	1.244	↓	73	60	22	155	194
2250	End sections, 8" diameter, 16 ga.	B-14	20	2.400	Ea.	44	112	10.80	166.80	227

508

33 42 11.40 Piping, Storm Drainage, Corrugated Metal		Crew	Daily Output	Labor-Hours	Unit	Material	2021 Bare Costs Labor	Equipment	Total	Total Incl O&P
2255	10" diameter, 16 ga.	B-14	20	2.400	Ea.	51.50	112	10.80	174.30	235
2260	12" diameter, 16 ga.		18	2.667		105	124	12	241	315
2265	15" diameter, 16 ga.		18	2.667		208	124	12	344	425
2270	18" diameter, 16 ga.	B-14	16	3	Ea.	242	140	13.50	395.50	490
2275	24" diameter, 16 ga.	B-13	16	3.500		320	169	36.50	525.50	645
2280	30" diameter, 16 ga.		14	4		520	193	42	755	910
2285	36" diameter, 14 ga.		14	4		650	193	42	885	1,050
2290	48" diameter, 14 ga.		10	5.600		1,875	270	58.50	2,203.50	2,525
2292	60" diameter, 14 ga.		6	9.333		1,850	450	98	2,398	2,825
2294	72" diameter, 14 ga.	B-13B	5	11.200		2,850	540	198	3,588	4,150
2300	Bends or elbows, 8" diameter	B-14	28	1.714		103	80	7.70	190.70	242
2320	10" diameter		25	1.920		128	89.50	8.65	226.15	284
2340	12" diameter, 16 ga.		23	2.087		151	97	9.40	257.40	320
2342	18" diameter, 16 ga.		20	2.400		216	112	10.80	338.80	415
2344	24" diameter, 14 ga.		16	3		360	140	13.50	513.50	620
2346	30" diameter, 14 ga.		15	3.200		440	149	14.40	603.40	725
2348	36" diameter, 14 ga.	B-13	15	3.733		575	180	39	794	945
2350	48" diameter, 12 ga.	"	12	4.667		740	225	49	1,014	1,200
2352	60" diameter, 10 ga.	B-13B	10	5.600		995	270	99	1,364	1,625
2354	72" diameter, 10 ga.	"	6	9.333		1,225	450	165	1,840	2,200
2360	Wyes or tees, 8" diameter	B-14	25	1.920		146	89.50	8.65	244.15	305
2380	10" diameter		21	2.286		182	106	10.30	298.30	370
2400	12" diameter, 16 ga.		19	2.526		208	118	11.35	337.35	420
2410	18" diameter, 16 ga.		16	3		289	140	13.50	442.50	545
2412	24" diameter, 14 ga.		16	3		525	140	13.50	678.50	800
2414	30" diameter, 14 ga.	B-13	12	4.667		685	225	49	959	1,150
2416	36" diameter, 14 ga.		11	5.091		795	246	53.50	1,094.50	1,300
2418	48" diameter, 12 ga.		10	5.600		1,075	270	58.50	1,403.50	1,650
2420	60" diameter, 10 ga.	B-13B	8	7		1,575	340	124	2,039	2,375
2422	72" diameter, 10 ga.	"	5	11.200		1,900	540	198	2,638	3,100
2500	Galvanized, uncoated, 20' lengths									
2520	8" diameter, 16 ga.	B-14	355	.135	L.F.	8.20	6.30	.61	15.11	19.05
2540	10" diameter, 16 ga.		280	.171		8.15	8	.77	16.92	22
2560	12" diameter, 16 ga.		220	.218		8.90	10.15	.98	20.03	26
2580	15" diameter, 16 ga.		220	.218		11.60	10.15	.98	22.73	29
2600	18" diameter, 16 ga.		205	.234		12.90	10.90	1.05	24.85	31.50
2610	21" diameter, 16 ga.		205	.234		19.25	10.90	1.05	31.20	38.50
2620	24" diameter, 14 ga.		175	.274		24.50	12.80	1.23	38.53	47
2630	27" diameter, 14 ga.		170	.282		24.50	13.15	1.27	38.92	48
2640	30" diameter, 14 ga.	B-13	130	.431		28	21	4.51	53.51	66.50
2660	36" diameter, 12 ga.		130	.431		32	21	4.51	57.51	71
2680	48" diameter, 12 ga.		110	.509		47.50	24.50	5.35	77.35	94.50
2690	60" diameter, 10 ga.	B-13B	78	.718		78	34.50	12.70	125.20	151
2695	72" diameter, 10 ga.	B-13B	60	.933	L.F.	90	45	16.50	151.50	184
2711	Bends or elbows, 12" diameter, 16 ga.	B-14	30	1.600	Ea.	123	74.50	7.20	204.70	254
2712	15" diameter, 16 ga.		25.04	1.917		156	89.50	8.65	254.15	315
2714	18" diameter, 16 ga.		20	2.400		175	112	10.80	297.80	370
2716	24" diameter, 14 ga.		16	3		265	140	13.50	418.50	515
2718	30" diameter, 14 ga.		15	3.200		355	149	14.40	518.40	630
2720	36" diameter, 14 ga.	B-13	15	3.733		475	180	39	694	830
2722	48" diameter, 12 ga.		12	4.667		640	225	49	914	1,100
2724	60" diameter, 10 ga.		10	5.600		995	270	58.50	1,323.50	1,575
2726	72" diameter, 10 ga.		6	9.333		1,250	450	98	1,798	2,150

33 42 11.40 Piping, Storm Drainage, Corrugated Metal	Crew	Daily Output	Labor-Hours	Unit	Material	2021 Bare Costs Labor	Equipment	Total	Total Incl O&P
2728	Wyes or tees, 12" diameter, 16 ga.	B-14	22.48	2.135	Ea.	158	99.50	9.60	267.10	335
2730	18" diameter, 16 ga.		15	3.200		245	149	14.40	408.40	505
2732	24" diameter, 14 ga.		15	3.200		425	149	14.40	588.40	710
2734	30" diameter, 14 ga.	▼	14	3.429		550	160	15.45	725.45	860
2736	36" diameter, 14 ga.	B-13	14	4		700	193	42	935	1,100
2738	48" diameter, 12 ga.		12	4.667		1,025	225	49	1,299	1,525
2740	60" diameter, 10 ga.		10	5.600		1,475	270	58.50	1,803.50	2,100
2742	72" diameter, 10 ga.	▼	6	9.333		1,775	450	98	2,323	2,725
2780	End sections, 8" diameter	B-14	35	1.371		102	64	6.15	172.15	214
2785	10" diameter		35	1.371		80.50	64	6.15	150.65	191
2790	12" diameter		35	1.371		96	64	6.15	166.15	207
2800	18" diameter	▼	30	1.600		112	74.50	7.20	193.70	242
2810	24" diameter	B-13	25	2.240		189	108	23.50	320.50	395
2820	30" diameter		25	2.240		340	108	23.50	471.50	560
2825	36" diameter		20	2.800		455	135	29.50	619.50	735
2830	48" diameter	▼	10	5.600		990	270	58.50	1,318.50	1,575
2835	60" diameter	B-13B	5	11.200		1,950	540	198	2,688	3,150
2840	72" diameter	"	4	14		2,250	675	248	3,173	3,750
2850	Couplings, 12" diameter					14.40			14.40	15.85
2855	18" diameter					17.80			17.80	19.55
2860	24" diameter					22.50			22.50	25
2865	30" diameter					27			27	29.50
2870	36" diameter					40.50			40.50	45
2875	48" diameter					59.50			59.50	65.50
2880	60" diameter					56.50			56.50	62
2885	72" diameter	▼			▼	64			64	70.50

33 42 11.50 Piping, Drainage & Sewage, Corrug. HDPE Type S

		Crew	Daily Output	Labor-Hours	Unit	Material	2021 Bare Costs Labor	Equipment	Total	Total Incl O&P
0010	**PIPING, DRAINAGE & SEWAGE, CORRUGATED HDPE TYPE S**									
0020	Not including excavation & backfill, bell & spigot									
1000	With gaskets, 4" diameter	B-20	425	.056	L.F.	.95	2.78		3.73	5.20
1010	6" diameter		400	.060		2.59	2.96		5.55	7.30
1020	8" diameter		380	.063		3.42	3.11		6.53	8.40
1030	10" diameter		370	.065		4.78	3.20		7.98	10.05
1040	12" diameter		340	.071		6.10	3.48		9.58	11.95
1050	15" diameter	▼	300	.080		9.75	3.94		13.69	16.65
1060	18" diameter	B-21	275	.102		12.85	5.20	.69	18.74	22.50
1070	24" diameter		250	.112		19.15	5.70	.76	25.61	30.50
1080	30" diameter		200	.140		24.50	7.15	.95	32.60	39
1090	36" diameter		180	.156		38	7.95	1.06	47.01	55
1100	42" diameter		175	.160		38	8.15	1.09	47.24	55
1110	48" diameter		170	.165		51.50	8.40	1.12	61.02	71
1120	54" diameter		160	.175		71.50	8.95	1.19	81.64	93.50
1130	60" diameter	▼	150	.187	▼	148	9.55	1.27	158.82	179
1135	Add 15% to material pipe cost for water tight connection bell & spigot									
1140	HDPE type S, elbows 12" diameter	B-20	11	2.182	Ea.	66.50	108		174.50	235
1150	15" diameter	"	9	2.667		102	131		233	310
1160	18" diameter	B-21	9	3.111		167	159	21	347	445
1170	24" diameter		9	3.111		355	159	21	535	650
1180	30" diameter		8	3.500		560	179	24	763	915
1190	36" diameter		8	3.500		725	179	24	928	1,100
1200	42" diameter		6	4.667		915	238	32	1,185	1,400
1220	48" diameter	▼	6	4.667		1,775	238	32	2,045	2,350

510

For customer support on your Site Work & Landscape Costs with RSMeans Data, call 800.448.8182.

33 42 Stormwater Conveyance

33 42 11 – Stormwater Gravity Piping

33 42 11.50 Piping, Drainage & Sewage, Corrug. HDPE Type S		Crew	Daily Output	Labor-Hours	Unit	Material	2021 Bare Costs Labor	Equipment	Total	Total Incl O&P
1240	HDPE type S, Tee 12" diameter	B-20	7	3.429	Ea.	100	169		269	365
1260	15" diameter	"	6	4		259	197		456	580
1280	18" diameter	B-21	6	4.667		380	238	32	650	805
1300	24" diameter		5	5.600		720	286	38	1,044	1,250
1320	30" diameter		5	5.600		1,225	286	38	1,549	1,825
1340	36" diameter		4	7		1,125	355	47.50	1,527.50	1,850
1360	42" diameter		4	7		1,300	355	47.50	1,702.50	2,025
1380	48" diameter	▼	4	7	▼	1,375	355	47.50	1,777.50	2,100
1400	Add to basic installation cost for each split coupling joint									
1402	HDPE type S, split coupling, 12" diameter	B-20	17	1.412	Ea.	10.80	69.50		80.30	116
1420	15" diameter		15	1.600		21	79		100	141
1440	18" diameter		13	1.846		29	91		120	168
1460	24" diameter		12	2		43	98.50		141.50	195
1480	30" diameter		10	2.400		81.50	118		199.50	267
1500	36" diameter		9	2.667		136	131		267	345
1520	42" diameter		8	3		162	148		310	400
1540	48" diameter	▼	8	3	▼	191	148		339	430

33 42 11.60 Sewage/Drainage Collection, Concrete Pipe

		Crew	Daily Output	Labor-Hours	Unit	Material	2021 Bare Costs Labor	Equipment	Total	Total Incl O&P
0010	**SEWAGE/DRAINAGE COLLECTION, CONCRETE PIPE**									
0020	Not including excavation or backfill									
0050	Box culvert, cast in place, 6' x 6'	C-15	16	4.500	L.F.	238	231		469	605
0060	8' x 8'		14	5.143		350	264		614	775
0070	12' x 12'	▼	10	7.200		690	370		1,060	1,300
0100	Box culvert, precast, base price, 8' long, 6' x 3'	B-69	140	.343		217	16.80	10.40	244.20	275
0150	6' x 7'		125	.384		296	18.80	11.65	326.45	365
0200	8' x 3'		113	.425		325	21	12.90	358.90	405
0250	8' x 8'		100	.480		380	23.50	14.60	418.10	470
0300	10' x 3'		110	.436		375	21.50	13.25	409.75	455
0350	10' x 8'		80	.600		525	29.50	18.25	572.75	640
0400	12' x 3'		100	.480		900	23.50	14.60	938.10	1,050
0450	12' x 8'	▼	67	.716	▼	895	35	22	952	1,050
0500	Set up charge at plant, add to base price				Job	6,350			6,350	6,975
0510	Inserts and keyway, add				Ea.	665			665	735
0520	Sloped or skewed end, add				"	1,125			1,125	1,250
1000	Non-reinforced pipe, extra strength, B&S or T&G joints									
1010	6" diameter	B-14	265.04	.181	L.F.	8.05	8.45	.81	17.31	22.50
1020	8" diameter		224	.214		8.90	10	.96	19.86	25.50
1030	10" diameter		216	.222		9.85	10.35	1	21.20	27.50
1040	12" diameter		200	.240		10.25	11.20	1.08	22.53	29
1050	15" diameter		180	.267		14.70	12.40	1.20	28.30	36
1060	18" diameter		144	.333		18.45	15.55	1.50	35.50	45
1070	21" diameter		112	.429		19.30	19.95	1.93	41.18	53.50
1080	24" diameter	▼	100	.480	▼	22.50	22.50	2.16	47.16	60.50
1560	Reinforced culvert, class 2, no gaskets									
1590	27" diameter	B-21	88	.318	L.F.	51.50	16.25	2.17	69.92	83.50
1592	30" diameter	B-13	80	.700		49.50	34	7.35	90.85	113
1594	36" diameter	"	72	.778	▼	79.50	37.50	8.15	125.15	152
2000	Reinforced culvert, class 3, no gaskets									
2010	12" diameter	B-14	150	.320	L.F.	16.50	14.90	1.44	32.84	41.50
2020	15" diameter		150	.320		20.50	14.90	1.44	36.84	46
2030	18" diameter		132	.364		25	16.95	1.64	43.59	55.50
2035	21" diameter		120	.400		28.50	18.65	1.80	48.95	61

33 42 11.60 Sewage/Drainage Collection, Concrete Pipe		Crew	Daily Output	Labor-Hours	Unit	Material	2021 Bare Costs Labor	Equipment	Total	Total Incl O&P
2040	24" diameter	B-14	100	.480	L.F.	27.50	22.50	2.16	52.16	66
2045	27" diameter	B-13	92	.609		46.50	29.50	6.40	82.40	103
2050	30" diameter		88	.636		44	30.50	6.65	81.15	102
2060	36" diameter	↓	72	.778		80	37.50	8.15	125.65	153
2070	42" diameter	B-13B	68	.824		104	40	14.55	158.55	190
2080	48" diameter		64	.875		108	42.50	15.45	165.95	199
2085	54" diameter		56	1		148	48.50	17.70	214.20	254
2090	60" diameter		48	1.167		229	56.50	20.50	306	360
2100	72" diameter		40	1.400		254	67.50	25	346.50	410
2120	84" diameter		32	1.750		410	84.50	31	525.50	615
2140	96" diameter	↓	24	2.333	↓	495	113	41.50	649.50	760
2200	With gaskets, class 3, 12" diameter	B-21	168	.167	L.F.	18.15	8.50	1.14	27.79	34
2220	15" diameter		160	.175		22.50	8.95	1.19	32.64	39.50
2230	18" diameter		152	.184		28	9.40	1.26	38.66	46
2235	21" diameter		152	.184		31	9.40	1.26	41.66	50
2240	24" diameter	↓	136	.206		34	10.50	1.40	45.90	54.50
2260	30" diameter	B-13	88	.636		52.50	30.50	6.65	89.65	111
2270	36" diameter	"	72	.778		90	37.50	8.15	135.65	163
2290	48" diameter	B-13B	64	.875		121	42.50	15.45	178.95	214
2310	72" diameter	"	40	1.400	↓	275	67.50	25	367.50	430
2330	Flared ends, 12" diameter	B-21	31	.903	Ea.	340	46	6.15	392.15	450
2340	15" diameter		25	1.120		395	57	7.65	459.65	530
2400	18" diameter		20	1.400		445	71.50	9.55	526.05	610
2420	24" diameter	↓	14	2		515	102	13.65	630.65	740
2440	36" diameter	B-13	10	5.600	↓	1,125	270	58.50	1,453.50	1,725
2500	Class 4									
2510	12" diameter	B-21	168	.167	L.F.	21	8.50	1.14	30.64	37.50
2512	15" diameter		160	.175		23	8.95	1.19	33.14	39.50
2514	18" diameter		152	.184		22	9.40	1.26	32.66	39.50
2516	21" diameter		144	.194		25.50	9.90	1.33	36.73	44.50
2518	24" diameter		136	.206		30	10.50	1.40	41.90	50
2520	27" diameter		120	.233		25.50	11.90	1.59	38.99	47.50
2522	30" diameter	B-13	88	.636		31	30.50	6.65	68.15	88
2524	36" diameter		72	.778		58.50	37.50	8.15	104.15	129
2528	42" diameter	↓	68	.824	↓	70	40	8.65	118.65	146
2600	Class 5									
2610	12" diameter	B-21	168	.167	L.F.	19.85	8.50	1.14	29.49	36
2612	15" diameter		160	.175		20.50	8.95	1.19	30.64	37
2614	18" diameter		152	.184		28.50	9.40	1.26	39.16	47
2616	21" diameter		144	.194		36	9.90	1.33	47.23	56
2618	24" diameter		136	.206		29	10.50	1.40	40.90	48.50
2620	27" diameter		120	.233		42	11.90	1.59	55.49	65.50
2622	30" diameter	B-13	88	.636		53	30.50	6.65	90.15	111
2624	36" diameter	"	72	.778		106	37.50	8.15	151.65	181
2800	Add for rubber joints 12" to 36" diameter					12%				
3080	Radius pipe, add to pipe prices, 12" to 60" diameter					50%				
3090	Over 60" diameter, add				↓	20%				
3500	Reinforced elliptical, 8' lengths, C507 class 3									
3520	14" x 23" inside, round equivalent 18" diameter	B-21	82	.341	L.F.	41	17.45	2.33	60.78	73.50
3530	24" x 38" inside, round equivalent 30" diameter	B-13	58	.966		70	46.50	10.10	126.60	158
3540	29" x 45" inside, round equivalent 36" diameter		52	1.077		110	52	11.30	173.30	211
3550	38" x 60" inside, round equivalent 48" diameter		38	1.474		178	71	15.45	264.45	320
3560	48" x 76" inside, round equivalent 60" diameter	B-13	26	2.154	L.F.	203	104	22.50	329.50	405

33 42 Stormwater Conveyance

33 42 11 – Stormwater Gravity Piping

33 42 11.60 Sewage/Drainage Collection, Concrete Pipe

		Crew	Daily Output	Labor-Hours	Unit	Material	2021 Bare Costs Labor	Equipment	Total	Total Incl O&P
3570	58" x 91" inside, round equivalent 72" diameter	B-13	22	2.545	L.F.	345	123	26.50	494.50	590
3780	Concrete slotted pipe, class 4 mortar joint									
3800	12" diameter	B-21	168	.167	L.F.	24.50	8.50	1.14	34.14	41
3840	18" diameter	"	152	.184	"	31	9.40	1.26	41.66	49.50
3900	Concrete slotted pipe, Class 4 O-ring joint									
3940	12" diameter	B-21	168	.167	L.F.	31.50	8.50	1.14	41.14	48.50
3960	18" diameter	"	152	.184	"	32	9.40	1.26	42.66	50.50
6200	Gasket, conc. pipe joint, 12"				Ea.	4.26			4.26	4.69
6220	24"					6.80			6.80	7.50
6240	36"					9.75			9.75	10.75
6260	48"					13.45			13.45	14.80
6270	60"					15.45			15.45	16.95
6280	72"					20.50			20.50	22.50

33 42 13 – Stormwater Culverts

33 42 13.13 Public Pipe Culverts

		Crew	Daily Output	Labor-Hours	Unit	Material	Labor	Equipment	Total	Total Incl O&P
0010	**PUBLIC PIPE CULVERTS**									
0020	Headwall, concrete									
0520	Precast, 12" diameter pipe	B-6	12	2	Ea.	2,350	96	18	2,464	2,750
0530	18" diameter pipe		10	2.400		2,350	115	21.50	2,486.50	2,775
0540	24" diameter pipe		10	2.400		3,175	115	21.50	3,311.50	3,700
0550	30" diameter pipe		8	3		2,475	144	27	2,646	2,950
0560	36" diameter pipe		8	3		3,275	144	27	3,446	3,850
0570	48" diameter pipe	B-69	7	6.857		6,925	335	208	7,468	8,350
0580	60" diameter pipe	"	6	8		11,500	390	243	12,133	13,600

33 42 13.15 Oval Arch Culverts

		Crew	Daily Output	Labor-Hours	Unit	Material	Labor	Equipment	Total	Total Incl O&P
0010	**OVAL ARCH CULVERTS**									
3000	Corrugated galvanized or aluminum, coated & paved									
3020	17" x 13", 16 ga., 15" equivalent	B-14	200	.240	L.F.	16.90	11.20	1.08	29.18	36.50
3040	21" x 15", 16 ga., 18" equivalent		150	.320		17.30	14.90	1.44	33.64	42.50
3060	28" x 20", 14 ga., 24" equivalent		125	.384		26.50	17.90	1.73	46.13	57.50
3080	35" x 24", 14 ga., 30" equivalent		100	.480		32.50	22.50	2.16	57.16	72
3100	42" x 29", 12 ga., 36" equivalent	B-13	100	.560		30	27	5.85	62.85	80
3120	49" x 33", 12 ga., 42" equivalent		90	.622		39	30	6.50	75.50	94.50
3140	57" x 38", 12 ga., 48" equivalent		75	.747		51.50	36	7.85	95.35	119
3160	Steel, plain oval arch culverts, plain									
3180	17" x 13", 16 ga., 15" equivalent	B-14	225	.213	L.F.	11.05	9.95	.96	21.96	28
3200	21" x 15", 16 ga., 18" equivalent		175	.274		14.40	12.80	1.23	28.43	36.50
3220	28" x 20", 14 ga., 24" equivalent		150	.320		20	14.90	1.44	36.34	45.50
3240	35" x 24", 14 ga., 30" equivalent	B-13	108	.519		26	25	5.45	56.45	72.50
3260	42" x 29", 12 ga., 36" equivalent	"	108	.519		46.50	25	5.45	76.95	94.50
3280	49" x 33", 12 ga., 42" equivalent	B-13	92	.609	L.F.	56	29.50	6.40	91.90	113
3300	57" x 38", 12 ga., 48" equivalent		75	.747	"	56.50	36	7.85	100.35	125
3320	End sections, 17" x 13"		22	2.545	Ea.	132	123	26.50	281.50	360
3340	42" x 29"		17	3.294	"	545	159	34.50	738.50	875
3360	Multi-plate arch, steel	B-20	1690	.014	Lb.	1.29	.70		1.99	2.47
8000	Bends or elbows, 6" diameter		30	.800	Ea.	71	39.50		110.50	138
8020	8" diameter		28	.857		94.50	42.50		137	168
8040	10" diameter		25	.960		110	47.50		157.50	192
8060	12" diameter	B-21	20	1.400		140	71.50	9.55	221.05	272
8100	Wyes or tees, 6" diameter	B-20	27	.889		108	44		152	184
8120	8" diameter		25	.960		136	47.50		183.50	220
8140	10" diameter		21	1.143		161	56.50		217.50	262

33 42 13 – Stormwater Culverts

33 42 13.15 Oval Arch Culverts	Crew	Daily Output	Labor-Hours	Unit	Material	2021 Bare Costs Labor	2021 Bare Costs Equipment	Total	Total Incl O&P
8160	12" diameter	B-21	18	1.556	Ea.	183	79.50	10.60	273.10	335

33 42 33 – Stormwater Curbside Drains and Inlets

33 42 33.13 Catch Basins

		Crew	Daily Output	Labor-Hours	Unit	Material	Labor	Equipment	Total	Total Incl O&P
0010	**CATCH BASINS**									
0011	Not including footing & excavation									
1510	Grates only, for pipe bells, plastic, 4" diameter pipe	1 Clab	50	.160	Ea.	3.73	7.10		10.83	14.70
1514	6" diameter pipe		50	.160		9.75	7.10		16.85	21.50
1516	8" diameter pipe	↓	50	.160		12.60	7.10		19.70	24.50
1520	For beehive type grate, add					60%				
1550	Gray iron, 4" diam. pipe, light duty	1 Clab	50	.160		27	7.10		34.10	40.50
1551	Heavy duty		50	.160		44	7.10		51.10	59
1552	6" diameter pipe, light duty		50	.160		39	7.10		46.10	53.50
1553	Heavy duty		50	.160		23	7.10		30.10	35.50
1554	8" diameter pipe, light duty		50	.160		55.50	7.10		62.60	71.50
1555	Heavy duty		50	.160		36.50	7.10		43.60	50.50
1556	10" diameter pipe, light duty		40	.200		80	8.90		88.90	101
1557	Heavy duty		40	.200		45.50	8.90		54.40	63.50
1558	12" diameter pipe, light duty		40	.200		109	8.90		117.90	132
1559	Heavy duty		40	.200		104	8.90		112.90	127
1560	15" diameter pipe, light duty		32	.250		162	11.10		173.10	195
1561	Heavy duty		32	.250		125	11.10		136.10	154
1562	18" diameter pipe, light duty		32	.250		172	11.10		183.10	206
1563	Heavy duty	↓	32	.250	↓	172	11.10		183.10	206
1568	For beehive type grate, add					60%				
1570	Covers only for pipe bells, gray iron, med. duty									
1571	4" diameter pipe	1 Clab	50	.160	Ea.	63.50	7.10		70.60	80.50
1572	6" diameter pipe		50	.160		110	7.10		117.10	133
1573	8" diameter pipe		50	.160		90.50	7.10		97.60	110
1574	10" diameter pipe		40	.200		218	8.90		226.90	253
1575	12" diameter pipe	1 Clab	40	.200	Ea.	106	8.90		114.90	129
1576	15" diameter pipe		32	.250		143	11.10		154.10	174
1577	18" diameter pipe	↓	32	.250		185	11.10		196.10	221
1578	24" diameter pipe	2 Clab	24	.667		263	29.50		292.50	335
1579	36" diameter pipe	"	24	.667	↓	340	29.50		369.50	415
1580	Curb inlet frame, grate, and curb box									
1582	Large 24" x 36" heavy duty	B-24	2	12	Ea.	870	605		1,475	1,850
1590	Small 10" x 21" medium duty	"	2	12		545	605		1,150	1,500
1600	Frames & grates, C.I., 24" square, 500 lb.	B-6	7.80	3.077		395	148	27.50	570.50	685
1700	26" D shape, 600 lb.		7	3.429		735	165	31	931	1,075
1800	Light traffic, 18" diameter, 100 lb.		10	2.400		216	115	21.50	352.50	435
1900	24" diameter, 300 lb.		8.70	2.759		209	133	25	367	455
2000	36" diameter, 900 lb.		5.80	4.138		720	199	37.50	956.50	1,125
2100	Heavy traffic, 24" diameter, 400 lb.		7.80	3.077		291	148	27.50	466.50	570
2200	36" diameter, 1,150 lb.		3	8		955	385	72	1,412	1,700
2300	Mass. State standard, 26" diameter, 475 lb.		7	3.429		695	165	31	891	1,050
2400	30" diameter, 620 lb.		7	3.429		380	165	31	576	695
2500	Watertight, 24" diameter, 350 lb.		7.80	3.077		430	148	27.50	605.50	725
2600	26" diameter, 500 lb.		7	3.429		445	165	31	641	770
2700	32" diameter, 575 lb.	↓	6	4	↓	915	192	36	1,143	1,325
2800	3 piece cover & frame, 10" deep,									
2900	1,200 lb., for heavy equipment	B-6	3	8	Ea.	1,075	385	72	1,532	1,850
3000	Raised for paving 1-1/4" to 2" high									

33 42 Stormwater Conveyance

33 42 33 – Stormwater Curbside Drains and Inlets

33 42 33.13 Catch Basins

		Crew	Daily Output	Labor-Hours	Unit	Material	2021 Bare Costs Labor	Equipment	Total	Total Incl O&P
3100	4 piece expansion ring									
3200	20" to 26" diameter	1 Clab	3	2.667	Ea.	223	118		341	425
3300	30" to 36" diameter	"	3	2.667	"	310	118		428	515
3320	Frames and covers, existing, raised for paving, 2", including									
3340	row of brick, concrete collar, up to 12" wide frame	B-6	18	1.333	Ea.	52.50	64	12	128.50	166
3360	20" to 26" wide frame		11	2.182		79	105	19.65	203.65	266
3380	30" to 36" wide frame		9	2.667		98	128	24	250	325
3400	Inverts, single channel brick	D-1	3	5.333		113	260		373	515
3500	Concrete		5	3.200		137	156		293	385
3600	Triple channel, brick		2	8		186	390		576	790
3700	Concrete		3	5.333		156	260		416	560

33 42 33.15 Remove and Replace Catch Basin Cover

		Crew	Daily Output	Labor-Hours	Unit	Material	2021 Bare Costs Labor	Equipment	Total	Total Incl O&P
0010	**REMOVE AND REPLACE CATCH BASIN COVER**									
0023	Remove catch basin cover	1 Clab	80	.100	Ea.		4.44		4.44	6.65
0033	Replace catch basin cover	"	80	.100	"		4.44		4.44	6.65

33 42 33.50 Stormwater Management

		Crew	Daily Output	Labor-Hours	Unit	Material	2021 Bare Costs Labor	Equipment	Total	Total Incl O&P
0010	**STORMWATER MANAGEMENT**									
0030	Add per S.F. of impervious surface	B-37	6000	.008	S.F.	3.19	.37	.04	3.60	4.12

33 46 Stormwater Management

33 46 23 – Modular Buried Stormwater Storage Units

33 46 23.10 HDPE Storm Water Infiltration Chamber

		Crew	Daily Output	Labor-Hours	Unit	Material	2021 Bare Costs Labor	Equipment	Total	Total Incl O&P
0010	**HDPE STORM WATER INFILTRATION CHAMBER**									
0020	Not including excavation or backfill									
0100	11" H x 16" W	2 Clab	300	.053	L.F.	6.75	2.37		9.12	11
0110	12" H x 22" W		300	.053		9.20	2.37		11.57	13.65
0120	12" H x 34" W		300	.053		10.90	2.37		13.27	15.50
0130	16" H x 34" W		300	.053		24	2.37		26.37	30
0140	30" H x 52" W		270	.059		37	2.63		39.63	45

33 52 Hydrocarbon Transmission and Distribution

33 52 13 – Liquid Hydrocarbon Piping

33 52 13.16 Gasoline Piping

		Crew	Daily Output	Labor-Hours	Unit	Material	2021 Bare Costs Labor	Equipment	Total	Total Incl O&P
0010	**GASOLINE PIPING**									
0020	Primary containment pipe, fiberglass-reinforced									
0030	Plastic pipe 15' & 30' lengths									
0040	2" diameter	Q-6	425	.056	L.F.	7.25	3.60		10.85	13.40
0050	3" diameter		400	.060		9.85	3.83		13.68	16.50
0060	4" diameter		375	.064		12.75	4.08		16.83	20
0100	Fittings									
0110	Elbows, 90° & 45°, bell ends, 2"	Q-6	24	1	Ea.	43.50	64		107.50	143
0120	3" diameter		22	1.091		57	69.50		126.50	167
0130	4" diameter		20	1.200		71.50	76.50		148	193
0200	Tees, bell ends, 2"		21	1.143		57.50	73		130.50	172
0210	3" diameter		18	1.333		65.50	85		150.50	199
0220	4" diameter		15	1.600		80.50	102		182.50	241
0230	Flanges bell ends, 2"		24	1		31	64		95	130
0240	3" diameter		22	1.091		38.50	69.50		108	147
0250	4" diameter		20	1.200		43	76.50		119.50	161

33 52 Hydrocarbon Transmission and Distribution

33 52 13 – Liquid Hydrocarbon Piping

33 52 13.16 Gasoline Piping		Crew	Daily Output	Labor-Hours	Unit	Material	2021 Bare Costs Labor	Equipment	Total	Total Incl O&P
0260	Sleeve couplings, 2"	Q-6	21	1.143	Ea.	11.95	73		84.95	122
0270	3" diameter		18	1.333		18.80	85		103.80	148
0280	4" diameter		15	1.600		21	102		123	175
0290	Threaded adapters, 2"		21	1.143		19.55	73		92.55	131
0300	3" diameter		18	1.333		25.50	85		110.50	155
0310	4" diameter		15	1.600		36.50	102		138.50	192
0320	Reducers, 2"		27	.889		26.50	56.50		83	114
0330	3" diameter		22	1.091		26.50	69.50		96	133
0340	4" diameter	▼	20	1.200	▼	35.50	76.50		112	153
1010	Gas station product line for secondary containment (double wall)									
1100	Fiberglass reinforced plastic pipe 25' lengths									
1120	Pipe, plain end, 3" diameter	Q-6	375	.064	L.F.	36	4.08		40.08	45.50
1130	4" diameter	Q-6	350	.069	L.F.	29.50	4.37		33.87	38.50
1140	5" diameter		325	.074		33	4.71		37.71	43
1150	6" diameter	▼	300	.080	▼	36.50	5.10		41.60	47.50
1200	Fittings									
1230	Elbows, 90° & 45°, 3" diameter	Q-6	18	1.333	Ea.	152	85		237	294
1240	4" diameter		16	1.500		183	95.50		278.50	345
1250	5" diameter		14	1.714		175	109		284	355
1260	6" diameter		12	2		218	128		346	430
1270	Tees, 3" diameter		15	1.600		189	102		291	360
1280	4" diameter		12	2		180	128		308	390
1290	5" diameter		9	2.667		235	170		405	515
1300	6" diameter		6	4		330	255		585	745
1310	Couplings, 3" diameter		18	1.333		53.50	85		138.50	186
1320	4" diameter		16	1.500		122	95.50		217.50	277
1330	5" diameter		14	1.714		217	109		326	400
1340	6" diameter		12	2		340	128		468	565
1350	Cross-over nipples, 3" diameter		18	1.333		10.20	85		95.20	138
1360	4" diameter		16	1.500		12.40	95.50		107.90	157
1370	5" diameter		14	1.714		17.20	109		126.20	182
1380	6" diameter		12	2		18	128		146	210
1400	Telescoping, reducers, concentric 4" x 3"		18	1.333		43.50	85		128.50	175
1410	5" x 4"		17	1.412		93	90		183	236
1420	6" x 5"	▼	16	1.500	▼	261	95.50		356.50	430

33 52 16 – Gas Hydrocarbon Piping

33 52 16.13 Steel Natural Gas Piping

		Crew	Daily Output	Labor-Hours	Unit	Material	2021 Bare Costs Labor	Equipment	Total	Total Incl O&P
0010	**STEEL NATURAL GAS PIPING**									
0020	Not including excavation or backfill, tar coated and wrapped									
4000	Pipe schedule 40, plain end									
4040	1" diameter	Q-4	300	.107	L.F.	4.95	6.85	.36	12.16	16.10
4080	2" diameter		280	.114		7.75	7.35	.38	15.48	19.95
4120	3" diameter	▼	260	.123		12.85	7.95	.41	21.21	26.50
4160	4" diameter	B-35	255	.220		16.75	12.45	3.59	32.79	41
4200	5" diameter		220	.255		24.50	14.45	4.17	43.12	52.50
4240	6" diameter		180	.311		29.50	17.65	5.10	52.25	64.50
4280	8" diameter		140	.400		47	22.50	6.55	76.05	92.50
4320	10" diameter		100	.560		118	32	9.15	159.15	187
4360	12" diameter		80	.700		131	39.50	11.45	181.95	216
4400	14" diameter		75	.747		140	42.50	12.20	194.70	230
4440	16" diameter		70	.800		153	45.50	13.10	211.60	250
4480	18" diameter	▼	65	.862	▼	196	49	14.10	259.10	305

516

For customer support on your Site Work & Landscape Costs with RSMeans Data, call 800.448.8182.

33 52 16.13 Steel Natural Gas Piping		Crew	Daily Output	Labor-Hours	Unit	Material	2021 Bare Costs Labor	Equipment	Total	Total Incl O&P
4520	20" diameter	B-35	60	.933	L.F.	305	53	15.30	373.30	430
4560	24" diameter	↓	50	1.120	↓	350	63.50	18.35	431.85	500
6000	Schedule 80, plain end									
6002	4" diameter	B-35	144	.389	L.F.	40.50	22	6.35	68.85	84.50
6004	5" diameter		140	.400		52	22.50	6.55	81.05	98
6006	6" diameter		126	.444		89	25	7.30	121.30	144
6008	8" diameter		108	.519		119	29.50	8.50	157	184
6010	10" diameter		90	.622		178	35.50	10.20	223.70	260
6012	12" diameter	↓	72	.778	↓	237	44	12.75	293.75	340
8008	Elbow, weld joint, standard weight									
8020	4" diameter	Q-16	6.80	3.529	Ea.	108	223	15.80	346.80	470
8022	5" diameter		5.40	4.444		219	281	19.90	519.90	685
8024	6" diameter		4.50	5.333		239	335	24	598	795
8026	8" diameter		3.40	7.059		450	445	31.50	926.50	1,200
8028	10" diameter		2.70	8.889		670	560	40	1,270	1,625
8030	12" diameter		2.30	10.435		780	660	47	1,487	1,900
8032	14" diameter		1.80	13.333		1,425	840	59.50	2,324.50	2,900
8034	16" diameter		1.50	16		1,950	1,000	71.50	3,021.50	3,725
8036	18" diameter		1.40	17.143		2,075	1,075	77	3,227	3,975
8038	20" diameter		1.20	20		3,375	1,275	89.50	4,739.50	5,675
8040	24" diameter	↓	1.02	23.529	↓	5,100	1,475	105	6,680	7,950
8100	Extra heavy									
8102	4" diameter	Q-16	5.30	4.528	Ea.	216	286	20.50	522.50	685
8104	5" diameter		4.20	5.714		440	360	25.50	825.50	1,050
8106	6" diameter		3.50	6.857		480	435	30.50	945.50	1,200
8108	8" diameter		2.60	9.231		675	585	41.50	1,301.50	1,650
8110	10" diameter		2.10	11.429		895	720	51	1,666	2,125
8112	12" diameter		1.80	13.333		1,050	840	59.50	1,949.50	2,475
8114	14" diameter		1.40	17.143		1,900	1,075	77	3,052	3,800
8116	16" diameter		1.20	20		2,600	1,275	89.50	3,964.50	4,850
8118	18" diameter		1.10	21.818		2,775	1,375	98	4,248	5,200
8120	20" diameter		.94	25.532		4,500	1,625	114	6,239	7,475
8122	24" diameter	↓	.80	30		6,800	1,900	134	8,834	10,400
8200	Malleable, standard weight									
8202	4" diameter	B-20	12	2	Ea.	855	98.50		953.50	1,100
8204	5" diameter		9	2.667		985	131		1,116	1,275
8206	6" diameter		8	3		1,250	148		1,398	1,600
8208	8" diameter		6	4		2,325	197		2,522	2,875
8210	10" diameter		5	4.800		2,225	237		2,462	2,800
8212	12" diameter	↓	4	6	↓	2,475	296		2,771	3,175
8300	Extra heavy									
8302	4" diameter	B-20	12	2	Ea.	1,700	98.50		1,798.50	2,025
8304	5" diameter	B-21	9	3.111		1,975	159	21	2,155	2,425
8306	6" diameter		8	3.500		2,350	179	24	2,553	2,875
8308	8" diameter		6	4.667		2,925	238	32	3,195	3,625
8310	10" diameter		5	5.600		3,550	286	38	3,874	4,375
8312	12" diameter	↓	4	7	↓	3,950	355	47.50	4,352.50	4,950
8500	Tee weld, standard weight									
8510	4" diameter	Q-16	4.50	5.333	Ea.	172	335	24	531	720
8512	5" diameter		3.60	6.667		285	420	30	735	980
8514	6" diameter		3	8		299	505	36	840	1,125
8516	8" diameter		2.30	10.435		520	660	47	1,227	1,600
8518	10" diameter	↓	1.80	13.333		1,025	840	59.50	1,924.50	2,450

For customer support on your Site Work & Landscape Costs with RSMeans Data, call 800.448.8182.

517

33 52 16 – Gas Hydrocarbon Piping

33 52 16.13 Steel Natural Gas Piping		Crew	Daily Output	Labor-Hours	Unit	Material	2021 Bare Costs Labor	Equipment	Total	Total Incl O&P
8520	12" diameter	Q-16	1.50	16	Ea.	1,425	1,000	71.50	2,496.50	3,150
8522	14" diameter		1.20	20		2,525	1,275	89.50	3,889.50	4,750
8524	16" diameter		1	24		2,825	1,525	108	4,458	5,500
8526	18" diameter		.90	26.667		4,425	1,675	119	6,219	7,500
8528	20" diameter		.80	30		7,000	1,900	134	9,034	10,700
8530	24" diameter	↓	.70	34.286	↓	9,025	2,175	154	11,354	13,300
8810	Malleable, standard weight									
8812	4" diameter	B-20	8	3	Ea.	1,450	148		1,598	1,825
8814	5" diameter	B-21	6	4.667		2,000	238	32	2,270	2,600
8816	6" diameter		5.30	5.283		1,725	270	36	2,031	2,350
8818	8" diameter		4	7		2,500	355	47.50	2,902.50	3,350
8820	10" diameter		3.30	8.485		3,900	435	58	4,393	5,025
8822	12" diameter	↓	2.70	10.370	↓	4,575	530	70.50	5,175.50	5,900
8900	Extra heavy									
8902	4" diameter	B-20	8	3	Ea.	2,175	148		2,323	2,625
8904	5" diameter	B-21	6	4.667		2,975	238	32	3,245	3,675
8906	6" diameter		5.30	5.283		2,600	270	36	2,906	3,300
8908	8" diameter		4	7		3,125	355	47.50	3,527.50	4,025
8910	10" diameter		3.30	8.485		3,900	435	58	4,393	5,025
8912	12" diameter	↓	2.70	10.370		4,575	530	70.50	5,175.50	5,900

33 52 16.20 Piping, Gas Service and Distribution, P.E.

		Crew	Daily Output	Labor-Hours	Unit	Material	2021 Bare Costs Labor	Equipment	Total	Total Incl O&P
0010	**PIPING, GAS SERVICE AND DISTRIBUTION, POLYETHYLENE**									
0020	Not including excavation or backfill									
1000	60 psi coils, compression coupling @ 100', 1/2" diameter, SDR 11	B-20A	608	.053	L.F.	.56	2.80		3.36	4.80
1010	1" diameter, SDR 11		544	.059		1.13	3.13		4.26	5.90
1040	1-1/4" diameter, SDR 11		544	.059		1.61	3.13		4.74	6.45
1100	2" diameter, SDR 11		488	.066		2.84	3.49		6.33	8.30
1160	3" diameter, SDR 11	↓	408	.078		5.05	4.17		9.22	11.75
1500	60 psi 40' joints with coupling, 3" diameter, SDR 11	B-21A	408	.098		7.50	5.35	1.06	13.91	17.40
1540	4" diameter, SDR 11		352	.114		14.50	6.25	1.23	21.98	26.50
1600	6" diameter, SDR 11		328	.122		37	6.70	1.32	45.02	52
1640	8" diameter, SDR 11	↓	272	.147	↓	62	8.05	1.59	71.64	82

33 52 16.23 Medium Density Polyethylene Piping

		Crew	Daily Output	Labor-Hours	Unit	Material	2021 Bare Costs Labor	Equipment	Total	Total Incl O&P
2010	**MEDIUM DENSITY POLYETHYLENE PIPING**									
2020	ASTM D2513, not including excavation or backfill									
2200	Butt fused pipe									
2205	80psi coils, butt fusion joint @ 100', 1/2" CTS diameter, SDR 7	B-22C	2050	.008	L.F.	.17	.40	.07	.64	.85
2210	1" CTS diameter, SDR 11.5		1900	.008		.24	.43	.07	.74	.98
2215	80psi coils, butt fusion joint @ 100', IPS 1/2" diameter, SDR 9.3		1950	.008		.25	.42	.07	.74	.98
2220	3/4" diameter, SDR 11		1950	.008		.32	.42	.07	.81	1.05
2225	1" diameter, SDR 11		1850	.009		.77	.44	.07	1.28	1.59
2230	1-1/4" diameter, SDR 11		1850	.009		.78	.44	.07	1.29	1.60
2235	1-1/2" diameter, SDR 11		1750	.009		.96	.46	.08	1.50	1.84
2240	2" diameter, SDR 11	↓	1750	.009		1.29	.46	.08	1.83	2.20
2245	80psi 40' lengths, butt fusion joint, IPS 3" diameter, SDR 11.5	B-22A	660	.061		2.48	3.08	1.22	6.78	8.65
2250	4" diameter, SDR 11.5	"	500	.080	↓	3.91	4.06	1.61	9.58	12.10
2400	Socket fused pipe									
2405	80psi coils, socket fusion coupling @ 100', 1/2" CTS diameter, SDR 7	B-20	1050	.023	L.F.	.19	1.13		1.32	1.90
2410	1" CTS diameter, SDR 11.5		1000	.024		.26	1.18		1.44	2.06
2415	80psi coils, socket fusion coupling @ 100', IPS 1/2" diameter, SDR 9.3		1000	.024		.27	1.18		1.45	2.06
2420	3/4" diameter, SDR 11		1000	.024		.34	1.18		1.52	2.14
2425	1" diameter, SDR 11	↓	950	.025		.79	1.25		2.04	2.73

33 52 16.23 Medium Density Polyethylene Piping	Crew	Daily Output	Labor-Hours	Unit	Material	2021 Bare Costs Labor	Equipment	Total	Total Incl O&P
2430 1-1/4" diameter, SDR 11	B-20	950	.025	L.F.	.80	1.25		2.05	2.74
2435 1-1/2" diameter, SDR 11		900	.027		.98	1.31		2.29	3.05
2440 2" diameter, SDR 11	↓	850	.028		1.32	1.39		2.71	3.53
2445 80psi 40' lengths, socket fusion coupling, IPS 3" diameter, SDR 11.5	B-21A	340	.118		2.88	6.45	1.27	10.60	14.15
2450 4" diameter, SDR 11.5	"	260	.154	↓	4.49	8.45	1.66	14.60	19.35
2600 Compression coupled pipe									
2605 80psi coils, compression coupling @ 100', 1/2" CTS diameter, SDR 7	B-20	2250	.011	L.F.	.30	.53		.83	1.12
2610 1" CTS diameter, SDR 11.5		2175	.011		.49	.54		1.03	1.34
2615 80psi coils, compression coupling @ 100', IPS 1/2" diameter, SDR 9.3		2175	.011		.48	.54		1.02	1.34
2620 3/4" diameter, SDR 11		2100	.011		.55	.56		1.11	1.45
2625 1" diameter, SDR 11	↓	2025	.012	↓	1.08	.58		1.66	2.06
3000 Fittings, butt fusion									
3010 SDR 11, IPS unless noted CTS									
3100 Caps, 3/4" diameter	B-22C	28.50	.561	Ea.	3.24	28.50	4.73	36.47	51.50
3105 1" diameter		27	.593		2.98	30	5	37.98	54
3110 1-1/4" diameter		27	.593		4.77	30	5	39.77	56
3115 1-1/2" diameter		25.50	.627		2.76	32	5.30	40.06	56.50
3120 2" diameter		24	.667		8.70	34	5.60	48.30	66.50
3125 3" diameter		24	.667		13.75	34	5.60	53.35	72
3130 4" diameter	↓	18	.889	↓	29	45	7.50	81.50	108
3200 Reducers, 1" x 3/4" diameters	B-22C	13.50	1.185	Ea.	13.60	60	10	83.60	116
3205 1-1/4" x 1" diameters		13.50	1.185		15.20	60	10	85.20	118
3210 1-1/2" x 3/4" diameters		12.75	1.255		19.85	63.50	10.60	93.95	129
3215 1-1/2" x 1" diameters		12.75	1.255		16.85	63.50	10.60	90.95	126
3220 1-1/2" x 1-1/4" diameters		12.75	1.255		16.85	63.50	10.60	90.95	126
3225 2" x 1" diameters		12	1.333		17.25	67.50	11.25	96	132
3230 2" x 1-1/4" diameters		12	1.333		17.25	67.50	11.25	96	132
3235 2" x 1-1/2" diameters		12	1.333		16	67.50	11.25	94.75	131
3240 3" x 2" diameters		12	1.333		20	67.50	11.25	98.75	135
3245 4" x 2" diameters		9	1.778		32	90	15	137	187
3250 4" x 3" diameters		9	1.778		29	90	15	134	183
3300 Elbows, 90°, 3/4" diameter		14.25	1.123		4.04	57	9.45	70.49	100
3302 1" diameter		13.50	1.185		4.94	60	10	74.94	106
3304 1-1/4" diameter		13.50	1.185		9.50	60	10	79.50	111
3306 1-1/2" diameter		12.75	1.255		8.25	63.50	10.60	82.35	116
3308 2" diameter		12	1.333		14.10	67.50	11.25	92.85	129
3310 3" diameter		12	1.333		28.50	67.50	11.25	107.25	144
3312 4" diameter		9	1.778		39	90	15	144	195
3350 45°, 3" diameter		12	1.333		31.50	67.50	11.25	110.25	148
3352 4" diameter		9	1.778		37	90	15	142	192
3400 Tees, 3/4" diameter		9.50	1.684		16.50	85.50	14.20	116.20	162
3405 1" diameter		9	1.778		4.97	90	15	109.97	157
3410 1-1/4" diameter		9	1.778		12.70	90	15	117.70	166
3415 1-1/2" diameter		8.50	1.882		18.40	95.50	15.85	129.75	181
3420 2" diameter		8	2		19.50	102	16.85	138.35	192
3425 3" diameter		8	2		35.50	102	16.85	154.35	210
3430 4" diameter	↓	6	2.667	↓	44.50	135	22.50	202	277
3500 Tapping tees, high volume, butt fusion outlets									
3505 1-1/2" punch, 2" x 2" outlet	B-22C	11	1.455	Ea.	110	74	12.25	196.25	246
3510 1-7/8" punch, 3" x 2" outlet		11	1.455		110	74	12.25	196.25	246
3515 4" x 2" outlet	↓	9	1.778		110	90	15	215	273
3550 For protective sleeves, add				↓	2.90			2.90	3.19
3600 Service saddles, saddle contour x outlet, butt fusion outlets									

33 52 16.23 Medium Density Polyethylene Piping	Crew	Daily Output	Labor-Hours	Unit	Material	2021 Bare Costs Labor	Equipment	Total	Total Incl O&P
3602	1-1/4" x 3/4" outlet	B-22C	17	.941	Ea.	14.15	48	7.95	70.10	96
3604	1-1/4" x 1" outlet		17	.941		6.65	48	7.95	62.60	87.50
3606	1-1/4" x 1-1/4" outlet		16	1		6.65	51	8.45	66.10	92.50
3608	1-1/2" x 3/4" outlet		16	1		6.65	51	8.45	66.10	92.50
3610	1-1/2" x 1-1/4" outlet		15	1.067		14.15	54	9	77.15	106
3612	2" x 3/4" outlet		12	1.333		6.65	67.50	11.25	85.40	121
3614	2" x 1" outlet		12	1.333		6.65	67.50	11.25	85.40	121
3616	2" x 1-1/4" outlet		11	1.455		6.65	74	12.25	92.90	132
3618	3" x 3/4" outlet		12	1.333		6.65	67.50	11.25	85.40	121
3620	3" x 1" outlet		12	1.333		6.65	67.50	11.25	85.40	121
3622	3" x 1-1/4" outlet		11	1.455		6.65	74	12.25	92.90	132
3624	4" x 3/4" outlet		10	1.600		6.65	81	13.50	101.15	144
3626	4" x 1" outlet		10	1.600		6.65	81	13.50	101.15	144
3628	4" x 1-1/4" outlet	↓	9	1.778		14.15	90	15	119.15	167
3685	For protective sleeves, 3/4" diameter outlets, add					.86			.86	.95
3690	1" diameter outlets, add					1.44			1.44	1.58
3695	1-1/4" diameter outlets, add				↓	2.48			2.48	2.73
3700	Branch saddles, contour x outlet, butt outlets, round base									
3702	2" x 2" outlet	B-22C	11	1.455	Ea.	17.30	74	12.25	103.55	144
3704	3" x 2" outlet		11	1.455		29	74	12.25	115.25	157
3706	3" x 3" outlet		11	1.455		27	74	12.25	113.25	154
3708	4" x 2" outlet		9	1.778		30	90	15	135	185
3710	4" x 3" outlet		9	1.778		50	90	15	155	207
3712	4" x 4" outlet		9	1.778		36	90	15	141	191
3750	Rectangular base, 2" x 2" outlet		11	1.455		17.30	74	12.25	103.55	144
3752	3" x 2" outlet		11	1.455		18.80	74	12.25	105.05	145
3754	3" x 3" outlet		11	1.455		27	74	12.25	113.25	154
3756	4" x 2" outlet		9	1.778		21.50	90	15	126.50	175
3758	4" x 3" outlet		9	1.778		25.50	90	15	130.50	180
3760	4" x 4" outlet	↓	9	1.778	↓	31	90	15	136	186
4000	Fittings, socket fusion									
4010	SDR 11, IPS unless noted CTS									
4100	Caps, 1/2" CTS diameter	B-20	30	.800	Ea.	2.37	39.50		41.87	61.50
4105	1/2" diameter		28.50	.842		2.37	41.50		43.87	64.50
4110	3/4" diameter		28.50	.842		3.63	41.50		45.13	66
4115	1" CTS diameter		28	.857		3.92	42.50		46.42	68
4120	1" diameter		27	.889		4.66	44		48.66	70.50
4125	1-1/4" diameter		27	.889		5.65	44		49.65	71.50
4130	1-1/2" diameter		25.50	.941		8.50	46.50		55	79
4135	2" diameter		24	1		6	49.50		55.50	80.50
4140	3" diameter		24	1		13.85	49.50		63.35	89.50
4145	4" diameter		18	1.333		20.50	65.50		86	121
4200	Reducers, 1/2" x 1/2" CTS diameters		14.25	1.684		6.15	83		89.15	131
4202	3/4" x 1/2" CTS diameters		14.25	1.684		5.70	83		88.70	130
4204	3/4" x 1/2" diameters		14.25	1.684		6.10	83		89.10	131
4206	1" CTS x 1/2" diameters		14	1.714		5.95	84.50		90.45	134
4208	1" CTS x 3/4" diameters		13.50	1.778		6	87.50		93.50	138
4210	1" x 1/2" CTS diameters		13.50	1.778		5.65	87.50		93.15	137
4212	1" x 1/2" diameters		13.50	1.778		6.30	87.50		93.80	138
4214	1" x 3/4" diameters		13.50	1.778		8.75	87.50		96.25	141
4216	1" x 1" CTS diameters		13.50	1.778		5.15	87.50		92.65	137
4218	1-1/4" x 1/2" CTS diameters		13.50	1.778		5.95	87.50		93.45	138
4220	1-1/4" x 1/2" diameters	↓	13.50	1.778		8.65	87.50		96.15	141

33 52 16.23 Medium Density Polyethylene Piping		Crew	Daily Output	Labor-Hours	Unit	Material	2021 Bare Costs Labor	Equipment	Total	Total Incl O&P
4222	1-1/4" x 3/4" diameters	B-20	13.50	1.778	Ea.	12.45	87.50		99.95	145
4224	1-1/4" x 1" CTS diameters		13.50	1.778		8.65	87.50		96.15	141
4226	1-1/4" x 1" diameters		13.50	1.778		10.60	87.50		98.10	143
4228	1-1/2" x 3/4" diameters		12.75	1.882		9.50	93		102.50	149
4230	1-1/2" x 1" diameters		12.75	1.882		8.05	93		101.05	148
4232	1-1/2" x 1-1/4" diameters		12.75	1.882		9.10	93		102.10	149
4234	2" x 3/4" diameters		12	2		9.95	98.50		108.45	159
4236	2" x 1" CTS diameters		12	2		9.80	98.50		108.30	159
4238	2" x 1" diameters		12	2		10.70	98.50		109.20	160
4240	2" x 1-1/4" diameters		12	2		9.75	98.50		108.25	159
4242	2" x 1-1/2" diameters		12	2		10.35	98.50		108.85	159
4244	3" x 2" diameters		12	2		11.75	98.50		110.25	161
4246	4" x 2" diameters		9	2.667		24.50	131		155.50	224
4248	4" x 3" diameters		9	2.667		28	131		159	228
4300	Couplings, 1/2" CTS diameter		15	1.600		1.80	79		80.80	120
4305	1/2" diameter		14.25	1.684		1.50	83		84.50	126
4310	3/4" diameter		14.25	1.684		1.68	83		84.68	126
4315	1" CTS diameter		14	1.714		2.03	84.50		86.53	129
4320	1" diameter		13.50	1.778		1.71	87.50		89.21	133
4325	1-1/4" diameter		13.50	1.778		2.06	87.50		89.56	133
4330	1-1/2" diameter		12.75	1.882		2.43	93		95.43	142
4335	2" diameter		12	2		2.59	98.50		101.09	151
4340	3" diameter		12	2		16.20	98.50		114.70	166
4345	4" diameter		9	2.667		23	131		154	223
4400	Elbows, 90°, 1/2" CTS diameter		15	1.600		3.41	79		82.41	122
4405	1/2" diameter		14.25	1.684		3.35	83		86.35	128
4410	3/4" diameter		14.25	1.684		3.96	83		86.96	128
4415	1" CTS diameter		14	1.714		3.81	84.50		88.31	131
4420	1" diameter		13.50	1.778		4.24	87.50		91.74	136
4425	1-1/4" diameter		13.50	1.778		4.04	87.50		91.54	135
4430	1-1/2" diameter		12.75	1.882		5.10	93		98.10	145
4435	2" diameter		12	2		4.16	98.50		102.66	153
4440	3" diameter		12	2		22	98.50		120.50	172
4445	4" diameter		9	2.667		57	131		188	260
4460	45°, 1" diameter		13.50	1.778		6.80	87.50		94.30	138
4465	2" diameter		12	2		9.10	98.50		107.60	158
4500	Tees, 1/2" CTS diameter		10	2.400		3.26	118		121.26	181
4505	1/2" diameter		9.50	2.526		2.89	125		127.89	189
4510	3/4" diameter		9.50	2.526		3.89	125		128.89	190
4515	1" CTS diameter		9.33	2.571		3.81	127		130.81	194
4520	1" diameter		9	2.667		4.75	131		135.75	202
4525	1-1/4" diameter		9	2.667		5.25	131		136.25	203
4530	1-1/2" diameter		8.50	2.824		8.70	139		147.70	218
4535	2" diameter		8	3		10.45	148		158.45	232
4540	3" diameter		8	3		32	148		180	256
4545	4" diameter		6	4		59.50	197		256.50	360
4600	Tapping tees, type II, 3/4" punch, socket fusion outlets									
4601	Saddle contour x outlet diameter, IPS unless noted CTS									
4602	1-1/4" x 1/2" CTS outlet	B-20	17	1.412	Ea.	11.95	69.50		81.45	117
4604	1-1/4" x 1/2" outlet		17	1.412		11.95	69.50		81.45	117
4606	1-1/4" x 3/4" outlet		17	1.412		11.95	69.50		81.45	117
4608	1-1/4" x 1" CTS outlet		17	1.412		13.05	69.50		82.55	118
4610	1-1/4" x 1" outlet		17	1.412		13.05	69.50		82.55	118

521

For customer support on your Site Work & Landscape Costs with RSMeans Data, call 800.448.8182.

33 52 16.23 Medium Density Polyethylene Piping	Crew	Daily Output	Labor-Hours	Unit	Material	2021 Bare Costs Labor	Equipment	Total	Total Incl O&P
4612	1-1/4" x 1-1/4" outlet	B-20	16	1.500	Ea.	22	74		96	135
4614	1-1/2" x 1/2" CTS outlet		16	1.500		11.95	74		85.95	124
4616	1-1/2" x 1/2" outlet		16	1.500		11.95	74		85.95	124
4618	1-1/2" x 3/4" outlet		16	1.500		11.95	74		85.95	124
4620	1-1/2" x 1" CTS outlet		16	1.500		13.05	74		87.05	125
4622	1-1/2" x 1" outlet		16	1.500		13.05	74		87.05	125
4624	1-1/2" x 1-1/4" outlet		15	1.600		22	79		101	142
4626	2" x 1/2" CTS outlet		12	2		11.95	98.50		110.45	161
4628	2" x 1/2" outlet		12	2		11.95	98.50		110.45	161
4630	2" x 3/4" outlet		12	2		11.95	98.50		110.45	161
4632	2" x 1" CTS outlet		12	2		13.05	98.50		111.55	162
4634	2" x 1" outlet		12	2		13.05	98.50		111.55	162
4636	2" x 1-1/4" outlet		11	2.182		22	108		130	185
4638	3" x 1/2" CTS outlet		12	2		11.95	98.50		110.45	161
4640	3" x 1/2" outlet		12	2		11.95	98.50		110.45	161
4642	3" x 3/4" outlet		12	2		11.95	98.50		110.45	161
4644	3" x 1" CTS outlet		12	2		13.05	98.50		111.55	162
4646	3" x 1" outlet		12	2		13.05	98.50		111.55	162
4648	3" x 1-1/4" outlet		11	2.182		22	108		130	185
4650	4" x 1/2" CTS outlet		10	2.400		11.95	118		129.95	190
4652	4" x 1/2" outlet		10	2.400		11.95	118		129.95	190
4654	4" x 3/4" outlet		10	2.400		11.95	118		129.95	190
4656	4" x 1" CTS outlet		10	2.400		13.05	118		131.05	191
4658	4" x 1" outlet		10	2.400		13.05	118		131.05	191
4660	4" x 1-1/4" outlet	▼	9	2.667		22	131		153	221
4685	For protective sleeves, 1/2" CTS to 3/4" diameter outlets, add					.86			.86	.95
4690	1" CTS & IPS diameter outlets, add					1.44			1.44	1.58
4695	1-1/4" diameter outlets, add				▼	2.48			2.48	2.73
4700	Tapping tees, high volume, socket fusion outlets									
4705	1-1/2" punch, 2" x 1-1/4" outlet	B-20	11	2.182	Ea.	110	108		218	282
4710	1-7/8" punch, 3" x 1-1/4" outlet		11	2.182		110	108		218	282
4715	4" x 1-1/4" outlet	▼	9	2.667		110	131		241	320
4750	For protective sleeves, add				▼	2.90			2.90	3.19
4800	Service saddles, saddle contour x outlet, socket fusion outlets									
4801	IPS unless noted CTS									
4802	1-1/4" x 1/2" CTS outlet	B-20	17	1.412	Ea.	3.70	69.50		73.20	108
4804	1-1/4" x 1/2" outlet		17	1.412		3.70	69.50		73.20	108
4806	1-1/4" x 3/4" outlet		17	1.412		3.70	69.50		73.20	108
4808	1-1/4" x 1" CTS outlet		17	1.412		6.10	69.50		75.60	111
4810	1-1/4" x 1" outlet		17	1.412		13.90	69.50		83.40	119
4812	1-1/4" x 1-1/4" outlet		16	1.500		9.80	74		83.80	122
4814	1-1/2" x 1/2" CTS outlet		16	1.500		3.70	74		77.70	115
4816	1-1/2" x 1/2" outlet		16	1.500		3.70	74		77.70	115
4818	1-1/2" x 3/4" outlet		16	1.500		12.65	74		86.65	125
4820	1-1/2" x 1-1/4" outlet		15	1.600		9.80	79		88.80	129
4822	2" x 1/2" CTS outlet		12	2		3.70	98.50		102.20	152
4824	2" x 1/2" outlet		12	2		3.70	98.50		102.20	152
4826	2" x 3/4" outlet		12	2		3.70	98.50		102.20	152
4828	2" x 1" CTS outlet		12	2		6.10	98.50		104.60	155
4830	2" x 1" outlet		12	2		6.10	98.50		104.60	155
4832	2" x 1-1/4" outlet		11	2.182		15.75	108		123.75	178
4834	3" x 1/2" CTS outlet		12	2		3.70	98.50		102.20	152
4836	3" x 1/2" outlet	▼	12	2		3.70	98.50		102.20	152

33 52 16.23 Medium Density Polyethylene Piping	Crew	Daily Output	Labor-Hours	Unit	Material	2021 Bare Costs Labor	Equipment	Total	Total Incl O&P
4838	3" x 3/4" outlet	B-20	12	2	Ea.	3.70	98.50		102.20	152
4840	3" x 1" CTS outlet		12	2		6.10	98.50		104.60	155
4842	3" x 1" outlet		12	2		6.10	98.50		104.60	155
4844	3" x 1-1/4" outlet		11	2.182		14.50	108		122.50	177
4846	4" x 1/2" CTS outlet		10	2.400		3.70	118		121.70	181
4848	4" x 1/2" outlet		10	2.400		3.70	118		121.70	181
4850	4" x 3/4" outlet		10	2.400		3.70	118		121.70	181
4852	4" x 1" CTS outlet		10	2.400		6.10	118		124.10	184
4854	4" x 1" outlet		10	2.400		6.10	118		124.10	184
4856	4" x 1-1/4" outlet		9	2.667		15.75	131		146.75	214
4885	For protective sleeves, 1/2" CTS to 3/4" diameter outlets, add					.86			.86	.95
4890	1" CTS & IPS diameter outlets, add					1.44			1.44	1.58
4895	1-1/4" diameter outlets, add					2.48			2.48	2.73
4900	Spigot fittings, tees, SDR 7, 1/2" CTS diameter	B-20	10	2.400		25.50	118		143.50	205
4901	SDR 9.3, 1/2" diameter		9.50	2.526		11.30	125		136.30	198
4902	SDR 10, 1-1/4" diameter		9	2.667		11.30	131		142.30	209
4903	SDR 11, 3/4" diameter		9.50	2.526		23.50	125		148.50	212
4904	2" diameter		8	3		21.50	148		169.50	245
4906	SDR 11.5, 1" CTS diameter		9.33	2.571		60	127		187	256
4907	3" diameter		8	3		61	148		209	288
4908	4" diameter		6	4		157	197		354	470
4921	90° elbows, SDR 7, 1/2" CTS diameter		15	1.600		13.45	79		92.45	133
4922	SDR 9.3, 1/2" diameter		14.25	1.684		5.80	83		88.80	130
4923	SDR 10, 1-1/4" diameter		13.50	1.778		5.90	87.50		93.40	138
4924	SDR 11, 3/4" diameter		14.25	1.684		6.60	83		89.60	131
4925	2" diameter		12	2		25	98.50		123.50	176
4927	SDR 11.5, 1" CTS diameter		14	1.714		23	84.50		107.50	152
4928	3" diameter		12	2		56.50	98.50		155	210
4929	4" diameter		9	2.667		121	131		252	330
4935	Caps, SDR 7, 1/2" CTS diameter		30	.800		4.46	39.50		43.96	64
4936	SDR 9.3, 1/2" diameter		28.50	.842		4.46	41.50		45.96	67
4937	SDR 10, 1-1/4" diameter		27	.889		4.62	44		48.62	70.50
4938	SDR 11, 3/4" diameter		28.50	.842		4.77	41.50		46.27	67.50
4939	1" diameter		27	.889		5.30	44		49.30	71.50
4940	2" diameter		24	1		5.70	49.50		55.20	80.50
4942	SDR 11.5, 1" CTS diameter		28	.857		18.70	42.50		61.20	84
4943	3" diameter		24	1		30.50	49.50		80	108
4944	4" diameter		18	1.333		29	65.50		94.50	131
4946	SDR 13.5, 4" diameter		18	1.333		73.50	65.50		139	179
4950	Reducers, SDR 10 x SDR 11, 1-1/4" x 3/4" diameters		13.50	1.778		15.25	87.50		102.75	148
4951	1-1/4" x 1" diameters		13.50	1.778		15.25	87.50		102.75	148
4952	SDR 10 x SDR 11.5, 1-1/4" x 1" CTS diameters		13.50	1.778		15.25	87.50		102.75	148
4953	SDR 11 x SDR 7, 3/4" x 1/2" CTS diameters		14.25	1.684		6.80	83		89.80	132
4954	SDR 11 x SDR 9.3, 3/4" x 1/2" diameters		14.25	1.684		6.80	83		89.80	132
4955	SDR 11 x SDR 10, 2" x 1-1/4" diameters		12	2		15.80	98.50		114.30	165
4956	SDR 11 x SDR 11, 1" x 3/4" diameters		13.50	1.778		7.85	87.50		95.35	140
4957	2" x 3/4" diameters		12	2		15.60	98.50		114.10	165
4958	2" x 1" diameters		12	2		15.70	98.50		114.20	165
4959	SDR 11 x SDR 11.5, 1" x 1" CTS diameters		13.50	1.778		8	87.50		95.50	140
4960	2" x 1" CTS diameters		12	2		15.60	98.50		114.10	165
4962	SDR 11.5 x SDR 11, 3" x 2" diameters		12	2		21	98.50		119.50	171
4963	4" x 2" diameters		9	2.667		32	131		163	232
4964	SDR 11.5 x SDR 11.5, 4" x 3" diameters		9	2.667		34.50	131		165.50	235

For customer support on your Site Work & Landscape Costs with RSMeans Data, call 800.448.8182.

523

33 52 16.23 Medium Density Polyethylene Piping	Crew	Daily Output	Labor-Hours	Unit	Material	2021 Bare Costs Labor	2021 Bare Costs Equipment	Total	Total Incl O&P
6100	Fittings, compression									
6101	MDPE gas pipe, ASTM D2513/ASTM F1924-98									
6102	Caps, SDR 7, 1/2" CTS diameter	B-20	60	.400	Ea.	12.85	19.70		32.55	43.50
6104	SDR 9.3, 1/2" IPS diameter		58	.414		22.50	20.50		43	55
6106	SDR 10, 1/2" CTS diameter		60	.400		11.35	19.70		31.05	42
6108	1-1/4" IPS diameter		54	.444		74.50	22		96.50	115
6110	SDR 11, 3/4" IPS diameter		58	.414		20	20.50		40.50	52.50
6112	1" IPS diameter		54	.444		32.50	22		54.50	68.50
6114	1-1/4" IPS diameter		54	.444		78	22		100	119
6116	1-1/2" IPS diameter		52	.462		95	23		118	139
6118	2" IPS diameter		50	.480		83.50	23.50		107	128
6120	SDR 11.5, 1" CTS diameter		56	.429		20.50	21		41.50	54.50
6122	SDR 12.5, 1" CTS diameter		56	.429		24	21		45	58
6202	Reducers, SDR 7 x SDR 10, 1/2" CTS x 1/2" CTS diameters		30	.800		17.35	39.50		56.85	78
6204	SDR 9.3 x SDR 7, 1/2" IPS x 1/2" CTS diameters		29	.828		50	41		91	116
6206	SDR 11 x SDR 7, 3/4" IPS x 1/2" CTS diameters		29	.828		49.50	41		90.50	116
6208	1" IPS x 1/2" CTS diameters		27	.889		60	44		104	132
6210	SDR 11 x SDR 9.3, 3/4" IPS x 1/2" IPS diameters		29	.828		49	41		90	115
6212	1" IPS x 1/2" IPS diameters		27	.889		62	44		106	134
6214	SDR 11 x SDR 10, 2" IPS x 1-1/4" IPS diameters		25	.960		95.50	47.50		143	176
6216	SDR 11 x SDR 11, 1" IPS x 3/4" IPS diameters		27	.889		60	44		104	132
6218	1-1/4" IPS x 1" IPS diameters		27	.889		57	44		101	128
6220	2" IPS x 1-1/4" IPS diameters		25	.960		95.50	47.50		143	176
6222	SDR 11 x SDR 11.5, 1" IPS x 1" CTS diameters		27	.889		46	44		90	117
6224	SDR 11 x SDR 12.5, 1" IPS x 1" CTS diameters		27	.889		56	44		100	128
6226	SDR 11.5 x SDR 7, 1" CTS x 1/2" CTS diameters		28	.857		34	42.50		76.50	101
6228	SDR 11.5 x SDR 9.3, 1" CTS x 1/2" IPS diameters		28	.857		39.50	42.50		82	107
6230	SDR 11.5 x SDR 11, 1" CTS x 3/4" IPS diameters		28	.857		55	42.50		97.50	124
6232	SDR 12.5 x SDR 7, 1" CTS x 1/2" CTS diameters		28	.857		36	42.50		78.50	103
6234	SDR 12.5 x SDR 9.3, 1" CTS x 1/2" IPS diameters		28	.857		40.50	42.50		83	108
6236	SDR 12.5 x SDR 11, 1" CTS x 3/4" IPS diameters		28	.857		34.50	42.50		77	102
6302	Couplings, SDR 7, 1/2" CTS diameter		30	.800		13.10	39.50		52.60	73.50
6304	SDR 9.3, 1/2" IPS diameter		29	.828		23	41		64	86.50
6306	SDR 10, 1/2" CTS diameter		30	.800		12.40	39.50		51.90	72.50
6308	SDR 11, 3/4" IPS diameter		29	.828		23.50	41		64.50	86.50
6310	1" IPS diameter		27	.889		31.50	44		75.50	100
6312	SDR 11.5, 1" CTS diameter		28	.857		24.50	42.50		67	90.50
6314	SDR 12.5, 1" CTS diameter	B-20	28	.857	Ea.	27.50	42.50		70	94
6402	Repair couplings, SDR 10, 1-1/4" IPS diameter		27	.889		71.50	44		115.50	144
6404	SDR 11, 1-1/4" IPS diameter		27	.889		73	44		117	146
6406	1-1/2" IPS diameter		26	.923		78.50	45.50		124	155
6408	2" IPS diameter		25	.960		91.50	47.50		139	172
6502	Elbows, 90°, SDR 7, 1/2" CTS diameter		30	.800		23	39.50		62.50	84.50
6504	SDR 9.3, 1/2" IPS diameter		29	.828		40	41		81	105
6506	SDR 10, 1-1/4" IPS diameter		27	.889		98	44		142	174
6508	SDR 11, 3/4" IPS diameter		29	.828		36.50	41		77.50	101
6510	1" IPS diameter		27	.889		36.50	44		80.50	106
6512	1-1/4" IPS diameter		27	.889		97.50	44		141.50	173
6514	1-1/2" IPS diameter		26	.923		146	45.50		191.50	229
6516	2" IPS diameter		25	.960		124	47.50		171.50	208
6518	SDR 11.5, 1" CTS diameter		28	.857		37	42.50		79.50	104
6520	SDR 12.5, 1" CTS diameter		28	.857		53.50	42.50		96	123
6602	Tees, SDR 7, 1/2" CTS diameter		20	1.200		40.50	59		99.50	133

33 52 16 – Gas Hydrocarbon Piping

33 52 16.23 Medium Density Polyethylene Piping

		Crew	Daily Output	Labor-Hours	Unit	Material	2021 Bare Costs Labor	2021 Bare Costs Equipment	Total	Total Incl O&P
6604	SDR 9.3, 1/2" IPS diameter	B-20	19.33	1.241	Ea.	45	61		106	141
6606	SDR 10, 1-1/4" IPS diameter		18	1.333		127	65.50		192.50	239
6608	SDR 11, 3/4" IPS diameter		19.33	1.241		56.50	61		117.50	154
6610	1" IPS diameter		18	1.333		62	65.50		127.50	167
6612	1-1/4" IPS diameter		18	1.333		104	65.50		169.50	214
6614	1-1/2" IPS diameter		17.33	1.385		216	68.50		284.50	340
6616	2" IPS diameter		16.67	1.440		143	71		214	264
6618	SDR 11.5, 1" CTS diameter		18.67	1.286		57	63.50		120.50	158
6620	SDR 12.5, 1" CTS diameter	▼	18.67	1.286	▼	59	63.50		122.50	160
8100	Fittings, accessories									
8105	SDR 11, IPS unless noted CTS									
8110	Protective sleeves, for high volume tapping tees, butt fusion outlets	1 Skwk	16	.500	Ea.	5.45	28.50		33.95	49
8120	For tapping tees, socket outlets, CTS, 1/2" diameter		18	.444		1.42	25.50		26.92	39.50
8122	1" diameter		18	.444		2.51	25.50		28.01	41
8124	IPS, 1/2" diameter		18	.444		1.96	25.50		27.46	40
8126	3/4" diameter		18	.444		2.51	25.50		28.01	41
8128	1" diameter		18	.444		3.42	25.50		28.92	42
8130	1-1/4" diameter		18	.444		4.31	25.50		29.81	42.50
8205	Tapping tee test caps, type I, yellow PE cap, "aldyl style"		45	.178		44	10.15		54.15	64
8210	Type II, yellow polyethylene cap		45	.178		44	10.15		54.15	64
8215	High volume, yellow polyethylene cap		45	.178		56.50	10.15		66.65	78
8250	Quick connector, female x female inlets, 1/4" N.P.T.		50	.160		18.30	9.15		27.45	34
8255	Test hose, 24" length, 3/8" ID, male outlets, 1/4" N.P.T.	▼	50	.160	▼	25	9.15		34.15	41.50
8260	Quick connector and test hose assembly	1 Skwk	50	.160	Ea.	43.50	9.15		52.65	61.50
8305	Purge point caps, butt fusion, SDR 10, 1-1/4" diameter	B-22C	27	.593		32	30	5	67	85.50
8310	SDR 11, 1-1/4" diameter		27	.593		32	30	5	67	85.50
8315	2" diameter		24	.667		32.50	34	5.60	72.10	92
8320	3" diameter		24	.667		45	34	5.60	84.60	106
8325	4" diameter		18	.889		44	45	7.50	96.50	124
8340	Socket fusion, SDR 11, 1-1/4" diameter		27	.593		8	30	5	43	59.50
8345	2" diameter		24	.667		9.60	34	5.60	49.20	67.50
8350	3" diameter		24	.667		58.50	34	5.60	98.10	121
8355	4" diameter	▼	18	.889		43	45	7.50	95.50	123
8360	Purge test quick connector, female x female inlets, 1/4" N.P.T.	1 Skwk	50	.160		18.30	9.15		27.45	34
8365	Purge test hose, 24" length, 3/8" ID, male outlets, 1/4" N.P.T.		50	.160		25	9.15		34.15	41.50
8370	Purge test quick connector and test hose assembly	▼	50	.160		43.50	9.15		52.65	61.50
8405	Transition fittings, MDPE x zinc plated steel, SDR 7, 1/2" CTS x 1/2" MPT	B-22C	30	.533		24.50	27	4.50	56	72.50
8410	SDR 9.3, 1/2" IPS x 3/4" MPT		28.50	.561		38.50	28.50	4.73	71.73	89.50
8415	SDR 10, 1-1/4" IPS x 1-1/4" MPT		27	.593		36.50	30	5	71.50	91
8420	SDR 11, 3/4" IPS x 3/4" MPT		28.50	.561		19.40	28.50	4.73	52.63	69
8425	1" IPS x 1" MPT		27	.593		23.50	30	5	58.50	76.50
8430	1-1/4" IPS x 1-1/4" MPT		17	.941		44.50	48	7.95	100.45	129
8435	1-1/2" IPS x 1-1/2" MPT		25.50	.627		45.50	32	5.30	82.80	103
8440	2" IPS x 2" MPT		24	.667		52	34	5.60	91.60	114
8445	SDR 11.5, 1" CTS x 1" MPT	▼	28	.571	▼	32	29	4.82	65.82	84

33 52 16.26 High Density Polyethylene Piping

		Crew	Daily Output	Labor-Hours	Unit	Material	2021 Bare Costs Labor	2021 Bare Costs Equipment	Total	Total Incl O&P
2010	**HIGH DENSITY POLYETHYLENE PIPING**									
2020	ASTM D2513, not including excavation or backfill									
2200	Butt fused pipe									
2205	125psi coils, butt fusion joint @ 100', 1/2" CTS diameter, SDR 7	B-22C	2050	.008	L.F.	.13	.40	.07	.60	.80
2210	160psi coils, butt fusion joint @ 100', IPS 1/2" diameter, SDR 9		1950	.008		.17	.42	.07	.66	.89
2215	3/4" diameter, SDR 11	▼	1950	.008	▼	.25	.42	.07	.74	.98

For customer support on your Site Work & Landscape Costs with RSMeans Data, call 800.448.8182.

525

33 52 16.26 High Density Polyethylene Piping	Crew	Daily Output	Labor-Hours	Unit	Material	2021 Bare Costs Labor	Equipment	Total	Total Incl O&P
2220 1" diameter, SDR 11	B-22C	1850	.009	L.F.	.41	.44	.07	.92	1.19
2225 1-1/4" diameter, SDR 11		1850	.009		.60	.44	.07	1.11	1.40
2230 1-1/2" diameter, SDR 11		1750	.009		1.13	.46	.08	1.67	2.02
2235 2" diameter, SDR 11		1750	.009		.92	.46	.08	1.46	1.79
2240 3" diameter, SDR 11		1650	.010		1.51	.49	.08	2.08	2.49
2245 160psi 40' lengths, butt fusion joint, IPS 3" diameter, SDR 11	B-22A	660	.061		1.64	3.08	1.22	5.94	7.75
2250 4" diameter, SDR 11		500	.080		2.71	4.06	1.61	8.38	10.80
2255 6" diameter, SDR 11		500	.080		5.85	4.06	1.61	11.52	14.25
2260 8" diameter, SDR 11		420	.095		9.95	4.83	1.91	16.69	20.50
2265 10" diameter, SDR 11		340	.118		23	5.95	2.36	31.31	36.50
2270 12" diameter, SDR 11		300	.133		32	6.75	2.68	41.43	48
2400 Socket fused pipe									
2405 125psi coils, socket fusion coupling @ 100', 1/2" CTS diameter, SDR 7	B-20	1050	.023	L.F.	.15	1.13		1.28	1.85
2410 160psi coils, socket fusion coupling @ 100', IPS 1/2" diameter, SDR 9		1000	.024		.19	1.18		1.37	1.98
2415 3/4" diameter, SDR 11		1000	.024		.27	1.18		1.45	2.06
2420 1" diameter, SDR 11		950	.025		.43	1.25		1.68	2.33
2425 1-1/4" diameter, SDR 11		950	.025		.62	1.25		1.87	2.54
2430 1-1/2" diameter, SDR 11		900	.027		1.15	1.31		2.46	3.24
2435 2" diameter, SDR 11		850	.028		.95	1.39		2.34	3.12
2440 3" diameter, SDR 11		850	.028		1.62	1.39		3.01	3.87
2445 160psi 40' lengths, socket fusion coupling, IPS 3" diameter, SDR 11	B-21A	340	.118		1.93	6.45	1.27	9.65	13.10
2450 4" diameter, SDR 11	"	260	.154		3.25	8.45	1.66	13.36	18
2600 Compression coupled pipe									
2610 160psi coils, compression coupling @ 100', IPS 1/2" diameter, SDR 9	B-20	1000	.024	L.F.	.40	1.18		1.58	2.21
2615 3/4" diameter, SDR 11		1000	.024		.48	1.18		1.66	2.30
2620 1" diameter, SDR 11		950	.025		.72	1.25		1.97	2.66
2625 1-1/4" diameter, SDR 11		950	.025		1.33	1.25		2.58	3.32
2630 1-1/2" diameter, SDR 11		900	.027		1.92	1.31		3.23	4.08
2635 2" diameter, SDR 11		850	.028		1.84	1.39		3.23	4.10
3000 Fittings, butt fusion									
3010 SDR 11, IPS unless noted CTS									
3105 Caps, 1/2" diameter	B-22C	28.50	.561	Ea.	6.60	28.50	4.73	39.83	55
3110 3/4" diameter		28.50	.561		3.43	28.50	4.73	36.66	51.50
3115 1" diameter		27	.593		3.74	30	5	38.74	54.50
3120 1-1/4" diameter		27	.593		6.10	30	5	41.10	57
3125 1-1/2" diameter		25.50	.627		4.49	32	5.30	41.79	58
3130 2" diameter		24	.667		8	34	5.60	47.60	65.50
3135 3" diameter		24	.667		16	34	5.60	55.60	74.50
3140 4" diameter		18	.889		22	45	7.50	74.50	100
3145 6" diameter	B-22A	18	2.222		61.50	113	44.50	219	286
3150 8" diameter		15	2.667		98	135	53.50	286.50	370
3155 10" diameter		12	3.333		330	169	67	566	685
3160 12" diameter		10.50	3.810		350	193	76.50	619.50	760
3205 Reducers, 1/2" x 1/2" CTS diameters	B-22C	14.25	1.123		12.25	57	9.45	78.70	109
3210 3/4" x 1/2" CTS diameters		14.25	1.123		8.80	57	9.45	75.25	106
3215 1" x 1/2" CTS diameters		13.50	1.185		13.65	60	10	83.65	116
3220 1" x 1/2" diameters		13.50	1.185		16.80	60	10	86.80	120
3225 1" x 3/4" diameters		13.50	1.185		13.60	60	10	83.60	116
3230 1-1/4" x 1" diameters		13.50	1.185		15.20	60	10	85.20	118
3232 1-1/2" x 3/4" diameters		12.75	1.255		7.90	63.50	10.60	82	116
3233 1-1/2" x 1" diameters		12.75	1.255		7.90	63.50	10.60	82	116
3234 1-1/2" x 1-1/4" diameters	B-22C	12.75	1.255	Ea.	7.90	63.50	10.60	82	116
3235 2" x 1" diameters		12	1.333		17.25	67.50	11.25	96	132

33 52 Hydrocarbon Transmission and Distribution

33 52 16 – Gas Hydrocarbon Piping

33 52 16.26 High Density Polyethylene Piping	Crew	Daily Output	Labor-Hours	Unit	Material	2021 Bare Costs Labor	Equipment	Total	Total Incl O&P
3240	2" x 1-1/4" diameters	B-22C	12	1.333	Ea.	17.25	67.50	11.25	96	132
3245	2" x 1-1/2" diameters		12	1.333		9.05	67.50	11.25	87.80	123
3250	3" x 2" diameters		12	1.333		10.55	67.50	11.25	89.30	125
3255	4" x 2" diameters		9	1.778		14.40	90	15	119.40	167
3260	4" x 3" diameters		9	1.778		15.45	90	15	120.45	169
3262	6" x 3" diameters	B-22A	9	4.444		35.50	226	89	350.50	470
3265	6" x 4" diameters		9	4.444		47.50	226	89	362.50	485
3270	8" x 6" diameters		7.50	5.333		81	271	107	459	610
3275	10" x 8" diameters		6	6.667		170	340	134	644	840
3280	12" x 8" diameters		5.25	7.619		250	385	153	788	1,025
3285	12" x 10" diameters		5.25	7.619		116	385	153	654	875
3305	Elbows, 90°, 3/4" diameter	B-22C	14.25	1.123		5.85	57	9.45	72.30	102
3310	1" diameter		13.50	1.185		5.75	60	10	75.75	107
3315	1-1/4" diameter		13.50	1.185		6.70	60	10	76.70	108
3320	1-1/2" diameter		12.75	1.255		7.60	63.50	10.60	81.70	116
3325	2" diameter		12	1.333		8.25	67.50	11.25	87	122
3330	3" diameter		12	1.333		16.05	67.50	11.25	94.80	131
3335	4" diameter		9	1.778		21	90	15	126	175
3340	6" diameter	B-22A	9	4.444		55	226	89	370	495
3345	8" diameter		7.50	5.333		164	271	107	542	705
3350	10" diameter		6	6.667		395	340	134	869	1,075
3355	12" diameter		5.25	7.619		370	385	153	908	1,150
3380	45°, 3/4" diameter	B-22C	14.25	1.123		6.50	57	9.45	72.95	103
3385	1" diameter		13.50	1.185		6.50	60	10	76.50	108
3390	1-1/4" diameter		13.50	1.185		6.65	60	10	76.65	108
3395	1-1/2" diameter		12.75	1.255		8.65	63.50	10.60	82.75	117
3400	2" diameter		12	1.333		9.70	67.50	11.25	88.45	124
3405	3" diameter		12	1.333		16.45	67.50	11.25	95.20	131
3410	4" diameter		9	1.778		20.50	90	15	125.50	174
3415	6" diameter	B-22A	9	4.444		55	226	89	370	495
3420	8" diameter		7.50	5.333		171	271	107	549	710
3425	10" diameter		6	6.667		375	340	134	849	1,075
3430	12" diameter		5.25	7.619		735	385	153	1,273	1,550
3505	Tees, 1/2" CTS diameter	B-22C	10	1.600		8	81	13.50	102.50	146
3510	1/2" diameter		9.50	1.684		8.80	85.50	14.20	108.50	153
3515	3/4" diameter		9.50	1.684		6.55	85.50	14.20	106.25	151
3520	1" diameter		9	1.778		7.05	90	15	112.05	159
3525	1-1/4" diameter		9	1.778		7.50	90	15	112.50	160
3530	1-1/2" diameter		8.50	1.882		12.35	95.50	15.85	123.70	174
3535	2" diameter		8	2		9.75	102	16.85	128.60	181
3540	3" diameter		8	2		18.05	102	16.85	136.90	190
3545	4" diameter		6	2.667		26.50	135	22.50	184	257
3550	6" diameter	B-22A	6	6.667		65.50	340	134	539.50	725
3555	8" diameter		5	8		264	405	161	830	1,075
3560	10" diameter		4	10		490	510	201	1,201	1,525
3565	12" diameter		3.50	11.429		650	580	229	1,459	1,825
3600	Tapping tees, type II, 3/4" punch, butt fusion outlets									
3601	Saddle contour x outlet diameter, IPS unless noted CTS									
3602	1-1/4" x 1/2" CTS outlet	B-22C	17	.941	Ea.	11.20	48	7.95	67.15	92.50
3604	1-1/4" x 1/2" outlet		17	.941		11.20	48	7.95	67.15	92.50
3606	1-1/4" x 3/4" outlet		17	.941		11.20	48	7.95	67.15	92.50
3608	1-1/4" x 1" outlet		17	.941		13.05	48	7.95	69	94.50
3610	1-1/4" x 1-1/4" outlet		16	1		23.50	51	8.45	82.95	111

For customer support on your Site Work & Landscape Costs with RSMeans Data, call 800.448.8182.

33 52 16.26 High Density Polyethylene Piping	Crew	Daily Output	Labor-Hours	Unit	Material	2021 Bare Costs Labor	Equipment	Total	Total Incl O&P
3612	1-1/2" x 1/2" CTS outlet	B-22C	16	1	Ea.	11.20	51	8.45	70.65	97.50
3614	1-1/2" x 1/2" outlet		16	1		11.20	51	8.45	70.65	97.50
3616	1-1/2" x 3/4" outlet		16	1		11.20	51	8.45	70.65	97.50
3618	1-1/2" x 1" outlet		16	1		13.05	51	8.45	72.50	99.50
3620	1-1/2" x 1-1/4" outlet		15	1.067		23.50	54	9	86.50	116
3622	2" x 1/2" CTS outlet		12	1.333		11.20	67.50	11.25	89.95	126
3624	2" x 1/2" outlet		12	1.333		11.20	67.50	11.25	89.95	126
3626	2" x 3/4" outlet		12	1.333		11.20	67.50	11.25	89.95	126
3628	2" x 1" outlet		12	1.333		13.05	67.50	11.25	91.80	128
3630	2" x 1-1/4" outlet		11	1.455		23.50	74	12.25	109.75	150
3632	3" x 1/2" CTS outlet		12	1.333		11.20	67.50	11.25	89.95	126
3634	3" x 1/2" outlet		12	1.333		11.20	67.50	11.25	89.95	126
3636	3" x 3/4" outlet		12	1.333		11.20	67.50	11.25	89.95	126
3638	3" x 1" outlet		12	1.333		13.05	67.50	11.25	91.80	128
3640	3" x 1-1/4" outlet		11	1.455		13.05	74	12.25	99.30	139
3642	4" x 1/2" CTS outlet		10	1.600		11.20	81	13.50	105.70	149
3644	4" x 1/2" outlet		10	1.600		11.20	81	13.50	105.70	149
3646	4" x 3/4" outlet		10	1.600		11.20	81	13.50	105.70	149
3648	4" x 1" outlet		10	1.600		13.05	81	13.50	107.55	151
3650	4" x 1-1/4" outlet	↓	9	1.778		23.50	90	15	128.50	177
3652	6" x 1/2" CTS outlet	B-22A	10	4		11.20	203	80.50	294.70	405
3654	6" x 1/2" outlet		10	4		11.20	203	80.50	294.70	405
3656	6" x 3/4" outlet		10	4		11.20	203	80.50	294.70	405
3658	6" x 1" outlet		10	4		13.05	203	80.50	296.55	410
3660	6" x 1-1/4" outlet		9	4.444		23.50	226	89	338.50	460
3662	8" x 1/2" CTS outlet		8	5		11.20	254	100	365.20	500
3664	8" x 1/2" outlet		8	5		11.20	254	100	365.20	500
3666	8" x 3/4" outlet		8	5		11.20	254	100	365.20	500
3668	8" x 1" outlet		8	5		13.05	254	100	367.05	505
3670	8" x 1-1/4" outlet	↓	7	5.714		23.50	290	115	428.50	585
3675	For protective sleeves, 1/2" to 3/4" diameter outlets, add					.86			.86	.95
3680	1" diameter outlets, add					1.44			1.44	1.58
3685	1-1/4" diameter outlets, add				↓	2.48			2.48	2.73
3700	Tapping tees, high volume, butt fusion outlets									
3705	1-1/2" punch, 2" x 2" outlet	B-22C	11	1.455	Ea.	110	74	12.25	196.25	246
3710	1-7/8" punch, 3" x 2" outlet		11	1.455		110	74	12.25	196.25	246
3715	4" x 2" outlet	↓	9	1.778		110	90	15	215	273
3720	6" x 2" outlet	B-22A	9	4.444		135	226	89	450	580
3725	8" x 2" outlet		7	5.714		110	290	115	515	680
3730	10" x 2" outlet		6	6.667		110	340	134	584	775
3735	12" x 2" outlet	↓	5	8		148	405	161	714	945
3750	For protective sleeves, add				↓	2.90			2.90	3.19
3800	Service saddles, saddle contour x outlet, butt fusion outlets									
3801	IPS unless noted CTS									
3802	1-1/4" x 3/4" outlet	B-22C	17	.941	Ea.	6.65	48	7.95	62.60	87.50
3804	1-1/4" x 1" outlet		16	1		6.65	51	8.45	66.10	92.50
3806	1-1/4" x 1-1/4" outlet		16	1		6.95	51	8.45	66.40	93
3808	1-1/2" x 3/4" outlet		16	1		14.15	51	8.45	73.60	101
3810	1-1/2" x 1-1/4" outlet		15	1.067		14.30	54	9	77.30	107
3812	2" x 3/4" outlet		12	1.333		14.15	67.50	11.25	92.90	129
3814	2" x 1" outlet		11	1.455		6.65	74	12.25	92.90	132
3816	2" x 1-1/4" outlet		11	1.455		6.95	74	12.25	93.20	132
3818	3" x 3/4" outlet	↓	12	1.333	↓	14.15	67.50	11.25	92.90	129

For customer support on your Site Work & Landscape Costs with RSMeans Data, call 800.448.8182.

33 52 16.26 High Density Polyethylene Piping		Crew	Daily Output	Labor-Hours	Unit	Material	2021 Bare Costs Labor	Equipment	Total	Total Incl O&P
3820	3" x 1" outlet	B-22C	11	1.455	Ea.	14.15	74	12.25	100.40	140
3822	3" x 1-1/4" outlet		11	1.455		6.95	74	12.25	93.20	132
3824	4" x 3/4" outlet		10	1.600		14.15	81	13.50	108.65	152
3826	4" x 1" outlet		9	1.778		14.15	90	15	119.15	167
3828	4" x 1-1/4" outlet		9	1.778		14.30	90	15	119.30	167
3830	6" x 3/4" outlet	B-22A	10	4		6.65	203	80.50	290.15	400
3832	6" x 1" outlet		9	4.444		6.65	226	89	321.65	440
3834	6" x 1-1/4" outlet		9	4.444		6.95	226	89	321.95	440
3836	8" x 3/4" outlet		8	5		14.15	254	100	368.15	505
3838	8" x 1" outlet		7	5.714		6.65	290	115	411.65	570
3840	8" x 1-1/4" outlet		7	5.714		6.95	290	115	411.95	570
3880	For protective sleeves, 3/4" diameter outlets, add					.86			.86	.95
3885	1" diameter outlets, add					1.44			1.44	1.58
3890	1-1/4" diameter outlets, add					2.48			2.48	2.73
3905	Branch saddles, contour x outlet, 2" x 2" outlet	B-22C	11	1.455		18.65	74	12.25	104.90	145
3910	3" x 2" outlet	B-22C	11	1.455	Ea.	18.65	74	12.25	104.90	145
3915	3" x 3" outlet		11	1.455		25	74	12.25	111.25	152
3920	4" x 2" outlet		9	1.778		18.65	90	15	123.65	172
3925	4" x 3" outlet		9	1.778		25	90	15	130	179
3930	4" x 4" outlet		9	1.778		34	90	15	139	189
3935	6" x 2" outlet	B-22A	9	4.444		18.65	226	89	333.65	455
3940	6" x 3" outlet		9	4.444		25	226	89	340	460
3945	6" x 4" outlet		9	4.444		34	226	89	349	470
3950	6" x 6" outlet		9	4.444		81.50	226	89	396.50	525
3955	8" x 2" outlet		7	5.714		18.65	290	115	423.65	580
3960	8" x 3" outlet		7	5.714		25	290	115	430	590
3965	8" x 4" outlet		7	5.714		34	290	115	439	600
3970	8" x 6" outlet		7	5.714		81.50	290	115	486.50	650
3980	Ball valves, full port, 3/4" diameter	B-22C	14.25	1.123		57.50	57	9.45	123.95	159
3982	1" diameter		13.50	1.185		57.50	60	10	127.50	164
3984	1-1/4" diameter		13.50	1.185		57.50	60	10	127.50	165
3986	1-1/2" diameter		12.75	1.255		84.50	63.50	10.60	158.60	200
3988	2" diameter		12	1.333		132	67.50	11.25	210.75	258
3990	3" diameter		12	1.333		310	67.50	11.25	388.75	455
3992	4" diameter		9	1.778		410	90	15	515	600
3994	6" diameter	B-22A	9	4.444		1,075	226	89	1,390	1,625
3996	8" diameter	"	7.50	5.333		1,750	271	107	2,128	2,450
4000	Fittings, socket fusion									
4010	SDR 11, IPS unless noted CTS									
4105	Caps, 1/2" CTS diameter	B-20	30	.800	Ea.	2.37	39.50		41.87	61.50
4110	1/2" diameter		28.50	.842		2.38	41.50		43.88	64.50
4115	3/4" diameter		28.50	.842		2.60	41.50		44.10	65
4120	1" diameter		27	.889		3.29	44		47.29	69
4125	1-1/4" diameter		27	.889		3.45	44		47.45	69.50
4130	1-1/2" diameter		25.50	.941		3.81	46.50		50.31	73.50
4135	2" diameter		24	1		4.01	49.50		53.51	78.50
4140	3" diameter		24	1		24.50	49.50		74	101
4145	4" diameter		18	1.333		40	65.50		105.50	143
4205	Reducers, 1/2" x 1/2" CTS diameters		14.25	1.684		6	83		89	131
4210	3/4" x 1/2" CTS diameters		14.25	1.684		6.15	83		89.15	131
4215	3/4" x 1/2" diameters		14.25	1.684		6.15	83		89.15	131
4220	1" x 1/2" CTS diameters		13.50	1.778		5.95	87.50		93.45	138
4225	1" x 1/2" diameters		13.50	1.778		6.30	87.50		93.80	138

529

33 52 16.26 High Density Polyethylene Piping		Crew	Daily Output	Labor-Hours	Unit	Material	2021 Bare Costs Labor	Equipment	Total	Total Incl O&P
4230	1" x 3/4" diameters	B-20	13.50	1.778	Ea.	5.15	87.50		92.65	137
4235	1-1/4" x 1/2" CTS diameters		13.50	1.778		5.95	87.50		93.45	138
4240	1-1/4" x 3/4" diameters		13.50	1.778		6.90	87.50		94.40	139
4245	1-1/4" x 1" diameters		13.50	1.778		5.65	87.50		93.15	137
4250	1-1/2" x 3/4" diameters		12.75	1.882		8.05	93		101.05	148
4252	1-1/2" x 1" diameters		12.75	1.882		8.05	93		101.05	148
4255	1-1/2" x 1-1/4" diameters		12.75	1.882		7.95	93		100.95	148
4260	2" x 3/4" diameters		12	2		9.90	98.50		108.40	159
4265	2" x 1" diameters		12	2		9.55	98.50		108.05	159
4270	2" x 1-1/4" diameters		12	2		10.35	98.50		108.85	159
4275	3" x 2" diameters		12	2		11.85	98.50		110.35	161
4280	4" x 2" diameters		9	2.667		28	131		159	228
4285	4" x 3" diameters		9	2.667		26	131		157	226
4305	Couplings, 1/2" CTS diameter		15	1.600		1.95	79		80.95	120
4310	1/2" diameter		14.25	1.684		1.95	83		84.95	126
4315	3/4" diameter		14.25	1.684		1.54	83		84.54	126
4320	1" diameter		13.50	1.778		1.52	87.50		89.02	133
4325	1-1/4" diameter		13.50	1.778		2.22	87.50		89.72	133
4330	1-1/2" diameter		12.75	1.882		2.42	93		95.42	142
4335	2" diameter		12	2		2.51	98.50		101.01	151
4340	3" diameter		12	2		11.45	98.50		109.95	161
4345	4" diameter		9	2.667		21.50	131		152.50	221
4405	Elbows, 90°, 1/2" CTS diameter		15	1.600		3.35	79		82.35	122
4410	1/2" diameter		14.25	1.684		3.38	83		86.38	128
4415	3/4" diameter		14.25	1.684		2.78	83		85.78	127
4420	1" diameter		13.50	1.778		3.36	87.50		90.86	135
4425	1-1/4" diameter		13.50	1.778		4.62	87.50		92.12	136
4430	1-1/2" diameter		12.75	1.882		6.75	93		99.75	146
4435	2" diameter		12	2		6.85	98.50		105.35	156
4440	3" diameter		12	2		25	98.50		123.50	176
4445	4" diameter		9	2.667		100	131		231	305
4450	45°, 3/4" diameter		14.25	1.684		6.80	83		89.80	132
4455	1" diameter		13.50	1.778		7.65	87.50		95.15	139
4460	1-1/4" diameter		13.50	1.778		6.45	87.50		93.95	138
4465	1-1/2" diameter		12.75	1.882		10.10	93		103.10	150
4470	2" diameter		12	2		7.50	98.50		106	156
4475	3" diameter		12	2		41	98.50		139.50	194
4480	4" diameter		9	2.667		53.50	131		184.50	256
4505	Tees, 1/2" CTS diameter		10	2.400		2.89	118		120.89	180
4510	1/2" diameter		9.50	2.526		2.87	125		127.87	189
4515	3/4" diameter		9.50	2.526		2.86	125		127.86	189
4520	1" diameter		9	2.667		3.53	131		134.53	201
4525	1-1/4" diameter		9	2.667		4.91	131		135.91	202
4530	1-1/2" diameter		8.50	2.824		8.45	139		147.45	217
4535	2" diameter		8	3		9.05	148		157.05	231
4540	3" diameter		8	3		37.50	148		185.50	262
4545	4" diameter	▼	6	4	▼	72.50	197		269.50	375
4600	Tapping tees, type II, 3/4" punch, socket fusion outlets									
4601	Saddle countour x outlet diameter, IPS unless noted CTS									
4602	1-1/4" x 1/2" CTS outlet	B-20	17	1.412	Ea.	11.20	69.50		80.70	116
4604	1-1/4" x 1/2" outlet		17	1.412		11.20	69.50		80.70	116
4606	1-1/4" x 3/4" outlet		17	1.412		11.20	69.50		80.70	116
4608	1-1/4" x 1" outlet	▼	17	1.412	▼	13.05	69.50		82.55	118

33 52 16.26 High Density Polyethylene Piping	Crew	Daily Output	Labor-Hours	Unit	Material	2021 Bare Costs Labor	Equipment	Total	Total Incl O&P
4610	1-1/4" x 1-1/4" outlet	B-20	16	1.500	Ea.	23.50	74		97.50	137
4612	1-1/2" x 1/2" CTS outlet		16	1.500		11.20	74		85.20	123
4614	1-1/2" x 1/2" outlet		16	1.500		11.20	74		85.20	123
4616	1-1/2" x 3/4" outlet		16	1.500		11.20	74		85.20	123
4618	1-1/2" x 1" outlet		16	1.500		13.05	74		87.05	125
4620	1-1/2" x 1-1/4" outlet		15	1.600		23.50	79		102.50	144
4622	2" x 1/2" CTS outlet		12	2		11.20	98.50		109.70	160
4624	2" x 1/2" outlet		12	2		11.20	98.50		109.70	160
4626	2" x 3/4" outlet		12	2		11.20	98.50		109.70	160
4628	2" x 1" outlet		12	2		13.05	98.50		111.55	162
4630	2" x 1-1/4" outlet		11	2.182		23.50	108		131.50	187
4632	3" x 1/2" CTS outlet		12	2		11.20	98.50		109.70	160
4634	3" x 1/2" outlet		12	2		11.20	98.50		109.70	160
4636	3" x 3/4" outlet		12	2		11.20	98.50		109.70	160
4638	3" x 1" outlet		12	2		13.05	98.50		111.55	162
4640	3" x 1-1/4" outlet		11	2.182		23.50	108		131.50	187
4642	4" x 1/2" CTS outlet		10	2.400		11.20	118		129.20	189
4644	4" x 1/2" outlet		10	2.400		11.20	118		129.20	189
4646	4" x 3/4" outlet		10	2.400		11.20	118		129.20	189
4648	4" x 1" outlet		10	2.400		13.05	118		131.05	191
4650	4" x 1-1/4" outlet		9	2.667		23.50	131		154.50	223
4652	6" x 1/2" CTS outlet	B-21A	10	4		11.20	219	43.50	273.70	385
4654	6" x 1/2" outlet	B-20	10	2.400		11.20	118		129.20	189
4656	6" x 3/4" outlet	B-21A	10	4		11.20	219	43.50	273.70	385
4658	6" x 1" outlet		10	4		13.05	219	43.50	275.55	385
4660	6" x 1-1/4" outlet		9	4.444		23.50	244	48	315.50	445
4662	8" x 1/2" CTS outlet		8	5		11.20	274	54	339.20	480
4664	8" x 1/2" outlet		8	5		11.20	274	54	339.20	480
4666	8" x 3/4" outlet		8	5		11.20	274	54	339.20	480
4668	8" x 1" outlet		8	5		13.05	274	54	341.05	485
4670	8" x 1-1/4" outlet		7	5.714		23.50	315	62	400.50	560
4675	For protective sleeves, 1/2" to 3/4" diameter outlets, add					.86			.86	.95
4680	1" diameter outlets, add				Ea.	1.44			1.44	1.58
4685	1-1/4" diameter outlets, add				"	2.48			2.48	2.73
4700	Tapping tees, high volume, socket fusion outlets									
4705	1-1/2" punch, 2" x 1-1/4" outlet	B-20	11	2.182	Ea.	110	108		218	282
4710	1-7/8" punch, 3" x 1-1/4" outlet		11	2.182		110	108		218	282
4715	4" x 1-1/4" outlet		9	2.667		110	131		241	320
4720	6" x 1-1/4" outlet	B-21A	9	4.444		110	244	48	402	540
4725	8" x 1-1/4" outlet		7	5.714		110	315	62	487	655
4730	10" x 1-1/4" outlet		6	6.667		110	365	72	547	745
4735	12" x 1-1/4" outlet		5	8		110	440	86.50	636.50	870
4750	For protective sleeves, add					2.90			2.90	3.19
4800	Service saddles, saddle contour x outlet, socket fusion outlets									
4801	IPS unless noted CTS									
4802	1-1/4" x 1/2" CTS outlet	B-20	17	1.412	Ea.	5.90	69.50		75.40	111
4804	1-1/4" x 1/2" outlet		17	1.412		5.90	69.50		75.40	111
4806	1-1/4" x 3/4" outlet		17	1.412		5.90	69.50		75.40	111
4808	1-1/4" x 1" outlet		16	1.500		6.95	74		80.95	119
4810	1-1/4" x 1-1/4" outlet		16	1.500		9.80	74		83.80	122
4812	1-1/2" x 1/2" CTS outlet		16	1.500		5.90	74		79.90	118
4814	1-1/2" x 1/2" outlet		16	1.500		5.90	74		79.90	118
4816	1-1/2" x 3/4" outlet		16	1.500		5.90	74		79.90	118

For customer support on your Site Work & Landscape Costs with RSMeans Data, call 800.448.8182.

531

33 52 16.26 High Density Polyethylene Piping		Crew	Daily Output	Labor-Hours	Unit	Material	2021 Bare Costs Labor	Equipment	Total	Total Incl O&P
4818	1-1/2" x 1-1/4" outlet	B-20	15	1.600	Ea.	9.80	79		88.80	129
4820	2" x 1/2" CTS outlet		12	2		5.90	98.50		104.40	155
4822	2" x 1/2" outlet		12	2		5.90	98.50		104.40	155
4824	2" x 3/4" outlet		12	2		5.90	98.50		104.40	155
4826	2" x 1" outlet		11	2.182		6.95	108		114.95	169
4828	2" x 1-1/4" outlet		11	2.182		9.80	108		117.80	172
4830	3" x 1/2" CTS outlet		12	2		5.90	98.50		104.40	155
4832	3" x 1/2" outlet		12	2		5.90	98.50		104.40	155
4834	3" x 3/4" outlet		12	2		5.90	98.50		104.40	155
4836	3" x 1" outlet		11	2.182		6.95	108		114.95	169
4838	3" x 1-1/4" outlet		11	2.182		9.80	108		117.80	172
4840	4" x 1/2" CTS outlet		10	2.400		5.90	118		123.90	184
4842	4" x 1/2" outlet		10	2.400		5.90	118		123.90	184
4844	4" x 3/4" outlet		10	2.400		5.90	118		123.90	184
4846	4" x 1" outlet		9	2.667		6.95	131		137.95	205
4848	4" x 1-1/4" outlet		9	2.667		9.80	131		140.80	208
4850	6" x 1/2" CTS outlet	B-21A	10	4		5.90	219	43.50	268.40	380
4852	6" x 1/2" outlet		10	4		5.90	219	43.50	268.40	380
4854	6" x 3/4" outlet		10	4		5.90	219	43.50	268.40	380
4856	6" x 1" outlet	B-21A	9	4.444	Ea.	6.95	244	48	298.95	425
4858	6" x 1-1/4" outlet		9	4.444		9.80	244	48	301.80	430
4860	8" x 1/2" CTS outlet		8	5		5.90	274	54	333.90	475
4862	8" x 1/2" outlet		8	5		5.90	274	54	333.90	475
4864	8" x 3/4" outlet		8	5		5.90	274	54	333.90	475
4866	8" x 1" outlet		7	5.714		7.05	315	62	384.05	540
4868	8" x 1-1/4" outlet		7	5.714		10	315	62	387	545
4880	For protective sleeves, 1/2" to 3/4" diameter outlets, add					.86			.86	.95
4885	1" diameter outlets, add					1.44			1.44	1.58
4890	1-1/4" diameter outlets, add					2.48			2.48	2.73
4900	Reducer tees, 1-1/4" x 3/4" x 3/4" diameters	B-20	9	2.667		7.50	131		138.50	205
4902	1-1/4" x 3/4" x 1" diameters		9	2.667		9.70	131		140.70	208
4904	1-1/4" x 3/4" x 1-1/4" diameters		9	2.667		9.70	131		140.70	208
4906	1-1/4" x 1" x 3/4" diameters		9	2.667		9.70	131		140.70	208
4908	1-1/4" x 1" x 1" diameters		9	2.667		7.40	131		138.40	205
4910	1-1/4" x 1" x 1-1/4" diameters		9	2.667		9.70	131		140.70	208
4912	1-1/4" x 1-1/4" x 3/4" diameters		9	2.667		7	131		138	205
4914	1-1/4" x 1-1/4" x 1" diameters		9	2.667		7.40	131		138.40	205
4916	1-1/2" x 3/4" x 3/4" diameters		8.50	2.824		12.30	139		151.30	222
4918	1-1/2" x 3/4" x 1" diameters		8.50	2.824		12.30	139		151.30	222
4920	1-1/2" x 3/4" x 1-1/4" diameters		8.50	2.824		12.30	139		151.30	222
4922	1-1/2" x 3/4" x 1-1/2" diameters		8.50	2.824		12.30	139		151.30	222
4924	1-1/2" x 1" x 3/4" diameters		8.50	2.824		12.30	139		151.30	222
4926	1-1/2" x 1" x 1" diameters		8.50	2.824		12.30	139		151.30	222
4928	1-1/2" x 1" x 1-1/4" diameters		8.50	2.824		12.30	139		151.30	222
4930	1-1/2" x 1" x 1-1/2" diameters		8.50	2.824		12.30	139		151.30	222
4932	1-1/2" x 1-1/4" x 3/4" diameters		8.50	2.824		10.35	139		149.35	219
4934	1-1/2" x 1-1/4" x 1" diameters		8.50	2.824		10.20	139		149.20	219
4936	1-1/2" x 1-1/4" x 1-1/4" diameters		8.50	2.824		10.35	139		149.35	219
4938	1-1/2" x 1-1/4" x 1-1/2" diameters		8.50	2.824		12.30	139		151.30	222
4940	1-1/2" x 1-1/2" x 3/4" diameters		8.50	2.824		9.65	139		148.65	219
4942	1-1/2" x 1-1/2" x 1" diameters		8.50	2.824		9.65	139		148.65	219
4944	1-1/2" x 1-1/2" x 1-1/4" diameters		8.50	2.824		9.65	139		148.65	219
4946	2" x 1-1/4" x 3/4" diameters		8	3		11.20	148		159.20	233

33 52 16.26 High Density Polyethylene Piping	Crew	Daily Output	Labor-Hours	Unit	Material	2021 Bare Costs Labor	2021 Bare Costs Equipment	Total	Total Incl O&P
4948	2" x 1-1/4" x 1" diameters	B-20	8	3	Ea.	10.85	148		158.85	233
4950	2" x 1-1/4" x 1-1/4" diameters		8	3		11.30	148		159.30	233
4952	2" x 1-1/2" x 3/4" diameters		8	3		11.20	148		159.20	233
4954	2" x 1-1/2" x 1" diameters		8	3		10.85	148		158.85	233
4956	2" x 1-1/2" x 1-1/4" diameters		8	3		10.85	148		158.85	233
4958	2" x 2" x 3/4" diameters		8	3		11.15	148		159.15	233
4960	2" x 2" x 1" diameters		8	3		10.95	148		158.95	233
4962	2" x 2" x 1-1/4" diameters	▼	8	3	▼	11.20	148		159.20	233
8100	Fittings, accessories									
8105	SDR 11, IPS unless noted CTS									
8110	Flange adapters, 2" x 6" long	1 Skwk	32	.250	Ea.	15.55	14.30		29.85	38.50
8115	3" x 6" long		32	.250		19.20	14.30		33.50	42.50
8120	4" x 6" long	▼	24	.333		24	19.05		43.05	55
8125	6" x 8" long	2 Skwk	24	.667		38	38		76	99.50
8130	8" x 9" long	"	20	.800		53.50	45.50		99	127
8135	Backup flanges, 2" diameter	1 Skwk	32	.250		19.80	14.30		34.10	43.50
8140	3" diameter		32	.250		26.50	14.30		40.80	51
8145	4" diameter	▼	24	.333		52.50	19.05		71.55	86
8150	6" diameter	2 Skwk	24	.667		54	38		92	117
8155	8" diameter	"	20	.800	▼	92	45.50		137.50	170
8200	Tapping tees, test caps									
8205	Type II, yellow polyethylene cap	1 Skwk	45	.178	Ea.	44	10.15		54.15	64
8210	High volume, black polyethylene cap		45	.178		56.50	10.15		66.65	78
8215	Quick connector, female x female inlets, 1/4" N.P.T.		50	.160		18.30	9.15		27.45	34
8220	Test hose, 24" length, 3/8" ID, male outlets, 1/4" N.P.T.		50	.160		25	9.15		34.15	41.50
8225	Quick connector and test hose assembly	▼	50	.160	▼	43.50	9.15		52.65	61.50
8300	Threaded transition fittings									
8302	HDPE x MPT zinc plated steel, SDR 7, 1/2" CTS x 1/2" MPT	B-22C	30	.533	Ea.	30	27	4.50	61.50	79
8304	SDR 9.3, 1/2" IPS x 3/4" MPT		28.50	.561		40	28.50	4.73	73.23	91.50
8306	SDR 11, 3/4" IPS x 3/4" MPT		28.50	.561		19.70	28.50	4.73	52.93	69
8308	1" IPS x 1" MPT		27	.593		26	30	5	61	79
8310	1-1/4" IPS x 1-1/4" MPT		27	.593		41	30	5	76	95.50
8312	1-1/2" IPS x 1-1/2" MPT		25.50	.627		47.50	32	5.30	84.80	106
8314	2" IPS x 2" MPT		24	.667		54.50	34	5.60	94.10	117
8322	HDPE x MPT 316 stainless steel, SDR 11, 3/4" IPS x 3/4" MPT		28.50	.561		26	28.50	4.73	59.23	76.50
8324	1" IPS x 1" MPT		27	.593		27	30	5	62	80
8326	1-1/4" IPS x 1-1/4" MPT		27	.593		33	30	5	68	87
8328	2" IPS x 2" MPT		24	.667		40	34	5.60	79.60	101
8330	3" IPS x 3" MPT		24	.667		79.50	34	5.60	119.10	144
8332	4" IPS x 4" MPT	▼	18	.889		109	45	7.50	161.50	196
8334	6" IPS x 6" MPT	B-22A	18	2.222		203	113	44.50	360.50	440
8342	HDPE x FPT 316 stainless steel, SDR 11, 3/4" IPS x 3/4" FPT	B-22C	28.50	.561		34	28.50	4.73	67.23	85
8344	1" IPS x 1" FPT		27	.593		45.50	30	5	80.50	101
8346	1-1/4" IPS x 1-1/4" FPT		27	.593		73.50	30	5	108.50	132
8348	1-1/2" IPS x 1-1/2" FPT	▼	25.50	.627	▼	70.50	32	5.30	107.80	131
8350	2" IPS x 2" FPT	B-22C	24	.667	Ea.	93	34	5.60	132.60	160
8352	3" IPS x 3" FPT		24	.667		180	34	5.60	219.60	255
8354	4" IPS x 4" FPT		18	.889		259	45	7.50	311.50	360
8362	HDPE x MPT epoxy carbon steel, SDR 11, 3/4" IPS x 3/4" MPT		28.50	.561		19.75	28.50	4.73	52.98	69
8364	1" IPS x 1" MPT		27	.593		22	30	5	57	74.50
8366	1-1/4" IPS x 1-1/4" MPT		27	.593		23	30	5	58	76
8368	1-1/2" IPS x 1-1/2" MPT		25.50	.627		25	32	5.30	62.30	81
8370	2" IPS x 2" MPT	▼	24	.667	▼	27.50	34	5.60	67.10	86.50

33 52 16 – Gas Hydrocarbon Piping

33 52 16.26 High Density Polyethylene Piping		Crew	Daily Output	Labor-Hours	Unit	Material	2021 Bare Costs Labor	Equipment	Total	Total Incl O&P
8372	3" IPS x 3" MPT	B-22C	24	.667	Ea.	42	34	5.60	81.60	103
8374	4" IPS x 4" MPT		18	.889		57	45	7.50	109.50	139
8382	HDPE x FPT epoxy carbon steel, SDR 11, 3/4" IPS x 3/4" FPT		28.50	.561		24	28.50	4.73	57.23	74
8384	1" IPS x 1" FPT		27	.593		26.50	30	5	61.50	79.50
8386	1-1/4" IPS x 1-1/4" FPT		27	.593		56.50	30	5	91.50	113
8388	1-1/2" IPS x 1-1/2" FPT		25.50	.627		57	32	5.30	94.30	116
8390	2" IPS x 2" FPT		24	.667		60.50	34	5.60	100.10	123
8392	3" IPS x 3" FPT		24	.667		103	34	5.60	142.60	170
8394	4" IPS x 4" FPT		18	.889		129	45	7.50	181.50	218
8402	HDPE x MPT poly coated carbon steel, SDR 11, 1" IPS x 1" MPT		27	.593		21.50	30	5	56.50	74
8404	1-1/4" IPS x 1-1/4" MPT		27	.593		28	30	5	63	81
8406	1-1/2" IPS x 1-1/2" MPT		25.50	.627		33	32	5.30	70.30	90
8408	2" IPS x 2" MPT		24	.667		37.50	34	5.60	77.10	97.50
8410	3" IPS x 3" MPT		24	.667		72.50	34	5.60	112.10	136
8412	4" IPS x 4" MPT		18	.889		113	45	7.50	165.50	200
8414	6" IPS x 6" MPT	B-22A	18	2.222		263	113	44.50	420.50	505
8602	Socket fused HDPE x MPT brass, SDR 11, 3/4" IPS x 3/4" MPT	B-20	28.50	.842		11.65	41.50		53.15	75
8604	1" IPS x 3/4" MPT		27	.889		14.10	44		58.10	81
8606	1" IPS x 1" MPT		27	.889		13.60	44		57.60	80.50
8608	1-1/4" IPS x 3/4" MPT		27	.889		14.60	44		58.60	81.50
8610	1-1/4" IPS x 1" MPT		27	.889		14.60	44		58.60	81.50
8612	1-1/4" IPS x 1-1/4" MPT		27	.889		21	44		65	88.50
8614	1-1/2" IPS x 1-1/2" MPT		25.50	.941		28.50	46.50		75	101
8616	2" IPS x 2" MPT		24	1		29	49.50		78.50	106
8618	Socket fused HDPE x FPT brass, SDR 11, 3/4" IPS x 3/4" FPT		28.50	.842		11.65	41.50		53.15	75
8620	1" IPS x 1/2" FPT		27	.889		11.65	44		55.65	78.50
8622	1" IPS x 3/4" FPT		27	.889		14.10	44		58.10	81
8624	1" IPS x 1" FPT		27	.889		13.60	44		57.60	80.50
8626	1-1/4" IPS x 3/4" FPT		27	.889		14.60	44		58.60	81.50
8628	1-1/4" IPS x 1" FPT		27	.889		14.10	44		58.10	81
8630	1-1/4" IPS x 1-1/4" FPT		27	.889		21	44		65	88.50
8632	1-1/2" IPS x 1-1/2" FPT		25.50	.941		29	46.50		75.50	102
8634	2" IPS x 1-1/2" FPT	B-20	24	1	Ea.	29.50	49.50		79	107
8636	2" IPS x 2" FPT	"	24	1	"	31	49.50		80.50	109

33 59 Hydrocarbon Utility Metering

33 59 33 – Natural-Gas Metering

33 59 33.10 Piping, Valves and Meters, Gas Distribution

		Crew	Daily Output	Labor-Hours	Unit	Material	2021 Bare Costs Labor	Equipment	Total	Total Incl O&P
0010	**PIPING, VALVES & METERS, GAS DISTRIBUTION**									
0020	Not including excavation or backfill									
0100	Gas stops, with or without checks									
0140	1-1/4" size	1 Plum	12	.667	Ea.	73.50	45		118.50	148
0180	1-1/2" size		10	.800		92	54		146	182
0200	2" size		8	1		148	67.50		215.50	263
0600	Pressure regulator valves, iron and bronze									
0640	1-1/2" diameter	1 Plum	13	.615	Ea.	173	41.50		214.50	253
0680	2" diameter	"	11	.727		276	49		325	380
0700	3" diameter	Q-1	13	1.231		630	75		705	800
0740	4" diameter	"	8	2		2,125	122		2,247	2,500
1000	Meter risers, ASTM D2513, 125 psi									
1002	MDPE x MPT epoxy coated carbon steel, fixed, anodeless									

534

33 59 Hydrocarbon Utility Metering

33 59 33 – Natural-Gas Metering

33 59 33.10 Piping, Valves and Meters, Gas Distribution	Crew	Daily Output	Labor-Hours	Unit	Material	2021 Bare Costs Labor	Equipment	Total	Total Incl O&P
1005 SDR 7, 1/2" CTS x 1/2" MPT, 22" x 24"	B-22C	30	.533	Ea.	67.50	27	4.50	99	119
1010 1/2" CTS x 1/2" MPT, 30" x 15"		30	.533		64	27	4.50	95.50	116
1015 1/2" CTS x 3/4" MPT, 22" x 24"		30	.533		35	27	4.50	66.50	84
1020 1/2" CTS x 3/4" MPT, 30" x 17"		30	.533		38	27	4.50	69.50	87.50
1025 1/2" CTS x 3/4" MPT, 30" x 28"		30	.533		24.50	27	4.50	56	72.50
1030 SDR 9.3, 1/2" IPS x 3/4" MPT, 30" x 17"		28.50	.561		39.50	28.50	4.73	72.73	91
1035 1/2" IPS x 3/4" MPT, 30" x 28"		28.50	.561		76.50	28.50	4.73	109.73	132
1040 SDR 10, 1-1/4" IPS x 1-1/4" MPT, 30" x 22"		27	.593		105	30	5	140	167
1045 SDR 11, 3/4" IPS x 3/4" MPT, 30" x 16"		28.50	.561		54.50	28.50	4.73	87.73	108
1050 3/4" IPS x 3/4" MPT, 36" x 23"		28.50	.561		59.50	28.50	4.73	92.73	113
1055 1" IPS x 1" MPT, 30" x 16"		27	.593		62.50	30	5	97.50	120
1060 1-1/4" IPS x 1-1/4" MPT, 30" x 22"		27	.593		107	30	5	142	169
1065 1-1/2" IPS x 1-1/2" MPT, 36" x 26"		25.50	.627		150	32	5.30	187.30	218
1070 2" IPS x 2" MPT, 36" x 36"		24	.667		194	34	5.60	233.60	271
1075 SDR 11.5, 1" CTS x 3/4" MPT, 30" x 16"	▼	28	.571	▼	57.50	29	4.82	91.32	112
1100 Compression joint x MPT stainless steel, HDPE or MDPE, flexible									
1105 SDR 7, 1/2" CTS x 1/2" MPT, 60" long	B-20	60	.400	Ea.	40.50	19.70		60.20	74
1110 1/2" CTS x 3/4" MPT, 60" long		60	.400		38.50	19.70		58.20	72
1115 SDR 11, 3/4" IPS x 3/4" MPT, 60" long		58	.414		61	20.50		81.50	97.50
1120 1" IPS x 1" MPT, 60" long		54	.444		134	22		156	180
1125 SDR 11.5, 1" CTS x 1" MPT, 60" long	▼	56	.429		78.50	21		99.50	118
2000 Lubricated semi-steel plug valve									
2040 3/4" diameter	1 Plum	16	.500	Ea.	124	34		158	187
2080 1" diameter		14	.571		130	38.50		168.50	201
2100 1-1/4" diameter		12	.667		154	45		199	237
2140 1-1/2" diameter		11	.727		149	49		198	238
2180 2" diameter	▼	8	1		615	67.50		682.50	780
2300 2-1/2" diameter	Q-1	5	3.200		675	195		870	1,025
2340 3" diameter	"	4.50	3.556	▼	415	217		632	785

33 61 Hydronic Energy Distribution

33 61 13 – Underground Hydronic Energy Distribution

33 61 13.10 Chilled/HVAC Hot Water Distribution

	Crew	Daily Output	Labor-Hours	Unit	Material	2021 Bare Costs Labor	Equipment	Total	Total Incl O&P
0010 **CHILLED/HVAC HOT WATER DISTRIBUTION**									
1005 Pipe, black steel w/2" polyurethane insul., 20' lengths									
1010 Align & tackweld on sleepers (NIC), 1-1/4" diameter	B-35	864	.065	L.F.	32.50	3.68	1.06	37.24	42.50
1020 1-1/2"		824	.068		34.50	3.86	1.11	39.47	45
1030 2"		680	.082		47	4.67	1.35	53.02	60
1040 2-1/2"		560	.100		51.50	5.70	1.64	58.84	67.50
1050 3"		528	.106		48	6	1.74	55.74	64
1060 4"		384	.146		53	8.30	2.39	63.69	73.50
1070 5"		360	.156		60.50	8.85	2.55	71.90	82.50
1080 6"		296	.189		85.50	10.75	3.10	99.35	113
1090 8"		264	.212		113	12.05	3.47	128.52	147
1100 12"		216	.259		176	14.70	4.24	194.94	221
1110 On trench bottom, 18" diameter	▼	176.13	.318		288	18.05	5.20	311.25	350
1120 24"	B-35A	145	.386		410	20.50	14.60	445.10	495
1130 30"		121	.463		665	24.50	17.50	707	785
1140 36"	▼	100	.560	▼	960	30	21	1,011	1,125
1150 Elbows, on sleepers, 1-1/2" diameter	Q-17	21	.762	Ea.	955	47	5.10	1,007.10	1,125
1160 3"	▼	9.36	1.709	▼	1,225	105	11.50	1,341.50	1,525

For customer support on your Site Work & Landscape Costs with RSMeans Data, call 800.448.8182.

535

33 61 13.10 Chilled/HVAC Hot Water Distribution		Crew	Daily Output	Labor-Hours	Unit	Material	2021 Bare Costs Labor	Equipment	Total	Total Incl O&P
1170	4"	Q-17	8	2	Ea.	1,475	123	13.45	1,611.45	1,825
1180	6"		6	2.667		1,800	164	17.90	1,981.90	2,250
1190	8"		4.64	3.448		2,400	212	23	2,635	3,000
1200	Tees, 1-1/2" diameter		17	.941		1,725	58	6.30	1,789.30	2,000
1210	3"		8.50	1.882		2,175	116	12.65	2,303.65	2,550
1220	4"		6	2.667		2,400	164	17.90	2,581.90	2,900
1230	6"	Q-17A	6.72	3.571		2,950	220	87	3,257	3,675
1240	8"	"	6.40	3.750		3,800	231	91	4,122	4,625
1250	Reducer, 3" diameter	Q-17	16	1		525	61.50	6.70	593.20	675
1260	4"		12	1.333		620	82	8.95	710.95	810
1270	6"		12	1.333		895	82	8.95	985.95	1,125
1280	8"		10	1.600		1,200	98.50	10.75	1,309.25	1,450
1290	Anchor, 4" diameter	Q-17A	12	2		1,375	123	48.50	1,546.50	1,750
1300	6"		10.50	2.286		1,675	141	55.50	1,871.50	2,125
1310	8"		10	2.400		2,000	148	58.50	2,206.50	2,475
1320	Cap, 1-1/2" diameter	Q-17	42	.381		169	23.50	2.56	195.06	223
1330	3"		14.64	1.093		220	67.50	7.35	294.85	350
1340	4"		16	1		253	61.50	6.70	321.20	375
1350	6"		16	1		435	61.50	6.70	503.20	580
1360	8"		13.50	1.185		560	73	7.95	640.95	735
1365	12"		11	1.455		855	89.50	9.75	954.25	1,075
1370	Elbow, fittings on trench bottom, 12" diameter	Q-17A	12	2		3,650	123	48.50	3,821.50	4,225
1380	18"		8	3		7,050	185	73	7,308	8,125
1390	24"		6	4		10,400	246	97	10,743	11,900
1400	30"		5.36	4.478		15,400	275	109	15,784	17,400
1410	36"		4	6		22,900	370	146	23,416	25,900
1420	Tee, 12" diameter		6.72	3.571		4,650	220	87	4,957	5,550
1430	18"		6	4		8,475	246	97	8,818	9,800
1440	24"		4.64	5.172		14,000	320	126	14,446	16,000
1450	30"		4.16	5.769		23,200	355	140	23,695	26,200
1460	36"		3.36	7.143		37,500	440	174	38,114	42,000
1470	Reducer, 12" diameter		10.64	2.256		1,475	139	55	1,669	1,900
1480	18"		9.68	2.479		2,575	152	60.50	2,787.50	3,125
1490	24"		8	3		4,200	185	73	4,458	4,950
1500	30"		7.04	3.409		7,000	210	83	7,293	8,100
1510	36"		5.36	4.478		11,100	275	109	11,484	12,700
1520	Anchor, 12" diameter		11	2.182		1,900	134	53	2,087	2,350
1530	18"		6.72	3.571		1,950	220	87	2,257	2,575
1540	24"		6.32	3.797		2,775	234	92.50	3,101.50	3,500
1550	30"		4.64	5.172		4,075	320	126	4,521	5,125
1560	36"		4	6		5,525	370	146	6,041	6,775
1565	Weld in place and install shrink collar									
1570	On sleepers, 1-1/2" diameter	Q-17A	18.50	1.297	Ea.	32	80	31.50	143.50	189
1580	3"		6.72	3.571		55	220	87	362	485
1590	4"		5.36	4.478		56	275	109	440	590
1600	6"		4	6		64.50	370	146	580.50	780
1610	8"		3.36	7.143		99.50	440	174	713.50	955
1620	12"		2.64	9.091		129	560	221	910	1,225
1630	On trench bottom, 18" diameter		2	12		194	740	292	1,226	1,625
1640	24"		1.36	17.647		258	1,075	430	1,763	2,375
1650	30"		1.04	23.077		320	1,425	560	2,305	3,100
1660	36"		1	24		385	1,475	585	2,445	3,275

For customer support on your Site Work & Landscape Costs with RSMeans Data, call 800.448.8182.

33 61 Hydronic Energy Distribution

33 61 13 – Underground Hydronic Energy Distribution

33 61 13.20 Pipe Conduit, Prefabricated/Preinsulated	Crew	Daily Output	Labor-Hours	Unit	Material	2021 Bare Costs Labor	Equipment	Total	Total Incl O&P
0010 **PIPE CONDUIT, PREFABRICATED/PREINSULATED**									
0020 Does not include trenching, fittings or crane.									
0300 For cathodic protection, add 12 to 14%									
0310 of total built-up price (casing plus service pipe)									
0580 Polyurethane insulated system, 250°F max. temp.									
0620 Black steel service pipe, standard wt., 1/2" insulation									
0660 3/4" diam. pipe size	Q-17	54	.296	L.F.	94.50	18.25	1.99	114.74	133
0670 1" diam. pipe size		50	.320		104	19.70	2.15	125.85	146
0680 1-1/4" diam. pipe size		47	.340		116	21	2.29	139.29	162
0690 1-1/2" diam. pipe size		45	.356		126	22	2.39	150.39	174
0700 2" diam. pipe size		42	.381		131	23.50	2.56	157.06	182
0710 2-1/2" diam. pipe size		34	.471		133	29	3.16	165.16	192
0720 3" diam. pipe size		28	.571		154	35	3.84	192.84	226
0730 4" diam. pipe size		22	.727		194	45	4.89	243.89	285
0740 5" diam. pipe size	↓	18	.889		247	54.50	5.95	307.45	360
0750 6" diam. pipe size	Q-18	23	1.043		287	66.50	4.67	358.17	420
0760 8" diam. pipe size		19	1.263		420	80.50	5.65	506.15	590
0770 10" diam. pipe size		16	1.500		530	95.50	6.70	632.20	735
0780 12" diam. pipe size		13	1.846		660	118	8.25	786.25	910
0790 14" diam. pipe size		11	2.182		735	139	9.75	883.75	1,025
0800 16" diam. pipe size		10	2.400		845	153	10.75	1,008.75	1,175
0810 18" diam. pipe size		8	3		975	191	13.45	1,179.45	1,375
0820 20" diam. pipe size		7	3.429		1,075	219	15.35	1,309.35	1,550
0830 24" diam. pipe size	↓	6	4		1,325	255	17.90	1,597.90	1,850
0900 For 1" thick insulation, add					10%				
0940 For 1-1/2" thick insulation, add					13%				
0980 For 2" thick insulation, add				↓	20%				
1500 Gland seal for system, 3/4" diam. pipe size	Q-17	32	.500	Ea.	1,100	31	3.36	1,134.36	1,250
1510 1" diam. pipe size		32	.500		1,100	31	3.36	1,134.36	1,250
1540 1-1/4" diam. pipe size		30	.533		1,175	33	3.58	1,211.58	1,325
1550 1-1/2" diam. pipe size		30	.533		1,175	33	3.58	1,211.58	1,325
1560 2" diam. pipe size		28	.571		1,375	35	3.84	1,413.84	1,575
1570 2-1/2" diam. pipe size		26	.615		1,500	38	4.14	1,542.14	1,700
1580 3" diam. pipe size		24	.667		1,600	41	4.48	1,645.48	1,850
1590 4" diam. pipe size		22	.727		1,900	45	4.89	1,949.89	2,175
1600 5" diam. pipe size	↓	19	.842		2,325	52	5.65	2,382.65	2,650
1610 6" diam. pipe size	Q-18	26	.923		2,475	59	4.14	2,538.14	2,825
1620 8" diam. pipe size		25	.960		2,875	61.50	4.30	2,940.80	3,275
1630 10" diam. pipe size		23	1.043		3,350	66.50	4.67	3,421.17	3,800
1640 12" diam. pipe size		21	1.143		3,700	73	5.10	3,778.10	4,200
1650 14" diam. pipe size		19	1.263		4,175	80.50	5.65	4,261.15	4,700
1660 16" diam. pipe size		18	1.333		4,925	85	5.95	5,015.95	5,525
1670 18" diam. pipe size	↓	16	1.500	↓	5,225	95.50	6.70	5,327.20	5,875
1680 20" diam. pipe size	Q-18	14	1.714	Ea.	5,925	109	7.70	6,041.70	6,700
1690 24" diam. pipe size	"	12	2	"	6,575	128	8.95	6,711.95	7,425
2000 Elbow, 45° for system									
2020 3/4" diam. pipe size	Q-17	14	1.143	Ea.	720	70.50	7.70	798.20	905
2040 1" diam. pipe size		13	1.231		740	75.50	8.25	823.75	930
2050 1-1/4" diam. pipe size		11	1.455		835	89.50	9.75	934.25	1,075
2060 1-1/2" diam. pipe size		9	1.778		840	109	11.95	960.95	1,100
2070 2" diam. pipe size		6	2.667		905	164	17.90	1,086.90	1,250
2080 2-1/2" diam. pipe size	↓	4	4		990	246	27	1,263	1,500

For customer support on your Site Work & Landscape Costs with RSMeans Data, call 800.448.8182.

537

33 61 Hydronic Energy Distribution

33 61 13 – Underground Hydronic Energy Distribution

33 61 13.20 Pipe Conduit, Prefabricated/Preinsulated	Crew	Daily Output	Labor-Hours	Unit	Material	2021 Bare Costs Labor	Equipment	Total	Total Incl O&P
2090	3" diam. pipe size	Q-17	3.50	4.571	Ea.	1,150	281	30.50	1,461.50	1,700
2100	4" diam. pipe size		3	5.333		1,325	330	36	1,691	2,000
2110	5" diam. pipe size		2.80	5.714		1,725	350	38.50	2,113.50	2,450
2120	6" diam. pipe size	Q-18	4	6		1,950	385	27	2,362	2,725
2130	8" diam. pipe size		3	8		2,825	510	36	3,371	3,900
2140	10" diam. pipe size		2.40	10		3,450	640	45	4,135	4,775
2150	12" diam. pipe size		2	12		4,525	765	54	5,344	6,200
2160	14" diam. pipe size		1.80	13.333		5,625	850	59.50	6,534.50	7,550
2170	16" diam. pipe size		1.60	15		6,700	955	67	7,722	8,875
2180	18" diam. pipe size		1.30	18.462		8,400	1,175	82.50	9,657.50	11,100
2190	20" diam. pipe size		1	24		10,600	1,525	108	12,233	14,000
2200	24" diam. pipe size		.70	34.286		13,300	2,175	154	15,629	18,000
2260	For elbow, 90°, add					25%				
2300	For tee, straight, add					85%	30%			
2340	For tee, reducing, add					170%	30%			
2380	For weldolet, straight, add					50%				

33 63 Steam Energy Distribution

33 63 13 – Underground Steam and Condensate Distribution Piping

33 63 13.10 Calcium Silicate Insulated System

		Crew	Daily Output	Labor-Hours	Unit	Material	2021 Bare Costs Labor	Equipment	Total	Total Incl O&P
0010	**CALCIUM SILICATE INSULATED SYSTEM**									
0011	High temp. (1200 degrees F)									
2840	Steel casing with protective exterior coating									
2850	6-5/8" diameter	Q-18	52	.462	L.F.	137	29.50	2.07	168.57	197
2860	8-5/8" diameter		50	.480		149	30.50	2.15	181.65	212
2870	10-3/4" diameter		47	.511		175	32.50	2.29	209.79	243
2880	12-3/4" diameter		44	.545		191	35	2.44	228.44	265
2890	14" diameter		41	.585		215	37.50	2.62	255.12	295
2900	16" diameter		39	.615		231	39.50	2.76	273.26	315
2910	18" diameter		36	.667		254	42.50	2.99	299.49	345
2920	20" diameter		34	.706		286	45	3.16	334.16	385
2930	22" diameter		32	.750		395	48	3.36	446.36	510
2940	24" diameter	Q-18	29	.828	L.F.	455	53	3.71	511.71	585
2950	26" diameter		26	.923		520	59	4.14	583.14	665
2960	28" diameter		23	1.043		640	66.50	4.67	711.17	810
2970	30" diameter		21	1.143		685	73	5.10	763.10	870
2980	32" diameter		19	1.263		770	80.50	5.65	856.15	975
2990	34" diameter		18	1.333		780	85	5.95	870.95	990
3000	36" diameter		16	1.500		835	95.50	6.70	937.20	1,075
3040	For multi-pipe casings, add					10%				
3060	For oversize casings, add					2%				
3400	Steel casing gland seal, single pipe									
3420	6-5/8" diameter	Q-18	25	.960	Ea.	2,100	61.50	4.30	2,165.80	2,400
3440	8-5/8" diameter		23	1.043		2,450	66.50	4.67	2,521.17	2,800
3450	10-3/4" diameter		21	1.143		2,725	73	5.10	2,803.10	3,125
3460	12-3/4" diameter		19	1.263		3,225	80.50	5.65	3,311.15	3,675
3470	14" diameter		17	1.412		3,575	90	6.30	3,671.30	4,100
3480	16" diameter		16	1.500		4,200	95.50	6.70	4,302.20	4,750
3490	18" diameter		15	1.600		4,650	102	7.15	4,759.15	5,250
3500	20" diameter		13	1.846		5,150	118	8.25	5,276.25	5,825
3510	22" diameter		12	2		5,775	128	8.95	5,911.95	6,550

33 63 13 – Underground Steam and Condensate Distribution Piping

33 63 13.10 Calcium Silicate Insulated System	Crew	Daily Output	Labor-Hours	Unit	Material	2021 Bare Costs Labor	Equipment	Total	Total Incl O&P
3520	24" diameter	Q-18	11	2.182	Ea.	6,500	139	9.75	6,648.75	7,375
3530	26" diameter		10	2.400		7,425	153	10.75	7,588.75	8,400
3540	28" diameter		9.50	2.526		8,375	161	11.30	8,547.30	9,450
3550	30" diameter		9	2.667		8,525	170	11.95	8,706.95	9,650
3560	32" diameter		8.50	2.824		9,575	180	12.65	9,767.65	10,800
3570	34" diameter		8	3		10,400	191	13.45	10,604.45	11,800
3580	36" diameter		7	3.429		11,100	219	15.35	11,334.35	12,500
3620	For multi-pipe casings, add					5%				
4000	Steel casing anchors, single pipe									
4020	6-5/8" diameter	Q-18	8	3	Ea.	1,875	191	13.45	2,079.45	2,375
4040	8-5/8" diameter		7.50	3.200		1,950	204	14.35	2,168.35	2,475
4050	10-3/4" diameter		7	3.429		2,550	219	15.35	2,784.35	3,150
4060	12-3/4" diameter		6.50	3.692		2,725	236	16.55	2,977.55	3,375
4070	14" diameter		6	4		3,225	255	17.90	3,497.90	3,950
4080	16" diameter		5.50	4.364		3,725	278	19.55	4,022.55	4,525
4090	18" diameter		5	4.800		4,225	305	21.50	4,551.50	5,125
4100	20" diameter		4.50	5.333		4,650	340	24	5,014	5,625
4110	22" diameter		4	6		5,175	385	27	5,587	6,300
4120	24" diameter		3.50	6.857		5,650	435	30.50	6,115.50	6,900
4130	26" diameter		3	8		6,375	510	36	6,921	7,825
4140	28" diameter		2.50	9.600		7,025	610	43	7,678	8,700
4150	30" diameter		2	12		7,450	765	54	8,269	9,400
4160	32" diameter		1.50	16		8,900	1,025	71.50	9,996.50	11,400
4170	34" diameter		1	24		10,000	1,525	108	11,633	13,400
4180	36" diameter		1	24		10,900	1,525	108	12,533	14,400
4220	For multi-pipe, add					5%	20%			
4800	Steel casing elbow									
4820	6-5/8" diameter	Q-18	15	1.600	Ea.	2,675	102	7.15	2,784.15	3,075
4830	8-5/8" diameter		15	1.600		2,850	102	7.15	2,959.15	3,275
4850	10-3/4" diameter		14	1.714		3,450	109	7.70	3,566.70	3,950
4860	12-3/4" diameter		13	1.846		4,075	118	8.25	4,201.25	4,650
4870	14" diameter		12	2		4,375	128	8.95	4,511.95	5,000
4880	16" diameter		11	2.182		4,775	139	9.75	4,923.75	5,475
4890	18" diameter		10	2.400		5,450	153	10.75	5,613.75	6,225
4900	20" diameter		9	2.667		5,800	170	11.95	5,981.95	6,675
4910	22" diameter		8	3		6,250	191	13.45	6,454.45	7,175
4920	24" diameter		7	3.429		6,875	219	15.35	7,109.35	7,925
4930	26" diameter		6	4		7,550	255	17.90	7,822.90	8,700
4940	28" diameter		5	4.800		8,150	305	21.50	8,476.50	9,425
4950	30" diameter		4	6		8,175	385	27	8,587	9,600
4960	32" diameter		3	8		9,300	510	36	9,846	11,000
4970	34" diameter		2	12		10,200	765	54	11,019	12,400
4980	36" diameter		2	12		10,900	765	54	11,719	13,200
5500	Black steel service pipe, std. wt., 1" thick insulation									
5510	3/4" diameter pipe size	Q-17	54	.296	L.F.	85.50	18.25	1.99	105.74	123
5540	1" diameter pipe size		50	.320		88	19.70	2.15	109.85	129
5550	1-1/4" diameter pipe size		47	.340		111	21	2.29	134.29	157
5560	1-1/2" diameter pipe size		45	.356		116	22	2.39	140.39	163
5570	2" diameter pipe size		42	.381		127	23.50	2.56	153.06	177
5580	2-1/2" diameter pipe size		34	.471		132	29	3.16	164.16	192
5590	3" diameter pipe size		28	.571		116	35	3.84	154.84	184
5600	4" diameter pipe size		22	.727		121	45	4.89	170.89	205
5610	5" diameter pipe size		18	.889		166	54.50	5.95	226.45	271

For customer support on your Site Work & Landscape Costs with RSMeans Data, call 800.448.8182.

539

33 63 13.10 Calcium Silicate Insulated System	Crew	Daily Output	Labor-Hours	Unit	Material	2021 Bare Costs Labor	Equipment	Total	Total Incl O&P
5620	6" diameter pipe size	Q-18	23	1.043	L.F.	179	66.50	4.67	250.17	300
6000	Black steel service pipe, std. wt., 1-1/2" thick insul.									
6010	3/4" diameter pipe size	Q-17	54	.296	L.F.	101	18.25	1.99	121.24	140
6040	1" diameter pipe size		50	.320		106	19.70	2.15	127.85	149
6050	1-1/4" diameter pipe size		47	.340		114	21	2.29	137.29	160
6060	1-1/2" diameter pipe size		45	.356		113	22	2.39	137.39	159
6070	2" diameter pipe size		42	.381		121	23.50	2.56	147.06	171
6080	2-1/2" diameter pipe size		34	.471		111	29	3.16	143.16	168
6090	3" diameter pipe size		28	.571		123	35	3.84	161.84	192
6100	4" diameter pipe size		22	.727		141	45	4.89	190.89	227
6110	5" diameter pipe size	Q-17	18	.889	L.F.	189	54.50	5.95	249.45	296
6120	6" diameter pipe size	Q-18	23	1.043		197	66.50	4.67	268.17	320
6130	8" diameter pipe size		19	1.263		253	80.50	5.65	339.15	405
6140	10" diameter pipe size		16	1.500		330	95.50	6.70	432.20	510
6150	12" diameter pipe size		13	1.846		375	118	8.25	501.25	595
6190	For 2" thick insulation, add					15%				
6220	For 2-1/2" thick insulation, add					25%				
6260	For 3" thick insulation, add					30%				
6800	Black steel service pipe, ex. hvy. wt., 1" thick insul.									
6820	3/4" diameter pipe size	Q-17	50	.320	L.F.	102	19.70	2.15	123.85	145
6840	1" diameter pipe size		47	.340		109	21	2.29	132.29	154
6850	1-1/4" diameter pipe size		44	.364		120	22.50	2.44	144.94	168
6860	1-1/2" diameter pipe size		42	.381		121	23.50	2.56	147.06	172
6870	2" diameter pipe size		40	.400		134	24.50	2.69	161.19	186
6880	2-1/2" diameter pipe size		31	.516		109	32	3.47	144.47	171
6890	3" diameter pipe size		27	.593		126	36.50	3.98	166.48	198
6900	4" diameter pipe size		21	.762		155	47	5.10	207.10	247
6910	5" diameter pipe size		17	.941		146	58	6.30	210.30	254
6920	6" diameter pipe size	Q-18	22	1.091		141	69.50	4.89	215.39	264
7400	Black steel service pipe, ex. hvy. wt., 1-1/2" thick insul.									
7420	3/4" diameter pipe size	Q-17	50	.320	L.F.	75.50	19.70	2.15	97.35	115
7440	1" diameter pipe size		47	.340		95	21	2.29	118.29	139
7450	1-1/4" diameter pipe size		44	.364		105	22.50	2.44	129.94	151
7460	1-1/2" diameter pipe size		42	.381		103	23.50	2.56	129.06	151
7470	2" diameter pipe size		40	.400		109	24.50	2.69	136.19	159
7480	2-1/2" diameter pipe size		31	.516		102	32	3.47	137.47	163
7490	3" diameter pipe size		27	.593		112	36.50	3.98	152.48	183
7500	4" diameter pipe size		21	.762		143	47	5.10	195.10	233
7510	5" diameter pipe size		17	.941		197	58	6.30	261.30	310
7520	6" diameter pipe size	Q-18	22	1.091		198	69.50	4.89	272.39	325
7530	8" diameter pipe size		18	1.333		325	85	5.95	415.95	490
7540	10" diameter pipe size		15	1.600		305	102	7.15	414.15	495
7550	12" diameter pipe size		13	1.846		365	118	8.25	491.25	585
7590	For 2" thick insulation, add					13%				
7640	For 2-1/2" thick insulation, add					18%				
7680	For 3" thick insulation, add					24%				

33 71 Electrical Utility Transmission and Distribution

33 71 16 – Electrical Utility Poles

33 71 16.20 Line Poles and Fixtures

		Crew	Daily Output	Labor-Hours	Unit	Material	2021 Bare Costs Labor	2021 Bare Costs Equipment	Total	Total Incl O&P
0010	**LINE POLES & FIXTURES**									
0200	Wood poles, material handling and spotting	R-7	6.49	7.396	Ea.		340	26	366	540
0220	Erect wood poles & backfill holes, in earth	R-5	6.77	12.999	"	1,850	725	177	2,752	3,325
0300	Crossarms for wood pole structure									
0310	Material handling and spotting	R-7	14.55	3.299	Ea.		151	11.55	162.55	240
0320	Install crossarms	R-5	11	8	"	610	445	109	1,164	1,450

33 71 16.23 Steel Electrical Utility Poles

		Crew	Daily Output	Labor-Hours	Unit	Material	2021 Bare Costs Labor	2021 Bare Costs Equipment	Total	Total Incl O&P
0010	**STEEL ELECTRICAL UTILITY POLES**									
6000	Digging holes in earth, average	R-5	25.14	3.500	Ea.		196	47.50	243.50	345
6010	In rock, average	"	4.51	19.512	"		1,100	265	1,365	1,925
6020	Formed plate pole structure									
6030	Material handling and spotting	R-7	2.40	20	Ea.		915	70	985	1,450
6040	Erect steel plate pole	R-5	1.95	45.128		11,500	2,525	615	14,640	17,200
6050	Guys, anchors and hardware for pole, in earth		7.04	12.500		700	700	170	1,570	2,000
6060	In rock		17.96	4.900		835	274	66.50	1,175.50	1,400
6070	Foundations for line poles									
6080	Excavation, in earth	R-5	135.38	.650	C.Y.		36.50	8.85	45.35	63.50
6090	In rock		20	4.400			246	60	306	430
6110	Concrete foundations		11	8		164	445	109	718	970

33 71 16.33 Wood Electrical Utility Poles

		Crew	Daily Output	Labor-Hours	Unit	Material	2021 Bare Costs Labor	2021 Bare Costs Equipment	Total	Total Incl O&P
0010	**WOOD ELECTRICAL UTILITY POLES**									
0011	Excludes excavation, backfill and cast-in-place concrete									
6200	Wood, class 3 Douglas Fir, penta-treated, 20'	R-3	3.10	6.452	Ea.	250	410	61.50	721.50	955
6600	30'		2.60	7.692		325	490	73.50	888.50	1,150
7000	40'		2.30	8.696		595	550	83	1,228	1,575
7200	45'		1.70	11.765		870	745	112	1,727	2,175
7400	Cross arms with hardware & insulators									
7600	4' long	1 Elec	2.50	3.200	Ea.	151	204		355	470
7800	5' long		2.40	3.333		166	212		378	495
8000	6' long		2.20	3.636		200	232		432	565
9000	Disposal of pole & hardware surplus material	R-7	20.87	2.300	Mile		105	8.05	113.05	168
9100	Disposal of crossarms & hardware surplus material	"	40	1.200	"		55	4.20	59.20	87

33 71 19 – Electrical Underground Ducts and Manholes

33 71 19.15 Underground Ducts and Manholes

		Crew	Daily Output	Labor-Hours	Unit	Material	2021 Bare Costs Labor	2021 Bare Costs Equipment	Total	Total Incl O&P
0010	**UNDERGROUND DUCTS AND MANHOLES**									
0011	Not incl. excavation, backfill, or concrete in slab and duct bank									
1000	Direct burial									
1010	PVC, schedule 40, w/coupling, 1/2" diameter	1 Elec	340	.024	L.F.	.18	1.50		1.68	2.43
1020	3/4" diameter		290	.028		.22	1.76		1.98	2.85
1030	1" diameter		260	.031		.32	1.96		2.28	3.26
1040	1-1/2" diameter		210	.038		.53	2.43		2.96	4.19
1050	2" diameter		180	.044		.64	2.83		3.47	4.91
1060	3" diameter	2 Elec	240	.067	L.F.	1.24	4.25		5.49	7.65
1070	4" diameter		160	.100		1.76	6.35		8.11	11.40
1080	5" diameter		120	.133		4.79	8.50		13.29	17.85
1090	6" diameter		90	.178		3.15	11.30		14.45	20.50
1110	Elbows, 1/2" diameter	1 Elec	48	.167	Ea.	.49	10.60		11.09	16.35
1120	3/4" diameter		38	.211		.55	13.40		13.95	20.50
1130	1" diameter		32	.250		.86	15.95		16.81	24.50
1140	1-1/2" diameter		21	.381		2.27	24.50		26.77	38.50
1150	2" diameter		16	.500		3.08	32		35.08	51
1160	3" diameter		12	.667		8.40	42.50		50.90	72.50

33 71 19.15 Underground Ducts and Manholes		Crew	Daily Output	Labor-Hours	Unit	Material	2021 Bare Costs Labor	Equipment	Total	Total Incl O&P
1170	4" diameter	1 Elec	9	.889	Ea.	17.05	56.50		73.55	103
1180	5" diameter		8	1		18.35	63.50		81.85	115
1190	6" diameter		5	1.600		36	102		138	191
1210	Adapters, 1/2" diameter		52	.154		.21	9.80		10.01	14.80
1220	3/4" diameter		43	.186		.35	11.85		12.20	18
1230	1" diameter		39	.205		.38	13.05		13.43	19.80
1240	1-1/2" diameter		35	.229		.81	14.55		15.36	22.50
1250	2" diameter		26	.308		1.02	19.60		20.62	30
1260	3" diameter		20	.400		2.13	25.50		27.63	40.50
1270	4" diameter		14	.571		4.33	36.50		40.83	59
1280	5" diameter		12	.667		8.65	42.50		51.15	72.50
1290	6" diameter		9	.889		10.30	56.50		66.80	95.50
1340	Bell end & cap, 1-1/2" diameter		35	.229		5.25	14.55		19.80	27.50
1350	Bell end & plug, 2" diameter		26	.308		6.50	19.60		26.10	36
1360	3" diameter		20	.400		7.95	25.50		33.45	47
1370	4" diameter		14	.571		9.35	36.50		45.85	64.50
1380	5" diameter		12	.667		14.20	42.50		56.70	78.50
1390	6" diameter		9	.889		16.15	56.50		72.65	102
1450	Base spacer, 2" diameter		56	.143		1.37	9.10		10.47	15
1460	3" diameter		46	.174		1.77	11.10		12.87	18.40
1470	4" diameter		41	.195		1.78	12.45		14.23	20.50
1480	5" diameter		37	.216		1.99	13.75		15.74	22.50
1490	6" diameter		34	.235		2.48	15		17.48	25
1550	Intermediate spacer, 2" diameter		60	.133		1.25	8.50		9.75	14
1560	3" diameter		46	.174		1.76	11.10		12.86	18.40
1570	4" diameter		41	.195		1.59	12.45		14.04	20
1580	5" diameter		37	.216		1.94	13.75		15.69	22.50
1590	6" diameter		34	.235		2.58	15		17.58	25.50
4010	PVC, schedule 80, w/coupling, 1/2" diameter		215	.037	L.F.	.97	2.37		3.34	4.59
4020	3/4" diameter		180	.044		1.20	2.83		4.03	5.55
4030	1" diameter		145	.055		1.79	3.51		5.30	7.15
4040	1-1/2" diameter		120	.067		2.94	4.25		7.19	9.55
4050	2" diameter	1 Elec	100	.080	L.F.	3.23	5.10		8.33	11.10
4060	3" diameter	2 Elec	130	.123		8.40	7.85		16.25	21
4070	4" diameter		90	.178		12	11.30		23.30	30
4080	5" diameter		70	.229		16.70	14.55		31.25	40
4090	6" diameter		50	.320		18.45	20.50		38.95	51
4110	Elbows, 1/2" diameter	1 Elec	29	.276	Ea.	2.11	17.55		19.66	28.50
4120	3/4" diameter		23	.348		5.70	22		27.70	39.50
4130	1" diameter		20	.400		4.60	25.50		30.10	43
4140	1-1/2" diameter		16	.500		6.60	32		38.60	55
4150	2" diameter		12	.667		7.95	42.50		50.45	72
4160	3" diameter		9	.889		21	56.50		77.50	107
4170	4" diameter		7	1.143		32	73		105	143
4180	5" diameter		6	1.333		81.50	85		166.50	216
4190	6" diameter		4	2		90.50	127		217.50	289
4210	Adapter, 1/2" diameter		39	.205		.21	13.05		13.26	19.65
4220	3/4" diameter		33	.242		.35	15.45		15.80	23.50
4230	1" diameter		29	.276		.38	17.55		17.93	26.50
4240	1-1/2" diameter		26	.308		.81	19.60		20.41	30
4250	2" diameter		23	.348		1.02	22		23.02	34
4260	3" diameter		18	.444		2.13	28.50		30.63	44.50
4270	4" diameter		13	.615		4.33	39		43.33	63.50

33 71 19.15 Underground Ducts and Manholes	Crew	Daily Output	Labor-Hours	Unit	Material	2021 Bare Costs Labor	Equipment	Total	Total Incl O&P
4280	5" diameter	1 Elec	11	.727	Ea.	8.65	46.50		55.15	78.50
4290	6" diameter		8	1		10.30	63.50		73.80	106
4310	Bell end & cap, 1-1/2" diameter		26	.308		5.25	19.60		24.85	35
4320	Bell end & plug, 2" diameter		23	.348		6.50	22		28.50	40
4330	3" diameter		18	.444		7.95	28.50		36.45	51
4340	4" diameter		13	.615		9.35	39		48.35	69
4350	5" diameter		11	.727		14.20	46.50		60.70	84.50
4360	6" diameter		8	1		16.15	63.50		79.65	112
4370	Base spacer, 2" diameter		42	.190		1.37	12.15		13.52	19.55
4380	3" diameter		33	.242		1.77	15.45		17.22	25
4390	4" diameter		29	.276		1.78	17.55		19.33	28
4400	5" diameter		26	.308		1.99	19.60		21.59	31
4410	6" diameter		25	.320		2.48	20.50		22.98	33
4420	Intermediate spacer, 2" diameter		45	.178		1.25	11.30		12.55	18.25
4430	3" diameter		34	.235		1.76	15		16.76	24.50
4440	4" diameter		31	.258		1.59	16.45		18.04	26.50
4450	5" diameter		28	.286		1.94	18.20		20.14	29
4460	6" diameter		25	.320		2.58	20.50		23.08	33.50

33 71 19.17 Electric and Telephone Underground

		Crew	Daily Output	Labor-Hours	Unit	Material	2021 Bare Costs Labor	Equipment	Total	Total Incl O&P
0010	**ELECTRIC AND TELEPHONE UNDERGROUND**									
0011	Not including excavation									
0200	backfill and cast in place concrete									
0250	For bedding, see Section 31 23 23.16									
0400	Hand holes, precast concrete, with concrete cover									
0600	2' x 2' x 3' deep	R-3	2.40	8.333	Ea.	530	530	79.50	1,139.50	1,450
0800	3' x 3' x 3' deep		1.90	10.526		690	670	101	1,461	1,850
1000	4' x 4' x 4' deep		1.40	14.286		1,550	905	136	2,591	3,200
1200	Manholes, precast with iron racks & pulling irons, C.I. frame									
1400	and cover, 4' x 6' x 7' deep	B-13	2	28	Ea.	2,950	1,350	293	4,593	5,600
1600	6' x 8' x 7' deep		1.90	29.474		3,325	1,425	310	5,060	6,125
1800	6' x 10' x 7' deep		1.80	31.111		3,725	1,500	325	5,550	6,700
4200	Underground duct, banks ready for concrete fill, min. of 7.5"									
4400	between conduits, center to center									
4580	PVC, type EB, 1 @ 2" diameter	2 Elec	480	.033	L.F.	.96	2.12		3.08	4.21
4600	2 @ 2" diameter		240	.067		1.91	4.25		6.16	8.40
4800	4 @ 2" diameter		120	.133		3.83	8.50		12.33	16.80
4900	1 @ 3" diameter		400	.040		1.62	2.55		4.17	5.55
5000	2 @ 3" diameter		200	.080		3.23	5.10		8.33	11.10
5200	4 @ 3" diameter		100	.160		6.45	10.20		16.65	22.50
5300	1 @ 4" diameter		320	.050		1.58	3.19		4.77	6.45
5400	2 @ 4" diameter		160	.100		3.16	6.35		9.51	12.90
5600	4 @ 4" diameter		80	.200		6.30	12.75		19.05	26
5800	6 @ 4" diameter		54	.296		9.45	18.85		28.30	38.50
5810	1 @ 5" diameter		260	.062		1.78	3.92		5.70	7.75
5820	2 @ 5" diameter		130	.123		3.56	7.85		11.41	15.55
5840	4 @ 5" diameter		70	.229		7.10	14.55		21.65	29.50
5860	6 @ 5" diameter		50	.320		10.65	20.50		31.15	42.50
5870	1 @ 6" diameter		200	.080		3.03	5.10		8.13	10.90
5880	2 @ 6" diameter		100	.160		6.05	10.20		16.25	22
5900	4 @ 6" diameter		50	.320		12.10	20.50		32.60	44
5920	6 @ 6" diameter		30	.533		18.15	34		52.15	70.50
6200	Rigid galvanized steel, 2 @ 2" diameter		180	.089		13.95	5.65		19.60	24

33 71 19.17 Electric and Telephone Underground		Crew	Daily Output	Labor-Hours	Unit	Material	2021 Bare Costs Labor	Equipment	Total	Total Incl O&P
6400	4 @ 2" diameter	2 Elec	90	.178	L.F.	28	11.30		39.30	47.50
6800	2 @ 3" diameter		100	.160		30.50	10.20		40.70	49
7000	4 @ 3" diameter		50	.320		61.50	20.50		82	98
7200	2 @ 4" diameter		70	.229		43.50	14.55		58.05	69
7400	4 @ 4" diameter		34	.471		86.50	30		116.50	140
7600	6 @ 4" diameter		22	.727		130	46.50		176.50	212
7620	2 @ 5" diameter		60	.267		94	17		111	128
7640	4 @ 5" diameter		30	.533		188	34		222	258
7660	6 @ 5" diameter	2 Elec	18	.889	L.F.	282	56.50		338.50	395
7680	2 @ 6" diameter		40	.400		114	25.50		139.50	164
7700	4 @ 6" diameter		20	.800		229	51		280	325
7720	6 @ 6" diameter		14	1.143		345	73		418	485
7800	For cast-in-place concrete, add									
7801	For cable, see Section 26 05 19.55									
7810	Under 1 C.Y.	C-6	16	3	C.Y.	199	138	3.39	340.39	430
7820	1 C.Y. to 5 C.Y.		19.20	2.500		180	115	2.83	297.83	375
7830	Over 5 C.Y.		24	2		149	92	2.26	243.26	305
7850	For reinforcing rods, add									
7860	#4 to #7	2 Rodm	1.10	14.545	Ton	1,250	855		2,105	2,650
7870	#8 to #14	"	1.50	10.667	"	1,250	630		1,880	2,325
8000	Fittings, PVC type EB, elbow, 2" diameter	1 Elec	16	.500	Ea.	15.45	32		47.45	64.50
8200	3" diameter		14	.571		17.80	36.50		54.30	73.50
8400	4" diameter		12	.667		36.50	42.50		79	104
8420	5" diameter		10	.800		113	51		164	200
8440	6" diameter		9	.889		217	56.50		273.50	325
8500	Coupling, 2" diameter					.90			.90	.99
8600	3" diameter					4.48			4.48	4.93
8700	4" diameter					4.79			4.79	5.25
8720	5" diameter					9.15			9.15	10.05
8740	6" diameter					25.50			25.50	28
8800	Adapter, 2" diameter	1 Elec	26	.308		.63	19.60		20.23	29.50
9000	3" diameter		20	.400		3.01	25.50		28.51	41.50
9200	4" diameter		16	.500		3.58	32		35.58	51.50
9220	5" diameter		13	.615		6.15	39		45.15	65.50
9240	6" diameter		10	.800		13.30	51		64.30	90
9400	End bell, 2" diameter		16	.500		2.30	32		34.30	50
9600	3" diameter		14	.571		4.07	36.50		40.57	58.50
9800	4" diameter		12	.667		4.62	42.50		47.12	68
9810	5" diameter		10	.800		8.10	51		59.10	84.50
9820	6" diameter		8	1		10.10	63.50		73.60	106
9830	5° angle coupling, 2" diameter		26	.308		6.95	19.60		26.55	36.50
9840	3" diameter		20	.400		25	25.50		50.50	65.50
9850	4" diameter		16	.500		20.50	32		52.50	70
9860	5" diameter		13	.615		17.90	39		56.90	78
9870	6" diameter		10	.800		27.50	51		78.50	106
9880	Expansion joint, 2" diameter		16	.500		35.50	32		67.50	86.50
9890	3" diameter		18	.444		74.50	28.50		103	124
9900	4" diameter		12	.667		100	42.50		142.50	173
9910	5" diameter		10	.800		181	51		232	275
9920	6" diameter	1 Elec	8	1	Ea.	172	63.50		235.50	284
9930	Heat bender, 2" diameter					515			515	565
9940	6" diameter					1,625			1,625	1,800
9950	Cement, quart					17.50			17.50	19.25

For customer support on your Site Work & Landscape Costs with RSMeans Data, call 800.448.8182.

33 81 Communications Structures

33 81 13 – Communications Transmission Towers

33 81 13.10 Radio Towers	Crew	Daily Output	Labor-Hours	Unit	Material	2021 Bare Costs Labor	Equipment	Total	Total Incl O&P
0010 **RADIO TOWERS**									
0020 Guyed, 50' H, 40 lb. section, 70 MPH basic wind speed	2 Sswk	1	16	Ea.	2,775	965		3,740	4,550
0100 Wind load 90 MPH basic wind speed	"	1	16		3,650	965		4,615	5,500
0300 190' high, 40 lb. section, wind load 70 MPH basic wind speed	K-2	.33	72.727		8,950	4,150	2,575	15,675	19,000
0400 200' high, 70 lb. section, wind load 90 MPH basic wind speed		.33	72.727		15,300	4,150	2,575	22,025	26,000
0600 300' high, 70 lb. section, wind load 70 MPH basic wind speed		.20	120		18,600	6,850	4,250	29,700	35,600
0700 270' high, 90 lb. section, wind load 90 MPH basic wind speed		.20	120		21,600	6,850	4,250	32,700	38,900
0800 400' high, 100 lb. section, wind load 70 MPH basic wind speed		.14	171		30,600	9,800	6,075	46,475	55,500
0900 Self-supporting, 60' high, wind load 70 MPH basic wind speed		.80	30		4,575	1,725	1,075	7,375	8,825
0910 60' high, wind load 90 MPH basic wind speed		.45	53.333		5,175	3,050	1,900	10,125	12,500
1000 120' high, wind load 70 MPH basic wind speed		.40	60		9,650	3,425	2,125	15,200	18,200
1200 190' high, wind load 90 MPH basic wind speed		.20	120		28,600	6,850	4,250	39,700	46,700
2000 For states west of Rocky Mountains, add for shipping					10%				

For customer support on your Site Work & Landscape Costs with RSMeans Data, call 800.448.8182.

545

Division Notes

	CREW	DAILY OUTPUT	LABOR-HOURS	UNIT	BARE COSTS				TOTAL INCL O&P
					MAT.	LABOR	EQUIP.	TOTAL	

Estimating Tips
34 11 00 Rail Tracks
This subdivision includes items that may involve either repair of existing or construction of new railroad tracks. Additional preparation work, such as the roadbed earthwork, would be found in Division 31. Additional new construction siding and turnouts are found in Subdivision 34 72. Maintenance of railroads is found under 34 01 23 Operation and Maintenance of Railways.

34 40 00 Traffic Signals
This subdivision includes traffic signal systems. Other traffic control devices such as traffic signs are found in Subdivision 10 14 53 Traffic Signage.

34 70 00 Vehicle Barriers
This subdivision includes security vehicle barriers, guide and guard rails, crash barriers, and delineators. The actual maintenance and construction of concrete and asphalt pavement are found in Division 32.

Reference Numbers
Reference numbers are shown at the beginning of some major classifications. These numbers refer to related items in the Reference Section. The reference information may be an estimating procedure, an alternate pricing method, or technical information.

Note: Not all subdivisions listed here necessarily appear. ∎

Same Data. Simplified.

Enjoy the convenience and efficiency of accessing your costs anywhere:

- **Skip the multiplier** by setting your location
- **Quickly search,** edit, favorite and share costs
- **Stay on top of price changes** with automatic updates

Discover more at rsmeans.com/online

---

34 01 13.10 Maintenance Grading of Roadways	Crew	Daily Output	Labor-Hours	Unit	Material	2021 Bare Costs Labor	Equipment	Total	Total Incl O&P
0010 **MAINTENANCE GRADING OF ROADWAYS**									
0100 Maintenance grading mob/demob of equipment per hour	B-11L	8	2	Hr.		103	134	237	300
0200 Maintenance grading of ditches one ditch slope only 1.0 MPH average		5	3.200	Mile		165	215	380	485
0210 1.5 MPH		7.50	2.133			110	143	253	320
0220 2.0 MPH		10	1.600			82.50	107	189.50	241
0230 2.5 MPH		12.50	1.280			66	86	152	193
0240 3.0 MPH		15	1.067			55	71.50	126.50	161
0300 Maintenance grading of roadway, 3 passes, 3.0 MPH average		5	3.200			165	215	380	485
0310 4 passes		3.80	4.211			218	282	500	635
0320 3 passes, 3.5 MPH average		5.80	2.759			143	185	328	415
0330 4 passes		4.40	3.636			188	244	432	550
0340 3 passes, 4.0 MPH average		6.70	2.388			123	160	283	360
0350 4 passes		5	3.200			165	215	380	485
0360 3 passes, 4.5 MPH average		7.50	2.133			110	143	253	320
0370 4 passes		5.60	2.857			148	192	340	430
0380 3 passes, 5.0 MPH average		8.30	1.928			99.50	129	228.50	291
0390 4 passes		6.30	2.540			131	170	301	385
0400 3 passes, 5.5 MPH average	B-11L	9.20	1.739	Mile		90	117	207	262
0410 4 passes		6.90	2.319			120	156	276	350
0420 3 passes, 6.0 MPH average		10	1.600			82.50	107	189.50	241
0430 4 passes		7.50	2.133			110	143	253	320
0440 3 passes, 6.5 MPH average		10.80	1.481			76.50	99.50	176	223
0450 4 passes		8.10	1.975			102	132	234	298
0460 3 passes, 7.0 MPH average		11.70	1.368			70.50	91.50	162	206
0470 4 passes		8.80	1.818			94	122	216	274
0480 3 passes, 7.5 MPH average		12.50	1.280			66	86	152	193
0490 4 passes		9.40	1.702			88	114	202	257
0500 3 passes, 8.0 MPH average		13.30	1.203			62	80.50	142.50	182
0510 4 passes		10	1.600			82.50	107	189.50	241
0520 3 passes, 8.5 MPH average		14.20	1.127			58.50	75.50	134	170
0530 4 passes		10.60	1.509			78	101	179	227
0540 3 passes, 9.0 MPH average		15	1.067			55	71.50	126.50	161
0550 4 passes		11.30	1.416			73	95	168	213
0560 3 passes, 9.5 MPH average		15.80	1.013			52.50	68	120.50	153
0570 4 passes		11.90	1.345			69.50	90	159.50	203
0600 Maintenance grading of roadway, 8' wide pass, 3.0 MPH average		15	1.067			55	71.50	126.50	161
0610 3.5 MPH average		17.50	.914			47.50	61.50	109	138
0620 4.0 MPH average		20	.800			41.50	53.50	95	121
0630 4.5 MPH average		22.50	.711			37	47.50	84.50	108
0640 5.0 MPH average		25	.640			33	43	76	96.50
0650 5.5 MPH average		27.50	.582			30	39	69	88
0660 6.0 MPH average		30	.533			27.50	36	63.50	80.50
0670 6.5 MPH average		32.50	.492			25.50	33	58.50	74.50
0680 7.0 MPH average		35	.457			23.50	30.50	54	68.50
0690 7.5 MPH average		37.50	.427			22	28.50	50.50	64.50
0700 8.0 MPH average		40	.400			20.50	27	47.50	60.50
0710 8.5 MPH average		42.50	.376			19.45	25.50	44.95	57
0720 9.0 MPH average		45	.356			18.40	24	42.40	53.50
0730 9.5 MPH average		47.50	.337			17.40	22.50	39.90	51

34 01 Operation and Maintenance of Transportation

34 01 23 – Operation and Maintenance of Railways

34 01 23.51 Maintenance of Railroads	Crew	Daily Output	Labor-Hours	Unit	Material	2021 Bare Costs Labor	Equipment	Total	Total Incl O&P
0010 **MAINTENANCE OF RAILROADS**									
0400 Resurface and realign existing track	B-14	200	.240	L.F.		11.20	1.08	12.28	17.90
0600 For crushed stone ballast, add	"	500	.096	"	17.75	4.47	.43	22.65	26.50

34 11 Rail Tracks

34 11 13 – Track Rails

34 11 13.23 Heavy Rail Track

	Crew	Daily Output	Labor-Hours	Unit	Material	2021 Bare Costs Labor	Equipment	Total	Total Incl O&P
0010 **HEAVY RAIL TRACK**	R347216-10								
1000 Rail, 100 lb. prime grade				L.F.	31.50			31.50	35
1500 Relay rail				L.F.	15.85			15.85	17.45

34 11 33 – Track Cross Ties

34 11 33.13 Concrete Track Cross Ties

	Crew	Daily Output	Labor-Hours	Unit	Material	2021 Bare Costs Labor	Equipment	Total	Total Incl O&P
0010 **CONCRETE TRACK CROSS TIES**									
1400 Ties, concrete, 8'-6" long, 30" OC	B-14	80	.600	Ea.	132	28	2.70	162.70	189

34 11 33.16 Timber Track Cross Ties

	Crew	Daily Output	Labor-Hours	Unit	Material	2021 Bare Costs Labor	Equipment	Total	Total Incl O&P
0010 **TIMBER TRACK CROSS TIES**									
1600 Wood, pressure treated, 6" x 8" x 8'-6", C.L. lots	B-14	90	.533	Ea.	56	25	2.40	83.40	101
1700 L.C.L. lots		90	.533		58.50	25	2.40	85.90	104
1900 Heavy duty, 7" x 9" x 8'-6", C.L. lots		70	.686		61.50	32	3.09	96.59	118
2000 L.C.L. lots		70	.686		61.50	32	3.09	96.59	118

34 11 33.17 Timber Switch Ties

	Crew	Daily Output	Labor-Hours	Unit	Material	2021 Bare Costs Labor	Equipment	Total	Total Incl O&P
0010 **TIMBER SWITCH TIES**									
1200 Switch timber, for a #8 switch, pressure treated	B-14	3.70	12.973	M.B.F.	3,350	605	58.50	4,013.50	4,675
1300 Complete set of timbers, 3.7 MBF for #8 switch	"	1	48	Total	12,900	2,225	216	15,341	17,800

34 11 93 – Track Appurtenances and Accessories

34 11 93.50 Track Accessories

	Crew	Daily Output	Labor-Hours	Unit	Material	2021 Bare Costs Labor	Equipment	Total	Total Incl O&P
0010 **TRACK ACCESSORIES**									
0020 Car bumpers, test	B-14	2	24	Ea.	4,150	1,125	108	5,383	6,350
0100 Heavy duty		2	24		7,850	1,125	108	9,083	10,400
0200 Derails hand throw (sliding)		10	4.800		1,625	224	21.50	1,870.50	2,150
0300 Hand throw with standard timbers, open stand & target		8	6		1,750	279	27	2,056	2,375
2400 Wheel stops, fixed		18	2.667	Pr.	975	124	12	1,111	1,275
2450 Hinged		14	3.429	"	1,325	160	15.45	1,500.45	1,700

34 11 93.60 Track Material

	Crew	Daily Output	Labor-Hours	Unit	Material	2021 Bare Costs Labor	Equipment	Total	Total Incl O&P
0010 **TRACK MATERIAL**									
0020 Track bolts				Ea.	3.64			3.64	4
0100 Joint bars				Pr.	86.50			86.50	95.50
0200 Spikes				Ea.	4			4	4.40
0300 Tie plates				"	14.50			14.50	15.95

For customer support on your Site Work & Landscape Costs with RSMeans Data, call 800.448.8182.

549

34 41 Roadway Signaling and Control Equipment

34 41 13 – Traffic Signals

34 41 13.10 Traffic Signals Systems

		Crew	Daily Output	Labor-Hours	Unit	Material	2021 Bare Costs Labor	Equipment	Total	Total Incl O&P
0010	**TRAFFIC SIGNALS SYSTEMS**									
0020	Component costs									
0600	Crew employs crane/directional driller as required									
1000	Vertical mast with foundation									
1010	Mast sized for single arm to 40'; no lighting or power function	R-11	.50	112	Signal	10,900	6,800	1,300	19,000	23,500
1100	Horizontal arm									
1110	Per linear foot of arm	R-11	50	1.120	Signal	234	68	12.90	314.90	375
1200	Traffic signal									
1210	Includes signal, bracket, sensor, and wiring	R-11	2.50	22.400	Signal	1,175	1,350	258	2,783	3,600
1300	Pedestrian signals and callers									
1310	Includes four signals with brackets and two call buttons	R-11	2.50	22.400	Signal	3,850	1,350	258	5,458	6,550
1400	Controller, design, and underground conduit									
1410	Includes miscellaneous signage and adjacent surface work	R-11	.25	224	Signal	26,200	13,600	2,575	42,375	52,000

34 71 Roadway Construction

34 71 13 – Vehicle Barriers

34 71 13.17 Security Vehicle Barriers

		Crew	Daily Output	Labor-Hours	Unit	Material	2021 Bare Costs Labor	Equipment	Total	Total Incl O&P
0010	**SECURITY VEHICLE BARRIERS**									
0020	Security planters excludes filling material									
0100	Concrete security planter, exposed aggregate 36" diam. x 30" high	B-11M	8	2	Ea.	540	103	29.50	672.50	780
0200	48" diam. x 36" high		8	2		655	103	29.50	787.50	905
0300	53" diam. x 18" high		8	2		830	103	29.50	962.50	1,100
0400	72" diam. x 18" high with seats		8	2		1,950	103	29.50	2,082.50	2,325
0450	84" diam. x 36" high		8	2		1,975	103	29.50	2,107.50	2,350
0500	36" x 36" x 24" high square		8	2		455	103	29.50	587.50	685
0600	36" x 36" x 30" high square		8	2		755	103	29.50	887.50	1,025
0700	48" L x 24" W x 30" H rectangle		8	2		470	103	29.50	602.50	705
0800	72" L x 24" W x 30" H rectangle		8	2		575	103	29.50	707.50	815
0900	96" L x 24" W x 30" H rectangle		8	2		840	103	29.50	972.50	1,100
0950	Decorative geometric concrete barrier, 96" L x 24" W x 36" H		8	2		845	103	29.50	977.50	1,125
1000	Concrete security planter, filling with washed sand or gravel <1 C.Y.		8	2		31	103	29.50	163.50	221
1050	2 C.Y. or less	↓	6	2.667		62.50	138	39	239.50	320
1200	Jersey concrete barrier, 10' L x 2' by 0.5' W x 30" H	B-21B	16	2.500		390	121	30	541	640
1300	10 or more same site		24	1.667		390	80.50	19.85	490.35	565
1400	10' L x 2' by 0.5' W x 32" H		16	2.500		1,200	121	30	1,351	1,550
1500	10 or more same site		24	1.667		1,200	80.50	19.85	1,300.35	1,475
1600	20' L x 2' by 0.5' W x 30" H		12	3.333		740	161	39.50	940.50	1,100
1700	10 or more same site		18	2.222		740	107	26.50	873.50	1,000
1800	20' L x 2' by 0.5' W x 32" H		12	3.333		875	161	39.50	1,075.50	1,250
1900	10 or more same site		18	2.222		875	107	26.50	1,008.50	1,150
2000	GFRC decorative security barrier per 10' section including concrete		4	10		2,975	480	119	3,574	4,125
2100	Per 12' section including concrete	↓	4	10	↓	3,575	480	119	4,174	4,775
2210	GFRC decorative security barrier will stop 30 MPH, 4,000 lb. vehicle									
2300	High security barrier base prep per 12' section on bare ground	B-11C	4	4	Ea.	23.50	207	54	284.50	395
2310	GFRC barrier base prep does not include haulaway of excavated matl.									
2400	GFRC decorative high security barrier per 12' section w/concrete	B-21B	3	13.333	Ea.	5,900	645	159	6,704	7,600
2410	GFRC decorative high security barrier will stop 50 MPH, 15,000 lb. vehicle									
2500	GFRC decorative impaler security barrier per 10' section w/prep	B-6	4	6	Ea.	2,275	289	54	2,618	3,025
2600	Per 12' section w/prep	"	4	6	"	2,750	289	54	3,093	3,525
2610	Impaler barrier should stop 50 MPH, 15,000 lb. vehicle w/some penetr.									
2700	Pipe bollards, steel, concrete filled/painted, 8' L x 4' D hole, 8" diam.	B-6	10	2.400	Ea.	890	115	21.50	1,026.50	1,175

34 71 Roadway Construction

34 71 13 – Vehicle Barriers

34 71 13.17 Security Vehicle Barriers

	Crew	Daily Output	Labor-Hours	Unit	Material	2021 Bare Costs Labor	Equipment	Total	Total Incl O&P
2710	Schedule 80 concrete bollards will stop 4,000 lb. vehicle @ 30 MPH								
2800	GFRC decorative jersey barrier cover per 10' section excludes soil	B-6	8	3	Ea.	1,000	144	27	1,171	1,350
2900	Per 12' section excludes soil		8	3		1,200	144	27	1,371	1,575
3000	GFRC decorative 8" diameter bollard cover		12	2		400	96	18	514	605
3100	GFRC decorative barrier face 10' section excludes earth backing		10	2.400		985	115	21.50	1,121.50	1,275
3200	Drop arm crash barrier, 15,000 lb. vehicle @ 50 MPH									
3205	Includes all material, labor for complete installation									
3210	12' width				Ea.				64,500	71,000
3310	24' width								86,000	95,000
3410	Wedge crash barrier, 10' width								98,000	108,000
3510	12.5' width								108,000	119,000
3520	15' width								118,500	130,000
3610	Sliding crash barrier, 12' width								178,500	196,000
3710	Sliding roller crash barrier, 20' width								198,000	217,500
3810	Sliding cantilever crash barrier, 20' width								198,000	217,500
3890	Note: Raised bollard crash barriers should be used w/tire shredders									
3910	Raised bollard crash barrier, 10' width				Ea.				38,800	42,900
4010	12' width								44,900	49,000
4110	Raised bollard crash barrier, 10' width, solar powered								46,900	52,000
4210	12' width								53,000	58,000
4310	In ground tire shredder, 16' width								43,900	47,900

34 71 13.26 Vehicle Guide Rails

		Crew	Daily Output	Labor-Hours	Unit	Material	2021 Bare Costs Labor	Equipment	Total	Total Incl O&P
0010	**VEHICLE GUIDE RAILS**									
0012	Corrugated stl., galv. stl. posts, 6'-3" OC - "W-Beam Guiderail"	B-80	850	.038	L.F.	12.10	1.84	1.24	15.18	17.40
0100	Double face		570	.056	"	36.50	2.74	1.85	41.09	46
0200	End sections, galvanized, flared		50	.640	Ea.	98.50	31	21	150.50	178
0300	Wrap around end		50	.640		141	31	21	193	225
0325	W-Beam, stl., galv. Terminal Connector, Concrete Mounted		8	4		249	195	132	576	710
0350	Anchorage units		15	2.133		1,200	104	70	1,374	1,550
0365	End section, flared	B-80A	78	.308		46	13.65	10.90	70.55	83
0370	End section, wrap-around, corrugated steel	"	78	.308		60	13.65	10.90	84.55	98.50
0410	Thrie Beam, stl., galv. stl. posts, 6'-3" OC	B-80	850	.038	L.F.	23.50	1.84	1.24	26.58	29.50
0440	Thrie Beam, stl., galv, Wrap Around End Section		50	.640	Ea.	90	31	21	142	168
0450	Thrie Beam, stl., galv. Terminal Connector, Concrete Mounted		8	4	"	274	195	132	601	735
0510	Rub Rail, stl, galv, steel posts (not included), 6'-3" OC		850	.038	L.F.	1.70	1.84	1.24	4.78	5.95
0600	Cable guide rail, 3 at 3/4" cables, steel posts, single face		900	.036	"	12.60	1.73	1.17	15.50	17.75
0650	Double face	B-80	635	.050	L.F.	23.50	2.46	1.66	27.62	31.50
0700	Wood posts		950	.034		13.30	1.64	1.11	16.05	18.25
0750	Double face		650	.049		23.50	2.40	1.62	27.52	31.50
0760	Breakaway wood posts		195	.164	Ea.	370	8	5.40	383.40	425
0800	Anchorage units, breakaway		15	2.133	"	1,550	104	70	1,724	1,925
0900	Guide rail, steel box beam, 6" x 6"		120	.267	L.F.	38.50	13	8.75	60.25	71
0950	End assembly	B-80A	48	.500	Ea.	1,550	22	17.70	1,589.70	1,750
1100	Median barrier, steel box beam, 6" x 8"	B-80	215	.149	L.F.	52.50	7.25	4.90	64.65	74
1120	Shop curved	B-80A	92	.261	"	71	11.60	9.25	91.85	105
1140	End assembly		48	.500	Ea.	2,575	22	17.70	2,614.70	2,875
1150	Corrugated beam		400	.060	L.F.	59	2.66	2.13	63.79	71.50
1400	Resilient guide fence and light shield, 6' high	B-2	130	.308	"	25	13.80		38.80	48
1500	Concrete posts, individual, 6'-5", triangular	B-80	110	.291	Ea.	68	14.20	9.55	91.75	107
1550	Square		110	.291		74.50	14.20	9.55	98.25	114
1600	Wood guide posts		150	.213		45	10.40	7	62.40	73
1650	Timber guide rail, 4" x 8" with 6" x 8" wood posts, treated		960	.033	L.F.	13.90	1.63	1.10	16.63	18.95

34 71 13 – Vehicle Barriers

34 71 13.26 Vehicle Guide Rails

		Crew	Daily Output	Labor-Hours	Unit	Material	2021 Bare Costs Labor	Equipment	Total	Total Incl O&P
2000	Median, precast concrete, 3'-6" high, 2' wide, single face	B-29	380	.147	L.F.	60.50	7.10	2.24	69.84	79.50
2200	Double face	"	340	.165		69.50	7.95	2.50	79.95	91
2300	Cast in place, steel forms	C-2	170	.282		49	15.05		64.05	76.50
2320	Slipformed	C-7	352	.205		43	9.80	3.10	55.90	65.50
2400	Speed bumps, thermoplastic, 10-1/2" x 2-1/4" x 48" long	B-2	120	.333	Ea.	144	14.95		158.95	182
3000	Energy absorbing terminal, 10 bay, 3' wide	B-80B	.10	320		46,000	15,100	2,375	63,475	75,500
3010	7 bay, 2'-6" wide		.20	160		35,900	7,550	1,200	44,650	52,000
3020	5 bay, 2' wide		.20	160		27,600	7,550	1,200	36,350	43,000
3030	Impact barrier, MUTCD, barrel type	B-16	30	1.067		395	49.50	19.30	463.80	530
3100	Wide hazard protection, foam sandwich, 7 bay, 7'-6" wide	B-80B	.14	229		36,700	10,800	1,700	49,200	58,500
3110	5' wide		.15	213		26,100	10,100	1,600	37,800	45,500
3120	3' wide		.18	178		24,800	8,400	1,325	34,525	41,300
5000	For bridge railing, see Section 32 34 10.10									

34 71 19 – Vehicle Delineators

34 71 19.13 Fixed Vehicle Delineators

		Crew	Daily Output	Labor-Hours	Unit	Material	2021 Bare Costs Labor	Equipment	Total	Total Incl O&P
0010	**FIXED VEHICLE DELINEATORS**									
0020	Crash barriers									
0100	Traffic channelizing pavement markers, layout only	A-7	2000	.012	Ea.		.71	.02	.73	1.09
0110	13" x 7-1/2" x 2-1/2" high, non-plowable, install	2 Clab	96	.167		30.50	7.40		37.90	44.50
0200	8" x 8" x 3-1/4" high, non-plowable, install		96	.167		29.50	7.40		36.90	43
0230	4" x 4" x 3/4" high, non-plowable, install		120	.133		2.66	5.90		8.56	11.80
0240	9-1/4" x 5-7/8" x 1/4" high, plowable, concrete pavmt.	A-2A	70	.343		4.70	15.75	4.45	24.90	33.50
0250	9-1/4" x 5-7/8" x 1/4" high, plowable, asphalt pavmt.	"	120	.200		3.83	9.15	2.59	15.57	21
0300	Barrier and curb delineators, reflectorized, 4" x 6"	2 Clab	150	.107	Ea.	3.31	4.74		8.05	10.70
0310	3" x 5"	"	150	.107	"	4.50	4.74		9.24	12
0500	Rumble strip, polycarbonate									
0510	24" x 3-1/2" x 1/2" high	2 Clab	50	.320	Ea.	10.10	14.20		24.30	32

34 72 16 – Railway Siding

34 72 16.50 Railroad Sidings

		Crew	Daily Output	Labor-Hours	Unit	Material	2021 Bare Costs Labor	Equipment	Total	Total Incl O&P
0010	**RAILROAD SIDINGS** R347216-10									
0800	Siding, yard spur, level grade									
0808	Wood ties and ballast, 80 lb. new rail	B-14	57	.842	L.F.	121	39	3.79	163.79	196
0809	80 lb. relay rail		57	.842		91	39	3.79	133.79	163
0812	90 lb. new rail		57	.842		128	39	3.79	170.79	204
0813	90 lb. relay rail		57	.842		91.50	39	3.79	134.29	164
0820	100 lb. new rail		57	.842		135	39	3.79	177.79	212
0822	100 lb. relay rail		57	.842		95.50	39	3.79	138.29	168
0830	110 lb. new rail		57	.842		141	39	3.79	183.79	219
0832	110 lb. relay rail		57	.842		100	39	3.79	142.79	173
1002	Steel ties in concrete, incl. fasteners & plates									
1003	80 lb. new rail	B-14	22	2.182	L.F.	182	102	9.80	293.80	365
1005	80 lb. relay rail		22	2.182		156	102	9.80	267.80	335
1012	90 lb. new rail		22	2.182		188	102	9.80	299.80	370
1015	90 lb. relay rail		22	2.182		159	102	9.80	270.80	340
1020	100 lb. new rail		22	2.182		194	102	9.80	305.80	375
1025	100 lb. relay rail		22	2.182		162	102	9.80	273.80	340
1030	110 lb. new rail		22	2.182		203	102	9.80	314.80	385
1035	110 lb. relay rail		22	2.182		166	102	9.80	277.80	345

34 72 Railway Construction

34 72 16 – Railway Siding

34 72 16.60 Railroad Turnouts		Crew	Daily Output	Labor-Hours	Unit	Material	2021 Bare Costs		Total	Total Incl O&P
							Labor	Equipment		
0010	**RAILROAD TURNOUTS**									
2200	Turnout, #8 complete, w/rails, plates, bars, frog, switch point,									
2250	timbers, and ballast to 6" below bottom of ties									
2280	90 lb. rails	B-13	.25	224	Ea.	39,500	10,800	2,350	52,650	62,000
2290	90 lb. relay rails		.25	224		26,500	10,800	2,350	39,650	47,900
2300	100 lb. rails		.25	224		44,000	10,800	2,350	57,150	67,000
2310	100 lb. relay rails		.25	224		29,600	10,800	2,350	42,750	51,000
2320	110 lb. rails		.25	224		48,500	10,800	2,350	61,650	72,000
2330	110 lb. relay rails		.25	224		32,600	10,800	2,350	45,750	54,500
2340	115 lb. rails		.25	224		53,000	10,800	2,350	66,150	77,000
2350	115 lb. relay rails		.25	224		34,600	10,800	2,350	47,750	56,500
2360	132 lb. rails		.25	224		60,500	10,800	2,350	73,650	85,500
2370	132 lb. relay rails	▼	.25	224	▼	40,200	10,800	2,350	53,350	63,000

Division Notes

		CREW	DAILY OUTPUT	LABOR-HOURS	UNIT	BARE COSTS				TOTAL INCL O&P
						MAT.	LABOR	EQUIP.	TOTAL	

Estimating Tips
35 01 50 Operation and Maintenance of Marine Construction
Includes unit price lines for pile cleaning and pile wrapping for protection.

35 20 16 Hydraulic Gates
This subdivision includes various types of gates that are commonly used in waterway and canal construction. Various earthwork and structural support items are found in Division 31, and concrete work is found in Division 3.

35 20 23 Dredging
This subdivision includes barge and shore dredging systems for rivers, canals, and channels.

35 31 00 Shoreline Protection
This subdivision includes breakwaters, bulkheads, and revetments for ocean and river inlets. Additional earthwork may be required from Division 31, as well as concrete work from Division 3.

35 41 00 Levees
Contains information on levee construction, including the estimated cost of clay cone material.

35 49 00 Waterway Structures
This subdivision includes breakwaters and bulkheads for canals.

35 51 00 Floating Construction
This section includes floating piers, docks, and dock accessories. Fixed Pier Timber Construction is found in Section 06 13 33. Driven piles are found in Division 31, as are sheet piling, cofferdams, and riprap.

Reference Numbers
Reference numbers are shown at the beginning of some major classifications. These numbers refer to related items in the Reference Section. The reference information may be an estimating procedure, an alternate pricing method, or technical information.

Note: Not all subdivisions listed here necessarily appear. ∎

Same Data. Simplified.
Enjoy the convenience and efficiency of accessing your costs anywhere:
- **Skip the multiplier** by setting your location
- **Quickly search,** edit, favorite and share costs
- **Stay on top of price changes** with automatic updates

Discover more at rsmeans.com/online

---

35 01 50.10 Cleaning of Marine Pier Piles		Crew	Daily Output	Labor-Hours	Unit	Material	Labor	Equipment	Total	Total Incl O&P
0010	**CLEANING OF MARINE PIER PILES**									
0015	Exposed piles pressure washing from boat									
0100	Clean wood piles from boat, 8" diam., 5' length	G	B-1D	17	.941	Ea.		42	12.65	54.65	76.50
0110	6-10' long	G		9	1.778			79	24	103	145
0120	11-15' long	G		6.50	2.462			109	33	142	200
0130	10" diam., 5' length	G		13.75	1.164			51.50	15.65	67.15	94
0140	6-10' long	G		7.25	2.207			98	29.50	127.50	179
0150	11-15' long	G		5	3.200			142	43	185	260
0160	12" diam., 5' length	G		11.50	1.391			62	18.70	80.70	113
0170	6-10' long	G		6	2.667			118	36	154	217
0180	11-15' long	G		4.25	3.765			167	50.50	217.50	305
0190	13" diam., 5' length	G		10.50	1.524			67.50	20.50	88	124
0200	6-10' long	G		5.50	2.909			129	39	168	236
0210	11-15' long	G		4	4			178	53.50	231.50	325
0220	14" diam., 5' length	G		9.75	1.641			73	22	95	133
0230	6-10' long	G		5.25	3.048			135	41	176	247
0240	11-15' long	G		3.75	4.267			189	57.50	246.50	345
0300	Clean concrete piles from boat, 12" diam., 5' length	G		14	1.143			50.50	15.35	65.85	92.50
0310	6-10' long	G		7.25	2.207			98	29.50	127.50	179
0320	11-15' long	G		5	3.200			142	43	185	260
0330	14" diam., 5' length	G		12	1.333			59	17.90	76.90	108
0340	6-10' long	G		6.25	2.560			114	34.50	148.50	208
0350	11-15' long	G		4.50	3.556			158	48	206	289
0360	16" diam., 5' length	G		10.50	1.524			67.50	20.50	88	124
0370	6-10' long	G		5.50	2.909			129	39	168	236
0380	11-15' long	G		4	4			178	53.50	231.50	325
0390	18" diam., 5' length	G		9.25	1.730			77	23	100	141
0400	6-10' long	G		5	3.200			142	43	185	260
0410	11-15' long	G		3.50	4.571			203	61.50	264.50	375
0420	20" diam., 5' length	G		8.50	1.882			83.50	25.50	109	153
0430	6-10' long	G		4.50	3.556			158	48	206	289
0440	11-15' long	G		3	5.333			237	71.50	308.50	435
0450	24" diam., 5' length	G		7	2.286			101	30.50	131.50	185
0460	6-10' long	G		3.75	4.267			189	57.50	246.50	345
0470	11-15' long	G		2.50	6.400			284	86	370	520
0480	36" diam., 5' length	G		4.75	3.368			150	45	195	273
0490	6-10' long	G		2.50	6.400			284	86	370	520
0500	11-15' long	G		1.75	9.143			405	123	528	740
0510	54" diam., 5' length	G		3	5.333			237	71.50	308.50	435
0520	6-10' long	G		1.75	9.143			405	123	528	740
0530	11-15' long	G		1.25	12.800			570	172	742	1,050
0540	66" diam., 5' length	G		2.50	6.400			284	86	370	520
0550	6-10' long	G		1.50	10.667			475	143	618	865
0560	11-15' long	G		1	16			710	215	925	1,275
0600	10" square pile, 5' length	G		13.25	1.208			53.50	16.20	69.70	98
0610	6-10' long	G		7	2.286			101	30.50	131.50	185
0620	11-15' long	G		4.75	3.368			150	45	195	273
0630	12" square pile, 5' length	G		11	1.455			64.50	19.55	84.05	118
0640	6-10' long	G		5.75	2.783			124	37.50	161.50	225
0650	11-15' long	G		4	4			178	53.50	231.50	325
0660	14" square pile, 5' length	G		9.50	1.684			75	22.50	97.50	137
0670	6-10' long	G		5	3.200			142	43	185	260
0680	11-15' long	G		3.50	4.571			203	61.50	264.50	375

35 01 50.10 Cleaning of Marine Pier Piles		Crew	Daily Output	Labor-Hours	Unit	Material	2021 Bare Costs Labor	Equipment	Total	Total Incl O&P
0690	16" square pile, 5' length G	B-1D	8.25	1.939	Ea.		86	26	112	157
0700	6-10' long G		4.25	3.765			167	50.50	217.50	305
0710	11-15' long G		3	5.333			237	71.50	308.50	435
0720	18" square pile, 5' length G		7.25	2.207			98	29.50	127.50	179
0730	6-10' long G		3.75	4.267			189	57.50	246.50	345
0740	11-15' long G		2.75	5.818			258	78	336	470
0750	20" square pile, 5' length G		6.50	2.462			109	33	142	200
0760	6-10' long G		3.50	4.571			203	61.50	264.50	375
0770	11-15' long G		2.50	6.400			284	86	370	520
0780	24" square pile, 5' length G		5.50	2.909			129	39	168	236
0790	6-10' long G		3	5.333			237	71.50	308.50	435
0800	11-15' long G		2	8			355	107	462	650
0810	14" diameter octagonal piles, 5' length G		12.50	1.280			57	17.20	74.20	104
0820	6-10' long G		6.50	2.462			109	33	142	200
0830	11-15' long G		4.50	3.556			158	48	206	289
0840	16" diameter octagonal piles, 5' length G		11	1.455			64.50	19.55	84.05	118
0850	6-10' long G		5.75	2.783			124	37.50	161.50	225
0860	11-15' long G		4	4			178	53.50	231.50	325
0870	18" diameter octagonal piles, 5' length G		9.75	1.641			73	22	95	133
0880	6-10' long G		5	3.200			142	43	185	260
0890	11-15' long G		3.50	4.571			203	61.50	264.50	375
0900	20" diameter octagonal piles, 5' length G		8.75	1.829			81	24.50	105.50	148
0910	6-10' long G		4.50	3.556			158	48	206	289
0920	11-15' long G		3.25	4.923			219	66	285	400
0930	24" diameter octagonal piles, 5' length G		7.25	2.207			98	29.50	127.50	179
0940	6-10' long G		3.75	4.267			189	57.50	246.50	345
0950	11-15' long G		2.75	5.818			258	78	336	470
1000	Clean steel piles from boat, 8" diam., 5' length G		14.50	1.103			49	14.80	63.80	89.50
1010	6-10' long G		7.75	2.065			91.50	27.50	119	168
1020	11-15' long G		5.50	2.909			129	39	168	236
1030	10" diam., 5' length G		11.50	1.391			62	18.70	80.70	113
1040	6-10' long G		6	2.667			118	36	154	217
1050	11-15' long G		4.50	3.556			158	48	206	289
1060	12" diam., 5' length G		9.50	1.684			75	22.50	97.50	137
1070	6-10' long G		5	3.200			142	43	185	260
1080	11-15' long G		3.75	4.267			189	57.50	246.50	345
1100	HP 8 x 36, 5' length G		7.50	2.133			94.50	28.50	123	173
1110	6-10' long G		4	4			178	53.50	231.50	325
1120	11-15' long G		3	5.333			237	71.50	308.50	435
1130	HP 10 x 42, 5' length G		6.25	2.560			114	34.50	148.50	208
1140	6-10' long G		3.25	4.923			219	66	285	400
1150	11-15' long G		2.30	6.957			310	93.50	403.50	565
1160	HP 10 x 57, 5' length G		6	2.667			118	36	154	217
1170	6-10' long G		3.25	4.923			219	66	285	400
1180	11-15' long G		2.30	6.957			310	93.50	403.50	565
1200	HP 12 x 53, 5' length G		5	3.200			142	43	185	260
1210	6-10' long G		2.75	5.818			258	78	336	470
1220	11-15' long G		2	8			355	107	462	650
1230	HP 12 x 74, 5' length G		5	3.200			142	43	185	260
1240	6-10' long G		2.75	5.818			258	78	336	470
1250	11-15' long G		1.90	8.421			375	113	488	685
1300	HP 14 x 73, 5' length G		4.25	3.765			167	50.50	217.50	305
1310	6-10' long G		2.33	6.867			305	92	397	555

For customer support on your Site Work & Landscape Costs with RSMeans Data, call 800.448.8182.

557

35 01 50 – Operation and Maintenance of Marine Construction

35 01 50.10 Cleaning of Marine Pier Piles		Crew	Daily Output	Labor-Hours	Unit	Material	2021 Bare Costs Labor	Equipment	Total	Total Incl O&P
1320	11-15' long	G	B-1D	1.60	10	Ea.		445	134	579	815
1330	HP 14 x 89, 5' length	G		4.25	3.765			167	50.50	217.50	305
1340	6-10' long	G		2.33	6.867			305	92	397	555
1350	11-15' long	G		1.60	10			445	134	579	815
1360	HP 14 x 102, 5' length	G		4.25	3.765			167	50.50	217.50	305
1370	6-10' long	G		2.33	6.867			305	92	397	555
1380	11-15' long	G		1.60	10			445	134	579	815
1385	HP 14 x 117, 5' length	G		4.25	3.765			167	50.50	217.50	305
1390	6-10' long	G		2.33	6.867			305	92	397	555
1395	11-15' long	G		1.60	10			445	134	579	815

35 01 50.20 Protective Wrapping of Marine Pier Piles

		Crew	Daily Output	Labor-Hours	Unit	Material	Labor	Equipment	Total	Total Incl O&P
0010	**PROTECTIVE WRAPPING OF MARINE PIER PILES**									
0015	Exposed piles wrapped using boat									
0020	Note: piles must be cleaned before wrapping									
0100	Wrap protective material on wood piles with nails 8" diameter	G	B-1G	162	.099	V.L.F.	18.65	4.39	.75	23.79	28
0110	10" diameter	G		130	.123		23.50	5.45	.93	29.88	34.50
0120	12" diameter	G		108	.148		28	6.60	1.12	35.72	42
0130	13" diameter	G		99	.162		30.50	7.20	1.22	38.92	45.50
0140	14" diameter	G		93	.172		32.50	7.65	1.30	41.45	49
0150	Wrap protective material on wood piles, straps, 8" diameter	G		180	.089		28	3.95	.67	32.62	37.50
0160	10" diameter	G		144	.111		35.50	4.93	.84	41.27	47.50
0170	12" diameter	G		120	.133		42.50	5.90	1.01	49.41	56.50
0180	13" diameter	G		110	.145		46	6.45	1.10	53.55	61.50
0190	14" diameter	G		103	.155		49.50	6.90	1.17	57.57	66
0200	Wrap protective material on concrete piles, straps, 12" diameter	G		120	.133		42.50	5.90	1.01	49.41	56.50
0210	14" diameter	G		103	.155		49.50	6.90	1.17	57.57	66
0220	16" diameter	G		90	.178		56.50	7.90	1.34	65.74	75.50
0230	18" diameter	G		80	.200		63.50	8.90	1.51	73.91	85
0240	20" diameter	G		72	.222		70.50	9.85	1.68	82.03	94
0250	24" diameter	G		60	.267		84.50	11.85	2.01	98.36	113
0260	36" diameter	G		40	.400		127	17.75	3.02	147.77	170
0270	54" diameter	G		27	.593		191	26.50	4.47	221.97	254
0280	66" diameter	G		22	.727		233	32.50	5.50	271	310
0300	Wrap protective material on concrete piles, straps, 10" square	G		113	.142		45	6.30	1.07	52.37	60
0310	12" square	G		94	.170		54	7.55	1.29	62.84	72
0320	14" square	G		81	.198		63	8.75	1.49	73.24	84
0330	16" square	G		71	.225		72	10	1.70	83.70	96
0340	18" square	G		63	.254		81	11.30	1.92	94.22	108
0350	20" square	G		63	.254		90	11.30	1.92	103.22	118
0360	24" square	G		47	.340		108	15.10	2.57	125.67	144
0400	Wrap protective material, concrete piles, straps, 14" octagon	G		107	.150		47.50	6.65	1.13	55.28	63.50
0410	16" octagon	G		94	.170		54.50	7.55	1.29	63.34	72
0420	18" octagon	G		83	.193		61	8.55	1.46	71.01	81.50
0430	20" octagon	G		75	.213		68	9.45	1.61	79.06	90.50
0440	24" octagon	G		62	.258		81.50	11.45	1.95	94.90	109
0500	Wrap protective material, steel piles, straps, HP 8 x 36	G		90	.178		72	7.90	1.34	81.24	92.50
0510	HP 10 x 42	G		72	.222		88.50	9.85	1.68	100.03	114
0520	HP 10 x 57	G		72	.222		90	9.85	1.68	101.53	116
0530	HP 12 x 53	G		60	.267		106	11.85	2.01	119.86	137
0540	HP 12 x 74	G		60	.267		108	11.85	2.01	121.86	139
0550	HP 14 x 73	G		52	.308		126	13.65	2.32	141.97	162
0560	HP 14 x 89	G		52	.308		128	13.65	2.32	143.97	163

558

For customer support on your Site Work & Landscape Costs with RSMeans Data, call 800.448.8182.

35 01 Operation and Maint. of Waterway & Marine Construction

35 01 50 – Operation and Maintenance of Marine Construction

35 01 50.20 Protective Wrapping of Marine Pier Piles		Crew	Daily Output	Labor-Hours	Unit	Material	2021 Bare Costs Labor	Equipment	Total	Total Incl O&P
0570	HP 14 x 102	G	B-1G	52	.308	V.L.F.	128	13.65	2.32	143.97	163
0580	HP 14 x 117	G	↓	52	.308	↓	130	13.65	2.32	145.97	165

35 22 Hydraulic Gates

35 22 26 – Sluice Gates

35 22 26.16 Hydraulic Sluice Gates

		Crew	Daily Output	Labor-Hours	Unit	Material	2021 Bare Costs Labor	Equipment	Total	Total Incl O&P
0010	**HYDRAULIC SLUICE GATES**									
0100	Heavy duty, self contained w/crank oper. gate, 18" x 18"	L-5A	1.70	18.824	Ea.	7,900	1,150	680	9,730	11,200
0110	24" x 24"	L-5A	1.20	26.667	Ea.	13,200	1,625	960	15,785	18,200
0120	30" x 30"		1	32		12,300	1,950	1,150	15,400	17,800
0130	36" x 36"		.90	35.556		14,200	2,175	1,275	17,650	20,300
0140	42" x 42"		.80	40		16,300	2,450	1,450	20,200	23,300
0150	48" x 48"		.50	64		18,900	3,900	2,300	25,100	29,400
0160	54" x 54"		.40	80		35,400	4,875	2,875	43,150	49,600
0170	60" x 60"		.30	107		28,100	6,525	3,850	38,475	45,100
0180	66" x 66"		.30	107		32,800	6,525	3,850	43,175	50,500
0190	72" x 72"		.20	160		36,300	9,775	5,775	51,850	61,500
0200	78" x 78"		.20	160		44,500	9,775	5,775	60,050	70,500
0210	84" x 84"	↓	.10	320		58,000	19,500	11,500	89,000	106,500
0220	90" x 90"	E-20	.30	213		69,000	12,700	7,100	88,800	103,500
0230	96" x 96"		.30	213		64,500	12,700	7,100	84,300	98,500
0240	108" x 108"		.20	320		73,500	19,100	10,600	103,200	122,000
0250	120" x 120"		.10	640		77,500	38,200	21,300	137,000	167,000
0260	132" x 132"	↓	.10	640	↓	100,000	38,200	21,300	159,500	192,000

35 22 63 – Through-Levee Access Gates

35 22 63.16 Canal Gates

		Crew	Daily Output	Labor-Hours	Unit	Material	2021 Bare Costs Labor	Equipment	Total	Total Incl O&P
0010	**CANAL GATES**									
0011	Cast iron body, fabricated frame									
0100	12" diameter	L-5A	4.60	6.957	Ea.	1,175	425	251	1,851	2,225
0110	18" diameter		4	8		1,375	490	289	2,154	2,575
0120	24" diameter		3.50	9.143		2,000	560	330	2,890	3,425
0130	30" diameter		2.80	11.429		2,700	700	410	3,810	4,500
0140	36" diameter		2.30	13.913		5,025	850	500	6,375	7,375
0150	42" diameter		1.70	18.824		5,725	1,150	680	7,555	8,825
0160	48" diameter		1.20	26.667		6,900	1,625	960	9,485	11,100
0170	54" diameter		.90	35.556		11,500	2,175	1,275	14,950	17,400
0180	60" diameter		.50	64		12,100	3,900	2,300	18,300	21,900
0190	66" diameter		.50	64		16,900	3,900	2,300	23,100	27,200
0200	72" diameter	↓	.40	80	↓	20,900	4,875	2,875	28,650	33,700

35 22 66 – Flap Gates, Hydraulic

35 22 66.16 Flap Gates

		Crew	Daily Output	Labor-Hours	Unit	Material	2021 Bare Costs Labor	Equipment	Total	Total Incl O&P
0010	**FLAP GATES**									
0100	Aluminum, 18" diameter	L-5A	5	6.400	Ea.	2,475	390	231	3,096	3,575
0110	24" diameter		4	8		2,725	490	289	3,504	4,075
0120	30" diameter		3.50	9.143		3,325	560	330	4,215	4,875
0130	36" diameter		2.80	11.429		4,250	700	410	5,360	6,200
0140	42" diameter		2.30	13.913		4,425	850	500	5,775	6,725
0150	48" diameter		1.70	18.824		6,050	1,150	680	7,880	9,200
0160	54" diameter		1.20	26.667		6,975	1,625	960	9,560	11,200
0170	60" diameter	↓	.80	40	↓	7,875	2,450	1,450	11,775	14,000

559

For customer support on your Site Work & Landscape Costs with RSMeans Data, call 800.448.8182.

35 22 Hydraulic Gates

35 22 66 – Flap Gates, Hydraulic

35 22 66.16 Flap Gates

35 22 66.16 Flap Gates		Crew	Daily Output	Labor-Hours	Unit	Material	2021 Bare Costs		Total	Total Incl O&P
							Labor	Equipment		
0180	66" diameter	L-5A	.50	64	Ea.	8,750	3,900	2,300	14,950	18,200
0190	72" diameter	↓	.40	80	↓	10,100	4,875	2,875	17,850	21,800

35 22 69 – Knife Gates, Hydraulic

35 22 69.16 Knife Gates

		Crew	Daily Output	Labor-Hours	Unit	Material	Labor	Equipment	Total	Total Incl O&P
0010	**KNIFE GATES**									
0100	Incl. handwheel operator for hub, 6" diameter	Q-23	7.70	3.117	Ea.	1,000	204	198	1,402	1,625
0110	8" diameter		7.20	3.333		1,500	218	212	1,930	2,200
0120	10" diameter		4.80	5		2,250	325	320	2,895	3,325
0130	12" diameter		3.60	6.667		4,375	435	425	5,235	5,950
0140	14" diameter		3.40	7.059		4,525	460	450	5,435	6,175
0150	16" diameter		3.20	7.500		5,425	490	475	6,390	7,225
0160	18" diameter		3	8		10,200	525	510	11,235	12,600
0170	20" diameter		2.70	8.889		8,475	580	565	9,620	10,800
0180	24" diameter		2.40	10		12,800	655	635	14,090	15,800
0190	30" diameter		1.80	13.333		31,000	875	850	32,725	36,300
0200	36" diameter	↓	1.20	20	↓	35,700	1,300	1,275	38,275	42,600

35 22 73 – Slide Gates, Hydraulic

35 22 73.16 Slide Gates

		Crew	Daily Output	Labor-Hours	Unit	Material	Labor	Equipment	Total	Total Incl O&P
0010	**SLIDE GATES**									
0100	Steel, self contained incl. anchor bolts and grout, 12" x 12"	L-5A	4.60	6.957	Ea.	4,925	425	251	5,601	6,325
0110	18" x 18"		4	8		5,450	490	289	6,229	7,075
0120	24" x 24"		3.50	9.143		7,100	560	330	7,990	9,025
0130	30" x 30"		2.80	11.429		6,800	700	410	7,910	9,000
0140	36" x 36"		2.30	13.913		7,225	850	500	8,575	9,800
0150	42" x 42"		1.70	18.824		8,750	1,150	680	10,580	12,100
0160	48" x 48"		1.20	26.667		7,875	1,625	960	10,460	12,200
0170	54" x 54"		.90	35.556		6,800	2,175	1,275	10,250	12,200
0180	60" x 60"		.55	58.182		9,225	3,550	2,100	14,875	17,900
0190	72" x 72"	↓	.36	88.889	↓	11,400	5,425	3,200	20,025	24,400

35 24 Dredging

35 24 13 – Suction Dredging

35 24 13.13 Cutter Suction Dredging

		Crew	Daily Output	Labor-Hours	Unit	Material	Labor	Equipment	Total	Total Incl O&P
0010	**CUTTER SUCTION DREDGING**									
0015	Add, Marine Equipment Rental, See Section 01 54 33.80									
1000	Hydraulic method, pumped 1,000' to shore dump, minimum	B-57	460	.104	B.C.Y.		5.30	4.01	9.31	12.30
1100	Maximum		310	.155			7.85	5.95	13.80	18.25
1400	Into scows dumped 20 miles, minimum		425	.113			5.75	4.34	10.09	13.30
1500	Maximum	↓	243	.198			10.05	7.60	17.65	23.50
1600	For inland rivers and canals in South, deduct				↓				30%	30%

35 24 23 – Clamshell Dredging

35 24 23.13 Mechanical Dredging

		Crew	Daily Output	Labor-Hours	Unit	Material	Labor	Equipment	Total	Total Incl O&P
0010	**MECHANICAL DREDGING**									
0015	Add, Marine Equipment Rental, See Section 01 54 33.80									
0020	Dredging mobilization and demobilization, add to below, minimum	B-8	.53	121	Total		6,175	5,450	11,625	15,200
0100	Maximum	"	.10	640	"		32,700	28,900	61,600	80,500
0300	Barge mounted clamshell excavation into scows									
0310	Dumped 20 miles at sea, minimum	B-57	310	.155	B.C.Y.		7.85	5.95	13.80	18.25
0400	Maximum	"	213	.225	"		11.45	8.65	20.10	26.50

35 24 Dredging

35 24 23 – Clamshell Dredging

35 24 23.13 Mechanical Dredging	Crew	Daily Output	Labor-Hours	Unit	Material	2021 Bare Costs Labor	Equipment	Total	Total Incl O&P
0500	Barge mounted dragline or clamshell, hopper dumped,								
0510	pumped 1,000' to shore dump, minimum	B-57	340	.141	B.C.Y.		7.15	5.40	12.55	16.65
0525	All pumping uses 2,000 gallons of water per cubic yard								
0600	Maximum	B-57	243	.198	B.C.Y.		10.05	7.60	17.65	23.50

35 31 Shoreline Protection

35 31 16 – Seawalls

35 31 16.13 Concrete Seawalls

		Crew	Daily Output	Labor-Hours	Unit	Material	2021 Bare Costs Labor	Equipment	Total	Total Incl O&P
0010	**CONCRETE SEAWALLS**									
0011	Reinforced concrete									
0015	include footing and tie-backs									
0020	Up to 6' high, minimum	C-17C	28	2.964	L.F.	62	171	19.55	252.55	345
0060	Maximum		24.25	3.423		99.50	197	22.50	319	430
0100	12' high, minimum		20	4.150		162	239	27.50	428.50	570
0160	Maximum		18.50	4.486		187	259	29.50	475.50	630
0180	Precast bulkhead, complete, including									
0190	vertical and battered piles, face panels, and cap									
0195	Using 16' vertical piles				L.F.				445	515
0196	Using 20' vertical piles				"				475	550

35 31 16.19 Steel Sheet Piling Seawalls

		Crew	Daily Output	Labor-Hours	Unit	Material	2021 Bare Costs Labor	Equipment	Total	Total Incl O&P
0010	**STEEL SHEET PILING SEAWALLS**									
0200	Steel sheeting, with 4' x 4' x 8" concrete deadmen, 10' OC									
0210	12' high, shore driven	B-40	27	2.370	L.F.	126	135	131	392	490
0260	Barge driven	B-76	15	4.800	"	190	274	250	714	905
6000	Crushed stone placed behind bulkhead by clam bucket	B-12H	120	.133	L.C.Y.	26.50	7.05	10.20	43.75	51

35 31 19 – Revetments

35 31 19.18 Revetments, Concrete

		Crew	Daily Output	Labor-Hours	Unit	Material	2021 Bare Costs Labor	Equipment	Total	Total Incl O&P
0010	**REVETMENTS, CONCRETE**									
0100	Concrete revetment matt 8' x 20' x 4-1/2" excluding site preparation									
0110	Includes all labor, material and equip. for complete installation	B-57	5	9.600	Ea.	2,725	485	370	3,580	4,125

35 49 Waterway Structures

35 49 13 – Floodwalls

35 49 13.30 Breakwaters, Bulkheads, Residential Canal

		Crew	Daily Output	Labor-Hours	Unit	Material	2021 Bare Costs Labor	Equipment	Total	Total Incl O&P
0010	**BREAKWATERS, BULKHEADS, RESIDENTIAL CANAL**									
0020	Aluminum panel sheeting, incl. concrete cap and anchor									
0030	Coarse compact sand, 4'-0" high, 2'-0" embedment	B-40	200	.320	L.F.	72	18.30	17.65	107.95	126
0040	3'-6" embedment		140	.457		84.50	26	25	135.50	161
0060	6'-0" embedment		90	.711		108	40.50	39	187.50	223
0120	6'-0" high, 2'-6" embedment		170	.376		92	21.50	21	134.50	157
0140	4'-0" embedment		125	.512		108	29	28	165	195
0160	5'-6" embedment		95	.674		143	38.50	37	218.50	257
0220	8'-0" high, 3'-6" embedment	B-40	140	.457	L.F.	124	26	25	175	204
0240	5'-0" embedment		100	.640		124	36.50	35.50	196	231
0420	Medium compact sand, 3'-0" high, 2'-0" embedment		235	.272		167	15.55	15	197.55	224
0440	4'-0" embedment		150	.427		205	24.50	23.50	253	289
0460	5'-6" embedment		115	.557		249	32	30.50	311.50	355
0520	5'-0" high, 3'-6" embedment		165	.388		213	22	21.50	256.50	292

35 49 Waterway Structures

35 49 13 – Floodwalls

35 49 13.30 Breakwaters, Bulkheads, Residential Canal	Crew	Daily Output	Labor-Hours	Unit	Material	2021 Bare Costs Labor	Equipment	Total	Total Incl O&P
0540	5'-0" embedment	B-40	120	.533	L.F.	246	30.50	29.50	306	350
0560	6'-6" embedment		105	.610		320	35	33.50	388.50	440
0620	7'-0" high, 4'-6" embedment		135	.474		285	27	26	338	385
0640	6'-0" embedment		110	.582		315	33	32	380	430
0720	Loose silty sand, 3'-0" high, 3'-0" embedment		205	.312		140	17.85	17.20	175.05	200
0740	4'-6" embedment		155	.413		168	23.50	23	214.50	246
0760	6'-0" embedment		125	.512		197	29	28	254	293
0820	4'-6" high, 4'-6" embedment		155	.413		204	23.50	23	250.50	285
0840	6'-0" embedment		125	.512		237	29	28	294	335
0860	7'-0" embedment		115	.557		286	32	30.50	348.50	400
0920	6'-0" high, 5'-6" embedment		130	.492		272	28	27	327	370
0940	7'-0" embedment		115	.557		300	32	30.50	362.50	415

35 51 Floating Construction

35 51 13 – Floating Piers

35 51 13.23 Floating Wood Piers

		Crew	Daily Output	Labor-Hours	Unit	Material	2021 Bare Costs Labor	Equipment	Total	Total Incl O&P
0010	**FLOATING WOOD PIERS**									
0020	Polyethylene encased polystyrene, no pilings included	F-3	330	.121	S.F.	32.50	6.80	1.44	40.74	47.50
0030	See Section 06 13 33.50 or Section 06 13 33.52 for fixed docks									
0200	Pile supported, shore constructed, minimum	F-3	130	.308	S.F.	29.50	17.25	3.66	50.41	62
0250	Maximum		120	.333		33	18.70	3.97	55.67	68.50
0400	Floating, small boat, prefab, no shore facilities, minimum		250	.160		27	8.95	1.90	37.85	45
0500	Maximum		150	.267		59	14.95	3.17	77.12	91
0700	Per slip, minimum (180 S.F. each)		1.59	25.157	Ea.	5,800	1,400	299	7,499	8,825
0800	Maximum		1.40	28.571	"	8,925	1,600	340	10,865	12,600

35 51 13.24 Jetties, Docks

		Crew	Daily Output	Labor-Hours	Unit	Material	2021 Bare Costs Labor	Equipment	Total	Total Incl O&P
0010	**JETTIES, DOCKS**									
0011	Floating including anchors									
0030	See Section 06 13 33.50 or Section 06 13 33.52 for fixed docks									
1000	Polystyrene flotation, minimum	F-3	200	.200	S.F.	34	11.20	2.38	47.58	56.50
1040	Maximum		135	.296	"	44	16.60	3.53	64.13	77.50
1100	Alternate method of figuring, minimum		1.13	35.398	Slip	6,125	1,975	420	8,520	10,100
1140	Maximum		.70	57.143	"	8,775	3,200	680	12,655	15,200
1200	Galv. steel frame and wood deck, 3' wide, minimum		320	.125	S.F.	18.25	7	1.49	26.74	32
1240	Maximum		200	.200		29.50	11.20	2.38	43.08	52
1300	4' wide, minimum		320	.125		22.50	7	1.49	30.99	37
1340	Maximum	F-3	200	.200	S.F.	36	11.20	2.38	49.58	59
1500	8' wide, minimum		250	.160		20.50	8.95	1.90	31.35	38
1540	Maximum		160	.250		31	14	2.98	47.98	58.50
1700	Treated wood frames and deck, 3' wide, minimum		250	.160		24	8.95	1.90	34.85	42
1740	Maximum		125	.320		83	17.95	3.81	104.76	123
2000	Polyethylene drums, treated wood frame and deck									
2100	6' wide, minimum	F-3	250	.160	S.F.	30.50	8.95	1.90	41.35	49
2140	Maximum		125	.320		41	17.95	3.81	62.76	76
2200	8' wide, minimum		233	.172		29	9.60	2.04	40.64	48.50
2240	Maximum		120	.333		40	18.70	3.97	62.67	76.50
2300	10' wide, minimum		200	.200		34	11.20	2.38	47.58	57
2340	Maximum		110	.364		47	20.50	4.33	71.83	87
2400	Concrete pontoons, treated wood frame and deck									
2500	Breakwater, concrete pontoon, 50' L x 8' W x 6.5' D	F-4	4	12	Ea.	45,700	665	248	46,613	52,000
2600	Inland breakwater, concrete pontoon, 50' L x 8' W x 4' D		4	12	"	45,900	665	248	46,813	52,000

562

35 51 Floating Construction

35 51 13 – Floating Piers

35 51 13.24 Jetties, Docks

		Crew	Daily Output	Labor-Hours	Unit	Material	2021 Bare Costs Labor	Equipment	Total	Total Incl O&P
2700	Docks, concrete pontoon w/treated wood deck	F-4	2400	.020	S.F.	108	1.11	.41	109.52	121
2800	Docks, concrete pontoons docks 400 S.F. minimum order									

35 51 13.28 Jetties, Floating Dock Accessories

		Crew	Daily Output	Labor-Hours	Unit	Material	2021 Bare Costs Labor	Equipment	Total	Total Incl O&P
0010	**JETTIES, FLOATING DOCK ACCESSORIES**									
0200	Dock connectors, stressed cables with rubber spacers									
0220	25" long, 3' wide dock	1 Clab	2	4	Joint	277	178		455	570
0240	5' wide dock		2	4		325	178		503	620
0400	38" long, 4' wide dock		1.75	4.571		305	203		508	640
0440	6' wide dock		1.45	5.517		345	245		590	745
1000	Gangway, aluminum, one end rolling, no hand rails									
1020	3' wide, minimum	1 Clab	67	.119	L.F.	285	5.30		290.30	325
1040	Maximum		32	.250		325	11.10		336.10	375
1100	4' wide, minimum		40	.200		300	8.90		308.90	345
1140	Maximum		24	.333		345	14.80		359.80	400
1180	For handrails, add					74.50			74.50	82
2000	Pile guides, beads on stainless cable	1 Clab	4	2	Ea.	87	89		176	229
2020	Rod type, 8" diameter piles, minimum		4	2		62.50	89		151.50	202
2040	Maximum		2	4		91	178		269	365
2100	10" to 14" diameter piles, minimum		3.20	2.500		88.50	111		199.50	264
2140	Maximum		1.75	4.571		155	203		358	475
2200	Roller type, 4 rollers, minimum		4	2		224	89		313	380
2240	Maximum		1.75	4.571		320	203		523	655

35 59 Marine Specialties

35 59 33 – Marine Bollards and Cleats

35 59 33.50 Jetties, Dock Accessories

		Crew	Daily Output	Labor-Hours	Unit	Material	2021 Bare Costs Labor	Equipment	Total	Total Incl O&P
0010	**JETTIES, DOCK ACCESSORIES**									
0100	Cleats, aluminum, "S" type, 12" long	1 Clab	8	1	Ea.	27	44.50		71.50	96.50
0140	10" long		6.70	1.194		65.50	53		118.50	151
0180	15" long		6	1.333		86	59		145	184
0400	Dock wheel for corners and piles, vinyl, 12" diameter		4	2		95	89		184	238
1000	Electrical receptacle with circuit breaker,									
1020	Pile mounted, double 30 amp, 125 volt	1 Elec	2	4	Unit	725	255		980	1,175
1060	Double 50 amp, 125/240 volt		1.60	5		980	320		1,300	1,550
1120	Free standing, add		4	2		217	127		344	430
1140	Double free standing, add		2.70	2.963		236	189		425	540
1160	Light, 2 louvered, with photo electric switch, add		8	1		204	63.50		267.50	320
1180	Telephone jack on stanchion		8	1		200	63.50		263.50	315
1300	Fender, vinyl, 4" high	1 Clab	160	.050	L.F.	16.20	2.22		18.42	21
1380	Corner piece		80	.100	Ea.	15.80	4.44		20.24	24
1400	Hose holder, cast aluminum		16	.500	"	47.50	22		69.50	85.50
1500	Ladder, aluminum, heavy duty									
1520	Crown top, 5 to 7 step, minimum	1 Clab	5.30	1.509	Ea.	203	67		270	325
1560	Maximum		2	4		305	178		483	600
1580	Bracket for portable clamp mounting		8	1		9.30	44.50		53.80	76.50
1800	Line holder, treated wood, small		16	.500		13.60	22		35.60	48
1840	Large		13.30	.602		23	26.50		49.50	65.50
2000	Mooring whip, fiberglass bolted to dock,									
2020	1,200 lb. boat	1 Clab	8.80	.909	Pr.	385	40.50		425.50	480
2040	10,000 lb. boat		6.70	1.194		775	53		828	935
2080	60,000 lb. boat		4	2		910	89		999	1,125

35 59 Marine Specialties

35 59 33 – Marine Bollards and Cleats

35 59 33.50 Jetties, Dock Accessories	Crew	Daily Output	Labor-Hours	Unit	Material	2021 Bare Costs Labor	Equipment	Total	Total Incl O&P
2400	Shock absorbing tubing, vertical bumpers									
2420	3" diameter, vinyl, white	1 Clab	80	.100	L.F.	5.95	4.44		10.39	13.20
2440	Polybutyl, clear		80	.100	"	6.60	4.44		11.04	13.90
2480	Mounts, polybutyl		20	.400	Ea.	6.95	17.75		24.70	34
2490	Deluxe		20	.400	"	15.55	17.75		33.30	43.50

Estimating Tips

Products such as conveyors, material handling cranes and hoists, and other items specified in this division require trained installers. The general contractor may not have any choice as to who will perform the installation or when it will be performed. Long lead times are often required for these products, making early decisions in purchasing and scheduling necessary. The installation of this type of equipment may require the embedment of mounting hardware during the construction of floors, structural walls, or interior walls/partitions. Electrical connections will require coordination with the electrical contractor.

Reference Numbers

Reference numbers are shown at the beginning of some major classifications. These numbers refer to related items in the Reference Section. The reference information may be an estimating procedure, an alternate pricing method, or technical information.

41 21 Conveyors

41 21 23 – Piece Material Conveyors

41 21 23.16 Container Piece Material Conveyors	Crew	Daily Output	Labor-Hours	Unit	Material	2021 Bare Costs Labor	Equipment	Total	Total Incl O&P
0010 **CONTAINER PIECE MATERIAL CONVEYORS**									
0020 Gravity fed, 2" rollers, 3" OC									
0050 10' sections with 2 supports, 600 lb. capacity, 18" wide				Ea.	530			530	580
0350 Horizontal belt, center drive and takeup, 60 fpm									
0400 16" belt, 26.5' length	2 Mill	.50	32	Ea.	3,675	1,875		5,550	6,750
0600 Inclined belt, 10' rise with horizontal loader and									
0620 End idler assembly, 27.5' length, 18" belt	2 Mill	.30	53.333	Ea.	7,750	3,125		10,875	13,100

41 22 Cranes and Hoists

41 22 13 – Cranes

41 22 13.10 Crane Rail

	Crew	Daily Output	Labor-Hours	Unit	Material	Labor	Equipment	Total	Total Incl O&P
0010 **CRANE RAIL**									
0020 Box beam bridge, no equipment included	E-4	3400	.009	Lb.	1.32	.57	.04	1.93	2.39
0200 Running track only, 104 lb. per yard	"	5600	.006	"	.66	.35	.03	1.04	1.30

41 22 23 – Hoists

41 22 23.10 Material Handling

	Crew	Daily Output	Labor-Hours	Unit	Material	Labor	Equipment	Total	Total Incl O&P
0010 **MATERIAL HANDLING**, cranes, hoists and lifts									
2100 Hoists, electric overhead, chain, hook hung, 15' lift, 1 ton cap.				Ea.	2,550			2,550	2,800
2200 3 ton capacity					3,950			3,950	4,350
2500 5 ton capacity					7,625			7,625	8,400

Estimating Tips

This division contains information about water and wastewater equipment and systems, which was formerly located in Division 44. The main areas of focus are total wastewater treatment plants and components of wastewater treatment plants. Also included in this section are oil/water separators for wastewater treatment.

Reference Numbers

Reference numbers are shown at the beginning of some major classifications. These numbers refer to related items in the Reference Section. The reference information may be an estimating procedure, an alternate pricing method, or technical information.

Same Data. Simplified.

Enjoy the convenience and efficiency of accessing your costs anywhere:

- **Skip the multiplier** by setting your location
- **Quickly search,** edit, favorite and share costs
- **Stay on top of price changes** with automatic updates

Discover more at rsmeans.com/online

46 07 53 – Packaged Wastewater Treatment Equipment

46 07 53.10 Biological Pkg. Wastewater Treatment Plants	Crew	Daily Output	Labor-Hours	Unit	Material	2021 Bare Costs Labor	Equipment	Total	Total Incl O&P
0010 **BIOLOGICAL PACKAGED WASTEWATER TREATMENT PLANTS**									
0011 Not including fencing or external piping									
0020 Steel packaged, blown air aeration plants									
0100 1,000 GPD				Gal.				55	60.50
0200 5,000 GPD								22	24
0300 15,000 GPD								22	24
0400 30,000 GPD								15.40	16.95
0500 50,000 GPD								11	12.10
0600 100,000 GPD								9.90	10.90
0700 200,000 GPD								8.80	9.70
0800 500,000 GPD				↓				7.70	8.45
1000 Concrete, extended aeration, primary and secondary treatment									
1010 10,000 GPD				Gal.				22	24
1100 30,000 GPD								15.40	16.95
1200 50,000 GPD								11	12.10
1400 100,000 GPD								9.90	10.90
1500 500,000 GPD				↓				7.70	8.45
1700 Municipal wastewater treatment facility									
1720 1.0 MGD				Gal.				11	12.10
1740 1.5 MGD								10.60	11.65
1760 2.0 MGD								10	11
1780 3.0 MGD								7.80	8.60
1800 5.0 MGD				↓				5.80	6.70
2000 Holding tank system, not incl. excavation or backfill									
2010 Recirculating chemical water closet	2 Plum	4	4	Ea.	575	271		846	1,025
2100 For voltage converter, add	"	16	1		288	67.50		355.50	415
2200 For high level alarm, add	1 Plum	7.80	1.026	↓	152	69.50		221.50	271

46 07 53.20 Wastewater Treatment System

	Crew	Daily Output	Labor-Hours	Unit	Material	Labor	Equipment	Total	Total Incl O&P
0010 **WASTEWATER TREATMENT SYSTEM**									
0020 Fiberglass, 1,000 gallon	B-21	1.29	21.705	Ea.	4,475	1,100	148	5,723	6,750
0100 1,500 gallon	"	1.03	27.184	"	9,600	1,400	185	11,185	12,900

46 23 Grit Removal And Handling Equipment

46 23 23 – Vortex Grit Removal Equipment

46 23 23.10 Rainwater Filters

	Crew	Daily Output	Labor-Hours	Unit	Material	Labor	Equipment	Total	Total Incl O&P
0010 **RAINWATER FILTERS**									
0100 42 gal./min	B-21	3.50	8	Ea.	41,400	410	54.50	41,864.50	46,300
0200 65 gal./min		3.50	8		41,500	410	54.50	41,964.50	46,300
0300 208 gal./min	↓	3.50	8	↓	41,500	410	54.50	41,964.50	46,400

46 25 Oil and Grease Separation and Removal Equipment

46 25 13 - Coalescing Oil-Water Separators

46 25 13.20 Oil/Water Separators

		Crew	Daily Output	Labor-Hours	Unit	Material	2021 Bare Costs Labor	2021 Bare Costs Equipment	Total	Total Incl O&P
0010	**OIL/WATER SEPARATORS**									
0020	Underground, tank only									
0030	Excludes excavation, backfill & piping									
0100	200 GPM	B-21	3.50	8	Ea.	41,200	410	54.50	41,664.50	46,000
0110	400 GPM		3.25	8.615		90,500	440	59	90,999	100,500
0120	600 GPM		2.75	10.182		91,000	520	69.50	91,589.50	101,000
0130	800 GPM		2.50	11.200		114,500	570	76.50	115,146.50	127,000
0140	1,000 GPM		2	14		127,500	715	95.50	128,310.50	141,500
0150	1,200 GPM		1.50	18.667		142,000	955	127	143,082	157,500
0160	1,500 GPM		1	28		161,500	1,425	191	163,116	180,000

46 25 16 - API Oil-Water Separators

46 25 16.10 Oil/Water Separators

		Crew	Daily Output	Labor-Hours	Unit	Material	2021 Bare Costs Labor	2021 Bare Costs Equipment	Total	Total Incl O&P
0010	**OIL/WATER SEPARATORS**									
0020	Complete system, not including excavation and backfill									
0100	Treated capacity 0.2 C.F./second	B-22	1	30	Ea.	9,675	1,550	287	11,512	13,200
0200	0.5 C.F./second	B-13	.75	74.667		12,800	3,600	785	17,185	20,300
0300	1.0 C.F./second		.60	93.333		17,100	4,500	980	22,580	26,600
0400	2.4 to 3.0 C.F./second		.30	187		28,600	9,000	1,950	39,550	47,100
0500	11 C.F./second		.17	329		23,800	15,900	3,450	43,150	53,500
0600	22 C.F./second		.10	560		39,600	27,000	5,875	72,475	90,500

46 51 Air and Gas Diffusion Equipment

46 51 13 - Floating Mechanical Aerators

46 51 13.10 Aeration Equipment

		Crew	Daily Output	Labor-Hours	Unit	Material	2021 Bare Costs Labor	2021 Bare Costs Equipment	Total	Total Incl O&P
0010	**AERATION EQUIPMENT**									
0020	Aeration equipment includes floats and excludes power supply and anchorage									
4320	Surface aerator, 50 lb. oxygen/hr, 30 HP, 900 RPM, inc floats, no anchoring	B-21B	1	40	Ea.	33,600	1,925	475	36,000	40,400
4322	Anchoring aerators per cell, 6 anchors per cell	B-6	.50	48		3,200	2,300	430	5,930	7,450
4324	Anchoring aerators per cell, 8 anchors per cell	"	.33	72.727		4,400	3,500	655	8,555	10,800

46 51 20 - Air and Gas Handling Equipment

46 51 20.10 Blowers and System Components

		Crew	Daily Output	Labor-Hours	Unit	Material	2021 Bare Costs Labor	2021 Bare Costs Equipment	Total	Total Incl O&P
0010	**BLOWERS AND SYSTEM COMPONENTS**									
0020	Rotary lobe blowers									
0030	Medium pressure (7 to 14 PSIG)									
0120	38 CFM, 2.1 BHP	Q-2	2.50	9.600	Ea.	1,575	605		2,180	2,625
0130	125 CFM, 5.5 BHP		2.40	10		1,750	630		2,380	2,875
0140	245 CFM, 10.2 BHP		2	12		2,025	760		2,785	3,375
0150	363 CFM, 14.5 BHP		1.80	13.333		2,950	840		3,790	4,500
0160	622 CFM, 24.5 BHP		1.40	17.143		4,900	1,075		5,975	7,000
0170	1,125 CFM, 42.8 BHP		1.10	21.818		7,800	1,375		9,175	10,600
0180	1,224 CFM, 47.4 BHP		1	24		8,500	1,525		10,025	11,600
0400	Prepackaged medium pressure (7 to 14 PSIG) blower									
0405	incl. 3 ph. motor, filter, silencer, valves, check valve, press. gage									
0420	38 CFM, 2.1 BHP	Q-2	2.50	9.600	Ea.	3,325	605		3,930	4,550
0430	125 CFM, 5.5 BHP		2.40	10		4,225	630		4,855	5,600
0440	245 CFM, 10.2 BHP		2	12		5,375	760		6,135	7,050
0450	363 CFM, 14.5 BHP		1.80	13.333		7,450	840		8,290	9,425
0460	521 CFM, 20 BHP		1.50	16		7,450	1,000		8,450	9,675
1010	Filters and silencers									
1100	Silencer with paper filter									

For customer support on your Site Work & Landscape Costs with RSMeans Data, call 800.448.8182.

569

46 51 20.10 Blowers and System Components	Crew	Daily Output	Labor-Hours	Unit	Material	2021 Bare Costs Labor	2021 Bare Costs Equipment	Total	Total Incl O&P
1105	1" connection	1 Plum	14	.571	Ea.	39	38.50		77.50	100
1110	1.5" connection		11	.727		43.50	49		92.50	121
1115	2" connection		9	.889		154	60		214	259
1120	2.5" connection		8	1		154	67.50		221.50	270
1125	3" connection	Q-1	8	2		159	122		281	355
1130	4" connection	"	5	3.200		320	195		515	645
1135	5" connection	Q-2	5	4.800		375	305		680	870
1140	6" connection	"	5	4.800		395	305		700	890
1300	Silencer with polyester filter									
1305	1" connection	1 Plum	14	.571	Ea.	44.50	38.50		83	107
1310	1.5" connection		11	.727		49	49		98	128
1315	2" connection		9	.889		12.95	60		72.95	104
1320	2.5" connection		8	1		25.50	67.50		93	129
1325	3" connection	Q-1	8	2		180	122		302	380
1330	4" connection	"	5	3.200		340	195		535	665
1335	5" connection	Q-2	5	4.800		430	305		735	930
1340	6" connection	"	5	4.800		475	305		780	980
1500	Chamber silencers									
1505	1" connection	1 Plum	14	.571	Ea.	91	38.50		129.50	158
1510	1.5" connection		11	.727		113	49		162	198
1515	2" connection		9	.889		139	60		199	243
1520	2.5" connection		8	1		225	67.50		292.50	350
1525	3" connection	Q-1	8	2	Ea.	320	122		442	530
1530	4" connection	"	5	3.200	"	410	195		605	745
1700	Blower couplings									
1701	Blower flexible coupling									
1710	1.5" connection	Q-1	71	.225	Ea.	33	13.75		46.75	57
1715	2" connection		67	.239		38	14.55		52.55	63.50
1720	2.5" connection		65	.246		45	15		60	72
1725	3" connection		64	.250		65	15.25		80.25	94
1730	4" connection		58	.276		65	16.80		81.80	96.50
1735	5" connection	Q-2	83	.289		75	18.25		93.25	110
1740	6" connection		79	.304		100	19.20		119.20	139
1745	8" connection		69	.348		180	22		202	231

		Crew	Daily Output	Labor-Hours	Unit	Material	2021 Bare Costs Labor	2021 Bare Costs Equipment	Total	Total Incl O&P
0010	**AERATION SYSTEM AIR PROCESS PIPING**									
0100	Blower pressure relief valves - adjustable									
0105	1" diameter	1 Plum	14	.571	Ea.	89.50	38.50		128	156
0110	2" diameter		9	.889		107	60		167	208
0115	2.5" diameter		8	1		131	67.50		198.50	245
0120	3" diameter	Q-1	8	2		169	122		291	370
0125	4" diameter	"	5	3.200		230	195		425	545
0200	Blower pressure relief valves - weight loaded									
0205	1" diameter	1 Plum	14	.571	Ea.	145	38.50		183.50	217
0210	2" diameter	"	9	.889		162	60		222	268
0220	3" diameter	Q-1	8	2		695	122		817	945
0225	4" diameter	"	5	3.200		920	195		1,115	1,300
0300	Pressure relief valves - preset high flow									
0310	2" diameter	1 Plum	9	.889	Ea.	335	60		395	460
0320	3" diameter	Q-1	8	2	"	745	122		867	1,000
1000	Check valves, wafer style									
1110	2" diameter	1 Plum	9	.889	Ea.	125	60		185	227

46 51 Air and Gas Diffusion Equipment

46 51 20 – Air and Gas Handling Equipment

46 51 20.20 Aeration System Air Process Piping

		Crew	Daily Output	Labor-Hours	Unit	Material	2021 Bare Costs Labor	Equipment	Total	Total Incl O&P
1120	3" diameter	Q-1	8	2	Ea.	185	122		307	385
1125	4" diameter	"	5	3.200		385	195		580	715
1130	5" diameter	Q-2	6	4		340	253		593	750
1135	6" diameter		5	4.800		370	305		675	860
1140	8" diameter	↓	4.50	5.333	↓	555	335		890	1,125
1200	Check valves, flanged steel									
1210	2" diameter	1 Plum	8	1	Ea.	230	67.50		297.50	355
1220	3" diameter	Q-1	4.50	3.556		175	217		392	520
1225	4" diameter	"	3	5.333		220	325		545	725
1230	5" diameter	Q-2	3	8		330	505		835	1,125
1235	6" diameter	"	3	8	↓	390	505		895	1,175
1240	8" diameter	Q-2	2.50	9.600	Ea.	1,450	605		2,055	2,500
2100	Butterfly valves, lever operated - wafer style									
2110	2" diameter	1 Plum	14	.571	Ea.	86.50	38.50		125	153
2120	3" diameter	Q-1	8	2		117	122		239	310
2125	4" diameter	"	5	3.200		148	195		343	455
2130	5" diameter	Q-2	6	4		177	253		430	570
2135	6" diameter		5	4.800		245	305		550	725
2140	8" diameter		4.50	5.333		345	335		680	885
2145	10" diameter	↓	4	6	↓	580	380		960	1,200
2200	Butterfly valves, gear operated - wafer style									
2210	2" diameter	1 Plum	14	.571	Ea.	106	38.50		144.50	175
2220	3" diameter	Q-1	8	2		195	122		317	395
2225	4" diameter	"	5	3.200		224	195		419	535
2230	5" diameter	Q-2	6	4		285	253		538	690
2235	6" diameter		5	4.800		370	305		675	860
2240	8" diameter		4.50	5.333		485	335		820	1,025
2245	10" diameter		4	6		595	380		975	1,225
2250	12" diameter	↓	3	8	↓	760	505		1,265	1,600

46 51 20.30 Aeration System Blower Control Panels

		Crew	Daily Output	Labor-Hours	Unit	Material	2021 Bare Costs Labor	Equipment	Total	Total Incl O&P
0010	**AERATION SYSTEM BLOWER CONTROL PANELS**									
0020	Single phase simplex									
0030	7 to 10 overload amp range	1 Elec	2.70	2.963	Ea.	580	189		769	920
0040	9 to 13 overload amp range		2.70	2.963		850	189		1,039	1,225
0050	12 to 18 overload amp range		2.70	2.963		795	189		984	1,150
0060	16 to 24 overload amp range		2.70	2.963		855	189		1,044	1,225
0070	23 to 32 overload amp range		2.70	2.963		700	189		889	1,050
0080	30 to 40 overload amp range	↓	2.70	2.963	↓	1,650	189		1,839	2,075
0090	Single phase duplex									
0100	7 to 10 overload amp range	1 Elec	2	4	Ea.	1,450	255		1,705	1,975
0110	9 to 13 overload amp range		2	4		875	255		1,130	1,350
0120	12 to 18 overload amp range		2	4		875	255		1,130	1,350
0130	16 to 24 overload amp range		2	4		875	255		1,130	1,350
0140	23 to 32 overload amp range		2	4		935	255		1,190	1,400
0150	30 to 40 overload amp range	↓	2	4	↓	1,150	255		1,405	1,650
0160	Three phase simplex									
0170	1.6 to 2.5 overload amp range	1 Elec	2.50	3.200	Ea.	975	204		1,179	1,375
0180	2.5 to 4 overload amp range		2.50	3.200		1,175	204		1,379	1,600
0190	4 to 6.3 overload amp range		2.50	3.200		1,175	204		1,379	1,600
0200	6 to 10 overload amp range		2.50	3.200		1,175	204		1,379	1,600
0210	9 to 14 overload amp range	↓	2.50	3.200	↓	1,350	204		1,554	1,775
0220	13 to 18 overload amp range	1 Elec	2.50	3.200	Ea.	1,225	204		1,429	1,650

46 51 20.30 Aeration System Blower Control Panels		Crew	Daily Output	Labor-Hours	Unit	Material	2021 Bare Costs Labor	2021 Bare Costs Equipment	Total	Total Incl O&P
0230	17 to 23 overload amp range	1 Elec	2.50	3.200	Ea.	1,025	204		1,229	1,425
0240	20 to 25 overload amp range		2.50	3.200		1,025	204		1,229	1,425
0250	23 to 32 overload amp range		2.50	3.200		1,175	204		1,379	1,600
0260	37 to 50 overload amp range		2.50	3.200		1,400	204		1,604	1,850
0270	Three phase duplex									
0280	1.6 to 2.5 overload amp range	1 Elec	1.90	4.211	Ea.	1,325	268		1,593	1,850
0290	2.5 to 4 overload amp range		1.90	4.211		1,325	268		1,593	1,850
0300	4 to 6.3 overload amp range		1.90	4.211		1,325	268		1,593	1,850
0310	6 to 10 overload amp range		1.90	4.211		1,325	268		1,593	1,850
0320	9 to 14 overload amp range		1.90	4.211		1,400	268		1,668	1,925
0330	13 to 18 overload amp range		1.90	4.211		1,425	268		1,693	1,975
0340	17 to 23 overload amp range		1.90	4.211		1,450	268		1,718	2,000
0350	20 to 25 overload amp range		1.90	4.211		1,450	268		1,718	2,000
0360	23 to 32 overload amp range		1.90	4.211		1,475	268		1,743	2,025
0370	37 to 50 overload amp range		1.90	4.211		2,200	268		2,468	2,825

46 51 36.10 Ceramic Disc Air Diffuser Systems		Crew	Daily Output	Labor-Hours	Unit	Material	2021 Bare Costs Labor	2021 Bare Costs Equipment	Total	Total Incl O&P
0010	**CERAMIC DISC AIR DIFFUSER SYSTEMS**									
0020	Price for air diffuser pipe system by cell size excluding concrete work									
0030	Price for air diffuser pipe system by cell size excluding air supply									
0040	Depth of 12' is the waste depth, not the cell dimensions									
0100	Ceramic disc air diffuser system, cell size 20' x 20' x 12'	2 Plum	.38	42.667	Ea.	10,900	2,900		13,800	16,300
0120	20' x 30' x 12'		.26	61.326		16,100	4,150		20,250	23,900
0140	20' x 40' x 12'		.20	80		21,300	5,425		26,725	31,600
0160	20' x 50' x 12'		.16	98.644		26,600	6,675		33,275	39,200
0180	20' x 60' x 12'		.14	117		31,800	7,950		39,750	46,900
0200	20' x 70' x 12'		.12	136		37,000	9,200		46,200	54,500
0220	20' x 80' x 12'		.10	155		42,200	10,500		52,700	62,000
0240	20' x 90' x 12'		.09	173		47,400	11,700		59,100	69,500
0260	20' x 100' x 12'		.08	192		52,500	13,000		65,500	77,500
0280	20' x 120' x 12'		.07	229		63,000	15,500		78,500	92,500
0300	20' x 140' x 12'		.06	267		73,500	18,100		91,600	108,000
0320	20' x 160' x 12'		.05	304		84,000	20,600		104,600	123,000
0340	20' x 180' x 12'		.05	341		94,500	23,100		117,600	138,500
0360	20' x 200' x 12'		.04	378		105,000	25,600		130,600	153,500
0380	20' x 250' x 12'		.03	472		131,000	32,000		163,000	191,500
0400	20' x 300' x 12'		.03	565		157,000	38,300		195,300	230,000
0420	20' x 350' x 12'		.02	658		183,000	44,600		227,600	268,000
0440	20' x 400' x 12'		.02	751		209,500	51,000		260,500	306,000
0460	20' x 450' x 12'	2 Plum	.02	847	Ea.	235,500	57,500		293,000	344,500
0480	20' x 500' x 12'	"	.02	941	"	261,500	63,500		325,000	382,500

572

For customer support on your Site Work & Landscape Costs with RSMeans Data, call 800.448.8182.

46 53 17 – Activated Sludge Treatment

46 53 17.10 Activated Sludge Treatment Cells	Crew	Daily Output	Labor-Hours	Unit	Material	2021 Bare Costs Labor	Equipment	Total	Total Incl O&P
0010 **ACTIVATED SLUDGE TREATMENT CELLS**									
0015 Price for cell construction excluding aerator piping & drain									
0020 Cell construction by dimensions									
0100 Treatment cell, end or single, 20' x 20' x 14' high (2' freeboard)	C-14D	.46	434	Ea.	20,900	23,600	975	45,475	59,500
0120 20' x 30' x 14' high		.35	568		27,100	30,900	1,275	59,275	77,500
0140 20' x 40' x 14' high		.29	702		33,400	38,100	1,575	73,075	95,500
0160 20' x 50' x 14' high		.24	835		39,600	45,400	1,875	86,875	113,000
0180 20' x 60' x 14' high		.21	969		45,900	52,500	2,175	100,575	131,500
0200 20' x 70' x 14' high		.18	1103		52,000	60,000	2,475	114,475	149,500
0220 20' x 80' x 14' high		.16	1237		58,500	67,000	2,775	128,275	167,500
0240 20' x 90' x 14' high		.15	1371		64,500	74,500	3,075	142,075	185,500
0260 20' x 100' x 14' high		.13	1504		71,000	81,500	3,375	155,875	203,500
0280 20' x 120' x 14' high		.11	1771		83,500	96,500	3,975	183,975	239,500
0300 20' x 140' x 14' high		.10	2039		96,000	111,000	4,575	211,575	276,000
0320 20' x 160' x 14' high		.09	2307		108,500	125,500	5,200	239,200	311,500
0340 20' x 180' x 14' high		.08	2574		121,000	140,000	5,800	266,800	348,000
0360 20' x 200' x 14' high		.07	2841		133,500	154,500	6,400	294,400	384,000
0380 20' x 250' x 14' high		.06	3509		164,500	190,500	7,900	362,900	474,000
0400 20' x 300' x 14' high		.05	4175		195,500	227,000	9,400	431,900	564,000
0420 20' x 350' x 14' high		.04	4843		227,000	263,000	10,900	500,900	654,000
0440 20' x 400' x 14' high		.04	5510		258,000	299,500	12,400	569,900	744,000
0460 20' x 450' x 14' high		.03	6192		289,500	336,500	13,900	639,900	836,000
0480 20' x 500' x 14' high		.03	6849		320,500	372,500	15,400	708,400	925,000
0500 Treatment cell, end or single, 20' x 20' x 12' high (2' freeboard)		.48	414		19,700	22,500	930	43,130	56,500
0520 20' x 30' x 12' high		.37	543		25,700	29,500	1,225	56,425	73,500
0540 20' x 40' x 12' high		.30	672		31,700	36,500	1,500	69,700	91,000
0560 20' x 50' x 12' high		.25	800		37,600	43,500	1,800	82,900	108,500
0580 20' x 60' x 12' high		.22	929		43,600	50,500	2,100	96,200	125,500
0600 20' x 70' x 12' high		.19	1058		49,500	57,500	2,375	109,375	142,500
0620 20' x 80' x 12' high		.17	1186		55,500	64,500	2,675	122,675	160,000
0640 20' x 90' x 12' high		.15	1315		61,500	71,500	2,950	135,950	177,500
0660 20' x 100' x 12' high		.14	1444		67,500	78,500	3,250	149,250	194,500
0680 20' x 120' x 12' high		.12	1701		79,500	92,500	3,825	175,825	229,000
0700 20' x 140' x 12' high		.10	1959		91,000	106,500	4,400	201,900	264,500
0720 20' x 160' x 12' high		.09	2215		103,000	120,500	4,975	228,475	298,500
0740 20' x 180' x 12' high	C-14D	.08	2472	Ea.	115,000	134,500	5,550	255,050	333,000
0760 20' x 200' x 12' high		.07	2732		127,000	148,500	6,150	281,650	368,000
0780 20' x 250' x 12' high		.06	3373		156,500	183,500	7,600	347,600	454,500
0800 20' x 300' x 12' high		.05	4016		186,500	218,500	9,025	414,025	540,500
0820 20' x 350' x 12' high		.04	4662		216,500	253,500	10,500	480,500	627,500
0840 20' x 400' x 12' high		.04	5305		246,000	288,500	11,900	546,400	713,500
0860 20' x 450' x 12' high		.03	5952		276,000	323,500	13,400	612,900	801,000
0880 20' x 500' x 12' high		.03	6601		305,500	358,500	14,900	678,900	887,500
1100 Treatment cell, connecting, 20' x 20' x 14' high (2' freeboard)		.57	351		16,700	19,100	790	36,590	47,800
1120 20' x 30' x 14' high		.45	443		20,800	24,100	995	45,895	60,000
1140 20' x 40' x 14' high		.37	535		25,000	29,100	1,200	55,300	72,000
1160 20' x 50' x 14' high		.32	627		29,200	34,100	1,400	64,700	84,500
1180 20' x 60' x 14' high		.28	719		33,300	39,100	1,625	74,025	97,000
1200 20' x 70' x 14' high		.25	811		37,400	44,100	1,825	83,325	108,500
1220 20' x 80' x 14' high		.22	903		41,500	49,100	2,025	92,625	121,000
1240 20' x 90' x 14' high		.20	995		45,700	54,000	2,250	101,950	133,500
1260 20' x 100' x 14' high		.18	1086		49,800	59,000	2,450	111,250	145,500
1280 20' x 120' x 14' high		.16	1271		58,000	69,000	2,850	129,850	170,000

46 53 Biological Treatment Systems

46 53 17 – Activated Sludge Treatment

46 53 17.10 Activated Sludge Treatment Cells		Crew	Daily Output	Labor-Hours	Unit	Material	2021 Bare Costs Labor	Equipment	Total	Total Incl O&P
1300	20' x 140' x 14' high	C-14D	.14	1455	Ea.	66,500	79,000	3,275	148,775	194,500
1320	20' x 160' x 14' high		.12	1638		74,500	89,000	3,675	167,175	219,000
1340	20' x 180' x 14' high		.11	1823		83,000	99,000	4,100	186,100	244,000
1360	20' x 200' x 14' high		.10	2006		91,000	109,000	4,525	204,525	268,000
1380	20' x 250' x 14' high		.08	2466		112,000	134,000	5,550	251,550	329,000
1400	20' x 300' x 14' high		.07	2924		132,500	159,000	6,575	298,075	390,500
1420	20' x 350' x 14' high		.06	3384		153,500	184,000	7,625	345,125	451,500
1440	20' x 400' x 14' high		.05	3846		174,000	209,000	8,650	391,650	513,000
1460	20' x 450' x 14' high		.05	4301		195,000	234,000	9,675	438,675	573,500
1480	20' x 500' x 14' high		.04	4762		215,500	259,000	10,700	485,200	635,000
1500	Treatment cell, connecting, 20' x 20' x 12' high (2' freeboard)		.60	336		15,800	18,300	755	34,855	45,400
1520	20' x 30' x 12' high		.47	425		19,800	23,100	955	43,855	57,500
1540	20' x 40' x 12' high		.39	515		23,800	28,000	1,150	52,950	69,000
1560	20' x 50' x 12' high		.33	604		27,800	32,800	1,350	61,950	81,000
1580	20' x 60' x 12' high		.29	694		31,800	37,700	1,550	71,050	92,500
1600	20' x 70' x 12' high		.26	783		35,800	42,600	1,750	80,150	105,000
1620	20' x 80' x 12' high		.23	873		39,800	47,400	1,975	89,175	116,500
1640	20' x 90' x 12' high		.21	962		43,800	52,500	2,175	98,475	128,500
1660	20' x 100' x 12' high		.19	1052		47,800	57,000	2,375	107,175	140,000
1680	20' x 120' x 12' high		.16	1230		56,000	67,000	2,775	125,775	164,000
1700	20' x 140' x 12' high		.14	1409		64,000	76,500	3,175	143,675	188,000
1720	20' x 160' x 12' high		.13	1589		72,000	86,500	3,575	162,075	212,000
1740	20' x 180' x 12' high	C-14D	.11	1767	Ea.	80,000	96,000	3,975	179,975	235,500
1760	20' x 200' x 12' high		.10	1946		88,000	105,500	4,375	197,875	259,000
1780	20' x 250' x 12' high		.08	2392		108,000	130,000	5,375	243,375	318,500
1800	20' x 300' x 12' high		.07	2841		127,500	154,500	6,400	288,400	378,000
1820	20' x 350' x 12' high		.06	3289		147,500	179,000	7,400	333,900	437,000
1840	20' x 400' x 12' high		.05	3738		167,500	203,000	8,400	378,900	497,000
1860	20' x 450' x 12' high		.05	4184		187,500	227,500	9,425	424,425	556,000
1880	20' x 500' x 12' high		.04	5571		207,500	303,000	12,500	523,000	694,000
1996	Price for aerator curb per cell excluding aerator piping & drain									
2000	Treatment cell, 20' x 20' aerator curb	F-7	1.88	17.067	Ea.	705	845		1,550	2,025
2010	20' x 30'		1.25	25.600		1,050	1,275		2,325	3,075
2020	20' x 40'		.94	34.133		1,400	1,700		3,100	4,075
2030	20' x 50'		.75	42.667		1,750	2,125		3,875	5,100
2040	20' x 60'		.63	51.200		2,125	2,525		4,650	6,100
2050	20' x 70'		.54	59.735		2,475	2,950		5,425	7,150
2060	20' x 80'		.47	68.259		2,825	3,375		6,200	8,150
2070	20' x 90'		.42	76.794		3,175	3,800		6,975	9,175
2080	20' x 100'		.38	85.333		3,525	4,225		7,750	10,200
2090	20' x 120'		.31	102		4,225	5,075		9,300	12,200
2100	20' x 140'		.27	119		4,925	5,925		10,850	14,300
2110	20' x 160'		.23	137		5,650	6,775		12,425	16,300
2120	20' x 180'		.21	154		6,350	7,600		13,950	18,400
2130	20' x 200'		.19	171		7,050	8,450		15,500	20,400
2160	20' x 250'		.15	213		8,800	10,600		19,400	25,500
2190	20' x 300'		.13	256		10,600	12,700		23,300	30,500
2220	20' x 350'		.11	299		12,300	14,800		27,100	35,700
2250	20' x 400'		.09	341		14,100	16,900		31,000	40,700
2280	20' x 450'		.08	384		15,900	19,000		34,900	45,800
2310	20' x 500'		.08	427		17,600	21,100		38,700	51,000

Assemblies Section

Table of Contents

Table of Contents

RSMeans data: Assemblies— How They Work

Assemblies estimating provides a fast and reasonably accurate way to develop construction costs. An assembly is the grouping of individual work items—with appropriate quantities— to provide a cost for a major construction component in a convenient unit of measure.

An assemblies estimate is often used during early stages of design development to compare

the cost impact of various design alternatives on total building cost.

Assemblies estimates are also used as an efficient tool to verify construction estimates.

Assemblies estimates do not require a completed design or detailed drawings. Instead, they are based on the general size of the structure and

other known parameters of the project. The degree of accuracy of an assemblies estimate is generally within +/- 15%.

Most assemblies consist of three major elements: a graphic, the system components, and the cost data itself. The **Graphic** is a visual representation showing the typical appearance of the assembly

❶ Unique 12-character Identifier

Our assemblies are identified by a **unique 12-character identifier**. The assemblies are numbered using UNIFORMAT II, ASTM Standard E1557. The first 5 characters represent this system to Level 3. The last 7 characters represent further breakdown in order to arrange items in understandable groups of similar tasks. Line numbers are consistent across all of our publications, so a line number in any assemblies data set will always refer to the same item.

❷ Narrative Descriptions

Our assemblies descriptions appear in two formats: narrative and table. **Narrative descriptions** are shown in a hierarchical structure to make them readable. In order to read a complete description, read up through the indents to the top of the section. Include everything that is above and to the left that is not contradicted by information below.

Narrative Format

G20 Site Improvements

G2040   Site Development

There are four basic types of Concrete Retaining Wall Systems: reinforced concrete with level backfill; reinforced concrete with sloped backfill or surcharge; unreinforced with level backfill; and unreinforced with sloped backfill or surcharge. System elements include: all necessary forms (4 uses); 3,000 p.s.i. concrete with an 8″ chute; all necessary reinforcing steel; and underdrain. Exposed concrete is patched and rubbed.

The Expanded System Listing shows walls that range in thickness from 10″ to 18″ for reinforced concrete walls with level backfill and 12″ to 24″ for reinforced walls with sloped backfill. Walls range from a height of 4′ to 20′. Unreinforced level and sloped backfill walls range from a height of 3′ to 10′.

System Components ❶	QUANTITY	UNIT	COST PER L.F. MAT.	COST PER L.F. INST.	COST PER L.F. TOTAL
SYSTEM G2040 210 1000					
CONC. RETAIN. WALL REINFORCED, LEVEL BACKFILL, 4' HIGH					
Forms in place, cont. wall footing & keyway, 4 uses	2.000	S.F.	5.64	10.30	15.94
Forms in place, retaining wall forms, battered to 8' high, 4 uses	8.000	SFCA	8.80	78.40	87.20
Reinforcing in place, walls, #3 to #7	.004	Ton	5.50	3.76	9.26
Concrete ready mix, regular weight, 3000 psi	.204	C.Y.	27.95		27.95
Placing concrete and vibrating footing con., shallow direct chute	.074	C.Y.		2.08	2.08
Placing concrete and vibrating walls, 8" thick, direct chute	.130	C.Y.		4.84	4.84
Pipe bedding, crushed or screened bank run gravel	1.000	L.F.	4.02	1.44	5.46
Pipe, subdrainage, corrugated plastic, 4" diameter	1.000	L.F.	.89	.88	1.77
Finish walls and break ties, patch walls	4.000	S.F.	.20	4.48	4.68
TOTAL			53	106.18	159.18

G2040 210	Concrete Retaining Walls	COST PER L.F. MAT.	COST PER L.F. INST.	COST PER L.F. TOTAL
1000	Conc. retain. wall, reinforced, level backfill, 4' high x 2'-2" base,10" th	53	106	159
1200	6' high x 3'-3" base, 10" thick	76	154	230
1400	8' high x 4'-3" base, 10" thick	99	201	300
1600	❷ 10' high x 5'-4" base, 13" thick	129	295	424
2200	16' high x 8'-6" base, 16" thick	241	480	721
2600	20' high x 10'-5" base, 18" thick	350	620	970
3000	Sloped backfill, 4' high x 3'-2" base, 12" thick	64	110	174
3200	6' high x 4'-6" base, 12" thick	90	158	248
3400	8' high x 5'-11" base, 12" thick	119	208	327
3600	10' high x 7'-5" base, 16" thick	170	310	480
3800	12' high x 8'-10" base, 18" thick	221	375	596
4200	16' high x 11'-10" base, 21" thick	370	530	900
4600	20' high x 15'-0" base, 24" thick	570	715	1,285
5000	Unreinforced, level backfill, 3'-0" high x 1'-6" base	28	72	100
5200	4'-0" high x 2'-0" base	43	95	138
5400	6'-0" high x 3'-0" base	80	147	227
5600	8'-0" high x 4'-0" base	126	196	322
5800	10'-0" high x 5'-0" base	186	305	491
7000	Sloped backfill, 3'-0" high x 2'-0" base	33.50	74.50	108
7200	4'-0" high x 3'-0" base	55	99	154
7400	6'-0" high x 5'-0" base	114	156	270
7600	8'-0" high x 7'-0" base	191	213	404
7800	10'-0" high x 9'-0" base	292	330	622

For supplemental customizable square foot estimating forms, visit: **RSMeans.com/2021books**

in question. It is frequently accompanied by additional explanatory technical information describing the class of items. The **System Components** is a listing of the individual tasks that make up the assembly, including the quantity and unit of measure for each item, along with the cost of material and installation. The **Assemblies**

data below lists prices for other similar systems with dimensional and/or size variations.

All of our assemblies costs represent the cost for the installing contractor. An allowance for profit has been added to all material, labor, and equipment rental costs. A markup for labor burdens, including workers' compensation, fixed

overhead, and business overhead, is included with installation costs.

The information in RSMeans cost data represents a "national average" cost. This data should be modified to the project location using the **City Cost Indexes** or **Location Factors** tables found in the Reference Section.

Table Format

A10 Foundations

A1010 Standard Foundations

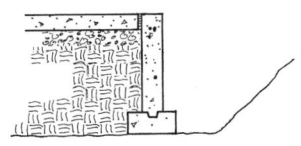

The Foundation Bearing Wall System includes: forms up to 6' high (four uses); 3,000 p.s.i. concrete placed and vibrated; and form removal with breaking form ties and patching walls. The wall systems list walls from 6" to 16" thick and are designed with minimum reinforcement.

Excavation and backfill are not included.

Please see the reference section for further design and cost information.

System Components			COST PER L.F.		
	QUANTITY	UNIT	MAT.	INST.	TOTAL
SYSTEM A1010 105 1500					
FOUNDATION WALL, CAST IN PLACE, DIRECT CHUTE, 4' HIGH, 6" THICK					
Formwork	8.000	SFCA	6.48	50.80	57.28
Reinforcing	3.300	Lb.	2.27	1.55	3.82
Unloading & sorting reinforcing	3.300	Lb.		.09	.09
Concrete, 3,000 psi	.074	C.Y.	10.14		10.14
Place concrete, direct chute	.074	C.Y.		2.75	2.75
Finish walls, break ties and patch voids, one side	4.000	S.F.	.20	4.48	4.68
TOTAL			19.09	59.67	78.76

A1010 105			Wall Foundations					
	WALL HEIGHT (FT.)	PLACING METHOD	CONCRETE (C.Y. per L.F.)	REINFORCING (LBS. per L.F.)	WALL THICKNESS (IN.)	COST PER L.F.		
						MAT.	INST.	TOTAL
1500	4'	direct chute	.074	3.3	6	19.10	59.50	78.60
1520			.099	4.8	8	23.50	61.50	85
1540			.123	6.0	10	27.50	62.50	90
1560			.148	7.2	12	32	64	96
1580			.173	8.1	14	36	65	101
1600			.197	9.44	16	40	66.50	106.50
1700	4'	pumped	.074	3.3	6	19.10	60.50	79.60
1720			.099	4.8	8	23.50	62.50	86
1740			.123	6.0	10	27.50	64	91.50
1760			.148	7.2	12	32	65.50	97.50
1780			.173	8.1	14	36	66.50	102.50
1800			.197	9.44	16	40	68	108

③ Unit of Measure

All RSMeans data: Assemblies include a typical **Unit of Measure** used for estimating that item. For instance, for continuous footings or foundation walls the unit is linear feet (L.F.). For spread footings the unit is each (Ea.). The estimator needs to take special care that the unit in the data matches the unit in the takeoff. Abbreviations and unit conversions can be found in the Reference Section.

④ System Components

System components are listed separately to detail what is included in the development of the total system price.

⑤ Table Descriptions

Table descriptions work similar to Narrative Descriptions, except that if there is a blank in the column at a particular line number, read up to the description above in the same column.

Sample Estimate

This sample demonstrates the elements of an estimate, including a tally of the RSMeans data lines. Published assemblies costs include all markups for labor burden and profit for the installing contractor. This estimate adds a summary of the markups applied by a general contractor on the installing contractor's work. These figures represent the total cost to the owner. The location factor with RSMeans data is applied at the bottom of the estimate to adjust the cost of the work to a specific location.

Project Name:	Interior Fit-out, ABC Office			
Location:	**Anywhere, USA**		**Date: 1/1/2021**	**STD**
Assembly Number	**Description**	**Qty.**	**Unit**	**Subtotal**
❶ C1010 124 1200	Wood partition, 2 x 4 @ 16" OC w/5/8" FR gypsum board	560	S.F.	$3,169.60
C1020 114 1800	Metal door & frame, flush hollow core, 3'-0" x 7'-0"	2	Ea.	$2,580.00
C3010 230 0080	Painting, brushwork, primer & 2 coats	1,120	S.F.	$1,512.00
C3020 410 0140	Carpet, tufted, nylon, roll goods, 12' wide, 26 oz	240	S.F.	$684.00
C3030 210 6000	Acoustic ceilings, 24" x 48" tile, tee grid suspension	200	S.F.	$1,812.00
D5020 125 0560	Receptacles incl plate, box, conduit, wire, 20 A duplex	8	Ea.	$2,448.00
D5020 125 0720	Light switch incl plate, box, conduit, wire, 20 A single pole	2	Ea.	$598.00
D5020 210 0560	Fluorescent fixtures, recess mounted, 20 per 1000 SF	200	S.F.	$2,414.00
	Assembly Subtotal			**$15,217.60**
	Sales Tax @ ❷		5 %	$ 380.44
	General Requirements @ ❸		7 %	$ 1,065.23
	Subtotal A			**$16,663.27**
	GC Overhead @ ❹		5 %	$ 833.16
	Subtotal B			**$17,496.44**
	GC Profit @ ❺		5 %	$ 874.82
	Subtotal C			**$18,371.26**
	Adjusted by Location Factor ❻	114.2		$ 20,979.98
	Architects Fee @ ❼		8 %	$ 1,678.40
	Contingency @ ❽		15 %	$ 3,147.00
	Project Total Cost			**$ 25,805.38**

This estimate is based on an interactive spreadsheet. You are free to download it and adjust it to your methodology. A copy of this spreadsheet is available at **RSMeans.com/2021books.**

1 Work Performed

The body of the estimate shows the RSMeans data selected, including line numbers, a brief description of each item, its takeoff quantity and unit, and the total installed cost, including the installing contractor's overhead and profit.

2 Sales Tax

If the work is subject to state or local sales taxes, the amount must be added to the estimate. In a conceptual estimate it can be assumed that one half of the total represents material costs. Therefore, apply the sales tax rate to 50% of the assembly subtotal.

3 General Requirements

This item covers project-wide needs provided by the general contractor. These items vary by project but may include temporary facilities and utilities, security, testing, project cleanup, etc. In assemblies estimates a percentage is used—typically between 5% and 15% of project cost.

4 General Contractor Overhead

This entry represents the general contractor's markup on all work to cover project administration costs.

5 General Contractor Profit

This entry represents the GC's profit on all work performed. The value included here can vary widely by project and is influenced by the GC's perception of the project's financial risk and market conditions.

6 Location Factor

RSMeans published data are based on national average costs. If necessary, adjust the total cost of the project using a location factor from the "Location Factor" table or the "City Cost Indexes" table found in the Reference Section. Use location factors if the work is general, covering the work of multiple trades. If the work is by a single trade (e.g., masonry) use the more specific data found in the City Cost Indexes.

To adjust costs by location factors, multiply the base cost by the factor and divide by 100.

7 Architect's Fee

If appropriate, add the design cost to the project estimate. These fees vary based on project complexity and size. Typical design and engineering fees can be found in the Reference Section.

8 Contingency

A factor for contingency may be added to any estimate to represent the cost of unknowns that may occur between the time that the estimate is performed and the time the project is constructed. The amount of the allowance will depend on the stage of design at which the estimate is done, as well as the contractor's assessment of the risk involved.

A10 Foundations

A1010 Standard Foundations

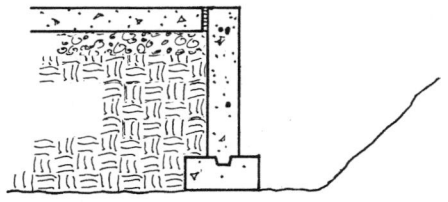

The Foundation Bearing Wall System includes: forms up to 6' high (four uses); 3,000 p.s.i. concrete placed and vibrated; and form removal with breaking form ties and patching walls. The wall systems list walls from 6" to 16" thick and are designed with minimum reinforcement.

Excavation and backfill are not included.

Please see the reference section for further design and cost information.

System Components	QUANTITY	UNIT	COST PER L.F.		
			MAT.	INST.	TOTAL
SYSTEM A1010 105 1500					
FOUNDATION WALL, CAST IN PLACE, DIRECT CHUTE, 4' HIGH, 6" THICK					
Formwork	8.000	SFCA	6.48	50.80	57.28
Reinforcing	3.300	Lb.	2.27	1.55	3.82
Unloading & sorting reinforcing	3.300	Lb.		.09	.09
Concrete, 3,000 psi	.074	C.Y.	10.14		10.14
Place concrete, direct chute	.074	C.Y.		2.75	2.75
Finish walls, break ties and patch voids, one side	4.000	S.F.	.20	4.48	4.68
TOTAL			19.09	59.67	78.76

A1010 105	Wall Foundations							
	WALL HEIGHT (FT.)	PLACING METHOD	CONCRETE (C.Y. per L.F.)	REINFORCING (LBS. per L.F.)	WALL THICKNESS (IN.)	COST PER L.F.		
						MAT.	INST.	TOTAL
1500	4'	direct chute	.074	3.3	6	19.10	59.50	78.60
1520			.099	4.8	8	23.50	61.50	85
1540			.123	6.0	10	27.50	62.50	90
1560			.148	7.2	12	32	64	96
1580			.173	8.1	14	36	65	101
1600			.197	9.44	16	40	66.50	106.50
1700	4'	pumped	.074	3.3	6	19.10	60.50	79.60
1720			.099	4.8	8	23.50	62.50	86
1740			.123	6.0	10	27.50	64	91.50
1760			.148	7.2	12	32	65.50	97.50
1780			.173	8.1	14	36	66.50	102.50
1800			.197	9.44	16	40	68	108

A10 Foundations

A1010 Standard Foundations

A1010 105				Wall Foundations				

	WALL HEIGHT (FT.)	PLACING METHOD	CONCRETE (C.Y. per L.F.)	REINFORCING (LBS. per L.F.)	WALL THICKNESS (IN.)	COST PER L.F.		
						MAT.	INST.	TOTAL
3000	6'	direct chute	.111	4.95	6	28.50	89.50	118
3020			.149	7.20	8	35.50	92	127.50
3040			.184	9.00	10	41.50	93.50	135
3060			.222	10.8	12	48	96	144
3080			.260	12.15	14	54	97.50	151.50
3100			.300	14.39	16	61	99.50	160.50
3200	6'	pumped	.111	4.95	6	28.50	91	119.50
3220			.149	7.20	8	35.50	94	129.50
3240			.184	9.00	10	41.50	96	137.50
3260			.222	10.8	12	48	98.50	146.50
3280			.260	12.15	14	54	100	154
3300			.300	14.39	16	61	103	164

For customer support on your Site Work & Landscape Costs with RSMeans Data, call 800.448.8182.

A10 Foundations

A1010 Standard Foundations

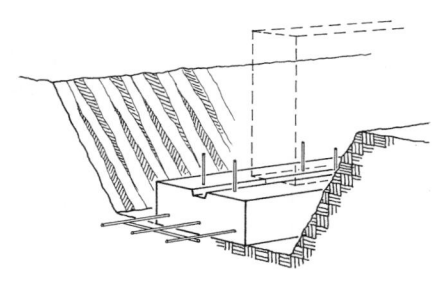

The Strip Footing System includes: excavation; hand trim; all forms needed for footing placement; forms for 2″ x 6″ keyway (four uses); dowels; and 3,000 p.s.i. concrete.

The footing size required varies for different soils. Soil bearing capacities are listed for 3 KSF and 6 KSF. Depths of the system range from 8″ and deeper. Widths range from 16″ and wider. Smaller strip footings may not require reinforcement.

Please see the reference section for further design and cost information.

System Components			COST PER L.F.		
	QUANTITY	UNIT	MAT.	INST.	TOTAL
SYSTEM A1010 110 2500					
STRIP FOOTING, LOAD 5.1 KLF, SOIL CAP. 3 KSF, 24″ WIDE X 12″ DEEP, REINF.					
Trench excavation	.148	C.Y.		1.45	1.45
Hand trim	2.000	S.F.		2.40	2.40
Compacted backfill	.074	C.Y.		.34	.34
Formwork, 4 uses	2.000	S.F.	5.64	10.30	15.94
Keyway form, 4 uses	1.000	L.F.	.58	1.31	1.89
Reinforcing, fy = 60000 psi	3.000	Lb.	2.13	2.01	4.14
Dowels	2.000	Ea.	2	5.88	7.88
Concrete, f'c = 3000 psi	.074	C.Y.	10.14		10.14
Place concrete, direct chute	.074	C.Y.		2.08	2.08
Screed finish	2.000	S.F.		.88	.88
TOTAL			20.49	26.65	47.14

A1010 110	Strip Footings	COST PER L.F.		
		MAT.	INST.	TOTAL
2100	Strip footing, load 2.6 KLF, soil capacity 3 KSF, 16″ wide x 8″ deep, plain	8.90	12.45	21.35
2300	Load 3.9 KLF, soil capacity 3 KSF, 24″ wide x 8″ deep, plain	11.20	14.10	25.30
2500	Load 5.1 KLF, soil capacity 3 KSF, 24″ wide x 12″ deep, reinf.	20.50	26.50	47
2700	Load 11.1 KLF, soil capacity 6 KSF, 24″ wide x 12″ deep, reinf.	20.50	26.50	47
2900	Load 6.8 KLF, soil capacity 3 KSF, 32″ wide x 12″ deep, reinf.	24.50	29	53.50
3100	Load 14.8 KLF, soil capacity 6 KSF, 32″ wide x 12″ deep, reinf.	24.50	29	53.50
3300	Load 9.3 KLF, soil capacity 3 KSF, 40″ wide x 12″ deep, reinf.	29	31.50	60.50
3500	Load 18.4 KLF, soil capacity 6 KSF, 40″ wide x 12″ deep, reinf.	29	31.50	60.50
3700	Load 10.1 KLF, soil capacity 3 KSF, 48″ wide x 12″ deep, reinf.	32	34	66
3900	Load 22.1 KLF, soil capacity 6 KSF, 48″ wide x 12″ deep, reinf.	34	36	70
4100	Load 11.8 KLF, soil capacity 3 KSF, 56″ wide x 12″ deep, reinf.	37.50	37.50	75
4300	Load 25.8 KLF, soil capacity 6 KSF, 56″ wide x 12″ deep, reinf.	40.50	40.50	81
4500	Load 10 KLF, soil capacity 3 KSF, 48″ wide x 16″ deep, reinf.	40.50	39.50	80
4700	Load 22 KLF, soil capacity 6 KSF, 48″ wide, 16″ deep, reinf.	41.50	40.50	82
4900	Load 11.6 KLF, soil capacity 3 KSF, 56″ wide x 16″ deep, reinf.	46	56.50	102.50
5100	Load 25.6 KLF, soil capacity 6 KSF, 56″ wide x 16″ deep, reinf.	48.50	58.50	107
5300	Load 13.3 KLF, soil capacity 3 KSF, 64″ wide x 16″ deep, reinf.	52.50	47	99.50
5500	Load 29.3 KLF, soil capacity 6 KSF, 64″ wide x 16″ deep, reinf.	56	50.50	106.50
5700	Load 15 KLF, soil capacity 3 KSF, 72″ wide x 20″ deep, reinf.	69.50	56.50	126
5900	Load 33 KLF, soil capacity 6 KSF, 72″ wide x 20″ deep, reinf.	73.50	60.50	134
6100	Load 18.3 KLF, soil capacity 3 KSF, 88″ wide x 24″ deep, reinf.	98	71.50	169.50
6300	Load 40.3 KLF, soil capacity 6 KSF, 88″ wide x 24″ deep, reinf.	106	79.50	185.50
6500	Load 20 KLF, soil capacity 3 KSF, 96″ wide x 24″ deep, reinf.	106	76	182
6700	Load 44 KLF, soil capacity 6 KSF, 96″ wide x 24″ deep, reinf.	113	82	195

A10 Foundations

A1010 Standard Foundations

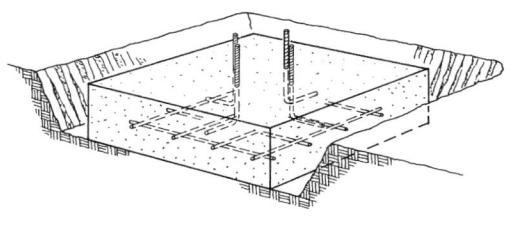

The Spread Footing System includes: excavation; backfill; forms (four uses); all reinforcement; 3,000 p.s.i. concrete (chute placed); and float finish.

Footing systems are priced per individual unit. The Expanded System Listing at the bottom shows various footing sizes. It is assumed that excavation is done by a truck mounted hydraulic excavator with an operator and oiler.

Backfill is with a dozer, and compaction by air tamp. The excavation and backfill equipment is assumed to operate at 30 C.Y. per hour.

Please see the reference section for further design and cost information.

System Components	QUANTITY	UNIT	COST EACH MAT.	COST EACH INST.	COST EACH TOTAL
SYSTEM A1010 210 2200					
SPREAD FOOTINGS, 3'-0" SQ. X 12" DEEP, 3 KSF SOIL CAP., 25K LOAD					
Excavation, Gradall, 30" bucket, 30 CY/Hr	.590	C.Y.		5.40	5.40
Trim sides & bottom	9.000	S.F.		10.80	10.80
Dozer backfill & roller compaction	.260	C.Y.		1.19	1.19
Forms in place, spread footing, 4 uses	12.000	S.F.	14.04	72	86.04
Reinforcing in place footings, #4 to #7	.006	Ton	8.25	8.10	16.35
Supports for dowels in footings, walls, and beams	6.000	L.F.	10.32	29.88	40.20
Concrete ready mix, regular weight, 3000 psi	.330	C.Y.	45.21		45.21
Place and vibrate concrete for spread footings over 5 CY, direct chute	.330	C.Y.		9.25	9.25
Finishing floor, bull float only	9.000	S.F.		3.96	3.96
TOTAL			77.82	140.58	218.40

A1010 210	Spread Footings	COST EACH MAT.	COST EACH INST.	COST EACH TOTAL
2200	Spread footings, 3'-0 square x 12" deep, 3 KSF soil cap., 25K load	78	141	219
2210	4'-0" sq. x 12" deep, 6 KSF soil cap., 75K load	133	208	341
2400	4'-6" sq. x 12" deep, 3 KSF soil capacity, 50K load	163	242	405
2600	15" deep, 6 KSF soil capacity, 100K load	199	284	483
2800	5'-0" sq. x 12" deep, 3 KSF soil capacity, 62K load	199	297	496
3000	16" deep, 6 KSF soil capacity, 125K load	284	355	639
3200	5'-6" sq. x 13" deep, 3 KSF soil capacity, 75K load	256	340	596
3400	18" deep, 6 KSF soil capacity, 150K load	340	430	770
3600	6'-0" sq. x 14" deep, 3 KSF soil capacity, 100K load	320	410	730
3800	20" deep, 6 KSF soil capacity, 200K load	440	530	970
4000	7'-6" sq. x 18" deep, 3 KSF soil capacity, 150K load	610	680	1,290
4200	25" deep, 6 KSF soil capacity, 300K load	830	875	1,705
4400	8'-6" sq. x 20" deep, 3 KSF soil capacity, 200K load	870	900	1,770
4600	27" deep, 6 KSF soil capacity, 400K load	1,150	1,150	2,300
4800	9'-6" sq. x 24" deep, 3 KSF soil capacity, 250K load	1,250	1,225	2,475
5000	30" deep, 6 KSF soil capacity, 500K load	1,575	1,500	3,075
5200	10'-6" sq. x 26" deep, 3 KSF soil capacity, 300K load	1,600	1,475	3,075
5400	33" deep, 6 KSF soil capacity, 600K load	2,125	1,900	4,025
5600	12'-0" sq. x 28" deep, 3 KSF soil capacity, 400K load	2,550	2,175	4,725
5800	38" deep, 6 KSF soil capacity, 800K load	3,075	2,575	5,650
6200	13'-6" sq. x 30" deep, 3 KSF soil capacity, 500K load	3,150	2,600	5,750
6400	42" deep, 6 KSF soil capacity, 1000K load	4,275	3,425	7,700

A1010 Standard Foundations

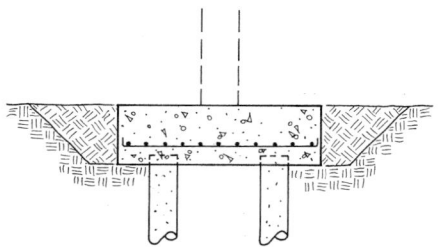

These pile cap systems include excavation with a truck mounted hydraulic excavator, hand trimming, compacted backfill, forms for concrete, templates for dowels or anchor bolts, reinforcing steel and concrete placed and floated.

Pile embedment is assumed as 6″. Design is consistent with the Concrete Reinforcing Steel Institute Handbook f'c = 3000 psi, fy = 60,000.

Please see the reference section for further design and cost information.

System Components	QUANTITY	UNIT	COST EACH MAT.	COST EACH INST.	COST EACH TOTAL
SYSTEM A1010 250 5100					
CAP FOR 2 PILES, 6'-6″X3'-6″X20″, 15 TON PILE, 8″ MIN. COL., 45K COL. LOAD					
Excavation, bulk, hyd excavator, truck mtd. 30″ bucket 1/2 CY	2.890	C.Y.		26.44	26.44
Trim sides and bottom of trench, regular soil	23.000	S.F.		27.60	27.60
Dozer backfill & roller compaction	1.500	C.Y.		6.89	6.89
Forms in place pile cap, square or rectangular, 4 uses	33.000	SFCA	52.80	214.50	267.30
Templates for dowels or anchor bolts	8.000	Ea.	13.76	39.84	53.60
Reinforcing in place footings, #8 to #14	.025	Ton	34.38	19.63	54.01
Concrete ready mix, regular weight, 3000 psi	1.400	C.Y.	191.80		191.80
Place and vibrate concrete for pile caps, under 5 CY, direct chute	1.400	C.Y.		52.02	52.02
Float finish	23.000	S.F.		10.12	10.12
TOTAL			292.74	397.04	689.78

A1010 250			Pile Caps					
	NO. PILES	SIZE FT-IN X FT-IN X IN	PILE CAPACITY (TON)	COLUMN SIZE (IN)	COLUMN LOAD (K)	COST EACH MAT.	COST EACH INST.	COST EACH TOTAL
5100	2	6-6x3-6x20	15	8	45	293	395	688
5150		26	40	8	155	360	485	845
5200		34	80	11	314	485	625	1,110
5250		37	120	14	473	520	665	1,185
5300	3	5-6x5-1x23	15	8	75	350	455	805
5350		28	40	10	232	390	515	905
5400		32	80	14	471	450	585	1,035
5450		38	120	17	709	520	670	1,190
5500	4	5-6x5-6x18	15	10	103	395	455	850
5550		30	40	11	308	570	655	1,225
5600		36	80	16	626	670	765	1,435
5650		38	120	19	945	705	800	1,505
5700	6	8-6x5-6x18	15	12	156	655	660	1,315
5750		37	40	14	458	1,050	975	2,025
5800		40	80	19	936	1,200	1,075	2,275
5850		45	120	24	1413	1,350	1,200	2,550
5900	8	8-6x7-9x19	15	12	205	985	890	1,875
5950		36	40	16	610	1,400	1,175	2,575
6000		44	80	22	1243	1,725	1,425	3,150
6050		47	120	27	1881	1,875	1,525	3,400

A10 Foundations

A1010 Standard Foundations

A1010 250	Pile Caps

	NO. PILES	SIZE FT-IN X FT-IN X IN	PILE CAPACITY (TON)	COLUMN SIZE (IN)	COLUMN LOAD (K)	COST EACH		
						MAT.	INST.	TOTAL
6100	10	11-6x7-9x21	15	14	250	1,375	1,100	2,475
6150		39	40	17	756	2,100	1,575	3,675
6200		47	80	25	1547	2,550	1,875	4,425
6250		49	120	31	2345	2,700	1,975	4,675
6300	12	11-6x8-6x22	15	15	316	1,750	1,325	3,075
6350		49	40	19	900	2,750	1,975	4,725
6400		52	80	27	1856	3,050	2,175	5,225
6450		55	120	34	2812	3,300	2,325	5,625
6500	14	11-6x10-9x24	15	16	345	2,275	1,625	3,900
6550		41	40	21	1056	2,950	2,025	4,975
6600		55	80	29	2155	3,850	2,575	6,425
6700	16	11-6x11-6x26	15	18	400	2,750	1,875	4,625
6750		48	40	22	1200	3,625	2,400	6,025
6800		60	80	31	2460	4,500	2,925	7,425
6900	18	13-0x11-6x28	15	20	450	3,125	2,050	5,175
6950		49	40	23	1349	4,225	2,675	6,900
7000		56	80	33	2776	4,975	3,150	8,125
7100	20	14-6x11-6x30	15	20	510	3,825	2,450	6,275
7150		52	40	24	1491	5,075	3,150	8,225

A10 Foundations

A1010 Standard Foundations

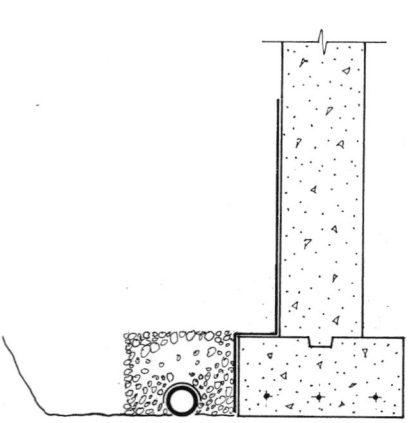

General: Footing drains can be placed either inside or outside of foundation walls depending upon the source of water to be intercepted. If the source of subsurface water is principally from grade or a subsurface stream above the bottom of the footing, outside drains should be used. For high water tables, use inside drains or both inside and outside.

The effectiveness of underdrains depends on good waterproofing. This must be carefully installed and protected during construction.

Costs below include the labor and materials for the pipe and 6″ of crushed stone around pipe. Excavation and backfill are not included.

System Components	QUANTITY	UNIT	COST PER L.F.		
			MAT.	INST.	TOTAL
SYSTEM A1010 310 1000					
FOUNDATION UNDERDRAIN, OUTSIDE ONLY, PVC, 4″ DIAM.					
PVC pipe 4″ diam. S.D.R. 35	1.000	L.F.	2.43	4.72	7.15
Crushed stone 3/4″ to 1/2″	.070	C.Y.	2	.92	2.92
TOTAL			4.43	5.64	10.07

A1010 310	Foundation Underdrain	COST PER L.F.		
		MAT.	INST.	TOTAL
1000	Foundation underdrain, outside only, PVC, 4″ diameter	4.43	5.65	10.08
1100	6″ diameter	7.25	6.25	13.50
1400	Perforated HDPE, 6″ diameter	4.57	2.36	6.93
1450	8″ diameter	8.95	2.95	11.90
1500	12″ diameter	11.30	7.30	18.60
1600	Corrugated metal, 16 ga. asphalt coated, 6″ diameter	9.55	5.85	15.40
1650	8″ diameter	14.65	6.25	20.90
1700	10″ diameter	18.05	8.25	26.30
3000	Outside and inside, PVC, 4″ diameter	8.85	11.25	20.10
3100	6″ diameter	14.50	12.45	26.95
3400	Perforated HDPE, 6″ diameter	9.15	4.72	13.87
3450	8″ diameter	17.85	5.90	23.75
3500	12″ diameter	22.50	14.60	37.10
3600	Corrugated metal, 16 ga., asphalt coated, 6″ diameter	19.15	11.70	30.85
3650	8″ diameter	29.50	12.45	41.95
3700	10″ diameter	36	16.45	52.45

A1010 Standard Foundations

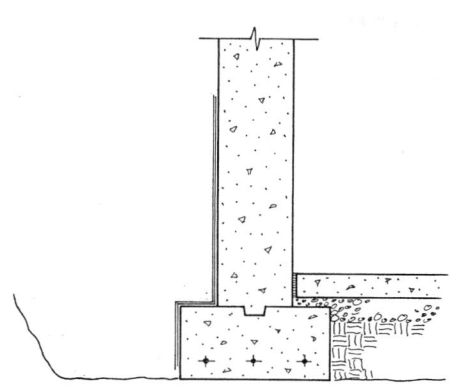

General: Apply foundation wall dampproofing over clean concrete giving particular attention to the joint between the wall and the footing. Use care in backfilling to prevent damage to the dampproofing.

Costs for four types of dampproofing are listed below.

System Components			COST PER L.F.		
	QUANTITY	UNIT	MAT.	INST.	TOTAL
SYSTEM A1010 320 1000					
FOUNDATION DAMPPROOFING, BITUMINOUS, 1 COAT, 4' HIGH					
Bituminous asphalt dampproofing brushed on below grade, 1 coat	4.000	S.F.	.96	3.76	4.72
Labor for protection of dampproofing during backfilling	4.000	S.F.		1.59	1.59
TOTAL			.96	5.35	6.31

A1010 320	Foundation Dampproofing	COST PER L.F.		
		MAT.	INST.	TOTAL
1000	Foundation dampproofing, bituminous, 1 coat, 4' high	.96	5.35	6.31
1400	8' high	1.92	10.70	12.62
1800	12' high	2.88	16.60	19.48
2000	2 coats, 4' high	1.92	6.60	8.52
2400	8' high	3.84	13.20	17.04
2800	12' high	5.75	20.50	26.25
3000	Asphalt with fibers, 1/16" thick, 4' high	1.84	6.60	8.44
3400	8' high	3.68	13.20	16.88
3800	12' high	5.50	20.50	26
4000	1/8" thick, 4' high	3.24	7.85	11.09
4400	8' high	6.50	15.65	22.15
4800	12' high	9.70	24	33.70
5000	Asphalt coated board and mastic, 1/4" thick, 4' high	5.70	7.15	12.85
5400	8' high	11.35	14.30	25.65
5800	12' high	17.05	22	39.05
6000	1/2" thick, 4' high	8.50	9.95	18.45
6400	8' high	17.05	19.85	36.90
6800	12' high	25.50	30.50	56
7000	Cementitious coating, on walls, 1/8" thick coating, 4' high	2.56	9.30	11.86
7400	8' high	5.10	18.65	23.75
7800	12' high	7.70	28	35.70
8000	Cementitious/metallic slurry, 4 coat, 1/2"thick, 2' high	1.59	10.10	11.69
8400	4' high	3.18	20	23.18
8800	6' high	4.77	30.50	35.27

A1020 Special Foundations

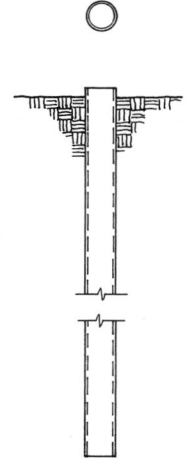

The Cast-in-Place Concrete Pile System includes: a defined number of 4,000 p.s.i. concrete piles with thin-wall, straight-sided, steel shells that have a standard steel plate driving point. An allowance for cutoffs is included.

The Expanded System Listing shows costs per cluster of piles. Clusters range from one pile to twenty piles. Loads vary from 50 Kips to 1,600 Kips. Both end-bearing and friction-type piles are shown.

Please see the reference section for cost of mobilization of the pile driving equipment and other design and cost information.

System Components	QUANTITY	UNIT	COST EACH		
			MAT.	INST.	TOTAL
SYSTEM A1020 110 2220					
CIP SHELL CONCRETE PILE, 25' LONG, 50K LOAD, END BEARING, 1 PILE					
7 Ga. shell, 12" diam.	27.000	V.L.F.	918	352.35	1,270.35
Steel pipe pile standard point, 12" or 14" diameter pile	1.000	Ea.	246	184	430
Pile cutoff, conc. pile with thin steel shell	1.000	Ea.		18.20	18.20
TOTAL			1,164	554.55	1,718.55

A1020 110	C.I.P. Concrete Piles	COST EACH		
		MAT.	INST.	TOTAL
2220	CIP shell concrete pile, 25' long, 50K load, end bearing, 1 pile	1,175	555	1,730
2240	100K load, end bearing, 2 pile cluster	2,075	925	3,000
2260	200K load, end bearing, 4 pile cluster	4,175	1,850	6,025
2280	400K load, end bearing, 7 pile cluster	7,275	3,250	10,525
2300	10 pile cluster	10,400	4,625	15,025
2320	800K load, end bearing, 13 pile cluster	21,600	7,600	29,200
2340	17 pile cluster	28,200	9,925	38,125
2360	1200K load, end bearing, 14 pile cluster	23,200	8,175	31,375
2380	19 pile cluster	31,500	11,100	42,600
2400	1600K load, end bearing, 19 pile cluster	31,500	11,100	42,600
2420	50' long, 50K load, end bearing, 1 pile	1,925	800	2,725
2440	Friction type, 2 pile cluster	3,700	1,600	5,300
2460	3 pile cluster	5,525	2,400	7,925
2480	100K load, end bearing, 2 pile cluster	3,850	1,600	5,450
2500	Friction type, 4 pile cluster	7,375	3,200	10,575
2520	6 pile cluster	11,100	4,800	15,900
2540	200K load, end bearing, 4 pile cluster	7,700	3,200	10,900
2560	Friction type, 8 pile cluster	14,800	6,425	21,225
2580	10 pile cluster	18,500	8,025	26,525
2600	400K load, end bearing, 7 pile cluster	13,500	5,625	19,125
2620	Friction type, 16 pile cluster	29,500	12,800	42,300
2640	19 pile cluster	35,100	15,200	50,300
2660	800K load, end bearing, 14 pile cluster	43,100	14,100	57,200
2680	20 pile cluster	61,500	20,200	81,700
2700	1200K load, end bearing, 15 pile cluster	46,100	15,100	61,200
2720	1600K load, end bearing, 20 pile cluster	61,500	20,200	81,700
3740	75' long, 50K load, end bearing, 1 pile	2,975	1,325	4,300
3760	Friction type, 2 pile cluster	5,725	2,625	8,350

A10 Foundations

A1020 Special Foundations

A1020 110	C.I.P. Concrete Piles	COST EACH		
		MAT.	INST.	TOTAL
3780	3 pile cluster	8,600	3,950	12,550
3800	100K load, end bearing, 2 pile cluster	5,950	2,625	8,575
3820	Friction type, 3 pile cluster	8,600	3,950	12,550
3840	5 pile cluster	14,300	6,575	20,875
3860	200K load, end bearing, 4 pile cluster	11,900	5,275	17,175
3880	6 pile cluster	17,900	7,900	25,800
3900	Friction type, 6 pile cluster	17,200	7,900	25,100
3910	7 pile cluster	20,000	9,225	29,225
3920	400K load, end bearing, 7 pile cluster	20,900	9,225	30,125
3930	11 pile cluster	32,800	14,500	47,300
3940	Friction type, 12 pile cluster	34,400	15,800	50,200
3950	14 pile cluster	40,100	18,400	58,500
3960	800K load, end bearing, 15 pile cluster	70,500	24,300	94,800
3970	20 pile cluster	93,500	32,400	125,900
3980	1200K load, end bearing, 17 pile cluster	79,500	27,500	107,000

For customer support on your Site Work & Landscape Costs with RSMeans Data, call 800.448.8182.

A1020 Special Foundations

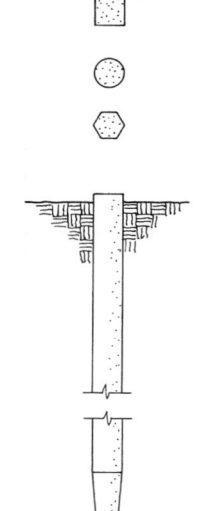

The Precast Concrete Pile System includes: pre-stressed concrete piles; standard steel driving point; and an allowance for cutoffs.

The Expanded System Listing shows costs per cluster of piles. Clusters range from one pile to twenty piles. Loads vary from 50 Kips to 1,600 Kips. Both end-bearing and friction type piles are listed.

Please see the reference section for cost of mobilization of the pile driving equipment and other design and cost information.

System Components	QUANTITY	UNIT	COST EACH		
			MAT.	INST.	TOTAL
SYSTEM A1020 120 2220					
PRECAST CONCRETE PILE, 50' LONG, 50K LOAD, END BEARING, 1 PILE					
Precast, prestressed conc. piles, 10" square, no mobil.	53.000	V.L.F.	845.35	591.48	1,436.83
Steel pipe pile standard point, 8" to 10" diameter	1.000	Ea.	56.50	81	137.50
Piling special costs cutoffs concrete piles plain	1.000	Ea.		126	126
TOTAL			901.85	798.48	1,700.33

A1020 120	Precast Concrete Piles	COST EACH		
		MAT.	INST.	TOTAL
2220	Precast conc pile, 50' long, 50K load, end bearing, 1 pile	900	800	1,700
2240	Friction type, 2 pile cluster	3,275	1,650	4,925
2260	4 pile cluster	6,525	3,300	9,825
2280	100K load, end bearing, 2 pile cluster	1,800	1,600	3,400
2300	Friction type, 2 pile cluster	3,275	1,650	4,925
2320	4 pile cluster	6,525	3,300	9,825
2340	7 pile cluster	11,400	5,800	17,200
2360	200K load, end bearing, 3 pile cluster	2,700	2,400	5,100
2380	4 pile cluster	3,600	3,200	6,800
2400	Friction type, 8 pile cluster	13,100	6,625	19,725
2420	9 pile cluster	14,700	7,450	22,150
2440	14 pile cluster	22,900	11,600	34,500
2460	400K load, end bearing, 6 pile cluster	5,400	4,800	10,200
2480	8 pile cluster	7,225	6,400	13,625
2500	Friction type, 14 pile cluster	22,900	11,600	34,500
2520	16 pile cluster	26,100	13,200	39,300
2540	18 pile cluster	29,400	14,900	44,300
2560	800K load, end bearing, 12 pile cluster	11,700	10,400	22,100
2580	16 pile cluster	14,400	12,800	27,200
2600	1200K load, end bearing, 19 pile cluster	54,000	18,600	72,600
2620	20 pile cluster	57,000	19,500	76,500
2640	1600K load, end bearing, 19 pile cluster	54,000	18,600	72,600
4660	100' long, 50K load, end bearing, 1 pile	1,725	1,375	3,100
4680	Friction type, 1 pile	3,125	1,425	4,550

A1020 Special Foundations

A1020 120	Precast Concrete Piles	COST EACH		
		MAT.	INST.	TOTAL
4700	2 pile cluster	6,225	2,850	9,075
4720	100K load, end bearing, 2 pile cluster	3,475	2,750	6,225
4740	Friction type, 2 pile cluster	6,225	2,850	9,075
4760	3 pile cluster	9,350	4,275	13,625
4780	4 pile cluster	12,500	5,700	18,200
4800	200K load, end bearing, 3 pile cluster	5,200	4,125	9,325
4820	4 pile cluster	6,925	5,525	12,450
4840	Friction type, 3 pile cluster	9,350	4,275	13,625
4860	5 pile cluster	15,600	7,125	22,725
4880	400K load, end bearing, 6 pile cluster	10,400	8,275	18,675
4900	8 pile cluster	13,900	11,000	24,900
4910	Friction type, 8 pile cluster	24,900	11,400	36,300
4920	10 pile cluster	31,200	14,300	45,500
4930	800K load, end bearing, 13 pile cluster	22,500	17,900	40,400
4940	16 pile cluster	27,700	22,100	49,800
4950	1200K load, end bearing, 19 pile cluster	104,000	32,400	136,400
4960	20 pile cluster	109,500	34,100	143,600
4970	1600K load, end bearing, 19 pile cluster	104,000	32,400	136,400

A10 Foundations

A1020 Special Foundations

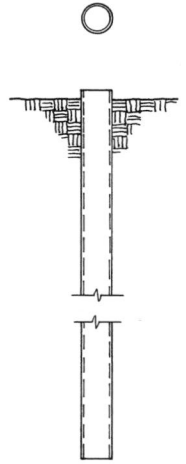

The Steel Pipe Pile System includes: steel pipe sections filled with 4,000 p.s.i. concrete; a standard steel driving point; splices when required and an allowance for cutoffs.

The Expanded System Listing shows costs per cluster of piles. Clusters range from one pile to twenty piles. Loads vary from 50 Kips to 1,600 Kips. Both end-bearing and friction-type piles are shown.

Please see the reference section for cost of mobilization of the pile driving equipment and other design and cost information.

System Components	QUANTITY	UNIT	COST EACH		
			MAT.	INST.	TOTAL
SYSTEM A1020 130 2220					
CONC. FILL STEEL PIPE PILE, 50' LONG, 50K LOAD, END BEARING, 1 PILE					
Piles, steel, pipe, conc. filled, 12" diameter	53.000	V.L.F.	2,173	999.05	3,172.05
Steel pipe pile, standard point, for 12" or 14" diameter pipe	1.000	Ea.	246	184	430
Pile cut off, concrete pile, thin steel shell	1.000	Ea.		18.20	18.20
TOTAL			2,419	1,201.25	3,620.25

A1020 130	Steel Pipe Piles	COST EACH		
		MAT.	INST.	TOTAL
2220	Conc. fill steel pipe pile, 50' long, 50K load, end bearing, 1 pile	2,425	1,200	3,625
2240	Friction type, 2 pile cluster	4,850	2,400	7,250
2250	100K load, end bearing, 2 pile cluster	4,850	2,400	7,250
2260	3 pile cluster	7,250	3,600	10,850
2300	Friction type, 4 pile cluster	9,675	4,800	14,475
2320	5 pile cluster	12,100	6,000	18,100
2340	10 pile cluster	24,200	12,000	36,200
2360	200K load, end bearing, 3 pile cluster	7,250	3,600	10,850
2380	4 pile cluster	9,675	4,800	14,475
2400	Friction type, 4 pile cluster	9,675	4,800	14,475
2420	8 pile cluster	19,400	9,600	29,000
2440	9 pile cluster	21,800	10,800	32,600
2460	400K load, end bearing, 6 pile cluster	14,500	7,200	21,700
2480	7 pile cluster	16,900	8,400	25,300
2500	Friction type, 9 pile cluster	21,800	10,800	32,600
2520	16 pile cluster	38,700	19,200	57,900
2540	19 pile cluster	46,000	22,800	68,800
2560	800K load, end bearing, 11 pile cluster	26,600	13,200	39,800
2580	14 pile cluster	33,900	16,800	50,700
2600	15 pile cluster	36,300	18,000	54,300
2620	Friction type, 17 pile cluster	41,100	20,400	61,500
2640	1200K load, end bearing, 16 pile cluster	38,700	19,200	57,900
2660	20 pile cluster	48,400	24,000	72,400
2680	1600K load, end bearing, 17 pile cluster	41,100	20,400	61,500
3700	100' long, 50K load, end bearing, 1 pile	4,725	2,350	7,075
3720	Friction type, 1 pile	4,725	2,350	7,075

For customer support on your Site Work & Landscape Costs with RSMeans Data, call 800.448.8182.

A1020 Special Foundations

A1020 130	Steel Pipe Piles	COST EACH		
		MAT.	INST.	TOTAL
3740	2 pile cluster	9,450	4,725	14,175
3760	100K load, end bearing, 2 pile cluster	9,450	4,725	14,175
3780	Friction type, 2 pile cluster	9,450	4,725	14,175
3800	3 pile cluster	14,200	7,075	21,275
3820	200K load, end bearing, 3 pile cluster	14,200	7,075	21,275
3840	4 pile cluster	18,900	9,425	28,325
3860	Friction type, 3 pile cluster	14,200	7,075	21,275
3880	4 pile cluster	18,900	9,425	28,325
3900	400K load, end bearing, 6 pile cluster	28,300	14,100	42,400
3910	7 pile cluster	33,100	16,500	49,600
3920	Friction type, 5 pile cluster	23,600	11,800	35,400
3930	8 pile cluster	37,800	18,900	56,700
3940	800K load, end bearing, 11 pile cluster	52,000	25,900	77,900
3950	14 pile cluster	66,000	33,000	99,000
3960	15 pile cluster	71,000	35,400	106,400
3970	1200K load, end bearing, 16 pile cluster	75,500	37,700	113,200
3980	20 pile cluster	94,500	47,100	141,600
3990	1600K load, end bearing, 17 pile cluster	80,500	40,100	120,600

For customer support on your Site Work & Landscape Costs with RSMeans Data, call 800.448.8182.

A1020 Special Foundations

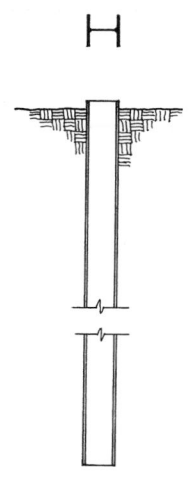

A Steel "H" Pile System includes: steel H sections; heavy duty driving point; splices where applicable and allowance for cutoffs.

The Expanded System Listing shows costs per cluster of piles. Clusters range from one pile to eighteen piles. Loads vary from 50 Kips to 2,000 Kips. All loads for Steel H Pile systems are given in terms of end bearing capacity.

Steel sections range from 10" x 10" to 14" x 14" in the Expanded System Listing. The 14" x 14" steel section is used for all H piles used in applications requiring a working load over 800 Kips.

Please see the reference section for cost of mobilization of the pile driving equipment and other design and cost information.

System Components	QUANTITY	UNIT	COST EACH MAT.	COST EACH INST.	COST EACH TOTAL
SYSTEM A1020 140 2220					
STEEL H PILES, 50' LONG, 100K LOAD, END BEARING, 1 PILE					
Steel H piles 10" x 10", 42 #/L.F.	53.000	V.L.F.	1,219	680.52	1,899.52
Heavy duty point, 10"	1.000	Ea.	206	187	393
Pile cut off, steel pipe or H piles	1.000	Ea.		36.50	36.50
TOTAL			1,425	904.02	2,329.02

A1020 140	Steel H Piles	COST EACH MAT.	COST EACH INST.	COST EACH TOTAL
2220	Steel H piles, 50' long, 100K load, end bearing, 1 pile	1,425	905	2,330
2260	2 pile cluster	2,850	1,800	4,650
2280	200K load, end bearing, 2 pile cluster	2,850	1,800	4,650
2300	3 pile cluster	4,275	2,700	6,975
2320	400K load, end bearing, 3 pile cluster	4,275	2,700	6,975
2340	4 pile cluster	5,700	3,625	9,325
2360	6 pile cluster	8,550	5,425	13,975
2380	800K load, end bearing, 5 pile cluster	7,125	4,525	11,650
2400	7 pile cluster	9,975	6,325	16,300
2420	12 pile cluster	17,100	10,800	27,900
2440	1200K load, end bearing, 8 pile cluster	11,400	7,225	18,625
2460	11 pile cluster	15,700	9,950	25,650
2480	17 pile cluster	24,200	15,400	39,600
2500	1600K load, end bearing, 10 pile cluster	18,700	9,450	28,150
2520	14 pile cluster	26,200	13,200	39,400
2540	2000K load, end bearing, 12 pile cluster	22,400	11,300	33,700
2560	18 pile cluster	33,700	17,000	50,700
3580	100' long, 50K load, end bearing, 1 pile	4,625	2,150	6,775
3600	100K load, end bearing, 1 pile	4,625	2,150	6,775
3620	2 pile cluster	9,275	4,275	13,550
3640	200K load, end bearing, 2 pile cluster	9,275	4,275	13,550
3660	3 pile cluster	13,900	6,425	20,325
3680	400K load, end bearing, 3 pile cluster	13,900	6,425	20,325
3700	4 pile cluster	18,500	8,575	27,075
3720	6 pile cluster	27,800	12,900	40,700
3740	800K load, end bearing, 5 pile cluster	23,200	10,700	33,900

A10 Foundations

A1020 Special Foundations

A1020 140	Steel H Piles	COST EACH		
		MAT.	INST.	TOTAL
3760	7 pile cluster	32,500	15,000	47,500
3780	12 pile cluster	55,500	25,700	81,200
3800	1200K load, end bearing, 8 pile cluster	37,100	17,100	54,200
3820	11 pile cluster	51,000	23,600	74,600
3840	17 pile cluster	79,000	36,400	115,400
3860	1600K load, end bearing, 10 pile cluster	46,400	21,400	67,800
3880	14 pile cluster	65,000	30,000	95,000
3900	2000K load, end bearing, 12 pile cluster	55,500	25,700	81,200
3920	18 pile cluster	83,500	38,600	122,100

For customer support on your Site Work & Landscape Costs with RSMeans Data, call 800.448.8182.

A10 Foundations

A1020 Special Foundations

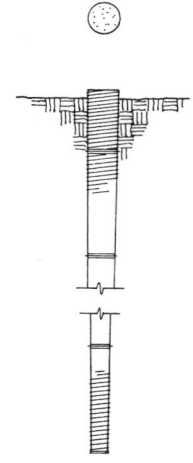

The Step Tapered Steel Pile System includes: step tapered piles filled with 4,000 p.s.i. concrete. The cost for splices and pile cutoffs is included.

The Expanded System Listing shows costs per cluster of piles. Clusters range from one pile to twenty-four piles. Both end bearing piles and friction piles are listed. Loads vary from 50 Kips to 1,600 Kips.

Please see the reference section for cost of mobilization of the pile driving equipment and other design and cost information.

System Components	QUANTITY	UNIT	COST EACH		
			MAT.	INST.	TOTAL
SYSTEM A1020 150 1000					
STEEL PILE, STEP TAPERED, 50' LONG, 50K LOAD, END BEARING, 1 PILE					
Steel shell step tapered conc. filled piles, 8" tip 60 ton capacity to 60'	53.000	V.L.F.	2,199.50	561.27	2,760.77
Pile cutoff, steel pipe or H piles	1.000	Ea.		36.50	36.50
TOTAL			2,199.50	597.77	2,797.27

A1020 150	Step-Tapered Steel Piles	COST EACH		
		MAT.	INST.	TOTAL
1000	Steel pile, step tapered, 50' long, 50K load, end bearing, 1 pile	2,200	600	2,800
1200	Friction type, 3 pile cluster	6,600	1,800	8,400
1400	100K load, end bearing, 2 pile cluster	4,400	1,200	5,600
1600	Friction type, 4 pile cluster	8,800	2,400	11,200
1800	200K load, end bearing, 4 pile cluster	8,800	2,400	11,200
2000	Friction type, 6 pile cluster	13,200	3,575	16,775
2200	400K load, end bearing, 7 pile cluster	15,400	4,175	19,575
2400	Friction type, 10 pile cluster	22,000	5,975	27,975
2600	800K load, end bearing, 14 pile cluster	30,800	8,375	39,175
2800	Friction type, 18 pile cluster	39,600	10,800	50,400
3000	1200K load, end bearing, 16 pile cluster	17,800	10,200	28,000
3200	Friction type, 21 pile cluster	23,400	13,400	36,800
3400	1600K load, end bearing, 18 pile cluster	20,000	11,500	31,500
3600	Friction type, 24 pile cluster	26,700	15,300	42,000
5000	100' long, 50K load, end bearing, 1 pile	6,650	1,175	7,825
5200	Friction type, 2 pile cluster	13,300	2,375	15,675
5400	100K load, end bearing, 2 pile cluster	13,300	2,375	15,675
5600	Friction type, 3 pile cluster	19,900	3,550	23,450
5800	200K load, end bearing, 4 pile cluster	26,600	4,750	31,350
6000	Friction type, 5 pile cluster	33,200	5,925	39,125
6200	400K load, end bearing, 7 pile cluster	46,500	8,300	54,800
6400	Friction type, 8 pile cluster	53,000	9,500	62,500
6600	800K load, end bearing, 15 pile cluster	99,500	17,800	117,300
6800	Friction type, 16 pile cluster	106,500	19,000	125,500
7000	1200K load, end bearing, 17 pile cluster	58,500	31,000	89,500
7200	Friction type, 19 pile cluster	44,000	23,500	67,500
7400	1600K load, end bearing, 20 pile cluster	46,400	24,700	71,100
7600	Friction type, 22 pile cluster	51,000	27,200	78,200

A1020 Special Foundations

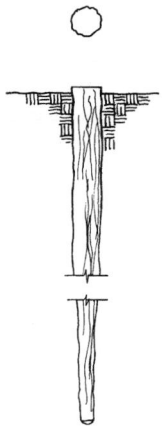

The Treated Wood Pile System includes: creosoted wood piles; a standard steel driving point; and an allowance for cutoffs.

The Expanded System Listing shows costs per cluster of piles. Clusters range from two piles to twenty piles. Loads vary from 50 Kips to 400 Kips. Both end-bearing and friction type piles are listed.

Please see the reference section for cost of mobilization of the pile driving equipment and other design and cost information.

System Components	QUANTITY	UNIT	COST EACH		
			MAT.	INST.	TOTAL
SYSTEM A1020 160 2220					
WOOD PILES, 25′ LONG, 50K LOAD, END BEARING, 3 PILE CLUSTER					
Wood piles, treated, 12″ butt, 8″ tip, up to 30′ long	81.000	V.L.F.	1,591.65	1,012.50	2,604.15
Point for pile tip, minimum	3.000	Ea.	156	103.50	259.50
Pile cutoff, wood piles	3.000	Ea.		54.60	54.60
TOTAL			1,747.65	1,170.60	2,918.25

A1020 160	Treated Wood Piles	COST EACH		
		MAT.	INST.	TOTAL
2220	Wood piles, 25′ long, 50K load, end bearing, 3 pile cluster	1,750	1,175	2,925
2240	Friction type, 3 pile cluster	1,600	1,075	2,675
2260	5 pile cluster	2,650	1,775	4,425
2280	100K load, end bearing, 4 pile cluster	2,325	1,550	3,875
2300	5 pile cluster	2,925	1,950	4,875
2320	6 pile cluster	3,500	2,350	5,850
2340	Friction type, 5 pile cluster	2,650	1,775	4,425
2360	6 pile cluster	3,175	2,125	5,300
2380	10 pile cluster	5,300	3,550	8,850
2400	200K load, end bearing, 8 pile cluster	4,650	3,125	7,775
2420	10 pile cluster	5,825	3,900	9,725
2440	12 pile cluster	7,000	4,675	11,675
2460	Friction type, 10 pile cluster	5,300	3,550	8,850
2480	400K load, end bearing, 16 pile cluster	9,325	6,250	15,575
2500	20 pile cluster	11,700	7,800	19,500
4520	50′ long, 50K load, end bearing, 3 pile cluster	4,925	1,725	6,650
4540	4 pile cluster	6,550	2,300	8,850
4560	Friction type, 2 pile cluster	3,125	1,050	4,175
4580	3 pile cluster	4,675	1,575	6,250
4600	100K load, end bearing, 5 pile cluster	8,200	2,875	11,075
4620	8 pile cluster	13,100	4,625	17,725
4640	Friction type, 3 pile cluster	4,675	1,575	6,250
4660	5 pile cluster	7,800	2,625	10,425
4680	200K load, end bearing, 9 pile cluster	14,700	5,200	19,900
4700	10 pile cluster	16,400	5,775	22,175
4720	15 pile cluster	24,600	8,650	33,250
4740	Friction type, 5 pile cluster	7,800	2,625	10,425
4760	6 pile cluster	9,350	3,150	12,500

A1020 Special Foundations

A1020 160	Treated Wood Piles	COST EACH		
		MAT.	INST.	TOTAL
4780	10 pile cluster	15,600	5,250	20,850
4800	400K load, end bearing, 18 pile cluster	29,500	10,400	39,900
4820	20 pile cluster	32,800	11,500	44,300
4840	Friction type, 9 pile cluster	14,000	4,725	18,725
4860	10 pile cluster	15,600	5,250	20,850
9000	Add for boot for driving tip, each pile	52	25.50	77.50

A10 Foundations

A1020 Special Foundations

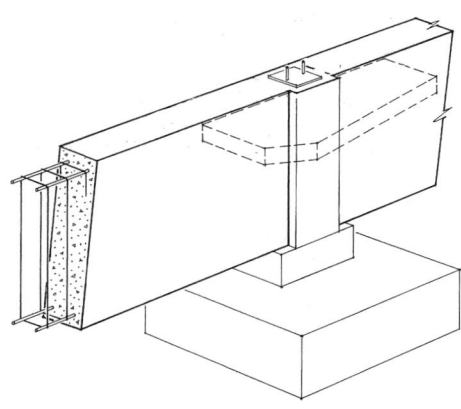

The Grade Beam System includes: excavation with a truck mounted backhoe; hand trim; backfill; forms (four uses); reinforcing steel; and 3,000 p.s.i. concrete placed from chute.

Superimposed loads vary in the listing from 1 Kip per linear foot (KLF) and above. In the Expanded System Listing, the span of the beams varies from 15' to 40'. Depth varies from 28" to 52". Width varies from 12" and wider.

Please see the reference section for further design and cost information.

System Components	QUANTITY	UNIT	MAT.	INST.	TOTAL
SYSTEM A1020 210 2220					
GRADE BEAM, 15' SPAN, 28" DEEP, 12" WIDE, 8 KLF LOAD					
Excavation, trench, hydraulic backhoe, 3/8 CY bucket	.260	C.Y.		2.54	2.54
Trim sides and bottom of trench, regular soil	2.000	S.F.		2.40	2.40
Backfill, by hand, compaction in 6" layers, using vibrating plate	.170	C.Y.		1.60	1.60
Forms in place, grade beam, 4 uses	4.700	SFCA	6.67	29.61	36.28
Reinforcing in place, beams & girders, #8 to #14	.019	Ton	26.13	19.95	46.08
Concrete ready mix, regular weight, 3000 psi	.090	C.Y.	12.33		12.33
Place and vibrate conc. for grade beam, direct chute	.090	C.Y.		2.02	2.02
TOTAL			45.13	58.12	103.25

A1020 210	Grade Beams	MAT.	INST.	TOTAL
2220	Grade beam, 15' span, 28" deep, 12" wide, 8 KLF load	45	58	103
2240	14" wide, 12 KLF load	46.50	58.50	105
2260	40" deep, 12" wide, 16 KLF load	46.50	72.50	119
2280	20 KLF load	53.50	77.50	131
2300	52" deep, 12" wide, 30 KLF load	62	98	160
2320	40 KLF load	75.50	108	183.50
3360	20' span, 28" deep, 12" wide, 2 KLF load	28.50	45.50	74
3380	16" wide, 4 KLF load	37.50	50.50	88
3400	40" deep, 12" wide, 8 KLF load	46.50	72.50	119
3440	14" wide, 16 KLF load	71.50	90	161.50
3460	52" deep, 12" wide, 20 KLF load	77	109	186
3480	14" wide, 30 KLF load	106	129	235
4540	30' span, 28" deep, 12" wide, 1 KLF load	30	46.50	76.50
4560	14" wide, 2 KLF load	47.50	63.50	111
4580	40" deep, 12" wide, 4 KLF load	56	76.50	132.50
4600	18" wide, 8 KLF load	79.50	92.50	172
4620	52" deep, 14" wide, 12 KLF load	102	126	228
4640	20" wide, 16 KLF load	124	137	261
4660	24" wide, 20 KLF load	152	156	308
5720	40' span, 40" deep, 12" wide, 1 KLF load	41	68.50	109.50
5740	2 KLF load	50.50	75.50	126
5760	52" deep, 12" wide, 4 KLF load	75.50	108	183.50
5780	20" wide, 8 KLF load	119	134	253
5800	28" wide, 12 KLF load	163	161	324

A1020 Special Foundations

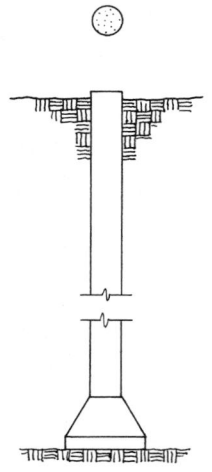

Caisson Systems are listed for three applications: stable ground, wet ground and soft rock. Concrete used is 3,000 p.s.i. placed from chute. Included are a bell at the bottom of the caisson shaft (if applicable) along with required excavation and disposal of excess excavated material up to two miles from job site.

The Expanded System lists cost per caisson. End-bearing loads vary from 200 Kips to 3,200 Kips. The dimensions of the caissons range from 2' x 50' to 7' x 200'.

Please see the reference section for further design and cost information.

System Components	QUANTITY	UNIT	COST EACH		
			MAT.	INST.	TOTAL
SYSTEM A1020 310 2200					
CAISSON, STABLE GROUND, 3000 PSI CONC., 10 KSF BRNG, 200K LOAD, 2'X50'					
Caissons, drilled, to 50', 24" shaft diameter, .116 C.Y./L.F.	50.000	V.L.F.	992.50	1,145	2,137.50
Reinforcing in place, columns, #3 to #7	.060	Ton	82.50	112.50	195
4' bell diameter, 24" shaft, 0.444 C.Y.	1.000	Ea.	61	217.50	278.50
Load & haul excess excavation, 2 miles	6.240	C.Y.		43.87	43.87
TOTAL			1,136	1,518.87	2,654.87

A1020 310	Caissons	COST EACH		
		MAT.	INST.	TOTAL
2200	Caisson, stable ground, 3000 PSI conc, 10 KSF brng, 200K load, 2'-0"x50'-0	1,125	1,525	2,650
2400	400K load, 2'-6"x50'-0"	1,850	2,425	4,275
2600	800K load, 3'-0"x100'-0"	5,150	5,700	10,850
2800	1200K load, 4'-0"x100'-0"	8,850	7,275	16,125
3000	1600K load, 5'-0"x150'-0"	19,900	11,400	31,300
3200	2400K load, 6'-0"x150'-0"	29,200	15,000	44,200
3400	3200K load, 7'-0"x200'-0"	52,500	21,900	74,400
5000	Wet ground, 3000 PSI conc., 10 KSF brng, 200K load, 2'-0"x50'-0"	985	2,250	3,235
5200	400K load, 2'-6"x50'-0"	1,625	3,800	5,425
5400	800K load, 3'-0"x100'-0"	4,475	10,200	14,675
5600	1200K load, 4'-0"x100'-0"	7,675	16,900	24,575
5800	1600K load, 5'-0"x150'-0"	17,100	36,400	53,500
6000	2400K load, 6'-0"x150'-0"	25,100	44,900	70,000
6200	3200K load, 7'-0"x200'-0"	45,000	72,000	117,000
7800	Soft rock, 3000 PSI conc., 10 KSF brng, 200K load, 2'-0"x50'-0"	985	18,900	19,885
8000	400K load, 2'-6"x50'-0"	1,625	21,500	23,125
8200	800K load, 3'-0"x100'-0"	4,475	56,500	60,975
8400	1200K load, 4'-0"x100'-0"	7,675	82,500	90,175
8600	1600K load, 5'-0"x150'-0"	17,100	169,500	186,600
8800	2400K load, 6'-0"x150'-0"	25,100	201,500	226,600
9000	3200K load, 7'-0"x200'-0"	45,000	321,000	366,000

A10 Foundations

A1020   Special Foundations

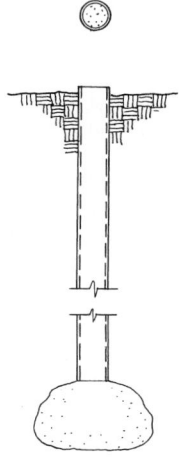

Pressure Injected Piles are usually uncased up to 25' and cased over 25' depending on soil conditions.

These costs include excavation and hauling of excess materials; steel casing over 25'; reinforcement; 3,000 p.s.i. concrete; plus mobilization and demobilization of equipment for a distance of up to fifty miles to and from the job site.

The Expanded System lists cost per cluster of piles. Clusters range from one pile to eight piles. End-bearing loads range from 50 Kips to 1,600 Kips.

Please see the reference section for further design and cost information.

System Components	QUANTITY	UNIT	COST EACH		
			MAT.	INST.	TOTAL
SYSTEM A1020 710 4200					
PRESSURE INJECTED FOOTING, END BEARING, 50' LONG, 50K LOAD, 1 PILE					
Pressure injected footings, cased, 30-60 ton cap., 12" diameter	50.000	V.L.F.	1,050	1,695	2,745
Pile cutoff, concrete pile with thin steel shell	1.000	Ea.		18.20	18.20
TOTAL			1,050	1,713.20	2,763.20

A1020 710	Pressure Injected Footings	COST EACH		
		MAT.	INST.	TOTAL
2200	Pressure injected footing, end bearing, 25' long, 50K load, 1 pile	490	1,200	1,690
2400	100K load, 1 pile	490	1,200	1,690
2600	2 pile cluster	980	2,375	3,355
2800	200K load, 2 pile cluster	980	2,375	3,355
3200	400K load, 4 pile cluster	1,950	4,750	6,700
3400	7 pile cluster	3,425	8,325	11,750
3800	1200K load, 6 pile cluster	4,125	9,025	13,150
4000	1600K load, 7 pile cluster	4,825	10,500	15,325
4200	50' long, 50K load, 1 pile	1,050	1,725	2,775
4400	100K load, 1 pile	1,825	1,725	3,550
4600	2 pile cluster	3,650	3,425	7,075
4800	200K load, 2 pile cluster	3,650	3,425	7,075
5000	4 pile cluster	7,300	6,850	14,150
5200	400K load, 4 pile cluster	7,300	6,850	14,150
5400	8 pile cluster	14,600	13,700	28,300
5600	800K load, 7 pile cluster	12,800	12,000	24,800
5800	1200K load, 6 pile cluster	11,700	10,300	22,000
6000	1600K load, 7 pile cluster	13,700	12,000	25,700

A1030 Slab on Grade

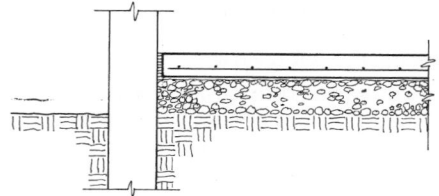

There are four types of Slab on Grade Systems listed: Non-industrial, Light industrial, Industrial and Heavy industrial. Each type is listed two ways: reinforced and non-reinforced. A Slab on Grade system includes three passes with a grader; 6″ of compacted gravel fill; polyethylene vapor barrier; 3500 p.s.i. concrete placed by chute; bituminous fibre expansion joint; all necessary edge forms (4 uses); steel trowel finish; and sprayed-on membrane curing compound.

The Expanded System Listing shows costs on a per square foot basis. Thicknesses of the slabs range from 4″ and above. Non-industrial applications are for foot traffic only with negligible abrasion. Light industrial applications are for pneumatic wheels and light abrasion. Industrial applications are for solid rubber wheels and moderate abrasion. Heavy industrial applications are for steel wheels and severe abrasion. All slabs are either shown unreinforced or reinforced with welded wire fabric.

System Components	QUANTITY	UNIT	COST PER S.F.		
			MAT.	INST.	TOTAL
SYSTEM A1030 120 2220					
SLAB ON GRADE, 4″ THICK, NON INDUSTRIAL, NON REINFORCED					
Fine grade, 3 passes with grader and roller	.110	S.Y.		.66	.66
Gravel under floor slab, 4″ deep, compacted	1.000	S.F.	.65	.36	1.01
Polyethylene vapor barrier, standard, 6 mil	1.000	S.F.	.08	.18	.26
Concrete ready mix, regular weight, 3500 psi	.012	C.Y.	1.67		1.67
Place and vibrate concrete for slab on grade, 4″ thick, direct chute	.012	C.Y.		.37	.37
Expansion joint, premolded bituminous fiber, 1/2″ x 6″	.220	L.F.	.10	.38	.48
Edge forms in place for slab on grade to 6″ high, 4 uses	.030	L.F.	.02	.12	.14
Cure with sprayed membrane curing compound	1.000	S.F.	.13	.11	.24
Finishing floor, monolithic steel trowel	1.000	S.F.		1.08	1.08
TOTAL			2.65	3.26	5.91

A1030 120	Plain & Reinforced	COST PER S.F.		
		MAT.	INST.	TOTAL
2220	Slab on grade, 4″ thick, non industrial, non reinforced	2.65	3.27	5.92
2240	Reinforced	2.75	3.56	6.31
2260	Light industrial, non reinforced	3.30	3.98	7.28
2280	Reinforced	3.46	4.38	7.84
2300	Industrial, non reinforced	4.31	8.20	12.51
2320	Reinforced	4.48	8.60	13.08
3340	5″ thick, non industrial, non reinforced	3.07	3.36	6.43
3360	Reinforced	3.23	3.76	6.99
3380	Light industrial, non reinforced	3.72	4.07	7.79
3400	Reinforced	3.89	4.47	8.36
3420	Heavy industrial, non reinforced	5.50	9.80	15.30
3440	Reinforced	5.60	10.25	15.85
4460	6″ thick, non industrial, non reinforced	3.62	3.28	6.90
4480	Reinforced	3.91	3.83	7.74
4500	Light industrial, non reinforced	4.29	3.99	8.28
4520	Reinforced	4.66	4.42	9.08
4540	Heavy industrial, non reinforced	6.05	9.95	16
4560	Reinforced	6.35	10.45	16.80

A20 Basement Construction

A2010 Basement Excavation

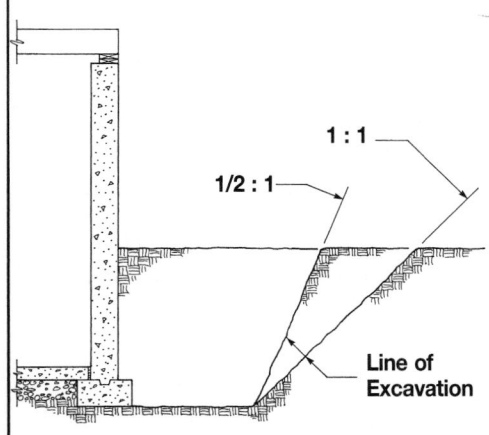

1:1

1/2:1

Line of Excavation

Pricing Assumptions: Two-thirds of excavation is by 2-1/2 C.Y. wheel mounted front end loader and one-third by 1-1/2 C.Y. hydraulic excavator.

Two-mile round trip haul by 12 C.Y. tandem trucks is included for excavation wasted and storage of suitable fill from excavated soil. For excavation in clay, all is wasted and the cost of suitable backfill with two-mile haul is included.

Sand and gravel assumes 15% swell and compaction; common earth assumes 25% swell and 15% compaction; clay assumes 40% swell and 15% compaction (non-clay).

In general, the following items are accounted for in the costs in the table below.

1. Excavation for building or other structure to depth and extent indicated.
2. Backfill compacted in place.
3. Haul of excavated waste.
4. Replacement of unsuitable material with bank run gravel.

Note: Additional excavation and fill beyond this line of general excavation for the building (as required for isolated spread footings, strip footings, etc.) are included in the cost of the appropriate component systems.

System Components	QUANTITY	UNIT	COST PER S.F.		
			MAT.	INST.	TOTAL
SYSTEM A2010 110 2280					
EXCAVATE & FILL, 1000 S.F., 8' DEEP, SAND, ON SITE STORAGE					
Excavating bulk shovel, 1.5 C.Y. bucket, 150 cy/hr	.262	C.Y.		.56	.56
Excavation, front end loader, 2-1/2 C.Y.	.523	C.Y.		1.36	1.36
Haul earth, 12 C.Y. dump truck	.341	L.C.Y.		2.37	2.37
Backfill, dozer bulk push 300', including compaction	.562	C.Y.		2.58	2.58
TOTAL				6.87	6.87

A2010 110	Building Excavation & Backfill	COST PER S.F.		
		MAT.	INST.	TOTAL
2220	Excav & fill, 1000 S.F., 4' sand, gravel, or common earth, on site storage		1.17	1.17
2240	Off site storage		1.72	1.72
2260	Clay excavation, bank run gravel borrow for backfill	3.25	2.59	5.84
2280	8' deep, sand, gravel, or common earth, on site storage		6.85	6.85
2300	Off site storage		14.65	14.65
2320	Clay excavation, bank run gravel borrow for backfill	13.20	12.10	25.30
2340	16' deep, sand, gravel, or common earth, on site storage		18.70	18.70
2350	Off site storage		36.50	36.50
2360	Clay excavation, bank run gravel borrow for backfill	36.50	31	67.50
3380	4000 S.F., 4' deep, sand, gravel, or common earth, on site storage		.63	.63
3400	Off site storage		1.18	1.18
3420	Clay excavation, bank run gravel borrow for backfill	1.66	1.32	2.98
3440	8' deep, sand, gravel, or common earth, on site storage		4.76	4.76
3460	Off site storage		8.20	8.20
3480	Clay excavation, bank run gravel borrow for backfill	6	7.20	13.20
3500	16' deep, sand, gravel, or common earth, on site storage		11.40	11.40
3520	Off site storage		22	22
3540	Clay, excavation, bank run gravel borrow for backfill	16.10	17.20	33.30
4560	10,000 S.F., 4' deep, sand, gravel, or common earth, on site storage		.36	.36
4580	Off site storage		.70	.70
4600	Clay excavation, bank run gravel borrow for backfill	1.01	.81	1.82
4620	8' deep, sand, gravel, or common earth, on site storage		4.09	4.09
4640	Off site storage		6.15	6.15
4660	Clay excavation, bank run gravel borrow for backfill	3.64	5.60	9.24
4680	16' deep, sand, gravel, or common earth, on site storage		9.20	9.20
4700	Off site storage		15.50	15.50
4720	Clay excavation, bank run gravel borrow for backfill	9.65	12.80	22.45

A20 Basement Construction

A2010 Basement Excavation

A2010 110	Building Excavation & Backfill	COST PER S.F.		
		MAT.	INST.	TOTAL
5740	30,000 S.F., 4' deep, sand, gravel, or common earth, on site storage		.20	.20
5760	Off site storage		.41	.41
5780	Clay excavation, bank run gravel borrow for backfill	.59	.47	1.06
5800	8' deep, sand, gravel, or common earth, on site storage		3.65	3.65
5820	Off site storage		4.79	4.79
5840	Clay excavation, bank run gravel borrow for backfill	2.05	4.50	6.55
5860	16' deep, sand, gravel, or common earth, on site storage		7.80	7.80
5880	Off site storage		11.25	11.25
5900	Clay excavation, bank run gravel borrow for backfill	5.35	9.85	15.20
6910	100,000 S.F., 4' deep, sand, gravel, or common earth, on site storage		.11	.11
6920	Off site storage		.23	.23
6930	Clay excavation, bank run gravel borrow for backfill	.29	.25	.54
6940	8' deep, sand, gravel, or common earth, on site storage		3.39	3.39
6950	Off site storage		4	4
6960	Clay excavation, bank run gravel borrow for backfill	1.11	3.85	4.96
6970	16' deep, sand, gravel, or common earth, on site storage		7.05	7.05
6980	Off site storage		8.85	8.85
6990	Clay excavation, bank run gravel borrow for backfill	2.89	8.15	11.04

For customer support on your Site Work & Landscape Costs with RSMeans Data, call 800.448.8182.

A2020 Basement Walls

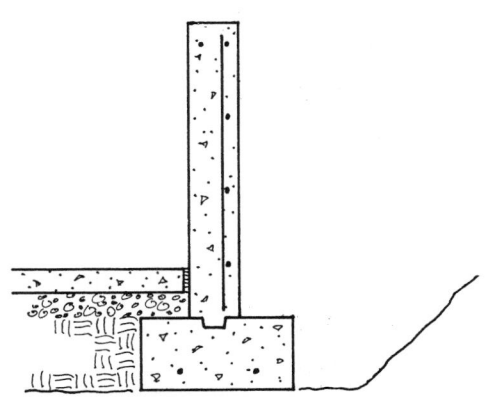

The Foundation Bearing Wall System includes: forms up to 16' high (four uses); 3,000 p.s.i. concrete placed and vibrated; and form removal with breaking form ties and patching walls. The wall systems list walls from 6" to 16" thick and are designed with minimum reinforcement.

Excavation and backfill are not included.

Please see the reference section for further design and cost information.

A2020 110						Walls, Cast in Place		
	WALL HEIGHT (FT.)	PLACING METHOD	CONCRETE (C.Y. per L.F.)	REINFORCING (LBS. per L.F.)	WALL THICKNESS (IN.)	COST PER L.F.		
						MAT.	INST.	TOTAL
5000	8'	direct chute	.148	6.6	6	38	119	157
5020			.199	9.6	8	47	125	172
5040			.250	12	10	56	125	181
5060			.296	14.39	12	64	128	192
5080			.347	16.19	14	66.50	128	194.50
5100			.394	19.19	16	80.50	133	213.50
5200	8'	pumped	.148	6.6	6	38	121	159
5220			.199	9.6	8	47	125	172
5240			.250	12	10	56	128	184
5260			.296	14.39	12	64	131	195
5280			.347	16.19	14	66.50	131	197.50
5300			.394	19.19	16	80.50	137	217.50
6020	10'	direct chute	.248	12	8	59	153	212
6040			.307	14.99	10	69	156	225
6060			.370	17.99	12	80	160	240
6080			.433	20.24	14	90	162	252
6100			.493	23.99	16	101	166	267
6220	10'	pumped	.248	12	8	59	157	216
6240			.307	14.99	10	69	160	229
6260			.370	17.99	12	80	164	244
6280			.433	20.24	14	90	166	256
6300			.493	23.99	16	101	171	272
7220	12'	pumped	.298	14.39	8	71	188	259
7240			.369	17.99	10	83	192	275
7260			.444	21.59	12	95.50	197	292.50
7280			.52	24.29	14	108	200	308
7300			.591	28.79	16	121	205	326
7420	12'	crane & bucket	.298	14.39	8	71	197	268
7440			.369	17.99	10	83	201	284
7460			.444	21.59	12	95.50	208	303.50
7480			.52	24.29	14	108	213	321
7500			.591	28.79	16	121	222	343
8220	14'	pumped	.347	16.79	8	82.50	219	301.50
8240			.43	20.99	10	96.50	224	320.50
8260			.519	25.19	12	112	230	342
8280			.607	28.33	14	126	233	359
8300			.69	33.59	16	141	239	380

A20 Basement Construction

A2020 Basement Walls

A2020 110				Walls, Cast in Place			

	WALL HEIGHT (FT.)	PLACING METHOD	CONCRETE (C.Y. per L.F.)	REINFORCING (LBS. per L.F.)	WALL THICKNESS (IN.)	COST PER L.F.		
						MAT.	INST.	TOTAL
8420	14'	crane & bucket	.347	16.79	8	82.50	229	311.50
8440			.43	20.99	10	96.50	234	330.50
8460			.519	25.19	12	112	243	355
8480			.607	28.33	14	126	248	374
8500			.69	33.59	16	141	256	397
9220	16'	pumped	.397	19.19	8	94.50	251	345.50
9240			.492	23.99	10	111	256	367
9260			.593	28.79	12	128	263	391
9280			.693	32.39	14	144	266	410
9300			.788	38.38	16	161	273	434
9420	16'	crane & bucket	.397	19.19	8	94.50	262	356.50
9440			.492	23.99	10	111	268	379
9460			.593	28.79	12	128	277	405
9480			.693	32.39	14	144	283	427
9500			.788	38.38	16	161	293	454

A2020 Basement Walls

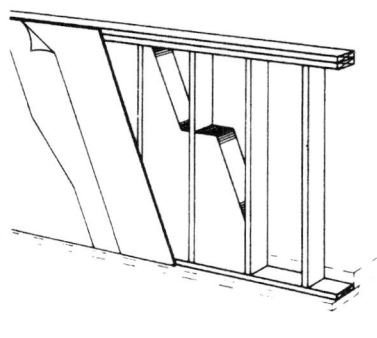

The Wood Wall Foundations System includes: all required pressure treated wood framing members; sheathing (treated and bonded with CDX-grade glue); asphalt paper; polyethylene vapor barrier; and 3-1/2″ R11 fiberglass insulation.

The Expanded System Listing shows various walls fabricated with 2″ x 4″, 2″ x 6″, 2″ x 8″, or 2″ x 10″ studs. Wall height varies from 2′ to 8′. The insulation is standard, fiberglass batt type for all walls listed.

System Components	QUANTITY	UNIT	COST PER S.F. MAT.	COST PER S.F. INST.	COST PER S.F. TOTAL
SYSTEM A2020 150 2000					
WOOD WALL FOUNDATION, 2″ X 4″ STUDS, 12″ O.C., 1/2″ SHEATHING, 2′ HIGH					
Framing plates & studs, 2″ x 4″ treated	.550	B.F.	2.32	2.91	5.23
Sheathing, 1/2″ thick CDX, treated	1.000	S.F.	1.40	1.16	2.56
Building paper, asphalt felt sheathing paper 15 lb	1.100	S.F.	.07	.19	.26
Polyethylene vapor barrier, standard, 6 mil	1.000	S.F.	.08	.18	.26
Fiberglass insulation, 3-1/2″ thick, 11″ wide	1.000	S.F.	.39	.57	.96
TOTAL			4.26	5.01	9.27

A2020 150	Wood Wall Foundations	COST PER S.F. MAT.	COST PER S.F. INST.	COST PER S.F. TOTAL
2000	Wood wall foundation, 2″ x 4″ studs, 12″ O.C., 1/2″ sheathing, 2′ high	4.26	5	9.26
2008	4′ high	3.60	4.18	7.78
2012	6′ high	3.38	3.91	7.29
2020	8′ high	3.27	3.77	7.04
2500	2″ x 6″ studs, 12″ O.C., 1/2″ sheathing, 2′ high	5.15	5.95	11.10
2508	4′ high	4.25	4.87	9.12
2512	6′ high	3.94	4.51	8.45
2520	8′ high	3.80	4.33	8.13
3000	2″ x 8″ studs, 12″ O.C., 1/2″ sheathing, 2′ high	6.80	5.35	12.15
3008	4′ high	5.45	4.42	9.87
3016	6′ high	4.96	4.11	9.07
3024	8′ high	4.74	3.96	8.70
3032	16″ O.C., 3/4″ sheathing, 2′ high	6.95	5.25	12.20
3040	4′ high	5.55	4.29	9.84
3048	6′ high	5.05	3.96	9.01
3056	8′ high	4.80	3.80	8.60
4000	2″ x 10″ studs, 12″ O.C., 1/2″ sheathing, 2′ high	7.15	5.45	12.60
4008	4′ high	5.65	4.50	10.15
4012	6′ high	5.20	4.19	9.39
4040	8′ high	4.94	4.03	8.97
4048	16″ O.C., 3/4″ sheathing, 2′ high	7.25	5.35	12.60
4056	4′ high	5.75	4.35	10.10
4070	6′ high	5.25	4.03	9.28
4080	8′ high	4.96	3.86	8.82

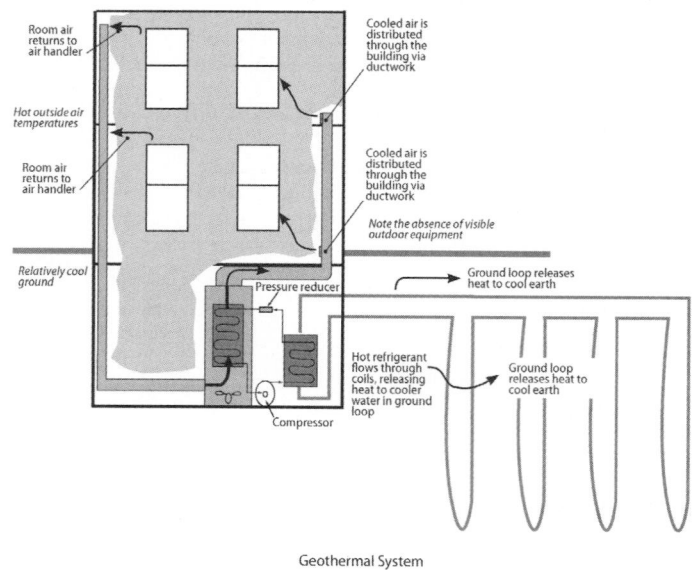

Geothermal System

System Components	QUANTITY	UNIT	COST EACH		
			MAT.	INST.	TOTAL
SYSTEM D3050 248 1000					
GEOTHERMAL HEAT PUMP SYSTEM 50 TON, VERTICAL LOOP, 200 LF PER TON					
Mobilization excavator	2.000	Ea.		1,772	1,772
Mobilization support crew and equipment	2.000	Ea.		700	700
Mobilization drill rig	2.000	Ea.		366	366
Drill wells 6" diameter	100.000	C.L.F.		55,700	55,700
Pipe loops 1 1/2" diameter	200.000	C.L.F.	45,400	45,200	90,600
Pipe headers 2" diameter	1600.000	L.F.	4,784	4,448	9,232
U-fittings for pipe loops	50.000	Ea.	392.50	997.50	1,390
Header tee fittings	100.000	Ea.	1,595	3,400	4,995
Header elbow fittings	10.000	Ea.	105	210	315
Excavate trench for pipe header	475.000	B.C.Y.		3,534	3,534
Backfill trench for pipe header	655.000	L.C.Y.		2,135.30	2,135.30
Compact trench for pipe header	475.000	E.C.Y.		1,344.25	1,344.25
Circulation pump 5 HP	1.000	Ea.	13,700	970	14,670
Pump control system	1.000	Ea.	1,425	1,225	2,650
Pump gauges	2.000	Ea.	48	51	99
Pump gauge fittings	2.000	Ea.	156	51	207
Pipe insulation for pump connection	12.000	L.F.	51.12	94.80	145.92
Pipe for pump connection	12.000	L.F.	204.60	441	645.60
Pipe fittings for pump connection	1.000	Ea.	25.50	224.90	250.40
Install thermostat wells	2.000	Ea.	18	137.30	155.30
Install gauge wells	2.000	Ea.	18	142.70	160.70
Thermometers, stem type, 9" case, 8" stem, 3/4" NPT	8.000	Ea.	344	899.60	1,243.60
Gauges, pressure or vacuum, 3-1/2" diameter dial	1.000	Ea.	1,150	325	1,475
Pipe strainer for pump	1.000	Ea.	266	325	591
Shut valve for pump	1.000	Ea.	815	350	1,165
Expansion joints for pump	2.000	Ea.	880	258	1,138
Heat pump 50 tons	1.000	Ea.	42,400	15,200	57,600
TOTAL			113,777.72	140,502.35	254,280.07

D30 HVAC

D3050 Terminal & Package Units

D3050 248	Geothermal Heat Pump System	COST EACH		
		MAT.	INST.	TOTAL
0990	Geothermal heat pump system, vertical loops 200' depth, 200 LF/4 gpm per ton			
1000	50 tons, vertical loops 200' depth, soft soil	114,000	140,500	254,500
1050	50 tons common soil	114,000	159,000	273,000
1075	50 tons dense soil	114,000	196,500	310,500
1100	40 tons, vertical loops 200' depth, soft soil	99,000	114,500	213,500
1150	40 tons common soil	108,500	153,000	261,500
1175	40 tons dense soil	99,000	159,000	258,000
1200	30 tons, vertical loops 200' depth, soft soil	75,000	90,500	165,500
1250	30 tons, common soil	75,000	101,500	176,500
1275	30 tons, dense soil	75,000	124,000	199,000
1300	25 tons, vertical loops 200' depth, soft soil	68,000	76,500	144,500
1350	25 tons, common soil	68,000	86,000	154,000
1375	25 tons, dense soil	68,000	104,500	172,500
1390	Geothermal heat pump system, vertical loops 250' depth, 250 LF/4 gpm per ton			
1400	50 tons, vertical loops 250' depth, soft soil	125,000	165,500	290,500
1450	50 tons common soil	125,000	189,000	314,000
1475	50 tons dense soil	125,000	235,500	360,500
1500	40 tons, vertical loops 250' depth, soft soil	108,500	134,500	243,000
1550	40 tons common soil	108,500	153,000	261,500
1575	40 tons dense soil	108,500	190,500	299,000
1600	30 tons, vertical loops 250' depth, soft soil	82,000	105,500	187,500
1650	30 tons, common soil	82,000	119,500	201,500
1675	30 tons, dense soil	82,000	147,500	229,500
1700	25 tons, vertical loops 250' depth, soft soil	73,500	89,500	163,000
1750	25 tons, common soil	73,500	101,000	174,500
1775	25 tons, dense soil	73,500	124,000	197,500
1998	Geothermal heat pump system, vertical loops 200' depth, 200 LF/4 gpm per ton			
1999	Includes pumping of wells and adding 1:1 cement grout mix with 5% waste factor			
2000	50 tons, vertical loops 200' depth, soft soil	148,000	165,500	313,500
2050	50 tons common soil	148,000	184,000	332,000
2075	50 tons dense soil	148,000	221,500	369,500
2100	40 tons, vertical loops 200' depth, soft soil	126,500	134,000	260,500
2150	40 tons common soil	135,500	173,000	308,500
2175	40 tons dense soil	126,500	179,000	305,500
2200	30 tons, vertical loops 200' depth, soft soil	95,500	105,500	201,000
2250	30 tons, common soil	95,500	116,500	212,000
2275	30 tons, dense soil	95,500	139,000	234,500
2300	25 tons, vertical loops 200' depth, soft soil	85,000	89,000	174,000
2350	25 tons, common soil	85,000	98,500	183,500
2375	25 tons, dense soil	85,000	117,000	202,000
2390	Geothermal heat pump system, vertical loops 250' depth, 250 LF/4 gpm per ton			
2391	Includes pumping of wells and adding 1:1 cement grout mix with 5% waste factor			
2400	50 tons, vertical loops 250' depth, soft soil	168,000	197,000	365,000
2450	50 tons common soil	168,000	220,000	388,000
2475	50 tons dense soil	168,000	266,500	434,500
2500	40 tons, vertical loops 250' depth, soft soil	142,500	159,500	302,000
2550	40 tons common soil	142,500	178,000	320,500
2575	40 tons dense soil	142,500	215,500	358,000
2600	30 tons, vertical loops 250' depth, soft soil	107,500	124,000	231,500
2650	30 tons, common soil	107,500	138,000	245,500
2675	30 tons, dense soil	107,500	165,500	273,000
2700	25 tons, vertical loops 250' depth, soft soil	95,000	105,000	200,000
2750	25 tons, common soil	95,000	116,500	211,500
2775	25 tons, dense soil	95,000	140,000	235,000
2998	Geothermal heat pump system, vertical loops 200' depth, 200 LF/4 gpm per ton			
2999	Includes pumping of wells and adding bentonite mix with 5% waste factor			
3000	50 tons, vertical loops 200' depth, soft soil	131,500	165,500	297,000
3050	50 tons common soil	131,500	184,000	315,500

D3050 Terminal & Package Units

D3050 248	Geothermal Heat Pump System	COST EACH		
		MAT.	INST.	TOTAL
3075	50 tons dense soil	131,500	221,500	353,000
3100	40 tons, vertical loops 200' depth, soft soil	113,500	134,000	247,500
3150	40 tons common soil	122,500	173,000	295,500
3175	40 tons dense soil	113,500	179,000	292,500
3200	30 tons, vertical loops 200' depth, soft soil	85,500	105,500	191,000
3250	30 tons, common soil	85,500	116,500	202,000
3275	30 tons, dense soil	85,500	139,000	224,500
3300	25 tons, vertical loops 200' depth, soft soil	76,500	89,000	165,500
3350	25 tons, common soil	76,500	98,500	175,000
3375	25 tons, dense soil	76,500	117,000	193,500
3390	Geothermal heat pump system, vertical loops 250' depth, 250 LF/4 gpm per ton			
3391	Includes pumping of wells and adding bentonite mix with 5% waste factor			
3400	50 tons, vertical loops 250' depth, soft soil	147,000	197,000	344,000
3450	50 tons common soil	147,000	220,000	367,000
3475	50 tons dense soil	147,000	266,500	413,500
3500	40 tons, vertical loops 250' depth, soft soil	126,000	159,500	285,500
3550	40 tons common soil	126,000	178,000	304,000
3575	40 tons dense soil	126,000	215,500	341,500
3600	30 tons, vertical loops 250' depth, soft soil	95,000	124,000	219,000
3650	30 tons, common soil	95,000	138,000	233,000
3675	30 tons, dense soil	95,000	165,500	260,500
3700	25 tons, vertical loops 250' depth, soft soil	84,500	105,000	189,500
3750	25 tons, common soil	84,500	116,500	201,000
3775	25 tons, dense soil	84,500	140,000	224,500

G1010 Site Clearing

System Components	QUANTITY	UNIT	COST PER ACRE		
			MAT.	INST.	TOTAL
SYSTEM G1010 120 1000					
REMOVE TREES & STUMPS UP TO 6″ DIA. BY CUT & CHIP & STUMP HAUL AWAY					
Cut & chip light trees to 6″ diameter	1.000	Acre		5,125	5,125
Grub stumps & remove, ideal conditions	1.000	Acre		1,990	1,990
TOTAL				7,115	7,115

G1010 120	Clear & Grub Site	COST PER ACRE		
		MAT.	INST.	TOTAL
0980	Clear & grub includes removal and haulaway of tree stumps			
1000	Remove trees & stumps up to 6 ″ in dia. by cut & chip & remove stumps		7,125	7,125
1100	Up to 12 ″ in diameter		11,300	11,300
1990	Removal of brush includes haulaway of cut material			
1992	Percentage of vegetation is actual plant coverage			
2000	Remove brush by saw, 4′ tall, 10 mile haul cycle		4,600	4,600

G10 Site Preparation

G1010 Site Clearing

Stripping and stock piling top soil can be done with a variety of equipment depending on the size of the job. For small jobs you can do by hand. Larger jobs can be done with a loader or skid steer. Even larger jobs can be done by a dozer.

System Components	QUANTITY	UNIT	COST PER S.Y.		
			MAT.	INST.	TOTAL
SYSTEM G1010 122 1100 **STRIP TOPSOIL AND STOCKPILE, 6 INCH DEEP BY SKID STEER** By skid steer, 6" deep, 100' haul, 101-500 S.Y.	1.000	S.Y.		3.84	3.84
TOTAL				3.84	3.84

G1010 122	Strip Topsoil	COST PER S.Y.		
		MAT.	INST.	TOTAL
1000	Strip topsoil & stockpile, 6 inch deep by hand, 50' haul		16.15	16.15
1100	100' haul, 101-500 S.Y., by skid steer		3.84	3.84
1200	100' haul, 501-900 S.Y.		2.13	2.13
1300	200' haul , 901-1100 S.Y.		2.98	2.98
1400	200' haul, 1101-4000 S.Y., by dozer		.66	.66

For customer support on your Site Work & Landscape Costs with RSMeans Data, call 800.448.8182.

G10 Site Preparation

G1020 Site Demolitions and Relocations

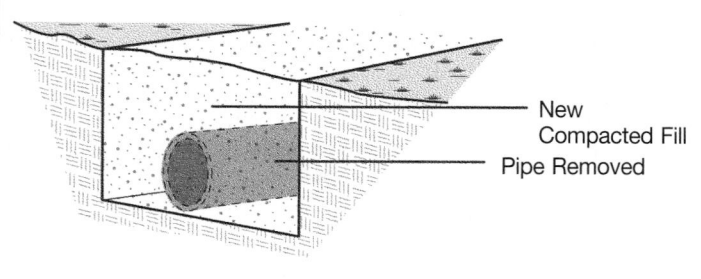

New Compacted Fill
Pipe Removed

Demolition of utility lines includes excavation, pipe removal, backfill, compaction and one day trucking cost for loading and disposal of pipe. No waste disposal fees are included.

Demolition of utility lines @ 6' depth include hydraulic shoring.

Demolition of utility lines @ 7' to 10' depth include a trench box.

Structural fill required after removal:

14" diameter pipe:	5.15 LCY per 100 LF
16" diameter pipe:	6.72 LCY per 100 LF
18" diameter pipe:	8.53 LCY per 100 LF
24" diameter pipe:	15.1 LCY per 100 LF

System Components			COST PER L.F.		
	QUANTITY	UNIT	MAT.	INST.	TOTAL
SYSTEM G1020 205 1000					
REMOVE 4″ D. I. WATER PIPE, 4′ DEPTH, EXCAVATE, BACKFILL, COMPACT& HAUL					
Excavate trench 4′ depth	.593	B.C.Y.		4.48	4.48
Backfill trench	.770	L.C.Y.		2.51	2.51
Compact trench	.593	E.C.Y.		3.25	3.25
Demolition of 4 inch D.I. pipe	1.000	L.F.		17.02	17.02
Truck for waste and hauling	.040	Hr.		6.24	6.24
TOTAL				33.50	33.50

G1020 205	Remove Underground Water Pipe Including Earthwork	COST PER L.F.		
		MAT.	INST.	TOTAL
1000	Remove D.I. pipe 4′ depth, excavate & backfill, hauling, no dump fee, 4″		33.50	33.50
1010	6″ to 12″ diameter		37	37
1020	Plastic pipe, 3/4″ to 4″ diameter		14.80	14.80
1030	6″ to 8″ diameter		16.65	16.65
1040	10″ to 12″ diameter		21	21
1050	Concrete water pipe, 4″ to 10″ diameter		23	23
1060	12″ diameter		28.50	28.50
1070	Welded steel water pipe, 4″ to 6″ diameter		30.50	30.50
1080	8″ to 10″ diameter		50.50	50.50
1090	Copper pipe, 3/4″ to 2″ diameter		15.65	15.65
1100	2-1/2″ to 3″ diameter		19.25	19.25
1110	4″ to 6″ diameter		23.50	23.50
1200	D.I. pipe 5′ depth, 4″ diameter		35.50	35.50
1210	6″ to 12″ diameter		39	39
1220	Plastic pipe, 3/4″ to 4″ diameter		16.80	16.80
1230	6″ to 8″ diameter		18.65	18.65
1240	10″ to 12″ diameter		23	23
1250	Concrete water pipe, 4″ to 10″ diameter		25	25
1260	12″ diameter		30.50	30.50
1270	Welded steel water pipe, 4″ to 6″ diameter		32.50	32.50
1280	8″ to 10″ diameter		52.50	52.50
1290	Copper pipe, 3/4″ to 2″ diameter		17.65	17.65
1300	2-1/2″ to 3″ diameter		21	21
1310	4″ to 6 ″ diameter		25.50	25.50
1400	D.I. pipe 6′ depth, 4″ diameter with hydraulic shoring	3.24	42.50	45.74
1410	6″ to 12″ diameter	3.24	46	49.24
1420	Plastic pipe, 3/4″ to 4″ diameter	3.24	24	27.24
1430	6″ to 8″ diameter	3.24	26	29.24
1440	10″ to 12″ diameter	3.24	30	33.24
1450	Concrete water pipe, 4″ to 10″ diameter	3.24	32	35.24

For customer support on your Site Work & Landscape Costs with RSMeans Data, call 800.448.8182.

G1020 Site Demolitions and Relocations

G1020 205	Remove Underground Water Pipe Including Earthwork	COST PER L.F.		
		MAT.	INST.	TOTAL
1460	12" diameter	3.24	37.50	40.74
1470	Welded steel water pipe, 4" to 6" diameter	3.24	39.50	42.74
1480	8" to 10" diameter	3.24	59.50	62.74
1490	Copper pipe, 3/4" to 2" diameter	3.24	25	28.24
1500	2-1/2" to 3" diameter	3.24	28.50	31.74
1510	4" to 6" diameter	3.24	33	36.24
1600	D.I. pipe, trench 7' deep by 6' wide, with trench box, 4" diameter pipe		44.50	44.50
1610	6" to 12" diameter		47.50	47.50
1620	Plastic pipe, 3/4" to 4" diameter		27.50	27.50
1630	6" to 8" diameter		28.50	28.50
1640	10" to 12" diameter		32	32
1650	Concrete water pipe, 4" to 10" diameter		34	34
1660	12" diameter		39.50	39.50
1670	Welded steel water pipe, 4" to 6" diameter		41	41
1680	8" to 10" diameter		61	61
1690	Copper pipe, 3/4" to 2" diameter		27.50	27.50
1700	2-1/2" to 3" diameter		30	30
1710	4" to 6" diameter		34.50	34.50
1800	D.I. pipe, trench 8' deep by 6' wide, with trench box, 4" diameter pipe		47.50	47.50
1810	6" to 12" diameter		50.50	50.50
1820	Plastic pipe, 3/4" to 4" diameter		28.50	28.50
1830	6" to 8" diameter		29.50	29.50
1840	10" to 12" diameter		35	35
1850	Concrete water pipe, 4" to 10" diameter		37	37
1860	12" diameter		42.50	42.50
1870	Welded steel water pipe, 4" to 6" diameter		44	44
1880	8" to 10" diameter		64	64
1890	Copper pipe, 3/4" to 2" diameter		28.50	28.50
1900	2-1/2" to 3" diameter		33	33
1910	4" to 6" diameter		37.50	37.50
2000	D.I. pipe, trench 9' deep by 6' wide, with trench box, 4" diameter pipe		49.50	49.50
2010	6" to 12" diameter		52.50	52.50
2020	Plastic pipe, 3/4" to 4" diameter		32.50	32.50
2030	6" to 8 " diameter		33.50	33.50
2040	10" to 12" diameter		37	37
2050	Concrete water pipe, 4" to 10" diameter		39	39
2060	12" diameter		44.50	44.50
2070	Welded steel water pipe, 4" to 6" diameter		46	46
2080	8" to 10" diameter		66	66
2090	Copper pipe, 3/4" to 2" diameter		32.50	32.50
2100	2-1/2" to 3" diameter		35	35
2110	4" to 6" diameter		39.50	39.50
2200	D.I. pipe, trench 10' deep by 6' wide, with trench box, 4" diameter pipe		52	52
2210	6" to 12" diameter		55.50	55.50
2220	Plastic pipe, 3/4" to 4" diameter		35.50	35.50
2230	6" to 8" diameter		36.50	36.50
2240	10" to 12" diameter		39.50	39.50
2250	Concrete water pipe, 4" to 10" diameter		42	42
2260	12" diameter		47.50	47.50
2270	Welded steel water pipe, 4" to 6" diameter		49	49
2280	8" to 10" diameter		69	69
2290	Copper pipe, 3/4" to 2" diameter		35.50	35.50
2300	2-1/2" to 3" diameter		38	38
2310	4" to 6" diameter		42.50	42.50
2400	Concrete water pipe 5' depth, 15" to 18" diameter		41	41
2410	21" to 24" diameter		49	49
2420	D.I. water pipe, 14" to 24" diameter		48	48
2430	Plastic water pipe, 14" to 18" diameter		20.50	20.50

G1020 Site Demolitions and Relocations

G1020 205	Remove Underground Water Pipe Including Earthwork	COST PER L.F.		
		MAT.	INST.	TOTAL
2440	20" to 24" diameter		25.50	25.50
2500	Concrete water pipe, 6' depth, 15" to 18" diameter with hydraulic shoring	3.24	47.50	50.74
2510	21" to 24" diameter	3.24	55.50	58.74
2520	D. l. water pipe, 14" to 24" diameter	3.24	54.50	57.74
2530	Plastic water pipe, 14" to 18" diameter	3.24	27	30.24
2540	20" to 24" diameter	3.24	32.50	35.74
2600	Concrete water pipe, 7' deep by 6' wide, w/trench box, 15" to 18" diameter		52	52
2610	21" to 24" diameter		60	60
2620	D. l. water pipe, 14" to 24" diameter		59	59
2630	Plastic water pipe, 14" to 18" inch diameter		31	31
2640	20" to 24" diameter		36.50	36.50
2700	Concrete water pipe, 8' deep by 6' wide, w/trench box, 15" to 18" diameter		55	55
2710	21" to 24" diameter		63	63
2720	D. l. water pipe, 14" to 24" diameter		62	62
2730	Plastic water pipe, 14" to 18" diameter		34	34
2740	20" to 24" diameter		39	39
2800	Concrete water pipe, 9' deep by 6' wide, w/trench box, 15" to 18" diameter		58	58
2810	21" to 24" diameter		66	66
2820	D. l. water pipe, 14" to 24" diameter		65	65
2830	Plastic water pipe, 14" to 18" diameter		37	37
2840	20" to 24" diameter		42	42
2900	Concrete water pipe, 10' deep by 6' wide, w/trench box, 15" to 18" diameter		60.50	60.50
2910	21" to 24" diameter		68.50	68.50
2920	D. l. water pipe, 14" to 24" diameter		67.50	67.50
2930	Plastic water pipe, 14" to 18" diameter		39.50	39.50
2940	20" to 24" diameter		45	45
3000	Import stuctural fill 10 miles round trip per cubic yard	33	12.60	45.60

For customer support on your Site Work & Landscape Costs with RSMeans Data, call 800.448.8182.

G1020 Site Demolitions and Relocations

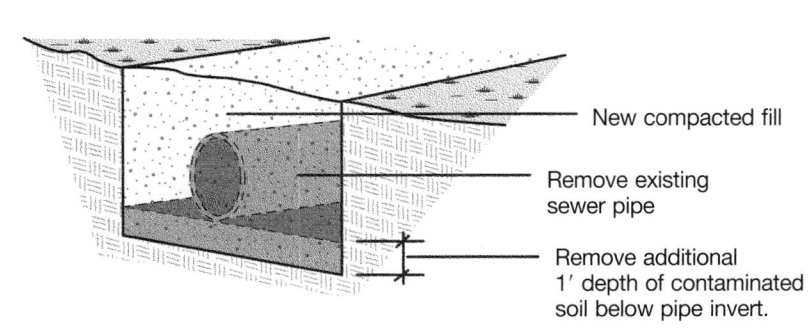

New compacted fill

Remove existing
sewer pipe

Remove additional
1' depth of contaminated
soil below pipe invert.

Demolition of utility lines includes excavation, pipe removal, backfill, compaction and one day trucking cost for loading and disposal of waste pipe and soil. No disposal fees are included.

Excavation is one foot deeper than pipe invert. For pipes to 12" diameter, two feet of excavated soil is waste. For pipes 15" to 18" diameter, 2.5 feet of excavated soil is waste. For pipes 21" to 24" diameter, 3 feet of excavated soil is waste.

Demolition of utility lines @ 5' to 6' depth includes hydraulic shoring.

Demolition of utility lines @ 7' to 10' depth includes trench box. Trench box requires 6' minimum trench width.

System Components			COST PER L.F.		
	QUANTITY	UNIT	MAT.	INST.	TOTAL
SYSTEM G1020 206 1000					
REMOVE 4 INCH PLASTIC SEWER PIPE, 4' DEPTH, EXCAVATE,					
BACKFILL, COMPACT & HAUL WASTE SOIL AND PIPE					
Excavate trench 5' depth (one foot over)	.741	B.C.Y.		5.02	5.02
Backfill trench	.963	L.C.Y.		3.14	3.14
Compact trench	.741	E.C.Y.		4.06	4.06
Demolition of 4 inch plastic sewer pipe	1.000	L.F.		2.80	2.80
Truck for waste hauling	.012	Hr.		1.87	1.87
TOTAL				16.89	16.89

G1020 206	Remove Underground Sewer Pipe Including Earthwork	COST PER L.F.		
		MAT.	INST.	TOTAL
1000	Remove plastic sewer, 4' depth, excavate, backfill, haul waste, 4" diameter		16.90	16.90
1010	6" to 8" diameter		18.65	18.65
1020	10" to 12" diameter		23	23
1030	Cast iron sewer, 4" diameter		25.50	25.50
1040	5" to 6" diameter		30	30
1050	8" to 12" diameter		34	34
1060	Concrete sewer, 4" to 10" diameter		25	25
1070	12" diameter		30.50	30.50
1100	Plastic sewer, 5' depth, with hydraulic shoring, 4" diameter	3.24	24	27.24
1110	6" to 8" diameter	3.24	26	29.24
1120	10" to 12" diameter	3.24	30	33.24
1130	Cast iron sewer, 4" diameter	3.24	33	36.24
1140	5" to 6" diameter	3.24	37	40.24
1150	8" to 12" diameter	3.24	41	44.24
1160	Concrete sewer, 4" to 10" inch diameter	3.24	32	35.24
1170	12" diameter	3.24	37.50	40.74
1200	Plastic sewer, 6' depth, with hydraulic shoring, 4" diameter	3.78	26	29.78
1210	6" to 8" diameter	3.78	27.50	31.28
1220	10" to 12" diameter	3.78	32	35.78
1230	Cast iron sewer, 4" diameter	3.78	34.50	38.28
1240	5" to 6" diameter	3.78	38.50	42.28
1250	8" to 12" diameter	3.78	43	46.78
1260	Concrete sewer, 4" to 10" diameter	3.78	34	37.78
1270	12" diameter	3.78	39.50	43.28
1300	Plastic sewer, 7' depth, with trench box, 4" diameter		29.50	29.50
1310	6" to 8" diameter		32	32

623

G1020 Site Demolitions and Relocations

G1020 206	Remove Underground Sewer Pipe Including Earthwork	COST PER L.F.		
		MAT.	INST.	TOTAL
1320	10" to 12" diameter		34	34
1330	Cast iron sewer, 4" diameter		36.50	36.50
1340	5" to 6" diameter		40.50	40.50
1350	8" to 12" diameter		45	45
1360	Concrete sewer, 4" to 10" diameter		36	36
1370	12" diameter		41.50	41.50
1400	Plastic sewer, 8' depth, with trench box, 4" diameter		32.50	32.50
1410	6" to 8" diameter		35	35
1420	10" to 12" diameter		39	39
1430	Cast iron sewer, 4" diameter		39.50	39.50
1440	5" to 6" diameter		43.50	43.50
1450	8" to 12" diameter		47.50	47.50
1460	Concrete sewer, 4" to 10" diameter		39	39
1470	12" diameter		44.50	44.50
1500	Plastic sewer, 9' depth, with trench box, 4" diameter		35.50	35.50
1510	6" to 8" diameter		38	38
1520	10" to 12" diameter		41.50	41.50
1530	Cast iron sewer, 4" diameter		42.50	42.50
1540	5" to 6" diameter		46.50	46.50
1550	8" to 12" diameter		50.50	50.50
1560	Concrete sewer, 4" to 10" diameter		42	42
1570	12" diameter		47.50	47.50
1600	Plastic sewer, 10' depth, with trench box, 4" diameter		38.50	38.50
1610	6" to 8" diameter		41	41
1620	10" to 12" diameter		44.50	44.50
1630	Cast iron sewer, 4" diameter		45.50	45.50
1640	5" to 6" diameter		49.50	49.50
1650	8" to 12" diameter		53.50	53.50
1660	Concrete sewer, 4" to 10" diameter		44.50	44.50
1670	12" diameter		50	50
1700	Concrete sewer, 5' depth, with hydraulic shoring, 15" to 18" diameter	3.24	51	54.24
1710	21" to 24" diameter	3.24	59	62.24
1720	Plastic sewer, 15" to 18" diameter	3.24	30	33.24
1730	21" to 24" diameter	3.24	35.50	38.74
1800	6' depth, 15" to 18" diameter	3.78	53	56.78
1810	21" to 24" diameter	3.78	62.50	66.28
1820	Plastic sewer, 15" to 18" diameter	3.78	33.50	37.28
1830	21" to 24" diameter	3.78	38.50	42.28
1900	7' depth, with trench box, 15" to 18" diameter		55	55
1910	21" to 24" diameter		63	63
1920	Plastic sewer, 15" to 18" diameter		34	34
1930	21" to 24" diameter		39	39
2000	8' depth, with trench box, 15" to 18" diameter		58	58
2010	21" to 24" diameter		66	66
2020	Plastic sewer, 15" to 18" diameter		37	37
2030	21" to 24" diameter		42	42
2100	9' depth, with trench box, 15" to 18" diameter		60.50	60.50
2110	21" to 24" diameter		68.50	68.50
2120	Plastic sewer, 15" to 18" diameter		39.50	39.50
2130	21" to 24" diameter		45	45
2200	10' depth, with trench box, 15" to 18" diameter		64.50	64.50
2210	21" to 24" diameter		72.50	72.50
2220	Plastic sewer, 15" to 18" diameter		43.50	43.50
2230	21" to 24" diameter		49	49

G10 Site Preparation

G1020 Site Demolitions and Relocations

G1020 210	Demolition Bituminous Sidewalk	COST PER S.Y.		
		MAT.	INST.	TOTAL
0990	Demolition of bituminous sidewalk & haulaway by truck			
1000	Sidewalk demolition, bituminous, 2 inch thick & trucking		9.15	9.15
1100	2-1/2 inch thick		9.85	9.85

G1020 212	Demolition Concrete Sidewalk	COST PER S.Y.		
		MAT.	INST.	TOTAL
0990	Demolition of concrete sidewalk & haulaway by truck			
1000	Sidewalk demolition, plain concrete, 4 inch thick & trucking		20	20
1100	Plain 5 inch thick		23	23
1200	Plain 6 inch thick		26.50	26.50
1300	Mesh reinforced, 4 inch thick		21.50	21.50
1400	Mesh reinforced, 5 inch thick		24.50	24.50
1500	Mesh reinforced, 6 inch thick		28.50	28.50

For customer support on your Site Work & Landscape Costs with RSMeans Data, call 800.448.8182.

G1030 Site Earthwork

The Cut and Fill Gravel System includes: moving gravel cut from an area above the specified grade to an area below the specified grade utilizing a bulldozer and/or scraper with the addition of compaction equipment, plus a water wagon to adjust the moisture content of the soil.

The Expanded System Listing shows Cut and Fill operations with hauling distances that vary from 50' to 5000'. Lifts for compaction in the filled area vary from 6" to 12". There is no waste included in the assumptions.

System Components			COST PER C.Y.		
	QUANTITY	UNIT	EQUIP.	LABOR	TOTAL
SYSTEM G1030 105 1000					
GRAVEL CUT & FILL, 80 HP DOZER & COMPACT, 50'HAUL, 6" LIFT, 2 PASSES					
Excavating, bulk, dozer, 80 HP, 50' haul sand and gravel	1.000	C.Y.	.97	2.10	3.07
Water wagon rent per day	.003	Hr.	.19	.23	.42
Backfill, dozer, 50' haul sand and gravel	1.150	C.Y.	.47	1.01	1.48
Compaction, vibrating roller, 6" lifts, 2 passes	1.000	C.Y.	.21	.32	.53
TOTAL			1.84	3.66	5.50

G1030 105	Cut & Fill Gravel	COST PER C.Y.		
		EQUIP.	LABOR	TOTAL
1000	Gravel cut & fill, 80HP dozer & roller compact, 50' haul, 6" lift, 2 passes	1.84	3.66	5.50
1050	4 passes	1.97	3.85	5.82
1100	12" lift, 2 passes	1.75	3.53	5.28
1150	4 passes	1.88	3.71	5.59
1200	150' haul, 6" lift, 2 passes	3.27	6.80	10.07
1250	4 passes	3.40	6.95	10.35
1300	12" lift, 2 passes	3.18	6.65	9.83
1350	4 passes	3.31	6.85	10.16
1400	300' haul, 6" lift, 2 passes	5.50	11.60	17.10
1450	4 passes	5.65	11.80	17.45
1500	12" lift, 2 passes	5.40	11.50	16.90
1550	4 passes	5.55	11.65	17.20
1600	105 HP dozer & roller compactor, 50' haul, 6" lift, 2 passes	2.01	2.76	4.77
1650	4 passes	2.14	2.95	5.09
1700	12" lift, 2 passes	1.92	2.63	4.55
1750	4 passes	2.05	2.81	4.86
1800	150' haul, 6" lift, 2 passes	3.88	5.35	9.23
1850	4 passes	4.01	5.55	9.56
1900	12" lift, 2 passes	3.79	5.20	8.99
1950	4 passes	3.92	5.40	9.32
2000	300' haul, 6" lift, 2 passes	7.20	9.85	17.05
2050	4 passes	7.35	10.05	17.40
2100	12' lift, 2 passes	7.10	9.70	16.80
2150	4 passes	7.25	9.90	17.15
2200	200 HP dozer & roller compactor, 150' haul, 6" lift, 2 passes	4.78	3.09	7.87
2250	4 passes	4.91	3.28	8.19
2300	12" lift, 2 passes	4.69	2.96	7.65
2350	4 passes	4.82	3.14	7.96

G10 Site Preparation

G1030 Site Earthwork

G1030 105	Cut & Fill Gravel	COST PER C.Y.		
		EQUIP.	LABOR	TOTAL
2600	300' haul, 6" lift, 2 passes	8.20	5.05	13.25
2650	4 passes	8.30	5.25	13.55
2700	12" lift, 2 passes	8.10	4.92	13.02
2750	4 passes	8.25	5.10	13.35
3000	300 HP dozer & roller compactors, 150' haul, 6" lift, 2 passes	3.56	2.11	5.67
3050	4 passes	3.69	2.30	5.99
3100	12" lift, 2 passes	3.47	1.98	5.45
3150	4 passes	3.60	2.16	5.76
3200	300' haul, 6" lift, 2 passes	6.10	3.36	9.46
3250	4 passes	6.20	3.55	9.75
3300	12" lift, 2 passes	6	3.23	9.23
3350	4 passes	6.15	3.41	9.56
4200	10 C.Y. elevating scraper & roller compacter, 1500' haul, 6" lift, 2 passes	4.45	4.58	9.03
4250	4 passes	4.87	5.10	9.97
4300	12" lift, 2 passes	4.32	4.47	8.79
4350	4 passes	4.48	4.58	9.06
4800	5000' haul, 6" lift, 2 passes	6	6.15	12.15
4850	4 passes	6.30	6.45	12.75
4900	12" lift, 2 passes	5.70	5.80	11.50
4950	4 passes	6.05	6.15	12.20
5000	15 C.Y. S.P. scraper & roller compactor, 1500' haul, 6" lift, 2 passes	4.95	3.38	8.33
5050	4 passes	5.20	3.69	8.89
5100	12" lift, 2 passes	4.81	3.25	8.06
5150	4 passes	4.97	3.38	8.35
5400	5000' haul, 6" lift, 2 passes	6.95	4.68	11.63
5450	4 passes	7.25	5	12.25
5500	12" lift, 2 passes	6.80	4.56	11.36
5550	4 passes	7	4.68	11.68
5600	21 C.Y. S.P. scraper & roller compactor, 1500' haul, 6" lift, 2 passes	4.13	2.70	6.83
5650	4 passes	4.40	3.02	7.42
5700	12" lift, 2 passes	4	2.57	6.57
5750	4 passes	4.15	2.71	6.86
6000	5000' haul, 6" lift, 2 passes	6.35	4.07	10.42
6050	4 passes	6.60	4.39	10.99
6100	12" lift, 2 passes	6.20	3.95	10.15
6150	4 passes	6.35	4.07	10.42

627

G1030 Site Earthwork

The Excavation of Sand and Gravel System balances the productivity of the Excavating equipment to the hauling equipment. It is assumed that the hauling equipment will encounter light traffic and will move up no considerable grades on the haul route. No mobilization cost is included. All costs given in these systems include a swell factor of 15%.

The Expanded System Listing shows excavation systems using backhoes ranging from 1/2 Cubic Yard capacity to 3-1/2 Cubic Yards. Power shovels indicated range from 1/2 Cubic Yard to 3 Cubic Yards. Dragline bucket rigs used range from 1/2 Cubic Yard to 3 Cubic Yards. Truck capacities indicated range from 8 Cubic Yards to 20 Cubic Yards. Each system lists the number of trucks involved and the distance (round trip) that each must travel.

System Components	QUANTITY	UNIT	COST PER C.Y. EQUIP.	LABOR	TOTAL
SYSTEM G1030 110 1000					
EXCAVATE SAND & GRAVEL, 1/2 CY BACKHOE, TWO 8 CY DUMP TRUCKS 2 MRT					
Excavating, bulk, hyd. backhoe, wheel mtd., 1-1/2 CY	1.000	B.C.Y.	1.57	3.95	5.52
Haul earth, 8 CY dump truck, 2 mile round trip, 2.6 loads/hr	1.150	L.C.Y.	4.30	5.87	10.17
Spotter at earth fill dump or in cut	.020	Hr.		1.20	1.20
TOTAL			5.87	11.02	16.89

G1030 110	Excavate and Haul Sand & Gravel	COST PER C.Y. EQUIP.	LABOR	TOTAL
1000	Excavate sand & gravel, 1/2 C.Y. backhoe, two 8 C.Y. dump trucks, 2 MRT	5.85	11	16.85
1200	Three 8 C.Y. dump trucks, 4 mile round trip	7.55	12.80	20.35
1400	Two 12 C.Y. dump trucks, 4 mile round trip	6.05	8.25	14.30
1600	3/4 C.Y. backhoe, four 8 C.Y. dump trucks, 2 mile round trip	5.20	8.50	13.70
1700	Six 8 C.Y. dump trucks, 4 mile round trip	7.30	11.40	18.70
1800	Three 12 C.Y. dump trucks, 4 mile round trip	6.05	7.15	13.20
1900	Two 16 C.Y. dump trailers, 3 mile round trip	5	5.75	10.75
2000	Two 20 C.Y. dump trailers, 4 mile round trip	4.68	5.40	10.08
2200	1-1/2 C.Y. backhoe, four 12 C.Y. dump trucks, 2 mile round trip	4.57	5.10	9.67
2300	Six 12 C.Y. dump trucks, 4 mile round trip	5.55	5.70	11.25
2400	Three 16 C.Y. dump trailers, 3 mile round trip	4.67	4.94	9.61
2600	Three 20 C.Y. dump trailers, 4 mile round trip	4.34	4.57	8.91
2800	2-1/2 C.Y. backhoe, six 12 C.Y. dump trucks, 2 mile round trip	4.80	5.25	10.05
2900	Eight 12 C.Y. dump trucks, 4 mile round trip	5.85	6.25	12.10
3000	Four 16 C.Y. dump trailers, 2 mile round trip	4.08	4.11	8.19
3100	Six 16 C.Y. dump trailers, 4 mile round trip	5.15	5.20	10.35
3200	Four 20 C.Y. dump trailers, 3 mile round trip	4.28	4.23	8.51
3400	3-1/2 C.Y. backhoe, eight 16 C.Y. dump trailers, 3 mile round trip	4.91	4.34	9.25
3500	Ten 16 C.Y. dump trailers, 4 mile round trip	5.20	4.54	9.74
3700	Six 20 C.Y. dump trailers, 2 mile round trip	3.58	3.13	6.71
3800	Nine 20 C.Y. dump trailers, 4 mile round trip	4.49	3.86	8.35
4000	1/2 C.Y. pwr. shovel, four 8 C.Y. dump trucks, 2 mile round trip	5.50	8.05	13.55
4100	Six 8 C.Y. dump trucks, 3 mile round trip	6.35	9.65	16
4200	Two 12 C.Y. dump trucks, 1 mile round trip	4.08	4.92	9
4300	Four 12 C.Y. dump trucks, 4 mile round trip	6.10	6.65	12.75
4400	Two 16 C.Y. dump trailers, 2 mile round trip	4.35	4.80	9.15
4500	Two 20 C.Y. dump trailers, 3 mile round trip	4.55	4.93	9.48
4600	Three 20 C.Y. dump trailers, 4 mile round trip	4.59	4.71	9.30
4800	3/4 C.Y. pwr. shovel, six 8 C.Y. dump trucks, 2 mile round trip	5.40	7.80	13.20
4900	Four 12 C.Y. dump trucks, 2 mile round trip	5	5.25	10.25
5000	Six 12 C.Y. dump trucks, 4 mile round trip	5.95	6.50	12.45
5100	Four 16 C.Y. dump trailers, 4 mile round trip	5.40	5.20	10.60
5200	Two 20 C.Y. dump trailers, 1 mile round trip	3.43	3.54	6.97
5400	1-1/2 C.Y. pwr. shovel, six 12 C.Y. dump trucks, 2 mile round trip	4.81	5.05	9.86

G10 Site Preparation

G1030 Site Earthwork

G1030 110	Excavate and Haul Sand & Gravel	COST PER C.Y.		
		EQUIP.	LABOR	TOTAL
5600	Four 16 C.Y. dump trailers, 1 mile round trip	3.56	3.30	6.86
5700	Six 16 C.Y. dump trailers, 3 mile round trip	4.73	4.54	9.27
5800	Six 20 C.Y. dump trailers, 4 mile round trip	4.40	4.18	8.58
6000	3 C.Y. pwr. shovel, eight 16 C.Y. dump trailers, 1 mile round trip	3.45	3.02	6.47
6200	Eight 20 C.Y. dump trailers, 2 mile round trip	3.37	2.96	6.33
6400	Twelve 20 C.Y. dump trailers, 4 mile round trip	4.30	3.71	8.01
6600	1/2 C.Y. dragline bucket, three 8 C.Y. dump trucks, 1 mile round trip	5.10	6.75	11.85
6700	Five 8 C.Y. dump trucks, 4 mile round trip	9.60	13.55	23.15
6800	Two 12 C.Y. dump trucks, 2 mile round trip	7.30	8.10	15.40
6900	Three 12 C.Y. dump trucks, 4 mile round trip	8.05	8.70	16.75
7000	Two 16 C.Y. dump trailers, 4 mile round trip	7.65	8.05	15.70
7200	3/4 C.Y. dragline bucket, four 8 C.Y. dump trucks, 1 mile round trip	5.45	7.20	12.65
7300	Six 8 C.Y. dump trucks, 3 mile round trip	7.60	10.75	18.35
7400	Four 12 C.Y. dump trucks, 4 mile round trip	7.30	8.25	15.55
7500	Two 16 C.Y. dump trailers, 2 mile round trip	5.70	6	11.70
7600	Three 16 C.Y. dump trailers, 4 mile round trip	6.60	6.65	13.25
7800	1-1/2 C.Y. dragline bucket, six 8 C.Y. dump trucks, 1 mile round trip	3.95	5.25	9.20
7900	Ten 8 C.Y. dump trucks, 3 mile round trip	7	9.40	16.40
8000	Six 12 C.Y. dump trucks, 2 mile round trip	5.35	5.55	10.90
8100	Eight 12 C.Y. dump trucks, 4 mile round trip	6.45	6.55	13
8200	Four 16 C.Y. dump trailers, 2 mile round trip	4.73	4.42	9.15
8300	Three 20 C.Y. dump trailers, 2 mile round trip	4.46	4.54	9
8400	3 C.Y. dragline bucket, eight 12 C.Y. dump trucks, 2 mile round trip	5.60	5.10	10.70
8500	Six 16 C.Y. dump trailers, 2 mile round trip	4.81	4.11	8.92
8600	Nine 16 C.Y. dump trailers, 4 mile round trip	5.80	4.97	10.77
8700	Eight 20 C.Y. dump trailers, 4 mile round trip	5.10	4.30	9.40

G1030 Site Earthwork

The Cut and Fill Common Earth System includes: moving common earth cut from an area above the specified grade to an area below the specified grade utilizing a bulldozer and/or scraper, with the addition of compaction equipment, plus a water wagon to adjust the moisture content of the soil.

The Expanded System Listing shows Cut and Fill operations with hauling distances that vary from 50' to 5000'. Lifts for compaction in the filled area vary from 4" to 8". There is no waste included in the assumptions.

System Components	QUANTITY	UNIT	COST PER C.Y.		
			EQUIP.	LABOR	TOTAL
SYSTEM G1030 115 1000					
EARTH CUT & FILL, 80 HP DOZER & COMPACTOR, 50' HAUL, 4" LIFT, 2 PASSES					
Excavating, bulk, dozer, 50' haul, common earth	1.000	C.Y.	1.59	3.44	5.03
Water wagon, rent per day	.008	Hr.	.51	.61	1.12
Backfill dozer, from existing stockpile, 75 H.P., 50' haul	1.310	C.Y.	.60	1.30	1.90
Compaction, roller, 4" lifts, 2 passes	1.000	C.Y.	.14	.48	.62
TOTAL			2.84	5.83	8.67

G1030 115	Cut & Fill Common Earth	COST PER C.Y.		
		EQUIP.	LABOR	TOTAL
1000	Earth cut & fill, 80 HP dozer & roller compact, 50' haul, 4" lift, 2 passes	2.85	5.85	8.70
1050	4 passes	2.52	5.30	7.80
1100	8" lift, 2 passes	2.38	4.81	7.20
1150	4 passes	2.52	5.30	7.80
1200	150' haul, 4" lift, 2 passes	4.57	9.35	13.90
1250	4 passes	4.82	10.20	15
1300	8" lift, 2 passes	4.57	9.35	13.90
1350	4 passes	4.82	10.20	15
1400	300' haul, 4" lift, 2 passes	8.20	16.85	25
1450	4 passes	8.60	18.30	27
1500	8" lift, 2 passes	8.20	16.85	25
1550	4 passes	8.60	18.30	27
1600	105 H.P. dozer and roller compactor, 50' haul, 4" lift, 2 passes	2.25	3.17	5.40
1650	4 passes	2.32	3.41	5.75
1700	8" lift, 2 passes	2.25	3.17	5.40
1750	4 passes	2.32	3.41	5.75
1800	150' haul, 4" lift, 2 passes	4.88	6.95	11.85
1850	4 passes	5.05	7.45	12.50
1900	8" lift, 2 passes	4.85	6.85	11.70
1950	4 passes	4.99	7.35	12.30
2000	300' haul, 4" lift, 2 passes	9.05	13.10	22
2050	4 passes	9.45	14.30	23.50
2100	8" lift, 2 passes	8.75	12.70	21.50
2150	4 passes	9.05	13.10	22
2200	200 H.P. dozer & roller compactor, 150' haul, 4" lift, 2 passes	5.60	3.66	9.30
2250	4 passes	5.70	3.90	9.60
2300	8" lift, 2 passes	5.60	3.54	9.10
2350	4 passes	5.60	3.66	9.30
2600	300' haul, 4" lift, 2 passes	9.90	6.60	16.55
2650	4 passes	10.05	7.10	17.15

G1030 Site Earthwork

G1030 115	Cut & Fill Common Earth	COST PER C.Y.		
		EQUIP.	LABOR	TOTAL
2700	8" lift, 2 passes	9.85	6.35	16.20
2750	4 passes	9.80	6.20	16
3000	300 H.P. dozer & roller compactor, 150' haul, 4" lifts, 2 passes	4.26	2.69	6.95
3050	4 passes	4.36	3.05	7.40
3100	8" lift, 2 passes	4.21	2.51	6.70
3150	4 passes	4.26	2.69	6.95
3200	300' haul, 4" lifts, 2 passes	7.30	4.28	11.60
3250	4 passes	7.40	4.64	12.05
3300	8" lifts, 2 passes	7.25	4.10	11.35
3350	4 passes	7.30	4.28	11.60
4200	10 C.Y. elevating scraper & roller compact, 1500' haul, 4" lifts, 2 passes	4.23	3.98	8.20
4250	4 passes	4.37	4.46	8.85
4300	8" lifts, 2 passes	4.16	3.74	7.90
4350	4 passes	4.23	3.98	8.20
4800	5000' haul, 4" lifts, 2 passes	5.55	5.05	10.60
4850	4 passes	5.70	5.55	11.20
4900	8" lifts, 2 passes	5.50	4.81	10.30
4950	4 passes	5.55	5.05	10.60
5000	15 C.Y. S.P. scraper & roller compact, 1500' haul, 4" lifts, 2 passes	5.35	3.28	8.60
5050	4 passes	5.40	3.58	9
5100	8" lifts, 2 passes	5.40	3.58	9
5150	4 passes	5.30	3.14	8.45
5400	5000' haul, 4" lifts, 2 passes	6.70	3.48	10.20
5450	4 passes	6.80	3.72	10.50
5500	8" lifts, 2 passes	7.30	4.10	11.40
5550	4 passes	7.30	4.04	11.35
5600	21 C.Y. S.P. scraper & roller compact, 1500' haul, 4" lifts, 2 passes	4.59	2.80	7.40
5650	4 passes	4.66	3.04	7.70
5700	8" lifts, 2 passes	4.79	2.95	7.75
5750	4 passes	4.82	3.07	7.90
6000	5000' haul, 4" lifts, 2 passes	6.55	3.55	10.10
6050	4 passes	6.65	3.91	10.55
6100	8" lifts, 2 passes	6.50	3.37	9.85
6150	4 passes	6.55	3.55	10.10

G1030 Site Earthwork

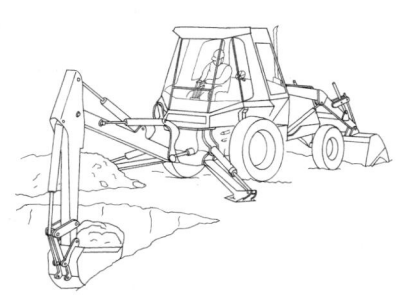

The Excavation of Common Earth System balances the productivity of the excavating equipment to the hauling equipment. It is assumed that the hauling equipment will encounter light traffic and will move up no considerable grades on the haul route. No mobilization cost is included. All costs given in these systems include a swell factor of 25% for hauling.

The Expanded System Listing shows Excavation systems using backhoes ranging from 1/2 Cubic Yard capacity to 3-1/2 Cubic Yards. Power shovels indicated range from 1/2 Cubic Yard to 3 Cubic Yards. Dragline bucket rigs range from 1/2 Cubic Yard to 3 Cubic Yards. Truck capacities range from 8 Cubic Yards to 20 Cubic Yards. Each system lists the number of trucks involved and the distance (round trip) that each must travel.

System Components	QUANTITY	UNIT	COST PER C.Y.		
			EQUIP.	LABOR	TOTAL
SYSTEM G1030 120 1000					
EXCAVATE COMMON EARTH, 1/2 CY BACKHOE, TWO 8 CY DUMP TRUCKS, 1 MRT					
Excavating, bulk hyd. backhoe wheel mtd., 1/2 C.Y.	.632	B.C.Y.	.99	2.50	3.49
Hauling, 8 CY truck, cycle 0.5 mile, 20 MPH, 15 min. wait/Ld./Uld.	1.280	L.C.Y.	2.47	3.38	5.85
Spotter at earth fill dump or in cut	.016	Hr.		1.06	1.06
TOTAL			3.46	6.94	10.40

G1030 120	Excavate and Haul Common Earth	COST PER C.Y.		
		EQUIP.	LABOR	TOTAL
1000	Excavate common earth, 1/2 C.Y. backhoe, two 8 C.Y. dump trucks, 1 MRT	3.46	6.95	10.41
1200	Three 8 C.Y. dump trucks, 3 mile round trip	7	11.95	18.95
1400	Two 12 C.Y. dump trucks, 4 mile round trip	6.70	9.10	15.80
1600	3/4 C.Y. backhoe, three 8 C.Y. dump trucks, 1 mile round trip	3.49	5.75	9.24
1700	Five 8 C.Y. dump trucks, 3 mile round trip	6.95	11.10	18.05
1800	Two 12 C.Y. dump trucks, 2 mile round trip	5.70	7.05	12.75
1900	Two 16 C.Y. dump trailers, 3 mile round trip	5.40	6	11.40
2000	Two 20 C.Y. dump trailers, 4 mile round trip	5.20	5.90	11.10
2200	1-1/2 C.Y. backhoe, eight 8 C.Y. dump trucks, 3 mile round trip	6.55	9.90	16.45
2300	Four 12 C.Y. dump trucks, 2 mile round trip	5.05	6.05	11.10
2400	Six 12 C.Y. dump trucks, 4 mile round trip	6.15	7	13.15
2500	Three 16 C.Y. dump trailers, 2 mile round trip	4.18	4.47	8.65
2600	Two 20 C.Y. dump trailers, 1 mile round trip	3.28	3.65	6.93
2700	Three 20 C.Y. dump trailers, 3 mile round trip	4.40	4.61	9.01
2800	2-1/2 C.Y. excavator, six 12 C.Y. dump trucks, 1 mile round trip	3.72	4.09	7.81
2900	Eight 12 C.Y. dump trucks, 3 mile round trip	5.35	5.75	11.10
3000	Four 16 C.Y. dump trailers, 1 mile round trip	3.75	3.72	7.47
3100	Six 16 C.Y. dump trailers, 3 mile round trip	5.05	5.05	10.10
3200	Six 20 C.Y. dump trailers, 4 mile round trip	4.69	4.67	9.36
3400	3-1/2 C.Y. backhoe, six 16 C.Y. dump trailers, 1 mile round trip	4.04	3.57	7.61
3600	Ten 16 C.Y. dump trailers, 4 mile round trip	5.75	5.10	10.85
3800	Eight 20 C.Y. dump trailers, 3 mile round trip	4.63	4.03	8.66
4000	1/2 C.Y. pwr. shovel, four 8 C.Y. dump trucks, 2 mile round trip	6.10	8.90	15
4100	Two 12 C.Y. dump trucks, 1 mile round trip	4.50	5.45	9.95
4200	Four 12 C.Y. dump trucks, 4 mile round trip	6.75	7.40	14.15
4300	Two 16 C.Y. dump trailers, 2 mile round trip	4.82	5.30	10.12
4400	Two 20 C.Y. dump trailers, 4 mile round trip	5.55	6.10	11.65
4800	3/4 C.Y. pwr. shovel, six 8 C.Y. dump trucks, 2 mile round trip	5.95	8.65	14.60
4900	Three 12 C.Y. dump trucks, 1 mile round trip	4.37	4.71	9.08
5000	Five 12 C.Y. dump trucks, 4 mile round trip	6.80	7.15	13.95
5100	Three 16 C.Y. dump trailers, 3 mile round trip	5.75	5.75	11.50
5200	Three 20 C.Y. dump trailers, 4 mile round trip	5.40	5.35	10.75

G1030 Site Earthwork

G1030 120	Excavate and Haul Common Earth	COST PER C.Y.		
		EQUIP.	LABOR	TOTAL
5400	1-1/2 C.Y. pwr. shovel, six 12 C.Y. dump trucks, 1 mile round trip	3.89	4.08	7.97
5500	Ten 12 C.Y. dump trucks, 4 mile round trip	6.35	6.50	12.85
5600	Six 16 C.Y. dump trailers, 3 mile round trip	5.25	5.05	10.30
5700	Four 20 C.Y. dump trailers, 2 mile round trip	3.85	3.85	7.70
5800	Six 20 C.Y. dump trailers, 4 mile round trip	4.88	4.66	9.54
6000	3 C.Y. pwr. shovel, ten 12 C.Y. dump trucks, 1 mile round trip	3.91	3.74	7.65
6200	Twelve 16 C.Y. dump trailers, 4 mile round trip	5.60	4.95	10.55
6400	Eight 20 C.Y. dump trailers, 2 mile round trip	3.73	3.26	6.99
6600	1/2 C.Y. dragline bucket, three 8 C.Y. dump trucks, 2 mile round trip	8.75	11.60	20.35
6700	Two 12 C.Y. dump trucks, 3 mile round trip	8.95	9.95	18.90
6800	Three 12 C.Y. dump trucks, 4 mile round trip	8.90	9.60	18.50
7000	Two 16 C.Y. dump trailers, 4 mile round trip	8.40	8.80	17.20
7200	3/4 C.Y. dragline bucket, four 8 C.Y. dump trucks, 1 mile round trip	4.81	6.40	11.21
7300	Six 8 C.Y. dump trucks, 3 mile round trip	8.40	11.85	20.25
7400	Two 12 C.Y. dump trucks, 1 mile round trip	6.10	6.80	12.90
7500	Four 12 C.Y. dump trucks, 4 mile round trip	8.10	8.55	16.65
7600	Two 16 C.Y. dump trailers, 2 mile round trip	6.30	6.65	12.95
7800	1-1/2 C.Y. dragline bucket, four 12 C.Y. dump trucks, 1 mile round trip	4.95	4.96	9.91
7900	Six 12 C.Y. dump trucks, 3 mile round trip	6.35	6.60	12.95
8000	Four 16 C.Y. dump trucks, 2 mile round trip	5.25	4.92	10.17
8200	Five 16 C.Y. dump trucks, 4 mile round trip	6.60	6.15	12.75
8400	3 C.Y. dragline bucket, eight 12 C.Y. dump trucks, 2 mile round trip	6.20	5.70	11.90
8500	Twelve 12 C.Y. dump trucks, 4 mile round trip	7.20	6.65	13.85
8600	Eight 16 C.Y. dump trailers, 3 mile round trip	6	5.15	11.15
8700	Six 20 C.Y. dump trailers, 2 mile round trip	4.52	3.78	8.30

G1030 Site Earthwork

The Cut and Fill Clay System includes: moving clay from an area above the specified grade to an area below the specified grade utilizing a bulldozer and compaction equipment; plus a water wagon to adjust the moisture content of the soil.

The Expanded System Listing shows Cut and Fill operations with hauling distances that vary from 50' to 5000'. Lifts for compaction in the filled areas vary from 4" to 8". There is no waste included in the assumptions.

System Components	QUANTITY	UNIT	COST PER C.Y.		
			EQUIP.	LABOR	TOTAL
SYSTEM G1030 125 1000					
CLAY CUT & FILL, 80 HP DOZER & COMPACTOR, 50' HAUL, 4" LIFTS, 2 PASSES					
Excavating , bulk, dozer, 50' haul, clay	1.000	C.Y.	1.79	3.87	5.66
Water wagon, rent per day	.004	Hr.	.26	.31	.57
Backfill dozer, from existing, stock pile, 75 H.P., 50' haul, clay	1.330	C.Y.	.70	1.52	2.22
Compaction, tamper, 4" lifts, 2 passes	245.000	S.F.	.62	.40	1.02
TOTAL			3.37	6.10	9.47

G1030 125	Cut & Fill Clay	COST PER C.Y.		
		EQUIP.	LABOR	TOTAL
1000	Clay cut & fill, 80 HP dozer & compactor, 50' haul, 4" lifts, 2 passes	3.37	6.10	9.47
1050	4 passes	4.16	6.75	10.91
1100	8" lifts, 2 passes	2.91	5.75	8.66
1150	4 passes	3.33	6.05	9.38
1200	150' haul, 4" lifts, 2 passes	5.85	11.50	17.35
1250	4 passes	6.65	12.15	18.80
1300	8" lifts, 2 passes	5.40	11.15	16.55
1350	4 passes	5.80	11.45	17.25
1400	300' haul, 4" lifts, 2 passes	9.75	20	29.75
1450	4 passes	10.55	20.50	31.05
1500	8" lifts, 2 passes	9.30	19.70	29
1550	4 passes	9.75	20	29.75
1600	105 HP dozer & sheepsfoot compactors, 50' haul, 4" lifts, 2 passes	3.56	4.39	7.95
1650	4 passes	4.34	5.05	9.39
1700	8" lifts, 2 passes	3.10	4.02	7.12
1750	4 passes	3.52	4.36	7.88
1800	150' haul, 4" lifts, 2 passes	6.75	8.75	15.50
1850	4 passes	7.50	9.40	16.90
1900	8" lifts, 2 passes	6.25	8.40	14.65
1950	4 passes	6.70	8.70	15.40
2000	300' haul, 4" lifts, 2 passes	10.45	13.90	24.35
2050	4 passes	11.25	14.55	25.80
2100	8" lifts, 2 passes	10	13.55	23.55
2150	4 passes	10.45	13.85	24.30
2200	200 HP dozer & sheepsfoot compactors, 150' haul, 4" lifts, 2 passes	8.30	5	13.30
2250	4 passes	9.10	5.65	14.75
2300	8" lifts, 2 passes	7.85	4.64	12.49
2350	4 passes	8.25	4.97	13.22
2600	300' haul, 4" lifts, 2 passes	14.10	8.35	22.45
2650	4 passes	14.90	9	23.90

G1030 Site Earthwork

G1030 125	Cut & Fill Clay	COST PER C.Y.		
		EQUIP.	LABOR	TOTAL
2700	8" lifts, 2 passes	13.65	8	21.65
2750	4 passes	14.05	8.35	22.40
3000	300 HP dozer & sheepsfoot compactors, 150' haul, 4" lifts, 2 passes	6.35	3.40	9.75
3050	4 passes	7.15	4.05	11.20
3100	8" lifts, 2 passes	5.90	3.04	8.94
3150	4 passes	6.30	3.37	9.67
3200	300' haul, 4" lifts, 2 passes	10.85	5.65	16.50
3250	4 passes	11.65	6.25	17.90
3300	8" lifts, 2 passes	10.40	5.25	15.65
3350	4 passes	10.80	5.60	16.40
4200	10 C.Y. elev. scraper & sheepsfoot rollers, 1500' haul, 4" lifts, 2 passes	10.55	7	17.55
4250	4 passes	10.80	7.35	18.15
4300	8" lifts, 2 passes	10.45	6.95	17.40
4350	4 passes	10.60	7.05	17.65
4800	5000' haul, 4" lifts, 2 passes	14.25	9.45	23.70
4850	4 passes	14.55	9.80	24.35
4900	8" lifts, 2 passes	14.20	9.40	23.60
4950	4 passes	14.35	9.45	23.80
5000	15 C.Y. SP scraper & sheepsfoot rollers, 1500' haul, 4" lifts, 2 passes	10.20	4.87	15.07
5050	4 passes	10.45	5.15	15.60
5100	8" lifts, 2 passes	10.10	4.75	14.85
5150	4 passes	10.25	4.87	15.12
5400	5000' haul, 4" lifts, 2 passes	14.55	6.85	21.40
5450	4 passes	14.85	7.20	22.05
5500	8" lifts, 2 passes	14.50	6.75	21.25
5550	4 passes	14.60	6.85	21.45
5600	21 C.Y. SP scraper & sheepsfoot rollers, 1500' haul, 4" lifts, 2 passes	8.40	3.84	12.24
5650	4 passes	8.65	4.15	12.80
5700	8" lifts, 2 passes	8.30	3.71	12.01
5750	4 passes	8.40	3.84	12.24
6000	5000' haul, 4" lifts, 2 passes	13.20	5.95	19.15
6050	4 passes	13.50	6.25	19.75
6100	8" lifts, 2 passes	13.15	5.85	19
6150	4 passes	13.25	5.95	19.20

G1030 Site Earthwork

The Excavation of Clay System balances the productivity of excavating equipment to hauling equipment. It is assumed that the hauling equipment will encounter light traffic and will move up no considerable grades on the haul route. No mobilization cost is included. All costs given in these systems include a swell factor of 40%.

The Expanded System Listing shows Excavation systems using backhoes ranging from 1/2 Cubic Yard capacity to 3-1/2 Cubic Yards. Power shovels indicated range from 1/2 Cubic Yard to 3 Cubic Yards. Dragline bucket rigs range from 1/2 Cubic Yard to 3 Cubic Yards. Truck capacities range from 8 Cubic Yards to 20 Cubic Yards. Each system lists the number of trucks involved and the distance (round trip) that each must travel.

System Components	QUANTITY	UNIT	COST PER C.Y.		
			EQUIP.	LABOR	TOTAL
SYSTEM G1030 130 1000					
EXCAVATE CLAY, 1/2 CY BACKHOE, TWO 8 CY DUMP TRUCKS, 2 MI ROUND TRIP					
Excavating bulk hyd. backhoe, wheel mtd. 1/2 C.Y.	1.000	B.C.Y.	1.57	3.95	5.52
Haul earth, 8 C.Y. dump truck, 2 mile round trip, 2.6 loads/hr	1.350	L.C.Y.	5.05	6.89	11.94
Spotter at earth fill dump or in cut	.023	Hr.		1.53	1.53
TOTAL			6.62	12.37	18.99

G1030 130	Excavate and Haul Clay	COST PER C.Y.		
		EQUIP.	LABOR	TOTAL
1000	Excavate clay, 1/2 C.Y. backhoe, two 8 C.Y. dump trucks, 2 mile round trip	6.60	12.35	18.95
1200	Three 8 C.Y. dump trucks, 4 mile round trip	8.90	15.20	24.10
1400	Two 12 C.Y. dump trucks, 4 mile round trip	7.15	9.85	17
1600	3/4 C.Y. backhoe, three 8 C.Y. dump trucks, 2 mile round trip	6.50	10.25	16.75
1700	Five 8 C.Y. dump trucks, 4 mile round trip	8.80	13.25	22.05
1800	Two 12 C.Y. dump trucks, 2 mile round trip	6.05	7.55	13.60
1900	Three 12 C.Y. dump trucks, 4 mile round trip	7.15	8.45	15.60
2000	1-1/2 C.Y. backhoe, eight 8 C.Y. dump trucks, 3 mile round trip	6.95	10.15	17.10
2100	Three 12 C.Y. dump trucks, 1 mile round trip	4.08	4.84	8.92
2200	Six 12 C.Y. dump trucks, 4 mile round trip	6.55	7.45	14
2300	Three 16 C.Y. dump trailers, 2 mile round trip	4.44	4.82	9.26
2400	Four 16 C.Y. dump trailers, 4 mile round trip	5.80	6	11.80
2600	Two 20 C.Y. dump trailers, 1 mile round trip	3.50	3.96	7.46
2800	2-1/2 C.Y. backhoe, eight 12 C.Y. dump trucks, 3 mile round trip	5.70	6.15	11.85
2900	Four 16 C.Y. dump trailers, 1 mile round trip	3.98	3.95	7.93
3000	Six 16 C.Y. dump trailers, 3 mile round trip	5.35	5.45	10.80
3100	Four 20 C.Y. dump trailers, 2 mile round trip	3.81	4.01	7.82
3200	Six 20 C.Y. dump trailers, 4 mile round trip	4.99	5	9.99
3400	3-1/2 C.Y. backhoe, ten 12 C.Y. dump trucks, 2 mile round trip	5.70	5.40	11.10
3500	Six 16 C.Y. dump trailers, 1 mile round trip	4.32	3.77	8.09
3600	Ten 16 C.Y. dump trailers, 4 mile round trip	6.10	5.35	11.45
3800	Eight 20 C.Y. dump trailers, 3 mile round trip	4.93	4.26	9.19
4000	1/2 C.Y. pwr. shovel, three 8 C.Y. dump trucks, 1 mile round trip	5.95	8.75	14.70
4100	Six 8 C.Y. dump trucks, 4 mile round trip	9	13.65	22.65
4200	Two 12 C.Y. dump trucks, 1 mile round trip	4.84	5.85	10.69
4400	Three 12 C.Y. dump trucks, 3 mile round trip	6.45	7.35	13.80
4600	3/4 C.Y. pwr. shovel, six 8 C.Y. dump trucks, 2 mile round trip	6.35	9.25	15.60
4700	Nine 8 C.Y. dump trucks, 4 mile round trip	8.80	12.60	21.40
4800	Three 12 C.Y. dump trucks, 1 mile round trip	4.68	5.10	9.78
4900	Four 12 C.Y. dump trucks, 3 mile round trip	6.45	6.90	13.35
5000	Two 16 C.Y. dump trailers, 1 mile round trip	4.80	4.99	9.79
5200	Three 16 C.Y. dump trailers, 3 mile round trip	6.15	6.20	12.35

G1030 Site Earthwork

G1030 130	Excavate and Haul Clay	COST PER C.Y.		
		EQUIP.	LABOR	TOTAL
5400	1-1/2 C.Y. pwr. shovel, six 12 C.Y. dump trucks, 2 mile round trip	5.65	6	11.65
5500	Four 16 C.Y. dump trailers, 1 mile round trip	4.20	3.93	8.13
5600	Seven 16 C.Y. dump trailers, 4 mile round trip	5.95	5.75	11.70
5800	Five 20 C.Y. dump trailers, 3 mile round trip	4.88	4.46	9.34
6000	3 C.Y. pwr. shovel, nine 16 C.Y. dump trailers, 2 mile round trip	4.59	4.02	8.61
6100	Twelve 16 C.Y. dump trailers, 4 mile round trip	5.95	5.25	11.20
6200	Six 20 C.Y. dump trailers, 1 mile round trip	3.65	3.20	6.85
6400	Nine 20 C.Y. dump trailers, 3 mile round trip	4.83	4.22	9.05
6600	1/2 C.Y. dragline bucket, two 8 C.Y. dump trucks, 1 mile round trip	9.70	12.85	22.55
6800	Four 8 C.Y. dump trucks, 4 mile round trip	12.35	16.45	28.80
7000	Two 12 C.Y. dump trucks, 3 mile round trip	9.65	10.75	20.40
7200	3/4 C.Y. dragline bucket, four 8 C.Y. dump trucks, 2 mile round trip	8.35	11.15	19.50
7300	Six 8 C.Y. dump trucks, 4 mile round trip	10.75	15.20	25.95
7400	Two 12 C.Y. dump trucks, 2 mile round trip	8.70	9.70	18.40
7500	Three 12 C.Y. dump trucks, 4 mile round trip	9.55	10.35	19.90
7600	Two 16 C.Y. dump trailers, 3 mile round trip	8.25	8.70	16.95
7800	1-1/2 C.Y. dragline bucket, five 12 C.Y. dump trucks, 3 mile round trip	7.30	7.25	14.55
7900	Three 16 C.Y. dump trailers, 1 mile round trip	5.30	4.99	10.29
8000	Five 16 C.Y. dump trailers, 4 mile round trip	7.10	6.55	13.65
8100	Three 20 C.Y. dump trailers, 2 mile round trip	5.30	5	10.30
8400	3 C.Y. dragline bucket, eight 16 C.Y. dump trailers, 3 mile round trip	6.20	5.30	11.50
8500	Six 20 C.Y. dump trailers, 2 mile round trip	4.86	4.09	8.95
8600	Eight 20 C.Y. dump trailers, 4 mile round trip	6.05	5.10	11.15

G1030 Site Earthwork

The Loading and Hauling of Rock System balances the productivity of loading equipment to hauling equipment. It is assumed that the hauling equipment will encounter light traffic and will move up no considerable grades on the haul route.

The Expanded System Listing shows Loading and Hauling systems that use either a track or wheel front-end loader. Track loaders indicated range from 1-1/2 Cubic Yards capacity to 4-1/2 Cubic Yards capacity. Wheel loaders range from 1-1/2 Cubic Yards to 5 Cubic Yards. Trucks for hauling range from 8 Cubic Yards capacity to 20 Cubic Yards capacity. Each system lists the number of trucks involved and the distance (round trip) that each must travel.

System Components	QUANTITY	UNIT	COST PER C.Y. EQUIP.	LABOR	TOTAL
SYSTEM G1030 150 1000					
LOAD & HAUL ROCK, 1-1/2 C.Y. TRACK LOADER, SIX 8 C.Y. TRUCKS, 1 MRT					
Excavating bulk, F.E. loader, track mtd., 1.5 C.Y.	1.000	B.C.Y.	1.12	1.73	2.85
8 C.Y. truck, cycle 2 miles	1.650	L.C.Y.	5.45	7.44	12.89
Spotter at earth fill dump or in cut	.014	Hr.		.93	.93
TOTAL			6.57	10.10	16.67

G1030 150	Load & Haul Rock	COST PER C.Y. EQUIP.	LABOR	TOTAL
1000	Load & haul rock, 1-1/2 C.Y. track loader, six 8 C.Y. trucks, 1 MRT	6.55	10.10	16.65
1200	Nine 8 C.Y. dump trucks, 3 mile round trip	8.85	13.30	22.15
1400	Six 12 C.Y. dump trucks, 4 mile round trip	8.45	9.70	18.15
1600	Three 16 C.Y. dump trucks, 2 mile round trip	5.90	6.55	12.45
2000	2-1/2 C.Y. track loader, twelve 8 C.Y. dump trucks, 3 mile round trip	9.10	12.60	21.70
2200	Five 12 C.Y. dump trucks, 1 mile round trip	5.45	5.55	11
2400	Eight 12 C.Y. dump trucks, 4 mile round trip	8.70	9.05	17.75
2600	Four 16 C.Y. dump trailers, 2 mile round trip	6.20	5.90	12.10
3000	3-1/2 C.Y. track loader, eight 12 C.Y. dump trucks, 2 mile round trip	7.05	7.10	14.15
3200	Five 16 C.Y. dump trucks, 1 mile round trip	5.35	4.79	10.14
3400	Seven 16 C.Y. dump trailers, 3 mile round trip	7.10	6.60	13.70
3600	Seven 20 C.Y. dump trailers, 4 mile round trip	6.60	6.10	12.70
4000	4-1/2 C.Y. track loader, nine 12 C.Y. dump trucks, 1 mile round trip	5.30	5.10	10.40
4200	Eight 16 C.Y. dump trailers, 2 mile round trip	5.85	5.20	11.05
4400	Eleven 16 C.Y. dump trailers, 4 mile round trip	7.50	6.65	14.15
4600	Seven 20 C.Y. dump trailers, 2 mile round trip	5.10	4.50	9.60
5000	1-1/2 C.Y. wheel loader, nine 8 C.Y. dump trucks, 2 mile round trip	6.90	10.65	17.55
5200	Four 12 C.Y. dump trucks, 1 mile round trip	4.95	5.95	10.90
5400	Seven 12 C.Y. dump trucks, 4 mile round trip	8.05	9.30	17.35
5600	Five 16 C.Y. dump trailers, 4 mile round trip	7.10	7.45	14.55
6000	3 C.Y. wheel loader, eight 12 C.Y. dump trucks, 2 mile round trip	6.35	6.85	13.20
6200	Five 16 C.Y. dump trailers, 1 mile round trip	4.61	4.52	9.13
6400	Seven 16 C.Y. dump trailers, 3 mile round trip	5.50	5.60	11.10
6600	Seven 20 C.Y. dump trailers, 4 mile round trip	5.85	5.80	11.65
7000	5 C.Y. wheel loader, twelve 12 C.Y. dump trucks, 1 mile round trip	4.74	4.79	9.53
7200	Nine 16 C.Y. dump trailers, 1 mile round trip	4.69	4.38	9.07
7400	Eight 20 C.Y. dump trailers, 1 mile round trip	4.03	3.73	7.76
7600	Twelve 20 C.Y. dump trailers, 3 mile round trip	5.45	4.98	10.43

G1030 Site Earthwork

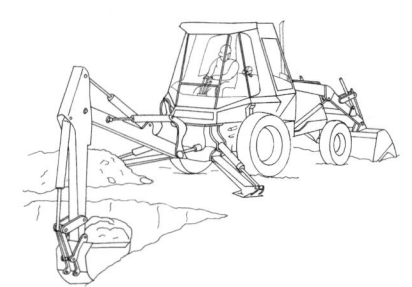

The Excavation of Sandy Clay/Loam System balances the productivity of the excavating equipment to the hauling equipment. It is assumed that the hauling equipment will encounter light traffic and will move up no considerable grades on the haul route. No mobilization cost is included. All costs given in these systems include a swell factor of 25% for hauling.

The Expanded System Listing shows Excavation systems using backhoes ranging from 1/2 Cubic Yard capacity to 3-1/2 Cubic Yards. Truck capacities range from 8 Cubic Yards to 20 Cubic Yards. Each system lists the number of trucks involved and the distance (round trip) that each must travel.

System Components	QUANTITY	UNIT	COST PER C.Y. EQUIP.	LABOR	TOTAL
SYSTEM G1030 160 1000					
EXCAVATE SANDY CLAY/LOAM, 1/2 CY BACKHOE, TWO 8 CY DUMP TRUCKS, 1 MRT					
Excavating, bulk hyd. backhoe wheel mtd., 1/2 C.Y.	1.000	B.C.Y.	1.43	3.59	5.02
8 C.Y. truck, cycle 2 miles	1.300	L.C.Y.	4.29	5.86	10.15
Spotter at earth fill dump or in cut	.016	Hr.		1.06	1.06
TOTAL			5.72	10.51	16.23

G1030 160	Excavate and Haul Sandy Clay/Loam	COST PER C.Y. EQUIP.	LABOR	TOTAL
1000	Excavate sandy clay/loam, 1/2 C.Y. backhoe, two 8 C.Y. dump trucks, 1 MRT	5.70	10.50	16.25
1200	Three 8 C.Y. dump trucks, 3 mile round trip	7	11.85	18.85
1400	Two 12 C.Y. dump trucks, 4 mile round trip	6.70	8.95	15.65
1600	3/4 C.Y. backhoe, three 8 C.Y. dump trucks, 1 mile round trip	5.20	8.10	13.30
1700	Five 8 C.Y. dump trucks, 3 mile round trip	6.95	11.10	18.05
1800	Two 12 C.Y. dump trucks, 2 mile round trip	5.65	6.95	12.55
1900	Two 16 C.Y. dump trailers, 3 mile round trip	5.40	5.90	11.30
2000	Two 20 C.Y. dump trailers, 4 mile round trip	5.15	5.80	10.95
2200	1-1/2 C.Y. excavator, eight 8 C.Y. dump trucks, 3 mile round trip	6.55	9.90	16.45
2300	Four 12 C.Y. dump trucks, 2 mile round trip	5.05	5.95	11.05
2400	Six 12 C.Y. dump trucks, 4 mile round trip	6.15	6.95	13.15
2500	Three 16 C.Y. dump trailers, 2 mile round trip	4.14	4.37	8.50
2600	Two 20 C.Y. dump trailers, 1 mile round trip	3.22	3.53	6.75
2700	Three 20 C.Y. dump trailers, 3 mile round trip	4.36	4.50	8.85
2800	2-1/2 C.Y. excavator, six 12 C.Y. dump trucks, 1 mile round trip	3.61	3.95	7.55
2900	Eight 12 C.Y. dump trucks, 3 mile round trip	5.25	5.65	10.90
3000	Four 16 C.Y. dump trailers, 1 mile round trip	3.61	3.54	7.15
3100	Six 16 C.Y. dump trailers, 3 mile round trip	4.95	4.92	9.85
3200	Six 20 C.Y. dump trailers, 4 mile round trip	4.58	4.51	9.10
3400	3-1/2 C.Y. excavator, six 16 C.Y. dump trailers, 1 mile round trip	3.88	3.49	7.35
3600	Ten 16 C.Y. dump trailers, 4 mile round trip	5.60	5.05	10.65
3800	Eight 20 C.Y. dump trailers, 3 mile round trip	4.48	3.97	8.45

G1030 Site Earthwork

The Gravel Backfilling System includes: a bulldozer to place and level backfill in specified lifts; compaction equipment; and a water wagon for adjusting moisture content.

The Expanded System Listing shows Gravel Backfilling operations with bulldozers ranging from 75 H.P. to 300 H.P. The maximum hauling distance ranges from 50' to 300'. Lifts for the compaction range from 6" to 12". There is no waste included in the assumptions.

System Components	QUANTITY	UNIT	COST PER C.Y.		
			EQUIP.	LABOR	TOTAL
SYSTEM G1030 205 1000					
GRAVEL BACKFILL, 80 HP DOZER & COMPACTORS, 50' HAUL, 6" LIFTS, 2 PASSES					
Backfilling, dozer, 80 H.P., 50' haul, sand and gravel, from stockpile	1.000	L.C.Y.	.41	.88	1.29
Water wagon rent per day	.003	Hr.	.19	.23	.42
Compaction, vibrating roller, 6" lifts, 2 passes	1.000	Hr.	.18	2.65	2.83
TOTAL			.78	3.76	4.54

G1030 205	Gravel Backfill	COST PER C.Y.		
		EQUIP.	LABOR	TOTAL
1000	Gravel backfill, 80 HP dozer & compactors, 50' haul, 6"lifts, 2 passes	.78	3.76	4.54
1050	4 passes	.98	5.05	6.03
1100	12" lifts, 2 passes	.80	1.91	2.71
1150	4 passes	1.25	3	4.25
1200	150' haul, 6" lifts, 2 passes	1.18	4.64	5.82
1250	4 passes	1.38	5.95	7.33
1300	12" lifts, 2 passes	1.05	4.49	5.54
1350	4 passes	1.25	5.80	7.05
1400	300' haul, 6" lifts, 2 passes	1.58	5.50	7.08
1450	4 passes	1.78	6.80	8.58
1500	12" lifts, 2 passes	1.60	3.65	5.25
1550	4 passes	2.05	4.74	6.79
1600	105 HP dozer & vibrating compactors, 50' haul, 6" lifts, 2 passes	.89	3.60	4.49
1650	4 passes	1.09	4.89	5.98
1700	12" lifts, 2 passes	.91	1.75	2.66
1750	4 passes	1.36	2.84	4.20
1800	150' haul, 6" lifts, 2 passes	1.42	4.33	5.75
1850	4 passes	1.62	5.60	7.22
1900	12" lifts, 2 passes	1.44	2.48	3.92
1950	4 passes	1.89	3.57	5.46
2000	300' haul, 6" lifts, 2 passes	1.89	4.96	6.85
2050	4 passes	2.09	6.25	8.34
2100	12" lifts, 2 passes	1.91	3.11	5.02
2150	4 passes	2.36	4.20	6.56
2200	200 HP dozer & roller compactors, 150' haul, 6" lifts, 2 passes	1.76	1.34	3.10
2250	4 passes	2.02	1.68	3.70
2300	12" lifts, 2 passes	1.54	1.06	2.60
2350	4 passes	1.80	1.39	3.19
2600	300' haul, 6" lifts, 2 passes	2.48	1.75	4.23
2650	4 passes	2.74	2.09	4.83
2700	12" lifts, 2 passes	2.26	1.47	3.73
2750	4 passes	2.52	1.80	4.32

G1030 Site Earthwork

G1030 205	Gravel Backfill	COST PER C.Y.		
		EQUIP.	LABOR	TOTAL
3000	300 HP dozer & roller compactors, 150' haul, 6" lifts, 2 passes	1.29	.99	2.28
3050	4 passes	1.55	1.33	2.88
3100	12" lifts, 2 passes	1.07	.71	1.78
3150	4 passes	1.33	1.04	2.37
3200	300' haul, 6" lifts, 2 passes	1.71	1.20	2.91
3250	4 passes	1.97	1.54	3.51
3300	12" lifts, 2 passes	1.49	.92	2.41
3350	4 passes	1.75	1.25	3

G1030 Site Earthwork

The Common Earth Backfilling System includes: a bulldozer to place and level backfill in specified lifts; compaction equipment; and a water wagon for adjusting moisture content.

The Expanded System Listing shows Common Earth Backfilling with bulldozers ranging from 80 H.P. to 300 H.P. The maximum distance ranges from 50' to 300'. Lifts for the compaction range from 4" to 8". There is no waste included in the assumptions.

System Components	QUANTITY	UNIT	COST PER C.Y. EQUIP.	LABOR	TOTAL
SYSTEM G1030 210 1000					
EARTH BACKFILL, 80 HP DOZER & ROLLER , 50′ HAUL, 4″ LIFTS, 2 PASSES					
Backfilling, dozer 80 H.P. 50′ haul, common earth, from stockpile	1.000	L.C.Y.	.46	.99	1.45
Water wagon, rent per day	.004	Hr.	.26	.31	.57
Compaction, roller, 4″ lifts, 2 passes	.035	Hr.	1.24	4.24	5.48
TOTAL			1.96	5.54	7.50

G1030 210	Common Earth Backfill	COST PER C.Y. EQUIP.	LABOR	TOTAL
1000	Earth backfill, 80 HP dozer & roller compactors, 50′ haul,4″ lifts,2 passes	1.96	5.55	7.51
1050	4 passes	3.20	9.75	12.95
1100	8″ lifts, 2 passes	1.36	3.47	4.83
1150	4 passes	1.96	5.55	7.51
1200	150′ haul, 4″ lifts, 2 passes	2.41	6.50	8.91
1250	4 passes	3.65	10.75	14.40
1300	8″ lifts, 2 passes	1.81	4.46	6.27
1350	4 passes	2.41	6.50	8.91
1400	300′ haul, 4″ lifts, 2 passes	2.85	7.45	10.30
1450	4 passes	4.09	11.70	15.79
1500	8″ lifts, 2 passes	2.25	5.40	7.65
1550	4 passes	2.85	7.45	10.30
1600	105 HP dozer & roller compactors, 50′ haul, 4″ lifts, 2 passes	2.08	5.35	7.43
1650	4 passes	3.32	9.55	12.87
1700	8″ lifts, 2 passes	1.48	3.27	4.75
1750	4 passes	2.08	5.35	7.43
1800	150′ haul, 4″ lifts, 2 passes	2.66	6.15	8.81
1850	4 passes	3.90	10.35	14.25
1900	8″ lifts, 2 passes	2.06	4.07	6.13
1950	4 passes	2.66	6.15	8.81
2000	300′ haul, 4″ lifts, 2 passes	3.20	6.85	10.05
2050	4 passes	4.44	11.10	15.54
2100	8″ lifts, 2 passes	2.60	4.81	7.41
2150	4 passes	3.20	6.85	10.05
2200	200 HP dozer & roller compactors, 150′ haul, 4″ lifts, 2 passes	2.46	3.61	6.07
2250	4 passes	3.14	6.05	9.19
2300	8″ lifts, 2 passes	2.12	2.40	4.52
2350	4 passes	2.46	3.61	6.07
2600	300′ haul, 4″ lifts, 2 passes	3.21	4.05	7.26
2650	4 passes	3.89	6.45	10.34
2700	8″ lifts, 2 passes	2.87	2.84	5.71
2750	4 passes	3.21	4.05	7.26
3000	300 HP dozer & roller compactors, 150′ haul, 4″ lift, 2 passes	1.95	3.23	5.18
3050	4 passes	2.63	5.65	8.28

G1030 Site Earthwork

G1030 210	Common Earth Backfill	COST PER C.Y.		
		EQUIP.	LABOR	TOTAL
3100	8″ lifts, 2 passes	1.61	2.02	3.63
3150	4 passes	1.95	3.23	5.18
3200	300′ haul, 4″ lifts, 2 passes	2.39	3.45	5.84
3250	4 passes	3.07	5.85	8.92
3300	8″ lifts, 2 passes	2.05	2.24	4.29
3350	4 passes	2.39	3.45	5.84

G1030 Site Earthwork

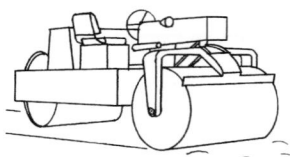

The Clay Backfilling System includes: a bulldozer to place and level backfill in specified lifts; compaction equipment; and a water wagon for adjusting moisture content.

The Expanded System Listing shows Clay Backfilling with bulldozers ranging from 75 H.P. to 300 H.P. The maximum distance ranges from 50′ to 300′. Lifts for the compaction range from 4″ to 8″. There is no waste included in the assumptions.

System Components	QUANTITY	UNIT	COST PER C.Y.		
			EQUIP.	LABOR	TOTAL
SYSTEM G1030 215 1000					
CLAY BACKFILL, 80 HP DOZER & TAMPER, 50″ HAUL, 4″ LIFTS, 2 PASSES					
Backfilling, dozer 80 H.P. 50′ haul, clay from stockpile	1.000	L.C.Y.	.53	1.14	1.67
Water wagon rent per day	.004	Hr.	.26	.31	.57
Compaction, tamper, 4″ lifts, 2 passes	1.000	Hr.	.15	1.36	1.51
TOTAL			.94	2.81	3.75

G1030 215	Clay Backfill	COST PER C.Y.		
		EQUIP.	LABOR	TOTAL
1000	Clay backfill, 80 HP dozer & tamper compactor, 50′ haul, 4″ lifts, 2 passes	.94	2.81	3.75
1050	4 passes	1.10	4.17	5.27
1100	8″ lifts, 2 passes	.87	2.13	3
1150	4 passes	.94	2.81	3.75
1200	150′ haul, 4″ lifts, 2 passes	1.46	3.95	5.41
1250	4 passes	1.62	5.30	6.92
1300	8″ lifts, 2 passes	1.39	3.27	4.66
1350	4 passes	1.46	3.95	5.41
1400	300′ haul, 4″ lifts, 2 passes	1.95	5	6.95
1450	4 passes	2.11	6.35	8.46
1500	8″ lifts, 2 passes	1.88	4.33	6.21
1550	4 Passes	1.95	5	6.95
1600	105 HP dozer & tamper compactors, 50′ haul, 4″ lifts, 2 passes	1.05	2.55	3.60
1650	4 passes	1.21	3.91	5.12
1700	8″ lifts, 2 passes	.98	1.87	2.85
1750	4 passes	1.05	2.55	3.60
1800	150′ haul, 4″ lifts, 2 passes	1.69	3.43	5.12
1850	4 passes	1.85	4.79	6.64
1900	8″ lifts, 2 passes	1.62	2.75	4.37
1950	4 passes	1.69	3.43	5.12
2000	300′ haul, 4″ lifts, 2 passes	2.32	4.29	6.61
2050	4 passes	2.48	5.65	8.13
2100	8″ lifts, 2 passes	2.25	3.61	5.86
2150	4 passes	2.32	4.29	6.61
2200	200 HP dozer & sheepsfoot compactors, 150′ haul, 5″ lifts, 2 passes	2.59	1.70	4.29
2250	4 passes	3.12	2.04	5.16
2300	8″ lifts, 2 passes	2.26	1.49	3.75
2350	4 passes	2.55	1.67	4.22
2600	300′ haul, 4″ lifts, 2 passes	3.41	2.18	5.59
2650	4 passes	3.94	2.52	6.46
2700	8″ lifts, 2 passes	3.08	1.97	5.05
2750	4 passes	3.37	2.15	5.52

G1030 Site Earthwork

G1030 215	Clay Backfill	COST PER C.Y.		
		EQUIP.	LABOR	TOTAL
3000	300 HP dozer & sheepsfoot compactors, 150' haul, 4" lifts, 2 passes	2.04	1.28	3.32
3050	4 passes	2.57	1.62	4.19
3100	8" lifts, 2 passes	1.71	1.07	2.78
3150	4 passes	2	1.25	3.25
3200	300' haul, 4" lifts, 2 passes	2.48	1.50	3.98
3250	4 passes	3.01	1.84	4.85
3300	8" lifts, 2 passes	2.15	1.29	3.44
3350	4 passes	2.44	1.47	3.91

G1030   Site Earthwork

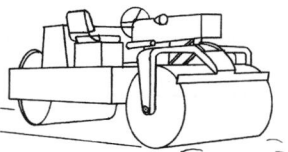

The Sandy Clay/Loam Backfilling System includes: a bulldozer to place and level backfill in specified lifts; compaction equipment; and a water wagon for adjusting moisture content.

The Expanded System Listing shows Common Earth Backfilling with bulldozers ranging from 80 H.P. to 300 H.P. The maximum distance ranges from 50′ to 300′. Lifts for the compaction range from 4″ to 8″. There is no waste included in the assumptions.

System Components	QUANTITY	UNIT	COST PER C.Y.		
			EQUIP.	LABOR	TOTAL
SYSTEM G1030 220 1000					
SANDY CLAY BACKFILL, 80 HP DOZER & ROLLER , 50′ HAUL, 4″LIFTS, 2 PASSES					
Backfilling, dozer 80 H.P. 50′ haul, common earth, from stockpile	1.000	L.C.Y.	.42	.90	1.32
Water wagon, rent per day	.004	Hr.	.26	.31	.57
Compaction, roller, 4″ lifts, 2 passes	.035	Hr.	1.24	4.24	5.48
TOTAL			1.92	5.45	7.37

G1030 220	Sandy Clay/Loam Backfill	COST PER C.Y.		
		EQUIP.	LABOR	TOTAL
0900	Backfill Sandy Clay/Loam using dozer and compactors			
1000	Backfill, 80 HP dozer & roller compactors, 50′ haul, 4″ lifts, 2 passes	1.92	5.45	7.35
1050	4 passes	3.16	9.70	12.85
1100	8″ lifts, 2 passes	1.32	3.38	4.70
1150	4 passes	1.92	5.45	7.35
1200	150′ haul, 4″ lifts, 2 passes	2.33	6.35	8.70
1250	4 passes	3.57	10.60	14.15
1300	8″ lifts, 2 passes	1.73	4.29	6
1350	4 passes	2.33	6.35	8.70
1400	300′ haul, 4″ lifts, 2 passes	2.74	7.25	9.95
1450	4 passes	3.98	11.45	15.45
1500	8″ lifts, 2 passes	2.14	5.15	7.30
1550	4 passes	2.74	7.25	9.95
1600	105 HP dozer & roller compactors, 50′ haul, 4″ lifts, 2 passes	2.03	5.25	7.30
1650	4 passes	3.27	9.50	12.80
1700	8″ lifts, 2 passes	1.43	3.21	4.64
1750	4 passes	2.03	5.25	7.30
1800	150′ haul, 4″ lifts, 2 passes	2.58	6	8.60
1850	4 passes	3.82	10.25	14.10
1900	8″ lifts, 2 passes	1.98	3.96	5.95
1950	4 passes	2.58	6	8.60
2000	300′ haul, 4″ lifts, 2 passes	3.05	6.65	9.70
2050	4 passes	4.29	10.90	15.20
2100	8″ lifts, 2 passes	2.45	4.61	7.05
2150	4 passes	3.05	6.65	9.70
2200	200 HP dozer & roller compactors, 150′ haul, 4″ lifts, 2 passes	2.33	3.54	5.85
2250	4 passes	3.01	5.95	8.95
2300	8″ lifts, 2 passes	1.99	2.33	4.31
2350	4 passes	2.33	3.54	5.85
2600	300′ haul, 4″ lifts, 2 passes	3.06	3.96	7
2650	4 passes	3.74	6.40	10.10
2700	8″ lifts, 2 passes	2.72	2.75	5.45

G1030 Site Earthwork

G1030 220	Sandy Clay/Loam Backfill	COST PER C.Y.		
		EQUIP.	LABOR	TOTAL
2750	4 passes	3.06	3.96	7
3000	300 HP dozer & roller compactors, 150' haul, 4" lifts, 2 passes	1.85	3.18	5
3050	4 passes	2.53	5.60	8.10
3100	8" lifts, 2 passes	1.51	1.97	3.47
3150	4 passes	1.85	3.18	5
3200	300' haul, 4" lifts, 2 passes	2.28	3.39	5.65
3250	4 passes	2.96	5.80	8.75
3300	8" lifts, 2 passes	1.94	2.18	4.11
3350	4 passes	2.28	3.39	5.65

For customer support on your Site Work & Landscape Costs with RSMeans Data, call 800.448.8182.

G1030 Site Earthwork

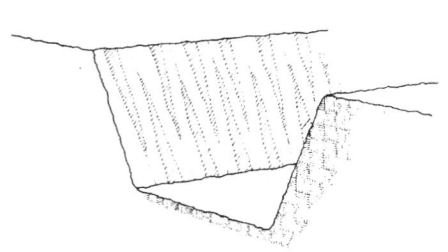

Trenching Systems are shown on a cost per linear foot basis. The systems include: excavation; backfill and removal of spoil; and compaction for various depths and trench bottom widths. The backfill has been reduced to accommodate a pipe of suitable diameter and bedding.

The slope for trench sides varies from none to 1:1.

The Expanded System Listing shows Trenching Systems that range from 2' to 12' in width. Depths range from 2' to 25'.

System Components	QUANTITY	UNIT	COST PER L.F.		
			EQUIP.	LABOR	TOTAL
SYSTEM G1030 805 1310					
TRENCHING, COMMON EARTH, NO SLOPE, 2' WIDE, 2' DP, 3/8 C.Y. BUCKET					
Excavation, trench, hyd. backhoe, track mtd., 3/8 C.Y. bucket	.148	B.C.Y.	.24	1.21	1.45
Backfill and load spoil, from stockpile	.153	L.C.Y.	.13	.37	.50
Compaction by vibrating plate, 6" lifts, 4 passes	.118	E.C.Y.	.03	.45	.48
Remove excess spoil, 8 C.Y. dump truck, 2 mile roundtrip	.040	L.C.Y.	.15	.20	.35
TOTAL			.55	2.23	2.78

G1030 805	Trenching Common Earth	COST PER L.F.		
		EQUIP.	LABOR	TOTAL
1310	Trenching, common earth, no slope, 2' wide, 2' deep, 3/8 C.Y. bucket	.54	2.24	2.78
1320	3' deep, 3/8 C.Y. bucket	.76	3.36	4.12
1330	4' deep, 3/8 C.Y. bucket	.98	4.48	5.46
1340	6' deep, 3/8 C.Y. bucket	1.28	5.80	7.08
1350	8' deep, 1/2 C.Y. bucket	1.66	7.70	9.36
1360	10' deep, 1 C.Y. bucket	3.61	9.20	12.81
1400	4' wide, 2' deep, 3/8 C.Y. bucket	1.28	4.45	5.73
1410	3' deep, 3/8 C.Y. bucket	1.71	6.70	8.41
1420	4' deep, 1/2 C.Y. bucket	1.94	7.45	9.39
1430	6' deep, 1/2 C.Y. bucket	3.20	12	15.20
1440	8' deep, 1/2 C.Y. bucket	6.75	15.50	22.25
1450	10' deep, 1 C.Y. bucket	8.20	19.25	27.45
1460	12' deep, 1 C.Y. bucket	10.55	24.50	35.05
1470	15' deep, 1-1/2 C.Y. bucket	7.70	22	29.70
1480	18' deep, 2-1/2 C.Y. bucket	13	30.50	43.50
1520	6' wide, 6' deep, 5/8 C.Y. bucket w/trench box	9.05	17.95	27
1530	8' deep, 3/4 C.Y. bucket	11.75	23.50	35.25
1540	10' deep, 1 C.Y. bucket	11.70	24.50	36.20
1550	12' deep, 1-1/2 C.Y. bucket	10.40	26	36.40
1560	16' deep, 2-1/2 C.Y. bucket	16.60	32.50	49.10
1570	20' deep, 3-1/2 C.Y. bucket	20.50	39	59.50
1580	24' deep, 3-1/2 C.Y. bucket	24.50	47	71.50
1640	8' wide, 12' deep, 1-1/2 C.Y. bucket w/trench box	14.50	33	47.50
1650	15' deep, 1-1/2 C.Y. bucket	18.80	43.50	62.30
1660	18' deep, 2-1/2 C.Y. bucket	24	43.50	67.50
1680	24' deep, 3-1/2 C.Y. bucket	33	60.50	93.50
1730	10' wide, 20' deep, 3-1/2 C.Y. bucket w/trench box	27	57.50	84.50
1740	24' deep, 3-1/2 C.Y. bucket	39.50	69.50	109
1780	12' wide, 20' deep, 3-1/2 C.Y. bucket w/trench box	42	73	115
1790	25' deep, bucket	52	93	145
1800	1/2 to 1 slope, 2' wide, 2' deep, 3/8 C.Y. bucket	.76	3.36	4.12
1810	3' deep, 3/8 C.Y. bucket	1.25	5.90	7.15
1820	4' deep, 3/8 C.Y. bucket	1.85	8.95	10.80
1840	6' deep, 3/8 C.Y. bucket	3.03	14.55	17.58

G1030 Site Earthwork

G1030 805	Trenching Common Earth	COST PER L.F.		
		EQUIP.	LABOR	TOTAL
1860	8' deep, 1/2 C.Y. bucket	4.78	23	27.78
1880	10' deep, 1 C.Y. bucket	12.45	32.50	44.95
2300	4' wide, 2' deep, 3/8 C.Y. bucket	1.49	5.55	7.04
2310	3' deep, 3/8 C.Y. bucket	2.20	9.20	11.40
2320	4' deep, 1/2 C.Y. bucket	2.72	11.35	14.07
2340	6' deep, 1/2 C.Y. bucket	5.35	21.50	26.85
2360	8' deep, 1/2 C.Y. bucket	13.10	31.50	44.60
2380	10' deep, 1 C.Y. bucket	18.15	44	62.15
2400	12' deep, 1 C.Y. bucket	24	59.50	83.50
2430	15' deep, 1-1/2 C.Y. bucket	22	64	86
2460	18' deep, 2-1/2 C.Y. bucket	45	99.50	144.50
2840	6' wide, 6' deep, 5/8 C.Y. bucket w/trench box	13.15	26.50	39.65
2860	8' deep, 3/4 C.Y. bucket	19.15	39.50	58.65
2880	10' deep, 1 C.Y. bucket	15.60	40.50	56.10
2900	12' deep, 1-1/2 C.Y. bucket	19.75	53	72.75
2940	16' deep, 2-1/2 C.Y. bucket	38	77.50	115.50
2980	20' deep, 3-1/2 C.Y. bucket	51.50	105	156.50
3020	24' deep, 3-1/2 C.Y. bucket	73	143	216
3100	8' wide, 12' deep, 1-1/2 C.Y. bucket w/trench box	24.50	60.50	85
3120	15' deep, 1-1/2 C.Y. bucket	35.50	87.50	123
3140	18' deep, 2-1/2 C.Y. bucket	52.50	104	156.50
3180	24' deep, 3-1/2 C.Y. bucket	81.50	157	238.50
3270	10' wide, 20' deep, 3-1/2 C.Y. bucket w/trench box	52.50	121	173.50
3280	24' deep, 3-1/2 C.Y. bucket	90	171	261
3370	12' wide, 20' deep, 3-1/2 C.Y. bucket w/trench box	75.50	140	215.50
3380	25' deep, 3-1/2 C.Y. bucket	105	200	305
3500	1 to 1 slope, 2' wide, 2' deep, 3/8 C.Y. bucket	.98	4.48	5.46
3520	3' deep, 3/8 C.Y. bucket	3.09	10.25	13.34
3540	4' deep, 3/8 C.Y. bucket	2.72	13.45	16.17
3560	6' deep, 1/2 C.Y. bucket	3.03	14.55	17.58
3580	8' deep, 1/2 C.Y. bucket	5.95	29	34.95
3600	10' deep, 1 C.Y. bucket	21.50	55.50	77
3800	4' wide, 2' deep, 3/8 C.Y. bucket	1.71	6.70	8.41
3820	3' deep, 3/8 C.Y. bucket	2.69	11.75	14.44
3840	4' deep, 1/2 C.Y. bucket	3.50	15.20	18.70
3860	6' deep, 1/2 C.Y. bucket	7.55	30.50	38.05
3880	8' deep, 1/2 C.Y. bucket	19.50	47	66.50
3900	10' deep, 1 C.Y. bucket	28	68.50	96.50
3920	12' deep, 1 C.Y. bucket	41.50	99.50	141
3940	15' deep, 1-1/2 C.Y. bucket	36	106	142
3960	18' deep, 2-1/2 C.Y. bucket	62	137	199
4030	6' wide, 6' deep, 5/8 C.Y. bucket w/trench box	17.30	35.50	52.80
4040	8' deep, 3/4 C.Y. bucket	25	49	74
4050	10' deep, 1 C.Y. bucket	22.50	59.50	82
4060	12' deep, 1-1/2 C.Y. bucket	30	81	111
4070	16' deep, 2-1/2 C.Y. bucket	59.50	123	182.50
4080	20' deep, 3-1/2 C.Y. bucket	83	171	254
4090	24' deep, 3-1/2 C.Y. bucket	121	240	361
4500	8' wide, 12' deep, 1-1/2 C.Y. bucket w/trench box	34.50	88	122.50
4550	15' deep, 1-1/2 C.Y. bucket	52	132	184
4600	18' deep, 2-1/2 C.Y. bucket	79.50	161	240.50
4650	24' deep, 3-1/2 C.Y. bucket	130	253	383
4800	10' wide, 20' deep, 3-1/2 C.Y. bucket w/trench box	78	184	262
4850	24' deep, 3-1/2 C.Y. bucket	138	268	406
4950	12' wide, 20' deep, 3-1/2 C.Y. bucket w/trench box	109	207	316
4980	25' deep, 3-1/2 C.Y. bucket	157	305	462

G1030 Site Earthwork

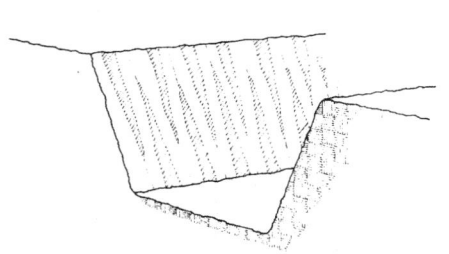

Trenching Systems are shown on a cost per linear foot basis. The systems include: excavation; backfill and removal of spoil; and compaction for various depths and trench bottom widths. The backfill has been reduced to accommodate a pipe of suitable diameter and bedding.

The slope for trench sides varies from none to 1:1.

The Expanded System Listing shows Trenching Systems that range from 2' to 12' in width. Depths range from 2' to 25'.

System Components	QUANTITY	UNIT	COST PER L.F.		
			EQUIP.	LABOR	TOTAL
SYSTEM G1030 806 1310					
TRENCHING, LOAM & SANDY CLAY, NO SLOPE, 2' WIDE, 2' DP, 3/8 C.Y. BUCKET					
Excavation, trench, hyd. backhoe, track mtd., 3/8 C.Y. bucket	.148	B.C.Y.	.22	1.12	1.34
Backfill and load spoil, from stockpile	.165	L.C.Y.	.14	.40	.54
Compaction by vibrating plate 18" wide, 6" lifts, 4 passes	.118	E.C.Y.	.03	.45	.48
Remove excess spoil, 8 C.Y. dump truck, 2 mile roundtrip	.042	L.C.Y.	.16	.21	.37
TOTAL			.55	2.18	2.73

G1030 806	Trenching Loam & Sandy Clay	COST PER L.F.		
		EQUIP.	LABOR	TOTAL
1310	Trenching, loam & sandy clay, no slope, 2' wide, 2' deep, 3/8 C.Y. bucket	.54	2.19	2.73
1320	3' deep, 3/8 C.Y. bucket	.81	3.59	4.40
1330	4' deep, 3/8 C.Y. bucket	.97	4.37	5.35
1340	6' deep, 3/8 C.Y. bucket	1.82	5.20	7.05
1350	8' deep, 1/2 C.Y. bucket	2.38	6.90	9.25
1360	10' deep, 1 C.Y. bucket	2.70	7.35	10.05
1400	4' wide, 2' deep, 3/8 C.Y. bucket	1.30	4.37	5.65
1410	3' deep, 3/8 C.Y. bucket	1.73	6.55	8.30
1420	4' deep, 1/2 C.Y. bucket	1.96	7.35	9.30
1430	6' deep, 1/2 C.Y. bucket	4.30	10.80	15.10
1440	8' deep, 1/2 C.Y. bucket	6.55	15.30	22
1450	10' deep, 1 C.Y. bucket	6.40	15.65	22
1460	12' deep, 1 C.Y. bucket	8.05	19.50	27.50
1470	15' deep, 1-1/2 C.Y. bucket	8.05	22.50	30.50
1480	18' deep, 2-1/2 C.Y. bucket	11.20	25	36
1520	6' wide, 6' deep, 5/8 C.Y. bucket w/trench box	8.70	17.60	26.50
1530	8' deep, 3/4 C.Y. bucket	11.30	23	34.50
1540	10' deep, 1 C.Y. bucket	10.70	23.50	34
1550	12' deep, 1-1/2 C.Y. bucket	10.25	26	36.50
1560	16' deep, 2-1/2 C.Y. bucket	16.20	33	49
1570	20' deep, 3-1/2 C.Y. bucket	19	39	58
1580	24' deep, 3-1/2 C.Y. bucket	24	47.50	71.50
1640	8' wide, 12' deep, 1-1/4 C.Y. bucket w/trench box	14.45	33.50	47.50
1650	15' deep, 1-1/2 C.Y. bucket	17.70	42.50	60
1660	18' deep, 2-1/2 C.Y. bucket	24.50	47.50	72
1680	24' deep, 3-1/2 C.Y. bucket	32	61.50	93.50
1730	10' wide, 20' deep, 3-1/2 C.Y. bucket w/trench box	32.50	62	94.50
1740	24' deep, 3-1/2 C.Y. bucket	40.50	76	117
1780	12' wide, 20' deep, 3-1/2 C.Y. bucket w/trench box	39	73.50	112
1790	25' deep, 3-1/2 C.Y. bucket	50.50	94.50	145
1800	1/2:1 slope, 2' wide, 2' deep, 3/8 C.Y. bucket	.76	3.28	4.04
1810	3' deep, 3/8 C.Y. bucket	1.24	5.75	7
1820	4' deep, 3/8 C.Y. bucket	1.83	8.75	10.60
1840	6' deep, 3/8 C.Y. bucket	4.37	13.05	17.40

G1030 Site Earthwork

G1030 806	Trenching Loam & Sandy Clay	COST PER L.F.		
		EQUIP.	LABOR	TOTAL
1860	8' deep, 1/2 C.Y. bucket	6.95	21	28
1880	10' deep, 1 C.Y. bucket	9.25	26	35.50
2300	4' wide, 2' deep, 3/8 C.Y. bucket	1.51	5.45	7
2310	3' deep, 3/8 C.Y. bucket	2.21	9	11.25
2320	4' deep, 1/2 C.Y. bucket	2.74	11.15	13.90
2340	6' deep, 1/2 C.Y. bucket	7.25	19.20	26.50
2360	8' deep, 1/2 C.Y. bucket	12.75	31	43.50
2380	10' deep, 1 C.Y. bucket	14.10	36	50
2400	12' deep, 1 C.Y. bucket	23.50	59.50	83
2430	15' deep, 1-1/2 C.Y. bucket	23	66	89
2460	18' deep, 2-1/2 C.Y. bucket	44.50	101	145
2840	6' wide, 6' deep, 5/8 C.Y. bucket w/trench box	12.70	26.50	39.50
2860	8' deep, 3/4 C.Y. bucket	18.35	39	57.50
2880	10' deep, 1 C.Y. bucket	19.20	44	63
2900	12' deep, 1-1/2 C.Y. bucket	20	53.50	73.50
2940	16' deep, 2-1/2 C.Y. bucket	37	78.50	116
2980	20' deep, 3-1/2 C.Y. bucket	50	106	156
3020	24' deep, 3-1/2 C.Y. bucket	70.50	145	216
3100	8' wide, 12' deep, 1-1/2 C.Y. bucket w/trench box	24	60.50	85
3120	15' deep, 1-1/2 C.Y. bucket	33	85	118
3140	18' deep, 2-1/2 C.Y. bucket	51	105	156
3180	24' deep, 3-1/2 C.Y. bucket	79	159	238
3270	10' wide, 20' deep, 3-1/2 C.Y. bucket w/trench box	63.50	129	192
3280	24' deep, 3-1/2 C.Y. bucket	87.50	174	261
3320	12' wide, 20' deep, 3-1/2 C.Y. bucket w/trench box	70	140	211
3380	25' deep, 3-1/2 C.Y. bucket w/trench box	95.50	188	284
3500	1:1 slope, 2' wide, 2' deep, 3/8 C.Y. bucket	.97	4.38	5.35
3520	3' deep, 3/8 C.Y. bucket	1.72	8.20	9.95
3540	4' deep, 3/8 C.Y. bucket	2.69	13.15	15.80
3560	6' deep, 1/2 C.Y. bucket	4.37	13.05	17.40
3580	8' deep, 1/2 C.Y. bucket	11.45	34.50	46
3600	10' deep, 1 C.Y. bucket	15.85	45	60.50
3800	4' wide, 2' deep, 3/8 C.Y. bucket	1.73	6.55	8.30
3820	3' deep, 1/2 C.Y. bucket	2.69	11.50	14.20
3840	4' deep, 1/2 C.Y. bucket	3.52	15	18.50
3860	6' deep, 1/2 C.Y. bucket	10.25	27.50	38
3880	8' deep, 1/2 C.Y. bucket	18.95	46.50	65.50
3900	10' deep, 1 C.Y. bucket	22	56	78
3920	12' deep, 1 C.Y. bucket	31.50	79.50	111
3940	15' deep, 1-1/2 C.Y. bucket	37.50	110	147
3960	18' deep, 2-1/2 C.Y. bucket	61	139	200
4030	6' wide, 6' deep, 5/8 C.Y. bucket w/trench box	16.70	35.50	52.50
4040	8' deep, 3/4 C.Y. bucket	25.50	55	80.50
4050	10' deep, 1 C.Y. bucket	27.50	64.50	92
4060	12' deep, 1-1/2 C.Y. bucket	30	81	111
4070	16' deep, 2-1/2 C.Y. bucket	58	124	182
4080	20' deep, 3-1/2 C.Y. bucket	81	173	254
4090	24' deep, 3-1/2 C.Y. bucket	118	243	360
4500	8' wide, 12' deep, 1-1/4 C.Y. bucket w/trench box	34	88	122
4550	15' deep, 1-1/2 C.Y. bucket	48	128	176
4600	18' deep, 2-1/2 C.Y. bucket	77.50	163	241
4650	24' deep, 3-1/2 C.Y. bucket	126	257	385
4800	10' wide, 20' deep, 3-1/2 C.Y. bucket w/trench box	94.50	196	291
4850	24' deep, 3-1/2 C.Y. bucket	134	272	405
4950	12' wide, 20' deep, 3-1/2 C.Y. bucket w/trench box	101	208	310
4980	25' deep, 3-1/2 C.Y. bucket	152	305	460

G1030 Site Earthwork

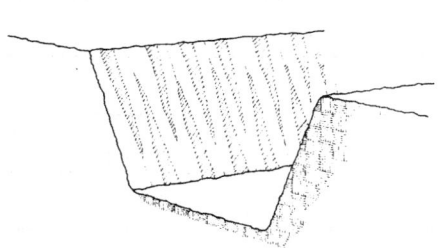

Trenching Systems are shown on a cost per linear foot basis. The systems include: excavation; backfill and removal of spoil; and compaction for various depths and trench bottom widths. The backfill has been reduced to accommodate a pipe of suitable diameter and bedding.

The slope for trench sides varies from none to 1:1.

The Expanded System Listing shows Trenching Systems that range from 2' to 12' in width. Depths range from 2' to 25'.

System Components	QUANTITY	UNIT	COST PER L.F.		
			EQUIP.	LABOR	TOTAL
SYSTEM G1030 807 1310					
TRENCHING, SAND & GRAVEL, NO SLOPE, 2' WIDE, 2' DEEP, 3/8 C.Y. BUCKET					
Excavation, trench, hyd. backhoe, track mtd., 3/8 C.Y. bucket	.148	B.C.Y.	.21	1.10	1.31
Backfill and load spoil, from stockpile	.140	L.C.Y.	.12	.34	.46
Compaction by vibrating plate 18" wide, 6" lifts, 4 passes	.118	E.C.Y.	.03	.45	.48
Remove excess spoil, 8 C.Y. dump truck, 2 mile roundtrip	.035	L.C.Y.	.13	.18	.31
TOTAL			.49	2.07	2.56

G1030 807	Trenching Sand & Gravel	COST PER L.F.		
		EQUIP.	LABOR	TOTAL
1310	Trenching, sand & gravel, no slope, 2' wide, 2' deep, 3/8 C.Y. bucket	.49	2.07	2.56
1320	3' deep, 3/8 C.Y. bucket	.75	3.45	4.20
1330	4' deep, 3/8 C.Y. bucket	.89	4.15	5.05
1340	6' deep, 3/8 C.Y. bucket	1.72	4.97	6.70
1350	8' deep, 1/2 C.Y. bucket	2.26	6.60	8.85
1360	10' deep, 1 C.Y. bucket	2.52	6.95	9.45
1400	4' wide, 2' deep, 3/8 C.Y. bucket	1.16	4.10	5.25
1410	3' deep, 3/8 C.Y. bucket	1.56	6.20	7.75
1420	4' deep, 1/2 C.Y. bucket	1.78	6.95	8.75
1430	6' deep, 1/2 C.Y. bucket	4.07	10.30	14.40
1440	8' deep, 1/2 C.Y. bucket	6.20	14.55	20.50
1450	10' deep, 1 C.Y. bucket	6	14.75	21
1460	12' deep, 1 C.Y. bucket	7.60	18.45	26
1470	15' deep, 1-1/2 C.Y. bucket	7.50	21.50	29
1480	18' deep, 2-1/2 C.Y. bucket	10.55	23.50	34
1520	6' wide, 6' deep, 5/8 C.Y. bucket w/trench box	8.25	16.65	25
1530	8' deep, 3/4 C.Y. bucket	10.75	22	32.50
1540	10' deep, 1 C.Y. bucket	10.05	22	32
1550	12' deep, 1-1/2 C.Y. bucket	9.60	24.50	34.50
1560	16' deep, 2 C.Y. bucket	15.50	31	46.50
1570	20' deep, 3-1/2 C.Y. bucket	17.90	36.50	54.50
1580	24' deep, 3-1/2 C.Y. bucket	22.50	44.50	67
1640	8' wide, 12' deep, 1-1/2 C.Y. bucket w/trench box	13.45	31.50	45
1650	15' deep, 1-1/2 C.Y. bucket	16.40	40	56
1660	18' deep, 2-1/2 C.Y. bucket	23	44.50	68
1680	24' deep, 3-1/2 C.Y. bucket	30.50	57.50	88
1730	10' wide, 20' deep, 3-1/2 C.Y. bucket w/trench box	30.50	58	88
1740	24' deep, 3-1/2 C.Y. bucket	38	71.50	109
1780	12' wide, 20' deep, 3-1/2 C.Y. bucket w/trench box	36.50	68.50	105
1790	25' deep, 3-1/2 C.Y. bucket	47.50	88.50	136
1800	1/2:1 slope, 2' wide, 2' deep, 3/8 C.Y. bucket	.69	3.11	3.80
1810	3' deep, 3/8 C.Y. bucket	1.14	5.45	6.60
1820	4' deep, 3/8 C.Y. bucket	1.68	8.35	10
1840	6' deep, 3/8 C.Y. bucket	4.16	12.45	16.60

G1030 807	Trenching Sand & Gravel	COST PER L.F.		
		EQUIP.	LABOR	TOTAL
1860	8' deep, 1/2 C.Y. bucket	6.60	19.90	26.50
1880	10' deep, 1 C.Y. bucket	8.65	24.50	33
2300	4' wide, 2' deep, 3/8 C.Y. bucket	1.36	5.15	6.50
2310	3' deep, 3/8 C.Y. bucket	2.01	8.55	10.55
2320	4' deep, 1/2 C.Y. bucket	2.49	10.60	13.05
2340	6' deep, 1/2 C.Y. bucket	6.95	18.35	25.50
2360	8' deep, 1/2 C.Y. bucket	12.05	29.50	41.50
2380	10' deep, 1 C.Y. bucket	13.30	34	47
2400	12' deep, 1 C.Y. bucket	22.50	56.50	78.50
2430	15' deep, 1-1/2 C.Y. bucket	21.50	62.50	84
2460	18' deep, 2-1/2 C.Y. bucket	42	95	137
2840	6' wide, 6' deep, 5/8 C.Y. bucket w/trench box	12.05	25.50	37.50
2860	8' deep, 3/4 C.Y. bucket	17.45	37	54.50
2880	10' deep, 1 C.Y. bucket	18.15	41.50	59.50
2900	12' deep, 1-1/2 C.Y. bucket	18.80	50.50	69.50
2940	16' deep, 2 C.Y. bucket	35.50	74	110
2980	20' deep, 3-1/2 C.Y. bucket	47.50	99.50	147
3020	24' deep, 3-1/2 C.Y. bucket	67	137	204
3100	8' wide, 12' deep, 1-1/4 C.Y. bucket w/trench box	22.50	57.50	80
3120	15' deep, 1-1/2 C.Y. bucket	31	80.50	111
3140	18' deep, 2-1/2 C.Y. bucket	48.50	99	148
3180	24' deep, 3-1/2 C.Y. bucket	75	150	224
3270	10' wide, 20' deep, 3-1/2 C.Y. bucket w/trench box	60	121	181
3280	24' deep, 3-1/2 C.Y. bucket	82.50	163	246
3370	12' wide, 20' deep, 3-1/2 C.Y. bucket w/trench box	67	134	201
3380	25' deep, 3-1/2 C.Y. bucket	96	188	284
3500	1:1 slope, 2' wide, 2' deep, 3/8 C.Y. bucket	1.76	5.25	7
3520	3' deep, 3/8 C.Y. bucket	1.58	7.80	9.40
3540	4' deep, 3/8 C.Y. bucket	2.48	12.50	15
3560	6' deep, 3/8 C.Y. bucket	4.16	12.45	16.65
3580	8' deep, 1/2 C.Y. bucket	10.95	33	44
3600	10' deep, 1 C.Y. bucket	14.80	42	57
3800	4' wide, 2' deep, 3/8 C.Y. bucket	1.56	6.20	7.75
3820	3' deep, 3/8 C.Y. bucket	2.45	10.90	13.35
3840	4' deep, 1/2 C.Y. bucket	3.21	14.20	17.40
3860	6' deep, 1/2 C.Y. bucket	9.80	26.50	36
3880	8' deep, 1/2 C.Y. bucket	17.95	44.50	62
3900	10' deep, 1 C.Y. bucket	20.50	53	73.50
3920	12' deep, 1 C.Y. bucket	30	75	105
3940	15' deep, 1-1/2 C.Y. bucket	35	104	139
3960	18' deep, 2-1/2 C.Y. bucket	57.50	131	188
4030	6' wide, 6' deep, 5/8 C.Y. bucket w/trench box	15.85	34	49.50
4040	8' deep, 3/4 C.Y. bucket	24	52.50	76.50
4050	10' deep, 1 C.Y. bucket	26	61	87
4060	12' deep, 1-1/2 C.Y. bucket	28	76.50	104
4070	16' deep, 2 C.Y. bucket	55.50	117	173
4080	20' deep, 3-1/2 C.Y. bucket	76.50	163	239
4090	24' deep, 3-1/2 C.Y. bucket	111	228	340
4500	8' wide, 12' deep, 1-1/2 C.Y. bucket w/trench box	32	83.50	115
4550	15' deep, 1-1/2 C.Y. bucket	45	121	166
4600	18' deep, 2-1/2 C.Y. bucket	73.50	153	227
4650	24' deep, 3-1/2 C.Y. bucket	119	241	360
4800	10' wide, 20' deep, 3-1/2 C.Y. bucket w/trench box	89.50	184	273
4850	24' deep, 3-1/2 C.Y. bucket	127	255	380
4950	12' wide, 20' deep, 3-1/2 C.Y. bucket w/trench box	95.50	195	290
4980	25' deep, 3-1/2 C.Y. bucket	144	288	430

G1030 Site Earthwork

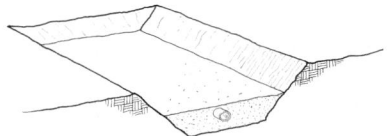

The Pipe Bedding System is shown for various pipe diameters. Compacted bank sand is used for pipe bedding and to fill 12″ over the pipe. No backfill is included. Various side slopes are shown to accommodate different soil conditions. Pipe sizes vary from 6″ to 84″ diameter.

System Components	QUANTITY	UNIT	COST PER L.F.		
			MAT.	INST.	TOTAL
SYSTEM G1030 815 1440					
PIPE BEDDING, SIDE SLOPE 0 TO 1, 1′ WIDE, PIPE SIZE 6″ DIAMETER					
Borrow, bank sand, 2 mile haul, machine spread	.086	C.Y.	1.75	.73	2.48
Compaction, vibrating plate	.086	C.Y.		.25	.25
TOTAL			1.75	.98	2.73

G1030 815	Pipe Bedding	COST PER L.F.		
		MAT.	INST.	TOTAL
1440	Pipe bedding, side slope 0 to 1, 1′ wide, pipe size 6″ diameter	1.75	.97	2.72
1460	2′ wide, pipe size 8″ diameter	3.79	2.11	5.90
1480	Pipe size 10″ diameter	3.87	2.15	6.02
1500	Pipe size 12″ diameter	3.96	2.20	6.16
1520	3′ wide, pipe size 14″ diameter	6.40	3.57	9.97
1540	Pipe size 15″ diameter	6.50	3.60	10.10
1560	Pipe size 16″ diameter	6.55	3.64	10.19
1580	Pipe size 18″ diameter	6.65	3.71	10.36
1600	4′ wide, pipe size 20″ diameter	9.45	5.25	14.70
1620	Pipe size 21″ diameter	9.55	5.30	14.85
1640	Pipe size 24″ diameter	9.75	5.45	15.20
1660	Pipe size 30″ diameter	9.95	5.55	15.50
1680	6′ wide, pipe size 32″ diameter	17	9.45	26.45
1700	Pipe size 36″ diameter	17.40	9.70	27.10
1720	7′ wide, pipe size 48″ diameter	27.50	15.40	42.90
1740	8′ wide, pipe size 60″ diameter	33.50	18.75	52.25
1760	10′ wide, pipe size 72″ diameter	47	26	73
1780	12′ wide, pipe size 84″ diameter	62	34.50	96.50
2140	Side slope 1/2 to 1, 1′ wide, pipe size 6″ diameter	3.27	1.82	5.09
2160	2′ wide, pipe size 8″ diameter	5.55	3.09	8.64
2180	Pipe size 10″ diameter	5.95	3.32	9.27
2200	Pipe size 12″ diameter	6.30	3.51	9.81
2220	3′ wide, pipe size 14″ diameter	9.10	5.05	14.15
2240	Pipe size 15″ diameter	9.35	5.20	14.55
2260	Pipe size 16″ diameter	9.55	5.30	14.85
2280	Pipe size 18″ diameter	10.05	5.60	15.65
2300	4′ wide, pipe size 20″ diameter	13.30	7.40	20.70
2320	Pipe size 21″ diameter	13.55	7.55	21.10
2340	Pipe size 24″ diameter	14.45	8.05	22.50
2360	Pipe size 30″ diameter	16	8.90	24.90
2380	6′ wide, pipe size 32″ diameter	23.50	13.15	36.65
2400	Pipe size 36″ diameter	25	13.95	38.95
2420	7′ wide, pipe size 48″ diameter	39	22	61
2440	8′ wide, pipe size 60″ diameter	49.50	27.50	77
2460	10′ wide, pipe size 72″ diameter	68	38	106
2480	12′ wide, pipe size 84″ diameter	89.50	49.50	139
2620	Side slope 1 to 1, 1′ wide, pipe size 6″ diameter	4.79	2.66	7.45
2640	2′ wide, pipe size 8″ diameter	7.40	4.10	11.50

G10 Site Preparation

G1030 Site Earthwork

G1030 815	Pipe Bedding	COST PER L.F.		
		MAT.	INST.	TOTAL
2660	Pipe size 10" diameter	8	4.46	12.46
2680	Pipe size 12" diameter	8.75	4.86	13.61
2700	3' wide, pipe size 14" diameter	11.85	6.60	18.45
2720	Pipe size 15" diameter	12.20	6.80	19
2740	Pipe size 16" diameter	12.60	7	19.60
2760	Pipe size 18" diameter	13.45	7.50	20.95
2780	4' wide, pipe size 20" diameter	17.10	9.50	26.60
2800	Pipe size 21" diameter	17.55	9.75	27.30
2820	Pipe size 24" diameter	19.05	10.60	29.65
2840	Pipe size 30" diameter	22	12.25	34.25
2860	6' wide, pipe size 32" diameter	30	16.80	46.80
2880	Pipe size 36" diameter	33	18.25	51.25
2900	7' wide, pipe size 48" diameter	50.50	28	78.50
2920	8' wide, pipe size 60" diameter	66	36.50	102.50
2940	10' wide, pipe size 72" diameter	89.50	50	139.50
2960	12' wide, pipe size 84" diameter	117	65	182

G2010 Roadways

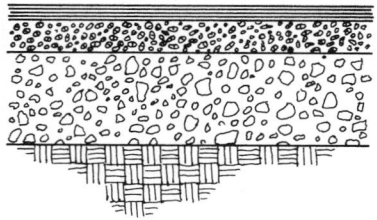

The Bituminous Roadway Systems are listed for pavement thicknesses between 3-I/2" and 7" and gravel bases from 3" to 22" in depth. Systems costs are expressed per linear foot for varying widths of two and multi-lane roads. Earth moving is not included. Granite curbs and line painting are added as required system components.

System Components	QUANTITY	UNIT	COST PER L.F.		
			MAT.	INST.	TOTAL
SYSTEM G2010 230 1050					
BITUM. ROADWAY, TWO LANES, 3-1/2" TH. PVMT., 3" TH. GRAVEL BASE, 24' WIDE					
Compact subgrade, 4 passes	.222	E.C.Y.		.54	.54
Bank gravel, 2 mi haul, dozer spread	.250	L.C.Y.	8.13	2.15	10.28
Compaction, granular material to 98%	.259	E.C.Y.		.18	.18
Grading, fine grade, 3 passes with grader	3.300	S.Y.		19.90	19.90
Bituminous paving, binder course, 2-1/2" thick	2.670	S.Y.	27.23	4.46	31.69
Bituminous paving, wearing course, 1" thick	2.670	S.Y.	10.81	2.62	13.43
Curbs, granite, split face, straight, 5" x 16"	2.000	L.F.	36.20	30.76	66.96
Thermoplastic, white or yellow, 4" wide, less than 6,000 L.F.	4.000	L.F.	2	1.24	3.24
TOTAL			84.37	61.85	146.22

G2010 230	Bituminous Roadways Gravel Base	COST PER L.F.		
		MAT.	INST.	TOTAL
1050	Bitum. roadway, two lanes, 3-1/2" th. pvmt., 3" th. gravel base, 24' wide	84.50	62	146.50
1100	28' wide	92.50	62.50	155
1150	32' wide	100	67	167
1200	4" th. gravel base, 24' wide	87.50	59	146.50
1210	28' wide	96	63.50	159.50
1220	32' wide	104	68	172
1222	36' wide	112	72.50	184.50
1224	40' wide	120	77	197
1230	6" th. gravel base, 24' wide	93	61.50	154.50
1240	28' wide	102	65.50	167.50
1250	32' wide	112	70.50	182.50
1252	36' wide	121	75.50	196.50
1254	40' wide	130	80	210
1256	8" th. gravel base, 24' wide	99	63	162
1258	28' wide	109	68	177
1260	32' wide	119	73	192
1262	36' wide	129	78	207
1264	40' wide	139	83.50	222.50
1300	9" th. gravel base, 24' wide	101	64	165
1350	28' wide	112	69.50	181.50
1400	32' wide	122	74.50	196.50
1410	10" th. gravel base, 24' wide	105	64.50	169.50
1412	28' wide	116	70	186
1414	32' wide	127	75.50	202.50
1416	36' wide	138	81	219
1418	40' wide	149	86.50	235.50
1420	12" thick gravel base, 24' wide	110	66.50	176.50
1422	28' wide	122	72.50	194.50
1424	32' wide	134	78	212
1426	36' wide	146	84	230

G2010 Roadways

G2010 230	Bituminous Roadways Gravel Base	COST PER L.F.		
		MAT.	INST.	TOTAL
1428	40' wide	158	89.50	247.50
1430	15" thick gravel base, 24' wide	119	69.50	188.50
1432	28' wide	132	75.50	207.50
1434	32' wide	146	82	228
1436	36' wide	159	88	247
1438	40' wide	173	94	267
1440	18" thick gravel base, 24' wide	127	72	199
1442	28' wide	142	79	221
1444	32' wide	157	85.50	242.50
1446	36' wide	172	92	264
1448	40' wide	187	99	286
1550	4" th. pvmt., 4" th. gravel base, 24' wide	93	59.50	152.50
1600	28' wide	102	64.50	166.50
1650	32' wide	112	69	181
1652	36' wide	120	73.50	193.50
1654	40' wide	130	78	208
1700	6" th. gravel base, 24' wide	99	61.50	160.50
1710	28' wide	109	66.50	175.50
1720	32' wide	119	71.50	190.50
1722	36' wide	129	76.50	205.50
1724	40' wide	139	81.50	220.50
1726	8" th. gravel base, 24' wide	104	63.50	167.50
1728	28' wide	115	68.50	183.50
1730	32' wide	127	74	201
1732	36' wide	137	79	216
1734	40' wide	149	84.50	233.50
1800	10" th. gravel base, 24' wide	110	65.50	175.50
1850	28' wide	122	69.50	191.50
1900	32' wide	134	76.50	210.50
1902	36' wide	146	82	228
1904	40' wide	151	85.50	236.50
2050	4" th. pvmt., 5" th. gravel base, 24' wide	96	63.50	159.50
2100	28' wide	106	69	175
2150	32' wide	116	74.50	190.50
2200	8" th. gravel base, 24' wide	105	66	171
2210	28' wide	116	72	188
2220	32' wide	128	78	206
2300	12" th. gravel base, 24' wide	118	67.50	185.50
2350	28' wide	131	73.50	204.50
2400	32' wide	144	79.50	223.50
2410	36' wide	157	85.50	242.50
2412	40' wide	171	91	262
2420	15" th. gravel base, 24' wide	120	68.50	188.50
2422	28' wide	141	76.50	217.50
2424	32' wide	156	83	239
2426	36' wide	170	89.50	259.50
2428	40' wide	185	96	281
2430	18" th. gravel base, 24' wide	135	73	208
2432	28' wide	151	80.50	231.50
2434	32' wide	167	87	254
2436	36' wide	183	93.50	276.50
2438	40' wide	199	100	299
2550	4-1/2" th. pvmt., 5" th. gravel base, 24' wide	102	64.50	166.50
2600	28' wide	112	70	182
2650	32' wide	123	75.50	198.50
2700	6" th. gravel base, 24' wide	105	65	170
2710	28' wide	116	71	187
2720	32' wide	127	76.50	203.50

G2010 Roadways

G2010 230	Bituminous Roadways Gravel Base	COST PER L.F.		
		MAT.	INST.	TOTAL
2800	13" th. gravel base, 24' wide	124	70.50	194.50
2850	28' wide	138	77	215
2900	32' wide	153	84	237
3050	5" th. pvmt., 6" th. gravel base, 24' wide	110	66	176
3100	28' wide	123	71.50	194.50
3150	32' wide	135	77.50	212.50
3300	14" th. gravel base, 24' wide	133	72.50	205.50
3350	28' wide	149	79	228
3400	32' wide	165	86	251
3550	5-1/2" th. pvmt., 7" th. gravel base, 24' wide	119	67.50	186.50
3600	28' wide	133	73.50	206.50
3650	32' wide	147	79.50	226.50
3800	17" th. gravel base, 24' wide	147	75	222
3850	28' wide	165	82.50	247.50
3900	32' wide	184	90	274
4050	Multi lane, 5" th. pvmt., 6" th. gravel base, 48' wide	184	101	285
4100	72' wide	258	135	393
4150	96' wide	330	170	500
4300	14" th. gravel base, 48' wide	227	113	340
4350	72' wide	320	155	475
4400	96' wide	415	196	611
4550	5-1/2" th. pvmt., 7" th. gravel base, 48' wide	201	104	305
4600	72' wide	284	140	424
4650	96' wide	365	176	541
4800	17" th. gravel base, 48' wide	257	119	376
4850	72' wide	365	163	528
4900	96' wide	475	208	683
5050	6" th. pvmt., 8" th. gravel base, 48' wide	217	106	323
5100	72' wide	305	143	448
5150	96' wide	400	181	581
5200	10" th. gravel base, 48' wide	229	109	338
5210	72' wide	325	148	473
5220	96' wide	420	187	607
5300	20" th. gravel base, 48' wide	283	125	408
5350	72' wide	405	171	576
5400	96' wide	530	219	749
5550	7" th. pvmt., 9" th. gravel base, 48' wide	246	110	356
5600	72' wide	350	149	499
5650	96' wide	455	189	644
5700	10" th. gravel base, 48' wide	253	112	365
5710	72' wide	360	152	512
5720	96' wide	470	192	662
5800	22" th. gravel base, 48' wide	320	130	450
5850	72' wide	460	180	640
5900	96' wide	600	230	830

G2010 Roadways

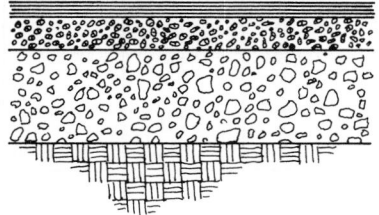

The Bituminous Roadway Systems are listed for pavement thicknesses between 3-1/2" and 7" and crushed stone bases from 3" to 22" in depth. Systems costs are expressed per linear foot for varying widths of two and multi-lane roads. Earth moving is not included. Granite curbs and line painting are added as required system components.

System Components	QUANTITY	UNIT	COST PER L.F.		
			MAT.	INST.	TOTAL
SYSTEM G2010 232 1050					
BITUMINOUS ROADWAY, TWO LANES, 3-1/2" PAVEMENT					
3" THICK CRUSHED STONE BASE, 24' WIDE					
Compact subgrade, 4 passes	.222	E.C.Y.		.54	.54
3/4" crushed stonel, 2 mi haul, dozer spread	.266	C.Y.	7.58	2.28	9.86
Compaction, granular material to 98%	.259	E.C.Y.		.18	.18
Grading, fine grade, 3 passes with grader	3.300	S.Y.		19.90	19.90
Bituminous paving, binder course, 2-1/2" thick	2.670	S.Y.	27.23	4.46	31.69
Bituminous paving, wearing course, 1" thick	2.670	S.Y.	10.81	2.62	13.43
Curbs, granite, split face, straight, 5" x 16"	2.000	L.F.	36.20	30.76	66.96
Thermoplastic, white or yellow, 4" wide, less than 6,000 L.F.	4.000	L.F.	2	1.24	3.24
TOTAL			83.82	61.98	145.80

G2010 232	Bituminous Roadways Crushed Stone	COST PER L.F.		
		MAT.	INST.	TOTAL
1050	Bitum. roadway, two lanes, 3-1/2" th. pvmt., 3" th. crushed stone, 24' wide	84	62	146
1100	28' wide	91.50	62.50	154
1150	32' wide	99	72	171
1200	4" thick crushed stone, 24' wide	86	59	145
1210	28' wide	94	63.50	157.50
1220	32' wide	102	68	170
1222	36' wide	110	72.50	182.50
1224	40' wide	118	77	195
1230	6" thick crushed stone, 24' wide	91	61	152
1240	28' wide	100	65.50	165.50
1250	32' wide	109	70.50	179.50
1252	36' wide	118	75.50	193.50
1254	40' wide	126	80	206
1256	8" thick crushed stone, 24' wide	96	63	159
1258	28' wide	106	68	174
1260	32' wide	116	73	189
1262	36' wide	125	78	203
1264	40' wide	135	83.50	218.50
1300	9" thick crushed stone, 24' wide	99	64	163
1350	28' wide	109	69.50	178.50
1400	32' wide	119	75	194
1410	10" thick crushed stone, 24' wide	101	64.50	165.50
1412	28' wide	112	70	182
1414	32' wide	122	75.50	197.50
1416	36' wide	133	81	214
1418	40' wide	143	86.50	229.50

659

G2010 Roadways

G2010 232	Bituminous Roadways Crushed Stone	COST PER L.F.		
		MAT.	INST.	TOTAL
1420	12" thick crushed stone, 24' wide	106	66.50	172.50
1422	28' wide	117	72.50	189.50
1424	32' wide	129	78	207
1426	36' wide	140	84	224
1428	40' wide	151	89.50	240.50
1430	15" thick crushed stone, 24' wide	114	69.50	183.50
1432	28' wide	126	75.50	201.50
1434	32' wide	139	82	221
1436	36' wide	151	88	239
1438	40' wide	164	94	258
1440	18" thick crushed stone, 24' wide	121	72	193
1442	28' wide	135	79	214
1444	32' wide	149	85.50	234.50
1446	36' wide	162	92	254
1448	40' wide	176	99	275
1550	4" thick pavement., 4" thick crushed stone, 24' wide	92	60	152
1600	28' wide	101	64.50	165.50
1650	32' wide	110	69	179
1652	36' wide	118	73.50	191.50
1654	40' wide	127	78	205
1700	6" thick crushed stone , 24' wide	96.50	61.50	158
1710	28' wide	106	66.50	172.50
1720	32' wide	116	71.50	187.50
1722	36' wide	126	76.50	202.50
1724	40' wide	136	81.50	217.50
1726	8" thick crushed stone , 24' wide	102	63.50	165.50
1728	28' wide	112	68.50	180.50
1730	32' wide	123	74	197
1732	36' wide	133	79	212
1734	40' wide	144	84.50	228.50
1800	10" thick crushed stone, 24' wide	107	65.50	172.50
1850	28' wide	118	69.50	187.50
1900	32' wide	130	76.50	206.50
1902	36' wide	141	82	223
1904	40' wide	146	85.50	231.50
2050	4" th. pavement, 5" thick crushed stone, 24' wide	94.50	64	158.50
2100	28' wide	104	69	173
2150	32' wide	114	74.50	188.50
2300	12" thick crushed stone, 24' wide	112	69.50	181.50
2350	28' wide	125	76	201
2400	32' wide	138	82.50	220.50
2550	4-1/2" thick pavement, 5" thick crushed stone, 24' wide	100	64.50	164.50
2600	28' wide	111	70	181
2650	32' wide	121	75.50	196.50
2800	13" thick crushed stone, 24' wide	120	71	191
2850	28' wide	134	77.50	211.50
2900	32' wide	148	84.50	232.50
3050	5" thick pavement, 6" thick crushed stone, 24' wide	108	65.50	173.50
3100	28' wide	120	71.50	191.50
3150	32' wide	132	77.50	209.50
3300	14" thick crushed stone, 24' wide	128	72.50	200.50
3350	28' wide	144	79	223
3400	32' wide	159	86	245
3550	5-1/2" thick pavement., 7" thick crushed stone, 24' wide	117	67.50	184.50
3600	28' wide	130	73.50	203.50
3650	32' wide	144	79.50	223.50
3800	17" thick crushed stone, 24' wide	142	75.50	217.50
3850	28' wide	174	87	261

G2010 Roadways

G2010 232	Bituminous Roadways Crushed Stone	COST PER L.F.		
		MAT.	INST.	TOTAL
3900	32' wide	206	99	305
4050	Multi lane, 5" thick pavement, 6" thick crushed stone, 48' wide	179	100	279
4100	72' wide	251	135	386
4150	96' wide	320	170	490
4300	14" thick crushed stone, 48' wide	220	113	333
4350	72' wide	310	155	465
4400	96' wide	405	196	601
4550	5-1/2" thick pavement, 7" thick crushed stone, 48' wide	197	103	300
4600	72' wide	277	140	417
4650	96' wide	360	176	536
4800	17" thick crushed stone, 48' wide	248	120	368
4850	72' wide	355	164	519
4900	96' wide	460	209	669
5050	6" thick pavement, 8" thick crushed stone, 48' wide	213	106	319
5100	72' wide	300	144	444
5150	96' wide	390	181	571
5300	20" thick crushed stone, 48' wide	274	126	400
5350	72' wide	390	173	563
5400	96' wide	510	221	731
5550	7" thick pavement, 9" thick crushed stone, 48' wide	241	110	351
5600	72' wide	345	150	495
5650	96' wide	445	190	635
5800	22" thick crushed stone, 48' wide	305	132	437
5850	72' wide	440	182	622
5900	96' wide	575	232	807

G2020 Parking Lots

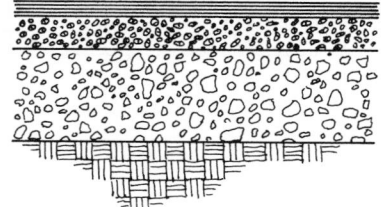

The Parking Lot System includes: compacted bank-run gravel; fine grading with a grader and roller; and bituminous concrete wearing course. All Parking Lot systems are on a cost per car basis. There are three basic types of systems: 90° angle, 60° angle, and 45° angle. The gravel base is compacted to 98%. Final stall design and lay-out of the parking lot with precast bumpers, sealcoating and white paint is also included.

The Expanded System Listing shows the three basic parking lot types with various depths of both gravel base and wearing course. The gravel base depths range from 6″ to 10″. The bituminous paving wearing course varies from a depth of 3″ to 6″.

System Components			COST PER CAR		
	QUANTITY	UNIT	MAT.	INST.	TOTAL
SYSTEM G2020 210 1500					
PARKING LOT, 90° ANGLE PARKING, 3″ BITUMINOUS PAVING, 6″ GRAVEL BASE					
Surveying crew for layout, 4 man crew	.020	Day		55.76	55.76
Borrow, bank run gravel, haul 2 mi., spread w/dozer, no compaction	7.223	L.C.Y.	234.75	61.90	296.65
Grading, fine grade 3 passes with motor grader	43.333	S.Y.		132.66	132.66
Compact w/vibrating plate, 8″ lifts, granular mat'l. to 98%	7.223	E.C.Y.		54.31	54.31
Binder course paving 2″ thick	43.333	S.Y.	353.16	64.13	417.29
Wear course paving 1″ thick	43.333	S.Y.	175.50	42.46	217.96
Seal coating, petroleum resistant under 1,000 S.Y.	43.333	S.Y.	72.37	66.73	139.10
Lines on pvmt., parking stall, paint, white, 4″ wide	1.000	Ea.	5.70	4.18	9.88
Precast concrete parking bar, 6″ x 10″ x 6′-0″	1.000	Ea.	63.50	22.50	86
TOTAL			904.98	504.63	1,409.61

G2020 210	Parking Lots Gravel Base	COST PER CAR		
		MAT.	INST.	TOTAL
1500	Parking lot, 90° angle parking, 3″ bituminous paving, 6″ gravel base	905	505	1,410
1520	8″ gravel base	985	545	1,530
1540	10″ gravel base	1,050	580	1,630
1560	4″ bituminous paving, 6″ gravel base	1,125	535	1,660
1580	8″ gravel base	1,200	570	1,770
1600	10″ gravel base	1,275	610	1,885
1620	6″ bituminous paving, 6″ gravel base	1,475	565	2,040
1640	8″ gravel base	1,550	605	2,155
1660	10″ gravel base	1,650	645	2,295
1800	60° angle parking, 3″ bituminous paving, 6″ gravel base	905	505	1,410
1820	8″ gravel base	985	545	1,530
1840	10″ gravel base	1,050	580	1,630
1860	4″ bituminous paving, 6″ gravel base	1,125	535	1,660
1880	8″ gravel base	1,200	570	1,770
1900	10″ gravel base	1,275	610	1,885
1920	6″ bituminous paving, 6″ gravel base	1,475	565	2,040
1940	8″ gravel base	1,550	605	2,155
1960	10″ gravel base	1,650	645	2,295
2200	45° angle parking, 3″ bituminous paving, 6″ gravel base	925	520	1,445
2220	8″ gravel base	1,000	560	1,560
2240	10″ gravel base	1,075	595	1,670
2260	4″ bituminous paving, 6″ gravel base	1,150	545	1,695
2280	8″ gravel base	1,225	585	1,810
2300	10″ gravel base	1,325	625	1,950
2320	6″ bituminous paving, 6″ gravel base	1,525	585	2,110
2340	8″ gravel base	1,600	620	2,220
2360	10″ gravel base	1,675	660	2,335

G2020 Parking Lots

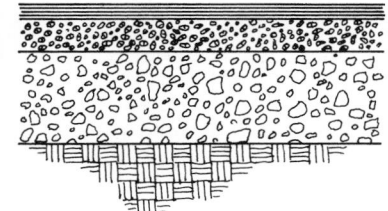

The Parking Lot System includes: compacted crushed stone; fine grading with a grader and roller; and bituminous concrete wearing course. All Parking Lot systems are on a cost per car basis. There are three basic types of systems: 90° angle, 60° angle, and 45° angle. The crushed stone is compacted to 98%. Final stall design and lay-out of the parking lot with precast bumpers, sealcoating and white paint is also included.

The Expanded System Listing shows the three basic parking lot types with various depths of both crushed stone and wearing course. The crushed stone depths range from 6" to 10". The bituminous paving wearing course varies from a depth of 3" to 6".

System Components	QUANTITY	UNIT	COST PER CAR		
			MAT.	INST.	TOTAL
SYSTEM G2020 212 1500					
PARKING LOT, 90° ANGLE PARKING, 3" BITUMINOUS PAVING, 6" CRUSHED STONE					
Surveying crew for layout, 4 man crew	.020	Day		55.76	55.76
Borrow, 3/4" crushed stone, haul 2 mi., spread w/dozer, no compaction	8.645	C.Y.	246.38	74.09	320.47
Grading, fine grade 3 passes with motor grader	43.333	S.Y.		261.30	261.30
Compact w/ vibrating plate, 8" lifts, granular mat'l. to 98%	7.222	E.C.Y.		54.31	54.31
Binder course paving, 2" thick	43.333	S.Y.	353.16	64.13	417.29
Wearing course, 1" thick	43.333	S.Y.	175.50	42.46	217.96
Seal coating, petroleum resistant under 1,000 S.Y.	43.333	S.Y.	72.37	66.73	139.10
Lines on pvmt., parking stall, paint, white, 4" wide	1.000	Ea.	5.70	4.18	9.88
Precast concrete parking bar, 6" x 10" x 6'-0"	1.000	Ea.	63.50	22.50	86
TOTAL			916.61	645.46	1,562.07

G2020 212	Parking Lots Crushed Stone	COST PER CAR		
		MAT.	INST.	TOTAL
1500	Parking lot, 90° angle parking, 3" bituminous paving, 6" crushed stone	915	645	1,560
1520	8" 3/4 inch crushed stone	1,000	690	1,690
1540	10" 3/4 inch crushed stone	1,075	730	1,805
1560	4" bituminous paving, 6" 3/4 inch crushed stone	1,150	675	1,825
1580	8" 3/4 inch crushed stone	1,225	715	1,940
1600	10" 3/4 inch crushed stone	1,300	760	2,060
1620	6" bituminous paving, 6" 3/4 inch crushed stone	1,500	710	2,210
1640	8" 3/4 inch crushed stone	1,575	750	2,325
1660	10" 3/4 inch crushed stone	1,650	795	2,445
1800	60° angle parking, 3" bituminous paving, 6" 3/4 inch crushed stone	915	645	1,560
1820	8" 3/4 inch crushed stone	1,000	690	1,690
1840	10" 3/4 inch crushed stone	1,075	730	1,805
1860	4" bituminous paving, 6" 3/4 inch crushed stone	1,150	675	1,825
1880	8" 3/4 inch crushed stone	1,225	715	1,940
1900	10" 3/4 inch crushed stone	1,300	760	2,060
1920	6" bituminous paving, 6" 3/4 inch crushed stone	1,525	710	2,235
1940	8" 3/4 inch crushed stone	1,575	750	2,325
1960	10" 3/4 inch crushed stone	1,650	795	2,445
2200	45° angle parking, 3" bituminous paving, 6" 3/4 inch crushed stone	760	615	1,375
2220	8" 3/4 inch crushed stone	1,025	705	1,730
2240	10" 3/4 inch crushed stone	1,100	750	1,850
2260	4" bituminous paving, 6" 3/4 inch crushed stone	1,175	690	1,865
2280	8" 3/4 inch crushed stone	1,250	735	1,985
2300	10" 3/4 inch crushed stone	1,325	775	2,100
2320	6" bituminous paving, 6" 3/4 inch crushed stone	1,525	725	2,250
2340	8" 3/4 inch crushed stone	1,625	770	2,395
2360	10" 3/4 inch crushed stone	1,700	810	2,510

G2020 Parking Lots

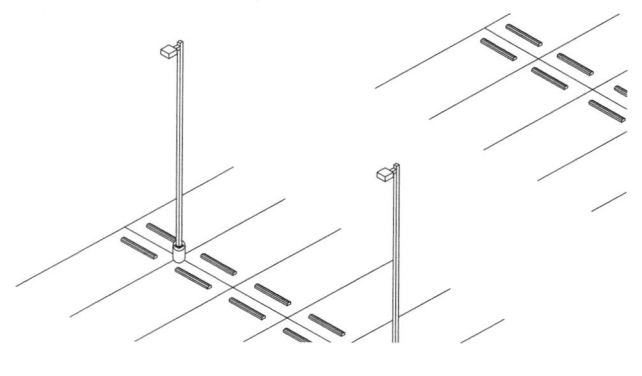

System Components	QUANTITY	UNIT	COST PER EACH		
			MAT.	INST.	TOTAL
SYSTEM G2020 214 1500					
PARKING LOT, 10 CARS WITH HANDICAP AND LIGHTING					
90 ° ANGLE PARKING, 3″ BITUMINOUS PAVING, 6″ CRUSHED STONE					
Surveying crew for layout, 4 man crew	.210	Day		585.48	585.48
Excavate for poles	4.740	B.C.Y.		630.42	630.42
Concrete pole bases	.940	C.Y.	215.26	292.42	507.68
Light poles	2.000	Ea.	1,330	1,445	2,775
Backfill around poles	3.800	L.C.Y.		182.40	182.40
Compact around poles	3.800	E.C.Y.		30.10	30.10
Power feed excavation	228.000	L.F.		239.40	239.40
Undergound conduit	228.000	L.F.	159.60	959.88	1,119.48
Power feed backfill	228.000	L.F.		168.72	168.72
Borrow, 3/4″ crushed stone, haul 2 mi., spread w/dozer, no compaction	91.000	C.Y.	2,593.50	779.87	3,373.37
Grading, fine grade 3 passes with motor grader	231.000	S.Y.		1,392.93	1,392.93
Compact crushed stone	75.840	E.C.Y.		415.60	415.60
Haul asphalt to site 5 miles	101.000	L.C.Y.		810.02	810.02
Binder course paving, 2″ thick	455.000	S.Y.	3,708.25	673.40	4,381.65
Wear course paving 1″ thick	455.000	S.Y.	1,842.75	445.90	2,288.65
Seal coating pavement	455.000	S.Y.	759.85	700.70	1,460.55
Precast concrete parking bar, 6″ x 10″ x 6′-0″	10.500	Ea.	666.75	236.25	903
Stall parking striping	10.500	Ea.	59.85	43.89	103.74
Handicap sign	1.000	Ea.	147	155.20	302.20
400 watt HPS luminaire	2.000	Ea.	1,690	690	2,380
Safety switch for lighting feed	1.000	Ea.	72	237	309
#8 wire feed for poles	4.560	C.L.F.	129.96	430.92	560.88
TOTAL			13,374.77	11,545.50	24,920.27

G2020 214	Parking Lots With Handicap & Lighting	COST PER EACH		
		MAT.	INST.	TOTAL
1490	Parking Lot, 90° Angle Parking, 3″ Bituminous Paving, 6″ Crushed Stone			
1500	10 cars with one handicap & lighting	13,400	11,500	24,900
1510	20 cars with one handicap	24,300	19,500	43,800
1520	30 cars with 2 handicap	37,400	30,000	67,400
1530	40 cars with 2 handicap	50,000	40,100	90,100
1540	50 cars with 3 handicap	61,500	48,500	110,000
1550	60 cars with 3 handicap	74,000	58,500	132,500
1560	70 cars with 4 handicap	87,000	69,000	156,000
1570	80 cars with 4 handicap	98,000	77,000	175,000
1580	90 cars with 5 handicap	111,500	87,500	199,000

G20 Site Improvements

G2020 Parking Lots

G2020 214	Parking Lots With Handicap & Lighting	COST PER EACH		
		MAT.	INST.	TOTAL
1590	100 cars with 5 handicap	122,000	95,500	217,500
1600	110 cars with 6 handicap	135,500	106,000	241,500
1610	120 cars with 6 handicap	146,000	114,000	260,000
1690	Parking Lot, 90° Angle Parking, 3" Bituminous Paving, 8" Crushed Stone			
1700	10 cars with one handicap & lighting	14,200	11,900	26,100
1710	20 cars with one handicap	25,900	20,200	46,100
1720	30 cars with 2 handicap	39,900	31,200	71,100
1730	40 cars with 2 handicap	53,500	41,600	95,100
1740	50 cars with 3 handicap	65,500	50,500	116,000
1750	60 cars with 3 handicap	79,000	61,000	140,000
1760	70 cars with 4 handicap	93,000	72,000	165,000
1770	80 cars with 4 handicap	105,000	80,000	185,000
1780	90 cars with 5 handicap	119,000	91,000	210,000
1790	100 cars with 5 handicap	130,500	99,500	230,000
1800	110 cars with 6 handicap	144,500	110,500	255,000
1810	120 cars with 6 handicap	156,500	119,000	275,500
1890	Parking Lot, 90° Angle Parking, 3" Bituminous Paving, 10" Crushed Stone			
1900	10 cars with one handicap & lighting	15,100	12,300	27,400
1910	20 cars with one handicap	27,600	21,000	48,600
1920	30 cars with 2 handicap	42,600	32,400	75,000
1930	40 cars with 2 handicap	57,000	43,200	100,200
1940	50 cars with 3 handicap	70,000	52,500	122,500
1950	60 cars with 3 handicap	84,000	63,000	147,000
1960	70 cars with 4 handicap	99,000	74,500	173,500
1970	80 cars with 4 handicap	111,500	83,500	195,000
1980	90 cars with 5 handicap	126,500	94,500	221,000
1990	100 cars with 5 handicap	139,000	103,500	242,500
2000	110 cars with 6 handicap	154,000	115,000	269,000
2010	120 cars with 6 handicap	166,500	123,500	290,000
2190	Parking Lot, 90° Angle Parking, 4" Bituminous Paving, 6" Crushed Stone			
2200	10 cars with one handicap & lighting	15,700	11,800	27,500
2210	20 cars with one handicap	28,800	20,100	48,900
2220	30 cars with 2 handicap	44,300	30,900	75,200
2230	40 cars with 2 handicap	59,000	41,200	100,200
2240	50 cars with 3 handicap	73,000	49,900	122,900
2250	60 cars with 3 handicap	87,500	60,500	148,000
2260	70 cars with 4 handicap	103,000	71,000	174,000
2270	80 cars with 4 handicap	116,000	79,500	195,500
2280	90 cars with 5 handicap	131,500	90,000	221,500
2290	100 cars with 5 handicap	145,000	98,500	243,500
2300	110 cars with 6 handicap	160,500	109,500	270,000
2310	120 cars with 6 handicap	173,500	117,500	291,000
2390	Parking Lot, 90° Angle Parking, 4" Bituminous Paving, 8" Crushed Stone			
2400	10 cars with one handicap & lighting	16,600	12,200	28,800
2410	20 cars with one handicap	30,400	20,800	51,200
2420	30 cars with 2 handicap	46,800	32,100	78,900
2430	40 cars with 2 handicap	62,500	42,800	105,300
2440	50 cars with 3 handicap	77,000	52,000	129,000
2450	60 cars with 3 handicap	92,500	62,500	155,000
2460	70 cars with 4 handicap	109,000	74,000	183,000
2470	80 cars with 4 handicap	123,000	82,500	205,500
2480	90 cars with 5 handicap	139,000	93,500	232,500
2490	100 cars with 5 handicap	153,000	102,500	255,500
2500	110 cars with 6 handicap	169,500	113,500	283,000
2510	120 cars with 6 handicap	183,500	122,500	306,000
2590	Parking Lot, 90° Angle Parking, 4" Bituminous Paving, 10" Crushed Stone			
2600	10 cars with one handicap & lighting	17,400	12,600	30,000
2610	20 cars with one handicap	32,200	21,600	53,800

G2020 Parking Lots

G2020 214	Parking Lots With Handicap & Lighting	COST PER EACH		
		MAT.	INST.	TOTAL
2620	30 cars with 2 handicap	49,400	33,300	82,700
2630	40 cars with 2 handicap	66,000	44,400	110,400
2640	50 cars with 3 handicap	81,500	54,000	135,500
2650	60 cars with 3 handicap	98,000	65,000	163,000
2660	70 cars with 4 handicap	115,000	76,500	191,500
2670	80 cars with 4 handicap	130,000	85,500	215,500
2680	90 cars with 5 handicap	147,000	97,000	244,000
2690	100 cars with 5 handicap	162,000	106,500	268,500
2700	110 cars with 6 handicap	179,000	118,000	297,000
2710	120 cars with 6 handicap	194,000	127,000	321,000
2790	Parking Lot, 90° Angle Parking, 5" Bituminous Paving, 6" Crushed Stone			
2800	10 cars with one handicap & lighting	17,600	12,000	29,600
2810	20 cars with one handicap	32,400	20,400	52,800
2820	30 cars with 2 handicap	49,800	31,500	81,300
2830	40 cars with 2 handicap	66,500	42,000	108,500
2840	50 cars with 3 handicap	82,000	51,000	133,000
2850	60 cars with 3 handicap	98,500	61,500	160,000
2860	70 cars with 4 handicap	116,000	72,500	188,500
2870	80 cars with 4 handicap	130,500	81,000	211,500
2880	90 cars with 5 handicap	148,000	92,000	240,000
2890	100 cars with 5 handicap	163,000	100,500	263,500
2900	110 cars with 6 handicap	180,500	111,500	292,000
2910	120 cars with 6 handicap	195,500	120,000	315,500
2990	Parking Lot, 90° Angle Parking, 5" Bituminous Paving, 8" Crushed Stone			
3000	10 cars with one handicap & lighting	18,400	12,400	30,800
3010	20 cars with one handicap	34,100	21,200	55,300
3020	30 cars with 2 handicap	52,500	32,600	85,100
3030	40 cars with 2 handicap	69,500	43,600	113,100
3040	50 cars with 3 handicap	86,000	53,000	139,000
3050	60 cars with 3 handicap	103,500	63,500	167,000
3060	70 cars with 4 handicap	122,000	75,000	197,000
3070	80 cars with 4 handicap	137,500	84,000	221,500
3080	90 cars with 5 handicap	155,500	95,500	251,000
3090	100 cars with 5 handicap	171,500	104,000	275,500
3100	110 cars with 6 handicap	189,500	116,000	305,500
3110	120 cars with 6 handicap	205,500	124,500	330,000
3190	Parking Lot, 90° Angle Parking, 5" Bituminous Paving, 10" Crushed Stone			
3200	10 cars with one handicap & lighting	19,300	12,800	32,100
3210	20 cars with one handicap	35,800	22,000	57,800
3220	30 cars with 2 handicap	55,000	33,800	88,800
3230	40 cars with 2 handicap	73,000	45,100	118,100
3240	50 cars with 3 handicap	90,500	55,000	145,500
3250	60 cars with 3 handicap	108,500	66,000	174,500
3260	70 cars with 4 handicap	128,000	78,000	206,000
3270	80 cars with 4 handicap	144,500	87,000	231,500
3280	90 cars with 5 handicap	163,500	99,000	262,500
3290	100 cars with 5 handicap	180,000	108,000	288,000
3300	110 cars with 6 handicap	199,000	120,000	319,000
3310	120 cars with 6 handicap	215,500	129,500	345,000
3390	Parking Lot, 90° Angle Parking, 6" Bituminous Paving, 6" Crushed Stone			
3400	10 cars with one handicap & lighting	19,400	12,200	31,600
3410	20 cars with one handicap	36,100	20,800	56,900
3420	30 cars with 2 handicap	55,500	32,000	87,500
3430	40 cars with 2 handicap	73,500	42,700	116,200
3440	50 cars with 3 handicap	91,000	51,500	142,500
3450	60 cars with 3 handicap	109,500	62,500	172,000
3460	70 cars with 4 handicap	128,500	73,500	202,000
3470	80 cars with 4 handicap	145,500	82,000	227,500

G2020 Parking Lots

G2020 214	Parking Lots With Handicap & Lighting	COST PER EACH		
		MAT.	INST.	TOTAL
3480	90 cars with 5 handicap	164,500	93,500	258,000
3490	100 cars with 5 handicap	181,000	102,000	283,000
3500	110 cars with 6 handicap	200,500	113,500	314,000
3510	120 cars with 6 handicap	217,000	122,000	339,000
3590	Parking Lot, 90° Angle Parking, 6" Bituminous Paving, 8" Crushed Stone			
3600	10 cars with one handicap & lighting	20,300	12,600	32,900
3610	20 cars with one handicap	37,700	21,500	59,200
3620	30 cars with 2 handicap	58,000	33,100	91,100
3630	40 cars with 2 handicap	77,000	44,200	121,200
3640	50 cars with 3 handicap	95,500	53,500	149,000
3650	60 cars with 3 handicap	114,500	64,500	179,000
3660	70 cars with 4 handicap	134,500	76,500	211,000
3670	80 cars with 4 handicap	152,000	85,500	237,500
3680	90 cars with 5 handicap	172,000	97,000	269,000
3690	100 cars with 5 handicap	189,500	106,000	295,500
3700	110 cars with 6 handicap	209,500	117,500	327,000
3710	120 cars with 6 handicap	227,000	126,500	353,500
3790	Parking Lot, 90° Angle Parking, 6" Bituminous Paving, 10" Crushed Stone			
3800	10 cars with one handicap & lighting	21,200	13,000	34,200
3810	20 cars with one handicap	39,500	22,300	61,800
3820	30 cars with 2 handicap	60,500	34,300	94,800
3830	40 cars with 2 handicap	80,500	45,800	126,300
3840	50 cars with 3 handicap	99,500	55,500	155,000
3850	60 cars with 3 handicap	119,500	67,000	186,500
3860	70 cars with 4 handicap	140,500	79,000	219,500
3870	80 cars with 4 handicap	159,000	88,500	247,500
3880	90 cars with 5 handicap	180,000	100,500	280,500
3890	100 cars with 5 handicap	198,000	110,000	308,000
3900	110 cars with 6 handicap	219,000	122,000	341,000
3910	120 cars with 6 handicap	237,500	131,500	369,000

G2030 Pedestrian Paving

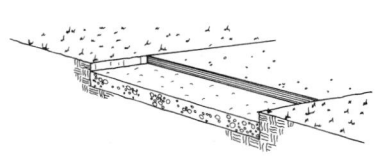

The Bituminous Sidewalk System includes: excavation; compacted gravel base (hand graded), bituminous surface; and hand grading along the edge of the completed walk.

The Expanded System Listing shows Bituminous Sidewalk systems with wearing course depths ranging from 1" to 2-1/2" of bituminous material. The gravel base ranges from 4" to 8". Sidewalk widths are shown ranging from 3' to 5'. Costs are on a linear foot basis.

System Components	QUANTITY	UNIT	COST PER L.F.		
			MAT.	INST.	TOTAL
SYSTEM G2030 110 1580					
BITUMINOUS SIDEWALK, 1" THICK PAVING, 4" GRAVEL BASE, 3' WIDTH					
Excavation, bulk, dozer, push 50'	.046	B.C.Y.		.11	.11
Borrow, bank run gravel, haul 2 mi., spread w/dozer, no compaction	.037	L.C.Y.	1.20	.32	1.52
Compact w/vib. plate, 8" lifts	.037	E.C.Y.		.11	.11
Fine grade, area to be paved, small area	.333	S.Y.		4.17	4.17
Asphaltic concrete, 2" thick	.333	S.Y.	1.33	.84	2.17
Backfill by hand, no compaction, light soil	.006	L.C.Y.		.23	.23
TOTAL			2.53	5.78	8.31

G2030 110	Bituminous Sidewalks	COST PER L.F.		
		MAT.	INST.	TOTAL
1580	Bituminous sidewalk, 1" thick paving, 4" gravel base, 3' width	2.53	5.75	8.28
1600	4' width	3.36	6.20	9.56
1620	5' width	4.27	6.50	10.77
1640	6" gravel base, 3' width	3.15	6	9.15
1660	4' width	4.17	6.55	10.72
1680	5' width	5.25	7.10	12.35
1700	8" gravel base, 3' width	3.73	6.30	10.03
1720	4' width	4.98	6.90	11.88
1740	5' width	6.25	7.55	13.80
1800	1-1/2" thick paving, 4" gravel base, 3' width	3.26	6	9.26
1820	4' width	4.31	6.60	10.91
1840	5' width	5.35	7.50	12.85
1860	6" gravel base, 3' width	3.88	6.25	10.13
1880	4' width	5.05	7.25	12.30
1900	5' width	6.40	7.65	14.05
1920	8" gravel base, 3' width	4.39	6.80	11.19
1940	4' width	5.95	7.30	13.25
1960	5' width	7.35	8.35	15.70
2120	2" thick paving, 4" gravel base, 3' width	3.85	6.85	10.70
2140	4' width	5.10	7.60	12.70
2160	5' width	6.45	8.35	14.80
2180	6" gravel base, 3' width	4.47	7.10	11.57
2200	4' width	5.95	7.95	13.90
2240	8" gravel base, 3' width	5.05	7.35	12.40
2280	5' width	8.45	9.20	17.65
2400	2-1/2" thick paving, 4" gravel base, 3' width	5.40	7.60	13
2420	4' width	7.20	8.50	15.70
2460	6" gravel base, 3' width	6.05	7.85	13.90
2500	5' width	10.10	9.95	20.05
2520	8" gravel base, 3' width	6.60	8.10	14.70
2540	4' width	8.80	9.20	18
2560	5' width	11.10	10.35	21.45

G20 Site Improvements

G2030 Pedestrian Paving

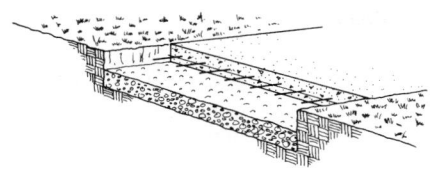

The Concrete Sidewalk System includes: excavation; compacted gravel base (hand graded); forms; welded wire fabric; and 3,000 p.s.i. air-entrained concrete (broom finish).

The Expanded System Listing shows Concrete Sidewalk systems with wearing course depths ranging from 4″ to 6″. The gravel base ranges from 4″ to 8″. Sidewalk widths are shown ranging from 3′ to 5′. Costs are on a linear foot basis.

System Components	QUANTITY	UNIT	COST PER L.F.		
			MAT.	INST.	TOTAL
SYSTEM G2030 120 1580					
CONCRETE SIDEWALK 4″ THICK, 4″ GRAVEL BASE, 3′ WIDE					
Excavation, box out with dozer	.100	B.C.Y.		.24	.24
Gravel base, haul 2 miles, spread with dozer	.037	L.C.Y.	1.20	.32	1.52
Compaction with vibrating plate	.037	E.C.Y.		.11	.11
Fine grade by hand	.333	S.Y.		4.23	4.23
Concrete in place including forms and reinforcing	.037	C.Y.	7.35	8.94	16.29
Backfill edges by hand	.010	L.C.Y.		.38	.38
TOTAL			8.55	14.22	22.77

G2030 120	Concrete Sidewalks	COST PER L.F.		
		MAT.	INST.	TOTAL
1580	Concrete sidewalk, 4″ thick, 4″ gravel base, 3′ wide	8.55	14.20	22.75
1600	4′ wide	11.40	17.40	28.80
1620	5′ wide	14.25	20.50	34.75
1640	6″ gravel base, 3′ wide	9.15	14.45	23.60
1660	4′ wide	12.20	17.70	29.90
1680	5′ wide	15.25	21	36.25
1700	8″ gravel base, 3′ wide	9.75	14.70	24.45
1720	4′ wide	13	18.05	31.05
1740	5′ wide	16.25	21.50	37.75
1800	5″ thick concrete, 4″ gravel base, 3′ wide	10.30	15.35	25.65
1820	4′ wide	13.75	18.85	32.60
1840	5′ wide	17.20	22.50	39.70
1860	6″ gravel base, 3′ wide	10.95	15.60	26.55
1880	4′ wide	14.55	19.20	33.75
1900	5′ wide	18.20	22	40.20
1920	8″ gravel base, 3′ wide	11.55	15.85	27.40
1940	4′ wide	15.40	20.50	35.90
1960	5′ wide	19.20	22.50	41.70
2120	6″ thick concrete, 4″ gravel base, 3′ wide	11.85	16.15	28
2140	4′ wide	15.80	19.90	35.70
2160	5′ wide	19.75	23.50	43.25
2180	6″ gravel base, 3′ wide	12.45	16.45	28.90
2200	4′ wide	16.60	20	36.60
2220	5′ wide	21	24	45
2240	8″ gravel base, 3′ wide	13.05	16.70	29.75
2260	4′ wide	17.40	20.50	37.90
2280	5′ wide	22	24.50	46.50

669

G2030 Pedestrian Paving

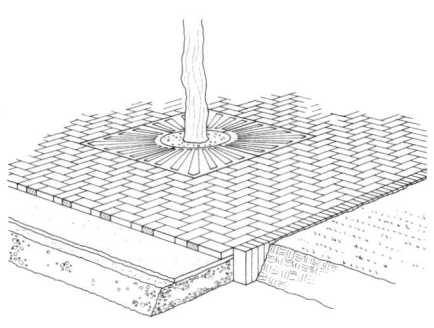

The Plaza Systems listed include several brick and tile paving surfaces on three different bases: gravel, slab on grade and suspended slab. The system cost includes this base cost with the exception of the suspended slab. The type of bedding for the pavers depends on the base being used, and alternate bedding may be desirable. Also included in the paving costs are edging and precast grating costs and where concrete bases are involved, expansion joints.

System Components	QUANTITY	UNIT	COST PER S.F. MAT.	COST PER S.F. INST.	COST PER S.F. TOTAL
SYSTEM G2030 150 2050					
PLAZA, BRICK PAVERS, 4″ X 8″ X 1-3/4″, GRAVEL BASE, STONE DUST BED					
Compact subgrade, static roller, 4 passes	.111	S.Y.		.05	.05
Bank gravel, 2 mi haul, dozer spread	.012	L.C.Y.	.40	.11	.51
Compact gravel bedding or base, vibrating plate	.012	E.C.Y.		.04	.04
Grading fine grade, 3 passes with grader	.111	S.Y.		.67	.67
Coarse washed sand bed, 1″ thick	.003	C.Y.	.06	.08	.14
Brick paver, 4″ x 8″ x 1-3/4″	4.150	Ea.	4.59	4.89	9.48
Brick edging, stood on end, 3 per L.F.	.060	L.F.	.37	.73	1.10
Concrete tree grate, 5′ square	.004	Ea.	1.83	.31	2.14
TOTAL			7.25	6.88	14.13

G2030 150	Brick & Tile Plazas	COST PER S.F. MAT.	COST PER S.F. INST.	COST PER S.F. TOTAL
1050	Plaza, asphalt pavers, 6″ x 12″ x 1-1/4″, gravel base, asphalt bedding	13.80	12.60	26.40
1100	Slab on grade, asphalt bedding	16.25	13.65	29.90
1150	Suspended slab, insulated & mastic bedding	18.45	16.75	35.20
1300	6″ x 12″ x 3″, gravel base, asphalt, bedding	19	13.05	32.05
1350	Slab on grade, asphalt bedding	20.50	14	34.50
1400	Suspended slab, insulated & mastic bedding	23	17.10	40.10
2050	Brick pavers, 4″ x 8″ x 1-3/4″, gravel base, stone dust bedding	7.25	6.85	14.10
2100	Slab on grade, asphalt bedding	9.80	9.85	19.65
2150	Suspended slab, insulated & no bedding	11.95	12.95	24.90
2300	4″ x 8″ x 2-1/4″, gravel base, stone dust bedding	6	6.85	12.85
2350	Slab on grade, asphalt bedding	8.50	9.85	18.35
2400	Suspended slab, insulated & no bedding	10.70	12.95	23.65
2550	Shale pavers, 4″ x 8″ x 2-1/4″, gravel base, stone dust bedding	6.80	7.80	14.60
2600	Slab on grade, asphalt bedding	9.30	10.80	20.10
2650	Suspended slab, insulated & no bedding	6.25	9.05	15.30
3050	Thin set tile, 4″ x 4″ x 3/8″, slab on grade	9.60	10.95	20.55
3300	4″ x 4″ x 3/4″, slab on grade	11.05	7.30	18.35
3550	Concrete paving stone, 4″ x 8″ x 2-1/2″, gravel base, sand bedding	5.30	4.85	10.15
3600	Slab on grade, asphalt bedding	7.25	7.20	14.45
3650	Suspended slab, insulated & no bedding	4.19	5.45	9.64
3800	4″ x 8″ x 3-1/4″, gravel base, sand bedding	5.30	4.85	10.15
3850	Slab on grade, asphalt bedding	6.55	5.60	12.15
3900	Suspended slab, insulated & no bedding	4.19	5.45	9.64
4050	Concrete patio blocks, 8″ x 16″ x 2″, gravel base, sand bedding	15.10	6.30	21.40
4100	Slab on grade, asphalt bedding	17.35	9.10	26.45
4150	Suspended slab, insulated & no bedding	14	6.90	20.90

G2030 Pedestrian Paving

G2030 150	Brick & Tile Plazas	COST PER S.F.		
		MAT.	INST.	TOTAL
4300	16" x 16" x 2", gravel base, sand bedding	20	5.05	25.05
4350	Slab on grade, asphalt bedding	22	6.30	28.30
4400	Suspended slab, insulated & no bedding	19.40	6.15	25.55
4550	24" x 24" x 2", gravel base, sand bedding	24.50	3.87	28.37
4600	Slab on grade, asphalt bedding	27	6.70	33.70
4650	Suspended slab, insulated & no bedding	24	4.95	28.95
5050	Bluestone flagging, 3/4" thick irregular, gravel base, sand bedding	15.40	14.90	30.30
5100	Slab on grade, mastic bedding	22	20.50	42.50
5300	1" thick, irregular, gravel base, sand bedding	13.50	16.35	29.85
5350	Slab on grade, mastic bedding	19.95	21.50	41.45
5550	Flagstone, 3/4" thick irregular, gravel base, sand bedding	25.50	14.20	39.70
5600	Slab on grade, mastic bedding	32	19.60	51.60
5800	1-1/2" thick, random rectangular, gravel base, sand bedding	27	16.35	43.35
5850	Slab on grade, mastic bedding	33.50	21.50	55
6050	Granite pavers, 3-1/2" x 3-1/2" x 3-1/2", gravel base, sand bedding	27	15	42
6100	Slab on grade, mortar bedding	30.50	24	54.50
6300	4" x 4" x 4", gravel base, sand bedding	28	14.60	42.60
6350	Slab on grade, mortar bedding	30.50	19.50	50
6550	4" x 12" x 4", gravel base, sand bedding	23	14.25	37.25
6600	Slab on grade, mortar bedding	25.50	19.15	44.65
6800	6" x 15" x 4", gravel base, sand bedding	13.75	13.45	27.20
6850	Slab on grade, mortar bedding	16.05	18.35	34.40
7050	Limestone, 3" thick, gravel base, sand bedding	14.05	18.55	32.60
7100	Slab on grade, mortar bedding	16.35	23.50	39.85
7300	4" thick, gravel base, sand bedding	17.60	19	36.60
7350	Slab on grade, mortar bedding	19.90	24	43.90
7550	5" thick, gravel base, sand bedding	21	19.50	40.50
7600	Slab on grade, mortar bedding	23.50	24.50	48
8050	Slate flagging, 3/4" thick, gravel base, sand bedding	13.85	15	28.85
8100	Slab on grade, mastic bedding	29	21.50	50.50
8300	1" thick, gravel base, sand bedding	15.60	16.05	31.65
8350	Slab on grade, mastic bedding	22.50	22	44.50

G2030 Pedestrian Paving

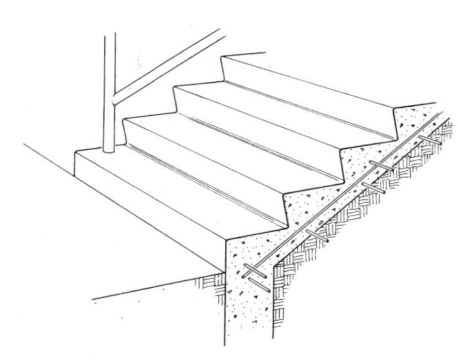

The Step System has three basic types: railroad tie, cast-in-place concrete or brick with a concrete base. System elements include: gravel base compaction; and backfill.

Wood Step Systems use 6" x 6" railroad ties that produce 3' to 6' wide steps that range from 2-riser to 5-riser configurations. Cast in Place Concrete Step Systems are either monolithic or aggregate finish. They range from 3' to 6' in width. Concrete Steps Systems are either in a 2-riser or 5-riser configuration. Precast Concrete Step Systems are listed for 4' to 7' widths with both 2-riser and 5-riser models. Brick Step Systems are placed on a 12" concrete base. The size range is the same as cost in place concrete. Costs are on a per unit basis. All systems are assumed to include a full landing at the top 4' long.

System Components			COST PER EACH		
	QUANTITY	UNIT	MAT.	INST.	TOTAL
SYSTEM G2030 310 0960					
STAIRS, RAILROAD TIES, 6" X 8", 3' WIDE, 2 RISERS					
Excavation, by hand, sandy soil	1.327	C.Y.		88.25	88.25
Borrow fill, bank run gravel	.664	C.Y.	19.59		19.59
Delivery charge	.664	C.Y.		19.75	19.75
Backfill compaction, 6" layers, hand tamp	.664	C.Y.		16.93	16.93
Railroad ties, wood, creosoted, 6" x 8" x 8'-6", C.L. lots	5.000	Ea.	307.50	198.20	505.70
Backfill by hand, no compaction, light soil	.800	C.Y.		30.40	30.40
Remove excess spoil, 12 C.Y. dump truck	1.200	L.C.Y.		8.34	8.34
TOTAL			327.09	361.87	688.96

G2030 310	Stairs	COST PER EACH		
		MAT.	INST.	TOTAL
0960	Stairs, railroad ties, 6" x 8", 3' wide, 2 risers	325	360	685
0980	5 risers	765	730	1,495
1000	4' wide, 2 risers	330	400	730
1020	5 risers	775	765	1,540
1040	5' wide, 2 risers	395	470	865
1060	5 risers	965	940	1,905
1080	6' wide, 2 risers	400	505	905
1100	5 risers	970	970	1,940
2520	Concrete, cast in place, 3' wide, 2 risers	248	515	763
2540	5 risers	330	730	1,060
2560	4' wide, 2 risers	305	675	980
2580	5 risers	405	940	1,345
2600	5' wide, 2 risers	365	840	1,205
2620	5 risers	470	1,125	1,595
2640	6' wide, 2 risers	420	980	1,400
2660	5 risers	545	1,300	1,845
2800	Exposed aggregate finish, 3' wide, 2 risers	251	530	781
2820	5 risers	335	730	1,065
2840	4' wide, 2 risers	310	675	985
2860	5 risers	410	940	1,350
2880	5' wide, 2 risers	375	840	1,215
2900	5 risers	475	1,125	1,600
2920	6' wide, 2 risers	425	980	1,405
2940	5 risers	555	1,300	1,855
3200	Precast, 4' wide, 2 risers	690	375	1,065
3220	5 risers	1,100	555	1,655

G2030 Pedestrian Paving

G2030 310	Stairs	COST PER EACH		
		MAT.	INST.	TOTAL
3240	5' wide, 2 risers	780	450	1,230
3260	5 risers	1,425	600	2,025
3280	6' wide, 2 risers	835	475	1,310
3300	5 risers	1,575	645	2,220
3320	7' wide, 2 risers	1,200	505	1,705
3340	5 risers	1,900	690	2,590
4100	Brick, incl 12" conc base, 3' wide, 2 risers	415	1,325	1,740
4120	5 risers	595	1,975	2,570
4140	4' wide, 2 risers	520	1,625	2,145
4160	5 risers	740	2,425	3,165
4180	5' wide, 2 risers	625	1,925	2,550
4200	5 risers	875	2,850	3,725
4220	6' wide, 2 risers	720	2,225	2,945
4240	5 risers	1,025	3,275	4,300

For customer support on your Site Work & Landscape Costs with RSMeans Data, call 800.448.8182.

G2040 Site Development

System Components	QUANTITY	UNIT	COST PER C.L.F.		
			MAT.	INST.	TOTAL
SYSTEM G2040 110 1000					
FENCE, CHAIN LINK, RESIDENTIAL 2-1/2″ TPOST, 1-5/8″ RAIL, 2-1/2″ MESH, 4′					
16 ga., steel, 2-1/2″ x 6′ x 0.065 wall, incl. post cap, excavation	2.000	Ea.	29.40	47.96	77.36
16 ga., steel, 1-5/8″ x 6′ x 0.065 wall, incl. post cap and excavation	12.000	Ea.	132.60	287.76	420.36
Fence, rail tubing, 16 ga., 1-3/8″ x 21′ x 0.065 wall incl. hardware	100.000	L.F.	134	952	1,086
Fence, fabric, steel, galv., 11-1/2 ga., 2-1/4″ mesh, 4′ H, incl. hardware	100.000	L.F.	294	952	1,246
Concrete, hand mix, bulk ingredients w/ gas powered cement mixer, 3000 psi	17.000	C.F.	88.40	87.21	175.61
Auger fence post hole, 3′ deep, medium soil, by hand	10.000	Ea.		176.50	176.50
TOTAL			678.40	2,503.43	3,181.83

G2040 110	Fences - Residential	COST PER C.L.F.		
		MAT.	INST.	TOTAL
1000	Fence, chain link, Residential 2-1/2″ post, 1-5/8″ rail, 2-1/2″ mesh, 4′	680	2,500	3,180
1050	Fence, chain link, Residential 2-1/2″ post, 1-5/8″ rail, 2-1/2″ mesh, 5′	775	2,525	3,300
1100	Fence, chain link, Residential 2-1/2″ post, 1-5/8″ rail, 2-1/2″ mesh, 6′	930	2,525	3,455

G2040 115	Fences - Internal	COST PER C.L.F.		
		MAT.	INST.	TOTAL
1000	Fence, chain link, Internal 2-1/2″ post, 1-5/8″ rail, 2″ mesh, 6′	1,050	2,275	3,325
1050	Fence, chain link, Internal 2-1/2″ post, 1-5/8″ rail, 2″ mesh, 8′	1,225	2,275	3,500
1100	Fence, chain link, Internal 4″ post, 1-5/8″ rail, 2″ mesh, 10′	2,500	2,275	4,775
1150	Fence, chain link, Internal 4″ post, 1-5/8″ rail, 2″ mesh, 12′	2,850	2,275	5,125

G2040 120	Fences - Industrial	COST PER C.L.F.		
		MAT.	INST.	TOTAL
1000	Fence, chain link, Industrial 4″ post, 2″ rail, 2″ mesh, 8′ high	3,150	2,650	5,800
1050	Fence, chain link, Industrial 4″ post, 2″ rail, 2″ mesh, 10′ high	3,525	2,700	6,225
1100	Fence, chain link, Industrial 4″ post, 2″ rail, 2″ mesh, 12′ high	4,375	2,750	7,125
1150	Fence, chain link, Industrial 4″ post, 2″ rail, 2″ mesh, 14′ high	4,450	2,775	7,225

G2040 125	Fences - Security	COST PER C.L.F.		
		MAT.	INST.	TOTAL
1000	Fence, chain link, Security 4″ post, 2″ rail, 1″ mesh, 8′ high	3,925	3,600	7,525
1050	Fence, chain link, Security 4″ post, 2″ rail, 1″ mesh, 10′ high	5,075	3,650	8,725
1100	Fence, chain link, Security 4″ post, 2″ rail, 1″ mesh, 12′ high	6,075	3,700	9,775

G2040 Site Development

There are four basic types of Concrete Retaining Wall Systems: reinforced concrete with level backfill; reinforced concrete with sloped backfill or surcharge; unreinforced with level backfill; and unreinforced with sloped backfill or surcharge. System elements include: all necessary forms (4 uses); 3,000 p.s.i. concrete with an 8″ chute; all necessary reinforcing steel; and underdrain. Exposed concrete is patched and rubbed.

The Expanded System Listing shows walls that range in thickness from 10″ to 18″ for reinforced concrete walls with level backfill and 12″ to 24″ for reinforced walls with sloped backfill. Walls range from a height of 4′ to 20′. Unreinforced level and sloped backfill walls range from a height of 3′ to 10′.

System Components	QUANTITY	UNIT	COST PER L.F.		
			MAT.	INST.	TOTAL
SYSTEM G2040 210 1000					
CONC. RETAIN. WALL REINFORCED, LEVEL BACKFILL, 4′ HIGH					
Forms in place, cont. wall footing & keyway, 4 uses	2.000	S.F.	5.64	10.30	15.94
Forms in place, retaining wall forms, battered to 8′ high, 4 uses	8.000	SFCA	8.80	78.40	87.20
Reinforcing in place, walls, #3 to #7	.004	Ton	5.50	3.76	9.26
Concrete ready mix, regular weight, 3000 psi	.204	C.Y.	27.95		27.95
Placing concrete and vibrating footing con., shallow direct chute	.074	C.Y.		2.08	2.08
Placing concrete and vibrating walls, 8″ thick, direct chute	.130	C.Y.		4.84	4.84
Pipe bedding, crushed or screened bank run gravel	1.000	L.F.	4.02	1.44	5.46
Pipe, subdrainage, corrugated plastic, 4″ diameter	1.000	L.F.	.89	.88	1.77
Finish walls and break ties, patch walls	4.000	S.F.	.20	4.48	4.68
TOTAL			53	106.18	159.18

G2040 210	Concrete Retaining Walls	COST PER L.F.		
		MAT.	INST.	TOTAL
1000	Conc. retain. wall, reinforced, level backfill, 4′ high x 2′-2″ base,10″ th	53	106	159
1200	6′ high x 3′-3″ base, 10″ thick	76	154	230
1400	8′ high x 4′-3″ base, 10″ thick	99	201	300
1600	10′ high x 5′-4″ base, 13″ thick	129	295	424
2200	16′ high x 8′-6″ base, 16″ thick	241	480	721
2600	20′ high x 10′-5″ base, 18″ thick	350	620	970
3000	Sloped backfill, 4′ high x 3′-2″ base, 12″ thick	64	110	174
3200	6′ high x 4′-6″ base, 12″ thick	90	158	248
3400	8′ high x 5′-11″ base, 12″ thick	119	208	327
3600	10′ high x 7′-5″ base, 16″ thick	170	310	480
3800	12′ high x 8′-10″ base, 18″ thick	221	375	596
4200	16′ high x 11′-10″ base, 21″ thick	370	530	900
4600	20′ high x 15′-0″ base, 24″ thick	570	715	1,285
5000	Unreinforced, level backfill, 3′-0″ high x 1′-6″ base	28	72	100
5200	4′-0″ high x 2′-0″ base	43	95	138
5400	6′-0″ high x 3′-0″ base	80	147	227
5600	8′-0″ high x 4′-0″ base	126	196	322
5800	10′-0″ high x 5′-0″ base	186	305	491
7000	Sloped backfill, 3′-0″ high x 2′-0″ base	33.50	74.50	108
7200	4′-0″ high x 3′-0″ base	55	99	154
7400	6′-0″ high x 5′-0″ base	114	156	270
7600	8′-0″ high x 7′-0″ base	191	213	404
7800	10′-0″ high x 9′-0″ base	292	330	622

G2040 Site Development

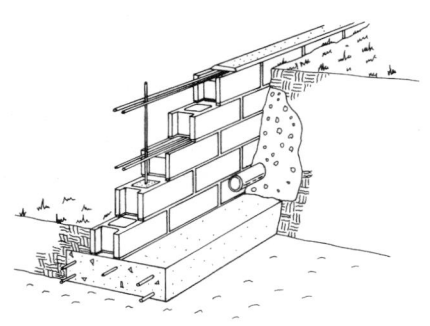

The Masonry Retaining Wall System includes: a reinforced concrete foundation footing for the wall with all necessary forms and 3,000 p.s.i. concrete; sand aggregate concrete blocks with steel reinforcing; solid grouting and underdrain.
The Expanded System Listing shows walls that range in thickness from 8″ to 12″ and in height from 3′-4″ to 8′.

System Components	QUANTITY	UNIT	COST PER L.F.		
			MAT.	INST.	TOTAL
SYSTEM G2040 220 1000					
REINF. CMU WALL, LEVEL FILL, 8″ THICK, 3′-4″ HIGH, 2′-4″ CONC. FTG.					
Forms in place, continuous wall footings 4 uses	1.670	S.F.	4.71	8.60	13.31
Joint reinforcing, #5 & #6 steel bars, vertical	5.590	Lb.	3.80	5.59	9.39
Concrete block, found. wall, cut joints, 8″ x 16″ x 8″ block	3.330	S.F.	12.19	23.48	35.67
Grout concrete block cores solid, 8″ thick, .258 CF/SF, pumped	3.330	S.F.	5.23	12.59	17.82
Concrete ready mix, regular wt., 3000 psi	.070	C.Y.	8.45		8.45
Pipe bedding, crushed or screened bank run gravel	1.000	L.F.	4.02	1.44	5.46
Pipe, subdrainage, corrugated plastic, 4″ diameter	1.000	L.F.	.89	.88	1.77
TOTAL			39.29	52.58	91.87

G2040 220	Masonry Retaining Walls	COST PER L.F.		
		MAT.	INST.	TOTAL
1000	Wall, reinforced CMU, level fill, 8″ thick, 3′-4″ high, 2′-4″ base	39.50	52.50	92
1200	4′-0″ high x 2′-9″ base	46	63.50	109.50
1400	4′-8″ high x 3′-3″ base	55	75	130
1600	5′-4″ high x 3′-8″ base	63.50	85.50	149
1800	6′-0″ high x 4′-2″ base	75	97.50	172.50
1840	10″ thick, 4′-0″ high x 2′-10″ base	47	64.50	111.50
1860	4′-8″ high x 3′-4″ base	56	76	132
1880	5′-4″ high x 3′-10″ base	65	86.50	151.50
1900	6′-0″ high x 4′-4″ base	76.50	99	175.50
1901	6′-0″ high x 4′-4″ base	76.50	99	175.50
1920	6′-8″ high x 4′-10″ base	84	109	193
1940	7′-4″ high x 5′-4″ base	100	123	223
2000	12″ thick, 5′-4″ high x 4′-0″ base	77	111	188
2200	6′-0″ high x 4′-6″ base	91	127	218
2400	6′-8″ high x 5′-0″ base	100	140	240
2600	7′-4″ high x 5′-6″ base	117	158	275
2800	8′-0″ high x 5′-11″ base	126	171	297

G2040 Site Development

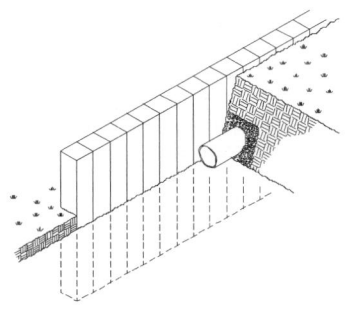

Wood Post Retaining Wall Systems are either constructed of redwood, cedar, pressure-treated lumber or creosoted lumber. System elements include wood posts installed side-by-side to various heights; and underdrain.

The Expanded System Listing shows a wide variety of configurations with each type of wood. Costs are on a linear foot basis. Each grouping in the listing shows walls with either 4" x 4", 6" x 6", 8" x 8", or random diameter lumber.

System Components	QUANTITY	UNIT	COST PER L.F.		
			MAT.	INST.	TOTAL
SYSTEM G2040 230 1000					
WOOD POST RETAINING WALL, SIDE-BY-SIDE 4 X 4 REDWOOD, 1' HIGH					
Redwood post, columns and girts, 4" x 4"	8.000	B.F.	74.40	20	94.40
Pipe bedding, crushed or screened bank run gravel	1.000	L.F.	1.10	.40	1.50
Pipe, subdrainage, corrugated plastic, 4" diameter	1.000	L.F.	.89	.88	1.77
TOTAL			76.39	21.28	97.67

G2040 230	Wood Post Retaining Walls	COST PER L.F.		
		MAT.	INST.	TOTAL
1000	Wood post retaining wall, side by side 4 x 4, redwood, 1' high	76.50	21.50	98
1200	2' high	152	41.50	193.50
1400	Cedar, 1' high	37	21.50	58.50
1600	2' high	72.50	41.50	114
1800	Pressure treated lumber, 1' high	16.80	18.65	35.45
2000	2' high	32.50	36.50	69
2200	Creosoted lumber, 1' high	14.95	16.50	31.45
2400	2' high	27	30	57
2600	6 x 6, redwood, 2' high	315	49.50	364.50
2800	4' high	625	98.50	723.50
3000	Cedar, 2' high	151	49.50	200.50
3200	4' high	300	98.50	398.50
3400	Pressure treated lumber, 2' high	56.50	40.50	97
3600	4' high	112	80.50	192.50
3800	Creosoted lumber, 2' high	45	32.50	77.50
4000	4' high	89.50	64	153.50
4200	8 x 8, redwood, 4' high	795	109	904
4400	6' high	1,500	204	1,704
4600	Cedar, 4' high	345	109	454
4800	6' high	650	204	854
5000	Pressure treated lumber, 4' high	207	126	333
5200	6' high	385	236	621
5400	Creosoted lumber, 4' high	168	102	270
5600	6' high	310	190	500
5800	Random diameter poles, cedar, 4' high	74.50	42.50	117
6000	6' high	111	63	174
6200	8' high	148	84	232
7000	Creosoted lumber, 4' high	34	23.50	57.50
7200	6' high	49	34	83
7400	8' high	67	46	113

G2040 Site Development

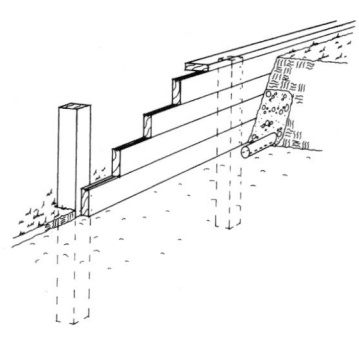

A Post and Board Retaining Wall System is constructed of one of four types of lumber: redwood, cedar, creosoted wood or pressure-treated lumber. The system includes all the elements that must go into a wall that resists lateral pressure. System elements include: posts and boards; cap; underdrain; and deadman where required.

The Expanded System Listing shows a wide variety of wall configurations with each type of lumber. The spacing of the deadman in the wall is indicated for each wall system. Wall heights vary from 3' to 10'.

System Components	QUANTITY	UNIT	COST PER L.F.		
			MAT.	INST.	TOTAL
SYSTEM G2040 240 1000					
POST & BOARD RETAINING WALL, 2" PLANKS, 4 X 4 POSTS, 3' HIGH					
Redwood post, 4" x 4"	2.000	B.F.	18.60	5	23.60
Redwood joist and cap, 2" x 10"	7.000	B.F.	52.40	5.96	58.36
Pipe bedding, crushed or screened bank run gravel	1.000	L.F.	2.01	.72	2.73
Pipe, subdrainage, corrugated plastic, 4" diameter	1.000	L.F.	.89	.88	1.77
TOTAL			73.90	12.56	86.46

G2040 240	Post & Board Retaining Walls	COST PER L.F.		
		MAT.	INST.	TOTAL
1000	Post & board retaining wall, 2" planking, redwood, 4 x 4 post, 4' spacing, 3' high	74	12.55	86.55
1200	2' spacing, 4' high	106	21.50	127.50
1400	6 x 6 post, 4' spacing, 5' high	174	26.50	200.50
1600	3' spacing, 6' high	246	37.50	283.50
1800	6 x 6 post with deadman, 3' spacing, 8' high	325	49	374
2000	8 x 8 post, 2' spacing, 8' high	715	98	813
2200	8 x 8 post with deadman, 2' spacing, 9' high	815	109	924
2400	Cedar, 4 x 4 post, 4' spacing, 3' high	36	13.15	49.15
2600	4 x 4 post with deadman, 2' spacing, 4' high	70.50	23	93.50
2800	6 x 6 post, 4' spacing, 4' high	73.50	22.50	96
3000	3' spacing, 5' high	105	32.50	137.50
3200	2' spacing, 6' high	167	52	219
3400	6 x 6 post with deadman, 3' spacing, 6' high	130	40	170
3600	8 x 8 post with deadman, 2' spacing, 8' high	345	101	446
3800	Pressure treated, 4 x 4 post, 4' spacing, 3' high	16.20	24.50	40.70
4000	3' spacing, 4' high	24.50	34.50	59
4200	4 x 4 post with deadman, 2' spacing, 5' high	48.50	47	95.50
4400	6 x 6 post, 3' spacing, 6' high	107	78.50	185.50
4600	2' spacing, 7' high	77.50	80	157.50
4800	6 x 6 post with deadman, 3' spacing, 8' high	79.50	77	156.50
5000	8 x 8 post with deadman, 2' spacing, 10' high	255	178	433
5200	Creosoted, 4 x 4 post, 4' spacing, 3' high	16.20	24.50	40.70
5400	3' spacing, 4' high	24.50	34.50	59
5600	4 x 4 post with deadman, 2' spacing, 5' high	48.50	47	95.50
5800	6 x 6 post, 3' spacing, 6' high	53	59.50	112.50
6000	2' spacing, 7' high	77.50	80	157.50
6200	6 x 6 post with deadman, 3' spacing 8' high	79.50	77	156.50
6400	8 x 8 post with deadman, 2' spacing, 10' high	255	178	433

G2040 Site Development

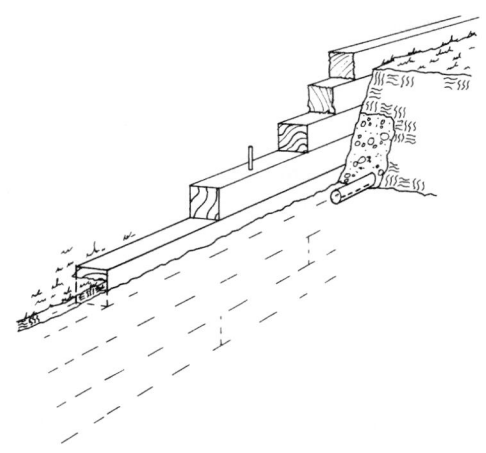

A Wood Tie Retaining Wall System is constructed of one of four types of lumber: redwood, cedar, creosoted wood or pressure-treated lumber. The system includes all the elements that must go into a wall that resists lateral pressure. System elements include: wood ties; threaded rod (1/2" diameter); underdrain; and deadman where required.

The Expanded System Listing shows a wide variety of wall configurations with each type of lumber. The spacing of the deadman in the wall is indicated for each wall system. Wall heights vary from 2' to 8'.

System Components	QUANTITY	UNIT	COST PER L.F.		
			MAT.	INST.	TOTAL
SYSTEM G2040 250 1000					
WOOD TIE WALL, ROD CONNECTOR AT 4'-0", REDWOOD, 6" X 6", 2'-0" HIGH					
Redwood post columns and girts, 6" x 6"	12.000	B.F.	154.80	24	178.80
Accessories, hangers, coil threaded rods, continuous, 1/2" diam.	.050	L.F.	.85		.85
Pipe bedding, crushed or screened bank run gravel	1.000	L.F.	2.01	.72	2.73
Pipe, subdrainage, corrugated plastic, 4" diameter	1.000	L.F.	.89	.88	1.77
TOTAL			158.55	25.60	184.15

G2040 250	Wood Tie Retaining Walls	COST PER L.F.		
		MAT.	INST.	TOTAL
1000	Wood tie wall, rod connector at 4'-0", redwood, 6" x 6", 2'-0" high	159	25.50	184.50
1400	Deadman at 6'-0", 4'-0" high	365	58.50	423.50
1600	6'-0" high	550	87	637
2000	8 x 8, deadman at 6'-0", 4'-0" high	550	75.50	625.50
2200	6'-0" high	830	114	944
2400	8'-0" high	1,100	150	1,250
3000	Cedar, 6 x 6, 2'-0" high	78	25.50	103.50
3400	Deadman at 6'-0", 4'-0" high	179	58.50	237.50
3600	6'-0" high	268	87	355
4000	8 x 8, deadman at 6'-0", 4'-0" high	240	75.50	315.50
4200	6'-0" high	365	114	479
4400	8'-0" high	480	150	630
7000	Pressure treated wood, 6 x 6, 2'-0" high	30.50	21	51.50
7400	Deadman at 6'-0", 4'-0" high	68.50	48	116.50
7600	6'-0" high	102	71.50	173.50
8000	8 x 8, deadman at 6'-0", 4'-0" high	144	87	231
8200	6'-0" high	218	131	349
8400	8'-0" high	288	173	461

G2040 Site Development

The Stone Retaining Wall System is constructed of one of four types of stone. Each of the four types is listed in terms of cost per ton. Construction is either dry set or mortar set. System elements include excavation; concrete base; crushed stone; underdrain; and backfill.

The Expanded System Listing shows five heights above grade for each type, ranging from 3' above grade to 12' above grade.

System Components	QUANTITY	UNIT	COST PER L.F.		
			MAT.	INST.	TOTAL
SYSTEM G2040 260 2400					
STONE RETAINING WALL, DRY SET, STONE AT $250/TON, 3' ABOVE GRADE					
Excavation, trench, hyd backhoe	.880	C.Y.		6.55	6.55
Strip footing, 36" x 12", reinforced	.111	C.Y.	20.31	16.27	36.58
Stone, wall material, type 1	6.550	C.F.	59.28		59.28
Setting stone wall, dry	6.550	C.F.		76.96	76.96
Stone borrow, delivered, 3/8", machine spread	.320	C.Y.	11.68	2.74	14.42
Piping, subdrainage, perforated PVC, 4" diameter	1.000	L.F.	2.43	4.72	7.15
Backfill with dozer, trench, up to 300' haul, no compaction	1.019	C.Y.		3	3
TOTAL			93.70	110.24	203.94

G2040 260	Stone Retaining Walls	COST PER L.F.		
		MAT.	INST.	TOTAL
2400	Stone retaining wall, dry set, stone at $250/ton, height above grade 3'	93.50	110	203.50
2420	Height above grade 4'	112	131	243
2440	Height above grade 6'	146	176	322
2460	Height above grade 8'	205	276	481
2480	Height above grade 10'	263	360	623
2500	Height above grade 12'	325	450	775
2600	$350/ton stone, height above grade 3'	118	110	228
2620	Height above grade 4'	143	131	274
2640	Height above grade 6'	191	176	367
2660	Height above grade 8'	272	276	548
2680	Height above grade 10'	355	360	715
2700	Height above grade 12'	440	450	890
2800	$450/ton stone, height above grade 3'	141	110	251
2820	Height above grade 4'	174	131	305
2840	Height above grade 6'	235	176	411
2860	Height above grade 8'	340	276	616
2880	Height above grade 10'	445	360	805
2900	Height above grade 12'	555	450	1,005
3000	$650/ton stone, height above grade 3'	188	110	298
3020	Height above grade 4'	235	131	366
3040	Height above grade 6'	325	176	501
3060	Height above grade 8'	470	276	746
3080	Height above grade 10'	620	360	980
3100	Height above grade 12'	780	450	1,230
5020	Mortar set, stone at $250/ton, height above grade 3'	93.50	97.50	191
5040	Height above grade 4'	112	115	227
5060	Height above grade 6'	146	153	299
5080	Height above grade 8'	205	237	442

G2040 Site Development

G2040 260	Stone Retaining Walls	COST PER L.F.		
		MAT.	INST.	TOTAL
5100	Height above grade 10'	263	310	573
5120	Height above grade 12'	325	385	710
5300	$350/ton stone, height above grade 3'	118	97.50	215.50
5320	Height above grade 4'	143	115	258
5340	Height above grade 6'	191	153	344
5360	Height above grade 8'	272	237	509
5380	Height above grade 10'	355	310	665
5400	Height above grade 12'	440	385	825
5500	$450/ton stone, height above grade 3'	141	97.50	238.50
5520	Height above grade 4'	174	115	289
5540	Height above grade 6'	235	153	388
5560	Height above grade 8'	340	237	577
5580	Height above grade 10'	445	310	755
5600	Height above grade 12'	555	385	940
5700	$650/ton stone, height above grade 3'	188	97.50	285.50
5720	Height above grade 4'	235	115	350
5740	Height above grade 6'	325	153	478
5760	Height above grade 8'	470	237	707
5780	Height above grade 10'	620	310	930
5800	Height above grade 12'	780	385	1,165

For customer support on your Site Work & Landscape Costs with RSMeans Data, call 800.448.8182.

G2040 Site Development

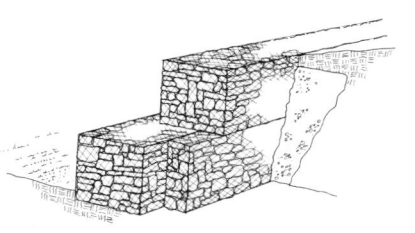

The Gabion Retaining Wall Systems list three types of surcharge conditions: level, sloped and highway, for two different facing configurations: stepped and straight with batter. Costs are expressed per L.F. for heights ranging from 6' to 18' for retaining sandy or clay soil. For protection against sloughing in wet clay materials counterforts have been added for these systems. Drainage stone has been added behind the walls to avoid any additional lateral pressures.

System Components			COST PER L.F.		
	QUANTITY	UNIT	MAT.	INST.	TOTAL
SYSTEM G2040 270 1000					
GABION RET. WALL, LEVEL BACKFILL, STEPPED FACE, 4' BASE, 6' HIGH					
3' x 3' cross section gabion	2.000	Ea.	80	172.89	252.89
3' x 1' cross section gabion	1.000	Ea.	21.67	10.37	32.04
Crushed stone drainage	.220	C.Y.	7.13	2.15	9.28
TOTAL			108.80	185.41	294.21

G2040 270	Gabion Retaining Walls	COST PER L.F.		
		MAT.	INST.	TOTAL
1000	Gabion ret. wall, lvl backfill, stepped face, 4' base, 6' high, sandy soil	109	185	294
1040	Clay soil with counterforts @ 16' O.C.	162	380	542
1100	5' base, 9' high, sandy soil	195	293	488
1140	Clay soil with counterforts @ 16' O.C.	330	585	915
1200	6' base, 12' high, sandy soil	279	470	749
1240	Clay soil with counterforts @ 16' O.C.	490	1,250	1,740
1300	7'-6" base, 15' high, sandy soil	395	655	1,050
1340	Clay soil with counterforts @ 16' O.C.	910	1,750	2,660
1400	9' base, 18' high, sandy soil	520	920	1,440
1440	Clay soil with counterforts @ 16' O.C.	1,050	2,075	3,125
2000	Straight face w/1:6 batter, 3' base, 6' high, sandy soil	94.50	177	271.50
2040	Clay soil with counterforts @ 16' O.C.	147	375	522
2100	4'-6" base, 9' high, sandy soil	157	289	446
2140	Clay soil with counterforts @ 16' O.C.	292	580	872
2200	6' base, 12' high, sandy soil	243	465	708
2240	Clay soil with counterforts @ 16' O.C.	455	1,250	1,705
2300	7'-6" base, 15' high, sandy soil	345	660	1,005
2340	Clay soil with counterforts @ 16' O.C.	685	1,375	2,060
2400	18' high, sandy soil	450	860	1,310
2440	Clay soil with counterforts @ 16' O.C.	740	1,000	1,740
3000	Backfill sloped 1-1/2:1, stepped face, 4'-6" base, 6' high, sandy soil	105	199	304
3040	Clay soil with counterforts @ 16' O.C.	158	395	553
3100	6' base, 9' high, sandy soil	187	370	557
3140	Clay soil with counterforts @ 16' O.C.	320	665	985
3200	7'-6" base, 12' high, sandy soil	289	570	859
3240	Clay soil with counterforts @ 16' O.C.	500	1,350	1,850
3300	9' base, 15' high, sandy soil	410	830	1,240
3340	Clay soil with counterforts @ 16' O.C.	925	1,925	2,850
3400	10'-6" base, 18' high, sandy soil	555	1,125	1,680
3440	Clay soil with counterforts @ 16' O.C.	1,100	2,275	3,375
4000	Straight face with 1:6 batter, 4'-6" base, 6' high, sandy soil	112	201	313
4040	Clay soil with counterforts @ 16' O.C.	165	395	560
4100	6' base, 9' high, sandy soil	197	375	572
4140	Clay soil with counterforts @ 16' O.C.	330	665	995

682

For customer support on your Site Work & Landscape Costs with RSMeans Data, call 800.448.8182.

G2040 Site Development

G2040 270	Gabion Retaining Walls	COST PER L.F.		
		MAT.	INST.	TOTAL
4200	7'-6" base, 12' high, sandy soil	300	575	875
4240	Clay soil with counterforts @ 16' O.C.	515	1,350	1,865
4300	9' base, 15' high, sandy soil	425	835	1,260
4340	Clay soil with counterforts @ 16' O.C.	940	1,925	2,865
4400	18' high, sandy soil	550	1,100	1,650
4440	Clay soil with counterforts @ 16' O.C.	1,100	2,250	3,350
5000	Highway surcharge, straight face, 6' base, 6' high, sandy soil	167	350	517
5040	Clay soil with counterforts @ 16' O.C.	220	545	765
5100	9' base, 9' high, sandy soil	290	610	900
5140	Clay soil with counterforts @ 16' O.C.	425	900	1,325
5200	12' high, sandy soil	545	905	1,450
5240	Clay soil with counterforts @ 16' O.C.	790	2,250	3,040
5300	12' base, 15' high, sandy soil	575	1,225	1,800
5340	Clay soil with counterforts @ 16' O.C.	915	1,925	2,840
5400	18' high, sandy soil	740	1,550	2,290
5440	Clay soil with counterforts @ 16' O.C.	1,275	2,725	4,000

G2040 Site Development

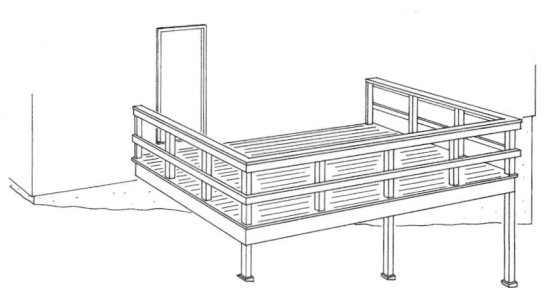

Wood Deck Systems are either constructed of pressure treated lumber or redwood lumber. The system includes: the deck, joists (16″ or 24″ on-center), girders, posts (8′ on-center) and railings. Decking is constructed of either 1″ x 4″ or 2″ x 6″ stock. Joists range from 2″ x 8″ to 2″ x 10″ depending on the size of the system. The size ranges hold true for both pressure treated lumber and redwood systems.

Costs are on a square foot basis.

System Components	QUANTITY	UNIT	COST PER S.F.		
			MAT.	INST.	TOTAL
SYSTEM G2040 910 1000					
WOOD DECK, TREATED LUMBER, 2″ X 8″ JOISTS @ 16″ O.C., 2″ X 6″ DECKING					
Decking, planks, fir 2″ x 6″ treated	2.080	B.F.	2.86	5.46	8.32
Framing, joists, fir 2″ x 8″ treated	1.330	B.F.	2.09	2.89	4.98
Framing, beams, fir 2″ x 10″ treated	.133	B.F.	.17	.26	.43
Framing, post, 4″ x 4″, treated	.333	B.F.	.61	.72	1.33
Framing, railing, 2″ x 4″ lumber, treated	.667	B.F.	.99	2.43	3.42
Excavating pits by hand, heavy soil or clay	.300	C.Y.		.40	.40
Spread footing, concrete	.200	C.Y.	.46	.62	1.08
TOTAL			7.18	12.78	19.96

G2040 910	Wood Decks	COST PER S.F.		
		MAT.	INST.	TOTAL
1000	Wood deck, treated lumber, 2″ x 8″ joists @ 16″ O.C., 2″ x 6″ decking	7.20	12.80	20
1004	2″ x 4″ decking	7.70	15.65	23.35
1008	1″ x 6″ decking	8.10	10.15	18.25
1012	1″ x 4″ decking	7.75	9.90	17.65
1500	2″ x 10″ joists @ 16″ O.C., 2″ x 6″ decking	7.70	13.50	21.20
1504	2″ x 4″ decking	8.25	16.40	24.65
1508	1″ x 6″ decking	8.65	10.85	19.50
1512	1″ x 4″ decking	8.30	10.60	18.90
1560	2″ x 10″ joists @ 24″ O.C., 2″ x 6″ decking	7.20	12.80	20
1564	2″ x 4″ decking	7.70	15.65	23.35
1568	1″ x 6″ decking	8.10	10.15	18.25
1572	1″ x 4″ decking	7.75	9.90	17.65
4000	Redwood lumber, 2″ x 8″ joists @ 16″ O.C., 2″ x 6″ decking	20	5.85	25.85
4004	2″ x 4″ decking	20.50	6	26.50
4008	1″ x 6″ decking	29	7.55	36.55
4012	1″ x 4″ decking	28.50	7.40	35.90
4500	2″ x 10″ joists @ 16″ O.C., 2″ x 6″ decking	33.50	6.40	39.90
4504	2″ x 4″ decking	32	6.80	38.80
4508	1″ x 6″ decking	32.50	7.85	40.35
4512	1″ x 4″ decking	32	7.70	39.70
4600	2″ x 10″ joists @ 24″ O.C., 2″ x 6″ decking	30	6.10	36.10
4612	2″ x 4″ decking	28.50	6.50	35
4616	1″ x 6″ decking	29	7.55	36.55
4620	1″ x 4″ decking	28.50	7.40	35.90

G20 Site Improvements

G2040 Site Development

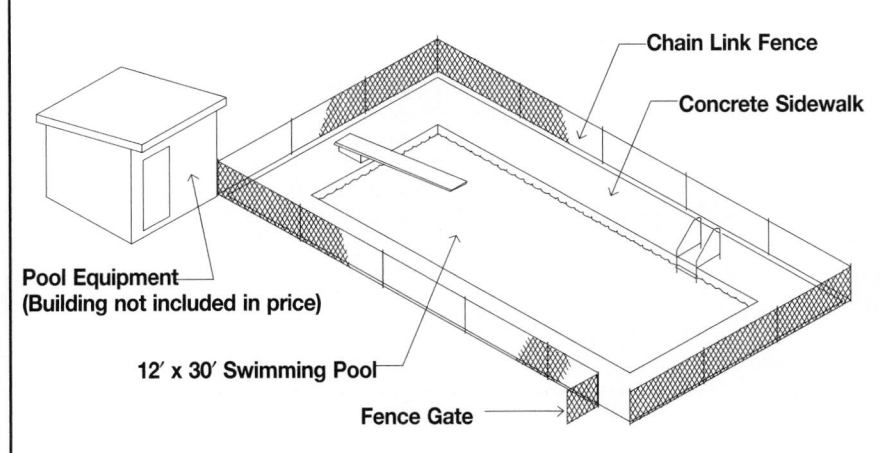

- Chain Link Fence
- Concrete Sidewalk
- Pool Equipment (Building not included in price)
- 12' x 30' Swimming Pool
- Fence Gate

The Swimming Pool System is a complete package. Everything from excavation to deck hardware is included in system costs. Below are three basic types of pool systems: residential, motel, and municipal. Systems elements include: excavation, pool materials, installation, deck hardware, pumps and filters, sidewalk, and fencing.

The Expanded System Listing shows three basic types of pools with a variety of finishes and basic materials. These systems are either vinyl lined with metal sides; gunite shell with a cement plaster finish or tile finish; or concrete sided with vinyl lining. Pool sizes listed here vary from 12' x 30' to 60' x 82.5'. All costs are on a per unit basis.

System Components	QUANTITY	UNIT	COST EACH		
			MAT.	INST.	TOTAL
SYSTEM G2040 920 1000					
SWIMMING POOL, RESIDENTIAL, CONC. SIDES, VINYL LINED, 12' X 30'					
Swimming pool, residential, in-ground including equipment	360.000	S.F.	11,700	5,832	17,532
4" thick reinforced concrete sidewalk, broom finish, no base	400.000	S.F.	980	1,192	2,172
Chain link fence, residential, 4' high	124.000	L.F.	762.60	596.44	1,359.04
Fence gate, chain link, 4' high	1.000	Ea.	104	192	296
TOTAL			13,546.60	7,812.44	21,359.04

G2040 920	Swimming Pools	COST EACH		
		MAT.	INST.	TOTAL
1000	Swimming pool, residential class, concrete sides, vinyl lined, 12' x 30'	13,500	7,800	21,300
1100	16' x 32'	15,400	8,825	24,225
1200	20' x 40'	19,800	11,300	31,100
1500	Tile finish, 12' x 30'	29,600	33,100	62,700
1600	16' x 32'	30,100	34,800	64,900
1700	20' x 40'	45,100	48,900	94,000
2000	Metal sides, vinyl lined, 12' x 30'	11,600	5,125	16,725
2100	16' x 32'	13,100	5,750	18,850
2200	20' x 40'	16,900	7,250	24,150
3000	Gunite shell, cement plaster finish, 12' x 30'	23,400	14,000	37,400
3100	16' x 32'	29,700	18,300	48,000
3200	20' x 40'	41,200	18,100	59,300
4000	Motel class, concrete sides, vinyl lined, 20' x 40'	28,400	15,500	43,900
4100	28' x 60'	39,900	21,700	61,600
4500	Tile finish, 20' x 40'	53,000	56,500	109,500
4600	28' x 60'	84,500	91,500	176,000
5000	Metal sides, vinyl lined, 20' x 40'	29,400	11,300	40,700
5100	28' x 60'	39,500	15,200	54,700
6000	Gunite shell, cement plaster finish, 20' x 40'	76,500	45,100	121,600
6100	28' x 60'	114,500	67,500	182,000
7000	Municipal class, gunite shell, cement plaster finish, 42' x 75'	305,000	158,000	463,000
7100	60' x 82.5'	392,000	203,000	595,000
7500	Concrete walls, tile finish, 42' x 75'	341,500	214,500	556,000
7600	60' x 82.5'	445,000	284,500	729,500
7700	Tile finish and concrete gutter, 42' x 75'	375,000	214,500	589,500
7800	60' x 82.5'	485,000	284,500	769,500
7900	Tile finish and stainless gutter, 42' x 75'	439,000	214,500	653,500
8000	60'x 82.5'	649,000	328,500	977,500

685

G2040 Site Development

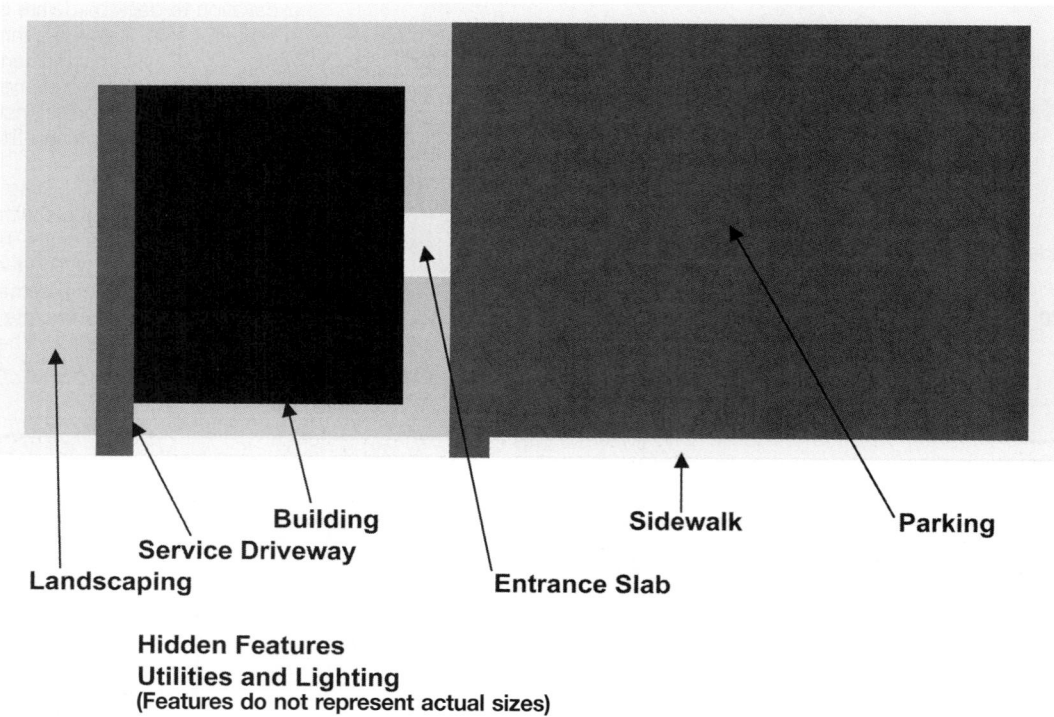

Building

Service Driveway

Landscaping

Sidewalk

Parking

Entrance Slab

Hidden Features
Utilities and Lighting
(Features do not represent actual sizes)

G2040 990	Site Development Components for Buildings	COST EACH		
		MAT.	INST.	TOTAL
0970	Assume minimal and balanced cut & fill, no rock, no demolition, no haz mat.			
0975	Lines can be adjusted linearly +/- 20% within same use & number of floors.			
0980	60,000 S.F. 2-Story Office Bldg on 3.3 Acres w/20% Green Space			
1000	Site Preparation	385	56,000	56,385
1002	Utilities	129,500	115,000	244,500
1004	Pavement	354,000	217,500	571,500
1006	Stormwater Management	402,500	70,000	472,500
1008	Sidewalks	18,800	21,800	40,600
1010	Exterior lighting	51,500	44,700	96,200
1014	Landscaping	76,000	58,500	134,500
1028	80,000 S.F. 4-Story Office Bldg on 3.5 Acres w/20% Green Space			
1030	Site preparation	395	61,000	61,395
1032	Utilities	122,500	104,500	227,000
1034	Pavement	423,500	260,500	684,000
1036	Stormwater Managmement	424,000	73,500	497,500
1038	Sidewalks	15,400	17,800	33,200
1040	Lighting	49,700	47,200	96,900
1042	Landscaping	78,000	61,000	139,000
1058	16,000 S.F. 1-Story Medical Office Bldg on 1.7 Acres w/25% Green Space			
1060	Site preparation	277	42,400	42,677
1062	Utilities	93,000	84,000	177,000
1064	Pavement	165,500	102,500	268,000
1066	Stormwater Management	196,000	34,100	230,100
1068	Sidewalks	13,700	15,900	29,600
1070	Lighting	32,400	25,900	58,300
1072	Landscaping	51,500	36,100	87,600
1098	36,000 S.F. 3-Story Apartment Bldg on 1.6 Acres w/25% Green Space			
1100	Site preparation	268	41,800	42,068

For customer support on your Site Work & Landscape Costs with RSMeans Data, call 800.448.8182.

G2040 Site Development

G2040 990	Site Development Components for Buildings	COST EACH		
		MAT.	INST.	TOTAL
1102	Utilities	87,500	76,500	164,000
1104	Pavement	167,000	103,000	270,000
1106	Stormwater Management	183,000	31,800	214,800
1108	Sidewalks	11,900	13,800	25,700
1110	Lighting	29,900	24,600	54,500
1112	Landscaping	49,200	34,300	83,500
1128	130,000 S.F. 3-Story Hospital Bldg on 5.9 Acres w/17% Green Space			
1130	Site preparation	515	80,500	81,015
1132	Utilities	175,000	150,500	325,500
1134	Pavement	720,000	442,500	1,162,500
1136	Stormwater Management	753,000	131,000	884,000
1138	Sidewalks	22,600	26,100	48,700
1140	Lighting	81,000	79,500	160,500
1142	Landscaping	109,000	93,000	202,000
1158	60,000 S.F. 1-Story Light Manufacturing Bldg on 3.9 Acres w/18% Green Space			
1160	Site preparation	420	60,000	60,420
1162	Utilities	149,500	143,000	292,500
1164	Pavement	346,500	189,500	536,000
1166	Stormwater Management	496,000	86,000	582,000
1168	Sidewalks	26,600	30,800	57,400
1170	Lighting	63,500	51,000	114,500
1172	Landscaping	82,000	66,000	148,000
1198	8,000 S.F. 1-Story Restaurant Bldg on 1.4 Acres w/26% Green Space			
1200	Site preparation	253	40,800	41,053
1202	Utilities	80,000	67,000	147,000
1204	Pavement	158,000	99,000	257,000
1206	Stormwater Management	161,000	28,000	189,000
1208	Sidewalks	9,725	11,200	20,925
1210	Lighting	26,300	22,700	49,000
1212	Landscaping	46,200	31,500	77,700
1228	20,000 S.F. 1-Story Retail Store Bldg on 2.1 Acres w/23% Green Space			
1230	Site preparation	310	49,700	50,010
1232	Utilities	103,000	93,500	196,500
1234	Pavement	214,500	132,500	347,000
1236	Stormwater Management	251,500	43,700	295,200
1238	Sidewalks	15,400	17,800	33,200
1240	Lighting	36,900	30,100	67,000
1242	Landscaping	58,500	42,500	101,000
1258	60,000 S.F. 1-Story Warehouse on 3.1 Acres w/19% Green Space			
1260	Site preparation	375	55,000	55,375
1262	Utilities	132,000	134,000	266,000
1264	Pavement	214,500	107,000	321,500
1266	Stormwater Management	386,500	67,000	453,500
1268	Sidewalks	26,600	30,800	57,400
1270	Lighting	55,000	39,800	94,800
1272	Landscaping	71,000	54,500	125,500

G20 Site Improvements

G2050 Landscaping

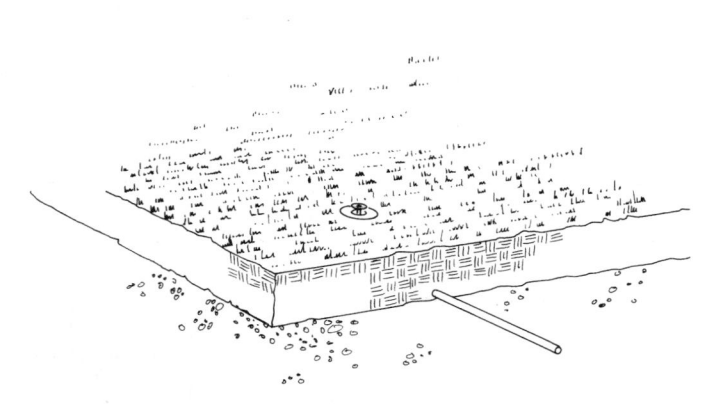

The Lawn Systems listed include different types of seeding, sodding and ground covers for flat and sloped areas. Costs are given per thousand square feet for different size jobs; residential, small commercial and large commercial. The size of the job relates to the type and productivity of the equipment being used. Components include furnishing and spreading screened loam, spreading fertilizer and limestone and mulching planted and seeded surfaces. Sloped surfaces include jute mesh or staking depending on the type of cover.

System Components	QUANTITY	UNIT	COST PER M.S.F. MAT.	COST PER M.S.F. INST.	COST PER M.S.F. TOTAL
SYSTEM G2050 410 1000					
LAWN, FLAT AREA, SEEDED, TURF MIX, RESIDENTIAL					
Scarify subsoil, residential, skid steer loader w/scarifiers, 50 HP	1.000	M.S.F.		29.70	29.70
Root raking and loading, residential, no boulders	1.000	M.S.F.		36.96	36.96
Spread topsoil, skid steer loader and hand dress	18.500	C.Y.	555	131.91	686.91
Spread ground limestone, push spreader	110.000	S.Y.	19.80	3.30	23.10
Spread fertilizer, push spreader	110.000	S.Y.	9.90	3.30	13.20
Till topsoil, 26" rototiller	110.000	S.Y.		64.90	64.90
By hand (raking)	1.000	M.S.F.		35.50	35.50
Rolling topsoil, hand push roller	18.500	C.Y.		3.15	3.15
Seeding, turf mix, push spreader	1.000	M.S.F.	14.05	66.50	80.55
Straw, mulch	110.000	S.Y.	74.80	123.20	198
TOTAL			673.55	498.42	1,171.97

G2050 410	Lawns & Ground Cover	COST PER M.S.F. MAT.	COST PER M.S.F. INST.	COST PER M.S.F. TOTAL
1000	Lawn, flat area, seeded, turf mix, residential	675	500	1,175
1040	Small commercial	675	545	1,220
1080	Large commercial	695	470	1,165
1200	Shade mix, residential	670	500	1,170
1240	Small commercial	670	545	1,215
1280	Large commercial	685	470	1,155
1400	Utility mix, residential	670	500	1,170
1440	Small commercial	670	545	1,215
1480	Large commercial	700	470	1,170
2000	Sod, bluegrass, residential	955	530	1,485
2040	Small commercial	1,025	695	1,720
2080	Large commercial	820	570	1,390
2200	Bentgrass, residential	1,000	520	1,520
2240	Small commercial	980	685	1,665
2280	Large commercial	950	585	1,535
2400	Ground cover, English ivy, residential	2,700	1,300	4,000
2440	Small commercial	2,700	1,525	4,225
2480	Large commercial	2,700	1,425	4,125
2600	Pachysandra, residential	3,675	6,850	10,525
2640	Small commercial	3,675	7,100	10,775
2680	Large commercial	3,700	6,975	10,675
2800	Vinca minor, residential	3,325	2,025	5,350

G2050 Landscaping

G2050 410	Lawns & Ground Cover	COST PER M.S.F.		
		MAT.	INST.	TOTAL
2840	Small commercial	3,325	2,275	5,600
2880	Large commercial	3,350	2,150	5,500
3000	Sloped areas, seeded, slope mix, residential	870	685	1,555
3040	Small commercial	870	680	1,550
3080	Large commercial	890	680	1,570
3200	Birdsfoot trefoil, residential	860	690	1,550
3240	Small commercial	860	680	1,540
3280	Large commercial	865	680	1,545
3400	Clover, residential	860	685	1,545
3440	Small commercial	860	680	1,540
3480	Large commercial	870	680	1,550
3600	Crown vetch, residential	1,025	685	1,710
3640	Small commercial	1,025	680	1,705
3680	Large commercial	1,050	680	1,730
3800	Wildflower, residential	860	685	1,545
3840	Small commercial	860	680	1,540
3880	Large commercial	865	680	1,545
4600	Sod, bluegrass, residential	1,025	1,100	2,125
4640	Small commercial	1,075	1,125	2,200
4680	Large commercial	880	1,025	1,905
4800	Bent grass, residential	1,075	595	1,670
4840	Small commercial	1,050	755	1,805
4880	Large commercial	1,000	735	1,735
5000	Ground cover, English ivy, residential	2,950	1,450	4,400
5040	Small commercial	2,950	1,675	4,625
5080	Large commercial	2,950	1,625	4,575
5200	Pachysandra, residential	4,025	7,550	11,575
5240	Small commercial	4,025	7,775	11,800
5280	Large commercial	4,025	7,750	11,775
5400	Vinca minor, residential	3,650	2,250	5,900
5440	Small commercial	2,900	2,450	5,350
5480	Large commercial	3,650	2,425	6,075

G2050 Landscaping

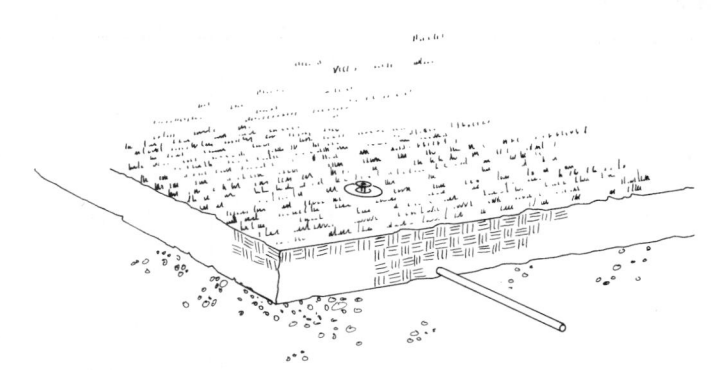

There are three basic types of Site Irrigation Systems: pop-up, riser mounted and quick coupling. Sprinkler heads are spray, impact or gear driven. Each system includes: the hardware for spraying the water; the pipe and fittings needed to deliver the water; and all other accessory equipment such as valves, couplings, nipples, and nozzles. Excavation heads and backfill costs are also included in the system.

The Expanded System Listing shows a wide variety of Site Irrigation Systems.

System Components	QUANTITY	UNIT	COST PER S.F.		
			MAT.	INST.	TOTAL
SYSTEM G2050 710 1000					
SITE IRRIGATION, POP UP SPRAY, PLASTIC, 10' RADIUS, 1000 S.F., PVC PIPE					
Excavation, chain trencher	54.000	L.F.		56.70	56.70
Pipe, fittings & nipples, PVC Schedule 40, 1" diameter	5.000	Ea.	4.60	127.50	132.10
Fittings, bends or elbows, 4" diameter	5.000	Ea.	47.25	287.50	334.75
Couplings PVC plastic, high pressure, 1" diameter	5.000	Ea.	4.60	162.50	167.10
Valves, bronze, globe, 125 lb. rising stem, threaded, 1" diameter	1.000	Ea.	298	42.50	340.50
Head & nozzle, pop-up spray, PVC plastic	5.000	Ea.	23.70	115	138.70
Backfill by hand with compaction	54.000	L.F.		39.96	39.96
Total cost per 1,000 S.F.			378.15	831.66	1,209.81
Total cost per S.F.			.38	.83	1.21

G2050 710	Site Irrigation	COST PER S.F.		
		MAT.	INST.	TOTAL
1000	Site irrigation, pop up spray, 10' radius, 1000 S.F., PVC pipe	.38	.83	1.21
1100	Polyethylene pipe	.42	.72	1.14
1200	14' radius, 8000 S.F., PVC pipe	.13	.36	.49
1300	Polyethylene pipe	.13	.33	.46
1400	18' square, 1000 S.F. PVC pipe	.39	.57	.96
1500	Polyethylene pipe	.39	.49	.88
1600	24' square, 8000 S.F., PVC pipe	.11	.23	.34
1700	Polyethylene pipe	.11	.22	.33
1800	4' x 30' strip, 200 S.F., PVC pipe	1.83	2.35	4.18
1900	Polyethylene pipe	1.82	2.18	4
2000	6' x 40' strip, 800 S.F., PVC pipe	.49	.78	1.27
2200	Economy brass, 11' radius, 1000 S.F., PVC pipe	.49	.80	1.29
2300	Polyethylene pipe	.48	.72	1.20
2400	14' radius, 8000 S.F., PVC pipe	.16	.36	.52
2500	Polyethylene pipe	.16	.33	.49
2600	3' x 28' strip, 200 S.F., PVC pipe	2.03	2.35	4.38
2700	Polyethylene pipe	2.02	2.18	4.20
2800	7' x 36' strip, 1000 S.F., PVC pipe	.45	.62	1.07
2900	Polyethylene pipe	.44	.57	1.01
3000	Hd brass, 11' radius, 1000 S.F., PVC pipe	.53	.80	1.33
3100	Polyethylene pipe	.53	.72	1.25
3200	14' radius, 8000 S.F., PVC pipe	.18	.36	.54
3300	Polyethylene pipe	.18	.33	.51
3400	Riser mounted spray, plastic, 10' radius, 1000 S.F., PVC pipe	.42	.80	1.22
3500	Polyethylene pipe	.42	.72	1.14
3600	12' radius, 5000 S.F., PVC pipe	.19	.47	.66

G20 Site Improvements

G2050 Landscaping

G2050 710	Site Irrigation	COST PER S.F.		
		MAT.	INST.	TOTAL
3700	Polyethylene pipe	.19	.44	.63
3800	19' square, 2000 S.F., PVC pipe	.19	.26	.45
3900	Polyethylene pipe	.19	.23	.42
4000	24' square, 8000 S.F., PVC pipe	.11	.23	.34
4100	Polyethylene pipe	.11	.22	.33
4200	5' x 32' strip, 300 S.F., PVC pipe	1.22	1.58	2.80
4300	Polyethylene pipe	1.22	1.46	2.68
4400	6' x 40' strip, 800 S.F., PVC pipe	.49	.78	1.27
4500	Polyethylene pipe	.49	.72	1.21
4600	Brass, 11' radius, 1000 S.F., PVC pipe	.46	.80	1.26
4700	Polyethylene pipe	.46	.72	1.18
4800	14' radius, 8000 S.F., PVC pipe	.15	.36	.51
4900	Polythylene pipe	.15	.33	.48
5000	Pop up gear drive stream type, plastic, 30' radius, 10,000 S.F., PVC pipe	.23	.69	.92
5020	Polyethylene pipe	.16	.21	.37
5040	40,000 S.F., PVC pipe	.22	.69	.91
5060	Polyethylene pipe	.12	.20	.32
5080	Riser mounted gear drive stream type, plas., 30' rad., 10,000 S.F.,PVC pipe	.22	.69	.91
5100	Polyethylene pipe	.15	.21	.36
5120	40,000 S.F., PVC pipe	.21	.69	.90
5140	Polyethylene pipe	.11	.20	.31
5200	Q.C. valve thread type w/impact head, brass, 75' rad, 20,000 S.F., PVC pipe	.09	.27	.36
5250	Polyethylene pipe	.06	.07	.13
5300	100,000 S.F., PVC pipe	.15	.43	.58
5350	Polyethylene pipe	.08	.10	.18
5400	Q.C. valve lug type w/impact head, brass, 75' rad, 20,000 S.F., PVC pipe	.09	.27	.36
5450	Polyethylene pipe	.06	.07	.13
5500	100,000 S.F. PVC pipe	.15	.43	.58
5550	Polyethylene pipe	.08	.10	.18
6000	Site irrigation, pop up impact type, plastic, 40' rad.,10000 S.F., PVC pipe	.13	.45	.58
6100	Polyethylene pipe	.09	.12	.21
6200	45,000 S.F., PVC pipe	.14	.49	.63
6300	Polyethylene pipe	.07	.13	.20
6400	High medium volume brass, 40' radius, 10000 S.F., PVC pipe	.14	.45	.59
6500	Polyethylene pipe	.09	.12	.21
6600	45,000 S.F. PVC pipe	.15	.49	.64
6700	Polyethylene pipe	.07	.13	.20
6800	High volume brass, 60' radius, 25,000 S.F., PVC pipe	.10	.35	.45
6900	Polyethylene pipe	.07	.09	.16
7000	100,000 S.F., PVC pipe	.13	.35	.48
7100	Polyethylene pipe	.07	.08	.15
7200	Riser mounted part/full impact type, plas., 40' rad., 10,000 S.F., PVC pipe	.13	.45	.58
7300	Polyethylene pipe	.08	.24	.32
7400	45,000 S.F., PVC pipe	.14	.49	.63
7500	Polyethylene pipe	.06	.13	.19
7600	Low medium volume brass, 40' radius, 10,000 S.F., PVC pipe	.14	.45	.59
7700	Polyethylene pipe	.09	.13	.22
7800	45,000 S.F., PVC pipe	.14	.49	.63
7900	Polyethylene pipe	.07	.13	.20
8000	Medium volume brass, 50' radius, 30,000 S.F., PVC pipe	.11	.37	.48
8100	Polyethylene pipe	.07	.10	.17
8200	70,000 S.F., PVC pipe	.18	.60	.78
8300	Polyethylene pipe	.09	.15	.24
8400	Riser mounted full only impact type, plas., 40' rad., 10,000 S.F., PVC pipe	.12	.45	.57
8500	Polyethylene pipe	.08	.13	.21
8600	45,000 S.F., PVC pipe	.13	.49	.62
8700	Polyethylene pipe	.06	.13	.19
8800	Low medium volume brass, 40' radius, 10,000 S.F., PVC pipe	.14	.45	.59

G20 Site Improvements

G2050 Landscaping

G2050 710	Site Irrigation	COST PER S.F.		
		MAT.	INST.	TOTAL
8900	Polyethylene pipe	.09	.13	.22
9000	45,000 S.F., PVC pipe	.15	.49	.64
9100	Polyethylene pipe	.07	.13	.20
9200	High volume brass, 80' radius, 75,000 S.F., PVC pipe	.09	.26	.35
9300	Polyethylene pipe	.01	.02	.03
9400	150,000 S.F., PVC pipe	.09	.24	.33
9500	Polyethylene pipe	.01	.02	.03
9600	Very high volume brass, 100' radius, 100,000 S.F., PVC pipe	.08	.20	.28
9700	Polyethylene pipe	.01	.02	.03
9800	200,000 S.F., PVC pipe	.08	.20	.28
9900	Polyethylene pipe	.01	.02	.03

For customer support on your Site Work & Landscape Costs with RSMeans Data, call 800.448.8182.

G2050 Landscaping

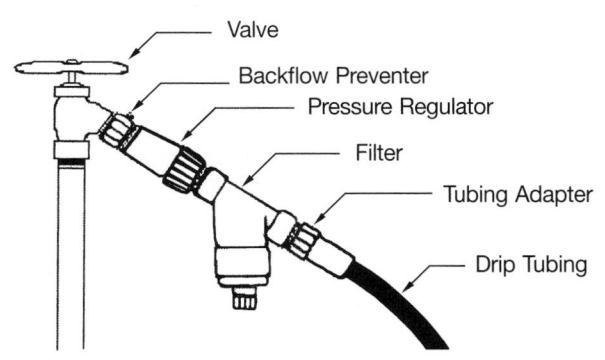

- Valve
- Backflow Preventer
- Pressure Regulator
- Filter
- Tubing Adapter
- Drip Tubing

Drip irrigation systems can be for either lawns or individual plants. Shown below are drip irrigation systems for lawns. These drip type systems are most cost effective where there are water restrictions on use or the cost of water is very high. The supply uses PVC piping with the laterals polyethylene drip lines.

System Components	QUANTITY	UNIT	COST PER S.F.		
			MAT.	INST.	TOTAL
SYSTEM G2050 720 1000					
SITE IRRIGATION, DRIP SYSTEM, 800 S.F. (20′ X 40′)					
Excavate supply and header lines 4″ x 18″ deep	50.000	L.F.		55.50	55.50
Excavate drip lines 4″ x 4″ deep	560.000	L.F.		935.20	935.20
Install supply manifold (stop valve, strainer, PRV, adapters)	1.000	Ea.	23.50	172	195.50
Install 3/4″ PVC supply line and risers to 4 ″ drip level	75.000	L.F.	24.75	171.75	196.50
Install 3/4″ PVC fittings for supply line	33.000	Ea.	60.39	567.60	627.99
Install PVC to drip line adapters	14.000	Ea.	26.18	240.80	266.98
Install 1/4 inch drip lines with hose clamps	560.000	L.F.	61.60	319.20	380.80
Backfill trenches	3.180	C.Y.		152.64	152.64
Cleanup area when completed	1.000	Ea.		133	133
Total cost per 800 S.F.			196.42	2,747.69	2,944.11
Total cost per S.F.			.25	3.43	3.68

G2050 720		Site Irrigation	COST PER S.F.		
			MAT.	INST.	TOTAL
1000	Site irrigation, drip lawn watering system, 800 S.F. (20′ x 40′ area)		.25	3.44	3.69
1100		1000 S.F. (20′ x 50′ area)	.21	3.09	3.30
1200		1600 S.F. (40′ x 40′ area), 2 zone	.26	3.30	3.56
1300		2000 S.F. (40′ x 50′ area), 2 zone	.23	2.98	3.21
1400		2400 S.F. (60′ x 40′ area), 3 zone	.25	3.26	3.51
1500		3000 S.F. (60′ x 50′ area), 3 zone	.22	2.93	3.15
1600		3200 S.F. (2 x 40′ x 40′), 4 zone	.26	3.24	3.50
1700		4000 S.F. (2 x 40′ x 50′), 4 zone	.22	2.93	3.15
1800		4800 S.F. (2 x 60′ x 40′), 6 zone with control	.27	3.22	3.49
1900		6000 S.F. (2 x 60′ x 50′), 6 zone with control	.23	2.90	3.13
2000		6400 S.F. (4 x 40′ x 40′), 8 zone with control	.27	3.26	3.53
2100		8000 S.F. (4 x 40′ x 50′), 8 zone with control	.23	2.95	3.18
2200		3200 S.F. (2 x 40′ x 40′), 4 zone with control	.28	3.31	3.59
2300		4000 S.F. (2 x 40′ x 50′), 4 zone with control	.24	2.99	3.23
2400		2400 S.F. (60′ x 40′), 3 zone with control	.28	3.36	3.64
2500		3000 S.F. (60′ x 50′), 3 zone with control	.24	3.01	3.25
2600		4800 S.F. (2 x 60′ x 40′), 6 zone, manual	.25	3.17	3.42
2700		6000 S.F. (2 x 60′ x 50′), 6 zone, manual	.21	2.86	3.07
2800		1000 S.F. (10′ x 100′ area)	.13	1.92	2.05
2900		2000 S.F. (2 x 10′ x 100′ area)	.15	1.85	2

693

G2050 Landscaping

The Tree Pit Systems are listed for different heights of deciduous and evergreen trees, from 4' and above. Costs are given showing the pit size to accommodate the soil ball and the type of soil being excavated. In clay soils drainage tile has been included in addition to gravel drainage. Woven drainage fabric is applied before bark mulch is spread, assuming runoff is not harmful to roots.

System Components	QUANTITY	UNIT	COST EACH MAT.	COST EACH INST.	COST EACH TOTAL
SYSTEM G2050 910 1220					
TREE PIT, 5' TO 6' TREE, DECIDUOUS, 2' X 1-1/4' DEEP PIT, CLAY SOIL					
Heavy soil or clay	.262	C.Y.		34.85	34.85
Fill in pit, screened gravel, 3/4" to 1/2"	.050	C.Y.	1.48	2.25	3.73
Perforated PVC, schedule 40, 4" diameter	3.000	L.F.	7.29	34.08	41.37
Mix planting soil, by hand, incl. loam, peat & manure	.304	C.Y.	15.81	5.37	21.18
Prepared planting mix, by hand	.304	C.Y.		13.38	13.38
Fabric, laid in trench, polypropylene, ideal conditions	.500	S.Y.	.86	.22	1.08
Bark mulch, hand spread, 3"	.346	C.Y.	1.53	1.83	3.36
TOTAL			26.97	91.98	118.95

G2050 910	Tree Pits	COST EACH MAT.	COST EACH INST.	COST EACH TOTAL
1000	Tree pit, 4' to 5' tree, evergreen, 2-1/2' x 1-1/2' deep pit, sandy soil	48	80	128
1020	Clay soil	47	117	164
1200	5' to 6' tree, deciduous, 2' x 1-1/4' deep pit, sandy soil	26	40.50	66.50
1220	Clay soil	27	92	119
1300	Evergreen, 3' x 1-3/4' deep pit, sandy soil	79	120	199
1320	Clay soil	72.50	173	245.50
1360	3-1/2' x 1-3/4' deep pit, sandy soil	100	150	250
1380	Clay soil	91	201	292
1400	6' to 7' tree, evergreen, 4' x 1-3/4' deep pit, sandy soil	121	175	296
1420	Clay soil	110	232	342
1600	7' to 9' tree, deciduous, 2-1/2' x 1-1/2' deep pit, sandy soil	48	99.50	147.50
1620	Clay soil	47	117	164
1800	8' to 10' tree, deciduous, 3' x 1-3/4' deep pit, sandy soil	79	120	199
1820	Clay soil	72.50	173	245.50
1900	Evergreen, 4-1/2' x 2' deep pit, sandy soil, smaller ball	179	252	431
1910	Larger ball	174	246	420
1920	Clay soil, smaller ball	156	300	456
1930	Larger ball	151	295	446
2000	10' to 12' tree, deciduous, 3-1/2' x 1-3/4' deep pit, sandy soil	100	148	248
2020	Clay soil	91	201	292

G2050 Landscaping

G2050 910	Tree Pits	COST EACH		
		MAT.	INST.	TOTAL
2060	4' x 1-3/4' deep pit, sandy soil	127	183	310
2080	Clay soil	117	248	365
2100	Evergreen, 5' x 2-1/4' deep pit, sandy soil, smaller ball	238	330	568
2110	Larger ball	230	320	550
2120	Clay soil, smaller ball	202	370	572
2130	Larger ball	194	360	554
2200	12' to 14' tree, deciduous, 4-1/2' x 2' deep pit, sandy soil, smaller ball	179	252	431
2210	Larger ball	174	246	420
2220	Clay soil, smaller ball	156	300	456
2230	Larger ball	151	295	446
2400	13' to 15' tree, evergreen, 6' x 2-1/2' deep pit, sandy soil	340	470	810
2420	Clay soil	281	490	771
2600	14' to 16' tree, deciduous, 5' x 2-1/4' deep pit, sandy soil, smaller ball	238	330	568
2610	Larger ball	230	320	550
2620	Clay soil, smaller ball	202	370	572
2630	Larger ball	194	360	554
2800	16' to 18' tree, deciduous, 6' x 2-1/2' deep pit, sandy soil	340	470	810
2820	Clay soil	281	490	771
2900	Evergreen, 7' x 3' deep pit, sandy soil	575	780	1,355
2920	Clay soil	465	760	1,225
3000	Over 18' tree, deciduous, 7' x 3' deep pit, sandy soil	575	780	1,355
3020	Clay soil	465	760	1,225
3100	Evergreen, 8' x 3-1/4' deep pit, sandy soil	800	1,075	1,875
3120	Clay soil	645	1,025	1,670

G3010 Water Supply

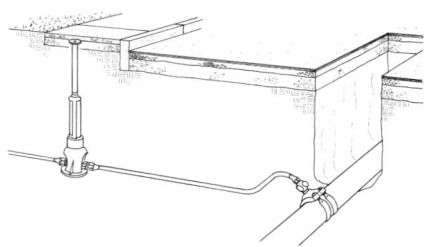

The Water Service Systems are for copper service taps from 1" to 2" diameter into pressurized mains from 6" to 8" diameter. Costs are given for two offsets with depths varying from 2' to 10'. Included in system components are excavation and backfill and required curb stops with boxes.

System Components	QUANTITY	UNIT	COST EACH		
			MAT.	INST.	TOTAL
SYSTEM G3010 121 1000					
WATER SERVICE, LEAD FREE, 6" MAIN, 1" COPPER SERVICE, 10' OFFSET, 2' DEEP					
Trench excavation, 1/2 C.Y. backhoe, 1 laborer	2.220	C.Y.		16.51	16.51
Drill & tap pressurized main, 6", 1" to 2" service	1.000	Ea.		485	485
Corporation stop, 1" diameter	1.000	Ea.	110	50.50	160.50
Saddles, 3/4" & 1" diameter, add	1.000	Ea.	73		73
Copper tubing, type K, 1" diameter	12.000	L.F.	144	54.60	198.60
Piping, curb stops, no lead, 1" diameter	1.000	Ea.	159	50.50	209.50
Curb box, cast iron, 1/2" to 1" curb stops	1.000	Ea.	44.50	67.50	112
Backfill by hand, no compaction, heavy soil	2.890	C.Y.		8.50	8.50
Compaction, vibrating plate	2.220	C.Y.		22.11	22.11
TOTAL			530.50	755.22	1,285.72

G3010 121	Water Service, Lead Free	COST EACH		
		MAT.	INST.	TOTAL
1000	Water service, Lead Free, 6" main, 1" copper service, 10' offset, 2' deep	530	755	1,285
1040	4' deep	530	800	1,330
1060	6' deep	530	1,025	1,555
1080	8' deep	530	1,125	1,655
1100	10' deep	530	1,250	1,780
1200	20' offset, 2' deep	530	800	1,330
1240	4' deep	530	895	1,425
1260	6' deep	530	1,350	1,880
1280	8' deep	530	1,550	2,080
1300	10' deep	530	1,775	2,305
1400	1-1/2" copper service, 10' offset, 2' deep	985	825	1,810
1440	4' deep	985	870	1,855
1460	6' deep	985	1,100	2,085
1480	8' deep	985	1,200	2,185
1500	10' deep	985	1,300	2,285
1520	20' offset, 2' deep	985	870	1,855
1530	4' deep	985	965	1,950
1540	6' deep	985	1,425	2,410
1560	8' deep	985	1,625	2,610
1580	10' deep	985	1,850	2,835
1600	2" copper service, 10' offset, 2' deep	1,300	855	2,155
1610	4' deep	1,275	905	2,180
1620	6' deep	1,275	1,125	2,400
1630	8' deep	1,275	1,225	2,500
1650	10' deep	1,275	1,350	2,625
1660	20' offset, 2' deep	1,275	905	2,180
1670	4' deep	1,275	1,000	2,275
1680	6' deep	1,275	1,450	2,725

G3010 Water Supply

G3010 121	Water Service, Lead Free	COST EACH		
		MAT.	INST.	TOTAL
1690	8' deep	1,275	1,650	2,925
1695	10' deep	1,275	1,875	3,150
1700	8" main, 1" copper service, 10' offset, 2' deep	530	800	1,330
1710	4' deep	530	845	1,375
1720	6' deep	530	1,075	1,605
1730	8' deep	530	1,175	1,705
1740	10' deep	530	1,275	1,805
1750	20' offset, 2' deep	530	845	1,375
1760	4' deep	530	940	1,470
1770	6' deep	530	1,400	1,930
1780	8' deep	530	1,600	2,130
1790	10' deep	530	1,825	2,355
1800	1-1/2" copper service, 10' offset, 2' deep	985	870	1,855
1810	4' deep	985	915	1,900
1820	6' deep	985	1,150	2,135
1830	8' deep	985	1,675	2,660
1840	10' deep	985	1,350	2,335
1850	20' offset, 2' deep	985	915	1,900
1860	4' deep	985	915	1,900
1870	6' deep	985	1,450	2,435
1880	8' deep	985	1,675	2,660
1890	10' deep	985	1,875	2,860
1900	2" copper service, 10' offset, 2' deep	1,300	900	2,200
1910	4' deep	1,275	950	2,225
1920	6' deep	1,275	1,175	2,450
1930	8' deep	1,275	1,275	2,550
1940	10' deep	1,275	1,375	2,650
1950	20' offset, 2' deep	1,275	950	2,225
1960	4' deep	1,275	1,050	2,325
1970	6' deep	1,275	1,500	2,775
1980	8' deep	1,275	1,700	2,975
1990	10' deep	1,275	1,925	3,200

G3010 Water Supply

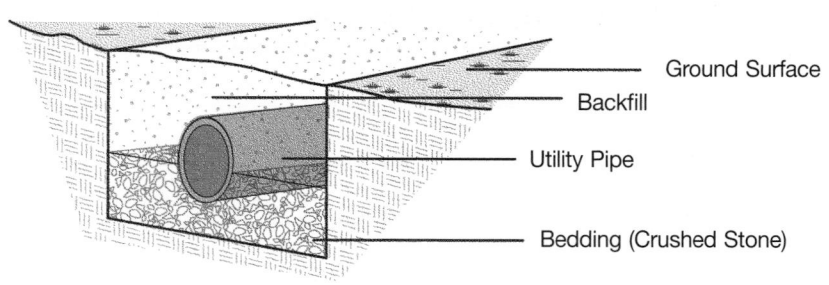

Ground Surface
Backfill
Utility Pipe
Bedding (Crushed Stone)

System Components	QUANTITY	UNIT	COST L.F.		
			MAT.	INST.	TOTAL
SYSTEM G3010 122 1000					
WATERLINE 4" DUCTILE IRON INCLUDING COMMON EARTH EXCAVATION,					
BEDDING, BACKFILL AND COMPACTION					
Excavate Trench	.296	B.C.Y.		2.24	2.24
Crushed stone bedding material	.203	L.C.Y.	5.79	2.65	8.44
Compact bedding	.173	E.C.Y.		1.09	1.09
Install 4" MJ ductile iron pipe	1.000	L.F.	51.50	11.75	63.25
Backfill trench	.160	L.C.Y.		.52	.52
Compact fill material in trench	.123	E.C.Y.		.68	.68
Dispose of excess fill material on-site	.225	L.C.Y.		2.94	2.94
TOTAL			57.29	21.87	79.16

G3010 122	Waterline (Common Earth Excavation)	COST L.F.		
		MAT.	INST.	TOTAL
0950	Waterline including common earth excavation, bedding, backfill and compaction			
1000	4" MJ ductile iron, 4' deep	57.50	22	79.50
1150	6 inch	66.50	25	91.50
1200	8 inch	61.50	28.50	90
1250	10 inch	79	32	111
1300	12 inch	105	34	139
1400	4" push joint ductile iron, 4' deep	32.50	16	48.50
1450	6 "	32	17.50	49.50
1500	8 "	43	22.50	65.50
1550	10 "	63.50	24	87.50
1600	12 "	67	27.50	94.50
1700	4" PVC Class 150, 4' deep	10.40	14.85	25.25
1750	6 "	12.70	16.10	28.80
1800	8 "	14.80	17.70	32.50
1900	4" PVC Class 200, 4' deep	10.35	14.85	25.20
1950	6 "	15.45	16.10	31.55
2000	8 "	16.35	17.70	34.05
2100	6" PVC C900, 4' deep	12.70	16.10	28.80
2150	8 "	14.80	19.80	34.60
2160	10 "	17.25	21.50	38.75
2200	12 "	22	24	46
2300	6" HDPE SDR 21 fusion, 4' deep	12.85	21	33.85
2350	8 "	18.05	23	41.05
2360	10 "	21.50	24	45.50
2400	12 "	27	26.50	53.50
2500	10", 1/4" wall black steel, 4' deep	38.50	44.50	83
2550	12"	43.50	46.50	90
3000	18" MJ ductile iron, 6' deep with trench box	175	76.50	251.50
3050	24"	194	94	288
3100	18" push joint ductile iron, 6' deep with trench box	90.50	66	156.50
3200	24"	123	75	198
3300	18" PVC C905, 6' deep with trench box	46	57	103

G30 Site Mechanical Utilities

G3010 Water Supply

G3010 122	Waterline (Common Earth Excavation)	MAT.	INST.	TOTAL
3350	24"	68	66	134
3400	18" HDPE fusion SDR 21, 6' deep with trench box	50.50	75.50	126
3450	24"	98.50	91	189.50
3500	24" Concrete (CCP) , 6' deep with trench box	125	80.50	205.50
3600	18", 1/4" wall black steel , 6' deep with trench box	108	81	189
3650	18", 5/16" wall	133	81	214
3700	18", 3/8" wall	255	199	454
3750	24", 3/8" wall	199	233	432
3800	36" Concrete (CCP) , 8' deep with trench box	197	122	319
3900	36" PVC C905 , 8' deep with trench box	156	97.50	253.50
4000	36" 5/16" wall black steel , 8' deep with trench box	189	305	494
4050	36" 1/2" wall black steel	370	330	700
5000	4" MJ ductile iron, 5' deep	57	23	80
5005	6"	66.50	26	92.50
5010	8"	61	29	90
5015	10"	78.50	32.50	111
5020	12"	104	34.50	138.50
5100	4" push on joint ductile iron, 5' deep	32	16.90	48.90
5105	6"	32	18.35	50.35
5110	8"	42.50	23.50	66
5115	10"	63	24.50	87.50
5120	12"	66	26.50	92.50
5200	4" PVC Class 150, 5' deep	10.25	15.80	26.05
5205	6"	12.40	16.95	29.35
5210	8"	14.35	18.35	32.70
5300	4" PVC Class 200, 5' deep	10.25	15.80	26.05
5305	6"	15.15	16.95	32.10
5310	8"	15.90	18.35	34.25
5405	6" PVC C900, 5' deep	12.40	16.95	29.35
5410	8"	14.35	20.50	34.85
5415	10"	17.25	22.50	39.75
5420	12"	21	24.50	45.50
5505	6" HDPE SDR 21 fusion, 5' deep	12.55	21.50	34.05
5510	8"	17.60	24	41.60
5515	10"	21.50	25	46.50
5520	12"	26	27	53
5615	10", 1/4" wall black steel, 5' deep	38	45	83
5620	12"	42.50	47	89.50
7000	4" MJ ductile iron, 7' deep	68.50	51.50	120
7005	6"	79	55	134
7010	8"	74.50	59	133.50
7015	10"	92.50	63	155.50
7020	12"	119	66	185
7100	4" push on joint ductile iron, 7' deep	43.50	45.50	89
7105	6"	44.50	47.50	92
7110	8"	56	53.50	109.50
7115	10"	77	55.50	132.50
7120	12"	81	58	139
7200	4" PVC Class 150, 7' deep	22	44.50	66.50
7205	6"	25	46	71
7210	8"	27.50	48.50	76
7300	4" PVC Class 200, 7' deep	22	44.50	66.50
7305	6"	27.50	46	73.50
7310	8"	29	48.50	77.50
7405	6" PVC C900, 7' deep	25	46	71
7410	8"	27.50	50.50	78
7415	10"	31.50	53	84.50
7420	12"	36	56	92

For customer support on your Site Work & Landscape Costs with RSMeans Data, call 800.448.8182.

G3010 Water Supply

G3010 122	Waterline (Common Earth Excavation)	COST L.F.		
		MAT.	INST.	TOTAL
7505	6" HDPE SDR 21 fusion, 7' deep	25	51	76
7510	8"	31	54	85
7515	10"	35.50	55.50	91
7520	12"	40.50	58.50	99

G30 Site Mechanical Utilities

G3010 Water Supply

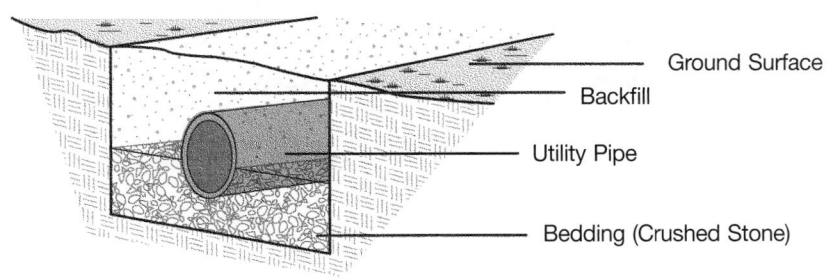

- Ground Surface
- Backfill
- Utility Pipe
- Bedding (Crushed Stone)

System Components	QUANTITY	UNIT	COST L.F.		
			MAT.	INST.	TOTAL
SYSTEM G3010 124 1000					
WATERLINE 4″ DUCTILE IRON INCLUDING LOAM & SANDY CLAY EXCAVATION,					
BEDDING, BACKFILL AND COMPACTION					
Excavate Trench	.296	B.C.Y.		2.06	2.06
Crushed stone bedding material	.203	L.C.Y.	5.79	2.65	8.44
Compact bedding	.173	E.C.Y.		1.09	1.09
Install 4″ MJ ductile iron pipe	1.000	L.F.	51.50	11.75	63.25
Backfill trench	.173	L.C.Y.		.56	.56
Compact fill material in trench	.123	E.C.Y.		.68	.68
Dispose of excess fill material on-site	.242	L.C.Y.		3.16	3.16
TOTAL			57.29	21.95	79.24

G3010 124	Waterline (Loam & Sandy Clay Excavation)	COST L.F.		
		MAT.	INST.	TOTAL
0950	Waterline including loam & sandy clay excavation, bedding, backfill & compaction			
1000	4″ MJ ductile iron, 4′ deep	57.50	22	79.50
1150	6 inch	66.50	25.50	92
1200	8 inch	61.50	28.50	90
1250	10 inch	79	32	111
1300	12 inch	105	34	139
1400	4″ push joint ductile iron, 4′ deep	32.50	16.10	48.60
1450	6 ″	32	17.60	49.60
1500	8 ″	43	22.50	65.50
1550	10 ″	63.50	24	87.50
1600	12 ″	67	26.50	93.50
1700	4″ PVC Class 150, 4′ deep	9.95	14.95	24.90
1750	6 ″	12.30	16.20	28.50
1800	8 ″	14.40	17.75	32.15
1900	4″ PVC Class 200, 4′ deep	9.95	14.95	24.90
1950	6 ″	15.05	16.20	31.25
2000	8 ″	15.95	17.75	33.70
2100	6″ PVC C900, 4′ deep	12.30	16.20	28.50
2150	8 ″	14.40	19.85	34.25
2200	12 ″	22	24.50	46.50
2300	6″ HDPE SDR 21 fusion, 4′ deep	12.45	21	33.45
2350	8 ″	17.65	23	40.65
2400	12 ″	26.50	27	53.50
2500	10″, 1/4″ wall black steel, 4′ deep	38.50	44.50	83
2550	12″	43.50	46.50	90
3000	18″ MJ ductile iron, 6′ deep with trench box	175	76.50	251.50
3050	24″	194	94.50	288.50
3100	18″ push joint ductile iron, 6′ deep with trench box	90.50	69.50	160
3200	24″	123	75.50	198.50
3300	18″ PVC C905, 6′ deep with trench box	45.50	57	102.50
3350	24″	67.50	66.50	134
3400	18″ HDPE fusion SDR 21, 6′ deep with trench box	50	75.50	125.50

701

G3010 Water Supply

G3010 124	Waterline (Loam & Sandy Clay Excavation)	COST L.F.		
		MAT.	INST.	TOTAL
3450	24"	98	91	189
3500	24" Concrete (CCP) , 6' deep with trench box	125	81	206
3600	18", 1/4" wall black steel , 6' deep with trench box	108	81.50	189.50
3650	18", 5/16" wall	133	81.50	214.50
3700	18", 3/8" wall	255	200	455
3750	24", 3/8" wall	199	233	432
3800	36" Concrete (CCP) , 8' deep with trench box	197	123	320
3900	36" PVC C905 , 8' deep with trench box	156	98.50	254.50
4000	36" 5/16" wall black steel , 8' deep with trench box	189	305	494
4050	36" 1/2" wall black steel	370	330	700
5000	4" MJ ductile iron, 5' deep	57	23	80
5005	6"	66.50	26	92.50
5010	8"	61	29	90
5015	10"	78.50	32.50	111
5020	12"	104	34.50	138.50
5100	4" push on joint ductile iron, 5' deep	32	17	49
5105	6"	32	18.40	50.40
5110	8"	42.50	23.50	66
5115	10"	63	24.50	87.50
5120	12"	66	26.50	92.50
5200	4" PVC Class 150, 5' deep	10.25	15.85	26.10
5205	6"	12.40	17	29.40
5210	8"	14.35	18.45	32.80
5300	4" PVC Class 200, 5' deep	10.25	15.85	26.10
5305	6"	15.15	17	32.15
5310	8"	15.90	18.45	34.35
5405	6" PVC C900, 5' deep	12.40	17	29.40
5410	8"	14.35	20.50	34.85
5415	10"	17.25	22.50	39.75
5420	12"	21	24.50	45.50
5505	6" HDPE SDR 21 fusion, 5' deep	12.55	21.50	34.05
5510	8"	17.60	24	41.60
5515	10"	21.50	25	46.50
5520	12"	26	27	53
5615	10", 1/4" wall black steel, 5' deep	38	45	83
5620	12"	42.50	47	89.50
6000	4" MJ ductile iron, 6' deep	68.50	47	115.50
6005	6"	79	51	130
6010	8"	74.50	55	129.50
6015	10"	92.50	58.50	151
6020	12"	119	61.50	180.50
6100	4" push on joint ductile iron, 6' deep	43.50	41	84.50
6105	6"	44.50	43.50	88
6110	8"	56	49	105
6115	10"	77	51	128
6120	12"	81	53.50	134.50
6200	4" PVC Class 150, 6' deep	22	40	62
6205	6"	25.50	42.50	68
6210	8"	27.50	44	71.50
6300	4" PVC Class 200, 6' deep	22	40	62
6305	6"	27.50	42	69.50
6310	8"	29	44	73
6405	6" PVC C900, 6' deep	25	42	67
6410	8"	27.50	46	73.50
6415	10"	31.50	49	80.50
6420	12"	36	52	88
6505	6" HDPE SDR 21 fusion, 6' deep	25	46.50	71.50
6510	8"	31	49.50	80.50

G30 Site Mechanical Utilities

G3010 Water Supply

G3010 124	Waterline (Loam & Sandy Clay Excavation)	COST L.F.		
		MAT.	INST.	TOTAL
6515	10"	35.50	52	87.50
6520	12"	40.50	54	94.50
6615	10", 1/4" wall black steel, 6' deep	52	71.50	123.50
6620	12"	57.50	74	131.50
7000	4" MJ ductile iron, 7' deep	68.50	50.50	119
7005	6"	79	54.50	133.50
7010	8"	74.50	58.50	133
7015	10"	92.50	62	154.50
7020	12"	119	65	184
7100	4" push on joint ductile iron, 7' deep	43.50	44.50	88
7105	6"	44.50	47	91.50
7110	8"	56	52.50	108.50
7115	10"	77	54.50	131.50
7120	12"	81	57	138
7200	4" PVC Class 150, 7' deep	22	43.50	65.50
7205	6"	25	45.50	70.50
7210	8"	27.50	47.50	75
7300	4" PVC Class 200, 7' deep	22	43.50	65.50
7305	6"	27.50	45.50	73
7310	8"	29	47.50	76.50
7405	6" PVC C900, 7' deep	25	45.50	70.50
7410	8"	27.50	49.50	77
7415	10"	31.50	52.50	84
7420	12"	36	55	91
7505	6" HDPE SDR 21 fusion, 7' deep	25	50	75
7510	8"	31	53	84
7515	10"	35.50	54.50	90
7520	12"	40.50	57.50	98

703

G3010 Water Supply

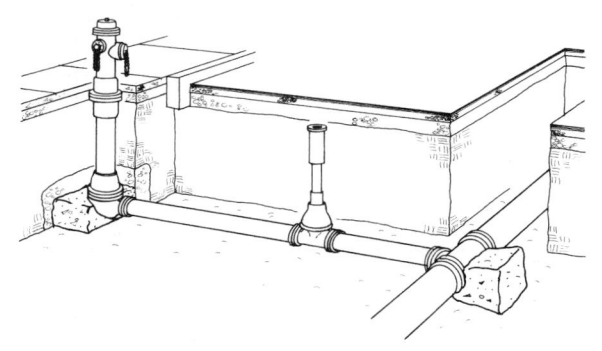

The Fire Hydrant Systems include: four different hydrants with three different lengths of offsets, at several depths of trenching. Excavation and backfill is included with each system as well as thrust blocks as necessary and pipe bedding. Finally spreading of excess material and fine grading complete the components.

System Components	QUANTITY	UNIT	COST PER EACH		
			MAT.	INST.	TOTAL
SYSTEM G3010 410 1500					
HYDRANT, 4-1/2″ VALVE SIZE, TWO WAY, 10′ OFFSET, 2′ DEEP					
Excavation, trench, 1 C.Y. hydraulic backhoe	3.556	C.Y.		32.11	32.11
Sleeve, 12″ x 6″	1.000	Ea.	3,475	130.65	3,605.65
Ductile iron pipe, 6″ diameter	8.000	L.F.	484	117.28	601.28
Gate valve, 6″	1.000	Ea.	3,600	326.50	3,926.50
Hydrant valve box, 6′ long	1.000	Ea.	256	88.50	344.50
4-1/2″ valve size, depth 2′-0″	1.000	Ea.	3,400	235	3,635
Thrust blocks, at valve, shoe and sleeve	1.600	C.Y.	366.40	497.73	864.13
Borrow, crushed stone, 3/8″	4.089	C.Y.	149.25	35.05	184.30
Backfill and compact, dozer, air tamped	4.089	C.Y.		85.06	85.06
Spread excess excavated material	.204	C.Y.		9.79	9.79
Fine grade, hand	400.000	S.F.		308	308
TOTAL			11,730.65	1,865.67	13,596.32

G3010 410	Fire Hydrants	COST PER EACH		
		MAT.	INST.	TOTAL
1000	Hydrant, 4-1/2″ valve size, two way, 0′ offset, 2′ deep	11,200	1,325	12,525
1100	4′ deep	11,500	1,350	12,850
1200	6′ deep	11,800	1,425	13,225
1300	8′ deep	11,900	1,475	13,375
1400	10′ deep	12,200	1,550	13,750
1500	10′ offset, 2′ deep	11,700	1,875	13,575
1600	4′ deep	12,000	2,200	14,200
1700	6′ deep	12,300	2,850	15,150
1800	8′ deep	12,400	3,800	16,200
1900	10′ deep	12,700	5,800	18,500
2000	20′ offset, 2′ deep	12,300	2,425	14,725
2100	4′ deep	12,600	2,975	15,575
2200	6′ deep	12,900	3,950	16,850
2300	8′ deep	13,000	5,325	18,325
2400	10′ deep	13,300	7,200	20,500
2500	Hydrant, 4-1/2″ valve size, three way, 0′ offset, 2′ deep	11,500	1,325	12,825
2600	4′ deep	11,800	1,350	13,150
2700	6′ deep	12,100	1,425	13,525
2800	8′ deep	12,200	1,500	13,700
2900	10′ deep	12,500	1,600	14,100
3000	10′ offset, 2′ deep	12,000	1,875	13,875
3100	4′ deep	12,300	2,225	14,525
3200	6′ deep	12,600	2,875	15,475
3300	8′ deep	12,700	3,825	16,525

G3010 Water Supply

G3010 410	Fire Hydrants	COST PER EACH		
		MAT.	INST.	TOTAL
3400	10' deep	13,000	5,825	18,825
3500	20' offset, 2' deep	12,600	2,450	15,050
3600	4' deep	12,900	3,000	15,900
3700	6' deep	13,200	3,975	17,175
3800	8' deep	13,300	5,350	18,650
3900	10' deep	13,600	7,225	20,825
5000	5-1/4" valve size, two way, 0' offset, 2' deep	11,500	1,325	12,825
5100	4' deep	11,900	1,350	13,250
5200	6' deep	12,200	1,425	13,625
5300	8' deep	12,600	1,475	14,075
5400	10' deep	13,000	1,550	14,550
5500	10' offset, 2' deep	12,000	1,875	13,875
5600	4' deep	12,400	2,200	14,600
5700	6' deep	12,700	2,850	15,550
5800	8' deep	13,100	3,800	16,900
5900	10' deep	13,500	5,800	19,300
6000	20' offset, 2' deep	12,600	2,425	15,025
6100	4' deep	13,000	2,975	15,975
6200	6' deep	13,300	3,950	17,250
6300	8' deep	13,700	5,325	19,025
6400	10' deep	14,100	7,200	21,300
6500	Three way, 0' offset, 2' deep	11,800	1,325	13,125
6600	4' deep	12,200	1,350	13,550
6700	6' deep	12,500	1,425	13,925
6800	8' deep	13,000	1,500	14,500
6900	10' deep	13,400	1,600	15,000
7000	10' offset, 2' deep	12,300	1,875	14,175
7100	4' deep	12,600	2,225	14,825
7200	6' deep	13,000	2,875	15,875
7300	8' deep	13,500	3,825	17,325
7400	10' deep	13,800	5,825	19,625
7500	20' offset, 2' deep	12,900	2,450	15,350
7600	4' deep	13,200	2,700	15,900
7700	6' deep	13,600	3,975	17,575

G3020 Sanitary Sewer

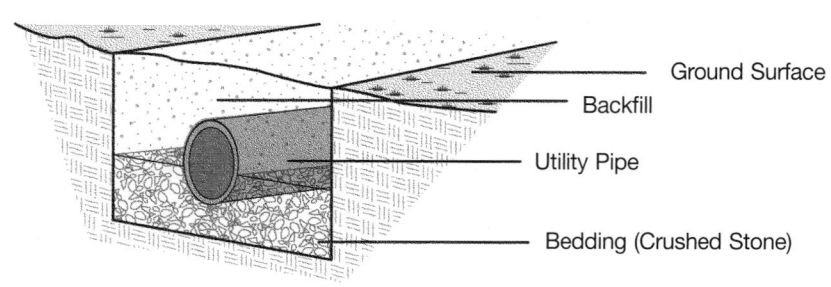

Ground Surface
Backfill
Utility Pipe
Bedding (Crushed Stone)

System Components			COST L.F.		
	QUANTITY	UNIT	MAT.	INST.	TOTAL
SYSTEM G3020 112 1000					
SEWERLINE 4″ B&S CAST IRON INCLUDING COMMON EARTH EXCAVATION,					
BEDDING, BACKFILL AND COMPACTION					
Excavate Trench	.296	B.C.Y.		2.24	2.24
Crushed stone bedding material	.203	L.C.Y.	5.79	2.65	8.44
Compact bedding	.173	E.C.Y.		1.09	1.09
Install 4″ B&S cast iron pipe	1.000	L.F.	27	22	49
Backfill trench	.160	L.C.Y.		.52	.52
Compact fill material in trench	.123	E.C.Y.		.68	.68
Dispose of excess fill material on-site	.225	L.C.Y.		2.94	2.94
TOTAL			32.79	32.12	64.91

G3020 112	Sewerline (Common Earth Excavation)	COST L.F.		
		MAT.	INST.	TOTAL
0950	Sewerline including common earth excavation, bedding, backfill and compaction			
1000	4″ B&S Cast iron, 4′ deep	33	32	65
1150	6 inch	52	37.50	89.50
1200	8 inch	80.50	55	135.50
1250	10 inch	131	57.50	188.50
1300	12 inch	185	65.50	250.50
1400	4″ HDPE type S watertight, 4′ deep	7	14.30	21.30
1450	6 ″	9.50	14.90	24.40
1500	8 ″	10.95	15.50	26.45
1550	10 ″	13.10	16.05	29.15
1600	12 ″	15.20	20.50	35.70
1700	4″ PVC, SDR 35, 20′ length , 4′ deep	8.20	14.85	23.05
1750	6 ″	10.95	15.55	26.50
1800	8 ″- 13′ length	16.35	16.15	32.50
1900	10″	20.50	18.40	38.90
1950	12 ″	26	18.90	44.90
2000	12 ″ Concrete w/gasket class 3	27.50	25.50	53
2100	12 ″ class 4	31	25.50	56.50
2200	12 ″ class 5	29.50	25.50	55
3050	24″	45.50	53.50	99
3100	18″ PVC, SDR 35, 13′ length, 6′ deep with trench box	58.50	54	112.50
3150	24″	90.50	57	147.50
3200	18″ concrete w/gasket class 3, 6′ deep with trench box	51	57.50	108.50
3250	24″ class 3	58.50	61.50	120
3300	18″ class 4	44.50	57.50	102
3350	24″ class 4	54	61.50	115.50
3400	18″ class 5	52	57.50	109.50
3450	24″ class 5	52.50	61.50	114
4000	30″ HDPE type S, watertight , 8′ deep with trench box	66	74.50	140.50
4050	36″	84	81.50	165.50
5000	30″ concrete w/gaskets class 3, 8′ deep with trench box	92.50	119	211.50
5050	36″ class 3	135	133	268

For customer support on your Site Work & Landscape Costs with RSMeans Data, call 800.448.8182.

G30 Site Mechanical Utilities

G3020 Sanitary Sewer

G3020 112	Sewerline (Common Earth Excavation)	COST L.F.		
		MAT.	INST.	TOTAL
5100	30" class 4	69.50	119	188.50
5150	36" class 4	101	133	234
5200	30" class 5	93	119	212
5250	36" class 5	152	133	285
5500	4" B&S Cast iron, 5' deep	32.50	33	65.50
5502	6"	52	38.50	90.50
5504	8"	80	55.50	135.50
5506	10"	130	58	188
5508	12"	184	66	250
5510	4" HDPE type S watertight, 5' deep	6.85	15.20	22.05
5512	6"	9.20	15.75	24.95
5514	8"	10.50	16.20	26.70
5516	10"	12.35	16.50	28.85
5518	12"	14.20	17.10	31.30
5520	4" PVC,SDR 35, 5' deep	8.10	15.75	23.85
5522	6"	10.60	16.35	26.95
5524	8"	15.85	16.85	32.70
5526	10"	19.55	18.85	38.40
5528	12"	25	19.25	44.25
5532	12" Concrete w/gasket class 3, 5' deep	26.50	26	52.50
5534	12" class 4	30	26	56
5536	12" class 5	28.50	26	54.50
6000	4" B&S Cast iron, 6' deep	44	59	103
6002	6"	64.50	66.50	131
6004	8"	93.50	83	176.50
6006	10"	144	86	230
6008	12"	199	94.50	293.50
6010	4" HDPE type S watertight, 6' deep	18.40	41	59.40
6012	6"	21.50	44	65.50
6014	8"	23.50	43.50	67
6016	10"	26.50	44.50	71
6018	12"	29	46	75
6020	4" PVC,SDR 35, 6' deep	19.60	41.50	61.10
6022	6"	23	44.50	67.50
6024	8"	29	44	73
6026	10"	33.50	47	80.50
6028	12"	40	48	88
6032	12" Concrete w/gasket class 3, 6' deep	41	54.50	95.50
6034	12" class 4	45	54.50	99.50
6036	12" class 5	43.50	54.50	98
6500	4" B&S Cast iron, 7' deep	44	61.50	105.50
6502	6"	64.50	69	133.50
6504	8"	93.50	85.50	179
6506	10"	144	89	233
6508	12"	199	97.50	296.50
6510	4" HDPE type S watertight, 7' deep	18.40	43.50	61.90
6512	6"	21.50	46.50	68
6514	8"	23.50	46	69.50
6516	10"	26.50	47	73.50
6518	12"	29	48.50	77.50
6520	4" PVC,SDR 35, 7' deep	19.60	44.50	64.10
6522	6"	23	47	70
6524	8"	29	47	76
6526	10"	33.50	49.50	83
6528	12"	40	50.50	90.50
6532	12" Concrete w/gasket class 3, 7' deep	41	57	98
6534	12" class 4	45	57	102
6536	12" class 5	43.50	57	100.50

G3020 Sanitary Sewer

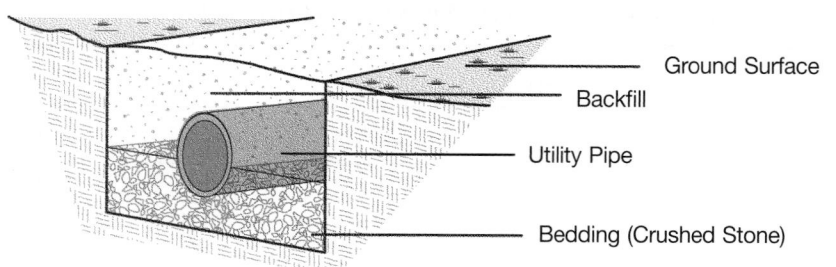

Ground Surface

Backfill

Utility Pipe

Bedding (Crushed Stone)

System Components			COST L.F.		
	QUANTITY	UNIT	MAT.	INST.	TOTAL
SYSTEM G3020 114 1000					
SEWERLINE 4″ B&S CAST IRON INCLUDING LOAM & SANDY CLAY EXCAVATION,					
BEDDING, BACKFILL AND COMPACTION					
Excavate Trench	.296	B.C.Y.		2.06	2.06
Crushed stone bedding material	.203	L.C.Y.	5.79	2.65	8.44
Compact bedding	.173	E.C.Y.		1.09	1.09
Install 4″ B&S cast iron pipe	1.000	L.F.	27	22	49
Backfill trench	.173	L.C.Y.		.56	.56
Compact fill material in trench	.123	E.C.Y.		.68	.68
Dispose of excess fill material on-site	.242	L.C.Y.		3.16	3.16
TOTAL			32.79	32.20	64.99

G3020 114	Sewerline (loam & sandy clay excavation)	COST L.F.		
		MAT.	INST.	TOTAL
0950	Sewerline incl. loam & sandy clay excavation, bedding, backfill and compaction			
1000	4″ B&S Cast iron, 4′ deep	33	32	65
1150	6 inch	52	37.50	89.50
1200	8 inch	80.50	55	135.50
1250	10 inch	131	58	189
1300	12 inch	185	65.50	250.50
1400	4″ HDPE type S watertight, 4′ deep	7	14.40	21.40
1450	6 ″	9.50	15	24.50
1500	8 ″	10.95	15.65	26.60
1550	10 ″	13.10	16.15	29.25
1600	12 ″	15.20	16.90	32.10
1700	4″ PVC, SDR 35, 20′ length , 4′ deep	8.20	14.95	23.15
1750	6 ″	10.95	15.65	26.60
1800	8 ″- 13′ length	16.35	16.25	32.60
1900	10″	20.50	18.50	39
1950	12 ″	26	19.05	45.05
2000	12 ″ Concrete w/gasket class 3	27.50	25.50	53
2100	12 ″ class 4	31	25.50	56.50
2200	12 ″ class 5	29.50	25.50	55
3000	18″ HDPE type S watertight, 6′ deep with trench box	36.50	51	87.50
3050	24″	45.50	54	99.50
3100	18″ PVC, SDR 35, 13′ length, 6′ deep with trench box	58.50	54.50	113
3150	24″	90.50	57.50	148
3200	18″ concrete w/gasket class 3, 6′ deep with trench box	51	58	109
3250	24″ class 3	58.50	62	120.50
3300	18″ class 4	44.50	58	102.50
3350	24″ class 4	54	62	116
3400	18″ class 5	52	58	110
3450	24″ class 5	52.50	62	114.50
4000	30″ HDPE type S, watertight , 8′ deep with trench box	66	78	144
4050	36″	84	82.50	166.50
5000	30″ concrete w/gaskets class 3, 8′ deep with trench box	92.50	120	212.50
5050	36″ class 3	135	134	269
5100	30″ class 4	69.50	120	189.50

G3020 Sanitary Sewer

G3020 114	Sewerline (loam & sandy clay excavation)	COST L.F.		
		MAT.	INST.	TOTAL
5150	36" class 4	101	134	235
5200	30" class 5	93	120	213
5250	36" class 5	152	134	286
5500	4" B&S Cast iron, 5' deep	32.50	33	65.50
5502	6"	52	38.50	90.50
5504	8"	80	55.50	135.50
5506	10"	130	58.50	188.50
5508	12"	184	66	250
5510	4" HDPE type S watertight, 5' deep	6.85	15.30	22.15
5512	6"	9.20	15.80	25
5514	8"	10.50	16.25	26.75
5516	10"	12.35	16.60	28.95
5518	12"	14.20	17.15	31.35
5520	4" PVC, SDR 35, 5' deep	8.10	15.85	23.95
5522	6"	10.60	16.40	27
5524	8"	15.85	16.90	32.75
5526	10"	19.55	18.95	38.50
5528	12"	25	19.35	44.35
5532	12" Concrete w/gasket class 3, 5' deep	26.50	26	52.50
5534	12" class 4	30	26	56
5536	12" class 5	28.50	26	54.50
6000	4" B&S Cast iron, 6' deep	44	59	103
6002	6"	64.50	66.50	131
6004	8"	93.50	83	176.50
6006	10"	144	86.50	230.50
6008	12"	199	95	294
6010	4" HDPE type S watertight, 6' deep	18.40	41.50	59.90
6012	6"	21.50	44	65.50
6014	8"	23.50	44	67.50
6016	10"	26.50	45	71.50
6018	12"	29	46	75
6020	4" PVC, SDR 35, 6' deep	19.60	42	61.60
6022	6"	23	44.50	67.50
6024	8"	29	44.50	73.50
6026	10"	33.50	47	80.50
6028	12"	40	48	88
6032	12" Concrete w/gasket class 3, 6' deep	41	55	96
6034	12" class 4	45	55	100
6036	12" class 5	43.50	55	98.50
6500	4" B&S Cast iron, 7' deep	44	61	105
6502	6"	64.50	68	132.50
6504	8"	93.50	84.50	178
6506	10"	144	88	232
6508	12"	199	96.50	295.50
6510	4" HDPE type S watertight, 7' deep	18.40	43	61.40
6512	6"	21.50	45.50	67
6514	8"	23.50	45.50	69
6516	10"	26.50	46.50	73
6518	12"	29	47.50	76.50
6520	4" PVC, SDR 35, 7' deep	19.60	43.50	63.10
6522	6"	23	46.50	69.50
6524	8"	29	46	75
6526	10"	33.50	49	82.50
6528	12"	40	50	90
6532	12" Concrete w/gasket class 3, 7' deep	41	56.50	97.50
6534	12" class 4	45	56.50	101.50
6536	12" class 5	43.50	56.50	100

G3020 Sanitary Sewer

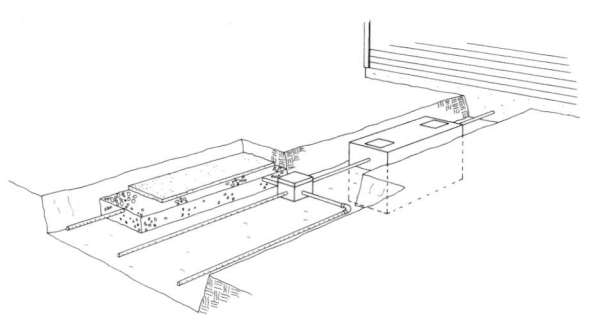

The Septic System includes: a septic tank; leaching field; concrete distribution boxes; plus excavation and gravel backfill.

The Expanded System Listing shows systems with tanks ranging from 1000 gallons to 2000 gallons. Tanks are either concrete or fiberglass. Cost is on a complete unit basis.

System Components	QUANTITY	UNIT	COST EACH		
			MAT.	INST.	TOTAL
SYSTEM G3020 302 0300					
1000 GALLON TANK, LEACHING FIELD, 600 S.F., 50' FROM BLDG.					
Mobilization	2.000	Ea.		1,772	1,772
Septic tank, precast, 1000 gallon	1.000	Ea.	1,100	293.50	1,393.50
Effluent filter	1.000	Ea.	39.50	86	125.50
Distribution box, precast, 5 outlets	1.000	Ea.	103	66.50	169.50
Flow leveler	3.000	Ea.	9.87	31.80	41.67
Sewer pipe, PVC, SDR 35, 4" diameter	160.000	L.F.	388.80	755.20	1,144
Tee	1.000	Ea.	21	86	107
Elbow	2.000	Ea.	18.90	115	133.90
Viewport cap	1.000	Ea.	7.80	28.50	36.30
Filter fabric	67.000	S.Y.	115.24	29.48	144.72
Detectable marking tape	1.600	C.L.F.	5.28	5.65	10.93
Excavation	160.000	C.Y.		2,240	2,240
Backfill	133.000	L.C.Y.		433.58	433.58
Spoil	55.000	L.C.Y.		405.35	405.35
Compaction	113.000	E.C.Y.		1,066.72	1,066.72
Stone fill, 3/4" to 1-1/2"	39.000	C.Y.	1,755	510.51	2,265.51
TOTAL			3,564.39	7,925.79	11,490.18

G3020 302	Septic Systems	COST EACH		
		MAT.	INST.	TOTAL
0300	1000 gal. concrete septic tank, 600 S.F. leaching field, 50 feet from bldg.	3,575	7,925	11,500
0301	200 feet from building	3,925	9,825	13,750
0305	750 S.F. leaching field, 50 feet from building	4,075	8,625	12,700
0306	200 feet from building	4,450	10,500	14,950
0400	1250 gal. concrete septic tank, 600 S.F. leaching field, 50 feet from bldg.	4,125	8,225	12,350
0401	200 feet from building	4,475	10,200	14,675
0405	750 S.F. leaching field, 50 feet from building	4,625	8,950	13,575
0406	200 feet from building	5,000	10,900	15,900
0410	1000 S.F. leaching field, 50 feet from building	5,525	10,200	15,725
0411	200 feet from building	5,900	12,200	18,100
0500	1500 gal. concrete septic tank, 600 S.F. leaching field, 50 feet from bldg.	4,325	8,525	12,850
0501	200 feet from building	4,675	10,400	15,075
0505	750 S.F. leaching field, 50 feet from building	4,850	9,250	14,100
0506	200 feet from building	5,225	12,200	17,425
0510	1000 S.F. leaching field, 50 feet from building	5,725	10,600	16,325
0511	200 feet from building	6,100	12,500	18,600
0515	1100 S.F. leaching field, 50 feet from building	6,075	11,000	17,075
0516	200 feet from building	6,450	13,100	19,550
0600	2000 gal concrete tank, 1200 S.F. leaching field, 50 feet from building	7,425	11,700	19,125
0601	200 feet from building	7,800	13,600	21,400

G3020 Sanitary Sewer

G3020 302	Septic Systems	COST EACH		
		MAT.	INST.	TOTAL
0605	1500 S.F. leaching field, 50 feet from building	8,475	13,200	21,675
0606	200 feet from building	8,850	15,100	23,950
0710	2500 gal concrete tank, 1500 S.F. leaching field, 50 feet from building	8,275	14,000	22,275
0711	200 feet from building	8,650	15,900	24,550
0715	1600 S.F. leaching field, 50 feet from building	8,650	14,500	23,150
0716	200 feet from building	9,025	16,400	25,425
0720	1800 S.F. leaching field, 50 feet from building	9,350	15,600	24,950
0721	200 feet from building	9,725	17,500	27,225
1300	1000 gal. 2 compartment tank, 600 S.F. leaching field, 50 feet from bldg.	3,250	7,925	11,175
1301	200 feet from building	3,600	9,825	13,425
1305	750 S.F. leaching field, 50 feet from building	3,750	8,625	12,375
1306	200 feet from building	4,125	10,500	14,625
1400	1250 gal. 2 compartment tank, 600 S.F. leaching field, 50 feet from bldg.	3,650	8,225	11,875
1401	200 feet from building	4,000	10,200	14,200
1405	750 S.F. leaching field, 50 feet from building	4,150	8,950	13,100
1406	200 feet from building	4,525	10,900	15,425
1410	1000 S.F. leaching field, 50 feet from building	5,050	10,200	15,250
1411	200 feet from building	5,425	12,200	17,625
1500	1500 gal. 2 compartment tank, 600 S.F. leaching field, 50 feet from bldg.	4,225	8,525	12,750
1501	200 feet from building	4,575	10,400	14,975
1505	750 S.F. leaching field, 50 feet from building	4,750	9,250	14,000
1506	200 feet from building	5,125	12,200	17,325
1510	1000 S.F. leaching field, 50 feet from building	5,625	10,600	16,225
1511	200 feet from building	6,000	12,500	18,500
1515	1100 S.F. leaching field, 50 feet from building	5,975	11,000	16,975
1516	200 feet from building	6,350	13,100	19,450
1600	2000 gal 2 compartment tank, 1200 S.F. leaching field, 50 feet from bldg.	7,450	11,700	19,150
1601	200 feet from building	7,825	13,600	21,425
1605	1500 S.F. leaching field, 50 feet from building	8,500	13,200	21,700
1606	200 feet from building	8,875	15,100	23,975
1710	2500 gal 2 compartment tank, 1500 S.F. leaching field, 50 feet from bldg.	9,225	14,000	23,225
1711	200 feet from building	9,600	15,900	25,500
1715	1600 S.F. leaching field, 50 feet from building	9,600	14,500	24,100
1716	200 feet from building	9,975	16,400	26,375
1720	1800 S.F. leaching field, 50 feet from building	10,300	15,600	25,900
1721	200 feet from building	10,700	17,500	28,200

G3020 Sanitary Sewer

Grass

Lagoon Liner

Gravel

System Components			COST EACH		
	QUANTITY	UNIT	MAT.	INST.	TOTAL
SYSTEM G3020 730 1000					
500,000 GPD, SECONDARY DOMESTIC SEWAGE LAGOON, COMMON EARTH					
Supervision of lagoon construction	60.000	Day		41,100	41,100
Quality Control of Earthwork per day	30.000	Day		15,900	15,900
Excavate lagoon	10780.000	B.C.Y.		29,645	29,645
Finish grade slopes for seeding	40.000	M.S.F.		1,508	1,508
Spread & grade top of lagoon banks for compaction	753.000	M.S.F.		60,616.50	60,616.50
Compact lagoon banks	10725.000	E.C.Y.		9,116.25	9,116.25
Install pond liner	202.000	M.S.F.	151,500	344,410	495,910
Provide water for compaction and dust control	2156.000	E.C.Y.	2,910.60	4,829.44	7,740.04
Dispose of excess material on-site	72.000	L.C.Y.		290.16	290.16
Seed non-lagoon side of banks	40.000	M.S.F.	1,240	1,232	2,472
Place gravel on top of banks	2311.000	S.Y.	12,710.50	3,697.60	16,408.10
Provide 6 ' high fence around lagoon	1562.000	L.F.	33,583	12,511.62	46,094.62
Provide 12' wide fence gate	3.000	Opng.	1,500	1,806	3,306
Temporary Fencing	1562.000	L.F.	5,248.32	4,139.30	9,387.62
TOTAL			208,692.42	530,801.87	739,494.29

G3020 730	Secondary Sewage Lagoon	COST EACH		
		MAT.	INST.	TOTAL
0980	Lagoons, Secondary, 10 day retention, 5.1 foot depth, no mechanical aeration			
1000	500,000 GPD Secondary Domestic Sewage Lagoon, common earth,10 day retention	208,500	531,000	739,500
1100	600,000 GPD	237,500	620,500	858,000
1200	700,000 GPD	265,000	801,000	1,066,000
1300	800,000 GPD	293,000	756,500	1,049,500
1400	900,000 GPD	320,500	838,500	1,159,000
1500	1 MGD	373,500	971,000	1,344,500
1600	2 MGD	624,000	1,699,500	2,323,500
1700	3 MGD	1,004,500	2,640,000	3,644,500
2000	500,000 GPD sandy clay & loam	208,500	532,500	741,000
2100	600,000 GPD	237,500	622,000	859,500
2200	700,000 GPD	265,000	679,000	944,000
2300	800,000 GPD	293,000	758,500	1,051,500
2400	900,000 GPD	320,500	840,500	1,161,000
2500	1 MGD	373,500	974,500	1,348,000
2600	2 MGD	624,000	1,701,000	2,325,000
2700	3 MGD	1,004,500	2,645,000	3,649,500
3000	500,000 GPD sand & gravel	208,500	521,000	729,500
3100	600,000 GPD	237,500	608,500	846,000
3200	700,000 GPD	265,000	664,000	929,000
3300	800,000 GPD	293,000	742,000	1,035,000
3400	900,000 GPD	320,500	822,000	1,142,500

G3020 Sanitary Sewer

G3020 730	Secondary Sewage Lagoon	COST EACH		
		MAT.	INST.	TOTAL
3500	1 MGD	373,500	952,500	1,326,000
3600	2 MGD	624,000	1,660,500	2,284,500
3700	3 MGD	1,004,500	2,582,000	3,586,500
6000	500,000 GPD, Domestic Sewage Sec. Lagoon, comm. earth, 10 day, no liner	48,900	184,000	232,900
6100	600,000 GPD	61,500	220,000	281,500
6200	700,000 GPD	65,500	230,500	296,000
6300	800,000 GPD	69,000	254,500	323,500
6400	900,000 GPD	72,500	274,500	347,000
6500	1 MGD	85,500	316,500	402,000
6600	2 MGD	114,000	540,000	654,000
6700	3 MGD	200,000	810,500	1,010,500
7000	500,000 GPD sandy clay & loam	57,000	188,000	245,000
7100	600,000 GPD	61,500	221,000	282,500
7200	700,000 GPD	65,500	225,500	291,000
7300	800,000 GPD	69,000	249,000	318,000
7400	900,000 GPD	72,500	276,500	349,000
7500	1 MGD	85,500	319,500	405,000
7600	2 MGD	114,000	541,500	655,500
7700	3 MGD	200,000	815,500	1,015,500

G3020 Sanitary Sewer

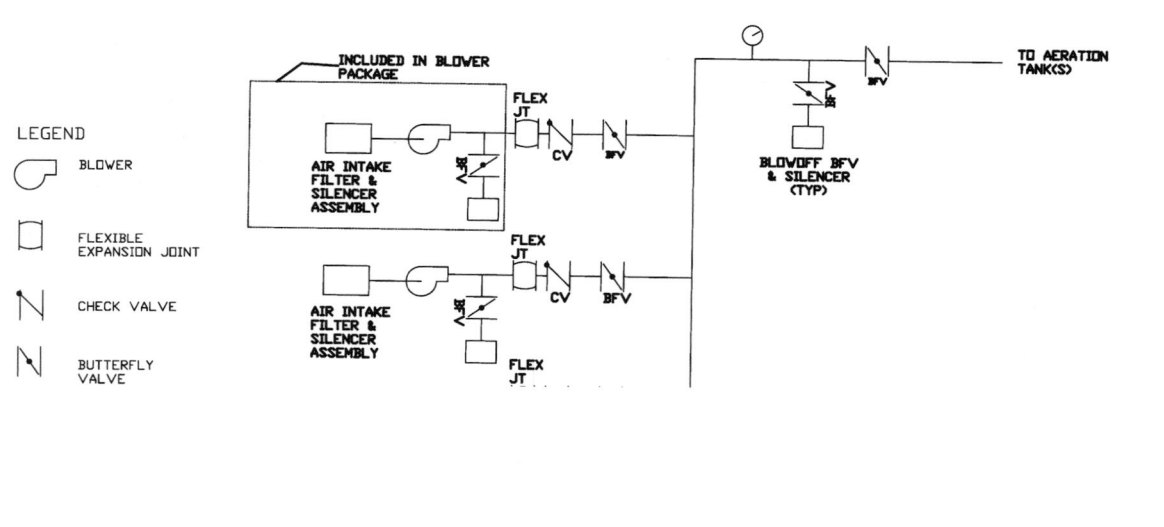

System Components	QUANTITY	UNIT	COST EACH		
			MAT.	INST.	TOTAL
SYSTEM G3020 740 1000					
WASTEWATER TREATMENT, AERATION, 75 CFM PREPACKAGED BLOWER SYSTEM					
Prepacked blower, 38 CFM	3.000	Ea.	10,950	2,715	13,665
Flexible coupling-2″ diameter	3.000	Ea.	126	64.50	190.50
Wafer style check valve-2″ diameter	3.000	Ea.	411	270	681
Wafer style butterfly valve-2″ diameter	3.000	Ea.	351	172.50	523.50
Wafer style butterfly valve-3″ diameter	1.000	Ea.	215	182	397
Blower Pressure Relief Valves-adjustable-3″ Diameter	1.000	Ea.	186	182	368
Silencer with polyester filter-3″ Connection	1.000	Ea.	198	182	380
Pipe-2″ diameter	30.000	L.F.	360	795	1,155
Pipe-3″ diameter	50.000	L.F.	1,075	1,450	2,525
3 x 3 x 2 Tee	3.000	Ea.	315	315	630
Pipe cap-3″ diameter	1.000	Ea.	67	48.50	115.50
Pressure gage	1.000	Ea.	24	25.50	49.50
Master control panel	1.000	Ea.	1,075	305	1,380
TOTAL			15,353	6,707	22,060

G3020 740	Blower System for Wastewater Aeration	COST EACH		
		MAT.	INST.	TOTAL
1000	75 CFM Prepackaged Blower System	15,400	6,700	22,100
1100	250 CFM	15,700	9,625	25,325
1200	450 CFM	27,700	10,200	37,900
1300	700 CFM	45,300	12,800	58,100
1400	1000 CFM	45,300	13,500	58,800

G3030 Storm Sewer

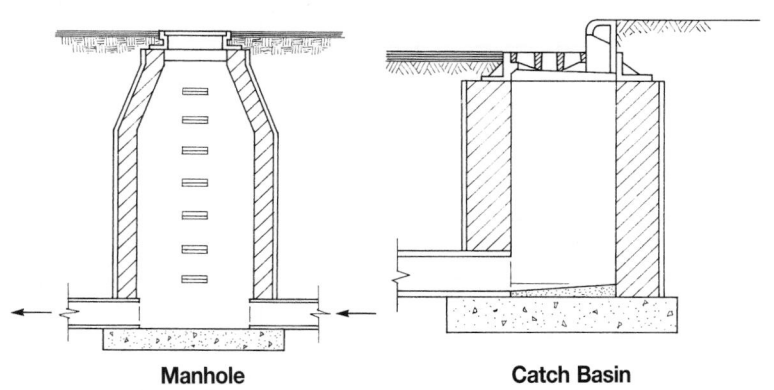

Manhole   Catch Basin

The Manhole and Catch Basin System includes: excavation with a backhoe; a formed concrete footing; frame and cover; cast iron steps and compacted backfill.

The Expanded System Listing shows manholes that have a 4', 5' and 6' inside diameter riser. Depths range from 4' to 14'. Construction material shown is either concrete, concrete block, precast concrete, or brick.

System Components	QUANTITY	UNIT	COST PER EACH		
			MAT.	INST.	TOTAL
SYSTEM G3030 210 1920					
MANHOLE/CATCH BASIN, BRICK, 4' I.D. RISER, 4' DEEP					
Excavation, hydraulic backhoe, 3/8 C.Y. bucket	14.815	B.C.Y.		133.78	133.78
Trim sides and bottom of excavation	64.000	S.F.		76.80	76.80
Forms in place, manhole base, 4 uses	20.000	SFCA	23.40	120	143.40
Reinforcing in place footings, #4 to #7	.019	Ton	26.13	25.65	51.78
Concrete, 3000 psi	.925	C.Y.	126.73		126.73
Place and vibrate concrete, footing, direct chute	.925	C.Y.		56.04	56.04
Catch basin or MH, brick, 4' ID, 4' deep	1.000	Ea.	770	1,175	1,945
Catch basin or MH steps; heavy galvanized cast iron	1.000	Ea.	19	16.20	35.20
Catch basin or MH frame and cover	1.000	Ea.	320	251.50	571.50
Fill, granular	12.954	L.C.Y.	427.48		427.48
Backfill, spread with wheeled front end loader	12.954	L.C.Y.		38.08	38.08
Backfill compaction, 12" lifts, air tamp	12.954	E.C.Y.		139.52	139.52
TOTAL			1,712.74	2,032.57	3,745.31

G3030 210	Manholes & Catch Basins	COST PER EACH		
		MAT.	INST.	TOTAL
1920	Manhole/catch basin, brick, 4' I.D. riser, 4' deep	1,725	2,025	3,750
1940	6' deep	2,450	2,825	5,275
1960	8' deep	3,300	3,875	7,175
1980	10' deep	4,025	4,800	8,825
3000	12' deep	5,150	5,250	10,400
3020	14' deep	6,475	7,325	13,800
3200	Block, 4' I.D. riser, 4' deep	1,425	1,650	3,075
3220	6' deep	1,975	2,325	4,300
3240	8' deep	2,650	3,200	5,850
3260	10' deep	3,150	3,975	7,125
3280	12' deep	4,075	5,050	9,125
3300	14' deep	5,150	6,175	11,325
4620	Concrete, cast-in-place, 4' I.D. riser, 4' deep	1,700	2,775	4,475
4640	6' deep	2,425	3,725	6,150
4660	8' deep	3,425	5,375	8,800
4680	10' deep	4,125	6,700	10,825
4700	12' deep	5,250	8,300	13,550
4720	14' deep	6,525	9,975	16,500
5820	Concrete, precast, 4' I.D. riser, 4' deep	1,950	1,500	3,450
5840	6' deep	2,525	2,025	4,550

G30 Site Mechanical Utilities

G3030 Storm Sewer

G3030 210	Manholes & Catch Basins	COST PER EACH		
		MAT.	INST.	TOTAL
5860	8' deep	3,975	2,850	6,825
5880	10' deep	4,550	3,500	8,050
5900	12' deep	5,525	4,250	9,775
5920	14' deep	6,700	5,500	12,200
6000	5' I.D. riser, 4' deep	3,100	1,625	4,725
6020	6' deep	5,025	2,275	7,300
6040	8' deep	5,050	3,025	8,075
6060	10' deep	6,775	3,850	10,625
6080	12' deep	8,675	4,925	13,600
6100	14' deep	10,700	6,025	16,725
6200	6' I.D. riser, 4' deep	3,525	2,125	5,650
6220	6' deep	6,775	2,825	9,600
6240	8' deep	8,300	3,950	12,250
6260	10' deep	10,000	5,000	15,000
6280	12' deep	12,000	6,325	18,325
6300	14' deep	14,100	7,675	21,775

G3030 Storm Sewer

The Headwall Systems are listed in concrete and different stone wall materials for two different backfill slope conditions. The backfill slope directly affects the length of the wing walls. Walls are listed for different culvert sizes starting at 30″ diameter. Excavation and backfill are included in the system components, and are figured from an elevation 2′ below the bottom of the pipe.

System Components	QUANTITY	UNIT	COST PER EACH MAT.	COST PER EACH INST.	COST PER EACH TOTAL
SYSTEM G3030 310 2000					
HEADWALL, C.I.P. CONCRETE FOR 30″ PIPE, 3′ LONG WING WALLS					
Excavation, hydraulic backhoe, 3/8 C.Y. bucket	28.000	B.C.Y.		252.84	252.84
Formwork, 2 uses	157.000	SFCA	573.71	2,170.99	2,744.70
Reinforcing in place, A615 Gr 60, longer and heavier dowels, add	45.000	Lb.	33.75	87.30	121.05
Concrete, 3000 psi	2.600	C.Y.	356.20		356.20
Place concrete, spread footings, direct chute	2.600	C.Y.		157.51	157.51
Backfill, dozer	28.000	L.C.Y.		61.60	61.60
TOTAL			963.66	2,730.24	3,693.90

G3030 310	Headwalls	MAT.	INST.	TOTAL
2000	Headwall, 1-1/2 to 1 slope soil, C.I.P. conc, 30″ pipe, 3′ long wing walls	965	2,725	3,690
2020	Pipe size 36″, 3′-6″ long wing walls	1,200	3,275	4,475
2040	Pipe size 42″, 4′ long wing walls	1,450	3,825	5,275
2060	Pipe size 48″, 4′-6″ long wing walls	1,750	4,475	6,225
2080	Pipe size 54″, 5′-0″ long wing walls	2,075	5,200	7,275
2100	Pipe size 60″, 5′-6″ long wing walls	2,400	5,925	8,325
2120	Pipe size 72″, 6′-6″ long wing walls	3,200	7,600	10,800
2140	Pipe size 84″, 7′-6″ long wing walls	4,050	9,400	13,450
2500	$250/ton stone, pipe size 30″, 3′ long wing walls	595	1,000	1,595
2520	Pipe size 36″, 3′-6″ long wing walls	765	1,200	1,965
2540	Pipe size 42″, 4′ long wing walls	960	1,425	2,385
2560	Pipe size 48″, 4′-6″ long wing walls	1,175	1,700	2,875
2580	Pipe size 54″, 5′ long wing walls	1,425	1,975	3,400
2600	Pipe size 60″, 5′-6″ long wing walls	1,700	2,275	3,975
2620	Pipe size 72″, 6′-6″ long wing walls	2,325	3,025	5,350
2640	Pipe size 84″, 7′-6″ long wing walls	3,075	3,875	6,950
3000	$350/ton stone, pipe size 30″, 3′ long wing walls	830	1,000	1,830
3020	Pipe size 36″, 3′-6″ long wing walls	1,075	1,200	2,275
3040	Pipe size 42″, 4′ long wing walls	1,350	1,425	2,775
3060	Pipe size 48″, 4′-6″ long wing walls	1,650	1,700	3,350
3080	Pipe size 54″, 5′ long wing walls	2,000	1,975	3,975
3100	Pipe size 60″, 5′-6″ long wing walls	2,375	2,275	4,650
3120	Pipe size 72″, 6′-6″ long wing walls	3,275	3,025	6,300
3140	Pipe size 84″, 7′-6″ long wing walls	4,325	3,875	8,200
3500	$450/ton stone, pipe size 30″, 3′ long wing walls	1,075	1,000	2,075
3520	Pipe size 36″, 3′-6″ long wing walls	1,375	1,200	2,575
3540	Pipe size 42″, 4′ long wing walls	1,725	1,425	3,150
3560	Pipe size 48″, 4′-6″ long wing walls	2,125	1,700	3,825
3580	Pipe size 54″, 5′ long wing walls	2,575	1,975	4,550
3600	Pipe size 60″, 5′-6″ long wing walls	3,050	2,275	5,325

G3030 Storm Sewer

G3030 310	Headwalls	COST PER EACH		
		MAT.	INST.	TOTAL
3620	Pipe size 72", 6'-6" long wing walls	4,200	3,025	7,225
3640	Pipe size 84", 7'-6" long wing walls	5,525	3,875	9,400
4000	$650/ton stone, pipe size 30", 3' long wing walls	1,525	1,000	2,525
4020	Pipe size 36", 3'-6" long wing walls	1,975	1,200	3,175
4040	Pipe size 42", 4' long wing walls	2,475	1,425	3,900
4060	Pipe size 48", 4'-6" long wing walls	3,050	1,700	4,750
4080	Pipe size 54", 5' long wing walls	3,700	1,975	5,675
4100	Pipe size 60", 5'-6" long wing walls	4,375	2,275	6,650
4120	Pipe size 72", 6'-6" long wing walls	6,050	3,025	9,075
4140	Pipe size 84", 7'-6" long wing walls	7,950	3,875	11,825
4500	2 to 1 slope soil, C.I.P. concrete, pipe size 30", 4'-3" long wing walls	1,150	3,175	4,325
4520	Pipe size 36", 5' long wing walls	1,450	3,850	5,300
4540	Pipe size 42", 5'-9" long wing walls	1,800	4,650	6,450
4560	Pipe size 48", 6'-6" long wing walls	2,150	5,375	7,525
4580	Pipe size 54", 7'-3" long wing walls	2,550	6,275	8,825
4600	Pipe size 60", 8'-0" long wing walls	2,975	7,175	10,150
4620	Pipe size 72", 9'-6" long wing walls	3,975	9,325	13,300
4640	Pipe size 84", 11'-0" long wing walls	5,125	11,600	16,725
5000	$250/ton stone, pipe size 30", 4'-3" long wing walls	720	1,150	1,870
5020	Pipe size 36", 5' long wing walls	945	1,425	2,370
5040	Pipe size 42", 5'-9" long wing walls	1,200	1,700	2,900
5060	Pipe size 48", 6'-6" long wing walls	1,475	2,025	3,500
5080	Pipe size 54", 7'-3" long wing walls	1,800	2,400	4,200
5100	Pipe size 60", 8'-0" long wing walls	2,125	2,800	4,925
5120	Pipe size 72", 9'-6" long wing walls	2,325	3,750	6,075
5140	Pipe size 84", 11'-0" long wing walls	3,075	4,900	7,975
5500	$350/ton stone, pipe size 30", 4'-3" long wing walls	1,000	1,150	2,150
5520	Pipe size 36", 5' long wing walls	1,325	1,425	2,750
5540	Pipe size 42", 5'-9" long wing walls	1,675	1,700	3,375
5560	Pipe size 48", 6'-6" long wing walls	2,075	2,025	4,100
5580	Pipe size 54", 7'-3" long wing walls	2,525	2,400	4,925
5600	Pipe size 60", 8'-0" long wing walls	3,000	2,800	5,800
5620	Pipe size 72", 9'-6" long wing walls	4,150	3,750	7,900
5640	Pipe size 84", 11'-0" long wing walls	5,550	4,900	10,450
6000	$450/ton stone, pipe size 30", 4'-3" long wing walls	1,300	1,150	2,450
6020	Pipe size 36", 5'-0" long wing walls	1,700	1,425	3,125
6040	Pipe size 42", 5'-9" long wing walls	2,150	1,700	3,850
6060	Pipe size 48", 6'-6" long wing walls	2,675	2,025	4,700
6080	Pipe size 54", 7'-3" long wing walls	3,225	2,400	5,625
6100	Pipe size 60", 8'-0" long wing walls	3,850	2,800	6,650
6120	Pipe size 72", 9'-6" long wing walls	5,325	3,750	9,075
6140	Pipe size 84", 11'-0" long wing walls	7,100	4,900	12,000
6500	$650/ton stone, pipe size 30", 4'-3" long wing walls	1,875	1,150	3,025
6520	Pipe size 36", 5' long wing walls	2,450	1,425	3,875
6540	Pipe size 42", 5'-9" long wing walls	3,100	1,700	4,800
6560	Pipe size 48", 6'-6" long wing walls	3,825	2,025	5,850
6580	Pipe size 54", 7'-3" long wing walls	4,650	2,400	7,050
6600	Pipe size 60", 8'-0" long wing walls	5,525	2,800	8,325
6620	Pipe size 72", 9'-6" long wing walls	7,675	3,750	11,425
6640	Pipe size 84", 11'-0" long wing walls	10,200	4,900	15,100

G3030 Storm Sewer

Most new project sites will require some degree of work to address the quantity and, sometimes, quality of the stormwater produced from that site. The extent of this work is determined by design professionals, based on local and federal guidelines. In general, the scope of stormwater work will be a function of the amount of impervious surface (roof and paved areas) to be constructed, the permeability of soil, and the sensitivity of the location to stormwater issues overall. Absent further information, an allowance of $3 per SF of impervious surface is advisable. Note that these costs rely on the inlet and piping of a traditional stormwater drainage system and are additive to that base work unless otherwise stated.

1050 Small Surface Retention

Small surface retention areas, or rain gardens, are typically 200 to 800 SF depressions that intercept surface storm run-off and retain and remediate (improve) water through filtration and absorption with specialized plants and fill material. These can be attractive if maintained and located in smaller areas to treat smaller flows.

1100 Large Surface Detention

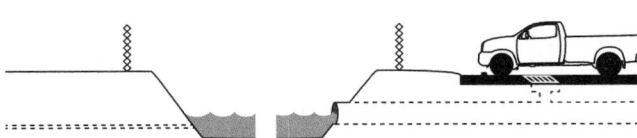

Large surface detention areas collect stormwater in open basins and release runoff at a set rate. These basins make up about 10% of the impervious area of a site and must be protected by a fence. This option is the least expensive per volume of capacity, but the results are usually not attractive and it is not always allowed by local ordinance.

1150 Sub-Surface Detention, Gravel

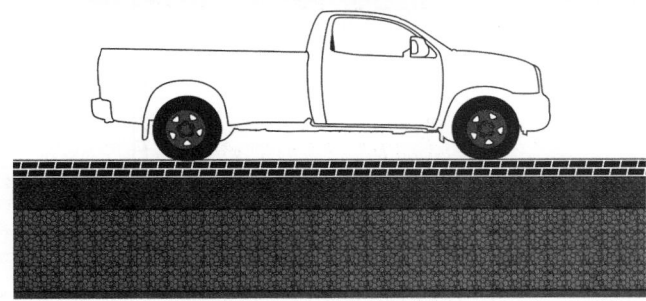

Sub-Surface detention systems (gravel) are large areas (typically located under surface parking) in which existing earth is replaced by gravel and the voids within are utilized as a temporary storage area for storm water. Water access can be by pervious pavers or a traditional inlet system.

1200 Sub-Surface Detention, Gravel & Void Structure

Sub-Surface detention (gravel and void structure) is a hybrid solution consisting of the gravel solution above with the presence of structures that create additional volumes of storage. These typically employ inlets which can be supplemented to provide improved water quality.

1250 Sub-Surface Detention, Structural Vaults

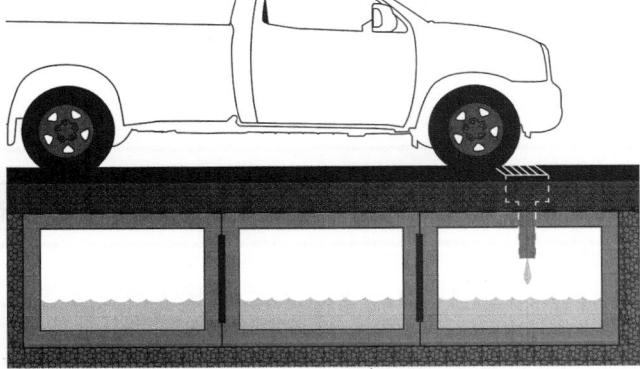

Sub-Surface detention systems (structural vaults) are large precast enclosures providing high-volume water storage. It is the most expensive system but requires the least amount of space for a given volume of capacity.

G3030 610	Stormwater Management (costs per CF of stormwater)	COST PER C.F.		
		MAT.	INST.	TOTAL
1050	Small Surface Retention	8.20	3.15	11.35
1100	Large Surface Detention	.28	.41	.69
1150	Sub-Surface Detention, Gravel	10.35	3.49	13.84
1200	Sub-Surface Detention, Gravel & Void Structure	9.30	1.79	11.09
1250	Sub-Surface Detention, Structural Vaults	21	2.07	23.07

For customer support on your Site Work & Landscape Costs with RSMeans Data, call 800.448.8182.

G3060 Fuel Distribution

System Components			COST L.F.		
	QUANTITY	UNIT	MAT.	INST.	TOTAL
SYSTEM G3060 112 1000					
GASLINE 1/2" PE, 60 PSI COILS, COMPRESSION COUPLING, INCLUDING COMMON					
EARTH EXCAVATION, BEDDING, BACKFILL AND COMPACTION					
Excavate Trench	.170	B.C.Y.		1.29	1.29
Sand bedding material	.090	L.C.Y.	1.85	1.18	3.03
Compact bedding	.070	E.C.Y.		.44	.44
Install 1/2" gas pipe	1.000	L.F.	.62	4.18	4.80
Backfill trench	.120	L.C.Y.		.39	.39
Compact fill material in trench	.100	E.C.Y.		.55	.55
Dispose of excess fill material on-site	.090	L.C.Y.		1.17	1.17
TOTAL			2.47	9.20	11.67

G3060 112	Gasline (Common Earth Excavation)	COST L.F.		
		MAT.	INST.	TOTAL
0950	Gasline, including common earth excavation, bedding, backfill and compaction			
1000	1/2" diameter PE, 60 psi coils, compression coupling, SDR 11, 2' deep	2.47	9.20	11.67
1010	4' deep	2.47	11.75	14.22
1020	6' deep	2.47	14.30	16.77
1050	1" diameter PE, 60 psi coils, compression coupling, SDR 11, 2' deep	3.09	9.70	12.79
1060	4' deep	3.09	12.25	15.34
1070	6' deep	3.09	14.80	17.89
1100	1 1/4" diameter PE, 60 psi coils, compression coupling, SDR 11, 2' deep	3.62	9.70	13.32
1110	4' deep	3.62	12.25	15.87
1120	6' deep	3.62	14.80	18.42
1150	2" diameter PE, 60 psi coils, compression coupling, SDR 11, 2' deep	4.97	10.20	15.17
1160	4' deep	4.97	12.80	17.77
1180	6' deep	4.97	15.35	20.32
1200	3" diameter PE, 60 psi coils, compression coupling, SDR 11, 2' deep	8.40	12.40	20.80
1210	4' deep	8.40	14.95	23.35
1220	6' deep	8.40	17.55	25.95

G3060 Fuel Distribution

G3060 112	Gasline (Common Earth Excavation)	COST L.F.		
		MAT.	INST.	TOTAL
1250	4" diameter PE, 60 psi 40' length, compression coupling, SDR 11, 2' deep	19.05	17.10	36.15
1260	4' deep	19.05	19.65	38.70
1270	6' deep	19.05	22	41.05
1300	6" diameter PE, 60 psi 40' length w/coupling, SDR 11, 2' deep	44	18.05	62.05
1310	4' deep	44	20.50	64.50
1320	6' deep	44	23	67
1350	8" diameter PE, 60 psi 40' length w/coupling, SDR 11, 4' deep	71.50	23.50	95
1360	6' deep	71.50	26	97.50
2000	1" diameter steel plain end, Schedule 40, 2' deep	7.30	15.65	22.95
2010	4' deep	7.30	18.25	25.55
2020	6' deep	7.30	21	28.30
2050	2" diameter, steel plain end, Schedule 40, 2' deep	10.40	16.45	26.85
2060	4' deep	10.40	19	29.40
2070	6' deep	10.40	21.50	31.90
2100	3" diameter, steel plain end, Schedule 40, 2' deep	17	18.55	35.55
2110	4' deep	17	21	38
2120	6' deep	17	23.50	40.50
2150	4" diameter, steel plain end, Schedule 40, 2' deep	21.50	29	50.50
2160	4' deep	21.50	31.50	53
2170	6' deep	21.50	34	55.50
2200	5" diameter, steel plain end, Schedule 40, 2' deep	29.50	32.50	62
2210	4' deep	29.50	35	64.50
2220	6' deep	29.50	37.50	67
2250	6" diameter, steel plain end, Schedule 40, 2' deep	36	39	75
2260	4' deep	36	41.50	77.50
2270	6' deep	36	44	80
2300	8" diameter, steel plain end, Schedule 40, 2' deep	55	48	103
2310	4' deep	55	50.50	105.50
2320	6' deep	55	53	108
2350	10" diameter, steel plain end, Schedule 40, 2' deep	132	64.50	196.50
2360	4' deep	132	67	199
2370	6' deep	132	69.50	201.50
2400	12" diameter, steel plain end, Schedule 40, 2' deep	147	78.50	225.50
2410	4' deep	147	81.50	228.50
2420	6' deep	147	84	231
2460	14" diameter, steel plain end, Schedule 40, 4' deep	157	87	244
2470	6' deep	157	90	247
2510	16" diameter, Schedule 40, 4' deep	173	93.50	266.50
2520	6' deep	173	96.50	269.50
2560	18" diameter, Schedule 40, 4' deep	221	101	322
2570	6' deep	221	104	325
2600	20" diameter, steel plain end, Schedule 40, 4' deep	340	109	449
2610	6' deep	340	112	452
2650	24" diameter, steel plain end, Schedule 40, 4' deep	390	130	520
2660	6' deep	390	133	523
3150	4" diameter, steel plain end, Schedule 80, 2' deep	47.50	46.50	94
3160	4' deep	47.50	49	96.50
3170	6' deep	47.50	51.50	99
3200	5" diameter, steel plain end, Schedule 80, 2' deep	60	47.50	107.50
3210	4' deep	60	50	110
3220	6' deep	60	52.50	112.50
3250	6" diameter, steel plain end, Schedule 80, 2' deep	101	52	153
3260	4' deep	101	54.50	155.50
3270	6' deep	101	57	158
3300	8" diameter, steel plain end, Schedule 80, 2' deep	134	60	194
3310	4' deep	134	62.50	196.50
3320	6' deep	134	65	199
3350	10" diameter, steel plain end, Schedule 80, 2' deep	199	70.50	269.50

G3060 Fuel Distribution

G3060 112	Gasline (Common Earth Excavation)	COST L.F.		
		MAT.	INST.	TOTAL
3360	4' deep	199	73	272
3370	6' deep	199	75.50	274.50
3400	12" diameter, steel plain end, Schedule 80, 2' deep	264	86.50	350.50
3410	4' deep	264	89	353
3420	6' deep	264	91.50	355.50

G4010 Electrical Distribution

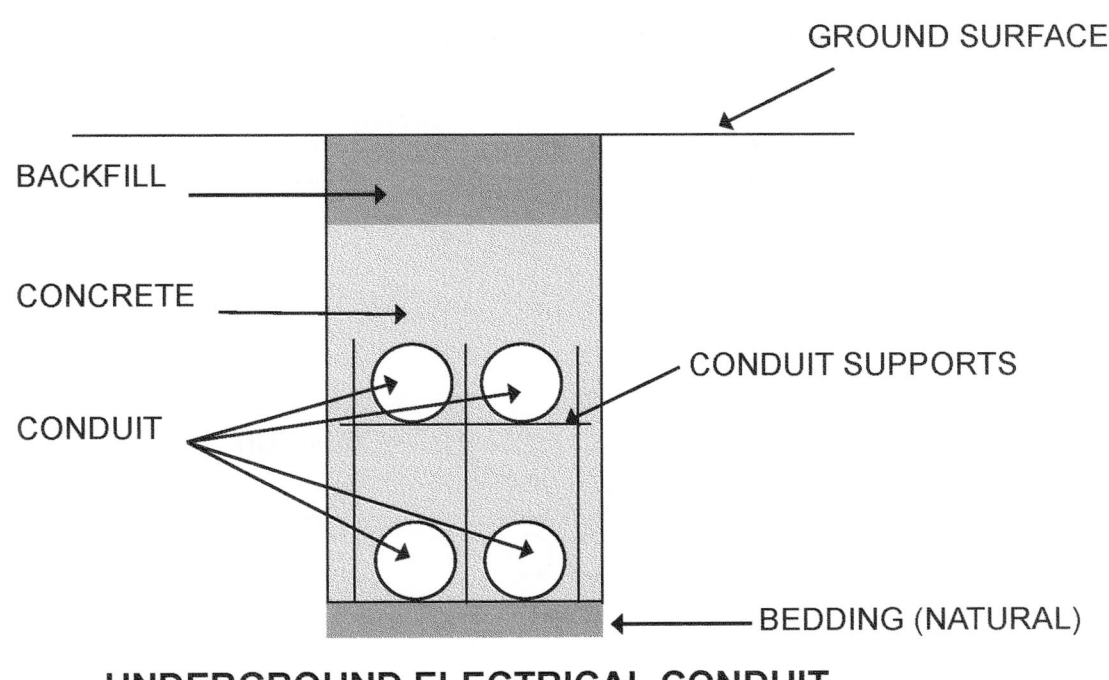

GROUND SURFACE

BACKFILL

CONCRETE

CONDUIT

CONDUIT SUPPORTS

BEDDING (NATURAL)

UNDERGROUND ELECTRICAL CONDUIT

System Components	QUANTITY	UNIT	COST EACH		
			MAT.	INST.	TOTAL
SYSTEM G4010 312 1000					
2000 AMP UNDERGROUND SERVICE INCLUDING COMMON EARTH EXCAVATION,					
CONCRETE, BACKFILL AND COMPACTION					
Excavate Trench	44.440	B.C.Y.		335.08	335.08
4 inch conduit bank	108.000	L.F.	1,123.20	3,024	4,147.20
4 inch fitting	8.000	Ea.	324	504	828
4 inch bells	4.000	Ea.	20.40	252	272.40
Concrete material	16.580	C.Y.	2,271.46		2,271.46
Concrete placement	16.580	C.Y.		464.24	464.24
Backfill trench	33.710	L.C.Y.		109.90	109.90
Compact fill material in trench	25.930	E.C.Y.		142.10	142.10
Dispose of excess fill material on-site	24.070	L.C.Y.		314.36	314.36
500 kcmil power cable	18.000	C.L.F.	24,750	8,550	33,300
Wire, 600 volt stranded copper wire, type THW, 1/0	6.000	C.L.F.	1,890	1,374	3,264
TOTAL			30,379.06	15,069.68	45,448.74

G4010 312	Underground Power Feed		COST EACH		
			MAT.	INST.	TOTAL
0900	Underground electrical				
0950	Including common earth excavation, concrete, backfill and compaction				
1000	2000 Amp service, 100' length, 4' depth		30,400	15,100	45,500
1100	1600 Amp service		23,800	11,600	35,400
1200	1200 Amp service		18,500	9,650	28,150
1300	1000 Amp service		21,700	8,825	30,525
1400	800 Amp service		10,000	8,275	18,275
1500	600 Amp service		8,100	7,700	15,800

G4010 Electrical Distribution

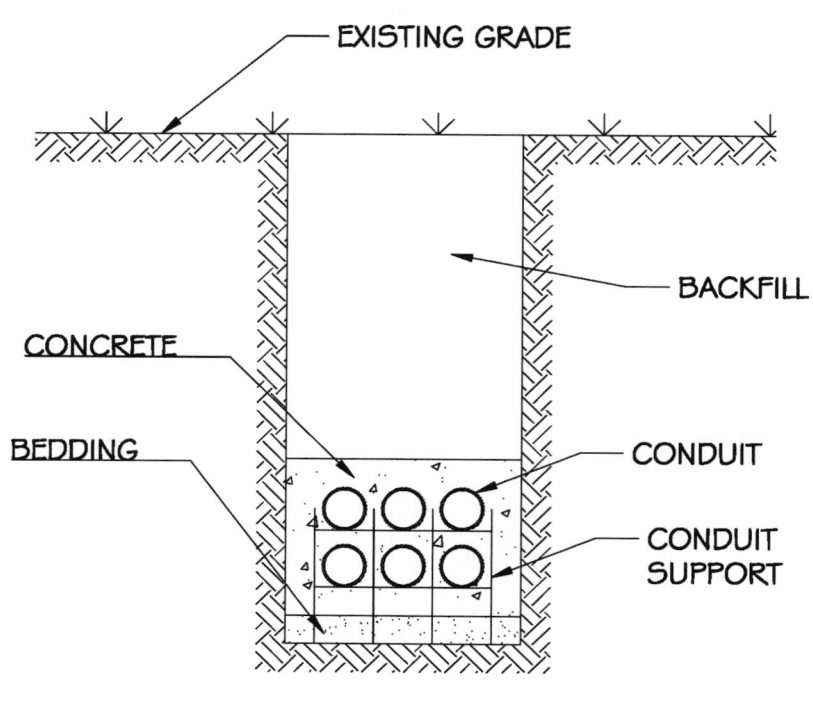

System Components	QUANTITY	UNIT	COST L.F.		
			MAT.	INST.	TOTAL
SYSTEM G4010 320 1012					
UNDERGROUND ELECTRICAL CONDUIT, 2″ DIAMETER, INCLUDING EXCAVATION,					
CONCRETE, BEDDING, BACKFILL AND COMPACTION					
Excavate Trench	.150	B.C.Y.		1.13	1.13
Base spacer	.200	Ea.	.30	2.70	3
Conduit	1.000	L.F.	.70	4.21	4.91
Concrete material	.050	C.Y.	6.85		6.85
Concrete placement	.050	C.Y.		1.41	1.41
Backfill trench	.120	L.C.Y.		.39	.39
Compact fill material in trench	.100	E.C.Y.		.55	.55
Dispose of excess fill material on-site	.120	L.C.Y.		1.56	1.56
Marking tape	.010	C.L.F.	.43	.04	.47
TOTAL			8.28	11.99	20.27

G4010 320	Underground Electrical Conduit	COST L.F.		
		MAT.	INST.	TOTAL
0950	Underground electrical conduit, including common earth excavation,			
0952	Concrete, bedding, backfill and compaction			
1012	2″ dia. Schedule 40 PVC, 2′ deep	8.30	12	20.30
1014	4′ deep	8.30	17	25.30
1016	6′ deep	8.30	22	30.30
1022	2 @ 2″ dia. Schedule 40 PVC, 2′ deep	9.30	18.90	28.20
1024	4′ deep	9.30	24	33.30
1026	6′ deep	9.30	28.50	37.80
1032	3 @ 2″ dia. Schedule 40 PVC, 2′ deep	10.30	26	36.30
1034	4′ deep	10.30	31	41.30
1036	6′ deep	10.30	35.50	45.80
1042	4 @ 2″ dia. Schedule 40 PVC, 2′ deep	13.95	32.50	46.45
1044	4′ deep	13.95	37.50	51.45
1046	6′ deep	13.95	42	55.95

G4010 Electrical Distribution

G4010 320	Underground Electrical Conduit	COST L.F.		
		MAT.	INST.	TOTAL
1062	6 @ 2" dia. Schedule 40 PVC, 2' deep	15.95	46	61.95
1064	4' deep	15.95	51	66.95
1066	6' deep	15.95	56	71.95
1082	8 @ 2" dia. Schedule 40 PVC, 2' deep	17.95	59.50	77.45
1084	4' deep	17.95	64.50	82.45
1086	6' deep	17.95	69.50	87.45
1092	9 @ 2" dia. Schedule 40 PVC, 2' deep	18.90	66	84.90
1094	4' deep	21.50	71	92.50
1096	6' deep	21.50	76	97.50
1212	3" dia. Schedule 40 PVC, 2' deep	10.40	14.75	25.15
1214	4' deep	10.40	19.80	30.20
1216	6' deep	10.40	24.50	34.90
1222	2 @ 3" dia. Schedule 40 PVC, 2' deep	10.80	24	34.80
1224	4' deep	10.80	29	39.80
1226	6' deep	10.80	34	44.80
1232	3 @ 3" dia. Schedule 40 PVC, 2' deep	12.55	33.50	46.05
1234	4' deep	12.55	38.50	51.05
1236	6' deep	12.55	43.50	56.05
1242	4 @ 3" dia. Schedule 40 PVC, 2' deep	18.40	43	61.40
1244	4' deep	18.40	48.50	66.90
1246	6' deep	18.40	53	71.40
1262	6 @ 3" dia. Schedule 40 PVC, 2' deep	20.50	62	82.50
1264	4' deep	20.50	67	87.50
1266	6' deep	20.50	72	92.50
1282	8 @ 3" dia. Schedule 40 PVC, 2' deep	24	81.50	105.50
1284	4' deep	24	86.50	110.50
1286	6' deep	24	92	116
1292	9 @ 3" dia. Schedule 40 PVC, 2' deep	28.50	92	120.50
1294	4' deep	30	96	126
1296	6' deep	30	101	131
1312	4" dia. Schedule 40 PVC, 2' deep	11	18.05	29.05
1314	4' deep	11	23	34
1316	6' deep	11	28	39
1322	2 @ 4" dia. Schedule 40 PVC, 2' deep	13.30	31	44.30
1324	4' deep	13.30	36.50	49.80
1326	6' deep	13.30	41	54.30
1332	3 @ 4" dia. Schedule 40 PVC, 2' deep	14.25	44	58.25
1334	4' deep	14.25	49	63.25
1336	6' deep	14.25	54	68.25
1342	4 @ 4" dia. Schedule 40 PVC, 2' deep	22	57.50	79.50
1344	4' deep	22	62.50	84.50
1346	6' deep	22	67.50	89.50
1362	6 @ 4" dia. Schedule 40 PVC, 2' deep	25.50	83.50	109
1364	4' deep	25.50	88.50	114
1366	6' deep	25.50	93.50	119
1382	8 @ 4" dia. Schedule 40 PVC, 2' deep	31	110	141
1384	4' deep	31	116	147
1386	6' deep	31	122	153
1392	9 @ 4" dia. Schedule 40 PVC, 2' deep	35	124	159
1394	4' deep	35	128	163
1396	6' deep	35	132	167
1412	5" dia. Schedule 40 PVC, 2' deep	14.35	21.50	35.85
1414	4' deep	14.35	26.50	40.85
1416	6' deep	14.35	31.50	45.85
1422	2 @ 5" dia. Schedule 40 PVC, 2' deep	20	38.50	58.50
1424	4' deep	20	43	63
1426	6' deep	20	48	68
1432	3 @ 5" dia. Schedule 40 PVC, 2' deep	25.50	55	80.50

G4010 Electrical Distribution

G4010 320	Underground Electrical Conduit	COST L.F.		
		MAT.	INST.	TOTAL
1434	4' deep	25.50	60.50	86
1436	6' deep	25.50	65.50	91
1442	4 @ 5" dia. Schedule 40 PVC, 2' deep	35.50	71.50	107
1444	4' deep	35.50	76.50	112
1446	6' deep	35.50	81	116.50
1462	6 @ 5" dia. Schedule 40 PVC, 2' deep	45.50	105	150.50
1464	4' deep	45.50	110	155.50
1466	6' deep	45.50	115	160.50
1482	8 @ 5" dia. Schedule 40 PVC, 2' deep	59.50	139	198.50
1484	4' deep	59.50	146	205.50
1486	6' deep	59.50	153	212.50
1492	9 @ 5" dia. Schedule 40 PVC, 2' deep	66.50	157	223.50
1494	4' deep	66.50	160	226.50
1496	6' deep	66.50	165	231.50
1512	6" dia. Schedule 40 PVC, 2' deep	14.05	26.50	40.55
1514	4' deep	14.05	31	45.05
1516	6' deep	14.05	36	50.05
1522	2 @ 6" dia. Schedule 40 PVC, 2' deep	16.70	47.50	64.20
1524	4' deep	16.70	52.50	69.20
1526	6' deep	16.70	57.50	74.20
1532	3 @ 6" dia. Schedule 40 PVC, 2' deep	22	69.50	91.50
1534	4' deep	22	75.50	97.50
1536	6' deep	22	81.50	103.50
1542	4 @ 6" dia. Schedule 40 PVC, 2' deep	29	89.50	118.50
1544	4' deep	29	94.50	123.50
1546	6' deep	29	99.50	128.50
1562	6 @ 6" dia. Schedule 40 PVC, 2' deep	38.50	133	171.50
1564	4' deep	38.50	139	177.50
1566	6' deep	38.50	145	183.50
1582	8 @ 6" dia. Schedule 40 PVC, 2' deep	50.50	177	227.50
1584	4' deep	50.50	185	235.50
1586	6' deep	50.50	192	242.50
1592	9 @ 6" dia. Schedule 40 PVC, 2' deep	56	199	255
1594	4' deep	56	203	259
1596	6' deep	56	208	264

G4020 Site Lighting

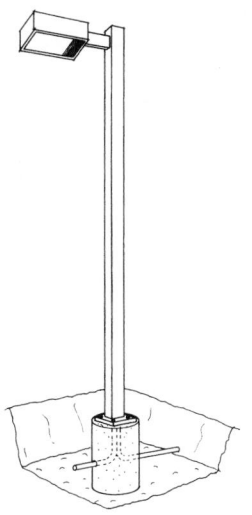

The Site Lighting System includes the complete unit from foundation to electrical fixtures. Each system includes: excavation; concrete base; backfill by hand; compaction with a plate compacter; pole of specified material; all fixtures; and lamps.

The Expanded System Listing shows Site Lighting Systems that use one of two types of lamps: high pressure sodium; and metal halide. Systems are listed for 400-watt and 1000-watt lamps. Pole height varies from 20' to 40'. There are four types of poles possibly listed: aluminum, fiberglass, steel and wood.

System Components	QUANTITY	UNIT	COST EACH		
			MAT.	INST.	TOTAL
SYSTEM G4020 110 5820					
SITE LIGHTING, METAL HALIDE, 400 WATT, ALUMINUM POLE, 20' HIGH					
Excavating, by hand, pits to 6' deep, heavy soil	2.368	C.Y.		314.94	314.94
Concrete in place incl. forms and reinf. stl. spread footings under 1 CY	.465	C.Y.	106.49	144.65	251.14
Backfill	1.903	C.Y.		91.34	91.34
Compact, vibrating plate	1.903	C.Y.		15.07	15.07
Aluminum light pole, 20', no concrete base	1.000	Ea.	665	722.50	1,387.50
Road area luminaire wire	.300	C.L.F.	8.55	28.35	36.90
Roadway area luminaire, metal halide 400 W	1.000	Ea.	705	345	1,050
TOTAL			1,485.04	1,661.85	3,146.89

G4020 110	Site Lighting	COST EACH		
		MAT.	INST.	TOTAL
2320	Site lighting, high pressure sodium, 400 watt, aluminum pole, 20' high	1,625	1,650	3,275
2340	30' high	2,725	2,025	4,750
2360	40' high	3,225	2,625	5,850
2520	Fiberglass pole, 20' high	2,025	1,450	3,475
2540	30' high	2,325	1,825	4,150
2560	40' high	3,500	2,325	5,825
2720	Steel pole, 20' high	2,200	1,750	3,950
2740	30' high	2,400	2,150	4,550
2760	40' high	2,675	2,800	5,475
2920	Wood pole, 20' high	1,225	1,625	2,850
2940	30' high	1,350	2,025	3,375
2960	40' high	1,675	2,500	4,175
3120	1000 watt, aluminum pole, 20' high	1,750	1,700	3,450
3140	30' high	2,825	2,075	4,900
3160	40' high	3,325	2,650	5,975
3320	Fiberglass pole, 20' high	2,150	1,500	3,650
3340	30' high	2,425	1,850	4,275
3360	40' high	3,600	2,350	5,950
3420	Steel pole, 20' high	2,325	1,775	4,100
3440	30' high	2,500	2,175	4,675

G40 Site Electrical Utilities

G4020 Site Lighting

G4020 110	Site Lighting	COST EACH		
		MAT.	INST.	TOTAL
3460	40' high	2,775	2,825	5,600
3520	Wood pole, 20' high	1,350	1,650	3,000
3540	30' high	1,450	2,075	3,525
3560	40' high	1,800	2,525	4,325
5820	Metal halide, 400 watt, aluminum pole, 20' high	1,475	1,650	3,125
5840	30' high	2,575	2,025	4,600
5860	40' high	3,075	2,625	5,700
6120	Fiberglass pole, 20' high	1,900	1,450	3,350
6140	30' high	2,175	1,825	4,000
6160	40' high	3,350	2,325	5,675
6320	Steel pole, 20' high	2,075	1,750	3,825
6340	30' high	2,250	2,150	4,400
6360	40' high	2,525	2,800	5,325
7620	1000 watt, aluminum pole, 20' high	1,625	1,700	3,325
7640	30' high	2,725	2,075	4,800
7660	40' high	3,225	2,650	5,875
7900	Fiberglass pole, 20' high	2,050	1,500	3,550
7920	30' high	2,325	1,850	4,175
7940	40' high	3,500	2,350	5,850
8120	Steel pole, 20' high	2,225	1,775	4,000
8140	30' high	2,400	2,175	4,575
8160	40' high	2,675	2,825	5,500

728

G4020   Site Lighting

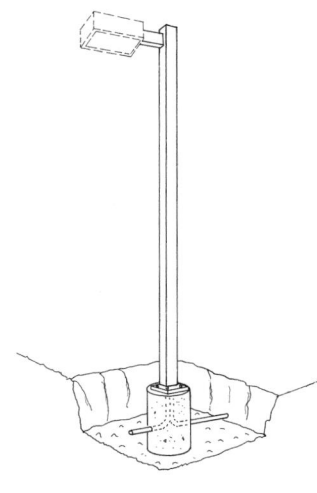

Table G4020 210 Procedure for Calculating Floodlights Required for Various Footcandles
Poles should not be spaced more than 4 times the fixture mounting height for good light distribution.

Estimating Chart
Select Lamp type.

Determine total square feet.

Chart will show quantity of fixtures to provide 1 footcandle initial, at intersection of lines. Multiply fixture quantity by desired footcandle level.

Chart based on use of wide beam luminaires in an area whose dimensions are large compared to mounting height and is approximate only.

To maintain 1 footcandle over a large area use these watts per square foot:

Incandescent	0.15
Metal Halide	0.032
Mercury Vapor	0.05
High Pressure Sodium	0.024

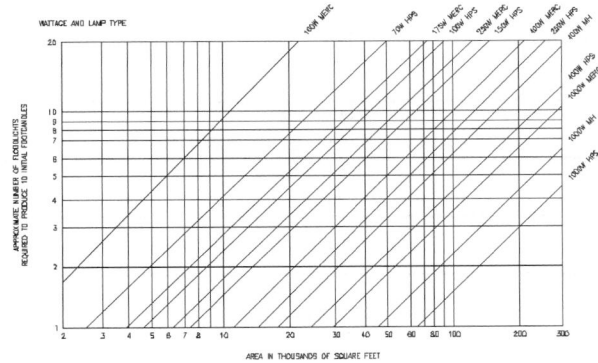

System Components	QUANTITY	UNIT	COST EACH		
			MAT.	INST.	TOTAL
SYSTEM G4020 210 0200					
LIGHT POLES, ALUMINUM, 20' HIGH, 1 ARM BRACKET					
Aluminum light pole, 20', no concrete base	1.000	Ea.	665	722.50	1,387.50
Bracket arm for aluminum light pole	1.000	Ea.	151	94.50	245.50
Excavation by hand, pits to 6' deep, heavy soil or clay	2.368	C.Y.		314.94	314.94
Footing, concrete incl forms, reinforcing, spread, under 1 C.Y.	.465	C.Y.	106.49	144.65	251.14
Backfill by hand	1.903	C.Y.		91.34	91.34
Compaction vibrating plate	1.903	C.Y.		15.07	15.07
TOTAL			922.49	1,383	2,305.49

G4020 210	Light Pole (Installed)	COST EACH		
		MAT.	INST.	TOTAL
0200	Light pole, aluminum, 20' high, 1 arm bracket	920	1,375	2,295
0240	2 arm brackets	1,075	1,375	2,450
0280	3 arm brackets	1,225	1,425	2,650
0320	4 arm brackets	1,375	1,450	2,825
0360	30' high, 1 arm bracket	2,000	1,750	3,750
0400	2 arm brackets	2,150	1,750	3,900
0440	3 arm brackets	2,300	1,800	4,100
0480	4 arm brackets	2,475	1,800	4,275
0680	40' high, 1 arm bracket	2,500	2,325	4,825
0720	2 arm brackets	2,650	2,325	4,975
0760	3 arm brackets	2,800	2,375	5,175
0800	4 arm brackets	2,975	2,400	5,375
0840	Steel, 20' high, 1 arm bracket	1,600	1,475	3,075
0880	2 arm brackets	1,675	1,475	3,150
0920	3 arm brackets	1,625	1,525	3,150
0960	4 arm brackets	1,725	1,525	3,250
1000	30' high, 1 arm bracket	1,775	1,850	3,625
1040	2 arm brackets	1,850	1,850	3,700
1080	3 arm brackets	1,800	1,900	3,700
1120	4 arm brackets	1,900	1,900	3,800
1320	40' high, 1 arm bracket	2,050	2,500	4,550
1360	2 arm brackets	2,125	2,500	4,625
1400	3 arm brackets	2,075	2,550	4,625
1440	4 arm brackets	2,175	2,550	4,725

G9020 Site Repair & Maintenance

G9020 100	Clean and Wrap Marine Piles	COST EACH		
		MAT.	INST.	TOTAL
1000	Clean & wrap 5 foot long, 8 inch diameter wood pile using nails from boat	103	113	216
1010	6 foot long	123	189	312
1020	7 foot long	144	196	340
1030	8 foot long	164	203	367
1040	9 foot long	185	211	396
1050	10 foot long	205	218	423
1060	11 foot long	226	281	507
1070	12 foot long	246	288	534
1075	13 foot long	267	295	562
1080	14 foot long	287	305	592
1090	15 foot long	310	310	620
1100	Clean & wrap 5 foot long, 10 inch diameter	128	140	268
1110	6 foot long	153	234	387
1120	7 foot long	179	243	422
1130	8 foot long	204	252	456
1140	9 foot long	230	261	491
1150	10 foot long	255	270	525
1160	11 foot long	281	360	641
1170	12 foot long	305	370	675
1175	13 foot long	330	380	710
1180	14 foot long	355	390	745
1190	15 foot long	385	395	780
1200	Clean & wrap 5 foot long, 12 inch diameter	155	168	323
1210	6 foot long	186	283	469
1220	7 foot long	217	294	511
1230	8 foot long	248	305	553
1240	9 foot long	279	315	594
1250	10 foot long	310	325	635
1260	11 foot long	340	425	765
1270	12 foot long	370	435	805
1275	13 foot long	405	450	855
1280	14 foot long	435	460	895
1290	15 foot long	465	470	935
1300	Clean & wrap 5 foot long, 13 inch diameter	168	184	352
1310	6 foot long	201	310	511
1320	7 foot long	235	320	555
1330	8 foot long	268	330	598
1340	9 foot long	300	345	645
1350	10 foot long	335	355	690
1360	11 foot long	370	455	825
1370	12 foot long	400	470	870
1375	13 foot long	435	480	915
1380	14 foot long	470	495	965
1390	15 foot long	505	505	1,010
1400	Clean & wrap 5 foot long, 14 inch diameter	180	197	377
1410	6 foot long	216	325	541
1420	7 foot long	252	335	587
1430	8 foot long	288	350	638
1440	9 foot long	325	360	685
1450	10 foot long	360	375	735
1460	11 foot long	395	485	880
1470	12 foot long	430	500	930
1475	13 foot long	470	515	985
1480	14 foot long	505	525	1,030
1490	15 foot long	540	540	1,080
1500	Clean & wrap 5 foot long, 8 inch diameter wood pile using straps from boat	155	110	265
1510	6 foot long	186	184	370
1520	7 foot long	217	191	408

G9020 Site Repair & Maintenance

G9020 100	Clean and Wrap Marine Piles	COST EACH		
		MAT.	INST.	TOTAL
1530	8 foot long	248	198	446
1540	9 foot long	279	204	483
1550	10 foot long	310	211	521
1560	11 foot long	340	273	613
1570	12 foot long	370	279	649
1575	13 foot long	405	286	691
1580	14 foot long	435	292	727
1590	15 foot long	465	299	764
1600	Clean & wrap 5 foot long, 10 inch diameter	195	136	331
1610	6 foot long	234	228	462
1620	7 foot long	273	236	509
1630	8 foot long	310	245	555
1640	9 foot long	350	253	603
1650	10 foot long	390	261	651
1660	11 foot long	430	350	780
1670	12 foot long	470	360	830
1675	13 foot long	505	365	870
1680	14 foot long	545	375	920
1690	15 foot long	585	385	970
1700	Clean & wrap 5 foot long, 12 inch diameter	233	162	395
1710	6 foot long	279	276	555
1720	7 foot long	325	286	611
1730	8 foot long	370	296	666
1740	9 foot long	420	305	725
1750	10 foot long	465	315	780
1760	11 foot long	510	415	925
1770	12 foot long	560	425	985
1775	13 foot long	605	435	1,040
1780	14 foot long	650	445	1,095
1790	15 foot long	700	455	1,155
1800	Clean & wrap 5 foot long, 13 inch diameter	253	178	431
1810	6 foot long	305	300	605
1820	7 foot long	355	310	665
1830	8 foot long	405	325	730
1840	9 foot long	455	335	790
1850	10 foot long	505	345	850
1860	11 foot long	555	445	1,000
1870	12 foot long	605	455	1,060
1875	13 foot long	655	465	1,120
1880	14 foot long	705	475	1,180
1890	15 foot long	760	485	1,245
1900	Clean & wrap 5 foot long, 14 inch diameter	273	191	464
1910	6 foot long	325	315	640
1920	7 foot long	380	330	710
1930	8 foot long	435	340	775
1940	9 foot long	490	350	840
1950	10 foot long	545	365	910
1960	11 foot long	600	475	1,075
1970	12 foot long	655	485	1,140
1975	13 foot long	710	495	1,205
1980	14 foot long	765	510	1,275
1990	15 foot long	820	520	1,340
2000	Clean & wrap 5 foot long, 12 inch diameter concrete pile from boat w/straps	233	142	375
2010	6 foot long	279	238	517
2020	7 foot long	325	248	573
2030	8 foot long	370	258	628
2040	9 foot long	420	268	688
2050	10 foot long	465	278	743

G9020 Site Repair & Maintenance

G9020 100	Clean and Wrap Marine Piles	COST EACH		
		MAT.	INST.	TOTAL
2060	11 foot long	510	370	880
2070	12 foot long	560	380	940
2080	13 foot long	605	390	995
2085	14 foot long	650	400	1,050
2090	15 foot long	700	410	1,110
2100	Clean & wrap 5 foot long, 14 inch diameter	273	166	439
2110	6 foot long	325	278	603
2120	7 foot long	380	289	669
2130	8 foot long	435	300	735
2140	9 foot long	490	310	800
2150	10 foot long	545	325	870
2160	11 foot long	600	415	1,015
2170	12 foot long	655	430	1,085
2180	13 foot long	710	440	1,150
2185	14 foot long	765	450	1,215
2190	15 foot long	820	460	1,280

Reference Section

All the reference information is in one section, making it easy to find what you need to know . . . and easy to use the data set on a daily basis. This section is visually identified by a vertical black bar on the page edges.

In this Reference Section, we've included Equipment Rental Costs, a listing of rental and operating costs; Crew Listings, a full listing of all crews and equipment, and their costs; Historical Cost Indexes for cost comparisons over time; City Cost Indexes and Location Factors for adjusting costs to the region you are in; Reference Tables, where you will find explanations, estimating information and procedures, or technical data; Change Orders, information on pricing changes to contract documents; and an explanation of all the Abbreviations in the data set.

Table of Contents

Estimating Tips

- This section contains the average costs to rent and operate hundreds of pieces of construction equipment. This is useful information when one is estimating the time and material requirements of any particular operation in order to establish a unit or total cost. Bare equipment costs shown on a unit cost line include, not only rental, but also operating costs for equipment under normal use.

Rental Costs

- Equipment rental rates are obtained from the following industry sources throughout North America: contractors, suppliers, dealers, manufacturers, and distributors.

- Rental rates vary throughout the country, with larger cities generally having lower rates. Lease plans for new equipment are available for periods in excess of six months, with a percentage of payments applying toward purchase.

- Monthly rental rates vary from 2% to 5% of the purchase price of the equipment depending on the anticipated life of the equipment and its wearing parts.

- Weekly rental rates are about 1/3 of the monthly rates, and daily rental rates are about 1/3 of the weekly rate.

- Rental rates can also be treated as reimbursement costs for contractor-owned equipment. Owned equipment costs include depreciation, loan payments, interest, taxes, insurance, storage, and major repairs.

Operating Costs

- The operating costs include parts and labor for routine servicing, such as the repair and replacement of pumps, filters, and worn lines. Normal operating expendables, such as fuel, lubricants, tires, and electricity (where applicable), are also included.

- Extraordinary operating expendables with highly variable wear patterns, such as diamond bits and blades, are excluded. These costs can be found as material costs in the Unit Price section.

- The hourly operating costs listed do not include the operator's wages.

Equipment Cost/Day

- Any power equipment required by a crew is shown in the Crew Listings with a daily cost.

- This daily cost of equipment needed by a crew includes both the rental cost and the operating cost and is based on dividing the weekly rental rate by 5 (the number of working days in the week), then adding the hourly operating cost multiplied by 8 (the number of hours in a day). This "Equipment Cost/Day" is shown in the far right column of the Equipment Rental section.

- If equipment is needed for only one or two days, it is best to develop your own cost by including components for daily rent and hourly operating costs. This is important when the listed Crew for a task does not contain the equipment needed, such as a crane for lifting mechanical heating/cooling equipment up onto a roof.

- If the quantity of work is less than the crew's Daily Output shown for a Unit Price line item that includes a bare unit equipment cost, the recommendation is to estimate one day's rental cost and operating cost for equipment shown in the Crew Listing for that line item.

- Please note, in some cases the equipment description in the crew is followed by a time period in parenthesis. For example: (daily) or (monthly). In these cases the equipment cost/day is calculated by adding the rental cost per time period to the hourly operating cost multiplied by 8.

Mobilization, Demobilization Costs

- The cost to move construction equipment from an equipment yard or rental company to the job site and back again is not included in equipment rental costs listed in the Reference Section. It is also not included in the bare equipment cost of any Unit Price line item or in any equipment costs shown in the Crew Listings.

- Mobilization (to the site) and demobilization (from the site) costs can be found in the Unit Price section.

- If a piece of equipment is already at the job site, it is not appropriate to utilize mobilization or demobilization costs again in an estimate. ■

Same Data. Simplified.

Enjoy the convenience and efficiency of accessing your costs anywhere:

- **Skip the multiplier** by setting your location
- **Quickly search,** edit, favorite and share costs
- **Stay on top of price changes** with automatic updates

Discover more at rsmeans.com/online

01 54 33 | Equipment Rental

		UNIT	HOURLY OPER. COST	RENT PER DAY	RENT PER WEEK	RENT PER MONTH	EQUIPMENT COST/DAY	
10	0010	**CONCRETE EQUIPMENT RENTAL** without operators							10
	0200	Bucket, concrete lightweight, 1/2 C.Y. [R015433-10]	Ea.	.76	38.50	116.16	350	29.30	
	0300	1 C.Y.		.99	63.50	190	570	45.90	
	0400	1-1/2 C.Y.		1.29	60.50	181.82	545	46.70	
	0500	2 C.Y.		1.40	74	222.23	665	55.65	
	0580	8 C.Y.		7.01	94.50	282.83	850	112.65	
	0600	Cart, concrete, self-propelled, operator walking, 10 C.F.		2.88	157	469.50	1,400	116.95	
	0700	Operator riding, 18 C.F.		4.85	167	500.50	1,500	138.95	
	0800	Conveyer for concrete, portable, gas, 16" wide, 26' long		10.71	162	484.86	1,450	182.70	
	0900	46' long		11.09	177	530.31	1,600	194.80	
	1000	56' long		11.26	194	580.82	1,750	206.25	
	1100	Core drill, electric, 2-1/2 H.P., 1" to 8" bit diameter		1.58	83.50	250	750	62.65	
	1150	11 H.P., 8" to 18" cores		5.45	130	390	1,175	121.60	
	1200	Finisher, concrete floor, gas, riding trowel, 96" wide		9.75	155	465.11	1,400	171.05	
	1300	Gas, walk-behind, 3 blade, 36" trowel		2.06	101	302.50	910	76.95	
	1400	4 blade, 48" trowel		3.10	116	347.50	1,050	94.30	
	1500	Float, hand-operated (Bull float), 48" wide		.08	13	39	117	8.45	
	1570	Curb builder, 14 H.P., gas, single screw		14.14	256	767.69	2,300	266.65	
	1590	Double screw		12.50	256	767.69	2,300	253.55	
	1600	Floor grinder, concrete and terrazzo, electric, 22" path		3.07	119	357.50	1,075	96.05	
	1700	Edger, concrete, electric, 7" path		1.19	57.50	172.50	520	44.05	
	1750	Vacuum pick-up system for floor grinders, wet/dry		1.63	103	309.38	930	74.95	
	1800	Mixer, powered, mortar and concrete, gas, 6 C.F., 18 H.P.		7.48	89	267.50	805	113.35	
	1900	10 C.F., 25 H.P.		9.08	124	372.50	1,125	147.15	
	2000	16 C.F.		9.44	146	436.88	1,300	162.90	
	2100	Concrete, stationary, tilt drum, 2 C.Y.		7.98	81	242.43	725	112.35	
	2120	Pump, concrete, truck mounted, 4" line, 80' boom		31.24	290	868.70	2,600	423.65	
	2140	5" line, 110' boom		39.99	290	868.70	2,600	493.65	
	2160	Mud jack, 50 C.F. per hr.		6.50	231	691.93	2,075	190.35	
	2180	225 C.F. per hr.		8.61	296	888.91	2,675	246.65	
	2190	Shotcrete pump rig, 12 C.Y./hr.		16.73	226	676.78	2,025	269.20	
	2200	35 C.Y./hr.		15.91	290	868.70	2,600	301.05	
	2600	Saw, concrete, manual, gas, 18 H.P.		5.59	115	345	1,025	113.70	
	2650	Self-propelled, gas, 30 H.P.		7.97	82	245.62	735	112.85	
	2675	V-groove crack chaser, manual, gas, 6 H.P.		1.65	100	300	900	73.25	
	2700	Vibrators, concrete, electric, 60 cycle, 2 H.P.		.47	70	210	630	45.80	
	2800	3 H.P.		.57	81	242.50	730	53.05	
	2900	Gas engine, 5 H.P.		1.56	17.05	51.21	154	22.70	
	3000	8 H.P.		2.10	17.25	51.74	155	27.15	
	3050	Vibrating screed, gas engine, 8 H.P.		2.83	108	325	975	87.65	
	3120	Concrete transit mixer, 6 x 4, 250 H.P., 8 C.Y., rear discharge		51.08	70.50	212.13	635	451.10	
	3200	Front discharge		59.30	136	409.10	1,225	556.20	
	3300	6 x 6, 285 H.P., 12 C.Y., rear discharge		58.56	167	500	1,500	568.50	
	3400	Front discharge		61.02	167	500	1,500	588.15	
20	0010	**EARTHWORK EQUIPMENT RENTAL** without operators							20
	0040	Aggregate spreader, push type, 8' to 12' wide [R015433-10]	Ea.	2.62	65	195	585	59.95	
	0045	Tailgate type, 8' wide		2.56	64	191.92	575	58.90	
	0055	Earth auger, truck mounted, for fence & sign posts, utility poles		13.95	152	454.55	1,375	202.55	
	0060	For borings and monitoring wells		42.95	84	252.53	760	394.15	
	0070	Portable, trailer mounted		2.32	92.50	277.50	835	74.05	
	0075	Truck mounted, for caissons, water wells		86.00	134	402	1,200	768.40	
	0080	Horizontal boring machine, 12" to 36" diameter, 45 H.P.		22.93	110	330	990	249.45	
	0090	12" to 48" diameter, 65 H.P.		31.47	130	390	1,175	329.75	
	0095	Auger, for fence posts, gas engine, hand held		.45	84.50	254.05	760	54.40	
	0100	Excavator, diesel hydraulic, crawler mounted, 1/2 C.Y. cap.		21.92	470	1,408.39	4,225	457	
	0120	5/8 C.Y. capacity		29.30	615	1,851.77	5,550	604.75	
	0140	3/4 C.Y. capacity		32.95	730	2,190.83	6,575	701.80	
	0150	1 C.Y. capacity		41.58	835	2,500	7,500	832.65	

For customer support on your Site Work & Landscape Costs with RSMeans Data, call 800.448.8182.

01 54 33 | Equipment Rental

		UNIT	HOURLY OPER. COST	RENT PER DAY	RENT PER WEEK	RENT PER MONTH	EQUIPMENT COST/DAY	
20	0200	1-1/2 C.Y. capacity	Ea.	49.02	505	1,518	4,550	695.80	20
	0300	2 C.Y. capacity		57.09	810	2,430	7,300	942.70	
	0320	2-1/2 C.Y. capacity		83.38	1,500	4,500	13,500	1,567	
	0325	3-1/2 C.Y. capacity		121.20	2,025	6,072	18,200	2,184	
	0330	4-1/2 C.Y. capacity		152.99	3,700	11,132	33,400	3,450	
	0335	6 C.Y. capacity		194.11	3,250	9,765.80	29,300	3,506	
	0340	7 C.Y. capacity		176.77	3,425	10,303.22	30,900	3,475	
	0342	Excavator attachments, bucket thumbs		3.43	261	783.90	2,350	184.20	
	0345	Grapples		3.17	225	674.15	2,025	160.20	
	0346	Hydraulic hammer for boom mounting, 4000 ft lb.		13.60	900	2,702.04	8,100	649.20	
	0347	5000 ft lb.		16.10	960	2,884.20	8,650	705.60	
	0348	8000 ft lb.		23.75	1,225	3,643.20	10,900	918.65	
	0349	12,000 ft lb.		25.95	1,125	3,373	10,100	882.20	
	0350	Gradall type, truck mounted, 3 ton @ 15' radius, 5/8 C.Y.		43.83	835	2,500	7,500	850.65	
	0370	1 C.Y. capacity		59.93	840	2,525.30	7,575	984.55	
	0400	Backhoe-loader, 40 to 45 H.P., 5/8 C.Y. capacity		12.01	286	857.50	2,575	267.55	
	0450	45 H.P. to 60 H.P., 3/4 C.Y. capacity		18.19	118	353.54	1,050	216.20	
	0460	80 H.P., 1-1/4 C.Y. capacity		20.54	118	353.54	1,050	235.05	
	0470	112 H.P., 1-1/2 C.Y. capacity		33.28	620	1,855.22	5,575	637.30	
	0482	Backhoe-loader attachment, compactor, 20,000 lb.		6.50	157	470.34	1,400	146.05	
	0485	Hydraulic hammer, 750 ft lb.		3.71	108	324.01	970	94.50	
	0486	Hydraulic hammer, 1200 ft lb.		6.61	207	621.89	1,875	177.25	
	0500	Brush chipper, gas engine, 6" cutter head, 35 H.P.		9.25	212	635	1,900	201	
	0550	Diesel engine, 12" cutter head, 130 H.P.		23.88	292	875	2,625	366.05	
	0600	15" cutter head, 165 H.P.		26.82	515	1,545	4,625	523.60	
	0750	Bucket, clamshell, general purpose, 3/8 C.Y.		1.41	92.50	277.78	835	66.85	
	0800	1/2 C.Y.		1.53	92.50	277.78	835	67.80	
	0850	3/4 C.Y.		1.65	92.50	277.78	835	68.80	
	0900	1 C.Y.		1.71	92.50	277.78	835	69.25	
	0950	1-1/2 C.Y.		2.81	92.50	277.78	835	78.05	
	1000	2 C.Y.		2.94	95	285	855	80.55	
	1010	Bucket, dragline, medium duty, 1/2 C.Y.		.83	92.50	277.78	835	62.15	
	1020	3/4 C.Y.		.79	92.50	277.78	835	61.85	
	1030	1 C.Y.		.80	92.50	277.78	835	62	
	1040	1-1/2 C.Y.		1.27	92.50	277.78	835	65.70	
	1050	2 C.Y.		1.30	92.50	277.78	835	65.95	
	1070	3 C.Y.		2.10	92.50	277.78	835	72.30	
	1200	Compactor, manually guided 2-drum vibratory smooth roller, 7.5 H.P.		7.28	181	542.50	1,625	166.75	
	1250	Rammer/tamper, gas, 8"		2.23	48.50	146.06	440	47	
	1260	15"		2.65	55.50	167.23	500	54.65	
	1300	Vibratory plate, gas, 18" plate, 3000 lb. blow		2.15	24.50	73.55	221	31.90	
	1350	21" plate, 5000 lb. blow		2.64	241	722.50	2,175	165.60	
	1370	Curb builder/extruder, 14 H.P., gas, single screw		12.50	256	767.69	2,300	253.55	
	1390	Double screw		12.50	256	767.69	2,300	253.55	
	1500	Disc harrow attachment, for tractor		.48	83.50	249.80	750	53.80	
	1810	Feller buncher, shearing & accumulating trees, 100 H.P.		43.75	465	1,393.97	4,175	628.80	
	1860	Grader, self-propelled, 25,000 lb.		33.65	1,125	3,366.73	10,100	942.50	
	1910	30,000 lb.		33.15	1,350	4,040.48	12,100	1,073	
	1920	40,000 lb.		52.35	1,325	4,000	12,000	1,219	
	1930	55,000 lb.		67.53	1,675	5,000	15,000	1,540	
	1950	Hammer, pavement breaker, self-propelled, diesel, 1000 to 1250 lb.		28.65	600	1,800	5,400	589.25	
	2000	1300 to 1500 lb.		43.19	1,025	3,051.51	9,150	955.80	
	2050	Pile driving hammer, steam or air, 4150 ft lb. @ 225 bpm		12.26	505	1,518	4,550	401.70	
	2100	8750 ft lb. @ 145 bpm		14.47	710	2,125.20	6,375	540.85	
	2150	15,000 ft lb. @ 60 bpm		14.81	845	2,530	7,600	624.45	
	2200	24,450 ft lb. @ 111 bpm		15.83	980	2,934.80	8,800	713.60	
	2250	Leads, 60' high for pile driving hammers up to 20,000 ft lb.		3.70	305	910.80	2,725	211.80	
	2300	90' high for hammers over 20,000 ft lb.		5.50	545	1,639.44	4,925	371.85	

01 54 33 | Equipment Rental

		UNIT	HOURLY OPER. COST	RENT PER DAY	RENT PER WEEK	RENT PER MONTH	EQUIPMENT COST/DAY	
20	2350	Diesel type hammer, 22,400 ft lb.	Ea.	17.98	495	1,489.40	4,475	441.70	20
	2400	41,300 ft lb.		25.91	625	1,881.35	5,650	583.55	
	2450	141,000 ft lb.		41.70	995	2,978.80	8,925	929.35	
	2500	Vib. elec. hammer/extractor, 200 kW diesel generator, 34 H.P.		41.74	725	2,168.78	6,500	767.70	
	2550	80 H.P.		73.68	1,050	3,135.58	9,400	1,217	
	2600	150 H.P.		136.35	2,000	6,035.99	18,100	2,298	
	2700	Hydro excavator w/ext boom 12 C.Y., 1200 gallons		38.15	1,625	4,857.60	14,600	1,277	
	2800	Log chipper, up to 22" diameter, 600 H.P.		45.00	325	975	2,925	555	
	2850	Logger, for skidding & stacking logs, 150 H.P.		43.84	940	2,813.18	8,450	913.35	
	2860	Mulcher, diesel powered, trailer mounted		20.06	310	924.26	2,775	345.35	
	2900	Rake, spring tooth, with tractor		14.85	375	1,123.58	3,375	343.50	
	3000	Roller, vibratory, tandem, smooth drum, 20 H.P.		7.88	325	978.60	2,925	258.75	
	3050	35 H.P.		10.22	275	825	2,475	246.80	
	3100	Towed type vibratory compactor, smooth drum, 50 H.P.		25.50	525	1,581.85	4,750	520.35	
	3150	Sheepsfoot, 50 H.P.		25.87	365	1,100	3,300	426.95	
	3170	Landfill compactor, 220 H.P.		70.51	1,675	5,035.45	15,100	1,571	
	3200	Pneumatic tire roller, 80 H.P.		13.04	410	1,228.10	3,675	349.90	
	3250	120 H.P.		19.56	565	1,700	5,100	496.50	
	3300	Sheepsfoot vibratory roller, 240 H.P.		62.77	1,425	4,303.41	12,900	1,363	
	3320	340 H.P.		84.58	2,200	6,565.78	19,700	1,990	
	3350	Smooth drum vibratory roller, 75 H.P.		23.55	660	1,982.18	5,950	584.80	
	3400	125 H.P.		27.86	750	2,247.17	6,750	672.35	
	3410	Rotary mower, brush, 60", with tractor		18.96	365	1,097.45	3,300	371.15	
	3420	Rototiller, walk-behind, gas, 5 H.P.		2.15	60	180	540	53.25	
	3422	8 H.P.		2.84	116	347.50	1,050	92.20	
	3440	Scrapers, towed type, 7 C.Y. capacity		6.50	129	386.72	1,150	129.30	
	3450	10 C.Y. capacity		7.27	172	517.37	1,550	161.65	
	3500	15 C.Y. capacity		7.46	199	595.76	1,775	178.85	
	3525	Self-propelled, single engine, 14 C.Y. capacity		134.48	2,250	6,739.92	20,200	2,424	
	3550	Dual engine, 21 C.Y. capacity		142.64	2,525	7,575.90	22,700	2,656	
	3600	31 C.Y. capacity		189.53	3,650	10,954.15	32,900	3,707	
	3640	44 C.Y. capacity		234.76	4,700	14,083.90	42,300	4,695	
	3650	Elevating type, single engine, 11 C.Y. capacity		62.42	935	2,800	8,400	1,059	
	3700	22 C.Y. capacity		115.65	1,625	4,850	14,600	1,895	
	3710	Screening plant, 110 H.P. w/5' x 10' screen		21.32	650	1,952.56	5,850	561.10	
	3720	5' x 16' screen		26.92	1,225	3,666.67	11,000	948.65	
	3850	Shovel, crawler-mounted, front-loading, 7 C.Y. capacity		220.61	3,975	11,915.21	35,700	4,148	
	3855	12 C.Y. capacity		339.92	5,500	16,514.06	49,500	6,022	
	3860	Shovel/backhoe bucket, 1/2 C.Y.		2.72	74	221.58	665	66.05	
	3870	3/4 C.Y.		2.69	83	248.76	745	71.25	
	3880	1 C.Y.		2.78	92	275.42	825	77.35	
	3890	1-1/2 C.Y.		2.98	108	324.01	970	88.65	
	3910	3 C.Y.		3.47	146	438.98	1,325	115.55	
	3950	Stump chipper, 18" deep, 30 H.P.		6.95	223	668	2,000	189.20	
	4110	Dozer, crawler, torque converter, diesel 80 H.P.		25.48	335	1,010.12	3,025	405.85	
	4150	105 H.P.		34.64	605	1,818.22	5,450	640.80	
	4200	140 H.P.		41.64	660	1,980	5,950	729.15	
	4260	200 H.P.		63.72	1,675	5,050.60	15,200	1,520	
	4310	300 H.P.		81.45	1,900	5,667.20	17,000	1,785	
	4360	410 H.P.		107.70	3,250	9,727.46	29,200	2,807	
	4370	500 H.P.		134.58	3,925	11,788.10	35,400	3,434	
	4380	700 H.P.		232.23	5,525	16,587.71	49,800	5,175	
	4400	Loader, crawler, torque conv., diesel, 1-1/2 C.Y., 80 H.P.		29.80	555	1,667.71	5,000	572	
	4450	1-1/2 to 1-3/4 C.Y., 95 H.P.		30.54	705	2,120	6,350	668.35	
	4510	1-3/4 to 2-1/4 C.Y., 130 H.P.		48.19	900	2,700	8,100	925.50	
	4530	2-1/2 to 3-1/4 C.Y., 190 H.P.		58.30	1,125	3,400	10,200	1,146	
	4560	3-1/2 to 5 C.Y., 275 H.P.		72.05	1,475	4,400	13,200	1,456	
	4610	Front end loader, 4WD, articulated frame, diesel, 1 to 1-1/4 C.Y., 70 H.P.		16.77	286	857.06	2,575	305.60	

For customer support on your Site Work & Landscape Costs with RSMeans Data, call 800.448.8182.

01 54 33 | Equipment Rental

		UNIT	HOURLY OPER. COST	RENT PER DAY	RENT PER WEEK	RENT PER MONTH	EQUIPMENT COST/DAY	
20	4620	1-1/2 to 1-3/4 C.Y., 95 H.P.	Ea.	20.17	465	1,400	4,200	441.40	**20**
	4650	1-3/4 to 2 C.Y., 130 H.P.		21.26	380	1,140	3,425	398.05	
	4710	2-1/2 to 3-1/2 C.Y., 145 H.P.		29.79	665	2,000	6,000	638.30	
	4730	3 to 4-1/2 C.Y., 185 H.P.		32.38	835	2,500	7,500	759	
	4760	5-1/4 to 5-3/4 C.Y., 270 H.P.		53.67	900	2,692.98	8,075	967.95	
	4810	7 to 9 C.Y., 475 H.P.		91.99	2,575	7,744.59	23,200	2,285	
	4870	9 to 11 C.Y., 620 H.P.		133.10	2,725	8,204.77	24,600	2,706	
	4880	Skid-steer loader, wheeled, 10 C.F., 30 H.P. gas		9.66	170	511.19	1,525	179.50	
	4890	1 C.Y., 78 H.P., diesel		18.60	495	1,487.50	4,475	446.30	
	4892	Skid-steer attachment, auger		.75	149	447.50	1,350	95.50	
	4893	Backhoe		.75	123	370.35	1,100	80.05	
	4894	Broom		.71	136	407.50	1,225	87.20	
	4895	Forks		.16	32	96	288	20.45	
	4896	Grapple		.73	89.50	267.93	805	59.45	
	4897	Concrete hammer		1.06	183	550	1,650	118.50	
	4898	Tree spade		.61	105	313.56	940	67.55	
	4899	Trencher		.66	102	305	915	66.25	
	4900	Trencher, chain, boom type, gas, operator walking, 12 H.P.		4.21	208	624.51	1,875	158.60	
	4910	Operator riding, 40 H.P.		16.83	525	1,580	4,750	450.70	
	5000	Wheel type, diesel, 4' deep, 12" wide		69.33	975	2,926.54	8,775	1,140	
	5100	6' deep, 20" wide		88.37	885	2,655	7,975	1,238	
	5150	Chain type, diesel, 5' deep, 8" wide		16.45	365	1,097.45	3,300	351.05	
	5200	Diesel, 8' deep, 16" wide		90.47	1,950	5,853.08	17,600	1,894	
	5202	Rock trencher, wheel type, 6" wide x 18" deep		47.46	91	272.73	820	434.20	
	5206	Chain type, 18" wide x 7' deep		105.44	286	858.60	2,575	1,015	
	5210	Tree spade, self-propelled		13.79	232	696.98	2,100	249.70	
	5250	Truck, dump, 2-axle, 12 ton, 8 C.Y. payload, 220 H.P.		24.16	355	1,071.50	3,225	407.60	
	5300	Three axle dump, 16 ton, 12 C.Y. payload, 400 H.P.		45.03	365	1,095.41	3,275	579.35	
	5310	Four axle dump, 25 ton, 18 C.Y. payload, 450 H.P.		50.45	585	1,750	5,250	753.60	
	5350	Dump trailer only, rear dump, 16-1/2 C.Y.		5.80	153	459.89	1,375	138.35	
	5400	20 C.Y.		6.26	172	517.37	1,550	153.55	
	5450	Flatbed, single axle, 1-1/2 ton rating		19.22	74.50	223.67	670	198.50	
	5500	3 ton rating		25.95	1,075	3,212.18	9,625	850.05	
	5550	Off highway rear dump, 25 ton capacity		63.42	1,525	4,600	13,800	1,427	
	5600	35 ton capacity		67.70	675	2,020.24	6,050	945.65	
	5610	50 ton capacity		84.88	2,175	6,500	19,500	1,979	
	5620	65 ton capacity		90.64	2,025	6,062.12	18,200	1,938	
	5630	100 ton capacity		122.70	2,975	8,936.41	26,800	2,769	
	6000	Vibratory plow, 25 H.P., walking	▼	6.84	305	909.11	2,725	236.55	
40	0010	**GENERAL EQUIPMENT RENTAL** without operators R015433 -10							**40**
	0020	Aerial lift, scissor type, to 20' high, 1200 lb. capacity, electric	Ea.	3.52	154	462.50	1,400	120.70	
	0030	To 30' high, 1200 lb. capacity		3.82	235	705	2,125	171.55	
	0040	Over 30' high, 1500 lb. capacity		5.20	288	865	2,600	214.55	
	0070	Articulating boom, to 45' high, 500 lb. capacity, diesel R015433 -15		10.04	250	750	2,250	230.30	
	0075	To 60' high, 500 lb. capacity		13.82	305	909.11	2,725	292.40	
	0080	To 80' high, 500 lb. capacity		16.24	995	2,987.50	8,975	727.45	
	0085	To 125' high, 500 lb. capacity		18.56	1,675	4,997.50	15,000	1,148	
	0100	Telescoping boom to 40' high, 500 lb. capacity, diesel		11.37	320	954.56	2,875	281.90	
	0105	To 45' high, 500 lb. capacity		12.66	325	974.77	2,925	296.20	
	0110	To 60' high, 500 lb. capacity		16.56	267	800	2,400	292.45	
	0115	To 80' high, 500 lb. capacity		21.52	360	1,077.80	3,225	387.75	
	0120	To 100' high, 500 lb. capacity		29.06	985	2,950	8,850	822.45	
	0125	To 120' high, 500 lb. capacity		29.51	1,525	4,545	13,600	1,145	
	0195	Air compressor, portable, 6.5 CFM, electric		.91	53.50	160	480	39.30	
	0196	Gasoline		.66	56	167.50	505	38.80	
	0200	Towed type, gas engine, 60 CFM		9.55	129	387.50	1,175	153.85	
	0300	160 CFM	▼	10.60	213	637.50	1,925	212.30	

01 54 | Construction Aids

01 54 33 | Equipment Rental

		UNIT	HOURLY OPER. COST	RENT PER DAY	RENT PER WEEK	RENT PER MONTH	EQUIPMENT COST/DAY	
0400	Diesel engine, rotary screw, 250 CFM	Ea.	12.23	175	525	1,575	202.85	40
0500	365 CFM		16.19	355	1,070	3,200	343.55	
0550	450 CFM		20.19	300	900	2,700	341.50	
0600	600 CFM		34.51	251	752.54	2,250	426.55	
0700	750 CFM		35.04	525	1,580	4,750	596.30	
0930	Air tools, breaker, pavement, 60 lb.		.57	81.50	245	735	53.60	
0940	80 lb.		.57	81	242.50	730	53.05	
0950	Drills, hand (jackhammer), 65 lb.		.68	64	192.50	580	43.90	
0960	Track or wagon, swing boom, 4" drifter		62.50	1,050	3,135.41	9,400	1,127	
0970	5" drifter		62.50	1,050	3,135.41	9,400	1,127	
0975	Track mounted quarry drill, 6" diameter drill		102.94	1,900	5,722.33	17,200	1,968	
0980	Dust control per drill		1.05	25.50	76.26	229	23.65	
0990	Hammer, chipping, 12 lb.		.61	46.50	140	420	32.85	
1000	Hose, air with couplings, 50' long, 3/4" diameter		.07	11	33	99	7.15	
1100	1" diameter		.08	12.35	37	111	8.05	
1200	1-1/2" diameter		.22	35	105	315	22.80	
1300	2" diameter		.24	45	135	405	28.95	
1400	2-1/2" diameter		.36	57.50	172.50	520	37.40	
1410	3" diameter		.42	58.50	175	525	38.35	
1450	Drill, steel, 7/8" x 2'		.09	13.05	39.12	117	8.50	
1460	7/8" x 6'		.12	19.85	59.58	179	12.85	
1520	Moil points		.03	7	21	63	4.40	
1525	Pneumatic nailer w/accessories		.48	39.50	119.19	360	27.70	
1530	Sheeting driver for 60 lb. breaker		.04	7.85	23.52	70.50	5.05	
1540	For 90 lb. breaker		.13	10.65	31.88	95.50	7.45	
1550	Spade, 25 lb.		.51	7.50	22.47	67.50	8.55	
1560	Tamper, single, 35 lb.		.60	58.50	175	525	39.75	
1570	Triple, 140 lb.		.90	62.50	187.09	560	44.60	
1580	Wrenches, impact, air powered, up to 3/4" bolt		.43	50	149.75	450	33.40	
1590	Up to 1-1/4" bolt		.58	67.50	202.50	610	45.15	
1600	Barricades, barrels, reflectorized, 1 to 99 barrels		.03	4	12	36	2.65	
1610	100 to 200 barrels		.03	4.46	13.38	40	2.90	
1620	Barrels with flashers, 1 to 99 barrels		.04	6.45	19.39	58	4.15	
1630	100 to 200 barrels		.03	5.15	15.52	46.50	3.35	
1640	Barrels with steady burn type C lights		.05	8.55	25.61	77	5.50	
1650	Illuminated board, trailer mounted, with generator		3.32	141	423.30	1,275	111.20	
1670	Portable barricade, stock, with flashers, 1 to 6 units		.04	6.45	19.34	58	4.15	
1680	25 to 50 units		.03	6	18.03	54	3.85	
1685	Butt fusion machine, wheeled, 1.5 HP electric, 2" - 8" diameter pipe		2.66	190	568.50	1,700	134.95	
1690	Tracked, 20 HP diesel, 4" - 12" diameter pipe		10.04	565	1,702.05	5,100	420.75	
1695	83 HP diesel, 8" - 24" diameter pipe		51.84	1,125	3,358.65	10,100	1,086	
1700	Carts, brick, gas engine, 1000 lb. capacity		2.98	65.50	196.97	590	63.20	
1800	1500 lb., 7-1/2' lift		2.95	70	210.50	630	65.70	
1822	Dehumidifier, medium, 6 lb./hr., 150 CFM		1.20	77.50	232.03	695	56.05	
1824	Large, 18 lb./hr., 600 CFM		2.22	585	1,750	5,250	367.75	
1830	Distributor, asphalt, trailer mounted, 2000 gal., 38 H.P. diesel		11.13	355	1,071.32	3,225	303.25	
1840	3000 gal., 38 H.P. diesel		13.02	385	1,149.71	3,450	334.10	
1850	Drill, rotary hammer, electric		1.13	72	216.17	650	52.25	
1860	Carbide bit, 1-1/2" diameter, add to electric rotary hammer		.03	41.50	125	375	25.25	
1865	Rotary, crawler, 250 H.P.		137.40	2,450	7,375	22,100	2,574	
1870	Emulsion sprayer, 65 gal., 5 H.P. gas engine		2.80	108	324.01	970	87.20	
1880	200 gal., 5 H.P. engine		7.31	181	543.50	1,625	167.20	
1900	Floor auto-scrubbing machine, walk-behind, 28" path		5.69	223	670	2,000	179.55	
1930	Floodlight, mercury vapor, or quartz, on tripod, 1000 watt		.46	36.50	110	330	25.70	
1940	2000 watt		.60	28.50	85.55	257	21.90	
1950	Floodlights, trailer mounted with generator, 1 - 300 watt light		3.59	79.50	238.30	715	76.35	
1960	2 - 1000 watt lights		4.54	88.50	265.48	795	89.40	
2000	4 - 300 watt lights		4.29	101	303.11	910	95	

01 54 33 | Equipment Rental

		UNIT	HOURLY OPER. COST	RENT PER DAY	RENT PER WEEK	RENT PER MONTH	EQUIPMENT COST/DAY	
40	2005	Foam spray rig, incl. box trailer, compressor, generator, proportioner	Ea.	25.76	540	1,620.05	4,850	530.15	**40**
	2015	Forklift, pneumatic tire, rough terr, straight mast, 5000 lb, 12' lift, gas		18.82	221	663.70	2,000	283.30	
	2025	8000 lb, 12' lift		22.95	365	1,097.45	3,300	403.10	
	2030	5000 lb, 12' lift, diesel		15.59	247	740.71	2,225	272.85	
	2035	8000 lb, 12' lift, diesel		16.90	280	839.82	2,525	303.15	
	2045	All terrain, telescoping boom, diesel, 5000 lb, 10' reach, 19' lift		17.40	200	600	1,800	259.25	
	2055	6600 lb, 29' reach, 42' lift		21.29	236	707.08	2,125	311.75	
	2065	10,000 lb, 31' reach, 45' lift		23.30	267	800	2,400	346.45	
	2070	Cushion tire, smooth floor, gas, 5000 lb capacity		8.32	262	785	2,350	223.60	
	2075	8000 lb capacity		11.47	320	957.50	2,875	283.25	
	2085	Diesel, 5000 lb capacity		7.82	257	770	2,300	216.55	
	2090	12,000 lb capacity		12.16	400	1,206.59	3,625	338.60	
	2095	20,000 lb capacity		17.41	800	2,405	7,225	620.30	
	2100	Generator, electric, gas engine, 1.5 kW to 3 kW		2.60	46.50	140	420	48.80	
	2200	5 kW		3.25	91	272.50	820	80.45	
	2300	10 kW		5.98	108	322.50	970	112.35	
	2400	25 kW		7.47	405	1,222.25	3,675	304.25	
	2500	Diesel engine, 20 kW		9.29	222	665	2,000	207.35	
	2600	50 kW		16.09	360	1,085	3,250	345.70	
	2700	100 kW		28.85	485	1,455	4,375	521.85	
	2800	250 kW		54.84	825	2,472.50	7,425	933.25	
	2850	Hammer, hydraulic, for mounting on boom, to 500 ft lb.		2.93	94.50	283.25	850	80.05	
	2860	1000 ft lb.		4.65	141	423.30	1,275	121.85	
	2900	Heaters, space, oil or electric, 50 MBH		1.48	47	141.42	425	40.10	
	3000	100 MBH		2.74	47	141.42	425	50.25	
	3100	300 MBH		8.00	136	409.10	1,225	145.80	
	3150	500 MBH		13.28	208	625	1,875	231.20	
	3200	Hose, water, suction with coupling, 20' long, 2" diameter		.02	5.65	17	51	3.55	
	3210	3" diameter		.03	14.15	42.50	128	8.75	
	3220	4" diameter		.03	28.50	85.01	255	17.25	
	3230	6" diameter		.11	41	122.96	370	25.50	
	3240	8" diameter		.27	54	161.92	485	34.55	
	3250	Discharge hose with coupling, 50' long, 2" diameter		.01	6.50	19.50	58.50	4	
	3260	3" diameter		.01	7.35	22	66	4.50	
	3270	4" diameter		.02	21	63.25	190	12.80	
	3280	6" diameter		.06	29.50	88.04	264	18.10	
	3290	8" diameter		.24	38	113.85	340	24.70	
	3295	Insulation blower		.84	118	353.54	1,050	77.40	
	3300	Ladders, extension type, 16' to 36' long		.18	41.50	125	375	26.45	
	3400	40' to 60' long		.64	123	368	1,100	78.75	
	3405	Lance for cutting concrete		2.22	66.50	200	600	57.80	
	3407	Lawn mower, rotary, 22", 5 H.P.		1.06	38.50	115	345	31.50	
	3408	48" self-propelled		2.92	160	481	1,450	119.60	
	3410	Level, electronic, automatic, with tripod and leveling rod		1.06	43	129.50	390	34.40	
	3430	Laser type, for pipe and sewer line and grade		2.20	117	350	1,050	87.55	
	3440	Rotating beam for interior control		.91	64	192.50	580	45.80	
	3460	Builder's optical transit, with tripod and rod		.10	43	129.50	390	26.70	
	3500	Light towers, towable, with diesel generator, 2000 watt		4.31	102	307.29	920	95.90	
	3600	4000 watt		4.55	142	425	1,275	121.40	
	3700	Mixer, powered, plaster and mortar, 6 C.F., 7 H.P.		2.08	86.50	260	780	68.60	
	3800	10 C.F., 9 H.P.		2.26	124	372.50	1,125	92.60	
	3850	Nailer, pneumatic		.48	30.50	91	273	22.10	
	3900	Paint sprayers complete, 8 CFM		.86	62.50	186.74	560	44.20	
	4000	17 CFM		1.61	111	333.84	1,000	79.70	
	4020	Pavers, bituminous, rubber tires, 8' wide, 50 H.P., diesel		32.31	575	1,724.57	5,175	603.40	
	4030	10' wide, 150 H.P.		96.77	1,975	5,905.34	17,700	1,955	
	4050	Crawler, 8' wide, 100 H.P., diesel		88.64	2,075	6,245.03	18,700	1,958	
	4060	10' wide, 150 H.P.		105.21	2,375	7,133.45	21,400	2,268	

01 54 33 | Equipment Rental

		UNIT	HOURLY OPER. COST	RENT PER DAY	RENT PER WEEK	RENT PER MONTH	EQUIPMENT COST/DAY	
40	4070	Concrete paver, 12' to 24' wide, 250 H.P.	Ea.	88.67	1,700	5,121.45	15,400	1,734	40
	4080	Placer-spreader-trimmer, 24' wide, 300 H.P.		118.92	2,575	7,760.56	23,300	2,503	
	4100	Pump, centrifugal gas pump, 1-1/2" diam., 65 GPM		3.97	54.50	164.10	490	64.60	
	4200	2" diameter, 130 GPM		5.04	44	132.50	400	66.80	
	4300	3" diameter, 250 GPM		5.18	55	165	495	74.40	
	4400	6" diameter, 1500 GPM		22.46	92.50	277.78	835	235.25	
	4500	Submersible electric pump, 1-1/4" diameter, 55 GPM		.40	35	105	315	24.25	
	4600	1-1/2" diameter, 83 GPM		.45	43.50	130	390	29.60	
	4700	2" diameter, 120 GPM		1.66	55	165	495	46.30	
	4800	3" diameter, 300 GPM		3.07	109	327.50	985	90.05	
	4900	4" diameter, 560 GPM		14.90	76	227.50	685	164.70	
	5000	6" diameter, 1590 GPM		22.30	65.50	196.97	590	217.80	
	5100	Diaphragm pump, gas, single, 1-1/2" diameter		1.14	38.50	116.16	350	32.35	
	5200	2" diameter		4.02	92.50	277.78	835	87.70	
	5300	3" diameter		4.09	86	258	775	84.35	
	5400	Double, 4" diameter		6.10	96	287.88	865	106.35	
	5450	Pressure washer 5 GPM, 3000 psi		3.92	110	330	990	97.35	
	5460	7 GPM, 3000 psi		5.00	90	270	810	93.95	
	5470	High pressure water jet 10 KSI		40.02	730	2,185.92	6,550	757.40	
	5480	40 KSI		28.21	990	2,975.28	8,925	820.75	
	5500	Trash pump, self-priming, gas, 2" diameter		3.85	109	328.29	985	96.50	
	5600	Diesel, 4" diameter		6.74	163	489.91	1,475	151.95	
	5650	Diesel, 6" diameter		17.02	163	489.91	1,475	234.15	
	5655	Grout pump		18.92	284	851.83	2,550	321.75	
	5700	Salamanders, L.P. gas fired, 100,000 Btu		2.92	57.50	172.50	520	57.85	
	5705	50,000 Btu		1.68	23	69	207	27.25	
	5720	Sandblaster, portable, open top, 3 C.F. capacity		.61	132	395	1,175	83.85	
	5730	6 C.F. capacity		1.02	133	399	1,200	87.90	
	5740	Accessories for above		.14	24	72.12	216	15.55	
	5750	Sander, floor		.78	73.50	220	660	50.20	
	5760	Edger		.52	35	105	315	25.20	
	5800	Saw, chain, gas engine, 18" long		1.77	63.50	190	570	52.20	
	5900	Hydraulic powered, 36" long		.79	59	176.77	530	41.65	
	5950	60" long		.79	65.50	196.97	590	45.70	
	6000	Masonry, table mounted, 14" diameter, 5 H.P.		1.34	76.50	230	690	56.70	
	6050	Portable cut-off, 8 H.P.		1.83	72.50	217.50	655	58.15	
	6100	Circular, hand held, electric, 7-1/4" diameter		.23	13.85	41.50	125	10.15	
	6200	12" diameter		.24	41	122.50	370	26.40	
	6250	Wall saw, w/hydraulic power, 10 H.P.		3.32	99.50	299	895	86.40	
	6275	Shot blaster, walk-behind, 20" wide		4.79	284	851.83	2,550	208.70	
	6280	Sidewalk broom, walk-behind		2.27	83.50	250.85	755	68.30	
	6300	Steam cleaner, 100 gallons per hour		3.38	83.50	250.85	755	77.20	
	6310	200 gallons per hour		4.38	101	303.11	910	95.70	
	6340	Tar kettle/pot, 400 gallons		16.65	128	383.85	1,150	209.95	
	6350	Torch, cutting, acetylene-oxygen, 150' hose, excludes gases		.46	15.50	46.51	140	12.95	
	6360	Hourly operating cost includes tips and gas		21.17	7.10	21.23	63.50	173.60	
	6410	Toilet, portable chemical		.13	23.50	70.03	210	15.05	
	6420	Recycle flush type		.16	29	86.75	260	18.65	
	6430	Toilet, fresh water flush, garden hose,		.20	34.50	103.47	310	22.25	
	6440	Hoisted, non-flush, for high rise		.16	28	84.66	254	18.20	
	6465	Tractor, farm with attachment		17.57	395	1,184.37	3,550	377.45	
	6480	Trailers, platform, flush deck, 2 axle, 3 ton capacity		1.71	96	287.41	860	71.15	
	6500	25 ton capacity		6.31	145	433.76	1,300	137.20	
	6600	40 ton capacity		8.14	206	616.66	1,850	188.45	
	6700	3 axle, 50 ton capacity		8.82	228	683.33	2,050	207.25	
	6800	75 ton capacity		11.21	305	907.63	2,725	271.20	
	6810	Trailer mounted cable reel for high voltage line work		5.94	28.50	85.86	258	64.75	
	6820	Trailer mounted cable tensioning rig		11.79	28.50	85.86	258	111.50	

For customer support on your Site Work & Landscape Costs with RSMeans Data, call 800.448.8182.

01 54 33 | Equipment Rental

		UNIT	HOURLY OPER. COST	RENT PER DAY	RENT PER WEEK	RENT PER MONTH	EQUIPMENT COST/DAY	
40								**40**
6830	Cable pulling rig	Ea.	74.51	28.50	85.86	258	613.30	
6850	Portable cable/wire puller, 8000 lb max pulling capacity		3.74	121	363.64	1,100	102.65	
6900	Water tank trailer, engine driven discharge, 5000 gallons		7.24	160	480.79	1,450	154.10	
6925	10,000 gallons		9.87	218	653.25	1,950	209.60	
6950	Water truck, off highway, 6000 gallons		72.61	845	2,529.89	7,600	1,087	
7010	Tram car for high voltage line work, powered, 2 conductor		6.95	29.50	88	264	73.20	
7020	Transit (builder's level) with tripod		.10	17.75	53.30	160	11.45	
7030	Trench box, 3000 lb., 6' x 8'		.57	97.50	293.15	880	63.15	
7040	7200 lb., 6' x 20'		.73	189	566.72	1,700	119.15	
7050	8000 lb., 8' x 16'		1.09	205	615	1,850	131.70	
7060	9500 lb., 8' x 20'		1.22	235	705.51	2,125	150.85	
7065	11,000 lb., 8' x 24'		1.28	221	663.70	2,000	142.95	
7070	12,000 lb., 10' x 20'		1.51	267	799.57	2,400	172	
7100	Truck, pickup, 3/4 ton, 2 wheel drive		9.35	62.50	187.09	560	112.20	
7200	4 wheel drive		9.60	167	500	1,500	176.75	
7250	Crew carrier, 9 passenger		12.81	109	328.29	985	168.15	
7290	Flat bed truck, 20,000 lb. GVW		15.44	134	402.40	1,200	204.05	
7300	Tractor, 4 x 2, 220 H.P.		22.52	218	653.25	1,950	310.80	
7410	330 H.P.		32.72	298	893.64	2,675	440.45	
7500	6 x 4, 380 H.P.		36.53	345	1,034.74	3,100	499.15	
7600	450 H.P.		44.76	420	1,254.23	3,775	608.95	
7610	Tractor, with A frame, boom and winch, 225 H.P.		25.04	296	886.76	2,650	377.65	
7620	Vacuum truck, hazardous material, 2500 gallons		12.94	315	940.67	2,825	291.70	
7625	5,000 gallons		13.18	445	1,332.62	4,000	371.95	
7650	Vacuum, HEPA, 16 gallon, wet/dry		.86	122	365	1,100	79.90	
7655	55 gallon, wet/dry		.79	25.50	76.50	230	21.60	
7660	Water tank, portable		.74	162	485.11	1,450	102.90	
7690	Sewer/catch basin vacuum, 14 C.Y., 1500 gallons		17.52	670	2,012	6,025	542.60	
7700	Welder, electric, 200 amp		3.86	33.50	101.01	305	51.10	
7800	300 amp		5.61	104	313.14	940	107.55	
7900	Gas engine, 200 amp		9.05	59	176.77	530	107.80	
8000	300 amp		10.25	111	333.96	1,000	148.75	
8100	Wheelbarrow, any size		.07	11.15	33.50	101	7.20	
8200	Wrecking ball, 4000 lb.		2.53	62	186	560	57.45	
50	**HIGHWAY EQUIPMENT RENTAL** without operators [R015433 -10]							**50**
0010								
0050	Asphalt batch plant, portable drum mixer, 100 ton/hr.	Ea.	89.47	1,550	4,677.24	14,000	1,651	
0060	200 ton/hr.		103.21	1,675	4,990.80	15,000	1,824	
0070	300 ton/hr.		121.29	1,950	5,853.08	17,600	2,141	
0100	Backhoe attachment, long stick, up to 185 H.P., 10-1/2' long		.37	26	77.34	232	18.45	
0140	Up to 250 H.P., 12' long		.42	29	86.75	260	20.70	
0180	Over 250 H.P., 15' long		.57	39.50	118.11	355	28.20	
0200	Special dipper arm, up to 100 H.P., 32' long		1.17	80.50	241.44	725	57.65	
0240	Over 100 H.P., 33' long		1.46	101	303.11	910	72.30	
0280	Catch basin/sewer cleaning truck, 3 ton, 9 C.Y., 1000 gal.		35.81	425	1,280.36	3,850	542.60	
0300	Concrete batch plant, portable, electric, 200 C.Y./hr.		24.47	565	1,698.44	5,100	535.50	
0520	Grader/dozer attachment, ripper/scarifier, rear mounted, up to 135 H.P.		3.19	64	192.32	575	63.95	
0540	Up to 180 H.P.		4.18	97	290.56	870	91.60	
0580	Up to 250 H.P.		5.92	155	465.11	1,400	140.40	
0700	Pvmt. removal bucket, for hyd. excavator, up to 90 H.P.		2.18	59	176.64	530	52.80	
0740	Up to 200 H.P.		2.33	75.50	225.76	675	63.80	
0780	Over 200 H.P.		2.55	92.50	276.98	830	75.80	
0900	Aggregate spreader, self-propelled, 187 H.P.		51.21	750	2,247.17	6,750	859.10	
1000	Chemical spreader, 3 C.Y.		3.20	99.50	299	895	85.40	
1900	Hammermill, traveling, 250 H.P.		68.03	520	1,565.69	4,700	857.40	
2000	Horizontal borer, 3" diameter, 13 H.P. gas driven		5.47	234	702.03	2,100	184.15	
2150	Horizontal directional drill, 20,000 lb. thrust, 78 H.P. diesel		27.86	535	1,606.09	4,825	544.10	
2160	30,000 lb. thrust, 115 H.P.		34.24	625	1,868.72	5,600	647.65	
2170	50,000 lb. thrust, 170 H.P.		49.09	720	2,156.61	6,475	824.05	

			UNIT	HOURLY OPER. COST	RENT PER DAY	RENT PER WEEK	RENT PER MONTH	EQUIPMENT COST/DAY	
50	2190	Mud trailer for HDD, 1500 gallons, 175 H.P., gas	Ea.	25.76	177	530.31	1,600	312.15	**50**
	2200	Hydromulcher, diesel, 3000 gallon, for truck mounting		17.61	193	580	1,750	256.85	
	2300	Gas, 600 gallon		7.57	96	287.88	865	118.15	
	2400	Joint & crack cleaner, walk behind, 25 H.P.		3.19	46	137.38	410	53	
	2500	Filler, trailer mounted, 400 gallons, 20 H.P.		8.43	173	517.50	1,550	170.95	
	3000	Paint striper, self-propelled, 40 gallon, 22 H.P.		6.83	123	368.69	1,100	128.35	
	3100	120 gallon, 120 H.P.		17.83	385	1,151.54	3,450	372.90	
	3200	Post drivers, 6" I-Beam frame, for truck mounting		12.54	325	969.72	2,900	294.25	
	3400	Road sweeper, self-propelled, 8' wide, 90 H.P.		36.34	720	2,164.75	6,500	723.65	
	3450	Road sweeper, vacuum assisted, 4 C.Y., 220 gallons		58.98	680	2,038.13	6,125	879.45	
	4000	Road mixer, self-propelled, 130 H.P.		46.79	835	2,508.46	7,525	876	
	4100	310 H.P.		75.91	2,175	6,558.59	19,700	1,919	
	4220	Cold mix paver, incl. pug mill and bitumen tank, 165 H.P.		96.11	2,350	7,015.87	21,000	2,172	
	4240	Pavement brush, towed		3.47	101	303.11	910	88.35	
	4250	Paver, asphalt, wheel or crawler, 130 H.P., diesel		95.37	2,300	6,898.28	20,700	2,143	
	4300	Paver, road widener, gas, 1' to 6', 67 H.P.		47.22	985	2,952.67	8,850	968.30	
	4400	Diesel, 2' to 14', 88 H.P.		57.06	1,175	3,501.40	10,500	1,157	
	4600	Slipform pavers, curb and gutter, 2 track, 75 H.P.		58.53	1,275	3,814.96	11,400	1,231	
	4700	4 track, 165 H.P.		36.11	855	2,560.72	7,675	801.05	
	4800	Median barrier, 215 H.P.		59.13	1,350	4,076.26	12,200	1,288	
	4901	Trailer, low bed, 75 ton capacity		10.84	286	857.06	2,575	258.10	
	5000	Road planer, walk behind, 10" cutting width, 10 H.P.		2.48	258	773	2,325	174.45	
	5100	Self-propelled, 12" cutting width, 64 H.P.		8.34	192	575.77	1,725	181.85	
	5120	Traffic line remover, metal ball blaster, truck mounted, 115 H.P.		47.03	1,025	3,100	9,300	996.25	
	5140	Grinder, truck mounted, 115 H.P.		51.41	1,025	3,100	9,300	1,031	
	5160	Walk-behind, 11 H.P.		3.59	143	430	1,300	114.75	
	5200	Pavement profiler, 4' to 6' wide, 450 H.P.		218.77	1,275	3,838.46	11,500	2,518	
	5300	8' to 10' wide, 750 H.P.		334.94	1,350	4,015.23	12,000	3,483	
	5400	Roadway plate, steel, 1" x 8' x 20'		.09	61.50	184.69	555	37.65	
	5600	Stabilizer, self-propelled, 150 H.P.		41.55	1,050	3,131.37	9,400	958.70	
	5700	310 H.P.		76.95	1,325	3,939.47	11,800	1,404	
	5800	Striper, truck mounted, 120 gallon paint, 460 H.P.		49.24	350	1,046	3,150	603.10	
	5900	Thermal paint heating kettle, 115 gallons		7.78	75	225	675	107.25	
	6000	Tar kettle, 330 gallon, trailer mounted		12.40	96	287.50	865	156.70	
	7000	Tunnel locomotive, diesel, 8 to 12 ton		30.11	625	1,881.35	5,650	617.20	
	7005	Electric, 10 ton		29.60	715	2,142.65	6,425	665.30	
	7010	Muck cars, 1/2 C.Y. capacity		2.33	27	81	243	34.80	
	7020	1 C.Y. capacity		2.54	35	105.56	315	41.45	
	7030	2 C.Y. capacity		2.69	39.50	118.11	355	45.10	
	7040	Side dump, 2 C.Y. capacity		2.90	49	146.33	440	52.50	
	7050	3 C.Y. capacity		3.90	53.50	160.96	485	63.40	
	7060	5 C.Y. capacity		5.69	69.50	207.99	625	87.10	
	7100	Ventilating blower for tunnel, 7-1/2 H.P.		2.16	53.50	159.91	480	49.30	
	7110	10 H.P.		2.45	55.50	167.23	500	53.05	
	7120	20 H.P.		3.58	72.50	217.40	650	72.15	
	7140	40 H.P.		6.21	96	287.43	860	107.20	
	7160	60 H.P.		8.79	103	308.33	925	132	
	7175	75 H.P.		10.49	160	480.79	1,450	180.10	
	7180	200 H.P.		21.03	315	945.90	2,850	357.45	
	7800	Windrow loader, elevating		54.49	1,700	5,125	15,400	1,461	
60	0010	**LIFTING AND HOISTING EQUIPMENT RENTAL** without operators	R015433 -10						**60**
	0150	Crane, flatbed mounted, 3 ton capacity	Ea.	14.58	203	610.30	1,825	238.75	
	0200	Crane, climbing, 106' jib, 6000 lb. capacity, 410 fpm	R312316 -45	40.13	2,675	8,034	24,100	1,928	
	0300	101' jib, 10,250 lb. capacity, 270 fpm		46.90	2,300	6,868.82	20,600	1,749	
	0500	Tower, static, 130' high, 106' jib, 6200 lb. capacity at 400 fpm		45.61	2,300	6,916	20,700	1,748	
	0520	Mini crawler spider crane, up to 24" wide, 1990 lb. lifting capacity		12.65	555	1,672.31	5,025	435.70	
	0525	Up to 30" wide, 6450 lb. lifting capacity		14.70	660	1,985.87	5,950	514.75	
	0530	Up to 52" wide, 6680 lb. lifting capacity		23.38	810	2,430.08	7,300	673.05	

01 54 33 | Equipment Rental

		UNIT	HOURLY OPER. COST	RENT PER DAY	RENT PER WEEK	RENT PER MONTH	EQUIPMENT COST/DAY
0535	Up to 55" wide, 8920 lb. lifting capacity	Ea.	26.10	895	2,691.37	8,075	747.10
0540	Up to 66" wide, 13,350 lb. lifting capacity		35.34	1,400	4,180.77	12,500	1,119
0600	Crawler mounted, lattice boom, 1/2 C.Y., 15 tons at 12' radius		37.34	855	2,558	7,675	810.30
0700	3/4 C.Y., 20 tons at 12' radius		54.72	960	2,874	8,625	1,013
0800	1 C.Y., 25 tons at 12' radius		68.10	1,025	3,038	9,125	1,152
0900	1-1/2 C.Y., 40 tons at 12' radius		66.98	1,150	3,476	10,400	1,231
1000	2 C.Y., 50 tons at 12' radius		89.66	1,375	4,120	12,400	1,541
1100	3 C.Y., 75 tons at 12' radius		65.65	2,400	7,210	21,600	1,967
1200	100 ton capacity, 60' boom		86.78	2,700	8,080.96	24,200	2,310
1300	165 ton capacity, 60' boom		107.18	3,025	9,091.08	27,300	2,676
1400	200 ton capacity, 70' boom		139.61	3,875	11,616.38	34,800	3,440
1500	350 ton capacity, 80' boom		184.04	4,200	12,626.50	37,900	3,998
1600	Truck mounted, lattice boom, 6 x 4, 20 tons at 10' radius		40.16	2,000	6,026	18,100	1,526
1700	25 tons at 10' radius		43.16	2,400	7,210	21,600	1,787
1800	8 x 4, 30 tons at 10' radius		54.71	2,575	7,725	23,200	1,983
1900	40 tons at 12' radius		54.71	2,825	8,446	25,300	2,127
2000	60 tons at 15' radius		54.23	1,675	5,000.09	15,000	1,434
2050	82 tons at 15' radius		60.03	1,800	5,404.14	16,200	1,561
2100	90 tons at 15' radius		67.06	1,950	5,883.95	17,700	1,713
2200	115 tons at 15' radius		75.66	2,200	6,591.03	19,800	1,923
2300	150 tons at 18' radius		81.91	2,775	8,343	25,000	2,324
2350	165 tons at 18' radius		87.91	2,825	8,498	25,500	2,403
2400	Truck mounted, hydraulic, 12 ton capacity		29.80	395	1,186.89	3,550	475.80
2500	25 ton capacity		36.72	490	1,464.67	4,400	586.70
2550	33 ton capacity		54.71	910	2,727.32	8,175	983.15
2560	40 ton capacity		54.71	910	2,727.32	8,175	983.15
2600	55 ton capacity		54.32	925	2,777.83	8,325	990.15
2700	80 ton capacity		71.13	1,475	4,444.53	13,300	1,458
2720	100 ton capacity		75.72	1,575	4,722.31	14,200	1,550
2740	120 ton capacity		106.34	1,850	5,555.66	16,700	1,962
2760	150 ton capacity		113.80	2,050	6,186.99	18,600	2,148
2800	Self-propelled, 4 x 4, with telescoping boom, 5 ton		15.29	435	1,298	3,900	381.95
2900	12-1/2 ton capacity		21.63	435	1,298	3,900	432.65
3000	15 ton capacity		34.77	455	1,363.66	4,100	550.85
3050	20 ton capacity		22.55	655	1,969.73	5,900	574.35
3100	25 ton capacity		37.06	1,425	4,293.01	12,900	1,155
3150	40 ton capacity		45.35	665	1,994.99	5,975	761.80
3200	Derricks, guy, 20 ton capacity, 60' boom, 75' mast		22.97	1,450	4,378	13,100	1,059
3300	100' boom, 115' mast		36.41	2,050	6,180	18,500	1,527
3400	Stiffleg, 20 ton capacity, 70' boom, 37' mast		25.66	635	1,906	5,725	586.50
3500	100' boom, 47' mast		39.71	685	2,060	6,175	729.70
3550	Helicopter, small, lift to 1250 lb. maximum, w/pilot		100.14	2,175	6,500.12	19,500	2,101
3600	Hoists, chain type, overhead, manual, 3/4 ton		.15	10.25	30.70	92	7.30
3900	10 ton		.80	6.25	18.78	56.50	10.10
4000	Hoist and tower, 5000 lb. cap., portable electric, 40' high		4.82	146	438	1,325	126.15
4100	For each added 10' section, add		.12	32.50	98	294	20.55
4200	Hoist and single tubular tower, 5000 lb. electric, 100' high		7.03	108	325	975	121.20
4300	For each added 6'-6" section, add		.21	39.50	118	355	25.25
4400	Hoist and double tubular tower, 5000 lb., 100' high		7.65	108	325	975	126.20
4500	For each added 6'-6" section, add		.23	42.50	128	385	27.45
4550	Hoist and tower, mast type, 6000 lb., 100' high		9.32	95.50	286.87	860	131.95
4570	For each added 10' section, add		.13	32.50	98	294	20.65
4600	Hoist and tower, personnel, electric, 2000 lb., 100' @ 125 fpm		17.67	25.50	75.76	227	156.55
4700	3000 lb., 100' @ 200 fpm		20.22	25.50	75.76	227	176.95
4800	3000 lb., 150' @ 300 fpm		22.44	25.50	75.76	227	194.70
4900	4000 lb., 100' @ 300 fpm		23.22	25.50	75.76	227	200.90
5000	6000 lb., 100' @ 275 fpm		24.95	25.50	75.76	227	214.75
5100	For added heights up to 500', add	L.F.	.01	3.33	10	30	2.10

01 54 33 | Equipment Rental

			UNIT	HOURLY OPER. COST	RENT PER DAY	RENT PER WEEK	RENT PER MONTH	EQUIPMENT COST/DAY	
60	5200	Jacks, hydraulic, 20 ton	Ea.	.05	19.85	59.60	179	12.30	60
	5500	100 ton		.40	26	78.50	236	18.95	
	6100	Jacks, hydraulic, climbing w/50' jackrods, control console, 30 ton cap.		2.19	32	96	288	36.70	
	6150	For each added 10' jackrod section, add		.05	5	15	45	3.40	
	6300	50 ton capacity		3.52	34.50	103	310	48.75	
	6350	For each added 10' jackrod section, add		.06	5	15	45	3.50	
	6500	125 ton capacity		9.20	53.50	160	480	105.55	
	6550	For each added 10' jackrod section, add		.62	5	15	45	8	
	6600	Cable jack, 10 ton capacity with 200' cable		1.60	36.50	110	330	34.80	
	6650	For each added 50' of cable, add		.22	15.35	46	138	11	
70	0010	**WELLPOINT EQUIPMENT RENTAL** without operators							70
	0020	Based on 2 months rental							
	0100	Combination jetting & wellpoint pump, 60 H.P. diesel	Ea.	15.83	305	922	2,775	311.05	
	0200	High pressure gas jet pump, 200 H.P., 300 psi	"	34.17	278	833.35	2,500	440.05	
	0300	Discharge pipe, 8" diameter	L.F.	.01	1.41	4.24	12.75	.90	
	0350	12" diameter		.01	2.09	6.26	18.80	1.35	
	0400	Header pipe, flows up to 150 GPM, 4" diameter		.01	.74	2.22	6.65	.50	
	0500	400 GPM, 6" diameter		.01	1.08	3.23	9.70	.70	
	0600	800 GPM, 8" diameter		.01	1.41	4.24	12.75	.95	
	0700	1500 GPM, 10" diameter		.01	1.75	5.25	15.75	1.15	
	0800	2500 GPM, 12" diameter		.03	2.09	6.26	18.80	1.45	
	0900	4500 GPM, 16" diameter		.03	2.42	7.27	22	1.70	
	0950	For quick coupling aluminum and plastic pipe, add		.03	9.60	28.84	86.50	6.05	
	1100	Wellpoint, 25' long, with fittings & riser pipe, 1-1/2" or 2" diameter	Ea.	.07	133	399	1,200	80.35	
	1200	Wellpoint pump, diesel powered, 4" suction, 20 H.P.		7.07	152	454.55	1,375	147.50	
	1300	6" suction, 30 H.P.		9.49	140	420	1,250	159.90	
	1400	8" suction, 40 H.P.		12.85	253	757.59	2,275	254.35	
	1500	10" suction, 75 H.P.		18.75	268	803.05	2,400	310.60	
	1600	12" suction, 100 H.P.		27.51	300	904.06	2,700	400.90	
	1700	12" suction, 175 H.P.		39.37	320	959.61	2,875	506.90	
80	0010	**MARINE EQUIPMENT RENTAL** without operators	Ea.						80
	0200	Barge, 400 Ton, 30' wide x 90' long		17.84	1,200	3,632.05	10,900	869.15	
	0240	800 Ton, 45' wide x 90' long		22.41	1,500	4,468.20	13,400	1,073	
	2000	Tugboat, diesel, 100 H.P.		29.92	240	721.18	2,175	383.65	
	2040	250 H.P.		58.10	435	1,306.49	3,925	726.10	
	2080	380 H.P.		126.49	1,300	3,919.48	11,800	1,796	
	3000	Small work boat, gas, 16-foot, 50 H.P.		11.48	48.50	145.01	435	120.85	
	4000	Large, diesel, 48-foot, 200 H.P.		75.57	1,375	4,154.64	12,500	1,436	

R015433 -10

R015433 -10

Crews - Standard

Crew A-1

Crew A-1	Hr.	Daily	Hr.	Daily	Bare Costs	Incl. O&P
1 Building Laborer	$44.40	$355.20	$66.25	$530.00	$44.40	$66.25
1 Concrete Saw, Gas Manual		113.70		125.07	14.21	15.63
8 L.H., Daily Totals		$468.90		$655.07	$58.61	$81.88

Crew A-1A

Crew A-1A	Hr.	Daily	Hr.	Daily	Bare Costs	Incl. O&P
1 Skilled Worker	$57.10	$456.80	$85.90	$687.20	$57.10	$85.90
1 Shot Blaster, 20"		208.70		229.57	26.09	28.70
8 L.H., Daily Totals		$665.50		$916.77	$83.19	$114.60

Crew A-1B

Crew A-1B	Hr.	Daily	Hr.	Daily	Bare Costs	Incl. O&P
1 Building Laborer	$44.40	$355.20	$66.25	$530.00	$44.40	$66.25
1 Concrete Saw		112.85		124.14	14.11	15.52
8 L.H., Daily Totals		$468.05		$654.13	$58.51	$81.77

Crew A-1C

Crew A-1C	Hr.	Daily	Hr.	Daily	Bare Costs	Incl. O&P
1 Building Laborer	$44.40	$355.20	$66.25	$530.00	$44.40	$66.25
1 Chain Saw, Gas, 18"		52.20		57.42	6.53	7.18
8 L.H., Daily Totals		$407.40		$587.42	$50.92	$73.43

Crew A-1D

Crew A-1D	Hr.	Daily	Hr.	Daily	Bare Costs	Incl. O&P
1 Building Laborer	$44.40	$355.20	$66.25	$530.00	$44.40	$66.25
1 Vibrating Plate, Gas, 18"		31.90		35.09	3.99	4.39
8 L.H., Daily Totals		$387.10		$565.09	$48.39	$70.64

Crew A-1E

Crew A-1E	Hr.	Daily	Hr.	Daily	Bare Costs	Incl. O&P
1 Building Laborer	$44.40	$355.20	$66.25	$530.00	$44.40	$66.25
1 Vibrating Plate, Gas, 21"		165.60		182.16	20.70	22.77
8 L.H., Daily Totals		$520.80		$712.16	$65.10	$89.02

Crew A-1F

Crew A-1F	Hr.	Daily	Hr.	Daily	Bare Costs	Incl. O&P
1 Building Laborer	$44.40	$355.20	$66.25	$530.00	$44.40	$66.25
1 Rammer/Tamper, Gas, 8"		47.00		51.70	5.88	6.46
8 L.H., Daily Totals		$402.20		$581.70	$50.27	$72.71

Crew A-1G

Crew A-1G	Hr.	Daily	Hr.	Daily	Bare Costs	Incl. O&P
1 Building Laborer	$44.40	$355.20	$66.25	$530.00	$44.40	$66.25
1 Rammer/Tamper, Gas, 15"		54.65		60.12	6.83	7.51
8 L.H., Daily Totals		$409.85		$590.12	$51.23	$73.76

Crew A-1H

Crew A-1H	Hr.	Daily	Hr.	Daily	Bare Costs	Incl. O&P
1 Building Laborer	$44.40	$355.20	$66.25	$530.00	$44.40	$66.25
1 Exterior Steam Cleaner		77.20		84.92	9.65	10.62
8 L.H., Daily Totals		$432.40		$614.92	$54.05	$76.86

Crew A-1J

Crew A-1J	Hr.	Daily	Hr.	Daily	Bare Costs	Incl. O&P
1 Building Laborer	$44.40	$355.20	$66.25	$530.00	$44.40	$66.25
1 Cultivator, Walk-Behind, 5 H.P.		53.25		58.58	6.66	7.32
8 L.H., Daily Totals		$408.45		$588.58	$51.06	$73.57

Crew A-1K

Crew A-1K	Hr.	Daily	Hr.	Daily	Bare Costs	Incl. O&P
1 Building Laborer	$44.40	$355.20	$66.25	$530.00	$44.40	$66.25
1 Cultivator, Walk-Behind, 8 H.P.		92.20		101.42	11.53	12.68
8 L.H., Daily Totals		$447.40		$631.42	$55.92	$78.93

Crew A-1M

Crew A-1M	Hr.	Daily	Hr.	Daily	Bare Costs	Incl. O&P
1 Building Laborer	$44.40	$355.20	$66.25	$530.00	$44.40	$66.25
1 Snow Blower, Walk-Behind		68.30		75.13	8.54	9.39
8 L.H., Daily Totals		$423.50		$605.13	$52.94	$75.64

Crew A-2

Crew A-2	Hr.	Daily	Hr.	Daily	Bare Costs	Incl. O&P
2 Laborers	$44.40	$710.40	$66.25	$1060.00	$45.87	$68.50
1 Truck Driver (light)	48.80	390.40	73.00	584.00		
1 Flatbed Truck, Gas, 1.5 Ton		198.50		218.35	8.27	9.10
24 L.H., Daily Totals		$1299.30		$1862.35	$54.14	$77.60

Crew A-2A

Crew A-2A	Hr.	Daily	Hr.	Daily	Bare Costs	Incl. O&P
2 Laborers	$44.40	$710.40	$66.25	$1060.00	$45.87	$68.50
1 Truck Driver (light)	48.80	390.40	73.00	584.00		
1 Flatbed Truck, Gas, 1.5 Ton		198.50		218.35		
1 Concrete Saw		112.85		124.14	12.97	14.27
24 L.H., Daily Totals		$1412.15		$1986.48	$58.84	$82.77

Crew A-2B

Crew A-2B	Hr.	Daily	Hr.	Daily	Bare Costs	Incl. O&P
1 Truck Driver (light)	$48.80	$390.40	$73.00	$584.00	$48.80	$73.00
1 Flatbed Truck, Gas, 1.5 Ton		198.50		218.35	24.81	27.29
8 L.H., Daily Totals		$588.90		$802.35	$73.61	$100.29

Crew A-3A

Crew A-3A	Hr.	Daily	Hr.	Daily	Bare Costs	Incl. O&P
1 Equip. Oper. (light)	$55.50	$444.00	$82.70	$661.60	$55.50	$82.70
1 Pickup Truck, 4x4, 3/4 Ton		176.75		194.43	22.09	24.30
8 L.H., Daily Totals		$620.75		$856.02	$77.59	$107.00

Crew A-3B

Crew A-3B	Hr.	Daily	Hr.	Daily	Bare Costs	Incl. O&P
1 Equip. Oper. (medium)	$59.00	$472.00	$87.90	$703.20	$55.15	$82.30
1 Truck Driver (heavy)	51.30	410.40	76.70	613.60		
1 Dump Truck, 12 C.Y., 400 H.P.		579.35		637.28		
1 F.E. Loader, W.M., 2.5 C.Y.		638.30		702.13	76.10	83.71
16 L.H., Daily Totals		$2100.05		$2656.22	$131.25	$166.01

Crew A-3C

Crew A-3C	Hr.	Daily	Hr.	Daily	Bare Costs	Incl. O&P
1 Equip. Oper. (light)	$55.50	$444.00	$82.70	$661.60	$55.50	$82.70
1 Loader, Skid Steer, 78 H.P.		446.30		490.93	55.79	61.37
8 L.H., Daily Totals		$890.30		$1152.53	$111.29	$144.07

Crew A-3D

Crew A-3D	Hr.	Daily	Hr.	Daily	Bare Costs	Incl. O&P
1 Truck Driver (light)	$48.80	$390.40	$73.00	$584.00	$48.80	$73.00
1 Pickup Truck, 4x4, 3/4 Ton		176.75		194.43		
1 Flatbed Trailer, 25 Ton		137.20		150.92	39.24	43.17
8 L.H., Daily Totals		$704.35		$929.35	$88.04	$116.17

Crew A-3E

Crew A-3E	Hr.	Daily	Hr.	Daily	Bare Costs	Incl. O&P
1 Equip. Oper. (crane)	$61.45	$491.60	$91.55	$732.40	$56.38	$84.13
1 Truck Driver (heavy)	51.30	410.40	76.70	613.60		
1 Pickup Truck, 4x4, 3/4 Ton		176.75		194.43	11.05	12.15
16 L.H., Daily Totals		$1078.75		$1540.43	$67.42	$96.28

Crew A-3F

Crew A-3F	Hr.	Daily	Hr.	Daily	Bare Costs	Incl. O&P
1 Equip. Oper. (crane)	$61.45	$491.60	$91.55	$732.40	$56.38	$84.13
1 Truck Driver (heavy)	51.30	410.40	76.70	613.60		
1 Pickup Truck, 4x4, 3/4 Ton		176.75		194.43		
1 Truck Tractor, 6x4, 380 H.P.		499.15		549.07		
1 Lowbed Trailer, 75 Ton		258.10		283.91	58.38	64.21
16 L.H., Daily Totals		$1836.00		$2373.40	$114.75	$148.34

For customer support on your Site Work & Landscape Costs with RSMeans Data, call 800.448.8182.

747

Crew No.	Bare Costs		Incl. Subs O&P		Cost Per Labor-Hour	
	Hr.	Daily	Hr.	Daily	Bare Costs	Incl. O&P
Crew A-3G						
1 Equip. Oper. (crane)	$61.45	$491.60	$91.55	$732.40	$56.38	$84.13
1 Truck Driver (heavy)	51.30	410.40	76.70	613.60		
1 Pickup Truck, 4x4, 3/4 Ton		176.75		194.43		
1 Truck Tractor, 6x4, 450 H.P.		608.95		669.85		
1 Lowbed Trailer, 75 Ton		258.10		283.91	65.24	71.76
16 L.H., Daily Totals		$1945.80		$2494.18	$121.61	$155.89
Crew A-3H	Hr.	Daily	Hr.	Daily	Bare Costs	Incl. O&P
1 Equip. Oper. (crane)	$61.45	$491.60	$91.55	$732.40	$61.45	$91.55
1 Hyd. Crane, 12 Ton (Daily)		733.15		806.47	91.64	100.81
8 L.H., Daily Totals		$1224.75		$1538.87	$153.09	$192.36
Crew A-3I	Hr.	Daily	Hr.	Daily	Bare Costs	Incl. O&P
1 Equip. Oper. (crane)	$61.45	$491.60	$91.55	$732.40	$61.45	$91.55
1 Hyd. Crane, 25 Ton (Daily)		810.50		891.55	101.31	111.44
8 L.H., Daily Totals		$1302.10		$1623.95	$162.76	$202.99
Crew A-3J	Hr.	Daily	Hr.	Daily	Bare Costs	Incl. O&P
1 Equip. Oper. (crane)	$61.45	$491.60	$91.55	$732.40	$61.45	$91.55
1 Hyd. Crane, 40 Ton (Daily)		1287.00		1415.70	160.88	176.96
8 L.H., Daily Totals		$1778.60		$2148.10	$222.32	$268.51
Crew A-3K	Hr.	Daily	Hr.	Daily	Bare Costs	Incl. O&P
1 Equip. Oper. (crane)	$61.45	$491.60	$91.55	$732.40	$56.98	$84.90
1 Equip. Oper. (oiler)	52.50	420.00	78.25	626.00		
1 Hyd. Crane, 55 Ton (Daily)		1377.00		1514.70		
1 P/U Truck, 3/4 Ton (Daily)		143.85		158.24	95.05	104.56
16 L.H., Daily Totals		$2432.45		$3031.34	$152.03	$189.46
Crew A-3L	Hr.	Daily	Hr.	Daily	Bare Costs	Incl. O&P
1 Equip. Oper. (crane)	$61.45	$491.60	$91.55	$732.40	$56.98	$84.90
1 Equip. Oper. (oiler)	52.50	420.00	78.25	626.00		
1 Hyd. Crane, 80 Ton (Daily)		2058.00		2263.80		
1 P/U Truck, 3/4 Ton (Daily)		143.85		158.24	137.62	151.38
16 L.H., Daily Totals		$3113.45		$3780.43	$194.59	$236.28
Crew A-3M	Hr.	Daily	Hr.	Daily	Bare Costs	Incl. O&P
1 Equip. Oper. (crane)	$61.45	$491.60	$91.55	$732.40	$56.98	$84.90
1 Equip. Oper. (oiler)	52.50	420.00	78.25	626.00		
1 Hyd. Crane, 100 Ton (Daily)		2253.00		2478.30		
1 P/U Truck, 3/4 Ton (Daily)		143.85		158.24	149.80	164.78
16 L.H., Daily Totals		$3308.45		$3994.93	$206.78	$249.68
Crew A-3N	Hr.	Daily	Hr.	Daily	Bare Costs	Incl. O&P
1 Equip. Oper. (crane)	$61.45	$491.60	$91.55	$732.40	$61.45	$91.55
1 Tower Crane (monthly)		1737.00		1910.70	217.13	238.84
8 L.H., Daily Totals		$2228.60		$2643.10	$278.57	$330.39
Crew A-3P	Hr.	Daily	Hr.	Daily	Bare Costs	Incl. O&P
1 Equip. Oper. (light)	$55.50	$444.00	$82.70	$661.60	$55.50	$82.70
1 A.T. Forklift, 31' reach, 45' lift		346.45		381.10	43.31	47.64
8 L.H., Daily Totals		$790.45		$1042.69	$98.81	$130.34
Crew A-3Q	Hr.	Daily	Hr.	Daily	Bare Costs	Incl. O&P
1 Equip. Oper. (light)	$55.50	$444.00	$82.70	$661.60	$55.50	$82.70
1 Pickup Truck, 4x4, 3/4 Ton		176.75		194.43		
1 Flatbed Trailer, 3 Ton		71.15		78.27	30.99	34.09
8 L.H., Daily Totals		$691.90		$934.29	$86.49	$116.79

Crew No.	Bare Costs		Incl. Subs O&P		Cost Per Labor-Hour	
	Hr.	Daily	Hr.	Daily	Bare Costs	Incl. O&P
Crew A-3R	Hr.	Daily	Hr.	Daily	Bare Costs	Incl. O&P
1 Equip. Oper. (light)	$55.50	$444.00	$82.70	$661.60	$55.50	$82.70
1 Forklift, Smooth Floor, 8,000 Lb.		283.25		311.57	35.41	38.95
8 L.H., Daily Totals		$727.25		$973.17	$90.91	$121.65
Crew A-4	Hr.	Daily	Hr.	Daily	Bare Costs	Incl. O&P
2 Carpenters	$54.70	$875.20	$81.65	$1306.40	$51.95	$77.40
1 Painter, Ordinary	46.45	371.60	68.90	551.20		
24 L.H., Daily Totals		$1246.80		$1857.60	$51.95	$77.40
Crew A-5	Hr.	Daily	Hr.	Daily	Bare Costs	Incl. O&P
2 Laborers	$44.40	$710.40	$66.25	$1060.00	$44.89	$67.00
.25 Truck Driver (light)	48.80	97.60	73.00	146.00		
.25 Flatbed Truck, Gas, 1.5 Ton		49.63		54.59	2.76	3.03
18 L.H., Daily Totals		$857.63		$1260.59	$47.65	$70.03
Crew A-6	Hr.	Daily	Hr.	Daily	Bare Costs	Incl. O&P
1 Instrument Man	$57.10	$456.80	$85.90	$687.20	$54.88	$82.15
1 Rodman/Chainman	52.65	421.20	78.40	627.20		
1 Level, Electronic		34.40		37.84	2.15	2.37
16 L.H., Daily Totals		$912.40		$1352.24	$57.02	$84.52
Crew A-7	Hr.	Daily	Hr.	Daily	Bare Costs	Incl. O&P
1 Chief of Party	$68.50	$548.00	$102.35	$818.80	$59.42	$88.88
1 Instrument Man	57.10	456.80	85.90	687.20		
1 Rodman/Chainman	52.65	421.20	78.40	627.20		
1 Level, Electronic		34.40		37.84	1.43	1.58
24 L.H., Daily Totals		$1460.40		$2171.04	$60.85	$90.46
Crew A-8	Hr.	Daily	Hr.	Daily	Bare Costs	Incl. O&P
1 Chief of Party	$68.50	$548.00	$102.35	$818.80	$57.73	$86.26
1 Instrument Man	57.10	456.80	85.90	687.20		
2 Rodmen/Chainmen	52.65	842.40	78.40	1254.40		
1 Level, Electronic		34.40		37.84	1.08	1.18
32 L.H., Daily Totals		$1881.60		$2798.24	$58.80	$87.44
Crew A-9	Hr.	Daily	Hr.	Daily	Bare Costs	Incl. O&P
1 Asbestos Foreman	$61.45	$491.60	$94.00	$752.00	$61.01	$93.34
7 Asbestos Workers	60.95	3413.20	93.25	5222.00		
64 L.H., Daily Totals		$3904.80		$5974.00	$61.01	$93.34
Crew A-10A	Hr.	Daily	Hr.	Daily	Bare Costs	Incl. O&P
1 Asbestos Foreman	$61.45	$491.60	$94.00	$752.00	$61.12	$93.50
2 Asbestos Workers	60.95	975.20	93.25	1492.00		
24 L.H., Daily Totals		$1466.80		$2244.00	$61.12	$93.50
Crew A-10B	Hr.	Daily	Hr.	Daily	Bare Costs	Incl. O&P
1 Asbestos Foreman	$61.45	$491.60	$94.00	$752.00	$61.08	$93.44
3 Asbestos Workers	60.95	1462.80	93.25	2238.00		
32 L.H., Daily Totals		$1954.40		$2990.00	$61.08	$93.44
Crew A-10C	Hr.	Daily	Hr.	Daily	Bare Costs	Incl. O&P
3 Asbestos Workers	$60.95	$1462.80	$93.25	$2238.00	$60.95	$93.25
1 Flatbed Truck, Gas, 1.5 Ton		198.50		218.35	8.27	9.10
24 L.H., Daily Totals		$1661.30		$2456.35	$69.22	$102.35

Crews - Standard

Crew No.	Bare Costs Hr.	Daily	Incl. Subs O&P Hr.	Daily	Cost Per Labor-Hour Bare Costs	Incl. O&P
Crew A-10D						
2 Asbestos Workers	$60.95	$975.20	$93.25	$1492.00	$58.96	$89.08
1 Equip. Oper. (crane)	61.45	491.60	91.55	732.40		
1 Equip. Oper. (oiler)	52.50	420.00	78.25	626.00		
1 Hydraulic Crane, 33 Ton		983.15		1081.46	30.72	33.80
32 L.H., Daily Totals		$2869.95		$3931.86	$89.69	$122.87
Crew A-11						
1 Asbestos Foreman	$61.45	$491.60	$94.00	$752.00	$61.01	$93.34
7 Asbestos Workers	60.95	3413.20	93.25	5222.00		
2 Chip. Hammers, 12 Lb., Elec.		65.70		72.27	1.03	1.13
64 L.H., Daily Totals		$3970.50		$6046.27	$62.04	$94.47
Crew A-12						
1 Asbestos Foreman	$61.45	$491.60	$94.00	$752.00	$61.01	$93.34
7 Asbestos Workers	60.95	3413.20	93.25	5222.00		
1 Trk-Mtd Vac, 14 CY, 1500 Gal.		542.60		596.86		
1 Flatbed Truck, 20,000 GVW		204.05		224.46	11.67	12.83
64 L.H., Daily Totals		$4651.45		$6795.31	$72.68	$106.18
Crew A-13						
1 Equip. Oper. (light)	$55.50	$444.00	$82.70	$661.60	$55.50	$82.70
1 Trk-Mtd Vac, 14 CY, 1500 Gal.		542.60		596.86		
1 Flatbed Truck, 20,000 GVW		204.05		224.46	93.33	102.66
8 L.H., Daily Totals		$1190.65		$1482.92	$148.83	$185.36
Crew B-1						
1 Labor Foreman (outside)	$46.40	$371.20	$69.25	$554.00	$45.07	$67.25
2 Laborers	44.40	710.40	66.25	1060.00		
24 L.H., Daily Totals		$1081.60		$1614.00	$45.07	$67.25
Crew B-1A						
1 Labor Foreman (outside)	$46.40	$371.20	$69.25	$554.00	$45.07	$67.25
2 Laborers	44.40	710.40	66.25	1060.00		
2 Cutting Torches		25.90		28.49		
2 Sets of Gases		347.20		381.92	15.55	17.10
24 L.H., Daily Totals		$1454.70		$2024.41	$60.61	$84.35
Crew B-1B						
1 Labor Foreman (outside)	$46.40	$371.20	$69.25	$554.00	$49.16	$73.33
2 Laborers	44.40	710.40	66.25	1060.00		
1 Equip. Oper. (crane)	61.45	491.60	91.55	732.40		
2 Cutting Torches		25.90		28.49		
2 Sets of Gases		347.20		381.92		
1 Hyd. Crane, 12 Ton		475.80		523.38	26.53	29.18
32 L.H., Daily Totals		$2422.10		$3280.19	$75.69	$102.51
Crew B-1C						
1 Labor Foreman (outside)	$46.40	$371.20	$69.25	$554.00	$45.07	$67.25
2 Laborers	44.40	710.40	66.25	1060.00		
1 Telescoping Boom Lift, to 60'		292.45		321.69	12.19	13.40
24 L.H., Daily Totals		$1374.05		$1935.69	$57.25	$80.65
Crew B-1D						
2 Laborers	$44.40	$710.40	$66.25	$1060.00	$44.40	$66.25
1 Small Work Boat, Gas, 50 H.P.		120.85		132.94		
1 Pressure Washer, 7 GPM		93.95		103.35	13.43	14.77
16 L.H., Daily Totals		$925.20		$1296.28	$57.83	$81.02

Crew No.	Bare Costs Hr.	Daily	Incl. Subs O&P Hr.	Daily	Cost Per Labor-Hour Bare Costs	Incl. O&P
Crew B-1E						
1 Labor Foreman (outside)	$46.40	$371.20	$69.25	$554.00	$44.90	$67.00
3 Laborers	44.40	1065.60	66.25	1590.00		
1 Work Boat, Diesel, 200 H.P.		1436.00		1579.60		
2 Pressure Washers, 7 GPM		187.90		206.69	50.75	55.82
32 L.H., Daily Totals		$3060.70		$3930.29	$95.65	$122.82
Crew B-1F						
2 Skilled Workers	$57.10	$913.60	$85.90	$1374.40	$52.87	$79.35
1 Laborer	44.40	355.20	66.25	530.00		
1 Small Work Boat, Gas, 50 H.P.		120.85		132.94		
1 Pressure Washer, 7 GPM		93.95		103.35	8.95	9.85
24 L.H., Daily Totals		$1483.60		$2140.68	$61.82	$89.19
Crew B-1G						
2 Laborers	$44.40	$710.40	$66.25	$1060.00	$44.40	$66.25
1 Small Work Boat, Gas, 50 H.P.		120.85		132.94	7.55	8.31
16 L.H., Daily Totals		$831.25		$1192.93	$51.95	$74.56
Crew B-1H						
2 Skilled Workers	$57.10	$913.60	$85.90	$1374.40	$52.87	$79.35
1 Laborer	44.40	355.20	66.25	530.00		
1 Small Work Boat, Gas, 50 H.P.		120.85		132.94	5.04	5.54
24 L.H., Daily Totals		$1389.65		$2037.34	$57.90	$84.89
Crew B-1J						
1 Labor Foreman (inside)	$44.90	$359.20	$67.00	$536.00	$44.65	$66.63
1 Laborer	44.40	355.20	66.25	530.00		
16 L.H., Daily Totals		$714.40		$1066.00	$44.65	$66.63
Crew B-1K						
1 Carpenter Foreman (inside)	$55.20	$441.60	$82.40	$659.20	$54.95	$82.03
1 Carpenter	54.70	437.60	81.65	653.20		
16 L.H., Daily Totals		$879.20		$1312.40	$54.95	$82.03
Crew B-2						
1 Labor Foreman (outside)	$46.40	$371.20	$69.25	$554.00	$44.80	$66.85
4 Laborers	44.40	1420.80	66.25	2120.00		
40 L.H., Daily Totals		$1792.00		$2674.00	$44.80	$66.85
Crew B-2A						
1 Labor Foreman (outside)	$46.40	$371.20	$69.25	$554.00	$45.07	$67.25
2 Laborers	44.40	710.40	66.25	1060.00		
1 Telescoping Boom Lift, to 60'		292.45		321.69	12.19	13.40
24 L.H., Daily Totals		$1374.05		$1935.69	$57.25	$80.65
Crew B-3						
1 Labor Foreman (outside)	$46.40	$371.20	$69.25	$554.00	$49.47	$73.84
2 Laborers	44.40	710.40	66.25	1060.00		
1 Equip. Oper. (medium)	59.00	472.00	87.90	703.20		
2 Truck Drivers (heavy)	51.30	820.80	76.70	1227.20		
1 Crawler Loader, 3 C.Y.		1146.00		1260.60		
2 Dump Trucks, 12 C.Y., 400 H.P.		1158.70		1274.57	48.01	52.82
48 L.H., Daily Totals		$4679.10		$6079.57	$97.48	$126.66
Crew B-3A						
4 Laborers	$44.40	$1420.80	$66.25	$2120.00	$47.32	$70.58
1 Equip. Oper. (medium)	59.00	472.00	87.90	703.20		
1 Hyd. Excavator, 1.5 C.Y.		695.80		765.38	17.40	19.13
40 L.H., Daily Totals		$2588.60		$3588.58	$64.72	$89.71

For customer support on your Site Work & Landscape Costs with RSMeans Data, call 800.448.8182.

749

Crew B-3B

Crew No.	Bare Costs Hr.	Bare Costs Daily	Incl. Subs O&P Hr.	Incl. Subs O&P Daily	Cost Per Labor-Hour Bare Costs	Cost Per Labor-Hour Incl. O&P
2 Laborers	$44.40	$710.40	$66.25	$1060.00	$49.77	$74.28
1 Equip. Oper. (medium)	59.00	472.00	87.90	703.20		
1 Truck Driver (heavy)	51.30	410.40	76.70	613.60		
1 Backhoe Loader, 80 H.P.		235.05		258.56		
1 Dump Truck, 12 C.Y., 400 H.P.		579.35		637.28	25.45	28.00
32 L.H., Daily Totals		$2407.20		$3272.64	$75.22	$102.27

Crew B-3C

Crew No.	Bare Costs Hr.	Bare Costs Daily	Incl. Subs O&P Hr.	Incl. Subs O&P Daily	Cost Per Labor-Hour Bare Costs	Cost Per Labor-Hour Incl. O&P
3 Laborers	$44.40	$1065.60	$66.25	$1590.00	$48.05	$71.66
1 Equip. Oper. (medium)	59.00	472.00	87.90	703.20		
1 Crawler Loader, 4 C.Y.		1456.00		1601.60	45.50	50.05
32 L.H., Daily Totals		$2993.60		$3894.80	$93.55	$121.71

Crew B-4

Crew No.	Bare Costs Hr.	Bare Costs Daily	Incl. Subs O&P Hr.	Incl. Subs O&P Daily	Cost Per Labor-Hour Bare Costs	Cost Per Labor-Hour Incl. O&P
1 Labor Foreman (outside)	$46.40	$371.20	$69.25	$554.00	$45.88	$68.49
4 Laborers	44.40	1420.80	66.25	2120.00		
1 Truck Driver (heavy)	51.30	410.40	76.70	613.60		
1 Truck Tractor, 220 H.P.		310.80		341.88		
1 Flatbed Trailer, 40 Ton		188.45		207.29	10.40	11.44
48 L.H., Daily Totals		$2701.65		$3836.78	$56.28	$79.93

Crew B-5

Crew No.	Bare Costs Hr.	Bare Costs Daily	Incl. Subs O&P Hr.	Incl. Subs O&P Daily	Cost Per Labor-Hour Bare Costs	Cost Per Labor-Hour Incl. O&P
1 Labor Foreman (outside)	$46.40	$371.20	$69.25	$554.00	$48.86	$72.86
4 Laborers	44.40	1420.80	66.25	2120.00		
2 Equip. Oper. (medium)	59.00	944.00	87.90	1406.40		
1 Air Compressor, 250 cfm		202.85		223.13		
2 Breakers, Pavement, 60 lb.		107.20		117.92		
2 -50' Air Hoses, 1.5"		45.60		50.16		
1 Crawler Loader, 3 C.Y.		1146.00		1260.60	26.82	29.50
56 L.H., Daily Totals		$4237.65		$5732.22	$75.67	$102.36

Crew B-5A

Crew No.	Bare Costs Hr.	Bare Costs Daily	Incl. Subs O&P Hr.	Incl. Subs O&P Daily	Cost Per Labor-Hour Bare Costs	Cost Per Labor-Hour Incl. O&P
1 Labor Foreman (outside)	$46.40	$371.20	$69.25	$554.00	$49.08	$73.22
6 Laborers	44.40	2131.20	66.25	3180.00		
2 Equip. Oper. (medium)	59.00	944.00	87.90	1406.40		
1 Equip. Oper. (light)	55.50	444.00	82.70	661.60		
2 Truck Drivers (heavy)	51.30	820.80	76.70	1227.20		
1 Air Compressor, 365 cfm		343.55		377.90		
2 Breakers, Pavement, 60 lb.		107.20		117.92		
8 -50' Air Hoses, 1"		64.40		70.84		
2 Dump Trucks, 8 C.Y., 220 H.P.		815.20		896.72	13.86	15.24
96 L.H., Daily Totals		$6041.55		$8492.58	$62.93	$88.46

Crew B-5B

Crew No.	Bare Costs Hr.	Bare Costs Daily	Incl. Subs O&P Hr.	Incl. Subs O&P Daily	Cost Per Labor-Hour Bare Costs	Cost Per Labor-Hour Incl. O&P
1 Powderman	$57.10	$456.80	$85.90	$687.20	$54.83	$81.97
2 Equip. Oper. (medium)	59.00	944.00	87.90	1406.40		
3 Truck Drivers (heavy)	51.30	1231.20	76.70	1840.80		
1 F.E. Loader, W.M., 2.5 C.Y.		638.30		702.13		
3 Dump Trucks, 12 C.Y., 400 H.P.		1738.05		1911.86		
1 Air Compressor, 365 cfm		343.55		377.90	56.66	62.33
48 L.H., Daily Totals		$5351.90		$6926.29	$111.50	$144.30

Crew B-5C

Crew No.	Bare Costs Hr.	Bare Costs Daily	Incl. Subs O&P Hr.	Incl. Subs O&P Daily	Cost Per Labor-Hour Bare Costs	Cost Per Labor-Hour Incl. O&P
3 Laborers	$44.40	$1065.60	$66.25	$1590.00	$51.09	$76.23
1 Equip. Oper. (medium)	59.00	472.00	87.90	703.20		
2 Truck Drivers (heavy)	51.30	820.80	76.70	1227.20		
1 Equip. Oper. (crane)	61.45	491.60	91.55	732.40		
1 Equip. Oper. (oiler)	52.50	420.00	78.25	626.00		
2 Dump Trucks, 12 C.Y., 400 H.P.		1158.70		1274.57		
1 Crawler Loader, 4 C.Y.		1456.00		1601.60		
1 S.P. Crane, 4x4, 25 Ton		1155.00		1270.50	58.90	64.79
64 L.H., Daily Totals		$7039.70		$9025.47	$110.00	$141.02

Crew B-5D

Crew No.	Bare Costs Hr.	Bare Costs Daily	Incl. Subs O&P Hr.	Incl. Subs O&P Daily	Cost Per Labor-Hour Bare Costs	Cost Per Labor-Hour Incl. O&P
1 Labor Foreman (outside)	$46.40	$371.20	$69.25	$554.00	$49.16	$73.34
4 Laborers	44.40	1420.80	66.25	2120.00		
2 Equip. Oper. (medium)	59.00	944.00	87.90	1406.40		
1 Truck Driver (heavy)	51.30	410.40	76.70	613.60		
1 Air Compressor, 250 cfm		202.85		223.13		
2 Breakers, Pavement, 60 lb.		107.20		117.92		
2 -50' Air Hoses, 1.5"		45.60		50.16		
1 Crawler Loader, 3 C.Y.		1146.00		1260.60		
1 Dump Truck, 12 C.Y., 400 H.P.		579.35		637.28	32.52	35.77
64 L.H., Daily Totals		$5227.40		$6983.10	$81.68	$109.11

Crew B-5E

Crew No.	Bare Costs Hr.	Bare Costs Daily	Incl. Subs O&P Hr.	Incl. Subs O&P Daily	Cost Per Labor-Hour Bare Costs	Cost Per Labor-Hour Incl. O&P
1 Labor Foreman (outside)	$46.40	$371.20	$69.25	$554.00	$49.16	$73.34
4 Laborers	44.40	1420.80	66.25	2120.00		
2 Equip. Oper. (medium)	59.00	944.00	87.90	1406.40		
1 Truck Driver (heavy)	51.30	410.40	76.70	613.60		
1 Water Tank Trailer, 5000 Gal.		154.10		169.51		
1 High Pressure Water Jet 40 KSI		820.75		902.83		
2 -50' Air Hoses, 1.5"		45.60		50.16		
1 Crawler Loader, 3 C.Y.		1146.00		1260.60		
1 Dump Truck, 12 C.Y., 400 H.P.		579.35		637.28	42.90	47.19
64 L.H., Daily Totals		$5892.20		$7714.38	$92.07	$120.54

Crew B-6

Crew No.	Bare Costs Hr.	Bare Costs Daily	Incl. Subs O&P Hr.	Incl. Subs O&P Daily	Cost Per Labor-Hour Bare Costs	Cost Per Labor-Hour Incl. O&P
2 Laborers	$44.40	$710.40	$66.25	$1060.00	$48.10	$71.73
1 Equip. Oper. (light)	55.50	444.00	82.70	661.60		
1 Backhoe Loader, 48 H.P.		216.20		237.82	9.01	9.91
24 L.H., Daily Totals		$1370.60		$1959.42	$57.11	$81.64

Crew B-6A

Crew No.	Bare Costs Hr.	Bare Costs Daily	Incl. Subs O&P Hr.	Incl. Subs O&P Daily	Cost Per Labor-Hour Bare Costs	Cost Per Labor-Hour Incl. O&P
.5 Labor Foreman (outside)	$46.40	$185.60	$69.25	$277.00	$50.64	$75.51
1 Laborer	44.40	355.20	66.25	530.00		
1 Equip. Oper. (medium)	59.00	472.00	87.90	703.20		
1 Vacuum Truck, 5000 Gal.		371.95		409.14	18.60	20.46
20 L.H., Daily Totals		$1384.75		$1919.35	$69.24	$95.97

Crew B-6B

Crew No.	Bare Costs Hr.	Bare Costs Daily	Incl. Subs O&P Hr.	Incl. Subs O&P Daily	Cost Per Labor-Hour Bare Costs	Cost Per Labor-Hour Incl. O&P
2 Labor Foremen (outside)	$46.40	$742.40	$69.25	$1108.00	$45.07	$67.25
4 Laborers	44.40	1420.80	66.25	2120.00		
1 S.P. Crane, 4x4, 5 Ton		381.95		420.14		
1 Flatbed Truck, Gas, 1.5 Ton		198.50		218.35		
1 Butt Fusion Mach., 4"-12" diam.		420.75		462.82	20.86	22.94
48 L.H., Daily Totals		$3164.40		$4329.32	$65.92	$90.19

Crew B-6C

Crew No.	Bare Costs Hr.	Bare Costs Daily	Incl. Subs O&P Hr.	Incl. Subs O&P Daily	Cost Per Labor-Hour Bare Costs	Cost Per Labor-Hour Incl. O&P
2 Labor Foremen (outside)	$46.40	$742.40	$69.25	$1108.00	$45.07	$67.25
4 Laborers	44.40	1420.80	66.25	2120.00		
1 S.P. Crane, 4x4, 12 Ton		432.65		475.92		
1 Flatbed Truck, Gas, 3 Ton		850.05		935.05		
1 Butt Fusion Mach., 8"-24" diam.		1086.00		1194.60	49.35	54.28
48 L.H., Daily Totals		$4531.90		$5833.57	$94.41	$121.53

Crew B-6D

Crew No.	Bare Costs Hr.	Bare Costs Daily	Incl. Subs O&P Hr.	Incl. Subs O&P Daily	Cost Per Labor-Hour Bare Costs	Cost Per Labor-Hour Incl. O&P
.5 Labor Foreman (outside)	$46.40	$185.60	$69.25	$277.00	$50.64	$75.51
1 Laborer	44.40	355.20	66.25	530.00		
1 Equip. Oper. (medium)	59.00	472.00	87.90	703.20		
1 Hydro Excavator, 12 C.Y.		1277.00		1404.70	63.85	70.23
20 L.H., Daily Totals		$2289.80		$2914.90	$114.49	$145.75

Crew B-7

Crew No.	Bare Costs Hr.	Bare Costs Daily	Incl. Subs O&P Hr.	Incl. Subs O&P Daily	Cost Per Labor-Hour Bare Costs	Cost Per Labor-Hour Incl. O&P
1 Labor Foreman (outside)	$46.40	$371.20	$69.25	$554.00	$47.17	$70.36
4 Laborers	44.40	1420.80	66.25	2120.00		
1 Equip. Oper. (medium)	59.00	472.00	87.90	703.20		
1 Brush Chipper, 12", 130 H.P.		366.05		402.65		
1 Crawler Loader, 3 C.Y.		1146.00		1260.60		
2 Chain Saws, Gas, 36" Long		83.30		91.63	33.24	36.56
48 L.H., Daily Totals		$3859.35		$5132.09	$80.40	$106.92

Crew B-7A

Crew No.	Bare Costs Hr.	Bare Costs Daily	Incl. Subs O&P Hr.	Incl. Subs O&P Daily	Cost Per Labor-Hour Bare Costs	Cost Per Labor-Hour Incl. O&P
2 Laborers	$44.40	$710.40	$66.25	$1060.00	$48.10	$71.73
1 Equip. Oper. (light)	55.50	444.00	82.70	661.60		
1 Rake w/Tractor		343.50		377.85		
2 Chain Saws, Gas, 18"		104.40		114.84	18.66	20.53
24 L.H., Daily Totals		$1602.30		$2214.29	$66.76	$92.26

Crew B-7B

Crew No.	Bare Costs Hr.	Bare Costs Daily	Incl. Subs O&P Hr.	Incl. Subs O&P Daily	Cost Per Labor-Hour Bare Costs	Cost Per Labor-Hour Incl. O&P
1 Labor Foreman (outside)	$46.40	$371.20	$69.25	$554.00	$47.76	$71.26
4 Laborers	44.40	1420.80	66.25	2120.00		
1 Equip. Oper. (medium)	59.00	472.00	87.90	703.20		
1 Truck Driver (heavy)	51.30	410.40	76.70	613.60		
1 Brush Chipper, 12", 130 H.P.		366.05		402.65		
1 Crawler Loader, 3 C.Y.		1146.00		1260.60		
2 Chain Saws, Gas, 36" Long		83.30		91.63		
1 Dump Truck, 8 C.Y., 220 H.P.		407.60		448.36	35.77	39.34
56 L.H., Daily Totals		$4677.35		$6194.05	$83.52	$110.61

Crew B-7C

Crew No.	Bare Costs Hr.	Bare Costs Daily	Incl. Subs O&P Hr.	Incl. Subs O&P Daily	Cost Per Labor-Hour Bare Costs	Cost Per Labor-Hour Incl. O&P
1 Labor Foreman (outside)	$46.40	$371.20	$69.25	$554.00	$47.76	$71.26
4 Laborers	44.40	1420.80	66.25	2120.00		
1 Equip. Oper. (medium)	59.00	472.00	87.90	703.20		
1 Truck Driver (heavy)	51.30	410.40	76.70	613.60		
1 Brush Chipper, 12", 130 H.P.		366.05		402.65		
1 Crawler Loader, 3 C.Y.		1146.00		1260.60		
2 Chain Saws, Gas, 36" Long		83.30		91.63		
1 Dump Truck, 12 C.Y., 400 H.P.		579.35		637.28	38.83	42.72
56 L.H., Daily Totals		$4849.10		$6382.97	$86.59	$113.98

Crew B-8

Crew No.	Bare Costs Hr.	Bare Costs Daily	Incl. Subs O&P Hr.	Incl. Subs O&P Daily	Cost Per Labor-Hour Bare Costs	Cost Per Labor-Hour Incl. O&P
1 Labor Foreman (outside)	$46.40	$371.20	$69.25	$554.00	$51.04	$76.15
2 Laborers	44.40	710.40	66.25	1060.00		
2 Equip. Oper. (medium)	59.00	944.00	87.90	1406.40		
1 Equip. Oper. (oiler)	52.50	420.00	78.25	626.00		
2 Truck Drivers (heavy)	51.30	820.80	76.70	1227.20		
1 Hyd. Crane, 25 Ton		586.70		645.37		
1 Crawler Loader, 3 C.Y.		1146.00		1260.60		
2 Dump Trucks, 12 C.Y., 400 H.P.		1158.70		1274.57	45.18	49.70
64 L.H., Daily Totals		$6157.80		$8054.14	$96.22	$125.85

Crew B-9

Crew No.	Bare Costs Hr.	Bare Costs Daily	Incl. Subs O&P Hr.	Incl. Subs O&P Daily	Cost Per Labor-Hour Bare Costs	Cost Per Labor-Hour Incl. O&P
1 Labor Foreman (outside)	$46.40	$371.20	$69.25	$554.00	$44.80	$66.85
4 Laborers	44.40	1420.80	66.25	2120.00		
1 Air Compressor, 250 cfm		202.85		223.13		
2 Breakers, Pavement, 60 lb.		107.20		117.92		
2 -50' Air Hoses, 1.5"		45.60		50.16	8.89	9.78
40 L.H., Daily Totals		$2147.65		$3065.22	$53.69	$76.63

Crew B-9A

Crew No.	Bare Costs Hr.	Bare Costs Daily	Incl. Subs O&P Hr.	Incl. Subs O&P Daily	Cost Per Labor-Hour Bare Costs	Cost Per Labor-Hour Incl. O&P
2 Laborers	$44.40	$710.40	$66.25	$1060.00	$46.70	$69.73
1 Truck Driver (heavy)	51.30	410.40	76.70	613.60		
1 Water Tank Trailer, 5000 Gal.		154.10		169.51		
1 Truck Tractor, 220 H.P.		310.80		341.88		
2 -50' Discharge Hoses, 3"		9.00		9.90	19.75	21.72
24 L.H., Daily Totals		$1594.70		$2194.89	$66.45	$91.45

Crew B-9B

Crew No.	Bare Costs Hr.	Bare Costs Daily	Incl. Subs O&P Hr.	Incl. Subs O&P Daily	Cost Per Labor-Hour Bare Costs	Cost Per Labor-Hour Incl. O&P
2 Laborers	$44.40	$710.40	$66.25	$1060.00	$46.70	$69.73
1 Truck Driver (heavy)	51.30	410.40	76.70	613.60		
2 -50' Discharge Hoses, 3"		9.00		9.90		
1 Water Tank Trailer, 5000 Gal.		154.10		169.51		
1 Truck Tractor, 220 H.P.		310.80		341.88		
1 Pressure Washer		97.35		107.08	23.80	26.18
24 L.H., Daily Totals		$1692.05		$2301.97	$70.50	$95.92

Crew B-9D

Crew No.	Bare Costs Hr.	Bare Costs Daily	Incl. Subs O&P Hr.	Incl. Subs O&P Daily	Cost Per Labor-Hour Bare Costs	Cost Per Labor-Hour Incl. O&P
1 Labor Foreman (outside)	$46.40	$371.20	$69.25	$554.00	$44.80	$66.85
4 Common Laborers	44.40	1420.80	66.25	2120.00		
1 Air Compressor, 250 cfm		202.85		223.13		
2 -50' Air Hoses, 1.5"		45.60		50.16		
2 Air Powered Tampers		79.50		87.45	8.20	9.02
40 L.H., Daily Totals		$2119.95		$3034.74	$53.00	$75.87

Crew B-9E

Crew No.	Bare Costs Hr.	Bare Costs Daily	Incl. Subs O&P Hr.	Incl. Subs O&P Daily	Cost Per Labor-Hour Bare Costs	Cost Per Labor-Hour Incl. O&P
1 Cement Finisher	$51.80	$414.40	$75.90	$607.20	$48.10	$71.08
1 Laborer	44.40	355.20	66.25	530.00		
1 Chip. Hammers, 12 Lb., Elec.		32.85		36.13	2.05	2.26
16 L.H., Daily Totals		$802.45		$1173.34	$50.15	$73.33

Crew B-10

Crew No.	Bare Costs Hr.	Bare Costs Daily	Incl. Subs O&P Hr.	Incl. Subs O&P Daily	Cost Per Labor-Hour Bare Costs	Cost Per Labor-Hour Incl. O&P
1 Equip. Oper. (medium)	$59.00	$472.00	$87.90	$703.20	$54.13	$80.68
.5 Laborer	44.40	177.60	66.25	265.00		
12 L.H., Daily Totals		$649.60		$968.20	$54.13	$80.68

Crew B-10A

Crew No.	Bare Costs Hr.	Bare Costs Daily	Incl. Subs O&P Hr.	Incl. Subs O&P Daily	Cost Per Labor-Hour Bare Costs	Cost Per Labor-Hour Incl. O&P
1 Equip. Oper. (medium)	$59.00	$472.00	$87.90	$703.20	$54.13	$80.68
.5 Laborer	44.40	177.60	66.25	265.00		
1 Roller, 2-Drum, W.B., 7.5 H.P.		166.75		183.43	13.90	15.29
12 L.H., Daily Totals		$816.35		$1151.63	$68.03	$95.97

For customer support on your Site Work & Landscape Costs with RSMeans Data, call 800.448.8182.

751

Left Column

Crew No.	Bare Costs Hr.	Daily	Incl. Subs O&P Hr.	Daily	Cost Per Labor-Hour Bare Costs	Incl. O&P
Crew B-10B	Hr.	Daily	Hr.	Daily	Bare Costs	Incl. O&P
1 Equip. Oper. (medium)	$59.00	$472.00	$87.90	$703.20	$54.13	$80.68
.5 Laborer	44.40	177.60	66.25	265.00		
1 Dozer, 200 H.P.		1520.00		1672.00	126.67	139.33
12 L.H., Daily Totals		$2169.60		$2640.20	$180.80	$220.02
Crew B-10C	Hr.	Daily	Hr.	Daily	Bare Costs	Incl. O&P
1 Equip. Oper. (medium)	$59.00	$472.00	$87.90	$703.20	$54.13	$80.68
.5 Laborer	44.40	177.60	66.25	265.00		
1 Dozer, 200 H.P.		1520.00		1672.00		
1 Vibratory Roller, Towed, 23 Ton		520.35		572.38	170.03	187.03
12 L.H., Daily Totals		$2689.95		$3212.59	$224.16	$267.72
Crew B-10D	Hr.	Daily	Hr.	Daily	Bare Costs	Incl. O&P
1 Equip. Oper. (medium)	$59.00	$472.00	$87.90	$703.20	$54.13	$80.68
.5 Laborer	44.40	177.60	66.25	265.00		
1 Dozer, 200 H.P.		1520.00		1672.00		
1 Sheepsft. Roller, Towed		426.95		469.64	162.25	178.47
12 L.H., Daily Totals		$2596.55		$3109.84	$216.38	$259.15
Crew B-10E	Hr.	Daily	Hr.	Daily	Bare Costs	Incl. O&P
1 Equip. Oper. (medium)	$59.00	$472.00	$87.90	$703.20	$54.13	$80.68
.5 Laborer	44.40	177.60	66.25	265.00		
1 Tandem Roller, 5 Ton		258.75		284.63	21.56	23.72
12 L.H., Daily Totals		$908.35		$1252.83	$75.70	$104.40
Crew B-10F	Hr.	Daily	Hr.	Daily	Bare Costs	Incl. O&P
1 Equip. Oper. (medium)	$59.00	$472.00	$87.90	$703.20	$54.13	$80.68
.5 Laborer	44.40	177.60	66.25	265.00		
1 Tandem Roller, 10 Ton		246.80		271.48	20.57	22.62
12 L.H., Daily Totals		$896.40		$1239.68	$74.70	$103.31
Crew B-10G	Hr.	Daily	Hr.	Daily	Bare Costs	Incl. O&P
1 Equip. Oper. (medium)	$59.00	$472.00	$87.90	$703.20	$54.13	$80.68
.5 Laborer	44.40	177.60	66.25	265.00		
1 Sheepsfoot Roller, 240 H.P.		1363.00		1499.30	113.58	124.94
12 L.H., Daily Totals		$2012.60		$2467.50	$167.72	$205.63
Crew B-10H	Hr.	Daily	Hr.	Daily	Bare Costs	Incl. O&P
1 Equip. Oper. (medium)	$59.00	$472.00	$87.90	$703.20	$54.13	$80.68
.5 Laborer	44.40	177.60	66.25	265.00		
1 Diaphragm Water Pump, 2"		87.70		96.47		
1 -20' Suction Hose, 2"		3.55		3.90		
2 -50' Discharge Hoses, 2"		8.00		8.80	8.27	9.10
12 L.H., Daily Totals		$748.85		$1077.38	$62.40	$89.78
Crew B-10I	Hr.	Daily	Hr.	Daily	Bare Costs	Incl. O&P
1 Equip. Oper. (medium)	$59.00	$472.00	$87.90	$703.20	$54.13	$80.68
.5 Laborer	44.40	177.60	66.25	265.00		
1 Diaphragm Water Pump, 4"		106.35		116.99		
1 -20' Suction Hose, 4"		17.25		18.98		
2 -50' Discharge Hoses, 4"		25.60		28.16	12.43	13.68
12 L.H., Daily Totals		$798.80		$1132.32	$66.57	$94.36
Crew B-10J	Hr.	Daily	Hr.	Daily	Bare Costs	Incl. O&P
1 Equip. Oper. (medium)	$59.00	$472.00	$87.90	$703.20	$54.13	$80.68
.5 Laborer	44.40	177.60	66.25	265.00		
1 Centrifugal Water Pump, 3"		74.40		81.84		
1 -20' Suction Hose, 3"		8.75		9.63		
2 -50' Discharge Hoses, 3"		9.00		9.90	7.68	8.45
12 L.H., Daily Totals		$741.75		$1069.57	$61.81	$89.13

Right Column

Crew No.	Bare Costs Hr.	Daily	Incl. Subs O&P Hr.	Daily	Cost Per Labor-Hour Bare Costs	Incl. O&P
Crew B-10K	Hr.	Daily	Hr.	Daily	Bare Costs	Incl. O&P
1 Equip. Oper. (medium)	$59.00	$472.00	$87.90	$703.20	$54.13	$80.68
.5 Laborer	44.40	177.60	66.25	265.00		
1 Centr. Water Pump, 6"		235.25		258.77		
1 -20' Suction Hose, 6"		25.50		28.05		
2 -50' Discharge Hoses, 6"		36.20		39.82	24.75	27.22
12 L.H., Daily Totals		$946.55		$1294.85	$78.88	$107.90
Crew B-10L	Hr.	Daily	Hr.	Daily	Bare Costs	Incl. O&P
1 Equip. Oper. (medium)	$59.00	$472.00	$87.90	$703.20	$54.13	$80.68
.5 Laborer	44.40	177.60	66.25	265.00		
1 Dozer, 80 H.P.		405.85		446.44	33.82	37.20
12 L.H., Daily Totals		$1055.45		$1414.64	$87.95	$117.89
Crew B-10M	Hr.	Daily	Hr.	Daily	Bare Costs	Incl. O&P
1 Equip. Oper. (medium)	$59.00	$472.00	$87.90	$703.20	$54.13	$80.68
.5 Laborer	44.40	177.60	66.25	265.00		
1 Dozer, 300 H.P.		1785.00		1963.50	148.75	163.63
12 L.H., Daily Totals		$2434.60		$2931.70	$202.88	$244.31
Crew B-10N	Hr.	Daily	Hr.	Daily	Bare Costs	Incl. O&P
1 Equip. Oper. (medium)	$59.00	$472.00	$87.90	$703.20	$54.13	$80.68
.5 Laborer	44.40	177.60	66.25	265.00		
1 F.E. Loader, T.M., 1.5 C.Y.		572.00		629.20	47.67	52.43
12 L.H., Daily Totals		$1221.60		$1597.40	$101.80	$133.12
Crew B-10O	Hr.	Daily	Hr.	Daily	Bare Costs	Incl. O&P
1 Equip. Oper. (medium)	$59.00	$472.00	$87.90	$703.20	$54.13	$80.68
.5 Laborer	44.40	177.60	66.25	265.00		
1 F.E. Loader, T.M., 2.25 C.Y.		925.50		1018.05	77.13	84.84
12 L.H., Daily Totals		$1575.10		$1986.25	$131.26	$165.52
Crew B-10P	Hr.	Daily	Hr.	Daily	Bare Costs	Incl. O&P
1 Equip. Oper. (medium)	$59.00	$472.00	$87.90	$703.20	$54.13	$80.68
.5 Laborer	44.40	177.60	66.25	265.00		
1 Crawler Loader, 3 C.Y.		1146.00		1260.60	95.50	105.05
12 L.H., Daily Totals		$1795.60		$2228.80	$149.63	$185.73
Crew B-10Q	Hr.	Daily	Hr.	Daily	Bare Costs	Incl. O&P
1 Equip. Oper. (medium)	$59.00	$472.00	$87.90	$703.20	$54.13	$80.68
.5 Laborer	44.40	177.60	66.25	265.00		
1 Crawler Loader, 4 C.Y.		1456.00		1601.60	121.33	133.47
12 L.H., Daily Totals		$2105.60		$2569.80	$175.47	$214.15
Crew B-10R	Hr.	Daily	Hr.	Daily	Bare Costs	Incl. O&P
1 Equip. Oper. (medium)	$59.00	$472.00	$87.90	$703.20	$54.13	$80.68
.5 Laborer	44.40	177.60	66.25	265.00		
1 F.E. Loader, W.M., 1 C.Y.		305.60		336.16	25.47	28.01
12 L.H., Daily Totals		$955.20		$1304.36	$79.60	$108.70
Crew B-10S	Hr.	Daily	Hr.	Daily	Bare Costs	Incl. O&P
1 Equip. Oper. (medium)	$59.00	$472.00	$87.90	$703.20	$54.13	$80.68
.5 Laborer	44.40	177.60	66.25	265.00		
1 F.E. Loader, W.M., 1.5 C.Y.		441.40		485.54	36.78	40.46
12 L.H., Daily Totals		$1091.00		$1453.74	$90.92	$121.15
Crew B-10T	Hr.	Daily	Hr.	Daily	Bare Costs	Incl. O&P
1 Equip. Oper. (medium)	$59.00	$472.00	$87.90	$703.20	$54.13	$80.68
.5 Laborer	44.40	177.60	66.25	265.00		
1 F.E. Loader, W.M., 2.5 C.Y.		638.30		702.13	53.19	58.51
12 L.H., Daily Totals		$1287.90		$1670.33	$107.33	$139.19

For customer support on your Site Work & Landscape Costs with RSMeans Data, call 800.448.8182.

Crew No.	Bare Costs		Incl. Subs O&P		Cost Per Labor-Hour	
Crew B-10U	**Hr.**	**Daily**	**Hr.**	**Daily**	**Bare Costs**	**Incl. O&P**
1 Equip. Oper. (medium)	$59.00	$472.00	$87.90	$703.20	$54.13	$80.68
.5 Laborer	44.40	177.60	66.25	265.00		
1 F.E. Loader, W.M., 5.5 C.Y.		967.95		1064.74	80.66	88.73
12 L.H., Daily Totals		$1617.55		$2032.94	$134.80	$169.41

Crew B-10V	Hr.	Daily	Hr.	Daily	Bare Costs	Incl. O&P
1 Equip. Oper. (medium)	$59.00	$472.00	$87.90	$703.20	$54.13	$80.68
.5 Laborer	44.40	177.60	66.25	265.00		
1 Dozer, 700 H.P.		5175.00		5692.50	431.25	474.38
12 L.H., Daily Totals		$5824.60		$6660.70	$485.38	$555.06

Crew B-10W	Hr.	Daily	Hr.	Daily	Bare Costs	Incl. O&P
1 Equip. Oper. (medium)	$59.00	$472.00	$87.90	$703.20	$54.13	$80.68
.5 Laborer	44.40	177.60	66.25	265.00		
1 Dozer, 105 H.P.		640.80		704.88	53.40	58.74
12 L.H., Daily Totals		$1290.40		$1673.08	$107.53	$139.42

Crew B-10X	Hr.	Daily	Hr.	Daily	Bare Costs	Incl. O&P
1 Equip. Oper. (medium)	$59.00	$472.00	$87.90	$703.20	$54.13	$80.68
.5 Laborer	44.40	177.60	66.25	265.00		
1 Dozer, 410 H.P.		2807.00		3087.70	233.92	257.31
12 L.H., Daily Totals		$3456.60		$4055.90	$288.05	$337.99

Crew B-10Y	Hr.	Daily	Hr.	Daily	Bare Costs	Incl. O&P
1 Equip. Oper. (medium)	$59.00	$472.00	$87.90	$703.20	$54.13	$80.68
.5 Laborer	44.40	177.60	66.25	265.00		
1 Vibr. Roller, Towed, 12 Ton		584.80		643.28	48.73	53.61
12 L.H., Daily Totals		$1234.40		$1611.48	$102.87	$134.29

Crew B-11A	Hr.	Daily	Hr.	Daily	Bare Costs	Incl. O&P
1 Equipment Oper. (med.)	$59.00	$472.00	$87.90	$703.20	$51.70	$77.08
1 Laborer	44.40	355.20	66.25	530.00		
1 Dozer, 200 H.P.		1520.00		1672.00	95.00	104.50
16 L.H., Daily Totals		$2347.20		$2905.20	$146.70	$181.57

Crew B-11B	Hr.	Daily	Hr.	Daily	Bare Costs	Incl. O&P
1 Equipment Oper. (light)	$55.50	$444.00	$82.70	$661.60	$49.95	$74.47
1 Laborer	44.40	355.20	66.25	530.00		
1 Air Powered Tamper		39.75		43.73		
1 Air Compressor, 365 cfm		343.55		377.90		
2 -50' Air Hoses, 1.5"		45.60		50.16	26.81	29.49
16 L.H., Daily Totals		$1228.10		$1663.39	$76.76	$103.96

Crew B-11C	Hr.	Daily	Hr.	Daily	Bare Costs	Incl. O&P
1 Equipment Oper. (med.)	$59.00	$472.00	$87.90	$703.20	$51.70	$77.08
1 Laborer	44.40	355.20	66.25	530.00		
1 Backhoe Loader, 48 H.P.		216.20		237.82	13.51	14.86
16 L.H., Daily Totals		$1043.40		$1471.02	$65.21	$91.94

Crew B-11J	Hr.	Daily	Hr.	Daily	Bare Costs	Incl. O&P
1 Equipment Oper. (med.)	$59.00	$472.00	$87.90	$703.20	$51.70	$77.08
1 Laborer	44.40	355.20	66.25	530.00		
1 Grader, 30,000 Lbs.		1073.00		1180.30		
1 Ripper, Beam & 1 Shank		91.60		100.76	72.79	80.07
16 L.H., Daily Totals		$1991.80		$2514.26	$124.49	$157.14

Crew B-11K	Hr.	Daily	Hr.	Daily	Bare Costs	Incl. O&P
1 Equipment Oper. (med.)	$59.00	$472.00	$87.90	$703.20	$51.70	$77.08
1 Laborer	44.40	355.20	66.25	530.00		
1 Trencher, Chain Type, 8' D		1894.00		2083.40	118.38	130.21
16 L.H., Daily Totals		$2721.20		$3316.60	$170.07	$207.29

Crew B-11L	Hr.	Daily	Hr.	Daily	Bare Costs	Incl. O&P
1 Equipment Oper. (med.)	$59.00	$472.00	$87.90	$703.20	$51.70	$77.08
1 Laborer	44.40	355.20	66.25	530.00		
1 Grader, 30,000 Lbs.		1073.00		1180.30	67.06	73.77
16 L.H., Daily Totals		$1900.20		$2413.50	$118.76	$150.84

Crew B-11M	Hr.	Daily	Hr.	Daily	Bare Costs	Incl. O&P
1 Equipment Oper. (med.)	$59.00	$472.00	$87.90	$703.20	$51.70	$77.08
1 Laborer	44.40	355.20	66.25	530.00		
1 Backhoe Loader, 80 H.P.		235.05		258.56	14.69	16.16
16 L.H., Daily Totals		$1062.25		$1491.76	$66.39	$93.23

Crew B-11N	Hr.	Daily	Hr.	Daily	Bare Costs	Incl. O&P
1 Labor Foreman (outside)	$46.40	$371.20	$69.25	$554.00	$52.47	$78.36
2 Equipment Operators (med.)	59.00	944.00	87.90	1406.40		
6 Truck Drivers (heavy)	51.30	2462.40	76.70	3681.60		
1 F.E. Loader, W.M., 5.5 C.Y.		967.95		1064.74		
1 Dozer, 410 H.P.		2807.00		3087.70		
6 Dump Trucks, Off Hwy., 50 Ton		11874.00		13061.40	217.35	239.08
72 L.H., Daily Totals		$19426.55		$22855.85	$269.81	$317.44

Crew B-11Q	Hr.	Daily	Hr.	Daily	Bare Costs	Incl. O&P
1 Equipment Operator (med.)	$59.00	$472.00	$87.90	$703.20	$54.13	$80.68
.5 Laborer	44.40	177.60	66.25	265.00		
1 Dozer, 140 H.P.		729.15		802.07	60.76	66.84
12 L.H., Daily Totals		$1378.75		$1770.27	$114.90	$147.52

Crew B-11R	Hr.	Daily	Hr.	Daily	Bare Costs	Incl. O&P
1 Equipment Operator (med.)	$59.00	$472.00	$87.90	$703.20	$54.13	$80.68
.5 Laborer	44.40	177.60	66.25	265.00		
1 Dozer, 200 H.P.		1520.00		1672.00	126.67	139.33
12 L.H., Daily Totals		$2169.60		$2640.20	$180.80	$220.02

Crew B-11S	Hr.	Daily	Hr.	Daily	Bare Costs	Incl. O&P
1 Equipment Operator (med.)	$59.00	$472.00	$87.90	$703.20	$54.13	$80.68
.5 Laborer	44.40	177.60	66.25	265.00		
1 Dozer, 300 H.P.		1785.00		1963.50		
1 Ripper, Beam & 1 Shank		91.60		100.76	156.38	172.02
12 L.H., Daily Totals		$2526.20		$3032.46	$210.52	$252.71

Crew B-11T	Hr.	Daily	Hr.	Daily	Bare Costs	Incl. O&P
1 Equipment Operator (med.)	$59.00	$472.00	$87.90	$703.20	$54.13	$80.68
.5 Laborer	44.40	177.60	66.25	265.00		
1 Dozer, 410 H.P.		2807.00		3087.70		
1 Ripper, Beam & 2 Shanks		140.40		154.44	245.62	270.18
12 L.H., Daily Totals		$3597.00		$4210.34	$299.75	$350.86

Crew B-11U	Hr.	Daily	Hr.	Daily	Bare Costs	Incl. O&P
1 Equipment Operator (med.)	$59.00	$472.00	$87.90	$703.20	$54.13	$80.68
.5 Laborer	44.40	177.60	66.25	265.00		
1 Dozer, 520 H.P.		3434.00		3777.40	286.17	314.78
12 L.H., Daily Totals		$4083.60		$4745.60	$340.30	$395.47

Crew B-11V

Crew No.	Bare Costs		Incl. Subs O&P		Cost Per Labor-Hour	
	Hr.	Daily	Hr.	Daily	Bare Costs	Incl. O&P
3 Laborers	$44.40	$1065.60	$66.25	$1590.00	$44.40	$66.25
1 Roller, 2-Drum, W.B., 7.5 H.P.		166.75		183.43	6.95	7.64
24 L.H., Daily Totals		$1232.35		$1773.43	$51.35	$73.89

Crew B-11W

Crew No.	Bare Costs		Incl. Subs O&P		Cost Per Labor-Hour	
	Hr.	Daily	Hr.	Daily	Bare Costs	Incl. O&P
1 Equipment Operator (med.)	$59.00	$472.00	$87.90	$703.20	$51.37	$76.76
1 Common Laborer	44.40	355.20	66.25	530.00		
10 Truck Drivers (heavy)	51.30	4104.00	76.70	6136.00		
1 Dozer, 200 H.P.		1520.00		1672.00		
1 Vibratory Roller, Towed, 23 Ton		520.35		572.38		
10 Dump Trucks, 8 C.Y., 220 H.P.		4076.00		4483.60	63.71	70.08
96 L.H., Daily Totals		$11047.55		$14097.18	$115.08	$146.85

Crew B-11Y

Crew No.	Bare Costs		Incl. Subs O&P		Cost Per Labor-Hour	
	Hr.	Daily	Hr.	Daily	Bare Costs	Incl. O&P
1 Labor Foreman (outside)	$46.40	$371.20	$69.25	$554.00	$49.49	$73.80
5 Common Laborers	44.40	1776.00	66.25	2650.00		
3 Equipment Operators (med.)	59.00	1416.00	87.90	2109.60		
1 Dozer, 80 H.P.		405.85		446.44		
2 Rollers, 2-Drum, W.B., 7.5 H.P.		333.50		366.85		
4 Vibrating Plates, Gas, 21"		662.40		728.64	19.47	21.42
72 L.H., Daily Totals		$4964.95		$6855.52	$68.96	$95.22

Crew B-12A

Crew No.	Bare Costs		Incl. Subs O&P		Cost Per Labor-Hour	
	Hr.	Daily	Hr.	Daily	Bare Costs	Incl. O&P
1 Equip. Oper. (crane)	$61.45	$491.60	$91.55	$732.40	$52.92	$78.90
1 Laborer	44.40	355.20	66.25	530.00		
1 Hyd. Excavator, 1 C.Y.		832.65		915.91	52.04	57.24
16 L.H., Daily Totals		$1679.45		$2178.32	$104.97	$136.14

Crew B-12B

Crew No.	Bare Costs		Incl. Subs O&P		Cost Per Labor-Hour	
	Hr.	Daily	Hr.	Daily	Bare Costs	Incl. O&P
1 Equip. Oper. (crane)	$61.45	$491.60	$91.55	$732.40	$52.92	$78.90
1 Laborer	44.40	355.20	66.25	530.00		
1 Hyd. Excavator, 1.5 C.Y.		695.80		765.38	43.49	47.84
16 L.H., Daily Totals		$1542.60		$2027.78	$96.41	$126.74

Crew B-12C

Crew No.	Bare Costs		Incl. Subs O&P		Cost Per Labor-Hour	
	Hr.	Daily	Hr.	Daily	Bare Costs	Incl. O&P
1 Equip. Oper. (crane)	$61.45	$491.60	$91.55	$732.40	$52.92	$78.90
1 Laborer	44.40	355.20	66.25	530.00		
1 Hyd. Excavator, 2 C.Y.		942.70		1036.97	58.92	64.81
16 L.H., Daily Totals		$1789.50		$2299.37	$111.84	$143.71

Crew B-12D

Crew No.	Bare Costs		Incl. Subs O&P		Cost Per Labor-Hour	
	Hr.	Daily	Hr.	Daily	Bare Costs	Incl. O&P
1 Equip. Oper. (crane)	$61.45	$491.60	$91.55	$732.40	$52.92	$78.90
1 Laborer	44.40	355.20	66.25	530.00		
1 Hyd. Excavator, 3.5 C.Y.		2184.00		2402.40	136.50	150.15
16 L.H., Daily Totals		$3030.80		$3664.80	$189.43	$229.05

Crew B-12E

Crew No.	Bare Costs		Incl. Subs O&P		Cost Per Labor-Hour	
	Hr.	Daily	Hr.	Daily	Bare Costs	Incl. O&P
1 Equip. Oper. (crane)	$61.45	$491.60	$91.55	$732.40	$52.92	$78.90
1 Laborer	44.40	355.20	66.25	530.00		
1 Hyd. Excavator, .5 C.Y.		457.00		502.70	28.56	31.42
16 L.H., Daily Totals		$1303.80		$1765.10	$81.49	$110.32

Crew B-12F

Crew No.	Bare Costs		Incl. Subs O&P		Cost Per Labor-Hour	
	Hr.	Daily	Hr.	Daily	Bare Costs	Incl. O&P
1 Equip. Oper. (crane)	$61.45	$491.60	$91.55	$732.40	$52.92	$78.90
1 Laborer	44.40	355.20	66.25	530.00		
1 Hyd. Excavator, .75 C.Y.		701.80		771.98	43.86	48.25
16 L.H., Daily Totals		$1548.60		$2034.38	$96.79	$127.15

Crew B-12G

Crew No.	Bare Costs		Incl. Subs O&P		Cost Per Labor-Hour	
	Hr.	Daily	Hr.	Daily	Bare Costs	Incl. O&P
1 Equip. Oper. (crane)	$61.45	$491.60	$91.55	$732.40	$52.92	$78.90
1 Laborer	44.40	355.20	66.25	530.00		
1 Crawler Crane, 15 Ton		810.30		891.33		
1 Clamshell Bucket, .5 C.Y.		67.80		74.58	54.88	60.37
16 L.H., Daily Totals		$1724.90		$2228.31	$107.81	$139.27

Crew B-12H

Crew No.	Bare Costs		Incl. Subs O&P		Cost Per Labor-Hour	
	Hr.	Daily	Hr.	Daily	Bare Costs	Incl. O&P
1 Equip. Oper. (crane)	$61.45	$491.60	$91.55	$732.40	$52.92	$78.90
1 Laborer	44.40	355.20	66.25	530.00		
1 Crawler Crane, 25 Ton		1152.00		1267.20		
1 Clamshell Bucket, 1 C.Y.		69.25		76.17	76.33	83.96
16 L.H., Daily Totals		$2068.05		$2605.78	$129.25	$162.86

Crew B-12I

Crew No.	Bare Costs		Incl. Subs O&P		Cost Per Labor-Hour	
	Hr.	Daily	Hr.	Daily	Bare Costs	Incl. O&P
1 Equip. Oper. (crane)	$61.45	$491.60	$91.55	$732.40	$52.92	$78.90
1 Laborer	44.40	355.20	66.25	530.00		
1 Crawler Crane, 20 Ton		1013.00		1114.30		
1 Dragline Bucket, .75 C.Y.		61.85		68.03	67.18	73.90
16 L.H., Daily Totals		$1921.65		$2444.74	$120.10	$152.80

Crew B-12J

Crew No.	Bare Costs		Incl. Subs O&P		Cost Per Labor-Hour	
	Hr.	Daily	Hr.	Daily	Bare Costs	Incl. O&P
1 Equip. Oper. (crane)	$61.45	$491.60	$91.55	$732.40	$52.92	$78.90
1 Laborer	44.40	355.20	66.25	530.00		
1 Gradall, 5/8 C.Y.		850.65		935.72	53.17	58.48
16 L.H., Daily Totals		$1697.45		$2198.11	$106.09	$137.38

Crew B-12K

Crew No.	Bare Costs		Incl. Subs O&P		Cost Per Labor-Hour	
	Hr.	Daily	Hr.	Daily	Bare Costs	Incl. O&P
1 Equip. Oper. (crane)	$61.45	$491.60	$91.55	$732.40	$52.92	$78.90
1 Laborer	44.40	355.20	66.25	530.00		
1 Gradall, 3 Ton, 1 C.Y.		984.55		1083.01	61.53	67.69
16 L.H., Daily Totals		$1831.35		$2345.41	$114.46	$146.59

Crew B-12L

Crew No.	Bare Costs		Incl. Subs O&P		Cost Per Labor-Hour	
	Hr.	Daily	Hr.	Daily	Bare Costs	Incl. O&P
1 Equip. Oper. (crane)	$61.45	$491.60	$91.55	$732.40	$52.92	$78.90
1 Laborer	44.40	355.20	66.25	530.00		
1 Crawler Crane, 15 Ton		810.30		891.33		
1 F.E. Attachment, .5 C.Y.		66.05		72.66	54.77	60.25
16 L.H., Daily Totals		$1723.15		$2226.39	$107.70	$139.15

Crew B-12M

Crew No.	Bare Costs		Incl. Subs O&P		Cost Per Labor-Hour	
	Hr.	Daily	Hr.	Daily	Bare Costs	Incl. O&P
1 Equip. Oper. (crane)	$61.45	$491.60	$91.55	$732.40	$52.92	$78.90
1 Laborer	44.40	355.20	66.25	530.00		
1 Crawler Crane, 20 Ton		1013.00		1114.30		
1 F.E. Attachment, .75 C.Y.		71.25		78.38	67.77	74.54
16 L.H., Daily Totals		$1931.05		$2455.07	$120.69	$153.44

Crew B-12N

Crew No.	Bare Costs		Incl. Subs O&P		Cost Per Labor-Hour	
	Hr.	Daily	Hr.	Daily	Bare Costs	Incl. O&P
1 Equip. Oper. (crane)	$61.45	$491.60	$91.55	$732.40	$52.92	$78.90
1 Laborer	44.40	355.20	66.25	530.00		
1 Crawler Crane, 25 Ton		1152.00		1267.20		
1 F.E. Attachment, 1 C.Y.		77.35		85.08	76.83	84.52
16 L.H., Daily Totals		$2076.15		$2614.68	$129.76	$163.42

Crew B-12O

Crew No.	Bare Costs		Incl. Subs O&P		Cost Per Labor-Hour	
	Hr.	Daily	Hr.	Daily	Bare Costs	Incl. O&P
1 Equip. Oper. (crane)	$61.45	$491.60	$91.55	$732.40	$52.92	$78.90
1 Laborer	44.40	355.20	66.25	530.00		
1 Crawler Crane, 40 Ton		1231.00		1354.10		
1 F.E. Attachment, 1.5 C.Y.		88.65		97.52	82.48	90.73
16 L.H., Daily Totals		$2166.45		$2714.01	$135.40	$169.63

For customer support on your Site Work & Landscape Costs with RSMeans Data, call 800.448.8182.

Crew No.	Bare Costs		Incl. Subs O&P		Cost Per Labor-Hour	

Crew B-12P	Hr.	Daily	Hr.	Daily	Bare Costs	Incl. O&P
1 Equip. Oper. (crane)	$61.45	$491.60	$91.55	$732.40	$52.92	$78.90
1 Laborer	44.40	355.20	66.25	530.00		
1 Crawler Crane, 40 Ton		1231.00		1354.10		
1 Dragline Bucket, 1.5 C.Y.		65.70		72.27	81.04	89.15
16 L.H., Daily Totals		$2143.50		$2688.77	$133.97	$168.05

Crew B-12Q	Hr.	Daily	Hr.	Daily	Bare Costs	Incl. O&P
1 Equip. Oper. (crane)	$61.45	$491.60	$91.55	$732.40	$52.92	$78.90
1 Laborer	44.40	355.20	66.25	530.00		
1 Hyd. Excavator, 5/8 C.Y.		604.75		665.23	37.80	41.58
16 L.H., Daily Totals		$1451.55		$1927.63	$90.72	$120.48

Crew B-12S	Hr.	Daily	Hr.	Daily	Bare Costs	Incl. O&P
1 Equip. Oper. (crane)	$61.45	$491.60	$91.55	$732.40	$52.92	$78.90
1 Laborer	44.40	355.20	66.25	530.00		
1 Hyd. Excavator, 2.5 C.Y.		1567.00		1723.70	97.94	107.73
16 L.H., Daily Totals		$2413.80		$2986.10	$150.86	$186.63

Crew B-12T	Hr.	Daily	Hr.	Daily	Bare Costs	Incl. O&P
1 Equip. Oper. (crane)	$61.45	$491.60	$91.55	$732.40	$52.92	$78.90
1 Laborer	44.40	355.20	66.25	530.00		
1 Crawler Crane, 75 Ton		1967.00		2163.70		
1 F.E. Attachment, 3 C.Y.		115.55		127.11	130.16	143.18
16 L.H., Daily Totals		$2929.35		$3553.20	$183.08	$222.08

Crew B-12V	Hr.	Daily	Hr.	Daily	Bare Costs	Incl. O&P
1 Equip. Oper. (crane)	$61.45	$491.60	$91.55	$732.40	$52.92	$78.90
1 Laborer	44.40	355.20	66.25	530.00		
1 Crawler Crane, 75 Ton		1967.00		2163.70		
1 Dragline Bucket, 3 C.Y.		72.30		79.53	127.46	140.20
16 L.H., Daily Totals		$2886.10		$3505.63	$180.38	$219.10

Crew B-12Y	Hr.	Daily	Hr.	Daily	Bare Costs	Incl. O&P
1 Equip. Oper. (crane)	$61.45	$491.60	$91.55	$732.40	$50.08	$74.68
2 Laborers	44.40	710.40	66.25	1060.00		
1 Hyd. Excavator, 3.5 C.Y.		2184.00		2402.40	91.00	100.10
24 L.H., Daily Totals		$3386.00		$4194.80	$141.08	$174.78

Crew B-12Z	Hr.	Daily	Hr.	Daily	Bare Costs	Incl. O&P
1 Equip. Oper. (crane)	$61.45	$491.60	$91.55	$732.40	$50.08	$74.68
2 Laborers	44.40	710.40	66.25	1060.00		
1 Hyd. Excavator, 2.5 C.Y.		1567.00		1723.70	65.29	71.82
24 L.H., Daily Totals		$2769.00		$3516.10	$115.38	$146.50

Crew B-13	Hr.	Daily	Hr.	Daily	Bare Costs	Incl. O&P
1 Labor Foreman (outside)	$46.40	$371.20	$69.25	$554.00	$48.28	$72.01
4 Laborers	44.40	1420.80	66.25	2120.00		
1 Equip. Oper. (crane)	61.45	491.60	91.55	732.40		
1 Equip. Oper. (oiler)	52.50	420.00	78.25	626.00		
1 Hyd. Crane, 25 Ton		586.70		645.37	10.48	11.52
56 L.H., Daily Totals		$3290.30		$4677.77	$58.76	$83.53

Crew B-13A	Hr.	Daily	Hr.	Daily	Bare Costs	Incl. O&P
1 Labor Foreman (outside)	$46.40	$371.20	$69.25	$554.00	$50.83	$75.85
2 Laborers	44.40	710.40	66.25	1060.00		
2 Equipment Operators (med.)	59.00	944.00	87.90	1406.40		
2 Truck Drivers (heavy)	51.30	820.80	76.70	1227.20		
1 Crawler Crane, 75 Ton		1967.00		2163.70		
1 Crawler Loader, 4 C.Y.		1456.00		1601.60		
2 Dump Trucks, 8 C.Y., 220 H.P.		815.20		896.72	75.68	83.25
56 L.H., Daily Totals		$7084.60		$8909.62	$126.51	$159.10

Crew B-13B	Hr.	Daily	Hr.	Daily	Bare Costs	Incl. O&P
1 Labor Foreman (outside)	$46.40	$371.20	$69.25	$554.00	$48.28	$72.01
4 Laborers	44.40	1420.80	66.25	2120.00		
1 Equip. Oper. (crane)	61.45	491.60	91.55	732.40		
1 Equip. Oper. (oiler)	52.50	420.00	78.25	626.00		
1 Hyd. Crane, 55 Ton		990.15		1089.17	17.68	19.45
56 L.H., Daily Totals		$3693.75		$5121.56	$65.96	$91.46

Crew B-13C	Hr.	Daily	Hr.	Daily	Bare Costs	Incl. O&P
1 Labor Foreman (outside)	$46.40	$371.20	$69.25	$554.00	$48.28	$72.01
4 Laborers	44.40	1420.80	66.25	2120.00		
1 Equip. Oper. (crane)	61.45	491.60	91.55	732.40		
1 Equip. Oper. (oiler)	52.50	420.00	78.25	626.00		
1 Crawler Crane, 100 Ton		2310.00		2541.00	41.25	45.38
56 L.H., Daily Totals		$5013.60		$6573.40	$89.53	$117.38

Crew B-13D	Hr.	Daily	Hr.	Daily	Bare Costs	Incl. O&P
1 Laborer	$44.40	$355.20	$66.25	$530.00	$52.92	$78.90
1 Equip. Oper. (crane)	61.45	491.60	91.55	732.40		
1 Hyd. Excavator, 1 C.Y.		832.65		915.91		
1 Trench Box		119.15		131.07	59.49	65.44
16 L.H., Daily Totals		$1798.60		$2309.38	$112.41	$144.34

Crew B-13E	Hr.	Daily	Hr.	Daily	Bare Costs	Incl. O&P
1 Laborer	$44.40	$355.20	$66.25	$530.00	$52.92	$78.90
1 Equip. Oper. (crane)	61.45	491.60	91.55	732.40		
1 Hyd. Excavator, 1.5 C.Y.		695.80		765.38		
1 Trench Box		119.15		131.07	50.93	56.03
16 L.H., Daily Totals		$1661.75		$2158.84	$103.86	$134.93

Crew B-13F	Hr.	Daily	Hr.	Daily	Bare Costs	Incl. O&P
1 Laborer	$44.40	$355.20	$66.25	$530.00	$52.92	$78.90
1 Equip. Oper. (crane)	61.45	491.60	91.55	732.40		
1 Hyd. Excavator, 3.5 C.Y.		2184.00		2402.40		
1 Trench Box		119.15		131.07	143.95	158.34
16 L.H., Daily Totals		$3149.95		$3795.86	$196.87	$237.24

Crew B-13G	Hr.	Daily	Hr.	Daily	Bare Costs	Incl. O&P
1 Laborer	$44.40	$355.20	$66.25	$530.00	$52.92	$78.90
1 Equip. Oper. (crane)	61.45	491.60	91.55	732.40		
1 Hyd. Excavator, .75 C.Y.		701.80		771.98		
1 Trench Box		119.15		131.07	51.31	56.44
16 L.H., Daily Totals		$1667.75		$2165.45	$104.23	$135.34

Crew B-13H	Hr.	Daily	Hr.	Daily	Bare Costs	Incl. O&P
1 Laborer	$44.40	$355.20	$66.25	$530.00	$52.92	$78.90
1 Equip. Oper. (crane)	61.45	491.60	91.55	732.40		
1 Gradall, 5/8 C.Y.		850.65		935.72		
1 Trench Box		119.15		131.07	60.61	66.67
16 L.H., Daily Totals		$1816.60		$2329.18	$113.54	$145.57

Crews - Standard

Crew B-13I

Crew No. B-13I	Hr.	Daily	Hr.	Daily	Bare Costs	Incl. O&P
1 Laborer	$44.40	$355.20	$66.25	$530.00	$52.92	$78.90
1 Equip. Oper. (crane)	61.45	491.60	91.55	732.40		
1 Gradall, 3 Ton, 1 C.Y.		984.55		1083.01		
1 Trench Box		119.15		131.07	68.98	75.88
16 L.H., Daily Totals		$1950.50		$2476.47	$121.91	$154.78

Crew B-13J

Crew No. B-13J	Hr.	Daily	Hr.	Daily	Bare Costs	Incl. O&P
1 Laborer	$44.40	$355.20	$66.25	$530.00	$52.92	$78.90
1 Equip. Oper. (crane)	61.45	491.60	91.55	732.40		
1 Hyd. Excavator, 2.5 C.Y.		1567.00		1723.70		
1 Trench Box		119.15		131.07	105.38	115.92
16 L.H., Daily Totals		$2532.95		$3117.17	$158.31	$194.82

Crew B-13K

Crew No. B-13K	Hr.	Daily	Hr.	Daily	Bare Costs	Incl. O&P
2 Equip. Opers. (crane)	$61.45	$983.20	$91.55	$1464.80	$61.45	$91.55
1 Hyd. Excavator, .75 C.Y.		701.80		771.98		
1 Hyd. Hammer, 4000 ft-lb		649.20		714.12		
1 Hyd. Excavator, .75 C.Y.		701.80		771.98	128.30	141.13
16 L.H., Daily Totals		$3036.00		$3722.88	$189.75	$232.68

Crew B-13L

Crew No. B-13L	Hr.	Daily	Hr.	Daily	Bare Costs	Incl. O&P
2 Equip. Opers. (crane)	$61.45	$983.20	$91.55	$1464.80	$61.45	$91.55
1 Hyd. Excavator, 1.5 C.Y.		695.80		765.38		
1 Hyd. Hammer, 5000 ft-lb		705.60		776.16		
1 Hyd. Excavator, .75 C.Y.		701.80		771.98	131.45	144.60
16 L.H., Daily Totals		$3086.40		$3778.32	$192.90	$236.15

Crew B-13M

Crew No. B-13M	Hr.	Daily	Hr.	Daily	Bare Costs	Incl. O&P
2 Equip. Opers. (crane)	$61.45	$983.20	$91.55	$1464.80	$61.45	$91.55
1 Hyd. Excavator, 2.5 C.Y.		1567.00		1723.70		
1 Hyd. Hammer, 8000 ft-lb		918.65		1010.52		
1 Hyd. Excavator, 1.5 C.Y.		695.80		765.38	198.84	218.72
16 L.H., Daily Totals		$4164.65		$4964.40	$260.29	$310.27

Crew B-13N

Crew No. B-13N	Hr.	Daily	Hr.	Daily	Bare Costs	Incl. O&P
2 Equip. Opers. (crane)	$61.45	$983.20	$91.55	$1464.80	$61.45	$91.55
1 Hyd. Excavator, 3.5 C.Y.		2184.00		2402.40		
1 Hyd. Hammer, 12,000 ft-lb		882.20		970.42		
1 Hyd. Excavator, 1.5 C.Y.		695.80		765.38	235.13	258.64
16 L.H., Daily Totals		$4745.20		$5603.00	$296.57	$350.19

Crew B-14

Crew No. B-14	Hr.	Daily	Hr.	Daily	Bare Costs	Incl. O&P
1 Labor Foreman (outside)	$46.40	$371.20	$69.25	$554.00	$46.58	$69.49
4 Laborers	44.40	1420.80	66.25	2120.00		
1 Equip. Oper. (light)	55.50	444.00	82.70	661.60		
1 Backhoe Loader, 48 H.P.		216.20		237.82	4.50	4.95
48 L.H., Daily Totals		$2452.20		$3573.42	$51.09	$74.45

Crew B-14A

Crew No. B-14A	Hr.	Daily	Hr.	Daily	Bare Costs	Incl. O&P
1 Equip. Oper. (crane)	$61.45	$491.60	$91.55	$732.40	$55.77	$83.12
.5 Laborer	44.40	177.60	66.25	265.00		
1 Hyd. Excavator, 4.5 C.Y.		3450.00		3795.00	287.50	316.25
12 L.H., Daily Totals		$4119.20		$4792.40	$343.27	$399.37

Crew B-14B

Crew No. B-14B	Hr.	Daily	Hr.	Daily	Bare Costs	Incl. O&P
1 Equip. Oper. (crane)	$61.45	$491.60	$91.55	$732.40	$55.77	$83.12
.5 Laborer	44.40	177.60	66.25	265.00		
1 Hyd. Excavator, 6 C.Y.		3506.00		3856.60	292.17	321.38
12 L.H., Daily Totals		$4175.20		$4854.00	$347.93	$404.50

Crew B-14C

Crew No. B-14C	Hr.	Daily	Hr.	Daily	Bare Costs	Incl. O&P
1 Equip. Oper. (crane)	$61.45	$491.60	$91.55	$732.40	$55.77	$83.12
.5 Laborer	44.40	177.60	66.25	265.00		
1 Hyd. Excavator, 7 C.Y.		3475.00		3822.50	289.58	318.54
12 L.H., Daily Totals		$4144.20		$4819.90	$345.35	$401.66

Crew B-14F

Crew No. B-14F	Hr.	Daily	Hr.	Daily	Bare Costs	Incl. O&P
1 Equip. Oper. (crane)	$61.45	$491.60	$91.55	$732.40	$55.77	$83.12
.5 Laborer	44.40	177.60	66.25	265.00		
1 Hyd. Shovel, 7 C.Y.		4148.00		4562.80	345.67	380.23
12 L.H., Daily Totals		$4817.20		$5560.20	$401.43	$463.35

Crew B-14G

Crew No. B-14G	Hr.	Daily	Hr.	Daily	Bare Costs	Incl. O&P
1 Equip. Oper. (crane)	$61.45	$491.60	$91.55	$732.40	$55.77	$83.12
.5 Laborer	44.40	177.60	66.25	265.00		
1 Hyd. Shovel, 12 C.Y.		6022.00		6624.20	501.83	552.02
12 L.H., Daily Totals		$6691.20		$7621.60	$557.60	$635.13

Crew B-14J

Crew No. B-14J	Hr.	Daily	Hr.	Daily	Bare Costs	Incl. O&P
1 Equip. Oper. (medium)	$59.00	$472.00	$87.90	$703.20	$54.13	$80.68
.5 Laborer	44.40	177.60	66.25	265.00		
1 F.E. Loader, 8 C.Y.		2285.00		2513.50	190.42	209.46
12 L.H., Daily Totals		$2934.60		$3481.70	$244.55	$290.14

Crew B-14K

Crew No. B-14K	Hr.	Daily	Hr.	Daily	Bare Costs	Incl. O&P
1 Equip. Oper. (medium)	$59.00	$472.00	$87.90	$703.20	$54.13	$80.68
.5 Laborer	44.40	177.60	66.25	265.00		
1 F.E. Loader, 10 C.Y.		2706.00		2976.60	225.50	248.05
12 L.H., Daily Totals		$3355.60		$3944.80	$279.63	$328.73

Crew B-15

Crew No. B-15	Hr.	Daily	Hr.	Daily	Bare Costs	Incl. O&P
1 Equipment Oper. (med.)	$59.00	$472.00	$87.90	$703.20	$52.51	$78.41
.5 Laborer	44.40	177.60	66.25	265.00		
2 Truck Drivers (heavy)	51.30	820.80	76.70	1227.20		
2 Dump Trucks, 12 C.Y., 400 H.P.		1158.70		1274.57		
1 Dozer, 200 H.P.		1520.00		1672.00	95.67	105.23
28 L.H., Daily Totals		$4149.10		$5141.97	$148.18	$183.64

Crew B-16

Crew No. B-16	Hr.	Daily	Hr.	Daily	Bare Costs	Incl. O&P
1 Labor Foreman (outside)	$46.40	$371.20	$69.25	$554.00	$46.63	$69.61
2 Laborers	44.40	710.40	66.25	1060.00		
1 Truck Driver (heavy)	51.30	410.40	76.70	613.60		
1 Dump Truck, 12 C.Y., 400 H.P.		579.35		637.28	18.10	19.92
32 L.H., Daily Totals		$2071.35		$2864.89	$64.73	$89.53

Crew B-17

Crew No. B-17	Hr.	Daily	Hr.	Daily	Bare Costs	Incl. O&P
2 Laborers	$44.40	$710.40	$66.25	$1060.00	$48.90	$72.97
1 Equip. Oper. (light)	55.50	444.00	82.70	661.60		
1 Truck Driver (heavy)	51.30	410.40	76.70	613.60		
1 Backhoe Loader, 48 H.P.		216.20		237.82		
1 Dump Truck, 8 C.Y., 220 H.P.		407.60		448.36	19.49	21.44
32 L.H., Daily Totals		$2188.60		$3021.38	$68.39	$94.42

Crew B-17A

Crew No. B-17A	Hr.	Daily	Hr.	Daily	Bare Costs	Incl. O&P
2 Labor Foremen (outside)	$46.40	$742.40	$69.25	$1108.00	$47.54	$71.08
6 Laborers	44.40	2131.20	66.25	3180.00		
1 Skilled Worker Foreman (out)	59.10	472.80	88.90	711.20		
1 Skilled Worker	57.10	456.80	85.90	687.20		
80 L.H., Daily Totals		$3803.20		$5686.40	$47.54	$71.08

Crew No.	Bare Costs		Incl. Subs O&P		Cost Per Labor-Hour	
Crew B-17B	Hr.	Daily	Hr.	Daily	Bare Costs	Incl. O&P
2 Laborers	$44.40	$710.40	$66.25	$1060.00	$48.90	$72.97
1 Equip. Oper. (light)	55.50	444.00	82.70	661.60		
1 Truck Driver (heavy)	51.30	410.40	76.70	613.60		
1 Backhoe Loader, 48 H.P.		216.20		237.82		
1 Dump Truck, 12 C.Y., 400 H.P.		579.35		637.28	24.86	27.35
32 L.H., Daily Totals		$2360.35		$3210.30	$73.76	$100.32
Crew B-18	Hr.	Daily	Hr.	Daily	Bare Costs	Incl. O&P
1 Labor Foreman (outside)	$46.40	$371.20	$69.25	$554.00	$45.07	$67.25
2 Laborers	44.40	710.40	66.25	1060.00		
1 Vibrating Plate, Gas, 21"		165.60		182.16	6.90	7.59
24 L.H., Daily Totals		$1247.20		$1796.16	$51.97	$74.84
Crew B-19	Hr.	Daily	Hr.	Daily	Bare Costs	Incl. O&P
1 Pile Driver Foreman (outside)	$57.90	$463.20	$89.50	$716.00	$57.11	$87.06
4 Pile Drivers	55.90	1788.80	86.40	2764.80		
2 Equip. Oper. (crane)	61.45	983.20	91.55	1464.80		
1 Equip. Oper. (oiler)	52.50	420.00	78.25	626.00		
1 Crawler Crane, 40 Ton		1231.00		1354.10		
1 Lead, 90' High		371.85		409.04		
1 Hammer, Diesel, 22k ft-lb		441.70		485.87	31.95	35.14
64 L.H., Daily Totals		$5699.75		$7820.60	$89.06	$122.20
Crew B-19A	Hr.	Daily	Hr.	Daily	Bare Costs	Incl. O&P
1 Pile Driver Foreman (outside)	$57.90	$463.20	$89.50	$716.00	$57.11	$87.06
4 Pile Drivers	55.90	1788.80	86.40	2764.80		
2 Equip. Oper. (crane)	61.45	983.20	91.55	1464.80		
1 Equip. Oper. (oiler)	52.50	420.00	78.25	626.00		
1 Crawler Crane, 75 Ton		1967.00		2163.70		
1 Lead, 90' High		371.85		409.04		
1 Hammer, Diesel, 41k ft-lb		583.55		641.90	45.66	50.23
64 L.H., Daily Totals		$6577.60		$8786.24	$102.78	$137.29
Crew B-19B	Hr.	Daily	Hr.	Daily	Bare Costs	Incl. O&P
1 Pile Driver Foreman (outside)	$57.90	$463.20	$89.50	$716.00	$57.11	$87.06
4 Pile Drivers	55.90	1788.80	86.40	2764.80		
2 Equip. Oper. (crane)	61.45	983.20	91.55	1464.80		
1 Equip. Oper. (oiler)	52.50	420.00	78.25	626.00		
1 Crawler Crane, 40 Ton		1231.00		1354.10		
1 Lead, 90' High		371.85		409.04		
1 Hammer, Diesel, 22k ft-lb		441.70		485.87		
1 Barge, 400 Ton		869.15		956.07	45.53	50.08
64 L.H., Daily Totals		$6568.90		$8776.67	$102.64	$137.14
Crew B-19C	Hr.	Daily	Hr.	Daily	Bare Costs	Incl. O&P
1 Pile Driver Foreman (outside)	$57.90	$463.20	$89.50	$716.00	$57.11	$87.06
4 Pile Drivers	55.90	1788.80	86.40	2764.80		
2 Equip. Oper. (crane)	61.45	983.20	91.55	1464.80		
1 Equip. Oper. (oiler)	52.50	420.00	78.25	626.00		
1 Crawler Crane, 75 Ton		1967.00		2163.70		
1 Lead, 90' High		371.85		409.04		
1 Hammer, Diesel, 41k ft-lb		583.55		641.90		
1 Barge, 400 Ton		869.15		956.07	59.24	65.17
64 L.H., Daily Totals		$7446.75		$9742.31	$116.36	$152.22
Crew B-20	Hr.	Daily	Hr.	Daily	Bare Costs	Incl. O&P
1 Labor Foreman (outside)	$46.40	$371.20	$69.25	$554.00	$49.30	$73.80
1 Skilled Worker	57.10	456.80	85.90	687.20		
1 Laborer	44.40	355.20	66.25	530.00		
24 L.H., Daily Totals		$1183.20		$1771.20	$49.30	$73.80

Crew No.	Bare Costs		Incl. Subs O&P		Cost Per Labor-Hour	
Crew B-20A	Hr.	Daily	Hr.	Daily	Bare Costs	Incl. O&P
1 Labor Foreman (outside)	$46.40	$371.20	$69.25	$554.00	$53.16	$79.34
1 Laborer	44.40	355.20	66.25	530.00		
1 Plumber	67.70	541.60	101.05	808.40		
1 Plumber Apprentice	54.15	433.20	80.80	646.40		
32 L.H., Daily Totals		$1701.20		$2538.80	$53.16	$79.34
Crew B-21	Hr.	Daily	Hr.	Daily	Bare Costs	Incl. O&P
1 Labor Foreman (outside)	$46.40	$371.20	$69.25	$554.00	$51.04	$76.34
1 Skilled Worker	57.10	456.80	85.90	687.20		
1 Laborer	44.40	355.20	66.25	530.00		
.5 Equip. Oper. (crane)	61.45	245.80	91.55	366.20		
.5 S.P. Crane, 4x4, 5 Ton		190.97		210.07	6.82	7.50
28 L.H., Daily Totals		$1619.97		$2347.47	$57.86	$83.84
Crew B-21A	Hr.	Daily	Hr.	Daily	Bare Costs	Incl. O&P
1 Labor Foreman (outside)	$46.40	$371.20	$69.25	$554.00	$54.82	$81.78
1 Laborer	44.40	355.20	66.25	530.00		
1 Plumber	67.70	541.60	101.05	808.40		
1 Plumber Apprentice	54.15	433.20	80.80	646.40		
1 Equip. Oper. (crane)	61.45	491.60	91.55	732.40		
1 S.P. Crane, 4x4, 12 Ton		432.65		475.92	10.82	11.90
40 L.H., Daily Totals		$2625.45		$3747.11	$65.64	$93.68
Crew B-21B	Hr.	Daily	Hr.	Daily	Bare Costs	Incl. O&P
1 Labor Foreman (outside)	$46.40	$371.20	$69.25	$554.00	$48.21	$71.91
3 Laborers	44.40	1065.60	66.25	1590.00		
1 Equip. Oper. (crane)	61.45	491.60	91.55	732.40		
1 Hyd. Crane, 12 Ton		475.80		523.38	11.90	13.08
40 L.H., Daily Totals		$2404.20		$3399.78	$60.10	$84.99
Crew B-21C	Hr.	Daily	Hr.	Daily	Bare Costs	Incl. O&P
1 Labor Foreman (outside)	$46.40	$371.20	$69.25	$554.00	$48.28	$72.01
4 Laborers	44.40	1420.80	66.25	2120.00		
1 Equip. Oper. (crane)	61.45	491.60	91.55	732.40		
1 Equip. Oper. (oiler)	52.50	420.00	78.25	626.00		
2 Cutting Torches		25.90		28.49		
2 Sets of Gases		347.20		381.92		
1 Lattice Boom Crane, 90 Ton		1713.00		1884.30	37.25	40.98
56 L.H., Daily Totals		$4789.70		$6327.11	$85.53	$112.98
Crew B-22	Hr.	Daily	Hr.	Daily	Bare Costs	Incl. O&P
1 Labor Foreman (outside)	$46.40	$371.20	$69.25	$554.00	$51.73	$77.35
1 Skilled Worker	57.10	456.80	85.90	687.20		
1 Laborer	44.40	355.20	66.25	530.00		
.75 Equip. Oper. (crane)	61.45	368.70	91.55	549.30		
.75 S.P. Crane, 4x4, 5 Ton		286.46		315.11	9.55	10.50
30 L.H., Daily Totals		$1838.36		$2635.61	$61.28	$87.85
Crew B-22A	Hr.	Daily	Hr.	Daily	Bare Costs	Incl. O&P
1 Labor Foreman (outside)	$46.40	$371.20	$69.25	$554.00	$50.75	$75.84
1 Skilled Worker	57.10	456.80	85.90	687.20		
2 Laborers	44.40	710.40	66.25	1060.00		
1 Equipment Operator, Crane	61.45	491.60	91.55	732.40		
1 S.P. Crane, 4x4, 5 Ton		381.95		420.14		
1 Butt Fusion Mach., 4"-12" diam.		420.75		462.82	20.07	22.07
40 L.H., Daily Totals		$2832.70		$3916.57	$70.82	$97.91

Crew No.	Bare Costs		Incl. Subs O&P		Cost Per Labor-Hour	

Crew B-22B

	Hr.	Daily	Hr.	Daily	Bare Costs	Incl. O&P
1 Labor Foreman (outside)	$46.40	$371.20	$69.25	$554.00	$50.75	$75.84
1 Skilled Worker	57.10	456.80	85.90	687.20		
2 Laborers	44.40	710.40	66.25	1060.00		
1 Equip. Oper. (crane)	61.45	491.60	91.55	732.40		
1 S.P. Crane, 4x4, 5 Ton		381.95		420.14		
1 Butt Fusion Mach., 8"-24" diam.		1086.00		1194.60	36.70	40.37
40 L.H., Daily Totals		$3497.95		$4648.35	$87.45	$116.21

Crew B-22C

	Hr.	Daily	Hr.	Daily	Bare Costs	Incl. O&P
1 Skilled Worker	$57.10	$456.80	$85.90	$687.20	$50.75	$76.08
1 Laborer	44.40	355.20	66.25	530.00		
1 Butt Fusion Mach., 2"-8" diam.		134.95		148.44	8.43	9.28
16 L.H., Daily Totals		$946.95		$1365.65	$59.18	$85.35

Crew B-23

	Hr.	Daily	Hr.	Daily	Bare Costs	Incl. O&P
1 Labor Foreman (outside)	$46.40	$371.20	$69.25	$554.00	$44.80	$66.85
4 Laborers	44.40	1420.80	66.25	2120.00		
1 Drill Rig, Truck-Mounted		768.40		845.24		
1 Flatbed Truck, Gas, 3 Ton		850.05		935.05	40.46	44.51
40 L.H., Daily Totals		$3410.45		$4454.30	$85.26	$111.36

Crew B-23A

	Hr.	Daily	Hr.	Daily	Bare Costs	Incl. O&P
1 Labor Foreman (outside)	$46.40	$371.20	$69.25	$554.00	$49.93	$74.47
1 Laborer	44.40	355.20	66.25	530.00		
1 Equip. Oper. (medium)	59.00	472.00	87.90	703.20		
1 Drill Rig, Truck-Mounted		768.40		845.24		
1 Pickup Truck, 3/4 Ton		112.20		123.42	36.69	40.36
24 L.H., Daily Totals		$2079.00		$2755.86	$86.63	$114.83

Crew B-23B

	Hr.	Daily	Hr.	Daily	Bare Costs	Incl. O&P
1 Labor Foreman (outside)	$46.40	$371.20	$69.25	$554.00	$49.93	$74.47
1 Laborer	44.40	355.20	66.25	530.00		
1 Equip. Oper. (medium)	59.00	472.00	87.90	703.20		
1 Drill Rig, Truck-Mounted		768.40		845.24		
1 Pickup Truck, 3/4 Ton		112.20		123.42		
1 Centr. Water Pump, 6"		235.25		258.77	46.49	51.14
24 L.H., Daily Totals		$2314.25		$3014.64	$96.43	$125.61

Crew B-24

	Hr.	Daily	Hr.	Daily	Bare Costs	Incl. O&P
1 Cement Finisher	$51.80	$414.40	$75.90	$607.20	$50.30	$74.60
1 Laborer	44.40	355.20	66.25	530.00		
1 Carpenter	54.70	437.60	81.65	653.20		
24 L.H., Daily Totals		$1207.20		$1790.40	$50.30	$74.60

Crew B-25

	Hr.	Daily	Hr.	Daily	Bare Costs	Incl. O&P
1 Labor Foreman (outside)	$46.40	$371.20	$69.25	$554.00	$48.56	$72.43
7 Laborers	44.40	2486.40	66.25	3710.00		
3 Equip. Oper. (medium)	59.00	1416.00	87.90	2109.60		
1 Asphalt Paver, 130 H.P.		2143.00		2357.30		
1 Tandem Roller, 10 Ton		246.80		271.48		
1 Roller, Pneum. Whl., 12 Ton		349.90		384.89	31.13	34.25
88 L.H., Daily Totals		$7013.30		$9387.27	$79.70	$106.67

Crew B-25B

	Hr.	Daily	Hr.	Daily	Bare Costs	Incl. O&P
1 Labor Foreman (outside)	$46.40	$371.20	$69.25	$554.00	$49.43	$73.72
7 Laborers	44.40	2486.40	66.25	3710.00		
4 Equip. Oper. (medium)	59.00	1888.00	87.90	2812.80		
1 Asphalt Paver, 130 H.P.		2143.00		2357.30		
2 Tandem Rollers, 10 Ton		493.60		542.96		
2 Roller, Pneum. Whl., 12 Ton		349.90		384.89	31.11	34.22
96 L.H., Daily Totals		$7732.10		$10361.95	$80.54	$107.94

Crew B-25C

	Hr.	Daily	Hr.	Daily	Bare Costs	Incl. O&P
1 Labor Foreman (outside)	$46.40	$371.20	$69.25	$554.00	$49.60	$73.97
3 Laborers	44.40	1065.60	66.25	1590.00		
2 Equip. Oper. (medium)	59.00	944.00	87.90	1406.40		
1 Asphalt Paver, 130 H.P.		2143.00		2357.30		
1 Tandem Roller, 10 Ton		246.80		271.48	49.79	54.77
48 L.H., Daily Totals		$4770.60		$6179.18	$99.39	$128.73

Crew B-25D

	Hr.	Daily	Hr.	Daily	Bare Costs	Incl. O&P
1 Labor Foreman (outside)	$46.40	$371.20	$69.25	$554.00	$49.82	$74.30
3 Laborers	44.40	1065.60	66.25	1590.00		
2.125 Equip. Oper. (medium)	59.00	1003.00	87.90	1494.30		
.125 Truck Driver (heavy)	51.30	51.30	76.70	76.70		
.125 Truck Tractor, 6x4, 380 H.P.		62.39		68.63		
.125 Dist. Tanker, 3000 Gallon		41.76		45.94		
1 Asphalt Paver, 130 H.P.		2143.00		2357.30		
1 Tandem Roller, 10 Ton		246.80		271.48	49.88	54.87
50 L.H., Daily Totals		$4985.06		$6458.35	$99.70	$129.17

Crew B-25E

	Hr.	Daily	Hr.	Daily	Bare Costs	Incl. O&P
1 Labor Foreman (outside)	$46.40	$371.20	$69.25	$554.00	$50.03	$74.61
3 Laborers	44.40	1065.60	66.25	1590.00		
2.250 Equip. Oper. (medium)	59.00	1062.00	87.90	1582.20		
.25 Truck Driver (heavy)	51.30	102.60	76.70	153.40		
.25 Truck Tractor, 6x4, 380 H.P.		124.79		137.27		
.25 Dist. Tanker, 3000 Gallon		83.53		91.88		
1 Asphalt Paver, 130 H.P.		2143.00		2357.30		
1 Tandem Roller, 10 Ton		246.80		271.48	49.96	54.96
52 L.H., Daily Totals		$5199.51		$6737.52	$99.99	$129.57

Crew B-26

	Hr.	Daily	Hr.	Daily	Bare Costs	Incl. O&P
1 Labor Foreman (outside)	$46.40	$371.20	$69.25	$554.00	$49.23	$73.32
6 Laborers	44.40	2131.20	66.25	3180.00		
2 Equip. Oper. (medium)	59.00	944.00	87.90	1406.40		
1 Rodman (reinf.)	58.90	471.20	88.05	704.40		
1 Cement Finisher	51.80	414.40	75.90	607.20		
1 Grader, 30,000 Lbs.		1073.00		1180.30		
1 Paving Mach. & Equip.		2503.00		2753.30	40.64	44.70
88 L.H., Daily Totals		$7908.00		$10385.60	$89.86	$118.02

Crew B-26A

	Hr.	Daily	Hr.	Daily	Bare Costs	Incl. O&P
1 Labor Foreman (outside)	$46.40	$371.20	$69.25	$554.00	$49.23	$73.32
6 Laborers	44.40	2131.20	66.25	3180.00		
2 Equip. Oper. (medium)	59.00	944.00	87.90	1406.40		
1 Rodman (reinf.)	58.90	471.20	88.05	704.40		
1 Cement Finisher	51.80	414.40	75.90	607.20		
1 Grader, 30,000 Lbs.		1073.00		1180.30		
1 Paving Mach. & Equip.		2503.00		2753.30		
1 Concrete Saw		112.85		124.14	41.92	46.11
88 L.H., Daily Totals		$8020.85		$10509.74	$91.15	$119.43

Crew B-26B

	Hr.	Daily	Hr.	Daily	Bare Costs	Incl. O&P
1 Labor Foreman (outside)	$46.40	$371.20	$69.25	$554.00	$50.04	$74.53
6 Laborers	44.40	2131.20	66.25	3180.00		
3 Equip. Oper. (medium)	59.00	1416.00	87.90	2109.60		
1 Rodman (reinf.)	58.90	471.20	88.05	704.40		
1 Cement Finisher	51.80	414.40	75.90	607.20		
1 Grader, 30,000 Lbs.		1073.00		1180.30		
1 Paving Mach. & Equip.		2503.00		2753.30		
1 Concrete Pump, 110' Boom		493.65		543.01	42.39	46.63
96 L.H., Daily Totals		$8873.65		$11631.82	$92.43	$121.16

For customer support on your Site Work & Landscape Costs with RSMeans Data, call 800.448.8182.

Crew No.	Bare Costs		Incl. Subs O&P		Cost Per Labor-Hour	

Crew B-26C

	Hr.	Daily	Hr.	Daily	Bare Costs	Incl. O&P
1 Labor Foreman (outside)	$46.40	$371.20	$69.25	$554.00	$48.25	$71.86
6 Laborers	44.40	2131.20	66.25	3180.00		
1 Equip. Oper. (medium)	59.00	472.00	87.90	703.20		
1 Rodman (reinf.)	58.90	471.20	88.05	704.40		
1 Cement Finisher	51.80	414.40	75.90	607.20		
1 Paving Mach. & Equip.		2503.00		2753.30		
1 Concrete Saw		112.85		124.14	32.70	35.97
80 L.H., Daily Totals		$6475.85		$8626.24	$80.95	$107.83

Crew B-27

	Hr.	Daily	Hr.	Daily	Bare Costs	Incl. O&P
1 Labor Foreman (outside)	$46.40	$371.20	$69.25	$554.00	$44.90	$67.00
3 Laborers	44.40	1065.60	66.25	1590.00		
1 Berm Machine		253.55		278.90	7.92	8.72
32 L.H., Daily Totals		$1690.35		$2422.91	$52.82	$75.72

Crew B-28

	Hr.	Daily	Hr.	Daily	Bare Costs	Incl. O&P
2 Carpenters	$54.70	$875.20	$81.65	$1306.40	$51.27	$76.52
1 Laborer	44.40	355.20	66.25	530.00		
24 L.H., Daily Totals		$1230.40		$1836.40	$51.27	$76.52

Crew B-29

	Hr.	Daily	Hr.	Daily	Bare Costs	Incl. O&P
1 Labor Foreman (outside)	$46.40	$371.20	$69.25	$554.00	$48.28	$72.01
4 Laborers	44.40	1420.80	66.25	2120.00		
1 Equip. Oper. (crane)	61.45	491.60	91.55	732.40		
1 Equip. Oper. (oiler)	52.50	420.00	78.25	626.00		
1 Gradall, 5/8 C.Y.		850.65		935.72	15.19	16.71
56 L.H., Daily Totals		$3554.25		$4968.11	$63.47	$88.72

Crew B-30

	Hr.	Daily	Hr.	Daily	Bare Costs	Incl. O&P
1 Equip. Oper. (medium)	$59.00	$472.00	$87.90	$703.20	$53.87	$80.43
2 Truck Drivers (heavy)	51.30	820.80	76.70	1227.20		
1 Hyd. Excavator, 1.5 C.Y.		695.80		765.38		
2 Dump Trucks, 12 C.Y., 400 H.P.		1158.70		1274.57	77.27	85.00
24 L.H., Daily Totals		$3147.30		$3970.35	$131.14	$165.43

Crew B-31

	Hr.	Daily	Hr.	Daily	Bare Costs	Incl. O&P
1 Labor Foreman (outside)	$46.40	$371.20	$69.25	$554.00	$46.86	$69.93
3 Laborers	44.40	1065.60	66.25	1590.00		
1 Carpenter	54.70	437.60	81.65	653.20		
1 Air Compressor, 250 cfm		202.85		223.13		
1 Sheeting Driver		7.45		8.20		
2 -50' Air Hoses, 1.5"		45.60		50.16	6.40	7.04
40 L.H., Daily Totals		$2130.30		$3078.69	$53.26	$76.97

Crew B-32

	Hr.	Daily	Hr.	Daily	Bare Costs	Incl. O&P
1 Laborer	$44.40	$355.20	$66.25	$530.00	$55.35	$82.49
3 Equip. Oper. (medium)	59.00	1416.00	87.90	2109.60		
1 Grader, 30,000 Lbs.		1073.00		1180.30		
1 Tandem Roller, 10 Ton		246.80		271.48		
1 Dozer, 200 H.P.		1520.00		1672.00	88.74	97.62
32 L.H., Daily Totals		$4611.00		$5763.38	$144.09	$180.11

Crew B-32A

	Hr.	Daily	Hr.	Daily	Bare Costs	Incl. O&P
1 Laborer	$44.40	$355.20	$66.25	$530.00	$54.13	$80.68
2 Equip. Oper. (medium)	59.00	944.00	87.90	1406.40		
1 Grader, 30,000 Lbs.		1073.00		1180.30		
1 Roller, Vibratory, 25 Ton		672.35		739.59	72.72	80.00
24 L.H., Daily Totals		$3044.55		$3856.28	$126.86	$160.68

Crew B-32B

	Hr.	Daily	Hr.	Daily	Bare Costs	Incl. O&P
1 Laborer	$44.40	$355.20	$66.25	$530.00	$54.13	$80.68
2 Equip. Oper. (medium)	59.00	944.00	87.90	1406.40		
1 Dozer, 200 H.P.		1520.00		1672.00		
1 Roller, Vibratory, 25 Ton		672.35		739.59	91.35	100.48
24 L.H., Daily Totals		$3491.55		$4347.98	$145.48	$181.17

Crew B-32C

	Hr.	Daily	Hr.	Daily	Bare Costs	Incl. O&P
1 Labor Foreman (outside)	$46.40	$371.20	$69.25	$554.00	$52.03	$77.58
2 Laborers	44.40	710.40	66.25	1060.00		
3 Equip. Oper. (medium)	59.00	1416.00	87.90	2109.60		
1 Grader, 30,000 Lbs.		1073.00		1180.30		
1 Tandem Roller, 10 Ton		246.80		271.48		
1 Dozer, 200 H.P.		1520.00		1672.00	59.16	65.08
48 L.H., Daily Totals		$5337.40		$6847.38	$111.20	$142.65

Crew B-33A

	Hr.	Daily	Hr.	Daily	Bare Costs	Incl. O&P
1 Equip. Oper. (medium)	$59.00	$472.00	$87.90	$703.20	$54.83	$81.71
.5 Laborer	44.40	177.60	66.25	265.00		
.25 Equip. Oper. (medium)	59.00	118.00	87.90	175.80		
1 Scraper, Towed, 7 C.Y.		129.30		142.23		
1.25 Dozers, 300 H.P.		2231.25		2454.38	168.61	185.47
14 L.H., Daily Totals		$3128.15		$3740.61	$223.44	$267.19

Crew B-33B

	Hr.	Daily	Hr.	Daily	Bare Costs	Incl. O&P
1 Equip. Oper. (medium)	$59.00	$472.00	$87.90	$703.20	$54.83	$81.71
.5 Laborer	44.40	177.60	66.25	265.00		
.25 Equip. Oper. (medium)	59.00	118.00	87.90	175.80		
1 Scraper, Towed, 10 C.Y.		161.65		177.82		
1.25 Dozers, 300 H.P.		2231.25		2454.38	170.92	188.01
14 L.H., Daily Totals		$3160.50		$3776.19	$225.75	$269.73

Crew B-33C

	Hr.	Daily	Hr.	Daily	Bare Costs	Incl. O&P
1 Equip. Oper. (medium)	$59.00	$472.00	$87.90	$703.20	$54.83	$81.71
.5 Laborer	44.40	177.60	66.25	265.00		
.25 Equip. Oper. (medium)	59.00	118.00	87.90	175.80		
1 Scraper, Towed, 15 C.Y.		178.85		196.74		
1.25 Dozers, 300 H.P.		2231.25		2454.38	172.15	189.37
14 L.H., Daily Totals		$3177.70		$3795.11	$226.98	$271.08

Crew B-33D

	Hr.	Daily	Hr.	Daily	Bare Costs	Incl. O&P
1 Equip. Oper. (medium)	$59.00	$472.00	$87.90	$703.20	$54.83	$81.71
.5 Laborer	44.40	177.60	66.25	265.00		
.25 Equip. Oper. (medium)	59.00	118.00	87.90	175.80		
1 S.P. Scraper, 14 C.Y.		2424.00		2666.40		
.25 Dozer, 300 H.P.		446.25		490.88	205.02	225.52
14 L.H., Daily Totals		$3637.85		$4301.27	$259.85	$307.23

Crew B-33E

	Hr.	Daily	Hr.	Daily	Bare Costs	Incl. O&P
1 Equip. Oper. (medium)	$59.00	$472.00	$87.90	$703.20	$54.83	$81.71
.5 Laborer	44.40	177.60	66.25	265.00		
.25 Equip. Oper. (medium)	59.00	118.00	87.90	175.80		
1 S.P. Scraper, 21 C.Y.		2656.00		2921.60		
.25 Dozer, 300 H.P.		446.25		490.88	221.59	243.75
14 L.H., Daily Totals		$3869.85		$4556.48	$276.42	$325.46

Crew No.	Bare Costs Hr.	Daily	Incl. Subs O&P Hr.	Daily	Cost Per Labor-Hour Bare Costs	Incl. O&P
Crew B-33F	Hr.	Daily	Hr.	Daily	Bare Costs	Incl. O&P
1 Equip. Oper. (medium)	$59.00	$472.00	$87.90	$703.20	$54.83	$81.71
.5 Laborer	44.40	177.60	66.25	265.00		
.25 Equip. Oper. (medium)	59.00	118.00	87.90	175.80		
1 Elev. Scraper, 11 C.Y.		1059.00		1164.90		
.25 Dozer, 300 H.P.		446.25		490.88	107.52	118.27
14 L.H., Daily Totals		$2272.85		$2799.78	$162.35	$199.98
Crew B-33G	Hr.	Daily	Hr.	Daily	Bare Costs	Incl. O&P
1 Equip. Oper. (medium)	$59.00	$472.00	$87.90	$703.20	$54.83	$81.71
.5 Laborer	44.40	177.60	66.25	265.00		
.25 Equip. Oper. (medium)	59.00	118.00	87.90	175.80		
1 Elev. Scraper, 22 C.Y.		1895.00		2084.50		
.25 Dozer, 300 H.P.		446.25		490.88	167.23	183.96
14 L.H., Daily Totals		$3108.85		$3719.38	$222.06	$265.67
Crew B-33H	Hr.	Daily	Hr.	Daily	Bare Costs	Incl. O&P
.5 Laborer	$44.40	$177.60	$66.25	$265.00	$54.83	$81.71
1 Equipment Operator (med.)	59.00	472.00	87.90	703.20		
.25 Equipment Operator (med.)	59.00	118.00	87.90	175.80		
1 S.P. Scraper, 44 C.Y.		4695.00		5164.50		
.25 Dozer, 410 H.P.		701.75		771.92	385.48	424.03
14 L.H., Daily Totals		$6164.35		$7080.43	$440.31	$505.74
Crew B-33J	Hr.	Daily	Hr.	Daily	Bare Costs	Incl. O&P
1 Equipment Operator (med.)	$59.00	$472.00	$87.90	$703.20	$59.00	$87.90
1 S.P. Scraper, 14 C.Y.		2424.00		2666.40	303.00	333.30
8 L.H., Daily Totals		$2896.00		$3369.60	$362.00	$421.20
Crew B-33K	Hr.	Daily	Hr.	Daily	Bare Costs	Incl. O&P
1 Equipment Operator (med.)	$59.00	$472.00	$87.90	$703.20	$54.83	$81.71
.25 Equipment Operator (med.)	59.00	118.00	87.90	175.80		
.5 Laborer	44.40	177.60	66.25	265.00		
1 S.P. Scraper, 31 C.Y.		3707.00		4077.70		
.25 Dozer, 410 H.P.		701.75		771.92	314.91	346.40
14 L.H., Daily Totals		$5176.35		$5993.63	$369.74	$428.12
Crew B-34A	Hr.	Daily	Hr.	Daily	Bare Costs	Incl. O&P
1 Truck Driver (heavy)	$51.30	$410.40	$76.70	$613.60	$51.30	$76.70
1 Dump Truck, 8 C.Y., 220 H.P.		407.60		448.36	50.95	56.05
8 L.H., Daily Totals		$818.00		$1061.96	$102.25	$132.75
Crew B-34B	Hr.	Daily	Hr.	Daily	Bare Costs	Incl. O&P
1 Truck Driver (heavy)	$51.30	$410.40	$76.70	$613.60	$51.30	$76.70
1 Dump Truck, 12 C.Y., 400 H.P.		579.35		637.28	72.42	79.66
8 L.H., Daily Totals		$989.75		$1250.89	$123.72	$156.36
Crew B-34C	Hr.	Daily	Hr.	Daily	Bare Costs	Incl. O&P
1 Truck Driver (heavy)	$51.30	$410.40	$76.70	$613.60	$51.30	$76.70
1 Truck Tractor, 6x4, 380 H.P.		499.15		549.07		
1 Dump Trailer, 16.5 C.Y.		138.35		152.19	79.69	87.66
8 L.H., Daily Totals		$1047.90		$1314.85	$130.99	$164.36
Crew B-34D	Hr.	Daily	Hr.	Daily	Bare Costs	Incl. O&P
1 Truck Driver (heavy)	$51.30	$410.40	$76.70	$613.60	$51.30	$76.70
1 Truck Tractor, 6x4, 380 H.P.		499.15		549.07		
1 Dump Trailer, 20 C.Y.		153.55		168.91	81.59	89.75
8 L.H., Daily Totals		$1063.10		$1331.57	$132.89	$166.45

Crew No.	Bare Costs Hr.	Daily	Incl. Subs O&P Hr.	Daily	Cost Per Labor-Hour Bare Costs	Incl. O&P
Crew B-34E	Hr.	Daily	Hr.	Daily	Bare Costs	Incl. O&P
1 Truck Driver (heavy)	$51.30	$410.40	$76.70	$613.60	$51.30	$76.70
1 Dump Truck, Off Hwy., 25 Ton		1427.00		1569.70	178.38	196.21
8 L.H., Daily Totals		$1837.40		$2183.30	$229.68	$272.91
Crew B-34F	Hr.	Daily	Hr.	Daily	Bare Costs	Incl. O&P
1 Truck Driver (heavy)	$51.30	$410.40	$76.70	$613.60	$51.30	$76.70
1 Dump Truck, Off Hwy., 35 Ton		945.65		1040.21	118.21	130.03
8 L.H., Daily Totals		$1356.05		$1653.82	$169.51	$206.73
Crew B-34G	Hr.	Daily	Hr.	Daily	Bare Costs	Incl. O&P
1 Truck Driver (heavy)	$51.30	$410.40	$76.70	$613.60	$51.30	$76.70
1 Dump Truck, Off Hwy., 50 Ton		1979.00		2176.90	247.38	272.11
8 L.H., Daily Totals		$2389.40		$2790.50	$298.68	$348.80
Crew B-34H	Hr.	Daily	Hr.	Daily	Bare Costs	Incl. O&P
1 Truck Driver (heavy)	$51.30	$410.40	$76.70	$613.60	$51.30	$76.70
1 Dump Truck, Off Hwy., 65 Ton		1938.00		2131.80	242.25	266.48
8 L.H., Daily Totals		$2348.40		$2745.40	$293.55	$343.18
Crew B-34I	Hr.	Daily	Hr.	Daily	Bare Costs	Incl. O&P
1 Truck Driver (heavy)	$51.30	$410.40	$76.70	$613.60	$51.30	$76.70
1 Dump Truck, 18 C.Y., 450 H.P.		753.60		828.96	94.20	103.62
8 L.H., Daily Totals		$1164.00		$1442.56	$145.50	$180.32
Crew B-34J	Hr.	Daily	Hr.	Daily	Bare Costs	Incl. O&P
1 Truck Driver (heavy)	$51.30	$410.40	$76.70	$613.60	$51.30	$76.70
1 Dump Truck, Off Hwy., 100 Ton		2769.00		3045.90	346.13	380.74
8 L.H., Daily Totals		$3179.40		$3659.50	$397.43	$457.44
Crew B-34K	Hr.	Daily	Hr.	Daily	Bare Costs	Incl. O&P
1 Truck Driver (heavy)	$51.30	$410.40	$76.70	$613.60	$51.30	$76.70
1 Truck Tractor, 6x4, 450 H.P.		608.95		669.85		
1 Lowbed Trailer, 75 Ton		258.10		283.91	108.38	119.22
8 L.H., Daily Totals		$1277.45		$1567.36	$159.68	$195.92
Crew B-34L	Hr.	Daily	Hr.	Daily	Bare Costs	Incl. O&P
1 Equip. Oper. (light)	$55.50	$444.00	$82.70	$661.60	$55.50	$82.70
1 Flatbed Truck, Gas, 1.5 Ton		198.50		218.35	24.81	27.29
8 L.H., Daily Totals		$642.50		$879.95	$80.31	$109.99
Crew B-34M	Hr.	Daily	Hr.	Daily	Bare Costs	Incl. O&P
1 Equip. Oper. (light)	$55.50	$444.00	$82.70	$661.60	$55.50	$82.70
1 Flatbed Truck, Gas, 3 Ton		850.05		935.05	106.26	116.88
8 L.H., Daily Totals		$1294.05		$1596.66	$161.76	$199.58
Crew B-34N	Hr.	Daily	Hr.	Daily	Bare Costs	Incl. O&P
1 Truck Driver (heavy)	$51.30	$410.40	$76.70	$613.60	$55.15	$82.30
1 Equip. Oper. (medium)	59.00	472.00	87.90	703.20		
1 Truck Tractor, 6x4, 380 H.P.		499.15		549.07		
1 Flatbed Trailer, 40 Ton		188.45		207.29	42.98	47.27
16 L.H., Daily Totals		$1570.00		$2073.16	$98.13	$129.57
Crew B-34P	Hr.	Daily	Hr.	Daily	Bare Costs	Incl. O&P
1 Pipe Fitter	$68.35	$546.80	$102.00	$816.00	$58.72	$87.63
1 Truck Driver (light)	48.80	390.40	73.00	584.00		
1 Equip. Oper. (medium)	59.00	472.00	87.90	703.20		
1 Flatbed Truck, Gas, 3 Ton		850.05		935.05		
1 Backhoe Loader, 48 H.P.		216.20		237.82	44.43	48.87
24 L.H., Daily Totals		$2475.45		$3276.07	$103.14	$136.50

For customer support on your Site Work & Landscape Costs with RSMeans Data, call 800.448.8182.

Crew No.	Bare Costs		Incl. Subs O&P		Cost Per Labor-Hour	
Crew B-34Q	Hr.	Daily	Hr.	Daily	Bare Costs	Incl. O&P
1 Pipe Fitter	$68.35	$546.80	$102.00	$816.00	$59.53	$88.85
1 Truck Driver (light)	48.80	390.40	73.00	584.00		
1 Equip. Oper. (crane)	61.45	491.60	91.55	732.40		
1 Flatbed Trailer, 25 Ton		137.20		150.92		
1 Dump Truck, 8 C.Y., 220 H.P.		407.60		448.36		
1 Hyd. Crane, 25 Ton		586.70		645.37	47.15	51.86
24 L.H., Daily Totals		$2560.30		$3377.05	$106.68	$140.71
Crew B-34R	Hr.	Daily	Hr.	Daily	Bare Costs	Incl. O&P
1 Pipe Fitter	$68.35	$546.80	$102.00	$816.00	$59.53	$88.85
1 Truck Driver (light)	48.80	390.40	73.00	584.00		
1 Equip. Oper. (crane)	61.45	491.60	91.55	732.40		
1 Flatbed Trailer, 25 Ton		137.20		150.92		
1 Dump Truck, 8 C.Y., 220 H.P.		407.60		448.36		
1 Hyd. Crane, 25 Ton		586.70		645.37		
1 Hyd. Excavator, 1 C.Y.		832.65		915.91	81.84	90.02
24 L.H., Daily Totals		$3392.95		$4292.97	$141.37	$178.87
Crew B-34S	Hr.	Daily	Hr.	Daily	Bare Costs	Incl. O&P
2 Pipe Fitters	$68.35	$1093.60	$102.00	$1632.00	$62.36	$93.06
1 Truck Driver (heavy)	51.30	410.40	76.70	613.60		
1 Equip. Oper. (crane)	61.45	491.60	91.55	732.40		
1 Flatbed Trailer, 40 Ton		188.45		207.29		
1 Truck Tractor, 6x4, 380 H.P.		499.15		549.07		
1 Hyd. Crane, 80 Ton		1458.00		1603.80		
1 Hyd. Excavator, 2 C.Y.		942.70		1036.97	96.51	106.16
32 L.H., Daily Totals		$5083.90		$6375.13	$158.87	$199.22
Crew B-34T	Hr.	Daily	Hr.	Daily	Bare Costs	Incl. O&P
2 Pipe Fitters	$68.35	$1093.60	$102.00	$1632.00	$62.36	$93.06
1 Truck Driver (heavy)	51.30	410.40	76.70	613.60		
1 Equip. Oper. (crane)	61.45	491.60	91.55	732.40		
1 Flatbed Trailer, 40 Ton		188.45		207.29		
1 Truck Tractor, 6x4, 380 H.P.		499.15		549.07		
1 Hyd. Crane, 80 Ton		1458.00		1603.80	67.05	73.75
32 L.H., Daily Totals		$4141.20		$5338.16	$129.41	$166.82
Crew B-34U	Hr.	Daily	Hr.	Daily	Bare Costs	Incl. O&P
1 Truck Driver (heavy)	$51.30	$410.40	$76.70	$613.60	$53.40	$79.70
1 Equip. Oper. (light)	55.50	444.00	82.70	661.60		
1 Truck Tractor, 220 H.P.		310.80		341.88		
1 Flatbed Trailer, 25 Ton		137.20		150.92	28.00	30.80
16 L.H., Daily Totals		$1302.40		$1768.00	$81.40	$110.50
Crew B-34V	Hr.	Daily	Hr.	Daily	Bare Costs	Incl. O&P
1 Truck Driver (heavy)	$51.30	$410.40	$76.70	$613.60	$56.08	$83.65
1 Equip. Oper. (crane)	61.45	491.60	91.55	732.40		
1 Equip. Oper. (light)	55.50	444.00	82.70	661.60		
1 Truck Tractor, 6x4, 450 H.P.		608.95		669.85		
1 Equipment Trailer, 50 Ton		207.25		227.97		
1 Pickup Truck, 4x4, 3/4 Ton		176.75		194.43	41.37	45.51
24 L.H., Daily Totals		$2338.95		$3099.84	$97.46	$129.16

Crew No.	Bare Costs		Incl. Subs O&P		Cost Per Labor-Hour	
Crew B-34W	Hr.	Daily	Hr.	Daily	Bare Costs	Incl. O&P
5 Truck Drivers (heavy)	$51.30	$2052.00	$76.70	$3068.00	$53.92	$80.50
2 Equip. Opers. (crane)	61.45	983.20	91.55	1464.80		
1 Equip. Oper. (mechanic)	61.50	492.00	91.65	733.20		
1 Laborer	44.40	355.20	66.25	530.00		
4 Truck Tractors, 6x4, 380 H.P.		1996.60		2196.26		
2 Equipment Trailers, 50 Ton		414.50		455.95		
2 Flatbed Trailers, 40 Ton		376.90		414.59		
1 Pickup Truck, 4x4, 3/4 Ton		176.75		194.43		
1 S.P. Crane, 4x4, 20 Ton		574.35		631.78	49.15	54.07
72 L.H., Daily Totals		$7421.50		$9689.01	$103.08	$134.57
Crew B-35	Hr.	Daily	Hr.	Daily	Bare Costs	Incl. O&P
1 Labor Foreman (outside)	$46.40	$371.20	$69.25	$554.00	$56.75	$84.76
1 Skilled Worker	57.10	456.80	85.90	687.20		
2 Welders	67.70	1083.20	101.05	1616.80		
1 Laborer	44.40	355.20	66.25	530.00		
1 Equip. Oper. (crane)	61.45	491.60	91.55	732.40		
1 Equip. Oper. (oiler)	52.50	420.00	78.25	626.00		
2 Welder, Electric, 300 amp		215.10		236.61		
1 Hyd. Excavator, .75 C.Y.		701.80		771.98	16.37	18.01
56 L.H., Daily Totals		$4094.90		$5754.99	$73.12	$102.77
Crew B-35A	Hr.	Daily	Hr.	Daily	Bare Costs	Incl. O&P
1 Labor Foreman (outside)	$46.40	$371.20	$69.25	$554.00	$53.42	$79.79
2 Laborers	44.40	710.40	66.25	1060.00		
1 Skilled Worker	57.10	456.80	85.90	687.20		
1 Welder (plumber)	67.70	541.60	101.05	808.40		
1 Equip. Oper. (crane)	61.45	491.60	91.55	732.40		
1 Equip. Oper. (oiler)	52.50	420.00	78.25	626.00		
1 Welder, Gas Engine, 300 amp		148.75		163.63		
1 Crawler Crane, 75 Ton		1967.00		2163.70	37.78	41.56
56 L.H., Daily Totals		$5107.35		$6795.32	$91.20	$121.35
Crew B-36	Hr.	Daily	Hr.	Daily	Bare Costs	Incl. O&P
1 Labor Foreman (outside)	$46.40	$371.20	$69.25	$554.00	$50.64	$75.51
2 Laborers	44.40	710.40	66.25	1060.00		
2 Equip. Oper. (medium)	59.00	944.00	87.90	1406.40		
1 Dozer, 200 H.P.		1520.00		1672.00		
1 Aggregate Spreader		59.95		65.94		
1 Tandem Roller, 10 Ton		246.80		271.48	45.67	50.24
40 L.H., Daily Totals		$3852.35		$5029.82	$96.31	$125.75
Crew B-36A	Hr.	Daily	Hr.	Daily	Bare Costs	Incl. O&P
1 Labor Foreman (outside)	$46.40	$371.20	$69.25	$554.00	$53.03	$79.05
2 Laborers	44.40	710.40	66.25	1060.00		
4 Equip. Oper. (medium)	59.00	1888.00	87.90	2812.80		
1 Dozer, 200 H.P.		1520.00		1672.00		
1 Aggregate Spreader		59.95		65.94		
1 Tandem Roller, 10 Ton		246.80		271.48		
1 Roller, Pneum. Whl., 12 Ton		349.90		384.89	38.87	42.76
56 L.H., Daily Totals		$5146.25		$6821.11	$91.90	$121.81

For customer support on your Site Work & Landscape Costs with RSMeans Data, call 800.448.8182.

761

Crew No.	Bare Costs		Incl. Subs O&P		Cost Per Labor-Hour	

Crew B-36B	Hr.	Daily	Hr.	Daily	Bare Costs	Incl. O&P
1 Labor Foreman (outside)	$46.40	$371.20	$69.25	$554.00	$52.81	$78.76
2 Laborers	44.40	710.40	66.25	1060.00		
4 Equip. Oper. (medium)	59.00	1888.00	87.90	2812.80		
1 Truck Driver (heavy)	51.30	410.40	76.70	613.60		
1 Grader, 30,000 Lbs.		1073.00		1180.30		
1 F.E. Loader, Crl, 1.5 C.Y.		668.35		735.18		
1 Dozer, 300 H.P.		1785.00		1963.50		
1 Roller, Vibratory, 25 Ton		672.35		739.59		
1 Truck Tractor, 6x4, 450 H.P.		608.95		669.85		
1 Water Tank Trailer, 5000 Gal.		154.10		169.51	77.53	85.28
64 L.H., Daily Totals		$8341.75		$10498.33	$130.34	$164.04

Crew B-36C	Hr.	Daily	Hr.	Daily	Bare Costs	Incl. O&P
1 Labor Foreman (outside)	$46.40	$371.20	$69.25	$554.00	$54.94	$81.93
3 Equip. Oper. (medium)	59.00	1416.00	87.90	2109.60		
1 Truck Driver (heavy)	51.30	410.40	76.70	613.60		
1 Grader, 30,000 Lbs.		1073.00		1180.30		
1 Dozer, 300 H.P.		1785.00		1963.50		
1 Roller, Vibratory, 25 Ton		672.35		739.59		
1 Truck Tractor, 6x4, 450 H.P.		608.95		669.85		
1 Water Tank Trailer, 5000 Gal.		154.10		169.51	107.33	118.07
40 L.H., Daily Totals		$6491.00		$7999.94	$162.28	$200.00

Crew B-36D	Hr.	Daily	Hr.	Daily	Bare Costs	Incl. O&P
1 Labor Foreman (outside)	$46.40	$371.20	$69.25	$554.00	$55.85	$83.24
3 Equip. Oper. (medium)	59.00	1416.00	87.90	2109.60		
1 Grader, 30,000 Lbs.		1073.00		1180.30		
1 Dozer, 300 H.P.		1785.00		1963.50		
1 Roller, Vibratory, 25 Ton		672.35		739.59	110.32	121.36
32 L.H., Daily Totals		$5317.55		$6546.98	$166.17	$204.59

Crew B-37	Hr.	Daily	Hr.	Daily	Bare Costs	Incl. O&P
1 Labor Foreman (outside)	$46.40	$371.20	$69.25	$554.00	$46.58	$69.49
4 Laborers	44.40	1420.80	66.25	2120.00		
1 Equip. Oper. (light)	55.50	444.00	82.70	661.60		
1 Tandem Roller, 5 Ton		258.75		284.63	5.39	5.93
48 L.H., Daily Totals		$2494.75		$3620.22	$51.97	$75.42

Crew B-37A	Hr.	Daily	Hr.	Daily	Bare Costs	Incl. O&P
2 Laborers	$44.40	$710.40	$66.25	$1060.00	$45.87	$68.50
1 Truck Driver (light)	48.80	390.40	73.00	584.00		
1 Flatbed Truck, Gas, 1.5 Ton		198.50		218.35		
1 Tar Kettle, T.M.		156.70		172.37	14.80	16.28
24 L.H., Daily Totals		$1456.00		$2034.72	$60.67	$84.78

Crew B-37B	Hr.	Daily	Hr.	Daily	Bare Costs	Incl. O&P
3 Laborers	$44.40	$1065.60	$66.25	$1590.00	$45.50	$67.94
1 Truck Driver (light)	48.80	390.40	73.00	584.00		
1 Flatbed Truck, Gas, 1.5 Ton		198.50		218.35		
1 Tar Kettle, T.M.		156.70		172.37	11.10	12.21
32 L.H., Daily Totals		$1811.20		$2564.72	$56.60	$80.15

Crew B-37C	Hr.	Daily	Hr.	Daily	Bare Costs	Incl. O&P
2 Laborers	$44.40	$710.40	$66.25	$1060.00	$46.60	$69.63
2 Truck Drivers (light)	48.80	780.80	73.00	1168.00		
2 Flatbed Trucks, Gas, 1.5 Ton		397.00		436.70		
1 Tar Kettle, T.M.		156.70		172.37	17.30	19.03
32 L.H., Daily Totals		$2044.90		$2837.07	$63.90	$88.66

Crew B-37D	Hr.	Daily	Hr.	Daily	Bare Costs	Incl. O&P
1 Laborer	$44.40	$355.20	$66.25	$530.00	$46.60	$69.63
1 Truck Driver (light)	48.80	390.40	73.00	584.00		
1 Pickup Truck, 3/4 Ton		112.20		123.42	7.01	7.71
16 L.H., Daily Totals		$857.80		$1237.42	$53.61	$77.34

Crew B-37E	Hr.	Daily	Hr.	Daily	Bare Costs	Incl. O&P
3 Laborers	$44.40	$1065.60	$66.25	$1590.00	$49.33	$73.62
1 Equip. Oper. (light)	55.50	444.00	82.70	661.60		
1 Equip. Oper. (medium)	59.00	472.00	87.90	703.20		
2 Truck Drivers (light)	48.80	780.80	73.00	1168.00		
4 Barrels w/ Flasher		16.60		18.26		
1 Concrete Saw		112.85		124.14		
1 Rotary Hammer Drill		52.25		57.48		
1 Hammer Drill Bit		25.25		27.77		
1 Loader, Skid Steer, 30 H.P.		179.50		197.45		
1 Conc. Hammer Attach.		118.50		130.35		
1 Vibrating Plate, Gas, 18"		31.90		35.09		
2 Flatbed Trucks, Gas, 1.5 Ton		397.00		436.70	16.68	18.34
56 L.H., Daily Totals		$3696.25		$5150.03	$66.00	$91.96

Crew B-37F	Hr.	Daily	Hr.	Daily	Bare Costs	Incl. O&P
3 Laborers	$44.40	$1065.60	$66.25	$1590.00	$45.50	$67.94
1 Truck Driver (light)	48.80	390.40	73.00	584.00		
4 Barrels w/ Flasher		16.60		18.26		
1 Concrete Mixer, 10 C.F.		147.15		161.87		
1 Air Compressor, 60 cfm		153.85		169.24		
1 -50' Air Hose, 3/4"		7.15		7.87		
1 Spade (Chipper)		8.55		9.40		
1 Flatbed Truck, Gas, 1.5 Ton		198.50		218.35	16.62	18.28
32 L.H., Daily Totals		$1987.80		$2758.98	$62.12	$86.22

Crew B-37G	Hr.	Daily	Hr.	Daily	Bare Costs	Incl. O&P
1 Labor Foreman (outside)	$46.40	$371.20	$69.25	$554.00	$46.58	$69.49
4 Laborers	44.40	1420.80	66.25	2120.00		
1 Equip. Oper. (light)	55.50	444.00	82.70	661.60		
1 Berm Machine		253.55		278.90		
1 Tandem Roller, 5 Ton		258.75		284.63	10.67	11.74
48 L.H., Daily Totals		$2748.30		$3899.13	$57.26	$81.23

Crew B-37H	Hr.	Daily	Hr.	Daily	Bare Costs	Incl. O&P
1 Labor Foreman (outside)	$46.40	$371.20	$69.25	$554.00	$46.58	$69.49
4 Laborers	44.40	1420.80	66.25	2120.00		
1 Equip. Oper. (light)	55.50	444.00	82.70	661.60		
1 Tandem Roller, 5 Ton		258.75		284.63		
1 Flatbed Truck, Gas, 1.5 Ton		198.50		218.35		
1 Tar Kettle, T.M.		156.70		172.37	12.79	14.07
48 L.H., Daily Totals		$2849.95		$4010.95	$59.37	$83.56

Crew No.		Bare Costs		Incl. Subs O&P		Cost Per Labor-Hour	

Crew B-37I

	Hr.	Daily	Hr.	Daily	Bare Costs	Incl. O&P
3 Laborers	$44.40	$1065.60	$66.25	$1590.00	$49.33	$73.62
1 Equip. Oper. (light)	55.50	444.00	82.70	661.60		
1 Equip. Oper. (medium)	59.00	472.00	87.90	703.20		
2 Truck Drivers (light)	48.80	780.80	73.00	1168.00		
4 Barrels w/ Flasher		16.60		18.26		
1 Concrete Saw		112.85		124.14		
1 Rotary Hammer Drill		52.25		57.48		
1 Hammer Drill Bit		25.25		27.77		
1 Air Compressor, 60 cfm		153.85		169.24		
1 -50' Air Hose, 3/4"		7.15		7.87		
1 Spade (Chipper)		8.55		9.40		
1 Loader, Skid Steer, 30 H.P.		179.50		197.45		
1 Conc. Hammer Attach.		118.50		130.35		
1 Concrete Mixer, 10 C.F.		147.15		161.87		
1 Vibrating Plate, Gas, 18"		31.90		35.09		
2 Flatbed Trucks, Gas, 1.5 Ton		397.00		436.70	22.33	24.56
56 L.H., Daily Totals		$4012.95		$5498.40	$71.66	$98.19

Crew B-37J

	Hr.	Daily	Hr.	Daily	Bare Costs	Incl. O&P
1 Labor Foreman (outside)	$46.40	$371.20	$69.25	$554.00	$46.58	$69.49
4 Laborers	44.40	1420.80	66.25	2120.00		
1 Equip. Oper. (light)	55.50	444.00	82.70	661.60		
1 Air Compressor, 60 cfm		153.85		169.24		
1 -50' Air Hose, 3/4"		7.15		7.87		
2 Concrete Mixers, 10 C.F.		294.30		323.73		
2 Flatbed Trucks, Gas, 1.5 Ton		397.00		436.70		
1 Shot Blaster, 20"		208.70		229.57	22.10	24.31
48 L.H., Daily Totals		$3297.00		$4502.70	$68.69	$93.81

Crew B-37K

	Hr.	Daily	Hr.	Daily	Bare Costs	Incl. O&P
1 Labor Foreman (outside)	$46.40	$371.20	$69.25	$554.00	$46.58	$69.49
4 Laborers	44.40	1420.80	66.25	2120.00		
1 Equip. Oper. (light)	55.50	444.00	82.70	661.60		
1 Air Compressor, 60 cfm		153.85		169.24		
1 -50' Air Hose, 3/4"		7.15		7.87		
2 Flatbed Trucks, Gas, 1.5 Ton		397.00		436.70		
1 Shot Blaster, 20"		208.70		229.57	15.97	17.57
48 L.H., Daily Totals		$3002.70		$4178.97	$62.56	$87.06

Crew B-38

	Hr.	Daily	Hr.	Daily	Bare Costs	Incl. O&P
1 Labor Foreman (outside)	$46.40	$371.20	$69.25	$554.00	$49.94	$74.47
2 Laborers	44.40	710.40	66.25	1060.00		
1 Equip. Oper. (light)	55.50	444.00	82.70	661.60		
1 Equip. Oper. (medium)	59.00	472.00	87.90	703.20		
1 Backhoe Loader, 48 H.P.		216.20		237.82		
1 Hyd. Hammer (1200 lb.)		177.25		194.97		
1 F.E. Loader, W.M., 4 C.Y.		759.00		834.90		
1 Pvmt. Rem. Bucket		63.80		70.18	30.41	33.45
40 L.H., Daily Totals		$3213.85		$4316.68	$80.35	$107.92

Crew B-39

	Hr.	Daily	Hr.	Daily	Bare Costs	Incl. O&P
1 Labor Foreman (outside)	$46.40	$371.20	$69.25	$554.00	$46.58	$69.49
4 Laborers	44.40	1420.80	66.25	2120.00		
1 Equip. Oper. (light)	55.50	444.00	82.70	661.60		
1 Air Compressor, 250 cfm		202.85		223.13		
2 Breakers, Pavement, 60 lb.		107.20		117.92		
2 -50' Air Hoses, 1.5"		45.60		50.16	7.41	8.15
48 L.H., Daily Totals		$2591.65		$3726.82	$53.99	$77.64

Crew B-40

	Hr.	Daily	Hr.	Daily	Bare Costs	Incl. O&P
1 Pile Driver Foreman (outside)	$57.90	$463.20	$89.50	$716.00	$57.11	$87.06
4 Pile Drivers	55.90	1788.80	86.40	2764.80		
2 Equip. Oper. (crane)	61.45	983.20	91.55	1464.80		
1 Equip. Oper. (oiler)	52.50	420.00	78.25	626.00		
1 Crawler Crane, 40 Ton		1231.00		1354.10		
1 Vibratory Hammer & Gen.		2298.00		2527.80	55.14	60.65
64 L.H., Daily Totals		$7184.20		$9453.50	$112.25	$147.71

Crew B-40B

	Hr.	Daily	Hr.	Daily	Bare Costs	Incl. O&P
1 Labor Foreman (outside)	$46.40	$371.20	$69.25	$554.00	$48.92	$72.97
3 Laborers	44.40	1065.60	66.25	1590.00		
1 Equip. Oper. (crane)	61.45	491.60	91.55	732.40		
1 Equip. Oper. (oiler)	52.50	420.00	78.25	626.00		
1 Lattice Boom Crane, 40 Ton		2127.00		2339.70	44.31	48.74
48 L.H., Daily Totals		$4475.40		$5842.10	$93.24	$121.71

Crew B-41

	Hr.	Daily	Hr.	Daily	Bare Costs	Incl. O&P
1 Labor Foreman (outside)	$46.40	$371.20	$69.25	$554.00	$45.91	$68.49
4 Laborers	44.40	1420.80	66.25	2120.00		
.25 Equip. Oper. (crane)	61.45	122.90	91.55	183.10		
.25 Equip. Oper. (oiler)	52.50	105.00	78.25	156.50		
.25 Crawler Crane, 40 Ton		307.75		338.52	6.99	7.69
44 L.H., Daily Totals		$2327.65		$3352.13	$52.90	$76.18

Crew B-42

	Hr.	Daily	Hr.	Daily	Bare Costs	Incl. O&P
1 Labor Foreman (outside)	$46.40	$371.20	$69.25	$554.00	$49.78	$74.67
4 Laborers	44.40	1420.80	66.25	2120.00		
1 Equip. Oper. (crane)	61.45	491.60	91.55	732.40		
1 Equip. Oper. (oiler)	52.50	420.00	78.25	626.00		
1 Welder	60.30	482.40	93.30	746.40		
1 Hyd. Crane, 25 Ton		586.70		645.37		
1 Welder, Gas Engine, 300 amp		148.75		163.63		
1 Horz. Boring Csg. Mch.		329.75		362.73	16.64	18.31
64 L.H., Daily Totals		$4251.20		$5950.52	$66.43	$92.98

Crew B-43

	Hr.	Daily	Hr.	Daily	Bare Costs	Incl. O&P
1 Labor Foreman (outside)	$46.40	$371.20	$69.25	$554.00	$48.92	$72.97
3 Laborers	44.40	1065.60	66.25	1590.00		
1 Equip. Oper. (crane)	61.45	491.60	91.55	732.40		
1 Equip. Oper. (oiler)	52.50	420.00	78.25	626.00		
1 Drill Rig, Truck-Mounted		768.40		845.24	16.01	17.61
48 L.H., Daily Totals		$3116.80		$4347.64	$64.93	$90.58

Crew B-44

	Hr.	Daily	Hr.	Daily	Bare Costs	Incl. O&P
1 Pile Driver Foreman (outside)	$57.90	$463.20	$89.50	$716.00	$56.10	$85.56
4 Pile Drivers	55.90	1788.80	86.40	2764.80		
2 Equip. Oper. (crane)	61.45	983.20	91.55	1464.80		
1 Laborer	44.40	355.20	66.25	530.00		
1 Crawler Crane, 40 Ton		1231.00		1354.10		
1 Lead, 60' High		211.80		232.98		
1 Hammer, Diesel, 15K ft.-lbs.		624.45		686.89	32.30	35.53
64 L.H., Daily Totals		$5657.65		$7749.57	$88.40	$121.09

Crew B-45

	Hr.	Daily	Hr.	Daily	Bare Costs	Incl. O&P
1 Equip. Oper. (medium)	$59.00	$472.00	$87.90	$703.20	$55.15	$82.30
1 Truck Driver (heavy)	51.30	410.40	76.70	613.60		
1 Dist. Tanker, 3000 Gallon		334.10		367.51		
1 Truck Tractor, 6x4, 380 H.P.		499.15		549.07	52.08	57.29
16 L.H., Daily Totals		$1715.65		$2233.38	$107.23	$139.59

Crew No.		Bare Costs		Incl. Subs O&P		Cost Per Labor-Hour	

Crew B-46	Hr.	Daily	Hr.	Daily	Bare Costs	Incl. O&P
1 Pile Driver Foreman (outside)	$57.90	$463.20	$89.50	$716.00	$50.48	$76.84
2 Pile Drivers	55.90	894.40	86.40	1382.40		
3 Laborers	44.40	1065.60	66.25	1590.00		
1 Chain Saw, Gas, 36" Long		41.65		45.81	.87	.95
48 L.H., Daily Totals		$2464.85		$3734.22	$51.35	$77.80

Crew B-47	Hr.	Daily	Hr.	Daily	Bare Costs	Incl. O&P
1 Blast Foreman (outside)	$46.40	$371.20	$69.25	$554.00	$48.77	$72.73
1 Driller	44.40	355.20	66.25	530.00		
1 Equip. Oper. (light)	55.50	444.00	82.70	661.60		
1 Air Track Drill, 4"		1127.00		1239.70		
1 Air Compressor, 600 cfm		426.55		469.20		
2 -50' Air Hoses, 3"		76.70		84.37	67.93	74.72
24 L.H., Daily Totals		$2800.65		$3538.88	$116.69	$147.45

Crew B-47A	Hr.	Daily	Hr.	Daily	Bare Costs	Incl. O&P
1 Drilling Foreman (outside)	$46.40	$371.20	$69.25	$554.00	$53.45	$79.68
1 Equip. Oper. (heavy)	61.45	491.60	91.55	732.40		
1 Equip. Oper. (oiler)	52.50	420.00	78.25	626.00		
1 Air Track Drill, 5"		1127.00		1239.70	46.96	51.65
24 L.H., Daily Totals		$2409.80		$3152.10	$100.41	$131.34

Crew B-47C	Hr.	Daily	Hr.	Daily	Bare Costs	Incl. O&P
1 Laborer	$44.40	$355.20	$66.25	$530.00	$49.95	$74.47
1 Equip. Oper. (light)	55.50	444.00	82.70	661.60		
1 Air Compressor, 750 cfm		596.30		655.93		
2 -50' Air Hoses, 3"		76.70		84.37		
1 Air Track Drill, 4"		1127.00		1239.70	112.50	123.75
16 L.H., Daily Totals		$2599.20		$3171.60	$162.45	$198.22

Crew B-47E	Hr.	Daily	Hr.	Daily	Bare Costs	Incl. O&P
1 Labor Foreman (outside)	$46.40	$371.20	$69.25	$554.00	$44.90	$67.00
3 Laborers	44.40	1065.60	66.25	1590.00		
1 Flatbed Truck, Gas, 3 Ton		850.05		935.05	26.56	29.22
32 L.H., Daily Totals		$2286.85		$3079.05	$71.46	$96.22

Crew B-47G	Hr.	Daily	Hr.	Daily	Bare Costs	Incl. O&P
1 Labor Foreman (outside)	$46.40	$371.20	$69.25	$554.00	$47.67	$71.11
2 Laborers	44.40	710.40	66.25	1060.00		
1 Equip. Oper. (light)	55.50	444.00	82.70	661.60		
1 Air Track Drill, 4"		1127.00		1239.70		
1 Air Compressor, 600 cfm		426.55		469.20		
2 -50' Air Hoses, 3"		76.70		84.37		
1 Gunite Pump Rig		321.75		353.93	61.00	67.10
32 L.H., Daily Totals		$3477.60		$4422.80	$108.68	$138.21

Crew B-47H	Hr.	Daily	Hr.	Daily	Bare Costs	Incl. O&P
1 Skilled Worker Foreman (out)	$59.10	$472.80	$88.90	$711.20	$57.60	$86.65
3 Skilled Workers	57.10	1370.40	85.90	2061.60		
1 Flatbed Truck, Gas, 3 Ton		850.05		935.05	26.56	29.22
32 L.H., Daily Totals		$2693.25		$3707.86	$84.16	$115.87

Crew B-48	Hr.	Daily	Hr.	Daily	Bare Costs	Incl. O&P
1 Labor Foreman (outside)	$46.40	$371.20	$69.25	$554.00	$49.86	$74.36
3 Laborers	44.40	1065.60	66.25	1590.00		
1 Equip. Oper. (crane)	61.45	491.60	91.55	732.40		
1 Equip. Oper. (oiler)	52.50	420.00	78.25	626.00		
1 Equip. Oper. (light)	55.50	444.00	82.70	661.60		
1 Centr. Water Pump, 6"		235.25		258.77		
1 -20' Suction Hose, 6"		25.50		28.05		
1 -50' Discharge Hose, 6"		18.10		19.91		
1 Drill Rig, Truck-Mounted		768.40		845.24	18.70	20.57
56 L.H., Daily Totals		$3839.65		$5315.98	$68.57	$94.93

Crew B-49	Hr.	Daily	Hr.	Daily	Bare Costs	Incl. O&P
1 Labor Foreman (outside)	$46.40	$371.20	$69.25	$554.00	$52.25	$78.46
3 Laborers	44.40	1065.60	66.25	1590.00		
2 Equip. Oper. (crane)	61.45	983.20	91.55	1464.80		
2 Equip. Oper. (oilers)	52.50	840.00	78.25	1252.00		
1 Equip. Oper. (light)	55.50	444.00	82.70	661.60		
2 Pile Drivers	55.90	894.40	86.40	1382.40		
1 Hyd. Crane, 25 Ton		586.70		645.37		
1 Centr. Water Pump, 6"		235.25		258.77		
1 -20' Suction Hose, 6"		25.50		28.05		
1 -50' Discharge Hose, 6"		18.10		19.91		
1 Drill Rig, Truck-Mounted		768.40		845.24	18.57	20.42
88 L.H., Daily Totals		$6232.35		$8702.15	$70.82	$98.89

Crew B-50	Hr.	Daily	Hr.	Daily	Bare Costs	Incl. O&P
2 Pile Driver Foremen (outside)	$57.90	$926.40	$89.50	$1432.00	$54.27	$82.68
6 Pile Drivers	55.90	2683.20	86.40	4147.20		
2 Equip. Oper. (crane)	61.45	983.20	91.55	1464.80		
1 Equip. Oper. (oiler)	52.50	420.00	78.25	626.00		
3 Laborers	44.40	1065.60	66.25	1590.00		
1 Crawler Crane, 40 Ton		1231.00		1354.10		
1 Lead, 60' High		211.80		232.98		
1 Hammer, Diesel, 15K ft.-lbs.		624.45		686.89		
1 Air Compressor, 600 cfm		426.55		469.20		
2 -50' Air Hoses, 3"		76.70		84.37		
1 Chain Saw, Gas, 36" Long		41.65		45.81	23.32	25.66
112 L.H., Daily Totals		$8690.55		$12133.37	$77.59	$108.33

Crew B-51	Hr.	Daily	Hr.	Daily	Bare Costs	Incl. O&P
1 Labor Foreman (outside)	$46.40	$371.20	$69.25	$554.00	$45.47	$67.88
4 Laborers	44.40	1420.80	66.25	2120.00		
1 Truck Driver (light)	48.80	390.40	73.00	584.00		
1 Flatbed Truck, Gas, 1.5 Ton		198.50		218.35	4.14	4.55
48 L.H., Daily Totals		$2380.90		$3476.35	$49.60	$72.42

Crew B-52	Hr.	Daily	Hr.	Daily	Bare Costs	Incl. O&P
1 Carpenter Foreman (outside)	$56.70	$453.60	$84.60	$676.80	$50.76	$75.55
1 Carpenter	54.70	437.60	81.65	653.20		
3 Laborers	44.40	1065.60	66.25	1590.00		
1 Cement Finisher	51.80	414.40	75.90	607.20		
.5 Rodman (reinf.)	58.90	235.60	88.05	352.20		
.5 Equip. Oper. (medium)	59.00	236.00	87.90	351.60		
.5 Crawler Loader, 3 C.Y.		573.00		630.30	10.23	11.26
56 L.H., Daily Totals		$3415.80		$4861.30	$61.00	$86.81

Crew B-53	Hr.	Daily	Hr.	Daily	Bare Costs	Incl. O&P
1 Equip. Oper. (light)	$55.50	$444.00	$82.70	$661.60	$55.50	$82.70
1 Trencher, Chain, 12 H.P.		158.60		174.46	19.82	21.81
8 L.H., Daily Totals		$602.60		$836.06	$75.33	$104.51

For customer support on your Site Work & Landscape Costs with RSMeans Data, call 800.448.8182.

Crew B-54

Crew No.	Bare Costs Hr.	Daily	Incl. Subs O&P Hr.	Daily	Cost Per Labor-Hour Bare Costs	Incl. O&P
1 Equip. Oper. (light)	$55.50	$444.00	$82.70	$661.60	$55.50	$82.70
1 Trencher, Chain, 40 H.P.		450.70		495.77	56.34	61.97
8 L.H., Daily Totals		$894.70		$1157.37	$111.84	$144.67

Crew B-54A

Crew No.	Bare Costs Hr.	Daily	Incl. Subs O&P Hr.	Daily	Cost Per Labor-Hour Bare Costs	Incl. O&P
.17 Labor Foreman (outside)	$46.40	$63.10	$69.25	$94.18	$57.17	$85.19
1 Equipment Operator (med.)	59.00	472.00	87.90	703.20		
1 Wheel Trencher, 67 H.P.		1140.00		1254.00	121.79	133.97
9.36 L.H., Daily Totals		$1675.10		$2051.38	$178.96	$219.16

Crew B-54B

Crew No.	Bare Costs Hr.	Daily	Incl. Subs O&P Hr.	Daily	Cost Per Labor-Hour Bare Costs	Incl. O&P
.25 Labor Foreman (outside)	$46.40	$92.80	$69.25	$138.50	$56.48	$84.17
1 Equipment Operator (med.)	59.00	472.00	87.90	703.20		
1 Wheel Trencher, 150 H.P.		1238.00		1361.80	123.80	136.18
10 L.H., Daily Totals		$1802.80		$2203.50	$180.28	$220.35

Crew B-54C

Crew No.	Bare Costs Hr.	Daily	Incl. Subs O&P Hr.	Daily	Cost Per Labor-Hour Bare Costs	Incl. O&P
1 Laborer	$44.40	$355.20	$66.25	$530.00	$51.70	$77.08
1 Equipment Operator (med.)	59.00	472.00	87.90	703.20		
1 Wheel Trencher, 67 H.P.		1140.00		1254.00	71.25	78.38
16 L.H., Daily Totals		$1967.20		$2487.20	$122.95	$155.45

Crew B-54D

Crew No.	Bare Costs Hr.	Daily	Incl. Subs O&P Hr.	Daily	Cost Per Labor-Hour Bare Costs	Incl. O&P
1 Laborer	$44.40	$355.20	$66.25	$530.00	$51.70	$77.08
1 Equipment Operator (med.)	59.00	472.00	87.90	703.20		
1 Rock Trencher, 6" Width		434.20		477.62	27.14	29.85
16 L.H., Daily Totals		$1261.40		$1710.82	$78.84	$106.93

Crew B-54E

Crew No.	Bare Costs Hr.	Daily	Incl. Subs O&P Hr.	Daily	Cost Per Labor-Hour Bare Costs	Incl. O&P
1 Laborer	$44.40	$355.20	$66.25	$530.00	$51.70	$77.08
1 Equipment Operator (med.)	59.00	472.00	87.90	703.20		
1 Rock Trencher, 18" Width		1015.00		1116.50	63.44	69.78
16 L.H., Daily Totals		$1842.20		$2349.70	$115.14	$146.86

Crew B-55

Crew No.	Bare Costs Hr.	Daily	Incl. Subs O&P Hr.	Daily	Cost Per Labor-Hour Bare Costs	Incl. O&P
2 Laborers	$44.40	$710.40	$66.25	$1060.00	$45.87	$68.50
1 Truck Driver (light)	48.80	390.40	73.00	584.00		
1 Truck-Mounted Earth Auger		394.15		433.57		
1 Flatbed Truck, Gas, 3 Ton		850.05		935.05	51.84	57.03
24 L.H., Daily Totals		$2345.00		$3012.62	$97.71	$125.53

Crew B-56

Crew No.	Bare Costs Hr.	Daily	Incl. Subs O&P Hr.	Daily	Cost Per Labor-Hour Bare Costs	Incl. O&P
1 Laborer	$44.40	$355.20	$66.25	$530.00	$49.95	$74.47
1 Equip. Oper. (light)	55.50	444.00	82.70	661.60		
1 Air Track Drill, 4"		1127.00		1239.70		
1 Air Compressor, 600 cfm		426.55		469.20		
1 -50' Air Hose, 3"		38.35		42.19	99.49	109.44
16 L.H., Daily Totals		$2391.10		$2942.69	$149.44	$183.92

Crew B-57

Crew No.	Bare Costs Hr.	Daily	Incl. Subs O&P Hr.	Daily	Cost Per Labor-Hour Bare Costs	Incl. O&P
1 Labor Foreman (outside)	$46.40	$371.20	$69.25	$554.00	$50.77	$75.71
2 Laborers	44.40	710.40	66.25	1060.00		
1 Equip. Oper. (crane)	61.45	491.60	91.55	732.40		
1 Equip. Oper. (light)	55.50	444.00	82.70	661.60		
1 Equip. Oper. (oiler)	52.50	420.00	78.25	626.00		
1 Crawler Crane, 25 Ton		1152.00		1267.20		
1 Clamshell Bucket, 1 C.Y.		69.25		76.17		
1 Centr. Water Pump, 6"		235.25		258.77		
1 -20' Suction Hose, 6"		25.50		28.05		
20 -50' Discharge Hoses, 6"		362.00		398.20	38.42	42.26
48 L.H., Daily Totals		$4281.20		$5662.40	$89.19	$117.97

Crew B-58

Crew No.	Bare Costs Hr.	Daily	Incl. Subs O&P Hr.	Daily	Cost Per Labor-Hour Bare Costs	Incl. O&P
2 Laborers	$44.40	$710.40	$66.25	$1060.00	$48.10	$71.73
1 Equip. Oper. (light)	55.50	444.00	82.70	661.60		
1 Backhoe Loader, 48 H.P.		216.20		237.82		
1 Small Helicopter, w/ Pilot		2101.00		2311.10	96.55	106.21
24 L.H., Daily Totals		$3471.60		$4270.52	$144.65	$177.94

Crew B-59

Crew No.	Bare Costs Hr.	Daily	Incl. Subs O&P Hr.	Daily	Cost Per Labor-Hour Bare Costs	Incl. O&P
1 Truck Driver (heavy)	$51.30	$410.40	$76.70	$613.60	$51.30	$76.70
1 Truck Tractor, 220 H.P.		310.80		341.88		
1 Water Tank Trailer, 5000 Gal.		154.10		169.51	58.11	63.92
8 L.H., Daily Totals		$875.30		$1124.99	$109.41	$140.62

Crew B-59A

Crew No.	Bare Costs Hr.	Daily	Incl. Subs O&P Hr.	Daily	Cost Per Labor-Hour Bare Costs	Incl. O&P
2 Laborers	$44.40	$710.40	$66.25	$1060.00	$46.70	$69.73
1 Truck Driver (heavy)	51.30	410.40	76.70	613.60		
1 Water Tank Trailer, 5000 Gal.		154.10		169.51		
1 Truck Tractor, 220 H.P.		310.80		341.88	19.37	21.31
24 L.H., Daily Totals		$1585.70		$2184.99	$66.07	$91.04

Crew B-60

Crew No.	Bare Costs Hr.	Daily	Incl. Subs O&P Hr.	Daily	Cost Per Labor-Hour Bare Costs	Incl. O&P
1 Labor Foreman (outside)	$46.40	$371.20	$69.25	$554.00	$51.45	$76.71
2 Laborers	44.40	710.40	66.25	1060.00		
1 Equip. Oper. (crane)	61.45	491.60	91.55	732.40		
2 Equip. Oper. (light)	55.50	888.00	82.70	1323.20		
1 Equip. Oper. (oiler)	52.50	420.00	78.25	626.00		
1 Crawler Crane, 40 Ton		1231.00		1354.10		
1 Lead, 60' High		211.80		232.98		
1 Hammer, Diesel, 15K ft.-lbs.		624.45		686.89		
1 Backhoe Loader, 48 H.P.		216.20		237.82	40.78	44.85
56 L.H., Daily Totals		$5164.65		$6807.40	$92.23	$121.56

Crew B-61

Crew No.	Bare Costs Hr.	Daily	Incl. Subs O&P Hr.	Daily	Cost Per Labor-Hour Bare Costs	Incl. O&P
1 Labor Foreman (outside)	$46.40	$371.20	$69.25	$554.00	$47.02	$70.14
3 Laborers	44.40	1065.60	66.25	1590.00		
1 Equip. Oper. (light)	55.50	444.00	82.70	661.60		
1 Cement Mixer, 2 C.Y.		112.35		123.58		
1 Air Compressor, 160 cfm		212.30		233.53	8.12	8.93
40 L.H., Daily Totals		$2205.45		$3162.72	$55.14	$79.07

Crew B-62

Crew No.	Bare Costs Hr.	Daily	Incl. Subs O&P Hr.	Daily	Cost Per Labor-Hour Bare Costs	Incl. O&P
2 Laborers	$44.40	$710.40	$66.25	$1060.00	$48.10	$71.73
1 Equip. Oper. (light)	55.50	444.00	82.70	661.60		
1 Loader, Skid Steer, 30 H.P.		179.50		197.45	7.48	8.23
24 L.H., Daily Totals		$1333.90		$1919.05	$55.58	$79.96

Crew No.	Bare Costs		Incl. Subs O&P		Cost Per Labor-Hour	

Crew B-62A

	Hr.	Daily	Hr.	Daily	Bare Costs	Incl. O&P
2 Laborers	$44.40	$710.40	$66.25	$1060.00	$48.10	$71.73
1 Equip. Oper. (light)	55.50	444.00	82.70	661.60		
1 Loader, Skid Steer, 30 H.P.		179.50		197.45		
1 Trencher Attachment		66.25		72.88	10.24	11.26
24 L.H., Daily Totals		$1400.15		$1991.93	$58.34	$83.00

Crew B-63

	Hr.	Daily	Hr.	Daily	Bare Costs	Incl. O&P
4 Laborers	$44.40	$1420.80	$66.25	$2120.00	$46.62	$69.54
1 Equip. Oper. (light)	55.50	444.00	82.70	661.60		
1 Loader, Skid Steer, 30 H.P.		179.50		197.45	4.49	4.94
40 L.H., Daily Totals		$2044.30		$2979.05	$51.11	$74.48

Crew B-63B

	Hr.	Daily	Hr.	Daily	Bare Costs	Incl. O&P
1 Labor Foreman (inside)	$44.90	$359.20	$67.00	$536.00	$47.30	$70.55
2 Laborers	44.40	710.40	66.25	1060.00		
1 Equip. Oper. (light)	55.50	444.00	82.70	661.60		
1 Loader, Skid Steer, 78 H.P.		446.30		490.93	13.95	15.34
32 L.H., Daily Totals		$1959.90		$2748.53	$61.25	$85.89

Crew B-64

	Hr.	Daily	Hr.	Daily	Bare Costs	Incl. O&P
1 Laborer	$44.40	$355.20	$66.25	$530.00	$46.60	$69.63
1 Truck Driver (light)	48.80	390.40	73.00	584.00		
1 Power Mulcher (small)		201.00		221.10		
1 Flatbed Truck, Gas, 1.5 Ton		198.50		218.35	24.97	27.47
16 L.H., Daily Totals		$1145.10		$1553.45	$71.57	$97.09

Crew B-65

	Hr.	Daily	Hr.	Daily	Bare Costs	Incl. O&P
1 Laborer	$44.40	$355.20	$66.25	$530.00	$46.60	$69.63
1 Truck Driver (light)	48.80	390.40	73.00	584.00		
1 Power Mulcher (Large)		345.35		379.88		
1 Flatbed Truck, Gas, 1.5 Ton		198.50		218.35	33.99	37.39
16 L.H., Daily Totals		$1289.45		$1712.23	$80.59	$107.01

Crew B-66

	Hr.	Daily	Hr.	Daily	Bare Costs	Incl. O&P
1 Equip. Oper. (light)	$55.50	$444.00	$82.70	$661.60	$55.50	$82.70
1 Loader-Backhoe, 40 H.P.		267.55		294.31	33.44	36.79
8 L.H., Daily Totals		$711.55		$955.90	$88.94	$119.49

Crew B-67

	Hr.	Daily	Hr.	Daily	Bare Costs	Incl. O&P
1 Millwright	$58.75	$470.00	$84.90	$679.20	$57.13	$83.80
1 Equip. Oper. (light)	55.50	444.00	82.70	661.60		
1 R.T. Forklift, 5,000 Lb., diesel		272.85		300.13	17.05	18.76
16 L.H., Daily Totals		$1186.85		$1640.93	$74.18	$102.56

Crew B-67B

	Hr.	Daily	Hr.	Daily	Bare Costs	Incl. O&P
1 Millwright Foreman (inside)	$59.25	$474.00	$85.60	$684.80	$59.00	$85.25
1 Millwright	58.75	470.00	84.90	679.20		
16 L.H., Daily Totals		$944.00		$1364.00	$59.00	$85.25

Crew B-68

	Hr.	Daily	Hr.	Daily	Bare Costs	Incl. O&P
2 Millwrights	$58.75	$940.00	$84.90	$1358.40	$57.67	$84.17
1 Equip. Oper. (light)	55.50	444.00	82.70	661.60		
1 R.T. Forklift, 5,000 Lb., diesel		272.85		300.13	11.37	12.51
24 L.H., Daily Totals		$1656.85		$2320.14	$69.04	$96.67

Crew B-68A

	Hr.	Daily	Hr.	Daily	Bare Costs	Incl. O&P
1 Millwright Foreman (inside)	$59.25	$474.00	$85.60	$684.80	$58.92	$85.13
2 Millwrights	58.75	940.00	84.90	1358.40		
1 Forklift, Smooth Floor, 8,000 Lb.		283.25		311.57	11.80	12.98
24 L.H., Daily Totals		$1697.25		$2354.78	$70.72	$98.12

Crew B-68B

	Hr.	Daily	Hr.	Daily	Bare Costs	Incl. O&P
1 Millwright Foreman (inside)	$59.25	$474.00	$85.60	$684.80	$62.79	$92.40
2 Millwrights	58.75	940.00	84.90	1358.40		
2 Electricians	63.70	1019.20	94.65	1514.40		
2 Plumbers	67.70	1083.20	101.05	1616.80		
1 R.T. Forklift, 5,000 Lb., gas		283.30		311.63	5.06	5.56
56 L.H., Daily Totals		$3799.70		$5486.03	$67.85	$97.96

Crew B-68C

	Hr.	Daily	Hr.	Daily	Bare Costs	Incl. O&P
1 Millwright Foreman (inside)	$59.25	$474.00	$85.60	$684.80	$62.35	$91.55
1 Millwright	58.75	470.00	84.90	679.20		
1 Electrician	63.70	509.60	94.65	757.20		
1 Plumber	67.70	541.60	101.05	808.40		
1 R.T. Forklift, 5,000 Lb., gas		283.30		311.63	8.85	9.74
32 L.H., Daily Totals		$2278.50		$3241.23	$71.20	$101.29

Crew B-68D

	Hr.	Daily	Hr.	Daily	Bare Costs	Incl. O&P
1 Labor Foreman (inside)	$44.90	$359.20	$67.00	$536.00	$48.27	$71.98
1 Laborer	44.40	355.20	66.25	530.00		
1 Equip. Oper. (light)	55.50	444.00	82.70	661.60		
1 R.T. Forklift, 5,000 Lb., gas		283.30		311.63	11.80	12.98
24 L.H., Daily Totals		$1441.70		$2039.23	$60.07	$84.97

Crew B-68E

	Hr.	Daily	Hr.	Daily	Bare Costs	Incl. O&P
1 Struc. Steel Foreman (inside)	$60.80	$486.40	$94.10	$752.80	$60.40	$93.46
3 Struc. Steel Workers	60.30	1447.20	93.30	2239.20		
1 Welder	60.30	482.40	93.30	746.40		
1 Forklift, Smooth Floor, 8,000 Lb.		283.25		311.57	7.08	7.79
40 L.H., Daily Totals		$2699.25		$4049.97	$67.48	$101.25

Crew B-68F

	Hr.	Daily	Hr.	Daily	Bare Costs	Incl. O&P
1 Skilled Worker Foreman (out)	$59.10	$472.80	$88.90	$711.20	$57.77	$86.90
2 Skilled Workers	57.10	913.60	85.90	1374.40		
1 R.T. Forklift, 5,000 Lb., gas		283.30		311.63	11.80	12.98
24 L.H., Daily Totals		$1669.70		$2397.23	$69.57	$99.88

Crew B-68G

	Hr.	Daily	Hr.	Daily	Bare Costs	Incl. O&P
2 Structural Steel Workers	$60.30	$964.80	$93.30	$1492.80	$60.30	$93.30
1 R.T. Forklift, 5,000 Lb., gas		283.30		311.63	17.71	19.48
16 L.H., Daily Totals		$1248.10		$1804.43	$78.01	$112.78

Crew B-69

	Hr.	Daily	Hr.	Daily	Bare Costs	Incl. O&P
1 Labor Foreman (outside)	$46.40	$371.20	$69.25	$554.00	$48.92	$72.97
3 Laborers	44.40	1065.60	66.25	1590.00		
1 Equip. Oper. (crane)	61.45	491.60	91.55	732.40		
1 Equip. Oper. (oiler)	52.50	420.00	78.25	626.00		
1 Hyd. Crane, 80 Ton		1458.00		1603.80	30.38	33.41
48 L.H., Daily Totals		$3806.40		$5106.20	$79.30	$106.38

Crew B-69A

	Hr.	Daily	Hr.	Daily	Bare Costs	Incl. O&P
1 Labor Foreman (outside)	$46.40	$371.20	$69.25	$554.00	$48.40	$71.97
3 Laborers	44.40	1065.60	66.25	1590.00		
1 Equip. Oper. (medium)	59.00	472.00	87.90	703.20		
1 Concrete Finisher	51.80	414.40	75.90	607.20		
1 Curb/Gutter Paver, 2-Track		1231.00		1354.10	25.65	28.21
48 L.H., Daily Totals		$3554.20		$4808.50	$74.05	$100.18

For customer support on your Site Work & Landscape Costs with RSMeans Data, call 800.448.8182.

Crew B-69B

Crew No.	Bare Costs Hr.	Daily	Incl. Subs O&P Hr.	Daily	Cost Per Labor-Hour Bare Costs	Incl. O&P
1 Labor Foreman (outside)	$46.40	$371.20	$69.25	$554.00	$48.40	$71.97
3 Laborers	44.40	1065.60	66.25	1590.00		
1 Equip. Oper. (medium)	59.00	472.00	87.90	703.20		
1 Cement Finisher	51.80	414.40	75.90	607.20		
1 Curb/Gutter Paver, 4-Track		801.05		881.15	16.69	18.36
48 L.H., Daily Totals		$3124.25		$4335.56	$65.09	$90.32

Crew B-70

Crew No.	Bare Costs Hr.	Daily	Incl. Subs O&P Hr.	Daily	Cost Per Labor-Hour Bare Costs	Incl. O&P
1 Labor Foreman (outside)	$46.40	$371.20	$69.25	$554.00	$50.94	$75.96
3 Laborers	44.40	1065.60	66.25	1590.00		
3 Equip. Oper. (medium)	59.00	1416.00	87.90	2109.60		
1 Grader, 30,000 Lbs.		1073.00		1180.30		
1 Ripper, Beam & 1 Shank		91.60		100.76		
1 Road Sweeper, S.P., 8' wide		723.65		796.01		
1 F.E. Loader, W.M. 1.5 C.Y.		441.40		485.54	41.60	45.76
56 L.H., Daily Totals		$5182.45		$6816.22	$92.54	$121.72

Crew B-70A

Crew No.	Bare Costs Hr.	Daily	Incl. Subs O&P Hr.	Daily	Cost Per Labor-Hour Bare Costs	Incl. O&P
1 Laborer	$44.40	$355.20	$66.25	$530.00	$56.08	$83.57
4 Equip. Oper. (medium)	59.00	1888.00	87.90	2812.80		
1 Grader, 40,000 Lbs.		1219.00		1340.90		
1 F.E. Loader, W.M., 2.5 C.Y.		638.30		702.13		
1 Dozer, 80 H.P.		405.85		446.44		
1 Roller, Pneum. Whl., 12 Ton		349.90		384.89	65.33	71.86
40 L.H., Daily Totals		$4856.25		$6217.15	$121.41	$155.43

Crew B-71

Crew No.	Bare Costs Hr.	Daily	Incl. Subs O&P Hr.	Daily	Cost Per Labor-Hour Bare Costs	Incl. O&P
1 Labor Foreman (outside)	$46.40	$371.20	$69.25	$554.00	$50.94	$75.96
3 Laborers	44.40	1065.60	66.25	1590.00		
3 Equip. Oper. (medium)	59.00	1416.00	87.90	2109.60		
1 Pvmt. Profiler, 750 H.P.		3483.00		3831.30		
1 Road Sweeper, S.P., 8' wide		723.65		796.01		
1 F.E. Loader, W.M. 1.5 C.Y.		441.40		485.54	83.00	91.30
56 L.H., Daily Totals		$7500.85		$9366.45	$133.94	$167.26

Crew B-72

Crew No.	Bare Costs Hr.	Daily	Incl. Subs O&P Hr.	Daily	Cost Per Labor-Hour Bare Costs	Incl. O&P
1 Labor Foreman (outside)	$46.40	$371.20	$69.25	$554.00	$51.95	$77.45
3 Laborers	44.40	1065.60	66.25	1590.00		
4 Equip. Oper. (medium)	59.00	1888.00	87.90	2812.80		
1 Pvmt. Profiler, 750 H.P.		3483.00		3831.30		
1 Hammermill, 250 H.P.		857.40		943.14		
1 Windrow Loader		1461.00		1607.10		
1 Mix Paver, 165 H.P.		2172.00		2389.20		
1 Roller, Pneum. Whl., 12 Ton		349.90		384.89	130.05	143.06
64 L.H., Daily Totals		$11648.10		$14112.43	$182.00	$220.51

Crew B-73

Crew No.	Bare Costs Hr.	Daily	Incl. Subs O&P Hr.	Daily	Cost Per Labor-Hour Bare Costs	Incl. O&P
1 Labor Foreman (outside)	$46.40	$371.20	$69.25	$554.00	$53.77	$80.16
2 Laborers	44.40	710.40	66.25	1060.00		
5 Equip. Oper. (medium)	59.00	2360.00	87.90	3516.00		
1 Road Mixer, 310 H.P.		1919.00		2110.90		
1 Tandem Roller, 10 Ton		246.80		271.48		
1 Hammermill, 250 H.P.		857.40		943.14		
1 Grader, 30,000 Lbs.		1073.00		1180.30		
.5 F.E. Loader, W.M. 1.5 C.Y.		220.70		242.77		
.5 Truck Tractor, 220 H.P.		155.40		170.94		
.5 Water Tank Trailer, 5000 Gal.		77.05		84.75	71.08	78.19
64 L.H., Daily Totals		$7990.95		$10134.29	$124.86	$158.35

Crew B-74

Crew No.	Bare Costs Hr.	Daily	Incl. Subs O&P Hr.	Daily	Cost Per Labor-Hour Bare Costs	Incl. O&P
1 Labor Foreman (outside)	$46.40	$371.20	$69.25	$554.00	$53.67	$80.06
1 Laborer	44.40	355.20	66.25	530.00		
4 Equip. Oper. (medium)	59.00	1888.00	87.90	2812.80		
2 Truck Drivers (heavy)	51.30	820.80	76.70	1227.20		
1 Grader, 30,000 Lbs.		1073.00		1180.30		
1 Ripper, Beam & 1 Shank		91.60		100.76		
2 Stabilizers, 310 H.P.		2808.00		3088.80		
1 Flatbed Truck, Gas, 3 Ton		850.05		935.05		
1 Chem. Spreader, Towed		85.40		93.94		
1 Roller, Vibratory, 25 Ton		672.35		739.59		
1 Water Tank Trailer, 5000 Gal.		154.10		169.51		
1 Truck Tractor, 220 H.P.		310.80		341.88	94.46	103.90
64 L.H., Daily Totals		$9480.50		$11773.83	$148.13	$183.97

Crew B-75

Crew No.	Bare Costs Hr.	Daily	Incl. Subs O&P Hr.	Daily	Cost Per Labor-Hour Bare Costs	Incl. O&P
1 Labor Foreman (outside)	$46.40	$371.20	$69.25	$554.00	$54.01	$80.54
1 Laborer	44.40	355.20	66.25	530.00		
4 Equip. Oper. (medium)	59.00	1888.00	87.90	2812.80		
1 Truck Driver (heavy)	51.30	410.40	76.70	613.60		
1 Grader, 30,000 Lbs.		1073.00		1180.30		
1 Ripper, Beam & 1 Shank		91.60		100.76		
2 Stabilizers, 310 H.P.		2808.00		3088.80		
1 Dist. Tanker, 3000 Gallon		334.10		367.51		
1 Truck Tractor, 6x4, 380 H.P.		499.15		549.07		
1 Roller, Vibratory, 25 Ton		672.35		739.59	97.83	107.61
56 L.H., Daily Totals		$8503.00		$10536.42	$151.84	$188.15

Crew B-76

Crew No.	Bare Costs Hr.	Daily	Incl. Subs O&P Hr.	Daily	Cost Per Labor-Hour Bare Costs	Incl. O&P
1 Dock Builder Foreman (outside)	$57.90	$463.20	$89.50	$716.00	$56.98	$86.98
5 Dock Builders	55.90	2236.00	86.40	3456.00		
2 Equip. Oper. (crane)	61.45	983.20	91.55	1464.80		
1 Equip. Oper. (oiler)	52.50	420.00	78.25	626.00		
1 Crawler Crane, 50 Ton		1541.00		1695.10		
1 Barge, 400 Ton		869.15		956.07		
1 Hammer, Diesel, 15K ft.-lbs.		624.45		686.89		
1 Lead, 60' High		211.80		232.98		
1 Air Compressor, 600 cfm		426.55		469.20		
2 -50' Air Hoses, 3"		76.70		84.37	52.08	57.29
72 L.H., Daily Totals		$7852.05		$10387.42	$109.06	$144.27

Crew B-76A

Crew No.	Bare Costs Hr.	Daily	Incl. Subs O&P Hr.	Daily	Cost Per Labor-Hour Bare Costs	Incl. O&P
1 Labor Foreman (outside)	$46.40	$371.20	$69.25	$554.00	$47.79	$71.29
5 Laborers	44.40	1776.00	66.25	2650.00		
1 Equip. Oper. (crane)	61.45	491.60	91.55	732.40		
1 Equip. Oper. (oiler)	52.50	420.00	78.25	626.00		
1 Crawler Crane, 50 Ton		1541.00		1695.10		
1 Barge, 400 Ton		869.15		956.07	37.66	41.42
64 L.H., Daily Totals		$5468.95		$7213.56	$85.45	$112.71

Crew B-77

Crew No.	Bare Costs Hr.	Daily	Incl. Subs O&P Hr.	Daily	Cost Per Labor-Hour Bare Costs	Incl. O&P
1 Labor Foreman (outside)	$46.40	$371.20	$69.25	$554.00	$45.68	$68.20
3 Laborers	44.40	1065.60	66.25	1590.00		
1 Truck Driver (light)	48.80	390.40	73.00	584.00		
1 Crack Cleaner, 25 H.P.		53.00		58.30		
1 Crack Filler, Trailer Mtd.		170.95		188.04		
1 Flatbed Truck, Gas, 3 Ton		850.05		935.05	26.85	29.54
40 L.H., Daily Totals		$2901.20		$3909.40	$72.53	$97.73

Crew No.	Bare Costs		Incl. Subs O&P		Cost Per Labor-Hour	
Crew B-78	**Hr.**	**Daily**	**Hr.**	**Daily**	**Bare Costs**	**Incl. O&P**
1 Labor Foreman (outside)	$46.40	$371.20	$69.25	$554.00	$45.47	$67.88
4 Laborers	44.40	1420.80	66.25	2120.00		
1 Truck Driver (light)	48.80	390.40	73.00	584.00		
1 Paint Striper, S.P., 40 Gallon		128.35		141.19		
1 Flatbed Truck, Gas, 3 Ton		850.05		935.05		
1 Pickup Truck, 3/4 Ton		112.20		123.42	22.72	24.99
48 L.H., Daily Totals		$3273.00		$4457.66	$68.19	$92.87

Crew B-78A	Hr.	Daily	Hr.	Daily	Bare Costs	Incl. O&P
1 Equip. Oper. (light)	$55.50	$444.00	$82.70	$661.60	$55.50	$82.70
1 Line Rem. (Metal Balls) 115 H.P.		996.25		1095.88	124.53	136.98
8 L.H., Daily Totals		$1440.25		$1757.47	$180.03	$219.68

Crew B-78B	Hr.	Daily	Hr.	Daily	Bare Costs	Incl. O&P
2 Laborers	$44.40	$710.40	$66.25	$1060.00	$45.63	$68.08
.25 Equip. Oper. (light)	55.50	111.00	82.70	165.40		
1 Pickup Truck, 3/4 Ton		112.20		123.42		
1 Line Rem.,11 H.P.,Walk Behind		114.75		126.22		
.25 Road Sweeper, S.P., 8' wide		180.91		199.00	22.66	24.92
18 L.H., Daily Totals		$1229.26		$1674.05	$68.29	$93.00

Crew B-78C	Hr.	Daily	Hr.	Daily	Bare Costs	Incl. O&P
1 Labor Foreman (outside)	$46.40	$371.20	$69.25	$554.00	$45.47	$67.88
4 Laborers	44.40	1420.80	66.25	2120.00		
1 Truck Driver (light)	48.80	390.40	73.00	584.00		
1 Paint Striper, T.M., 120 Gal.		603.10		663.41		
1 Flatbed Truck, Gas, 3 Ton		850.05		935.05		
1 Pickup Truck, 3/4 Ton		112.20		123.42	32.61	35.87
48 L.H., Daily Totals		$3747.75		$4979.89	$78.08	$103.75

Crew B-78D	Hr.	Daily	Hr.	Daily	Bare Costs	Incl. O&P
2 Labor Foremen (outside)	$46.40	$742.40	$69.25	$1108.00	$45.24	$67.53
7 Laborers	44.40	2486.40	66.25	3710.00		
1 Truck Driver (light)	48.80	390.40	73.00	584.00		
1 Paint Striper, T.M., 120 Gal.		603.10		663.41		
1 Flatbed Truck, Gas, 3 Ton		850.05		935.05		
3 Pickup Trucks, 3/4 Ton		336.60		370.26		
1 Air Compressor, 60 cfm		153.85		169.24		
1 -50' Air Hose, 3/4"		7.15		7.87		
1 Breaker, Pavement, 60 lb.		53.60		58.96	25.05	27.56
80 L.H., Daily Totals		$5623.55		$7606.78	$70.29	$95.08

Crew B-78E	Hr.	Daily	Hr.	Daily	Bare Costs	Incl. O&P
2 Labor Foremen (outside)	$46.40	$742.40	$69.25	$1108.00	$45.10	$67.31
9 Laborers	44.40	3196.80	66.25	4770.00		
1 Truck Driver (light)	48.80	390.40	73.00	584.00		
1 Paint Striper, T.M., 120 Gal.		603.10		663.41		
1 Flatbed Truck, Gas, 3 Ton		850.05		935.05		
4 Pickup Trucks, 3/4 Ton		448.80		493.68		
2 Air Compressors, 60 cfm		307.70		338.47		
2 -50' Air Hoses, 3/4"		14.30		15.73		
2 Breakers, Pavement, 60 lb.		107.20		117.92	24.28	26.71
96 L.H., Daily Totals		$6660.75		$9026.26	$69.38	$94.02

Crew No.	Bare Costs		Incl. Subs O&P		Cost Per Labor-Hour	
Crew B-78F	**Hr.**	**Daily**	**Hr.**	**Daily**	**Bare Costs**	**Incl. O&P**
2 Labor Foremen (outside)	$46.40	$742.40	$69.25	$1108.00	$45.00	$67.16
11 Laborers	44.40	3907.20	66.25	5830.00		
1 Truck Driver (light)	48.80	390.40	73.00	584.00		
1 Paint Striper, T.M., 120 Gal.		603.10		663.41		
1 Flatbed Truck, Gas, 3 Ton		850.05		935.05		
7 Pickup Trucks, 3/4 Ton		785.40		863.94		
3 Air Compressors, 60 cfm		461.55		507.70		
3 -50' Air Hoses, 3/4"		21.45		23.59		
3 Breakers, Pavement, 60 lb.		160.80		176.88	25.74	28.31
112 L.H., Daily Totals		$7922.35		$10692.58	$70.74	$95.47

Crew B-79	Hr.	Daily	Hr.	Daily	Bare Costs	Incl. O&P
1 Labor Foreman (outside)	$46.40	$371.20	$69.25	$554.00	$45.68	$68.20
3 Laborers	44.40	1065.60	66.25	1590.00		
1 Truck Driver (light)	48.80	390.40	73.00	584.00		
1 Paint Striper, T.M., 120 Gal.		603.10		663.41		
1 Heating Kettle, 115 Gallon		107.25		117.97		
1 Flatbed Truck, Gas, 3 Ton		850.05		935.05		
2 Pickup Trucks, 3/4 Ton		224.40		246.84	44.62	49.08
40 L.H., Daily Totals		$3612.00		$4691.28	$90.30	$117.28

Crew B-79A	Hr.	Daily	Hr.	Daily	Bare Costs	Incl. O&P
1.5 Equip. Oper. (light)	$55.50	$666.00	$82.70	$992.40	$55.50	$82.70
.5 Line Remov. (Grinder) 115 H.P.		515.50		567.05		
1 Line Rem. (Metal Balls) 115 H.P.		996.25		1095.88	125.98	138.58
12 L.H., Daily Totals		$2177.75		$2655.32	$181.48	$221.28

Crew B-79B	Hr.	Daily	Hr.	Daily	Bare Costs	Incl. O&P
1 Laborer	$44.40	$355.20	$66.25	$530.00	$44.40	$66.25
1 Set of Gases		173.60		190.96	21.70	23.87
8 L.H., Daily Totals		$528.80		$720.96	$66.10	$90.12

Crew B-79C	Hr.	Daily	Hr.	Daily	Bare Costs	Incl. O&P
1 Labor Foreman (outside)	$46.40	$371.20	$69.25	$554.00	$45.31	$67.64
5 Laborers	44.40	1776.00	66.25	2650.00		
1 Truck Driver (light)	48.80	390.40	73.00	584.00		
1 Paint Striper, T.M., 120 Gal.		603.10		663.41		
1 Heating Kettle, 115 Gallon		107.25		117.97		
1 Flatbed Truck, Gas, 3 Ton		850.05		935.05		
3 Pickup Trucks, 3/4 Ton		336.60		370.26		
1 Air Compressor, 60 cfm		153.85		169.24		
1 -50' Air Hose, 3/4"		7.15		7.87		
1 Breaker, Pavement, 60 lb.		53.60		58.96	37.71	41.48
56 L.H., Daily Totals		$4649.20		$6110.76	$83.02	$109.12

Crew B-79D	Hr.	Daily	Hr.	Daily	Bare Costs	Incl. O&P
2 Labor Foremen (outside)	$46.40	$742.40	$69.25	$1108.00	$45.45	$67.84
5 Laborers	44.40	1776.00	66.25	2650.00		
1 Truck Driver (light)	48.80	390.40	73.00	584.00		
1 Paint Striper, T.M., 120 Gal.		603.10		663.41		
1 Heating Kettle, 115 Gallon		107.25		117.97		
1 Flatbed Truck, Gas, 3 Ton		850.05		935.05		
4 Pickup Trucks, 3/4 Ton		448.80		493.68		
1 Air Compressor, 60 cfm		153.85		169.24		
1 -50' Air Hose, 3/4"		7.15		7.87		
1 Breaker, Pavement, 60 lb.		53.60		58.96	34.75	38.22
64 L.H., Daily Totals		$5132.60		$6788.18	$80.20	$106.07

For customer support on your Site Work & Landscape Costs with RSMeans Data, call 800.448.8182.

Crew No.	Bare Costs		Incl. Subs O&P		Cost Per Labor-Hour	
Crew B-79E	**Hr.**	**Daily**	**Hr.**	**Daily**	**Bare Costs**	**Incl. O&P**
2 Labor Foremen (outside)	$46.40	$742.40	$69.25	$1108.00	$45.24	$67.53
7 Laborers	44.40	2486.40	66.25	3710.00		
1 Truck Driver (light)	48.80	390.40	73.00	584.00		
1 Paint Striper, T.M., 120 Gal.		603.10		663.41		
1 Heating Kettle, 115 Gallon		107.25		117.97		
1 Flatbed Truck, Gas, 3 Ton		850.05		935.05		
5 Pickup Trucks, 3/4 Ton		561.00		617.10		
2 Air Compressors, 60 cfm		307.70		338.47		
2 -50' Air Hoses, 3/4"		14.30		15.73		
2 Breakers, Pavement, 60 lb.		107.20		117.92	31.88	35.07
80 L.H., Daily Totals		$6169.80		$8207.66	$77.12	$102.60
Crew B-80	**Hr.**	**Daily**	**Hr.**	**Daily**	**Bare Costs**	**Incl. O&P**
1 Labor Foreman (outside)	$46.40	$371.20	$69.25	$554.00	$48.77	$72.80
1 Laborer	44.40	355.20	66.25	530.00		
1 Truck Driver (light)	48.80	390.40	73.00	584.00		
1 Equip. Oper. (light)	55.50	444.00	82.70	661.60		
1 Flatbed Truck, Gas, 3 Ton		850.05		935.05		
1 Earth Auger, Truck-Mtd.		202.55		222.81	32.89	36.18
32 L.H., Daily Totals		$2613.40		$3487.46	$81.67	$108.98
Crew B-80A	**Hr.**	**Daily**	**Hr.**	**Daily**	**Bare Costs**	**Incl. O&P**
3 Laborers	$44.40	$1065.60	$66.25	$1590.00	$44.40	$66.25
1 Flatbed Truck, Gas, 3 Ton		850.05		935.05	35.42	38.96
24 L.H., Daily Totals		$1915.65		$2525.05	$79.82	$105.21
Crew B-80B	**Hr.**	**Daily**	**Hr.**	**Daily**	**Bare Costs**	**Incl. O&P**
3 Laborers	$44.40	$1065.60	$66.25	$1590.00	$47.17	$70.36
1 Equip. Oper. (light)	55.50	444.00	82.70	661.60		
1 Crane, Flatbed Mounted, 3 Ton		238.75		262.63	7.46	8.21
32 L.H., Daily Totals		$1748.35		$2514.22	$54.64	$78.57
Crew B-80C	**Hr.**	**Daily**	**Hr.**	**Daily**	**Bare Costs**	**Incl. O&P**
2 Laborers	$44.40	$710.40	$66.25	$1060.00	$45.87	$68.50
1 Truck Driver (light)	48.80	390.40	73.00	584.00		
1 Flatbed Truck, Gas, 1.5 Ton		198.50		218.35		
1 Manual Fence Post Auger, Gas		54.40		59.84	10.54	11.59
24 L.H., Daily Totals		$1353.70		$1922.19	$56.40	$80.09
Crew B-81	**Hr.**	**Daily**	**Hr.**	**Daily**	**Bare Costs**	**Incl. O&P**
1 Laborer	$44.40	$355.20	$66.25	$530.00	$51.57	$76.95
1 Equip. Oper. (medium)	59.00	472.00	87.90	703.20		
1 Truck Driver (heavy)	51.30	410.40	76.70	613.60		
1 Hydromulcher, T.M., 3000 Gal.		256.85		282.54		
1 Truck Tractor, 220 H.P.		310.80		341.88	23.65	26.02
24 L.H., Daily Totals		$1805.25		$2471.22	$75.22	$102.97
Crew B-81A	**Hr.**	**Daily**	**Hr.**	**Daily**	**Bare Costs**	**Incl. O&P**
1 Laborer	$44.40	$355.20	$66.25	$530.00	$46.60	$69.63
1 Truck Driver (light)	48.80	390.40	73.00	584.00		
1 Hydromulcher, T.M., 600 Gal.		118.15		129.97		
1 Flatbed Truck, Gas, 3 Ton		850.05		935.05	60.51	66.56
16 L.H., Daily Totals		$1713.80		$2179.02	$107.11	$136.19
Crew B-82	**Hr.**	**Daily**	**Hr.**	**Daily**	**Bare Costs**	**Incl. O&P**
1 Laborer	$44.40	$355.20	$66.25	$530.00	$49.95	$74.47
1 Equip. Oper. (light)	55.50	444.00	82.70	661.60		
1 Horiz. Borer, 6 H.P.		184.15		202.57	11.51	12.66
16 L.H., Daily Totals		$983.35		$1394.17	$61.46	$87.14

Crew No.	Bare Costs		Incl. Subs O&P		Cost Per Labor-Hour	
Crew B-82A	**Hr.**	**Daily**	**Hr.**	**Daily**	**Bare Costs**	**Incl. O&P**
2 Laborers	$44.40	$710.40	$66.25	$1060.00	$49.95	$74.47
2 Equip. Opers. (light)	55.50	888.00	82.70	1323.20		
2 Dump Truck, 8 C.Y., 220 H.P.		815.20		896.72		
1 Flatbed Trailer, 25 Ton		137.20		150.92		
1 Horiz. Dir. Drill, 20k lb. Thrust		544.10		598.51		
1 Mud Trailer for HDD, 1500 Gal.		312.15		343.37		
1 Pickup Truck, 4x4, 3/4 Ton		176.75		194.43		
1 Flatbed Trailer, 3 Ton		71.15		78.27		
1 Loader, Skid Steer, 78 H.P.		446.30		490.93	78.21	86.04
32 L.H., Daily Totals		$4101.25		$5136.34	$128.16	$160.51
Crew B-82B	**Hr.**	**Daily**	**Hr.**	**Daily**	**Bare Costs**	**Incl. O&P**
2 Laborers	$44.40	$710.40	$66.25	$1060.00	$49.95	$74.47
2 Equip. Opers. (light)	55.50	888.00	82.70	1323.20		
2 Dump Truck, 8 C.Y., 220 H.P.		815.20		896.72		
1 Flatbed Trailer, 25 Ton		137.20		150.92		
1 Horiz. Dir. Drill, 30k lb. Thrust		647.65		712.41		
1 Mud Trailer for HDD, 1500 Gal.		312.15		343.37		
1 Pickup Truck, 4x4, 3/4 Ton		176.75		194.43		
1 Flatbed Trailer, 3 Ton		71.15		78.27		
1 Loader, Skid Steer, 78 H.P.		446.30		490.93	81.45	89.59
32 L.H., Daily Totals		$4204.80		$5250.24	$131.40	$164.07
Crew B-82C	**Hr.**	**Daily**	**Hr.**	**Daily**	**Bare Costs**	**Incl. O&P**
2 Laborers	$44.40	$710.40	$66.25	$1060.00	$49.95	$74.47
2 Equip. Opers. (light)	55.50	888.00	82.70	1323.20		
2 Dump Truck, 8 C.Y., 220 H.P.		815.20		896.72		
1 Flatbed Trailer, 25 Ton		137.20		150.92		
1 Horiz. Dir. Drill, 50k lb. Thrust		824.05		906.46		
1 Mud Trailer for HDD, 1500 Gal.		312.15		343.37		
1 Pickup Truck, 4x4, 3/4 Ton		176.75		194.43		
1 Flatbed Trailer, 3 Ton		71.15		78.27		
1 Loader, Skid Steer, 78 H.P.		446.30		490.93	86.96	95.66
32 L.H., Daily Totals		$4381.20		$5444.28	$136.91	$170.13
Crew B-82D	**Hr.**	**Daily**	**Hr.**	**Daily**	**Bare Costs**	**Incl. O&P**
1 Equip. Oper. (light)	$55.50	$444.00	$82.70	$661.60	$55.50	$82.70
1 Mud Trailer for HDD, 1500 Gal.		312.15		343.37	39.02	42.92
8 L.H., Daily Totals		$756.15		$1004.97	$94.52	$125.62
Crew B-83	**Hr.**	**Daily**	**Hr.**	**Daily**	**Bare Costs**	**Incl. O&P**
1 Tugboat Captain	$59.00	$472.00	$87.90	$703.20	$51.70	$77.08
1 Tugboat Hand	44.40	355.20	66.25	530.00		
1 Tugboat, 250 H.P.		726.10		798.71	45.38	49.92
16 L.H., Daily Totals		$1553.30		$2031.91	$97.08	$126.99
Crew B-84	**Hr.**	**Daily**	**Hr.**	**Daily**	**Bare Costs**	**Incl. O&P**
1 Equip. Oper. (medium)	$59.00	$472.00	$87.90	$703.20	$59.00	$87.90
1 Rotary Mower/Tractor		371.15		408.26	46.39	51.03
8 L.H., Daily Totals		$843.15		$1111.46	$105.39	$138.93
Crew B-85	**Hr.**	**Daily**	**Hr.**	**Daily**	**Bare Costs**	**Incl. O&P**
3 Laborers	$44.40	$1065.60	$66.25	$1590.00	$48.70	$72.67
1 Equip. Oper. (medium)	59.00	472.00	87.90	703.20		
1 Truck Driver (heavy)	51.30	410.40	76.70	613.60		
1 Telescoping Boom Lift, to 80'		387.75		426.52		
1 Brush Chipper, 12", 130 H.P.		366.05		402.65		
1 Pruning Saw, Rotary		26.40		29.04	19.50	21.46
40 L.H., Daily Totals		$2728.20		$3765.02	$68.20	$94.13

Crew B-86

Crew No.	Bare Hr.	Bare Daily	Incl. Subs O&P Hr.	Incl. Subs O&P Daily	Cost Per Labor-Hour Bare Costs	Cost Per Labor-Hour Incl. O&P
1 Equip. Oper. (medium)	$59.00	$472.00	$87.90	$703.20	$59.00	$87.90
1 Stump Chipper, S.P.		189.20		208.12	23.65	26.02
8 L.H., Daily Totals		$661.20		$911.32	$82.65	$113.92

Crew B-86A

Crew No.	Bare Hr.	Bare Daily	Incl. Subs O&P Hr.	Incl. Subs O&P Daily	Bare Costs	Incl. O&P
1 Equip. Oper. (medium)	$59.00	$472.00	$87.90	$703.20	$59.00	$87.90
1 Grader, 30,000 Lbs.		1073.00		1180.30	134.13	147.54
8 L.H., Daily Totals		$1545.00		$1883.50	$193.13	$235.44

Crew B-86B

Crew No.	Bare Hr.	Bare Daily	Incl. Subs O&P Hr.	Incl. Subs O&P Daily	Bare Costs	Incl. O&P
1 Equip. Oper. (medium)	$59.00	$472.00	$87.90	$703.20	$59.00	$87.90
1 Dozer, 200 H.P.		1520.00		1672.00	190.00	209.00
8 L.H., Daily Totals		$1992.00		$2375.20	$249.00	$296.90

Crew B-87

Crew No.	Bare Hr.	Bare Daily	Incl. Subs O&P Hr.	Incl. Subs O&P Daily	Bare Costs	Incl. O&P
1 Laborer	$44.40	$355.20	$66.25	$530.00	$56.08	$83.57
4 Equip. Oper. (medium)	59.00	1888.00	87.90	2812.80		
2 Feller Bunchers, 100 H.P.		1257.60		1383.36		
1 Log Chipper, 22" Tree		555.00		610.50		
1 Dozer, 105 H.P.		640.80		704.88		
1 Chain Saw, Gas, 36" Long		41.65		45.81	62.38	68.61
40 L.H., Daily Totals		$4738.25		$6087.35	$118.46	$152.18

Crew B-88

Crew No.	Bare Hr.	Bare Daily	Incl. Subs O&P Hr.	Incl. Subs O&P Daily	Bare Costs	Incl. O&P
1 Laborer	$44.40	$355.20	$66.25	$530.00	$56.91	$84.81
6 Equip. Oper. (medium)	59.00	2832.00	87.90	4219.20		
2 Feller Bunchers, 100 H.P.		1257.60		1383.36		
1 Log Chipper, 22" Tree		555.00		610.50		
2 Log Skidders, 50 H.P.		1826.70		2009.37		
1 Dozer, 105 H.P.		640.80		704.88		
1 Chain Saw, Gas, 36" Long		41.65		45.81	77.17	84.89
56 L.H., Daily Totals		$7508.95		$9503.13	$134.09	$169.70

Crew B-89

Crew No.	Bare Hr.	Bare Daily	Incl. Subs O&P Hr.	Incl. Subs O&P Daily	Bare Costs	Incl. O&P
1 Equip. Oper. (light)	$55.50	$444.00	$82.70	$661.60	$52.15	$77.85
1 Truck Driver (light)	48.80	390.40	73.00	584.00		
1 Flatbed Truck, Gas, 3 Ton		850.05		935.05		
1 Concrete Saw		112.85		124.14		
1 Water Tank, 65 Gal.		102.90		113.19	66.61	73.27
16 L.H., Daily Totals		$1900.20		$2417.98	$118.76	$151.12

Crew B-89A

Crew No.	Bare Hr.	Bare Daily	Incl. Subs O&P Hr.	Incl. Subs O&P Daily	Bare Costs	Incl. O&P
1 Skilled Worker	$57.10	$456.80	$85.90	$687.20	$50.75	$76.08
1 Laborer	44.40	355.20	66.25	530.00		
1 Core Drill (Large)		121.60		133.76	7.60	8.36
16 L.H., Daily Totals		$933.60		$1350.96	$58.35	$84.44

Crew B-89B

Crew No.	Bare Hr.	Bare Daily	Incl. Subs O&P Hr.	Incl. Subs O&P Daily	Bare Costs	Incl. O&P
1 Equip. Oper. (light)	$55.50	$444.00	$82.70	$661.60	$52.15	$77.85
1 Truck Driver (light)	48.80	390.40	73.00	584.00		
1 Wall Saw, Hydraulic, 10 H.P.		86.40		95.04		
1 Generator, Diesel, 100 kW		521.85		574.03		
1 Water Tank, 65 Gal.		102.90		113.19		
1 Flatbed Truck, Gas, 3 Ton		850.05		935.05	97.58	107.33
16 L.H., Daily Totals		$2395.60		$2962.92	$149.72	$185.18

Crew B-89C

Crew No.	Bare Hr.	Bare Daily	Incl. Subs O&P Hr.	Incl. Subs O&P Daily	Bare Costs	Incl. O&P
1 Cement Finisher	$51.80	$414.40	$75.90	$607.20	$51.80	$75.90
1 Masonry cut-off saw, gas		58.15		63.97	7.27	8.00
8 L.H., Daily Totals		$472.55		$671.16	$59.07	$83.90

Crew B-90

Crew No.	Bare Hr.	Bare Daily	Incl. Subs O&P Hr.	Incl. Subs O&P Daily	Bare Costs	Incl. O&P
1 Labor Foreman (outside)	$46.40	$371.20	$69.25	$554.00	$49.15	$73.35
3 Laborers	44.40	1065.60	66.25	1590.00		
2 Equip. Oper. (light)	55.50	888.00	82.70	1323.20		
2 Truck Drivers (heavy)	51.30	820.80	76.70	1227.20		
1 Road Mixer, 310 H.P.		1919.00		2110.90		
1 Dist. Truck, 2000 Gal.		303.25		333.57	34.72	38.19
64 L.H., Daily Totals		$5367.85		$7138.88	$83.87	$111.54

Crew B-90A

Crew No.	Bare Hr.	Bare Daily	Incl. Subs O&P Hr.	Incl. Subs O&P Daily	Bare Costs	Incl. O&P
1 Labor Foreman (outside)	$46.40	$371.20	$69.25	$554.00	$53.03	$79.05
2 Laborers	44.40	710.40	66.25	1060.00		
4 Equip. Oper. (medium)	59.00	1888.00	87.90	2812.80		
2 Graders, 30,000 Lbs.		2146.00		2360.60		
1 Tandem Roller, 10 Ton		246.80		271.48		
1 Roller, Pneum. Whl., 12 Ton		349.90		384.89	48.98	53.87
56 L.H., Daily Totals		$5712.30		$7443.77	$102.01	$132.92

Crew B-90B

Crew No.	Bare Hr.	Bare Daily	Incl. Subs O&P Hr.	Incl. Subs O&P Daily	Bare Costs	Incl. O&P
1 Labor Foreman (outside)	$46.40	$371.20	$69.25	$554.00	$52.03	$77.58
2 Laborers	44.40	710.40	66.25	1060.00		
3 Equip. Oper. (medium)	59.00	1416.00	87.90	2109.60		
1 Roller, Pneum. Whl., 12 Ton		349.90		384.89		
1 Road Mixer, 310 H.P.		1919.00		2110.90	47.27	52.00
48 L.H., Daily Totals		$4766.50		$6219.39	$99.30	$129.57

Crew B-90C

Crew No.	Bare Hr.	Bare Daily	Incl. Subs O&P Hr.	Incl. Subs O&P Daily	Bare Costs	Incl. O&P
1 Labor Foreman (outside)	$46.40	$371.20	$69.25	$554.00	$50.45	$75.28
4 Laborers	44.40	1420.80	66.25	2120.00		
3 Equip. Oper. (medium)	59.00	1416.00	87.90	2109.60		
3 Truck Drivers (heavy)	51.30	1231.20	76.70	1840.80		
3 Road Mixers, 310 H.P.		5757.00		6332.70	65.42	71.96
88 L.H., Daily Totals		$10196.20		$12957.10	$115.87	$147.24

Crew B-90D

Crew No.	Bare Hr.	Bare Daily	Incl. Subs O&P Hr.	Incl. Subs O&P Daily	Bare Costs	Incl. O&P
1 Labor Foreman (outside)	$46.40	$371.20	$69.25	$554.00	$49.52	$73.89
6 Laborers	44.40	2131.20	66.25	3180.00		
3 Equip. Oper. (medium)	59.00	1416.00	87.90	2109.60		
3 Truck Drivers (heavy)	51.30	1231.20	76.70	1840.80		
3 Road Mixers, 310 H.P.		5757.00		6332.70	55.36	60.89
104 L.H., Daily Totals		$10906.60		$14017.10	$104.87	$134.78

Crew B-90E

Crew No.	Bare Hr.	Bare Daily	Incl. Subs O&P Hr.	Incl. Subs O&P Daily	Bare Costs	Incl. O&P
1 Labor Foreman (outside)	$46.40	$371.20	$69.25	$554.00	$50.26	$74.96
4 Laborers	44.40	1420.80	66.25	2120.00		
3 Equip. Oper. (medium)	59.00	1416.00	87.90	2109.60		
1 Truck Driver (heavy)	51.30	410.40	76.70	613.60		
1 Road Mixer, 310 H.P.		1919.00		2110.90	26.65	29.32
72 L.H., Daily Totals		$5537.40		$7508.10	$76.91	$104.28

Crew B-91

Crew No.	Bare Hr.	Bare Daily	Incl. Subs O&P Hr.	Incl. Subs O&P Daily	Bare Costs	Incl. O&P
1 Labor Foreman (outside)	$46.40	$371.20	$69.25	$554.00	$52.81	$78.76
2 Laborers	44.40	710.40	66.25	1060.00		
4 Equip. Oper. (medium)	59.00	1888.00	87.90	2812.80		
1 Truck Driver (heavy)	51.30	410.40	76.70	613.60		
1 Dist. Tanker, 3000 Gallon		334.10		367.51		
1 Truck Tractor, 6x4, 380 H.P.		499.15		549.07		
1 Aggreg. Spreader, S.P.		859.10		945.01		
1 Roller, Pneum. Whl., 12 Ton		349.90		384.89		
1 Tandem Roller, 10 Ton		246.80		271.48	35.77	39.34
64 L.H., Daily Totals		$5669.05		$7558.35	$88.58	$118.10

For customer support on your Site Work & Landscape Costs with RSMeans Data, call 800.448.8182.

Crews - Standard

Crew No.	Bare Costs		Incl. Subs O&P		Cost Per Labor-Hour	

Crew B-91B

	Hr.	Daily	Hr.	Daily	Bare Costs	Incl. O&P
1 Laborer	$44.40	$355.20	$66.25	$530.00	$51.70	$77.08
1 Equipment Oper. (med.)	59.00	472.00	87.90	703.20		
1 Road Sweeper, Vac. Assist.		879.45		967.39	54.97	60.46
16 L.H., Daily Totals		$1706.65		$2200.59	$106.67	$137.54

Crew B-91C

	Hr.	Daily	Hr.	Daily	Bare Costs	Incl. O&P
1 Laborer	$44.40	$355.20	$66.25	$530.00	$46.60	$69.63
1 Truck Driver (light)	48.80	390.40	73.00	584.00		
1 Catch Basin Cleaning Truck		542.60		596.86	33.91	37.30
16 L.H., Daily Totals		$1288.20		$1710.86	$80.51	$106.93

Crew B-91D

	Hr.	Daily	Hr.	Daily	Bare Costs	Incl. O&P
1 Labor Foreman (outside)	$46.40	$371.20	$69.25	$554.00	$51.23	$76.42
5 Laborers	44.40	1776.00	66.25	2650.00		
5 Equip. Oper. (medium)	59.00	2360.00	87.90	3516.00		
2 Truck Drivers (heavy)	51.30	820.80	76.70	1227.20		
1 Aggreg. Spreader, S.P.		859.10		945.01		
2 Truck Tractors, 6x4, 380 H.P.		998.30		1098.13		
2 Dist. Tankers, 3000 Gallon		668.20		735.02		
2 Pavement Brushes, Towed		176.70		194.37		
2 Rollers Pneum. Whl., 12 Ton		699.80		769.78	32.71	35.98
104 L.H., Daily Totals		$8730.10		$11689.51	$83.94	$112.40

Crew B-92

	Hr.	Daily	Hr.	Daily	Bare Costs	Incl. O&P
1 Labor Foreman (outside)	$46.40	$371.20	$69.25	$554.00	$44.90	$67.00
3 Laborers	44.40	1065.60	66.25	1590.00		
1 Crack Cleaner, 25 H.P.		53.00		58.30		
1 Air Compressor, 60 cfm		153.85		169.24		
1 Tar Kettle, T.M.		156.70		172.37		
1 Flatbed Truck, Gas, 3 Ton		850.05		935.05	37.92	41.72
32 L.H., Daily Totals		$2650.40		$3478.96	$82.83	$108.72

Crew B-93

	Hr.	Daily	Hr.	Daily	Bare Costs	Incl. O&P
1 Equip. Oper. (medium)	$59.00	$472.00	$87.90	$703.20	$59.00	$87.90
1 Feller Buncher, 100 H.P.		628.80		691.68	78.60	86.46
8 L.H., Daily Totals		$1100.80		$1394.88	$137.60	$174.36

Crew B-94A

	Hr.	Daily	Hr.	Daily	Bare Costs	Incl. O&P
1 Laborer	$44.40	$355.20	$66.25	$530.00	$44.40	$66.25
1 Diaphragm Water Pump, 2"		87.70		96.47		
1 -20' Suction Hose, 2"		3.55		3.90		
2 -50' Discharge Hoses, 2"		8.00		8.80	12.41	13.65
8 L.H., Daily Totals		$454.45		$639.17	$56.81	$79.90

Crew B-94B

	Hr.	Daily	Hr.	Daily	Bare Costs	Incl. O&P
1 Laborer	$44.40	$355.20	$66.25	$530.00	$44.40	$66.25
1 Diaphragm Water Pump, 4"		106.35		116.99		
1 -20' Suction Hose, 4"		17.25		18.98		
2 -50' Discharge Hoses, 4"		25.60		28.16	18.65	20.52
8 L.H., Daily Totals		$504.40		$694.12	$63.05	$86.77

Crew B-94C

	Hr.	Daily	Hr.	Daily	Bare Costs	Incl. O&P
1 Laborer	$44.40	$355.20	$66.25	$530.00	$44.40	$66.25
1 Centrifugal Water Pump, 3"		74.40		81.84		
1 -20' Suction Hose, 3"		8.75		9.63		
2 -50' Discharge Hoses, 3"		9.00		9.90	11.52	12.67
8 L.H., Daily Totals		$447.35		$631.37	$55.92	$78.92

Crew B-94D

	Hr.	Daily	Hr.	Daily	Bare Costs	Incl. O&P
1 Laborer	$44.40	$355.20	$66.25	$530.00	$44.40	$66.25
1 Centr. Water Pump, 6"		235.25		258.77		
1 -20' Suction Hose, 6"		25.50		28.05		
2 -50' Discharge Hoses, 6"		36.20		39.82	37.12	40.83
8 L.H., Daily Totals		$652.15		$856.64	$81.52	$107.08

Crew C-1

	Hr.	Daily	Hr.	Daily	Bare Costs	Incl. O&P
3 Carpenters	$54.70	$1312.80	$81.65	$1959.60	$52.13	$77.80
1 Laborer	44.40	355.20	66.25	530.00		
32 L.H., Daily Totals		$1668.00		$2489.60	$52.13	$77.80

Crew C-2

	Hr.	Daily	Hr.	Daily	Bare Costs	Incl. O&P
1 Carpenter Foreman (outside)	$56.70	$453.60	$84.60	$676.80	$53.32	$79.58
4 Carpenters	54.70	1750.40	81.65	2612.80		
1 Laborer	44.40	355.20	66.25	530.00		
48 L.H., Daily Totals		$2559.20		$3819.60	$53.32	$79.58

Crew C-2A

	Hr.	Daily	Hr.	Daily	Bare Costs	Incl. O&P
1 Carpenter Foreman (outside)	$56.70	$453.60	$84.60	$676.80	$52.83	$78.62
3 Carpenters	54.70	1312.80	81.65	1959.60		
1 Cement Finisher	51.80	414.40	75.90	607.20		
1 Laborer	44.40	355.20	66.25	530.00		
48 L.H., Daily Totals		$2536.00		$3773.60	$52.83	$78.62

Crew C-3

	Hr.	Daily	Hr.	Daily	Bare Costs	Incl. O&P
1 Rodman Foreman (outside)	$60.90	$487.20	$91.05	$728.40	$55.10	$82.31
4 Rodmen (reinf.)	58.90	1884.80	88.05	2817.60		
1 Equip. Oper. (light)	55.50	444.00	82.70	661.60		
2 Laborers	44.40	710.40	66.25	1060.00		
3 Stressing Equipment		56.85		62.53		
.5 Grouting Equipment		123.33		135.66	2.82	3.10
64 L.H., Daily Totals		$3706.57		$5465.79	$57.92	$85.40

Crew C-4

	Hr.	Daily	Hr.	Daily	Bare Costs	Incl. O&P
1 Rodman Foreman (outside)	$60.90	$487.20	$91.05	$728.40	$59.40	$88.80
3 Rodmen (reinf.)	58.90	1413.60	88.05	2113.20		
3 Stressing Equipment		56.85		62.53	1.78	1.95
32 L.H., Daily Totals		$1957.65		$2904.14	$61.18	$90.75

Crew C-4A

	Hr.	Daily	Hr.	Daily	Bare Costs	Incl. O&P
2 Rodmen (reinf.)	$58.90	$942.40	$88.05	$1408.80	$58.90	$88.05
4 Stressing Equipment		75.80		83.38	4.74	5.21
16 L.H., Daily Totals		$1018.20		$1492.18	$63.64	$93.26

Crew C-5

	Hr.	Daily	Hr.	Daily	Bare Costs	Incl. O&P
1 Rodman Foreman (outside)	$60.90	$487.20	$91.05	$728.40	$58.64	$87.58
4 Rodmen (reinf.)	58.90	1884.80	88.05	2817.60		
1 Equip. Oper. (crane)	61.45	491.60	91.55	732.40		
1 Equip. Oper. (oiler)	52.50	420.00	78.25	626.00		
1 Hyd. Crane, 25 Ton		586.70		645.37	10.48	11.52
56 L.H., Daily Totals		$3870.30		$5549.77	$69.11	$99.10

Crew C-6

	Hr.	Daily	Hr.	Daily	Bare Costs	Incl. O&P
1 Labor Foreman (outside)	$46.40	$371.20	$69.25	$554.00	$45.97	$68.36
4 Laborers	44.40	1420.80	66.25	2120.00		
1 Cement Finisher	51.80	414.40	75.90	607.20		
2 Gas Engine Vibrators		54.30		59.73	1.13	1.24
48 L.H., Daily Totals		$2260.70		$3340.93	$47.10	$69.60

For customer support on your Site Work & Landscape Costs with RSMeans Data, call 800.448.8182.

771

Crew No.	Bare Costs		Incl. Subs O&P		Cost Per Labor-Hour	
Crew C-6A	Hr.	Daily	Hr.	Daily	Bare Costs	Incl. O&P
2 Cement Finishers	$51.80	$828.80	$75.90	$1214.40	$51.80	$75.90
1 Concrete Vibrator, Elec, 2 HP		45.80		50.38	2.86	3.15
16 L.H., Daily Totals		$874.60		$1264.78	$54.66	$79.05
Crew C-7	Hr.	Daily	Hr.	Daily	Bare Costs	Incl. O&P
1 Labor Foreman (outside)	$46.40	$371.20	$69.25	$554.00	$47.97	$71.39
5 Laborers	44.40	1776.00	66.25	2650.00		
1 Cement Finisher	51.80	414.40	75.90	607.20		
1 Equip. Oper. (medium)	59.00	472.00	87.90	703.20		
1 Equip. Oper. (oiler)	52.50	420.00	78.25	626.00		
2 Gas Engine Vibrators		54.30		59.73		
1 Concrete Bucket, 1 C.Y.		45.90		50.49		
1 Hyd. Crane, 55 Ton		990.15		1089.17	15.14	16.66
72 L.H., Daily Totals		$4543.95		$6339.78	$63.11	$88.05
Crew C-7A	Hr.	Daily	Hr.	Daily	Bare Costs	Incl. O&P
1 Labor Foreman (outside)	$46.40	$371.20	$69.25	$554.00	$46.38	$69.24
5 Laborers	44.40	1776.00	66.25	2650.00		
2 Truck Drivers (heavy)	51.30	820.80	76.70	1227.20		
2 Conc. Transit Mixers		1176.30		1293.93	18.38	20.22
64 L.H., Daily Totals		$4144.30		$5725.13	$64.75	$89.46
Crew C-7B	Hr.	Daily	Hr.	Daily	Bare Costs	Incl. O&P
1 Labor Foreman (outside)	$46.40	$371.20	$69.25	$554.00	$47.79	$71.29
5 Laborers	44.40	1776.00	66.25	2650.00		
1 Equipment Operator, Crane	61.45	491.60	91.55	732.40		
1 Equipment Oiler	52.50	420.00	78.25	626.00		
1 Conc. Bucket, 2 C.Y.		55.65		61.22		
1 Lattice Boom Crane, 165 Ton		2403.00		2643.30	38.42	42.26
64 L.H., Daily Totals		$5517.45		$7266.92	$86.21	$113.55
Crew C-7C	Hr.	Daily	Hr.	Daily	Bare Costs	Incl. O&P
1 Labor Foreman (outside)	$46.40	$371.20	$69.25	$554.00	$48.30	$72.04
5 Laborers	44.40	1776.00	66.25	2650.00		
2 Equipment Operators (med.)	59.00	944.00	87.90	1406.40		
2 F.E. Loaders, W.M., 4 C.Y.		1518.00		1669.80	23.72	26.09
64 L.H., Daily Totals		$4609.20		$6280.20	$72.02	$98.13
Crew C-7D	Hr.	Daily	Hr.	Daily	Bare Costs	Incl. O&P
1 Labor Foreman (outside)	$46.40	$371.20	$69.25	$554.00	$46.77	$69.77
5 Laborers	44.40	1776.00	66.25	2650.00		
1 Equip. Oper. (medium)	59.00	472.00	87.90	703.20		
1 Concrete Conveyer		206.25		226.88	3.68	4.05
56 L.H., Daily Totals		$2825.45		$4134.07	$50.45	$73.82
Crew C-8	Hr.	Daily	Hr.	Daily	Bare Costs	Incl. O&P
1 Labor Foreman (outside)	$46.40	$371.20	$69.25	$554.00	$48.89	$72.53
3 Laborers	44.40	1065.60	66.25	1590.00		
2 Cement Finishers	51.80	828.80	75.90	1214.40		
1 Equip. Oper. (medium)	59.00	472.00	87.90	703.20		
1 Concrete Pump (Small)		423.65		466.01	7.57	8.32
56 L.H., Daily Totals		$3161.25		$4527.61	$56.45	$80.85
Crew C-8A	Hr.	Daily	Hr.	Daily	Bare Costs	Incl. O&P
1 Labor Foreman (outside)	$46.40	$371.20	$69.25	$554.00	$47.20	$69.97
3 Laborers	44.40	1065.60	66.25	1590.00		
2 Cement Finishers	51.80	828.80	75.90	1214.40		
48 L.H., Daily Totals		$2265.60		$3358.40	$47.20	$69.97

Crew No.	Bare Costs		Incl. Subs O&P		Cost Per Labor-Hour	
Crew C-8B	Hr.	Daily	Hr.	Daily	Bare Costs	Incl. O&P
1 Labor Foreman (outside)	$46.40	$371.20	$69.25	$554.00	$47.72	$71.18
3 Laborers	44.40	1065.60	66.25	1590.00		
1 Equip. Oper. (medium)	59.00	472.00	87.90	703.20		
1 Vibrating Power Screed		87.65		96.42		
1 Roller, Vibratory, 25 Ton		672.35		739.59		
1 Dozer, 200 H.P.		1520.00		1672.00	57.00	62.70
40 L.H., Daily Totals		$4188.80		$5355.20	$104.72	$133.88
Crew C-8C	Hr.	Daily	Hr.	Daily	Bare Costs	Incl. O&P
1 Labor Foreman (outside)	$46.40	$371.20	$69.25	$554.00	$48.40	$71.97
3 Laborers	44.40	1065.60	66.25	1590.00		
1 Cement Finisher	51.80	414.40	75.90	607.20		
1 Equip. Oper. (medium)	59.00	472.00	87.90	703.20		
1 Shotcrete Rig, 12 C.Y./hr		269.20		296.12		
1 Air Compressor, 160 cfm		212.30		233.53		
4 -50' Air Hoses, 1"		32.20		35.42		
4 -50' Air Hoses, 2"		115.80		127.38	13.11	14.43
48 L.H., Daily Totals		$2952.70		$4146.85	$61.51	$86.39
Crew C-8D	Hr.	Daily	Hr.	Daily	Bare Costs	Incl. O&P
1 Labor Foreman (outside)	$46.40	$371.20	$69.25	$554.00	$49.52	$73.53
1 Laborer	44.40	355.20	66.25	530.00		
1 Cement Finisher	51.80	414.40	75.90	607.20		
1 Equipment Oper. (light)	55.50	444.00	82.70	661.60		
1 Air Compressor, 250 cfm		202.85		223.13		
2 -50' Air Hoses, 1"		16.10		17.71	6.84	7.53
32 L.H., Daily Totals		$1803.75		$2593.65	$56.37	$81.05
Crew C-8E	Hr.	Daily	Hr.	Daily	Bare Costs	Incl. O&P
1 Labor Foreman (outside)	$46.40	$371.20	$69.25	$554.00	$47.82	$71.10
3 Laborers	44.40	1065.60	66.25	1590.00		
1 Cement Finisher	51.80	414.40	75.90	607.20		
1 Equipment Oper. (light)	55.50	444.00	82.70	661.60		
1 Shotcrete Rig, 35 C.Y./hr.		301.05		331.15		
1 Air Compressor, 250 cfm		202.85		223.13		
4 -50' Air Hoses, 1"		32.20		35.42		
4 -50' Air Hoses, 2"		115.80		127.38	13.58	14.94
48 L.H., Daily Totals		$2947.10		$4129.89	$61.40	$86.04
Crew C-9	Hr.	Daily	Hr.	Daily	Bare Costs	Incl. O&P
1 Cement Finisher	$51.80	$414.40	$75.90	$607.20	$49.02	$72.78
2 Laborers	44.40	710.40	66.25	1060.00		
1 Equipment Oper. (light)	55.50	444.00	82.70	661.60		
1 Grout Pump, 50 C.F./hr.		190.35		209.38		
1 Air Compressor, 160 cfm		212.30		233.53		
2 -50' Air Hoses, 1"		16.10		17.71		
2 -50' Air Hoses, 2"		57.90		63.69	14.90	16.38
32 L.H., Daily Totals		$2045.45		$2853.11	$63.92	$89.16
Crew C-10	Hr.	Daily	Hr.	Daily	Bare Costs	Incl. O&P
1 Laborer	$44.40	$355.20	$66.25	$530.00	$49.33	$72.68
2 Cement Finishers	51.80	828.80	75.90	1214.40		
24 L.H., Daily Totals		$1184.00		$1744.40	$49.33	$72.68
Crew C-10B	Hr.	Daily	Hr.	Daily	Bare Costs	Incl. O&P
3 Laborers	$44.40	$1065.60	$66.25	$1590.00	$47.36	$70.11
2 Cement Finishers	51.80	828.80	75.90	1214.40		
1 Concrete Mixer, 10 C.F.		147.15		161.87		
2 Trowels, 48" Walk-Behind		188.60		207.46	8.39	9.23
40 L.H., Daily Totals		$2230.15		$3173.72	$55.75	$79.34

For customer support on your Site Work & Landscape Costs with RSMeans Data, call 800.448.8182.

Crew No.	Bare Costs		Incl. Subs O&P		Cost Per Labor-Hour	
Crew C-10C	**Hr.**	**Daily**	**Hr.**	**Daily**	**Bare Costs**	**Incl. O&P**
1 Laborer	$44.40	$355.20	$66.25	$530.00	$49.33	$72.68
2 Cement Finishers	51.80	828.80	75.90	1214.40		
1 Trowel, 48" Walk-Behind		94.30		103.73	3.93	4.32
24 L.H., Daily Totals		$1278.30		$1848.13	$53.26	$77.01
Crew C-10D	**Hr.**	**Daily**	**Hr.**	**Daily**	**Bare Costs**	**Incl. O&P**
1 Laborer	$44.40	$355.20	$66.25	$530.00	$49.33	$72.68
2 Cement Finishers	51.80	828.80	75.90	1214.40		
1 Vibrating Power Screed		87.65		96.42		
1 Trowel, 48" Walk-Behind		94.30		103.73	7.58	8.34
24 L.H., Daily Totals		$1365.95		$1944.55	$56.91	$81.02
Crew C-10E	**Hr.**	**Daily**	**Hr.**	**Daily**	**Bare Costs**	**Incl. O&P**
1 Laborer	$44.40	$355.20	$66.25	$530.00	$49.33	$72.68
2 Cement Finishers	51.80	828.80	75.90	1214.40		
1 Vibrating Power Screed		87.65		96.42		
1 Cement Trowel, 96" Ride-On		171.05		188.16	10.78	11.86
24 L.H., Daily Totals		$1442.70		$2028.97	$60.11	$84.54
Crew C-10F	**Hr.**	**Daily**	**Hr.**	**Daily**	**Bare Costs**	**Incl. O&P**
1 Laborer	$44.40	$355.20	$66.25	$530.00	$49.33	$72.68
2 Cement Finishers	51.80	828.80	75.90	1214.40		
1 Telescoping Boom Lift, to 60'		292.45		321.69	12.19	13.40
24 L.H., Daily Totals		$1476.45		$2066.09	$61.52	$86.09
Crew C-11	**Hr.**	**Daily**	**Hr.**	**Daily**	**Bare Costs**	**Incl. O&P**
1 Struc. Steel Foreman (outside)	$62.30	$498.40	$96.40	$771.20	$59.78	$91.78
6 Struc. Steel Workers	60.30	2894.40	93.30	4478.40		
1 Equip. Oper. (crane)	61.45	491.60	91.55	732.40		
1 Equip. Oper. (oiler)	52.50	420.00	78.25	626.00		
1 Lattice Boom Crane, 150 Ton		2324.00		2556.40	32.28	35.51
72 L.H., Daily Totals		$6628.40		$9164.40	$92.06	$127.28
Crew C-12	**Hr.**	**Daily**	**Hr.**	**Daily**	**Bare Costs**	**Incl. O&P**
1 Carpenter Foreman (outside)	$56.70	$453.60	$84.60	$676.80	$54.44	$81.22
3 Carpenters	54.70	1312.80	81.65	1959.60		
1 Laborer	44.40	355.20	66.25	530.00		
1 Equip. Oper. (crane)	61.45	491.60	91.55	732.40		
1 Hyd. Crane, 12 Ton		475.80		523.38	9.91	10.90
48 L.H., Daily Totals		$3089.00		$4422.18	$64.35	$92.13
Crew C-13	**Hr.**	**Daily**	**Hr.**	**Daily**	**Bare Costs**	**Incl. O&P**
1 Struc. Steel Worker	$60.30	$482.40	$93.30	$746.40	$58.43	$89.42
1 Welder	60.30	482.40	93.30	746.40		
1 Carpenter	54.70	437.60	81.65	653.20		
1 Welder, Gas Engine, 300 amp		148.75		163.63	6.20	6.82
24 L.H., Daily Totals		$1551.15		$2309.63	$64.63	$96.23
Crew C-14	**Hr.**	**Daily**	**Hr.**	**Daily**	**Bare Costs**	**Incl. O&P**
1 Carpenter Foreman (outside)	$56.70	$453.60	$84.60	$676.80	$53.39	$79.54
5 Carpenters	54.70	2188.00	81.65	3266.00		
4 Laborers	44.40	1420.80	66.25	2120.00		
4 Rodmen (reinf.)	58.90	1884.80	88.05	2817.60		
2 Cement Finishers	51.80	828.80	75.90	1214.40		
1 Equip. Oper. (crane)	61.45	491.60	91.55	732.40		
1 Equip. Oper. (oiler)	52.50	420.00	78.25	626.00		
1 Hyd. Crane, 80 Ton		1458.00		1603.80	10.13	11.14
144 L.H., Daily Totals		$9145.60		$13057.00	$63.51	$90.67

Crew No.	Bare Costs		Incl. Subs O&P		Cost Per Labor-Hour	
Crew C-14A	**Hr.**	**Daily**	**Hr.**	**Daily**	**Bare Costs**	**Incl. O&P**
1 Carpenter Foreman (outside)	$56.70	$453.60	$84.60	$676.80	$54.68	$81.58
16 Carpenters	54.70	7001.60	81.65	10451.20		
4 Rodmen (reinf.)	58.90	1884.80	88.05	2817.60		
2 Laborers	44.40	710.40	66.25	1060.00		
1 Cement Finisher	51.80	414.40	75.90	607.20		
1 Equip. Oper. (medium)	59.00	472.00	87.90	703.20		
1 Gas Engine Vibrator		27.15		29.86		
1 Concrete Pump (Small)		423.65		466.01	2.25	2.48
200 L.H., Daily Totals		$11387.60		$16811.88	$56.94	$84.06
Crew C-14B	**Hr.**	**Daily**	**Hr.**	**Daily**	**Bare Costs**	**Incl. O&P**
1 Carpenter Foreman (outside)	$56.70	$453.60	$84.60	$676.80	$54.57	$81.36
16 Carpenters	54.70	7001.60	81.65	10451.20		
4 Rodmen (reinf.)	58.90	1884.80	88.05	2817.60		
2 Laborers	44.40	710.40	66.25	1060.00		
2 Cement Finishers	51.80	828.80	75.90	1214.40		
1 Equip. Oper. (medium)	59.00	472.00	87.90	703.20		
1 Gas Engine Vibrator		27.15		29.86		
1 Concrete Pump (Small)		423.65		466.01	2.17	2.38
208 L.H., Daily Totals		$11802.00		$17419.08	$56.74	$83.75
Crew C-14C	**Hr.**	**Daily**	**Hr.**	**Daily**	**Bare Costs**	**Incl. O&P**
1 Carpenter Foreman (outside)	$56.70	$453.60	$84.60	$676.80	$52.29	$77.96
6 Carpenters	54.70	2625.60	81.65	3919.20		
2 Rodmen (reinf.)	58.90	942.40	88.05	1408.80		
4 Laborers	44.40	1420.80	66.25	2120.00		
1 Cement Finisher	51.80	414.40	75.90	607.20		
1 Gas Engine Vibrator		27.15		29.86	.24	.27
112 L.H., Daily Totals		$5883.95		$8761.86	$52.54	$78.23
Crew C-14D	**Hr.**	**Daily**	**Hr.**	**Daily**	**Bare Costs**	**Incl. O&P**
1 Carpenter Foreman (outside)	$56.70	$453.60	$84.60	$676.80	$54.35	$81.07
18 Carpenters	54.70	7876.80	81.65	11757.60		
2 Rodmen (reinf.)	58.90	942.40	88.05	1408.80		
2 Laborers	44.40	710.40	66.25	1060.00		
1 Cement Finisher	51.80	414.40	75.90	607.20		
1 Equip. Oper. (medium)	59.00	472.00	87.90	703.20		
1 Gas Engine Vibrator		27.15		29.86		
1 Concrete Pump (Small)		423.65		466.01	2.25	2.48
200 L.H., Daily Totals		$11320.40		$16709.48	$56.60	$83.55
Crew C-14E	**Hr.**	**Daily**	**Hr.**	**Daily**	**Bare Costs**	**Incl. O&P**
1 Carpenter Foreman (outside)	$56.70	$453.60	$84.60	$676.80	$53.34	$79.52
2 Carpenters	54.70	875.20	81.65	1306.40		
4 Rodmen (reinf.)	58.90	1884.80	88.05	2817.60		
3 Laborers	44.40	1065.60	66.25	1590.00		
1 Cement Finisher	51.80	414.40	75.90	607.20		
1 Gas Engine Vibrator		27.15		29.86	.31	.34
88 L.H., Daily Totals		$4720.75		$7027.86	$53.64	$79.86
Crew C-14F	**Hr.**	**Daily**	**Hr.**	**Daily**	**Bare Costs**	**Incl. O&P**
1 Labor Foreman (outside)	$46.40	$371.20	$69.25	$554.00	$49.56	$73.02
2 Laborers	44.40	710.40	66.25	1060.00		
6 Cement Finishers	51.80	2486.40	75.90	3643.20		
1 Gas Engine Vibrator		27.15		29.86	.38	.41
72 L.H., Daily Totals		$3595.15		$5287.06	$49.93	$73.43

Crew No.	Bare Costs		Incl. Subs O&P		Cost Per Labor-Hour	
Crew C-14G	**Hr.**	**Daily**	**Hr.**	**Daily**	**Bare Costs**	**Incl. O&P**
1 Labor Foreman (outside)	$46.40	$371.20	$69.25	$554.00	$48.91	$72.19
2 Laborers	44.40	710.40	66.25	1060.00		
4 Cement Finishers	51.80	1657.60	75.90	2428.80		
1 Gas Engine Vibrator		27.15		29.86	.48	.53
56 L.H., Daily Totals		$2766.35		$4072.67	$49.40	$72.73

Crew C-14H	Hr.	Daily	Hr.	Daily	Bare Costs	Incl. O&P
1 Carpenter Foreman (outside)	$56.70	$453.60	$84.60	$676.80	$53.53	$79.68
2 Carpenters	54.70	875.20	81.65	1306.40		
1 Rodman (reinf.)	58.90	471.20	88.05	704.40		
1 Laborer	44.40	355.20	66.25	530.00		
1 Cement Finisher	51.80	414.40	75.90	607.20		
1 Gas Engine Vibrator		27.15		29.86	.57	.62
48 L.H., Daily Totals		$2596.75		$3854.67	$54.10	$80.31

Crew C-14L	Hr.	Daily	Hr.	Daily	Bare Costs	Incl. O&P
1 Carpenter Foreman (outside)	$56.70	$453.60	$84.60	$676.80	$51.19	$76.28
6 Carpenters	54.70	2625.60	81.65	3919.20		
4 Laborers	44.40	1420.80	66.25	2120.00		
1 Cement Finisher	51.80	414.40	75.90	607.20		
1 Gas Engine Vibrator		27.15		29.86	.28	.31
96 L.H., Daily Totals		$4941.55		$7353.06	$51.47	$76.59

Crew C-14M	Hr.	Daily	Hr.	Daily	Bare Costs	Incl. O&P
1 Carpenter Foreman (outside)	$56.70	$453.60	$84.60	$676.80	$53.08	$79.03
2 Carpenters	54.70	875.20	81.65	1306.40		
1 Rodman (reinf.)	58.90	471.20	88.05	704.40		
2 Laborers	44.40	710.40	66.25	1060.00		
1 Cement Finisher	51.80	414.40	75.90	607.20		
1 Equip. Oper. (medium)	59.00	472.00	87.90	703.20		
1 Gas Engine Vibrator		27.15		29.86		
1 Concrete Pump (Small)		423.65		466.01	7.04	7.75
64 L.H., Daily Totals		$3847.60		$5553.88	$60.12	$86.78

Crew C-15	Hr.	Daily	Hr.	Daily	Bare Costs	Incl. O&P
1 Carpenter Foreman (outside)	$56.70	$453.60	$84.60	$676.80	$51.31	$76.28
2 Carpenters	54.70	875.20	81.65	1306.40		
3 Laborers	44.40	1065.60	66.25	1590.00		
2 Cement Finishers	51.80	828.80	75.90	1214.40		
1 Rodman (reinf.)	58.90	471.20	88.05	704.40		
72 L.H., Daily Totals		$3694.40		$5492.00	$51.31	$76.28

Crew C-16	Hr.	Daily	Hr.	Daily	Bare Costs	Incl. O&P
1 Labor Foreman (outside)	$46.40	$371.20	$69.25	$554.00	$48.89	$72.53
3 Laborers	44.40	1065.60	66.25	1590.00		
2 Cement Finishers	51.80	828.80	75.90	1214.40		
1 Equip. Oper. (medium)	59.00	472.00	87.90	703.20		
1 Gunite Pump Rig		321.75		353.93		
2 -50' Air Hoses, 3/4"		14.30		15.73		
2 -50' Air Hoses, 2"		57.90		63.69	7.03	7.74
56 L.H., Daily Totals		$3131.55		$4494.94	$55.92	$80.27

Crew C-16A	Hr.	Daily	Hr.	Daily	Bare Costs	Incl. O&P
1 Laborer	$44.40	$355.20	$66.25	$530.00	$51.75	$76.49
2 Cement Finishers	51.80	828.80	75.90	1214.40		
1 Equip. Oper. (medium)	59.00	472.00	87.90	703.20		
1 Gunite Pump Rig		321.75		353.93		
2 -50' Air Hoses, 3/4"		14.30		15.73		
2 -50' Air Hoses, 2"		57.90		63.69		
1 Telescoping Boom Lift, to 60'		292.45		321.69	21.45	23.59
32 L.H., Daily Totals		$2342.40		$3202.64	$73.20	$100.08

Crew No.	Bare Costs		Incl. Subs O&P		Cost Per Labor-Hour	
Crew C-17	**Hr.**	**Daily**	**Hr.**	**Daily**	**Bare Costs**	**Incl. O&P**
2 Skilled Worker Foremen (out)	$59.10	$945.60	$88.90	$1422.40	$57.50	$86.50
8 Skilled Workers	57.10	3654.40	85.90	5497.60		
80 L.H., Daily Totals		$4600.00		$6920.00	$57.50	$86.50

Crew C-17A	Hr.	Daily	Hr.	Daily	Bare Costs	Incl. O&P
2 Skilled Worker Foremen (out)	$59.10	$945.60	$88.90	$1422.40	$57.55	$86.56
8 Skilled Workers	57.10	3654.40	85.90	5497.60		
.125 Equip. Oper. (crane)	61.45	61.45	91.55	91.55		
.125 Hyd. Crane, 80 Ton		182.25		200.47	2.25	2.48
81 L.H., Daily Totals		$4843.70		$7212.02	$59.80	$89.04

Crew C-17B	Hr.	Daily	Hr.	Daily	Bare Costs	Incl. O&P
2 Skilled Worker Foremen (out)	$59.10	$945.60	$88.90	$1422.40	$57.60	$86.62
8 Skilled Workers	57.10	3654.40	85.90	5497.60		
.25 Equip. Oper. (crane)	61.45	122.90	91.55	183.10		
.25 Hyd. Crane, 80 Ton		364.50		400.95		
.25 Trowel, 48" Walk-Behind		23.57		25.93	4.73	5.21
82 L.H., Daily Totals		$5110.98		$7529.98	$62.33	$91.83

Crew C-17C	Hr.	Daily	Hr.	Daily	Bare Costs	Incl. O&P
2 Skilled Worker Foremen (out)	$59.10	$945.60	$88.90	$1422.40	$57.64	$86.68
8 Skilled Workers	57.10	3654.40	85.90	5497.60		
.375 Equip. Oper. (crane)	61.45	184.35	91.55	274.65		
.375 Hyd. Crane, 80 Ton		546.75		601.42	6.59	7.25
83 L.H., Daily Totals		$5331.10		$7796.07	$64.23	$93.93

Crew C-17D	Hr.	Daily	Hr.	Daily	Bare Costs	Incl. O&P
2 Skilled Worker Foremen (out)	$59.10	$945.60	$88.90	$1422.40	$57.69	$86.74
8 Skilled Workers	57.10	3654.40	85.90	5497.60		
.5 Equip. Oper. (crane)	61.45	245.80	91.55	366.20		
.5 Hyd. Crane, 80 Ton		729.00		801.90	8.68	9.55
84 L.H., Daily Totals		$5574.80		$8088.10	$66.37	$96.29

Crew C-17E	Hr.	Daily	Hr.	Daily	Bare Costs	Incl. O&P
2 Skilled Worker Foremen (out)	$59.10	$945.60	$88.90	$1422.40	$57.50	$86.50
8 Skilled Workers	57.10	3654.40	85.90	5497.60		
1 Hyd. Jack with Rods		36.70		40.37	.46	.50
80 L.H., Daily Totals		$4636.70		$6960.37	$57.96	$87.00

Crew C-18	Hr.	Daily	Hr.	Daily	Bare Costs	Incl. O&P
.125 Labor Foreman (outside)	$46.40	$46.40	$69.25	$69.25	$44.62	$66.58
1 Laborer	44.40	355.20	66.25	530.00		
1 Concrete Cart, 10 C.F.		116.95		128.65	12.99	14.29
9 L.H., Daily Totals		$518.55		$727.89	$57.62	$80.88

Crew C-19	Hr.	Daily	Hr.	Daily	Bare Costs	Incl. O&P
.125 Labor Foreman (outside)	$46.40	$46.40	$69.25	$69.25	$44.62	$66.58
1 Laborer	44.40	355.20	66.25	530.00		
1 Concrete Cart, 18 C.F.		138.95		152.85	15.44	16.98
9 L.H., Daily Totals		$540.55		$752.10	$60.06	$83.57

Crew C-20	Hr.	Daily	Hr.	Daily	Bare Costs	Incl. O&P
1 Labor Foreman (outside)	$46.40	$371.20	$69.25	$554.00	$47.40	$70.54
5 Laborers	44.40	1776.00	66.25	2650.00		
1 Cement Finisher	51.80	414.40	75.90	607.20		
1 Equip. Oper. (medium)	59.00	472.00	87.90	703.20		
2 Gas Engine Vibrators		54.30		59.73		
1 Concrete Pump (Small)		423.65		466.01	7.47	8.21
64 L.H., Daily Totals		$3511.55		$5040.15	$54.87	$78.75

For customer support on your Site Work & Landscape Costs with RSMeans Data, call 800.448.8182.

Crew No.	Bare Costs		Incl. Subs O&P		Cost Per Labor-Hour	

Crew C-21

	Hr.	Daily	Hr.	Daily	Bare Costs	Incl. O&P
1 Labor Foreman (outside)	$46.40	$371.20	$69.25	$554.00	$47.40	$70.54
5 Laborers	44.40	1776.00	66.25	2650.00		
1 Cement Finisher	51.80	414.40	75.90	607.20		
1 Equip. Oper. (medium)	59.00	472.00	87.90	703.20		
2 Gas Engine Vibrators		54.30		59.73		
1 Concrete Conveyer		206.25		226.88	4.07	4.48
64 L.H., Daily Totals		$3294.15		$4801.01	$51.47	$75.02

Crew C-22

	Hr.	Daily	Hr.	Daily	Bare Costs	Incl. O&P
1 Rodman Foreman (outside)	$60.90	$487.20	$91.05	$728.40	$59.19	$88.47
4 Rodmen (reinf.)	58.90	1884.80	88.05	2817.60		
.125 Equip. Oper. (crane)	61.45	61.45	91.55	91.55		
.125 Equip. Oper. (oiler)	52.50	52.50	78.25	78.25		
.125 Hyd. Crane, 25 Ton		73.34		80.67	1.75	1.92
42 L.H., Daily Totals		$2559.29		$3796.47	$60.94	$90.39

Crew C-23

	Hr.	Daily	Hr.	Daily	Bare Costs	Incl. O&P
2 Skilled Worker Foremen (out)	$59.10	$945.60	$88.90	$1422.40	$57.48	$86.30
6 Skilled Workers	57.10	2740.80	85.90	4123.20		
1 Equip. Oper. (crane)	61.45	491.60	91.55	732.40		
1 Equip. Oper. (oiler)	52.50	420.00	78.25	626.00		
1 Lattice Boom Crane, 90 Ton		1713.00		1884.30	21.41	23.55
80 L.H., Daily Totals		$6311.00		$8788.30	$78.89	$109.85

Crew C-23A

	Hr.	Daily	Hr.	Daily	Bare Costs	Incl. O&P
1 Labor Foreman (outside)	$46.40	$371.20	$69.25	$554.00	$49.83	$74.31
2 Laborers	44.40	710.40	66.25	1060.00		
1 Equip. Oper. (crane)	61.45	491.60	91.55	732.40		
1 Equip. Oper. (oiler)	52.50	420.00	78.25	626.00		
1 Crawler Crane, 100 Ton		2310.00		2541.00		
3 Conc. Buckets, 8 C.Y.		337.95		371.75	66.20	72.82
40 L.H., Daily Totals		$4641.15		$5885.15	$116.03	$147.13

Crew C-24

	Hr.	Daily	Hr.	Daily	Bare Costs	Incl. O&P
2 Skilled Worker Foremen (out)	$59.10	$945.60	$88.90	$1422.40	$57.48	$86.30
6 Skilled Workers	57.10	2740.80	85.90	4123.20		
1 Equip. Oper. (crane)	61.45	491.60	91.55	732.40		
1 Equip. Oper. (oiler)	52.50	420.00	78.25	626.00		
1 Lattice Boom Crane, 150 Ton		2324.00		2556.40	29.05	31.95
80 L.H., Daily Totals		$6922.00		$9460.40	$86.53	$118.26

Crew C-25

	Hr.	Daily	Hr.	Daily	Bare Costs	Incl. O&P
2 Rodmen (reinf.)	$58.90	$942.40	$88.05	$1408.80	$47.42	$73.15
2 Rodmen Helpers	35.95	575.20	58.25	932.00		
32 L.H., Daily Totals		$1517.60		$2340.80	$47.42	$73.15

Crew C-27

	Hr.	Daily	Hr.	Daily	Bare Costs	Incl. O&P
2 Cement Finishers	$51.80	$828.80	$75.90	$1214.40	$51.80	$75.90
1 Concrete Saw		112.85		124.14	7.05	7.76
16 L.H., Daily Totals		$941.65		$1338.54	$58.85	$83.66

Crew C-28

	Hr.	Daily	Hr.	Daily	Bare Costs	Incl. O&P
1 Cement Finisher	$51.80	$414.40	$75.90	$607.20	$51.80	$75.90
1 Portable Air Compressor, Gas		38.80		42.68	4.85	5.34
8 L.H., Daily Totals		$453.20		$649.88	$56.65	$81.23

Crew C-29

	Hr.	Daily	Hr.	Daily	Bare Costs	Incl. O&P
1 Laborer	$44.40	$355.20	$66.25	$530.00	$44.40	$66.25
1 Pressure Washer		97.35		107.08	12.17	13.39
8 L.H., Daily Totals		$452.55		$637.09	$56.57	$79.64

Crew C-30

	Hr.	Daily	Hr.	Daily	Bare Costs	Incl. O&P
1 Laborer	$44.40	$355.20	$66.25	$530.00	$44.40	$66.25
1 Concrete Mixer, 10 C.F.		147.15		161.87	18.39	20.23
8 L.H., Daily Totals		$502.35		$691.87	$62.79	$86.48

Crew C-31

	Hr.	Daily	Hr.	Daily	Bare Costs	Incl. O&P
1 Cement Finisher	$51.80	$414.40	$75.90	$607.20	$51.80	$75.90
1 Grout Pump		321.75		353.93	40.22	44.24
8 L.H., Daily Totals		$736.15		$961.13	$92.02	$120.14

Crew C-32

	Hr.	Daily	Hr.	Daily	Bare Costs	Incl. O&P
1 Cement Finisher	$51.80	$414.40	$75.90	$607.20	$48.10	$71.08
1 Laborer	44.40	355.20	66.25	530.00		
1 Crack Chaser Saw, Gas, 6 H.P.		73.25		80.58		
1 Vacuum Pick-Up System		74.95		82.44	9.26	10.19
16 L.H., Daily Totals		$917.80		$1300.22	$57.36	$81.26

Crew D-1

	Hr.	Daily	Hr.	Daily	Bare Costs	Incl. O&P
1 Bricklayer	$53.70	$429.60	$80.90	$647.20	$48.70	$73.38
1 Bricklayer Helper	43.70	349.60	65.85	526.80		
16 L.H., Daily Totals		$779.20		$1174.00	$48.70	$73.38

Crew D-2

	Hr.	Daily	Hr.	Daily	Bare Costs	Incl. O&P
3 Bricklayers	$53.70	$1288.80	$80.90	$1941.60	$50.15	$75.50
2 Bricklayer Helpers	43.70	699.20	65.85	1053.60		
.5 Carpenter	54.70	218.80	81.65	326.60		
44 L.H., Daily Totals		$2206.80		$3321.80	$50.15	$75.50

Crew D-3

	Hr.	Daily	Hr.	Daily	Bare Costs	Incl. O&P
3 Bricklayers	$53.70	$1288.80	$80.90	$1941.60	$49.94	$75.20
2 Bricklayer Helpers	43.70	699.20	65.85	1053.60		
.25 Carpenter	54.70	109.40	81.65	163.30		
42 L.H., Daily Totals		$2097.40		$3158.50	$49.94	$75.20

Crew D-4

	Hr.	Daily	Hr.	Daily	Bare Costs	Incl. O&P
1 Bricklayer	$53.70	$429.60	$80.90	$647.20	$49.15	$73.83
2 Bricklayer Helpers	43.70	699.20	65.85	1053.60		
1 Equip. Oper. (light)	55.50	444.00	82.70	661.60		
1 Grout Pump, 50 C.F./hr.		190.35		209.38	5.95	6.54
32 L.H., Daily Totals		$1763.15		$2571.78	$55.10	$80.37

Crew D-5

	Hr.	Daily	Hr.	Daily	Bare Costs	Incl. O&P
1 Bricklayer	53.70	429.60	80.90	647.20	53.70	80.90
8 L.H., Daily Totals		$429.60		$647.20	$53.70	$80.90

Crew D-6

	Hr.	Daily	Hr.	Daily	Bare Costs	Incl. O&P
3 Bricklayers	$53.70	$1288.80	$80.90	$1941.60	$48.94	$73.71
3 Bricklayer Helpers	43.70	1048.80	65.85	1580.40		
.25 Carpenter	54.70	109.40	81.65	163.30		
50 L.H., Daily Totals		$2447.00		$3685.30	$48.94	$73.71

Crew D-7

	Hr.	Daily	Hr.	Daily	Bare Costs	Incl. O&P
1 Tile Layer	$51.70	$413.60	$75.55	$604.40	$46.65	$68.17
1 Tile Layer Helper	41.60	332.80	60.80	486.40		
16 L.H., Daily Totals		$746.40		$1090.80	$46.65	$68.17

Crew D-8

	Hr.	Daily	Hr.	Daily	Bare Costs	Incl. O&P
3 Bricklayers	$53.70	$1288.80	$80.90	$1941.60	$49.70	$74.88
2 Bricklayer Helpers	43.70	699.20	65.85	1053.60		
40 L.H., Daily Totals		$1988.00		$2995.20	$49.70	$74.88

For customer support on your Site Work & Landscape Costs with RSMeans Data, call 800.448.8182.

775

Crew No.	Bare Costs		Incl. Subs O&P		Cost Per Labor-Hour	
Crew D-9	**Hr.**	**Daily**	**Hr.**	**Daily**	**Bare Costs**	**Incl. O&P**
3 Bricklayers	$53.70	$1288.80	$80.90	$1941.60	$48.70	$73.38
3 Bricklayer Helpers	43.70	1048.80	65.85	1580.40		
48 L.H., Daily Totals		$2337.60		$3522.00	$48.70	$73.38

Crew D-10	Hr.	Daily	Hr.	Daily	Bare Costs	Incl. O&P
1 Bricklayer Foreman (outside)	$55.70	$445.60	$83.90	$671.20	$53.64	$80.55
1 Bricklayer	53.70	429.60	80.90	647.20		
1 Bricklayer Helper	43.70	349.60	65.85	526.80		
1 Equip. Oper. (crane)	61.45	491.60	91.55	732.40		
1 S.P. Crane, 4x4, 12 Ton		432.65		475.92	13.52	14.87
32 L.H., Daily Totals		$2149.05		$3053.51	$67.16	$95.42

Crew D-11	Hr.	Daily	Hr.	Daily	Bare Costs	Incl. O&P
1 Bricklayer Foreman (outside)	$55.70	$445.60	$83.90	$671.20	$51.03	$76.88
1 Bricklayer	53.70	429.60	80.90	647.20		
1 Bricklayer Helper	43.70	349.60	65.85	526.80		
24 L.H., Daily Totals		$1224.80		$1845.20	$51.03	$76.88

Crew D-12	Hr.	Daily	Hr.	Daily	Bare Costs	Incl. O&P
1 Bricklayer Foreman (outside)	$55.70	$445.60	$83.90	$671.20	$49.20	$74.13
1 Bricklayer	53.70	429.60	80.90	647.20		
2 Bricklayer Helpers	43.70	699.20	65.85	1053.60		
32 L.H., Daily Totals		$1574.40		$2372.00	$49.20	$74.13

Crew D-13	Hr.	Daily	Hr.	Daily	Bare Costs	Incl. O&P
1 Bricklayer Foreman (outside)	$55.70	$445.60	$83.90	$671.20	$52.16	$78.28
1 Bricklayer	53.70	429.60	80.90	647.20		
2 Bricklayer Helpers	43.70	699.20	65.85	1053.60		
1 Carpenter	54.70	437.60	81.65	653.20		
1 Equip. Oper. (crane)	61.45	491.60	91.55	732.40		
1 S.P. Crane, 4x4, 12 Ton		432.65		475.92	9.01	9.91
48 L.H., Daily Totals		$2936.25		$4233.52	$61.17	$88.20

Crew D-14	Hr.	Daily	Hr.	Daily	Bare Costs	Incl. O&P
3 Bricklayers	$53.70	$1288.80	$80.90	$1941.60	$51.20	$77.14
1 Bricklayer Helper	43.70	349.60	65.85	526.80		
32 L.H., Daily Totals		$1638.40		$2468.40	$51.20	$77.14

Crew E-1	Hr.	Daily	Hr.	Daily	Bare Costs	Incl. O&P
1 Welder Foreman (outside)	$62.30	$498.40	$96.40	$771.20	$59.37	$90.80
1 Welder	60.30	482.40	93.30	746.40		
1 Equip. Oper. (light)	55.50	444.00	82.70	661.60		
1 Welder, Gas Engine, 300 amp		148.75		163.63	6.20	6.82
24 L.H., Daily Totals		$1573.55		$2342.82	$65.56	$97.62

Crew E-2	Hr.	Daily	Hr.	Daily	Bare Costs	Incl. O&P
1 Struc. Steel Foreman (outside)	$62.30	$498.40	$96.40	$771.20	$59.64	$91.34
4 Struc. Steel Workers	60.30	1929.60	93.30	2985.60		
1 Equip. Oper. (crane)	61.45	491.60	91.55	732.40		
1 Equip. Oper. (oiler)	52.50	420.00	78.25	626.00		
1 Lattice Boom Crane, 90 Ton		1713.00		1884.30	30.59	33.65
56 L.H., Daily Totals		$5052.60		$6999.50	$90.22	$124.99

Crew E-3	Hr.	Daily	Hr.	Daily	Bare Costs	Incl. O&P
1 Struc. Steel Foreman (outside)	$62.30	$498.40	$96.40	$771.20	$60.97	$94.33
1 Struc. Steel Worker	60.30	482.40	93.30	746.40		
1 Welder	60.30	482.40	93.30	746.40		
1 Welder, Gas Engine, 300 amp		148.75		163.63	6.20	6.82
24 L.H., Daily Totals		$1611.95		$2427.63	$67.16	$101.15

Crew E-3A	Hr.	Daily	Hr.	Daily	Bare Costs	Incl. O&P
1 Struc. Steel Foreman (outside)	$62.30	$498.40	$96.40	$771.20	$60.97	$94.33
1 Struc. Steel Worker	60.30	482.40	93.30	746.40		
1 Welder	60.30	482.40	93.30	746.40		
1 Welder, Gas Engine, 300 amp		148.75		163.63		
1 Telescoping Boom Lift, to 40'		281.90		310.09	17.94	19.74
24 L.H., Daily Totals		$1893.85		$2737.72	$78.91	$114.07

Crew E-4	Hr.	Daily	Hr.	Daily	Bare Costs	Incl. O&P
1 Struc. Steel Foreman (outside)	$62.30	$498.40	$96.40	$771.20	$60.80	$94.08
3 Struc. Steel Workers	60.30	1447.20	93.30	2239.20		
1 Welder, Gas Engine, 300 amp		148.75		163.63	4.65	5.11
32 L.H., Daily Totals		$2094.35		$3174.03	$65.45	$99.19

Crew E-5	Hr.	Daily	Hr.	Daily	Bare Costs	Incl. O&P
2 Struc. Steel Foremen (outside)	$62.30	$996.80	$96.40	$1542.40	$60.03	$92.24
5 Struc. Steel Workers	60.30	2412.00	93.30	3732.00		
1 Equip. Oper. (crane)	61.45	491.60	91.55	732.40		
1 Welder	60.30	482.40	93.30	746.40		
1 Equip. Oper. (oiler)	52.50	420.00	78.25	626.00		
1 Lattice Boom Crane, 90 Ton		1713.00		1884.30		
1 Welder, Gas Engine, 300 amp		148.75		163.63	23.27	25.60
80 L.H., Daily Totals		$6664.55		$9427.13	$83.31	$117.84

Crew E-6	Hr.	Daily	Hr.	Daily	Bare Costs	Incl. O&P
3 Struc. Steel Foremen (outside)	$62.30	$1495.20	$96.40	$2313.60	$59.96	$92.17
9 Struc. Steel Workers	60.30	4341.60	93.30	6717.60		
1 Equip. Oper. (crane)	61.45	491.60	91.55	732.40		
1 Welder	60.30	482.40	93.30	746.40		
1 Equip. Oper. (oiler)	52.50	420.00	78.25	626.00		
1 Equip. Oper. (light)	55.50	444.00	82.70	661.60		
1 Lattice Boom Crane, 90 Ton		1713.00		1884.30		
1 Welder, Gas Engine, 300 amp		148.75		163.63		
1 Air Compressor, 160 cfm		212.30		233.53		
2 Impact Wrenches		90.30		99.33	16.91	18.60
128 L.H., Daily Totals		$9839.15		$14178.39	$76.87	$110.77

Crew E-7	Hr.	Daily	Hr.	Daily	Bare Costs	Incl. O&P
1 Struc. Steel Foreman (outside)	$62.30	$498.40	$96.40	$771.20	$60.03	$92.24
4 Struc. Steel Workers	60.30	1929.60	93.30	2985.60		
1 Equip. Oper. (crane)	61.45	491.60	91.55	732.40		
1 Equip. Oper. (oiler)	52.50	420.00	78.25	626.00		
1 Welder Foreman (outside)	62.30	498.40	96.40	771.20		
2 Welders	60.30	964.80	93.30	1492.80		
1 Lattice Boom Crane, 90 Ton		1713.00		1884.30		
2 Welder, Gas Engine, 300 amp		297.50		327.25	25.13	27.64
80 L.H., Daily Totals		$6813.30		$9590.75	$85.17	$119.88

Crew E-8	Hr.	Daily	Hr.	Daily	Bare Costs	Incl. O&P
1 Struc. Steel Foreman (outside)	$62.30	$498.40	$96.40	$771.20	$59.73	$91.67
4 Struc. Steel Workers	60.30	1929.60	93.30	2985.60		
1 Welder Foreman (outside)	62.30	498.40	96.40	771.20		
4 Welders	60.30	1929.60	93.30	2985.60		
1 Equip. Oper. (crane)	61.45	491.60	91.55	732.40		
1 Equip. Oper. (oiler)	52.50	420.00	78.25	626.00		
1 Equip. Oper. (light)	55.50	444.00	82.70	661.60		
1 Lattice Boom Crane, 90 Ton		1713.00		1884.30		
4 Welder, Gas Engine, 300 amp		595.00		654.50	22.19	24.41
104 L.H., Daily Totals		$8519.60		$12072.40	$81.92	$116.08

Crew No.	Bare Costs		Incl. Subs O&P		Cost Per Labor-Hour	

Crew E-9	Hr.	Daily	Hr.	Daily	Bare Costs	Incl. O&P
2 Struc. Steel Foremen (outside)	$62.30	$996.80	$96.40	$1542.40	$59.96	$92.17
5 Struc. Steel Workers	60.30	2412.00	93.30	3732.00		
1 Welder Foreman (outside)	62.30	498.40	96.40	771.20		
5 Welders	60.30	2412.00	93.30	3732.00		
1 Equip. Oper. (crane)	61.45	491.60	91.55	732.40		
1 Equip. Oper. (oiler)	52.50	420.00	78.25	626.00		
1 Equip. Oper. (light)	55.50	444.00	82.70	661.60		
1 Lattice Boom Crane, 90 Ton		1713.00		1884.30		
5 Welder, Gas Engine, 300 amp		743.75		818.13	19.19	21.11
128 L.H., Daily Totals		$10131.55		$14500.03	$79.15	$113.28

Crew E-10	Hr.	Daily	Hr.	Daily	Bare Costs	Incl. O&P
1 Welder Foreman (outside)	$62.30	$498.40	$96.40	$771.20	$61.30	$94.85
1 Welder	60.30	482.40	93.30	746.40		
1 Welder, Gas Engine, 300 amp		148.75		163.63		
1 Flatbed Truck, Gas, 3 Ton		850.05		935.05	62.42	68.67
16 L.H., Daily Totals		$1979.60		$2616.28	$123.72	$163.52

Crew E-11	Hr.	Daily	Hr.	Daily	Bare Costs	Incl. O&P
2 Painters, Struc. Steel	$47.20	$755.20	$75.80	$1212.80	$48.58	$75.14
1 Building Laborer	44.40	355.20	66.25	530.00		
1 Equip. Oper. (light)	55.50	444.00	82.70	661.60		
1 Air Compressor, 250 cfm		202.85		223.13		
1 Sandblaster, Portable, 3 C.F.		83.85		92.23		
1 Set Sand Blasting Accessories		15.55		17.11	9.45	10.39
32 L.H., Daily Totals		$1856.65		$2736.88	$58.02	$85.53

Crew E-11A	Hr.	Daily	Hr.	Daily	Bare Costs	Incl. O&P
2 Painters, Struc. Steel	$47.20	$755.20	$75.80	$1212.80	$48.58	$75.14
1 Building Laborer	44.40	355.20	66.25	530.00		
1 Equip. Oper. (light)	55.50	444.00	82.70	661.60		
1 Air Compressor, 250 cfm		202.85		223.13		
1 Sandblaster, Portable, 3 C.F.		83.85		92.23		
1 Set Sand Blasting Accessories		15.55		17.11		
1 Telescoping Boom Lift, to 60'		292.45		321.69	18.58	20.44
32 L.H., Daily Totals		$2149.10		$3058.57	$67.16	$95.58

Crew E-11B	Hr.	Daily	Hr.	Daily	Bare Costs	Incl. O&P
2 Painters, Struc. Steel	$47.20	$755.20	$75.80	$1212.80	$46.27	$72.62
1 Building Laborer	44.40	355.20	66.25	530.00		
2 Paint Sprayer, 8 C.F.M.		88.40		97.24		
1 Telescoping Boom Lift, to 60'		292.45		321.69	15.87	17.46
24 L.H., Daily Totals		$1491.25		$2161.74	$62.14	$90.07

Crew E-12	Hr.	Daily	Hr.	Daily	Bare Costs	Incl. O&P
1 Welder Foreman (outside)	$62.30	$498.40	$96.40	$771.20	$58.90	$89.55
1 Equip. Oper. (light)	55.50	444.00	82.70	661.60		
1 Welder, Gas Engine, 300 amp		148.75		163.63	9.30	10.23
16 L.H., Daily Totals		$1091.15		$1596.43	$68.20	$99.78

Crew E-13	Hr.	Daily	Hr.	Daily	Bare Costs	Incl. O&P
1 Welder Foreman (outside)	$62.30	$498.40	$96.40	$771.20	$60.03	$91.83
.5 Equip. Oper. (light)	55.50	222.00	82.70	330.80		
1 Welder, Gas Engine, 300 amp		148.75		163.63	12.40	13.64
12 L.H., Daily Totals		$869.15		$1265.63	$72.43	$105.47

Crew E-14	Hr.	Daily	Hr.	Daily	Bare Costs	Incl. O&P
1 Welder Foreman (outside)	$62.30	$498.40	$96.40	$771.20	$62.30	$96.40
1 Welder, Gas Engine, 300 amp		148.75		163.63	18.59	20.45
8 L.H., Daily Totals		$647.15		$934.83	$80.89	$116.85

Crew E-16	Hr.	Daily	Hr.	Daily	Bare Costs	Incl. O&P
1 Welder Foreman (outside)	$62.30	$498.40	$96.40	$771.20	$61.30	$94.85
1 Welder	60.30	482.40	93.30	746.40		
1 Welder, Gas Engine, 300 amp		148.75		163.63	9.30	10.23
16 L.H., Daily Totals		$1129.55		$1681.22	$70.60	$105.08

Crew E-17	Hr.	Daily	Hr.	Daily	Bare Costs	Incl. O&P
1 Struc. Steel Foreman (outside)	$62.30	$498.40	$96.40	$771.20	$61.30	$94.85
1 Structural Steel Worker	60.30	482.40	93.30	746.40		
16 L.H., Daily Totals		$980.80		$1517.60	$61.30	$94.85

Crew E-18	Hr.	Daily	Hr.	Daily	Bare Costs	Incl. O&P
1 Struc. Steel Foreman (outside)	$62.30	$498.40	$96.40	$771.20	$60.44	$92.84
3 Structural Steel Workers	60.30	1447.20	93.30	2239.20		
1 Equipment Operator (med.)	59.00	472.00	87.90	703.20		
1 Lattice Boom Crane, 20 Ton		1526.00		1678.60	38.15	41.97
40 L.H., Daily Totals		$3943.60		$5392.20	$98.59	$134.81

Crew E-19	Hr.	Daily	Hr.	Daily	Bare Costs	Incl. O&P
1 Struc. Steel Foreman (outside)	$62.30	$498.40	$96.40	$771.20	$59.37	$90.80
1 Structural Steel Worker	60.30	482.40	93.30	746.40		
1 Equip. Oper. (light)	55.50	444.00	82.70	661.60		
1 Lattice Boom Crane, 20 Ton		1526.00		1678.60	63.58	69.94
24 L.H., Daily Totals		$2950.80		$3857.80	$122.95	$160.74

Crew E-20	Hr.	Daily	Hr.	Daily	Bare Costs	Incl. O&P
1 Struc. Steel Foreman (outside)	$62.30	$498.40	$96.40	$771.20	$59.72	$91.59
5 Structural Steel Workers	60.30	2412.00	93.30	3732.00		
1 Equip. Oper. (crane)	61.45	491.60	91.55	732.40		
1 Equip. Oper. (oiler)	52.50	420.00	78.25	626.00		
1 Lattice Boom Crane, 40 Ton		2127.00		2339.70	33.23	36.56
64 L.H., Daily Totals		$5949.00		$8201.30	$92.95	$128.15

Crew E-22	Hr.	Daily	Hr.	Daily	Bare Costs	Incl. O&P
1 Skilled Worker Foreman (out)	$59.10	$472.80	$88.90	$711.20	$57.77	$86.90
2 Skilled Workers	57.10	913.60	85.90	1374.40		
24 L.H., Daily Totals		$1386.40		$2085.60	$57.77	$86.90

Crew E-24	Hr.	Daily	Hr.	Daily	Bare Costs	Incl. O&P
3 Structural Steel Workers	$60.30	$1447.20	$93.30	$2239.20	$59.98	$91.95
1 Equipment Operator (med.)	59.00	472.00	87.90	703.20		
1 Hyd. Crane, 25 Ton		586.70		645.37	18.33	20.17
32 L.H., Daily Totals		$2505.90		$3587.77	$78.31	$112.12

Crew E-25	Hr.	Daily	Hr.	Daily	Bare Costs	Incl. O&P
1 Welder Foreman (outside)	$62.30	$498.40	$96.40	$771.20	$62.30	$96.40
1 Cutting Torch		12.95		14.24	1.62	1.78
8 L.H., Daily Totals		$511.35		$785.45	$63.92	$98.18

Crew E-26	Hr.	Daily	Hr.	Daily	Bare Costs	Incl. O&P
1 Struc. Steel Foreman (outside)	$62.30	$498.40	$96.40	$771.20	$61.64	$94.84
1 Struc. Steel Worker	60.30	482.40	93.30	746.40		
1 Welder	60.30	482.40	93.30	746.40		
.25 Electrician	63.70	127.40	94.65	189.30		
.25 Plumber	67.70	135.40	101.05	202.10		
1 Welder, Gas Engine, 300 amp		148.75		163.63	5.31	5.84
28 L.H., Daily Totals		$1874.75		$2819.03	$66.96	$100.68

For customer support on your Site Work & Landscape Costs with RSMeans Data, call 800.448.8182.

777

Crews - Standard

Crew E-27	Hr.	Daily	Hr.	Daily	Bare Costs	Incl. O&P
1 Struc. Steel Foreman (outside)	$62.30	$498.40	$96.40	$771.20	$59.72	$91.59
5 Struc. Steel Workers	60.30	2412.00	93.30	3732.00		
1 Equip. Oper. (crane)	61.45	491.60	91.55	732.40		
1 Equip. Oper. (oiler)	52.50	420.00	78.25	626.00		
1 Hyd. Crane, 12 Ton		475.80		523.38		
1 Hyd. Crane, 80 Ton		1458.00		1603.80	30.22	33.24
64 L.H., Daily Totals		$5755.80		$7988.78	$89.93	$124.82

Crew F-3	Hr.	Daily	Hr.	Daily	Bare Costs	Incl. O&P
4 Carpenters	$54.70	$1750.40	$81.65	$2612.80	$56.05	$83.63
1 Equip. Oper. (crane)	61.45	491.60	91.55	732.40		
1 Hyd. Crane, 12 Ton		475.80		523.38	11.90	13.08
40 L.H., Daily Totals		$2717.80		$3868.58	$67.94	$96.71

Crew F-4	Hr.	Daily	Hr.	Daily	Bare Costs	Incl. O&P
4 Carpenters	$54.70	$1750.40	$81.65	$2612.80	$55.46	$82.73
1 Equip. Oper. (crane)	61.45	491.60	91.55	732.40		
1 Equip. Oper. (oiler)	52.50	420.00	78.25	626.00		
1 Hyd. Crane, 55 Ton		990.15		1089.17	20.63	22.69
48 L.H., Daily Totals		$3652.15		$5060.36	$76.09	$105.42

Crew F-5	Hr.	Daily	Hr.	Daily	Bare Costs	Incl. O&P
1 Carpenter Foreman (outside)	$56.70	$453.60	$84.60	$676.80	$55.20	$82.39
3 Carpenters	54.70	1312.80	81.65	1959.60		
32 L.H., Daily Totals		$1766.40		$2636.40	$55.20	$82.39

Crew F-6	Hr.	Daily	Hr.	Daily	Bare Costs	Incl. O&P
2 Carpenters	$54.70	$875.20	$81.65	$1306.40	$51.93	$77.47
2 Building Laborers	44.40	710.40	66.25	1060.00		
1 Equip. Oper. (crane)	61.45	491.60	91.55	732.40		
1 Hyd. Crane, 12 Ton		475.80		523.38	11.90	13.08
40 L.H., Daily Totals		$2553.00		$3622.18	$63.83	$90.55

Crew F-7	Hr.	Daily	Hr.	Daily	Bare Costs	Incl. O&P
2 Carpenters	$54.70	$875.20	$81.65	$1306.40	$49.55	$73.95
2 Building Laborers	44.40	710.40	66.25	1060.00		
32 L.H., Daily Totals		$1585.60		$2366.40	$49.55	$73.95

Crew G-1	Hr.	Daily	Hr.	Daily	Bare Costs	Incl. O&P
1 Roofer Foreman (outside)	$50.20	$401.60	$81.35	$650.80	$44.99	$72.92
4 Roofers Composition	48.20	1542.40	78.15	2500.80		
2 Roofer Helpers	35.95	575.20	58.25	932.00		
1 Application Equipment		194.80		214.28		
1 Tar Kettle/Pot		209.95		230.94		
1 Crew Truck		168.15		184.97	10.23	11.25
56 L.H., Daily Totals		$3092.10		$4713.79	$55.22	$84.17

Crew G-2	Hr.	Daily	Hr.	Daily	Bare Costs	Incl. O&P
1 Plasterer	$49.85	$398.80	$74.25	$594.00	$46.27	$68.95
1 Plasterer Helper	44.55	356.40	66.35	530.80		
1 Building Laborer	44.40	355.20	66.25	530.00		
1 Grout Pump, 50 C.F./hr.		190.35		209.38	7.93	8.72
24 L.H., Daily Totals		$1300.75		$1864.18	$54.20	$77.67

Crew G-2A	Hr.	Daily	Hr.	Daily	Bare Costs	Incl. O&P
1 Roofer Composition	$48.20	$385.60	$78.15	$625.20	$42.85	$67.55
1 Roofer Helper	35.95	287.60	58.25	466.00		
1 Building Laborer	44.40	355.20	66.25	530.00		
1 Foam Spray Rig, Trailer-Mtd.		530.15		583.16		
1 Pickup Truck, 3/4 Ton		112.20		123.42	26.76	29.44
24 L.H., Daily Totals		$1670.75		$2327.78	$69.61	$96.99

Crew G-3	Hr.	Daily	Hr.	Daily	Bare Costs	Incl. O&P
2 Sheet Metal Workers	$65.45	$1047.20	$98.70	$1579.20	$54.92	$82.47
2 Building Laborers	44.40	710.40	66.25	1060.00		
32 L.H., Daily Totals		$1757.60		$2639.20	$54.92	$82.47

Crew G-4	Hr.	Daily	Hr.	Daily	Bare Costs	Incl. O&P
1 Labor Foreman (outside)	$46.40	$371.20	$69.25	$554.00	$45.07	$67.25
2 Building Laborers	44.40	710.40	66.25	1060.00		
1 Flatbed Truck, Gas, 1.5 Ton		198.50		218.35		
1 Air Compressor, 160 cfm		212.30		233.53	17.12	18.83
24 L.H., Daily Totals		$1492.40		$2065.88	$62.18	$86.08

Crew G-5	Hr.	Daily	Hr.	Daily	Bare Costs	Incl. O&P
1 Roofer Foreman (outside)	$50.20	$401.60	$81.35	$650.80	$43.70	$70.83
2 Roofers Composition	48.20	771.20	78.15	1250.40		
2 Roofer Helpers	35.95	575.20	58.25	932.00		
1 Application Equipment		194.80		214.28	4.87	5.36
40 L.H., Daily Totals		$1942.80		$3047.48	$48.57	$76.19

Crew G-6A	Hr.	Daily	Hr.	Daily	Bare Costs	Incl. O&P
2 Roofers Composition	$48.20	$771.20	$78.15	$1250.40	$48.20	$78.15
1 Small Compressor, Electric		39.30		43.23		
2 Pneumatic Nailers		55.40		60.94	5.92	6.51
16 L.H., Daily Totals		$865.90		$1354.57	$54.12	$84.66

Crew G-7	Hr.	Daily	Hr.	Daily	Bare Costs	Incl. O&P
1 Carpenter	$54.70	$437.60	$81.65	$653.20	$54.70	$81.65
1 Small Compressor, Electric		39.30		43.23		
1 Pneumatic Nailer		27.70		30.47	8.38	9.21
8 L.H., Daily Totals		$504.60		$726.90	$63.08	$90.86

Crew H-1	Hr.	Daily	Hr.	Daily	Bare Costs	Incl. O&P
2 Glaziers	$52.65	$842.40	$78.40	$1254.40	$56.48	$85.85
2 Struc. Steel Workers	60.30	964.80	93.30	1492.80		
32 L.H., Daily Totals		$1807.20		$2747.20	$56.48	$85.85

Crew H-2	Hr.	Daily	Hr.	Daily	Bare Costs	Incl. O&P
2 Glaziers	$52.65	$842.40	$78.40	$1254.40	$49.90	$74.35
1 Building Laborer	44.40	355.20	66.25	530.00		
24 L.H., Daily Totals		$1197.60		$1784.40	$49.90	$74.35

Crew H-3	Hr.	Daily	Hr.	Daily	Bare Costs	Incl. O&P
1 Glazier	$52.65	$421.20	$78.40	$627.20	$47.35	$71.03
1 Helper	42.05	336.40	63.65	509.20		
16 L.H., Daily Totals		$757.60		$1136.40	$47.35	$71.03

Crew H-4	Hr.	Daily	Hr.	Daily	Bare Costs	Incl. O&P
1 Carpenter	$54.70	$437.60	$81.65	$653.20	$51.44	$77.05
1 Carpenter Helper	42.05	336.40	63.65	509.20		
.5 Electrician	63.70	254.80	94.65	378.60		
20 L.H., Daily Totals		$1028.80		$1541.00	$51.44	$77.05

Crew No.	Bare Costs		Incl. Subs O&P		Cost Per Labor-Hour	

Crew J-1	Hr.	Daily	Hr.	Daily	Bare Costs	Incl. O&P
3 Plasterers	$49.85	$1196.40	$74.25	$1782.00	$47.73	$71.09
2 Plasterer Helpers	44.55	712.80	66.35	1061.60		
1 Mixing Machine, 6 C.F.		113.35		124.69	2.83	3.12
40 L.H., Daily Totals		$2022.55		$2968.28	$50.56	$74.21

Crew J-2	Hr.	Daily	Hr.	Daily	Bare Costs	Incl. O&P
3 Plasterers	$49.85	$1196.40	$74.25	$1782.00	$48.79	$72.41
2 Plasterer Helpers	44.55	712.80	66.35	1061.60		
1 Lather	54.10	432.80	79.00	632.00		
1 Mixing Machine, 6 C.F.		113.35		124.69	2.36	2.60
48 L.H., Daily Totals		$2455.35		$3600.28	$51.15	$75.01

Crew J-3	Hr.	Daily	Hr.	Daily	Bare Costs	Incl. O&P
1 Terrazzo Worker	$51.75	$414.00	$75.60	$604.80	$47.65	$69.63
1 Terrazzo Helper	43.55	348.40	63.65	509.20		
1 Floor Grinder, 22" Path		96.05		105.66		
1 Terrazzo Mixer		162.90		179.19	16.18	17.80
16 L.H., Daily Totals		$1021.35		$1398.85	$63.83	$87.43

Crew J-4	Hr.	Daily	Hr.	Daily	Bare Costs	Incl. O&P
2 Cement Finishers	$51.80	$828.80	$75.90	$1214.40	$49.33	$72.68
1 Laborer	44.40	355.20	66.25	530.00		
1 Floor Grinder, 22" Path		96.05		105.66		
1 Floor Edger, 7" Path		44.05		48.45		
1 Vacuum Pick-Up System		74.95		82.44	8.96	9.86
24 L.H., Daily Totals		$1399.05		$1980.95	$58.29	$82.54

Crew J-4A	Hr.	Daily	Hr.	Daily	Bare Costs	Incl. O&P
2 Cement Finishers	$51.80	$828.80	$75.90	$1214.40	$48.10	$71.08
2 Laborers	44.40	710.40	66.25	1060.00		
1 Floor Grinder, 22" Path		96.05		105.66		
1 Floor Edger, 7" Path		44.05		48.45		
1 Vacuum Pick-Up System		74.95		82.44		
1 Floor Auto Scrubber		179.55		197.51	12.33	13.56
32 L.H., Daily Totals		$1933.80		$2708.46	$60.43	$84.64

Crew J-4B	Hr.	Daily	Hr.	Daily	Bare Costs	Incl. O&P
1 Laborer	$44.40	$355.20	$66.25	$530.00	$44.40	$66.25
1 Floor Auto Scrubber		179.55		197.51	22.44	24.69
8 L.H., Daily Totals		$534.75		$727.51	$66.84	$90.94

Crew J-6	Hr.	Daily	Hr.	Daily	Bare Costs	Incl. O&P
2 Painters	$46.45	$743.20	$68.90	$1102.40	$48.20	$71.69
1 Building Laborer	44.40	355.20	66.25	530.00		
1 Equip. Oper. (light)	55.50	444.00	82.70	661.60		
1 Air Compressor, 250 cfm		202.85		223.13		
1 Sandblaster, Portable, 3 C.F.		83.85		92.23		
1 Set Sand Blasting Accessories		15.55		17.11	9.45	10.39
32 L.H., Daily Totals		$1844.65		$2626.47	$57.65	$82.08

Crew J-7	Hr.	Daily	Hr.	Daily	Bare Costs	Incl. O&P
2 Painters	$46.45	$743.20	$68.90	$1102.40	$46.45	$68.90
1 Floor Belt Sander		50.20		55.22		
1 Floor Sanding Edger		25.20		27.72	4.71	5.18
16 L.H., Daily Totals		$818.60		$1185.34	$51.16	$74.08

Crew No.	Bare Costs		Incl. Subs O&P		Cost Per Labor-Hour	

Crew K-1	Hr.	Daily	Hr.	Daily	Bare Costs	Incl. O&P
1 Carpenter	$54.70	$437.60	$81.65	$653.20	$51.75	$77.33
1 Truck Driver (light)	48.80	390.40	73.00	584.00		
1 Flatbed Truck, Gas, 3 Ton		850.05		935.05	53.13	58.44
16 L.H., Daily Totals		$1678.05		$2172.26	$104.88	$135.77

Crew K-2	Hr.	Daily	Hr.	Daily	Bare Costs	Incl. O&P
1 Struc. Steel Foreman (outside)	$62.30	$498.40	$96.40	$771.20	$57.13	$87.57
1 Struc. Steel Worker	60.30	482.40	93.30	746.40		
1 Truck Driver (light)	48.80	390.40	73.00	584.00		
1 Flatbed Truck, Gas, 3 Ton		850.05		935.05	35.42	38.96
24 L.H., Daily Totals		$2221.25		$3036.66	$92.55	$126.53

Crew L-1	Hr.	Daily	Hr.	Daily	Bare Costs	Incl. O&P
1 Electrician	$63.70	$509.60	$94.65	$757.20	$65.70	$97.85
1 Plumber	67.70	541.60	101.05	808.40		
16 L.H., Daily Totals		$1051.20		$1565.60	$65.70	$97.85

Crew L-2	Hr.	Daily	Hr.	Daily	Bare Costs	Incl. O&P
1 Carpenter	$54.70	$437.60	$81.65	$653.20	$48.38	$72.65
1 Carpenter Helper	42.05	336.40	63.65	509.20		
16 L.H., Daily Totals		$774.00		$1162.40	$48.38	$72.65

Crew L-3	Hr.	Daily	Hr.	Daily	Bare Costs	Incl. O&P
1 Carpenter	$54.70	$437.60	$81.65	$653.20	$59.64	$89.16
.5 Electrician	63.70	254.80	94.65	378.60		
.5 Sheet Metal Worker	65.45	261.80	98.70	394.80		
16 L.H., Daily Totals		$954.20		$1426.60	$59.64	$89.16

Crew L-3A	Hr.	Daily	Hr.	Daily	Bare Costs	Incl. O&P
1 Carpenter Foreman (outside)	$56.70	$453.60	$84.60	$676.80	$59.62	$89.30
.5 Sheet Metal Worker	65.45	261.80	98.70	394.80		
12 L.H., Daily Totals		$715.40		$1071.60	$59.62	$89.30

Crew L-4	Hr.	Daily	Hr.	Daily	Bare Costs	Incl. O&P
2 Skilled Workers	$57.10	$913.60	$85.90	$1374.40	$52.08	$78.48
1 Helper	42.05	336.40	63.65	509.20		
24 L.H., Daily Totals		$1250.00		$1883.60	$52.08	$78.48

Crew L-5	Hr.	Daily	Hr.	Daily	Bare Costs	Incl. O&P
1 Struc. Steel Foreman (outside)	$62.30	$498.40	$96.40	$771.20	$60.75	$93.49
5 Struc. Steel Workers	60.30	2412.00	93.30	3732.00		
1 Equip. Oper. (crane)	61.45	491.60	91.55	732.40		
1 Hyd. Crane, 25 Ton		586.70		645.37	10.48	11.52
56 L.H., Daily Totals		$3988.70		$5880.97	$71.23	$105.02

Crew L-5A	Hr.	Daily	Hr.	Daily	Bare Costs	Incl. O&P
1 Struc. Steel Foreman (outside)	$62.30	$498.40	$96.40	$771.20	$61.09	$93.64
2 Structural Steel Workers	60.30	964.80	93.30	1492.80		
1 Equip. Oper. (crane)	61.45	491.60	91.55	732.40		
1 S.P. Crane, 4x4, 25 Ton		1155.00		1270.50	36.09	39.70
32 L.H., Daily Totals		$3109.80		$4266.90	$97.18	$133.34

Crew No.	Bare Costs		Incl. Subs O&P		Cost Per Labor-Hour	

Crew L-5B	Hr.	Daily	Hr.	Daily	Bare Costs	Incl. O&P
1 Struc. Steel Foreman (outside)	$62.30	$498.40	$96.40	$771.20	$62.33	$94.01
2 Structural Steel Workers	60.30	964.80	93.30	1492.80		
2 Electricians	63.70	1019.20	94.65	1514.40		
2 Steamfitters/Pipefitters	68.35	1093.60	102.00	1632.00		
1 Equip. Oper. (crane)	61.45	491.60	91.55	732.40		
1 Equip. Oper. (oiler)	52.50	420.00	78.25	626.00		
1 Hyd. Crane, 80 Ton		1458.00		1603.80	20.25	22.27
72 L.H., Daily Totals		$5945.60		$8372.60	$82.58	$116.29

Crew L-6	Hr.	Daily	Hr.	Daily	Bare Costs	Incl. O&P
1 Plumber	$67.70	$541.60	$101.05	$808.40	$66.37	$98.92
.5 Electrician	63.70	254.80	94.65	378.60		
12 L.H., Daily Totals		$796.40		$1187.00	$66.37	$98.92

Crew L-7	Hr.	Daily	Hr.	Daily	Bare Costs	Incl. O&P
2 Carpenters	$54.70	$875.20	$81.65	$1306.40	$53.04	$79.11
1 Building Laborer	44.40	355.20	66.25	530.00		
.5 Electrician	63.70	254.80	94.65	378.60		
28 L.H., Daily Totals		$1485.20		$2215.00	$53.04	$79.11

Crew L-8	Hr.	Daily	Hr.	Daily	Bare Costs	Incl. O&P
2 Carpenters	$54.70	$875.20	$81.65	$1306.40	$57.30	$85.53
.5 Plumber	67.70	270.80	101.05	404.20		
20 L.H., Daily Totals		$1146.00		$1710.60	$57.30	$85.53

Crew L-9	Hr.	Daily	Hr.	Daily	Bare Costs	Incl. O&P
1 Labor Foreman (inside)	$44.90	$359.20	$67.00	$536.00	$50.19	$75.58
2 Building Laborers	44.40	710.40	66.25	1060.00		
1 Struc. Steel Worker	60.30	482.40	93.30	746.40		
.5 Electrician	63.70	254.80	94.65	378.60		
36 L.H., Daily Totals		$1806.80		$2721.00	$50.19	$75.58

Crew L-10	Hr.	Daily	Hr.	Daily	Bare Costs	Incl. O&P
1 Struc. Steel Foreman (outside)	$62.30	$498.40	$96.40	$771.20	$61.35	$93.75
1 Structural Steel Worker	60.30	482.40	93.30	746.40		
1 Equip. Oper. (crane)	61.45	491.60	91.55	732.40		
1 Hyd. Crane, 12 Ton		475.80		523.38	19.82	21.81
24 L.H., Daily Totals		$1948.20		$2773.38	$81.17	$115.56

Crew L-11	Hr.	Daily	Hr.	Daily	Bare Costs	Incl. O&P
2 Wreckers	$44.40	$710.40	$67.40	$1078.40	$51.44	$77.26
1 Equip. Oper. (crane)	61.45	491.60	91.55	732.40		
1 Equip. Oper. (light)	55.50	444.00	82.70	661.60		
1 Hyd. Excavator, 2.5 C.Y.		1567.00		1723.70		
1 Loader, Skid Steer, 78 H.P.		446.30		490.93	62.92	69.21
32 L.H., Daily Totals		$3659.30		$4687.03	$114.35	$146.47

Crew M-1	Hr.	Daily	Hr.	Daily	Bare Costs	Incl. O&P
3 Elevator Constructors	$90.30	$2167.20	$133.80	$3211.20	$85.79	$127.11
1 Elevator Apprentice	72.25	578.00	107.05	856.40		
5 Hand Tools		50.50		55.55	1.58	1.74
32 L.H., Daily Totals		$2795.70		$4123.15	$87.37	$128.85

Crew M-3	Hr.	Daily	Hr.	Daily	Bare Costs	Incl. O&P
1 Electrician Foreman (outside)	$65.70	$525.60	$97.65	$781.20	$67.62	$100.41
1 Common Laborer	44.40	355.20	66.25	530.00		
.25 Equipment Operator (med.)	59.00	118.00	87.90	175.80		
1 Elevator Constructor	90.30	722.40	133.80	1070.40		
1 Elevator Apprentice	72.25	578.00	107.05	856.40		
.25 S.P. Crane, 4x4, 20 Ton		143.59		157.95	4.22	4.65
34 L.H., Daily Totals		$2442.79		$3571.75	$71.85	$105.05

Crew M-4	Hr.	Daily	Hr.	Daily	Bare Costs	Incl. O&P
1 Electrician Foreman (outside)	$65.70	$525.60	$97.65	$781.20	$66.92	$99.38
1 Common Laborer	44.40	355.20	66.25	530.00		
.25 Equipment Operator, Crane	61.45	122.90	91.55	183.10		
.25 Equip. Oper. (oiler)	52.50	105.00	78.25	156.50		
1 Elevator Constructor	90.30	722.40	133.80	1070.40		
1 Elevator Apprentice	72.25	578.00	107.05	856.40		
.25 S.P. Crane, 4x4, 40 Ton		190.45		209.50	5.29	5.82
36 L.H., Daily Totals		$2599.55		$3787.09	$72.21	$105.20

Crew Q-1	Hr.	Daily	Hr.	Daily	Bare Costs	Incl. O&P
1 Plumber	$67.70	$541.60	$101.05	$808.40	$60.92	$90.92
1 Plumber Apprentice	54.15	433.20	80.80	646.40		
16 L.H., Daily Totals		$974.80		$1454.80	$60.92	$90.92

Crew Q-1A	Hr.	Daily	Hr.	Daily	Bare Costs	Incl. O&P
.25 Plumber Foreman (outside)	$69.70	$139.40	$104.00	$208.00	$68.10	$101.64
1 Plumber	67.70	541.60	101.05	808.40		
10 L.H., Daily Totals		$681.00		$1016.40	$68.10	$101.64

Crew Q-1C	Hr.	Daily	Hr.	Daily	Bare Costs	Incl. O&P
1 Plumber	$67.70	$541.60	$101.05	$808.40	$60.28	$89.92
1 Plumber Apprentice	54.15	433.20	80.80	646.40		
1 Equip. Oper. (medium)	59.00	472.00	87.90	703.20		
1 Trencher, Chain Type, 8' D		1894.00		2083.40	78.92	86.81
24 L.H., Daily Totals		$3340.80		$4241.40	$139.20	$176.72

Crew Q-2	Hr.	Daily	Hr.	Daily	Bare Costs	Incl. O&P
2 Plumbers	$67.70	$1083.20	$101.05	$1616.80	$63.18	$94.30
1 Plumber Apprentice	54.15	433.20	80.80	646.40		
24 L.H., Daily Totals		$1516.40		$2263.20	$63.18	$94.30

Crew Q-3	Hr.	Daily	Hr.	Daily	Bare Costs	Incl. O&P
1 Plumber Foreman (inside)	$68.20	$545.60	$101.80	$814.40	$64.44	$96.17
2 Plumbers	67.70	1083.20	101.05	1616.80		
1 Plumber Apprentice	54.15	433.20	80.80	646.40		
32 L.H., Daily Totals		$2062.00		$3077.60	$64.44	$96.17

Crew Q-4	Hr.	Daily	Hr.	Daily	Bare Costs	Incl. O&P
1 Plumber Foreman (inside)	$68.20	$545.60	$101.80	$814.40	$64.44	$96.17
1 Plumber	67.70	541.60	101.05	808.40		
1 Welder (plumber)	67.70	541.60	101.05	808.40		
1 Plumber Apprentice	54.15	433.20	80.80	646.40		
1 Welder, Electric, 300 amp		107.55		118.31	3.36	3.70
32 L.H., Daily Totals		$2169.55		$3195.91	$67.80	$99.87

Crew Q-5	Hr.	Daily	Hr.	Daily	Bare Costs	Incl. O&P
1 Steamfitter	$68.35	$546.80	$102.00	$816.00	$61.52	$91.83
1 Steamfitter Apprentice	54.70	437.60	81.65	653.20		
16 L.H., Daily Totals		$984.40		$1469.20	$61.52	$91.83

Crews - Standard

Crew No.	Bare Costs Hr.	Daily	Incl. Subs O&P Hr.	Daily	Cost Per Labor-Hour Bare Costs	Incl. O&P
Crew Q-6	Hr.	Daily	Hr.	Daily	Bare Costs	Incl. O&P
2 Steamfitters	$68.35	$1093.60	$102.00	$1632.00	$63.80	$95.22
1 Steamfitter Apprentice	54.70	437.60	81.65	653.20		
24 L.H., Daily Totals		$1531.20		$2285.20	$63.80	$95.22
Crew Q-7	Hr.	Daily	Hr.	Daily	Bare Costs	Incl. O&P
1 Steamfitter Foreman (inside)	$68.85	$550.80	$102.75	$822.00	$65.06	$97.10
2 Steamfitters	68.35	1093.60	102.00	1632.00		
1 Steamfitter Apprentice	54.70	437.60	81.65	653.20		
32 L.H., Daily Totals		$2082.00		$3107.20	$65.06	$97.10
Crew Q-8	Hr.	Daily	Hr.	Daily	Bare Costs	Incl. O&P
1 Steamfitter Foreman (inside)	$68.85	$550.80	$102.75	$822.00	$65.06	$97.10
1 Steamfitter	68.35	546.80	102.00	816.00		
1 Welder (steamfitter)	68.35	546.80	102.00	816.00		
1 Steamfitter Apprentice	54.70	437.60	81.65	653.20		
1 Welder, Electric, 300 amp		107.55		118.31	3.36	3.70
32 L.H., Daily Totals		$2189.55		$3225.51	$68.42	$100.80
Crew Q-9	Hr.	Daily	Hr.	Daily	Bare Costs	Incl. O&P
1 Sheet Metal Worker	$65.45	$523.60	$98.70	$789.60	$58.90	$88.83
1 Sheet Metal Apprentice	52.35	418.80	78.95	631.60		
16 L.H., Daily Totals		$942.40		$1421.20	$58.90	$88.83
Crew Q-10	Hr.	Daily	Hr.	Daily	Bare Costs	Incl. O&P
2 Sheet Metal Workers	$65.45	$1047.20	$98.70	$1579.20	$61.08	$92.12
1 Sheet Metal Apprentice	52.35	418.80	78.95	631.60		
24 L.H., Daily Totals		$1466.00		$2210.80	$61.08	$92.12
Crew Q-11	Hr.	Daily	Hr.	Daily	Bare Costs	Incl. O&P
1 Sheet Metal Foreman (inside)	$65.95	$527.60	$99.50	$796.00	$62.30	$93.96
2 Sheet Metal Workers	65.45	1047.20	98.70	1579.20		
1 Sheet Metal Apprentice	52.35	418.80	78.95	631.60		
32 L.H., Daily Totals		$1993.60		$3006.80	$62.30	$93.96
Crew Q-12	Hr.	Daily	Hr.	Daily	Bare Costs	Incl. O&P
1 Sprinkler Installer	$66.50	$532.00	$99.35	$794.80	$59.85	$89.42
1 Sprinkler Apprentice	53.20	425.60	79.50	636.00		
16 L.H., Daily Totals		$957.60		$1430.80	$59.85	$89.42
Crew Q-13	Hr.	Daily	Hr.	Daily	Bare Costs	Incl. O&P
1 Sprinkler Foreman (inside)	$67.00	$536.00	$100.10	$800.80	$63.30	$94.58
2 Sprinkler Installers	66.50	1064.00	99.35	1589.60		
1 Sprinkler Apprentice	53.20	425.60	79.50	636.00		
32 L.H., Daily Totals		$2025.60		$3026.40	$63.30	$94.58
Crew Q-14	Hr.	Daily	Hr.	Daily	Bare Costs	Incl. O&P
1 Asbestos Worker	$60.95	$487.60	$93.25	$746.00	$54.85	$83.90
1 Asbestos Apprentice	48.75	390.00	74.55	596.40		
16 L.H., Daily Totals		$877.60		$1342.40	$54.85	$83.90
Crew Q-15	Hr.	Daily	Hr.	Daily	Bare Costs	Incl. O&P
1 Plumber	$67.70	$541.60	$101.05	$808.40	$60.92	$90.92
1 Plumber Apprentice	54.15	433.20	80.80	646.40		
1 Welder, Electric, 300 amp		107.55		118.31	6.72	7.39
16 L.H., Daily Totals		$1082.35		$1573.11	$67.65	$98.32

Crew No.	Bare Costs Hr.	Daily	Incl. Subs O&P Hr.	Daily	Cost Per Labor-Hour Bare Costs	Incl. O&P
Crew Q-16	Hr.	Daily	Hr.	Daily	Bare Costs	Incl. O&P
2 Plumbers	$67.70	$1083.20	$101.05	$1616.80	$63.18	$94.30
1 Plumber Apprentice	54.15	433.20	80.80	646.40		
1 Welder, Electric, 300 amp		107.55		118.31	4.48	4.93
24 L.H., Daily Totals		$1623.95		$2381.51	$67.66	$99.23
Crew Q-17	Hr.	Daily	Hr.	Daily	Bare Costs	Incl. O&P
1 Steamfitter	$68.35	$546.80	$102.00	$816.00	$61.52	$91.83
1 Steamfitter Apprentice	54.70	437.60	81.65	653.20		
1 Welder, Electric, 300 amp		107.55		118.31	6.72	7.39
16 L.H., Daily Totals		$1091.95		$1587.51	$68.25	$99.22
Crew Q-17A	Hr.	Daily	Hr.	Daily	Bare Costs	Incl. O&P
1 Steamfitter	$68.35	$546.80	$102.00	$816.00	$61.50	$91.73
1 Steamfitter Apprentice	54.70	437.60	81.65	653.20		
1 Equip. Oper. (crane)	61.45	491.60	91.55	732.40		
1 Hyd. Crane, 12 Ton		475.80		523.38		
1 Welder, Electric, 300 amp		107.55		118.31	24.31	26.74
24 L.H., Daily Totals		$2059.35		$2843.28	$85.81	$118.47
Crew Q-18	Hr.	Daily	Hr.	Daily	Bare Costs	Incl. O&P
2 Steamfitters	$68.35	$1093.60	$102.00	$1632.00	$63.80	$95.22
1 Steamfitter Apprentice	54.70	437.60	81.65	653.20		
1 Welder, Electric, 300 amp		107.55		118.31	4.48	4.93
24 L.H., Daily Totals		$1638.75		$2403.51	$68.28	$100.15
Crew Q-19	Hr.	Daily	Hr.	Daily	Bare Costs	Incl. O&P
1 Steamfitter	$68.35	$546.80	$102.00	$816.00	$62.25	$92.77
1 Steamfitter Apprentice	54.70	437.60	81.65	653.20		
1 Electrician	63.70	509.60	94.65	757.20		
24 L.H., Daily Totals		$1494.00		$2226.40	$62.25	$92.77
Crew Q-20	Hr.	Daily	Hr.	Daily	Bare Costs	Incl. O&P
1 Sheet Metal Worker	$65.45	$523.60	$98.70	$789.60	$59.86	$89.99
1 Sheet Metal Apprentice	52.35	418.80	78.95	631.60		
.5 Electrician	63.70	254.80	94.65	378.60		
20 L.H., Daily Totals		$1197.20		$1799.80	$59.86	$89.99
Crew Q-21	Hr.	Daily	Hr.	Daily	Bare Costs	Incl. O&P
2 Steamfitters	$68.35	$1093.60	$102.00	$1632.00	$63.77	$95.08
1 Steamfitter Apprentice	54.70	437.60	81.65	653.20		
1 Electrician	63.70	509.60	94.65	757.20		
32 L.H., Daily Totals		$2040.80		$3042.40	$63.77	$95.08
Crew Q-22	Hr.	Daily	Hr.	Daily	Bare Costs	Incl. O&P
1 Plumber	$67.70	$541.60	$101.05	$808.40	$60.92	$90.92
1 Plumber Apprentice	54.15	433.20	80.80	646.40		
1 Hyd. Crane, 12 Ton		475.80		523.38	29.74	32.71
16 L.H., Daily Totals		$1450.60		$1978.18	$90.66	$123.64
Crew Q-22A	Hr.	Daily	Hr.	Daily	Bare Costs	Incl. O&P
1 Plumber	$67.70	$541.60	$101.05	$808.40	$56.92	$84.91
1 Plumber Apprentice	54.15	433.20	80.80	646.40		
1 Laborer	44.40	355.20	66.25	530.00		
1 Equip. Oper. (crane)	61.45	491.60	91.55	732.40		
1 Hyd. Crane, 12 Ton		475.80		523.38	14.87	16.36
32 L.H., Daily Totals		$2297.40		$3240.58	$71.79	$101.27

Crew No.	Bare Costs		Incl. Subs O&P		Cost Per Labor-Hour	
	Hr.	Daily	Hr.	Daily	Bare Costs	Incl. O&P
Crew Q-23						
1 Plumber Foreman (outside)	$69.70	$557.60	$104.00	$832.00	$65.47	$97.65
1 Plumber	67.70	541.60	101.05	808.40		
1 Equip. Oper. (medium)	59.00	472.00	87.90	703.20		
1 Lattice Boom Crane, 20 Ton		1526.00		1678.60	63.58	69.94
24 L.H., Daily Totals		$3097.20		$4022.20	$129.05	$167.59
Crew R-1	Hr.	Daily	Hr.	Daily	Bare Costs	Incl. O&P
1 Electrician Foreman	$64.20	$513.60	$95.40	$763.20	$59.53	$88.46
3 Electricians	63.70	1528.80	94.65	2271.60		
2 Electrician Apprentices	50.95	815.20	75.70	1211.20		
48 L.H., Daily Totals		$2857.60		$4246.00	$59.53	$88.46
Crew R-1A	Hr.	Daily	Hr.	Daily	Bare Costs	Incl. O&P
1 Electrician	$63.70	$509.60	$94.65	$757.20	$57.33	$85.17
1 Electrician Apprentice	50.95	407.60	75.70	605.60		
16 L.H., Daily Totals		$917.20		$1362.80	$57.33	$85.17
Crew R-1B	Hr.	Daily	Hr.	Daily	Bare Costs	Incl. O&P
1 Electrician	$63.70	$509.60	$94.65	$757.20	$55.20	$82.02
2 Electrician Apprentices	50.95	815.20	75.70	1211.20		
24 L.H., Daily Totals		$1324.80		$1968.40	$55.20	$82.02
Crew R-1C	Hr.	Daily	Hr.	Daily	Bare Costs	Incl. O&P
2 Electricians	$63.70	$1019.20	$94.65	$1514.40	$57.33	$85.17
2 Electrician Apprentices	50.95	815.20	75.70	1211.20		
1 Portable cable puller, 8000 lb.		102.65		112.92	3.21	3.53
32 L.H., Daily Totals		$1937.05		$2838.51	$60.53	$88.70
Crew R-2	Hr.	Daily	Hr.	Daily	Bare Costs	Incl. O&P
1 Electrician Foreman	$64.20	$513.60	$95.40	$763.20	$59.81	$88.90
3 Electricians	63.70	1528.80	94.65	2271.60		
2 Electrician Apprentices	50.95	815.20	75.70	1211.20		
1 Equip. Oper. (crane)	61.45	491.60	91.55	732.40		
1 S.P. Crane, 4x4, 5 Ton		381.95		420.14	6.82	7.50
56 L.H., Daily Totals		$3731.15		$5398.55	$66.63	$96.40
Crew R-3	Hr.	Daily	Hr.	Daily	Bare Costs	Incl. O&P
1 Electrician Foreman	$64.20	$513.60	$95.40	$763.20	$63.45	$94.33
1 Electrician	63.70	509.60	94.65	757.20		
.5 Equip. Oper. (crane)	61.45	245.80	91.55	366.20		
.5 S.P. Crane, 4x4, 5 Ton		190.97		210.07	9.55	10.50
20 L.H., Daily Totals		$1459.97		$2096.67	$73.00	$104.83
Crew R-4	Hr.	Daily	Hr.	Daily	Bare Costs	Incl. O&P
1 Struc. Steel Foreman (outside)	$62.30	$498.40	$96.40	$771.20	$61.38	$94.19
3 Struc. Steel Workers	60.30	1447.20	93.30	2239.20		
1 Electrician	63.70	509.60	94.65	757.20		
1 Welder, Gas Engine, 300 amp		148.75		163.63	3.72	4.09
40 L.H., Daily Totals		$2603.95		$3931.22	$65.10	$98.28

Crew No.	Bare Costs		Incl. Subs O&P		Cost Per Labor-Hour	
	Hr.	Daily	Hr.	Daily	Bare Costs	Incl. O&P
Crew R-5						
1 Electrician Foreman	$64.20	$513.60	$95.40	$763.20	$55.87	$83.45
4 Electrician Linemen	63.70	2038.40	94.65	3028.80		
2 Electrician Operators	63.70	1019.20	94.65	1514.40		
4 Electrician Groundmen	42.05	1345.60	63.65	2036.80		
1 Crew Truck		168.15		184.97		
1 Flatbed Truck, 20,000 GVW		204.05		224.46		
1 Pickup Truck, 3/4 Ton		112.20		123.42		
.2 Hyd. Crane, 55 Ton		198.03		217.83		
.2 Hyd. Crane, 12 Ton		95.16		104.68		
.2 Earth Auger, Truck-Mtd.		40.51		44.56		
1 Tractor w/Winch		377.65		415.42	13.59	14.95
88 L.H., Daily Totals		$6112.55		$8658.52	$69.46	$98.39
Crew R-6	Hr.	Daily	Hr.	Daily	Bare Costs	Incl. O&P
1 Electrician Foreman	$64.20	$513.60	$95.40	$763.20	$55.87	$83.45
4 Electrician Linemen	63.70	2038.40	94.65	3028.80		
2 Electrician Operators	63.70	1019.20	94.65	1514.40		
4 Electrician Groundmen	42.05	1345.60	63.65	2036.80		
1 Crew Truck		168.15		184.97		
1 Flatbed Truck, 20,000 GVW		204.05		224.46		
1 Pickup Truck, 3/4 Ton		112.20		123.42		
.2 Hyd. Crane, 55 Ton		198.03		217.83		
.2 Hyd. Crane, 12 Ton		95.16		104.68		
.2 Earth Auger, Truck-Mtd.		40.51		44.56		
1 Tractor w/Winch		377.65		415.42		
3 Cable Trailers		194.25		213.68		
.5 Tensioning Rig		55.75		61.33		
.5 Cable Pulling Rig		306.65		337.32	19.91	21.91
88 L.H., Daily Totals		$6669.20		$9270.84	$75.79	$105.35
Crew R-7	Hr.	Daily	Hr.	Daily	Bare Costs	Incl. O&P
1 Electrician Foreman	$64.20	$513.60	$95.40	$763.20	$45.74	$68.94
5 Electrician Groundmen	42.05	1682.00	63.65	2546.00		
1 Crew Truck		168.15		184.97	3.50	3.85
48 L.H., Daily Totals		$2363.75		$3494.17	$49.24	$72.80
Crew R-8	Hr.	Daily	Hr.	Daily	Bare Costs	Incl. O&P
1 Electrician Foreman	$64.20	$513.60	$95.40	$763.20	$56.57	$84.44
3 Electrician Linemen	63.70	1528.80	94.65	2271.60		
2 Electrician Groundmen	42.05	672.80	63.65	1018.40		
1 Pickup Truck, 3/4 Ton		112.20		123.42		
1 Crew Truck		168.15		184.97	5.84	6.42
48 L.H., Daily Totals		$2995.55		$4361.59	$62.41	$90.87
Crew R-9	Hr.	Daily	Hr.	Daily	Bare Costs	Incl. O&P
1 Electrician Foreman	$64.20	$513.60	$95.40	$763.20	$52.94	$79.24
1 Electrician Lineman	63.70	509.60	94.65	757.20		
2 Electrician Operators	63.70	1019.20	94.65	1514.40		
4 Electrician Groundmen	42.05	1345.60	63.65	2036.80		
1 Pickup Truck, 3/4 Ton		112.20		123.42		
1 Crew Truck		168.15		184.97	4.38	4.82
64 L.H., Daily Totals		$3668.35		$5379.98	$57.32	$84.06
Crew R-10	Hr.	Daily	Hr.	Daily	Bare Costs	Incl. O&P
1 Electrician Foreman	$64.20	$513.60	$95.40	$763.20	$60.17	$89.61
4 Electrician Linemen	63.70	2038.40	94.65	3028.80		
1 Electrician Groundman	42.05	336.40	63.65	509.20		
1 Crew Truck		168.15		184.97		
3 Tram Cars		219.60		241.56	8.08	8.89
48 L.H., Daily Totals		$3276.15		$4727.73	$68.25	$98.49

Crew R-11

Crew No.	Hr.	Daily	Hr.	Daily	Bare Costs	Incl. O&P
1 Electrician Foreman	$64.20	$513.60	$95.40	$763.20	$60.69	$90.26
4 Electricians	63.70	2038.40	94.65	3028.80		
1 Equip. Oper. (crane)	61.45	491.60	91.55	732.40		
1 Common Laborer	44.40	355.20	66.25	530.00		
1 Crew Truck		168.15		184.97		
1 Hyd. Crane, 12 Ton		475.80		523.38	11.50	12.65
56 L.H., Daily Totals		$4042.75		$5762.74	$72.19	$102.91

Crew R-12

Crew No.	Hr.	Daily	Hr.	Daily	Bare Costs	Incl. O&P
1 Carpenter Foreman (inside)	$55.20	$441.60	$82.40	$659.20	$51.90	$77.75
4 Carpenters	54.70	1750.40	81.65	2612.80		
4 Common Laborers	44.40	1420.80	66.25	2120.00		
1 Equip. Oper. (medium)	59.00	472.00	87.90	703.20		
1 Steel Worker	60.30	482.40	93.30	746.40		
1 Dozer, 200 H.P.		1520.00		1672.00		
1 Pickup Truck, 3/4 Ton		112.20		123.42	18.55	20.40
88 L.H., Daily Totals		$6199.40		$8637.02	$70.45	$98.15

Crew R-13

Crew No.	Hr.	Daily	Hr.	Daily	Bare Costs	Incl. O&P
1 Electrician Foreman	$64.20	$513.60	$95.40	$763.20	$61.55	$91.52
3 Electricians	63.70	1528.80	94.65	2271.60		
.25 Equip. Oper. (crane)	61.45	122.90	91.55	183.10		
1 Equipment Oiler	52.50	420.00	78.25	626.00		
.25 Hydraulic Crane, 33 Ton		245.79		270.37	5.85	6.44
42 L.H., Daily Totals		$2831.09		$4114.27	$67.41	$97.96

Crew R-15

Crew No.	Hr.	Daily	Hr.	Daily	Bare Costs	Incl. O&P
1 Electrician Foreman	$64.20	$513.60	$95.40	$763.20	$62.42	$92.78
4 Electricians	63.70	2038.40	94.65	3028.80		
1 Equipment Oper. (light)	55.50	444.00	82.70	661.60		
1 Telescoping Boom Lift, to 40'		281.90		310.09	5.87	6.46
48 L.H., Daily Totals		$3277.90		$4763.69	$68.29	$99.24

Crew R-15A

Crew No.	Hr.	Daily	Hr.	Daily	Bare Costs	Incl. O&P
1 Electrician Foreman	$64.20	$513.60	$95.40	$763.20	$55.98	$83.32
2 Electricians	63.70	1019.20	94.65	1514.40		
2 Common Laborers	44.40	710.40	66.25	1060.00		
1 Equip. Oper. (light)	55.50	444.00	82.70	661.60		
1 Telescoping Boom Lift, to 40'		281.90		310.09	5.87	6.46
48 L.H., Daily Totals		$2969.10		$4309.29	$61.86	$89.78

Crew R-18

Crew No.	Hr.	Daily	Hr.	Daily	Bare Costs	Incl. O&P
.25 Electrician Foreman	$64.20	$128.40	$95.40	$190.80	$55.89	$83.05
1 Electrician	63.70	509.60	94.65	757.20		
2 Electrician Apprentices	50.95	815.20	75.70	1211.20		
26 L.H., Daily Totals		$1453.20		$2159.20	$55.89	$83.05

Crew R-19

Crew No.	Hr.	Daily	Hr.	Daily	Bare Costs	Incl. O&P
.5 Electrician Foreman	$64.20	$256.80	$95.40	$381.60	$63.80	$94.80
2 Electricians	63.70	1019.20	94.65	1514.40		
20 L.H., Daily Totals		$1276.00		$1896.00	$63.80	$94.80

Crew R-21

Crew No.	Hr.	Daily	Hr.	Daily	Bare Costs	Incl. O&P
1 Electrician Foreman	$64.20	$513.60	$95.40	$763.20	$63.71	$94.67
3 Electricians	63.70	1528.80	94.65	2271.60		
.1 Equip. Oper. (medium)	59.00	47.20	87.90	70.32		
.1 S.P. Crane, 4x4, 25 Ton		115.50		127.05	3.52	3.87
32.8 L.H., Daily Totals		$2205.10		$3232.17	$67.23	$98.54

Crew R-22

Crew No.	Hr.	Daily	Hr.	Daily	Bare Costs	Incl. O&P
.66 Electrician Foreman	$64.20	$338.98	$95.40	$503.71	$58.30	$86.62
2 Electricians	63.70	1019.20	94.65	1514.40		
2 Electrician Apprentices	50.95	815.20	75.70	1211.20		
37.28 L.H., Daily Totals		$2173.38		$3229.31	$58.30	$86.62

Crew R-30

Crew No.	Hr.	Daily	Hr.	Daily	Bare Costs	Incl. O&P
.25 Electrician Foreman (outside)	$65.70	$131.40	$97.65	$195.30	$51.98	$77.40
1 Electrician	63.70	509.60	94.65	757.20		
2 Laborers (Semi-Skilled)	44.40	710.40	66.25	1060.00		
26 L.H., Daily Totals		$1351.40		$2012.50	$51.98	$77.40

Crew R-31

Crew No.	Hr.	Daily	Hr.	Daily	Bare Costs	Incl. O&P
1 Electrician	$63.70	$509.60	$94.65	$757.20	$63.70	$94.65
1 Core Drill, Electric, 2.5 H.P.		62.65		68.92	7.83	8.61
8 L.H., Daily Totals		$572.25		$826.12	$71.53	$103.26

Crew W-41E

Crew No.	Hr.	Daily	Hr.	Daily	Bare Costs	Incl. O&P
.5 Plumber Foreman (outside)	$69.70	$278.80	$104.00	$416.00	$58.78	$87.72
1 Plumber	67.70	541.60	101.05	808.40		
1 Laborer	44.40	355.20	66.25	530.00		
20 L.H., Daily Totals		$1175.60		$1754.40	$58.78	$87.72

For customer support on your Site Work & Landscape Costs with RSMeans Data, call 800.448.8182.

783

Historical Cost Indexes

The table below lists both the RSMeans® historical cost index based on Jan. 1, 1993 = 100 as well as the computed value of an index based on Jan. 1, 2021 costs. Since the Jan. 1, 2021 figure is estimated, space is left to write in the actual index figures as they become available through the quarterly *RSMeans Construction Cost Indexes*.

To compute the actual index based on Jan. 1, 2021 = 100, divide the historical cost index for a particular year by the actual Jan. 1, 2021 construction cost index. Space has been left to advance the index figures as the year progresses.

Year	Historical Cost Index Jan. 1, 1993 = 100		Current Index Based on Jan. 1, 2021 = 100		Year	Historical Cost Index Jan. 1, 1993 = 100	Current Index Based on Jan. 1, 2021 = 100		Year	Historical Cost Index Jan. 1, 1993 = 100	Current Index Based on Jan. 1, 2021 = 100	
	Est.	Actual	Est.	Actual		Actual	Est.	Actual		Actual	Est.	Actual
Oct 2021					July 2004	143.7	60.7		1984	82.0	34.6	
July 2021					2003	132.0	55.8		1983	80.2	33.9	
April 2021					2002	128.7	54.4		1982	76.1	32.2	
Jan 2021	236.7		100.0		2001	125.1	52.9		1981	70.0	29.6	
2020		234.6	99.1		2000	120.9	51.1		1980	62.9	26.6	
2019		232.2	98.1		1999	117.6	49.7		1979	57.8	24.4	
2018		222.9	94.2		1998	115.1	48.6		1978	53.5	22.6	
2017		213.6	90.2		1997	112.8	47.7		1977	49.5	20.9	
2016		207.3	87.6		1996	110.2	46.6		1976	46.9	19.8	
2015		206.2	87.1		1995	107.6	45.5		1975	44.8	18.9	
2014		204.9	86.6		1994	104.4	44.1		1974	41.4	17.5	
2013		201.2	85.0		1993	101.7	43.0		1973	37.7	15.9	
2012		194.6	82.2		1992	99.4	42.0		1972	34.8	14.7	
2011		191.2	80.8		1991	96.8	40.9		1971	32.1	13.6	
2010		183.5	77.5		1990	94.3	39.8		1970	28.7	12.1	
2009		180.1	76.1		1989	92.1	38.9		1969	26.9	11.4	
2008		180.4	76.2		1988	89.9	38.0					
2007		169.4	71.6		1987	87.7	37.0					
2006		162.0	68.4		1986	84.2	35.6					
2005		151.6	64.0		1985	82.6	34.9					

Adjustments to Costs

The "Historical Cost Index" can be used to convert national average building costs at a particular time to the approximate building costs for some other time.

Example:

Estimate and compare construction costs for different years in the same city.

To estimate the national average construction cost of a building in 1970, knowing that it cost $900,000 in 2021:

INDEX in 1970 = 28.7

INDEX in 2021 = 236.7

Note: The city cost indexes for Canada can be used to convert U.S. national averages to local costs in Canadian dollars.

Example:

To estimate and compare the cost of a building in Toronto, ON in 2021 with the known cost of $600,000 (US$) in New York, NY in 2021:

INDEX Toronto = 112.5

INDEX New York = 134.1

$$\frac{\text{INDEX Toronto}}{\text{INDEX New York}} \times \text{Cost New York} = \text{Cost Toronto}$$

$$\frac{112.5}{134.1} \times \$600,000 = .839 \times \$600,000 = \$503,400$$

The construction cost of the building in Toronto is $503,400 (CN$).

Time Adjustment Using the Historical Cost Indexes:

$$\frac{\text{Index for Year A}}{\text{Index for Year B}} \times \text{Cost in Year B} = \text{Cost in Year A}$$

$$\frac{\text{INDEX 1970}}{\text{INDEX 2021}} \times \text{Cost 2021} = \text{Cost 1970}$$

$$\frac{28.7}{236.7} \times \$900,000 = .121 \times \$900,000 = \$108,900$$

The construction cost of the building in 1970 was $108,900.

*Historical Cost Index updates and other resources are provided on the following website:
rsmeans.com/2021books

For customer support on your Site Work & Landscape Costs with RSMeans Data, call 800.448.8182.

How to Use the City Cost Indexes

What you should know before you begin

RSMeans City Cost Indexes (CCI) are an extremely useful tool for when you want to compare costs from city to city and region to region.

This publication contains average construction cost indexes for 731 U.S. and Canadian cities covering over 930 three-digit zip code locations, as listed directly under each city.

Keep in mind that a City Cost Index number is a percentage ratio of a specific city's cost to the national average cost of the same item at a stated time period.

In other words, these index figures represent relative construction factors (or, if you prefer, multipliers) for material and installation costs, as well as the weighted average for Total In Place costs for each CSI MasterFormat division. Installation costs include both labor and equipment rental costs. When estimating equipment rental rates only for a specific location, use 01 54 33 EQUIPMENT RENTAL COSTS in the Reference Section.

The 30 City Average Index is the average of 30 major U.S. cities and serves as a national average. This national average represents the baseline from which all locations are calculated.

Index figures for both material and installation are based on the 30 major city average of 100 and represent the cost relationship as of July 1, 2020. The index for each division is computed from representative material and labor quantities for that division. The weighted average for each city is a weighted total of the components listed above it. It does not include relative productivity between trades or cities.

As changes occur in local material prices, labor rates, and equipment rental rates (including fuel costs), the impact of these changes should be accurately measured by the change in the City Cost Index for each particular city (as compared to the 30 city average).

Therefore, if you know (or have estimated) building costs in one city today, you can easily convert those costs to expected building costs in another city.

In addition, by using the Historical Cost Index, you can easily convert national average building costs at a particular time to the approximate building costs for some other time. The City Cost Indexes can then be applied to calculate the costs for a particular city.

Quick calculations

Location Adjustment Using the City Cost Indexes:

$$\frac{\text{Index for City A}}{\text{Index for City B}} \times \text{Cost in City B} = \text{Cost in City A}$$

**Time Adjustment for the National Average
Using the Historical Cost Index:**

$$\frac{\text{Index for Year A}}{\text{Index for Year B}} \times \text{Cost in Year B} = \text{Cost in Year A}$$

Adjustment from the National Average:

$$\frac{\text{Index for City A}}{100} \times \text{National Average Cost} = \text{Cost in City A}$$

Since each of the other RSMeans data sets contains many different items, any *one* item multiplied by the particular city index may give incorrect results. However, the larger the number of items compiled, the closer the results should be to actual costs for that particular city.

The City Cost Indexes for Canadian cities are calculated using Canadian material and equipment prices and labor rates in Canadian dollars. When compared to the baseline mentioned previously, the resulting index enables the user to convert baseline costs to local costs. Therefore, indexes for Canadian cities can be used to convert U.S. national average prices to local costs in Canadian dollars.

How to use this section

1. Compare costs from city to city.

In using the RSMeans Indexes, remember that an index number is not a fixed number but a ratio: It's a percentage ratio of a building component's cost at any stated time to the national average cost of that same component at the same time period. Put in the form of an equation:

$$\frac{\text{Specific City Cost}}{\text{National Average Cost}} \times 100 = \text{City Index Number}$$

Therefore, when making cost comparisons between cities, do not subtract one city's index number from the index number of another city and read the result as a percentage difference. Instead, divide one city's index number by that of the other city. The resulting number may then be used as a multiplier to calculate cost differences from city to city.

The formula used to find cost differences between cities for the purpose of comparison is as follows:

$$\frac{\text{City A Index}}{\text{City B Index}} \times \text{City B Cost (Known)} = \text{City A Cost (Unknown)}$$

In addition, you can use RSMeans CCI to calculate and compare costs division by division between cities using the same basic formula. (Just be sure that you're comparing similar divisions.)

2. Compare a specific city's construction costs with the national average.

When you're studying construction location feasibility, it's advisable to compare a prospective project's cost index with an index of the national average cost.

For example, divide the weighted average index of construction costs of a specific city by that of the 30 City Average, which = 100.

$$\frac{\text{City Index}}{100} = \text{\% of National Average}$$

As a result, you get a ratio that indicates the relative cost of construction in that city in comparison with the national average.

3. Convert U.S. national average to actual costs in Canadian City.

$$\frac{\text{Index for Canadian City}}{100} \times \text{National Average Cost} = \text{Cost in Canadian City in \$ CAN}$$

Adjust construction cost data based on a national average.
When you use a source of construction cost data which is based on a national average (such as RSMeans cost data), it is necessary to adjust those costs to a specific location.

$$\frac{\text{City Index}}{100} \times \frac{\text{Cost Based on}}{\text{National Average Costs}} = \frac{\text{City Cost}}{\text{(Unknown)}}$$

When applying the City Cost Indexes to demolition projects, use the appropriate division installation index. For example, for removal of existing doors and windows, use the Division 8 (Openings) index.

What you might like to know about how we developed the Indexes

The information presented in the CCI is organized according to the Construction Specifications Institute (CSI) MasterFormat 2018 classification system.

To create a reliable index, RSMeans researched the building type most often constructed in the United States and Canada. Because it was concluded that no one type of building completely represented the building construction industry, nine different types of buildings were combined to create a composite model.

The exact material, labor, and equipment quantities are based on detailed analyses of these nine building types, and then each quantity is weighted in proportion to expected usage. These various material items, labor hours, and equipment rental rates are thus combined to form a composite building representing as closely as possible the actual usage of materials, labor, and equipment in the North American building construction industry.

The following structures were chosen to make up that composite model:

1. Factory, 1 story
2. Office, 2–4 stories
3. Store, Retail
4. Town Hall, 2–3 stories
5. High School, 2–3 stories
6. Hospital, 4–8 stories
7. Garage, Parking
8. Apartment, 1–3 stories
9. Hotel/Motel, 2–3 stories

For the purposes of ensuring the timeliness of the data, the components of the index for the composite model have been streamlined. They currently consist of:

specific quantities of 66 commonly used construction materials;

specific labor-hours for 21 building construction trades; and

specific days of equipment rental for 6 types of construction equipment (normally used to install the 66 material items by the 21 trades.) Fuel costs and routine maintenance costs are included in the equipment cost.

Material and equipment price quotations are gathered quarterly from cities in the United States and Canada. These prices and the latest negotiated labor wage rates for 21 different building trades are used to compile the quarterly update of the City Cost Index.

The 30 major U.S. cities used to calculate the national average are:

Atlanta, GA	Memphis, TN
Baltimore, MD	Milwaukee, WI
Boston, MA	Minneapolis, MN
Buffalo, NY	Nashville, TN
Chicago, IL	New Orleans, LA
Cincinnati, OH	New York, NY
Cleveland, OH	Philadelphia, PA
Columbus, OH	Phoenix, AZ
Dallas, TX	Pittsburgh, PA
Denver, CO	St. Louis, MO
Detroit, MI	San Antonio, TX
Houston, TX	San Diego, CA
Indianapolis, IN	San Francisco, CA
Kansas City, MO	Seattle, WA
Los Angeles, CA	Washington, DC

What the CCI does not indicate

The weighted average for each city is a total of the divisional components weighted to reflect typical usage. It does not include the productivity variations between trades or cities.

In addition, the CCI does not take into consideration factors such as the following:

- managerial efficiency
- competitive conditions
- automation
- restrictive union practices
- unique local requirements
- regional variations due to specific building codes

UNITED STATES / ALABAMA

	DIVISION	30 CITY AVERAGE MAT.	INST.	TOTAL	ANNISTON 362 MAT.	INST.	TOTAL	BIRMINGHAM 350-352 MAT.	INST.	TOTAL	BUTLER 369 MAT.	INST.	TOTAL	DECATUR 356 MAT.	INST.	TOTAL	DOTHAN 363 MAT.	INST.	TOTAL
015433	CONTRACTOR EQUIPMENT		100.0	100.0		100.5	100.5		102.7	102.7		98.2	98.2		100.5	100.5		98.2	98.2
0241, 31 - 34	SITE & INFRASTRUCTURE, DEMOLITION	100.0	100.0	100.0	89.8	87.0	87.9	88.9	90.9	90.3	101.8	83.2	89.0	81.6	86.8	85.1	99.5	83.1	88.2
0310	Concrete Forming & Accessories	100.0	100.0	100.0	88.9	66.7	70.0	94.3	67.4	71.4	84.5	67.1	71.4	94.5	65.7	70.0	94.3	67.3	71.3
0320	Concrete Reinforcing	100.0	100.0	100.0	88.6	69.5	79.4	92.3	69.5	81.3	93.7	71.5	83.0	86.4	68.4	77.7	93.7	69.6	82.1
0330	Cast-in-Place Concrete	100.0	100.0	100.0	83.3	67.0	77.1	101.0	68.9	88.9	81.2	67.3	76.0	93.8	67.0	83.7	81.2	67.3	76.0
03	CONCRETE	100.0	100.0	100.0	91.4	69.0	81.4	97.0	69.8	84.8	92.5	70.5	82.6	90.0	68.3	80.3	91.8	69.4	81.7
04	MASONRY	100.0	100.0	100.0	96.7	62.5	75.8	89.8	62.6	73.2	101.5	62.5	77.7	87.1	62.6	72.1	102.9	62.5	78.3
05	METALS	100.0	100.0	100.0	101.4	94.5	99.3	98.7	92.8	96.8	100.3	95.9	99.0	100.9	93.8	98.7	100.4	95.1	98.8
06	WOOD, PLASTICS & COMPOSITES	100.0	100.0	100.0	85.3	67.5	76.0	93.4	67.7	80.0	79.6	70.1	74.6	97.4	65.6	80.7	92.5	67.5	79.4
07	THERMAL & MOISTURE PROTECTION	100.0	100.0	100.0	95.0	62.3	80.9	94.7	66.1	82.4	95.0	65.7	82.4	93.6	65.3	81.5	95.0	64.7	82.0
08	OPENINGS	100.0	100.0	100.0	94.4	68.2	88.0	102.3	68.7	94.1	94.5	70.6	88.6	106.7	67.3	97.1	94.5	68.6	88.2
0920	Plaster & Gypsum Board	100.0	100.0	100.0	86.0	67.2	73.7	97.5	67.2	77.6	84.0	69.9	74.8	96.9	65.2	76.1	93.4	67.2	76.2
0950, 0980	Ceilings & Acoustic Treatment	100.0	100.0	100.0	85.8	67.2	74.1	96.2	67.2	78.0	85.8	69.9	75.8	94.3	65.2	76.1	85.8	67.2	74.1
0960	Flooring	100.0	100.0	100.0	83.5	68.4	79.1	95.8	68.4	87.8	88.2	68.4	82.4	89.0	68.4	83.0	93.3	68.4	86.0
0970, 0990	Wall Finishes & Painting/Coating	100.0	100.0	100.0	87.4	53.7	67.2	91.6	53.7	68.9	87.4	45.5	62.3	81.3	60.5	68.8	87.4	78.1	81.8
09	FINISHES	100.0	100.0	100.0	84.4	65.7	74.3	94.3	65.9	78.9	87.4	66.3	76.0	89.3	65.3	76.3	89.9	68.4	78.3
COVERS	DIVS. 10 - 14, 25, 28, 41, 43, 44, 46	100.0	100.0	100.0	100.0	84.1	96.2	100.0	84.7	96.4	100.0	85.2	96.5	100.0	83.9	96.2	100.0	85.0	96.5
21, 22, 23	FIRE SUPPRESSION, PLUMBING & HVAC	100.0	100.0	100.0	102.3	51.9	82.0	100.8	65.0	86.4	98.5	64.4	84.7	100.8	64.9	86.4	98.5	63.8	84.5
26, 27, 3370	ELECTRICAL, COMMUNICATIONS & UTIL.	100.0	100.0	100.0	99.2	57.7	78.6	98.6	65.4	82.2	101.0	57.8	79.7	94.2	65.2	79.9	99.9	73.3	86.7
MF2018	WEIGHTED AVERAGE	100.0	100.0	100.0	97.1	66.6	83.9	98.2	70.9	86.4	97.1	69.7	85.2	96.9	70.2	85.4	97.2	71.6	86.1

ALABAMA

	DIVISION	EVERGREEN 364 MAT.	INST.	TOTAL	GADSDEN 359 MAT.	INST.	TOTAL	HUNTSVILLE 357-358 MAT.	INST.	TOTAL	JASPER 355 MAT.	INST.	TOTAL	MOBILE 365-366 MAT.	INST.	TOTAL	MONTGOMERY 360-361 MAT.	INST.	TOTAL
015433	CONTRACTOR EQUIPMENT		98.2	98.2		100.5	100.5		100.5	100.5		100.5	100.5		98.2	98.2		103.5	103.5
0241, 31 - 34	SITE & INFRASTRUCTURE, DEMOLITION	102.3	83.2	89.1	87.5	87.0	87.2	81.3	86.7	85.0	87.4	87.0	87.1	95.1	83.2	86.9	93.5	92.2	92.6
0310	Concrete Forming & Accessories	80.9	67.0	69.0	86.0	67.2	70.0	94.4	65.3	69.7	91.7	67.1	70.8	93.6	67.0	71.0	97.8	67.4	71.9
0320	Concrete Reinforcing	93.8	71.5	83.0	91.6	69.5	80.9	86.4	72.8	79.8	86.4	69.5	78.2	91.3	71.5	81.8	99.5	69.6	85.1
0330	Cast-in-Place Concrete	81.2	67.2	75.9	93.8	67.2	83.8	91.3	66.7	82.1	104.0	67.2	90.1	85.2	67.2	78.4	85.1	68.9	79.0
03	CONCRETE	92.9	69.5	82.4	94.3	69.3	83.1	88.8	68.8	79.9	97.7	69.3	84.9	87.9	69.6	79.7	90.8	69.8	81.4
04	MASONRY	101.5	62.5	77.7	85.8	62.5	71.6	88.5	62.0	72.3	83.5	62.5	70.7	100.5	62.5	77.3	99.7	62.6	77.0
05	METALS	100.4	95.7	98.9	98.7	95.0	97.5	100.9	95.5	99.2	98.6	94.9	97.5	102.5	95.8	100.4	101.5	92.8	98.8
06	WOOD, PLASTICS & COMPOSITES	75.5	67.5	71.3	87.0	67.5	76.8	97.4	65.3	80.6	94.4	67.5	80.3	91.1	67.5	78.8	95.4	67.7	80.9
07	THERMAL & MOISTURE PROTECTION	95.0	64.7	82.0	93.9	65.4	81.6	93.6	65.0	81.3	93.9	65.2	81.6	94.6	64.7	81.7	93.3	66.5	81.8
08	OPENINGS	94.5	69.1	88.3	103.4	68.6	94.9	106.4	68.2	97.1	103.3	68.6	94.9	97.0	69.1	90.2	95.7	68.7	89.2
0920	Plaster & Gypsum Board	83.5	67.2	72.8	89.3	67.2	74.8	96.9	65.0	76.0	93.3	67.2	76.2	90.6	67.2	75.3	93.0	67.2	76.1
0950, 0980	Ceilings & Acoustic Treatment	85.8	67.2	74.1	87.9	67.2	74.9	95.7	65.0	76.4	87.9	67.2	74.9	92.9	67.2	76.8	94.8	67.2	77.5
0960	Flooring	86.1	68.4	80.9	93.8	68.4	86.4	89.0	68.4	83.0	87.4	68.4	81.9	92.8	68.4	85.7	92.5	68.4	85.5
0970, 0990	Wall Finishes & Painting/Coating	87.4	45.5	62.3	81.3	53.7	64.7	81.3	60.6	68.9	81.3	53.7	64.7	90.5	45.5	63.5	92.1	53.7	69.1
09	FINISHES	86.7	64.8	74.9	86.0	65.7	75.0	89.6	65.1	76.3	87.1	65.7	75.5	90.7	64.8	76.7	93.1	65.9	78.3
COVERS	DIVS. 10 - 14, 25, 28, 41, 43, 44, 46	100.0	84.9	96.4	100.0	84.1	96.3	100.0	83.7	96.2	100.0	84.1	96.2	100.0	84.9	96.5	100.0	84.7	96.4
21, 22, 23	FIRE SUPPRESSION, PLUMBING & HVAC	98.5	58.1	82.2	103.7	65.5	88.3	100.8	63.8	85.9	103.7	64.9	88.1	100.6	60.4	84.4	100.6	63.3	85.6
26, 27, 3370	ELECTRICAL, COMMUNICATIONS & UTIL.	98.6	57.8	78.4	94.2	65.4	80.0	95.0	63.4	79.4	93.9	57.7	76.0	101.6	60.3	81.2	101.5	73.2	87.5
MF2018	WEIGHTED AVERAGE	96.8	67.8	84.3	97.2	70.8	85.8	96.9	69.9	85.2	97.6	69.6	85.5	97.8	68.7	85.2	98.0	71.8	86.6

ALABAMA / ALASKA

	DIVISION	PHENIX CITY 368 MAT.	INST.	TOTAL	SELMA 367 MAT.	INST.	TOTAL	TUSCALOOSA 354 MAT.	INST.	TOTAL	ANCHORAGE 995-996 MAT.	INST.	TOTAL	FAIRBANKS 997 MAT.	INST.	TOTAL	JUNEAU 998 MAT.	INST.	TOTAL
015433	CONTRACTOR EQUIPMENT		98.2	98.2		98.2	98.2		100.5	100.5		107.6	107.6		111.2	111.2		107.6	107.6
0241, 31 - 34	SITE & INFRASTRUCTURE, DEMOLITION	105.9	83.1	90.2	99.3	83.1	88.2	81.8	87.0	85.4	124.0	115.8	118.4	123.2	121.2	121.8	139.7	115.8	123.3
0310	Concrete Forming & Accessories	88.7	68.6	71.6	85.9	67.2	70.0	94.4	67.1	71.2	112.9	113.4	113.3	120.7	112.7	113.9	124.0	113.4	115.0
0320	Concrete Reinforcing	93.7	68.2	81.4	93.7	69.5	82.1	86.4	69.5	78.2	143.1	118.2	131.1	145.0	118.2	132.1	146.2	118.2	132.7
0330	Cast-in-Place Concrete	81.2	66.9	75.8	81.2	67.2	75.9	95.1	67.2	84.6	117.0	114.9	116.2	116.0	112.4	114.6	116.9	114.9	116.2
03	CONCRETE	95.8	66.9	84.1	91.3	69.3	81.4	90.6	69.3	81.0	112.6	114.1	113.3	105.1	113.1	108.7	119.7	114.1	117.2
04	MASONRY	101.5	61.9	77.3	105.9	62.5	79.4	87.4	62.5	72.2	185.2	116.7	143.4	187.5	115.0	143.3	171.5	116.7	138.0
05	METALS	100.3	94.6	98.5	100.3	95.0	98.7	100.1	95.0	98.5	126.3	103.8	119.3	122.7	105.5	117.3	121.4	103.8	115.9
06	WOOD, PLASTICS & COMPOSITES	84.9	69.6	76.9	81.5	67.5	74.2	97.4	67.5	81.7	102.3	111.1	106.9	117.3	111.0	114.0	117.0	111.1	113.9
07	THERMAL & MOISTURE PROTECTION	95.4	65.4	82.5	94.8	65.3	82.2	93.7	65.4	81.5	177.8	113.6	150.2	188.1	112.6	155.7	191.7	113.6	158.2
08	OPENINGS	94.4	69.4	88.3	94.4	68.6	88.1	106.4	68.6	97.2	131.0	113.6	126.8	133.2	112.9	128.2	131.7	113.6	127.3
0920	Plaster & Gypsum Board	87.4	69.4	75.6	85.4	67.2	73.5	96.9	67.2	77.4	139.6	111.3	121.0	165.5	111.3	129.9	141.3	111.3	121.6
0950, 0980	Ceilings & Acoustic Treatment	85.8	69.4	75.6	85.8	67.2	74.1	95.7	67.2	77.8	120.9	111.3	114.9	107.6	111.3	109.9	116.2	111.3	113.1
0960	Flooring	90.0	68.4	83.7	88.5	68.4	82.7	89.0	68.4	83.0	114.5	116.7	115.1	117.5	116.7	117.3	126.7	116.7	123.8
0970, 0990	Wall Finishes & Painting/Coating	87.4	79.8	82.8	87.4	53.7	67.2	81.3	53.7	64.7	108.6	117.6	114.0	109.0	113.4	111.6	108.9	117.6	114.1
09	FINISHES	88.8	69.6	78.4	87.4	65.7	75.6	89.6	65.7	76.7	123.7	114.5	118.7	125.5	113.5	119.0	125.5	114.5	119.5
COVERS	DIVS. 10 - 14, 25, 28, 41, 43, 44, 46	100.0	84.8	96.4	100.0	84.1	96.3	100.0	84.1	96.3	100.0	110.1	102.4	100.0	109.3	102.2	100.0	110.1	102.4
21, 22, 23	FIRE SUPPRESSION, PLUMBING & HVAC	98.5	63.0	84.2	98.5	63.8	84.5	100.8	64.3	86.1	101.2	104.1	102.4	101.2	106.5	103.3	101.8	104.1	102.7
26, 27, 3370	ELECTRICAL, COMMUNICATIONS & UTIL.	100.5	68.2	84.5	99.6	73.3	86.6	94.7	65.4	80.2	115.8	107.9	111.9	125.5	107.9	116.8	109.2	107.9	108.5
MF2018	WEIGHTED AVERAGE	97.7	70.8	86.1	96.9	71.2	85.8	96.9	70.5	85.5	119.7	110.4	115.7	120.2	111.0	116.2	119.7	110.4	115.7

ALASKA / ARIZONA

		ALASKA			ARIZONA													
		KETCHIKAN			CHAMBERS			FLAGSTAFF			GLOBE			KINGMAN			MESA/TEMPE		
	DIVISION	999			865			860			855			864			852		
		MAT.	INST.	TOTAL	MAT.	INST.	TOTAL	MAT.	INST.	TOTAL	MAT.	INST.	TOTAL	MAT.	INST.	TOTAL	MAT.	INST.	TOTAL
015433	CONTRACTOR EQUIPMENT		111.2	111.2		85.9	85.9		85.9	85.9		87.5	87.5		85.9	85.9		87.5	87.5
0241, 31 - 34	SITE & INFRASTRUCTURE, DEMOLITION	176.0	121.3	138.4	69.8	87.5	82.0	89.3	87.6	88.2	103.1	88.7	93.2	69.8	87.6	82.0	93.2	88.9	90.2
0310	Concrete Forming & Accessories	112.4	113.3	113.2	96.6	68.8	73.0	102.5	68.9	74.0	86.2	69.0	71.6	94.7	64.9	69.3	89.0	70.1	72.9
0320	Concrete Reinforcing	108.2	118.2	113.0	99.5	73.6	87.0	99.3	73.6	86.9	104.3	73.6	89.5	99.6	73.6	87.0	105.0	73.6	89.8
0330	Cast-in-Place Concrete	234.6	113.8	189.0	88.4	67.6	80.6	88.5	67.8	80.7	79.6	67.4	75.0	88.1	67.8	80.4	80.3	67.6	75.5
03	CONCRETE	177.3	113.8	148.8	93.4	69.2	82.6	114.1	69.3	94.0	98.0	69.3	85.1	93.1	67.5	81.6	89.7	69.9	80.8
04	MASONRY	194.7	116.6	147.1	94.9	60.0	73.6	95.0	60.7	74.1	100.5	59.9	75.8	94.9	60.1	73.6	100.7	60.0	75.9
05	METALS	122.8	105.4	117.4	101.2	70.2	91.5	101.7	70.7	92.1	102.3	71.4	92.7	101.9	70.6	92.2	102.7	71.8	93.0
06	WOOD, PLASTICS & COMPOSITES	107.8	111.0	109.5	97.3	70.4	83.3	104.0	70.4	86.4	80.2	70.6	75.2	92.3	65.0	78.0	83.8	72.0	77.6
07	THERMAL & MOISTURE PROTECTION	194.3	112.5	159.2	99.3	70.0	86.7	101.2	70.4	88.0	99.0	68.7	86.0	99.2	69.6	86.5	98.7	69.1	86.0
08	OPENINGS	128.9	113.5	125.2	105.8	66.8	96.3	105.9	70.0	97.2	95.3	66.9	88.4	106.0	65.4	96.1	95.4	67.9	88.7
0920	Plaster & Gypsum Board	148.8	111.3	124.2	94.9	70.0	78.6	98.4	70.0	79.8	85.5	70.0	75.3	87.4	64.4	72.3	88.9	71.5	77.5
0950, 0980	Ceilings & Acoustic Treatment	101.1	111.3	107.5	116.5	70.0	87.4	117.1	70.0	87.6	103.7	70.0	82.6	117.1	64.4	84.1	103.7	71.5	83.5
0960	Flooring	117.2	116.7	117.1	89.1	61.5	81.1	91.6	62.7	83.2	100.0	61.5	88.7	87.7	61.5	80.1	101.7	62.7	90.3
0970, 0990	Wall Finishes & Painting/Coating	109.0	117.6	114.1	84.7	56.9	68.0	84.7	56.9	68.0	86.9	56.9	68.9	84.7	56.9	68.0	86.9	56.9	68.9
09	FINISHES	125.7	114.4	119.6	95.2	66.0	79.4	98.3	66.3	80.9	98.1	66.2	80.8	93.8	62.8	77.1	97.9	67.2	81.3
COVERS	DIVS. 10 - 14, 25, 28, 41, 43, 44, 46	100.0	109.8	102.3	100.0	84.0	96.2	100.0	84.0	96.2	100.0	84.4	96.3	100.0	83.4	96.1	100.0	84.5	96.4
21, 22, 23	FIRE SUPPRESSION, PLUMBING & HVAC	98.6	104.1	100.8	98.5	75.7	89.3	101.4	76.1	91.2	96.3	75.8	88.0	98.5	76.1	89.5	101.5	75.8	91.1
26, 27, 3370	ELECTRICAL, COMMUNICATIONS & UTIL.	125.5	107.9	116.8	101.5	64.1	83.0	100.6	60.8	80.9	97.3	64.6	81.1	101.5	64.6	83.2	95.4	64.6	80.1
MF2018	WEIGHTED AVERAGE	130.1	110.9	121.8	98.2	70.4	86.2	102.4	70.4	88.6	98.2	70.7	86.3	98.2	69.8	85.9	98.0	71.1	86.4

ARIZONA / ARKANSAS

		ARIZONA												ARKANSAS					
		PHOENIX			PRESCOTT			SHOW LOW			TUCSON			BATESVILLE			CAMDEN		
	DIVISION	850,853			863			859			856 - 857			725			717		
		MAT.	INST.	TOTAL	MAT.	INST.	TOTAL	MAT.	INST.	TOTAL	MAT.	INST.	TOTAL	MAT.	INST.	TOTAL	MAT.	INST.	TOTAL
015433	CONTRACTOR EQUIPMENT		94.1	94.1		85.9	85.9		87.5	87.5		87.5	87.5		86.8	86.8		86.8	86.8
0241, 31 - 34	SITE & INFRASTRUCTURE, DEMOLITION	92.8	96.1	95.0	77.0	87.7	84.3	105.4	88.7	94.0	88.6	88.9	88.8	71.6	82.8	79.3	77.4	82.8	81.1
0310	Concrete Forming & Accessories	92.8	70.9	74.1	98.4	71.0	75.1	92.7	69.0	72.6	89.4	77.4	79.2	85.5	59.9	63.7	77.9	60.4	63.0
0320	Concrete Reinforcing	99.7	73.6	87.1	99.3	73.6	86.9	105.0	73.6	89.8	87.1	73.7	80.6	83.0	67.4	75.5	91.3	68.5	80.3
0330	Cast-in-Place Concrete	84.5	69.9	79.0	88.4	67.9	80.7	79.7	67.4	75.0	82.6	67.6	77.0	68.7	73.2	70.4	74.9	73.3	74.3
03	CONCRETE	92.6	71.0	82.9	99.0	70.3	86.1	100.2	69.3	86.3	88.2	73.2	81.5	71.2	66.6	69.1	75.0	67.0	71.4
04	MASONRY	93.7	60.6	73.5	95.0	60.1	73.7	100.6	59.9	75.8	87.5	60.6	71.1	96.4	61.0	74.8	102.4	61.8	77.6
05	METALS	104.2	72.4	94.3	101.8	70.9	92.2	102.1	71.4	92.5	103.4	72.0	93.6	95.5	75.7	89.4	101.7	76.0	93.7
06	WOOD, PLASTICS & COMPOSITES	88.5	72.3	80.0	98.9	73.2	85.5	87.8	70.6	78.8	84.0	81.8	82.9	92.8	61.3	76.3	86.3	61.8	73.5
07	THERMAL & MOISTURE PROTECTION	99.2	69.9	86.6	99.8	70.5	87.3	99.3	68.7	86.1	99.6	70.3	87.0	104.4	61.1	85.8	98.1	61.8	82.5
08	OPENINGS	101.2	70.6	93.7	105.9	68.3	96.7	94.7	68.5	88.3	92.0	76.3	88.1	99.8	59.4	90.0	103.1	60.0	92.6
0920	Plaster & Gypsum Board	96.9	71.5	80.2	95.1	72.9	80.5	90.9	70.0	77.2	93.2	81.6	85.6	89.0	60.8	70.5	82.7	61.3	68.7
0950, 0980	Ceilings & Acoustic Treatment	116.2	71.5	88.2	115.2	72.9	88.7	103.7	70.0	82.6	105.0	81.6	90.3	92.4	60.8	72.6	90.9	61.3	72.4
0960	Flooring	102.6	62.7	90.9	90.1	64.0	82.5	103.6	61.5	91.3	93.3	65.3	85.1	84.9	70.4	80.7	85.7	71.8	81.7
0970, 0990	Wall Finishes & Painting/Coating	94.3	56.7	71.8	84.7	56.9	68.0	86.9	56.9	68.0	87.5	56.9	69.2	88.6	52.4	66.9	92.7	54.0	69.5
09	FINISHES	102.7	67.7	83.8	95.6	68.2	80.8	100.1	66.2	81.8	96.3	73.5	84.0	83.6	60.9	71.3	83.8	61.7	71.8
COVERS	DIVS. 10 - 14, 25, 28, 41, 43, 44, 46	100.0	85.6	96.6	100.0	84.3	96.3	100.0	84.4	96.3	100.0	85.6	96.6	100.0	78.8	95.0	100.0	78.8	95.0
21, 22, 23	FIRE SUPPRESSION, PLUMBING & HVAC	96.7	76.9	88.7	101.4	75.8	91.1	96.3	75.8	88.0	101.4	73.7	90.2	97.3	51.7	78.9	97.1	57.2	81.0
26, 27, 3370	ELECTRICAL, COMMUNICATIONS & UTIL.	97.6	62.7	80.3	100.4	64.6	82.7	95.1	64.6	80.0	97.2	59.7	78.6	96.2	58.2	77.4	94.1	56.0	75.2
MF2018	WEIGHTED AVERAGE	98.4	72.1	87.0	99.9	71.1	87.5	98.4	70.8	86.5	96.9	71.8	86.1	92.4	62.9	79.6	94.1	64.1	81.1

ARKANSAS

		FAYETTEVILLE			FORT SMITH			HARRISON			HOT SPRINGS			JONESBORO			LITTLE ROCK		
	DIVISION	727			729			726			719			724			720 - 722		
		MAT.	INST.	TOTAL	MAT.	INST.	TOTAL	MAT.	INST.	TOTAL	MAT.	INST.	TOTAL	MAT.	INST.	TOTAL	MAT.	INST.	TOTAL
015433	CONTRACTOR EQUIPMENT		86.8	86.8		86.8	86.8		86.8	86.8		86.8	86.8		109.9	109.9		92.6	92.6
0241, 31 - 34	SITE & INFRASTRUCTURE, DEMOLITION	71.1	82.8	79.1	76.5	82.6	80.7	76.1	82.8	80.7	80.3	82.7	81.9	95.2	99.4	98.1	83.6	91.3	88.9
0310	Concrete Forming & Accessories	80.5	60.4	63.4	103.6	59.9	66.4	90.9	60.3	64.8	75.1	59.6	62.1	89.2	60.3	64.6	101.6	60.5	66.6
0320	Concrete Reinforcing	83.0	65.1	74.4	84.0	64.9	74.8	82.6	68.5	75.8	89.6	68.4	79.4	80.2	66.9	73.8	85.2	67.5	76.7
0330	Cast-in-Place Concrete	68.7	73.3	70.4	78.5	74.2	76.9	76.2	73.2	75.0	76.6	72.6	75.1	74.8	74.2	74.6	76.2	76.1	76.2
03	CONCRETE	70.9	66.4	68.9	78.2	66.5	72.9	77.5	66.9	72.7	78.0	66.5	72.8	75.2	68.0	71.9	79.7	67.7	74.3
04	MASONRY	87.2	61.8	71.7	94.6	60.6	73.9	96.5	61.8	75.3	76.7	58.3	65.5	88.6	60.6	71.5	91.6	61.5	73.3
05	METALS	95.5	74.8	89.1	97.9	74.2	90.5	96.7	75.8	90.2	101.7	75.5	93.5	92.4	90.0	91.7	98.0	74.2	90.6
06	WOOD, PLASTICS & COMPOSITES	88.2	61.8	74.4	114.0	61.8	86.7	99.8	61.8	79.9	83.2	61.8	72.0	97.2	62.1	78.8	107.3	62.0	83.6
07	THERMAL & MOISTURE PROTECTION	105.2	61.8	86.6	105.8	60.9	86.5	104.7	61.8	86.3	98.3	59.8	81.8	110.5	60.9	89.2	100.1	62.4	83.9
08	OPENINGS	99.8	59.1	89.9	101.8	58.6	91.3	100.6	59.9	90.7	103.1	59.5	92.4	105.4	59.8	94.3	95.5	59.1	86.6
0920	Plaster & Gypsum Board	88.4	61.3	70.7	94.6	61.3	72.8	93.7	61.3	72.5	81.0	61.3	68.1	102.6	61.3	75.5	109.0	61.3	77.7
0950, 0980	Ceilings & Acoustic Treatment	92.4	61.3	72.9	94.3	61.3	73.7	94.3	61.3	73.7	90.9	61.3	72.4	97.0	61.3	74.7	97.4	61.3	74.8
0960	Flooring	81.9	71.8	79.0	91.9	70.4	85.6	87.5	71.8	82.9	84.6	67.7	79.7	61.3	70.4	64.0	93.8	80.4	89.9
0970, 0990	Wall Finishes & Painting/Coating	88.6	54.0	67.9	88.6	51.6	66.5	88.6	54.0	67.9	92.7	51.9	68.3	78.4	53.0	63.2	91.9	53.0	68.6
09	FINISHES	82.8	61.7	71.3	87.1	61.0	73.0	85.8	61.7	72.7	83.6	60.3	71.0	82.4	61.4	71.0	95.4	63.4	78.1
COVERS	DIVS. 10 - 14, 25, 28, 41, 43, 44, 46	100.0	78.8	95.0	100.0	78.7	95.0	100.0	78.8	95.0	100.0	78.5	94.9	100.0	77.6	94.7	100.0	79.4	95.1
21, 22, 23	FIRE SUPPRESSION, PLUMBING & HVAC	97.4	61.6	83.0	101.2	48.3	79.9	97.3	49.5	78.0	97.1	49.0	77.7	101.8	50.8	81.2	100.8	48.4	79.7
26, 27, 3370	ELECTRICAL, COMMUNICATIONS & UTIL.	91.2	51.9	71.7	94.0	58.3	76.3	95.0	55.4	75.4	95.8	62.8	79.5	99.8	60.0	80.1	102.3	59.6	81.2
MF2018	WEIGHTED AVERAGE	91.3	64.2	79.6	95.1	61.9	80.8	93.7	62.3	80.1	93.4	62.5	80.1	94.8	65.7	82.2	96.0	63.5	81.9

For customer support on your Site Work & Landscape Costs with RSMeans Data, call 800.448.8182.

ARKANSAS / CALIFORNIA

DIVISION		PINE BLUFF 716 MAT.	INST.	TOTAL	RUSSELLVILLE 728 MAT.	INST.	TOTAL	TEXARKANA 718 MAT.	INST.	TOTAL	WEST MEMPHIS 723 MAT.	INST.	TOTAL	ALHAMBRA 917-918 MAT.	INST.	TOTAL	ANAHEIM 928 MAT.	INST.	TOTAL
015433	CONTRACTOR EQUIPMENT		86.8	86.8		86.8	86.8		88.0	88.0		109.9	109.9		92.5	92.5		97.5	97.5
0241, 31 - 34	SITE & INFRASTRUCTURE, DEMOLITION	82.8	82.6	82.7	72.8	82.6	79.5	93.4	84.6	87.4	101.1	99.2	99.8	99.1	101.8	101.0	100.7	102.6	102.1
0310	Concrete Forming & Accessories	74.8	60.1	62.3	86.6	59.6	63.6	81.6	60.2	63.4	95.6	60.1	65.4	115.5	135.6	132.6	102.0	139.4	133.8
0320	Concrete Reinforcing	91.2	67.4	79.7	83.5	68.3	76.2	90.8	68.5	80.0	80.2	60.4	70.6	98.9	133.2	115.4	93.2	133.1	112.5
0330	Cast-in-Place Concrete	76.6	73.0	75.2	71.9	72.5	72.2	83.5	73.0	79.6	78.4	73.7	76.6	82.3	126.4	99.0	88.3	129.7	104.0
03	CONCRETE	78.7	66.6	73.3	74.0	66.3	70.5	77.3	66.8	72.6	81.4	66.5	74.7	92.6	130.8	109.8	96.9	133.8	113.5
04	MASONRY	109.6	60.6	79.7	93.9	58.3	72.2	89.2	60.8	71.9	77.5	58.2	65.8	113.1	137.8	128.2	77.2	136.8	113.6
05	METALS	102.5	75.3	94.0	95.5	75.4	89.2	94.8	75.8	88.9	91.4	87.4	90.2	79.2	113.8	90.0	107.0	115.7	109.7
06	WOOD, PLASTICS & COMPOSITES	82.8	61.8	71.8	94.6	61.8	77.4	91.2	61.8	75.8	104.4	62.1	82.3	101.6	133.4	118.2	96.3	138.5	118.4
07	THERMAL & MOISTURE PROTECTION	98.4	61.2	82.4	105.5	59.8	85.9	99.1	61.0	82.8	111.1	59.8	89.1	103.4	128.8	114.3	109.4	132.9	119.5
08	OPENINGS	104.3	59.2	93.3	99.8	59.5	90.0	109.1	60.0	97.1	102.9	57.8	91.9	88.1	133.7	98.9	93.0	136.2	111.8
0920	Plaster & Gypsum Board	80.7	61.3	68.0	89.0	61.3	70.8	84.2	61.3	69.2	104.6	61.3	76.2	96.7	134.7	121.6	111.5	139.7	130.0
0950, 0980	Ceilings & Acoustic Treatment	90.9	61.3	72.4	92.4	61.3	72.9	96.6	61.3	74.5	95.6	61.3	74.1	115.1	134.7	127.4	106.8	139.7	127.4
0960	Flooring	84.4	71.8	80.7	84.4	67.7	79.5	86.5	70.8	81.9	63.8	67.7	65.0	104.7	119.9	109.1	98.9	119.9	105.0
0970, 0990	Wall Finishes & Painting/Coating	92.7	51.6	68.1	88.6	51.9	66.6	92.7	54.0	69.5	78.4	51.9	62.5	102.8	119.8	113.0	90.1	119.8	107.9
09	FINISHES	83.5	61.3	71.5	83.7	60.3	71.1	86.5	61.4	72.9	83.7	60.6	71.2	103.9	130.6	118.4	100.3	133.8	118.4
COVERS	DIVS. 10 - 14, 25, 28, 41, 43, 44, 46	100.0	78.7	95.0	100.0	78.5	94.9	100.0	78.7	95.0	100.0	79.3	95.1	100.0	116.6	103.9	100.0	117.7	104.2
21, 22, 23	FIRE SUPPRESSION, PLUMBING & HVAC	101.0	51.7	81.1	97.4	48.5	77.7	101.0	54.9	82.4	98.0	61.5	83.3	96.9	128.4	109.6	100.8	128.4	112.0
26, 27, 3370	ELECTRICAL, COMMUNICATIONS & UTIL.	94.2	57.6	76.1	94.0	55.4	74.9	95.7	57.6	76.9	101.1	62.8	82.1	121.7	131.9	126.7	91.7	114.7	103.1
MF2018	WEIGHTED AVERAGE	96.2	62.8	81.8	92.5	61.3	79.0	95.1	63.7	81.5	94.2	67.5	82.7	97.4	127.0	110.2	99.5	125.9	110.9

CALIFORNIA

DIVISION		BAKERSFIELD 932-933 MAT.	INST.	TOTAL	BERKELEY 947 MAT.	INST.	TOTAL	EUREKA 955 MAT.	INST.	TOTAL	FRESNO 936-938 MAT.	INST.	TOTAL	INGLEWOOD 903-905 MAT.	INST.	TOTAL	LONG BEACH 906-908 MAT.	INST.	TOTAL
015433	CONTRACTOR EQUIPMENT		100.3	100.3		98.8	98.8		95.3	95.3		95.5	95.5		95.6	95.6		95.6	95.6
0241, 31 - 34	SITE & INFRASTRUCTURE, DEMOLITION	100.0	108.7	106.0	115.4	103.4	107.2	112.0	100.4	104.1	102.2	100.9	101.3	87.5	99.7	95.9	94.5	99.7	98.1
0310	Concrete Forming & Accessories	100.8	138.4	132.8	115.8	173.3	164.7	111.2	160.0	152.7	100.3	159.0	150.3	105.7	136.0	131.5	99.8	136.0	130.6
0320	Concrete Reinforcing	95.8	133.1	113.8	87.7	134.6	110.4	101.3	134.7	117.4	80.1	133.7	106.0	99.7	133.2	115.9	98.9	133.2	115.5
0330	Cast-in-Place Concrete	84.7	130.1	101.9	110.8	136.5	120.5	95.9	132.9	109.9	91.7	132.5	107.1	81.8	129.1	99.6	93.1	129.1	106.7
03	CONCRETE	90.9	133.4	109.9	107.8	151.7	127.5	107.5	144.6	124.1	93.6	143.8	116.1	90.5	132.1	109.2	99.9	132.1	114.3
04	MASONRY	91.2	136.3	118.7	120.6	152.8	140.3	99.7	157.9	135.2	98.9	143.6	126.2	72.7	137.9	112.5	81.7	137.9	116.0
05	METALS	101.8	113.6	105.5	107.1	119.4	110.9	106.7	119.1	110.6	102.1	117.7	107.0	87.5	116.3	96.5	87.4	116.3	96.4
06	WOOD, PLASTICS & COMPOSITES	92.7	137.5	116.1	109.5	180.0	146.4	110.1	164.4	138.5	98.7	164.4	133.1	99.0	133.8	117.2	91.7	133.8	113.7
07	THERMAL & MOISTURE PROTECTION	108.0	123.5	114.6	114.9	155.3	132.2	112.9	153.3	130.3	98.9	133.7	113.8	105.2	129.9	115.8	105.5	129.9	116.0
08	OPENINGS	93.1	135.2	103.4	90.7	165.9	109.0	103.3	146.8	113.9	95.0	150.6	108.6	87.5	134.0	98.8	87.4	134.0	98.8
0920	Plaster & Gypsum Board	97.5	138.4	124.4	110.3	181.9	157.3	116.3	166.2	149.0	92.4	166.2	140.8	103.4	134.7	123.9	98.6	134.7	122.3
0950, 0980	Ceilings & Acoustic Treatment	109.8	138.4	127.7	105.7	181.9	153.4	111.3	166.2	145.7	102.6	166.2	142.4	102.5	134.7	122.7	102.5	134.7	122.7
0960	Flooring	99.9	119.9	105.7	121.5	149.0	129.5	102.5	149.0	116.1	99.3	122.1	106.0	113.7	119.9	115.5	110.4	119.9	113.2
0970, 0990	Wall Finishes & Painting/Coating	92.6	108.4	102.1	108.4	168.0	144.1	91.7	135.6	118.0	97.9	122.3	112.5	108.9	119.8	115.4	108.9	119.8	115.4
09	FINISHES	100.1	131.9	117.3	109.8	169.8	142.3	105.5	157.2	133.5	97.8	150.6	126.4	105.1	131.0	119.1	104.0	131.0	118.6
COVERS	DIVS. 10 - 14, 25, 28, 41, 43, 44, 46	100.0	127.0	106.4	100.0	132.5	107.7	100.0	129.6	107.0	100.0	129.5	107.0	100.0	117.6	104.1	100.0	117.6	104.1
21, 22, 23	FIRE SUPPRESSION, PLUMBING & HVAC	100.9	127.1	111.5	97.6	167.8	125.9	97.1	133.2	111.7	100.9	131.4	113.2	96.1	128.5	109.2	96.1	127.5	108.8
26, 27, 3370	ELECTRICAL, COMMUNICATIONS & UTIL.	105.5	109.7	107.6	100.1	155.8	127.7	98.1	130.8	114.3	96.2	108.6	102.3	99.5	131.9	115.5	99.3	131.9	115.4
MF2018	WEIGHTED AVERAGE	99.0	124.8	110.1	103.3	151.9	124.3	102.6	137.6	117.7	98.6	131.2	112.7	93.6	127.4	108.2	95.2	127.2	109.0

CALIFORNIA

DIVISION		LOS ANGELES 900-902 MAT.	INST.	TOTAL	MARYSVILLE 959 MAT.	INST.	TOTAL	MODESTO 953 MAT.	INST.	TOTAL	MOJAVE 935 MAT.	INST.	TOTAL	OAKLAND 946 MAT.	INST.	TOTAL	OXNARD 930 MAT.	INST.	TOTAL
015433	CONTRACTOR EQUIPMENT		104.7	104.7		95.3	95.3		95.3	95.3		95.5	95.5		98.8	98.8		94.5	94.5
0241, 31 - 34	SITE & INFRASTRUCTURE, DEMOLITION	94.0	110.4	105.3	108.4	100.0	102.6	103.7	100.4	101.4	94.8	100.6	98.8	121.8	103.4	109.2	102.4	98.8	99.9
0310	Concrete Forming & Accessories	103.2	140.0	134.5	100.4	159.0	150.2	96.1	159.3	149.9	109.7	138.1	133.8	103.2	173.4	162.9	101.9	139.4	133.8
0320	Concrete Reinforcing	98.1	131.4	114.2	101.3	133.5	116.9	105.0	133.8	118.9	96.3	133.0	114.0	89.7	137.2	112.6	94.5	133.1	113.1
0330	Cast-in-Place Concrete	84.6	131.3	102.2	107.1	132.4	116.7	95.9	132.6	109.7	79.3	128.8	98.0	105.1	136.5	117.0	92.4	129.1	106.2
03	CONCRETE	96.2	134.4	113.3	108.1	143.7	124.0	99.1	143.9	119.2	85.7	132.9	106.9	108.9	152.2	128.3	93.5	133.6	111.5
04	MASONRY	88.1	138.0	118.6	100.7	141.0	125.3	99.7	144.6	127.1	97.0	136.3	120.9	129.2	152.8	143.6	99.8	136.5	122.2
05	METALS	95.8	116.4	102.2	106.2	116.6	109.4	102.9	117.2	107.4	99.4	114.8	104.2	102.5	120.4	108.1	97.3	115.4	103.0
06	WOOD, PLASTICS & COMPOSITES	107.3	139.0	123.9	96.4	164.4	132.0	91.2	164.4	129.5	98.7	137.4	118.9	95.6	180.0	139.8	93.3	138.6	117.0
07	THERMAL & MOISTURE PROTECTION	101.7	132.5	114.9	112.3	137.8	123.3	111.9	139.8	123.9	104.8	121.8	112.1	113.2	155.5	131.3	108.1	131.7	118.2
08	OPENINGS	98.2	137.6	107.8	102.6	148.4	113.8	101.4	150.9	113.4	90.3	135.2	101.3	90.8	166.5	109.2	92.8	136.3	103.4
0920	Plaster & Gypsum Board	98.4	139.7	125.5	109.4	166.2	146.6	111.5	166.2	147.4	102.3	138.4	126.0	104.5	181.9	155.3	96.0	139.7	124.7
0950, 0980	Ceilings & Acoustic Treatment	109.6	139.7	128.5	110.6	166.2	145.4	106.8	166.2	144.0	104.9	138.4	125.9	107.7	181.9	154.2	104.7	139.7	126.6
0960	Flooring	110.9	119.9	113.5	98.1	114.0	102.7	98.4	133.0	108.5	100.7	115.5	105.0	114.4	149.0	124.5	93.1	119.9	100.9
0970, 0990	Wall Finishes & Painting/Coating	108.1	119.8	115.1	91.7	135.6	118.0	91.7	135.6	118.0	84.5	108.8	99.0	108.4	168.0	144.1	84.5	114.6	102.5
09	FINISHES	105.8	134.1	121.2	102.6	150.7	128.6	101.6	154.0	129.9	98.7	131.1	116.3	107.9	170.0	141.5	95.9	133.3	116.1
COVERS	DIVS. 10 - 14, 25, 28, 41, 43, 44, 46	100.0	118.9	104.5	100.0	129.6	107.0	100.0	129.6	107.0	100.0	113.0	103.1	100.0	132.5	107.7	100.0	117.9	104.2
21, 22, 23	FIRE SUPPRESSION, PLUMBING & HVAC	100.4	128.6	111.8	97.1	126.1	108.8	100.8	131.4	113.2	97.0	127.0	109.1	101.3	167.9	128.2	100.8	128.5	112.0
26, 27, 3370	ELECTRICAL, COMMUNICATIONS & UTIL.	98.1	131.9	114.8	95.3	116.3	105.6	97.4	110.2	103.7	94.8	109.7	102.2	99.4	163.0	130.9	100.6	117.4	108.9
MF2018	WEIGHTED AVERAGE	98.4	129.3	111.8	101.8	130.7	114.3	100.8	132.1	114.3	95.6	123.6	107.7	103.9	153.2	125.2	98.2	125.8	110.1

CALIFORNIA

DIVISION		PALM SPRINGS 922			PALO ALTO 943			PASADENA 910 - 912			REDDING 960			RICHMOND 948			RIVERSIDE 925		
		MAT.	INST.	TOTAL	MAT.	INST.	TOTAL	MAT.	INST.	TOTAL	MAT.	INST.	TOTAL	MAT.	INST.	TOTAL	MAT.	INST.	TOTAL
015433	CONTRACTOR EQUIPMENT		96.3	96.3		98.8	98.8		92.5	92.5		95.3	95.3		98.8	98.8		96.3	96.3
0241, 31 - 34	SITE & INFRASTRUCTURE, DEMOLITION	91.8	100.8	98.0	111.6	103.4	106.0	95.9	101.8	100.0	124.5	100.2	107.8	120.5	103.4	108.8	99.2	100.8	100.3
0310	Concrete Forming & Accessories	98.4	132.1	127.0	101.1	163.9	154.6	102.8	135.6	130.7	102.8	155.1	147.3	119.0	173.0	165.0	102.4	139.4	133.9
0320	Concrete Reinforcing	106.5	132.9	119.3	87.7	136.9	111.4	99.8	133.2	115.9	134.5	133.8	134.2	87.7	136.9	111.4	103.5	133.1	117.8
0330	Cast-in-Place Concrete	84.2	129.6	101.3	93.7	136.4	109.8	78.1	126.4	96.3	117.0	132.6	122.9	107.8	136.4	118.6	91.6	129.7	106.0
03	CONCRETE	91.3	130.4	108.9	97.6	147.8	120.2	88.3	130.8	107.4	118.6	142.0	129.1	110.9	152.0	129.3	97.3	133.8	113.7
04	MASONRY	75.2	134.1	111.1	102.6	149.2	131.0	99.1	137.8	122.7	120.2	144.6	135.1	120.4	149.2	138.0	76.2	136.5	113.0
05	METALS	107.6	115.2	110.0	100.0	119.7	106.2	79.2	113.8	90.0	103.2	117.3	107.6	100.1	119.8	106.2	107.0	115.7	109.7
06	WOOD, PLASTICS & COMPOSITES	90.6	129.0	110.7	92.8	167.7	132.0	86.2	133.4	110.9	104.7	158.6	132.9	113.2	180.0	148.2	96.3	138.5	118.4
07	THERMAL & MOISTURE PROTECTION	108.7	129.3	117.5	112.4	149.9	128.5	102.9	128.3	113.8	126.7	139.9	132.4	113.1	152.8	130.1	109.6	132.8	119.6
08	OPENINGS	100.0	131.3	107.6	90.8	155.9	106.7	88.1	133.7	99.2	116.4	147.7	124.0	114.7	155.1	126.1	106.2	136.2	110.8
0920	Plaster & Gypsum Board	106.1	129.9	121.7	102.8	169.2	146.4	91.4	134.7	119.8	120.9	160.2	146.7	110.7	181.9	157.4	110.6	139.7	129.7
0950, 0980	Ceilings & Acoustic Treatment	103.6	129.9	120.1	106.3	169.2	145.7	115.1	134.7	127.4	138.0	160.2	151.9	106.3	181.9	153.7	111.5	139.7	129.1
0960	Flooring	101.0	115.5	105.2	112.9	149.0	123.5	98.2	119.9	104.6	88.8	126.7	99.9	124.1	149.0	131.3	102.4	119.9	107.5
0970, 0990	Wall Finishes & Painting/Coating	88.7	114.6	104.2	108.4	168.0	144.1	102.8	119.8	113.0	101.6	135.6	122.0	108.4	168.0	144.1	88.7	119.8	107.3
09	FINISHES	98.7	126.7	113.9	106.3	162.6	136.7	101.1	130.7	117.1	108.9	149.7	131.0	111.2	169.8	142.9	102.0	133.8	119.2
COVERS	DIVS. 10 - 14, 25, 28, 41, 43, 44, 46	100.0	116.7	103.9	100.0	131.2	107.4	100.0	116.6	103.9	100.0	129.0	106.8	100.0	132.5	107.6	100.0	120.4	104.8
21, 22, 23	FIRE SUPPRESSION, PLUMBING & HVAC	97.1	125.9	108.7	97.6	164.8	124.7	96.9	128.4	109.6	101.1	131.4	113.3	97.6	162.4	123.7	100.8	128.4	112.0
26, 27, 3370	ELECTRICAL, COMMUNICATIONS & UTIL.	94.4	106.7	100.5	99.3	170.0	134.3	118.7	131.9	125.2	99.5	121.7	110.5	99.8	140.3	119.8	91.5	116.0	103.6
MF2018	WEIGHTED AVERAGE	97.3	121.9	108.0	99.4	150.8	121.6	95.4	127.0	109.0	107.8	132.6	118.6	102.8	148.2	122.4	99.4	126.0	110.9

CALIFORNIA

DIVISION		SACRAMENTO 942,956 - 958			SALINAS 939			SAN BERNARDINO 923 - 924			SAN DIEGO 919 - 921			SAN FRANCISCO 940 - 941			SAN JOSE 951		
		MAT.	INST.	TOTAL	MAT.	INST.	TOTAL	MAT.	INST.	TOTAL	MAT.	INST.	TOTAL	MAT.	INST.	TOTAL	MAT.	INST.	TOTAL
015433	CONTRACTOR EQUIPMENT		97.6	97.6		95.5	95.5		96.3	96.3		103.0	103.0		111.2	111.2		97.0	97.0
0241, 31 - 34	SITE & INFRASTRUCTURE, DEMOLITION	99.6	108.5	105.7	115.6	100.9	105.5	78.2	100.8	93.7	104.5	109.5	107.9	121.3	114.9	116.9	135.4	95.7	108.2
0310	Concrete Forming & Accessories	101.9	157.3	149.1	105.9	157.9	150.1	106.5	139.4	134.5	104.0	129.4	125.6	107.4	168.5	159.4	102.5	173.6	163.0
0320	Concrete Reinforcing	83.3	133.8	107.7	95.0	134.3	113.9	103.5	133.1	117.8	96.0	131.1	112.9	100.8	135.5	117.5	92.7	137.4	114.3
0330	Cast-in-Place Concrete	88.9	133.2	105.6	91.2	132.9	107.0	63.3	129.7	88.4	89.5	125.9	103.3	117.4	133.9	123.6	111.4	135.9	120.6
03	CONCRETE	100.4	143.0	119.5	103.5	143.5	121.5	72.1	133.8	99.8	97.1	127.6	110.8	120.5	149.1	133.3	105.0	152.5	126.3
04	MASONRY	103.5	144.6	128.6	96.9	150.1	129.4	83.0	136.5	115.6	90.7	132.6	116.3	136.8	153.9	147.2	128.4	156.1	145.3
05	METALS	98.2	112.2	102.6	102.0	118.3	107.1	107.0	115.7	109.7	95.8	115.0	101.8	107.2	124.6	112.7	101.1	125.7	108.8
06	WOOD, PLASTICS & COMPOSITES	90.8	161.6	127.8	98.7	161.4	131.5	100.4	138.5	120.3	95.5	126.9	111.9	101.5	174.3	139.5	103.1	179.8	143.2
07	THERMAL & MOISTURE PROTECTION	122.2	141.7	130.6	105.5	145.8	122.8	107.8	132.8	118.6	108.1	120.8	113.5	119.0	156.1	134.9	108.5	160.6	130.8
08	OPENINGS	103.6	149.3	114.7	93.9	155.7	108.9	100.0	137.8	109.2	100.1	127.5	106.8	99.4	162.2	114.7	92.6	164.4	110.6
0920	Plaster & Gypsum Board	98.9	163.1	141.0	97.2	163.1	140.4	112.3	139.7	130.3	97.5	127.4	117.1	100.4	175.9	149.9	107.1	181.9	156.2
0950, 0980	Ceilings & Acoustic Treatment	106.3	163.1	141.9	104.9	163.1	141.4	106.8	139.7	127.4	126.2	127.4	127.0	119.8	175.9	154.9	107.6	181.9	154.2
0960	Flooring	112.7	126.7	116.8	95.0	142.9	109.0	104.6	119.9	109.0	108.0	119.9	111.5	111.2	142.9	120.5	92.1	149.0	108.7
0970, 0990	Wall Finishes & Painting/Coating	105.2	135.6	123.4	85.4	168.0	134.9	88.7	119.8	107.3	103.5	119.8	113.3	109.1	176.8	149.7	92.0	168.0	137.5
09	FINISHES	104.6	151.3	129.9	98.3	157.7	130.4	100.2	133.8	118.4	108.8	126.5	118.4	109.1	165.9	139.8	100.9	170.0	138.3
COVERS	DIVS. 10 - 14, 25, 28, 41, 43, 44, 46	100.0	129.8	107.0	100.0	129.3	106.9	100.0	117.7	104.2	100.0	116.1	103.8	100.0	128.9	106.8	100.0	132.1	107.6
21, 22, 23	FIRE SUPPRESSION, PLUMBING & HVAC	101.2	130.5	113.0	97.0	136.9	113.1	97.1	128.4	109.7	100.7	127.0	111.3	101.2	179.4	132.8	100.8	173.4	130.1
26, 27, 3370	ELECTRICAL, COMMUNICATIONS & UTIL.	95.3	121.6	108.3	95.7	132.6	114.0	94.4	111.9	103.0	105.3	103.3	104.3	100.0	186.3	142.7	100.5	188.8	144.2
MF2018	WEIGHTED AVERAGE	101.0	133.1	114.9	99.2	137.8	115.9	95.0	125.4	108.1	100.3	121.3	109.4	107.6	159.0	129.8	103.0	158.4	126.9

CALIFORNIA

DIVISION		SAN LUIS OBISPO 934			SAN MATEO 944			SAN RAFAEL 949			SANTA ANA 926 - 927			SANTA BARBARA 931			SANTA CRUZ 950		
		MAT.	INST.	TOTAL	MAT.	INST.	TOTAL	MAT.	INST.	TOTAL	MAT.	INST.	TOTAL	MAT.	INST.	TOTAL	MAT.	INST.	TOTAL
015433	CONTRACTOR EQUIPMENT		95.5	95.5		98.8	98.8		98.0	98.0		96.3	96.3		95.5	95.5		97.0	97.0
0241, 31 - 34	SITE & INFRASTRUCTURE, DEMOLITION	107.4	100.6	102.7	118.1	103.5	108.1	111.8	108.2	109.4	90.2	100.8	97.5	102.4	100.6	101.2	135.0	95.6	107.9
0310	Concrete Forming & Accessories	111.6	139.5	135.4	108.1	173.7	163.9	114.7	173.0	164.3	107.0	139.4	134.5	102.5	139.4	133.9	102.6	158.0	149.7
0320	Concrete Reinforcing	96.3	133.1	114.0	87.7	137.6	111.8	88.4	134.8	110.8	107.1	133.1	119.7	94.5	133.1	113.1	114.4	134.3	124.0
0330	Cast-in-Place Concrete	98.1	129.0	109.8	104.3	136.7	116.5	121.4	135.5	126.7	80.8	129.7	99.3	92.0	129.0	106.0	110.7	134.6	119.7
03	CONCRETE	101.1	133.6	115.7	107.3	152.5	127.6	131.4	151.0	140.2	88.9	133.8	109.1	93.4	133.6	111.4	107.6	144.4	124.1
04	MASONRY	98.5	135.2	120.9	120.1	159.0	143.8	97.5	155.8	133.0	72.5	136.8	111.8	97.2	135.2	120.4	132.2	150.2	143.2
05	METALS	100.1	115.4	104.9	99.9	121.2	106.5	101.2	115.6	105.7	107.0	115.7	109.7	97.9	115.3	103.3	108.6	122.6	112.9
06	WOOD, PLASTICS & COMPOSITES	100.9	138.6	120.6	101.7	180.0	142.7	101.5	179.8	142.5	102.4	138.5	121.3	93.3	138.6	117.0	103.1	161.5	133.6
07	THERMAL & MOISTURE PROTECTION	105.6	130.8	116.4	112.8	160.0	133.1	117.9	155.8	134.2	109.1	132.9	119.3	105.0	131.0	116.2	108.0	148.5	125.4
08	OPENINGS	92.1	135.8	102.8	90.8	166.5	109.2	101.3	165.1	116.8	99.3	136.2	108.3	93.6	137.8	104.4	93.9	155.7	108.9
0920	Plaster & Gypsum Board	102.8	139.7	127.0	108.7	181.9	156.7	111.1	181.9	157.5	113.4	139.7	130.7	96.0	139.7	124.7	114.8	163.1	146.5
0950, 0980	Ceilings & Acoustic Treatment	104.9	139.7	126.7	106.3	181.9	153.7	115.9	181.9	157.2	106.8	139.7	127.4	104.7	139.7	126.6	108.1	163.1	142.6
0960	Flooring	101.5	119.9	106.9	116.9	149.0	126.3	129.4	142.9	133.3	105.2	119.9	109.5	94.1	119.9	101.7	96.2	142.9	109.8
0970, 0990	Wall Finishes & Painting/Coating	84.5	114.6	102.5	108.4	168.0	144.1	104.6	168.0	142.6	88.7	119.8	107.3	84.5	114.6	102.5	108.4	168.0	137.5
09	FINISHES	100.1	133.3	118.0	108.7	170.2	142.0	113.0	168.6	143.1	101.7	133.8	119.0	96.3	133.3	116.3	103.3	157.8	132.8
COVERS	DIVS. 10 - 14, 25, 28, 41, 43, 44, 46	100.0	126.9	106.3	100.0	132.6	107.7	100.0	131.9	107.5	100.0	117.7	104.2	100.0	117.9	104.2	100.0	129.6	107.0
21, 22, 23	FIRE SUPPRESSION, PLUMBING & HVAC	97.0	128.5	109.7	97.6	169.3	126.5	97.6	179.6	130.6	97.1	128.4	109.7	100.8	128.5	112.0	100.8	137.0	115.4
26, 27, 3370	ELECTRICAL, COMMUNICATIONS & UTIL.	94.8	113.6	104.1	99.3	178.8	138.6	96.2	125.7	110.8	94.4	114.7	104.4	93.9	111.7	102.7	99.7	132.6	116.0
MF2018	WEIGHTED AVERAGE	98.4	125.5	110.1	101.9	156.6	125.5	105.0	150.3	124.6	97.1	125.7	109.5	97.4	125.0	109.4	104.9	138.1	119.2

CALIFORNIA / COLORADO

DIVISION		SANTA ROSA 954 MAT.	INST.	TOTAL	STOCKTON 952 MAT.	INST.	TOTAL	SUSANVILLE 961 MAT.	INST.	TOTAL	VALLEJO 945 MAT.	INST.	TOTAL	VAN NUYS 913-916 MAT.	INST.	TOTAL	ALAMOSA 811 MAT.	INST.	TOTAL
015433	CONTRACTOR EQUIPMENT		95.8	95.8		95.3	95.3		95.3	95.3		98.0	98.0		92.5	92.5		87.8	87.8
0241, 31-34	SITE & INFRASTRUCTURE, DEMOLITION	104.1	100.3	101.5	103.4	100.4	101.3	131.6	100.0	109.9	99.9	108.5	105.8	113.5	101.8	105.5	135.8	81.2	98.3
0310	Concrete Forming & Accessories	99.5	172.4	161.5	100.8	157.1	148.7	104.3	150.2	143.4	103.8	172.1	161.9	110.3	135.6	131.8	100.4	68.5	73.3
0320	Concrete Reinforcing	102.2	134.8	117.9	105.0	133.8	118.9	134.5	133.6	134.1	89.5	134.7	111.3	99.8	133.2	115.9	107.3	66.6	87.7
0330	Cast-in-Place Concrete	105.2	134.1	116.1	93.4	132.5	108.2	106.4	132.5	116.3	96.7	134.7	111.1	82.4	126.4	99.0	98.6	70.9	88.1
03	CONCRETE	108.0	150.6	127.1	98.2	142.9	118.3	120.9	139.8	129.4	105.0	150.3	125.3	102.2	130.8	115.0	112.5	69.6	93.2
04	MASONRY	98.8	154.8	132.9	99.7	144.6	127.1	118.6	141.0	132.3	75.8	154.8	124.0	113.1	137.8	128.2	128.3	58.5	85.7
05	METALS	107.4	120.3	111.4	103.2	117.2	107.5	102.2	116.7	106.7	101.1	115.1	105.5	78.4	113.8	89.4	103.2	77.4	95.2
06	WOOD, PLASTICS & COMPOSITES	92.3	179.6	138.0	97.4	161.4	130.9	106.6	152.4	130.6	89.3	179.8	136.6	95.6	133.4	115.4	93.9	71.0	81.9
07	THERMAL & MOISTURE PROTECTION	109.5	155.2	129.1	112.3	138.9	123.7	129.0	136.4	132.2	115.8	155.0	132.6	104.2	128.3	114.5	108.4	66.6	90.5
08	OPENINGS	100.9	165.7	116.7	101.4	149.2	113.0	117.4	141.8	123.3	102.9	164.4	117.9	88.0	133.7	99.1	96.4	69.9	89.9
0920	Plaster & Gypsum Board	108.3	181.9	156.6	111.5	163.1	145.4	121.4	153.8	142.7	105.4	181.9	155.6	94.9	134.7	121.0	83.8	70.3	75.0
0950, 0980	Ceilings & Acoustic Treatment	106.8	181.9	153.9	114.7	163.1	145.0	128.9	153.8	144.5	117.3	181.9	157.8	112.6	134.7	126.4	112.0	70.3	85.9
0960	Flooring	101.5	136.3	111.7	98.4	133.0	108.5	89.3	126.7	100.2	122.7	149.0	130.4	101.6	119.9	107.0	106.8	66.9	95.1
0970, 0990	Wall Finishes & Painting/Coating	88.7	168.0	136.2	91.7	132.0	115.8	101.6	135.6	122.0	105.6	168.0	143.0	102.8	119.8	113.0	97.5	76.1	84.7
09	FINISHES	100.9	167.2	136.8	103.4	151.9	129.6	108.0	145.9	128.5	109.3	169.6	141.9	103.3	130.7	118.1	102.6	69.0	84.4
COVERS	DIVS. 10-14, 25, 28, 41, 43, 44, 46	100.0	131.1	107.3	100.0	129.2	106.9	100.0	128.3	106.7	100.0	131.6	107.4	100.0	116.6	103.9	100.0	84.8	96.4
21, 22, 23	FIRE SUPPRESSION, PLUMBING & HVAC	97.1	179.8	130.4	100.8	131.4	113.2	97.3	131.4	111.0	101.3	147.6	120.0	96.9	128.4	109.6	97.1	69.8	86.1
26, 27, 3370	ELECTRICAL, COMMUNICATIONS & UTIL.	94.7	125.7	110.0	97.4	112.8	105.0	99.8	121.7	110.6	92.9	126.7	109.6	118.6	131.9	125.2	95.5	60.9	78.4
MF2018	WEIGHTED AVERAGE	101.2	149.8	122.2	101.0	131.9	114.3	107.3	131.0	117.5	100.5	143.4	119.0	98.5	127.0	110.8	103.3	69.2	88.6

COLORADO

DIVISION		BOULDER 803 MAT.	INST.	TOTAL	COLORADO SPRINGS 808-809 MAT.	INST.	TOTAL	DENVER 800-802 MAT.	INST.	TOTAL	DURANGO 813 MAT.	INST.	TOTAL	FORT COLLINS 805 MAT.	INST.	TOTAL	FORT MORGAN 807 MAT.	INST.	TOTAL
015433	CONTRACTOR EQUIPMENT		89.3	89.3		87.3	87.3		96.4	96.4		87.8	87.8		89.3	89.3		89.3	89.3
0241, 31-34	SITE & INFRASTRUCTURE, DEMOLITION	94.7	87.8	90.0	97.2	83.4	87.7	101.1	97.6	98.7	129.1	81.3	96.3	107.4	87.9	94.1	97.0	87.3	90.4
0310	Concrete Forming & Accessories	103.7	69.1	74.3	93.6	68.5	72.3	102.6	68.7	73.8	106.6	68.3	74.0	101.6	68.4	73.4	104.2	68.3	73.6
0320	Concrete Reinforcing	102.3	66.8	85.2	101.5	69.1	85.8	100.3	69.1	85.2	107.3	66.6	87.7	102.4	66.8	85.2	102.6	66.8	85.3
0330	Cast-in-Place Concrete	113.5	71.4	97.6	116.6	71.5	99.6	118.1	72.5	100.9	113.4	70.9	97.4	128.2	71.0	106.6	111.4	70.2	95.8
03	CONCRETE	107.8	70.0	90.8	111.4	70.2	92.9	115.2	70.6	95.1	114.2	69.5	94.1	119.2	69.6	96.9	106.2	69.2	89.6
04	MASONRY	104.1	63.9	79.6	107.5	62.1	79.8	113.5	62.7	82.5	116.0	59.5	81.6	120.9	61.9	84.9	119.6	63.7	85.5
05	METALS	93.9	76.5	88.5	97.1	77.6	91.0	99.6	77.3	92.6	103.2	77.4	95.2	95.2	76.5	89.4	93.6	76.6	88.3
06	WOOD, PLASTICS & COMPOSITES	102.9	70.5	86.0	91.8	70.7	80.7	107.2	70.7	88.1	103.4	71.0	86.4	100.2	70.5	84.7	102.9	70.5	86.0
07	THERMAL & MOISTURE PROTECTION	106.3	71.0	91.2	107.3	70.2	91.4	104.5	72.1	90.6	108.3	66.9	90.3	106.7	70.2	91.0	106.3	69.3	90.4
08	OPENINGS	95.0	69.7	88.8	99.1	70.3	92.1	102.4	70.3	94.6	102.9	69.9	94.9	95.0	69.7	88.8	95.0	69.7	88.8
0920	Plaster & Gypsum Board	123.3	70.3	88.5	108.5	70.3	83.5	119.1	70.3	87.1	97.1	70.3	79.5	116.5	70.3	86.2	123.3	70.3	88.5
0950, 0980	Ceilings & Acoustic Treatment	97.4	70.3	80.4	107.6	70.3	84.2	111.0	70.3	85.5	112.0	70.3	85.9	97.4	70.3	80.4	97.4	70.3	80.4
0960	Flooring	107.5	73.2	97.5	98.7	71.3	90.7	103.3	74.7	95.0	112.3	65.8	98.7	103.5	74.7	95.1	108.0	73.2	97.8
0970, 0990	Wall Finishes & Painting/Coating	96.2	76.1	84.2	95.9	80.7	86.8	105.9	74.7	87.2	97.5	76.1	84.7	96.2	76.1	84.2	96.2	76.1	84.2
09	FINISHES	102.3	70.4	85.1	100.2	70.1	83.9	104.6	70.2	86.0	105.2	68.9	85.6	100.7	70.2	84.2	102.4	69.9	84.8
COVERS	DIVS. 10-14, 25, 28, 41, 43, 44, 46	100.0	84.5	96.4	100.0	84.1	96.3	100.0	84.3	96.3	100.0	84.7	96.4	100.0	83.8	96.2	100.0	83.8	96.2
21, 22, 23	FIRE SUPPRESSION, PLUMBING & HVAC	97.2	71.9	87.0	101.1	71.5	89.2	100.9	74.7	90.4	97.1	56.8	80.9	101.0	71.5	89.1	97.2	70.9	86.6
26, 27, 3370	ELECTRICAL, COMMUNICATIONS & UTIL.	99.0	79.8	89.5	102.1	73.9	88.1	103.7	79.8	91.8	95.1	53.5	74.5	99.0	79.8	89.5	99.3	75.2	87.4
MF2018	WEIGHTED AVERAGE	99.2	73.7	88.2	101.8	72.3	89.1	104.0	75.1	91.5	103.6	65.5	87.1	102.8	73.2	90.0	99.9	72.5	88.0

COLORADO

DIVISION		GLENWOOD SPRINGS 816 MAT.	INST.	TOTAL	GOLDEN 804 MAT.	INST.	TOTAL	GRAND JUNCTION 815 MAT.	INST.	TOTAL	GREELEY 806 MAT.	INST.	TOTAL	MONTROSE 814 MAT.	INST.	TOTAL	PUEBLO 810 MAT.	INST.	TOTAL
015433	CONTRACTOR EQUIPMENT		90.9	90.9		89.3	89.3		90.9	90.9		89.3	89.3		89.4	89.4		87.8	87.8
0241, 31-34	SITE & INFRASTRUCTURE, DEMOLITION	145.1	88.7	106.3	107.6	87.6	93.9	129.0	89.1	101.6	94.0	87.8	89.8	138.5	85.4	102.0	120.8	81.5	93.8
0310	Concrete Forming & Accessories	97.0	68.1	72.4	96.3	68.7	72.8	105.4	69.3	74.7	99.2	68.4	73.0	96.6	68.3	72.5	102.8	68.8	73.9
0320	Concrete Reinforcing	106.1	66.8	87.2	102.6	66.8	85.3	106.5	66.6	87.3	102.3	66.8	85.2	106.0	66.8	87.1	102.6	69.0	86.4
0330	Cast-in-Place Concrete	98.6	69.9	87.7	111.5	71.4	96.4	109.2	72.2	95.2	107.1	71.0	93.5	98.6	70.4	88.0	97.9	71.9	88.1
03	CONCRETE	117.9	69.1	96.0	117.2	69.8	95.9	110.7	70.4	92.6	102.7	69.6	87.9	108.4	69.3	90.9	100.8	70.5	87.2
04	MASONRY	102.2	63.6	78.6	122.7	62.7	86.1	136.2	61.7	90.8	113.7	63.2	82.9	108.4	63.6	81.1	98.7	60.9	75.6
05	METALS	102.9	77.3	94.9	93.8	76.5	88.4	104.6	77.3	96.1	95.2	76.6	89.4	102.0	77.6	94.4	106.4	78.9	97.8
06	WOOD, PLASTICS & COMPOSITES	88.9	70.7	79.4	94.1	70.5	81.8	101.4	70.7	85.3	97.2	70.5	83.2	89.7	70.8	79.8	96.6	71.0	83.2
07	THERMAL & MOISTURE PROTECTION	108.2	68.0	90.9	107.5	70.6	91.6	107.2	69.3	90.9	105.9	70.5	90.7	108.4	68.0	91.0	106.8	68.6	90.4
08	OPENINGS	101.9	69.8	94.1	95.0	69.7	88.8	102.6	69.8	94.6	95.0	69.7	88.8	103.1	69.9	95.0	98.1	70.5	91.4
0920	Plaster & Gypsum Board	125.3	70.3	89.3	113.4	70.3	85.1	138.6	70.3	93.8	115.1	70.3	85.7	82.7	70.3	74.6	88.1	70.3	76.4
0950, 0980	Ceilings & Acoustic Treatment	111.4	70.3	85.7	97.4	70.3	80.4	111.4	70.3	85.7	97.4	70.3	80.4	112.0	70.3	85.9	119.7	70.3	88.8
0960	Flooring	105.8	73.2	96.3	101.0	73.2	92.9	111.6	69.6	99.3	102.3	72.6	93.7	109.5	69.8	97.9	108.1	72.6	97.7
0970, 0990	Wall Finishes & Painting/Coating	97.4	76.1	84.7	96.2	76.1	84.2	97.4	76.1	84.7	96.2	76.1	84.2	97.5	76.1	84.7	97.5	76.1	84.7
09	FINISHES	108.5	70.1	87.7	100.3	70.1	84.0	110.2	69.9	88.4	99.4	69.8	83.4	103.2	69.5	85.0	103.6	70.1	85.5
COVERS	DIVS. 10-14, 25, 28, 41, 43, 44, 46	100.0	84.1	96.3	100.0	84.1	96.3	100.0	84.9	96.5	100.0	83.8	96.2	100.0	84.4	96.3	100.0	84.8	96.4
21, 22, 23	FIRE SUPPRESSION, PLUMBING & HVAC	97.1	56.7	80.8	97.2	71.3	86.7	100.9	73.3	89.8	101.0	71.4	89.1	97.1	56.8	80.9	100.9	71.6	89.1
26, 27, 3370	ELECTRICAL, COMMUNICATIONS & UTIL.	93.0	53.5	73.4	99.3	77.6	88.5	94.8	52.5	73.9	99.0	79.8	89.5	94.8	53.5	74.3	95.5	63.3	79.6
MF2018	WEIGHTED AVERAGE	103.6	66.6	87.6	101.5	73.0	89.2	105.6	70.0	90.2	99.9	73.3	88.4	102.3	66.3	86.7	101.6	70.7	88.2

DIVISION		COLORADO SALIDA 812			CONNECTICUT BRIDGEPORT 066			BRISTOL 060			HARTFORD 061			MERIDEN 064			NEW BRITAIN 060		
		MAT.	INST.	TOTAL	MAT.	INST.	TOTAL	MAT.	INST.	TOTAL	MAT.	INST.	TOTAL	MAT.	INST.	TOTAL	MAT.	INST.	TOTAL
015433	CONTRACTOR EQUIPMENT		89.4	89.4		94.2	94.2		94.2	94.2		100.4	100.4		94.6	94.6		94.2	94.2
0241, 31 - 34	SITE & INFRASTRUCTURE, DEMOLITION	128.8	85.0	98.7	107.9	95.6	99.4	107.1	95.6	99.2	102.3	104.7	103.9	104.5	96.2	98.8	107.2	95.6	99.2
0310	Concrete Forming & Accessories	105.4	68.4	74.0	104.3	115.7	114.0	104.3	117.0	115.1	102.9	117.4	115.2	104.0	117.0	115.0	104.8	117.0	115.2
0320	Concrete Reinforcing	105.7	66.6	86.9	116.0	144.7	129.8	116.0	144.7	129.8	110.4	144.7	126.9	116.0	144.7	129.8	116.0	144.7	129.8
0330	Cast-in-Place Concrete	113.0	70.5	97.0	99.8	126.6	110.0	93.5	126.6	106.0	95.3	128.4	107.8	90.0	126.6	103.8	95.1	126.6	107.0
03	CONCRETE	109.0	69.4	91.3	101.9	123.9	111.8	99.0	124.5	110.5	100.2	125.2	111.4	97.3	124.5	109.5	99.7	124.5	110.9
04	MASONRY	136.5	59.5	89.5	111.8	130.5	123.2	103.7	130.5	120.1	107.3	130.6	121.5	103.4	130.5	119.9	105.6	130.5	120.8
05	METALS	101.7	77.5	94.1	103.5	118.7	108.2	103.5	118.7	108.2	108.2	117.6	111.1	100.3	118.6	106.0	99.4	118.7	105.4
06	WOOD, PLASTICS & COMPOSITES	97.9	70.8	83.8	106.2	112.7	109.6	106.2	114.5	110.6	97.6	114.8	106.6	106.2	114.5	110.6	106.2	114.5	110.6
07	THERMAL & MOISTURE PROTECTION	107.2	66.9	89.9	98.9	121.6	108.7	99.1	120.6	108.3	103.5	121.6	111.3	99.1	120.6	108.3	99.1	120.5	108.3
08	OPENINGS	96.4	69.9	90.0	96.6	119.9	102.3	96.6	121.6	102.7	98.3	121.7	104.0	98.8	120.9	104.2	96.6	121.6	102.7
0920	Plaster & Gypsum Board	82.9	70.3	74.7	122.9	112.9	116.3	122.9	114.8	117.6	110.1	114.8	113.2	125.4	114.8	118.4	122.9	114.8	117.6
0950, 0980	Ceilings & Acoustic Treatment	112.0	70.3	85.9	102.9	112.9	109.2	102.9	114.8	110.3	103.3	114.8	110.5	111.0	114.8	113.4	102.9	114.8	110.3
0960	Flooring	115.2	66.7	101.1	96.3	123.6	104.2	96.3	123.6	104.2	100.3	126.0	107.8	96.3	118.8	102.8	96.3	123.6	104.2
0970, 0990	Wall Finishes & Painting/Coating	97.5	76.1	84.7	90.7	126.8	112.3	90.7	130.8	114.7	98.9	130.8	118.0	90.7	126.8	112.3	90.7	130.8	114.7
09	FINISHES	103.9	68.8	84.9	100.3	117.7	109.8	100.4	119.2	110.6	101.8	119.9	111.6	102.7	118.0	111.0	100.4	119.2	110.6
COVERS	DIVS. 10 - 14, 25, 28, 41, 43, 44, 46	100.0	84.5	96.3	100.0	112.2	102.9	100.0	112.4	102.9	100.0	113.1	103.1	100.0	112.4	102.9	100.0	112.4	102.9
21, 22, 23	FIRE SUPPRESSION, PLUMBING & HVAC	97.1	70.9	86.6	101.0	119.0	108.3	101.0	119.0	108.3	100.9	119.0	108.2	97.2	119.0	106.0	101.0	119.0	108.3
26, 27, 3370	ELECTRICAL, COMMUNICATIONS & UTIL.	95.0	60.9	78.1	94.1	105.2	99.6	94.1	103.9	99.0	97.1	110.8	103.9	94.1	101.9	98.0	94.2	103.9	99.0
MF2018	WEIGHTED AVERAGE	102.9	69.8	88.6	100.9	116.7	107.7	100.1	116.9	107.4	101.7	118.7	109.0	98.8	116.4	106.4	99.7	116.9	107.1

DIVISION		CONNECTICUT NEW HAVEN 065			NEW LONDON 063			NORWALK 068			STAMFORD 069			WATERBURY 067			WILLIMANTIC 062		
		MAT.	INST.	TOTAL	MAT.	INST.	TOTAL	MAT.	INST.	TOTAL	MAT.	INST.	TOTAL	MAT.	INST.	TOTAL	MAT.	INST.	TOTAL
015433	CONTRACTOR EQUIPMENT		94.6	94.6		94.6	94.6		94.2	94.2		94.2	94.2		94.2	94.2		94.2	94.2
0241, 31 - 34	SITE & INFRASTRUCTURE, DEMOLITION	107.2	96.2	99.7	98.6	96.2	97.0	107.7	95.6	99.4	108.4	95.6	99.6	107.6	95.6	99.3	107.8	95.4	99.3
0310	Concrete Forming & Accessories	104.0	116.9	115.0	104.0	116.9	115.0	104.3	115.7	114.0	104.3	116.0	114.3	104.3	117.0	115.1	104.3	116.9	115.0
0320	Concrete Reinforcing	116.0	144.7	129.8	90.9	144.7	116.9	116.0	144.6	129.8	116.0	144.7	129.8	116.0	144.7	129.8	116.0	144.7	129.8
0330	Cast-in-Place Concrete	96.7	124.7	107.3	82.4	124.7	98.3	98.2	125.8	108.7	99.8	126.0	109.7	99.8	126.6	110.0	93.2	124.7	105.1
03	CONCRETE	114.3	123.8	118.6	87.5	123.8	103.8	101.2	123.7	111.3	101.9	123.9	111.8	101.9	124.5	112.1	98.9	123.8	110.0
04	MASONRY	104.0	130.5	120.2	102.3	130.5	119.5	103.4	129.7	119.4	104.3	129.7	119.8	104.3	130.5	120.3	103.6	130.5	120.0
05	METALS	99.7	118.6	105.6	99.3	118.6	105.3	103.5	118.6	108.2	103.5	119.0	108.3	103.5	118.6	108.2	103.2	118.5	108.0
06	WOOD, PLASTICS & COMPOSITES	106.2	114.5	110.6	106.2	114.5	110.6	106.2	112.7	109.6	106.2	112.7	109.6	106.2	114.5	110.6	106.2	114.5	110.6
07	THERMAL & MOISTURE PROTECTION	99.2	119.4	107.9	99.0	120.3	108.1	99.2	121.3	108.7	99.1	121.3	108.6	99.1	119.6	107.9	99.3	120.0	108.2
08	OPENINGS	96.6	120.9	102.5	98.9	120.1	104.1	96.6	119.9	102.3	96.6	119.9	102.3	96.6	120.9	102.5	98.9	120.9	104.3
0920	Plaster & Gypsum Board	122.9	114.8	117.6	122.9	114.8	117.6	122.9	112.9	116.3	122.9	112.9	116.3	122.9	114.8	117.6	122.9	114.8	117.6
0950, 0980	Ceilings & Acoustic Treatment	102.9	114.8	110.3	101.4	114.8	109.8	102.9	112.9	109.2	102.9	112.9	109.2	102.9	114.8	110.3	101.4	114.8	109.8
0960	Flooring	96.3	123.6	104.2	96.3	121.2	103.5	96.3	123.6	104.2	96.3	123.6	104.2	96.3	123.6	104.2	96.3	115.5	101.9
0970, 0990	Wall Finishes & Painting/Coating	90.7	124.4	110.9	90.7	126.8	112.3	90.7	126.8	112.3	90.7	126.8	112.3	90.7	126.8	112.3	90.7	126.8	112.3
09	FINISHES	100.4	118.5	110.2	99.5	118.4	109.7	100.4	117.7	109.8	100.5	117.7	109.8	100.3	118.8	110.3	100.2	117.4	109.5
COVERS	DIVS. 10 - 14, 25, 28, 41, 43, 44, 46	100.0	112.4	102.9	100.0	112.4	102.9	100.0	112.2	102.9	100.0	112.4	102.9	100.0	112.4	102.9	100.0	112.4	102.9
21, 22, 23	FIRE SUPPRESSION, PLUMBING & HVAC	101.0	119.0	108.3	97.2	119.0	106.0	101.0	119.0	108.3	101.0	119.0	108.3	101.0	119.0	108.3	101.0	118.9	108.2
26, 27, 3370	ELECTRICAL, COMMUNICATIONS & UTIL.	94.1	106.2	100.1	91.9	103.9	97.8	94.1	107.1	100.6	94.1	156.4	125.0	93.8	107.3	100.4	94.1	107.1	100.5
MF2018	WEIGHTED AVERAGE	101.5	117.0	108.2	96.7	116.6	105.3	100.4	116.9	107.5	100.5	123.9	110.6	100.5	117.2	107.7	100.3	116.9	107.5

DIVISION		D.C. WASHINGTON 200 - 205			DELAWARE DOVER 199			NEWARK 197			WILMINGTON 198			FLORIDA DAYTONA BEACH 321			FORT LAUDERDALE 333		
		MAT.	INST.	TOTAL	MAT.	INST.	TOTAL	MAT.	INST.	TOTAL	MAT.	INST.	TOTAL	MAT.	INST.	TOTAL	MAT.	INST.	TOTAL
015433	CONTRACTOR EQUIPMENT		104.0	104.0		115.6	115.6		117.0	117.0		115.8	115.8		98.2	98.2		91.4	91.4
0241, 31 - 34	SITE & INFRASTRUCTURE, DEMOLITION	105.0	95.4	98.4	102.7	103.6	103.3	102.0	106.0	104.7	98.9	104.5	102.8	117.0	82.6	93.4	94.9	72.4	79.4
0310	Concrete Forming & Accessories	100.6	73.0	77.1	100.0	100.2	100.2	96.2	99.9	99.4	98.0	100.4	100.0	98.6	61.6	67.1	93.8	67.7	71.6
0320	Concrete Reinforcing	115.4	90.3	103.3	100.7	116.3	108.2	98.8	116.3	107.2	96.7	116.4	106.2	90.5	61.5	76.5	88.7	57.6	73.7
0330	Cast-in-Place Concrete	98.3	79.7	91.3	106.2	108.3	107.0	88.6	106.3	95.3	103.4	108.3	105.2	85.5	65.0	77.8	88.0	72.3	82.1
03	CONCRETE	103.7	79.2	92.7	98.6	106.8	102.3	92.1	106.2	98.4	96.6	106.9	101.2	88.7	64.5	77.8	90.6	69.1	81.0
04	MASONRY	99.1	87.1	91.8	103.9	101.7	102.5	102.6	101.6	102.0	99.8	101.7	100.9	87.0	60.0	70.5	90.9	72.8	79.9
05	METALS	104.5	96.2	101.9	105.3	123.5	111.0	107.4	125.8	113.1	105.5	123.6	111.2	100.0	89.2	96.7	96.8	87.7	94.0
06	WOOD, PLASTICS & COMPOSITES	100.7	71.4	85.4	96.1	98.1	97.1	91.0	97.8	94.6	89.5	98.1	94.0	93.5	60.3	76.1	79.1	62.3	70.3
07	THERMAL & MOISTURE PROTECTION	100.8	86.0	94.5	105.4	112.4	108.4	109.9	111.2	110.5	104.9	112.4	108.1	100.8	64.5	85.2	105.5	68.1	89.5
08	OPENINGS	100.8	74.8	94.5	91.1	109.9	95.7	91.1	109.7	95.6	88.6	109.9	93.8	92.9	59.8	84.8	94.3	59.9	85.9
0920	Plaster & Gypsum Board	118.0	70.6	86.9	103.9	97.9	100.0	99.8	97.9	98.6	100.9	97.9	99.0	90.9	59.8	70.5	111.4	61.8	78.9
0950, 0980	Ceilings & Acoustic Treatment	108.3	70.6	84.7	105.5	97.9	100.7	100.8	97.9	99.0	101.9	97.9	99.4	86.1	59.8	69.7	94.4	61.8	74.0
0960	Flooring	94.7	76.7	89.5	91.5	106.8	95.9	91.9	106.8	96.2	99.5	106.8	101.6	104.4	61.0	91.8	102.1	61.0	90.1
0970, 0990	Wall Finishes & Painting/Coating	101.5	70.0	82.6	91.4	117.9	107.3	84.1	117.9	104.4	94.5	117.9	108.5	99.0	62.1	76.9	90.6	58.4	71.3
09	FINISHES	101.8	72.3	85.8	97.1	102.3	99.9	92.3	102.1	97.6	98.6	102.3	100.6	94.9	60.9	76.5	97.1	64.8	79.6
COVERS	DIVS. 10 - 14, 25, 28, 41, 43, 44, 46	100.0	95.5	99.0	100.0	100.0	100.0	100.0	99.3	99.8	100.0	106.3	101.5	100.0	82.0	95.8	100.0	88.0	97.2
21, 22, 23	FIRE SUPPRESSION, PLUMBING & HVAC	100.8	89.4	96.2	99.4	120.9	108.0	99.6	120.8	108.2	100.7	120.9	108.9	100.7	73.7	89.8	100.7	76.0	90.7
26, 27, 3370	ELECTRICAL, COMMUNICATIONS & UTIL.	96.4	103.9	100.1	98.2	109.9	104.0	98.0	109.9	103.9	97.4	109.9	103.6	98.0	61.4	79.9	95.1	67.5	81.4
MF2018	WEIGHTED AVERAGE	101.3	88.0	95.5	99.5	110.4	104.2	98.7	110.6	103.8	99.1	110.7	104.1	97.2	69.0	85.0	96.5	72.2	86.0

		FLORIDA																
	DIVISION	FORT MYERS			GAINESVILLE			JACKSONVILLE			LAKELAND			MELBOURNE			MIAMI		
		339,341			326,344			320,322			338			329			330 - 332,340		
		MAT.	INST.	TOTAL	MAT.	INST.	TOTAL	MAT.	INST.	TOTAL	MAT.	INST.	TOTAL	MAT.	INST.	TOTAL	MAT.	INST.	TOTAL
015433	CONTRACTOR EQUIPMENT		98.2	98.2		98.2	98.2		98.2	98.2		98.2	98.2		98.2	98.2		95.7	95.7
0241, 31 - 34	SITE & INFRASTRUCTURE, DEMOLITION	106.7	82.7	90.2	125.2	82.4	95.8	117.0	82.2	93.1	108.7	82.7	90.9	124.6	82.6	95.7	94.8	80.7	85.1
0310	Concrete Forming & Accessories	89.8	62.0	66.1	93.1	55.5	61.1	98.5	61.8	67.3	86.0	63.2	66.6	94.6	62.3	67.1	101.3	64.3	69.9
0320	Concrete Reinforcing	89.7	75.0	82.6	96.0	59.8	78.5	90.5	59.7	75.7	91.9	75.1	83.8	91.5	67.1	79.7	98.0	57.6	78.5
0330	Cast-in-Place Concrete	91.8	64.1	81.3	98.2	64.6	85.5	86.4	64.5	78.1	93.9	65.1	83.0	103.1	60.0	88.7	85.2	69.6	79.4
03	CONCRETE	91.0	66.7	80.1	99.5	61.3	82.3	89.1	64.0	77.9	92.6	67.6	81.4	99.7	65.8	84.5	89.6	66.5	79.2
04	MASONRY	85.7	57.3	68.4	99.7	59.2	75.0	86.8	57.7	69.1	101.1	60.0	76.0	84.3	60.0	69.4	94.7	62.6	75.1
05	METALS	98.8	93.7	97.2	98.8	87.5	95.3	98.6	87.0	95.0	98.7	94.2	97.3	108.4	91.4	103.1	97.0	85.5	93.4
06	WOOD, PLASTICS & COMPOSITES	75.9	61.0	68.1	86.6	52.7	68.9	93.5	61.3	76.6	71.2	62.7	66.8	88.6	61.0	74.2	94.1	62.5	77.6
07	THERMAL & MOISTURE PROTECTION	105.3	61.3	86.4	101.2	61.8	84.3	101.1	62.3	84.5	105.2	62.2	86.7	101.3	63.3	85.0	104.8	64.8	87.6
08	OPENINGS	95.7	63.3	87.8	92.6	55.2	83.5	92.9	59.9	84.9	95.6	64.2	88.0	92.2	61.5	84.7	96.7	60.1	87.8
0920	Plaster & Gypsum Board	107.3	60.6	76.6	87.9	52.0	64.4	90.9	60.8	71.1	105.0	62.3	77.0	87.9	60.6	70.0	98.6	61.8	74.5
0950, 0980	Ceilings & Acoustic Treatment	89.2	60.6	71.3	80.5	52.0	62.7	86.1	60.8	70.3	89.2	62.3	72.3	85.5	60.6	69.9	95.1	61.8	74.2
0960	Flooring	99.0	73.7	91.6	101.9	59.8	89.6	104.4	61.0	91.8	97.1	61.0	86.6	102.2	61.0	90.2	100.9	61.6	89.5
0970, 0990	Wall Finishes & Painting/Coating	95.1	62.5	75.5	99.0	62.5	77.1	99.0	62.5	77.1	95.1	62.5	75.5	99.0	79.5	87.3	96.8	58.4	73.8
09	FINISHES	96.6	63.9	78.9	93.4	56.2	73.3	95.0	61.5	76.8	95.8	62.3	77.7	94.4	63.2	77.5	96.9	62.6	78.3
COVERS	DIVS. 10 - 14, 25, 28, 41, 43, 44, 46	100.0	80.8	95.5	100.0	79.3	95.1	100.0	80.3	95.4	100.0	82.2	95.8	100.0	82.1	95.8	100.0	85.2	96.5
21, 22, 23	FIRE SUPPRESSION, PLUMBING & HVAC	97.6	57.0	81.2	97.8	60.6	82.8	100.7	64.0	85.9	97.6	58.4	81.8	100.7	74.1	90.0	96.3	67.9	84.8
26, 27, 3370	ELECTRICAL, COMMUNICATIONS & UTIL.	96.3	64.2	80.4	98.0	56.4	77.4	97.7	64.4	81.3	94.8	56.4	75.8	99.0	66.5	82.9	97.7	77.1	87.5
MF2018	WEIGHTED AVERAGE	96.4	66.7	83.6	98.3	63.8	83.4	97.0	66.9	84.0	97.1	66.3	83.8	99.9	70.6	87.2	96.2	70.4	85.1

		FLORIDA																
	DIVISION	ORLANDO			PANAMA CITY			PENSACOLA			SARASOTA			ST. PETERSBURG			TALLAHASSEE		
		327 - 328,347			324			325			342			337			323		
		MAT.	INST.	TOTAL	MAT.	INST.	TOTAL	MAT.	INST.	TOTAL	MAT.	INST.	TOTAL	MAT.	INST.	TOTAL	MAT.	INST.	TOTAL
015433	CONTRACTOR EQUIPMENT		103.5	103.5		98.2	98.2		98.2	98.2		98.2	98.2		98.2	98.2		103.5	103.5
0241, 31 - 34	SITE & INFRASTRUCTURE, DEMOLITION	114.1	91.6	98.7	130.1	82.6	97.5	129.9	82.6	97.4	116.8	82.7	93.4	110.9	82.6	91.5	110.1	91.9	97.6
0310	Concrete Forming & Accessories	103.9	61.6	67.9	97.6	64.3	69.2	95.6	61.5	66.6	95.9	63.0	67.9	93.8	61.1	66.0	101.8	62.1	68.1
0320	Concrete Reinforcing	96.5	65.0	81.3	94.5	65.8	80.7	96.9	66.7	82.4	92.3	72.6	82.8	91.9	75.0	83.8	103.1	59.9	82.3
0330	Cast-in-Place Concrete	105.9	66.6	91.0	90.8	64.9	81.0	112.2	64.7	94.3	101.2	65.0	87.5	94.9	64.9	83.6	88.6	66.5	80.3
03	CONCRETE	99.2	65.5	84.0	98.1	66.4	83.9	107.2	65.3	88.4	94.9	67.0	82.4	94.3	66.6	81.8	93.2	64.8	80.4
04	MASONRY	93.2	60.0	73.0	91.2	59.4	71.8	109.8	58.9	78.7	91.4	60.0	72.3	139.1	57.9	89.6	91.5	59.5	72.0
05	METALS	95.2	88.0	93.0	99.7	90.5	96.8	100.7	90.7	97.6	102.2	92.5	99.1	99.6	93.8	97.8	99.9	86.1	95.6
06	WOOD, PLASTICS & COMPOSITES	94.0	60.6	76.5	92.2	64.3	77.6	90.7	61.3	75.3	93.5	62.7	77.4	80.7	60.0	69.9	98.6	61.5	79.2
07	THERMAL & MOISTURE PROTECTION	108.2	65.6	89.9	101.5	62.5	84.7	101.4	61.6	84.3	100.3	62.2	83.9	105.5	61.1	86.4	96.2	63.0	82.0
08	OPENINGS	97.7	60.7	88.7	91.1	63.0	84.2	91.1	61.5	83.9	97.7	63.2	89.3	94.5	62.7	86.7	97.8	60.1	88.6
0920	Plaster & Gypsum Board	92.7	59.8	71.1	90.1	63.9	72.9	98.9	60.8	73.9	98.3	62.3	74.6	109.3	59.6	76.6	101.5	60.8	74.8
0950, 0980	Ceilings & Acoustic Treatment	99.0	59.8	74.5	85.5	63.9	72.0	85.5	60.8	70.0	91.1	62.3	73.0	90.6	59.6	71.1	99.9	60.8	75.4
0960	Flooring	98.8	61.0	87.7	104.1	72.0	94.7	99.9	61.0	88.6	105.4	52.4	89.9	101.0	57.4	88.3	101.4	60.1	89.4
0970, 0990	Wall Finishes & Painting/Coating	94.5	62.1	75.1	99.0	62.5	77.1	99.0	62.5	77.1	95.1	62.5	75.6	95.1	62.5	75.5	99.0	62.5	77.4
09	FINISHES	98.0	61.1	78.0	96.0	65.1	79.3	95.7	61.2	77.0	98.5	60.8	78.1	98.0	60.0	77.4	100.0	61.4	79.1
COVERS	DIVS. 10 - 14, 25, 28, 41, 43, 44, 46	100.0	82.6	95.9	100.0	79.4	95.2	100.0	78.7	95.0	100.0	80.9	95.5	100.0	78.7	95.0	100.0	78.9	95.0
21, 22, 23	FIRE SUPPRESSION, PLUMBING & HVAC	96.4	56.0	80.1	100.7	64.2	86.0	100.7	62.0	85.1	100.3	56.2	82.5	100.7	58.4	83.6	96.5	64.2	83.5
26, 27, 3370	ELECTRICAL, COMMUNICATIONS & UTIL.	98.3	63.2	80.9	96.9	56.4	76.9	100.2	52.3	76.5	97.5	56.4	77.1	95.3	59.6	77.6	103.3	56.4	80.1
MF2018	WEIGHTED AVERAGE	97.9	66.3	84.3	98.7	67.2	85.1	101.2	65.3	85.7	99.1	65.3	84.5	100.4	65.8	85.4	98.1	66.7	84.6

		FLORIDA						GEORGIA											
	DIVISION	TAMPA			WEST PALM BEACH			ALBANY			ATHENS			ATLANTA			AUGUSTA		
		335 - 336,346			334,349			317,398			306			300 - 303,399			308 - 309		
		MAT.	INST.	TOTAL	MAT.	INST.	TOTAL	MAT.	INST.	TOTAL	MAT.	INST.	TOTAL	MAT.	INST.	TOTAL	MAT.	INST.	TOTAL
015433	CONTRACTOR EQUIPMENT		98.2	98.2		91.4	91.4		92.3	92.3		91.0	91.0		97.5	97.5		91.0	91.0
0241, 31 - 34	SITE & INFRASTRUCTURE, DEMOLITION	111.3	84.6	93.0	91.1	72.4	78.2	104.9	74.5	84.0	101.9	88.8	92.9	99.5	97.2	97.9	95.4	89.4	91.3
0310	Concrete Forming & Accessories	97.0	63.4	68.4	97.7	67.4	71.9	90.2	67.7	71.0	91.9	42.0	49.5	97.2	75.8	78.9	93.1	73.6	76.5
0320	Concrete Reinforcing	88.7	75.2	82.2	91.2	54.8	73.7	91.8	71.8	82.1	92.3	60.2	76.8	94.5	71.8	83.6	92.6	70.8	82.1
0330	Cast-in-Place Concrete	92.8	65.2	82.4	83.7	72.2	79.4	86.0	69.7	79.8	103.9	69.5	90.9	107.3	72.4	94.1	98.1	70.8	87.8
03	CONCRETE	93.0	67.8	81.7	87.5	68.5	79.0	84.0	70.8	78.1	97.8	56.1	79.1	100.4	74.4	88.7	90.9	72.8	82.8
04	MASONRY	91.4	60.0	72.3	90.4	70.0	78.0	98.5	69.3	80.7	80.0	68.2	72.8	92.1	70.3	78.8	94.1	69.4	79.0
05	METALS	98.8	94.4	97.4	95.7	86.3	92.8	105.9	98.3	103.5	95.3	77.3	89.7	96.3	83.2	92.2	95.0	83.4	91.4
06	WOOD, PLASTICS & COMPOSITES	85.1	62.7	73.4	84.1	62.3	72.7	80.5	68.0	73.9	94.7	34.5	63.2	104.6	78.6	91.0	96.2	76.2	85.7
07	THERMAL & MOISTURE PROTECTION	105.8	62.2	87.1	105.2	68.0	89.2	99.5	68.8	86.3	94.3	65.5	82.0	95.9	73.9	86.5	94.0	71.0	84.1
08	OPENINGS	95.6	64.2	88.0	93.8	59.3	85.4	85.4	70.3	81.7	91.4	49.5	81.2	99.6	76.7	94.0	91.4	75.1	87.5
0920	Plaster & Gypsum Board	112.2	62.3	79.5	114.9	61.8	80.1	101.3	67.7	79.3	92.6	33.3	53.7	99.3	78.3	85.5	93.4	76.1	82.0
0950, 0980	Ceilings & Acoustic Treatment	94.4	62.3	74.3	89.2	61.8	72.0	83.6	67.7	73.7	102.6	33.3	59.2	93.9	78.3	84.1	103.3	76.1	86.2
0960	Flooring	102.1	61.0	90.1	104.0	52.4	88.9	103.5	71.1	94.1	93.9	71.1	87.3	95.6	72.5	88.9	94.1	71.1	87.4
0970, 0990	Wall Finishes & Painting/Coating	95.1	62.5	75.5	90.6	57.8	70.9	85.1	95.0	91.0	89.7	89.5	89.6	96.3	96.7	96.5	89.7	88.4	89.0
09	FINISHES	99.7	62.3	79.5	96.6	63.0	78.4	94.1	70.7	81.4	95.5	50.2	71.0	95.4	77.5	85.7	95.2	74.8	84.1
COVERS	DIVS. 10 - 14, 25, 28, 41, 43, 44, 46	100.0	81.0	95.5	100.0	87.9	97.2	100.0	84.3	96.3	100.0	80.5	95.4	100.0	86.3	96.8	100.0	85.4	96.6
21, 22, 23	FIRE SUPPRESSION, PLUMBING & HVAC	100.7	58.5	83.6	97.6	66.1	84.9	100.7	69.4	88.1	97.2	63.1	83.5	101.0	72.0	89.3	101.0	66.4	87.1
26, 27, 3370	ELECTRICAL, COMMUNICATIONS & UTIL.	95.0	63.2	79.3	95.6	64.9	80.4	95.2	62.0	78.8	99.2	61.5	80.6	98.9	72.7	85.9	99.7	68.5	84.3
MF2018	WEIGHTED AVERAGE	98.0	67.4	84.8	95.1	69.0	83.8	96.5	72.2	86.0	95.9	63.8	82.0	98.7	76.7	89.2	96.4	73.5	86.5

DIVISION		GEORGIA																	
		COLUMBUS			DALTON			GAINESVILLE			MACON			SAVANNAH			STATESBORO		
		318 - 319			307			305			310 - 312			313 - 314			304		
		MAT.	INST.	TOTAL	MAT.	INST.	TOTAL	MAT.	INST.	TOTAL	MAT.	INST.	TOTAL	MAT.	INST.	TOTAL	MAT.	INST.	TOTAL
015433	CONTRACTOR EQUIPMENT		92.3	92.3		106.1	106.1		91.0	91.0		101.8	101.8		97.7	97.7		94.0	94.0
0241, 31 - 34	SITE & INFRASTRUCTURE, DEMOLITION	104.8	74.8	84.2	101.6	93.8	96.2	101.6	88.7	92.7	106.3	88.8	94.3	106.2	83.8	90.8	102.8	75.0	83.7
0310	Concrete Forming & Accessories	90.1	70.3	73.3	83.5	62.2	65.4	96.0	39.9	48.3	89.8	69.2	72.3	96.2	71.6	75.3	78.1	50.2	54.3
0320	Concrete Reinforcing	91.7	71.8	82.1	91.8	57.7	75.3	92.1	60.2	76.7	92.9	71.8	82.7	97.8	70.9	84.8	91.4	67.5	79.8
0330	Cast-in-Place Concrete	85.7	70.0	79.8	100.9	68.0	88.5	109.2	69.0	94.0	84.5	70.9	79.4	86.8	71.3	81.0	103.7	68.6	90.5
03	CONCRETE	83.8	72.2	78.6	97.0	65.3	82.8	99.6	55.0	79.6	83.4	72.0	78.3	84.5	72.8	79.2	97.2	61.9	81.4
04	MASONRY	98.2	69.4	80.7	81.1	66.6	72.2	88.5	67.3	75.5	112.4	68.8	85.8	92.4	70.3	78.9	83.2	68.1	74.0
05	METALS	105.5	99.0	103.5	96.2	92.0	94.9	94.5	77.3	89.2	100.7	99.2	100.2	101.8	95.8	99.9	99.8	96.9	98.9
06	WOOD, PLASTICS & COMPOSITES	80.5	71.4	75.7	77.4	63.0	69.9	99.0	32.0	64.0	85.3	70.1	77.3	92.0	73.3	82.2	70.3	45.5	57.4
07	THERMAL & MOISTURE PROTECTION	99.4	70.1	86.8	96.1	65.7	83.0	94.3	63.4	81.0	98.0	71.2	86.5	97.4	70.6	85.9	94.6	62.8	81.0
08	OPENINGS	85.4	72.8	82.3	92.9	64.5	86.0	91.4	48.1	80.8	85.2	72.0	82.0	94.9	73.6	89.7	93.9	57.4	85.0
0920	Plaster & Gypsum Board	101.3	71.3	81.6	82.3	62.5	69.3	94.3	30.7	52.6	104.0	69.9	81.6	103.3	72.9	83.4	84.2	44.6	58.2
0950, 0980	Ceilings & Acoustic Treatment	83.6	71.3	75.9	118.4	62.5	83.4	102.6	30.7	57.6	79.2	69.9	73.4	98.2	72.9	82.4	113.4	44.6	70.3
0960	Flooring	103.5	71.1	94.1	94.8	71.1	87.9	95.5	71.1	88.4	81.4	71.1	78.4	97.9	72.5	90.5	111.4	71.1	99.6
0970, 0990	Wall Finishes & Painting/Coating	85.1	85.9	85.6	80.3	66.7	72.2	89.7	89.5	89.6	87.1	95.0	91.8	86.9	85.3	85.9	88.1	66.7	75.3
09	FINISHES	94.0	71.7	81.9	104.6	64.0	82.7	96.0	48.6	70.3	83.4	71.8	77.1	96.9	73.1	84.0	107.4	54.2	78.6
COVERS	DIVS. 10 - 14, 25, 28, 41, 43, 44, 46	100.0	84.7	96.4	100.0	81.0	95.5	100.0	80.0	95.3	100.0	84.4	96.3	100.0	85.4	96.6	100.0	81.6	95.7
21, 22, 23	FIRE SUPPRESSION, PLUMBING & HVAC	100.7	66.3	86.8	97.3	57.9	81.4	97.2	62.6	83.3	100.7	71.9	89.1	100.8	64.3	86.1	98.0	66.3	85.2
26, 27, 3370	ELECTRICAL, COMMUNICATIONS & UTIL.	95.3	70.8	83.2	108.0	59.9	84.2	99.2	70.7	85.1	93.9	62.8	78.5	99.2	63.8	81.7	99.7	63.4	81.7
MF2018	WEIGHTED AVERAGE	96.4	73.4	86.4	97.7	68.0	84.9	96.5	64.3	82.6	95.3	74.4	86.3	97.3	72.8	86.7	97.9	67.1	84.6

DIVISION		GEORGIA						HAWAII									IDAHO		
		VALDOSTA			WAYCROSS			HILO			HONOLULU			STATES & POSS., GUAM			BOISE		
		316			315			967			968			969			836 - 837		
		MAT.	INST.	TOTAL	MAT.	INST.	TOTAL	MAT.	INST.	TOTAL	MAT.	INST.	TOTAL	MAT.	INST.	TOTAL	MAT.	INST.	TOTAL
015433	CONTRACTOR EQUIPMENT		92.3	92.3		92.3	92.3		96.0	96.0		100.7	100.7		159.0	159.0		92.3	92.3
0241, 31 - 34	SITE & INFRASTRUCTURE, DEMOLITION	114.8	74.5	87.1	111.1	74.1	85.7	146.1	100.4	114.7	153.4	108.5	122.6	192.5	97.3	127.1	84.6	88.8	87.5
0310	Concrete Forming & Accessories	80.0	44.0	49.4	81.9	61.5	64.5	106.1	124.1	121.4	123.2	124.2	124.0	108.4	63.3	70.1	97.3	80.4	82.9
0320	Concrete Reinforcing	93.8	60.7	77.8	93.8	56.6	75.8	140.0	122.0	131.3	160.1	122.1	141.7	251.6	43.8	151.4	101.2	76.0	89.0
0330	Cast-in-Place Concrete	84.2	69.5	78.7	95.1	68.5	85.1	196.9	125.0	169.7	153.7	126.2	143.3	170.6	104.3	145.6	89.2	93.5	90.8
03	CONCRETE	88.2	58.1	74.7	91.1	64.8	79.3	153.2	123.2	139.7	146.4	123.5	136.1	156.1	74.9	119.7	97.6	84.3	91.7
04	MASONRY	104.4	69.3	83.0	105.1	68.1	82.5	140.4	124.4	130.7	132.4	124.5	127.6	200.5	48.8	108.0	125.0	84.2	100.1
05	METALS	105.0	93.4	101.4	104.0	86.8	98.7	110.0	107.8	109.4	124.9	106.4	119.1	141.9	81.4	123.0	107.7	80.4	99.2
06	WOOD, PLASTICS & COMPOSITES	67.9	36.3	51.3	69.6	61.2	65.2	108.3	124.7	116.9	135.2	124.8	129.8	119.7	63.6	90.4	89.5	79.4	84.2
07	THERMAL & MOISTURE PROTECTION	99.7	65.4	84.9	99.4	64.9	84.6	127.6	118.4	123.6	144.9	119.6	134.1	151.5	70.0	116.5	99.1	85.4	93.2
08	OPENINGS	82.2	50.5	74.5	82.5	60.7	77.2	114.7	122.3	116.6	128.7	122.4	127.2	118.9	52.8	102.8	97.1	74.0	91.4
0920	Plaster & Gypsum Board	94.2	35.1	55.4	94.2	60.8	72.3	122.9	125.3	124.5	169.2	125.3	140.4	249.0	53.0	120.4	94.3	79.1	84.3
0950, 0980	Ceilings & Acoustic Treatment	82.3	35.1	52.7	80.9	60.8	68.3	125.1	125.3	125.2	132.7	125.3	128.1	231.1	53.0	119.5	112.5	79.1	91.6
0960	Flooring	96.9	71.1	89.4	98.3	71.1	90.4	105.9	140.4	116.0	122.8	140.4	127.9	123.3	47.3	101.1	91.8	86.5	90.3
0970, 0990	Wall Finishes & Painting/Coating	85.1	43.1	89.9	85.1	66.7	74.1	96.1	137.0	120.6	109.0	137.0	125.8	101.6	39.6	64.5	89.3	45.7	63.2
09	FINISHES	91.8	51.9	70.2	91.4	63.4	76.3	111.9	129.0	121.1	126.8	129.0	128.0	186.6	58.2	117.1	96.1	77.7	86.1
COVERS	DIVS. 10 - 14, 25, 28, 41, 43, 44, 46	100.0	80.9	95.5	100.0	81.2	95.6	100.0	112.1	102.9	100.0	112.3	102.9	100.0	92.9	98.3	100.0	86.9	96.9
21, 22, 23	FIRE SUPPRESSION, PLUMBING & HVAC	100.7	69.1	88.0	98.4	60.5	83.1	101.1	108.9	104.2	101.1	108.9	104.3	104.1	52.5	83.2	101.2	71.5	89.2
26, 27, 3370	ELECTRICAL, COMMUNICATIONS & UTIL.	93.8	56.7	75.5	97.1	63.4	80.4	105.8	122.0	113.8	106.7	122.0	114.3	150.4	49.2	100.3	97.2	69.3	83.4
MF2018	WEIGHTED AVERAGE	96.7	65.4	83.2	96.6	66.9	83.8	115.6	117.2	116.3	120.3	117.8	119.2	138.0	63.4	105.7	100.9	78.2	91.1

DIVISION		IDAHO															ILLINOIS		
		COEUR D'ALENE			IDAHO FALLS			LEWISTON			POCATELLO			TWIN FALLS			BLOOMINGTON		
		838			834			835			832			833			617		
		MAT.	INST.	TOTAL	MAT.	INST.	TOTAL	MAT.	INST.	TOTAL	MAT.	INST.	TOTAL	MAT.	INST.	TOTAL	MAT.	INST.	TOTAL
015433	CONTRACTOR EQUIPMENT		87.3	87.3		92.3	92.3		87.3	87.3		92.3	92.3		92.3	92.3		98.8	98.8
0241, 31 - 34	SITE & INFRASTRUCTURE, DEMOLITION	84.4	83.1	83.5	83.3	88.6	86.9	91.3	84.4	86.5	86.1	88.8	87.9	93.0	88.6	90.0	94.4	92.2	92.9
0310	Concrete Forming & Accessories	103.6	78.3	82.1	90.7	75.5	77.8	109.1	81.3	85.5	97.4	79.9	82.5	98.5	79.8	82.6	80.3	111.8	107.1
0320	Concrete Reinforcing	109.3	99.8	104.7	103.2	74.2	89.2	109.3	100.0	104.8	101.6	76.1	89.3	103.6	77.0	90.7	95.4	97.5	96.4
0330	Cast-in-Place Concrete	96.5	81.5	90.8	85.0	80.1	83.1	100.3	81.9	93.4	91.7	93.3	92.3	94.2	81.6	89.4	94.4	109.3	100.0
03	CONCRETE	103.8	83.3	94.6	89.5	77.2	84.0	107.5	84.8	97.3	96.7	84.0	91.0	104.6	80.1	93.6	91.5	109.3	99.5
04	MASONRY	126.7	83.6	100.4	120.2	84.2	98.2	127.1	85.3	101.6	122.5	87.2	101.0	125.4	81.8	98.8	115.8	116.7	116.4
05	METALS	101.4	88.5	97.4	115.9	79.1	104.4	100.8	89.4	97.4	116.0	80.1	104.8	116.0	80.0	104.8	93.0	121.1	101.7
06	WOOD, PLASTICS & COMPOSITES	90.7	77.1	83.6	82.7	73.6	78.0	96.9	80.1	88.1	89.5	79.4	84.2	90.7	79.4	84.7	76.5	109.2	93.6
07	THERMAL & MOISTURE PROTECTION	152.8	80.4	121.7	97.7	72.2	86.8	153.1	82.5	122.8	98.3	74.7	88.2	99.2	80.7	91.2	95.5	107.9	100.8
08	OPENINGS	114.1	75.0	104.6	99.8	68.6	92.2	107.0	81.1	100.7	97.7	68.2	90.5	100.5	64.3	91.7	91.4	114.3	96.9
0920	Plaster & Gypsum Board	170.7	76.8	109.1	83.2	73.2	76.6	172.4	79.9	111.7	85.0	79.1	81.1	87.2	79.1	81.9	85.4	109.8	101.4
0950, 0980	Ceilings & Acoustic Treatment	146.0	76.8	102.6	113.9	73.2	88.4	146.0	79.9	104.6	119.7	79.1	94.2	116.5	79.1	93.0	84.1	109.8	100.2
0960	Flooring	129.9	73.5	113.5	91.5	73.5	86.2	133.5	78.0	117.3	95.2	78.0	90.2	96.5	82.3	92.3	85.5	115.3	94.2
0970, 0990	Wall Finishes & Painting/Coating	106.4	70.3	84.7	89.3	37.6	58.3	106.4	72.1	85.8	89.2	40.7	60.2	89.3	39.9	59.7	85.6	128.3	111.2
09	FINISHES	159.1	76.3	114.3	94.3	71.2	81.8	160.4	79.2	116.5	97.5	75.8	85.7	98.1	76.6	86.4	86.1	114.2	101.3
COVERS	DIVS. 10 - 14, 25, 28, 41, 43, 44, 46	100.0	88.2	97.2	100.0	86.3	96.8	100.0	97.4	99.4	100.0	86.9	96.9	100.0	86.9	96.9	100.0	105.5	101.3
21, 22, 23	FIRE SUPPRESSION, PLUMBING & HVAC	100.6	81.5	92.9	102.3	78.2	92.6	102.4	85.6	95.6	100.9	73.0	89.7	100.9	71.5	89.0	96.8	100.4	98.2
26, 27, 3370	ELECTRICAL, COMMUNICATIONS & UTIL.	88.4	78.5	83.5	88.0	68.2	78.2	86.8	84.7	85.8	94.3	67.5	81.1	89.1	70.7	80.0	94.7	87.0	90.8
MF2018	WEIGHTED AVERAGE	108.1	81.5	96.6	100.2	76.9	90.1	108.4	84.9	98.2	101.8	77.8	91.4	102.9	76.9	91.7	94.8	105.4	99.4

ILLINOIS

DIVISION		CARBONDALE 629 MAT.	INST.	TOTAL	CENTRALIA 628 MAT.	INST.	TOTAL	CHAMPAIGN 618-619 MAT.	INST.	TOTAL	CHICAGO 606-608 MAT.	INST.	TOTAL	DECATUR 625 MAT.	INST.	TOTAL	EAST ST. LOUIS 620-622 MAT.	INST.	TOTAL
015433	CONTRACTOR EQUIPMENT		107.2	107.2		107.2	107.2		99.7	99.7		97.0	97.0		99.7	99.7		107.2	107.2
0241, 31 - 34	SITE & INFRASTRUCTURE, DEMOLITION	103.0	92.8	96.0	103.4	94.7	97.4	103.2	93.2	96.3	105.6	100.0	101.7	96.7	93.4	94.4	106.0	94.3	97.9
0310	Concrete Forming & Accessories	88.6	104.8	102.4	90.5	109.9	107.0	86.9	112.4	108.6	100.5	159.0	150.3	91.5	116.6	112.9	86.0	112.8	108.8
0320	Concrete Reinforcing	90.9	98.5	94.6	90.9	99.1	94.8	95.4	96.8	96.1	104.8	151.0	127.1	89.1	96.9	92.9	90.8	105.1	97.7
0330	Cast-in-Place Concrete	92.1	98.4	94.5	92.5	114.7	100.9	109.5	106.3	108.3	120.6	154.2	133.3	100.5	113.9	105.6	94.1	117.7	103.0
03	CONCRETE	82.4	103.1	91.7	82.9	111.3	95.6	103.5	108.3	105.7	112.4	155.4	131.7	92.9	112.9	101.9	83.9	114.7	97.7
04	MASONRY	81.6	106.8	97.0	81.6	117.8	103.7	141.6	120.2	128.5	102.7	164.4	140.4	77.1	120.0	103.3	81.9	121.5	106.1
05	METALS	98.5	127.7	107.6	98.6	130.8	108.7	93.0	117.8	100.7	95.7	147.2	111.8	102.5	119.3	107.8	99.7	134.5	110.6
06	WOOD, PLASTICS & COMPOSITES	85.6	102.2	94.3	88.4	107.4	98.3	83.4	110.1	97.4	108.0	158.3	134.3	87.7	116.1	102.6	82.6	109.5	96.6
07	THERMAL & MOISTURE PROTECTION	90.6	96.7	93.2	90.7	106.5	97.5	96.3	110.4	102.3	94.7	149.2	118.1	95.9	111.7	102.7	90.7	109.1	98.6
08	OPENINGS	87.1	112.0	93.2	87.1	114.8	93.9	91.9	113.4	97.1	101.2	168.5	117.6	97.9	116.8	102.5	87.2	115.3	94.0
0920	Plaster & Gypsum Board	86.8	102.6	97.2	87.9	107.9	101.0	87.1	110.7	102.6	96.0	159.9	138.0	89.8	116.9	107.7	85.4	110.1	101.6
0950, 0980	Ceilings & Acoustic Treatment	79.6	102.6	94.0	79.6	107.9	97.4	84.1	110.7	100.8	89.6	159.9	133.7	86.6	116.9	105.6	79.6	110.1	98.7
0960	Flooring	116.3	111.3	114.8	117.3	111.3	115.5	88.9	111.3	95.4	95.4	161.3	114.6	104.6	114.0	107.3	115.2	113.5	114.7
0970, 0990	Wall Finishes & Painting/Coating	99.9	99.9	99.9	99.9	106.3	103.7	85.6	109.4	99.9	93.9	162.4	135.0	92.8	112.2	104.4	99.9	110.9	106.5
09	FINISHES	91.4	106.2	99.4	91.9	109.8	101.6	88.0	112.2	101.1	94.8	160.6	130.4	92.0	116.2	105.1	91.0	112.6	102.7
COVERS	DIVS. 10 - 14, 25, 28, 41, 43, 44, 46	100.0	101.3	100.3	100.0	100.2	100.1	100.0	106.1	101.4	100.0	127.2	106.4	100.0	106.2	101.5	100.0	107.4	101.8
21, 22, 23	FIRE SUPPRESSION, PLUMBING & HVAC	96.7	101.7	98.7	96.7	93.2	95.3	96.8	101.9	98.9	100.4	134.9	114.3	100.5	95.9	98.7	100.5	96.7	99.0
26, 27, 3370	ELECTRICAL, COMMUNICATIONS & UTIL.	91.9	103.8	97.8	92.9	102.3	97.6	97.4	91.9	94.6	96.1	136.4	116.0	94.9	99.9	97.4	92.6	105.6	99.1
MF2018	WEIGHTED AVERAGE	92.7	105.2	98.1	92.9	106.7	98.9	98.4	106.3	101.8	100.5	144.4	119.5	96.7	107.6	101.4	94.1	109.8	100.9

ILLINOIS

DIVISION		EFFINGHAM 624 MAT.	INST.	TOTAL	GALESBURG 614 MAT.	INST.	TOTAL	JOLIET 604 MAT.	INST.	TOTAL	KANKAKEE 609 MAT.	INST.	TOTAL	LA SALLE 613 MAT.	INST.	TOTAL	NORTH SUBURBAN 600-603 MAT.	INST.	TOTAL
015433	CONTRACTOR EQUIPMENT		99.7	99.7		98.8	98.8		90.4	90.4		90.4	90.4		98.8	98.8		90.4	90.4
0241, 31 - 34	SITE & INFRASTRUCTURE, DEMOLITION	101.0	92.0	94.8	96.7	92.1	93.5	101.7	94.8	97.0	94.9	93.1	93.7	96.1	92.9	93.9	101.0	93.9	96.1
0310	Concrete Forming & Accessories	95.9	110.3	108.2	86.8	112.8	108.9	100.1	157.8	149.2	92.5	139.9	132.8	101.1	120.7	117.7	99.2	152.6	144.6
0320	Concrete Reinforcing	92.1	89.3	90.7	94.9	100.6	97.6	111.9	138.0	124.5	112.6	128.7	120.4	95.1	126.5	110.3	111.9	142.0	126.4
0330	Cast-in-Place Concrete	100.2	105.4	102.1	97.2	104.9	100.1	109.6	146.6	123.5	102.1	130.3	112.8	97.1	116.9	104.6	109.6	146.8	123.7
03	CONCRETE	93.6	105.6	99.0	94.4	108.8	100.9	104.1	149.8	124.6	97.9	134.3	114.2	95.3	121.3	106.9	104.1	148.2	123.9
04	MASONRY	85.7	110.5	100.8	116.0	117.4	116.8	101.8	159.4	136.9	98.3	140.9	124.3	116.0	124.3	121.1	98.8	152.2	131.4
05	METALS	99.5	112.4	103.6	93.0	123.5	102.5	93.8	138.8	107.8	93.8	130.6	105.3	93.0	139.3	107.5	94.8	138.9	108.6
06	WOOD, PLASTICS & COMPOSITES	90.3	110.1	100.6	83.3	111.4	98.0	101.1	158.5	131.1	93.0	139.8	117.5	99.4	119.0	109.6	100.0	152.9	127.7
07	THERMAL & MOISTURE PROTECTION	95.4	104.2	99.2	95.7	105.1	99.7	99.3	145.0	118.9	98.4	133.2	113.3	95.9	117.0	105.0	99.7	141.0	117.4
08	OPENINGS	93.1	110.3	97.3	91.4	112.9	96.6	99.2	163.6	114.9	92.5	150.5	106.6	91.4	128.4	100.4	99.2	161.8	114.5
0920	Plaster & Gypsum Board	89.3	110.7	103.3	87.1	112.1	103.5	90.0	160.4	136.2	89.4	141.2	123.4	94.5	119.9	111.1	94.3	154.7	133.9
0950, 0980	Ceilings & Acoustic Treatment	79.6	110.7	99.1	84.1	112.1	101.6	89.3	160.4	133.9	89.3	141.2	121.9	84.1	119.9	106.5	89.3	154.7	130.3
0960	Flooring	105.9	111.3	107.4	88.6	115.3	96.4	94.2	154.7	111.9	91.1	143.8	106.5	95.5	119.1	102.4	94.6	152.2	111.4
0970, 0990	Wall Finishes & Painting/Coating	92.8	104.1	99.6	85.6	93.8	90.5	83.9	169.6	135.2	83.9	129.2	111.0	85.6	126.7	110.2	85.4	156.1	127.8
09	FINISHES	90.6	110.5	101.4	87.3	111.7	100.5	91.4	159.6	128.3	90.1	140.1	117.1	90.3	119.8	106.3	92.1	153.8	125.5
COVERS	DIVS. 10 - 14, 25, 28, 41, 43, 44, 46	100.0	104.7	101.1	100.0	105.7	101.3	100.0	123.1	105.4	100.0	111.6	102.7	100.0	106.7	101.6	100.0	117.7	104.2
21, 22, 23	FIRE SUPPRESSION, PLUMBING & HVAC	96.7	99.5	97.9	96.8	100.7	98.4	100.4	131.4	112.9	96.6	123.4	107.4	96.8	122.3	107.1	100.3	129.7	112.2
26, 27, 3370	ELECTRICAL, COMMUNICATIONS & UTIL.	93.3	103.7	98.5	95.4	83.2	89.4	95.2	133.5	114.2	91.1	131.9	111.3	93.0	131.9	112.3	95.1	130.5	112.6
MF2018	WEIGHTED AVERAGE	95.3	104.9	99.4	95.5	104.8	99.5	98.5	140.3	116.5	95.1	129.5	110.0	95.8	122.2	107.2	98.5	137.3	115.3

ILLINOIS

DIVISION		PEORIA 615-616 MAT.	INST.	TOTAL	QUINCY 623 MAT.	INST.	TOTAL	ROCK ISLAND 612 MAT.	INST.	TOTAL	ROCKFORD 610-611 MAT.	INST.	TOTAL	SOUTH SUBURBAN 605 MAT.	INST.	TOTAL	SPRINGFIELD 626-627 MAT.	INST.	TOTAL
015433	CONTRACTOR EQUIPMENT		98.8	98.8		99.7	99.7		98.8	98.8		98.8	98.8		90.4	90.4		105.4	105.4
0241, 31 - 34	SITE & INFRASTRUCTURE, DEMOLITION	97.7	93.0	94.5	99.7	92.4	94.7	94.9	90.9	92.1	97.1	94.1	95.0	101.0	93.8	96.0	102.5	102.4	102.4
0310	Concrete Forming & Accessories	90.6	117.9	113.8	93.5	113.8	110.8	88.4	95.3	94.3	94.3	130.0	124.6	99.2	152.6	144.6	90.9	119.9	115.6
0320	Concrete Reinforcing	92.5	103.8	98.0	91.7	82.1	87.0	94.9	94.2	94.6	87.6	132.9	109.4	111.9	142.0	126.4	88.3	101.9	94.9
0330	Cast-in-Place Concrete	94.3	114.9	102.1	100.4	102.1	101.0	95.1	94.2	94.8	96.5	129.2	108.8	109.6	146.7	123.6	95.6	113.1	102.2
03	CONCRETE	91.7	115.2	102.3	93.2	104.9	98.5	92.3	96.0	94.0	92.3	130.9	109.6	104.1	148.1	123.8	89.3	114.9	100.8
04	MASONRY	114.6	122.5	119.4	109.1	108.5	108.7	115.8	96.1	103.8	89.4	142.0	121.5	98.8	152.1	131.3	89.0	124.3	110.5
05	METALS	95.7	125.1	104.8	99.6	110.1	102.9	93.0	115.9	100.2	95.7	144.2	110.8	94.8	138.7	108.5	100.1	120.3	106.4
06	WOOD, PLASTICS & COMPOSITES	91.8	115.9	104.4	87.5	116.1	102.5	85.0	94.2	89.8	91.7	126.6	110.0	100.0	152.9	127.7	87.1	119.0	103.8
07	THERMAL & MOISTURE PROTECTION	96.4	112.7	103.4	95.4	102.6	98.5	95.7	93.9	94.9	99.1	129.3	112.1	99.7	141.0	117.4	98.1	116.3	105.9
08	OPENINGS	96.6	120.9	102.5	93.8	111.3	98.1	91.4	99.5	93.3	96.6	138.1	106.7	99.2	161.8	114.5	98.8	119.9	103.9
0920	Plaster & Gypsum Board	91.9	116.6	108.1	87.9	116.9	106.9	87.1	94.4	91.9	91.9	127.6	115.3	94.3	154.7	133.9	90.4	119.5	109.5
0950, 0980	Ceilings & Acoustic Treatment	89.2	116.6	106.4	79.6	116.9	103.0	84.1	94.4	90.5	89.2	127.6	113.3	89.3	154.7	130.3	88.5	119.5	108.0
0960	Flooring	92.3	121.0	100.7	104.6	108.4	105.7	89.8	90.3	89.9	92.3	124.0	101.5	94.6	152.2	111.4	105.3	128.7	112.1
0970, 0990	Wall Finishes & Painting/Coating	85.6	140.2	118.3	92.8	106.2	100.8	85.6	91.0	88.8	85.6	144.2	120.8	85.4	156.1	127.8	98.4	114.0	107.7
09	FINISHES	90.3	121.0	106.9	90.0	113.2	102.5	87.6	93.7	90.9	90.3	130.0	111.8	92.1	153.8	125.5	95.5	121.7	109.7
COVERS	DIVS. 10 - 14, 25, 28, 41, 43, 44, 46	100.0	106.8	101.6	100.0	104.5	101.1	100.0	92.3	98.2	100.0	113.3	103.1	100.0	117.7	104.2	100.0	107.9	101.9
21, 22, 23	FIRE SUPPRESSION, PLUMBING & HVAC	100.6	102.2	101.2	96.7	98.0	97.2	96.8	92.9	95.2	100.7	115.5	106.7	100.3	129.7	112.2	100.5	102.5	101.3
26, 27, 3370	ELECTRICAL, COMMUNICATIONS & UTIL.	96.3	93.5	94.9	91.5	79.2	85.4	88.9	89.7	89.3	96.5	128.3	112.3	95.1	130.5	112.6	100.9	91.4	96.2
MF2018	WEIGHTED AVERAGE	97.4	109.9	102.8	96.1	101.1	98.3	94.5	95.5	94.9	96.3	126.3	109.3	98.5	137.2	115.3	97.8	110.4	103.2

INDIANA

DIVISION		ANDERSON 460			BLOOMINGTON 474			COLUMBUS 472			EVANSVILLE 476 - 477			FORT WAYNE 467 - 468			GARY 463 - 464		
		MAT.	INST.	TOTAL	MAT.	INST.	TOTAL	MAT.	INST.	TOTAL	MAT.	INST.	TOTAL	MAT.	INST.	TOTAL	MAT.	INST.	TOTAL
015433	CONTRACTOR EQUIPMENT		92.7	92.7		79.6	79.6		79.6	79.6		108.7	108.7		92.7	92.7		92.7	92.7
0241, 31 - 34	SITE & INFRASTRUCTURE, DEMOLITION	99.7	86.5	90.6	87.3	85.1	85.8	83.6	85.1	84.6	92.8	112.6	106.4	100.5	86.4	90.9	100.3	90.2	93.4
0310	Concrete Forming & Accessories	94.5	78.9	81.3	103.2	80.5	83.9	96.8	79.5	82.1	95.6	80.1	82.4	92.9	73.8	76.7	94.6	109.1	107.0
0320	Concrete Reinforcing	102.1	83.9	93.3	89.6	84.9	87.3	90.0	84.9	87.6	98.1	80.4	89.6	102.1	77.2	90.1	102.1	112.5	107.1
0330	Cast-in-Place Concrete	102.2	75.4	92.1	99.8	75.1	90.5	99.4	73.3	89.5	95.4	83.2	90.8	108.6	74.3	95.7	106.8	109.8	107.9
03	CONCRETE	94.9	79.2	87.8	99.2	79.1	90.2	98.4	78.0	89.2	99.3	81.5	91.3	97.8	75.3	87.7	97.0	109.9	102.8
04	MASONRY	83.7	75.3	78.6	90.4	73.1	79.8	90.2	73.1	79.7	85.8	78.7	81.4	87.6	71.2	77.6	85.1	107.9	99.0
05	METALS	102.9	89.4	98.7	100.8	75.6	92.9	100.8	75.1	92.8	93.4	84.6	90.7	102.9	86.8	97.8	102.9	106.8	104.1
06	WOOD, PLASTICS & COMPOSITES	92.8	79.4	85.8	110.3	81.3	95.1	104.0	80.1	91.5	91.9	79.4	85.4	92.7	74.1	83.0	90.6	107.5	99.4
07	THERMAL & MOISTURE PROTECTION	107.9	75.3	93.9	95.3	77.3	87.6	94.7	76.6	86.9	99.8	81.7	92.0	107.6	76.0	94.0	106.5	103.5	105.2
08	OPENINGS	92.9	77.4	89.1	97.3	78.9	92.8	93.8	78.2	90.0	91.7	76.9	88.1	92.9	71.3	87.6	92.9	112.6	97.7
0920	Plaster & Gypsum Board	100.4	79.3	86.5	95.4	81.7	86.4	92.3	80.4	84.5	90.9	78.7	82.9	99.8	73.8	82.8	94.2	108.2	103.3
0950, 0980	Ceilings & Acoustic Treatment	96.0	79.3	85.5	78.5	81.7	80.5	78.5	80.4	79.7	82.1	78.7	80.0	96.0	73.8	82.1	96.0	108.2	103.6
0960	Flooring	94.1	76.0	88.8	100.5	81.8	95.1	95.2	81.8	91.3	95.4	72.3	88.6	94.1	70.6	87.2	94.1	107.6	98.0
0970, 0990	Wall Finishes & Painting/Coating	92.0	67.9	77.5	84.7	79.7	81.7	84.7	79.7	81.7	90.9	83.6	86.5	92.0	71.0	79.4	92.0	117.3	107.1
09	FINISHES	92.8	77.2	84.3	91.3	80.8	85.6	89.3	80.1	84.3	90.2	78.8	84.1	92.6	72.9	81.9	91.8	109.5	101.4
COVERS	DIVS. 10 - 14, 25, 28, 41, 43, 44, 46	100.0	89.9	97.6	100.0	89.3	97.5	100.0	89.2	97.5	100.0	91.6	98.0	100.0	90.2	97.7	100.0	102.9	100.7
21, 22, 23	FIRE SUPPRESSION, PLUMBING & HVAC	100.6	76.7	90.9	100.2	77.6	91.1	96.4	76.9	88.6	100.5	78.2	91.5	100.6	71.6	88.9	100.6	103.1	101.6
26, 27, 3370	ELECTRICAL, COMMUNICATIONS & UTIL.	89.4	84.9	87.2	100.2	85.0	92.7	99.6	87.1	93.4	96.2	83.2	89.7	90.1	75.0	82.6	100.6	108.1	104.3
MF2018	WEIGHTED AVERAGE	96.8	80.4	89.7	98.2	79.7	90.2	96.5	79.4	89.1	96.0	83.2	90.5	97.5	75.9	88.2	98.3	105.8	101.5

INDIANA

DIVISION		INDIANAPOLIS 461 - 462			KOKOMO 469			LAFAYETTE 479			LAWRENCEBURG 470			MUNCIE 473			NEW ALBANY 471		
		MAT.	INST.	TOTAL	MAT.	INST.	TOTAL	MAT.	INST.	TOTAL	MAT.	INST.	TOTAL	MAT.	INST.	TOTAL	MAT.	INST.	TOTAL
015433	CONTRACTOR EQUIPMENT		87.9	87.9		92.7	92.7		79.6	79.6		98.9	98.9		91.0	91.0		88.9	88.9
0241, 31 - 34	SITE & INFRASTRUCTURE, DEMOLITION	99.5	93.7	95.5	95.4	86.5	89.3	84.3	85.1	84.8	81.6	99.5	93.8	87.1	85.5	86.0	79.5	87.3	84.9
0310	Concrete Forming & Accessories	97.4	85.0	86.8	97.6	77.3	80.3	93.8	81.9	83.7	92.7	77.6	79.8	93.1	78.5	80.7	92.1	76.2	78.6
0320	Concrete Reinforcing	101.0	85.3	93.4	92.3	85.2	88.9	89.6	85.2	87.4	88.8	78.0	83.6	99.1	83.8	91.7	90.2	80.4	85.5
0330	Cast-in-Place Concrete	99.3	85.1	94.0	101.2	81.0	93.6	99.9	78.4	91.8	93.5	72.7	85.6	104.9	74.6	93.5	96.5	72.6	87.5
03	CONCRETE	97.7	84.5	91.8	91.7	80.6	86.7	98.6	80.9	90.6	91.8	76.6	85.0	97.5	78.7	89.0	97.2	76.1	87.7
04	MASONRY	94.5	79.0	85.0	83.4	76.0	78.8	95.3	76.2	83.6	75.0	73.0	73.8	92.1	75.4	81.9	81.4	70.3	74.7
05	METALS	99.7	75.4	92.1	99.2	89.9	96.3	99.1	75.7	91.8	95.5	86.1	92.6	102.5	89.3	98.4	97.6	82.0	92.8
06	WOOD, PLASTICS & COMPOSITES	96.5	85.6	90.8	96.0	76.7	85.9	100.7	83.1	91.5	89.9	77.9	83.6	101.3	79.0	89.6	92.5	77.1	84.4
07	THERMAL & MOISTURE PROTECTION	97.6	81.3	90.6	106.9	76.4	93.8	94.7	78.9	87.9	99.7	75.9	89.5	97.7	76.7	88.7	88.3	73.3	81.9
08	OPENINGS	103.5	81.2	98.1	88.2	76.3	85.3	92.3	79.9	89.3	93.5	74.1	88.8	90.8	77.2	87.4	91.3	73.2	86.9
0920	Plaster & Gypsum Board	87.9	85.3	86.2	105.1	76.5	86.3	89.9	83.6	85.7	70.8	78.0	75.5	90.9	79.3	83.3	88.3	77.1	81.0
0950, 0980	Ceilings & Acoustic Treatment	97.6	85.3	89.9	96.6	76.5	84.0	74.7	83.6	80.2	86.5	78.0	81.2	78.5	79.3	79.0	82.1	77.1	79.0
0960	Flooring	95.9	81.8	91.8	98.2	87.9	95.2	94.0	81.8	90.4	69.7	81.8	73.2	94.8	76.0	89.3	92.6	51.5	80.6
0970, 0990	Wall Finishes & Painting/Coating	99.6	79.7	87.7	92.0	67.2	77.2	84.7	82.3	83.2	85.5	71.8	77.3	84.7	67.9	74.6	90.9	64.4	75.0
09	FINISHES	94.1	84.1	88.7	94.6	78.1	85.6	87.7	82.1	84.7	80.7	77.9	79.2	88.7	76.8	82.3	89.3	70.1	78.9
COVERS	DIVS. 10 - 14, 25, 28, 41, 43, 44, 46	100.0	92.3	98.2	100.0	89.7	97.6	100.0	89.4	97.5	100.0	88.8	97.4	100.0	88.9	97.4	100.0	88.6	97.3
21, 22, 23	FIRE SUPPRESSION, PLUMBING & HVAC	100.5	78.5	91.6	96.8	77.1	88.9	96.4	77.9	88.9	97.4	73.8	87.9	100.2	76.5	90.7	96.7	76.7	88.7
26, 27, 3370	ELECTRICAL, COMMUNICATIONS & UTIL.	102.5	87.1	94.9	93.3	77.5	85.5	99.1	80.3	89.8	94.2	73.9	84.2	92.2	75.7	84.0	94.9	74.8	84.9
MF2018	WEIGHTED AVERAGE	99.5	82.9	92.3	94.9	79.8	88.4	96.2	79.9	89.1	93.0	78.3	86.6	96.6	78.9	89.0	94.3	76.3	86.5

INDIANA / IOWA

DIVISION		SOUTH BEND 465 - 466			TERRE HAUTE 478			WASHINGTON 475			BURLINGTON 526			CARROLL 514			CEDAR RAPIDS 522 - 524		
		MAT.	INST.	TOTAL	MAT.	INST.	TOTAL	MAT.	INST.	TOTAL	MAT.	INST.	TOTAL	MAT.	INST.	TOTAL	MAT.	INST.	TOTAL
015433	CONTRACTOR EQUIPMENT		108.4	108.4		108.7	108.7		108.7	108.7		97.0	97.0		97.0	97.0		93.8	93.8
0241, 31 - 34	SITE & INFRASTRUCTURE, DEMOLITION	100.0	95.3	96.8	94.3	112.8	107.0	93.8	113.1	107.1	98.1	88.8	91.7	86.9	89.6	88.7	100.4	89.1	92.6
0310	Concrete Forming & Accessories	96.2	77.8	80.5	96.6	78.6	81.3	97.7	81.6	84.0	98.0	90.7	91.8	83.6	80.8	81.2	104.8	85.9	88.7
0320	Concrete Reinforcing	100.3	82.9	91.9	98.1	84.0	91.3	90.8	85.0	88.0	94.3	93.2	93.8	95.0	81.8	88.6	95.0	99.3	97.1
0330	Cast-in-Place Concrete	102.3	80.1	93.9	92.4	78.9	87.3	100.5	85.8	95.0	105.0	52.5	85.2	105.0	78.9	95.1	105.3	84.4	97.4
03	CONCRETE	101.5	80.9	92.2	102.3	80.0	92.3	108.0	83.9	97.2	96.3	78.4	88.2	94.9	81.1	88.7	96.4	88.4	92.8
04	MASONRY	81.6	75.4	77.8	93.6	75.4	82.5	85.9	78.3	81.2	101.6	68.9	81.7	103.1	69.4	82.6	107.5	82.7	92.4
05	METALS	106.8	102.5	105.5	94.1	86.1	91.6	88.6	87.1	88.1	86.0	97.3	89.5	86.1	93.7	88.4	88.5	102.5	92.9
06	WOOD, PLASTICS & COMPOSITES	97.3	77.1	86.7	94.5	78.4	86.1	94.8	80.9	87.5	93.7	94.8	94.3	77.6	85.4	81.7	102.0	85.6	93.4
07	THERMAL & MOISTURE PROTECTION	102.2	79.6	92.5	99.9	79.9	91.3	99.8	82.7	92.5	104.1	73.6	91.0	104.3	75.3	91.9	105.2	81.4	95.0
08	OPENINGS	92.9	76.5	88.9	92.2	76.6	88.4	89.1	80.3	87.0	94.3	92.3	93.8	98.5	81.1	94.3	99.0	87.8	96.3
0920	Plaster & Gypsum Board	88.4	76.6	80.7	90.9	77.7	82.2	90.9	80.2	83.9	99.9	95.0	96.7	95.9	85.4	89.0	105.4	85.7	92.5
0950, 0980	Ceilings & Acoustic Treatment	96.2	76.6	83.9	82.1	77.7	79.3	77.6	80.2	79.2	103.1	95.0	98.0	103.1	85.4	92.0	105.6	85.7	93.1
0960	Flooring	92.3	87.3	90.8	95.4	76.6	89.9	96.4	81.8	92.1	95.3	65.9	86.7	89.3	75.9	85.4	109.7	87.3	103.3
0970, 0990	Wall Finishes & Painting/Coating	95.9	83.4	88.4	90.9	79.1	83.8	90.9	86.5	88.3	88.7	80.6	83.8	88.7	80.6	83.8	90.6	72.3	79.6
09	FINISHES	93.1	79.8	85.9	90.2	78.0	83.6	89.7	81.9	85.5	95.9	86.2	90.7	92.1	80.3	85.7	101.4	84.7	92.4
COVERS	DIVS. 10 - 14, 25, 28, 41, 43, 44, 46	100.0	92.0	98.1	100.0	91.1	97.9	100.0	91.8	98.1	100.0	93.8	98.5	100.0	89.7	97.6	100.0	93.1	98.4
21, 22, 23	FIRE SUPPRESSION, PLUMBING & HVAC	100.5	75.9	90.6	100.5	78.1	91.5	96.7	80.0	90.0	96.9	79.8	90.0	96.9	74.1	87.7	100.7	84.4	94.1
26, 27, 3370	ELECTRICAL, COMMUNICATIONS & UTIL.	103.5	88.8	96.2	94.7	85.3	90.0	95.1	83.8	89.5	99.7	69.6	84.8	100.3	74.1	87.3	97.7	80.2	89.1
MF2018	WEIGHTED AVERAGE	99.6	83.4	92.6	96.9	82.9	90.8	95.1	84.8	90.7	95.8	81.1	89.5	95.5	79.3	88.5	98.3	86.6	93.2

City Cost Indexes - V2

IOWA

DIVISION		COUNCIL BLUFFS 515 MAT.	INST.	TOTAL	CRESTON 508 MAT.	INST.	TOTAL	DAVENPORT 527-528 MAT.	INST.	TOTAL	DECORAH 521 MAT.	INST.	TOTAL	DES MOINES 500-503,509 MAT.	INST.	TOTAL	DUBUQUE 520 MAT.	INST.	TOTAL
015433	CONTRACTOR EQUIPMENT		93.2	93.2		97.0	97.0		97.0	97.0		97.0	97.0		103.5	103.5		92.7	92.7
0241, 31 - 34	SITE & INFRASTRUCTURE, DEMOLITION	105.5	86.2	92.2	92.0	90.6	91.0	98.8	92.7	94.6	96.7	88.6	91.1	98.1	101.0	100.1	98.2	86.5	90.2
0310	Concrete Forming & Accessories	82.7	76.4	77.3	77.2	83.5	82.6	104.3	100.5	101.1	95.1	69.1	73.0	96.1	87.1	88.4	84.5	83.2	83.4
0320	Concrete Reinforcing	96.9	84.5	91.0	94.0	81.8	88.1	95.0	99.9	97.4	94.3	81.2	88.0	103.3	99.6	101.5	93.7	80.8	87.5
0330	Cast-in-Place Concrete	109.4	88.2	101.4	107.8	82.2	98.1	101.4	96.5	99.6	102.2	72.9	91.1	91.9	92.7	92.2	103.1	83.8	95.8
03	CONCRETE	98.3	82.9	91.4	95.5	83.5	90.1	94.6	99.6	96.8	94.2	73.7	85.0	95.4	91.5	93.6	93.0	83.8	88.9
04	MASONRY	107.3	78.0	89.4	104.4	76.5	87.4	103.1	95.1	98.3	123.7	66.3	88.7	89.6	87.2	88.1	108.6	70.0	85.1
05	METALS	93.1	96.0	94.0	86.4	94.3	88.9	88.5	108.4	94.7	86.1	92.5	88.1	92.0	96.4	93.4	87.1	93.8	89.2
06	WOOD, PLASTICS & COMPOSITES	76.1	75.1	75.6	69.3	85.4	77.7	102.0	100.3	101.1	90.1	68.5	78.8	90.4	85.6	87.9	78.2	84.2	81.3
07	THERMAL & MOISTURE PROTECTION	104.4	75.6	92.1	106.0	78.5	94.2	104.6	93.8	100.0	104.3	66.8	88.2	98.6	85.5	92.9	104.7	76.9	92.8
08	OPENINGS	98.1	80.0	93.7	107.9	82.8	101.8	99.0	101.9	99.7	97.2	75.0	91.8	100.6	87.8	97.5	98.1	84.2	94.7
0920	Plaster & Gypsum Board	95.9	75.0	82.2	91.6	85.4	87.5	105.4	100.7	102.3	98.8	62.9	78.5	84.6	85.4	85.1	95.9	84.3	88.3
0950, 0980	Ceilings & Acoustic Treatment	103.1	75.0	85.5	88.4	85.4	86.5	105.6	100.7	102.5	103.1	67.9	81.0	91.0	85.4	87.5	103.1	84.3	91.3
0960	Flooring	88.3	85.3	87.4	79.3	65.9	75.4	97.8	90.6	95.7	94.8	65.9	86.4	92.2	92.0	92.2	100.5	71.3	92.0
0970, 0990	Wall Finishes & Painting/Coating	85.6	57.0	68.4	82.2	80.6	81.2	88.7	93.0	91.3	88.7	80.6	83.8	93.1	82.9	87.0	89.8	79.3	83.5
09	FINISHES	93.0	75.6	83.6	83.5	80.1	81.7	98.0	97.9	97.9	95.6	69.5	81.5	90.3	87.4	88.7	96.7	80.6	88.0
COVERS	DIVS. 10 - 14, 25, 28, 41, 43, 44, 46	100.0	90.5	97.8	100.0	92.1	98.1	100.0	97.3	99.4	100.0	89.2	97.5	100.0	94.7	98.8	100.0	91.9	98.1
21, 22, 23	FIRE SUPPRESSION, PLUMBING & HVAC	100.7	79.6	92.2	96.8	76.0	88.4	100.7	95.6	98.6	96.9	70.9	86.4	100.5	85.0	94.3	100.7	76.3	90.9
26, 27, 3370	ELECTRICAL, COMMUNICATIONS & UTIL.	102.1	82.0	92.2	93.3	74.1	83.8	95.9	88.1	92.0	97.7	45.5	71.9	103.8	91.4	97.6	100.9	77.0	89.1
MF2018	WEIGHTED AVERAGE	98.8	81.9	91.5	95.2	81.1	89.1	97.3	96.6	97.0	97.3	71.0	85.6	97.3	90.0	94.2	97.3	80.6	90.1

IOWA

DIVISION		FORT DODGE 505 MAT.	INST.	TOTAL	MASON CITY 504 MAT.	INST.	TOTAL	OTTUMWA 525 MAT.	INST.	TOTAL	SHENANDOAH 516 MAT.	INST.	TOTAL	SIBLEY 512 MAT.	INST.	TOTAL	SIOUX CITY 510-511 MAT.	INST.	TOTAL
015433	CONTRACTOR EQUIPMENT		97.0	97.0		97.0	97.0		92.7	92.7		93.2	93.2		97.0	97.0		97.0	97.0
0241, 31 - 34	SITE & INFRASTRUCTURE, DEMOLITION	99.5	87.5	91.2	99.5	88.5	92.0	98.1	83.7	88.2	103.2	85.6	91.1	108.1	88.5	94.7	110.2	90.7	96.8
0310	Concrete Forming & Accessories	77.8	75.1	75.5	82.2	68.7	70.7	92.7	83.9	85.2	84.7	74.3	75.8	85.6	36.4	43.7	104.8	75.2	79.6
0320	Concrete Reinforcing	94.0	81.1	87.8	93.9	81.2	87.8	94.3	93.3	93.8	96.9	81.7	89.6	96.9	81.1	89.3	95.0	97.9	96.4
0330	Cast-in-Place Concrete	101.2	39.9	78.0	101.2	68.3	88.7	105.7	61.2	88.9	105.8	79.3	95.8	103.7	53.1	84.6	104.4	87.3	97.9
03	CONCRETE	91.1	64.9	79.4	91.4	71.9	82.6	95.7	78.4	88.0	95.9	78.3	88.0	94.9	51.9	75.6	95.8	84.3	90.7
04	MASONRY	103.4	49.8	70.7	117.4	65.1	85.5	104.3	52.9	73.0	106.8	71.4	85.2	126.4	50.0	79.8	101.1	68.4	81.2
05	METALS	86.5	92.2	88.3	86.5	92.5	88.4	86.0	98.0	89.7	92.1	93.8	92.6	86.2	91.6	87.9	88.5	101.2	92.4
06	WOOD, PLASTICS & COMPOSITES	69.8	85.4	78.0	74.4	68.5	71.3	86.8	94.6	90.9	78.2	75.8	77.0	79.2	32.9	55.0	102.0	74.6	87.7
07	THERMAL & MOISTURE PROTECTION	105.4	62.2	86.8	104.9	67.3	88.8	104.9	67.1	88.7	103.7	70.0	89.3	104.0	52.7	82.0	104.6	71.6	90.4
08	OPENINGS	102.1	73.1	95.1	94.6	75.0	89.9	98.5	89.4	96.3	90.2	73.6	86.2	95.4	44.3	82.9	99.0	80.9	94.6
0920	Plaster & Gypsum Board	91.6	85.4	87.5	91.6	67.9	76.1	96.2	95.0	95.4	95.9	75.7	82.6	95.9	31.4	53.6	105.4	74.2	84.9
0950, 0980	Ceilings & Acoustic Treatment	88.4	85.4	86.5	88.4	67.9	75.6	103.1	95.0	98.0	103.1	75.7	85.9	103.1	31.4	58.2	105.6	74.2	86.0
0960	Flooring	80.5	65.9	76.3	82.6	65.9	77.7	103.7	65.9	92.7	89.1	70.5	83.7	90.2	65.9	83.2	97.9	71.3	90.1
0970, 0990	Wall Finishes & Painting/Coating	82.2	77.4	79.3	82.2	80.6	81.2	89.8	80.6	84.3	85.6	80.6	82.6	88.7	79.6	83.2	88.7	67.8	76.1
09	FINISHES	85.3	74.9	79.7	85.8	69.2	76.8	97.7	81.5	88.9	93.1	74.0	82.8	95.4	44.5	67.8	99.5	73.1	85.2
COVERS	DIVS. 10 - 14, 25, 28, 41, 43, 44, 46	100.0	85.6	96.6	100.0	88.9	97.4	100.0	87.3	97.0	100.0	83.7	96.2	100.0	80.0	95.3	100.0	90.5	97.8
21, 22, 23	FIRE SUPPRESSION, PLUMBING & HVAC	96.8	68.0	85.2	96.8	74.0	87.6	96.9	69.6	85.9	96.9	79.9	90.0	96.9	68.1	85.3	100.7	78.8	91.9
26, 27, 3370	ELECTRICAL, COMMUNICATIONS & UTIL.	98.7	66.8	82.9	97.9	45.5	72.0	99.6	68.3	84.1	97.7	75.5	86.7	97.7	45.5	71.9	97.7	74.1	86.0
MF2018	WEIGHTED AVERAGE	95.0	71.0	84.6	95.0	71.2	84.7	96.4	75.7	87.5	96.1	78.7	88.6	96.9	60.1	81.0	98.0	80.3	90.3

IOWA / KANSAS

DIVISION		SPENCER 513 MAT.	INST.	TOTAL	WATERLOO 506-507 MAT.	INST.	TOTAL	BELLEVILLE 669 MAT.	INST.	TOTAL	COLBY 677 MAT.	INST.	TOTAL	DODGE CITY 678 MAT.	INST.	TOTAL	EMPORIA 668 MAT.	INST.	TOTAL
015433	CONTRACTOR EQUIPMENT		97.0	97.0		97.0	97.0		100.6	100.6		100.6	100.6		100.6	100.6		98.8	98.8
0241, 31 - 34	SITE & INFRASTRUCTURE, DEMOLITION	108.1	87.3	93.9	105.1	90.4	95.0	110.4	88.3	95.2	103.8	88.8	93.5	106.8	89.1	94.6	102.5	85.9	91.1
0310	Concrete Forming & Accessories	92.5	36.1	44.5	94.4	70.1	73.7	90.9	53.1	58.7	93.9	59.1	64.3	87.1	60.8	64.7	82.1	67.5	69.7
0320	Concrete Reinforcing	96.9	81.0	89.3	94.6	81.1	88.1	96.1	99.7	97.8	96.5	99.7	98.0	94.0	99.5	96.7	94.8	99.8	97.2
0330	Cast-in-Place Concrete	103.7	62.8	88.3	108.4	84.8	99.5	113.4	81.1	101.2	108.8	83.5	99.2	110.7	87.9	102.1	109.6	83.6	99.8
03	CONCRETE	95.4	55.1	77.3	97.0	78.2	88.6	107.6	72.4	91.8	104.3	76.0	91.6	105.5	78.2	93.3	100.3	79.8	91.1
04	MASONRY	126.4	50.0	79.8	104.1	76.4	87.2	91.2	58.0	70.9	97.2	61.9	75.6	111.1	59.8	79.8	96.9	63.3	76.4
05	METALS	86.2	91.3	87.8	88.8	94.5	90.6	93.4	97.6	94.7	90.6	98.3	93.1	92.2	98.0	94.0	93.1	98.6	94.8
06	WOOD, PLASTICS & COMPOSITES	86.7	32.9	58.6	89.3	65.8	77.0	85.8	49.3	66.7	92.8	55.3	73.2	84.5	56.2	69.7	76.8	66.4	71.4
07	THERMAL & MOISTURE PROTECTION	104.9	53.2	82.7	105.2	77.7	93.4	91.3	60.3	78.0	97.1	62.5	82.2	97.0	66.5	83.9	89.4	71.9	81.9
08	OPENINGS	106.1	44.3	91.0	95.1	74.5	90.0	93.7	60.6	85.6	97.5	63.9	89.3	97.5	64.3	89.4	91.8	72.0	87.0
0920	Plaster & Gypsum Board	96.2	31.4	53.7	99.5	65.2	77.0	85.3	48.3	61.0	97.1	54.4	69.1	90.6	55.3	67.4	82.2	65.9	71.5
0950, 0980	Ceilings & Acoustic Treatment	103.1	31.4	58.2	89.7	65.2	74.3	82.2	48.3	60.9	79.7	54.4	63.9	79.7	55.3	64.4	82.2	65.9	71.9
0960	Flooring	93.1	65.9	85.2	87.9	79.9	85.5	89.6	65.2	82.5	87.7	65.2	81.1	83.8	66.5	78.7	85.0	65.2	79.2
0970, 0990	Wall Finishes & Painting/Coating	88.7	54.5	68.2	82.2	84.6	83.6	89.0	98.1	94.4	89.0	98.1	94.4	94.9	98.1	96.8	89.0	98.1	94.4
09	FINISHES	96.2	40.6	66.1	89.0	72.4	80.0	86.5	58.6	71.4	86.6	63.1	73.9	84.6	64.3	73.7	83.7	69.6	76.1
COVERS	DIVS. 10 - 14, 25, 28, 41, 43, 44, 46	100.0	80.0	95.3	100.0	90.6	97.8	100.0	79.5	95.2	100.0	81.5	95.6	100.0	85.5	96.6	100.0	81.3	95.6
21, 22, 23	FIRE SUPPRESSION, PLUMBING & HVAC	96.9	68.1	85.3	100.6	81.0	92.7	96.7	68.0	85.1	96.7	67.7	85.0	100.5	76.3	90.7	96.7	70.0	85.9
26, 27, 3370	ELECTRICAL, COMMUNICATIONS & UTIL.	99.2	45.5	72.6	94.9	64.2	79.7	99.8	60.4	80.3	96.8	65.4	81.2	94.6	71.2	83.0	97.4	64.4	81.1
MF2018	WEIGHTED AVERAGE	98.3	59.9	81.7	96.5	78.4	88.7	96.9	69.1	84.9	96.5	71.7	85.7	98.0	74.8	88.0	95.2	74.0	86.3

For customer support on your Site Work & Landscape Costs with RSMeans Data, call 800.448.8182.

797

City Cost Indexes - V2

KANSAS

DIVISION		FORT SCOTT 667			HAYS 676			HUTCHINSON 675			INDEPENDENCE 673			KANSAS CITY 660 - 662			LIBERAL 679		
		MAT.	INST.	TOTAL	MAT.	INST.	TOTAL	MAT.	INST.	TOTAL	MAT.	INST.	TOTAL	MAT.	INST.	TOTAL	MAT.	INST.	TOTAL
015433	CONTRACTOR EQUIPMENT		99.7	99.7		100.6	100.6		100.6	100.6		100.6	100.6		97.3	97.3		100.6	100.6
0241, 31 - 34	SITE & INFRASTRUCTURE, DEMOLITION	99.2	86.0	90.1	108.8	88.6	94.9	88.5	88.9	88.8	107.2	89.0	94.7	94.0	86.8	89.1	108.5	88.6	94.8
0310	Concrete Forming & Accessories	99.4	79.8	82.7	91.4	57.6	62.7	81.6	54.7	58.7	102.5	65.4	70.9	95.9	99.3	98.8	87.5	56.8	61.4
0320	Concrete Reinforcing	94.1	96.0	95.0	94.0	99.7	96.8	94.0	99.7	96.7	93.5	96.0	94.7	91.3	104.3	97.6	95.4	99.4	97.3
0330	Cast-in-Place Concrete	101.6	78.9	93.1	85.8	81.5	84.2	79.5	83.4	81.0	111.1	83.5	100.7	87.1	96.7	90.7	85.8	80.1	83.7
03	CONCRETE	96.0	83.1	90.2	96.7	74.6	86.8	80.2	73.9	77.4	106.7	78.2	93.9	88.6	99.7	93.6	98.5	73.7	87.4
04	MASONRY	99.4	53.8	71.6	108.5	58.1	77.8	97.2	61.9	75.6	94.6	61.9	74.6	98.5	98.1	98.3	109.0	51.4	73.8
05	METALS	93.1	96.9	94.3	90.2	98.3	92.8	90.0	97.8	92.5	89.9	97.0	92.1	101.0	105.9	102.5	90.5	97.5	92.7
06	WOOD, PLASTICS & COMPOSITES	96.1	87.4	91.5	89.8	55.3	71.8	79.2	49.5	63.7	103.9	63.5	82.8	92.1	99.8	96.1	85.1	55.3	69.5
07	THERMAL & MOISTURE PROTECTION	90.5	69.5	81.5	97.4	61.0	81.8	95.8	61.6	81.1	97.2	71.5	86.1	89.9	97.8	93.3	97.6	58.2	80.7
08	OPENINGS	91.8	82.7	89.6	97.4	63.9	89.3	97.4	60.7	88.5	95.6	67.5	88.7	93.0	98.5	94.3	97.5	63.9	89.3
0920	Plaster & Gypsum Board	87.9	87.4	87.5	94.9	54.4	68.3	89.2	48.5	62.5	106.2	62.9	77.8	81.9	100.1	93.9	91.5	54.4	67.2
0950, 0980	Ceilings & Acoustic Treatment	82.2	87.4	85.4	79.7	54.4	63.9	79.7	48.5	60.1	79.7	62.9	69.2	82.2	100.1	93.4	79.7	54.4	63.9
0960	Flooring	99.0	63.8	88.7	86.4	65.2	80.2	80.7	65.2	76.2	92.1	63.8	83.9	79.7	96.0	84.4	84.0	65.2	78.5
0970, 0990	Wall Finishes & Painting/Coating	90.5	73.8	80.5	94.9	98.1	96.8	94.9	98.1	96.8	94.9	98.1	96.8	96.5	104.4	101.2	94.9	98.1	96.8
09	FINISHES	88.7	77.4	82.6	86.4	62.1	73.3	82.1	59.7	70.0	89.3	68.1	77.8	84.2	99.3	92.3	85.5	61.6	72.6
COVERS	DIVS. 10 - 14, 25, 28, 41, 43, 44, 46	100.0	82.2	95.8	100.0	80.2	95.3	100.0	80.8	95.5	100.0	82.4	95.9	100.0	95.6	99.0	100.0	79.6	95.2
21, 22, 23	FIRE SUPPRESSION, PLUMBING & HVAC	96.7	64.2	83.6	96.7	65.3	84.0	96.7	67.7	85.0	96.7	67.1	84.8	100.4	103.0	101.4	96.7	66.0	84.3
26, 27, 3370	ELECTRICAL, COMMUNICATIONS & UTIL.	96.7	65.4	81.2	96.0	65.4	80.9	92.5	58.7	75.7	94.0	70.1	82.2	101.6	97.1	99.4	94.6	65.4	80.1
MF2018	WEIGHTED AVERAGE	95.2	73.9	86.0	96.0	70.4	84.9	91.9	69.7	82.3	96.4	73.6	86.6	96.4	99.2	97.6	96.0	69.5	84.6

KANSAS / KENTUCKY

DIVISION		SALINA 674			TOPEKA 664 - 666			WICHITA 670 - 672			ASHLAND 411 - 412			BOWLING GREEN 421 - 422			CAMPTON 413 - 414		
		MAT.	INST.	TOTAL	MAT.	INST.	TOTAL	MAT.	INST.	TOTAL	MAT.	INST.	TOTAL	MAT.	INST.	TOTAL	MAT.	INST.	TOTAL
015433	CONTRACTOR EQUIPMENT		100.6	100.6		104.7	104.7		106.1	106.1		95.7	95.7		88.9	88.9		95.0	95.0
0241, 31 - 34	SITE & INFRASTRUCTURE, DEMOLITION	97.5	88.7	91.5	98.6	96.9	97.4	95.0	98.7	97.6	112.9	76.0	87.6	80.0	87.5	85.1	87.9	87.8	87.8
0310	Concrete Forming & Accessories	83.6	61.6	64.9	91.8	68.0	71.6	91.0	57.3	62.3	87.2	86.3	86.5	87.8	81.4	82.3	89.9	78.4	80.1
0320	Concrete Reinforcing	93.5	91.6	92.5	92.7	101.1	96.8	87.4	101.4	94.1	91.8	91.3	91.5	89.0	79.2	84.3	89.8	88.7	89.2
0330	Cast-in-Place Concrete	96.1	84.2	91.6	96.5	88.8	93.6	89.9	78.8	85.7	87.8	90.9	89.0	87.1	68.3	80.0	97.2	65.4	85.2
03	CONCRETE	93.3	77.4	86.1	91.4	82.0	87.2	87.6	73.6	81.3	92.9	90.0	91.6	91.5	76.9	84.9	94.9	76.1	86.5
04	MASONRY	126.2	55.1	82.8	94.2	65.1	76.4	95.3	53.2	69.6	92.5	88.2	89.9	94.7	69.6	79.4	91.3	52.4	67.6
05	METALS	92.0	98.6	94.1	97.3	98.6	97.7	94.1	96.6	94.9	96.7	106.5	99.8	98.4	85.1	94.2	97.6	88.6	94.8
06	WOOD, PLASTICS & COMPOSITES	80.8	60.7	70.3	90.6	66.7	78.1	96.3	55.1	74.8	74.3	84.7	79.7	86.3	83.8	85.0	84.5	85.3	84.9
07	THERMAL & MOISTURE PROTECTION	96.5	61.8	81.6	94.1	84.1	89.8	95.3	60.9	80.5	90.5	85.1	88.2	88.3	77.9	83.9	99.9	64.8	84.8
08	OPENINGS	97.4	67.3	90.1	103.2	74.8	96.3	100.9	64.6	92.1	90.5	86.1	89.4	91.3	79.5	88.4	92.5	84.1	90.4
0920	Plaster & Gypsum Board	89.2	59.9	70.0	94.3	65.8	75.6	93.9	53.9	67.7	56.9	84.7	75.2	84.1	84.0	84.0	84.1	84.7	84.5
0950, 0980	Ceilings & Acoustic Treatment	79.7	59.9	67.3	95.5	65.8	76.9	91.9	53.9	68.1	77.8	84.7	82.1	82.1	84.0	83.3	82.1	84.7	83.7
0960	Flooring	82.1	66.5	77.6	92.3	66.5	84.7	93.0	66.9	85.4	74.8	77.5	75.6	90.4	60.3	81.6	92.8	61.4	83.6
0970, 0990	Wall Finishes & Painting/Coating	94.9	98.1	96.8	95.3	98.1	97.0	96.4	57.3	73.0	91.6	86.1	88.3	90.9	66.9	76.5	90.9	51.3	67.2
09	FINISHES	83.3	65.3	73.6	93.6	69.5	80.6	92.7	57.6	73.7	76.9	84.6	81.1	88.0	76.0	81.5	88.8	72.9	80.2
COVERS	DIVS. 10 - 14, 25, 28, 41, 43, 44, 46	100.0	85.5	96.6	100.0	85.9	96.7	100.0	85.4	96.6	100.0	83.6	96.1	100.0	89.3	97.5	100.0	87.6	97.1
21, 22, 23	FIRE SUPPRESSION, PLUMBING & HVAC	100.5	69.4	87.9	100.6	74.5	90.1	100.3	70.6	88.3	96.4	81.0	90.2	100.5	77.9	91.4	96.7	72.8	87.1
26, 27, 3370	ELECTRICAL, COMMUNICATIONS & UTIL.	94.4	73.4	84.0	102.0	72.0	87.2	99.5	73.4	86.6	92.7	85.0	88.9	95.2	76.5	85.9	92.9	84.9	89.0
MF2018	WEIGHTED AVERAGE	96.8	73.3	86.6	98.1	78.0	89.4	96.5	72.2	86.0	93.6	86.3	90.5	95.2	78.4	87.9	94.9	76.3	86.9

KENTUCKY

DIVISION		CORBIN 407 - 409			COVINGTON 410			ELIZABETHTOWN 427			FRANKFORT 406			HAZARD 417 - 418			HENDERSON 424		
		MAT.	INST.	TOTAL	MAT.	INST.	TOTAL	MAT.	INST.	TOTAL	MAT.	INST.	TOTAL	MAT.	INST.	TOTAL	MAT.	INST.	TOTAL
015433	CONTRACTOR EQUIPMENT		95.0	95.0		98.9	98.9		88.9	88.9		99.9	99.9		95.0	95.0		108.7	108.7
0241, 31 - 34	SITE & INFRASTRUCTURE, DEMOLITION	92.1	88.2	89.4	83.1	99.8	93.9	73.8	86.5	82.5	90.3	97.6	95.3	85.7	88.8	87.8	82.2	111.3	102.2
0310	Concrete Forming & Accessories	86.2	73.1	75.1	85.1	68.2	70.7	81.8	69.7	71.5	101.4	75.9	79.7	85.9	79.3	80.3	93.3	74.2	77.0
0320	Concrete Reinforcing	87.1	87.9	87.5	88.5	74.9	81.9	89.4	79.8	84.7	103.5	80.1	92.2	90.2	88.2	89.2	89.1	80.3	84.9
0330	Cast-in-Place Concrete	90.3	69.6	82.5	93.0	78.2	87.4	78.7	64.9	73.5	90.5	74.1	84.3	93.4	66.9	83.4	77.0	81.1	78.5
03	CONCRETE	84.3	75.0	80.2	93.2	73.8	84.5	83.5	70.4	77.6	88.8	76.4	83.2	91.7	76.9	85.1	88.6	78.1	83.9
04	MASONRY	90.6	57.5	70.4	105.7	69.0	83.3	78.7	59.2	66.8	88.7	71.2	78.0	90.2	54.4	68.4	98.1	75.5	84.3
05	METALS	92.5	87.7	91.0	95.5	86.5	92.7	97.5	84.0	93.3	95.2	84.2	91.7	97.6	88.6	94.8	88.3	85.3	87.4
06	WOOD, PLASTICS & COMPOSITES	74.1	74.9	74.5	81.5	65.9	73.3	80.3	71.4	75.7	100.2	75.0	87.0	80.7	85.3	83.1	89.1	73.2	80.8
07	THERMAL & MOISTURE PROTECTION	104.5	66.4	88.1	99.9	69.5	86.9	87.7	65.4	78.2	102.1	72.9	89.5	99.8	66.4	85.4	99.1	79.0	90.5
08	OPENINGS	87.1	67.9	82.4	94.3	69.5	88.2	91.3	70.3	86.2	98.7	75.7	93.1	92.8	84.0	90.6	89.5	75.6	86.1
0920	Plaster & Gypsum Board	93.4	74.1	80.7	67.4	65.8	66.3	83.0	71.2	75.3	99.3	74.1	82.7	83.0	84.7	84.1	87.2	72.3	77.4
0950, 0980	Ceilings & Acoustic Treatment	79.5	74.1	76.1	86.5	65.8	73.5	82.1	71.2	75.3	96.5	74.1	82.4	82.1	84.7	83.7	77.6	72.3	74.3
0960	Flooring	88.3	61.4	80.4	67.0	79.6	70.7	87.5	70.4	82.5	95.8	75.0	89.7	90.9	61.4	82.3	94.4	72.6	88.0
0970, 0990	Wall Finishes & Painting/Coating	85.4	59.0	69.6	85.5	67.7	74.8	90.9	64.9	75.3	96.1	87.4	90.9	90.9	51.3	67.2	90.9	82.6	85.9
09	FINISHES	85.6	69.6	76.9	79.6	69.8	74.3	86.6	68.7	76.9	95.7	77.0	85.6	88.0	73.4	80.1	87.9	74.8	80.8
COVERS	DIVS. 10 - 14, 25, 28, 41, 43, 44, 46	100.0	89.0	97.4	100.0	87.8	97.1	100.0	86.1	96.7	100.0	91.1	97.9	100.0	88.3	97.2	100.0	52.8	88.9
21, 22, 23	FIRE SUPPRESSION, PLUMBING & HVAC	97.2	70.6	86.5	97.5	72.4	87.3	96.9	74.3	87.8	100.7	77.2	91.2	96.8	73.9	87.5	96.9	75.6	88.3
26, 27, 3370	ELECTRICAL, COMMUNICATIONS & UTIL.	89.3	84.9	87.1	96.0	69.8	83.0	92.8	75.0	84.0	100.1	75.0	87.7	92.9	84.9	89.0	94.7	75.0	84.9
MF2018	WEIGHTED AVERAGE	91.8	75.0	84.6	94.9	75.0	86.3	91.8	73.4	83.8	96.8	78.6	88.9	94.3	77.1	86.9	92.7	78.7	86.7

For customer support on your Site Work & Landscape Costs with RSMeans Data, call 800.448.8182.

KENTUCKY

DIVISION		LEXINGTON 403 - 405			LOUISVILLE 400 - 402			OWENSBORO 423			PADUCAH 420			PIKEVILLE 415 - 416			SOMERSET 425 - 426		
		MAT.	INST.	TOTAL	MAT.	INST.	TOTAL	MAT.	INST.	TOTAL	MAT.	INST.	TOTAL	MAT.	INST.	TOTAL	MAT.	INST.	TOTAL
015433	CONTRACTOR EQUIPMENT		95.0	95.0		95.3	95.3		108.7	108.7		108.7	108.7		95.7	95.7		95.0	95.0
0241, 31 - 34	SITE & INFRASTRUCTURE, DEMOLITION	94.6	90.7	91.9	88.4	97.1	94.4	92.7	112.7	106.5	84.9	111.5	103.2	124.2	75.3	90.6	78.2	88.3	85.1
0310	Concrete Forming & Accessories	100.4	75.0	78.8	101.1	81.5	84.4	91.5	78.4	80.4	89.1	78.2	79.8	97.5	82.4	84.7	87.1	73.8	75.8
0320	Concrete Reinforcing	95.5	86.8	91.3	87.9	86.7	87.3	89.1	79.8	84.6	89.6	78.5	84.3	92.3	91.2	91.8	89.4	88.1	88.8
0330	Cast-in-Place Concrete	92.4	83.9	89.2	87.0	69.7	80.5	89.8	83.9	87.6	82.0	77.8	80.4	96.5	85.8	92.5	77.0	84.4	79.8
03	CONCRETE	87.4	80.6	84.3	85.1	78.5	82.1	100.4	81.0	91.7	93.0	78.5	86.5	106.4	86.5	97.5	78.7	80.5	79.5
04	MASONRY	89.2	71.8	78.6	87.3	70.6	77.1	90.5	80.0	84.1	93.3	75.9	82.7	89.7	79.8	83.6	84.9	60.6	70.1
05	METALS	94.9	88.7	93.0	97.1	86.1	93.6	89.9	87.3	89.1	86.8	85.8	86.5	96.6	106.3	99.6	97.5	88.3	94.7
06	WOOD, PLASTICS & COMPOSITES	92.1	73.2	82.2	97.8	84.1	90.6	86.7	77.5	81.9	84.1	78.7	81.3	85.8	84.7	85.2	81.4	74.9	78.0
07	THERMAL & MOISTURE PROTECTION	104.9	78.0	93.4	101.8	75.1	90.3	99.8	82.0	92.2	99.2	78.9	90.5	91.4	76.9	85.2	99.1	68.7	86.0
08	OPENINGS	87.4	76.6	84.8	88.7	79.1	86.3	89.5	78.4	86.8	88.8	78.1	86.2	91.0	82.9	89.1	91.9	74.2	87.6
0920	Plaster & Gypsum Board	102.8	72.3	82.8	94.0	84.0	87.4	85.8	76.7	79.8	85.2	77.9	80.4	60.6	84.7	76.4	83.0	74.1	77.1
0950, 0980	Ceilings & Acoustic Treatment	82.0	72.3	75.9	99.0	84.0	89.6	77.6	76.7	77.1	77.6	77.9	77.8	77.8	84.7	82.1	82.1	74.1	77.1
0960	Flooring	93.4	63.8	84.8	96.2	60.1	85.6	93.8	60.3	84.0	92.7	72.6	86.8	79.2	61.4	74.0	91.2	61.4	82.5
0970, 0990	Wall Finishes & Painting/Coating	85.4	80.5	82.5	93.7	66.8	77.6	90.9	86.0	87.9	90.9	69.6	78.2	91.6	87.4	89.0	90.9	64.9	75.3
09	FINISHES	89.1	72.8	80.3	95.2	76.1	84.9	88.1	75.3	81.1	87.3	76.2	81.3	79.5	79.3	79.4	87.3	70.3	78.1
COVERS	DIVS. 10 - 14, 25, 28, 41, 43, 44, 46	100.0	90.7	97.8	100.0	89.8	97.6	100.0	93.4	98.4	100.0	86.4	96.8	100.0	45.6	87.2	100.0	89.2	97.5
21, 22, 23	FIRE SUPPRESSION, PLUMBING & HVAC	101.0	76.9	91.2	100.7	78.4	91.7	100.5	77.9	91.4	96.9	74.5	87.9	96.4	78.4	89.2	96.9	70.5	86.2
26, 27, 3370	ELECTRICAL, COMMUNICATIONS & UTIL.	91.5	76.3	84.0	95.5	76.5	86.1	94.7	73.8	84.4	96.7	74.4	85.7	95.3	85.0	90.2	93.3	84.9	89.2
MF2018	WEIGHTED AVERAGE	94.2	78.8	87.5	94.9	79.6	88.3	95.3	81.8	89.4	93.0	80.0	87.4	96.1	82.1	90.1	92.1	76.6	85.4

LOUISIANA

DIVISION		ALEXANDRIA 713 - 714			BATON ROUGE 707 - 708			HAMMOND 704			LAFAYETTE 705			LAKE CHARLES 706			MONROE 712		
		MAT.	INST.	TOTAL	MAT.	INST.	TOTAL	MAT.	INST.	TOTAL	MAT.	INST.	TOTAL	MAT.	INST.	TOTAL	MAT.	INST.	TOTAL
015433	CONTRACTOR EQUIPMENT		88.0	88.0		92.5	92.5		86.6	86.6		86.6	86.6		86.1	86.1		88.0	88.0
0241, 31 - 34	SITE & INFRASTRUCTURE, DEMOLITION	99.2	84.7	89.2	101.4	92.8	95.5	99.5	82.6	87.9	101.1	84.5	89.7	101.8	84.2	89.7	99.2	84.6	89.2
0310	Concrete Forming & Accessories	77.1	61.1	63.5	94.2	71.5	74.9	74.4	54.3	57.3	91.4	65.5	69.4	91.9	69.1	72.5	76.7	60.6	63.0
0320	Concrete Reinforcing	92.6	53.8	73.9	88.3	53.8	71.7	86.3	51.8	69.6	87.5	51.8	70.3	87.5	53.8	71.3	91.6	53.8	73.3
0330	Cast-in-Place Concrete	87.2	67.1	79.6	92.2	69.5	83.6	87.0	65.6	78.9	86.6	67.5	79.3	91.1	68.4	82.6	87.2	66.3	79.3
03	CONCRETE	82.0	62.6	73.3	89.2	68.3	79.8	84.3	58.6	72.8	85.3	64.3	75.9	87.5	66.6	78.1	81.8	62.2	73.0
04	MASONRY	106.0	64.5	80.7	86.4	62.3	71.7	90.4	62.2	73.2	90.4	65.3	75.1	89.9	66.6	75.8	100.9	63.2	77.9
05	METALS	93.5	70.3	86.2	93.7	73.6	87.4	85.6	68.0	80.1	84.9	68.3	79.7	84.9	69.2	80.0	93.4	70.3	86.2
06	WOOD, PLASTICS & COMPOSITES	86.2	60.1	72.6	96.1	75.0	85.0	75.4	52.3	63.3	95.7	65.7	80.0	94.0	69.5	81.2	85.6	60.1	72.2
07	THERMAL & MOISTURE PROTECTION	99.6	66.2	85.2	98.0	68.1	85.2	97.3	63.3	82.7	97.9	67.1	84.7	97.6	68.2	85.0	99.5	65.6	85.0
08	OPENINGS	110.5	59.5	98.1	97.9	71.9	91.5	96.2	52.8	85.7	99.6	59.8	89.9	99.6	63.2	90.8	110.5	58.0	97.7
0920	Plaster & Gypsum Board	81.3	59.6	67.1	98.4	74.6	82.8	100.4	51.6	68.4	109.1	65.3	80.4	109.1	69.3	83.0	81.0	59.6	67.0
0950, 0980	Ceilings & Acoustic Treatment	92.2	59.6	71.8	100.3	74.6	84.2	103.0	51.6	70.8	99.8	65.3	78.2	100.5	69.3	81.0	92.2	59.6	71.8
0960	Flooring	84.6	67.6	79.6	93.6	67.6	86.0	92.7	67.6	85.3	101.3	67.6	91.5	101.3	67.6	91.5	84.2	67.6	79.4
0970, 0990	Wall Finishes & Painting/Coating	92.7	60.4	73.3	95.8	60.4	74.6	94.9	57.0	72.2	94.9	56.9	72.1	94.9	60.4	74.2	92.7	62.2	74.4
09	FINISHES	84.9	61.9	72.5	96.6	70.1	82.3	95.8	56.4	74.4	98.7	65.0	80.5	98.9	68.1	82.2	84.8	61.8	72.3
COVERS	DIVS. 10 - 14, 25, 28, 41, 43, 44, 46	100.0	82.7	95.9	100.0	85.3	96.5	100.0	82.3	95.8	100.0	84.8	96.4	100.0	85.7	96.6	100.0	82.3	95.8
21, 22, 23	FIRE SUPPRESSION, PLUMBING & HVAC	101.0	63.0	85.6	100.9	61.5	85.0	97.3	59.8	82.2	101.1	61.2	85.0	101.1	64.2	86.2	101.0	61.9	85.2
26, 27, 3370	ELECTRICAL, COMMUNICATIONS & UTIL.	92.7	54.9	74.0	103.0	59.3	81.4	95.9	67.0	81.6	96.8	61.4	79.3	96.5	66.1	81.4	94.2	58.5	76.5
MF2018	WEIGHTED AVERAGE	96.1	64.7	82.5	97.0	68.3	84.6	93.2	63.4	80.3	95.1	65.9	82.5	95.3	68.4	83.7	96.0	64.6	82.4

LOUISIANA / MAINE

DIVISION		NEW ORLEANS 700 - 701			SHREVEPORT 710 - 711			THIBODAUX 703			AUGUSTA 043			BANGOR 044			BATH 045		
		MAT.	INST.	TOTAL	MAT.	INST.	TOTAL	MAT.	INST.	TOTAL	MAT.	INST.	TOTAL	MAT.	INST.	TOTAL	MAT.	INST.	TOTAL
015433	CONTRACTOR EQUIPMENT		88.6	88.6		94.3	94.3		86.6	86.6		99.3	99.3		94.2	94.2		94.2	94.2
0241, 31 - 34	SITE & INFRASTRUCTURE, DEMOLITION	102.0	93.6	96.2	101.8	94.0	96.5	101.9	84.3	89.8	90.4	101.7	98.2	92.8	93.0	92.9	90.1	91.5	91.1
0310	Concrete Forming & Accessories	94.6	69.8	73.5	93.5	62.4	67.1	85.2	62.3	65.7	101.2	78.6	82.0	95.5	78.4	81.0	90.7	78.1	80.0
0320	Concrete Reinforcing	87.7	53.2	71.0	93.0	52.8	73.6	86.3	51.8	69.7	97.7	80.9	89.6	88.8	80.9	85.0	87.9	80.6	84.4
0330	Cast-in-Place Concrete	83.7	71.5	79.1	90.3	68.3	82.0	93.5	65.4	82.9	90.3	111.5	98.3	71.4	110.3	86.1	71.4	110.4	86.1
03	CONCRETE	88.9	67.4	79.3	88.9	63.4	77.4	89.0	62.2	77.0	96.1	90.9	93.8	87.5	90.5	88.9	87.5	90.4	88.8
04	MASONRY	92.9	63.2	74.8	88.9	63.6	73.5	112.5	61.7	81.5	93.0	93.1	93.1	108.3	92.7	98.7	114.5	90.6	99.9
05	METALS	95.9	61.8	85.2	95.5	68.7	87.1	85.6	68.2	80.2	115.0	92.3	107.9	98.7	93.8	97.2	97.0	93.2	95.8
06	WOOD, PLASTICS & COMPOSITES	94.8	72.6	83.2	97.7	62.3	79.2	82.7	63.3	72.5	98.5	75.9	86.7	94.6	75.9	84.8	87.7	75.9	81.5
07	THERMAL & MOISTURE PROTECTION	96.1	68.6	84.3	98.0	67.1	84.8	97.5	64.3	83.3	110.2	100.7	106.1	107.7	100.0	104.4	107.6	99.0	103.9
08	OPENINGS	98.7	65.0	90.5	104.5	57.6	93.1	100.5	54.8	89.4	102.9	79.1	97.1	96.6	82.0	93.1	96.6	78.1	92.1
0920	Plaster & Gypsum Board	101.0	72.1	82.0	97.8	61.6	74.0	102.1	62.9	76.3	107.5	75.0	86.2	110.3	75.0	87.2	105.2	75.0	85.4
0950, 0980	Ceilings & Acoustic Treatment	95.8	72.1	80.9	100.9	61.6	76.3	103.0	62.9	77.9	97.1	75.0	83.3	80.2	75.0	77.0	79.5	75.0	76.7
0960	Flooring	107.9	68.9	96.6	92.9	67.6	85.5	98.4	67.0	89.3	91.0	109.8	96.5	84.2	109.8	91.7	82.6	105.5	89.3
0970, 0990	Wall Finishes & Painting/Coating	99.9	59.9	75.9	94.6	62.2	75.2	95.3	57.0	72.4	95.3	89.8	92.0	89.1	98.4	94.7	89.1	84.7	86.5
09	FINISHES	100.1	69.0	83.3	95.3	63.2	77.9	97.8	62.6	78.8	96.6	84.6	90.1	88.9	85.4	87.0	87.5	83.1	85.1
COVERS	DIVS. 10 - 14, 25, 28, 41, 43, 44, 46	100.0	85.4	96.6	100.0	83.2	96.0	100.0	83.3	96.1	100.0	100.7	100.2	100.0	100.3	100.1	100.0	96.2	99.1
21, 22, 23	FIRE SUPPRESSION, PLUMBING & HVAC	101.0	61.4	85.0	100.8	62.5	85.3	97.3	59.6	82.1	101.2	75.0	90.6	101.3	74.7	90.6	97.5	74.9	88.4
26, 27, 3370	ELECTRICAL, COMMUNICATIONS & UTIL.	100.3	69.4	85.0	101.4	65.6	83.7	94.8	67.0	81.0	99.2	78.4	88.9	98.4	70.9	84.8	96.8	78.4	87.7
MF2018	WEIGHTED AVERAGE	97.7	68.3	85.0	97.7	66.8	84.4	95.5	65.0	82.3	101.6	86.0	94.9	97.5	84.5	91.9	96.2	84.6	91.2

		MAINE																	
		HOULTON			KITTERY			LEWISTON			MACHIAS			PORTLAND			ROCKLAND		
DIVISION		047			039			042			046			040 - 041			048		
		MAT.	INST.	TOTAL	MAT.	INST.	TOTAL	MAT.	INST.	TOTAL	MAT.	INST.	TOTAL	MAT.	INST.	TOTAL	MAT.	INST.	TOTAL
015433	CONTRACTOR EQUIPMENT		94.2	94.2		94.2	94.2		94.2	94.2		94.2	94.2		94.2	94.2		94.2	94.2
0241, 31 - 34	SITE & INFRASTRUCTURE, DEMOLITION	91.9	91.5	91.7	81.0	91.5	88.2	90.1	93.0	92.1	91.3	91.5	91.5	90.6	102.0	98.4	87.7	91.5	90.3
0310	Concrete Forming & Accessories	100.0	78.1	81.3	89.1	78.3	79.9	101.8	78.4	81.9	96.4	78.1	80.8	102.8	78.5	82.2	97.8	78.1	81.1
0320	Concrete Reinforcing	88.8	80.6	84.9	87.5	80.7	84.2	109.1	80.9	95.5	88.8	80.6	84.9	108.3	80.9	95.1	88.8	80.6	84.9
0330	Cast-in-Place Concrete	71.4	109.3	85.7	71.9	110.4	86.4	72.9	110.3	87.0	71.4	110.3	86.1	86.7	111.4	96.0	72.9	110.4	87.0
03	CONCRETE	88.6	90.0	89.2	81.9	90.5	85.8	87.5	90.5	88.9	88.0	90.4	89.1	95.9	90.8	93.6	85.3	90.4	87.6
04	MASONRY	91.5	90.6	91.0	111.6	90.6	98.8	92.2	92.7	92.5	91.5	90.6	91.0	98.3	92.7	94.9	86.2	90.6	88.9
05	METALS	97.3	93.1	96.0	91.6	93.4	92.1	102.4	93.9	99.8	97.3	93.2	96.0	108.6	92.4	103.5	97.1	93.2	95.9
06	WOOD, PLASTICS & COMPOSITES	99.2	75.9	87.0	89.7	75.9	82.4	101.5	75.9	88.1	95.6	75.9	85.3	101.7	76.0	88.2	96.7	75.9	85.8
07	THERMAL & MOISTURE PROTECTION	107.8	99.0	104.0	108.1	99.0	104.2	107.5	100.0	104.3	107.7	99.0	104.0	110.7	101.0	106.6	107.4	99.0	103.8
08	OPENINGS	96.7	78.1	92.2	96.6	81.5	92.9	99.6	82.0	95.3	96.7	78.1	92.2	97.0	82.0	93.4	96.6	78.1	92.1
0920	Plaster & Gypsum Board	112.0	75.0	87.8	100.3	75.0	83.7	115.4	75.0	88.9	110.9	75.0	87.4	107.5	75.0	86.2	110.9	75.0	87.4
0950, 0980	Ceilings & Acoustic Treatment	79.5	75.0	76.7	90.6	75.0	80.9	89.7	75.0	80.5	79.5	75.0	76.7	96.1	75.0	82.9	79.5	75.0	76.7
0960	Flooring	85.3	105.5	91.2	89.0	105.5	93.8	86.8	108.9	93.5	84.7	105.5	90.8	90.7	109.8	96.3	85.0	105.5	91.0
0970, 0990	Wall Finishes & Painting/Coating	89.1	84.7	86.5	79.1	97.3	90.0	89.1	98.4	94.7	89.1	84.7	86.5	95.4	98.4	97.2	89.1	84.7	86.5
09	FINISHES	89.3	83.1	86.0	91.9	84.5	87.9	92.3	85.4	88.5	88.9	83.1	85.8	95.2	85.5	89.9	88.7	83.1	85.7
COVERS	DIVS. 10 - 14, 25, 28, 41, 43, 44, 46	100.0	93.1	98.4	100.0	96.2	99.1	100.0	100.3	100.1	100.0	93.1	98.4	100.0	100.6	100.1	100.0	96.2	99.1
21, 22, 23	FIRE SUPPRESSION, PLUMBING & HVAC	97.5	74.9	88.4	97.6	74.9	88.4	101.3	74.8	90.6	97.5	74.9	88.4	100.7	74.8	90.3	97.5	74.9	88.4
26, 27, 3370	ELECTRICAL, COMMUNICATIONS & UTIL.	100.1	78.4	89.4	93.6	78.4	86.1	100.1	73.9	87.1	100.1	78.4	89.4	103.2	73.8	88.7	100.0	78.4	89.3
MF2018	WEIGHTED AVERAGE	95.9	84.4	91.0	94.4	84.9	90.3	98.0	84.9	92.4	95.8	84.5	90.9	100.6	85.6	94.1	95.0	84.6	90.5

		MAINE			MARYLAND														
		WATERVILLE			ANNAPOLIS			BALTIMORE			COLLEGE PARK			CUMBERLAND			EASTON		
DIVISION		049			214			210 - 212			207 - 208			215			216		
		MAT.	INST.	TOTAL	MAT.	INST.	TOTAL	MAT.	INST.	TOTAL	MAT.	INST.	TOTAL	MAT.	INST.	TOTAL	MAT.	INST.	TOTAL
015433	CONTRACTOR EQUIPMENT		94.2	94.2		105.0	105.0		102.6	102.6		106.0	106.0		100.4	100.4		100.4	100.4
0241, 31 - 34	SITE & INFRASTRUCTURE, DEMOLITION	91.8	91.5	91.6	98.7	94.3	95.7	99.8	95.2	96.7	101.7	90.8	94.2	89.9	87.3	88.1	96.5	85.6	89.0
0310	Concrete Forming & Accessories	90.1	78.1	79.9	102.6	75.4	79.5	103.2	75.3	79.5	84.9	73.6	75.3	94.6	80.5	82.6	92.3	71.4	74.5
0320	Concrete Reinforcing	88.8	80.6	84.9	104.8	90.4	97.9	113.9	90.5	102.6	110.6	97.8	104.4	95.2	87.4	91.4	94.3	83.1	88.9
0330	Cast-in-Place Concrete	71.4	110.3	86.1	114.5	77.1	100.3	123.0	78.2	106.1	103.7	73.5	92.3	95.0	85.2	91.3	105.5	63.2	89.5
03	CONCRETE	89.1	90.4	89.6	103.1	79.9	92.7	113.1	79.9	98.2	103.1	79.4	92.5	90.3	84.6	87.8	98.3	72.1	86.5
04	MASONRY	101.8	90.6	95.0	103.3	73.4	85.1	106.0	75.2	87.2	110.9	71.6	86.9	103.3	88.9	94.5	118.6	56.5	80.7
05	METALS	97.2	93.2	96.0	104.8	102.3	104.0	103.2	97.6	101.4	90.7	110.5	96.9	100.2	103.7	101.3	100.4	100.9	100.6
06	WOOD, PLASTICS & COMPOSITES	87.0	75.9	81.2	99.2	74.7	86.4	105.6	74.5	89.4	76.9	73.1	74.9	89.3	78.2	83.5	86.6	78.1	82.1
07	THERMAL & MOISTURE PROTECTION	107.7	99.0	104.0	99.9	80.7	91.7	99.6	81.4	91.7	101.9	79.2	92.2	97.6	81.2	90.6	97.8	71.7	86.6
08	OPENINGS	96.7	78.1	92.2	102.1	81.0	97.0	108.0	81.0	96.0	92.7	82.1	90.1	97.7	82.5	94.0	96.1	81.0	92.4
0920	Plaster & Gypsum Board	105.2	75.0	85.4	102.9	74.2	84.1	104.9	73.9	84.6	109.2	72.6	85.2	108.0	78.1	88.4	108.0	77.9	88.3
0950, 0980	Ceilings & Acoustic Treatment	79.5	75.0	76.7	94.3	74.2	81.7	106.9	73.9	86.2	113.8	72.6	88.0	107.3	78.1	89.0	107.3	77.9	88.9
0960	Flooring	82.3	105.5	89.1	94.7	75.9	89.2	96.5	75.9	90.5	88.1	74.8	84.2	89.4	94.3	90.8	88.5	72.2	83.7
0970, 0990	Wall Finishes & Painting/Coating	89.1	84.7	86.5	96.0	69.4	80.1	97.7	69.4	80.7	100.3	69.4	81.8	91.8	80.6	85.1	91.8	69.4	78.4
09	FINISHES	87.6	83.1	85.2	95.2	74.3	83.9	99.8	74.2	85.9	99.0	72.6	84.7	98.5	83.0	90.1	98.7	71.5	83.9
COVERS	DIVS. 10 - 14, 25, 28, 41, 43, 44, 46	100.0	93.0	98.3	100.0	89.5	97.5	100.0	89.7	97.6	100.0	82.3	95.8	100.0	91.4	98.0	100.0	83.8	96.2
21, 22, 23	FIRE SUPPRESSION, PLUMBING & HVAC	97.5	74.9	88.4	100.8	82.9	93.6	100.8	81.6	93.0	97.1	83.4	91.6	96.8	70.7	86.3	96.9	64.6	85.8
26, 27, 3370	ELECTRICAL, COMMUNICATIONS & UTIL.	100.1	78.4	89.4	101.1	85.6	93.4	96.7	86.9	91.8	96.4	103.9	100.1	97.2	79.9	88.6	96.7	60.0	78.6
MF2018	WEIGHTED AVERAGE	96.2	84.5	91.2	101.4	83.4	93.6	102.3	83.1	94.0	97.5	85.8	92.5	97.1	83.1	91.0	98.9	72.6	87.5

		MARYLAND															MASSACHUSETTS		
		ELKTON			HAGERSTOWN			SALISBURY			SILVER SPRING			WALDORF			BOSTON		
DIVISION		219			217			218			209			206			020 - 022, 024		
		MAT.	INST.	TOTAL	MAT.	INST.	TOTAL	MAT.	INST.	TOTAL	MAT.	INST.	TOTAL	MAT.	INST.	TOTAL	MAT.	INST.	TOTAL
015433	CONTRACTOR EQUIPMENT		100.4	100.4		100.4	100.4		100.4	100.4		97.9	97.9		97.9	97.9		105.8	105.8
0241, 31 - 34	SITE & INFRASTRUCTURE, DEMOLITION	84.4	86.1	85.6	88.9	87.5	87.9	96.4	85.3	88.8	88.8	82.9	84.8	95.0	82.8	86.6	92.0	105.4	101.2
0310	Concrete Forming & Accessories	99.5	89.0	90.6	93.6	78.0	80.3	109.4	48.7	57.8	93.3	72.9	75.9	101.4	72.8	77.1	104.7	136.7	131.9
0320	Concrete Reinforcing	94.3	111.8	102.8	95.2	87.4	91.4	94.3	62.1	78.8	109.4	97.7	103.7	110.1	97.7	104.1	117.1	149.5	132.8
0330	Cast-in-Place Concrete	85.4	69.5	79.4	90.5	85.3	88.5	105.5	61.1	88.7	106.2	73.9	94.0	118.9	73.7	101.8	98.2	142.4	114.9
03	CONCRETE	83.2	87.3	85.0	86.8	83.5	85.3	99.4	57.5	80.6	100.9	79.0	91.1	111.1	78.9	96.6	103.4	140.4	120.0
04	MASONRY	103.2	64.3	79.5	109.7	88.9	97.0	117.7	53.1	78.3	110.2	71.9	86.9	94.8	71.9	80.8	107.9	144.2	130.1
05	METALS	100.5	112.7	104.3	100.3	103.9	101.4	100.5	91.6	97.7	95.2	106.5	98.7	95.2	106.0	98.6	102.4	134.6	112.4
06	WOOD, PLASTICS & COMPOSITES	95.0	97.9	96.5	88.2	74.4	81.0	107.9	49.6	77.4	84.3	72.4	78.1	92.4	72.4	81.9	104.9	135.7	121.0
07	THERMAL & MOISTURE PROTECTION	97.3	77.5	88.8	97.8	84.5	92.1	98.2	67.0	84.8	104.3	83.9	95.5	104.9	83.9	95.9	108.4	138.0	121.1
08	OPENINGS	96.1	99.5	96.9	96.1	79.8	92.1	96.3	59.6	87.4	84.8	81.7	84.1	85.4	81.7	84.5	100.3	143.3	110.8
0920	Plaster & Gypsum Board	110.0	98.3	102.3	108.0	74.2	85.8	118.0	48.7	72.5	116.0	72.6	87.5	118.5	72.6	88.4	105.5	136.3	125.7
0950, 0980	Ceilings & Acoustic Treatment	107.3	98.3	101.6	110.1	74.2	87.6	107.3	48.7	70.6	121.4	72.6	90.9	121.4	72.6	90.9	94.0	136.3	120.5
0960	Flooring	90.9	72.2	85.4	88.9	94.3	90.5	94.5	72.2	88.0	94.0	74.8	88.4	97.6	74.8	91.0	96.8	162.4	115.9
0970, 0990	Wall Finishes & Painting/Coating	91.8	69.4	78.4	91.8	69.4	78.4	91.8	69.4	78.4	108.0	69.4	84.9	108.0	69.4	84.9	101.6	154.9	133.5
09	FINISHES	98.8	85.0	91.4	98.8	79.5	88.4	101.8	53.9	75.9	99.6	72.0	84.7	101.3	72.1	85.5	98.5	143.8	123.0
COVERS	DIVS. 10 - 14, 25, 28, 41, 43, 44, 46	100.0	53.7	89.1	100.0	91.0	97.9	100.0	79.6	95.2	100.0	80.8	95.5	100.0	78.8	95.0	100.0	117.4	104.1
21, 22, 23	FIRE SUPPRESSION, PLUMBING & HVAC	96.8	75.1	88.0	100.6	88.3	95.6	96.8	67.5	85.0	97.1	83.8	91.7	97.1	83.8	91.7	96.7	127.1	109.0
26, 27, 3370	ELECTRICAL, COMMUNICATIONS & UTIL.	98.2	83.7	91.0	97.0	79.9	88.5	95.7	57.9	77.0	93.8	103.9	98.8	91.8	103.9	97.8	98.3	129.3	113.7
MF2018	WEIGHTED AVERAGE	96.1	83.0	90.5	97.7	86.2	92.8	99.3	65.0	84.5	96.7	84.9	91.6	97.4	84.8	92.0	100.2	132.9	114.3

MASSACHUSETTS

DIVISION		BROCKTON 023 MAT.	INST.	TOTAL	BUZZARDS BAY 025 MAT.	INST.	TOTAL	FALL RIVER 027 MAT.	INST.	TOTAL	FITCHBURG 014 MAT.	INST.	TOTAL	FRAMINGHAM 017 MAT.	INST.	TOTAL	GREENFIELD 013 MAT.	INST.	TOTAL
015433	CONTRACTOR EQUIPMENT		97.1	97.1		97.1	97.1		98.0	98.0		94.2	94.2		95.9	95.9		94.2	94.2
0241, 31 - 34	SITE & INFRASTRUCTURE, DEMOLITION	91.0	96.4	94.7	80.6	96.0	91.2	90.2	96.5	94.5	82.8	95.9	91.8	79.6	95.4	90.4	86.5	94.5	92.0
0310	Concrete Forming & Accessories	101.9	122.8	119.7	99.4	122.0	118.7	101.9	122.5	119.4	94.5	115.2	112.1	102.5	122.7	119.7	92.6	119.0	115.1
0320	Concrete Reinforcing	105.1	144.2	124.0	84.3	121.3	102.1	105.1	121.3	112.9	85.2	135.8	109.6	85.2	144.0	113.5	88.8	120.8	104.3
0330	Cast-in-Place Concrete	84.5	133.0	102.8	70.2	132.7	93.8	81.7	133.3	101.2	77.1	132.7	98.1	77.1	132.9	98.2	79.2	117.8	93.8
03	CONCRETE	91.6	129.8	108.7	77.7	125.4	99.1	89.0	125.8	105.5	78.6	124.6	99.2	81.3	129.7	103.0	82.0	118.5	98.4
04	MASONRY	104.3	136.3	123.8	96.8	136.3	120.9	104.9	136.2	124.0	98.7	128.1	116.7	105.3	132.1	121.7	103.3	118.6	112.6
05	METALS	100.9	129.6	109.9	95.5	119.2	102.9	100.9	119.8	106.8	100.3	121.8	107.0	100.3	128.7	109.2	102.9	113.1	106.1
06	WOOD, PLASTICS & COMPOSITES	101.3	122.2	112.2	97.4	122.2	110.4	101.3	122.4	112.4	95.2	112.6	104.3	102.9	122.0	112.9	92.8	122.6	108.4
07	THERMAL & MOISTURE PROTECTION	105.5	127.4	114.9	104.4	124.1	112.8	105.4	124.0	113.3	102.4	120.0	109.9	102.6	125.1	112.2	102.5	109.9	105.7
08	OPENINGS	97.8	130.5	105.8	93.9	119.4	100.1	97.8	120.5	103.4	99.6	123.0	105.3	90.2	130.3	99.9	99.8	121.0	105.0
0920	Plaster & Gypsum Board	92.9	122.6	112.4	87.2	122.6	110.5	92.9	122.6	112.4	108.7	112.8	111.4	112.4	122.6	119.1	109.7	123.0	118.4
0950, 0980	Ceilings & Acoustic Treatment	98.6	122.6	113.7	80.4	122.6	106.9	98.6	122.6	113.7	86.4	122.8	103.0	86.4	122.6	109.1	94.7	123.0	112.5
0960	Flooring	89.9	158.8	110.0	87.4	158.8	108.2	88.7	158.8	109.1	89.1	158.8	109.4	90.7	158.8	110.6	88.3	136.7	102.5
0970, 0990	Wall Finishes & Painting/Coating	87.0	138.9	118.1	87.0	138.9	118.1	87.0	138.9	118.1	84.9	138.9	117.2	85.8	138.9	117.6	84.9	114.9	102.9
09	FINISHES	92.8	131.5	113.7	86.3	131.5	110.7	92.5	131.6	113.7	90.1	125.8	109.5	90.9	131.3	112.8	92.4	123.0	108.9
COVERS	DIVS. 10 - 14, 25, 28, 41, 43, 44, 46	100.0	112.8	103.0	100.0	112.8	103.0	100.0	113.3	103.1	100.0	106.3	101.5	100.0	112.4	102.9	100.0	105.4	101.3
21, 22, 23	FIRE SUPPRESSION, PLUMBING & HVAC	101.7	104.3	102.8	97.2	99.5	98.1	101.7	104.0	102.6	97.6	100.5	98.8	97.6	116.3	105.2	97.6	98.2	97.8
26, 27, 3370	ELECTRICAL, COMMUNICATIONS & UTIL.	100.8	99.8	100.3	98.0	96.4	97.2	100.6	101.2	100.9	99.8	100.2	100.0	97.0	119.8	108.3	99.8	101.5	100.7
MF2018	WEIGHTED AVERAGE	98.9	117.8	107.1	93.3	114.1	102.3	98.5	116.0	106.1	95.4	113.2	103.1	94.9	122.4	106.8	96.7	109.5	102.3

MASSACHUSETTS

DIVISION		HYANNIS 026 MAT.	INST.	TOTAL	LAWRENCE 019 MAT.	INST.	TOTAL	LOWELL 018 MAT.	INST.	TOTAL	NEW BEDFORD 027 MAT.	INST.	TOTAL	PITTSFIELD 012 MAT.	INST.	TOTAL	SPRINGFIELD 010 - 011 MAT.	INST.	TOTAL
015433	CONTRACTOR EQUIPMENT		97.1	97.1		96.6	96.6		94.2	94.2		98.0	98.0		94.2	94.2		94.2	94.2
0241, 31 - 34	SITE & INFRASTRUCTURE, DEMOLITION	87.4	96.2	93.4	92.0	96.1	94.8	91.1	96.3	94.7	88.6	96.5	94.0	92.1	94.6	93.8	91.5	95.0	93.9
0310	Concrete Forming & Accessories	92.7	122.5	118.0	103.8	123.2	120.3	100.4	125.2	121.5	101.9	122.6	119.5	100.3	106.6	105.6	100.6	120.3	117.4
0320	Concrete Reinforcing	84.3	121.3	102.2	105.5	141.8	123.0	106.3	141.6	123.3	105.1	121.3	112.9	88.2	117.5	102.3	106.3	120.9	113.3
0330	Cast-in-Place Concrete	77.1	132.9	98.1	89.2	133.1	105.8	81.1	136.1	101.8	72.0	133.3	95.1	88.5	116.0	98.9	84.4	119.7	97.7
03	CONCRETE	83.3	125.6	102.3	94.9	129.6	110.5	87.0	131.3	106.9	85.8	125.8	103.8	88.0	111.6	98.6	88.6	119.7	102.6
04	MASONRY	103.4	136.3	123.4	110.9	138.8	127.9	97.8	139.9	123.5	103.3	134.8	122.5	98.4	113.3	107.5	98.1	121.9	112.6
05	METALS	97.1	119.7	104.1	103.3	128.9	111.3	103.3	125.6	110.2	100.9	119.8	106.8	103.0	111.3	105.6	106.3	113.2	108.4
06	WOOD, PLASTICS & COMPOSITES	89.7	122.2	106.7	103.7	122.2	113.4	102.6	122.2	112.8	101.3	122.4	112.4	102.6	107.1	104.9	102.6	122.6	113.0
07	THERMAL & MOISTURE PROTECTION	104.9	126.5	114.2	103.2	128.1	113.9	102.9	129.3	114.2	105.3	123.6	113.1	103.0	106.3	104.4	102.9	111.4	106.6
08	OPENINGS	94.4	120.6	100.8	93.7	129.7	102.5	100.9	129.7	107.9	97.8	124.9	104.4	100.9	111.6	103.5	100.9	121.0	105.8
0920	Plaster & Gypsum Board	83.2	122.6	109.1	115.1	122.6	120.0	115.1	122.6	120.0	92.9	122.6	112.4	115.1	107.1	109.8	115.1	123.0	120.3
0950, 0980	Ceilings & Acoustic Treatment	89.7	122.6	110.3	96.2	122.6	112.8	96.2	122.6	112.8	98.6	122.6	113.7	96.2	107.1	103.0	96.2	123.0	113.0
0960	Flooring	84.8	158.8	106.4	91.3	158.8	111.0	91.3	158.8	111.0	88.7	158.8	109.1	91.6	131.5	103.3	90.7	136.7	104.1
0970, 0990	Wall Finishes & Painting/Coating	87.0	138.9	118.1	85.0	138.9	117.3	84.9	138.9	117.2	87.0	138.9	118.1	84.9	114.9	102.9	85.7	114.9	103.2
09	FINISHES	87.6	131.5	111.3	94.5	131.4	114.5	94.5	132.8	115.2	92.4	131.6	113.6	94.6	112.2	104.1	94.4	123.8	110.3
COVERS	DIVS. 10 - 14, 25, 28, 41, 43, 44, 46	100.0	112.8	103.0	100.0	112.9	103.0	100.0	114.6	103.4	100.0	113.3	103.1	100.0	102.9	100.7	100.0	106.5	101.5
21, 22, 23	FIRE SUPPRESSION, PLUMBING & HVAC	101.7	108.6	104.5	101.0	120.3	108.8	101.0	126.6	111.3	101.7	104.0	102.6	101.0	95.2	98.7	101.0	99.9	100.6
26, 27, 3370	ELECTRICAL, COMMUNICATIONS & UTIL.	98.8	101.2	100.0	99.0	125.6	112.2	99.4	124.4	111.8	101.6	103.2	102.4	99.4	101.5	100.5	99.5	102.9	101.2
MF2018	WEIGHTED AVERAGE	96.0	116.9	105.1	99.4	124.9	110.4	98.4	126.4	110.5	98.1	116.3	106.0	98.6	105.1	101.4	99.1	110.8	104.2

MASSACHUSETTS / MICHIGAN

DIVISION		WORCESTER 015 - 016 MAT.	INST.	TOTAL	ANN ARBOR 481 MAT.	INST.	TOTAL	BATTLE CREEK 490 MAT.	INST.	TOTAL	BAY CITY 487 MAT.	INST.	TOTAL	DEARBORN 481 MAT.	INST.	TOTAL	DETROIT 482 MAT.	INST.	TOTAL
015433	CONTRACTOR EQUIPMENT		94.2	94.2		106.7	106.7		94.4	94.4		106.7	106.7		106.7	106.7		97.5	97.5
0241, 31 - 34	SITE & INFRASTRUCTURE, DEMOLITION	91.4	95.9	94.5	82.0	90.0	87.5	93.9	79.0	83.7	72.7	88.3	83.4	81.7	90.2	87.5	100.3	102.4	101.8
0310	Concrete Forming & Accessories	101.0	124.0	120.5	98.0	105.4	104.3	100.4	79.4	82.6	98.1	79.2	82.0	97.9	106.1	104.9	98.6	104.1	103.3
0320	Concrete Reinforcing	106.3	148.7	126.8	99.0	104.8	101.8	90.4	81.5	86.1	99.0	103.9	101.3	99.0	106.1	102.4	96.9	102.4	99.6
0330	Cast-in-Place Concrete	83.9	132.8	102.4	85.2	99.1	90.5	83.2	93.3	87.0	81.6	85.2	83.0	83.4	100.0	89.7	101.5	99.3	100.7
03	CONCRETE	88.4	130.9	107.4	88.0	104.0	95.2	84.4	84.4	84.4	86.3	87.1	86.7	87.1	104.9	95.1	99.6	101.5	100.5
04	MASONRY	97.7	134.7	120.3	96.0	100.1	98.5	93.9	77.4	83.8	95.6	79.8	85.9	95.9	123.5	112.7	101.0	98.3	99.4
05	METALS	106.3	127.9	113.1	99.3	118.8	105.4	105.5	83.2	98.5	99.9	114.3	104.4	99.4	120.7	106.1	100.8	93.0	98.4
06	WOOD, PLASTICS & COMPOSITES	103.1	124.1	114.1	90.9	107.1	99.3	93.0	78.6	85.5	90.9	79.1	84.7	90.9	107.1	99.3	93.7	106.0	100.1
07	THERMAL & MOISTURE PROTECTION	102.9	123.0	111.6	104.6	100.1	102.6	96.3	78.8	88.8	102.1	82.1	93.5	103.1	107.8	105.1	102.2	103.7	102.9
08	OPENINGS	100.9	132.7	108.6	94.1	102.1	96.1	86.7	75.5	84.0	94.1	83.6	91.6	94.1	102.1	96.1	97.0	100.9	97.9
0920	Plaster & Gypsum Board	115.1	124.6	121.3	93.9	107.1	102.6	87.6	75.0	79.3	93.9	78.4	83.7	93.9	107.1	102.6	91.5	106.1	101.1
0950, 0980	Ceilings & Acoustic Treatment	96.2	124.6	114.0	80.2	107.1	97.1	87.9	75.0	79.8	80.9	78.4	79.3	80.2	107.1	97.1	84.2	106.1	97.9
0960	Flooring	91.3	160.0	111.4	92.8	110.8	98.0	92.1	67.2	84.8	92.8	85.5	90.7	92.0	104.9	95.8	94.9	104.8	97.8
0970, 0990	Wall Finishes & Painting/Coating	84.9	138.9	117.2	83.4	100.4	93.6	86.5	72.6	78.2	83.4	82.4	82.8	83.4	98.6	92.5	94.8	97.3	96.3
09	FINISHES	94.5	132.8	115.2	90.5	106.0	98.9	88.7	76.0	81.8	90.2	80.0	84.7	90.3	105.2	98.3	94.0	103.9	99.4
COVERS	DIVS. 10 - 14, 25, 28, 41, 43, 44, 46	100.0	107.6	101.8	100.0	101.8	100.4	100.0	90.2	97.7	100.0	95.1	98.8	100.0	102.2	100.5	100.0	101.0	100.2
21, 22, 23	FIRE SUPPRESSION, PLUMBING & HVAC	101.0	105.1	102.7	100.7	94.1	98.0	100.6	79.9	92.3	100.7	80.2	92.4	100.7	103.1	101.6	100.5	101.5	100.9
26, 27, 3370	ELECTRICAL, COMMUNICATIONS & UTIL.	99.5	106.4	102.9	97.2	105.3	101.2	95.6	74.8	85.3	96.3	79.4	88.0	97.2	100.1	98.6	99.8	101.6	100.7
MF2018	WEIGHTED AVERAGE	99.1	118.6	107.5	96.3	102.0	98.7	95.7	79.4	88.6	95.7	85.3	91.2	96.1	106.0	100.4	99.5	100.9	100.1

For customer support on your Site Work & Landscape Costs with RSMeans Data, call 800.448.8182.

801

MICHIGAN

DIVISION		FLINT 484 - 485			GAYLORD 497			GRAND RAPIDS 493,495			IRON MOUNTAIN 498 - 499			JACKSON 492			KALAMAZOO 491		
		MAT.	INST.	TOTAL	MAT.	INST.	TOTAL	MAT.	INST.	TOTAL	MAT.	INST.	TOTAL	MAT.	INST.	TOTAL	MAT.	INST.	TOTAL
015433	CONTRACTOR EQUIPMENT		106.7	106.7		101.9	101.9		98.3	98.3		89.3	89.3		101.9	101.9		94.4	94.4
0241, 31 - 34	SITE & INFRASTRUCTURE, DEMOLITION	70.3	88.9	83.1	87.6	76.8	80.2	93.3	87.7	89.5	96.4	85.6	89.0	111.0	78.4	88.6	94.2	79.0	83.8
0310	Concrete Forming & Accessories	101.5	84.9	87.4	98.6	71.4	75.4	97.0	78.1	80.9	92.0	75.8	78.3	95.2	79.4	81.8	100.4	79.8	82.8
0320	Concrete Reinforcing	99.0	104.2	101.5	84.3	88.6	86.4	93.6	81.5	87.8	84.3	80.9	82.7	81.9	103.9	92.5	90.4	80.0	85.4
0330	Cast-in-Place Concrete	85.8	88.7	86.9	83.0	75.3	80.1	87.5	91.2	88.9	98.3	64.9	85.7	82.9	90.0	85.6	84.9	93.4	88.1
03	CONCRETE	88.5	90.9	89.6	81.2	77.5	79.5	93.3	83.1	88.7	89.5	73.7	82.4	76.6	88.7	82.0	87.4	84.4	86.1
04	MASONRY	96.1	89.0	91.7	104.3	69.7	83.2	92.1	76.8	82.8	90.8	76.4	82.0	84.9	83.3	83.9	92.6	81.2	85.6
05	METALS	99.4	114.9	104.2	107.2	109.2	107.9	102.5	82.5	96.2	106.6	89.5	101.3	107.5	112.5	109.0	105.5	83.1	98.5
06	WOOD, PLASTICS & COMPOSITES	94.9	84.1	89.2	86.6	70.9	78.4	93.2	76.7	84.6	84.7	76.2	80.3	85.3	77.1	81.0	93.0	78.6	85.5
07	THERMAL & MOISTURE PROTECTION	102.4	86.2	95.4	95.0	71.4	84.9	98.0	73.4	87.4	98.4	71.4	86.8	94.3	84.6	90.1	96.3	80.1	89.4
08	OPENINGS	94.1	85.5	92.0	86.0	75.8	83.5	101.8	76.2	95.6	92.5	67.3	86.3	85.3	83.7	84.9	86.7	76.7	84.3
0920	Plaster & Gypsum Board	95.6	83.5	87.7	87.1	69.4	75.5	95.3	73.4	80.9	52.9	76.3	68.3	85.7	75.7	79.1	87.6	75.0	79.3
0950, 0980	Ceilings & Acoustic Treatment	80.2	83.5	82.3	86.0	69.4	75.6	98.2	73.4	82.7	85.8	76.3	79.9	86.0	75.7	79.5	87.9	75.0	79.8
0960	Flooring	92.8	88.8	91.6	85.5	81.7	84.4	94.8	78.6	90.0	102.3	86.8	97.8	84.0	75.1	81.4	92.1	73.5	86.6
0970, 0990	Wall Finishes & Painting/Coating	83.4	80.7	81.8	82.6	77.7	79.7	97.1	77.7	85.5	99.1	66.3	79.4	82.6	89.3	86.6	86.5	78.7	81.8
09	FINISHES	89.8	84.6	87.0	87.8	73.2	79.9	95.4	77.6	85.8	88.8	76.8	82.3	88.9	78.6	83.4	88.7	78.0	82.9
COVERS	DIVS. 10 - 14, 25, 28, 41, 43, 44, 46	100.0	97.2	99.3	100.0	94.4	98.7	100.0	100.9	100.2	100.0	93.0	98.4	100.0	97.7	99.5	100.0	101.8	100.4
21, 22, 23	FIRE SUPPRESSION, PLUMBING & HVAC	100.7	85.9	94.7	97.0	75.3	88.3	100.6	81.8	93.0	96.9	80.0	90.1	97.0	81.7	90.9	100.6	80.5	92.5
26, 27, 3370	ELECTRICAL, COMMUNICATIONS & UTIL.	97.2	92.8	95.0	93.8	73.5	83.7	100.7	86.4	93.6	99.4	77.7	88.7	97.2	99.2	98.2	95.4	77.2	86.4
MF2018	WEIGHTED AVERAGE	95.9	90.9	93.8	94.6	78.1	87.5	98.9	82.2	91.6	96.5	78.9	88.9	94.1	88.0	91.4	96.0	80.9	89.5

MICHIGAN / MINNESOTA

DIVISION		LANSING 488 - 489			MUSKEGON 494			ROYAL OAK 480,483			SAGINAW 486			TRAVERSE CITY 496			BEMIDJI 566		
		MAT.	INST.	TOTAL	MAT.	INST.	TOTAL	MAT.	INST.	TOTAL	MAT.	INST.	TOTAL	MAT.	INST.	TOTAL	MAT.	INST.	TOTAL
015433	CONTRACTOR EQUIPMENT		110.1	110.1		94.4	94.4		87.2	87.2		106.7	106.7		89.3	89.3		95.8	95.8
0241, 31 - 34	SITE & INFRASTRUCTURE, DEMOLITION	92.8	97.2	95.8	92.0	78.9	83.0	86.2	88.5	87.8	73.7	88.3	83.7	81.9	84.6	83.8	95.2	92.3	93.2
0310	Concrete Forming & Accessories	95.5	82.2	84.2	101.0	78.5	81.9	93.7	104.1	102.6	98.0	82.3	84.6	92.0	69.8	73.1	87.3	84.8	85.1
0320	Concrete Reinforcing	100.8	103.9	102.3	91.1	81.6	86.5	89.7	99.8	94.5	99.0	103.9	101.3	85.6	76.5	81.2	97.1	100.7	98.8
0330	Cast-in-Place Concrete	94.0	88.4	91.9	82.9	91.6	86.2	74.6	93.5	81.7	84.2	85.2	84.6	76.7	77.0	76.8	99.7	95.0	97.9
03	CONCRETE	94.2	89.4	92.0	82.9	83.5	83.2	75.7	99.0	86.1	87.5	88.5	88.0	73.8	74.4	74.0	91.3	92.5	91.9
04	MASONRY	89.3	85.2	86.8	91.4	78.0	83.2	89.8	95.4	93.2	97.6	79.8	86.7	89.1	71.7	78.5	100.5	103.0	102.0
05	METALS	98.3	111.4	102.4	103.1	83.5	97.0	102.8	90.0	98.8	99.4	114.1	104.0	106.5	87.3	100.5	89.2	118.8	98.4
06	WOOD, PLASTICS & COMPOSITES	89.7	80.4	84.8	90.9	77.4	83.8	86.4	107.1	97.2	88.0	83.5	85.6	84.7	69.5	76.7	70.2	80.1	75.4
07	THERMAL & MOISTURE PROTECTION	101.0	85.2	94.2	95.3	72.9	85.7	100.6	96.8	99.0	103.2	82.6	94.3	97.4	71.1	86.1	106.3	90.0	99.3
08	OPENINGS	102.5	83.4	97.9	86.0	76.7	83.8	93.9	99.8	95.4	92.3	86.0	90.8	92.5	63.0	85.3	99.4	103.7	100.4
0920	Plaster & Gypsum Board	86.8	79.8	82.2	69.5	73.8	72.3	91.0	107.1	101.6	93.9	82.9	86.7	52.9	69.4	63.7	98.3	80.1	86.3
0950, 0980	Ceilings & Acoustic Treatment	84.2	79.8	81.5	87.9	73.8	79.0	79.6	107.1	96.9	80.2	82.9	81.9	85.8	69.4	75.5	131.6	80.1	99.3
0960	Flooring	97.2	93.6	96.2	90.7	84.1	88.8	89.8	101.8	93.3	92.8	85.5	90.7	102.3	81.7	96.3	92.2	104.1	95.7
0970, 0990	Wall Finishes & Painting/Coating	96.8	77.7	85.4	84.7	79.0	81.3	84.9	89.3	87.6	83.4	82.4	82.8	99.1	37.6	62.2	85.6	96.5	92.1
09	FINISHES	91.9	83.3	87.2	85.2	79.2	82.0	89.4	102.5	96.5	90.2	82.6	86.1	87.7	68.2	77.1	101.1	89.0	94.5
COVERS	DIVS. 10 - 14, 25, 28, 41, 43, 44, 46	100.0	97.0	99.3	100.0	101.5	100.3	100.0	95.3	98.9	100.0	95.6	99.0	100.0	91.6	98.0	100.0	97.5	99.4
21, 22, 23	FIRE SUPPRESSION, PLUMBING & HVAC	100.5	85.9	94.6	100.4	81.5	92.8	97.1	96.6	96.9	100.7	79.7	92.2	96.9	75.2	88.1	97.1	83.6	91.7
26, 27, 3370	ELECTRICAL, COMMUNICATIONS & UTIL.	101.3	94.5	97.9	95.8	70.2	83.2	99.0	99.3	99.2	95.6	85.2	90.5	95.3	73.4	84.5	103.3	96.9	100.2
MF2018	WEIGHTED AVERAGE	98.1	90.5	94.8	94.5	79.7	88.1	94.1	96.9	95.3	95.6	86.7	91.8	93.5	75.2	85.6	96.8	94.6	95.8

MINNESOTA

DIVISION		BRAINERD 564			DETROIT LAKES 565			DULUTH 556 - 558			MANKATO 560			MINNEAPOLIS 553 - 555			ROCHESTER 559		
		MAT.	INST.	TOTAL	MAT.	INST.	TOTAL	MAT.	INST.	TOTAL	MAT.	INST.	TOTAL	MAT.	INST.	TOTAL	MAT.	INST.	TOTAL
015433	CONTRACTOR EQUIPMENT		98.4	98.4		95.8	95.8		104.2	104.2		98.4	98.4		108.3	108.3		99.5	99.5
0241, 31 - 34	SITE & INFRASTRUCTURE, DEMOLITION	95.6	97.0	96.6	93.4	92.5	92.8	100.9	103.6	102.8	92.2	97.9	96.1	96.0	107.6	104.0	98.0	96.1	96.7
0310	Concrete Forming & Accessories	88.3	85.4	85.8	83.8	80.4	81.0	100.3	100.0	100.0	97.8	101.7	101.1	102.4	116.9	114.7	104.9	105.9	105.7
0320	Concrete Reinforcing	95.9	100.8	98.3	97.1	100.6	98.8	108.1	101.2	104.7	95.8	110.2	102.7	94.0	108.1	100.8	100.3	110.5	105.2
0330	Cast-in-Place Concrete	108.4	99.7	105.1	96.8	98.1	97.3	102.2	103.0	102.5	99.8	110.4	103.8	90.7	118.0	101.0	97.5	102.5	99.4
03	CONCRETE	95.1	94.4	94.8	88.9	91.6	90.1	102.0	102.3	102.1	90.9	107.2	98.2	95.9	116.4	105.1	93.3	106.5	99.2
04	MASONRY	127.3	112.3	118.2	126.0	108.9	115.6	100.4	112.3	107.7	114.9	125.9	121.6	109.4	121.3	116.7	101.3	114.3	109.2
05	METALS	90.2	118.7	99.1	89.2	118.0	98.2	97.5	120.1	104.6	90.1	123.8	100.6	96.1	125.4	105.2	96.6	126.9	106.1
06	WOOD, PLASTICS & COMPOSITES	84.4	77.1	80.6	66.9	71.9	69.5	97.6	97.7	97.6	95.5	92.2	93.7	103.6	114.6	109.3	104.7	104.6	104.6
07	THERMAL & MOISTURE PROTECTION	103.8	98.2	101.4	106.1	94.0	100.9	98.0	106.4	101.6	104.3	100.4	102.6	100.1	118.1	107.8	104.4	98.6	101.9
08	OPENINGS	86.3	102.0	90.1	99.4	99.2	99.3	103.1	108.5	104.4	90.7	111.8	95.8	98.9	123.7	104.9	98.2	120.4	103.6
0920	Plaster & Gypsum Board	85.8	77.2	80.2	98.0	71.7	80.7	93.3	98.0	96.4	89.8	92.7	91.7	96.1	115.0	108.5	101.9	105.3	104.1
0950, 0980	Ceilings & Acoustic Treatment	60.2	77.2	70.8	131.6	71.7	94.0	92.0	98.0	95.8	60.2	92.7	80.5	101.2	115.0	109.8	94.5	105.3	101.3
0960	Flooring	90.9	104.1	94.8	90.9	81.6	88.2	94.8	103.5	97.4	93.0	88.8	91.8	101.9	121.5	107.6	94.9	88.8	93.1
0970, 0990	Wall Finishes & Painting/Coating	80.1	96.5	90.0	85.6	79.4	81.9	91.4	108.9	95.6	90.8	98.8	95.6	99.5	127.3	116.2	84.4	102.5	95.2
09	FINISHES	81.5	89.0	85.6	100.6	79.8	89.3	93.3	101.5	97.8	83.2	98.8	91.6	98.7	118.8	109.6	92.8	102.8	98.2
COVERS	DIVS. 10 - 14, 25, 28, 41, 43, 44, 46	100.0	99.3	99.8	100.0	98.5	99.6	100.0	101.2	100.3	100.0	105.4	101.3	100.0	106.7	101.6	100.0	102.0	100.5
21, 22, 23	FIRE SUPPRESSION, PLUMBING & HVAC	96.2	87.4	92.7	97.1	86.4	92.8	100.5	96.4	98.8	96.2	95.5	96.0	100.6	113.3	105.7	100.6	101.7	101.1
26, 27, 3370	ELECTRICAL, COMMUNICATIONS & UTIL.	101.4	101.9	101.7	103.1	68.4	86.0	101.4	101.9	101.7	106.9	99.5	103.2	104.2	115.4	109.8	101.4	99.5	100.4
MF2018	WEIGHTED AVERAGE	95.5	97.9	96.5	97.6	90.3	94.4	99.9	103.9	101.6	95.5	105.0	99.6	99.7	116.6	107.0	98.4	106.1	101.7

MINNESOTA / MISSISSIPPI

DIVISION		SAINT PAUL 550-551			ST. CLOUD 563			THIEF RIVER FALLS 567			WILLMAR 562			WINDOM 561			BILOXI 395		
		MAT.	INST.	TOTAL	MAT.	INST.	TOTAL	MAT.	INST.	TOTAL	MAT.	INST.	TOTAL	MAT.	INST.	TOTAL	MAT.	INST.	TOTAL
015433	CONTRACTOR EQUIPMENT		104.2	104.2		98.4	98.4		95.8	95.8		98.4	98.4		98.4	98.4		98.7	98.7
0241, 31 - 34	SITE & INFRASTRUCTURE, DEMOLITION	97.6	104.3	102.2	91.4	98.3	96.2	94.1	92.1	92.8	90.3	97.0	94.9	84.1	96.2	92.4	101.8	83.7	89.4
0310	Concrete Forming & Accessories	102.2	120.4	117.7	85.2	119.7	114.5	88.3	84.1	84.7	84.9	85.2	85.2	89.9	81.2	82.5	86.3	68.0	70.7
0320	Concrete Reinforcing	105.3	110.8	108.0	95.9	110.5	102.9	97.4	100.6	98.9	95.5	110.1	102.6	95.5	109.2	102.1	89.6	64.5	77.5
0330	Cast-in-Place Concrete	102.5	119.7	109.0	95.6	117.5	103.9	98.8	92.0	96.2	97.1	78.5	90.1	84.0	82.8	83.5	121.8	66.9	101.0
03	CONCRETE	101.6	119.0	109.4	86.8	117.9	100.7	90.1	91.1	90.5	86.7	88.7	87.6	78.0	88.2	82.6	98.8	68.6	85.3
04	MASONRY	98.7	123.0	113.5	110.2	119.5	115.8	100.5	103.0	102.0	113.7	112.3	112.9	125.9	89.1	103.5	90.1	63.9	74.1
05	METALS	97.4	126.6	106.5	90.9	125.4	101.7	89.3	117.8	98.2	90.0	123.4	100.4	89.9	121.5	99.8	90.1	90.6	90.3
06	WOOD, PLASTICS & COMPOSITES	101.3	118.2	110.1	81.1	117.8	100.3	71.6	80.1	76.1	80.7	77.2	78.9	86.1	77.2	81.4	83.0	68.6	75.5
07	THERMAL & MOISTURE PROTECTION	101.5	119.2	109.1	104.0	112.7	107.7	107.2	87.9	98.9	103.7	97.9	101.2	103.7	81.5	94.2	101.4	63.5	85.1
08	OPENINGS	97.9	127.9	105.2	91.1	127.7	100.0	99.4	103.7	100.4	88.2	101.1	91.3	91.6	101.1	93.9	96.1	63.1	88.1
0920	Plaster & Gypsum Board	92.5	119.0	109.9	85.8	119.0	107.6	98.3	80.1	86.3	85.8	77.3	80.2	85.8	77.3	80.2	112.4	68.3	83.5
0950, 0980	Ceilings & Acoustic Treatment	93.3	119.0	109.4	60.2	119.0	97.1	131.6	80.1	99.3	60.2	77.3	70.9	60.2	77.3	70.9	90.4	68.3	76.6
0960	Flooring	93.5	121.5	101.7	87.7	119.6	97.0	91.9	81.6	88.9	89.2	81.6	87.0	91.8	81.6	88.8	93.8	64.3	85.2
0970, 0990	Wall Finishes & Painting/Coating	90.6	131.3	115.0	90.8	127.3	112.7	85.6	79.4	81.9	85.6	79.4	81.9	85.6	98.8	93.5	80.7	48.3	61.3
09	FINISHES	93.1	121.8	108.6	80.9	120.7	102.5	101.0	83.2	91.3	81.0	83.3	82.2	81.2	82.6	82.0	92.5	65.4	77.8
COVERS	DIVS. 10 - 14, 25, 28, 41, 43, 44, 46	100.0	107.0	101.6	100.0	106.1	101.4	100.0	97.4	99.4	100.0	99.2	99.8	100.0	96.2	99.1	100.0	82.9	96.0
21, 22, 23	FIRE SUPPRESSION, PLUMBING & HVAC	100.5	117.9	107.5	100.0	109.3	103.8	97.1	83.3	91.5	96.2	100.6	98.0	96.2	80.7	90.0	101.1	56.6	83.1
26, 27, 3370	ELECTRICAL, COMMUNICATIONS & UTIL.	102.1	119.9	110.9	101.4	119.9	110.6	100.9	68.4	84.8	101.4	80.6	91.1	106.9	99.4	103.2	100.2	56.1	78.4
MF2018	WEIGHTED AVERAGE	99.3	119.2	107.9	94.9	115.9	104.0	96.4	89.4	93.4	93.7	96.6	94.9	94.0	91.7	93.0	97.1	66.6	83.9

MISSISSIPPI

DIVISION		CLARKSDALE 386			COLUMBUS 397			GREENVILLE 387			GREENWOOD 389			JACKSON 390 - 392			LAUREL 394		
		MAT.	INST.	TOTAL	MAT.	INST.	TOTAL	MAT.	INST.	TOTAL	MAT.	INST.	TOTAL	MAT.	INST.	TOTAL	MAT.	INST.	TOTAL
015433	CONTRACTOR EQUIPMENT		98.7	98.7		98.7	98.7		98.7	98.7		98.7	98.7		103.9	103.9		98.7	98.7
0241, 31 - 34	SITE & INFRASTRUCTURE, DEMOLITION	99.6	82.0	87.5	100.1	82.8	88.2	105.8	83.7	90.6	102.6	81.8	88.3	97.2	92.6	94.0	105.3	82.1	89.4
0310	Concrete Forming & Accessories	83.7	42.6	48.7	75.8	44.7	49.4	80.2	62.0	64.7	94.0	42.7	50.4	87.3	67.4	70.3	75.9	57.6	60.4
0320	Concrete Reinforcing	99.6	63.4	82.1	96.0	63.5	80.3	100.1	63.7	82.5	99.6	63.5	82.1	100.9	64.6	83.4	96.6	31.2	65.1
0330	Cast-in-Place Concrete	100.3	57.4	84.1	124.2	59.6	99.8	103.3	66.1	89.2	107.8	57.2	88.7	105.0	67.6	90.9	121.5	58.8	97.8
03	CONCRETE	93.3	53.5	75.4	100.3	55.3	80.1	98.6	65.5	83.8	99.6	53.5	78.9	96.4	68.4	83.8	102.3	55.3	81.2
04	MASONRY	91.9	50.0	66.4	114.1	52.7	76.6	137.0	63.5	92.2	92.5	49.9	66.5	97.1	64.3	77.1	110.5	52.6	75.2
05	METALS	92.3	86.7	90.6	87.2	89.2	87.8	93.4	90.3	92.4	92.3	86.5	90.5	96.6	88.6	94.1	87.3	75.3	83.6
06	WOOD, PLASTICS & COMPOSITES	78.8	42.9	59.9	69.9	43.7	56.2	74.8	60.7	67.4	92.4	42.9	66.5	88.3	67.8	77.5	70.9	62.3	66.4
07	THERMAL & MOISTURE PROTECTION	97.7	50.1	77.3	101.3	53.5	80.8	98.1	62.7	83.0	98.2	52.3	78.4	99.8	64.7	84.7	101.5	55.6	81.8
08	OPENINGS	94.9	46.4	83.1	95.8	46.8	83.8	94.6	56.3	85.3	94.9	46.4	83.1	99.2	60.4	89.8	93.1	49.9	82.6
0920	Plaster & Gypsum Board	91.5	41.9	59.0	101.8	42.7	63.0	91.0	60.2	70.8	101.7	41.9	62.5	95.9	67.2	77.1	101.8	61.9	75.6
0950, 0980	Ceilings & Acoustic Treatment	89.2	41.9	59.5	84.0	42.7	58.1	91.3	60.2	71.8	89.2	41.9	59.5	93.6	67.2	77.1	84.0	61.9	70.1
0960	Flooring	99.6	64.3	89.3	87.5	64.3	80.7	98.0	64.3	88.2	105.4	64.3	93.4	91.5	65.6	84.0	86.2	64.3	79.9
0970, 0990	Wall Finishes & Painting/Coating	90.9	44.7	63.2	80.7	44.7	59.1	90.9	56.2	70.1	90.9	44.7	63.2	87.0	56.2	68.5	80.7	44.7	59.1
09	FINISHES	92.8	46.5	67.7	87.7	47.6	66.0	93.3	61.6	76.1	96.2	46.5	69.3	92.5	66.0	78.2	87.9	58.6	72.0
COVERS	DIVS. 10 - 14, 25, 28, 41, 43, 44, 46	100.0	45.4	87.2	100.0	46.4	87.4	100.0	81.9	95.7	100.0	45.4	87.1	100.0	83.2	96.0	100.0	32.4	84.1
21, 22, 23	FIRE SUPPRESSION, PLUMBING & HVAC	99.1	48.9	78.8	99.0	50.8	79.5	100.9	56.5	83.0	99.1	49.2	79.0	101.2	60.0	84.6	99.0	45.4	77.4
26, 27, 3370	ELECTRICAL, COMMUNICATIONS & UTIL.	97.4	40.1	69.0	98.0	53.0	75.7	97.4	56.1	77.0	97.4	37.5	67.8	102.6	56.1	79.6	99.3	55.8	77.8
MF2018	WEIGHTED AVERAGE	95.7	53.8	77.6	96.7	57.1	79.6	99.4	65.2	84.6	97.0	53.5	78.2	98.6	67.9	85.3	96.8	56.4	79.4

MISSISSIPPI / MISSOURI

DIVISION		MCCOMB 396			MERIDIAN 393			TUPELO 388			BOWLING GREEN 633			CAPE GIRARDEAU 637			CHILLICOTHE 646		
		MAT.	INST.	TOTAL	MAT.	INST.	TOTAL	MAT.	INST.	TOTAL	MAT.	INST.	TOTAL	MAT.	INST.	TOTAL	MAT.	INST.	TOTAL
015433	CONTRACTOR EQUIPMENT		98.7	98.7		98.7	98.7		98.7	98.7		104.0	104.0		104.0	104.0		98.6	98.6
0241, 31 - 34	SITE & INFRASTRUCTURE, DEMOLITION	93.0	81.9	85.4	97.4	83.8	88.0	97.2	81.9	86.7	88.2	86.1	86.8	90.2	86.3	87.5	101.7	84.8	90.1
0310	Concrete Forming & Accessories	75.8	43.9	48.6	73.3	68.1	68.9	80.7	44.1	49.6	91.6	90.7	90.8	84.2	81.4	81.8	77.7	92.3	90.1
0320	Concrete Reinforcing	97.2	32.7	66.1	96.0	64.6	80.8	97.4	63.3	81.0	100.6	92.0	96.4	101.9	84.9	93.7	100.7	97.8	99.3
0330	Cast-in-Place Concrete	107.8	57.0	88.6	115.3	66.9	97.1	100.3	58.3	84.4	82.7	91.5	86.0	81.8	84.7	82.9	97.5	82.5	91.8
03	CONCRETE	89.9	48.6	71.3	94.4	68.7	82.8	92.8	54.5	75.6	87.9	92.7	90.1	87.1	85.0	86.1	96.7	90.7	94.0
04	MASONRY	115.5	49.5	75.3	89.7	63.9	74.0	126.1	51.0	80.3	111.9	93.8	100.9	109.4	80.1	91.5	101.3	88.4	93.4
05	METALS	87.4	74.3	83.3	88.4	90.8	89.1	92.2	86.2	90.4	88.9	115.3	97.1	90.0	111.5	96.7	82.3	108.6	90.5
06	WOOD, PLASTICS & COMPOSITES	69.9	45.2	57.0	67.6	68.6	68.1	75.4	43.7	58.8	90.7	91.2	90.9	83.0	79.6	81.2	80.0	93.7	87.2
07	THERMAL & MOISTURE PROTECTION	100.8	52.3	80.0	101.0	63.9	85.1	97.7	50.5	77.4	96.6	94.7	95.8	96.1	83.0	90.5	94.1	88.5	91.7
08	OPENINGS	95.8	39.7	82.1	95.5	64.2	87.9	94.9	49.5	83.8	98.2	94.6	97.3	98.2	78.0	93.3	86.4	92.6	87.9
0920	Plaster & Gypsum Board	101.8	44.3	64.1	101.8	68.3	79.8	91.0	42.7	59.3	95.9	91.5	93.0	95.4	79.6	85.0	103.7	93.7	97.1
0950, 0980	Ceilings & Acoustic Treatment	84.0	44.3	59.1	84.7	68.3	74.4	89.2	42.7	60.1	89.6	91.5	90.8	89.6	79.6	83.3	88.5	93.7	91.7
0960	Flooring	87.5	64.3	80.7	86.1	64.3	79.8	98.2	64.3	88.3	95.6	92.4	94.7	92.1	84.6	89.9	92.4	95.1	93.2
0970, 0990	Wall Finishes & Painting/Coating	80.7	44.7	59.1	80.7	56.2	66.0	90.9	43.1	62.3	95.1	101.2	98.7	95.1	76.7	84.1	90.4	98.3	95.1
09	FINISHES	87.1	47.7	65.8	87.2	66.3	75.9	92.3	47.4	68.0	96.5	91.9	94.0	95.4	80.8	87.5	95.9	93.7	94.7
COVERS	DIVS. 10 - 14, 25, 28, 41, 43, 44, 46	100.0	48.2	87.8	100.0	82.9	96.0	100.0	46.4	87.4	100.0	96.8	99.3	100.0	95.6	99.0	100.0	96.6	99.2
21, 22, 23	FIRE SUPPRESSION, PLUMBING & HVAC	99.0	48.3	78.6	101.1	60.1	84.5	99.2	50.2	79.4	96.9	94.8	96.1	100.7	98.2	99.7	97.3	95.2	96.4
26, 27, 3370	ELECTRICAL, COMMUNICATIONS & UTIL.	96.8	53.9	75.6	99.3	58.1	78.9	97.2	53.0	75.3	99.4	74.0	86.8	99.4	94.8	97.1	94.2	73.0	83.7
MF2018	WEIGHTED AVERAGE	95.1	53.8	77.3	95.4	67.8	83.5	97.2	56.4	79.6	95.6	92.3	94.2	96.4	90.6	93.9	93.7	90.7	92.4

For customer support on your Site Work & Landscape Costs with RSMeans Data, call 800.448.8182.

803

DIVISION		MISSOURI																
		COLUMBIA 652			FLAT RIVER 636			HANNIBAL 634			HARRISONVILLE 647			JEFFERSON CITY 650 - 651			JOPLIN 648		
		MAT.	INST.	TOTAL	MAT.	INST.	TOTAL	MAT.	INST.	TOTAL	MAT.	INST.	TOTAL	MAT.	INST.	TOTAL	MAT.	INST.	TOTAL
015433	CONTRACTOR EQUIPMENT		107.2	107.2		104.0	104.0		104.0	104.0		98.6	98.6		112.3	112.3		102.0	102.0
0241, 31 - 34	SITE & INFRASTRUCTURE, DEMOLITION	93.0	91.4	91.9	90.7	85.8	87.3	86.0	85.9	85.9	93.6	85.9	88.3	93.0	99.4	97.4	103.0	89.7	93.9
0310	Concrete Forming & Accessories	77.7	89.1	87.4	98.4	85.0	87.0	89.9	79.6	81.1	75.1	95.5	92.5	90.8	78.5	80.4	88.0	75.8	77.6
0320	Concrete Reinforcing	87.1	100.0	93.3	101.9	98.0	100.0	100.1	91.9	96.1	100.3	105.7	102.9	90.9	89.1	90.0	104.0	102.3	103.2
0330	Cast-in-Place Concrete	83.7	87.0	84.9	85.3	86.7	85.8	78.3	89.5	82.5	99.9	95.7	98.3	88.8	85.6	87.6	105.6	74.7	93.9
03	CONCRETE	79.4	91.9	86.0	90.8	89.5	90.2	84.5	87.0	85.6	93.2	98.2	95.4	85.6	84.6	85.2	96.3	81.2	89.5
04	MASONRY	146.8	91.4	113.0	109.3	73.2	87.3	102.9	90.6	95.4	95.6	95.3	95.4	99.2	91.5	94.5	94.5	80.6	86.0
05	METALS	94.1	119.5	102.0	88.8	116.6	97.5	88.9	114.8	97.0	82.7	113.1	92.2	93.6	112.1	99.4	85.2	106.6	91.8
06	WOOD, PLASTICS & COMPOSITES	71.4	86.7	79.4	99.3	86.5	92.6	88.8	78.3	83.3	77.0	95.4	86.6	89.1	73.1	80.7	90.2	75.2	82.3
07	THERMAL & MOISTURE PROTECTION	92.1	89.7	91.0	96.8	86.8	92.5	96.4	89.2	93.3	93.3	96.2	94.5	99.2	88.9	94.8	93.7	80.7	88.2
08	OPENINGS	94.3	88.7	93.0	98.2	93.8	97.1	98.2	80.6	93.9	86.1	99.4	89.4	94.5	78.4	90.6	87.4	78.5	85.2
0920	Plaster & Gypsum Board	72.2	86.7	81.7	101.9	86.7	91.9	95.7	78.3	84.3	100.0	95.3	96.9	80.3	72.4	75.1	110.4	74.6	86.9
0950, 0980	Ceilings & Acoustic Treatment	79.8	86.7	84.1	89.6	86.7	87.8	89.6	78.3	82.5	88.5	95.3	92.8	84.2	72.4	76.8	89.1	74.6	81.2
0960	Flooring	87.3	95.9	89.8	98.9	80.7	93.6	94.8	92.4	94.1	87.9	96.1	90.3	93.5	68.2	86.1	116.9	71.5	103.6
0970, 0990	Wall Finishes & Painting/Coating	94.1	82.0	86.8	95.1	68.3	79.1	95.1	89.6	91.8	94.6	102.7	99.4	95.1	82.0	87.2	90.0	72.2	79.3
09	FINISHES	80.9	89.1	85.4	98.4	81.7	89.4	96.1	81.9	88.4	93.7	96.2	95.1	88.4	76.1	81.7	102.5	74.7	87.4
COVERS	DIVS. 10 - 14, 25, 28, 41, 43, 44, 46	100.0	96.9	99.3	100.0	94.7	98.8	100.0	94.4	98.7	100.0	98.0	99.5	100.0	97.6	99.4	100.0	93.4	98.4
21, 22, 23	FIRE SUPPRESSION, PLUMBING & HVAC	100.9	101.2	101.0	96.9	92.2	95.0	96.9	93.1	95.4	97.2	96.1	96.7	101.0	95.6	98.8	101.1	69.9	88.5
26, 27, 3370	ELECTRICAL, COMMUNICATIONS & UTIL.	94.2	82.5	88.4	103.4	94.8	99.1	98.3	74.0	86.3	99.9	97.5	98.7	100.9	82.5	91.8	92.5	74.7	83.7
MF2018	WEIGHTED AVERAGE	95.7	94.5	95.2	96.6	90.6	94.0	94.5	88.5	91.9	93.2	97.4	95.0	95.7	90.0	93.2	95.2	80.1	88.7

DIVISION		MISSOURI																
		KANSAS CITY 640 - 641			KIRKSVILLE 635			POPLAR BLUFF 639			ROLLA 654 - 655			SEDALIA 653			SIKESTON 638		
		MAT.	INST.	TOTAL	MAT.	INST.	TOTAL	MAT.	INST.	TOTAL	MAT.	INST.	TOTAL	MAT.	INST.	TOTAL	MAT.	INST.	TOTAL
015433	CONTRACTOR EQUIPMENT		105.1	105.1		94.1	94.1		96.4	96.4		107.2	107.2		97.3	97.3		96.4	96.4
0241, 31 - 34	SITE & INFRASTRUCTURE, DEMOLITION	95.5	100.3	98.8	89.1	81.1	83.6	76.6	84.6	82.1	91.4	90.2	90.6	89.1	85.5	86.6	79.9	85.2	83.6
0310	Concrete Forming & Accessories	87.2	100.5	98.5	82.2	75.6	76.6	82.5	76.3	77.2	84.9	90.5	89.6	82.9	75.4	76.5	83.2	76.4	77.4
0320	Concrete Reinforcing	92.2	107.0	99.4	100.8	79.4	90.5	103.9	72.4	88.2	87.6	88.9	88.2	86.4	105.1	95.4	103.2	72.4	88.3
0330	Cast-in-Place Concrete	103.6	100.5	102.4	85.3	78.7	82.8	65.6	79.8	71.0	85.6	90.6	87.5	89.5	77.9	85.1	70.1	79.8	73.8
03	CONCRETE	94.3	102.1	97.8	102.5	78.6	91.8	78.9	78.1	78.6	81.1	91.8	85.9	94.1	82.7	89.0	82.3	78.2	80.4
04	MASONRY	101.2	100.7	100.9	116.0	81.3	94.8	107.7	71.0	85.3	117.2	81.8	95.6	124.2	78.2	96.2	107.1	71.0	85.1
05	METALS	91.9	110.2	97.6	88.6	98.7	91.7	89.1	95.3	91.0	93.4	113.4	99.7	92.4	110.5	98.1	89.5	95.4	91.3
06	WOOD, PLASTICS & COMPOSITES	85.1	100.4	93.1	76.7	73.9	75.2	75.7	77.2	76.5	78.4	92.3	85.7	73.0	74.1	73.5	77.4	77.2	77.3
07	THERMAL & MOISTURE PROTECTION	92.2	103.0	96.9	101.9	86.7	95.4	100.5	79.2	91.4	92.3	88.7	90.8	96.9	83.8	91.3	100.7	78.4	91.1
08	OPENINGS	96.9	100.2	97.7	103.7	75.9	96.9	104.6	72.4	96.8	94.3	87.7	92.7	99.4	83.8	95.6	104.6	72.4	96.8
0920	Plaster & Gypsum Board	110.5	100.5	103.9	91.1	73.7	79.7	91.4	77.2	82.1	73.6	92.4	86.0	68.5	73.7	71.9	93.4	77.2	82.7
0950, 0980	Ceilings & Acoustic Treatment	90.7	100.5	96.9	88.4	73.7	79.2	89.6	77.2	81.8	79.8	92.4	87.7	79.8	73.7	76.0	89.6	77.2	81.8
0960	Flooring	93.1	99.9	95.1	73.8	92.0	79.1	87.1	80.7	85.2	91.0	92.0	91.3	71.2	69.6	70.7	87.7	80.7	85.6
0970, 0990	Wall Finishes & Painting/Coating	96.2	111.5	105.4	90.9	74.6	81.1	90.2	63.9	74.4	94.1	85.9	89.2	94.1	98.3	96.6	90.2	63.9	74.4
09	FINISHES	96.8	101.5	99.4	95.1	77.8	85.7	94.9	75.2	84.2	82.3	90.3	86.6	80.2	75.9	77.9	95.6	75.5	84.7
COVERS	DIVS. 10 - 14, 25, 28, 41, 43, 44, 46	100.0	99.4	99.9	100.0	93.9	98.6	100.0	93.0	98.4	100.0	96.8	99.2	100.0	92.0	98.1	100.0	93.0	98.4
21, 22, 23	FIRE SUPPRESSION, PLUMBING & HVAC	100.8	101.4	101.0	97.0	92.5	95.2	97.0	90.4	94.3	97.1	93.9	95.8	97.0	90.4	94.4	97.0	90.4	94.3
26, 27, 3370	ELECTRICAL, COMMUNICATIONS & UTIL.	101.7	99.7	100.7	98.3	73.9	86.2	98.6	94.8	96.7	93.0	76.4	84.8	93.8	97.4	95.6	97.8	94.8	96.3
MF2018	WEIGHTED AVERAGE	97.5	101.8	99.3	98.0	83.6	91.8	94.4	84.3	90.0	93.4	90.6	92.2	95.6	88.1	92.4	94.9	84.4	90.4

DIVISION		MISSOURI									MONTANA								
		SPRINGFIELD 656 - 658			ST. JOSEPH 644 - 645			ST. LOUIS 630 - 631			BILLINGS 590 - 591			BUTTE 597			GREAT FALLS 594		
		MAT.	INST.	TOTAL	MAT.	INST.	TOTAL	MAT.	INST.	TOTAL	MAT.	INST.	TOTAL	MAT.	INST.	TOTAL	MAT.	INST.	TOTAL
015433	CONTRACTOR EQUIPMENT		99.7	99.7		98.6	98.6		107.9	107.9		96.1	96.1		95.8	95.8		95.8	95.8
0241, 31 - 34	SITE & INFRASTRUCTURE, DEMOLITION	91.6	89.3	90.0	97.5	84.2	88.3	96.9	98.8	98.2	91.9	90.6	91.0	98.2	90.3	92.8	101.9	90.6	94.1
0310	Concrete Forming & Accessories	90.7	78.3	80.2	86.8	93.4	92.4	94.6	102.5	101.3	99.1	69.4	73.8	85.7	69.1	71.6	99.4	69.2	73.7
0320	Concrete Reinforcing	84.0	99.6	91.5	97.5	108.6	102.9	91.4	112.4	101.5	90.1	81.0	85.7	97.7	80.6	89.4	90.1	81.0	85.7
0330	Cast-in-Place Concrete	90.9	77.0	85.7	98.1	96.5	97.5	92.4	104.0	97.1	112.2	70.3	96.4	123.5	70.1	103.3	130.5	71.1	108.1
03	CONCRETE	90.5	82.7	87.0	92.3	98.0	94.8	92.0	106.0	98.3	95.7	72.6	85.4	99.0	72.3	87.1	104.1	72.8	90.0
04	MASONRY	92.8	84.5	87.7	96.4	93.2	94.5	87.6	107.7	99.8	131.1	83.1	101.8	126.6	76.1	95.8	131.5	80.3	100.3
05	METALS	98.5	104.9	100.5	88.5	113.9	96.4	94.3	121.5	102.8	104.6	89.4	99.8	98.9	88.5	95.7	102.1	89.4	98.1
06	WOOD, PLASTICS & COMPOSITES	80.0	76.3	78.1	90.2	93.5	91.9	92.1	100.1	96.3	89.4	66.5	77.5	75.0	66.5	70.6	90.9	66.5	78.1
07	THERMAL & MOISTURE PROTECTION	96.1	78.5	88.6	93.6	92.7	93.2	95.3	105.6	99.7	108.7	72.2	93.0	108.3	69.8	91.7	108.9	72.1	93.1
08	OPENINGS	101.7	88.2	98.4	90.3	99.9	92.7	100.0	106.8	101.7	98.4	66.4	90.6	96.8	66.4	89.4	99.6	71.0	92.6
0920	Plaster & Gypsum Board	74.8	76.0	75.6	111.9	93.4	99.8	98.1	100.5	99.7	117.1	66.1	83.7	116.3	66.1	83.4	126.4	66.1	86.9
0950, 0980	Ceilings & Acoustic Treatment	79.8	76.0	77.4	94.9	93.4	94.0	86.3	100.5	95.2	110.0	66.1	82.5	117.4	66.1	85.3	119.3	66.1	86.0
0960	Flooring	90.3	71.5	84.8	97.5	99.9	98.2	97.5	97.1	97.4	90.6	77.1	86.7	87.9	79.4	85.4	94.6	77.1	89.5
0970, 0990	Wall Finishes & Painting/Coating	88.5	104.5	98.1	90.4	103.0	97.9	99.3	106.8	103.8	87.6	88.4	88.1	86.1	64.5	73.2	86.1	91.1	89.1
09	FINISHES	84.4	80.0	82.0	99.4	95.7	97.4	97.2	101.5	99.5	97.8	72.2	84.0	98.3	70.1	83.0	102.4	72.3	86.1
COVERS	DIVS. 10 - 14, 25, 28, 41, 43, 44, 46	100.0	94.3	98.7	100.0	97.3	99.4	100.0	101.8	100.4	100.0	92.3	98.2	100.0	92.3	98.2	100.0	92.0	98.1
21, 22, 23	FIRE SUPPRESSION, PLUMBING & HVAC	100.9	74.0	90.1	101.1	90.6	96.8	100.7	104.5	102.3	100.8	74.4	90.2	100.8	68.3	87.7	100.8	74.0	90.0
26, 27, 3370	ELECTRICAL, COMMUNICATIONS & UTIL.	97.4	67.1	82.4	99.9	80.0	90.1	103.2	96.7	100.0	99.2	72.4	85.9	105.1	70.5	88.0	98.7	73.4	86.1
MF2018	WEIGHTED AVERAGE	96.5	81.4	90.0	96.0	93.3	94.9	97.5	104.6	100.6	101.4	77.2	90.9	101.3	74.4	89.7	102.8	77.1	91.7

MONTANA

DIVISION		HAVRE 595			HELENA 596			KALISPELL 599			MILES CITY 593			MISSOULA 598			WOLF POINT 592		
		MAT.	INST.	TOTAL	MAT.	INST.	TOTAL	MAT.	INST.	TOTAL	MAT.	INST.	TOTAL	MAT.	INST.	TOTAL	MAT.	INST.	TOTAL
015433	CONTRACTOR EQUIPMENT		95.8	95.8		100.9	100.9		95.8	95.8		95.8	95.8		95.8	95.8		95.8	95.8
0241, 31 - 34	SITE & INFRASTRUCTURE, DEMOLITION	105.1	90.1	94.8	92.1	98.1	96.2	87.8	90.2	89.5	94.1	90.1	91.4	81.0	90.3	87.4	111.6	90.1	96.8
0310	Concrete Forming & Accessories	78.0	67.7	69.2	102.5	68.9	74.0	89.3	68.8	71.9	97.3	67.9	72.3	89.3	69.3	72.2	89.4	67.8	71.1
0320	Concrete Reinforcing	98.5	75.1	87.2	101.6	80.6	91.5	100.3	84.7	92.8	98.1	75.2	87.0	99.4	84.5	92.2	99.5	74.5	87.5
0330	Cast-in-Place Concrete	133.0	68.3	108.6	95.2	71.2	86.1	107.2	69.5	93.0	117.4	68.3	98.9	91.0	70.1	83.1	131.5	67.3	107.3
03	CONCRETE	106.6	70.2	90.3	96.9	72.5	86.0	89.2	72.7	81.8	96.1	70.3	84.5	78.4	73.1	76.0	110.1	69.8	92.0
04	MASONRY	127.6	77.2	96.9	115.5	75.1	90.9	125.5	79.2	97.3	133.6	77.2	99.2	153.1	75.9	106.0	134.8	77.2	99.7
05	METALS	95.1	86.4	92.4	100.9	86.7	96.5	95.0	90.2	93.5	94.3	86.6	91.9	95.5	89.9	93.8	94.4	86.3	91.9
06	WOOD, PLASTICS & COMPOSITES	65.7	66.5	66.2	94.6	66.8	80.1	78.8	66.5	72.4	87.5	66.5	76.5	78.8	66.5	72.4	77.9	66.5	72.0
07	THERMAL & MOISTURE PROTECTION	108.6	63.7	89.3	104.8	70.4	90.0	107.8	70.7	91.9	108.2	65.0	89.7	107.4	71.9	92.2	109.2	64.9	90.2
08	OPENINGS	97.3	65.1	89.5	96.1	66.6	88.9	97.3	67.3	90.0	96.8	65.1	89.1	96.8	67.3	89.6	96.8	65.0	89.1
0920	Plaster & Gypsum Board	111.7	66.1	81.8	114.1	66.1	82.6	116.3	66.1	83.4	124.7	66.1	86.3	116.3	66.1	83.4	118.8	66.1	84.2
0950, 0980	Ceilings & Acoustic Treatment	117.4	66.1	85.3	122.3	66.1	87.1	117.4	66.1	85.3	112.9	66.1	83.6	117.4	66.1	85.3	112.9	66.1	83.6
0960	Flooring	85.5	86.8	85.9	97.2	79.4	92.0	89.6	86.8	88.8	94.5	86.8	92.2	89.6	79.4	86.7	91.0	86.8	89.8
0970, 0990	Wall Finishes & Painting/Coating	86.1	64.5	73.2	93.9	57.2	71.9	86.1	86.7	86.4	86.1	64.5	73.2	86.1	86.7	86.4	86.1	64.5	73.2
09	FINISHES	97.7	70.9	83.2	105.0	69.3	85.6	98.2	73.8	85.0	100.2	70.9	84.3	97.7	72.5	84.1	99.7	70.9	84.1
COVERS	DIVS. 10 - 14, 25, 28, 41, 43, 44, 46	100.0	84.7	96.4	100.0	92.6	98.2	100.0	85.4	96.6	100.0	84.7	96.4	100.0	92.2	98.2	100.0	84.7	96.4
21, 22, 23	FIRE SUPPRESSION, PLUMBING & HVAC	97.0	65.6	84.3	100.9	67.9	87.6	97.0	66.1	84.5	97.0	71.7	86.8	100.8	68.4	87.7	97.0	71.6	86.8
26, 27, 3370	ELECTRICAL, COMMUNICATIONS & UTIL.	98.7	66.7	82.8	105.3	70.9	88.3	102.3	65.6	84.1	98.7	71.7	85.3	103.2	69.1	86.3	98.7	71.7	85.3
MF2018	WEIGHTED AVERAGE	100.2	72.5	88.3	101.2	74.7	89.7	98.0	74.1	87.7	99.1	74.6	88.5	98.8	74.8	88.5	101.3	74.5	89.7

NEBRASKA

DIVISION		ALLIANCE 693			COLUMBUS 686			GRAND ISLAND 688			HASTINGS 689			LINCOLN 683 - 685			MCCOOK 690		
		MAT.	INST.	TOTAL	MAT.	INST.	TOTAL	MAT.	INST.	TOTAL	MAT.	INST.	TOTAL	MAT.	INST.	TOTAL	MAT.	INST.	TOTAL
015433	CONTRACTOR EQUIPMENT		92.2	92.2		98.8	98.8		98.8	98.8		98.8	98.8		104.9	104.9		98.8	98.8
0241, 31 - 34	SITE & INFRASTRUCTURE, DEMOLITION	97.6	92.3	94.0	100.4	86.8	91.1	105.6	87.3	93.0	103.8	86.8	92.1	95.3	96.6	96.2	100.5	86.8	91.1
0310	Concrete Forming & Accessories	83.9	53.2	57.8	95.7	72.2	75.7	95.2	67.4	71.5	98.9	69.9	74.2	96.8	73.9	77.3	89.3	53.8	59.0
0320	Concrete Reinforcing	110.1	83.2	97.1	100.7	81.9	91.6	100.1	80.4	90.6	100.1	72.5	86.8	95.3	80.7	88.3	102.7	72.7	88.2
0330	Cast-in-Place Concrete	104.2	78.3	94.4	104.6	78.3	94.7	110.7	75.6	97.5	110.7	70.8	95.6	87.5	81.9	85.4	112.6	70.7	96.8
03	CONCRETE	113.9	68.0	93.3	100.2	77.2	89.9	104.8	73.8	90.9	105.1	71.9	90.2	93.8	78.9	87.1	103.2	64.6	85.9
04	MASONRY	109.4	74.2	87.9	115.8	74.2	90.4	109.1	73.0	87.1	119.3	70.9	89.8	98.4	77.2	85.5	103.6	74.2	85.7
05	METALS	96.7	83.0	92.4	89.2	95.8	91.3	91.0	94.8	92.2	91.7	91.5	91.6	98.6	93.4	97.0	91.7	91.5	91.6
06	WOOD, PLASTICS & COMPOSITES	78.2	47.0	61.9	92.3	72.3	81.8	91.4	64.0	77.0	95.5	69.6	81.9	100.2	72.6	85.7	85.8	47.9	66.0
07	THERMAL & MOISTURE PROTECTION	101.1	72.3	88.7	101.4	76.7	90.8	101.5	77.2	91.1	101.6	74.1	89.8	100.3	79.1	91.1	96.1	72.6	86.0
08	OPENINGS	90.9	55.9	82.4	91.7	71.0	86.7	91.7	66.3	85.5	91.7	67.2	85.8	105.8	68.3	96.7	91.3	53.7	82.1
0920	Plaster & Gypsum Board	75.5	45.9	56.1	84.4	71.8	76.2	83.5	63.3	70.3	85.2	69.1	74.6	90.3	71.8	78.2	87.1	46.8	60.7
0950, 0980	Ceilings & Acoustic Treatment	88.5	45.9	61.8	84.8	71.8	76.7	84.8	63.3	71.3	84.8	69.1	74.9	106.7	71.8	84.8	84.1	46.8	60.7
0960	Flooring	90.9	83.3	88.7	83.7	83.3	83.6	83.4	75.3	81.1	84.8	77.5	82.7	94.9	82.6	91.3	89.6	83.3	87.8
0970, 0990	Wall Finishes & Painting/Coating	156.3	49.2	92.1	74.2	57.2	64.0	74.2	59.5	65.4	74.2	57.2	64.0	94.7	75.6	83.2	85.6	42.9	60.0
09	FINISHES	91.0	56.8	72.5	85.1	72.5	78.3	85.1	67.1	75.4	85.8	69.7	77.1	96.5	75.4	85.1	88.2	56.6	71.1
COVERS	DIVS. 10 - 14, 25, 28, 41, 43, 44, 46	100.0	85.1	96.5	100.0	88.0	97.2	100.0	88.1	97.2	100.0	87.7	97.1	100.0	89.7	97.6	100.0	85.3	96.5
21, 22, 23	FIRE SUPPRESSION, PLUMBING & HVAC	97.1	71.4	86.7	97.0	71.7	86.8	100.8	80.2	92.5	97.0	71.0	86.5	100.6	80.3	92.4	96.8	71.7	86.7
26, 27, 3370	ELECTRICAL, COMMUNICATIONS & UTIL.	92.7	63.2	78.1	96.1	78.4	87.3	94.9	66.0	80.6	94.3	76.5	85.5	105.4	66.0	85.9	94.4	63.3	79.0
MF2018	WEIGHTED AVERAGE	98.4	70.4	86.3	96.0	77.7	88.1	97.4	76.2	88.2	97.1	75.2	87.7	99.8	79.3	90.9	95.9	70.2	84.8

NEBRASKA / NEVADA

DIVISION		NORFOLK 687			NORTH PLATTE 691			OMAHA 680 - 681			VALENTINE 692			CARSON CITY 897			ELKO 898		
		MAT.	INST.	TOTAL	MAT.	INST.	TOTAL	MAT.	INST.	TOTAL	MAT.	INST.	TOTAL	MAT.	INST.	TOTAL	MAT.	INST.	TOTAL
015433	CONTRACTOR EQUIPMENT		88.5	88.5		98.8	98.8		94.2	94.2		91.9	91.9		95.7	95.7		92.3	92.3
0241, 31 - 34	SITE & INFRASTRUCTURE, DEMOLITION	82.6	85.9	84.9	102.3	86.6	91.5	89.9	95.2	93.6	85.3	91.3	89.4	80.5	93.7	89.6	65.3	88.0	80.9
0310	Concrete Forming & Accessories	82.3	71.4	73.0	91.8	72.8	75.6	92.3	78.3	80.4	80.6	51.4	55.7	100.8	84.7	87.1	105.3	92.7	94.6
0320	Concrete Reinforcing	100.8	63.3	82.7	102.2	77.7	90.4	103.7	81.0	92.7	102.7	63.2	83.6	116.4	117.8	117.1	117.6	109.1	113.5
0330	Cast-in-Place Concrete	105.3	68.1	91.2	112.6	58.0	92.0	86.7	79.6	84.0	99.4	51.6	81.4	97.1	84.3	92.3	90.9	69.3	82.8
03	CONCRETE	98.8	69.4	85.6	103.3	69.7	88.2	94.2	79.5	87.6	100.8	54.7	80.1	106.5	90.2	99.2	97.2	87.2	92.7
04	MASONRY	123.9	74.1	93.5	93.1	73.7	81.2	96.8	81.6	87.5	104.8	73.6	85.8	116.8	67.2	86.5	123.4	64.9	87.7
05	METALS	92.5	78.1	88.0	91.0	93.4	91.8	98.6	84.8	94.3	102.5	77.5	94.7	109.5	95.4	105.1	113.1	91.6	106.4
06	WOOD, PLASTICS & COMPOSITES	76.8	71.8	74.2	87.9	74.2	80.8	92.6	78.0	85.0	73.6	45.7	59.0	87.1	85.6	86.3	96.2	98.8	97.6
07	THERMAL & MOISTURE PROTECTION	100.3	74.7	89.3	96.0	74.8	86.9	95.9	82.2	90.0	95.8	70.6	85.0	114.3	80.1	99.6	110.4	70.2	93.2
08	OPENINGS	92.9	66.1	86.4	90.7	70.6	85.8	99.7	79.0	94.7	92.7	51.4	82.7	101.1	83.8	96.9	102.4	89.2	99.2
0920	Plaster & Gypsum Board	84.4	71.8	76.1	87.1	73.9	78.4	92.2	77.9	82.9	88.7	45.0	60.0	98.5	85.2	89.7	103.2	99.0	100.5
0950, 0980	Ceilings & Acoustic Treatment	99.4	71.8	82.1	84.1	73.9	77.7	93.6	77.9	83.8	98.4	45.0	64.9	107.1	85.2	93.3	105.3	99.0	101.4
0960	Flooring	102.7	83.3	97.0	90.8	75.3	86.3	95.0	90.4	93.7	115.7	83.3	106.3	99.0	64.1	88.8	99.1	64.1	88.9
0970, 0990	Wall Finishes & Painting/Coating	133.5	57.2	87.8	85.6	56.1	67.9	103.2	58.1	76.1	154.7	58.1	96.8	91.6	85.1	87.7	87.4	75.6	80.3
09	FINISHES	101.7	72.2	85.7	88.6	71.7	79.5	96.7	78.2	86.7	108.1	56.8	80.4	98.3	80.4	88.7	95.4	86.8	90.8
COVERS	DIVS. 10 - 14, 25, 28, 41, 43, 44, 46	100.0	86.9	96.9	100.0	87.9	97.2	100.0	89.3	97.5	100.0	83.9	96.2	100.0	106.3	101.5	100.0	86.2	96.8
21, 22, 23	FIRE SUPPRESSION, PLUMBING & HVAC	96.6	70.8	86.2	100.6	73.4	89.6	100.6	83.9	93.9	96.3	70.1	85.7	100.9	81.4	93.0	99.0	73.3	88.6
26, 27, 3370	ELECTRICAL, COMMUNICATIONS & UTIL.	95.2	78.4	86.8	92.9	66.0	79.6	102.9	83.1	93.1	90.6	63.2	77.1	100.9	92.0	96.5	97.5	85.0	91.3
MF2018	WEIGHTED AVERAGE	97.4	74.5	87.5	96.1	74.9	86.9	98.6	83.0	91.9	98.0	67.4	84.8	103.2	85.6	95.6	101.5	81.8	93.0

For customer support on your Site Work & Landscape Costs with RSMeans Data, call 800.448.8182.

805

NEVADA / NEW HAMPSHIRE

DIVISION		ELY 893 MAT.	INST.	TOTAL	LAS VEGAS 889-891 MAT.	INST.	TOTAL	RENO 894-895 MAT.	INST.	TOTAL	CHARLESTON 036 MAT.	INST.	TOTAL	CLAREMONT 037 MAT.	INST.	TOTAL	CONCORD 032-033 MAT.	INST.	TOTAL
015433	CONTRACTOR EQUIPMENT		92.3	92.3		92.3	92.3		92.3	92.3		94.2	94.2		94.2	94.2		98.6	98.6
0241, 31 - 34	SITE & INFRASTRUCTURE, DEMOLITION	70.7	89.1	83.3	73.8	92.7	86.8	70.9	89.8	83.9	81.6	93.4	89.7	75.5	93.4	87.8	91.0	102.3	98.7
0310	Concrete Forming & Accessories	97.9	97.8	97.8	99.1	112.9	110.9	94.5	84.5	85.9	86.6	82.9	83.5	93.3	83.0	84.5	99.3	94.4	95.1
0320	Concrete Reinforcing	116.2	109.3	112.9	107.1	125.5	116.0	109.9	123.8	116.6	87.5	84.4	86.0	87.5	84.4	86.0	93.9	84.5	89.4
0330	Cast-in-Place Concrete	97.6	91.3	95.2	94.5	109.4	100.1	103.2	82.2	95.3	86.0	111.8	95.7	79.0	111.8	91.4	104.7	113.8	108.2
03	CONCRETE	106.0	97.3	102.1	101.0	113.3	106.5	105.1	90.5	98.6	90.3	93.5	91.7	83.0	93.5	87.7	102.2	99.3	100.9
04	MASONRY	129.0	71.7	94.1	115.5	105.1	109.2	122.4	67.1	88.7	92.3	95.4	94.2	93.1	95.4	94.5	105.2	98.9	101.4
05	METALS	113.0	94.4	107.2	122.1	105.3	116.8	114.8	98.7	109.8	99.2	90.8	96.6	99.2	90.8	96.6	106.7	89.8	101.4
06	WOOD, PLASTICS & COMPOSITES	86.5	100.9	94.0	85.0	112.0	99.1	81.0	85.3	83.2	87.6	80.6	83.9	94.7	80.6	87.3	98.2	94.4	96.2
07	THERMAL & MOISTURE PROTECTION	110.9	88.0	101.1	125.1	103.2	115.7	110.4	78.5	96.7	107.6	102.2	105.3	107.4	102.2	105.2	111.3	107.1	109.5
08	OPENINGS	102.3	90.3	99.4	101.3	115.0	104.7	100.2	84.9	96.5	96.6	79.8	92.5	97.6	79.8	93.3	97.4	87.4	95.0
0920	Plaster & Gypsum Board	98.1	101.1	100.1	92.4	112.5	105.6	87.8	85.2	86.1	100.3	79.9	86.9	100.8	79.9	87.1	100.2	94.1	96.2
0950, 0980	Ceilings & Acoustic Treatment	105.3	101.1	102.7	111.7	112.5	112.2	108.5	85.2	93.9	90.6	79.9	83.9	90.6	79.9	83.9	92.6	94.1	93.5
0960	Flooring	96.6	64.1	87.1	88.5	102.0	92.4	93.3	64.1	84.8	87.5	107.9	93.4	90.0	107.9	95.2	96.0	107.9	99.5
0970, 0990	Wall Finishes & Painting/Coating	87.4	106.7	99.0	89.9	118.5	107.1	87.4	85.1	86.1	79.1	85.7	83.1	79.1	85.7	83.1	95.0	85.7	89.4
09	FINISHES	94.4	93.1	93.7	93.0	112.2	103.4	92.6	80.2	85.9	90.6	87.2	88.8	90.9	87.2	88.9	95.0	95.9	95.5
COVERS	DIVS. 10 - 14, 25, 28, 41, 43, 44, 46	100.0	67.0	92.2	100.0	106.4	101.5	100.0	105.6	101.3	100.0	84.4	96.3	100.0	84.4	96.3	100.0	106.8	101.6
21, 22, 23	FIRE SUPPRESSION, PLUMBING & HVAC	99.0	91.6	96.0	101.1	103.6	102.1	100.9	81.4	93.0	97.6	78.3	89.8	97.6	78.3	89.8	101.2	85.4	94.8
26, 27, 3370	ELECTRICAL, COMMUNICATIONS & UTIL.	97.8	89.5	93.7	102.1	109.9	105.9	98.1	92.0	95.1	94.6	72.4	83.7	94.6	72.4	83.7	95.9	72.4	84.3
MF2018	WEIGHTED AVERAGE	102.9	89.5	97.1	104.2	107.0	105.4	102.8	85.6	95.4	95.6	85.6	91.3	94.8	85.7	90.8	100.8	91.3	96.7

NEW HAMPSHIRE / NEW JERSEY

DIVISION		KEENE 034 MAT.	INST.	TOTAL	LITTLETON 035 MAT.	INST.	TOTAL	MANCHESTER 031 MAT.	INST.	TOTAL	NASHUA 030 MAT.	INST.	TOTAL	PORTSMOUTH 038 MAT.	INST.	TOTAL	ATLANTIC CITY 082,084 MAT.	INST.	TOTAL
015433	CONTRACTOR EQUIPMENT		94.2	94.2		94.2	94.2		99.1	99.1		94.2	94.2		94.2	94.2		91.8	91.8
0241, 31 - 34	SITE & INFRASTRUCTURE, DEMOLITION	89.5	93.4	92.2	75.6	92.4	87.1	87.4	102.4	97.7	91.2	93.6	92.8	85.0	94.3	91.4	92.0	97.2	95.6
0310	Concrete Forming & Accessories	91.8	83.2	84.5	103.8	77.3	81.3	100.2	94.8	95.6	100.8	94.7	95.7	88.2	94.3	93.4	112.0	145.5	140.5
0320	Concrete Reinforcing	87.5	84.4	86.0	88.2	84.4	86.4	110.5	84.6	98.0	109.3	84.6	97.4	87.5	84.6	86.1	82.0	137.7	108.9
0330	Cast-in-Place Concrete	86.4	111.9	96.0	77.5	103.4	87.3	98.7	114.3	104.6	81.8	113.4	93.7	77.5	113.6	91.2	79.8	135.5	100.8
03	CONCRETE	90.0	93.7	91.6	82.4	88.0	85.0	101.6	99.7	100.7	90.3	99.4	94.4	82.5	99.4	90.1	87.8	139.2	110.9
04	MASONRY	94.9	95.4	95.2	105.1	80.6	90.1	100.7	98.9	99.6	97.7	98.8	98.4	93.2	98.3	96.3	106.3	140.0	126.9
05	METALS	99.9	91.2	97.2	99.9	90.8	97.1	107.4	90.3	102.0	106.0	91.8	101.5	101.5	93.2	98.9	104.0	116.0	107.7
06	WOOD, PLASTICS & COMPOSITES	93.0	80.6	86.5	105.7	80.6	92.6	100.4	94.5	97.3	104.3	94.4	99.1	89.1	94.4	91.9	117.5	147.1	133.0
07	THERMAL & MOISTURE PROTECTION	108.2	102.2	105.6	107.6	95.5	102.4	112.5	107.1	110.2	108.7	106.1	107.6	108.1	105.7	107.1	100.7	136.4	116.0
08	OPENINGS	95.3	83.1	92.3	98.5	79.8	94.0	96.3	92.5	95.4	99.5	90.1	97.2	100.0	81.1	95.4	96.7	142.0	107.7
0920	Plaster & Gypsum Board	100.5	79.9	87.0	115.8	79.9	92.2	99.2	94.1	95.8	110.7	94.1	99.8	100.3	94.1	96.2	109.6	148.2	134.9
0950, 0980	Ceilings & Acoustic Treatment	90.6	79.9	83.9	90.6	79.9	83.9	89.3	94.1	92.3	100.9	94.1	96.6	91.4	94.1	93.1	89.1	148.2	126.1
0960	Flooring	89.6	107.9	95.0	100.4	107.9	102.6	95.0	110.1	99.4	93.3	110.1	98.2	87.7	110.1	94.2	102.7	158.5	119.0
0970, 0990	Wall Finishes & Painting/Coating	79.1	98.4	90.7	79.1	85.7	83.1	96.3	118.0	109.3	79.1	118.0	102.4	79.1	97.4	90.1	77.1	147.2	119.1
09	FINISHES	92.5	88.6	90.4	96.0	83.5	89.2	94.3	99.9	97.3	97.4	99.9	98.7	91.5	97.4	94.7	94.4	149.6	124.3
COVERS	DIVS. 10 - 14, 25, 28, 41, 43, 44, 46	100.0	90.3	97.7	100.0	91.1	97.9	100.0	107.0	101.7	100.0	106.9	101.6	100.0	106.6	101.5	100.0	116.5	103.9
21, 22, 23	FIRE SUPPRESSION, PLUMBING & HVAC	97.6	78.4	89.8	97.6	70.6	86.7	101.2	85.5	94.9	101.4	85.5	95.0	101.4	85.0	94.8	100.4	130.0	112.4
26, 27, 3370	ELECTRICAL, COMMUNICATIONS & UTIL.	94.6	72.4	83.7	95.4	47.3	71.6	97.4	78.5	88.0	96.3	78.5	87.5	95.0	77.6	86.4	92.2	138.0	114.9
MF2018	WEIGHTED AVERAGE	96.1	86.2	91.8	96.1	77.7	88.1	100.5	93.0	97.3	99.3	92.3	96.3	96.5	91.5	94.3	97.8	132.6	112.8

NEW JERSEY

DIVISION		CAMDEN 081 MAT.	INST.	TOTAL	DOVER 078 MAT.	INST.	TOTAL	ELIZABETH 072 MAT.	INST.	TOTAL	HACKENSACK 076 MAT.	INST.	TOTAL	JERSEY CITY 073 MAT.	INST.	TOTAL	LONG BRANCH 077 MAT.	INST.	TOTAL
015433	CONTRACTOR EQUIPMENT		91.8	91.8		94.2	94.2		94.2	94.2		94.2	94.2		91.8	91.8		91.4	91.4
0241, 31 - 34	SITE & INFRASTRUCTURE, DEMOLITION	92.8	97.5	96.0	107.6	98.9	101.6	113.1	98.9	103.3	108.7	98.9	101.9	98.3	98.8	98.6	103.5	97.6	99.5
0310	Concrete Forming & Accessories	101.8	146.6	139.9	96.0	148.3	140.5	109.0	148.4	142.5	96.0	148.1	140.3	100.0	148.3	141.1	100.6	134.2	129.2
0320	Concrete Reinforcing	107.7	137.9	122.2	72.3	152.4	110.9	72.3	152.4	110.9	72.3	152.4	110.9	93.8	152.4	122.1	72.3	151.9	110.7
0330	Cast-in-Place Concrete	77.4	135.7	99.4	82.7	135.9	102.8	71.1	137.8	96.2	80.8	137.7	102.3	64.5	135.9	91.5	71.7	128.9	93.3
03	CONCRETE	88.5	139.8	111.5	85.9	143.3	111.7	83.0	144.0	110.4	84.3	143.8	111.0	80.6	143.2	108.7	84.6	134.2	106.9
04	MASONRY	95.5	138.5	121.8	98.0	141.7	124.6	115.6	141.7	131.5	102.2	141.7	126.3	92.5	141.7	122.5	107.4	129.8	121.1
05	METALS	110.3	116.4	112.2	101.5	125.2	108.9	103.1	125.3	110.0	101.6	125.0	108.9	107.9	122.3	112.4	101.6	120.6	107.6
06	WOOD, PLASTICS & COMPOSITES	104.7	149.8	128.3	93.1	149.8	122.8	109.1	149.8	130.4	93.1	149.8	122.8	94.2	149.8	123.3	95.4	134.7	115.9
07	THERMAL & MOISTURE PROTECTION	100.5	135.9	115.7	102.2	140.6	118.6	102.5	140.8	118.9	102.0	133.1	115.3	101.7	140.6	118.4	101.8	126.2	112.3
08	OPENINGS	98.8	143.4	109.7	102.7	144.8	113.0	101.2	144.8	111.8	100.6	144.8	111.4	99.2	144.8	110.3	95.3	134.1	104.7
0920	Plaster & Gypsum Board	105.8	151.0	135.4	118.5	151.0	139.8	125.5	151.0	142.2	118.5	151.0	139.8	121.8	151.0	140.9	120.4	135.5	130.3
0950, 0980	Ceilings & Acoustic Treatment	98.1	151.0	131.2	83.8	151.0	125.9	85.2	151.0	126.4	83.8	151.0	125.9	93.5	151.0	129.5	83.8	135.5	116.2
0960	Flooring	98.9	158.5	116.3	82.4	183.6	112.0	87.4	183.6	115.5	82.4	183.6	112.0	83.3	183.6	112.6	83.6	168.8	108.5
0970, 0990	Wall Finishes & Painting/Coating	77.1	147.2	119.1	82.6	142.9	118.7	82.6	142.9	118.7	82.6	142.9	118.7	82.7	142.9	118.8	82.7	147.2	121.4
09	FINISHES	94.8	150.4	124.9	90.5	154.7	125.2	93.8	155.2	127.1	90.4	154.7	125.2	93.3	155.2	126.8	91.5	142.7	119.2
COVERS	DIVS. 10 - 14, 25, 28, 41, 43, 44, 46	100.0	115.8	103.7	100.0	125.9	106.1	100.0	125.9	106.1	100.0	125.9	106.1	100.0	125.9	106.1	100.0	113.2	103.1
21, 22, 23	FIRE SUPPRESSION, PLUMBING & HVAC	100.7	132.5	113.6	100.4	135.3	114.5	100.7	135.2	114.6	100.4	135.3	114.5	100.7	135.3	114.7	100.4	125.0	110.3
26, 27, 3370	ELECTRICAL, COMMUNICATIONS & UTIL.	95.8	141.5	118.4	94.7	144.4	119.3	95.2	144.4	119.5	94.7	143.1	118.7	98.7	143.1	120.6	94.5	125.1	109.6
MF2018	WEIGHTED AVERAGE	98.9	133.7	114.0	97.6	137.5	114.8	98.9	137.6	115.6	97.4	137.1	114.6	97.7	137.1	114.7	97.1	126.6	109.9

NEW JERSEY

DIVISION		NEW BRUNSWICK 088-089			NEWARK 070-071			PATERSON 074-075			POINT PLEASANT 087			SUMMIT 079			TRENTON 085-086		
		MAT.	INST.	TOTAL	MAT.	INST.	TOTAL	MAT.	INST.	TOTAL	MAT.	INST.	TOTAL	MAT.	INST.	TOTAL	MAT.	INST.	TOTAL
015433	CONTRACTOR EQUIPMENT		91.4	91.4		99.5	99.5		94.2	94.2		91.4	91.4		94.2	94.2		95.5	95.5
0241, 31 - 34	SITE & INFRASTRUCTURE, DEMOLITION	105.8	98.0	100.4	114.9	107.9	110.1	110.5	98.8	102.5	107.2	97.6	100.6	110.3	98.9	102.4	90.8	103.4	99.5
0310	Concrete Forming & Accessories	105.4	147.6	141.3	99.8	148.4	141.2	98.2	148.0	140.6	99.0	134.2	129.0	98.9	148.2	140.9	98.9	133.8	128.6
0320	Concrete Reinforcing	83.0	151.9	116.2	97.7	152.4	124.1	93.8	152.4	122.1	83.0	151.9	116.2	72.3	152.4	110.9	107.7	110.7	109.2
0330	Cast-in-Place Concrete	98.6	132.0	111.3	93.5	138.9	110.6	82.2	137.6	103.1	98.6	130.8	110.8	68.7	137.7	94.8	94.5	129.4	107.7
03	CONCRETE	104.2	141.4	120.9	93.7	144.3	116.4	88.7	143.8	113.4	103.8	134.9	117.7	80.2	143.9	108.8	95.4	127.2	109.7
04	MASONRY	104.4	135.2	123.2	102.5	141.7	126.4	98.5	141.7	124.8	92.6	129.8	115.3	101.0	141.1	125.8	102.5	129.9	119.2
05	METALS	104.1	121.0	109.3	109.6	123.9	114.0	102.9	125.0	109.8	104.1	120.7	109.2	101.5	125.2	108.9	109.6	106.4	108.6
06	WOOD, PLASTICS & COMPOSITES	110.5	149.7	131.0	96.4	149.9	124.4	95.9	149.8	124.1	101.9	134.7	119.1	97.2	149.8	124.7	96.7	134.7	116.6
07	THERMAL & MOISTURE PROTECTION	101.0	136.2	116.1	104.1	141.9	120.3	102.3	133.1	115.5	101.0	128.4	112.8	102.5	140.8	119.0	100.9	132.6	114.5
08	OPENINGS	92.1	144.8	104.9	101.2	144.9	111.8	105.7	144.8	115.2	93.8	138.5	104.7	106.6	144.8	115.9	97.2	125.2	104.1
0920	Plaster & Gypsum Board	108.2	151.0	136.3	112.0	151.0	137.6	121.8	151.0	140.9	102.8	135.5	124.3	120.4	151.0	140.5	97.7	135.5	122.5
0950, 0980	Ceilings & Acoustic Treatment	89.1	151.0	127.9	92.7	151.0	129.2	93.5	151.0	129.5	89.1	135.5	118.2	83.8	151.0	125.9	92.7	135.5	119.5
0960	Flooring	100.3	183.6	124.6	97.6	183.6	122.7	83.3	183.6	112.6	97.6	158.5	115.4	83.6	183.6	112.8	104.3	168.8	123.2
0970, 0990	Wall Finishes & Painting/Coating	77.1	142.9	116.5	96.1	142.9	124.2	82.6	142.9	118.7	77.1	147.2	119.1	82.6	142.9	118.7	92.4	147.2	125.3
09	FINISHES	94.8	154.7	127.2	98.0	155.3	129.0	93.4	154.7	126.6	93.3	141.0	119.1	91.5	154.7	125.7	97.0	142.8	121.8
COVERS	DIVS. 10 - 14, 25, 28, 41, 43, 44, 46	100.0	125.7	106.1	100.0	126.2	106.2	100.0	125.9	106.1	100.0	105.9	101.4	100.0	125.9	106.1	100.0	113.3	103.1
21, 22, 23	FIRE SUPPRESSION, PLUMBING & HVAC	100.4	127.8	111.5	100.8	134.2	114.3	100.7	135.1	114.6	100.4	125.0	110.3	100.4	132.3	113.3	100.8	124.6	110.4
26, 27, 3370	ELECTRICAL, COMMUNICATIONS & UTIL.	92.7	128.7	110.5	103.3	143.1	123.0	98.7	144.4	121.3	92.2	125.1	108.5	95.2	144.4	119.5	100.5	123.9	112.1
MF2018	WEIGHTED AVERAGE	99.7	132.2	113.8	101.8	137.9	117.4	99.3	137.3	115.7	99.1	126.5	110.9	97.7	136.9	114.6	100.5	124.4	110.8

NEW JERSEY / NEW MEXICO

DIVISION		NEW JERSEY VINELAND 080,083			ALBUQUERQUE 870-872			CARRIZOZO 883			CLOVIS 881			FARMINGTON 874			GALLUP 873		
		MAT.	INST.	TOTAL	MAT.	INST.	TOTAL	MAT.	INST.	TOTAL	MAT.	INST.	TOTAL	MAT.	INST.	TOTAL	MAT.	INST.	TOTAL
015433	CONTRACTOR EQUIPMENT		91.8	91.8		105.2	105.2		105.2	105.2		105.2	105.2		105.2	105.2		105.2	105.2
0241, 31 - 34	SITE & INFRASTRUCTURE, DEMOLITION	96.7	97.3	97.1	88.4	94.4	92.5	106.6	94.4	98.2	93.9	94.4	94.2	94.6	94.4	94.5	103.2	94.4	97.1
0310	Concrete Forming & Accessories	95.7	133.1	127.5	103.1	65.0	70.7	95.9	65.0	69.6	95.9	64.9	69.5	103.2	65.0	70.7	103.2	65.0	70.7
0320	Concrete Reinforcing	82.0	125.1	102.8	96.3	69.5	83.4	109.2	69.5	90.0	110.4	69.5	90.7	105.3	69.5	88.1	100.8	69.5	85.7
0330	Cast-in-Place Concrete	86.0	129.1	102.3	90.4	68.9	82.2	93.4	68.9	84.1	93.3	68.8	84.1	91.2	68.9	82.8	85.8	68.9	79.4
03	CONCRETE	92.2	129.2	108.8	94.1	68.5	82.6	116.6	68.5	95.0	104.3	68.4	88.2	97.6	68.5	84.5	104.4	68.5	88.3
04	MASONRY	94.6	129.9	116.1	105.7	62.4	79.3	107.4	62.4	80.0	107.4	62.4	80.0	112.7	62.4	82.0	101.8	62.4	77.8
05	METALS	103.9	111.5	106.3	106.2	89.8	101.1	104.7	89.8	100.0	104.3	89.6	99.7	103.9	89.8	99.5	103.0	89.8	98.9
06	WOOD, PLASTICS & COMPOSITES	98.3	133.3	116.6	104.5	64.9	83.8	89.5	64.9	76.6	89.5	64.9	76.6	104.6	64.9	83.8	104.6	64.9	83.8
07	THERMAL & MOISTURE PROTECTION	100.4	131.1	113.6	98.9	72.0	87.4	103.9	72.0	90.2	102.5	72.0	89.4	99.1	72.0	87.4	100.3	72.0	88.1
08	OPENINGS	93.4	131.8	102.8	98.9	65.1	90.7	96.5	65.1	88.9	96.7	65.1	89.0	101.1	65.1	92.3	101.1	65.1	92.3
0920	Plaster & Gypsum Board	101.1	134.0	122.7	126.4	63.9	85.4	82.9	63.9	70.5	82.9	63.9	70.5	111.9	63.9	80.4	111.9	63.9	80.4
0950, 0980	Ceilings & Acoustic Treatment	89.1	134.0	117.3	111.4	63.9	81.6	112.0	63.9	81.9	112.0	63.9	81.9	108.5	63.9	80.6	108.5	63.9	80.6
0960	Flooring	96.9	158.5	114.9	88.4	65.0	81.6	96.5	65.0	87.3	96.5	65.0	87.3	89.8	65.0	82.6	89.8	65.0	82.6
0970, 0990	Wall Finishes & Painting/Coating	77.1	147.2	119.1	92.0	51.1	67.5	89.3	51.1	66.4	89.3	51.1	66.4	86.4	51.1	65.3	86.4	51.1	65.3
09	FINISHES	92.0	140.1	118.0	99.0	63.3	79.6	97.5	63.3	78.9	96.2	63.3	78.4	96.6	63.3	78.5	97.8	63.3	79.1
COVERS	DIVS. 10 - 14, 25, 28, 41, 43, 44, 46	100.0	113.2	103.1	100.0	86.4	96.8	100.0	86.4	96.8	100.0	86.4	96.8	100.0	86.4	96.8	100.0	86.4	96.8
21, 22, 23	FIRE SUPPRESSION, PLUMBING & HVAC	100.4	124.7	110.2	101.4	68.1	87.9	98.7	68.1	86.3	98.7	68.1	86.2	101.2	68.1	87.9	98.9	68.1	86.5
26, 27, 3370	ELECTRICAL, COMMUNICATIONS & UTIL.	92.2	138.0	114.9	86.6	69.5	78.1	89.7	69.5	79.7	87.8	69.5	78.7	84.9	69.5	77.3	84.3	69.5	77.0
MF2018	WEIGHTED AVERAGE	97.2	126.6	109.9	98.9	71.6	87.1	101.4	71.6	88.5	99.1	71.5	87.2	99.3	71.6	87.3	99.2	71.6	87.3

NEW MEXICO

DIVISION		LAS CRUCES 880			LAS VEGAS 877			ROSWELL 882			SANTA FE 875			SOCORRO 878			TRUTH/CONSEQUENCES 879		
		MAT.	INST.	TOTAL	MAT.	INST.	TOTAL	MAT.	INST.	TOTAL	MAT.	INST.	TOTAL	MAT.	INST.	TOTAL	MAT.	INST.	TOTAL
015433	CONTRACTOR EQUIPMENT		81.1	81.1		105.2	105.2		105.2	105.2		109.7	109.7		105.2	105.2		81.1	81.1
0241, 31 - 34	SITE & INFRASTRUCTURE, DEMOLITION	95.6	74.7	81.2	93.4	94.4	94.1	96.5	94.4	95.0	95.6	102.8	100.5	90.0	94.4	93.0	108.8	74.7	85.4
0310	Concrete Forming & Accessories	92.5	64.0	68.3	103.2	65.0	70.7	95.9	65.0	69.6	101.9	65.1	70.6	103.2	65.0	70.7	100.9	64.0	69.5
0320	Concrete Reinforcing	106.0	69.3	88.3	102.6	69.5	86.6	110.4	69.5	90.7	102.1	69.5	86.4	104.5	69.5	87.6	98.0	69.4	84.2
0330	Cast-in-Place Concrete	88.2	62.1	78.3	88.7	68.9	81.2	93.3	68.9	84.1	94.2	70.0	85.1	86.9	68.9	80.1	95.3	62.1	82.7
03	CONCRETE	82.0	65.2	74.5	95.2	68.5	83.2	105.2	68.5	88.7	99.1	68.8	85.5	94.1	68.5	82.6	87.3	65.2	77.4
04	MASONRY	103.0	60.7	77.2	102.1	62.4	77.9	118.1	62.4	84.2	98.8	62.5	76.7	102.0	62.4	77.9	98.6	62.1	76.4
05	METALS	103.1	82.1	96.5	102.7	89.8	98.7	105.6	89.8	100.7	101.0	87.7	96.8	103.0	89.8	98.9	102.5	82.1	96.2
06	WOOD, PLASTICS & COMPOSITES	79.9	63.9	71.5	104.6	64.9	83.8	89.5	64.9	76.6	101.3	64.9	82.3	104.6	64.9	83.8	96.9	64.9	79.6
07	THERMAL & MOISTURE PROTECTION	91.8	67.2	81.2	98.6	72.0	87.2	102.7	72.0	87.9	100.5	73.4	88.8	98.6	72.0	87.2	89.2	67.6	79.9
08	OPENINGS	92.1	64.6	85.4	97.7	65.1	89.8	96.5	65.1	88.8	99.1	65.2	90.9	97.6	65.1	89.7	91.1	64.6	84.6
0920	Plaster & Gypsum Board	82.0	63.9	70.1	111.9	63.9	80.4	82.9	63.9	70.5	125.2	63.9	85.0	111.9	63.9	80.4	113.4	63.9	80.9
0950, 0980	Ceilings & Acoustic Treatment	96.8	63.9	76.2	108.5	63.9	80.6	112.0	63.9	81.9	109.4	63.9	80.9	108.5	63.9	80.6	98.5	63.9	76.9
0960	Flooring	127.1	65.0	109.0	89.8	65.0	82.6	96.5	65.0	87.3	98.9	65.0	89.0	89.8	65.0	82.6	118.3	65.0	102.8
0970, 0990	Wall Finishes & Painting/Coating	78.5	51.1	62.1	86.4	51.1	65.3	89.3	51.1	66.4	96.7	51.1	69.4	86.4	51.1	65.3	79.1	51.1	62.3
09	FINISHES	104.6	62.5	81.8	96.4	63.3	78.5	96.3	63.3	78.4	103.8	63.3	81.9	96.3	63.3	78.4	107.6	62.5	83.2
COVERS	DIVS. 10 - 14, 25, 28, 41, 43, 44, 46	100.0	84.1	96.3	100.0	86.4	96.8	100.0	86.4	96.8	100.0	86.5	96.8	100.0	86.4	96.8	100.0	84.1	96.2
21, 22, 23	FIRE SUPPRESSION, PLUMBING & HVAC	101.7	67.8	88.0	98.9	68.1	86.5	100.9	68.1	87.7	101.2	68.1	87.8	98.9	68.1	86.5	98.8	67.8	86.3
26, 27, 3370	ELECTRICAL, COMMUNICATIONS & UTIL.	90.0	80.7	85.4	86.2	69.5	77.9	88.9	69.5	79.3	100.7	69.4	85.2	84.7	69.5	77.2	88.1	69.4	78.9
MF2018	WEIGHTED AVERAGE	96.8	70.0	85.2	97.4	71.6	86.3	100.7	71.6	88.1	100.5	72.2	88.3	97.1	71.6	86.1	96.8	68.6	84.6

DIVISION		NEW MEXICO TUCUMCARI 884			NEW YORK ALBANY 120-122			BINGHAMTON 137-139			BRONX 104			BROOKLYN 112			BUFFALO 140-142		
		MAT.	INST.	TOTAL	MAT.	INST.	TOTAL	MAT.	INST.	TOTAL	MAT.	INST.	TOTAL	MAT.	INST.	TOTAL	MAT.	INST.	TOTAL
015433	CONTRACTOR EQUIPMENT		105.2	105.2		117.0	117.0		115.8	115.8		102.9	102.9		107.5	107.5		101.3	101.3
0241, 31-34	SITE & INFRASTRUCTURE, DEMOLITION	93.5	94.4	94.1	78.5	107.4	98.4	94.1	87.0	89.2	98.8	107.1	104.5	117.8	117.5	117.6	99.2	103.8	102.4
0310	Concrete Forming & Accessories	95.9	64.9	69.5	99.3	104.4	103.7	100.5	92.2	93.4	94.9	183.3	170.1	106.9	183.3	171.9	103.4	115.3	113.6
0320	Concrete Reinforcing	108.2	69.5	89.5	99.2	112.1	105.4	96.5	106.7	101.4	95.6	173.5	133.2	97.9	231.2	162.2	95.5	113.4	104.2
0330	Cast-in-Place Concrete	93.3	68.8	84.1	76.1	115.5	91.0	105.2	103.6	104.6	82.3	167.1	114.3	107.2	165.5	129.2	117.3	122.0	119.1
03	CONCRETE	103.6	68.4	87.8	89.6	110.6	99.0	94.1	101.0	97.2	86.3	175.5	126.3	105.6	183.9	140.7	107.1	116.9	111.5
04	MASONRY	119.1	62.4	84.5	97.2	116.3	108.9	109.7	103.6	106.0	88.5	184.6	147.1	119.1	184.6	159.1	110.8	123.5	118.5
05	METALS	104.3	89.6	99.7	103.1	123.4	109.4	96.0	135.0	108.2	92.9	173.2	117.9	103.9	173.4	125.6	93.9	107.2	98.0
06	WOOD, PLASTICS & COMPOSITES	89.5	64.9	76.6	97.4	100.6	99.1	102.1	89.0	95.2	93.5	182.4	140.0	105.3	182.1	145.5	100.9	114.0	107.8
07	THERMAL & MOISTURE PROTECTION	102.5	72.0	89.4	107.7	109.4	108.6	110.1	93.1	102.8	102.8	163.8	129.0	110.3	163.2	133.0	104.0	111.2	107.1
08	OPENINGS	96.4	65.1	88.8	96.4	101.9	97.7	90.5	94.1	91.4	92.5	191.0	116.5	87.9	190.9	113.0	100.0	109.5	102.3
0920	Plaster & Gypsum Board	82.9	63.9	70.5	105.1	100.4	102.0	113.6	88.6	97.2	92.2	184.6	152.8	106.0	184.6	157.6	119.9	114.3	116.2
0950, 0980	Ceilings & Acoustic Treatment	112.0	63.9	81.9	97.3	100.4	99.3	102.4	88.6	93.8	80.3	184.6	145.7	96.1	184.6	151.6	102.1	114.3	109.7
0960	Flooring	96.5	65.0	87.3	92.6	107.6	97.0	103.0	97.8	101.5	99.4	182.7	123.7	110.8	182.7	131.2	94.1	119.1	101.4
0970, 0990	Wall Finishes & Painting/Coating	89.3	51.1	66.4	93.9	99.4	97.2	84.5	101.3	94.5	104.7	164.2	140.4	109.6	164.2	142.3	93.8	114.4	106.1
09	FINISHES	96.1	63.3	78.3	94.8	103.9	99.7	96.7	93.4	94.9	93.8	181.8	141.4	107.8	181.6	147.7	101.6	116.4	109.6
COVERS	DIVS. 10-14, 25, 28, 41, 43, 44, 46	100.0	86.4	96.8	100.0	100.9	100.2	100.0	98.0	99.5	100.0	137.9	108.9	100.0	137.2	108.8	100.0	106.3	101.5
21, 22, 23	FIRE SUPPRESSION, PLUMBING & HVAC	98.7	67.7	86.2	101.2	108.0	103.9	101.6	98.6	100.4	100.4	171.5	129.1	100.5	171.5	129.1	100.9	102.0	101.3
26, 27, 3370	ELECTRICAL, COMMUNICATIONS & UTIL.	89.7	69.5	79.7	101.8	107.2	104.5	100.2	97.3	98.7	93.6	183.2	137.9	100.2	183.2	141.3	99.9	104.3	102.1
MF2018	WEIGHTED AVERAGE	99.7	71.5	87.6	98.3	109.4	103.1	98.5	100.5	99.3	94.8	171.2	127.8	102.7	173.1	133.1	100.9	109.4	104.8

DIVISION		NEW YORK ELMIRA 148-149			FAR ROCKAWAY 116			FLUSHING 113			GLENS FALLS 128			HICKSVILLE 115,117,118			JAMAICA 114		
		MAT.	INST.	TOTAL	MAT.	INST.	TOTAL	MAT.	INST.	TOTAL	MAT.	INST.	TOTAL	MAT.	INST.	TOTAL	MAT.	INST.	TOTAL
015433	CONTRACTOR EQUIPMENT		118.7	118.7		107.5	107.5		107.5	107.5		110.8	110.8		107.5	107.5		107.5	107.5
0241, 31-34	SITE & INFRASTRUCTURE, DEMOLITION	97.9	87.0	90.4	120.6	117.5	118.5	120.6	117.5	118.5	70.0	96.4	88.1	111.0	116.0	114.4	114.9	117.5	116.7
0310	Concrete Forming & Accessories	83.6	94.4	92.8	92.0	183.3	169.7	96.2	183.3	170.3	82.5	96.2	94.2	88.1	153.9	144.1	96.2	183.3	170.3
0320	Concrete Reinforcing	97.6	105.5	101.4	97.9	231.2	162.2	99.6	231.2	163.1	95.2	107.4	101.1	97.9	169.7	132.5	97.9	231.2	162.2
0330	Cast-in-Place Concrete	103.2	102.9	103.1	116.0	165.5	134.7	116.0	165.5	134.7	73.3	107.2	86.1	98.4	155.6	120.0	107.2	165.5	129.2
03	CONCRETE	92.8	101.7	96.8	111.7	183.9	144.1	112.2	183.9	144.4	78.7	103.3	89.7	97.4	156.5	123.9	104.9	183.9	140.4
04	MASONRY	106.2	103.1	104.3	124.2	184.6	161.0	117.4	184.6	158.4	102.3	107.9	105.7	114.2	168.6	147.4	122.3	184.6	160.3
05	METALS	92.4	136.3	106.1	104.0	173.4	125.6	104.0	173.4	125.6	96.4	122.5	104.6	105.5	147.9	118.7	104.0	173.4	125.6
06	WOOD, PLASTICS & COMPOSITES	81.9	92.7	87.6	86.5	182.1	136.5	91.8	182.1	139.0	83.3	92.9	88.3	82.6	151.9	118.9	91.8	182.1	139.0
07	THERMAL & MOISTURE PROTECTION	110.8	92.9	103.1	110.2	163.4	133.0	110.2	163.4	133.1	101.0	102.3	101.5	109.8	155.1	129.2	110.0	163.4	132.9
08	OPENINGS	96.8	95.8	96.6	86.7	190.9	112.1	86.7	190.9	112.1	90.2	95.4	91.5	87.1	159.2	104.7	86.7	190.9	112.1
0920	Plaster & Gypsum Board	110.2	92.6	98.6	92.7	184.6	153.0	95.6	184.6	154.0	90.9	92.8	92.1	92.5	153.6	132.6	95.6	184.6	154.0
0950, 0980	Ceilings & Acoustic Treatment	109.1	92.6	98.7	83.4	184.6	146.8	83.4	184.6	146.8	88.5	92.8	91.2	82.6	153.6	127.1	83.4	184.6	146.8
0960	Flooring	86.8	97.8	90.0	104.5	182.7	127.3	106.1	182.7	128.4	83.8	102.8	89.4	103.4	180.3	125.9	106.1	182.7	128.4
0970, 0990	Wall Finishes & Painting/Coating	90.7	88.9	89.6	109.6	164.2	142.3	109.6	164.2	142.3	85.3	99.4	93.8	109.6	164.2	142.3	109.6	164.2	142.3
09	FINISHES	97.2	94.1	95.5	101.7	181.6	144.9	102.5	181.6	145.3	86.6	97.1	92.3	100.3	159.5	132.3	102.1	181.6	145.1
COVERS	DIVS. 10-14, 25, 28, 41, 43, 44, 46	100.0	98.2	99.6	100.0	137.2	108.8	100.0	137.2	108.8	100.0	91.9	98.1	100.0	131.3	107.4	100.0	137.2	108.8
21, 22, 23	FIRE SUPPRESSION, PLUMBING & HVAC	97.3	94.4	96.1	96.6	171.5	126.8	96.6	171.5	126.8	97.5	104.1	100.1	100.5	155.3	122.6	96.6	171.5	126.8
26, 27, 3370	ELECTRICAL, COMMUNICATIONS & UTIL.	97.6	100.3	98.9	106.4	183.2	144.4	106.4	183.2	144.4	95.8	101.1	98.4	99.7	145.8	122.5	98.9	183.2	140.6
MF2018	WEIGHTED AVERAGE	96.9	100.4	98.4	102.7	173.1	133.1	102.5	173.1	133.0	92.8	103.2	97.3	100.5	151.7	122.6	100.9	173.1	132.1

DIVISION		NEW YORK JAMESTOWN 147			KINGSTON 124			LONG ISLAND CITY 111			MONTICELLO 127			MOUNT VERNON 105			NEW ROCHELLE 108		
		MAT.	INST.	TOTAL	MAT.	INST.	TOTAL	MAT.	INST.	TOTAL	MAT.	INST.	TOTAL	MAT.	INST.	TOTAL	MAT.	INST.	TOTAL
015433	CONTRACTOR EQUIPMENT		89.2	89.2		107.5	107.5		107.5	107.5		107.5	107.5		102.9	102.9		102.9	102.9
0241, 31-34	SITE & INFRASTRUCTURE, DEMOLITION	99.4	87.2	91.0	137.1	112.4	120.1	119.0	117.5	118.0	132.6	112.4	118.7	104.6	102.9	103.5	103.8	102.9	103.2
0310	Concrete Forming & Accessories	83.7	86.7	86.2	83.9	126.0	119.7	101.0	183.3	171.0	92.0	126.0	120.9	84.9	134.4	127.0	99.9	131.0	126.4
0320	Concrete Reinforcing	97.8	105.9	101.7	95.6	152.8	123.2	97.9	231.2	162.2	94.8	152.8	122.8	94.5	172.2	132.0	94.6	172.1	132.0
0330	Cast-in-Place Concrete	107.1	101.1	104.8	99.0	140.7	114.7	110.6	165.5	131.3	92.6	140.8	110.8	91.9	145.2	112.1	91.9	144.9	111.9
03	CONCRETE	95.9	95.3	95.6	98.8	135.6	115.3	108.1	183.9	142.1	94.1	135.6	112.7	95.0	145.5	117.7	94.5	143.8	116.6
04	MASONRY	116.0	100.3	106.4	118.4	148.9	137.0	115.6	184.6	157.7	110.9	148.9	134.1	94.5	151.4	129.2	94.5	151.4	129.2
05	METALS	90.0	100.3	93.2	104.8	133.9	113.9	103.9	173.4	125.6	104.8	134.0	113.9	92.7	168.0	116.1	92.9	167.2	116.1
06	WOOD, PLASTICS & COMPOSITES	80.7	83.3	82.1	84.7	119.9	103.1	98.6	182.1	142.3	92.9	119.8	107.0	83.3	128.0	106.7	100.9	124.4	113.2
07	THERMAL & MOISTURE PROTECTION	110.3	91.1	102.0	123.0	139.6	130.2	110.2	163.4	133.0	122.7	139.6	130.0	103.6	141.4	119.8	103.8	139.3	118.9
08	OPENINGS	96.7	89.6	95.0	92.4	137.8	103.5	86.7	190.9	112.1	88.5	137.7	100.5	92.5	162.7	109.6	92.5	160.7	109.2
0920	Plaster & Gypsum Board	98.3	82.9	88.2	91.3	120.7	110.6	100.9	184.6	155.8	91.9	120.7	110.8	88.3	128.7	114.8	100.4	125.0	116.5
0950, 0980	Ceilings & Acoustic Treatment	104.6	82.9	91.0	75.7	120.7	103.9	83.4	184.6	146.8	75.7	120.7	103.9	78.4	128.7	109.9	78.4	125.0	107.6
0960	Flooring	89.2	97.8	91.7	100.9	153.2	116.2	107.9	182.7	129.7	103.4	153.2	117.9	90.8	171.4	114.3	98.9	156.5	115.7
0970, 0990	Wall Finishes & Painting/Coating	92.2	95.8	94.4	111.2	124.7	119.3	109.6	164.2	142.3	111.2	117.5	115.0	103.2	164.2	139.8	103.2	164.2	139.8
09	FINISHES	95.3	88.9	91.8	98.5	130.1	115.6	103.6	181.6	145.8	98.9	129.2	115.3	90.9	143.1	119.1	94.8	137.9	118.1
COVERS	DIVS. 10-14, 25, 28, 41, 43, 44, 46	100.0	91.9	98.1	100.0	115.1	103.6	100.0	137.2	108.8	100.0	115.1	103.6	100.0	123.2	105.5	100.0	108.7	101.8
21, 22, 23	FIRE SUPPRESSION, PLUMBING & HVAC	97.1	89.2	93.9	97.4	128.3	109.8	100.5	171.5	129.1	97.4	132.3	111.5	96.8	146.3	116.7	96.8	146.0	116.3
26, 27, 3370	ELECTRICAL, COMMUNICATIONS & UTIL.	96.6	91.4	94.0	97.2	112.0	104.5	99.3	183.2	140.8	97.2	112.0	104.5	91.7	164.2	127.6	91.7	136.4	113.9
MF2018	WEIGHTED AVERAGE	97.1	92.3	95.0	101.2	128.8	113.1	102.2	173.1	132.8	99.8	129.5	112.7	94.9	147.0	117.4	95.3	141.4	115.2

For customer support on your Site Work & Landscape Costs with RSMeans Data, call 800.448.8182.

NEW YORK

DIVISION		NEW YORK 100 - 102			NIAGARA FALLS 143			PLATTSBURGH 129			POUGHKEEPSIE 125 - 126			QUEENS 110			RIVERHEAD 119		
		MAT.	INST.	TOTAL	MAT.	INST.	TOTAL	MAT.	INST.	TOTAL	MAT.	INST.	TOTAL	MAT.	INST.	TOTAL	MAT.	INST.	TOTAL
015433	CONTRACTOR EQUIPMENT		105.6	105.6		89.2	89.2		92.5	92.5		107.5	107.5		107.5	107.5		107.5	107.5
0241, 31 - 34	SITE & INFRASTRUCTURE, DEMOLITION	106.6	112.7	110.8	101.5	88.1	92.3	107.7	93.5	98.0	133.6	111.4	118.3	114.2	117.5	116.5	112.0	115.7	114.6
0310	Concrete Forming & Accessories	103.6	186.4	174.0	83.6	108.9	105.1	88.0	88.0	88.0	83.9	161.5	149.9	88.3	183.3	169.1	93.3	154.5	145.4
0320	Concrete Reinforcing	107.3	175.3	140.1	96.5	106.7	101.4	99.5	106.7	103.0	95.6	152.4	123.0	99.6	231.2	163.1	99.7	241.9	168.3
0330	Cast-in-Place Concrete	95.3	168.6	123.0	110.8	119.4	114.0	89.6	96.6	92.2	95.8	130.2	108.8	101.8	165.5	125.9	100.1	155.7	121.1
03	CONCRETE	99.3	177.7	134.5	98.2	111.8	104.3	91.1	94.2	92.5	96.2	147.9	119.4	100.6	183.9	138.0	98.0	169.2	130.0
04	MASONRY	99.0	186.2	152.2	123.1	119.9	121.1	96.8	93.3	94.7	110.0	133.7	124.5	109.4	184.6	155.3	119.6	168.6	149.5
05	METALS	105.2	174.6	126.8	92.5	100.8	95.0	100.7	97.6	99.7	104.9	133.6	113.8	103.9	173.4	125.6	106.0	174.2	127.3
06	WOOD, PLASTICS & COMPOSITES	99.3	186.1	144.7	80.6	104.6	93.2	90.5	85.2	87.7	84.7	173.3	131.1	82.8	182.1	134.7	88.5	151.9	121.7
07	THERMAL & MOISTURE PROTECTION	105.0	168.7	132.3	110.4	105.1	108.1	116.3	92.8	106.2	123.0	138.8	129.8	109.7	163.4	132.7	111.3	154.4	129.8
08	OPENINGS	96.2	197.3	120.8	96.7	100.8	97.7	97.9	91.2	96.0	92.5	164.3	110.0	86.7	190.9	112.1	87.1	177.0	109.0
0920	Plaster & Gypsum Board	99.4	188.3	157.7	98.3	104.8	102.6	111.6	84.5	93.8	91.3	175.6	146.6	92.5	184.6	152.9	93.8	153.6	133.1
0950, 0980	Ceilings & Acoustic Treatment	98.1	188.3	154.6	104.6	104.8	104.7	104.9	84.5	92.1	75.7	175.6	138.3	83.4	184.6	146.8	83.3	153.6	127.4
0960	Flooring	99.8	184.4	124.5	89.2	106.9	94.4	106.3	100.2	104.5	100.9	151.4	115.6	103.4	182.7	126.6	104.5	180.3	126.6
0970, 0990	Wall Finishes & Painting/Coating	104.0	169.0	143.0	92.2	106.1	100.6	106.7	90.6	97.1	111.2	117.5	115.0	109.6	164.2	142.3	109.6	164.2	142.3
09	FINISHES	99.6	185.1	145.9	95.4	108.5	102.5	98.2	89.8	93.7	98.3	157.8	130.5	100.6	181.6	144.4	100.9	159.5	132.6
COVERS	DIVS. 10 - 14, 25, 28, 41, 43, 44, 46	100.0	141.5	109.8	100.0	97.5	99.4	100.0	87.9	97.2	100.0	110.5	102.5	100.0	137.2	108.8	100.0	128.5	106.7
21, 22, 23	FIRE SUPPRESSION, PLUMBING & HVAC	99.4	174.6	129.7	97.1	100.2	98.4	97.4	95.6	96.6	97.4	114.6	104.3	100.5	171.5	129.1	100.8	157.4	123.6
26, 27, 3370	ELECTRICAL, COMMUNICATIONS & UTIL.	99.3	185.3	141.9	95.4	95.2	95.3	93.4	88.3	90.9	97.2	116.7	106.9	99.7	183.2	141.0	101.1	146.0	123.3
MF2018	WEIGHTED AVERAGE	100.3	174.2	132.2	98.0	103.4	100.3	97.7	92.8	95.6	100.3	131.7	113.9	100.5	173.1	131.8	101.3	156.9	125.3

NEW YORK

DIVISION		ROCHESTER 144 - 146			SCHENECTADY 123			STATEN ISLAND 103			SUFFERN 109			SYRACUSE 130 - 132			UTICA 133 - 135		
		MAT.	INST.	TOTAL	MAT.	INST.	TOTAL	MAT.	INST.	TOTAL	MAT.	INST.	TOTAL	MAT.	INST.	TOTAL	MAT.	INST.	TOTAL
015433	CONTRACTOR EQUIPMENT		117.0	117.0		110.8	110.8		102.9	102.9		102.9	102.9		110.8	110.8		110.8	110.8
0241, 31 - 34	SITE & INFRASTRUCTURE, DEMOLITION	89.1	104.7	99.8	79.9	97.2	91.8	109.3	107.1	107.8	100.9	100.7	100.8	92.9	96.5	95.4	71.8	96.0	88.4
0310	Concrete Forming & Accessories	103.4	98.9	99.6	100.9	104.1	103.6	84.3	183.5	168.7	93.5	135.6	129.4	99.6	93.3	94.2	100.5	92.8	93.9
0320	Concrete Reinforcing	96.8	105.7	101.1	94.2	112.0	102.8	95.6	205.9	148.8	94.6	144.8	118.8	97.5	106.8	102.0	97.5	105.7	101.5
0330	Cast-in-Place Concrete	99.2	107.0	102.1	88.5	113.5	97.9	91.9	167.1	120.3	89.1	133.7	106.0	97.7	107.2	101.3	89.3	106.6	95.9
03	CONCRETE	98.0	104.1	100.8	91.9	109.9	100.0	96.7	180.6	134.4	91.9	136.1	111.7	97.0	101.9	99.2	95.1	101.2	97.8
04	MASONRY	104.3	106.9	105.9	99.6	116.2	109.7	100.6	184.6	151.8	94.3	136.2	119.9	100.9	108.0	105.2	92.5	107.5	101.6
05	METALS	101.2	121.3	107.4	100.7	125.7	108.5	91.0	173.5	116.7	91.0	130.1	103.2	99.6	121.8	106.5	97.6	121.3	105.0
06	WOOD, PLASTICS & COMPOSITES	104.7	97.0	100.6	104.9	100.3	102.5	82.0	182.4	134.5	93.0	137.7	116.4	98.9	89.3	93.9	98.9	88.4	93.4
07	THERMAL & MOISTURE PROTECTION	117.5	102.1	110.9	102.6	108.5	105.2	103.1	163.8	129.2	103.6	135.8	117.4	105.3	99.4	102.8	94.9	99.1	96.7
08	OPENINGS	100.6	98.4	100.1	95.3	101.7	96.9	92.5	191.0	116.5	92.5	145.7	105.5	92.5	92.4	92.5	95.1	91.7	94.2
0920	Plaster & Gypsum Board	106.4	96.8	100.1	102.1	100.4	101.0	88.5	184.6	151.5	91.4	138.7	122.4	102.4	89.1	93.7	102.4	88.1	93.0
0950, 0980	Ceilings & Acoustic Treatment	104.8	96.8	99.8	99.2	100.4	100.0	80.3	184.6	145.7	78.4	138.7	116.2	102.4	89.1	94.0	102.4	88.1	93.5
0960	Flooring	92.2	107.5	96.7	91.4	107.6	96.2	94.5	182.7	120.2	94.6	163.7	114.8	91.9	96.7	93.3	89.6	98.1	92.1
0970, 0990	Wall Finishes & Painting/Coating	93.7	99.7	97.3	85.3	99.4	93.8	104.7	164.2	140.4	103.2	120.9	113.8	87.1	102.6	96.4	81.0	102.6	93.9
09	FINISHES	98.8	100.4	99.7	93.6	103.7	99.0	92.8	181.8	141.0	92.1	140.1	118.1	95.3	94.1	94.6	93.6	93.9	93.8
COVERS	DIVS. 10 - 14, 25, 28, 41, 43, 44, 46	100.0	99.7	99.9	100.0	100.1	100.0	100.0	137.9	108.9	100.0	108.9	102.1	100.0	98.5	99.7	100.0	98.0	99.5
21, 22, 23	FIRE SUPPRESSION, PLUMBING & HVAC	100.9	91.1	96.9	101.2	107.9	103.9	100.4	171.5	129.1	96.8	117.8	105.2	101.2	95.8	99.0	101.2	101.7	101.4
26, 27, 3370	ELECTRICAL, COMMUNICATIONS & UTIL.	102.6	96.5	99.6	99.8	107.2	103.5	93.6	183.2	137.9	97.3	110.4	103.8	100.1	106.8	103.4	98.0	106.8	102.4
MF2018	WEIGHTED AVERAGE	100.8	101.2	101.0	97.9	108.6	102.5	96.5	171.9	129.1	94.9	125.4	108.1	98.7	101.6	99.9	96.8	102.5	99.3

DIVISION		NEW YORK									NORTH CAROLINA								
		WATERTOWN 136			WHITE PLAINS 106			YONKERS 107			ASHEVILLE 287 - 288			CHARLOTTE 281 - 282			DURHAM 277		
		MAT.	INST.	TOTAL	MAT.	INST.	TOTAL	MAT.	INST.	TOTAL	MAT.	INST.	TOTAL	MAT.	INST.	TOTAL	MAT.	INST.	TOTAL
015433	CONTRACTOR EQUIPMENT		110.8	110.8		102.9	102.9		102.9	102.9		97.0	97.0		100.2	100.2		102.2	102.2
0241, 31 - 34	SITE & INFRASTRUCTURE, DEMOLITION	79.3	96.3	90.9	99.0	103.0	101.7	106.6	103.0	104.1	96.5	76.0	82.4	99.5	83.5	88.5	101.8	84.5	89.9
0310	Concrete Forming & Accessories	83.9	96.0	94.2	98.8	145.4	138.4	99.0	145.5	138.6	93.8	59.6	64.7	98.3	64.7	69.8	101.5	59.6	65.9
0320	Concrete Reinforcing	98.2	106.8	102.3	94.6	172.3	132.1	98.4	172.3	134.0	93.6	62.9	78.8	96.6	68.3	83.0	95.4	63.6	80.1
0330	Cast-in-Place Concrete	104.1	109.4	106.1	81.6	145.4	105.7	91.3	145.4	111.8	102.3	70.1	90.1	104.4	72.6	92.4	115.8	70.1	98.5
03	CONCRETE	107.5	103.8	105.8	86.0	150.5	114.9	94.6	150.6	119.7	89.6	65.6	78.8	92.6	69.5	82.3	99.0	65.7	84.1
04	MASONRY	93.7	111.5	104.8	93.9	151.4	129.0	96.6	151.4	130.0	85.3	62.6	71.4	87.3	65.2	73.8	86.9	62.6	72.0
05	METALS	97.7	121.6	105.1	92.5	168.3	116.2	101.3	168.4	122.3	101.0	89.1	97.3	101.9	89.1	97.9	117.4	89.3	108.6
06	WOOD, PLASTICS & COMPOSITES	78.9	91.9	85.7	98.9	142.5	121.7	98.8	142.5	121.7	91.4	57.0	73.4	92.6	63.2	77.3	96.9	57.0	76.1
07	THERMAL & MOISTURE PROTECTION	95.2	101.2	97.8	103.4	144.0	120.8	103.8	144.6	121.3	98.7	61.5	82.7	91.9	66.2	80.9	101.7	61.4	84.4
08	OPENINGS	95.1	95.4	95.1	92.5	170.7	111.6	96.2	170.7	114.3	89.3	57.6	81.6	99.2	61.5	90.1	95.9	57.8	86.6
0920	Plaster & Gypsum Board	93.1	91.7	92.2	94.8	143.6	126.8	99.7	143.6	128.5	108.8	55.9	74.1	100.2	62.3	75.3	95.8	55.9	69.6
0950, 0980	Ceilings & Acoustic Treatment	102.4	91.7	95.7	78.4	143.6	119.3	98.1	143.6	126.6	89.6	55.9	68.5	92.1	62.3	73.4	98.1	55.9	71.6
0960	Flooring	82.6	98.1	87.1	97.1	180.3	121.4	96.5	182.7	121.7	98.2	63.7	88.1	97.2	68.4	88.8	104.5	63.7	92.6
0970, 0990	Wall Finishes & Painting/Coating	81.0	101.0	93.0	103.2	164.2	139.8	103.2	164.2	139.8	98.6	54.4	72.1	95.0	66.8	78.1	100.1	54.4	72.7
09	FINISHES	90.9	96.4	93.9	92.9	153.5	125.7	98.7	153.9	128.6	96.0	59.3	76.2	95.1	65.3	79.0	96.7	59.3	76.5
COVERS	DIVS. 10 - 14, 25, 28, 41, 43, 44, 46	100.0	99.6	99.9	100.0	127.5	106.5	100.0	134.8	108.2	100.0	84.2	96.3	100.0	85.0	96.5	100.0	84.2	96.3
21, 22, 23	FIRE SUPPRESSION, PLUMBING & HVAC	101.2	89.9	96.6	101.0	146.4	119.3	101.0	146.4	119.3	101.8	58.8	84.5	101.0	63.4	85.8	101.8	58.8	84.5
26, 27, 3370	ELECTRICAL, COMMUNICATIONS & UTIL.	100.0	88.7	94.4	91.7	164.2	127.6	97.4	164.2	130.5	99.9	55.7	78.0	101.6	84.2	93.0	95.4	55.2	75.5
MF2018	WEIGHTED AVERAGE	98.4	98.9	98.6	94.9	149.8	118.6	99.1	150.1	121.1	97.0	64.6	83.0	98.4	72.1	87.0	101.2	65.2	85.6

For customer support on your Site Work & Landscape Costs with RSMeans Data, call 800.448.8182.

809

City Cost Indexes - V2

NORTH CAROLINA

DIVISION		ELIZABETH CITY 279			FAYETTEVILLE 283			GASTONIA 280			GREENSBORO 270,272 - 274			HICKORY 286			KINSTON 285		
		MAT.	INST.	TOTAL	MAT.	INST.	TOTAL	MAT.	INST.	TOTAL	MAT.	INST.	TOTAL	MAT.	INST.	TOTAL	MAT.	INST.	TOTAL
015433	CONTRACTOR EQUIPMENT		106.2	106.2		102.2	102.2		97.0	97.0		102.2	102.2		102.2	102.2		102.2	102.2
0241, 31 - 34	SITE & INFRASTRUCTURE, DEMOLITION	105.9	85.8	92.1	95.7	84.2	87.8	96.2	76.0	82.3	101.6	84.5	89.8	94.9	84.2	87.6	93.7	84.2	87.2
0310	Concrete Forming & Accessories	84.6	61.5	64.9	93.0	58.1	63.3	101.1	59.8	66.0	101.3	59.6	65.8	89.0	59.7	64.1	84.9	58.1	62.1
0320	Concrete Reinforcing	93.4	66.5	80.4	97.2	63.6	81.0	93.9	63.7	79.3	94.2	63.6	79.5	93.6	63.6	79.1	93.1	63.6	78.9
0330	Cast-in-Place Concrete	116.0	70.1	98.7	107.3	68.4	92.6	100.0	70.8	89.0	114.9	70.7	98.2	102.3	69.9	90.1	98.8	67.8	87.0
03	CONCRETE	98.8	67.0	84.5	91.6	64.4	79.4	88.5	66.0	78.4	98.4	65.9	83.8	89.3	65.6	78.7	86.5	64.2	76.5
04	MASONRY	99.1	58.5	74.4	88.2	58.5	70.1	89.4	62.6	73.0	83.7	62.6	70.8	75.0	62.5	67.4	81.5	58.5	67.5
05	METALS	103.3	91.4	99.6	121.6	89.4	111.5	101.7	89.5	97.9	109.7	89.4	103.4	101.1	88.7	97.2	99.9	89.3	96.6
06	WOOD, PLASTICS & COMPOSITES	77.7	61.2	69.1	89.8	57.0	72.7	100.8	57.0	77.9	96.6	57.0	75.9	84.9	57.0	70.3	81.0	57.0	68.4
07	THERMAL & MOISTURE PROTECTION	100.7	59.8	83.2	98.0	59.6	81.5	99.0	61.4	82.9	101.3	61.4	84.2	99.1	61.4	82.9	98.9	59.6	82.0
08	OPENINGS	93.3	60.5	85.4	89.4	57.8	81.7	92.5	57.8	84.1	95.9	57.8	86.6	89.4	57.8	81.7	89.4	57.8	81.7
0920	Plaster & Gypsum Board	89.7	59.6	70.0	112.1	55.9	75.2	114.7	55.9	76.1	96.8	55.9	70.0	108.8	55.9	74.1	108.7	55.9	74.1
0950, 0980	Ceilings & Acoustic Treatment	98.1	59.6	74.0	92.8	55.9	69.7	94.7	55.9	70.4	98.1	55.9	71.6	89.6	55.9	68.5	94.7	55.9	70.4
0960	Flooring	96.3	63.7	86.8	98.3	63.7	88.2	101.6	63.7	90.5	104.5	63.7	92.6	98.1	63.7	88.1	95.2	63.7	86.0
0970, 0990	Wall Finishes & Painting/Coating	100.1	54.4	72.7	98.6	54.4	72.1	98.6	54.4	72.1	100.1	54.4	72.7	98.6	54.4	72.1	98.6	54.4	72.1
09	FINISHES	93.8	60.9	76.0	97.2	58.3	76.2	98.9	59.3	77.5	96.8	59.3	76.6	96.2	59.3	76.2	96.3	58.3	75.8
COVERS	DIVS. 10 - 14, 25, 28, 41, 43, 44, 46	100.0	85.3	96.5	100.0	82.9	96.0	100.0	84.3	96.3	100.0	84.2	96.3	100.0	84.3	96.3	100.0	82.8	96.0
21, 22, 23	FIRE SUPPRESSION, PLUMBING & HVAC	97.9	55.8	80.9	101.6	56.7	83.5	101.8	57.8	84.1	101.7	58.8	84.4	98.0	57.8	81.8	98.0	55.6	80.9
26, 27, 3370	ELECTRICAL, COMMUNICATIONS & UTIL.	95.1	66.6	81.0	99.9	55.2	77.8	99.4	84.2	91.9	94.7	55.7	75.4	97.9	84.2	91.1	97.7	53.7	76.0
MF2018	WEIGHTED AVERAGE	98.1	66.6	84.5	100.5	64.0	84.7	97.7	68.5	85.1	99.7	65.3	84.8	95.2	69.0	83.9	94.9	63.5	81.3

NORTH CAROLINA / NORTH DAKOTA

DIVISION		MURPHY 289			RALEIGH 275 - 276			ROCKY MOUNT 278			WILMINGTON 284			WINSTON-SALEM 271			BISMARCK 585		
		MAT.	INST.	TOTAL	MAT.	INST.	TOTAL	MAT.	INST.	TOTAL	MAT.	INST.	TOTAL	MAT.	INST.	TOTAL	MAT.	INST.	TOTAL
015433	CONTRACTOR EQUIPMENT		97.0	97.0		106.9	106.9		102.2	102.2		97.0	97.0		102.2	102.2		100.9	100.9
0241, 31 - 34	SITE & INFRASTRUCTURE, DEMOLITION	97.1	75.7	82.4	101.1	93.2	95.7	103.8	84.1	90.3	97.5	75.7	82.5	102.0	84.5	90.0	101.0	99.3	99.8
0310	Concrete Forming & Accessories	101.9	57.9	64.5	99.4	59.3	65.3	92.2	59.0	64.0	95.3	58.1	63.6	103.1	59.6	66.1	107.3	78.9	83.1
0320	Concrete Reinforcing	93.1	61.3	77.8	95.7	63.6	80.2	93.4	63.6	79.0	94.3	63.6	79.5	94.2	63.6	79.5	94.4	100.2	97.2
0330	Cast-in-Place Concrete	105.9	67.7	91.5	119.2	70.7	100.9	113.5	69.1	96.7	101.9	67.8	89.0	117.8	70.1	99.8	99.4	89.1	95.5
03	CONCRETE	92.5	63.7	79.6	99.4	65.6	84.2	99.3	65.0	83.9	89.6	64.2	78.2	99.9	65.7	84.5	96.5	86.6	92.1
04	MASONRY	78.1	58.5	66.2	85.1	61.4	70.6	76.9	61.2	67.3	75.8	58.5	65.2	84.0	62.6	70.9	102.0	84.1	91.1
05	METALS	98.8	88.5	95.6	102.0	87.0	97.3	102.5	88.5	98.1	100.6	89.3	97.1	106.8	89.3	101.4	93.6	94.5	93.8
06	WOOD, PLASTICS & COMPOSITES	101.6	56.9	78.2	93.7	57.2	74.6	86.5	57.0	71.1	93.2	57.0	74.3	96.6	57.0	75.9	103.2	75.4	88.6
07	THERMAL & MOISTURE PROTECTION	99.0	59.6	82.1	94.6	62.0	80.6	101.3	60.9	83.9	98.7	59.6	81.9	101.3	61.4	84.2	107.9	86.4	98.6
08	OPENINGS	89.3	57.2	81.5	98.0	57.9	88.2	92.6	57.8	84.1	89.5	57.8	81.7	95.9	57.8	86.6	103.4	84.6	98.8
0920	Plaster & Gypsum Board	113.3	55.7	75.5	91.5	55.9	68.1	90.9	55.9	67.9	110.2	55.9	74.6	96.8	55.9	70.0	91.1	74.9	80.5
0950, 0980	Ceilings & Acoustic Treatment	89.6	55.7	68.4	97.4	55.9	71.4	94.9	55.9	70.5	92.8	55.9	69.7	98.1	55.9	71.6	113.7	74.9	89.4
0960	Flooring	101.9	63.7	90.8	96.2	63.7	86.7	99.9	63.7	89.4	98.9	63.7	88.6	104.5	63.7	92.6	88.9	54.9	79.0
0970, 0990	Wall Finishes & Painting/Coating	98.6	54.4	72.1	96.7	54.4	71.4	100.1	54.4	72.7	98.6	54.4	72.1	100.1	54.4	72.7	89.1	61.0	72.3
09	FINISHES	97.9	58.2	76.4	95.1	59.1	75.6	94.4	59.0	75.3	97.1	58.3	76.1	96.8	59.3	76.6	97.5	73.2	84.3
COVERS	DIVS. 10 - 14, 25, 28, 41, 43, 44, 46	100.0	82.8	96.0	100.0	84.1	96.3	100.0	83.8	96.2	100.0	82.8	96.0	100.0	84.2	96.3	100.0	94.3	98.6
21, 22, 23	FIRE SUPPRESSION, PLUMBING & HVAC	98.0	55.6	80.9	100.9	58.2	83.7	97.9	57.0	81.4	101.8	56.7	83.6	101.7	58.8	84.4	100.5	78.1	91.5
26, 27, 3370	ELECTRICAL, COMMUNICATIONS & UTIL.	100.6	53.7	77.4	100.4	54.2	77.5	96.7	55.2	76.2	100.2	53.7	77.2	94.7	55.7	75.4	98.8	76.2	87.6
MF2018	WEIGHTED AVERAGE	96.1	62.6	81.6	99.0	65.3	84.4	97.1	64.5	83.0	96.6	63.1	82.1	99.5	65.3	84.7	99.1	83.1	92.2

NORTH DAKOTA

DIVISION		DEVILS LAKE 583			DICKINSON 586			FARGO 580 - 581			GRAND FORKS 582			JAMESTOWN 584			MINOT 587		
		MAT.	INST.	TOTAL	MAT.	INST.	TOTAL	MAT.	INST.	TOTAL	MAT.	INST.	TOTAL	MAT.	INST.	TOTAL	MAT.	INST.	TOTAL
015433	CONTRACTOR EQUIPMENT		95.8	95.8		95.8	95.8		100.9	100.9		95.8	95.8		95.8	95.8		95.8	95.8
0241, 31 - 34	SITE & INFRASTRUCTURE, DEMOLITION	104.7	90.7	95.1	113.1	90.6	97.7	101.2	99.0	99.7	109.3	91.4	97.0	103.8	90.7	94.8	106.7	91.2	96.0
0310	Concrete Forming & Accessories	106.2	69.3	74.8	93.9	69.1	72.8	96.6	69.8	73.8	98.7	76.2	79.6	95.7	69.2	73.2	93.5	69.4	73.0
0320	Concrete Reinforcing	97.8	100.2	99.0	98.7	100.1	99.4	97.4	100.8	99.1	96.3	100.3	98.2	98.4	100.7	99.5	99.7	100.1	99.9
0330	Cast-in-Place Concrete	116.6	78.4	102.2	105.5	78.3	95.3	96.6	84.9	92.2	105.5	84.9	97.7	115.2	78.4	101.3	105.5	78.6	95.4
03	CONCRETE	101.9	78.6	91.5	101.2	78.5	91.0	98.7	81.1	90.8	98.7	84.1	92.1	100.4	78.7	90.7	97.2	78.8	88.9
04	MASONRY	115.6	82.4	95.4	116.1	79.8	93.9	98.2	91.2	94.0	109.1	85.6	94.7	128.5	91.1	105.7	109.0	79.1	90.8
05	METALS	92.6	95.3	93.5	92.5	94.7	93.2	98.0	95.1	97.1	92.6	96.1	93.7	92.6	95.9	93.6	92.9	95.6	93.7
06	WOOD, PLASTICS & COMPOSITES	98.8	65.2	81.2	84.4	65.2	74.4	91.8	65.5	78.0	89.6	72.6	80.7	86.6	65.2	75.4	84.0	65.2	74.2
07	THERMAL & MOISTURE PROTECTION	106.3	81.0	95.4	106.9	81.4	95.9	104.3	85.9	96.4	106.5	84.2	96.9	106.1	83.8	96.5	106.2	81.9	95.8
08	OPENINGS	100.6	79.0	95.3	100.6	79.0	95.3	100.8	79.1	95.5	99.2	83.0	95.3	100.6	79.0	95.3	99.4	79.0	94.4
0920	Plaster & Gypsum Board	102.2	64.8	77.6	93.1	64.8	74.5	87.1	64.8	72.4	94.3	72.3	79.9	94.0	64.8	74.8	93.1	64.8	74.5
0950, 0980	Ceilings & Acoustic Treatment	109.4	64.8	81.4	109.4	64.8	81.4	97.9	64.8	77.1	109.4	72.3	86.2	109.4	64.8	81.4	109.4	64.8	81.4
0960	Flooring	96.3	54.9	84.2	89.3	54.9	79.2	100.3	54.9	87.1	91.2	54.9	80.6	90.0	54.9	79.8	89.0	54.9	79.0
0970, 0990	Wall Finishes & Painting/Coating	83.5	53.7	65.6	83.5	53.7	65.6	94.7	64.1	76.4	83.5	62.9	71.2	83.5	53.7	65.6	83.5	54.7	66.3
09	FINISHES	97.1	65.2	79.8	94.7	65.2	78.7	97.0	66.6	80.5	94.9	71.3	82.1	94.1	65.2	78.5	93.8	65.3	78.4
COVERS	DIVS. 10 - 14, 25, 28, 41, 43, 44, 46	100.0	82.8	95.9	100.0	82.7	95.9	100.0	91.8	98.1	100.0	90.8	97.8	100.0	82.7	95.9	100.0	91.2	97.9
21, 22, 23	FIRE SUPPRESSION, PLUMBING & HVAC	96.9	74.8	88.0	96.9	69.9	86.0	100.6	70.7	88.5	100.7	78.0	91.5	96.9	68.9	85.6	100.7	68.8	87.8
26, 27, 3370	ELECTRICAL, COMMUNICATIONS & UTIL.	94.2	70.0	82.2	100.7	65.3	83.2	100.6	66.8	83.9	96.8	70.1	83.6	94.2	65.7	80.1	99.2	69.9	84.7
MF2018	WEIGHTED AVERAGE	98.6	77.8	89.6	99.2	75.7	89.0	99.6	78.9	90.6	98.8	81.0	91.1	98.7	76.9	89.3	98.7	76.5	89.1

For customer support on your Site Work & Landscape Costs with RSMeans Data, call 800.448.8182.

City Cost Indexes - V2

DIVISION		NORTH DAKOTA WILLISTON 588			OHIO AKRON 442 - 443			OHIO ATHENS 457			OHIO CANTON 446 - 447			OHIO CHILLICOTHE 456			OHIO CINCINNATI 451 - 452		
		MAT.	INST.	TOTAL	MAT.	INST.	TOTAL	MAT.	INST.	TOTAL	MAT.	INST.	TOTAL	MAT.	INST.	TOTAL	MAT.	INST.	TOTAL
015433	CONTRACTOR EQUIPMENT		95.8	95.8		87.0	87.0		83.5	83.5		87.0	87.0		94.0	94.0		96.6	96.6
0241, 31 - 34	SITE & INFRASTRUCTURE, DEMOLITION	106.6	91.5	96.2	96.4	91.2	92.8	108.7	82.7	90.8	96.5	91.1	92.8	95.0	92.1	93.0	93.3	98.5	96.9
0310	Concrete Forming & Accessories	100.7	76.2	79.8	102.6	79.9	83.3	94.3	76.1	78.8	102.6	72.0	76.6	97.7	78.6	81.4	100.4	74.9	78.7
0320	Concrete Reinforcing	100.6	100.1	100.4	92.2	87.2	89.8	93.1	85.0	89.2	92.2	71.5	82.2	90.0	84.6	87.4	95.5	83.7	89.8
0330	Cast-in-Place Concrete	105.5	84.9	97.7	107.2	85.5	99.0	111.7	92.1	104.3	108.2	84.5	99.2	101.4	88.8	96.6	96.8	76.5	89.1
03	CONCRETE	98.6	84.0	92.1	101.3	82.8	93.0	103.4	83.1	94.3	101.8	76.2	90.3	101.8	83.7	93.7	96.0	77.2	87.6
04	MASONRY	103.2	82.2	90.4	99.8	85.7	91.2	79.7	93.2	87.9	100.5	78.5	87.1	86.3	86.7	86.6	91.7	79.2	84.1
05	METALS	92.7	95.8	93.7	94.3	79.0	89.5	102.3	79.1	95.1	94.3	72.4	87.5	94.1	88.4	92.3	96.3	81.9	91.8
06	WOOD, PLASTICS & COMPOSITES	91.5	72.6	81.6	104.1	78.3	90.6	87.1	71.5	78.9	104.5	69.3	86.1	99.3	75.3	86.8	101.3	73.6	86.8
07	THERMAL & MOISTURE PROTECTION	106.5	83.3	96.5	104.1	88.1	97.3	102.9	87.4	96.2	105.0	84.9	96.4	104.3	84.4	95.8	102.1	80.9	93.0
08	OPENINGS	100.7	83.0	96.4	107.3	80.0	100.6	97.4	73.0	91.5	101.1	67.2	92.9	88.4	75.2	85.2	98.6	73.2	92.4
0920	Plaster & Gypsum Board	94.3	72.3	79.9	98.1	77.9	84.8	94.2	70.8	78.8	99.2	68.6	79.1	97.5	75.2	82.9	96.5	73.2	81.2
0950, 0980	Ceilings & Acoustic Treatment	109.4	72.3	86.2	93.2	77.9	83.6	102.8	70.8	82.7	93.2	68.6	77.8	95.5	75.2	82.6	88.5	73.2	78.9
0960	Flooring	92.3	54.9	81.4	91.6	79.5	88.1	124.9	71.8	109.4	91.8	70.7	85.6	101.5	71.8	92.8	101.5	79.3	95.0
0970, 0990	Wall Finishes & Painting/Coating	83.5	60.4	69.7	97.9	86.5	91.1	104.9	84.9	92.9	97.9	77.3	85.5	102.2	84.9	91.8	101.5	73.0	84.4
09	FINISHES	95.1	71.2	82.1	95.5	80.3	87.3	102.9	75.6	88.1	95.7	71.5	82.6	99.4	77.2	87.4	97.2	75.1	85.3
COVERS	DIVS. 10 - 14, 25, 28, 41, 43, 44, 46	100.0	93.0	98.4	100.0	91.6	98.0	100.0	90.3	97.7	100.0	90.2	97.7	100.0	89.4	97.5	100.0	88.1	97.2
21, 22, 23	FIRE SUPPRESSION, PLUMBING & HVAC	96.9	74.7	87.9	100.5	83.7	93.7	96.7	77.3	88.9	100.5	75.5	90.4	97.2	88.0	93.5	100.4	78.1	91.4
26, 27, 3370	ELECTRICAL, COMMUNICATIONS & UTIL.	97.1	75.1	86.2	99.2	78.8	89.1	100.7	86.4	93.6	98.5	81.9	90.3	99.8	86.4	93.2	99.1	74.0	86.7
MF2018	WEIGHTED AVERAGE	97.8	80.7	90.4	99.7	83.0	92.5	99.2	81.8	91.7	99.2	77.5	89.8	96.3	85.3	91.5	98.0	79.2	89.9

DIVISION		OHIO CLEVELAND 441			OHIO COLUMBUS 430 - 432			OHIO DAYTON 453 - 454			OHIO HAMILTON 450			OHIO LIMA 458			OHIO LORAIN 440		
		MAT.	INST.	TOTAL	MAT.	INST.	TOTAL	MAT.	INST.	TOTAL	MAT.	INST.	TOTAL	MAT.	INST.	TOTAL	MAT.	INST.	TOTAL
015433	CONTRACTOR EQUIPMENT		91.8	91.8		93.7	93.7		87.4	87.4		94.0	94.0		86.8	86.8		87.0	87.0
0241, 31 - 34	SITE & INFRASTRUCTURE, DEMOLITION	95.5	95.2	95.3	102.9	94.1	96.8	91.4	91.5	91.5	91.4	91.7	91.6	102.5	82.4	88.7	95.7	91.4	92.7
0310	Concrete Forming & Accessories	101.3	89.9	91.6	100.6	76.3	80.0	99.8	74.8	78.5	99.9	74.6	78.4	94.3	73.7	76.7	102.6	72.0	76.6
0320	Concrete Reinforcing	92.8	90.7	91.8	104.6	78.8	92.2	95.5	75.4	85.8	95.5	73.6	84.9	93.1	75.6	84.6	92.2	87.5	89.9
0330	Cast-in-Place Concrete	106.9	95.9	102.7	102.9	86.0	96.5	86.8	77.9	83.5	93.1	79.5	87.9	102.6	86.1	96.4	100.8	88.3	96.8
03	CONCRETE	101.7	91.7	97.2	100.1	80.2	91.2	89.3	76.0	83.4	92.2	76.8	85.3	96.3	78.5	88.3	98.9	80.3	90.5
04	MASONRY	104.6	97.3	100.1	87.9	83.9	85.5	85.4	74.1	78.5	85.7	77.8	80.9	108.9	74.9	88.2	96.4	91.0	93.1
05	METALS	95.9	83.8	92.1	103.2	79.6	95.9	95.6	75.5	89.4	95.7	84.3	92.1	102.3	79.0	95.1	94.9	79.8	90.2
06	WOOD, PLASTICS & COMPOSITES	98.4	87.9	92.9	100.3	75.1	87.1	103.2	74.5	88.2	102.2	74.5	87.7	87.0	72.6	79.4	104.1	67.7	85.0
07	THERMAL & MOISTURE PROTECTION	100.9	96.7	99.1	92.7	83.8	88.9	108.1	76.3	94.4	104.4	76.8	92.5	102.5	79.9	92.8	101.4	87.6	97.5
08	OPENINGS	100.2	86.0	96.7	97.8	73.6	91.9	95.1	72.2	89.5	93.0	72.3	87.9	97.4	71.0	91.0	101.1	74.2	94.6
0920	Plaster & Gypsum Board	98.1	87.5	91.1	92.7	74.5	80.7	99.0	74.3	82.8	99.0	74.3	82.8	94.2	71.9	79.5	98.1	66.9	77.6
0950, 0980	Ceilings & Acoustic Treatment	91.2	87.5	88.8	88.8	74.5	79.8	95.7	74.3	82.3	95.0	74.3	82.1	102.8	71.9	83.4	93.2	66.9	76.7
0960	Flooring	92.5	87.3	91.0	96.0	75.3	89.9	105.2	69.0	94.7	102.5	74.9	94.5	124.2	73.6	109.4	91.8	87.3	90.5
0970, 0990	Wall Finishes & Painting/Coating	101.9	89.3	94.3	99.5	77.7	86.4	102.2	67.4	81.3	102.2	67.9	81.6	104.9	72.8	85.7	97.9	86.5	91.1
09	FINISHES	95.4	89.2	92.0	93.2	75.7	83.8	100.6	72.6	85.5	99.6	73.9	85.7	102.3	73.0	86.4	95.6	75.4	84.7
COVERS	DIVS. 10 - 14, 25, 28, 41, 43, 44, 46	100.0	97.4	99.4	100.0	88.7	97.3	100.0	87.2	97.0	100.0	87.1	97.0	100.0	91.6	98.0	100.0	86.5	96.8
21, 22, 23	FIRE SUPPRESSION, PLUMBING & HVAC	100.5	91.3	96.8	100.4	89.8	96.1	101.2	77.8	91.7	101.0	71.4	89.1	96.7	87.4	93.0	100.5	84.6	94.1
26, 27, 3370	ELECTRICAL, COMMUNICATIONS & UTIL.	99.1	92.0	95.6	101.2	81.0	91.2	97.6	72.9	85.4	97.8	72.7	85.4	101.0	72.9	87.1	98.6	75.3	87.1
MF2018	WEIGHTED AVERAGE	99.4	91.5	96.0	99.2	83.2	92.3	96.9	76.6	88.1	96.9	76.7	88.2	99.6	79.0	90.7	98.7	81.8	91.4

DIVISION		OHIO MANSFIELD 448 - 449			OHIO MARION 433			OHIO SPRINGFIELD 455			OHIO STEUBENVILLE 439			OHIO TOLEDO 434 - 436			OHIO YOUNGSTOWN 444 - 445		
		MAT.	INST.	TOTAL	MAT.	INST.	TOTAL	MAT.	INST.	TOTAL	MAT.	INST.	TOTAL	MAT.	INST.	TOTAL	MAT.	INST.	TOTAL
015433	CONTRACTOR EQUIPMENT		87.0	87.0		87.4	87.4		87.4	87.4		91.1	91.1		91.1	91.1		87.0	87.0
0241, 31 - 34	SITE & INFRASTRUCTURE, DEMOLITION	91.5	91.1	91.2	94.9	90.0	91.9	91.7	91.5	91.6	139.6	95.6	109.4	99.5	88.3	91.8	96.2	91.0	92.6
0310	Concrete Forming & Accessories	90.5	70.9	73.8	98.8	76.5	79.8	99.8	74.8	78.5	98.9	76.3	79.7	103.1	81.7	84.9	102.6	74.0	78.3
0320	Concrete Reinforcing	84.0	75.3	79.8	98.8	75.3	87.4	95.5	75.4	85.8	96.1	95.5	95.9	107.2	78.9	93.5	92.2	84.2	88.3
0330	Cast-in-Place Concrete	99.2	84.3	93.6	89.0	84.7	87.4	89.2	77.9	84.9	96.7	86.0	92.7	97.5	86.8	93.4	106.1	83.9	97.7
03	CONCRETE	93.3	76.3	85.7	88.1	79.1	84.1	90.4	76.0	84.0	91.9	82.8	87.8	96.1	83.2	90.3	100.8	79.1	91.0
04	MASONRY	98.8	85.2	90.5	90.5	86.1	87.8	85.6	74.1	78.5	78.7	87.3	83.9	96.6	88.0	91.3	100.2	83.0	89.7
05	METALS	95.2	74.6	88.7	102.2	77.5	94.5	95.6	75.5	89.3	98.4	82.6	93.5	103.0	84.2	97.1	94.4	77.6	89.2
06	WOOD, PLASTICS & COMPOSITES	89.7	67.7	78.2	95.0	74.9	84.5	104.4	74.5	88.8	89.8	73.8	81.5	99.8	81.1	90.0	104.1	72.1	87.3
07	THERMAL & MOISTURE PROTECTION	103.3	85.4	95.7	90.0	84.9	87.8	108.0	76.3	94.4	101.6	84.2	94.2	90.8	87.1	89.3	105.2	84.8	96.4
08	OPENINGS	102.3	68.4	94.1	90.9	72.4	86.4	93.4	71.9	88.2	91.1	76.7	87.6	93.4	78.3	89.7	101.1	77.7	95.4
0920	Plaster & Gypsum Board	91.2	66.9	75.2	91.7	74.5	80.4	99.0	74.3	82.8	89.5	72.8	78.6	93.7	80.8	85.2	98.1	71.4	80.6
0950, 0980	Ceilings & Acoustic Treatment	93.9	66.9	77.0	95.0	74.5	82.1	95.7	74.3	82.3	92.0	72.8	80.0	95.0	80.8	86.1	93.2	71.4	79.6
0960	Flooring	86.4	88.4	87.0	96.4	88.4	94.1	105.2	69.0	94.7	125.5	90.8	115.4	96.9	90.1	94.9	91.8	85.9	90.1
0970, 0990	Wall Finishes & Painting/Coating	97.9	74.1	83.6	101.1	74.1	84.9	102.2	67.4	81.3	113.5	82.9	95.2	101.1	82.9	90.2	97.9	76.2	84.9
09	FINISHES	92.9	73.6	82.4	95.1	78.0	85.9	100.0	72.6	85.5	111.4	78.8	93.9	95.9	83.1	89.0	95.6	75.7	84.8
COVERS	DIVS. 10 - 14, 25, 28, 41, 43, 44, 46	100.0	89.8	97.6	100.0	88.4	97.3	100.0	87.1	97.0	100.0	91.8	98.1	100.0	86.7	96.9	100.0	90.0	97.6
21, 22, 23	FIRE SUPPRESSION, PLUMBING & HVAC	96.7	83.3	91.3	96.6	88.7	93.4	101.2	77.8	91.7	97.1	87.1	93.1	100.4	89.2	95.9	100.5	81.7	92.9
26, 27, 3370	ELECTRICAL, COMMUNICATIONS & UTIL.	96.6	87.4	92.0	97.1	87.4	92.3	97.6	77.2	87.5	91.5	102.5	96.9	102.3	99.6	101.0	98.6	71.6	85.2
MF2018	WEIGHTED AVERAGE	96.6	81.1	89.9	95.5	83.6	90.3	96.9	77.2	88.4	97.1	87.4	92.9	98.9	87.7	94.1	99.1	79.7	90.7

For customer support on your Site Work & Landscape Costs with RSMeans Data, call 800.448.8182.

811

DIVISION		OHIO ZANESVILLE 437-438			OKLAHOMA ARDMORE 734			CLINTON 736			DURANT 747			ENID 737			GUYMON 739		
		MAT.	INST.	TOTAL	MAT.	INST.	TOTAL	MAT.	INST.	TOTAL	MAT.	INST.	TOTAL	MAT.	INST.	TOTAL	MAT.	INST.	TOTAL
015433	CONTRACTOR EQUIPMENT		87.4	87.4		77.7	77.7		76.9	76.9		76.9	76.9		76.9	76.9		76.9	76.9
0241, 31 - 34	SITE & INFRASTRUCTURE, DEMOLITION	98.0	87.9	91.0	94.4	89.1	90.8	95.7	87.7	90.2	92.9	85.0	87.5	97.9	87.7	90.9	99.9	86.8	90.9
0310	Concrete Forming & Accessories	95.2	75.5	78.5	90.9	53.4	59.0	89.5	53.4	58.8	83.1	52.9	57.4	93.5	53.6	59.6	97.0	53.1	59.7
0320	Concrete Reinforcing	98.2	88.2	93.4	88.2	66.3	77.6	88.7	66.3	77.9	92.1	60.6	76.9	88.1	66.3	77.6	88.7	60.3	75.0
0330	Cast-in-Place Concrete	93.8	83.1	89.8	93.3	67.5	83.5	90.1	67.5	81.6	90.3	67.3	81.6	90.1	67.6	81.6	90.2	66.9	81.4
03	CONCRETE	91.7	80.5	86.7	86.4	60.5	74.5	85.9	60.5	74.5	85.3	59.2	73.6	86.4	60.7	74.8	89.0	59.1	75.6
04	MASONRY	88.9	83.0	85.3	96.7	56.7	72.3	123.2	56.7	82.7	88.5	59.6	70.8	104.6	56.7	75.4	98.4	53.1	70.8
05	METALS	103.7	84.7	97.8	98.0	61.0	86.5	98.1	61.0	86.6	93.2	58.6	82.4	99.6	61.2	87.6	98.6	57.3	85.8
06	WOOD, PLASTICS & COMPOSITES	89.7	74.9	82.0	99.0	52.5	74.7	98.0	52.5	74.2	86.4	52.5	68.7	101.9	52.5	76.1	106.1	52.5	78.1
07	THERMAL & MOISTURE PROTECTION	90.1	81.2	86.3	101.2	62.7	84.7	101.3	62.7	84.7	97.7	62.0	82.4	101.5	62.7	84.8	101.8	59.4	83.6
08	OPENINGS	90.9	77.3	87.6	103.4	53.3	91.2	103.4	53.3	91.2	95.9	51.9	85.2	104.6	54.5	92.4	103.6	51.9	91.0
0920	Plaster & Gypsum Board	87.5	74.5	78.9	89.9	51.9	65.0	89.6	51.9	64.9	79.2	51.9	61.3	90.5	51.9	65.2	90.5	51.9	65.2
0950, 0980	Ceilings & Acoustic Treatment	95.0	74.5	82.1	93.5	51.9	67.5	93.5	51.9	67.5	87.1	51.9	65.1	93.5	51.9	67.5	93.5	51.9	67.5
0960	Flooring	94.5	71.8	87.9	86.9	54.7	77.5	85.8	54.7	76.7	92.3	49.3	79.8	87.5	71.6	82.9	89.1	54.7	79.0
0970, 0990	Wall Finishes & Painting/Coating	101.1	84.9	91.4	84.3	43.5	59.9	84.3	43.5	59.9	91.5	43.5	62.8	84.3	43.5	59.9	84.3	40.2	57.9
09	FINISHES	94.2	75.1	83.9	88.1	51.4	68.2	88.0	51.4	68.2	87.3	50.3	67.3	88.6	54.9	70.4	89.5	52.3	69.3
COVERS	DIVS. 10 - 14, 25, 28, 41, 43, 44, 46	100.0	87.5	97.1	100.0	77.4	94.7	100.0	77.4	94.7	100.0	77.4	94.7	100.0	77.4	94.7	100.0	77.4	94.7
21, 22, 23	FIRE SUPPRESSION, PLUMBING & HVAC	96.6	85.6	92.2	97.3	65.5	84.4	97.3	65.5	84.4	97.4	65.4	84.5	101.1	65.5	86.7	97.3	65.2	84.3
26, 27, 3370	ELECTRICAL, COMMUNICATIONS & UTIL.	97.3	86.4	91.9	95.2	70.0	82.7	96.1	70.0	83.2	97.6	66.6	82.3	96.1	70.0	83.2	97.5	63.2	80.5
MF2018	WEIGHTED AVERAGE	96.1	83.0	90.4	95.9	63.8	82.0	97.3	63.7	82.8	93.8	62.7	80.4	97.8	64.2	83.3	97.0	61.7	81.7

DIVISION		OKLAHOMA LAWTON 735			MCALESTER 745			MIAMI 743			MUSKOGEE 744			OKLAHOMA CITY 730 - 731			PONCA CITY 746		
		MAT.	INST.	TOTAL	MAT.	INST.	TOTAL	MAT.	INST.	TOTAL	MAT.	INST.	TOTAL	MAT.	INST.	TOTAL	MAT.	INST.	TOTAL
015433	CONTRACTOR EQUIPMENT		77.7	77.7		76.9	76.9		88.0	88.0		88.0	88.0		87.3	87.3		76.9	76.9
0241, 31 - 34	SITE & INFRASTRUCTURE, DEMOLITION	94.0	89.1	90.7	86.3	86.9	86.7	87.6	84.5	85.5	88.7	84.5	85.8	94.3	99.0	97.5	93.5	87.3	89.2
0310	Concrete Forming & Accessories	96.7	53.6	60.0	81.0	39.8	46.0	95.5	53.1	59.5	100.4	56.6	63.1	99.6	62.7	68.2	90.4	53.3	58.8
0320	Concrete Reinforcing	88.3	66.3	77.7	91.7	60.4	76.6	90.3	66.3	78.7	91.2	65.7	78.9	98.4	66.4	83.0	91.2	66.3	79.2
0330	Cast-in-Place Concrete	87.2	67.5	79.7	79.2	66.8	74.5	83.0	68.3	77.4	84.0	68.9	78.3	91.6	72.8	84.5	92.8	67.4	83.2
03	CONCRETE	83.0	60.6	73.0	76.5	53.1	66.0	81.0	61.7	72.3	82.6	63.3	74.0	90.4	66.7	79.8	87.4	60.5	75.3
04	MASONRY	98.4	56.7	73.0	106.3	56.6	76.0	91.3	56.8	70.2	108.2	48.3	71.6	98.8	57.5	73.6	83.8	56.7	67.3
05	METALS	103.4	61.2	90.3	93.2	57.4	82.0	93.1	75.7	87.7	94.7	74.9	88.5	96.6	64.0	86.4	93.1	60.9	83.1
06	WOOD, PLASTICS & COMPOSITES	105.0	52.5	77.6	83.7	34.9	58.2	100.4	52.6	75.4	105.5	56.8	80.0	100.9	64.2	81.7	95.3	52.5	72.9
07	THERMAL & MOISTURE PROTECTION	101.2	62.7	84.7	97.3	59.3	81.0	97.8	62.0	82.5	98.1	59.9	81.7	93.9	65.8	81.8	97.9	62.9	82.9
08	OPENINGS	106.4	54.5	93.8	95.9	42.2	82.8	94.9	53.4	85.5	97.1	55.7	87.0	99.8	60.9	90.3	95.9	53.3	85.5
0920	Plaster & Gypsum Board	92.8	51.9	66.0	78.7	33.8	49.2	84.9	51.9	63.3	87.8	56.2	67.1	93.9	63.7	74.1	83.8	51.9	62.9
0950, 0980	Ceilings & Acoustic Treatment	102.4	51.9	70.8	87.1	33.8	53.7	87.1	51.9	65.1	97.9	56.2	71.8	90.7	63.7	73.8	87.1	51.9	65.1
0960	Flooring	89.3	71.6	84.2	91.2	54.7	80.5	98.6	49.3	84.2	101.2	31.7	80.9	89.6	71.6	84.3	95.6	54.7	83.7
0970, 0990	Wall Finishes & Painting/Coating	84.3	43.5	59.9	91.5	40.2	60.8	91.5	40.2	60.8	91.5	40.2	60.8	90.1	43.5	62.2	91.5	43.5	62.8
09	FINISHES	91.1	54.9	71.5	86.4	40.7	61.6	89.3	50.0	68.0	93.0	49.5	69.5	91.1	61.9	75.3	89.0	52.3	69.1
COVERS	DIVS. 10 - 14, 25, 28, 41, 43, 44, 46	100.0	77.4	94.7	100.0	75.5	94.2	100.0	77.8	94.8	100.0	78.7	95.0	100.0	79.5	95.2	100.0	77.4	94.7
21, 22, 23	FIRE SUPPRESSION, PLUMBING & HVAC	101.1	65.5	86.7	97.4	60.5	82.5	97.4	60.5	82.5	101.2	62.1	85.4	101.0	65.9	86.8	97.4	60.6	82.5
26, 27, 3370	ELECTRICAL, COMMUNICATIONS & UTIL.	97.5	66.6	82.2	96.2	64.5	80.5	97.4	64.5	81.2	95.9	86.7	91.4	102.4	70.0	86.3	95.9	63.2	79.7
MF2018	WEIGHTED AVERAGE	98.1	63.9	83.3	93.2	58.3	78.1	93.5	63.0	80.3	96.1	65.7	83.0	97.6	67.8	84.8	93.9	61.8	80.0

DIVISION		OKLAHOMA POTEAU 749			SHAWNEE 748			TULSA 740 - 741			WOODWARD 738			OREGON BEND 977			EUGENE 974		
		MAT.	INST.	TOTAL	MAT.	INST.	TOTAL	MAT.	INST.	TOTAL	MAT.	INST.	TOTAL	MAT.	INST.	TOTAL	MAT.	INST.	TOTAL
015433	CONTRACTOR EQUIPMENT		86.8	86.8		76.9	76.9		88.0	88.0		76.9	76.9		95.5	95.5		95.5	95.5
0241, 31 - 34	SITE & INFRASTRUCTURE, DEMOLITION	74.4	80.3	78.4	96.4	87.3	90.1	94.7	84.2	87.5	96.1	87.7	90.3	112.5	95.7	100.9	102.9	95.7	98.0
0310	Concrete Forming & Accessories	88.1	52.9	58.1	83.0	53.3	57.7	100.2	56.0	63.4	89.6	42.1	49.2	103.1	101.8	102.0	99.5	102.1	101.7
0320	Concrete Reinforcing	92.2	66.2	79.7	91.2	62.4	77.3	91.4	66.3	79.3	88.1	66.3	77.6	90.5	114.3	102.0	94.4	114.4	104.0
0330	Cast-in-Place Concrete	83.0	68.2	77.4	95.7	67.4	85.0	91.5	70.8	83.7	90.1	67.4	81.6	118.3	101.0	111.8	114.6	101.1	109.5
03	CONCRETE	83.0	61.5	73.3	88.9	59.8	75.9	87.6	64.3	77.1	86.1	55.4	72.3	110.9	103.2	107.5	102.0	103.4	102.6
04	MASONRY	91.5	56.8	70.3	107.7	56.7	76.6	92.2	57.8	71.2	92.3	56.7	70.6	101.7	102.7	102.3	98.6	102.7	101.1
05	METALS	93.2	75.4	87.6	93.1	59.5	82.6	97.9	76.1	91.1	98.2	60.9	86.6	107.3	98.1	104.4	108.0	98.5	105.0
06	WOOD, PLASTICS & COMPOSITES	91.5	52.6	71.2	86.2	52.5	68.6	104.6	56.8	79.6	98.1	37.3	66.3	95.9	102.1	99.1	91.5	102.1	97.0
07	THERMAL & MOISTURE PROTECTION	97.9	62.0	82.5	97.9	61.6	82.3	98.1	64.9	83.9	101.4	61.2	84.2	118.6	102.9	111.8	117.7	105.5	112.5
08	OPENINGS	95.9	53.4	85.5	95.9	52.4	85.3	98.8	56.9	88.6	103.4	44.9	89.2	97.0	105.0	99.0	97.3	105.0	99.2
0920	Plaster & Gypsum Board	81.8	51.9	62.2	79.2	51.9	61.3	87.8	56.2	67.0	89.6	36.3	54.6	116.1	102.2	107.0	113.9	102.2	106.2
0950, 0980	Ceilings & Acoustic Treatment	87.1	51.9	65.1	87.1	51.9	65.1	97.9	56.2	71.8	93.5	36.3	57.7	90.0	102.2	97.6	90.7	102.2	97.9
0960	Flooring	94.9	49.3	81.6	92.3	54.7	81.3	100.0	62.4	89.0	85.8	52.1	76.0	101.3	106.4	102.8	99.7	106.4	101.7
0970, 0990	Wall Finishes & Painting/Coating	91.5	43.5	62.8	91.5	40.2	60.8	91.5	54.9	69.6	84.3	43.5	59.9	93.9	76.5	83.5	93.9	66.6	77.6
09	FINISHES	87.0	50.4	67.2	87.5	51.0	67.8	92.9	57.0	73.5	88.0	42.0	63.1	100.7	99.9	100.3	99.0	98.8	98.9
COVERS	DIVS. 10 - 14, 25, 28, 41, 43, 44, 46	100.0	77.7	94.7	100.0	77.4	94.7	100.0	78.6	95.0	100.0	75.8	94.3	100.0	102.5	100.6	100.0	102.5	100.6
21, 22, 23	FIRE SUPPRESSION, PLUMBING & HVAC	97.4	60.6	82.5	97.4	65.4	84.5	101.2	61.3	85.1	97.3	65.5	84.4	97.2	102.3	99.3	101.2	117.3	107.7
26, 27, 3370	ELECTRICAL, COMMUNICATIONS & UTIL.	96.0	64.5	80.4	97.6	70.0	83.9	97.6	69.9	83.9	97.4	70.0	83.8	98.3	96.1	97.2	97.6	96.1	96.8
MF2018	WEIGHTED AVERAGE	93.0	62.6	79.9	95.3	63.3	81.5	96.9	65.5	83.3	95.9	61.2	80.9	102.3	100.5	101.5	101.6	103.7	102.5

For customer support on your Site Work & Landscape Costs with RSMeans Data, call 800.448.8182.

OREGON

DIVISION		KLAMATH FALLS 976			MEDFORD 975			PENDLETON 978			PORTLAND 970 - 972			SALEM 973			VALE 979		
		MAT.	INST.	TOTAL	MAT.	INST.	TOTAL	MAT.	INST.	TOTAL	MAT.	INST.	TOTAL	MAT.	INST.	TOTAL	MAT.	INST.	TOTAL
015433	CONTRACTOR EQUIPMENT		95.5	95.5		95.5	95.5		93.1	93.1		95.5	95.5		101.4	101.4		93.1	93.1
0241, 31 - 34	SITE & INFRASTRUCTURE, DEMOLITION	116.6	95.7	102.2	110.8	95.7	100.4	110.6	89.2	95.9	105.6	95.7	98.8	98.5	104.0	102.3	97.5	89.2	91.8
0310	Concrete Forming & Accessories	96.1	101.4	100.6	94.9	101.6	100.6	97.2	101.7	101.1	101.0	102.0	101.9	99.2	102.2	101.7	103.8	100.4	100.9
0320	Concrete Reinforcing	90.5	114.2	101.9	92.0	114.3	102.7	89.7	114.4	101.6	95.1	114.5	104.4	102.2	114.5	108.1	87.6	114.2	100.4
0330	Cast-in-Place Concrete	118.4	95.7	109.8	118.3	100.9	111.7	119.2	96.2	110.5	117.8	101.1	111.5	108.3	103.0	106.3	93.7	96.4	94.7
03	CONCRETE	113.8	101.2	108.1	108.3	103.1	106.0	94.7	101.5	97.8	103.7	103.4	103.6	99.7	104.0	101.6	78.8	100.9	88.7
04	MASONRY	114.7	102.7	107.4	95.2	102.7	99.8	105.6	104.9	105.2	100.5	104.8	103.1	101.3	104.9	103.5	103.5	104.9	104.3
05	METALS	107.3	97.8	104.3	107.6	97.9	104.5	115.5	98.7	110.2	109.4	98.6	106.0	116.4	97.2	110.4	115.3	97.2	109.7
06	WOOD, PLASTICS & COMPOSITES	86.2	102.1	94.5	85.1	102.1	94.0	88.7	102.2	95.8	92.7	102.1	97.6	85.5	102.4	94.3	97.7	102.2	100.1
07	THERMAL & MOISTURE PROTECTION	118.8	98.4	110.0	118.4	98.0	109.7	110.5	95.7	104.1	117.7	103.4	111.6	114.0	104.8	110.0	109.9	91.8	102.1
08	OPENINGS	97.0	105.0	99.0	99.6	105.0	100.9	93.1	105.0	96.0	95.3	105.0	97.7	102.7	105.1	103.3	93.1	92.3	92.9
0920	Plaster & Gypsum Board	110.2	102.2	104.9	109.6	102.2	104.7	96.3	102.2	100.2	113.5	102.2	106.1	109.6	102.2	104.8	103.4	102.2	102.6
0950, 0980	Ceilings & Acoustic Treatment	97.6	102.2	100.5	102.6	102.2	102.3	65.8	102.2	88.6	92.8	102.2	98.7	105.1	102.2	103.3	65.8	102.2	88.6
0960	Flooring	98.3	106.4	100.7	97.8	106.4	100.3	67.3	106.4	78.7	97.6	106.4	101.0	100.1	106.4	101.9	69.4	106.4	80.2
0970, 0990	Wall Finishes & Painting/Coating	93.9	76.8	83.7	93.9	76.8	83.7	85.0	76.8	80.1	93.8	76.8	83.6	95.5	74.5	82.9	85.0	76.5	79.9
09	FINISHES	101.3	99.9	100.6	101.4	99.9	100.6	71.3	100.0	86.8	98.9	99.9	99.5	100.6	99.9	100.2	71.9	100.0	87.1
COVERS	DIVS. 10 - 14, 25, 28, 41, 43, 44, 46	100.0	102.4	100.6	100.0	102.4	100.6	100.0	105.2	101.2	100.0	102.6	100.6	100.0	103.1	100.7	100.0	102.6	100.6
21, 22, 23	FIRE SUPPRESSION, PLUMBING & HVAC	97.2	102.3	99.2	101.2	108.1	104.0	100.0	108.8	103.5	101.2	111.1	105.2	100.2	108.2	103.4	100.0	69.8	87.8
26, 27, 3370	ELECTRICAL, COMMUNICATIONS & UTIL.	97.2	79.6	88.5	100.6	79.6	90.2	90.3	93.3	91.8	97.9	110.5	104.1	101.5	96.1	98.8	90.3	65.5	78.0
MF2018	WEIGHTED AVERAGE	103.3	97.8	100.9	103.1	99.3	101.4	98.3	101.0	99.5	102.0	104.7	103.2	103.3	102.8	103.0	96.0	87.8	92.4

PENNSYLVANIA

DIVISION		ALLENTOWN 181			ALTOONA 166			BEDFORD 155			BRADFORD 167			BUTLER 160			CHAMBERSBURG 172		
		MAT.	INST.	TOTAL	MAT.	INST.	TOTAL	MAT.	INST.	TOTAL	MAT.	INST.	TOTAL	MAT.	INST.	TOTAL	MAT.	INST.	TOTAL
015433	CONTRACTOR EQUIPMENT		110.8	110.8		110.8	110.8		108.0	108.0		110.8	110.8		110.8	110.8		110.0	110.0
0241, 31 - 34	SITE & INFRASTRUCTURE, DEMOLITION	91.6	95.0	93.9	94.9	95.1	95.0	102.8	91.1	94.8	90.1	92.9	92.0	85.8	94.8	92.0	87.6	92.2	90.7
0310	Concrete Forming & Accessories	98.9	109.3	107.7	82.5	87.6	86.9	81.1	78.4	78.8	85.0	93.1	91.9	83.9	92.0	90.8	87.4	74.0	76.0
0320	Concrete Reinforcing	97.5	112.8	104.9	94.5	109.6	101.8	93.7	100.7	97.1	96.5	100.8	98.6	95.1	121.2	107.7	98.9	106.9	102.7
0330	Cast-in-Place Concrete	88.5	103.6	94.2	98.6	90.6	95.6	107.2	82.3	97.8	94.2	87.1	91.5	87.1	97.0	90.8	93.8	90.0	92.4
03	CONCRETE	91.8	108.8	99.4	87.2	93.8	90.2	100.9	85.1	93.8	93.5	93.4	93.5	79.4	100.0	88.6	97.4	87.0	92.7
04	MASONRY	96.6	98.8	97.9	100.3	89.8	93.9	119.2	76.9	93.4	97.9	78.6	86.1	102.6	91.8	96.0	98.3	75.8	84.6
05	METALS	99.8	122.3	106.8	93.8	118.2	101.4	98.8	111.2	102.7	97.5	111.5	101.9	93.5	123.0	102.7	99.2	116.8	104.7
06	WOOD, PLASTICS & COMPOSITES	98.2	110.8	104.8	75.5	86.2	81.1	78.1	78.9	78.5	81.3	97.8	90.0	77.1	90.9	84.3	83.6	73.0	78.0
07	THERMAL & MOISTURE PROTECTION	105.3	112.6	108.4	103.3	95.0	99.7	99.6	85.1	93.4	105.1	85.4	96.6	102.7	94.0	99.0	96.8	79.3	89.3
08	OPENINGS	92.5	108.5	96.4	86.3	89.6	87.1	93.6	83.6	91.1	92.4	93.0	92.5	86.3	99.3	89.4	89.0	78.4	86.4
0920	Plaster & Gypsum Board	99.7	111.2	107.3	91.2	85.9	87.7	98.2	78.5	85.3	91.6	97.9	95.7	91.2	90.7	90.9	115.9	72.4	87.3
0950, 0980	Ceilings & Acoustic Treatment	92.2	111.2	104.1	95.4	85.9	89.4	113.3	78.5	91.5	95.5	97.9	97.0	96.1	90.7	92.7	96.9	72.4	81.5
0960	Flooring	91.9	97.7	93.6	85.4	103.7	90.7	92.2	98.9	94.1	86.1	98.9	89.8	86.4	100.5	90.5	91.4	74.4	86.4
0970, 0990	Wall Finishes & Painting/Coating	87.1	108.2	99.7	82.8	112.0	100.3	89.7	100.6	96.2	87.1	100.6	95.2	82.8	101.1	93.8	87.1	95.9	92.4
09	FINISHES	92.4	107.2	100.4	90.4	92.5	91.5	101.6	83.9	92.0	90.4	95.4	93.1	90.3	94.0	92.3	93.8	75.7	84.0
COVERS	DIVS. 10 - 14, 25, 28, 41, 43, 44, 46	100.0	103.5	100.8	100.0	98.6	99.7	100.0	94.9	98.8	100.0	97.7	99.5	100.0	99.7	99.9	100.0	92.6	98.3
21, 22, 23	FIRE SUPPRESSION, PLUMBING & HVAC	101.2	116.9	107.5	100.5	90.6	96.5	93.1	80.5	88.0	97.4	86.3	92.9	96.7	92.5	95.0	95.1	85.5	91.2
26, 27, 3370	ELECTRICAL, COMMUNICATIONS & UTIL.	99.4	99.5	99.5	90.7	113.3	101.9	94.7	113.3	103.9	94.1	113.3	103.6	91.2	103.7	97.4	93.6	82.7	88.2
MF2018	WEIGHTED AVERAGE	97.5	108.1	102.1	94.2	97.5	95.6	98.0	90.1	94.6	95.6	95.0	95.3	92.1	98.7	95.0	95.4	86.0	91.3

PENNSYLVANIA

DIVISION		DOYLESTOWN 189			DUBOIS 158			ERIE 164 - 165			GREENSBURG 156			HARRISBURG 170 - 171			HAZLETON 182		
		MAT.	INST.	TOTAL	MAT.	INST.	TOTAL	MAT.	INST.	TOTAL	MAT.	INST.	TOTAL	MAT.	INST.	TOTAL	MAT.	INST.	TOTAL
015433	CONTRACTOR EQUIPMENT		89.5	89.5		108.0	108.0		110.8	110.8		108.0	108.0		112.2	112.2		110.8	110.8
0241, 31 - 34	SITE & INFRASTRUCTURE, DEMOLITION	104.5	82.9	89.6	107.6	91.5	96.6	91.8	94.9	93.9	98.7	93.2	94.9	88.7	98.1	95.1	84.6	93.9	91.0
0310	Concrete Forming & Accessories	81.3	122.1	116.1	80.5	80.8	80.7	98.1	87.4	89.0	87.9	86.7	86.8	100.5	89.6	91.2	79.0	85.6	84.6
0320	Concrete Reinforcing	94.3	143.7	118.1	93.0	113.7	103.0	96.5	109.3	102.7	93.0	120.9	106.5	103.9	112.4	108.0	94.7	107.7	101.0
0330	Cast-in-Place Concrete	83.6	124.7	99.1	103.3	93.2	99.5	96.9	91.0	94.7	99.4	96.4	98.3	92.5	102.4	96.2	83.6	91.8	86.7
03	CONCRETE	87.3	126.3	104.8	102.9	92.2	98.1	86.4	93.8	89.7	96.1	97.3	96.6	92.6	99.2	95.6	84.4	93.0	88.3
04	MASONRY	100.2	129.3	117.9	120.1	89.5	101.4	88.2	90.7	89.7	130.3	88.0	104.5	93.7	92.9	93.2	110.1	85.8	95.3
05	METALS	97.3	121.9	105.0	98.8	116.7	104.4	94.0	117.3	101.3	98.8	120.7	105.6	106.0	118.6	109.9	99.5	117.8	105.2
06	WOOD, PLASTICS & COMPOSITES	76.8	121.6	100.2	76.9	78.9	78.0	94.9	85.7	90.1	85.0	84.1	84.5	100.3	86.4	93.0	75.1	84.1	79.8
07	THERMAL & MOISTURE PROTECTION	101.9	125.8	112.2	99.9	91.2	96.2	103.6	90.7	98.1	99.5	91.8	96.2	98.9	108.2	102.9	104.5	96.2	100.9
08	OPENINGS	94.9	128.3	103.0	93.6	86.6	91.9	86.4	90.7	87.5	93.5	95.6	94.0	100.6	87.8	97.4	93.0	89.9	92.2
0920	Plaster & Gypsum Board	90.1	122.2	111.2	96.8	78.5	84.8	99.7	85.4	90.3	98.6	83.9	89.0	122.6	85.9	98.5	90.4	83.7	86.0
0950, 0980	Ceilings & Acoustic Treatment	91.6	122.2	110.8	113.3	78.5	91.5	92.2	85.4	87.9	112.7	83.9	94.6	102.3	85.9	92.0	92.9	83.7	87.2
0960	Flooring	75.9	132.9	92.6	91.9	98.9	93.9	92.5	103.7	95.8	95.9	74.4	89.6	95.2	94.1	94.9	83.1	85.2	83.7
0970, 0990	Wall Finishes & Painting/Coating	86.7	136.8	116.7	89.7	101.1	96.6	92.9	93.5	93.3	89.7	101.1	96.6	93.7	89.5	91.2	87.1	99.1	94.3
09	FINISHES	84.1	125.4	106.4	101.8	85.3	92.9	93.5	90.5	91.9	102.3	85.2	93.0	99.2	89.6	94.0	88.3	86.6	87.4
COVERS	DIVS. 10 - 14, 25, 28, 41, 43, 44, 46	100.0	114.9	103.5	100.0	96.7	99.2	100.0	98.9	99.7	100.0	98.8	99.7	100.0	98.6	99.7	100.0	98.5	99.6
21, 22, 23	FIRE SUPPRESSION, PLUMBING & HVAC	96.7	126.3	108.7	93.1	83.4	89.2	100.5	93.8	97.8	93.1	84.7	89.7	96.4	98.5	97.2	97.4	91.0	94.8
26, 27, 3370	ELECTRICAL, COMMUNICATIONS & UTIL.	93.5	127.5	110.3	95.2	113.3	104.1	92.2	90.5	91.4	95.2	113.3	104.2	100.6	89.0	94.9	94.8	86.7	90.8
MF2018	WEIGHTED AVERAGE	94.7	122.6	106.7	98.5	94.0	96.6	94.0	94.6	94.3	98.0	95.8	97.1	98.5	97.0	97.9	95.1	92.5	94.0

PENNSYLVANIA

DIVISION		INDIANA 157			JOHNSTOWN 159			KITTANNING 162			LANCASTER 175 - 176			LEHIGH VALLEY 180			MONTROSE 188		
		MAT.	INST.	TOTAL	MAT.	INST.	TOTAL	MAT.	INST.	TOTAL	MAT.	INST.	TOTAL	MAT.	INST.	TOTAL	MAT.	INST.	TOTAL
015433	CONTRACTOR EQUIPMENT		108.0	108.0		108.0	108.0		110.8	110.8		110.0	110.0		110.8	110.8		110.8	110.8
0241, 31 - 34	SITE & INFRASTRUCTURE, DEMOLITION	96.9	92.1	93.6	103.3	93.4	96.5	88.3	94.7	92.7	79.7	94.3	89.7	88.4	94.3	92.4	87.2	94.0	91.8
0310	Concrete Forming & Accessories	81.7	89.8	88.6	80.5	87.3	86.3	83.9	86.8	86.3	89.5	87.8	88.0	91.8	106.3	104.1	80.0	86.1	85.2
0320	Concrete Reinforcing	92.4	121.2	106.3	93.7	121.1	106.9	95.1	121.0	107.6	98.5	112.5	105.2	94.7	102.9	98.6	99.2	110.7	104.7
0330	Cast-in-Place Concrete	97.5	93.6	96.0	108.1	89.8	101.2	90.5	94.0	91.8	79.8	96.7	86.2	90.5	95.3	92.3	88.7	89.3	88.9
03	CONCRETE	93.7	97.7	95.5	101.9	95.3	98.9	81.9	96.6	88.5	85.7	96.6	90.6	90.8	102.8	96.2	89.4	92.9	91.0
04	MASONRY	115.5	89.6	99.7	116.4	89.2	99.8	104.9	85.1	92.8	104.0	89.5	95.1	96.6	92.5	94.1	96.5	87.7	91.1
05	METALS	98.9	120.5	105.6	98.9	120.6	105.6	93.6	121.8	102.4	99.2	121.3	106.1	99.4	117.1	104.9	97.6	118.5	104.1
06	WOOD, PLASTICS & COMPOSITES	78.9	90.8	85.1	76.9	86.1	81.7	77.1	86.3	81.9	86.8	86.2	86.5	88.8	108.8	99.2	76.0	84.1	80.2
07	THERMAL & MOISTURE PROTECTION	99.3	92.5	96.4	99.6	91.9	96.3	102.9	90.9	97.7	96.1	105.1	100.0	105.0	106.9	105.8	104.6	86.9	97.0
08	OPENINGS	93.6	94.9	93.9	93.5	92.3	93.2	86.3	96.8	88.8	89.0	87.8	88.7	92.9	102.7	95.3	89.8	88.5	89.5
0920	Plaster & Gypsum Board	98.5	90.7	93.4	96.7	85.9	89.6	91.2	86.0	87.8	118.7	85.9	97.2	92.4	109.1	103.3	91.1	83.7	86.3
0950, 0980	Ceilings & Acoustic Treatment	113.3	90.7	99.1	112.7	85.9	95.9	96.1	86.0	89.8	96.9	85.9	90.0	92.9	109.1	103.1	95.5	83.7	88.1
0960	Flooring	92.7	98.9	94.5	91.9	103.7	95.3	86.4	98.9	90.0	92.5	94.4	93.0	88.8	89.8	89.1	83.8	98.9	88.2
0970, 0990	Wall Finishes & Painting/Coating	89.7	101.1	96.6	89.7	112.0	103.1	82.8	101.1	93.8	87.1	89.5	88.6	87.1	96.2	92.5	87.1	99.1	94.3
09	FINISHES	101.4	92.2	96.4	101.2	92.2	96.4	90.5	89.6	90.0	93.9	88.3	90.9	90.5	102.3	96.9	89.5	89.0	89.2
COVERS	DIVS. 10 - 14, 25, 28, 41, 43, 44, 46	100.0	98.0	99.5	100.0	98.2	99.6	100.0	97.8	99.5	100.0	96.5	99.2	100.0	102.7	100.6	100.0	99.1	99.8
21, 22, 23	FIRE SUPPRESSION, PLUMBING & HVAC	93.1	81.5	88.4	93.1	88.3	91.1	96.7	86.6	92.6	95.1	96.0	95.5	97.4	112.0	103.3	97.4	90.3	94.5
26, 27, 3370	ELECTRICAL, COMMUNICATIONS & UTIL.	95.2	113.3	104.2	95.2	113.3	104.2	90.7	113.3	101.9	94.8	96.4	95.6	94.8	124.6	109.5	94.1	91.7	92.9
MF2018	WEIGHTED AVERAGE	96.8	96.2	96.5	98.0	97.2	97.7	92.6	96.6	94.3	94.2	96.5	95.2	95.6	107.5	100.8	94.6	93.3	94.0

PENNSYLVANIA

DIVISION		NEW CASTLE 161			NORRISTOWN 194			OIL CITY 163			PHILADELPHIA 190 - 191			PITTSBURGH 150 - 152			POTTSVILLE 179		
		MAT.	INST.	TOTAL	MAT.	INST.	TOTAL	MAT.	INST.	TOTAL	MAT.	INST.	TOTAL	MAT.	INST.	TOTAL	MAT.	INST.	TOTAL
015433	CONTRACTOR EQUIPMENT		110.8	110.8		95.3	95.3		110.8	110.8		100.4	100.4		96.9	96.9		110.0	110.0
0241, 31 - 34	SITE & INFRASTRUCTURE, DEMOLITION	86.2	95.3	92.5	94.7	91.9	92.8	84.8	93.1	90.5	97.9	102.1	100.8	102.7	93.3	96.3	82.6	92.7	89.5
0310	Concrete Forming & Accessories	83.9	95.9	94.1	81.6	120.7	114.9	83.9	89.4	88.6	95.6	140.6	133.9	98.9	99.1	99.1	80.3	84.4	83.8
0320	Concrete Reinforcing	94.1	99.4	96.6	97.4	143.7	119.7	95.1	90.4	92.8	107.8	151.5	128.9	92.6	125.8	108.6	97.7	108.7	103.0
0330	Cast-in-Place Concrete	87.9	96.9	91.3	83.6	122.4	98.3	85.4	90.6	87.4	93.9	133.6	108.9	106.7	100.9	104.5	85.0	92.5	87.8
03	CONCRETE	79.7	98.0	87.9	86.6	124.8	103.8	78.2	91.5	84.1	100.3	138.9	117.6	101.5	104.3	102.8	89.1	93.0	90.8
04	MASONRY	102.7	94.8	97.9	111.2	125.7	120.0	101.0	85.0	91.3	99.9	135.2	121.4	105.5	100.7	102.6	97.5	83.9	89.2
05	METALS	93.6	115.1	100.3	102.9	122.0	108.8	93.6	112.7	99.5	106.5	124.1	112.0	100.3	108.9	103.0	99.4	118.5	105.4
06	WOOD, PLASTICS & COMPOSITES	77.1	96.4	87.2	74.4	121.5	99.1	77.1	90.9	84.3	93.8	142.0	119.0	100.3	99.2	99.7	75.6	82.3	79.1
07	THERMAL & MOISTURE PROTECTION	102.8	93.4	98.8	108.8	124.3	115.4	102.6	87.6	96.2	103.7	137.2	118.1	99.9	99.6	99.8	96.3	95.2	95.8
08	OPENINGS	86.3	97.0	88.8	86.6	128.3	96.7	86.6	92.3	87.7	97.7	143.0	108.8	97.0	104.7	98.9	89.0	89.6	89.2
0920	Plaster & Gypsum Board	91.2	96.4	94.6	87.2	122.2	110.2	91.2	90.7	90.9	103.0	143.0	129.3	100.4	99.1	99.5	113.3	81.9	92.7
0950, 0980	Ceilings & Acoustic Treatment	96.1	96.4	96.3	95.1	122.2	112.1	96.1	90.7	92.7	99.7	143.0	126.9	106.5	99.1	101.9	96.9	81.9	87.5
0960	Flooring	86.4	103.4	91.3	87.1	132.9	100.5	86.4	98.9	90.0	98.4	149.0	113.2	99.8	106.7	101.8	88.3	98.9	91.4
0970, 0990	Wall Finishes & Painting/Coating	82.8	109.2	98.6	85.2	136.8	116.2	82.8	101.1	93.8	95.6	168.5	139.3	95.3	110.6	104.5	87.1	99.1	94.3
09	FINISHES	90.4	98.7	94.9	87.4	124.4	107.4	90.2	92.3	91.4	98.7	145.6	124.1	102.5	101.3	101.9	92.1	87.7	89.7
COVERS	DIVS. 10 - 14, 25, 28, 41, 43, 44, 46	100.0	100.3	100.1	100.0	113.6	103.2	100.0	98.2	99.6	100.0	119.6	104.6	100.0	101.4	100.3	100.0	98.3	99.6
21, 22, 23	FIRE SUPPRESSION, PLUMBING & HVAC	96.7	96.0	96.4	95.7	124.4	107.3	96.7	89.8	93.9	96.4	136.9	112.7	96.5	98.6	97.3	95.1	91.2	93.5
26, 27, 3370	ELECTRICAL, COMMUNICATIONS & UTIL.	91.2	100.8	95.9	93.4	136.7	114.9	92.7	103.7	98.2	99.2	157.6	128.1	96.8	111.5	104.1	93.2	87.9	90.6
MF2018	WEIGHTED AVERAGE	92.2	98.9	95.1	95.1	123.4	107.3	92.1	94.4	93.1	99.7	136.9	115.8	99.3	102.6	100.7	94.0	92.5	93.3

PENNSYLVANIA

DIVISION		READING 195 - 196			SCRANTON 184 - 185			STATE COLLEGE 168			STROUDSBURG 183			SUNBURY 178			UNIONTOWN 154		
		MAT.	INST.	TOTAL	MAT.	INST.	TOTAL	MAT.	INST.	TOTAL	MAT.	INST.	TOTAL	MAT.	INST.	TOTAL	MAT.	INST.	TOTAL
015433	CONTRACTOR EQUIPMENT		117.0	117.0		110.8	110.8		110.0	110.0		110.8	110.8		110.8	110.8		108.0	108.0
0241, 31 - 34	SITE & INFRASTRUCTURE, DEMOLITION	99.9	105.0	103.4	92.1	94.6	93.8	82.5	93.8	90.3	86.2	94.2	91.7	94.0	93.5	93.7	97.4	93.0	94.4
0310	Concrete Forming & Accessories	98.4	89.2	90.6	99.0	89.9	91.3	84.0	88.0	87.4	85.8	87.2	87.0	92.3	81.7	83.2	74.5	91.8	89.2
0320	Concrete Reinforcing	98.8	147.8	122.4	97.5	112.5	104.8	95.8	109.9	102.6	97.8	110.9	104.1	100.5	106.9	103.6	93.0	121.3	106.7
0330	Cast-in-Place Concrete	74.9	97.3	83.4	92.4	93.5	92.8	89.1	90.8	89.7	87.1	90.8	88.5	92.9	91.1	92.2	97.5	96.3	97.0
03	CONCRETE	85.9	103.6	93.8	93.6	96.4	94.9	93.7	94.1	93.9	88.2	94.0	90.8	92.7	90.9	91.9	93.3	99.6	96.1
04	MASONRY	99.8	92.0	95.0	97.0	94.1	95.2	102.4	92.9	96.6	94.4	92.3	93.1	98.1	77.6	85.6	133.8	94.1	109.6
05	METALS	103.2	135.3	113.2	101.9	120.9	107.8	97.4	118.8	104.1	99.5	119.0	105.6	99.2	117.4	104.9	98.6	121.2	105.6
06	WOOD, PLASTICS & COMPOSITES	94.0	86.1	89.9	98.2	88.5	93.1	83.4	86.2	84.8	82.3	84.1	83.2	84.9	82.3	83.5	70.4	90.8	81.1
07	THERMAL & MOISTURE PROTECTION	109.9	106.2	108.3	105.1	92.7	99.8	104.3	105.9	105.0	104.7	87.2	97.2	97.3	88.5	93.6	99.2	94.6	97.2
08	OPENINGS	90.8	103.1	93.8	92.5	92.7	92.5	89.7	89.6	89.6	93.0	91.1	92.5	89.1	86.6	88.5	93.5	97.4	94.4
0920	Plaster & Gypsum Board	96.3	85.9	89.5	102.4	88.2	93.1	92.2	85.9	88.1	90.9	83.7	86.2	112.5	81.9	92.4	94.7	90.7	92.1
0950, 0980	Ceilings & Acoustic Treatment	86.7	85.9	86.2	102.4	88.2	93.5	92.3	85.9	88.3	91.6	83.7	86.7	93.7	81.9	86.3	112.7	90.7	98.9
0960	Flooring	91.9	94.4	92.6	91.9	96.9	93.4	89.3	96.4	91.4	86.6	89.8	87.5	89.2	98.9	92.1	88.7	105.0	93.5
0970, 0990	Wall Finishes & Painting/Coating	84.1	108.2	98.5	87.1	108.2	99.7	87.1	112.0	102.0	87.1	99.1	94.3	87.1	99.1	94.3	89.7	109.2	101.4
09	FINISHES	88.4	90.7	89.7	95.2	92.6	93.8	89.7	91.3	90.6	89.2	88.2	88.7	92.9	86.1	89.2	89.6	95.3	97.3
COVERS	DIVS. 10 - 14, 25, 28, 41, 43, 44, 46	100.0	98.9	99.8	100.0	99.6	99.9	100.0	96.9	99.3	100.0	100.1	100.0	100.0	94.2	98.6	100.0	99.9	99.9
21, 22, 23	FIRE SUPPRESSION, PLUMBING & HVAC	101.1	114.5	106.5	101.2	96.0	99.1	97.4	96.3	97.0	97.4	93.3	95.7	95.1	83.2	90.3	93.1	88.9	91.4
26, 27, 3370	ELECTRICAL, COMMUNICATIONS & UTIL.	99.5	96.5	98.0	99.5	92.0	95.8	93.3	113.3	103.2	94.8	136.5	115.4	93.6	88.5	91.1	92.7	113.3	102.9
MF2018	WEIGHTED AVERAGE	97.3	104.7	100.5	98.3	96.8	97.7	95.2	99.1	96.9	95.0	100.9	97.6	95.0	89.3	92.5	97.1	99.2	98.0

PENNSYLVANIA

DIVISION		WASHINGTON 153			WELLSBORO 169			WESTCHESTER 193			WILKES-BARRE 186-187			WILLIAMSPORT 177			YORK 173-174		
		MAT.	INST.	TOTAL	MAT.	INST.	TOTAL	MAT.	INST.	TOTAL	MAT.	INST.	TOTAL	MAT.	INST.	TOTAL	MAT.	INST.	TOTAL
015433	CONTRACTOR EQUIPMENT		108.0	108.0		110.8	110.8		95.3	95.3		110.8	110.8		110.8	110.8		110.0	110.0
0241, 31 - 34	SITE & INFRASTRUCTURE, DEMOLITION	97.5	93.9	95.0	93.5	93.4	93.5	100.7	92.8	95.3	84.2	94.6	91.4	85.2	94.4	91.5	84.7	94.3	91.3
0310	Concrete Forming & Accessories	81.9	98.1	95.7	84.2	81.0	81.5	88.6	121.9	117.0	88.8	89.9	89.7	88.6	86.9	87.2	84.0	87.3	86.8
0320	Concrete Reinforcing	93.0	121.5	106.8	95.8	110.6	102.9	96.5	143.7	119.3	96.5	112.5	104.2	99.7	112.5	105.9	100.5	112.5	106.3
0330	Cast-in-Place Concrete	97.5	96.9	97.3	93.4	84.6	90.1	92.7	124.5	104.7	83.6	93.4	87.3	78.6	91.6	83.5	85.4	96.6	89.6
03	CONCRETE	93.8	102.8	97.8	95.9	89.0	92.8	94.2	126.1	108.5	85.3	96.4	90.3	80.9	94.5	87.0	90.3	96.4	93.0
04	MASONRY	114.3	96.7	103.5	102.6	78.0	87.6	105.6	129.3	120.1	110.4	94.1	100.5	89.7	91.9	91.1	99.5	89.5	93.4
05	METALS	98.6	122.4	106.0	97.5	117.7	103.8	102.9	122.0	108.8	97.6	120.9	104.9	99.2	120.9	106.0	100.9	121.2	107.2
06	WOOD, PLASTICS & COMPOSITES	79.0	98.8	89.3	80.7	81.5	81.1	81.9	121.5	102.6	85.1	88.5	86.8	81.0	86.2	83.7	79.4	85.7	82.7
07	THERMAL & MOISTURE PROTECTION	99.3	96.9	98.3	105.4	82.1	95.4	109.3	121.4	114.5	104.5	92.7	99.4	96.6	91.3	94.3	96.4	105.0	100.1
08	OPENINGS	93.5	103.6	95.9	92.3	86.7	91.0	86.6	128.3	96.7	89.8	92.7	90.5	89.1	91.4	89.7	89.0	87.4	88.6
0920	Plaster & Gypsum Board	98.4	98.9	98.7	90.8	81.1	84.4	87.8	122.2	110.4	91.9	88.2	89.5	113.3	85.9	95.3	113.3	85.4	95.0
0950, 0980	Ceilings & Acoustic Treatment	112.7	98.9	104.1	92.3	81.1	85.3	95.1	122.2	112.1	95.5	88.2	90.9	96.9	85.9	90.0	96.2	85.4	89.4
0960	Flooring	92.8	105.0	96.4	85.6	98.9	89.5	90.0	132.9	102.5	87.5	96.9	90.3	88.1	94.4	89.9	89.8	94.4	91.1
0970, 0990	Wall Finishes & Painting/Coating	89.7	109.2	101.4	87.1	99.1	94.3	85.2	136.8	116.2	87.1	108.2	99.7	87.1	108.2	99.7	87.1	89.5	88.6
09	FINISHES	101.3	100.3	100.7	89.8	86.1	87.8	88.8	125.3	108.6	90.4	92.6	91.6	92.7	90.0	91.2	92.4	88.0	90.0
COVERS	DIVS. 10 - 14, 25, 28, 41, 43, 44, 46	100.0	100.4	100.1	100.0	94.1	98.6	100.0	114.7	103.5	100.0	99.6	99.9	100.0	96.3	99.1	100.0	96.4	99.2
21, 22, 23	FIRE SUPPRESSION, PLUMBING & HVAC	93.1	92.5	92.8	97.4	83.2	91.7	95.7	126.3	108.0	97.4	96.0	96.8	95.1	93.4	94.4	100.9	96.0	98.9
26, 27, 3370	ELECTRICAL, COMMUNICATIONS & UTIL.	94.7	113.3	103.9	94.1	91.7	92.9	93.3	127.5	110.2	94.8	92.0	93.4	94.1	85.1	89.6	95.6	83.9	89.8
MF2018	WEIGHTED AVERAGE	96.6	101.9	98.9	96.2	89.3	93.2	96.1	123.2	107.8	94.9	96.8	95.7	92.8	94.2	93.4	96.2	94.6	95.5

DIVISION		PUERTO RICO SAN JUAN 009			RHODE ISLAND NEWPORT 028			PROVIDENCE 029			SOUTH CAROLINA AIKEN 298			BEAUFORT 299			CHARLESTON 294		
		MAT.	INST.	TOTAL	MAT.	INST.	TOTAL	MAT.	INST.	TOTAL	MAT.	INST.	TOTAL	MAT.	INST.	TOTAL	MAT.	INST.	TOTAL
015433	CONTRACTOR EQUIPMENT		83.7	83.7		96.9	96.9		102.4	102.4		101.8	101.8		101.8	101.8		101.8	101.8
0241, 31 - 34	SITE & INFRASTRUCTURE, DEMOLITION	129.7	83.1	97.7	86.0	95.7	92.7	89.3	104.5	99.8	133.3	83.3	99.0	128.2	83.3	97.4	112.3	83.5	92.5
0310	Concrete Forming & Accessories	86.6	21.0	30.8	101.8	120.8	118.0	101.6	121.0	118.1	97.7	62.5	67.8	96.3	38.1	46.8	95.2	62.9	67.7
0320	Concrete Reinforcing	179.7	18.0	101.7	105.1	121.3	112.9	99.2	121.3	109.9	94.4	57.8	76.8	93.5	63.6	79.1	93.4	64.2	79.3
0330	Cast-in-Place Concrete	97.1	31.1	72.2	68.7	116.8	86.9	94.2	118.3	103.3	89.6	65.3	80.4	89.5	65.3	80.4	105.1	65.6	90.2
03	CONCRETE	97.5	25.0	65.0	84.3	119.2	99.9	98.1	119.7	107.8	99.0	64.3	83.4	96.5	54.3	77.5	93.4	65.7	81.0
04	MASONRY	87.9	23.3	48.5	97.5	121.7	112.3	106.6	121.8	115.9	85.7	62.1	71.3	102.1	62.1	77.7	103.5	65.6	80.4
05	METALS	127.6	38.7	99.9	100.9	115.9	105.6	105.3	114.5	108.2	102.4	87.6	97.8	102.4	89.7	98.4	104.4	90.8	100.2
06	WOOD, PLASTICS & COMPOSITES	84.8	20.0	50.9	101.2	119.8	110.9	103.2	120.1	112.0	94.6	65.4	78.9	92.7	31.3	60.6	91.2	64.5	77.3
07	THERMAL & MOISTURE PROTECTION	136.9	26.4	89.5	105.0	117.2	110.2	109.2	118.1	113.1	96.7	61.5	81.6	96.3	58.5	80.1	95.3	63.5	81.7
08	OPENINGS	155.1	18.6	121.8	97.8	120.6	103.4	99.4	120.7	104.6	94.1	60.2	85.8	94.1	43.2	81.7	97.6	62.7	89.1
0920	Plaster & Gypsum Board	157.4	18.0	66.0	91.6	120.2	110.4	105.2	120.2	115.0	91.2	63.6	73.1	95.8	29.5	52.3	96.5	63.6	74.9
0950, 0980	Ceilings & Acoustic Treatment	226.3	18.0	95.8	84.2	120.2	106.8	96.9	120.2	111.5	80.1	63.6	69.8	85.8	29.5	50.5	85.8	63.6	71.9
0960	Flooring	212.2	28.4	158.5	88.7	125.9	99.5	89.1	125.9	99.8	101.5	85.7	96.9	103.0	72.3	94.1	102.7	80.0	96.1
0970, 0990	Wall Finishes & Painting/Coating	183.5	23.4	87.6	87.0	116.8	104.9	93.8	116.8	107.6	90.9	62.9	74.1	90.9	54.2	68.9	90.9	68.0	77.1
09	FINISHES	201.9	22.4	104.8	88.9	121.8	106.7	95.1	122.0	109.6	93.4	66.9	79.0	95.4	44.1	67.6	93.4	66.7	78.9
COVERS	DIVS. 10 - 14, 25, 28, 41, 43, 44, 46	99.3	25.1	81.8	100.0	107.8	101.8	100.0	108.3	101.9	100.0	83.3	96.1	100.0	76.1	94.4	100.0	83.3	96.1
21, 22, 23	FIRE SUPPRESSION, PLUMBING & HVAC	117.7	18.0	77.5	100.9	111.4	105.1	101.2	111.4	105.3	97.8	52.6	79.6	97.8	58.2	81.8	101.6	58.3	84.1
26, 27, 3370	ELECTRICAL, COMMUNICATIONS & UTIL.	117.5	24.1	71.3	101.3	97.6	99.4	102.1	97.6	99.8	95.2	59.6	77.6	98.3	63.9	81.3	96.9	56.8	77.1
MF2018	WEIGHTED AVERAGE	124.9	28.3	83.1	97.0	112.6	103.8	101.0	113.3	106.3	98.1	65.2	83.9	98.9	61.5	82.7	99.4	67.0	85.4

DIVISION		SOUTH CAROLINA COLUMBIA 290 - 292			FLORENCE 295			GREENVILLE 296			ROCK HILL 297			SPARTANBURG 293			SOUTH DAKOTA ABERDEEN 574		
		MAT.	INST.	TOTAL	MAT.	INST.	TOTAL	MAT.	INST.	TOTAL	MAT.	INST.	TOTAL	MAT.	INST.	TOTAL	MAT.	INST.	TOTAL
015433	CONTRACTOR EQUIPMENT		105.4	105.4		101.8	101.8		101.8	101.8		101.8	101.8		101.8	101.8		95.8	95.8
0241, 31 - 34	SITE & INFRASTRUCTURE, DEMOLITION	112.3	92.3	98.6	122.4	83.2	95.5	117.2	83.5	94.0	114.1	83.2	92.9	117.0	83.5	94.0	100.1	90.9	93.8
0310	Concrete Forming & Accessories	98.1	62.9	68.1	81.5	62.9	65.6	95.0	63.0	67.7	92.9	62.2	66.8	98.3	63.0	68.2	100.0	66.5	71.5
0320	Concrete Reinforcing	95.8	64.2	80.6	93.0	64.2	79.1	92.9	61.5	77.7	93.7	62.7	78.8	92.9	64.2	79.1	92.9	70.1	81.9
0330	Cast-in-Place Concrete	106.4	66.0	91.2	89.5	65.3	80.4	89.5	65.6	80.5	89.5	64.8	80.2	89.5	65.6	80.5	111.2	75.5	97.7
03	CONCRETE	93.4	65.7	81.0	91.4	65.5	79.8	90.4	65.3	79.2	88.6	64.9	78.0	90.6	65.8	79.5	100.0	71.2	87.1
04	MASONRY	95.8	65.6	77.4	85.8	65.5	73.4	83.5	65.6	72.5	109.3	61.3	80.0	103.8	65.6	73.5	118.1	69.8	88.6
05	METALS	101.5	88.3	97.4	103.2	90.0	99.1	103.2	90.5	99.3	102.4	89.3	98.3	103.2	90.9	99.4	91.7	81.5	88.6
06	WOOD, PLASTICS & COMPOSITES	95.4	64.5	79.3	75.9	64.5	69.9	91.1	64.5	77.2	89.2	64.5	76.3	95.5	64.5	79.3	98.9	62.7	79.9
07	THERMAL & MOISTURE PROTECTION	91.1	64.3	79.6	95.6	63.5	81.9	95.6	63.5	81.8	95.5	56.7	78.8	95.7	63.5	81.9	102.9	73.5	90.3
08	OPENINGS	99.8	62.7	90.7	94.2	62.7	86.5	94.1	62.4	86.4	94.1	61.3	86.1	94.1	62.7	86.5	96.9	55.4	86.8
0920	Plaster & Gypsum Board	94.4	63.6	74.2	85.3	63.6	71.1	89.8	63.6	72.6	89.2	63.6	72.4	92.0	63.6	73.4	99.6	62.2	75.1
0950, 0980	Ceilings & Acoustic Treatment	88.5	63.6	72.9	81.4	63.6	70.2	80.1	63.6	69.8	80.1	63.6	69.8	80.1	63.6	69.8	103.6	62.2	77.7
0960	Flooring	92.8	80.0	89.1	93.7	80.0	89.7	100.3	80.0	94.4	99.2	72.3	91.3	101.7	80.0	95.4	93.9	41.8	78.7
0970, 0990	Wall Finishes & Painting/Coating	92.4	68.0	77.7	90.9	68.0	77.1	90.9	68.0	77.1	90.9	62.9	74.1	90.9	68.0	77.1	83.5	31.8	52.5
09	FINISHES	92.2	66.7	78.4	89.4	66.7	77.1	91.2	66.7	77.9	90.5	64.4	76.4	91.9	66.7	78.3	94.1	59.6	75.4
COVERS	DIVS. 10 - 14, 25, 28, 41, 43, 44, 46	100.0	83.3	96.1	100.0	83.3	96.1	100.0	83.4	96.1	100.0	83.1	96.0	100.0	83.4	96.1	100.0	86.0	96.7
21, 22, 23	FIRE SUPPRESSION, PLUMBING & HVAC	101.0	56.5	83.0	101.6	56.5	83.4	101.6	55.5	83.0	97.8	52.3	79.4	101.6	55.5	83.0	100.6	51.4	80.8
26, 27, 3370	ELECTRICAL, COMMUNICATIONS & UTIL.	100.3	60.4	80.5	95.2	60.4	78.0	97.0	83.5	90.3	97.0	83.5	90.3	97.0	83.5	90.3	99.8	64.8	82.4
MF2018	WEIGHTED AVERAGE	98.8	67.6	85.3	97.4	67.0	84.2	97.5	70.0	85.6	97.3	68.2	84.7	97.7	70.2	85.8	99.1	66.7	85.1

SOUTH DAKOTA

DIVISION		MITCHELL 573 MAT.	INST.	TOTAL	MOBRIDGE 576 MAT.	INST.	TOTAL	PIERRE 575 MAT.	INST.	TOTAL	RAPID CITY 577 MAT.	INST.	TOTAL	SIOUX FALLS 570-571 MAT.	INST.	TOTAL	WATERTOWN 572 MAT.	INST.	TOTAL
015433	CONTRACTOR EQUIPMENT		95.8	95.8		95.8	95.8		100.9	100.9		95.8	95.8		102.0	102.0		95.8	95.8
0241, 31 - 34	SITE & INFRASTRUCTURE, DEMOLITION	96.3	90.2	92.1	96.3	90.2	92.1	99.5	98.4	98.7	98.4	90.5	92.9	93.7	101.1	98.8	96.2	90.2	92.1
0310	Concrete Forming & Accessories	98.9	42.8	51.2	87.8	43.2	49.9	99.0	46.2	54.1	108.3	55.8	63.7	101.8	84.8	87.4	83.7	72.1	73.8
0320	Concrete Reinforcing	92.4	67.3	80.3	94.8	67.2	81.5	94.8	97.5	96.1	86.9	97.6	92.0	97.3	99.8	98.5	89.8	67.2	78.9
0330	Cast-in-Place Concrete	108.0	50.1	86.2	108.0	74.2	95.3	112.8	77.8	99.6	107.2	75.4	95.2	88.9	82.6	86.5	108.0	74.1	95.2
03	CONCRETE	97.7	51.2	76.8	97.3	59.7	80.4	102.3	67.3	86.6	97.2	70.9	85.4	94.3	86.9	91.0	96.4	72.7	85.8
04	MASONRY	104.5	74.2	86.1	112.4	69.8	86.4	105.3	74.4	86.5	111.7	72.7	87.9	106.1	79.2	89.7	139.5	70.6	97.5
05	METALS	90.7	80.4	87.5	90.8	80.7	87.6	94.2	88.6	92.4	93.4	90.7	92.6	94.8	93.0	94.3	90.7	80.6	87.6
06	WOOD, PLASTICS & COMPOSITES	97.7	32.5	63.6	84.4	32.2	57.1	103.8	33.0	66.8	104.4	46.3	74.0	98.4	84.0	90.9	79.7	71.7	75.5
07	THERMAL & MOISTURE PROTECTION	102.7	70.1	88.7	102.6	71.7	89.4	103.0	72.7	90.0	103.3	77.2	92.1	103.3	86.1	95.9	102.4	73.2	89.8
08	OPENINGS	96.1	38.2	82.0	98.6	37.5	83.7	99.5	62.4	90.5	100.6	69.8	93.1	103.8	90.9	100.6	96.1	59.2	87.1
0920	Plaster & Gypsum Board	98.5	31.1	54.3	92.8	30.9	52.2	95.0	31.4	53.3	99.4	45.3	63.9	91.5	83.8	86.5	90.8	71.4	78.1
0950, 0980	Ceilings & Acoustic Treatment	100.4	31.1	57.0	103.6	30.9	58.0	109.6	31.4	60.6	104.9	45.3	67.6	105.6	83.8	91.9	100.4	71.4	82.3
0960	Flooring	93.5	74.8	88.1	89.0	44.3	76.0	99.9	29.8	79.4	93.2	69.2	86.2	95.8	77.2	90.3	87.6	41.8	74.2
0970, 0990	Wall Finishes & Painting/Coating	83.5	34.8	54.3	83.5	35.7	54.8	93.6	118.9	108.8	83.5	118.9	104.7	96.5	118.9	109.9	83.5	31.8	52.5
09	FINISHES	92.9	46.2	67.6	91.6	42.0	64.8	99.8	50.0	72.8	94.1	64.2	77.9	97.4	87.0	91.8	90.1	64.3	76.1
COVERS	DIVS. 10 - 14, 25, 28, 41, 43, 44, 46	100.0	77.8	94.8	100.0	77.7	94.7	100.0	85.8	96.7	100.0	86.1	96.7	100.0	91.4	98.0	100.0	81.4	95.6
21, 22, 23	FIRE SUPPRESSION, PLUMBING & HVAC	96.8	46.7	76.6	96.8	65.8	84.3	100.6	75.1	90.3	100.6	74.2	90.0	100.5	72.2	89.1	96.8	49.6	77.8
26, 27, 3370	ELECTRICAL, COMMUNICATIONS & UTIL.	98.3	61.0	79.9	99.8	38.3	69.3	102.9	45.9	74.7	96.7	45.9	71.6	102.6	66.7	84.8	97.6	61.0	79.5
MF2018	WEIGHTED AVERAGE	96.6	59.6	80.6	97.1	60.7	81.4	100.2	68.9	86.7	98.7	71.0	86.7	99.3	82.1	91.8	97.7	66.6	84.3

TENNESSEE

DIVISION		CHATTANOOGA 373-374 MAT.	INST.	TOTAL	COLUMBIA 384 MAT.	INST.	TOTAL	COOKEVILLE 385 MAT.	INST.	TOTAL	JACKSON 383 MAT.	INST.	TOTAL	JOHNSON CITY 376 MAT.	INST.	TOTAL	KNOXVILLE 377-379 MAT.	INST.	TOTAL
015433	CONTRACTOR EQUIPMENT		103.4	103.4		98.2	98.2		98.2	98.2		104.2	104.2		97.4	97.4		97.4	97.4
0241, 31 - 34	SITE & INFRASTRUCTURE, DEMOLITION	106.1	91.8	96.3	89.8	82.3	84.6	95.5	79.0	84.2	99.4	91.6	94.0	112.8	81.3	91.2	91.9	82.1	85.2
0310	Concrete Forming & Accessories	101.1	62.0	67.9	80.5	60.3	63.3	80.7	31.2	38.6	87.8	39.7	46.9	85.3	59.8	63.6	99.9	63.1	68.6
0320	Concrete Reinforcing	96.3	68.3	82.8	87.4	61.9	75.1	87.4	61.6	75.0	87.4	69.2	78.6	96.9	64.3	81.2	96.3	67.8	82.5
0330	Cast-in-Place Concrete	98.2	63.8	85.2	89.4	63.2	79.5	101.3	56.2	84.3	98.9	57.7	83.3	79.0	58.0	71.1	92.3	65.1	82.0
03	CONCRETE	95.3	65.4	81.9	91.9	63.3	79.1	101.8	47.6	77.5	92.6	53.3	75.0	104.8	61.7	85.5	92.8	66.2	80.9
04	MASONRY	100.5	56.9	73.9	115.5	53.2	77.5	110.6	37.8	66.2	115.8	43.8	71.9	115.3	43.9	71.7	78.5	56.3	65.0
05	METALS	94.4	90.6	93.2	93.2	87.6	91.5	93.3	86.6	91.2	95.8	90.3	94.1	91.7	88.7	90.8	94.8	90.0	93.3
06	WOOD, PLASTICS & COMPOSITES	108.9	62.7	84.7	67.8	61.2	64.4	68.0	28.4	47.3	82.4	38.3	59.3	80.3	65.3	72.4	97.4	63.3	79.5
07	THERMAL & MOISTURE PROTECTION	100.7	62.2	84.2	95.5	60.3	80.4	96.0	50.5	76.5	97.7	52.4	78.2	95.9	57.1	79.3	93.8	62.4	80.4
08	OPENINGS	101.7	61.3	91.8	91.0	51.3	81.3	91.0	34.2	77.1	97.4	43.8	84.3	98.0	62.1	89.3	95.5	56.6	86.0
0920	Plaster & Gypsum Board	81.2	62.3	68.8	88.2	60.8	70.2	88.2	27.1	48.1	90.2	37.2	55.4	101.5	65.0	77.5	108.9	62.9	78.7
0950, 0980	Ceilings & Acoustic Treatment	99.0	62.3	76.0	80.9	60.8	68.3	80.9	27.1	47.1	88.7	37.2	56.4	94.8	65.0	76.1	96.1	62.9	75.3
0960	Flooring	98.9	61.4	87.9	83.3	51.5	74.0	83.4	47.7	73.0	82.7	56.1	75.0	92.8	56.1	82.1	98.3	56.1	86.0
0970, 0990	Wall Finishes & Painting/Coating	92.3	56.7	71.0	81.9	53.8	65.0	81.9	53.8	65.0	84.0	60.6	70.0	89.1	56.7	69.7	89.1	56.7	69.7
09	FINISHES	94.8	61.0	76.5	87.8	57.7	71.5	88.3	34.7	59.3	88.1	43.5	64.0	98.2	58.9	76.9	92.3	60.9	75.3
COVERS	DIVS. 10 - 14, 25, 28, 41, 43, 44, 46	100.0	81.3	95.6	100.0	81.0	95.5	100.0	73.5	93.8	100.0	75.0	94.1	100.0	78.1	94.9	100.0	82.1	95.8
21, 22, 23	FIRE SUPPRESSION, PLUMBING & HVAC	101.1	56.7	83.2	98.3	70.2	87.0	98.3	63.9	84.5	101.0	58.6	83.9	100.8	53.0	81.5	100.8	62.9	85.5
26, 27, 3370	ELECTRICAL, COMMUNICATIONS & UTIL.	101.8	79.8	90.9	94.9	48.7	72.0	96.3	59.0	77.8	101.1	63.3	82.4	92.9	50.5	72.0	97.9	57.7	78.0
MF2018	WEIGHTED AVERAGE	99.0	68.7	85.9	95.2	64.6	82.0	96.6	56.1	79.1	98.1	60.0	81.6	99.2	60.5	82.5	95.7	66.0	82.9

TENNESSEE / TEXAS

DIVISION		MCKENZIE 382 MAT.	INST.	TOTAL	MEMPHIS 375,380-381 MAT.	INST.	TOTAL	NASHVILLE 370-372 MAT.	INST.	TOTAL	ABILENE 795-796 MAT.	INST.	TOTAL	AMARILLO 790-791 MAT.	INST.	TOTAL	AUSTIN 786-787 MAT.	INST.	TOTAL
015433	CONTRACTOR EQUIPMENT		98.2	98.2		99.2	99.2		106.9	106.9		88.0	88.0		94.3	94.3		92.5	92.5
0241, 31 - 34	SITE & INFRASTRUCTURE, DEMOLITION	95.1	79.2	84.2	90.0	91.3	90.9	103.9	99.0	100.5	93.8	84.9	87.7	93.9	94.4	94.2	97.0	91.0	92.9
0310	Concrete Forming & Accessories	89.4	33.4	41.8	98.0	72.1	76.0	102.1	68.6	73.6	99.0	60.4	66.2	99.3	54.0	60.8	98.7	54.8	61.3
0320	Concrete Reinforcing	87.6	62.5	75.5	104.0	71.5	88.3	102.4	69.0	86.3	97.1	50.5	74.6	97.8	50.4	74.9	92.2	45.2	69.5
0330	Cast-in-Place Concrete	99.1	57.3	83.3	93.9	69.8	84.8	91.2	70.5	83.4	86.0	66.6	78.7	83.9	67.6	77.8	93.6	67.1	83.6
03	CONCRETE	100.6	49.2	77.5	96.7	72.0	85.6	94.5	70.4	83.7	86.5	61.6	75.3	93.3	59.0	77.9	89.2	58.2	75.3
04	MASONRY	114.4	43.2	71.0	93.6	60.0	73.1	92.1	63.9	74.9	101.3	63.3	78.1	98.4	61.7	76.0	98.8	63.3	77.1
05	METALS	93.3	87.2	91.4	84.5	83.1	84.1	107.0	85.7	100.4	108.3	70.2	96.4	103.5	68.6	92.6	102.5	65.4	91.0
06	WOOD, PLASTICS & COMPOSITES	77.8	30.6	53.1	100.2	73.0	86.0	105.3	67.9	85.7	103.7	62.1	81.9	107.1	53.2	78.9	96.9	54.8	74.9
07	THERMAL & MOISTURE PROTECTION	96.1	51.4	76.9	93.2	67.3	82.1	99.2	68.9	86.2	100.4	64.7	85.1	102.3	63.8	85.8	96.3	65.1	82.9
08	OPENINGS	91.0	35.8	77.5	99.2	66.9	91.4	100.8	66.5	92.5	101.7	57.9	91.0	103.1	53.0	90.9	101.3	49.9	88.8
0920	Plaster & Gypsum Board	91.0	29.3	50.5	95.7	72.4	80.4	90.3	66.9	75.0	101.7	61.6	75.4	113.8	52.3	73.4	88.4	53.9	65.8
0950, 0980	Ceilings & Acoustic Treatment	80.9	29.3	48.6	101.5	72.4	83.2	89.3	66.9	75.3	97.3	61.6	74.9	95.4	52.3	68.4	99.4	53.9	70.9
0960	Flooring	86.2	51.5	76.1	98.1	56.1	85.8	96.8	64.1	87.2	96.5	71.2	89.1	96.3	67.8	88.0	94.5	60.9	84.7
0970, 0990	Wall Finishes & Painting/Coating	81.9	42.0	58.0	90.6	74.8	81.1	97.7	67.9	79.9	91.5	57.0	70.8	94.5	54.3	70.4	94.3	42.1	63.0
09	FINISHES	89.4	36.0	60.5	97.5	69.4	82.3	95.3	67.3	80.1	92.2	62.1	75.9	97.7	56.1	75.2	94.8	54.1	72.8
COVERS	DIVS. 10 - 14, 25, 28, 41, 43, 44, 46	100.0	74.3	94.0	100.0	85.3	96.5	100.0	85.1	96.5	100.0	79.6	95.2	100.0	79.4	95.1	100.0	79.1	95.1
21, 22, 23	FIRE SUPPRESSION, PLUMBING & HVAC	98.3	55.8	81.2	100.7	74.2	90.0	101.1	82.1	93.4	101.2	50.4	80.7	100.9	52.4	81.4	100.8	59.6	84.2
26, 27, 3370	ELECTRICAL, COMMUNICATIONS & UTIL.	96.1	53.3	74.9	101.6	64.8	83.4	99.6	62.0	81.0	97.6	55.2	76.6	100.3	60.6	80.6	99.4	59.4	79.6
MF2018	WEIGHTED AVERAGE	96.7	54.7	78.6	96.5	72.5	86.1	100.0	74.5	89.0	99.0	61.6	82.9	99.9	61.8	83.5	98.6	62.3	82.9

For customer support on your Site Work & Landscape Costs with RSMeans Data, call 800.448.8182.

TEXAS

DIVISION		BEAUMONT 776-777			BROWNWOOD 768			BRYAN 778			CHILDRESS 792			CORPUS CHRISTI 783-784			DALLAS 752-753		
		MAT.	INST.	TOTAL	MAT.	INST.	TOTAL	MAT.	INST.	TOTAL	MAT.	INST.	TOTAL	MAT.	INST.	TOTAL	MAT.	INST.	TOTAL
015433	CONTRACTOR EQUIPMENT		92.1	92.1		88.0	88.0		92.1	92.1		88.0	88.0		97.9	97.9		107.6	107.6
0241, 31-34	SITE & INFRASTRUCTURE, DEMOLITION	90.8	88.8	89.4	102.1	84.8	90.3	81.3	88.6	86.3	104.0	84.9	90.8	141.2	79.9	99.1	108.1	99.1	101.9
0310	Concrete Forming & Accessories	102.4	57.2	63.9	97.6	58.5	64.4	80.7	56.1	59.8	97.5	58.7	64.5	106.9	54.1	62.0	96.4	62.9	67.9
0320	Concrete Reinforcing	93.9	59.3	77.2	84.6	48.7	67.3	96.3	48.7	73.3	97.3	48.7	73.8	82.4	50.1	66.8	103.3	51.8	78.5
0330	Cast-in-Place Concrete	98.6	63.2	85.2	107.8	56.3	88.3	79.0	56.8	70.6	88.2	56.4	76.2	114.2	67.3	96.5	100.1	69.8	88.7
03	CONCRETE	95.5	60.6	79.8	95.6	56.9	78.3	79.7	56.1	69.1	95.0	57.0	77.9	96.2	59.9	79.9	97.2	64.7	82.6
04	MASONRY	99.2	63.9	77.7	126.7	58.2	84.9	139.3	58.1	89.8	104.9	57.8	76.1	81.5	61.7	69.4	102.7	60.0	76.6
05	METALS	101.1	74.6	92.8	103.1	68.7	92.3	101.0	69.9	91.3	105.5	68.8	94.0	98.8	83.9	94.1	103.4	80.7	96.3
06	WOOD, PLASTICS & COMPOSITES	111.4	57.3	83.1	102.8	60.9	80.9	76.5	57.2	66.4	103.0	60.9	81.0	121.9	53.6	86.1	97.9	64.6	80.5
07	THERMAL & MOISTURE PROTECTION	92.8	64.7	80.7	92.9	61.1	79.2	86.2	62.1	75.8	101.0	61.0	83.8	99.3	63.1	83.8	90.4	67.8	80.7
08	OPENINGS	94.6	55.8	85.1	100.6	55.7	89.6	95.4	53.9	85.3	96.1	55.7	86.3	101.5	51.5	89.3	96.6	60.0	87.7
0920	Plaster & Gypsum Board	106.6	56.8	73.9	96.9	60.5	73.0	91.6	56.7	68.7	100.7	60.5	74.3	90.6	52.8	65.8	93.3	63.6	73.8
0950, 0980	Ceilings & Acoustic Treatment	116.3	56.8	79.0	92.5	60.5	72.4	105.7	56.7	75.0	93.5	60.5	72.8	103.2	52.8	71.6	109.9	63.6	80.9
0960	Flooring	120.5	77.1	107.8	76.0	52.1	69.0	89.7	63.9	82.1	94.7	52.1	82.3	111.3	73.7	100.3	97.7	69.8	89.5
0970, 0990	Wall Finishes & Painting/Coating	95.9	51.3	69.2	89.5	49.2	65.3	93.2	54.7	70.1	91.5	49.2	66.1	106.8	45.6	70.1	102.3	54.3	73.5
09	FINISHES	104.1	60.0	80.3	84.0	56.4	69.1	90.6	56.9	72.4	91.9	56.4	72.7	105.7	56.4	79.0	101.5	63.2	80.8
COVERS	DIVS. 10-14, 25, 28, 41, 43, 44, 46	100.0	80.3	95.4	100.0	79.1	95.1	100.0	79.8	95.2	100.0	79.1	95.1	100.0	78.4	94.9	100.0	81.3	95.6
21, 22, 23	FIRE SUPPRESSION, PLUMBING & HVAC	101.0	60.8	84.8	97.0	47.3	77.0	97.2	59.5	82.0	97.4	51.6	78.9	101.0	61.4	84.8	101.0	62.1	85.3
26, 27, 3370	ELECTRICAL, COMMUNICATIONS & UTIL.	99.6	61.0	80.5	99.4	44.4	72.2	98.0	61.0	79.7	97.6	53.7	75.9	96.7	66.1	81.6	98.2	60.5	79.5
MF2018	WEIGHTED AVERAGE	99.2	64.9	84.4	99.1	57.2	81.0	96.3	62.4	81.7	98.7	59.4	81.7	100.2	64.6	84.8	100.1	67.4	86.0

TEXAS

DIVISION		DEL RIO 788			DENTON 762			EASTLAND 764			EL PASO 798-799,885			FORT WORTH 760-761			GALVESTON 775		
		MAT.	INST.	TOTAL	MAT.	INST.	TOTAL	MAT.	INST.	TOTAL	MAT.	INST.	TOTAL	MAT.	INST.	TOTAL	MAT.	INST.	TOTAL
015433	CONTRACTOR EQUIPMENT		87.2	87.2		97.4	97.4		88.0	88.0		94.3	94.3		94.3	94.3		104.9	104.9
0241, 31-34	SITE & INFRASTRUCTURE, DEMOLITION	120.3	82.8	94.6	101.8	79.3	86.3	104.9	84.8	91.1	89.7	94.0	92.7	97.7	94.4	95.4	108.5	87.3	93.9
0310	Concrete Forming & Accessories	102.5	49.3	57.2	106.3	58.9	65.9	98.7	58.6	64.6	100.5	57.8	64.2	97.1	60.7	66.1	89.8	53.8	59.2
0320	Concrete Reinforcing	83.0	45.3	64.8	86.1	48.9	68.1	84.9	48.7	67.4	99.2	50.5	75.7	93.6	51.4	73.2	95.7	56.9	77.0
0330	Cast-in-Place Concrete	122.9	55.8	97.6	83.3	57.5	73.5	113.9	56.4	92.2	76.1	66.4	72.4	96.9	67.4	85.7	104.9	57.8	87.1
03	CONCRETE	115.8	51.8	87.1	75.3	58.5	67.8	100.3	57.0	80.9	90.0	60.2	76.6	89.5	62.1	77.2	96.9	57.8	79.3
04	MASONRY	95.8	58.0	72.8	135.3	60.1	89.4	95.1	58.2	72.6	96.8	63.3	76.3	93.8	61.2	73.9	96.0	58.2	72.9
05	METALS	98.6	65.4	88.2	102.6	84.1	96.9	102.9	68.7	92.2	103.7	67.8	92.5	105.4	69.1	94.1	102.6	89.1	98.4
06	WOOD, PLASTICS & COMPOSITES	103.1	48.6	74.6	115.7	61.0	87.1	109.5	60.9	84.1	94.1	58.4	75.4	96.5	62.3	78.6	92.5	53.9	72.3
07	THERMAL & MOISTURE PROTECTION	96.4	60.4	81.0	91.4	62.4	79.0	93.2	61.1	79.4	94.6	64.6	81.7	90.9	65.1	79.8	85.6	62.4	75.7
08	OPENINGS	96.7	46.0	84.4	118.8	55.4	103.4	70.6	55.7	67.0	96.3	53.7	85.9	101.0	58.2	90.6	99.4	53.9	88.3
0920	Plaster & Gypsum Board	86.9	47.8	61.3	101.8	60.5	74.7	96.9	60.5	73.0	112.0	57.6	76.3	106.5	61.6	77.1	99.1	53.2	69.0
0950, 0980	Ceilings & Acoustic Treatment	98.6	47.8	66.8	96.3	60.5	73.8	92.5	60.5	72.4	92.4	57.6	70.6	100.9	61.6	76.3	110.8	53.2	74.7
0960	Flooring	95.7	52.1	83.0	72.0	52.1	66.2	95.3	52.1	82.7	97.6	70.6	89.7	95.3	61.5	85.5	106.0	63.9	93.7
0970, 0990	Wall Finishes & Painting/Coating	95.1	40.5	62.4	100.1	46.5	68.0	90.8	49.2	65.9	98.2	49.2	68.8	100.8	54.4	73.0	106.0	51.1	73.1
09	FINISHES	98.2	48.2	71.2	82.8	56.2	68.4	89.9	56.4	71.8	96.8	59.1	76.4	97.6	60.0	77.3	101.0	54.6	75.9
COVERS	DIVS. 10-14, 25, 28, 41, 43, 44, 46	100.0	77.2	94.6	100.0	79.4	95.1	100.0	79.1	95.1	100.0	78.9	95.0	100.0	80.2	95.3	100.0	79.7	95.2
21, 22, 23	FIRE SUPPRESSION, PLUMBING & HVAC	97.1	56.6	80.8	97.0	53.3	79.4	97.0	47.3	77.0	100.9	63.7	85.9	100.8	56.8	83.1	97.1	59.6	82.0
26, 27, 3370	ELECTRICAL, COMMUNICATIONS & UTIL.	98.6	57.6	78.3	101.7	56.5	79.3	99.3	56.4	78.1	100.4	49.6	75.3	100.8	59.8	80.5	99.5	61.1	80.5
MF2018	WEIGHTED AVERAGE	100.7	58.2	82.3	98.9	61.6	82.8	95.7	58.9	79.8	98.3	63.3	83.2	99.0	63.9	83.9	98.8	64.0	83.8

TEXAS

DIVISION		GIDDINGS 789			GREENVILLE 754			HOUSTON 770-772			HUNTSVILLE 773			LAREDO 780			LONGVIEW 756		
		MAT.	INST.	TOTAL	MAT.	INST.	TOTAL	MAT.	INST.	TOTAL	MAT.	INST.	TOTAL	MAT.	INST.	TOTAL	MAT.	INST.	TOTAL
015433	CONTRACTOR EQUIPMENT		87.2	87.2		97.5	97.5		102.2	102.2		92.1	92.1		87.2	87.2		88.8	88.8
0241, 31-34	SITE & INFRASTRUCTURE, DEMOLITION	105.5	83.1	90.1	97.4	81.8	86.7	107.7	94.9	98.9	97.2	88.6	91.3	99.5	83.3	88.3	95.1	87.7	90.0
0310	Concrete Forming & Accessories	100.2	49.4	57.0	87.0	55.5	60.2	94.8	59.5	64.8	88.7	52.2	57.7	102.5	53.7	61.0	82.4	58.7	62.2
0320	Concrete Reinforcing	83.7	46.2	65.6	103.8	48.9	77.3	95.5	50.6	73.8	96.5	48.7	73.5	83.0	50.1	67.1	102.7	48.4	76.5
0330	Cast-in-Place Concrete	104.1	56.1	86.0	102.5	57.4	85.5	97.5	68.6	86.6	108.7	56.7	89.1	87.8	62.8	78.4	119.3	56.3	95.5
03	CONCRETE	93.4	52.2	74.9	93.2	56.8	76.9	94.9	62.2	80.2	103.4	54.3	81.3	87.8	57.2	74.1	109.5	56.8	85.9
04	MASONRY	103.2	58.1	75.7	163.6	60.1	100.5	94.2	63.7	75.6	136.9	58.1	88.8	89.3	62.4	72.9	159.1	56.8	96.7
05	METALS	98.0	66.4	88.2	100.8	82.5	95.1	105.6	75.0	96.1	100.9	69.8	91.2	101.2	68.5	91.0	93.6	67.4	85.4
06	WOOD, PLASTICS & COMPOSITES	102.4	48.6	74.3	84.2	56.4	69.7	98.0	59.8	78.0	86.5	52.1	68.5	103.1	53.5	77.1	77.8	61.0	69.0
07	THERMAL & MOISTURE PROTECTION	96.9	61.1	81.5	89.4	61.3	77.3	87.8	68.4	79.5	87.3	61.5	76.2	95.7	63.2	81.7	90.6	60.2	77.6
08	OPENINGS	95.9	46.8	83.9	92.3	53.2	82.8	106.3	56.9	94.3	95.4	50.5	84.4	96.9	50.4	85.5	83.5	55.3	76.6
0920	Plaster & Gypsum Board	86.1	47.8	61.0	79.3	55.6	63.7	100.1	58.7	72.9	95.5	51.5	66.6	103.1	52.8	64.9	77.5	60.5	66.3
0950, 0980	Ceilings & Acoustic Treatment	98.6	47.8	66.8	103.5	55.6	73.4	103.7	58.7	75.5	105.7	51.5	71.7	103.0	52.8	71.6	96.3	60.5	73.8
0960	Flooring	95.4	52.1	82.8	92.3	52.1	80.5	105.4	75.2	96.6	94.3	52.1	82.0	95.6	60.9	85.5	98.2	52.1	84.7
0970, 0990	Wall Finishes & Painting/Coating	95.1	42.1	63.3	93.4	49.2	66.9	106.5	58.0	77.4	93.2	54.7	70.1	95.1	42.1	63.3	84.8	46.5	61.9
09	FINISHES	96.9	48.4	70.6	95.3	53.8	72.9	105.6	62.0	82.0	93.3	51.9	70.9	97.7	53.3	73.7	99.7	56.2	76.1
COVERS	DIVS. 10-14, 25, 28, 41, 43, 44, 46	100.0	77.5	94.7	100.0	79.0	95.1	100.0	82.1	95.8	100.0	79.2	95.1	100.0	78.0	94.8	100.0	75.1	94.1
21, 22, 23	FIRE SUPPRESSION, PLUMBING & HVAC	97.2	58.5	81.6	97.3	55.4	80.4	101.0	63.8	85.9	97.2	59.5	82.0	100.9	56.8	83.1	97.2	55.0	80.2
26, 27, 3370	ELECTRICAL, COMMUNICATIONS & UTIL.	95.9	53.5	74.9	95.4	56.5	76.1	102.0	65.1	83.7	98.0	56.9	77.7	98.7	59.1	79.1	95.5	49.0	72.5
MF2018	WEIGHTED AVERAGE	97.3	58.3	80.5	99.6	61.3	83.1	101.3	67.3	86.6	99.9	60.7	83.0	97.6	60.9	81.8	99.8	59.2	82.3

TEXAS

DIVISION		LUBBOCK 793 - 794			LUFKIN 759			MCALLEN 785			MCKINNEY 750			MIDLAND 797			ODESSA 797		
		MAT.	INST.	TOTAL	MAT.	INST.	TOTAL	MAT.	INST.	TOTAL	MAT.	INST.	TOTAL	MAT.	INST.	TOTAL	MAT.	INST.	TOTAL
015433	CONTRACTOR EQUIPMENT		100.1	100.1		88.8	88.8		98.0	98.0		97.5	97.5		100.1	100.1		88.0	88.0
0241, 31 - 34	SITE & INFRASTRUCTURE, DEMOLITION	118.4	83.8	94.7	90.0	87.1	88.0	145.4	80.0	100.5	93.8	81.8	85.5	120.9	83.8	95.4	94.0	85.0	87.8
0310	Concrete Forming & Accessories	99.3	53.1	60.0	85.7	52.4	57.4	106.9	49.9	58.4	86.0	55.5	60.1	103.8	60.4	66.9	99.0	60.4	66.2
0320	Concrete Reinforcing	98.3	50.4	75.2	104.6	63.6	84.8	82.6	50.0	66.9	103.8	48.9	77.3	99.4	50.2	75.7	97.1	50.3	74.5
0330	Cast-in-Place Concrete	86.2	66.7	78.8	106.7	55.7	87.4	124.0	57.8	99.0	96.1	57.4	81.5	91.6	64.3	81.3	86.0	65.7	78.4
03	CONCRETE	85.3	59.3	73.6	101.4	56.3	81.2	103.7	54.6	81.7	88.3	56.8	74.2	89.4	61.7	77.0	86.5	61.3	75.2
04	MASONRY	100.6	61.6	76.9	120.5	58.0	82.3	96.1	61.7	75.1	176.4	60.1	105.4	117.8	59.2	82.1	101.3	61.6	77.1
05	METALS	112.0	84.9	103.5	100.9	71.4	91.7	98.5	83.2	93.7	100.7	82.5	95.0	110.1	84.7	102.1	107.5	69.9	95.8
06	WOOD, PLASTICS & COMPOSITES	104.8	52.1	77.2	84.8	52.2	67.8	120.0	48.7	82.8	83.0	56.4	69.1	110.2	62.2	85.1	103.7	62.1	81.9
07	THERMAL & MOISTURE PROTECTION	90.3	63.4	78.7	90.4	59.6	77.2	99.2	62.4	83.4	89.2	61.3	77.2	90.5	63.4	78.9	100.4	63.6	84.6
08	OPENINGS	108.7	52.0	94.9	63.4	54.4	61.2	100.6	47.1	87.6	92.3	53.2	82.8	110.2	57.8	97.4	101.7	57.7	91.0
0920	Plaster & Gypsum Board	101.9	51.3	68.7	76.1	51.6	60.0	91.5	47.8	62.9	79.3	55.6	63.7	103.0	61.6	75.9	101.7	61.6	75.4
0950, 0980	Ceilings & Acoustic Treatment	97.9	51.3	68.7	91.2	51.5	66.3	103.0	47.8	68.4	103.3	55.6	73.4	94.7	61.6	74.0	97.3	61.6	74.9
0960	Flooring	90.1	67.7	83.6	131.5	52.1	108.3	110.8	73.7	100.0	91.7	52.1	80.2	91.3	60.9	82.4	96.5	66.5	87.7
0970, 0990	Wall Finishes & Painting/Coating	102.4	52.2	72.3	84.8	49.2	63.5	106.8	40.5	67.1	93.4	49.2	66.9	102.4	49.2	70.5	91.5	52.2	67.9
09	FINISHES	94.7	55.2	73.3	107.5	51.7	77.3	106.1	53.0	77.4	95.0	53.8	72.7	94.8	59.3	75.6	92.2	60.7	75.2
COVERS	DIVS. 10 - 14, 25, 28, 41, 43, 44, 46	100.0	78.9	95.0	100.0	79.5	95.2	100.0	77.7	94.7	100.0	79.0	95.1	100.0	80.0	95.3	100.0	79.7	95.2
21, 22, 23	FIRE SUPPRESSION, PLUMBING & HVAC	100.5	50.6	80.4	97.2	55.5	80.4	97.2	47.1	77.0	97.3	57.5	81.2	96.7	50.6	78.1	101.2	50.6	80.8
26, 27, 3370	ELECTRICAL, COMMUNICATIONS & UTIL.	96.4	52.2	74.5	96.7	59.1	78.1	96.5	30.4	63.8	95.5	56.5	76.2	96.4	52.6	74.7	97.7	52.5	75.3
MF2018	WEIGHTED AVERAGE	100.4	60.7	83.3	96.6	60.5	81.0	100.9	55.2	81.2	99.5	61.8	83.2	100.9	61.7	84.0	98.9	60.8	82.5

TEXAS

DIVISION		PALESTINE 758			SAN ANGELO 769			SAN ANTONIO 781 - 782			TEMPLE 765			TEXARKANA 755			TYLER 757		
		MAT.	INST.	TOTAL	MAT.	INST.	TOTAL	MAT.	INST.	TOTAL	MAT.	INST.	TOTAL	MAT.	INST.	TOTAL	MAT.	INST.	TOTAL
015433	CONTRACTOR EQUIPMENT		88.8	88.8		88.0	88.0		95.0	95.0		88.0	88.0		88.8	88.8		88.8	88.8
0241, 31 - 34	SITE & INFRASTRUCTURE, DEMOLITION	95.3	87.7	90.1	98.5	84.9	89.1	97.5	95.9	96.4	87.3	84.4	85.3	84.7	87.8	86.8	94.3	87.8	89.8
0310	Concrete Forming & Accessories	76.2	58.7	61.3	97.9	49.7	56.9	101.3	55.7	62.5	101.9	49.3	57.2	93.9	59.0	64.2	87.6	58.8	63.1
0320	Concrete Reinforcing	102.0	48.7	76.3	84.5	50.2	67.9	89.1	47.3	68.9	84.7	48.4	67.2	101.8	51.2	77.4	102.7	51.2	77.9
0330	Cast-in-Place Concrete	97.5	56.3	81.9	101.7	62.4	86.8	83.8	69.4	78.3	83.3	55.9	72.9	98.2	64.8	85.6	117.2	56.3	94.2
03	CONCRETE	102.4	56.9	82.0	91.2	55.3	75.1	86.3	59.6	74.3	77.8	52.5	66.5	94.1	60.4	79.0	109.0	57.4	85.8
04	MASONRY	115.0	56.8	79.5	123.2	58.2	83.6	89.0	63.4	73.4	134.3	58.1	87.8	181.0	56.8	105.3	169.7	56.8	100.9
05	METALS	100.7	67.5	90.3	103.3	69.4	92.7	103.4	64.2	91.2	102.9	67.2	91.8	93.5	68.8	85.8	100.5	68.5	90.5
06	WOOD, PLASTICS & COMPOSITES	75.0	61.0	67.7	103.1	48.6	74.6	103.8	54.9	78.2	112.5	48.6	79.1	91.3	61.0	75.5	86.7	61.0	73.3
07	THERMAL & MOISTURE PROTECTION	90.9	60.4	77.8	92.7	60.9	79.0	88.7	67.6	79.6	92.3	60.1	78.5	90.3	61.9	78.1	90.8	60.4	77.7
08	OPENINGS	63.3	55.7	61.5	100.6	49.3	88.1	102.3	51.7	90.0	67.3	48.9	62.8	83.5	56.3	76.8	63.3	56.3	61.6
0920	Plaster & Gypsum Board	75.6	60.5	65.7	96.9	47.8	64.7	92.1	53.9	67.0	96.9	47.8	64.7	82.6	60.5	68.1	76.1	60.5	65.8
0950, 0980	Ceilings & Acoustic Treatment	91.2	60.5	71.9	92.5	47.8	64.5	93.2	53.9	68.5	92.5	47.8	64.5	96.3	60.5	73.8	91.2	60.5	71.9
0960	Flooring	121.6	52.1	101.3	76.1	52.1	69.1	105.7	73.9	96.4	96.9	52.1	83.8	107.4	60.9	93.8	134.1	52.1	110.2
0970, 0990	Wall Finishes & Painting/Coating	84.8	49.2	63.5	89.5	49.2	65.3	101.1	45.6	67.8	90.8	42.1	61.6	84.8	49.2	63.5	84.8	49.2	63.5
09	FINISHES	105.2	56.5	78.8	83.8	49.2	65.1	106.7	57.7	80.2	89.1	48.4	67.1	102.5	58.2	78.5	108.6	56.5	80.4
COVERS	DIVS. 10 - 14, 25, 28, 41, 43, 44, 46	100.0	79.2	95.1	100.0	77.5	94.7	100.0	79.7	95.2	100.0	77.8	94.8	100.0	79.2	95.1	100.0	79.2	95.1
21, 22, 23	FIRE SUPPRESSION, PLUMBING & HVAC	97.2	54.1	79.9	97.0	49.8	78.0	100.9	60.0	84.4	97.0	50.6	78.3	97.2	53.0	79.4	97.2	55.6	80.4
26, 27, 3370	ELECTRICAL, COMMUNICATIONS & UTIL.	93.3	46.6	70.2	103.2	52.6	78.1	100.2	61.5	81.0	100.4	50.1	75.5	96.6	52.5	74.8	95.5	51.7	73.9
MF2018	WEIGHTED AVERAGE	95.9	58.9	80.0	98.7	57.3	80.8	98.9	63.8	83.7	94.1	56.4	77.8	99.1	60.4	82.4	100.1	60.2	82.8

TEXAS / UTAH

DIVISION		VICTORIA 779			WACO 766 - 767			WAXAHACHIE 751			WHARTON 774			WICHITA FALLS 763			LOGAN 843		
		MAT.	INST.	TOTAL	MAT.	INST.	TOTAL	MAT.	INST.	TOTAL	MAT.	INST.	TOTAL	MAT.	INST.	TOTAL	MAT.	INST.	TOTAL
015433	CONTRACTOR EQUIPMENT		103.5	103.5		88.0	88.0		97.5	97.5		104.9	104.9		88.0	88.0		91.5	91.5
0241, 31 - 34	SITE & INFRASTRUCTURE, DEMOLITION	113.6	85.0	93.9	96.0	84.6	88.2	95.6	81.7	86.0	119.1	87.0	97.1	96.8	85.0	88.7	97.2	86.8	90.0
0310	Concrete Forming & Accessories	89.8	49.6	55.6	100.2	60.4	66.3	86.0	58.9	63.0	84.0	51.3	56.1	100.2	60.5	66.4	100.2	70.4	74.9
0320	Concrete Reinforcing	91.9	48.5	71.0	84.3	47.7	66.7	103.8	48.9	77.3	95.6	48.5	72.9	84.3	50.4	68.0	101.6	87.8	94.9
0330	Cast-in-Place Concrete	117.8	57.7	95.1	90.1	65.7	80.9	101.5	57.2	84.8	121.1	56.8	96.8	96.3	71.1	86.7	85.1	75.0	81.3
03	CONCRETE	104.9	54.4	82.2	84.2	60.8	73.7	92.2	58.3	77.0	108.9	54.8	84.6	87.1	63.2	76.4	105.3	75.5	91.9
04	MASONRY	113.9	58.2	79.9	93.2	62.4	74.4	164.3	60.0	100.7	97.3	58.1	73.4	93.7	62.0	74.4	108.5	63.1	80.8
05	METALS	101.0	85.7	96.2	105.4	68.1	93.7	100.8	82.2	95.0	102.6	85.2	97.2	105.3	70.2	94.4	107.5	84.3	100.3
06	WOOD, PLASTICS & COMPOSITES	95.1	48.7	70.9	110.5	62.1	85.2	83.0	61.1	71.6	85.0	50.8	67.1	110.5	62.1	85.2	83.0	70.7	76.5
07	THERMAL & MOISTURE PROTECTION	88.6	59.5	76.1	93.1	64.9	81.0	89.3	61.8	77.5	85.9	62.0	75.7	93.1	63.8	80.5	100.9	68.7	87.0
08	OPENINGS	99.2	49.1	87.0	78.2	57.5	73.2	92.3	55.8	83.4	99.4	50.3	87.4	78.2	57.9	73.2	92.4	73.1	87.7
0920	Plaster & Gypsum Board	94.7	47.8	63.9	97.0	61.6	73.8	79.8	60.5	67.1	94.3	50.0	65.2	97.0	61.6	73.8	83.4	70.2	74.7
0950, 0980	Ceilings & Acoustic Treatment	112.1	47.8	71.8	93.1	61.6	73.4	105.2	60.5	77.2	110.8	50.0	72.7	93.1	61.6	73.4	113.9	70.2	86.5
0960	Flooring	104.3	52.1	89.0	96.3	69.9	88.6	91.7	52.1	80.2	103.0	63.5	91.5	96.9	74.3	90.3	96.5	62.9	86.7
0970, 0990	Wall Finishes & Painting/Coating	106.2	54.7	75.3	90.8	57.0	70.6	93.4	49.2	66.9	106.0	53.1	74.3	92.8	54.3	69.7	89.3	60.8	72.3
09	FINISHES	98.3	49.9	72.1	89.5	61.8	74.5	95.7	56.6	74.5	100.1	52.9	74.6	89.9	62.4	75.0	97.3	67.7	81.3
COVERS	DIVS. 10 - 14, 25, 28, 41, 43, 44, 46	100.0	79.1	95.1	100.0	79.6	95.2	100.0	79.5	95.2	100.0	79.4	95.1	100.0	79.6	95.2	100.0	85.6	96.6
21, 22, 23	FIRE SUPPRESSION, PLUMBING & HVAC	97.2	59.4	81.9	100.8	59.7	84.2	97.3	55.4	80.4	97.1	59.0	81.7	100.8	54.3	82.1	100.9	74.6	90.3
26, 27, 3370	ELECTRICAL, COMMUNICATIONS & UTIL.	104.2	52.0	78.3	103.3	55.4	79.6	95.4	56.5	76.2	103.2	57.0	80.3	105.3	54.7	80.3	94.6	70.9	82.9
MF2018	WEIGHTED AVERAGE	101.0	60.8	83.6	95.8	63.2	81.7	99.5	62.0	83.3	101.0	62.1	84.2	96.5	62.5	81.8	100.7	74.0	89.2

UTAH / VERMONT

DIVISION		OGDEN 842,844 MAT.	INST.	TOTAL	PRICE 845 MAT.	INST.	TOTAL	PROVO 846-847 MAT.	INST.	TOTAL	SALT LAKE CITY 840-841 MAT.	INST.	TOTAL	BELLOWS FALLS 051 MAT.	INST.	TOTAL	BENNINGTON 052 MAT.	INST.	TOTAL
015433	CONTRACTOR EQUIPMENT		91.5	91.5		90.6	90.6		90.6	90.6		91.5	91.5		94.2	94.2		94.2	94.2
0241, 31-34	SITE & INFRASTRUCTURE, DEMOLITION	85.1	86.8	86.3	94.4	84.4	87.5	93.5	85.3	87.9	84.9	86.7	86.1	86.8	93.3	91.3	86.2	93.3	91.1
0310	Concrete Forming & Accessories	100.2	70.4	74.9	102.9	63.3	69.2	102.1	70.4	75.2	102.6	70.4	75.2	99.6	98.4	98.6	96.7	98.2	98.0
0320	Concrete Reinforcing	101.3	87.8	94.7	108.9	80.0	95.0	109.9	87.8	99.2	103.4	87.8	95.9	82.2	82.8	82.5	82.2	82.8	82.5
0330	Cast-in-Place Concrete	86.4	75.0	82.1	85.2	70.8	79.7	85.2	75.0	81.3	94.4	75.0	87.1	86.5	111.7	96.0	86.5	111.6	96.0
03	CONCRETE	94.3	75.5	85.9	106.5	69.4	89.9	104.9	75.5	91.7	114.1	75.5	96.8	86.9	100.2	92.9	86.7	100.1	92.7
04	MASONRY	102.8	63.1	78.6	114.0	66.3	84.9	114.2	63.1	83.0	116.8	63.1	84.0	98.8	91.9	94.6	108.9	91.9	98.5
05	METALS	108.1	84.3	100.7	104.6	80.6	97.1	105.6	84.3	98.9	111.8	84.3	103.2	100.2	90.8	97.2	100.1	90.6	97.2
06	WOOD, PLASTICS & COMPOSITES	83.0	70.7	76.5	86.4	61.8	73.6	84.7	70.7	77.4	85.0	70.7	77.5	106.0	103.2	104.5	102.9	103.2	103.1
07	THERMAL & MOISTURE PROTECTION	99.7	68.7	86.4	102.7	67.0	87.4	102.8	68.7	88.1	107.6	68.7	90.9	101.4	87.7	95.5	101.3	87.7	95.5
08	OPENINGS	92.4	73.1	87.7	96.1	73.4	90.5	96.1	73.1	90.5	94.2	73.1	89.0	100.6	94.0	99.0	100.6	94.0	99.0
0920	Plaster & Gypsum Board	83.4	70.2	74.7	86.6	61.0	69.8	84.3	70.2	75.0	92.8	70.2	77.9	113.9	103.1	106.8	113.0	103.1	106.5
0950, 0980	Ceilings & Acoustic Treatment	113.9	70.2	86.5	113.9	61.0	80.8	113.9	70.2	86.5	106.8	70.2	83.8	81.4	103.1	95.0	81.4	103.1	95.0
0960	Flooring	94.6	62.9	85.3	97.7	53.5	84.8	97.5	62.9	87.4	98.6	62.9	88.1	90.3	110.7	96.3	89.4	110.7	95.6
0970, 0990	Wall Finishes & Painting/Coating	89.3	60.8	72.3	89.3	57.8	70.4	89.3	60.8	72.3	92.4	62.1	74.2	82.8	96.7	91.1	82.8	96.7	91.1
09	FINISHES	95.5	67.7	80.4	98.5	60.2	77.8	98.0	67.7	81.6	97.0	67.8	81.2	88.5	101.2	95.4	88.1	101.2	95.2
COVERS	DIVS. 10-14, 25, 28, 41, 43, 44, 46	100.0	85.6	96.6	100.0	84.6	96.4	100.0	85.6	96.6	100.0	85.6	96.6	100.0	100.5	100.1	100.0	100.4	100.1
21, 22, 23	FIRE SUPPRESSION, PLUMBING & HVAC	100.9	74.6	90.3	98.9	62.2	84.1	100.9	74.6	90.3	101.2	74.6	90.5	97.6	83.5	91.9	97.6	83.5	91.9
26, 27, 3370	ELECTRICAL, COMMUNICATIONS & UTIL.	95.0	70.9	83.1	98.9	66.2	82.7	95.1	71.6	83.4	97.5	68.6	83.2	104.3	76.6	90.6	104.3	52.4	78.6
MF2018	WEIGHTED AVERAGE	98.6	74.0	88.0	101.1	68.6	87.1	101.1	74.0	88.0	103.3	73.7	90.5	97.1	90.6	94.3	97.5	87.2	93.0

VERMONT

DIVISION		BRATTLEBORO 053 MAT.	INST.	TOTAL	BURLINGTON 054 MAT.	INST.	TOTAL	GUILDHALL 059 MAT.	INST.	TOTAL	MONTPELIER 056 MAT.	INST.	TOTAL	RUTLAND 057 MAT.	INST.	TOTAL	ST. JOHNSBURY 058 MAT.	INST.	TOTAL
015433	CONTRACTOR EQUIPMENT		94.2	94.2		99.4	99.4		94.2	94.2		99.4	99.4		94.2	94.2		94.2	94.2
0241, 31-34	SITE & INFRASTRUCTURE, DEMOLITION	87.7	93.3	91.6	91.6	103.4	99.7	85.8	93.0	90.7	90.0	102.5	98.6	90.6	94.3	93.2	85.8	92.4	90.3
0310	Concrete Forming & Accessories	99.9	98.4	98.6	98.8	81.9	84.5	96.0	92.7	93.2	101.0	99.2	99.5	100.2	81.8	84.6	93.2	92.7	92.8
0320	Concrete Reinforcing	81.3	82.8	82.0	82.2	82.8	82.5	82.9	82.8	82.8	96.1	82.8	89.7	102.8	82.8	93.1	81.3	82.7	82.0
0330	Cast-in-Place Concrete	89.3	111.7	97.7	107.3	113.9	109.8	83.9	103.5	91.3	99.6	113.9	105.0	84.8	112.9	95.4	83.9	103.5	91.3
03	CONCRETE	88.9	100.2	93.9	104.9	93.4	99.8	84.4	94.8	89.1	100.2	101.2	100.6	89.7	93.1	91.2	84.0	94.8	88.8
04	MASONRY	108.6	91.9	98.4	109.6	94.2	100.2	108.9	77.8	90.0	108.7	94.2	99.9	89.7	94.1	92.4	136.9	77.8	100.9
05	METALS	100.1	90.8	97.2	108.3	89.2	102.4	100.2	90.5	97.1	106.7	89.2	101.3	106.5	90.6	101.6	100.2	90.4	97.1
06	WOOD, PLASTICS & COMPOSITES	106.4	103.2	104.7	99.7	79.8	89.3	101.1	103.2	102.2	101.3	103.3	102.3	106.6	79.7	92.5	95.1	103.2	99.4
07	THERMAL & MOISTURE PROTECTION	101.5	87.7	95.6	103.7	96.8	102.8	101.2	81.3	92.6	110.1	99.3	105.5	101.6	95.8	99.1	101.0	81.3	92.6
08	OPENINGS	100.6	94.2	99.1	103.1	81.5	97.9	100.6	94.4	99.1	103.6	94.5	101.4	103.7	81.5	98.3	100.6	94.4	99.1
0920	Plaster & Gypsum Board	113.9	103.1	106.8	115.4	79.0	91.5	120.4	103.1	109.0	117.5	103.1	108.0	114.4	79.0	91.1	122.4	103.1	109.7
0950, 0980	Ceilings & Acoustic Treatment	81.4	103.1	95.0	93.3	79.0	84.3	81.4	103.1	95.0	83.9	103.1	95.9	87.6	79.0	82.2	81.4	103.1	95.0
0960	Flooring	90.6	110.7	96.4	96.9	110.7	100.9	93.4	110.7	98.4	97.4	110.7	101.3	90.3	110.7	96.3	96.7	110.7	100.8
0970, 0990	Wall Finishes & Painting/Coating	82.8	96.7	91.1	95.3	104.6	100.9	82.8	104.6	95.9	98.3	104.6	102.1	82.8	104.6	95.9	82.8	104.6	95.9
09	FINISHES	88.7	101.2	95.4	97.0	88.9	92.6	90.2	98.5	94.7	95.4	102.7	99.4	90.2	88.8	89.4	91.5	98.5	95.3
COVERS	DIVS. 10-14, 25, 28, 41, 43, 44, 46	100.0	100.5	100.1	100.0	98.9	99.7	100.0	95.7	99.0	100.0	101.4	100.3	100.0	98.6	99.7	100.0	95.7	99.0
21, 22, 23	FIRE SUPPRESSION, PLUMBING & HVAC	97.6	83.5	91.9	101.2	68.1	87.8	97.6	59.5	82.3	97.4	68.1	85.6	101.4	68.0	87.9	97.6	59.5	82.3
26, 27, 3370	ELECTRICAL, COMMUNICATIONS & UTIL.	104.3	76.6	90.6	104.0	56.4	80.5	104.3	52.4	78.6	102.8	56.4	79.9	104.3	56.4	80.6	104.3	52.4	78.6
MF2018	WEIGHTED AVERAGE	97.8	90.7	94.7	103.1	82.3	94.1	97.3	79.2	89.5	101.2	86.0	94.6	99.4	81.6	91.7	98.7	79.2	90.3

VERMONT / VIRGINIA

DIVISION		WHITE RIVER JCT. 050 MAT.	INST.	TOTAL	ALEXANDRIA 223 MAT.	INST.	TOTAL	ARLINGTON 222 MAT.	INST.	TOTAL	BRISTOL 242 MAT.	INST.	TOTAL	CHARLOTTESVILLE 229 MAT.	INST.	TOTAL	CULPEPER 227 MAT.	INST.	TOTAL
015433	CONTRACTOR EQUIPMENT		94.2	94.2		103.4	103.4		102.2	102.2		102.2	102.2		106.2	106.2		102.2	102.2
0241, 31-34	SITE & INFRASTRUCTURE, DEMOLITION	90.2	92.5	91.8	114.5	86.0	94.9	125.0	83.8	96.7	108.9	84.0	91.8	113.5	85.7	94.4	112.0	83.5	92.4
0310	Concrete Forming & Accessories	93.7	93.2	93.2	94.9	72.8	76.1	93.1	61.8	66.5	87.7	59.2	63.4	85.9	60.9	64.6	82.9	68.7	70.8
0320	Concrete Reinforcing	82.2	82.7	82.5	86.9	88.5	87.7	98.0	86.2	92.3	98.0	70.4	84.7	97.4	71.9	85.1	98.0	88.5	93.4
0330	Cast-in-Place Concrete	89.3	104.3	94.9	107.2	76.7	95.7	104.3	75.2	93.3	103.9	49.2	83.3	108.1	73.3	94.9	106.8	74.5	94.6
03	CONCRETE	90.5	95.2	92.6	99.4	78.3	89.9	103.8	72.4	89.7	100.2	59.5	81.9	100.7	68.9	86.5	97.9	75.6	87.9
04	MASONRY	122.1	79.1	95.9	91.9	72.8	80.3	105.7	71.4	84.8	95.2	55.0	70.7	120.7	55.3	80.8	108.0	70.4	85.0
05	METALS	100.2	90.4	97.1	105.9	102.2	104.8	104.4	101.1	103.4	103.2	93.4	100.2	103.5	96.4	101.3	103.6	101.1	102.8
06	WOOD, PLASTICS & COMPOSITES	99.2	103.2	101.3	94.6	71.7	82.6	90.7	57.7	73.4	81.8	57.7	69.2	79.9	58.3	68.6	78.3	67.7	72.8
07	THERMAL & MOISTURE PROTECTION	101.5	81.9	93.1	100.6	79.9	91.7	102.5	77.4	91.8	101.8	60.7	84.1	101.4	68.1	87.1	101.6	77.6	91.3
08	OPENINGS	100.6	94.4	99.1	95.5	76.0	90.8	93.8	67.8	87.4	96.4	64.7	88.7	94.8	65.3	87.6	95.1	73.8	89.9
0920	Plaster & Gypsum Board	111.6	103.1	106.0	101.6	71.0	81.5	98.1	56.6	70.9	94.2	56.6	69.5	94.2	56.6	69.5	94.4	66.9	76.3
0950, 0980	Ceilings & Acoustic Treatment	81.4	103.1	95.0	99.0	71.0	81.5	96.5	56.6	71.5	95.8	56.6	71.2	95.8	56.6	71.2	96.5	66.9	77.9
0960	Flooring	88.4	110.7	94.9	90.9	75.3	92.1	97.5	74.8	90.8	94.1	55.0	82.7	92.5	55.0	81.6	92.5	71.0	86.2
0970, 0990	Wall Finishes & Painting/Coating	82.8	104.6	95.9	111.0	70.1	86.5	111.0	70.1	86.5	97.7	75.0	84.1	97.7	59.1	74.6	111.0	69.8	86.4
09	FINISHES	88.0	98.8	93.9	98.3	72.2	84.2	98.0	63.5	79.3	94.7	59.8	75.8	94.3	58.2	74.8	94.9	68.4	80.6
COVERS	DIVS. 10-14, 25, 28, 41, 43, 44, 46	100.0	96.1	99.1	100.0	88.8	97.4	100.0	79.8	95.2	100.0	82.2	95.8	100.0	83.8	96.2	100.0	87.6	97.1
21, 22, 23	FIRE SUPPRESSION, PLUMBING & HVAC	97.6	60.2	82.5	101.5	86.1	95.3	101.5	85.6	95.1	97.7	49.7	78.3	97.7	86.0	93.0	97.7	83.2	91.8
26, 27, 3370	ELECTRICAL, COMMUNICATIONS & UTIL.	104.3	52.3	78.6	95.4	100.8	98.0	93.4	102.2	97.7	95.0	35.0	65.3	95.0	72.0	83.6	97.0	97.3	97.1
MF2018	WEIGHTED AVERAGE	98.7	79.6	90.4	100.0	84.8	93.4	101.0	81.6	92.6	98.5	59.4	81.6	99.8	74.2	88.7	99.1	82.0	91.7

For customer support on your Site Work & Landscape Costs with RSMeans Data, call 800.448.8182.

819

VIRGINIA

DIVISION		FAIRFAX 220 - 221			FARMVILLE 239			FREDERICKSBURG 224 - 225			GRUNDY 246			HARRISONBURG 228			LYNCHBURG 245		
		MAT.	INST.	TOTAL	MAT.	INST.	TOTAL	MAT.	INST.	TOTAL	MAT.	INST.	TOTAL	MAT.	INST.	TOTAL	MAT.	INST.	TOTAL
015433	CONTRACTOR EQUIPMENT		102.2	102.2		106.2	106.2		102.2	102.2		102.2	102.2		102.2	102.2		102.2	102.2
0241, 31 - 34	SITE & INFRASTRUCTURE, DEMOLITION	123.2	83.8	96.1	109.8	84.6	92.5	111.5	83.4	92.2	106.7	82.5	90.1	120.0	83.8	95.2	107.5	84.0	91.4
0310	Concrete Forming & Accessories	86.3	61.6	65.3	102.5	60.2	66.5	86.3	63.3	66.7	91.5	58.3	63.2	81.7	61.2	64.3	87.7	60.6	64.6
0320	Concrete Reinforcing	98.0	88.5	93.4	95.4	64.9	80.7	98.7	77.4	88.5	96.7	42.2	70.4	98.0	80.5	89.6	97.4	70.7	84.5
0330	Cast-in-Place Concrete	104.3	75.3	93.4	107.7	83.8	98.6	105.9	72.5	93.3	103.9	49.3	83.3	104.3	75.5	93.4	103.9	74.1	92.7
03	CONCRETE	103.4	72.7	89.6	102.5	70.9	88.3	97.6	70.6	85.5	98.8	53.9	78.7	101.2	71.3	87.8	98.7	68.9	85.3
04	MASONRY	105.6	72.0	85.1	105.6	56.1	75.4	106.9	81.6	91.5	97.1	56.0	72.0	104.5	71.4	84.3	112.5	55.3	77.6
05	METALS	103.7	101.8	103.1	101.3	89.6	97.7	103.6	97.9	101.8	103.2	74.2	94.2	103.5	99.0	102.1	103.4	95.4	100.9
06	WOOD, PLASTICS & COMPOSITES	81.8	57.1	68.9	98.5	58.3	77.4	81.8	62.4	71.7	85.4	57.7	70.9	77.0	57.9	67.0	81.8	57.9	69.3
07	THERMAL & MOISTURE PROTECTION	102.2	77.2	91.5	103.8	69.7	89.1	101.6	79.5	92.1	101.8	59.1	83.4	102.0	76.8	91.2	101.6	68.2	87.3
08	OPENINGS	93.8	68.0	87.5	94.3	53.2	84.3	94.8	67.5	88.2	96.4	44.4	83.7	95.1	67.0	88.2	95.1	64.8	87.7
0920	Plaster & Gypsum Board	94.4	56.0	69.2	106.4	56.6	73.7	94.4	61.5	72.8	94.2	56.6	69.5	94.2	56.8	69.6	94.2	56.8	69.6
0950, 0980	Ceilings & Acoustic Treatment	96.5	56.0	71.1	90.1	56.6	69.1	96.5	61.5	74.5	95.8	56.6	71.2	95.8	56.8	71.4	95.8	56.8	71.4
0960	Flooring	94.3	76.4	89.1	95.3	55.0	83.5	94.3	69.1	86.9	95.3	55.0	83.5	92.3	76.9	87.8	94.1	55.0	82.7
0970, 0990	Wall Finishes & Painting/Coating	111.0	70.1	86.5	93.3	54.5	70.0	111.0	54.3	77.0	97.7	29.7	57.0	111.0	58.9	79.8	97.7	75.0	84.1
09	FINISHES	96.5	63.6	78.7	94.4	58.0	74.7	95.4	62.4	77.5	94.9	54.9	73.2	95.4	62.6	77.7	94.5	59.6	75.6
COVERS	DIVS. 10 - 14, 25, 28, 41, 43, 44, 46	100.0	87.0	96.9	100.0	83.7	96.2	100.0	74.3	93.9	100.0	82.4	95.8	100.0	82.4	95.9	100.0	82.0	95.8
21, 22, 23	FIRE SUPPRESSION, PLUMBING & HVAC	97.7	84.0	92.2	98.0	52.9	79.8	97.7	76.1	89.0	97.7	63.1	83.7	97.7	75.0	88.5	97.7	86.0	93.0
26, 27, 3370	ELECTRICAL, COMMUNICATIONS & UTIL.	95.9	102.2	99.0	89.8	69.9	79.9	93.6	102.1	97.8	95.0	69.8	82.6	95.2	78.3	86.8	95.9	65.3	80.7
MF2018	WEIGHTED AVERAGE	99.9	81.7	92.0	98.5	66.0	84.5	98.6	79.9	90.5	98.4	63.2	83.2	99.4	75.6	89.1	99.1	73.1	87.9

VIRGINIA

DIVISION		NEWPORT NEWS 236			NORFOLK 233 - 235			PETERSBURG 238			PORTSMOUTH 237			PULASKI 243			RICHMOND 230 - 232		
		MAT.	INST.	TOTAL	MAT.	INST.	TOTAL	MAT.	INST.	TOTAL	MAT.	INST.	TOTAL	MAT.	INST.	TOTAL	MAT.	INST.	TOTAL
015433	CONTRACTOR EQUIPMENT		106.3	106.3		107.9	107.9		106.2	106.2		106.2	106.2		102.2	102.2		108.0	108.0
0241, 31 - 34	SITE & INFRASTRUCTURE, DEMOLITION	109.0	85.8	93.0	109.6	92.1	97.6	112.4	85.7	94.1	107.7	85.4	92.4	106.1	82.8	90.1	101.8	92.3	95.3
0310	Concrete Forming & Accessories	101.5	60.5	66.6	103.2	60.2	66.6	93.8	61.0	65.9	89.0	60.2	64.5	91.5	59.0	63.8	96.3	60.9	66.2
0320	Concrete Reinforcing	95.3	66.8	81.5	105.0	66.8	86.5	94.9	71.9	83.8	94.9	66.8	81.3	96.7	80.0	88.6	104.0	71.9	88.5
0330	Cast-in-Place Concrete	104.6	76.1	93.8	116.2	77.0	101.4	111.1	76.3	98.0	103.6	64.7	88.9	103.9	80.3	95.0	88.4	77.3	84.2
03	CONCRETE	99.6	68.8	85.8	104.1	68.8	88.2	105.0	70.0	89.3	98.2	64.7	83.2	98.8	71.7	86.7	90.5	70.1	81.4
04	MASONRY	100.4	55.3	72.9	104.4	55.3	74.5	114.9	55.3	78.5	106.8	55.2	75.3	91.0	53.5	68.1	103.3	55.3	74.1
05	METALS	103.8	94.0	100.7	105.2	91.0	100.8	101.4	96.4	99.9	102.7	93.6	99.9	103.3	94.8	100.6	105.9	93.6	102.0
06	WOOD, PLASTICS & COMPOSITES	96.9	58.5	76.8	99.2	58.3	77.8	86.7	58.3	71.8	82.2	58.3	69.7	85.4	57.7	70.9	96.5	58.3	76.5
07	THERMAL & MOISTURE PROTECTION	103.7	66.3	87.6	101.2	68.9	87.3	103.7	68.5	88.6	103.7	66.6	87.8	101.8	68.6	87.5	98.8	69.7	86.3
08	OPENINGS	94.6	63.8	87.1	95.3	64.2	87.7	94.0	65.3	87.0	94.7	63.7	87.2	96.4	66.0	86.6	101.6	65.3	92.8
0920	Plaster & Gypsum Board	107.9	56.8	74.3	99.3	56.8	71.4	100.4	56.6	71.6	101.4	56.6	72.0	94.2	56.6	69.5	98.9	56.8	71.3
0950, 0980	Ceilings & Acoustic Treatment	95.8	56.8	71.4	90.7	56.8	69.4	92.0	56.6	69.8	95.8	56.6	71.2	95.8	56.6	71.2	93.6	56.8	70.5
0960	Flooring	95.3	55.0	83.5	95.9	55.0	83.9	91.3	55.0	80.7	88.3	55.0	78.6	95.3	55.0	83.5	91.1	55.0	80.6
0970, 0990	Wall Finishes & Painting/Coating	93.3	58.0	72.1	97.7	58.9	74.4	93.3	59.1	72.8	93.3	58.9	72.7	97.7	46.0	66.7	92.5	59.1	72.5
09	FINISHES	98.5	58.2	75.4	94.5	58.2	74.9	93.0	58.2	74.2	92.9	58.2	74.1	94.9	56.0	73.8	93.7	58.2	74.5
COVERS	DIVS. 10 - 14, 25, 28, 41, 43, 44, 46	100.0	83.8	96.2	100.0	83.4	96.1	100.0	83.8	96.2	100.0	81.5	95.6	100.0	81.5	95.7	100.0	83.5	96.1
21, 22, 23	FIRE SUPPRESSION, PLUMBING & HVAC	101.8	65.5	87.1	101.1	65.0	86.5	98.0	86.0	93.1	101.8	65.5	87.1	97.7	63.9	84.1	101.1	86.0	95.0
26, 27, 3370	ELECTRICAL, COMMUNICATIONS & UTIL.	91.9	72.1	82.1	95.9	62.3	79.3	92.0	72.0	82.1	90.6	62.3	76.6	95.0	80.3	87.7	98.6	72.0	85.5
MF2018	WEIGHTED AVERAGE	99.5	69.5	86.5	100.7	68.3	86.7	99.4	74.4	88.6	99.0	67.4	85.3	98.1	69.7	85.8	99.6	74.6	88.8

VIRGINIA / WASHINGTON

DIVISION		ROANOKE 240 - 241			STAUNTON 244			WINCHESTER 226			CLARKSTON 994			EVERETT 982			OLYMPIA 985		
		MAT.	INST.	TOTAL	MAT.	INST.	TOTAL	MAT.	INST.	TOTAL	MAT.	INST.	TOTAL	MAT.	INST.	TOTAL	MAT.	INST.	TOTAL
015433	CONTRACTOR EQUIPMENT		102.2	102.2		106.2	106.2		102.2	102.2		88.4	88.4		97.7	97.7		100.2	100.2
0241, 31 - 34	SITE & INFRASTRUCTURE, DEMOLITION	106.9	84.1	91.2	110.1	84.4	92.5	118.7	83.5	94.5	102.2	83.1	89.1	93.2	103.3	100.1	94.8	105.9	102.4
0310	Concrete Forming & Accessories	99.0	60.8	66.5	91.1	60.1	64.8	84.5	63.9	67.0	102.5	62.0	68.0	114.5	105.5	106.8	98.0	106.0	105.0
0320	Concrete Reinforcing	97.7	70.7	84.7	97.4	69.3	83.8	97.4	76.4	87.3	106.6	99.4	103.1	107.5	113.7	110.5	120.0	113.5	116.9
0330	Cast-in-Place Concrete	118.0	83.8	105.1	108.1	83.6	98.8	104.3	61.2	88.0	87.0	79.6	84.2	100.9	111.7	105.0	88.8	114.3	98.5
03	CONCRETE	103.3	72.4	89.4	100.2	71.5	87.3	100.7	66.8	85.5	90.0	75.1	83.3	94.5	108.6	100.8	93.0	109.7	100.5
04	MASONRY	98.4	57.1	73.2	107.8	56.0	76.2	101.8	66.8	80.5	100.6	87.8	92.8	109.1	104.3	106.2	106.7	102.2	104.0
05	METALS	105.7	95.3	102.4	103.5	90.8	99.5	103.6	96.1	101.3	92.9	88.6	91.6	116.2	99.5	111.0	117.2	98.0	111.2
06	WOOD, PLASTICS & COMPOSITES	95.7	57.9	75.9	85.4	58.3	71.2	79.9	62.6	70.9	95.5	56.5	75.1	112.1	105.1	108.4	92.4	105.4	99.2
07	THERMAL & MOISTURE PROTECTION	101.6	70.4	88.2	101.4	69.6	87.7	102.1	74.2	90.1	157.0	79.1	123.6	113.5	107.0	110.7	109.4	107.5	108.6
08	OPENINGS	95.5	64.8	88.0	95.1	54.1	85.1	96.5	65.8	89.0	116.6	64.4	103.9	105.6	107.8	106.2	109.3	108.0	109.0
0920	Plaster & Gypsum Board	101.6	56.8	72.2	94.2	56.6	69.5	94.4	61.6	72.9	143.7	55.3	85.7	109.5	105.5	106.9	105.5	105.5	105.5
0950, 0980	Ceilings & Acoustic Treatment	99.0	56.8	72.5	95.8	56.6	71.2	96.5	61.6	74.6	91.6	55.3	68.9	104.4	105.5	105.1	114.7	105.5	108.9
0960	Flooring	99.0	56.7	86.7	94.9	55.0	83.2	93.7	71.0	87.1	85.8	73.9	82.3	109.8	101.4	107.4	99.9	91.6	97.5
0970, 0990	Wall Finishes & Painting/Coating	97.7	75.0	84.1	97.7	29.2	56.7	111.0	73.9	88.8	83.5	68.7	74.6	95.4	92.6	93.7	92.6	94.3	93.6
09	FINISHES	97.3	60.3	77.2	94.7	55.3	73.4	95.9	65.6	79.5	105.6	63.1	82.6	106.2	103.0	104.5	102.6	101.5	102.0
COVERS	DIVS. 10 - 14, 25, 28, 41, 43, 44, 46	100.0	82.3	95.8	100.0	83.8	96.2	100.0	78.4	94.9	100.0	93.5	98.5	100.0	102.5	100.6	100.0	103.6	100.9
21, 22, 23	FIRE SUPPRESSION, PLUMBING & HVAC	101.5	67.2	87.6	97.7	56.8	81.2	97.7	77.5	89.6	97.5	76.8	89.2	101.2	102.5	101.7	101.1	108.1	103.9
26, 27, 3370	ELECTRICAL, COMMUNICATIONS & UTIL.	95.0	59.7	77.5	94.1	77.8	86.0	93.9	88.4	91.2	91.6	92.1	91.8	105.7	105.9	105.8	103.5	105.6	104.6
MF2018	WEIGHTED AVERAGE	100.5	69.1	87.0	99.0	67.8	85.5	99.2	76.4	89.3	99.9	79.5	91.1	104.5	104.2	104.4	103.9	105.2	104.5

WASHINGTON

DIVISION		RICHLAND 993			SEATTLE 980-981,987			SPOKANE 990-992			TACOMA 983-984			VANCOUVER 986			WENATCHEE 988		
		MAT.	INST.	TOTAL	MAT.	INST.	TOTAL	MAT.	INST.	TOTAL	MAT.	INST.	TOTAL	MAT.	INST.	TOTAL	MAT.	INST.	TOTAL
015433	CONTRACTOR EQUIPMENT		88.4	88.4		100.2	100.2		88.4	88.4		97.7	97.7		94.1	94.1		97.7	97.7
0241, 31 - 34	SITE & INFRASTRUCTURE, DEMOLITION	104.7	83.9	90.5	97.8	105.0	102.8	104.2	83.9	90.3	96.3	103.3	101.1	106.3	90.8	95.6	104.9	99.9	101.4
0310	Concrete Forming & Accessories	102.7	82.2	85.2	110.0	105.8	106.4	108.0	81.7	85.7	104.5	106.2	105.9	104.9	97.1	98.3	105.9	73.7	78.5
0320	Concrete Reinforcing	102.2	99.3	100.8	103.9	111.8	107.7	102.8	100.1	101.5	106.2	113.7	109.8	107.1	113.7	110.3	107.1	91.4	99.5
0330	Cast-in-Place Concrete	87.2	85.3	86.5	106.5	110.7	108.1	90.6	85.1	88.5	103.8	112.4	107.0	115.9	100.3	110.0	105.9	87.4	98.9
03	CONCRETE	89.5	86.2	88.0	101.4	108.1	104.4	91.7	86.1	89.2	96.3	109.1	102.0	106.2	100.9	103.8	104.1	81.8	94.1
04	MASONRY	101.7	85.8	92.0	114.3	102.8	107.3	102.3	100.4	101.2	108.1	105.5	106.5	108.7	104.8	106.3	111.4	90.2	98.5
05	METALS	93.3	89.6	92.2	115.8	100.2	110.9	95.4	90.3	93.8	118.1	99.5	112.3	115.6	100.0	110.7	115.4	86.1	106.2
06	WOOD, PLASTICS & COMPOSITES	95.7	80.5	87.8	108.8	105.2	106.9	104.2	80.5	91.8	100.4	105.1	102.9	93.9	96.2	95.1	102.3	70.8	85.8
07	THERMAL & MOISTURE PROTECTION	158.7	87.1	128.0	112.4	106.5	109.9	155.4	88.0	126.5	113.3	108.5	111.2	113.7	100.9	108.2	112.9	84.3	100.6
08	OPENINGS	114.7	76.6	105.4	105.7	106.1	106.0	115.3	77.4	106.0	106.4	107.8	106.7	102.5	101.2	102.2	105.9	70.3	97.2
0920	Plaster & Gypsum Board	143.7	80.0	101.9	104.6	105.5	105.2	132.3	80.0	98.0	106.7	105.5	105.9	104.8	96.6	99.4	110.3	70.3	84.1
0950, 0980	Ceilings & Acoustic Treatment	98.0	80.0	86.7	107.2	105.5	106.1	94.4	80.0	85.4	107.6	105.5	106.3	106.2	96.6	100.2	100.4	70.3	81.5
0960	Flooring	86.1	81.8	84.9	106.6	98.7	104.3	85.3	101.4	90.0	102.0	101.4	101.9	107.8	102.2	106.2	105.2	73.9	96.1
0970, 0990	Wall Finishes & Painting/Coating	83.5	75.8	78.9	108.1	91.2	98.0	83.7	79.1	80.9	95.4	92.6	93.7	97.6	79.1	86.5	95.4	67.5	78.7
09	FINISHES	107.4	80.5	92.8	106.5	102.3	104.2	104.8	84.8	94.0	104.5	103.3	103.8	103.0	95.7	99.1	105.2	71.9	87.2
COVERS	DIVS. 10 - 14, 25, 28, 41, 43, 44, 46	100.0	97.3	99.4	100.0	102.7	100.6	100.0	97.2	99.3	100.0	102.9	100.7	100.0	98.1	99.5	100.0	95.3	98.9
21, 22, 23	FIRE SUPPRESSION, PLUMBING & HVAC	101.4	111.4	105.4	101.2	117.2	107.6	101.3	84.2	94.4	101.3	108.1	104.0	101.3	107.5	103.8	97.5	87.6	93.5
26, 27, 3370	ELECTRICAL, COMMUNICATIONS & UTIL.	89.1	101.3	95.2	103.1	114.5	108.7	88.0	83.5	85.8	105.5	105.7	105.6	111.7	109.1	110.4	106.3	91.0	98.7
MF2018	WEIGHTED AVERAGE	100.7	93.0	97.4	105.4	108.4	106.7	101.0	86.8	94.9	104.9	105.7	105.2	106.2	102.2	104.4	105.0	85.6	96.6

DIVISION		WASHINGTON YAKIMA 989			WEST VIRGINIA BECKLEY 258 - 259			BLUEFIELD 247 - 248			BUCKHANNON 262			CHARLESTON 250 - 253			CLARKSBURG 263 - 264		
		MAT.	INST.	TOTAL	MAT.	INST.	TOTAL	MAT.	INST.	TOTAL	MAT.	INST.	TOTAL	MAT.	INST.	TOTAL	MAT.	INST.	TOTAL
015433	CONTRACTOR EQUIPMENT		97.7	97.7		102.2	102.2		102.2	102.2		102.2	102.2		107.1	107.1		102.2	102.2
0241, 31 - 34	SITE & INFRASTRUCTURE, DEMOLITION	98.7	101.7	100.8	100.1	84.5	89.4	100.6	84.5	89.5	106.8	84.7	91.6	99.1	94.7	96.1	107.5	84.7	91.8
0310	Concrete Forming & Accessories	104.4	102.1	102.5	85.3	84.3	84.5	88.1	84.1	84.7	87.4	87.3	87.3	97.9	88.8	90.2	84.5	87.4	87.0
0320	Concrete Reinforcing	106.7	100.2	103.6	95.7	85.1	90.6	96.3	78.6	87.8	96.9	86.3	91.8	101.7	85.3	93.8	96.9	96.2	96.6
0330	Cast-in-Place Concrete	110.8	87.3	101.9	100.0	89.1	95.9	101.6	89.0	96.8	101.3	90.4	97.3	96.8	90.5	94.4	111.0	86.6	101.8
03	CONCRETE	100.9	96.1	98.7	91.9	87.3	89.8	94.8	86.0	90.8	97.6	89.4	94.0	91.2	89.7	90.5	101.6	89.7	96.3
04	MASONRY	101.5	102.1	101.9	91.7	85.4	87.8	92.6	85.4	88.2	104.0	88.3	94.5	87.2	90.5	89.2	108.0	88.3	96.0
05	METALS	116.1	91.1	108.3	98.0	102.6	99.5	103.5	100.1	102.4	103.7	103.3	103.6	96.8	100.9	98.1	103.7	106.8	104.7
06	WOOD, PLASTICS & COMPOSITES	100.8	105.1	103.0	80.4	84.4	82.5	84.0	84.4	84.2	83.1	86.8	85.0	92.2	88.6	90.3	79.1	86.8	83.1
07	THERMAL & MOISTURE PROTECTION	113.5	95.8	105.9	102.6	83.1	94.2	101.5	83.1	93.6	101.8	86.7	95.3	97.9	87.6	93.5	101.7	86.4	95.2
08	OPENINGS	105.9	100.8	104.6	94.4	83.1	91.7	96.9	81.6	93.2	96.9	84.7	94.0	95.0	85.4	92.6	96.9	86.9	94.5
0920	Plaster & Gypsum Board	106.3	105.5	105.8	86.6	84.1	85.0	93.3	84.1	87.2	93.9	86.5	89.1	92.6	88.1	89.7	91.7	86.5	88.3
0950, 0980	Ceilings & Acoustic Treatment	101.9	105.5	104.1	81.0	84.1	82.9	93.3	84.1	87.5	95.8	86.5	90.0	90.6	88.1	89.1	95.8	86.5	90.0
0960	Flooring	103.1	81.8	96.9	89.1	98.4	91.8	91.7	98.4	93.7	91.4	94.7	92.4	93.4	98.4	94.9	90.3	94.7	91.6
0970, 0990	Wall Finishes & Painting/Coating	95.4	79.1	85.6	90.2	87.7	88.7	97.7	85.5	90.4	97.7	89.0	92.5	92.3	89.2	90.4	97.7	89.0	92.5
09	FINISHES	103.6	95.9	99.4	86.6	87.4	87.0	92.7	87.2	89.7	93.8	88.9	91.2	92.6	90.7	91.6	93.1	88.9	90.8
COVERS	DIVS. 10 - 14, 25, 28, 41, 43, 44, 46	100.0	100.4	100.1	100.0	91.6	98.0	100.0	91.6	98.0	100.0	92.9	98.3	100.0	93.4	98.5	100.0	92.9	98.3
21, 22, 23	FIRE SUPPRESSION, PLUMBING & HVAC	101.3	113.5	106.2	98.0	89.4	94.6	97.7	80.1	90.6	97.7	91.1	95.0	101.1	91.3	97.1	97.7	91.3	95.1
26, 27, 3370	ELECTRICAL, COMMUNICATIONS & UTIL.	108.2	101.3	104.8	91.9	81.0	86.5	94.2	81.0	87.7	95.3	90.7	93.0	98.9	86.4	92.7	95.3	90.7	93.0
MF2018	WEIGHTED AVERAGE	105.0	101.6	103.5	95.2	87.7	91.9	97.3	85.2	92.1	98.7	90.5	95.1	96.6	91.1	94.2	99.3	91.0	95.7

WEST VIRGINIA

DIVISION		GASSAWAY 266			HUNTINGTON 255 - 257			LEWISBURG 249			MARTINSBURG 254			MORGANTOWN 265			PARKERSBURG 261		
		MAT.	INST.	TOTAL	MAT.	INST.	TOTAL	MAT.	INST.	TOTAL	MAT.	INST.	TOTAL	MAT.	INST.	TOTAL	MAT.	INST.	TOTAL
015433	CONTRACTOR EQUIPMENT		102.2	102.2		102.2	102.2		102.2	102.2		102.2	102.2		102.2	102.2		102.2	102.2
0241, 31 - 34	SITE & INFRASTRUCTURE, DEMOLITION	104.2	84.7	90.8	105.0	85.9	91.9	116.4	84.5	94.5	103.8	84.9	90.9	101.5	85.8	90.7	110.4	85.8	93.5
0310	Concrete Forming & Accessories	86.7	84.5	84.8	98.2	88.2	89.7	84.8	84.1	84.2	85.3	76.5	77.8	84.8	87.7	87.3	89.4	89.3	89.3
0320	Concrete Reinforcing	96.9	87.3	92.3	97.1	90.3	93.8	96.9	78.6	88.1	95.7	90.6	93.2	96.9	96.3	96.6	96.3	86.2	91.4
0330	Cast-in-Place Concrete	106.0	88.9	99.6	109.0	94.2	103.4	101.7	89.0	96.9	104.7	84.8	97.2	101.3	92.9	98.1	103.5	92.8	99.5
03	CONCRETE	98.1	87.6	93.4	97.1	91.7	94.7	104.2	86.0	96.0	95.3	83.1	89.8	94.5	92.0	93.4	99.7	91.0	95.8
04	MASONRY	108.8	88.2	96.2	89.7	92.1	91.2	96.4	85.4	89.7	92.9	80.5	85.3	126.8	88.3	103.4	82.5	85.5	84.3
05	METALS	103.6	103.3	103.5	100.5	105.2	102.0	103.6	100.2	102.5	98.4	102.5	99.7	103.7	107.1	104.8	104.4	103.2	104.0
06	WOOD, PLASTICS & COMPOSITES	81.9	83.2	82.6	92.8	87.3	89.9	79.5	84.4	82.1	80.4	75.9	78.1	79.5	86.8	83.3	83.5	89.2	86.5
07	THERMAL & MOISTURE PROTECTION	101.5	86.3	95.0	102.8	86.9	96.0	102.5	83.1	94.2	102.8	77.1	91.8	101.5	87.3	95.4	101.8	86.8	95.4
08	OPENINGS	95.3	82.9	92.3	93.7	85.8	91.8	96.9	81.6	93.2	96.1	71.1	90.0	98.0	86.9	95.3	95.9	83.4	92.8
0920	Plaster & Gypsum Board	92.8	82.8	86.3	86.9	87.0	88.7	91.7	84.1	86.7	86.6	75.3	79.2	91.7	86.5	88.3	94.2	89.0	90.8
0950, 0980	Ceilings & Acoustic Treatment	95.8	82.8	87.7	81.0	87.0	84.8	95.8	84.1	88.5	81.0	75.3	77.4	95.8	86.5	90.0	95.8	89.0	91.5
0960	Flooring	91.3	98.4	93.3	96.7	100.4	97.8	90.5	98.4	92.8	89.1	94.4	90.6	90.5	94.7	91.7	94.4	98.4	95.6
0970, 0990	Wall Finishes & Painting/Coating	97.7	89.2	92.6	90.2	89.2	89.6	97.7	61.4	75.9	90.2	79.7	83.9	97.7	89.0	92.5	97.7	89.2	92.6
09	FINISHES	93.3	87.6	90.2	89.5	90.5	90.1	94.2	84.5	89.0	86.8	79.7	83.0	92.7	88.9	90.7	94.8	91.1	92.8
COVERS	DIVS. 10 - 14, 25, 28, 41, 43, 44, 46	100.0	92.5	98.2	100.0	93.1	98.4	100.0	91.6	98.0	100.0	77.1	94.6	100.0	92.9	98.3	100.0	93.1	98.4
21, 22, 23	FIRE SUPPRESSION, PLUMBING & HVAC	97.7	81.5	91.2	101.8	91.6	97.7	97.7	89.4	94.3	98.0	77.1	89.6	97.7	91.3	95.1	101.4	91.2	97.3
26, 27, 3370	ELECTRICAL, COMMUNICATIONS & UTIL.	95.3	86.4	90.9	95.0	89.8	92.4	92.1	81.0	86.6	96.8	74.0	85.5	95.5	90.7	93.1	95.4	85.0	90.3
MF2018	WEIGHTED AVERAGE	98.7	87.3	93.7	97.8	91.7	95.1	99.0	86.8	93.8	96.5	80.8	89.7	99.3	91.4	95.9	98.9	90.0	95.1

WEST VIRGINIA / WISCONSIN

DIVISION		PETERSBURG 268 MAT.	INST.	TOTAL	ROMNEY 267 MAT.	INST.	TOTAL	WHEELING 260 MAT.	INST.	TOTAL	BELOIT 535 MAT.	INST.	TOTAL	EAU CLAIRE 547 MAT.	INST.	TOTAL	GREEN BAY 541 - 543 MAT.	INST.	TOTAL
015433	CONTRACTOR EQUIPMENT		102.2	102.2		102.2	102.2		102.2	102.2		96.6	96.6		98.1	98.1		95.8	95.8
0241, 31 - 34	SITE & INFRASTRUCTURE, DEMOLITION	100.8	85.6	90.4	103.7	85.6	91.3	111.0	85.7	93.7	96.1	98.3	97.6	96.4	97.7	97.3	99.9	93.7	95.7
0310	Concrete Forming & Accessories	88.6	85.3	85.8	84.0	85.2	85.0	91.5	87.1	87.7	99.0	93.1	94.0	100.2	98.1	98.4	112.4	102.8	104.3
0320	Concrete Reinforcing	96.3	87.3	91.9	96.9	90.9	94.0	95.7	96.3	96.0	102.7	136.5	119.0	93.7	113.9	103.5	91.9	113.9	102.5
0330	Cast-in-Place Concrete	101.3	90.7	97.3	106.0	82.4	97.1	103.5	93.0	99.6	101.9	96.5	99.8	97.8	103.1	99.8	101.1	103.0	101.8
03	CONCRETE	94.6	88.6	91.9	97.9	86.3	92.7	99.7	91.8	96.2	98.8	102.0	100.2	93.6	102.8	97.7	96.9	105.0	100.5
04	MASONRY	99.1	88.3	92.5	96.8	88.2	91.6	107.0	86.8	94.7	95.1	109.8	104.0	92.1	111.9	104.1	122.5	111.0	115.5
05	METALS	103.8	103.3	103.6	103.8	104.4	104.0	104.5	107.1	105.3	96.7	113.6	102.0	91.7	108.5	96.9	94.3	108.9	98.8
06	WOOD, PLASTICS & COMPOSITES	84.3	84.4	84.4	78.5	84.4	81.6	85.4	86.8	86.1	95.7	90.9	93.2	101.8	96.5	99.0	112.8	102.9	107.6
07	THERMAL & MOISTURE PROTECTION	101.6	81.9	93.2	101.7	81.8	93.1	102.0	86.0	95.1	105.0	93.5	100.1	103.7	105.4	104.5	106.2	105.9	106.0
08	OPENINGS	98.0	80.5	93.8	98.0	81.4	93.9	96.6	86.9	94.2	97.9	105.5	99.7	103.0	96.1	101.3	98.8	105.8	100.5
0920	Plaster & Gypsum Board	93.9	84.1	87.5	91.4	84.1	86.6	94.2	86.5	89.2	90.9	91.3	91.2	99.7	96.9	97.9	98.3	103.5	101.7
0950, 0980	Ceilings & Acoustic Treatment	95.8	84.1	88.5	95.8	84.1	88.5	95.8	86.5	90.0	87.4	91.3	89.9	94.1	96.9	95.9	86.3	103.5	97.1
0960	Flooring	92.3	94.7	93.0	90.2	94.4	91.5	95.3	94.7	95.1	95.4	113.0	100.6	82.3	117.9	92.7	97.7	117.9	103.6
0970, 0990	Wall Finishes & Painting/Coating	97.7	89.0	92.5	97.7	89.0	92.5	97.7	89.0	92.5	88.9	98.7	94.7	80.1	83.6	82.2	87.9	106.2	98.8
09	FINISHES	93.5	87.5	90.3	92.8	87.5	89.9	95.0	88.5	91.5	92.5	96.7	94.8	89.0	99.9	94.9	92.4	106.0	99.7
COVERS	DIVS. 10 - 14, 25, 28, 41, 43, 44, 46	100.0	92.5	98.2	100.0	92.5	98.2	100.0	91.6	98.0	100.0	98.2	99.6	100.0	96.7	99.2	100.0	98.1	99.5
21, 22, 23	FIRE SUPPRESSION, PLUMBING & HVAC	97.7	81.9	91.3	97.7	81.6	91.2	101.5	90.5	97.1	98.4	91.2	95.5	100.6	90.3	96.5	100.9	97.4	99.5
26, 27, 3370	ELECTRICAL, COMMUNICATIONS & UTIL.	98.0	74.0	86.1	97.4	74.0	85.8	93.2	90.7	91.9	98.8	81.1	90.1	102.4	102.9	102.6	97.9	102.9	100.4
MF2018	WEIGHTED AVERAGE	98.3	85.6	92.8	98.5	85.3	92.8	100.1	90.9	96.1	97.8	97.3	97.6	97.4	100.4	98.7	99.4	103.1	101.0

WISCONSIN

DIVISION		KENOSHA 531 MAT.	INST.	TOTAL	LA CROSSE 546 MAT.	INST.	TOTAL	LANCASTER 538 MAT.	INST.	TOTAL	MADISON 537 MAT.	INST.	TOTAL	MILWAUKEE 530,532 MAT.	INST.	TOTAL	NEW RICHMOND 540 MAT.	INST.	TOTAL
015433	CONTRACTOR EQUIPMENT		94.6	94.6		98.1	98.1		96.6	96.6		101.6	101.6		90.8	90.8		98.4	98.4
0241, 31 - 34	SITE & INFRASTRUCTURE, DEMOLITION	103.0	96.2	98.3	90.1	97.7	95.3	94.8	97.7	96.8	94.9	107.7	103.7	91.0	97.2	95.3	93.0	96.7	95.5
0310	Concrete Forming & Accessories	110.2	113.8	113.3	84.6	98.0	96.0	98.2	92.0	92.9	103.6	106.2	105.8	102.6	116.9	114.8	93.8	89.2	89.9
0320	Concrete Reinforcing	102.5	113.9	108.0	93.5	113.9	103.3	103.9	102.1	103.0	103.4	114.0	108.5	102.0	116.7	109.1	91.0	109.7	100.0
0330	Cast-in-Place Concrete	110.9	105.5	108.8	87.9	103.0	93.6	101.3	96.0	99.3	93.7	108.9	99.4	94.5	118.1	103.4	101.8	99.0	100.7
03	CONCRETE	103.7	110.6	106.8	85.0	102.8	92.9	98.5	95.4	97.1	99.9	108.2	103.6	100.0	116.2	107.3	91.1	96.6	93.6
04	MASONRY	93.0	115.8	106.9	91.3	111.9	103.8	95.1	109.8	104.1	94.5	119.5	109.7	97.4	121.2	111.9	117.0	109.9	112.7
05	METALS	97.8	106.1	100.4	91.6	108.4	96.8	94.0	100.2	95.9	100.8	104.6	102.0	95.7	98.9	96.7	92.0	105.5	96.2
06	WOOD, PLASTICS & COMPOSITES	105.7	115.2	110.7	84.0	96.5	90.5	94.8	90.9	92.8	98.9	103.0	101.0	102.3	115.6	109.3	89.7	87.1	88.4
07	THERMAL & MOISTURE PROTECTION	105.7	109.9	107.5	103.1	104.8	103.8	104.8	93.2	99.8	106.0	111.2	108.2	104.4	115.3	109.1	104.0	100.1	102.3
08	OPENINGS	92.2	112.7	97.2	102.9	96.2	101.3	93.8	88.5	92.6	101.5	105.7	102.5	102.5	113.5	105.2	88.5	87.0	88.1
0920	Plaster & Gypsum Board	78.8	116.3	103.4	95.1	96.9	96.3	89.5	91.3	90.7	100.5	103.5	102.5	92.9	116.0	108.0	85.8	87.5	86.9
0950, 0980	Ceilings & Acoustic Treatment	87.4	116.3	105.5	92.9	96.9	95.4	82.3	91.3	88.0	95.7	103.5	100.6	88.8	116.0	105.8	58.9	87.5	76.8
0960	Flooring	113.2	117.9	114.6	76.3	117.9	88.4	95.1	105.9	98.2	95.0	117.9	101.7	101.5	118.1	106.4	91.2	109.3	96.5
0970, 0990	Wall Finishes & Painting/Coating	99.7	122.4	113.3	80.1	81.1	80.7	88.9	93.0	91.4	90.9	104.7	99.2	100.8	122.2	113.6	90.8	77.8	83.0
09	FINISHES	96.7	116.2	107.2	85.9	99.7	93.4	90.9	94.5	92.9	95.8	108.0	102.4	96.3	118.1	108.1	81.9	91.4	87.0
COVERS	DIVS. 10 - 14, 25, 28, 41, 43, 44, 46	100.0	99.8	100.0	100.0	96.7	99.2	100.0	98.2	99.6	100.0	103.4	100.8	100.0	104.2	101.0	100.0	90.3	97.7
21, 22, 23	FIRE SUPPRESSION, PLUMBING & HVAC	100.8	97.3	99.4	100.6	90.3	96.5	94.7	85.0	90.8	96.1	100.6	97.9	96.1	117.1	104.6	96.2	84.8	91.6
26, 27, 3370	ELECTRICAL, COMMUNICATIONS & UTIL.	100.0	102.9	101.4	102.7	102.9	102.8	98.5	81.7	90.2	98.9	102.8	100.9	97.8	102.8	100.3	100.8	81.7	91.4
MF2018	WEIGHTED AVERAGE	99.2	106.2	102.2	95.7	100.3	97.7	95.8	92.9	94.6	98.6	106.4	101.9	97.9	111.7	103.9	94.9	92.9	94.0

WISCONSIN

DIVISION		OSHKOSH 549 MAT.	INST.	TOTAL	PORTAGE 539 MAT.	INST.	TOTAL	RACINE 534 MAT.	INST.	TOTAL	RHINELANDER 545 MAT.	INST.	TOTAL	SUPERIOR 548 MAT.	INST.	TOTAL	WAUSAU 544 MAT.	INST.	TOTAL
015433	CONTRACTOR EQUIPMENT		95.8	95.8		96.6	96.6		96.6	96.6		95.8	95.8		98.4	98.4		95.8	95.8
0241, 31 - 34	SITE & INFRASTRUCTURE, DEMOLITION	91.1	92.9	92.3	85.7	98.6	94.5	96.7	99.8	98.8	103.5	92.4	95.9	89.7	96.4	94.3	86.9	93.5	91.5
0310	Concrete Forming & Accessories	92.5	92.4	92.4	89.7	92.7	92.2	99.4	110.8	109.1	89.7	91.5	91.2	91.6	85.6	86.5	91.9	94.3	94.0
0320	Concrete Reinforcing	92.0	105.5	98.6	104.0	102.2	103.1	102.7	113.9	108.1	92.2	102.3	97.1	91.0	109.5	99.9	92.2	113.7	102.6
0330	Cast-in-Place Concrete	93.7	95.7	94.5	86.9	94.7	89.9	99.9	105.3	101.9	106.3	94.4	101.8	95.9	94.7	95.5	87.4	89.2	88.1
03	CONCRETE	86.7	96.2	91.0	86.2	95.5	90.4	97.9	109.2	103.0	98.4	94.8	96.8	85.9	93.5	89.3	81.9	96.3	88.4
04	MASONRY	105.8	111.3	109.1	94.0	109.8	103.6	95.3	115.8	107.8	125.9	109.1	115.7	116.3	107.6	111.0	105.3	109.1	107.6
05	METALS	92.2	104.0	95.9	94.7	104.3	97.7	98.3	106.1	100.8	92.0	102.8	95.4	93.0	105.8	97.0	91.9	107.6	96.8
06	WOOD, PLASTICS & COMPOSITES	89.0	91.0	90.1	83.8	90.9	87.5	96.1	111.2	104.0	85.9	91.0	88.6	87.8	82.9	85.2	88.4	93.9	91.3
07	THERMAL & MOISTURE PROTECTION	105.1	85.3	96.6	104.2	99.4	102.1	105.1	109.1	106.8	106.0	83.6	96.4	103.7	94.7	99.8	104.9	95.2	100.7
08	OPENINGS	95.4	92.8	94.8	94.0	96.0	94.5	97.9	110.4	101.0	95.4	88.5	93.8	87.9	91.0	88.6	95.6	96.3	95.8
0920	Plaster & Gypsum Board	86.1	91.3	89.5	83.2	91.3	88.5	90.9	112.1	104.8	86.1	91.3	89.5	86.1	83.1	84.1	86.1	94.3	91.5
0950, 0980	Ceilings & Acoustic Treatment	86.3	91.3	89.4	86.1	91.3	89.4	87.4	112.1	102.9	86.3	91.3	89.4	60.2	83.1	74.5	86.3	94.3	91.3
0960	Flooring	89.2	109.5	95.1	91.1	109.3	96.4	95.4	117.9	102.0	88.5	109.3	94.6	92.3	117.4	99.7	89.1	109.3	95.0
0970, 0990	Wall Finishes & Painting/Coating	85.6	106.2	98.0	88.9	93.0	91.4	88.8	121.1	108.2	85.6	76.2	79.9	80.1	100.7	92.5	85.6	79.3	81.8
09	FINISHES	87.5	97.3	92.8	89.2	95.4	92.6	92.5	113.6	103.9	88.3	93.4	91.1	81.5	92.5	87.5	87.2	95.4	91.7
COVERS	DIVS. 10 - 14, 25, 28, 41, 43, 44, 46	100.0	81.9	95.7	100.0	98.3	99.6	100.0	99.4	99.9	100.0	96.0	99.1	100.0	89.1	97.4	100.0	96.4	99.1
21, 22, 23	FIRE SUPPRESSION, PLUMBING & HVAC	97.1	84.4	92.0	94.7	91.3	93.4	100.6	99.2	100.0	97.1	84.4	92.0	96.2	86.8	92.5	97.1	85.0	92.2
26, 27, 3370	ELECTRICAL, COMMUNICATIONS & UTIL.	101.3	81.2	91.3	101.8	90.5	96.2	99.6	102.9	101.2	100.8	75.2	88.1	105.0	94.9	100.0	102.3	75.2	88.9
MF2018	WEIGHTED AVERAGE	95.2	92.7	94.1	94.2	96.5	95.2	98.5	106.2	101.8	98.0	91.1	95.0	94.6	94.6	94.6	94.5	92.9	93.8

WYOMING

DIVISION		CASPER 826 MAT.	INST.	TOTAL	CHEYENNE 820 MAT.	INST.	TOTAL	NEWCASTLE 827 MAT.	INST.	TOTAL	RAWLINS 823 MAT.	INST.	TOTAL	RIVERTON 825 MAT.	INST.	TOTAL	ROCK SPRINGS 829-831 MAT.	INST.	TOTAL
015433	CONTRACTOR EQUIPMENT		98.5	98.5		92.3	92.3		92.3	92.3		92.3	92.3		92.3	92.3		92.3	92.3
0241, 31 - 34	SITE & INFRASTRUCTURE, DEMOLITION	98.8	96.8	97.4	91.5	87.2	88.5	83.3	87.2	86.0	97.1	87.2	90.3	90.7	87.2	88.3	87.0	87.2	87.1
0310	Concrete Forming & Accessories	99.6	64.4	69.7	103.4	63.8	69.7	92.8	64.2	68.5	97.4	64.2	69.1	91.6	64.1	68.2	99.2	64.0	69.2
0320	Concrete Reinforcing	105.9	81.4	94.1	97.3	81.5	89.7	104.5	81.6	93.5	104.2	81.6	93.3	105.2	81.6	93.8	105.2	80.9	93.5
0330	Cast-in-Place Concrete	104.7	79.4	95.1	98.7	78.0	90.9	99.7	78.0	91.5	99.8	78.0	91.5	99.7	77.9	91.5	99.7	77.9	91.4
03	CONCRETE	109.2	73.0	93.0	101.7	72.3	88.6	101.8	72.5	88.7	116.6	72.5	96.9	110.7	72.4	93.5	102.3	72.3	88.8
04	MASONRY	96.8	65.0	77.4	98.5	66.5	79.0	95.5	68.0	78.7	95.5	68.0	78.7	95.5	68.0	78.7	152.7	61.1	96.8
05	METALS	101.5	79.0	94.5	103.8	80.4	96.5	100.0	80.5	93.9	100.0	80.5	93.9	100.1	80.3	93.9	100.9	79.7	94.2
06	WOOD, PLASTICS & COMPOSITES	94.7	62.2	77.7	94.4	61.4	77.2	83.1	61.9	72.0	87.7	61.9	74.2	81.8	61.9	71.4	92.4	61.9	76.4
07	THERMAL & MOISTURE PROTECTION	109.2	67.6	91.3	105.5	67.5	89.2	106.5	67.2	89.6	108.0	67.2	90.5	107.4	70.4	91.5	106.6	68.3	90.2
08	OPENINGS	109.2	67.1	99.0	107.0	66.7	97.2	111.0	65.5	99.9	110.6	65.5	99.5	110.8	65.5	99.8	111.4	66.3	100.4
0920	Plaster & Gypsum Board	96.9	61.1	73.4	85.8	60.6	69.3	82.8	61.1	68.6	83.1	61.1	68.7	82.8	61.1	68.6	94.7	61.1	72.7
0950, 0980	Ceilings & Acoustic Treatment	119.8	61.1	83.0	107.8	60.6	78.2	110.7	61.1	79.6	110.7	61.1	79.6	110.7	61.1	79.6	110.7	61.1	79.6
0960	Flooring	103.9	72.8	94.8	102.9	72.8	94.1	96.6	67.6	88.1	99.7	67.6	90.4	95.9	67.6	87.7	102.4	58.3	89.5
0970, 0990	Wall Finishes & Painting/Coating	98.3	58.1	74.2	97.7	58.1	74.0	94.3	74.6	82.5	94.3	74.6	82.5	94.3	57.6	72.3	94.3	74.6	82.5
09	FINISHES	103.7	64.6	82.6	99.4	64.1	80.3	95.0	65.2	78.9	97.2	65.2	79.9	95.6	63.3	78.1	98.3	63.3	79.4
COVERS	DIVS. 10 - 14, 25, 28, 41, 43, 44, 46	100.0	88.6	97.3	100.0	87.9	97.2	100.0	98.4	99.6	100.0	98.4	99.6	100.0	87.2	97.0	100.0	85.0	96.5
21, 22, 23	FIRE SUPPRESSION, PLUMBING & HVAC	100.9	74.6	90.3	101.2	74.6	90.5	99.1	71.8	88.1	99.1	71.8	88.1	99.1	71.8	88.1	101.1	71.8	89.3
26, 27, 3370	ELECTRICAL, COMMUNICATIONS & UTIL.	97.0	62.0	79.7	95.2	67.1	81.6	94.0	60.0	77.2	94.0	60.0	77.2	94.0	64.2	79.3	92.7	64.8	78.9
MF2018	WEIGHTED AVERAGE	102.5	72.3	89.4	101.0	72.4	88.7	99.4	71.3	87.3	101.9	71.3	88.7	100.8	71.4	88.1	103.3	70.6	89.1

WYOMING / CANADA

DIVISION		SHERIDAN 828 MAT.	INST.	TOTAL	WHEATLAND 822 MAT.	INST.	TOTAL	WORLAND 824 MAT.	INST.	TOTAL	YELLOWSTONE NAT'L PA 821 MAT.	INST.	TOTAL	BARRIE, ONTARIO MAT.	INST.	TOTAL	BATHURST, NEW BRUNSWICK MAT.	INST.	TOTAL
015433	CONTRACTOR EQUIPMENT		92.3	92.3		92.3	92.3		92.3	92.3		92.3	92.3		97.9	97.9		97.4	97.4
0241, 31 - 34	SITE & INFRASTRUCTURE, DEMOLITION	91.1	87.2	88.4	87.6	87.2	87.3	84.9	87.2	86.5	85.0	87.2	86.5	116.9	91.6	99.5	104.6	87.4	92.8
0310	Concrete Forming & Accessories	101.8	64.0	69.7	94.8	63.9	68.5	94.9	64.2	68.8	95.0	64.1	68.7	125.7	80.9	87.6	104.9	56.8	63.9
0320	Concrete Reinforcing	105.2	81.6	93.8	104.5	80.8	93.1	105.2	81.6	93.8	107.0	79.0	93.5	171.9	84.6	129.8	137.9	56.0	98.4
0330	Cast-in-Place Concrete	103.1	77.9	93.6	104.1	77.9	94.2	99.7	78.0	91.5	99.7	77.9	91.5	155.6	80.1	127.1	112.6	54.5	90.6
03	CONCRETE	110.8	72.4	93.6	107.0	72.2	94.2	102.0	72.6	88.8	102.2	72.0	88.7	141.5	81.8	114.7	116.6	56.9	89.8
04	MASONRY	95.8	68.0	78.9	95.8	56.3	71.7	95.5	68.0	78.7	95.5	68.0	78.8	169.0	87.4	119.2	164.9	56.2	98.6
05	METALS	103.6	80.3	96.4	99.9	79.3	93.5	100.1	80.5	94.0	100.7	79.0	94.0	111.5	90.6	105.0	113.4	72.2	100.6
06	WOOD, PLASTICS & COMPOSITES	96.1	61.9	78.2	85.2	61.9	73.0	85.3	61.9	73.1	85.3	61.9	73.1	116.6	79.7	97.3	95.0	57.0	75.2
07	THERMAL & MOISTURE PROTECTION	107.9	68.0	90.8	106.7	63.9	88.3	106.5	67.2	89.6	106.0	67.2	89.3	114.2	83.8	101.1	110.4	56.8	87.4
08	OPENINGS	111.6	65.5	100.4	109.6	65.5	98.8	111.2	65.5	100.1	104.3	64.6	94.6	90.4	79.6	87.8	84.2	50.3	75.9
0920	Plaster & Gypsum Board	104.5	61.1	76.0	82.8	61.1	68.6	82.8	61.1	68.6	83.0	61.1	68.6	151.7	79.4	104.2	119.4	56.0	77.8
0950, 0980	Ceilings & Acoustic Treatment	114.9	61.1	81.2	110.7	61.1	79.6	110.7	61.1	79.6	111.3	61.1	79.8	95.4	79.4	85.3	115.9	56.0	78.4
0960	Flooring	101.0	57.5	88.3	98.2	67.6	89.3	98.2	67.6	89.3	98.2	67.6	89.3	121.5	85.1	110.8	101.2	40.5	83.4
0970, 0990	Wall Finishes & Painting/Coating	96.6	58.1	73.5	94.3	55.6	71.1	94.3	74.6	82.5	94.3	57.6	72.3	102.9	81.6	90.1	108.6	46.6	71.5
09	FINISHES	103.0	61.8	80.7	95.7	63.1	78.1	95.4	65.2	79.1	95.6	63.3	78.1	112.9	81.9	96.1	109.1	52.7	78.6
COVERS	DIVS. 10 - 14, 25, 28, 41, 43, 44, 46	100.0	87.9	97.2	100.0	87.9	97.2	100.0	98.4	99.6	100.0	98.4	99.6	139.3	62.8	121.3	131.2	55.8	113.4
21, 22, 23	FIRE SUPPRESSION, PLUMBING & HVAC	99.1	71.8	88.1	99.1	71.8	88.1	99.1	71.8	88.1	99.1	71.8	88.1	106.2	91.1	100.1	106.2	63.0	88.8
26, 27, 3370	ELECTRICAL, COMMUNICATIONS & UTIL.	95.9	64.2	80.2	94.0	64.2	79.3	94.0	64.2	79.3	93.2	64.2	78.9	117.2	81.6	99.6	117.0	55.0	86.3
MF2018	WEIGHTED AVERAGE	102.4	71.1	88.9	100.1	69.9	87.0	99.5	71.9	87.0	98.9	71.4	87.0	118.0	85.2	103.8	112.7	60.8	90.3

CANADA

DIVISION		BRANDON, MANITOBA MAT.	INST.	TOTAL	BRANTFORD, ONTARIO MAT.	INST.	TOTAL	BRIDGEWATER, NOVA SCOTIA MAT.	INST.	TOTAL	CALGARY, ALBERTA MAT.	INST.	TOTAL	CAP-DE-LA-MADELEINE, QUEBEC MAT.	INST.	TOTAL	CHARLESBOURG, QUEBEC MAT.	INST.	TOTAL
015433	CONTRACTOR EQUIPMENT		99.6	99.6		97.9	97.9		97.3	97.3		122.7	122.7		98.1	98.1		98.1	98.1
0241, 31 - 34	SITE & INFRASTRUCTURE, DEMOLITION	124.4	89.9	100.7	116.1	91.9	99.4	100.2	89.3	92.7	126.4	112.5	116.9	96.2	90.6	92.4	96.2	90.6	92.4
0310	Concrete Forming & Accessories	149.5	64.9	77.5	129.0	87.6	93.8	97.4	66.7	71.3	125.3	92.0	96.9	136.3	77.8	86.5	136.3	77.8	86.5
0320	Concrete Reinforcing	167.0	53.1	112.0	160.7	83.3	123.3	136.9	46.4	93.2	133.9	78.7	107.3	136.9	70.5	104.8	136.9	70.5	104.8
0330	Cast-in-Place Concrete	108.0	68.4	93.0	129.0	99.6	117.9	132.5	65.1	107.0	134.5	101.3	122.0	104.1	85.4	97.0	104.1	85.4	97.0
03	CONCRETE	125.5	65.1	98.4	128.4	91.4	111.8	126.4	63.6	98.3	136.8	93.7	117.5	114.1	79.8	98.7	114.1	79.8	98.7
04	MASONRY	221.6	59.4	122.6	173.0	91.5	123.3	167.9	64.2	104.6	231.4	86.3	142.9	169.1	74.8	111.5	169.1	74.8	111.5
05	METALS	126.0	77.2	110.8	111.4	91.2	105.1	110.6	74.9	99.5	128.4	98.4	119.1	109.7	84.0	101.7	109.7	84.0	101.7
06	WOOD, PLASTICS & COMPOSITES	150.7	65.9	106.3	120.4	86.6	102.7	86.3	66.5	75.9	99.8	91.9	95.7	132.8	78.0	104.1	132.8	78.0	104.1
07	THERMAL & MOISTURE PROTECTION	126.8	66.8	101.0	120.1	89.0	106.8	114.8	65.8	93.8	136.1	94.8	118.4	113.9	81.9	100.2	113.9	81.9	100.2
08	OPENINGS	100.2	58.9	90.1	87.7	85.3	87.1	82.7	60.3	77.3	82.4	81.2	82.1	89.2	71.1	84.8	89.2	71.1	84.8
0920	Plaster & Gypsum Board	114.9	64.8	82.0	114.7	86.4	96.1	120.4	65.7	84.5	132.6	90.9	105.2	143.9	77.4	100.3	143.9	77.4	100.3
0950, 0980	Ceilings & Acoustic Treatment	122.1	64.8	86.2	102.7	86.4	92.5	102.7	65.7	79.5	146.1	90.9	111.5	102.7	77.4	86.9	102.7	77.4	86.9
0960	Flooring	133.4	60.4	112.1	115.3	85.1	106.5	96.7	57.4	85.2	124.1	81.3	111.5	115.3	83.8	106.1	115.3	83.8	106.1
0970, 0990	Wall Finishes & Painting/Coating	114.4	52.6	77.4	107.2	89.5	96.6	107.2	57.7	77.6	107.8	101.3	103.9	107.2	81.4	91.7	107.2	81.4	91.7
09	FINISHES	124.1	63.1	91.1	109.7	87.6	97.8	104.5	64.3	82.8	127.8	91.5	108.2	113.1	79.6	95.0	113.1	79.6	95.0
COVERS	DIVS. 10 - 14, 25, 28, 41, 43, 44, 46	131.2	58.0	114.0	131.2	64.7	115.5	131.2	58.7	114.1	131.2	88.4	121.1	131.2	73.1	117.5	131.2	73.1	117.5
21, 22, 23	FIRE SUPPRESSION, PLUMBING & HVAC	106.5	76.8	94.5	106.2	93.8	101.2	106.2	77.4	94.6	106.6	86.0	98.3	106.6	82.0	96.7	106.6	82.0	96.7
26, 27, 3370	ELECTRICAL, COMMUNICATIONS & UTIL.	116.4	62.1	89.5	114.5	81.1	98.0	118.8	57.7	88.6	114.0	90.4	102.3	113.2	64.2	89.0	113.2	64.2	89.0
MF2018	WEIGHTED AVERAGE	122.8	68.9	99.5	115.2	88.8	103.8	113.3	68.7	94.0	123.3	91.7	109.7	112.8	78.2	97.8	112.8	78.2	97.8

CANADA

DIVISION		CHARLOTTETOWN, PRINCE EDWARD ISLAND			CHICOUTIMI, QUEBEC			CORNER BROOK, NEWFOUNDLAND			CORNWALL, ONTARIO			DALHOUSIE, NEW BRUNSWICK			DARTMOUTH, NOVA SCOTIA		
		MAT.	INST.	TOTAL	MAT.	INST.	TOTAL	MAT.	INST.	TOTAL	MAT.	INST.	TOTAL	MAT.	INST.	TOTAL	MAT.	INST.	TOTAL
015433	CONTRACTOR EQUIPMENT		115.1	115.1		97.9	97.9		97.7	97.7		97.9	97.9		98.0	98.0		96.8	96.8
0241, 31 - 34	SITE & INFRASTRUCTURE, DEMOLITION	132.1	101.4	111.0	101.7	90.5	94.0	127.6	87.3	99.9	114.3	91.4	98.6	99.9	87.8	91.6	116.0	88.9	97.4
0310	Concrete Forming & Accessories	125.5	52.1	63.0	138.3	89.8	97.0	122.9	74.1	81.4	126.3	81.0	87.8	104.3	57.0	64.0	111.1	66.7	73.3
0320	Concrete Reinforcing	153.8	45.3	101.4	104.0	93.7	99.0	153.1	47.5	102.1	160.7	83.0	123.2	140.0	56.0	99.5	160.1	46.4	105.2
0330	Cast-in-Place Concrete	150.0	56.3	114.6	106.0	93.8	101.4	125.7	61.8	101.5	116.0	90.7	106.4	109.0	54.7	88.5	121.3	65.0	100.0
03	CONCRETE	147.7	54.0	105.6	107.7	92.0	100.6	160.3	65.9	118.0	122.2	85.2	105.6	120.0	57.1	91.7	140.5	63.6	106.0
04	MASONRY	189.1	53.8	106.6	167.5	89.8	120.1	216.3	72.6	128.7	172.0	83.8	118.2	170.1	56.2	100.6	231.9	64.2	129.6
05	METALS	133.6	75.8	115.6	113.2	91.5	106.4	126.4	73.9	110.0	111.2	89.9	104.6	104.6	72.8	94.6	126.8	74.6	110.5
06	WOOD, PLASTICS & COMPOSITES	102.2	51.7	75.8	133.0	90.7	110.9	120.8	80.1	101.8	118.2	80.5	98.5	93.2	57.1	74.3	112.1	66.4	88.2
07	THERMAL & MOISTURE PROTECTION	135.4	56.3	101.5	111.0	95.6	104.4	130.8	64.1	102.2	119.9	83.9	104.5	119.6	56.8	92.6	128.1	65.8	101.3
08	OPENINGS	85.8	45.0	75.9	87.9	76.3	85.1	106.0	65.3	96.1	89.2	79.3	86.8	85.7	50.3	77.1	91.6	60.3	83.9
0920	Plaster & Gypsum Board	126.1	49.9	76.1	138.9	90.5	107.2	148.0	79.7	103.1	172.0	80.2	111.7	127.7	56.0	80.6	144.2	65.7	92.7
0950, 0980	Ceilings & Acoustic Treatment	131.4	49.9	80.3	114.6	90.5	99.5	122.7	79.7	95.7	106.6	80.2	90.0	105.0	56.0	74.3	131.0	65.7	90.1
0960	Flooring	119.9	53.9	100.6	117.8	83.8	107.9	114.3	48.3	95.0	115.3	83.9	106.1	103.2	61.9	91.1	108.9	57.4	93.9
0970, 0990	Wall Finishes & Painting/Coating	108.7	38.8	66.8	108.6	103.3	105.4	114.3	55.0	78.7	107.2	83.4	93.0	110.3	46.6	72.2	114.3	57.7	80.4
09	FINISHES	123.0	51.3	84.2	116.0	91.2	102.6	123.8	68.7	94.0	118.7	82.0	98.8	109.3	57.1	81.1	122.4	64.2	90.9
COVERS	DIVS. 10 - 14, 25, 28, 41, 43, 44, 46	131.2	56.4	113.6	131.2	78.2	118.7	131.2	58.4	114.1	131.2	62.4	115.0	131.2	55.9	113.4	131.2	58.7	114.1
21, 22, 23	FIRE SUPPRESSION, PLUMBING & HVAC	106.5	57.4	86.7	106.2	80.3	95.7	106.5	64.1	89.4	106.6	91.8	100.7	106.3	63.1	88.8	106.5	77.4	94.7
26, 27, 3370	ELECTRICAL, COMMUNICATIONS & UTIL.	119.8	46.7	83.6	113.5	84.3	99.0	113.5	50.7	82.4	114.6	82.1	98.5	117.6	52.0	85.1	117.6	57.7	88.0
MF2018	WEIGHTED AVERAGE	124.1	58.7	95.8	112.5	86.9	101.4	127.3	66.7	101.1	115.3	85.5	102.4	112.4	61.0	90.2	124.0	68.6	100.1

CANADA

DIVISION		EDMONTON, ALBERTA			FORT MCMURRAY, ALBERTA			FREDERICTON, NEW BRUNSWICK			GATINEAU, QUEBEC			GRANBY, QUEBEC			HALIFAX, NOVA SCOTIA		
		MAT.	INST.	TOTAL	MAT.	INST.	TOTAL	MAT.	INST.	TOTAL	MAT.	INST.	TOTAL	MAT.	INST.	TOTAL	MAT.	INST.	TOTAL
015433	CONTRACTOR EQUIPMENT		123.6	123.6		100.2	100.2		115.0	115.0		98.1	98.1		98.1	98.1		112.9	112.9
0241, 31 - 34	SITE & INFRASTRUCTURE, DEMOLITION	126.6	114.2	118.1	120.8	92.5	101.4	116.1	102.5	106.8	96.0	90.6	92.3	96.4	90.6	92.4	101.8	102.3	102.1
0310	Concrete Forming & Accessories	128.7	92.0	97.4	126.2	85.9	91.9	128.1	57.7	68.2	136.3	77.7	86.4	136.3	77.6	86.4	125.4	85.1	91.1
0320	Concrete Reinforcing	133.5	78.7	107.1	148.7	78.6	114.9	138.3	56.2	98.7	144.7	70.5	108.9	144.7	70.5	108.9	153.3	75.3	115.7
0330	Cast-in-Place Concrete	143.8	101.3	127.8	171.2	95.1	142.5	111.1	56.8	90.6	102.7	85.4	96.1	106.1	85.3	98.3	97.9	82.3	92.0
03	CONCRETE	141.3	93.7	119.9	146.3	88.3	120.3	127.7	58.6	96.7	114.5	79.7	98.9	116.1	79.7	99.7	122.5	83.4	105.0
04	MASONRY	230.8	86.3	142.7	214.4	83.6	134.6	187.9	57.6	108.4	168.9	74.8	111.5	169.2	74.8	111.6	184.4	83.7	122.9
05	METALS	131.9	98.3	121.4	138.5	89.6	123.2	133.8	81.1	117.4	109.7	83.9	101.6	109.9	83.8	101.8	133.5	95.0	121.5
06	WOOD, PLASTICS & COMPOSITES	99.1	91.9	95.3	114.1	85.6	99.1	104.4	57.5	79.8	132.8	78.0	104.1	132.8	78.0	104.1	101.6	85.6	93.2
07	THERMAL & MOISTURE PROTECTION	141.3	94.8	121.3	128.8	89.4	111.9	135.5	58.3	102.4	113.9	81.9	100.2	113.9	80.5	99.5	137.8	85.9	115.5
08	OPENINGS	81.1	81.2	81.1	89.2	77.7	86.4	88.1	49.5	78.7	89.2	66.8	83.8	89.2	66.8	83.8	89.1	77.0	86.2
0920	Plaster & Gypsum Board	131.9	90.9	105.0	115.2	85.0	95.4	128.8	56.9	81.0	114.1	77.4	90.0	114.1	77.4	90.0	122.7	85.0	97.9
0950, 0980	Ceilings & Acoustic Treatment	141.7	90.9	109.9	111.7	85.0	95.0	132.7	56.0	84.6	102.7	77.4	86.9	102.7	77.4	86.9	129.1	85.0	101.4
0960	Flooring	126.3	81.3	113.1	115.3	81.3	105.4	120.7	64.6	104.3	115.3	83.8	106.1	115.3	83.8	106.1	113.9	78.1	103.4
0970, 0990	Wall Finishes & Painting/Coating	110.4	101.3	105.0	107.3	85.9	94.5	109.9	59.4	79.7	107.2	81.4	91.7	107.2	81.4	91.7	109.5	87.5	96.3
09	FINISHES	128.6	91.5	107.7	113.1	85.4	98.1	126.3	59.3	90.1	108.9	79.6	93.0	108.9	79.6	93.0	121.1	84.7	101.4
COVERS	DIVS. 10 - 14, 25, 28, 41, 43, 44, 46	131.2	89.6	121.4	131.2	86.5	120.7	131.2	56.8	113.7	131.2	73.1	117.5	131.2	73.1	117.5	131.2	65.6	115.7
21, 22, 23	FIRE SUPPRESSION, PLUMBING & HVAC	106.4	86.0	98.2	106.7	90.7	100.2	106.7	71.8	92.6	106.6	82.0	96.7	106.2	82.0	96.4	106.5	78.9	95.4
26, 27, 3370	ELECTRICAL, COMMUNICATIONS & UTIL.	123.4	90.4	107.1	108.5	76.3	92.6	122.1	68.5	95.6	113.2	64.2	89.0	113.9	64.2	89.3	123.7	87.2	105.6
MF2018	WEIGHTED AVERAGE	125.3	91.9	110.9	123.9	86.2	107.6	121.9	67.8	98.5	112.5	78.0	97.6	112.7	77.9	97.7	120.5	84.9	105.1

CANADA

DIVISION		HAMILTON, ONTARIO			HULL, QUEBEC			JOLIETTE, QUEBEC			KAMLOOPS, BRITISH COLUMBIA			KINGSTON, ONTARIO			KITCHENER, ONTARIO		
		MAT.	INST.	TOTAL	MAT.	INST.	TOTAL	MAT.	INST.	TOTAL	MAT.	INST.	TOTAL	MAT.	INST.	TOTAL	MAT.	INST.	TOTAL
015433	CONTRACTOR EQUIPMENT		110.1	110.1		98.1	98.1		98.1	98.1		101.3	101.3		100.1	100.1		99.4	99.4
0241, 31 - 34	SITE & INFRASTRUCTURE, DEMOLITION	105.3	100.8	102.2	96.0	90.6	92.3	96.5	90.6	92.5	119.7	94.1	102.1	114.3	95.0	101.0	93.6	96.1	95.3
0310	Concrete Forming & Accessories	132.8	91.0	97.2	136.3	77.7	86.4	136.3	77.8	86.5	123.0	80.8	87.1	126.5	81.1	87.8	118.4	84.7	89.7
0320	Concrete Reinforcing	134.4	98.4	117.0	144.7	70.5	108.9	136.9	70.5	104.8	107.3	73.8	91.1	160.7	83.0	123.2	94.7	98.2	96.4
0330	Cast-in-Place Concrete	106.1	97.9	103.0	102.7	85.4	96.1	106.9	85.4	96.8	92.6	89.6	91.5	116.0	90.7	106.4	110.3	93.3	103.9
03	CONCRETE	120.1	95.2	108.9	114.5	79.7	98.9	115.4	79.8	99.4	126.4	83.2	107.0	124.3	85.3	106.8	102.2	90.4	96.9
04	MASONRY	189.2	96.3	132.5	168.9	74.8	111.5	169.3	74.8	111.6	177.4	82.2	119.3	179.6	83.9	121.2	150.0	94.5	116.1
05	METALS	132.0	102.4	122.7	109.9	83.9	101.8	109.9	84.0	101.8	111.9	86.2	103.9	112.6	89.9	105.6	122.8	96.1	114.5
06	WOOD, PLASTICS & COMPOSITES	110.7	90.7	100.2	132.8	78.0	104.1	132.8	78.0	104.1	100.5	79.6	89.5	118.2	80.6	98.6	109.9	83.0	95.8
07	THERMAL & MOISTURE PROTECTION	132.8	97.6	117.7	113.9	81.9	100.1	113.9	81.9	100.1	129.8	79.5	108.2	119.9	85.0	104.9	115.3	94.3	106.3
08	OPENINGS	85.5	89.8	86.5	89.2	66.8	83.8	89.2	71.1	84.8	86.3	77.2	84.1	89.2	79.0	86.7	80.3	83.9	81.2
0920	Plaster & Gypsum Board	124.4	90.0	101.8	114.1	77.4	90.0	143.9	77.4	100.3	100.1	78.8	86.2	175.4	80.3	113.0	104.8	82.7	90.3
0950, 0980	Ceilings & Acoustic Treatment	137.4	90.0	107.7	102.7	77.4	86.9	102.7	77.4	86.9	102.7	78.8	87.8	119.3	80.3	94.8	106.7	82.7	91.7
0960	Flooring	121.1	91.0	112.3	115.3	83.8	106.1	115.3	83.8	106.1	114.8	48.0	95.3	115.3	83.9	106.1	102.9	91.1	99.5
0970, 0990	Wall Finishes & Painting/Coating	107.2	99.2	102.4	107.2	81.4	91.7	107.2	81.4	91.7	107.2	74.4	87.5	107.2	77.1	89.2	101.7	89.8	94.6
09	FINISHES	123.1	91.8	106.1	108.9	79.6	93.0	113.1	79.6	95.0	109.3	74.3	90.4	122.2	81.3	100.1	103.2	85.9	93.9
COVERS	DIVS. 10 - 14, 25, 28, 41, 43, 44, 46	131.2	87.6	120.9	131.2	73.1	117.5	131.2	73.1	117.5	131.2	81.6	119.5	131.2	62.4	115.0	131.2	85.2	120.3
21, 22, 23	FIRE SUPPRESSION, PLUMBING & HVAC	106.8	87.7	99.1	106.2	82.0	96.4	106.2	82.0	96.4	106.2	85.2	97.7	106.6	91.9	100.7	105.5	87.7	98.3
26, 27, 3370	ELECTRICAL, COMMUNICATIONS & UTIL.	115.5	99.8	107.7	115.3	64.2	90.0	113.9	64.2	89.3	117.3	73.0	95.4	114.6	80.8	97.9	111.3	97.5	104.5
MF2018	WEIGHTED AVERAGE	119.2	94.5	108.6	112.6	78.0	97.7	113.0	78.2	97.9	115.6	81.6	100.9	116.5	85.5	103.1	109.9	91.2	101.8

City Cost Indexes - V2

CANADA

DIVISION		LAVAL, QUEBEC MAT.	INST.	TOTAL	LETHBRIDGE, ALBERTA MAT.	INST.	TOTAL	LLOYDMINSTER, ALBERTA MAT.	INST.	TOTAL	LONDON, ONTARIO MAT.	INST.	TOTAL	MEDICINE HAT, ALBERTA MAT.	INST.	TOTAL	MONCTON, NEW BRUNSWICK MAT.	INST.	TOTAL
015433	CONTRACTOR EQUIPMENT		98.1	98.1		100.2	100.2		100.2	100.2		113.0	113.0		100.2	100.2		97.4	97.4
0241, 31 - 34	SITE & INFRASTRUCTURE, DEMOLITION	96.4	90.7	92.5	113.8	93.1	99.6	113.8	92.5	99.2	104.4	101.5	102.4	112.6	92.6	98.8	104.1	89.2	93.8
0310	Concrete Forming & Accessories	136.5	78.4	87.0	128.1	86.0	92.3	125.8	76.9	84.2	131.9	85.8	92.7	128.0	76.8	84.4	104.9	63.8	70.0
0320	Concrete Reinforcing	144.7	71.2	109.2	148.7	78.6	114.9	148.7	78.5	114.9	124.4	97.2	111.3	148.7	78.5	114.9	137.9	66.7	103.6
0330	Cast-in-Place Concrete	106.1	86.0	98.5	128.4	95.1	115.9	119.2	91.7	108.8	118.5	97.3	110.5	119.2	91.6	108.8	108.4	65.3	92.1
03	CONCRETE	116.1	80.3	100.0	126.5	88.3	109.4	122.0	83.0	104.5	124.5	92.6	110.1	122.2	82.9	104.6	114.6	66.0	92.8
04	MASONRY	169.2	75.4	112.0	188.1	83.6	124.4	168.3	77.6	113.0	200.2	95.4	136.3	168.3	77.6	113.0	164.6	71.7	107.9
05	METALS	109.8	84.2	101.8	132.7	89.6	119.3	111.9	89.4	104.9	132.0	104.1	123.3	112.1	89.4	105.0	113.4	85.0	104.5
06	WOOD, PLASTICS & COMPOSITES	132.9	78.6	104.5	117.9	85.6	101.0	114.1	76.4	94.4	113.7	83.9	98.1	117.9	76.4	96.2	95.0	63.0	78.3
07	THERMAL & MOISTURE PROTECTION	114.7	82.4	100.0	126.1	89.4	110.4	122.4	85.2	106.4	128.3	95.3	114.1	128.5	85.2	109.9	114.6	69.3	95.2
08	OPENINGS	89.2	67.5	83.9	89.2	77.7	86.4	89.2	72.7	85.2	81.8	85.0	82.6	89.2	72.7	85.2	84.2	59.8	78.3
0920	Plaster & Gypsum Board	114.4	78.1	90.6	106.2	85.0	92.3	101.6	75.6	84.6	128.9	82.9	98.7	103.9	75.6	85.4	119.4	62.2	81.9
0950, 0980	Ceilings & Acoustic Treatment	102.7	78.1	87.3	111.7	85.0	95.0	102.7	75.6	85.8	135.0	82.9	102.4	102.7	75.6	85.8	115.9	62.2	82.2
0960	Flooring	115.3	84.7	106.4	115.3	81.3	105.4	115.3	81.3	105.4	115.7	91.1	108.5	115.3	81.3	105.4	101.2	63.3	90.1
0970, 0990	Wall Finishes & Painting/Coating	107.2	82.1	92.2	107.2	93.8	99.2	107.3	73.1	86.8	109.6	96.1	101.5	107.2	73.1	86.8	108.6	79.9	91.4
09	FINISHES	108.9	80.2	93.4	111.0	86.3	97.6	108.3	77.1	91.4	123.5	87.5	104.0	108.5	77.1	91.5	109.1	65.1	85.3
COVERS	DIVS. 10 - 14, 25, 28, 41, 43, 44, 46	131.2	73.7	117.6	131.2	85.3	120.4	131.2	83.5	119.9	131.2	87.1	120.8	131.2	82.3	119.7	131.2	58.2	114.0
21, 22, 23	FIRE SUPPRESSION, PLUMBING & HVAC	105.8	82.7	96.5	106.6	87.6	98.9	106.6	87.6	98.9	106.2	86.3	98.4	106.2	84.5	97.4	106.2	70.5	91.8
26, 27, 3370	ELECTRICAL, COMMUNICATIONS & UTIL.	114.3	64.8	89.8	110.1	76.3	93.4	107.7	76.3	92.2	110.1	98.2	104.2	107.7	76.3	92.2	120.9	85.5	103.4
MF2018	WEIGHTED AVERAGE	112.6	78.6	97.9	118.9	85.7	104.6	113.6	82.6	100.2	119.2	92.9	107.8	113.7	81.9	100.0	113.0	73.2	95.8

CANADA

DIVISION		MONTREAL, QUEBEC MAT.	INST.	TOTAL	MOOSE JAW, SASKATCHEWAN MAT.	INST.	TOTAL	NEW GLASGOW, NOVA SCOTIA MAT.	INST.	TOTAL	NEWCASTLE, NEW BRUNSWICK MAT.	INST.	TOTAL	NORTH BAY, ONTARIO MAT.	INST.	TOTAL	OSHAWA, ONTARIO MAT.	INST.	TOTAL
015433	CONTRACTOR EQUIPMENT		116.2	116.2		96.5	96.5		96.8	96.8		97.4	97.4		97.4	97.4		99.4	99.4
0241, 31 - 34	SITE & INFRASTRUCTURE, DEMOLITION	114.3	102.2	106.0	113.6	87.1	95.4	109.9	88.9	95.5	104.6	87.4	92.8	125.0	90.6	101.4	103.8	96.3	98.6
0310	Concrete Forming & Accessories	132.9	90.6	96.9	108.5	53.6	61.8	111.0	66.7	73.3	104.9	57.0	64.1	150.5	78.6	89.4	124.9	88.2	93.6
0320	Concrete Reinforcing	124.6	93.8	109.8	105.0	59.8	83.1	153.1	46.4	101.6	137.9	56.0	98.4	181.1	82.6	133.5	150.0	99.2	125.5
0330	Cast-in-Place Concrete	122.7	97.5	113.2	115.6	62.0	95.4	121.3	65.0	100.0	112.6	54.6	90.7	116.8	78.2	102.3	127.5	102.1	118.0
03	CONCRETE	126.5	94.2	112.0	106.5	58.7	85.0	139.6	63.6	105.5	116.6	57.0	89.8	141.6	79.8	113.8	124.7	95.2	111.5
04	MASONRY	185.6	89.9	127.2	166.9	55.2	98.8	215.9	64.2	123.4	164.9	56.2	98.6	222.8	80.2	135.8	152.8	97.7	119.2
05	METALS	142.7	101.5	129.8	108.6	74.0	97.8	124.4	74.6	108.8	113.4	72.5	100.6	125.2	89.4	114.0	113.1	96.8	108.0
06	WOOD, PLASTICS & COMPOSITES	109.7	91.4	100.1	97.1	52.4	73.7	112.1	66.4	88.2	95.0	57.0	75.2	154.0	79.2	114.8	118.1	86.7	101.7
07	THERMAL & MOISTURE PROTECTION	126.1	97.1	113.7	113.1	58.4	89.6	128.1	65.8	101.3	114.6	56.8	89.8	134.2	80.5	111.1	116.2	99.6	109.1
08	OPENINGS	87.6	78.6	85.4	85.5	50.5	77.0	91.6	60.3	83.9	84.2	50.3	75.9	98.4	77.0	93.2	85.1	87.2	85.6
0920	Plaster & Gypsum Board	126.6	90.5	102.9	97.5	51.3	67.2	141.8	65.7	91.9	119.4	56.0	77.8	138.3	78.8	99.3	108.5	86.6	94.1
0950, 0980	Ceilings & Acoustic Treatment	139.7	90.5	108.9	102.7	51.3	70.5	122.1	65.7	86.8	115.9	56.0	78.4	122.1	78.8	95.0	102.2	86.6	92.4
0960	Flooring	118.9	86.8	109.5	105.0	52.3	89.6	108.9	57.4	93.9	101.2	61.9	89.7	133.4	83.9	119.0	105.9	93.2	102.2
0970, 0990	Wall Finishes & Painting/Coating	115.1	103.3	108.0	107.2	59.4	78.6	114.3	57.7	80.4	108.6	46.6	71.5	114.3	82.8	95.4	101.7	103.0	102.5
09	FINISHES	124.7	92.2	107.1	104.6	53.7	77.0	119.9	64.2	89.8	109.1	57.1	80.9	127.2	80.4	101.8	104.1	90.2	96.6
COVERS	DIVS. 10 - 14, 25, 28, 41, 43, 44, 46	131.2	79.8	119.1	131.2	55.5	113.4	131.2	58.7	114.1	131.2	55.8	113.4	131.2	61.2	114.7	131.2	85.9	120.5
21, 22, 23	FIRE SUPPRESSION, PLUMBING & HVAC	106.8	80.5	96.2	106.6	68.4	91.2	106.5	77.4	94.7	106.2	63.0	88.8	106.5	90.0	99.8	105.5	87.8	98.3
26, 27, 3370	ELECTRICAL, COMMUNICATIONS & UTIL.	118.0	84.3	101.4	116.3	55.4	86.2	113.9	57.7	86.1	116.4	55.0	86.0	115.2	82.0	98.7	112.2	100.3	106.3
MF2018	WEIGHTED AVERAGE	122.1	89.4	108.0	111.0	62.4	90.0	121.9	68.6	98.9	112.8	61.4	90.5	125.0	83.4	107.0	112.4	93.6	104.3

CANADA

DIVISION		OTTAWA, ONTARIO MAT.	INST.	TOTAL	OWEN SOUND, ONTARIO MAT.	INST.	TOTAL	PETERBOROUGH, ONTARIO MAT.	INST.	TOTAL	PORTAGE LA PRAIRIE, MANITOBA MAT.	INST.	TOTAL	PRINCE ALBERT, SASKATCHEWAN MAT.	INST.	TOTAL	PRINCE GEORGE, BRITISH COLUMBIA MAT.	INST.	TOTAL
015433	CONTRACTOR EQUIPMENT		112.7	112.7		97.9	97.9		97.9	97.9		100.1	100.1		96.5	96.5		101.3	101.3
0241, 31 - 34	SITE & INFRASTRUCTURE, DEMOLITION	108.0	101.0	103.2	116.9	91.5	99.4	116.1	91.3	99.1	114.6	90.2	97.8	109.6	87.2	94.2	122.9	94.1	103.1
0310	Concrete Forming & Accessories	126.4	85.0	91.2	125.7	77.3	84.5	129.0	79.6	87.0	128.2	64.5	74.0	108.5	53.4	61.6	114.6	76.1	81.9
0320	Concrete Reinforcing	134.5	97.2	116.5	171.9	84.6	129.7	160.7	83.1	123.2	148.7	53.1	102.6	109.6	59.7	85.5	107.3	73.8	91.1
0330	Cast-in-Place Concrete	120.6	99.1	112.5	155.6	74.7	125.1	129.0	79.8	110.4	129.2	67.9	99.8	104.8	61.9	88.6	116.0	89.7	106.1
03	CONCRETE	126.4	92.9	111.4	141.5	78.3	113.1	128.4	80.9	107.1	115.1	69.4	92.5	102.1	58.5	82.5	136.7	81.1	111.8
04	MASONRY	189.1	95.7	132.1	169.0	85.3	117.9	173.0	86.2	120.1	171.7	58.5	102.7	166.2	55.2	98.5	179.3	82.2	120.1
05	METALS	132.7	105.0	124.0	111.5	90.4	104.9	111.4	90.1	104.7	112.1	77.4	101.3	108.7	73.8	97.8	111.9	86.3	103.9
06	WOOD, PLASTICS & COMPOSITES	106.3	82.7	94.0	116.6	76.2	95.5	120.4	78.0	98.2	117.9	65.9	90.7	97.1	52.4	73.7	100.5	73.1	86.1
07	THERMAL & MOISTURE PROTECTION	141.4	95.9	121.9	114.2	81.3	100.0	120.1	85.8	105.4	113.6	66.4	93.4	113.0	57.4	89.1	123.8	78.8	104.5
08	OPENINGS	90.5	84.3	89.0	90.4	76.4	87.0	87.7	78.6	85.5	89.2	58.9	81.8	84.5	50.5	76.2	86.3	73.7	83.2
0920	Plaster & Gypsum Board	124.0	81.8	96.3	151.7	75.8	101.9	114.7	77.6	90.4	103.6	64.8	78.2	97.5	51.3	67.2	100.1	72.2	81.8
0950, 0980	Ceilings & Acoustic Treatment	145.7	81.8	105.6	95.4	75.8	83.1	102.7	77.6	87.0	102.7	64.8	79.0	102.7	51.3	70.5	102.7	72.2	83.6
0960	Flooring	114.4	86.8	106.4	121.5	85.1	110.8	115.3	83.9	106.1	115.3	60.4	99.3	105.0	52.3	89.6	110.5	65.7	97.4
0970, 0990	Wall Finishes & Painting/Coating	109.6	91.3	98.6	102.9	81.6	90.1	107.2	84.8	93.8	107.3	52.6	74.6	107.2	50.6	73.3	107.2	74.4	87.5
09	FINISHES	126.0	85.5	104.1	112.9	79.3	94.7	109.7	81.0	94.2	108.4	62.8	83.8	104.6	52.7	76.5	108.1	73.6	89.4
COVERS	DIVS. 10 - 14, 25, 28, 41, 43, 44, 46	131.2	85.0	120.3	139.3	61.7	121.0	131.2	62.6	115.0	131.2	57.7	113.9	131.2	55.5	113.4	131.2	80.9	119.3
21, 22, 23	FIRE SUPPRESSION, PLUMBING & HVAC	106.7	87.9	99.1	106.2	89.9	99.7	106.2	93.2	101.0	106.2	76.3	94.1	106.6	61.7	88.5	106.2	85.3	97.8
26, 27, 3370	ELECTRICAL, COMMUNICATIONS & UTIL.	111.3	98.1	104.8	118.5	80.8	99.9	114.5	81.6	98.2	116.0	53.6	85.1	116.3	55.4	86.2	114.5	73.0	94.0
MF2018	WEIGHTED AVERAGE	120.6	93.0	108.7	118.1	83.5	103.2	115.2	85.2	102.3	113.6	67.5	93.6	110.2	60.7	88.8	116.5	81.0	101.2

For customer support on your Site Work & Landscape Costs with RSMeans Data, call 800.448.8182.

825

CANADA

DIVISION		QUEBEC CITY, QUEBEC			RED DEER, ALBERTA			REGINA, SASKATCHEWAN			RIMOUSKI, QUEBEC			ROUYN-NORANDA, QUEBEC			SAINT HYACINTHE, QUEBEC		
		MAT.	INST.	TOTAL	MAT.	INST.	TOTAL	MAT.	INST.	TOTAL	MAT.	INST.	TOTAL	MAT.	INST.	TOTAL	MAT.	INST.	TOTAL
015433	CONTRACTOR EQUIPMENT		117.5	117.5		100.2	100.2		124.0	124.0		98.1	98.1		98.1	98.1		98.1	98.1
0241, 31 - 34	SITE & INFRASTRUCTURE, DEMOLITION	113.4	102.2	105.7	112.6	92.6	98.8	126.0	113.0	117.1	96.3	90.6	92.4	96.0	90.6	92.3	96.4	90.6	92.4
0310	Concrete Forming & Accessories	129.6	90.7	96.5	140.1	76.8	86.3	131.7	88.2	94.7	136.3	89.8	96.8	136.3	77.7	86.4	136.3	77.7	86.4
0320	Concrete Reinforcing	121.9	93.9	108.4	148.7	78.5	114.9	121.5	84.6	103.7	103.3	93.7	98.7	144.7	70.5	108.9	144.7	70.5	108.9
0330	Cast-in-Place Concrete	129.8	97.6	117.6	119.2	91.6	108.8	151.6	94.6	130.1	108.0	93.8	102.7	102.7	85.4	96.1	106.1	85.4	98.3
03	CONCRETE	129.2	94.4	113.6	122.9	82.9	105.0	141.3	90.6	118.5	111.4	92.0	102.7	114.5	79.7	98.9	116.1	79.7	99.8
04	MASONRY	189.0	89.9	128.5	168.3	77.6	113.0	221.2	83.1	137.0	168.7	89.8	120.6	168.9	74.8	111.5	169.2	74.8	111.6
05	METALS	136.5	103.3	126.2	112.1	89.4	105.0	137.4	97.7	125.0	109.4	91.6	103.9	109.9	83.9	101.8	109.9	83.9	101.8
06	WOOD, PLASTICS & COMPOSITES	112.5	91.4	101.4	117.9	76.4	96.2	105.7	89.2	97.1	132.8	90.7	110.8	132.8	78.0	104.1	132.8	78.0	104.1
07	THERMAL & MOISTURE PROTECTION	123.0	97.1	111.9	138.2	85.2	115.5	148.5	83.4	120.5	113.9	95.6	106.0	113.9	81.9	100.1	114.4	81.9	100.4
08	OPENINGS	88.5	85.7	87.9	89.2	72.7	85.2	88.2	77.0	85.4	88.8	76.3	85.8	89.2	66.8	83.8	89.2	66.8	83.8
0920	Plaster & Gypsum Board	123.0	90.5	101.7	103.9	75.6	85.4	142.0	88.1	106.6	143.7	90.5	108.8	113.9	77.4	90.0	113.9	77.4	90.0
0950, 0980	Ceilings & Acoustic Treatment	134.2	90.5	106.8	102.7	75.6	85.8	156.4	88.1	113.6	102.1	90.5	94.8	102.1	77.4	86.6	102.1	77.4	86.6
0960	Flooring	115.6	86.8	107.2	118.3	81.3	107.5	125.4	91.1	115.4	116.4	83.8	106.9	115.3	83.8	106.1	115.3	83.8	106.1
0970, 0990	Wall Finishes & Painting/Coating	115.9	103.3	108.3	107.2	73.1	86.8	109.6	85.1	95.0	110.0	103.3	105.9	107.2	81.4	91.7	107.2	81.4	91.7
09	FINISHES	123.5	92.2	106.6	109.4	77.1	91.9	134.2	89.4	109.9	113.5	91.2	101.4	108.7	79.6	93.0	108.7	79.6	93.0
COVERS	DIVS. 10 - 14, 25, 28, 41, 43, 44, 46	131.2	79.8	119.1	131.2	82.3	119.7	131.2	67.6	116.2	131.2	78.3	118.7	131.2	73.1	117.5	131.2	73.1	117.5
21, 22, 23	FIRE SUPPRESSION, PLUMBING & HVAC	106.5	80.5	96.0	106.2	84.5	97.4	106.4	82.8	96.9	106.2	80.3	95.8	106.2	82.0	96.4	102.4	82.0	94.1
26, 27, 3370	ELECTRICAL, COMMUNICATIONS & UTIL.	124.2	84.3	104.5	107.7	76.3	92.2	121.0	88.1	104.7	113.9	84.3	99.2	113.9	64.2	89.3	114.5	64.2	89.6
MF2018	WEIGHTED AVERAGE	122.2	89.9	108.2	114.2	81.9	100.2	127.0	88.5	110.3	112.3	86.9	101.4	112.4	78.0	97.6	111.8	78.0	97.2

CANADA

DIVISION		SAINT JOHN, NEW BRUNSWICK			SARNIA, ONTARIO			SASKATOON, SASKATCHEWAN			SAULT STE MARIE, ONTARIO			SHERBROOKE, QUEBEC			SOREL, QUEBEC		
		MAT.	INST.	TOTAL	MAT.	INST.	TOTAL	MAT.	INST.	TOTAL	MAT.	INST.	TOTAL	MAT.	INST.	TOTAL	MAT.	INST.	TOTAL
015433	CONTRACTOR EQUIPMENT		97.4	97.4		97.9	97.9		97.0	97.0		97.9	97.9		98.1	98.1		98.1	98.1
0241, 31 - 34	SITE & INFRASTRUCTURE, DEMOLITION	104.8	89.2	94.1	114.7	91.4	98.7	111.8	90.0	96.8	105.1	91.0	95.4	96.4	90.6	92.4	96.5	90.6	92.5
0310	Concrete Forming & Accessories	128.1	61.3	71.2	127.5	86.1	92.3	108.3	87.3	90.4	115.0	83.4	88.1	136.3	77.7	86.4	136.3	77.8	86.5
0320	Concrete Reinforcing	137.9	66.7	103.6	114.6	84.4	100.1	112.2	84.4	98.8	103.1	83.2	93.5	144.7	70.5	108.9	136.9	70.5	104.8
0330	Cast-in-Place Concrete	110.9	65.3	93.7	119.0	92.1	108.8	112.0	89.5	103.5	106.9	78.5	96.2	106.1	85.4	98.3	106.9	85.4	98.8
03	CONCRETE	117.5	64.8	93.8	117.5	88.3	104.4	106.6	87.8	98.1	102.6	82.3	93.5	116.1	79.7	99.8	115.4	79.8	99.4
04	MASONRY	188.3	71.5	117.1	185.9	88.7	126.6	179.7	83.0	120.7	169.9	89.2	120.7	169.2	74.8	111.6	169.3	74.8	111.6
05	METALS	113.3	84.9	104.5	111.3	90.5	104.8	105.4	87.1	99.7	110.5	93.9	105.3	109.7	83.9	101.6	109.9	84.0	101.8
06	WOOD, PLASTICS & COMPOSITES	120.8	59.5	88.7	119.3	85.6	101.7	93.8	88.3	90.9	105.1	84.9	94.5	132.8	78.0	104.1	132.8	78.0	104.1
07	THERMAL & MOISTURE PROTECTION	115.0	69.2	95.3	120.2	89.1	106.8	116.7	81.0	101.4	118.9	85.1	104.4	113.9	81.9	100.2	113.9	81.9	100.1
08	OPENINGS	84.1	56.9	77.5	90.7	82.2	88.6	85.5	76.5	83.3	82.5	83.3	82.7	89.2	66.8	83.8	89.2	71.1	84.8
0920	Plaster & Gypsum Board	133.4	58.5	84.3	139.9	85.4	104.1	116.5	88.1	97.9	106.4	84.7	92.1	113.9	77.4	90.0	143.7	77.4	100.2
0950, 0980	Ceilings & Acoustic Treatment	120.4	58.5	81.6	107.8	85.4	93.8	124.2	88.1	101.6	102.7	84.7	91.4	102.1	77.4	86.6	102.1	77.4	86.6
0960	Flooring	113.3	63.3	98.7	115.3	91.8	108.4	108.0	91.1	103.1	107.9	88.7	102.3	115.3	83.8	106.1	115.3	83.8	106.1
0970, 0990	Wall Finishes & Painting/Coating	108.6	79.9	91.4	107.2	95.9	100.4	110.3	85.1	95.2	107.2	88.9	96.2	107.2	81.4	91.7	107.2	81.4	91.7
09	FINISHES	116.9	63.1	87.2	114.5	88.5	100.5	114.4	88.7	100.5	105.7	85.3	94.6	108.7	79.6	93.0	112.9	79.6	94.9
COVERS	DIVS. 10 - 14, 25, 28, 41, 43, 44, 46	131.2	58.1	114.0	131.2	63.9	115.3	131.2	65.9	115.8	131.2	83.8	120.0	131.2	73.1	117.5	131.2	73.1	117.5
21, 22, 23	FIRE SUPPRESSION, PLUMBING & HVAC	106.2	72.0	92.4	106.2	98.9	103.2	106.0	82.6	96.5	106.2	87.8	98.8	106.6	82.0	96.7	102.4	82.0	94.4
26, 27, 3370	ELECTRICAL, COMMUNICATIONS & UTIL.	123.5	85.5	104.7	117.6	83.6	100.8	118.6	88.0	103.5	116.2	82.0	99.3	113.9	64.2	89.3	113.9	64.2	89.3
MF2018	WEIGHTED AVERAGE	115.6	72.9	97.1	115.5	89.4	104.2	112.0	85.1	100.4	110.6	86.4	100.1	112.8	78.0	97.7	112.9	78.2	97.9

CANADA

DIVISION		ST. CATHARINES, ONTARIO			ST JEROME, QUEBEC			ST. JOHN'S, NEWFOUNDLAND			SUDBURY, ONTARIO			SUMMERSIDE, PRINCE EDWARD ISLAND			SYDNEY, NOVA SCOTIA		
		MAT.	INST.	TOTAL	MAT.	INST.	TOTAL	MAT.	INST.	TOTAL	MAT.	INST.	TOTAL	MAT.	INST.	TOTAL	MAT.	INST.	TOTAL
015433	CONTRACTOR EQUIPMENT		97.3	97.3		98.1	98.1		120.5	120.5		97.3	97.3		96.8	96.8		96.8	96.8
0241, 31 - 34	SITE & INFRASTRUCTURE, DEMOLITION	93.8	92.8	93.1	96.0	90.6	92.3	113.1	111.0	111.6	93.9	92.4	92.8	119.4	86.4	96.7	106.1	88.9	94.3
0310	Concrete Forming & Accessories	116.1	90.6	94.4	136.3	77.7	86.4	127.4	83.3	89.8	111.3	85.6	89.4	111.3	51.5	60.4	111.0	66.7	73.3
0320	Concrete Reinforcing	95.5	98.3	96.8	144.7	70.5	108.9	168.7	80.9	126.4	96.2	96.8	96.5	151.1	45.2	100.0	153.1	46.4	101.6
0330	Cast-in-Place Concrete	105.3	95.0	101.4	102.7	85.4	96.1	135.2	96.3	120.5	106.2	91.5	100.7	115.2	53.1	91.7	93.6	65.0	82.8
03	CONCRETE	99.9	93.7	97.1	114.5	79.7	98.9	143.9	88.2	118.9	100.1	89.9	95.5	149.5	52.1	105.8	126.7	63.0	98.4
04	MASONRY	149.6	97.0	117.5	168.9	74.8	111.5	212.1	88.8	136.9	149.7	93.2	115.2	215.2	53.7	116.7	213.6	64.2	122.5
05	METALS	113.0	96.0	107.7	109.9	83.9	101.8	139.5	95.2	125.7	112.5	95.4	107.1	124.4	68.0	106.8	124.4	74.6	108.8
06	WOOD, PLASTICS & COMPOSITES	107.2	90.4	98.4	132.8	78.0	104.1	105.1	81.6	92.8	102.4	84.9	93.2	112.8	51.1	80.5	112.1	66.4	88.2
07	THERMAL & MOISTURE PROTECTION	115.3	97.6	107.7	113.9	81.9	100.1	137.8	92.7	118.5	114.5	92.6	105.1	127.6	55.6	96.7	128.1	65.8	101.3
08	OPENINGS	79.8	88.3	81.9	89.2	66.8	83.8	85.6	72.7	82.5	80.4	83.7	81.2	102.3	44.7	88.3	91.6	60.3	83.9
0920	Plaster & Gypsum Board	98.8	90.3	93.2	113.9	77.4	90.0	146.3	80.6	103.2	97.6	84.7	89.1	143.0	49.9	81.9	141.8	65.7	91.9
0950, 0980	Ceilings & Acoustic Treatment	102.2	90.3	94.8	102.1	77.4	86.6	144.0	80.6	104.3	97.1	84.7	89.3	122.1	49.9	76.9	122.1	65.7	86.8
0960	Flooring	101.7	88.1	97.8	115.3	83.8	106.1	118.7	50.0	98.7	99.9	88.7	96.6	109.0	53.9	92.9	108.9	57.4	93.9
0970, 0990	Wall Finishes & Painting/Coating	101.7	99.6	100.4	107.2	81.4	91.7	111.8	96.7	102.8	101.7	90.7	95.2	114.3	38.8	69.0	114.3	57.7	80.4
09	FINISHES	101.0	91.2	95.7	108.7	79.6	93.0	131.0	78.4	102.5	99.0	86.5	92.2	121.1	50.8	83.1	119.9	64.2	89.8
COVERS	DIVS. 10 - 14, 25, 28, 41, 43, 44, 46	131.2	66.2	115.9	131.2	73.1	117.5	131.2	65.4	115.7	131.2	85.1	120.3	131.2	55.0	113.2	131.2	58.7	114.1
21, 22, 23	FIRE SUPPRESSION, PLUMBING & HVAC	105.5	87.1	98.1	106.2	82.0	96.4	106.3	81.4	96.3	105.9	85.4	97.7	106.5	57.3	86.6	106.5	77.4	94.7
26, 27, 3370	ELECTRICAL, COMMUNICATIONS & UTIL.	118.9	98.5	105.7	114.6	64.2	89.6	122.9	78.3	100.8	110.7	98.3	104.6	112.8	46.7	80.1	113.9	57.7	86.1
MF2018	WEIGHTED AVERAGE	108.0	92.1	101.1	112.5	78.0	97.6	126.2	85.2	108.5	107.7	90.3	100.2	124.4	56.4	95.0	120.1	68.6	97.9

CANADA

DIVISION		THUNDER BAY, ONTARIO			TIMMINS, ONTARIO			TORONTO, ONTARIO			TROIS RIVIERES, QUEBEC			TRURO, NOVA SCOTIA			VANCOUVER, BRITISH COLUMBIA		
		MAT.	INST.	TOTAL	MAT.	INST.	TOTAL	MAT.	INST.	TOTAL	MAT.	INST.	TOTAL	MAT.	INST.	TOTAL	MAT.	INST.	TOTAL
015433	CONTRACTOR EQUIPMENT		97.3	97.3		97.9	97.9		112.2	112.2		97.6	97.6		97.3	97.3		130.1	130.1
0241, 31 - 34	SITE & INFRASTRUCTURE, DEMOLITION	98.3	92.7	94.4	116.1	91.0	98.8	107.8	101.6	103.6	106.3	90.3	95.3	100.4	89.3	92.7	115.2	116.1	115.8
0310	Concrete Forming & Accessories	124.9	89.4	94.7	129.0	78.7	86.2	131.3	98.5	103.4	160.0	77.8	90.1	97.4	66.7	71.3	132.4	89.3	95.7
0320	Concrete Reinforcing	85.3	98.1	91.5	160.7	82.6	123.0	134.4	100.9	118.2	153.1	70.5	113.3	136.9	46.4	93.2	130.1	91.4	111.4
0330	Cast-in-Place Concrete	116.0	94.3	107.8	129.0	78.3	109.8	106.9	109.2	107.8	96.9	85.3	92.5	133.9	65.1	107.9	121.8	98.7	113.0
03	CONCRETE	107.6	92.8	101.0	128.4	79.8	106.6	120.3	103.1	112.6	129.2	79.7	107.0	127.1	63.6	98.6	130.1	94.0	113.9
04	MASONRY	150.2	96.7	117.6	173.0	80.2	116.4	189.2	104.2	137.4	218.0	74.8	130.6	168.0	64.2	104.7	180.3	93.1	127.1
05	METALS	113.0	95.2	107.5	111.4	89.7	104.6	132.5	106.3	124.3	123.6	83.7	111.2	110.6	74.9	99.5	132.2	106.9	124.3
06	WOOD, PLASTICS & COMPOSITES	118.1	88.6	102.7	120.4	79.2	98.9	113.5	97.0	104.9	170.1	77.9	121.9	86.3	66.5	75.9	108.4	88.8	98.1
07	THERMAL & MOISTURE PROTECTION	115.6	95.2	106.9	120.1	80.5	103.1	139.9	105.1	125.0	127.1	81.9	107.7	114.8	65.8	93.8	145.8	90.2	121.9
08	OPENINGS	79.1	86.9	81.0	87.7	77.0	85.1	84.1	95.7	87.0	100.2	71.1	93.1	82.7	60.3	77.3	84.7	86.0	85.0
0920	Plaster & Gypsum Board	123.9	88.5	100.7	114.7	78.8	91.2	123.9	96.6	106.0	171.2	77.4	109.7	120.4	65.7	84.5	135.1	87.5	103.9
0950, 0980	Ceilings & Acoustic Treatment	97.1	88.5	91.7	102.7	78.8	87.8	143.2	96.6	114.0	120.8	77.4	93.6	102.7	65.7	79.5	147.4	87.5	109.9
0960	Flooring	105.9	94.3	102.5	115.3	83.9	106.1	115.9	96.6	110.3	133.4	83.8	118.9	96.7	57.4	85.2	130.0	84.8	116.8
0970, 0990	Wall Finishes & Painting/Coating	101.7	92.0	95.9	107.2	82.8	92.6	106.0	103.0	104.2	114.3	81.4	94.5	107.2	57.7	77.6	105.8	95.5	99.6
09	FINISHES	104.8	90.5	97.1	109.7	80.4	93.8	123.6	98.7	110.1	130.7	79.6	103.0	104.5	64.3	82.8	131.7	89.0	108.6
COVERS	DIVS. 10 - 14, 25, 28, 41, 43, 44, 46	131.2	66.3	115.9	131.2	61.3	114.7	131.2	90.4	121.6	131.2	73.0	117.5	131.2	58.7	114.1	131.2	86.6	120.7
21, 22, 23	FIRE SUPPRESSION, PLUMBING & HVAC	105.5	87.2	98.1	106.2	90.0	99.7	106.8	95.0	102.0	106.5	82.0	96.6	106.2	77.4	94.6	106.8	86.4	98.6
26, 27, 3370	ELECTRICAL, COMMUNICATIONS & UTIL.	111.3	97.8	104.6	116.2	82.0	99.3	113.7	100.0	106.9	114.2	64.2	89.5	113.6	57.7	85.9	118.8	79.5	99.4
MF2018	WEIGHTED AVERAGE	109.3	91.6	101.6	115.4	83.4	101.6	119.3	99.9	111.0	122.8	78.1	103.5	112.8	68.7	93.7	121.7	91.7	108.7

CANADA

DIVISION		VICTORIA, BRITISH COLUMBIA			WHITEHORSE, YUKON			WINDSOR, ONTARIO			WINNIPEG, MANITOBA			YARMOUTH, NOVA SCOTIA			YELLOWKNIFE, NWT		
		MAT.	INST.	TOTAL	MAT.	INST.	TOTAL	MAT.	INST.	TOTAL	MAT.	INST.	TOTAL	MAT.	INST.	TOTAL	MAT.	INST.	TOTAL
015433	CONTRACTOR EQUIPMENT		104.8	104.8		139.6	139.6		97.3	97.3		122.1	122.1		96.8	96.8		126.5	126.5
0241, 31 - 34	SITE & INFRASTRUCTURE, DEMOLITION	123.6	99.0	106.7	134.2	120.5	124.8	90.4	92.7	92.0	112.7	109.6	110.6	109.7	88.9	95.4	143.2	116.8	125.0
0310	Concrete Forming & Accessories	114.0	86.6	90.7	136.3	56.3	68.2	124.9	87.2	92.8	131.3	63.3	73.4	111.0	66.7	73.3	138.9	74.7	84.3
0320	Concrete Reinforcing	109.7	91.2	100.8	164.6	61.7	114.9	93.6	97.1	95.3	125.0	58.4	92.9	153.1	46.4	101.6	136.0	64.1	101.3
0330	Cast-in-Place Concrete	116.0	94.5	107.9	150.5	72.3	121.0	107.9	95.7	103.3	141.6	70.1	114.6	120.0	65.0	99.2	172.6	86.5	140.1
03	CONCRETE	138.8	90.4	117.1	156.1	64.8	115.2	101.4	92.2	97.2	137.1	66.2	105.3	139.0	63.6	105.1	162.8	78.1	124.8
04	MASONRY	182.8	93.0	128.0	256.6	56.9	134.8	149.8	96.0	116.9	211.7	62.8	120.9	215.8	64.2	123.3	244.6	67.6	136.6
05	METALS	106.7	91.9	102.1	143.5	93.1	127.8	113.0	95.5	107.6	139.8	84.8	122.7	124.4	74.6	108.8	138.9	90.0	123.7
06	WOOD, PLASTICS & COMPOSITES	98.2	86.1	91.9	143.9	55.1	86.3	118.1	86.0	101.3	105.3	64.1	83.8	129.9	76.2	101.8	129.7	76.2	101.8
07	THERMAL & MOISTURE PROTECTION	126.3	87.3	109.6	143.9	62.8	109.1	115.4	94.7	106.5	140.8	67.9	109.5	128.1	65.8	101.3	142.4	77.5	114.5
08	OPENINGS	87.0	80.6	85.4	99.6	52.4	88.1	79.0	85.8	80.6	87.8	57.8	80.5	91.6	60.3	83.9	94.1	64.6	86.9
0920	Plaster & Gypsum Board	109.6	85.5	93.8	180.6	52.4	96.5	109.0	85.8	93.8	138.3	62.3	88.4	141.8	65.7	91.9	189.8	74.7	114.2
0950, 0980	Ceilings & Acoustic Treatment	104.4	85.5	92.6	170.0	52.4	96.3	97.1	85.8	90.0	148.7	62.3	94.6	122.1	65.7	86.8	168.4	74.7	109.6
0960	Flooring	113.6	65.7	99.7	124.5	53.8	103.8	105.9	91.8	101.8	121.0	65.6	104.8	108.9	57.4	93.9	124.3	80.9	111.6
0970, 0990	Wall Finishes & Painting/Coating	110.3	95.5	101.5	117.1	51.7	77.9	101.7	92.5	96.2	106.0	51.1	73.1	114.3	57.7	80.4	120.3	73.9	92.5
09	FINISHES	111.8	84.1	96.8	148.1	55.1	97.7	102.4	88.5	94.9	131.5	62.8	94.3	119.9	64.2	89.8	150.9	75.3	110.0
COVERS	DIVS. 10 - 14, 25, 28, 41, 43, 44, 46	131.2	63.2	115.2	131.2	59.8	114.4	131.2	65.7	115.8	131.2	61.0	114.6	131.2	58.7	114.1	131.2	62.0	114.9
21, 22, 23	FIRE SUPPRESSION, PLUMBING & HVAC	106.3	84.7	97.6	106.9	69.9	92.0	105.5	87.6	98.3	106.8	61.8	88.7	106.5	77.4	94.7	107.6	86.2	99.0
26, 27, 3370	ELECTRICAL, COMMUNICATIONS & UTIL.	116.5	79.0	98.0	138.0	56.2	97.5	116.0	98.1	107.1	123.2	61.6	92.7	113.9	57.7	86.1	132.2	76.3	104.5
MF2018	WEIGHTED AVERAGE	116.8	86.4	103.7	136.0	68.6	106.9	108.6	91.2	101.1	125.8	68.4	101.0	121.9	68.6	98.8	135.0	81.1	111.7

Location Factors - Commercial - V2

Costs shown in RSMeans cost data publications are based on national averages for materials and installation. To adjust these costs to a specific location, simply multiply the base cost by the factor and divide by 100 for that city. The data is arranged alphabetically by state and postal zip code numbers. For a city not listed, use the factor for a nearby city with similar economic characteristics.

STATE/ZIP	CITY	MAT.	INST.	TOTAL
ALABAMA				
350-352	Birmingham	98.2	70.9	86.4
354	Tuscaloosa	96.9	70.5	85.5
355	Jasper	97.6	69.6	85.5
356	Decatur	96.9	70.2	85.4
357-358	Huntsville	96.9	69.9	85.2
359	Gadsden	97.2	70.8	85.8
360-361	Montgomery	98.0	71.8	86.6
362	Anniston	97.1	66.6	83.9
363	Dothan	97.2	71.6	86.1
364	Evergreen	96.8	67.8	84.3
365-366	Mobile	97.8	68.7	85.2
367	Selma	96.9	71.2	85.8
368	Phenix City	97.7	70.8	86.1
369	Butler	97.1	69.7	85.2
ALASKA				
995-996	Anchorage	119.7	110.4	115.7
997	Fairbanks	120.2	111.0	116.2
998	Juneau	119.7	110.4	115.7
999	Ketchikan	130.1	110.9	121.8
ARIZONA				
850,853	Phoenix	98.4	72.1	87.0
851,852	Mesa/Tempe	98.0	71.1	86.4
855	Globe	98.2	70.7	86.3
856-857	Tucson	96.9	71.8	86.1
859	Show Low	98.4	70.8	86.5
860	Flagstaff	102.4	70.4	88.6
863	Prescott	99.9	71.1	87.5
864	Kingman	98.2	69.8	85.9
865	Chambers	98.2	70.4	86.2
ARKANSAS				
716	Pine Bluff	96.2	62.8	81.8
717	Camden	94.1	64.1	81.1
718	Texarkana	95.1	63.7	81.5
719	Hot Springs	93.4	62.5	80.1
720-722	Little Rock	96.0	63.5	81.9
723	West Memphis	94.2	67.5	82.7
724	Jonesboro	94.8	65.7	82.2
725	Batesville	92.4	62.9	79.6
726	Harrison	93.7	62.3	80.1
727	Fayetteville	91.3	64.2	79.6
728	Russellville	92.5	61.3	79.0
729	Fort Smith	95.1	61.9	80.8
CALIFORNIA				
900-902	Los Angeles	98.4	129.3	111.8
903-905	Inglewood	93.6	127.4	108.2
906-908	Long Beach	95.2	127.2	109.0
910-912	Pasadena	95.4	127.0	109.0
913-916	Van Nuys	98.5	127.0	110.8
917-918	Alhambra	97.4	127.0	110.2
919-921	San Diego	100.3	121.3	109.4
922	Palm Springs	97.3	121.9	108.0
923-924	San Bernardino	95.0	124.4	108.1
925	Riverside	99.4	126.0	110.9
926-927	Santa Ana	97.1	125.7	109.5
928	Anaheim	99.5	125.9	110.9
930	Oxnard	98.2	125.8	110.1
931	Santa Barbara	97.4	125.0	109.4
932-933	Bakersfield	99.0	124.8	110.1
934	San Luis Obispo	98.4	125.5	110.1
935	Mojave	95.6	123.6	107.7
936-938	Fresno	98.6	131.2	112.7
939	Salinas	99.2	137.8	115.9
940-941	San Francisco	107.6	159.0	129.8
942,956-958	Sacramento	101.0	133.1	114.9
943	Palo Alto	99.4	150.8	121.6
944	San Mateo	101.9	156.6	125.5
945	Vallejo	100.5	143.4	119.0
946	Oakland	103.9	153.2	125.2
947	Berkeley	103.3	151.9	124.3
948	Richmond	102.8	148.2	122.4
949	San Rafael	105.0	150.3	124.6
950	Santa Cruz	104.9	138.1	119.2

STATE/ZIP	CITY	MAT.	INST.	TOTAL
CALIFORNIA (CONT'D)				
951	San Jose	103.0	158.4	126.9
952	Stockton	101.0	131.9	114.3
953	Modesto	100.8	132.1	114.3
954	Santa Rosa	101.2	149.8	122.2
955	Eureka	102.6	137.6	117.7
959	Marysville	101.8	130.7	114.3
960	Redding	107.8	132.6	118.6
961	Susanville	107.3	131.0	117.5
COLORADO				
800-802	Denver	104.0	75.1	91.5
803	Boulder	99.2	73.7	88.2
804	Golden	101.5	73.0	89.2
805	Fort Collins	102.8	73.2	90.0
806	Greeley	99.9	73.3	88.4
807	Fort Morgan	99.9	72.5	88.0
808-809	Colorado Springs	101.8	72.3	89.1
810	Pueblo	101.6	70.7	88.2
811	Alamosa	103.3	69.2	88.6
812	Salida	102.9	69.8	88.6
813	Durango	103.6	65.5	87.1
814	Montrose	102.3	66.3	86.7
815	Grand Junction	105.6	70.0	90.2
816	Glenwood Springs	103.6	66.6	87.6
CONNECTICUT				
060	New Britain	99.7	116.9	107.1
061	Hartford	101.7	118.7	109.0
062	Willimantic	100.3	116.9	107.5
063	New London	96.7	116.6	105.3
064	Meriden	98.8	116.4	106.4
065	New Haven	101.5	117.0	108.2
066	Bridgeport	100.9	116.7	107.7
067	Waterbury	100.5	117.2	107.7
068	Norwalk	100.4	116.9	107.5
069	Stamford	100.5	123.9	110.6
D.C.				
200-205	Washington	101.3	88.0	95.5
DELAWARE				
197	Newark	98.7	110.6	103.8
198	Wilmington	99.1	110.7	104.1
199	Dover	99.5	110.4	104.2
FLORIDA				
320,322	Jacksonville	97.0	66.9	84.0
321	Daytona Beach	97.2	69.0	85.0
323	Tallahassee	98.1	66.7	84.6
324	Panama City	98.7	67.2	85.1
325	Pensacola	101.2	65.3	85.7
326,344	Gainesville	98.3	63.8	83.4
327-328,347	Orlando	97.9	66.3	84.3
329	Melbourne	99.9	70.6	87.2
330-332,340	Miami	96.2	70.4	85.1
333	Fort Lauderdale	96.5	72.2	86.0
334,349	West Palm Beach	95.1	69.0	83.8
335-336,346	Tampa	98.0	67.4	84.8
337	St. Petersburg	100.4	65.8	85.4
338	Lakeland	97.1	66.3	83.8
339,341	Fort Myers	96.4	66.7	83.6
342	Sarasota	99.1	65.3	84.5
GEORGIA				
300-303,399	Atlanta	98.7	76.7	89.2
304	Statesboro	97.9	67.1	84.6
305	Gainesville	96.5	64.3	82.6
306	Athens	95.9	63.8	82.0
307	Dalton	97.7	68.0	84.9
308-309	Augusta	96.4	73.5	86.5
310-312	Macon	95.3	74.4	86.3
313-314	Savannah	97.3	72.8	86.7
315	Waycross	96.6	66.9	83.8
316	Valdosta	96.7	65.4	83.2
317,398	Albany	96.5	72.2	86.0
318-319	Columbus	96.4	73.4	86.4

STATE/ZIP	CITY	MAT.	INST.	TOTAL
HAWAII				
967	Hilo	115.6	117.2	116.3
968	Honolulu	120.3	117.8	119.2
STATES & POSS.				
969	Guam	138.0	63.4	105.7
IDAHO				
832	Pocatello	101.8	77.8	91.4
833	Twin Falls	102.9	76.9	91.7
834	Idaho Falls	100.2	76.9	90.1
835	Lewiston	108.4	84.9	98.2
836-837	Boise	100.9	78.2	91.1
838	Coeur d'Alene	108.1	81.5	96.6
ILLINOIS				
600-603	North Suburban	98.5	137.3	115.3
604	Joliet	98.5	140.3	116.5
605	South Suburban	98.5	137.2	115.3
606-608	Chicago	100.5	144.4	119.5
609	Kankakee	95.1	129.5	110.0
610-611	Rockford	96.3	126.3	109.3
612	Rock Island	94.5	95.5	94.9
613	La Salle	95.8	122.2	107.2
614	Galesburg	95.5	104.8	99.5
615-616	Peoria	97.4	109.9	102.8
617	Bloomington	94.8	105.4	99.4
618-619	Champaign	98.4	106.3	101.8
620-622	East St. Louis	94.1	109.8	100.9
623	Quincy	96.1	101.1	98.3
624	Effingham	95.3	104.9	99.4
625	Decatur	96.7	107.6	101.4
626-627	Springfield	97.8	110.4	103.2
628	Centralia	92.9	106.7	98.9
629	Carbondale	92.7	105.2	98.1
INDIANA				
460	Anderson	96.8	80.4	89.7
461-462	Indianapolis	99.5	82.9	92.3
463-464	Gary	98.3	105.8	101.5
465-466	South Bend	99.6	83.4	92.6
467-468	Fort Wayne	97.5	75.9	88.2
469	Kokomo	94.9	79.8	88.4
470	Lawrenceburg	93.0	78.3	86.6
471	New Albany	94.3	76.3	86.5
472	Columbus	96.5	79.4	89.1
473	Muncie	96.6	78.9	89.0
474	Bloomington	98.2	79.7	90.2
475	Washington	95.1	84.8	90.7
476-477	Evansville	96.0	83.2	90.5
478	Terre Haute	96.9	82.9	90.8
479	Lafayette	96.2	79.9	89.1
IOWA				
500-503,509	Des Moines	97.3	90.0	94.2
504	Mason City	95.0	71.2	84.7
505	Fort Dodge	95.0	71.0	84.6
506-507	Waterloo	96.5	78.4	88.7
508	Creston	95.2	81.1	89.1
510-511	Sioux City	98.0	80.3	90.3
512	Sibley	96.9	60.1	81.0
513	Spencer	98.3	59.9	81.7
514	Carroll	95.5	79.3	88.5
515	Council Bluffs	98.8	81.9	91.5
516	Shenandoah	96.1	78.7	88.6
520	Dubuque	97.3	80.6	90.1
521	Decorah	96.7	71.0	85.6
522-524	Cedar Rapids	98.3	86.6	93.2
525	Ottumwa	96.4	75.7	87.5
526	Burlington	95.8	81.1	89.5
527-528	Davenport	97.3	96.6	97.0
KANSAS				
660-662	Kansas City	96.4	99.2	97.6
664-666	Topeka	98.1	78.0	89.4
667	Fort Scott	95.2	73.9	86.0
668	Emporia	95.2	74.0	86.0
669	Belleville	96.9	69.1	84.9
670-672	Wichita	96.5	72.2	86.0
673	Independence	96.4	73.6	86.6
674	Salina	96.8	73.3	86.6
675	Hutchinson	91.9	69.7	82.3
676	Hays	96.0	70.4	84.9
677	Colby	96.5	71.7	85.7

STATE/ZIP	CITY	MAT.	INST.	TOTAL
KANSAS (CONT'D)				
678	Dodge City	98.0	74.8	88.0
679	Liberal	96.0	69.5	84.6
KENTUCKY				
400-402	Louisville	94.9	79.6	88.3
403-405	Lexington	94.2	78.8	87.5
406	Frankfort	96.8	78.6	88.9
407-409	Corbin	91.8	75.0	84.6
410	Covington	94.9	75.0	86.3
411-412	Ashland	93.6	86.3	90.5
413-414	Campton	94.9	76.3	86.9
415-416	Pikeville	96.1	82.1	90.1
417-418	Hazard	94.3	77.1	86.9
420	Paducah	93.0	80.0	87.4
421-422	Bowling Green	95.2	78.4	87.9
423	Owensboro	95.3	81.8	89.4
424	Henderson	92.7	78.7	86.7
425-426	Somerset	92.1	76.6	85.4
427	Elizabethtown	91.8	73.4	83.8
LOUISIANA				
700-701	New Orleans	97.7	68.3	85.0
703	Thibodaux	95.5	65.0	82.3
704	Hammond	93.2	63.4	80.3
705	Lafayette	95.1	65.9	82.5
706	Lake Charles	95.3	68.4	83.7
707-708	Baton Rouge	97.0	68.3	84.6
710-711	Shreveport	97.7	66.8	84.4
712	Monroe	96.0	64.6	82.4
713-714	Alexandria	96.1	64.7	82.5
MAINE				
039	Kittery	94.4	84.9	90.3
040-041	Portland	100.6	85.6	94.1
042	Lewiston	98.0	84.9	92.4
043	Augusta	101.6	86.0	94.9
044	Bangor	97.5	84.5	91.9
045	Bath	96.2	84.6	91.2
046	Machias	95.8	84.5	90.9
047	Houlton	95.9	84.4	91.0
048	Rockland	95.0	84.6	90.5
049	Waterville	96.2	84.5	91.2
MARYLAND				
206	Waldorf	97.4	84.8	92.0
207-208	College Park	97.5	85.8	92.5
209	Silver Spring	96.7	84.9	91.6
210-212	Baltimore	102.3	83.1	94.0
214	Annapolis	101.4	83.4	93.6
215	Cumberland	97.1	83.1	91.0
216	Easton	98.9	72.6	87.5
217	Hagerstown	97.7	86.2	92.8
218	Salisbury	99.3	65.0	84.5
219	Elkton	96.1	83.0	90.5
MASSACHUSETTS				
010-011	Springfield	99.1	110.8	104.2
012	Pittsfield	98.6	105.1	101.4
013	Greenfield	96.7	109.5	102.3
014	Fitchburg	95.4	113.2	103.1
015-016	Worcester	99.1	118.6	107.5
017	Framingham	94.9	122.4	106.8
018	Lowell	98.4	126.4	110.5
019	Lawrence	99.4	124.9	110.4
020-022, 024	Boston	100.2	132.9	114.3
023	Brockton	98.9	117.8	107.1
025	Buzzards Bay	93.3	114.1	102.3
026	Hyannis	96.0	116.9	105.1
027	New Bedford	98.1	116.3	106.0
MICHIGAN				
480,483	Royal Oak	94.1	96.9	95.3
481	Ann Arbor	96.3	102.0	98.7
482	Detroit	99.5	100.9	100.1
484-485	Flint	95.9	90.9	93.8
486	Saginaw	95.6	86.7	91.8
487	Bay City	95.7	85.3	91.2
488-489	Lansing	98.1	90.5	94.8
490	Battle Creek	95.7	79.4	88.6
491	Kalamazoo	96.0	80.9	89.5
492	Jackson	94.1	88.0	91.4
493,495	Grand Rapids	98.9	82.2	91.6
494	Muskegon	94.5	79.7	88.1

Location Factors - Commercial - V2

STATE/ZIP	CITY	MAT.	INST.	TOTAL
MICHIGAN (CONT'D)				
496	Traverse City	93.5	75.2	85.6
497	Gaylord	94.6	78.1	87.5
498-499	Iron Mountain	96.5	78.9	88.9
MINNESOTA				
550-551	Saint Paul	99.3	119.2	107.9
553-555	Minneapolis	99.7	116.6	107.0
556-558	Duluth	99.9	103.9	101.6
559	Rochester	98.4	106.1	101.7
560	Mankato	95.5	105.0	99.6
561	Windom	94.0	91.7	93.0
562	Willmar	93.7	96.6	94.9
563	St. Cloud	94.9	115.9	104.0
564	Brainerd	95.5	97.9	96.5
565	Detroit Lakes	97.6	90.3	94.4
566	Bemidji	96.8	94.6	95.8
567	Thief River Falls	96.4	89.4	93.4
MISSISSIPPI				
386	Clarksdale	95.7	53.8	77.6
387	Greenville	99.4	65.2	84.6
388	Tupelo	97.2	56.4	79.6
389	Greenwood	97.0	53.5	78.2
390-392	Jackson	98.6	67.9	85.3
393	Meridian	95.4	67.8	83.5
394	Laurel	96.8	56.4	79.4
395	Biloxi	97.1	66.6	83.9
396	McComb	95.1	53.8	77.3
397	Columbus	96.7	57.1	79.6
MISSOURI				
630-631	St. Louis	97.5	104.6	100.6
633	Bowling Green	95.6	92.3	94.2
634	Hannibal	94.5	88.5	91.9
635	Kirksville	98.0	83.6	91.8
636	Flat River	96.6	90.6	94.0
637	Cape Girardeau	96.4	90.6	93.9
638	Sikeston	94.9	84.4	90.4
639	Poplar Bluff	94.4	84.3	90.0
640-641	Kansas City	97.5	101.8	99.3
644-645	St. Joseph	96.0	93.3	94.9
646	Chillicothe	93.7	90.7	92.4
647	Harrisonville	93.2	97.4	95.0
648	Joplin	95.2	80.1	88.7
650-651	Jefferson City	95.7	90.0	93.2
652	Columbia	95.7	94.5	95.2
653	Sedalia	95.6	88.1	92.4
654-655	Rolla	93.4	90.6	92.2
656-658	Springfield	96.5	81.4	90.0
MONTANA				
590-591	Billings	101.4	77.2	90.9
592	Wolf Point	101.3	74.5	89.7
593	Miles City	99.1	74.6	88.5
594	Great Falls	102.8	77.1	91.7
595	Havre	100.2	72.5	88.3
596	Helena	101.2	74.7	89.7
597	Butte	101.3	74.4	89.7
598	Missoula	98.8	74.8	88.5
599	Kalispell	98.0	74.1	87.7
NEBRASKA				
680-681	Omaha	98.6	83.0	91.9
683-685	Lincoln	99.8	79.3	90.9
686	Columbus	96.0	77.7	88.1
687	Norfolk	97.4	74.5	87.5
688	Grand Island	97.4	76.2	88.2
689	Hastings	97.1	75.2	87.7
690	McCook	95.9	70.2	84.8
691	North Platte	96.1	74.9	86.9
692	Valentine	98.0	67.4	84.8
693	Alliance	98.4	70.4	86.3
NEVADA				
889-891	Las Vegas	104.2	107.0	105.4
893	Ely	102.9	89.5	97.1
894-895	Reno	102.8	85.6	95.4
897	Carson City	103.2	85.6	95.6
898	Elko	101.5	81.8	93.0
NEW HAMPSHIRE				
030	Nashua	99.3	92.3	96.3
031	Manchester	100.5	93.0	97.3

STATE/ZIP	CITY	MAT.	INST.	TOTAL
NEW HAMPSHIRE (CONT'D)				
032-033	Concord	100.8	91.3	96.7
034	Keene	96.1	86.2	91.8
035	Littleton	96.1	77.7	88.1
036	Charleston	95.6	85.6	91.3
037	Claremont	94.8	85.7	90.8
038	Portsmouth	96.5	91.5	94.3
NEW JERSEY				
070-071	Newark	101.8	137.9	117.4
072	Elizabeth	98.9	137.6	115.6
073	Jersey City	97.7	137.1	114.7
074-075	Paterson	99.3	137.3	115.7
076	Hackensack	97.4	137.1	114.6
077	Long Branch	97.1	126.6	109.9
078	Dover	97.6	137.5	114.8
079	Summit	97.7	136.9	114.6
080,083	Vineland	97.2	126.6	109.9
081	Camden	98.9	133.7	114.0
082,084	Atlantic City	97.8	132.6	112.8
085-086	Trenton	100.5	124.4	110.8
087	Point Pleasant	99.1	126.5	110.9
088-089	New Brunswick	99.7	132.2	113.8
NEW MEXICO				
870-872	Albuquerque	98.9	71.6	87.1
873	Gallup	99.2	71.6	87.3
874	Farmington	99.3	71.6	87.3
875	Santa Fe	100.5	72.2	88.3
877	Las Vegas	97.4	71.6	86.3
878	Socorro	97.1	71.6	86.1
879	Truth/Consequences	96.8	68.6	84.6
880	Las Cruces	96.8	70.0	85.2
881	Clovis	99.1	71.5	87.2
882	Roswell	100.7	71.6	88.1
883	Carrizozo	101.4	71.6	88.5
884	Tucumcari	99.7	71.5	87.6
NEW YORK				
100-102	New York	100.3	174.2	132.2
103	Staten Island	96.5	171.9	129.1
104	Bronx	94.8	171.2	127.8
105	Mount Vernon	94.9	147.0	117.4
106	White Plains	94.9	149.8	118.6
107	Yonkers	99.1	150.1	121.1
108	New Rochelle	95.3	141.4	115.2
109	Suffern	94.9	125.4	108.1
110	Queens	100.5	173.1	131.8
111	Long Island City	102.2	173.1	132.8
112	Brooklyn	102.7	173.1	133.1
113	Flushing	102.5	173.1	133.0
114	Jamaica	100.9	173.1	132.1
115,117,118	Hicksville	100.5	151.7	122.6
116	Far Rockaway	102.7	173.1	133.1
119	Riverhead	101.3	156.9	125.3
120-122	Albany	98.3	109.4	103.1
123	Schenectady	97.9	108.6	102.5
124	Kingston	101.2	128.8	113.1
125-126	Poughkeepsie	100.3	131.7	113.9
127	Monticello	99.8	129.5	112.7
128	Glens Falls	92.8	103.2	97.3
129	Plattsburgh	97.7	92.8	95.6
130-132	Syracuse	98.7	101.6	99.9
133-135	Utica	96.8	102.5	99.3
136	Watertown	98.4	98.9	98.6
137-139	Binghamton	98.5	100.5	99.3
140-142	Buffalo	100.9	109.8	104.8
143	Niagara Falls	98.0	103.4	100.3
144-146	Rochester	100.8	101.2	101.0
147	Jamestown	97.1	92.3	95.0
148-149	Elmira	96.9	100.4	98.4
NORTH CAROLINA				
270,272-274	Greensboro	99.7	65.3	84.8
271	Winston-Salem	99.5	65.3	84.7
275-276	Raleigh	99.0	65.3	84.4
277	Durham	101.2	65.2	85.6
278	Rocky Mount	97.1	64.5	83.0
279	Elizabeth City	98.1	66.6	84.5
280	Gastonia	97.7	68.5	85.1
281-282	Charlotte	98.4	72.1	87.0
283	Fayetteville	100.5	64.0	84.7
284	Wilmington	96.6	63.1	82.1
285	Kinston	94.9	63.5	81.3

For customer support on your Site Work & Landscape Costs with RSMeans Data, call 800.448.8182.

STATE/ZIP	CITY	MAT.	INST.	TOTAL
NORTH CAROLINA (CONT'D)				
286	Hickory	95.2	69.0	83.9
287-288	Asheville	97.0	64.6	83.0
289	Murphy	96.1	62.6	81.6
NORTH DAKOTA				
580-581	Fargo	99.6	78.9	90.6
582	Grand Forks	98.8	81.0	91.1
583	Devils Lake	98.6	77.8	89.6
584	Jamestown	98.7	76.9	89.3
585	Bismarck	99.1	83.1	92.2
586	Dickinson	99.2	75.7	89.0
587	Minot	98.7	76.5	89.1
588	Williston	97.8	80.7	90.4
OHIO				
430-432	Columbus	99.2	83.2	92.3
433	Marion	95.5	83.6	90.3
434-436	Toledo	98.9	87.7	94.1
437-438	Zanesville	96.1	83.0	90.4
439	Steubenville	97.1	87.4	92.9
440	Lorain	98.7	81.8	91.4
441	Cleveland	99.4	91.5	96.0
442-443	Akron	99.7	83.0	92.5
444-445	Youngstown	99.1	79.7	90.7
446-447	Canton	99.2	77.5	89.8
448-449	Mansfield	96.6	81.1	89.9
450	Hamilton	96.9	76.7	88.2
451-452	Cincinnati	98.0	79.2	89.9
453-454	Dayton	96.9	76.6	88.1
455	Springfield	96.9	77.2	88.4
456	Chillicothe	96.3	85.3	91.5
457	Athens	99.2	81.8	91.7
458	Lima	99.6	79.0	90.7
OKLAHOMA				
730-731	Oklahoma City	97.6	67.8	84.8
734	Ardmore	95.9	63.8	82.0
735	Lawton	98.1	63.9	83.3
736	Clinton	97.3	63.7	82.8
737	Enid	97.8	64.2	83.3
738	Woodward	95.9	61.2	80.9
739	Guymon	97.0	61.7	81.7
740-741	Tulsa	96.9	65.5	83.3
743	Miami	93.5	63.0	80.3
744	Muskogee	96.1	65.7	83.0
745	McAlester	93.2	58.3	78.1
746	Ponca City	93.9	61.8	80.0
747	Durant	93.8	62.7	80.4
748	Shawnee	95.3	63.3	81.5
749	Poteau	93.0	62.6	79.9
OREGON				
970-972	Portland	102.0	104.7	103.2
973	Salem	103.3	102.8	103.0
974	Eugene	101.6	103.7	102.5
975	Medford	103.1	99.3	101.4
976	Klamath Falls	103.3	97.8	100.9
977	Bend	102.3	100.5	101.5
978	Pendleton	98.3	101.0	99.5
979	Vale	96.0	87.8	92.4
PENNSYLVANIA				
150-152	Pittsburgh	99.3	102.6	100.7
153	Washington	96.6	101.9	98.9
154	Uniontown	97.1	99.2	98.0
155	Bedford	98.0	90.1	94.6
156	Greensburg	98.0	95.8	97.1
157	Indiana	96.8	96.2	96.5
158	Dubois	98.5	94.0	96.6
159	Johnstown	98.0	97.2	97.7
160	Butler	92.1	98.7	95.0
161	New Castle	92.2	98.9	95.1
162	Kittanning	92.6	96.6	94.3
163	Oil City	92.1	94.4	93.1
164-165	Erie	94.0	94.6	94.3
166	Altoona	94.2	97.5	95.6
167	Bradford	95.6	95.0	95.3
168	State College	95.2	99.1	96.9
169	Wellsboro	96.2	89.3	93.2
170-171	Harrisburg	98.5	97.0	97.9
172	Chambersburg	95.4	86.0	91.3
173-174	York	96.2	94.6	95.5
175-176	Lancaster	94.2	96.5	95.2

STATE/ZIP	CITY	MAT.	INST.	TOTAL
PENNSYLVANIA (CONT'D)				
177	Williamsport	92.8	94.2	93.4
178	Sunbury	95.0	89.3	92.5
179	Pottsville	94.0	92.5	93.3
180	Lehigh Valley	95.6	107.5	100.8
181	Allentown	97.5	108.1	102.1
182	Hazleton	95.1	92.5	94.0
183	Stroudsburg	95.0	100.9	97.6
184-185	Scranton	98.3	96.8	97.7
186-187	Wilkes-Barre	94.9	96.8	95.7
188	Montrose	94.6	93.3	94.0
189	Doylestown	94.7	122.6	106.7
190-191	Philadelphia	99.7	136.9	115.8
193	Westchester	96.1	123.2	107.8
194	Norristown	95.1	123.4	107.3
195-196	Reading	97.3	104.7	100.5
PUERTO RICO				
009	San Juan	124.9	28.3	83.1
RHODE ISLAND				
028	Newport	97.0	112.6	103.8
029	Providence	101.0	113.3	106.3
SOUTH CAROLINA				
290-292	Columbia	98.8	67.6	85.3
293	Spartanburg	97.7	70.2	85.8
294	Charleston	99.4	67.0	85.4
295	Florence	97.4	67.0	84.2
296	Greenville	97.5	70.0	85.6
297	Rock Hill	97.3	68.2	84.7
298	Aiken	98.1	65.2	83.9
299	Beaufort	98.9	61.5	82.7
SOUTH DAKOTA				
570-571	Sioux Falls	99.3	82.1	91.8
572	Watertown	97.7	66.6	84.3
573	Mitchell	96.6	59.6	80.6
574	Aberdeen	99.1	66.7	85.1
575	Pierre	100.2	68.9	86.7
576	Mobridge	97.1	60.7	81.4
577	Rapid City	98.7	71.0	86.7
TENNESSEE				
370-372	Nashville	100.0	74.5	89.0
373-374	Chattanooga	99.0	68.7	85.9
375,380-381	Memphis	96.5	72.5	86.1
376	Johnson City	99.2	60.5	82.5
377-379	Knoxville	95.7	66.0	82.9
382	McKenzie	96.7	54.7	78.6
383	Jackson	98.1	60.0	81.6
384	Columbia	95.2	64.6	82.0
385	Cookeville	96.6	56.1	79.1
TEXAS				
750	McKinney	99.5	61.8	83.2
751	Waxahachie	99.5	62.0	83.3
752-753	Dallas	100.1	67.4	86.0
754	Greenville	99.6	61.3	83.1
755	Texarkana	99.1	60.4	82.4
756	Longview	99.8	59.2	82.3
757	Tyler	100.1	60.2	82.8
758	Palestine	95.9	58.9	80.0
759	Lufkin	96.6	60.5	81.0
760-761	Fort Worth	99.0	63.9	83.9
762	Denton	98.9	61.6	82.8
763	Wichita Falls	96.5	62.5	81.8
764	Eastland	95.7	58.9	79.8
765	Temple	94.1	56.4	77.8
766-767	Waco	95.8	63.2	81.7
768	Brownwood	99.1	57.2	81.0
769	San Angelo	98.7	57.3	80.8
770-772	Houston	101.3	67.3	86.6
773	Huntsville	99.9	60.7	83.0
774	Wharton	101.0	62.1	84.2
775	Galveston	98.8	64.0	83.8
776-777	Beaumont	99.2	64.9	84.4
778	Bryan	96.3	62.4	81.7
779	Victoria	101.0	60.8	83.6
780	Laredo	97.6	60.9	81.8
781-782	San Antonio	98.9	63.8	83.7
783-784	Corpus Christi	100.2	64.6	84.8
785	McAllen	100.9	55.2	81.2
786-787	Austin	98.6	62.3	82.9

STATE/ZIP	CITY	MAT.	INST.	TOTAL
TEXAS (CONT'D)				
788	Del Rio	100.7	58.2	82.3
789	Giddings	97.3	58.3	80.5
790-791	Amarillo	99.9	61.8	83.5
792	Childress	98.7	59.4	81.7
793-794	Lubbock	100.4	60.7	83.3
795-796	Abilene	99.0	61.6	82.9
797	Midland	100.9	61.7	84.0
798-799,885	El Paso	98.3	63.3	83.2
UTAH				
840-841	Salt Lake City	103.3	73.7	90.5
842,844	Ogden	98.6	74.0	88.0
843	Logan	100.7	74.0	89.2
845	Price	101.1	68.6	87.1
846-847	Provo	101.1	74.0	89.4
VERMONT				
050	White River Jct.	98.7	79.6	90.4
051	Bellows Falls	97.1	90.6	94.3
052	Bennington	97.5	87.2	93.0
053	Brattleboro	97.8	90.7	94.7
054	Burlington	103.1	82.3	94.1
056	Montpelier	101.2	86.0	94.6
057	Rutland	99.4	81.6	91.7
058	St. Johnsbury	98.7	79.2	90.3
059	Guildhall	97.3	79.2	89.5
VIRGINIA				
220-221	Fairfax	99.9	81.7	92.0
222	Arlington	101.0	81.6	92.6
223	Alexandria	100.0	84.8	93.4
224-225	Fredericksburg	98.6	79.9	90.5
226	Winchester	99.2	76.4	89.3
227	Culpeper	99.1	82.0	91.7
228	Harrisonburg	99.4	75.6	89.1
229	Charlottesville	99.8	74.2	88.7
230-232	Richmond	99.6	74.6	88.8
233-235	Norfolk	100.7	68.3	86.7
236	Newport News	99.5	69.5	86.5
237	Portsmouth	99.0	67.4	85.3
238	Petersburg	99.4	74.4	88.6
239	Farmville	98.5	66.0	84.5
240-241	Roanoke	100.5	69.1	87.0
242	Bristol	98.5	59.4	81.6
243	Pulaski	98.1	69.7	85.8
244	Staunton	99.0	67.8	85.5
245	Lynchburg	99.1	73.1	87.9
246	Grundy	98.4	63.2	83.2
WASHINGTON				
980-981,987	Seattle	105.4	108.4	106.7
982	Everett	104.5	104.2	104.4
983-984	Tacoma	104.9	105.7	105.2
985	Olympia	103.9	105.2	104.5
986	Vancouver	106.2	102.2	104.4
988	Wenatchee	105.0	85.6	96.6
989	Yakima	105.0	101.6	103.5
990-992	Spokane	101.0	86.8	94.9
993	Richland	100.7	93.0	97.4
994	Clarkston	99.9	79.5	91.1
WEST VIRGINIA				
247-248	Bluefield	97.3	85.2	92.1
249	Lewisburg	99.0	86.8	93.8
250-253	Charleston	96.6	91.1	94.2
254	Martinsburg	96.5	80.8	89.7
255-257	Huntington	97.8	91.7	95.1
258-259	Beckley	95.2	87.7	91.9
260	Wheeling	100.1	90.9	96.1
261	Parkersburg	98.9	90.0	95.1
262	Buckhannon	98.7	90.5	95.1
263-264	Clarksburg	99.3	91.0	95.7
265	Morgantown	99.3	91.4	95.9
266	Gassaway	98.7	87.3	93.7
267	Romney	98.5	85.3	92.8
268	Petersburg	98.3	85.6	92.8
WISCONSIN				
530,532	Milwaukee	97.9	111.7	103.9
531	Kenosha	99.2	106.2	102.2
534	Racine	98.5	106.2	101.8
535	Beloit	97.8	97.3	97.6
537	Madison	98.6	106.4	101.9

STATE/ZIP	CITY	MAT.	INST.	TOTAL
WISCONSIN (CONT'D)				
538	Lancaster	95.8	92.9	94.6
539	Portage	94.2	96.5	95.2
540	New Richmond	94.9	92.9	94.0
541-543	Green Bay	99.4	103.1	101.0
544	Wausau	94.5	92.9	93.8
545	Rhinelander	98.0	91.1	95.0
546	La Crosse	95.7	100.3	97.7
547	Eau Claire	97.4	100.4	98.7
548	Superior	94.6	94.6	94.6
549	Oshkosh	95.2	92.7	94.1
WYOMING				
820	Cheyenne	101.0	72.4	88.7
821	Yellowstone Nat'l Park	98.9	71.4	87.0
822	Wheatland	100.1	69.9	87.0
823	Rawlins	101.9	71.3	88.7
824	Worland	99.5	71.9	87.6
825	Riverton	100.8	71.4	88.1
826	Casper	102.5	72.3	89.4
827	Newcastle	99.4	71.3	87.3
828	Sheridan	102.4	71.1	88.9
829-831	Rock Springs	103.3	70.6	89.1
CANADIAN FACTORS (reflect Canadian currency)				
ALBERTA				
	Calgary	123.3	91.7	109.7
	Edmonton	125.3	91.9	110.9
	Fort McMurray	123.9	86.2	107.6
	Lethbridge	118.9	85.7	104.6
	Lloydminster	113.6	82.6	100.2
	Medicine Hat	113.7	81.9	100.0
	Red Deer	114.2	81.9	100.2
BRITISH COLUMBIA				
	Kamloops	115.6	81.6	100.9
	Prince George	116.5	81.0	101.2
	Vancouver	121.7	91.7	108.7
	Victoria	116.8	86.4	103.7
MANITOBA				
	Brandon	122.8	68.9	99.5
	Portage la Prairie	113.6	67.5	93.6
	Winnipeg	125.8	68.4	101.0
NEW BRUNSWICK				
	Bathurst	112.7	60.8	90.3
	Dalhousie	112.4	61.0	90.2
	Fredericton	121.9	67.8	98.5
	Moncton	113.0	73.2	95.8
	Newcastle	112.8	61.4	90.5
	St. John	115.6	72.9	97.1
NEWFOUNDLAND				
	Corner Brook	127.3	66.7	101.1
	St. Johns	126.2	85.2	108.5
NORTHWEST TERRITORIES				
	Yellowknife	135.0	81.1	111.7
NOVA SCOTIA				
	Bridgewater	113.3	68.7	94.0
	Dartmouth	124.0	68.6	100.1
	Halifax	120.5	84.9	105.1
	New Glasgow	121.9	68.6	98.9
	Sydney	120.1	68.6	97.9
	Truro	112.8	68.7	93.7
	Yarmouth	121.9	68.6	98.8
ONTARIO				
	Barrie	118.0	85.2	103.8
	Brantford	115.2	88.8	103.8
	Cornwall	115.3	85.5	102.4
	Hamilton	119.2	94.5	108.6
	Kingston	116.5	85.5	103.1
	Kitchener	109.9	91.2	101.8
	London	119.2	92.9	107.8
	North Bay	125.0	83.4	107.0
	Oshawa	112.4	93.6	104.3
	Ottawa	120.6	93.0	108.7
	Owen Sound	118.1	83.5	103.2
	Peterborough	115.2	85.2	102.3
	Sarnia	115.5	89.4	104.2

STATE/ZIP	CITY	MAT.	INST.	TOTAL
ONTARIO (CONT'D)				
	Sault Ste. Marie	110.6	86.4	100.1
	St. Catharines	108.0	92.1	101.1
	Sudbury	107.7	90.3	100.2
	Thunder Bay	109.3	91.6	101.6
	Timmins	115.4	83.4	101.6
	Toronto	119.3	99.9	111.0
	Windsor	108.6	91.2	101.1
PRINCE EDWARD ISLAND				
	Charlottetown	124.1	58.7	95.8
	Summerside	124.4	56.4	95.0
QUEBEC				
	Cap-de-la-Madeleine	112.8	78.2	97.8
	Charlesbourg	112.8	78.2	97.8
	Chicoutimi	112.5	86.9	101.4
	Gatineau	112.5	78.0	97.6
	Granby	112.7	77.9	97.7
	Hull	112.6	78.0	97.7
	Joliette	113.0	78.2	97.9
	Laval	112.6	78.6	97.9
	Montreal	122.1	89.4	108.0
	Quebec City	122.2	89.9	108.2
	Rimouski	112.3	86.9	101.4
	Rouyn-Noranda	112.4	78.0	97.6
	Saint-Hyacinthe	111.8	78.0	97.2
	Sherbrooke	112.8	78.0	97.7
	Sorel	112.9	78.2	97.9
	Saint-Jerome	112.5	78.0	97.6
	Trois-Rivieres	122.8	78.1	103.5
SASKATCHEWAN				
	Moose Jaw	111.0	62.4	90.0
	Prince Albert	110.2	60.7	88.8
	Regina	127.0	88.5	110.3
	Saskatoon	112.0	85.1	100.4
YUKON				
	Whitehorse	136.0	68.6	106.9

R011105-05 Tips for Accurate Estimating

1. Use pre-printed or columnar forms for orderly sequence of dimensions and locations and for recording telephone quotations.

2. Use only the front side of each paper or form except for certain pre-printed summary forms.

3. Be consistent in listing dimensions: For example, length x width x height. This helps in rechecking to ensure that, the total length of partitions is appropriate for the building area.

4. Use printed (rather than measured) dimensions where given.

5. Add up multiple printed dimensions for a single entry where possible.

6. Measure all other dimensions carefully.

7. Use each set of dimensions to calculate multiple related quantities.

8. Convert foot and inch measurements to decimal feet when listing. Memorize decimal equivalents to .01 parts of a foot (1/8″ equals approximately .01′).

9. Do not "round off" quantities until the final summary.

10. Mark drawings with different colors as items are taken off.

11. Keep similar items together, different items separate.

12. Identify location and drawing numbers to aid in future checking for completeness.

13. Measure or list everything on the drawings or mentioned in the specifications.

14. It may be necessary to list items not called for to make the job complete.

15. Be alert for: Notes on plans such as N.T.S. (not to scale); changes in scale throughout the drawings; reduced size drawings; discrepancies between the specifications and the drawings.

16. Develop a consistent pattern of performing an estimate. For example:
    a. Start the quantity takeoff at the lower floor and move to the next higher floor.
    b. Proceed from the main section of the building to the wings.
    c. Proceed from south to north or vice versa, clockwise or counterclockwise.
    d. Take off floor plan quantities first, elevations next, then detail drawings.

17. List all gross dimensions that can be either used again for different quantities, or used as a rough check of other quantities for verification (exterior perimeter, gross floor area, individual floor areas, etc.).

18. Utilize design symmetry or repetition (repetitive floors, repetitive wings, symmetrical design around a center line, similar room layouts, etc.). Note: Extreme caution is needed here so as not to omit or duplicate an area.

19. Do not convert units until the final total is obtained. For instance, when estimating concrete work, keep all units to the nearest cubic foot, then summarize and convert to cubic yards.

20. When figuring alternatives, it is best to total all items involved in the basic system, then total all items involved in the alternates. Therefore you work with positive numbers in all cases. When adds and deducts are used, it is often confusing whether to add or subtract a portion of an item; especially on a complicated or involved alternate.

R011105-50   Metric Conversion Factors

Description: This table is primarily for converting customary U.S. units in the left hand column to SI metric units in the right hand column. In addition, conversion factors for some commonly encountered Canadian and non-SI metric units are included.

	If You Know		Multiply By		To Find
Length	Inches	x	25.4[a]	=	Millimeters
	Feet	x	0.3048[a]	=	Meters
	Yards	x	0.9144[a]	=	Meters
	Miles (statute)	x	1.609	=	Kilometers
Area	Square inches	x	645.2	=	Square millimeters
	Square feet	x	0.0929	=	Square meters
	Square yards	x	0.8361	=	Square meters
Volume (Capacity)	Cubic inches	x	16,387	=	Cubic millimeters
	Cubic feet	x	0.02832	=	Cubic meters
	Cubic yards	x	0.7646	=	Cubic meters
	Gallons (U.S. liquids)[b]	x	0.003785	=	Cubic meters[c]
	Gallons (Canadian liquid)[b]	x	0.004546	=	Cubic meters[c]
	Ounces (U.S. liquid)[b]	x	29.57	=	Milliliters[c, d]
	Quarts (U.S. liquid)[b]	x	0.9464	=	Liters[c, d]
	Gallons (U.S. liquid)[b]	x	3.785	=	Liters[c, d]
Force	Kilograms force[d]	x	9.807	=	Newtons
	Pounds force	x	4.448	=	Newtons
	Pounds force	x	0.4536	=	Kilograms force[d]
	Kips	x	4448	=	Newtons
	Kips	x	453.6	=	Kilograms force[d]
Pressure, Stress, Strength (Force per unit area)	Kilograms force per square centimeter[d]	x	0.09807	=	Megapascals
	Pounds force per square inch (psi)	x	0.006895	=	Megapascals
	Kips per square inch	x	6.895	=	Megapascals
	Pounds force per square inch (psi)	x	0.07031	=	Kilograms force per square centimeter[d]
	Pounds force per square foot	x	47.88	=	Pascals
	Pounds force per square foot	x	4.882	=	Kilograms force per square meter[d]
Flow	Cubic feet per minute	x	0.4719	=	Liters per second
	Gallons per minute	x	0.0631	=	Liters per second
	Gallons per hour	x	1.05	=	Milliliters per second
Bending Moment Or Torque	Inch-pounds force	x	0.01152	=	Meter-kilograms force[d]
	Inch-pounds force	x	0.1130	=	Newton-meters
	Foot-pounds force	x	0.1383	=	Meter-kilograms force[d]
	Foot-pounds force	x	1.356	=	Newton-meters
	Meter-kilograms force[d]	x	9.807	=	Newton-meters
Mass	Ounces (avoirdupois)	x	28.35	=	Grams
	Pounds (avoirdupois)	x	0.4536	=	Kilograms
	Tons (metric)	x	1000	=	Kilograms
	Tons, short (2000 pounds)	x	907.2	=	Kilograms
	Tons, short (2000 pounds)	x	0.9072	=	Megagrams[e]
Mass per Unit Volume	Pounds mass per cubic foot	x	16.02	=	Kilograms per cubic meter
	Pounds mass per cubic yard	x	0.5933	=	Kilograms per cubic meter
	Pounds mass per gallon (U.S. liquid)[b]	x	119.8	=	Kilograms per cubic meter
	Pounds mass per gallon (Canadian liquid)[b]	x	99.78	=	Kilograms per cubic meter
Temperature	Degrees Fahrenheit	(F-32)/1.8		=	Degrees Celsius
	Degrees Fahrenheit	(F+459.67)/1.8		=	Degrees Kelvin
	Degrees Celsius	C+273.15		=	Degrees Kelvin

[a]The factor given is exact
[b]One U.S. gallon = 0.8327 Canadian gallon
[c]1 liter = 1000 milliliters = 1000 cubic centimeters
  1 cubic decimeter = 0.001 cubic meter

[d]Metric but not SI unit
[e]Called "tonne" in England and
  "metric ton" in other metric countries

R011105-60 Weights and Measures

Measures of Length
1 Mile = 1760 Yards = 5280 Feet
1 Yard = 3 Feet = 36 inches
1 Foot = 12 Inches
1 Mil = 0.001 Inch
1 Fathom = 2 Yards = 6 Feet
1 Rod = 5.5 Yards = 16.5 Feet
1 Hand = 4 Inches
1 Span = 9 Inches
1 Micro-inch = One Millionth Inch or 0.000001 Inch
1 Micron = One Millionth Meter + 0.00003937 Inch

Surveyor's Measure
1 Mile = 8 Furlongs = 80 Chains
1 Furlong = 10 Chains = 220 Yards
1 Chain = 4 Rods = 22 Yards = 66 Feet = 100 Links
1 Link = 7.92 Inches

Square Measure
1 Square Mile = 640 Acres = 6400 Square Chains
1 Acre = 10 Square Chains = 4840 Square Yards = 43,560 Sq. Ft.
1 Square Chain = 16 Square Rods = 484 Square Yards = 4356 Sq. Ft.
1 Square Rod = 30.25 Square Yards = 272.25 Square Feet = 625 Square Lines
1 Square Yard = 9 Square Feet
1 Square Foot = 144 Square Inches
An Acre equals a Square 208.7 Feet per Side

Cubic Measure
1 Cubic Yard = 27 Cubic Feet
1 Cubic Foot = 1728 Cubic Inches
1 Cord of Wood = 4 x 4 x 8 Feet = 128 Cubic Feet
1 Perch of Masonry = 16½ x 1½ x 1 Foot = 24.75 Cubic Feet

Avoirdupois or Commercial Weight
1 Gross or Long Ton = 2240 Pounds
1 Net or Short Ton = 2000 Pounds
1 Pound = 16 Ounces = 7000 Grains
1 Ounce = 16 Drachms = 437.5 Grains
1 Stone = 14 Pounds

Power
1 British Thermal Unit per Hour = 0.2931 Watts
1 Ton (Refrigeration) = 3.517 Kilowatts
1 Horsepower (Boiler) = 9.81 Kilowatts
1 Horsepower (550 ft-lb/s) = 0.746 Kilowatts

Shipping Measure
For Measuring Internal Capacity of a Vessel:
   1 Register Ton = 100 Cubic Feet

For Measurement of Cargo:
   Approximately 40 Cubic Feet of Merchandise is considered a Shipping Ton, unless that bulk would weigh more than 2000 Pounds, in which case Freight Charge may be based upon weight.

40 Cubic Feet = 32.143 U.S. Bushels = 31.16 Imp. Bushels

Liquid Measure
1 Imperial Gallon = 1.2009 U.S. Gallon = 277.42 Cu. In.
1 Cubic Foot = 7.48 U.S. Gallons

R011110-10 Architectural Fees

Tabulated below are typical percentage fees by project size, for good professional architectural service. Fees may vary from those listed depending upon degree of design difficulty and economic conditions in any particular area.

Rates can be interpolated horizontally and vertically. Various portions of the same project requiring different rates should be adjusted proportionately. For alterations, add 50% to the fee for the first $500,000 of project cost and add 25% to the fee for project cost over $500,000.

Architectural fees tabulated below include Structural, Mechanical and Electrical Engineering Fees. They do not include the fees for special consultants such as kitchen planning, security, acoustical, interior design, etc.

Civil Engineering fees are included in the Architectural fee for project sites requiring minimal design such as city sites. However, separate Civil Engineering fees must be added when utility connections require design, drainage calculations are needed, stepped foundations are required, or provisions are required to protect adjacent wetlands.

Building Types	Total Project Size in Thousands of Dollars						
	100	250	500	1,000	5,000	10,000	50,000
Factories, garages, warehouses, repetitive housing	9.0%	8.0%	7.0%	6.2%	5.3%	4.9%	4.5%
Apartments, banks, schools, libraries, offices, municipal buildings	12.2	12.3	9.2	8.0	7.0	6.6	6.2
Churches, hospitals, homes, laboratories, museums, research	15.0	13.6	12.7	11.9	9.5	8.8	8.0
Memorials, monumental work, decorative furnishings	—	16.0	14.5	13.1	10.0	9.0	8.3

R011110-30   Engineering Fees

Typical **Structural Engineering Fees** based on type of construction and total project size. These fees are included in Architectural Fees.

Type of Construction	Total Project Size (in thousands of dollars)			
	$500	$500-$1,000	$1,000-$5,000	Over $5000
Industrial buildings, factories & warehouses	Technical payroll times 2.0 to 2.5	1.60%	1.25%	1.00%
Hotels, apartments, offices, dormitories, hospitals, public buildings, food stores		2.00%	1.70%	1.20%
Museums, banks, churches and cathedrals		2.00%	1.75%	1.25%
Thin shells, prestressed concrete, earthquake resistive		2.00%	1.75%	1.50%
Parking ramps, auditoriums, stadiums, convention halls, hangars & boiler houses		2.50%	2.00%	1.75%
Special buildings, major alterations, underpinning & future expansion	▼	Add to above 0.5%	Add to above 0.5%	Add to above 0.5%

For complex reinforced concrete or unusually complicated structures, add 20% to 50%.

Typical **Mechanical and Electrical Engineering Fees** are based on the size of the subcontract. The fee structure for both is shown below. These fees are included in Architectural Fees.

Type of Construction	Subcontract Size							
	$25,000	$50,000	$100,000	$225,000	$350,000	$500,000	$750,000	$1,000,000
Simple structures	6.4%	5.7%	4.8%	4.5%	4.4%	4.3%	4.2%	4.1%
Intermediate structures	8.0	7.3	6.5	5.6	5.1	5.0	4.9	4.8
Complex structures	10.1	9.0	9.0	8.0	7.5	7.5	7.0	7.0

For renovations, add 15% to 25% to applicable fee.

R012153-60   Security Factors

Contractors entering, working in, and exiting secure facilities often lose productive time during a normal workday. The recommended allowances in this section are intended to provide for the loss of productivity by increasing labor costs. Note that different costs are associated with searches upon entry only and searches upon entry and exit. Time spent in a queue is unpredictable and not part of these allowances. Contractors should plan ahead for this situation.

Security checkpoints are designed to reflect the level of security required to gain access or egress. An extreme example is when contractors, along with any materials, tools, equipment, and vehicles, must be physically searched and have all materials, tools, equipment, and vehicles inventoried and documented prior to both entry and exit.

Physical searches without going through the documentation process represent the next level and take up less time.

Electronic searches—passing through a detector or x-ray machine with no documentation of materials, tools, equipment, and vehicles—take less time than physical searches.

Visual searches of materials, tools, equipment, and vehicles represent the next level of security.

Finally, access by means of an ID card or displayed sticker takes the least amount of time.

Another consideration is if the searches described above are performed each and every day, or if they are performed only on the first day with access granted by ID card or displayed sticker for the remainder of the project. The figures for this situation have been calculated to represent the initial check-in as described and subsequent entry by ID card or displayed sticker for up to 20 days on site. For the situation described above, where the time period is beyond 20 days, the impact on labor cost is negligible.

There are situations where tradespeople must be accompanied by an escort and observed during the work day. The loss of freedom of movement will slow down productivity for the tradesperson. Costs for the observer have not been included. Those costs are normally born by the owner.

R012153-65   Infectious Disease Precautions

Contractors entering and working on job sites may be required to take precautions to prevent the spread of infectious diseases. Those precautions will reduce the amount of productive time during a normal workday. The recommended allowances in this section are intended to provide for the loss of productive time by increasing labor costs.

The estimator should be aware that one, many or none of these precautions apply to the line items in an estimate. Job site requirements and sound judgement must be applied.

Labor cost implications are based upon:

Temperature checks	Once per day
Donning/doffing masks and gloves	Four times per day (2 times each)
Washing hands (additional over normal)	Eight times per day
Informational meetings	Once per day
Maintaining social distance	Throughout the day
Disinfecting tools or equipment	Throughout the day

R012909-80   Sales Tax by State

State sales tax on materials is tabulated below (5 states have no sales tax). Many states allow local jurisdictions, such as a county or city, to levy additional sales tax.

Some projects may be sales tax exempt, particularly those constructed with public funds.

State	Tax (%)	State	Tax (%)	State	Tax (%)	State	Tax (%)
Alabama	4	Illinois	6.25	Montana	0	Rhode Island	7
Alaska	0	Indiana	7	Nebraska	5.5	South Carolina	6
Arizona	5.6	Iowa	6	Nevada	6.85	South Dakota	4.5
Arkansas	6.5	Kansas	6.5	New Hampshire	0	Tennessee	7
California	7.25	Kentucky	6	New Jersey	6.625	Texas	6.25
Colorado	2.9	Louisiana	4.45	New Mexico	5.125	Utah	6.10
Connecticut	6.35	Maine	5.5	New York	4	Vermont	6
Delaware	0	Maryland	6	North Carolina	4.75	Virginia	5.3
District of Columbia	6	Massachusetts	6.25	North Dakota	5	Washington	6.5
Florida	6	Michigan	6	Ohio	5.75	West Virginia	6
Georgia	4	Minnesota	6.875	Oklahoma	4.5	Wisconsin	5
Hawaii	4	Mississippi	7	Oregon	0	Wyoming	4
Idaho	6	Missouri	4.225	Pennsylvania	6	Average	5.11 %

Sales Tax by Province (Canada)

GST - a value-added tax, which the government imposes on most goods and services provided in or imported into Canada. PST - a retail sales tax, which five of the provinces impose on the prices of most goods and some

services. QST - a value-added tax, similar to the federal GST, which Quebec imposes. HST - Three provinces have combined their retail sales taxes with the federal GST into one harmonized tax.

Province	PST (%)	QST(%)	GST(%)	HST(%)
Alberta	7	0	5	0
British Columbia	7	0	5	0
Manitoba	7	0	5	0
New Brunswick	0	0	0	15
Newfoundland	0	0	0	15
Northwest Territories	0	0	5	0
Nova Scotia	0	0	0	15
Ontario	0	0	0	13
Prince Edward Island	0	0	0	15
Quebec	9.975	0	5	0
Saskatchewan	6	0	5	0
Yukon	0	0	5	0

R012909-85   Unemployment Taxes and Social Security Taxes

State unemployment tax rates vary not only from state to state, but also with the experience rating of the contractor. The federal unemployment tax rate is 6.2% of the first $7,000 of wages. This is reduced by a credit of up to 5.4% for timely payment to the state. The minimum federal unemployment tax is 0.6% after all credits.

Social security (FICA) for 2021 is estimated at time of publication to be 7.65% of wages up to $137,700.

R012909-86   Unemployment Tax by State

Information is from the U.S. Department of Labor, state unemployment tax rates.

State	Tax (%)	State	Tax (%)	State	Tax (%)	State	Tax (%)
Alabama	6.80	Illinois	6.93	Montana	6.30	Rhode Island	9.49
Alaska	5.4	Indiana	7.4	Nebraska	5.4	South Carolina	5.46
Arizona	12.76	Iowa	7.5	Nevada	5.4	South Dakota	9.35
Arkansas	14.3	Kansas	7.6	New Hampshire	7.5	Tennessee	10.0
California	6.2	Kentucky	9.3	New Jersey	5.8	Texas	6.5
Colorado	8.15	Louisiana	6.2	New Mexico	5.4	Utah	7.1
Connecticut	6.8	Maine	5.46	New York	9.1	Vermont	7.7
Delaware	8.20	Maryland	7.50	North Carolina	5.76	Virginia	6.21
District of Columbia	7	Massachusetts	12.65	North Dakota	10.74	Washington	7.73
Florida	5.4	Michigan	10.3	Ohio	9	West Virginia	8.5
Georgia	5.4	Minnesota	9.1	Oklahoma	5.5	Wisconsin	12.0
Hawaii	5.6	Mississippi	5.6	Oregon	5.4	Wyoming	8.8
Idaho	5.4	Missouri	8.37	Pennsylvania	11.03	Median	7.40%

R012909-90   Overtime

One way to improve the completion date of a project or eliminate negative float from a schedule is to compress activity duration times. This can be achieved by increasing the crew size or working overtime with the proposed crew.

To determine the costs of working overtime to compress activity duration times, consider the following examples. Below is an overtime efficiency and cost chart based on a five, six, or seven day week with an eight through twelve hour day. Payroll percentage increases for time and one half and double times are shown for the various working days.

Days per Week	Hours per Day	Production Efficiency					Payroll Cost Factors	
		1st Week	2nd Week	3rd Week	4th Week	Average 4 Weeks	@ 1-1/2 Times	@ 2 Times
	8	100%	100%	100%	100%	100%	1.000	1.000
	9	100	100	95	90	96	1.056	1.111
5	10	100	95	90	85	93	1.100	1.200
	11	95	90	75	65	81	1.136	1.273
	12	90	85	70	60	76	1.167	1.333
	8	100	100	95	90	96	1.083	1.167
	9	100	95	90	85	93	1.130	1.259
6	10	95	90	85	80	88	1.167	1.333
	11	95	85	70	65	79	1.197	1.394
	12	90	80	65	60	74	1.222	1.444
	8	100	95	85	75	89	1.143	1.286
	9	95	90	80	70	84	1.183	1.365
7	10	90	85	75	65	79	1.214	1.429
	11	85	80	65	60	73	1.240	1.481
	12	85	75	60	55	69	1.262	1.524

R013113-40   Builder's Risk Insurance

Builder's risk insurance is insurance on a building during construction. Premiums are paid by the owner or the contractor. Blasting, collapse and underground insurance would raise total insurance costs.

For customer support on your Site Work & Landscape Costs with RSMeans Data, call 800.448.8182.

83

R013113-50   General Contractor's Overhead

There are two distinct types of overhead on a construction project: Project overhead and main office overhead. Project overhead includes those costs at a construction site not directly associated with the installation of construction materials. Examples of project overhead costs include the following:

1. Superintendent
2. Construction office and storage trailers
3. Temporary sanitary facilities
4. Temporary utilities
5. Security fencing
6. Photographs
7. Cleanup
8. Performance and payment bonds

The above project overhead items are also referred to as general requirements and therefore are estimated in Division 1. Division 1 is the first division listed in the CSI MasterFormat but it is usually the last division estimated. The sum of the costs in Divisions 1 through 49 is referred to as the sum of the direct costs.

All construction projects also include indirect costs. The primary components of indirect costs are the contractor's main office overhead and profit. The amount of the main office overhead expense varies depending on the following:

1. Owner's compensation
2. Project managers' and estimators' wages
3. Clerical support wages
4. Office rent and utilities
5. Corporate legal and accounting costs
6. Advertising
7. Automobile expenses
8. Association dues
9. Travel and entertainment expenses

These costs are usually calculated as a percentage of annual sales volume. This percentage can range from 35% for a small contractor doing less than $500,000 to 5% for a large contractor with sales in excess of $100 million.

R013113-55   Installing Contractor's Overhead

Installing contractors (subcontractors) also incur costs for general requirements and main office overhead.

Included within the total incl. overhead and profit costs is a percent mark-up for overhead that includes:

1. Compensation and benefits for office staff and project managers
2. Office rent, utilities, business equipment, and maintenance
3. Corporate legal and accounting costs

4. Advertising
5. Vehicle expenses (for office staff and project managers)
6. Association dues
7. Travel, entertainment
8. Insurance
9. Small tools and equipment

R013113-60   Workers' Compensation Insurance Rates by Trade

The table below tabulates the national averages for workers' compensation insurance rates by trade and type of building. The average "Insurance Rate" is multiplied by the "% of Building Cost" for each trade. This produces the "Workers' Compensation" cost by % of total labor cost, to be added for each trade by building type to determine the weighted average workers' compensation rate for the building types analyzed.

Trade	Insurance Rate (% Labor Cost) Range			Average	% of Building Cost Office Bldgs.	Schools & Apts.	Mfg.	Workers' Compensation Office Bldgs.	Schools & Apts.	Mfg.
Excavation, Grading, etc.	2.1 %	to	16.5%	9.3%	4.8%	4.9%	4.5%	0.45%	0.46%	0.42%
Piles & Foundations	3.1	to	28.0	15.6	7.1	5.2	8.7	1.11	0.81	1.36
Concrete	2.5	to	24.9	13.7	5.0	14.8	3.7	0.69	2.03	0.51
Masonry	3.2	to	53.5	28.4	6.9	7.5	1.9	1.96	2.13	0.54
Structural Steel	3.3	to	30.8	17.1	10.7	3.9	17.6	1.83	0.67	3.01
Miscellaneous & Ornamental Metals	2.3	to	22.8	12.5	2.8	4.0	3.6	0.35	0.50	0.45
Carpentry & Millwork	2.6	to	29.1	15.8	3.7	4.0	0.5	0.58	0.63	0.08
Metal or Composition Siding	4.2	to	135.3	69.7	2.3	0.3	4.3	1.60	0.21	3.00
Roofing	4.2	to	105.1	54.7	2.3	2.6	3.1	1.26	1.42	1.70
Doors & Hardware	3.1	to	29.1	16.1	0.9	1.4	0.4	0.14	0.23	0.06
Sash & Glazing	2.6	to	20.5	11.5	3.5	4.0	1.0	0.40	0.46	0.12
Lath & Plaster	2.1	to	33.7	17.9	3.3	6.9	0.8	0.59	1.24	0.14
Tile, Marble & Floors	1.9	to	17.4	9.7	2.6	3.0	0.5	0.25	0.29	0.05
Acoustical Ceilings	1.7	to	21.5	11.6	2.4	0.2	0.3	0.28	0.02	0.03
Painting	2.8	to	37.7	20.3	1.5	1.6	1.6	0.30	0.32	0.32
Interior Partitions	2.6	to	29.1	15.8	3.9	4.3	4.4	0.62	0.68	0.70
Miscellaneous Items	1.6	to	97.7	49.7	5.2	3.7	9.7	2.58	1.84	4.82
Elevators	1.0	to	8.9	5.0	2.1	1.1	2.2	0.11	0.06	0.11
Sprinklers	1.4	to	14.2	7.8	0.5	—	2.0	0.04	—	0.16
Plumbing	1.4	to	13.2	7.3	4.9	7.2	5.2	0.36	0.53	0.38
Heat., Vent., Air Conditioning	2.8	to	15.8	9.3	13.5	11.0	12.9	1.26	1.02	1.20
Electrical	1.2	to	10.6	5.9	10.1	8.4	11.1	0.60	0.50	0.65
Total	1.0 %	to	135.3%	68.1	100.0%	100.0%	100.0%	17.36%	16.05%	19.81%
						Overall Weighted Average	17.74%			

Workers' Compensation Insurance Rates by States

The table below lists the weighted average Workers' Compensation base rate for each state with a factor comparing this with the national average of 15.1%.

State	Weighted Average	Factor	State	Weighted Average	Factor	State	Weighted Average	Factor
Alabama	17.2%	159	Kentucky	17.5%	162	North Dakota	8.3%	77
Alaska	11.9	110	Louisiana	25.5	236	Ohio	6.4	59
Arizona	11.1	103	Maine	13.5	125	Oklahoma	13.3	123
Arkansas	7.3	68	Maryland	15.2	141	Oregon	10.8	100
California	32.6	302	Massachusetts	12.0	111	Pennsylvania	26.1	242
Colorado	8.1	75	Michigan	5.8	54	Rhode Island	11.0	102
Connecticut	19.8	183	Minnesota	20.7	192	South Carolina	24.6	228
Delaware	18.2	169	Mississippi	13.7	127	South Dakota	11.3	105
District of Columbia	11.3	105	Missouri	21.4	198	Tennessee	8.8	81
Florida	12.5	116	Montana	12.2	113	Texas	8.0	74
Georgia	50.0	463	Nebraska	16.6	154	Utah	9.2	85
Hawaii	14.0	130	Nevada	12.0	111	Vermont	13.6	126
Idaho	12.9	119	New Hampshire	12.7	118	Virginia	9.9	92
Illinois	27.9	258	New Jersey	21.5	199	Washington	10.4	96
Indiana	4.9	45	New Mexico	20.7	192	West Virginia	6.7	62
Iowa	15.1	140	New York	25.9	240	Wisconsin	16.0	148
Kansas	8.8	81	North Carolina	17.4	161	Wyoming	7.0	65
			Weighted Average for U.S. is	15.1% of payroll = 100%				

The weighted average skilled worker rate for 35 trades is 17.74%. For bidding purposes, apply the full value of Workers' Compensation directly to total labor costs, or if labor is 38%, materials 42% and overhead and profit 20% of total cost, carry 38/80 x 17.74% = 8.43% of cost (before overhead and profit) into overhead. Rates vary not only from state to state but also with the experience rating of the contractor.

Rates are the most current available at the time of publication.

R015423-10   Steel Tubular Scaffolding

On new construction, tubular scaffolding is efficient up to 60′ high or five stories. Above this it is usually better to use a hung scaffolding if construction permits. Swing scaffolding operations may interfere with tenants. In this case, the tubular is more practical at all heights.

In repairing or cleaning the front of an existing building the cost of tubular scaffolding per S.F. of building front increases as the height increases above the first tier. The first tier cost is relatively high due to leveling and alignment.

The minimum efficient crew for erecting and dismantling is three workers. They can set up and remove 18 frame sections per day up to 5 stories high. For 6 to 12 stories high, a crew of four is most efficient. Use two or more on top and two on the bottom for handing up or hoisting. They can

also set up and remove 18 frame sections per day. At 7′ horizontal spacing, this will run about 800 S.F. per day of erecting and dismantling. Time for placing and removing planks must be added to the above. A crew of three can place and remove 72 planks per day up to 5 stories. For over 5 stories, a crew of four can place and remove 80 planks per day.

The table below shows the number of pieces required to erect tubular steel scaffolding for 1000 S.F. of building frontage. This area is made up of a scaffolding system that is 12 frames (11 bays) long by 2 frames high.

For jobs under twenty-five frames, add 50% to rental cost. Rental rates will be lower for jobs over three months duration. Large quantities for long periods can reduce rental rates by 20%.

Description of Component	Number of Pieces for 1000 S.F. of Building Front	Unit
5′ Wide Standard Frame, 6′-4″ High	24	Ea.
Leveling Jack & Plate	24	
Cross Brace	44	
Side Arm Bracket, 21″	12	
Guardrail Post	12	
Guardrail, 7′ section	22	
Stairway Section	2	
Stairway Starter Bar	1	
Stairway Inside Handrail	2	
Stairway Outside Handrail	2	
Walk-Thru Frame Guardrail	2	↓

Scaffolding is often used as falsework over 15′ high during construction of cast-in-place concrete beams and slabs. Two foot wide scaffolding is generally used for heavy beam construction. The span between frames depends upon the load to be carried with a maximum span of 5′.

Heavy duty shoring frames with a capacity of 10,000#/leg can be spaced up to 10′ O.C. depending upon form support design and loading.

Scaffolding used as horizontal shoring requires less than half the material required with conventional shoring.

On new construction, erection is done by carpenters.

Rolling towers supporting horizontal shores can reduce labor and speed the job. For maintenance work, catwalks with spans up to 70′ can be supported by the rolling towers.

R015433-10   Contractor Equipment

Rental Rates shown elsewhere in the data set pertain to late model high quality machines in excellent working condition, rented from equipment dealers. Rental rates from contractors may be substantially lower than the rental rates from equipment dealers depending upon economic conditions; for older, less productive machines, reduce rates by a maximum of 15%. Any overtime must be added to the base rates. For shift work, rates are lower. Usual rule of thumb is 150% of one shift rate for two shifts; 200% for three shifts.

For periods of less than one week, operated equipment is usually more economical to rent than renting bare equipment and hiring an operator.

Costs to move equipment to a job site (mobilization) or from a job site (demobilization) are not included in rental rates, nor in any Equipment costs on any Unit Price line items or crew listings. These costs can be found elsewhere. If a piece of equipment is already at a job site, it is not appropriate to utilize mob/demob costs in an estimate again.

Rental rates vary throughout the country with larger cities generally having lower rates. Lease plans for new equipment are available for periods in excess of six months with a percentage of payments applying toward purchase.

Rental rates can also be treated as reimbursement costs for contractor-owned equipment. Owned equipment costs include depreciation, loan payments, interest, taxes, insurance, storage, and major repairs.

Monthly rental rates vary from 2% to 5% of the cost of the equipment depending on the anticipated life of the equipment and its wearing parts. Weekly rates are about 1/3 the monthly rates and daily rental rates are about 1/3 the weekly rates.

The hourly operating costs for each piece of equipment include costs to the user such as fuel, oil, lubrication, normal expendables for the equipment, and a percentage of the mechanic's wages chargeable to maintenance. The hourly operating costs listed do not include the operator's wages.

The daily cost for equipment used in the standard crews is figured by dividing the weekly rate by five, then adding eight times the hourly operating cost to give the total daily equipment cost, not including the operator. This figure is in the right hand column of the Equipment listings under Equipment Cost/Day.

Pile Driving rates shown for pile the hammer and extractor do not include leads, cranes, boilers or compressors. Vibratory pile driving requires an added field specialist during set-up and pile driving operation for the electric model. The hydraulic model requires a field specialist for set-up only. Up to 125 reuses of sheet piling are possible using vibratory drivers. For normal conditions, crane capacity for hammer type and size is as follows.

Crane Capacity	Hammer Type and Size		
	Air or Steam	Diesel	Vibratory
25 ton	to 8,750 ft.-lb.		70 H.P.
40 ton	15,000 ft.-lb.	to 32,000 ft.-lb.	170 H.P.
60 ton	25,000 ft.-lb.		300 H.P.
100 ton		112,000 ft.-lb.	

Cranes should be specified for the job by size, building and site characteristics, availability, performance characteristics, and duration of time required.

Backhoes & Shovels rent for about the same as equivalent size cranes but maintenance and operating expenses are higher. The crane operator's rate must be adjusted for high boom heights. Average adjustments: for 150' boom add 2% per hour; over 185', add 4% per hour; over 210', add 6% per hour; over 250', add 8% per hour and over 295', add 12% per hour.

Tower Cranes of the climbing or static type have jibs from 50' to 200' and capacities at maximum reach range from 4,000 to 14,000 pounds. Lifting capacities increase up to maximum load as the hook radius decreases.

Typical rental rates, based on purchase price, are about 2% to 3% per month. Erection and dismantling run between 500 and 2000 labor hours. Climbing operation takes 10 labor hours per 20' climb. Crane dead time is about 5 hours per 40' climb. If crane is bolted to side of the building add cost of ties and extra mast sections. Climbing cranes have from 80' to 180' of mast while static cranes have 80' to 800' of mast.

Truck Cranes can be converted to tower cranes by using tower attachments. Mast heights over 400' have been used.

A single 100' high material **Hoist and Tower** can be erected and dismantled in about 400 labor hours; a double 100' high hoist and tower in about 600 labor hours. Erection times for additional heights are 3 and 4 labor hours per vertical foot respectively up to 150', and 4 to 5 labor hours per vertical foot over 150' high. A 40' high portable Buck hoist takes about 160 labor hours to erect and dismantle. Additional heights take 2 labor hours per vertical foot to 80' and 3 labor hours per vertical foot for the next 100'. Most material hoists do not meet local code requirements for carrying personnel.

A 150' high **Personnel Hoist** requires about 500 to 800 labor hours to erect and dismantle. Budget erection time at 5 labor hours per vertical foot for all trades. Local code requirements or labor scarcity requiring overtime can add up to 50% to any of the above erection costs.

Earthmoving Equipment: The selection of earthmoving equipment depends upon the type and quantity of material, moisture content, haul distance, haul road, time available, and equipment available. Short haul cut and fill operations may require dozers only, while another operation may require excavators, a fleet of trucks, and spreading and compaction equipment. Stockpiled material and granular material are easily excavated with front end loaders. Scrapers are most economically used with hauls between 300' and 1-1/2 miles if adequate haul roads can be maintained. Shovels are often used for blasted rock and any material where a vertical face of 8' or more can be excavated. Special conditions may dictate the use of draglines, clamshells, or backhoes. Spreading and compaction equipment must be matched to the soil characteristics, the compaction required and the rate the fill is being supplied.

R015433-15   Heavy Lifting

Hydraulic Climbing Jacks

The use of hydraulic heavy lift systems is an alternative to conventional type crane equipment. The lifting, lowering, pushing, or pulling mechanism is a hydraulic climbing jack moving on a square steel jackrod from 1-5/8" to 4" square, or a steel cable. The jackrod or cable can be vertical or horizontal, stationary or movable, depending on the individual application. When the jackrod is stationary, the climbing jack will climb the rod and push or pull the load along with itself. When the climbing jack is stationary, the jackrod is movable with the load attached to the end and the climbing jack will lift or lower the jackrod with the attached load. The heavy lift system is normally operated by a single control lever located at the hydraulic pump.

The system is flexible in that one or more climbing jacks can be applied wherever a load support point is required, and the rate of lift synchronized.

Economic benefits have been demonstrated on projects such as: erection of ground assembled roofs and floors, complete bridge spans, girders and trusses, towers, chimney liners and steel vessels, storage tanks, and heavy machinery. Other uses are raising and lowering offshore work platforms, caissons, tunnel sections and pipelines.

R015436-50   Mobilization

Costs to move rented construction equipment to a job site from an equipment dealer's or contractor's yard (mobilization) or off the job site (demobilization) are not included in the rental or operating rates, nor in the equipment cost on a unit price line or in a crew listing. These costs can be found consolidated in the Mobilization section of the data and elsewhere in particular site work sections. If a piece of equipment is already on the job site, it is not appropriate to include mob/demob costs in a new estimate that requires use of that equipment. The following table identifies approximate sizes of rented construction equipment that would be hauled on a towed trailer. Because this listing is not all-encompassing, the user can infer as to what size trailer might be required for a piece of equipment not listed.

3-ton Trailer	20-ton Trailer	40-ton Trailer	50-ton Trailer
20 H.P. Excavator	110 H.P. Excavator	200 H.P. Excavator	270 H.P. Excavator
50 H.P. Skid Steer	165 H.P. Dozer	300 H.P. Dozer	Small Crawler Crane
35 H.P. Roller	150 H.P. Roller	400 H.P. Scraper	500 H.P. Scraper
40 H.P. Trencher	Backhoe	450 H.P. Art. Dump Truck	500 H.P. Art. Dump Truck

R024119-10 Demolition Defined

Whole Building Demolition - Demolition of the whole building with no concern for any particular building element, component, or material type being demolished. This type of demolition is accomplished with large pieces of construction equipment that break up the structure, load it into trucks and haul it to a disposal site, but disposal or dump fees are not included. Demolition of below-grade foundation elements, such as footings, foundation walls, grade beams, slabs on grade, etc., is not included. Certain mechanical equipment containing flammable liquids or ozone-depleting refrigerants, electric lighting elements, communication equipment components, and other building elements may contain hazardous waste, and must be removed, either selectively or carefully, as hazardous waste before the building can be demolished.

Foundation Demolition - Demolition of below-grade foundation footings, foundation walls, grade beams, and slabs on grade. This type of demolition is accomplished by hand or pneumatic hand tools, and does not include saw cutting, or handling, loading, hauling, or disposal of the debris.

Gutting - Removal of building interior finishes and electrical/mechanical systems down to the load-bearing and sub-floor elements of the rough building frame, with no concern for any particular building element, component, or material type being demolished. This type of demolition is accomplished by hand or pneumatic hand tools, and includes loading into trucks, but not hauling, disposal or dump fees, scaffolding, or shoring. Certain mechanical equipment containing flammable liquids or ozone-depleting refrigerants, electric lighting elements, communication equipment components, and other building elements may contain hazardous waste, and must be removed, either selectively or carefully, as hazardous waste, before the building is gutted.

Selective Demolition - Demolition of a selected building element, component, or finish, with some concern for surrounding or adjacent elements, components, or finishes (see the first Subdivision (s) at the beginning of appropriate Divisions). This type of demolition is accomplished by hand or pneumatic hand tools, and does not include handling, loading,

storing, hauling, or disposal of the debris, scaffolding, or shoring. "Gutting" methods may be used in order to save time, but damage that is caused to surrounding or adjacent elements, components, or finishes may have to be repaired at a later time.

Careful Removal - Removal of a piece of service equipment, building element or component, or material type, with great concern for both the removed item and surrounding or adjacent elements, components or finishes. The purpose of careful removal may be to protect the removed item for later re-use, preserve a higher salvage value of the removed item, or replace an item while taking care to protect surrounding or adjacent elements, components, connections, or finishes from cosmetic and/or structural damage. An approximation of the time required to perform this type of removal is 1/3 to 1/2 the time it would take to install a new item of like kind. This type of removal is accomplished by hand or pneumatic hand tools, and does not include loading, hauling, or storing the removed item, scaffolding, shoring, or lifting equipment.

Cutout Demolition - Demolition of a small quantity of floor, wall, roof, or other assembly, with concern for the appearance and structural integrity of the surrounding materials. This type of demolition is accomplished by hand or pneumatic hand tools, and does not include saw cutting, handling, loading, hauling, or disposal of debris, scaffolding, or shoring.

Rubbish Handling - Work activities that involve handling, loading or hauling of debris. Generally, the cost of rubbish handling must be added to the cost of all types of demolition, with the exception of whole building demolition.

Minor Site Demolition - Demolition of site elements outside the footprint of a building. This type of demolition is accomplished by hand or pneumatic hand tools, or with larger pieces of construction equipment, and may include loading a removed item onto a truck (check the Crew for equipment used). It does not include saw cutting, hauling or disposal of debris, and, sometimes, handling or loading.

R024119-20 Dumpsters

Dumpster rental costs on construction sites are presented in two ways.

The cost per week rental includes the delivery of the dumpster; its pulling or emptying once per week, and its final removal. The assumption is made that the dumpster contractor could choose to empty a dumpster by simply bringing in an empty unit and removing the full one. These costs also include the disposal of the materials in the dumpster.

The Alternate Pricing can be used when actual planned conditions are not approximated by the weekly numbers. For example, these lines can be used when a dumpster is needed for 4 weeks and will need to be emptied 2 or 3 times per week. Conversely the Alternate Pricing lines can be used when a dumpster will be rented for several weeks or months but needs to be emptied only a few times over this period.

R024119-30 Rubbish Handling Chutes

To correctly estimate the cost of rubbish handling chute systems, the individual components must be priced separately. First choose the size of the system; a 30-inch diameter chute is quite common, but the sizes range from 18 to 36 inches in diameter. The 30-inch chute comes in a standard weight and two thinner weights. The thinner weight chutes are sometimes chosen for cost savings, but they are more easily damaged.

There are several types of major chute pieces that make up the chute system. The first component to consider is the top chute section (top intake hopper) where the material is dropped into the chute at the highest point. After determining the top chute, the intermediate chute pieces called the regular chute sections are priced. Next, the number of chute control door sections (intermediate intake hoppers) must be determined. In the more complex systems, a chute control door section is provided at each floor level. The last major component to consider is bolt down frames; these are usually provided at every other floor level.

There are a number of accessories to consider for safe operation and control. There are covers for the top chute and the chute door sections. The top

chute can have a trough that allows for better loading of the chute. For the safest operation, a chute warning light system can be added that will warn the other chute intake locations not to load while another is being used. There are dust control devices that spray a water mist to keep down the dust as the debris is loaded into a Dumpster. There are special breakaway cords that are used to prevent damage to the chute if the Dumpster is removed without disconnecting from the chute. There are chute liners that can be installed to protect the chute structure from physical damage from rough abrasive materials. Warning signs can be posted at each floor level that is provided with a chute control door section.

In summary, a complete rubbish handling chute system will include one top section, several intermediate regular sections, several intermediate control door (intake hopper) sections and bolt down frames at every other floor level starting with the top floor. If so desired, the system can also include covers and a light warning system for a safer operation. The bottom of the chute should always be above the Dumpster and should be tied off with a breakaway cord to the Dumpster.

R026510-20   Underground Storage Tank Removal

Underground Storage Tank Removal can be divided into two categories: Non-Leaking and Leaking. Prior to removing an underground storage tank, tests should be made, with the proper authorities present, to determine whether a tank has been leaking or the surrounding soil has been contaminated.

To safely remove Liquid Underground Storage Tanks:

1. Excavate to the top of the tank.
2. Disconnect all piping.
3. Open all tank vents and access ports.
4. Remove all liquids and/or sludge.
5. Purge the tank with an inert gas.
6. Provide access to the inside of the tank and clean out the interior using proper personal protective equipment (PPE).
7. Excavate soil surrounding the tank using proper PPE for on-site personnel.
8. Pull and properly dispose of the tank.
9. Clean up the site of all contaminated material.
10. Install new tanks or close the excavation.

R031113-10   Wall Form Materials

Aluminum Forms

Approximate weight is 3 lbs. per S.F.C.A. Standard widths are available from 4" to 36" with 36" most common. Standard lengths of 2', 4', 6' to 8' are available. Forms are lightweight and fewer ties are needed with the wider widths. The form face is either smooth or textured.

Metal Framed Plywood Forms

Manufacturers claim over 75 reuses of plywood and over 300 reuses of steel frames. Many specials such as corners, fillers, pilasters, etc. are available. Monthly rental is generally about 15% of purchase price for first month and 9% per month thereafter with 90% of rental applied to purchase for the first month and decreasing percentages thereafter. Aluminum framed forms cost 25% to 30% more than steel framed.

After the first month, extra days may be prorated from the monthly charge. Rental rates do not include ties, accessories, cleaning, loss of hardware or freight in and out. Approximate weight is 5 lbs. per S.F. for steel; 3 lbs. per S.F. for aluminum.

Forms can be rented with option to buy.

Plywood Forms, Job Fabricated

There are two types of plywood used for concrete forms.

1. Exterior plyform is completely waterproof. This is face oiled to facilitate stripping. Ten reuses can be expected with this type with 25 reuses possible.
2. An overlaid type consists of a resin fiber fused to exterior plyform. No oiling is required except to facilitate cleaning. This is available in both high density (HDO) and medium density overlaid (MDO). Using HDO, 50 reuses can be expected with 200 possible.

Plyform is available in 5/8" and 3/4" thickness. High density overlaid is available in 3/8", 1/2", 5/8" and 3/4" thickness.

5/8" thick is sufficient for most building forms, while 3/4" is best on heavy construction.

Plywood Forms, Modular, Prefabricated

There are many plywood forming systems without frames. Most of these are manufactured from 1-1/8" (HDO) plywood and have some hardware attached. These are used principally for foundation walls 8' or less high. With care and maintenance, 100 reuses can be attained with decreasing quality of surface finish.

Steel Forms

Approximate weight is 6-1/2 lbs. per S.F.C.A. including accessories. Standard widths are available from 2" to 24", with 24" most common. Standard lengths are from 2' to 8', with 4' the most common. Forms are easily ganged into modular units.

Forms are usually leased for 15% of the purchase price per month prorated daily over 30 days.

Rental may be applied to sale price, and usually rental forms are bought. With careful handling and cleaning 200 to 400 reuses are possible.

Straight wall gang forms up to 12' x 20' or 8' x 30' can be fabricated. These crane handled forms usually lease for approx. 9% per month.

Individual job analysis is available from the manufacturer at no charge.

For customer support on your Site Work & Landscape Costs with RSMeans Data, call 800.448.8182.

847

R031113-40   Forms for Reinforced Concrete

Design Economy

Avoid many sizes in proportioning beams and columns.

From story to story avoid changing column dimensions. Gain strength by adding steel or using a richer mix. If a change in size of column is necessary, vary one dimension only to minimize form alterations. Keep beams and columns the same width.

From floor to floor in a multi-story building vary beam depth, not width, as that will leave the slab panel form unchanged. It is cheaper to vary the strength of a beam from floor to floor by means of a steel area than by 2″ changes in either width or depth.

Cost Factors

Material includes the cost of lumber, cost of rent for metal pans or forms if used, nails, form ties, form oil, bolts and accessories.

Labor includes the cost of carpenters to make up, erect, remove and repair, plus common labor to clean and move. Having carpenters remove forms minimizes repairs.

Improper alignment and condition of forms will increase finishing cost. When forms are heavily oiled, concrete surfaces must be neutralized before finishing. Special curing compounds will cause spillages to spall off in first frost. Gang forming methods will reduce costs on large projects.

Materials Used

Boards are seldom used unless their architectural finish is required. Generally, steel, fiberglass and plywood are used for contact surfaces. Labor on plywood is 10% less than with boards. The plywood is backed up with

2 x 4's at 12″ to 32″ O.C. Walers are generally 2 - 2 x 4's. Column forms are held together with steel yokes or bands. Shoring is with adjustable shoring or scaffolding for high ceilings.

Reuse

Floor and column forms can be reused four or possibly five times without excessive repair. Remember to allow for 10% waste on each reuse.

When modular sized wall forms are made, up to twenty uses can be expected with exterior plyform.

When forms are reused, the cost to erect, strip, clean and move will not be affected. 10% replacement of lumber should be included and about one hour of carpenter time for repairs on each reuse per 100 S.F.

The reuse cost for certain accessory items normally rented on a monthly basis will be lower than the cost for the first use.

After the fifth use, new material required plus time needed for repair prevent the form cost from dropping further; it may go up. Much depends on care in stripping, the number of special bays, changes in beam or column sizes and other factors.

Costs for multiple use of formwork may be developed as follows:

2 Uses

$$\frac{(\text{1st Use + Reuse})}{2} = \text{avg. cost/2 uses}$$

3 Uses

$$\frac{(\text{1st Use + 2 Reuses})}{3} = \text{avg. cost/3 uses}$$

4 Uses

$$\frac{(\text{1st use + 3 Reuses})}{4} = \text{avg. cost/4 uses}$$

R031113-60   Formwork Labor-Hours

Item	Unit	Hours Required			Total Hours	Multiple Use		
		Fabricate	Erect & Strip	Clean & Move	1 Use	2 Use	3 Use	4 Use
Beam and Girder, interior beams, 12" wide	100 S.F.	6.4	8.3	1.3	16.0	13.3	12.4	12.0
Hung from steel beams		5.8	7.7	1.3	14.8	12.4	11.6	11.2
Beam sides only, 36" high		5.8	7.2	1.3	14.3	11.9	11.1	10.7
Beam bottoms only, 24" wide		6.6	13.0	1.3	20.9	18.1	17.2	16.7
Box out for openings		9.9	10.0	1.1	21.0	16.6	15.1	14.3
Buttress forms, to 8' high		6.0	6.5	1.2	13.7	11.2	10.4	10.0
Centering, steel, 3/4" rib lath			1.0		1.0			
3/8" rib lath or slab form			0.9		0.9			
Chamfer strip or keyway	100 L.F.		1.5		1.5	1.5	1.5	1.5
Columns, fiber tube 8" diameter			20.6		20.6			
12"			21.3		21.3			
16"			22.9		22.9			
20"			23.7		23.7			
24"			24.6		24.6			
30"			25.6		25.6			
Columns, round steel, 12" diameter			22.0		22.0	22.0	22.0	22.0
16"			25.6		25.6	25.6	25.6	25.6
20"			30.5		30.5	30.5	30.5	30.5
24"			37.7		37.7	37.7	37.7	37.7
Columns, plywood 8" x 8"	100 S.F.	7.0	11.0	1.2	19.2	16.2	15.2	14.7
12" x 12"		6.0	10.5	1.2	17.7	15.2	14.4	14.0
16" x 16"		5.9	10.0	1.2	17.1	14.7	13.8	13.4
24" x 24"		5.8	9.8	1.2	16.8	14.4	13.6	13.2
Columns, steel framed plywood 8" x 8"			10.0	1.0	11.0	11.0	11.0	11.0
12" x 12"			9.3	1.0	10.3	10.3	10.3	10.3
16" x 16"			8.5	1.0	9.5	9.5	9.5	9.5
24" x 24"			7.8	1.0	8.8	8.8	8.8	8.8
Drop head forms, plywood		9.0	12.5	1.5	23.0	19.0	17.7	17.0
Coping forms		8.5	15.0	1.5	25.0	21.3	20.0	19.4
Culvert, box			14.5	4.3	18.8	18.8	18.8	18.8
Curb forms, 6" to 12" high, on grade		5.0	8.5	1.2	14.7	12.7	12.1	11.7
On elevated slabs		6.0	10.8	1.2	18.0	15.5	14.7	14.3
Edge forms to 6" high, on grade	100 L.F.	2.0	3.5	0.6	6.1	5.6	5.4	5.3
7" to 12" high	100 S.F.	2.5	5.0	1.0	8.5	7.8	7.5	7.4
Equipment foundations		10.0	18.0	2.0	30.0	25.5	24.0	23.3
Flat slabs, including drops		3.5	6.0	1.2	10.7	9.5	9.0	8.8
Hung from steel		3.0	5.5	1.2	9.7	8.7	8.4	8.2
Closed deck for domes		3.0	5.8	1.2	10.0	9.0	8.7	8.5
Open deck for pans		2.2	5.3	1.0	8.5	7.9	7.7	7.6
Footings, continuous, 12" high		3.5	3.5	1.5	8.5	7.3	6.8	6.6
Spread, 12" high		4.7	4.2	1.6	10.5	8.7	8.0	7.7
Pile caps, square or rectangular		4.5	5.0	1.5	11.0	9.3	8.7	8.4
Grade beams, 24" deep		2.5	5.3	1.2	9.0	8.3	8.0	7.9
Lintel or Sill forms		8.0	17.0	2.0	27.0	23.5	22.3	21.8
Spandrel beams, 12" wide		9.0	11.2	1.3	21.5	17.5	16.2	15.5
Stairs			25.0	4.0	29.0	29.0	29.0	29.0
Trench forms in floor		4.5	14.0	1.5	20.0	18.3	17.7	17.4
Walls, Plywood, at grade, to 8' high		5.0	6.5	1.5	13.0	11.0	9.7	9.5
8' to 16'		7.5	8.0	1.5	17.0	13.8	12.7	12.1
16' to 20'		9.0	10.0	1.5	20.5	16.5	15.2	14.5
Foundation walls, to 8' high		4.5	6.5	1.0	12.0	10.3	9.7	9.4
8' to 16' high		5.5	7.5	1.0	14.0	11.8	11.0	10.6
Retaining wall to 12' high, battered		6.0	8.5	1.5	16.0	13.5	12.7	12.3
Radial walls to 12' high, smooth		8.0	9.5	2.0	19.5	16.0	14.8	14.3
2' chords		7.0	8.0	1.5	16.5	13.5	12.5	12.0
Prefabricated modular, to 8' high		—	4.3	1.0	5.3	5.3	5.3	5.3
Steel, to 8' high		—	6.8	1.2	8.0	8.0	8.0	8.0
8' to 16' high		—	9.1	1.5	10.6	10.3	10.2	10.2
Steel framed plywood to 8' high		—	6.8	1.2	8.0	7.5	7.3	7.2
8' to 16' high		—	9.3	1.2	10.5	9.5	9.2	9.0

For customer support on your Site Work & Landscape Costs with RSMeans Data, call 800.448.8182.

84●

R032110-10   Reinforcing Steel Weights and Measures

Bar Designation No.**	Nominal Weight Lb./Ft.	U.S. Customary Units			SI Units			
		Nominal Dimensions*			Nominal Dimensions*			
		Diameter in.	Cross Sectional Area, in.²	Perimeter in.	Nominal Weight kg/m	Diameter mm	Cross Sectional Area, cm²	Perimeter mm
3	.376	.375	.11	1.178	.560	9.52	.71	29.9
4	.668	.500	.20	1.571	.994	12.70	1.29	39.9
5	1.043	.625	.31	1.963	1.552	15.88	2.00	49.9
6	1.502	.750	.44	2.356	2.235	19.05	2.84	59.8
7	2.044	.875	.60	2.749	3.042	22.22	3.87	69.8
8	2.670	1.000	.79	3.142	3.973	25.40	5.10	79.8
9	3.400	1.128	1.00	3.544	5.059	28.65	6.45	90.0
10	4.303	1.270	1.27	3.990	6.403	32.26	8.19	101.4
11	5.313	1.410	1.56	4.430	7.906	35.81	10.06	112.5
14	7.650	1.693	2.25	5.320	11.384	43.00	14.52	135.1
18	13.600	2.257	4.00	7.090	20.238	57.33	25.81	180.1

* The nominal dimensions of a deformed bar are equivalent to those of a plain round bar having the same weight per foot as the deformed bar.
** Bar numbers are based on the number of eighths of an inch included in the nominal diameter of the bars.

R032110-70   Bend, Place and Tie Reinforcing

Placing and tying by rodmen for footings and slabs run from nine hrs. per ton for heavy bars to fifteen hrs. per ton for light bars. For beams, columns, and walls, production runs from eight hrs. per ton for heavy bars to twenty hrs. per ton for light bars. The overall average for typical reinforced concrete buildings is about fourteen hrs. per ton. These production figures include the time for placing accessories and usual inserts, but not their material cost (allow 15% of the cost of delivered bent rods). Equipment handling is necessary for the larger-sized bars so that installation costs for the very heavy bars will not decrease proportionately.

Installation costs for splicing reinforcing bars include allowance for equipment to hold the bars in place while splicing as well as necessary scaffolding for iron workers.

R032110-80   Shop-Fabricated Reinforcing Steel

The material prices for reinforcing, shown in the unit cost sections of the data set, are for 50 tons or more of shop-fabricated reinforcing steel and include:

1. Mill base price of reinforcing steel
2. Mill grade/size/length extras
3. Mill delivery to the fabrication shop
4. Shop storage and handling
5. Shop drafting/detailing
6. Shop shearing and bending
7. Shop listing
8. Shop delivery to the job site

Both material and installation costs can be considerably higher for small jobs consisting primarily of smaller bars, while material costs may be slightly lower for larger jobs.

R032205-30　Common Stock Styles of Welded Wire Fabric

This table provides some of the basic specifications, sizes, and weights of welded wire fabric used for reinforcing concrete.

New Designation		Old Designation		Steel Area per Foot				Approximate Weight per 100 S.F.	
Spacing — Cross Sectional Area (in.) — (Sq. in. 100)		Spacing — Wire Gauge (in.) — (AS & W)		Longitudinal		Transverse			
				in.	cm	in.	cm	lbs	kg
Rolls	6 x 6 — W1.4 x W1.4	6 x 6 — 10 x 10		.028	.071	.028	.071	21	9.53
	6 x 6 — W2.0 x W2.0	6 x 6 — 8 x 8	1	.040	.102	.040	.102	29	13.15
	6 x 6 — W2.9 x W2.9	6 x 6 — 6 x 6		.058	.147	.058	.147	42	19.05
	6 x 6 — W4.0 x W4.0	6 x 6 — 4 x 4		.080	.203	.080	.203	58	26.91
	4 x 4 — W1.4 x W1.4	4 x 4 — 10 x 10		.042	.107	.042	.107	31	14.06
	4 x 4 — W2.0 x W2.0	4 x 4 — 8 x 8	1	.060	.152	.060	.152	43	19.50
	4 x 4 — W2.9 x W2.9	4 x 4 — 6 x 6		.087	.227	.087	.227	62	28.12
	4 x 4 — W4.0 x W4.0	4 x 4 — 4 x 4		.120	.305	.120	.305	85	38.56
Sheets	6 x 6 — W2.9 x W2.9	6 x 6 — 6 x 6		.058	.147	.058	.147	42	19.05
	6 x 6 — W4.0 x W4.0	6 x 6 — 4 x 4		.080	.203	.080	.203	58	26.31
	6 x 6 — W5.5 x W5.5	6 x 6 — 2 x 2	2	.110	.279	.110	.279	80	36.29
	4 x 4 — W1.4 x W1.4	4 x 4 — 4 x 4		.120	.305	.120	.305	85	38.56

NOTES: 1. Exact W—number size for 8 gauge is W2.1
　　　　2. Exact W—number size for 2 gauge is W5.4

The above table was compiled with the following excerpts from the WRI Manual of Standard Practices, 7th Edition, Copyright 2006. Reproduced with permission of the Wire Reinforcement Institute, Inc.:

1. Chapter 3, page 7, Table 1 Common Styles of Metric Wire Reinforcement (WWR) With Equivalent US Customary Units
2. Chapter 6, page 19, Table 5 Customary Units
3. Chapter 6, Page 23, Table 7 Customary Units (in.) Welded Plain Wire Reinforcement
4. Chapter 6, Page 25 Table 8 Wire Size Comparison
5. Chapter 9, Page 30, Table 9 Weight of Longitudinal Wires Weight (Mass) Estimating Tables
6. Chapter 9, Page 31, Table 9M Weight of Longitudinal Wires Weight (Mass) Estimating Tables
7. Chapter 9, Page 32, Table 10 Weight of Transverse Wires Based on 62" lengths of transverse wire (60" width plus 1" overhand each side)
8. Chapter 9, Page 33, Table 10M Weight of Transverse Wires

For customer support on your Site Work & Landscape Costs with RSMeans Data, call 800.448.8182.

851

Reference Tables

R033053-60 Maximum Depth of Frost Penetration in Inches

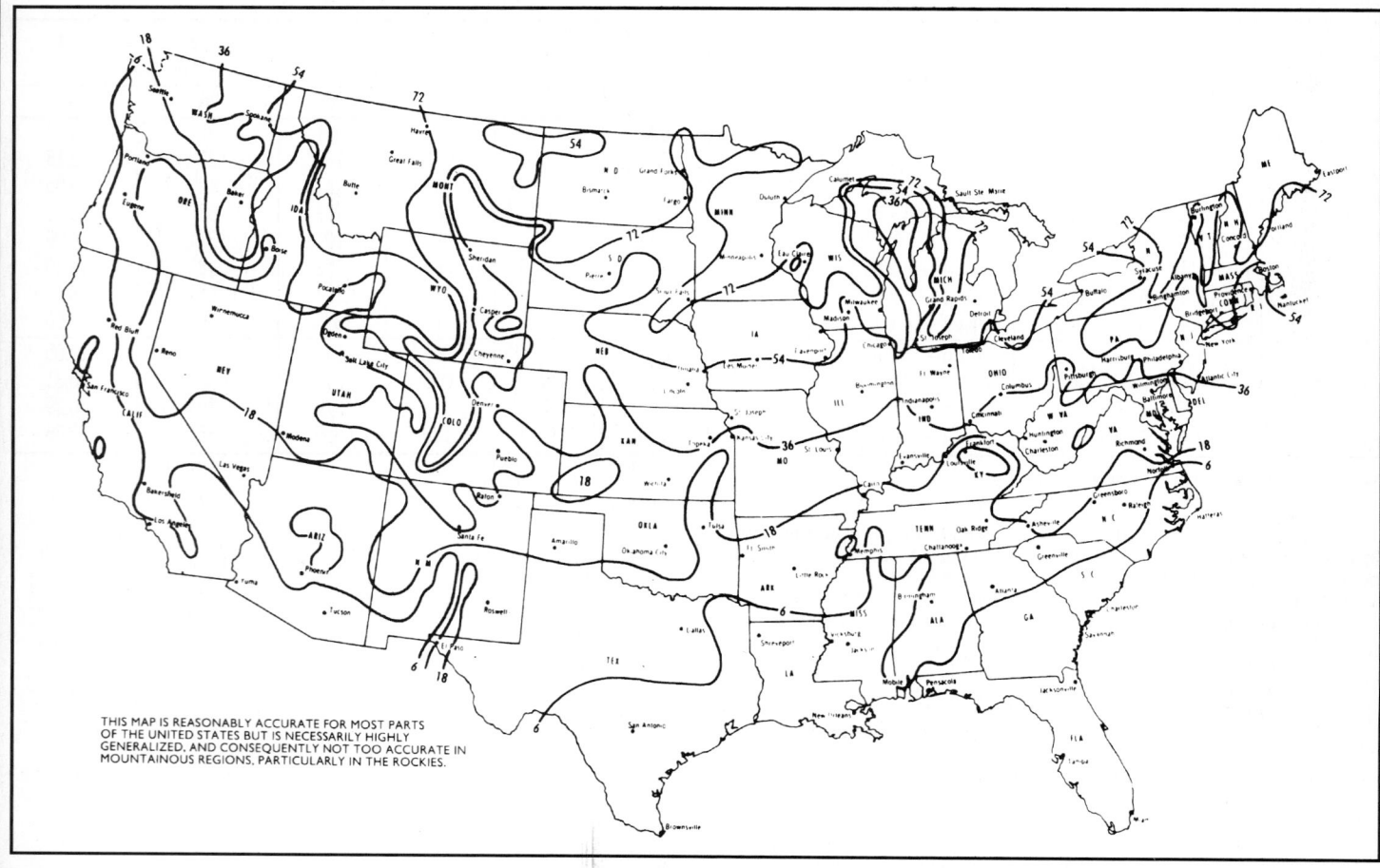

THIS MAP IS REASONABLY ACCURATE FOR MOST PARTS
OF THE UNITED STATES BUT IS NECESSARILY HIGHLY
GENERALIZED, AND CONSEQUENTLY NOT TOO ACCURATE IN
MOUNTAINOUS REGIONS, PARTICULARLY IN THE ROCKIES.

852

For customer support on your Site Work & Landscape Costs with RSMeans Data, call 800.448.8182.

R033105-20   Materials for One C.Y. of Concrete

This is an approximate method of figuring quantities of cement, sand and coarse aggregate for a field mix with waste allowance included.

With crushed gravel as coarse aggregate, to determine barrels of cement required, divide 10 by total mix; that is, for 1:2:4 mix, 10 divided by 7 = 1-3/7 barrels.

If the coarse aggregate is crushed stone, use 10-1/2 instead of 10 as given for gravel.

To determine tons of sand required, multiply barrels of cement by parts of sand and then by 0.2; that is, for the 1:2:4 mix, as above, 1-3/7 x 2 x .2 = .57 tons.

Tons of crushed gravel are in the same ratio to tons of sand as parts in the mix, or 4/2 x .57 = 1.14 tons.

1 bag cement = 94#	1 C.Y. sand or crushed gravel = 2700#	1 C.Y. crushed stone = 2575#
4 bags = 1 barrel	1 ton sand or crushed gravel = 20 C.F.	1 ton crushed stone = 21 C.F.

Average carload of cement is 692 bags; of sand or gravel is 56 tons.

Do not stack stored cement over 10 bags high.

R033105-65   Field-Mix Concrete

Presently most building jobs are built with ready-mixed concrete except at isolated locations and some larger jobs requiring over 10,000 C.Y. where land is readily available for setting up a temporary batch plant.

The most economical mix is a controlled mix using local aggregate proportioned by trial to give the required strength with the least cost of material.

R033105-70   Placing Ready-Mixed Concrete

For ground pours allow for 5% waste when figuring quantities.

Prices in the front of the data set assume normal deliveries. If deliveries are made before 8 A.M. or after 5 P.M. or on Saturday afternoons add 30%. Negotiated discounts for large volumes are not included in prices in front of the data set.

For the lower floors without truck access, concrete may be wheeled in rubber-tired buggies, conveyer handled, crane handled or pumped. Pumping is economical if there is top steel. Conveyers are more efficient for thick slabs.

At higher floors the rubber-tired buggies may be hoisted by a hoisting tower and wheeled to the location. Placement by a conveyer is limited to three floors and is best for high-volume pours. Pumped concrete is best when the building has no crane access. Concrete may be pumped directly as high as thirty-six stories using special pumping techniques. Normal maximum height is about fifteen stories.

The best pumping aggregate is screened and graded bank gravel rather than crushed stone.

Pumping downward is more difficult than pumping upward. The horizontal distance from pump to pour may increase preparation time prior to pour. Placing by cranes, either mobile, climbing or tower types, continues as the most efficient method for high-rise concrete buildings.

R040513-10 Cement Mortar (material only)

Type N - 1:1:6 mix by volume. Use everywhere above grade except as noted below. - 1:3 mix using conventional masonry cement which saves handling two separate bagged materials.

Type M - 1:1/4:3 mix by volume, or 1 part cement, 1/4 (10% by wt.) lime, 3 parts sand. Use for heavy loads and where earthquakes or hurricanes may occur. Also for reinforced brick, sewers, manholes and everywhere below grade.

Mix Proportions by Volume and Compressive Strength of Mortar

Where Used	Mortar Type	Allowable Proportions by Volume				Compressive Strength @ 28 days
		Portland Cement	Masonry Cement	Hydrated Lime	Masonry Sand	
Plain Masonry	M	1	1	—	6	2500 psi
		1	—	1/4	3	
	S	1/2	1	—	4	1800 psi
		1	—	1/4 to 1/2	4	
	N	—	1	—	3	750 psi
		1	—	1/2 to 1-1/4	6	
	O	—	1	—	3	350 psi
		1	—	1-1/4 to 2-1/2	9	
	K	1	—	2-1/2 to 4	12	75 psi
Reinforced Masonry	PM	1	1	—	6	2500 psi
	PL	1	—	1/4 to 1/2	4	2500 psi

Note: The total aggregate should be between 2.25 to 3 times the sum of the cement and lime used.

The labor cost to mix the mortar is included in the productivity and labor cost of unit price lines in unit cost sections for brickwork, blockwork and stonework.

The material cost of mixed mortar is included in the material cost of those same unit price lines and includes the cost of renting and operating a 10 C.F. mixer at the rate of 200 C.F. per day.

There are two types of mortar color used. One type is the inert additive type with about 100 lbs. per M brick as the typical quantity required. These colors are also available in smaller-batch-sized bags (1 lb. to 15 lb.) which can be placed directly into the mixer without measuring. The other type is premixed and replaces the masonry cement. Dark green color has the highest cost.

R040519-50 Masonry Reinforcing

Horizontal joint reinforcing helps prevent wall cracks where wall movement may occur and in many locations is required by code. Horizontal joint reinforcing is generally not considered to be structural reinforcing and an unreinforced wall may still contain joint reinforcing.

Reinforcing strips come in 10' and 12' lengths and in truss and ladder shapes, with and without drips. Field labor runs between 2.7 to 5.3 hours per 1000 L.F. for wall thicknesses up to 12".

The wire meets ASTM A82 for cold drawn steel wire and the typical size is 9 ga. sides and ties with 3/16" diameter also available. Typical finish is mill galvanized with zinc coating at .10 oz. per S.F. Class I (.40 oz. per S.F.) and Class III (.80. oz. per S.F.) are also available, as is hot dipped galvanizing at 1.50 oz. per S.F.

R042110-10   Economy in Bricklaying

Have adequate supervision. Be sure bricklayers are always supplied with materials so there is no waiting. Place experienced bricklayers at corners and openings.

Use only screened sand for mortar. Otherwise, labor time will be wasted picking out pebbles. Use seamless metal tubs for mortar as they do not leak or catch the trowel. Locate stack and mortar for easy wheeling.

Have brick delivered for stacking. This makes for faster handling, reduces chipping and breakage, and requires less storage space. Many dealers will deliver select common in 2' x 3' x 4' pallets or face brick packaged. This affords quick handling with a crane or forklift and easy tonging in units of ten, which reduces waste.

Use wider bricks for one wythe wall construction. Keep scaffolding away from the wall to allow mortar to fall clear and not stain the wall.

On large jobs develop specialized crews for each type of masonry unit.

Consider designing for prefabricated panel construction on high rise projects.

Avoid excessive corners or openings. Each opening adds about 50% to the labor cost for area of opening.

Bolting stone panels and using window frames as stops reduce labor costs and speed up erection.

R042110-20   Common and Face Brick

Common building brick manufactured according to ASTM C62 and facing brick manufactured according to ASTM C216 are the two standard bricks available for general building use.

Building brick is made in three grades: SW, where high resistance to damage caused by cyclic freezing is required; MW, where moderate resistance to cyclic freezing is needed; and NW, where little resistance to cyclic freezing is needed. Facing brick is made in only the two grades SW and MW. Additionally, facing brick is available in three types: FBS, for general use; FBX, for general use where a higher degree of precision and lower permissible variation in size than FBS are needed; and FBA, for general use to produce characteristic architectural effects resulting from non-uniformity in size and texture of the units.

In figuring the material cost of brickwork, an allowance of 25% mortar waste and 3% brick breakage was included. If bricks are delivered palletized

with 280 to 300 per pallet, or packaged, allow only 1-1/2% for breakage. Packaged or palletized delivery is practical when a job is big enough to have a crane or other equipment available to handle a package of brick. This is so on all industrial work but not always true on small commercial buildings.

The use of buff and gray face is increasing, and there is a continuing trend to the Norman, Roman, Jumbo and SCR brick.

Common red clay brick for backup is not used that often. Concrete block is the most usual backup material with occasional use of sand lime or cement brick. Building brick is commonly used in solid walls for strength and as a fire stop.

Brick panels built on the ground and then crane erected to the upper floors have proven to be economical. This allows the work to be done under cover and without scaffolding.

R042110-50   Brick, Block & Mortar Quantities

Type Brick	Nominal Size (incl. mortar) L	H	W	Modular Coursing	Number of Brick per S.F.	3/8" Joint	1/2" Joint	Bond Type	Description	Factor
Standard	8	2-2/3	4	3C=8"	6.75	8.1	10.3	Common	full header every fifth course	+20%
Economy	8	4	4	1C=4"	4.50	9.1	11.6		full header every sixth course	+16.7%
Engineer	8	3-1/5	4	5C=16"	5.63	8.5	10.8	English	full header every second course	+50%
Fire	9	2-1/2	4-1/2	2C=5"	6.40	550 # Fireclay	—	Flemish	alternate headers every course	+33.3%
Jumbo	12	4	6 or 8	1C=4"	3.00	22.5	29.2		every sixth course	+5.6%
Norman	12	2-2/3	4	3C=8"	4.50	11.2	14.3	Header = W x H exposed		+100%
Norwegian	12	3-1/5	4	5C=16"	3.75	11.7	14.9	Rowlock = H x W exposed		+100%
Roman	12	2	4	2C=4"	6.00	10.7	13.7	Rowlock stretcher = L x W exposed		+33.3%
SCR	12	2-2/3	6	3C=8"	4.50	21.8	28.0	Soldier = H x L exposed		—
Utility	12	4	4	1C=4"	3.00	12.3	15.7	Sailor = W x L exposed		-33.3%

Running Bond — Number of Brick per S.F. of Wall - Single Wythe with 3/8" Joints; C.F. of Mortar per M Bricks, Waste Included; For Other Bonds Standard Size Add to S.F. Quantities in Table to Left

Concrete Blocks Nominal Size	Standard	Lightweight	Blocks per 100 S.F.	Partitions	Back up
2" x 8" x 16"	20 PSF	15 PSF	113	27 C.F.	36 C.F.
4"	30	20		41	51
6"	42	30		56	66
8"	55	38		72	82
10"	70	47		87	97
12"	85	55		102	112

Approximate Weight per S.F.; Mortar per M block, waste included

Brick & Mortar Quantities
©Brick Industry Association. 2009 Feb. Technical Notes on Brick Construction 10:
    Dimensioning and Estimating Brick Masonry. Reston (VA): BIA. Table 1 Modular Brick Sizes and Table 4 Quantity Estimates for Brick Masonry.

R042210-20 Concrete Block

The material cost of special block such as corner, jamb and head block can be figured at the same price as ordinary block of equal size. Labor on specials is about the same as equal-sized regular block.

Bond beams and 16″ high lintel blocks are more expensive than regular units of equal size. Lintel blocks are 8″ long and either 8″ or 16″ high.

Use of a motorized mortar spreader box will speed construction of continuous walls.

Hollow non-load-bearing units are made according to ASTM C129 and hollow load-bearing units according to ASTM C90.

R050516-30 Coating Structural Steel

On field-welded jobs, the shop-applied primer coat is necessarily omitted. All painting must be done in the field and usually consists of red oxide rust inhibitive paint or an aluminum paint. The table below shows paint coverage and daily production for field painting.

See Division 09 97 13.23 for hot-dipped galvanizing and for field-applied cold galvanizing and other paints and protective coatings.

See Division 05 01 10.51 for steel surface preparation treatments such as wire brushing, pressure washing and sand blasting.

Type Construction	Surface Area per Ton	Coat	One Gallon Covers		In 8 Hrs. Person Covers		Average per Ton Spray	
			Brush	Spray	Brush	Spray	Gallons	Labor-hours
Light Structural	300 S.F. to 500 S.F.	1st	500 S.F.	455 S.F.	640 S.F.	2000 S.F.	0.9 gals.	1.6 L.H.
		2nd	450	410	800	2400	1.0	1.3
		3rd	450	410	960	3200	1.0	1.0
Medium	150 S.F. to 300 S.F.	All	400	365	1600	3200	0.6	0.6
Heavy Structural	50 S.F. to 150 S.F.	1st	400	365	1920	4000	0.2	0.2
		2nd	400	365	2000	4000	0.2	0.2
		3rd	400	365	2000	4000	0.2	0.2
Weighted Average	225 S.F.	All	400	365	1350	3000	0.6	0.6

R050521-20 Welded Structural Steel

Usual weight reductions with welded design run 10% to 20% compared with bolted or riveted connections. This amounts to about the same total cost compared with bolted structures since field welding is more expensive than bolts. For normal spans of 18' to 24' figure 6 to 7 connections per ton.

Trusses — For welded trusses add 4% to weight of main members for connections. Up to 15% less steel can be expected in a welded truss compared to one that is shop bolted. Cost of erection is the same whether shop bolted or welded.

General — Typical electrodes for structural steel welding are E6010, E6011, E60T and E70T. Typical buildings vary between 2# to 8# of weld rod per

ton of steel. Buildings utilizing continuous design require about three times as much welding as conventional welded structures. In estimating field erection by welding, it is best to use the average linear feet of weld per ton to arrive at the welding cost per ton. The type, size and position of the weld will have a direct bearing on the cost per linear foot. A typical field welder will deposit 1.8# to 2# of weld rod per hour manually. Using semiautomatic methods can increase production by as much as 50% to 75%.

R050523-10 High Strength Bolts

Common bolts (A307) are usually used in secondary connections (see Division 05 05 23.10).

High strength bolts (A325 and A490) are usually specified for primary connections such as column splices, beam and girder connections to columns, column bracing, connections for supports of operating equipment or of other live loads which produce impact or reversal of stress, and in structures carrying cranes of over 5-ton capacity.

Allow 20 field bolts per ton of steel for a 6 story office building, apartment house or light industrial building. For 6 to 12 stories allow 25 bolts per ton, and above 12 stories, 30 bolts per ton. On power stations, 20 to 25 bolts per ton are needed.

R051223-10 Structural Steel

The bare material prices for structural steel, shown in the unit cost sections of the data set, are for 100 tons of shop-fabricated structural steel and include:
1. Mill base price of structural steel
2. Mill scrap/grade/size/length extras
3. Mill delivery to a metals service center (warehouse)
4. Service center storage and handling
5. Service center delivery to a fabrication shop
6. Shop storage and handling
7. Shop drafting/detailing
8. Shop fabrication
9. Shop coat of primer paint
10. Shop listing
11. Shop delivery to the job site

In unit cost sections of the data set that contain items for field fabrication of steel components, the bare material cost of steel includes:
1. Mill base price of structural steel
2. Mill scrap/grade/size/length extras
3. Mill delivery to a metals service center (warehouse)
4. Service center storage and handling
5. Service center delivery to the job site

Reference Tables

R051223-20   Steel Estimating Quantities

One estimate on erection is that a crane can handle 35 to 60 pieces per day. Say the average is 45. With usual sizes of beams, girders, and columns, this would amount to about 20 tons per day. The type of connection greatly affects the speed of erection. Moment connections for continuous design slow down production and increase erection costs.

Short open web bar joists can be set at the rate of 75 to 80 per day, with 50 per day being the average for setting long span joists.

After main members are calculated, add the following for usual allowances: base plates 2% to 3%; column splices 4% to 5%; and miscellaneous details 4% to 5%, for a total of 10% to 13% in addition to main members.

The ratio of column to beam tonnage varies depending on type of steels used, typical spans, story heights and live loads.

It is more economical to keep the column size constant and to vary the strength of the column by using high strength steels. This also saves floor space. Buildings have recently gone as high as ten stories with 8″ high strength columns. For light columns under W8X31 lb. sections, concrete filled steel columns are economical.

High strength steels may be used in columns and beams to save floor space and to meet head room requirements. High strength steels in some sizes sometimes require long lead times.

Round, square and rectangular columns, both plain and concrete filled, are readily available and save floor area, but are higher in cost per pound than rolled columns. For high unbraced columns, tube columns may be less expensive.

Below are average minimum figures for the weights of the structural steel frame for different types of buildings using A36 steel, rolled shapes and simple joints. For economy in domes, rise to span ratio = .13. Open web joist framing systems will reduce weights by 10% to 40%. Composite design can reduce steel weight by up to 25% but additional concrete floor slab thickness may be required. Continuous design can reduce the weights up to 20%. There are many building codes with different live load requirements and different structural requirements, such as hurricane and earthquake loadings, which can alter the figures.

Structural Steel Weights per S.F. of Floor Area									
Type of Building	No. of Stories	Avg. Spans	L.L. #/S.F.	Lbs. Per S.F.	Type of Building	No. of Stories	Avg. Spans	L.L. #/S.F.	Lbs. Per S.F.
Steel Frame Mfg.	1	20′x20′ 30′x30′ 40′x40′	40	8 13 18	Apartments	2-8 9-25	20′x20′	40	8 14
Parking garage	4	Various	80	8.5	Office	to 10 20 30 over 50	Various	80	10 18 26 35
Domes (Schwedler)*	1	200′ 300′	30	10 15					

R051223-25 Common Structural Steel Specifications

ASTM A992 (formerly A36, then A572 Grade 50) is the all-purpose carbon grade steel widely used in building and bridge construction.

The other high-strength steels listed below may each have certain advantages over ASTM A992 structural carbon steel, depending on the application. They have proven to be economical choices where, due to lighter members, the reduction of dead load and the associated savings in shipping cost can be significant.

ASTM A588 atmospheric weathering, high-strength, low-alloy steels can be used in the bare (uncoated) condition, where exposure to normal atmosphere causes a tightly adherant oxide to form on the surface, protecting the steel from further oxidation. ASTM A242 corrosion-resistant, high-strength, low-alloy steels have enhanced atmospheric corrosion resistance of at least two times that of carbon structural steels with copper, or four times that of carbon structural steels without copper. The reduction or elimination of maintenance resulting from the use of these steels often offsets their higher initial cost.

Steel Type	ASTM Designation	Minimum Yield Stress in KSI	Shapes Available
Carbon	A36	36	All structural shape groups, and plates & bars up through 8" thick
	A529	50	Structural shape group 1, and plates & bars up through 2" thick
High-Strength Low-Alloy Quenched & Self-Tempered	A913	50	All structural shape groups
		60	
		65	
		70	
High-Strength Low-Alloy Columbium-Vanadium	A572	42	All structural shape groups, and plates & bars up through 6" thick
		50	All structural shape groups, and plates & bars up through 4" thick
		55	Structural shape groups 1 & 2, and plates & bars up through 2" thick
		60	Structural shape groups 1 & 2, and plates & bars up through 1-1/4" thick
		65	Structural shape group 1, and plates & bars up through 1-1/4" thick
High-Strength Low-Alloy Columbium-Vanadium	A992	50	All structural shape groups
Weathering High-Strength Low-Alloy	A242	42	Structural shape groups 4 & 5, and plates & bars over 1-1/2" up through 4" thick
		46	Structural shape group 3, and plates & bars over 3/4" up through 1-1/2" thick
		50	Structural shape groups 1 & 2, and plates & bars up through 3/4" thick
Weathering High-Strength Low-Alloy	A588	42	Plates & bars over 5" up through 8" thick
		46	Plates & bars over 4" up through 5" thick
		50	All structural shape groups, and plates & bars up through 4" thick
Quenched and Tempered	A852	70	Plates & bars up through 4" thick
Low-Alloy Quenched and Tempered Alloy	A514	90	Plates & bars over 2-1/2" up through 6" thick
		100	Plates & bars up through 2-1/2" thick

R051223-30 High Strength Steels

The mill price of high strength steels may be higher than A992 carbon steel, but their proper use can achieve overall savings through total reduced weights. For columns with L/r over 100, A992 steel is best; under 100, high strength steels are economical. For heavy columns, high strength steels are economical when cover plates are eliminated. There is no economy using high strength steels for clip angles or supports or for beams where deflection governs. Thinner members are more economical than thick.

The per ton erection and fabricating costs of the high strength steels will be higher than for A992 since the same number of pieces, but less weight, will be installed.

For customer support on your Site Work & Landscape Costs with RSMeans Data, call 800.448.8182.

859

R051223-35   Common Steel Sections

The upper portion of this table shows the name, shape, common designation and basic characteristics of commonly used steel sections. The lower portion explains how to read the designations used for the above illustrated common sections.

Shape & Designation	Name & Characteristics	Shape & Designation	Name & Characteristics
W	W Shape — Parallel flange surfaces	MC	Miscellaneous Channel — Infrequently rolled by some producers
S	American Standard Beam (I Beam) — Sloped inner flange	L	Angle — Equal or unequal legs, constant thickness
M	Miscellaneous Beams — Cannot be classified as W, HP or S; infrequently rolled by some producers	T	Structural Tee — Cut from W, M or S on center of web
C	American Standard Channel — Sloped inner flange	HP	Bearing Pile — Parallel flanges and equal flange and web thickness

Common drawing designations follow:

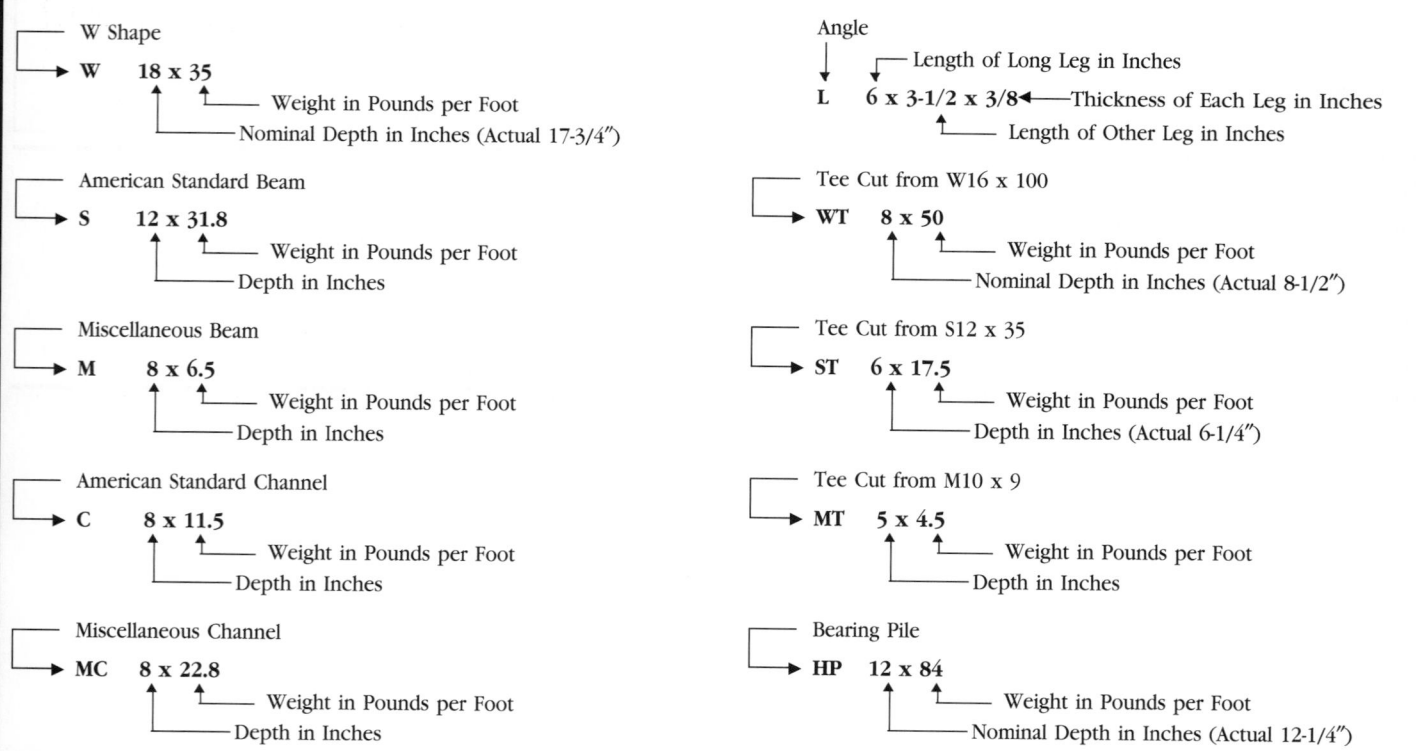

W Shape
W 18 x 35
— Weight in Pounds per Foot
— Nominal Depth in Inches (Actual 17-3/4″)

American Standard Beam
S 12 x 31.8
— Weight in Pounds per Foot
— Depth in Inches

Miscellaneous Beam
M 8 x 6.5
— Weight in Pounds per Foot
— Depth in Inches

American Standard Channel
C 8 x 11.5
— Weight in Pounds per Foot
— Depth in Inches

Miscellaneous Channel
MC 8 x 22.8
— Weight in Pounds per Foot
— Depth in Inches

Angle
L 6 x 3-1/2 x 3/8
— Length of Long Leg in Inches
— Thickness of Each Leg in Inches
— Length of Other Leg in Inches

Tee Cut from W16 x 100
WT 8 x 50
— Weight in Pounds per Foot
— Nominal Depth in Inches (Actual 8-1/2″)

Tee Cut from S12 x 35
ST 6 x 17.5
— Weight in Pounds per Foot
— Depth in Inches (Actual 6-1/4″)

Tee Cut from M10 x 9
MT 5 x 4.5
— Weight in Pounds per Foot
— Depth in Inches

Bearing Pile
HP 12 x 84
— Weight in Pounds per Foot
— Nominal Depth in Inches (Actual 12-1/4″)

R051223-45 Installation Time for Structural Steel Building Components

The following tables show the expected average installation times for various structural steel shapes. Table A presents installation times for columns, Table B for beams, Table C for light framing and bolts, and Table D for structural steel for various project types.

Table A

Description	Labor-Hours	Unit
Columns		
Steel, Concrete Filled		
3-1/2" Diameter	.933	Ea.
6-5/8" Diameter	1.120	Ea.
Steel Pipe		
3" Diameter	.933	Ea.
8" Diameter	1.120	Ea.
12" Diameter	1.244	Ea.
Structural Tubing		
4" x 4"	.966	Ea.
8" x 8"	1.120	Ea.
12" x 8"	1.167	Ea.
W Shape 2 Tier		
W8 x 31	.052	L.F.
W8 x 67	.057	L.F.
W10 x 45	.054	L.F.
W10 x 112	.058	L.F.
W12 x 50	.054	L.F.
W12 x 190	.061	L.F.
W14 x 74	.057	L.F.
W14 x 176	.061	L.F.

Table B

Description	Labor-Hours	Unit	Labor-Hours	Unit
Beams, W Shape				
W6 x 9	.949	Ea.	.093	L.F.
W10 x 22	1.037	Ea.	.085	L.F.
W12 x 26	1.037	Ea.	.064	L.F.
W14 x 34	1.333	Ea.	.069	L.F.
W16 x 31	1.333	Ea.	.062	L.F.
W18 x 50	2.162	Ea.	.088	L.F.
W21 x 62	2.222	Ea.	.077	L.F.
W24 x 76	2.353	Ea.	.072	L.F.
W27 x 94	2.581	Ea.	.067	L.F.
W30 x 108	2.857	Ea.	.067	L.F.
W33 x 130	3.200	Ea.	.071	L.F.
W36 x 300	3.810	Ea.	.077	L.F.

Table C

Description	Labor-Hours	Unit
Light Framing		
Angles 4" and Larger	.055	lbs.
Less than 4"	.091	lbs.
Channels 8" and Larger	.048	lbs.
Less than 8"	.072	lbs.
Cross Bracing Angles	.055	lbs.
Rods	.034	lbs.
Hanging Lintels	.069	lbs.
High Strength Bolts in Place		
3/4" Bolts	.070	Ea.
7/8" Bolts	.076	Ea.

Table D

Description	Labor-Hours	Unit	Labor-Hours	Unit
Apartments, Nursing Homes, etc.				
1-2 Stories	4.211	Piece	7.767	Ton
3-6 Stories	4.444	Piece	7.921	Ton
7-15 Stories	4.923	Piece	9.014	Ton
Over 15 Stories	5.333	Piece	9.209	Ton
Offices, Hospitals, etc.				
1-2 Stories	4.211	Piece	7.767	Ton
3-6 Stories	4.741	Piece	8.889	Ton
7-15 Stories	4.923	Piece	9.014	Ton
Over 15 Stories	5.120	Piece	9.209	Ton
Industrial Buildings				
1 Story	3.478	Piece	6.202	Ton

R051223-80   Dimensions and Weights of Sheet Steel

Gauge No.	Approximate Thickness				Weight		
	Inches (in fractions)	Inches (in decimal parts)		Millimeters			per Square Meter in Kg.
	Wrought Iron	Wrought Iron	Steel	Steel	per S.F. in Ounces	per S.F. in Lbs.	
0000000	1/2"	.5	.4782	12.146	320	20.000	97.650
000000	15/32"	.46875	.4484	11.389	300	18.750	91.550
00000	7/16"	.4375	.4185	10.630	280	17.500	85.440
0000	13/32"	.40625	.3886	9.870	260	16.250	79.330
000	3/8"	.375	.3587	9.111	240	15.000	73.240
00	11/32"	.34375	.3288	8.352	220	13.750	67.130
0	5/16"	.3125	.2989	7.592	200	12.500	61.030
1	9/32"	.28125	.2690	6.833	180	11.250	54.930
2	17/64"	.265625	.2541	6.454	170	10.625	51.880
3	1/4"	.25	.2391	6.073	160	10.000	48.820
4	15/64"	.234375	.2242	5.695	150	9.375	45.770
5	7/32"	.21875	.2092	5.314	140	8.750	42.720
6	13/64"	.203125	.1943	4.935	130	8.125	39.670
7	3/16"	.1875	.1793	4.554	120	7.500	36.320
8	11/64"	.171875	.1644	4.176	110	6.875	33.570
9	5/32"	.15625	.1495	3.797	100	6.250	30.520
10	9/64"	.140625	.1345	3.416	90	5.625	27.460
11	1/8"	.125	.1196	3.038	80	5.000	24.410
12	7/64"	.109375	.1046	2.657	70	4.375	21.360
13	3/32"	.09375	.0897	2.278	60	3.750	18.310
14	5/64"	.078125	.0747	1.897	50	3.125	15.260
15	9/128"	.0713125	.0673	1.709	45	2.813	13.730
16	1/16"	.0625	.0598	1.519	40	2.500	12.210
17	9/160"	.05625	.0538	1.367	36	2.250	10.990
18	1/20"	.05	.0478	1.214	32	2.000	9.765
19	7/160"	.04375	.0418	1.062	28	1.750	8.544
20	3/80"	.0375	.0359	.912	24	1.500	7.324
21	11/320"	.034375	.0329	.836	22	1.375	6.713
22	1/32"	.03125	.0299	.759	20	1.250	6.103
23	9/320"	.028125	.0269	.683	18	1.125	5.490
24	1/40"	.025	.0239	.607	16	1.000	4.882
25	7/320"	.021875	.0209	.531	14	.875	4.272
26	3/160"	.01875	.0179	.455	12	.750	3.662
27	11/640"	.0171875	.0164	.417	11	.688	3.357
28	1/64"	.015625	.0149	.378	10	.625	3.052

R053100-10   Decking Descriptions

General - All Deck Products

A steel deck is made by cold forming structural grade sheet steel into a repeating pattern of parallel ribs. The strength and stiffness of the panels are the result of the ribs and the material properties of the steel. Deck lengths can be varied to suit job conditions, but because of shipping considerations, are usually less than 40 feet. Standard deck width varies with the product used but full sheets are usually 12″, 18″, 24″, 30″, or 36″. The deck is typically furnished in a standard width with the ends cut square. Any cutting for width, such as at openings or for angular fit, is done at the job site.

The deck is typically attached to the building frame with arc puddle welds, self-drilling screws, or powder or pneumatically driven pins. Sheet to sheet fastening is done with screws, button punching (crimping), or welds.

Composite Floor Deck

After installation and adequate fastening, a floor deck serves several purposes. It (a) acts as a working platform, (b) stabilizes the frame, (c) serves as a concrete form for the slab, and (d) reinforces the slab to carry the design loads applied during the life of the building. Composite decks are distinguished by the presence of shear connector devices as part of the deck. These devices are designed to mechanically lock the concrete and deck together so that the concrete and the deck work together to carry subsequent floor loads. These shear connector devices can be rolled-in embossments, lugs, holes, or wires welded to the panels. The deck profile can also be used to interlock concrete and steel.

Composite deck finishes are either galvanized (zinc coated) or phosphatized/painted. Galvanized deck has a zinc coating on both the top and bottom surfaces. The phosphatized/painted deck has a bare (phosphatized) top surface that will come in contact with the concrete. This bare top surface can be expected to develop rust before the concrete is placed. The bottom side of the deck has a primer coat of paint.

A composite floor deck is normally installed so the panel ends do not overlap on the supporting beams. Shear lugs or panel profile shapes often prevent a tight metal to metal fit if the panel ends overlap; the air gap caused by overlapping will prevent proper fusion with the structural steel supports when the panel end laps are shear stud welded.

Adequate end bearing of the deck must be obtained as shown on the drawings. If bearing is actually less in the field than shown on the drawings, further investigation is required.

Roof Deck

A roof deck is not designed to act compositely with other materials. A roof deck acts alone in transferring horizontal and vertical loads into the building frame. Roof deck rib openings are usually narrower than floor deck rib openings. This provides adequate support of the rigid thermal insulation board.

A roof deck is typically installed to endlap approximately 2″ over supports. However, it can be butted (or lapped more than 2″) to solve field fit problems. Since designers frequently use the installed deck system as part of the horizontal bracing system (the deck as a diaphragm), any fastening substitution or change should be approved by the designer. Continuous perimeter support of the deck is necessary to limit edge deflection in the finished roof and may be required for diaphragm shear transfer.

Standard roof deck finishes are galvanized or primer painted. The standard factory applied paint for roof decks is a primer paint and is not intended to weather for extended periods of time. Field painting or touching up of abrasions and deterioration of the primer coat or other protective finishes is the responsibility of the contractor.

Cellular Deck

A cellular deck is made by attaching a bottom steel sheet to a roof deck or composite floor deck panel. A cellular deck can be used in the same manner as a floor deck. Electrical, telephone, and data wires are easily run through the chase created between the deck panel and the bottom sheet.

When used as part of the electrical distribution system, the cellular deck must be installed so that the ribs line up and create a smooth cell transition at abutting ends. The joint that occurs at butting cell ends must be taped or otherwise sealed to prevent wet concrete from seeping into the cell. Cell interiors must be free of welding burrs, or other sharp intrusions, to prevent damage to wires.

When used as a roof deck, the bottom flat plate is usually left exposed to view. Care must be maintained during erection to keep good alignment and prevent damage.

A cellular deck is sometimes used with the flat plate on the top side to provide a flat working surface. Installation of the deck for this purpose requires special methods for attachment to the frame because the flat plate, now on the top, can prevent direct access to the deck material that is bearing on the structural steel. It may be advisable to treat the flat top surface to prevent slipping.

A cellular deck is always furnished galvanized or painted over galvanized.

Form Deck

A form deck can be any floor or roof deck product used as a concrete form. Connections to the frame are by the same methods used to anchor floor and roof decks. Welding washers are recommended when welding a deck that is less than 20 gauge thickness.

A form deck is furnished galvanized, prime painted, or uncoated. A galvanized deck must be used for those roof deck systems where a form deck is used to carry a lightweight insulating concrete fill.

R061110-30   Lumber Product Material Prices

The price of forest products fluctuates widely from location to location and from season to season depending upon economic conditions. The bare material prices in the unit cost sections of the data set show the National Average material prices in effect Jan. 1 of this data year. It must be noted that lumber prices in general may change significantly during the year.

Availability of certain items depends upon geographic location and must be checked prior to firm-price bidding.

R061636-20   Plywood

There are two types of plywood used in construction: interior, which is moisture-resistant but not waterproofed, and exterior, which is waterproofed.

The grade of the exterior surface of the plywood sheets is designated by the first letter: A, for smooth surface with patches allowed; B, for solid surface with patches and plugs allowed; C, which may be surface plugged or may have knot holes up to 1″ wide; and D, which is used only for interior type plywood and may have knot holes up to 2-1/2″ wide. "Structural Grade" is specifically designed for engineered applications such as box beams. All CC & DD grades have roof and floor spans marked on them.

Underlayment-grade plywood runs from 1/4″ to 1-1/4″ thick. Thicknesses 5/8″ and over have optional tongue and groove joints which eliminate the need for blocking the edges. Underlayment 19/32″ and over may be referred to as Sturd-i-Floor.

The price of plywood can fluctuate widely due to geographic and economic conditions.

Typical uses for various plywood grades are as follows:

AA-AD Interior — cupboards, shelving, paneling, furniture

BB Plyform — concrete form plywood

CDX — wall and roof sheathing

Structural — box beams, girders, stressed skin panels

AA-AC Exterior — fences, signs, siding, soffits, etc.

Underlayment — base for resilient floor coverings

Overlaid HDO — high density for concrete forms & highway signs

Overlaid MDO — medium density for painting, siding, soffits & signs

303 Siding — exterior siding, textured, striated, embossed, etc.

R081313-20   Steel Door Selection Guide

Standard steel doors are classified into four levels, as recommended by the Steel Door Institute in the chart below. Each of the four levels offers a range of construction models and designs to meet architectural requirements for preference and appearance, including full flush, seamless, and stile & rail. Recommended minimum gauge requirements are also included.

For complete standard steel door construction specifications and available sizes, refer to the Steel Door Institute Technical Data Series, ANSI A250.8-98 (SDI-100), and ANSI A250.4-94 Test Procedure and Acceptance Criteria for Physical Endurance of Steel Door and Hardware Reinforcements.

Level		Model	Construction	For Full Flush or Seamless		
				Min. Gauge	Thickness (in)	Thickness (mm)
I	Standard Duty	1	Full Flush			
		2	Seamless	20	0.032	0.8
II	Heavy Duty	1	Full Flush			
		2	Seamless	18	0.042	1.0
III	Extra Heavy Duty	1	Full Flush			
		2	Seamless			
		3	*Stile & Rail	16	0.053	1.3
IV	Maximum Duty	1	Full Flush			
		2	Seamless	14	0.067	1.6

*Stiles & rails are 16 gauge; flush panels, when specified, are 18 gauge

R092910-10   Levels of Gypsum Board Finish

In the past, contract documents often used phrases such as "industry standard" and "workmanlike finish" to specify the expected quality of gypsum board wall and ceiling installations. The vagueness of these descriptions led to unacceptable work and disputes.

In order to resolve this problem, four major trade associations concerned with the manufacture, erection, finish, and decoration of gypsum board wall and ceiling systems developed an industry-wide *Recommended Levels of Gypsum Board Finish.*

The finish of gypsum board walls and ceilings for specific final decoration is dependent on a number of factors. A primary consideration is the location of the surface and the degree of decorative treatment desired. Painted and unpainted surfaces in warehouses and other areas where appearance is normally not critical may simply require the taping of wallboard joints and 'spotting' of fastener heads. Blemish-free, smooth, monolithic surfaces often intended for painted and decorated walls and ceilings in habitated structures, ranging from single-family dwellings through monumental buildings, require additional finishing prior to the application of the final decoration.

Other factors to be considered in determining the level of finish of the gypsum board surface are (1) the type of angle of surface illumination (both natural and artificial lighting), and (2) the paint and method of application or the type and finish of wallcovering specified as the final decoration. Critical lighting conditions, gloss paints, and thin wall coverings require a higher level of gypsum board finish than heavily textured surfaces which are subsequently painted or surfaces which are to be decorated with heavy grade wall coverings.

The following descriptions were developed by the Association of the Wall and Ceiling Industries-International (AWCI), Ceiling & Interior Systems Construction Association (CISCA), Gypsum Association (GA), and Painting and Decorating Contractors of America (PDCA) as a guide.

Level 0: Used in temporary construction or wherever the final decoration has not been determined. Unfinished. No taping, finishing or corner beads are required. Also could be used where non-predecorated panels will be used in demountable-type partitions that are to be painted as a final finish.

Level 1: Frequently used in plenum areas above ceilings, in attics, in areas where the assembly would generally be concealed, or in building service corridors and other areas not normally open to public view. Some degree of sound and smoke control is provided; in some geographic areas, this level is referred to as "fire-taping," although this level of finish does not typically meet fire-resistant assembly requirements. Where a fire resistance rating is required for the gypsum board assembly, details of construction should be in accordance with reports of fire tests of assemblies that have met the requirements of the fire rating acceptable.

All joints and interior angles shall have tape embedded in joint compound. Accessories are optional at specifier discretion in corridors and other areas with pedestrian traffic. Tape and fastener heads need not be covered with joint compound. Surface shall be free of excess joint compound. Tool marks and ridges are acceptable.

Level 2: It may be specified for standard gypsum board surfaces in garages, warehouse storage, or other similar areas where surface appearance is not of primary importance.

All joints and interior angles shall have tape embedded in joint compound and shall be immediately wiped with a joint knife or trowel, leaving a thin coating of joint compound over all joints and interior angles. Fastener heads and accessories shall be covered with a coat of joint compound. Surface shall be free of excess joint compound. Tool marks and ridges are acceptable.

Level 3: Typically used in areas receiving heavy texture (spray or hand applied) finishes before final painting, or where commercial-grade (heavy duty) wall coverings are to be applied as the final decoration. This level of finish should not be used where smooth painted surfaces or where lighter weight wall coverings are specified. The prepared surface shall be coated with a drywall primer prior to the application of final finishes.

All joints and interior angles shall have tape embedded in joint compound and shall be immediately wiped with a joint knife or trowel, leaving a thin coating of joint compound over all joints and interior angles. One additional coat of joint compound shall be applied over all joints and interior angles. Fastener heads and accessories shall be covered with two separate coats of joint compound. All joint compounds shall be smooth and free of tool marks and ridges. The prepared surface shall be covered with a drywall primer prior to the application of the final decoration.

Level 4: This level should be used where residential grade (light duty) wall coverings, flat paints, or light textures are to be applied. The prepared surface shall be coated with a drywall primer prior to the application of final finishes. Release agents for wall coverings are specifically formulated to minimize damage if coverings are subsequently removed.

The weight, texture, and sheen level of the wall covering material selected should be taken into consideration when specifying wall coverings over this level of drywall treatment. Joints and fasteners must be sufficiently concealed if the wall covering material is lightweight, contains limited pattern, has a glossy finish, or has any combination of these features. In critical lighting areas, flat paints applied over light textures tend to reduce joint photographing. Gloss, semi-gloss, and enamel paints are not recommended over this level of finish.

All joints and interior angles shall have tape embedded in joint compound and shall be immediately wiped with a joint knife or trowel, leaving a thin coating of joint compound over all joints and interior angles. In addition, two separate coats of joint compound shall be applied over all flat joints and one separate coat of joint compound applied over interior angles. Fastener heads and accessories shall be covered with three separate coats of joint compound. All joint compounds shall be smooth and free of tool marks and ridges. The prepared surface shall be covered with a drywall primer like Sheetrock® First Coat prior to the application of the final decoration.

Level 5: The highest quality finish is the most effective method to provide a uniform surface and minimize the possibility of joint photographing and of fasteners showing through the final decoration. This level of finish is required where gloss, semi-gloss, or enamel is specified; when flat joints are specified over an untextured surface; or where critical lighting conditions occur. The prepared surface shall be coated with a drywall primer prior to the application of the final decoration.

All joints and interior angles shall have tape embedded in joint compound and be immediately wiped with a joint knife or trowel, leaving a thin coating of joint compound over all joints and interior angles. Two separate coats of joint compound shall be applied over all flat joints and one separate coat of joint compound applied over interior angles. Fastener heads and accessories shall be covered with three separate coats of joint compound.

A thin skim coat of joint compound shall be trowel applied to the entire surface. Excess compound is immediately troweled off, leaving a film or skim coating of compound completely covering the paper. As an alternative to a skim coat, a material manufactured especially for this purpose may be applied such as Sheetrock® Tuff-Hide primer surfacer. The surface must be smooth and free of tool marks and ridges. The prepared surface shall be covered with a drywall primer prior to the application of the final decoration.

R131113-20   Swimming Pools

Pool prices given per square foot of surface area include pool structure, filter and chlorination equipment, pumps, related piping, ladders/steps, maintenance kit, skimmer and vacuum system. Decks and electrical service to equipment are not included.

Residential in-ground pool construction can be divided into two categories: vinyl lined and gunite. Vinyl lined pool walls are constructed of different materials including wood, concrete, plastic or metal. The bottom is often graded with sand over which the vinyl liner is installed. Vermiculite or soil cement bottoms may be substituted for an added cost.

Gunite pool construction is used both in residential and municipal installations. These structures are steel reinforced for strength and finished with a white cement limestone plaster.

Municipal pools will have a higher cost because plumbing codes require more expensive materials, chlorination equipment and higher filtration rates.

Municipal pools greater than 1,800 S.F. require gutter systems to control waves. This gutter may be formed into the concrete wall. Often a vinyl/stainless steel gutter or gutter/wall system is specified, which will raise the pool cost.

Competition pools usually require tile bottoms and sides with contrasting lane striping, which will also raise the pool cost.

R133113-10   Air Supported Structures

Air supported structures are made from fabrics that can be classified into two groups: temporary and permanent. Temporary fabrics include nylon, woven polyethylene, vinyl film, and vinyl coated dacron. These have lifespans that range from five to fifteen plus years. The cost per square foot includes a fabric shell, tension cables, primary and back-up inflation systems and doors. The lower cost structures are used for construction shelters, bulk storage and pond covers. The more expensive are used for recreational structures and warehouses.

Permanent fabrics are teflon coated fiberglass. The life of this structure is twenty plus years. The high cost limits its application to architectural designed structures which call for a clear span covered area, such as stadiums and convention centers. Both temporary and permanent structures are available in translucent fabrics which eliminate the need for daytime lighting.

Areas to be covered vary from 10,000 S.F. to any area up to 1000 feet wide by any length. Height restrictions range from a maximum of 1/2 of the

width to a minimum of 1/6 of the width. Erection of even the largest of the temporary structures requires no more than a week.

Centrifugal fans provide the inflation necessary to support the structure during application of live loads. Airlocks are usually used at large entrances to prevent loss of static pressure. Some manufacturers employ propeller fans which generate sufficient airflow (30,000 CFM) to eliminate the need for airlocks. These fans may also be automatically controlled to resist high wind conditions, regulate humidity (air changes), and provide cooling and heat.

Insulation can be provided with the addition of a second or even third interior liner, creating a dead air space with an "R" value of four to nine. Some structures allow for the liner to be collapsed into the outer shell to enable the internal heat to melt accumulated snow. For cooling or air conditioning, the exterior face of the liner can be aluminized to reflect the sun's heat.

R133419-10   Pre-Engineered Steel Buildings

These buildings are manufactured by many companies and normally erected by franchised dealers throughout the U.S. The four basic types are: Rigid Frames, Truss type, Post and Beam and the Sloped Beam type. The most popular roof slope is a low pitch of 1" in 12". The minimum economical area of these buildings is about 3000 S.F. of floor area. Bay sizes are usually 20' to 24' but can go as high as 30' with heavier girts and purlins. Eave heights are usually 12' to 24' with 18' to 20' most typical.

Material prices shown in the Unit Price section are bare costs for the building shell only and do not include floors, foundations, anchor bolts, interior finishes or utilities. Costs assume at least three bays of 24' each, a 1" in 12" roof slope, and they are based on a 30 psf roof load and a 20 psf wind

load and no unusual requirements. Wind load is a function of wind speed, building height, and terrain characteristics; this should be determined by a registered structural engineer. Costs include the structural frame, 26 ga. non-insulated colored corrugated or ribbed roofing and siding panels, fasteners, closures, trim and flashing but no allowance for insulation, doors, windows, skylights, gutters or downspouts. Very large projects would generally cost less for materials than the prices shown. For roof panel substitutions and wall panel substitutions, see appropriate Unit Price sections.

Conditions at the site, weather, shape and size of the building, and labor availability will affect the erection cost of the building.

R133423-30   Dome Structures

Steel — The four types are Lamella, Schwedler, Arch and Geodesic. For maximum economy, the rise should be about 15 to 20% of the diameter. Most common diameters are in the 200' to 300' range. Lamella domes weigh about 5 P.S.F. of floor area less than Schwedler domes. The Schwedler dome weight in lbs. per S.F. approaches .046 times the diameter. Domes below 125' in diameter weigh .07 times the diameter and the cost per ton of steel is higher. See R051223-20 for estimating weight.

Wood — Small domes are of sawn lumber. Larger ones are laminated. In larger sizes, triaxial and triangular cost about the same; radial domes cost more. Radial domes are economical in the 60' to 70' diameter range. The most economical range of all types is 80' to 200' diameters. Diameters can

run over 400'. All costs are quoted above the foundation. Prices include 2" decking and a tension tie ring in place.

Plywood — Stock prefab geodesic domes are available with diameters from 24' to 60'.

Fiberglass — Aluminum framed translucent sandwich panels with spans from 5' to 45' are commercially available.

Aluminum — Stressed skin aluminum panels form geodesic domes with spans ranging from 82' to 232'. An aluminum space truss, triangulated or nontriangulated, with aluminum or clear acrylic closure panels can be used for clear spans of 40' to 415'.

R220102-20 Labor Adjustment Factors

Labor Adjustment Factors are provided for Divisions 21, 22, and 23 to assist the mechanical estimator account for the various complexities and special conditions of any particular project. While a single percentage has been entered on each line of Division 22 01 02.20, it should be understood that these are just suggested midpoints of ranges of values commonly used by mechanical estimators. They may be increased or decreased depending on the severity of the special conditions.

The group for "existing occupied buildings" has been the subject of requests for explanation. Actually there are two stages to this group: buildings that are existing and "finished" but unoccupied, and those that also are occupied. Buildings that are "finished" may result in higher labor costs due to the

workers having to be more careful not to damage finished walls, ceilings, floors, etc. and may necessitate special protective coverings and barriers. Also corridor bends and doorways may not accommodate long pieces of pipe or larger pieces of equipment. Work above an already hung ceiling can be very time consuming. The addition of occupants may force the work to be done on premium time (nights and/or weekends), eliminate the possible use of some preferred tools such as pneumatic drivers, powder charged drivers, etc. The estimator should evaluate the access to the work area and just how the work is going to be accomplished to arrive at an increase in labor costs over "normal" new construction productivity.

R220523-80 Valve Materials

VALVE MATERIALS

Bronze:
Bronze is one of the oldest materials used to make valves. It is most commonly used in hot and cold water systems and other non-corrosive services. It is often used as a seating surface in larger iron body valves to ensure tight closure.

Carbon Steel:
Carbon steel is a high strength material. Therefore, valves made from this metal are used in higher pressure services, such as steam lines up to 600 psi at 850° F. Many steel valves are available with butt-weld ends for economy and are generally used in high pressure steam service as well as other higher pressure non-corrosive services.

Forged Steel:
Valves from tough carbon steel are used in service up to 2000 psi and temperatures up to 1000° F in gate, globe and check valves.

Iron:
Valves are normally used in medium to large pipe lines to control non-corrosive fluid and gases, where pressures do not exceed 250 psi at 450° F or 500 psi cold water, oil or gas.

Stainless Steel:
Developed steel alloys can be used in over 90% corrosive services.

Plastic PVC:
This is used in a great variety of valves generally in high corrosive services with lower temperatures and pressures.

VALVE SERVICE PRESSURES

Pressure ratings on valves provide an indication of the safe operating pressure for a valve at some elevated temperature. This temperature is dependent upon the materials used and the fabrication of the valve. When specific data is not available, a good "rule-of-thumb" to follow is the temperature of saturated steam on the primary rating indicated on the valve body. Example: The valve has the number 150S printed on the side indicating 150 psi and hence, a maximum operating temperature of 367° F (temperature of saturated steam and 150 psi).

DEFINITIONS

1. "WOG" – Water, oil, gas (cold working pressures).
2. "SWP" – Steam working pressure.
3. 100% area (full port) – means the area through the valve is equal to or greater than the area of standard pipe.
4. "Standard Opening" – means that the area through the valve is less than the area of standard pipe and therefore these valves should be used only where restriction of flow is unimportant.
5. "Round Port" – means the valve has a full round opening through the plug and body, of the same size and area as standard pipe.
6. "Rectangular Port" – valves have rectangular shaped ports through the plug body. The area of the port is either equal to 100% of the area of standard pipe, or restricted (standard opening). In either case it is clearly marked.
7. "ANSI" – American National Standards Institute.

R220523-90 Valve Selection Considerations

INTRODUCTION: In any piping application, valve performance is critical. Valves should be selected to give the best performance at the lowest cost.

The following is a list of performance characteristics generally expected of valves.
1. Stopping flow or starting it.
2. Throttling flow (Modulation).
3. Flow direction changing.
4. Checking backflow (Permitting flow in only one direction).
5. Relieving or regulating pressure.

In order to properly select the right valve, some facts must be determined.

A. What liquid or gas will flow through the valve?

B. Does the fluid contain suspended particles?

C. Does the fluid remain in liquid form at all times?

D. Which metals does fluid corrode?

E. What are the pressure and temperature limits? (As temperature and pressure rise, so will the price of the valve.)

F. Is there constant line pressure?

G. Is the valve merely an on-off valve?

H. Will checking of backflow be required?

I. Will the valve operate frequently or infrequently?

Valves are classified by design type into such classifications as gate, globe, angle, check, ball, butterfly and plug. They are also classified by end connection, stem, pressure restrictions and material such as bronze, cast iron, etc. Each valve has a specific use. A quality valve used correctly will provide a lifetime of trouble-free service, but a high quality valve installed in the wrong service may require frequent attention.

For customer support on your Site Work & Landscape Costs with RSMeans Data, call 800.448.8182.

86

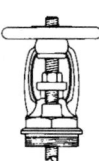

STEM TYPES
(O.S. & Y)—Rising Stem-Outside Screw and Yoke

Offers a visual indication of whether the valve is open or closed. Recommended where high temperatures, corrosives, and solids in the line might cause damage to inside-valve stem threads. The stem threads are engaged by the yoke bushing so the stem rises through the hand wheel as it is turned.

(R.S.)—Rising Stem-Inside Screw

Adequate clearance for operation must be provided because both the hand wheel and the stem rise.
The valve wedge position is indicated by the position of the stem and hand wheel.

(N.R.S.)—Non-Rising Stem-Inside Screw

A minimum clearance is required for operating this type of valve. Excessive wear or damage to stem threads inside the valve may be caused by heat, corrosion, and solids. Because the hand wheel and stem do not rise, wedge position cannot be visually determined.

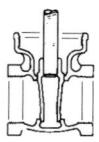

VALVE TYPES
Gate Valves

Provide full flow, minute pressure drop, minimum turbulence and minimum fluid trapped in the line.
They are normally used where operation is infrequent.

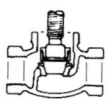

Globe Valves

Globe valves are designed for throttling and/or frequent operation with positive shut-off. Particular attention must be paid to the several types of seating materials available to avoid unnecessary wear. The seats must be compatible with the fluid in service and may be composition or metal. The configuration of the globe valve opening causes turbulence which results in increased resistance. Most bronze globe valves are rising stem-inside screw, but they are also available on O.S. & Y.

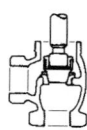

Angle Valves

The fundamental difference between the angle valve and the globe valve is the fluid flow through the angle valve. It makes a 90° turn and offers less resistance to flow than the globe valve while replacing an elbow. An angle valve thus reduces the number of joints and installation time.

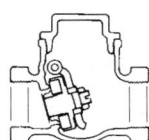

Check Valves

Check valves are designed to prevent backflow by automatically seating when the direction of fluid is reversed.

Swing check valves are generally installed with gate valves, as they provide comparable full flow. Usually recommended for lines where flow velocities are low and should not be used on lines with pulsating flow. Recommended for horizontal installation or in vertical lines only where flow is upward.

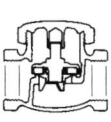 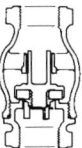

Lift Check Valves

These are commonly used with globe and angle valves since they have similar diaphragm seating arrangements and are recommended for preventing backflow of steam, air, gas and water, and on vapor lines with high flow velocities. For horizontal lines, horizontal lift checks should be used and vertical lift checks for vertical lines.

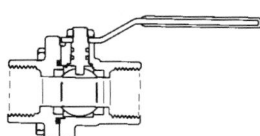

Ball Valves

Ball valves are light and easily installed, yet because of modern elastomeric seats, provide tight closure. Flow is controlled by rotating up to 90° a drilled ball which fits tightly against resilient seals. This ball seats with flow in either direction, and valve handle indicates the degree of opening. Recommended for frequent operation readily adaptable to automation, ideal for installation where space is limited.

Butterfly Valves

Butterfly valves provide bubble-tight closure with excellent throttling characteristics. They can be used for full-open, closed and for throttling applications.

The butterfly valve consists of a disc within the valve body which is controlled by a shaft. In its closed position, the valve disc seals against a resilient seat. The disc position throughout the full 90° rotation is visually indicated by the position of the operator.

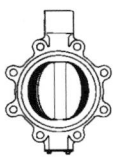

A butterfly valve is only a fraction of the weight of a gate valve and requires no gaskets between flanges in most cases. Recommended for frequent operation and adaptable to automation where space is limited.

Wafer and lug type bodies, when installed between two pipe flanges, can be easily removed from the line. The pressure of the bolted flanges holds the valve in place.
Locating lugs makes installation easier.

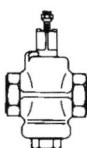

Plug Valves

Lubricated plug valves, because of the wide range of services to which they are adapted, may be classified as all purpose valves. They can be safely used at all pressure and vacuums, and at all temperatures up to the limits of available lubricants. They are the most satisfactory valves for the handling of gritty suspensions and many other destructive, erosive, corrosive and chemical solutions.

R221113-50   Pipe Material Considerations

1. Malleable fittings should be used for gas service.
2. Malleable fittings are used where there are stresses/strains due to expansion and vibration.
3. Cast fittings may be broken as an aid to disassembling heating lines frozen by long use, temperature and minerals.
4. A cast iron pipe is extensively used for underground and submerged service.
5. Type M (light wall) copper tubing is available in hard temper only and is used for nonpressure and less severe applications than K and L.

6. Type L (medium wall) copper tubing, available hard or soft for interior service.
7. Type K (heavy wall) copper tubing, available in hard or soft temper for use where conditions are severe. For underground and interior service.
8. Hard drawn tubing requires fewer hangers or supports but should not be bent. Silver brazed fittings are recommended, but soft solder is normally used.
9. Type DMV (very light wall) copper tubing designed for drainage, waste and vent plus other non-critical pressure services.

Domestic/Imported Pipe and Fittings Costs

The prices shown in this publication for steel/cast iron pipe and steel, cast iron, and malleable iron fittings are based on domestic production sold at the normal trade discounts. The above listed items of foreign manufacture may be available at prices 1/3 to 1/2 of those shown. Some imported items after minor machining or finishing operations are being sold as domestic to further complicate the system.

Caution: Most pipe prices in this data set also include a coupling and pipe hangers which for the larger sizes can add significantly to the per foot cost and should be taken into account when comparing "book cost" with the quoted supplier's cost.

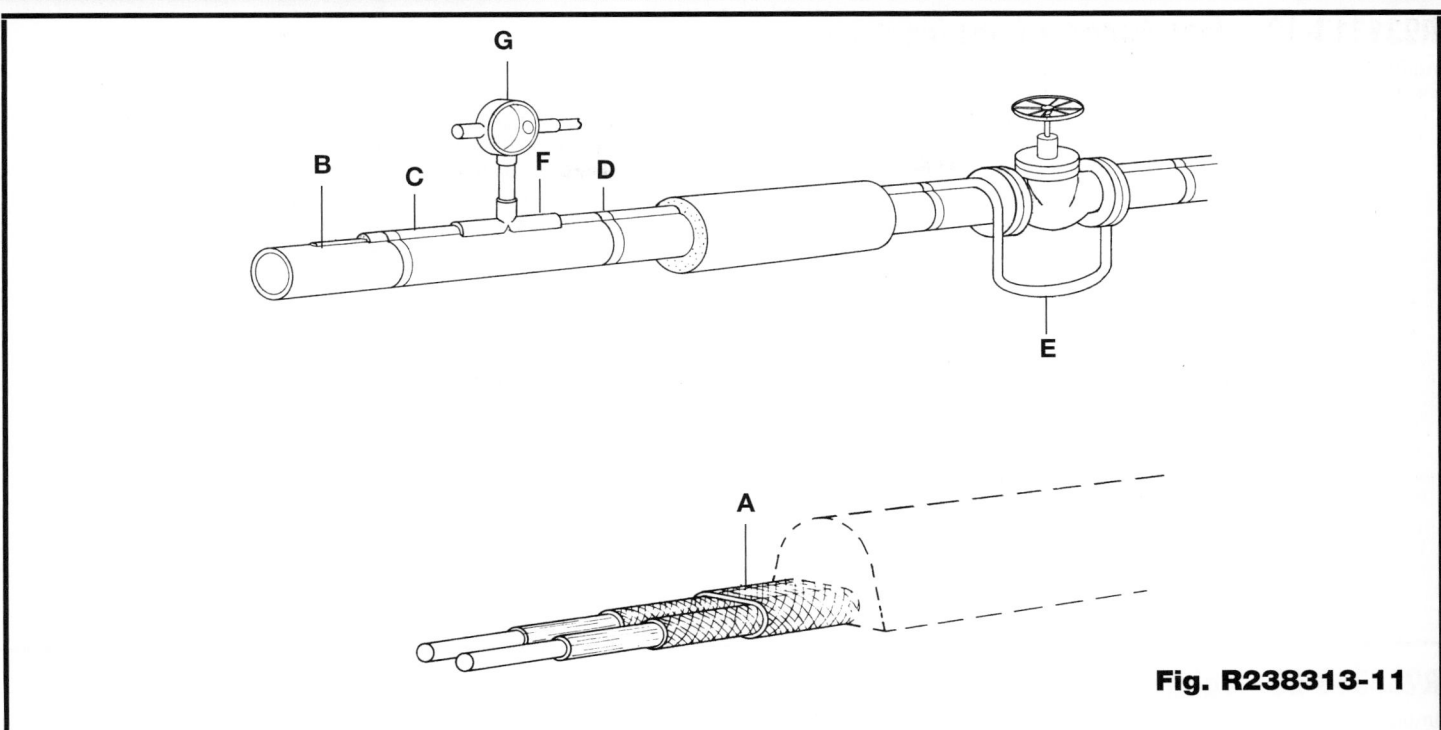

Fig. R238313-11

R238313-10 Heat Trace Systems

Before you can determine the cost of a HEAT TRACE installation
the method of attachment must be established. There are four (4) common
methods:

1. Cable is simply attached to the pipe with polyester tape
   every 12′.
2. Cable is attached with a continuous cover of 2″ wide
   aluminum tape.
3. Cable is attached with factory extruded heat transfer cement
   and covered with metallic raceway with clips every 10′.
4. Cable is attached between layers of pipe insulation using either clips or
   polyester tape.

Example: Components for method 3 must include:

A. Heat trace cable by voltage and watts per linear foot.
B. Heat transfer cement, 1 gallon per 60 linear feet of cover.
C. Metallic raceway by size and type.
D. Raceway clips by size of pipe.

When taking off linear foot lengths of cable add the following for each valve
in the system. (E)

In all of the above methods each component of the system must be priced
individually.

SCREWED OR WELDED VALVE:			FLANGED VALVE:			BUTTERFLY VALVES:		
1/2″	=	6″	1/2″	=	1′ -0″	1/2″	=	0′
3/4″	=	9″	3/4″	=	1′ -6″	3/4″	=	0′
1″	=	1′ -0″	1″	=	2′ -0″	1″	=	1′ -0″
1-1/2″	=	1′ -6″	1-1/2″	=	2′ -6″	1-1/2″	=	1′ -6″
2″	=	2′	2″	=	2′ -6″	2″	=	2′ -0″
2-1/2″	=	2′ -6″	2-1/2″	=	3′ -0″	2-1/2″	=	2′ -6″
3″	=	2′ -6″	3″	=	3′ -6″	3″	=	2′ -6″
4″	=	4′ -0″	4″	=	4′ -0″	4″	=	3′ -0″
6″	=	7′ -0″	6″	=	8′ -0″	6″	=	3′ -6″
8″	=	9′ -6″	8″	=	11′ -0″	8″	=	4′ -0″
10″	=	12′ -6″	10″	=	14′ -0″	10″	=	4′ -0″
12″	=	15′ -0″	12″	=	16′ -6″	12″	=	5′ -0″
14″	=	18′ -0″	14″	=	19′ -6″	14″	=	5′ -6″
16″	=	21′ -6″	16″	=	23′ -0″	16″	=	6′ -0″
18″	=	25′ -6″	18″	=	27′ -0″	18″	=	6′ -6″
20″	=	28′ -6″	20″	=	30′ -0″	20″	=	7′ -0″
24″	=	34′ -0″	24″	=	36′ -0″	24″	=	8′ -0″
30″	=	40′ -0″	30″	=	42′ -0″	30″	=	10′ -0″

For customer support on your Site Work & Landscape Costs with RSMeans Data, call 800.448.8182.

871

R238313-10   Heat Trace Systems (cont.)

Add the following quantities of heat transfer cement to linear foot totals for each valve:

Nominal Valve Size	Gallons of Cement per Valve
1/2"	0.14
3/4"	0.21
1"	0.29
1-1/2"	0.36
2"	0.43
2-1/2"	0.70
3"	0.71
4"	1.00
6"	1.43
8"	1.48
10"	1.50
12"	1.60
14"	1.75
16"	2.00
18"	2.25
20"	2.50
24"	3.00
30"	3.75

The following must be added to the list of components to accurately price HEAT TRACE systems:

1. Expediter fitting and clamp fasteners (F)
2. Junction box and nipple connected to expediter fitting (G)
3. Field installed terminal blocks within junction box
4. Ground lugs
5. Piping from power source to expediter fitting
6. Controls
7. Thermostats
8. Branch wiring
9. Cable splices
10. End of cable terminations
11. Branch piping fittings and boxes

Deduct the following percentages from labor if cable lengths in the same area exceed:

150' to 250'	10%	351' to 500'	20%
251' to 350'	15%	Over 500'	25%

Add the following percentages to labor for elevated installations:

15' to 20' high	10%	31' to 35' high	40%
21' to 25' high	20%	36' to 40' high	50%
26' to 30' high	30%	Over 40' high	60%

R238313-20   Spiral-Wrapped Heat Trace Cable (Pitch Table)

In order to increase the amount of heat, occasionally a heat trace cable is wrapped in a spiral fashion around a pipe, increasing the number of feet of heater cable per linear foot of pipe.

Engineers first determine the heat loss per foot of pipe (based on the insulating material, its thickness, and the temperature differential across it). A ratio is then calculated by the formula:

$$\text{Feet of Heat Trace per Foot of Pipe} = \frac{\text{Watts/Foot of Heat Loss}}{\text{Watts/Foot of the Cable}}$$

The linear distance between wraps (pitch) is then taken from a chart or table. Generally, the pitch is listed on a drawing leaving the estimator to calculate the total length of heat tape required. An approximation may be taken from this table.

Feet of Heat Trace Per Foot of Pipe

Pitch In Inches	Nominal Pipe Size in Inches															
	1	1¼	1½	2	2½	3	4	6	8	10	12	14	16	18	20	24
3.5	1.80															
4	1.65															
5	1.46	1.60	1.80													
6	1.34	1.45	1.55	1.75												
7	1.25	1.35	1.43	1.57	1.75											
8	1.20	1.28	1.34	1.45	1.60	1.80										
9	1.16	1.23	1.28	1.37	1.51	1.68										
10	1.13	1.19	1.24	1.32	1.44	1.57	1.82									
15	1.06	1.08	1.10	1.15	1.21	1.29	1.42	1.78								
20	1.04	1.05	1.06	1.08	1.13	1.17	1.25	1.49	1.73							
25		1.04	1.04	1.06	1.08	1.11	1.17	1.33	1.51	1.72						
30				1.04	1.05	1.07	1.12	1.24	1.37	1.54	1.70	1.80				
35					1.06	1.06	1.09	1.17	1.28	1.42	1.54	1.64	1.78			
40						1.05	1.07	1.14	1.22	1.33	1.44	1.52	1.64	1.75		
50							1.05	1.09	1.15	1.22	1.29	1.35	1.44	1.53	1.64	1.83
60								1.06	1.11	1.16	1.21	1.25	1.31	1.39	1.46	1.62
70								1.05	1.08	1.12	1.17	1.19	1.24	1.30	1.35	1.47
80									1.06	1.09	1.13	1.15	1.19	1.24	1.28	1.38
90									1.04	1.06	1.10	1.13	1.16	1.19	1.23	1.32
100										1.05	1.08	1.10	1.13	1.15	1.19	1.23

Note: Common practice would normally limit the lower end of the table to 5% of additional heat. Above 80% an engineer would likely opt for two (2) parallel cables.

872

For customer support on your Site Work & Landscape Costs with RSMeans Data, call 800.448.8182.

R260519-93 Metric Equivalent, Wire

United States		European	
Size AWG or kcmil	Area Cir. Mils. (cmil) mm²	Size mm²	Area Cir. Mils.
18	1620/.82	.75	1480
16	2580/1.30	1.0	1974
14	4110/2.08	1.5	2961
12	6530/3.30	2.5	4935
10	10,380/5.25	4	7896
8	16,510/8.36	6	11,844
6	26,240/13.29	10	19,740
4	41,740/21.14	16	31,584
3	52,620/26.65	25	49,350
2	66,360/33.61	–	–
1	83,690/42.39	35	69,090
1/0	105,600/53.49	50	98,700
2/0	133,100/67.42	–	–
3/0	167,800/85.00	70	138,180
4/0	211,600/107.19	95	187,530
250	250,000/126.64	120	236,880
300	300,000/151.97	150	296,100
350	350,000/177.30	–	–
400	400,000/202.63	185	365,190
500	500,000/253.29	240	473,760
600	600,000/303.95	300	592,200
700	700,000/354.60	–	–
750	750,000/379.93	–	–

U.S. vs. European Wire – Approximate Equivalents

R260533-22 Conductors in Conduit

The table below lists the maximum number of conductors for various sized conduit using THW, TW or THWN insulations.

Copper Wire Size	1/2"			3/4"			1"			1-1/4"			1-1/2"			2"			2-1/2"			3"		3-1/2"		4"	
	TW	THW	THWN	TW	THW	THWN	TW	THW	THWN	TW	THW	THWN	TW	THW	THWN	TW	THW	THWN	TW	THW	THWN	THW	THWN	THW	THWN	THW	THWN
#14	9	6	13	15	10	24	25	16	39	44	29	69	60	40	94	99	65	154	142	93		143		192			
#12	7	4	10	12	8	18	19	13	29	35	24	51	47	32	70	78	53	114	111	76	164	117		157			
#10	5	4	6	9	6	11	15	11	18	26	19	32	36	26	44	60	43	73	85	61	104	95	160	127		163	
#8	2	1	3	4	3	5	7	5	9	12	10	16	17	13	22	28	22	36	40	32	51	49	79	66	106	85	136
#6		1	1		2	4		4	6		7	11		10	15		16	26		23	37	36	57	48	76	62	98
#4		1	1		1	2		3	4		5	7		7	9		12	16		17	22	27	35	36	47	47	60
#3		1	1		1	1		2	3		4	6		6	8		10	13		15	19	23	29	31	39	40	51
#2		1	1		1	1		2	3		4	5		5	7		9	11		13	16	20	25	27	33	34	43
#1					1	1		1	1		3	3		4	5		6	8		9	12	14	18	19	25	25	32
1/0					1	1		1	1		2	3		3	4		5	7		8	10	12	15	16	21	21	27
2/0					1	1		1	1		1	2		3	3		5	6		7	8	10	13	14	17	18	22
3/0					1	1		1	1		1	1		2	3		4	5		6	7	9	11	12	14	15	18
4/0						1		1	1		1	1		1	2		3	4		5	6	7	9	10	12	13	15
250 kcmil								1	1		1	1		1	1		2	3		4	4	6	7	8	10	10	12
300								1	1		1	1		1	1		2	3		3	4	5	6	7	8	9	11
350									1		1	1		1	1		1	2		3	3	4	5	6	7	8	9
400											1	1		1	1		1	1		2	3	4	5	5	6	7	8
500											1	1		1	1		1	1		1	2	3	4	4	5	6	7
600												1		1	1		1	1		1	1	3	3	4	4	5	5
700														1	1		1	1		1	1	2	3	3	4	4	5
750														1	1		1	1		1	1	2	2	3	3	4	4

For customer support on your Site Work & Landscape Costs with RSMeans Data, call 800.448.8182.

873

R260590-05 Typical Overhead Service Entrance

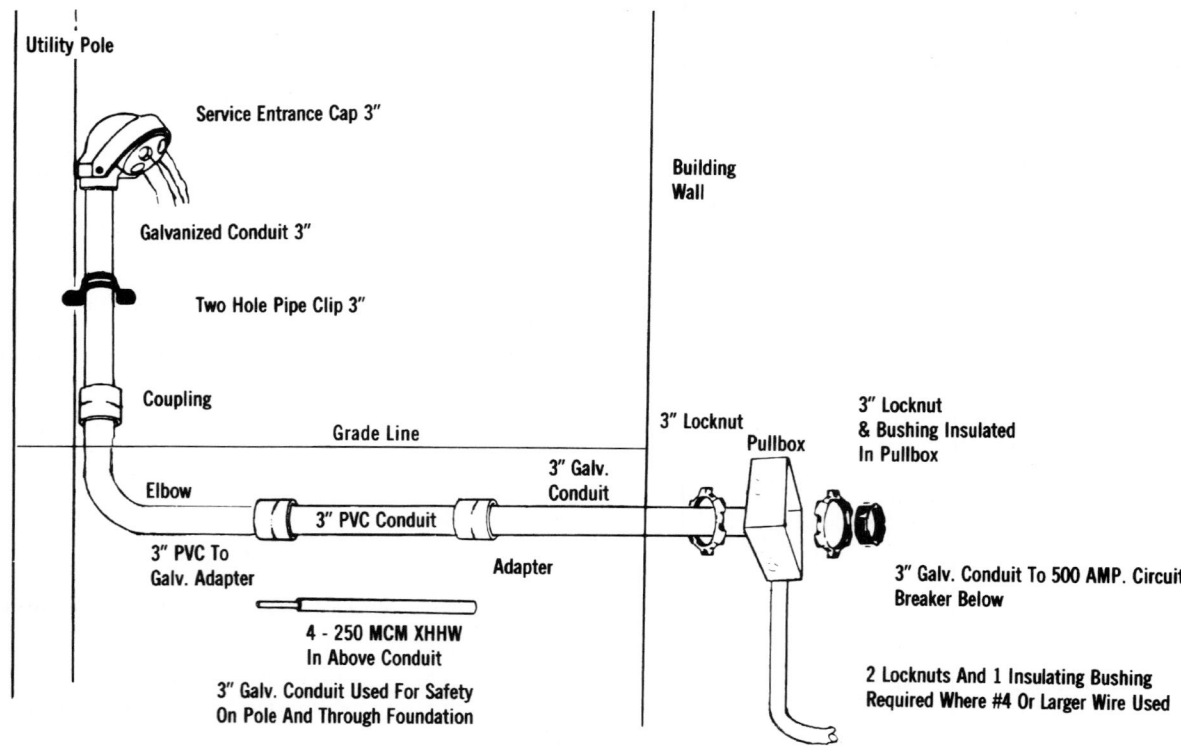

Utility Pole

Service Entrance Cap 3″

Galvanized Conduit 3″

Two Hole Pipe Clip 3″

Coupling

Grade Line

Building Wall

Elbow

3″ Galv. Conduit

3″ PVC Conduit

3″ PVC To Galv. Adapter

Adapter

4 - 250 MCM XHHW In Above Conduit

3″ Galv. Conduit Used For Safety On Pole And Through Foundation

3″ Locknut

Pullbox

3″ Locknut & Bushing Insulated In Pullbox

3″ Galv. Conduit To 500 AMP. Circuit Breaker Below

2 Locknuts And 1 Insulating Bushing Required Where #4 Or Larger Wire Used

874

For customer support on your Site Work & Landscape Costs with RSMeans Data, call 800.448.8182.

R262716-40   Standard Electrical Enclosure Types

NEMA Enclosures

Electrical enclosures serve two basic purposes: they protect people from accidental contact with enclosed electrical devices and connections, and they protect the enclosed devices and connections from specified external conditions. The National Electrical Manufacturers Association (NEMA) has established the following standards. Because these descriptions are not intended to be complete representations of NEMA listings, consultation of NEMA literature is advised for detailed information.

The following definitions and descriptions pertain to NONHAZARDOUS locations.

NEMA   Type 1: General purpose enclosures intended for use indoors, primarily to prevent accidental contact of personnel with the enclosed equipment in areas that do not involve unusual conditions.

NEMA   Type 2: Dripproof indoor enclosures intended to protect the enclosed equipment against dripping noncorrosive liquids and falling dirt.

NEMA   Type 3: Dustproof, raintight and sleet-resistant (ice-resistant) enclosures intended for use outdoors to protect the enclosed equipment against wind-blown dust, rain, sleet, and external ice formation.

NEMA   Type 3R: Rainproof and sleet-resistant (ice-resistant) enclosures which are intended for use outdoors to protect the enclosed equipment against rain. These enclosures are constructed so that the accumulation and melting of sleet (ice) will not damage the enclosure and its internal mechanisms.

NEMA   Type 3S: Enclosures intended for outdoor use to provide limited protection against wind-blown dust, rain, and sleet (ice) and to allow operation of external mechanisms when ice-laden.

NEMA   Type 4: Watertight and dust-tight enclosures intended for use indoors and out – to protect the enclosed equipment against splashing water, seepage of water, falling or hose-directed water, and severe external condensation.

NEMA   Type 4X: Watertight, dust-tight, and corrosion-resistant indoor and outdoor enclosures featuring the same provisions as Type 4 enclosures, plus corrosion resistance.

NEMA   Type 5: Indoor enclosures intended primarily to provide limited protection against dust and falling dirt.

NEMA   Type 6: Enclosures intended for indoor and outdoor use – primarily to provide limited protection against the entry of water during occasional temporary submersion at a limited depth.

NEMA   Type 6R: Enclosures intended for indoor and outdoor use – primarily to provide limited protection against the entry of water during prolonged submersion at a limited depth.

NEMA   Type 11: Enclosures intended for indoor use – primarily to provide, by means of oil immersion, limited protection to enclosed equipment against the corrosive effects of liquids and gases.

NEMA   Type 12: Dust-tight and driptight indoor enclosures intended for use indoors in industrial locations to protect the enclosed equipment against fibers, flyings, lint, dust, and dirt, as well as light splashing, seepage, dripping, and external condensation of noncorrosive liquids.

NEMA   Type 13: Oil-tight and dust-tight indoor enclosures intended primarily to house pilot devices, such as limit switches, foot switches, push buttons, selector switches, and pilot lights, and to protect these devices against lint and dust, seepage, external condensation, and sprayed water, oil, and noncorrosive coolant.

The following definitions and descriptions pertain to HAZARDOUS, or CLASSIFIED, locations.

NEMA   Type 7: Enclosures intended for use in indoor locations classified as Class 1, Groups A, B, C, or D, as defined in the National Electrical Code.

NEMA   Type 9: Enclosures intended for use in indoor locations classified as Class 2, Groups E, F, or G, as defined in the National Electrical Code.

For customer support on your Site Work & Landscape Costs with RSMeans Data, call 800.448.8182.

875

R312316-40   Excavating

The selection of equipment used for structural excavation and bulk excavation or for grading is determined by the following factors.

1. Quantity of material
2. Type of material
3. Depth or height of cut
4. Length of haul
5. Condition of haul road
6. Accessibility of site
7. Moisture content and dewatering requirements
8. Availability of excavating and hauling equipment

Some additional costs must be allowed for hand trimming the sides and bottom of concrete pours and other excavation below the general excavation.

Number of B.C.Y. per truck = 1.5 C.Y. bucket x 8 passes = 12 loose C.Y.

$$= 12 \times \frac{100}{118} = 10.2 \text{ B.C.Y. per truck}$$

Truck Haul Cycle:

Load truck, 8 passes	=	4 minutes
Haul distance, 1 mile	=	9 minutes
Dump time	=	2 minutes
Return, 1 mile	=	7 minutes
Spot under machine	=	1 minute
		23 minute cycle

Add the mobilization and demobilization costs to the total excavation costs. When equipment is rented for more than three days, there is often no mobilization charge by the equipment dealer. On larger jobs outside of urban areas, scrapers can move earth economically provided a dump site or fill area and adequate haul roads are available. Excavation within sheeting bracing or cofferdam bracing is usually done with a clamshell and production

When planning excavation and fill, the following should also be considered.

1. Swell factor
2. Compaction factor
3. Moisture content
4. Density requirements

A typical example for scheduling and estimating the cost of excavation of a 15' deep basement on a dry site when the material must be hauled off the site is outlined below.

Assumptions:

1. Swell factor, 18%
2. No mobilization or demobilization
3. Allowance included for idle time and moving on job
4. No dewatering, sheeting, or bracing
5. No truck spotter or hand trimming

Fleet Haul Production per day in B.C.Y.

$$4 \text{ trucks} \times \frac{50 \text{ min. hour}}{23 \text{ min. haul cycle}} \times 8 \text{ hrs.} \times 10.2 \text{ B.C.Y.}$$

$$= 4 \times 2.2 \times 8 \times 10.2 = 718 \text{ B.C.Y./day}$$

is low, since the clamshell may have to be guided by hand between the bracing. When excavating or filling an area enclosed with a wellpoint system, add 10% to 15% to the cost to allow for restricted access. When estimating earth excavation quantities for structures, allow work space outside the building footprint for construction of the foundation and a slope of 1:1 unless sheeting is used.

R312316-45 Excavating Equipment

The table below lists theoretical hourly production in C.Y./hr. bank measure for some typical excavation equipment. Figures assume 50 minute hours, 83% job efficiency, 100% operator efficiency, 90° swing and properly sized hauling units, which must be modified for adverse digging and loading conditions. Actual production costs in the front of the data set average about 50% of the theoretical values listed here.

Equipment	Soil Type	B.C.Y. Weight	% Swell	1 C.Y.	1-1/2 C.Y.	2 C.Y.	2-1/2 C.Y.	3 C.Y.	3-1/2 C.Y.	4 C.Y.
Hydraulic Excavator	Moist loam, sandy clay	3400 lb.	40%	85	125	175	220	275	330	380
"Backhoe"	Sand and gravel	3100	18	80	120	160	205	260	310	365
15' Deep Cut	Common earth	2800	30	70	105	150	190	240	280	330
	Clay, hard, dense	3000	33	65	100	130	170	210	255	300
Power Shovel Optimum Cut (Ft.)	Moist loam, sandy clay	3400	40	170 (6.0)	245 (7.0)	295 (7.8)	335 (8.4)	385 (8.8)	435 (9.1)	475 (9.4)
	Sand and gravel	3100	18	165 (6.0)	225 (7.0)	275 (7.8)	325 (8.4)	375 (8.8)	420 (9.1)	460 (9.4)
	Common earth	2800	30	145 (7.8)	200 (9.2)	250 (10.2)	295 (11.2)	335 (12.1)	375 (13.0)	425 (13.8)
	Clay, hard, dense	3000	33	120 (9.0)	175 (10.7)	220 (12.2)	255 (13.3)	300 (14.2)	335 (15.1)	375 (16.0)
Drag Line Optimum Cut (Ft.)	Moist loam, sandy clay	3400	40	130 (6.6)	180 (7.4)	220 (8.0)	250 (8.5)	290 (9.0)	325 (9.5)	385 (10.0)
	Sand and gravel	3100	18	130 (6.6)	175 (7.4)	210 (8.0)	245 (8.5)	280 (9.0)	315 (9.5)	375 (10.0)
	Common earth	2800	30	110 (8.0)	160 (9.0)	190 (9.9)	220 (10.5)	250 (11.0)	280 (11.5)	310 (12.0)
	Clay, hard, dense	3000	33	90 (9.3)	130 (10.7)	160 (11.8)	190 (12.3)	225 (12.8)	250 (13.3)	280 (12.0)

Equipment	Soil Type	B.C.Y. Weight	% Swell	Wheel Loaders				Track Loaders		
				3 C.Y.	4 C.Y.	6 C.Y.	8 C.Y.	2-1/4 C.Y.	3 C.Y.	4 C.Y.
Loading Tractors	Moist loam, sandy clay	3400	40	260	340	510	690	135	180	250
	Sand and gravel	3100	18	245	320	480	650	130	170	235
	Common earth	2800	30	230	300	460	620	120	155	220
	Clay, hard, dense	3000	33	200	270	415	560	110	145	200
	Rock, well-blasted	4000	50	180	245	380	520	100	130	180

R312319-90   Wellpoints

A single stage wellpoint system is usually limited to dewatering an average 15′ depth below normal ground water level. Multi-stage systems are employed for greater depth with the pumping equipment installed only at the lowest header level. Ejectors with unlimited lift capacity can be economical when two or more stages of wellpoints can be replaced or when horizontal clearance is restricted, such as in deep trenches or tunneling projects, and where low water flows are expected. Wellpoints are usually spaced on 2-1/2′ to 10′ centers along a header pipe. Wellpoint spacing, header size, and pump size are all determined by the expected flow as dictated by soil conditions.

In almost all soils encountered in wellpoint dewatering, the wellpoints may be jetted into place. Cemented soils and stiff clays may require sand wicks about 12″ in diameter around each wellpoint to increase efficiency and eliminate weeping into the excavation. These sand wicks require 1/2 to 3 C.Y. of washed filter sand and are installed by using a 12″ diameter steel casing and hole puncher jetted into the ground 2′ deeper than the wellpoint. Rock may require predrilled holes.

Labor required for the complete installation and removal of a single stage wellpoint system is in the range of 3/4 to 2 labor-hours per linear foot of header, depending upon jetting conditions, wellpoint spacing, etc.

Continuous pumping is necessary except in some free draining soil where temporary flooding is permissible (as in trenches which are backfilled after each day's work). Good practice requires provision of a stand-by pump during the continuous pumping operation.

Systems for continuous trenching below the water table should be installed three to four times the length of expected daily progress to ensure uninterrupted digging, and header pipe size should not be changed during the job.

For pervious free draining soils, deep wells in place of wellpoints may be economical because of lower installation and maintenance costs. Daily production ranges between two to three wells per day, for 25′ to 40′ depths, to one well per day for depths over 50′.

Detailed analysis and estimating for any dewatering problem is available at no cost from wellpoint manufacturers. Major firms will quote "sufficient equipment" quotes or their affiliates will offer lump sum proposals to cover complete dewatering responsibility.

Description for 200′ System with 8″ Header		Quantities
Equipment & Material	Wellpoints 25′ long, 2″ diameter @ 5′ O.C.	40 Each
	Header pipe, 8″ diameter	200 L.F.
	Discharge pipe, 8″ diameter	100 L.F.
	8″ valves	3 Each
	Combination jetting & wellpoint pump (standby)	1 Each
	Wellpoint pump, 8″ diameter	1 Each
	Transportation to and from site	1 Day
	Fuel for 30 days x 60 gal./day	1800 Gallons
	Lubricants for 30 days x 16 lbs./day	480 Lbs.
	Sand for points	40 C.Y.
Labor	Technician to supervise installation	1 Week
	Labor for installation and removal of system	300 Labor-hours
	4 Operators straight time 40 hrs./wk. for 4.33 wks.	693 Hrs.
	4 Operators overtime 2 hrs./wk. for 4.33 wks.	35 Hrs.

R312323-30   Compacting Backfill

Compaction of fill in embankments, around structures, in trenches, and under slabs is important to control settlement. Factors affecting compaction are:

1. Soil gradation
2. Moisture content
3. Equipment used
4. Depth of fill per lift
5. Density required

Production Rate:

$$\frac{1.75′ \text{ plate width x 50 F.P.M. x 50 min./hr. x .67′ lift}}{27 \text{ C.F. per C.Y.}} = 108.5 \text{ C.Y./hr.}$$

Production Rate for 4 Passes:

$$\frac{108.5 \text{ C.Y.}}{4 \text{ passes}} = 27.125 \text{ C.Y./hr. x 8 hrs.} = 217 \text{ C.Y./day}$$

Example:

Compact granular fill around a building foundation using a 21″ wide x 24″ vibratory plate in 8″ lifts. Operator moves at 50 F.P.M. working a 50 minute hour to develop 95% Modified Proctor Density with 4 passes.

R314116-40   Wood Sheet Piling

Wood sheet piling may be used for depths to 20' where there is no ground water. If moderate ground water is encountered Tongue & Groove

sheeting will help to keep it out. When considerable ground water is present, steel sheeting must be used.

For estimating purposes on trench excavation, sizes are as follows:

Depth	Sheeting	Wales	Braces	B.F. per S.F.
To 8'	3 x 12's	6 x 8's, 2 line	6 x 8's, @ 10'	4.0 @ 8'
8' x 12'	3 x 12's	10 x 10's, 2 line	10 x 10's, @ 9'	5.0 average
12' to 20'	3 x 12's	12 x 12's, 3 line	12 x 12's, @ 8'	7.0 average

Sheeting to be toed in at least 2' depending upon soil conditions. A five person crew with an air compressor and sheeting driver can drive and brace 440 SF/day at 8' deep, 360 SF/day at 12' deep, and 320 SF/day at 16' deep.

For normal soils, piling can be pulled in 1/3 the time to install. Pulling difficulty increases with the time in the ground. Production can be increased by high pressure jetting.

R314116-45   Steel Sheet Piling

Limiting weights are 22 to 38#/S.F. of wall surface with 27#/S.F. average for usual types and sizes. (Weights of piles themselves are from 30.7#/L.F. to 57#/L.F. but they are 15" to 21" wide.) Lightweight sections 12" to 28" wide from 3 ga. to 12 ga. thick are also available for shallow excavations. Piles may be driven two at a time with an impact or vibratory hammer (use vibratory to pull) hung from a crane without leads. A reasonable estimate of the life of steel sheet piling is 10 uses with up to 125 uses possible if a vibratory hammer is used. Used piling costs from 50% to 80% of new piling depending on location and market conditions. Sheet piling and H piles

can be rented for about 30% of the delivered mill price for the first month and 5% per month thereafter. Allow 1 labor-hour per pile for cleaning and trimming after driving. These costs increase with depth and hydrostatic head. Vibratory drivers are faster in wet granular soils and are excellent for pile extraction. Pulling difficulty increases with the time in the ground and may cost more than driving. It is often economical to abandon the sheet piling, especially if it can be used as the outer wall form. Allow about 1/3 additional length or more for toeing into ground. Add bracing, waler and strut costs. Waler costs can equal the cost per ton of sheeting.

R314513-90   Vibroflotation and Vibro Replacement Soil Compaction

Vibroflotation is a proprietary system of compacting sandy soils in place to increase relative density to about 70%. Typical bearing capacities attained will be 6000 psf for saturated sand and 12,000 psf for dry sand. Usual range is 4000 to 8000 psf capacity. Costs in the front of the data set are for a vertical foot of compacted cylinder 6' to 10' in diameter.

Vibro replacement is a proprietary system of improving cohesive soils in place to increase bearing capacity. Most silts and clays above or below the water table can be strengthened by installation of stone columns.

The process consists of radial displacement of the soil by vibration. The created hole is then backfilled in stages with coarse granular fill which is thoroughly compacted and displaced into the surrounding soil in the form of a column.

The total project cost would depend on the number and depth of the compacted cylinders. The installing company guarantees relative soil density of the sand cylinders after compaction and the bearing capacity of the soil after the replacement process. Detailed estimating information is available from the installer at no cost.

R316326-60 Caissons

The three principal types of caissons are:

(1) Belled Caissons, which except for shallow depths and poor soil conditions, are generally recommended. They provide more bearing than shaft area. Because of its conical shape, no horizontal reinforcement of the bell is required.

(2) Straight Shaft Caissons are used where relatively light loads are to be supported by caissons that rest on high value bearing strata. While the shaft is larger in diameter than for belled types this is more than offset by the savings in time and labor.

(3) Keyed Caissons are used when extremely heavy loads are to be carried. A keyed or socketed caisson transfers its load into rock by a combination of end-bearing and shear reinforcing of the shaft. The most economical shaft often consists of a steel casing, a steel wide flange core and concrete. Allowable compressive stresses of .225 f'c for concrete, 16,000 psi for the wide flange core, and 9,000 psi for the steel casing are commonly used. The usual range of shaft diameter is 18" to 84". The number of sizes specified for any one project should be limited due to the problems of casing and auger storage. When hand work is to be performed, shaft diameters should not be less than 32". When inspection of borings is required a minimum shaft diameter of 30" is recommended. Concrete caissons are intended to be poured against earth excavation so permanent forms, which add to cost, should not be used if the excavation is clean and the earth is sufficiently impervious to prevent excessive loss of concrete.

Soil Conditions for Belling		
Good	**Requires Handwork**	**Not Recommended**
Clay	Hard Shale	Silt
Sandy Clay	Limestone	Sand
Silty Clay	Sandstone	Gravel
Clayey Silt	Weathered Mica	Igneous Rock
Hard-pan		
Soft Shale		
Decomposed Rock		

R329219-50 Seeding

The type of grass is determined by light, shade and moisture content of soil plus intended use. Fertilizer should be disked 4" before seeding. For steep slopes disk five tons of mulch and lay two tons of hay or straw on surface per acre after seeding. Surface mulch can be staked, lightly disked or tar emulsion sprayed. Material for mulch can be wood chips, peat moss, partially rotted hay or straw, wood fibers and sprayed emulsions. Hemp seed blankets with fertilizer are also available. For spring seeding, watering is necessary. Late fall seeding may have to be reseeded in the spring. Hydraulic seeding, power mulching, and aerial seeding can be used on large areas.

R329343-10   Plant Spacing Chart

This chart may be used when plants are to be placed equidistant from each other, staggering their position in each row.

Plant Spacing (Inches)	Row Spacing (Inches)	Plants Per (CSF)	Plant Spacing (Feet)	Row Spacing (Feet)	Plants Per (MSF)
			4	3.46	72
6	5.20	462	5	4.33	46
8	6.93	260	6	5.20	32
10	8.66	166	8	6.93	18
12	10.39	115	10	8.66	12
15	12.99	74	12	10.39	8
18	15.59	51.32	15	12.99	5.13
21	18.19	37.70	20	17.32	2.89
24	20.78	28.87	25	21.65	1.85
30	25.98	18.48	30	25.98	1.28
36	31.18	12.83	40	34.64	0.72

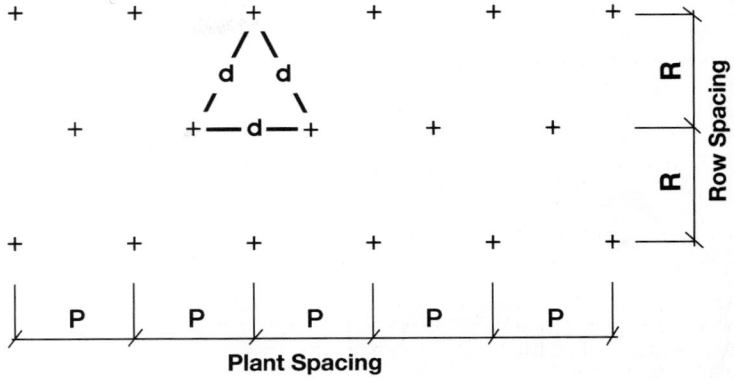

R329343-20   Trees and Plants by Environment and Purposes

Dry, Windy, Exposed Areas
Barberry
Junipers, all varieties
Locust
Maple
Oak
Pines, all varieties
Poplar, Hybrid
Privet
Spruce, all varieties
Sumac, Staghorn

Lightly Wooded Areas
Dogwood
Hemlock
Larch
Pine, White
Rhododendron
Spruce, Norway
Redbud

Total Shade Areas
Hemlock
Ivy, English
Myrtle
Pachysandra
Privet
Spice Bush
Yews, Japanese

Cold Temperatures of Northern U.S. and Canada
Arborvitae, American
Birch, White
Dogwood, Silky
Fir, Balsam
Fir, Douglas
Hemlock
Juniper, Andorra
Juniper, Blue Rug
Linden, Little Leaf
Maple, Sugar
Mountain Ash
Myrtle
Olive, Russian

Pine, Mugho
Pine, Ponderosa
Pine, Red
Pine, Scotch
Poplar, Hybrid
Privet
Rosa Rugosa
Spruce, Dwarf Alberta
Spruce, Black Hills
Spruce, Blue
Spruce, Norway
Spruce, White, Engelman
Yellow Wood

Wet, Swampy Areas
American Arborvitae
Birch, White
Black Gum
Hemlock
Maple, Red
Pine, White
Willow

Poor, Dry, Rocky Soil
Barberry
Crownvetch
Eastern Red Cedar
Juniper, Virginiana
Locust, Black
Locust, Bristly
Locust, Honey
Olive, Russian
Pines, all varieties
Privet
Rosa Rugosa
Sumac, Staghorn

Seashore Planting
Arborvitae, American
Juniper, Tamarix
Locust, Black
Oak, White
Olive, Russian
Pine, Austrian
Pine, Japanese Black

Pine, Mugho
Pine, Scotch
Privet, Amur River
Rosa Rugosa
Yew, Japanese

City Planting
Barberry
Fir, Concolor
Forsythia
Hemlock
Holly, Japanese
Ivy, English
Juniper, Andorra
Linden, Little Leaf
Locust, Honey
Maple, Norway, Silver
Oak, Pin, Red
Olive, Russian
Pachysandra
Pine, Austrian
Pine, White
Privet
Rosa Rugosa
Sumac, Staghorn
Yew, Japanese

Bonsai Planting
Azaleas
Birch, White
Ginkgo
Junipers
Pine, Bristlecone
Pine, Mugho
Spruce, Engelmann
Spruce, Dwarf Alberta

Street Planting
Linden, Little Leaf
Oak, Pin
Ginkgo

Fast Growth
Birch, White
Crownvetch
Dogwood, Silky

Fir, Douglas
Juniper, Blue Pfitzer
Juniper, Blue Rug
Maple, Silver
Olive, Autumn
Pines, Austrian, Ponderosa, Red Scotch and White
Poplar, Hybrid
Privet
Spruce, Norway
Spruce, Serbian
Texus, Cuspidata, Hicksi
Willow

Dense, Impenetrable Hedges
Field Plantings:
  Locust, Bristly,
  Olive, Autumn
  Sumac
Residential Area:
  Barberry, Red or Green
  Juniper, Blue Pfitzer
  Rosa Rugosa

Food for Birds
Ash, Mountain
Barberry
Bittersweet
Cherry, Manchu
Dogwood, Silky
Honeysuckle, Rem Red
Hawthorn
Oaks
Olive, Autumn, Russian
Privet
Rosa Rugosa
Sumac

Erosion Control
Crownvetch
Locust, Bristly
Willow

For customer support on your Site Work & Landscape Costs with RSMeans Data, call 800.448.8182.

881

R329343-30 Zones of Plant Hardiness

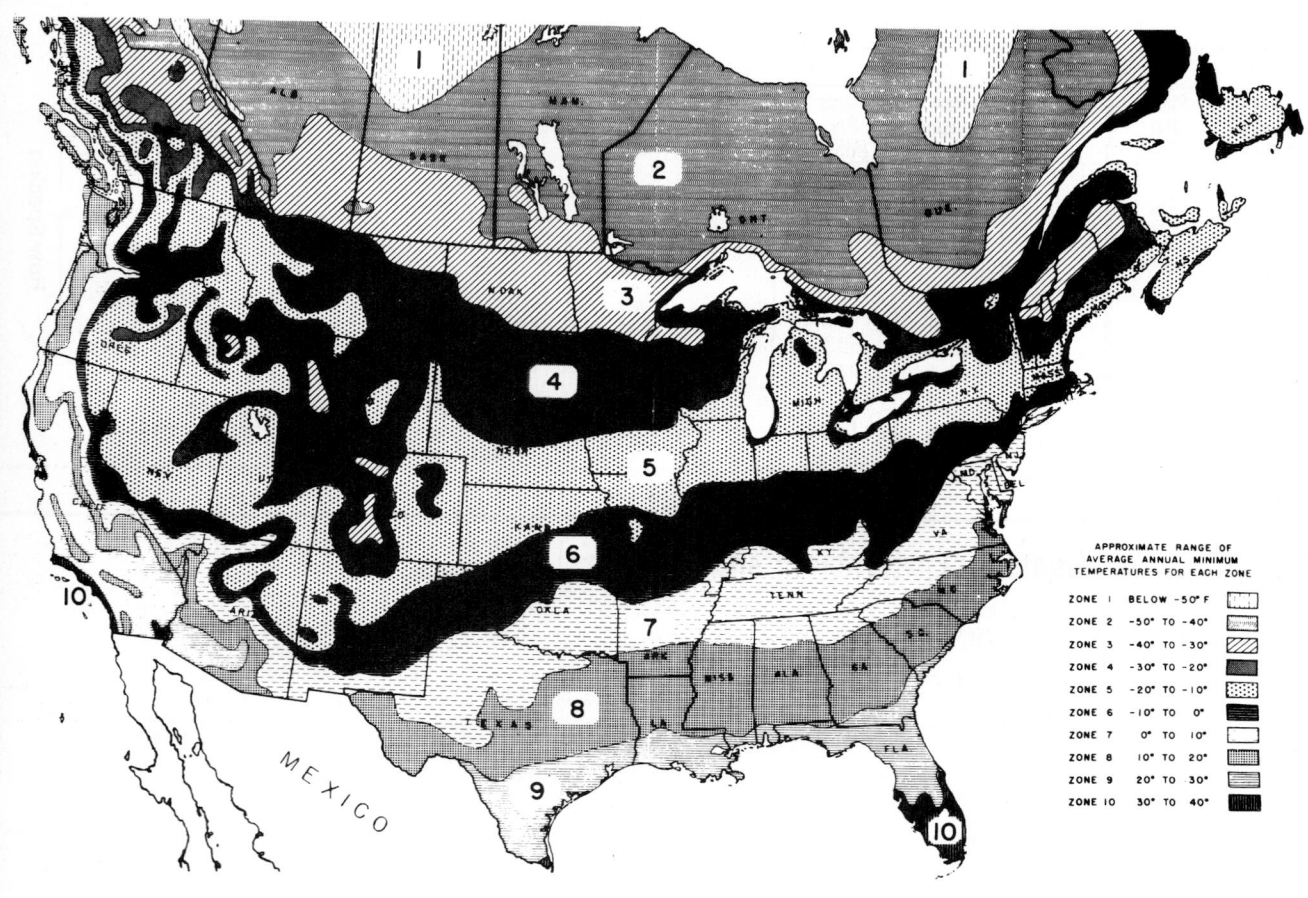

R331113-80 Piping Designations

There are several systems currently in use to describe pipe and fittings. The following paragraphs will help to identify and clarify classifications of piping systems used for water distribution.

Piping may be classified by schedule. Piping schedules include 5S, 10S, 10, 20, 30, Standard, 40, 60, Extra Strong, 80, 100, 120, 140, 160 and Double Extra Strong. These schedules are dependent upon the pipe wall thickness. The wall thickness of a particular schedule may vary with pipe size.

Ductile iron pipe for water distribution is classified by Pressure Classes such as Class 150, 200, 250, 300 and 350. These classes are actually the rated water working pressure of the pipe in pounds per square inch (psi). The pipe in these pressure classes is designed to withstand the rated water working pressure plus a surge allowance of 100 psi.

The American Water Works Association (AWWA) provides standards for various types of **plastic pipe**. C-900 is the specification for polyvinyl chloride (PVC) piping used for water distribution in sizes ranging from 4" through 12". C-901 is the specification for polyethylene (PE) pressure pipe, tubing and fittings used for water distribution in sizes ranging from 1/2" through 3". C-905 is the specification for PVC piping sizes 14" and greater.

PVC pressure-rated pipe is identified using the standard dimensional ratio (SDR) method. This method is defined by the American Society for Testing and Materials (ASTM) Standard D 2241. This pipe is available in SDR numbers 64, 41, 32.5, 26, 21, 17, and 13.5. A pipe with an SDR of 64 will have the thinnest wall while a pipe with an SDR of 13.5 will have the thickest wall. When the pressure rating (PR) of a pipe is given in psi, it is based on a line supplying water at 73 degrees F.

The National Sanitation Foundation (NSF) seal of approval is applied to products that can be used with potable water. These products have been tested to ANSI/NSF Standard 14.

Valves and strainers are classified by American National Standards Institute (ANSI) Classes. These Classes are 125, 150, 200, 250, 300, 400, 600, 900, 1500 and 2500. Within each class there is an operating pressure range dependent upon temperature. Design parameters should be compared to the appropriate material dependent, pressure-temperature rating chart for accurate valve selection.

Utilities

R3371 Elec. Utility Transmission & Distribution

R337119-30 Concrete for Conduit Encasement

The table below lists C.Y. of concrete for 100 L.F. of trench. Conduits separation center to center should meet 7.5" (N.E.C.).

Number of Conduits	1	2	3	4	6	8	9	Number of Conduits
Trench Dimension	11.5" x 11.5"	11.5" x 19"	11.5" x 27"	19" x 19"	19" x 27"	19" x 38"	27" x 27"	Trench Dimension
Conduit Diameter 2.0"	3.29	5.39	7.64	8.83	12.51	17.66	17.72	Conduit Diameter 2.0"
2.5"	3.23	5.29	7.49	8.62	12.19	17.23	17.25	2.5"
3.0"	3.15	5.13	7.24	8.29	11.71	16.59	16.52	3.0"
3.5"	3.08	4.97	7.02	7.99	11.26	15.98	15.84	3.5"
4.0"	2.99	4.80	6.76	7.65	10.74	15.30	15.07	4.0"
5.0"	2.78	4.37	6.11	6.78	9.44	13.57	13.12	5.0"
6.0"	2.52	3.84	5.33	5.74	7.87	11.48	10.77	6.0"

For customer support on your Site Work & Landscape Costs with RSMeans Data, call 800.448.8182.

883

R347216-10   Single Track R.R. Siding

The costs for a single track RR siding in the Unit Price section include the components shown in the table below.

Description of Component	Qty. per L.F. of Track	Unit
Ballast, 1-1/2" crushed stone	.667	C.Y.
6" x 8" x 8'-6" Treated timber ties, 22" O.C.	.545	Ea.
Tie plates, 2 per tie	1.091	Ea.
Track rail	2.000	L.F.
Spikes, 6", 4 per tie	2.182	Ea.
Splice bars w/ bolts, lock washers & nuts, @ 33' O.C.	.061	Pair
Crew B-14 @ 57 L.F./Day	.018	Day

R347216-20   Single Track, Steel Ties, Concrete Bed

The costs for a R.R. siding with steel ties and a concrete bed in the Unit Price section include the components shown in the table below.

Description of Component	Qty. per L.F. of Track	Unit
Concrete bed, 9' wide, 10" thick	.278	C.Y.
Ties, W6x16 x 6'-6" long, @ 30" O.C.	.400	Ea.
Tie plates, 4 per tie	1.600	Ea.
Track rail	2.000	L.F.
Tie plate bolts, 1", 8 per tie	3.200	Ea.
Splice bars w/bolts, lock washers & nuts, @ 33' O.C.	.061	Pair
Crew B-14 @ 22 L.F./Day	.045	Day

Change Orders

Change Order Considerations

A change order is a written document usually prepared by the design professional and signed by the owner, the architect/engineer, and the contractor. A change order states the agreement of the parties to: an addition, deletion, or revision in the work; an adjustment in the contract sum, if any; or an adjustment in the contract time, if any. Change orders, or "extras", in the construction process occur after execution of the construction contract and impact architects/ engineers, contractors, and owners.

Change orders that are properly recognized and managed can ensure orderly, professional, and profitable progress for everyone involved in the project. There are many causes for change orders and change order requests. In all cases, change orders or change order requests should be addressed promptly and in a precise and prescribed manner. The following paragraphs include information regarding change order pricing and procedures.

The Causes of Change Orders

Reasons for issuing change orders include:

- Unforeseen field conditions that require a change in the work
- Correction of design discrepancies, errors, or omissions in the contract documents
- Owner-requested changes, either by design criteria, scope of work, or project objectives
- Completion date changes for reasons unrelated to the construction process
- Changes in building code interpretations, or other public authority requirements that require a change in the work
- Changes in availability of existing or new materials and products

Procedures

Properly written contract documents must include the correct change order procedures for all parties—owners, design professionals, and contractors—to follow in order to avoid costly delays and litigation.

Being "in the right" is not always a sufficient or acceptable defense. The contract provisions requiring notification and documentation must be adhered to within a defined or reasonable time frame.

The appropriate method of handling change orders is by a written proposal and acceptance by all parties involved. Prior to starting work on a project, all parties should identify their

authorized agents who may sign and accept change orders, as well as any limits placed on their authority.

Time may be a critical factor when the need for a change arises. For such cases, the contractor might be directed to proceed on a "time and materials" basis, rather than wait for all paperwork to be processed—a delay that could impede progress. In this situation, the contractor must still follow the prescribed change order procedures including, but not limited to, notification and documentation.

Lack of documentation can be very costly, especially if legal judgments are to be made, and if certain field personnel are no longer available. For time and material change orders, the contractor should keep accurate daily records of all labor and material allocated to the change.

Owners or awarding authorities who do considerable and continual building construction (such as the federal government) realize the inevitability of change orders for numerous reasons, both predictable and unpredictable. As a result, the federal government, the American Institute of Architects (AIA), the Engineers Joint Contract Documents Committee (EJCDC), and other contractor, legal, and technical organizations have developed standards and procedures to be followed by all parties to achieve contract continuance and timely completion, while being financially fair to all concerned.

Pricing Change Orders

When pricing change orders, regardless of their cause, the most significant factor is when the change occurs. The need for a change may be perceived in the field or requested by the architect/engineer *before* any of the actual installation has begun, or may evolve or appear *during* construction when the item of work in question is partially installed. In the latter cases, the original sequence of construction is disrupted, along with all contiguous and supporting systems. Change orders cause the greatest impact when they occur *after* the installation has been completed and must be uncovered, or even replaced. Post-completion changes may be caused by necessary design changes, product failure, or changes in the owner's requirements that are not discovered until the building or the systems begin to function.

Specified procedures of notification and record keeping must be adhered to and enforced regardless of the stage of construction: *before*, *during*, or *after* installation. Some bidding documents anticipate change orders by requiring that unit prices including overhead and profit percentages—for additional as well as deductible changes—be listed. Generally these unit prices do not fully take into account the ripple effect, or impact on other trades, and should be used for general guidance only.

When pricing change orders, it is important to classify the time frame in which the change occurs. There are two basic time frames for change orders: *pre-installation change orders*, which occur before the start of construction, and *post-installation change orders*, which involve reworking after the original installation. Change orders that occur between these stages may be priced according to the extent of work completed using a combination of techniques developed for pricing *pre-* and *post-installation* changes.

Factors To Consider When Pricing Change Orders

As an estimator begins to prepare a change order, the following questions should be reviewed to determine their impact on the final price.

General

- *Is the change order work* pre-installation *or* post-installation?

  Change order work costs vary according to how much of the installation has been completed. Once workers have the project scoped in their minds, even though they have not started, it can be difficult to refocus. Consequently they may spend more than the normal amount of time understanding the change. Also, modifications to work in place, such as trimming or refitting, usually take more time than was initially estimated. The greater the amount of work in place, the more reluctant workers are to change it. Psychologically they may resent the change and as a result the rework takes longer than normal. Post-installation change order estimates must include demolition of existing work as required to accomplish the change. If the work is performed at a later time, additional obstacles, such as building finishes, may be present which must be protected. Regardless of whether the change occurs

For customer support on your Site Work & Landscape Costs with RSMeans Data, call 800.448.8182.

897

pre-installation or post-installation, attempt to isolate the identifiable factors and price them separately. For example, add shipping costs that may be required pre-installation or any demolition required post-installation. Then analyze the potential impact on productivity of psychological and/or learning curve factors and adjust the output rates accordingly. One approach is to break down the typical workday into segments and quantify the impact on each segment.

Change Order Installation Efficiency

The labor-hours expressed (for new construction) are based on average installation time, using an efficiency level. For change order situations, adjustments to this efficiency level should reflect the daily labor-hour allocation for that particular occurrence.

- *Will the change substantially delay the original completion date?*

A significant change in the project may cause the original completion date to be extended. The extended schedule may subject the contractor to new wage rates dictated by relevant labor contracts. Project supervision and other project overhead must also be extended beyond the original completion date. The schedule extension may also put installation into a new weather season. For example, underground piping scheduled for October installation was delayed until January. As a result, frost penetrated the trench area, thereby changing the degree of difficulty of the task. Changes and delays may have a ripple effect throughout the project. This effect must be analyzed and negotiated with the owner.

- *What is the net effect of a deduct change order?*

In most cases, change orders resulting in a deduction or credit reflect only bare costs. The contractor may retain the overhead and profit based on the original bid.

Materials

- *Will you have to pay more or less for the new material, required by the change order, than you paid for the original purchase?*

The same material prices or discounts will usually apply to materials purchased for change orders as new construction. In some instances, however, the contractor may forfeit the advantages of competitive pricing for change orders. Consider the following example:

A contractor purchased over $20,000 worth of fan coil units for an installation and obtained the maximum discount. Some time later it was determined the project required an additional matching unit. The contractor has to purchase this unit from the original supplier to ensure a match. The supplier at this time may not discount the unit because of the small quantity, and he is no longer in a competitive situation. The impact of quantity on purchase can add between 0% and 25% to material prices and/or subcontractor quotes.

- *If materials have been ordered or delivered to the job site, will they be subject to a cancellation charge or restocking fee?*

Check with the supplier to determine if ordered materials are subject to a cancellation charge. Delivered materials not used as a result of a change order may be subject to a restocking fee if returned to the supplier. Common restocking charges run between 20% and 40%. Also, delivery charges to return the goods to the supplier must be added.

Labor

- *How efficient is the existing crew at the actual installation?*

Is the same crew that performed the initial work going to do the change order? Possibly the change consists of the installation of a unit identical to one already installed; therefore, the change should take less time. Be sure to consider this potential productivity increase and modify the productivity rates accordingly.

- *If the crew size is increased, what impact will that have on supervision requirements?*

Under most bargaining agreements or management practices, there is a point at which a working foreman is replaced by a nonworking foreman. This replacement increases project overhead by adding a nonproductive worker. If additional workers are added to accelerate the project or to perform changes while maintaining the schedule, be sure to add additional supervision time if warranted. Calculate the hours involved and the additional cost directly if possible.

- *What are the other impacts of increased crew size?*

The larger the crew, the greater the potential for productivity to decrease. Some of the factors that cause this productivity loss are: overcrowding (producing restrictive conditions in the working space) and possibly a shortage of any special tools and equipment required. Such factors affect not only the crew working on the elements directly involved in the change order, but other crews whose movements may also be hampered. As the crew increases, check its basic composition for changes by the addition or deletion of apprentices or nonworking foreman, and quantify the potential effects of equipment shortages or other logistical factors.

- *As new crews, unfamiliar with the project, are brought onto the site, how long will it take them to become oriented to the project requirements?*

The orientation time for a new crew to become 100% effective varies with the site and type of project. Orientation is easiest at a new construction site and most difficult at existing, very restrictive renovation sites. The type of work also affects orientation time. When all elements of the work are exposed, such as concrete or masonry work, orientation is decreased. When the work is concealed or less visible, such as existing electrical systems, orientation takes longer. Usually orientation can be accomplished in one day or less. Costs for added orientation should be itemized and added to the total estimated cost.

- *How much actual production can be gained by working overtime?*

Short term overtime can be used effectively to accomplish more work in a day. However, as overtime is scheduled to run beyond several weeks, studies have shown marked decreases in output. The following chart shows the effect of long term overtime on worker efficiency. If the anticipated change requires extended overtime to keep the job on schedule, these factors can be used as a guide to predict the impact on time and cost. Add project overhead, particularly supervision, that may also be incurred.

For customer support on your Site Work & Landscape Costs with RSMeans Data, call 800.448.8182.

Days per Week	Hours per Day	Production Efficiency					Payroll Cost Factors	
		1st Week	2nd Week	3rd Week	4th Week	Average 4 Weeks	@ 1-1/2 Times	@ 2 Times
5	8	100%	100%	100%	100%	100%	100%	100%
	9	100	100	95	90	96	1.056	1.111
	10	100	95	90	85	93	1.100	1.200
	11	95	90	75	65	81	1.136	1.273
	12	90	85	70	60	76	1.167	1.333
6	8	100	100	95	90	96	1.083	1.167
	9	100	95	90	85	93	1.130	1.259
	10	95	90	85	80	88	1.167	1.333
	11	95	85	70	65	79	1.197	1.394
	12	90	80	65	60	74	1.222	1.444
7	8	100	95	85	75	89	1.143	1.286
	9	95	90	80	70	84	1.183	1.365
	10	90	85	75	65	79	1.214	1.429
	11	85	80	65	60	73	1.240	1.481
	12	85	75	60	55	69	1.262	1.524

Effects of Overtime

Caution: Under many labor agreements, Sundays and holidays are paid at a higher premium than the normal overtime rate.

The use of long-term overtime is counterproductive on almost any construction job; that is, the longer the period of overtime, the lower the actual production rate. Numerous studies have been conducted, and while they have resulted in slightly different numbers, all reach the same conclusion. The figure above tabulates the effects of overtime work on efficiency.

As illustrated, there can be a difference between the *actual* payroll cost per hour and the *effective* cost per hour for overtime work. This is due to the reduced production efficiency with the increase in weekly hours beyond 40. This difference between actual and effective cost results from overtime work over a prolonged period. Short-term overtime work does not result in as great a reduction in efficiency and, in such cases, effective cost may not vary significantly from the actual payroll cost. As the total hours per week are increased on a regular basis, more time is lost due to fatigue, lowered morale, and an increased accident rate.

As an example, assume a project where workers are working 6 days a week, 10 hours per day. From the figure above (based on productivity studies), the average effective productive hours over a 4-week period are:

$$0.875 \times 60 = 52.5$$

Depending upon the locale and day of week, overtime hours may be paid at time and a half or double time. For time and a half, the overall (average) *actual* payroll cost (including regular and overtime hours) is determined as follows:

$$\frac{40 \text{ reg. hrs.} + (20 \text{ overtime hrs.} \times 1.5)}{60 \text{ hrs.}} = 1.167$$

Based on 60 hours, the payroll cost per hour will be 116.7% of the normal rate at 40 hours per week. However, because the effective production (efficiency) for 60 hours is reduced to the equivalent of 52.5 hours, the effective cost of overtime is calculated as follows:

For time and a half:

$$\frac{40 \text{ reg. hrs.} + (20 \text{ overtime hrs.} \times 1.5)}{52.5 \text{ hrs.}} = 1.33$$

The installed cost will be 133% of the normal rate (for labor).

Thus, when figuring overtime, the actual cost per unit of work will be higher than the apparent overtime payroll dollar increase, due to the reduced productivity of the longer work week. These efficiency calculations are true only for those cost factors determined by hours worked. Costs that are applied weekly or monthly, such as equipment rentals, will not be similarly affected.

Equipment

- *What equipment is required to complete the change order?*

Change orders may require extending the rental period of equipment already on the job site, or the addition of special equipment brought in to accomplish the change work. In either case, the additional rental charges and operator labor charges must be added.

Summary

The preceding considerations and others you deem appropriate should be analyzed and applied to a change order estimate. The impact of each should be quantified and listed on the estimate to form an audit trail.

Change orders that are properly identified, documented, and managed help to ensure the orderly, professional, and profitable progress of the work. They also minimize potential claims or disputes at the end of the project.

Back by customer demand!

You asked and we listened. For customer convenience and estimating ease, we have made the 2021 Project Costs available for download at **RSMeans.com/2021books**. You will also find sample estimates, an RSMeans data overview video, and a book registration form to receive quarterly data updates throughout 2021.

Estimating Tips

- The cost figures available in the download were derived from hundreds of projects contained in the RSMeans database of completed construction projects. They include the contractor's overhead and profit. The figures have been adjusted to January of the current year.

- These projects were located throughout the U.S. and reflect a tremendous variation in square foot (S.F.) costs. This is due to differences, not only in labor and material costs, but also in individual owners' requirements. For instance, a bank in a large city would have different features than one in a rural area. This is true of all the different types of buildings analyzed. Therefore, caution should be exercised when using these Project Costs. For example, for courthouses, costs in the database are local courthouse costs and will not apply to the larger, more elaborate federal courthouses.

- None of the figures "go with" any others. All individual cost items were computed and tabulated separately. Thus, the sum of the median figures for plumbing, HVAC, and electrical will not normally total up to the total mechanical and electrical costs arrived at by separate analysis and tabulation of the projects.

- Each building was analyzed as to total and component costs and percentages. The figures were arranged in ascending order with the results tabulated as shown. The 1/4 column shows that 25% of the projects had lower costs and 75% had higher. The 3/4 column shows that 75% of the projects had lower costs and 25% had higher. The median column shows that 50% of the projects had lower costs and 50% had higher.

- Project Costs are useful in the conceptual stage when no details are available. As soon as details become available in the project design, the square foot approach should be discontinued and the project should be priced as to its particular components. When more precision is required, or for estimating the replacement cost of specific buildings, the current edition of *Square Foot Costs with RSMeans data* should be used.

- In using the figures in this section, it is recommended that the median column be used for preliminary figures if no additional information is available. The median figures, when multiplied by the total city construction cost index figures (see City Cost Indexes) and then multiplied by the project size modifier at the end of this section, should present a fairly accurate base figure, which would then have to be adjusted in view of the estimator's experience, local economic conditions, code requirements, and the owner's particular requirements. There is no need to factor in the percentage figures, as these should remain constant from city to city.

- The editors of this data would greatly appreciate receiving cost figures on one or more of your recent projects, which would then be included in the averages for next year. All cost figures received will be kept confidential, except that they will be averaged with other similar projects to arrive at square foot cost figures for next year.

See the website above for details and the discount available for submitting one or more of your projects.

Same Data. Simplified.

Enjoy the convenience and efficiency of accessing your costs anywhere:

- **Skip the multiplier** by setting your location
- **Quickly search,** edit, favorite and share costs
- **Stay on top of price changes** with automatic updates

Discover more at rsmeans.com/online

50 17 00	Project Costs		UNIT COSTS			% OF TOTAL			
		UNIT	1/4	MEDIAN	3/4	1/4	MEDIAN	3/4	
01 0000	**Auto Sales with Repair**	S.F.							**01**
0100	Architectural		107	120	130	58%	64%	67%	
0200	Plumbing		8.95	9.40	12.55	4.84%	5.20%	6.80%	
0300	Mechanical		12	16.10	17.75	6.40%	8.70%	10.15%	
0400	Electrical		18.45	23	28.50	9.05%	11.70%	15.90%	
0500	Total Project Costs		180	187	193				
02 0000	**Banking Institutions**	S.F.							**02**
0100	Architectural		161	198	241	59%	65%	69%	
0200	Plumbing		6.50	9.10	12.60	2.12%	3.39%	4.19%	
0300	Mechanical		12.95	17.90	21	4.41%	5.10%	10.75%	
0400	Electrical		31.50	38	58.50	10.45%	13.05%	15.90%	
0500	Total Project Costs		268	300	370				
03 0000	**Court House**	S.F.							**03**
0100	Architectural		85	167	167	54.50%	58.50%	58.50%	
0200	Plumbing		3.22	3.22	3.22	2.07%	2.07%	2.07%	
0300	Mechanical		20	20	20	12.95%	12.95%	12.95%	
0400	Electrical		26	26	26	16.60%	16.60%	16.60%	
0500	Total Project Costs		155	286	286				
04 0000	**Data Centers**	S.F.							**04**
0100	Architectural		192	192	192	68%	68%	68%	
0200	Plumbing		10.55	10.55	10.55	3.71%	3.71%	3.71%	
0300	Mechanical		27	27	27	9.45%	9.45%	9.45%	
0400	Electrical		25.50	25.50	25.50	9%	9%	9%	
0500	Total Project Costs		284	284	284				
05 0000	**Detention Centers**	S.F.							**05**
0100	Architectural		178	188	200	52%	53%	60.50%	
0200	Plumbing		18.80	23	27.50	5.15%	7.10%	7.25%	
0300	Mechanical		24	34	41	7.55%	9.50%	13.80%	
0400	Electrical		39	46.50	60.50	10.90%	14.85%	17.95%	
0500	Total Project Costs		300	320	375				
06 0000	**Fire Stations**	S.F.							**06**
0100	Architectural		101	130	190	46%	54.50%	61.50%	
0200	Plumbing		10.25	14.05	18.35	4.59%	5.60%	6.30%	
0300	Mechanical		15.10	22	29.50	6.05%	8.25%	10.20%	
0400	Electrical		23.50	30	40	10.75%	12.55%	14.95%	
0500	Total Project Costs		211	241	335				
07 0000	**Gymnasium**	S.F.							**07**
0100	Architectural		89	118	118	57%	64.50%	64.50%	
0200	Plumbing		2.19	7.15	7.15	1.58%	3.48%	3.48%	
0300	Mechanical		3.35	30	30	2.42%	14.65%	14.65%	
0400	Electrical		11	21.50	21.50	7.95%	10.35%	10.35%	
0500	Total Project Costs		139	206	206				
08 0000	**Hospitals**	S.F.							**08**
0100	Architectural		108	178	193	43%	47.50%	48%	
0200	Plumbing		7.95	15.15	33	6%	7.45%	7.65%	
0300	Mechanical		52.50	59	77	14.20%	17.95%	23.50%	
0400	Electrical		24	48	62	10.95%	13.75%	16.85%	
0500	Total Project Costs		253	375	405				
09 0000	**Industrial Buildings**	S.F.							**09**
0100	Architectural		59	116	235	52%	56%	82%	
0200	Plumbing		1.70	4.62	13.35	1.57%	2.11%	6.30%	
0300	Mechanical		4.86	9.25	44	4.77%	5.55%	14.80%	
0400	Electrical		7.40	8.45	70.50	7.85%	13.55%	16.20%	
0500	Total Project Costs		88	133	435				
10 0000	**Medical Clinics & Offices**	S.F.							**10**
0100	Architectural		90.50	122	163	48.50%	55.50%	62.50%	
0200	Plumbing		9	13.50	21.50	4.47%	6.65%	8.75%	
0300	Mechanical		14.60	22.50	41.50	7.80%	10.90%	16.05%	
0400	Electrical		19.90	24.50	38	8.90%	11.40%	14.05%	
0500	Total Project Costs		168	224	295				

		50 17 00	Project Costs	UNIT	UNIT COSTS			% OF TOTAL		
				1/4	MEDIAN	3/4	1/4	MEDIAN	3/4
11	0000	**Mixed Use**	S.F.							11
	0100	Architectural		92	130	198	45.50%	52.50%	61.50%
	0200	Plumbing		6.25	11.45	12.15	3.31%	3.47%	4.18%
	0300	Mechanical		15.25	25	46	4.68%	13.60%	17.05%
	0400	Electrical		16.15	36	53.50	8.30%	11.40%	15.65%
	0500	Total Project Costs		190	335	340			
12	0000	**Multi-Family Housing**	S.F.							12
	0100	Architectural		77.50	105	155	54.50%	61.50%	66.50%
	0200	Plumbing		6.90	11.10	15.10	5.30%	6.85%	8%
	0300	Mechanical		7.15	9.55	27.50	4.92%	6.90%	10.40%
	0400	Electrical		10	15.70	22.50	6.20%	8%	10.25%
	0500	Total Project Costs		128	210	253			
13	0000	**Nursing Home & Assisted Living**	S.F.							13
	0100	Architectural		72.50	94.50	119	51.50%	55.50%	63.50%
	0200	Plumbing		7.80	11.75	12.90	6.25%	7.40%	8.80%
	0300	Mechanical		6.40	9.45	18.50	4.04%	6.70%	9.55%
	0400	Electrical		10.60	16.70	23.50	7%	10.75%	13.10%
	0500	Total Project Costs		123	161	194			
14	0000	**Office Buildings**	S.F.							14
	0100	Architectural		93	130	179	54.50%	61%	69%
	0200	Plumbing		5.15	8.10	15.15	2.70%	3.78%	5.85%
	0300	Mechanical		10.10	17.15	26.50	5.60%	8.20%	11.10%
	0400	Electrical		12.90	22	34	7.75%	10%	12.70%
	0500	Total Project Costs		159	202	285			
15	0000	**Parking Garage**	S.F.							15
	0100	Architectural		32	39	40.50	70%	79%	88%
	0200	Plumbing		1.05	1.10	2.06	2.05%	2.70%	2.83%
	0300	Mechanical		.82	1.25	4.76	2.11%	3.62%	3.81%
	0400	Electrical		2.80	3.08	6.45	5.30%	6.35%	7.95%
	0500	Total Project Costs		39	47.50	51.50			
16	0000	**Parking Garage/Mixed Use**	S.F.							16
	0100	Architectural		103	113	115	61%	62%	65.50%
	0200	Plumbing		3.33	4.36	6.65	2.47%	2.72%	3.66%
	0300	Mechanical		14.20	16	23	7.80%	13.10%	13.60%
	0400	Electrical		14.90	21.50	22	8.20%	12.65%	18.15%
	0500	Total Project Costs		169	176	182			
17	0000	**Police Stations**	S.F.							17
	0100	Architectural		117	131	165	49%	56.50%	61%
	0200	Plumbing		15.45	18.55	18.65	5.05%	5.55%	9.05%
	0300	Mechanical		35	48.50	50.50	13%	14.55%	16.55%
	0400	Electrical		26.50	29	30.50	9.15%	12.10%	14%
	0500	Total Project Costs		219	270	305			
18	0000	**Police/Fire**	S.F.							18
	0100	Architectural		114	114	350	55.50%	66%	68%
	0200	Plumbing		9.15	9.45	35	5.45%	5.50%	5.55%
	0300	Mechanical		13.95	22	80	8.35%	12.70%	12.80%
	0400	Electrical		15.90	20.50	91.50	9.50%	11.75%	14.55%
	0500	Total Project Costs		167	173	630			
19	0000	**Public Assembly Buildings**	S.F.							19
	0100	Architectural		116	160	240	57.50%	61.50%	66%
	0200	Plumbing		6.15	9	13.50	2.60%	3.36%	4.79%
	0300	Mechanical		12.95	23	35.50	6.55%	8.95%	12.45%
	0400	Electrical		19.15	26	41.50	8.60%	10.75%	13%
	0500	Total Project Costs		186	255	375			
20	0000	**Recreational**	S.F.							20
	0100	Architectural		111	176	238	49.50%	59%	65.50%
	0200	Plumbing		8.60	14.85	25.50	3.12%	4.76%	7.40%
	0300	Mechanical		13.25	20	32	4.98%	6.60%	11.55%
	0400	Electrical		19.15	27.50	40	7.15%	8.95%	10.75%
	0500	Total Project Costs		198	296	445			

For customer support on your Site Work & Landscape Costs with RSMeans Data, call 800.448.8182.

89

		50 17 00	Project Costs	UNIT	UNIT COSTS			% OF TOTAL			
					1/4	MEDIAN	3/4	1/4	MEDIAN	3/4	
21	0000	**Restaurants**		S.F.							21
	0100	Architectural			127	195	250	59%	60%	63.50%	
	0200	Plumbing			13.90	32	40.50	7.35%	7.75%	8.95%	
	0300	Mechanical			15	17.70	37.50	6.50%	8.15%	11.15%	
	0400	Electrical			14.90	24.50	48.50	7.10%	10.30%	11.60%	
	0500	Total Project Costs			210	305	420				
22	0000	**Retail**		S.F.							22
	0100	Architectural			56.50	88.50	182	54.50%	60%	64.50%	
	0200	Plumbing			7.05	9.85	12.25	4.18%	5.45%	8.45%	
	0300	Mechanical			6.60	9.35	17.20	4.98%	6.15%	7.05%	
	0400	Electrical			10.50	21.50	31.50	7.90%	11.25%	12.45%	
	0500	Total Project Costs			86	153	292				
23	0000	**Schools**		S.F.							23
	0100	Architectural			98	127	173	51%	56%	60.50%	
	0200	Plumbing			7.75	10.70	15.65	3.66%	4.59%	7%	
	0300	Mechanical			18.70	26.50	38.50	8.95%	12%	14.55%	
	0400	Electrical			18.45	25.50	33	9.45%	11.25%	13.30%	
	0500	Total Project Costs			178	227	310				
24	0000	**University, College & Private School Classroom & Admin Buildings**		S.F.							24
	0100	Architectural			126	160	194	50.50%	55%	59.50%	
	0200	Plumbing			8.50	11.50	19.90	3.02%	4.46%	7.35%	
	0300	Mechanical			25	37.50	46.50	9.95%	11.95%	14.70%	
	0400	Electrical			20	28.50	35.50	7.70%	9.90%	11.55%	
	0500	Total Project Costs			219	293	380				
25	0000	**University, College & Private School Dormitories**		S.F.							25
	0100	Architectural			81.50	143	152	53%	61.50%	68%	
	0200	Plumbing			10.80	15.25	17.50	6.35%	6.65%	8.95%	
	0300	Mechanical			4.84	20.50	32.50	4.13%	9%	11.80%	
	0400	Electrical			5.75	19.95	30.50	4.75%	7.35%	10.60%	
	0500	Total Project Costs			121	229	271				
26	0000	**University, College & Private School Science, Eng. & Lab Buildings**		S.F.							26
	0100	Architectural			140	166	195	48.50%	56.50%	58%	
	0200	Plumbing			9.65	14.65	24.50	3.29%	3.77%	5%	
	0300	Mechanical			37	69	71	11.70%	19.40%	23.50%	
	0400	Electrical			29	38	39	9%	12.05%	13.15%	
	0500	Total Project Costs			293	315	370				
27	0000	**University, College & Private School Student Union Buildings**		S.F.							27
	0100	Architectural			111	292	292	54.50%	54.50%	59.50%	
	0200	Plumbing			16.80	16.80	25	3.13%	4.27%	11.45%	
	0300	Mechanical			32	51.50	51.50	9.60%	9.60%	14.55%	
	0400	Electrical			28	48.50	48.50	9.05%	12.80%	13.15%	
	0500	Total Project Costs			219	535	535				
28	0000	**Warehouses**		S.F.							28
	0100	Architectural			47.50	72.50	132	60.50%	67%	71.50%	
	0200	Plumbing			2.48	5.30	10.20	2.82%	3.72%	5%	
	0300	Mechanical			2.93	16.70	26	4.56%	8.15%	10.70%	
	0400	Electrical			6.15	20	33.50	7.75%	10.10%	18.30%	
	0500	Total Project Costs			71	113	228				

For customer support on your Site Work & Landscape Costs with RSMeans Data, call 800.448.8182.

Square Foot Project Size Modifier

One factor that affects the S.F. cost of a particular building is the size. In general, for buildings built to the same specifications in the same locality, the larger building will have the lower S.F. cost. This is due mainly to the decreasing contribution of the exterior walls plus the economy of scale usually achievable in larger buildings. The Area Conversion Scale shown below will give a factor to convert costs for the typical size building to an adjusted cost for the particular project.

The Square Foot Base Size lists the median costs, most typical project size in our accumulated data, and the range in size of the projects.

The Size Factor for your project is determined by dividing your project area in S.F. by the typical project size for the particular Building Type. With this factor, enter the Area Conversion Scale at the appropriate Size Factor and determine the appropriate Cost Multiplier for your building size.

Example: Determine the cost per S.F. for a 107,200 S.F. Multi-family housing.

$$\frac{\text{Proposed building area} = 107{,}200 \text{ S.F.}}{\text{Typical size from below} = 53{,}600 \text{ S.F.}} = 2.00$$

Enter Area Conversion Scale at 2.0, intersect curve, read horizontally the appropriate cost multiplier of .94. Size adjusted cost becomes .94 x $210.00 = $197.40 based on national average costs.

Note: For Size Factors less than .50, the Cost Multiplier is 1.1
For Size Factors greater than 3.5, the Cost Multiplier is .90

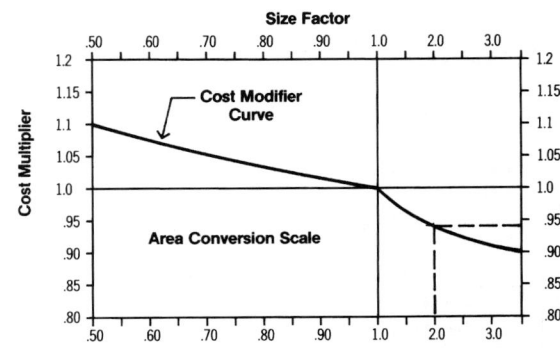

System	Median Cost (Total Project Costs)	Typical Size Gross S.F. (Median of Projects)	Typical Range (Low – High) (Projects)
Auto Sales with Repair	$187.00	24,900	4,680 – 29,253
Banking Institutions	300.00	9,300	3,267 – 38,148
Court House	286.00	47,600	24,680 – 70,500
Data Centers	284.00	14,400	14,369 – 14,369
Detention Centers	320.00	37,800	12,257 – 183,339
Fire Stations	241.00	12,500	6,300 – 49,577
Gymnasium	206.00	52,400	22,844 – 81,992
Hospitals	375.00	87,100	22,428 – 410,273
Industrial Buildings	133.00	21,100	5,146 – 200,625
Medical Clinics & Offices	224.00	22,500	2,273 – 327,000
Mixed Use	335.00	28,500	7,200 – 188,944
Multi-Family Housing	210.00	53,600	2,472 – 1,161,450
Nursing Home & Assisted Living	161.00	38,200	1,515 – 242,555
Office Buildings	202.00	20,600	1,115 – 930,000
Parking Garage	47.5	151,800	99,884 – 287,040
Parking Garage/Mixed Use	176.00	254,200	5,302 – 318,033
Police Stations	270.00	28,500	15,377 – 88,625
Police/Fire	173.00	44,300	8,600 – 50,279
Public Assembly Buildings	255.00	21,000	2,232 – 235,301
Recreational	296.00	27,700	1,000 – 223,787
Restaurants	305.00	6,000	3,975 – 42,000
Retail	153.00	26,000	4,035 – 84,270
Schools	227.00	69,900	1,344 – 410,848
University, College & Private School Classroom & Admin Buildings	293.00	48,200	8,806 – 196,187
University, College & Private School Dormitories	229.00	39,600	1,500 – 126,889
University, College & Private School Science, Eng. & Lab Buildings	315.00	60,000	5,311 – 117,643
University, College & Private School Student Union Buildings	535.00	47,300	42,075 – 50,000
Warehouses	113.00	11,100	640 – 303,750

For customer support on your Site Work & Landscape Costs with RSMeans Data, call 800.448.8182.

893

A	Area Square Feet; Ampere
AAFES	Army and Air Force Exchange Service
ABS	Acrylonitrile Butadiene Stryrene; Asbestos Bonded Steel
A.C., AC	Alternating Current; Air-Conditioning; Asbestos Cement; Plywood Grade A & C
ACI	American Concrete Institute
ACR	Air Conditioning Refrigeration
ADA	Americans with Disabilities Act
AD	Plywood, Grade A & D
Addit.	Additional
Adh.	Adhesive
Adj.	Adjustable
af	Audio-frequency
AFFF	Aqueous Film Forming Foam
AFUE	Annual Fuel Utilization Efficiency
AGA	American Gas Association
Agg.	Aggregate
A.H., Ah	Ampere Hours
A hr.	Ampere-hour
A.H.U., AHU	Air Handling Unit
A.I.A.	American Institute of Architects
AIC	Ampere Interrupting Capacity
Allow.	Allowance
alt., alt	Alternate
Alum.	Aluminum
a.m.	Ante Meridiem
Amp.	Ampere
Anod.	Anodized
ANSI	American National Standards Institute
APA	American Plywood Association
Approx.	Approximate
Apt.	Apartment
Asb.	Asbestos
A.S.B.C.	American Standard Building Code
Asbe.	Asbestos Worker
ASCE	American Society of Civil Engineers
A.S.H.R.A.E.	American Society of Heating, Refrig. & AC Engineers
ASME	American Society of Mechanical Engineers
ASTM	American Society for Testing and Materials
Attchmt.	Attachment
Avg., Ave.	Average
AWG	American Wire Gauge
AWWA	American Water Works Assoc.
Bbl.	Barrel
B&B, BB	Grade B and Better; Balled & Burlapped
B&S	Bell and Spigot
B.&W.	Black and White
b.c.c.	Body-centered Cubic
B.C.Y.	Bank Cubic Yards
BE	Bevel End
B.F.	Board Feet
Bg. cem.	Bag of Cement
BHP	Boiler Horsepower; Brake Horsepower
B.I.	Black Iron
bidir.	bidirectional
Bit., Bitum.	Bituminous
Bit., Conc.	Bituminous Concrete
Bk.	Backed
Bkrs.	Breakers
Bldg., bldg	Building
Blk.	Block
Bm.	Beam
Boil.	Boilermaker
bpm	Blows per Minute
BR	Bedroom
Brg., brng.	Bearing
Brhe.	Bricklayer Helper
Bric.	Bricklayer

Brk., brk	Brick
brkt	Bracket
Brs.	Brass
Brz.	Bronze
Bsn.	Basin
Btr.	Better
BTU	British Thermal Unit
BTUH	BTU per Hour
Bu.	Bushels
BUR	Built-up Roofing
BX	Interlocked Armored Cable
°C	Degree Centigrade
c	Conductivity, Copper Sweat
C	Hundred; Centigrade
C/C	Center to Center, Cedar on Cedar
C-C	Center to Center
Cab	Cabinet
Cair.	Air Tool Laborer
Cal.	Caliper
Calc	Calculated
Cap.	Capacity
Carp.	Carpenter
C.B.	Circuit Breaker
C.C.A.	Chromate Copper Arsenate
C.C.F.	Hundred Cubic Feet
cd	Candela
cd/sf	Candela per Square Foot
CD	Grade of Plywood Face & Back
CDX	Plywood, Grade C & D, exterior glue
Cefi.	Cement Finisher
Cem.	Cement
CF	Hundred Feet
C.F.	Cubic Feet
CFM	Cubic Feet per Minute
CFRP	Carbon Fiber Reinforced Plastic
c.g.	Center of Gravity
CHW	Chilled Water; Commercial Hot Water
C.I., CI	Cast Iron
C.I.P., CIP	Cast in Place
Circ.	Circuit
C.L.	Carload Lot
CL	Chain Link
Clab.	Common Laborer
Clam	Common Maintenance Laborer
C.L.F.	Hundred Linear Feet
CLF	Current Limiting Fuse
CLP	Cross Linked Polyethylene
cm	Centimeter
CMP	Corr. Metal Pipe
CMU	Concrete Masonry Unit
CN	Change Notice
Col.	Column
CO_2	Carbon Dioxide
Comb.	Combination
comm.	Commercial, Communication
Compr.	Compressor
Conc.	Concrete
Cont., cont	Continuous; Continued, Container
Corkbd.	Cork Board
Corr.	Corrugated
Cos	Cosine
Cot	Cotangent
Cov.	Cover
C/P	Cedar on Paneling
CPA	Control Point Adjustment
Cplg.	Coupling
CPM	Critical Path Method
CPVC	Chlorinated Polyvinyl Chloride
C.Pr.	Hundred Pair
CRC	Cold Rolled Channel
Creos.	Creosote
Crpt.	Carpet & Linoleum Layer
CRT	Cathode-ray Tube
CS	Carbon Steel, Constant Shear Bar Joist

Csc	Cosecant
C.S.F.	Hundred Square Feet
CSI	Construction Specifications Institute
CT	Current Transformer
CTS	Copper Tube Size
Cu	Copper, Cubic
Cu. Ft.	Cubic Foot
cw	Continuous Wave
C.W.	Cool White; Cold Water
Cwt.	100 Pounds
C.W.X.	Cool White Deluxe
C.Y.	Cubic Yard (27 cubic feet)
C.Y./Hr.	Cubic Yard per Hour
Cyl.	Cylinder
d	Penny (nail size)
D	Deep; Depth; Discharge
Dis., Disch.	Discharge
Db	Decibel
Dbl.	Double
DC	Direct Current
DDC	Direct Digital Control
Demob.	Demobilization
d.f.t.	Dry Film Thickness
d.f.u.	Drainage Fixture Units
D.H.	Double Hung
DHW	Domestic Hot Water
DI	Ductile Iron
Diag.	Diagonal
Diam., Dia	Diameter
Distrib.	Distribution
Div.	Division
Dk.	Deck
D.L.	Dead Load; Diesel
DLH	Deep Long Span Bar Joist
dlx	Deluxe
Do.	Ditto
DOP	Dioctyl Phthalate Penetration Test (Air Filters)
Dp., dp	Depth
D.P.S.T.	Double Pole, Single Throw
Dr.	Drive
DR	Dimension Ratio
Drink.	Drinking
D.S.	Double Strength
D.S.A.	Double Strength A Grade
D.S.B.	Double Strength B Grade
Dty.	Duty
DWV	Drain Waste Vent
DX	Deluxe White, Direct Expansion
dyn	Dyne
e	Eccentricity
E	Equipment Only; East; Emissivity
Ea.	Each
EB	Encased Burial
Econ.	Economy
E.C.Y	Embankment Cubic Yards
EDP	Electronic Data Processing
EIFS	Exterior Insulation Finish System
E.D.R.	Equiv. Direct Radiation
Eq.	Equation
EL	Elevation
Elec.	Electrician; Electrical
Elev.	Elevator; Elevating
EMT	Electrical Metallic Conduit; Thin Wall Conduit
Eng.	Engine, Engineered
EPDM	Ethylene Propylene Diene Monomer
EPS	Expanded Polystyrene
Eqhv.	Equip. Oper., Heavy
Eqlt.	Equip. Oper., Light
Eqmd.	Equip. Oper., Medium
Eqmm.	Equip. Oper., Master Mechanic
Eqol.	Equip. Oper., Oilers
Equip.	Equipment
ERW	Electric Resistance Welded

E.S.	Energy Saver	H	High Henry	Lath.	Lather
Est.	Estimated	HC	High Capacity	Lav.	Lavatory
esu	Electrostatic Units	H.D., HD	Heavy Duty; High Density	lb.; #	Pound
E.W.	Each Way	H.D.O.	High Density Overlaid	L.B., LB	Load Bearing; L Conduit Body
EWT	Entering Water Temperature	HDPE	High Density Polyethylene Plastic	L. & E.	Labor & Equipment
Excav.	Excavation	Hdr.	Header	lb./hr.	Pounds per Hour
excl	Excluding	Hdwe.	Hardware	lb./L.F.	Pounds per Linear Foot
Exp., exp	Expansion, Exposure	H.I.D., HID	High Intensity Discharge	lbf/sq.in.	Pound-force per Square Inch
Ext., ext	Exterior; Extension	Help.	Helper Average	L.C.L.	Less than Carload Lot
Extru.	Extrusion	HEPA	High Efficiency Particulate Air	L.C.Y.	Loose Cubic Yard
f.	Fiber Stress		Filter	Ld.	Load
F	Fahrenheit; Female; Fill	Hg	Mercury	LE	Lead Equivalent
Fab., fab	Fabricated; Fabric	HIC	High Interrupting Capacity	LED	Light Emitting Diode
FBGS	Fiberglass	HM	Hollow Metal	L.F.	Linear Foot
F.C.	Footcandles	HMWPE	High Molecular Weight	L.F. Hdr	Linear Feet of Header
f.c.c.	Face-centered Cubic		Polyethylene	L.F. Nose	Linear Foot of Stair Nosing
f'c.	Compressive Stress in Concrete;	HO	High Output	L.F. Rsr	Linear Foot of Stair Riser
	Extreme Compressive Stress	Horiz.	Horizontal	Lg.	Long; Length; Large
F.E.	Front End	H.P., HP	Horsepower; High Pressure	L & H	Light and Heat
FEP	Fluorinated Ethylene Propylene	H.P.F.	High Power Factor	LH	Long Span Bar Joist
	(Teflon)	Hr.	Hour	L.H.	Labor Hours
F.G.	Flat Grain	Hrs./Day	Hours per Day	L.L., LL	Live Load
F.H.A.	Federal Housing Administration	HSC	High Short Circuit	L.L.D.	Lamp Lumen Depreciation
Fig.	Figure	Ht.	Height	lm	Lumen
Fin.	Finished	Htg.	Heating	lm/sf	Lumen per Square Foot
FIPS	Female Iron Pipe Size	Htrs.	Heaters	lm/W	Lumen per Watt
Fixt.	Fixture	HVAC	Heating, Ventilation & Air-	LOA	Length Over All
FJP	Finger jointed and primed		Conditioning	log	Logarithm
Fl. Oz.	Fluid Ounces	Hvy.	Heavy	L-O-L	Lateralolet
Flr.	Floor	HW	Hot Water	long.	Longitude
Flrs.	Floors	Hyd.; Hydr.	Hydraulic	L.P., LP	Liquefied Petroleum; Low Pressure
FM	Frequency Modulation;	Hz	Hertz (cycles)	L.P.F.	Low Power Factor
	Factory Mutual	I.	Moment of Inertia	LR	Long Radius
Fmg.	Framing	IBC	International Building Code	L.S.	Lump Sum
FM/UL	Factory Mutual/Underwriters Labs	I.C.	Interrupting Capacity	Lt.	Light
Fdn.	Foundation	ID	Inside Diameter	Lt. Ga.	Light Gauge
FNPT	Female National Pipe Thread	I.D.	Inside Dimension; Identification	L.T.L.	Less than Truckload Lot
Fori.	Foreman, Inside	I.F.	Inside Frosted	Lt. Wt.	Lightweight
Foro.	Foreman, Outside	I.M.C.	Intermediate Metal Conduit	L.V.	Low Voltage
Fount.	Fountain	In.	Inch	M	Thousand; Material; Male;
fpm	Feet per Minute	Incan.	Incandescent		Light Wall Copper Tubing
FPT	Female Pipe Thread	Incl.	Included; Including	M²CA	Meters Squared Contact Area
Fr	Frame	Int.	Interior	m/hr.; M.H.	Man-hour
F.R.	Fire Rating	Inst.	Installation	mA	Milliampere
FRK	Foil Reinforced Kraft	Insul., insul	Insulation/Insulated	Mach.	Machine
FSK	Foil/Scrim/Kraft	I.P.	Iron Pipe	Mag. Str.	Magnetic Starter
FRP	Fiberglass Reinforced Plastic	I.P.S., IPS	Iron Pipe Size	Maint.	Maintenance
FS	Forged Steel	IPT	Iron Pipe Threaded	Marb.	Marble Setter
FSC	Cast Body; Cast Switch Box	I.W.	Indirect Waste	Mat; Mat'l.	Material
Ft., ft	Foot; Feet	J	Joule	Max.	Maximum
Ftng.	Fitting	J.I.C.	Joint Industrial Council	MBF	Thousand Board Feet
Ftg.	Footing	K	Thousand; Thousand Pounds;	MBH	Thousand BTU's per hr.
Ft lb.	Foot Pound		Heavy Wall Copper Tubing, Kelvin	MC	Metal Clad Cable
Furn.	Furniture	K.A.H.	Thousand Amp. Hours	MCC	Motor Control Center
FVNR	Full Voltage Non-Reversing	kcmil	Thousand Circular Mils	M.C.F.	Thousand Cubic Feet
FVR	Full Voltage Reversing	KD	Knock Down	MCFM	Thousand Cubic Feet per Minute
FXM	Female by Male	K.D.A.T.	Kiln Dried After Treatment	M.C.M.	Thousand Circular Mils
Fy.	Minimum Yield Stress of Steel	kg	Kilogram	MCP	Motor Circuit Protector
g	Gram	kG	Kilogauss	MD	Medium Duty
G	Gauss	kgf	Kilogram Force	MDF	Medium-density fibreboard
Ga.	Gauge	kHz	Kilohertz	M.D.O.	Medium Density Overlaid
Gal., gal.	Gallon	Kip	1000 Pounds	Med.	Medium
Galv., galv	Galvanized	KJ	Kilojoule	MF	Thousand Feet
GC/MS	Gas Chromatograph/Mass	K.L.	Effective Length Factor	M.F.B.M.	Thousand Feet Board Measure
	Spectrometer	K.L.F.	Kips per Linear Foot	Mfg.	Manufacturing
Gen.	General	Km	Kilometer	Mfrs.	Manufacturers
GFI	Ground Fault Interrupter	KO	Knock Out	mg	Milligram
GFRC	Glass Fiber Reinforced Concrete	K.S.F.	Kips per Square Foot	MGD	Million Gallons per Day
Glaz.	Glazier	K.S.I.	Kips per Square Inch	MGPH	Million Gallons per Hour
GPD	Gallons per Day	kV	Kilovolt	MH, M.H.	Manhole; Metal Halide; Man-Hour
gpf	Gallon per Flush	kVA	Kilovolt Ampere	MHz	Megahertz
GPH	Gallons per Hour	kVAR	Kilovar (Reactance)	Mi.	Mile
gpm, GPM	Gallons per Minute	KW	Kilowatt	MI	Malleable Iron; Mineral Insulated
GR	Grade	KWh	Kilowatt-hour	MIPS	Male Iron Pipe Size
Gran.	Granular	L	Labor Only; Length; Long;	mj	Mechanical Joint
Grnd.	Ground		Medium Wall Copper Tubing	m	Meter
GVW	Gross Vehicle Weight	Lab.	Labor	mm	Millimeter
GWB	Gypsum Wall Board	lat	Latitude	Mill.	Millwright
				Min., min.	Minimum, Minute

Abbreviations

Misc.	Miscellaneous	PCM	Phase Contrast Microscopy	SBS	Styrene Butadiere Styrene
ml	Milliliter, Mainline	PDCA	Painting and Decorating	SC	Screw Cover
M.L.F.	Thousand Linear Feet		Contractors of America	SCFM	Standard Cubic Feet per Minute
Mo.	Month	P.E., PE	Professional Engineer;	Scaf.	Scaffold
Mobil.	Mobilization		Porcelain Enamel;	Sch., Sched.	Schedule
Mog.	Mogul Base		Polyethylene; Plain End	S.C.R.	Modular Brick
MPH	Miles per Hour	P.E.C.I.	Porcelain Enamel on Cast Iron	S.D.	Sound Deadening
MPT	Male Pipe Thread	Perf.	Perforated	SDR	Standard Dimension Ratio
MRGWB	Moisture Resistant Gypsum	PEX	Cross Linked Polyethylene	S.E.	Surfaced Edge
	Wallboard	Ph.	Phase	Sel.	Select
MRT	Mile Round Trip	P.I.	Pressure Injected	SER, SEU	Service Entrance Cable
ms	Millisecond	Pile.	Pile Driver	S.F.	Square Foot
M.S.F.	Thousand Square Feet	Pkg.	Package	S.F.C.A.	Square Foot Contact Area
Mstz.	Mosaic & Terrazzo Worker	Pl.	Plate	S.F. Flr.	Square Foot of Floor
M.S.Y.	Thousand Square Yards	Plah.	Plasterer Helper	S.F.G.	Square Foot of Ground
Mtd., mtd., mtd	Mounted	Plas.	Plasterer	S.F. Hor.	Square Foot Horizontal
Mthe.	Mosaic & Terrazzo Helper	plf	Pounds Per Linear Foot	SFR	Square Feet of Radiation
Mtng.	Mounting	Pluh.	Plumber Helper	S.F. Shlf.	Square Foot of Shelf
Mult.	Multi; Multiply	Plum.	Plumber	S4S	Surface 4 Sides
MUTCD	Manual on Uniform Traffic Control	Ply.	Plywood	Shee.	Sheet Metal Worker
	Devices	p.m.	Post Meridiem	Sin.	Sine
M.V.A.	Million Volt Amperes	Pntd.	Painted	Skwk.	Skilled Worker
M.V.A.R.	Million Volt Amperes Reactance	Pord.	Painter, Ordinary	SL	Saran Lined
MV	Megavolt	pp	Pages	S.L.	Slimline
MW	Megawatt	PP, PPL	Polypropylene	Sldr.	Solder
MXM	Male by Male	P.P.M.	Parts per Million	SLH	Super Long Span Bar Joist
MYD	Thousand Yards	Pr.	Pair	S.N.	Solid Neutral
N	Natural; North	P.E.S.B.	Pre-engineered Steel Building	SO	Stranded with oil resistant inside
nA	Nanoampere	Prefab.	Prefabricated		insulation
NA	Not Available; Not Applicable	Prefin.	Prefinished	S-O-L	Socketolet
N.B.C.	National Building Code	Prop.	Propelled	sp	Standpipe
NC	Normally Closed	PSF, psf	Pounds per Square Foot	S.P.	Static Pressure; Single Pole; Self-
NEMA	National Electrical Manufacturers	PSI, psi	Pounds per Square Inch		Propelled
	Assoc.	PSIG	Pounds per Square Inch Gauge	Spri.	Sprinkler Installer
NEHB	Bolted Circuit Breaker to 600V.	PSP	Plastic Sewer Pipe	spwg	Static Pressure Water Gauge
NFPA	National Fire Protection Association	Pspr.	Painter, Spray	S.P.D.T.	Single Pole, Double Throw
NLB	Non-Load-Bearing	Psst.	Painter, Structural Steel	SPF	Spruce Pine Fir; Sprayed
NM	Non-Metallic Cable	P.T.	Potential Transformer		Polyurethane Foam
nm	Nanometer	P. & T.	Pressure & Temperature	S.P.S.T.	Single Pole, Single Throw
No.	Number	Ptd.	Painted	SPT	Standard Pipe Thread
NO	Normally Open	Ptns.	Partitions	Sq.	Square; 100 Square Feet
N.O.C.	Not Otherwise Classified	Pu	Ultimate Load	Sq. Hd.	Square Head
Nose.	Nosing	PVC	Polyvinyl Chloride	Sq. In.	Square Inch
NPT	National Pipe Thread	Pvmt.	Pavement	S.S.	Single Strength; Stainless Steel
NQOD	Combination Plug-on/Bolt on	PRV	Pressure Relief Valve	S.S.B.	Single Strength B Grade
	Circuit Breaker to 240V.	Pwr.	Power	sst, ss	Stainless Steel
N.R.C., NRC	Noise Reduction Coefficient/	Q	Quantity Heat Flow	Sswk.	Structural Steel Worker
	Nuclear Regulator Commission	Qt.	Quart	Sswl.	Structural Steel Welder
N.R.S.	Non Rising Stem	Quan., Qty.	Quantity	St.; Stl.	Steel
ns	Nanosecond	Q.C.	Quick Coupling	STC	Sound Transmission Coefficient
NTP	Notice to Proceed	r	Radius of Gyration	Std.	Standard
nW	Nanowatt	R	Resistance	Stg.	Staging
OB	Opposing Blade	R.C.P.	Reinforced Concrete Pipe	STK	Select Tight Knot
OC	On Center	Rect.	Rectangle	STP	Standard Temperature & Pressure
OD	Outside Diameter	recpt.	Receptacle	Stpi.	Steamfitter, Pipefitter
O.D.	Outside Dimension	Reg.	Regular	Str.	Strength; Starter; Straight
ODS	Overhead Distribution System	Reinf.	Reinforced	Strd.	Stranded
O.G.	Ogee	Req'd.	Required	Struct.	Structural
O.H.	Overhead	Res.	Resistant	Sty.	Story
O&P	Overhead and Profit	Resi.	Residential	Subj.	Subject
Oper.	Operator	RF	Radio Frequency	Subs.	Subcontractors
Opng.	Opening	RFID	Radio-frequency Identification	Surf.	Surface
Orna.	Ornamental	Rgh.	Rough	Sw.	Switch
OSB	Oriented Strand Board	RGS	Rigid Galvanized Steel	Swbd.	Switchboard
OS&Y	Outside Screw and Yoke	RHW	Rubber, Heat & Water Resistant;	S.Y.	Square Yard
OSHA	Occupational Safety and Health		Residential Hot Water	Syn.	Synthetic
	Act	rms	Root Mean Square	S.Y.P.	Southern Yellow Pine
Ovhd.	Overhead	Rnd.	Round	Sys.	System
OWG	Oil, Water or Gas	Rodm.	Rodman	t.	Thickness
Oz.	Ounce	Rofc.	Roofer, Composition	T	Temperature; Ton
P.	Pole; Applied Load; Projection	Rofp.	Roofer, Precast	Tan	Tangent
p.	Page	Rohe.	Roofer Helpers (Composition)	T.C.	Terra Cotta
Pape.	Paperhanger	Rots.	Roofer, Tile & Slate	T & C	Threaded and Coupled
P.A.P.R.	Powered Air Purifying Respirator	R.O.W.	Right of Way	T.D.	Temperature Difference
PAR	Parabolic Reflector	RPM	Revolutions per Minute	TDD	Telecommunications Device for
P.B., PB	Push Button	R.S.	Rapid Start		the Deaf
Pc., Pcs.	Piece, Pieces	Rsr	Riser	T.E.M.	Transmission Electron Microscopy
P.C.	Portland Cement; Power Connector	RT	Round Trip	temp	Temperature, Tempered, Temporary
P.C.F.	Pounds per Cubic Foot	S.	Suction; Single Entrance; South	TFFN	Nylon Jacketed Wire

896

Abbreviations

TFE	Tetrafluoroethylene (Teflon)	U.L., UL	Underwriters Laboratory	w/	With
T. & G.	Tongue & Groove;	Uld.	Unloading	W.C., WC	Water Column; Water Closet
	Tar & Gravel	Unfin.	Unfinished	W.F.	Wide Flange
Th., Thk.	Thick	UPS	Uninterruptible Power Supply	W.G.	Water Gauge
Thn.	Thin	URD	Underground Residential	Wldg.	Welding
Thrded	Threaded		Distribution	W. Mile	Wire Mile
Tilf.	Tile Layer, Floor	US	United States	W-O-L	Weldolet
Tilh.	Tile Layer, Helper	USGBC	U.S. Green Building Council	W.R.	Water Resistant
THHN	Nylon Jacketed Wire	USP	United States Primed	Wrck.	Wrecker
THW.	Insulated Strand Wire	UTMCD	Uniform Traffic Manual For Control	WSFU	Water Supply Fixture Unit
THWN	Nylon Jacketed Wire		Devices	W.S.P.	Water, Steam, Petroleum
T.L., TL	Truckload	UTP	Unshielded Twisted Pair	WT., Wt.	Weight
T.M.	Track Mounted	V	Volt	WWF	Welded Wire Fabric
Tot.	Total	VA	Volt Amperes	XFER	Transfer
T-O-L	Threadolet	VAT	Vinyl Asbestos Tile	XFMR	Transformer
tmpd	Tempered	V.C.T.	Vinyl Composition Tile	XHD	Extra Heavy Duty
TPO	Thermoplastic Polyolefin	VAV	Variable Air Volume	XHHW	Cross-Linked Polyethylene Wire
T.S.	Trigger Start	VC	Veneer Core	XLPE	Insulation
Tr.	Trade	VDC	Volts Direct Current	XLP	Cross-linked Polyethylene
Transf.	Transformer	Vent.	Ventilation	Xport	Transport
Trhv.	Truck Driver, Heavy	Vert.	Vertical	Y	Wye
Trlr	Trailer	V.F.	Vinyl Faced	yd	Yard
Trlt.	Truck Driver, Light	V.G.	Vertical Grain	yr	Year
TTY	Teletypewriter	VHF	Very High Frequency	Δ	Delta
TV	Television	VHO	Very High Output	%	Percent
T.W.	Thermoplastic Water Resistant	Vib.	Vibrating	~	Approximately
	Wire	VLF	Vertical Linear Foot	Ø	Phase; diameter
UCI	Uniform Construction Index	VOC	Volatile Organic Compound	@	At
UF	Underground Feeder	Vol.	Volume	#	Pound; Number
UGND	Underground Feeder	VRP	Vinyl Reinforced Polyester	<	Less Than
UHF	Ultra High Frequency	W	Wire; Watt; Wide; West	>	Greater Than
U.I.	United Inch			Z	Zone

For customer support on your Site Work & Landscape Costs with RSMeans Data, call 800.448.8182.

Index

Index

For customer support on your Site Work & Landscape Costs with RSMeans Data, call 800.448.8182.

Index

For customer support on your Site Work & Landscape Costs with RSMeans Data, call 800.448.8182.

Index

Index

For customer support on your Site Work & Landscape Costs with RSMeans Data, call 800.448.8182.

Index

S

Index

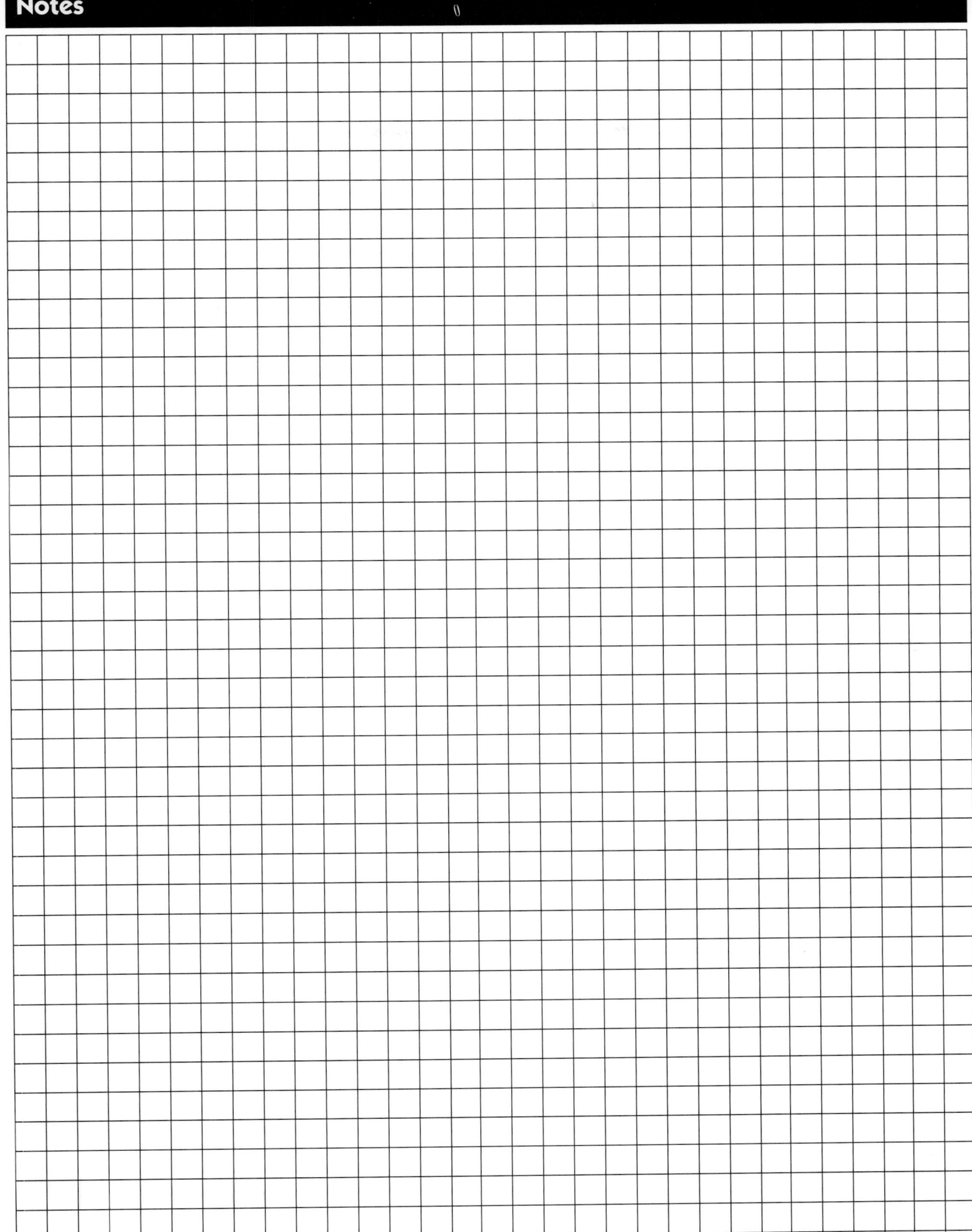

Division Notes

	CREW	DAILY OUTPUT	LABOR-HOURS	UNIT	BARE COSTS				TOTAL INCL O&P
					MAT.	LABOR	EQUIP.	TOTAL	

Division Notes

	CREW	DAILY OUTPUT	LABOR-HOURS	UNIT	BARE COSTS				TOTAL INCL O&P
					MAT.	LABOR	EQUIP.	TOTAL	

Division Notes

	CREW	DAILY OUTPUT	LABOR-HOURS	UNIT	BARE COSTS				TOTAL INCL O&P
					MAT.	LABOR	EQUIP.	TOTAL	

Division Notes

	CREW	DAILY OUTPUT	LABOR-HOURS	UNIT	BARE COSTS				TOTAL INCL O&P
					MAT.	LABOR	EQUIP.	TOTAL	

Division Notes

	CREW	DAILY OUTPUT	LABOR-HOURS	UNIT	BARE COSTS				TOTAL INCL O&P
					MAT.	LABOR	EQUIP.	TOTAL	

Division Notes

	CREW	DAILY OUTPUT	LABOR-HOURS	UNIT	BARE COSTS				TOTAL INCL O&P
					MAT.	LABOR	EQUIP.	TOTAL	

Cost Data Selection Guide

The following table provides definitive information on the content of each cost data publication. The number of lines of data provided in each unit price or assemblies division, as well as the number of crews, is listed for each data set. The presence of other elements such as reference tables, square foot models, equipment rental costs, historical cost indexes, and city cost indexes, is also indicated. You can use the table to help select the RSMeans data set that has the quantity and type of information you most need in your work.

Unit Cost Divisions	Building Construction	Mechanical	Electrical	Commercial Renovation	Square Foot	Site Work Landsc.	Green Building	Interior	Concrete Masonry	Open Shop	Heavy Construction	Light Commercial	Facilities Construction	Plumbing	Residential
1	621	454	475	576	0	543	200	377	505	620	560	322	1092	462	219
2	825	347	157	781	0	972	181	466	221	824	739	550	1268	355	340
3	1745	341	232	1265	0	1537	1043	355	2274	1745	1930	538	2028	317	445
4	960	22	0	920	0	724	180	613	1158	928	614	532	1175	0	446
5	1895	158	155	1099	0	858	1793	1106	735	1895	1031	985	1912	204	752
6	2468	18	18	2127	0	110	589	1544	281	2464	123	2157	2141	22	2677
7	1593	215	128	1633	0	580	761	532	523	1590	26	1326	1693	227	1046
8	2140	80	3	2733	0	255	1138	1813	105	2142	0	2328	2966	0	1552
9	2125	86	45	1943	0	313	464	2216	424	2062	15	1779	2379	54	1544
10	1088	17	10	684	0	232	32	898	136	1088	34	588	1179	237	224
11	1095	199	166	539	0	135	56	923	29	1062	0	229	1115	162	108
12	539	0	2	297	0	219	147	1552	14	506	0	272	1571	23	216
13	745	149	157	252	0	370	124	250	77	721	271	109	761	115	103
14	273	36	0	223	0	0	0	257	0	273	0	12	293	16	6
21	127	0	41	37	0	0	0	293	0	127	0	121	665	685	259
22	1165	7543	160	1226	0	2010	1061	849	20	1154	2109	875	7505	9400	719
23	1263	7010	639	1033	0	250	864	775	38	1246	191	980	5244	2011	579
25	0	0	14	14	0	0	0	0	0	0	0	0	0	0	0
26	1513	491	10473	1293	0	859	645	1160	55	1439	648	1360	10254	399	636
27	95	0	467	105	0	0	0	71	0	95	45	67	389	0	56
28	143	79	223	124	0	0	28	97	0	127	0	70	209	57	41
31	1511	733	610	807	0	3261	284	7	1216	1456	3280	607	1568	660	616
32	906	49	8	944	0	4523	419	418	367	877	1942	496	1803	140	544
33	1319	1100	563	310	0	3142	33	0	253	596	3277	135	1790	2165	161
34	112	0	47	9	0	195	0	0	36	67	226	0	141	0	0
35	18	0	0	0	0	327	0	0	0	18	442	0	84	0	0
41	63	0	0	34	0	8	0	22	0	62	31	0	69	14	0
44	75	79	0	0	0	0	0	0	0	0	0	0	75	75	0
46	23	16	0	0	0	274	261	0	0	23	264	0	33	33	0
48	8	0	36	2	0	0	21	0	0	8	15	0	21	0	8
Totals	26453	19222	14829	21010	0	21697	10324	16594	8467	25215	17813	16446	51423	17833	13297

Assem Div	Building Construction	Mechanical	Electrical	Commercial Renovation	Square Foot	Site Work Landscape	Assemblies	Green Building	Interior	Concrete Masonry	Heavy Construction	Light Commercial	Facilities Construction	Plumbing	Asm Div	Residential
A		15	0	190	165	579	599	0	0	537	573	155	24	0	1	378
B		0	0	848	2567	0	5673	56	329	1979	368	2108	174	0	2	211
C		0	0	647	956	0	1338	0	1645	146	0	846	251	0	3	591
D		1060	945	712	1866	72	2544	330	827	0	0	1353	1108	1090	4	851
E		0	0	85	262	0	302	0	5	0	0	259	5	0	5	391
F		0	0	0	114	0	143	0	0	0	0	114	0	0	6	357
G		527	447	318	312	3378	792	0	0	535	1349	205	293	677	7	307
															8	760
															9	80
															10	0
															11	0
															12	0
Totals		1602	1392	2800	6242	4029	11391	386	2806	3197	2290	5040	1855	1767		3926

Reference Section	Building Construction Costs	Mechanical	Electrical	Commercial Renovation	Square Foot	Site Work Landscape	Assem.	Green Building	Interior	Concrete Masonry	Open Shop	Heavy Construction	Light Commercial	Facilities Construction	Plumbing	Resi.
Reference Tables	yes	yes	yes	yes	no	yes	yes	yes	yes	yes	yes	yes	yes	yes	yes	yes
Models					111		25						50			28
Crews	584	584	584	564		584		584	584	584	562	584	562	564	584	560
Equipment Rental Costs	yes	yes	yes	yes		yes		yes	yes	yes	yes	yes	yes	yes	yes	yes
Historical Cost Indexes	yes	yes	yes	yes	yes	yes	yes	yes	yes	yes	yes	yes	yes	yes	yes	no
City Cost Indexes	yes	yes	yes	yes	yes	yes	yes	yes	yes	yes	yes	yes	yes	yes	yes	yes

2021 Training Class Options   📞 877.620.6245

RSMeans offers training classes in a variety of formats—eLearning training modules, instructor-led classes, on-site training, and virtual training classes. Due to our current Covid-19 situation, RSMeans has adjusted our current training offerings and instructor-led classes for 2021 have not been determined. Please visit our website at https://www.rsmeans.com/products/training/seminars for our current schedule of class offerings.

Training classes that are offered in one or more of our formats are: Building Systems and the Construction Process; Construction Cost Estimating: Concepts & Practice; Introduction to Estimating; Scope of Work for Facilities Estimating; Facilities Construction Estimating; Maintenance & Repair Estimating for Facilities; Mechanical & Electrical Estimating; RSMeans CostWorks CD; and RSMeans Online Training.

If you have any questions or you would like to register for any class or purchase a training module, call us at 877-620.6245.

Facilities Construction Estimating

In this two-day course, professionals working in facilities management can get help with their daily challenges to establish budgets for all phases of a project.

Some of what you'll learn:
- Determining the full scope of a project
- Understanding of RSMeans data and what is included in prices
- Identifying appropriate factors to be included in your estimate
- Creative solutions to estimating issues
- Organizing estimates for presentation and discussion
- Special estimating techniques for repair/remodel and maintenance projects
- Appropriate use of contingency, city cost indexes, and reference notes
- Techniques to get to the correct estimate quickly

Who should attend: facility managers, engineers, contractors, facility tradespeople, planners, and project managers.

Construction Cost Estimating: Concepts and Practice

This one- or two-day introductory course to improve estimating skills and effectiveness starts with the details of interpreting bid documents and ends with the summary of the estimate and bid submission.

Some of what you'll learn:
- Using the plans and specifications to create estimates
- The takeoff process—deriving all tasks with correct quantities
- Developing pricing using various sources; how subcontractor pricing fits in
- Summarizing the estimate to arrive at the final number
- Formulas for area and cubic measure, adding waste and adjusting productivity to specific projects
- Evaluating subcontractors' proposals and prices
- Adding insurance and bonds
- Understanding how labor costs are calculated
- Submitting bids and proposals

Who should attend: project managers, architects, engineers, owners' representatives, contractors, and anyone who's responsible for budgeting or estimating construction projects.

Assessing Scope of Work for Facilities Construction Estimating

This two-day practical training program addresses the vital importance of understanding the scope of projects in order to produce accurate cost estimates for facilities repair and remodeling.

Some of what you'll learn:
- Discussions of site visits, plans/specs, record drawings of facilities, and site-specific lists
- Review of CSI divisions, including means, methods, materials, and the challenges of scoping each topic
- Exercises in scope identification and scope writing for accurate estimating of projects
- Hands-on exercises that require scope, takeoff, and pricing

Who should attend: corporate and government estimators, planners, facility managers, and others who need to produce accurate project estimates.

Maintenance & Repair Estimating for Facilities

This two-day course teaches attendees how to plan, budget, and estimate the cost of ongoing and preventive maintenance and repair for existing buildings and grounds.

Some of what you'll learn:
- The most financially favorable maintenance, repair, and replacement scheduling and estimating
- Auditing and value engineering facilities
- Preventive planning and facilities upgrading
- Determining both in-house and contract-out service costs
- Annual, asset-protecting M&R plan

Who should attend: facility managers, maintenance supervisors, buildings and grounds superintendents, plant managers, planners, estimators, and others involved in facilities planning and budgeting.

Mechanical & Electrical Estimating

This two-day course teaches attendees how to prepare more accurate and complete mechanical/electrical estimates, avoid the pitfalls of omission and double-counting, and understand the composition and rationale within the RSMeans mechanical/electrical database.

Some of what you'll learn:
- The unique way mechanical and electrical systems are interrelated
- M&E estimates—conceptual, planning, budgeting, and bidding stages
- Order of magnitude, square foot, assemblies, and unit price estimating
- Comparative cost analysis of equipment and design alternatives

Who should attend: architects, engineers, facilities managers, mechanical and electrical contractors, and others who need a highly reliable method for developing, understanding, and evaluating mechanical and electrical contracts.

Building Systems and the Construction Process

This one-day course was written to assist novices and those outside the industry in obtaining a solid understanding of the construction process – from both a building systems and construction administration approach.

Some of what you'll learn:
- Various systems used and how components come together to create a building
- Start with foundation and end with the physical systems of the structure such as HVAC and Electrical
- Focus on the process from start of design through project closeout

This training session requires you to bring a laptop computer to class.

Who should attend: building professionals or novices to help make the crossover to the construction industry; suited for anyone responsible for providing high level oversight on construction projects.

Practical Project Management for Construction Professionals

In this two-day course you will acquire the essential knowledge and develop the skills to effectively and efficiently execute the day-to-day responsibilities of the construction project manager.

Some of what you'll learn:
- General conditions of the construction contract
- Contract modifications: change orders and construction change directives
- Negotiations with subcontractors and vendors
- Effective writing: notification and communications
- Dispute resolution: claims and liens

Who should attend: architects, engineers, owners' representatives, and project managers.

RSMeans data Training

Training for our Online Estimating Solution

Construction estimating is vital to the decision-making process at each state of every project. Our online solution works the way you do. It's systematic, flexible, and intuitive. In this one-day class you will see how you can estimate any phase of any project faster and better.

Some of what you'll learn:
- Customizing our online estimating solution
- Making the most of RSMeans "Circle Reference" numbers
- How to integrate your cost data
- Generating reports, exporting estimates to MS Excel, sharing, collaborating, and more

Also offered as a self-paced or on-site training program!

Training for our CD Estimating Solution

This one-day course helps users become more familiar with the functionality of the CD. Each menu, icon, screen, and function found in the program is explained in depth. Time is devoted to hands-on estimating exercises.

Some of what you'll learn:
- Searching the database using all navigation methods
- Exporting RSMeans data to your preferred spreadsheet format
- Viewing crews, assembly components, and much more
- Automatically regionalizing the database

This training session requires you to bring a laptop computer to class.

When you register for this course you will receive an outline for your laptop requirements.

Also offered as a self-paced or on-site training program!

Site Work Estimating with RSMeans data

This one-day program focuses directly on site work costs. Accurately scoping, quantifying, and pricing site preparation, underground utility work, and improvements to exterior site elements are often the most difficult estimating tasks on any project. Some of what you'll learn:
- Evaluation of site work and understanding site scope including: site clearing, grading, excavation, disposal and trucking of materials, backfill and compaction, underground utilities, paving, sidewalks, and seeding & planting.
- Unit price site work estimates—Correct use of RSMeans site work cost data to develop a cost estimate.
- Using and modifying assemblies—Save valuable time when estimating site work activities using custom assemblies.

Who should attend: engineers, contractors, estimators, project managers, owners' representatives, and others who are concerned with the proper preparation and/or evaluation of site work estimates.

Please bring a laptop with capability to access the internet.

Facilities Estimating Using the CD

This two-day class combines hands-on skill-building with best estimating practices and real-life problems. You will learn key concepts, tips, pointers, and guidelines to save time and avoid cost oversights and errors.

Some of what you'll learn:
- Estimating process concepts
- Customizing and adapting RSMeans cost data
- Establishing scope of work to account for all known variables
- Budget estimating: when, why, and how
- Site visits: what to look for and what you can't afford to overlook
- How to estimate repair and remodeling variables

This training session requires you to bring a laptop computer to class.

Who should attend: facility managers, architects, engineers, contractors, facility tradespeople, planners, project managers, and anyone involved with JOC, SABRE, or IDIQ.

Registration Information

How to register

By Phone
Register by phone at 877.620.6245

Online
Register online at
RSMeans.com/products/services/training

Note: Purchase Orders or Credits Cards are required to register.

Two-day seminar registration fee - $1,300*.

One-Day Construction Cost Estimating or Building Systems and the Construction Process - $825*.

Two-day virtual training classes - $825*.

Three-day virtual training classes - $995*.

Instructor-led Government pricing

All federal government employees save off the regular seminar price. Other promotional discounts cannot be combined with the government discount. Call 781.422.5115 for government pricing.

CANCELLATION POLICY FOR INSTRUCTOR-LED CLASSES:

If you are unable to attend a seminar, substitutions may be made at any time before the session starts by notifying the seminar registrar at 781.422.5115 or your sales representative.

If you cancel twenty-one (21) days or more prior to the seminar, there will be no penalty and your registration fees will be refunded. These cancellations must be received by the seminar registrar or your sales representative and will be confirmed to be eligible for cancellation.

If you cancel fewer than twenty-one (21) days prior to the seminar, you will forfeit the registration fee.

In the unfortunate event of an RSMeans cancellation, RSMeans will work with you to reschedule your attendance in the same seminar at a later date or will fully refund your registration fee. RSMeans cannot be responsible for any non-refundable travel expenses incurred by you or another as a result of your registration, attendance at, or cancellation of an RSMeans seminar.

Any on-demand training modules are not eligible for cancellation, substitution, transfer, return, or refund.

AACE approved courses

Many seminars described and offered here have been approved for 14 hours (1.4 recertification credits) of credit by the AACE International Certification Board toward meeting the continuing education requirements for recertification as a Certified Cost Engineer/Certified Cost Consultant.

AIA Continuing Education

We are registered with the AIA Continuing Education System (AIA/CES) and are committed to developing quality learning activities in accordance with the CES criteria. Many seminars meet the AIA/CES criteria for Quality Level 2. AIA members may receive 14 learning units (LUs) for each two-day RSMeans course.

Daily course schedule

The first day of each seminar session begins at 8:30 a.m. and ends at 4:30 p.m. The second day begins at 8:00 a.m. and ends at 4:00 p.m. Participants are urged to bring a hand-held calculator since many actual problems will be worked out in each session.

Continental breakfast

Your registration includes the cost of a continental breakfast and a morning and afternoon refreshment break. These informal segments allow you to discuss topics of mutual interest with other seminar attendees. (You are free to make your own lunch and dinner arrangements.)

Hotel/transportation arrangements

We arrange to hold a block of rooms at most host hotels. To take advantage of special group rates when making your reservation, be sure to mention that you are attending the RSMeans Institute data seminar. You are, of course, free to stay at the lodging place of your choice. (Hotel reservations and transportation arrangements should be made directly by seminar attendees.)

Important

Class sizes are limited, so please register as soon as possible.

*Note: Pricing subject to change.